# PRE-INDUSTRIAL IRON
# Its Technology and Ethnology

by

William Rostoker

and

Bennet Bronson

ARCHEOMATERIALS MONOGRAPH No. 1
PHILADELPHIA, PENNSYLVANIA
1990

**Library of Congress Cataloging-in-Publication Data**

Rostoker, William.

Pre-industrial iron : its technology and ethnology / by William Rostoker and Bennet Bronson.
p. cm.—Archeomaterials monographs ; no. 1)
Includes bibliographical references and index.
ISBN 0-924171-03-0 : $55.00
1. Iron—Metallurgy—History. 2. Iron industry and trade--History. I. Bronson, Bennet, 1938- . II. Title. III. Series.
TN703.R67 1990
669′.141′09—dc20 90-48198
CIP

ISBN 0-924171-03-0

# Table of Contents

**LIST OF FIGURES**

## LIST OF TABLES

# Introduction

The contemporary world is built on iron in its various forms. Our structures, machines, tools, vehicles, and weapons are, for the most part, constructed of iron. From natural resources and recycled scrap, the world produces about half a billion tons of new iron every year—more by weight than any other material—so this commodity shapes our way of life. Production of iron has been escalating for more than 3500 years, the greater part of written history.

The ironmaking industry was created by the imagination and skill of many men, most of whom are anonymous. Until the twentieth century, ironmaking represented a triumph of empiricism. Since history is our record of trends and events, the scientific study of the development of ironmaking technology must be taken into account. The difficulty is that historians and anthropologists have little knowledge of science, so that aspect of history is often treated with vagueness and simplicity.

Ironmaking is far more than an evolving pattern of technology. It is as much a mistake to isolate the technology of ironmaking from the mainstream of historical studies as to reduce it to insubstantial footnotes. The possession of ironmaking capacity has always been a concern at the highest level of society. Peoples have risen, become dominant, and fallen because of their decisions whether to make or buy commodities, to use local or imported craftsmen, to remain static technologically or exploit new concepts and developments.

The authors represent two different disciplines—technology and anthropology. Together we have tried to treat the subject of iron and ironmaking in a way which encompasses both of our perspectives on what is important. It has been an experiment in interdisciplinary synthesis, out of which has come a mutual understanding of our disparate fields. In order to converse constructively, we have had to abandon individual jargon which so often disguises unclear thought. This exercise became an education in itself. The fact that we have come this far shows the potential for such collaboration.

We are not trying to revise history, but rather to construct a stronger foundation upon which better history can be written by those professionally trained to do so. There is nonetheless a great deal of historical content inherent in the subject—because the interplay of technology and ethnology is the theme, because a purely technological explanation of some phenomena may be unpalatable to some, and because events and technology influenced each other.

Should the historian and anthropologist perceive unintended condescension, let it be said that the modern technological perspective is ill-equipped to rationalize and explain those methods of processing and fabrication which have long since ceased to be used. The modern metallurgist will probably find this subject unfamiliar territory. It is not that the scientific principles are inadequate, but that the means by which the principles were exercised are buried in the literature or often lost completely, and can be recovered only with considerable effort.

The explanation of early technology is like working a puzzle. For example, not until the nineteenth century was phosphorus identified, and its influences on the properties of iron recognized and to some extent moderated. Yet this element has always been a hazard to the successful fabrication of iron. The early craftsman knew nothing about phosphorus. How then did he deal with the problems of the peculiar variations in the formability of phosphorus-bearing iron and its propensity to shatter under impact? Did he simply discontinue the use of the iron ore from which the unsatisfactory metal derived? Or did he learn to segregate the high formability iron from the poorer grades? Were there ways, empirically discovered, by which the problems caused by unseen phosphorus could be ameliorated? Would it have been simpler to import metal that was free of the undesirable characteristics? Answers to these and other questions may explain successes and failures throughout history and in turn the events where iron performance might have been crucial. The literature of the last 40 years is of little assistance. The effect of phosphorus on iron is understood in a general way, but the details are omitted because modern irons do not contain significant levels of phosphorus. The problems of eliminating it from iron were solved in the late nineteenth and early twentieth centuries, and by now, the subject is little discussed and largely forgotten.

The term *pre-industrial* is used here to describe the period up to the advent of the use of steam and electrical power for machines. At that point, scale of production and unit costs of products abruptly changed. This event occurred in the early nineteenth century in Europe and the technology spread from there, although in many parts of the world it still has not arrived. We have therefore studied recent ironmaking and iron usage in some remote parts of the world. These examples, when well documented, show what is being done without industrial power sources and scientific knowledge. In the absence of details about ironmaking in early history, we suggest logical options and determine their feasibility using ethnographic and textual evidence, even when it is derived from a different period. This pragmatic approach assumes that what sensible people did in one set of circumstances, other sensible people may also have done at another place and time. This premise at least provides a starting point in the search for evidence.

What emerges from this study of iron and ironmaking is that the use of the term "backward" must be re-examined. It is evident that high levels of sophistication were achieved in some unlikely places and that both scale of output and quality of product were often reconciled with regional needs. The absence of large-scale commercial activity does not necessarily mean that a good product could not be made.

The reader may approach this book on various levels. The first two chapters explain the performance qualities of irons that make them pre-eminent and also the many features by which functional values are achieved or obstructed. Chapters IX to XV are necessarily technical so that the reader can see the differences in process basics and product control that the empirical or Edisonian ironmaster could exploit. These chapters highlight the special features of processes with local origins. Throughout we attempt to explain superstitions and misconceptions which have survived so long that their origins have become lost.

In Chapter XVI the authors reflect on topics which are broader in scope than the individual chapters. There are many anecdotes which support the general arguments but would distract the reader if they were incorporated into the main text. These illustrations may help clarify points which seem obscure to the reader with an elementary understanding of the subject.

While we have assembled a large bibliography, it will be clear from the text that some sources were consulted more frequently than others. Most of the authors are well known to students of ancient technology, but the extensive quotation from the neglected works of Frederick Overman may come as a surprise. Since we believe that he was the first American metallurgist, we feel that his career should be appreciated by a larger audience. Therefore, we have summarized what we have been able to discover about this remarkable man in Appendix D. Also cited are chronicles by travellers of the nineteenth century whose detailed and often astute observations of Asian and African ironmaking, published in what are now obscure works, provide the factual bases for many of our explanations of what could be done without modern scientific knowledge and facilities.

William Rostoker
Bennet Bronson

June 1986

# 1. Iron As A Material

## 1. INTRODUCTION

Iron is one among many useful materials, metallic and nonmetallic. Its physical and mechanical properties make it ideal for some purposes and adequate for others. Even now, although better materials may exist, iron is often preferred because it is widely available and, priced in cash or labor, inexpensive. Before the modern period, there was no better material for a surprisingly wide range of applications, because iron is itself a family of materials in which the members may differ more from each other than from other metals in their range of properties and the ways they may be processed. These differences are important in understanding the history of metallurgical technology.

The craftsman, artisan, farmer, shipwright, and soldier of the past viewed metals, like other materials, in practical terms, considering qualities like strength, durability, and ease of fabrication, as well as cost. Although until the nineteenth century scientific understanding of the subject was limited, perhaps especially in the case of iron, there was probably a fair empirical knowledge of ore sources, how to convert ores into metal, and which were best for certain purposes. Materials and techniques were chosen pragmatically. It is certain that the pre-industrial metalworker was as conscious as we of the inevitable trade-off between performance and price.

## 2. MECHANICAL PROPERTIES OF METALLIC MATERIALS

While individual attributes of nonmetallic materials may be equal to or even better than those of metals, a combination of properties such as strength, toughness, and ease of shaping can be achieved only in metals. Any material can be evaluated in terms of functional attributes. Although most of these are now known to be related, if sometimes in a nonspecific way, they often must have seemed to vary independently of one another. Attributes that were relevant in pre-industrial contexts include:

—resistance to bending or breaking under transverse loads, needed in beams and lintels
—resistance to fracture under compression, needed in pillars and pedestals
—resistance to buckling, needed in nails
—resistance to twisting, needed in drills and axles
—resistance to stretching or breaking under tension, needed in cables, chains, and ropes
—resistance of cutting edge to dulling, needed in knives, files, and saws
—resistance to abrasive wear, needed in plow tips and bearings
—resistance to blunting, needed in spears and pickaxes
—resistance to penetration, needed in armor
—resistance to shattering when struck sharply, needed in anvils and arrowheads
—resistance to thermal shock, needed in cooking pots
—ability to conduct heat, also needed in cooking pots
—resistance to chemical damage, also desirable in storage vessels
—ability to withstand high temperature, needed in stoves and furnaces
—ability to store elastic energy, needed in bows and springs

Deformability, one of the more distinctive traits of metals and their alloys, was a crucial property in the development of many fabrication processes in the past. While stone, wood, and bone usually had to be shaped by removing material in the form of chips and fragments, metal could be worked by deformation, and hot and cold forging without removal of material. Most of the metals known in antiquity could also be shaped by chip removal: that is, by milling, grinding, and other forms of machining. Many could also be cast by being melted to low viscosity liquids and then poured into molds. Moreover, waste from machining or casting could be recovered by remelting.

Metals have a wide range of useful properties, their major disadvantages being that they are generally poor insulators and offer only fair resistance to abrasive wear and high temperature. Certain pre-industrial metals in their unalloyed state—gold, silver, antimony, tin, lead, and zinc—had limited applications. Copper, on the other hand, was highly versatile, and alloying with other metals imparted to it functional attributes that few nonmetallic materials had. Yet even copper and its alloys took second place to iron. For many tasks, members of the iron family equalled or surpassed all other pre-modern materials.

Under tension, or bending, iron has variable strengths depending on its carbon content (see Table 1.1 for comparison of tensile strengths of irons and other materials). A high carbon steel, which fractures at a stress of almost 170,000 pounds per square inch (i.e., 1172.4 megapascals), is extremely strong. Several organic fibers, especially hemp and silk, are comparable in strength. On the other hand, irons at both ends of the spectrum of carbon content are mediocre. Low carbon

TABLE 1.1. Tensile strengths of selected materials

| Material | Strength* | Additional information |
|---|---|---|
| *Iron & Steel* | | |
| Cast iron, white | 25 | 3–4% C |
| Cast iron, gray | 25–40 | 3–4% C |
| Wrought (or bloom) iron | 46 (30) | |
| Steel, 0.05–0.10% | 40–60 (25–45) | Plain; equivalent to bloom iron |
| Cast iron, malleable | 52 (33) | Ca. 3% C |
| Steel, 0.3% C, soft | 67 (50) | Plain; annealed to softest state |
| 0.8% C, soft | 89 (55) | Plain; annealed |
| 0.3% C, tempered | 92 (75) | Plain; heat treated (538°C temper) |
| 0.5% C, soft | 92 (53) | Plain; annealed |
| Cast iron, spheroid graphite | 94 (74) | Grade SNG 47/2; ca. 3.5% C |
| Steel, 0.5% C, tempered | 133 (88) | Plain; heat treated (538°C temper) |
| 0.8% C, tempered | 169 (122) | Plain; heat treated (538°C temper) |
| 0.8% C, cold drawn | 300 (280) | Plain; cold worked |
| *Other Metals* | | |
| Lead | 2.5 (1.9) | Pure |
| Pewter, traditional | 6.8 (–) | Alloy; tin with 30% Pb |
| Pewter, modern | 8.6 (–) | Alloy; Britannia metal |
| Gold | 18 (~0) | Pure |
| Copper | 34 (10) | Hot rolled, tough pitch (0.04%) |
| Silver, sterling | 45 (–) | Alloy; 7.5% Cu |
| Bronze, 5% Sn, annealed | 47 (19) | Alloy; rest is Cu |
| 11% Sn, as cast | 55 (29) | |
| 8% Sn, annealed | 60 (24) | |
| 10% Sn, annealed | 66 (28) | |
| 5% Sn, cold worked | 81 (75) | |
| 8% Sn, cold worked | 93 (72) | |
| 10% Sn, cold worked | 100 (–) | |
| *Stone & Ceramics* | | |
| Concrete, reinforced | 0.9 | |
| Glass | 2 | Minimum |
| Sandstone | 4.8–34 | |
| Limestone | 20–30 | |
| Granite | 25–35 | |
| Marble | 30–35 | |
| Glass, filament | 99 | |
| *Natural Polymers* | | |
| Wood | 1.2–1.9 | |
| Leather | 4–6.5 | |
| Paper | 5–15 | |
| Bone | 17.5–21.6 | |
| Wool fiber | 25 | |
| Bamboo fiber | 50 | |
| Catgut | 60 | |
| Cotton fiber | 75 | |
| Silk fiber | 100 | |
| Hemp fiber | 140 | |

*For comparative purposes on the same scale. ( ) = yield strength. Data taken from standard engineering manuals.

steels and wrought irons, and also high carbon cast irons, have less tensile strength than bronze.

An object is said to be brittle if, like glass, it breaks without visible distortion when subjected to increasing stress. It is tough if it bends or stretches before breaking. The range of the toughest to the most brittle materials covers several orders of magnitude (Table 1.2). Heat treated steels and all low carbon irons are very tough, but high carbon cast iron is only moderately tough, ranking well below most steels but far above materials such as ceramics and plate glass.

TABLE 1.2. Selected materials representing a scale of toughness

| Material | Toughness |
|---|---|
| Heat treated steel | 200 |
| Soft, low carbon, constructional steel | 120 |
| Softened, medium carbon steel | 70 |
| Gray cast iron | 50 |
| Dense (void free), crystalline ceramic | 4 |
| Plate glass | 0.7 |

Resistance to abrasion is a desirable property in objects like plow shares, wheel rims, axles, and pivots, and is indispensable in tools like files and grindstones which are meant to cut or abrade other materials. Most members of the iron family have only limited uses as abrasives, although high carbon steel makes good files, which have an abrading as well as a cutting action and are used (in cutlers' "steels") to finish the sharpening of the edges of steel cutting tools. However, no type of iron can replace the harder minerals—diamond, corundum, emery, or sandstone—as simple abrasives.

Iron is particularly desirable where it is necessary to combine abrasion resistance with strength and toughness. Quartz and flint, for instance, despite their great hardness, make poor plow tips; these materials may resist wear in rocky soils but they shatter so readily that early peoples preferred tips made of soft metal or even wood. Cast iron plow shares, on the other hand, have long been used with success in both Europe and China. Here, the choice of high carbon iron is a compromise between the toughness of softer metals and the abrasion resistance of brittle stone. Cast iron plows remained popular on American farms until tractors came into use; high plowing speeds necessitated yet another compromise, and plows made of heat-treated steel, which is even more shock-resistant, were developed. The resistance to abrasion and the carbon contents of iron alloys are directly correlated (Fig. 1.1). Bloom iron is the softest and contains the least carbon. The hardest is heat-treated white cast iron.

Resistance to blunting is related both to hardness and toughness. A pointed object designed to penetrate a workpiece, rather than to cut it, becomes blunt not because of abrasion but because of shattering or deformation. A good spearhead or nail must be strong and tough, since a penetrating implement works by retaining its pointed shape and its ability to focus kinetic energy long enough to perforate a target material. Bronze is better than obsidian for spearheads because it does not shatter and better than copper because it has a much higher resistance to deformation. Low carbon iron is inferior to bronze for this purpose because it is soft and cast irons are too brittle. Steels, however, can be superior to both bronze and other irons if suitably treated.

Abrasion, deformation, and corrosion can dull a cutting edge. Shattering is a problem only if the cutting action involves impact, as with saw teeth, or twisting and prying, as with chisels and axes. Razors and swords have frequently been made of high carbon steel, although it is brittle, because a smooth-edged cutting implement can function well if carefully used. American cabinetmakers now often employ traditional Japanese chisels, which, because of their hardness and abrasion resistance, stay sharp longer than European chisels. However, the high carbon Japanese blades are delicate, and will snap if, like European chisels, they are used to pry as well as cut.

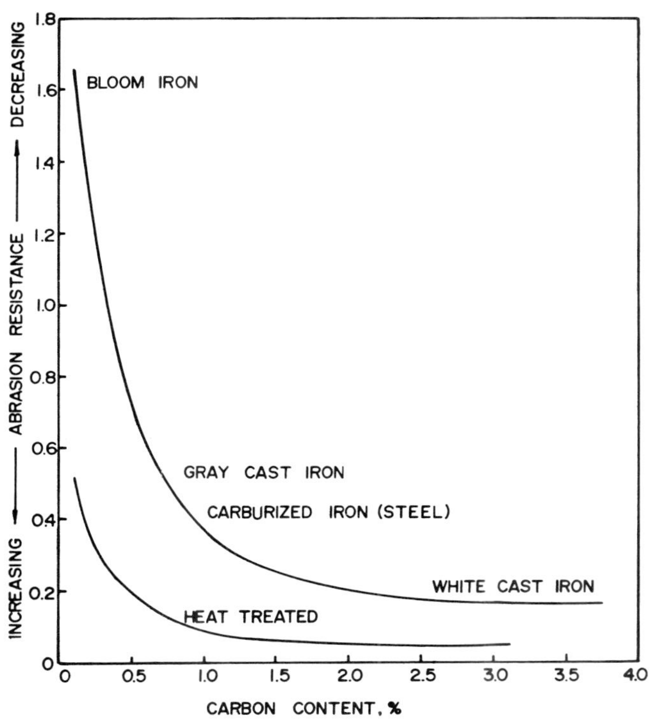


Fig. 1.1 *The Relationship of Abrasion Resistance to Carbon Content.* Note that superior wear resistance is achieved by a combination of high carbon content and hardening by heat treatment. (Adapted from ASM *Metals Handbook*, 9th edition, 1978: 608)

A major shortcoming of thin cutting edges made of ordinary steel is the propensity to rust. Many modern cooks recognize, for example, that a carbon steel kitchen knife is easier to sharpen to a fine edge and that it cuts more effectively than its equivalent in stainless steel, yet most prefer the latter because they are easier to care for. The carbon steel blade should be dried and oiled after each use, whereas the stainless steel blade will retain its edge after months of careless handling. A similar preference for corrosion resistance over other qualities must have sometimes influenced the choice of bronze rather than iron in antiquity. In both the Mediterranean and the Far East, bronze swords and knives continued to be used and even preferred until late in the Iron Age.

In general, however, points will be most resistant to blunting and edges to dulling if made from steels with carbon contents between about 0.4 and 2.0 percent. The first crucial development in iron technology was the ability to produce medium or high carbon steel; next was the knowledge of how to form it into useful shapes, weld it to softer but tougher iron, and in the final step of production, harden it through one of several techniques. Those who became expert in the handling of steel, an achievement rare in pre-industrial societies, produced artifacts—usually weapons—that could become legendary. Those who were not as skilled technically often made steel even though the performance potential of their metal could not be fully exploited: this paradox is a recurring theme in the history of ironworking.

Another attribute of steel and low carbon iron is their high elastic modulus or stiffness, which even today makes them the materials of choice for springs and related devices that store and release mechanical energy. Cold worked bronze is reasonably stiff and thus performs adequately in coil springs that, like those incorporated in the safety pin-like fibulae of Bronze and Iron Age Europe, carry only a moderate load. Both iron and steel have a higher elastic modulus than bronze. Steels also have a very high elastic limit or point at which they lose the ability to recover from bending. A steel spring stores almost twice as much elastic energy as one made of bronze and at least fifteen times more than springs of wood or bone.

European makers of crossbows in the fourteenth and fifteenth centuries exploited the elasticity of iron. At that time, according to the traditional explanation (e.g., Payne-Gallwey 1958), personal armor had developed so that arrows propelled by wooden bows were no longer effective against a fully equipped knight. Military designers therefore had to experiment with new weapons, including bows made of iron or steel, which were rapidly perfected. Once devices for cocking bows with a pull of many hundreds of pounds had been developed, the "steel" crossbow (which may often have been made of low carbon iron) became a weapon with as much penetrating capability as a modern handgun.

It seems possible, in fact, that by the sixteenth century the use of iron or steel crossbows in warfare was limited more by expense than by any technical superiority of firearms. Good steel springs like those that formed the arms of crossbows had to be made of a uniform and slag-free metal which was, until the early nineteenth century, difficult and costly to produce. High performance springs used in crossbows and clocks were as near the forefront of pre-industrial technology as sword and razor steels. Most of the spring-powered flintlock mechanisms in non-European guns of the eighteenth and nineteenth centuries were imported from Europe, and as late as the 1780s Chinese craftsmen, although able to make all other parts of a watch or clock, still could not fabricate the springs (Cipolla 1965: 160). The spring steel for clocks as well as the strings of musical instruments (Goodway 1984) required metal with few and only very small nonmetallic inclusions. This was an English development of the late eighteenth century.

## 3. STRENGTHENING IRON

There are four general methods by which iron can be made stronger: alloying, cold working, reducing the size of its crystals, and inducing a change of state. These techniques may be used alone or in combination.

### 3a. *Alloying*

Among the alloying elements of iron used in antiquity, carbon was by far the most important. Molten iron can dissolve about 4.5 percent carbon. Small increases in carbon content are accompanied by impressive gains in hardness; a high purity iron has only one-third the hardness of a 1.5 percent carbon-iron alloy. Of other elements that form alloys with iron, those with deleterious effects, such as sulfur and phosphorus, played a larger role in the past than beneficial additives like manganese and nickel. In pre-modern times iron-manganese and iron-nickel alloys were occasionally produced from mixed ores or (in the case of iron-nickel) recovered from meteorites (see Chapter 2.4a-4c). However, until the late nineteenth century iron-carbon was the only ferrous alloy to be regularly made and effectively used.

### 3b. *Cold Working*

Cold working, strain hardening, and work hardening are synonymous terms to a metallurgist. All three refer to the strengthening of a metal by plastic deformation at normal or slightly elevated temperatures. Cold deformation yields an improvement in the performance of most metals, including copper, bronze, and any iron with a carbon content of less than about 1.5 percent (Fig. 1.2). Bronzes respond well to such treatment, increasing in strength by a factor of five. Low carbon iron, on the other hand, is not only less strong than a

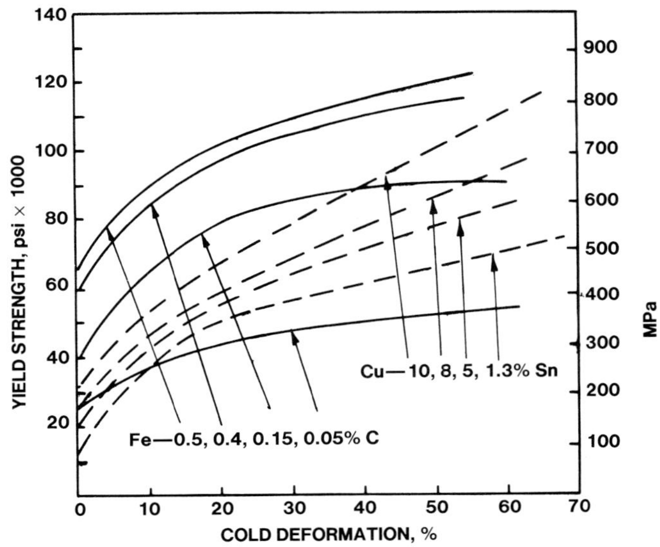


Fig. 1.2 *Cold Deformation and Strength in Bronze and Steel.* Alloys of both copper-tin and iron-carbon respond to cold deformation—that is, hammering or rolling at room temperature—by becoming harder and stronger. Low carbon steel is less responsive than low tin bronze; high carbon steel doubles in strength as deformation proceeds, whereas the strength of high tin bronze can be more than tripled.

high tin bronze in its undeformed state, but is less improved by low temperature deformation. Since early iron in the form of uncarburized blooms was mostly of the low carbon type, the reasons for the long coexistence of bronze with iron are clear. Until methods for producing medium carbon iron alloys—steels—were discovered, many bronzes were more serviceable than iron.

Cold working is also a method of fabrication: stretching, bending, twisting, or hammering at normal temperatures results not only in strain hardening but also in a change of shape, both of which are often desirable. A good example is the drawing of wire. A thin bar of iron or bronze pulled through a die that reduces its diameter by 30 percent is not only doubled in length, but its yield strength increases by a factor of two. The chief disadvantage of cold working is that a progressive loss of ductility accompanies it. Even after the first pass through the die, wire becomes more likely to break if bent at a sharp angle and is also markedly more resistant to drawing through a second die with a smaller diameter. Several passages through a set of successively smaller die holes make it almost impossible to draw the wire further without snapping. Cold hammered strips react the same way; as hammering proceeds and the strip becomes thinner, it also becomes harder and more likely to crack when bent.

This problem can be overcome because in metals strain hardening can be reversed by heating to temperatures well below the melting point, which softens the metal and restores the original ductility. This medium temperature heat treatment is called *annealing*. Thus, a smith may thin a plate by hammering on an anvil until it becomes too hard to deform further or until he suspects that cracking is about to occur. He must then anneal it in the forge for a few minutes. After the plate has cooled again, it has returned to its original soft and ductile state while retaining the shape and dimensions it had before annealing. A repetitive sequence of cold working and annealing can be used to produce highly complex shapes and significant changes of dimensions in metals.

Hot forging is easier, but it is impractical when the workpiece is so thin that it loses most of its heat as soon as it is removed from the forge. We therefore assume that most early wire and sheet metal were produced by repeated cold working and annealing, and that annealing was known to many of the peoples in the past who used native or imported metals but had no knowledge of smelting or casting. The Andaman Islanders of the Indian Ocean are reputed to have once made all their spears from shipwreck iron, as are the Northwest Coast Indians of Washington and British Columbia and the Hopewell peoples of the American Midwest. The Hopewells, who flourished in the early first millennium A.D., are among the true masters of cold metalworking. Made from the abundant native copper of the Lake Superior region, the delicate copper bells and intricate ear ornaments of Hopewell rank with the finest products of ancient metallurgy.

### 3c. *Altering Crystal Size*

Sophisticated cold working of iron must sometimes have made smiths aware of the benefits of controlling the temperature at which annealing occurs. Iron will anneal at any temperature above about 400°C. At higher temperatures, annealing proceeds more rapidly. Although the process is slower at lower temperatures, it produces an iron structure with a smaller crystal size which can be a significant factor in strength. A piece of iron annealed at 600° to 700°C has an average crystal size of more than 1 mm. An identical piece annealed at very close to 400°C will be composed of crystals less than 0.12 mm. in diameter, giving it a yield strength at least one-third higher than the other piece.

### 3d. *Inducing a Change of State*

The term *heat treatment* is often loosely used to denote a method of hardening and strengthening by subjecting a piece of metal to a particular series of thermal processing steps: cooling it very rapidly from forging temperatures, reheating it to a lower temperature than that required for forging, and then letting it cool again at a convenient rate. In the heat treatment of steels, the first step is called *quenching* and the last two *tempering*; the whole hardening process is often described as a *quench and temper* procedure. Steels particularly benefit from being handled in this way.

The hard but brittle metal produced by quenching can be manipulated by tempering to yield a remarkable range of trade-offs between strength and toughness. Since steel objects may differ in carbon contents and since moderate changes in carbon content can have dramatic effects on the hardness and toughness of the final product, the total range of combinations becomes great (Fig. 1.3).

## 4. FABRICATION BY HOT FORGING

For most cultures in most periods, the common method of converting iron to a useful shape has been hot forging—hammering the metal while it is at a high temperature. The technique involves, essentially, the use of brute force which, in order to be successful, depends on delivering to the point of impact a quantity of energy that overwhelms the resistance of the material to squeezing (compression) and lateral flow.

It is fundamental to the forging of all deformable metals that the force required to effect deformation is reduced as the temperature rises (Fig. 1.4). At 1200°C the deformation resistance of iron falls to less than one-third of its value at normal temperatures, and even a moderate degree of heating will noticeably decrease the effort needed; however, the curve of resistance only begins to drop sharply at temperatures above 750°C, when the workpiece begins to glow red. Blacksmiths may choose to work at cherry red or orange heats, in the 700° to 900°C range (see Table 1.5). Most, however, prefer to leave workpieces in the forge until they begin

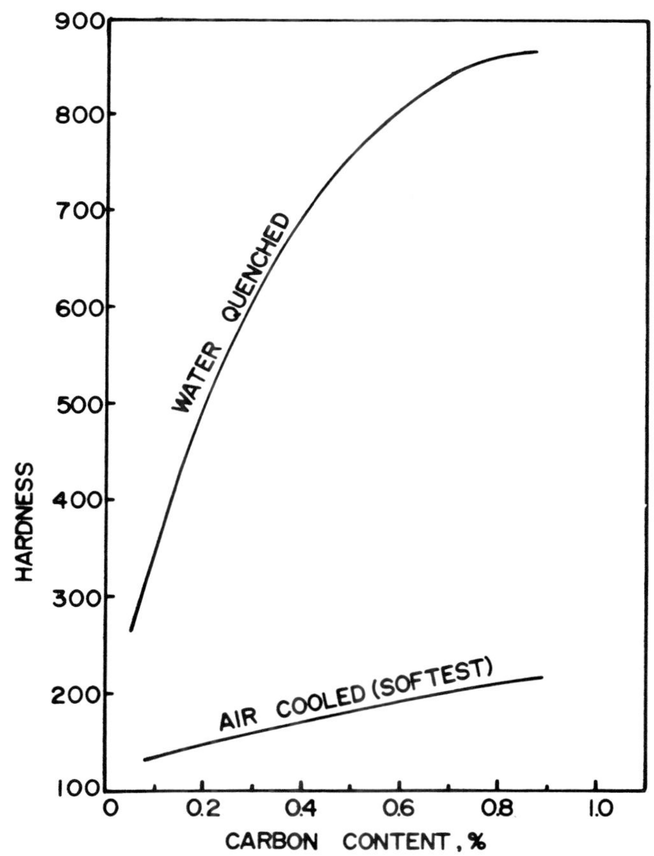


Fig. 1.3 *The Capacity of Steels for Hardening by Heat Treatment.* The upper curve represents the maximum hardness obtainable by quenching plain steels of varying carbon content, and the lower the minimum obtained through slow cooling. By quenching and subsequent tempering, a steel can be made to fall between these curves.

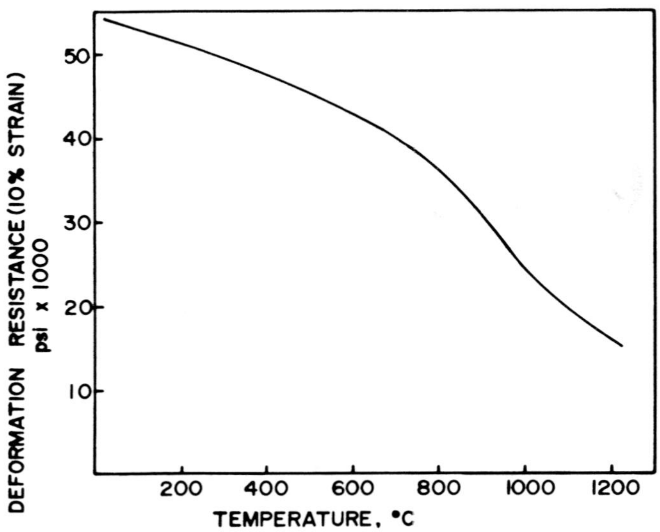


Fig. 1.4 *The Deformation Resistance of Low Carbon Iron as a Function of Temperature.* Iron objects are usually forged at or above yellow heat (ca. 1000°C) because flow stress is influenced by deformation rate (that is, by the velocity with which the deforming force is applied). The curve has been adjusted to a rate appropriate for hammering.

to glow light yellow or white and a temperature above 1000°C is reached, because they feel that the extra expenditure of fuel is repaid by the conservation of physical labor. There are other advantages to forging at "hot working" temperatures: the metal does not strain harden regardless of the degree of deformation, and levels of alloying have little effect on resistance to deformation, so that at white heat both iron and steel are forged with almost equal ease.

An iron alloy that contains appreciable amounts of sulfur is an important exception. Above a certain level, sulfur can cause an otherwise highly forgeable metal to become severely embrittled and to crack under the hammer. Such a metal is termed *hot short* (Figure 1.5 illustrates the effect of attempting to forge iron rods with a high sulfur content at hot working temperatures). The cracking constitutes serious and irreversible damage and in a blacksmith's shop, rods damaged in this way would have to be discarded. Hot shortness is not related to ductility at normal temperatures.

The hot forging qualities of iron made in early workshops contrasted clearly with those of copper alloys. Most bronzes can be worked only at relatively low temperatures. At high temperatures they tend to crack badly, so that any fabrication by hammering must be done with moderate heating or, preferably, with no heating (Chadwick 1939). The effort required to induce deformation in a very hot piece of iron is much less than that needed to work bronze at any temperature below its hot shortness limit.

Two factors control the hot working of iron objects—the rates at which the workpiece cools and at which it oxidizes. Because of the restrictions imposed by both of these factors, the workpiece should be of substantial size but not so large that it cannot be easily manipulated.

The limitations caused by cooling rates can be best understood by visualizing a blacksmith shaping a piece of iron. He places the piece in the forge and waits until it begins to glow orange or yellow. He then pulls it out and begins hammering immediately. He can hammer for only a short time, because it will cool and deformation resistance will increase to a point where further hammering produces no impression; the normal limit is at about 600° to 700°C (Fig. 1.4). He then must return the piece to the forge and repeat the process again.

The effectiveness of this process depends on the rate at which the piece cools, and this in turn depends on its size and shape. Small, thin pieces cool quickly. Once a thin sheet or rod has been reduced to a diameter of a few millimeters, it becomes too cool almost before it reaches the anvil. One solution to this problem is to have a gang of hammer men who can deliver a large number of blows to the workpiece in a brief span of time; another is cold working. Even today, products made of wire and sheet are usually fabricated at room temperature, with periodic annealing heat treatments.

Fig. 1.5 *Hot Shortness in Steel.* Cracking and fracturing are caused by the presence of sulfur in steel forged at white heat. Oxidation, which occurs quickly at elevated temperatures, makes it impossible to heal the cracks through continued hammering. One end of the longer rod has been flattened by hammering at room temperature, showing that sulfur-containing steel can be tough and ductile when cool.

Fig. 1.6a *A Water Powered Tilt Hammer.* This large tilt hammer with a heavy metal head is actuated by a rotating cam shaft connected to a waterwheel. The hammer arm or helve pivots at one end and is raised and then released by the cams at a point midway between the pivot and the head. The hammer appears to be heavier than the one shown in Figure 1.6b. (Illustration of the late eighteenth century from *Encyclopédie Méthodique* by Denis Diderot, copied from Gillespie v.1, 1959: pl. 97)

Fig. 1.6b *A Water Powered Rocking Hammer.* Here the pivot is between the hammer head and the cam actuation point. The cams on the wheel shaft drive the hammer arm down against a heavy wood block which, by recoil, accelerates the hammer against the workpiece resting on the anvil and thereby increases the energy imparted to it. The heads of such hammers weighed 300 to 400 pounds and the cams produced about 140 strokes per minute. (Nineteenth-century woodcut from Osborn 1869: 754)

Because forges are usually open furnaces fueled with charcoal or coal and powered by a bellows, oxidation occurs. The combustion gases of these furnaces are necessarily oxidizing, so that with each immersion in the forge the surface of the workpiece acquires a layer of iron oxide similar in physical character to the original ore. The oxide layer or "hammer scale" is knocked off each time on the anvil, and the quantity of iron in the piece is thus progressively reduced as forging and reheating proceed. In a large or medium-sized workpiece this loss is tolerable, although it leaves the surface rough and makes control of dimensions difficult. In a small workpiece, however, the loss of metal may be so serious as to preclude hot working.

Very large workpieces were unforgeable in antiquity, not because of oxidation and cooling rates, but because their mass was beyond the physical capacity of even a hammer gang. It was not until the invention of the waterwheel-operated rocking hammer (Fig. 1.6) in the late Middle Ages that single pieces larger than twenty or thirty kilograms could be easily forged. Water powered hammers for processing grain had long been used in several parts of East and Southeast Asia, but only in Europe were such devices adapted for use by smiths. The invention and development of the water powered hammer meant that large pieces of iron could be worked and thus provided an important element in the technological system that preceded the Industrial Revolution.

## 5. FABRICATION BY COLD WORKING

Cold deformation is the only method available for shaping thin sheets and rods of small diameter. It can also be used as a finishing procedure. At ambient temperatures, detailing of smooth or figured surfaces can be done in a leisurely fashion without the time pressure of hot working.

Shaping procedures carried out at room temperature have their own lexicon of terms (Table 1.3) In practice, finishing by cold deformation is usually accompanied by filing, incising, cutting, and other methods that entail removal rather than simple displacement of metal. Excellent results, however, can be obtained by cold deformation alone. As the surfaces of struck coins show, the finish achieved on an irregular surface can be better than that on the finest casting.

## 6. FABRICATION BY CASTING

Melting bloom iron and steel was not generally feasible until the nineteenth century. The chief exception was in India and the Middle East, where high carbon steel was traditionally made by liquefying pieces of iron blooms (Chapter 11). Otherwise, the only irons to be regularly melted during the pre-industrial period were "cast irons." With carbon contents in the 3 to 4 percent range, they could become completely liquid below 1275°C, a temperature which could be reached in shaft furnaces of moderate size. Many of these furnaces, which were used in several parts of the world (Chapter

*TABLE 1.3. Specialized shaping operations usually done at normal temperatures*

| | |
|---|---|
| Upsetting, heading | By compression forces gathering metal rod into a head as for nails, rivets, pins, bolts. |
| Forming | Thinning and flexing sheet to develop curved bowl shapes. |
| Beading | Developing a rolled-over edge to provide stiffness. |
| Embossing | Raising a design on a flat plate or ribbon using a tool replicating the design. |
| Coining | Embossing on both sides of a blank. |
| Dimpling | Bulging small areas of a larger plate or sheet. |
| Blanking | Cutting a flat plate or sheet to a special edge profile or contour. |

3), served the dual functions of smelting the original pig iron from ore and remelting it for casting.

Any metal that melts at a temperature within an achievable range can be cast into near-finished shapes. High carbon cast iron is a good casting metal, with viscosity and surface tension low enough for it to flow readily into complex molds. The surfaces of most cast iron objects tend to be somewhat rough, but pieces have been produced (for instance, in the Ironbridge Gorge Museum at Coalbrookdale in England) which are as smooth as cast bronze.

Most European and Chinese molds for iron casting were made from a clay based material and usually consisted of an assembly of blocks defining a cavity. The mold was filled with metal which was allowed to cool, thus producing a metal replica of the cavity (Fig. 1.7). The lost wax method of making molds was undoubtedly also used in some areas but is rarely reported in the literature.

## 7. TEMPERATURE MEASUREMENT IN EARLY IRONWORKING

Some method of judging temperatures was necessary in several pre-industrial metallurgical processes—heat treatment, annealing and forging, and perhaps melting—but appropriate temperature gauges did not exist until recently. John Percy's encyclopedic *Metallurgy*, which contains the most advanced information on ironmaking in the 1860s, gives no scaled measurements of temperature, although he clearly recognized the importance of estimating and regulating it. Like many ironmasters and blacksmiths before him, he discussed the subject by reference to an unquantified but reliable indicator of temperature, the color of a radiant surface.

Hot materials will radiate light in the visible range of the spectrum. Fortunately for those who work it, hot

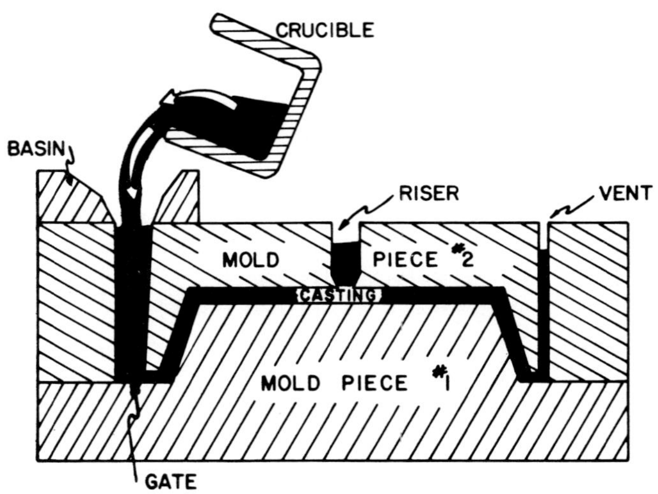


Fig. 1.7 *Mold Assembly for Casting Metal.* The vent allows air pushed ahead by the liquid metal front to escape; otherwise the air might be trapped as a compressed pocket and prevent complete filling of the mold. The riser serves to show the foundrymen that the mold has been filled and also, together with the basin, provides a pool of metal which stays liquid longer and thus can feed extra metal into the space caused by shrinkage as the liquid within the mold solidifies.

*TABLE 1.4. Color-temperature scale*

| | |
|---|---|
| Lowest red visible in the dark | 470°C |
| Dark red, blood-red | 550 |
| Dark cherry | 625 |
| Cherry red | 700 |
| Light red | 850 |
| Orange | 900 |
| Light orange | 950 |
| Yellow | 1000 |
| Light yellow | 1050 |
| White | 1150 up |

iron radiates an unusually broad range of colors at temperatures relevant to processing (Table 1.4). Although the literature on traditional ironworking shows that there were differences in nomenclature and in the number of separate colors distinguished, estimating heat by radiant color was widely used, objective, and reasonably satisfactory as a production guide. The main defect was insensitivity to temperatures above 1150°C, within the actual smelting range, which undoubtedly contributed to the severe problems in controlling blast furnace operations before reliable measuring devices were invented.

A second visual temperature gauge was provided by temper colors. If a clean iron surface is polished to a fairly bright finish, heating causes the formation of an oxide film with thickness proportional to temperature. This film filters reflected light so that it provides a range of colors. The temper colors of steel extend from light yellow to black, representing temperatures between 200° and 370°C. This is the full span of temperatures relevant to tempering, the reheating of quenched steel objects to render them less hard and tougher (Table 1.5). By reheating to a deep straw color, for instance, a smith could have created a tool which combined reasonable hardness with reasonable toughness, able to cut stone, wood, bone, and metal without cracking or dulling too rapidly.

*TABLE 1.5. Colors and temperatures for tempering nonalloy steels*

| Heat to: | At color: | To temper (or "draw the temper from"): |
|---|---|---|
| 200°C | faint straw | milling cutters, gear tooth cutters, tools to cut hard cast iron |
| 230°C | straw yellow | dies, razors, drills for tempered saws |
| 240°C ? | "yellow" | pen knives, screw tops, dies, and taps |
| 250°C | deep straw | taps, cutting edges of drills, chipping chisels, butchers' knives, tools to cut soft cast iron |
| 275°C | bronze | axes |
| 290°C | peacock | stone cutting chisels, springs, axes, tools to cut wrought iron |
| 310°C | full blue | drill stems, saws, tools to cut soft metals, axes |
| 340°C | light blue | (needles, hacksaws, screwdrivers) |
| 370°C | black | |

All data except centigrade equivalents have been compiled from notes chosen by M. T. Richardson for inclusion in his *Practical Blacksmithing* (1889–91, 3: 134–152; 4: 186–209). Most of the notes, written by and for practicing smiths, appeared originally in pages of *Blacksmith and Wheelwright* during the 1880s. They represent a consensus of opinion among ordinary (if literate) American users of iron and steel during the few decades that lay between the discovery of the main outlines of ferrous chemistry and the extinction of the traditional forge. Note that the objects heated to cherry red and then quenched are essentially untempered; they are thus even harder than the objects tempered by reheating to a light straw color.

Other uncalibrated methods of temperature measurement, although rarely recorded, were undoubtedly used by ironworkers. One method was reported in the nineteenth century by Baden-Powell (1872, v.2: 146), who observed that smiths in Sialkot, northwestern India, tempered steel by:

> placing the article to be tempered in a vessel containing oil along with some alloy, the melting point of which answers to the temper required. Springs, if small or if equal in thickness, are blazed off in the usual way; but if otherwise they are roasted in flaming oil in order to render unequal thicknesses alike in temper. The alloy consists of lead, tin and bismuth, melted together in various proportions to suit the tempers required in cutlery.

A related method involved placing a metal object with a specific melting temperature in a furnace and then observing it through a hole while the heat was gradually raised. If the metal object was of copper, for instance, it would liquefy at a temperature of 1083°C. If it was of silver, it would melt at 960°C. A furnace operator would therefore have found it possible in principle to hold the peak temperature within the 120°C range bracketed by the melting points of the two metals. In reality, however, the technique of temperature measurement using indicator metals was hampered by a lack of standardization. For true accuracy, measurements would have to have been based either on perfectly pure metals or on samples from the same batches. Moreover, as most potential indicator metals are chemically reactive and thus likely to change their composition while in the furnace, they needed to be protected from the furnace atmosphere.

In the eighteenth century, Josiah Wedgwood, the noted English potter, conceived the idea of measuring temperature not with metals but with a series of ceramic blocks with differing compositions. By inserting these blocks into a kiln and withdrawing them at intervals as it grew hotter, he was able to derive a temperature scale based on their measured shrinkage (Wedgwood 1782; Rostoker and Rostoker 1989). In the late eighteenth and early nineteenth centuries, the Wedgwood system was widely used by European metallurgists like Pearson (1795) in his pioneering studies of Indian wootz and Osborn (1869: 876–877) in his recommended method for controlling the carburization of blister steel.

A further improvement on this method, and one which has survived to modern times, is the Seger cone system. In 1886, Hermann Seger at the Royal Porcelain Factory in Berlin devised a set of carefully formulated ceramics, each of which was sensitive to a specific temperature within the range important for kiln firing. These ceramics were made into long tetrahedral shapes or cones which slumped when heated beyond the specified temperature. A series of such cones placed in a kiln made it possible to control peak temperatures with fair accuracy. The fact that these temperatures could not be measured on an absolute scale was of little importance. As in the case of color "temperatures" in metals, what mattered was that the cones gave reproducible results which could be empirically related to the quality of a product.

## 8. IRON VERSUS BRONZE

In the past, archaeologists believed that iron was technically superior to bronze. They were especially taken with the military consequences of the introduction of iron, postulating technologically superior armies (of Dorians, Celts, Qin Chinese, and unspecified Indo-Europeans) wielding iron weapons to sweep away the hapless armies of Bronze Age peoples. Now it is known that the changeover from bronze to iron proceeded at a very moderate pace in most parts of the world, and that iron weapons were often not the first iron implements to be adopted.

The best bronzes were alloys of either copper and arsenic or copper and tin. The latter was a more versatile material, containing tin in amounts ranging from 3 to 10 percent for forgeable metal and as high as 25 percent for castings. The earliest iron was usually an almost pure metal, with a carbon content that was generally less than 0.1 percent. Larger amounts of carbon were sometimes present because of uncontrollable variations in the smelting process. The bulk of the early iron was therefore very soft. If the ores used were high in phosphorus, the iron produced could have been harder—perhaps as hard as a medium carbon steel before heat treatment. However, a low carbon iron high in phosphorus would have been too brittle for use except in very warm climates (Chapter 2).

Bronze, although also soft when in an as-cast or annealed condition, becomes substantially stronger and harder when subjected to cold working—to hammering or deforming while at room temperature. In contrast, low carbon iron will harden only slightly by deformation. Iron with 0.05 percent carbon is stronger than a low tin bronze when both are in a fully annealed state, but the iron is quickly surpassed in strength by the bronze when both are reduced in thickness through rolling or hammering (Fig. 1.2). Even a mild steel, which when annealed is stronger than even a high tin bronze and which has a great initial gain in strength when lightly cold worked, is not as strong as bronze when both are deformed to the same extent.

The fact that bronze can be greatly hardened by working is sometimes overlooked. Thus, for example, one specialist on the raw materials of antiquity writes, "Iron is so much harder than copper and bronze that it replaced those metals for most types of tool, in spite of the very laborious forging and tempering processes" (Rosenfeld 1965: 68). The statement is true only for metals that left the workshop in a fully annealed condition. If they had been even moderately deformed without subsequent heating, the bronze would probably have been harder and stronger than the iron.

Work hardenability explains why a bronze sword might have been preferred to an iron one. Unless a bronze sword were annealed after shaping, it would have been superior in terms of strength and hardness because of the hammering needed to finish a tapered edge. It might also have been tougher than an iron sword, since the forgeable bronzes as a class are very tough, and therefore more resistant to chipping and notching. Swords of low carbon iron and tempered steel can also be tough, but if occluded slag is too close to the edge, breakage will occur, because slag is a glass. Furthermore, swords might contain patches of high carbon iron, which can be very brittle, and those with too much phosphorus might crumble under impact on a cold day.

Both bronze and iron have advantages with respect to fabrication. A bronze sword can be cast into semifinished form; an iron sword cannot. On the other hand, a bronze with a tin content appropriate for weapons is much more difficult to hot forge than either iron or steel. Medium and high tin bronzes cannot be forge welded. In addition, since they resist deformation until heated to a temperature at which they crack, they must be cold forged with considerable effort and with constant interruptions for annealing. Even though bloom iron could not be cast to shape, it may have sometimes been preferable to bronze because it is easily forge welded and relatively plastic at high temperatures, and that containing low sulfur does not crack when white hot. Iron would have been much easier to work when minor adjustments and repairs had to be made in the field. Repairing a broken bronze helmet meant either rivetting or melting more bronze in a crucible in order to cast on a patch. An iron helmet can be repaired simply by heating the broken edges and forge welding them together with a hammer.

However, it is doubtful that ease of repair explains the historical trend toward iron. The real advantage of this metal in early periods was that it was more inexpensive than copper or bronze, whether the cost was measured in money or labor. Workable iron deposits exist almost everywhere, while copper and especially tin are not only scarce but distributed unevenly over the surface of the earth. Iron, therefore, has a great advantage in terms of transportation costs and predictability of supply. Iron ores are often found on or close to the surface, thus minimizing mining costs. They are easier to smelt and require less fuel than the sulfide ores that constitute the great bulk of the world's copper reserves. As a result, in those pre-industrial economies for which records of prices exist, a kilogram of iron costs between one-fifth and one-twentieth as much as a kilogram of bronze. This discrepancy explains why bronze, although superior to early irons in terms of mechanical properties, was eventually used more for luxury and special purpose items.

# 2. The Varieties of Iron

## 1. BLOOM AND WROUGHT IRON

By far the most important type of iron before the nineteenth century was a soft, ductile metal with a low carbon content. This metal is called bloom iron when produced directly from the smelting furnace and wrought iron when produced indirectly by removing carbon from cast iron. The difference in manufacturing methods should not obscure the essential identity of the products. They usually cannot be distinguished either by their behavior on the anvil or by metallographic examination. Many writers call both wrought iron, but here we use a terminology that reflects production methods, calling a metal *bloom* iron when it is made by the direct or bloomery process, reserving the term *wrought* for iron produced by a multi-stage indirect process.

The very low carbon content of bloom and wrought irons means that they do not melt below about 1500°C. Although temperatures very close to this were possible in traditional ceramic kilns in the Far East (Leach 1940: 178–192), low carbon irons could not be liquefied in metallurgical furnaces before the late nineteenth century. It is probable that the crucible hearths used to make wootz steel in India sometimes reached 1400°C, for melting was essential to the process and the product is known (Zschokke 1924: 656) to have sometimes contained as little as 1.5 percent carbon. However, there is no evidence that either low carbon iron or steel was ever shaped by casting during the pre-industrial period; only machining and forging could have been used.

Wrought and bloom irons were too soft and easily bent for high performance cutting tools or weaponry, but both were well suited to the manufacture of such items as body armor, nails, chains, wheel rims, tongs, farm tools, harness hardware, and ship fittings. Any metal tool that did not require a sturdy point or cutting edge could have been made of low carbon iron, especially if it needed reasonable tensile strength, toughness under impact, and enough ductility to minimize the cracking caused by repeated bending and straightening (e.g., Table 2.1). It is a common misconception that wrought and bloom irons are more rust resistant than steel or cast iron,[1] but the slight differences in the corrosion rates of wrought iron and carbon steel (Table 2.2) are not significant. Before the invention of chromium- and nickel-containing stainless steels in the late nineteenth century, all forms of iron were vulnerable to the attack of oxygen from the atmosphere and other sources. Only two types of rustproofing were available: the application of protective oils, waxes, and resins, and bluing, the controlled oxidation of polished surfaces either by heating to about 250°C or by treatment with a chemical solution. Neither method was permanent.

Bloom and wrought irons can be worked both hot and cold, and the same impurities that cause hot and cold shortness in other kinds of iron—sulfur and phosphorus—also affect them. Pre-industrial bloom iron was usually freer of both impurities than wrought iron, because charcoal was used instead of coal in most bloomery furnaces and these furnaces were operated at lower temperatures. The cast iron from a blast furnace was likely to contain substantial amounts of both elements and to pass these on to the wrought iron made from it, unless it was smelted from a low phosphorus ore using charcoal or low sulfur coal.

The presence of occluded slag, retained in significant amounts (sometimes as much as several percent by volume) from the smelting of bloom iron and the puddling or fining of wrought iron, may have been advantageous in hot forging. Slag inhibits oxidation at high temperatures and, when molten, serves as a valuable flux cover that facilitates hammer or forge welding, but it hinders coldworking. Its brittle, glass-like character means that it shatters and eventually powders, leaving cracks and voids in the metal that can be closed only by heating and forge welding.

Nearly carbon-free iron is very tough and ductile at normal temperatures; it is the most cold formable member of the iron family. Moreover, the fact that it strain hardens only moderately means that it remains soft longer under the hammer, requiring only occasional annealing.

Even though wrought and bloom irons could be worked into jewelry and other small objects, they corrode quickly and in an unsightly fashion when neglected. However, during the early Iron Age in the eastern Mediterranean, iron was often fabricated into luxury furnishings and jewelry such as fibulae (Snodgrass 1980) and bracelets (Pigott et al. 1982), presumably because it was an expensive novelty. In Europe and the Middle East, iron became less fashionable for personal adornment as the metal became more common, but it continued to be appreciated elsewhere, as in East Africa (Fig. 2.1) and Japan (Fig. 2.2). The black

---

[1] Rollason (1973: 154) indicates that Staffordshire (but not Swedish) wrought iron "resists atmospheric corrosion about 20 percent better than mild steel." Considering the normal variability of corrosion measurements on commercially produced metals, a 20 percent difference is too small to be significant. In their definitive study of the corrosion resistance of wrought iron, Chilton and Evans (1955) conclude that known variations in piled macrostructure, in sulfide content, and in the degree of segregation of minor components are sufficient to render inappropriate any generalization about corrosion in wrought irons as a class.

*TABLE 2.1. Compilation of iron and steel products identified with 9th–13th century A.D. Russia**

| | | | | | |
|---|---|---|---|---|---|
| *Farming Related Tools* | | | | | |
| colter | sickle | tree-climbing hook | fish hook | horse harness | |
| plow tip | scythe | harpoon | crow bar | horseshoe | |
| spade | shears | boat hook | axe | | |
| hoe | knives | fish spear | cattle bell | | |
| *Craftsmen Tools* | | | | | |
| sledge hammer | punches | files | saw blade | knives | |
| peen hammer | chaser | wire draw die | wedges | awl | |
| tongs | pincer | soldering iron | wood cutters | pick | |
| chisels | metal shears | adze | plane | anvils | |
| *Hunting and Fighting Weapons* | | | | | |
| spear | chain mail | crossbow | | | |
| saber | arrow tip | shield face | | | |
| lance | axe | shield clamp | | | |
| helmet | mace | swords | | | |
| *Household Implements* | | | | | |
| scissors | needles | hinges | chain | latches | hasp |
| frying pans | nails | handles | bucket | padlock | springs |
| balance | rivets | kettle | hoops | key | pots |

*Kolchin 1967: 29–30

finish of the metal may, however, have been considered as desirable as its strength and workability in those societies in which it was frequently used for furnishings and architectural decorations such as lamps, gates, grills, hinges, and locks (e.g., Fig. 2.3).

A wide range of surface hues can be induced on polished iron surfaces by heating them until the temper colors (Table 1.5) begin to appear; these are retained after cooling although they are susceptible to oxidation and abrasion. Other adherent surface colors can be formed by chemical treatment: by bluing through oxidation in low melting nitrate salts, as on gun barrels; by browning with nitric acid and other ingredients, again as on gun barrels; and by applying various sulfides and sulfates to produce dark grays and yellows, as on Asian knives and swords.[2] The blacking of old-fashioned iron stoves is achieved by heating an oiled surface or by quenching a heated object in oil—in both cases the oil decomposes to produce a dark glossy finish.

Iron has also been an important medium for engraving, etching, and inlay. The great armorers of Renaissance Europe, like their predecessors and contemporaries in the Islamic world, did much of their best work on plates of bloom or wrought iron, and—less often—of low carbon steel (Williams 1977; 1978). Soft irons and steels have good forming properties and can hold finely incised detail (e.g., Fig. 2.4). A related development was the production of printing plates in Europe, believed to have originated in the inked paper squeezes made by armorers to check or record engraved designs. Not only were the first known etchings prepared from iron plates, but steel "again became the most common material for engraving in the nineteenth century, when it displaced copper, partly because of the finer detail that could be cut into it, but principally because of its superior wearing qualities" (Smith 1960: 11).

## 2. STEELS

### 2a. *General Character*

Modern metallurgists define steel as iron having a carbon content between about 0.05 and 2.0 percent, usually between 0.1 and 1.2 percent. Earlier, however, steel was defined in a variety of ways. Laymen in most places probably used the term to refer to any valued iron alloy, while metalworking specialists focused on technical traits.

Many peoples have long recognized that a distinction between iron and steel exists, although they undoubtedly knew the two were related. Most European languages have a separate word for steel, and so do Arabic, Persian, Chinese, Japanese, and the majority of the tongues of India and Southeast Asia. The terms for steel connoted desirability, indicating that some of it was a specialty commodity which few producers could make reliably and which therefore sold for a higher price than ordinary iron (Chapter 16). Another consequence was that the making and working of steel were shrouded in superstition.

Iron may often have been called steel to enhance its value. On the other hand, many steels appear to have

*TABLE 2.2. Relative corrosion rates* of wrought iron and plain carbon steel***

| Medium | Duration (yr) | Wrought iron | Plain carbon steel |
|---|---|---|---|
| Sea Water | 1 | 10 | 13 |
| | 3 | 7 | 11 |
| | 5 | 5 | 5 |
| Brackish Water | 1 | 13 | 13 |
| | 3 | 12 | 13 |
| | 5 | 7 | 8 |
| Fresh Water | 1 | 8 | 7 |
| | 3 | 5 | 5 |
| | 5 | 4 | 4 |

*Corrosion rate in mils (or thousandths of an inch) per year.
**Data taken from *Metals Handbook*, vol. 1. 1964: 263, fig. 8.

[2] A summary of such procedures as practiced within recent years is given in *Metals Handbook* (1948: 730–732). More exotic treatments from other sources include: 1) Holland's (1831, v.2: 107) recipe for browning gun barrels, using a mixture of nitric acid, sweet spirit of niter, spirit of wine, blue vitriol, and "tincture of steel"; 2) the classic technique for yellowing and etching damascened swords through the use of a ferric sulfate solution (Allan 1979: 81–82); and 3) the traditional method of darkening and etching the blades of *keris* daggers in Indonesia with a mixture of arsenic sulfide and lime juice (Solyom and Solyom 1978; Groneman 1905).

Fig. 2.1. *Masai Bloom Iron Jewelry*. Tanzania, East Africa, nineteenth century. Made from locally smelted iron, apparently drawn through a die. (From Stuhlmann 1910: 74)

Fig. 2.2. *Japanese Iron Sword Guards*. Whereas *tsuba* of nonferrous metal were usually made by other craftsmen, swordsmiths often made iron *tsuba* themselves. A wide variety of ironworking techniques is represented. The diameter of b (top center) is 8 cm. All are in the Gunsaulus Collection at the Field Museum. Top row, left to right: Quatrefoil with overall cherry blossom decoration. Choshu or Bushu school, unsigned, nineteenth century. The design is chiselled out in medium relief. FM 235499; Chrysanthemum. Ito style, unsigned, eighteenth-nineteenth century. Although similar *tsuba* are said to have been made by piercing and sawing with an abrasive-loaded wire, in this piece the fins are joined to the hub and rim by brazing, apparently with a low melting iron (-arsenic?) alloy. FM 130940; Lobed rosette in form of helmet. Myochin school, signed "Munenori" and dated 1861. Formed from 26 trapezoidal plates, each forged to a shallow S-curve and fastened to the next with three iron rivets, a technique favored by Japanese helmet makers. FM 131258. Second row, left to right: Weathered wood or *mokumé* pattern. Myochin school, unsigned, probably eighteenth century. Made from a stack of thin, distorted steel plates that were forge welded together, ground flat on a bias, and then etched to show the wavy laminations. FM 130860; Skull in grass on rugged surface. Kaneiye school and signed "Kaneiye," possibly late sixteenth century. The texture resembles partly melted wax. Although this suggests lost wax casting followed by malleabilization, the same effect could be achieved in wrought iron with hammers and gravers. FM 130783; Eggplant on vine. Satsuma style, unsigned, early eighteenth century. Like e, this gives a strong impression of having been cast to shape, but close examination shows thin slag ribbons in several places. It is therefore forged of wrought or bloom iron. FM 131100

been treated like ordinary iron and presumably were regarded as such. Failure to recognize steel must have been common where all iron was made in bloomeries, which, unlike fineries, often yielded a heterogeneous product with steely patches resulting from the pickup of carbon from charcoal or liquid cast iron in the hearth (Chapter 9). Examples of steels that were used like any bloom iron often appear in the ethnographic and archaeological literature (see especially Piaskowski 1983). The properties of some blooms of this kind were, however, appreciated. As in the case of the famous steel of Noricum during Roman times, an unevenly carburized bloom could be converted to a metal with a homogeneous and significant carbon content through repeated forging.

Various methods, all more or less laborious, were used during the pre-industrial period for making steel deliberately: running a bloomery to produce a medium carbon metal directly without intermediate steps, augmenting the carbon content of a bloom or wrought iron, or removing some of the carbon from a cast iron. In many cases it is not obvious why such methods were even tried. They were costly, time-consuming, and so technically difficult that they often failed. The fabrication of steel objects was even more susceptible to failure. Few ancient smiths could effectively exploit their raw material, but the demand for steel in ancient societies was often great. Thermal and mechanical treatments can give it a broad range of properties, which are caused by iron carbide (cementite) crystals that can be induced to alter their shape, size, and spatial relationship with associated crystals of carbon-free iron (ferrite). The higher the carbon content, the greater the proportion of carbide crystals and the more the resulting aggregate differs from pure iron in its behavior. The most important differences are that like bronze, steel can be strengthened by cold working, and that unlike any other pre-industrial metal, it can also be strengthened by heat treatment.

## 2b. *The Cold Working of Steel*

A difference between bloom and wrought iron, and steel is that the latter becomes much harder and stronger when cold worked. At high temperatures (900°C and above) steel deforms almost as easily as pure iron. At lower temperatures, however, steel is difficult to shape with a hammer or a drawing die, needing frequent annealing if it is not to be too hard and brittle to work. Great strength accompanies hardness. A cold drawn steel wire containing about 0.8 percent carbon can reach yield strengths in excess of 2069 megapascals, or 300,000 pounds per square inch (Table 1.1). Few other materials, including modern alloys, can surpass this.

The strain hardenability of iron must have been

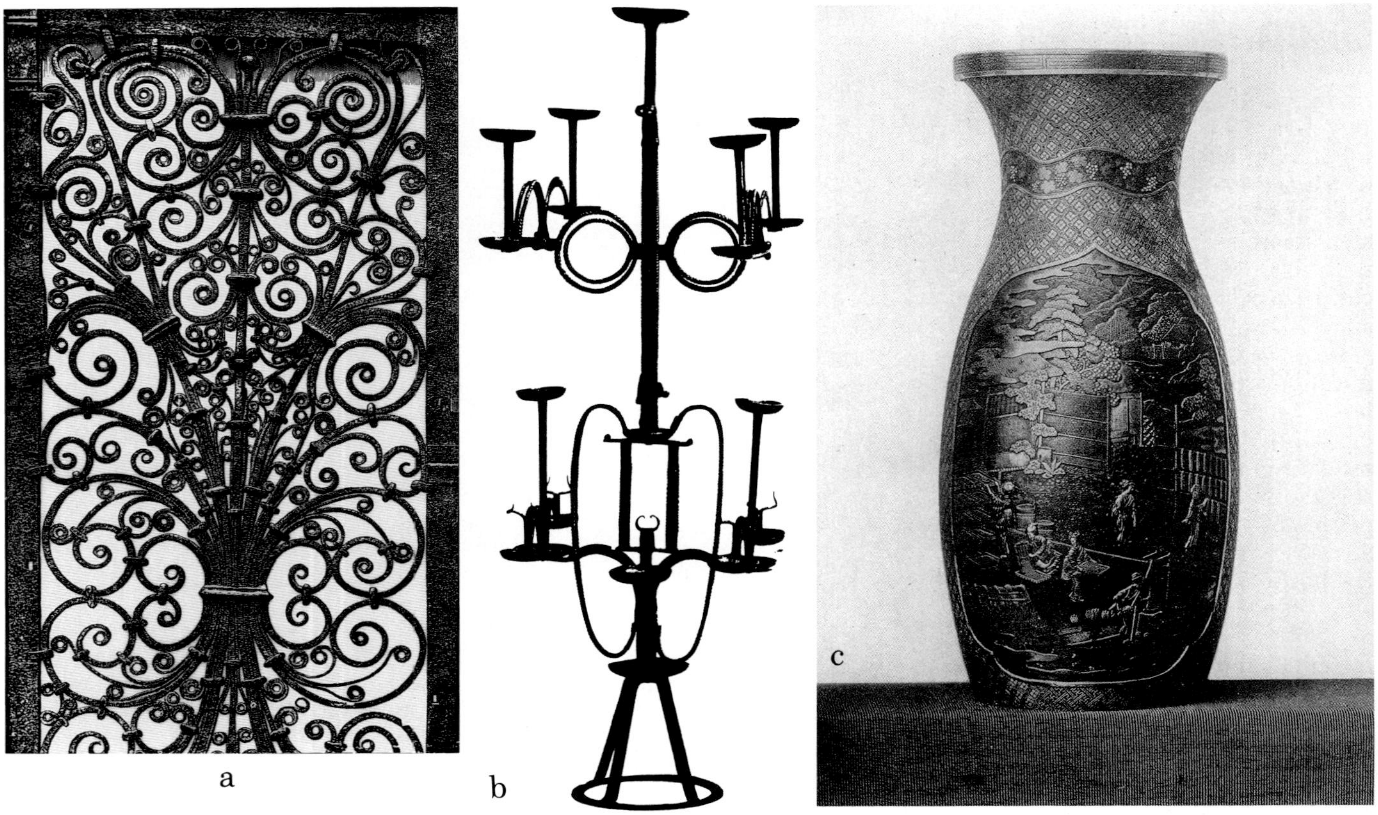


Fig. 2.3. *Furnishings of Soft Iron.* Blacksmiths' methods are used in many regions to produce decorative furnishings. a. Iron gate or panel. From an abbey in Picardy, northern France, dated to the early thirteenth century. Forged from bloom iron. (From Gardner 1914); b. Wrought iron lamp stand with cattle effigies. Betsileo tribe, Malagasy Republic (Madagascar). Nineteenth century. Forged from bloom iron, 120 cm. high, and said to be former royal property. FM 187258; c. Cast iron vase with decarburized surface. Japan, nineteenth century. Japanese craftsmen were skilled at decarburizing portions of white iron castings to facilitate engraving and inlaying. (From Rein 1889: fig. XVII)

recognized at an early date, for cold working had long been used to strengthen bronze. A cold worked steel with a carbon content in the 0.1 to 0.5 percent range would have been moderately tough and at least as hard and wear resistant as bronze, and far superior in these respects to bloom or wrought iron, but a very high carbon steel would have been too brittle.

### 2c. *The Heat Treatment of Steel*

Heat treating steel may be even more important than cold working. Modern heat treatment has three essential steps: heating, quenching, and tempering.

In the first, a steel object is heated above a temperature which ranges from 723° to 850°C, depending on carbon content. Without measuring instruments, a skilled blacksmith can judge temperature by the color of the radiant emission of the hot steel (Table 1.5). If the threshold color (somewhere between cherry red and light orange) is not reached, the piece will lack the full hardness conferred by quenching.

Quenching is accomplished by sudden immersion in a liquid that can transfer heat efficiently—usually water or oil.[3] The speed and completeness of heat transfer are

Fig. 2.4. *Engraved and Embossed Steel Helmet.* Burgundy, eastern France, sixteenth century. The period was a high point in the history of European decorative armor. (From Beck 1893–95, v.5: 360)

[3]Considerable mystery has always been associated with quenching fluids. Sixteenth- and seventeenth-century manuscripts (Smith 1968: 11–12, 56–61) advise the use of blood, urine, "dew from the month of May," and other more exotic liquids. However, clean, cold water serves best for most purposes, although the addition of common salt can yield a small improvement in quench severity. Oil is often used in quenching modern alloy steels. Cold mercury, which conducts heat better than water, has been used to provide a quench that is exceptionally severe.

critical. To achieve effective hardening, the entire piece must lose as much as 400° to 600°C within about 2 seconds. This means that the quenching medium should be in motion (a powerful jet of water works well) and that the piece being quenched has to be agitated rapidly. Even then, sufficiently high cooling rates occur only near the surface of the steel. Farther in the rate of heat transfer is limited by the thermal diffusivity of the steel itself, rendering the effectiveness of the quenching fluid irrelevant. Thus, a piece of ordinary carbon steel can be hardened throughout only if its cross section is small—7 to 10 mm. or less.

In the as-quenched condition, the metal is at maximum hardness. However, unless its carbon content is less than about 0.15 percent, it is also so brittle that it is unusable. Therefore, quenched steel is tempered by reheating to a temperature well below the previous threshold, this time using the oxide-film temper colors (Table 1.5) as a guide. After the desired point is reached and the object has cooled again, it will have lost some of the hardness and brittleness acquired at the quenching stage. This loss and a corresponding gain in toughness are proportional to the temperature reached. Heating to only a faint straw color yields a steel which is still extremely hard but, if used carefully, tough enough to cut other metals. Heating to a light blue temperature would make the same steel much softer but also much tougher, and thus suitable for axles and other machinery parts in which brittleness is undesirable.

A great range of properties can be obtained by controlling the trade-off between hardness with strength on the one hand and toughness with ductility on the other (Tables 1.5, 2.3). Of particular importance are tempering levels that make a steel both ductile and strong; in contrast, cold worked bronzes and steels lose ductility as they gain strength. Another advantage of heat treatment is that objects of any shape can be hardened, while cold working is only practical for thin pieces like sheets and rods. A third advantage is its reversibility. A tool can be heat treated to a soft and ductile state, shaped while in that condition, and then heat treated again for hardness and strength. If the results are unsatisfactory, the process can be repeated a second or third time.

The ability of steel to respond to quenching and tempering is crucial in modern manufacturing, but the archaeological evidence proves that full heat treatment was uncommon in antiquity. Quenching was early and widespread; numerous steel artifacts show signs of having been cooled from a high temperature after forging.[4] Some may have been thrown into water to make them safe to handle, but there can be no doubt that quenching of others was intended to enhance the properties of the steel. Sixty-nine percent of a sample of iron and steel tools from late Iron Age Celtic sites in Czechoslovakia which contained enough carbon to respond to heat treatment were quenched (Pleiner 1982). Some

*TABLE 2.3. The influence of tempering temperature on the strength, ductility, and fracture energy of a 0.5% carbon steel in the form of a 0.5 inch diameter round bar**

| Tempering temperature °C | Strengths, psi × 1000 | | Ductility, % reduction in area | Elongation | Toughness, ft-lbs |
|---|---|---|---|---|---|
| | yield | tensile | | | |
| 260 | 115 | 160 | 30 | 10 | 6 |
| 370 | 110 | 150 | 40 | 15 | 30 |
| 540 | 99 | 134 | 54 | 20 | 42 |
| 590 | 88 | 119 | 60 | 22 | 51 |
| 650 | 86 | 110 | 61 | 25 | 66 |

*Water quenched from 830°C

knife and sword blades of that period not only had steel edges welded to iron bodies, but were placed in the quenching bath so that only the edges and tips were hardened (Pleiner 1980: 405). European smiths of the last few centuries B.C. evidently understood the connection between steels and their capacity to harden.

Tempering, however, was rare in antiquity. Coghlan (1977: 95) knows of only four examples of tempered steel objects from the entire Roman period in Europe, and there is no evidence that tempering was regularly practiced in any part of the world until the second millennium A.D. Sixteenth-century writers like Biringuccio (Smith and Gnudi 1942: 308) make it clear that the technique was widely used by then in western Europe, and Kolchin (1967: 60–61) implies that tempering was not uncommon in Russia in the eleventh and twelfth centuries. Yet steel workers in most other regions seem to have done relatively little tempering until modern times. For example, even now Japanese swordsmiths do not usually temper the meticulously quenched edges of their blades.

The brittleness induced by severe quenching could have been counteracted by means other than tempering. One was to use a steel, however adventitiously, with a low carbon content, in the 0.1 to 0.2 percent range. This amount of carbon hardened the metal somewhat without making it excessively brittle in the as-quenched state. A second was to quench only the working edge of a steel tool, protecting the rest with a coating of clay or other insulating material. A third was to mitigate the severity of quenching, either by using a liquid less heat conductive than water or by interrupting the quenching process.

Interrupted quenching is well attested in pre-industrial settings (e.g., Smith and Gnudi 1942: 371; Wertime 1962: 18), although it is rarely used now in the processing of non-alloy steels. As its name implies, the technique involves withdrawing the hot steel from the quenching bath as soon as it has passed the point where the hardening transformation occurs (at ca. 400°C for non-alloy steels), leaving it to finish cooling at a slower rate in room temperature air. The "self-tempered" structure that results is called bainite, which is significantly tougher and less hard than the martensite produced by a full quench. A different type of self-tempering occurs when only part of the implement is quenched, so that the residual high temperature of the rest can, if the smith judges properly, raise the martensite of the quenched portion back to a tempering heat.

[4]As examples see: for Jordan, Smith et al. (1984); for Celtic Europe, Pleiner (1982); for western Russia, Gurin (1980); for Syria, Wheeler et al. (1976); for Renaissance Europe, Smith (1968); for China, Li (1975); and for Cyprus, Tholander (1971).

*TABLE 2.4. Physical and mechanical properties of cast irons found in artifacts*

| Cast iron type | Microstructure | Yield strength, psi | Tensile strength, psi |
|---|---|---|---|
| White | interconnecting networks of iron carbide with enclosed zones of steel crystals | none | 20,000 |
| Gray | matrix of steel crystals with a distributed population of graphite platelets or flakes | none | 25,000–40,000 |
| Malleable | matrix of steel crystals with a distributed population of graphite in rough nodular form. | 32,000–40,000 | 50,000–60,000 |

Because temper colors cannot be used as a guide,[5] it is difficult to gauge precisely when the hardening transformation occurs so that the object can be removed from the bath almost immediately. A full quench to maximum hardness can be completed in much less than a minute, so an interrupted quench is much harder to control than ordinary quenching and tempering.

Tempering must have been practiced by some ancient ironworkers unless the early tempered pieces—like the four Roman objects mentioned above—are spurious or accidental. Perhaps tempering was known only to a minority of smiths. The idea could not have spread easily by stimulus diffusion, since even modern blacksmiths find that detailed written instructions or actual demonstrations are needed before control by temper colors can be mastered.

Tempering is a difficult process to master and problems of control are inherent in the interrupted quenching technique, but ancient iron workers did not have to depend entirely on heat treatment to take advantage of the potential of medium carbon steels. These steels respond well to cold working, reaching levels of hardness and tensile strength which are close to those of the same metals when quenched and tempered. It is possible that swordsmiths in some early cultures preferred cold working to heat treatment. The most important qualities were sharpness and resistance to wear, which could have been obtained in a medium carbon content steel with a high degree of strain hardening at the edge.

## 3. CAST IRON

### 3a. *General Character*

There are several different types of cast iron with slightly different properties (Table 2.4), but they all have a similar technological background. The melting temperature of pure iron is 1550°C. With the progressive addition of carbon, this falls to a minimum of 1143°C at the 4.3 percent carbon level. Cast irons therefore liquefy at a temperature only slightly higher than the melting point of pure copper, 1083°C. The fluidity of liquid iron of high carbon content is comparable to that of bronze. In the hands of a skilled foundryman, castings made with it can be almost as thin and detailed as those in bronze. The Chinese bells and statues cast in white iron during the Tang through Qin periods often have exceptionally sharp detailing (e.g., Fig. 2.5).

Castable iron is different from wrought iron and steel, principally because high carbon irons usually are

Fig. 2.5. *Cast Iron Guardian Figure.* From an arsenal building near Zhengchou, Henan province, northern central China. Ming period (fifteenth-seventeenth century). White cast iron, about 2 m. high. The detail photograph shows a raised inscription on the same statue, illustrating the ability of cast iron to replicate fine mold details. The raised vertical and horizontal lines represent gaps between pieces of the original mold into which liquid metal has penetrated. These were too hard to be filed off.

[5]Modern jewelers and other users of small steel punches and engraving tools do use temper colors in an interrupted quenching process sometimes called "point hardening." In Untracht's (1975: 447) words, "The tool is heated to cherry red, and the point is dipped into the quenching liquid. While being held with tongs, the point is polished with a piece of emery cloth held on a piece of wood. As the remainder of the tool has retained its heat, the polished portion is watched until the desired tempering color has been reached. The whole tool is then quenched immediately." It will be observed that this process combines hardening and tempering in one operation. There is no evidence that the key step, polishing the piece immediately to make the temper colors visible, was understood in ancient times.

not hot or cold forgeable. Some can be straightened by bending, but most are too brittle. Despite their excellent compression strength, they are much less tough and, with only a few exceptions, lower in bending and tensile strength than other types of iron. On the other hand, cast iron became the cheapest form of that metal once the technology of smelting in blast furnaces had been developed.

Most ancient peoples seem not to have known how to handle cast iron. Before the Industrial Revolution, it was made and used regularly in only three parts of the world: western Europe, the Middle East, and the Far East, with chronological primacy in the last. Well-cast iron implements in China are as early as the middle of the first millennium B.C.

### 3b. *White Cast Iron*

White cast iron or "white iron" is characterized by extreme hardness and a bright, shiny fracture that seems whiter than that of most other metals. Both the color and hardness result from the presence of abundant carbon in the form of iron carbide. Metallurgical theory and archaeological evidence suggest that white iron was the earliest form of cast iron. Most or all of the oldest analyzed specimens from China, the cradle of cast iron technology, are white. This is to be expected, since the white form of cast iron would be the usual result of smelting at the lowest temperatures at which liquid iron can exist and the appropriate type of slag can form.

White cast iron not only is the weakest of the irons in terms of tensile strength and shock resistance, but also its great hardness, which increases with rising carbon content (Fig. 2.6), makes the metal nearly impossible to file or chisel. Even today white iron is regarded as unmachineable.

The problems caused by extreme hardness can be appreciated by considering an iron casting operation of the traditional Chinese type. A typical casting mold (Fig. 1.7) can produce a complex semi-finished shape with a relatively smooth surface, but it will necessarily have unwanted parts: lumps and rods attached to the surface which represent the partially filled gating, feeding, and venting apertures of the original mold. If they are white cast iron, the only way they can be removed is by breaking, since they are too hard to be cut with a chisel or file. Chinese ironworkers, who clearly could not do this with very massive castings, cut the redundant parts off the body of the casting before or just after the temperature had dropped below the complete solidification point (Hommel 1937: 30). However, this procedure was never adopted by Western founders, who have avoided making very large objects from white cast iron.

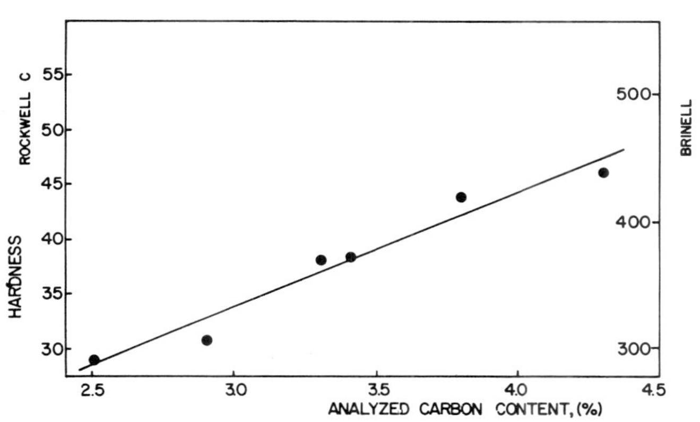


Fig. 2.6. *Hardness and Carbon Content in White Cast Iron.* The test samples on which this graph is based were cooled at a rate consistent with that of a large casting allowed to become cold while still in the mold.

The chief advantage of white iron is its great resistance to abrasion. For this reason it was used until recently in plows and is still used, with alloying modifications, in rock crushing and grinding equipment. In earlier times, white iron may have served a wider range of functions, for in reality it is not so brittle and intractable that it cannot be formed into serviceable cutting implements. Tests of experimental tools made of white iron show that it is a superior metal for graving tools and arrow points and that it is adequate even for axeheads and heavy chopping blades (Rostoker 1987).

Ancient Chinese foundrymen seem to have been more aware than their later Western counterparts of the utility of unmodified white iron, since they made many early cutting and digging tools from it. From a modern standpoint, however, the most unexpected application of white iron by the Chinese was in making bells. Gray cast iron does have an exceptional capacity for damping vibrations and is now much used in beds for heavy machinery, but the white variety of cast iron actually has a low damping capacity. Like steel and high tin bronze but unlike most other cast irons, white iron when struck will ring with a clear and sustained tone. Bells made from it perform almost as well as bells of bronze (Rostoker et al. 1984).

### 3c. *Gray Cast Iron*

Gray cast iron is named for the dull, nonreflective appearance of its fracture surface. It is softer and less brittle than white iron. The gray color is due to the presence in its microstructure of petal-shaped flakes of black graphite. The graphite forms during solidification, aggregating into a pattern of dispersed flaws within an otherwise steel-like structure. The fact that this structure is continuous while the graphite flakes are discontinuous, which limits the growth of cracks, accounts for the relative toughness of gray iron. In white cast iron a small crack encounters no obstacles as it grows along the linked boundaries of brittle iron carbide crystals. Accordingly, gray cast iron has greater tensile strength than white cast iron, and can even exceed that of wrought iron or low carbon steel. By lowering the carbon content, and hence the volume of graphite, the strength of gray iron is increased.

Gray cast iron is relatively soft, and can be filed and cut with hardened steel tools. In modern manufacturing it is regarded as an ideally machineable metal, partly because the graphite in it serves as a lubricant during cutting. Its self-lubricating qualities also make it a good material for bearings. Like steel, its properties can be modified by quenching and tempering.

### 3d. *Malleablized and Mottled Cast Iron*

Malleablized cast iron, or in modern terminology *malleable iron*, was known in China as early as the Han dynasty (ca. 200 B.C. to A.D. 200). It was developed independently in England in the seventeenth century (Tylecote 1976: 93), in the eighteenth century in France (by Réaumur—Sisco and Smith 1956: 320–339), and in the United States in the early nineteenth century (Davis 1899). It is formed by heat treating white cast iron for 24 to 72 hours at 900° to 1000°C, which transforms the hard, brittle carbide into graphite. Because the graphite accumulates within the metal as spheroidal nodules separated by a steel-like matrix, the malleablized metal has special properties. Its composite structure is stronger than gray iron and is comparatively deformable, soft, and tough. Iron that had been both malleablized and decarburized was used extensively by the ancient Chinese when steel might otherwise have been necessary.[6] Ordinary malleablized iron can be hardened in much the same way as steel, with a quench-and-temper heat treatment.

There are two types of malleable iron, whiteheart and blackheart. The former was used by the ancient Chinese and recently by the English and French; the latter is mainly American. The chief difference between the two is that whiteheart malleable has a decarburized structure like pure iron on its surface but a graphitized interior, whereas the blackheart variety is graphitized throughout. Although both types were in commercial use during the nineteenth century, the blackheart product ultimately dominated the malleable iron industry.

The terms *mottled cast iron* and *mottled iron* denote a variable mixture of white and gray cast iron. These hybrids are now considered to lack the good qualities of either of the parent varieties, and therefore to be undesirable, but objects made of them were not uncommon in the past. While they must often have resulted from poor control of manufacturing processes, there is reason to believe that in some cases mottled iron was made deliberately.

Early writers on the making of cast iron cannon, for instance, often seem to specify a metal that is neither white nor gray. Mottled iron may actually have been superior for this purpose, since it is more machinable than white iron and less vulnerable to failure at high temperature and pressure than the gray variety (Rostoker 1987). Modern metallurgists generally think of mottled iron as useless, but John Percy, a noted nineteenth-century authority on iron and steel, had a high opinion of some kinds of mottled iron. In his discussion of the iron used for casting cannon at the arsenal of Finspong in Sweden, he describes a metal that "consists essentially of a delicate white network of white-iron, the meshes being more or less circular, and filled with grey-iron." This metal was "of great strength and well adapted for cannon" (Percy 1864: 553–554).

### 3e. *The Chemistry of Cast Iron*

The same furnace may yield either white or gray cast iron, depending primarily on the presence or absence of silicon and sulfur and on the rate at which the solid iron freezes from the melt. The chief difference in performance between the two metals results from the state of the carbon they contain—in white iron, the carbon is in hard carbide crystals; in gray iron, it is largely dispersed in soft graphite flakes. Silicon is essential to the formation of graphite during solidification. Thus, at silicon levels below about 0.2 percent, the cast iron that solidifies from the melt will always be white. When silicon exceeds 1.5 percent, it will always be gray. At intermediate silicon levels, the nature of the product is controlled by freezing rates: the iron is white when the liquid iron freezes quickly and gray when the transition from liquid to solid is prolonged (Fig. 2.7).

The silicon derives from silica, a major component of all slags and refractories. Silica itself has no effect on graphite formation, so the furnace must be able to reach temperatures sufficient to reduce silica to elemental silicon before gray iron will form. This occurs above about 1300°C. Because the silicon thus produced is highly soluble in iron, the cast iron produced in a hotter furnace is more likely to be gray.

Sulfur has an effect opposite to and even stronger than that of silicon. Liquid iron with a sulfur content greater than about 0.5 percent will always solidify as white cast iron, no matter how great the silicon content and how slowly it is cooled. A high sulfur content usually indicates that the metal has been smelted with coal; high sulfur is not common in charcoal smelted irons.

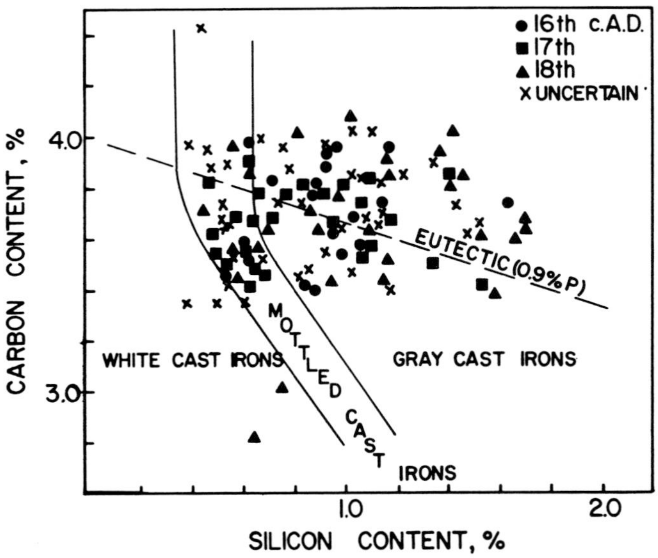


Fig. 2.7. *Structure, Silicon, and Carbon in Historical Cast Irons.* The data points represent dated specimens from Belgium, Germany, and France (Doat and Evrard 1951). The scarcity of white cast iron in this group is noteworthy.

[6]Discussions of early Chinese malleabilizing techniques may be found in Li 1975, Wang 1982: 129, and Rostoker et al. 1983.

If we assume that early blast furnaces tended to operate at temperatures no higher than necessary to liquefy the iron and that most of them were charcoal fueled, a historical sequence of cast iron types can be proposed. White cast iron should have come first. In the beginning it might have been little used except for castings where strength was unimportant, and only later served as a raw material in the manufacture of wrought iron and of malleable iron castings. Gray iron should have come next as furnaces were made larger and thus became hotter. Since gray iron and white iron are interchangeable for some purposes, the differences might not have been discernible. Gray iron might not have been made deliberately until tougher, more machineable castings were needed. Still later, interest in white iron would have revived as fossil fuels began to replace charcoal in blast furnaces and as levels of sulfur in the iron began to rise.

Cast iron production in China follows this sequence. The earliest iron castings in China are almost all white irons, presumably because the metal was smelted with a low sulfur fuel (charcoal) at a temperature too low to reduce silica to silicon. Gray iron appeared about 200 B.C. and continued to be used with white iron until about A.D. 1000. It may have been used simply because it was easy to make in furnaces that ran at temperatures high enough to reduce silicon; Chinese ironmasters of the Han period, who had already discovered how to malleablize white cast iron, had no real need for the special qualities of gray iron. After about A.D. 1000, as coal became widely used for smelting, the rise in sulfur levels in smelted iron caused a return to the making of irons that were mostly white. So much sulfur was present in cast irons of this later period that even very massive and slow-freezing castings such as giant bells came out white rather than gray.

The much shorter history of cast iron in Europe is similar, beginning with white iron, proceeding to gray iron as the need for machineable castings increased, and ending in the eighteenth and nineteenth centuries when ironmakers, forced to switch to coal as fuel, had difficulty in producing gray iron of good quality instead of white iron. Perhaps the major difference between the Chinese and European experiences with cast iron lies in the early Chinese mastery of malleablizing techniques which Europeans did not use extensively until the middle of the nineteenth century.

## 4. MINOR CONSTITUENTS IN BLOOM IRON, STEEL, AND CAST IRON

### 4a. *Adventitious Alloys and Impurities*

Iron produced by any smelting method is likely to acquire small amounts of other elements from the ore that are reducible under the same conditions as iron. This represents a type of alloying that was both unintentional and unsuspected before the rise of modern chemistry. In many cases the presence of these elements had no influence on the properties of the metal product. In other cases, however, good or bad effects were substantial. We now know how to control undesirable impurities by management of the slag and furnace conditions or by counteracting the effect of those impurities through the addition of other elements as antidotes. Earlier ironmakers lacked these options. Either they managed to avoid ores and fuels that contained the more deleterious impurities, or they learned to live with them.

Furnaces that are designed to reduce iron oxides will reduce the minerals of many other metallic elements as well, and most of these, at least in small amounts, are soluble in solid and liquid iron. The main exceptions to this rule are metals such as sodium, potassium, calcium, magnesium, and aluminum that bond strongly with oxygen. Such oxides, commonly present in the rocks and clays associated with iron minerals, are not reducible at iron smelting temperatures. They therefore exist in the finished iron only as slag inclusions which, although chemically inert, may cause physical problems by stimulating metal fracture.

### 4b. *Manganese*

Modern carbon steels intended to be heat treatable usually contain small amounts of one or more of four metallic elements—manganese, chromium, nickel, and molybdenum. Their presence in moderate quantities increases the capacity of the steel to harden deeply through quenching or to achieve full hardening at slower cooling rates, as when a tool is quenched in oil or air rather than in water. The ability to achieve an increased depth of hardening with slower quenching rates is called *hardenability* in the modern literature. Manganese is one of the most significant contributors to hardenability. A bar of alloy steel 15 centimeters in diameter with about 0.6 percent carbon and about 6 percent of other metals, including a substantial proportion of manganese, hardens to the center when cooled only in air. In contrast, when a plain steel bar of the same size with a similar carbon content is quenched, even in very cold water, the hardened zone will probably be no deeper than a few millimeters.[7]

The mineral forms of manganese often occur together with iron ore (Chapter 4). Natural associations of iron with nickel are much less common, and associations of iron with significant quantities of chromium or molybdenum are almost nonexistent. Thus, manganese is more likely than the other beneficial alloying metals to have been present in the iron ores used by pre-modern smelters.

A bloomery run at a minimum temperature for smelting iron will not reduce manganese oxides. Any manganese present in the ore will therefore wind up in the slag, where its only effect will be to replace iron oxides as a fluxing material and thus to increase the percentage of iron metal recovered from the ore. Bloomery

[7]It may be useful at this point to distinguish "hardenability" from "hardness capacity." The latter term denotes the maximum hardness that a steel with a given carbon content can reach under ideally rapid quenching. The former term refers to the ease with which that maximum hardness can be achieved and the depth to which it occurs. Manganese and other alloying elements, including carbon, contribute to hardenability but only carbon content regulates the achievable hardness level.

slags containing substantial amounts of manganese are often observed in archaeological contexts (Tylecote 1976: table 36; Piaskowski 1983: table 9). Percy (1864: 301–308) notes that the iron ores smelted in the Catalan hearths at Ariège in the French Pyrenees contained between 3 and 6 percent manganese, but the finished iron had only 0.52 percent of that metal. The slag held about 13 percent of manganese oxide which represents the bulk of that metal in the ore. Percy concludes that manganese "is not sensibly reduced in the Catalan furnace." Tylecote (1962: 191) reaches a similar conclusion about the smelting of manganiferous bog ores in ancient Britain, stating that "in all cases, at least 99 percent of the manganese in the ore finds its way into the slag."

Some early bloomeries, however, were evidently capable of reducing manganese oxides. Klevenhüller and his colleagues (1961) present analyses of three iron blooms of the early Roman period from Magdalensberg in Carinthia, the ancient Noricum. One of these contained 1.13 percent manganese, and another 0.57 percent. Coghlan (1977: 83) cites a description of a late Hallstatt or La Tène Period sword from Scandinavia, which contained about 1 percent manganese, enough to effect a marked improvement in the hardenability of the 0.5 percent carbon steel. Substantial quantities of manganese exist in a number of Austrian and Swedish iron deposits (Walzel 1951; Wertime 1962: 192). Ore from such deposits must have been the source of the metals described by Klevenhüller and Coghlan's source.

In post-medieval Europe, the art of extracting manganese in bloomeries was most highly developed in Austria, where the *Stücköfen* furnaces of Carinthia and Styria supplied a steel that was long considered the best in Europe. Although these furnaces were bloomeries in that they yielded solid lumps of metal, they were run at temperatures high enough to produce small quantities of liquid iron as well (Percy 1864: 329). Such temperatures would have reduced manganese minerals efficiently. It seems probable that other bloomeries known to have produced a natural iron-manganese alloy (e.g., at Jabalpur in India—Fermor 1909: 595–597) did so because the furnaces could achieve temperatures above the minimum needed to extract iron from ore.

The temperatures reached by blast furnaces for making cast iron are more than adequate to reduce manganese oxides. Significant amounts of manganese are often present in early steels found in China, where most if not all smelting was done in blast furnaces. Three of six analyzed swords of the Han period (ca. 200 B.C.-A.D. 200) have manganese contents in the 0.5 percent range (Li 1975: 17). In the sixteenth through eighteenth centuries, the blast furnace iron of Dannemora in Sweden was famous for its capacity to be converted to steel, a property which probably derived from the high manganese content of the magnetite ores from which it was made (Percy 1864: 199).

### 4c. *Nickel*

Nickel can be reduced at bloomery temperatures, but it is associated with meteoritic iron more often than with terrestrial iron ores. Nevertheless, nickel ascribed to nonmeteoritic origins has been detected in chemical analyses of early irons. One early composite-welded iron axe from Sweden had thin bands of iron containing as much as 6 percent nickel (Hermelin, Tholander, and Blomgren 1979). Although the scholars who studied this axe say the iron is nonmeteoritic, one of them (Blomgren 1980) notes that there are no appropriate iron-nickel ores in the region where the axe was discovered. Seven Greek iron artifacts dating to the early first millennium B.C. have nickel contents ranging from 1.40 to 10.77 percent (Varoufakis 1982: 316). These are believed to have been made from garnierite ores that occur on the island of Euboea and at other localities in Greece and Turkey. In addition, a strong case can be made that smelted nickel-bearing iron from deposits in Sulawesi was regularly incorporated into *keris* and other traditional Indonesian weapons during the eighteenth and nineteenth centuries (Bronson 1987).

Other early nickel-containing irons might have been smelted from ores rather than forged from meteorites.[8] In many, the nickel content is low, rarely exceeding 1 percent (Tylecote 1962: 42). Steel made from iron with this amount of nickel would have much less hardenability than one with a similar amount of manganese, although streaks of even a low nickel steel would, as in the pattern-welded *keris* blades of Indonesia and Malaysia, have a decorative effect if the metal surface were colored and etched.

### 4d. *Copper*

While published analyses of ancient iron artifacts rarely indicate whether they contain copper, cupriferous iron objects may have been relatively common in some areas. For example, a number of objects from Early Iron Age Britain contain between 0.01 and 0.2 percent copper (Haldane 1970). Iron and copper ores are often associated, especially in gossan deposits like those of Huelva in southern Spain. The copper in such an ore could be easily reduced at temperatures sufficient to smelt iron. In the sixteenth century, Biringuccio (Smith and Gnudi 1942: 66) was already conscious of this possibility, warning against the use of iron ores that show traces of the colors denoting copper minerals.

Copper will alloy with bloom or cast iron during smelting and will remain with it even if converted to steel. It has a strengthening effect on the finished metal, and, in modern formulations under such trade names as Corten, it also renders steel more corrosion resistant (McGannon 1964: 1089). Furthermore, minor additions of copper to iron can be decorative. Traditional Japanese swordsmiths sometimes added copper to their steel, presumably to achieve a well-defined grain pattern on the polished surface of the blades (Smith 1960: 52).

From the standpoint of an early smith, however, these beneficial effects would have been nullified by the

[8]See for example Piaskowski 1969–70; Tylecote 1962: table 30; Piaskowski and Hensel 1979, as well as several of the sources cited by Tylecote n.d. We know of no data showing that smelted iron-nickel alloys existed in China, India, or Africa.

substantial hot shortness that copper can impart to iron. As Biringuccio's comment shows, the general undesirability of copper in iron has long been recognized and is still of concern to the modern scrap steel industry. There was and is no practical method for removing copper from iron. Its presence in ores might or might not have been detectable by the blue and green colors of copper carbonates. Where it was not detectable, experience would usually have taught smelters that certain ores were best avoided.

### 4e. *Sulfur*

The iron sulfide mineral, pyrite, occurs in many iron ores. Sulfur is commonly present in coal as well, partly as disseminated pyrite particles and partly bound in the molecular structure of the coal itself. Both sulfur and sulfides will dissolve readily at high temperatures in liquid and solid iron. Thus, the product of smelting, whether bloom or cast iron, is bound to pick up much of the sulfur in an ore. When coal or coke is used as fuel, sulfur is absorbed from that source as well.

Sulfur in iron and steel reduces forgeability by rendering the metal piece hot short—that is, subject to cracking under the blows of the hammer at high temperatures. When a piece of steel or iron cracks during forging, the freshly formed surface of the crack immediately reacts with air to form an oxide film of sufficient thickness to prevent subsequent rewelding of the surfaces when forging is resumed. Large cracks formed during forging cause iron or steel to fragment immediately. Small cracks remain as hidden flaws that can fracture when the finished product is put into service.

The hot shortness of sulfur-contaminated iron depends on the level of sulfur and the forging temperature. Even very low levels of sulfur are potentially dangerous because at temperatures above 988°C the sulfur promotes melting of the iron, which forms pockets of liquid within the solid metal structure. These pockets, in turn, cause irreversible cracking. Iron containing as little as 0.03 percent sulfur is considered to be at hazard by modern metallurgists.

There are two possible remedies. One is to forge only at temperatures below 988°C, which is feasible but costly in both labor and fuel. Because 988°C is not far above the point where resistance to deformation rises sharply, a smith would have to strike the iron harder and reheat it more frequently (Fig. 1.4). He would also have to pay close attention to the color of his workpiece to guard against ruining it through overheating.

The modern remedy is to add manganese in the proportion of at least five times the sulfur content. By now this procedure is so well established as a precaution that its origin has disappeared from contemporary texts, and hot shortness in steels has become a rare occurrence. In pre-industrial settings, however, very few irons or steels could have contained enough manganese to neutralize even moderate amounts of sulfur.

Many historical iron samples contain ratios of manganese to sulfur that signify excessive sulfur and therefore diminished forgeability (Fig. 2.8). Since the samples did come from finished implements, it is clear that such

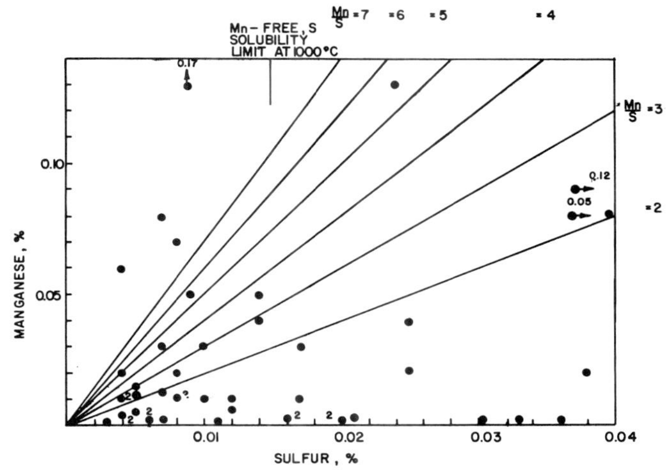


Fig. 2.8. *Manganese and Sulfur in Historical Wrought Irons.* The analyses are plotted to show their conformity with various Mn/S ratios. By modern standards any point with a Mn/S ratio less than 5 and a sulfur content greater than 0.03 percent is potentially hot short, yet the data points all represent successfully forged iron products. (Sources: Coghlan 1956; Schubert 1957; Morton 1963; Tylecote 1976; Samuels 1980a)

metals could be successfully forged, but we believe that the procedure must have been difficult to manage. Smiths may have forged at lower temperatures if they had encountered cracking while at white and yellow heats. Or perhaps the abundant slag inclusions in traditional bloom and wrought irons, by acting as flux, helped in rewelding minor cracks, which may have made such metals more sulfur-tolerant. Yet hot shortness was evidently a major problem for pre-industrial ironworkers. A preference for iron that was not hot short helps to explain the premium prices paid in sixteenth- through eighteenth-century Europe for the low sulfur wrought and bloom irons of Sweden, Spain, and Russia.

Sulfur plays a different role in the solidification of cast iron, and it is a powerful factor in inducing white rather than gray iron to form (Fig. 2.9). Not much more than 0.5 percent sulfur is enough to counteract the factors—high silicon content and slow freezing rate—that would otherwise force the metal to solidify as gray. Minimizing sulfur content was therefore important to foundrymen producing castings that had to be finished by machining.

Traditional iron founders often must have been aware of the differing qualities of high and low sulfur cast iron even though they did not recognize the presence of sulfur itself. Tegengren (1923: 339), for example, notes that in Hunan, southern China, "the [cast] iron produced with charcoal is stated to be suitable for pans, while that blown with coke is harder and used for bells, chimes and agricultural implements." This implies that the pans, because they could not be brittle, had to be made of a low sulfur and thus gray or mottled cast iron. The agricultural implements could presumably be of either gray or white iron. The bells and chimes, on the other hand, had to be of white iron and thus were made from the high sulfur product of furnaces fired with coke.

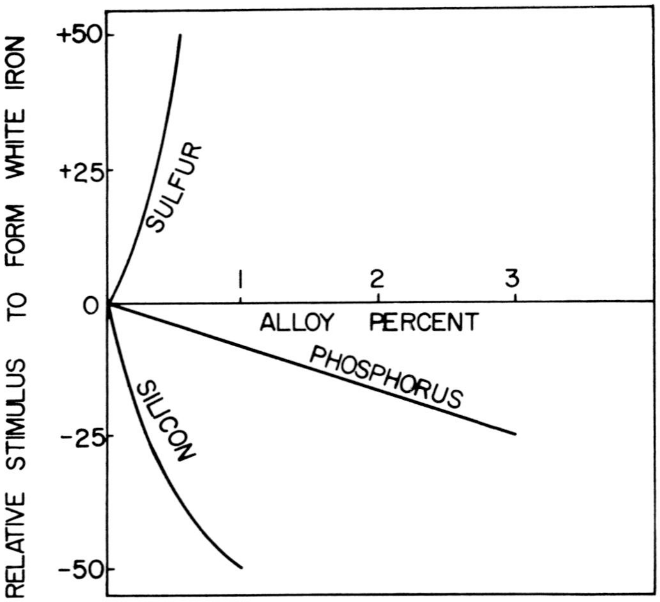


Fig. 2.9. *Effect on Structure of Sulfur, Silicon, and Phosphorus in Cast Iron.* Silicon and phosphorus both predispose liquid cast iron to solidify in the gray state. Sulfur, however, has an overriding effect. In amounts of more than 0.5 percent it will force the metal to solidify as white iron even when silicon and phosphorus are high and the freezing rate low. (Adapted from Massari 1938)

High sulfur iron must sometimes have been preferred for its superior casting qualities. While there appears to be no published information on the subject, our own experiments indicate that sulfur reduces the surface tension of liquid cast iron,[9] a trait of importance in making finely detailed castings. It is possible that Chinese foundrymen regularly chose high sulfur metals for making cast iron statues. Many of these statues contain very large amounts of sulfur (Pinel et al. 1938: 178).

### 4f. *Phosphorus*

In small amounts phosphorus greatly strengthens bloom iron and steel. Modern steelmakers sometimes add phosphorus within the carefully controlled range of about 0.07 to 0.12 percent, resulting in a gain in yield strength of up to 30 percent over normal steels (*Metals Handbook*, v.1, 9th ed. 1980: 162).

On the other hand, in large amounts phosphorus imparts a strong tendency toward brittleness. While strength increases as more phosphorus is added, this is accompanied by a disastrous drop in toughness and ability to withstand repeated bending (Table 2.5). These deficiencies reveal themselves to the blacksmith when high phosphorus wrought iron and steel are hammered or bent at ambient temperatures; such metals are traditionally said to be "cold short." To the user, embrittlement by phosphorus might not become apparent until the weather turns cold. Sword blades and spear points, serviceable enough in summer, might shatter upon impact during a cold spell, or an anchor chain might suddenly snap during a winter storm.

Phosphorus in iron and steel behaves in some ways like carbon because it strengthens iron and confers high work hardenability. A high phosphorus wrought iron can be hardened through cold working to the same level as any bronze (Tylecote 1981). Phosphorus by itself, at levels of about 0.5 percent, can make plain wrought iron as hard and strong as a medium carbon steel (compare Fig. 1.3). Yet it imparts such brittleness to the metal that modern ironmakers consider a 0.05 percent phosphorus level unacceptable. Ancient ironmakers evidently did not share this view, however. They often smelted ores with a seemingly unacceptable phosphorus content and made extensive use of the resulting high phosphorus metal. Although this metal would seem more suitable for implements that did not involve impact, archaeological and historical weapons with very high phosphorus contents are known from several parts of the world.

Phosphorus also hardens and embrittles steels. It significantly increases the capacity of steel to harden in depth when quenched. On the other hand, it reduces the solubility of carbon in iron at high (austenitizing) temperatures. Under comparable conditions pure iron can be carburized up to a 0.945 percent carbon level, whereas iron containing 0.6 to 1.2 percent phosphorus is held to a maximum of 0.6 to 0.53 percent carbon (Stead 1918). This differential probably accounts for the strong preference shown by makers of historical English blister steel for low phosphorus Spanish and Swedish iron (Chapter 11). These steels were made by heating iron bars in sealed, carbon-filled boxes. Achieving the highest possible carbon content was critical to this steelmaking method, so phosphorus-containing iron had to be avoided.

In cast irons, phosphorus strongly depresses solidification temperatures. In the early part of this century, additions of up to 1 percent phosphorus were made to cast irons in order to increase the time they would stay liquid when poured into a mold. The addition of 1 percent phosphorus to a normal gray iron can double the distance that the iron can flow before freezing (Heine et al. 1967). It was once believed that ancient Chinese foundrymen added even greater amounts of mineral phosphorus, perhaps as much as 7 percent, to liquid cast iron to enhance fluidity (Read 1913: 27). However, more recent analyses of early Chinese cast iron samples have not shown unusually high phosphorus levels (e.g., Pinel et al. 1938).

### 4g. *Silicon*

When temperatures rise above ca. 1300°C, conditions in a blast furnace will not only reduce iron oxides to iron but also to some extent silica ($SiO_2$, quartz) to elemental silicon. Silica is almost always present in ore, and silicon is very soluble in liquid cast iron. European

[9]The experiments involved measuring the surface tension of 20-gram buttons of cast iron partially melted by an electric arc in vacuum on a water-cooled hearth. During melting, the liquid metal rested on a plate of still-solid metal. Surface tension caused this liquid portion to assume a domed form. Measurement of the angle between the dome and shell after freezing made it possible to estimate the influence of sulfur. A button of sulfur-free cast iron with 4 percent carbon yielded an angle of about 92°. A button with the same carbon content but with 0.5 percent sulfur added had an angle of only 62°.

*TABLE 2.5. Effect of phosphorus content on tensile test, bend, and impact fracture properties**

| Phosphorus, % | Yield strength, psi | Tensile strength, psi | Elongation, % | Repeated reverse bends, number | Impact fracture toughness, ft-lb |
|---|---|---|---|---|---|
| 0.06 | 24,800 | 42,000 | 30 | 96 | 43.5 |
| 0.10 | 26,000 | 43,900 | 22.5 | 82 | 31 |
| 0.19 | 44,500 | 56,100 | 30 | 57 | 36.5 |
| 0.29 | 40,000 | 47,700 | 30 | 61 | 4 |
| 0.49 | 69,200 | 77,000 | 21 | 40 | 1 |
| 0.76 | 73,100 | 83,200 | 20 | 11 | 1 |
| 1.07 | 72,400 | 85,500 | 15 | 10 | 1 |

Other elements: 0.015% C, 0.055% Mn, 0.045% Si, 0.03% S
Annealed 720°C—8 hours after hot forging and hot rolling
Reversed 90° bends over 0.25 in. radius
Charpy impact test—probably larger notch radius than present convention
*Lorig and Krause 1936: 12

cast iron artifacts of the late pre-industrial period, the sixteenth through eighteenth centuries, may contain silicon in amounts up to about 1.5 percent (Fig. 2.7). Its presence is a strong factor in causing liquid cast iron to solidify in the gray form (Fig. 2.10).

Silicon pickup did not usually occur in bloom iron production because the furnaces involved ran at temperatures below 1300°C. However, temperatures in the hottest zones of operating bloomeries cannot have been uniform, so droplets of silicon-containing cast iron must often have formed within the furnace shaft. Any elemental silicon of this sort would usually have been reoxidized when it entered the iron oxide-rich slag pool.

The presence of silicon in cast iron made it difficult to convert to wrought iron in a puddling furnace or finery.[10] Once the conversion was completed, however, the finished wrought iron would have inherited little or no elemental silicon from the parent cast iron. Analyses of historical wrought irons do sometimes show high silicon contents, but this is almost certainly because of error in analytical procedure. The analyst in such cases has failed to distinguish between alloyed silicon and the silica in occluded slag.

### 4h. *Nitrogen*

Nitrogen acts very much like carbon in its alloying behavior with iron. It is more effective than carbon as a strengthening agent in pure iron and steel, and in the latter increases both hardening capacity and hardenability. Unlike carbon, nitrogen enrichment at the surface of a steel can effect a modest improvement in wear resistance even without heat treatment.

Nitrogen is less dominant than carbon in iron chemistry because it is not as easily absorbed. Nitrogen molecules as they occur in air and combustion gases must be dissociated to the monatomic form before they can enter the structure of either solid or liquid iron, and temperatures sufficient to cause this dissociation were seldom achieved in early furnaces. Embrittlement associated with nitrogen alloying did become a problem after the introduction of the Bessemer process in the mid-nineteenth century. Before then the only practical way to add nitrogen to iron or steel was to expose the metal to compounds that already contained monatomic nitrogen (i.e., ammonia, amines, or cyanogens) at temperatures between 600° and 900°C. These processes are now called carbonitriding and nitriding.

When the hardness of artifacts is higher than might be expected from carbon content, alloyed nitrogen might be the cause. However, because phosphorus can have a similar hardening effect, more specific evidence for nitriding must be sought microscopically. Alloying with nitrogen, unlike phosphorus, can be observed in low carbon iron through metallographic examination (Smith 1960: 4; Pigott et al. 1982; Piaskowski 1983).

Some early formulas for carburizing iron may, in fact, have incorporated ingredients capable of diffusing both carbon and nitrogen into the metal. In 1739, Cramer recommended a blend of charcoal with bones, horn, leather, hoofs, and hair as a carburizing compound (Smith 1968: 111; Wertime 1962: 201–202). All these substances contain amines with monatomic nitrogen. In the sixteenth century, Biringuccio described a supposed Middle Eastern practice as follows:

> They say that they file it [the iron], knead it with a certain meal, make little cakes of it, and feed these to geese. They collect the

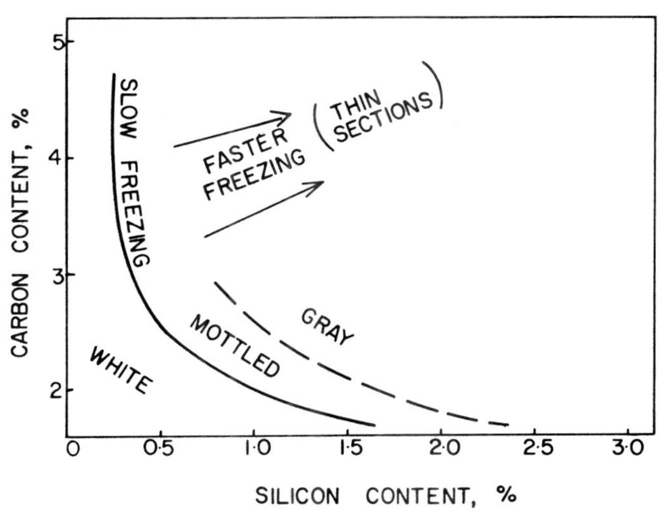


Fig. 2.10. *Effect on Structure of Freezing Rate, Silicon, and Carbon in Cast Iron.* It will be seen that iron with high silicon tends to be gray no matter what the carbon content except when freezing is very fast. The diagram does not take sulfur into account; it has an opposite effect (Fig. 2.9). (Adapted from Heine, Loper, and Rosenthal 1967: fig. 18.2)

[10]The problems caused by silicon during the fining process are discussed in Chapter 13.

*TABLE 2.6. Summary of influence of minor elements on iron, steel, and cast irons*

| Element | Level, % | Iron | Behavior as alloyed in iron |
|---|---|---|---|
| Manganese | 0.2–1 | LCI | Improves hot forgeability by sulfur control |
| | | ST | Same as above plus considerable increase in hardenability |
| Phosphorus | <1 | LCI | Increases strength but also raises the temperature at which embrittlement occurs |
| | | ST | Increases hardenability |
| | | CI | Increases liquid metal fluidity but at some cost in strength |
| Silicon | <2 | CI | Predisposes solidification to gray cast iron |
| Sulfur | <0.5 | LCI, ST | Induces hot shortness in forging |
| | | CI | Predisposes solidification to white cast iron |
| Copper | <0.5 | LCI, ST | Modest strengthening but can induce hot shortness. Can improve corrosion resistance |
| Nitrogen | <0.1 | LCI | Increases strength but also low temperature brittleness |
| | | ST | Increases hardenability |
| | >0.1 | LCI, ST | Increases surface wear resistance |

LCI = low carbon iron (bloom or wrought); ST = steel; CI = cast iron

> dung of these geese when they wish, shrink it with fire, and convert it into steel. I do not much believe this, but I think that whatever they do is by virtue of the tempering (quenching), if not by virtue of the iron itself (Smith and Gnudi 1942: 70).

An almost identical method of making steel from iron was attributed to Wayland, a legendary English smith of the Dark Ages (Straker 1931: 178–179; Willen et al. 1976; Martin and Foley 1979). Biringuccio's skepticism about the details of this particular procedure was probably well founded. On the other hand, the association of amine-rich goose dung with the hardening of iron may be significant. "Shrinking with fire" might carbonitride iron filings. If these were then consolidated by sintering and hot forging, the resulting metal would have many of the qualities of steel.[11]

### 4i. *Other Metals*

Arsenic, tin, bismuth, and antimony are all known to occur in pre-modern irons as deliberate or inadvertent additives. European cannonball founders of the fifteenth century made wrought iron fusible (and presumably increased the fusibility of cast iron) by adding large quantities of these metals to it (Wertime 1962: 23). Biringuccio (Smith and Gnudi 1942: 321) confirms this, adding that it is an undesirable practice.

Archaeological irons made from ores that contain arsenic are not rare (Tylecote and Thomsen 1973). Arsenic appears in amounts up to 0.25 percent in medieval bloomery irons from Bulgaria (Piaskowski 1982b). Even larger amounts of arsenic—as much as 7 percent—are present in alloys made from arseniferous iron ores in nineteenth-century Thailand (Hogboom 1913–14: 78). High arsenic iron has a low melting point but is too brittle for most purposes. Arsenic in the 0.25 percent range would somewhat increase the hardness and brittleness of steel. The fact that arsenic tends to segregate during forging, producing enriched zones along the margins of pile-welded strips, suggests that it may have helped to produce a visible and possibly decorative banded structure in finished iron objects.

### 4j. *Summary*

Thus, the subject of the influence of minor elements on the properties of the various types of iron is important and complex. Certain alloying elements did not occur in iron made before the modern period. However, a surprising number were present during earlier times, either as deliberate or inadvertent additives, and had significant effects on the fabrication and use qualities of irons and steels (Table 2.6). It should be remembered that the elements involved may interact in ways that are often not obvious even to professional metallurgists.

[11]Confirming evidence appears in Martin and Foley 1979. Other instances of possible carbonitriding are suggested by della Porta's (1658: 307–308) instructions for the hardening of files. In one recipe he recommends heating the files in a sealed iron chest filled with a powder composed of ox hoofs, salt, crushed glass, and chimney soot. In another he recommends using the powdered cores of goat horns without any other admixture.

# 3. Smelting Furnaces and Furnace Systems

## 1. THE RAW MATERIALS

Iron is chemically too reactive to exist as a native metal in usable quantities, except in meteorites and in one or two terrestrial deposits (Chapter 4). The great bulk of the iron in the earth's crust occurs in combined or mineral form. As with other metallic minerals, iron is usually mixed with gangue—rock that is not reducible to its component elements at ordinary iron smelting temperatures. A mixture of iron mineral and gangue that is worth exploiting is called an ore, which is categorized as rich or lean on the basis of the relative amounts of recoverable metal it contains. Ores as they come from the mine usually must be crushed and sorted, and other preliminary processing may be needed before they are ready for the smelting furnace.

Fuels known to have been used alone or in combination for smelting iron in the pre-industrial world include a wide variety of carbon-containing substances: coke, coal, charcoal, peat, wood, grass, and dung. Some have a higher calorific value than others. Charcoal is chemically pure and thus unlikely to add undesirable elements to the smelted product. Any fuel for smelting is expensive because of the large quantities needed for the process and, throughout the history of the industry, fuel costs have been the dominant factor in the economics of iron production in most areas (Chapter 6).

## 2. THE BASICS OF A FURNACE SYSTEM

A smelting furnace has two main functions. It must produce and contain high temperatures, and it must maintain combustion gases in a reducing condition, thus insuring that smelting does occur and that the smelted metal does not revert to a metal oxide in the furnace.

An open fire, unless it is of large dimensions, cannot heat objects to smelting temperatures. A deep bed of burning fuel improves the heating ability of a fire if the object to be heated is buried in the middle, because much of the radiated heat is reflected toward the center by the outer layers of glowing coals. Although a large open fire itself provides a measure of containment, it is still not thermally efficient. Fuel costs will be high unless radiant and convective heat loss can be controlled.

To improve thermal efficiency it is necessary to enclose the area where heat is generated. Enclosing walls provide surfaces that radiate heat back toward the center of the combustion space and also restrict the loss of heat through convection and conduction. The radiating wall surfaces have the same effect as the outer layer of glowing coals in an open fire but require a much smaller expenditure of fuel. The wall surfaces may be made of any refractory material that can withstand the temperatures involved without melting or crumbling (Chapter 5). Those best suited for furnace building also have low thermal conductivity, but it is often necessary to use layers of different refractories to achieve the dual goals of mechanical integrity and heat containment.

It is not enough, however, for a smelting furnace only to generate and retain heat, since the release of metal from the mineral state requires that certain specific chemical reactions occur. The most critical of these is the reducing reaction that separates an oxide into elemental metal and an oxygen-containing gas. Separation is accomplished by a reagent with an affinity for oxygen strong enough to pull oxygen away from metal oxides when temperatures are high. One of the few suitable reagents is carbon monoxide. This gas, along with carbon dioxide, is the coincidental product of burning carbonaceous fuel, but it makes the smelting of metals possible.

The central problem in designing a smelting furnace is that fires do not necessarily produce much carbon monoxide. The purpose of an ordinary stove, kiln, or heating furnace is to make something—food, pottery, air—hot. Such a device can be designed to maximize the efficiency with which fuel is burned, so all the carbon present is converted to carbon dioxide. In smelting, however, the fuel must provide both heat and abundant carbon monoxide, which means that efficient burning is counterproductive. Hence, a smelting furnace must be designed so that it can reach smelting heat but keep the air supply to a minimum. Walls, air inlets, fuel, and ore must be positioned so that the fire is regulated with just enough injected air to force the fire to high enough temperatures.

This contradiction can be understood in the context of the chemistry of combustion. If one kilogram of carbon is burned with 2.67 kg. of oxygen (about 10.67 kg. of air), all of it becomes carbon dioxide and 7900 kilocalories of heat are generated. The carbon dioxide ($CO_2$) itself as well as the residual nitrogen, the ore, the slag, and the furnace walls are all raised in temperature. But no iron is produced, because the reduction process requires that carbon monoxide (CO) be present:

$$Fe_2O_3 + 3CO \rightarrow 2Fe + 3CO_2$$

If, on the other hand, one kilogram of carbon is burned with only 1.33 kg. of oxygen, the product is wholly carbon monoxide, but the heat produced is only 2250 kilocalories of heat, about 28.5 percent of the thermal potential of the fuel. Obviously the use of carbon as a fuel conflicts with its use as a reducing agent.

Hence, it is essential for air input into a smelting furnace to be controlled so that an appropriate ratio of CO to $CO_2$ is maintained. This ratio can vary within limits. A low $CO/CO_2$ ratio increases the thermal efficiency of combustion but lessens the efficiency of metal recovery; a high ratio has the reverse effect, using more fuel but economizing on ore. When the metal being smelted is iron, variations in the $CO/CO_2$ ratio also affect the nature of the metal produced. In general terms, a relatively low ratio of monoxide to dioxide will result in a low carbon bloom iron with much loss of metal to the slag, whereas a high $CO/CO_2$ ratio will yield a high carbon cast iron and a slag that is almost iron-free.

In all furnace systems there is a continuous input of oxygen and a continuous expulsion of combustion gases. In most pre-industrial furnaces, the air comes in cold and exits hot. This is fundamentally inefficient; energy expended in heating the nitrogen and other inert components of air may greatly exceed that used in the actual reduction process. Among the most important features of modern furnace design are devices for recycling heat by using exhaust gases to preheat the incoming air. An important difference between shaft and bowl furnaces is that in the former waste gases can be used to preheat the fuel and ore, so the bowl furnace is generally less fuel-efficient. However, the thermal efficiency of a bowl furnace can be improved by heaping fuel and ore well above the rim, and/or by preheating the air intake.

The preheating of air with recycled heat was not scientifically understood before the nineteenth century, but it was accomplished in some traditional smelting systems by exposing the air to burning fuel before injecting it into the furnace—for instance, by feeding the air through a tuyere that projected well into the fire (e.g., as practiced by the Haya of East Africa—Schmidt and Avery 1978). The high temperatures attained may have had a significant effect on the reactions within the charge.

Using exhaust gases to preheat the fuel and ore also has great potential benefit. This particular economy is inherent in any shaft-type furnace in which solids are fed to the reaction zone through the same vertical stack that serves as a chimney. There is a temperature gradient within a shaft furnace (Fig. 3.1), so a large proportion of the energy produced is utilized before the exhaust gases leave the furnace. Even in furnaces without shafts, such as Catalan hearths, a certain amount of heat was sometimes recovered by piling fuel and ore at the edge of the furnace so that the hot exhaust gases could pass through or over them.

When preheated fuel and/or preheated air are brought into contact, the combustion temperatures are raised proportionately. Two possible results of preheating are that satisfactory operating temperatures can be obtained using less fuel, or that the temperature at which thermochemical reactions occur can be deliberately raised. Thus the thermal efficiency of a furnace can be

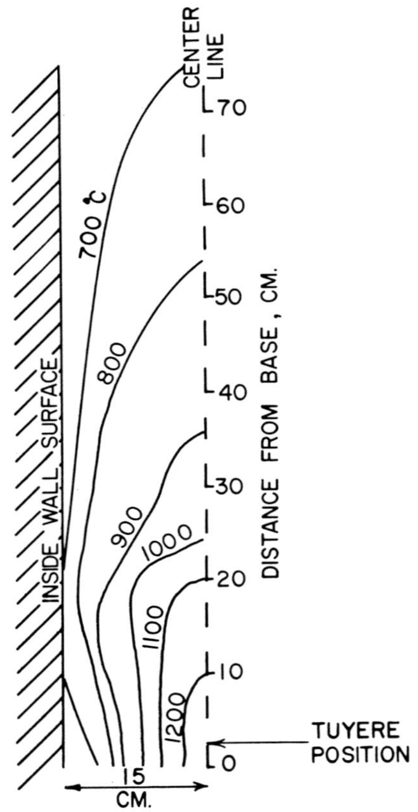


Fig. 3.1. *Temperatures within a Furnace Shaft.* Combustion occurs near the base of the shaft, just above the tuyere. The flame temperature there may be much higher than 1200°C, but the solids—ore, fuel, and furnace walls—never get that hot. Instead, energy is transferred upward and outward. This transfer creates the temperature gradient shown here. Much of the energy is returned later as the heated charge settles downward toward the combustion zone. (Based on the data of Tylecote et al. 1971)

understood in two ways: the economical achievement of a desired temperature or maximization of available energy.

## 3. INDIRECT HEATING FURNACES

In antiquity, it was not known that carbon was both reagent and fuel in smelting, yet some pre-industrial smelting systems seem to have been designed to separate reduction from combustion. As a result, the iron ore and a carbonaceous reagent could be raised to high temperatures without direct contact with the flame. The heat might have been transferred by conduction, as in a crucible furnace, or by radiation from hot furnace surfaces, as in a reverberatory furnace.

The main advantage of using crucibles to separate combustion from reduction is greater freedom in the choice of fuel. Because the heating element can occupy as large a space as needed, it can burn fuels with a relatively low calorific value, such as wood and low grade coal. Because the fuel need not be densely packed, air can be supplied at low pressure, perhaps by labor-efficient natural draft. And because the combustion gases have no contact with the crucible contents, the heating fuel can contain large amounts of sulfur, which would in an open system contaminate the finished iron. The reducing agent has to supply only carbon, not heat. Because less of it is needed than of the heating fuels, it can be selected for efficacy and chemical purity rather than for cost efficiency. Moreover, close regulation of the proportions of ore and reducing agent allows, in principle, the direct smelting of a metal with any carbon content desired—that is, bloom iron, cast iron, or steel can be produced in a crucible.

Iron smelting crucible furnaces were used in Shanxi in northern China until at least the late 1930s (Fig. 3.2; see also Chapter 10).[1] They were characterized by crucibles as long as 1.2 m., low three-sided stall furnaces which held the crucibles, and anthracite and bituminous coal as fuel. These furnaces often had no bellows attached, being supplied with air solely by natural draft. Most could produce several hundred kilograms of iron per day with relative efficiency in terms of labor and other input costs (Tables 14.7, 14.9a, b). A smelting run lasted between one and three days, after which the crucibles were broken and a slaggy mass of iron extracted. Variants of these furnaces not only smelted but also refined the smelted product to a serviceable cast iron.

Outside China, the use of crucibles for making or processing iron is documented only in western Africa, southern India and Sri Lanka, the Middle East, and western Europe. Although several modern commentators claim that a Shanxi-like process was also known in Japan, their information derives partly from Hulme's forced interpretation (1940–41: 27–28) of a passage in Kaempfer's seventeenth-century *History of Japan*. The original states only that iron "is refined upon the spot and cast into staffs and cylinders," and no mention is made of crucibles.

The case for crucible ironmaking in Africa is not well substantiated. One instance seems clear—that cited by Cline (1937: 35, 55) in the western Sudan, where a single crucible was used in the smelting of especially high quality iron, but he only supposes that in other places in the Ivory Coast and elsewhere in West Africa crucibles must have been employed in iron casting. Crucibles in South Asia and the Middle East seem to have been not primarily for smelting but for converting bloom (and sometimes, cast) iron to steel—the steel called *wootz* by Europeans. Most witnesses to the wootz process (e.g., Buchanan 1807; Heyne 1814; Heath 1839) emphasize the small scale, simple equipment, and the high quality of the product. In Central India at least, the material placed in the crucible was reportedly not metallic iron but unrefined ore (Schwartz 1899); if this is correct, Indians sometimes used crucibles for smelting as well as for steelmaking.

Early European ironworkers did not recognize the potential applications of crucibles in smelting, steelmaking, or even simple melting. The only crucible-like

Fig. 3.2. *Shanxi Crucible Smelting Furnace.* A diagrammatic representation of the smelting equipment used in Shanxi, northern China. The crucibles are filled with ore and coal, placed within a crude stall, and then covered with more coal. Firing is often by natural draft. See also Figure 10.6 and the description in Tegengren (1923–24: 323–325). (Adapted from Shockley 1904)

process widely used in Europe before the eighteenth century was the pack or cementation method of carburizing iron to "blister" steel. Patented in England by Ellyott and Meysey in 1614, and probably introduced in Germany and Holland at about the same time (Schubert 1957: 323), pack carburizing took place in large clay chests charged with bar iron and charcoal and covered with a lid. The chests were heated with coal in a reverberatory furnace until the bars became partially carburized. Since the temperature was never high enough to liquefy the metal, the product was inhomogeneous, but when pack forged it produced a usable steel (Chapter 11).

Benjamin Huntsman is believed to have been the first European to have actually melted iron in a crucible. In the 1740s he developed a method for rendering carburized (blister) steel homogeneous by liquefying it in a small crucible subjected to intense heat (Schubert 1957: 330). The first crucibles for making, rather than simply for homogenizing, steel did not appear in Europe until 1800, when David Mushet received a patent for a process involving small crucibles and a charge of charcoal with either bar iron or rich iron ore. Percy (1864: 776) makes it clear that he regards Mushet's process as a borrowing from India, not as an original invention. Crucible smelting was included in Mushet's patent but was never practiced in Europe on a production basis. In general, European ironmakers have followed Huntsman's lead in using crucibles mainly for improving steel made by other proceses and, since the late nineteenth century, for making special alloys.

Workers in nonferrous metals used reverberatory furnaces long before the introduction of pack cementation. Biringuccio, in the 1530s, described an Italian reverberatory furnace for remelting bronze (Smith and Gnudi 1942: 286, 287), but the antiquity of the process is unknown. The only archaeological example is possibly a clay foundation at the well-known ironworking site of Tieshenggou in Henan Province, central China, dating to the end of the first millennium B.C. According

[1] For example, by Shockley (1904), Richtofen (1882; 1907), Read (1913; 1934), Tegengren (1923–24), Wang (1923–24), Foster (1926), and Wang and Wang (1930). As late as the 1920s the process was carried on in at least four widely separated parts of Shanxi. In about 1930, Wang and Wang (1930: 110) estimated that the eastern part of the province alone still produced about 30,000 tons of crucible iron per year.

to the excavators, this device operated in a reverberatory fashion and was definitely used to smelt iron (Henan Archaeological Team 1962).

There are several possible ethnographic examples. A side-blown, horizontal smelting hearth was operating in Iran in the 1830s (Hulme 1937–38: 186), and a sloping hearth functioned as a finery in Anatolia in the mid-nineteenth century (Beck 1891, v.1: 288). Two horizontal furnaces are reported to have been used in India for smelting (Jacob 1843) and forging iron blooms (Franklin 1971). None of the descriptions, however, says clearly that the ore or iron was placed out of direct contact with the flame, so it is uncertain if any of these furnaces were reverberatory.

The reverberatory furnace did not become essential in European ironmaking until the development of the system for converting cast to wrought iron known as the puddling process, patented by Henry Cort in 1783–84 (Fig. 3.3). The operation of a puddling furnace resembled that of a pottery kiln, in which direct contact of the work with flame and fuel is generally undesirable. The fuel was burned in a chamber separate from the material to be heated; heat was transferred by radiation as the hot combustion gases heated a roof which then radiated to the metal in the hearth below. Most potential contaminants in the fuel therefore passed off without coming in contact with the metal, which was further protected by a covering of molten slag. This made it possible to use fuels which were impure and inexpensive like uncoked bituminous coal, as Cort himself did.

A further benefit of the reverberatory design was the lack of obstruction to the flow of air and combustion gases, so that natural draft could be used rather than bellows. The draft pressure generated by tall chimneys was strong enough to permit the development of new methods of recycling waste heat, building on Neilson's demonstration in the late 1820s of the advantages of preheating the air used in smelting. The double-ended puddling furnace, with preheating chambers warmed by waste heat, that evolved in the 1830s and 1840s was in turn the forerunner of the open hearth furnace that dominated large volume steelmaking for much of the subsequent nineteenth and twentieth centuries.

Because the puddling furnace was not fully developed until the Industrial Age, it is discussed only briefly here. The nature of its evolution may, however, have earlier, undocumented parallels since technological advance in one field may cause similar movement in

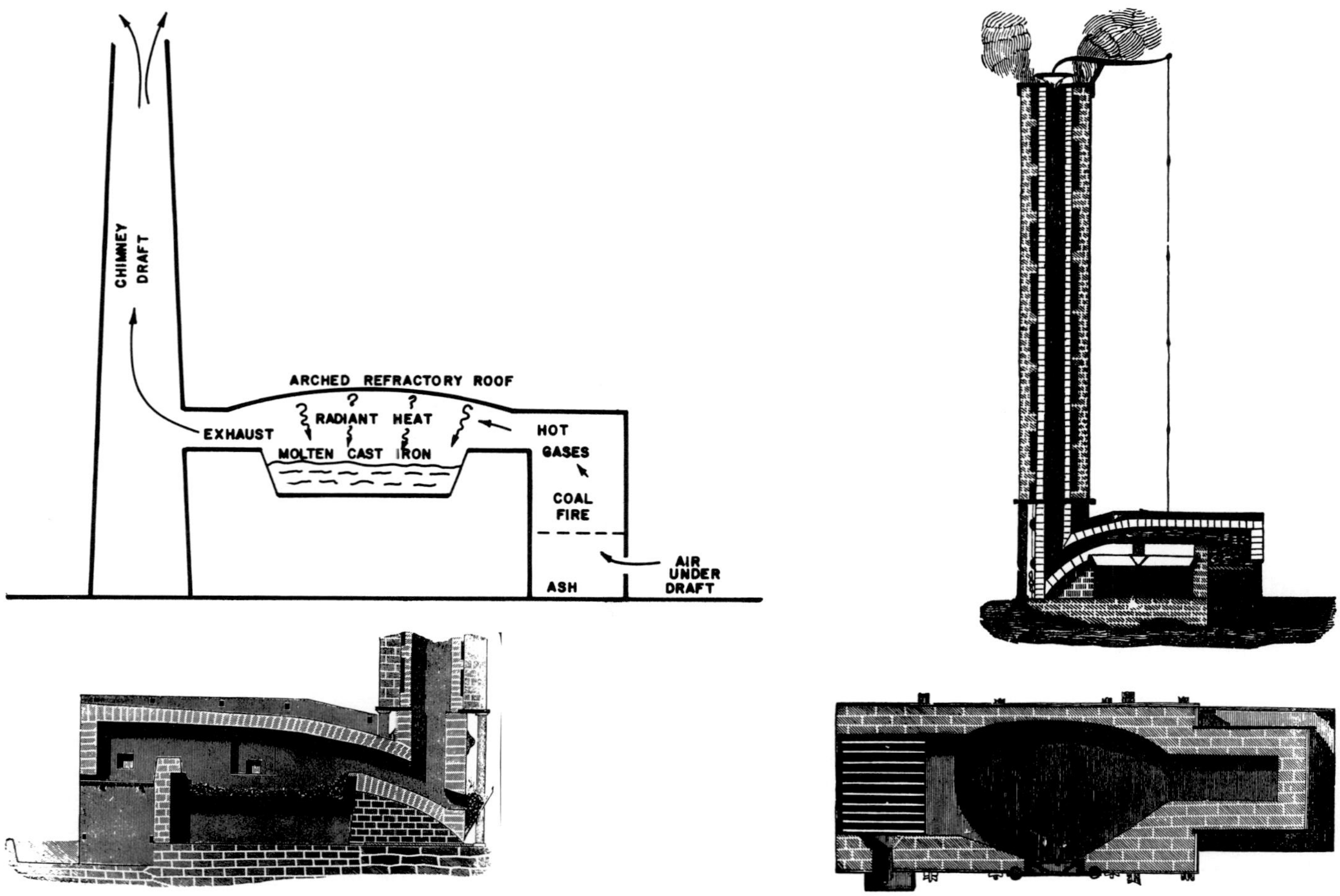


Fig. 3.3. *Reverberatory Puddling Furnace.* When introduced in Europe during the early nineteenth century, such furnaces were a major advance. Because heating of the charge was indirect, any fuel—including high sulfur coal—could be used. Energy from the fuel burned in a separate fire chamber was transferred to the charge primarily by radiation from the roof. (Adapted from Osborne 1869: 660–662)

another. The simultaneous appearance of a series of inventions in the period 1780 to 1840 set the stage for the beginning of the Industrial Age, in that its hallmarks—the puddling furnace, the steam engine, and new textile machinery—were developed. Synergistic linkages between inventions are predictable features of any period of technological progress.

## 4. DIRECT HEATING FURNACES

The bowl furnace, the shaft furnace, and their many variants have predominated in iron smelting since its beginnings (Figs. 3.4, 3.5). In both, the fuel and ore are mixed or packed in alternating layers, so the fuel burns while in direct contact with the ore. All such furnaces must be supplied with air at a rate appropriate to the desired reaction, which is dictated largely by temperature and by the critical $CO/CO_2$ ratio. Air supply is introduced into a furnace by one or more tuyeres, tubes of metal or fired clay. These project partly into the furnace space and are tilted slightly downward so they do not become plugged by accumulations of semi-solid materials and droplets of slag. A tuyere must be located far enough above the bottom of the furnace to leave room for a growing pool of slag and metal. Air input in a bowl furnace has to be supplied by a bellows or other mechanical device; in a shaft furnace, natural draft is sometimes sufficient. The pressure of the air blast must be strong enough to project across the chamber space, otherwise unreduced ore and fuel will remain against the opposite wall and restrict the available reaction area.

When air blast pressure is limited, as was the case in all pre-industrial smelting operations, complete reduction can be insured in only two ways: either the furnace must be small in cross section or it must be furnished with more than one tuyere. Before the development of steam power and the iron blowing cylinder, furnaces could sustain relatively large capacities by having multiple tuyeres, arranged either radially around the circumference of a wider furnace or in a line along the sides of a furnace with an elongated cross section (Fig. 3.6). Circular arrays of tuyeres are found in many traditional African smelting furnaces, where they may function by either natural or forced draft, and in most modern blast furnaces, which have powerful blowers. Linear tuyere arrays used for smelting are much less common; examples are those in the Japanese tatara furnace, the low Kui bloomery of Cambodia, and perhaps some of the bank furnaces of southern India.

Immediately above the tuyere in any furnace is an open space created by the continuing combustion of the fuel as it comes in contact with the incoming air. The upper extremity of this space is arch-shaped, formed by a partially melted aggregate of ore particles, metal, slag, and fuel. The most important thermochemical reactions between the combustion flame and the hot solids occur at this interface; the products then rain down through the combustion space and collect in the basin below. Because the interface is porous, the combustion gases can percolate upward through the bed of fuel and ore, finally exiting at the top opening. In the course of this upward motion, the gases transfer heat to the

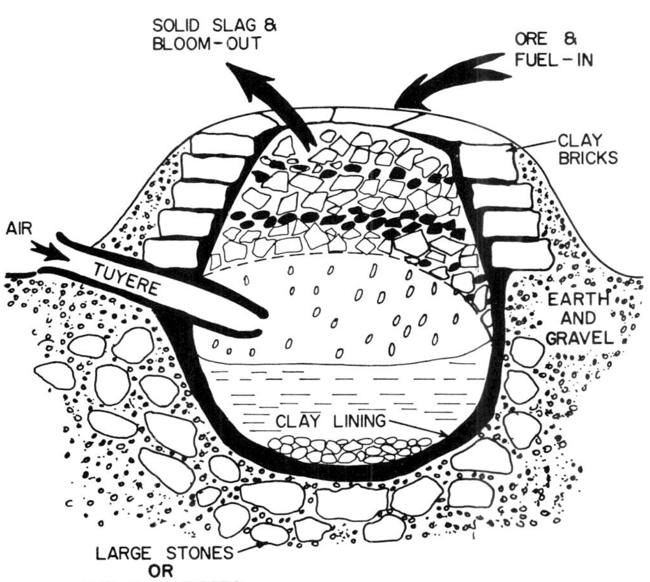


Fig. 3.4. *Generalized Bowl Furnace.* Although inexpensive and easy to build, bowl furnaces have various limitations. There is no chimney effect, so all air has to be supplied by bellows. The shortness of the column of fuel and ore means that a large proportion of combustion heat is lost through the opening at the top. Many, though not all, bowl furnaces lack provisions for tapping slag, which make it necessary to shut the furnace down every few hours due to the rising level of the slag pool.

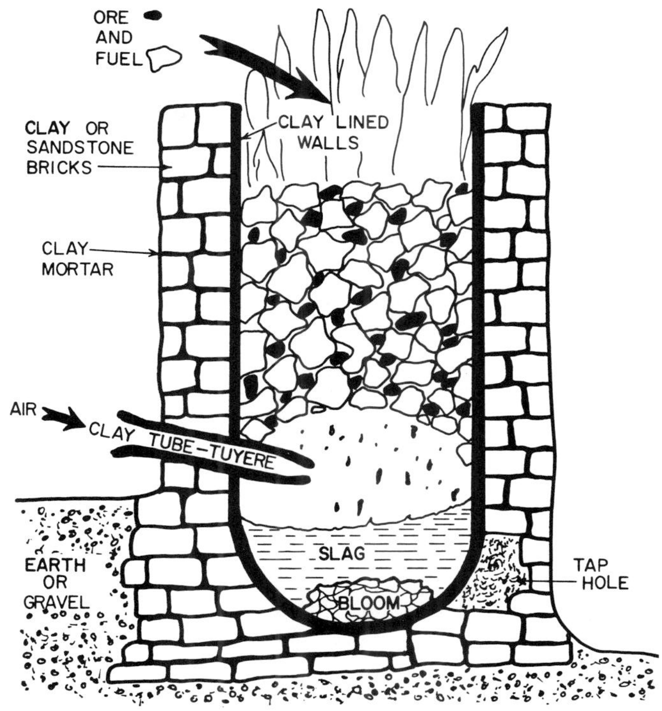


Fig. 3.5. *Generalized Shaft Furnace.* The crucial feature of such a furnace is that the chimney also serves as the charging chute, making it possible to use waste heat to raise the charge to a high temperature before it reaches the arched combustion zone. Most (but not all) shaft furnaces can be tapped during smelting. Here, the furnace wall would be periodically pierced at the level of the basin to allow the slag to drain out. Extracting the solid bloom would mean shutting the furnace down and breaking a larger hole in the wall.

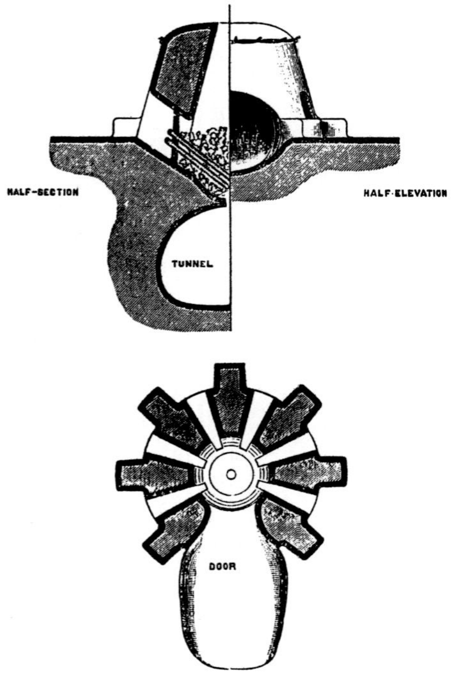


Fig. 3.6. *Yoruba Multi-Tuyere Domed Furnace.* The six radially arranged double tuyeres permit the furnace to operate by natural draft. (From Bellamy 1904: figs. 6, 7)

downward moving bed of solids so that it is progressively heated while the gases themselves cool. Certain reducing reactions may therefore begin far up the stack during preheating of this kind. Hence, the height of a shaft furnace relates directly to the efficiency with which heat is transferred to the descending charge and with which the ore lumps are reduced.

For sustained operation, a furnace must be tapped periodically so that the pool of molten slag does not rise to the level of the tuyere(s) and choke off the air blast. Because furnaces are likely to produce a greater volume of slag than of metal, this must be done frequently. Many bowl furnaces were not designed to be tapped (e.g., Fig. 3.4), and could be run for only a limited time before charging had to be stopped, the blast discontinued, the furnace cooled, and the slag and metal removed as a lump. A bowl furnace, which is by design a hole in the ground, can be modified for tapping by locating it on a slope or by digging a deeper hole next to it. Furthermore, non-tapping shaft furnaces exist, some of considerable size. A traditional furnace of the Fipa of Tanzania is 2.5 m. wide at the base and more than 3 m. high; the smelted iron was removed after each 24-hour run by waiting several days for it to cool and then climbing inside (Wembah-Raschid 1973).

The shaft furnace is usually designed so that the furnace basin can be broken open at an appropriate level in order to drain off part of the slag (Fig. 3.5). The hole can be restoppered with a lump of clay and the cycle of opening, draining, and plugging can then be repeated. By breaking the furnace open at a lower level the metal can be tapped or raked out, depending on whether it is cast or bloom iron. In this way the furnace can operate continuously, stopping only when its walls have eroded to a point where further operation would be hazardous. Continuous operation is an important factor in fuel efficiency and total output. If, to recover the metal, the furnace must be shut down, allowed to cool, and then broken apart, much fuel must later be expended to reheat the rebuilt furnace. Output will be limited by long unproductive periods spent in cooling and rebuilding.

Ironmakers not using semipermanent European- or Chinese-style systems may, however, erect a one time or short term furnace quickly and cheaply. Many clays are sufficiently refractory when simply air dried, making it possible to build a furnace quickly, using only mud formed into bricks or applied as a cement-like paste. The furnace basin of a short shaft furnace used by the Haya in western Tanzania (Avery and Schmidt 1979) was constructed two days before smelting, with slag fragments from previous operations and earth from termite mounds. One day before smelting, a wood fire was built in the basin to roast the ore. The roasted ore was then removed and, on the day of smelting, a short (1.25 m.) shaft quickly constructed, again of slag fragments and termite earth. Six to eight hours elapsed from the time smelting began until shutdown. The furnace was then knocked apart and the bloom extracted. In all, three days were sufficient for a complete cycle, which perhaps could have been shortened to two days if roasting, as is normal in most ironmaking systems, had been carried out concurrently with one of the other operations.

A similar smelting operation was observed among the Dayak of southeastern Borneo in the 1840s (Percy 1864: 273–275; Harrison and O'Connor 1969, v.2: 394–397). The furnace was also of the short shaft variety, 1.2 m. high, built of clay over a substantial basin-like hearth; it produced a single lump of iron each day. Here, however, the furnace had thick and durable walls as well as a tap hole for draining slag. A portion of the furnace front was designed to be opened for removal of the day's production and was easily resealed for the next day's run. A reusable smelting apparatus like this might seem much more economical than that of the Haya, yet building a Dayak furnace reportedly took more than a month, because of the time spent waiting for the thick clay walls to dry.

Perhaps the best known type of bowl furnace is the Catalan hearth, varieties of which were once used over much of southern Europe, but which by the nineteenth century was confined to the neighborhood of the Pyrenees in southern France and northern Spain, and also parts of North America. By then the Catalan hearth had been developed into a system that could surpass many of the earlier blast furnaces and fineries of northern Europe in terms of efficiency and product quality (Chapter 13). The classic description in English of the Catalan process is by Percy (1864: 278–314), who summarizes exceptionally detailed data assembled in the early 1840s. A typical Catalan furnace was built of stones set in clay. The hearth itself, partly lined with iron plates, had a rectangular plan and walls that sloped out toward the top, giving it the appearance of a hopper or funnel (Fig. 3.7). Apart from the more sophisticated stone and iron construction, the layout is similar to that of some Indian furnaces.

Air was supplied to the hearth through a tuyere set about 0.5 m. above the bottom. Fuel and ore were piled as much as 1 m. higher than the tuyere level so that the preheating of the charge was effective. While the slag could be tapped from the bottom of the hearth, the metal bloom had to be hooked and lifted out from the top. Despite this awkward arrangement, Catalan furnaces sustained operation over periods as long as several months and yielded several hundred kilograms of iron per day. The quality of the saleable bar iron that could be produced in a single operation compensated for the energy lost through the open mouth, whereas traditional blast furnace-and-finery systems required two or more fuel-intensive processing steps before the iron was ready for the market.

Perhaps the most efficient version of the Catalan hearth was the late nineteenth-century Champlain forge of upper New York State. Equipped with a hot blast and two metal tuyeres, they produced one ton of finished iron from two tons of ore with a 60 to 65 percent iron content, while consuming only about two tons of charcoal (Maynard 1874: 131–132). It seems probable that no other ironmaking system of the period could produce low carbon iron with such efficiency of ore and fuel (Tables 14.5a, 14.9b). Labor costs were high but so was the quality of the product. Its extremely low phosphorus and sulfur contents enabled it to be competitive with iron made in larger operations, so it sold at a good profit to manufacturers of crucible steel and boiler plate.

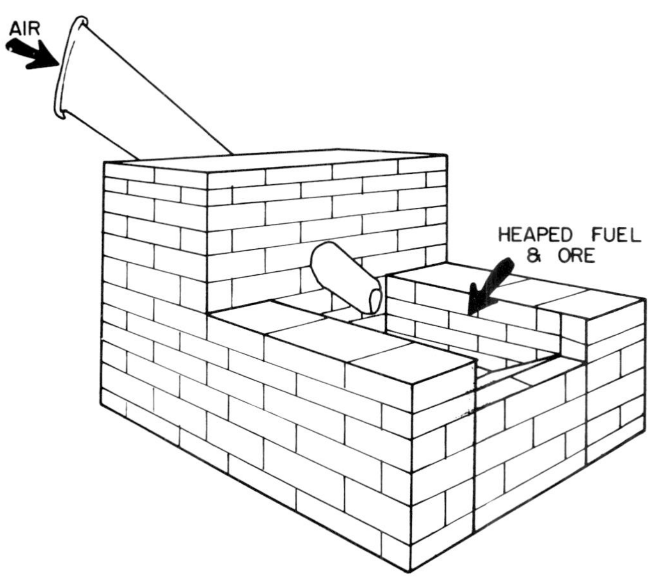


Fig. 3.7. *Catalan Hearth.* Successful operation depended on a large volume of air through the tuyere and careful placement of heaped-up fuel and ore. Such furnaces are only called Catalan hearths when used as bloomeries for smelting, but traditional European fining hearths and even blacksmiths' forges often had similar designs. (After Percy's description [1864: 278–285])

## 5. FACTORS THAT DETERMINE FURNACE DIMENSIONS

The height of a shaft furnace affects its efficiency. Maximizing the height, and hence the length of the column within which heat transfer and thermochemical reactions can occur, might seem desirable, but limits are imposed by the possibility of structural deterioration and the need to maintain gas permeability within a charged shaft.

The advantage of the slag- and iron-tapping shaft furnace is that it can be operated continuously over extended periods, thus economizing on fuel and raising total output. Pre-modern ironmasters in both China and Europe usually tried to keep their furnaces in blast for at least several months. However, the length of a smelting campaign depended on the continued physical integrity of the furnace structure, which because of the nature of ironmaking thermochemistry was difficult to maintain.

The intense heat will crack, decompose, or melt many structural materials. Moreover, liquid slag is a corrosive fluid that will eventually dissolve almost any clay or rock material used to line the shaft and hearth. This is not surprising, for the primary function of slag is to act as a solvent for the fuel ash and the nonreducible fraction of the ore. Only a very small number of substances can resist rapid dissolution by liquid slag. Materials that combine this quality with resistance to melting, spalling, softening, and calcining, and that also have a useful load-bearing capacity are scarce, which must have been a significant obstacle to the construction of larger and more permanent furnaces. Without a suitable structural material, the investment in large, tall furnaces was not likely to have been returned.

Despite the rarity of these construction materials, tall furnaces were built, using many methods of reinforcement to improve the structural stability of furnace walls: iron hoops and bars in China (e.g., Wagner 1985: 14); timber props and cribwork in Africa (e.g., Cline 1937: 35); and construction against or into hillsides and artificial mounds. Among the tallest recorded non-Western furnaces were those in western China during the nineteenth century. All had reinforced structures and were as much as 9 m. high (e.g., the furnaces at Tonghai in Yunnan [Fig. 3.8] and at Huangnipu in Sichuan [Tegengren 1923–24: 343], which were constructed of wood-framed clay or brick and of wood-framed sandstone respectively). Few tall furnaces equalled these during the pre-industrial period. Some furnaces in western Europe may have reached 11 m. before the mid-eighteenth century, and by the 1770s and 1780s furnaces in the Urals were 15 m. tall (Wertime 1962: 104). In contrast, no known Indian examples exceeded 3 m., and none in Africa was taller than about 6 m. (Cline 1937).

The height of a shaft furnace was limited by three factors: the structural stability of its walls; the ability of the air blast to overcome the back pressure produced by a high column of packed fuel and ore; and the ability of the fuel to resist crushing under the weight of the column of solids above. Coarse fragments of fuel and

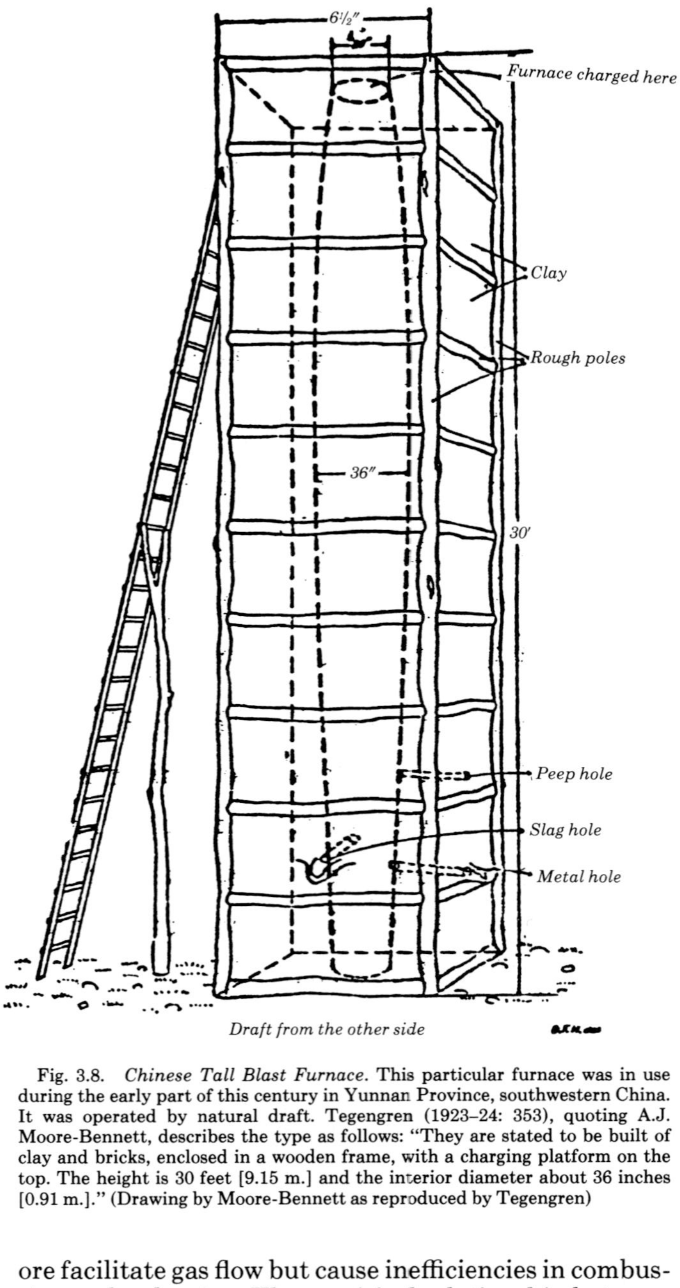


Fig. 3.8. *Chinese Tall Blast Furnace.* This particular furnace was in use during the early part of this century in Yunnan Province, southwestern China. It was operated by natural draft. Tegengren (1923–24: 353), quoting A.J. Moore-Bennett, describes the type as follows: "They are stated to be built of clay and bricks, enclosed in a wooden frame, with a charging platform on the top. The height is 30 feet [9.15 m.] and the interior diameter about 36 inches [0.91 m.]." (Drawing by Moore-Bennett as reproduced by Tegengren)

ore facilitate gas flow but cause inefficiencies in combustion and reduction. The empirical relationship between furnace height and production volume is, despite irregularities, fairly close (Table 3.1). Furnace height also correlates closely with labor productivity, moderately well with efficiency in ore reduction, and—perhaps surprisingly—less well with fuel efficiency.

The resistance of fuel to crushing is important in charcoal fueled furnaces, since even the most powerfully driven furnace with an optimally sized charge will become choked when the weight of the column exceeds the very limited load-bearing capacity of charcoal. The maximum height thus imposed has been estimated at about 7.5 m. (Schubert 1951–53: 75), but this figure is too low. Charcoal fueled blast furnaces in Russia sometimes reached 15 m. (Table 3.2). Many of the American charcoal furnaces (catalogued by Dunlap in 1874, in *Wiley's American Iron Trade Manual*) were more than 10 m. high, and a few (e.g., in Georgia—Dunlap 1874: 99–101) exceeded 18 m. Yet the latter, perhaps the tallest charcoal furnaces on record, seem to have been no more productive than furnaces half their size. It is clear that some limit had been exceeded.

In addition, the internal diameter of a furnace, although less important in functional terms than its height, was also a significant factor in efficient fuel use and ore reduction. One reason is that in a wider furnace more reacting material is present at the level of the combustion zone; another is that the center of such a furnace can operate at a higher temperature because it is thermally contained by the outer portions of the bed, rather than by the furnace wall. Since the rate of any thermochemical reaction is always substantially increased by even small rises in temperature, a wide furnace will operate more efficiently than a narrow one if it is supplied with sufficient air throughout the internal area.

This may explain the popularity of smelting furnaces with short shafts and large diameters in Chinese metallurgical tradition (Fig. 3.9). These furnaces offered less back pressure and thus permitted the use of the high output but relatively low pressure box bellows that is peculiar to China. Similar furnaces date back to at least the Han period, indicating a continuing effort by Chinese ironmakers to reach a compromise between the limitations of available air delivery systems and the desire to expand production per unit.

## 6. THE DEVELOPMENT OF THE EUROPEAN BLAST FURNACE

In fifteenth-century Europe the shaft furnaces that had recently begun to produce cast iron were made larger as ironmasters sought to achieve efficiencies of scale. The trend toward increasing furnace height and diameter led to changes in structural materials and shape, and may also have been a major force behind the development of techniques for smelting with more crush resistant fuels like coke rather than charcoal. By then the superior efficiency of the very large European-type blast furnace was recognized by ironmakers everywhere, but it was not universally adopted until recently in some places. As late as the mid-nineteenth century, few furnaces outside northwestern Europe and the eastern United States were more than 10 m. in height even when anthracite or coke was the fuel.

Because of the high temperatures reached in taller furnaces by the sixteenth century, it was necessary to abandon traditional construction with unfired clay and to begin building furnaces of refractory sandstone. Because of the difficulties of stone cutting and fitting, such furnaces were often designed with rectangular rather than the earlier circular (and more efficient) cross sections.[2] However, when prefired refractory bricks came into use a century or two later, the design became more flexible, and the cylindrical form that offered better

[2]The relationship between structural materials and furnace form in the European context is discussed by many authorities, including Morton (1963), anonymous (1964), Pfannenschmidt (1977), Linsley and Heatherington (1978), and White (1981).

*TABLE 3.1. Shaft furnace height (m.) and volume of production (charcoal fuel)*

| Place/group | Nation | Date | Process* | Height | Kg/day | Reference |
|---|---|---|---|---|---|---|
| (Koi) | E India | 1920s | d | 0.8 | 13.6 | Cammiade 1931 |
| Magadi | S India | 1800s | d | 0.9 | 33.0 | Buchanan 1807,1: 175 |
| Sukur | Nigeria | 1962 | d | 0.9 | 20.3 | Sassoon 1964 |
| (Agaria) | E India | 1870s | d | 0.9 | 14.1 | Ball 1880: 122 |
| (Kol) | E India | 1900s | d | 0.9 | 19.5 | Watson 1910: 325 |
| Narbada Valley | W India | 1850s | d | 1.1 | 5.5 | Kennedy 1855 |
| (Agaria) | C India | 1940s | d | 1.2 | 12.4 | Swarup & Misra 1944: 128 |
| Mirjati | E India | 1920 | d | 1.2 | 14.4 | McWilliam 1920 |
| Tendukhera | C India | 1900 | d | 1.2 | 60.0 | Bartlett 1900: 150–153 |
| Sisala | Ghana | 1971 | d | 1.3 | 4.0 | Pole 1974: 29 |
| Rajodha | C India | 1890s | d | 1.4 | 34.4 | Turner 1893: 166 |
| Salem | S India | 1880s | d | 1.5 | 32.4 | Holland 1892: 145 |
| Birbhum | E India | 1840– | i | 1.5 | 232.5 | Sanyal 1968: 103 |
| (Chinese) | Philippines | 1900s | i | 1.8 | 499.0 | McCaskey 1903b |
| Malabar | S India | 1880s | d | 1.9 | 90.4 | Holland 1892: 145 |
| Velatar | S India | 1800s | d | 2.0 | 145.0 | Buchanan 1807, 2: 439 |
| South Henan | C China | 1960s | i | 2.0 | 720.0 | Wagner 1984 |
| Letchemporam | C India | 1810 | d | 2.1 | 50.7 | Heyne 1814: 222 |
| Mailiban | SW China | 1900s | i | 2.2 | 88.5 | Coggin-Brown 1923: 264 |
| Jabalpur | C India | 1880s | d | 2.3 | 72.5 | Balfour 1885, 2: 373 |
| Xinyangshen | C China | 1910s | i | 2.5 | 544.0 | Tegengren 1923–24: 334 |
| Lufengxian | SW China | 1870s | i | 2.5 | 700.0 | Tegengren 1923–24: 350 |
| Elburz Area | N Iran | 1930s | i | 3.0 | 90.0 | Böhne 1928 |
| Basari | Togo | 1890s | d | 3.5 | 11.0 | Cline 1937: 36 |
| Eisenerz | Austria | 1770s | d | 4.9 | 1205.0 | Percy 1864: 329 |
| Baojing area | C China | 1910s | i | 5.5 | 1225.0 | Tegengren 1923–24: 339 |
| Zhongjing area | W China | 1890s | i | 6.0 | 2718.0 | Tegengren 1923–24: 343 |
| Puppadaung | C Burma | 1860s | d | 6.1 | 40.0 | Blanford in Percy 1864: 272 |
| (Uppland?) | Sweden | 1660s | i | 7.2 | 1280.8 | Pipping 1982: 224 |
| (Uppland?) | Sweden | 1660s | i | 10.9 | 2120.0 | Pipping 1982: 219 |
| Urals | Russia | 1790s | i | 15 | 13,000. | Wertime 1962: 104 and Johannsen 1953: 269 |

*i = indirect; the product is cast rather than bloom iron.

thermal uniformity and fewer obstructions to the downward motion of the charge reappeared.

What may have been the first successful production blast furnace in the American colonies was built at Saugus, Massachusetts, in the early 1640s, based on a contemporary English design. It was first operated in 1645 and ran, with periodic rebuildings, for 30 years (Hartley 1957) (Fig. 3.10). The massive stone buttresses, which are characteristic of the period, provided not only structural support but also thermal insulation.

By the middle of the eighteenth century the height from the bottom of the hearth to the furnace mouth of a typical western European blast furnace had increased to an average of 7 to 8 m. (compare Pfannenschmidt 1977: 219–222; Gillespie 1959: pl. 87; Percy 1864: 260–264). There were also changes in the design of the internal walls and hearth (Figs. 3.11, 3.12). Below the level where the charge began to melt, the walls of the furnace converged to form an inverted cone, which in some cases terminated in a narrow cylindrical well. These designs created a deep pool at the bottom of the hearth, where the resulting hydrostatic pressure helped in the draining of liquid iron and slag. The pool also allowed the lighter slag to separate more cleanly from the liquid metal. Inverted conic/cylindrical hearth designs of this kind are from at least as early as the seventeenth century (Pfannenschmidt 1977).

*TABLE 3.2. Some names associated with shaft furnaces in Europe*

| Name(s)-origin | Product | Approx. heights, m. | Sources |
|---|---|---|---|
| Rennofen (Germany, Austria) | Bloom | 1–2 | Wertime 1962: 60 |
| Blasshutten<br>Blaseofen (Germany)<br>Bauerofen | Bloom | 2 | Percy 1864: 32<br>Wertime 1962: 58, 60, 76 |
| Osmond (Sweden) | Bloom<br>cast iron | 2–2.5<br>3–5 | Wertime 1962: 77<br>Percy 1864: 321 |
| Hohe Ofen (Germany)<br>Stuckofen (Austria)<br>High Bloomery (England)<br>Haute Fourneau (France)<br>Canet (Italy) | either | 4–5 | Wertime 1962: 60<br>Percy 1864: 326–329<br>Johannsen 1953: 133 |
| Flossofen (Germany)<br>Blauofen | cast iron | 5+ | Wertime 1962: 60, 72<br>Johannsen 1953: 139, 141 |

The upper part of the shaft, in which the ore and fuel were packed, occupied the greatest portion of the furnace height. This zone was in the form of an erect truncated cone or pyramid, while the original cylindrical shaft with vertical walls disappeared from use. The term *bosh*, used even today, is applied to the plane of jjncture between the inverted conical hearth and the erect conical shaft. By the eighteenth century the average bosh diameter was 1.5 to 2 m. For comparison, a typical blast furnace of the 1960s has the following characteristics: output per day—4200 tons; overall height—43 m.; and bosh diameter—ca. 10 m. Some recent blast furnaces of Japanese design have about twice this capacity.

## 7. THE DEVELOPMENT OF OTHER SMELTING FURNACES

Western European blast furnaces have had a fairly consistent pattern of growth in size during the past

Fig. 3.9. *Chinese Smelting and "Puddling" Furnaces.* A woodblock illustration from the seventeenth-century technical work, *Tiangong Kaiwu* [T'ien-kung K'ai-wu]. It shows a short shaft smelting furnace equipped with a box bellows. The liquid cast iron flows into a basin-like holding furnace and then into a rectangular apparatus in which, according to the accompanying text, the cast iron is converted to forgeable iron or steel by sprinkling on a wet earthy powder while stirring with willow poles. None of the many commentators on the drawing has felt that the system would work as described: liquid iron in an open puddling hearth would certainly freeze before significant decarburization occurred. (From T'ao Hsiung's 1929 edition, based on a version in the Imperial Library, Tokyo; compare Sun and Sun 1966)

three hundred years. As noted above, discussions of pre-industrial technology in general, and of metallurgy in particular, usually treat this trend in evolutionary terms. It is usually assumed that the smallest and simplest techniques are the earliest and that a fairly regular pattern of increase in size and complexity characterizes the rest of the history of the exploitation of iron.

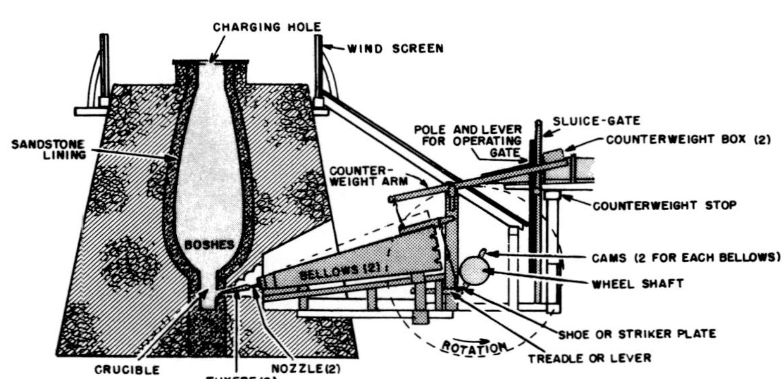


Fig. 3.10 *Blast Furnace at Saugus, Massachusetts.* A schematic section of the Hammersmith furnace, operated in the latter half of the seventeenth century. The two side by side bellows were driven by a water wheel. Liquid cast iron ran out through a tap hole at the base of the "crucible" (i.e., deep hearth). According to Hartley (1957: 172), the exterior measured 7.9 m. square at the base and was 6.4 m. high. (After McGannon 1964: 6)

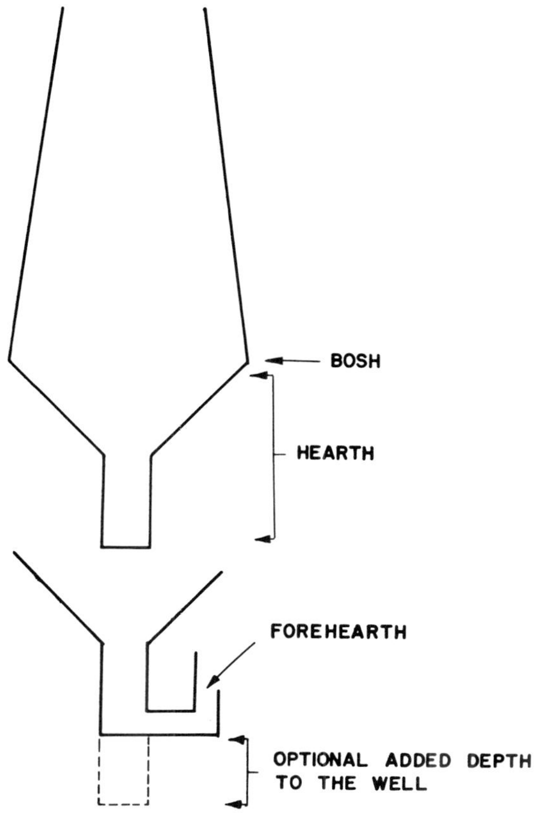


Fig. 3.11. *Inner Profile of a European Blast Furnace.* The biconical form, already used in the late seventeenth century, has survived to the present day, as have the terms *bosh* and *hearth*.

While modern ironmaking technologies are more complex than ancient ones, it is doubtful that the difference has much value for explaining why a given system was used in a particular place, and why it survived or was eventually replaced. First, the archaeological and historical record is not clear about the chronological priority of basic types of furnaces. Second, there is no theoretical basis for assuming that smaller furnaces predate larger, since economic and sociocultural factors have always controlled technological change. It is meaningless to classify technical phenomena like furnace designs in terms of primitiveness or sophistication; what matters is how well these served the needs of the society in which they functioned. There may have been a tendency in any technological tradition to improve quality, increase output, and optimize the efficient use of outputs, but this was not necessary. Currently, conflicts exist among purely technical goals, for instance between output quality and production efficiency, and between these and goals of other kinds. Further, processes are selected on the basis of resource availability, capital cost, market opportunity, and cultural outlook, as well as their technological feasibility.

The inhabitants of sub-Saharan Africa, for example, had many ironmaking methods. Although these pre-modern Africans may have been capable of building large furnaces with large outputs, their efforts would have been wasted. Local markets were limited, while supralocal commerce, for political rather than economic reasons, was subject to many restrictions. Consumers had the iron they needed, and it is possible that per capita consumption of iron in parts of Africa exceeded that of most of pre-industrial Europe. The multiplicity of small furnaces in operation provided employment, and some of them turned reasonable profits.

The Bassari of Togo operated one of the most productive African ironmaking systems. At the end of the nineteenth century, they were exporting 500 to 1000 tons of iron each year to other parts of Togo, Ghana, and Dahomey (Cornevin 1962). This level of production was reached by running several hundred shaft furnaces between 3 and 3.5 m. high, each capable of smelting 30 kg. of iron every three days. The Bassari furnaces were not exceptionally efficient in the use of fuel or ore (Table 14.3a), although the labor efficiency of such furnaces, which did not require bellows, may have been high. The African blacksmiths who bought the iron appreciated the quality of the product (Hupfeld 1899), since it sold at prices high enough to make the Bassari a prosperous tribe.

There are few aspects of this system which could have been improved. Increasing total production would have been pointless, since the market for premium grade iron, already exported distances of several hundred kilometers, could not have been enlarged because of transportation difficulties and competition from the many other ironmaking groups in the region. The use of water powered machinery was impractical, since the area is dry. Building larger capacity furnaces would have called for manual bellows and thus a much greater input of manpower. The technology involved in the Bassari system is primitive only if advancement is equated with mechanized, large volume production.

Ironmaking in pre-modern India, despite its dense population, large markets, and developed economic institutions, always remained at a small or medium scale. Perhaps because of the excellence of its economic organization which coordinated the output of large numbers of cottage-level producers, Indian steel dominated much of the international market for more than a thousand years. It is possible that there was little incentive for local capitalists and inventors to interfere with a system that worked well.

## 8. THE CLASSIFICATION OF SMELTING FURNACES

Attempts have been made to classify furnace designs according to logical evolutionary schemes, influential among them being those proposed by Klusemann (1924), Coghlan (1956: 86), Cleere (1972), and Tylecote (1980). None is completely satisfactory, yet no significant improvement has been formulated. A meaningful classification scheme cannot be constructed even if developmental aspects are not considered.

Some have sorted furnaces according to cross section, dividing them into bowl and shaft furnaces, and making further subdivisions. Others have recognized that certain design features unrelated to overall shape, such as whether slag can be tapped from the furnace or whether

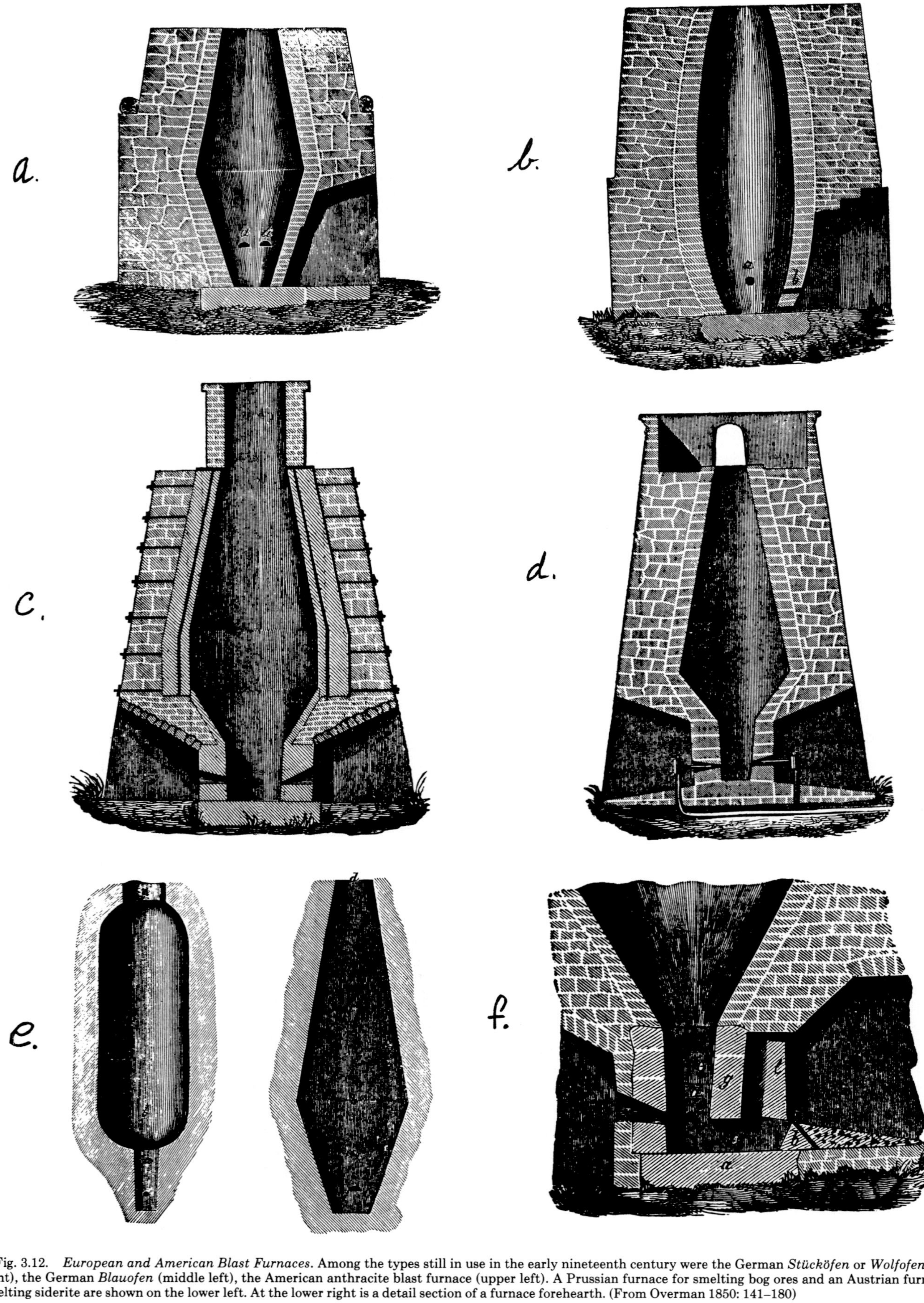


Fig. 3.12. *European and American Blast Furnaces.* Among the types still in use in the early nineteenth century were the German *Stücköfen* or *Wolfofen* (upper right), the German *Blauofen* (middle left), the American anthracite blast furnace (upper left). A Prussian furnace for smelting bog ores and an Austrian furnace for smelting siderite are shown on the lower left. At the lower right is a detail section of a furnace forehearth. (From Overman 1850: 141–180)

it has single or multiple tuyeres, can be equally important. The second approach is preferable, but there are other design features that could have as much functional significance as overall shape or provisions for tapping slag. These include features related to:

- Charging—can it be charged continuously or can only a single charge of fixed volume be smelted during one run?
- Metal removal—can the smelted product be extracted from the furnace without shutting it down?
- Preheating—is the charge or the incoming air, or both, significantly preheated before reaching the combustion zone?
- Air supply—can natural draft be used? Does the design require the use of bellows too large for manual operation?
- Separation of heating and reduction with the flame out of contact with the ore

Other features which could be considered in the classification of furnace systems include durability, insulating qualities, ability to separate metal cleanly from slag, ability to char or coke raw fuel within the furnace itself, and the presence or absence of a charcoal-covered furnace bottom. Yet another relevant feature is simply size, for it is conceivable that two identically designed furnaces, one of them twice as large the other, would operate differently.

If all the significant features are considered, the combinations are too numerous to be useful. If, on the other hand, the majority of these features are arbitrarily excluded from the classification scheme, the resulting system is meaningless in technological and economic terms. It is therefore best to avoid systematic taxonomy and to use a simple classification based on those distinctions already recognized in the literature: the differences between indirect (reverberatory and crucible) and direct (shaft and bowl) heating furnaces. In the pre-industrial period indirect heating furnaces were not often used for ironmaking, so subdivision is unnecessary. Bowl furnaces, however, were common. The main problem in their classification is that most of the formal and operational traits are distributed continuously. The developed Catalan hearth is an exception and, with related smelting hearths in other parts of Europe, constitutes a distinctive family of furnace types. But otherwise, each bowl furnace within a village may differ considerably, and those within a given region may cover a broad range of shape, insulation, slag-tapping capability, position relative to the surface of the ground, number of bellows and tuyeres, and so forth (for an example of one kind, see Fig. 3.13). Therefore, no attempt has been made here to classify bowl furnaces.

However, one unusual variety of bowl furnace is the *tatara*-like hearth of the Kui of northern Cambodia, observed in the 1880s (Moura 1882; 1883: 43–46; Beck 1891, v.1: 1009–1011). The hearth itself stands on an earthen platform 80 cm. high; its interior is 90 cm. wide, 250 cm. long, and 40 cm. deep, with walls of fire resistant earth. Each of the long sides is pierced by 26 holes, furnished with a tuyere connected to a bamboo pipe. All the pipes on each side are connected to a single large drum bellows of the Indian type, depressed by the feet of three operators and raised by a cord attached to an overhead bamboo spring—a unique device in Southeast Asia, where metalworkers traditionally used mainly vertical piston bellows, with some Chinese box bellows and simple bag bellows made of skins. The daily yield was about 10 to 15 kg. of notably good iron. The unusual shape, together with the unique blowing system, suggests that this is a distinct bowl furnace type.

The only general class of furnace that can be readily divided into subgroups is the shaft furnace, which includes a variety of distinctive, geographically limited designs that might be considered types. The traditional nomenclature of shaft furnace types in Europe is complex and not entirely consistent, being based not only on formal and functional traits but also on idiosyncratic local preferences, so that types similar in appearance and operation appear under different names. Many of these names are useful guides to their ancestry, however, and others recur frequently in older writings on historical metallurgy (Table 3.2; Fig. 3.12).

Distinctive types of non-European shaft furnaces also exist, along with many that are too variable or too poorly described to be easily classified. Some of the more historically important and technically interesting are discussed below. Unless otherwise stated, all are

Fig. 3.13. *Malagasy Bowl Furnace.* As used on Madagascar during the mid-nineteenth century, such furnaces were covered by a low mound but provided with a short vent pipe. The air supply was from pairs of manually operated air pistons of the Southeast Asian type (see also Fig. 7.3). To the right of the furnace mound is a forging operation using a pit hearth that is an open-topped version of the bowl smelting unit. (From Ellis 1858: 264)

made of clay, have some provision for slag tapping, and produce bloom iron.

*The West African high natural draft furnace.* Hupfeld's (1899) description is the best available, but there are many other references in the Africanist literature (see Cline 1937). The furnaces are most common in the dry Sahel region of Senegal, Mali, and Upper Volta. Features include a cylindrical shape, a height of several meters, relative permanence, radially arranged tuyeres, and an absence of bellows. Output is about 10 kg. per day, with runs that last several days. No operational examples are extant. A few shaft furnaces using natural draft are also known from southern Central Africa, in the area between Lake Nyasa and Lake Tanganyika; these resemble the West African variety but are usually shorter and less permanent. The Fipa furnace (Wembah-Raschid 1973) appears to have been broken up after each smelting run.

*The Indian bank furnace* (Fig. 3.14; Buchanan 1807, v.1: 171–174; v.2: 436–440; Franklin 1971; Heyne 1814: 219–223). Percy's (1864: 255–570) tripartite classification of Indian smelting furnaces is inadequate. There are actually many kinds of furnaces in India. The common characteristic is that the furnace is built into a manmade earthen bank so that three sides are supported by earth while the fourth side is covered by only a thin screen of clay. Several furnaces are often built in a row along the same bank. The tops of the shafts stand free on all four sides, like chimneys. Most shafts are slot-like, rectangular to horizontal in section, and much less deep than wide. All are blown with bellows. Size and capacity are variable; the largest (in Malabar—modern Kerala) were about 3 m. high and had a daily output of about 100 kg. Widespread in the central and southwestern parts of India until the mid-nineteenth century, the type is now extinct.

*The Indian miniature tube furnace* (Ahmed 1967; Blanford in Percy 1864: 261–263; Buchanan 1807: 601–603; Schwartz 1901: 279–280; Watson 1910: 324–326). As employed by "tribal" smelters in eastern and central

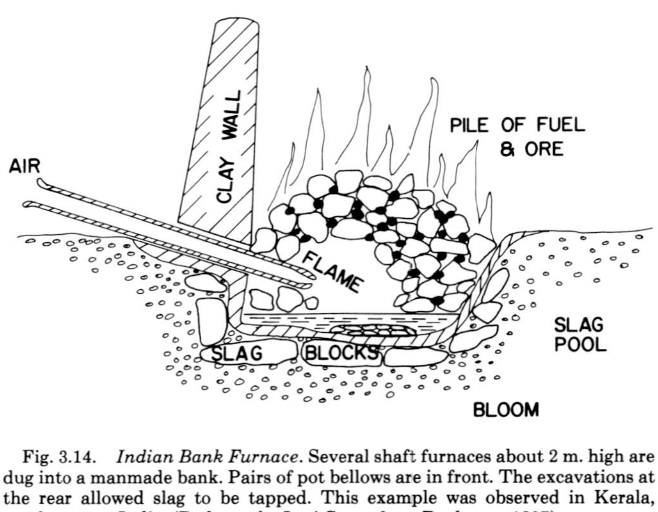


Fig. 3.14. *Indian Bank Furnace.* Several shaft furnaces about 2 m. high are dug into a manmade bank. Pairs of pot bellows are in front. The excavations at the rear allowed slag to be tapped. This example was observed in Kerala, southwestern India. (Redrawn by Lori Grove from Buchanan 1807)

India, these furnaces are about 1.5 m. in height and have inside shaft diameters that may be as small as 20 cm.; daily output is about 10 kg. They are shaped like a freestanding cylinder or elongated cone, and often are furnished with a wide tray-like appendage at the lip that holds the finely divided charcoal and ore which is pushed into the furnace continuously during the several hours needed to form a bloom of several kilograms. All employ bellows, usually of the foot-operated drum variety, and can be worked by one individual, who stands on the bellows while raking fuel and ore into the furnace mouth with a long handled implement.

*The Chinese "cupola" smelting furnace.* The characteristic small furnace of China, attested in the central and southern parts of that country and also in the Philippines, resembles a cupola furnace for remelting cast iron, and may have sometimes filled that role. The inside height is 2 to 3 m. and the diameter between 0.5 and 1 m., tapering from a wider top to a narrower bottom. The outside is often supported by iron bands. In Hunan and Guizhou Provinces, the whole furnace is mounted on trunnions, so that the molten iron is removed by tilting rather than tapping. Blowing is usually done manually, using box bellows. The fuel is often charcoal but may also be either anthracite or coke. Daily output is several hundred kilograms of cast iron, and the furnaces can be operated continuously for periods of several months. Wagner (1985), in the most detailed description, translates and comments on official reports made during the "backyard furnace" campaign of the Great Leap Forward period of the 1960s (see also Tegengren 1923–24: 335, 339; Johannsen 1953: 22). In the early twentieth century, this type of furnace was operated by Chinese located in the Philippines (McCaskey 1903a; 1903b).

*The Chinese high furnace* (Fig. 3.8; Tegengren 1923–24: 339, 342–346, 352). This type was seen by European observers only in the west central and southwestern provinces of China. The furnaces are constructed of firebrick or stone, with an internal height ranging from 6 to 9 m. The diameter at the midpoint is about 2 m.; the shaft is a smoothly curving ellipse that is sharply constricted at the mouth and at the base. Fuel may be charcoal, coke, or anthracite. Daily output ranges from 1 to more than 2 tons of liquid cast iron. All except the tall, narrow Yunnan version reported by Moore-Bennett (Tegengren 1923–24: 352) are blown by large water-powered box bellows; Moore-Bennett says that the furnaces he saw used natural draft.

*The Japanese tatara furnace* (Fig. 12.1). This yields a complex product composed of bloom iron, cast iron, and steel. The term *tatara* may also refer to the horizontal fan bellows with which these elongated multiple tuyere furnaces were usually, if not invariably, equipped. The Koreans used similar furnaces. The Japanese had several other smelting devices, including conventional cylindrical shaft furnaces (Geerts 1875: 11–12) and the small *noro* pit furnaces operated by traditional swordmakers (Nishio 1913: 63; Tanimura 1980: 666–667).

There are also a few minor types that should be included more for their technical interest than their general historical or economic importance.

*The Nigerian vertical tuyere furnace* (Fig. 3.15; for good recent descriptions, see Sassoon 1964; Nadel 1942: 259-265; and Gardi 1969: 15–22). This unique design may still be in use in central and eastern Nigeria and adjacent parts of the Cameroons. The "shaft" is about 1 m. high, 40 cm. wide, and 30 cm. deep, with a large tuyere driven by two small drum bellows projecting downward from the top; this preheats the incoming air producing relatively high smelting temperatures. About 10 kg. of bloom iron are produced each day in a series of very short smelting runs, each lasting an hour or less and producing a 2 kg. bloom.

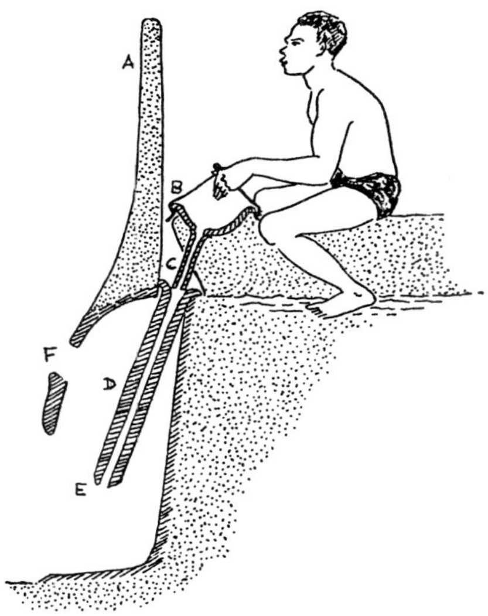


Fig. 3.15. *Nigerian Vertical Tuyere Furnace.* An unusual hybrid of bowl and shaft designs. A. Wall to protect bellows operator. B. Pot bellows. C. Air tube. D. Tuyere. E. Combustion zone. F. Charging hole. A tuyere in this position seems capable of substantially preheating the incoming air. (From Sassoon 1964: fig. 6)

*The Ghanaian leaning furnace.* The chief authority is Pole (1974; 1975), who has traced the distribution of this furnace in northern Ghana with exemplary completeness. It was known to have been used in ten iron-making villages scattered over an area 400 km. by 100 km., belonging to a number of different ethnic groups. It is cylindrical, about 130 cm. high, 20 cm. in internal diameter, inclined at an angle of about 20 degrees from the vertical, and blown by pot bellows that often sit on stands in front of the furnace. Blooms in the 5 to 10 kg. range are the normal result of a one day furnace run. In most of the villages studied by Pole, the furnaces had been in use within the last few decades.

*The Yoruba multiple tuyere domed furnace* (Fig. 3.6). Such furnaces were seen in use by Bellamy (1904) in the Yoruba village of Ola-Igbi in southwestern Nigeria and owe their fame to his notably good description. As very little iron was still smelted in the Yoruba area at the time of Bellamy's visit, the Ola-Igbi examples may be the only such furnaces ever recorded, although slightly similar structures from Mali are described by Gardi (1969) and others. The key characteristics of the Yoruba furnace are its low proportions, with an interior about 2.2 m. wide and 1.1 m. high, and operation entirely by natural draft, supplied through bundles of radially arranged tuyeres.

*The Burmese natural draft furnace.* This furnace is known from only a single locality, Mount Popa in central Burma, yet it is nonetheless frequently discussed in the literature. Interpretations of the shape of the Mount Popa furnaces differ (Blanford 1862; Percy 1864: 270–273; Chhibber 1926). Despite Blanford's opinion to the contrary, the overall design of the furnaces looks Indian: the plan is rectangular, and the entire shaft is dug into a natural slope so that three sides are supported by earth while the relatively thin clay front is held in place with a wooden framework. Each furnace was said to produce about 20 kg. of iron per day.

# 4. Mineral Iron

## 1. INTRODUCTION

Iron ores have sufficient concentrations of minerals for conversion to metallic iron. The availability and character of ores must have influenced the various paths chosen in the history of ironmaking.

## 2. NATIVE IRON

Metals rarely occur in their metallic or "native" form. The chief exceptions are chemically inert elements like gold, platinum, silver, and copper. A few that are more reactive, such as antimony, bismuth, arsenic, lead, mercury, and iron, may also on occasion be discovered in an uncombined state; that is, not associated with other elements. Native gold and copper, used in remote antiquity and conspicuous in nature, may have shown ancient man that metals are distinctive among natural raw materials. Several native alloys may also have been known early: electrum (gold alloyed with silver) and platinum group alloys (platinum with iridium, rhodium, palladium, and osmium).

Native iron of terrestrial origin is extremely scarce. The only reliably reported occurrence of terrestrial native iron in exploitable amounts is on Disko Island off the coast of West Greenland, a deposit mined by local Eskimos (Coghlan 1977: 1). Its nonmeteoritic character has been securely established (e.g., by Bird and Weathers 1977). Very small quantities of native iron dispersed in basalts exist in several other areas, but there is no evidence that these were detected before modern times.[1]

Meteoritic native iron is more common. Approximately 6 percent of all meteorites contain metallic iron, always alloyed with significant amounts of nickel. These extraterrestrial objects weigh from a few pounds to as much as forty tons. It has been estimated that over the last 250,000 years about ten million tons of metallic meteorites have struck dry land—many of them as dust, but still a great quantity (Maddin 1975).

Because nickel is a prominent contaminant of the iron in meteorites, ancient iron artifacts that contain nickel are usually assumed to have been made of meteoritic iron (see, e.g., Gettens et al. 1971; Waldbaum 1980). Those meteorites that consist almost entirely of iron are forgeable, so the fabrication of knives and other objects from them is not difficult. Since the early nineteenth century, for example, pile-welded Indonesian *keris* daggers have been made from an iron-nickel alloy hacked from the large Prambanan meteorite, currently kept at the sultan's palace in Surakarta on Java.[2] The idea that nickel content always indicates meteoritic origin has been challenged by the results of studies of an ancient Swedish axe pile-welded from a high nickel-iron alloy and a smelted iron of low carbon content (Hermelin, Tholander, and Blomgren 1979). The iron-nickel alloy is said to be terrestrial in origin because of its low phosphorus content (about 0.25 percent), which is allegedly atypical of meteorites. However, the chemical analyses of 100 iron meteorites (Moore, Lewis, and Nava 1969) show the following ranges: nickel—5.52 to 18.57 percent; cobalt—0.32 to 1.03 percent; and phosphorus—0.01 to 1.13 percent. It is therefore clear that the phosphorus content of iron meteorites can be low. The problem of recognizing meteoritic iron through chemical analysis, not yet completely resolved, is of importance in historical metallurgy (see Appendix C).

## 3. OTHER IRON MINERALS

All other iron minerals are chemical compounds, which often are closely associated with noniron minerals from which they cannot easily be separated by physical methods. Mixtures of valued and unwanted minerals are called ores. Some of the components of an ore, like manganese, nickel, and phosphorus, can be reduced along with the iron in a smelting furnace; if this happens, the product of smelting is an alloyed iron. The remaining components are driven off as gases or absorbed by the slag. The proportion of iron- (and manganese-) bearing minerals as opposed to slag forming minerals (i.e., gangue) is a crucial factor in evaluating an ore. If the iron mineral content is low and the gangue content high, more ore and fuel will be needed. The furnace will also have to be larger and will probably have a shorter life.

However, an ore is more than a mixture of extractable metals and slag forming rock. The geological processes by which an ore body was formed have an important effect on the ease with which it can be mined, on the possibility of separating unwanted rock before smelting, on the rate at which smelting reactions occur, and on the impurities that may appear in the smelted metal.

From a purely chemical standpoint, the great bulk of past and present iron production relied on two oxides,

---

[1] As early as 1854, Overman (1854: 2) observed that "Native iron has been found in Canaan, Conn., in a vein or plate two inches thick; it is sufficiently ductile to be wrought into nails by a blacksmith. It was found in a mica slate rock, upon a primitive mountain, and very much intermixed with plumbago." This has been overlooked by twentieth-century scholars.

[2] For utilization of the Prambanan meteorite, see Groneman 1910, and Solyom and Solyom 1978. For its date, see Bronson 1987. It has been seen in Solo within the last few years by both Garrett Solyom and Edward Frey (personal communication). For the opinions of President Thomas Jefferson and other historical insights on iron meteorites, see Mehl 1938.

one variable hydroxide, and a carbonate (Table 4.1). Theoretically, all except the sulfide can be reduced to metal in a single step by heating in the presence of carbon monoxide, but actual mineralogy makes successful smelting more complicated. Each compound can have a number of physical forms with different properties caused by variable degrees of hydration. The limited but significant mobility of water-soluble iron has created a series of replacement and residual deposits where iron minerals—particularly hematites—have collected in or have wholly replaced materials like limestone and chert.

The great variety of shapes and textures among the iron minerals may not have confused early miners. They would have been guided by the red, orange, and yellow colors of many hematite and limonite ores, by the gunmetal luster of magnetite and specular hematite, and by the fact that even the less distinctive iron compounds are heavier than most rocks. A few other metal-bearing minerals, including those of mercury and arsenic, may be as brightly colored, and some—like manganese, tin, and lead ores—may closely resemble the darker colored iron ores in appearance and weight. But iron is so much more common than any of these that mistaken identifications cannot have been frequent even in the early Iron Age. Any reddish, yellowish, or lustrous gray-black rock that was also heavy was likely to contain appreciable amounts of iron. The main difficulty was whether that particular rock could be economically smelted: whether it was a usable or "payable" ore rather than an attractive mineral specimen.

## 4. VARIETIES OF IRON ORE

Ores are often given the same names as the minerals which ideally (but not necessarily) form their main components. In order to emphasize the difference between ores and minerals and to introduce the vocabulary used in this book, we list here the principal ores recognized by Western ironmakers. It should be noted that this traditional nomenclature, while by no means comprehensive or consistent, is still more widely used by miners and smelters than the chemically defined mineral names (Table 4.1).

*Bog ore.* A limonite ore formed in a bog, often under peat. Many writers use *bog ore* as a general term to describe all recently formed iron concretions, including lake ores and the nodules in waterlogged tropical clays. Defined in this way the bog ores are the most ubiquitous of all iron ores. The fact that they occur near the surface in otherwise oreless areas explains their significance to early smelters. Bog ores have not been important in recent years except in Japan, where an unusual variety thought to have been formed by deposition from hot waters in volcanic areas was exploited extensively in the 1930s and 1940s (Mitsuchi 1952: 542–546).

*Banded iron formation.* Now a technical term (cf. Trendall and Morris 1983) designating the group of critically important deposits also known as BIF ores, BHQ (banded hematite quartzite) ores, itabirites, taconites, or jaspillites. The great extent of these deposits, along with the fact that their iron contents often exceed 60 percent while gangue contents are in the 1 to 2 percent range, make them the premium ores of the present day. Earlier, however, they were probably not extensively used because many of the principal banded iron formations are too deep or too distant from early ironmaking centers. Among those known to have been mined and smelted in the pre-industrial period are the deposits of Krivoi Rog in the Ukraine, Bihar and Orissa in eastern India, and the Liberian shield in West Africa. The BIF ores of Bihar and Orissa average 60 to 69 percent Fe, 1 to 4 percent $SiO_2$, and 1.5 to 3 percent $Al_2O_3$ with very low phosphorus and sulfur (Krishnan 1964: 12). Many others are as rich.

*Blackband.* A siderite ore that occurs in strata associated with coal deposits. As the name indicates, blackband ores contain at least some coal as well as considerable sulfur. Very important in early nineteenth-century Britain and Germany (e.g., Ure 1856: 1103) and used to some extent in southern China (Tegengren 1923–24: 9), these ores were not generally favored by pre-industrial ironmakers.

*TABLE 4.1. Selected iron minerals*

| Name | Chemical formula | Iron content | Description |
|---|---|---|---|
| Magnetite | $Fe_3O_4$ | 72.35 | Black to grey. Magnetic. May be associated with ilmenite and thus titaniferous. Density 5.2. Hardness 6 (Mohs' scale)*. |
| Hematite | $Fe_2O_3$ | 70.0 | Orange through black but most often red or brown color. The most common. Density 5.2. Hardness 5.5 to 6.5; weathered specimens softer. |
| Specularite | $Fe_2O_3$ | 70.0 | Lustrous gray-black. Important to rock collectors and to ancient Mexican and Iraqi lapidaries but of minor significance as an ore. |
| Goethite | FeO(OH) | 62.85 | Mostly yellow or orange but colors range to dark brown. Density <4. Hardness 5 to 5.5 |
| Limonite | $FeO(OH).nH_2O$ | <62.85 | Usual name for impure goethite. Next to hematite in importance among ores. Often quite soft. |
| Siderite | $FeCO_3$ | 48.21 | Yellow to brown color. Density 3 to 3.8. Hardness 3.5 to 4. Third in importance to traditional European ironmakers but little used elsewhere. |
| Pyrite | $FeS_2$ | 46.6 | Metallic silver-gold color. Density 5.2. Hardness 6 to 6½ (Mohs). Used briefly as an ore in 19th-century Britain (see Section 3) |
| Greenalite | $(Fe,Mg)_6Si_4\text{-}O_{10}(OH)$ | 33.8–37.5** | Often present in significant quantities in Lake Superior-type ores but not usually considered an ore mineral. |
| Chamosite | $(Fe,Mg)_{3\text{-}6}(O,OH)$ | 28.5–37.3** | Green to gray. Density 3 to 3.4. A minor ore mineral in 19th-century Europe. |

*Mohs' scale of hardness for minerals: diamond (10), corundum (9), topaz (8), quartz (7), feldspar (6), apatite (5), fluorite (4), calcite (3), gypsum (2), talc (1).
** = empirical measurements. For greenalite, see Klein 1983: 428; for chamosite, see Amstutz 1971: 155.

*Chamosite.* A complex hydrous iron-magnesium basic oxide of a greenish color, known to have been mined mainly at Nucitz in Bohemia during the eighteenth and nineteenth centuries and in England in the early twentieth century. A roasted Nucitz chamosite contains about 44 percent iron (Beyschlag, Vogt, and Krusch 1916: 1042). Chamosites and thuringites accounted for about 1 percent of world iron production in the early 1900s (Beyschlag, Vogt, and Krusch 1916: 1094).

*Clay ironstone.* An impure siderite ore which, like blackband, is associated with deposits of coal. The term *ironstone* without a "clay" prefix has a broader meaning (see below).

*Fayalite.* A simple iron silicate mineral which, although present in some magnetite ores that are currently smelted (cf. Klein 1983: 452–453), is unimportant in its natural form. It is included here because it is a major component of bloomery slags. When, as often was the case, these slags were later resmelted in blast furnaces, the constituent fayalite (as well as the wustite, magnetite, and the "glassy phase") became the equivalent of an ore. In modern idiom, bloomery slags are often called fayalite slags even though the fayalite mineral is only one of several components.

*Goethite.* A mineral rather than an ore, though sometimes used as a synonym for limonite in metallurgical contexts (Table 4.1). A constituent of bog and lake ore.

*Gossan.* The weathered near-surface portion of many pyrite/chalcopyrite deposits. It has a high content of iron in oxide form, the other metals and the sulfides having been leached out and deposited in deeper levels. Residues of sulfide and copper may make some iron-rich gossans undesirable as ores, but they were exploited in some areas. LeClère (1901: 354), for example, saw gossan hematites being smelted for iron in Yunnan, southwestern China.

*Greenalite.* Also a mineral rather than an ore (Table 4.1).

*Hematite.* Ores at present called hematites are likely to be from one of the Precambrian deposits called *banded iron formations*. However, the term can be applied to any ore in which most of the iron present is in the form of the nonhydrated sesquioxide. In the earlier literature, the designation of an ore as hematite may have meant only that it was red or brown and not conspicuously nodular in form. One of the most historically significant hematite deposits is on Elba in Italy, a mine renowned for the purity of its ore and thought to have been worked continuously from Etruscan times to the present.

*Iron formation.* A common but easily misunderstood synonym of banded iron formation.

*Ironstone.* An imprecise term used by some authorities to mean clay ironstone, by others to mean a dense weathering product derived from softer underlying ores, and by still others to mean any hard, dark iron ore that is otherwise unidentifiable.

*Lake ores.* Limonite-hematites that occur as concretions on freshwater lake bottoms. Although formed in recent times (and probably still forming), they closely resemble oolitic ores. Lake and bog ores were widely collected and smelted in northern Europe and eastern North America through the late nineteenth century. Although low in iron content, they were readily available and had a porous structure that made them easy to smelt. Concretions of a similar type occur in lakes and deltas in tropical areas; these too are known to have been exploited by early ironmakers (e.g., Harrison and O'Connor 1969).

*Laterite.* A residual hematite-limonite ore formed in ferruginous tropical soils through the action of abundant rainfall, which washes most other soil components such as silicates and alumina down to deeper levels and leaves the iron oxides behind. This ongoing process, called laterization, is a major problem for tropical farmers, but the iron oxide concentrates have often been smelted in Africa and India. Some are rich; Krishnan (1954: 14), for instance, cites an analysis of 57.58 percent Fe and 7.28 percent $SiO_2$ for a laterite in the important traditional smelting area of Jabalpur in Madhya Pradesh, Central India. Although most laterites are much poorer, they exist in enormous quantities. Many authorities agree that laterite ores comprise the bulk of the world's reserves of potentially recoverable iron.

*Limonite.* The term applied to any ore in which limonite—amorphous hydrated iron oxide—is the chief iron mineral. It is loosely used to designate almost any mixture of hydrated oxides (perhaps with some nonhydrated sesquioxides as well) that has a yellowish rather than reddish color, including most bog and certain oolitic ores. The majority of the ores called limonites occur in the temperate zones, where they are abundant.

*Magnetite.* This mineral ($Fe_3O_4$) can be considered an ore when it occurs in relatively pure form, as in massive deposits associated with intrusions of magma (e.g., Kiruna in Sweden) or as placer sands (e.g., in Japan, India, China, and West Africa). Because of its density, it is relatively easy to separate hydraulically from the silica in the sands. Like siderites, many magnetite ores were highly regarded by pre-industrial steelmakers. However, the hard and compact nature of the more massive magnetite deposits made them difficult to mine and smelt.

*Minette ore.* Oolitic limonites in Lorraine and elsewhere on the French-German border, rarely worked until the nineteenth century and conspicuously low in grade.

*Oolitic limonite.* An ore that occurs in nodules of "fish-egg" size, very widespread. The minette ores are the best-known example.

*Pisolitic ore.* "Pea"-sized ores, either limonites or hematites. Essentially a size variant of oolitic ores.

*Pyrite.* Large deposits of pyrite ores are not common, and most of those worked in early times were exploited for copper (in the form of chalcopyrite) and other nonferrous metals. During the nineteenth century, intensive efforts were made to extract a usable iron from the pyrite ores mined at Huelva in southern Spain and shipped to England because of their sulfur and copper contents; methods of smelting the iron-rich residues were eventually developed (Checkland 1967). However, such ores are not known to have been used earlier for iron production, although dispersed pyrite mineral was undoubtedly a minor component of many other iron ores.

*Siderite.* The term as applied to an ore often implies not only that iron carbonate is the principal iron mineral present, but also that the ore is not from a coal-associated sedimentary deposit. Nonsedimentary siderite ores occur in veins and larger ore bodies in complexly mineralized regions. Important examples include the iron mountains at Eisenerz and Hüttenberg in Austria and the "lode" deposits of Siegerland in northern Germany. Such ores are moderately abundant in those countries and have played a prominent role in the manufacture of premium grade iron and steel since ancient times, but they appear to have been much less important in other regions. In India, for example, the 113 ore deposits reported to have supplied traditional bloomery furnaces include a few siderite-containing black band and clay ironstone ores, but not a single nonsedimentary siderite ore of the German-Austrian type (LaTouche 1918, v.2: 231–281).

*Spathic ore.* A synonym for siderite.

*Thuringite.* An ore formerly mined at Schmiedefeld in the Thuringerwald, central Germany. Although a silicate, in its oolitic structure and low iron content it resembles minette.

## 5. THE QUALITY OF IRON ORES

Academic writings on metallurgy often characterize the quality of an ore by calculating the percentage of metallic iron it contains. The famous itabirite ores of Minas Gerais in Brazil are said, for example, to contain an average of 60 percent iron or the minette ores of eastern France 25 to 30 percent. This is, however, an unsatisfactory index to the value of an ore in terms of actual mining and smelting, since other features may be at least as important as overall iron content.

Ease of mining is one. During the pre-industrial period, before the advent of explosives, miners preferred soft and easily excavated ore, which often caused them to neglect rich deposits in favor of softer deposits with a lower iron content, or to dig substantial underground galleries even though large amounts of high grade ore were still present on the surface. About 1910, Coggin-Brown (1923: 88) observed a large surface outcrop of high grade hematite above an operating underground iron mine in southwestern China:

> It would be expected that the miners would prefer to win rich ore from the outcropping portion, rather than to mine it. But the tools at their disposal are of poor quality and gunpowder does not appear to be used locally, so that it is really easier to make small tunnels in soft portions of the ore body, than to attempt to smash up the hard portion which weathering has left on the outcrop.

On the other hand, some easily obtained ores may not have been used because they produced unsatisfactory metal; it is likely that a number of phosphorus- and sulfur-containing deposits fell into this category. Other high grade ores—such as the Lake Superior-type banded hematites of eastern India and the nearly pure magnetite in the mountains at Kiruna and other places in northern Sweden—were probably also avoided because of the difficulty of mining and smelting them. One Western ironmaster who worked in India for many years was struck by this neglect of high grade ores: "Even in Chanda District [in central India], lateritic ores were preferred by the natives to the many rich magnetites and specular hematites that occur there in large quantities, for the lateritic ores were easier for the natives to mine and to smelt" (Schwartz 1901: 338).

Not all ores that were difficult to handle were rejected. In India, Japan, China, and Africa there was widespread use of magnetite sands even though the recovery and concentration of magnetite from them by hydraulic methods was a labor-intensive process. Smelting these sands was also difficult, and some Western students of metallurgy (e.g., Richardson 1934) have regarded magnetite sands as unsmeltable. Yet a marked preference for such ores existed even in places where, as in many parts of India, alternative supplies of good ores were readily available. In at least some instances, as at Sikhbar near Darjeeling, magnetites were explicitly recognized as being more difficult to smelt but were used because they "produced iron of superior hardness" (LaTouche 1918, v.2: 239).

Pure magnetite and hematite are high in iron content, while pure limonite and siderite are relatively low (Table 4.1). For this reason, the former may seem preferable to the latter since they require less fuel to be smelted and give higher yields of iron, but in reality it makes little practical difference whether the amount of oxygen and carbonates in chemical combination with the iron is large or small. What matters most in smelting an actual ore, as opposed to a textbook mineral, is the percentage of gangue it contains.

The proportion of nonreducible gangue minerals like lime and aluminum silicate (expressed as percentages of CaO, $Al_2O_3$, and $SiO_2$) is markedly variable (Table 4.2). A high proportion of gangue is an especially serious problem in bloomery smelting, in which fuel costs are closely dependent on the amount of slag produced. Energy is required to liquefy the slag in any furnace, but in a bloomery much of the energy used in reducing the iron is also lost, because slag fluidity at bloom-forming temperatures can be achieved only by using iron oxides as a flux. No less than 2.3 percent of the available iron is lost to the slag for every 1 percent of silica in the ore. Very large quantities of fuel must therefore be expended before any recoverable iron appears in the bloom.

Hence, the gangue content of an ore and the type of iron mineral are likely to be much more important to a

*TABLE 4.2. Some characteristics of iron ores*

| Location | Major iron minerals | Iron content, % | Major non-iron minerals,* % |
|---|---|---|---|
| Hupei, China | hematite | 60–62 | 0.2–5 Mn oxides; 3–5 $SiO_2$; 0.05–0.25 P; 0.05–0.12 S; some Cu[1] |
| Chitaldroog, India | hematite | 54–61 | up to 3.5 Mn oxides; 1–3 $SiO_2$[2] |
| Togo, West Africa | laterite | 69 | 1.5 $SiO_2$; 0.02 P[3] |
| Elba, Italy | specularite, hematite limonite | 50–68 | 6 Mn (local); low P, 10 $SiO_2$[4] chalcopyrite, pyrite[5] |
| Minette, France | oolitic limonite | 25–45 | 0.3 Mn, 0.7–0.9 P[6] 12–30 CaO, 5–10 $SiO_2$, 2–10 $Al_2O_3$[7] |
| Ironbands, Britain | siderites, limonites | 25–35 (dried) | 0.3–1 Mn oxides; 1.4 $P_2O_5$[8] 9 $Al_2O_3$; 6 CaO; 12 $SiO_2$; 4 MgO[9] |
| Siegerland, Germany | siderite | 35 50 (roasted) | up to 10 Mn; up to 1 CaO; up to 0.2 S[10] $<$ 0.3 P, up to 0.6 Cu[11] |
| Krivoi Rog, Russia | specularite, hematite | 50–69 | 1–7.5 $SiO_2$; $<$ 0.02 P[12] |
| Erzberg, Austria | siderite | 38–40 | 2 Mn; 0.03 P; 0.04 S[13] |
| Bilbao, Spain | siderite, hematite, limonite | 39–54 | $<$ 1 Mn; $<$ 0.02 P; $<$ 0.27 S; 1–2 $Al_2O_3$; 1–5 CaO; 5–10 $SiO_2$[14] |

[1]Tegengren 1923–24: 185–190, [2] Sambhasiva Iyer 1901, [3]Hupfeld 1899, [4]Beyschlag, Vogt, and Krusch 1914, 1: 178, 369–372, [5]Bray 1942: 20, [6]Beyschlag, Vogt, and Krusch 1916, 2: 1009, [7]Bray 1942: 21, [8]Bray 1942: 21, [9]Beyschlag, Vogt, and Krusch 1916, 2: 1024, [10]Bray 1942: 20, [11]Beyschlag, Vogt, and Krusch 1916, 2: 803–804, [12]Beyschlag, Vogt, and Krusch 1916, 2: 1059, [13]Beyschlag, Vogt, and Krusch 1916, 2: 817–820, [14]Beyschlag, Vogt, and Krusch 1916, 2: 832.
*Reported as percentages; sometimes as a mineral or a slag component and sometimes as an element; sometimes specific analyses and sometimes as ranges.

bloomery operator than total iron content. For example, an ore that contained 50 percent iron, 30 percent silica, and 70 percent magnetite would be unsmeltable by bloomery methods because there would be too little iron to flux the slag. A second ore, also with a 50 percent iron content but composed of 95 percent siderite and 5 percent gangue, would yield substantial amounts of metal with a minimal consumption of fuel. A third ore, with 95 percent magnetite and 5 percent gangue, would be only slightly preferable to the second. Although a kilogram of the magnetite ore would contain a third more iron than the siderite ore, thus lowering transportation costs from the mine, the two ores would perform about the same in the furnace. The finished metal from the magnetite ore might have a slightly lower fuel cost, but the difference would not be significant.[3]

## 6. BENEFICIAL ELEMENTS IN ORES

Several metals sometimes associated with iron in ores could have been reduced in early iron smelting furnaces. However, only nickel and manganese are likely to have improved the quality of the finished metal.

Iron ores that contain nickel are not especially rare. Many are iron pyrites with admixed nickel sulfides, which are unlikely to have been smelted in antiquity. Others, however, are lateritic in origin, consisting of varying amounts of iron oxides and a few percent of the mineral garnierite, a nickel-magnesium silicate derived by weathering from underlying rocks with a dispersed nickel content. Although a number of the principal deposits of garnierite-containing iron ores are in places like Cuba and New Caledonia, where iron was unknown before the arrival of Europeans, a few were accessible to pre-industrial peoples who smelted iron. These include the ores of Frankenstein in Silesia, Poland, and of Soroako on Sulawesi, Indonesia. Both deposits were worked during the eighteenth and nineteenth centuries.

The Frankenstein ores are for the most part poor in iron, with a maximum content of 8 percent (Beyschlag, Vogt, and Krusch 1916, v.2: 951), too low for ancient smelting unless the deposit once had portions with a much higher iron content. The Soroako deposits, on the other hand, contain up to 55 percent iron in the form of limonite, along with 1 to 2 percent nickel. An ore of this quality might have been smelted at an early date for the iron alone, even if the unusual characteristics of the metal were unnoticed. By the mid-nineteenth century, however, Soroako iron appears to have been regarded as different. It was exported to several other islands and recognized explicitly by the Javanese as a material that behaved in the same way as the meteoritic nickel-iron from the Prambanan meteorite (Groneman 1910; Bronson 1987).

No smelting experiments have yet been carried out on nickel-containing ores. If a sample of Soroako ore with a 2 percent nickel content were smelted in an ordinary bloomery, an iron that contained some nickel might be produced, but it is far from certain that the amount of nickel present would be enough to affect the performance of an object made from this iron. Iron-nickel alloys with less than 5 percent nickel are not much harder or stronger than ordinary iron (Chapter 2). The main effect of adding small amounts of nickel would be to render the metal more resistant to etching by acids; Indonesian *keris* makers exploited this decorative potential. Although none of the known iron-nickel ores contains enough nickel to yield a noticeably superior metal, the nickel content could have significantly affected the performance if the iron were to be converted to steel.

Manganese-containing iron ores, on the other hand, may have been important in several regions before modern times. Manganese is perhaps the single most beneficial alloying element besides carbon that can enter iron by accident during the smelting process. The manganese content of an iron deposit can range from only a few percent to over 30. An ore with 30 percent

[3]The siderite ore would contain 0.083 kg. of gangue per kg. of iron and the magnetite ore only 0.05 kg., giving the latter a slight advantage in terms of fuel consumed per unit of product. On the other hand, the more porous structure of a roasted siderite would make it easier to smelt and thus would offset much of this advantage.

manganese and 30 percent iron would now be termed a manganese ore, but it likely would formerly have been treated as an iron ore by bloomery operators. It would yield useful quantities of iron despite its manganese content because the manganese oxide would serve as an effective flux for the slag.

Many European limonites and siderites contain appreciable amounts of manganese, the oxide of which behaves much like the oxides of iron in terms of solubilization, transport while in solution, and precipitation through geological processes. Twenty-five percent of the iron ores in Germany bear from 12 to 30 percent manganese, and the famous manganiferous siderites of Styria and Carinthia in Austria may have even more (Beyschlag, Vogt, and Krusch 1916, v.2: 814). High manganese values also exist in many lake and bog ores, and in a number of the metasomatic and sedimentary limonites and siderites of Sweden, Norway, Yugoslavia, Finland, and the Soviet Union. Low manganese values characterize most magnetites (with the major exception of those at Dannemora in Sweden), some oolitic limonites (e.g., the minette ores of France and Belgium), and many of the hematites that come from contact deposits and "iron formations."

In sub-Saharan Africa and in Asia other than in Siberia, manganiferous iron ores are much less common. A number of hematites in India have fairly high manganese contents, and scattered occurrences of similar ores are reported in East Asia and Africa, but manganiferous hematites and magnetites seem no more common there than in Europe, and manganiferous limonites and siderites are almost entirely absent. It is possible that this partly reflects incomplete exploration, since Asia and Africa are not as well known geologically as Europe. But the scarcity could be real, and manganese-containing iron ores are less common in tropical and semitropical regions. A verified interregional difference of this kind would have far-reaching effects on comparative studies of European and non-European traditional ironmaking.

## 7. UNDESIRABLE ELEMENTS IN ORES

Some iron ores contain significant amounts of disseminated pyrite, $FeS_2$ (see Tylecote 1976: table 28). In the normal course of passing through an iron smelting shaft furnace, some of the sulfur in an ore will be absorbed by the metal product, whether bloom or cast iron. Ores containing pyrite are uncommon enough in most regions that they could be easily avoided. Those that could not may have been usable anyway, because roasting ores before smelting them, often done by primitive smelters to break the ores up and make them more porous, would have had the secondary effect of driving off most of the sulfur.

Phosphorus presented more serious difficulties, because many iron ores contain substantial quantities of phosphates (Table 4.2). At iron smelting temperatures, these phosphates are reducible to elemental phosphorus in vapor form which is readily absorbed by all types of metallic iron. If the metal product is bloom iron, phosphorus absorption above certain levels will cause it to have poor bending qualities at room temperature, and cracks will form after even minor deformation.

The problems of making good iron from high phosphorus ores could not be solved in the pre-industrial context. No preliminary treatment, such as hand sorting or roasting, could remove mineral phosphates and prevent the production of an unsatisfactory metal. The early ironmaster had to learn from experience not to use ore from certain deposits, no matter how attractive they might appear superficially. For example, some rich iron ores in Germany and France that are high in phosphorus were traditionally considered inferior. Not until the twentieth century, when the technique of phosphorus absorption into the slag had been mastered, did these ores became valuable.

Oolitic limonites and bog ores are particularly variable in phosphorus content, which ranges from less than 0.04 percent to as high as 3.6 percent. This is undoubtedly because ores of this kind were originally formed by precipitation in aqueous environments where phosphorus-containing organisms were abundant. Biogenic phosphates may be occluded during the sedimentation process (as in the Newfoundland oolitic ores—Hayes 1915). Phosphorus ions in solution may be bound into the mineral crystal linkages at the moment of precipitation. Because such ores are widely distributed, readily accessible, and—due to their porous nature—

*TABLE 4.3. Phosphorus and sulfur levels in limonite ores of Europe**

| Iron, % | Phosphorus, % | Count | Sulfur % | Count |
|---|---|---|---|---|
| 60–70 | 0–0.2 | 2 | 0–0.1 | 3 |
| | 0.2–0.4 | 0 | 0.1–0.2 | 0 |
| | 0.4–0.6 | 0 | 0.2–0.3 | 0 |
| | 0.6–0.8 | 0 | 0.3–0.4 | 0 |
| | >0.8 | 22 | 0.4–0.5 | 0 |
| | | | >0.5 | 0 |
| 50–60 | 0–0.2 | 0 | 0–0.1 | 23 |
| | 0.2–0.4 | 10 | 0.1–0.2 | 0 |
| | 0.4–0.6 | 5 | 0.2–0.3 | 0 |
| | 0.6–0.8 | 3 | 0.3–0.4 | 1 |
| | >0.8 | 12 | 0.4–0.5 | 0 |
| | | | >0.5 | 2 |
| 40–50 | 0–0.2 | 20 | 0–0.1 | 20 |
| | 0.2–0.4 | 21 | 0.1–0.2 | 2 |
| | 0.4–0.6 | 11 | 0.2–0.3 | 1 |
| | 0.6–0.8 | 6 | 0.3–0.4 | 1 |
| | >0.8 | 24 | 0.4–0.5 | 0 |
| | | | >0.5 | 1 |
| 30–40 | 0–0.2 | 7 | 0–0.1 | 9 |
| | 0.2–0.4 | 7 | 0.1–0.2 | 0 |
| | 0.4–0.6 | 0 | 0.2–0.3 | 0 |
| | 0.6–0.8 | 0 | 0.3–0.4 | 0 |
| | >0.8 | 29 | 0.4–0.5 | 0 |
| | | | >0.5 | 0 |
| Totals | | 179 | | 63 |

Note: In cast iron smelting, the recoveries of both iron and phosphorus are nearly 100%. In bloom iron smelting, the recoveries of iron depend on the silica content in the ore and the fuel-to-ore ratios but were rarely above 50% (see Tables 9.2 and 9.3). The recoveries of phosphorus cannot be calculated and from the meager data collected in Table 10.4 are highly variable but are probably less than 50%.

Recoveries of sulfur are likely to be very high in bloom iron smelting if the ore was not previously roasted. With roasted ores the recovery should be very low. In cast iron smelting with charcoal the recovery should be very high without roasting. When coke or coal was used for cast iron smelting, the resultant sulfur content in the metal is completely governed by the sulfur content of the fuel.

*Abstracted from analyses in Doelter and Leitmeyer 1926.

easily smelted, they were often exploited. Their phosphorus content is undoubtedly responsible for the deficient cold formability of some kinds of iron.

Analyses of more than 200 limonite ores from all parts of Europe (Doelter and Leitmeyer 1926: 680–714) show the extent of the problems caused by phosphorus and sulfur in early smelting and fining (Table 4.3). The sulfur content of most limonites is evidently low, but the phosphorus content is often high. Cast iron gains in fluidity by the addition of phosphorus, but low carbon iron and steel are rendered more or less cold short by the amount of phosphorus in many limonites. Roughly 30 percent of the phosphorus present in an ore is absorbed by the metal when smelted in a bloomery. The indirect process, because of the higher temperatures reached in blast furnaces, extracts nearly 100 percent of the phosphorus present; it is absorbed by the cast iron and passed on to the wrought iron at the fining stage. European sources of the sixteenth through nineteenth centuries often state that wrought iron was more likely to be cold short than bloom iron.

It seems likely that 0.5 percent phosphorus is the level above which cold shortness is a serious problem (Table 2.5). Therefore, more than 43 percent of European limonite ores were too high in phosphorus to yield a good iron when smelted in a bloomery, while 60 percent of them were too high when the iron was produced by the blast furnace-finery method (Table 4.3). Even 0.5 percent phosphorus might have been too much in the colder areas of Europe. Under subarctic conditions, cold formability in the workshop and toughness in use might be measurably decreased at phosphorus levels as low as 0.2 to 0.3 percent. Theoretically, then, limonite ores should have been more popular in warm climates than cold ones, and bloomeries should have persisted in cold regions after nearby, but warmer, regions had converted fully to the indirect process.

## 8. MINING AND PREPARATION OF ORE

Most pre-industrial iron mining was carried out at or near the surface, because the value of the ore was rarely great enough to repay the cost of digging the deep shafts and galleries that were necessary, even in early periods, for mining copper and silver. Underground iron mines with galleries several hundred meters long did exist in Germany in the eighteenth century (Gleichmann and Gleichmann 1983: 15–17) and China in the nineteenth century (Tegengren 1923–24: 164–168), but they were unusual. More often ore was quarried from exposed faces or pits where the overburden was not too deep, or simply collected from the surface (Figs. 4.1a, b).

What may have been the most common method of ore recovery in areas with mature but minimally capitalized iron industries is represented at one African mine (Fig. 4.2), a circular pit perhaps 10 m. deep. Layers of iron concretions and bedded ores that are softer and thus more easily worked are often found within a few

Fig. 4.1a. *Collecting Bog Ore in China, Seventeenth Century.* The concretions of limonite are being turned up with a plow. (From the *Tiangong Kaiwu;* compare Sun and Sun 1966: fig. 14–8)

Fig. 4.1b. *Collecting Lake Ore in France, Eighteenth Century.* Limonite nodules could also be dredged up from lake bottoms. (From the *Encyclopedia* of Diderot; Gillespie 1959: pl. 83)

meters of the surface. Before the advent of modern earthmoving equipment, the most economical way to mine these deposits was to pockmark the surface with large numbers of small bell pits, bottle-shaped excavations with narrow entrances that widened once they reached the ore-bearing levels.[4]

Relatively little is known about the extent to which pre-industrial peoples understood geological differences. It seems likely that many had a practical knowledge of stratigraphy, that they could recognize and trace the common kinds of deposits, and that they could make a reasonably good guess about the location of usable concentrations of ore before they dug into the ground. All iron miners in a part of Shanxi in China recognized by name the series of horizontally bedded strata (Table 4.4); they dug vertical shafts from the surface and then followed the richer layers with horizontal galleries, often to a distance of a hundred meters or so (Shockley 1904). A British deposit was mined in a similar way in the 1920s (Fig. 4.3).

Some of the theoretical and technical knowledge needed for early iron mining may have derived from nonferrous metal industries, in which mining operations sometimes attained a high level of technological sophistication. In the first millennia B.C. and A.D., the great copper and silver mines of Spain, Greece, and China already had elaborate manually powered systems for pumping air and water, along with complex arrangements for surveying and timbering shafts and galleries. Some of these methods may have been used occasionally in iron mines, either because certain deposits were of high enough quality to repay the expense or because, in complexly mineralized zones, the miners were also looking for other metals.[5] Generally, however, the low per unit value of iron ore precluded the use of mining equipment with a high capital or operating cost. No iron ore was valuable enough to justify the expense of using long chains of hand pumps (Fig. 4.4) and there are no reports of waterwheel operated pumps in iron mines in either China or Europe.

Tools and other equipment were simple. Those used for the actual excavation of ore did not change greatly between the Early Bronze Age and the time that gunpowder began to be used as a mining explosive in Europe, probably in the fifteenth or sixteenth century A.D. Ore was loosened with picks, chisels, and sledgehammers, pried out with picks and crowbars, piled with shovels and hoes, and carried out in baskets. Fire setting—building fires against the ore face and then quenching the face with cold water—was used in several parts of Eurasia but does not seem to have been very common, perhaps because of both the expense and the danger of asphyxiation.

Fig. 4.2. *Pit Mining in Burkina Faso, West Africa, Twentieth Century.* The pit passes through overburden to reach a stratum of laterite several meters below the surface. Gardi (1968) states that similar excavations in the northern Ivory Coast may be as much as 15 m. deep. (From Forbes 1933: 232)

Hydraulic methods of mining and concentrating iron ores were widespread. They were generally used to process magnetite sands, either concentrated in alluvial deposits or dispersed in an earthy or soft rock matrix. A common technique for working matrix-bound iron sand deposits was practiced by the Khasi of northeastern India in the early nineteenth century:

> A small stream of water is conducted along the slope to the site of the present excavation. The excavators standing on one side of their work, poke out the soil from between the boulders with long poles terminating in iron spikes. The loosened soil tumbles into the stream, and is carried by it violently down a narrow channel to a point 200 yards distant and about eighty feet perpendicularly below. Here a little post is fixed on each side of the stream, and against the upper side of these poles, little bits of stick are laid, so as to form a kind of dam, which stops the

*TABLE 4.4. Layered (vein) structure of an iron ore body in Alabama**

| Layer material | Layer thickness, inches | Depth from surface, inches |
|---|---|---|
| soil and clay | 72 | 72 |
| sandstone | 36 | 108 |
| clay, sandstone, ore | 34 | 142 |
| ore | 10 | 152 |
| clay, ore, slate | 11 | 163 |
| ore | 16 | 179 |
| slate, ore, clay | 98 | 277 |
| good ore | 120 | 397 |
| poor (lean) ore | 144 | 541 |

*Phillips 1898: 38–39C

A somewhat more colorful description of the layers in a vein of iron ore in southeastern Shansi, China, has been assembled by Shockley 1904: 851.

| | | |
|---|---|---|
| Upper 5 layers, . | Pai Kung,<br>Wu Hua Kung,<br>Huang Mien Kung,<br>Pa I Kung,<br>Ti Kung, | First layer ore.<br>Five flower ore.<br>Yellow face ore.<br>Eight good friends ore.<br>Earth ore. |
| Boundary, . . | Huo Shih, | Flint. |
| Lower 5 layers, . | Sha Kung,<br>Huang Ni Kung,<br>Feng O Kung,<br>Pei Kung, | Sand ore.<br>Yellow mud ore.<br>Wind goose ore.<br>?? " |
| Below this, . . | *Water reaching to the sea; full of dragons and dangerous to disturb!* | |

[4] Similar mining techniques were used in England during Roman and post-Roman times. "The ore was worked by bell-pits, of about 6 feet diameter at the top, widening toward the bottom. They were generally shallow, rarely more than 20 feet deep; sometimes they were connected by levels" (Straker 1931: 101). Forbes (1933: 232) notes that pits of this kind in West Africa were often connected by tunnels.

[5] Among the best known of these complexly mineralized deposits are the siderite lodes of Siegerland in western Germany, where high grade ores of iron, copper, lead-zinc, and cobalt occur in very close association (Beyschlag, Vogt, and Krusch 1916, v.2: 792–805).

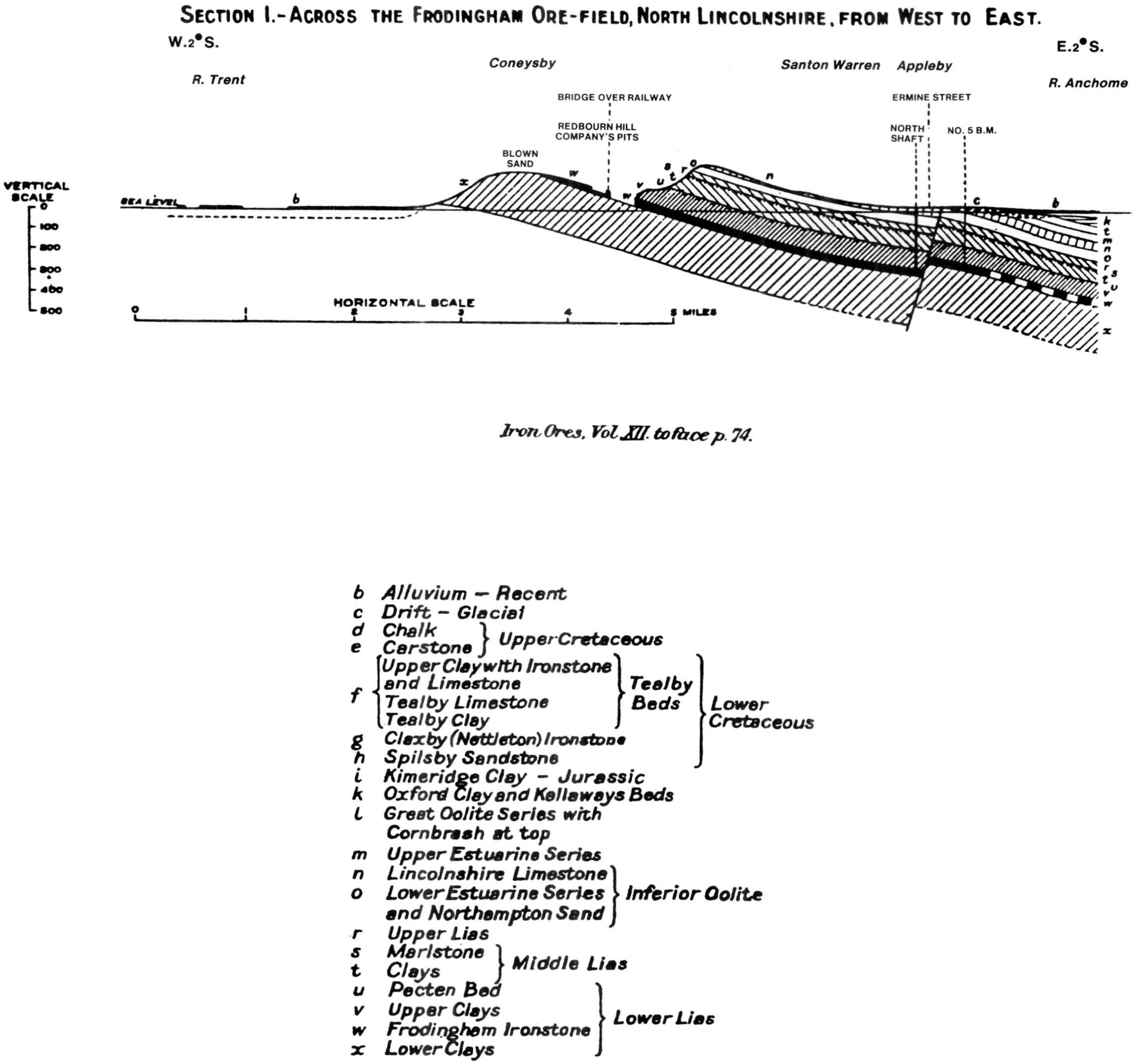


Fig. 4.3. *Iron Deposit in Cross Section.* The stratum of ore emerges onto the surface at one point, which is where it could be recognized and where mining could begin. (From *Geol. Surv. Great Britain, Mem.* 1920: pl. 4)

> heavy particles of iron, whilst the lighter grains of soil are carried off by the rapid stream bounding over the obstacle. As the iron accumulates, sticks are added to heighten the dam, and when this is nearly as high as the bank, (about one foot), the ore, a fine black sand is taken out, the dam lowered, and the process repeated (Yule 1842: 854).

Similar methods of mining with sluices were used in Japan. A nineteenth-century Japanese woodcut shows soil containing iron sand being dug from a hillside and thrown into an artificially directed stream, where the iron sand was caught in straw mats (Fig. 4.5). In 1878, Lyman (1879: 75–76) observed a more elaborate operation of this kind at Minari in Izumo Province, western Honshu:

> The washing is done in a sluice about forty fathoms [73 m.] long in eight parts of unequal length, sloping gently with the ground of a small valley but with a fall of a couple of feet between the different parts. The width of the sluice is in general 1 1/2 or two

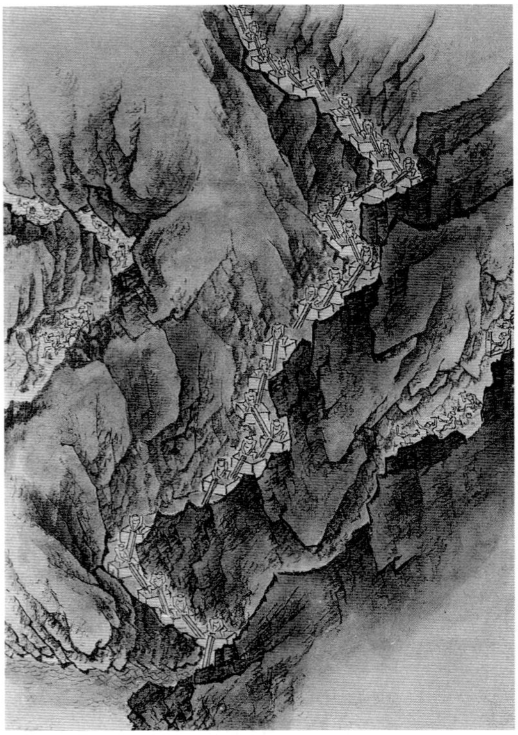

Fig. 4.4. *Deep Mining in Japan, Eighteenth Century.* A contemporary woodblock illustration of a copper mine on Shikoku controlled by the Sumitomo family. The chain of 130 hand pumps serves to raise water 220 m. to the surface. (From Tsuna and Smith 1983: 5, 35)

> feet, but in one part is four feet. The bottom is of boards and the sides partly of boards but mostly of flat stakes.

The Japanese also employed sluices to separate iron sands recovered from beaches and stream beds, perhaps by passing them down the chute repeatedly until the heavy magnetite fraction was as concentrated as possible, as was done with gold-bearing sands during the nineteenth-century California Gold Rush (Fig. 4.6).

The separation of iron sands by simpler hydraulic methods—panning or using a short trough—is still done in West Africa. In a typical operation of this kind in northeastern Nigeria:

> Women are entirely responsible for the collection and preparation of the ore. During the latter part of the rainy season, July-September, they go to the sandy stream beds in between heavy storms and wash and collect the black sands, which being heavier than the quartz sand have been left behind on the surface by the storm water. These sands which are primarily composed of magnetite are stored in a large pot sealed with clay until the next smelting season, April-May. A sloping trough is then dug inside the threshing area and enough magnetite sand for a day's smelting is washed repeatedly to purify and concentrate it and get rid of the quartz and other impurities (Sassoon 1964: 174–175).

The technique of concentrating iron sands in short troughs is also reported in Mysore, southern India (Primrose 1900–1901: 213) and in Henan, central China (Wagner 1985: 28–32). In Henan, 2 kg. of concentrated sand yielded only 1 kg. of iron when smelted in a blast furnace, indicating that much gangue remained in the concentrate.

A concern with the size as well as the composition of

LAVAGE DU MINÊRAI D'OXYDE DE FER MAGNÊTIQUE OU DU FER TITANÊ MAGNÊTIQUE SABLONNEUX SUR DES NATTES DANS UNE FOSSE A EAU COURANTE.

Fig. 4.5. *Sluicing Iron Sand in Japan, Nineteenth Century.* Earth containing magnetite sand is dug from the hillside and dumped into a fast-moving stream. The magnetite settles out and is caught on reed mats placed on the stream bed. (From a woodblock print reproduced in Geerts 1883: 527)

A—SLUICE. B—BOX. C—BOTTOM OF INVERTED BOX. D—OPEN PART OF IT. E—IRON HOE. F—RIFFLES. G—SMALL LAUNDER. H—BOWL WITH WHICH SETTLINGS ARE TAKEN AWAY. I—BLACK BOWL IN WHICH THEY ARE WASHED.

Fig. 4.6. *The Wood Construction of a Sluice.* The channel is called a launder. Raised plates on the floor of the launder, riffles, provide local turbulence, allowing denser particles to settle and be trapped by the ledges of the riffles. (From Agricola; Hoover and Hoover 1912: 323)

ore fragments is evident in many recorded smelting traditions. Ore is usually mined not as fine sand but as lumps which require further processing. This is especially true if they are to be smelted in a shaft furnace, for there is an optimum size of ore fragment for each furnace. If the fragments are too small relative to the height of the furnace shaft, they will increase the resistance of the bed to the upward flow of air and combustion gases, causing a rise in fuel consumption and perhaps smothering the furnace entirely (Chapter 7). On the other hand, if the fragments are too large or the furnace shaft too short, they will not be completely reduced by the time they reach the combustion zone and pass into the slag, where they are lost to the metal recovery process. A bowl furnace or crucible hearth might operate efficiently with heterogeneously sized ores, but for a shaft furnace to burn efficiently, the ore must be broken and sifted into uniform pieces which are of the appropriate size for the furnace shaft.

In experiments with bloom iron smelting, Tylecote and his co-workers (1971) tried different ores crushed and screened to three size fractions—12 to 25 mm., 3 to 12 mm., and less than 3 mm. The largest size gave poor metal recovery for each ore type, indicating that the reduction reaction was not able to penetrate to the centers of the ore fragments in the time it took them to descend through a one meter-high shaft furnace. Optimum metal recovery from the two smaller size ranges depended on the microporosity of the ore. Porous minerals such as limonite (bog ore) allowed hot reducing gases to penetrate the fragment, so there was an increase in the surface area of the solid available for reaction. Because of their high surface to volume ratio, fragments in the 3 to 12 mm. range were reduced efficiently; their size was sufficient to keep back pressure in the furnace from building up to a significant extent. Nonporous hematites, on the other hand, could not be completely reduced unless broken into fragments less than 3 mm. in diameter.

Breaking up large lumps of ore also allows for beneficiation—the upgrading of ore quality by separating fragments with a high iron content from those that are mostly gangue. This can be done by hydraulic methods if the ore is soft enough to be finely crushed by sledge hammers or water powered stamp mills. A stamp mill (Fig. 4.7), described by Agricola in the sixteenth century (Hoover and Hoover 1912: 284), was used in conjunction with sluices to upgrade nonferrous ores. The same principle is employed today in the beneficiation of the low grade taconite iron ores of Minnesota and Labrador through the use of massive grinding machines followed by magnetic separation or flotation. Crushed laterite ores are still separated by panning in Mali, West Africa (Gardi 1969: 30). In the nineteenth century, South Indians used a separation method that combined trough or pan concentration with winnowing: after the crushed and washed ore had dried, it was poured from a basket while a strong wind was blowing; the fraction that drifted farthest was discarded, while the rest was washed once again (Holland 1892: 145–146).

Where the ore was too hard to be crushed by available equipment to sizes appropriate for separation by wind or water, it was sorted by hand. The importance of hiring skillful sorters is emphasized by Biringuccio:

> But whoever wishes to make the iron good and soft by virtue of the ore itself, in addition to the method and the charcoal, must first provide an experienced and intelligent sorter who will carefully sort the pure iron from the impure and will separate them according to the indications of their appearance and by breaking them. Then . . . [after roasting] he must look them over piece by piece before they are brought to the furnace in order to see whether some trace of another metal has appeared (Smith and Gnudi 1942: 64–65).

A—MORTAR. B—UPRIGHT POSTS. C—CROSS-BEAMS. D—STAMPS. E—THEIR HEADS. F—AXLE (CAM-SHAFT). G—TOOTH OF THE STAMP (TAPPET). H—TEETH OF AXLE (CAMS).

Fig. 4.7. *Ore Crushing Mill in Czechoslovakia, Sixteenth Century.* A woodcut illustration from Agricola's *De Re Metallica*, showing a water powered stamp mill. The hammer heads are of iron or of wood faced with iron, affixed to the ends of long posts that are alternately lifted and dropped by the cams on the wheel shaft. (From Hoover and Hoover 1912: 284)

Comparable procedures seem to have been nearly universal among pre-industrial ironmakers. Much iron must have been thrown away during selection, which may help to account for the high ore costs sometimes reported by early writers like Cronström in the seventeenth century (Pipping 1982). However, as Biringuccio makes clear, hand sorting had advantages. It allowed a sharp upgrading of ore quality in terms of iron versus gangue content and enabled sizing to be done simultaneously with upgrading. Moreover, unlike modern beneficiation systems, hand sorting permitted the separation of significant percentages of unwanted components, like iron pyrites and copper minerals, that cannot be removed by other processing methods.

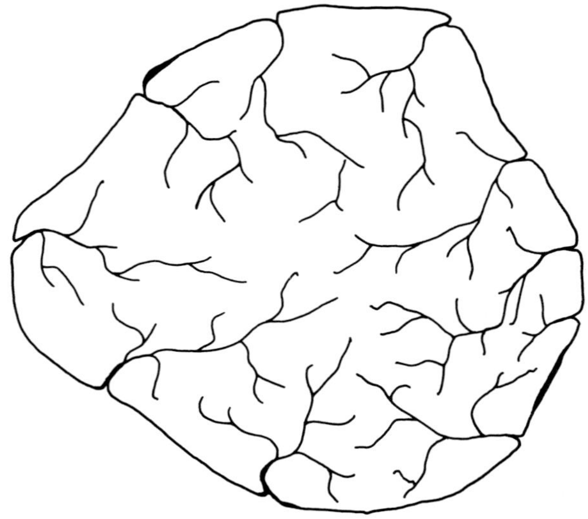

Fig. 4.8. *Effects of Roasting on an Ore Fragment.* The roasting operation creates pockets of gas under pressure and thus causes the ore to crack. The network of cracks allows reducing gases to penetrate the fragment when it is placed in the smelting furnace.

## 9. ROASTING IRON ORE

The roasting of iron ores is known to have been practiced in Europe in ancient times and was very common there during the seventeenth century (Tylecote 1976: 118). It was also done in Africa (Bellamy 1904: 102; Avery and Schmidt 1979), China (Tegengren 1923–24: 338, 343, 351), Sri Lanka (Coomeraswamy 1956), and Southeast Asia (Percy 1864: 274). In most cases the ore was roasted to dry it, to break it into smaller pieces, and to make these more porous.

Porosity is advantageous in an ore because, even when the lumps are large, the hot gases in the reduction reaction have access to a greater area of contact with the mineral. Although some ores, like bog ores, have this attribute naturally, others, like some siderite and limonite ores, are dense. By roasting them at a temperature sufficient to drive various chemically combined or structurally associated substances off as gases, they too can be made porous. The most important of these gases are carbon dioxide, which comes from the breakdown of the carbonates in siderite ($FeCO_3$), and the water vapor released when limonite ($FeO(OH).nH_2O$) loses its water of hydration. The pressure from the gases causes microcracking within the ore fragments, as do the volumetric contractions that occur when crystalline structures are reorganized. Hence, the result of roasting is a group of ore fragments riddled with cracks (Fig. 4.8). If desired, these can be easily crushed to a finer size, although it may be unnecessary, since even the larger fragments will have a high surface to volume area and will be accessible to reducing gases during smelting. A roasted siderite or limonite is as easily reduced in a furnace as a naturally porous bog or lake ore .

To roast an iron ore, it has to be heated to a temperature of about 600°C. No special atmospheric controls are necessary, since the process involves only drying and driving off chemically bound water and carbon dioxide. Therefore roasting was traditionally done in heaps, in crudely enclosed stalls (Fig. 4.9), or in recesses dug into hillsides. The fuel was usually wood (although coal would have worked equally well) stacked with the ore lumps so that combustion could be sustained by natural draft. Because high quality fuels were unnecessary at this stage, charcoal, which was expensive, was rarely used.

Roasting was practiced as early as the Bronze Age, because it is a necessary step in converting the sulfides in copper ores to oxides prior to smelting. The roasting of iron ores must often have had a similar effect, changing any iron sulfides to oxides. This was desirable, although probably unintentional, since there are no indications in the ethnometallurgical literature that traditional ironmakers are aware that roasting improves the quality of iron. They roasted ores to make them easier to crush and smelt, not to remove sulfur or to make iron that was less hot short.

A—Area. B—Wood. C—Ore. D—Cone-shaped piles. E—Canal.

Fig. 4.9. *Roasting Stall, Czechoslovakia, Sixteenth Century.* The equipment needed for roasting is simple—in this case, a low three-sided enclosure that serves mainly to regulate air access during burning. (From Hoover and Hoover 1912: 274, 275)

Nonetheless, roasting is an efficient desulfurizing method. The sulfur is eliminated by any of several reactions. In an unrestricted air supply, pyrite oxidizes readily at temperatures above ca. 500°C, so that the sulfur is driven off in the form of sulfur dioxide gas:

$$3FeS_2 + 8O_2 \rightarrow Fe_3O_4 + 6SO_2.$$

The magnetite produced may then turn to hematite:

$$4Fe_3O_4 + O_2 \rightarrow 6Fe_2O_3.$$

If oxygen is deficient or absent, other reactions can occur:

$$5FeS_2 \rightarrow 2FeS + S_2(\text{vapor}), \text{ or}$$

$$FeS_2 + 3O_2 \rightarrow FeSO_4 + SO_2.$$

The ferrous sulfate is harmless to the final product, as it will dissociate to sulfur trioxide gas when heated to a moderate temperature in the smelting furnace:

$$FeSO_4 \rightarrow FeO + SO_3.$$

Thus, the complete dissociation of $FeS_2$ to $Fe_2O_3$, $Fe_3O_4$, $SO_2$, and $SO_3$ eliminates sulfur from an ore. However, dissociation to the lower sulfide of iron (FeS) mitigates but does not necessarily solve the problem. If the FeS comes directly in contact with an iron oxide during roasting, a reaction producing sulfur dioxide will occur:

$$FeS + (Fe_2O_3 \text{ or } Fe_3O_4 \text{ or } FeO) \rightarrow Fe + SO_2 \text{ (unbalanced)}.$$

On the other hand, if the FeS becomes embedded in certain other minerals in the ore, such as silica, clay, or limestone, part of the sulfur will pass through the roasting process and into the smelting furnace. Although this will be partly eliminated during smelting by the reaction,

$$FeS_2 + 2Fe_2O_3 \rightarrow 5Fe + 2SO_3,$$

a small fraction of the sulfur originally present in the ore will enter the finished iron.

Because all the above reactions probably took place when ore was roasted in a large heap with nonuniform temperatures and air supply, traditional roasting methods were effective in removing sulfur from ore. Iron smelted from roasted ores in charcoal furnaces therefore had a low sulfur content and was rarely hot short. Some ancient iron samples do contain substantial amounts of sulfur, probably because they were either smelted with coal or made from ores that had not been roasted before smelting.

Since roasting is so simple and economical, it is surprising that pyrite-containing ores were not always pretreated in this way. However, it must be remembered that the role of sulfur in iron was not understood and that roasting was usually, if not always, intended solely to improve the structure of an ore. From the point of view of an early smelter, certain types of ores would not have benefited from roasting. The denser magnetites and hematites do not contain anything which can be gasified at roasting temperatures, and so are unlikely to crack and gain in porosity. Bog and lake ores are often already porous enough in the as-mined state. It is true that all these types of ores can contain excessive sulfur, but thus far there is no evidence that any were roasted during pre-industrial times.

## 10. THE SUPERIORITY OF AN IRON ORE

Early writers on ironmaking often praised the ores from specific districts and mines: the Dauphiné and Ariège in France, Dannemora in Sweden, Siegerland in Germany, Bilbao in Spain, Elba in Italy, Eisenerz and Hüttenberg in Austria, Banyeri in Togo, Jabalpur and Salem in central and southern India, Izumo in western Japan, Daye and Shanxi in central and northern China, and others. Some of these deposits were famous for the same reasons that Lake Superior, Minais Gerais, and Krivoi Rog became well known in later times: because the ores were abundant, accessible, high in iron content, and free from phosphorus and sulfur. Elba, for instance, owed a reputation that lasted more than two millennia to large reserves of high grade, but otherwise ordinary, hematite.

The fame of other deposits had a different basis, for the ores involved were neither rich nor especially free from impurities. In Shanxi, for instance, the hematite and limonite ores were only medium grade and fairly high in phosphorus and sulfur, so the success of ironmaking there derived from factors other than ore quality. The ore was widely distributed and accessible, and a good mineral fuel, anthracite coal, was present in the same places as the ore. Much of the metal produced was cast iron and thus not sensitive to high phosphorus and sulfur contents, and local specialists had developed a series of unique and labor-efficient methods centering on the use of crucibles for smelting and secondary processing. The result was that Shanxi became one of the largest iron producers in pre-industrial Asia.

A similar situation existed during the early nineteenth century in the "blackband" districts of northern Britain. The blackband siderite ores were celebrated even though they were low in iron and very high in sulfur and phosphorus. Their great advantage was that they occurred in veins that alternated with layers of bituminous coal, making them exceptionally inexpensive to mine and smelt (Ure 1856: 1103). Some of the blackband ironstones in Scotland contained so much carbonaceous matter that they could be smelted without any added fuel (Richardson 1934: 562). Ores of this particular kind would not have been valued anywhere (except perhaps in China) before the Industrial Age because they were low in iron and high in detrimental impurities, but they show that some notable ores of earlier periods might have had little to recommend them other than the low cost of mining and smelting.

A third group of deposits owed their reputations to the special character of the metal made from them. Eisenerz and Hüttenberg in Austria are notable examples; from the Iron Age through the nineteenth century, their siderite ores were known as *steel ores* because they yielded an iron that could be converted to what many specialists deemed the best steel in Europe. The magnetite sands of southern India and western Honshu had a similar local reputation as sources of steel of unmatched quality. Some modern writers (e.g., Wertime 1962: 92–93) have suggested that the mineralogical character of these ores—the fact that the chief iron minerals involved were siderites and magnetites—may have made them especially suitable for steel production.

However, in theory, no iron mineral is better than any other in terms of the quality of the steel that can be made from it. Certain iron ores, on the other hand, do have characteristics that would make them suitable for steelmaking. The most important of these characteristics is the presence of nonferrous elements that could be passed on to the steel after smelting and converting.

Several elements that can improve steel sometimes occur in iron ore. Those reducible at bloom iron smelting temperatures include phosphorus, copper, and manganese. The benefits conferred by phosphorus and copper are mixed, but alloying iron with manganese has no undesirable side effects, and a steel made from such an alloy is distinctly superior to ordinary steels in terms of the balance between hardness and strength. Further, some early irons and steels from China (Li 1975: 17) and Austria, ancient Noricum (Klevenhüller et al. 1961), contain significant quantities of manganese. It is virtually certain that the often-praised steel of the Roman province of Noricum (i.e., modern Carinthia) was smelted from the high manganese siderites and limonites of the Hüttenberg area.

Thus, it seems likely that some ore deposits were famous because of the presence of manganese. Among the ores known to contain substantial percentages of this metal are Eisenerz, Hüttenberg, Siegerland, Ariège, Dannemora, and Jabalpur. A Jabalpur hematite ore with 46.43 percent iron and 12.96 percent manganese was smelted in a local bloomery to give an "iron" called *kheri* with a 0.296 percent manganese and, presumably before partial decarburization in the smithy, a 1.214 percent carbon content. Although Fermor (1909: 595–597) concludes that "the superior hardness of the kheri over the soft iron can hardly be attributed to the 0.3 percent of manganese it contains," that amount of manganese can significantly improve the hardenability of a steel and might explain the reputation of Jabalpur *kheri*. It is possible, moreover, that the sample Fermor analyzed is atypical. A bloomery run hot enough to produce a 1.2 percent carbon steel is also capable of extracting much of the manganese from a manganiferous iron ore.

Not all "steel ores" contain manganese. The magnetite sands used in steelmaking in India and Japan, and some of the well-known hematites of Spain and Italy, usually did not. Traditional makers and consumers of steel strongly favored these ores, yet there is no apparent scientific reason they should have been preferred to any other ore of good quality.

## 11. TYPES AND DISTRIBUTIONS OF ORES

The nonferrous ores have a patchy or coarse-grained world distribution, while iron ores are distributed in a fine-grained pattern; that is, iron ores are more common than nonferrous ones. Most of the world's best iron ores are confined to a relatively small number of large deposits, chiefly the hematite "iron formations," but smaller deposits of high grade iron ore are widespread. Moreover, there are types of medium and low grade iron ores that have no real analogues among the ores of other metals: the tropical laterites and the lake/bog ores and related concretions. These are often found in places without any nonferrous metals. Laterites are extremely common in tropical areas, and recently formed iron concretions exist everywhere except in the Arctic. Consequently, iron in at least moderate quantities is widely available. An ancient bronzemaker might have had to obtain copper from sources a few hundred kilometers away, and tin might have travelled a thousand kilometers or more. Iron, on the other hand, could have been found within a few hours' walk from virtually any place. That so many areas have at least enough ore for a small scale smelting industry has had a profound effect on the history of iron use. The frequency of marginal iron deposits explains the rapid diffusion of ironmaking in antiquity and its low price in later times, because a monopoly could not be maintained over a material within access of all.

The availability of large, good quality ore sources, though less critical than the availability of fuel and (where appropriate) water power, played an increasingly important role as regional economies expanded. Marginal deposits were sufficient while production was oriented toward consumption by a few households or villages in the immediate neighborhood of the producer. When, however, ironmakers began to sell to a wider market and to compete with more distant ironmakers, it became necessary to look for deposits of greater size and better quality. In economically developed societies, like Roman Italy and Han China, small producers using minor deposits mostly ceased operations, while limited numbers of major ironmaking centers evolved in places with large iron ore reserves.

For an ore to be exceptional for pre-industrial usage, it had to be accessible by simple mining techniques and capable of good yields when smelted in simple furnaces. Large deposits of such ores were neither numerous nor evenly distributed: none of the ancient empires probably contained more than twenty or thirty. Deposits like those of Elba and Carinthia were rare enough to be considered important national resources, and their locations influenced strategic planning and the positions of cities and trade routes. Moreover, their size and consequent ability to be worked for a long time increased their importance. Although the major copper and silver producing areas of antiquity were often mined out within a century or so, the major iron deposits were, in effect, inexhaustible.

Not all major deposits of excellent ores were intensively exploited. Areas with access to banded hematite deposits of the Krivoi Rog-Lake Superior type could have become great ironmaking centers in antiquity, since those ores are now considered to be among the world's best in terms of purity and iron content. Yet few of the areas near deposits of this kind—in eastern India, western Africa, Manchuria, the Ukraine—were important iron producers before modern times. The banded hematites may have been difficult for ancient smelters to handle, or they may have been too far from populations large enough to support an iron industry of significant size.

# 5. Refractories

## 1. INTRODUCTION

Furnace walls, hearths, tuyeres, and crucibles must be made of materials which will maintain structural integrity at high temperatures; they should neither melt, soften, sag, nor collapse under their own weight. They must also support the additional burden of a column of ore and fuel or a stack of crucibles loaded with ore and reagents. Construction materials having these qualities are called refractories.

Several additional requirements must be met for refractories to be useful in metal processing. They must withstand the corrosive behavior of liquid slags. In a furnace operation, slags should be fluid and refractories solid. The purpose of a slag is to dissolve all extraneous rock and fuel ash, providing a covering liquid with a low melting temperature through which the reduced metal can settle; it also dissolves the furnace walls and tuyeres in the process. A good refractory will dissolve into the slag at a tolerably slow rate and so prolongs furnace life because it maintains its structural integrity even in an extremely corrosive environment. In general, the metallic products of iron smelting do not attack or interact with refractories.

Refractories are oxides, and so in principle are reducible. Their reduction to elemental or even sub-oxide form by high temperatures, combined with the strongly reducing nature of the gases within the smelting environment, must be avoided. The simple oxides of silicon (silica), aluminum (alumina), calcium (lime), and magnesium (magnesia) function well under furnace conditions, but silica is susceptible to the highly reducing atmospheres of a cast iron smelting furnace at temperatures above 1300°C. Even so, the rate of reduction is slow enough that a furnace constructed of sandstone (silica) blocks should withstand the charge intact long enough to justify the investment in the furnace. The other oxides can be regarded as nonreducible in the context of iron smelting. Note that reducibility and resistance to melting are different. In fact, some mixtures of these oxides can have low melting temperatures and thus serve as slags. Pure alumina is a very high melting, refractory material, yet when associated with silica and lime, a low melting liquid slag may form (Appendix A). However, it cannot be reduced to aluminum metal in the environment of an iron smelting process.

Oxide refractories conduct heat poorly, and their thermal conductivities are smaller than those of any metals or alloys. For example, the thermal conductivity of magnesia is only about 1/1800 that of iron, high for a ceramic material. Poor or low thermal conductivity is useful for containing the heat generated by combustion within the reaction zone of a smelting unit. This thermal containment is vital to fuel efficiency. The refractories most resistant to melting, sagging, and other problems are not as insulating as the less refractory oxide aggregates. Thus, the walls of a furnace were often built of a refractory clay or stone, and sand, gravel, and even earth were mounded around the outside (Figs. 3.2, 3.3, 3.5). The great porosity of a loose aggregate reduces thermal conductivity. In modern furnace design, "refractory" brick is used for areas in direct contact with the highest temperatures, and is backed with "insulating" brick at that point in the wall where the temperature is low enough for more heat sensitive materials to be used.

Poor thermal conductivity has some hazards. A furnace which is at normal ambient temperatures must be brought up to the operating temperature very slowly, and it must be shut down by cooling very slowly. If neither precaution is taken, the refractory construction will be damaged by the stresses generated in the body of the ceramic materials. Thermally induced stresses result from temperature gradients which are steeper in poorly conductive substances. All ceramic materials expand when heated and contract when cooled. Therefore, a temperature gradient can be translated directly into an expansion gradient, and into a pattern of tensile and compressive stresses of remarkably high magnitude. Since ceramic materials are notably weak in tension and cannot deform to relieve thermal stresses, they crack. These cracks combine to produce fractures. Cracks forming in ceramics branch as they grow and after only a few cycles of heating and cooling, a refractory can break up and fall away. The destructive action of thermal and other internal stresses can be reduced by altering the formula of the refractory.

## 2. CLAY MINERALS

The oxides silica and alumina have very small but finite solubilities in water. Over long periods of time, dilute solutions were formed as water drained over weathered rocks and into streams and rivers; they finally collected in ponds and lakes. With changes of temperature, the solutions became supersaturated and precipitated the excess solute. The precipitates were aluminum silicates compounded with hydroxyl (OH) groups to form tiny crystals (visible only at magnifications of several thousand) which settled to the bottom of the body of water. Over millions of years, deep beds formed and remained unchanged even after the lakes had dried up. The consolidations of these tiny crystals of hydroxy-alumino-silicates are called clay minerals, or simply clays. Natural clay deposits occluded larger particles of sand and gravel from the silt brought down by rivers. When the water dried up, vegetation took root in the clay, and thus the upper surface of an old clay bed is usually permeated with roots, humus, and other vegetable matter.

The variety in clay minerals derives from the interaction of the clay with other soluble oxides, such as soda, potash, and magnesia. Special clay formulations which include these oxides are different in their atomic and crystalline structures and in their physical properties. Other oxides, carbonates, and hydroxides can be deposited simultaneously with the clay minerals. The most important of these are limestone and the hydrated iron oxides. Clay and limestone mixtures are called *marls*, and the red clays which exist in many parts of the world owe their color to iron oxide. Both of these naturally occurring mixtures are used for pottery manufacture. A marl clay can be fabricated into blocks, as other clays can, dried, and used in furnace construction, with the constraint that the initial heating rate must be very slow. The fired iron-rich clays are not sufficiently refractory for furnace construction, and some clays, such as the laterites, contain so much iron that they can be smelted into metal (Chapter 4).

With movements of the earth's crust, clay beds were folded under the surface and subjected to high pressures and temperatures, conditions which can cause a metamorphosis in physical nature without changing chemical composition. Thus, while natural clays are friable, mudstone and shale are hard and fracture in planar patterns. Micas, a species of modified clays, have been heated sufficiently under geological conditions to grow into large crystals that cleave into thin, translucent leaves.

Natural clay, which has not been altered by geological processes, has remarkable properties when mixed with water. Depending on the amount of water added, a variety of textures can be formed, from a malleable paste which can be modeled to a more fluid slip which can be used for trowelling and painting; between the two there is a great range of consistencies. Although clays are generally plastic, some cracking can occur, especially when the material is bent, but this is easily repaired using only finger pressure and moisture. Other ceramic powders ground to the same size do not have the same strength and resistance to slumping. Clay and water mixtures can be molded because of the tabular crystal shapes of the clay particles which developed as a result of precipitation from aqueous solutions.

Nonuniform cooling and nonuniform drying act in the same way to produce cracks. One way to reduce the risk of cracks without significantly increasing the drying time is to temper the clay. Fired clay, mudstone, or shale can be crushed into small fragments and blended with the clay. The fired pottery temper does not absorb much water and will not shrink in the drying cycle. Although it does not prevent differential shrinkage during drying, it creates impassable blocks so that the cracks which do form cannot propagate and cause the body to shatter. Tempered pottery has a network of invisible microcracks rather than large yawning cracks. The sum of all the spaces in the microcracks is the same as the space created by one large crack. One large crack is destructive, but the microcrack assembly is not.

To form a furnace, clay paste can be poured into molds, and left until the clay has dried sufficiently to support its own weight; then the molds are removed to permit the water to evaporate more quickly. The principle is similar to that of modern poured concrete construction, although the hardening processes are completely different. Blocks of sun-dried clay mortared with a paste of the same clay can also be used to build a furnace. The load-bearing capacity of sun-dried brick is great enough that the furnace would be structurally stable without forms even before baking and firing (Rostoker and Gebhard 1981b).

Clay paste was used for the hearth but more often it was only a binder for a charcoal base. The *tatara* furnace and some African furnaces had thick charcoal bottoms produced by burning and charring wood in situ. The bottoms were then pounded into position. If they had not been bonded with clay, the charcoal would have floated up through the slag and the metal pools which rested upon them.

Other sources of internal stress become apparent when the dried clay is fired. One is the thermal expansion associated with nonuniform heating, a serious problem because furnace walls could be several inches thick. The thermal conductivity of the wall could be increased, and the thermal gradients made less steep by blending clay with ground charcoal (later coke). Charcoal is an effective temper, and also provides higher thermal conductivity which reduces the thermal stresses.

Since clay beds usually have an overburden of decaying plant matter, any vegetable particles occluded in the clay mix will decompose into water vapor and hydrocarbon and oxycarbon gases during firing. If the gases are trapped within the clay, they will build up pressures that can cause the body to explode. It is necessary, therefore, to process the raw clay before using it. It must be washed, so that clay floats away from the gravel, and then washed again so that the plant debris separates from the clay. Clay from termite mounds was preferred by Africans because it was already processed and immediately usable.

The choice of clays suitable for furnace construction is limited. Those with appreciable amounts of potash, soda, and/or iron oxides have unacceptably low softening and melting temperatures. The term *fireclay* was coined to describe the soda- and potash-free clays of which kaolinite is the most common. The alumino-silicate products of firing are naturally more refractory than any others. Fireclays can be made more refractory by blending them with quartz (silica) sand. Corundum (alumina), where available, could be substituted for quartz to produce a superior refractory material. Modern fireclays contain, in their finished condition for refractory use, 35 to 55 percent alumina, but the addition either of silica or alumina (not both) will raise the softening and fusion temperature above a base of about 1580°C.

## 3. STONE REFRACTORIES

As taller furnaces were developed and higher smelting temperatures achieved, the structural capacities of fireclays were inadequate. Most rocks are poor refractories, because they have intrinsically low melting temperatures (granites and igneous rock) or they decompose on heating and crumble (hydrated minerals or carbonates),[1] but sandstone is not. The term *sandstone* denotes a family of sedimentary rocks which are bonded aggregates of silica particles. Silica is a refractory ceramic material because it retains both strength and integrity at very high temperatures. Sandstone is a porous aggregate and therefore has a lower thermal conductivity and an increased resistance to the fragmentation or spalling caused by thermal stresses. It can be easily quarried and dressed into large, well-faced blocks.

Not all sandstones are equally suited to the construction of high temperature furnaces. The distinctions rest in the nature of the bond between quartz particles. The best for furnace construction has a quartzitic, or recrystallized, bond; this is a pressure bond between quartz surfaces in direct contact with each other. Sandstones with a calcium carbonate bond would be unsuitable because there is a large gap between the temperature at which the carbonate dissociates to lime and the temperature at which the lime can react with the silica to form a new ceramic bond. In that interval, the quartz particle bonds would, for the most part, be lost. In contrast, common red sandstone has an iron carbonate bond. The FeO residue from the carbonate dissociation is very active and would create a glass bond at a relatively low temperature which would evolve from a transient liquid to a solid bond by silica diffusion within a very narrow range of temperatures.[2]

Sandstone furnace construction is known primarily from Europe, but there is documentary evidence that it was also used in China. Tegengren (1923-24: 338, 360, 362) quotes an earlier source that in Hunan, "The blast furnaces, built of sandstone slabs with a wooden frame and clay filling in the intervening space, measure some 20 ft [6 m.] in height." Plans of furnaces in Yunnan which designated sandstone blocks in both the hearth and the walls were copied.

Although they were not widely used, other naturally occurring rock materials could have been suitable refractories. Common stone and clay faced with slabs of quartzose schist luted with a "compost of chaff and clay" have been cited as examples (Atkinson 1890: 269). More specifically, unusual rocks were used in the Catalan furnaces.

> The bottom of the furnace is formed of a refractory stone, such as granite, mica-schist, sandstone or even limestone. It must not be too large to admit of its being taken out and replaced by a fresh one. Its upper surface would be flat or, by preference, somewhat concave. These stones last from 6 hours to 3 or 6 months, and sometimes longer; but this depends on the skill of the furnace man... The bottom stone rests upon a bed of pounded slags and clay brasque [paste], from 0.40 to 0.58 m in thickness; and below this is a large stone, generally an old mill-stone (Percy 1864: 284).

The surprise expressed by "even limestone" derives from the fact that it would not have been an effective material for walls, since it decomposes to lime and then erodes rapidly. However, as a hearthstone it would have created a certain amount of physical stability, and its reaction with the slag pool could have cemented the fragments back into place.

Cutting and shaping flat faces on sandstone are difficult but feasible, and forming curved surfaces is even harder. The internal profiles of furnaces changed accordingly from circular to rectangular. Large slabs of sandstone, which were used until recently, made both refractory and structurally strong hearth bases. In the early nineteenth century, refractory bricks were manufactured by blending fireclay and corundum (mineral alumina). Since bricks can be molded with either flat or curved faces, furnace design was no longer dependent on stone cutting. Nevertheless, construction of rectangular furnace shafts with sandstone persisted well into the nineteenth century.

## 4. CRUCIBLES AND TUYERES

Furnace walls and hearths can be so thick that their structural strength at high temperatures is not likely to be challenged, but wall thicknesses of crucibles and tuyeres are limited. A crack in one of these is more serious than a crack in a furnace wall and is very difficult to detect at white heat levels within the furnace. Even if the crack is seen, it is impossible to counteract the damage. Cracks in crucibles obviously prevent them from containing liquid iron and from isolating the ore/carbon mixture from the flame. Early nineteenth-century articles on wootz steel production in India report that a significant fraction of the liquid

---

[1]Osborn (1869: 534–535) lists sandstone, granite, clay-slate, soapstone, mica-slate, and gneiss as refractory stones used in nineteenth-century American furnaces. This is essentially a statement after Overman 1854: 453.

[2]A quotation from Overman (1854: 158) is informative:

> The construction of the hearth [blast furnace] is a business in which the founder himself takes an active part. Still, as this is governed by general rules, we shall give a statement, sufficient to serve our present purpose, of the principles by which we should be guided. It is a mistaken notion that every sandstone which resists fire will make a good hearthstone. It is not the heat which destroys the hearth, but the chemical action of the materials in the furnace. The durability of a hearth is determined by the manipulation of the founder and the keeper. Any refractory material constitutes hearthstone, particulary silex [silica], clay and lime; but a mixture of these three substances must not be applied. The form of the aggregation has considerable influence; but of this we shall speak hereafter. Sandstones are most commonly used in this country, while in other countries, the material varies according to ore and fuel. Limestone, sandstone, gneiss, granite, plastic clay, or fire clay is employed, as circumstances require. But sandstone will answer in all cases, if the ore and fuel are properly prepared. Any sandstone which is free of iron, or from lime, or from matter which towards silex [silica] acts as a strong alkali, may be used for hearthstones. Its refractory quality must be proved by some previous test. This test consists of drying a fragment of the rock in question by a very low heat on the top of a stove, or near a fire grate, for twelve or twenty-four hours; and then exposing it to the gradual heat of a blacksmith's fire. If the stone is friable, after a good red or white heat, or if it falls to pieces by being moistened with water, we may conclude that the rock contains lime, and that it is not good for hearthstones. But if the fragment resists the first heat well, and if it is still hard and compact, we may expose it to a welding heat in the blacksmith's fire, urging the bellows strongly for half or three quarters of an hour. If the stone resists this heat, and if its color is not altered to brown, we may conclude that it is perfectly safe to construct a hearth of it.

metal mixture in crucibles was lost in every run through cracks in some crucibles (Bronson 1986c).

The clay-carbon mixes played an important role in crucible fabrication. Blending powdered charcoal, powdered coke, or mineral graphite with a refractory clay has several advantages. The temper limits the extent of microcracking and significantly raises the thermal conductivity of the clay structure. It also acts as a reducing agent on the inside of the crucible, enhancing the efficiency of smelting, and on the outside it lessens the corrosiveness of the slagged fuel ash by reducing the FeO content. The carbon in the fireclay slowly oxidizes during furnace operation so its presence in the clay blend may be undetectable after use.

The modern equivalent, the "plumbago" crucible, is a blend of fireclay and mineral graphite (Percy 1861). The thermal conductivity of graphite is superior to that of either charcoal or coke and its use represents a sophisticated technology. Burnt clay, fired as a loose, purified powder, is an alternative to pottery fragment temper. Although it improves the crack resistance of fireclays in the drying stage, crucibles with burnt clay temper are not so long lasting as the plumbago crucibles.

High performance crucibles are heavily documented, but very little has been written about tuyeres, perhaps because they fit loosely into the furnace wall and could have been easily replaced during furnace operation. Therefore, they could have been made of lower quality materials and frequently replaced from a large stockpile. Tuyeres were subjected to conditions likely to cause damage, since they protruded directly into the combustion zone where the temperatures in the furnace were highest. The outer surface, which projected beyond the inner furnace wall, was heated directly by the flame while the inner surface was bathed by the injection of relatively cold air. The outer surface was attacked by corrosive slag, and loose pieces of incompletely reduced ore fragments would periodically fall on it.

There are a few brief references to tuyere fabrication in Africa (Cline 1937: 41–42, 105). In addition,

> The central pipe or tuyere is made by the smelter or his assistants from specially selected clay moulded around a stick and formed into a massive tube with an internal diameter of two inches, an external diameter of five inches, and a total length of two and a half feet. The top of the pipe is fixed into a hole immediately behind the base of the guard wall with stones and building earth (Sassoon 1964).

The descriptions omit the fact that the molded tuyere is sundried after the stick has been withdrawn and before installation into the furnace. Also, nothing is said about the nature of the clay, except that "Tuyeres in Africa are almost exclusively made of baked clay with heavy quartz temper" (van der Merwe 1980: 490). Presumably "heavy" means a large amount.

## 5. METAL AS A REFRACTORY

A metal plate, well cooled on one side, can withstand the direct impingement of a flame on the other side without melting. Metals provide strength and wear resistance at elevated temperatures unequalled by ceramic refractories. Moreover, they can be cast or forged, have few shape or size limitations, and are not dependent on unique mineral deposits. A coating of slag freezes on them in a furnace, and thus the direct effects of flame are intercepted by a stable, renewable coating. For these reasons, the use of metals is an important advance in furnace containment. Air-cooled walls and water-cooled hearth plates of cast iron were English innovations, first applied to the fining furnace and later to the puddling furnace. The Chinese used a cast iron pan filled with water as a furnace roof (Tegengren 1923-24: 359). In both cases, the melting point of the cast iron is well below the furnace temperatures, indicating that despite the internal temperature, thermal conduction and the cooling system can keep the iron surface cool enough to be stable. Modern iron and steel producing furnaces make extensive use of water-cooled (water-jacketed) metal plates as walls and hearths.

In the Catalan furnace, it was necessary to change the inclination of the tuyere in order to maximize performance. Ceramic materials were too fragile for this, so copper tuyeres were used (Percy 1864: 285; Overman 1854: 251–253). Iron tuyeres were introduced for the long-running blast furnaces of the eighteenth century (Sanders and Gould 1976: 30). The metal tuyere did not melt because it was cooled on the inside bore by the steady flow of air from a double bellows powered by a waterwheel.

# 6. Fuels

## 1. INTRODUCTION

Until recently, the combustion of carbonaceous substances supplied the heat for iron processing (Table 6.1). The common ironmaking processes that required the burning of at least some carbon varied in terms of both the quantity and quality of fuel. Some, like fining, needed more fuel per unit weight of metal than did smelting. Others, like roasting ore and baking refractories, consumed only moderate amounts of fuel. Certain processes also required that the fuel have special qualities such as chemical purity, high caloric value, mechanical strength, and uniform fragment size. Fuels with these properties are usually rare and expensive, so historical ironmakers have tried to devise methods in which fuels of less than the best quality could be utilized. In densely populated, industrially developed regions, it was necessary to economize on fuel, especially wood and charcoal. Frequent charcoal shortages had great impact on the development of metallurgy.

The history of ironmaking has been affected not only by wood supply, but also by competition between metal producers and nonmetallurgical industries which also consume wood, as demonstrated by an acrimonious debate within the English defense industry in the sixteenth and seventeenth centuries. The manufacturers of the iron cannon which were a decisive factor in the great Elizabethan naval victories were accused of making charcoal from the oak trees needed to build the ships that carried the cannon. Although both industries preferred oak, the issue was eventually resolved by improving forest management methods and by enforcing laws that forbade the burning of large trees for charcoal (Schubert 1942; 1957: 218–222).

## 2. WOOD, GRASS, AND ANIMAL DUNG

Wood and similar fuels have a high cellulose content and a low heating value per unit volume, and are also inferior in heating value and calorific density compared with charcoal, coal, and coke (Table 6.2). A large amount of the heat produced by burning wood is dissipated in driving off water and breaking down cellulose molecules. If the fuel pieces are thick enough, more efficient combustion may occur once the remaining hydrocarbons have been largely converted to charcoal, but the heat of any wood or dung fire in ordinary circumstances is still limited and unlikely to exceed 1000°C.

Very large fires, such as the firestorms sometimes caused by earthquakes and incendiary bombs, are exceptions. Fires can also reach high temperatures in specially constructed furnaces which are generally unsuitable for metallurgical processes. Temperatures in the 1250° to 1300°C range were regularly achieved in a wood-fired climbing pottery kiln at St. Ives (Leach 1976: 187). Even this small and efficient device, modeled after a traditional Japanese kiln, took 35 hours to reach maximum temperature; similar but larger Japanese kilns needed two weeks (Leach 1976: 187). Wood-burning ceramic furnaces in the Far East have much larger fireboxes than direct heating metallurgical furnaces. The need to generate large amounts of heat within a small combustion space is common to many metal processing operations and is presumably why, even in China and Japan, ironmakers rarely used kiln-like furnaces.

Despite the low calorific density of unconverted plant fuels, they were important in traditional iron manufacturing. Preliminary roasting of an ore (Fig. 4.9) was almost always done with a wood fire. A typical operation of this kind was observed among the Yoruba of Nigeria in the early twentieth century: "Having taken this [ore] out in pieces weighing three to five pounds, they roast it all night over fires of green timber, placing about a barrow full of ore on each fire" (Bellamy 1904). In pre-industrial Europe, the roasting of one metric ton of ore over an open fire required 100 to 200 kg. of wood, but a roasting kiln could function with only 30 kg. of fuel per ton of ore (Tylecote 1976: 119). Two cases of wood being used in the startup of furnaces have been documented in African iron smelting (Cline 1937: 44, 49), while most early European descriptions of starting up furnaces specify charcoal with wood for kindling. Wood or grass was also good for baking freshly dried furnace walls, an operation that required slowly increasing amounts of moderate heat.

Although wood is usually relegated to ancillary operations in ironmaking, in principle it is possible to generate charcoal from wood within the smelting furnace itself, with the waste heat contained in the off gases. Thus, for example, if wood were stacked in the upper part of the shaft of a blast furnace, conversion to charcoal would occur before the charge had descended as far as the combustion zone. Although lifting so much wood to the top of the furnace might mean higher labor costs and create problems in efficiently layering ore and fuel within the charge column, a wood fueled smelting furnace seems feasible.

An early nineteenth-century blast furnace (4 m. tall) in central Turkey produced cast iron and was reportedly fueled only with wood (Beck 1891, v.1: 286–290). A tall blast furnace in Yunnan province, southwestern China, was charged every two hours with 120 kg. of ore,

*TABLE 6.1. Processes involved in ironmaking that require fuel*

| |
|---|
| Roasting the ore |
| Smelting to bloom iron or cast iron |
| Fining and puddling cast iron |
| Heating and reheating for consolidation to bar and plate |
| Heating and reheating for forging to finished products |
| Heating and reheating for annealing |
| Heating for carburization, malleabilizing, quench and temper hardening |
| Baking of casting molds and furnace walls, hearth |
| Remelting for casting liquid iron |

*TABLE 6.2. Heating values of solid fuels*

| Fuel | Heating value, Kcal per kg | Density, kg/liter | Heating value, Kcal/liter |
|---|---|---|---|
| Air dried wood* | | | |
| yellow birch | $2.9 \times 10^3$ | 0.50 | $1.45 \times 10^3$ |
| hickory | $3.0 \times 10^3$ | 0.57 | $1.71 \times 10^3$ |
| white oak | $3.1 \times 10^3$ | 0.54 | $1.67 \times 10^3$ |
| white elm | $3.2 \times 10^3$ | 0.39 | $1.25 \times 10^3$ |
| willow | $3.3 \times 10^3$ | 0.29 | $0.96 \times 10^3$ |
| Charcoal* | | | |
| oak | $6.95 \times 10^3$ | 0.57 | $3.96 \times 10^3$ |
| pine | average | 0.28–0.44 | $1.95–3.06 \times 10^3$ |
| Anthracite coal** | $7.17 \times 10^3$ | 1.4–1.8 | $10.04–12.9 \times 10^3$ |
| Bituminous coal** | $6.0–8.0 \times 10^3$ | 1.2–1.5 | $7.2–12.0 \times 10^3$ |
| Coke** | $6.89 \times 10^3$ | 1.3 | $8.96 \times 10^3$ |
| Air dried peat* | $4.3 \times 10^3$ | — | — |

**Chemical Engineers' Handbook.* 2nd ed. McGraw-Hill. 1941: 2341-2342.
***Handbook of Tables for Applied Engineering Science.* The Chemical Rubber Co. 1970: 134, 308.

300 kg. of wood, and 30 kg. of charcoal. Every four hours 100 to 150 kg. of liquid iron were tapped for a total of 600 to 900 kg. of metal per day; after ten days of operation, daily output rose to 1200 to 4000 kg. (Yang 1982: 187–188). The greater part of the energy input into this system was from the wood rather than the charcoal. It is probable that all the wood was converted to charcoal soon after entering the furnace shaft.

Grass, leaves, and other dried herbaceous materials were widely used in African smelting. In Tanzania, swamp reeds were packed into the bowl of a furnace recessed into the ground. Then they were burned until the bowl was completely filled with reed charcoal. The furnace superstructure was built above this, so that the bloom and slag produced by smelting rested upon the charcoal base (Avery and Schmidt 1979). In this case, the reed charcoal acted more as a refractory than as a fuel. Plantain and umbrella leaves were also used in African furnace operation. They were laid against the freshly dried clay lining of furnace walls to provide a degree of temporary insulation against the direct radiant heat of the burning charcoal (Cline 1937: 39). A refractory charcoal furnace base is also employed in the Japanese *tatara* iron smelting process.

In India, wood chips or leaves were an essential part of the charge in crucibles for making wootz steel, while charcoal seems to have been used only as an external fuel. Inside the crucibles, along with pieces of low carbon bloom iron, the chief carburizing agents were chips of dried wood and a variety of green leaves.[1] The specific function of these herbaceous additives is not clear. The wood chips, of course, provided charcoal, and the leafy material may have also, but it was a finely particulate and active form of charcoal that carburized more rapidly.

Dried cow and camel dung are common fireplace and cooking fuels in some regions, and could have also had metallurgical applications. Dung was used to start up one smelting unit in northwestern India, where it was piled in the furnace basin to a depth of 12 inches, but still below the tuyeres; the actual ignition and combustion were provided by charcoal above the tuyeres (Percy 1864: 259). The result was probably a bed of charred dung in the basin that could have functioned like the bed of reed charcoal in the Tanzanian furnaces.

In the Middle East, dung fuels were preferred for some secondary ironworking processes such as tempering damascened swords at Shiraz in Persia. "After the object is forged, it is placed for six or eight days in the furnace of a hot bath, where the greatest attention has to be paid to the even heating of the article. The bath is heated with the dried dung of cows and other animals, which gives a steady and not very intense heat, and is supposed to contain the salts necessary for the formation of true damascene" (Egerton 1896: 59, after Murdock Smith).

## 3. WOOD CHARCOAL

It is generally accepted that wood charcoal was the most common fuel for smelting iron and for most subsequent processing. Charcoal, which can burn at temperatures reaching to well above 1300°C (Fig. 6.1), is particularly suited for metallurgical use. If it is kept dry, it can be stored indefinitely; it is also easily ignited, rapid and even burning, noncaking and nondecrepitating, thus allowing a good draught in a narrow throated furnace, with less sulfur and other contaminants than fossil fuels. However, it also has shortcomings. Its heating value is only moderate relative to its bulk, it has poor resistance to crushing, it does not transport well, and, most importantly, it is expensive.

Charcoal is difficult to transport because vibration causes it to abrade, resulting in a great quantity of fine particles. These particles cannot be used in a furnace charge because they pack tightly and increase resistance to air flow. It has been estimated that in India at least 10 to 15 percent of the charcoal carried over a moderate distance had to be thrown away (Oldham 1859); a similar loss ratio has been suggested for other regions.[2]

The cost of charcoal in relation to the total cost of iron production can be illustrated by figures cited in a manuscript written in 1704 describing operations at the blast furnace and fining forge at Pontypool in South Wales. Costs are in pounds sterling: cost of charcoal at the smelting site—890; total cost of smelting for 400

[1]The wood is usually identified as *Cassia auriculata* and the green leaves as *Convolvulus laurifolia* (Heyne 1814: 358), *Asclepias gigantea* (Heath 1839: 390), *Calatropis gigantea* (Holland 1893), *Jatropha curcas* (Leschenault de la Tour 1820: 334), or an *Ipomoea* species (Buchanan 1807, v.2: 19).

[2]The poor transportability of charcoal is noted by Schubert (1957: 224), Hyde (1977: 10–11), and Hammersley (1973: 606), who say that in Britain an acceptable distance for carrying charcoal to a furnace was considered to be 3 to 5 miles. However, a charcoal furnace in Bulacan province of the Philippines gets all its charcoal from Bicol province—a distance of several hundred kilometers.

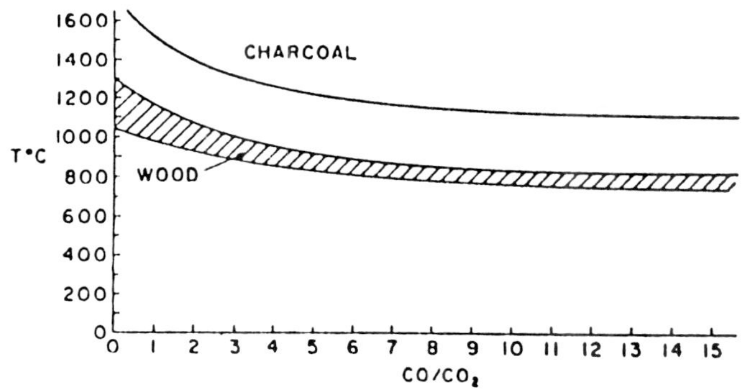


Fig. 6.1. *$CO/CO_2$ Ratio and Peak Temperature.* The graph shows estimates of the maximum temperatures achievable by burning wood and charcoal with air supply rates adjusted to given ratios of carbon monoxide to carbon dioxide. The estimates are based on the assumption that no heat is lost to the surroundings; in real furnaces, only flame reaches such temperatures. (Based on data from Avery and Schmidt 1979, van der Merwe 1982, and Avery 1982)

[long] tons of pig iron—1425; cost of charcoal at the finery—875; total cost of fining 400 tons of pig to 300 tons of wrought iron—1530; charcoal as percentage of pig iron cost—62.5 percent; charcoal as percentage of overall wrought iron cost—59.7 percent (Schubert 1957: 423–426). These figures are not unusual; it has been estimated that in England in the sixteenth through eighteenth centuries charcoal accounted for 58 to 75 percent of the total cost of ironmaking (Hammersley 1973: 610). Although wood was scarce and expensive in England, these data should not be discounted as unrepresentative of the industry in other areas, since charcoal was expensive in most parts of the world, because of the labor costs alone. In China, India, and Iran, the man days of labor needed for one day of smelting were primarily spent in the preparation and transportation of charcoal, rather than in furnace operation or the mining and preparation of ore (Table 6.3). Sometimes the labor cost of charcoal was greater than that of all other ironmaking operations combined.

The production of charcoal involves the thermal decomposition of wood, which is predominantly a mixture of cellulose polymers. The cellulose molecules begin to pyrolysize at temperatures of 250° to 300°C, breaking down into elemental carbon, water vapor, and hydrooxycarbon gases.

The first step in the conversion of wood to charcoal is reducing the water content of the freshly cut raw material. Much of the water in green wood evaporates over a period of months if the logs are allowed to stand or lie in a reasonably well-ventilated location. The weight loss by water evaporation at this "seasoning" stage can vary from 15 to 32 percent (*Chem. Eng. Handbook* 1941: 2342). Hardwoods fall at the lower end of this range. Since all the original water must be evaporated before pyrolysis can begin, attempts to accelerate the seasoning process by heating or charring the green wood would be expensive.

The next step, charring or charking, has been described in observations of metallurgy among the "wild inhabitants" of the Tributary Mehals (now part of Orissa) in eastern India:

> Only logs of one size are employed, namely, about 12–18 inches in diameter. Trees of larger dimensions cannot be broken up by the tools the natives possess, and the trunks are left to lie in the forest. The smaller branches are perhaps used for firewood; they are never, as far as I am aware, used for making charcoal. The logs are cut into pieces about 5–6 ft long which are piled on each other in layers crosswise, and then the small stack of logs, without any other preparation, is set fire to. When well lighted, water is thrown on, and the crusts of charcoal formed are knocked off each log (Oldham 1859: 14, after Blanford).

These particular charcoal producers were considered exceptional by one expert: "They seem to exceed all others that I have heard of, in the ingenuity with which they seem to have adopted a process for obtaining the very smallest possible amount of useful fuel from the greatest amount of timber" (Oldham 1859).

A more efficient technique was recommended by Biringuccio in 1540:

> Four large poles are set up in the middle of a square, or three in a triangle, which make an empty space of little less than half "braccio" [58.4 cm]. Then proceed around these, covering them upwards, circle by circle, with all your cut wood and with chumps split into smaller pieces, building it in the form of a round pyramid or haystack. . . until you have arranged the width and height that you wish the charcoal pile to be. The middle space between the poles is left empty up to the top. When this pile has been made, it is well covered all over the outside with fern leaves and broom and then, on top of these, it is plastered well from the top with good, firm earth, as dry as it is dug up. This plastering is. . . sealed so that it may not breathe except at the top where ten or twelve vents are left as outlets for the smoke and humidity contained in the wood and earth. This done, the fire is thrown to the bottom of the little hole that you left in the middle between the poles. . . Then this opening is also closed over with earth and only the vents are left open. In six or eight days the whole charcoal pile gradually takes fire and bakes. When it is seen that the vents cease to give off strong

*TABLE 6.3. Labor costs of fuel, ore, and furnace work in smelting operations*

| | | Man days/day | | | | |
|---|---|---|---|---|---|---|
| Country | Area | Fuel | Ore | Furnace | Output/day | Reference |
| China, C | South Henan | 34 | 32 | 9 | 720* | Wagner 1984 |
| India, C | Madhya Pradesh | 10 | 2 | 2 | 23 | Hole 1898 |
| India, C | Madhya Pradesh | 5 | 5 | 4 | 60 | Bartlett 1900: 150-153 |
| India, E | Orissa | 2 | 2 | 3 | 22 | Ahmed 1967: 17-44 |
| India, S | Mysore | 9 | 1 | 4 | 18 | Buchanan 1807, 2: 35 |
| India, S | Mysore | 10 | 8 | 4 | 22 | Buchanan 1807, 2: 139 |
| India, S | Mysore | 4 | 2 | 4 | 33 | Buchanan 1807, 1: 175 |
| India, S | Mysore | 7 | 4 | 4 | 27 | Buchanan 1807, 2: 17 |
| India, S | Mysore | 9 | 2 | 3 | 31 | Buchanan 1807, 3: 362 |
| Iran, N | Mazanderan | 5 | 4 | 4 | 90* | Böhne 1928 |

Output is in kg; * = cast iron, other output is bloom iron.

Fig. 6.2. *European Charcoal Making Pile, Eighteenth Century.* Cordwood is tightly stacked into a conical pile and covered with earth to render the sides impermeable to air. The projecting log at the apex will later be pulled out to provide a vent for combustion gases. (From Beck 1897, v.3: 294)

> exhalations it is to be believed that it is baked. Then, with the same kind of earth, seal the top and all around and in every place so that none of the vents can breathe at all, and the fire inside immediately suffocates and dies when it finds itself without respiration. Thus your wood is converted into wholly spent charcoal without ashes or any humidity (Smith and Gnudi 1942: 177–178).

Biringuccio does not imply that there was anything new about this system in his day; similar structures (Fig. 6.2) were still common in western and southern Europe during the nineteenth century, although many attempts were being made to promote improved techniques (Percy 1875: 366–414). The better traditional systems can attain respectable yields when properly managed, ranging from 15 to 28 percent by weight (Percy 1875: 366–414). Comparable wood-charcoal conversion figures are given by Osborn (1869: 221–223). The charcoal produced is not pure carbon, about 30 percent of its weight being residual hydrocarbon and ash.

Therefore, almost any traditional charcoal making system should perform adequately as long as its operators are diligent in regulating vents and patching leaks that might reduce efficiency of recovery. The one possible improvement that could have been made in traditional methods would have been to prolong the process, by keeping the vents closed as much as possible; quick charring, which takes only four to five days, produces half as much charcoal by weight as when the process lasts for ten to twelve days (Percy 1875: 366–414).

It is not clear whether supplies of charcoal were sufficient to meet the large and increasing demands of metallurgical industries. Some scholars have suggested that the forests of Europe and several parts of Asia were formerly denuded by their iron industries, causing acute industrial and domestic fuel shortages. Such a shortage has been documented for Britain (Ashton 1924; Nef 1932; Schubert 1957), and for Europe as a whole; the traditional view is that fuel crises worsened in the sixteenth through eighteenth centuries (Braudel 1967: 269–271). Similar deforestation caused by the metallurgical industry may have occurred in northern China (Hartwell 1963: 95) and northeastern Iran (Horne 1982).

These shortages may have been exaggerated, and perhaps nonexistent (e.g., Hammersley 1973). Data show that forest management and regulated timber cropping existed in many areas of Britain by the early seventeenth century, that these were potentially very efficient, and that, despite high prices, there is no evidence of actual shortages of metallurgical charcoal. It is estimated that in eighteenth-century England about 2800 hectares of carefully coppiced forest were enough to provide a continuous supply of charcoal for an average smelting furnace.[3] An estimate made for early nineteenth-century France was that less than 200 square miles (about 500 km.$^2$ or 50,000 hectares) of coppice would serve the whole iron industry (Wertime 1962: 118). In contrast, a Scottish iron furnace in the eighteenth century would have needed about 4000 hectares of coppiced forest land (Lindsay 1975: 287).

In China, systems of selective cutting and planned regrowth had been developed by at least the eleventh century A.D. (Wagner 1985: 33–37); these practices are still employed by charcoal makers in modern Henan. In the nineteenth century, British geological surveyors assessed the potential of various districts in India as production sites for iron and estimated sustainable fuel supplies based on rates of timber regrowth (e.g., Oldham 1859; Holland 1892). Before the middle of the nineteenth century, several careful measurements of timber regrowth rates were made in continental Europe (Table 6.4; Percy 1875). These efforts helped balance the annual use of charcoal for a given rate of iron production with a sustainable rate of cropping wood.

When estimates of annual rates of timber cropping are combined with figures on the acreage of forest needed to maintain the known rates of iron production from 1540 to 1750, it becomes clear that the British iron industry had never been in danger of running out of fuel and timber (Hammersley 1973). Instead, the real problem was the steadily rising cost of labor to cut the wood, make it into charcoal, and transport it to the furnace (Fig. 6.3). The best strategy for decreasing fuel

[3] Coppicing is a forest management system whereby, when mature timber is cut, the stumps are left to send up shoots. The shoots are pruned to preserve the best which, using the existing root system, grow at a much faster rate than a seedling. The regrowth can be cropped about every 20 years.

*TABLE 6.4. Estimates of annual timber crop per acre of forest land*

| Weight of timber, lbs. | Location |
|---|---|
| 9400 | India[1] |
| 3370 (oak) | Black Forest, Germany[2] |
| 3470 (ash, med. altitude) | Black Forest, Germany[2] |
| 2990 (ash, high altitude) | Black Forest, Germany[2] |
| 2580 (hornbeam) | Black Forest, Germany[2] |
| 3940 (silver fir) | Black Forest, Germany[2] |
| 3250 (scotch fir) | Black Forest, Germany[2] |
| 3270 average* | Black Forest, Germany[2] |
| 4600** | Britain[3] |

*Corrected to green timber using an average dry weight loss of 24%
**Assuming oak timber at an average density of 46 lb/cu. ft.
Note: To these numbers may be added a modern estimate (Clawson 1975: 79) of the United States average natural regrowth of hardwoods—including all land-water-climate production levels—of 4875 lb. per acre. An average density of 65 lb./cu ft. (Percy 1875: 194) has been used to convert data given in cubic feet per acre.
[1]Oldham 1859, [2]Percy 1875, [3]Hammersley 1973.

costs was to find a substitute for charcoal. The obvious candidate was mineral coal, already used as a fuel for centuries in both Europe and China; in the form of coke, coal was superior to charcoal in some ways. Because labor costs and the prices charged by the owners of woodlands were much higher in England than in areas like Sweden, the change from charcoal to coke in Britain was motivated by the need to meet foreign competition rather than by a scarcity of wood (Hammersley 1973: 610).

## 4. PEAT

Peat, also called "turf" in some parts of Britain, is a carbonaceous material composed of subfossil plant remains preserved under anaerobic conditions in swamps. It occurs in many parts of the world, including tropical areas. Beds of peat may be 20 m. thick, but more commonly are 2 to 3 m. Because peat lies close to the surface and thus is easily removed, it has been widely exploited as a household and industrial fuel where coal and wood are scarce. Its use in ironmaking has been limited, but its abundance and low cost have often

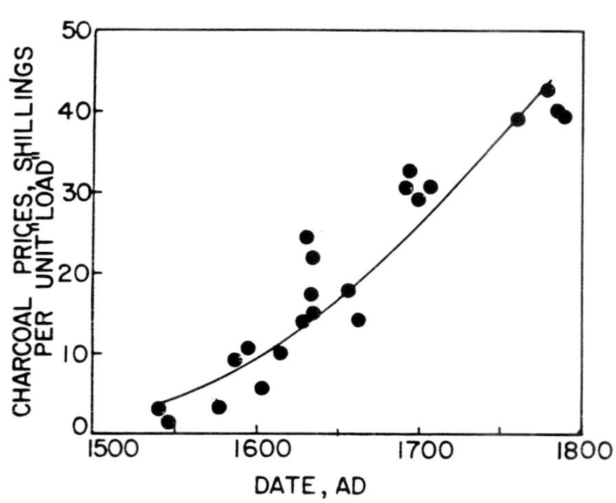


Fig. 6.3. *British Charcoal Prices, 1550–1800.* Although some inflation occurred during this period, the trend of rising costs is nonetheless real. It undoubtedly provided a strong incentive for ironmakers to look at alternative fuels. (Data from Hammersley 1973)

attracted smelters, smiths, and other manufacturers (Percy 1875: 200–226, 498–515).

Peat is a loosely knit material of clearly recognizable botanical origin. It can be cut with a spade and is usually excavated in blocks. Because it is about 25 percent water, it makes a poor fuel unless it is dried. Air drying will reduce the water content to an acceptable 10 percent and kiln drying will remove an even larger percentage. Air-dried peat has densities which vary between about 250 to 450 kg/m$^3$ and a calorific value about two-thirds that of bituminous coal. The organic fraction of dried peat contains about 50 to 60 percent carbon, 5 to 7 percent hydrogen, 30 to 50 percent oxygen, and 1 to 2 percent nitrogen. The inorganic ash content can be as high as 22 percent but is more commonly about 5 percent. The composition of the ash is variable but can be averaged to 1.8 percent potash, 3 percent soda, 30 percent lime, 12 percent magnesia, 9 percent alumina, 8 percent silica, 26 percent iron, 4 percent residual carbon, 6 percent sulfur, and 1.7 percent phosphorus. The sulfur content, of special concern to ironmakers, averages only 0.3 percent of the combined organic and inorganic fractions (Percy 1875: 200–226, 498–515). This is low compared to most bituminous coals.

Peat can be converted to charcoal, but the process is much more difficult than that for wood. The charcoal yield is about 30 percent of the dry weight of the peat. Because peat charcoal is very powdery and readily crushed, it has to be made in kilns or pits rather than piles, usually mixed or "co-charked" with wood to provide added structural support (Schubert 1957: 18; Percy 1875: 514). Some peat charcoals can be used in lump form, but most must first be made into briquettes. The calorific value of peat charcoal appears to be similar to that of wood charcoal.[4]

There are fragmentary references to the use of both air-dried and charked peat in iron manufacturing. In northern England and Scotland, smelting was fueled by peat charcoal from "time immemorial" (Schubert 1957: 18). In England in the sixteenth century great efforts were made to substitute peat for a portion of the wood charcoal (Schubert 1957: 224–226). The extent of this practice is not well known, but there is evidence that it survived into the late seventeenth century (Sturdie 1693; Powle 1678). Some commentators, it should be noted, believe that claims about the use and value of peat are exaggerated (Schubert 1957: 18).

Peat charcoal had industrial applications in the nineteenth century. Equal weights of peat and wood charcoal were used in several tall blast furnaces in the Erzgebirge on the German-Austrian border and in blast furnaces 10.8 m. high in Josefstadt, Austria. In Britain, briquetted peat charcoal alone was reportedly successful in an experimental cupola remelting of pig iron and in pig iron smelting in blast furnaces 7.8 m. high (Percy 1875: 512–514). The qualities of peat charcoal were

[4]Overman (1854: 87) says, "If turf is dug for the purposes of charring, it is advisable to employ a good strong press. Turf, thus pressed, chars excellently, and yields a charcoal as hard again as the best sugar maple, or hickory coal."

appreciated: "Turf charcoal is an excellent fuel, but expensive; it burns freely, and produces a fine heat. In Styria, sheet iron, and reheating furnaces are heated by it: and in Bohemia, Bavaria and Russia, it is extensively used in the blast furnaces, and produces in most cases, very liquid, lively iron. Good turf coal is superior to charcoal in the blacksmith's fire" (Overman 1854: 85). The implications are that peat possessed high calorific value, but, despite its efficiency as a fuel and the lower labor costs of excavating it, its ultimate cost—which includes the charking process—precluded wider use. There is no literary evidence which indicates whether peat, either in dried or charked form, was ever part of non-European ironmaking traditions.

*TABLE 6.5. Classes of coal*

| |
|---|
| Anthracite |
| Bituminous |
| Sub-bituminous |
| Brown coal or lignite |
| Peat |

Note: Schubert 1957: 224-226 summarizes the unsuccessful trials and unsubstantiated claims on the use of peat in part or the whole of the fuel for iron smelting in England from the 16th to as late as the 18th centuries.

## 5. COAL

The history of coal in minor iron processing activities is long but obscure. It was in regular but limited use in Europe by the thirteenth century (Mott, in Read 1939–40) and in China by the fourth century (Needham et al. 1980: 520). Archaeological finds of high sulfur iron strongly suggest the sporadic employment of coal by ironmakers as early as the second century A.D. in China (Read 1939–40) and in Roman Britain (Webster 1955; Coghlan 1977). These areas have been more thoroughly studied by archaeometallurgists than most, so it is probable that evidence of an ancient connection between coal and metal will eventually appear in other areas.

Coal, because it is easily identified and obtained, is likely to have been used even earlier in nonmetallurgical processes. The fact that it burns could have been discovered independently in several places, but the disadvantages of coal as a domestic fuel would have been immediately obvious. It is hard to ignite, and emits a sulfurous smoke which would give food directly heated by it a bad smell and taste. In ironmaking it can cause hot shortness due to sulfur contamination. Yet for those living near an exposed coal seam or a beach covered with coal, fossil carbon could have been acquired with much less labor than an equivalent weight of charcoal, and therefore a lower quality product may have been acceptable because of the lower labor cost and ease of exploitation.

Coal is the generic name for a family of mineral fuels that differ considerably in the amount of carbon, hydrocarbons, and water they contain (Table 6.5). The hydrocarbons in coal may range from methane to heavy tars. Thus, if coal is heated, gases are evolved over a wide range of temperatures up to 600°C. Some of these are condensable to liquids (hence the term coal oil) and some to tarry semi-solids. When both the water and hydrocarbons have been pyrolysized, the residuum consists largely of carbon, which is called coke. The higher the temperature at which the coke is formed, the harder and stronger it will be.

When burned in a furnace, lignite and bituminous coals differ from charcoal in that they undergo some degree of fusion or softening when heated (*Chem. Eng. Handbook* 1941: 2333). This partial melting is not a sharp transition nor does it progress beyond a very viscous state, but it is enough to cause fragments of coal to cake together and thus to increase the resistance of a furnace charge to gas flow. Because the softening of coal occurs at 330° to 440°C—well below the temperature in the combustion zone of any shaft furnace, and therefore well above that zone in a physical sense—the resulting back pressure will tend to choke off combustion. The softer coals also contain a high proportion of hydrocarbons that can recondense at a relatively low temperature. The resulting tars collect near the top of the furnace and add to the choking effect.

The problem of back pressure therefore limits the height of a shaft in which lignite or bituminous coal is burned. These fuels are not practical except in a shallow or short furnace structure where the top becomes sufficiently hot to prevent the condensation of tars and where the column of fuel and ore is not heavy enough to compress the softened coal.

These difficulties would not have been encountered in early fineries and forges, where the bed depth of fuel did not exceed a few feet. Even smelting would have been possible if the furnace were short and wide and if the bellows provided sufficient blast pressure. The low Chinese smelting furnaces pictured in the seventeenth-century *Tiangong Kaiwu* (Sun and Sun 1966: 250) (Fig. 3.9) had enormous windbox bellows and top openings wider than their height. Since coal became common in iron smelting in China 500 years before these pictures were drawn, it is plausible that such back pressure minimizing furnace designs were developed in response to the need to use coal directly in smelting.

The special physical characteristics of uncoked mineral carbon therefore did not pose insoluble problems in the pre-industrial period. In fact, it is likely that lignite and bituminous coals were occasionally used by ancient ironmakers in regions where they occur as surface outcroppings or colluvial deposits. The Romano-British iron objects with high sulfur contents should not necessarily be considered relics of isolated experiments. It is possible that there was an irregular but routine utilization of a familiar resource, even though the awareness that raw coal produces inferior but sometimes usable iron must have existed.

Coal smelted iron would have been inferior because mineral coal, unlike charcoal, contains substantial amounts of sulfur, ranging from about 0.4 to 3.5 percent. In the furnace, only 13 to 24 percent of the sulfur will pass off as the gases $SO_2$ and $H_2S$ (Percy 1875: 275). The rest will be retained in the ash, where it is harmful to iron. The fine particles of newly reduced iron come into contact with the sulfide in the ash long before they enter the protective pool of slag. The result is that the finished metal, whether bloom or cast iron, will have absorbed sulfur in substantial amounts—as much as several percent in cast iron. Little of this sulfur is lost during subsequent processing. Wrought iron made from high sulfur cast iron, for instance, will be hot short. Wrought iron made from the same high sulfur cast iron in a finery that is also fueled with coal will be unusable.

The presence of sulfur in coal was not an obstacle to early Chinese foundrymen making or remelting white pig iron for casting, since sulfur helps to insure that a white iron structure develops even at less than optimal freezing rates. Moreover, sulfur in coal would not cause serious problems for any ironworkers, in China or elsewhere, engaged in the solid state processing of their metal: roasting, annealing, carburizing, or smithing. The fact that solid iron is almost invulnerable to sulfur contamination (except superficially) helps to explain why coal was used in European chaferies or reheating furnaces long before Dud Dudley and his rivals in the seventeenth century started testing it as a smelting fuel (Braudel 1967: 272).

The poor forgeability of iron smelted with coal remained a major problem until well into the nineteenth century, when desulfurization methods involving high lime slags and manganese alloying were discovered. Before then, since sulfur was not generally recognized as the origin of the problem, coal smelted iron would have been subject to marked and seemingly random variations in quality. Coal from certain sources was known to produce better iron than coal from elsewhere, yet rarely could the product compete with the better grades of charcoal smelted iron. The dominant role of fuel costs in ironmaking had long been understood; even earlier it was recognized that coal would burn at smelting temperatures and was also inexpensive. Nevertheless, because coal often made iron hot short, the shift to cheap coal from expensive charcoal was slow.

## 6. COKE

Many of the drawbacks of coal can be eliminated by converting it to coke. The coking process is similar to charcoal making because in both the coal or wood is heated to a high temperature in the absence of air, thus driving off the volatile constituents of the parent material and pyrolysizing the rest. The volatile products evolved include noncondensable hydrocarbon gases such as methane and propane, condensable light oils, condensable heavy hydrocarbons or tars, and ammonia. Although these substances became the bases of several new industries in the late nineteenth century, until that time they were discarded as waste products. The nonvolatile residue is a very porous carbon sponge which is much stronger than charcoal. Because of this strength, it can support a much taller fuel and ore column without being crushed and obstructing air flow.[5] The bearing strength of coke can be further increased by raising the peak coking temperature. A modern coke oven delivers finished coke at temperatures much higher than would be needed simply to eliminate volatile hydrocarbons.

Coking does not, however, reduce the coal's sulfur content. Despite a widely held misconception (Braudel 1967: 273; Hyde 1977: 84), coked coal retains most of the sulfur that was originally present. Rosenquist (1974: 227) concludes: "Of the total sulfur content in the coal about one-third is expelled in the gas and two-thirds is retained in the coke. Since the weight of the coke is about two-thirds of the weight of the coal this means that the sulfur percentage in the coke is about the same as in the coal." Thus, before desulfurizing slags had been developed, coke was as potentially hazardous as coal to the smelted iron.

In pre-industrial Europe, a standard coke pile was similar to the wood charking pile described by Biringuccio—a domed heap with a central chimney of unmortared brick and an airtight covering of wetted coal dust rather than clay. A typical pile had a diameter of 9 m. and a height of 1.5 m., with a base built of larger coal pieces to provide for the ingress of air. After ignition, the normal coking time was three or four days (Percy 1861: 144). The yield varied from 30 to 65 percent of the weight of the original coal and had a heating value which was about the same as that of an equal weight of bituminous coal of average quality (Table 6.2). The energy lost in the conversion was therefore considerable. Metallurgically, however, the benefits more than outweighed the losses. Coke is more porous than coal and so combusts more rapidly, has no tendency to soften at any temperature, and is more resistant to crushing and abrasion, all of which allow a tall furnace to function more efficiently. Moreover, coke has a greater energy density. Although its heating value per unit weight is similar to that of charcoal, its heating value per unit volume is more than twice as high. This is of major importance in a thermochemical process which must operate within the restricted space of a furnace shaft.

In both China and Europe, coke was first used as a cooking fuel, not in metallurgy. The earliest reference to coke in Chinese annals is a brief account by a wealthy resident of Loyang, who in 874–879 A.D. had his food

[5]Not all bituminous coals produce coke of equal resistance to crushing. This became evident in Britain at the beginning of the nineteenth century. As blast furnaces began to reach heights of 15 or more meters, certain coals came to be preferred for coke production. Gale (1966: 103) suggests that the absence of coal of this kind helps to explain the decline of iron production in the "Black Country" of England, noting that blast furnaces there had to remain short to accommodate the coke produced from local coal.

*TABLE 6.6. Coke and charcoal in British ironmaking*

| | 1740 | 1760 | 1788 | 1806 |
|---|---|---|---|---|
| Imported charcoal iron, tons | 30,000 | — | 51,716 | 17,015 |
| Domestic charcoal iron | 17,350 | 25,600 | 13,100 | 7,800 |
| Domestic coke iron | 1,000? | 9,800 | 48,200 | 227,156 |
| Domestic charcoal furnaces | 59 | 64 | 24 | 11 |
| Domestic coke furnaces | 2? | 14 | 53 | 186 |

From Scrivenor 1854: 57, 58, 87, 99, 137. The figures for coke iron and furnaces in 1740 are interpolations; those for 1760 are based on the estimates of Hyde 1977: 67.

cooked with coal that had been preburned to remove its bad smell (Hartwell 1966: 68, after Ichisada). One of the earliest records of coke in Europe describes its introduction in brewing in Derbyshire in 1642: "Almost immediately the brewers in the region began to use coke instead of straw and ordinary coal for drying and heating the malt. 'This new fuel will give Derby beer the paleness and sweetness which will make its reputation, ridding it of the unpleasant smell of ordinary coal.' It duly became the leading beer in England" (Braudel 1967: 273). Although Braudel implies that this effect was due to a significant removal of sulfur through the coking process, it seems more likely that the improvement in smell reflected only the differences between the two sulfur gases, $H_2S$ and $SO_2$. The former smells and tastes like rotten eggs. The latter adds little to the flavor, and is now used to bleach various foodstuffs. In the coking of coal, much of the small percentage of sulfur removed is in the form of $H_2S$ associated with other gaseous hydrogen compounds.

While there may have been experiments with coke in iron manufacture during the late seventeenth century, it was not until about 1718 that the first coke fueled blast furnace came into operation. Abraham Darby, the proprietor of that furnace, ran one of the few British iron firms to specialize in foundry production. His fame rests not only on his development of the first successful European method of smelting with coke, but also on his invention of molding and pouring techniques. These techniques were used to make the large cast iron "missionary-cooking" pots of Coalbrookdale that were famous in places as distant as Pennsylvania and Africa. A high sulfur white iron was presumably well suited for casting them.

Darby claimed that his cost per ton was only half that of iron smelted with charcoal (Hyde 1977: 24), yet other British ironmakers and processors did not rush to take advantage of these impressive savings, undoubtedly because high sulfur iron like that made by Darby was of little interest to the finery and forge operators who were their principal customers. It took about fifty years for the output of coke furnaces in Britain to equal that of charcoal furnaces, and even after that Britain still consumed more charcoal iron than coke iron (Table 6.6). Coke smelting was not adopted as readily in other European countries, and it was not until almost a hundred years after Darby's discovery that any place other than Britain was substantially dependent on iron smelted with coke.

Coke in ironmaking is mainly a phenomenon of the Industrial Revolution. In the pre-industrial period, there was a limited market for coke among secondary processors since it could be used only in foundries and perhaps a few fineries. Coke would have been desirable because of its price and mechanical strength, but it was only a remedy and not itself a sufficient cause for an industrial revolution. Its potential could not have been realized until a whole range of other developments had been made: air preheating, steam powered air blast machines, mechanical charge-conveying systems and, most importantly, reverberatory refining and puddling furnaces with desulfurizing slags. Without these inventions, the coke fueled blast furnace would have remained an essentially marginal device.

In China, where coke smelting is much older than in Europe, the total pig iron production in 1916 was estimated at 370,000 tons, of which 199,000 were smelted by recently introduced European techniques, 8000 by undescribed techniques, and about 163,000 by methods that were fully traditional. The European-type furnaces burned coke, but the traditional furnaces usually did not. More than half burned charcoal and about 40 percent (all of them crucible furnaces), anthracite. Only a handful used coke (Tables 6.7, 6.8).

*TABLE 6.7. Chinese production of pig iron by traditional methods, ca. 1916*

| Province | | Furnace | Number | Iron (Tons) | Tegengren 1923-24: |
|---|---|---|---|---|---|
| Manchuria | coke | small blast | some | 1,700 | 316 |
| | coke | crucible | some | | 317 |
| Shansi | anthracite | crucible | many | 70,000 | 324-326 |
| Honan | charcoal | small blast | many | 14,400 | 334 |
| Fukien | charcoal, some anthracite | large blast | some | 2,600 | 337 |
| | charcoal | small blast | some | | 337 |
| Hunan | charcoal, some coke | large blast | many | 35,000 | 339 |
| | coke or anthracite | small blast | many | | 339 |
| | coal | large blast | one+ | | 338 |
| Kueichou | charcoal | small blast | some | 6,000 | 341 |
| Szechuan | charcoal | large blast | many | 23,700 | 343-345 |
| Yunnan | charcoal | large blast | many | 9,300 | 350-363 |
| | charcoal | small blast | many | | 354-363 |
| | coke & ore-coal mix | large blast | some | | 354 |
| | charred wood | large blast or natural draft | some | | 352 |

*TABLE 6.8. Iron in China: 1916 production and estimated fuel use*

| | | |
|---|---|---|
| Max. coke-smelted | = | 4,000 tons (2.4%) |
| Max. anthracite-smelted | = | 72,000 tons (44.2%) |
| Max. raw coal-smelted | = | 3,000 tons (1.8%) |
| Min. charcoal-smelted | = | 83,700 tons (51.4%) |

Table 6.8 summarizes the data presented in Table 6.7. The totals are rough estimates made by the authors, not Tegengren. We have taken the gross figures for iron produced in each province and attempted to subdivide these according to Tegengren's unquantified but often precise statements about the relative importance of the smelting systems used in that province. Although a number of arbitrary choices had to be made in cases where several fuels were reported to be used, much of Tegengren's data comes from single-fuel centers or provinces.

Even though coke had been known in China for a thousand years, it had not inspired an industrial revolution.[6] If, indeed, coked bituminous coal was a principal metallurgical fuel in northern central China during the eleventh and twelfth centuries (Hartwell 1966, 1967), there must not have been better alternatives such as charcoal or anthracite available. When superior fuels could be obtained, Chinese ironmasters returned to them. By the last half of the nineteenth century, coke was not commonly used as fuel despite China's substantial coal reserves and the tremendous deforestation that resulted from the use of charcoal.

## 7. ANTHRACITE

Although anthracite is, in its native state, similar to coke in low ash and volatile hydrocarbon contents, it played only a small role in pre-industrial ironmaking. It is an excellent fuel because it is noncaking, strong, dry, often very low in sulfur content, and extremely high in heating value (Table 6.2). Some types of anthracite, moreover, make excellent iron. In Pennsylvania, more than fifty anthracite fueled furnaces were in operation by 1840 (Scrivenor 1854: 267); in 1872, there were 152, many with an average annual capacity of between 10,000 and 15,000 tons (Dunlap 1874: 55–60).

In China, the main metallurgical use of anthracite was as a fuel and reagent in the Shanxi crucible process, even though its unique characteristics were not crucial to success. Either raw bituminous coal or charcoal would have served the Shanxi ironmasters as well. In Europe, anthracite was almost ignored, despite substantial deposits in several countries, such as Britain and France, that were most affected by the high price of charcoal.

In 1831, the unpopularity of anthracite as a blast furnace fuel during the pre-industrial period was explained as follows:

> I will now speak of some experiments made with a curious, combustible substance, called anthracite,—a kind of pure carbon, without any mixture of bitumen—compact, igniting with great difficulty, and giving out such a heat, when once in a state of combustion, that it is very difficult to procure materials for the construction of the blast furnace which will not melt. It has been determined that cast iron cannot be made with anthracite, except by excessive care. . . . (M. A. Perdonnet, as quoted by Scrivenor 1854: 265).

A complete account of the French experiments performed at Vizille in 1827–28 indicates that the main cause of failure was the tendency of that particular type of anthracite to splinter at high temperatures. When only coke was burned in the furnace, a blast pressure of 1.6 psi was adequate. At 70 percent anthracite and 30 percent coke, pressure had to be increased to 3 psi and the yield declined. At 80 percent anthracite even 4 psi was not enough, while all output ceased when the anthracite was raised to 90 percent (Johnson 1841: 14–26).

The hot blast system which had been recently invented in Scotland was found to be the key to success in using anthracite in blast furnaces, and was exploited almost simultaneously in both Wales and the United States during the 1830s. By 1850, numerous hot blast anthracite furnaces were in operation in Wales and eastern Pennsylvania (Johnson 1841: 34; Hyde 1977: 159). They were so successful in the U.S. that the introduction of coke into American ironworks was substantially retarded. Coke furnaces in the United States were not adopted on a significant scale until the 1850s and 1860s, when ironmakers began to move away from the Pennsylvania-New Jersey anthracite sources and when the scale of production had begun to outstrip the capacity of the highly developed American types of charcoal furnaces (Osborn 1869: 198–199).[7]

[6]It is not clear that coke was in fact used for ironmaking in China as early as the twelfth century. Hartwell (1966, 1967) is cautious but concludes that it was. McNeill (1982: 32–33) is certain about it. Needham (1958: 18; 1966: 234) gives no references earlier than the late eighteenth century. Tegengren (1923–24: 310, 317, 354) too quotes only relatively recent references.

[7]The use of coke in the U.S. had been tested but rejected in at least one major trial (Johnson 1841: 6–8). For the efficiency of anthracite furnaces, see Johnson (1841: 30, 31, 76) and Hyde (1977: 159); both cite a fuel:iron ratio of 1.35–1.84: 1, whereas the average coke furnace of the period had a fuel:ore ratio of about 2.3: 1 (Scrivenor 1854: 260). On the long survival of charcoal furnaces in the U.S., see Schallenberg and Ault (1977).

# 7. Air Supply

## 1. INTRODUCTION

An adequate supply of air is necessary for successful furnace operation. There must be enough air to support combustion and it must be distributed throughout the furnace so that it reaches most of the bed of fuel and ore. The distribution of air can be controlled by regulating containment conditions and the size, number, and position of the tuyeres. Systems requiring only a low air pressure can and have been sustained by natural draft, but most have utilized mechanical air pumping devices.

The air supply required for a bloomery furnace can be calculated on the basis of observed parameters. About 9.8 $m^3$ of air is consumed for each kilogram of iron produced. In experiments with a replica of an early British shaft furnace, Tylecote (1971: 348) measured the empirical relationship between rate of air input and temperature (Fig. 7.1). A rise in temperature of only 100°C meant air input had to be tripled, from 0.12 to 0.34 $m^3$ per minute. A small, traditional cast iron producing furnace in central China, which had to be run at higher temperatures than Tylecote's bloomery to produce cast rather than bloom iron, consumed 1.6 $m^3$/minute (Wagner 1985: 17).

Empirical data on the relationship of total air input to yield and of air input rate to temperature in preindustrial furnaces are scarce. By taking into account differences in furnace capacity and in the nature of raw materials and products, we can, however, estimate that an average of 10 $m^3$ of air is required to produce a kilogram of smelted iron, and that many small smelting furnaces of the past operated at input rates of 0.2–2 $m^3$/minute. The low end of this range was probably close to the lower limit at which smelting temperatures can be attained.

However, not all furnaces with about the same yields need air supply systems of a similar capacity. Indirect heating reverberatory or crucible furnaces, in which combustion is physically separated from the reactions, may require similar amounts of air per unit of product and may also be relatively insensitive to the rate at which this air is supplied. Direct heating furnaces such as bowl furnaces and smelting forges, because of their inherently poor heat containment, need a high rate of air input to reach initial working temperatures and will not operate well unless this input rate can be maintained. Shaft furnaces vary in their air supply requirements. One may need air at a relatively high pressure while another with the same shape and dimensions may work well with a low pressure air supply.

The main reason for variability among shaft furnaces is that their contents offer differing degrees of resistance to the upward flow of combustion gases. The amount of resistance or back pressure is inversely proportional to the size of the spaces between fragments of fuel and ore, and thus to the size of the fragments themselves. A charge composed entirely of large fragments creates little back pressure but causes inefficient operation, since a large piece of ore will not be completely reduced by the time it reaches the combustion zone and so falls through into the slag at the bottom of the furnace. A charge of small fragments, on the other hand, will resist the passage of effluent gases. An air blast at high pressure may be needed to overcome this resistance.

A furnace operator must consider several factors before choosing an air supply system. The first is the quantity of iron which has to be produced within a given period, for this is closely related to the total amount of air that must be delivered within that time. High production from a single furnace will almost certainly require a high pressure air supply system. If yield per day is not crucial, however, other factors may influence the furnace operator's choice.

In indirect heating furnaces, the only constraint other than the desired yield per day is the maintenance of smelting temperatures. This is not difficult in a properly constructed reverberatory furnace or enclosed crucible oven, so a low pressure air delivery system will work. Almost all bowl furnaces require air delivery at high pressure because reaching and maintaining adequate temperatures are difficult. A low pressure system may work in shaft furnaces in which fragment sizes are large or the bed shallow, although smelting will be slow. Shaft furnaces with more finely divided ore and fuel or with a higher charge column need high pressure systems; early in ironmaking history, column heights were limited to several meters because the necessary air pressures could not be attained.

## 2. NATURAL DRAFT

Indirect heating furnaces often make use of natural draft, the flow of oxygen-containing air that feeds any fire as the hot gases at its center move upward. In an open fire, the effect of natural draft is limited. The only documented iron smelting process employing an open fire without bellows is the Shanxi crucible system (Chapter 3), in which the draft created by large fires together with partial containment by the sides of the crucibles themselves was sufficient to achieve smelting temperatures.

Enclosed fires provided with chimneys can reach

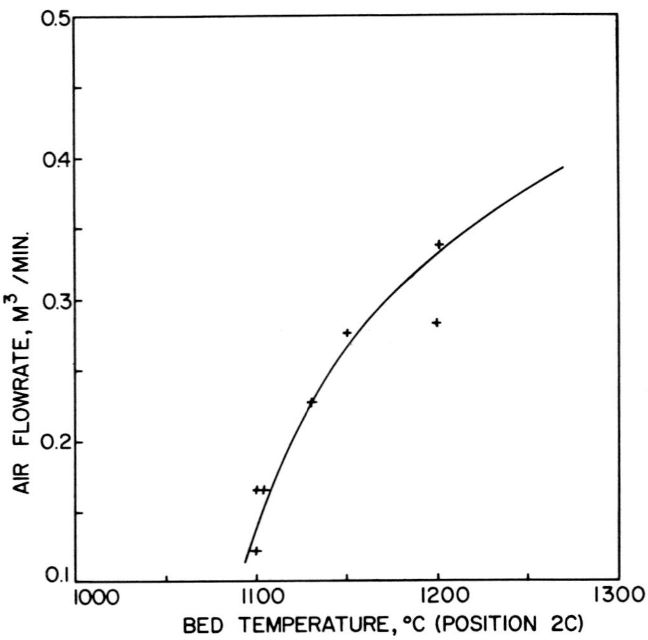


Fig. 7.1. *Air Flow Rate and Temperature in Bloomery Furnace.* Temperatures measured at a position above but close to the tuyere are plotted as a function of the air input rate to the combustion zone of an experimental shaft furnace. At air input rates above 0.3 $m^3$ per minute, cast iron was produced. (Data from Tylecote, Austin, and Wraith 1971: 348)

high temperatures fairly easily, as in European puddling furnaces of the late eighteenth and nineteenth centuries. The fuel in these furnaces can be placed so that it offers little resistance to air flow; under such conditions even a short chimney can draw so much air that temperature rise must be restrained by dampers in the chimney or at the intake to the fire chamber. Higher chimneys will cause greater air intake rates, as the traditional Chinese "dragon" or climbing kiln used in porcelain manufacture shows. Built in the form of a tunnel running up the side of a hill, this furnace is in effect a sloping chimney in which the natural draft is proportionate to total vertical height. The temperatures reached are greater than those in most or all iron smelting furnaces of the pre-industrial era. Large climbing kilns could not only vitrify high alumina porcelain, but are said to have sometimes melted it, indicating temperatures in excess of 1450°C.

Reaching high temperatures in a direct heating shaft furnace, where the "chimney" is filled with fuel and ore, is more difficult. Tylecote and Owles's (1960) calculations of the theoretical capabilities of shaft furnaces blown by natural draft compare achievable intake air pressures with the back pressure caused by charcoal-ore beds of varying depth and permeability. As in the chimney of a house, the difference in density between the hot column of combustion gases inside the furnace and the cold (normal) air outside creates a natural pressure that displaces the hot gases upward at a calculable rate and forms a vacuum that sucks in cold air. However, the obstruction caused by the furnace charge creates a need for additional draft, which increases sharply as the bed becomes higher or as it becomes more tightly packed. Since a certain amount of packing is inevitable if the ore fragments are small enough to smelt, the only viable way to produce a satisfactory amount of iron in this type of furnace is either to supplement the natural draft with bellows or to use a tall furnace that is only partially filled. Adequate natural draft requires about 2 m. of empty shaft for each 1 m. of bed depth. In short furnaces, the draft can be augmented by orienting the air intake toward prevailing winds.

In actual practice, shaft furnaces using natural draft have no regular pattern of distribution. The best-known archaeological examples are a group of furnaces discovered in the Swiss Jura in the 1860s. These date roughly in the period of 800 to 500 BC. Gowland (1899: 316) describes furnaces with "a curious similarity" to these, found in Carinthia, the ancient Noricum: "Vertical holes excavated near the face of a steep bank, with a lateral hole near their base to which the wind was admitted by a horizontal channel in the front of the bank, lined with stone. According to Munichsdorfer the blast for these was obtained from the natural wind alone." In the reconstruction (Fig. 7.2), the vertical component of the furnace is not near the face of the bank, which might have caused difficulties by the hazards of wall collapse if too close to the face. The long horizontal hole would have been used for both the entry of air and the extraction of slag and iron.

Other furnaces that may have been operated without bellows occur elsewhere in Europe. Polish experiments with replicas of the ancient 1.2 m. tall shaft furnaces found in the Holy Cross Mountains near Kielce did produce "smelting conditions" using only natural draft (Coghlan 1977: 35). On the other hand, the short, second-century A.D. shaft furnace at Ashwicken in Britain (Tylecote and Owles 1960) when replicated and run by natural draft produced no metal (Tylecote et al. 1971). The differences between the two seem to reflect the measure of success or failure that the experimenters

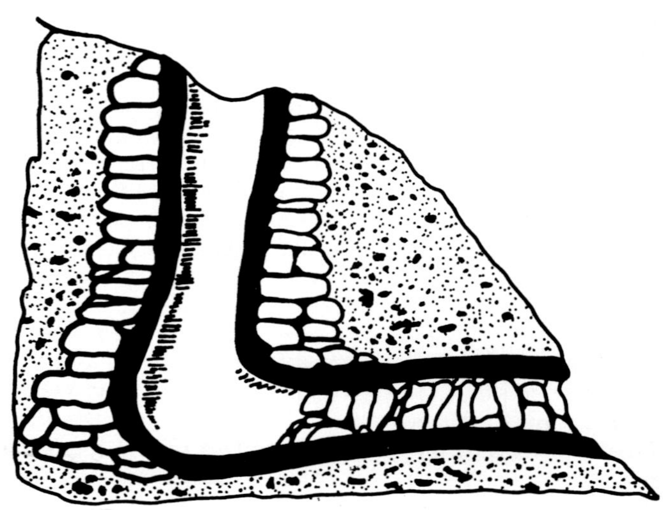

Fig. 7.2. *An Early Swiss Natural Draft Furnace?* Quiquerez's (1871) reconstruction of a furnace in the Jura. (From Gowland 1899)

chose to use. The evidences of natural draft furnaces in iron smelting in Europe are not convincing.

Little is known about Asian natural draft furnaces. The best documented are those of the Mount Popa or "Puppadaung" type in Burma (Chapter 3), in which the absence of bellows is indicated only by the geologist Blanford (1862: 218–219; Percy 1864: 270–272). While Blanford was normally a good observer, his account of the 20-tuyere shaft furnaces, which has become a standard in the ethnometallurgical literature, is completely unlike the series of bowl furnaces described by the only other technically oriented visitor to Mount Popa, Chhibber (1926). Smelting without bellows was also observed at Tonghai in Yunnan province in southwestern China, a few hundred kilometers from Mount Popa (Tegengren 1923–24: 352). The furnaces were tall structures, 9.15 m. high, built of clay and brick, with an internal diameter of 0.9 m. at the waist and only 0.5 m. at the mouth and hearth (Fig. 3.9). Since ore was smelted to cast iron in these furnaces, they must have been able to generate temperatures in excess of about 1250°C.

Much more is known about natural draft furnaces in Africa, because over the past two centuries European travellers have reported the smelting of iron in tall, narrow shaft furnaces that are entirely wind blown. The classic technical account of these furnaces is Hupfeld's description (1899) of ironmaking among the Bassari of Togo. Detailed descriptions of shorter natural draft furnaces are given by Bellamy (1904)—extensively quoted by Cline (1937: 31–32) and van der Merwe (1980: 491–492)—focusing on a Yoruba ironworking village in western Nigeria, and by Todd and Charles (1978) who worked with Dimi smelters in eastern Ethiopia.

The important point is that, since natural draft smelting was common in one area, it could have been elsewhere in the pre-industrial world. Cline's exhaustive tabulation of sub-Saharan smelting methods, which covers 106 published descriptions and a minimum of 98 separate tribes or subtribes, shows that furnaces worked without bellows are not unusual in Africa. He concludes (1937: 52) that iron smelting shaft furnaces which were tall enough not to require bellows, and are known often to have been used without them, were concentrated in three areas at the time of European contact: the western Sudan, including Ghana, Togo, and Nigeria; the "Katanga-Nyasa" region of Zaire, Zambia, and Malawi; and the Horn of Africa. Many of the furnaces, at least in the first two areas, had circular plans with four, eight, or more equally spaced tuyeres or bundles of tuyeres. The use of several tuyeres, one above the other, provided a larger total opening for air intake so that a wider tube of fragile ceramic was not needed. Although Cline emphasizes the connection between natural draft and tall furnaces, this is not supported by the research of Todd and Charles (1978) and of van der Merwe and Avery (1982), who describe natural draft furnaces that are no more than a meter or two in height. This could be taken to indicate that taller furnaces could have been filled only part way. However, Wembah-Raschid (1973: 29) describes the experimental run of a 3 meter high furnace in Tanzania that was filled to the top with charcoal, ore, and wood. This work demonstrated that if the fuel and ore fragment sizes are large enough, a fully packed furnace of considerable height will function with natural draft.

Using natural draft is not a fast way to make iron: smelting times of two to three days and furnace cycle times of a week or more are common. However, in the absence of powered blowing machinery, natural draft seems far more labor-efficient than bellows. It may be no accident that the more professional and commercially oriented of the African ironmakers, like the Yoruba and Bassari, often preferred the natural draft system, while their part time counterparts, like the Pangwe of Equatorial Guinea (Tessmann 1913), operated even tall furnaces with very large numbers of bellows.

## 3. FORCED AIR DEVICES

In most regions the development of air blowing devices must have played an important role in the evolution of iron metallurgy since many smelting and smithing processes are impossible without bellows. A fairly close correlation exists between the capacity of bellows, the size of furnaces, and the efficiency of the ironmaking process. Larger furnaces produce more metal per unit of furnace space, and are more economical in labor and—perhaps—in fuel input.

Most authorities[1] agree that the leather bag bellows and the skin covered drum (or bowl or pot) bellows are the earliest practical varieties, that the piston bellows and the accordion bellows came next, and that rotary blowers were not used anywhere until the modern period (Fig. 7.3; Foy 1909). Although little can be gained by assigning bellows types to particular developmental stages, this order is followed below.

## 4. BAG BELLOWS

Whole animal skins must have been used as bellows almost everywhere in the Old World, but, because they are perishable, they do not survive in archaeological contexts. In India, bellows of buffalo hide (Rice 1877), goat skin (Heath 1839), and bullock skin (Fig. 7.4a; Voysey 1832) were recorded (Holland 1892: pl. 1; Buchanan 1807). Malcolmson (1836) describes their use: "The fire is kept up by a blast from four bullock skins, placed at right angles to each other, the muzzles resting on the upper edge of the furnace so as to force the blast downwards." Early representations of Indian animal skin bellows are in Abu 'l-Fazl's biography of the Emperor Akbar, dating to the sixteenth century (Blochmann 1927: pl. 1). The largest bag bellows could achieve very high temperatures in small furnaces; sets of four bullock hide bellows like those Voysey or Malcolmson describe, each with an air capacity that must exceed a cubic meter, were regularly employed in heating small

[1] Writers on the distribution and history of bellows include Foy 1909; von Luschan 1909; Stuhlmann 1910; van Gennep 1911; and Forbes 1964, v.8: 111–119.

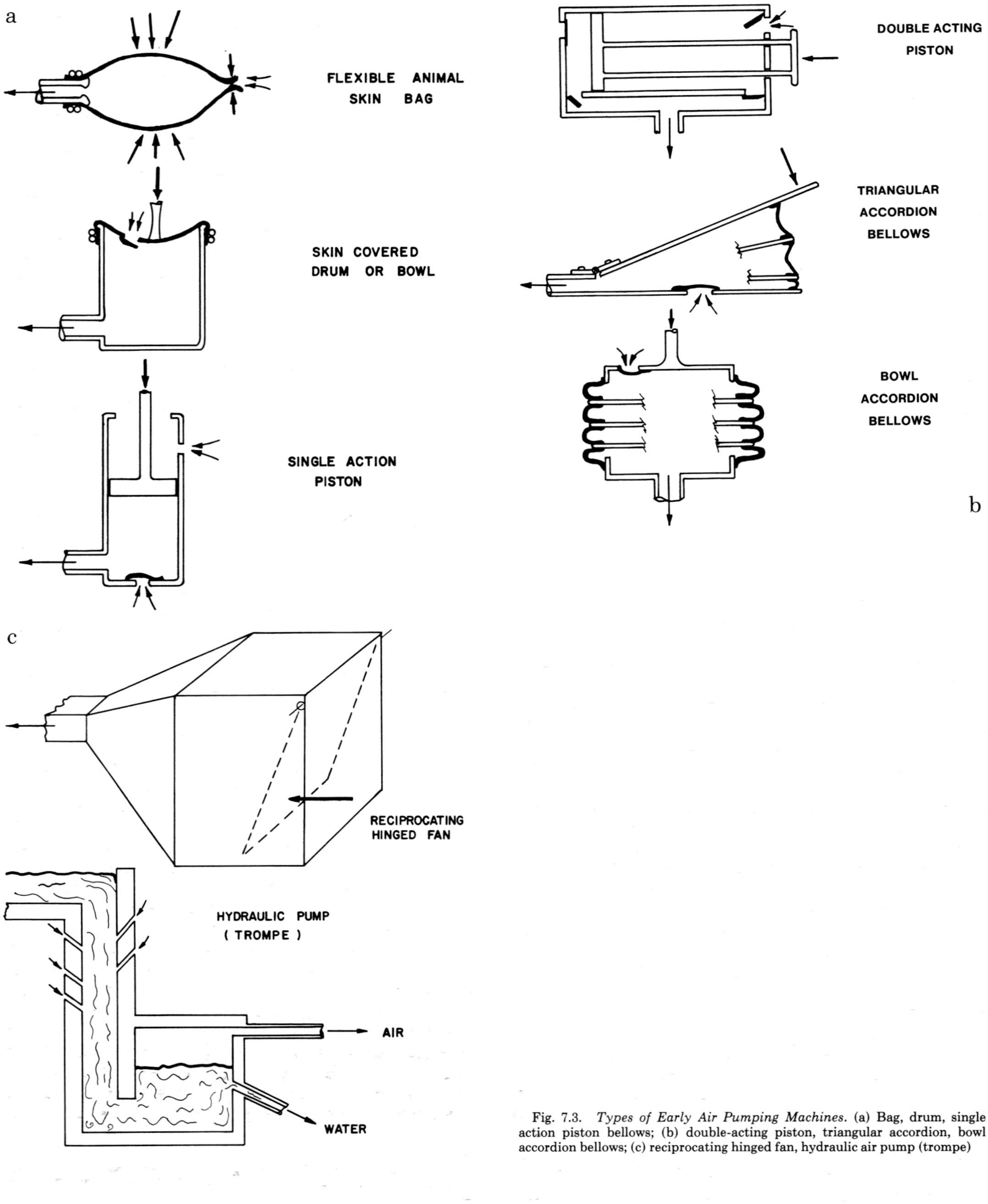


Fig. 7.3. *Types of Early Air Pumping Machines.* (a) Bag, drum, single action piston bellows; (b) double-acting piston, triangular accordion, bowl accordion bellows; (c) reciprocating hinged fan, hydraulic air pump (trompe)

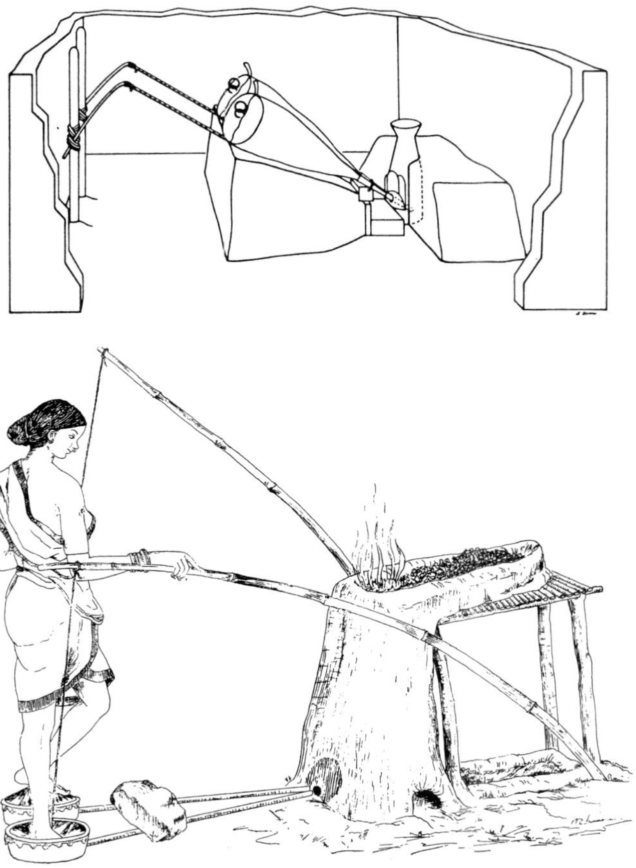

Fig. 7.4. *Indian Bag and Drum Bellows.* (above) A pair of large bag bellows made from whole bullock hides, used to drive a small bloomery in Mysore, southern India. The spring poles help to pull the bellows back after each stroke. (Redrawn by L. Grove from Buchanan 1807); (below) a pair of bowl bellows used with a bloomery furnace in Orissa, eastern India. The operator steps alternately on the bellows; the spring pole pulls the diaphragm up when the weight is released. (From *Survey of Traditional Crafts—Orissa* 1961: 19–21)

batches of tiny crucibles for making wootz steel, in a hearth not more than 25 cm. wide.

Bag bellows, often made of goat hide, were widely used with small smelting furnaces in northern and eastern Africa (Stuhlmann 1910: 62; Cline 1937: 108; van der Merwe 1980: 489); because of their large volumetric displacement, they generally occurred in pairs rather than in the sets of twenty or more as did African bowl bellows. They appear sporadically in Southeast Asia (Marschall 1968: map 2), and seem to be the only type of bellows known in much of Southwest and Central Asia. Only in Korea, Japan, eastern China, northern Europe, and southwestern Africa does the bag bellows seem to have been uncommon. It was unquestionably the most widespread of all air supply devices.

The bag bellows is inefficient in terms of the volume of air delivered per unit of labor because the simplicity of its intake valve (many varieties are filled with air simply by opening and closing one end with the fingers) means that it can only be pumped slowly. Its advantages are that it is portable, has a relatively large capacity, and is inexpensive and easy to make. The preference for these bellows among wootz makers shows that high temperatures that can be reached in furnaces equipped with them.

## 5. DRUM BELLOWS

One of the earliest representations of a drum bellows is in an Egyptian wall painting of a bronze workshop, dating to the fifteenth century B.C. (Davies 1943: pl.17). Archaeological remains of drum, or pot, bellows are known from the Near East, the Levant, Anatolia, and Cyprus (Davey 1979). The drum or pot bellows during recent times is used chiefly in Africa, where its distribution is even more extensive than that of the bag bellows (Stuhlmann 1910: 62; Cline 1937: 107–108), and in central and eastern India (Ball 1880: 668; Beck 1891: 235–237; Cammiade 1931; Swarup and Misra 1944), where some measured more than a meter in diameter (Ball 1877: pl. 5). In Southeast Asia the only reported drum bellows were used by the nineteenth-century Kui of Cambodia to supply air to the multiple tuyeres of their *tatara*-style smelting furnaces (Moura 1882). These bellows were large like those of eastern India.

Almost always employed in pairs and pumped with either the hands (in sub-Saharan Africa) or the feet (in India and ancient Egypt), a drum bellows delivers only a small volume of air at each stroke but can be worked with great rapidity. If provided with a rod attached to its center, as it was in some parts of Africa, or with a cord connecting the diaphragm to a bamboo spring pole, as it was in India (Fig. 7.4b), it would have a fast cycle time—perhaps faster than any other blowing device before the invention of the rotary fan or turbine.

The steadiness of the blast provided by sets of drum bellows, and the consequent high temperatures achieved, may account for the production of steel rather than carbon-free iron in some African furnaces, like those of the Haya (Avery and Schmidt 1979). During smelting, the Dimi of Ethiopia use three pairs of hand operated drum bellows simultaneously (Todd and Charles 1978). With each chamber pumping at a rate of 47 to 60 strokes per minute, the bloom produced averaged 0.4 percent carbon. Larger batteries of drum bellows are known among the BaSakata of Zaire who used up to 16 drum chambers with relatively tall (2.5 m.) shaft furnaces, while the Manyema sometimes had as many as 24 with large (2 × 1.3 m.) pit furnaces (Cline 1937: 52). The Pangwe of Equatorial Guinea smelted in a shaft furnace with 32 drum chambers worked by 16 bellows operators (Tessmann 1913: 230).

## 6. ACCORDION BELLOWS

The accordion bellows is a bag bellows with rigid sides or ends. In Europe and much of the Mediterranean littoral the triangular variety, the type commonly found near the fireplaces of American homes, became the norm. Introduced during the Roman period, the triangular bellows was extensively developed during the Middle Ages and Renaissance. For example, the bellows of a blast furnace operating in the Italian Alps in 1460 was 3.7 m. long (Schubert 1957: 199). Both Agricola and Biringuccio, writing in the sixteenth century, illustrate only triangular accordion bellows; they usually occur in pairs and were similar in size to the Alpine bellows. Jenkins (1925–26: 59) mentions an English, seventeenth-century blast furnace with a bellows of 5.5 × 1.2 m. By the eighteenth century they had become even larger; contemporary illustrations (e.g., in Diderot's *Encyclopedia*—Gillespie 1959: pl. 24) show water powered furnace bellows that are at least 7 m. long.

Other varieties of accordion bellows are known. A bowl or cylindrical accordion bellows (Fig. 7.3b) appears similar in design to the large, football-shaped blowing devices shown in Chinese art of the Han period. A Han relief (Needham 1965: fig. 13) depicts one suspended from ropes in a horizontal position and manually operated; its capacity is approximately 1/3 $m^3$. Liu (1978: 21) believes that these bellows, assumed to be air bags with solid ends strengthened with hoops, were used singly as blowers for the four tuyeres of the large smelting furnaces excavated during the 1960s and 1970s in Hebixian and Gongxian in northern Henan province.

A bellows resembling the Han type, although it is less complex, existed in western and northern India. In Kathiawar, western India, these bellows were oriented vertically, made of "bullock hides sewn round hoops in vertical rings and worked alternately by downward pressure" (Jacob 1843: 990). In the 1880s, smelters in the Kumaon district of the Himalayan foothills used a similar, but horizontal, set of bellows which were 3 to 4 m. long, cornucopia-shaped, furnished with wooden hoops, and made of whole cowskins (Atkinson 1973: 269).

Cylindrical accordion bellows are known in sixteenth-century Portugal where they were used horizontally and in twentieth-century Morocco (Paris and Ferriol 1922: 341). A type of upright accordion bellows that resembles a drum bellows with the diaphragm extended upward over a series of hoops also occurred in Morocco and among several groups in Togo (Cline 1937: 105–106).

An accordion bellows of any type is an improvement over the simple bag bellows because it is easier to mechanize and can be made much larger. However, when compared with the piston bellows, possibly known in principle to Europeans long before its first appearance there in the seventeenth century (Sanders and Gould 1976: 30) and its eventual wide adoption by ironmakers at the beginning of the nineteenth century, the advantages of the accordion bellows are less obvious. It is much more susceptible to leakage and heat damage, and delivers a more intermittent blast. It wastes more energy because of the greater mass of the top side of the accordion bellows as compared with a piston face.

## 7. PISTON BELLOWS

The single-acting piston bellows (Fig. 7.3a) has as limited a pre-modern distribution as the accordion bellows, being confined to Europe (in recent times), Japan (for single-acting Japanese box bellows, see Hommel 1937: 20), and all of Southeast Asia, including Taiwan, Madagascar, northwestern New Guinea, and northeastern India (Marschall 1968: map 2). The earliest evidence for the characteristic Southeast Asian form is in the twelfth-century temple of Candi Suku in Java (Kempers 1959: pl. 334). Limited numbers of bag bellows and Chinese double-acting piston bellows were also used in many parts of the region without displacing the dominant type (Marschall 1968: 194–203, map 3).

Mallet (1877) describes a single-acting piston bellows in Assam, northeastern India, at the western edge of the traditional range of such devices:

> The cylinder, about 3 feet long by 10 inches in diameter and three-quarters of an inch thick, is cut out of a solid piece of wood. . . Into the ends, disks of wood are fitted by rough dove-tailing, the circular joints between the disks and cylinder being rendered air-tight with clay. In the centre of one disk is a small hole in which the piston-rod works. The latter is generally made from a slip of bamboo split from a large-sized piece and has a diameter of perhaps three-eighths of an inch. At one end is a cross-handle; at the other, the piston: this is a disk of wood somewhat smaller in diameter than the cylinder and having a groove cut in the circumference, into which strips of skin from the necks of cocks are sewn by twine passing through holes for the purpose in the edge of the piston. The long, soft feathers fill up the space between the piston and the cylinder, and produce but little friction, while not allowing much air to pass through... In both of the end discs there is a double orifice fitted with a valve opening inwards; the latter made of leather or a couple of folds of stout paper, fastened at one side by a slip of wood nailed to the disc.

Because much of Southeast Asia had switched to imported European or Chinese iron by the time technically oriented Western observers appeared, published descriptions of smelting with these bellows are rare. A particularly large piston bellows in use in the 1840s consisted of a 5 feet, 5 inch high wooden cylinder, open at the top but closed at the bottom, 3 feet in circumference (Harrison and O'Connor 1969: 344–349, 395). This may be the largest of its type ever recorded. Smelters with piston bellows were still operated by the Shan of eastern Burma in the 1890s (Scott and Hardiman 1900, v.2: 299–300), and a similarly equipped bowl furnace at Soroako in central Sulawesi may have been used to smelt the nickeliferous iron ores of that area, thus producing an iron-nickel alloy (Abendanon 1917–18: 1684–1686).[2]

The capabilities of the Southeast Asian piston bellows can be learned from Burns's account (1849: 151–152) of smelting among the Kayan in the interior of Borneo. He states that the villagers customarily made iron in pits in the ground equipped with as many as 10 or 12 piston bellows, each 2 m. long and 0.2 m. in diameter, with counterweighted piston rods. Taking a conservative estimate of 30 strokes per minute (Bronson has timed a similar but noncounterweighted bellows used by a blacksmith in Sumatra at 45 to 50 strokes per minute) and assuming that the piston trav-

[2]For nickel-iron smelting in Sulawesi, see Chapter 1 and Bronson 1987; for other examples of smelting with piston bellows, see Bronson and Charoenwongsa 1985.

els only half the length of the cylinder, an array of twelve such bellows would deliver a minimum of 8.13 $m^3$ of air per minute, which is more than sufficient to compensate for any inadequacy in furnace design. In ironmaking systems where bellows capacity is large enough, details of furnace design cease to be critical. Recent measurements of the air supply to a small traditional smelting furnace in Henan in central China (Wagner 1985: 17) provide a helpful comparison. The Chinese furnace, blown by a double-acting piston bellows, received 1.6 $m^3$ of air per minute and produced a minimum of 30 kg. of cast iron per hour. The Kayan furnace with its 10 to 12 piston bellows received a blast at least five times as great. Since the metal outputs were comparable, air leakages in the ducting can apparently be as important to iron smelting as the bellows design.

## 8. DOUBLE-ACTING PISTON BELLOWS

If the Southeast Asian type of piston bellows has ever been used in the ethnically Chinese parts of China, it has long since been replaced by the double-acting piston bellows or "windbox," said by Hommel (1937: 18) to have "the merit of surpassing in efficiency any other bellows made before the advent of machinery." Needham (1965: 19) suggests that the windbox (Fig. 7.3b) first appeared "perhaps in the 4th century BC, but more probably in the 3rd century AD." By the early twentieth century, it could be found in a wide variety of shapes (Barnard 1961; Hommel 1937: 18–20, 28–30) and had become a major element in the unique, small batch cast iron technology of China. The seventeenth-century work, *Tiangong Kaiwu*, illustrates windboxes in service in twelve separate metallurgical processes (Sun and Sun 1966). The device is not, however, exclusively Chinese, since it is found in both Korea and Japan; the Japanese version is sometimes single-acting. Windboxes also occur in Southeast Asia as far south as Java and Borneo (Marschall 1968: 188–189, 193–194), where they are often used by metalworkers who are not immigrants from China. In the isolated Land Dayak area of Sarawak in Borneo, a contemporary swordsmith has one beside his forge (Neimitz and Neimitz 1975), and iron smelters in Upper Assam in northeastern India once used them (Mallet 1877).

The double-acting piston is advantageous because it provides positive air pressure on both strokes, whereas most other air pumping systems require two complementary units to maintain a reasonably constant blast. Some small, but efficient windboxes could be pumped by one person, which explains how, until recently, travelling Chinese tinkers could mend cast iron pans by patching them with liquid iron melted with a small brazier and a portable windbox (Hommel 1937: 31–33). However, even a double-acting piston bellows does not have as steady a blast as a rotary blower. In a technical study conducted in Hubei in 1959 (Wagner 1985: 55), a windbox attached to a smelting furnace produced a blast pressure that fluctuated from about 220 millibars near the end of the forward stroke to about 29 millibars at the pause between the forward and backward strokes. When used in pairs, other types of bellows should be able to do as well.

Tegengren (1923–24: 346) describes a water powered windbox in Sichuan Province with a cylindrical cross section, 1 m. in diameter and 3.5 m. long, which was supposedly large enough for a blast furnace with a daily output of at least 1.3 tons of cast iron. Windboxes with rectangular cross sections that are even larger are illustrated in the *Tiangong Kaiwu*; they are manually operated by only two laborers and have an estimated capacity of as much as 10 $m^3$ per stroke. It is unlikely however, that these devices functioned as described. Tegengren (1923–24: 351) also observed cylindrical windboxes in Yunnan Province only 2.5 m. long and 0.45 m. in diameter (and hence with a capacity of only 0.5 $m^3$ per stroke) which needed four men to operate them if the waterwheels were stopped. It is doubtful that a windbox twenty times as large could be worked by half as many men. Perhaps the illustrator of the *Tiangong Kaiwu* has simply enlarged the small rectangular bellows used by blacksmiths and tinkers rather than depicting a type actually used in larger ironworks.

The Chinese did not use the double-acting piston bellows to the exclusion of other types. Woodcuts and paintings by Ming Dynasty illustrators of technology (Needham 1965: pl. 19, 20; Liu 1978: 20–21), made long after the invention of the windbox, often show metalworkers employing muscle or water powered hinged fans rather than windboxes, especially with large furnaces. The complexity of the windbox and perhaps the difficulty of achieving good piston sealing when the design was scaled up beyond a moderate size may explain why other types were sometimes preferred. Large windboxes may have been even more common before Tegengren's day (1920s), when Western methods had already been adopted by the larger and less remote Chinese ironworks, but they must have been delicate contrivances, especially since they were generally made of wood. The usual feather or paper piston seals would not have worked well in supporting a heavy horizontal piston, and the basic unit could tolerate little warping or shrinkage without serious loss of efficiency. For these reasons, Chinese ironmasters would have turned to other kinds of blowers.

## 9. HINGED FAN BELLOWS

The main alternative to the windbox in China was the fan bellows (Fig. 7.3c), which had become comparatively rare by the time Westerners began writing about Chinese metallurgy. A hinged fan was, however, still in use during the 1920s in Shanxi Province, at a crucible stall furnace used to convert raw iron sponge to cast iron (Foster 1926: 176). The oldest existing representation of an operating Chinese blast furnace, in a book written in 1331 by Chu Chun, includes a pair of hinged fans (Needham 1965: 64, fig. 25). These devices must each contain a single leaf or blade (presumably with flap-covered holes to ease the return stroke) of about 1 × 2 m., swinging through an arc about 1 m. long. The estimated air capacity is about 1 cubic meter per stroke.

Since the design is relatively friction-free, a pair of

these bellows should have a fast enough cycle rate to deliver a considerable quantity of air. They were not much less efficient than windboxes of similar capacity. Moreover, they may have been cheaper, especially when very large bellows were needed. The Chinese hinged fan design is simple and sturdy: four walls built of brick or even pounded earth, a wooden fan leaf with a flap valve perhaps of leather or treated cloth, and a pushrod attached to the blade which slides through a bushing in the rear wall. This device is not portable, unlike the windbox. On the other hand, because its major moving part—the suspended fan leaf—is basically weightless, it is not hard to work.

The tatara bellows of Japan is a horizontally hinged version of the Chinese fan bellows, powered not by pushing a rod but by stepping on and off the fan leaf. The two known principal versions are the simple *fumi-fuigo*, where a single leaf rocks seesaw-fashion (Fig. 12.2) over a central axle (Gowland 1899: 307–308), and the balance bellows, or *tembin-fuigo*, with two opposed leaves hinged at opposite ends and connected by rods to an overhead balance beam (Tanimura 1980: 71; Ledebur 1901: 844). The second type, invented in the seventeenth or early eighteenth century, significantly improved labor efficiency, since it required two rather than eight operators (Nishio 1913: 76; Takahashi 1986: 76). Lyman (1879: 64) notes that even the balance bellows required substantial human effort: "The work is so severe that the men are relieved about once an hour, and in the twenty-four hours there are three sets of men or six men in all." It is curious that the Japanese, who knew (and reprinted) the *Tiangong Kaiwu* and other Chinese works that describe waterwheel driven blowing devices, used only manpower to drive their tatara furnaces.

## 10. HYDRAULIC AIR PUMPS

An air pumping device without moving parts, known in English by its French name, *trompe*, first appeared in southern Europe during the seventeenth century; according to Percy (1864: 285), it is supposed to have been invented in Italy in 1640.[3] At the end of the eighteenth century on the threshold of the Industrial Revolution, the trompe had displaced the accordion bellows in many mountainous regions of Europe where the large amounts of hydraulic power required to run it were available. Despite the competition of water and steam powered piston bellows, the trompe survived in isolated areas of the Pyrenees and Tennessee until late in the nineteenth century (Swank 1892: 298–299). The trompe had been used with most kinds of smelting furnaces, but late in its history it was usually associated with Catalan-type hearths.

Percy (1864: 286–92) cites studies carried out in the 1840s showing that a well-constructed trompe of the type used with a Catalan hearth in Ariège could develop an air pressure of 140 millibars or 2 psi, which would be adequate even for a moderately tall blast furnace. Later versions of the trompe had barometer-like attachments for gauging the air blast. Percy (1864: 289) notes that "no machine produces a blast of such constant tension as the trompe, the mercury appearing as though it were congealed in the gauge."

The trompe worked because the air entrained within a falling column of water retains the velocity of the fall after the column reaches the bottom; this air can be caught and piped directly into the tuyere of a furnace. The essential elements of a successful trompe (Fig. 7.3c) are a water column of sufficient height and an air catching enclosure—what Percy calls a "wind-chest"—designed to keep as much spray as possible out of the air pipe that leads to the furnace. In the 1840s, the height of trompe columns in Ariège ranged from ten to a minimum of seven meters. This may seem wasteful of power by the standards of an age when centimeters of hydraulic head were the equivalent of kilowatts and when waterwheels 4 or 5 m. high could drive the machinery of entire factories, but waste was not an issue in the places where trompes existed, for many had more water power than they could possibly use.

The need for water confined use of the trompe to areas with tremendous resources. A second and more important disadvantage was the wetness of the blast it produced, the evaporation of which meant a modest but possibly significant waste of fuel. Its advantages were low maintenance cost, inexpensive operation, and the power and constancy of the blast. It was the only truly continuous air delivery system invented during the pre-industrial period.

## 11. WATER POWERED MECHANISMS

Connecting waterwheels to mechanical blowing devices antedates the invention of the trompe by many centuries. The first water powered, mechanism driven bellows in Europe appeared in France and Germany during the twelfth century, perhaps at ironworks operated by Carthusian or Cistercian monks (Wertime 1961: 61–63). In China and probably also in Korea, the accordion and/or hinged fan bellows may have been operating with attached waterwheels as early as the first century A.D. (Wertime 1961: 53). Only one installation is known in the Middle East: the small cast iron producing blast furnaces of the Amol area in Mazanderan Province in Iran (Böhne 1928), where triangular accordion bellows were driven by horizontal waterwheels.[4] The combination of a Mediterranean-style blowing device with an East or Central Asian type of waterwheel may indicate that the Amol system was locally invented, perhaps long before Böhne's visit in the 1870s.

The water powered bellows was important in European and Chinese ironmaking. A manually blown furnace could be costly to operate because it required so much manpower, so the adoption of water power would sharply reduce the size of the work force at the smelting

---

[3]Good discussions of the trompe appear in Percy 1864: 286–292; Beck 1897, v.3: 119–121, 311–315; and Sanders and Gould 1976: 3; the classic illustrations are those in Diderot's *Encyclopedia* (Gillespie 1963: pl. 88). The trompe in America is described briefly in Swank 1892: 298–299.

[4]Wertime's (1962: 52) description of these as "bag bellows" has misled others. The blowing devices illustrated by Böhne are triangular bellows, set on edge and driven by pushrods coming directly from the (crankshaft equipped?) waterwheel.

site. Moreover, furnaces of greatly increased capacity could be built, since a well-designed waterwheel would deliver much more power than could be generated by even the largest teams of workers. Only one type of manually operated furnace, the Japanese tatara, is known to have been able to produce more than 100 to 200 kg. of iron per day.[5] In contrast, daily yields of 1000 kg. were often achieved by early water powered furnaces. Hartwell (1963: 65) estimates that furnaces which made at least 1.5 tons of iron per day already existed in China by the eleventh century. Europe had many furnaces of equivalent capacity before 1700; by the end of the eighteenth century, there were water powered charcoal furnaces in Siberia that were more than 12 m. high, with capacities of more than 20,000 kg. (Tables 14.3b, 14.4).

Drive mechanisms for linking waterwheels with bellows were relatively complex. Gearing was sometimes used, and devices for converting rotary to reciprocating linear motion—cranks or cams—were indispensable. Many devices of this kind were illustrated in Chinese technical writings (e.g., Needham and Wang 1965) and in European works like those of Agricola and Biringuccio.[6]

An extensive account of the history of the mechanics of either area is not appropriate here, but it should be noted that, at least in Europe, most of the devices necessary for applying the power of the waterwheel to furnace operation—including counterweights, springs, rockers, and camshafts—were in existence long before the first water powered bellows appeared. Why it should have taken so many centuries to make the connection between water power and furnaces is not clear. Perhaps here as elsewhere what was technologically possible was less important than perceived need. Like the Japanese, a number of peoples who possessed the technology for hitching waterwheels to bellows did not do it, perhaps because of conservatism or the lack of need for very large quantities of iron.

In the eighteenth century, increased European interest in the efficient use of water power led to the development of two other blowing devices: the wooden box bellows (Fig. 7.5) and the wooden or cast iron cylinder blower or piston compressor (Fig. 7.6). The first resembled the wood and leather triangular bellows but operated more like a tatara bellows, with tight-fitting wooden sides that slid past each other. Although invented much earlier—in Nürnberg in 1550, according to Smith and Forbes (1957: 32)—the box bellows was not adopted by British ironmakers until shortly after 1700 (Schubert 1957: 208). The second was a cylindrical device with a single-acting piston similar in form and concept to the traditional Southeast Asian piston bellows, but much larger and stronger. The first cylinder blowers were installed at the Carron Iron Works of Scotland in the 1760s (Scrivenor 1854: 83). Driven by large waterwheels, these machines produced the powerful blast important to the success of that enterprise.

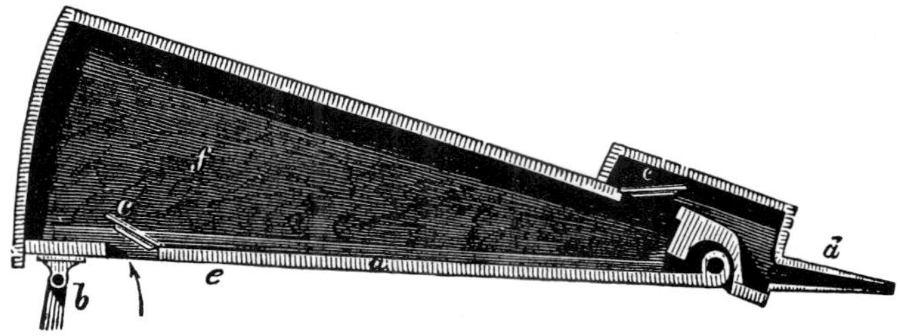

Fig. 7.5. *European Wooden Box Bellows.* Although shaped like a triangular accordion bellows, this is made entirely of wood and functions like a Japanese tatara fan bellows. Such devices were widely used in Europe and North America during the seventeenth through nineteenth centuries. (From Osborn 1869: 543–544)

## 12. STEAM POWER AND THE HOT BLAST

Several developments in blowing devices after the Industrial Revolution did not affect pre-industrial ironmaking directly, but some principles are general to all processes.

The steam engine was first used for pumping air into furnaces in the late 1770s. By the early nineteenth century, it had become the primary power source for the rapidly proliferating coke furnaces that were transforming ironmaking. The most important effect of steam power was to sever the longstanding bond between ironworks and sources of water power, making it possible to establish furnaces without regard to locations of streams. A secondary effect was almost as important. Perhaps because both were made of machined cast iron, the first blowing devices powered by steam engines were the newly introduced cylinders rather than the traditional triangular bellows. The steam driven piston

Fig. 7.6. *European Piston Blowing Cylinders.* The two pistons, operating in tandem, could be driven by either a water wheel or a steam engine. The cylinders and pistons in this case are both of wood. Nineteenth century. (From Osborn 1869: 545)

[5]For tatara yields, see Lyman 1879: 61–66. Indian furnaces with capacities greater than 100 kg./day existed at Velatar in Kerala (Buchanan 1807, v.2: 439), in Kumaon in the Himalayan foothills (Atkinson 1973: 270–271), and at Birbhum in West Bengal (Sanyal 1968: 105). We know of no traditional furnaces with such capacities in Africa.

[6]For more details on drive mechanisms and power outputs, consult Landels 1978 and Johnson 1841.

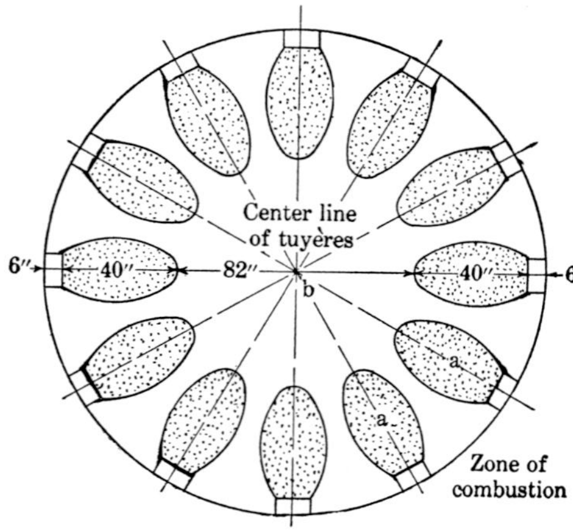


Fig. 7.7. *Radial Penetration of the Flame.* Measurements made by Kinney, Royster, and Joseph (1927) of the spaces defined by the flame zones of a 12-tuyere blast furnace of the early twentieth century. The momentum of the gases carry the reducing action further, but if they do not reach the center, a pillar of unreacted coke, ore, and limestone remains. Then considerable ore is dumped into the slag pool.

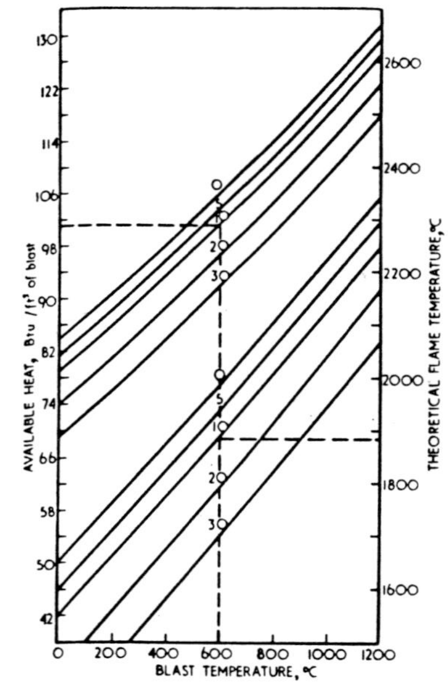


Fig. 7.9. *The Relationship of Preheat Temperature and Furnace Temperature.* It will be seen that the relationship is directly additive; the water content of the input air absorbs heat and hence opposes the benefits of preheating. None of the contents of the furnace ever reaches the same temperature as the flame. (From Grieve 1960)

and the air compressing piston were closely associated through the early twentieth century, a combination which allowed an increase in furnace size.

Until about 1800, furnaces in most parts of Europe were limited in height because of back pressure and the weakness of available blowing devices. The most powerful bellows in use before the 1770s probably could not deliver an air blast at pressures much greater than about 2 psi, inadequate for furnaces larger than about ten to twelve meters. The 15 meter furnaces of the 1820s and the 25 meter furnaces of the 1900s were possible largely because steam engines could be used to drive large piston (and later, turbine) blowers. Measurements made in an early twentieth-century blast furnace show the limitations imposed by air insert pressure, since furnace heights of around 17 or 18 meters required blast pressures as great as 16 psi (Figs. 7.7, 7.8).

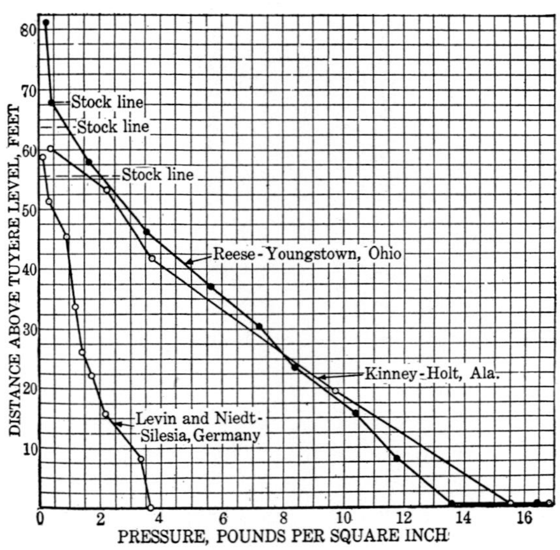


Fig. 7.8. *Gas Pressure Gradients in Twentieth-Century Blast Furnaces.* The top of the burden of fuel and ore must always be at or below the stock line, the point in the shaft at which gas pressure drops below one psi above atmospheric pressure. The two American furnaces here operate at much higher pressures than the one in Silesia (now in Poland). (From Kinney, Royster, and Joseph 1927)

Regular preheating of the blast was also an important development in modern ironmaking. Earlier, in several African systems, tuyeres were arranged to provide some preheating to the air before it entered the combustion zone (Chapter 3). A preheated blast may also have been used in some Chinese furnaces (e.g., Rocher 1880), where tuyeres were placed so that the air in them was exposed to furnace flames for a significant time before reaching the tuyere mouth. In Europe, preheating was not practiced until the nineteenth century, when Neilson's experiments in the 1820s (Roberts-Austen 1899) created a controversy over the relative merits of hot and cold blast iron that was to last for more than a hundred years. Neilson demonstrated satisfactorily that the hot blast saved fuel even when no effort was made to recycle waste heat from the furnace mouth. There were benefits other than fuel conservation, since the anthracite in Wales and Pennsylvania required a hot blast to be used in successful smelting. Moreover, furnace temperatures could now be easily raised, for a given quantity of heat applied to the input air increases the temperature of the reaction zone by almost exactly that amount (Fig. 7.9). The improvement in fuel efficiency was quantified by Routledge (1891: 33): "Thus, at the Clyde Ironworks, in 1829, the production of each ton of iron required 153 cwts of coal, and 10 1/2 tons of limestone; whereas, in 1833, after the hot blast had been adopted, 59 cwts of raw coal, and 7 cwts of limestone, sufficed for the same work." The ability to make furnaces run at temperatures much higher than could be achieved by ordinary combustion ultimately made it possible to melt and cast steel and to create many of the common modern alloys.

# 8. Slags

## 1. PHYSICAL NATURE

All iron ores are mixtures of reducible minerals and rocks, or gangue, that cannot be reduced at normal smelting temperatures. The gangue must be eliminated, so that the product removed from the furnace is not a large glassy lump filled with tiny bits of dispersed iron. The gangue minerals can be separated from the metal by creating conditions in which they form a slag. Ironmakers are likely to pay as much attention to slag as to iron during smelting, and may in fact be most concerned with keeping the slag in a suitably liquid condition, believing that the iron will take care of itself.

Slags have several positive functions. In theory, they absorb the gangue and the nonreducible part of the fuel ash into a homogeneous liquid. By forming a pool with high fluidity, they allow the newly smelted metal to settle out and collect in a single mass. They also protect the metal from reoxidation, and with adjustments to composition, remove impurities from the iron.

On the other hand, making a slag consumes fuel. An ore with a high gangue content will produce much more slag than is actually needed to form a thin cover over the metal and so may require a great deal of fuel (see Bray 1942: 156–157 for the relevant figures). In a modern blast furnace, 1 kg. of coke is needed for each 4 kg. of slag. The rich ores of the Mesabi Range in the United States produce 550 to 600 kg. of slag per metric ton of iron; the lower grade ores of Britain yield between 1600 and 1800 kg. per ton. Thus, to make a ton of iron from British ore, the slag alone will use up 400 to 450 kg. of fuel—about half of the total fuel consumed in an efficient modern smelting process. Earlier, the loss of energy to slag formation would have been proportionally greater.

## 2. THE SETTLING PROCESS

Physically, slags are similar to glasses. Because liquid or solid iron is much heavier than liquid slag, in a properly regulated furnace it settles through the slag pool and collects beneath it. Slags have densities of 3 to 4 grams per cubic centimeter, while iron has a density of 7.8 g/cc.

A clean separation of metal from slag should therefore be easy to achieve, but even the most fluid slag is viscous compared to water or liquid cast iron. Moreover, minimal viscosities in slags exist only in a narrow range of temperatures and compositions (Fig. 8.1). If the viscosity is much above the minimum, the settling rate of iron particles through the slag will be reduced, slag-metal separation will be poor, and metal recovery will be low.

The importance of settling rate can be illustrated by calculations which include density, particle diameter, and viscosity expressed as representative values (Appendix B). The time needed for droplets of cast iron to settle through a slag pool only 6 cm. deep can vary from 750 to 24,000 seconds—from 12.5 minutes to almost 7 hours. This very wide range would be even greater if turbulence and rising bubbles within the slag pool had been considered. It is clear that the settling time must be minimized in order to recover useful concentrations of metal, which, in turn, means minimizing viscosity through close control of slag composition.

## 3. COMPONENTS OF IRON SMELTING SLAGS

The gangue in an ore may consist of a wide variety of minerals, many of them highly complex geologically. Because actual mineralogical identity is inconsequential to the smelting process, it is customary among metallurgists to refer to all gangue minerals as the equivalent oxides. These include silica ($SiO_2$), alumina ($Al_2O_3$), lime (CaO), magnesia (MgO), potash ($K_2O$), soda ($Na_2O$), wustite (FeO), and hematite ($Fe_2O_3$). For example, an ore that actually consists of limestone (calcium carbonate, $CaCO_3$), dolomite (magnesium carbonate, $MgCO_3$), assorted clays (complex aluminum silicates), quartz (silicon dioxide, $SiO_2$), and hematite might nonetheless be described as containing only CaO, MgO, $Al_2O_3$, $SiO_2$, and $Fe_2O_3$. During smelting, the more complex minerals are converted to these simple oxides.

The most common noniron oxides in both ores and slags are $SiO_2$, $Al_2O_3$, and CaO (Table 8.1). All three have melting points well above the temperatures reached by smelting furnaces ($SiO_2$ at 1728°C, $Al_2O3$ at 2015°C, and CaO at 2580°C), but mixtures of these oxides in certain proportions melt at much lower temperatures. For instance, about 10 percent $Al_2O_3$ added to pure $SiO_2$ causes it to melt at 1587°C, and adding about 20 percent CaO as well brings the melting point down below 1350°C. The phase diagrams (Appendix A) can be consulted to clarify the crucial importance of the melting points of mixtures of compounds.

A modern slag contains not only gangue from the ore and ash from the fuel, but also minerals that are deliberately added to adjust the melting point and viscosity. Those additives which lower temperature and viscosity are called fluxes or fluxing agents. In pre-industrial times, the functions of fluxing entities were most likely

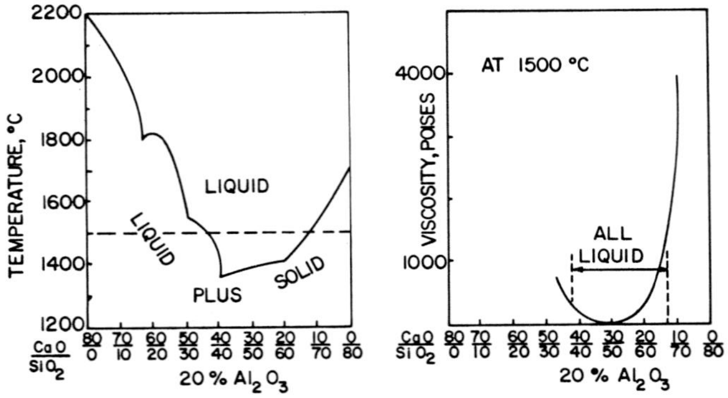


Fig. 8.1. *Composition, Temperature, and Viscosity in Slags.* Silica-lime-alumina slags, typical of blast furnace operation, not only have a high melting point but can be melted at achievable furnace temperatures only within a narrow range of compositions. Note that without a change of temperature but with a shift in composition, slags can change from fluid to viscous. (Graph on the left adapted from Muan and Osborn 1965; the one on the right uses data from Machin and Yue, given in Geiger and Poirier 1973: 28)

not understood and therefore the presence of fluxing agents was usually adventitious. Potash and soda are important fluxing agents, as are lime and iron oxide.

One last slag component that should be mentioned is dissolved refractory material (mostly silica and alumina) from the walls and other parts of furnaces, since molten slags are excellent solvents for most other ceramic substances. Usually, however, the rate of solution of the furnace walls is slow enough that slag composition is little affected. Only in the Japanese *tatara* hearth (Chapter 12) are structural parts of the furnace known to enter the slag in significant proportions.

## 4. CONTRIBUTIONS TO SLAG FROM ORE AND FUEL

Ore components other than iron are reducible in the smelting furnace, including phosphorus, which as phosphate or hydroxyphosphate of iron or calcium is present in many ores (Table 8.1). Most of the reduced elemental phosphorus is likely to be picked up immediately by the metal, but some of it, along with unreduced phosphate, goes into the slag. Certain slag compositions, however, can inhibit the reduction of phosphorus or even cause it to be extracted from liquid iron.

Sulfur in the form of sulfides also occurs in some ores. Much of it was probably eliminated in preliminary roasting, but any sulfides that remained passed to the metal or slag. As in the case of phosphorus, certain slags have the ability to absorb and thus neutralize large amounts of sulfur. These important neutralizing capabilities of slags are discussed further in later chapters.

When burned, all solid fuels leave an ash composed of a variety of minerals (see, e.g., Table 8.2 for published analyses of wood ash). The ash minerals are in effect concentrated when the wood is made into charcoal. Percy (1875: 358) suggests that the amount of ash in charcoal can be estimated by dividing the ash content of the wood by 0.23, an average yield for wood-charcoal conversion. Thus, for instance, charcoal made from oak with a 1.65 percent ash content will contain 7.16 percent ash. Applying Percy's calculation to the mean ash content of various wood species (Table 8.3) shows that most kinds of charcoal contribute significant quantities of ash to the smelting process.

Coal and coke also contain ash, but the minerals involved are different from those in wood and charcoal. Silica tends to predominate, with considerable alumina, lime, and iron oxide, but usually with little soda or potash (Table 8.4). Sulfur is often present in coal in substantial quantities. The ash content in raw coal ranges from about 3 percent to as much as 14 percent (Butts 1943: 75–76; Percy 1875: 349–352). Coking increases the ash content by 45 to 50 percent (Bray 1942: 61; Butts 1943: 61–68). Some cokes contain so much ash, in fact, that they can contribute more mineral to the slag than the ore does.

## 5. INTRINSIC OR SELF-FLUXING SLAGS

Although not used in modern ironmaking, slags without fluxing additives—i.e., composed only of gangue from the ore and ash from the fuel—must have been almost universal in pre-industrial production. Some furnace operators may have learned to add separate fluxes when the slag began to thicken or the furnace to choke, but most probably controlled their slags simply by choosing particular ores, by varying the proportion of ore to fuel, and by increasing or decreasing the force of the blast.

There are two groups of smelting slags. One, with silica and iron oxides, becomes liquid at a relatively low

*TABLE 8.1. Selection of published analyses of noniron minerals in iron ores*

| $SiO_2$ % | CaO % | $Al_2O_3$ % | MgO % | P %* | Origin | Source |
|---|---|---|---|---|---|---|
| 12–25 | 4–10 | 5–10 | — | — | Lorraine, France | Bray 1942: 21 |
| 5–10 | 12–20 | 2–10 | — | — | Lorraine, France | Bray 1942: 21 |
| 10 av | — | — | — | 0.07 | Elba, Italy | Bray 1942: 20 |
| 8 av | — | — | — | 0.05 | Krivoi Rog, USSR | Bray 1942: 22 |
| 5–10 | 1–6 | 3–6 | — | 0.5/0.9 | USA | Butts 1943: 125–127 |
| 3–20 | <2 | <10 | <1 | 0.03/0.3 | China | Tegengren 1923–24 |
| 9 | 1 | 0.3 | 0.7 | 0.01 | Japan | Tylecote 1976: 43 |
| 1–13 | 0.25/4 | <3 | <2 | 0.02/1 | Britain | Tylecote 1976: 43 |
| 7 | — | 1 | — | 0.05 | Labrador, Canada | Tylecote 1971: 347 |
| 14 | 0.5 | 0.5 | — | 0.03 | West Africa | Bellamy 1904 |
| 37 | 3.6 | 10 | 2.2 | 0.2 | Britain | Morton and Wingrove 1972 |
| 14.4 | 0.5 | 8.5 | 4.7 | 0.06 | Britain | Morton and Wingrove 1972 |
| 16.7 | 1.15 | 8.6 | 3.5 | 0.2 | Britain | Morton and Wingrove 1972 |

*Phosphorus in iron ores can exist as calcium or iron phosphates, dispersed atomically in the iron oxide crystal structure.

*TABLE 8.2. Analyses of wood ash**

| Wood species | Potash % | Soda % | Lime % | Magnesia % | Silica % | Source |
|---|---|---|---|---|---|---|
| Beech | 8.2 | 7.4 | 27.7 | 7.7 | 2.5 | Percy 1875: 188, 190 |
| Pine | 7.9 | 4.4 | 28.5 | 5.6 | 13.4 | Percy 1875: 188, 190 |
| Silver Fir | — | — | 35.7 | 1.9 | 10.3 | Percy 1875: 188, 190 |
| Oak | 8.4 | 5.7 | 75.5 | 4.5 | 0.8 | Percy 1875: 188, 190 |
| Elm | 21.9 | 13.7 | 47.8 | 7.7 | 3.1 | Percy 1875: 188, 190 |
| Beech | 22.8 | 9.5 | 24.5 | 11.6 | 1.9 | Percy 1875: 188, 190 |
| Silver Fir | 16.9 | — | 34.6 | 6.8 | 2.4 | Pfannenschmidt 1977: 60 |
| Birch | 18.6 | — | 30.8 | 10.3 | 1.9 | Pfannenschmidt 1977: 60 |

*For additional figures, see Osborn 1869: 180–182.

temperature and is characteristic of bloomery operation. Its main disadvantage is the waste of iron, since a substantial part of the metal in the ore is necessarily lost to the slag. The other type of slag is composed of silica, alumina, and lime; it melts at a higher temperature and is the usual product of a blast furnace, but a high temperature is required to liquefy it. Both slag systems can be formed without other additives to the furnace. Here we discuss how these slags derive from the assortment of minerals naturally present in ores and fuels. (See Chapters 9 and 10 for examples of these slags and the technical and economic consequences of using them.)

The control of a hypothetical bloomery slag has been analyzed (Table 8.5). Here we have calculated in terms of a mildly reducing condition typical of many somewhat inefficient smelting operations; this is defined in terms of 18 percent silica content in the slag. The melting temperature, about 1150°C,[1] is controlled by the composition, and is depressed further by soda and potash. In this case, the sum of the components in the fuel, the ore, and the unreduced iron oxides produces a self-fluxing slag, and no extra additions are required in the charge to insure that the slag can be tapped from the furnace. This particular slag is in fact relatively economical in terms of the iron lost to the slag and fuel expended. Although a 62.7 percent FeO content in the slag indicates a degree of metal recovery that is not ideal, the returns on the fuel and ore are reasonable; many actual smelting systems had even higher losses of iron to the slag.

In a second calculation (Table 8.6), for a slag associated with smelting to cast iron, only the nonreducible components in the ore and the fuel ash combine to make the slag, which is composed of silica, lime, magnesia (which functions much like lime), and alumina. The melting temperature is deduced to be about 1250°C (see Fig. A.11), well within the range of an early blast furnace. The real melting temperature of this hypothetical slag might be even lower when the small amounts of potash, soda, and wustite are considered, but the iron produced would certainly be cast iron. Although it is conceivable that some cast iron could come from a process using an iron silicate slag system like that in bloomeries, it is not possible for a low carbon iron to be produced with a silica-lime-alumina slag.

[1]Percentages of pairs of oxides with similar behaviors have been added: i.e., %CaO + %MgO and %K2O + %$Na_2O$. Melting or liquidus temperature has been estimated from the ternary silica-wustite-anorthite phase diagram (Fig. A.10).

*TABLE 8.3. Average ash contents from various wood species**

| Wood species | Mean ash content % |
|---|---|
| Willow | 2.00 |
| Aspen | 1.73 |
| Oak | 1.65 |
| Hornbeam | 1.62 |
| Alder | 1.38 |
| Beech | 1.06 |
| Scotch Fir | 1.04 |
| Silver Fir | 1.02 |
| Birch | 0.85 |

Note: These averaged data probably represent large ranges in values because in other tables (Percy 1875: 184, 185, 186, 188) some very different numbers appear. For example:

| | |
|---|---|
| Beech | 0.38–2.15 |
| Oak | 1.45–2.43 |
| Birch | 0.60–1.54 |
| Poplar, Aspen | 1.86–2.98 |
| Willow | 3.67–4.57 |
| Hornbeam | 0.87 |
| Scotch Fir | 0.63 |

*From Percy 1875: 187

## 6. ADDITION OF FLUXES

In bloom iron smelting, slags can have a low melting temperature over a wide range of FeO content because of the nature of the iron silicate system. They therefore react little to minor variations in temperature and do not benefit from the addition of small amounts of fluxes. However, when FeO is only a minor component of the slag, as in cast iron smelting, the nonreducible components are less likely to be self-fluxing. This can be corrected by adding larger lumps of ore or lime.

Although the conditions for smelting to cast iron are so strongly reducing that theoretically no oxide of iron can pass through to the slag, this is true only if the ore remains above the combustion zone long enough. When the added ore lumps are sufficiently large, however, some of the oxide remains unreduced and thus enters the slag pool, where it lowers the fusion temperature of the slag and thereby decreases its viscosity.[2] A second approach is to add lime in quantities up to 30 to 45 percent, which greatly lowers the melting point, as the ternary diagram for the system silica-lime-alumina (Fig.

[2]The authors witnessed this in 1987 during a visit to the traditional blast furnaces at Angat in the northern Philippines. The furnaces are run without added flux. Each time the very viscous slag is to be tapped, the operator throws a shovelful of crushed ore through the open furnace front directly into the slag pool. The slag quickly becomes fluid, so it can be tapped.

*TABLE 8.4. Analyses of coal ash**

| Silica % | Alumina % | Wustite %** | Lime % | Magnesia % | Potash + soda % |
|---|---|---|---|---|---|
| 35 | 26 | 20 | 6 | 2 | 3 |
| 35 | 28 | 19 | 5 | 2 | 4 |
| 36 | 28 | 26 | 2 | 2 | 2 |
| 52 | 14 | 27 | 2 | 1 | — |
| 36 | 41 | 11 | 3 | 5 | — |
| 24 | 21 | 26 | 9 | 9 | — |
| 38 | 38 | 15 | 3 | 3 | — |
| 40 | 39 | 12 | 2 | 3 | — |
| 44† | 39 | 8 | 6 | 3 | — |
| 53† | 37 | 6 | 3 | 1 | — |
| 59 | 35 | 2 | 1 | 1 | 1 |
| 55 | 27 | 13 | 2 | 0 | 1 |
| 49 | 39 | 4 | 5 | 0 | 1 |
| 45 | 41 | 7 | 4 | 1 | 2 |
| 37 | 41 | 10 | 6 | 2 | 1 |

*Percy 1875: 349–353
**And other iron oxides
†Anthracites, all the others are bituminous
Note: Figures have been rounded off to the nearest whole number. The geographic origins include Britain, France, America.

A.11) shows. Of the three principal slag components, lime is most likely to be lacking in the ore as mined.

Adding limestone to the charge of cast iron smelting furnaces was common in England by at least the seventeenth century (Schubert 1957: 232; Tylecote 1976: 108), when the proportions by weight of limestone to iron ore ranged from about 1:8 to 1:17. By the late eighteenth century in England larger additions of limestone, close to a ratio of 1 to 3, caused the sulfur content in the cast iron to be reduced. This innovation marks the beginning of modern steelmaking. Also associated with high lime slags was the need to run the furnace at a much higher temperature. Cast iron is tapped from modern blast furnaces, for example, above 1500°C.

It is not easy to judge from a slag analysis whether flux additions were made. Iron oxides and limestone are expected to occur in every slag because of the charge composition, but not necessarily in ideal proportions. The same is true of potash and soda which are effective in reducing the melting temperatures of slags. However, since they are always present where charcoal is used as fuel, slag analysis is unlikely to reveal that mineral feldspar or extra wood ash was added to flux. In contrast, the deliberate addition of calcium fluoride (i.e., fluorspar), a potentially effective flux,[3] could be recognized if laboratory methods for detecting fluorine were used. At present, nothing in either the historical or archaeometallurgical literature points to the presence (or absence) of fluorine in ancient slags. Another possible, but undocumented, smelting flux is the mineral borax, used in several regions as a flux in welding. Since light elements such as boron and fluorine are difficult to detect by spectroscopic methods, they have rarely been sought in slag analyses; they can, however, be determined by other analytical techniques.

[3]The efficacy of calcium fluoride additions to ironmaking slags is well established. For example, the addition of only 4 percent to a reasonably typical blast furnace slag reduces the viscosity at 1300°C by one-half, from 40 to 20 poises (Richardson 1974, v.1: 109)

*TABLE 8.5. Calculation of a representative slag yield and composition in bloom iron smelting using charcoal fuel*

Use data from Tylecote et al. 1971: 353, 354 for a furnacing run involving ca. 2 hrs. heating up time and ca. 2 hrs. smelting.

| Run no. | Total ore, kg. | Total charcoal, kg. | Bloom wt., kg. |
|---|---|---|---|
| 19 | 27.2 | 37 | 5.8 |
| 21 | 18.2 | 45 | 6.4 |
| 27 | 22.2 | 38 | 6.5 |
| totals | 68.1 | 120 | 18.7 |

Use an averaged ore analysis from Tylecote et al. 1971: 347. 12% silica, 2% lime; 2% magnesia; 4% alumina

| | silica | lime | magnesia | alumina | potash, soda |
|---|---|---|---|---|---|
| 68.1 kg. ore, kg. = | 8.2 | 1.4 | 1.4 | 2.7 | — |
| 120 kg. charcoal @ ca. 3% ash * = | 0.2 | 1.7 | 0.4 | — | 1.3 |
| totals, kg. | 8.4 | 3.1 | 1.8 | 2.7 | 1.3 |

Let slag = ca. 18% silica (typical)
slag yield, kg. = 47

slag composition: 18% silica; 10.5% lime + magnesia; 6% alumina; 2.8% potash, soda; rest wustite

convert to anorthite (see Figure A.10) = 19% anorthite; 13% silica; 68% wustite

from which the all-liquid-state (liquidus) temperature (without factoring in the 2.8% potash, soda which are strong melting depressants) = ca. 1150°C

*typical ash composition (Percy 1861: 109): 4.5% silica; 47% lime; 12% magnesia; 21% potash, soda

*TABLE 8.6. Calculation of a representative slag yield and composition in cast iron smelting with coke*

Use a coal of 5% ash by weight which converts 80% to coke providing about 6% ash per unit weight of coke.

Use an average ash composition: 44% silica; 33% alumina; 4% lime; 2% magnesia; 1.4% potash, soda (taken from data in Percy 1875: 352–353)

Use an ore/coke ratio of 1.2 which is an average of 0.66 (Tegengren 1923–24: 387) and 1.8 (Butts 1943: 109)

Use an ore analysis: 12% silica; 2% lime; 2% magnesia 4% alumina (Tylecote et al. 1971: 347)

| | silica | lime | magnesia | alumina |
|---|---|---|---|---|
| 100 kg. ore provides, kg. | 12 | 2 | 2 | 4 |
| 83 kg. coke yields 5 kg. ash: kg. | 2 | 0.2 | 0.1 | 2 |
| totals kg. | 14 | 2.2 | 2.1 | 6 |
| slag yield = 24 kg. | 60% | 8% | 8% | 24% |

This assumes little wustite, potash, soda, all of which act to lower the slag melting temperature.
Using lime plus magnesia jointly, from the ternary liquidus surface for the system lime-alumina-silica (Figure A.11)
The melting temperature of the slag = ca. 1250°C

## 7. OTHER SLAG FUNCTIONS

Besides collecting the rock and ash in a convenient liquid form and protecting the reduced metal from continued reaction with combustion gases during the smelting operation, slags have other important functions. The carbon dissolved in iron metal can react with iron oxides in the slag to change the composition of both, a critical factor in several historical processes for decarburizing cast iron to wrought iron. In the Walloon fining process, for example, cast iron was drip melted in an oxidizing atmosphere which removed part of the carbon. The rest was eliminated by reaction with high FeO slags after the droplets of iron reached the bottom of the fining hearth. High FeO slags are efficient decarburizing agents when they act on iron in a liquid or finely divided solid state. Such slags also remain fluid over a wide range of composition, from about 60 to 85 percent FeO at temperatures lower than 1250°C (Fig. A.8). In a finery operation the FeO would soon be reduced to elemental iron, allowing the slag to become viscous again. This could be corrected simply by shoveling in additional $Fe_3O_4$ in the form of hammer scale from the forge.[4]

An appropriate slag has a dephosphorizing and decarburizing effect on cast iron as it is converted to wrought iron in a puddling hearth. Phosphorus in liquid iron can be oxidized by contact with $Fe_3O_4$ in slag to phosphorus pentoxide ($P_2O_5$), a compound of phosphorus soluble in slags, particularly those with high lime content. Thus, the addition of large amounts of hammer scale or iron ore to the contents of a puddling furnace causes the elimination of phosphorus and carbon simultaneously. After the mid-nineteenth century, the development of the "basic" puddling process using a high lime refractory brick, which was introduced several decades after Cort's puddling furnace, made possible the production of a wrought iron with mechanical properties equal to those of the better quality bloom irons.

[4]Hammer scale is the product of oxidation during forging. A smith's workpiece reacts with the high carbon dioxide content of the forge gases, converting the metal at the surface back to a layer of almost pure $Fe_3O_4$. Blows from the forge hammer shatter this oxide layer, which collects on the floor of the forge shop. Hammer scale must have been collected carefully for use in fining furnaces and other industrial operations.

Sulfur can be removed from cast iron in similar ways, since the sulfur in the metal reacts with slags having significant amounts of lime, soda, and/or potash, producing a calcium, sodium, or potassium sulfate, each of which is soluble in slag. The precise conditions for considerably reducing sulfur through the use of basic slags were discovered in the late nineteenth century, and since then have been important in the world's iron and steel industries.

The high FeO slag trapped in bloom and wrought irons performed another function—that of a welding flux in the forge welding process. Until recently, smaller or thinner pieces of iron had to be hammered together at high temperatures to form larger ones (a technique known in the nineteenth century as *piling*). However, welding temperatures quickly oxidize metal surfaces, and even a very thin film of oxide can prevent adhesion. If pieces of slag-containing iron are heated above the melting temperature of the slag, it will ooze out onto the surfaces, inhibiting further oxidation and dissolving existing oxide. (Slags as welding fluxes are discussed further in Chapter 15.)

## 8. SLAG IDENTIFICATION AND SURVIVAL

Residual slags are often more informative about ancient processing technologies than the metal artifacts themselves. Rusting sometimes destroys iron artifacts, leaving only solid pieces of rust that may have retained their original shapes. Slags, on the other hand, are relatively resistant to environmental conditions produced by air, water, and ambient heat, so their composition may be precisely the same as when they were made. For this reason, slag study has been of major importance in the development of archaeometallurgy.[5]

The melting temperature of a well-preserved slag can be determined in the laboratory, so that the minimum temperature of the smelting operation can be established. The minor elements in the slag such as $P_2O_5$ and MnO give some idea of the reducing conditions in the smelting operation, and also help identify the ore and its possible provenance. Metal droplets or fragments

[5]A useful book in English on the analysis of archaeological slags is by Bachmann (1982); he includes an excellent bibliography.

suspended in the slag can be completely preserved (Fig. 8.2), if the slag has not been modified by the environment and the passage of time. Often small pieces of charcoal are embedded in slag; they probably came out of the furnace when the slag was tapped. This charcoal can be used for radiocarbon dating of the furnace operation.

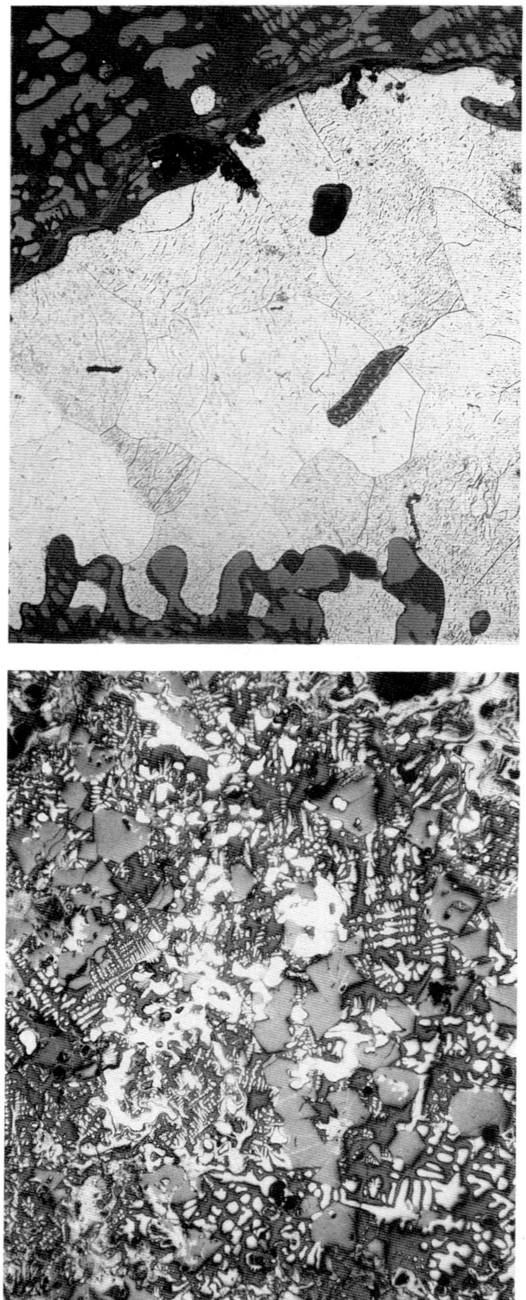

Fig. 8.2. *Microstructure of a Thai Bloomery Slag.* The low magnification view (above, ×50) of the slag shows prills or metal inclusions in iron. That at higher magnification (×200) is of the metal which is very low carbon iron. The precipitates in the iron are carbides or nitrides, whose well-formed shapes denotes very slow cooling as would take place in a small bowl furnace smelting run. The metal hardness is only 108 BHN. (Micrographs by James Dvorak; specimens provided through the courtesy of the National Geographic Society and Vincent Pigott)

Fig. 8.3. *Microstructure of a Canadian Bloomery Slag.* Recovered from the Baffin Island area of Arctic Canada (Rostoker and Dvorak 1986), this bloom may date to the sixteenth century A.D. The white strings of globules and crosses (dendrites) are wustite (FeO) crystals; the gray background is the iron silicate called fayalite; the light gray angular crystals are an iron-alumina-silicate called hercynite; and the black matrix is a lime-alumina-silicate called anorthite. The overall or average composition is about 54 percent FeO, 16 percent $SiO_2$, 9 percent MnO, 6 percent $P_2O_5$, and 3 percent $Al_2O_3$. ×400. (Photomicrograph by James Dvorak)

Just as prills of metal entrapped in residual slag are valid samples of the smelting product, so too can slag particles embedded in bloom or wrought iron help determine the nature and details of the smelting or refining process. Slag inclusions trapped in uncorroded bloom iron are representative of the smelting process; that is, they are unchanged by later heating and forging in the production of an object. Only a carburization heat treatment might reduce the FeO content of the slag inclusions, and it would affect only the carburized zone. Slag inclusions in wrought iron indicate production by the fining or puddling process, and they are also accurate guides to the original method of manufacture, except when carburization or decarburization has occurred.[6]

The identity of a slag is most simply established by chemical analysis. This gives the weight percentages of the heavy elements present, which can then be converted by calculation to the weight percentages of the corresponding oxides (see Tables 9.1 and 10.2 for examples of analyses; for more details on analyses and their

[6]Examples of the use of slag inclusion analyses to interpret iron processing are given by Hedges and Salter (1978–79), Gordon (1983), Gordon and van der Merwe (1984), Rostoker et al. (1985), and Killick and Gordon (1987).

relationships to the behavior of slags, see White 1980 and Appendix A). In addition, slags, like metals, have a characteristic appearance when polished and examined under the microscope, because crystalline and glassy entities which have specific mineral identities and shapes form during solidification. Microstructural characteristics can be used to identify the slag almost as unambiguously as chemical analysis, especially when the slag is occluded in the metal as, for example, in a raw bloom of iron (Fig. 8.3; for and detailed discussions of microstructural identification procedures, see Morton and Wingrove 1969, 1972; Wingrove 1970; Bachmann 1982). Even without a chemical analysis, it can be determined whether the smelting operation was wasteful of iron, since the slag would contain significant numbers of wustite crystals (FeO) and probably magnetite crystals ($Fe_3O_4$). In the microscope, they are distinguished by their shapes, shades of white, gray, or black, and arrangements with respect to other entities in the microstructure. They can be more specifically identified by micro-beam, X-ray diffraction analysis. Magnetite in a slag is easily detected by a simple permanent magnet, especially if a sample of the slag is first crushed to powder.

Without additional information, an iron silicate slag cannot be taken as positive evidence of a bloom iron smelting operation. The smelting of both copper and lead involves the same slag formulation to eliminate silica and other rock components of the ores, and iron ore was added to produce the slag. Such slags can be distinguished from iron smelting slags by the presence of small quantities of copper or lead, which show in chemical analysis and sometimes as entrapped metal globules, or "prills." In both cases, the copper or lead values converted by calculation to their oxides will amount to more than 1 percent up to as much as 10 percent (Rostoker and Gebhard 1981a). On the other hand, iron smelting slags would be expected to contain less than 0.1 percent of either lead or copper.

Gangue components such as alumina, lime, magnesia, and potash are present only in the primary smelting slag. Secondary smelting slags which have been in contact only with fuel ash (as in fining) are likely to be very low in alumina which is not a major component of wood ash. When secondary smelting does not involve contact with fuel ash, as in puddling, potash is absent (Killick and Gordon 1987).

Iron smelting slags from the bloomery process can react to the environment. When bloom iron slags solidify, there are four products into which the FeO may segregate—FeO crystals called wustite, magnetite, an iron silicate called fayalite, and a glassy iron silicate with a higher proportion of silica. Of these, only wustite reacts with water to form a product similar to iron rust. Instead of being dense and hard, the slag becomes loose and crumbly, rather like a bog ore, a process which takes several hundred years in a ground water environment. All the other components are unaffected; they remain disseminated in the degraded products of the wustite rust and may be recovered and identified (Rostoker and Gebhard 1981a). Slags which have absorbed a high content of $P_2O_5$ or potash and soda are also slowly soluble in water, producing a rust-like residue.[7] When slags have degraded, they reveal less about the smelting operation in which they were produced (see Rostoker and Dvorak 1986).

[7]For the influence of potash and soda on the survival of glasses in a ground water environment, see Frank 1982: 12–13.

# 9. Smelting to Bloom Iron

## 1. INTRODUCTION

Furnaces formerly used for converting iron ore directly into low carbon iron are known generically as *bloomeries* because they yield an unfused solid mass, or bloom. Because the solid metal has a low carbon content, they are distinguished from the other type of pre-industrial smelting device, the blast furnace, which produces high carbon liquid iron.

Many of the bloomeries described in the archaeological and ethnohistorical literature are shaft furnaces, with an elongated vertical section that serves as both a chimney and a charging chute. Since the operation of these furnaces is well understood, they are treated as models in the discussions that follow. Other types of bloomeries would have worked differently, but most of the discussion here is applicable to the operation of any device used for the direct production of low carbon iron from ore.

In Chapter 3, the basic operation of the shaft furnace was outlined. The charge, a column composed of a mixture of fuel and ore, moves downward through the shaft and becomes steadily hotter as it approaches the combustion zone at the level of the tuyeres. The combustion zone is an open space between an arch formed of the partially melted charge above a pool of slag composed mainly of silica and incompletely reduced iron oxides below.[1] Particles of iron, probably protected by a coating of slag, fall through the combustion zone into the slag pool. If the pool is sufficiently fluid, the iron particles, being much denser than the slag, settle to the bottom and aggregate into a spongy bloom. If the slag pool is very viscous or incompletely melted, the metal does not segregate but remains suspended as particles of microscopic size, inseparably mixed with the bulk of slag, so no metal is recovered.

When the furnace has been run long enough that the slag level begins to reach the tuyere, the slag must be drained off so the tuyere opening is not blocked. In order to do this, the air blast is stopped and the plug removed from the tap hole. When enough molten slag has poured out, the tap hole is replugged and the blast started again. This cycle may be repeated several times before the desired amount of metallic iron accumulates. Because the iron forms a solid mass, charging and blowing must be halted while it is extracted. Thus, part or all of a shaft furnace may be demolished, while a bowl furnace has to cool sufficiently to be emptied. Many shaft furnaces, however, are designed to permit removal of the bloom through a temporary hole in the side, so smelting can be resumed while the furnace is still hot.

[1]The form and dynamics of the combustion zone in modern blast furnaces are described by Kinney et al. (1927) and by Elliott et al. (1952).

## 2. THE THERMOCHEMISTRY OF BLOOMERY OPERATION

The reduction of ore to iron begins high in the furnace shaft, and the process is basically complete by the time a given unit of the charge reaches the arch of the combustion zone. Furnace gases are meanwhile moving in the opposite direction. At the moment they meet the descending charge they consist of pure $CO_2$—the product of combustion at the tuyeres—mixed with nitrogen. As soon as they penetrate the charge, however, they come into contact with unburned fuel, which converts much of the $CO_2$ into CO, making the gases strongly reducing. They continue to rise and soon encounter unreduced iron oxides, which are converted to metal while reconverting some of the CO to $CO_2$. The $CO/CO_2$ ratio changes as the gases move upward, and it does not stabilize until the gases reach either the furnace mouth or a point within the shaft where the temperature is too low for oxygen to be extracted from ore.

The temperature and the $CO/CO_2$ ratio control conditions in the furnace (see Fig. A.2, where the ratio is expressed as the percentage of $CO_2$ in all the carbon oxide gases). Reduction does not occur at temperatures below about 700°C. Because no slag will melt at less than 1000° to 1100°C, these lower reduction temperatures are irrelevant in practice. Reduction also does not occur unless the percentage of $CO_2$ is lower than 20 to 40, depending on the temperature—that is, unless the $CO/CO_2$ ratio exceeds 4:1–3:2. The $CO_2$ content must be above 3 percent (and the $CO/CO_2$ ratio below 97:3) if the product is to be bloom iron. When it is lower than 1 percent, the result is cast iron. In theory, steel will be made if the $CO_2$ is held within the 1 to 3 percent range, but such precise control of furnace atmosphere is almost impossible to achieve.

Ores such as hematites or roasted limonites and siderites are oxides of iron that are predominantly $Fe_2O_3$. When the oxide is exposed to a heated gas which is mostly CO, it goes through a series of reduction steps:

$$3Fe_2O_3 + CO \rightarrow 2Fe_3O_4 + CO_2$$

$$Fe_3O_4 + CO \rightarrow 3FeO + CO_2$$

$$FeO + CO \rightarrow Fe + CO_2$$

Magnetite ores, since they are already $Fe_3O_4$, do not have to be converted to that oxide; similarly any fayalite ($2FeO.SiO_2$) that may be present (as when old slags are added to the charge) is not altered until the final step, which all ores must undergo. This is the critical reaction for metal production, requiring the greatest temperature and the highest ratio of $CO/CO_2$.

The reduction relationship between temperature and carbon monoxide (Fig. 9.1) shows that the heat required for successful reduction is not necessarily very great. When the iron oxide becomes associated with the silicate in slag, a larger CO/CO2 ratio is needed, which can be obtained by raising the temperature. This increase in heat is not an inefficient use of fuel, because reaction speed is greater and thus a larger output per unit of time can be gained. In addition, much of the iron oxide is inaccessible to the reducing gases because ore chunks vary in size and permeability, which means that differential reduction occurs during the descent of the ore through the furnace shaft. Thus, the efficiency of metal recovery depends on the reduction rate, which is determined by temperature.

In the reduction of a typical hematite fragment (Fig. 9.2), the CO from the burning fuel reacts only when it is in direct contact with the iron oxide. The first stage of reduction occurs quickly as a thin film of metallic iron forms on the surface of the fragment. As soon as the film becomes continuous, however, the CO is excluded from contact with the underlying oxides, so that further reduction cannot occur more rapidly than oxygen can diffuse outward through the iron film to pass off as $CO_2$. The second stage therefore proceeds at a much slower rate. Eventually the fragment becomes layered with oxides in the following order: FeO on the outside, $Fe_3O_4$ next, and $Fe_2O_3$ at the center. The boundaries between the layers move slowly inward as reduction proceeds. In time the $Fe_2O_3$ core is reduced, then the $Fe_3O_4$ layer,

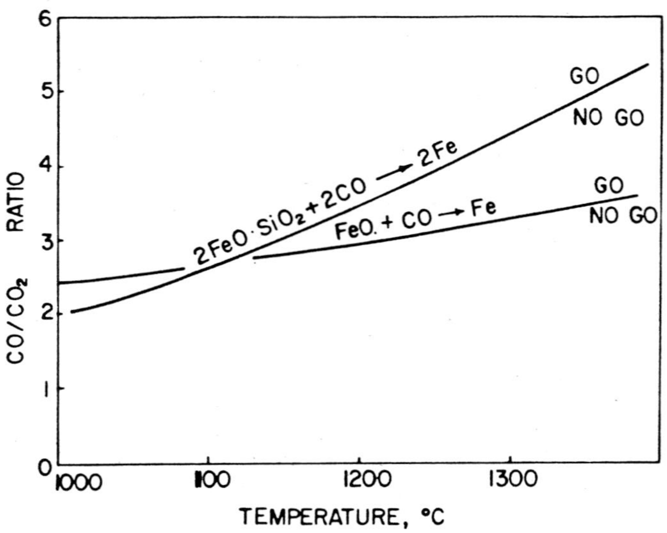


Fig. 9.1. *Reduction Thresholds for Iron Oxides in Ore and Slag.* The minimum $CO/CO_2$ ratio needed to reduce FeO (and the other iron oxides in ore) rises slowly with increasing temperature. When, however, the FeO has dissolved in the slag, taking the form of fayalite or $2FeO.SiO_2$, the minimum necessary $CO/CO_2$ ratio rises so sharply that reduction is unlikely except under blast furnace conditions. Therefore, iron-rich slags generally cannot be resmelted in bloomeries.

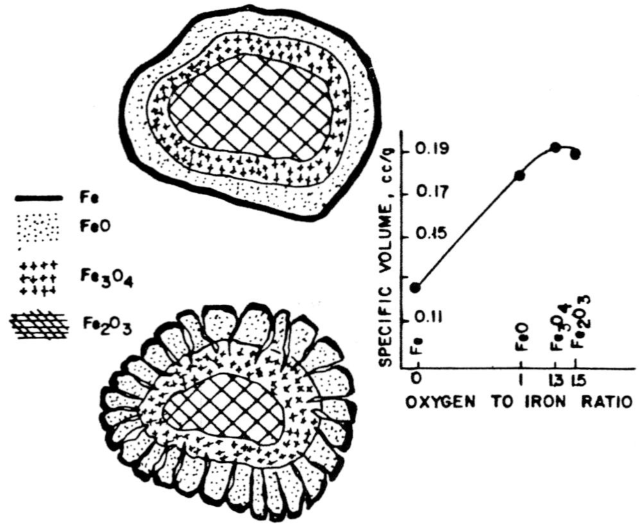


Fig. 9.2. *Stages in the Reduction of a Hematite Fragment.* The reduction process begins at the surface and proceeds stepwise toward the center. The hematite ($Fe_2O_3$) is first reduced to magnetite ($Fe_3O_4$), then to wustite (FeO), and finally to iron. The cross section of the fragment exhibits these stages in inverse order. Because the more reduced stages have smaller specific volumes, the outer layers shrink and crack, which exposes the interior and accelerates reduction. (Drawing based on photomicrographs in von Bogdandy and Engell 1971 and Türkdoğan 1978)

and finally the remaining FeO. At that point, the fragment is entirely converted to metal.

The difficulty with this model is that it depends entirely on oxygen diffusion rates in solid iron, which are so slow that under normal furnacing conditions even fragments as small as a centimeter might be incompletely reduced by the time they descend through the shaft. Fortunately, the volume changes which accompany the formation of oxide layers cause stretching and rupturing of the overlying strata of Fe and FeO; the cracks then propagate down through the brittle $Fe_2O_3$ (Fig. 9.2). The reduction process is accelerated considerably because the increase in the surface area of the ore fragment at the cracks makes it more accessible to reducing gases. It will be even faster if naturally porous or finely divided ores are used, or if cracks are induced in other ores by roasting, by heating and then cooling them rapidly with water, or by crushing them with a hammer before they go into the furnace. Macrocracking of an ore fragment by prior treatment and microcracking by surface reduction within the furnace improve the rate at which smelting proceeds.

In bloomery smelting, dense ores like hematite and magnetite must be in finely particulate form to be reduced within an economical length of time. Small particles pack tightly and thus inhibit the flow of gases within the furnace shaft. These ores—particularly magnetite sands—were, however, widely used by traditional Asian and African smelters, who compensated for tight packing by keeping the column short and the fuel to ore ratio high. In contrast, European bloomery operators preferred bog ores and siderites which were lower in iron content but could be deeply cracked by roasting. The subsequent development of the high fur-

*TABLE 9.1a. Composition of typical bloom iron smelting slags*

| Origin | FeO | $Fe_2O_3$ | $SiO_2$ | CaO | MnO | $Al_2O_3$ | MgO | $P_2O_5$ | $K_2O$ | $Na_2O$ |
|---|---|---|---|---|---|---|---|---|---|---|
| 1. Britain, 2 c. A.D. | 59.9 | 1.7 | 28.9 | 2.5 | 1.0 | 5.3 | 0.2 | 0.2 | ND | ND |
| 2. Poland, 1 c. B.C.? | 59.1 | 4.4 | 20.0 | 2.8 | 8.8 | 1.2 | 0.4 | 2.8 | ND | ND |
| 3. USSR, 5 c. B.C. | 56.6 | 13.8 | 18.7 | 3.5 | 0.1 | 3.3 | ND | 0.6 | ND | ND |
| 4. Turkey, 6 c. B.C. | 55.7 | 14.0 | 8.6 | 5.2 | 7.2 | 1.9 | 4.9 | 2.0 | ND | ND |
| 5. Austria, 5 c. B.C. | 55.3 | 12.6 | 24.5 | 2.0 | 2.4 | 2.5 | 1.4 | 0.1 | ND | ND |
| 6. India, 20 c. A.D. | 54.3 | 5.0 | 25.9 | 2.1 | 1.9 | 7.5 | 1.9 | 0.4 | ND | ND |
| 7. India, 20 c. A.D. | 53.4 | 10.0 | 18.0 | 0.4 | 4.7 | 9.0 | 2.2 | 0.5 | ND | ND |
| 8. Britain, 1 c. A.D. | 53.0 | 22.9 | 16.0 | 2.8 | ND | 1.5 | 0.5 | 0.4 | ND | ND |
| 9. Nigeria, 5 c. B.C. | 46.7 | 10.7 | 25.5 | 1.4 | 2.4 | 8.9 | 0.9 | 1.0 | 0.6 | 0.05 |
| 10. France (Roman) | 46.1 | 4.8 | 31.8 | 2.1 | 2.2 | 9.9 | 0.8 | 0.3 | 0.5 | 0.5 |
| 11. Greece, 5 c. B.C. | 42.5 | 33.0 | 18.9 | ND | ND | 3.1 | 1.3 | 0.4 | ND | ND |
| 12. Sudan, 2–6 c. A.D. | 42.0 | 20.8 | 20.7 | 2.4 | 2.4 | 6.7 | 0.9 | 1.0 | 0.5 | 0.1 |
| 13. Denmark, 2–4 c. A.D. | 41.2 | 3.6 | 22.7 | 1.4 | 16.8 | 1.0 | 1.1 | 2.2 | ND | ND |
| 14. Britain, 15 c. A.D. | 39.0 | 5.0 | 29.4 | 5.5 | 2.5 | 9.3 | 4.9 | 0.4 | ND | ND |
| 15. Britain, 13 c. A.D. | 31.0 | 26.4 | 20.3 | 2.7 | 13.1 | 2.4 | 1.9 | ND | ND | ND |

(ND = not determined)
Sources: (1) Tylecote 1969; (2) Piaskowski 1983: 221; (3) Kolchin 1953: 11; (4) (5) (8) (9) (10) (12) (13) (14) (15) Tylecote 1980: 219; (6) Swarup and Misra 1944: 129; (7) McWilliam 1920: 163; (11) Rostoker and Gebhard 1981a: 42

nace for making cast iron allowed the reduction of larger chunks of dense ores, because of higher operating temperatures and a longer time available for the charge to descend through the shaft. Increased size of the ore chunks meant that gases could pass more freely through the bed of ore. Even in modern furnaces, the recovery of metal is less efficient if the ore is extremely hard and dense.

### 3. BLOOMERY SLAG

Many ores consist almost entirely of silica and various iron oxides, which are converted to ferrous oxide during smelting. In theory, therefore, the slags would have only two components, FeO and $SiO_2$. The binary FeO-$SiO_2$ ceramic system has a minimum melting point of about 1178°C; at slightly above 1250°C it has good fluidity between about 60 and 80 percent FeO (Fig. A.8). Actual slags, however, tend to have more than two significant components. Most commonly, they also contain $Fe_2O_3$ and $Fe_3O_4$, derived from pieces of ore that have not been completely reduced by the time they descend through the shaft and reach the slag pool. The majority of bloomery slags can therefore be regarded as belonging to the ternary system SiO2-FeO-$Fe_2O_3$ (Fig. A.9).[2] These ternary slags work better than simple binary slags because the presence of the $Fe_2O_3$ and $Fe_3O_4$ widens the range of compositions that will be highly fluid at or even below 1200°C.

Representative analyses of normal or typical slags produced by bloomery operations are given in Table 9.1a; they are similar to perhaps 90 percent of published analyses. The behavior of these slags can be understood in terms of the $SiO_2$-FeO-$Fe_2O_3$ ternary diagram; most fall within the lowest temperature melting zone (the 1200°C contour) (Fig. A.9). Their melting points are influenced by components other than silica and iron compounds, but those do not occur in quantities great enough to alter the behavior of the slags. Alumina, lime, and magnesia might raise or lower the melting point slightly, and manganese oxide would have the same effect as FeO. Potash and soda have usually not been determined in analyses; both strongly depress the melting point in slag (Chapter 8).

The densities of all the slags listed in Table 9.1a probably lie between 4.3 to 4.5 g./cc., which is much less than the 7.87 g./cc. of pure iron. They would therefore have satisfactory melting points, and also allow a good separation of iron within the slag pool. As long as the pool remained fluid and turbulence was minimal, particles of iron would settle through it without difficulty.

A randomly selected group of less typical bloomery slags have melting points that seem improbably high or unpredictable on the basis of the $SiO_2$-FeO-$Fe_2O_3$ ternary diagram (Table 9.1b). Some are very high or low in iron and silicon oxides, while the rest contain unusual amounts of other components. Most can be understood by consulting graphs that show the relationship among more than three components, like the quaternary diagram (Fig. A.10). For example, the melting points of the Ethiopian and Ivory Coast samples (Table 9.1b) were estimated to be about 1150°C; the Ivory Coast sample actually melted at this temperature (Zacharias and Bachmann 1983: 4).

On the other hand, some atypical slags from early sites have melting points that seem extremely high regardless of the number of constituent components considered. If the slag is from a bloomery and the analysis is correct, the disparity between the predicted and actual melting points could result from the selection for analysis of an infusible fraction of a more normal slag, or the disappearance or omission of one or more components that can affect melting points. There are several melting point depressants—potash, soda, fluorine, boron—likely to be present in certain slags, but which are rarely sought by laboratory analysts. Since charcoal fuel always contains at least some and often much potash and soda,[3] the habitual omission of

[2]See Muan and Osborn (1965: 113) for phase diagrams and Osann (1959: 1209) for measured melting temperatures of actual ancient slags.

[3]Percy (1875: 358) implies that a 3 percent ash content in charcoal is a low average. The tables he presents (1875: 188–189) indicate that hardwood (and hardwood charcoal) ash averages between 15 and 20 percent potash, plus a few percent of soda. Frank (1982: 75), discussing potash and soda as used in glassmaking, emphasizes that the composition of a given batch of plant ash depends not only on the species involved but also on the soil in which the plant is grown.

*TABLE 9.1b. Composition of unusual bloom iron smelting slags*

| Origin | FeO | $Fe_2O_3$ | $SiO_2$ | CaO | MnO | $Al_2O_3$ | MgO | $P_2O_5$ | $K_2O$ | $Na_2O$ |
|---|---|---|---|---|---|---|---|---|---|---|
| 1. India, 19 c. | 74.0 | 8.1 | 10.3 | 2.5 | 0.2 | 1.9 | 1.1 | 0.4 | ND | ND |
| 2. Poland, 2-1 c. B.C. | 72.9 | 14.1 | 8.6 | 1.2 | 0.2 | 1.4 | 0.1 | 1.2 | ND | ND |
| 3. Britain (Roman) | 62.1 | 7.7 | 21.2 | 0.4 | 0.5 | 3.2 | 1.4 | 1.7 | ND | ND |
| 4. Britain (Roman?) | 61.5 | 19.7 | 10.2 | 5.7 | tr | 1.1 | 0.4 | 1.0 | ND | ND |
| 5. Poland, 2-1 c. B.C. | 54.5 | 10.0 | 19.1 | 2.5 | 1.2 | 1.8 | 0.02 | 9.3 | ND | ND |
| 6. Britain (Mediev.) | 47.5 | 6.9 | 24.2 | 1.1 | 2.5 | 14.5 | 0.3 | 1.6 | ND | ND |
| 7. Ethiopia, 20 c. A.D. | 43.1 | * | 34.8 | 1.1 | 1.6 | 13.6 | ND? | 1.0 | 1.4 | ND |
| 8. France, 19 c. | 41.8 | * | 33.5 | 7.2 | 13.0 | 3.7 | 2.4 | ND | ND | ND |
| 9. Poland, 2-1 c. B.C. | 39.0 | 9.9 | 35.6 | 6.3 | 0.3 | 4.4 | 1.2 | 1.9 | ND | ND |
| 10. Nigeria, 20 c. | 33.7 | 6.1 | 52.0 | 3.7 | ND | 2.5 | 0.7 | 0.2 | ND | ND |
| 11. Britain (Roman?) | 24.4 | 37.5 | 31.5 | 1.8 | tr | 5.8 | 1.1 | ND | ND | ND |
| 12. Ivory Coast, 20 c. | 21.6 | * | 41.1 | 22.2 | 0.4 | 10.2 | 0.4 | ND | 0.5 | ND |
| 13. Britain (Roman?) | 18.1 | 3.8 | 59.7 | ND? | 0.2 | 13.4 | ND? | 0.5 | ND | ND |
| 14. Palestine, 13 c. B.C. | 4.1 | 4.6 | 53.5 | 14.0 | 0.1 | 13.0 | 3.5 | 1.3 | 2.1 | 0.5 |

(* = $Fe_2O_3$ may be included in FeO. ND = not determined)
Sources: (1) Turner 1893: 168; (2) (5) (9) Piaskowski 1983: 221; (3) (6) Morton and Wingrove 1969: 56; (4) (13) Coghlan 1977: 20; (7) Todd and Charles 1978: 71; (8) Morton 1963, citing Percy 1864; (10) Bellamy 1904: 23; (11) anon. 1965: 4; (12) Zacharias and Bachmann 1983: 4; (14) Liebowitz and Folk 1984: 272.

$K_2O$ and $Na_2O$ from published analyses is especially unfortunate. In addition, both compounds tend to leach out of glassy structures (Frank 1982: 12–22, 73–77), so it is probable that many ancient slags had actual melting points lower than analysis indicates.

Bloomery slags are distinguished from blast furnace slags primarily by substantial quantities of iron oxides, which are in fact exceptionally good slag-forming substances. In antiquity, the ores of other metals, especially copper and lead, were often fluxed with iron ore because it can form slags at low temperatures. Because of the strongly reducing nature of blast furnace gases, these slags are very low in iron oxide and thus are of the $SiO_2$-$Al_2O_3$-CaO type. Cast iron smelting slags like these are not fluid below about 1350°C (Fig. A.11), which is 100° to 150°C higher than the melting range of bloomery slags that belong to the FeO-$Fe_2O_3$-$SiO_2$ system.

The price for using a low melting slag is a loss of iron. Much of the available metal in the ore will necessarily wind up in the slag (see, e.g., Table 9.2), because a substantial percentage of the iron oxides present must be utilized in lowering the temperature at which silica will form a slag with good fluidity. If the silica and other gangue materials do not form a slag, no metal recovery can occur. If the slag is not very fluid, the particles of metallic iron will settle out slowly or not at all, so that metal recovery is minimal. Experimental smelts in replicas of ancient furnaces have often produced some metal which, because of problems with slag fluidity, had to be collected by crushing the slag and pulling tiny particles of iron out with a magnet (Tylecote et al. 1971).

A large part of the iron oxide in the ore must therefore be committed to fluxing the silica and other nonreducible oxides. The amount of iron needed is directly proportional to the amount of these gangue materials present; for example, an ore in which the ratio of iron (and manganese) oxides to gangue falls below about 50 to 60 percent is not smeltable in a traditional bloomery.[4] With iron oxide contents of 70 to 80 percent, the loss of metal to the slag, although still great, was evidently tolerable to smelters before cast iron was produced regularly. To recover more iron, the combustion zone temperature would have to have been increased in order to obtain a slag with the same fluidity. While possible in a well-insulated furnace with a good air supply system, raising the temperature would have required a greater amount of charcoal which was, in most areas, more costly than ore. In addition, the product would often have been cast iron.

When figures on relative weights of slag and metal produced in a typical smelting operation are converted to volumes (Table 9.2), the quantitative difference between slag and iron becomes even more striking. At average densities of 7.87 g./cc. for iron and 4.4 g./cc. for slag (*Handbook of Chemistry and Physics* 1969: B214), the volume ratios of slag to iron become 1545:166 (for the Roman furnace) and 1368:360 (for the British furnace). Depending on the quality of the ore and the efficiency of the smelting system, an ordinary pre-industrial bloomery would have yielded between 4 and 10 liters of slag for each liter of iron. The ratio may be lowered to 2:1 by controlling the amount of ore and furnace conditions carefully. Such a ratio was achieved, for example, in the most efficient recorded bloomery furnace, the late nineteenth-century hot blast Champlain forge (Maynard 1874).

Because a bloomery produced a generally large volume of slag, tapping had to be frequent. In furnaces without tap holes, the amount of iron that could be made during one production run was severely limited, since the furnace would quickly fill with slag past the level of the tuyere. Once it had cooled, the slag could be removed only by chipping it out. The process was so labor-intensive that proprietors of non-tapping fur-

[4]The ternary diagram in Figure A.9 indicates that an iron oxide content between 65 and 60 percent is the lower limit for a slag usable in a bloomery. This limit, however, might be pushed downward by a few percent if the diagram included the effects of potash and soda, present in significant quantities in all charcoal ash. The strongly fluxing effect of these compounds would help to keep the slag liquid while in the furnace and would thus decrease the wastage of metal as FeO and $Fe_2O_3$. The limit would also be pushed downward if a higher melting temperature slag chemistry could be applied. At high bloomery temperatures in the 1250° to 1300°C range, the loss of iron will be greatly reduced if the gangue and fuel ash are such that a quaternary FeO-$SiO_2$-CaO-$Al_2O_3$ slag is evolved. Slags of this kind can have a melting point well below that of normal blast furnace slags while—due to their low iron oxide content—allowing a fairly efficient recovery of metal.

*TABLE 9.2. Calculations of iron recovery in a bloom iron smelting operation*

Assume a hematite ore (Tylecote et al. 1971):
68.6% $Fe_2O_3$, 17% $SiO_2$, 2% $Al_2O_3$, 3% CaO, 0.5% MgO
Assume a charcoal to iron ratio (by weight) of 10:1 (Tylecote et al. 1971)
Assume charcoal ash (3% of charcoal weight; Percy 1864) composition:
16% $K_2O$, 53% CaO, 12% MgO, 5% $SiO_2$, 14% $P_2O_5$

(a) Assume a Roman slag (Tylecote 1976: 56):
26% $SiO_2$, 66% FeO, 5% $Al_2O_3$, 2% CaO, 1% MgO
Using for calculation: 10 kg. ore and 48 kg. charcoal
Total $SiO_2$ into slag = 10 × 0.17 + 48 × 0.03 × 0.05 = 1.77 kg.
1.77 kg. $SiO_2$ (into slag) ≡ 26% (in slag) = 6.8 kg. slag
66% FeO (in the slag) = 4.5 kg. FeO = 3.5 kg. Fe
10 kg. ore ≡ 6.86 kg. $Fe_2O_3$ = 4.8 kg. Fe
Metal recovery in raw bloom = [(4.8 − 3.5)/4.8]100 = 27%

(b) Assume a slag from Britain (15 c. A.D.) (Tylecote 1981)
39% FeO, 5% $Fe_2O_3$, 29.4% $SiO_2$
1.77 kg. $SiO_2$ (into slag) ≡ 29.4% (in slag) = 6.0 kg. slag
39% FeO, 5% $Fe_2O_3$ (in the slag) = 2 kg. Fe in the slag
Metal recovery in raw bloom = [(4.8 − 2.0)/4.8]100 = 58%

Note: The only really effective way to reduce the iron loss to the slag is to use a richer ore; that is, one with less silica, lime, and alumina.

Compare with the estimate below taken from Tylecote 1962: 289
Input, Output and Yield at Byrkeknot Bloomery
A.D. 1408–1409

| | |
|---|---|
| Total ore supplied for one year | 170 tons* |
| Iron produced in one year | 24.3 tons |
| Weekly output | 1200 lbs. |
| Size of bloom | 195 lbs. |
| Overall yield; lbs. of iron/lb. of ore | 14.3% |
| Actual yield; lbs. of iron/lb. of iron in the ore** | 33.0% |
| Charcoal consumption (2 hearths) per week | 12.9 tons |
| Charcoal rate; per pound of iron smelted | 12 lbs. |
| Total weight of slag produced (ash content of charcoal = 3%) per year | 155 tons |

A notably good confirmation of iron recovery figures is given by Hupfeld 1899: 184 for a 19th-century bloomery in the German colony of Togo in West Africa.

*Input*
20 kg. ore (54.9% Fe) ≃ 11 kg. Fe.

*Yield*
11.8 kg. of "pure" slag (49.6% Fe)
1.5 kg. cake slag (15.4% Fe)—probably accretions formed by reaction with the furnace wall
1.5 kg. slag trapped in the bloom (ca. 50% Fe)
3.1 kg. iron in the bloom

Recovery
lbs. of iron in the ore to lbs. of iron in the bloom = 28.2%

Note: This is one of the few records of smelting iron where the slag entrapped in the bloom is accounted for. Bloom weights which take no account of this provide overestimates of the iron recovery. Using a density of 3.5 g/cc. for the slag, the bloom is about 52 percent by volume of slag.

*Based on a fother of 2400 lbs. This figure is a revision of the estimate given in the author's paper, reference 1.
**Based on analysis of ore from Harthope Mill.

naces often chose either to move the upper part of the furnace or to build a new one after each smelting run.[5]

Residual slag had little value. A small but important part remained in the bloom and in the finished iron. In a few areas, limited quantities of high iron slag were set aside and fed into the furnace during the next run. The Romans used slag for building roads (e.g., Straker 1931: 28). In general, it was a waste material, produced and discarded in large quantities. Unfortunately for the industrial archaeologist, old bloomery slags eventually became such an important feedstock for blast furnaces that most of the ancient slags in western Europe have disappeared. In his classic account of the ironworks of the Forest of Dean in western Britain, Powle (1678: 932) notes that ancient "cinder" still available "in inexhaustible quantities" was being added (in a proportion of 1:1) to the ore in contemporary blast furnaces. According to Jenkins (1925–26: 45), the known cinder deposits of the Forest continued to be exploited until the early nineteenth century when they seemingly became exhausted. Their reputation among ironmakers appears to have survived even longer since as late as 1926 a newly discovered deposit of Forest cinder was bought by a blast furnace operator in South Wales (Jenkins 1925–26: 45).

## 4. THE FORMATION OF A BLOOM

By the time it reaches the arch of the combustion zone, the furnace charge has become a porous aggregate

[5]Small bowl furnaces dug into the ground, like those of the Kikuyu of Kenya (Routledge and Routledge 1910: 97–105), were often used only once, as the substantial shaft furnaces of the Fipa of Tanzania may have been (Wembah-Raschid 1973: 32; Cline 1937: 42). Furnaces whose upper portions are picked up and moved after each smelting run are known ethnographically from Nigeria and Niger in West Africa (Krieger 1963: 324); possible archaeological examples include the "slag pit" furnaces of the Roman and early Migration periods in eastern Britain, northern Germany, Denmark, and perhaps Poland (Tylecote 1980: 216–218).

of charcoal, unreduced oxides, slag, and particles of iron metal. The slag, when the system functions well, develops as droplets which in moving down touch, join, and grow like water condensing on a window pane. When the droplets become large enough, they fall through the combustion space, collecting in a pool below the tuyere in the furnace basin. Good sidewall insulation and radiant heat transfer from above keep the pool molten. If the force of the blast is too weak to propel the air across the combustion space to the opposite furnace wall, the radiant heat source and thus the slag pool will be limited in area. Any liquid trickling toward the margin of the heated zone freezes, forming a glassy shell around the pool.

Nonmetallurgists often describe the formation of a bloom in terms of melting and later solidification at the furnace bottom. It should be reiterated that iron never becomes liquid in a bloomery operated at normal temperatures. Instead, the reduced metal takes the form of tiny flakes of solid iron that are enveloped and dragged downward by the slag drops through the arch of the combustion zone and the empty space underneath to fall into the slag pool. There the iron particles remain in unstable suspension as long as the pool is agitated. When it becomes quiescent, they begin to settle slowly toward the bottom.

At some point these settling particles, even while still in motion, start to touch one another. At elevated temperatures, film-free surfaces of iron particles that are in point contact will quickly develop area contact, or sinter. Small groups of particles will then cluster into larger ones as free energy driving forces seek to minimize total surface area.

If there were no slag pool at the bottom of the furnace, sintering would have only a limited effect. In a laboratory oven, a loose bed of iron particles heated to 1000°C in a nonoxidizing atmosphere will become a coherent aggregate within three or four hours. Although the particles in such an aggregate largely lose their separate identity, their interconnections are weak so the aggregate shatters if forged.

In normal sintering among solid particles the initial bonding is moderately rapid, but the advance of densification is slow and it proceeds at a steadily decreasing rate. This is because the iron atoms can move only from one surface site to another through the body of solid particles (represented by the arrows with broken lines, Fig. 9.3). The rate at which this movement occurs is governed by the capacity of iron to diffuse within itself, and the volume diffusion coefficient involved is very small.[6] Substantial consolidation beyond a fragile aggregate is therefore not likely to occur within a reasonable span of time; that is, hours rather than days.

Many blooms are dense. Figure 9.4 shows examples of pieces of raw low carbon metal, not yet consolidated in the forge and with no sign of having once been melted, which are nonetheless markedly solid and compact.

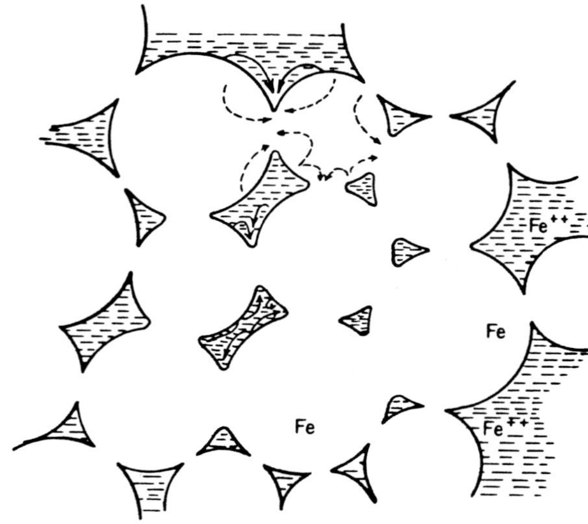


Fig. 9.3. *Solid Phase and Liquid Phase Sintering.* Sintering occurs as point contacts between particles become area contacts due to the movement of atoms between surfaces. Solid phase sintering is slowed because iron atoms diffuse only with difficulty through solid iron (solid arrows). Liquid phase sintering, whereby iron atoms travel along a gradient of ions in a suitable liquid—perhaps a slag—is much faster (dashed arrows). At a certain point in the sintering process, individual particles lose their identity and the aggregate assumes properties approaching those of a dense body.

The most likely explanation is that the immersion of the bed of iron particles in liquid slag causes sintering to be accelerated. Since iron ions are in the majority in the slag, iron atoms must be diffusing through the liquid (illustrated by the solid arrows, Fig. 9.3). Because diffusion in liquids can take place at a far greater rate than in solids, the acceleration of the sintering process is enormous. This special phenomenon is called *liquid phase sintering.* Although metallurgists usually associate the term with transport through liquid metals, there is no fundamental reason why liquid phase sintering cannot also function when molten slag serves as the conduit for iron atoms.

## 5. CONSOLIDATING THE BLOOM

No bloom directly from the smelting furnace is solid enough to be forged immediately into finished products. All blooms require preliminary processing to expel most of the occluded and adhering slag, causing a reduction in volume of at least 50 percent. If the bloom is still hot when withdrawn from the furnace, the first stage of consolidation occurs then; the bloom is placed on an anvil and hammered until it grows cool. In smelting systems where the furnace must be cooled before the iron can be extracted, nothing can be done immediately except to knock away the excess slag from the surface of the bloom or to break up the slaggy mass at the furnace bottom so the bits of metal can be picked out.

Except in smelting devices also used as forges—the Catalan hearth and the German *Rennofen* are examples—most of the consolidation work has to be done at separate furnaces. Often they were specially designed, as in India and parts of Europe where shaft furnaces

[6] The volume coefficient for the self-diffusion of iron at 1200°C is only $5.7 \times 10^{-11}$ cm.$^2$/sec. (Elliott et al. 1963). In contrast, diffusivities in liquids are in the range of $1\text{-}10^{-5}$ cm$^2$/sec.; a million-fold difference.

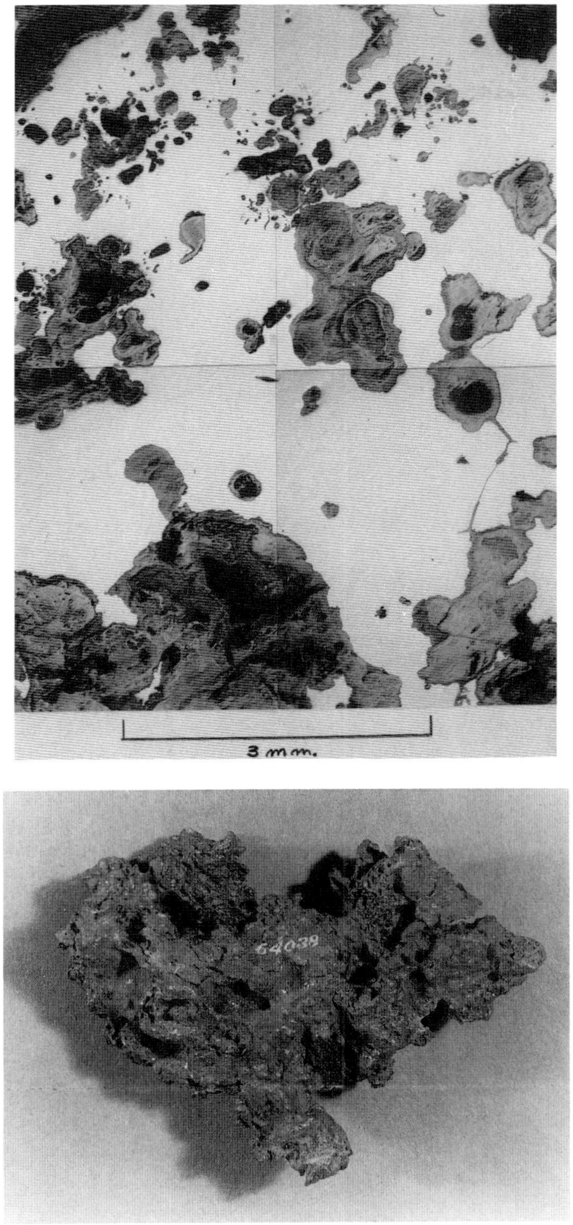


Fig. 9.4. (a) *Canadian Bloom.* Low magnification view of the cross section of one of the "Frobisher" blooms, found near Baffin Island and dated by radiocarbon to the twelfth-thirteenth century A.D. (Sayre et al. 1982). The dark portions are slag-filled pores, representing about 50 percent of the total volume. (Photograph by James Dvorak); (b) *African Bloom.* Produced by Ovimbundu smelters in Angola during the 1890s. The slaggy, porous appearance is typical of blooms before consolidation. This example has a maximum diameter of 10 cm. and weighs about 210 grams. (Field Museum cat. no. 64038)

were used as bloomeries. Otherwise, the bloom was worked up in an ordinary blacksmith's forge, as it was in some areas of Africa, for example.

The product of a finery—a furnace for converting cast iron into wrought iron—was treated similarly and was often called a bloom. In the developed European tradition, the equipment used for consolidating finery blooms sometimes included a special furnace, called a *chafery*. This term was once applied to any furnace used in the working up of blooms, whether from a finery or a bloomery. Modern historians, however, define a chafery as a furnace attached to a finery, and more specifically, to the type of finery known as a Walloon hearth.

The first hammer blows given to the hot or reheated bloom are critical. Observers of traditional ironmaking often noted that light hammers were used at this stage and that the strokes were very rapid, apparently to flatten projections and tack together separated fragments of metal. In India, where blooms were often sold to secondary processors in only partially consolidated form, "Care was taken not to hammer out too much of the slag as the iron is sold by weight. . . The smelters said that on being refined this would yield half its weight of *pucca* [real, high quality] iron" (Watson 1910: 325). Except in situations like this,[7] the light initial strokes are normally followed by more vigorous hammering to squeeze out slag and begin the transformation of the spongy mass into a solid plate or bar of metal.

In some cases, much of the slag is allowed to drain out of the bloom while it rests within the hearth used for consolidation, as it was in India (Atkinson 1890: 270): "The fire being lit, a mixture of [pieces of]. . . bloom metal is placed on the hearth opposite the bellows, with the larger pieces nearest the fire. The blast having commenced, in a quarter of an hour the slag begins to flow, and in another quarter of an hour the metal (now a porous, pasty mass) is taken out of the fire and subjected to the blows of two or more sledge-hammers." Slag that drains off by itself like this collects at the bottom of the hearth, and resembles deposits from the bottoms of smelting furnaces. Such deposits are seldom reported (see, however, Rostoker and Gebhard 1981a), probably because they are difficult to distinguish or the majority of ironmakers depended more on muscle (or water) power than gravity to get the slag out of the bloom. Hammering the slag out rather than waiting for it to drain out saves fuel and time. It also minimizes the changes in carbon content that result from prolonged heating at high temperatures.

Bloom consolidation can be understood by comparing it with the squeezing of a sponge filled with a liquid, the viscosity of which lies somewhere between that of water and heavy grease. Like a sponge, a heated bloom squirts or oozes liquid when deformed (Fig. 9.5). After several squeezings (i.e., cycles of heating and hammering), the bloom becomes relatively liquid-free. However, a bloom reacts differently than a sponge in one important respect: it undergoes plastic instead of elastic deformation, meaning that the squeezed-out pores remain closed rather than springing open again after pressure is removed. Iron surfaces are forced against iron surfaces; since the slag has removed any obstruct-

[7]An interesting variant of the practice of leaving as much slag as possible in the worked-up bloom is reported by Franklin (1971: 240–241). Describing ironmaking at Jabalpur in central India, he complains that the "refiners" deliberately add lumps of raw bloom to the finished iron, "mixing all these crude pieces with the other so that the cheat cannot be detected except on trial; thus they not only shorten the time of the operation, and thereby use less fuel, but contrive by this nefarious practice to sell a large part of their crude iron at the same price as the malleable."

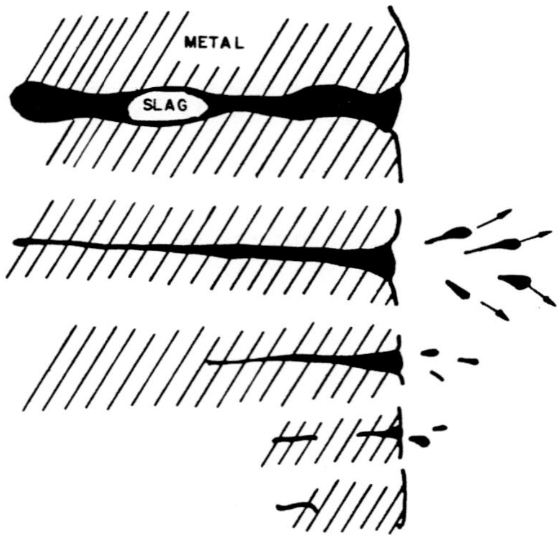


Fig. 9.5. *Expulsion of Slag.* The bloom is a sponge represented by two interpenetrating networks, one of iron and one of slag. When the slag is molten, it is retained in the composite by hydrostatic pressure under the slag pool and by surface tension when removed and before cooling. When reheated in a forging preheat furnace, some of the slag seeps out because of gravity and collects at the bottom of the furnace. Some of it remains occluded. When at a temperature of about 1200°C the bloom is removed and rapidly hammered on an anvil, the slag is viscous but the force of the hammer ejects it like toothpaste squeezed from a tube. As the pores close they also elongate. Repeated heating and hammering expel most of the slag, and the pore channels lose their continuity with the external surface.

ing films, the surfaces weld to one another, forming a unified mass in which the original separations often cannot be recognized even by careful metallography.

It is most unlikely that all the slag will be expelled. Some is bound to be trapped in pores that are welded shut before all of it escapes. When the bloom is subsequently worked down to a shape like a plate or bar, the pockets of trapped slag show how working took place. Iron forged into a plate will contain slag ribbons that are readily observed at low power magnification under a metallurgical microscope, and iron forged to round bar has elongated, wire-like slag inclusions.

Iron consolidated until it is reasonably free of slag is as deformable as metal shaped by the more conventional modern processes of casting and forging. It can be hammered to complex shapes at normal temperatures, with periodic recrystallization anneals to restore ductility. However, both bloom and wrought iron are easily distinguished from the modern low carbon iron called *mild steel* which is free of slag stringer inclusions. If a segment of wrought iron bar is notched on one side with a chisel and bent to open the notch, the fracture that develops has a splintered appearance like wood broken in a similar manner (Fig. 9.6).

Raw blooms were normally consolidated at the smelting site, but not always. Iron sold in the form of unworked blooms was common in India, and is also known in Africa. Among the Bassari of Togo, for instance, the raw blooms produced at the great smelting center of Banyeri were carried to forging centers as much as fifty kilometers distant (Cornevin 1962: 88–91). Elsewhere in Africa, dispersed blooms were broken up and the resulting iron fragments, with minimal further working, divided among the furnace laborers (Tessmann 1913:

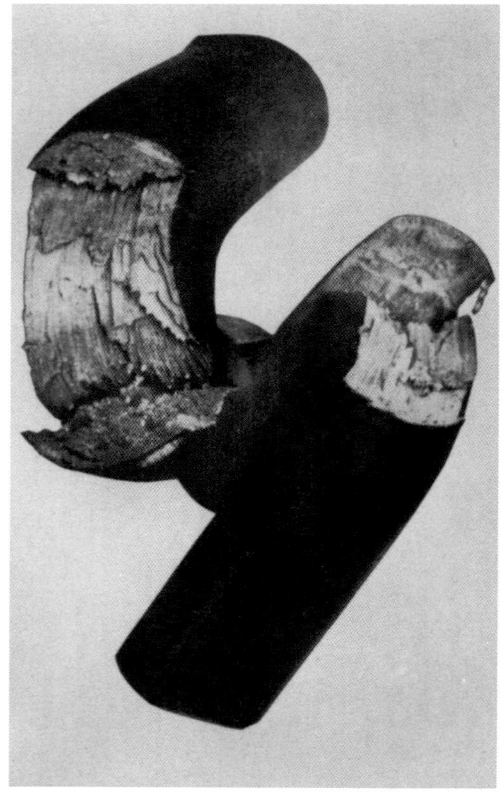

Fig. 9.6. *Woody Fracture.* Normally a steel bar notched and bent to fracture will show a fracture path which runs in a simple plane across the section. In a woody fracture the crack path is diverted by branches which follow the brittle slag ribbons. In some respects wrought iron was safer than the mild steel which had supplanted it by the late nineteenth century because fracture in bending often did not lead to complete separation into two pieces. (From Aston and Story 1957: 47)

232) or between the furnace operator and the owner of the ore (Sassoon 1964: 176). The slaggy fragments of bloom might be traded or kept by their owners until it was time to get a new tool forged.

In other parts of the world smelters themselves usually worked the iron into a consolidated, marketable form—what in Europe came to be called *merchant iron* or simply *bar iron.* Most often, in Asia and Africa as well as Europe, bloom iron was marketed in the shape of bars or flattened balls. Other shapes included plates and roughed-out tool blanks; in some cases the form of these blanks had become symbolic, as they were too small to be used in manufacturing the tool they represented. In a number of areas, merchant iron was marked with an indication of quality. The most common and practical mark was a simple split, made with an axe, that revealed the metal in the interior of the piece. Conventional trade marks made by striking bar iron with a die were widely used in Europe from the Middle Ages onward and are also known in China (see, for example, Barraclough 1984: 173–175).

## 6. FUEL TO ORE RATIOS

The composition of the furnace charge is a primary controlling factor in the smelting process. In many ironmaking traditions this has been conceived in terms of ratios between the weight or volume of fuel to ore. Western ironmakers have generally used weight ratios for the past century, but the volumetric ratios of earlier systems probably worked well enough. Either provided a degree of control and made it possible to estimate production cost and efficiency.

Tylecote and his associates conducted an important series of experiments using charges of different compositions in a reconstructed Romano-British bloomery (Tylecote et al. 1971: table VII). Lowering the fuel to ore ratio from 2:1 to 1:4 resulted in the increase of the percentage of iron (as oxide) in the slag. It also decreased the carbon content of the iron metal, the $CO/CO_2$ ratio in the furnace gases, and the overall metal yield. (For a summary of the relevant data, see Table 9.3.)

In general, raising the fuel to ore ratio in smelting has several consequences: the temperature within the furnace rises, more air is needed for complete combustion, and gases with a higher $CO/CO_2$ ratio reach farther up the shaft from the combustion zone. The result is that the metal is more effectively reduced and partially carburized. If carburization proceeds to the saturation point, the iron melts, a different slag chemistry is introduced, and recovery efficiencies of metal from ore improve dramatically. If this happens all across the hearth, the furnace is no longer technically a bloomery since its product is cast rather than bloom iron.

The experiments led Tylecote and his associates (1971) to conclude that:

> . . .most of the charcoal was used for filling and preheating the furnace before the first charge of ore was made. Thereafter the charcoal consumption was directly related to the fuel to ore ratio. The burning rate of charcoal varied between 2.6 and 3.5 kg/hr once the optimum conditions were established. For a furnace of given size this related to the air rate. The charcoal consumption during smelting is only about a third of the total; so the fuel to ore ratio does not have much influence on it. It was to be expected that the total fuel to ore ratio would vary with the bloom size. The ratio varied from 60:1 for blooms containing 1 kg of metal to as low as 4:1 for the larger blooms. An allowance made for reworking blooms into forged bars would double the ratio.

*TABLE 9.3. Effects of fuel-to-ore ratio (by weight) on process factors**

| Fuel/ore | $CO/CO + CO_2$ | % C in metal | % Fe in slag | % Metal yield |
|---|---|---|---|---|
| 2.1 | 0.75, 0.72 | 1.8, 1.7 | 8.0, 14.0 | 78, 80 |
| 1.5 | 0.65 | 0.5 | 16.0 | (27) |
| 1.0 | 0.50 | 0.6 | 18.0 | (93) |
| 0.75 | 0.57, 0.62 | 0.2, 0.4 | 35.8, 18.9 | (29), (24) |
| 0.50 | 0.48 | 0.07 | 39.0 | 48 |
| 0.25 | 0.67 | 0.03 | 44.1 | 49 |

( ) = Questionable because of uncertainty on the amount of slag trapped in the bloom.

Other citations on fuel-to-ore ratios and metal yield:

| | | |
|---|---|---|
| Lyman 1879: 63 | fuel-to-ore ratio ~1.0 | (Tatara furnace) |
| Atkinson 1890: 270–271 | fuel-to-ore ratio ~0.33<br>metal yield ~60% | (India) |
| Sambhasiva Iyer 1898–99: 102–116 | fuel-to-ore ratio ~1.0 | (India) |
| Krishnan 1954: 80–81 | fuel-to-ore ratio ~0.9<br>metal yield ~25% | (India) |
| Sassoon 1964: 178 | fuel-to-ore ratio ~1.12 | (Africa) |

A useful summation of fuel use and metal recoveries given by Bell 1884: 30–31 reads as follows: "In producing the bloom, six times its weight of charcoal is used, and half the iron contained in the ore is lost. In reheating the crude lump for forging into a bar, one half its weight is wasted and again there is considerable expenditure of charcoal. We are probably within the mark in accepting the fact, that ten times as much fuel and three times as much ore are required, for every unit of metal produced, as is consumed in the blast furnace and Bessemer furnace together." The loss of 50 percent of the bloom in forging to bar probably reflects the expulsion of slag as well as the oxidation of metal in the reheating furnace.

*Taken from experiments published in Tylecote et al. 1971

An average of the more successful furnace runs in this experiment gives a fuel:iron ratio of about 10:1, or about 20:1 allowing for reworking. This is a good rate of recovery when compared to the efficiency of other bloom furnaces (cf. Tables 14.3a, 14.5a).

Most estimates of the inputs and outputs of traditional bloomery smelting are vague about the inclusion of the fuel needed for reheating and consolidating the semi-finished blooms. Among the observers who are clear on the subject is Atkinson (1890: 270–271). Describing the ironmaking system used by highly skilled smelters in Kumaon in the Himalayan foothills, he noted that 930 kg. of ore and 340 kg. of charcoal yielded 327 kg. of raw bloom, which was converted to 82 kg. of bar iron by expending another 667 kg. of charcoal. The indicated fuel to ore ratio for the full process, to finished bar iron, is slightly better than 1.1:1, the fuel to metal ratio is 12.3:1, and the ore to metal ratio is about 11.3:1.

Such ratios cannot serve as absolute measures of efficiency without considering data on the richness of the ore. The most useful measure of ore quality is not the percentage of iron present but the ratio of iron oxides to gangue. This is of particular concern to operators of bloomeries, for in these furnaces a considerable fraction not only of fuel but also of iron from the ore must be expended in converting the gangue into molten slag. Twenty percent silica in the slag requires a substantial amount of fuel, and most of the iron originally present in the ore is needed to flux it (Table 9.2).

## 7. ORIGIN OF CARBON ENRICHMENT IN BLOOMS

Blooms of raw iron and artifacts made from them often have a heterogeneous carbon content. Although the direct or bloomery process can theoretically produce a metal with carbon in the 0.01 percent range, many pieces of ethnographic and archaeological bloom iron have zones with carbon contents as high as 0.4 percent. Sometimes these zones can be securely attributed to forge carburization, either of blooms or artifacts, especially when they occur on the working edges of implements or when they are arranged in the continuous bands that would result from the pile-welding of carburized strips. In other cases, however, the zones of higher carbon content are randomly distributed, suggesting that carburization took place while the bloom was forming within the furnace.

There are two possible ways to enrich the carbon content of a bloom before it leaves the furnace. The first is carburization through contact with solid carbon (Chapter 11). Few ironmakers except the Japanese are known to have done this deliberately to produce a high carbon metal, but many must have been aware that parts of a bloom with unburned charcoal in them were likely to be hard, especially when the bellows had been worked with unusual vigor.

The second carbon-enriching mechanism depends on heterogeneous conditions in the combustion zone (Fig. 9.7). In an ordinary bloomery furnace, the fuel to ore ratio is kept relatively low and air is supplied through the tuyeres at a moderate rate, which keeps the $CO/CO_2$ ratio well within the range for production of a low carbon iron. However, the fuel to ore and carbon monoxide to dioxide ratios can easily, perhaps inadvertently, increase to levels at which some of the newly reduced iron particles in the area just above the arch of the combustion zone become carburized and form droplets of liquid cast iron. While this might happen throughout the cross section of the furnace (which then would be operating as a blast furnace), the limits on the amount of blast pressure likely to be available in a device designed as a bloomery make it probable that only a small part of the cross section will be affected. The portions of the combustion zone arch at some distance from the tuyere will not become hot enough or have a sufficiently high $CO/CO_2$ ratio for iron particles to be liquefied; those areas will function like a normal bloomery, shedding solid iron particles down to form a bloom. Only the portions very near the tuyere can produce cast iron, and they may do so only intermittently during a smelting run, but the liquid high carbon metal also rains down. When the droplets hit the porous bloom, they spread out and soak in. Within a minute or two the carbon in them begins to diffuse into the solid particles of bloom iron. The droplets solidify as their carbon content falls to the level of a steel.

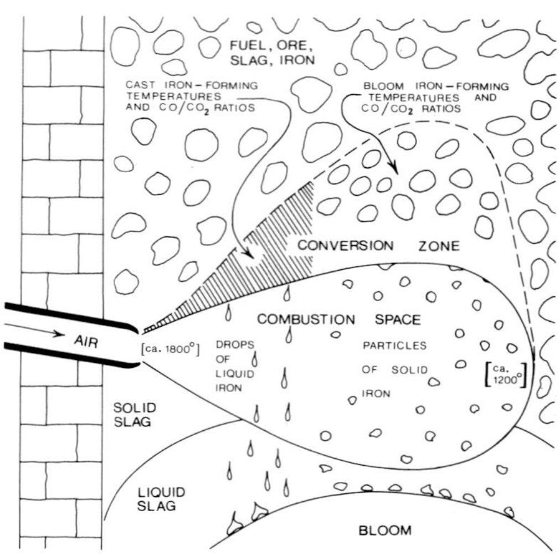


Fig. 9.7. *Heterogeneous Conditions in the Combustion Zone.* In some circumstances zones of high carbon content can occur in a bloom. The illustration suggests that both the temperature and the $CO/CO_2$ ratio are higher than elsewhere immediately above the tuyere orifice inside the furnace. According to fuel/ore ratio and air input rate, the conditions for smelting to cast iron could be achieved locally. Droplets of liquid cast iron settle through the slag pool and wet the iron bloom immediately. The high carbon in the cast iron diffuses into the solid iron, carburizing it in zones to medium carbon steel; that is, 0.3 percent to 0.6 percent carbon. The bloom here is below the slag pool, but in some modern replications of the operation of a short shaft furnace (e.g., Tylecote et al. 1971), the bloom was discovered hanging from the furnace wall above the slag pool and opposite the tuyere. The positions of these blooms were probably caused by the force of the blast of the modern air fans, which are stronger and more constant than ancient bellows; in addition, the tuyeres in the replication experiments were probably more firmly sealed in the furnace wall. In larger shaft furnaces, the bloom was resting on the hearth, to be hooked out after the slag had been tapped.

The appearance of the finished bloom depends on the amount of cast iron that has fallen on it. When the temperature and $CO/CO_2$ ratio have been only slightly above the cast iron threshold, the blooms should have only patches of harder metal. Very hard patches are cast iron droplets that fell late in the process of bloom formation, while less hard patches represent droplets that fell earlier and have lost more of their carbon to the surrounding metal. On the other hand, in cases where a large portion of the furnace cross section was pushed above the cast iron threshold for a considerable part of the smelting run, the bloom might contain a large amount of steel and even cast iron.

Heterogeneous blooms of this sort were the usual product of the *Stücköfen* of Styria and Carinthia in southern Austria. Percy (1864: 328–329) quotes Jars as saying that these furnaces produced liquid cast iron simultaneously with solid, steely blooms and that these blooms sometimes had undergone "incipient fusion," making them extremely difficult to extract. Overman (1854: 142–143) confirms the presence of cast iron in Stückofen, adding that it is converted on the spot to wrought iron or steel by fining in the same hearth used for consolidating the *Stückwolf* or bloom. Osborn (1869: 639–643) gives a similar description of the product of Catalan forges in the Lake Champlain area of New York and Vermont: "a mixture of fibrous iron, cast iron, and steel—an aggregation of unavoidable irregularities." Blooms containing steel and cast iron are also characteristic of the Japanese *tatara* furnace (Chapter 12).

## 8. CONTAMINANTS

In contrast to blast furnaces, bloomeries have long been known to produce a very pure iron free of most contaminants because of the temperatures at which it is smelted and the use of charcoal as a fuel (Chapters 2, 6). Unlike coal and coke, charcoal contains little or no sulfur. The relatively small quantities of sulfur and phosphorus in iron produced by the direct process often made it a premium material for high stress applications in which possible hot or cold shortness was unacceptable. For this reason, the direct process survived in competition with the indirect, although the latter was more efficient on a large scale.

A small furnace needing only a few tons of fuel per year could continue to use an expensive fuel like charcoal when operators of large furnaces were switching to coal and coke. During the nineteenth century, after large European ironworks had already made the change, writers on metallurgical topics often were amazed at the purity of bloom iron produced in less industrialized regions; by pure they usually mean that the iron had, by their standards, a low sulfur content. The British admiralty, for example, continued to prefer Spanish bloom iron even after British wrought iron had been developed (Chapter 13). The low temperature at which the average bloomery operates also offers the product limited protection from contamination by sulfur in the ore or by phosphorus in the ore and fuel. Pyrite ($FeS_2$), by far the most common of the sulfur-containing minerals, is often present in hard iron ores. It becomes liquid at iron smelting temperatures, melting at 1171°C. If while in liquid form it comes in contact with particles of reduced iron in the combustion zone, alloying will be very rapid. Pyrite droplets that reach the slag pool without coming in contact with metal are probably not deleterious, for they are likely to remain suspended in the slag. The density of $FeS_2$ is 5.0 g./cc., which is very close to that of a liquid $FeO$-$Fe_2O_3$-$SiO_2$ slag.

At about 700°C, $FeS_2$ will slowly undergo thermal dissociation to FeS with the evolution of sulfur vapor. In furnaces to which no ore is added after smelting begins, and the ore is therefore heated gradually, there is probably time for most of the pyrite to be converted and for the sulfur vapor to be taken off with the exit gases. Although the remaining FeS, which melts just below 1200°C, would behave in the same way as $FeS_2$, descending with the charge and alloying readily with any metallic iron it touches, the sulfur content has been decreased by 50 percent so much less is available to enter the product. In shaft furnaces designed so charging can continue after the furnace is lighted, the sulfur passes more readily into the metal. The rapid heating undergone by ore put into a shaft already at smelting heat would insure that most of the $FeS_2$ would melt and begin to alloy with iron before conversion to FeS occurred.

Sulfate minerals, which tend to replace sulfides in bog ores and perhaps in extensively weathered hard-rock ores as well, pose few problems in a bloomery (or a blast furnace). Since they dissociate to gaseous $SO_2$ and $SO_3$ at a temperature of about 600°C, they are unlikely to be absorbed by the iron. Sulfides can also be converted to sulfates by roasting the ore (Chapter 4). Roasting is usually done before smelting begins by heating the ore in piles to 400° to 500°C in air, but some furnace designs may allow substantial roasting to occur as an integral part of the smelting process. If the sulfides in the charge can be brought up to smelting heat slowly enough and if the air supply is adequate, most will change first to ferrous sulfate and then to sulfur dioxide and trioxide gases. These reactions will not take place in a shaft furnace, whether run as a bloomery or a blast furnace, because of the absence of free oxygen in the upper part of the shaft. However, in bowl furnaces where it is customary to preheat the charge by stacking it around the furnace edge, like the Catalan forge, substantial roasting of sulfides probably occurs.

Phosphorus, which can occur in iron ores as phosphates of iron or calcium, can also be deleterious. The mineral vivianite, $Fe_3(PO_4)_2.8H_2O$, is an easily reducible additive used intentionally by Chinese foundrymen to increase the fluidity of cast iron (Read 1913). Magnetite ores commonly contain the less reducible mineral apatite, $Ca_5(PO_4)_3.(OH,F,Cl)$. Charcoal ash can also be a source of phosphorus. According to Percy (1875: 188–190), the phosphoric acid content of wood ash can be as high as 10 percent but is more generally between 1 and 3 percent. Analytical reports usually do not distinguish among the phosphates of sodium, potassium, calcium, and magnesium, but all can be reduced to elemental phosphorus at smelting temperatures. The most refractory of this group of compounds is calcium phosphate, the reduction of which by the reaction

$$Ca_3(PO_4)_2 + 5CO \rightarrow 3CaO + 2P + 5CO_2$$

can occur at temperatures not far above 1000°C. Elemental phosphorus is a vapor at those temperatures and is readily absorbed by metallic iron.

Several factors control the actual reduction of phosphorus in ore and its subsequent deposition in the slag or iron. Some of the phosphorus which is difficult to reduce will pass through to the slag pool and can be detected by analysis as $P_2O_5$ (Tables 9.1a, 1b). Some reduced phosphorus will become immobilized in slag while still in the furnace shaft, but the reason for this is not completely clear.[8] Temperature influences the amount of phosphorus which will enter the iron, since the higher the temperature, the greater the proportion of phosphorus. A limited sample of smelting experiments (Table 10.4) indicates that the amount of phosphorus entering the iron is very low at minimum smelting temperatures, but around 50 percent when a bloomery is run at near-blast furnace heat. In a true blast furnace, phosphorus recovery by the iron approaches 100 percent.

## 9. THE REDUCTION OF MANGANESE

Manganese is a minor component in many iron ores. Since manganese greatly alleviates hot shortness and increases hardenability in iron-carbon alloys, it is important to understand the efficiency of its reduction in bloomeries. In general, manganese in amounts of about five times the sulfur content will eliminate hot shortness and any manganese in excess of that amount will increase hardenability proportionately.

Theoretically, manganese cannot be reduced from MnO by the action of carbon monoxide below about 1400°C (Alcock 1976: 199–201), but some early bloom irons and steels contain up to 0.5 percent manganese and occasionally even more (see, for example, Klevenhüller et al. 1961), amounts which have a significant effect on the properties of the finished product. Kinney (1926: 11) found in a study of the reactions in blast furnaces that "approximately one-third of the manganese had entered the metal while it was in a sponge-iron state"—that is, solid iron particles below 1130°C. How then does manganese reduction and alloying occur at these temperatures?

[8]The ubiquity of phosphorus compounds in slag shows that all the phosphorus does not enter the iron. Because the Thomas process dephosphorized the metal in puddling furnaces by using a secondary slag high in FeO (Chapter 13), we know that a slag could affect phosphorus levels in iron. Although bloomery temperatures are probably too low for the formation of the type of slag utilized in this process, there are a number of slag-like materials which can immobilize phosphorus mechanically. Among these are several Chinese ceramic glazes with structures and compositions similar to those of slags which contain substantial amounts of phosphorus in the form of "millions of bubbles. . . that are reluctant to disperse even with longer soaking or greater heat" (Wood 1978: 39).

The presence of manganese in iron may result from reactions forced by the absorption of vapor. When the product of any reaction is quickly absorbed, the reaction will proceed even though the elementary thermodynamics indicate that it should not. Manganese has a vapor pressure of 1 mm. at about 1250°C, and in vapor form is easily absorbed by iron particles. Vapor absorption is not an efficient mechanism for transferring manganese from ore to iron, but it could account for a manganese content in the 0.1 to 0.5 percent range.

In addition, many bloomeries were customarily run so that parts of the charge could be reduced to liquid cast iron. In some instances, the temperature at localized hot spots must have been high enough to reduce manganese under negative free energy conditions. Spathic ores containing high manganese contents were used in the Stücköfen of Styria and Carinthia, and in some regions of Germany. These areas were famous as regions of "steel ores" because of the quality of their steel and the direct production of steel as a bloom (Wertime 1962: 196; Smith 1968: 116). We believe that these Stücköfen were run hotter than usual bloomeries, so that the ore was reduced to cast iron droplets in some zones. Under such conditions manganese oxide is more readily reducible and absorbed by the liquid cast iron. The droplets, when absorbed by the bloom, impart to it both carbon and manganese, producing a steel of considerable hardenability and therefore more responsive to hardening heat treatments even if they are poorly controlled.

# 10. Smelting to Cast Iron

## 1. INTRODUCTION

Although the Chinese had mastered the smelting of cast iron and the art of making things from it by 400 B.C., it was in Europe that the most important advances in technology eventually occurred. In the late fourteenth century A.D. Europeans began deliberately to build their furnaces higher and to run them hot enough to produce a liquid alloy of iron and carbon that could be cast to final form in a mold.

The development of this new technology in the West was sporadic and slow—the true blast furnace was still unknown to Biringuccio in the most generally scientifically advanced nation in Europe, almost two centuries after the device had first appeared in Belgium and northwestern Germany. After another two centuries, by the 1770s or 1780s, Europe had equalled or even surpassed China in the production of iron castings. The succeeding century saw the heyday of cast iron, an era of very large furnaces, enormous production volumes, and ingenuity in finding uses for a metal that in retrospect seems an unlikely raw material for bridges, ocean liners, statues, and exposition halls. The "Age of Cast Iron" in Europe lasted no more than a hundred years and, in the United States, less than forty.[1] Cast iron started to be replaced when wrought iron from puddling furnaces and improved rolling mills became inexpensive enough to be used instead of cast iron in rails and structural applications (Temin 1964: 25–34). The age ended with the introduction of Bessemer's converter and Siemens's open hearth, which made both cast and rolled steel much less expensively than ever before.

Most of the cast iron in America or Europe visible today is in a sense archaeological, since it was cast fifty or a hundred years ago into fire escape weights, manhole covers, and architectural moldings so massive that they are virtually indestructible. Newly made consumer goods of cast iron are scarce in most modern countries; if it were not still used in the engine blocks of cars, underground water pipes, and fire hydrants, Americans might never see it at all. Yet very large amounts of cast iron continue to be made. It is an important material for industrial machinery and is still the almost universal feedstock in steelmaking. Therefore, modern nations produce and consume it in quantities unimagined by the most visionary industrialist of Victorian times. Modern blast furnaces are often huge, heavily instrumented, and computer controlled. With daily capacities as high as 20,000 tons, some can make more iron in a single day than all the furnaces of Elizabethan England could in a year.[2]

Such a furnace would undoubtedly impress a Victorian or Elizabethan ironmaster, although the principle behind its operation would not be difficult for either to understand. The Victorian would recognize a modern blast furnace immediately as an enlarged version of the smelting devices of his own time. The Elizabethan would be briefly puzzled until he saw the furnace operating, examined its bottle-shaped interior, and heard what its various parts—bosh, tuyere, and so forth—are called. Both would easily grasp the basic processes involved, as would a Chinese ironmaster of the Han Dynasty. They would be familiar with the chief inputs and outputs, the principles of furnace operation, and the overall economics of cast iron production. Once they had become accustomed to the vast size of modern ironmaking equipment, they might even maintain that the blast furnace has undergone little essential change in the past few centuries.

Thus, the roots of the modern steel industry lie in the early history of cast iron production. The direct ancestor of the present-day blast furnace was in Europe where, at some time between the late fourteenth and the mid-sixteenth centuries, the techniques crucial in smelting iron to a liquid form, casting it into useful shapes, and converting it into wrought iron were borrowed or discovered. These developments were predicated on a sound knowledge of water power, a habit of using relatively tall shaft furnaces, a tradition of extensive utilization of iron, and a moderately developed economy with a fairly dense population. Even though each of these factors was in place by A.D. 1100, they were not effectively combined until a special need arose.

The most likely reason that cast iron was made in Europe was the discovery that the material was suited to a new technology of war. Cast iron was made into projectiles as soon as it was introduced, and the subsequent increase in its production closely parallels the growing use of artillery in European warfare. This could be a coincidence, but some urgent need must have justified the substantial capital investment involved, and in its early years cast iron was used for little else in

[1]Temin (1964: 28) shows that in 1830 only 15 percent of the pig iron produced in the United States was used to make castings. In 1850 this had risen to 55 percent but by 1870—before the introduction of the Bessemer converter and the open hearth—it had fallen again to 45 percent. Considering that during this period the United States was still producing a moderate amount of bloom iron by the direct process (in 1831, about 8000 out of a total 90,000 tons of bar iron—Temin 1964: 25), it seems probable that castings were the main product of the American iron industry for only a decade or two.

[2]Scrivenor (1854: 57) estimates the total annual production of England and Wales to have been only 17,350 tons as late as 1740.

Europe. Later, when it was converted to wrought iron, it could have been consumed in large quantities, but this is an expensive way to obtain an inferior product. Before cast iron had wide application, smelting and fining it to wrought iron must have been subsidized for military purposes (Chapter 14). It may have taken several centuries for the blast furnace and finery to become commercially viable in a purely civilian economy, assuming that such has ever existed.

## 2. OVERVIEW

The smelting of cast iron is different from that of bloom iron. The furnaces tend to be larger and must be run at higher temperatures; blowing machines are almost always employed; the slags produced are different in composition; and the yield of metal from ore is good. The product, along with a large amount of dissolved carbon, is likely also to contain substantial quantities of other elements such as phosphorus, silicon, sulfur, and manganese.

Among the less obvious differences between cast and bloom iron is the fact that the former is substantially slag-free, because it collects at the bottom of the furnace basin, as a dense liquid underneath a pool of less dense liquid slag. The combined effects of surface tension, the low viscosity typical of liquid metals, and the great density differences allow a clean separation of slag and iron. The slag pool floats above the cast iron pool in a substantially quiescent manner so that there is almost no mixing.

Another difference is that slag and iron can both be taken from the furnace as liquids, making continuous operation over a period of weeks or months easy—and almost inevitable. In the great majority of the world's bloomery systems, furnace runs lasted no longer than a single day; by then enough solid iron had accumulated so that the furnace had to be shut down. In shaft furnaces making cast iron, on the other hand, one day runs are unknown, and most furnaces were regularly kept in blast for periods of no less than a week or two.

Tapping can be done through one hole or two. If there are two, one is positioned above the other so that the lighter slag flows from the upper hole and the heavier iron from the one below. If there is a single tap hole, the stream of liquid that pours out initially is iron. After the metal/slag interface has dropped to the tap hole level, liquid slag begins to emerge and the stream is diverted to a slag dump. Whether the one or two tap hole method is used depends on the relative volumes of iron and slag that accumulate in the basin and the amount of metal required from a single tapping. Having two tap holes makes it possible to collect larger amounts of liquid metal in the furnace, for slag can be removed at frequent intervals to make room for more iron.

The ability to match the volume of metal available at one pouring time with a specific casting objective is an important feature in systems where furnaces are tapped directly into molds for making finished objects. Most traditional Chinese and recent Western ironmakers have preferred to run the iron from the furnace into pigs and later to remelt these for casting into molds. Remelting was desirable because it yielded a stronger and more predictable cast iron with a lower carbon content. In the early sixteenth century, Western furnace operators went to considerable lengths to avoid having to melt the same metal twice, presumably to save on fuel. European furnaces of the sixteenth and seventeenth centuries were therefore designed for control of the amount of metal that could be poured in a single tapping. When the volume of metal from one furnace was not enough to fill a mold, such as those for the larger types of cannon, double furnaces from which iron could be tapped simultaneously were built. A sixteenth-century English example has twin furnace shafts, separated by a common wall, and each has a single large bellows activated by its own water wheel which was supplied with water from a shared dam (Schubert 1957: 248).

In later European smelting plants the newly made iron was usually run into shallow, open molds of sand or loam to solidify in the form of stick- or log-shaped pigs, which were then sent to either fineries or foundries, depending on their composition. Pigs of high carbon white iron that were relatively low in silicon, phosphorus, and sulfur almost certainly went to fineries, since high quality finery pig or forge pig brought a high price and since foundries in Western countries rarely used white iron. Pig iron of intermediate quality and composition could go to either place. Foundries preferred pig that was likely to solidify as gray iron when cast. Thus, good foundry pig had a low sulfur content and fairly high silicon and phosphorus contents. It too sold for a premium price.

## 3. REDUCTION CONDITIONS FOR CAST IRON SMELTING

The conditions for producing a carbon saturated iron in a smelting furnace are defined in Appendix A (Fig. A.2). At any temperature higher than about 1150°C, the furnace gases must have a $CO/CO_2$ ratio above 90 percent. This condition exists over a large part of the interior of a modern blast furnace (Fig. 10.1). Slag under such strongly reducing conditions has a melting temperature much higher than that of bloomery slag. Furnaces must therefore be operated at temperatures above 1350°C even though reduction occurs and cast iron becomes liquid at much lower temperatures—as in the case of bloomery smelting, the operating temperature of a blast furnace is dictated by slag characteristics rather than by the smelting reaction itself. A saturated liquid iron contains about 4.8 percent carbon at 1350°C (Fig. A.1).

Theoretically, the fuel to ore and fuel to metal ratios in making cast iron should be higher than in making bloom iron. A greater consumption of carbon would appear to be necessary to keep the $CO/CO_2$ ratio high and to achieve higher temperatures in the combustion zone. In practice, this is not the case. The empirical evidence is ambiguous about fuel to ore ratios. The five bloom iron smelting systems listed in the lower part of Table 9.3 have a median charcoal to ore ratio of about 1:1; the median for the seven cast iron systems shown in

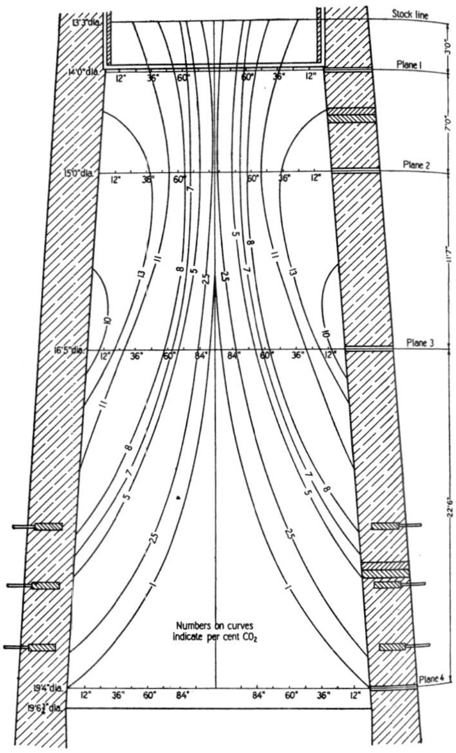


Fig. 10.1. *Patterns of Carbon Dioxide Content.* Measurements made at various points within an early twentieth-century blast furnace about 8 meters high. The horizontal line at the bottom represents the level of the tuyeres; the one at the top is the stock line, the upper limit of the column of fuel and ore. The proportion of carbon dioxide is less than 1 percent (that is, almost totally carbon monoxide) throughout a triangular zone that reaches from the tuyeres halfway to the stock line. Conditions within this zone are appropriate for reduction to cast iron. (From Kinney et al. 1927)

Table 10.1 is about the same. We have better information on recorded fuel to iron ratios. Tables 14.3a, 14.3b, and 14.4 present relevant data on 17 bloomeries and 44 blast furnaces, all of the pre-industrial period. The bloomeries have a median fuel to iron ratio of more than 5:1, but only one-tenth of the blast furnaces have fuel to iron ratios as high as this.

Admittedly, these data cannot fully resolve the issue of the fuel use efficiency of blast furnaces as opposed to bloomeries. The sample is nonrandom and the comparison is not direct, since the types of iron produced are

*TABLE 10.1. Some published fuel (charcoal)-to-ore ratios in cast iron smelting**

| Location | Period | Ratio by weight | Source |
|---|---|---|---|
| Britain | 16th c. | 1.5 to 2.5 | Hammersley 1973: 604–606 |
| | 17th, 18th c. | 1.25 | |
| Britain | 17th c. | 1.0 | Schubert 1957: 232 |
| India | 20th c. (estimated) | 0.5 | Richards 1918: 273, 274 |
| China | 20th c. | 3.0 | Tegengren 1923–24: 334 |
| Sweden | 20th c. | 0.5 | Tegengren 1923–24 |
| China | | 3.5 | Wagner 1984: Pt. I: 40 |
| China | | | Pt. II: 17–19 |

Note that these ratios vary widely partly reflecting the thermal efficiencies of larger furnaces (Britain) and partly reflecting the grade of the ore. For example the ratio of 3.0 (Tegengren 1923) represents smelting an ore whose iron content was only 40% whereas a pure hematite ore is 70 percent iron. Low grade ores use a great deal of fuel to melt the large volume of slag produced.
*For a larger compendium, see Tables 14.3b and 14.4.

different. The efficiency of the blast furnace seems much less impressive when the fuel cost of converting the cast iron to a forgeable metal is included (Chapter 13). However, the data discussed above do not seriously misrepresent the situation. The ordinary blast furnace of the pre-modern period produced a kilogram of iron with a minimum expenditure of fuel as efficiently as the average bloomery. Even though fuel costs climbed steeply when blast furnace pig was fed into a finery, the blast furnace itself was reasonably economical.

One reason for the fuel efficiency of blast furnaces was that they could be operated for long periods of time without interruption. As modern smelting experiments show (Tylecote, Austin, and Wraith 1971), a large proportion of the total fuel used in the first day of smelting was spent in heating up the furnace. Fuel consumption dropped sharply on the second day and thereafter, provided that charging and blowing did not have to be stopped more than momentarily for the removal of metal and slag.

Another reason was that blast furnaces could extract more metal from a given unit of ore, an advantage when comparing fuel costs per unit weight of metal recovered. In bloom iron smelting, a high proportion of the iron available had to go into the slag, whereas cast iron smelting slags can be completely iron-free. This difference is illustrated by figures for iron recovery from bloomeries (Table 9.2). In the examples, 10 kg. of ore containing 4.8 kg. of iron yielded between 1.3 and 2.8 kg. of iron as metal. However, in some blast furnaces (Table 14.9a, bottom), the same amount of the same ore would produce between 3.75 and 4.75 kg. of metal. Recovery efficiencies in the 90 to 95 percent range were possible even when the blast furnaces were simple and small. Relatively low fuel costs reflect both increased metal recovery and decreased slag volume.

Although blast furnace slag can have a low iron content, it will not always, but two factors favor the production of a low iron slag. First, blast furnaces are usually taller and always hotter than bloomeries. The temperature gradient within the shaft is therefore lengthened, giving more time for ore fragments in the descending charge to become completely reduced. Second, the ore fragments which do reach the slag pool in an only partially reduced condition are still likely to become converted to iron because carbon in liquid cast iron can reduce FeO when that oxide is present in the slag in appreciable concentrations. Under certain conditions this mechanism allows the recovery of iron from unreduced ore in slag to approach 100 percent.

The costs of fuel, ore, and other inputs in blast furnace smelting and the comparative economic efficiency of bloomery and blast furnace-finery systems in making low carbon iron are discussed below (Chapter 14). Here it should be emphasized that historically, regardless of the ironmaking system in use, fuel costs played a major role in technological decisions before the introduction of coke. In Europe, the perceived (if not necessarily real) fuel shortage of the sixteenth to eighteenth centuries was the stimulus for important technical innovations in the eighteenth and early nineteenth centuries, and it is likely that many changes in other ironmaking traditions have been similarly motivated. Crises in ore supply, in contrast, were unusual or unknown in most parts of the world. Few ironmasters of the pre-industrial period economized on ore, so the superior iron recovery of the blast furnace was irrelevant. Hyde's (1977: 30–31) discussion of a typical British blast furnace helps to put fuel costs into perspective:

> The Bank furnace, for example, produced 360 tons of pig iron in 1721 with operating costs of 2054 pounds [Sterling]. Of this total, 1459 pounds was spent on charcoal, 313 pounds on iron ore, and 282 pounds on all other expenses. Half of the 282 pounds went to rent (40 pounds), the clerk's salary (40 pounds), and furnace labor (61 pounds). Replacement of the hearth and bellows, plus general repairs cost 63 pounds or roughly 3% of the operating costs.

## 4. SLAGS ASSOCIATED WITH CAST IRON SMELTING

The fluidity of bloomery slags derives chiefly from the low melting temperature of the $FeO$-$Fe_2O_3$-$SiO_2$ system (Chapter 9.3; Appendix A). Magnesia ($MgO$), alumina ($Al_2O_3$), and lime ($CaO$) in combined amounts of up to 20 percent can be tolerated in these slags without raising the liquidus point much above 1200°C (Morton and Wingrove 1969, 1972; see also Figure A.11).

However, the more strongly reducing conditions needed to produce cast iron—that is, the higher temperatures and $CO/CO_2$ ratios, as well as the greater carbon content of the metal itself—limit the amount of FeO that can survive in the slag. Some will undoubtedly enter the slag pool as incompletely reduced ore or as iron that has reoxidized in the combustion zone, but it will remain in oxide form there only with difficulty because of the reducing action of the carbon in cast iron. The rate of reduction of iron oxides in blast furnace slags is limited only by sluggish rates of approach to equilibrium at the slag/metal interface. If held long enough in the furnace, all iron oxides in the slag will disappear. Hence, adequate slag fluidity depends on the polycomponent behavior of the residual oxides ($SiO_2$, $CaO$, $Al_2O_3$, and $MgO$) that derive from the gangue in the ore.

The melting characteristics of these residual oxides can be visualized by examining a phase diagram of the ternary system $SiO_2$-$CaO$-$Al_2O_3$ (Fig. A.11). A slag formed by those three elements in certain proportions can be liquid at temperatures that are only moderately high. A wide zone of ternary compositions exists within which good fluidity can be attained at about 1350°C. The addition of up to 10 percent MgO does not constrict this zone. The presence either of FeO from the ore or of $Na_2O$ and $K_2O$ from the fuel serves to widen it, increasing the probability that a given mixture of silica, lime, alumina, and magnesia will melt at about 1350°C, and lowering the melting point of some mixtures to 1300°C.

Table 10.2a summarizes the compositions of slags from 22 European and North American furnaces of the sixteenth through nineteenth centuries. The two groups of nineteenth-century slags have fairly uniform melting points; most of them cluster around 1350°C, despite marked differences in the compositions of the two groups—one comes from coke-burning furnaces and has a moderately high lime content; the other, from charcoal furnaces, has more silica and less lime. Evidently, the art of mixing ores and adding separate fluxes had advanced considerably by the time these slags were produced.

A model for slag composition during the early industrial period is based on E. Riley's elaborate analyses carried out at Dowlais in Wales in the mid-1850s. Percy (1864: 499) reproduces one of Riley's tables of the composition of slags, collected according to a careful sampling scheme, from twelve furnaces operating over a period of one week. All furnaces used coke, a mixture of several Welsh and English ores, and a hot blast; all produced white cast iron. It is unlikely that the slag composition given below would contain so much lime without a deliberate addition of limestone to the charge. The slags from the furnaces had the following composition by percent:

| | |
|---|---|
| Silica ($SiO_2$) | 41.96 ± 2.77 |
| Lime ($CaO$) | 30.83 ± 2.52 |
| Alumina ($Al_2O_3$) | 14.70 ± 1.43 |
| Magnesia ($MgO$) | 4.54 ± 0.46 |
| Iron Oxide ($FeO$) | 2.79 ± 1.42 |
| Manganese Oxide ($MnO$) | 1.21 ± 0.23 |
| Potash ($K_2O$) | 1.89 ± 0.21 |
| Sulfur | 0.92 ± 0.22 |
| Phosphorus ($P_2O_5$) | 0.15 ± 0.14 |

Most of the slags within one standard deviation of the Dowlais mean would also have had melting temperatures near 1300°C. The consistency achieved in the operation of the Dowlais furnaces undoubtedly meant that less fuel had to be used, as compared with other systems.

Older Chinese slags (200 B.C.-1150 A.D.) constitute a small and perhaps unrepresentative sample from only two furnaces (Table 10.2b). Nonetheless, they have a fairly consistent composition and closely resemble nineteenth-century European and North American slags. The lime content of the Chinese slags is as high as that of the eighteenth- and nineteenth-century European examples. It is unlikely that all this lime was present naturally in the ore or fuel, so it must have been added deliberately, either to lower the melting point of the slag or to aid in lowering the sulfur content of the metal produced (Section 6, this chapter).

The presence of unlisted fluxes would often explain the improbable slag analyses, of both European and Chinese samples, that appear in archaeological reports. Like bloomery slags, the melting point of a blast furnace slag can be significantly lowered by the presence of potash or soda, or less commonly boron (as oxides) and fluorine (as fluorides). Iron and manganese oxides also have strong fluxing effects at the temperatures used to make cast iron.

Other nineteenth-century slags are unusual in composition when compared with those of Dowlais and other

*TABLE 10.2a. Analyses of European and American blast furnace slags*

| Century | $SiO_2$ | CaO | $Al_2O_3$ | MgO | FeO | MnO | Other | Melting Point |
|---|---|---|---|---|---|---|---|---|
| | | | Charcoal Furnaces, 16th–18th c. | | | | | |
| 1. 16–17th | 49.7 | 11.9 | 23.2 | 7.2 | 4.4 | 3.3 | | 1400°C |
| 2. 16–18th | 45.3 | 22.8 | 22.5 | 3.7 | * | 1.2 | 3.7 $Fe_2O_3$* | 1375°C |
| 3. 17th | 49.7 | 14.2 | 18.2 | 4.8 | 6.1 | 4.8 | | 1325°C |
| 4. 17th | 49.3 | 22.8 | 11.4 | 12.0 | 2.7 | 0.8 | 2.0 $K_2O$ | 1325°C |
| 5. 17th | 62.8 | 15.9 | 7.3 | 8.4 | 4.8 | 0.4 | | 1400°C |
| 6. 17th | 47.8 | 7.6 | 24.0 | 3.9 | * | 0.6 | 9.3 $Fe_2O_3$* | 1525°C |
| 7. 17–18th | 45.9 | 18.4 | 19.1 | 9.2 | * | 3.0 | 2.4 $Fe_2O_3$* | 1325°C |
| 8. 18th | 41.6 | 14.1 | 22.7 | 14.2 | * | 3.0 | 2.6 $Fe_2O_3$* | 1400°C |
| 9. 18th | 57.8 | 0.7 | 18.6 | — | 16.2 | nd? | 2.7 $K_2O$ | >1600°C |
| 10. 18th | 52.5 | 17.0 | 20.2 | 4.6 | * | 1.9 | 4.3 $Fe_2O_3$* | 1325°C |
| 11. 18th | 61.4 | 22.2 | 7.4 | 0.9 | 2.9 | 1.2 | | 1400°C |
| 12. 18–19th | 56.4 | 14.6 | 12.4 | 3.6 | 2.6 | 9.8 | | 1325°C |
| | | | Charcoal Furnaces, 19th c. (U.S.) | | | | | |
| 13. 19th | 51.3 | 16.0 | 9.9 | 17.3 | 4.0 | 0.6 | | 1350°C |
| 14. 19th | 52.0 | 18.0 | 12.0 | 14.9 | 3.0 | 0.6 | | 1350°C |
| 15. 19th | 54.7 | 19.0 | 18.2 | 1.9 | 3.1 | 1.9 | | 1350°C |
| 16. 19th? | 48.5 | 26.5 | 16.0 | 3.7 | 1.0 | 1.7 | | <1300°C |
| 17. 19th | 65.1 | 15.6 | 7.4 | 4.9 | 3.8 | 0.2 | 3.1 $K_2O$ | 1450°C |
| | | | Coal and Coke Furnaces, 18th–19th c. | | | | | |
| 18. 18th | 38.1 | 41.1 | 11.2 | 2.5 | — | 1.9 | | 1375°C |
| 19. 19th | 39.5 | 32.5 | 15.1 | 3.5 | 2.0 | 2.9 | 1.1 $K_2O$ | 1325°C |
| 20. 19th | 44.6 | 29.2 | 18.1 | 1.7 | 2.6 | 2.6 | | 1325°C |
| 21. 19th | 46.0 | 25.3 | 17.0 | 4.0 | 2.0 | 1.5 | | 1350°C |
| 22. 19th | 43.6 | 28.9 | 20.4 | 1.1 | 3.7 | 0.3 | | 1325°C |

* = iron oxide as $Fe_2O_3$

Melting points inferred on basis of $SiO_2$-$Al_2O_3$-MgO-CaO content.

*Ores with $Fe_2O_3$ content: The use of "$Fe_2O_3$" in specifying the iron oxide content of slag conforms to British conventions; in most cases this should be interpreted as the equivalent amount of FeO.

Names of Furnaces:

1. Cannock
2. Rievaulx
3. Saugus
4. Sharpley Pool
5. Coed Ithil
6. Himley
7. Rockley
8. Melbourne
9. Low Mill
10. Charlcot
11. Courrance
12. Duddon
13. Hopewell
14. Joanna
15. Willroy
16. Trumbull
17. Hammond
18. Bradley
19. Tipton
20. Dowlais
21. Decazeville
22. Ebbw Vale

Sources: Morton and Wingrove (1969: 55, 57)—1, 2, 7, 10, 12, 19, 22; Tylecote (1976: 82, 109)—4, 5, 6, 8, 9, 18, 20, 21; White (1980: 60)—3, 13–16; Mutton (1976: 65)—11; Maynard (1874: 120)—17.

large English ironworks of the period (Table 10.2c). Several are from furnaces that used manganiferous siderite ores and produced *Spiegeleisen*, cast iron with a high content of manganese. These slags are in a sense intermediate between bloomery slags and Dowlais-type blast furnace slags. Although all the furnaces were true blast furnaces (and not *Stückofen*), their operating temperatures were not high enough to avoid the loss of much manganese to the slag. The presence of manganese oxides meant that the usual constraints on slag composition were substantially lessened. As the examples from Siegen and Styria show, lime could be either dispensed with entirely or added in moderate amounts in order to improve the yield of iron.

A reasonably complete understanding of slags was not gained until near the middle of the twentieth cen-

*TABLE 10.2b. Analyses of Chinese blast furnace slags, 200 B.C.–A.D. 1150*

| Period | $SiO_2$ | CaO | $Al_2O_3$ | MgO | FeO | MnO |
|---|---|---|---|---|---|---|
| 23. Han | 54.4 | 24.0 | 13.2 | 2.1 | 2.4 | 0.3 |
| 24. Han | 53.7 | 22.7 | 12.1 | 2.5 | 3.7 | 0.6 |
| 25. Han | 51.5 | 27.5 | 11.8 | 3.3 | 1.5 | 0.4 |
| 26. Han | 49.2 | 25.6 | 11.0 | 2.6 | 2.9 | 0.4 |
| 27. Song | 48.7 | 23.2 | n/a | 9.6 | nd? | 0.2 |
| 28. Song | 52.8 | 24.5 | n/a | 9.3 | nd? | 0.3 |
| 29. Song | 43.8 | 19.6 | n/a | 10.9 | nd? | nd? |
| 30. Song | 54.7 | 25.2 | n/a | 11.7 | nd? | nd? |

Key: n/a = not available, nd = not determined
Sources: 23–26—Henan Provincial Museum 1978: 9, 27–30—Yang 1982: 172.

*TABLE 10.2c. Analyses of contemporary blast furnace slags from 19th-century sources*

| | $SiO_2$ | CaO | $Al_2O_3$ | MgO | FeO | MnO |
|---|---|---|---|---|---|---|
| Overman (1850: 251) | | | | | | |
| United States[1] | 51.8 | 21.8 | 15.2 | 4.8 | 3.7 | 1.1 |
| United States[2] | 63.6 | 24.0 | 3.8 | 1.2 | 1.7 | 3.9 |
| Sweden | 31.1 | 14.1 | 8.9 | 34.2 | 1.0 | 4.4 |
| France | 52.0 | 30.2 | 5.0 | 5.2 | 1.6 | 4.7 |
| Savoy (Italy)[3] | 71.0 | 7.2 | 2.5 | 5.2 | 5.0 | 6.5 |
| Styria[4] | 37.8 | — | 2.1 | 8.6 | 21.5 | 29.2 |
| Wales[5] | 40.6 | 32.2 | 16.8 | — | 10.4 | — |
| Percy (1864: 500–503) | | | | | | |
| Rothau, Alsace[6] | 70.2 | 20.4 | 6.4 | — | 0.2 | 2.7 |
| Rothau | 35.8 | 2.4 | 4.2 | 4.1 | 21.2 | 29.6 |
| Rothau | 54.2 | 24.6 | 6.8 | 0.3 | 9.2 | 4.8 |
| Peitz, Saxony[7] | 70.1 | 19.7 | 6.3 | 0.7 | 1.5 | 1.4 |
| Sjogren, Sweden[8] | 61.1 | 19.8 | 5.4 | 7.1 | 3.3 | 2.6 |
| Siegen, Westphalia[9] | 49.6 | — | 9.0 | 15.2 | 0.4 | 25.8 |
| Siegen[10] | 48.4 | — | 6.7 | 10.2 | 0.1 | 34.0 |
| Siegen[11] | 43.4 | 18.1 | 12.2 | 4.5 | 5.8 | 4.0 |
| Eisenerz, Styria[12] | 40.6 | 11.9 | 9.8 | 2.4 | 25.6 | 8.9 |
| Edsken, Sweden[13] | 46.4 | 38.6 | 4.3 | 7.4 | 1.0 | 1.9 |

All analyses are from charcoal furnaces except number 5 below.
[1]"Average cinder of good ore"; [2]"Derived from a furnace smelting bog ore"; [3]"The result of bad work at the furnace"; [4]Overman says this is from Siegen but has confused with it another analysis; Bauerman (1882: 155) cites von Mayrhofer as the source, noting that it is Styrian and from the smelting of white cast iron. [5]An aberrant slag by contemporary Welsh standards, but Overman makes no comment. [6]The three Rothau analyses are Klasek's; the ores are sphaerosiderites. [7]Analysis by Karsten; bog ore. [8]Analysis by Berthier; magnetite ore. [9] By Karsten; "from spathic and brown iron ore; furnace working very hot; grey iron." [10]By Karsten. "Furnace working very well; the product spiegeleisen. [11] Perhaps by Lindnauer; gray iron. [12]By Mrazek; from spathic ore. [13]By Ullgren; "excellent pig-iron suitable for making Bessemer steel."
Overman took his analyses from J. H. Alexander's *Report on the Manufacture of Iron* (Baltimore 1840s). Percy took the ones shown here from Lindnauer's *Compendium der Hütten-Chemie* (Prague 1861).

tury, but we should not assume that every aspect is recent. A scientific approach to slag composition dates to the early nineteenth century in Western countries, but practical ironmakers evidently had some conscious control over slag behavior before then. Overman (1850: 227) clearly summarizes his understanding of the behavior of a polycomponent slag:

> Silex [silica], in combination with clay or magnesia, will not melt, but a mixture of a given amount of alkali [potash, soda], magnesia, and clay with a given amount of silex or acid, is fusible. If to the above two silicates we add a third silicate in itself either infusible or strongly refractory, say silicate of lime, the whole mixture is melted at a lower temperature than that at which any two of them will melt.

Overman knew about the fluxing actions of various substances, particularly iron oxides which he naturally wanted to minimize in slag in order to increase the yield of metal. He recommended strongly that furnace operators attempt to achieve a free running slag by mixing lime and clay (or preferably calcareous and clayey ores) with siliceous ores rather than by adding large numbers of separate ingredients (Overman 1850: 226):

> In most cases, artificial fluxes are too expensive for use. In fact, those fluxes are unnecessary, because we can produce almost any degree of fusibility we desire by means of lime, clay and silex. All other materials which serve as fluxes are too small to be entitled to notice, and impracticable for general application; such as soda, potash, manganese and magnesia.

Much of what Overman and his contemporaries knew about slags must have been understood by the eighteenth century or even earlier. The deliberate addition of lime to the charge of blast furnaces was already widespread in Europe by the seventeenth century and in China long before that. The addition of bloomery slags (i.e., of iron oxides plus iron silicate) was also common in Europe by the seventeenth century and must often have been beneficial to slag quality and thus to the yield of metal.

Sometimes a single ore smelted with a given type of charcoal or coal would have fortuitously yielded a slag with a workably low melting point. Figure 10.2 shows types of ore and coal ash plotted on a ternary $SiO_2$-CaO-$Al_2O_3$ diagram. One or two of the ores involved would be fully "self-fluxing"—that is, they would form a free running slag at blast furnace temperatures without any additives. Others would have a low enough melting point if mixed with a second type of ore.

There are four simple and direct ways to adjust the fluidity of a blast furnace slag that is too high in alumina and silica. The first, adding limestone or other lime-containing materials, has been mentioned. It is the best solution where calcareous materials are available, although instances are known—for instance, in Angat in the Philippines (McCaskey 1903b)—where limestone is abundant and yet local furnace operators did not use it. The second is to add large amounts of $K_2O$ or $Na_2O$ in the form of plant ash or minerals like feldspar; this was rarely done, possibly because of the corrosive effects of the slags on furnace walls. The third is to add manganese oxide or a manganiferous iron ore. Unless furnace temperatures are very high, most of the manganese will enter the slag in the form of MnO, which has a fluxing effect similar to that of iron. In many cases, the reason for adding manganese must have been to produce Spiegeleisen, manganiferous cast iron, but some blast furnace slags contained enough MnO to affect their melting points strongly (Table 10.2c).

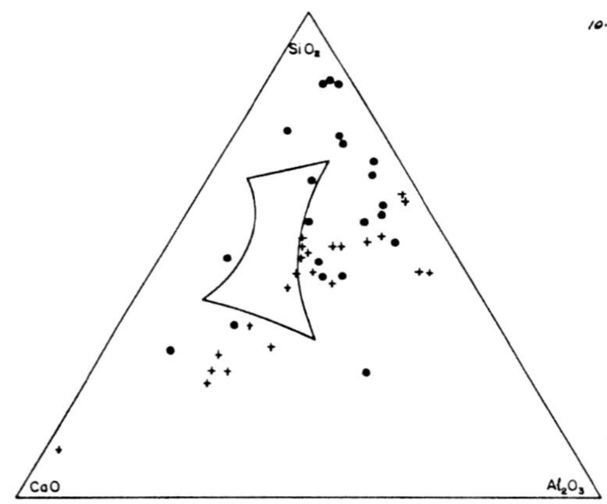


Fig. 10.2. *Melting Points of Ore and Coal Ash.* Various types of iron ore (solid circles) and coal ash (crosses), plotted onto a triangular composition scale. The three corners of the diagram represent alumina ($Al_2O_3$), lime (CaO), and silica ($SiO_2$), and only the 1350°C contour has been drawn. Certain ores and coals would produce a "naturally" low melting slag if used together. The majority, however, contain too much $SiO_2$ and $Al_2O_3$; these would have to be fluxed either with CaO or with FeO unless the furnace could reach temperatures in the 1400° to 1600°C range.

The fourth method of improving the fluidity of a slag is to increase the fragment size of the ore in the charge, so that more FeO than usual is carried past the combustion zone into the slag pool. This may have been the first fluidity improving technique in cast iron smelting. It continued to be used fairly frequently, for blast furnace slags with a significant FeO content were relatively common through the nineteenth century. Charging large ore fragments is such an easy way to solve the slag fluidity problem that many furnace operators who used limestone might have deliberately not broken up the ore as finely as would otherwise have been desirable. A reduction in metal yield by a few percent would have been preferable to having viscous slag block the furnace. In furnaces where no limestone was added, the technique of using large fragments of ore had the advantages of dependability and of low temperature operation, since FeO-rich slags remain liquid at temperatures below the lower limit for normal $SiO_2$-CaO-$Al_2O_3$ slags.

The only disadvantages of using a FeO-rich blast furnace slag are the loss of metal and the need to replenish the iron oxide frequently so that its content in the slag does not become significantly reduced by reaction with the carbon at the slag/metal interface. McCaskey's (1903b) description of smelting at Angat provides relevant figures on the metal loss:

| | $Al_2O_3$ | MgO | $SiO_2$ | $Fe_2O_3$ | Fe |
|---|---|---|---|---|---|
| Ore: | 2–12% | 0.1–2% | 6–14% | 50–90% | |
| Slag: | 24% | 2% | 42% | 27% | + 1.7% |

The ores analyzed by McCaskey seem to have been those regularly used at the furnaces, and he states that they are "self-fluxing"; the slag was tapped every two or three minutes. The furnace produced perhaps half a ton of iron per day, which was cast directly from the furnace to form points and shares for plows. McCaskey implies that the operation, performed by Chinese workers using traditional methods, was still moderately profitable in the early 1900s because good ore and charcoal were inexpensive in that part of Bulacan.[3]

Ironmaking operations producing similar slags may have been fairly widespread earlier in East and Southeast Asia—furnaces like the one at Bulacan are still used in several parts of China (Wagner 1985), though they are often run in more conventional ways and presumably produce more conventional slags. High iron slags from these furnaces would be easy to mistake for slags from bloomeries, as a comparison of McCaskey's Bulacan slag sample with one of the slags from a furnace of the Dimi in Ethiopia shows (Todd and Charles 1978: 75):

| | $Al_2O_3$ | CaO | MgO | $SiO_2$ | $Fe_2O_3$ | FeO |
|---|---|---|---|---|---|---|
| Dimi: | 16% | 3% | - | 48% | - | 27% |
| Bulacan: | 24% | nd | 2% | 42% | 27% | - |

The "$Fe_2O_3$" in the Bulacan sample is a convention in reports by mining oriented laboratories; in fact, the iron oxides were most likely present as FeO or $Fe_3O_4$.

It is not always easy to draw conclusions from analyses because slags from a bloomery may be almost identical in composition with slags from a blast furnace, even when neither of the furnaces is unusual in design or in any sense intermediate in type. The Dimi furnace is an ordinary East African bloomery, consisting of a pit about 40 cm. deep with a 60 cm. shaft placed on top, whereas the Bulacan device, more than 2 meters high, belongs to a common class of Chinese trumpet blast furnaces which are known to produce only cast iron. A high FeO content suggests that the slag came from a bloomery, but it does not prove it since true blast furnaces can be run so that they produce similar slags.

## 5. WHITE AND GRAY CAST IRONS

Carbon in cast iron can take the form of either iron carbide (cementite) or graphite, and the proportions of these determine whether the metal is white, mottled, or gray (Chapter 2). When carbon contents are between three and four percent, which is the most common range, the metal can be in any form between gray and white, depending on cooling rate and on the presence or absence of sulfur, phosphorus, and silicon.

The phosphorus and silicon play a double role. First, they limit the solubility of carbon in iron while smelting is under way. A cast iron with about 1.5 percent phosphorus plus silicon will not have a carbon content higher than about 3.8 percent. Second, the two elements act in opposition to the effects of sulfur and a fast cooling rate. When freezing is very fast or the sulfur content above about 0.5 percent, white iron will always be produced. If freezing is slow and the sulfur content lower and if enough silicon and phosphorus are present, the iron will freeze as mottled or gray.

Foundrymen may have chosen casting metals because of phosphorus as well as silicon content, not on the basis of chemical analysis but rather empirical knowledge of the behavior of irons made in certain ways from certain ores. In Figure 10.3a, securely dated iron castings are plotted according to carbon contents on one axis and combined silicon and phosphorus contents on the other. A comparison of this graph with Figure 2.7 shows that adding phosphorus to silicon gives a tighter clustering of data points than using silicon alone. Figure 10.3b helps to understand how a composite of factors controls the structure and properties of a cast iron.

## 6. REDUCTION TO SILICON

The high ratios of CO to $CO_2$ and the high temperatures in the combustion zone of a cast iron smelting unit create appropriate conditions for the reduction of oxides other than those of iron. The most ubiquitous of these is silicon dioxide or silica, $SiO_2$. Its reduction yields silicon, an element that is readily soluble in both liquid and solid iron.

Because of the high temperatures required to reduce silica, it is likely to be already part of a slag when reduction begins. The silica from the ore begins to form a slag with iron oxides at about 1200°C. As the temperature rises, other oxides—including alumina, lime, and magnesia—dissolve into the slag. It becomes fully fluid at about 1300° to 1400°C, depending on the proportions of other oxides, and starts to drip downward through the charge. As it encounters particles of charcoal or coal, it forms a thin film over them. The physical contact between solid carbon and liquid slag sets up conditions for reduction to elemental silicon.

This happens only slowly at first. But above 1350°C, silicon begins to be produced in measurable quantities and is rapidly dissolved by any liquid iron with which it comes in contact as the charge continues to descend through the shaft. At temperatures in the 1600°C range of modern blast furnaces, enough silicon is reduced for the iron to reach silicon levels of several percent while still in the furnace shaft. At lower temperatures, however, most silicon reduction occurs subsequently at the interface between the pools of liquid slag and iron at the bottom of the furnace. There, silica is converted to silicon by reaction with the dissolved carbon in the iron. If tapping is delayed long enough, the iron can pick up considerable amounts of silicon even in furnaces where combustion zone temperatures do not rise above 1350° or 1400°C.

It might seem that approximate levels of silicon in

[3]Smelting and casting iron in Bulacan province are still profitable. The authors, during their visit in 1987, saw two ironworks in the town of Angat where blast furnaces were in operation. As in McCaskey's day, the smelted iron is poured directly into molds for plow points that continue to be popular with local farmers. The furnaces have capacities of about 1 ton/day.

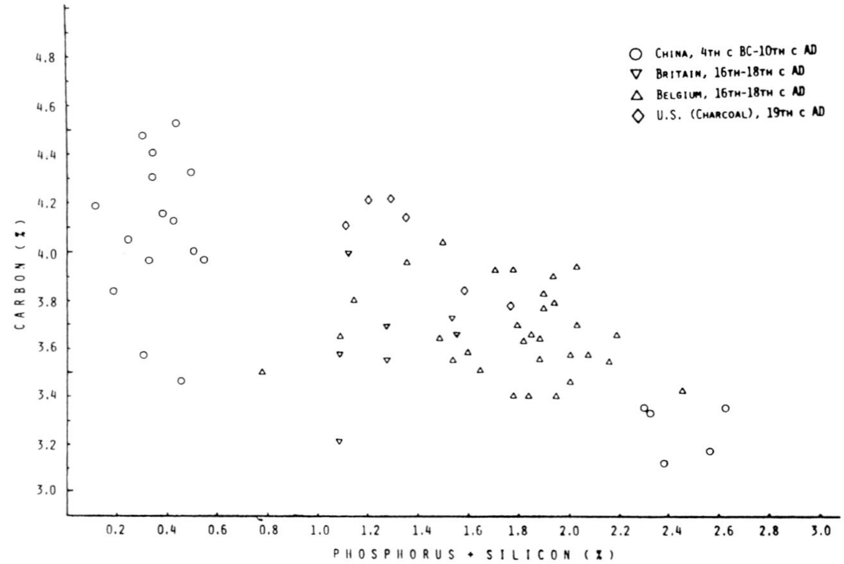


Fig. 10.3a. *Analyses of Dated Iron Castings.* The horizontal axis represents added percentages of silicon and phosphorus; the vertical axis carbon content. Irons at the far left are white and those at the far right, gray. All castings shown have been fairly securely dated; examples listed in the sources as "18th century?" and so forth were omitted from the diagram. The diagonal alignment shows the approximate position of the carbon-silicon-phosphorus eutectic, which roughly defines the carbon saturation limit in liquid iron. Sources: China, 4th c. B.C.–1st c. A.D.: (Henan Provincial Museum 1978: 7–10; Li 1975: 8; Henger 1970: 47); China, 6th–10 c. A.D.: (Pinel et al. 1938: 178); Belgium and Luxembourg, 16th–18th c. A.D. (Evrard and Descy 1948: 203–204); Britain, 16th–18th c. (Schubert 1957: 398); United States, 19th c. [charcoal] (Schenck 1983: 109)

early cast iron could be inferred solely from data on the relevant slags, but this is impossible, for there is no way of estimating one of the four variables involved. Three of them can be derived from slag analysis: the furnace temperature, the silica content of the slag, and its lime-to-silica ratio. The lime content of a slag affects the reducibility of silica to silicon (Fig. 10.4); all else being equal, a lower lime-to-silica ratio means more silicon in the iron. The problem is that the fourth variable, the time that the slag and iron remain in contact, depends on the rate of descent of the charge through the shaft and the length of the interval between the times the iron is tapped. These cannot be estimated through examining a sample of slag.

An example can clarify the complex relationship among the factors controlling silicon content. Assuming that a slag has a lime-to-silica ratio of 0.28:1, an overall silica content of about 50 percent, and a melting point of about 1350°C, the ratio between the percentage of silicon in the iron to the percentage of silica in the slag is about 0.064 at 1350°C (Fig. 10.4). Multiplying this by the 50 percent silica content of the slag in the example gives a potential total silicon in iron content of 3.2 percent, which would be high even in a gray cast iron of the modern period. Dwell time becomes a factor at this point, for few early ironmasters would have left their furnaces untapped long enough for silicon content to

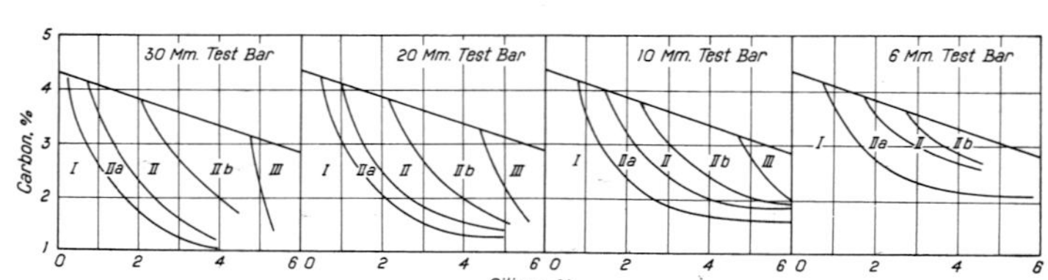


Fig. 10.3b. *Relationship between Thickness (and Cooling Rate) on the Structure of Cast Iron.* A present-day correlation between three cast iron production factors—carbon content, silicon content, and the casting thickness—and the resultant solidification structures developed in sand molds. The structure designation codes are as follows: I = white iron; IIa = mottled iron; II = pearlitic gray iron; IIb = mixed pearlitic-ferritic gray iron; III = ferritic gray iron. (Charts from LaPlanche 1947)

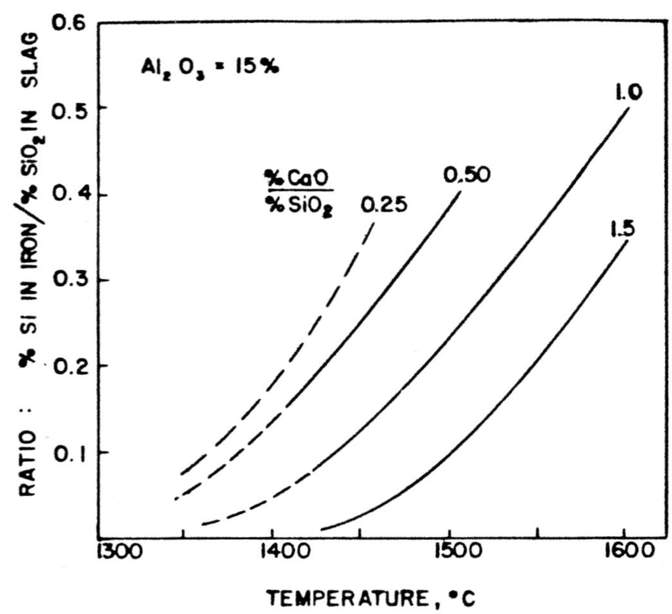


Fig. 10.4. *Lime in the Slag and the Silicon Content of Iron.* The presence of lime complicates the silicon reduction process. Low and medium lime slags, in which the $CaO/SiO_2$ ratio is in the 25 to 50 percent range, will inhibit silicon reduction at the temperatures reached by most early furnaces. High lime slags will have a strongly inhibiting effect unless the furnace is run at 1500°C or above. (From McGannon 1964: fig. 12–142)

approach its potential maximum. Even at 1400° to 1500°C, nearly 24 hours are needed to reach maximum levels of silicon in cast iron (McGannon 1964: fig. 12–144). At 1350°C, with the furnace tapped at normal six to eight hour intervals, the amount in the iron actually produced will be well below the upper limit.

In general, most modern gray cast irons contain about one to two percent silicon (in Table 10.3 are analyses of cast irons with the highest and lowest silicon contents given in published data). It is generally believed that the introduction of coke as a smelting fuel was the principal cause of the increase in the amount of silicon in European and North American cast irons of the nineteenth century—before 1700, known examples range from 0.5 percent to a high of less than 2 percent, whereas coke smelted irons made after 1800 rarely have below 1 percent and often as high as 3 to 4 percent.[4] During the nineteenth century, charcoal pig averaged only 0.39 percent silicon, whereas coke pig averaged about 2.0 percent (Morton 1966).

Although this average for nineteenth-century charcoal smelted iron seems low, there is no doubt that the use of coke is closely associated with an increase in silicon content. However, several facts suggest that this relationship is not direct and not in itself causal. First, high silicon charcoal irons did exist, and so did coke irons with a fairly low silicon content (Table 10.3). Second, the trend toward high silicon cast iron had already begun in Europe well before coke came into use for smelting, possibly as early as the sixteenth century (Fig. 2.7). European ironmakers may have been becoming more aware of the desirability of (necessarily high silicon) gray irons for the manufacture of those types of castings which, like cannon, had to be both strong and machinable. Finally, the Chinese experience seems not to parallel the European one. Analyses of datable Chinese cast irons are still scarce, and none indicates a clear correlation between the presumed use of coal or coke in smelting and the production of high silicon cast iron.[5] Perhaps because they made more white iron, and perhaps also because they used short as well as tall shaft furnaces, Chinese ironmakers succeeded in producing low silicon pig iron with coke and charcoal pig containing almost no silicon (Fig. 10.3a).

Therefore, we think that long term changes in the silicon content of European cast irons, to the extent that such changes are not an illusion caused by a small and biased sample, can be explained by the interaction of factors other than the choice of fuel: 1) an increasing need for gray iron, with its greater strength and good machineability, 2) an improved ability to control the smelting process, 3) a tendency to build taller and hotter furnaces, in a conscious effort to improve efficiency, and 4) the use of the hot blast. The introduction of coke and anthracite as furnace fuels occurs in the context of these factors. Both fuels undoubtedly could have produced a metal with a low silicon content if burned in a sixteenth-century European furnace or in a Chinese-type short shaft furnace. Similarly, charcoal burned in a tall, high temperature furnace like those used in eighteenth-century Russia and nineteenth-century America would have yielded a cast iron with silicon levels within the range of the cold-blast coke pig of the same period.

On the other hand, there is no doubt that in western Europe the coincidence of large furnaces, heated blasts, and mineral fuels created a problem for ironmakers of the nineteenth century. Contemporary furnace operators, most of whose customers needed a low silicon white cast iron suitable for conversion to wrought iron in the finery and puddling furnace, would have considered a high silicon content a disadvantage. The sulfur in the first coke smelted irons must sometimes have been present in sufficient quantities to overcome the influence of the silicon, forcing the metal to solidify as white iron, but that made the product unsuitable for fining and puddling. The European world greeted Abraham Darby's discovery of smelting with coke unenthusiastically partly because it it did not know what to do with cast iron high in both sulfur and silicon. His firm's initial solution, foreshadowed by many centuries of Chinese experience, was to specialize in the production of castings for undemanding domestic applications like cooking, where white or mottled iron served as well as gray. Later, the Darbys and others worked toward the removal of silicon and sulfur from the iron, efforts which stimulated some of the principal innovations in eighteenth- and nineteenth-century ironmaking.

## 7. SULFUR PARTITION BETWEEN METAL AND SLAG

Although sulfur has no recognized effect on the fluidity of cast iron, it does play an important role during solidification, encouraging the formation of white iron rather than gray (Fig. 2.9). Even more important are the limits sulfur in cast iron impose on the forgeability of wrought irons and steels. With charcoal fueled smelting furnaces, the chief sources of sulfur were the iron ores themselves, from which it could be largely eliminated if the ores were roasted (Chapter 4). With the advent of coke, however, a major new source of sulfur was introduced to the ironmaking process, and the sulfur content of ordinary cast iron increased markedly in those parts of Europe and North America where mineral fuels were used. This was not extremely deleterious because, fortunately, lime in significant quantities had already become an ingredient of the normal blast furnace charge.

In the smelting furnace, sulfur can partition to both the metal and the slag. The proportion that goes to the slag correlates with the amount of lime and magnesia present. Both of these substances react with sulfur to

[4]For silicon in early European cast irons, see Schubert 1957: 398, Tylecote 1976: 84, Pfannenschmidt 1977: 215, Evrard and Descy 1948: 198–255, and Figure 2.8 here. For cast irons of the nineteenth century, see the analyses in Percy 1864: 532–552. Almost all the analyses of earlier objects are of firebacks and stove plates. Those reported by Pfannenschmidt include several hundred pieces of varying date; only a handful of these have silicon contents higher than 1.4 percent. Over a hundred dated examples are reported by Evrard and Descy, including early eighteenth-century German stove plates with 1.6 to 1.7 percent silicon.

[5]The thirteen analyses of Zhou and Han period cast irons (Henan Provincial Museum 1978: 7–10; Li 1975: 8) show silicon contents between 0.04 and 0.28 percent. The eight analyses in Pinel et al. (1938: 178) include five examples in the 2.0 to 2.42 percent range, but all of these are types of objects which would perform better if made of gray iron.

*TABLE 10.3. High and low silicon contents of cast irons*

| | (Place, Date) | Silicon | Carbon | Sulfur | Phos-phorus | Man-ganese |
|---|---|---|---|---|---|---|
| High | | | | | | |
| Coal/Coke | | | | | | |
| Hot Blast | (Britain, 19th c.) | 5.73 | 4.41 | 0.12 | 0.13 | 1.33[1] |
| Cold Blast | (Britain, 19th c.) | 3.45 | 1.70 | 0.44 | 1.30 | tr[2] |
| Charcoal | | | | | | |
| Hot Blast | (Germany, 19th c.) | 3.21 | 4.15 | tr | 1.22 | tr[3] |
| Cold Blast | (Germany, 18th c.) | 1.70 | 3.65 | 0.06 | 0.85 | 0.15[4] |
| Fuel not known | | | | | | |
| Cold Blast | (China, 5th c. A.D.) | 2.42 | 3.35 | 0.13 | 0.21 | 0.13[5] |
| Low | | | | | | |
| Coal/Coke | | | | | | |
| Hot Blast | (U.S., 19th c.) | 0.83 | 2.89 | 0.01 | — | nd[6] |
| Cold Blast | (Britain, 19th c.) | 0.57 | 2.81 | 0.06 | 0.29 | 0.13[7] |
| Crucible | (China, 16th c.) | 0.33 | 4.7 | 0.50 | 0.34 | nd[8] |
| Charcoal | | | | | | |
| Cold Blast | (China, 20th c.) | 0.01 | 3.02 | 0.04 | 0.42 | nd[9] |
| Cold Blast | (Germany, 16th c.?) | 0.38 | 3.96 | nd? | 0.22 | 1.59[10] |
| Fuel not known | | | | | | |
| Cold Blast | (China, 1st c. B.C.) | 0.02 | 4.05 | 0.06 | 0.22 | 0.03[11] |

[1]Percy (1864: 535): Tow Law "glazed pig," suited for Bessemer Process. [2]Percy (1864: 540): made at Wellingborough with uncoked Yorkshire coal. [3]Overman (1850: 485), gray pig iron smelted with hard charcoal and hot blast of 400°C. [4]Evrard and Descy (1948: 224), stove plate from Geislautern in the Palatinate. [5]Pinel et al. (1938: 178): gray iron statue with cast-on date of A.D. 550. [6]Percy (1864: 539), after Svanberg: anthracite smelted pig iron from Williamsburg, Pennsylvania. [7]Tylecote (1976: 109), after Percy: cold blast forge pig. [8]Analysis for present authors by Michael Notis, 1985; white iron bell from Pingyang in Shanxi with cast-on date, almost certainly smelted with anthracite in a crucible. [9]Wagner (1985: 52–57): demonstration smelt at traditional furnace in Henan Province; manually operated bellows; hearth temperature 1570°C and slag temperature 1400°C. [10]Pfannenschmidt (1977: 37), after Rathgen: cannon chamber found in Friedberg, Oberhessen. [11]Li (1975: 8): western Han Period, white iron, from Mancheng.

form calcium and magnesium sulfides or sulfates which are soluble in slag and which in that form will not dissolve in iron. If the lime-magnesia content of the slag is high enough, it can even extract sulfur from the droplets of liquid metal falling through the pool of slag and from the cast iron pool underneath. In this respect a blast furnace has a greater potential to produce a low sulfur iron than a bloomery, for low temperature slags of the $FeO$-$Fe_2O_3$-$SiO_2$ type will not dissolve any of the sulfur that might be present, dumping it entirely into the metal.

The "high lime" slag is an essential feature of modern blast furnace technology (see Figure 10.5 for the effect on the levels of sulfur in the carbon saturated, liquid metal). As a basis for comparison, a normal composition for an early blast furnace slag, estimated from Table 10.2a-c, is 50 percent silica, 19 percent lime, 15 percent alumina, and 7 percent magnesia, while a modern blast furnace might contain 35 percent silica, 41 percent lime, 13 percent alumina, and 8 percent magnesia. The significant difference is the relative proportion of lime plus magnesia to silica; in modern slags this may exceed 1.5:1. Such a slag will force the sulfur content of a cast iron down to less than 0.01 percent (Fig. 10.5), whereas a typical slag of the seventeenth through mid-nineteenth centuries, with a lime-silica ratio of about 1:2, cannot reduce the sulfur content of the metal below about 0.06 percent. Even that degree of sulfur reduction will not be reached unless the liquid slag and metal remain in contact at active temperatures for a relatively long time. This is why many European cast irons made when coke was first adopted have sulfur contents between 0.10 and 0.40 percent.[6] Although most were smelted with a limestone flux, not enough lime was used for all the sulfur present to pass to the slag before the iron was tapped.

Sulfur contents in pre-modern cast irons vary widely from as high as 3.75 percent (Overman 1850: 487) to as low as 0.01 percent (Evrard and Descy 1948: 204). In general the higher sulfur levels reflect the use of coal or coke and the lower levels, that of wood charcoal as the smelting fuel in the blast furnace. There are, however, a few coals as low in sulfur as 0.03 percent[7] (Percy 1865: 550), and some charcoal pig iron ran sulfur levels as high as 0.40 percent (Evrard and Descy 1948: 203).

The pyrite content of the iron ore is another factor affecting the sulfur content of the cast iron product despite the use of low sulfur fuel. We rarely have data on ores to help explain the sulfur levels in the product metal. Moreover, only occasionally is there information about whether the iron ore was roasted before smelting. Roasting induces thermal decomposition of the $FeS_2$ to $FeS$ and the evolution of sulfur vapor before it can be absorbed by reduced iron.

Limestone additions to the furnace charge caused the removal of sulfur into the slag by the solubilization of calcium sulfide or sulfate. While the sulfur absorption capacity can be approximated from the slag composition, the action of an efficient slag reaction leaves no trace in the metal. So a low sulfur cast iron can mean a low sulfur coal/coke and a low sulfur iron ore or a lime/magnesia level in the slag which acts strongly to desulfurize the metal. There is no indication that manufacturers of iron before the early nineteenth century knew that lime and magnesia in the blast furnace slag

[6]See, e.g., Percy 1864: 540, 542, 547, etc.; the cokes of South Wales seem to have been especially high in sulfur.

[7]Analyses of 193 German stove plates presented by Pfannenschmidt (1977: 215) show a mean sulfur content of only about 0.03 percent. Over fifty firebacks and stove plates from Belgium, Luxembourg, and France were analyzed by Evrard and Descy (1948: 203–204); they have a similar average sulfur content. The thirteen Belgian and French cast iron pots for which the same authors give sulfur analyses have a mean content of about 0.06 percent (Evrard and Descy 1948: 234). Thus, at 0.08 percent, Schubert's Sussex firebacks are high in sulfur.

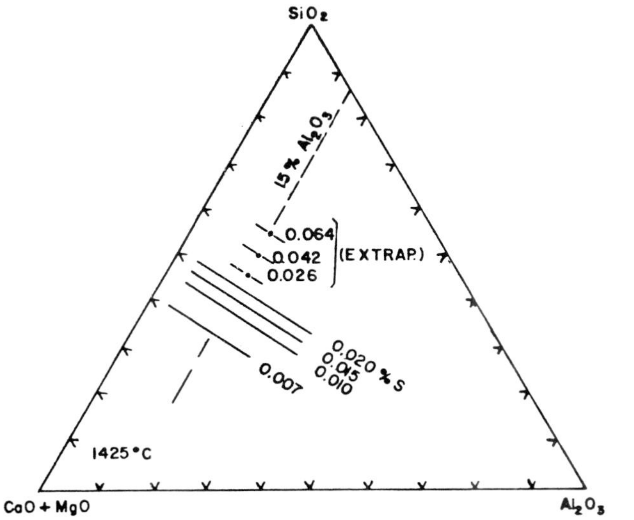


Fig. 10.5. *Lime in the Slag and the Sulfur Content of Iron.* The solid lines indicate experimentally established saturation sulfur levels in modern pig irons; the dashed lines, sulfur levels that have been extrapolated in order to match the lower lime/silica ratios of early blast furnace slags. Thus, for example, with the furnace running at 1425°C, a silica content of 55 percent, and an alumina content of 15 percent in the slag, the expected sulfur content of the metal could reach 0.064 percent. Lime and magnesia are considered additive in their influence on sulfur. (From McGannon 1964: fig. 12–164)

reduced sulfur,[8] since lime was usually added to lower the melting temperature of the slag.

We do not know if a high lime/magnesia content in the slag was intrinsic to the coal and iron ore or represented a deliberate addition of limestone or dolomite. In general if the lime plus magnesia is higher than about 30 percent of the slag, it is likely that the addition was deliberate, but below 30 percent we cannot tell. Soda and potash are also strong sulfide formers and the sodium and potassium sulfides or sulfates are absorbed by the slag at the furnace temperature. In the amounts typical of coal or coke, they desulfurize little. Alkali oxides must exceed about 5 percent to have an effect, so amounts above that must have been deliberately added.

A fairly high sulfur content was, however, acceptable for many of the uses to which cast iron was put. In China, the longstanding existence of a large domestic and foreign market for cast iron pans for cooking and industrial processing made it feasible to utilize ironmaking systems which, like the Shanxi crucible process, yielded a metal with so much sulfur that it almost invariably solidified as white cast iron. In Europe, where the potential consumption of white iron was more limited, a substantial market existed for hot short wrought iron which could be worked with fair success by most blacksmiths. Only late in the nineteenth century did it become both possible and desirable to remove virtually all sulfur from most irons and steels.

Earlier, when a metal with no hot shortness was required, it would have been much better to use charcoal in smelting. Bar iron made from charcoal pig commanded higher prices than its coke smelted equivalent until the beginning of the twentieth century, when the development of high lime slags and other sulfur removal techniques—basic linings in open hearths and Bessemer converters, and manganese additions to liquid steel—allowed the manufacture of a virtually sulfur free iron. As a result, the traditional prejudice of secondary processors and industrial consumers against wrought iron and steel made with coke gradually disappeared.

Excessive sulfur content could be removed by post-treating pig iron in crucibles. This system was tried experimentally in Britain during the eighteenth century, with the explicit goal of improving the forgeability of coke smelted iron. Morton and Gould (1967) have described in twentieth-century terms how the processes patented by John Wood in 1761 and 1763 in Britain actually worked. In the so-called *potting process*, the high sulfur cast iron, granulated or in small pieces, was heated with scrap iron in closed crucibles placed in a coal fired reverberatory furnace. The other ingredients of the crucible charge were mill scale (i.e., iron oxide), old slag, and ashes of kelp, a soda rich substance which at the time was used extensively in the glass industry. The kelp ash was the desulfurizing ingredient, forming strongly bonded alkali sulfides and sulfates that dissolved in the slag rather than the metal. The process was repeated if the iron produced in the first try was not easier to forge. Wood's process apparently worked but, because of its high cost, was not used for more than a few decades. It was less expensive for British steelmakers to import Swedish and Russian charcoal iron than to remove the sulfur from the local coke smelted metal.

## 8. PHOSPHORUS IN CAST IRON SMELTING

Phosphorus in iron ore is substantially reduced into the pig iron during smelting with deleterious results to the mechanical properties of the wrought iron subsequently produced by fining and puddling. In cast iron, phosphorus is less harmful. It increases casting fluidity and the likelihood of solidifying as gray iron, reduces strength when present in amounts of about 1 percent or more and the level of carbon saturation in the liquid iron (as silicon does), and in very large amounts lowers the melting point of the metal.

The reduction of ore in a modern blast furnace entails converting virtually all the phosphates present to elemental phosphorus and dissolving it, in the gas phase, into the liquid iron (McGannon 1964: 418). Earlier blast furnaces were probably less efficient at extracting phosphorus because of the lower temperatures at which they operated, but any blast furnace would have produced an iron with a higher phosphorus content than a bloomery using the same ore. Schubert (1957: 152–153) states that one of the reasons for the long survival of the

[8]Twentieth-century blast furnace slags contain 1 to 4 percent sulfur (Bray 1942: 157; McGannon 1964: 431), and some nineteenth-century slags have similar amounts (Tylecote 1976: 109; Percy 1864: 497–503). However, this does not prove that the nineteenth-century slags were being consciously formulated for desulfurizing; indeed, the fact that almost all of them have CaO + MgO contents of well below the 40 to 50 percent of modern slags indicates that they were not optimally efficient at sulfur removal. White (1980) has discussed at length the question of slag composition for optimum desulfurizing qualities, in the context of iron manufacture in the eastern United States during the eighteenth and nineteenth centuries. He concludes that none of the 27 slags whose analyses he considers "came anywhere close to being effective" from the standpoint of the ability to remove sulfur from iron.

direct process in making low carbon iron was "the unsuitability of early blast furnaces for smelting high-phosphorus ores." Morton and Gould (1967) provide examples of "tough" pig iron, at 0.124 percent phosphorus, and "cold short" pig iron, at 1.01 percent; both terms refer to qualities of those metals after conversion to wrought iron.

Some cast irons with unusually high and low phosphorus contents are shown in Table 10.4. The content of the Shanxi example reported by Tegengren is extraordinary (6.95 percent), but cast irons with one to two percent phosphorus are relatively common in pre-industrial and early industrial contexts, produced by furnaces using either charcoal or coke. As the table shows, low phosphorus cast irons were also known. The average pig as it came from the furnace, in both China and Europe, fell into an intermediate range of 0.50 to 1.50 percent.[9] Although not of such high quality as steels of the time (and unacceptable compared to modern steels), intermediate pig irons were used extensively in finery and foundry.

Irons known to make good castings may often have contained amounts of phosphorus in excess of about 0.5 percent. Arguing from the indirect evidence of the phosphorus content of modern ores and comments about "highly fluid" irons in historical sources, Wertime (1962: 175–176) proposes that in sixteenth- through eighteenth-century Europe there was not only a direct correlation between phosphorus and success in the foundry business, but also an inverse correlation with the production of forgeable bar iron:

> As suggested by the case of Périgord, the presence of phosphorus in ores otherwise generally free of sulfur appears, contrariwise, to have been a critical factor in the success of the first major cast-iron foundries which spread through Europe. Barring Scandinavia, wherever there was successful casting, bar iron was also generally described as cold-short.

The very high phosphorus contents of 12 cast irons from Pingting in Shanxi—between 0.46 and 6.95 percent (mean = 2.06±1.97 percent) phosphorus—caused Read (1913: 27–28), a student of ancient Chinese cast irons, to observe that "it is not at all improbable that . . . [this] has been the chief factor in the development of the industry" there, and that the presence of phosphorus in the coal as well as the ore explains its very high concentration in the metal. He (1934: 551) later retracted his statement that the making of a high phosphorus metal with such good fluidity and with such a low melting point (i.e., 980°C at 6.7 percent P) could have induced the Chinese to adopt cast iron 2000 years ago, but the connection between phosphorus and success in iron founding still seems valid. It is possible that Chinese as well as European foundries specialized in the production of iron castings because their local coal and ore had unusually high phosphorus contents.

No method for removing most of the phosphorus from cast iron (or from wrought iron and steel made of that cast iron) existed until the late 1870s, when the Thomas-Gilchrist process was introduced. The essence of Thomas's concept was the use of a basic magnesia-lime furnace lining and the highly oxidizing atmosphere of a Bessemer converter, both of which were unavailable in the pre-industrial period. Some earlier systems, such as puddling with a slag high in FeO (Percy 1864: 663), did partially remove the phosphorus, and several of the recorded European fining systems may also have reduced phosphorus contents to some extent (Chapter 13). In general, however, phosphorus proved more difficult to control than sulfur. The only method certain to produce a metal without undesirable levels of phosphorus was to begin with a phosphorus free fuel and a very low phosphorus ore.

*TABLE 10.4. High and low phosphorus contents of cast irons*

| | Phosphorus | Carbon | Silicon | Sulfur | Manganese |
|---|---|---|---|---|---|
| High | | | | | |
| Coal/Coke | | | | | |
| China, 20th c. | 6.95 | n/a | n/a | 0.22 | n/a[1] |
| France, 19th c. | 2.30 | 1.40 | 4.10 | 0.30 | n/a[2] |
| Charcoal | | | | | |
| Britain, 17–18th c. | 1.43 | 2.97 | 0.93 | 0.17 | 1.64[3] |
| Low | | | | | |
| Coal/Coke | | | | | |
| Britain, 19th c. | 0.03 | 2.21 | 2.63 | 0.10 | 0.07[4] |
| Charcoal | | | | | |
| Germany, 19th c. | 0.06 | 4.32 | 1.00 | 0.01 | 10.71[5] |

[1]Tegengren (1923–24: 325): crucible cast iron from Dayang, Shanxi Province. [2]Overman (1850: 488): from Firmy, "very brittle." [3]Schubert (1957: 398): fireback from Sussex. [4]Percy (1865: 536): "No. 3 foundry-pig," Whitehaven, Cumberland; smelted with hot blast from [imported?] red hematite with 98.71 $Fe_2O_3$ content. [5]Percy (1865: 532): spiegeleisen from Musen.

## 9. MANGANESE IN CAST IRON SMELTING

In a bloomery, a third or less of the manganese present in ore will enter the finished iron (Kinney 1926), considerably less than can be expected in a blast furnace. If the pig iron from the blast furnace is made into steel, the manganese will improve hardenability and reduce hot shortness. If instead the pig iron is poured into a mold, the manganese content can still confer useful properties. It will reduce the ability of sulfur to force the molten metal to freeze as white cast iron—according to Massari (1938), the influence of sulfur is entirely negated by manganese in the amount of twice the sulfur content plus an additional 0.3 percent. Any manganese above that level will counteract silicon and phosphorus in inducing gray iron formation.

When smelted in the blast furnace, manganiferous ores yielded a cast iron that contained at least several percent manganese. This product was usually called *Spiegeleisen* (mirror-iron), a term still recognized as a synonym for the ferromanganese used to make additions of manganese to steels. According to Percy (1864: 782, 796), Spiegeleisen was the main factor in the traditional Siegen process of Germany, which converted pig iron by fining or puddling to a noted type of steel. Beck (1899, v.4: 78) attributes Krupp's success in making large steel castings (for cannon barrels) during the

[9]The median for the sixteenth- to eighteenth-century cast iron objects from Belgium, Luxembourg, and France analyzed by Evrard and Descy (1948) is about 1.0 percent phosphorus. The 230 German stove plates and firebacks studied by Pfannenschmidt (1977: 215) have a median phosphorus content of about 0.8 percent, with several examples that may be as high as 2 percent.

1850s to the high manganese cast iron of the Siegen area. He traces the production of manganese steel and of Spiegeleisen in that area (i.e., the Ruhr), and seemingly the use of the term as well, back to the sixteenth century (Beck 1893–95, v.2: 183).

## 10. CRUCIBLE SMELTING

Whereas pig iron was smelted almost exclusively in blast furnaces in Europe (from the fifteenth century) and North America (from the seventeenth century), a system of production in crucibles was used concurrently in Eastern Asia, which competed effectively with conventional blast furnaces through the early 1900s. The main features were crucibles, open furnaces with almost no insulation, and coal as the sole fuel and reducing agent.

Chinese crucible smelting continued after the mid-nineteenth century mainly in Shanxi Province,[10] where it was observed by foreign and native travelers. Because several (Richtofen, Shockley, Read, Tegengren, Foster, and Wang) were mining geologists, mining engineers, or metallurgists, and because they often wrote extensively on the subject, the Shanxi process is among the best-recorded of all pre-industrial smelting systems. The age of the process is controversial. The blast furnace appears to predate it, but the crucible system has been so perfected and so well adapted to its environment that it must be very old.

There are two main varieties; the first has relatively short crucibles and small stall furnaces and is found mainly in southeastern Shanxi, and the other, in eastern Shanxi, employs 200 or more crucibles of over a meter in length placed in a single furnace (Fig. 10.6). The former were all equipped with large manually operated fan bellows, while the latter were sometimes operated by natural draft. The fuel was not, as has been stated, invariably anthracite. Shockley (1904: 867–868) mentions two areas in eastern Shanxi where smelters used ordinary bituminous coal. Read (1934: 550) observed that not only ore and coal went into the crucibles, but also a third material called *heitu*, "black earth." As this contained "abundant crystals" of vivianite (iron phosphate), he concluded that it might be the source of the large amounts of phosphorus in the iron. Tegengren (1923–24: 323), on the other hand, maintained that *heitu* functioned mainly as a flux due to its high ash content; his analysis of *heitu* showed a $P_2O_5$ content of only 0.07 percent.

All agree that the iron produced from the smelting crucibles was not immediately usable. Read (1913: 27) calls three of the twelve samples he analyzed "malleable," two others "slightly malleable," and the other seven "brittle" or "hard"; he does not report carbon contents. He says that the individual "blooms" (the term and quotation marks are his) as they come from the crucible are heterogeneous, consisting of "irregular fragments of iron, earthy residues, and a certain amount of coke. The bloom is sold to the makers of wrought iron, the small pieces of iron are sold to the makers of cast-iron, and the coke is used in the manufacture of crucibles." Tegengren (1923–24: 324), who visited two centers specializing in the "refining" of bar iron, describes the raw iron cakes shipped there from the smelting sites:

> The resulting raw iron cakes at Ta-Yang weigh about 8–11 catties (4.8–6.6 kilos) each. The product is not quite similar to what we generally mean by the term pig iron. First, it is extremely impure, containing earthy residues, slag and a certain amount of coke. Secondly its composition is different in as much as the iron contains rather little carbon and is malleable.

He gives one analysis which shows 2.27 percent carbon and 9.3l percent silicon; much of the latter, he says, is probably occluded slag.

The cakes became bar iron only after extensive processing. Tegengren (1923–24: 327–330) describes an elaborate multi-stage procedure that involved three separate types of furnace. Other commentators seem to imply that conversion to wrought iron in the areas they visited was simpler, but none says that the cakes could be used in the forge as they came from the crucible.

Further processing was also required to convert the cakes to usable cast iron. Several observers say that this was done by reheating the cakes with more coal in a second type of crucible, often smaller and perhaps subjected to a higher temperature than for smelting. In

Fig. 10.6. *Crucible Smelting in Shanxi.* The upper photograph shows the crucibles, more than a meter long, being made; the lower, the arrangement of the crucibles within the stall furnace. A diagramatic illustration of such a crucible is in Figure 3.3. (Photographs from Shockley 1904)

[10] Iron smelting crucible furnaces are known archaeologically from Shandong Province in eastern China (Yang 1982: 77) as well as from Shanxi. Nineteenth-century travelers observed them in Shanxi, Liaoning in northeastern China (Richtofen 1882: 96; 1907: 498–499), Hunan in south central China (photograph by Knapp in *Archives, Verein Deutscher Eisenhüttenleute*, 1936), and Xinjiang in northwestern China (Davidov, "Montanindustrie an der Grenze Chinas," *Berg- und Hüttenmännischen Zeitung* 31 [1872]: 394–400). We are indebted to Donald Wagner for the last two references. Additional sources are: Shockley 1904: 855–859, Tegengren 1923–24: 319ff., Wang and Wang 1930: 110, Wang 1923–24: 443, Licent 1924, v.2: 620–625).

the one existing description of crucible smelting outside Shanxi (in the Manchurian province of Liaoning, in northeastern China—see Richtofen 1882: 96), the conversion of the cakes to cast iron took place in a small blast furnace.

In the absence of metallographic studies and more chemical analyses of the crucible cakes, perhaps they should be characterized as a complex mixture of bloom and cast iron. The mechanisms by which the cakes were formed within the crucible are, however, reasonably clear. Their spongy form and the slag occluded by, rather than cleanly separated from, the iron suggest that reduction took place at a low temperature in a highly reducing atmosphere for most of the furnace run. Since the metal did not liquefy completely, the condition was one in which gas composition constantly changed from almost pure CO to one with a considerable proportion of $CO_2$. Early in the run, cast iron would have been made but later bloom iron was. The cast iron would carburize the bloom iron to some extent, but the metal product was always largely a solid framework entrapping the slag.

As late as the early twentieth century, more than one-third of all the iron made in China by traditional means was still smelted in crucibles, which demonstrates that the Shanxi system was economically competitive. The antiquity of crucible smelting has been linked to the first documented use of coal in the fourth century A.D. (Needham 1958: 144), but the process clearly does not require coal as a fuel. Since the crucibles of Shanxi clearly could and often did yield slaggy lumps of metal with carbon contents low enough to classify them as directly smelted steels or wrought irons, it cannot be stated unequivocally that the bloomery process was unknown in China. Even cakes with carbon contents in the cast iron range were different from an ordinary cast iron pig.

The hypothesis that the early Chinese preferred to shape iron by casting rather than forging because most crucible made iron contained excessive amounts of sulfur and phosphorus was tenable only in light of early analyses of traditional irons (Read 1913: 27). Wrought iron products are rarer in ancient China than in Europe, but the analyses of early cast irons now available show that they do not have greater sulfur and phosphorus contents than pre-industrial European cast irons. In addition, there are many blast furnaces in late Zhou and Han smelting sites (ca. 400 B.C.-A.D. 200). Such conventional smelting devices played a larger role in early Chinese ironmaking than in the nineteenth century, and there is no reason why crucible furnaces had to utilize high sulfur and high phosphorus ore or coal. Loaded with good ore, good charcoal, and an appropriate flux, and then heated to the right temperature, a Shanxi-type crucible could produce cast iron (or bloom iron, or direct process steel) that would equal the best.

## 11. IRON CASTINGS AS COMMODITIES

At present, groups of identical iron objects can be made in a variety of ways: conventional casting, hot forging in die blocks that have complementary cavities to replicate a desired shape, cold drawing between conforming dies, stamping from continuous coils of metal sheet, and pressure die casting, among others. Any can be used in mass production, and most involve the direct application of great force, so powerful machines, heavy dies made from special steels, and equipment for temperature control are crucial.

In the past, mass production of iron objects was more limited, but the majority of modern force-fabricating techniques do have early analogues. Stamps and swages have long been part of the smith's toolkit, as have heavy hammers and dies. However, only rarely could the available techniques be controlled well enough to produce large numbers of identical or similar objects. Any method of metal forming that depended on force was inevitably incremental, labor-intensive, costly, and slow.

The early fabrication procedure best suited to mass production was casting. With minimal organization, a group of craftsmen who could melt metal could produce large numbers of simple mold made objects efficiently. The results are described vividly in Biringuccio's account of a visit to a brass foundry in Milan, probably in A.D. 1507. He talks about the master craftsmen "who did nothing but mould in lute and form an infinite number of moulds of all those little objects of everyday use which can be made from brass by casting," and about the sheer quantity of objects produced: "Pondering on this I thought that this shop alone was sufficient to furnish not only Milan but also the whole of Italy, and surely it seemed to me a fine and splendid undertaking... and surely it pleased me to see so many things being moulded continuously, and continuously being cast" (Smith and Gnudi 1942: 74). He reacts to this with Renaissance enthusiasm:

> Whoever entered that shop and saw the activity of so many persons would, I think, believe that he had entered an Inferno, nay, on the contrary, a Paradise, where there was a mirror in which sparkled all the beauty of genius and the power of art... In that place I turned my eyes nowhere but that I saw some ingenious novelty and beauty of craftsmanship, and thinking of the arrangement and the greatness of the new things that were shown to me, I was sometimes stupefied (Smith and Gnudi 1942: 72).

Biringuccio's Milanese brass foundry probably had no iron casting counterpart in Europe until the eighteenth century. Earlier iron foundries were small, focused mainly on military markets, and able to make, or at least to sell, a limited variety of goods.

This was not true in China, however. By the early Han Dynasty in the second and first centuries B.C., Chinese iron founders had already begun to operate on a fully industrial scale, casting many kinds of objects in large numbers and in the process becoming wealthy.[11] Analyzed objects include a cooking pot, a tripod brazier, a lamp, and a heavy adze or trenching tool (Fig.

[11]The great Han period historian Sima Qian (Ssu-ma Ch'ien) enumerated nine families as illustrations of "the ways in which some of the worthy men of the present age, working within an area of a thousand miles, have managed to acquire wealth." Five of the nine were merchants or money lenders. The other four were all smelters and casters of iron (Watson 1969: 352–355).

Fig. 10.7. *(a) Cast Iron Cooking Bowl.* Designed to fit into a circular hole on the top of a stove. The outer rim has a diameter of about 35 cm. Its date is uncertain. Berthold Laufer, who acquired it in Xian (Sian), believed it to be from the Han period (2nd c. B.C.–2nd c. A.D.), but one Chinese archaeologist has suggested that it may be as late as the twelfth century. This later date would fit well with the idea that the Chinese did not make extensive use of coal for smelting until the tenth or eleventh century. The microstructure is a white cast iron with a considerable dispersion of sulfides, signifying that the metal was smelted with coal. (Field Museum cat. no. 120989) *(b) Cast Iron Cooking Pan or Brazier.* The style is that of the late Han period (1st–2nd c. A.D.). The microstructure is white cast iron and there are almost no sulfide inclusions visible. This would signify the use of charcoal for smelting. (Field Museum cat. no. 120990) *(c) Cast Iron Lamp or Lamp Stand.* The central spike suggests that it may have supported a candle. It is stylistically dated to the Han through Sui Dynasties (2nd c. B.C.– 6th c. A.D.). (Field Museum cat. no. 120998) *(d) Heavy Cast Iron Trenching Tool or Adze.* Sturdy enough for industrial use, it represents a type of tool that was widely used from the late Warring States period to the beginning of the Sui Dynasty, between perhaps 300 B.C. and A.D. 589. The microstructure shows that it is a white cast iron which has been partially decarburized to a low carbon steel. (Field Museum cat. no. 127393)

10.7). All are white cast iron, though the surface of one adze has been decarburized to low carbon steel. All are utilitarian and without nonfunctional ornament, presumably intended for purchase and use by those who were not among the social elite. The cost of production and retail price could not have been high.

A plow point (Fig. 10.8) is one of the many types of cast iron agricultural implements made by the Han Dynasty Chinese. Similar plow points continued to be used throughout East and Southeast Asia until modern times, and even in the 1900s they still were usually imported from China or made by locally based Chinese workmen. The small blast furnaces at Bulacan in the Philippines, discussed above in this chapter, specialized in the production of plow shares and points, cast directly from the furnace into three-part molds. Larger than the Han example shown here, each point made at Bulacan weighed about 2.2 kg. and each share about the same. McCaskey (1903a: 56–61) estimates that eight points and shares could be poured every two or three hours, for a daily production of 70 to 80 sets and a monthly production of 2000 to 3000 sets during the four month smelting season. About fourteen men were needed to keep the furnace at full output. Much of their labor went into operating the manual windbox bellows. They also mined ore and made charcoal during the months when the furnace was not in blast.

For fourteen men to make about 330 kg. of plow points per day may seem a modest level of productivity. If, however, these figures are compared with those for the making of low carbon iron in some other bloomeries and fineries (Table 14.9), the production of Bulacan is impressive. A combined bloomery and forge without water powered bellows and hammers has a maximum productivity of about 5 kg. of bar iron per man day; most fall below 2.5 kg. per man day. If we assume that a smith with a hand hammer needs about a day to convert 5 kg. of bar iron into finished goods, a bloomery based manufacturing system would have a maximum net productivity of 2.5 kg. per man day of plow points or the equivalent. The Bulacan system, simple though it was, produced 23.5 kg. per man day. The manufacture

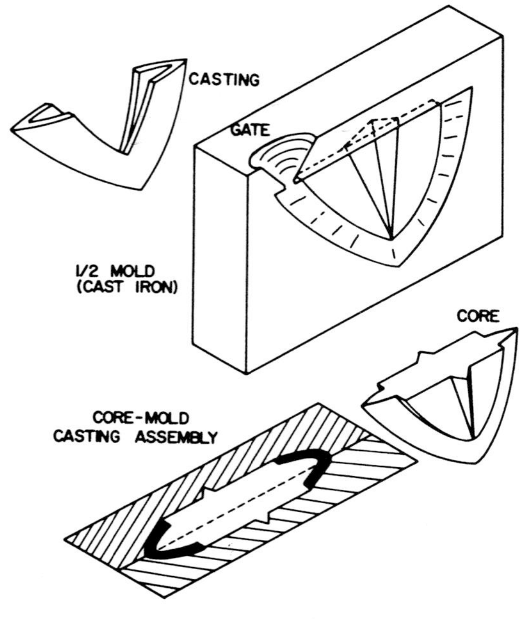


Fig. 10.8. *Cast Iron Mold.* It is demountable and "permanent," used in the casting of iron tips for wooden plows (Rostoker et al. 1983); such a mold was capable of producing hundreds of castings before it succumbed to thermal fatigue cracking. The earliest iron plow tips in China appear late in the Warring States period, in the third century B.C. The mold is an assembly of three cast iron units, two of which are mirror images using the same pattern. The third, the core, has a shape complementary to the other two pieces. All three parts would have been cast in baked clay molds (Zhengzhou 1978). Comparable permanent mold systems made of cast iron are illustrated in Needham 1958: pl. 3, Barnard 1961: pl. 8, and Henan Provincial Museum 1978: 14.

of consumer goods from cast iron could clearly revolutionize the economics of ironmaking.

Operations in an early Chinese foundry do not appear to have been greatly different from those today. Through the use of master patterns, large numbers of molds could be made quickly on an assembly line. Archaeological evidence from the site of Wafangzhuang (Hua 1983), a Han period foundry in Henan Province, shows that the casting system was complex enough and large enough in scale to indicate a considerable division of labor. Complex stacked molds were made on the spot, and it is thought that these had to be used while still red hot from the kiln. No smelting was done there; pigs and scrap iron were remelted in either crucibles (Yang 1982: 126) or shaft furnaces (Hua 1983: 126).

Several teams with specialized functions would therefore have been needed: one to obtain and prepare the clay based molding material; a second to make the master patterns and to form the component parts of the mold assemblies; a third to carry the molds to drying areas, operate the kilns used for baking them, and then bring them immediately to the casting floor; and a fourth to operate the melting furnaces and, using ladles or small crucibles, to pour the liquid iron into rows of high mold stacks. A fifth team might have been assigned to break the molds open, extract the castings, remove from them redundant parts such as iron filled gates or sprues, and prepare the finished products for sale.

This system had extensive variations in the larger casting shops of the Han period. Stacked molds were, for example, used, so that a number of mold cavities could be filled in a single pouring. Hua (1983: 126) describes how as many as 120 small buckles were cast in a single assembly of molds of this kind. Another refinement, still frequently found in Western foundries, is the reusable, "permanent" mold which is itself made of cast iron (Fig. 10.8). Its chief advantage was the labor saved in mold making and the speed with which castings could be turned out, since the iron mold would quickly freeze the liquid metal, making it possible to separate the mold pieces and dump out a hot but solid casting within a minute or so after pouring. Disadvantages included the substantial cost of replacing the molds, which had a lifetime of only several hundred castings (Rostoker et al. 1983), and the limitations on the range of their shapes. Yet labor productivity would have been high, and might have exceeded that of workshops employing conventional stone or clay molds.

The Chinese made such extensive use of iron castings in part because the metal was much cheaper than any copper alloy. It is also a good casting material, easy to handle, and sufficiently fluid to replicate fine details, thin sections, and sharp corners in the interior of molds. The extraordinary thinness of traditionally cast Chinese pans impressed foreign observers. Foster (1926: 177, 220), an experienced foundryman, says that ordinary pans made in Shanxi have a maximum thickness of 1/16 inch when three feet in diameter; he illustrates a broken fragment which is about 1/32 inch (0.8 mm.) thick, extremely thin by the standards of pre-industrial Europe.[12] A preference for very thin walled pans appears to have been shared by the manufacturers, to whom raw material and shipping costs were important, and by domestic and foreign consumers, who liked cooking wares that would heat up with a minimal expenditure of fuel. In addition, salt makers and other processors of corrosive substances must have considered the resistance of cast iron to chemical attack as important as its low price.

We do not know why Europeans took so long to appreciate the usefulness of cast iron. It could have been produced in Greek and Roman (or for that matter, Celtic and German) furnaces, and there is good evidence that the Romans not only made cast iron occasionally but even experimented with fashioning finished objects from it (Tylecote 1976: 57). This technology lag may have resulted from a lack of skill in working with multi-part sectional molds. While the Chinese, from the Bronze Age onward, had generally used molds that fitted together in sections for casting complex shapes, the earlier bronze casting industries of Europe and the Middle East were strongly oriented toward one-piece molds made by the lost wax process. The tradition of pouring all except the simplest castings into such molds continued unbroken to the fourteenth and fifteenth centuries when Europeans first began to use cast iron,

[12]For comparison, Maréchal (1955: 141) describes castings made at Liège in Belgium which have walls 2 to 4 mm. thick; he feels that such thinness "denotes a rare mastery of the foundry trade."

even though the lost wax process is expensive and one- or two-part sectional molds limit the cast shapes attainable.

Europeans may also have had difficulty working white cast iron because of its extreme hardness and brittleness. Perhaps as much as 2000 years ago, the Chinese discovered techniques for malleabilizing and solid state decarburizing; they could adjust the performance of their cast iron to meet a wide range of functional criteria (e.g., the adze in Fig. 10.7d). The Europeans evidently could not. Although some practical ironworkers may have known how to soften and toughen cast iron at a relatively early date, it was not until Réaumur's experiments of the eighteenth century that these methods began to be worked out, and not until the nineteenth century that they were widely applied.

Whatever the explanation, Europeans of the Renaissance were slow in adopting cast iron for domestic use. At Rievaulx, as at other English furnaces that did not make cannon and shot, castings never exceeded 5 percent of total output during the sixteenth and seventeenth centuries (Schubert 1957: 246); the other 95 percent went to fineries to be converted to wrought iron. Following in the left-hand column are the earliest known dates for English nonmilitary items made of cast iron (Schubert 1957: 257–268); the dates in the right-hand column are for Germany (Johannsen 1953: 203–211; Pfannenschmidt 1977: 45–56):

| | | |
|---|---|---|
| Bells | | 1434 |
| Water pipes | | 1444–45 |
| Pots | 1543 | 1477 |
| Firebacks/stove plates | 1548 | 1474 |
| Branding irons | | 1549 |
| Grave slabs | 1550s | 1516 |
| Plow shares | | 1556 |
| Firedogs (andirons) | 1573 | 1444 |
| Anvils | | 1591 |
| Wagon bolts | 1650s | |
| Sugar cane rollers | 1663 | |
| Garden rollers | 1663 | |
| Large boilers | 1706 | |
| Railings | 1700s | |

Cast iron military items came before, with the earliest cast iron cannon on the Continent dating to the 1390s. Iron cannon balls were routinely cast in Germany and France by the early fifteenth century and in Britain by the 1490s (Schubert 1942: 132); by 1514 the French were making them in molds that were also of cast iron (Johannsen 1953: 205). The cast iron grenade was known by 1467 (Johansen 1953: 206). Biringuccio, writing in the 1530s, was familiar with bombs and cannon balls of cast iron, but appears to be unaware of other applications for it.

We should not assume that the use of cast iron in an area means that blast furnaces were also in production there. Furnaces for remelting scrap iron are known to have existed in the early sixteenth century. Pfannenschmidt (1977: 36) cites a description of one operating in 1500, a small shaft furnace 1.5 ells high by 0.5 ells wide (0.74–1.17 × 0.25–0.39 m.), with two tuyeres and charged exclusively with "old iron." The metal may have been scrap cast iron, but similar furnaces were also used to make cast iron by what may be called the indirect process, by the carburization of bloom iron. This was commonplace to Biringuccio, who recommended that the converting furnace be charged only with bloom iron and charcoal, and disparaged those who added antimony to their iron in order to create an alloy with a lower melting point. A system in which bloom iron was carburized must have been employed at Spanish gun and ammunition foundries. Cipolla (1965: 35) says that cast iron manufactures were established in the Bilbao area, the center of the already famous Spanish bloom iron industry, in the 1620s. More than 200 years elapsed before the first blast furnaces were built there or elsewhere in Spain (Alzola 1896: 14–15).

Promoting the conversion of bloom iron to castable metal with additives is a poorly documented practice, but it may have been widespread. Tylecote (1976: 69) cites Johannsen as saying that arsenical cast irons were used in making cannon balls in the mid-fifteenth century. Percy (1864: 76–77) gives analyses of nineteenth-century cast iron cannonballs from two parts of the Islamic world, Algeria and Turkey, that have arsenic contents of 16.2 and 27.0 percent. Since most of the iron made then in those areas came from bloomeries, the original material may have been bloom rather than blast furnace made pig iron. Wertime (1962: 68) suggests that additives were common in fifteenth-century Germany,[13] noting that the *Feuerwerksbuch* of 1454 prescribes several melting devices (including a cupola-like furnace) and the admixture of either powdered glass or *speiss* (an arsenide byproduct of copper smelting) and of bismuth or tin. The use of bismuth is curious because that element does not alloy with iron and has a low boiling temperature.

The conversion of bloom to cast iron required a furnace separate from that in which the bloom was produced. Remelting converted cast iron in special furnaces is generally thought to have started in Europe only in the early eighteenth century (Beck 1897, v.3: 163–175; Schubert 1957: 269; Evrard and Descy 1948: 214; Mott 1957–59: 76), but the remelting of cast iron scrap and of excess metal set aside in the form of pigs must be as old as the blast furnace itself. When scrap iron had to be melted, a second run through the blast furnace might have served, but often a secondary heating device, like the small shaft furnace mentioned above or even a crucible set in the fire of a forge, would have been more practical. Beck, although he may have originated the idea that before 1700 iron casting was done "almost entirely direct from the blast furnace," was also aware of the existence of an old but small scale scrap remelting industry. He notes, for instance, Swedenborg's observation of 1734 that remelting furnaces had long been used for the casting of small objects in

[13]The use of cast iron containing other metals that lowered its melting point continued in Germany long after the fifteenth century. Evrard and Descy (1948: 243) cite Le Verrier's observation in the 1890s that the finely detailed decorative castings then called "bijoux de Berlin" were made of a natural iron-arsenic alloy from an ore mined in Silesia.

urban foundries in France and Italy (Beck 1897, v.3: 166).

What Beck and other historians mean when they say that remelting furnaces did not exist before 1700 is that this date marks the approximate beginning of their regular use in large foundries and of the realization among European ironmakers that a second melting can improve the quality of cast iron objects. Remelted cast iron can produce a stronger gray iron by a lowered carbon content, which was achieved by altering temperatures and amounts of fuel, and thus the CO/CO2 ratio of the furnace atmosphere, or by mixing different sorts of pig iron with wrought and cast iron scrap. In the mid-eighteenth century, Grignon (1807: 438–440) recommended, and apparently practiced, a less conventional technique for improving the quality of iron for cannon castings by adding saltpeter (i.e., potassium nitrate), a violent oxidizer, to the liquid metal in the well of the remelting furnace. This was potentially very effective in lowering the carbon content of a cast iron—so much so, in fact, that Grignon was conscious of the risk that the metal might turn to steel, causing "vexatious embarrassment and the ruin of the furnace." [14]

In Europe, any remelting furnace is likely to be called a *cupola*. Some writers use the term to include shaft furnaces, often small (Fig. 10.9), that look and operate like blast furnaces, while others reserve it for reverberatory remelting devices, like those introduced in Britain at the beginning of the eighteenth century (Schubert 1957: 269) and adopted in many other parts of Europe within the next hundred years (Pfannenschmidt 1977: 152–164). The term is also applied by others to the full range of indirectly and directly heated furnaces in European foundries between the mid-eighteenth and mid-nineteenth centuries (for illustrations see Beck 1897, v.3: 765–768 and 1899, v.4: 92–102).

The smallest of all "cupolas" may be the foot-high portable cylinder furnace of the Chinese cast iron pot mender (Hommel 1937: 31–32; Lockhart in Percy 1864: 747–748, Foster 1926: 134). The furnace encloses a clay crucible several inches deep and wide, charged with a few bits of old cast iron. The pot mender heats this in the blast of a small windbox bellows. The iron becomes liquid in about 15 minutes. The mender pours liquid metal from the crucible onto a piece of felt which he holds in his hand, and pushing it into the break, he pats the molten iron into place from the other side with a piece of rolled-up cotton, also held in the hand.

Fig. 10.9. *Scrap Iron Melting Furnace.* These furnaces were used in the early eighteenth century by itinerant foundrymen, who collected cast iron scrap and made it into various household utensils. This one has two parts, a demountable shaft and a hearth or basin which can be lifted and used as a pouring ladle. (From Beck 1897, v.3: 171, after Swedenborg, *Regnum Subterraneum sive Minerale de Ferro*, 1734)

Most authorities agree that military applications of cast iron predated domestic ones, and that they were quantitatively more important during the fourteenth to eighteenth centuries. Nef (1963: 81), however, as part of his generally convincing effort to show that war has contributed less to human progress than is often thought, has claimed a major role for civilian castings:

> What was done in Britain with the iron and other metals which did not go into guns and ammunition? There was a remarkable increase in the domestic demand for metals, to equip the new industrial enterprises, to supply new conveyances, and to provide commodities for ordinary living and even for dying. Sugar refining, brewing, soap boiling, the dyeing and carding of wool and cloth—all manufactures which grew rapidly—required metal parts such as rollers, boilers, stoves, and tools.

Nef's case is not strong. There is no evidence that civilian demand for cast iron in Britain (or on the Continent) increased rapidly during the sixteenth and seventeenth centuries, or that there was even much of a civilian market. No early furnace records that we know of show more than a very small percentage of production devoted to pots, firebacks, and similar items. Without the demand for munitions, and the large profits to be made from defense contracts, iron casting in Britain would have remained a minor enterprise for the first four hundred years after the introduction of the blast furnace. Only in the late eighteenth century did consumer interest in domestic and industrial wares made

---

[14] This is the principle behind the Bessemer converter as well as the oxygen lance used in modern steelmaking systems. Grignon's (1807: 439) description reads:

> It is necessary at three different times, that is, when the bath [of liquid iron in the well of the remelting furnace] is half, three- quarters and completely full, to introduce into it a tube of wood with a cast iron rod for a handle. This tube should contain saltpeter purified of its sea salt and without humidity. One moves the tube around into all parts of the bath, so that a lively effervescence is caused by the "deflagration" of the saltpeter.

He says that eight ounces of saltpeter is enough for a hundred pounds (a *quintal*) of good-quality iron. Although only a few tenths of one percent of carbon could be removed by such an amount, it might be enough to effect a noticeable improvement in the iron. Adding much more saltpeter would increase the risk Grignon mentions of having the metal "come to nature," freezing in the furnace basin as a solid block of steel.

of cast iron begin to reach the level that it had in China 2000 years previously.

Abraham Darby, at Coalbrookdale in Shropshire, is said to have been the first British ironmaker to dedicate the production of an entire blast furnace to making commercial, nonmilitary castings. Mott (1957–59) attributes Darby's success in doing this not only to his use of coke as a smelting fuel, but also to his innovative complex sand molds. Mott sees a connection between these factors, suggesting that the high silicon content resulting from coke smelting and the high phosphorus content of local ores helped make better and thinner castings; he may, however, be wrong about the role of silicon. Three analyses of cast iron found at Coalbrookdale—one of charcoal smelted metal with a seventeenth-century (and pre-Darby) date and the other two of metal made in the eighteenth century by the Darby firm (Mott 1957–59: 71)—show that the differences in silicon contents are not statistically significant. The sample of charcoal iron contains about 1.0 percent silicon, while the two coke iron samples contain 1.50 percent. Mott may be correct about the phosphorus, because Darby's iron, with about 0.50 percent, would have had good casting fluidity, although it is not exceptionally phosphorus-rich (compare with the irons cited in Tables 10.3 and 10.4). The seventeenth-century sample of charcoal smelted iron from Coalbrookdale contains 0.55 percent phosphorus, which is actually higher than the phosphorus content of one of the two coke smelted samples.

According to Mott's evidence, Darby's castings were as thin and almost as strong if made with charcoal smelted iron. The chief and perhaps sole advantage of using coke was that the iron could be made and poured at a higher temperature,[15] and it was less expensive. The cost advantage enjoyed by Darby in the early decades of operations at Coalbrookdale was slight or even nonexistent, but by the 1750s it cost half as much to make iron with coke as with charcoal (Mott 1957–59: 68–73). This and the superior casting techniques used at Coalbrookdale are sufficient to explain the success of Darby and his heirs in the foundry business.

Even the Darbys did not find that the market for iron castings was open-ended. In the first year of operation for which good records exist, 1709, Darby sold 22.8 tons of castings and 58.3 tons of raw pig iron; in the period 1718 to 1737, the quantity of castings sold by the firm reached a plateau of 250 to 300 tons per year, with surplus iron—usually about 100 to 150 tons per year—being sold as pig (Mott 1957–59: 69). The limited demand for civilian castings may be the reason why this Quaker firm not only entered the gun trade for a time,[16] but also made repeated efforts to branch out into the production of wrought iron. During the 1720s and 1730s, fineries at Coalbrookdale, at first using an mixture of charcoal pig and later their own coke pig, produced 100 to 150 tons of cold short wrought iron per year. In the 1750s the Darbys built several new furnaces at Horsehay and Ketley, where ore with a lower phosphorus content was available. Virtually all the pig iron produced by these furnaces went to fineries rather than foundries (Mott 1957–59: 82–83).

Not until the 1760s and 1770s did the Darbys or anyone else in Europe succeed in finding a nonmilitary market for large quantities of iron castings. One outlet was bridge building, inspired by the success of the structure erected by the Darby firm with its own funds at Ironbridge, a few kilometers from Coalbrookdale (for historical details, see Trinder 1973: 186). Another was the casting of parts for heavy machinery, including cylinders and other pieces for Watt's new steam engines. This developing industrial market grew almost explosively. European production of nonmilitary iron castings probably passed that of China shortly after the beginning of the nineteenth century.[17] At some point in the first quarter of that century, citizens of European countries must have begun to use as much cast iron in their homes and on their farms as the Chinese had for the past two millennia.

[15]While much has been said about the influence of composition on the fluidity of a liquid metal, the degree of superheat can also be a dominating factor. If any metal is heated well above the temperature at which it is completely liquid, it will have sufficient fluidity to fill intricate, thin molds. However, the fuel costs to superheat metal as well as difficulties with controlling the metal composition are deterrents to superheat procedures. Modern casting practices call for minimum superheat temperatures to insure mold filling according to the complexity of the casting. Naturally the increment of temperature called superheat is less if, because of its composition, the liquid metal is very fluid.

[16]According to Raistrick (1953: 66–67), about 2000 cannon and 182 tons of cannon parts—a substantial percentage of the total cast iron output of Coalbrookdale (at that time, about 600 tons/year)—were produced between 1740 and 1748; gun casting was stopped shortly after that, when Abraham Darby II perfected his method for making finery pig with coke.

[17]On the assumptions that: 1) Chinese iron production in 1800 totalled about 170,000 tons (the amount still made by traditional methods in the 1900s—Tegengren 1923–24: 396), 2) 75 percent of this was used in castings, 3) 25 percent of the total European pig iron production went to nonmilitary foundries, and 4) Europe produced about 500,000 tons of pig in 1800 (Scrivenor 1854: 150, 179; Beck 1899, v.4: 165).

# 11. Steelmaking by Carburization

## 1. INTRODUCTION

*Steel* is used here to mean any iron-carbon alloy with a carbon content between 0.1 and about 1.8 percent. Although significant amounts of manganese and/or phosphorus were not uncommon in ancient steel, their presence would not have altered the basic nature of the transformations that could occur in the metal or the microstructures that were developed.

In the past, as today, reliable grades of steel were rarely the direct product of smelting (for exceptions, see Chapter 12), because of the behavior of slag in the furnace. There are two fundamental types of slag which, in combination with the furnace temperature and the $CO/CO_2$ ratio of the furnace gases, define the carbon content of the iron produced. Slags high in FeO, such as those typical of bloomery operations, are likely to oxidize out any carbon present in the metal, and therefore, under normal conditions the bloom produced should have a carbon content well below 0.1 percent. In contrast, the silica-lime-alumina slags necessary for furnaces run at higher temperatures and with higher $CO/CO_2$ ratios do not remove carbon from the metal as it forms. Hence, cast iron as it comes from the blast furnace is a more or less saturated solution of carbon in iron, often with a 4 to 5 percent carbon content.

Smelting directly and homogeneously to steel, with a carbon content intermediate between those appropriate to the two types of slag, requires a precise control of the $CO/CO_2$ ratio. Since this was generally not feasible in the past, direct process steel was rarely produced. Often portions of a bloom contained much carbon, especially when the furnace was run at temperatures higher than those of an ordinary bloomery. Occasionally, as in some Catalan hearth operations, the blooms became sufficiently carburized to be considered steel. But in many smelting traditions, heterogeneous mixtures of soft and steely iron were not considered better than the soft iron usually produced, and may not even have been recognized. Much of the carbon in this type of mixture might be lost through decarburization during the working up of the bloom, especially at the surfaces of the forged bars. Therefore, the majority of ancient processes for making a predictable and good quality steel were indirect, and worked by either augmenting the carbon content of a low carbon iron or reducing the carbon content of a cast iron. The latter process was not so common as the former in the preindustrial world, perhaps due to the difficulty of halting a decarburizing reaction before the metal lost almost all its carbon and thus became a soft wrought iron (Chapter 13).

Increasing the amount of carbon in a low carbon bloom or wrought iron was more usual. Carbon could be added by cementation, the reaction of solid wrought iron with carbon in gaseous form; by allowing the solid metal to come in contact with liquid cast iron; or by melting the wrought iron in a crucible in the presence of charcoal or another form of solid carbon. The first two methods yielded a product with a carbon content that tended to be nonuniform, so further processing was necessary to produce a useful homogeneous steel. Heterogeneous blooms, for example, were usually pile forged.

## 2. THE THEORY OF CARBURIZATION BY GAS/METAL REACTION

In principle, solid wrought iron will absorb carbon when placed in contact with charcoal or graphite, and ultimately will attain a level of saturation which varies according to the temperature of the iron (Fig. A.1). In practice, however, contact with solid carbon is not an effective means for enriching iron which is at a temperature lower than its melting point. Solid carbon can diffuse into solid iron only at points where the two substances touch each other. Even if the carbon is powdered and the iron buried in it, the total area of these contact points is only a small proportion of the overall surface area of the metal. Although the iron can take up carbon at the contact points readily, and although this will diffuse into the interior of the metal with moderate speed (Table 11.1), the carbon solids themselves have poor diffusivity—that is, carbon atoms move slowly within a solid matrix of other carbon atoms. Once the few carbon atoms at the contact points have entered the iron, more carbon atoms move slowly to take their places. The carburizing reaction therefore effectively stops.

The restraints imposed by carbon-within-carbon diffusion can be overcome by liquefying the iron or by making the carbon available in the form of a gas. Almost any gaseous, carbonaceous compound capable of giving up some of its constituent carbon could be used, but carbon monoxide is most effective. At high temperatures it is readily converted to the dioxide and to elemental carbon,

$$2\,CO \rightarrow CO_2 + C,$$

or preferably,

$$2\,CO \rightarrow CO_2 + [C]_{Fe},$$

indicating that the carbon atoms are deposited on the surface of the hot iron and promptly absorbed.

*TABLE 11.1. Diffusivities (D) associated with carburization**

| Interdiffusion system | Temperature | D(cm²/sec.) |
|---|---|---|
| Carbon Self Diffusion | 1000°C | $5 \times 10^{-28}$** |
| γ-Iron Self Diffusion | 1000°C | $1.1 \times 10^{-12}$ |
| γ-Iron/Carbon | 1000°C | $2.2 \times 10^{-7}$ |
| α-Iron/Carbon | 1000°C | $2.0 \times 10^{-5}$* |
| | 850°C | $4.4 \times 10^{-6}$ |

**Handbook of Chemistry and Physics*. 60th ed., CRC Press 1980
**Extrapolated

A continued supply of carbon monoxide molecules can be provided in an ordinary forge. However, the monoxide supply can be more easily controlled in a closed container, either by feeding in a continuous stream from an outside source (which is the method used in several modern carburizing systems) or by promoting a restoration reaction within the container, so that there is enough carbon dioxide to initiate the conversion of elemental carbon to the monoxide:

$$CO_2 + C \rightarrow 2\,CO$$

If a piece of iron within a closed container is embedded in carbon particles, furnished with a small initial source of $CO_2$, and then heated to an appropriate temperature, a continuous transport process is established:

$$CO_2 + C \rightarrow 2\,CO$$
$$2\,CO \rightarrow CO_2 + [C]_{Fe}$$
$$CO_2 + C \rightarrow 2\,CO, \qquad \text{and so on.}$$

This is the principle behind the pack carburization system. The same reactions also occur in open furnaces, where their monoxide restoring character is less critical. These furnaces will achieve carburizing conditions, if only intermittently.

## 3. CARBURIZATION IN THE OPEN FORGE

The discovery that the properties of solid iron could be altered in open furnaces must have been made repeatedly since the start of the Iron Age. Iron is worked hot, and is usually heated by burying the workpiece in a bed of burning charcoal or mineral carbon. The combustion is maintained by an air blast through a tuyere projecting into the fuel bed. To heat the workpiece quickly, a smith strengthens the air blast, thus maximizing calorific output from the fuel by burning it completely to carbon dioxide. When the piece is sufficiently heated, however, he must shut off the blast to avoid wasting fuel. The carbon dioxide trapped between the charcoal fragments then quickly converts to the monoxide. At this point the bed will carburize the surface of any iron object within it (Chapter 3).

Most smiths probably recognized that leaving a piece of consolidated bloom iron in a quiescent bed of incandescent charcoal created a product that could be made harder and sharper than ordinary iron. The technique, usually called *cementation*, *case hardening*, or *forge carburization* ("forge" used in the sense of the heating system for iron products), was probably the earliest method of producing steel. Until the mid-nineteenth century, it accounted for the bulk of the world's steel.

Forge cementation had disadvantages as a production technique. The carburized layer of steel on the surface of the iron was likely to be thin and easily removed if the finished implement were resharpened or abraded in use. In addition, the process was difficult to control. The bed of fuel would soon cool down and carburization cease unless the bellows were periodically activated. But whenever this was done, the $CO_2$ produced could decarburize the piece. Although a skillful smith might succeed in adjusting the blast to hold the temperature and to maintain the necessary $CO/CO_2$ ratio, there was no visual indication that the critical gas ratio had been achieved. Forge cementation had another disadvantage: carburization is an intrinsically slow process, so it is tedious to make a steel layer of reasonable depth.

Carbon is only slightly soluble in ferrite, the form that unalloyed iron takes at low temperatures. Essentially no carburization occurs while the iron remains below 723°C (Fig. A.1). Carbon is soluble in austenite, the high temperature form of iron, when heated above 910°C. In the intermediate temperature range between 723° and 910°C, carburization can proceed but at a reduced rate.

The $CO/CO_2$ ratio of the gases in the forge also affects the level of carbon saturation in iron (Figs. A.2, 11.1).

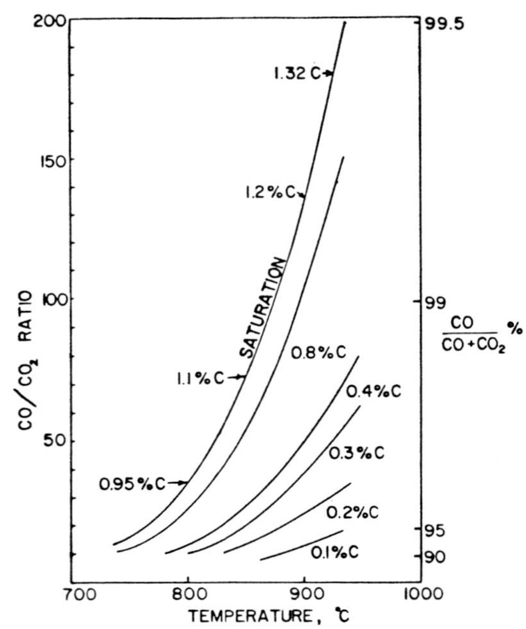


Fig. 11.1. *Conditions for Carburizing and Decarburizing.* This is an expansion of the section of Figure A.2 that relates changes in carbon content to temperature and the percent of carbon monoxide in an atmosphere containing both carbon oxide gases. Any point in the diagram defines a carbon content in equilibrium with a gas of specified carbon monoxide content and at a specified temperature. If any of these magnitudes changes, the carbon content of the steel will shift: first at the surface of the metal and progressing inward with time. For example, if a steel containing 1.1 percent carbon was placed in an atmosphere whose $CO/CO_2$ ratio was 20 at a temperature of 850°C, the composition would slowly decrease to about 0.4 percent carbon.

An atmosphere where the ratio is in the 100:1 range can produce a carbon level of greater than 1 percent at the surface of a piece of iron even when the temperature is relatively low. On the other hand, when the monoxide/dioxide ratio drops only slightly, the achievable carbon level plummets. When there are still ten parts of CO to each part of CO2, this level is only about 0.1 percent even at temperatures close to the melting point of iron, and those temperatures are hard to maintain in an open forge without pumping the bellows so hard that the monoxide/dioxide ratio drops far below 10:1. Decarburization and even oxidation would be the result.

Saturation is the upper limit to carbon absorption and is likely to be stabilized only at the surface of an iron object. The actual carbon content of the enriched zone decreases inward from the surface. This gradient of decreasing carbon content represents a dynamic balance between the rates of carbon ingress at the surface and of diffusion into the interior. Unless the iron object is very thin in cross section or unless carburizing conditions can be maintained for a period of many hours or days, the inner part of the object is not likely to receive any carbon from the surface. The technical term for those outer parts with increased carbon levels is *case*; the deeper unaffected part is called the *core*.

The growth of the case depth proceeds slowly. The difficulties inherent in forge cementation can be seen in Figures 11.1 and 11.2, the latter showing the rate of change as a function of temperature and time. An acceptable $CO/CO_2$ ratio is hard to maintain while keeping the air blast strong enough to hold the temperature above 900°C, but a temperature of at least 950°C is necessary to attain a case depth greater than a few millimeters within a few days. Clearly, the physical endurance of the smith and the bellows operators limits what can be done. No piece of iron more than a centimeter thick can be feasibly converted to a homogeneous steel in an open forge, and it is unlikely that the real case depth will exceed a small fraction of a millimeter.

On the other hand, homogeneity is not always necessary. When a hard penetrating point or a sharp cutting edge resistant to dulling is needed, the surface carburization of a larger object to depths of a millimeter or even less will be adequate. Repeated sharpening will ultimately penetrate below the high carbon case, but it is still possible to re-establish the case by returning the object to the forge and recarburizing. There has been no research on the possible procedures used, but we believe that carburizing in an open forge furnace is not likely to produce more than very superficial carburization.

Where it is necessary to impart some of the qualities of steel to the interior of an object—perhaps because its intended use requires greater elastic deflection than low carbon iron has—it is also possible to weld several surface carburized pieces together, creating a composite of layers with alternating high and low carbon contents. This technique is often called *pile welding* or *faggoting*.

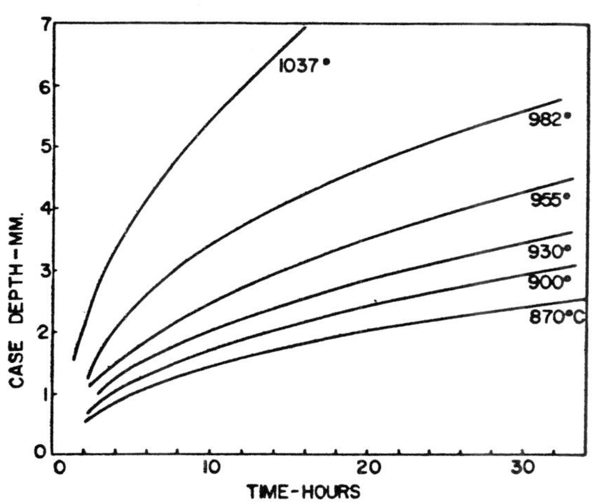


Fig. 11.2. *Penetration of a Carbon Enriched Zone.* The rate at which the case thickens or deepens is an important factor in the control of surface hardening of penetrating and cutting tools or weapons. The chart shows, for instance, that a case depth of 5 mm. requires about twenty hours at 1000°C in a carburizing atmosphere. (Adapted from *Metals Handbook* 1964, v.2: 93–114)

## 4. CARBURIZATION IN CLOSED CONTAINERS

The contradiction between the need for a stronger air blast for efficient combustion and the maintenance of carburizing conditions seems to have been recognized empirically long before the cementation process was understood. Smiths must have noticed that prolonged direct contact between the metal and vigorously burning coals yielded a steel of uneven hardness and that the fire would soften the edges of steel artifacts. The obvious solution was to protect the surface of the metal from the oxidizing flame while keeping it in contact with a substance capable of emitting carburizing gases. As in smelting in crucibles and reverberatory furnaces, the essential principle was to separate the carbon used as an energy source from the carbon used as a reagent: that is, to package the carburizing system and to immerse the package in a heating system.

The simplest way of accomplishing this was to coat the workpiece first with a carbonaceous material and then with a protective coating that was both refractory and airtight. Finely divided clay, borax, or even powdered glass worked well enough for the outer coating, while popular substances for the inner coating were powdered plant charcoal or charred hoofs, horns, or leather. For "tempering" (presumably steeling) files, Biringuccio recommends covering them with "soot of rams' horns or of ox hoofs, crushed glass, and common salt, all moistened with vinegar" (Smith and Gnudi 1942: 372). Theophilus suggests a similar recipe, without glass and vinegar, to be sprinkled on the file when it was red hot (Hawthorne and Smith 1963: 93–94). His second method—smearing the files with pig fat and wrapping them in leather strips tied with linen thread—would presumably have case hardened the objects more effectively (Hawthorne and Smith 1963: 94–95). For case hardening iron bars and forged articles, traditional

Russian smiths used "leather, shavings of horn and hoofs, etc., placed in a vessel or covered with clay and then put into a smith forge" (Kolchin 1967: 27). As late as the 1890s, American blacksmiths employed similar formulas for "steelifying" iron: mixtures of leather, hoofs, and salt; or of salt, sal soda, pulverized charcoal, manganese oxide, black rosin, and raw limestone (Richardson 1978: 257–261).

To proceed from a clay covering to a closed clay pot seems an obvious step, and the ideas must have developed at about the same time. The simplest receptacle is a clay box with a clay lid, air-dried and lightly baked. The clay might have a large volume percentage of charcoal powder as temper. The lid would be of a similar material and luted with a thin paste of the same clay after the box is loaded. The iron piece(s) would be packed with charcoal and/or other organic and inorganic substances (limestone is common). At least one of these must break down when heated to provide modest amounts of carbon dioxide, which is necessary to start the carburizing reaction.

The box can be placed directly in the forging furnace and buried in the charcoal bed. Although its temperature can be only roughly regulated (by observation of radiant colors—see Table 1.5), achieving a precise heat is much less critical than in open forge carburizing operations. Not only does the box buffer its contents against rapid temperature fluctuations but it protects them from inadvertent decarburization when the blast is applied too vigorously, for the large quantity of charcoal and the small amount of starter in the box insure a carburizing condition at all times.

This "pack" carburizing technique has survived to modern times. Sometimes known as *box hardening* to American blacksmiths, it is now performed in iron boxes or segments of pipe, but otherwise the practice is much the same. Richardson (1978: 263) states that a case depth of an eighth of an inch can be obtained by exposing the charcoal filled box or pipe to a cherry red heat for 24 hours and that 4 or 5 hours are enough "to make a good surface of steel." He considers the process useful mainly in case hardening artifacts that have already been given their finished shape, emphasizing that they should be cleaned and polished before they go into the hardening box. Richardson naturally knew that the actual hardening would be accomplished by quenching and tempering, but earlier the carburized surfaces of iron objects were often hardened by cold working. Livadefs (1956), for instance, has presented evidence that this was the procedure followed in making the iron clamps used to fasten building stones together in classical Greece.

From the seventeenth century onward, pack carburizing of bar iron was carried out on an industrial scale in northern Europe. In England, France, Germany, and Poland, most of the steel was produced by this technique, and it could be further processed into what by the standards of the day was moderately good metal for a variety of industrial purposes. A type of kiln-like "chest converter" furnace was commonly used for this purpose in England until the late nineteenth century (Fig. 11.3). The design is reverberatory; the long bars of iron were packed with charcoal into two large rectangular chests made of fireclay and heated in them for as long as eight days.

The immediate product of such a furnace was known as *blister* steel. The term describes the appearance of the carburized rods or plates: on removal from the furnace, their surfaces were covered with blister-like swellings, which were the result of a thermochemical reaction between dissolved carbon and the high FeO slag inclusions present in all wrought iron—bloom, fined, or puddled—of the period. As the carbon penetrated the surface of the bars, it reduced the ferrous oxide of the slag inclusions, so that carbon monoxide evolved at sufficient pressure to deform the steel and to force bubbles above its surface.[1] Although an unworked bar of blister steel might have looked like a product unlikely to be useable, the bubbles did not hamper performance, since they could easily have been welded shut.

The bars as they came from the furnace had a heterogeneous carbon content, and it was evidently difficult to form them into a finished product which had the higher carbon portions at the working edges, or faces, and the lower carbon portions at the back of the implement or in other places where hardness was less important. For this reason, it was customary to pile weld blister steel bars to even out the carbon content,[2] so that "shear" and "double shear" steels were made. Even these were unsatisfactory for some high performance applications. In the making of files, for instance, these steels "when ground ready for cutting would present no outward appearance of flaws; but the cutter's chisels would often penetrate parts which were unsound, occasioned apparently by a white powder embedded in the steel. . . It was a source of great annoyance to the file-makers" (Percy 1864: 838, quoting an anonymous contemporary). The problem, as Percy suggests, was the presence in the metal of "cinder"—that is, slag. Crucible steel processing in England was undertaken to homogenize blister steel and to rid it of the occluded slag that was present in even the finest "double-shear."

Seventeenth- to nineteenth-century literature on iron-

[1]One of the first to demonstrate this conclusively was John Percy, who persuaded the owner of a blister steel plant in Sheffield to melt a bar of Swedish wrought iron in a crucible and then to subject the resulting slag free ingot to the normal chest conversion process. Percy states that the metal produced "showed a fracture at each end characteristic of converted steel, but there was not the slightest indication of a blister" (Percy 1877: 462).

[2]Barraclough (1967) gives a comprehensive description of the developed cementation process for blister steel. Wrought iron bars, about 3 by 3/4 inches in section, were packed with charcoal in sandstone chests sealed with cement and heated to 1000° to 1100°C for ten to as many as fourteen days. According to the time and temperature involved, as well as the rate of cooling which could last up to ten days, the carburized bars attained average carbon contents ranging from 0.5 to 1.8 percent. For better homogeneity, the bars of blister steel were cut to about 18 inch lengths, stacked and wired in bundles of six to eight pieces, and hot forged to a single bar of about 2 by 3/4 inch cross section. This product was called *shear steel*, and was used to manufacture shears and other types of medium quality cutlery. Samuels (1980: 185–187, figs. 53.1–3) gives micrographs of typical blister, shear, and crucible steels. Contemporary descriptions of the developed blister steel process are given by Percy (1864: 768–773), Overman (1854, Steel: 120–133), and Osborn (1869: 878–879).

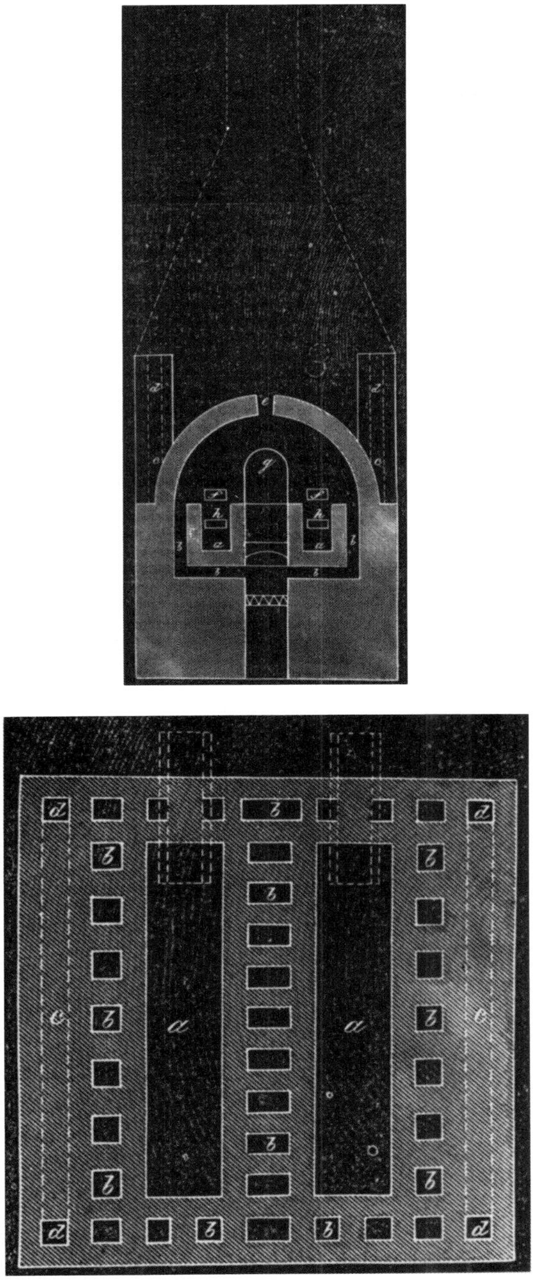

Fig. 11.3. *Eighteenth-Century Blister Steel Furnace.* Two chests (aa, width and depth at 66/91 cm. and length, 244/457 cm.) were built over a fire chamber. They were filled with wrought iron bars or plates separated by a mixture of charcoal, salt, and ash. The top of each chest was sealed with a mixture of moist sand and crushed, old brick to a depth of 15 cm. and luted with clay. The furnace was run for six to eight days for cutting grade steel and for less time for less demanding uses such as saws and springs. (After Osborn 1869: 873–878)

making contains statements indicating that only certain types of bar iron could be carburized satisfactorily (see, e.g., Overman 1854, Steel: 155–156; Barraclough 1984: 57). English steelmakers found it necessary to import all the bar iron as raw material from Sweden, and the French had little success in their longstanding attempts to modify the pack carburizing process so that it could function with the bar iron produced in France. Since this particular problem was eliminated in the later nineteenth century with the development of crucible, Bessemer, and open hearth steelmaking processes, it was never studied by scientific metallurgists, so explanations for it must be largely conjectural.

There are clues as to why some kinds of iron, satisfactory in other respects, made bad blister steel. An eighteenth-century Swedish manuscript (Barraclough 1984: app. 11) describes the appearance of the fractures of bar irons that have good or poor carburizing capabilities. Iron favorable for conversion to blister steel had a fracture surface with few "fibers" (slag ribbons or stringers) and a "fine" as opposed to a "coarse" grain size. This assessment is consistent with Richardson's (1978: 264) opinion, published in 1891, that iron suitable for case hardening should have a very fine grain size, and be ductile after quenching and free from "ash holes" (slag).

It is possible that the good characteristics are related partly to the absence of phosphorus. Bar iron that contains significant quantities of it has a characteristically shiny, faceted fracture as opposed to the matte appearance of the fracture in low phosphorus iron. Phosphorus also can decrease the solid solubility of carbon in iron at elevated temperatures. At about 950°C, for instance, a 0.6 percent phosphorus content reduces carbon uptake from about 1.4 percent to a maximum of 0.6 percent (Stead 1918) and, more seriously, causes a marked decrease in the rate at which carbon diffuses from the surface into the interior. Bar iron with too much phosphorus would therefore be inadequately carburized after lying in the furnace for the length of time sufficient to make a purer iron into steel. Both Richardson and the Swedish source make it clear that an excessive slag content, revealed by the degree of fibrous or woody fracture, was not favorable to making steel. This seems reasonable, for the slag ribbons in the bar would certainly act as barriers to the inward diffusion of carbon.

## 5. CARBURIZATION BY LIQUID CAST IRON

In principle, carbon dissolved in a liquid would be a much more efficient carburizing agent than carbon in solid or gaseous form. Almost all carbonaceous (hydrocarbon) liquids, however, will turn to gases long before carburizing temperatures are reached. The chief exception is the solution of elemental carbon in its principal solvent—that is, cast iron. A piece of bloom or wrought iron coated with molten cast iron will take up carbon readily because all parts of its surface are in contact with the liquid. The carbon in that liquid—unlike carbon in a solid—has excellent diffusivity, and the temperature of melted cast iron is higher than that generally used for pack carburizing. These advantages could not have been recognized in those terms by pre-industrial steelmakers, yet a number of early steelmaking processes exploited the principle of wetting solid low carbon iron with liquid high carbon iron.

One is the Japanese *tatara* process, in which a very large bloom absorbed liquid cast iron, so that its middle portion was converted to a much-prized but heterogeneous steel (Rostoker, Bronson, and Dvorak 1989; see Chapter 12 here). Another example, which according to al-Biruni originated in Sind, Pakistan (Allen 1979: 79), is the practice of sprinkling liquid *dus* onto steel sword blades while they are being forged. As al-Biruni states elsewhere that *dus* in the Herat area in Persia is mixed with ordinary iron in crucibles to make steel, Allen (1979: 75) reasonably interprets the term to mean cast iron.

This heating of the mixture in crucibles is a third method of carburization through the use of cast iron; al-Biruni's statement shows that it was known in the Middle East by at least the eleventh century. Steel made in crucibles charged with both wrought and cast iron was a traditional specialty of Hoki and Mimasaka Provinces in Japan (Geerts 1883: 541). The technique may also have been used in India since one of the recorded systems for making wootz steel employed cast iron rather than dried wood as the carburizing agent in the crucibles.

A closely related method, called by Needham (1958: 5–6) *cofusion*, is thought to have considerable antiquity in China, where clay coverings were used. The classic description of the Chinese process is included in the seventeenth-century *Tiangong Kaiwu* (Sun and Sun 1966: 250). Bundles of iron strips (about 13 mm. wide × 28 mm. long) were initially packed with pieces of cast iron and encased in a mixture of clay and straw. After the clay coating was dry, these packages were heated in a furnace to a temperature sufficient to melt the cast iron and held at that temperature for some time. Upon their removal from the furnace, the iron bundles had become welded lumps which had heterogeneous carbon distribution. These were then hot forged into strips, cut up, stacked, packaged in clay, and reheated, again to a temperature above the melting point of cast iron. The product of several repetitions of this cycle was a relatively homogeneous steel. It is not absolutely clear that the cast iron always became hot enough to melt. Whitaker and Williams (1969) have shown experimentally that crushed or powdered cast iron placed between layers of wrought iron and heated to 900° to 1000°C will be fully absorbed within about eight hours.

A similar steelmaking process is described by Biringuccio (Smith and Gnudi 1942: 67–69), who implies that it was a standard method in Italy during the sixteenth century. According to him, a round, deep receptacle, made of one-third clay and two-thirds charcoal dust, was formed within the bed of a forge. Small pieces of cast iron were mixed with powdered marble, crushed slag, "or other fusible and nonearthy stones" and placed inside. These ingredients were then covered with charcoal and ignited, the tuyere of the forge being adjusted so that the blast struck the center of the receptacle. When the metal had melted into a puddle of cast iron, partially consolidated blooms of low carbon iron were placed in the puddle or "bath" and kept there between four and six hours, being turned frequently "so that the solid iron may take into its pores those subtle substances that are found in the melted iron." Next, the blooms were extracted, flattened under the hammer, and tested by quenching them in cold water, breaking them into pieces and checking to see whether each fragment had the right appearance and, presumably, whether all of them had hardened, showing that they had been entirely converted to steel. They were then returned to the bath, withdrawn again, and hammered into bars that once more were vigorously quenched. The transformation that occurred was evidently striking. Whereas the original metal "resembled only a lump of lead or wax, in this way it is now made so hard that it surpasses almost every other hard thing. It also becomes very white, much more so than is the nature of the iron in it; indeed it is almost like silver" (Smith and Gnudi 1942: 69–70). Biringuccio's description is so clear that only a few details need comment. First, it is possible that the metal initially placed in the container was a low rather than a high carbon iron and that this was subsequently converted to cast iron *in situ* through the action of a powerful blast under strongly reducing conditions. This possibility is raised by the fact that Biringuccio in other translations (e.g., Agricola as quoted by Lister 1693: 571; Percy 1864: 808) seems to say that the initial metal and the blooms added later were the same kind of iron. Second, Biringuccio's idea that the solid iron is taking up through its pores a substance that transforms it shows remarkable prevision. The ability of low carbon sponge iron to aspirate liquid cast iron is also the basis of the traditional Japanese *tatara* steelmaking process (Chapter 12). Third, the explicit testing procedures are noteworthy. The idea of quenching a steel, breaking it, and judging its quality by the appearance of the fracture is ingenious and still used. A hard, high carbon, martensitic structure would have a specular—white and shining—fracture. Inadequately carburized zones would be indicated by the matte fibrous fracture characteristic of a soft pearlitic steel.

Biringuccio's method (also known as the Brescian steel process) appears to have died out in Europe within a century after his own death. As Lister shows, the idea of making steel by "boiling iron in its own sow metal" was considered a lost art by the English as early as the 1690s, and Réaumur's experiments of a decade or two later (Percy 1864: 806), though successful in producing steel by this technique, did not interest professional metalmakers. This is puzzling since, assuming that a suitably low sulfur and low phosphorus cast iron could be found, "iron boiling" should produce a metal that is better and cheaper than blister steel. The time required—between four and six hours as opposed to the ten or even twenty days needed by the chest conversion process—should have been by itself a decisive factor in favor of carburization with liquid cast iron. We can offer only one explanation for its disappearance in Europe: that it was not easily applicable to the tonnage production which was rapidly becoming a primary objective.

## 6. WOOTZ OR INDIAN CRUCIBLE STEEL

*Wootz*, a word of uncertain etymology,[3] has since the early nineteenth century been the standard term in Western metallurgical literature for the crucible steel of India. The steel itself had been known long before in the Middle East and the Mediterranean. Although some of the more extreme statements about its antiquity and excellence may be discounted, there can be no doubt that by the late first millennium A.D. India was exporting substantial quantities of this steel to the West. Wootz was made in crucibles, sold in "cakes" that were shaped like the crucibles, and forged—in Iran and elsewhere in the Middle East as well as in India—into sword blades, some of which had a characteristic pattern called by Europeans *damasked* or *damascene* from its supposed origin at Damascus in Syria (Bronson 1986a).

The steel was often forged to final shape in the Middle East and was sometimes manufactured there as well,[4] but by the time Europeans began to write about the material, India had a near-monopoly on it. The metal that Europeans associated with India made good cutting tools but was extremely hard to work, as noted by della Porta (1658/1588) in the sixteenth century, Moxon in the seventeenth century (Smith 1960: 24), and Réaumur (Sisco and Smith 1956: 176) in the early eighteenth century. Like the early English experimenters of the 1790s, these three accounts imply that the metal had to be forged at a dark or cherry red temperature to avoid cracking, which suggests that the Indian steel known to della Porta and his successors was similar to what later came to be called wootz.

Metallurgically, wootz is a macro-homogeneous steel containing more than 1 percent carbon. Swords made of it are thick, inclined to fracture with small flexure, and very hard and sharp. They also can have surfaces which when polished (and usually, lightly etched) show the damascene structure: a pattern of markings often compared with fish roe or watered silk (Fig. 11.4). This structure was long considered a guarantee of superior performance and has attracted much scientific attention since the early nineteenth century—the damascus sword is one of the classic puzzles in the history of scientific metallurgy. Several reliable observers saw wootz being made during the nineteenth century. The secondary literature focuses on only a few of the reports, not always the most reliable, and usually attempts to rationalize their descriptions in terms of each other, combining them to produce a synthesis of *the* wootz manufacturing process. This approach is based on the questionable assumption that each observer is attempting with varying success to describe the same process. A careful reading of the sources shows that, in reality, Indian steelmakers employed crucibles in a variety of ways.

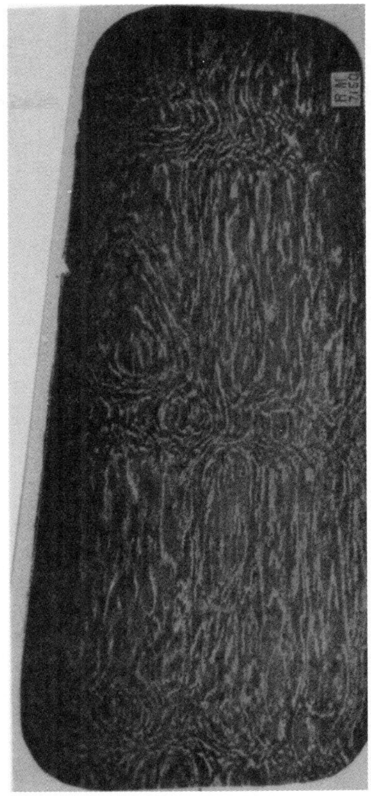

Fig. 11.4. *Damascene Pattern of a Sword Blade.* The white (bright) lines represent bands of iron carbide standing against a dark background of deeply etched steel of a high carbon content. The carbide bands are inherited from the solidified structure. (From Sherby and Wadsworth 1983–84: 29)

The common denominator of all true wootz making systems (some so-called wootz was made in other ways[5]) was the use of small crucibles, typically between 10 and 20 cm. high, 5 or 6 cm. in external diameter, and more or

[3]*Wootz* as a term does not seem to have ever been common in India. The word first appears in a letter written in January 1794 by Dr. Helenus Scott, a physician in Bombay, to Joseph Banks, then president of the Royal Society in London: "I enclose in one of the boxes a specimen of a kind of steel which is called wootz and is in high esteem among Indians" (Franklin 1971: 270). The first appearance of wootz in print was in a 1795 report of a lecture by George Pearson, who as one of the distinguished scientists chosen by Banks to investigate the specimen had access to Scott's unpublished letter. By the mid-nineteenth century the term was so well known as to be a standard entry in encyclopedias. Whether Scott heard and transcribed it correctly and which Indian language it might have come from are still controversial issues. Several South Indian languages call steel "ukku" or "wukku." Yule and Burnell (1903) favor a Canarese origin for the word transcribed by Scott; Krishnan (1952: 505) prefers Telingana.

[4]Early Middle Eastern writers describe crucible steelmaking in some detail. These include al-Biruni in the eleventh century (Allan 1979: 75) and al-Tarsusi in the twelfth century (Cahen 1947–48: 127–128), both of whom make it clear that the processes were widely used and well understood in various parts of the Islamic world.

[5]T.H. Holland (1892: 147–148) discovered that all steel made in the Salem district near Madras was in his day produced by an unusual noncrucible process: by collecting the cast iron "shot" that accumulated in bloomery slag (the bloomeries in question were presumably run at temperatures high enough to yield more than the very small amounts that are normally present in such furnaces); by cleaning this through pounding it in a mortar; by throwing it on top of burning charcoal in a small pit furnace; and by applying a decarburizing blast. After a half-hour, the blast was stopped and a cake of steel was removed. Holland states that "the workers are well aware of the fact that if they continued the process too long, the resulting product would be of no more use than ordinary wrought iron." The steel cakes were apparently sold as wootz and in 1890 were the only type of steel made or remembered to have been made anywhere in Salem. Holland pointedly does not comment on the fact that Heath (1839), in his famous description of the crucible process for making wootz, implies that this was in common use in Salem, where Heath lived in the 1830s. Holland seems to suspect that Heath was guilty either of carelessness or of the lesser sin of presenting hearsay in the guise of firsthand observation.

less conical in shape. The clay, often tempered with charcoal and numerous other ingredients, had to be sufficiently refractory to survive heating to at least 1400°C. The crucibles when charged were sealed with either a wad of moist clay or a clay lid luted with a clay slurry. Sometimes it was perforated with a small hole.

The crucible charge was usually bloom iron forged into bars, presumably to eliminate as much slag as possible, and then cut into pieces small enough to be packed into the crucibles. Schwartz and Hunter report that the charge was composed of ore rather than smelted iron, while Voysey, who probably saw more wootz being made over a longer period than any other observer, says that two kinds of iron, one hard and the other soft, were placed in the crucible. He claims that nothing except slag was added, an indication that he might be describing carburization by cast iron.[6] Most other sources state that carbonaceous substances were also added: charcoal, wood, and—an especially famous ingredient—green leaves. The fact that several commentators copy from one another the botanical names of the plants used does not inspire confidence in the independence of their data, but there can be no doubt that leaves were often included, especially in Mysore (Buchanan 1807, v.1: 174 and Heyne 1814: 358, 361), and that dry wood was generally preferred to charcoal.

Later commentators have often observed, and probably rightly so, that herbaceous matter serves no purpose in this process. Wootz making is, however, essentially a pack carburization process carried out at an unusually high temperature and so is subject to the same requirement as ordinary pack carburization in a sealed container—a small quantity of carbon dioxide "starter" must be present to initiate the conversion reaction. A gram or two of the wood or leaves in the charge, or some hair and chaff in the crucible wall, could have done this.

From five to as many as fifty crucibles were loaded into the furnace for a production run. Some furnaces were simple firepits, but others designed to handle large numbers of crucibles may have been built with substantial enclosing walls. The crucibles were stacked or arranged in rows or in a dome-like pattern above the firepit. The fuel was charcoal and there was an air blast through a tuyere. The bellows attached to this tuyere could be large. Voysey and Schwartz both say that in Hyderabad four bellows made of entire bullock hides were used; in Mysore and Tamilnadu, pairs of bellows were more typical.

As heating proceeded, the contents of the crucible underwent what amounted to pack carburization except that the temperature and carbon present were sufficient to bring the average composition of the steel to at least 1.5 percent carbon. As the temperature was above the liquidus for this steel, it melted. By the time the crucibles were removed from the furnace or the furnace was shut down with the crucibles still inside, they contained a pool of liquid with a homogeneous carbon content. Several sources (Heyne, Coomeraswamy, Wood, Holland) state that complete liquefaction was an explicit goal of the wootz makers, who would place individual crucibles directly in the air blast until, by lifting and shaking them with iron tongs, the motion of a liquid could be felt. "C.V.B." decribes an operation where as many as 54 crucibles were stacked in one furnace. This was then heated for about nine hours, after which the crucibles were extracted and, when cooled, cracked open. The twenty to thirty that contained a fully fused lump of steel were considered finished. The others were reloaded into new crucibles and returned to the furnace for another run. (For an extended and detailed discussion of the making and marketing of wootz steel, see Bronson 1986a.)

## 7. ENGLISH EXPERIMENTS WITH WOOTZ STEEL

European experiments with "damascus" steels go back to at least the sixteenth century, but it was not until the 1790s that laboratory researchers began to work with steels that were specifically known to be Indian and that can be securely identified as wootz. Beginning with tests carried out on a small number of samples shipped from Bombay to the Royal Society in London in 1794 (Pearson 1795), work on the material increased rapidly. Within twenty-five years wootz had attracted the interest of many leading scientists and had inspired patents and commercial products, among them David Mushet's crucible steels (Mushet 1805; Percy 1864: 778), Stodart's surgical instruments—which were actually made of imported wootz (Stodart 1818)—and the "aluminum" and silver steels of Faraday and Stodart (Faraday 1819; Stodart and Faraday 1820).[7]

Two problems troubled British experimenters: how best to work imported wootz and how to reproduce wootz and the damascene structure. The literature of the period shows no evidence that a real effort was made to prove that wootz, whether imported or homemade, was a better steel than that already being made in Europe. The formidable reputation of Indian steels and Persian swords was enough for most researchers

[6] "The materials used in the preparation of the steel are two different kinds of iron; one from Mirtpalli—the other from Kondapur, in the proportion of three parts of the former to two of the latter. The Mirtpalli iron is derived originally from the iron sand already mentioned, and is sent in the state of large amorphous masses of a reddish grey color and of an extremely porous texture. The Kondapur iron is procured from an ore found amongst the iron clay, at a place about 20 miles distant. It [the ore] is said to be of a dirty brown color, and very frangible. The iron, however, is moderately compact and of a brilliant white fracture" (Voysey 1832: 246–247). Whether this second metal is cast iron is not clear, but it sounds like it. However, Voysey was technologically knowledgeable, so probably could have identified cast iron and recognized that its presence in an indigenous Indian ironmaking process was worthy of comment. That he does not make the identification therefore suggests that the second metal may have been only a steely bloom iron. As Section 5 of this chapter indicates, the "brilliant white fracture" could stem from a high phosphorus content.

[7] J. Holland (1831, v.1: 248–249), writing in Lardner's *Cabinet Encyclopaedia*, gives an amusing account of the marketing of the numerous patent metals inspired by the fad in the 1820s for exotic steels. Although products called *silver steel* and *damascus steel* were widely promoted (the former by Michael Faraday), Holland notes that these terms had disadvantages: "Silver, however, is an ingredient within the reach of every manufacturer who chooses to put it into his melting pot; and Damascus being a somewhat equivocal epithet, at best applying rather to a mode than a material of steel-making, a third party procures from abroad a few lumps of the celebrated Indian *wootz*, a substance which not many persons have seen and fewer still thoroughly understand, and under these auspices comes forth *Peruvian steel*. . ." Silver does not alloy with iron either in the liquid or the solid state.

who continued to try to discover the secret, and in the process, become famous and rich.

The difficulty of working wootz is a recurring theme. Like della Porta and Réaumur before them, English experimenters found that the metal could not be heated much above a color that they describe variously as "a slight red heat" (Scott in Franklin 1971: 270), "a pale red color" (Pearson 1795: 323), and "cherry red" (Stodart 1818: 571). Stodart comments on the "unmanageable shape" of the small crucible-shaped wootz cakes and on their often nonhomogeneous texture. In his cutlery business he seems to have been obliged to remelt most wootz in order to achieve a sufficiently uniform raw material, although he continued to claim that wootz steel, when worked by the proper techniques, was "vastly superior to the best cast steel of Europe."

The requirement that forging be conducted at or below the cherry red limit meant more work for the forgeman, because steel is less easily deformable at that temperature and the workpiece would have cooled quickly, meaning that it had to be returned to the forge more often during the shaping process. The explanation for this is not entirely clear. A very high carbon steel will crack during forging if the temperature reaches its melting point. A homogeneous 1.5 percent steel will begin to melt at about 1250°C: that is, above the lower limit of white heat (Table 1.5). A steel with an even higher carbon content might melt while still at a light yellow color, at about 1150°C. However, this is still much hotter than the temperature indicated in Scott's letter to the Royal Society, which Pearson (1795: 322) quotes as saying "that when wootz is heated above a slight red heat, part of the mass seems to run, and the whole is lost, as if it consisted of metals of different degrees of fusibility." Scott must be mistaken about this. Even if the wootz in question contained substantial amounts of cast iron with 4 to 5 percent carbon, the indicated melting point would still not be below 1143°C—far above the 800° to 850°C indicated by any color that could be called a "slight red."

The reason for the sensitivity of wootz to high forging temperatures must lie in the phenomenon of carbide network formation. In a very high carbon steel heated to temperatures in the orange range, about 900°C, brittle carbide networks will form at crystal boundaries as it cools. Forging would go well for the first few strokes but then, as the temperature dropped, the structure might suddenly become brittle. Stodart's cautions against temperatures higher than cherry red (about 700°C) make sense because, if the steel is kept below that temperature, carbide networks do not form around grain boundaries on cooling.

Only in the first quarter of the twentieth century was an adequate theory developed for the deformation processing of steels with very high carbon levels. Significantly, the pioneers in this—Belaiew (1912, 1918), and Desch and Roberts (1923)—were engaged in research on the damascus problem. Their theoretical explanations centered on the spheroidization of the carbides present through a long-time anneal at 750°C prior to hot forging at a temperature not much higher than that. This previous heat treatment creates a spheroidal carbide structure, permitting extensive cold deformation without cracking, as in the rolling of razor blades (Fig. 11.5).

The excellence of the cutting edge of a sword produced from wootz steel derives from its high carbon content. However, the fact that it can be successfully forged and that it has moderate toughness afterward is possibly due to a spheroidizing anneal, a process discovered originally by Indian and Iranian craftsmen and later by European manufacturers of razors and similar implements. Belaiew (1918: 433) suggests that, in the case of wootz, the crucial annealing step was performed immediately after the steel was made, citing the "annealing" operations described by Voysey and other observers of wootz manufacture. If certain of the craftsmen who fabricated the wootz into swords did not understand this lengthy annealing, then they would have had to be extremely careful about how high they heated the metal; once they allowed it to reach a yellow heat, the globular carbides would start to dissolve in the iron, making the workpiece brittle as it cooled.

Some swordmakers evidently did know how to re-anneal a wootz blade. Egerton (1880: 59) quotes Murdoch Smith as saying that craftsmen in Shiraz and several other Persian cities would place the forged object "for six or eight days in the furnace of a hot bath,

Fig. 11.5. *Microstructures of Very High Carbon Steel.* The micrographs illustrate the carbide arrangements reputed to be crucial to the workability of wootz steel, and are actually a modern razor blade steel containing more than 1 percent carbon. The globular dispersion of carbides permits both hot and cold working. The carbide network makes deformation at any temperature difficult. Both ×800

where the greatest attention has to be paid to the even heating of the article," using dried dung as fuel because of "the steady and not very intense heat" it gives. If the "bath" was a container filled with a dry material like sand rather than water, the treatment would have produced a structure of spheroidized carbides. The temperatures might have been close to the 750°C specified by Desch and Roberts (1923) for the production of an optimum "globular" structure.

The toughness that can be gained from such treatment should not, however, be exaggerated, as General Anosoff, the great Russian scientist, who experimented with damasacus steel in the 1830s and 1840s, may have when he claimed:

> A dagger-blade of good damascene steel, properly polished and correspondingly hardened, cannot be broken by bending, but can yet be bent to such an extent as to lose its elasticity. When bent in the usual fashion, the blade flies back and retains its original shape. When bent more forcibly (say to a right angle) it does not break or lose its original elasticity after straightening again (Belaiew 1918: 424).[8]

In reality, as arms experts (Abbott 1847: 421; Egerton 1880: 57) have maintained on the basis of experience, and as Zschokke (1924) has shown through testing in the laboratory, neither Indian nor Persian swords were especially tough or flexible in comparison with European swords. Weapons made of wootz could be extraordinarily hard and sharp, and some of them may have approached the limit of hardness possible in a carbon steel. However, without sophisticated alloying and the "stressed quench" techniques used in making Japanese swords, wootz weapons could not have been tough. While the use of a spheroidizing anneal made them workable and, if they were thick, usable, they were not superweapons. The nineteenth-century Europeans who attempted to duplicate wootz steel and damascene swords made notable contributions to the development of scientific metallurgy, but their efforts were misdirected. The high performance steels of the future would be developed in other ways.

## 8. THE ORIGIN OF THE DAMASCENE PATTERN

The legendary damascus effect stems from the presence of crystallized carbides, which have a freeform pattern that can be revealed by etching a carefully polished blade of forged wootz steel (Fig. 11.6). The etchant can be a dilute inorganic acid, a solution of copper sulfate (as suggested by Wilkinson 1837), an organic acid or, classically, a hot solution of ferrous sulfate or "zag" (Smith 1960). A typical damascene pattern (Fig. 11.4), developed by differential attack of the etchant on the steel, is composed of alternating dark and shiny bands which metallographically are zones of high and low populations of crystallized carbide. The banded effect can be induced only in certain wootz-type steels. Other steels of comparable carbon content do not have it, despite many nineteenth-century efforts to reproduce it.

Modern understanding of the damascening effect began when Wilkinson, the sword manufacturer (1837: 191–193), became the first European to make an explicit correlation between the visible crystals on the surface of an unforged cake of wootz steel and the patterning on the finished sword. His own forging experiments with wootz cakes showed that some yielded the damascene pattern and others did not. The latter, he noted, were the cakes that "gave only slight indications of a pattern, the crystals being very small." He suggested that the complexity of the crystalline pattern would be increased if, as was necessary when an actual sword was being made, several cakes were welded together by folding and reforging.

Belaiew (1918), combining his own work with that of two earlier Russian investigators, Anosoff and Tchernoff, further clarified the issue. He demonstrated that the etched pattern reflected bands with high and low populations of coarse, globular iron carbide which had been twisted and turned during the deformation processes. The filed and etched surface of a blade showed these bands in planar section. Like Wilkinson, he associated the bands with the large crystals evident on the surface of solidified cakes of wootz, attributing their size to slow freezing in the crucible.

As indicated above, the time necessary for freezing might not have been lengthy. Belaiew thought in terms of days, which is why he attempted to show, contrary to the ethnographic evidence, that the wootz makers described by Voysey and others cooled their furnaces down slowly. Understanding of the freezing transition has advanced since Belaiew's day. It is now clear that the solidifying metal had to be held within the transition range of 1350° to 1150°C for only an hour or so, which would have been easy in an Indian furnace where small crucibles could be manipulated frequently within the furnace, but difficult when, as in English crucible processes like those of Huntsman and Mushet, the steel was melted in large immobile crucibles and then poured into ingot molds where freezing would proceed rapidly.

In theory, the segregation of carbides is associated with the formation of dendrites during the freezing process. The dendrites themselves are relatively low in carbon, but as they form, the liquid in their immediate neighborhood becomes abnormally carbon-rich, and solidifies in bands around the dendrites during the last stage of freezing. The larger the crystals formed by freezing are, the thicker the bands and the greater the space separating them—both features that contribute to the visibility of the bands when etched. Similar carbide patterns are in fact present even in crucible

[8]General Anosoff's opinion evidently is based on the damascene swords he himself manufactured in the 1830s and 1840s in the Russian government arms factory he directed at Zlatoust in the Urals. Abbott (1847: 420–421) not only visited Zlatoust but also investigated damascene swordmaking in India, at Jalalabad in Gujarat. The Gujarati smiths forged swords with a good damascene pattern from cakes of wootz; the methods used by Anosoff have never been clearly understood. While Abbott is enthusiastic about the Russian blades, he has serious reservations about the Indian ones. With regard to a sword made by Anosoff, he comments: "In appearance it differs from the Jullalabad blades chiefly in the greater uniformity of the interlaced streaks; attributable probably to a better figure in the mass of steel from which it was forged. It is perfectly elastic. The simple damask of Jullalabad being tempered in oil, has little elasticity, and the makers will not warrant it to undergo any proof. It is liable both to bend without recovery and to snap short on concussion."

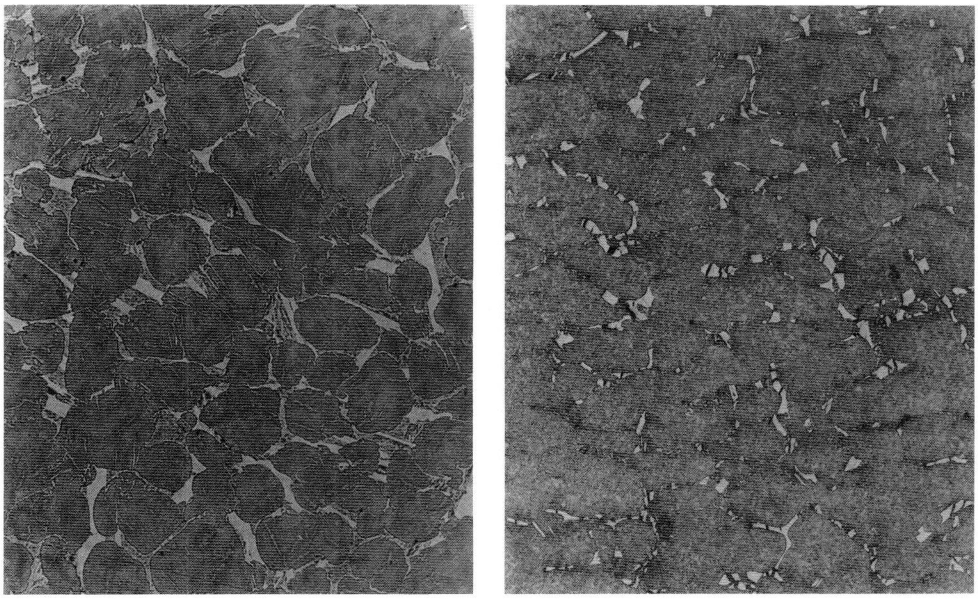

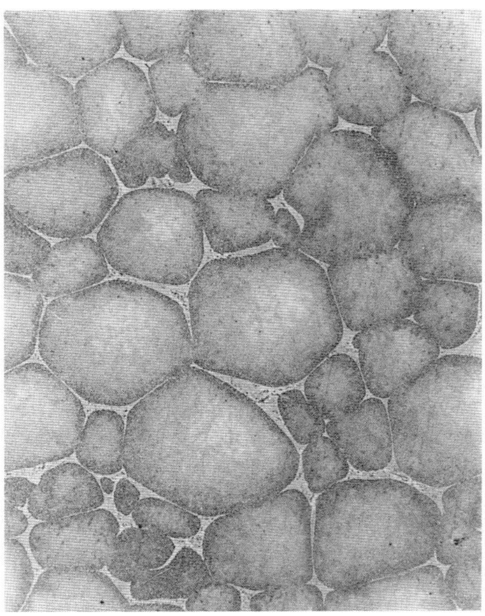

Fig. 11.6. *Carbide Networks Produced by Solidification.* The steel contains 1.3 percent carbon. Left: structure of a slowly solidified ingot showing a network of massive carbide (the bright line pattern in Fig. 11.4). Top left: after hot rolling, the massive carbides are cracked but the cracks are sessile and the piece shows no surface cracks. Top right: the grains of austenite are enlarged by heating to the beginning of melting regenerating the carbide network. The temperatures of melting and hot working are very close. All ×50

steels that have been rapidly solidified, but these are visible only under a microscope.

"Micro-damascene" steels would have properties identical to those with fully developed damascene figures on their surfaces. While the emphasis placed on the macroscopic patterns by weapon makers and users may have resulted in part from the prestige of owning a weapon easily recognized as costly and rare, there could have been other reasons for their desirability. A sword made of a crucible steel without a visible damascene pattern might have been as good, but macroscopic watering may have been a hallmark of the quality of the blade that was difficult to fake.

Whether these patterns can be attributed entirely to crystallization effects is also not clear. The free form of the watering on a damascene blade is consistent with the irregular three-dimensional grids that represent the carbon segregations of the contiguous crystals in a solidified steel. Not only is each grid unlike others in principal dimensions but each is oriented with respect to the rest in a random manner. Thus, the production of the pattern on a finished blade may require only that the swordsmith preserve a pre-existing configuration through the steps of fabrication. However, Wilkinson, a skilled swordmaker, felt that a certain amount of piling and folding was involved. Smith (1960) has argued that the intrinsically three-dimensional carbide segregation patterns in coarse grained cast steel structures cannot fully account for the damascening effect, maintaining that some intermediate cutting or engraving stage must be introduced so that desired patterns appear in the final forging. Since other specialists are convinced that this is not necessarily true, the issue has not been resolved. Even though we seem to have reasonable explanations for the origin of the damascene pattern, no one has yet replicated the true Damascus sword.

A final problem in reproducing a damascene pattern is the choice of etchants. If the bands of carbide crystals are to be distinguished visually, they must react with the etching solution at a markedly different rate from the steel matrix. While carbide reacts more slowly than the steel matrix to ordinary modern etchants, the differential is brought out by very short times of immersion in the solutions and is apparent only under the microscope. For macroscopically visible differences to develop, the chemical process must be slow enough that it can be controlled. Moreover, the carbide itself must be almost totally resistant to the etchant, so the steel can be etched to a lower level than the polished carbides, resulting in the characteristic relief. This effect is less dramatic when inorganic acids are used instead of some of the naturally occurring organic acids like those in citrus fruits and the metal sulfates such as the *zag* employed by Persian weaponsmiths. As with other proprietary processes in making specialty irons and steels, the etching of damascus blades required not only skill, but also knowledge of formulations that were kept secret from the outside world.

## 9. EUROPEAN CRUCIBLE STEELMAKING

The first European to use crucibles in the production of steel is usually said to have been Benjamin Huntsman, at an uncertain date in the mid-eighteenth century.[9] He was not trying to convert iron to steel, but rather, in seeking a better metal for clock springs, he devised a system for melting blister steel in crucibles and pouring this into ingots (Carr and Taplin 1962: 4), which naturally had a more homogeneous composition than even double-shear steel and were largely free of slag occlusions. This metal served his purpose well: like music wire (Chapter 13.10), a thin ribbon spring must have a uniform carbon content throughout its length and few if any nonmetallic inclusions that might cause fractures. In Huntsman's melting system (Tylecote 1976: ill. on 127), crucibles were buried in coke on a small, horizontal hearth. A chimney about 15 meters high provided a natural draft in the manner of the chest converting furnaces used in making blister steel. Melting took as much as six hours.

The first European to employ a crucible to melt steel was not a colleague of Huntsman but a scientist on the Royal Society team that carried out the initial experiments with wootz. David Mushet patented his process in 1800, four years after the Royal Society project began, and promptly sold his patent for the substantial sum of 4000 pounds. His first publication on wootz did not appear until 1805. Although he implicitly denied that he knew about the use of crucibles by Indians, Percy (1864: 778) does not doubt where Mushet got the idea: "It is curious that Mushet's process, as far as relates to the use of malleable iron in the production of cast-steel, should in principle, and I may add even in practice too, be identical with that by which the Hindoos from ancient times prepared their wootz. I cannot discover any essential difference between the two." By the middle of the nineteenth century, Huntsman's process too had been adapted to making steel from wrought iron. Writing in the 1880s, Greenwood (n.d.: 229) recommends that for the harder kinds of steel the crucible should contain pieces of bar iron, powdered charcoal, and fragments of *Spiegeleisen*, manganiferous cast iron. For milder steels, the necessary carbon would be supplied by the walls of a "plumbago" crucible made by mixing refractory clay with coke, charcoal, or graphite.

## 10. COMPOSITE STEELS

Welding is usually a fabrication technique: a method for joining pieces of iron or steel either by local fusion or by hot forging with sufficient deformation to effect an interfacial bond between solids without fusion. In some cases, forge welding can also cause a significant improvement in the properties of the metal, rendering it denser and more homogeneous.

Welding intended to promote homogenization was once routine in many steel manufacturing traditions. The process, known as piling or faggotting, entailed the bundling of several unevenly carburized bars and forging the bundle down to the dimensions of a bar again. The potential of this simple method for homogenizing the structure of a metal can be appreciated by considering the working up of the steels used in the sides, the back, and especially the edges of Japanese swords. The steels as they were received by the swordsmith were pieces of the large blooms made in tatara furnaces and were often unevenly carburized. However, smiths routinely subjected them to repeated pile welding, forging them out, folding them, and forging them out again. The process was normally repeated between fifteen and twenty times, but thirty or more repetitions were not unknown (e.g., Joly and Inada 1913: 84).

How this treatment effects a microscopic-level rearrangement of structure can be understood by visualizing a bar two centimeters thick, heavily carburized on one side and uncarburized on the other. If this bar were subjected to hot forging and folding only twenty-five times, and if it were possible to forge each layer perfectly parallel with the others, the resulting structure would be a sandwich of layers each 1/3355 of a micron thick, with the carbon at least as evenly distributed through it as if the steel had been melted in a crucible. Although in reality the results would be less perfect than this, the Japanese system, simple and mechanical though it seems, can produce steels that have a composition as uniform as any made by the wootz process.

Other pile welding systems tended to be less elaborate. Typical of many of these is the one used in eighteenth- and nineteenth-century England to work

[9]It is difficult to believe that earlier European ironworkers did not sometimes use the crucible melting technique which was widely practiced in the Middle East and North Africa (Chapter 11.6). Smith and Forbes (1957: 35) imply that Europeans probably did, citing the fact that both Joseph Moxon in the sixteenth century and Robert Hooke in the seventeenth appear to have been familiar enough with cast steel to mention it casually.

up bars of blister steel as they came from the converting furnace:

> Bars of blister-steel are broken into pieces about 16 in. long, and drawn out at a red heat under the hammer to the width of about 1 ½ in., and the thickness of about ½ in. Several bars thus obtained are piled on one another, and the end of the pile is firmly wedged into an iron hoop with a handle attached to it. The free end of the pile is raised to a welding heat in a hollow coke fire, during the process being dusted over with a finely-pounded clay, welded under the hammer, and reduced to about 2 in. square. The hoop is now detached from the other end of the bar, and this end is treated in exactly the same manner as the first end. In this state it is termed "single shear-steel." It may be broken into two equal parts, which may again be welded together and drawn out to any required dimensions to form "double shear-steel" (Percy 1864: 859).

The layers within a bar of double shear steel were not thin, but the steel itself would have achieved reasonable uniformity in composition through carbon diffusion. English cutlers considered double shear steel to be of high quality. A moderate level of homogenization was evidently enough for most of the purposes to which pre-industrial steels were put.

In Germany, though perhaps not in England, greater uniformity was achieved by numerous repetitions of the doubling process. As early as the sixteenth century the welded bundles of bars, usually between eight and ten in number and carefully selected to have a similar steel content, could be folded and reforged four or more times, yielding a rod of *Garbstahl* (= shear steel) composed of 160 or more layers (Beck 1893–95: 490–492). Steel of this quality was used for springs and coinmaking dies, while steel with fewer laminations served for cutlery. In Japan, too, the number of times a bundle was folded depended on the use to which the steel would be put. The encyclopediast Ono Ranzan (quoted by Geerts 1883: 542) recommended fifteen times for sabers and swords; thirteen for razors; eleven for the blades of carpenter's planes; five for kitchen knives; and four for ordinary knives.

In many circumstances a highly homogeneous structure in finished implements was not desirable. No premodern steel of uniform composition could match the range of properties attainable in a composite of steel and iron or of high and low carbon steel. The fourteenth-century Spanish writer Ibn Hodeil, in comparing the homogeneous wootz ("hindy") swords used by his Moorish fellows with the composite blades of their Frankish (that is, western European Christian) enemies, states:

> If the two edges of the saber are of steel while the blade itself is of iron, the saber is called *mudakkar* (steeled), which is typical of Frankish swords. The Arabs claim that the fabrication of these is the work of Genies. They are more resistant to the blows one gives with them even during cold weather, while the *hindy* saber often breaks when the weather is cold and shows itself better [than the Frankish sword] when the weather is warm (Mercier 1924: 231).

Here, the advantage of the composite blade is said to lie in the protection it gives against cold shortness, presumably an effect of an excessive level of phosphorus in the steel. If one of the metals were iron or low carbon steel, such a structure would also help to alleviate the normal brittleness of a steel with a relatively high carbon content: the high carbon component would give it strength and hardness, while the low carbon one would make it tough. (The subject of welded heterogeneous structures is discussed further in Chapter 15.)

The more complex steelmaking methods of the pre-industrial period show that early craftsmen possessed considerable technological skill. However, the most elegant solutions, then as now, were not invariably the best. While creations such as wootz and the true damascene sword may be justly admired, it is important to remember that simpler methods requiring great force often worked, and worked well. Pile welding is one of these simpler techniques which enabled smiths in many parts of the world to produce serviceable tools and weapons. Japanese smiths, with this intrinsically simple technology, produced items that may have performed better than any piece of steel made until the twentieth century.

# 12. Carbon Steel as a Smelting Product

## 1. INTRODUCTION

So far, we have discussed two forms of iron that can be taken directly from the furnace, one which is almost free of carbon and the other saturated with it. Between the two, however, is a third form—steel—with an intermediate carbon level of roughly 0.1 to 1.8 percent. It is difficult to produce directly in the smelting furnace (Chapter 11.1).

## 2. INADVERTENT CARBURIZING

Ancient bloomeries produced carbon steel. Pleiner (1980: 397–398) describes a pre-Roman Celtic bloom iron smelting furnace which had a deep hearth filled initially with brushwood. In the early stages of the smelting campaign, the bloom and slag were supported by the packed brushwood at a relatively high level within the furnace. As the brushwood slowly burned, the liquid slag began to run to the bottom, baring the bloom of iron to the atmosphere of the combustion zone. Depending on the $CO/CO_2$ ratio in this zone, the exposed part of the bloom might have been carburized from an initially low carbon state or decarburized from an initially high carbon state. Or, to look at this process in another way, the brushwood was converted to a bed of charcoal upon which the bloom rested (covered with slag). Physical contact with an essentially unlimited supply of carbon could carburize the bloom wherever contact was made. Whatever the mechanism, a bloom of inhomogeneously carburized steel resulted, so the process has been taken to account for forged artifacts with highly variable carbon contents in no definable patterns (Williams 1977; Tylecote 1976: 55; Pleiner 1980). In the bloomery, the atmosphere must be reducing in order to produce metal; it needs to be only a little more so to be carburizing. On the other hand, the forging preheat hearth is intended to heat as quickly as possible, which demands that fuel combustion be as efficient as possible and therefore that the atmosphere be nearly pure $CO_2$. Moreover, hammer scale was an invariable product of bloom preheating. Carburization and the generation of hammer scale cannot take place at the same time or even in sequence, because the rates of oxide film formation are always higher than the rates of carburization.

Tylecote and his associates (1971) in their experiments with a model bloom smelting furnace found that twenty of thirty-two runs produced blooms, ten of which had carbon contents in the 0.2 to 1.8 percent range. Similar replicative bloom iron smelting in Austria had the same results (Vetters 1966). Since the associated slags were high in FeO and $Fe_2O_3$, usually a characteristic of slags from low carbon iron smelting, the experimenters felt that it was necessary to explain such high carbon levels in the product metal. They offered models for the occurrence of high carbon levels in metal immersed in a slag pool high in iron oxides:

a) carburization by charcoal fragments occluded in the bloom;

b) exposure of reduced iron particles in the shaft to carburizing gas conditions followed by rapid oxidation just above the flame zone so that the composite particles falling rapidly into the slag pool partition the oxides to the slag and the high carbon, iron kernels to the bloom;

c) a modification of b in which the high carbon particles become coated with slag reduced to a low iron oxide level of higher melting point so that the particle of metal is protected from the oxidizing blast.

Models b and c are not unlikely, but they depend on the precision of nonequilibrium events. Model a is clearly feasible.

There is at least one other mechanism by which inhomogeneous carbon contents can be obtained in iron blooms (Chapter 9.7). The tuyere level of the shaft furnace is highly nonuniform both in temperature and $CO/CO_2$ gas ratio. The combustion space projects partially across the diameter of the furnace. Within that space and immediately above it, the temperature is much higher than elsewhere, but in it the combustion gas is almost pure carbon dioxide (Kinney, Royster, and Joseph 1927; Elliott, Buchanan, and Wagstaff 1952). Only a short distance above and laterally away from the combustion space, the temperature is slightly lower and the gases have reverted largely to carbon monoxide. In that narrow band of the charge bed around the combustion space, it is possible to have the conditions for reduction to cast iron. Thus while the bulk of the reduction zone rains down fragments of nearly carbon-free iron, a local portion releases droplets of high carbon cast iron (Fig. 9.7). Since they all settle into the same bed below the slag pool but at different locations, localized areas of higher carbon occur.

Figure A.2 shows the relationship (ignoring nitrogen) between the percentage of carbon monoxide in the carbonaceous gas mixture and the carbon content of iron at various specific temperatures. Note that for the carbon content of an iron to exceed 0.1 percent above 1150°C, the proportion of carbon monoxide must not be less than about 95 percent. At about 99 percent carbon monoxide (less than 1 percent carbon dioxide), the iron will saturate with carbon to over 4 percent and be liquid cast iron. Accordingly, to produce directly a steel with a carbon content between 0.1 and 1.8 percent, the carbon dioxide content of the combustion gases must be between 1 and 5 percent. This range is so narrow that the direct reduction of ore to a metal with the carbon content of steels would only be fortuitous and difficult to control. Therefore, in seeking methods by which early metallurgists could deliberately smelt to steel, it is more logical to examine the possibilities for carbon modification of the reduced product after it has collected in the furnace basin below the tuyere.[1]

## 3. THE *TATARA* FURNACE

This unique Japanese smelting furnace produced a heterogeneous bloom with a vertical carbon gradient from high carbon (>0.8 percent) at the bottom to an almost pure iron at the top—the whole resting on a pool of cast iron (see, e.g., Geerts 1883: 528–532; Tanimura 1980). The entire metal deposit was covered by a slag pool (Fig. 12.1). Lyman (1879: 62) gives a detailed account:

> The height is only 3.7 feet, and the width at top 3.2 feet outside or 2.4 feet inside, but the length is 9.5 feet outside. The outer sides are battering [inclined], and the inner ones still more so near the bottom, so that only half a foot in width along the middle of the bottom is flat. The vertical section crosswise differs then from that of the blast furnace in being widest at the top instead of near the middle. In the course of the operation, however, the inner sides rapidly wear away toward the bottom so that the section becomes much wider there than at first. The extreme lowness of the furnace can be accounted for by the fact that the minuteness of the particles of the ore [magnetite sands] enable them to be deoxidized in a very short descent. The great length of the furnace is made up for by a row of nineteen tweers [tuyeres] on each side, for which there are elliptical holes 0.2 feet high and 0.7 feet above the bottom outside and 0.5 feet inside; so that the heat is kept up throughout the whole length. The progress of the smelting can also be viewed through the holes, at the side of the nozzles, and any obstructing slag can be poked away with a small rod inserted there; and at each end of the furnace at the bottom there is a hole 0.4 foot in diameter for drawing off slag and pig iron, and there likewise the smelting can be watched. At the end of the first twenty four hours that hole is closed and two other similar ones are opened on either side of it near the corners, as the side walls have already become thin. Below the bottom, which is on a level with the ground, there is a layer 0.8 foot thick of ashes of the maki, a conifer (Podocarpus macrophylla); and below them there are 4.2 feet in depth by 3.5 feet in width and 9.5 feet in length of hot coals; and the five feet together are called the fukutoko (or hearth). Below it is a depth of seven feet filled with red and black clay; from the bottom of which there is a horizontal drain to a neighboring hillside. The whole depth of twelve feet is walled about with half a foot or more in thickness (granite and other kinds). The fukutoko with the parts below is permanent; and at some places is not allowed to cool down for a hundred years. The furnace proper is built of wet clay in a day, and dried with a wood fire in the following night.

The name *tatara* refers to the air blowing machine (Fig. 12.2) which was a two-chamber version of the reciprocating, hinged fan (Fig. 7.3c). The fan was rocked on an axle between the two chambers by foot power so that it produced a continuous air supply by pushing air into one of the two chambers at each cycle of rocking. Many of the details are corroborated by Joly and Inada (1913), who also note that more than one tatara air blast machine was sometimes used and that the smelting could last for 168 hours.

The base of the furnace wall was unusually thick. Gowland (1899: 306–307) augments Lyman's account, stating it was constructed of a mixture of weathered granite and clay. The feldspar in the granite would have provided an important flux for the slag. This appears to have been one of its intended functions, as over a 58 to 68 hour run of the furnace much of the wall dissolved while a low melting slag composition was maintained. When it thinned to the point of potential perforation, the furnace had to be shut down, the metal removed by knocking down the walls, and all the furnace above ground rebuilt.

For data on furnace output we turn again to Lyman (1879):

> Throughout the operation, the slag is running out at the end holes of the furnace, and some pig iron too, called dzuku, which, though variable, amounts to perhaps 180 kamme (1500 lbs) in all. On the fourth morning at three o'clock (that is at the end of 72 hours), after having charged in all about 3800 kamme of charcoal and 3600 kamme (30,000 lbs) of ore, they stop the blast (which has been continuous) and take away the nozzles. Then they break up the furnace, brush off the coals that remain on the mass of metal that has been produced, and let it cool. The mass is steel, but the outside of it is bad steel called kera; below it there is melted pig iron, or dzuku, which on the removal of the steel mass cools and becomes solid at once. The steel mass weighs about 540 kamme (4500 lbs) of which two-thirds are good steel and the rest kera. The two are separated with a hammer. The dzuku weighs about 360 kamme (3000 lbs), or together with the previous 180 kamme (4500 lbs) in all; making the whole product then about 1080 kamme (9000 lbs or about four tons) or 30% of the ore. With a larger furnace sometimes 1200 kamme are produced.

In order to understand the descriptions of the tatara process, we must recognize that the bulk of the product metal is originally sponge or bloom iron, and that the hearth is a deep bed of charcoal which never cools down. The early bloom iron deposit on the charcoal hearth is carburized to saturation which, at the temperatures attained, means that it becomes liquid cast iron. The further deposit of bloom iron as a sponge aspirates this high carbon liquid metal and is thereby itself carburized. The height to which liquid cast iron can rise through the capillary channels in the bloom is limited by wettability of the liquid metal on the solid metal surfaces. The physics of capillary rise set a limit of height above which the molten cast iron cannot rise and where, above that point, a relatively low carbon metal can consolidate.

Thus, it is a secondary process of carburization that yields a mass composed of liquid cast iron with a thick,

[1]Despite the difficulties in the control of reduction potential and temperature, steel was made in a bloomery smelting operation. Overman (1854: 466) describes a process called "German" steel: "This steel is made in two different ways: either directly from the ore, or by converting the ore into pig metal, and then into steel. The steel manufactured by the first method is generally crude and irregular, and therefore this method is seldom practised." Further, he wrote: "The stuckofen, or wulf's-oven, as well as the Catalan forge, is one of the furnaces employed in the manufacture of steel from ore. In making steel, the blast is directed more upon the fuel than upon the iron. The tuyere is level. The iron is impregnated with carbon. In the blue (blau) oven, a kind of pig metal is frequently made, which is almost pure steel, but it is coarse, and never, even after the best refining, makes a good finished product. All attempts to make steel directly from ore are therefore unprofitable."

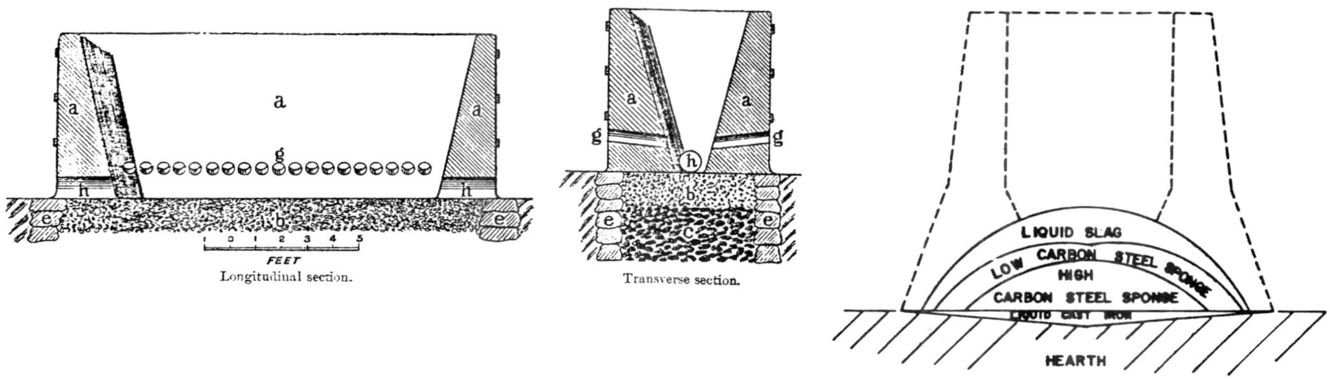


Fig. 12.1. *The Japanese Tatara Furnace.* It succeeded in making steel during a bloom smelting operation. Left: the furnace had a long rectangular inner cross section with opposing lines of tuyeres projecting air blast across the short dimension (Gowland 1899). An important feature is the construction of the furnace superstucture on a foundation of a deep bed of charcoal (Lyman 1879). Although the furnace structure was destroyed each time the product metal was removed, the foundation of charcoal was replenished and was a permanent feature of each successive reconstruction. Right: the metal product was heterogeneous in carbon content and massive (Tanimura 1980). The liquid slag cover was drained off but the sponge had to be dragged off the pool of liquid cast iron. The latter promptly solidified as a plate-shaped piece which was broken up for remelting in another furnace either for castings or for fining to wrought iron (Lyman 1879: 760). The iron and steel were segregated by breaking the sponge mass into small pieces and hand sorting (Dodd 1978).

high to low carbon steel crust. The primary product is ordinary sponge iron, converted through the action of an unusual hearth bottom capable of carburizing what rests upon it.

## 4. AFRICAN COUNTERPARTS TO THE TATARA PROCESS

Some East African ironmakers traditionally packed the bottoms of their furnaces with dried grass. These include several Ugandan groups: the Banyankole, Baganda, and Labwor or Karamoja (Cline 1937: 43, 47), and Haya (Avery and Schmidt 1979; Schmidt and Avery 1978). All use low furnaces with multiple bellows and deep pits underneath; at least two, the Labwor and Haya, produced a nonhomogeneous steel.

Fig. 12.2. *The Tatara Air Blast System.* An early Japanese woodcut (Gowland 1899) shows the air blast system from which the furnace derives its name. It is a double-acting version of the hinged fan in Figure 7.3. There are two chambers, each connected to the banks of tuyeres. By the rocking motion, air is forced into the tuyeres from one or the other chamber. It was a labor-intensive function since six men are needed to run the air blast and there must be a relief team since the furnace run occupies many hours. See also Swank (1892: 70) for a description of the same air blast system used on a Korean blast furnace.

The Haya operation, studied in detail by Avery and Schmidt in the 1970s, is like a small scale tatara process. The furnace pit was first filled with a grass charcoal made by burning dried reeds *in situ*. An iron ore and charcoal charge was placed on top of this and, after ignition, the fire was blown by tuyeres, driven by drum bellows, which projected well within the furnace so that the blast was substantially preheated. A bloom and slag cover were generated which rested upon the bed of filamentary charcoal fragments. While it is not apparent from the description that the bloom itself became hot enough to develop a superficial film of cast iron, the slag effervesced because of the large surface area of the contact zone between the slag and the charcoal. The result was a bloom with a heterogeneous carbon content, ranging from very little to more than 0.8 percent. While Schmidt and Avery stressed the very high temperatures reached through preheating the air, it is clear that the charcoal bed beneath the combustion zone was also important in this process. As in the tatara method, the hot charcoal served to carburize a nearly pure iron to a steel.

## 5. DIRECT STEELMAKING BY SECONDARY DECARBURIZATION

Other direct steelmaking methods seem to reverse the procedure, so that the metal produced is initially cast iron but is decarburized to steel before smelting is complete. Cases of smelting which regularly produced steely iron are common in the literature on nineteenth- and early twentieth-century traditional metallurgy both in Africa (e.g., Todd and Charles 1978; Routledge and Routledge 1910; see also the discussion in Cline 1937: 53–55) and in South Asia (e.g., Percy 1864: 279; Hunter 1875: 149; Swarup and Misra 1944; Campbell 1971;

McWilliam 1920: 165–166). Most of the reports are not detailed enough to determine exactly how the processes worked.

An exception to the usual laconic account is Bellamy's (1904) of smelting by the Yoruba, in the hinterland of Nigeria. They produced a high carbon steel bloom or, perhaps more properly, lump. Natural draft was supplied by a large number of tuyeres disposed radially around the circumference of the furnace shaft (Fig. 3.6). The furnace was filled with charcoal, ignited, and run with only fuel additions for about two hours. After that, both a specially selected slag and charcoal were added at intervals in the proportions of about 5 lbs. of slag to 2 bushels of charcoal. The smelted slag was drawn off and charges of ore, slag, and charcoal were made at ten intervals. The quantity of ore in each charge was gradually increased from about 5 to as much as 30 lbs. by the eighth charge. Slag was drawn off as judged necessary by the head smelter, and slag from the last three ore additions was set aside for the next smelting run. The smelting run from start to metal extraction lasted about 36 hours. The metal was recovered as a solid, dense lump with an average weight of 75 lbs. The ratio of metal recovered to ore weight added was about 0.46.

The specimens which Bellamy brought back to England were examined by F.W. Harbord, whose report is appended to Bellamy's paper. The smelted metal was shown to contain 1.67 percent carbon, 0.252 percent silicon, 0.006 percent sulfur, and 0.010 percent phosphorus—a very high carbon steel. It would be difficult to forge at these carbon levels, and Bellamy notes that in subsequent processing by the smith the metal was decarburized in the forging furnace to 1.022 percent carbon, whereupon it was rendered into useful shapes such as the chisel analyzed by Harbord. The slags used as flux and those discarded as unsuitable for that purpose were also studied. The former contained 38 percent $SiO_2$, 38.43 percent FeO, and 10.0 percent $Fe_2O_3$; the latter had 52.0 percent $SiO_2$, 33.68 percent FeO, and 6.14 percent $Fe_2O_3$.

According to the $SiO_2$-FeO-$Fe_2O_3$ melting diagram (Muan and Osborn 1965: 54), both of these slags should be very viscous, having been formed well below their full fusion temperatures. However, the "preferred" slag which was drawn off after the large ore additions near the end of the smelting run contained 10 percent of lime, alumina, and magnesia as well as potash (not analyzed) from the fuel, so it could have been adequately fluid. The addition of the high iron slag undoubtedly contributed to the high ore/iron recovery rate of the Yoruba process—46 lbs. of iron for each 100 lbs. of 60 percent ore, indicating a loss of only 25 percent, which is remarkably efficient for a direct steelmaking process.

As Harbord recognized, the shape and slag-free character of the "bloom" indicate that it was actually a lump of somewhat decarburized cast iron made in the following manner. The high iron flux was added every two hours. During the period between flux additions, there was a continuing reaction between the iron oxide-rich slag and the carbon dissolved in the metal pool below and in contact with the slag pool. The slag/metal reaction slowly removed the carbon present near the surface of the gradually accreting lump of iron. As the lump was decarburized, the iron oxide content of the slag was simultaneously reduced to iron, which helped to increase the metal yield. Since this decarburization reaction generated no heat, fuel had to be added with the flux to keep both slag and metal molten. The slag which Bellamy calls dross was lower in iron oxide and likely to be very viscous and difficult to reduce, so it was reasonably "thrown to spoil."

Hence, the function of the preferred slag was to assist in the partial decarburization of an iron in which the carbon content would otherwise be unmananageably high due to the large amounts of fuel used and the high and only somewhat controllable temperatures reached in a windblown shaft furnace. Whether similar procedures were followed elsewhere in West Africa, a region where tall shaft furnaces easily capable of smelting to cast iron were used, is not known.

## 6. SUMMARY

Thus, it is clear that steelmaking in the smelting furnace was actually a two-stage operation. In the Japanese tatara furnace and the Celtic pit hearth (Pleiner 1980), a bloom iron smelting product was carburized to steel in the hearth. In the African case, the cast iron product of smelting was decarburized by an iron-oxide-rich slag to a steel. In each case, the steel product was largely a very high carbon steel which, although difficult to forge, produced a serviceable metal. Direct smelting to steel probably happened often during the transition from bloom smelting to cast iron smelting, but the difficulty of producing it consistently meant that it could not be routine. Fining, puddling, and ultimately converting by bottom air blast proved more efficient and reproducible.

# 13. Conversion of Cast Iron to Wrought Iron or Steel

## 1. INTRODUCTION

Cast irons usually cannot be forged at either normal or high temperatures, but they can be converted to forgeable iron or steel by a secondary processing step. This two-stage route to a forgeable metal has been termed *indirect*. *Wrought iron*, a term generally applied to forgeable, low carbon iron, is a holdover from the nineteenth century. It is presently used to describe a very formable iron, although traditional wrought iron ceased to be made in the early 1930s. What is used now is an extra low carbon iron which is called *mild steel*. Mild steel, in general, has a somewhat higher carbon content than wrought iron, but actually has the same formability. The real distinction between wrought iron and mild steel is the absence of long, thick slag inclusions in the latter. Wrought iron is a first cousin of bloom iron and is often not distinguishable from it. Steel from the indirect process does not carry the adjective "wrought" although it too possesses the large slag inclusion population which modern steels do not.

The conversion of cast iron to wrought iron involves the reduction of the high carbon content (usually 3 to 5 percent) to the much lower carbon levels characteristic of iron and steel. This is accomplished by reacting the carbon alloyed in the cast iron with oxygen in some form to produce a gas which escapes. If the cast iron is liquid, the gas can form at the surface, or in the interior and come to the surface by the action buoyancy of gas bubbles. If the cast iron is in solid form, the carbon atoms must diffuse to the surface of the piece of metal where they can be oxidized and released as carbon dioxide or monoxide. Because of the differences in carbon diffusion rates, the solid state decarburization process is much slower, but it has some advantages (see below). The production of steel by the indirect process involves simply the cessation of the decarburizing action when the metal has reached a carbon content appropriate to a steel (Chapter 12). In practice, stopping the process at the right carbon level was often too difficult to be done on a regular basis so that, for many years, it was common to decarburize to wrought iron and then to recarburize back to steel (Chapter 11).

The oxygen for decarburization could be delivered in several different forms which, in many ways, dictated the nature of the decarburization process. These include: a) oxygen in air; b) oxygen in carbon dioxide—but not carbon monoxide; c) oxygen in solid, mineral oxides as found in ores; d) oxygen in hammer (mill) scale which is predominantly $Fe_3O_4$; and e) oxygen in iron smelting slags which are rich in FeO. The means of transferring these sources of available oxygen to the carbon alloyed in the cast iron are described below.

These decarburization methods have various complications. If the cast iron is initially liquid, the reduction of the carbon level raises the freezing temperatures (Fig. A.3). In order to keep the metal molten, the temperature in the furnace must be continuously raised during the progress of the conversion to iron or steel. This was not always possible, so a solid or semi-solid product much like bloom iron was formed. If the oxygen source is air, it is theoretically possible to keep the metal molten as decarburization proceeds, but the necessary technology was not exploited until the development of the steam powered air compressor permitted the invention of the Bessemer process. The reason is that the formation of carbon dioxide or monoxide by the direct reaction between carbon and gaseous oxygen produces the heat to raise temperatures in exactly the same way as burning a fuel. However, the oxidation of carbon by carbon dioxide or by any of the iron oxide forms absorbs heat and so provides a cooling action. If this occurred, the molten state could not be maintained by furnacing systems of the pre-industrial period.

## 2. THE FINING PROCESS

In Europe, the earliest system for converting cast to forgeable iron was the cluster of related techniques known as *fining*. They appeared during the fifteenth century and survived in one form or another until the middle of the nineteenth century when they were gradually supplanted by the puddling process. Numerous variations of furnace design and operating procedures were devised, including one or two hearths for fining and forge reheating, one or two tuyeres, one or two meltings, manipulation of the slag and metal with an iron bar, and use of white or gray pig iron (or mottled). Percy (1861: 579–620), in naming some of these methods—Walloon, Franche-Comté, Styrian, Carinthian, Tyrol, Siegen, Lombardy, Salzberg, Bohemian, and Rohnitz—illustrates the highly regional nature of experimentation with cast iron conversion. The diversity suggests that none of the experimentation was successful enough to establish one method as superior to the others.

In each process, cast iron in the form of pigs is drip melted, the droplets descending through a bed of charcoal which is in vigorous combustion by the action of forced air delivered through a tuyere. Efficient combustion to carbon dioxide not only maximizes heat output, but also provides a gaseous environment which is both decarburizing and oxidizing. The droplets, having percolated through the hot charcoal bed, settle through a high FeO slag and agglomerate to a lump or bloom on the furnace bottom (Fig. 13.1).

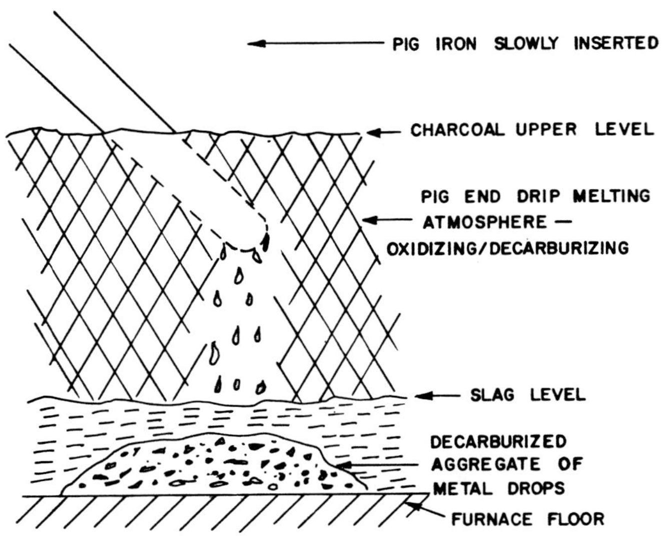


Fig. 13.1. *Diagrammatic Illustration of Fining.* A stick of pig iron was drip melted in a charcoal fire combusted with an excess air supply so that the highest rate of heating and an oxidizing gas environment are achieved. During the descent through the combusting bed, the cast iron loses silicon and carbon by oxidation. The silica produced by oxidation of the pig iron was an expensive nuisance because it formed a film over the iron that retarded the loss of carbon.

The formation of droplets of liquid cast iron is a natural consequence of the fact that a large, solid piece of the metal resting on top of a bed of very hot coals melts only gradually. Most of the time the bulk of the piece is solid because of the thermal gradient that exists from surface to center. A thin film of liquid metal forms at the surface, runs by gravity to the lowest position, and forms droplets, much like the drip melting of icicles during a thaw. When sufficiently heavy, these droplets break away and begin a descent through the bed of packed pieces of fuel. The droplets, as they ooze through the bed, offer a high ratio of surface to volume which greatly increases their ability to react rapidly with the surrounding gases.

Most commonly, the pigs of cast iron perched on the edge of the fining furnace were cigar-shaped, as much as 5 to 6 m. long, and tapered at both ends. They were cast in sand beds so that the upper surface was flat and the lower convex. In some cases, however, broken pieces of pig iron were piled around the rim of the fining furnace, which was covered with fuel. As they became adequately preheated by the exiting hot gases, they were tipped one by one into the column of fuel in combustion and drip melted as the fuel bed descended. The pig iron might also be piled against the furnace wall opposite the tuyere. The furnace was filled with fuel and combustion started. The blast from the tuyere was directed against this pile of pig iron so that a strongly oxidizing gas stream of oxygen and carbon dioxide decarburized the drops of iron running down the pile. In both these cases, the decarburized metal collected on the hearth under a slag pool of high FeO slag. The slag itself provided a decarburizing environment as the bloom of wrought iron collected, and iron was reduced from the slag to join the bloom. From time to time it was necessary to add hammer scale to restore the FeO content to a level consistent with good slag fluidity.

Because of its low silicon content, white iron was preferred. However, with the advent of the use of coke in blast furnaces, the higher smelting temperatures meant a greater silicon content in the pigs and a slower rate of fining (Tylecote 1976: 107). Experimental data on the effect of silicon on the rate of decarburization (Heine 1951; see Fig. 13.2) show that at a pool temperature of 1300°C the silicon oxidizes while the carbon content remains unchanged until the silicon has been almost completely eliminated. Moreover, the oxidation of the silicon forms an interfering scum over the liquid iron surface through which the silicon but not the carbon can continue to diffuse to the reacting gases (Heine 1951). Similarly, as iron droplets produced in the fining process descend through the hot bed of coals, the rate of decarburization should be retarded in proportion to the silicon content of the cast iron.

Fining was performed in a shallow hearth, like a Catalan hearth, but modified with a chimney-shaped superstructure to draw off hot gases and to protect the workmen from the heat and showers of sparks. Descriptions of finery hearths show that sometimes the combusting fuel, the accumulated metal, and the slag were contained by air or water cooled cast iron plates used as the floor and walls. The tuyere projected into the combustion zone through a hole in one of the wall plates. In some furnace systems, two tuyeres were positioned opposite each other in the walls. The interior faces of the finery hearth were commonly constructed of refractory brick or stone. The inside dimensions of an average hearth were approximately 68 × 45 × 22 cm. depth

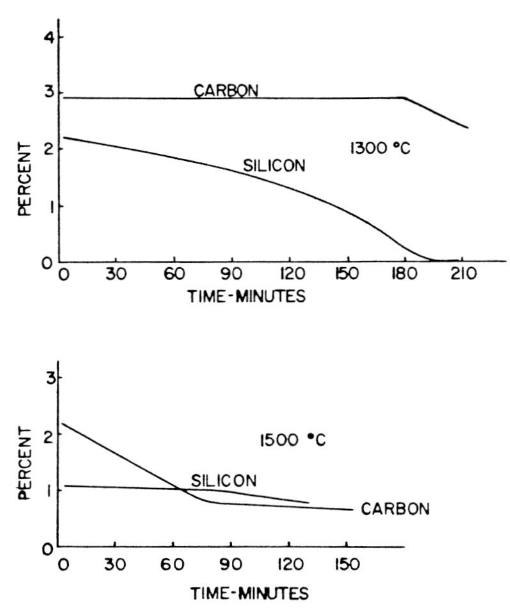


Fig. 13.2. *Silicon Interference to Carbon Removal in Fining.* The silica film retarding carbon oxidation can be overcome only at very high furnace temperatures, meaning higher fuel costs. The long-lasting preference for charcoal smelted pig iron resulted from the lower smelting temperatures and the consequent lower silicon content in the metal product of the blast furnace. (Adapted from Heine 1951)

below the tuyere; to assist the workmen, the bottom was raised about 50 cm. above floor level (Fig. 13.3).

Air and water cooled iron plates were major innovations which prolonged the life of a furnace, since high FeO slags are corrosive to any ceramic wall or basin. The cooled metal plates created and maintained a frozen layer of the slag separating the metal from the liquid slag pool. This precluded reaction by the slag with the furnace structure as long as the cooling of the iron plates was maintained. Frozen films of slag were later used in the puddling furnace and are integral parts of the modern blast furnace.

Around 1760, Pashley vividly described the fining process (Schubert 1951–53: 72):

> The hearth being filled with charcoal considerably above its top, and the fire blown up by a pair of pretty large bellows and more charcoal being from time to time supplied as the occasion requires; the finer lays upon the fire, of the produce of the last process [cast iron smelting] about 100 pounds with a covering of charcoal over it. The metal, as the operation advances, sinks lower, becomes hot, goes into fusion and drops through the fire to the bottom of the hearth; where escaping the violence of the heat, it grows cooler and resumes in some degree its solid state. As soon as all the metal is gone down, it is by the help of an iron crow called by the workmen, a fourgon, raised again to the top of the fire and is melted down again to the bottom of the hearth. If the metal had no bad quality which made it require a longer continuance in the fire, and if the operation has been properly conducted, the process is now finished, all but what belongs to the hammer. But that it may be better understood, I shall describe the effects in three different stages.
>
> The first fusion separates from the metal a small proportion of a very corrosive scoria [slag] and renders the metal much more difficult to flux [melt] again, and it therefore after the first elevation goes down the second time in a less fluid state, when the quantity of scoria, or cinder as they call it, is also considerably increased. And at the third time it is melted down, it drops through the fire so fluid as bricklayer's mortar, during which operation a large proportion of cinder is separated from it, which is let off, on a redundance, through the hole in the iron plate mentioned above; and the redundance is constantly known by the snuffling of the blast, occasioned by the interruption it meets with from the cinder at the nose of the tuiron [tuyere].

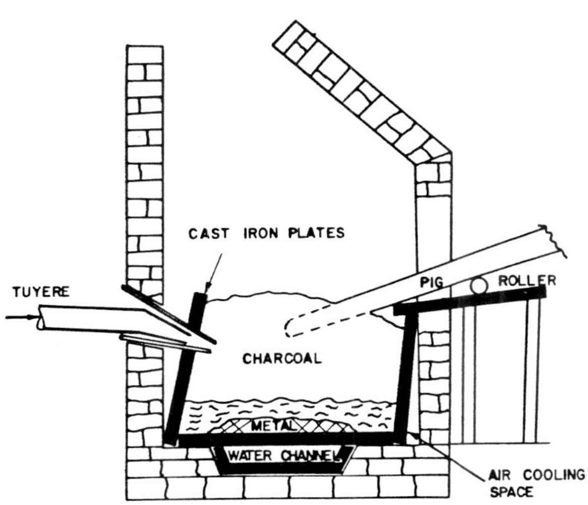


Fig. 13.3. *Improved Fining Furnace*. The use of water and air cooled cast iron plates was a major factor in improving (raising) the furnace temperatures and furnace longevity. Frozen coatings of slag provided thermal insulation sufficient to keep the cast iron from melting superficially. (Adapted from Schubert 1951–53 and Maréchal 1955)

> The change made in the cinder, in the progress of the whole operation, is no less remarkable and thence the finer forms his judgement of the success of it. That which was at the beginning exceeding corrosive and fumed and spattered like liquors in fermentation [carbon boil], at the last becomes mild and flows like wax. The mass at the bottom of the hearth, instead of the fragile reguline substance which, so far from being tractable with the hammer, is become a tolerable connected lump of malleable iron having the interstices filled with the cinder above described.

Clearly, the first drip melting step may not have adequately reduced the carbon content, since it was still a low melting metal with more than 2 percent carbon. It is therefore necessary to recycle the mass accumulated at the bottom of the hearth. According to Pashley, after the second drip melting the product is usually forgeable; that is, the carbon content has dropped well below 2 percent.

Much charcoal was consumed in all fining operations. According to Hammersley (1973: 604), in the sixteenth century consumption was as much as 5 long tons of charcoal per long ton of bar iron produced; better controls brought this down to about 2.5 tons per ton of bar iron by the early eighteenth century. These figures are in close agreement with seventeenth-century records cited by Morton and Gould (1967). Throughout, the amount of fuel needed for fining was often greater than that for smelting the same quantity of iron.

A fluid slag was necessary to permit a reasonably rapid gravity separation of the metal droplets to a metal pool, as it was in bloom iron smelting. This was the high Fe0 slag called "redundance" by Pashley, some of which was often added to the initial charge of the next run. During the fining operation, the slag tended to become viscous partly because of the input of oxidized silicon from the pig iron and partly because the metal accumulation retained enough carbon to reduce FeO out of the slag. In the early stages of the operation, mill or hammer scale, rich in FeO, had to be added to keep the slag fluid. Later, when the carbon content of the metal dropped low enough, fluidity was maintained by the incorporation of re-oxidized metal into the slag. Actually much more iron was lost to slag than would be expected from the elimination of silicon and carbon. Both Hammersley (1973: 603) and Morton and Gould (1967) cite a ratio of 26 to 29 cwt. pig iron to 20 cwt. bar iron, indicating only a 70 to 77 percent recovery of forgeable metal. Hyde (1977: 79–80) gives similar figures for a sample of fourteen fineries in eighteenth-century Britain: the range was from 1.26:1 to 1.49:1, representing recoveries of 67 to 80 percent.

At about 1200°C, carbon in the metal can react with and reduce the FeO and higher iron oxides in the molten slag that covers the metal accumulation at the bottom of the hearth. Decarburization is thus accomplished partly by the carbon dioxide and oxygen in the combustion zone but also partly by reaction with the iron oxides dissolved in the slag. Historically, metal produced by fining was usually very low in carbon. There was no reliable method of process control that would allow stopping the conversion when the carbon content was at an intermediate level. Thus, in its low carbon content and in other respects, a lump of fined iron was much like bloom iron.

Both Pashley and later commentators attest that the silicon oxidation out of the pig iron interfered with fining. The silicon dioxide (silica) scum over the liquid metal droplets shielded the latter from the decarburizing gases. As the metal descended though the slag pool, the viscosity of the pool was raised as the silica scum dissolved into it. Both of these circumstances slowed the process and increased the fuel costs, so a preliminary step of desiliconizing aided the fining operation. In the South Wales process, which operated into the middle of the nineteenth century, the pig iron was premelted in a shaft furnace much like the modern cupola using a highly oxidizing combustion which converted the silicon in the pig iron to a slag, the desiliconized metal being drained at a trickle into a fining furnace (Percy 1864: 581–591).

As was the product of the bloomery furnace, that of the fining furnace was permeated with slag which had to be expelled by draining (sometimes called *sweating*) in a reheat furnace and then forging to squeeze out most of the liquid slag that remained. The resulting forged iron bar still retained numerous slag particles in the form of thin ribbons, indistinguishable in number and shape from the slag in bloom iron. The slag composition is basically the same including the potash and other minor oxides contributed by the charcoal ash.

Charcoal rather than coal or coke was always used for the fining process because any sulfur in the fuel would have been efficiently picked up by the liquid iron droplets, making the decarburized metal likely to be hot short. The slags in the fining operation were not able to reduce the quantity of sulfur that might be in the metal. Fossil fuels could be burned in the chafery or reheating furnace in which the fined lump was prepared for forging because sulfur penetration into a massive solid piece is slow. After coke began to be used in blast furnaces, pig iron of the older charcoal smelted type continued to be preferred in fineries, because the pig iron was lower in sulfur and silicon.

We do not know how fining was invented or discovered, but a reconstruction can be suggested. Because the demands for castings were not as steady as the output of the blast furnace, it was necessary from time to time to pour off liquid iron into pigs which were later remelted for casting. The remelting would have taken place in a smaller version of the smelting furnace (now called a *cupola furnace*). Remelting requires much less fuel than smelting because the combustion can be conducted more efficiently, but careful control is necessary. If the combustion is carried to a point at which equal proportions of carbon monoxide and carbon dioxide exist, the liquid cast iron emerges with a lower carbon content. However, if the combustion takes place with enough air to go completely to carbon dioxide, the carbon content of the pig iron is drastically reduced and it can collect at the bottom of the furnace as a bloom of iron, or steel, or some mixture of the two. Whatever the outcome, the furnace would not have poured, but the extracted lump would have been sufficiently forgeable to warrant further investigation. In due course it would have been clear that if the lump were recycled through the melting furnace, its forgeability would improve. Ultimately, convenience caused the remelting furnace to be adapted to the forging operation so the lump could be extracted from the open top of the furnace.

## 3. THE EUROPEAN PUDDLING PROCESS

The coke fueled blast furnace, despite its cost effectiveness and high productivity, had disadvantages: it made the metal high in silicon, sulfur, and phosphorus. (The interference of silicon in the decarburization process has been discussed above in this chapter, and the serious degradation of the forging performance imparted by sulfur and phosphorus in Chapter 2.) All the sulfur and much of the phosphorus in the pig iron carried over to the wrought iron. With the demand for forgeable iron steadily increasing in the eighteenth century, much effort was devoted to eliminating the undesirable elements in the cast iron product of the coke fueled blast furnace. Because the charcoal furnaces of Sweden and Russia produced cheaper bar iron, the British had to develop a higher level of technology or see their iron industry decline.

The solution was the reverberatory furnace. It was a flat, shallow hearth, walled on all sides and roofed with refractory brick. The heat was supplied by a contiguous fire chamber in which any coal could be burned. The air supply was by natural draft raised by a tall chimney stack also contiguous with the hearth (Fig. 3.3). The chimney draft was so strong that dampers had to be used to control the temperature. The hot combustion gases moved from the fire chamber across the hearth, impinging on the refractory brick roof above, which in turn radiated heat directly onto the charge on the hearth. This radiant heat transfer was called *reverberation* at the time—hence the name of the furnace. Although Biringuccio indicates that a primitive form of the fundamental principle of indirect heating was employed during the sixteenth century (Smith and Gnudi 1942: 281–288), it was not until the middle of the eighteenth century that it was applied to the refining of pig iron. It was first used to remove most of the silicon which retarded the fining operation, a shift away from the South Wales process (Chapter 13.2). The pigs were stacked around the edge of the shallow hearth, drip melted, and the liquid metal collected as a pool in the hearth. The high surface-to-volume ratio of the thin rivulets of flowing iron allowed oxidization of the silicon by the carbon dioxide and excess oxygen in the combustion gases moving across the hearth to the chimney. The metal in the hearth became nearly free of silicon, although the carbon content remained largely unchanged, and it solidified as white cast iron. Much high silica slag was produced. The furnace was allowed to cool until a pool of white iron could solidify. The iron was pulled from the "refining" furnace in large plate-like fragments and broken (stamped) into smaller pieces for the decarburization treatment. At this point, it could have been fined in the usual furnace, but the "potting" process came into use and ultimately set British steelmaking on a new course.

In the potting process, fireclay crucibles were filled with a charge of refined white iron, iron ore, or mill scale as a source of oxygen, and, most importantly, alkali oxides in the form of kelp ash and limestone. Each crucible was covered and sealed. A large number of these were stacked on the hearth of a reverberatory furnace and heated to well above the melting temperature of the white iron. The lime and alkali absorbed sulfur from liquid iron to form strongly bonded sulfides which are soluble in the slag. The iron oxides provided the basis for decarburization and in each of the "pots" a sponge iron-slag aggregate—like a bloom—was formed. Many of the pots cracked, probably because of the corrosive effect of the slag on the fireclay, and the small blooms could be raked together and pulled out as a ball to be forged to bar iron in the established manner. The major importance of potting was that it was the first chemical desulfurization process for iron. The potting process was pre-eminent for only about 30 years (Morton and Mutton 1967), probably because it did not allow the high productivity that was demanded. However, it did mean that coke fueled blast furnace cast iron could yield a forgeable (low sulfur) iron, which probably explains the subsequent rapid decline of the charcoal fueled blast furnace in Britain.

In the 1780s, a major advance in processing cast iron to wrought iron—the "puddling" process[1]—was associated with the name of Henry Cort. For a considerable time, fining and puddling co-existed in Europe, and both had advantages.[2] However, the lower costs of puddling and the improved technology that evolved tipped the balance.

In the Cort patent, the hearth of the reverberatory furnace was deepened so that a pool or "puddle" of cast iron could be formed. The name *reverberatory* was replaced by *puddling furnace*. The charge, in the early days of the process, was refined cast iron. The pool (puddle) of molten cast iron was stirred with a *rabble*, a long, heavy iron pole with a hoe-shaped end, until it was turbulent. By moving the rabble back and forth while partially submerged in the pool, oxidizing gases reached the carbon in the metal and decarburization proceeded rapidly without interference from floating slag.

As the cast iron decarburized, crystals of iron formed (Fig. A.3). The crystals were agglomerated by the turbulence and these aggregations grew to substantial wads, occluding lots of slag. The rabble itself was an efficient collector because crystals and wads clung to the iron pole and formed crude balls which, when large enough, were lifted out, pried off the pole, and moved immediately to a forge or a special squeezing-action rolling mill so that much of the occluded slag was expelled before the ball cooled and the slag lost its fluidity.

A complete description of the "hand" puddling furnace operation is given by McGannon (1964: 12):

> With the hearth properly built up or repaired and the furnace in good working order and at a proper temperature, about 500 pounds of pig iron were charged by hand through the door. Following this operation, occupying 2 to 3 minutes, the purification of the pig iron and the process of puddling advanced by stages known as Melting, Clearing, Boiling, Balling and Drawing. To achieve quick melting, the door and other openings were closed, the furnace was fired vigorously, and the pigs turned over once or twice by the puddler or his helper. In this way the charge was usually melted in 20 to 25 minutes after charging. The molten metal, covered with a thin layer of slag, was then stirred or rabbled by the puddler to hasten the oxidation of the silicon, manganese, and a part of the phosphorus, an operation known as clearing [refining] and requiring 8 to 10 minutes. As soon as the metal had cleared, as revealed by a change in its appearance, the puddler endeavored to bring on the boil by raising the temperature of the furnace, charging some dry roll scale (iron-oxide scale detached from bars in rolling) and stirring the bath vigorously. After some 8 to 10 minutes of strenuous effort, the oxidation of the silicon and manganese was brought to a point where the carbon could also be oxidized.
>
> As the product of this reaction was the gas, CO, and since the slag was somewhat viscous, the action caused the latter to foam or boil and rise in the furnace. At this point the slag was permitted to flow from the furnace freely unless it was desired to hold the phosphorus high in the iron, when a little coal was added from time to time, and as much of the slag as possible was held in the furnace. As the elimination of the carbon became more rapid, the gas [CO] would escape in larger bubbles and burst into flame at the surface of the slag to form small flames called "puddler's candles". With the disappearance of the candles, the puddler increased the stirring of the bath during the lowering of the heat [temperature] until the metal, in terms of the puddler, would 'come to nature'. In this phenomenon, most characteristic of puddling, the metal appeared in small globules, like butter in churned milk, each globule representing a portion of the iron that had become decarburized. As this reaction neared completion, the bath became pasty and very hard to work. This change occurred because the high-carbon pig iron, which was molten at that temperature, was converted to low-carbon iron, which is solid (though pasty) at the same temperature. The charge progressed rapidly, lasting only 6 to 8 minutes, so that in some 30 to 35 minutes after clearing the metal was ready for balling.
>
> The globules agglomerated by the rabbling tended to collect in sponge-like clusters on the bottom; these clusters had to be raised constantly and exposed to the heat to prevent them from freezing to the bottom [of the hearth]. So the temperature of the

---

[1] For a more detailed account of the story of Henry Cort and his patents, see Dickinson 1940–41, and Mott and Singer 1983. Percy (1864: 627–639), in an extended digression from technical matters, wrote an impassioned account of Cort's martyrdom.

[2] Overman (1854: 280, 282) later provided a retrospective on the ultimate dominance of the puddling process over the finery:

> It is undeniable that charcoal forge iron is, in many respects, superior to puddled iron. For all the purposes for which wrought iron is applied, it is more malleable, compact and durable. The puddling process is conducted on more philosophical principles than the charcoal forge, and in the course of time may be brought to such a state of perfection as to supercede the latter altogether. But this is not the case at present; and the charcoal forge will be needed so long as the puddling process does not furnish a quality of iron equal to it. Another reason why the former will command precedence for some time to come, is that it is less expensive than the more complicated puddling establishments, and permits the manufacture of iron on a small scale without serious disadvantages. Iron works, situated at remote places in the country, frequently find a favorable market for a limited quantity or iron; while an increase of that quantity would not prove profitable. Such cases are very common in the farming districts of the interior of the country which are not easily accessible, as well as in the growing Western States. The same remark is applicable to the new States and territories. It is questionable whether the charcoal forges of the West, and even in the heart of the anthracite and bituminous basin, do not yield larger profits than the puddling forges and rolling mills; at least, an investment in charcoal establishments may be considered quite as safe as in those of stone coal, at the present time. . . .
>
> This method [puddling] of converting cast iron into malleable iron is designed to supercede every other method by which that result is effected. But, thus far, the quality of puddled iron has been such that we have been unable entirely to dispense with the charcoal forge. Still, this quality would be much improved if better metal was generally employed. The nature of the puddling process is such, as we have elsewhere stated, that we are enabled by it to employ inferior metals to a great degree. Thus, blast furnaces have been erected at places where charcoal furnaces would not have flourished. Inferior pig iron answers tolerably well for the puddling furnace. Metal perfectly useless in the charcoal fire will, in this furnace, produce a very good article. In fact, every kind of pig iron, however bad in quality, may, by the puddling process, be advantageously worked.

> furnace was raised as high as possible, and the metal was worked into a mass which was next separated into three parts or balls of about 150 pounds each, a size convenient for handling with tongs. This operation required about 15 minutes.

Similar descriptions are given by Scrivenor (1854: 118), Percy (1864: 654–656), Ward (1972), and Morton and Birt (1974); the last is less graphic but more detailed.

At this stage of the process development, the refining and decarburization steps had been combined into one continuous furnacing operation. The oxidized silicon created a high silica, viscous slag which could be thinned by the addition of hammer scale or raked off and replaced by high FeO slag from previous operations. This was only one of the improvements in puddling, a process that evolved rapidly over a 60 year period. It could be run 24 hours a day, being stopped only for repairs to the hearth and less frequent maintenance of the refractory brick structure. While there was some increase in the capacity of the furnaces, production was largely expanded by having more furnaces in one plant. The real economy of scale came when rolling mills to squeeze and convert the puddled balls to bar and plate were associated with the furnaces (Hyde 1977: 125). The charging of solid pigs was abandoned in favor of pouring molten pig iron tapped from a nearby blast furnace. Thus, the concept of the integrated steel mill was inaugurated.

The structure of the hearth was changed over time. The original puddling furnace hearth was a deep bed of silica sand bonded by sintering. It was attacked by iron oxides during puddling and slagged away rapidly; production time was lost to frequent patching. The process associated with the deep sand hearth was called *dry puddling*. This hearth system was, in due course, replaced by a water or air cooled cast iron plate floor upon which a sand base was piled which became sintered into position by the liquid slag. The cast iron plate understructure extended the time between major furnace repairs (Tylecote 1976: 111). Later, a deep bed of old slag, rich in iron oxides, replaced the sand. The slag bottom was a source of oxygen for the decarburizing boil and obviated the need for rabbling and mill scale additions. The hearth was quickly refurbished after each balling operation. Sometimes lump ore was used instead of or in addition to slag (Maynard 1874). An important improvement was that the refining or desiliconizing and the decarburizing could be done simultaneously, with the liquid pig iron at the top of the pool reacting with the oxidizing combustion gases to remove silicon and manganese, and at the bottom of the pool with the oxidizing, iron oxide-rich slag to decarburize and dephosphorize. The slag hearth-bottom system was called *wet puddling*.

Dry puddling drew oxygen for decarburization from the hot combustion gases introduced by mechanical stirring. Wet puddling produced a reaction between carbon dissolved in the liquid metal and oxygen in the iron silicate slag sintered to the hearth below the liquid pool. It was the gas generated by the reaction below the surface of the pool of metal that produced the carbon boil. Dry puddling without the turbulence of the carbon boil was wasteful because of the large fraction of metal which was oxidized along with the carbon, and which wound up in the slag floating over the pool. The later wet puddling was considerably more economical in metal loss. About 40 cwt. of pig iron were needed to produce one ton (20 cwt.) of bar iron (ca. 50 percent recovery) by the dry puddling process (Gale 1969: 48). In comparison the wet puddling process consumed only 21 to 27 cwt. of pig iron per ton of bar iron (ca. 83 percent recovery), but various sources unaccountably differ on this figure (Gale 1969: 48; Mott 1957–59: 80; Schubert 1957: 291; and Roberts 1983: 41). Roberts adds that the two-stage process combining the refining (silicon removal) operation with the puddling process gave about a 70 percent iron recovery although the fuel costs for double melting must have been high.

As the carbon content of the puddled iron ball fell, the hearth temperature had to be raised or the metal became too stiff to be manageable (Fig. A.1). Accurate temperature measurements were not possible and control was difficult, so the carbon content was highly variable if attempts were made to produce a steel instead of a near-zero carbon iron (Tylecote 1976: 127). On the other hand, the near-zero carbon level appropriate to wrought iron could be judged by the cessation of the carbon boil and the "puddler's candles."

As in the case of the lumps or loupes from the finery and the blooms from the direct reduction process, the ball of forgeable iron as it came from the puddling furnace occluded a great deal of slag which was essentially an iron oxide-rich iron silicate. The ball was hammered or squeezed to eliminate most of it as a viscous liquid, but a significant amount remained as elongated, ribbon-like inclusions that were the trademark of wrought iron when it ultimately competed with low carbon steel from the Bessemer and open hearth processes of the later nineteenth century. The distinction could be readily made by notching a bar and bending it so that the notched side was in tension until it fractured. The fracture of wrought iron under those circumstances looked like that of a splintered wooden board—hence the term of *woody* fracture (Fig. 9.6)—which the materials called mild steels did not show.

## 4. PHOSPHORUS IN FINED AND PUDDLED IRON

From its inception, fined iron and its product—wrought or puddled iron—did not compete well with conventional bloom iron, so the fact that it ultimately almost completely supplanted the bloom iron process clearly suggests a complex development (Chapters 14, 16). Here it is sufficient to note that, since iron made by the indirect process was usually higher both in phosphorus and sulfur, it could be both cold and hot short. Wrought iron could be entirely satisfactory in both respects, but often it was not, and precise control of the process took a long time to perfect. Note that bloom iron could also be cold short although not often hot short depending on the ore used.

The problem with phosphorus originated in the blast

furnace, because many iron ores contain considerable amounts of it. While, theoretically, the conditions in any smelting furnace can reduce mineral-bound phosphorus to elemental phosphorus which can be absorbed by any iron, the actual efficiency of conversion depends on the temperature and the reducing conditions (the data are sparse; see Table 14.1 for figures based on production at Gittelde in Germany). Where the product is low carbon bloom iron, the phosphorus absorbed by the metal can be much less than is in the ore. On the other hand, as Kinney (1926: 12) states: "It is generally conceded that all of the phosphorus entering the furnace finds its way to the metal, unless the [blast] furnace is working abnormally." [3] Given that cast iron efficiently absorbs phosphorus from the ore during smelting, the question is to what extent the phosphorus was oxidized out again in the conversion to forgeable metal. There is little information on this topic, but we do know that the British navy, for a long time, specified Spanish or Swedish bar iron for anchors or chains because fined iron was unreliable. Since these fittings were sometimes exposed to very cold water, this preference probably reflects an awareness of the extreme brittleness of iron containing even small amounts of phosphorus at low temperatures. Some of the Swedish and Russian iron ores were notably low in phosphorus (and sulfur) and were not as subject to low temperature embrittlement. The puddling operation (and also the fining process) could significantly reduce the phosphorus content of the metal, because of the slag characteristics at the time that the iron was "brought to nature." Two types of slag allow absorption of phosphorus from the metal: the first, an iron oxide-rich, low silica slag, and, the second, a high lime-magnesia slag (other major components being silica and alumina) in an oxidizing atmosphere with a total of lime and magnesia in excess of 50 percent. That the first slag could be maintained throughout most of the puddling operation was the key to reliable removal of phosphorus because there was enough time and turbulence for the dephosphorizing process to approach completion. The second slag condition was too corrosive to silica based refractory bricks and could be exploited only later, after the development of basic (high lime) refractory ceramics. Percy (1864: 663–664) emphasizes that the puddling process was very effective, as compared with the later Bessemer process, in removing the phosphorus brought in with the pig iron.[4] The Bessemer process supplanted puddling but the use of high phosphorus pig iron had to be abandoned in the conversion to mild steel. Percy also notes that by the expedient of "double puddling" which involved the reconversion of scrap wrought iron to cast iron by melting with coke and repuddling to wrought iron, the final product was almost completely free of phosphorus (by the analytical chemistry techniques of the time). The double puddling system was not cost effective except for making steel by the cementation process, according to Percy.

Percy also provides data that suggest why the fined iron was less reliable in cold working and in cold service temperatures as compared with puddled iron in its most sophisticated state. He notes (1864: 619) that the mean percentage of silica in finery slags was about 30 percent, whereas the average silica content of six compositions (p. 668) of puddling slags was only 16 percent. In the absence of good analyses of iron oxides, the silica analysis (which is easier to perform) has been taken as an indication of the iron oxide content of the slag, i.e., low silica signifies high FeO. McGannon (1964: 350) states that the dephosphorization of iron and steel is proportionate to the iron content of the slag. Therefore, fining slags seem to have reduced some phosphorus in the metal (Morton 1963), but puddling slags, because their iron oxide contents were higher, were more efficient in this respect.[5]

## 5. SULFUR IN FINED AND PUDDLED IRON

The fining and puddling processes used to produce wrought iron did not remove much of the sulfur brought in with the pig iron (Bray 1942: 207). The iron silicate slags, neither high in silica, as at the refining stage, nor high in iron oxide, as at the end of the process, had little capacity to absorb sulfur, a regression from the successful potting process. Large additions of limestone and alkali-rich ashes were tolerable in the closed crucible which, in any event, was not reused, but these ingredients, so important to desulfurizing, attacked the furnace construction too aggressively.

The ironmasters of the early nineteenth century were aware of the difficulties in forging caused by the high sulfur in the wrought iron which came from coke fueled blast furnace pig iron. They also knew that the puddling process did not solve the problem, while the potting process, which it replaced, efficiently desulfurized the wrought iron. How, then, did they did they deal with these difficulties, since basic refractories and the concomitant use of high lime slags did not become normal procedures until the last quarter of the nineteenth century?

An acceptable sulfur level could have been achieved in two ways: selective use of low sulfur pig iron pro-

[3] According to Bell's calculations (1884: 357) which incorporate the sources of phosphorus (iron ore, limestone, and coke), 86 percent of the total available phosphorus is absorbed by the liquid iron and 14 percent by the slag. In comparison, for the same charge of ore, limestone, and coke, the sulfur distribution was 2.09 percent to the metal, 82.64 percent to the slag, and 15.27 percent to the gases carried off at the top of the furnace.

[4] The original Bessemer process used silica brick in contact with the molten metal. This was resistant to the high silica slag produced but this slag could not absorb phosphorus. Later in the nineteenth century, the use of high lime/magnesia bricks for furnace lining made it possible to operate with a high lime slag which could absorb phosphorus from the metal.

[5] Bell (1884: 359–361) cites the following data on phosphorus removal at the operations of the Bowling Iron Works in England:

| | carbon | silicon | sulfur | phosphorus |
|---|---|---|---|---|
| cold blast iron | 3.656% | 1.255% | 0.033% | 0.565% |
| as-refined metal | 3.342 | 0.130 | 0.025 | 0.490 |
| finished-puddled iron | 0.226 | 0.109 | 0.012 | 0.064 |

The puddled slag from this operation had the following composition: silica, 12.75 percent; alumina, 2.35; lime, 0.18; magnesia, 0.18; $Fe_3O_4$, 14.39; FeO, 67.09 (total iron 62.25 percent); MnO, 0.55; $P_2O_5$, 3.12; and sulfur 0.05.

When the refining was combined with the puddling operation, the early forming slag was too high in silica and was drawn off "by making it foam" (McGannon 1964: 12). The later forming slag was much lower in silica and very high in iron oxides which permitted the absorption of phosphorus, as the analyses demonstrate.

duced by coke unusually low in sulfur or by unusual smelting practices; or importation of charcoal pig iron from Sweden, Spain, and/or Russia, augmented by locally produced pig iron from charcoal blast furnaces (if they were still in production—see Riden 1977: table 2) to be blended with high sulfur, coke pig iron. Blending of pig irons was certainly an established practice:

> the great concentration of forges of the Stour Valley was based on the blending, in the forge, of Forest of Dean pig iron (which made Merchant or 'tough' iron) and the pig iron from the coal fields which made a cold short (brittle when cold) iron, the blend ('Best Merchant Bar' or 'Ordinary Mill Bar' according to the quantity of tough iron) being suitable for making nails, though for certain purposes the tough 'Forest' or Furness iron was required (Mott 1957–59: 53; see also Trinder 1982: 27).

This description refers to blending to reduce the average phosphorus content, but the same method dilutes the sulfur content also. Percy (1864: 658) distinguishes between gray pig iron and white pig iron from the coke fueled blast furnace as representing low sulfur and high sulfur, respectively (Fig. 2.9). Thus, it was possible to judge sulfur levels in pig iron by fracture appearance long before modern techniques of chemical analysis were developed.

Using British low sulfur pig and imported charcoal pig iron to dilute high sulfur pig iron does not seem to be a tenable procedure. Foreign iron imports declined from about 52,000 tons in the last decade of the eighteenth century to less than 20,000 tons after 1815 (mostly bar iron), while at the same time British production of coke pig iron rose to over 500,000 tons per annum (Scrivenor 1854: 136–137). In the same period, locally produced charcoal pig iron had fallen to less than 10,000 tons per annum (Hyde 1977: 67). The available low sulfur charcoal pig iron could not have made a significant reduction in the sulfur content of coke pig iron even if all of it had been committed to this task.

What seems clear is that substantial quantities of low sulfur coke were available. The few published data on the sulfur contents of coke pig irons of the late seventeenth and early eighteenth centuries indicate that in Shropshire there were extensive workings of low sulfur coal: "Shropshire coal seams produced good coke with high reactivity and low sulfur" (Tylecote 1976: 105–106). At Coalmoor, Shropshire, the coal contained 0.50 to 0.55 percent sulfur, in contrast to the Forest of Dean (1.13 to 1.95 percent) and Bristol (0.95 to 2.07 percent) (Mott 1957–59: 49). Contra Hyde (1977: 84), coking does not significantly change the sulfur content of coal (Chapter 6).

High lime slags could have reduced the sulfur content of the pig iron product, but the extent to which they were used is unknown. Mott (1957–59: 69) cites blast furnace charges for the period of 1718 to 1737 as follows: 9 tons coke, 3.3 tons ore, 0.7 tons limestone, 0.5 tons coal. A blast furnace slag analysis from an early coke blast furnace shows 1.37 percent sulfur and 43.62 percent lime plus magnesia (Morton 1966). The use of high lime slags could not have been widespread, since only one of six blast furnace slag samples (Tylecote 1976: 109), ranging in date from 1650 to 1850, has enough lime plus magnesia to have been effective in desulfurizing pig iron.

Fortunately, because of the indirect heating system in the puddling furnace, the sulfur content of the fuel added little sulfur to the metal. Fired in a remote grate, even low grade bituminous coal could be used. The volatile hydrocarbons were burned completely along with the mineral carbon, and the condensation of tars caused no blockage as was likely in the blast furnace. The sulfur dioxide effluent came off with the other combustion gases; while this contributed to atmospheric pollution, it was not picked up by the molten iron in significant quantities.

The difficulties created by sulfur may, however, have been exaggerated. It has been shown above that many forged and formed iron artifacts have sulfur levels without the compensating manganese that would make a modern carbon steel impossible to hot forge (Fig. 2.8). However, bloom, fined, and wrought irons were different from modern "mild" steel in one important respect—the high FeO slag present as ribbon inclusions. Small pieces of these irons can be consolidated by forge welding because, it is argued, the slag inclusions provide a welding flux. Perhaps, therefore, hot cracking—if it is not too severe—can be closed and welded shut by continued hammering because of these slag inclusions. Modern mild steel—in fact, any steel from the Bessemer period on—since it is almost free from these slag inclusions, is not similarly amenable to forge welding. In the nineteenth and twentieth centuries, before fusion welding techniques were widely practiced, a borax powder was used as a flux in forge welding mild steel.

## 6. FINING AND PUDDLING TO STEEL

Overman in his *Manufacture of Steel* (1854: 81–114) speaks of *German* or *natural* steel, a product of fining, which was a metal with significant but highly variable carbon content. He recommended against its use for high performance tools because it often cracked during the quenching stage of heat treatment. The input material was a moderately high silicon pig iron. The maximum temperature of the fining furnace was high—ca. 1500°C, at which silicon and carbon are oxidized at about the same rate. The fining was a one-step process, in which the presence of silica films over the droplets of metal retarded the rate of decarburization. The bloom was heterogeneous with carbon ranging from 0.3 to 0.8 percent. Overman remarks that the German steel process was best applied to "sparry carbonate of iron" which is the manganese-bearing ore of Germany and Austria. Presumably there was sufficient manganese left in the fined metal to make the bar steel product hardenable.

Although puddling to the carbon levels for steel (greater than 0.2 percent) was difficult to control, procedures were developed sometime before about 1840 (Barraclough 1971). Certain features of steel puddling are different from wrought iron puddling. The most important is that the whole process of oxidation is slowed down. While wrought iron puddling took from 73 to 85 minutes, steel puddling was protracted to as much as 120 minutes (Barraclough 1971). As a consequence the output was less—1600 to 1900 lbs. of steel bar versus 2500 to 2700 lbs. of wrought iron bar in a twelve hour period, and there was a commensurate increase in fuel costs (Barraclough 1971). A high manganese, low sulfur pig iron was used. Presumably, although most of the manganese was oxidized into the slag, enough remained to combine with the residual alloyed sulfur to prevent hot shortness. Oxidation was slowed by shutting down the air intake at certain times, a process which required a delicate balance between limiting air into the space above the hearth and keeping the temperature up by the necessary combustion of coal in the forechamber.

Data from 1875 on the rates of oxidation of the principal elements in the metal have been reduced to fractional decreases to show more clearly the sequence of events (Fig. 13.4). Note that silicon and manganese are most effectively removed before much change has occurred in the carbon content. Slower operation is also better in reducing the phosphorus and sulfur levels. In comparison to the puddling furnace for producing wrought iron, that for steel has a higher ratio of grate-to-hearth area so that the need for higher temperature could be met more rapidly. The hearth was deeper and the bridges higher to permit a more effective slag cover. The roof was higher to keep the air in the gases more remote from the metal. In a similar furnace, high manganese pig iron created a more fluid slag (with MnO instead of FeO) which presumably provided a better protection against uncontrolled oxidation of carbon by the burned gases (Greenwood n.d.: 226–228)

## 7. FUEL EFFICIENCIES IN THE PUDDLING FURNACE

The invention of puddling, despite its problems, meant that, for the first time in European history, coal and its coke derivative could be used in all stages of ironmaking. While charcoal continued to be preferred by specialty steel producers and blacksmiths until the end of the nineteenth century, it was no longer a dominating factor in metallurgical economics.

The layout of the puddling hearth readily permitted effective air preheating by the combustion of waste gases (Chapter 7) or the checker-brick scheme. The latter was composed of two identical forechambers filled loosely with bricks to store heat, one of which was heated by exhaust gases before they vented up the chimney, while the other, already hot from a previous cycle, preheated the intake air for the combustion chamber. When the first chamber was hot and the other had begun to cool, the routing of the intake air was switched. The result was an efficient recycling of energy, so that the temperature of the combustion gas was raised by the preheated air with no additional fuel cost. Although similar heat regenerating schemes were used in blast furnaces earlier in the nineteenth century, the checker-brick method pioneered in the puddling hearths was arguably the key to the development of the open hearth steelmaking process, in which furnace temperatures were sufficient to produce fully molten steel from molten cast iron so that liquid steel could be poured into ingots of sizes and shapes more adaptable to large scale hot working into bar and plate.

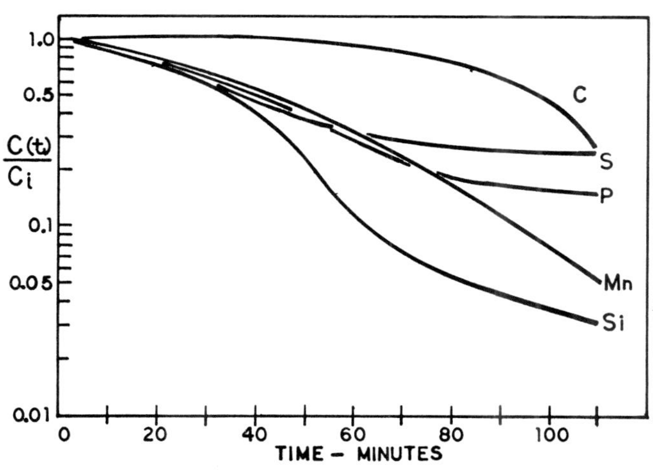


Fig. 13.4. *Rates of Elimination of Various Elements during Puddling.* In this instance the processes are conducted slowly enough that the operation can be stopped at some carbon level appropriate to a useful steel. (From Barraclough 1971)

## 8. THE CHINESE PUDDLING PROCESS

Li Chung (1975) explicitly uses the term *puddling* in the English abstract of an article on the archaeometallurgy of ancient China:

> Wrought iron and wrought steel were produced by puddling with cast iron as raw material. The method was discovered probably in the 1st century BC, since remains of puddling furnaces of the West Han period have been found along with pieces of steel containing 1.3% carbon and of wrought iron. . . .The evidences indicate that the puddling method was used for producing either wrought iron of low carbon content or steel of medium and high carbon contents by arresting the decarburization process when the desired carbon content had been reached.

We use the term *explicit* here because none of the facts provided about these furnaces makes it clear that a process similar to European puddling was in fact used. It may be that the translator chose the term, puddling, as generic to the wrought iron product.

Further, Cheng (1978) observes: "A furnace for steelmaking from molten pig iron was unearthed at the site of Honan No. 3 in Kunghsien County. Discovery of this method was a milestone in mankind's use of iron." No further details on these ancient Chinese puddling operations have been found in more recent literature, but the techniques were definitely known later. An account of an iron refining operation in the nineteenth century (Rocher, as quoted by Tegengren 1923–24: 359) included a recognizable puddling process; both the location—the remote province of Yunnan—and the date—sometime in the 1860s—make Western influence improbable (see also Wagner 1984: figs. 15–18).

The *Tiangong Kaiwu* of Sung Ying-Hsing published in 1637 (Sun and Sun 1966: 250) is important historically and puzzling technically. One illustration depicts what appears to be a puddling operation in its left-hand panel (Fig. 3.9) and the accompanying text reads as follows:

> If the pig iron is intended for casting purposes the molten iron is discharged directly into pig beds molded in the form of long bars and round pieces. If it is intended for the making of wrought iron, the molten iron is run into a square ditch built a few inches below the furnace and a few feet away from it, surrounded by a low wall, where a few persons holding willow sticks and a certain amount of dry and finely screened earth stand ready. When the molten metal flows into the square ditch, one man quickly throws the earthen material into the molten mass, while the others vigorously stir it with the willow sticks, and soon wrought iron is obtained. Two or three inches of each willow stick will be burnt off at the end of an operation so that new ones will have to be used at the next turn. When the wrought iron is somewhat cooled (and becomes pasty) it is either cut into square blocks in the ditch, or lifted and hammered into round pieces before it is sent to market. This method, however, is not used in the refineries of Liu-Yang.

The translators (Sun and Sun 1966: 258) suggest that the "earthen materials" consisted largely of iron oxide-rich silicate slag which decarburized the liquid iron and added to the resultant low carbon iron. This slag would have to derive from some other smelting or fining operation. But since bloom iron smelting does not appear to have been practiced at the time in China, and since fining is an alternative to puddling, the source of such a high iron oxide slag is unknown. It is possible that the translators have inferred wrongly, and the "dry and finely screened earth" was really iron ore, which in fact can be a richer source of oxygen than any slag.

The *Tiangong Kaiwu* description has other technical anomalies. In the illustration (Fig. 3.9), the "ditch" is about 5 ft. square. A simple calculation shows that the iron weight per inch of depth of liquid metal pool is about 1000 lbs. In such an open ditch, the cast iron would have frozen almost immediately. To retard total freezing, the pool of molten metal would have to have been several inches deep. Even then, it is doubtful whether the uncovered pool would stay liquid long enough for much decarburization to occur. From descriptions of the European puddling processes, more than thirty minutes are needed for removing the carbon in the presence of a radiantly hot roof. Further, a deep, 5 foot square pool of iron would be difficult to handle when solidified. The wrought iron product is "cut into square blocks in the ditch, or lifted and hammered into round pieces before it is sent to market" (Sun and Sun ibid). This would require manipulators and hammers capable of handling a piece of hot iron weighing several thousand pounds.

It seems clear that the *Tiangong Kaiwu* depicts a true puddling process, although the description is incorrect. A possible explanation is that the seventeenth-century author and artist had not personally observed a puddling hearth and so omitted certain important details such as a heated roof and walls over and around the "ditch." Another interpretation is suggested by the description of a Chinese double-furnace system observed in the nineteenth century by Julien (Julien and Champion 1869: 55):

> Sometimes the two furnaces are placed side by side, in such a manner that the cast iron from the first can trickle into the second, with a little silica added as need be. This combination is very advantageous in the economy of fuel and moreover the cast iron is decarburized easily by the action of the air as the liquid metal is flowing.

The flow through a short, shallow trough exposed to the air would cause some decarburization, and trickling through the combusting fuel bed of the second furnace completed the process. Lantenois (1907: 418) described a similar two-stage process, which is also like the South Wales process (section 3, this chapter), in which a refining furnace was tapped into a fining furnace.

In the illustration (Fig. 3.9), the first furnace for smelting is being drained into a trough with a holding pool to regulate the rate of flow into what must be a second furnace, although no box bellows is shown. It may be that natural draft is being used, the openings for which would be obscured by the platform, or it may be that the artist simply omitted the second bellows. The "earth" being scattered into the furnace is silica or an easily fused clay, and the trickle of cast iron is being spread over the area of combusting charcoal with willow sticks. If this is, indeed, what is being shown, it is a fining rather than a puddling process. This is an understandable mistake by the translators who, taking the Chinese text to read "wrought iron," thought that the process had to be "puddling." The Chinese characters on the holding pool refer to droplets of steel. What is shown is consistent with the expectation that the open air decarburization of spills over the sides of the pool would produce a metal less completely reduced in carbon, that is, steel.

Processes similar to puddling existed in Japan (Geerts 1874: 13) as well as China, and in East Asia other decarburization methods, including European-style fining, were known (e.g., Lyman 1879: 76–77). These methods seem to have included techniques for controlling the removal of carbon from iron while in a solid state.

## 9. DECARBURIZATION OF SOLID CAST IRON TO IRON AND STEEL

A wrought iron product from the forging of a fined or puddled ball of raw metal is first made into a semi-finished bar or plate. It can then be converted to a finished product by a further series of operations. The Chinese discovered (as the Europeans did in the seventeenth century) that decarburizing a casting as a solid, shaped piece eliminated the need for expensive forging and machining. The casting approach to shape-forming is less expensive and more amenable to mass production, and shapes can be made which are impossible to forge.

It is feasible to decarburize unmelted white cast iron to the carbon levels of iron or steel and thereby convert the metal from a brittle to a formable material. This is true only of white, not gray, cast iron. There is good metallographic evidence that the decarburization of white iron castings was practiced in China by the fourth century B.C. (Li Chung 1975).[6] It seems possible that the early development of methods for solid state decarburization of thin cast iron plates by the Chinese made bloom iron smelting unnecessary (e.g., BUIST 1981), and the archaeological record contains no indisputable examples of early Chinese direct process iron.

There is potential confusion in the interpretation of the ductility and toughness acquired by white iron castings heat treated at temperatures below their melting point. In a neutral environment, the massive crystals of iron carbide in the white iron will decompose to graphite in about 24 hours at 800°C. The resultant metal is called malleablized iron and is capable of modest bending. A similar heat treatment performed in an oxidizing environment will produce a shallow surface layer of pure iron or steel with an interior of malleable iron (Fig. 13.5). Because the surface layer takes much of the stress when a metal is flexed, iron thus treated has an even greater bending capability. The latter is called the *white heart malleable iron* process in the twentieth-century literature.

Many of the farm implement castings manufactured in China from the early Han period in the second and first centuries B.C. were sufficiently thin that they were decarburized to steel throughout. A product suitable for use in swords could be made by decarburizing still thinner plates of white iron to a homogeneous high carbon steel and then forging stacks of these into thicker bars. Since the solid state decarburization process involves some hours at high temperature, it is relatively easy to stop the exposure of the piece of metal when the carbon content has been reduced to levels appropriate to high carbon steel.

Solid state decarburization can be followed by forging to a wrought iron or steel. In one account from nineteenth-century China, the low carbon (2.27 percent), poorly fused, slag-riddled iron product of a crucible smelting process was further processed by heating and forging (Tegengren 1923–24: 324–329). This low carbon content is inappropriate for a cast iron product of smelting, since it is actually a mixture of cast iron and bloom iron averaging 2.27 percent carbon. The cakes of raw iron are first simply heated on a grate covered with anthracite so that much of the slag drains off, as when a raw bloom is given a preliminary heating. Then the hot lumps are quickly drawn out and forged to short bars. The bars are reheated and reforged several times until finished bar dimensions are attained. The reheating is done in an oxidizing combustion atmosphere so the finished product has a carbon content of only 0.34 percent. The sheath of oxide on the bars acts as the source of oxygen for decarburization.

Any atmosphere containing less than 90 percent carbon monoxide will decarburize to a steel structure of some specific carbon level (Fig. 11.1). For example, when the carbon monoxide content is only 75 percent, the equilibrium product will be pure iron with an extremely small carbon content. However, below 31 percent, the atmosphere is oxidizing and decarburizing so there is a major loss of metal. Tegengren (1923–24: 324–329) estimates that the metal lost in the repeated heating in the forge furnace amounted to close to 50 percent.

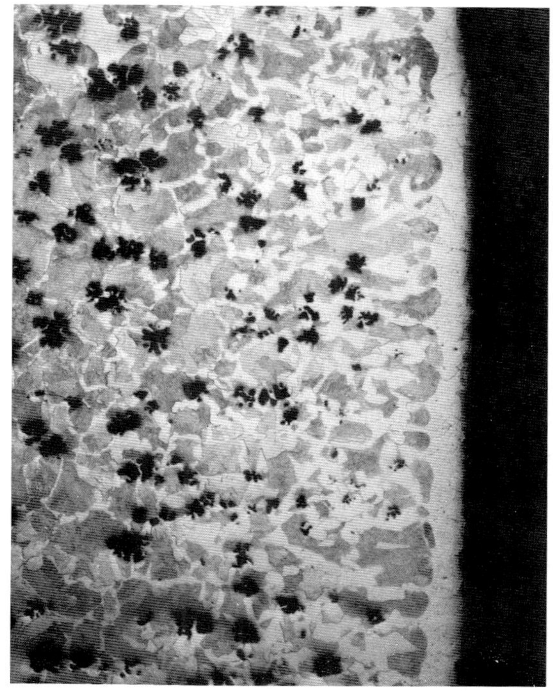

Fig. 13.5. *Decarburization of White Cast Iron.* The historical technique (shown here) was to pack the white iron casting in mill scale and heat to temperatures of about 1000°C for at least 24 hours. The micrograph (×50) shows the pure iron surface layer backed by a higher carbon, steel structure and further back (not illustrated) the original white iron structure. Black nodules of graphite are the product of simultaneous malleablization. Carbon is lost from the carbon nodules much more slowly. Plates of white iron 6 mm. thick can be decarburized completely to pure iron without graphite formation. (See Rostoker et al. 1985)

Although the use of a controlled atmosphere for deep carbon augmentation was not practical until modern times, a similar result was achieved in Europe by pack carburizing with solid carbon in a more or less airtight container (Chapter 11). The same is true of decarburizing. The pieces or castings of white iron had to be pack decarburized in order to minimize the loss of metal by oxidation. Just as solid carbon was the preferred packing medium for pack carburization, the agent for decarburization had to be oxidizing. The most common was the suboxide $Fe_3O_4$, which is the major component of mill or hammer scale. This was the basis for the more successful of the pack decarburization processes developed in France and England during the eighteenth and nineteenth centuries. For many years there was uncertainty about the proper mix for the pack. Réaumur in the 1720s (Sisco and Smith 1956: 292–359) experimented with all sorts of substances: chalk, sand, bone

[6]Joly and Inada (1913: 80) quote from a Japanese book published in 1829 that in the early fifteenth century, before the development of the tatara process for making steel, steel swords were made from cast iron; that is, the sword was cast as cast iron. The description goes no further. Taken literally, the statement is feasible since a casting of white iron can be converted to a steel of any carbon content by solid state decarburization. No other approach would convert a cast iron object to steel.

ash, wood ash, potash, charcoal, calcined oyster shells. Most of these had no effect on decarburization except the chalk which would release carbon dioxide at elevated temperatures.

The speed of the puddling process spelled the end of pack decarburization as a method for producing bar iron, but it survived for many years for applications that involved the conversion of semi-finished white iron castings. Packing such castings in iron ore and limestone and annealing for five to six days formed the basis of an English patent to Samuel Lucas in 1804 (Percy 1864: 804–805). The process, although not immediately successful, eventually came to be widely used in the cutlery industry. Ede (1873: 64) describes how low priced tableware was made by the developed process:

> Cast iron cutlery is enclosed in boxes and cemented with some substance containing oxygen such as poor iron ores, free from sulfur, the scales from the smith's anvil, and various other absorbents of carbon. The boxes are luted in a similar manner as the boxes when annealing steel or case-hardening iron; they are afterwards placed in suitable furnaces and the cast articles are kept in a state little short of fusion for two or three days; they are then found to possess a considerable degree of malleability, and can be readily bent and slightly forged.

The white heart malleable iron product survived into the twentieth century (Angus 1976: 30, 85). As it is understood today, the bulk of the packing should be of $Fe_3O_4$, the primary component of mill or forging scale, to which some limestone has been added to provide carbon dioxide as a gas transfer starter.

European consumers were doubtful about the quality of malleabilized semi-finished cast iron objects, regarding them as inferior substitutes for higher priced prototypes made of forged or cast steel. This was not so in pre-industrial Japan, however, where softened cast iron reached its uncontested peak in the manufacture of the extraordinary teapots employed in the tea ceremony. Rein (1889: 433–435) describes a process in which articles of white cast iron were decarburized by packing them with charcoal in a container which allowed the aspiration of air so that, while being heated externally, a gas atmosphere high in $CO_2$ was maintained in the interior. The surface was decarburized to sufficient softness to allow engraving and inlay work (Fig. 2.4).

## 10. MUSIC WIRE

The diameter of wire in long lengths is reduced by pulling the tapered end through a conical orifice with an exit diameter less than the entry diameter of the wire. This hole is pierced through a plate or plug of a hard, wear-resistant material. A series of openings with increasingly smaller diameters allows wire to be produced in a succession of reductions to a desired fineness. When all the orifices for conversion of wire from some larger diameter to a smaller one are pierced in one steel plate, the tool is referred to as a *draw plate*. Wire drawing is a cold working operation, so intermediate annealing heat treatments are necessary. The reductions in the cross section area of the wire per draw are small so that the pulling or draw force does not stretch or, more critically, rupture the wire as it exits the die.

From as early as the fourteenth century until the eighteenth century, the specialized wire producers in and around Nürnberg and in the German state of Westphalia dominated the European market (von Stromer 1977). By progressive innovation of the machinery and tools for wire making, the Nürnberg industries in particular maintained a nearly unchallenged superiority in wire quality. The product was so important that the technology used to produce it was uncovered by the English partly through industrial espionage and partly through the importation of Westphalian craftsmen by English industrialists (Schubert 1957: 297–302).

Fine wire, like that used for stringed musical instruments, was particularly difficult to fabricate in iron or steel. Both bloom iron and fined iron of the period were populated by many thick ribbons of glassy slag (Fig. 13.6). As the wire was drawn to smaller (finer) diameters, the probability that a fragment of these inclusions occupied a considerable portion of the wire cross section increased, so that beyond some critical wire diameter a particular batch of iron would repeatedly break during the drawing operation. Even if the drawing to final gauge was successful, the fatigue fracture resistance of the wire would lead to frequent breakage during musical performances because of the vibratory (fluctuating stress) nature of the sound generation.

By the sixteenth century, Nürnberg alone was producing iron wire of consistently good quality which was less than 0.5 mm. in diameter. Metallurgical examination of

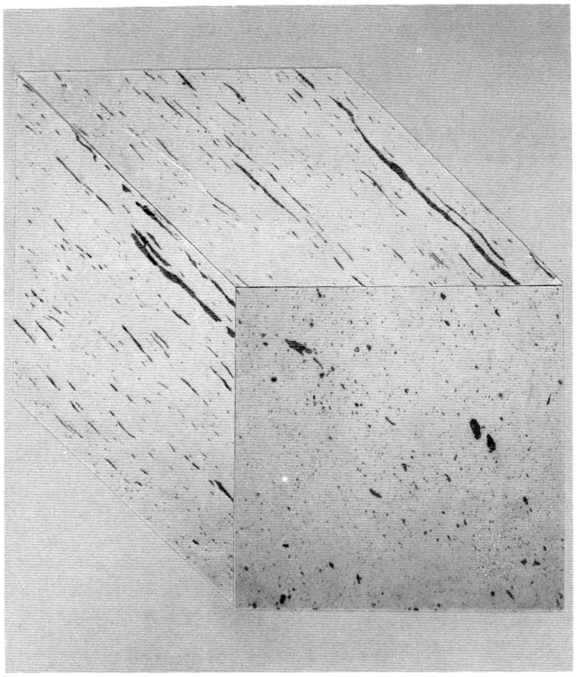

Fig. 13.6. *Spatial View of Slag Stringers.* A montage of microstructures (×50) taken at three orthogonal planes of section through a round bar of early twentieth-century wrought iron helps to visualize the ribbon shapes of slag inclusions. (Photomicrographs courtesy of James Dvorak)

harpsichord wire from instruments of the seventeenth and eighteenth centuries shows a very low carbon iron which is as free of nonmetallic inclusions as the modern steel product called music wire (Goodway and Odell 1984). The hardnesses of these wire samples were higher than pure iron usually is, probably because of the presence of as little as 0.1 percent phosphorus.

The Westphalian process for producing nearly inclusion-free iron was a special variant on fining. We do not know whether the Nürnbergers adopted the Westphalian process or developed a similar one. Schubert (1957: 299–302) unfortunately calls the Westphalian process the Osmond iron process, which is confusing because the term *Osmond* has also been applied to a Swedish design of a bloom iron smelting furnace and more generally to Swedish iron exports (Tholander 1975). It may be that "Osmond" denoted a high quality product. In the Westphalian process, as in fining, the pig of cast iron was drip-melted beneath a bed of charcoal in full combustion. Instead of the liquid metal drops descending through the bed into the slag pool and collecting as a slag-riddled aggregate, they were caught on the end of an iron bar slowly rotated by the furnace operator. The drops spread over the end of the bar forming a thin film which decarburized rapidly in the combustion atmosphere. The accumulation of such droplets generated a ball of nearly carbon-free iron. If, from time to time, the oxidizing environment went too far and created an oxide film over the growing ball, the operator would dip the ball into a molten slag pool which would quickly dissolve the iron oxide. The rod and ball would be withdrawn, cooled until the slag was solid, rapped on rock to shatter off the slag, and then the drip collection would proceed. When it attained some convenient size, the ball would be removed to be forged, both hot and then cold, to bar to provide the input material for the wire drawing process (Schubert 1957: fig. 25 on 295). Since the fined product was collected above the slag pool, it did not occlude much slag.

Modern music wire is a high carbon steel whereas early music wire was almost carbon-free (probably less than 0.01 percent). Sound production demands not only the capacity for tensioning, but also other properties such as fatigue resistance—in which copper alloys are deficient compared to iron—and creep resistance—polymeric materials have difficulty sustaining tension. Iron and steel have a superior combination of properties which meet all these demands. High carbon steel has taken over from low carbon iron partly because it can be adjusted to higher tension and partly because it has higher fatigue fracture resistance. The high strength metal wires will not relax tension (creep) much at normal temperatures.

The early nineteenth-century Indian use of wootz steel for the manufacture of fine wire is mentioned in communications from the surgeon assigned to the Resident in Mysore (Buchanan 1807, v.1: 151–152):

> The mode of preparing country steel before it is drawn into wire, is by taking any quantity, and heating it in a charcoal fire until it be red hot; when it is taken out, beaten into a long thin plate upwards of an inch in breadth, and rolled up into an oval or round form, leaving a small space between each of the folds. It is then put into the fire again. well heated, and hammered out as before. This process is repeated eight times, by which the weight of the steel is reduced to one fifth of the original quantity.
>
> When this is done, it is ready for being formed into wire, and is again heated, and beat into small square rods. It is heated again, and drawn through a small hole, in a plate of common steel, into wire by means of pincers. In this plate there are several holes, of various dimensions, for the purpose of gradually reducing the wire to the size required.
>
> After it has been once drawn, it is necessary to heat it again, before it can be drawn a second time, which is done through a hole somewhat smaller than the former one. It afterwards requires no further heat; but is drawn eight or ten times more until it be sufficiently fine; and this is partly ascertained by the sound it gives, when struck by the finger on being stretched out. At the time of drawing it through plates, a small quantity of oil is applied to it, to make it pass easily. The length of take up in making four pieces of wire, nearly of the same weight, and as fine as the larger of the specimens (its weight 169 grains, its fineness 2 grains per foot) from the time of commencing the preparation of the steel, will occupy one man four days.

The folding and forging eight to ten times decarburize to lower carbon levels so that the drawing process is made easier. The space between the folds and the loss of weight testify to the oxidizing and decarburizing nature of the heating process. Like the music wire of Nürnberg and Westphalia, the wire is a cold drawn, low carbon iron. Also, the "heating" associated with hammering implies a hot working operation, whereas "heating" between stages of drawing the wire implies an annealing procedure to soften the iron so that it may be drawn further to finer diameters without breaking or wearing out the die holes.

## 11. PILING

Whether done to a bloom of iron or a puddled ball, hot forging both expelled a great deal of slag and consolidated the metal. The remainder of the slag was occluded in the consolidated metal and, as forging developed an extended shape such as a bar or plate, slag inclusions also became narrow ribbon shapes within the metal body. The inclusions did not impede hot forging, as they did any subsequent cold forming. If rod stock was to be fabricated into nails or rivets, short lengths were upset to form the flat or mushroom heads, in which cracks or splits commonly developed. Cracking—edge splits or laminations—could also occur in cold formed plate or sheet, as in plywood which has been soaked and quickly dried. The problem was caused by the long, thick, glass-like slag inclusions lying in the cut faces, which cracked in cold forming and the cracks propagated into the metal. At hot forming temperatures the slag was either liquid or solid deformable as are other glasses.

If the inclusions could be rendered thinner and shorter, there was less hazard of splitting but this meant deformation to smaller dimensions of thickness or diameter than were necessary. The splitting problem was solved in Britain late in the eighteenth or early in the nineteenth century by forge welding stacks of bars and reforging them hot back to the bar sizes (Gale 1966: 45–46).

This reworking greatly thinned the slag ribbons without loss of control of the desired thickness of the metal piece itself. As the ribbons were elongated during hot working, they also tended to separate into a line of shorter inclusion ribbons. This combination of thinning and shortening reduced the incidence of splitting. Making plate by piling in a cross pattern was done in the hope that the finished product might be deformed in any direction without splitting. The achievement of such dual directionality or isotropic deformability remains a problem in contemporary fabrication.

# 14. The Transition from the Bloomery to the Finery in Europe

## 1. INTRODUCTION

The transition from the bloomery to the finery in Europe took place in less than two centuries beginning sometime in the sixteenth century. By 1600, the iron production of Europe had increased several times over, and some of its iron goods, among them cast iron cannon and steel swords, were exported in large numbers to places as distant and as metallurgically sophisticated as southern and eastern Asia. An expanded use of water power and machinery, new attitudes toward science and technology, and major alterations in political, military, and economic institutions contributed to the success of this industrial aspect of the Renaissance. From a strictly metallurgical point of view, however, one innovation stands out. Between 1500 and 1700 in much of northwestern Europe, the bloomery was almost completely replaced by a more complex smelting system based on the large scale production of cast iron in blast furnaces and the subsequent conversion of part of this to wrought iron in fineries.

The true blast furnace and various sorts of finery had been known in the Far East for two millennia. The actual European invention or borrowing of the indirect process is not as important as the fact that the historical records available make it possible to trace its causes and effects in detail. Some of the results were far-reaching, since the modern steel industry is the direct descendant of Renaissance ironmaking systems. They also influenced the progress—or lack thereof—of ironmaking elsewhere as the new techniques and their products were distributed worldwide in the ships of European empire-builders, merchants, and colonists. By the beginning of the Industrial Revolution in the late eighteenth century, inexpensive—although not necessarily superior—European iron was making inroads into the traditional markets of African and Asian ironmakers. By the late nineteenth century, most non-European ironmaking systems had disappeared, and virtually all iron produced in the last hundred years has been made by methods derived from the indirect process developed in northwestern Europe during the sixteenth and seventeenth centuries.

Here the effects are not the concern, but rather the causes, for these are less obvious than they may seem. The overwhelming success of the indirect process has led some historians to regard it as inevitable, in that the new technology was instantly perceived as superior and spread as quickly as competitors could master it. In reality, for several centuries this superiority was not apparent to most ironmakers, and in some ways the new technology may have been inferior, because it cost more and yielded a poorer iron, except in unusually favorable circumstances.

## 2. THE NEW IRONMAKING OF THE RENAISSANCE

Early in the fifteenth century, most of Europe had begun to recover from the devastating mortality of the first great outbreak of the Black Death. People in the south, in the midst of an economic boom, felt the stirrings of a new intellectual self-confidence that soon caused them to boast that they were the equals of the ancients. The north—or at least the western part of the north—though still medieval in terms of institutions and artistic outlook, was not especially backward technologically. Workers in France, the Low Countries, Germany, and even Britain were skilled in mining, used water powered bellows, tilt hammers, and stamping mills, were able bronze founders, and could make both guns and gunpowder. Moreover, with the spread of new ideas and institutions from Italy, it is not surprising that ironmaking, like many other arts, was due for a change.

Ironmakers could have concentrated on perfecting the old bloomery processes, which since at least Roman times had been producing good iron in amounts adequate for the domestic needs of substantial populations and for the military requirements of large states. For over a millennium there had been fluctuations in demand and in the size of furnaces and workshops, but the only real trend was to making larger batches of iron. At the end of the Roman period the average bloom size was only about 5 kg., while in the fifteenth century blooms of 100 kg. were common (Schubert 1957: 122–145). The significance of this is hard to judge in the absence of statistically meaningful samples and information about how long it took to make these blooms. Water powered hammers made handling large blooms easier (Schubert 1957: 122–145), and the larger furnaces were more efficient in their use of fuel. In due course the fully developed Catalan hearth and the Stückofen showed that a carefully designed bloomery, well equipped with powered hammers and blowing devices, could achieve significant efficiencies in production.

Another alternative was to abandon the one-stage process altogether and, as the Chinese had already done, build furnaces so large that the production of cast iron was consistent. The cast iron was then converted to the desired forgeable metal by a second processing stage. This approach had more potential for the future than the perfection of the bloomery process, and eventually led to ironmaking methods of great efficiency which produced metal in much larger quantities and at much lower prices than before. This was, of course, unforeseeable in the fifteenth or even seventeenth century, because the indirect process was initially inferior to the old one in some ways, and, as understood by Europeans, took a relatively long time to perfect. The advantages of the blast furnace-finery method were not obvious to the majority of ironmakers in the first few centuries after its introduction in the early fifteenth century, to the extent that a nineteenth-century historian commented (Beck 1893–95, v.2: 186; translation Bronson). : "this improvement, which had first appeared in the Rhineland, spread with the most extreme slowness, so that by the first half of the sixteenth century it was still confined to western Germany and eastern France."

More recent research (Wertime 1962: 72–78) has shown that the area where blast furnaces, and presumably fineries, were known at least by 1500 to 1550 also included parts of England and Sweden, and perhaps a few places in western Czechoslovakia and Poland. Bloomeries were still used throughout Europe until the mid-seventeenth century, and most of the southern and eastern parts of the continent did not begin the shift to blast furnaces until the late seventeenth, eighteenth, or even nineteenth century.

It is not likely that the widespread reluctance to adopt the new techniques was due to scientific backwardness or lack of information, because several of the areas which did not, including southern Germany, Austria, and northern Italy, were among the most metallurgically advanced of the time. Those ironworkers who did could not have foreseen the consequences.

Schubert (1957: 152–153) highlights the fact that the early European indirect process had deficiencies, citing the superior quality of the iron produced in bloomeries, the unsuitability of early blast furnaces for smelting high phosphorus ores, and the relative costliness of the furnace and other necessary equipment. For him, the advantages of the indirect process lay in the better yield of iron from ore, the lower per unit consumption of charcoal, and the much higher potential volume of output.

Morelli (1982: 24–27) also treats the issue that the bloom iron producing "low furnace" survived in eighteenth-century Tuscany side by side with the cast iron producing "high furnace" only because of differences in ores and, perhaps, because of conservative attitudes among some contemporary Italian ironmakers. Pointing out that both types of furnace made use of ore from the same Elban mines, Morelli concludes that the main disadvantages of the indirect process were the larger investment required, the need for a market able to absorb a continuous production of high volume, and the relatively fixed location of high furnaces and the consequent difficulty of moving them as the availability of resources changed. She enumerates several counterbalancing advantages such as the high quality (i.e., freedom from slag) of decarburized cast as opposed to bloom iron, the shortness of the time of smelting in a high furnace, and the saving of fuel because such furnaces can be operated continuously without breaks for extracting the metal.

We believe that Schubert and Morelli are incorrect on some points. Wrought iron was no more slag-free than bloom iron, and the indirect process was less fuel efficient than the direct process for several centuries. On the other hand, both authors correctly consider economic as well as technical explanations. In order to understand the rise of cast and wrought ironmaking, a complex model must be constructed with technological, economic, and sociopolitical factors.

## 3. THE MARKETPLACE

Until recently the bulk of the iron sold in and exported from Europe was "bar iron"—a low carbon forgeable metal that formed the main input for most secondary manufacturing procedures. The direct and indirect processes were in head to head competition in this market. Recorded monetary costs indicate that through the seventeenth or eighteenth century indirectly produced wrought iron was rarely a cost effective alternative to directly produced bloomery iron, but from an early date it occupied a major place in European commerce.

A manuscript written in 1670 and ascribed to the Swedish ironmaster Cronström has provided valuable information about production costs of the two different processes (Pipping 1982: 163, 178, 182, 189). The component costs of Catalan hearth (i.e., direct process) production in Spain, Sweden's chief competitor in producing premium bar iron for the international market, amounted to 15 daler per skippund of 195 kg. In comparison, four times in the manuscript the cost of the same weight of bar iron produced by the Swedish indirect process is estimated at 21 to 29 daler (the manuscript implies an efficiency of conversion of pig iron to bar iron of 75 percent). These figures make the indirect process appear uncompetitive, but Cronström shows that it need not have been. In estimating the cost of bar iron made by the indirect process in Sweden if the smelting furnace and the fining-forging hearths were on the same site, he finds that the cost of one skippund of bar iron would be only about 15.5 daler—almost the same as the cost of Spanish bloom iron. The effect on the cost by the integration of the production operations is not explained, but it may be due in part to transportation expenses.

The records for indirect process bar iron manufacture at Pontypool in Wales in 1704 (Schubert 1957: 426) show that it was an outstandingly efficient operation which may be atypical. An annual pig iron production of 400 long tons at a total cost of £1425 was valued at £2400, a net profit of 40.6 percent, but demand was too limited to justify running the furnaces full time. The pig iron was not sold, but converted to 300 tons of bar iron at an additional cost of £1530. Since the sale price came to £4500, the real overall profit amounted to 34.5 percent—a good margin in a period when blue-chip investments often yielded only a few percent per annum. In considering the Pontypool figures, several points must be borne in mind. First, the bar iron did sell on the market and therefore was competitive with bloomery made bar iron for quality and price. Second, the demand for bar iron was so great that furnaces could run full time. In addition, like almost all financial records of ironmaking during the pre-industrial period, those of Pontypool are clearly incomplete, making no allowance for initial construction costs, maintenance costs, depreciation, or interest on fixed and operating capital.

In a discussion of the economics of the Catalan forges that survived in southern France until the 1860s, Percy (1864: 280, 310, 311) compares the costs of the direct and indirect processes, and finds the direct process more expensive. In 1840, a ton of bar from three representative forges in Ariège sold for an average price of £18.4.0 per ton, for a profit of only 6.1 percent. This contrasts markedly with the average price in England for bar iron made by the puddling process, which sold in Percy's day for only £6 per ton. The comparison is unfair since the French product was a better grade of iron, and the English was not only of an ordinary grade but also produced with coke, a much cheaper fuel. The charcoal consumed in the French furnace accounted for most of the production cost—£10 for each £18 in sales—while the coke needed to make a ton of English iron probably cost no more than a pound or two. Even if the fuel costs had been the same, the English blast furnace/puddling hearth combination still produced cheaper bar iron.

The most convincing comparison of the two routes to forgeable iron is made by Beck (1897, v.3: 882) on the basis of records preserved at Gittelde in the Harz Mountains of northern Germany, where the dukes of Brunswick established a major metallurgical center in the sixteenth century. The Gittelde enterprise simultaneously produced wrought iron in blast furnaces/fineries and bloom iron in a type of furnace locally called a *Zerrennherd*. Both systems, which utilized local charcoal, ores, and labor, were operated side by side for more than one hundred years. Comprehensive records, kept for the duration, preserve unique data from which the average production cost and sales price for iron made by the two systems can be figured (Table 14.1). The indirect process obviously became more profitable as the evidence shows that revenues from the bloomeries decreased while those from the fineries rose. At the beginning of production in 1590, still more than a century after the introduction of the blast furnace in Germany, the bloomery not only was more cost-efficient but also had a higher margin of profit than the finery. By the mid-seventeenth century, however, the economic status of the bloomery had deteriorated to the point that it was less profitable in relative terms and sometimes actually ran at a loss. Gittelde's managers seem to have remained optimistic about bloomeries, continuing to build new ones until after the middle of the seventeenth century, but the cost was evidently too great. The last Zerrennherd was closed in 1689.

*TABLE 14.1. Bloom iron vs. wrought iron, cost and price per ton at Gittelde, 1590-1678*

| Bloomery bar iron | | | Finery bar iron | | |
|---|---|---|---|---|---|
| Dates | Cost | Price | Dates | Cost | Price |
| 1590 | 116.36 | 155.55 | 1573-90 | 151.33 | 186.36 |
| 1616-25 | 178.96 | 228.50 | 1617 | 165.35 | 170.00 |
| 1647-54 | 186.90 | 170.00 | 1647-54 | 109.71 | 170.60 |
| 1678-89 | 145.00 | 167.00 | 1678-95 | 100.00 | 165.40 |

Ton = metric ton. Cost and price are converted where necessary to marks. Finery costs include the cost of making the pig iron at the blast furnace. Source: Beck 1897: 882.

The reasons for the shutdown are not completely clear. The price at which metal was sold seems to have fluctuated violently, undoubtedly because Brunswick was embroiled in the Thirty Years War in the early seventeenth century. The political situation may help explain the apparently erratic behavior of the prices of wrought iron vis-à-vis bloom iron, but it makes comparison difficult (see Table 14.2 for data on costs). The chief reason for the poor performance of the bloomery was a sharp increase in labor costs after 1647, a period when expenditures for labor at the blast furnace and finery stayed the same or even declined. There is no obvious technological explanation for the shift to the finery, since no major labor-saving innovations in fining operations were introduced during the mid-seventeenth century.

## 4. FUEL CONSUMPTION

The data in Tables 14.3a and b indicate that, at the smelting stage, the indirect process appears to be more fuel-efficient than the direct. However, the comparison is deceptive because the data are for blast furnaces which yielded only raw cast iron, and until well after 1800 it was not often a final product. In almost all pre-industrial economies, those of the Far East being notable exceptions, the market demanded a forgeable iron low in carbon. Raw iron as extracted from the bloomery was already in a semi-marketable condition; that from the blast furnace was not. The conversion of raw cast iron to a saleable low carbon iron required large quantities of fuel. Small furnaces should be more wasteful of fuel than larger ones, because, in principle,

*TABLE 14.2. Costs of fuel, ore, and labor at Gittelde*

| | Fuel loads | Fuel cost | Ore loads | Ore cost | Labor cost |
|---|---|---|---|---|---|
| Bloomery (Zerrennherd) | | | | | |
| 1590 | 13.9 | 49.0% | 8.8 | 29.6% | 21.4% |
| 1613-25 | 20.4 | 58.1% | 11.4 | 30.1% | 11.8% |
| 1647-54 | 30.5 | 42.6% | 9.4 | 20.0% | 37.4% |
| 1678-89 | 27.4 | 44.8% | 7.6 | 14.8% | 40.4% |
| Blast Furnace and Finery | | | | | |
| 1573-90 | 22.3 | 60.3% | 7.7 | 19.8% | 19.9% |
| 1617 | 37.1 | 59.3% | 10.3 | 24.7% | 16.0% |
| 1647-54 | 18.1 | 47.3% | 4.4 | 18.2% | 34.5% |
| 1674-95 | 26.9 | 43.4% | 9.6 | 35.2% | 21.3% |

Loads = fuder; a load of ore and a load of fuel are not stated to weigh the same. Source: Beck 1897: 882.

*TABLE 14.3a. Fuel use efficiency—first stage, direct process*[1]

| Place | Date | Kg. fuel per 100 kg. iron | Furnace Type* | Furnace Height** | Furnace Capacity*** | Reference |
|---|---|---|---|---|---|---|
| Jefisi, Ghana, W Africa | 1971 | 1727 | SS | 1.3 | 4 | Pole 1974: 29 |
| Sukur, Nigeria, W Africa | 1962 | 1406 | SS | — | 16 | Sassoon 1964: 216 |
| Byrkeknott, Britain | 1408 | 1200 | BF | — | 77 | Tylecote 1962: 289 |
| Basari, Togo, W Africa | 1890s | 1050 | HS | 3.5 | 10 | Hupfeld 1899: 185–188 |
| Tendukhera, C India | 1990 | 630 | SS | 1.2 | 60 | Bartlett 1900: 150–153 |
| Kol, Bengal, E India | 1900s | 610 | SS | 0.9 | 19 | Watson 1910: 325 |
| Velatar, Malabar, S India | 1800s | 590 | SS | 2.0 | 145 | Buchanan 1807, v. 2: 439 |
| Schmalkalden, Germany | 1790s?[2] | 573 | HS | 4.9 | 750 | Percy 1864: 329 |
| Sassello, Sardinia | 1840s | 530 | CA | — | 527 | Percy 1864: 314 |
| Puppadaung, C Burma | 1860s | 490 | HS | 6.1 | 40 | Blanford in Percy 1864 |
| Narbada Valley, W India | 1850s | 470 | SS | 1.1 | 5 | Krishnan 1954: 75–76 |
| Tripalda, Sardinia | 1840s | 317 | CA | — | 500+ | Percy 1864: 344 |
| Rajodha, C India | 1890s | 270 | SS | 1.4 | 34 | Turner 1893 |
| Chitaldrug, S India | 1900 | 230 | SS | 1.7 | 11 | Sambhasiva Iyer 1901: 106 |
| Tripalda, Sardinia | 1840s | 178 | CA | — | — | Percy 1864: 315 |
| Jabalpur, C India | 1880s | 120 | SS | 2.0 | 72 | Balfour 1885, v. 2: 373 |
| Mirjati, Orissa, E India | 1920 | 110 | SS | 1.2 | 14 | McWilliam 1920 |
| Kumaon, N India | 1890s[3] | 104 | BF | 0.6 | 140 | Atkinson 1890: 270–271 |

*Furnace Types: BF = Bowl Furnace, CA = Catalan/Corsican Hearth, HS = High Shaft, SS = Short Shaft.
**m.
***kg./day
[1]Table 14.3 presents an extensive compilation of data focusing on fuel inputs. In cases where fuel consumption is reported in terms of volume and where no weight equivalences are suggested in the sources, this has been converted to kilograms by using Percy's (1875: 401) data on the weight of Pyrenees beech charcoal, at 225 kg. to the cubic meter. We have also used Hammersley's estimate that the English "load" equals 1550 lb.
In order to insure the independence of at least some of the cases, the data on fuel consumption presented here have been drawn from as broad a sample as possible. Information about Africa, Asia, and the more remote parts of Europe is from the nineteenth and twentieth centuries, as no earlier quantified descriptions are available and, with the exception of recent European and North American examples, is representative of pre-industrial technologies and economic contexts that have not been influenced by the European Industrial Revolution. Later non-European technological systems may not be especially primitive and are not necessarily "survivals" of more ancient ironmaking methods. However, they do represent to some degree the great range of possibilities open to the world's ironmakers in pre-industrial times.
[2]A Stückofen. The product may include some steel in the bloom or even some cast iron.
[3]As shown by a 4:1 reduction at the "refining" stage, a very impure metal.

a large body mass relative to surface area retains heat better.

Except for the enormous, efficient charcoal furnaces of Siberia, for a long time the largest in the world, the correlation between fuel efficiency and furnace height or capacity is, however, weak (see Table 14.4 after Beck 1899, v.4: 81). This discrepancy may be attributed to differences in the methods of furnace construction, in

*TABLE 14.3b. Fuel use efficiency—first stage, indirect process*

| Place | Date | Kg. fuel per 100 kg iron | Furnace Type* | Furnace Height** | Furnace Capacity*** | Reference |
|---|---|---|---|---|---|---|
| Ferghana, Central Asia | 1880s | 1500 | SS | — | — | Lansdell 1885, v.1: 198 |
| Glasgow, Scotland | 1840s (a) | 575 | HS | 13.7 | — | Ure 1856: 1100 |
| Unspecified place, Sweden | 1660s | 520 | SS | 7.2 | 1280 | Pipping 1982: 163 |
| Pingding, Shanxi, China | 1926 (b) | 488 | CR | — | 800 | Wang & Wang 1930: 109 |
| Unspecified place, Sweden | 1660s | 470 | HS | 10.1 | 2120 | Pipping 1982: 224 |
| Dowlais, Wales, Britain | 1820s | 450 | HS | 15.8 | 13,500 | Beck 1899: 320 |
| Dayang, Shanxi, N China | 1910s (b) | 390 | CR | 2.7 | 347 | Tegengren 1923–24: 334–335 |
| Cannock Chase, Britain | 1588 | 258 | HS | — | ? | Nef 1932, v.1: 193–194 |
| Baojing etc., C China | 1910s (c) | 370 | SS | 1.5 | 453 | Tegengren 1923–24: 339 |
| Tomsk, Siberia | 1790s | 370 | HS | 6.5 | 2021 | Beck 1897: 1146 |
| Unspecified place, Japan | 1890s (d) | 360 | TA | 1.2 | 1238 | Ledebur 1901 |
| Amegawa, Idzumo, Japan | 1870s (d) | 350 | TA | 1.1 | 1360 | Lyman 1879: 61–66 |
| Amol, Elburz, N Iran | 1920s | 350 | SS | 3.0 | 90 | Böhne 1928 |
| Baojing etc., C China | 1910s | 310 | HS | 5.5 | 1225 | Tegengren 1923–24: 338–339 |
| South Henan, C China | 1960s | 300 | SS | 2.0 | 720 | Wagner 1985 |
| Merthyr, Wales | ca. 1845 (a) | 280 | HS | 13.7 | 6000+ | Ure 1856: 110 |
| Dauphiné avg., SE France | 1778 | 273 | HS | 6.7 | — | Léon 1961: 57 |
| Allevard, Dauphiné, France | 1778 (e) | 262 | HS | 7.0 | 1779 | Léon 1961: 110 |
| Staffordshire avg., Brit. | 1840s (f) | 205 | HS | 14.3 | 6300 | Ure 1856: 1073 |
| Yangchuan, N China | 1926 (g) | 185 | HS | 13.8 | 15,300 | Wang & Wang 1930: 111 |
| Northern France avg. | 1770s | 175 | HS | — | 2000? | Léon 1961: 56 |
| Sulzbach, Germany | 1750s? (f) | 163 | HS | — | 800 | Beck 1897: 365–366 |
| Allentown, Pennsylvania | 1839 (h) | 139 | HS | 12.2 | 7000 | Johnson 1841: 28 |
| Pontypool, Wales | 1706 | 120 | HS | — | 2410 | Schubert 1957: 424 |
| Pottsville, Pennsylvania | 1838 | 111 | HS | 10.7 | 4000 | Johnson 1841: 28 |
| Zhongjing area, W China | 1890s | 110 | HS | 6.0 | 3000+ | Tegengren 1923–24: 343–344 |
| Monceau, Belgium | 1850s (g) | 105 | HS | 15.3 | 22,000 | Beck 1899: 977–998 |
| Recent USA | 1940s (g) | 85 | HS | 20+ | 50,000+ | Butts 1943: 102 |

*Furnace Types: CR = Crucible, HS = High Shaft, SS = Short Shaft; TA = Tatara.
**m.
***kg./day
(a) In coal needed to make the coke used in the furnace; (b) Anthracite; (c) Coke and anthracite; (d) *Tatara* furnaces produced lumps composed of wrought iron, steel, and cast iron; (e) Average of 8 experimental runs; (f) Coke, coal:coke conversion ratios are between 2:1 and 3:2; (g) Coke and hot blast; (h) Anthracite and hot blast.

*TABLE 14.4. Fuel use for cast iron in charcoal and coke blast furnaces*

| Place | Fuel for 100 kg. iron | Furnace height, m. | Output in kg/ day |
|---|---|---|---|
| Charcoal Furnaces[1] | | | |
| Libethen, Hungary | 549 | 7.2 | 2721 |
| Rohnitz, Hungary | 404 | 7.6 | 3175 |
| Laurwig, Norway | 357? | 9.0 | 1270 |
| Ruschwinski, Siberia | 321 | 7.3 | 5625 |
| St. Gertrud, Carniola | 299 | 6.1 | 1587 |
| Neuberg, Styria | 221 | 5.1 | 1452 |
| Mitterwald, Lower Hungary | 201 | 9.1 | 8119 |
| Johann-Georgenstadt, Saxony | 195 | 6.0 | 2223 |
| Horzowitz, Bohemia | 184 | 9.8 | 3492 |
| Vordernberg, Carniola | 184 | 5.8 | 3629 |
| Eisenerz, Styria | 182 | 9.1 | 4990 |
| Petrokamensk, Siberia | 160 | 10.8 | 18,824 |
| Mossinz, Carniola | 150 | 8.2 | 3583 |
| Vordernberg, Styria | 147 | 5.2 | 2268 |
| v. Rauscher, Carniola | 137 | 5.5 | 2359 |
| Newiansk, Siberia | 115 | 12.5 | 28,577 |
| v. Rauscher, Carniola | 95 | 9.1 | 3402 |
| Coke Furnaces[2] | | | |
| Creuzot, France | 230 | 11.6 | 2500 |
| Glamorganshire, Wales | 260 | 15.7–20.4 | 5650–7045 |
| Gleiwitz, Silesia | 350 | 11.6–11.9 | 2100 |
| Shropshire, England | 350 | 12.6–15.7 | 5000–6000 |
| Clyde Ironworks, Scotland | 350 | 9.7 | 2500 |

[1]Charcoal: Hassenfratz's data, quoted by Beck 1899: 81. Centners converted to kilograms at rate for English cwts.
[2]Coke: Beck 1899: 91, data apparently assembled by him.

the iron content of the ores,[1] in local ore and labor expenses, or even in the ironmaster's expertise.

The direct process is substantially more fuel efficient than the indirect, a conclusion further substantiated by the greater amount of charcoal needed to make bar iron by the indirect process in England between the sixteenth and early eighteenth centuries (Tables 14.5b, 14.6), and in Germany and France in the eighteenth century (Table 14.7) than for the direct process systems (Table 14.5a). The discrepancy results from the great amount of fuel consumed in the fining stage of the indirect process. In most of the cases examined, these fuel requirements are more than enough to offset whatever economies might result from the efficiency of the blast furnaces used in the smelting stage. The average finery of the pre-industrial period was not very different from a bloomery, in that it needed as much fuel and operated in an inefficient, small-batch mode. It is evident that fining and puddling furnaces required as much fuel to make a ton of bar from cast iron as did a Catalan hearth to make the same amount from the original ore (Table 14.8).

In evaluating the fuel efficiency of systems for making low carbon iron, a convenient reference point may be set at the 4:1 ratio between total fuel used and finished iron produced. Many direct process ironmaking systems, such as almost all Catalan hearths and at least two Indian shaft furnace-forge combinations, are even more efficient. Several other Indian and African examples, for which only data on the fuel required to make raw blooms have been recorded (Table 14.3a), almost certainly would have been well below 4:1 even if the fuel cost of working up the blooms into finished iron had been considered. In contrast, very few indirect ironmaking processes reached this level of efficiency until modern times. The chief exceptions are the charcoal furnaces of Llanelly and Pontypool (Table 14.5b) and of the Foley Works (Table 14.6) around 1700 and the coke fueled furnaces and hearths at Charleroi in the 1850s. There is little evidence that fuel/iron ratios better than 4:1 were regularly achieved by blast furnace-finery combinations before the nineteenth century even in the most technologically advanced European countries. In France in the 1790s a minimum of 200 kg. of charcoal was needed for each 100 kg. of blast furnace pig iron, and refining this to wrought iron usually required at least another 400 kg. (Grignon in Léon 1961: 57; Courtviron and Bouchu in Beck 1897: 331).

By the second quarter of the nineteenth century, a 1.5:1 ratio of fuel to cast iron was reached and even surpassed in most large European blast furnaces. Although fuels like coke and anthracite with a high caloric value per unit weight may partly account for this, charcoal alone was used with equal success in Russia. However, European ironmakers were unable to improve significantly the fuel efficiency of methods for the subsequent conversion of cast to wrought iron. In the early 1800s there may even have been a reversal of the long-term trend toward fuel economy, due to the tremendous consumption of coal and coke in early puddling hearths (Table 14.8; compare Beck 1897: 708). Because these fuels were inexpensive, ironmakers of the early Industrial Age may not have been careful in using techniques that were comparatively efficient. This casual attitude did not last long, and by the mid-nineteenth century commentators like Scrivenor and Percy were expressing interest in economizing on fuel. Not until after 1850 did the modern system of ironmaking begin to reach a reasonable level of efficiency in energy consumption.

## 5. EFFICIENCY OF REDUCTION FROM ORE

When the ore/metal ratios in the manufacture of raw iron and finished low carbon bar iron produced by the indirect and direct processes are compared, it is clear that the indirect process is moderately more efficient in recovering metal from ore (Table 14.9). Since the extraction of the iron present in an ore is of concern here, the data used for the comparison have been culled from sources which are geologically reliable or which report ore samples analyzed in a laboratory. The indirect process has a moderate advantage in the recovery of metal from ore, which might be greater at the raw metal stage if the weights of blooms were corrected for occluded slag. Atypically low metal recoveries from the cast iron smelting operations may reflect loss of metal to a high viscosity slag, but more probably indicate a high gangue content in the ore. The usual European blast furnace of the pre-industrial period recovered perhaps 80 percent of the iron in the ore and between 25 and 33 percent of this was lost during the fining process (Hammersley 1973: 603; Morton and Gould 1967; Hyde

[1]Remember that the fuel requirements to melt a slag are a major factor in smelting. When there is a large percentage of gangue in the iron ore, there must be a correspondingly large slag volume. Unfortunately the iron contents of the ores used are rarely given.

*TABLE 14.5a. Fuel use efficiency[1]—all stages, direct process*

| Place | Date | Kg. fuel per 100 kg. iron | Furnace | | Reference |
|---|---|---|---|---|---|
| | | | Types* | Capacity** | |
| Kamar, Orissa, E. India | 1961 | 1500 (a) | SS + FH | 13 | Ahmed 1967: 17–44 |
| Bilbao area?, N Spain | 1660s | 1120 (b) | CA | 150 | Pipping 1982: 163 |
| Madras area, S India | 1842 | 870 | SS + FH | 15 | Campbell 1971: 258 |
| Jabalpur, C India | 1898 | 820 (c) | SS + FH | 24 | Hole 1898 |
| Kumaon, N India | 1890s | 810 (d) | BF + FH | 41 | Atkinson 1890: 271–272 |
| Unspecified place, Corsica | 1840s | 800 (e) | CA | 203 | Percy 1864: 319 |
| Sassello, Sardinia | 1840s | 665 (f) | CA + FH | 528 | Percy 1864: 313 |
| Ariège, SW France | 1667 | 590 (g) | CA | ? | Beck 1893–95: 969 |
| Tripalda, Sardinia | 1840s | 430 (f) | CA + FH | 520 | Percy 1864: 313 |
| Echeaux, B.-Navarre, France | 1780s | 423 (h) | CA | 527 | Beck 1897: 660 |
| Chica-Bayly, Mysore, India | 1800s | 400 | SS + FH | 31 | Buchanan 1807, v. 3: 362 |
| Ressecq, Ariège, SW France | 1830s | 359 (c) | CA | 606 | Percy 1864: 298 |
| Jabalpur, C India | 1900 | 320 | SS + FH | 34 | Fernandez 1900: 158–160 |
| Vicdessos, Ariège, SW France | 1840 | 310 | CA | 612 | Percy 1864: 311 |
| Celles, Ariège, SW France | 1840s | 282 (a) | CA | 665 | Percy 1864: 310 |
| Vicdessos, Ariège, SW France | 1789 | 280 (i) | CA | 750 | Beck 1897: 658 |
| Foix, SW France | 1789 | 270 (j) | CA | 650 | Beck 1897: 1000 |
| Ariège, SW France | 1840s | 185 (k) | CA | 652 | Percy 1864: 309 |
| Navarra, N Spain | 1716 | 100 (l) | CA | @600 | Beck 1893–95: 867 |

[1]Fuel use includes smelting and finishing stages
*Furnace types: BF = Bowl Furnace, CA = Catalan/Corsican Hearth, FH = Finishing Hearth (at bloomery), SS = Short Shaft
**kg./day
(a) all fuel-iron ratios are for a form of iron said by the source to be ready for sale to blacksmiths; some may still contain appreciable slag.
(b) a very high fuel:iron ratio for the Catalan process. See below.
(c) 2.1:1 at forge + 5.4:1 at furnace.
(d) smelting and forging by independent units at one site.
(e) considered "enormous" fuel consumption by Percy.
(f) with traditional charge: ca. 85% ore, 5% wrought iron scrap, 10% cast iron; data from Balduacco.
(g) no source cited by Beck.
(h) conversions as above.
(i) weights converted at 1 Marc pound = 2.04 kg.; 1 centner = 1 quintal = 50 kg.
(j) production of a hearth going well; charcoal and iron figures are averages for 38 hearths in Foix and neighborhood.
(k) experiment by François with hematite and refuse manganese ore.
(l) conversions as above; data from Réaumur; leather bellows, not trompe.

1977: 79–80). The average yield of the blast furnace-finery combination was therefore about 50 to 60 percent of the iron originally present.

Such a recovery percentage is better than most bloomeries could achieve. Because iron oxides serve as the main fluxing component of the slag, a certain percentage of the iron must be lost in a direct process furnace. Less was lost in Catalan hearths because they were charged with high quality ore and were extremely efficient (Table 14.9b). The amount lost is roughly proportional to the amount of gangue present; unless the ore is almost free of silica, the iron lost as flux can exceed 75 percent even in the best run bloomeries. On the other hand, there is a similar loss of metal at the

*TABLE 14.5b. Fuel use efficiency[1]—all stages, indirect process*

| Place | Date | Kg. fuel per 100 kg. iron | Furnace | | Reference |
|---|---|---|---|---|---|
| | | | Types* | Capacity** | |
| Unspecified place, Britain | 1790s | 1000 (a) | HS + PU | 4000+ | Beck 1897: 708 |
| Unspecified place, Sweden | 1660s | 990 (b) | HS + FI | 200? | Pipping 1982: 203 |
| Kisuki, Idzumo, Japan | 1878 | 906 (c) | TA + FI | 151 | Lyman 1879: 77 |
| Nancun, Shanxi, N China | 1910s | 890 (d) | CR + FI | 30 | Tegengren 1923–24: 329 |
| Gr. Chartreuse, S. France | 1778 | 867 (e) | HS + FI | — | Léon 1961: 57 |
| S Wales, Britain | 1820s | 796 (f) | HS + PU | — | Beck 1899: 322 |
| Cannock Chase, Britain | 1588 | 775 (g) | HS + FI | — | Nef 1932, v. 1: 193–194 |
| Champagne, E France | 1750s? | 750 (h) | HS + FI | — | Beck 1897: 331 |
| Avg. ironworks, N France | 1778 | 500 (e) | HS + FI | — | Léon 1961: 57 |
| S Henan, C China | 1960s | 370 (i) | SS + FI | 619 | Wagner 1985 |
| Llanelly, Wales, Britain | 1704 | 340 (j) | HS + FI | 2000? | Schubert 1957: 426–428 |
| Pontypool, Wales, Britain | 1704 | 240 (k) | HS + FI | 2000? | Schubert 1957: 424–426 |
| Monceau, Belgium | 1850s | 208 | HS + PU | 10,000+ | Beck 1899: 977–978 |

[1]Fuel use includes smelting, fining/puddling and finishing stages
*Furnace types: CR = Crucible, FH = Finishing Heart (at bloomery), FI = Finery, HS = High Shaft, SS = Short Shaft; TA = Tatara; PU = Puddling Hearth.
**kg./day
(a) 5:1 (coal, coked) at blast furnace; ca. 5:1 (mostly coal) for 3 stages of puddling process.
(b) 4.3:1 at blast furnace, 5.6:1 at finery.
(c) Only 1.5:1 at the 2-stage finery, which Lyman implies may be too good to be true.
(d) Includes fuel cost of crucible smelting at Dayang (see above).
(e) Grignon's figures; fuel costs for local blast furnaces and fineries are based on detailed records; others are then-current estimates.
(f) Coste and Perdonnet's estimate. The stages of the puddling process here use only 2278 kg. of fuel for 1000 kg. of bar.
(g) Loads charcoal converted at 1550 lbs/load.
(h) Courtviron & Bouchu's figures for an unsatisfactory but average operation.
(i) Includes 3.0:1 smelting fuel consumption; fuel use in Henan refining is very low.
(j) 1.2:1 at blast furnace; 1550 lbs/load charcoal.
(k) 0.9:1 at blast furnace; charcoal converted at 1550 lbs/load.

*TABLE 14.6. Charcoal consumption for bar iron made by the indirect process, Britain, 16th–18th c.**

| Place | Date | Fuel:Iron |
|---|---|---|
| Newbridge | 1539 | 12.4:1 |
| Robertsbridge | 1546 | 11.6:1 |
| Glamorgan | 1568 | 8.5:1 |
| Cannock | 1580s | 8.5:1 |
| Middleton | 1590s | 5.6:1 |
| Oakamoor | 1590s | 5.6:1 |
| Lismore | 1607–1608 | 6.6:1 |
| Lismore | 1616 | 6.2:1 |
| Dean | 1654–60 | 4.8:1 |
| Dean | 1661–62 | 4.4:1 |
| Dean | 1691 | 4.3:1 |
| Lydney | 1699 | 4.6:1 |
| Lanelthy | 1704 | 3.0:1 |
| Foley Works | 1688–1717 | 3.3:1 |
| Yorks. Partnership | 1701–20 | 4.4:1 |

*Hammersley 1973: 604. Loads fuel converted at 1550 lbs/load.

fining stage of the indirect process. The 50 to 55 percent recovery that seems to have been about average for the blast furnace-finery combination was certainly possible for some bloomeries, including non-European types. For example, the Yoruba of the Oyo region of Nigeria and the smelters of Chitaldrug in southern India would probably have exceeded the 50 percent level if their product had been weighed after rather than before final forging (Table 14.9a). Moreover, some indirect process operations recover less than 50 to 60 percent of the metal, including most if not all traditional Japanese wrought ironmaking operations.

Thus, the superiority of the indirect process in recovering metal is far from overwhelming despite the opinions of Victorian travellers. Many non-European bloomeries could probably extract as much forgeable iron from ore of a given quantity and quality as the average pre-industrial blast furnace-finery. Moreover, efficiency in the use of ore cannot have been of great importance to both Europeans and non-Europeans before smelting with coke. The monetary or labor cost of ore has rarely been a dominating factor in the economics of ironmaking. Data on costs of fuel and ore show that ore was rarely as costly as fuel, in most cases the latter exceeding the former by a factor of three or four (Table 14.10).

We believe that fuel was normally more expensive than ore during the pre-industrial period in most ironmaking regions. There may be a few exceptions where ores of specially valued types were controlled by a small number of owners or where, as in England during the eighteenth century, mineral fuel was cheap. Enterprising furnace operators in pre-industrial settings made strenuous efforts to conserve charcoal, and cared little that much of the iron in the ore was lost with the slag.

*TABLE 14.7. Charcoal per 100 pounds of bar iron, indirect process, continental Europe, late 18th c.**

| | |
|---|---|
| France | 500–800 lb. |
| Department of Lot | 1000–1400 lb. |
| Lauchhammer | 629 lb. |
| Malapane | 645 lb. |
| Peitz | 672 lb. |
| Burghammer in Lausitz | 712 lb. |
| Schmeideberg near Dresden | 770 lb. |
| Turrach | 573 lb. |
| Horzowitz in Bohemia | 354 lb. |
| Siberia | 493 lb. |

*Hassenfratz, quoted by Beck 1897: 681

*TABLE 14.8. Fuel use for wrought iron at finery only*

| Method | Fuel Type | Kg. Fuel per 1000 kg. iron | Capacity |
|---|---|---|---|
| | 17th c.[1] | | |
| Unspecified fining forge | Charcoal | 2800 | — |
| | 18th c.[2] | | |
| Walloon Hearth, Styria | Charcoal | 3199 | 291 |
| Loschfeuer, Schmalkalden | Charcoal | 4789 | 460 |
| German hearth | Charcoal | 2520 | 362 |
| Kaltfrischfeuer | Charcoal | 3247 | 227 |
| Carniola system (3 stages) | Charcoal | 2583 | 411 |
| Osmund hearth in Mark | Charcoal | 3010 | 257 |
| Coke hearth in Altwied | Coke | 2454 | — |
| | 19th c.[3] | | |
| German hearth, cold blast | Charcoal | 4497 | 145 |
| German hearth, hot blast | Charcoal | 2249 | 145 |
| Lancashire hearth, Sweden | Charcoal | 1870 | 2177 |
| Puddling hearth, Wales | Coke | 8500 | — |
| Puddling hearth, England | Coke | 4171 | — |
| Puddling hearth, Wales | Coke | 3020 | — |

[1]Morton and Gould 1967; [2]Beck 1897: 681; [3]den Ouden 1981: 67, 73; Hyde 1977: 142

## 6. LABOR EFFICIENCY

The contribution of labor cost to total cost varies widely among the world's ironmaking systems. Usually labor costs were not too high for the pre-industrial ironmaster. In a few cases, as in southern India where wages were paid in shares of the iron produced, labor at the furnace and forge seems to have been more costly than either fuel or ore.

The question of labor productivity is of particular interest, for this is an area in which the indirect process might be thought to have an indisputable advantage over the direct process. As expected, blast furnaces producing cast iron were highly labor efficient in both China and Europe (Table 14.11). Even the tatara furnaces of Japan and the short shaft furnaces of China and the Philippines, all of which used manually powered bellows and thus required relatively large numbers of workers, were still more productive in terms of kilograms of iron per man day than any bloomery except the water powered Stückofen of Austria.

The relative productivity of labor in the indirect process decreases, however, when the finery stage is included. Fining methods were highly labor-intensive in both Europe and eastern Asia (Table 14.12). As late as the nineteenth century, output per man day at a finery was similar to that of a Corsican or Catalan hearth, even though the latter began with raw ore while the finery utilized a manufactured product incorporating an additional increment of labor. Very small bloomeries with small finishing forges seem to be less labor efficient than the blast furnace-finery combination, but this is at least partly because none of the smaller bloomeries considered here (Tables 14.11, 14.12) was equipped with water powered hammers and blowing devices. When similarly mechanized and worked on a similar scale, as in the Pyrenees and Corsica, the

*TABLE 14.9a. Iron recovery efficiency—ore-to-iron ratios and percentage of iron recovered at first stage, direct and indirect processes*

| Place | Date | Ore: iron ratio | Recovery pct. | Furnace Type* | Furnace Height** | Furnace Capacity*** | Reference |
|---|---|---|---|---|---|---|---|
| | | | Direct Process: Raw Bloom Iron | | | | |
| Sukur, Nigeria, W Africa | 1962 (a) | 12.5 | 12.0% | SS | 0.9 | 16 | Sassoon 1964: 216 |
| Byrkeknott, Britain | 1408 | 6.9 | 33.9% | BF | — | 78 | Tylecote 1962: 289 |
| Tendukhera, C India | 1900 | 6.3 | 36.2% | SS | 1.2 | 60 | Bartlett 1900: 150 |
| Jefisi (Sisala), Ghana | 1971 | 6.1 | 32.7% | SS | 1.3 | 4 | Pole 1974: 29 |
| Chitaldrug, S India | 1900 (b) | 2.4 | 72.6% | SS | 1.8 | 8 | Sambhasiva Iyer 1902 |
| Oyo (Yoruba), Nigeria | 1900 | 2.2 | 75.0% | SS | — | 34 | Bellamy 1904 |
| | | | Indirect Process: Cast Iron | | | | |
| Unspecified place, Japan | 1880s | 3.6 | 46.2% | TA | 2.0 | 1310 | Ledebur 1901 |
| Une, Idzumo, Japan | 1878 (c) | 3.8 | 43.9% | TA | 2.0 | 1096 | Lyman 1879: 68 |
| Amegawa, Idzumo, Japan | 1878 (c) | 3.3 | 55.5% | TA | 2.0 | 1360 | Lyman 1879: 62 |
| Pingding, Shanxi, N China | 1920s (k) | 3.3 | 64.0% | CR | — | 488 | Wang & Wang 1930 |
| S Henan, C China | 1960s | 2.0 | 77.5% | SS | 2.0 | 720 | Wagner 1985 |
| Coalbrookdale, Britain | 1755–1806 | 3.0–5.0 | 57.0–95.0% | HS | 6.0 | 400–600 | Tylecote 1976: 106 |
| Staffordshire avg., Britain | 1840s (l) | 3.4 | 99.0%? | HS | 7.0 | 6000 | Ure 1856: 110 |

*Furnace Types: BF = Bowl Furnace, CR = Crucible Furnace, HS = High Shaft, SS = Short Shaft, TA = Tatara.
**m.
***kg./day
(a) Sukur: smelter's estimates; assume 65% Fe in well-washed magnetite sand.
(b) Chitaldrug: ore data from Sambhasiva Iyer 1901: 102.
(c) Une & Amegawa: assume 60% Fe in well-washed magnetite sand; product is mixed cast iron and steel.
(d) Ressecq: from Percy, based on detailed measurements by Richard.
(e) Ressecq etc. 3:1, assume Ariège average 43.8% ore.
(f) Vicdessos?: François' experimental runs using ca. 10% refuse manganese ore
(g) Celles: avg. for 3 hearths: Niaux-Vieux, Ramade & Celles, assume avg. for Ariège ore; Percy's data are from François.
(h) Sasello: assume 68.5% avg. Fe for Elba ore
(i) Tripalda: experimental process with waste heat recycling. Elba ore.
(j) Kikusi: 37.5% metal loss in refining.
(k) Pingding: product is malleable spongy mixture, perhaps more like bloom than cast iron.
(l) Staffordshire: according to contemporary methods of calculating the iron content of ores, well-run blast furnaces were almost 100% efficient in extracting the iron that existed in the ore.

direct process appears to have been at least as productive in terms of yield per man day as any indirect iron-making system used before the middle of the nineteenth century.

Ironmasters of the past undoubtedly based their decisions about new processes on a more practical criterion of labor efficiency, that of labor cost—the number of personnel was less important than what they cost collectively per unit of output. It is here that the indirect process may have had the advantage, for there are indications that a major shortcoming of many of the old bloom furnaces was their relative isolation and the consequent difficulty of controlling a labor force that tended to be divided into numerous widely separated units of five or ten men.

This may have been the case at Gittelde. As noted above (Table 14.1), the sharp rise in labor costs at the Zerrennherden after the mid-seventeenth century was not accompanied by similar increases in wages at the blast furnaces and fineries, suggesting that the Zerrennherd possessed some quality that encouraged independent attitudes on the part of its workers. Because operating a finery and a bloomery would seem to require similar levels of skill, and thus would not justify differential wage rates, it is possible that supervisory difficulties arising from isolation are the real explanation of this seemingly arbitrary differential in labor costs. Percy (1864: 294) criticizes laborers at Catalan hearths in the French Pyrenees in the 1840s for widespread featherbedding and laziness.

*TABLE 14.9b. Iron recovery efficiency—ore-to-iron ratios and percentage of iron recovered by all stages, direct and indirect processes*

| Place | Date | Ore: iron ratio | Recovery pct. | Furnace Types* | Furnace Capacity** | Reference |
|---|---|---|---|---|---|---|
| | | | Direct Process: Finished Bloom Iron | | | |
| Kamar, Orissa E India | 1961 | 15.0 | 14.8 | SS + FH | 13 | Ahmed 1967: 17–44 |
| Basari, Togo, W Africa | 1890s | 6.0 | 24.0 | HS + FH | 7 | Luschan 1909: 40–43 |
| Unspecified, Corsica | 1820s | 3.9 | 56.4 | CA | 149 | Percy 1864: 319 |
| Ressecq, Ariège, S France | 1830s (d)† | 3.2 | 69.5 | CA | 456 | Percy 1864: 298 |
| Vicdessos, S France | 1840 (e) | 3.1 | 72.1 | CA | 612 | Percy 1864: 311 |
| Vicdessors?, S France | 1840s (f) | 2.6 | 87.4 | CA | 652 | Percy 1864: 309 |
| Celles etc., S France | 1841 (g) | 3.0 | 74.2 | CA | 664 | Percy 1864: 310 |
| Sassello, Sardinia | 1840s (h) | 2.1 | 65.7 | CA | 501 | Percy 1864: 314 |
| Tripalda, Sardinia | 1840s (i) | 1.5 | 86.9 | CA | 700? | Percy 1864: 315 |
| | | | Indirect Process: Finished Wrought Iron | | | |
| Kikusi, Idzumo, Japan | 1877 (j) | 5.2 | 20.8 | TA + FI | 151 | Lyman 1879: 77 |
| Staffordshire etc. | 1840s | 4.8 | 70.0 | HS + PU | — | Ure 1856: 110 |

*Furnace Types: CA = Catalan/Corsican Hearth, FH = Finishing Hearth (at bloomery), FI = Finery, HS = High Shaft, SS = Short Shaft, TA = Tatara, PU = Puddling Hearth
**kg./day
†Letters refer to notes in Table 14.9a.

*TABLE 14.10. Fuel and ore expenditures for finished bar iron as a percentage of total monetary production costs*[1]

| Place | Date | Fuel | Ore | Furance* | Reference |
|---|---|---|---|---|---|
| | | Direct Process Bloom Iron | | | |
| Andhra Pradesh, C India | 1795 | 65.8% | 13.2% | SS + FH (a) | Heyne 1971: 218 |
| Vicdessos, Ariège | 1840s | 60.9% | 15.3% | CA | Percy 1864: 311 |
| Vordenberg, Austria | 1550 | 56.0% | 35.0% | HS + FI (b) | Johannsen 1953: 244 |
| Tendukhera, C India | 1829 | 46.4% | 7.6% | SS + FH (c) | Franklin 1971: 244 |
| Oberpfalz, Germany | 1450 | 40.7% | 25.6% | BF + FH | Johannsen 1953: 122 |
| Malabar, S India | 1803 | 35.8% | 35.8% | SS + FH (d) | Buchanan 1807, v. 2: 440 |
| Mysore, S India | 1803 | 30.3% | 18.1% | SS + FH (e) | Buchanan 1807, v. 2: 17 |
| Mysore, S India | 1803 | 23.8% | 12.1% | SS + FH (e) | Buchanan 1807, v. 2: 35 |
| Mysore, S India | 1803 | 22.0% | 8.5% | SS + FH (e) | Buchanan 1807, v. 3: 362 |
| | | Indirect Process Wrought Iron | | | |
| Dillenberg, Nassau | 1563 | 72.7% | 17.8% | HS + FI (f) | Beck 1893-95: 718–719 |
| Llanelly, Wales | 1704 | 66.2% | 9.4% | HS + FI (g) | Schubert 1957: 426–428 |
| Nancun, Shanxi | 1910s | 62.1% | 13.5% | CR + FI (h) | Tegengren 1923–24: 329 |
| Pontypool, Wales | 1704 | 57.6% | 13.4% | HS + FI (i) | Schubert 1957: 424–426 |
| Cannock Chase, England | 1588 | 55.6% | 27.2% | HS + FI (j) | Nef 1932, v.1: 193–194 |
| Kikusi, W Japan | 1878 | 53.8% | 20.1% | TA + FI (k) | Lyman 1879 |
| Durbon, Dauphiné | 1696 | 47.8% | 24.5% | HS + FI (l) | Gille 1947: 140 |

*BF = Bowl Furnace, CA = Catalan Hearth, CR = Crucible, FH = Finishing Hearth (at bloomery), FI = Finery, HS = High Shaft, SS = Short Shaft, TA = Tatara.
(a) Earliest available quantitative Indian furnace data; the iron produced is not fully refined. 16 annas = 48 dubs.
(b) Loss-making operation due to unusually high charcoal costs; relatively high ore costs may have been the rule in Styria.
(c) Experimental runs with 4 traditional furnaces for 5 weeks.
(d) Costs and wages cited by source in monetary terms.
(e) Costs and wages in source given as metal pieces. Although simple in technological terms, these are fully commercial operations; as often in Europe, fuel and ore costs are mainly for labor, the materials themselves being either free or subject to very moderate taxes and fees.
(f) At blast furnace, fuel = 66.2% and ore = 29.7% of total pig iron cost.
(g) At blast furnace, fuel = 66% and ore = 22% of total pig iron cost.
(h) Labor at finery may be substantially underestimated. "Price" of iron at finery set equal to production cost of raw iron at crucible furnace. Fuel is anthracite (at crucib. furn.) and anthracite & wood (at finery).
(i) At blast furnace, fuel = 61% and ore = 28% of total pig iron cost.
(j) At blast furnace, fuel = 55.4% and ore = 40.6% of total pig cost.
(k) Setting cost of pig at finery equal to production cost at Amegawa and counting only half of refining cost as fuel.
(l) Gille calculations 50% for raw material and 25% for fuel, but he is not including the fuel cost of the pig as it enters the finery. The percentages given here do not make allowance for loss of iron during refining.
[1]Table 14.10 presents a selection of data showing the costs of fuel and ore as percentages of the total cost of making a batch of iron. In only a few of these cases (e.g., Idzumo in Japan, Shanxi in China and Malabar in India) were both fuel and ore bought on the open market. In the others, either or both inputs were collected by the smelting enterprise itself; here, "cost" represents the wages of miners, charcoal makers and haulers plus whatever fees were owed for the use of forest and mine.

We do not believe that the managerial problems inherent in an industry composed of dispersed units were insoluble or that these problems can have been a common major cause of the changeover from the direct to the indirect process. On the other hand, we do maintain there is no reason to think that the indirect process was markedly superior to its predecessor in terms of any of the common measures of efficiency. It was slightly more efficient in extracting iron from ore, at most equally efficient in utilizing labor, and much

*TABLE 14.11. Labor productivity at furnace—direct and indirect processes, first stage*

| Place | Date | Kg iron per man day | Furnace Type* | Furnace Height** | Furnace Capacity*** | Reference |
|---|---|---|---|---|---|---|
| | | | Direct Process, Raw Bloom Iron | | | |
| Sisala, Ghana, W Africa | 1971 | 1.0 | SS | 1.3 | 4.0 | Pole 1974: 29 |
| Kol, Bengal, E India | 1807 | 1.4 | SS | 1.1 | 4.2 | Martin 1838, v.2: 35 |
| Sukur, Nigeria, W Africa | 1962 | 6.8 | SS | 0.9 | 20.3 | Sassoon 1964 |
| Motta Betta, S India | 1800s | 5.5 | SS | — | 22.0 | Buchanan 1807, v.2:13 |
| Magadi, Mysore, S India | 1800s | 8.4 | SS | 0.9 | 33.0 | Buchanan 1807, v.1: 175 |
| Puppadaung, C Burma | 1860s | 13.3 | HS | 6.1 | 40.0 | Blandford in Percy 1864: 272 |
| Tendukhera, C India | 1900 | 15.0 | SS | 1.2 | 60.0 | Bartlett 1900: 150–153 |
| Velatar, Malabar, S India | 1800s | 24.0 | SS | 2.0 | 145.0 | Buchanan 1807, v.2: 439 |
| Kumaon, N India | 1890s | 36.8 | BF | 0.6 | 276.0 | Atkinson 1973: 270–271 |
| Eisenerz, Styria, Austria | 1770s | 121.0 | HS | 4.9 | 1205.0 | Percy 1864: 329 |
| | | | Indirect Process, Raw Cast Iron | | | |
| Amol, Elburz, N Iran | 1930s | 22.5 | SS | 3.0 | 90.0 | Böhne 1928 |
| Amegawa, Idzumo, Japan | 1870s | 41.0 | TA | 1.1 | 1360.0 | Lyman 1879: 61–66 |
| Unspecified place, Japan | 1880s | 47.0 | TA | 1.2 | 1238.0 | Ledebur 1901 |
| Sanchuan, N China | 1930 | 52.0 | CR | — | 311.0 | Wang & Wang 1930: 109 |
| Dayang, N China | 1910s | 58.0 | CR | 2.7 | 347.9 | Tegengren 1923–24: 324 |
| Luzon, Philippines | 1900s | 62.2 | SS | 1.8 | 499.0 | McCaskey 1903 |
| South Henan, C China | 1960s | 80.0 | SS | 2.0 | 720.0 | Wagner 1985 |
| Provotière, Brittany | 1660s | 102.9 | HS | — | 720.0 | Pipping 1982: 232–234 |
| Zhongjing area W China | 1890s | 151.0 | HS | 6.0 | 2718.0 | Tegengren 1923–24: 343 |
| Unspecified place, Sweden | 1660s | 241.5 | HS | 7.2 | 1280.8 | Pipping 1982: 163 |
| Unspecified place, Sweden | 1660s | 282.7 | HS | 10.1 | 2120.0 | Pipping 1982: 224 |
| Pontypool, Wales | 1706 | 229.5 | HS | — | 2410.0 | Percy 1964: 329 |

*Furnace types: BF = Bowl Furnace, CR = Crucible, HS = High Shaft, SS = Short Shaft, TA = Tatara.
**m.
***kg./day

*TABLE 14.12. Labor productivity at furnace—direct and indirect processes, all stages*

| Place | Date | Kg. iron per man day | Furnace Type* | Furnace Capacity** | Reference |
|---|---|---|---|---|---|
| | | Direct Process, Finished Bloom Iron | | | |
| Doray Guda, Mysore, India | 1803 | 1.8 | SS + FH | 18 | Buchanan 1807, v.2: 35 |
| Chinnarayan, Mysore, India | 1803 | 2.7 | SS + FH | 27 | Buchanan 1807, v.2: 17 |
| Chica-Bayly, Mysore, India | 1803 | 2.8 | SS + FH | 31 | Buchanan 1807, v.3: 362 |
| (Kamar), Orissa, E India | 1961 | 4.1 | SS + FH | 22 | Ahmed 1967: 17–44 |
| Jabalpur, C India | 1898 | 5.1 | SS + FH | 24 | Hole 1898 |
| Oberpfalz, Germany | 14th c. | 18.2 | BF(+FH?) | 128 | Johannsen 1953: 122 |
| Ressecq, Ariège, S France | 1840s | 45.6 | CA | 456 | Percy 1864: 294, 298 |
| Place not stated, Corsica | 1840s | 50.0 | CA | 200 | Percy 1864: 329 |
| | | Indirect Process, Finished Wrought Iron | | | |
| Nancun, Shanxi, N China | 1910s | 2.7 | CR + FI | 302 | Tegengren 1923–23: 329 |
| Stara Hut, Bohemia | 1575 | 8.3 | HS + FI | 220 | Kellenbenz 1974: 210 |
| Pontypool, Wales | 1704 | 22.0 | HS + FI | 1000 | Schubert 1957: 424 |
| South Henan, C China | 1960s | 34.4 | HS + FI | 619 | Wagner 1985 |
| A Walloon forge, Sweden | | 38.3 | HS + FI | 1125 | Ouden 1981: 67–69 |
| A German hearth, Sweden | 1800s | 39.3 | HS + FI | 432 | Ouden 1981: 67 |
| A Lancashire hearth, Sweden | 1890s | 87.7 | HS + FI | 2149 | Ouden 1981: 73 |

*Furnace Types: BF = Bowl Furnace, CA = Catalan/Corsican Hearth, CR = Crucible, FH = Finishing Hearth (at bloomery), FI = Finery, HS = High Shaft, SS = Short Shaft
**kg./day

less efficient in the amount of forgeable iron that could be made from a given quantity of fuel. All else being equal, the indirect process should never have been adopted in Europe. It, of course, was, but we think this happened for reasons unrelated to efficiency that are in part noneconomic.

## 7. THE RISE OF THE INDIRECT PROCESS

Three key elements were present in early fifteenth-century Europe: a rising demand for iron, an increasing use of water power, and a widespread interest in an imperfect but rapidly improving weapon, the cannon. The spread of the blast furnace and finery is closely linked with these.

The development of the first stage of the indirect process, a method of making cast iron in smelting furnaces, seems a natural consequence of the use of bellows driven by water wheels and of an expanding market for low carbon iron. Furnaces grew wider and taller. By the late fourteenth century, bloomeries several meters in height furnished with water powered bellows had appeared in central Germany, Flanders, and northeastern France. It seems likely that most high bloomeries of this kind, unless very carefully controlled, sometimes produced cast rather than bloom iron.

The theoretical explanation for this is straightforward. In a furnace with a larger inside diameter, more fuel enters the reactive zone, requiring a greater volume of air. If the air input system falls behind, the $CO/CO_2$ ratio goes up. The width of the furnace cuts down radial heat loss through its walls. The greater height of the ore-plus-fuel column leads to a more efficient preheating of the charge. All of this means a higher peak temperature just above the tuyeres. If the gas mixture is allowed to become more reducing, these higher peak temperatures will create the conditions for the formation of liquid cast iron. Avoiding this in a tall, hot furnace cannot have been easy. Any slowdown in the air blast was likely to yield a stream of molten iron instead of the expected slag when tapped.

Cast iron was undesirable in a traditional bloomery because it was not forgeable. Moreover, gradual accumulations of it in the bottom of a furnace could cause long shutdowns while the sometimes massive floors of high carbon iron were removed. At first, it may have been only the presence of this unwanted substance that caused some ironmakers to search for a practical application. Eventually it was discovered that cast iron was a substitute for the various materials used in making cannon and cannon balls.

The use of cannon had begun in Europe during the late thirteenth century, and by the late fourteenth century, they had some military importance. Their chief function in those days was in siege operations, for it had become clear that the stone fortifications of the Middle Ages were vulnerable to breaching by bombardment. The tubes of even the earliest siege cannon were much too large to fabricate as one- or two-piece forgings, but they could be built up by assembling wrought iron bars (Fig. 14.1) or could be cast from bronze, since cannon-sized objects were within the scope of contemporary foundry technology.

Manufacturing satisfactory ammunition was as much a challenge as making the guns themselves. The ideal gunpowder-propelled projectile had to be heavy, strong, and—because expended—inexpensive. Bronze was far too costly to be regularly used in cannonballs, although it would have been effective. Lead had the right weight and a low price, but a lead cannonball would flatten against stone walls, dissipating its energy over too large an area to be destructive. Welded wrought iron cannonballs performed effectively but were expensive to fabricate. Most early artillery men therefore had to settle for ammunition made of stone.

An obvious alternative existed: balls and cannon made of cast iron. How the art of casting them was initially discovered is not known, but it seems a comparatively easy step for anyone familiar with bronze founding and with the behavior of the obviously fluid iron that by then must have been a common byproduct of high bloomery furnaces. Wertime (1962: 64–68) sug-

Fig. 14.1. *Earliest Cannon Construction.* These were fabricated from wrought iron square bars arranged like the staves of a barrel and reinforced with heavy iron rings. The rings were trapped by bending over the bars at the muzzle and breech end. The breech was a separate device that hooked onto the back. Sometimes propellant pressures were sustained by sealing the gaps between the bars with molten lead. This Indian gun was 7.163 m. long and weighed 40,000 kg. The projectile was a carved stone of 63.5 cm. diameter weighing 350 kg. (Rathgen 1925)

gests that cannon were the first objects in Europe to be cast from iron, quoting Rathgen's report of iron cannon casting at Frankfurt am Main in 1390, while records indicate that iron cannonballs were not regularly used until after the fifteenth century (Wertime citing Johannsen 1919: 1463–1466).

The apparent chronological priority of cast iron cannon over cannonballs is unexpected, for the latter seem in principle much easier to make. A satisfactory cannon must have considerable strength and requires a bore that can be machined to a reasonable smoothness. This requires gray or mottled rather than white cast iron, and the problem of making this consistently is not thought to have been solved until the 1550s or 1560s. An adequate cast iron cannonball, on the other hand, only needs to be inexpensive to make and easy to cast into a spherical form. Projectile iron could therefore be produced in two ways. First, the high carbon waste product of bloomery accidents could be utilized: it made little difference whether this would cool as white or gray iron. Second, wrought iron, presumably in the form of very thin bars, could be carburized in crucibles or cupolas until melted. The *Feuerwerksbuch* of 1454 prescribes the addition of 5 to 6 percent of bismuth to crucible-carburized iron in order to increase its fluidity (Wertime 1962: 68), although bismuth is actually insoluble in both liquid and solid iron. Biringuccio (Smith and Gnudi 1942: 320–321) rightly condemns "corrupted" cast iron of this kind since it would not be good for cannon, but there is no reason why it could not have been used for cannonballs.

The issue of whether cast iron cannon or cannonballs were first cannot at present be resolved. What matters is that both came into existence at or slightly before the time when blast furnaces first appeared and cast iron cannonballs came to play a major role in European warfare. After the middle of the fifteenth century, most cannon seem to have used cast iron ammunition, no matter what metal they themselves were made of, and in northern Europe the iron used in most cannonballs came from true blast furnaces.

Parenthetically, cast iron entered Europe as the product of a large volume process for making military matériel, so it was therefore not a local industry. Cast iron manufacturing became centralized at almost the moment it started, usually under the aegis of a great military power. A state iron foundry equipped with tall blast furnaces could achieve single-plant outputs that had never been within the practical reach of most bloomery operations. This statist and warlike orientation may explain why cast iron manufacturing in Europe never took the form, long familiar in China, of a village industry serving strictly local markets with civilian goods.

The first European cast iron products intended for nonmilitary use appear as sidelines of the iron gun casting industry during in the late fifteenth and early sixteenth centuries. Without knowledge of Chinese malleabilizing methods and with less skill in producing thin and inexpensive castings, the gunfounders of Europe evidently had difficulty making products that civilians would buy, but the impetus to develop them existed. Since excess liquid metal is always produced in a smelting or foundry operation, 20 percent of the liquid metal might have been left after pouring off a gun casting. This was too little even for a small cannon, so it was economical to use the overrun for a purpose other than stockpiling.

Even more important was the irregular demand inherent in defense industries. During war, business was brisk, and a munitions manufacturer might be able to sell as much he could make, gaining large profits in the process. In peacetime, demand for cannon was likely to be so slack that many foundries were forced to close, but a complete shutdown of all furnaces was undesirable for governments and foundrymen alike, both of whom were interested in being ready for the next war. Some ironworks could have been kept open by producing guns and ammunition for government stockpiles, but excess capacity inevitably existed. It was only natural that foundry owners looked to civilian markets.

In the beginning, nonmilitary production seems to have focused on cast iron gravestones, firebacks, and decorative panels (Fig. 14.2). Most such castings were simple, flat designs made in open molds which had images and inscriptions impressed on their surfaces (Schubert 1957: 256–266; Browne 1967; Pfannenschmidt 1977). It is surprising to learn that these objects serving marginal functions found—or rather created—a substantial market; many sixteenth-century firebacks still survive. Eventually, as skills improved, European foundrymen began to experiment with more useful objects like stoves, plow shares, and kettles, but it was not until almost the end of the pre-industrial period that these products were more than a sideline for the average European iron foundry. As late as the early eighteenth century, Abraham Darby appears to have been unusual among iron founders in specializing in kettles and similar domestic and agricultural implements, while most of his competitors still depended heavily on defense contracts (Chapter 10.11).

In a sense, wrought iron made by the indirect process may also be considered a spin-off of military cast iron production. When the *Feuerwerksbuch* of 1454 was written, cast iron in Germany was still being made from bloomery iron, and Biringuccio (Smith and Gnudi 1942: 66) indicates that indirectly made cast iron continued in Italy until the 1540s. By then there were parts of Europe where the order of manufacture was reversed—cast iron was reprocessed by fining to wrought iron. When and where this first happened are not known (see, e.g., Crossley 1981: 39–41). Wertime (1962: 57) declares that there seems to be no one place of origin, that it emerged almost simultaneously in northern Italy, western Germany, and eastern France. In the first literary mention of cast iron reprocessing, Filarete puts its beginnings in Italy in the latter half of the fifteenth century (Spencer 1963). On the other hand, the name "Walloon" finery process and the pictorial work of the Flemish artist Henri Blès (Schubert 1957: 195) may indicate an origin in southern Belgium sometime in the second half of the fifteenth century. It is likely that the real discoverers of the indirect method of making low carbon iron were not acknowledged by contemporaries; then as now, inventors must often have not been recognized while entrepreneurs received the credit for new processes.

Fig. 14.2. *Cast Iron Fireback.* Decorative flat plates were used for superfluous cast iron from smelting and remelting furnaces built for the cannon industry. These plates served as grave markers, firebacks, ovens, and stoves. This fireback was probably produced in the seventeenth century. (Jenkins 1920)

The circumstances of the discovery itself may be linked to the nature of the casting procedure. All castings in their raw form have redundant metal in the form of gates, risers, runners, and entire defective castings. These accumulate and are remelted for new castings. If the redundant pieces are broken up and fed as a scrap charge with ore into a tall smelting furnace run at a high rate of air blast to achieve melting before the weight of the iron metal crushed the charcoal, the oxidizing nature of the ore and the excess oxygen in the gases would decarburize and desiliconize the iron to a point where a bloom-like mass would form in the hearth. The mass might be too large to remove without breaching the furnace wall and thus jeopardizing its structural stability. However, after the iron mass could be successfully extracted, the material would be recognized by any experienced ironworker. Forgeability tests would confirm that it was identical to the iron made in bloomeries.

While this seems a plausible reconstruction of the initial discovery of the fining process, using tall shaft furnaces as fineries on a regular basis would present difficulties. Shorter furnaces were needed in order to reprocess cast iron in more manageable batches and to give easier access to the accumulating bloom. The obvious solution was to adapt a second type of traditional smelting device, the bowl furnace or "forge" of the Catalan type. Almost all parts of Europe already used these furnaces as low volume bloomeries; most could probably function as fineries without significant changes in design. The first experiments with forge-like fineries must have occurred soon after it was discovered that cast iron could be decarburized on a production basis. Much wrought iron was fined in this way in northwestern Europe by the middle of the sixteenth century.

It was rapidly apparent that this new type of forgeable iron unfortunately had a broad range of forging qualities in comparison with the old bloom iron, mostly because greater amounts of phosphorus were present. While bloomeries were run at temperatures too low to extract much of the phosphorus that might be present in the ore, blast furnaces extracted it very efficiently, and whatever phosphorus the cast iron contained was inevitably passed on to the wrought iron. A finery fed with cast iron that came from a low phosphorus ore might yield an excellent iron, such as the premium

Osmund wrought iron of Sweden. But the average finery product was likely to be more cold short than the bloomery products with which it competed in the marketplace.

## 8. WHY THE INDIRECT PROCESS WON

By the sixteenth century, the bloom iron industry had slowed in growth; by the seventeenth century, it was declining sharply. By the beginning of the eighteenth century, bloomeries had disappeared in much of northern and western Europe, and the indirect process dominated the industry even though the innovations of the Industrial Revolution that would give wrought iron an advantage in quality and price were in the future. Cast iron was clearly a better material for making cannon and a few other things, so it would seem reasonable to have blast furnaces for those special purposes and bloomeries for ordinary low carbon iron production. Fineries nonetheless prevailed because ironmakers had to have the equipment to produce cast iron and, once this investment was made, they found it economical to turn part of their product into wrought iron even if it sometimes had to be sold at a loss (Chapter 10).

The indirect process in Europe survived losses because it was ultimately linked to the cast iron industry. The immediate product of the blast furnace, cast iron, had so solid a market base in ordnance production that it was profitable enough to support civilian product lines like firebacks and gravestones which had a limited market and which must often have been produced mainly in order to keep plants open during peacetime. We believe that a similar rationale lies behind the switch to wrought iron manufacture. The iron from the blast furnace and finery may have been poorer and more expensive to make than bloomery iron, but it kept the furnaces going and ready for the next war.

The idea that wrought iron was subsidized by cast iron during the pre-industrial period in Europe is in a sense supported by the absence of the indirect process in most of Asia and by its long dominance in China, Korea, and Japan. Cast iron was in fact known before the modern period in many parts of Asia other than the Far East: Turkey (Beck 1891, v.1: 286–290), Persia (Böhne 1928), Soviet Central Asia (Lansdell 1885, v.1: 198; Küssmaul and Schäfer 1973), the Philippines (McCaskey 1903a), Thailand (Crawfurd 1971), Vietnam (Gourou 1956, v.2: 559), and Indonesia (Raffles 1817, v.1: 192). In most of these places it was used to make either cannon or plows and cooking pots in small quantities and did not serve as an input to a finery. Forgeable iron still came from bloomery furnaces everywhere except in the Far East, where all of it was from fineries, and in Thailand and Turkey, where fined and bloomery iron both were made (Bronson 1986b; Hamilton 1837: 47).

Since the market for cast iron in most of Asia was small, there was little blast furnace smelting and consequently even less fining. In most of the countries listed above, the demand for iron pots or locally made iron cannon was too small to subsidize the production of other iron goods. In the Far East, however, iron founders produced an enormous range of domestic and industrial implements which, due to the precision of casting methods and the use of malleabilizing processes, filled many of the functions that in the West were served by forgeable iron. This large volume of production of cast iron for civilian use arguably had the same effect as military production in Europe: it lowered the cost of castable iron so that bloomeries could not compete.

# 15. The Fabrication of Composites from Iron and Steel

## 1. INTRODUCTION

There are iron products which are composites of separate pieces of steel or iron. Before the nineteenth century, when fusion welding—the joining of even high melting-point metals by heating adjacent surfaces until they liquefy—became known, the only way to create a metallurgical union of two wrought iron or steel objects was with a low melting braze or solder, in effect a metallic glue, or by a process called *forge welding*.

Forge welding was far more important to iron fabricators than brazing or soldering.[1] Forge welding is different from fusion welding. The pieces being joined are generally hot, but never enough to liquefy. They can be joined so closely that there is no boundary between them and they become a single piece of metal, but this is done entirely by solid state mechanisms, using deformation and heat.

The art of forge welding must have been developed in the early stages of iron metallurgy, because the furnaces produced bloom iron only in very small fragments, which then had to be consolidated to make a piece large enough for a tool. On the other hand, the concept of forge welding may not have been obvious to copper and bronze workers. Although those metals could have been forge welded with a suitable flux, they were rarely, if ever, joined in that way, presumably because other methods like casting and casting-on, which were not appropriate for low carbon irons, could be used to make small metal pieces into large ones.

## 2. MECHANISMS IN FORGE WELDING

A good weld, whether created by fusion or forging, is not visible, because the original interfaces have completely disappeared. Although forge welding has almost disappeared in Western countries as an ironworking technique, it continues to be used for working other metals such as the clad aluminum used for aircraft plate and sheet (Tylecote 1968). This bonding process has two components, deformation and heating, which usually are applied simultaneously. Deformation causes the metal to extend and rupture the oxide films at the interface between the work pieces, allowing oxide-free metal to make contact. Pressure forces this contact to be close enough for the fresh surfaces to begin to exchange atoms—that is, to form microwelds.

Oxide films prevent the formation of good welds, so they must be fragmented and cleaned off carefully. For metals that oxidize slowly, it may be feasible to abrade the surfaces to be joined down to bright metal, place them in contact, and heat and hammer them as soon as possible after that, but usually chemical methods for eliminating oxides at the interfaces of work pieces to be integrated, combined with hot hammering, are more practical. If an oxide solvent is liquid at forging temperatures, it can be spread over the interface and then squeezed out as the now-clean metal surfaces are forced together, so that no new oxidation can take place and the surfaces remain as open to microweld formation as if they were in a vacuum.

Welding, brazing, and soldering fluxes all function in this way. In bloom or wrought iron, or most sorts of steel made from them, the numerous slag inclusions already present provide an excellent flux (Fig. 15.1). The oxide-dissolving properties of slag explain why bloom and wrought iron have traditionally been considered amenable to forge welding. On the other hand, steels made by fusion in crucibles as well as those processed by frequent folding and forging are likely to be relatively slag-free and thus difficult to weld. The interface of these steels must be coated with a separate flux—often powdered borax, though crushed slag or other substances[2] would work equally well—before welding.

## 3. FORGE WELDING SMALL INTO LARGE PIECES

The discovery of forge welding is likely to have been incidental to forging individual blooms into a cluster. The same actions deform and shape the blooms, expel excess slag, and weld the individual blooms into a bar, so it follows that these bars can be bonded together

[1]Brazing as a method of joining pieces of wrought iron or steel appears sporadically, and appears to have been most common in the Far East and Europe. In the Far East, cast iron as well as copper alloys seem to have been used as brazing compounds. We have seen Chinese and Japanese objects with parts that are apparently joined by a cast iron braze: e.g., the wheel-like Japanese sword guard of wrought iron with thin spokes that are joined to its hub and rim in such a way that they cannot have been forge welded on (Fig. 2.2b). Examination with a microscope shows that the joints are embedded in a material that oxidizes to the same color as the rest of the sword guard and is almost certainly iron of some kind. The material may be an iron-arsenic alloy, since only this alloy has adequate toughness and a melting point (as low as 835°C with 30 percent As) in the temperature range of brazing compounds suitable for such delicate work. Iron-sulfur (988°C at 31 percent S) and iron-antimony (1002°C at ca. 52 percent Sb) are too brittle. Iron-phosphorus (1050°C at 10.5 percent P) and ordinary cast iron (1153°C at 4.25 percent C) are tough enough but melt at temperatures that would make brazing such a small object difficult.

[2]Correspondents of Richardson (1978: 101–105), writing in the magazine *Practical Blacksmithing* in the 1880s, stated that they used a number of substances other than pure borax as fluxes for welding steel. Among the recipes suggested are: 1) copperas (ferrous sulfate), saltpeter, sal ammoniac (ammonium chloride), salt, and sand; 2) copperas, salt, and white sand; 3) borax and glass; 4) copperas, saltpeter, salt, black oxide of manganese, and prussiate (cyanide) of potash; 5) lime and river sand; 6) carbonate of iron and borax; 7) rock saltpeter, oil of vitriol (sulfuric acid), and rainwater.

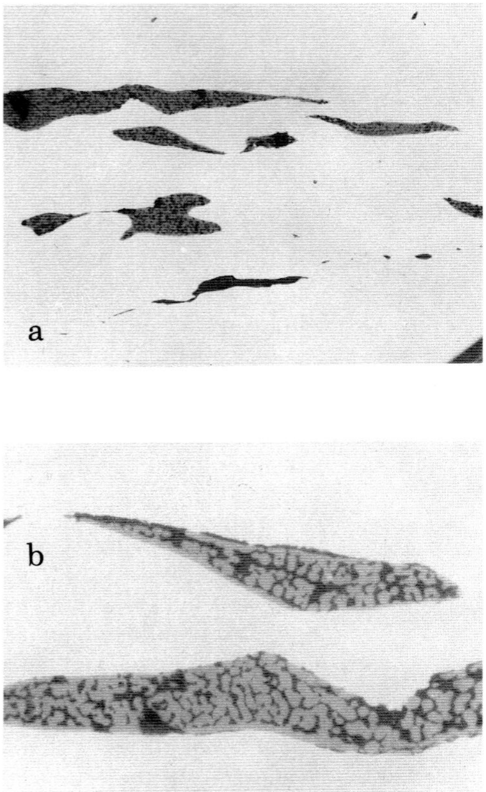


Fig. 15.1. *Slag Inclusions in Bloom Iron.* a) (×22) a cluster of slag inclusions. The elongated shapes reflect the deformation imposed on the original bloom. b) (×1000) the structure of the inclusions as a high percentage of wustite (FeO) phase in a matrix of what must be fayalite and possibly glass. The high content of the wustite phase is evidence of the poor recovery of metal in the smelting operation.

because slag will continue to eliminate solid oxide films that prevent good welding.

Forge welding can form shapes too complex to be easily forged from a single workpiece, increase the homogeneity of a piece of metal, create a heterogeneous structure of more than one metal or more than one type of the same metal, and assemble smaller units into larger ones. The formation of complex shapes is now the most familiar application. Earlier, however, much and perhaps most of the welding performed had other purposes. Forge welding to homogenize steels is one (Chapter 11); by dividing a piece, forging those pieces together, and repeating these operations a number of times, the distances over which carbon has to diffuse can be so reduced that reasonably complete homogenization can be achieved. The final product, as shown by traditional Japanese steels, can be as uniform in composition and properties as steel processed in crucibles.

The assembly of large iron pieces by forge welding has long been an important process in India, where massive iron pillars dating to the first millennium A.D. are known. The famous pillar at Kutb Minar in Delhi (Fig. 15.2a) weighs approximately six tons, and the one at Dhar about seven tons. In some areas, as in coastal Orissa, large iron beams were customary elements in temple construction. Beams in the Sun Temple at Kon-

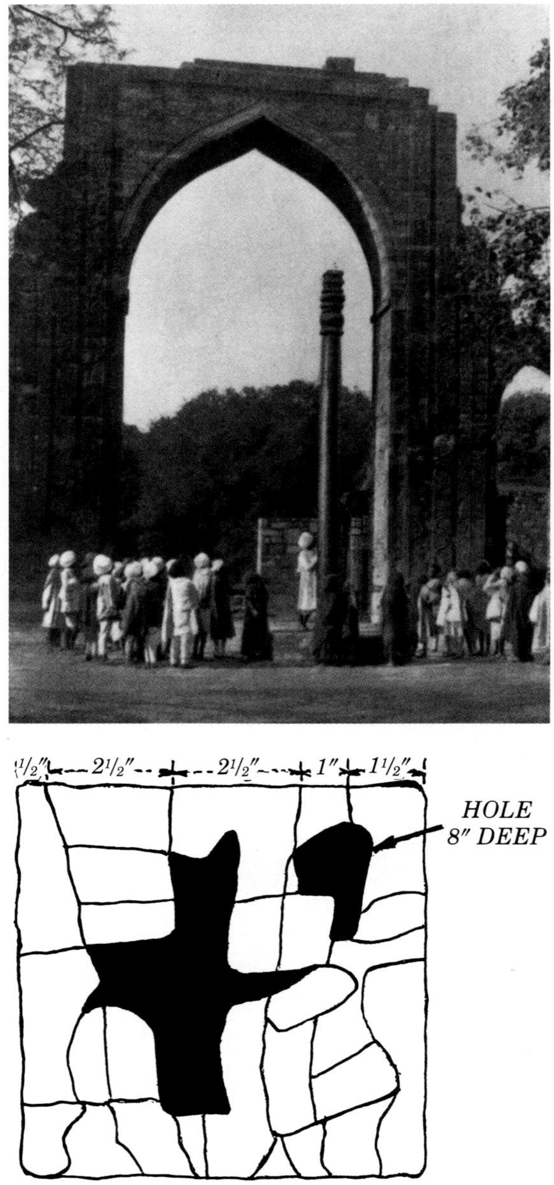


Fig. 15.2. *Massive Indian Iron Composite Castings.* above) The iron pillar of Delhi is nearly 8 m. high and ranges from 30 to 42 cm. in diameter, and its weight has been estimated to be about 6 metric tons (Graves 1912). (From Johannsen 1953: 17); below) The fracture appearance of a broken iron beam from the Sun Temple at Konarak in eastern India. It shows by the opened weld seams that the beam was produced by the composite welding of small blooms. (From Graves 1912, after the 1904 *Annual Report* of the Archaeological Survey of India)

arak, and perhaps some of those in other Orissan temples, exceed four tons in weight. In the sixteenth and seventeenth centuries, Indian craftsmen also forge welded the gigantic cannon that were fashionable in several parts of the subcontinent. It is unclear whether these weapons were single units rather than loose assemblies of bars and hoops, but they could be the most massive of all pre-modern welded structures. The cannon known as "Raja Gopal," preserved at Tanjore in southern India, weighs at least 35 tons, and "Landa Kasub," at Bijapur in central India, 47 tons (Krishnan

1954: 63–67; Rathgen 1925: 208). European welded objects were usually much smaller. Perhaps the biggest wrought iron cannon made in Europe, "Dolle Griet" in Ghent, Belgium (Rathgen 1925: 208), weighs less than a fifth as much. The heaviest European anchors of the pre-industrial period, all forged from low carbon iron, weighed about four tons (Beck 1897, v.3: 225).

The chief technical problem which had to be overcome in making the Indian pieces, as well as the much smaller, but still substantial, iron beams employed by the Romans (e.g., in Britain—Wright 1972; Tylecote 1962: 238), was simply the conversion of a large number of small blooms into a single bar, rather than homogenizing the composition. The surface of the Delhi pillar still seems smooth and dense, but its interior probably resembles that of early Indian pillars and beams which have fallen and broken, so that their fracture surfaces can be examined.

A number of the beams at Konarak were composites of many small blooms (Graves 1912), which were fitted and welded casually, as shown by the presence of hollow spaces in the interior of the beams. Graves's sketch of the structure of one fracture surface (Fig. 15.2b) suggests that the beam was fabricated by adding blooms end to end. Manipulating the growing mass of these blooms must have been difficult, but the manner in which it was done was simple. Only the end of the beam had to be kept hot. This must have been accomplished by laying the beam horizontally in a large, but open, hearth so that the end being worked could be heated sufficiently for welding and so that relays of workers bringing yellow or white hot blooms from separate hearths had easy access to the heated end. The blooms would be held one at a time to the working face and hammered on. The awkwardness of this arrangement and the problem of keeping the beam face and the blooms hot enough explain why many of the forge welds were imperfect, so that the beam eventually fractured along the weld lines.

A sounder structure would have resulted if the blooms had first been formed into long bars and then welded together side by side. The great expenditure of fuel needed to keep the entire length of the beam and the bars at welding heat might seem prohibitive, but a similar method would have been employed in making wrought iron cannon in both India and Europe if the longitudinal bars forming their barrels were interwelded and not simply held in place by iron hoops. Moreover, satisfactory anchors could only be made by forge welding parallel bars, since their successful function depended on great longitudinal tensile strength so welds that ran across the anchor shaft had to be avoided. Iron anchors of substantial size were made by manually operated hammers in China (see, e.g., the *Tiangong Kaiwu*—Sun and Sun 1966: 196, fig. 10.2). This was also the standard way in which anchors were made in Europe until water powered hammers came into use late in the seventeenth century (Beck 1897: 252).

Massive equipment was developed for anchor manufacture in eighteenth-century France (Fig. 15.3) to accommodate hammers weighing as much as 800 pounds

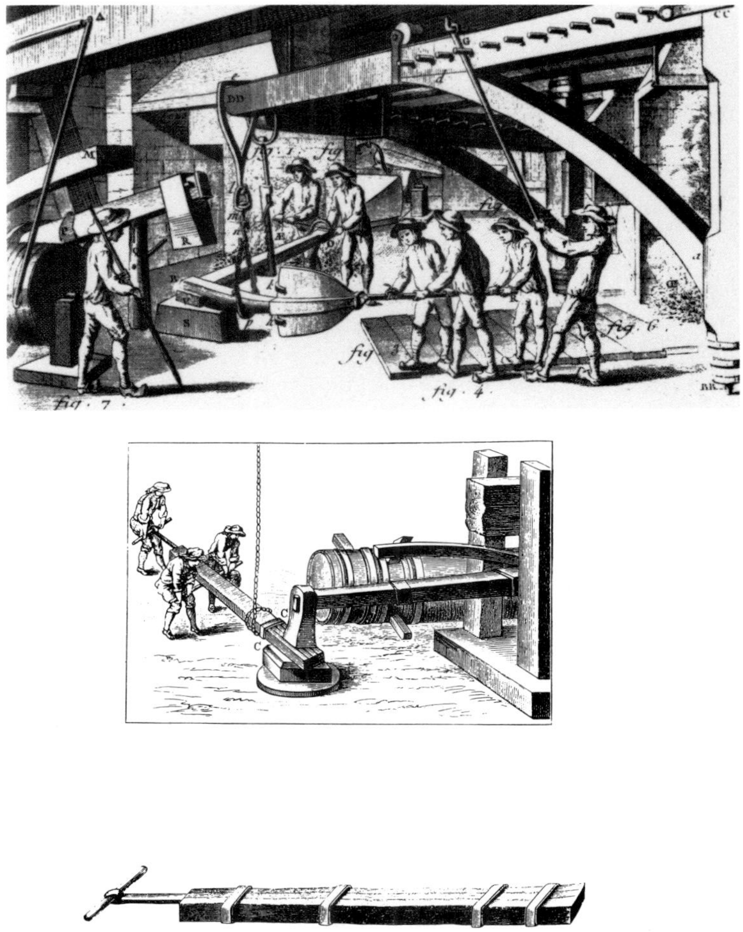

Fig. 15.3. *Stages of Anchor Manufacture.* above) The forge welding of one arm to the main shaft of the anchor using a heavy, water powered hammer. The preheating furnace is visible at the right. (From the eighteenth-century *Encyclopedia* of Diderot reproduced by Gillespie 1959, v.1: pl. 104); below) The anchor shaft itself being produced by the forge welding of a stack of iron bars or plates. A T-shaped handle has been welded onto one of the plates to facilitate manipulation. (From Beck 1897, v.3: 171, after Duhamel de Monceau's *Artes et Mêtiers*)

(Beck 1897: 252). A metallographic study (Samuels 1980b) of an anchor which was produced during this period, from James Cook's famous bark *Endeavour*, shows that it was excellently crafted. Its selection is a credit to Captain Cook's reputation as one of the most technically knowledgeable sailors of his day.

## 4. SIMPLE COMPOSITES OF IRON AND HIGH CARBON STEEL

An equally important function of forge welding was the production of heterogeneous structures in which components with differing properties complemented one another. For example, most of the body of a sword or knife must be tough and moderately stiff; it is the edge that needs the hardness and resistance to dulling that characterize medium and high carbon steels. To make the entire blade of high carbon steel would be expensive, and would probably result in an unsatisfactory sword. Edged weapons made entirely of high carbon steel, such as damascus swords, were, however, used in some regions, and performed well enough when fabricated with expertise and handled with care. Most of these blades, although their cutting ability was far superior to that of blades of low carbon iron, were likely to break when struck against a hard object.

The obvious solution was to make the body of the blade from soft low carbon iron and the working edge or face from a more hardenable metal. An easy way to do this was to carburize the edge or the entire outer surface by placing the blade in a converting chest (Chapter 11) or in the carbon monoxide-rich zone of a forging hearth. However, the effectiveness of pack carburization is limited because of the time required to raise near-surface carbon contents to a useful level and because of the case depth that can be attained. If a deeply carburized cutting edge that can be resharpened repeatedly without grinding away all the harder metal is needed, it is often more practical to weld a separately made piece of high carbon steel onto the body of the sword.

The steel could come from a variety of sources. For example, in the manufacture of steel-clad composites in Karelia, Finland, during the early nineteenth century, the steel was chiseled in thin flakes from the surface of carburized blooms until enough flakes had accumulated to make a thick covering on the surface of a second bloom which then was presumably forged into cutting implements (Kolchin 1967: 26, citing Fullon). In this context, however, the source of the steel is not important, but rather that iron-carbon alloys with differing properties can be combined by welding, and that the combination may have properties superior to those of a metal piece with a single composition.

Iron-steel composites appear to have been invented independently in several ironworking centers. The history of the technique is especially well documented in central Europe, where iron-to-steel welding was common by the late first millennium B.C. For example, a large percentage of chisels and axes made during the Late Hallstatt and Early La Tène periods have welded-on steel edges (Pleiner 1980: 389). One of the earliest Mediterranean objects known to have been made of steel welded onto iron is an adze of the late fifth or early fourth century B.C. from Al Mina in northern Syria. The steel for the edge and working face of this adze may have been obtained, as among the Karelians, by carburizing a portion of a bloom similar to the one from which the iron body was made (Muhly et al. 1977: 158).

The manufacture of composite tools with iron bodies and forged-on steel edges had become widespread by the end of the pre-industrial period. We have seen numerous eighteenth- and nineteenth-century examples from Europe, North America, and East and Southeast Asia; our impression is that the technique was less commonly used in South and Southwest Asia and in Africa. In the United States, not only steel but also cast iron was welded to wrought iron (Overman 1854: 122–125; Richardson 1978, v.3: 71). In Burma (Bell 1907: 17–19) even ordinary iron work knives were routinely given welded-on steel edges.

With the rise of modern alloy steels in the early twentieth century, many composite tool types disappeared, perhaps because they were more expensive to produce than the homogeneous single metal alloy tools, especially since the new alloys make superior tools. About the Finch Brothers Foundry on Dartmoor in southwestern England, Barron (n.d.: 33) comments that in the early twentieth century:

> [S]teels were developed which were claimed to be suitable for making cutting tools of a single homogeneous piece of metal. . . There is little doubt that tools made of this supposedly all-purpose steel do not hold their edge or withstand repeated sharpening, nor the shock of hard work, as did the older composite tools. I have been shown Finch tools made in the old way, described by the farmer owner as 'the best tools on the farm' and bought many years before by his grandfather.

The Finch Foundry may have been one of the last in an industrialized Western country to make such tools, but they are still widely used in Japan.

The ways in which a high carbon edge can be joined to the body of a cutting implement are limited by its shape: it can be butt welded directly against the body, inserted into it, or bent into a U or V shape in order to surround it (Fig. 15.4). The last assemblage, forged down so that the steel case is well bonded to the iron core, would be especially suited to a tool like a plane blade, where hardness and smoothness are required on the sides as well as on the cutting edge. Japanese sword specialists call this particular configuration *kobushi* (Fig. 15.8). The side and edge of one of these swords are strikingly dissimilar in appearance because of differential quenching and polishing; the grain of the surface shows that the edge and both sides were shaped from a single piece of steel, wrapped around a hidden core of what is presumably a much softer iron.

## 5. COMPLEXLY STRUCTURED COMPOSITES

Most tool functions seem to be adequately filled by simple composites like those described in the preceding

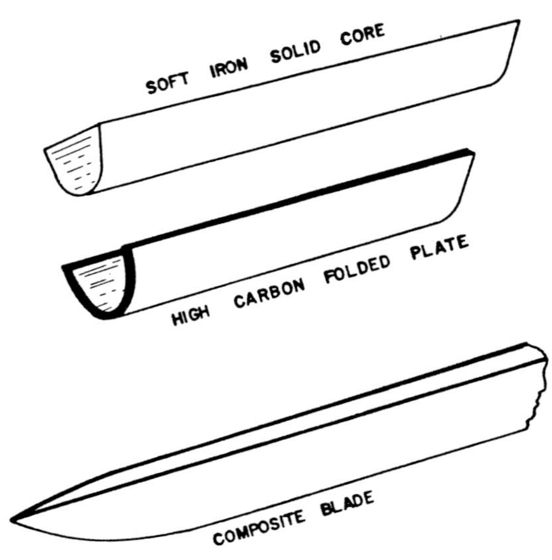


Fig. 15.4. *Composite of Steel Welded onto an Iron Core.* This diagrammatic illustration shows the assembly by which a steel-edged sword or knife could be produced by forge welding. To protect the steel from decarburization during heating in the forge furnace, it was commonly coated with clay.

paragraph, in which no more than two pieces of metal form the finished implement. However, much more complex composites are common in ethnographic and archaeological contexts. The intention of the makers may have been purely decorative or to some extent functional. Like the Japanese swords discussed below, certain multiple-component welded structures may perform better than those with only two components. Further, the appearance of these structures can be highly distinctive—an important quality in artifacts which, in the case of weapons, functioned as status symbols.

The most basic of the complex composites is a laminated structure of rods or plates, welded parallel so that the laminations extend along the length of a blade. The idea may derive from the practice of forging a group of small blooms into bars before welding them side by side into a larger unit. An object assembled in this way can show a visible pattern of longitudinal lines. If the component bars had inadvertently become surface carburized during forging, the result might not only look like an intentionally produced iron-steel composite but might even perform like one, being more resistant to bending and fracturing than a blade made from a single piece of metal.

It is only a short step from this to the deliberate creation of laminated composites of separately prepared pieces of iron and steel or of low and high carbon steel (Fig. 15.5). If forged and filed into the form of a cutting implement, this particular structure would behave in much the same way as an iron blade with welded-on steel edges. If polished and treated with a suitable etchant, the two metals would appear as slightly wavy longitudinal bands of alternating darker and lighter colors. If there were enough of these bands and if they were not straight, the appearance of the finished blade might be decorative.

Examples of implements—generally weapons—with a visibly laminated structure are often found in northern and eastern European archaeological sites of Iron Age, Roman, and early medieval date. In many cases the laminations seem to be no more than a byproduct of pile forging, but in others it is clear that the layered patterns were made to be decorative. Pattern welded weapons appeared by the third century A.D. and persisted, in eastern Europe at least, until the thirteenth or fourteenth century. Experiments have been conducted to replicate the characteristic patterns of fifth- through ninth-century Merovingian and Carolingian swords from France (France-Lanord 1948) and of English and Scandinavian swords of roughly the same period (Maryon 1948; 1960; Anstee and Biek 1961). Most of the weapons in both groups were manufactured by forming twisted ribbons of low carbon iron and steel, forging these together into a bar with a rectangular cross section, and welding narrow strips of more homogeneous steel onto either side to serve as cutting edges (Maryon 1960: 32). The ordinary iron and steel in the central portion of the weapon could be etched so that they showed "good contrasts in brilliance" (Maryon 1960: 28).

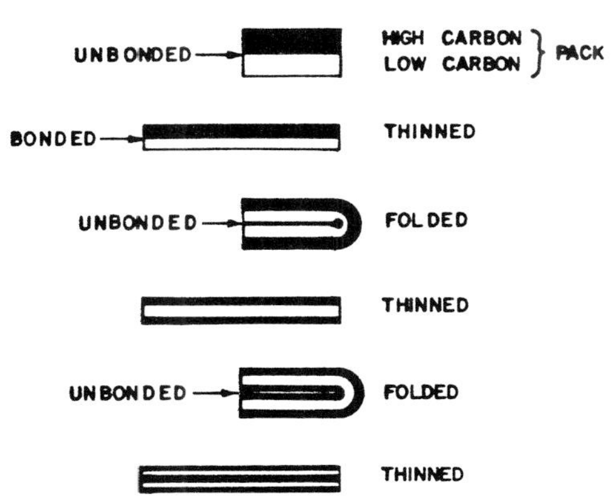


Fig. 15.5. *Generation of a Multi-Laminate of Two Grades of Steel.* The procedure shown could be used to homogenize the carbon content of two batches of steel, and also to create a duplex structure which would become visible by polishing and etching the finished piece.

Weapons with simple longitudinally piled structures include Iron Age swords from Europe (as well as the contemporary objects called "currency bars," probably sword blanks—Tylecote 1962: 205–208), a socketed spearhead of the seventh or sixth century B.C. from Syria (Coghlan 1977: 84), and many eighteenth- and nineteenth-century swords from Indonesia, western China, and Tibet. It is often not easy to tell whether these structures represent two separately prepared metals or whether the linear patterns thus formed had aesthetic value for their manufacturers and owners. The Tibetan swords are definitely composed of two metals, and the coarse longitudinal bands formed by these metals are clearly meant to be decorative; the banding is intentionally accentuated by mechanical abrasion (Smith 1960: 8–9). Thirty-seven of the 42 Tibetan swords in the collection of the Field Museum have been treated in this way.

A distinctive type of pattern welded structure occurs on the blades of *keris* and other weapons (mainly swords and spearheads) from Indonesia, Malaysia, and the southern parts of Thailand and the Philippines. It was usually made by forging two or more types of iron into thin plates, placing these symmetrically around a wider central plate of steel, and forging together the resulting sandwich. The decorative patterning is developed either by creating undulations in the plates and then filing or forging them flat again so that the surface shows a wood grain-like effect, or by preparing the upper plates from thin rods of contrasting types of metal, folded or twisted, and then forged together. Southeast Asian pattern welding is different from the earlier European version in that the patterning rarely extends through the blade but instead is laid out above and below the central plate, which not only forms the edges of the weapon but carries the entire structural load when the weapon is in use; and that the patterns are often developed by combining soft iron and steel with an iron-nickel alloy called *pamor*. This alloy was sometimes meteoritic in origin but, in the majority of cases, it derived from the smelting of a nickeliferous ore found in Sulawesi (Chapter 4).

There are two principal variants of the pattern welded keris: the Javanese (and Malay, etc.) type where the structure is revealed in three dimensions through deep etching, and the Balinese type where the surface is polished flat and etched only enough to blacken the iron or steel matrix, revealing the pamor as a network of silvery lines (Fig. 15.6). In neither case does the pamor contribute significantly to the performance of the blade, because the nickel content, rarely above a few percent, is too low to strengthen the iron much and—in the non-Balinese blades, at least—is confined to the non-

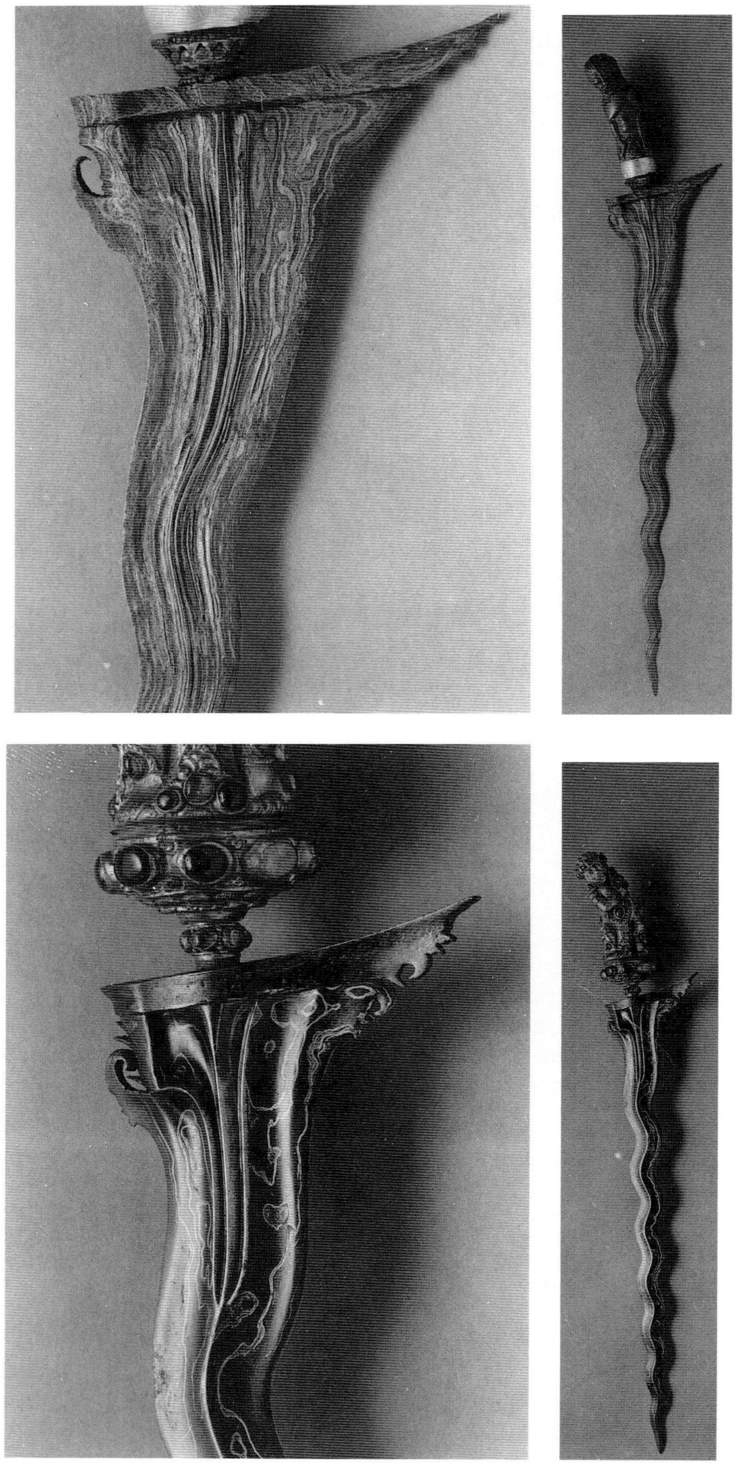

Fig. 15.6. *Decorative Laminations in Indonesian Keris Blades.* Both examples are composed of multiple layers of ordinary steel interleaved with an iron alloy which by preliminary tests has a 2 to 3 percent nickel content. The patterns have been developed by forging to bend over as in Figure 15.5 and twisting, then forging back to a flat. Above)—(Field Museum cat.no. 162547) This keris was probably fabricated in central Java. The driftwood effect was produced by deep etching so that the nickel-iron alloy stood in relief; below)—(FM cat.no. 111003) This keris from Bali has had a different treatment, having been filed and polished to a smooth surface, then etched only enough to highlight the shiny streaks against the dark matrix.

load-bearing surface layers. Moreover, the severe etching necessary to bring out three-dimensional structural patterns clearly weakens many blades. It is evident that the metallurgical aesthetic involved has long since become divorced from considerations of practical function.

Several other important pattern welding traditions should be mentioned. One has existed for perhaps the last thousand years in the Middle East, where a fashion for iron-in-steel or steel-in-steel ornament may have developed because of the traditional appreciation of damascus swords and of the "watered" patterns developed in the wootz steel from which the swords were made. Although the watered pattern of true damascus is formed by carbide networks that crystallize around steel dendrites when the wootz solidifies in the crucible, similar effects were often produced mechanically by welding together bits of different iron and steel in semi-random patterns (see, e.g., Smith 1960: 17). Crystallized structures are not imitated in many pattern welded objects from South Asia and the Middle East, including swords, other cutting implements, and the striking gun barrels made at least as early as the seventeenth century in Iran, Kabul in Afghanistan, Sind in Pakistan, Kashmir, and Istanbul (Smith 1960: 31–32; Egerton 1880: 60–62).

European interest in Asian crucible steels led, in the late eighteenth century, to the development of numerous semi-secret "damascening" processes, the majority of which were based on the pattern welding of various metals. Increased Western understanding of pattern welding influenced the design of the barrels of shotguns and pistols. Asian-type gun barrels were imitated in the West by the 1730s.[3] Smith (1960: 31) notes that "all the best European sporting guns were of this type from the mid-eighteenth until near the end of the nineteenth century." The imported Asian techniques were a commercial success, and the pattern welded "twist" shotguns made by the Mantons and their peers in early nineteenth-century Britain must be among the finest objects ever crafted from iron or steel.

Pattern welding was also practiced in Japan, where it appears on swords and on *tsuba,* or sword guards. Two different techniques were used. Although the visual effects are distinct, they (and sometimes also the techniques) are often called by the same name, *mokume* or "wood grain." The more common technique, which some specialists call *masame-tsukuri* (Robinson 1961: 26), produces the thin, seemingly random lines that form swirls on the polished surfaces of most old swords. The swordsmith apparently achieved this effect by the deliberate retention of small amounts of interlamellar flux when subjecting the steel to the last few cycles of homogenization through folding and forging. The grain pattern produced depends on the direction in which the semi-finished steel billet is forged:sideways folding and hammering yield a grain that runs parallel to the edge, while hammering at an angle gives configurations that look like wood burls or swirling water. Most of these patterns are delicate and difficult to see when, as on the edge of a sword, the surface is not smoothly and brightly polished.

The second technique is like the one used to make keris in Southeast Asia, and is sometimes called *hada-gitai* (Robinson 1961: 26). During the last stages of working up the steel, a smith selects two pieces which differ in appearance (perhaps because of different carbon contents), forges them together, folds and reforges them a few times, and incorporates the resulting layered plates into the blade. When the shaping of the blade is almost complete, he makes a series of parallel indentations or broad cuts down through the upper layers. After the surface is forged or planed flat again, the deeper layers are revealed as a series of concentric curves arranged regularly along a line parallel to the edge. The effect is highly visible even when the surface is only polished and not chemically etched, the technique which Japanese metalworkers apparently applied. The conspicuousness of the patterns may help to explain why blades made in this way have not been popular among smiths and sword fanciers. Although some of these swords[4] have been widely admired, the Japanese prefer more subtle decoration.

The use of *mokume,* or weathered wood patterns, in sword guards is relatively common, both in nonferrous metals (Notis 1979) and in iron (Smith 1960: figs. 43–49; Fig. 2.2d here). The layering on the sword guards may be thick, and more deeply etched than any mokume blade would be. (See also Smith 1960: figs. 31–34 and Robinson 1961: figs. 24a-d.)

The success of the decorative aspect of pattern welding depended in part on a knowledge of suitable etchants, many of which could be derived from natural sources. The most widely used were the vitriols—the sulfates of copper, iron, and zinc. Each of these rapidly attacks the low carbon forms of iron but acts slowly on higher carbon steels. The low carbon areas would be not only blackened but also partly dissolved, leaving the lighter portions standing in relief and making it possible to increase the contrast by carefully polishing the higher and more silvery steel. The nickel-containing pamor in Southeast Asian blades is brought out not by vitriol but by mixtures of arsenic sulfide and lime juice. Numerous other recipes for increasing the contrast between the elements of welded composites are known from other parts of the world. While many of these formulations appear to have worked, few are easy to use under laboratory conditions, but modern etchants also often do not produce satisfactory results. Those who

[3]Blair's edition of *Pollard* (1983: 150) and Smith's *History of Metallography* (1960: 30) agree that the first published description of the making of pattern welded or "damascened" gun barrels by Europeans is Wasström's, which appeared in 1773, but there are earlier European guns of this sort. Bailey (1975: 58) illustrates a pattern welded flintlock made by the English gunmaker Griffin in about 1740, and it seems probable that others were produced before that in both England and Europe. The fashion for such guns and perhaps also the technique for making them may have come from either India or, more probably, Turkey. On the other hand, Johannsen (1953: 170) says that "damask guns," forged from bundles of wire, were made in Liège during the mid-sixteenth century. Although it is not easy to show that this tradition stayed alive until the mid-eighteenth century, it too must be considered a possible source of the developed European pattern welded handgun.

[4]Two schools of the Koto or Old Sword period, the Namonihira and the Gwassan, are especially known for their use of *ayasugi* grain (Robinson 1961: 47, 50). This was presumably produced by the hada-gitai technique. The ayasugi swords made by certain swordmakers of the Shin-Shinto or Modern period, particularly those by Gwassan Sadayoshi, are often illustrated in books and articles on Japanese swords.

experiment with the coloring of composite blades soon learn to accept as possible even the most bizarre and laborious of traditional etching systems.

## 6. THE STRUCTURE OF THE JAPANESE SWORD

The vast majority of forge welded iron objects made outside Japan fall into one of two structural classes: they are either simple composites, containing two elements of which each is a metal with differing properties, or compound composites in which multiple elements, again of two different metals, are piled together or otherwise arranged in patterns. A third structural class also exists: one whose members contain more than two elements, each of which may be of a separate type of iron or steel. Most objects of this type are Japanese in origin and include many of the swords and related weapons (daggers, *naginata* halberds, spears) made during the last few centuries.

Five-element *shihozume* and four-element *oriawase-sanmai* types have been standard Japanese designs for good blades in recent centuries (Fig. 15.7a), but there is no consensus about which structure is best. Excellent swords can be made with more or fewer elements, including the ancient and still common two-element type called *kobushi* or *kofuse* (Fig. 15.8). The only kinds that are considered intrinsically inferior are one-element swords like those made in factories in the late 1930s and early 1940s, and perhaps also the two-element swords with steel edges inserted into an iron back.

Three structural features seem to be common to good swords and must therefore be essential to achieving desired performance: the soft iron core or "heart," the high carbon steel edge, and the sides made of medium or high carbon steel, either applied separately or formed as an integral part of the edge. The functional roles of the core and edge may not be greatly different from those of the iron backs and steel cutting edges of composite tools in other cultures—the one provides toughness and the other hardness. However, the function of the steel sides, which are unusual in the composite blades of other cultures but common in those of Japan, is not clear. Most specialists seem to regard the steel sides as aesthetic, added to increase the subtle metallographic features visible on polished metal. There is, however, a technological explanation for the steel sides. Steel is used to control the stresses generated when the blade is quenched. Although most swords are

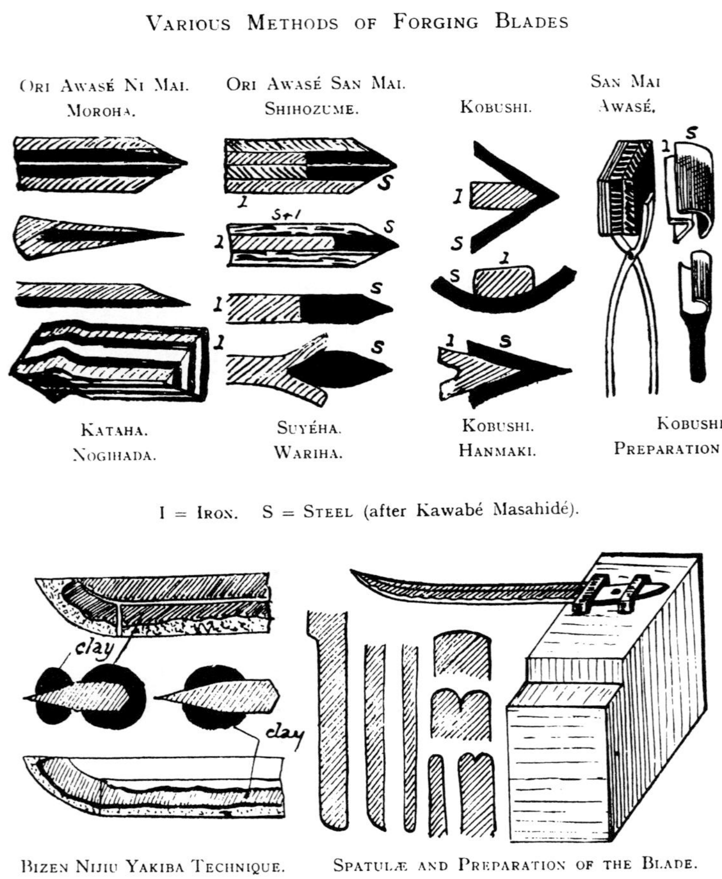


Fig. 15.7. *Composite Nature of Japanese Sword Blades.* Above) Various assemblies of high carbon steel (black) and lower carbon steels (hatched). The assemblies Masahide called *moroha* and *kataha* are more suitable for woodworking implements than for swords. Below) The preparation for heat treatment. The essential idea was to leave the blade itself bare but to cover the adjacent flanks partially with clay. When quenched the clay covering produces a differential transformation at the boundary betwen high and low carbon steel that is responsive to etching. The clay coating also helps to generate a high level of residual stress that places the blade edge in compression so that it can have a superlative cutting ability without being fragile. (After Joly and Inada 1913: 84, 91, based on sketches by the swordsmith Masahide)

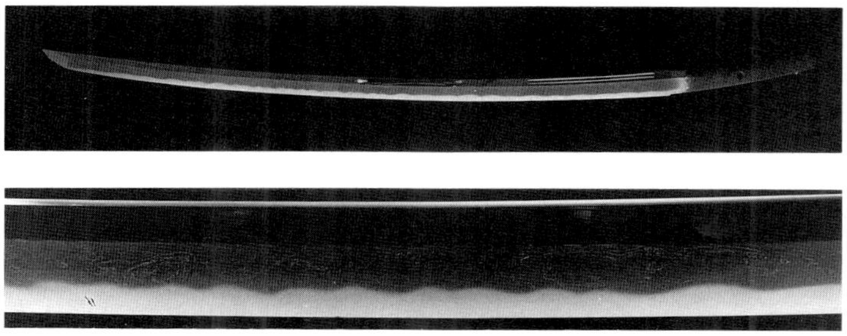

Fig. 15.8. *Japanese Sword Blade.* Signed by Suyehiro, a smith of the Osafuna School of Bizen Province. The tang is inscribed with a date equivalent to A.D. 1392. The overall view shows the strong backward curve that an already slightly curved blade acquires when quenched under the differential cooling rates imposed by the clay cover. The detail view shows the light colored graining produced by thin lines of retained, interlaminar welding flux. The visual contrast between the pearlitic transformation structure of the flanks and the martensitic edge, the *yakiba*, marks the line where the protective clay coating was trimmed away prior to quenching. (Field Museum cat.no. 131638)

quenched during manufacture, only Japanese swords are quenched so that stresses actually distort their structure, but the technique appears to be essential to good performance.

In quench-stressing, the back and sides of the blade are covered with a thick layer of clay, charcoal, and other ingredients (Fig. 15.7b). The smith carefully trims this layer to expose the full length of the edge but only about a centimeter of its width. He then heats the blade to a temperature in the light yellow range[5] and plunges it into water. This transforms the high carbon edge into a predominantly martensitic structure, while the clay-insulated portions of the blade cool more slowly and so assume a fine pearlitic form. The steel along the line where the insulating envelope has previously been trimmed away has a complex structure transitional between the hard martensite and the softer pearlite. After polishing, the pearlitic sides take on a mirror-like finish, while the edge looks frosted. The transition zone contains various dark and shiny flecks that connoisseurs appreciate. The effect of the differential quenching treatment on performance has only recently begun to be understood, thanks to engineering analyses of blades (Gee 1985). The crucial point is that the difference in quenching rate between the edge and sides sets up a pattern of strong internal stress, measured at about 6000 pounds to the square inch (Gee 1985). Martensite crystals occupy more space than pearlite and ferrite crystals. At the moment of quenching, the edge therefore attempts to expand longitudinally but is restrained by the pearlitic sides and ferritic core to which it is bonded. As a result, the edge comes under strong compressive stress while the sides and core are in tension. Thus, the shape of the blade distorts, bowing backward away from the edge, and the martensite of the edge, which normally would be almost glass-like in its brittleness, becomes fracture resistant.

The principle is similar to that of surface-stressed safety glass, where the normal tensile stresses generated by impact are reduced by compressive stresses at the surface (Gee 1985). A Japanese sword thus treated needs no tempering. It is as "tough" as a good European sword that has been quenched and then tempered in the normal way, but it is much harder and can take a sharper and more durable edge.

It is not entirely clear when Japanese swordsmiths first discovered how to use residual and internal stresses to circumvent the traditional problem of having either toughness with softness on the one hand or hardness with brittleness on the other. Japanese swords were already famous in China during the eleventh century, and extant tenth-century blades usually show the visible signs of the quench-stressing treatment—that is, the differentiated surface appearance of the frosted edge and the shiny sides. It is of course possible that some of these early swords have been subjected to later heating and quenching, for many and perhaps most have been repolished during later centuries, and more fundamental alterations may have occurred. It is even possible that some of the swords considered to be the earliest are forgeries, but by the twelfth or thirteenth centuries the tradition of exploiting the residual stresses caused by quenching was well established; there are too many contemporary documents and surviving swords for one to doubt that by then, and probably a century or two earlier, this kind of a sword could be regularly produced.[6]

## 7. JOINING NONFERROUS METALS TO IRON

In general, other metals are rarely joined to iron or steel by true forge welding, although such welds can be made if the surfaces are clean and an appropriate flux is

[5]Joly and Inada (1913: 91–92) say that swords are quenched from one of three (radiant, not temper) colors: dark red, cherry red, or yellow "like summer moonlight." A cherry red temperature would be expected to produce acceptable hardness with better toughness after quenching, while higher austenitizing temperatures would lead to dangerous brittleness no matter how carefully the quench-stressing of the edge is carried out. However, those higher temperatures are certainly often used with success and apparently without subsequent tempering. Joly and Inada may not be correct about the use of dark red temperatures. As Table 1.5 shows, a cherry red corresponds to the minimum temperature needed to produce hardness.

[6]There are many other aspects of the Japanese sword relevant to the history of metallurgy and of iron. A good discussion of grisly but systematic testing procedures is in Joly and Inada (1913), as well as a partial translation of the remarkably objective and analytical published opinions of the great swordsmith Masahide. The best explanations of the close relationship between traditional connoisseurship and modern metallurgical understanding are by C.S. Smith (e.g., 1960: 40–57). Useful analyses and micrographs are presented by Tanimura (1980). Interestingly, good photographs of swords are not as easy to find. Those in Smith (1960) and Robinson (1961) are helpful, but, for real insights, the swords themselves should be examined. The best collections outside Japan are in the Museum of Fine Arts in Boston and the Victoria and Albert Museum in London.

used, but it is better and/or easier to make the joins in other ways. Historically, the metals most often joined to iron seem to have been gold and silver. They were simply applied in the form of leaf, or chemically deposited as in mercury gilding. They can also be inlaid into iron by hammering a piece of gold or silver (or brass) wire into a groove; the groove may or may not be undercut to hold the inlaid metal more securely. Or they can be overlaid on it; the iron substrate is first prepared by scoring with crosshatched parallel lines, after which fragments of gold or silver are beaten in so that they adhere to the crosshatching. This technique, now generally known by the Indian term *kuftgari*,[7] is less secure than inlaying. Neither technique depends primarily on the establishment of actual welds across metal to metal interfaces, though it seems likely that a minor amount of microwelding does occur when the grooves or scoring lines are freshly made before the gold or silver is applied. Both techniques are widespread, reaching a peak of intricacy in the "damascene" arms and armor of the medieval Islamic world. Kuftgari work has long been known in Europe as well. Theophilus, writing in the early twelfth century, not only regards it as a standard technique, but even describes a geared machine for automating the production of the crosshatched lines (Hawthorne and Smith 1963: 184).

In theory it is possible to bond most metals to a hot, clean iron surface by heating them until they melt; the resulting joins involve actual diffusion across interfaces and so can be considered true welds. In practice this particular form of fusion welding appears to have been used most often in joining iron to copper alloys and to tin.

Tin plating on iron dates back much further than the sealed food containers that, legend says, were first used by Napoleon's army. Theophilus states that iron objects can be tinned simply by dropping them into a bath of the molten metal, emphasizing that surfaces to be tinned must first be freshly filed (Hawthorne and Smith 1963: 187). Other documentary mentions of tin-plated objects, as well the objects themselves, occur in medieval European contexts. Hedges's (1964: 108) list of early tinned iron objects includes spurs, nails, hinges, keys, drinking vessels, and, in Chaucer's *House of Fame*, an allegorical pillar "that was of tynned yron clere." We do not know if other cultures practiced tin plating at such an early date, or if other methods of tin to iron welding were used during the pre-industrial period. In Islamic cultures, in which much copper was tin-plated, tin must also have been put on iron occasionally. But compound objects with welded three-dimensional tin and iron parts are rare. We know of no examples even in places like the Malay Peninsula, where tin has historically been abundant and inexpensive.

The fusion welding of iron to copper alloys in the form of brazing compounds is an old and important technique. Most brazing served an intermediate role, facilitating the bonding of a third piece to a piece of iron. Often that third piece was itself iron, with the braze being used to join parts of objects, such as tubes and lock mechanisms, which could not easily be fabricated by forge welding. However, the cast-on metal could also act as a substrate for other metals or, in its own right, as a part of the finished artifact. One sword guard (Fig. 2.2a) illustrates the former approach: small bits of a copper alloy, seemingly cast in, have been covered with gold leaf. Examples of the latter include the various bimetallic weapons of bronze cast onto iron known from early Iron Age contexts at sites in the Middle East, the Mediterranean, China, and Thailand.[8]

Less often cited but equally interesting are the weapons of the Sahel of West Africa. There, brass hilts appear to have been cast onto iron swords during most of this century; more recently, aluminum has been substituted for brass. Several peoples, such as the Nupe and Hausa of Nigeria and the Tuareg of Mali, make these weapons. In fact, the Tuareg are among the modern world's leading exponents of polymetallic decoration; riveted or soldered objects combining three or four metals are not uncommon in collections of Tuareg artifacts.

[7] *Kuftgari* or *koftgari* has been a standard term applied by writers to Indian metalworking since at least the third quarter of the nineteenth century (see, e.g., the descriptions by Hunter [1875: 150] and Egerton [1880: 69]). Hendley (1892: 5–11) uses it but prefers the confusing synonym, "false damascening." Untracht (1975: 153–158) is among the modern specialists who take kuftgari in a general, not specifically Indian sense. Excellent macrophotographs of kuftgari work appear in Bauer (1979).

[8] See, for instance, the references in Wertime and Muhly (1980): for Cyprus and Palestine (p. 84), for Iran (p. 451), and for China (p. 512). For Thailand, see Stech Wheeler and Maddin (1976: 45–46) and White (1982: 46).

# 16. Economics and Geography

## 1. INTRODUCTION

Here we attempt to fit the technical data presented throughout this book into a broader explanatory framework, since the scientific explanations need to be integrated with causal factors of a nontechnological nature. We do not claim to understand all the possible relationships among a major technological system like ironmaking and the socioeconomic, cultural, and environmental systems within which it is necessarily embedded, but if enough examples are considered, patterns may be discerned.

## 2. IRON AS A MARKET COMMODITY

Metals have certain desirable properties which nonmetallic materials lack: they can be shaped by deformation and casting as well as by cutting, and they combine the qualities of toughness, hardness, and flexibility in ways not equalled by organic materials or stone. The superiority of iron would not necessarily have been obvious, since it is more difficult to form and less aesthetically appealing than gold and silver, and until the techniques of steeling were developed, copper alloys were better than iron for some purposes. Iron is not sharper or tougher than cold worked bronze and is less resistant to corrosion. Modern steels are indeed supermaterials, but ferrous metals initially became popular because they are easily obtainable and thus relatively inexpensive.

In antiquity, exploitable deposits of iron ore were more widely available than those of other metals, and in most places more than any other economic mineral except clay. Even flint and other stones suitable for flaking into tools are less common than smeltable iron ore (Chapter 4). Iron not only occurs more frequently than any other metal except aluminum, but it is also more widely distributed because some of its ores, such as bog ores, clay ironstones, and laterites, are formed secondarily within sedimentary deposits. Copper and other metal deposits occur only in rocks created or affected by vulcanism and in secondary deposits directly derived from them, but iron appears in massive quantities in areas in which there is no geological evidence for volcanic activity. Moreover, iron ores, unlike many copper ores, do not require complex processing to yield a useful metal, since the most common iron minerals, simple oxides and carbonates, are relatively easy to smelt.

Iron ore was available in great quantities in some places, unlike the readily mined deposits of silver, tin, and copper which can be soon exhausted. Investors and treasury officials (among others) have always known that the supplies of nonferrous metals can be quickly mined out. In contrast, some iron deposits must have seemed almost inexhaustible to people using early mining techniques; such ancient iron mining centers as Elba in Italy, Eisenerz in Austria, Canigou in the French Pyrenees, Santander-Vizcaya in Spain, Kielce-Radom in Poland, Krivoi Rog in the Soviet Union, Bihar and Salem-Trichinopoly in India, and Daye in China have continued producing ore to the present.

The existence of such vast reserves must have simplified economic planning: the output of iron could be expanded indefinitely when the need arose. The quantities available could have fostered visions of abundant iron implements for armies, homes, and farms, but the potential for enormous increases in per capita production may have been realized only once before the modern era—during the brief industrial florescence in Song period China in the eleventh century—because of social and political factors rather than limits imposed by ore and fuel resources.

Centralized control of primary iron production was impossible because ore deposits were large, widely distributed, and easily smelted, but the secondary processing of iron—raising and lowering carbon content and effective heat treatment—is difficult in theory and practice, so trade secrets could have developed. No other pre-industrial metal can be so greatly improved by secondary processing. Only gold was surrounded by such elaborate ritual and pseudo-science as iron. Although these rituals must have stemmed from the desire to control phenomena which were not understood, the mystery surrounding iron manufacture probably served the more practical ends of discouraging competitors and persuading consumers to pay a higher price.

Iron, as other metals, is relatively more valuable than other commodities. For example, in Calcutta in the early nineteenth century,[1] copper by weight was as valuable as cloves, and even ordinary iron almost as costly as pepper (Table 16.1). The relative value of metals would be even more striking if measured by volume, a matter of as much concern as weight in water-borne commerce. It is clear that a cubic meter of iron was worth at least several times as much as an equivalent volume of spice. In fact, few pre-industrial commodities except precious stones, dyes, and drugs

[1]Prices for locally made Indian iron and steel are not included on Milburn's (1813, v.2: 155–156) list but would presumably be about the same as those of equivalent English products; native iron remained competitive with English iron until the 1820s in the coastal parts of India and, as data cited by Hoey (1880) show, in interior cities like Lucknow as late as the 1880s.

*TABLE 16.1. Calcutta market prices, 1809–1811*[1]

| Commodity | Source | Price sold |
|---|---|---|
| Rice, Wheat | India | 1–2 Rs.*/Md.** |
| Pepper | Indonesia | 10 Rs./Md. |
| Cloves | Indonesia | 55 Rs./Md. |
| Cinnamon | Sri Lanka | 90 Rs./Md. |
| Nutmeg | Indonesia | 320 Rs./Md. |
| Iron | England | 9 Rs./Md. |
| Lead | England | 13 Rs./Md. |
| Steel | England | 14 Rs./Md. |
| Tin | Malaysia | 29 Rs./Md. |
| Zinc | China | 35 Rs./Md. |
| Copper | Japan | 55 Rs./Md. |

*Rs. = sicca rupees, about 2 s, 6 d. in contemporary English money.
**Md. = (Bengal) maund = ca. 37.3 kg.
[1]Milburn 1813, v.2: 155–156

were as valuable on a volumetric basis as ordinary metals.

Iron was generally less expensive than other metals (Table 16.1), except for steel. Medium and high carbon steels are unrivalled for military purposes and their fabrication technology is similar to that of ordinary iron. Its outstanding performance made it worth the effort and cost to manufacture weapons which were both decorative and functional. While the stories about large sums paid for steel swords and armor may not always be credible, some of these objects were worth as much as a silver or gold object of equivalent mass. Therefore, the high market value of steel commodities encouraged manufacturing specialization and trade.

## 3. THE DISTRIBUTION OF IRONMAKING

Historical iron production does not follow a regular geographical pattern, being absent from some regions, concentrated in a few parts of others, and spread evenly over still others. One of the greatest contrasts is between western Europe and Southwest Asia from 1000 to 1900. Most county-sized districts in medieval and Renaissance Europe had at least one iron mine and bloomery. The Middle East, in contrast, has made little iron during the past thousand years. Despite the antiquity of ironworking there, as well as its scientific sophistication and developed economy, the Middle East from the medieval period onward imported most of its iron from South and Central Asia and from Europe.[2]

Many other examples of the uneven distribution of ironmaking exist. In China during the eighteenth and nineteenth centuries, blast furnaces and other establishments for smelting iron were present in every province and virtually every prefecture. During the same period ironmaking was dispersed more widely in India, with tens of thousands of minuscule producing units scattered evenly over the southern and central portions of the subcontinent and along the Himalayan fringe. In insular Southeast Asia, on the other hand, a region with close and longstanding contacts with both India and China, furnaces for smelting iron were rare. Major populous Asian areas such as Java, the Malay Peninsula, and most of Luzon and Sumatra have made little if any iron since the sixteenth century.[3]

Similar contrasts existed within regions. In South Asia, no iron was made in the densely populated and economically developed Ganges-Jamuna interfluve and Brahmaputra delta in the north and northeast; both have been entirely dependent on imports from other subregions and from foreign countries during the past few hundred years. In Vietnam, the lowland, ethnically Vietnamese have consistently for at least several centuries imported most of their iron from China, highland ethnic groups in Yunnan, and Thailand. Historically, little iron has been smelted in Mesopotamia and Egypt, and the same appears to be true of the Levant—coastal Israel, Lebanon, and Syria—and much of the Arabian Peninsula, whereas areas such as Iran, Anatolia, and Yemen were self-sufficient or even net exporters. In sub-Saharan Africa, smelting localities may have been more widely and evenly distributed than in any other region of comparable size in the world, but there are several areas, including Mozambique and the southern Ivory Coast, where ironmaking has not been common in recent centuries. In Europe, where iron is as important to the economy as in Africa, there are regions, such as eastern England, southern Italy, northern Poland, Greece, and most of the Netherlands, in which iron was not made during the last few hundred years of the pre-industrial period.

Ironmaking is much more unevenly distributed than are the necessary natural resources. An early smelter wishing to produce a few hundred kilograms of iron per year and willing to expend the labor would probably have been able to find sufficient ore within twenty kilometers of his home. Fuel would also have been available since enough vegetation to sustain a small scale smelting industry existed almost everywhere (Chapter 6). Horne (1982) has shown, for instance, that even in the semi-desert of northeastern Iran an annual yield of about 4000 kg. of charcoal—enough for at least 400 kg. of bloom iron—can be sustained indefinitely by 1000 hectares of land. Therefore, in principle, every district or even every village could have had its own smelting furnace. Most larger regions could usually support, in terms of resources (ore, fuel, and water power), larger scale production enterprises proportionate to their ability to commit capital wealth.

Price, or rather cost, is the crucial factor. The relative cost of extracting and processing is more likely to explain the absence of ironmaking than a lack of resources. Even in early societies decisions about making iron must have been based on factors besides the physical effort needed, or perceived to be needed. These included the availability of and other uses for capital, the presence of basic technical knowledge, the willing-

[2]For imports of metals to the Middle East from India and Spain in the twelfth century, see Goitein 1980. For earlier and later references, see Bronson 1986c.

[3]To our knowledge, smelting in insular Southeast Asia is recorded only in the interior of Borneo, central Sulawesi, single locations in Sumatra, Luzon, and Sumbawa, and the small islands of Billiton and Karimata. It may also have occurred in a few other places too remote for foreign observers to notice it. However, Java, with an old and literate civilization and a population of several million, has been well known to Europeans since the sixteenth century, and there is no reference to Javanese smelting in any European or Indonesian historical source; Raffles (1817) is explicit that in his day Java imported all its iron.

ness of some individuals to undertake the work, the time needed for developing expertise, and—perhaps most important—the existence of alternative sources of raw or fabricated metal.

## 4. DOMESTIC AND COMMERCIAL PRODUCTION

In many societies, dispersed production patterns existed for making pottery and woven cloth, both of which require about as much technical knowledge and capital investment as simple smelting, and iron implements were regularly fabricated by blacksmiths in villages in numerous pre-modern societies, including those of Europe and the United States. Village smelting can be done with no more than a general knowledge of smelting methods, an amount of capital no greater than that needed for a pottery kiln or blacksmith's forge, and an intravillage economic system in which one or two households could specialize part time in smelting iron which would be exchanged for other necessities.

It is likely that village-level ironmaking took place, but there is no archaeological and ethnographic proof. It may not be possible to obtain the necessary evidence by excavating; few archaeological research programs can be complete enough to show that every settlement possessed a furnace and that each had only enough smelting capacity for its own needs. Ethnographic research, on the other hand, should be able to demonstrate this easily, but published descriptions concentrate on single villages and are vague about the distribution of ironmaking establishments throughout the broader region.

Village smiths often smelted iron. In Africa this occurred among the BaNyankole, the Pangwe, the Ewe (Thurnwald 1932: 125–133), the Fouta, the BaSakata, the Bambara, and the Ovimbundu (Cline 1937: 114–126). Most tribal smiths seem to have lived and worked separately in ordinary farming villages rather than in communities of ironworking specialists. The Pangwe of Equatorial Guinea have solved the village smith's problem of finding the skill, labor, and capital needed to mine and smelt his own iron (Tessman 1913). A Pangwe smith first recruits a group of about five fellow villagers who contribute the capital needed to build a furnace and to purchase the necessary but costly smelting "medicines"; these associates receive substantial shares of the smelted iron. He then gathers a larger number of laborers, also from his fellow villagers, to help with mining and charcoal making and to operate the bellows, of which there are as many as seventeen pairs for one furnace; these laborers are paid with small shares of the iron. Finally he finds, apparently in another village, a smelting expert, who is also a shaman, to direct the endeavor. If the expert also provides the smelting medicines, he may get more than a third of all iron made; his salary for serving as a foreman is a share slightly smaller than that of a member of the investment syndicate.

All of those involved must abstain from sexual relations for two months while ore is fetched and magical preparations made. Finally, smelting can begin. The furnace, of small but unspecified size, is filled with ore and charcoal. It is then lit, and no further ore or fuel is added. Urged on by the chants and shouts of the medicine man, the seventeen bellows operators pump furiously for nine or ten hours until the smelt is completed. All go home to bed to wait for the furnace to cool enough for someone to climb inside to extract the bloom. A second smelting run is then started. This continues for as many times as there are investors, for each is entitled to the proceeds of one smelting run. The blooms are then broken up, the workers' and medicine man's shares distributed, and the furnace abandoned until the next year. Tessmann does not report quantities, but we estimate that the total amount of iron produced did not exceed one or two hundred kilograms.

The Pangwe ironmaking system is one of the most labor-intensive and onerous on record, but some general conclusions can be drawn. First, complex organizational and financial arrangements are possible even within simple ironmaking traditions. Second, since few descriptions of African smelting give details about organization and finance, it is possible that similarly elaborate systems may have been common. Third, the Pangwe case represents what is probably a typical compromise between domestic and export-oriented iron production. The bellows operators apparently use their iron themselves, taking it to the village blacksmith for forging when they need a new hoe or knife. The investors, however, intend to sell their iron; Tessmann notes that it goes not only to neighboring villages but also to other tribes; the medicine man, being an outsider, presumably sells his share. Last, the Pangwe show why a village may elect not to smelt even when it knows how. Tessmann reports that some villages had furnaces which were rarely used because the inhabitants were reluctant to become bellows operators, and implies that the low wages were an obstacle.

Evidence for true village by village ironmaking is strongest not in Africa but in Borneo, where several nineteenth-century observers noted that smiths among the Kayan and Kenyah of the northern interior knew how to smelt and that there was at least one smith in most communities (Harrison and O'Connor 1969: 345–347). Nieuwenhuis (1907, v.2: 198) confirms this, adding that most Kayan and Bahau adult males were able to forge iron themselves. Most sources are unclear about village self-sufficiency, although forgeable iron is rarely reported as an important trade item in the interior of Borneo. Robert Burns, an official of Rajah Brooke's administration in Sarawak in the early 1850s, says, however, that "in every Kayan village there is a place for smelting iron, in all the processes of which the community mutually partakes" (Harrison and O'Connor 1969: 345). Generally, societies where smelting is so dispersed are rarely mentioned in the ethnographic and historical literature, probably because there really are few of them. In most of Asia, the various stages of ironworking were traditionally carried out by separate groups or communities of specialists. Smelters in Japan and China sold most or all of what they made to outsiders; they often did their own mining but rarely worked their iron up into finished implements. The situation in India was similar, although miners and smelters usually belonged to castes or ethnic groups distinct from those of blacksmiths. While smiths were attached to individual villages or groups of villages, smelters sold their product to wholesalers who carried it to a wide market.

Specialized smelting communities also existed in Africa, and separate communities of "smiths" who carried out the full sequence of ironworking processes from mining to forging were common. Many were strongly commercial in orientation, as was the entrepreneurial village of Sukur in northeastern Nigeria, said by Sassoon (1964: 174) not only to have contained as many as fifty smelting furnaces as late as the 1940s, but also to have decided in the early 1960s "that the iron imported from Europe made such poor tools that it would pay them to revive the smelting of their locally produced ores."[4]

Southeast Asia too had commercial ironmakers among its tribal peoples. Hendriks (1842: 6) lists seven "Dusun" (Maanyan or Ot Danum) Dayak villages that supplied a celebrated natural steel to the arms manufacturing and exporting center at Negara in the Malay kingdom of Banjermasin on the south coast of Borneo. Other Southeast Asians who formerly smelted iron for export rather than local use include the Minangkabau of Selimpawang in Sumatra (Burger 1836: 194–200; Horner 1861: 371–373), the Lawa of Thailand (Hutchinson 1934), the West Toraja of Sulawesi (Kruyt 1901), and the Kui of Cambodia (Moura 1882; 1883: 43–46). The Selimpawang and apparently some of the Kui villages did not engage in smithing. As was the case with the Bassari of West Africa, the entire production of these villages was sold to smiths in other settlements to be worked up into finished implements.

Even in Siberia, where the environment enforced an especially high level of village self-sufficiency and where most peoples had long known the use of iron, traditional smelting was geographically concentrated. The great majority of Siberian peoples bought their iron, either from traders carrying Chinese and Russian metal or from a small number of tribal groups who smelted on a commercial basis. These groups included certain of the Kirgiz, the Yakut of the far north (Okladnikov 1970: 197–200), and the highly specialized Shor on the northern edge of the Altai, who were requested by the Russian government in 1641 "not to sell any breastplates or iron helmets or spears or hunting-spears or any armor to the Black and White Kalmyks or Kirgiz or Sayan peoples." The Shor refused, stating "that is how we Yasak people live" (Levin and Potapov 1964: 257, 445).

One type of village-level commercial production involves specialized mining and smelting camps attended for a few months each year by smiths from a number of separate communities (Krieger 1963). These camps were the chief source of metal in several provinces of northern Nigeria during pre- and early colonial times. In the Hausa emirates some smelting camps were large: a hundred or more smiths from at least five districts in Sokoto province went annually to smelt ore from the deposits at Tureta (Krieger 1963: 318). Several hundred are said to have attended the camp at Gidan Dan Tsofuwa in northern Sokoto (Krieger 1963: 326), and even more temporary and resident ironmakers may have sometimes worked at Kuriga in Zaria Province.

The identities of those who worked at the camps are not made clear. Some evidently came from villages that specialized in forging iron implements and some may have been peddlers of blacksmith's iron. Others, however, could have been village smiths from farming communities. If so, the Nigerian smelting camp system represents an unusual mixture of commercial and domestic production. The fact that camps like Kuriga also had resident ironmakers, and that such camps seem usually to have been established under the aegis of (and taxed by) a local chief, suggests one possible model for the development of larger industrial centers. It is not hard to imagine a situation in which independent laborers working on their own account could gradually become employees, and a chief or an enterprising individual could turn into an industrialist.

We do not know what percentage of commercially oriented ironmaking centers can be described as specialized, in the sense that the extraction and processing of metal constituted the chief livelihood for the majority of residents. Most smelters probably did some farming. Many were primarily farmers who mined and smelted iron during idle seasons, as was the case in much of medieval Europe and Song Dynasty China (Hartwell 1967). There are also reports of very small scale smelting carried out by full time professionals; most of the ironmaking localities visited by Buchanan (1807) in southern India appear to have been in operation for 300 or more days per year.

We also do not know whether commercial ironmaking in specialist villages was the usual pattern of production. At some sites so much iron was made that the bulk of it must have been exported. Iron Age sites with multiple furnaces, at least some of which are thought to have been used at the same time, have been found in several parts of Europe: northwestern Germany, western Jutland, Silesia, Little Poland, and the Carpathian Basin (Pleiner 1977). Iron was used but not made at many other sites in Europe and elsewhere, since no slag or other smelting debris are reported. As far as is now known, the great bulk of early iron was smelted in a minority of settlements and supplied to the rest through trade and other modes of exchange.

## 5. THE TRADE IN IRON

Early trade was mostly in luxury items such as silks and spices, but tons of nonprecious metals also moved along the commercial routes of the pre-industrial world, often being shipped as far as types of goods usually considered more exotic. The metals were frequently those such as tin and copper which are rarities, if not luxuries. Certain types of iron and steel objects also

[4]Several sources speak of the revival of smelting in West Africa during and after World War II. Froelich et al. (1963: 29), for instance, note that this occurred at Banjeli (Banyeri), the chief ironmaking center of the Bassari of Togo. Jonkers (1979: 111) refers to the same phenomenon among the Minyanka of Mali.

were valuable because of their singularity: Persian "bintie" and Japanese swords imported to China; Chinese cast iron pots exported to Southeast Asia; wootz steel shipped to Persia from India and onward in the form of Damascus swords to much of the Middle East and Mediterranean; Toledo swords; English cannon; Nürnberg piano wire; Sheffield shears; Connecticut axes. Imported objects of this kind were unique enough that consumers were willing to pay large prices for them and thus justify long distance transportation.

Much commodity-grade iron was also handled by international traders of the pre-industrial period: from Spain and from the Levant to Rome in the first century A.D.; from Spain to England in the thirteenth century; from Spain and India to Egypt in the twelfth; from India to Java and the Middle East in the sixteenth. Ordinary iron has been shipped over long distances as often as many more exotic commodities, only partly because iron is a convenient material for shipping, being resistant to breakage and heavy enough to be a useful ballast in ships. The main reason iron was frequently traded was because, like most metals in ancient economies, it was relatively high in value and low in bulk.

It is easy to see how a pattern of commercial production by certain villages could develop. Smelting could be developed by a small group within a single village and the technology kept secret from neighboring villages long enough to establish a local monopoly. Alternatively, specialized centers could evolve because of differences in efficiency among ironmakers in neighboring villages. Some might eventually find that they were producing iron with less labor than their competitors. Since even the smallest furnace if run for a month or so can produce more iron than a single village could easily consume, it would have made sense for competent manufacturers to begin trading their surplus.

The process of centralization might be accelerated if the iron in question possessed, or was thought to possess, unusual qualities as a material for implements with high cultural and market value, like weapons. Long before the Bronze and Iron Ages began, many parts of the world witnessed the development of extensive interlocal trade in cutting implements made of stone with special functional traits, such as jade, certain flints, and obsidian: evidently, some ancient peoples often felt that the quality of such implements was important enough to obtain them from distant specialist producers.

It is worth emphasizing the difference in commercial potential between commodity and specialty irons. Ordinary commodity iron would not travel far when communications were poor and markets small: the value added by smelting and fabrication into a bar or rude implement is too low to justify the cost of more than a few days' transportation in an area where consumers can easily begin production themselves. However, specialty iron in the form of weapons or unwrought metal ready for making weapons or high performance tools has a much greater potential value, with a selling price proportional to consumers' perceptions of the superiority, uniqueness, and indispensability of the product. The margin of actual rather than perceived superiority does not need to be great. A sword that is ten percent better could be valued at many times the price of ordinary swords and might therefore be traded great distances from its point of origin.

We therefore expect two kinds of iron trade to coexist in regions with nonindustrial economies: commodity irons traded within the neighborhood and specialty irons traded throughout an area that might be very large. The corresponding production centers would also be of two types: one with small numbers of ordinary bloomeries serving markets that comprise a few tens of other settlements, and a second type, not necessarily larger, which is oriented to a wider market and which may employ special equipment for smelting or subsequent processing. The second type of center might also make some commodity iron for local use. The trading networks involved, however, might be separate, because specialty ironmakers could ignore local markets entirely.

A relevant example of trans-local orientation in small scale ironmaking is the village of Konasamundran in Hyderabad, Central India, where an especially high grade of wootz steel was produced. Judging by Voysey's (1832) description, the organization and equipment were no more elaborate than those of other Indian wootz making villages, most of which sold their steel in nearby manufacturing towns. The steel made in Konasamundram, however, had an exclusively foreign market. It was monopolized by Persian merchants, who bought it in the village and exported it all to Iran.

The same trans-local orientation in small scale smelting can be cited in West African groups like the Bassari, whose production was geared to a regional economy not much less integrated than that of medieval Europe. In addition, some historical societies with undeveloped economies had only slightly less extreme levels of specialization and export orientation in ironmaking. The "Dusun" Dayak groups, who supplied steel to Negara in Borneo, were headhunters at the time Hendriks (1842) visited the area, without chiefdoms or other modes of supra-village sociopolitical organization. Other examples include the equally stateless and warlike West Toraja and Tobada of Central Sulawesi, who were significant exporters of weapons in the eighteenth century (Kruyt 1901: 148–191); the Shor of Siberia, who were nomadic pastoralists when not making iron and armaments; and the impoverished Jur of the southern Sudan, who supplied much of the iron used in that region (Cline 1937: 25).

## 6. MARKET PATTERNS

In undeveloped economies, only two types of marketing patterns for iron seem to have been common: short-range networks for distributing commodity iron and long-range networks for specialty iron. Neither of these is well documented in the ethnographic and historical literature.

Commodity iron networks in undeveloped economies were not only small but simply organized, so that the product passed through very few hands—network

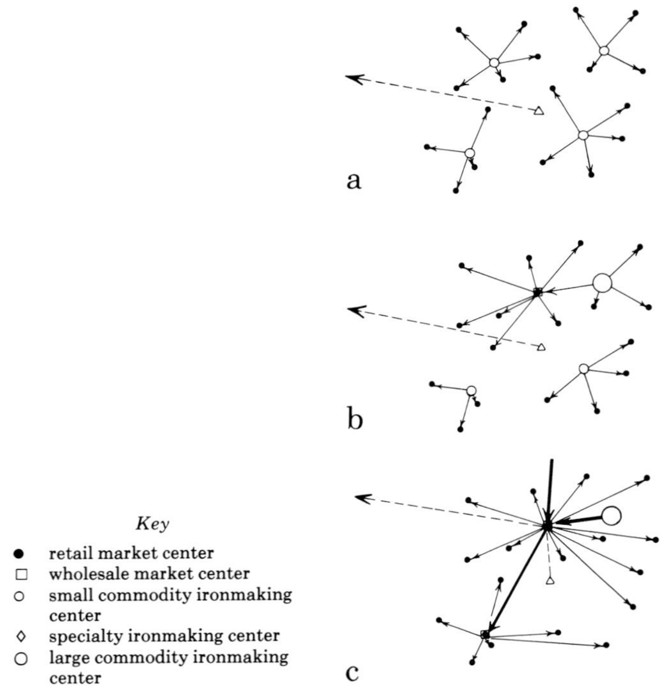


Fig. 16.1. *Regional Iron Production and Marketing Patterns in the Pre-industrial Period.* a) An undeveloped pattern. Commodity iron is made by numerous small producers and shipped to retail centers and consumers in the immediate neighborhood. Only the specialty ironmaking centers have a wider market; b) A partially developed pattern. Some commodity ironmaking centers continue to produce for strictly local markets but others have begun to operate on a larger scale. Wholesale trading centers have begun to develop, and small ironmaking centers near these have dropped out. Specialty ironmakers are not affected by these trends; c) A fully developed pattern. A few large centers now dominate the production of commodity iron, selling their metal through a hierarchy of wholesale trading centers. The specialty ironmakers, although their products and volume of production may not have changed greatly, now make use of the same marketing network.

nodes—on its way from the mine to the consumer. In many instances only two nodes were involved: the smelter himself and a blacksmith, whose shop was unlikely to be more than a day's journey distant. The smelter might have bought his ore or charcoal from others, and one or two peddlers or shopkeepers might have handled transportation between the smelter and the smith, but the maximum number of network nodes was small (see Fig. 16.1 for the spatial patterns that several hypothetical networks of this kind would form in a schematized subregion).

Actual subregions with iron trading networks of the undeveloped short-range type are often encountered in the earlier ethnographic literature. Unexpectedly, in western North Carolina and eastern Tennessee in the early 1880s, forgeable iron was supplied by more than three dozen Catalan hearths, mostly equipped with trompes (Swank 1892: 298–299):

> They were fitfully operated, as the wants of their owners and of the neighboring farmers and blacksmiths required, or as the supply of water for the forges and hammers would permit. . . . They furnished their respective neighborhoods with iron for horseshoes, wagon-tires and harrow-teeth. . . . Iron is taken in exchange for shoes, coffee, sugar, calico, salt and other articles used by the people of the country. It is considered a legal tender in the settlement of all debts and liabilities. This bar iron, after being collected by the merchants, is sent out and sold in Knoxville, Bristol and other points affording a market.

Thus, part of the production of these hearths went directly to consumers, while part passed through the hands of local storekeepers and their contacts in larger market centers. No more than three or four network nodes were involved.

Although Swank does not say so, we may be certain that in the 1880s a significant proportion of the iron used in North Carolina and Tennessee came from outside suppliers in the form of steel bars and finished steel articles like axes, cutlery, and guns. These represented a separate and larger network for the distribution of specialty as opposed to commodity iron. Networks for the distribution of specialty iron, attested in many parts of the world, tend to cover greater distances than the commodity iron networks with which they are partially congruous. They also usually have a more complex structure and closer integration with regional and national economies. The specialty iron- and steelmakers whose products reached rural parts of the United States during the third quarter of the nineteenth century included Sheffield in Britain, Solingen in Germany, and several Swedish centers, as well as Collinsville and other toolmaking cities in Connecticut.

Greater complexity characterized iron production and exchange in the more developed pre-modern economies. Historical details for very early periods are generally lacking, but a model that should be applicable can be derived from Gille's (1947) classification of makers of wrought iron in France during the mid-eighteenth century. He suggests that in terms of market orientation these ironmaking firms fall into four main types, ranging from those that served largely local markets to those engaged only in national-level "grande commerce" (Table 16.2). The "forge" most oriented toward supra-local, national-level sales—Clairvaux—is neither the largest producer nor the one with the most distant regular customers; in fact, the relative absence of direct

*TABLE 16.2. Ironmaking centers classified by type of market served: France, eighteenth century*[1]

| | La Frette (Perche) | St.-Baudel (Berry) | Mareuil (Berry) | Charenton (Berry) | Clairvaux (Champagne) |
|---|---|---|---|---|---|
| Type | 1st | 2nd | 3rd | 3rd | 4th |
| Sold Locally | 50% | * | 10% | 10% | 0% |
| Non-local Markets | 1? | 5 | 11 | 9 | 6 |
| Farthest Market[2] | 50 km. | 250 km. | 700 km. | 250 km. | 200 km. |
| Sales (Livres)[3] × 1000 | 500 | >125 | >4,000 | >2,000 | 400 |

*59 of 69 customers.
[1]All data from Gille (1947: 101–102).
[2]Distances are calculated by us, not Gille.
[3]An extraordinary sale by Charenton to Douai is not included.

sales to small local users (who are in this case retailers and minor manufacturers—wire drawers, cutlers, nail and lock makers—rather than households) does not appear to be an especially good criterion for the importance of a manufacturing enterprise. Significant divisions within the wrought iron industry clearly did exist in mid-eighteenth-century France. The four types discerned by Gille can be combined into two major types of wrought iron producers:

> 1. Small to medium-sized firms whose iron either passed directly to small industrial consumers or indirectly to them through one or two levels of jobbers and wholesalers, most of them located in minor commercial centers within the immediate neighborhood.
>
> 2. Medium to large firms most of which had some local sales but the bulk of whose product either went directly to big, nationally oriented consumers (such as arsenals and shipyards) or indirectly to large and small customers through the several levels of wholesalers who constituted the national market of the period.

We add a third type of firm not considered by Gille: makers of specialty iron, as illustrated by the celebrated steel forges of Rives in the Dauphiné, southeastern France. Grignon and Binelli reported that this ancient industrial center in the 1770s, despite its organizational fragmentation and archaic technology, was still the only large volume producer of high quality steel in France (Léon 1961). In 1778, the 27 or 28 active forges at Rives produced 1,758,000 marc pounds of steel (Léon 1961: 19, 64). At the time most of this went to the royal arsenal at Saint-Etienne, about 200 kilometers distant, but Rives had regular customers as far away as the Near East (Léon 1961: 20): its market radius was much larger than that of any French maker of commodity iron. Its individual forges were small (with an average of only four workers and an output of about 30 tons per year) and were independently owned, although a guild-like cooperation in negotiating with raw material (i.e., cast iron) suppliers and with government customers seems often to have enabled the Rives forges to appear like a single corporation to outsiders. Rives was therefore in its own way as much a representative of the pre-industrial *grande commerce* as Mareuil, but even in the 1770s Rives continued to supply surrounding farmers with iron shovels, bill hooks, and pitchforks.

Thus, on the eve of the Industrial Revolution, the overall French system for making and marketing iron was complex, and its component subsystems were varied. In some cases iron may have reached the consumer by paths as short as those in primitive economies; in others it may have passed through several nodes between furnace and fabricator and through several more before it reached the peddler's pack or retail shop. Indeed, commercial organization in the eighteenth-century French iron trade was as varied in its structural aspects as the iron trade of the present day. In the more developed economies of the pre-industrial period, the small volume of total production and small size of individual enterprises do not imply simplicity in the corresponding commercial systems.

Ironmaking centers in the pre-industrial period can be placed within a regional landscape in one of three distinct patterns.

1. In an undeveloped economy, commodity iron is traded only within the immediate neighborhood of an ironmaking center, while specialty iron may be shipped fairly long distances (Fig. 16.1a). Assuming that supplies of ore and fuel are adequate, the size and production of a center specializing in commodity iron are roughly proportional to the density of population within a radius set by transportation costs. In a region without political barriers, the radius of such a market area might be as large as several days' travel by land and even larger if water transport is available. The existence of hindrances—tolls, banditry, local hostility—will act to increase the number of market areas and iron producing centers. Within limits, the number of locally oriented centers is independent of regional population size. The number of producing units will rise and fall as demand within the market area expands and contracts, but the number of centers is determined largely by transportation costs and therefore should remain more or less constant until demand vanishes almost entirely or until it overwhelms either the resource base or the organizational capacity of the center. As centers that make commodity iron for local markets are likely to be small, they can respond to the depletion of resources by shifting from one place to another within their market region. In contrast, a specialty ironmaking center is fixed to a particular location to the extent that the quality of its product depends or is thought to depend on using a certain kind of ore, fuel, or flowing water resource.

2. In a partially developed economy, like those of India and West Africa during the seventeenth and eighteenth centuries or Europe in the Roman period and Renaissance, centers producing entirely for local markets will still exist in isolated districts (Fig. 16.1b). But there will also be centers that make commodity iron in fairly large quantities and distribute it over a sizeable region; while transportation methods may not be significantly more efficient than in less developed economies, a decrease in the number of political barriers results in the expansion of marketing radii.

An instructive comparison of locally and subregionally oriented centers occurs in a description of the historical iron trade of the Lake Kivu area in eastern Central Africa (Newbury 1980: 11–12):

> Trade in iron products was especially important within Rwanda. During the early 19th century, one major source for hoes used in Rwanda was the small state of Kayonza, located north of Rwanda... West of the lake, however, neither the sources of iron ore nor the skills of ironmaking were sufficiently restricted to promote a regional trade in iron products such as occurred in Rwanda. Instead, iron products tended to radiate out from many local centers, wherever families or villages engaged in smelting and forging.

We doubt that differences in ores and skills explain the difference in trade patterns, since Ruanda was the home of a relatively well-organized state, while the area west of the lake was not. Yet the strong contrast between the two areas shows that different systems for commerce in iron can coexist in adjacent districts, apparently for fairly long periods of time.

In more complex economies, the growth of individual producers is facilitated by the development of market institutions, which provide credit and help during short-term fluctuations in demand. Producers who have grown to middle size may increase output by building more furnaces or bigger ones; the former strategy may be less risky and the latter more economical. Such producers need larger quantities of ore and fuel as well as better access to communication routes. They are therefore less mobile than small producers; they may find it difficult to survive in areas where the resource base is severely limited. On the other hand, they are highly competitive in areas with adequate resources and transportation. The result is a fall in the total number of producers while net regional output rises.

Makers of specialty iron and steel are not necessarily affected by these trends. Those with a genuinely superior product can remain competitive while staying small; as in the case of wootz making, a scaling up of processes might entail an unacceptable loss in quality. Further, the size of the regional and supra-regional markets to which the specialty manufacturers are already oriented may not increase simply because the subregion or region has become politically and economically unified. On the other hand, when (as in the case of Styrian-Carinthian steel) product quality depends less on special processes than on special raw materials, larger producing units may prove desirable. Some specialty producers may therefore evolve in the same way as makers of commodity iron, although the organization and technology used (as in Styria and Rives during the eighteenth century) may still be conservative.

3. Since only two examples of a developed pre-industrial economy exist—China at various periods in its history and Europe during the sixteenth to eighteenth centuries—model building is simplified (Fig. 16.1c). Two of the three earlier types of centers would have persisted under such conditions. While locally oriented small producers will die out, medium-sized makers of commodity iron and specialty producers will survive everywhere except in the immediate hinterland of large cities and industrial centers, perhaps with little change in scale or technology. In the metropolitan areas, however, a fourth type of producer appears: Gille's Types 3 and 4, oriented toward *grande commerce* and with markets as large as entire regions. Size here is economical: hence large furnaces and large integrated enterprises begin to appear. The low price of the iron produced, as well the manufacturers' ability to coerce traders and government regulators, hastens the demise of small local ironworks. In a few places competition among large and medium-sized firms might be intense enough to lower the price of iron to a point where consumption in the household and the farm begins to rise.

## 7. THE LOCATIONS OF FURNACES

Students of European ironmaking have long accepted the axiom that "ore moves to coal," reasoning that more fuel than ore is needed to make metal and that the lowest-cost location for a smelting plant is therefore as close as possible to the coal mine or other fuel source. Hartshorne (1928) was one of the first to challenge this notion, pointing out that in pig iron manufacture during the early twentieth century the ore generally outweighed the coal, and that the average ironmaker was more interested in proximity to markets than to iron and coal mines. Citing examples like Bilbao in Spain, which exported ore while making iron for local markets with imported Welsh coal, Hartshorne makes a convincing case that actual locational decisions in modern ironmaking are much more complex than "ore moves to coal" implies. Not only is market access of primary importance but other factors—among them, differential freight rates and government taxes and subsidies—can on occasion be decisive in determining the location of a modern ironworking enterprise.

Access to fuel supplies might have been a strong motivation, since a great deal of fuel was often needed, the fuel was generally charcoal, and charcoal is a fragile material which does not travel well. Yet smelting did take place for several centuries at or very near iron mines in the Bassari area of Togo, Kielce-Radom in Poland, Hüttenberg and Eisenerz in Austria, and Ariège in France. In addition, government and market factors probably had no less influence in the past than at present. Tolls and other fees imposed at various governmental levels may sometimes have determined plant locations even in primitive economies, and early ironmakers must have been at least as conscious as their modern counterparts of the need to be near their markets.

It is not possible to predict exactly where an early ironmaking center should be located even if there are detailed data on all the factors involved: mines, forests, water, labor, communications, and markets, among others. On the basis of the models of ironmaking enterprises outlined in the previous section, however, we can suggest how different kinds of centers will be affected by local constraints.

*Specialty iron producing centers* are likely to stay fixed for long periods of time. The markup on the product is high enough that premium prices for transportation and fuel can be paid, so they need not be located near their principal markets, and they are thus able to survive local fuel shortages that might adversely affect the manufacturers of other kinds of iron. Since they do not depend on nearby customers, specialty ironworks in one place may not affect the existence of others.

Examples include Rives, Sheffield, and Solingen-Remscheid in Europe as well as the wootz makers of the Hyderabad and Salem-Trichinopoly areas in India. All these traditional steelmaking centers were stable over time and were composed of large numbers of small independent enterprises. The European works are in places with good water power, but the fact that there are many locations with fast-running streams and very few with clusters of small steel mills suggests that choices of site were made for other reasons.

*Locally oriented smelting centers*, due to their low capitalization and the availability of usable quantities of ore and fuel almost anywhere, are likely to have been moved from time to time. The main locational constraints are proximity to customers and distance from other producers of similar iron. The archetypal examples are the forest settlements of the Kol and Agaria in central and eastern India: dispersed hamlets of professional smelters, semi-itinerant but tied to a particular area by traditional relationships with certain smiths and moneylenders.[5] With yields per furnace of no more than 1000 kg. per year, the strain on resources is not excessive, although conflicts over woodcutting can develop where the forests are near densely populated areas. The part time smelting operations of African groups like the Pangwe are less mobile, because a large labor force is needed and smelting is only done for a few days per year. Therefore, furnaces must be located in or near farming villages; if those villages sometimes are moved, it is for reasons unrelated to the supply of fuel and ore.

*Subregionally oriented centers* will be located to minimize the transportation costs of fuel and ore while remaining a reasonable distance from markets. The availability of labor will not usually be an important factor in determining the location of a center at this level; it should offer steady enough work to attract specialists and laborers from a considerable distance. Access to water power, if relevant, will determine the exact siting of the center but not its general location, since usable streams are likely to be frequent enough that an appropriate mill site can be found within a short distance of the sources of ore and fuel.

The question of whether an early ironmaker preferred sites nearer the fuel source than the mine must remain open. It seems reasonable that he would choose to minimize the distance that charcoal must be carried, but the fact that many historical smelting furnaces were situated close to mines indicates that this was not the case. The charcoal used in a given furnace may have come from several more or less widely separated places, and harvestable parts of forests may have varied from year to year, so it may have been simpler to locate the furnace near the mine, despite the difficulties of transporting fuel. In addition, in many parts of the world the makers and carriers of charcoal belonged to underprivileged classes or ethnic groups,[6] so ironmasters may not have been concerned that the location of the furnace was inconvenient to them.

The furnaces of a subregional center may range in size from small to medium, with daily outputs in the 10 to 1000 kg. range. Examples of medium-sized furnaces in subregional centers include many of the Stückofen and early blast furnaces of northern and central Europe, the Catalan and Corsican hearths of southern Europe, the shaft furnaces of Kerala and Bihar in India which produced over 100 kg. per day, and many of the provincial Chinese blast furnaces described by Tegengren (1923–24). Subregional centers containing numerous small furnaces are also known; the Bassari of Togo, who operated more than 500 furnaces in three localities, produced between 250 and 500 tons of iron per year, and sold this to markets within a 200 km. radius. Other centers of this kind include the Kui area of northern Cambodia, Tendukhera in central India and, in all probability, the more important iron producing localities of the Roman Empire. Few Roman furnaces can have had capacities greater than about 20 kg. of iron per day.

*Regionally oriented centers* differ from subregional centers chiefly because they function within a more widely integrated economy. The only example outside Europe of an economic system developed enough to sustain regionally oriented ironmaking centers is China. The scale of production there was much larger than in, for instance, the contemporary Roman Empire. Archaeological finds of the Han period (200 B.C.-A.D. 200) at Tieshenggou, Wafangzhuang, and Guxingzhen in Henan Province include groups of blast furnaces two or more meters in diameter with associated roasting ovens, cupola and crucible furnaces (for melting rather than smelting iron), and fining hearths.[7] (The organization of these large enterprises is discussed in the next section.) A model for the way in which either large or medium-sized enterprises could combine to serve extensive markets is provided by the nineteenth-century Shanxi system. The province contained about a thousand ironmaking firms, some independent and some linked by common ownership or by indebtedness to the same iron merchants. Each operated a crucible furnace with a daily capacity that averaged about 500 kg. The bulk of the 50,000 tons produced each year was bought on the spot by wholesalers and shipped by donkey train and boat to most other provinces of northern China (Tegengren 1923–24: 320–321).

Centers for making specialty iron often served regions as large as this, but few non-Chinese makers of commodity iron reached as wide a market until shortly before the Industrial Revolution. The first European producer of ordinary iron to become regionally oriented may have been in the Basque-speaking area of northern Spain, perhaps in the thirteenth or fourteenth century. Next was Sweden in the sixteenth century, followed by Russia and Britain in the early and very late eighteenth centuries. In Spain, producing units had daily capacities in the same range as the furnaces of Shanxi; Swedish units could produce one to two tons per day, about average for European (and Chinese) blast furnaces of the time; Russian and British production by 1800 had grown much larger, sometimes reaching a daily output of ten to twenty tons.

All such regional centers appear to be located in areas with considerable resources of ore and fuel, as well as of water power. Proximity to markets is of less importance than at lower levels; none of the centers cited above was near a major city when it was established, although they

[5]Fully itinerant smelters are also known, like the Hungana of western Zaire who carry on their trade while travelling through the territories of their customers, the Mbala, Pende, Ngongo, Pindi, Yansi, Mbunda, and Kwese (Sousberghe 1955).

[6]The low status of woodsmen and charcoal burners is evident in much European folklore—for instance, in Grimm's fairy tales. In Yunnan, southwestern China, Rocher (1880) notes that charcoal for smelting furnaces was usually supplied at prices greatly below urban market rates by impoverished Yi tribesmen.

[7]The main source is Archaeological Team 1962. Further details and opinions appear in Li 1975, Henan Provincial Museum 1978, and Wang 1982: 125–128.

eventually attracted numerous secondary industries and a large local population. Several such centers may exist in a given region and may even be near one another, but they are strong competitors for centers at lower levels. Where communications are good, they will squeeze out locally and subregionally oriented ironmaking centers even when these are distant.

## 8. THE SIZE OF ENTERPRISES

Iron was once a relatively high-priced commodity but it was not a conspicuous one in most places until the nineteenth century. Modern historians and early records agree in treating the long-distance trade in iron as an unexciting phenomenon; few early sources devote much space to it, and no poems have been written about iron-laden barks from exotic shores. This is not because iron is an intrinsically unromantic substance. The Victorians found it exciting enough, while many equally utilitarian materials have aroused their share of enthusiasm among businessmen and poets. The real reason why iron was inconspicuous is that, despite its fairly high unit value, it brought only modest profits.

Iron is simply too common. Not only are iron ores widespread, but the knowledge of basic smelting and fabricating methods was also. True expertise in ironmaking was undoubtedly rare, but the inhabitants of any part of the Old World were capable of learning to make a good product if forced to do so by high prices or genuine scarcity. The vast size of this normally marginal reserve production capacity must have acted as a severe restraint on profits in ancient iron industries and on long-term upward movements of prices. It almost precluded oligopolistic market manipulations: any attempt by established ironmakers to increase profit margins by raising prices would have been defeated within a short time by the appearance in the marketplace of numerous new producers.

The history of the copper industry is an instructive contrast about potentials for market manipulation, price fixing, and other methods for achieving large profits. Copper ores are not only scarcer than those of iron but in most regions are distributed in a less dispersed pattern: in all Europe, for instance, there have never been more than a few hundred significant occurrences of copper, and no more than a handful of these are large and accessible enough to have yielded significant quantities of metal using pre-industrial mining and smelting technologies. Controlling all European sources of copper was technically feasible, as shown by the temporary success of sixteenth-century financiers like Christoph Fuerer of Nürnberg (Kellenbenz 1974: 261), either in cornering the copper market or creating monopolistic cartels of copper producers.

Moreover, extracting and smelting copper ores were capital-intensive. Most of the European deep mining of the fifteenth through eighteenth centuries was for copper or copper-silver deposits. Agricola clearly illustrates the complexity of techniques and mechanisms used, and the scale of operations was similar. A shaft at the Roherbuhl copper-silver mine in the Tirol in southern Austria reached a depth of 886 meters in 1597, when the mine employed two thousand workers (Kellenbenz 1974: 201–202). In the 1550s a single smelting works at Neusohl in Slovakia had sixteen shaft furnaces, fifty roasting hearths, and an annual output of a half-million kilograms of copper (Kellenbenz 1974: 208).

Europe was not alone in having enormous coppermaking enterprises during the pre-industrial period; works on a similar scale existed in both China and Japan. In the early 1800s, before the introduction of Western technology to Japan but well after that country had passed its peak as a world copper exporter, the great mine at Ijo on the island of Shikoku still employed 10,000 men and produced about 480,000 kilograms of metal per year (Burger 1832: 10, 14). Western travelers in Yunnan in southwestern China during the late nineteenth and early twentieth centuries recorded numerous traditional iron and copper works. The largest of the former may have employed a hundred men in mining and smelting, but the work force at the latter was much greater. Brown (1923: 103–122) lists four Yunnanese copper works which in the early twentieth century employed more than a thousand men. The mine at Laochang included several galleries more than 1500 meters long; that at Baopingchang in the early nineteenth century is said, presumably on the basis of government records, to have produced over 1785 tons annually—about 1.6 million kilograms—and to have employed between five and six thousand men.

The consequence of the linked factors of intrinsic scarcity and operational scale was to make copper attractive to both industrialists and merchants. The Neusohl works were owned by the Fuggers, the greatest bankers of sixteenth-century Europe. A large proportion of Japanese copper production during the eighteenth and nineteenth centuries was controlled by the Sumitomo family, still among the world's principal industrialists. Governmental and private agencies regularly achieved monopolistic control of prices in several regions, and in periods when prices were inflated copper became a prime speculative commodity. The main European copper markets in Hamburg and Amsterdam appear to have undergone substantial fluctuations due to speculation and war-induced shifts in supply and demand. During the seventeenth century, copper prices were sometimes highly volatile, so that opportunities for large speculative profits existed (Cipolla 1965: 157).

In contrast, iron is a dull product from a commercial standpoint: difficult to monopolize, low-priced under ordinary conditions, and so easy to make and transport that demand rarely runs far ahead of supply. The present depressed state of the world's iron and steel industries represents what may well be the historic norm: fierce competition, more or less permanent overcapacity, unenterprising management, and modest profits. It is not hard to see why the Fuggers and Sumitomos of the pre-industrial period were more interested in metals other than iron.

The mid-eighteenth through mid-twentieth centuries seem in fact to be a historical anomaly: a period in which the making of ordinary iron was profitable and, briefly at least, at the forefront of the development of materials technology. Otherwise, except for several periods in Chinese history there are few indications that the making of commodity iron was regarded as exciting. The smelting and fabrication of iron have been by and large conservative, routine, and small scale.

The principal exceptions are in defense-related iron industries. One of the best documented examples is the cast iron used for making cannon in Europe during the Renaissance (Chapters 13, 14). As a speculative commodity, the cast iron cannon was unlike ordinary iron, in that it had volatile and periodically urgent demand, relatively fixed inventories, hoarding value for consumers, and distinct restrictions on market entry because of the substantial technical skills and capitalization needed to cast a satisfactory cannon tube. The result was that the iron cannon business became the main (and virtually the only) branch of pre-industrial European ironmaking to be economically concentrated. The De Geers and Trips in Holland and Sweden, the Brownes in southern England, and the Marselis family in Russia—all cannon-makers—were big businessmen in their day, controlling large segments of the markets and probably employing in their branches as many workers as a good-sized copper mine.

Military needs also gave rise to a number of other types of manufacturing concerns that sometimes grew to substantial size. In the post-medieval period, the state arsenals of Venice and Turkey, the personally owned enterprises of a few northern European rulers (e.g., the mining and manufacturing organization belonging to the Duke of Brunswick, which controlled the ironworks at Gittelde), the *kharkhana* or palace workshop complexes of Persia and Mughal India, and government arms making centers in China are all known to have each employed more than a thousand men; the Arsenale at Venice is often said to have been the largest industrial plant in Europe before the nineteenth century. However, all of these were general manufacturing organizations which often did not mine or smelt metals and which employed only a moderate number of armorers and smiths. Except for brief periods of military emergency, it seems unlikely that any of these organizations retained more than a few hundred employees whose jobs were directly connected with the processing of iron.

Individual plants for making and working iron were small in both the public and private sectors, and even cannon foundries did not grow significantly. As late as 1695 the famous De Geer-founded gun factory at Finspong in Sweden employed only 37 persons at the factory itself along with 21 miners (Cipolla 1965: 153) and about a hundred woodcutters and charcoal burners—a tiny operation compared with Roherbuhl or Neusohl. Ironworks in Europe did not begin to approach copperworks in size and capitalization until the beginning of the Industrial Revolution. The first to employ a thousand men may have been Le Creusot in France, in the 1770s. The first to do so without massive state subventions must have been somewhere in England; in 1806 the giant Cyfartha works of South Wales is said to have had two thousand workers engaged in making iron and mining ore and coal (Beck 1899: 154).

Pre-modern ironworks and ironmaking enterprises were small in relation to plants and firms in other sectors in most parts of the world. In India it was rare for a "firm" to be much larger than the personnel needed to supply and operate a single furnace and secondary hearth. Of the eight South Indian ironmaking villages described in detail by Buchanan (1807), the furnaces in six employed between 18 and 22 men, including miners and charcoal makers; in the seventh, groups of only four or five operated small furnaces cooperatively; and in the eighth, where several larger furnaces seem to have been run by one entrepreneur, fewer employees were needed because both ore and charcoal were purchased from independent suppliers. Buchanan's figures seem to be about average for fully commercial ironmaking in South Asia. The largest enterprises on which we have found employment data were in the ironworking center of Tendukhera in Central India; each unit seems to have operated a single furnace with between twenty-four and twenty-seven operators, miners, and charcoal makers.

African and Southeast Asian ironmaking was organized on an even smaller scale. Although in some cases large numbers of people are reported to have helped casually during a smelting run, the majority of furnaces in both areas had ten or fewer full time workers. The village of productive and fully professional ironmakers in southwestern Nigeria that Bellamy (1904) studied had eleven furnaces and a total population of only 125 people. Unless the entire village was a single firm, these individual enterprises must not have been large. The same is probably true of village-level ironmaking operations in northern and southwestern Asia. While little is known about their organization, there is no reason to believe that any nonstate smelting or smithing enterprise in those regions employed more than a handful of workers.

In Japan, on the other hand, ironmaking was organized in larger units, at least partly because of the great labor demands of the *tatara* process. Lyman (1879: 65) reports that a 32-man crew was needed to operate an average-sized tatara furnace at Amegawa in Izumo, most of them occupied in treading the bellows and breaking up pigs of iron by hand. At this particular furnace, ore and fuel were both purchased, but we believe that furnaces existed where these were extracted by regular employees; in such cases the number of workers attached to a single furnace could approach two hundred—about the size of a similarly integrated ironmaking operation in sixteenth-through eighteenth-century Europe.

China is the principal exception to the rule that all pre-industrial ironmaking was organized on a small scale. Needham (1958) notes that outstandingly wealthy private ironmakers began to emerge in the late Zhou period, in the fourth and third centuries B.C. Building on Zhou traditions of government licensing and oversight of certain industries, the early Han emperors gradually consolidated control over all iron production until, during the reign of Wu Di in the late second century, iron and salt became government "monopolies" and major sources of revenue. In theory this meant that the state manufactured and distributed all iron. In practice, much (and usually most) iron was made by selected private contractors working under the close supervision of special government offices. Some of these built up firms of substantial but undetermined size.

The theoretical government monopoly on iron in China survived the fall of the Han Dynasty, so large firms with good official connections continued to be favored and small bootleg smelting enterprises suppressed wherever the authority of national and provincial rulers was strong. Individual firms may have reached their maximum size during the Song period in the eleventh century. A single smelting plant at Qicunzhen in southern Hebei Province had between 35 and 112 furnaces, produced an estimated 14,000 tons of pig iron per year, and employed about 3000 workers; a plant at Guzhenzhen produced about 13,000 tons per year (Hartwell 1967). At the end of the tenth century, the two largest government armories employed 8000 laborers (Hartwell 1963). Due to a copper shortage in the late eleventh century, three imperial mints in the Kaifeng area began making cast iron coins. In 1078 they cast 1350 tons of coins and employed as many as 7000 workers; several decades later all three came under the control of a single intendant of iron mining and mints (Hartwell 1967). Song entrepreneurs in the private (and perhaps the public) sector acquired vast fortunes, while per capita consumption of iron in the zone around the capital at Kaifeng reached the astonishing total of 6 kg. per year (Hartwell 1963). Probably no European population of similar size consumed as much iron before the nineteenth century.

A comparison of Chinese and other iron industries suggests why most siderurgical traditions have been conservative, even backward, in trying new methods of making and working iron. The Chinese industry was not conservative. Even after the great inventions that were made during the late Zhou and early Han periods in the fifth to first centuries B.C., new techniques continued to be introduced: puddling and solid state fining in the first to third centuries A.D., crucible smelting in perhaps the fourth century A.D., large-scale iron casting techniques and the coking of coal in the eighth or ninth century, and the reciprocating piston and fan bellows as well as the extensive application of water power at some point between the third and twelfth centuries. From a commercial point of view, the Chinese iron industry had other unusual features: it produced for a large and relatively prosperous population; producers could reach most of this population because of good communications and well-organized commercial institutions; and individual firms often became large, encouraged by government policies and the size of potential markets.

There may be a connection between an innovative attitude toward productive technology and commercial success, but—this would not at first seem to be the case—causation runs from the latter to the former, as the Chinese evidence suggests. There are a few other examples in the pre-industrial world with markets as large and communications as good, such as the Roman Empire, but they did not advance in ironmaking as the Chinese did. The size and wealth of ironmaking enterprises made China unique, which may be the real explanation for the difference. Just as the European and East Asian copper industries, Chinese ironworks could be large and profitable enough to attract major concentrations of capital, making it possible to experiment with new techniques.

Economies of scale are then of considerable importance, but the relationship between the size of furnaces and savings in fuel, ore, and labor costs is far from clear. Large furnaces do seem to have a higher output per unit of labor, but it cannot be shown that they are superior in terms of fuel to iron or ore to iron ratios. Furthermore, we have almost no information on one important set of variables, the real capital costs of inventory and plant.

Yet size, in furnaces and enterprises, has unquestionably been important in the recent history of ironmaking. It has been crucial to Western ironmakers since the late eighteenth century, a source of pride to owners and managers, a major goal to researchers, and to lay contemporaries the keystone of the Industrial Revolution. Many inventions have been made to increase furnace size, in the confident expectation that this will reduce the unit cost of iron. Managers of American steel corporations in the twentieth century have consistently preferred larger furnaces, but there is little good evidence that these produce significant savings in costs (Boylan 1975). Although a major reduction in the real cost of iron was effected between 1900 and 1970, Boylan demonstrates convincingly that only a modest decrease in labor costs can be definitely attributed to the enlargement of furnaces; the rest of the savings came from size-independent technical improvements in the use of energy and materials.

We do not maintain that Boylan's conclusions can be extended back to the beginning of the Iron Age. Scale economies have clearly sometimes been significant, and it is possible that the search for them has led to important if somewhat accidental technical advances—conceivably, the first Japanese tatara furnace and the first European blast furnace both were results of attempts to enlarge earlier types of bloomeries. Like their Victorian predecessors, modern students of ancient and non-Western ironmaking methods are usually impressed by bigness, although the evidence does not justify it. The real reason European iron eventually became inexpensive in world markets was not the use of giant blast furnaces but skill in employing inexpensive fuel and the development of efficient methods of converting cast iron to wrought iron and steel. As late as the 1920s, Tegengren found that cast iron made in Shanxi crucibles was sold in China for about the same price as blast furnace-made cast iron in Europe.

Moreover, some of the other pre-industrial smelting methods must not have been inefficient in comparison with modern blast furnaces and basic oxygen process converters. We are reasonably confident, for instance, that with certain modifications we could operate a traditional Catalan hearth at a profit. We also are among those who have begun to suspect that the giant blast furnace of the 1960s and 1970s is already a dinosaur and that the days of the indirect process itself are numbered. The future may belong to smelting devices that resemble pre-industrial ones: small furnaces which, as in traditional Africa, can smelt directly to steel or which, as in several parts of China, can feed frequently or continuously tapped cast iron into a finery-like converter to be turned to steel. As the front end of a continuous processing system like that of the modern mini-mill, such direct process smelting devices will, we think, eventually become the foundation of a steel industry that is radically transformed. In some ways these future smelting systems would be more familiar to a small seventeenth- or eighteenth-century ironmaker in Ariège, Bihar, or Togo than to the prophets of the first two centuries of the Industrial Age.

# Appendix A
# The Phase Relationships Involved In Ancient Iron Processing

Given here are selected phase diagrams that relate to discussions in the text. These diagrams represent equilibrium conditions which are not always met in practice. Nevertheless, they define the processes and transformations which will occur and the direction they will take.

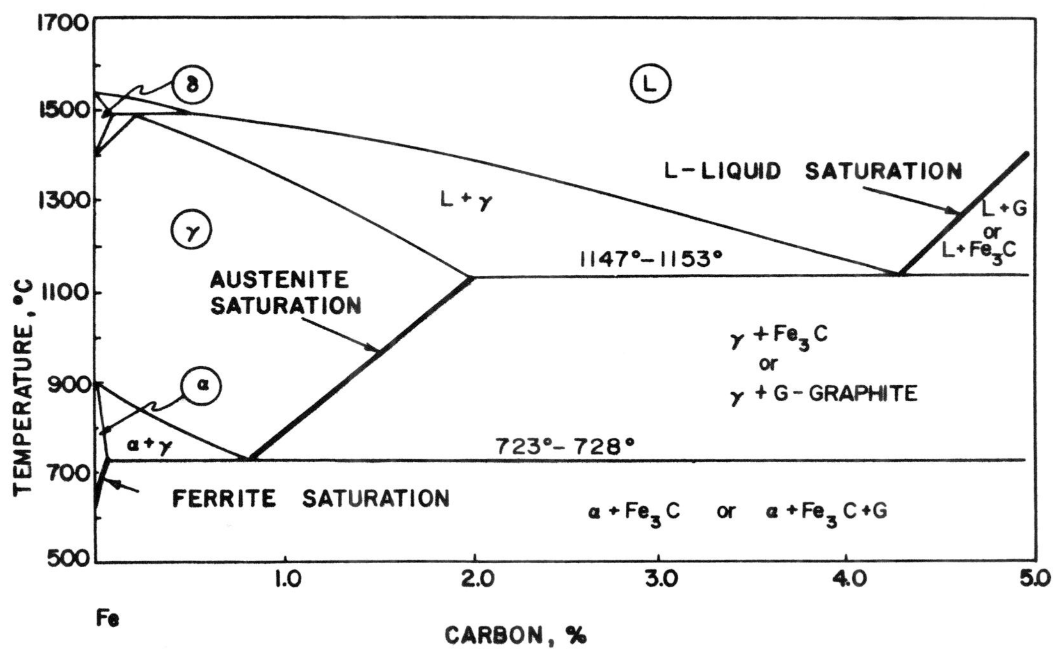


Fig. A.1. Reconstruction of the iron-carbon phase diagram, with heavy lines indicating the carbon saturation over the whole range of important temperatures. Note that these lines represent saturation in the presence of an infinite source of carbon. When the source of carbon is limited as in Figure A.2, the saturation is correspondingly less.

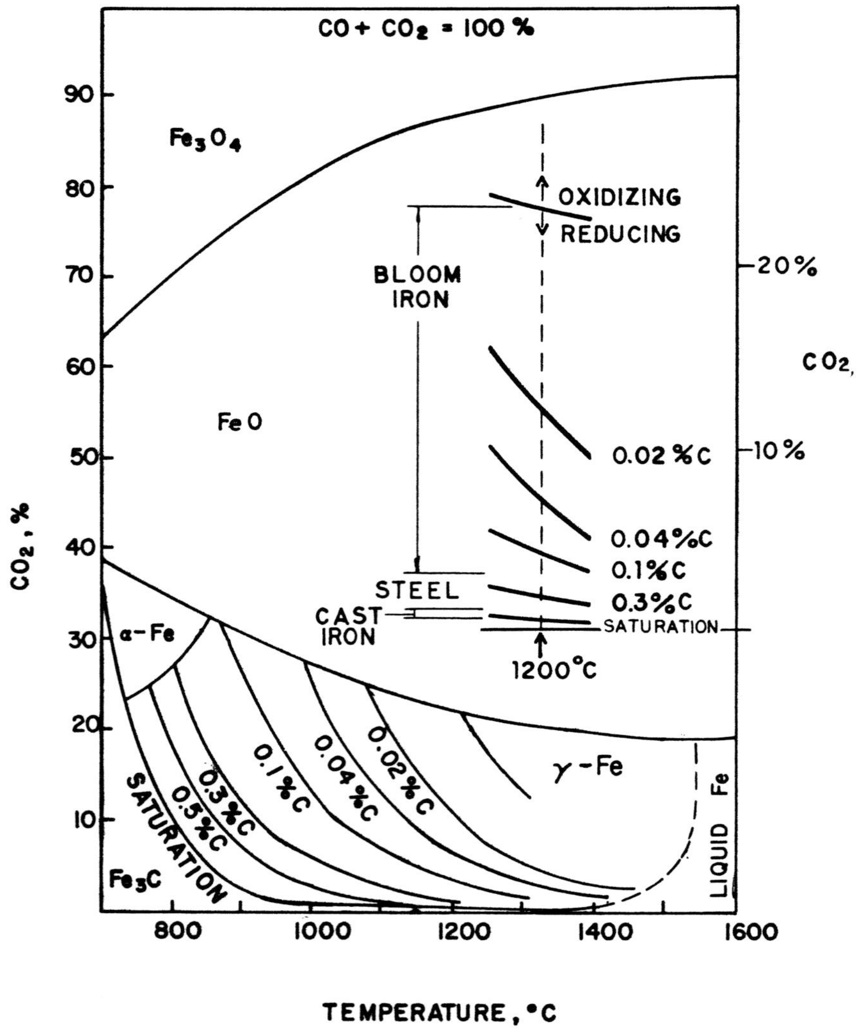


Fig. A.2. When the carbon supply is a mixture of carbon monoxide (CO) and carbon dioxide ($CO_2$) gases, the carbon saturation is reduced in proportion to the ratio of $CO/CO_2$. At a given ratio, a steel may gain or lose carbon according to the temperature in order to achieve an equilibrium with the gas atmosphere. Note that the percentage of $CO_2$ can be estimated from the relation: $1/1+R \times 100$ where R is the $CO/CO_2$ ratio. As an example of how to use this diagram, let us take a temperature of 900°C and an atmosphere of 5 percent $CO_2$. A steel containing 0.1 percent carbon will gain carbon up to about 0.6 percent, which represents saturation under the stated conditions. In comparison, a 0.8 percent carbon steel will lose carbon until the 0.6 percent saturation level is reached. The influence of the $CO/CO_2$ ratio is independent of dilution by a neutral gas such as nitrogen. (Adapted from *Metals Handbook* 1964, v.2)

The inset presents the equilibrium levels of carbon close to the temperature of 1200°C against an expanded scale of $CO_2$ content in the gas mixture of CO and $CO_2$. It demonstrates that under the reducing conditions of a shaft or blast furnace the conditions for direct smelting to steel are extremely narrow.

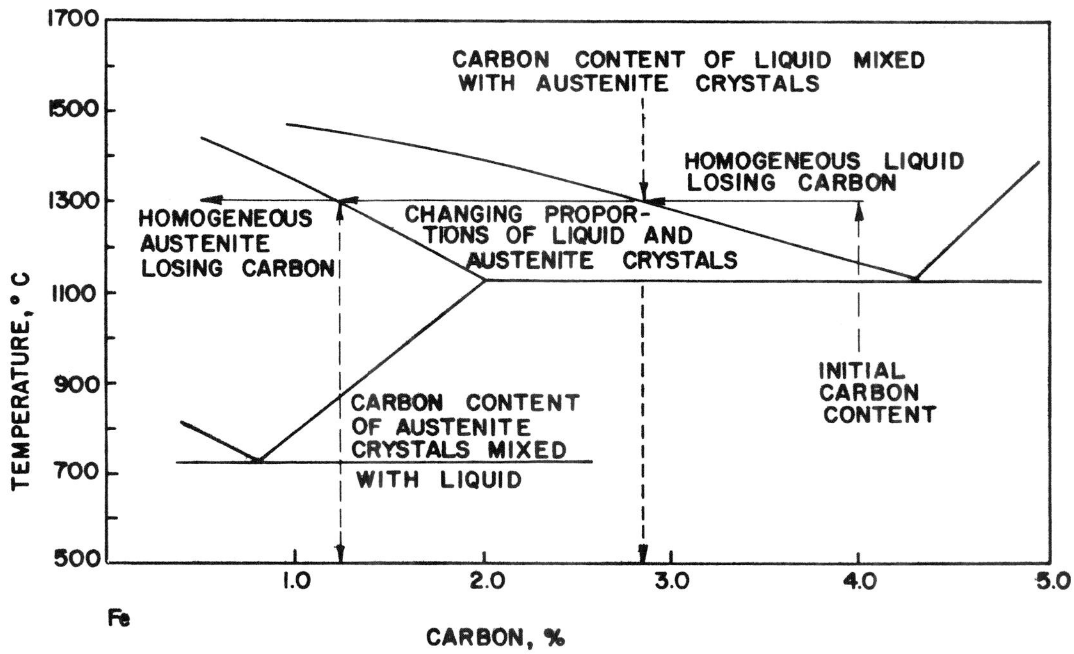


Fig. A.3. If a liquid iron of the carbon content indicated at 1300°C is covered with a flowing gas—either oxygen, air, water vapor, or carbon dioxide—the carbon in the liquid iron will react with the atmosphere and be progressively oxidized out of the metal in the form of CO or $CO_2$. When the carbon content in the liquid iron has been reduced to about 3 percent, crystals of the austenitic form of iron will form and will be suspended in the liquid. These crystals have a carbon content slightly above 1 percent. As carbon is continuously extracted, more austenite crystals form and the volume of residual liquid decreases. Ultimately all the liquid converts to these crystals and the pool of iron becomes a solid piece. Note that solidification can occur not only by cooling but also by change in composition. The solid can also continue to lose carbon in the presence of this atmosphere but at a slower rate. This is the basic principle of the fining or puddling of cast iron in which the sources of oxygen can also be liquid slag or solid iron ore. The carbon content of the iron crystals which form is also usually much lower because of the nature of the oxygen sources.

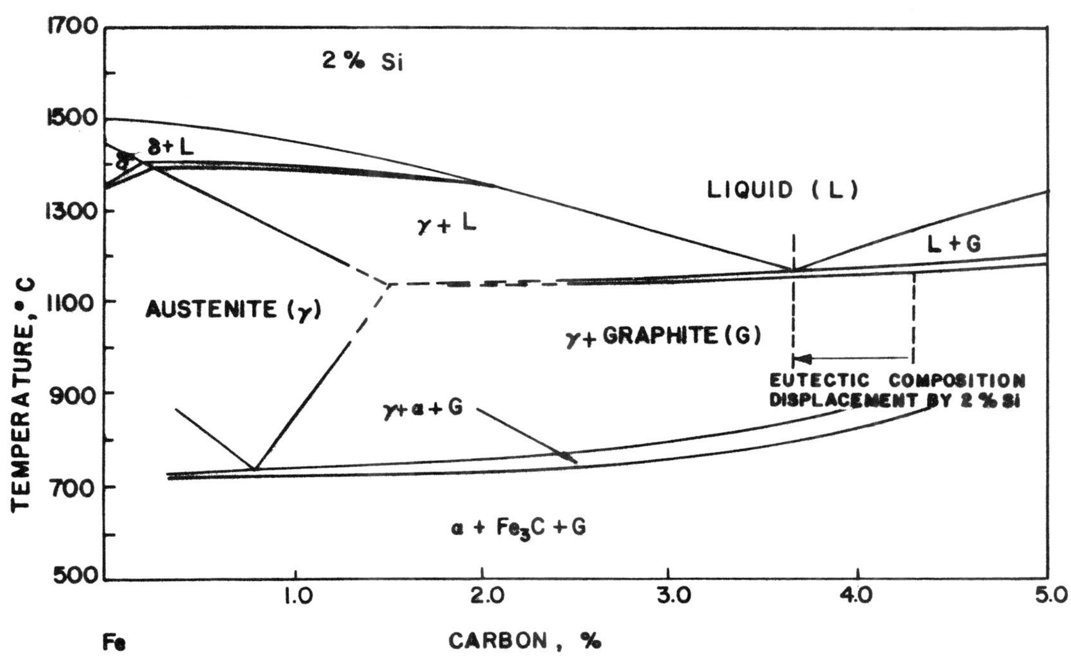


Fig. A.4. Reconstruction of the iron-carbon phase diagram with a constant level of 2 percent silicon. This involves only a few modifications of the simple iron-carbon diagram. One of these is the shift of the eutectic composition from 4.3 percent to about 3.6 percent. Another is that the liquid saturation curve beyond the eutectic composition is less steep. Most important is that only the crystallization of graphite occurs during eutectic freezing. The cast irons which this diagram represents solidify completely as gray iron. The suppression of the tendency to crystallize as white iron can be reversed by the addition of a fourth element such as sulfur. See Greiner, Marsh, and Stoughton 1933, and Massari 1938.

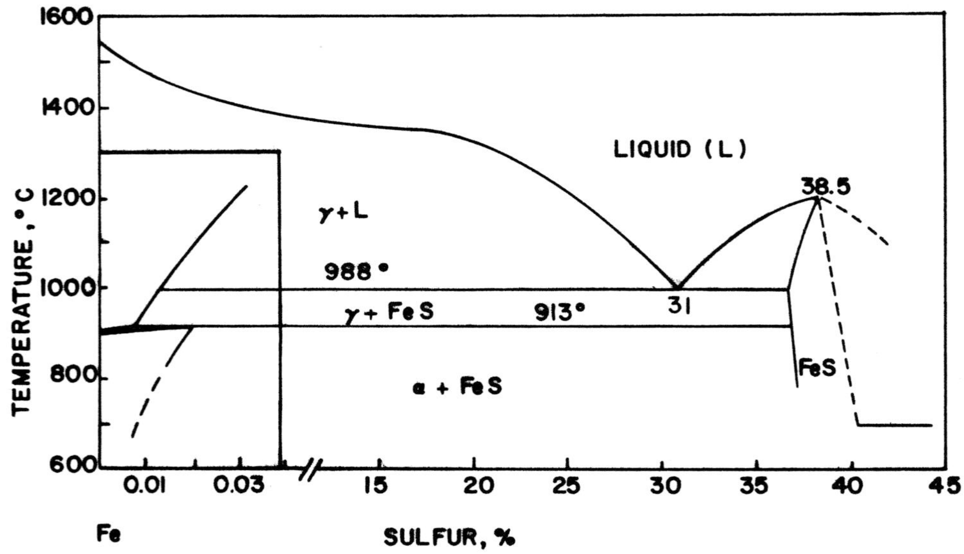


Fig. A.5. The phase diagram of the iron-sulfur system has some unusual and important features. Sulfur lowers the melting temperature of iron to as low as 988°C. The solid solubilities of sulfur in any of the forms of iron are very low. In the hot forging range—about 1000° to 1200°C—the solid solubilities of sulfur are only 0.01 to 0.03 percent. If the metal contains more than these limits, liquid forms between the crystals of the structure and makes the metal difficult to forge. (Adapted from Hansen 1958: 705 and McGannon 1964: 408)

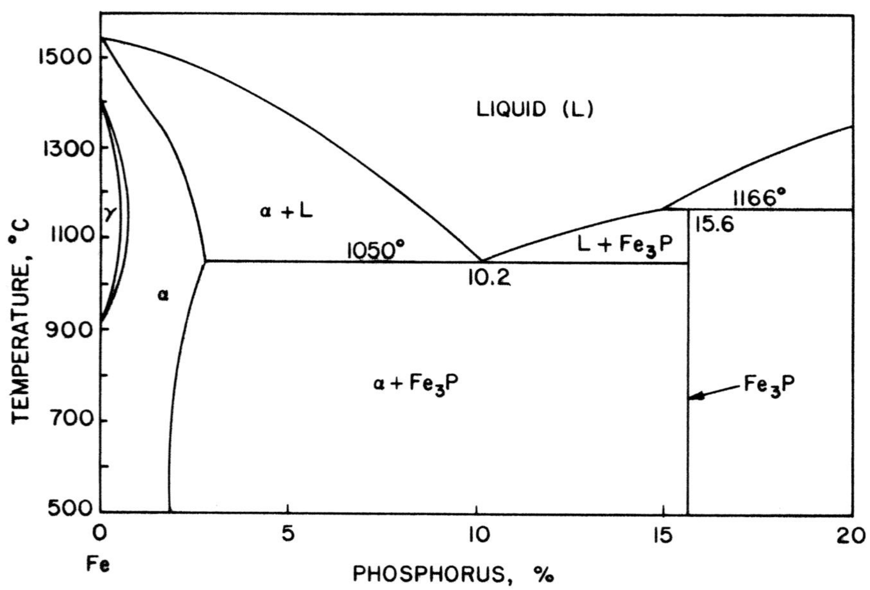


Fig. A.6. The phase diagram of the iron-phosphorus system, like that of the iron-sulfur system, shows that the melting temperature of iron can be greatly lowered by the addition of sufficient phosphorus. However, this does not affect the hot forgeability of iron because the solid solubilities of phosphorus in iron are at least 2.5 percent in the forging temperature range, and the phosphorus pickup during smelting is likely to be much less than 1 percent.

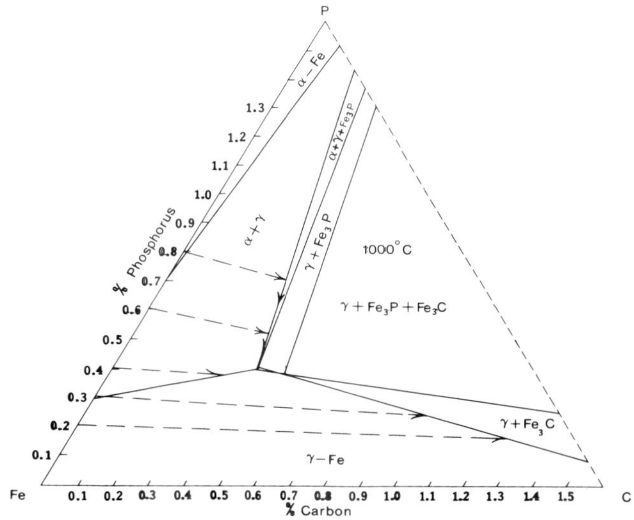


Fig. A.7. Hypothetical view of the alloy system iron-phosphorus-carbon at an isothermal section for 1000°C where carburizing might be performed. The dashed arrows indicate the the limits of carbon saturation imposed by different phosphorus levels. In all cases, reduced carbon solid solubility is to be expected when phosphorus is introduced into iron during smelting.

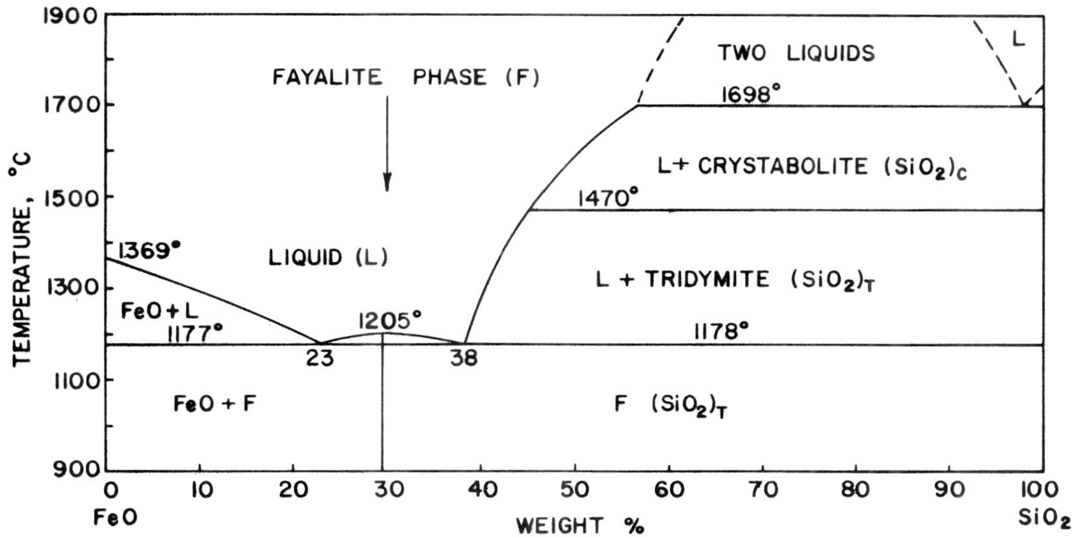


Fig. A.8. The wustite-silica phase diagram is fundamental to an understanding of bloom iron smelting, fining, and puddling. The silica which is the major rock constituent in iron ore is a very high melting substance. The addition of the suboxide of iron, FeO or wustite, lowers the melting temperature dramatically. There is a wide range of compositions between about 60 to 85 percent FeO where the melting temperatures are below 1250°C. When the slag composition is in this range, reduced iron metal particles can settle and aggregate to a bloom, while being protected from re-oxidation by the slag cover. When slags within this composition range solidify, they crystallize predominantly in the form of an interceramic compound, $2FeO.SiO_2$ or $Fe_2SiO_4$, designated as *fayalite*. These are commonly called fayalite slags although either glassy silica or crystalline wustite, FeO, co-exists in the solidified structure (Fig. 15.1). Actual bloomery slags contain considerable amounts of alumina, magnesia, and trivalent iron ions. The fayalite slag is also capable of oxidizing phosphorus out of liquid cast iron and absorbing it during the puddling operation. (Adapted from Levin, Robbins, and McMurdie 1964, and Morton and Wingrove 1969)

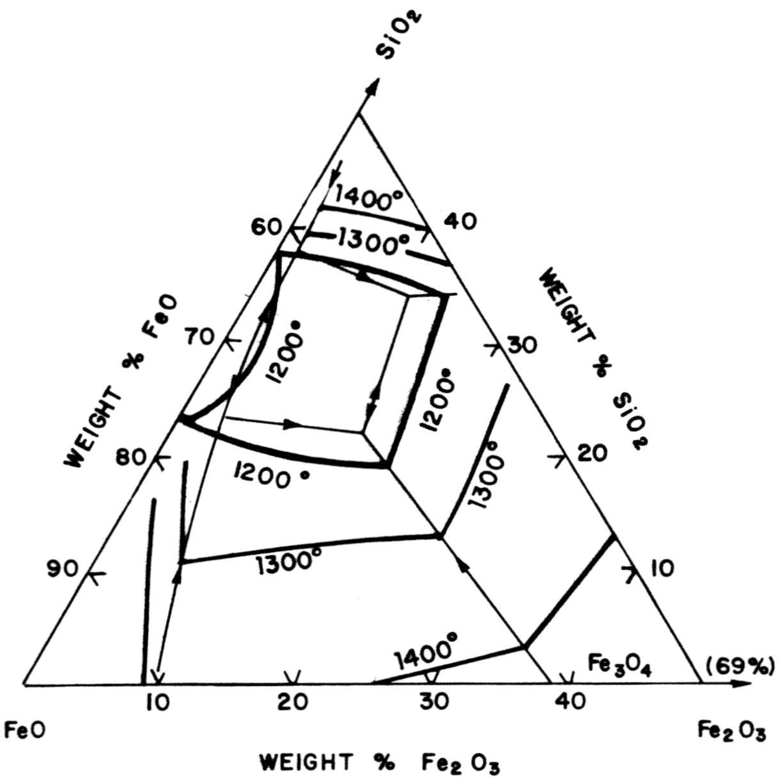


Fig. A.9. This portion of the ternary phase diagram shows that trivalent iron residual from incompletely reduced iron ore particles has an important influence in widening the range of slag compositions with very low melting temperatures. The triangular composition grid is used here to define a projection of the melting (liquidus) surface. This surface has valleys designated by the thin lines with arrows. The heavy isochronal lines are contours of constant temperature of melting. The loop identified as 1200°C encloses a slag region of compositions that melt below 1200°C. While the simple binary FeO-$SiO_2$ has narrow ranges of composition capable of such low melting temperatures, the addition of relatively small amounts of the higher oxides of iron make it much simpler to attain that desired state in a smelting operation. (Adapted from Muan and Osborn 1965: 161)

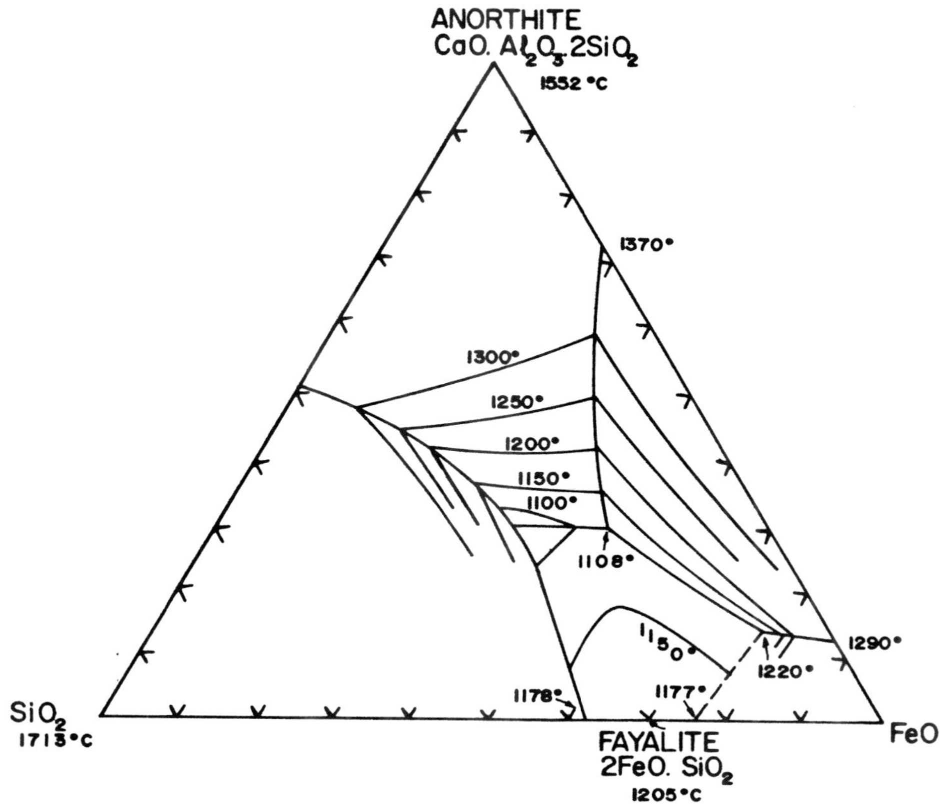


Fig. A.10. Real slags contain many more than three oxide components so that it becomes difficult to describe graphically the trends in melting temperatures. This is a segment of a quaternary phase diagram of the system silica-wustite-lime-alumina which is defined by an arbitrary triangle of compositions in terms of the components: silica, wustite, and an interceramic compound called anorthite ($CaO.Al_2O_3.2SiO_2$). It shows another zone of low melting slags formed from lime, silica, and clay, all of which can occur in iron ores or the ash from coal. (From Levin, Robbins, and McMurdie 1974: 288, fig. 869)

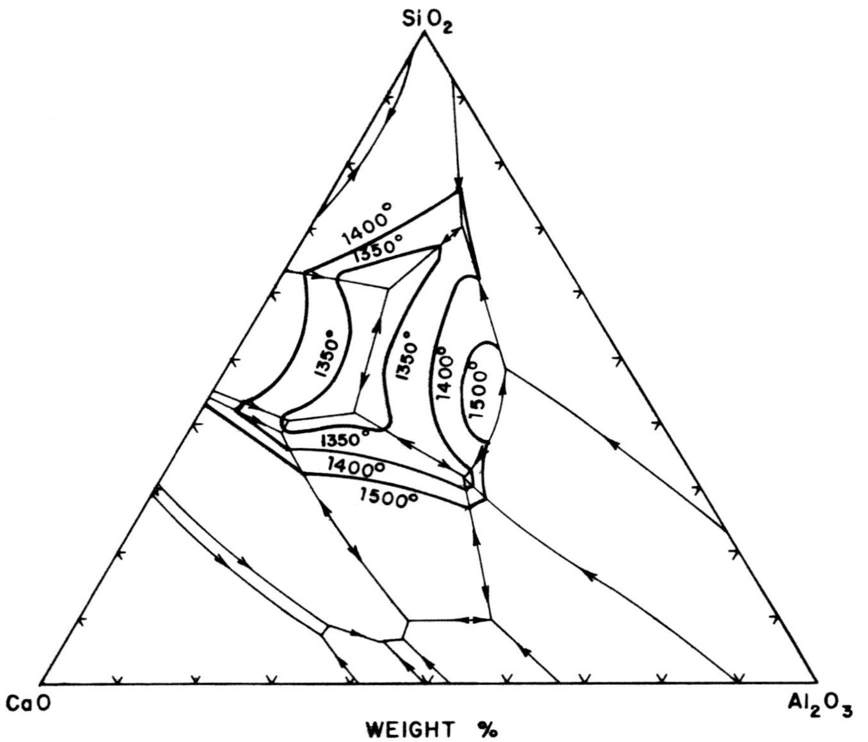


Fig. A.11. In the absence of a significant content of FeO, low melting slags can be achieved in a moderately broad range of compositions in the ternary system silica-alumina-lime. The melting surface (liquidus) is projected on the composition triangle and the zone of melting below 1350°C is outlined as a closed loop. This ceramic system is important because in the absence of FeO, these three oxides are the major components of the ore, coal, and charcoal that contribute to the slag composition. Real slags may have lower melting temperatures because of relatively small amounts of FeO, $K_2O$, and $Na_2O$. (Adapted from Muan and Osborn 1965: 95)

# Appendix B
# Settling Rates of Iron Particles in a Slag Pool

The settling or sedimentation velocity of (ideally spherical) solid particles in a fluid medium which is itself motionless is given by a form of Stoke's Law (Geiger and Poirier 1973: 71):

$$V_{max} = \frac{2D^2(\rho_S - \rho_L)g}{36\eta}$$

where

$V_{max}$ = maximum or terminal velocity of the settling particles, cm/sec
D = diameter of the particles assuming a spherical shape, cm.
g = gravitational acceleration, cm/sec/sec
$\eta$ = viscosity of the fluid, poises
$\rho_S$ = density of the solid particles, g/cc
$\rho_L$ = density of the liquid, g/cc

Note: $\eta$, $—_S$, $\rho_L$ are at the temperature of the settling process.

The actual settling rates may be considerably less if there are bubbles of gas rising through the fluid or if there are convection currents caused by thermal gradients. Settling rates are increased if the falling particles touch each other and join, as is to be expected when the iron particles are molten droplets.

Unfortunately, data on densities of molten slags are not readily available, so calculations of settling rates of iron in slag must be based on several assumptions. Let us assume therefore that:

solids density = 7.3 g/cc,
liquid density = 3.5 g/cc, and
particle diameter = 0.1 mm, which gives:
$V_{max}$ = $0.02/\eta$ cm/sec.

Using the isoviscosity lines for compositions in the ternary system $SiO_2$-CaO-$Al_2O_3$ at 1500°C (Machin and Yee as shown in Geiger and Poirier 1973: 28), we see that blast furnace slags would have viscosities as low as 2.5 poises and as high as 100 poises. If the slag pool is 6 cm. deep, this range of viscosities would provide the following maximum settling velocities and times:

| $\eta$, poises | $V_{max}$, cm/sec | Time, sec (6 cm) |
|---|---|---|
| 2.5 | $8 \times 10^{-3}$ | 750 |
| 5 | $4 \times 10^{-3}$ | 1,500 |
| 10 | $2 \times 10^{-3}$ | 3,000 |
| 20 | $1 \times 10^{-3}$ | 6,000 |
| 40 | $5 \times 10^{-4}$ | 12,000 |
| 80 | $2.5 \times 10^{-4}$ | 24,000 |

These figures indicate that a slag pool should be kept shallow by frequent draining and at a viscosity of less than about 20 poises.

The control of viscosity rests both with slag composition and temperature. The basic idea is that the furnace temperature must be above that of the liquidus for a good slag to form. Below the liquidus, even a sparse suspension of crystallized particles will radically increase viscosity. This is demonstrated by comparison of viscosity measurements with a vertical section of the ternary phase diagram (Fig. 8.1).

# Appendix C
# Iron Meteorites

Metallic meteorites have characteristic features. They are iron-nickel alloys with nickel contents that range from slightly below 5 percent to slightly above 35 percent, with probably 90 percent having between 5 and 15 percent (Fig. C.1). They have transformation structures of the kind caused by very slow cooling and solidification: giant crystals of the high temperature austenite phase from which the ferritic phase has subsequently precipitated as massive plate-like crystals, forming a lattice that cross hatches each original crystal of the high temperature phase. These structures are so large that they may be seen, with proper polishing and etching, without a microscope. Similar structures in man-made alloys are only visible at magnifications of at least 100x.

Iron meteorites consistently contain other elements. Those present in percentages from 0.01 to 1 (rarely above 1) include phosphorus, cobalt, carbon, and sulfur. Elements present in lower concentrations measured in parts per million (ppm) include copper, zinc, gallium, germanium, iridium, and platinum. Still other elements have been detected in amounts of less than one ppm. The many analyses given by Buchwald (1975: v.1–3), for example, show that phosphorus content ranges from 0.02 to 0.9 percent (Fig. C.2), and is systematically correlated with nickel content (Fig. C.3). Cobalt content usually lies between 0.3 and 0.8 percent (Fig. C.4) but—unlike phosphorus—has no apparent correlation with nickel content. Carbon content is generally very low. At least 80 percent of iron meteorites have less than 0.03 percent carbon, and very few have more than 0.1 percent (Moore, Lewis, and Nava 1969).

Because most iron meteorites are cold and hot forgeable, they could have been used by man before techniques for smelting iron had been developed, and many are known to have been. Unless forging and heating have been minimal, however, the structure of meteoritic iron is not distinguishable from that of a smelted or melt-alloyed product of the same composition. Yet making this distinction is a matter of considerable historical importance.

Nickel content is a clue. Because no iron meteorite contains less than about 5 percent Ni, a metal sample in good condition with a lower Ni level is probably not meteoritic. Similarly, few smelted Fe-Ni ores could yield an alloy with more than a few percent Ni, so a sample with a Ni content greater than 10 percent is probably meteoritic.

Contrary to what has sometimes been claimed, meteoritic iron cannot be securely identified on the basis of the presence or absence of Co, P, S, or Cu. Natural alloys made from ores can contain Co, and meteoritic

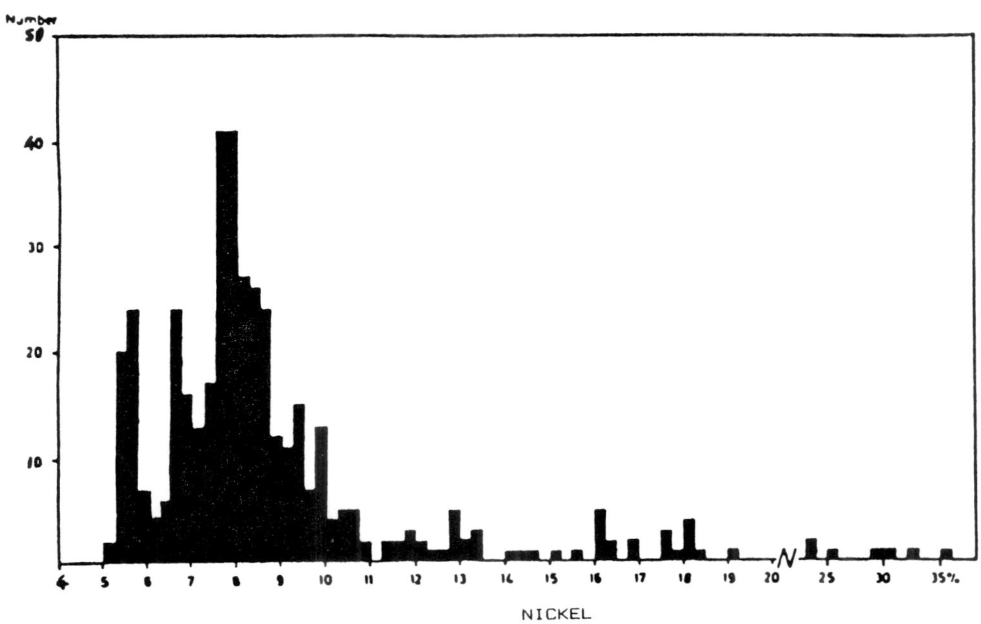


Fig. C.1. Frequency (ordinate) of occurrence of nickel levels (abscissa) in iron meteorites. No iron meteorite contains less than 5 percent Ni, and only seven contain more than 20 percent (Buchwald 1975, v.1: 77).

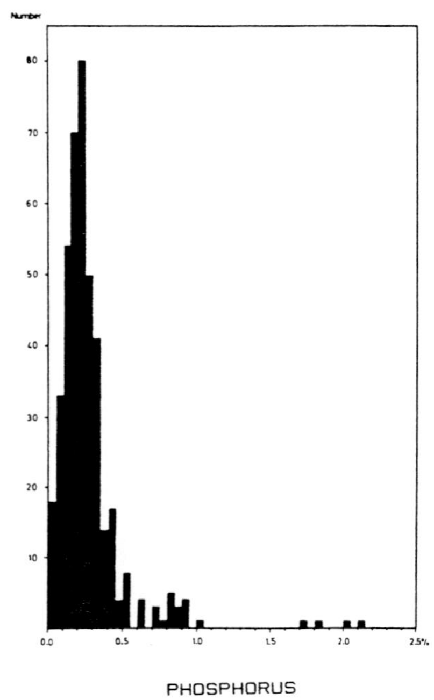


Fig. C.2. Frequency of occurrence of phosphorus in iron meteorites (Buchwald 1975, v.1: 80)

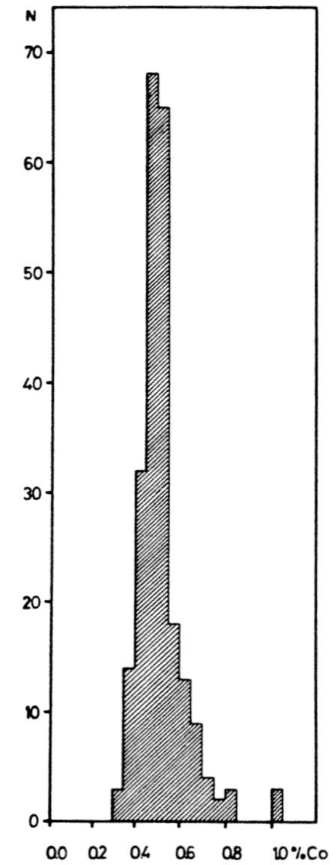


Fig. C.4. Frequency of occurrence of cobalt in iron meteorites (Buchwald 1975, v.1: 78)

iron often includes moderate amounts of P, S and Cu. However, there are two elements that occur systematically in minor levels in iron meteorites but are often absent from terrestrial iron minerals and their smelted products: germanium (Ge) and gallium (Ga). The average value of the former in meteoritic iron is 150 ppm and of the latter, 50 ppm (Buchwald 1975, v.1: 86). The absence of these elements from some terrestrial deposits is shown by analyses of several North American smelted nickel samples in which there were as many as sixteen elements at the level of 100 ppm or less, but none of them were germanium or gallium (Rengstorff 1966: 31). Admittedly, there may be nickel or iron ores in some regions that contain Ge and Ga. However, enough data exist to justify the staement that an iron-nickel alloy artifact lacking germanium and gallium in detectable amounts is probably not of meteoritic origin.

The feasibility of producing a smelted iron-nickel alloy in simple furnaces from nickeliferous iron ore is now

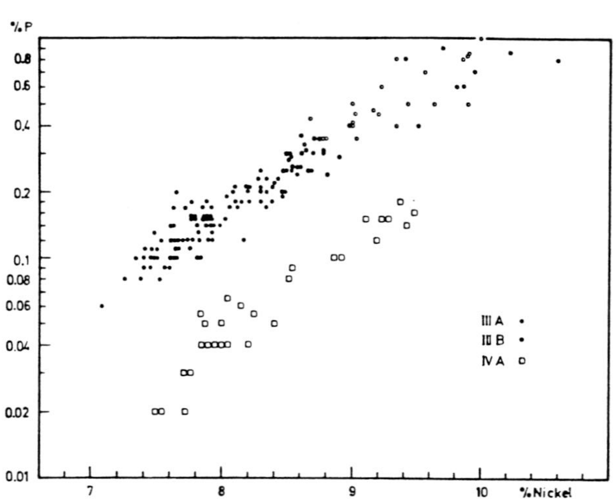


Fig. C.3. Correlations between the nickel and phosphorus contents in iron meteorites (Buchwald 1975, v.1: 81)

*Table C.1. Nickeliferous ores exploited in historical iron producing regions*

| Region | Nature of ore | Source |
|---|---|---|
| Greenland | | |
| (Disko Island) | native iron* | BVK(1): 341** |
| Norway | | |
| (Evje) | nickel-pyrrhotite | BVK(1): 281, 294 |
| Sweden | | |
| (Klefva) | " | BVK(1): 297–298 |
| Italy | | |
| (Piedmont) | " | BVK(1): 298 |
| Saxony/Bohemia | " | BVK(1): 299 |
| Hungary | | |
| (Dobschau) | nickel arsenides/siderite | BVK(2):947 |
| Switzerland | | |
| (Valais) | " | BVK(2): 949 |
| Silesia | | |
| (Frankenstein) | garnierite | BVK(2): 958 |
| Greece | | |
| (Larymna) | laterites | Boldt: 74*** |
| Serbia | | |
| (Golesh mtns) | " | Boldt: 75 |
| Indonesia | | |
| (Sulawesi-Celebes) | " | Boldt: 65 |
| Philippines | | |
| (Mindanao) | " | Boldt: 67 |
| Russia | | |
| (Urals, Ukraine) | " | Boldt: 69 |

* = with high carbon content.
**BVK = Beyschlag, Vogt, and Krusch 1914–16 (2 vols.)
***Boldt = Boldt 1967

well established (Piaskowski 1960, 1969–70; Hermelin, Tholander, and Blomgren 1979). Nickel-bearing minerals capable of conversion to high nickel irons include pyrrhotite and pendlandite (nickel sulfides), garnierite (an iron-nickel-magnesium-silicate), and nickelite (a nickel arsenide). Ores containing all these minerals have been exploited for their nickel content in recent times (Table C.1), and some of them are in areas where they could have been used in antiquity. The formation of an alloy is reasonably likely to occur if the mineral contains both nickel and iron, less so if nickel mineral is admixed with iron mineral. The product of the latter may be a physical mixture of nickel and iron metal particles in which only limited alloying by diffusion occurred during the smelting and forging operations.

The garnierites tend to be associated with laterites of an intermediate but sometimes high iron content. Such ores, which would behave like ordinary iron ores in the furnace, are believed to have been smelted during the eighteenth and nineteenth centuries by traditional iron-makers in Sulawesi, Indonesia (Bronson 1987). The nickel-pyrrhotites are often admixed with copper sulfides such as chalcopyrite. If the copper content of such an ore is high, it is an unlikely ancient source of iron-nickel alloys. Others, however, could conceivably be exploited for their iron content if the sulfides in them were first dead roasted to oxides.

Nickel arsenides too can be dead roasted to oxides of iron and nickel, which can then easily be reduced to metal (Gowland 1921: 546–547). Piaskowski (1982a) reports that arsenic contents from 0.24 to 1.15 percent were identified in the iron-nickel alloy bands in the composite structure of an ancient socketed axe from Poland, and that nickel arsenide-siderite ores exist in Hungary (Table C.1). If such ores had been smelted, residual arsenic from incompletely roasted ore particles would have entered the metal. The presence of as much arsenic as Piaskowski reports is a certain indicator of terrestrial origin. According to Buchwald (1975, v.1: 84), the arsenic content of iron meteorites is always less than 0.003 percent.

# Appendix D
# Frederick Overman — An Appreciation

Blast furnaces, fining operations, forges, and foundries were all established in the American Colonies by the middle of the seventeenth century. These operations were run by ironmasters who were really craftsmen—they learned by example and experience and only rarely innovated. Certainly none of them was sufficiently educated and even knowledgable enough to write about contemporary technology. In this context, Frederick Overman must be regarded as the first American metallurgist.

In the remarkably short time of less than five years, Overman wrote six books, each rich in the details of existing technology in the United States around 1850. He had first-hand experience with all aspects of ironmaking and a strong background in contemporary European ironmaking, so he could and did weave comparisons with American technology into his texts. His books are a major resource for early American ironmaking practices, but they are not widely recognized. Overman himself is almost unknown to metallurgists today, although judging by the number of reprintings that two of his books enjoyed, he must have been frequently consulted in his own time (see his bibliography at the end of this Appendix).

In 1869, Henry Stafford Osborn (who was the first Professor of Mining and Metallurgy at Lafayette College, in Easton, Pennsylvania) published a 972-page treatise on *Metallurgy of Iron and Steel.* In his preface, Osborn recounts that his original intention was to re-edit Overman's *Manufacture of Iron* and bring it up to date. He then stated that so much had happened in the intervening eighteen years that extensive revision was necessary. Nevertheless, a large part of Osborn's book was direct quotation from Overman (not always acknowledged).

Overman must have been a remarkable person as the preface to *A Treatise on Metallurgy* by John A. Roebling attests. Roebling was the distinguished civil engineer who designed machinery for the production of wire rope and cable, designed the Brooklyn bridge with its steel cable suspension system, and founded J.A. Roebling & Sons to manufacture steel cable. Roebling's long and sometimes emotional tribute to Overman tells us that he was born in Germany in the town of Elberfeldt and died in Philadelphia on 7 January 1852 at the age of 49 of arsene poisoning contracted during laboratory analytical procedures. The *Dictionary of National Biography* (v.7, 1934) notes that had he survived many more years he would have been a "leading figure in the development of metallurgy in the United States."

He came of "humble" parents who could give him only an elementary education. He was first apprenticed to learn a mercantile trade, but on his release he became an apprentice to a cabinet maker. He went to Berlin to work and attracted the attention of the superintendent of the new and unique Royal Polytechnic Institute. He was enrolled there as a student for four years and also attended lectures at the Royal University.

Following graduation, he entered industry where he was responsible for several process improvements in ironmaking, particularly in puddling. He was in charge of plant design and operations of a chemical works at Chemnitz in Saxony, and for a while was employed by the Austrian government under the personal direction of Prince Metternich for projects on national planning. He married in 1829 (*DAB*).

From Ludwig Beck's monumental *Geschichte des Eisens* (1891–99), we learn that Overman in 1838 wrote his first book, *The Fining of Pig Iron*, which was published in Brunn (now Brno, Czechoslovakia, but then in the Austro-Hungarian Empire). Further, he was managing director for the firm of Blansko in Mahren. In his citation of *Manufacture of Iron*, Beck identifies the place of publication as London which shows that his subsequent works were also available to a European audience.

Apparently Overman developed a distaste for the political climate of central Europe and emigrated to the United States in 1842. There can be no doubt of his genuine affection for his adopted country for he writes about "we" as himself and other Americans whereas "they" are the Europeans—Germans, British, and Swedes. He appears to have found employment as a consultant in American ironworking industries, and the details of his writings show that he must have been familiar with works as far west as Pittsburgh. He devoted the last three or four years of his life primarily to the writing of books. Retailed in Europe by the publisher H.C. Baird of Philadelphia, his *Manufacture of Iron* was reviewed glowingly by the *London Mining Journal*.

Roebling's tribute comes across as a combination of respect, admiration, and affection, although he alludes to some human frailties. Apparently, Overman was likely to be blunt and therefore probably managed to alienate some people. However, his first-hand knowledge of so many details of operations suggests that he got along well with operators of iron and steel plants at the floor level.

He had an uncommon scope of expertise. In the *Manufacture of Iron*, he writes authoritatively on geology, mining, smelting, refining, and metalworking. In the *Manufacture of Steel*, he describes heat treament and provides critical commentary on the details that only one who had done or supervised such procedures could know. His *Treatise on Metallurgy* deals heavily with the geology, mining, and smelting of nonferrous metals.

A matter of some curiosity is his relative lack of recognition by later writers on the subject of the metallurgy of iron and steel. Apart from H.S. Osborn and the *London Mining Journal*, he received no accolades although his books on iron and steel went through several printings and revisions. John Percy, who wrote the justifiably famous treatise *Metallurgy: Iron and Steel* in 1864, does not mention his name although the text is replete with many others. J.M. Swank in his *History of the Manufacture of Iron in All Ages (and particularly in the United States)*, first published in 1884, makes no mention of him. However, in the second edition (1892), Swank included five references to him. He is not mentioned in Tylecote's *History of Metallurgy* (1976), C.A. Sanders and D.C. Gould's *History Cast in Metal* (1976), or K.C. Barracough's *Blister Steel* (1984). There are two short references in T.A. Wertime's *The Coming of the Age of Steel* (1962) and another in Barraclough's *Crucible Steel* (1984).

A comparison between Percy's and Overman's styles is striking. Percy was very much the university scientist. Although he visited many iron and steel plants, his descriptions were those of a visiting observer culling bits of interesting facts to rationalize and weave into an emerging understanding of metallurgical fundamentals. Overman, on the other hand, was the consummate engineer—strong on operational details, willing to use contemporary science but in the last analysis content with the empirical approach. Both had impressive analytical abilities, but Overman comes to strong positive, sometimes dogmatic conclusions appropriate to the demands of decision-making whereas Percy very often took an ambiguous position as befits one who was conscious that the facts were not all in place and that final judgment should be reserved until more information was available.

Overman's books were written at a time of impending transition in steelmaking. The Industrial Revolution had begun in Britain but was by no means in full swing. In much of the rest of Europe and in the United States it was only beginning. Steam power had not completely replaced water power; mineral coal and coke had not by any means displaced charcoal. Even the ancient bloomery, in a sophisticated form, was still in use because its bar iron was superior in many ways, and so commanded a premium price.

In the United States, European developments were adopted at a judicious pace. The vast forests made the continued use of charcoal for smelting and fining feasible and desirable. Anthracite, for much of the first half of the nineteenth century, was the major mineral coal used. The development of the hot blast in conjunction with the use of anthracite for blast furnace operation was probably as much an American as a Welsh accomplishment. Puddling was not as quickly adopted in the United States as it was in Britain and Germany. However, the bar iron produced in the United States was often deficient in ductility, both cold and hot. It was still necessary to import premium grades from Sweden and Russia. In fact there was some need for protectionism to safeguard the local iron industries.

As Overman pointed out, it was in the growing recognition of the wide advantages of the use of steel that American industries were most deficient. The tonnage conversion of simple bar iron to medium and high carbon steel by the blister steelmaking process grew slowly. In particular, the melting and casting of very high carbon steels using blister steel in the crucible steel process constituted a small industry beset by poor quality control with the result that this expensive material was being largely imported.

These were the process and product that most exercised Overman. He argued that the potential for American self-sufficiency was limited largely by the ignorance of the steelmakers themselves. His book *Manufacture of Steel* was in many ways a call for national progress as well as a technical treatise. He wrote with obvious persuasion that all the resources for steelmaking were available in the United States except the expertise.

To read Overman's *Manufacture of Steel* is to experience the scientific ferment of the day expressed in the contemporary idiom. The ideas of the atom and the atom assemblies of the solid state were emerging with the normal degree of confusion attendant on very new concepts. It was appropriate for a thoughtful engineer such as Overman to try to fit them to the facts about steel as they seemed to be and test their ability to rationalize and predict. Matters have not changed all that much.

Following are the titles of his books: *The Manufacture of Iron* published first in 1850 (the third reprinting was a revision published in 1854), ca. 500 pages; *Practical Mineralogy, Assaying and Mining* (1851); *The Moulder's and Founder's Pocket Guide* (1851), 252 pages; *Mechanics for the Millwright, Machinist, Engineer, Civil Engineer, Architect and Student* (1851); *Manufacture of Steel* (1854), 226 pages; *A Treatise on Metallurgy* (1852—this text enjoyed at least six reprintings), ca. 700 pages.

## *APPENDIX E* Benchmarks in Ironmaking

Despite our resolve not to write a history of iron and ironmaking, the complexity of events and the lack of a chronological perspective in this book require some ordered historical view. For the most part, we have made no judgments about the validity of dates. A lack of notations for large spans of time can be explained alternatively by the absence of literature known to us or of trends in technology.

| | EASTERN MEDITERRANEAN (Greece, Anatolia, Levant, Cyprus, Egypt) | MIDDLE EAST (Iran, Iraq, Arabian Peninsula, Transcaucasia) | SOUTH ASIA (India, Pakistan, Bangla Desh, Sri Lanka) | FAR EAST (China, Korea, Mongolia, Manchuria) | WESTERN MEDITERRANEAN (Italy, Spain, and North Africa) | |
|---|---|---|---|---|---|---|
| **1500 B.C.** | Iron a precious metal reserved for ornaments and 1500 B.C.—prestige gifts[1] | | | | | **1500 B.C.** |
| **1000 B.C.** | Early use of iron for knives, daggers, and swords although bronze still the most common choice[2]; iron bars (spits) used in Greece for currency[3]; after ca. 700 B.C., iron weapons, tools, implements, nails became relatively common[4]; large slag piles indicate major smelting activity in Nubia from 6th c. B.C. to ca. A.D. 320 | Iron artifacts, daggers, spear points, ornaments, tools, implements in western Iran—Azerbaijan, Kurdistan[9]; evidence of iron manufacture in southern Arabia[10]; Persian bloom iron produced in bowl furnaces using either natural or forced draft; natural draft forge preheating furnaces[11] | Iron enters the material culture of India diffusing from Persia and Media[14] | Earliest indications of substantial ironmaking in China, primarily cast iron,[20] using a shaft furnace[21] | Egypt uses iron technologies from Greece, Assyria; Etruria and Southern Italy produce iron; Phoenician and Greek settlements in North Africa and Spain use iron[40] | **1000 B.C.** |
| **500 B.C.** | Carburization of bloom iron, quench and temper hardening of steel seen in artifacts[5] | | Large numbers and varieties of iron artifacts—tools, implements, utensils—suggest availability; products mostly from low carbon bloom iron.[14] Crucible steel (wootz) process probably developed toward the end of the pre-Christian era | Production expansion in China sufficient for widespread use of cast tools, utensils, hardware[22] Rapid expansion of Chinese iron production over the span of the Han dynasty using shaft and crucible | Celtic iron smelting technology brought to Spain[41] | **500 B.C.** |
| **B.C./A.D.** | Indian wootz steel supplied to Persia to fabricate ornate sword blades[6] | | Iron production sufficient for an export trade in finished products; iron pillars consolidated from small blooms to weights of 6 tons[15] | Furnace smelting[23] with water-powered air supply.[24] Manufacturing techniques in China included stack molds,[25] iron permanent molds,[26] solid state decarburization of white cast iron.[27] Cast iron technology diffuses from China to Korea[28] | Spanish iron noted for weapons and major source for Roman Empire[42] | **B.C./A.D.** |
| **A.D. 500** | | Wootz steel crafted in Persia with damascene etch patterns[12] | Sinhalese (Sri Lanka) technology includes carburizing and quench hardening of tools[16]; wootz and Sri Lankan steel exported to Arabia and Persia[17] Iron smelting remains as multiple small furnace smelting to bloom iron until the 19th century[18] Indian iron and steel export continues to the 19th century[11] | Iron technology diffuses from China and Korea to Japan, earliest sword making[29]; smelting in bowl furnaces[30]; inlaid iron[31] | Berbers of North Africa continue primitive bowl smelting[43] | **A.D. 500** |
| **A.D. 1000** | Damascus steel, named for the distribution point, appears in Europe[7] | Wootz steelmaking process copied or duplicated on the Arabian peninsula[13] | | Industrial use of coal in China[32]; cast iron technology diffuses from China to Manchuria[33] and Mongolia[34]; annual cast iron production in China reaches 150,000 tons[35]; Japanese samurai sword decor and styles begin[36] | Milanese etched and engraved ironwork acclaimed[44]; Spanish iron and steel exported to France, England, Low Countries[45]; cast iron smelting, water powered bellows in Italy[46] | **A.D. 1000** |

*APPENDIX E* **Benchmarks in Ironmaking** (Continued)

| | **EASTERN MEDITERRANEAN** (Greece, Anatolia, Levant, Cyprus, Egypt) | **MIDDLE EAST** (Iran, Iraq, Arabian Peninsula, Transcaucasia) | **SOUTH ASIA** (India, Pakistan, Bangla Desh, Sri Lanka) | **FAR EAST** (China, Korea, Mongolia, Manchuria) | **WESTERN MEDITERRANEAN** (Italy, Spain, and North Africa) | |
|---|---|---|---|---|---|---|
| **A.D. 1500** | Ironmaking technology static until blast furnaces and fineries adopted in Turkey[8] | Ironmaking technology essentially static until the modern era | | Carburizing wrought iron with liquid cast iron[37]; iron technology in China essentially static into the 19th century; invention of the box bellows[38]; Japanese tatara smelting system predates late 18th century[39] | Catalan forge with trompe air blast operating in Spain[47]; Brescian steelmaking process in Italy[48] | **A.D. 1500** |
| **A.D. 1600** | | | | | | **A.D. 1600** |
| **A.D. 1700** | | | | | Spanish iron widely exported[49] | **A.D. 1700** |
| **A.D. 1800** | | | | | First blast furnace in Spain[50] | **A.D. 1800** |
| **A.D. 1900** | | | | | | **A.D. 1900** |

[1]Pleiner 1980; Waldbaum 1980: 69–98; [2]Pleiner 1969; Snodgrass 1980: 335–374; [3]Snodgrass 1980: 335–374; [4]Pleiner 1969; Snodgrass 1980: 335–374; van der Merwe 1980: 463–506; [5]Pleiner 1969; Williams and Maxwell-Hyslop 1976; Wheeler, Muhly, and Maddin 1976; [6]Maryon 1960: 63; [7]Belaiew 1918; [8]Beck 1891, v.1: 288; [9]Pigott 1980: 417–462; [10]van der Merwe 1980: 475; [11]Hulme 1937–38; [12]Allan 1979; [13]Cahen 1947–48; [14]Pleiner 1971; [15]Graves 1912; [16]Hadfield 1912; [17]Allan 1979; [18]Holland 1892; Buchanan 1807, v.1: 174; Hoey 1880; [19]La Touche 1917–18: 271; [20]Needham 1958: 1–4; [21]Liu 1978; [22]Needham 1958: 5–10; [23]Cheng 1978; Yang 1982: 72; Hartwell 1967; [24]Liu 1978; [25]Hua 1983; [26]Zhengzhao 1978; [27]Li 1975; [28]Sasaki 1982; [29]Tanimura 1980; [30]Nishio 1913; [31]Rein 1889: 434; [32]Hartwell 1967; [33]Len'kov 1974; [34]Terekhova 1974; [35]Hartwell 1966; [36]Tanimura 1980; [37]Sun and Sun 1966: 250; [38]Liu 1978; [39]Lyman 1879; [40]Snodgrass 1980: 335–374; [41]Pleiner 1980: 382; Forbes 1956; [42]Scrivenor 1854; Schubert 1957: 54; Swank 1892: 16–17; [43]Wertime 1962: 83; [44]Schubert 1957: 302; [45]Schubert 1957: 110; Crossley 1981: 35; Swank 1884: 16–17; [46]Wertime 1962: 105; [47]Wertime 1962: 84; [48]Smith and Gnudi 1942: 67–70; [49]Scrivenor 1854; Alzola 1896; Wertime 1962: 101; [50]Alzola 1896; Chilcote 1968: 14–17; [51]van der Merwe 1980: 463–506; Tylecote 1975; [52]Tylecote 1975; [53]van der Merwe 1980: 463–506; [54]Cline 1937; [55]Bellamy 1904; [56]Kolchin 1953; [57]Piaskowski 1960; [58]Kolchin 1953; [59]Crossley 1981: 33; [60]Wertime 1962: 79, 91–95; [61]Johannsen 1953: 267–269; [62]Rubstov

*APPENDIX E* **Benchmarks in Ironmaking** (Continued)

| | **SUB-SAHARAN AFRICA** | **EASTERN EUROPE** (Poland, Czechoslovakia, Hungary, Balkans, Soviet Union) | **CENTRAL AND NORTHERN EUROPE** (Germany, Austria, Scandinavia) | **WESTERN EUROPE** (British Isles, France, Low Countries) | **NORTH AMERICA** | |
|---|---|---|---|---|---|---|
| **1500 B.C.** | | | | | | **1500 B.C.** |
| **1000 B.C.** | | Bowl furnaces producing 2 kg. blooms in Ukraine[56] | Small bloomeries in Celtic Austria[71] | | | **1000 B.C.** |
| **500 B.C.** | Ironmaking begins in West Africa without any prior bronze technology[51]; short shaft smelting furnaces[52] | Bloom iron smelting, pattern welding, steelmaking in Poland[57] | Spread of ironmaking technology with eastern and western migrations of the Celts[71] | Bloom iron technology diffuses from Central Europe[87] | | **500 B.C.** |
| **B.C./A.D.** | Ironmaking diffused both east and south across Africa by A.D. 300[53] | Bloom sizes 5 kg. in Ukraine[58]; iron smelting in Moravia[59] | Further spread of bloom iron technology to Scandinavia and Eastern Europe[71]; centralization of iron production by the Roman Empire[72] | Romans organize bloom iron production centers and uniform technology[88] | | **B.C./A.D.** |
| **A.D. 500** | Forged iron in the forms of broad-blade arrow heads, razors, and rings[53] | | Pattern welded sword technology in northern Europe[73] | Bloom iron technology static but slow growth of bloom sizes[89]; pile welding in use[90] | | **A.D. 500** |

1975; [63]Wertime 1962: 11; [64]Rubstov 1975; [65]Pipping 1982: 39–42; [66]Scrivenor 1854: 160–161; [67]Johannsen 1953: 267–269; [68]Tylecote 1976: 113; [69]Esper 1982; [70]Tylecote 1976: 113; [71]Pleiner 1980: 375–415;[72]Tylecote 1976: 53; [73]Tylecote 1974: 64; [74]Crossley 1981: 35–37; [75]Tylecote 1974: 78; [76]Crossley 1981: 35–37; [77]Wertime 1962: 57–63; [78]Wertime 1962: 70–72; [79]Schubert 1957: 146–149; [80]Tylecote 1974: 90; [81]Johannsen 1953: 228–233; [82]Goodway and Odell 1984; [83]Pipping 1982: 39–42; [84]Gille 1947: appended chart; [85]Wertime 1962: 130–135; [86]van den Ouden 1981; Walzell 1951; [87]Pleiner 1980: 375–415; [88]Tylecote 1974: 53; [89]Schubert 1957: 93–111; [90]Schubert 1957: 73–77; [91]Schubert 1957: 110; Crossley 1981: 35–36; [92]Crossley 1981: 35–36; [93]Wertime 1962: 72–75; [94]Crossley 1981: 35–36; [95]Schubert 1942; [96]Hammersley 1973; [97]Schubert 1957: 304; [98]Jenkins 1938–39; Doat and Evrard 1951; [99]Tylecote 1974: 91; [100]Wertime 1962: 124–127; [101]Tylecote 1974: 91; [102]Scrivenor 1854: 141, 169, 172; [103]Tylecote 1974: 127; [104]Percy 1864: 627; [105]Percy 1864: 466–468; [106]Roberts-Austen 1899; [107]Scrivenor 1854: 141, 169, 172; [108]Percy 1864: 310–315; [109]Sayre et al. 1982; [110]Hartley 1957; [111]Scrivenor 1854: 69; [112]Bruce 1930: 12; [113]Scrivenor 1854: 73; Bruce 1930: 71; [114]Bruce 1930: 71; [115]Scrivenor 1854: 69, 73; 1[16]Taussig 1892: 55; [117]Taussig 1892: 133; Overman 1854: 179–193; [118]Taussig 1892: 133; [119]Taussig 1892: 128; [120]Swank 1892: 297–300; [121]Overman 1854: 245–249; [122]Dunn 1980.

*APPENDIX E* **Benchmarks in Ironmaking** (Continued)

| | SUB-SAHARAN AFRICA | EASTERN EUROPE (Poland, Czechoslovakia, Hungary, Balkans, Soviet Union) | CENTRAL AND NORTHERN EUROPE (Germany, Austria, Scandinavia) | WESTERN EUROPE (British Isles, France, Low Countries) | NORTH AMERICA | |
|---|---|---|---|---|---|---|
| **A.D. 1000** | Iron production remained a local activity until the modern era; during the 18th and early 19th centuries a large number of furnace variants, both bowl and shaft types, have been described; many had multiple tuyeres, both natural and forced draft air supply; all these are primitive enough to have had early origins[54] | Advanced iron production in Hungary, Bulgaria with German, Italian workers[60]; German workers to Russia [61]; cast iron smelting and casting diffuses from Mongol kingdoms[62] | Sweden exports iron to Baltic and North Sea neighbors[74]; tinned iron plate[75]; water power technology in smelting and forging[76]; cast iron smelting in Austria, Germany, and Sweden[77] | Much iron imported from Spain, Sweden[91]; water power for bellows and hammers in France[92]; cast iron smelting moves from Central Europe to France[93] | Evidence of iron smelting by Norse visitors in the 13th century[109] | **A.D. 1000** |
| **A.D. 1500** | | Water power for bellows and hammers in Poland[63]; cast iron cannon produced in Russia[64]; blister steel a major industry in Danzig[65] | Cast iron cannon, fining cast iron to wrought iron[78]; high bloomery furnaces make 150 kg. blooms[79]; slitting mill at Liège[80] | Continues to Low Countries and to England[93]; water power in general use[94]; England a major exporter of cast iron cannon[95]; fining begins to compete with bloom iron production[96] | | **A.D. 1500** |
| **A.D. 1600** | | | Development of the iron castings industry[81]; monopolies on quality iron wire, in Nürnberg, Siegen[82]; blister steel production[83] | Slitting mill developed in Liège[97]; iron casting industry grows[98]; blister steel industry in England[99] | First integrated iron mill (blast furnace, finery, tilt hammer) near Boston; later other mills in Connecticut, Rhode Island, New Jersey[110] | **A.D. 1600** |
| **A.D. 1700** | | Polish and German workers imported to establish blast furnaces and fineries in eastern Russia[66]; Russia one of largest bar iron exporters in world[67]; Russian charcoal blast furnace largest in the world[68] | Sweden becomes a major exporter of bar iron to Europe[84] | Coke for cast iron smelting in England[100]; potting/stamping process in England[101]; bar iron imported from Russia, Sweden, Spain[102]; crucible melting of steel in England[103] | Cast iron smelting in Virginia, Maryland, Pennsylvania[111]; Virginia 'iron plantations' used slave labor throughout the century[112]; Britain lifted ban on import of American pig iron but maintained ban on export of slitting and rolling mills for years[113]; first cannon cast in Virginia[114]; first iron smelting in French Canada in 1736 but lasted only 30 years[115] | **A.D. 1700** |
| **A.D. 1800** | | Serf labor discontinued in Russia[69]; charcoal still the primary smelting fuel in Russia[70] | Slow adoption of use of coke[85]; Swedish exports decline[86]; English puddling process adopted[87] | Puddling process developed in England[104]; anthracite use for smelting in Wales[105]; preheated air for smelting in England[106]; imports of bar iron from Russia, Sweden, Spain decline[107]; some Catalan bloomeries still operating in Pyrenees[108] | United States remained a net importer of plate and bar iron[116]; anthracite-fueled blast furnaces coupled with hot blast briefly the dominant development[117]; charcoal blast furnaces persisted, coke taking over in latter half of century[118]; puddling furnaces operating in the first quarter but finery persisted to middle of century[119]; Catalan furnaces with trompe operate until late in century[120]; Catalan furnace produced steel blooms[121]; first blast furnace in Upper Canada[122] | **A.D. 1800** |
| **A.D. 1900** | Considerable evidence of medium carbon steel produced directly in the smelting furnace[55] | | | | | **A.D. 1900** |

# Glossary

*Air Blast.* In most kinds of smelting furnaces the air necessary for combustion must be forced into the combustion zone under pressure. This pressurized air, supplied by bellows or other blowing devices, is called the air blast. See also *hot blast.*

*Anneal.* In metallurgical usage, both a noun and a transitive verb meaning to subject a piece of metal to a thermal excursion—a cycle of heating, holding at temperature and cooling—for the sake of modifying its properties. A second, prefixed noun can be used to designate specific objectives. If it is done to soften a piece after cold working, it can be called a recrystallization anneal; if to relax internal stresses, a stress relief anneal; and if to induce a transformation, pearlitic or normalizing anneal.

*Anthracite.* A hard, dense coal with a high energy content, often called a "natural *coke*." Used for smelting iron in some parts of the world.

*Austenite.* Iron and iron alloys can exist in two different atomic arrangements or crystal forms. One of these occurs normally at elevated temperatures and transforms to the lower temperature form on cooling. The high temperature form is called austenite. Its capacity for dissolving more carbon than the low temperature form, *ferrite,* is responsible for the heat treatment capability of steels. Under appropriate conditions and levels of alloying, some of the austenite can remain stable after cooling to room temperature. Nickel and manganese are the only austenite stabilizers likely to be present in a pre-industrial iron alloy.

*Bainite.* When the austenitic condition of a steel is cooled at a specific rate, one of the transformation products is an arrangement of *ferrite* and *cementite* that is identifiable by its unique appearance under the microscope. Bainite is associated with considerable but not the highest strength attainable by heat treatment. It has rarely been identified in pre-industrial steels because of the very narrow range of cooling rates necessary to produce it and the difficulty in recognizing it in mixed transformation structures.

*Bar Iron.* The traditional name given in English to the marketed form of wrought iron produced by a *bloomery, finery,* or *puddling* operation. The term could be applied to flattened, plate-like shapes as well as to elongated bars with round or square cross sections. Smiths wishing to fabricate larger objects from these could forge weld bundles of them into thicker rods or plates, a process called *piling.*

*Bear.* A mass of iron, slag, and infusible material that could form in the bottom of smelting furnaces. Bears were sometimes the result of accidents at blast furnaces, where the pool of cast iron solidified before it could be tapped. They may also represent accumulations of metal that have leaked through the bottom of a blast furnace, bloomery, or other smelting device. Many synonyms have been used, including *salamander.*

*Bituminous Coal.* The most common type of coal, intermediate in hardness, density, and chemical purity between *anthracite* and *lignite.* Bituminous suited for conversion to coke is called "coking coal."

*Blackband.* A *siderite* ore associated with coal.

*Blackheart Malleable.* See *malleable iron.*

*Blast Furnace.* A shaft furnace used for smelting metals. In ironmaking parlance, a blast furnace is one whose product is liquid cast iron rather than a solid bloom.

*Blister Steel.* When bars or plates of wrought iron were pack carburized in a special, coal fired furnace, the carbon enriched product frequently exhibited surface blisters. This was taken as evidence that the steel had been carburized as advertised. Such blisters were formed by the reaction of the infusing carbon with the internal slag inclusions, producing carbon monoxide gas at sufficient pressure to bulge the metal near the surface. Since the blisters could easily be closed by subsequent forging, they constituted no hazard to later performance of the steel thus produced.

*Bloom.* A spongy aggregate of metal formed under the slag pool in a smelting or fining furnace. As it comes from the furnace, a typical iron bloom contains much slag. Most of this is expelled by draining and squeezing when the bloom is consolidated by subsequent forging, but finished wrought or bloom iron still contains visible slag inclusions. The term comes from the Anglo-Saxon "bloma," meaning a lump of metal.

*Bloom Iron.* The solid product of the direct reduction of ore in a smelting furnace. Most bloom iron has a very low carbon content, although partially carburized metal can be produced under some conditions. Bloom iron is often called *wrought iron* when the emphasis is on properties rather than on the way it was produced.

*Bog Ore.* A *limonite* ore formed in swamps.

*Bosh.* In a blast furnace, the juncture between the erect conical superstructure and the inverted-conical hearth. The *tuyeres* project into the combustion zone near but below the bosh.

*Brasque.* A paste of clay and charcoal used for the lining of a furnace or crucible.

*Brazing.* A method of joining iron pieces by fusion welding them to each other using an intermediate layer of brass or a similarly low-melting copper alloy. Copper alloys are still used by blacksmiths for fabricating rings, tubes, and other delicate iron shapes.

*Brittle.* The term describes a material which (a) shows no capacity for shape change by deformation, (b) absorbs little energy or work in fracturing, and (c) fractures readily under impact. White and gray cast iron are considered brittle by comparison with wrought iron and steel.

*Calcining.* As applied to ores, a synonym for *roasting.*

*Carbonic Acid, Carbonic Anhydride.* Carbon dioxide.

*Carbonic Oxide.* The modern term is carbon monoxide.

*Carburet, Carburette.* When suffixed to "iron" or "manganese," an obsolete term for the carbides of those metals.

*Carburization.* Process by which carbon is deposited on the surface of an iron object at a temperature at which the carbon diffuses inward and alloys with the iron to form a steel. The temperature must be high enough for the iron to have transformed to the *austenite* form.

*Case.* Usually, the carbon-enriched zone near the surface of a piece of iron produced by a carburization process. The unenriched interior is referred to as the "core." In some circumstances, a carbon-depleted surface zone is also called a case.

*Cast Iron.* Iron containing between 2 percent and 4.5 percent of alloyed carbon. Because it has the ability to become liquid at temperatures well below the melting points of wrought iron and steel, it is well suited to casting.

*Catalan Hearth, Forge, or Furnace.* A smelting device that has the form of a bowl, often rectangular, built into the top of a waist-high platform and often supplied with air by a *trompe* rather than a conventional bellows. Although "Catalan" refers to a language and ethnic group of northeastern Spain, the Catalan hearth has its historical center of distribution in the Basque-speaking areas of northern Spain and southwestern France. The term has been used in the literature to designate a very wide variety of forge-like bloomeries in places as far apart as India, China, northern Europe, and North America.

*Cementation.* An obsolete term for *carburization.*

*Cement.* The formulations of charcoal powder, wood ash, and other additives used for *carburization* in the *blister steelmaking process.*

*Cementite.* The iron carbide in the microstructure of steel and cast iron. It has the nominal formula $Fe_3C$ and an atomic arrangement quite unlike that of *ferrite* or *austenite.* The hardness of a steel or cast iron is proportional to the number and close spacing of the cementite crystals present.

*Chafery.* The reheating furnace in a two-stage *finery* system. In a single-stage finery, the same facility served for decarburizing the cast iron and for reheating the resulting wrought iron so as to forge it into a marketable form. In two-stage (i.e., Walloon, etc.) systems, different furnaces were employed for fining and for reheating. The use of a chafery for the latter function had the important advantage that inexpensive coal rather than charcoal could serve as fuel. Since the iron during this second stage of working did not leave the solid state, it could not become contaminated by sulfur from the coal.

*Chalcopyrite.* An iron-copper sulfide mineral.

*Charge.* (1) The *ore*, fuel, and *flux* added to a smelting furnace. (2) To put these into a furnace.

*Cinder.* Obsolete synonym for *slag.*

*Coke.* A fuel made by roasting *bituminous coal* in order to drive off volatile components. It has a high energy density and is still used extensively in the smelting and further processing of metals.

*Cold Short.* The tendency for a metal to crack when hammered or bent at ambient temperatures. This tendency is exacerbated when the temperatures are low, as in winter. Alloyed phosphorus is the chief historical cause of cold shortness in iron and steel.

*Copperas, Couperose.* Obsolete synonyms for ferrous sulfate.

*Coppicing.* A system of forest management that involves leaving the stumps of cut trees undisturbed until they throw out shoots, pruning all except the sturdiest of these and harvesting them after they have grown to sapling size. The per-hectare yield of wood or charcoal under a coppicing system may be more than triple that of an ordinary forest. In pre-industrial Europe, coppiced forests supplied the fuel needs of a number of ironmaking areas.

*Creep.* Slow, permanent deformation under constant stretching, squeezing, or twisting forces applied over a period of time. Creep is the cause of slackening in taut wires and of sagging in loaded beams.

*Cupola.* In the context of ironmaking, a shaft or reverberatory furnace used to remelt cast iron for pouring into molds.

*Cuprous.* In nontechnical usage, the term means simply "pertaining to copper." In scientific terminology, however, it denotes specifically the divalent form of copper in chemical compounds. The distinction parallels the two meanings of *ferrous.*

*Damascene.* A distinctive pattern that appears on the etched surfaces of swords and other implements made of *wootz* steel, sometimes compared to the appearance of silk moiré. The name and its variants (damasked, damascened, damascus, etc.) reflect the European belief that such swords were made at Damascus in Syria, although it seems probable that the great majority were fabricated in Iran and India. The true damascus pattern derives from the crystal arrangements of the metal and should be distinguished from the somewhat similar patterns produced by (a) *pattern welding* and (b) inlaying or *kuftgari* work.

*Decarburizing.* Process events by which the proportion of carbon in a high carbon steel is decreased. Decarburization of cast iron in the liquid state is called *fining, puddling,* or converting, depending on the methods used. Solid state decarburization is also possible, as in the annealing of blooms with an excessive carbon content by African smelters and the conversion of cast iron swords to steel by the ancient Chinese.

*Die.* Any of several different types of device used in forming metals, usually by cold working. (1) A plate with holes in it through which wire is drawn. (2) A combined guide and cutter for making the threading on screws, bolts, etc. (3) A shallow mold or stamp of very hard material, often high carbon steel, into which a softer metal is forced or which is forced into that metal by use of a heavy hammer or press.

*Direct Process.* The production of solid, low carbon iron from ore in a single operation, as in a bloomery or the Krupp-Renn and Chenot processes. To be distinguished from the multistage indirect process.

*Draft.* The pressurized air that feeds combustion in a furnace. Natural or induced draft is generated by the density difference between the hot gases in a stack or chimney and the cold air on the outside. Forced draft utilizes machines to pump the air. See *air blast.*

*Draw.* (1) A synonym for tempering (a steel), as in "to draw a freshly quenched blade." (2) The action of pulling a wire, bar, or tube through a hole in a die plate in order to reduce its diameter.

*Ductility.* The capacity for deformation without cracking or tearing. A very ductile material can be hammered, bent, twisted, or drawn to almost any target dimension. Bloom and wrought irons and steels are quite ductile at ambient and elevated temperatures, though much less so than gold or silver. Cast irons are not usually considered to be ductile except when *malleabilized*; even then their ductility is limited.

*Endurance Limit.* See *fatigue*.

*Eutectic.* Pertaining to (1) an alloy that solidifies through the simultaneous crystallization of two or more metal or alloy components or (2) the conditions under which this occurs. A eutectic alloy solidifies at a lower temperature, the eutectic point, than any of its component entities. Iron forms eutectic alloys with carbon (at 4.3 percent C), phosphorus (at 10.2 percent P) and many other elements. Any element that generates a eutectic alloy will decrease the incipient melting temperatures even when present in smaller amounts.

*Eutectoid.* In iron metallurgy, the term denotes the carbon content of a steel at which the temperature of solid state transformation of *austenite* to *ferrite* and *cementite* is minimized. In pure iron, this transformation occurs at about 910°C; at the eutectoid composition, with 0.8 percent carbon, it occurs at only 723°C. The eutectoid temperature is significant in the heat treatment of steels. If they are not heated initially past this point, no structural changes can occur. See also *pearlite* and *bainite*.

*Faggotting.* A traditional English term referring to the practice of further improving bar iron after initial forging has consolidated and removed most of the slag from the metal as it comes from the finery or puddling furnace (see *muck bar*). The bars are cut into short lengths, stacked, bound with wire into bundles or "faggots," and then heated and forged down to a single bar again. This procedure renders the composition more homogeneous and reduces the size of slag inclusions. "Bushelling" is an equally obsolete synonym; *piling* includes faggotting but has a wider area of reference.

*Fatigue.* A form of brittle fracture associated with fluctuating stress. All metals and polymers are subject to this form of failure. Some metals, notably iron and steel, can withstand fatigue indefinitely if the stresses do not exceed a given level, the *endurance limit*.

*Fayalite.* An iron silicate mineral, usually but not always artificial in origin.

*Fayalite Slag.* A *slag* which solidifies substantially to a fayalite structure. Chemical analyses of such slags, usually reported in terms of silica and iron oxide rather than iron silicate content, often show more than 60 percent FeO. They are fluid at temperatures in the 1200°–1300°C range and are associated with smelting in bloomeries and with fining and puddling. Blast furnace slags do not have a high fayalite content.

*Feeding.* Supplying liquid metal to a casting which is solidifying in a mold. As the metal freezes in the mold, a reduction in volume occurs, leaving voids and pores in the casting. Such defects can be largely prevented by careful control of feeding.

*Ferrite.* The low temperature crystal form of iron. In the pure metal, ferrite transforms to *austenite* when heated above 910°C. It can dissolve much phosphorus but very little carbon, which is why many steels viewed under the microscope show distinct zones of cementite or iron carbide and of pure ferrite.

*Ferrous.* Pertaining to iron in general or to the divalent form of iron in chemical compounds. See *cuprous*.

*Fettling.* Repairing holes in the hearth of a puddling furnace by filling them with hammer scale and iron-rich slag.

*Finery.* (1) A furnace in which *fining* is performed. (2) An ironworks at which such furnaces are present.

*Fining.* Any of a number of extinct processes for converting cast iron to wrought iron, usually in a furnace that somewhat resembles a smith's forge or a Catalan hearth.

*Fire Clay.* A natural clay that has a high *kaolin* content and is substantially free of minerals with a fluxing effect such as alkali and iron oxides. When mixed with a silica or charcoal temper, fire clay forms a refractory solid suitable for furnace linings, bricks, tuyeres, and crucibles.

*Flux.* (1) A substance which is capable of lowering the fusion temperature of a *slag* or other ceramic. The oxides of iron, manganese, sodium, potassium, and calcium, as well as the alkaline earth oxides and calcium fluoride, are all effective in this role. (2) A substance which when molten can dissolve the oxide films that form on most metal surfaces and can prevent their reoxidation, making it possible to join them by welding, soldering, or brazing.

*Forge.* (1) To shape a piece of metal by hammering. (2) A facility where forging is done. (3) In France, the term was extended to include integrated operations that might contain blast furnaces, steelmaking furnaces, and even foundries in addition to (or instead of) forges in the strict sense.

*Gangue.* The nonmetal-bearing and nonvolatile component of an *ore*. As much gangue as possible is removed before smelting. What remains becomes part of the slag.

*Gate.* The basin and channel system by which liquid metal is conducted into a mold. After the casting has solidified, a portion of the metal-filled basin and channel remains attached. This too is called a gate; it is usually broken or cut off and later returned to the melting furnace.

*Goethite.* A mineral, essentially a hydrated ferrous oxide. It is the modern geologists' name for what metallurgists continue to call *limonite*.

*Hammer Scale.* A surface deposit of oxide, primarily *magnetite* and *wustite*, formed while a piece of iron is heated for forging. This scales off during hammering or rolling. Hammer scale was formerly used as a flux for finery slags, an oxidizing agent in puddling hearths, and an additive to the charge in smelting furnaces. *Mill scale* is a synonym.

*Hardness.* The resistance of a solid material to indentation or scratching. In metals, though not always in other materials, measured hardness correlates reasonably well with strength.

*Hardenability.* When used in connection with steel, the term does not refer to the capacity for achieving hardness but rather to the depth of hardening that can be attained through quenching. Simple iron-carbon alloys can become very hard but only to a depth of a few millimeters beneath the surface; their hardenability is therefore low. The addition of certain alloying elements, on the other hand, can result in a steel with very high hardenability: the hardened layer is thick, and quenching need not be as severe. Among the potential alloying elements of this kind, manganese was the only one to be widely available in pre-industrial times.

*Haute Fourneau, Hochofen.* The French or German "high furnace" is a more or less strict equivalent of the English term, "blast furnace." Not all furnaces that were physically high were hautes fourneaux or hochofen.

*Heat Treatment.* In metallurgical jargon, the term denotes the full cycle of thermal processes whereby steel is rendered both hard and tough: heating until the steel takes the form of homogeneous *austenite*, *quenching* back to room temperature, and then reheating or *tempering* to remove some of the hardness and brittleness induced by quenching. More generally, any thermal cycle of heating and cooling for the purpose of modifying the properties of a metal.

*Hot Blast.* The preheating of air before it is forced through the tuyere into the furnace. The resulting increase in the temperatures in and around the combustion zone causes smelting to proceed at a faster rate, yields significant economies in fuel use, and facilitates a number of reactions, including the reduction of manganese and silicon.

*Hot Shortness.* The propensity of a metal to crack during hot forging, often caused in iron by the presence of sulfur. Red Short is a synonym, although actually hot shortness is not exhibited at temperatures below about 988°C, when the color of the metal is bright orange verging on straw yellow.

*Helve.* The shaft of a water wheel-powered *tilt hammer*. When the cam of the wheel shaft lifts the helve at its middle, the device is called a "belly helve hammer."

*Hematite.* A mineral ($Fe_2O_3$) and class of ore.

*Hypereutectoid, Hypoeutectoid.* Refers to steels that are higher or lower in carbon than the *eutectoid* composition. In general, hypoeutectoid steels when heat treated are less wear resistant but tougher, whereas hypereutectoid steels are harder and more brittle. When applied to cast iron the terms "hypereutectic" and "hypoeutectic" have the same meanings with reference to the eutectic composition of 4.3 percent carbon.

*Inclusions.* Nonmetallic zones or particles of slag and perhaps of infusible substances as well in the body of a metal. Even after extensive forging, the volume of a piece of either bloom or wrought iron comprises about 5 percent of slag inclusions.

*Indirect Process.* The production of low carbon iron and steel from ore in two or more steps: by smelting to cast iron in a blast furnace and then *decarburizing* to wrought iron or steel in a *finery, puddling furnace,* or other device.

*Kaolin.* The most refractory of the clay minerals, a hydroxysilicate of alumina. A high proportion of kaolin is necessary in a good *fireclay*.

*Keris, Kris.* A type of dagger and sword traditionally used in Indonesia, Malaysia, the Philippines, and Thailand. Many examples have wavy blades. Most are decorated by *pattern welding*.

*Kuftgari.* A technique for decorating iron and other metals that resembles inlaying. The technique has no common Western name, but the Indian term now seems to be in the process of incorporation into European languages.

*Laterite.* A residual hematite-limonite deposit formed in tropical soils. Some laterites contain enough iron to be used as ores.

*Lignite.* A soft, brown-colored coal with a low energy content, rarely used in ironmaking.

*Limonite.* (1) The mineral now usually called goethite, a hydrated ferrous oxide; (2) An ore, much of whose iron content is in the form of *goethite*.

*Liquidus.* Most alloys and ceramics melt progressively over a span of temperatures on heating and solidify progressively on cooling over the same span. The upper limit of this span is called the liquidus or the liquidus temperature. Above it, the substance is all liquid; between it and the lower limit, the "solidus," the substance is a mushy mixture of solids and liquids. To produce a sound casting, it is important to pour at a temperature well above the liquidus of the alloy involved. Good casting alloys have a low liquidus and a narrow range between the liquidus and the solidus. *Slags,* like other ceramics, may have a rather wide liquidus-solidus range.

*Loam.* In metal casters' terminology, a naturally occurring mixture of clay and silica sand used for making molds. European (though not Asian) foundrymen traditionally preferred such mixtures for their resistance to cracking and shrinking during drying. Casting loams are no longer used, having been replaced by mixtures of selected grades of sand and bentonite clay.

*Loupe, Luppe.* See *bloom*

*Magnetite.* (1) The mineral with the formula $Fe_3O_4$; (2) An ore predominantly of magnetite; (3) The mineral associated with "loadstone".

*Malleable.* An obsolete term for *ductile*.

*Malleable Iron, Malleabilized Iron.* The product of annealing a white cast iron at a temperature of about 1000°C for periods of up to 5–6 days. This treatment decomposes much of the *cementite* present to graphite and *ferrite*. The resulting metal is much softer and less brittle than cast iron and is capable of undergoing limited deformation. There are two basic malleabilizing processes, one yielding "whiteheart" and the other "blackheart" malleable iron. The former has a decarburized, ferritic surface and an interior composed mainly of ferrite and pearlite with interspersed graphite nodules. The latter, having been heated in a neutral atmosphere, still has all of its carbon, and has the same internal structure but without the decarburized *case*.

*Martensite.* The hardest and most brittle crystal form of steel. When hot steel is cooled slowly past the transformation temperature, its *austenite* structure is altered to *ferrite* and *pearlite*. When cooling occurs quickly, however, the transformation product takes the form of martensite. The hardness of martensite increases with carbon content but so also does *brittleness*. *Tempering* has the effect of transforming martensite to a softer and tougher condition. The choice of tempering temperature and time allows for a wide choice of strength-toughness combinations.

*Mill Scale.* See *hammer scale.*

*Minette.* An abundant but low grade iron ore found in the French-German border area.

*Muck Bar.* Iron forged directly from a consolidated lump of wrought iron as it came from the puddling furnace. As the term implies, the product was not highly regarded, due to its excessive slag content. Muck bar had to be subjected to *faggotting* before it was considered usable by blacksmiths.

*Ore.* A natural deposit of metal-bearing minerals in economic concentrations.

*Pamor.* In most Indonesian languages, the etchant-resisting iron used to form welded patterns in the blades of *keris* and other weapons. Some pamor is known to have been meteoritic iron. Much of it contained nickel, apparently derived from smelting nickeliferous *laterite* ores.

*Pattern Welding.* An archaeometallurgist's term referring to the use of *pile welding* to produce visible patterns of iron or steel laminations, with an at least partly decorative intent.

*Pearlite.* One of the transformation products of slowly cooled *austenite*: a structure of interleaving crystals of *cementite* and *ferrite*. As a pearlitic steel is relatively easy to machine and to deform on an anvil or in a die, most steel products are given their semi-final shape while in this condition. They can subsequently be reheated and quenched to produce a harder and stronger *martensitic* structure.

*Peroxide Of Iron.* Obsolete synonym for $Fe_2O_3$, *hematite.*

*Pig Iron.* Raw cast iron as it comes from the smelting furnace. The name derives from the former practice of allowing the tapped iron to run into a series of open molds arranged along a central channel. The arrangement reminded observers of rows of suckling piglets. Although iron pigs are now cast quite differently, the name has remained.

*Pile Welding, Piling.* The most general term for processes that involve forge welding several plates or bars together in order to create more homogeneous structures, improved properties or decorative effects. See *pattern welding* and *faggotting.*

*Plumbago.* Obsolete synonym for graphite. In ironmaking, the term was used mainly with reference to crucibles formed from a mixture of graphite and refractory clay.

*Potash.* An old-fashioned term for potassium carbonate, especially that derived from wood ashes. Also formerly used to denote potassium oxide and elemental potassium. It has a strongly fluxing effect in slags, other ceramics, and glass.

*Potting Process.* A crucible refining system used in Britain and perhaps elsewhere in Europe during the late eighteenth and early nineteenth centuries for desulfurizing and simultaneously *decarburizing* cast iron. Lime, kelp ash, and mill scale all formed part of the crucible charge.

*Protoxide Of Iron.* Obsolete synonym for FeO, *wustite.*

*Puddling.* A method of decarburizing cast iron to wrought iron characterized by the use of a *reverberatory furnace* and of *hammer scale* or *fayalite slag* as an oxidizing agent. In dry puddling, these oxidizers are simply thrown into the furnace along with the cast iron. In wet puddling, they are bonded to the furnace bottom before the operation begins.

*Puddlers' Candles.* Blue flames of burning carbon monoxide that bob to the surface of the liquid iron as bubbles during wet puddling. The appearance and character of these flames served as a guide to the furnace operator in judging the progress of decarburization.

*Pyrite.* An iron sulfide mineral, $FeS_2$.

*Pyrolysis.* The breaking down of large molecules into smaller and simpler ones by the application of heat. The pyrolysis of wood produces charcoal and the pyrolysis of coal, coke.

*Quenching.* The second step in the heat treatment of steel: the piece, heated to *austenite* temperatures, is quickly cooled by plunging it into water. Much of the austenite thereby takes the form of hard and brittle *martensite.*

*Rabbling.* A raking or stirring action performed during the *puddling* of cast iron and the *roasting* of ores. A rabbling tool is a long iron rod with a hoe-like end.

*Red Short.* See *hot short*

*Refractory.* (1) The property of resisting softening or melting at high temperatures. (2) A ceramic material that has this property. To be useful, a metallurgical refractory must also be chemically inert, and thus able to resist the corrosive action of slags. Certain functions may also require a refractory to have low thermal conductivity.

*Reverberatory Furnace.* A furnace in which the hearth, where the charge is placed, is separated from but connected to the firebox where the fuel is burned. As used in the remelting of cast iron and in puddling, such furnaces generally took the form of short horizontal tunnels with flat shallow hearths, equipped with a firebox at one end and a chimney at the other. Flames from the burning fuel heated the roof of the tunnel which in turn radiated ("reverberated") the heat downward onto the charge.

*Riser.* Secondary vertical channels in a mold assembly. Liquid metal is added through the main channel, the *gate.* The risers fill with metal as the mold fills but feed metal back into the mold as shrinkage occurs during solidification. By observing the level of metal in the risers, the foundryman can judge the progress of pouring and freezing.

*Roasting.* Heating ore in air as a preliminary to smelting in order to dry it, to crack it, to decompose carbonates, hydroxides, sulfates, and sulfides, and to drive off their volatile components such as sulfur dioxide, carbon dioxide, and water vapor.

*Salamander.* Synonym for *bear.*

*Scaling.* The forming of oxide crusts on metal surfaces exposed at high temperatures to air, water vapor or carbon dioxide. In the case of iron, these crusts are primarily of *magnetite*, $Fe_3O_4$. See *hammer scale.*

*Scoria.* Obsolete synonym for *slag.*

*Sea Coal.* Early synonym for mineral coal. In Britain, it was often found washed up on beaches.

*Sesquioxide Of Iron.* Obsolete synonym for *magnetite*, $Fe_3O_4$.

*Shear Steel. Blister steel* that has been homogenized by *piling.* When the bundling and forging down of bars was repeated a second time, the product was known as "double shear" steel. The name derives from the use of this metal in the cutlery trades.

*Shingling.* The initial forging given to a bloom or puddled ball of iron as it came from the furnace. Shingling served to consolidate the metal, to remove much of the slag and to produce *muck bar.*

*Siderite.* An iron carbonate ore and mineral, important in pre-modern European steelmaking.

*Silex.* Obsolete synonym for silica: $SiO_2$, silicon dioxide or quartz.

*Silicate.* Slags and glasses with a significant content of *silica.* Term for a compound mineral such as iron silicate or *fayalite.* Analyses of slags are conventionally reported in terms of their component oxides such as silica, lime, alumina, etc. although when solidified they are constituted of silicates, aluminates, etc.

*Sintering.* The process whereby microwelded bonds develop at contact points between solid particles at elevated temperatures in the absence of pressure or deformation.

*Slag.* A ceramic or glass composed of *gangue* and metal oxides derived from *ore,* fuel ash, separately added *fluxes*, and sometimes portions of furnace linings. Slag is not only the residue but also an active component of the smelting process.

*Sluicing.* A method for concentrating an ore that employs running water in a channel to separate out the lighter gangue particles from heavy metallic minerals from a sand or ground *ore* aggregate.

*Smelting.* The use of thermochemical reactions to decompose ore minerals to metals. In traditional iron smelting, the key reaction involves the reduction of iron oxides by carbon monoxide at high temperatures.

*Soda.* The sodium-based equivalent of *potash.* It had similar uses during the pre-industrial period.

*Solidus.* The temperature below which an alloy is completely solid and above which melting beings. See *liquidus.*

*Sow.* An obsolete term for a large, long cast iron pig. See *pig iron.*

*Spathic Ore.* Synonym for *siderite* ore.

*Spiegeleisen.* A type of cast iron with an appreciable but very variable manganese content, usually derived from a natural iron-manganese ore. It served as a raw material for steelmaking and as a desulfurizing additive to ordinary cast iron in fining and puddling operations. Its name, "mirror-iron," comes from the white, shiny appearance of its fracture surface.

*Stall Furnace.* A furnace open at the top and in front.

*Stamping.* A British, eighteenth century de-siliconizing process preliminary to *potting.* By one method, pigs were packed around the edge of a reverberatory furnace and drip-melted so that liquid collected in a shallow pool in the hearth. Silicon was effectively oxidized out and the cast iron was pulled out in thick plates to be broken up in stamping mills for further treatment on the way to the production of wrought iron.

*Stamp mill.* A heavy hammer repeatedly lifted vertically and dropped on to an anvil. Usually powered by a waterwheel. Used to crush ore lumps for hand sorting or *sluicing*; also to break up plates of white iron from the *stamping* process.

*Stückofen.* A type of tall bloomery used in Germany and Austria, often regarded as the ancestor of the blast furnace.

*Tatara.* (1) A Japanese horizontal fan bellows; (2) The type of elongated, multiple-tuyere smelting furnace peculiar to Japan.

*Temper.* (1) Mineral or organic matter added to wet clay in order to reduce cracking during drying and baking. Tempers known to have been used in making furnaces, crucibles, etc. include sand, crushed pottery, crushed slag, charcoal, coal, chaff, straw, hair and manure. (2) To modify the quenched condition of a steel, as below and the selected hardness and toughness resulting.

*Temper Colors.* The colors developed in oxide films formed on the polished surface of iron when heated to temperatures in the range 200°–350°C. These serve as a guide in the tempering of steels.

*Tempering.* The last stage of the heat treatment of steel: through reheating to a temperature between about 200° and 400°C and judging this by observing the temper colors, brittleness can be decreased and various combinations of hardness and toughness achieved. Nonmetallurgists and early metallurgical writers often use "tempering" to denote the full heat treatment process, including initial heating to austenitizing temperatures, quenching, and reheating.

*Tilt Hammer.* A heavy fixed hammer pivoting on a horizontal shaft, driven by human, animal, or water power. The later European and North American water-powered tilt hammers weighed as much as 300–400 kilograms and could deliver as many as 250 blows per minute.

*Toughness.* In metals, the ability to resist cracking and to resist the growth of cracks or flaws already present.

*Tough Pig.* Pig iron which can be converted to a wrought iron that is tough and deformable at ambient temperatures. The major factor is a low phosphorus content.

*Trompe.* An air pumping system that makes use of the fact that large amounts of air are entrained by falling water. Physically, a trompe consists of a wide vertical tube through which the water falls and a box at the bottom that separates the entrained air, compresses it, and directs it to an air pipe that may be connected to a tuyere. Historically, the trompe seems usually but not invariably to have been used with the *Catalan hearth.* It has no moving parts and, although wasteful of water power, can provide an exceptionally constant air blast.

*Troostite.* Synonym for *bainite,* more often used in Europe than North America.

*Tuyere, Twyer, Etc.* A tube that projects into the combustion area of a furnace or forge, through which air is delivered for combustion. Most tuyeres of the pre-industrial period were of refractory ceramic. Those of the modern period are often air-or water-cooled and made of metal.

*Vent.* A small channel in the wall of a mold to allow the escape of air and other gases as liquid metal is poured in through the *gate.*

*Vitriol.* An obsolete term for the sulfates of iron, copper, zinc, and aluminum.

*Welding.* Iron pieces can be bonded at touching surfaces by melting them together—by fusion welding—or by inducing micro-welds to form between adjacent surfaces while these remain in a solid state—by forge welding.

*Whiteheart Malleable.* See *malleable iron.*

*Woody Fracture.* The term describes the appearance of a wrought or bloom iron bar that has been notched and then bent until it fractures. The presence of numerous ribbons of slag makes this look somewhat like a piece of broken wood.

*Wootz.* A crucible steel with a very high carbon content, traditionally made in India and used for the manufacture of *damascene* blades. The term is of uncertain derivation.

*Wrought iron.* The low carbon metal produced by fineries and puddling furnaces. It is essentially identical in appearance and properties to bloom iron. Its name derives not from the way it is made but the way it is often used: it can be readily worked or wrought by forging.

*Wustite.* A mineral, FeO, that is rarely found in nature but is a transition oxide in the reduction of iron *ore.* It is also a significant component of some slags. It plays a major role in the thermochemistry of the decarburization of liquid and solid cast iron.

*Yield Strength.* A measure of the ability to withstand permanent deformation under stress. Low carbon irons and especially steels behave elastically when subjected to moderate force: they bend and then snap back once the force is removed. However, as the force increases a point is eventually reached where some of the bend becomes irreversible. The force (stress) necessary to produce this effect equals the yield strength of the material involved.

*Zag.* The Persian name for the etchant used in developing the *damascene* pattern on the surface of steel. It is thought to have been a somewhat variable mixture of the sulfates of iron and perhaps other metals, more or less equivalent to the European term, *vitriol.* In solution, such mixtures would attack iron aggressively.

# Bibliography

Abbreviations used for journal identifications. Prepositions and articles have been omitted.

| | | | |
|---|---|---|---|
| Acad. | = Academy | Mem. | = Memoirs, Mémoires |
| Am. | = American | Met. | = Metallurgy, Metallurgical |
| Assoc. | = Association | Mitt. | = Mitteilungen |
| Ber. | = Bericht | Phil. | = Philosophical |
| Br. | = British | Proc. | = Proceedings |
| Bull. | = Bulletin | Rev. | = Review, Revue |
| Bur. | = Bureau | Sci. | = Science, Scientific |
| Econ. | = Economics | Soc. | = Society, Société |
| Eng. | = Engineers, Engineering | Sociol. | = Sociological |
| Geogr. | = Geographical | Surv. | = Survey |
| Geol. | = Geology, Geological | Trans. | = Transactions |
| Hist. | = History, Historical | Univ. | = University |
| Inst. | = Institute, Institution | Z. | = Zeitschrift |
| J. | = Journal | | |


Abbott, J.
1847 Process of Working the Damascus Blade of Goojarat. *J. Asiatic Soc. Bengal* 16.1: 417–423.

Abendanon, E. C.
1917–18 *Midden-Celebes-Expeditie; Geologische en Geographische Doorkruisingen Van Midden-Celebes (1909–10)*, vol. 1. Brill, Leiden.

Ahmed, M.
1967 *Survey of Traditional Crafts, Orissa*. Census of India 1961, vol. 12, pt. 7-A. Manager of Publications, Delhi.

Alcock, C. B.
1976 *Principles of Pyrometallurgy*. Academic Press, New York.

Allan, J. W.
1979 *Persian Metal Technology 700–1300 A.D.* Oxford Oriental Monographs No. 2. Ithaca Press, London.

de Alzola, P.
1896 The Iron and Steel Industry of Spain. *J. Iron and Steel Inst.* 52.2: 5–35.

Amstutz, G. C.
1971 *Glossary of Mining Geology*. Ferdinand Enke, Stuttgart.

Angus, H. T.
1976 *Cast Iron: Physical and Engineering Properties*, 2nd ed. Butterworths, London.

Anonymous
1964 Blast Furnaces. *Bull. Hist. Met. Group* 1.3: 1–8.

Anonymous
1965 Blast Furnaces. *Bull. Hist. Met. Group* 1.4: 1–5.

Anstee, J. W. and L. Biek
1961 A Study in Pattern Welding. *Mediaeval Arch.* 5: 71–93.

Archaeological Team, Honan Province
1962 *T'ieh Sheng Kou, Kung Hsein*. Inst. of Archaeology, Academia Sinica, Peking. [in Chinese]

Ashton, T. S.
1924 *Iron and Steel in the Industrial Revolution*. University of Manchester, Econ. Hist. Series No. 2.

Aston, J. and E. B. Story
1957 *Wrought Iron In America*. Byers, Pittsburgh, Pennsylvania.

Atkinson, E. F. T.
1890 *The Himalayan Gazetteer*. 1973 Reprint, Cosmo, New Delhi.

Avery D. H. and P. Schmidt
1979 A Metallurgical Study of the Iron Bloomery, Particularly as Practised in Buhaya (Tanzania). *J. Metals* (Oct): 14–20.

Avery, D. H.
1982 The Iron Bloomery. In *The Evolution of the First Fire-Using Industries*, edited by T.A. Wertime and S.F. Wertime, pp. 205–214. Smithsonian Institution Press, Washington, D.C.

Bachmann, H.-G.
1982 *The Identification of Slags from Archaeological Sites*. University of London Institute of Archaeology Occasional Paper 6. Institute of Archaeology, London.

Baden-Powell, B. H.
1872 *Hand-book of the Manufactures and Arts of the Punjab*. 2 vols. Punjab Printing Co., Lahore.

Bailey, D.
1975 *Rare and Beautiful Firearms*. Gahahas, New York.

Balfour, E.
1855 *On the Iron Ores, the Manufacture of Iron and Steel and the Coals of the Madras Presidency*. Madras.
1885 *The Cyclopaedia of India and of Eastern and Southern Asia*. Quartich, London.

Ball, V.
1877 Geology of the Rajmehal Hills. *Mem. Geol. Surv. India* 13.2: 1–72.
1880 On the Aurunga and Hutar Coal Fields and the Iron Ores of Palamow and Tores. *Mem. Geol. Surv. India* 1: 1–128.

Barnard, N.
1961 *Bronze Casting and Bronze Alloys in Ancient China*. Monograph 14, Australian National University and Monumenta Sinica.

Barnard, N. and T. Sato
1975 *Metallurgical Remains of Ancient China*. Nichiosha, Tokyo.

Barraclough, K. C.
1967 The Cementation and Crucible Steelmaking Processes. *Bull. Hist. Met. Group* 8: 24–34.
1971 "Puddled Steel": The Technology. *J. Iron and Steel Inst.* 209: 952–957.

Barraclough, K. C.
1984 *Steelmaking Before Bessemer, I—Blister Steel*. The Metals Inst., London.

Barron, R. A.
n.d. [1980?] *The Finch Foundry Trust and Sticklepath Museum of Rural Industry*. Finch Foundry, Okehampton, Devon.

Bartlett. H. E.
1900 Letters (1899). *The Agricultural Ledger* 14: 150–153. Calcutta.

Bauer, W. P.
1979 A Scientific Examination of the Applied Decoration on Two Indian Swords. In *Islamic Arms and Armor*, edited by R. Elgood, pp. 1–4. Scolar Press, London.

Bauerman, H.
1882 *A Treatise on the Metallurgy of Iron*. Crosby Lockwood, London.

Beck, L.
1891–99 *Dei Geschichte des Eisens in technischer und kulturgeschichtlicher Beziehung*, vols. 1–4. F. Vieweg, Brunswick.

Belaiew, N.
1912 Sur la cristallisation et structure des aciers refroidis. *Rev. Mét.* 9: 321–342.
1918 Damascene Steel. *J. Iron and Steel Inst.* 97.1: 419–439.

Bell, E. N.
1907 *A Monograph on Iron and Steel Work in Burma*. Superintendent of Government Printing, Rangoon.

Bell, I. L.
1884 *Principles of the Manufacture of Iron and Steel.* George Routledge, London.

Bellamy, C. B.
1904 A West African Smelting House, with Analyses of Specimens by F.W. Harbord. *J. Iron and Steel Inst.* 66.2: 99–126.

Beyschlag, F., J. H. L. Vogt, and P. Krusch
1914–16 *The Deposits of the Useful Minerals and Rocks.* Macmillan, London.

Bird, J. M. and M. S. Weathers
1977 Native Iron Occurrences of Disko Island, Greenland. *J. Geol.* 85: 359–371.

Blair, C., ed.
1983 *Pollard's History of Firearms.* MacMillan, New York.

Blanford, W. T.
1862 Account of a Visit to Puppadoung, an Extinct Volcano in Upper Burma. *J. Asiatic Soc. Bengal* 31.3: 217–226.

Blochmann, H., trans. and ed.
1927 *The A'In-I-Akhbari of Abu 'L Fazl,* 2nd. ed. Asiatic Soc. of Bengal, Calcutta.

Blomgren, S.
1980 The Possibilities of Producing Iron Nickel Alloys in Prehistoric Times. *J. Hist. Met. Soc.* 14.2: 103–104.

von Bogdandy, L. and H. S. Engell
1971 *The Reduction of Iron Ores.* Springer, Berlin.

Böhne, E.
1928 Die Eisenindustrie Masenderans. *Stahl und Eisen* 48.45: 1577–1580.

Boldt, J. R.
1967 *The Winning of Nickel—Its Geology, Mining and Extractive Metallurgy.* Van Nostrand, Princeton.

Boylan, M. G.
1975 *Economic Effects of Scale Increases in the Steel Industry, The Case of U.S. Blast Furnaces.* Praeger, New York.

Braudel, F.
1967 *Capitalism and Material Life: 1400–1800.* Harper Colophon Books, New York.

Bray J. L.
1942 *Ferrous Production Metallurgy.* Wiley, New York.

Bronson, B.
1986a Patterns in the Early Southeast Asia Trade. *Research Conference On Early Southeast Asia, Bangkok and Nakhon Pathom,* 58–121. Br. Inst. in Southeast Asia, Bangkok.
1986b Notes on the History of Iron in Thailand. *J. Siam Soc.* 73: 205–225.
1986c The Making and Selling of Wootz. A Crucible Steel of India. *Archeomaterials* 1: 13–51.
1987 Terrestrial and Meteoritic Iron in the Indonesian Kris. *J. Hist. Met. Soc.* 21.1: 8–15.

Bronson, B. and P. Chareonwongsa
1985 *Eyewitness Accounts of Traditional Mining and Smelting in Mainland Southeast Asia.* Thailand Academic Publishers, Bangkok.

Brown, J. C.
1923 The Mines and Mineral Resources of Yunnan with Short Accounts of its Agricultural Products and Trade. *Mem. Geol. Surv. India* 47.

Browne, P. J.
1967 Sussex Firebacks—History of an Early Application of Cast Iron. *Foundry Trade J.* (July 27): 109–111.

Bruce, K.
1930 *Virginia Iron Manufactures in the Slave Era.* A.M. Kelley, New York.

Buchanan, F.
1807 *A Journey from Madras Through the Countries of Mysore, Canara and Malabar.* 3 vols. T. Cadell & W. Davies and Black, Parry and Kingsbury, London.

Buchwald, V. F.
1975 *Handbook of Iron Meteorites.* 3 vols. Univ. California Press, Los Angeles.

BUIST
1981 Iron Works (He-1) of Han Dynasty at Guxing, Henan. Text of a lecture presented by the staffs of the City Museum of Zhangzhou and the Beijing University of Iron and Steel Technology (BUIST) at the Symposium on Archeometallurgy, Beijing, Oct. 1981.

Burger, H.
1832 Beschrijving der Japansche Kopermijnen en der Bereiding van het Koper. *Verhandelingen van Het Bataviaasch Genootschap* 16, sect. 1: 1–28.
1836 Aanmerkingen Gehouden op eene Reize door eenige Districten der Padangsche Bovenlanden. *Verhandelingen van Het Bataviaasch Genootschap* 16, sect. 5: 157–226.

Burns, R.
1849 The Kayans of the North-West Borneo. *J. Indian Archipelago and Eastern Asia* 3: 140–152.

Butts, A.
1943 *Metallurgical Problems.* McGraw-Hill, New York.

C.V.B.
1820 Methods of Making Steel in Mysore. *Asiatic J.* 9: 441–442.

Cahen, C.
1947–48 Un Traité d'Armurerie Composé pour Saladin. *Bull. d'Etudes Orientales* 12: 103–163.

Cammiade, L. A.
1931 Iron Smelting by Kois, a Jungle Tribe in the Eastern Ghats of India. *Man* 31: 66–67.

Campbell J.
1971 [1842] Manufacture of Bar Iron in Southern India. In *Indian Science in the Eighteenth Century, Some Contemporary Accounts,* edited by Dharampal, pp. 249–262. Impex, Delhi.

Carr, J. C. and W. Taplin
1962 *History of the British Steelmaking Industry.* Harvard Univ. Press, Cambridge, Massachusetts.

*Cast Metals Handbook.*
1944 Am. Foundryman's Soc., Chicago.

Chadwick, R.
1939 The Effect of Composition and Constitution on the Working and on Some Physical Properties of the Tin Bronzes. *J. Inst. Met.* 64: 331–346.

Chakrabarti, D. K.
1976 The Beginning of Iron in India. *Antiquity* 50.198: 114–128.

Chakrabarti, D. K.
1977 Distribution of Iron Ores and the Archeological Evidence of Early Iron in India. *J. Econ. and Soc. Hist. of the Orient* 20.2: 166–184.

Checkland, S. G.
1967 *The Mines of Tharsis.* Allen and Unwin, London.

*Chemical Engineering Handbook*
1941 2nd ed., edited by J.H. Perry. McGraw-Hill, New York.

Cheng, S.-P.
1978 An Iron and Steel Works of 2000 Years Ago. *China Reconstructs* 27.1: 32–34.

Chhibber, H. L.
1926 The Extinct Iron Industry of the Neighborhood of Mount Popa, Upper Burma. *J. Asiatic Soc. Bengal* n.s. 22: 219–232.

Chilcote, R. H.
1968 *Spain's Iron and Steel Industry.* Research Monograph 32, University of Texas at Austin.

Chilton, J. P. and U. R. Evans
1955 The Corrosion Resistance of Wrought Iron. *J. Iron and Steel Inst.* 181: 113–122.

Cipolla, C. M.
1965 *Guns, Sails and Empires.* Minerva Press, Funk & Wagnall's, New York.

Clawson, M.
1975 *Forests for Whom and for What?* Johns Hopkins University Press, Baltimore.

Cleere, H. F.
1972 The Classification of Early Iron-Smelting Furnaces. *Antiquaries J.* 52: 8–23.

Cline, W.
1937 *Mining and Metallurgy in Negro Africa.* General Series in Anthropology No. 5. George Banta, Manasha, Wisconsin.

Coggin-Brown, J.
1923 The Mines and Mineral Resources of Yunnan.... *Mem. Geol. Surv. India* 47.

Coghlan, H. H.
1956, 1977 *Notes on Prehistoric Iron in the Old World*, 1st and 2nd eds. Occasional Papers in Technology 8. Pitt Rivers Museum, Oxford.

Coomaraswamy, A. K.
1956 [c.a. 1905] *Mediaeval Sinhalese Art*. Pantheon, New York.

Cornevin, R.
1962 *Les Bassari du Nord Togo*. Berger-Levrault, Paris.

Crawfurd, J.
1971 [1856] *A Descriptive Dictionary of the Indian Islands and Adjacent Countries*. Oxford University Press, Kuala Lumpur.

Crossley, D. W.
1981 Medieval Iron Smelting. In *Mediaeval Industry*, edited by D.W. Crossley, pp. 29–41. Council for British Archaeology Research Report No. 40.

Davis, G. C.
1899 Malleable Cast Iron—Its History in the United States. *J. Franklin Inst.* 148: 134–144, 181–195.

Day, St.-J. V.
1875 *On the High Antiquity of Iron and Steel*. W.H. Guest, London.

Desch, C. H. and A. T. Roberts
1923 Some Properties of Steel Containing Globular Cementite. *J. Iron and Steel Inst.* 107.1: 249–259, 260–266.

Dickinson, H. W.
1940–41 Henry Cort's Bicentenary. *Trans. Newcomen Soc.* 21: 31–46.

Diderot, D.
1783 *Encyclopédie Méthodique—Recueil des Planches* Ser. 9—*Artes et Métiers*, vol. 2. Panckouke, Paris.
1813 *Encylopédie Méthodique—Recueil des Planches* Ser. 15—*Chimie et Métallurgie*. Agasse, Paris.

Doat, R. and R. Evrard
1951 Three Centuries of Cast Iron Metallurgy. *Am. Foundryman* (Aug): 58–60.

Dodd, B.
1978 The Making of Old Japanese Swords. *J. Mechanical Working Technology* 2: 75–84.

Doelter, C. and H. Leitmeier
1926 *Handbuch der Mineralchemie*, vol. 3, pt. 2. T. Steinkopf, Dresden.

Dunlap, T., ed.
1874 *Wiley's American Iron Trade Manual*. Wiley, New York.

Dunn, A. D.
1980a The Marmora Iron Works in Upper Canada; The First Phase 1819 to 1826. *J. Hist. Met. Soc.* 14.2: 74–79.

Egerton, W.
1880 *An Illustrated Handbook of Indian Arms*. London.
1896 *A Description of Indian and Oriental Armour*. 2nd ed. London.

Ede, G.
1873 *The Management of Steel*. 2nd ed. Tweede, London.

Elliot, J. F., R. A. Buchanan, and J. B. Wagstaff
1952 Physical Conditions in the Combustion and Smelting Zones of a Blast Furnace. *Trans. Am. Inst. Met. Eng.* 194.7: 709–717.

Elliot, J. F., M. Gleiser, and V. Ramakrishna
1963 *Thermochemistry of Steelmaking*, vol. 2. Addison-Wesley. Reading, MA.

Ellis, W.
1858 *Three Visits to Madagascar During the Years 1853–1854–1856*. John Murray, London.

Esper, T.
1982 Industrial Serfdom and Metallurgical Technology in 19th-Century Russia. *Technology and Culture* 23: 583–608.

Evrard, R. and A. Descy
1948 *Histoire de l'Usine des Vennes*. Editions Solédi, Liège.

Faraday, M.
1819 An Analysis of Wootz or Indian Steel. *Quarterly J. Sci., Literature, and Arts* 1: 319–330.

Fermor, L. L.
1909 The Manganese-Ore Deposits of India. *Mem. Geol. Surv. India* 37: 1–610.

Fernandez, E. F.
1900 Letter. *The Agricultural Ledger 1898* 17: 158–160. Calcutta.

Forbes, R. H.
1933 The Black Man's Industries. *The Geographical Review* 23: 230–247.

Forbes, R. J.
1956 Metallurgy. In *A History of Technology*, vol. 2, edited by C. J. Singer, E. J. Holmyard, A. R. Hall, and T. J. Williams, pp. 41–80. Clarendon Press, Oxford.

Forbes, R. J.
1964 *Studies in Ancient Technology*, vol. 8. R. J. Brill, Leiden.

Foster, F. A.
1926 Ancient Chinese Excelled in Making Delicate Art Castings. *The Foundry* 54 (Feb.): 129–134, 146.

Foy, W.
1909 Zur Geschichte der Eisentechnik, inbesondere des Gebläses. *Ethnologica* 1: 185–222.

France-Lanord, A.
1948 La fabrication des épées damassés aux époques Merovingienne et Carolingienne. *Pays Gaumais* 10: 1–3.

Frank, S.
1982 *Glass and Archeology*. Academic Press, London.

Franklin, J.
1971 [1829] The Mode of Manufacturing Iron in Central India. In *Indian Science in the Eighteenth Century, Some Contemporary Accounts*, edited by Dharampal, pp. 220–248. Impex India, Delhi.

Froelich, J.-C., P. Alexandre, and R. Cornevin
1963 *Les populations du Nord-Togo*. Presses Universitaires de France, Paris.

Gale, W. K. V.
1966 *The Black Country Iron Industry, A Technical History*. Iron and Steel Inst., London.
1969 *Iron and Steel*. Longmans, Green, London.

Gardi, R.
1969 *African Crafts and Craftsmen*. Van Nostrand-Reinhold, New York.

Gardner, J. S.
1914 *Ironwork, Part I—From the Earliest Times to the End of the Mediaeval Period*. H.M. Stationery Office, London.

Gaur, R. C.
1983 *Excavations at Atranjikhera*. Motilal Banarsidaaa, Delhi.

Gee, D. R.
1985 Metallurgical Analysis of a 16th Century Japanese Sword. Unpublished Winning Paper in the Am. Soc. Metals International Student Paper Contest.

Geerts, A. J. C.
1883 *Les produits de la nature Japonaise et Chinois*, pt. 2. Lévy & Salabelle, Yokohama.
1874 Useful Minerals and Metallurgy of the Japanese. *Trans. Asiatic Soc. Japan* 3.1: 85–93.
1875 Useful Minerals and Metallurgy of the Japanese. *Trans. Asiatic Soc. Japan* 3.1: 85–93.

Geiger, G. H. and D. R. Poirier
1973 *Transport Phenomena in Metallurgy*. Addison-Wesley, Reading, Massachusetts.

van Gennep, A.
1911 Etudes d'ethnographie algeriennes. *Revue d'Ethnologie et de Sociologie* 2: 265–280.

Geol. Surv. Br.
1920 Special Reports on the Mineral Resources of Great Britain. *Mem. Geol. Surv.* (BR) 12.

Gettens, R. J., R. S. Clarke, and W. T. Chase
1971 *Two Early Chinese Bronze Weapons with Meteoric Iron Blades*. Freer Gallery of Art Occasional Papers, vol. 4.1. Washington, D.C.

Ghosh, A. K. and P. K. Chattopadhyay
1982 An Early Steel Implement from Barudih, Bihar Province, India: Metallurgical Studies. *MASCA J.* 2.2: 63–64.

Gille, B.
1947 *Les Origines de la grande industrie métallurgique en France*. Domat-Montchrestien, Paris.

Gilles, J. W.
1958 Versuchsschmelze in einem vorgeschichtlichen Rennofen. *Stahl und Eisen* 78: 1690–1695.

Gillespie, C. C.
1959 *Pictorial Encyclopedia of Trades and Industry by Denis Diderot*, vol. 1. Dover, New York.

Gleichmann, H. D. and J. Gleichmann
1983 Der Hollerten Zug—ein wichtiges Eisensteinwerk im Siegerland. *Der Anschnitt* 35: 12–21.

Goitein, S. D.
1980 From Aden to India, Specimens of the Correspondence of India Traders of the Twelfth Century. *J. Econ. and Soc. Hist. of the Orient* 23: 43–66.

Goodway, M. and J. S. Odell
1984 *The Metallurgy of 17th and 18th Century Music Wire.* Hubbard Memorial Series no. 3. Pendragon Press, Stuyvesant, New York.

Gordon, R. B.
1983 English Iron for American Arms: Laboratory Evidence on the Source of Iron Used at the Springfield Armory in 1860. *J. Hist. Met. Soc.* 17: 91–98.

Gordon, R. B. and N. J. van der Merwe
1984 Metallographic Study of Iron Artifacts from the Eastern Transvaal. *Archaeometry* 26.4: 108–127.

Gourou, P.
1956 *The Peasants of the Tonkin Delta.* Human Relations Area Files Press, New Haven, Connecticut.

Gowland, W.
1899 The Early Metallurgy of Copper, Tin and Iron in Europe as Illustrated by Ancient Remains and the Primitive Processes Surviving in Japan. *Archeologia* 56: 267–322.
1921 *The Metallurgy of the Non-Ferrous Metals.* Griffin, London.

Grandidier, G.
1928 *Histoire physique, naturelle et politique de Madagascar, IV: Ethnographie de Madagascar.* Libraire Hachette, Paris.

Graves, H. R.
1912 Further Notes on the Early Use of Iron in India. *J. Iron and Steel Inst.* 85.1: 187–202.

Greenwood, W. H.
n.d. [ca. 1883] *A Manual of Metallurgy*, vol. 1. 2nd ed. G.P. Putnam, New York.

Greiner, E. S., J. S. Marsh, and B. Stoughton
1933 *The Alloys of Iron and Silicon.* McGraw-Hill, New York.

Grieve, A.
1960 The Effects of Humidified Blast. *J. Iron & Steel Inst.* 194: 37–43.

Grignon, P. C.
1807 *L'Art de fabriquer le fer.* Auguste Delalain, Paris.

Groneman, J.
1905 Het Njirami de Jaarlijksche Reiniging van de Erfwapens en Andere Poesakas in Midden-Java. *Internationales Archiv für Ethnographie* 17: 81–90.
1910 Der Kris der Javaner. *Internationales Archiv für Ethnographie* 19: 90–109, 123–61, 178–211.

Gurin, M. F.
1980 Metallographic Investigations of Iron Artifacts from Abidnaya, Belorussia. *Sovetskaya Archeologica* 4: 251–259. [in Russian]

Hadfield, R.
1912 Sinhalese Iron and Steel of Ancient Origin. *J. Iron and Steel Inst.* 85.1: 134–186.

Haldane, W.
1970 A Study of the Chemical Composition of Pre-Roman Ironwork from Somerset. *Bull. Hist. Met. Group* 4.2: 53–66.

Hamilton, W. I.
1837 Extracts from Notes Made on a Journey in Central Asia. *J. Royal Geogr. Soc.* 27: 34–61.

Hammersley, G.
1973 The Charcoal Iron Industry and its Fuel, 1540–1750. *Econ. Hist. Rev.* 26.3: 593–613.

*Handbook of Chemistry and Physics.* 50th ed.
1970 CRC Publishing Co., Cleveland, Ohio.

*Handbook of Tables for Applied Engineering Science.* Edited by R. E. Bolz and G. L. Tove. CRC Publishing Co., Cleveland, Ohio.

Hansen, M.
1958 *Constitution of Binary Alloys.* McGraw-Hill, New York.

Harrison, T. and S. J. O'Connor
1969 *Excavations of the Prehistoric Iron Industry in West Borneo.* 2 Vols. Data Paper 72, Southeast Asia Program, Cornell Univ., Ithaca, New York.

Hartley, E. N.
1957 *Ironworks on the Saugus.* Univ. of Oklahoma Press, Norman, Oklahoma.

Hartshorne, R.
1928 Location Factors In the Iron and Steel Industry. *Econ. Geogr.* 4.2: 241–252.

Hartwell, R. M.
1962 A Revolution in the Chinese Iron and Coal Industries during the Sung, 960–1126 A.D. *J. Asian Studies* 21: 153–62.
1963 Iron and Early Industrialism in Eleventh Century China. PhD. Dissertation, Univ. of Chicago.
1966 Markets, Technology and the Structure of Enterprise in the Development of the 11th Century Chinese Iron and Steel Industry. *J. Econ. Hist.* 26: 29–58.
1967 A Cycle of Economic Change in Imperial China: Coal and Iron in Northeast China, 750–1350. *J. Econ. and Soc. Hist. Orient* 10: 102–159.

Hawthorne, J. G. and C. S. Smith, tr. and eds.
1963 *On Divers Arts; the Treatise of Theophilus.* Univ. of Chicago Press, Chicago.

Hayes, A. O.
1915 Waban Iron Ore of Newfoundland. *Canada Geol. Surv.*, Mem. No. 78.

Heath, J. M.
1839 On Indian Iron and Steel. *J. Royal Asiatic Society* 5: 390–393.

Hedges, E. S.
1964 *Tin in Social and Economic History.* St. Martin's Press, New York.

Hedges, R. E. M. and C. J. Salter
1978-79 Source Determination of Iron Currency Bars through Analysis of Slag Inclusions. *Archaeometry* 21.2: 161–175.

Heine, R. J.
1951 Oxidation-Reduction Principles Controlling the Composition of Molten Cast Irons. *Trans. Am. Foundrymen's Soc.* 59: 121–137.

Heine, R. W., C. R. Loper and P. C. Rosenthal
1967 *Principles of Metal Casting.* 2nd ed. McGraw-Hill, New York.

Henan Archaeological Team
1962 *Report on an Archaeological Site with Ironworking Facilities.* Wen Wu Press, Beijing.

Henan Provincial Museum
1978 Iron and Steel Making Techniques of the Han Dynasty in Henan. *Kaogu Xuebao* 1978.1: 1–26. [in Chinese]

Hendley, T. H.
1892 *Damascening of Steel or Iron, as Practised in India.* W. Griggs, London.

Hendriks, A.
1842 Iets over de Wapenfabrikatie op Borneo. *Verhandelingen van het Bataviaasch Genootschap* 18.2: 1–30.

Henger, G. W.
1970 The Metallography and Chemical Analyses of Iron-Base Samples Dating form Antiquity to Modern Times. *Bull. Hist. Met. Group.* 4.2: 45–52.

Hermelin, E., E. Tholander, and S. Blomgren
1979 A Prehistoric Nickel-Alloyed Iron Axe. *J. Hist. Met. Soc.* 13.2: 69–94.

Heyne, B.
1814 *Tracts, Historical and Statistical, on India.* London.
1971 [1795] Iron Works at Ramanakapettah. In *Indian Science and Technology in the Eighteenth Century*, edited by Dharampal, pp. 215–219. Impex India, Delhi.

*History of Metallurgy in China*
1978 By the staff in Archeometallurgy at the Beijing University of Iron and Steel Technology. Science Publications, Beijing.

Hoey, W.
1880 *A Monograph on Trade and Manufactures in Northern India.* American Methodist Mission, Lucknow.

Hogboom, B.
1913-14 Contributions to the Geology and Morphology of Siam. *Bull. Geol. Inst. Univ. Uppsala* 12: 65–128.

Hole, R. S.
1898 Iron Industry in the Jabalpur District. *The Agricultural Ledger* 17: 8–27. Calcutta.

Holland, T. H.
1892 Preliminary Report on the Iron-Ores and Iron-Industries of the Salem District. *Records, Geol. Surv. India* 25.3: 135–159.
1893 Iron Resources and Iron Industries of the Southern Districts of Madras Presidency. *Imperial Inst. Handbook of Commercial Products*, No. 8. Government Press, Calcutta.

Holland, J.
1831 A Treatise on the Progressive Improvement and Present State of the Manufacture of Metal, vol. 1: Iron and Steel. In *The Cabinet Encyclopaedia*, edited by D. Lardner. Longmans, Rees, Orme, Brown, and Greene, London.

Hommel, R. P.
1937 *China at Work.* John Day, New York.

Hoover, H. C. and L. H. Hoover, trans.
1912 *De Re Metallica by Georgius Agricola.* The Mining Magazine. Reprinted in 1950 by Dover Publications, New York.

Horne, L.
1982 Fuel for the Metal Worker: the Role of Charcoal and Charcoal Production in Ancient Metallurgy. *Expedition* 25.1: 6–13.

Horner, L.
1861 Reizen over Sumatra. *Tijdschrift Voor Indische Tal-, Land-en Volkenkunde* 10 (5th ser., 1): 322–376.

Hua, J.-M.
1983 The Mass Production of Iron Castings in Ancient China. *Sci. Am.* 248.1: 106–114.

Hulme, E. W.
1937–38 Prehistoric and Primitive Iron-Smelting. *Trans. Newcomen Soc.* 18: 181–192.
1940–41 Prehistoric and Primitive Iron Smelting Pt. 2, The Crucible Processes of the East. *Trans. Newcomen Soc.* 21: 23–30.

Hunter, A.
1875 Metal-Work among the Hindoos. *Art J.* 14: 45–48, 149–151.

Hupfeld, F.
1899 Die Eisenindustrie in Togo. *Mitteilungen aus den Deutschen Schutzgebieten* 12.4: 175–194.

Hutchinson, E. W.
1934 The Lawa of Northern Siam. *J. Siam Soc.* 27: 153–182.

Hyde, C. K.
1977 *Technological Change and the British Iron Industry, 1700–1870.* Princeton Univ. Press, Princeton, New Jersey.

Jacob, G. Le G.
1843 Report on the Iron of Kattywar; its Comparative Value with British Metal, the Mines and Modes of Smelting the Ore. *J. Royal Asiatic Soc.* 7: 98–104.

Jenkins, R.
1920 The Rise and Fall of the Sussex Iron Industry. *Trans. Newcomen Soc.* 1: 16–33.
1925–26 Iron-Making in the Forest of Dean. *Trans. Newcomen Soc.* 6: 43–65.
1938–39 Ironfounding in England, 1490–1603. *Trans. Newcomen Soc.* 19: 35–49.

Johannsen, O.
1919 Die Erfindung der Eisengusstechnik. *Stahl und Eisen* 39: 1457–1466, 1625–1629.
1953 *Geschichte des Eisens.* Verlag Stahleisen, Düsseldorf.

Johnson, W. R.
1841 *Notes on Use of Anthracite in the Manufacture of Iron.* Charles C. Little and James Brown, Boston.

Joly, H. L. and H. Inada
1913 *The Sword Book and the Book of Same.* Privately printed, London.

Jonkers, D.
1979 Forgeron, forge et métaux en pays minyanka. *J. des Africanistes* 50.2: 7–30.

Julien, S. and P. Champion
1869 *Industries anciennes et modernes de l'empire Chinois.* Eugene La Croix, Paris.

Kellenbenz, H.
1974 Technology in the Age of Scientific Revolution, 1500–1700. In *Fontana Economic History of Europe, 16th and 17th Centuries*, edited by C.M. Cipolla, pp. 177–270. Fontana, London.

Kempers, A. J. B.
1959 *Ancient Indonesian Art.* Van der Peet, Amsterdam.

Kennedy, J. P.
1855 Abstracted by Krishnan (1964): 75.

Kenny, M. G.
1979 Pre-Colonial Trade in Eastern Lake Victoria. *Azania* 14: 97–107.

Killick, D. and R. B. Gordon
1987 Microstructures of Puddling Slags from Fontley, England and Roxbury, Connecticut. *J. Hist. Met. Soc.* 21.1: 28–36.

Kinney, S. P.
1926 Composition of Materials from Various Elevations in a Blast Furnace. *U.S. Bureau Mines.* Technical Paper 387.

Kinney, S. P., P. H. Royster, and T. L. Joseph
1927 Iron Blast Furnace Reactions. *U.S. Bureau Mines.* Technical Paper 391.

Klein, C.
1983 Diagenesis and Metamorphism of Precambrian Banded Iron-Formations. In *Iron-Formation: Facts and Problems*, edited by A.F. Trendall and R.C. Morris, pp. 445–460. Elsevier, Amsterdam.

Klevenhüller, G., R. Mitsche, and F. Trojer
1961 Aufbau dreier römischer Eisenluppen und der damit Verbundenen Schlackenbestandteile vom Magdelensberg, Kärnten. *Carinthia II* 71: 81–87.

Klusemann, K.
1924 Die Entwicklung der Eisengewinnung in Africa und Europe. *Mitt. Anthropologischen Gesellschaft in Wien* 54: 120–140.

Kolchin, B. A.
1967 [1953] *Metallurgy and Metalworking in Ancient Russia.* Israel Program for Scientific Translation, Jerusalem.

Krieger, K.
1963 Notizen zur Eisengewinnung der Hausa. *Z. für Ethnologie* 88: 318–331.

Krishnan, M. S.
1952 The Iron Ores of India. In *Sym. sur les gisements de fer du monde*, vol. 1, edited by F. Blondel and L. Marvier, pp. 502–532. XIXe Congrès Géologique International, Algiers.

Krishnan, M. S.
1954 Iron Ore, Iron and Steel. *Bull. Geol. Surv. India, Series A—Economic Geology*, no. 8: 1–236.
1964 *Iron-Ores in India.* 22nd Session, International Geological Congress, India. New Delhi.

Kruyt, A. C.
1901 Het Ijzer in Midden-Celebes. *Bijdragen Voor Taal-, Land-En Volkenkunde* 53: 148–160.

Küssmaul, F. and F. Schäfer
1973 Tadschiken (Afghanistan, Badakhshan) Formen und Eisengiessen. E748/1965, *Encyclopaedia Cinematographica*:3–33. Göttingen.

Landels, J. G.
1978 *Engineering in the Ancient World.* Univ. California Press, Berkeley.

Lansdell, H.
1885 *Russian Central Asia.* Sampson Low, Marston, Searle and Rivington, London.

Lantenois, N. H.
1907 Résultats de la mission géologique et minière du Yunnan Méridional. *Annales des Mines*, 10th ser., Mém., vol. 11: 385–503.

LaPlanche, H.
1947 A New Structural Diagram for Cast Iron. *Metals Progress* 52: 991–993.

LaTouche, T. H. D.
1917–18 *A Bibliography of Indian Geology and Physical Geography with Annotated Index.* 2 vols. Geological Survey of India, Calcutta.

Leach, B.
1940 *A Potter's Book.* Faber & Faber, London. Reprinted in 1976 by Transatlantic Arts Inc.

Ledebur, A.
1901 Über de Japanischen Eisenhüttenbetrieb. *Stahl und Eisen* 21.16: 841–850.

LeClère, M. A.
1901 Etude géologique et minière des provinces Chinois voisines du Tonkin. *Annales des Mines*, 9th ser., Mém. vol. 20: 287–492.

Len'kov, V. D.
1974 *Metallurgy and Metal Working of the Jurchens in the 12th Century A.D.* Novosibirsk. [in Russian]

Léon, P.
1961 *Les Techniques métallurgiques Dauphinoises au dix-huitieme siècle.* Histoire de la Pensée 5. Hermann, Paris.

Leschenault de la Tour
1820 Rélation d'un Voyage à Karikal et Salem. *Mém. du Musée d'Histoire Naturelle* 6: 329–348.

Levin, E. M., C. R. Robbins, and H. T. McMurdie
1964 *Phase Diagrams for Ceramists.* American Ceramic Soc., Columbus, Ohio.

Levin, M. G. and L. P. Potapov, eds.
1964 *The Peoples of Siberia.* Univ. of Chicago Press, Chicago.

Li Chung
1975 The Development of Iron and Steel Technology in China. *Kaogu Xuebao* 1975.2: 1–22. [in Chinese]

Li, T.-M.
1979 Innovative Technology Opens New Reserves at Soroako. *Mining Eng.* (July): 780.

Licent, E.
1924 *Dix années (1914–1923) dans le bassin du fleuve Jaune et autres tributaires du Golfe du Pei Tcheu Ly.* Libraire Française, Tientsin.

Liebowitz, H. and R. Folk
1984 The Dawning of Iron Smelting in Palestine: The Late Bronze Age. *World Archaeology* 11.3: 265–280.

Lindsay, J. M.
1975 Charcoal Smelting and its Fuel Supply; The Example of Lorn Furnace, Argyllshire, 1753–1872. *J. Hist. Geogr.* 1.3: 283–298.

Linsley, S. and R. Hetherington
1978 A Seventeenth Century Blast Furnace at Allensford, Northumberland. *J. Hist. Met. Soc.* 12.1: 1–11.

Lister, M.
1693 The Manner of Making and Tempering, with a Conjecture at the Method the Ancients Used to Steel their Picks for Cutting Porphyry. *Phil. Trans. Royal Soc.* 17.203: 865–869.

Liu, Y.-T.
1978 The Evolution of the Furnace in Ancient China. *Wen Wu* 1978.2: 18–27. [in Chinese]

Livadefs, C. J.
1956 The Structural Iron of the Parthenon. *J. Iron and Steel Inst.* 182: 46–66.

Lorig, C. H. and D. E. Krause
1936 Phosphorus as an Alloying Element in Low Carbon, Low Alloy Steels. *Metals and Alloys* 7: 9–13.

von Luschan, F.
1909 Eisentechnik in Afrika. *Z. für Ethnographie* 41: 22–59.

Lyman, B. S.
1879 *Geological Survey of Japan: Reports of Progress for 1878 and 1879.* Tookei.

Maddin, R.
1975 Early Iron Metallurgy in the Near East. *Trans. Iron and Steel Inst. Japan* 15: 59–68.

Malcolmson, J. G.
1836 Notes on Geological Specimens from the Country between Hyderabad and Nagpur. *J. Asiatic Soc. Bengal* 5: 96–122.

Mallet, E. R.
1877 Blowing Machines Used by the Smiths of Upper Assam. *Records, Geol. Surv. India* 10.3: 152–154.

Maréchal, J. R.
1955 Evolution de la fabrication de la fonte en Europe et ses relations avec la méthode wallone d'affinage. *Techniques et Civilisations* 4.4: 129–143.

Marschall, W.
1968 Metallurgie und frühe Beschiedlungeschichte Indonesiens. *Ethnologica* n.s. 4: 29–263.

Martin, J. M.
1838 *The History, Antiquities, Topography and Statistics of Eastern India.* 3 vols. London.

Martin, G. and V. Foley
1979 Weyland the Smith: Some Findings. *J. Hist. Met. Soc.* 13.1: 38–39.

Maryon, H.
1948 A Sword of the Nydam Type from Ely Fields Farm, near Ely. *Cambridge Antiquarian Soc. J.* 41: 73–76.
1960 Pattern Welding and Damascening of Sword Blades—Part Two, The Damascene Process. *Studies in Conservation, IIC* 5: 52–59.

Massalski, Capt.
1841 Préparation de l'acier damassé en Perse. *Annuaire du Journal des Mines de Russie* 1841: 297–308.

Massari, S. C.
1938 The Properties and Uses of Chilled Iron. *Proc. Am. Soc. Testing Materials* 38, pt. 2: 217–232.

Maynard, G. W.
1874 The Iron Ores of Lake Champlain, United States of America. *J. Iron & Steel Inst.* 1: 109–136.

McCaskey, H. D.
1903a Iron Metallurgy in the Philippines. *The Engineering and Mining J.* 76: 780.
1903b Report on a Geological Reconnaissance of the Iron Region of Angat, Bulacan. *Philippines Mining Bureau, Bull. #3.* Manilla.

McGannon, H. E., ed.
1964 *Making, Shaping and Treating of Steel.* 8th edition. United States Steel Corp., Columbus, Ohio. See also 9th edition, 1971.

McNeill, W. H.
1982 *The Pursuit of Power.* Univ. of Chicago Press, Chicago.

McWilliam, A.
1920 Indian Iron Making at Mirjati, Chota Nagpur. *J. Iron and Steel Inst.* 102.2: 159–170.

Mehl, R. F.
1938 Iron from Heaven. *The Carnegie Magazine* 11: 241–246.

Mercier, L., trans.
1924 *Aly Ben 'Abderrahman ibn Hodeil, La Parure des Chevaliers et l'Insigne des Preux.* Paris.

Van der Merwe, N. J.
1980 The Advent of Iron in Africa. In Wertime and Muhly 1980: 463–506.

Van der Merwe, N. J. and D. H. Avery
1982 Pathways to Steel. *Am. Scientist* 70.2: 146–154.

*Metallurgy of Ancient China*
1978 By the Archeometallurgy Group of the Beijing Univ. of Iron and Steel Technology. Cultural Publishing House, Beijing.

*Metals Handbook*
1948 7th ed.; 1964 8th ed., vol. 2; 1978, 9th ed., vol. 1. Am. Soc. Metals, Metals Park, Ohio.

Milburn, W.
1813 *Oriental Commerce.* 2 vols. Black, Parry & Co., London.

Mitsuchi, T.
1952 Iron Deposits in Japan. In *Sym. sur les gisements de fer du monde*, vol. 1, edited by F. Blondel and L. Marvier, pp. 537–560. XIXe Congrès Géologique International, Algiers.

Moore, C. B., C. F. Lewis, and D. Nava
1969 Superior Analyses of Iron Meteorites. *Meteorite Research; Proc. Sym. in Austria, 1968*, pp. 738–762. Springer Verlag, New York.

Morelli, R.
1982 Swedenborg and Iron Founding in Italy. *J. Hist. Met. Soc.* 16.1: 24–28.

Morton, G. R.
1963 The Products of Nibthwaite Iron Works. *The Metallurgist* 2.11: 259–268.
1966 The Early Coke Era. *Bull. Hist. Met. Group* 1.6: 49–60.

Morton, G. R. and R. G. Birt
1974 The Present Day Production of Wrought Iron. *J. Met. Soc.* 8.2: 96–102.

Morton, G. R. and J. Gould
1967 Little Aston Forge: 1574–1798. *J. Iron and Steel Inst.* 205: 237–244.

Morton, G. R. and N. Mutton
1967 The Transition to Cort's Puddling Process. *J. Iron and Steel Inst.* 205: 722–728.

Morton, G. R. and J. Wingrove
1969 Constitution of Bloomery Slags: Part I—Roman. *J. Iron and Steel Inst.* 207: 1556–1564.
1972 Constitution of Bloomery Slags: Part II—Medieval. *J. Iron and Steel Inst.* 210: 478–488.

Mott, R. A.
1957–9 Abraham Darby (I and II) and the Coal-Iron Industry. *Trans. Newcomen Soc.* 31: 49–93.

Mott, R. A. and P. Singer, eds.
1983 *Henry Cort: The Great Finer.* The Metals Soc., London.

Moura, J.
1882 Fabrication du fer chez les Cuois du Cumpong-Soai. *Rev. d'Ethnographie* 1: 435–437.
1883 *Le Royaume du Cambodge.* E Leroux, Paris.

Muan, A. and E. F. Osborn
1965 *Phase Equilibria Among Oxides in Steelmaking.* Addison-Wesley, Reading, Massachusetts.

Muhly, J. D., T. S. Wheeler, and R. Maddin
1977 An Iron Adze of the Fifth-Fourth Centuries BC from Al-Mina. *Levant* 9: 156–161.

Mushet, D.
1805 Experiments on Wootz. *Phil. Trans. Royal Soc.* 95: 322–346.

Mutton, N.
1966 Charlcot Furnace 1733–1779. *Bull. Hist. Met. Group* 1.6.
1976 Charcoal Iron Making in the Department of Lot-et-Garonne, France. *J. Hist. Met. Soc.* 10.2: 64–69.

Nadel, S. F.
1942 *A Black Byzantium: The Kingdom of Nupe in Nigeria.* Oxford Univ. Press, London.

Needham, J.
1958 *The Development of Iron and Steel Technology in China.* The Newcomen Soc., London.

Needham, J. and Wang L.
1965 *Science and Civilization in China,* vol. 4. Physics and Physical Technology, pt. 2. Cambridge Univ. Press.

Needham, J., P. Y. Ho, and G. D. Lu
1976 *Science and Civilization in China,* vol. 5. Chemistry and Chemical Technology, pt. 3. Cambridge Univ. Press.

Needham, J., P. Y. Ho, G. D. Lu, and N. Sivin
1980 *Science and Civilization in China,* vol. 5. Chemistry and Chemical Technology, pt. 4. Cambridge Univ. Press.

Nef, J. U.
1932 *The Rise of the British Coal Industry.* 2 vols. Routledge, London.
1963 *War and Human Progress.* Norton, New York.

Neimitz, C. and I. Neimitz
1975 The Forging of a Ritual Knife (Pendat) by Land Dayaks in Sarawak, Borneo. *Sarawak Museum J.* 23.44: 243–257.

Newbury, D. S.
1980 Lake Kiva, Regional Trade in the Nineteenth Century. *J. Africanistes* 50.2: 7–30.

Nieuwenhuis, A. W.
1904–7 *Quer durch Borneo,* 2 vols. E.J. Brill, Leiden.

Nishio, K.
1913 The Mining Industry of Japan. *Trans. Am. Inst. Mining Eng.* 43: 54–98.

Notis, M. R.
1979 Study of Japanese Mokumé Technique by Electron Micro-Probe Analysis. *MASCA J.* 1.3: 38–47.

Okladnikov, A. P.
1970 *Yakutia Before its Incorporation into the Russian State.* Edited by H. N. Michael. McGill-Queens Univ. Press, Montreal.

Oldham, T.
1859 Preliminary Notice on the Coal and Iron of Talcheer in the Tributary Mehals of Cuttack. *Mem., Geol. Surv. India* 1: 1–32.

Osann, B.
1959 Eisenhüttenmanorische Aussagen der Rennofen-Rennstahl- und Rennschlackfünde von Salzgitter-Lobmachtersen. *Stahl und Eisen* 79: 1206–1211.

Osborn, H. S.
1869 *The Metallurgy of Iron and Steel.* H.C. Baird, Philadelphia.

den Ouden, A.
1981 The Production of Wrought Iron in Finery Hearths, Pt. 1: The Finery Process and its Development. *J. Hist. Met. Soc.* 15.2: 63–87.

Overman, F.
1850 –1854 *The Manufacture of Iron,* 2nd and 3rd eds. H.C. Baird, Philadelphia.

Paris, A. and F. Ferriol
1922 L'industrie du fer chez les Berbères du Maroc. *Hespéris* 2: 339–344.

Payne-Gallwey, R.
1958 [1907] *The Crossbow.* Bramhall House, New York.

Pearson, G.
1795 Experiments and Observations to Investigate the Nature of a Kind of Steel... Called Wootz. *Phil. Trans. Royal Soc.* 1795: 322–346.

Percy, J.
1861 *Metallurgy—Fuel, Fireclays, Copper, Zinc, Brass.* John Murray, London.
1864 *Metallurgy—Iron and Steel.* John Murray, London.
1875 *Metallurgy—Introduction, Refractory Metals and Fuels.* John Murray, London.
1877 On the Cause of the Blisters on "Blister Steel." *J. Iron and Steel Inst.* 2: 460–463.

Pfannenschmidt, C. W.
1977 *Die Anwendung des Holzkohlenhochofens Seit ende des 16. Jahrhunderts zur Erzeugung von Gusswaren Erster Schmeltzung und die Spätere Zweite Schmelzung in Flamm-und Kupolofen Bis Mitte des 19. Jahr-Hunderts.* Verein Deutscher Eisenhüttenleute Fachausschussbericht 9.007, Düsseldorf.

Phillips, W. B.
1898 *Iron Making in Alabama.* 2nd ed. Alabama Geol. Surv., Montgomery, Alabama.

Piaskowski, J.
1960 An Interesting Example of Early Technology. *J. Iron and Steel Inst.* 194: 336–340.
1969–70 On the Production of Iron with High Nickel Content in Antiquity. *ACTA Archeologica Carpathica* 11: 319–328. [in Polish]
1982a A Study of the Origin of the Ancient High-Nickel Iron Generally Regarded as Meteorites. In *The Evolution of the First Fire-Using Industries,* edited by T.A. Wertime and S.F. Wertime, pp. 237–243. Smithsonian Inst. Press, Washington, D.C.
1982b The Occurrence of Arsenic in Ancient and Mediaeval Iron Implements Made of Bloomery Iron. *Kwartalnik Historii Nauki I Techniki* 27.2: 397–410. [in Polish]
1983 Metallographic Examinations of Iron Objects of the Late La Tène and Roman Periods from the Northeastern Areas of Poland. *Materialy Starozytne I Wczesnosredniowieczne* 5: 197–228. Panstwowe Muzeum Archeologiczne, Warsaw. [in Polish]

Piaskowski, J. and Z. Hensel
1979 Metallographic Investigations of Iron Objects from a Cremation Cemetery at Brzezce near Bialobrzegi, Province of Radom. *Sprowozdania Archeologiczne* 31: 141–150. [in Polish]

Pigott, V. C.
1980 The Iron Age in Western Iran. In Wertime and Muhly 1980: 417–464.

Pigott, V. C., P. E. McGovern, and M. R. Notis
1982 The Earliest Steel from Transjordan. *MASCA J.* 2.2: 35–39.

Pinel, M. L., T. T. Read, and T. A. Wright
1938 Composition and Microstructure of Ancient Chinese Iron Castings. *Trans. Am. Inst. Mining Met. and Eng.* 131: 174–184.

Pipping, G., ed.
1982 *Iron and Steel on the European Market in the 17th Century.* The Hist. Met. Group of the Swedish Ironmaster's Association, Stockholm.

Pleiner, R.
1971 The Problem of the Beginning of the Iron Age in India. *ACTA Praehistorica et Archeologica* 2: 5–36.
1977 Extensive Iron Production Centers in Germany. In *Symposium Ausland der la Tène-Civilisation und Anfange der Germanische Beseidlung in Mittleren Donaugebiet*, pp. 297–305. Bratislava.
1980 Early Iron Metallurgy in Europe. In Wertime and Muhly 1980: 375–415.
1982 Investigation into the Blacksmith's Technology on Celtic Oppida. *Pamatky Archeologicke* 73: 86–177. [in Czech]

Pole, L. M.
1974 Iron Smelting Procedures in the Upper Region of Ghana. *J. Hist. Met. Soc.* 8.1: 21–31.
1975 Iron Working Apparatus and Techniques: Upper Region of Ghana. *West African J. Arch.* 5: 11–39.

della Porta, G. B.
1658 [1588] *Natural Magick.* Reprinted in 1957 by Basic Books, New York.

Powle, H.
1678 An Account of the Iron Works in the Forest of Dean. *Phil. Trans. Royal Soc.* 12: 931–935.

Primrose, A.
1900–1901 Report on Economic Mineral Products of the Bangalore, Kolar, Tumkur and Mysor Districts. *Records, Mysore Geol. Department* 3: 204–239.

Raffles, T. S.
1817 *The History of Java.* 2 vols. London.

Raistrick, A.
1953 *Dynasty of Ironfounders: The Darbys and Coalbrookdale.* Longmans, Green, London.
1972 *Industrial Archeology.* Methuen, London.

Rao, K. N. P., J. K. Mukherjee, and A. K. Lahiri
1970 Some Observations on the Structure of Ancient Steel from South India. *Bull. Hist. Met. Group* 4.1: 12–17.

Rathgen, B.
1925 Die Pulverwaffe in Indien. *Ostasiatische Z.* 12: 196–217.

Read, T. T.
1913 The Mineral Production and Resources of China. *Trans. Am. Inst. Mining Met. and Eng.* 43: 3–53.
1934 The Early Casting of Iron. *Geogr. Rev.* 24: 544–554.
1937 Chinese Iron—A Puzzle. *Harvard J. Asiatic Studies* 2 (Dec): 398–407.
1939–40 The Earliest Industrial Use of Coal. *Trans. Newcomen Soc.* 20: 119–133.

Rein, J. J.
1889 *The Industries of Japan.* A.C. Armstrong, New York.

Renstorff, G. W. P.
1966 *High Purity Metals.* DMIC Report 222, Defense Metals Information Center, Columbus, Ohio.

Rice, B. L.
1877 *Mysore and Coorg.* A Gazetteer Compiled for the Government of India. Bangalore.

Richards, R. J.
1918 *Salem,* vol. 1, pt. 1. Madras District Gazeteers. Government Press, Madras.

Richardson, F. D.
1974 *Physical Chemistry of Melts in Metallurgy.* 2 vols. Academic Press, London.

Richardson, H. C.
1934 Iron, Prehistoric and Ancient. *Am. J. Arch.* 38.4: 555–583.

Richardson, M. T.
1978 [1889–91] *Practical Blacksmithing.* Weathervane Books, New York.

von Richtofen, F.
1882 *China. Ergebnisse Eigener Reisen und Darauf Gegrundeter Studien.* Dietrich Reimer, Berlin.
1907 *Ferdinand von Richtofen's Tagebucher Aus China.* Edited by E. Tiessen. Dietrich Reimer, Berlin.

Rickard, T. A.
1929 Iron in Antiquity. *J. Iron and Steel Inst.* 120: 323–342.

Riden, P.
1977 The Output of the British Iron Industry Before 1870. *Economic History Review,* series 2, 30: 442–459.

Roberts, C. W.
1983 *A Legacy from Victorian Enterprise.* A. Sutton, Gloucester, UK.

Roberts-Austen, W.
1899 Presidential Address. *J. Iron & Steel Inst.* 55: 17–43.

Robinson, B. W.
1961 *The Arts of the Japanese Sword.* Tuttle, Rutland, Vermont.

Rocher, E.
1880 *La Province Chinoise du Yün-Nan.* E. Leroux, Paris.

Rollason, E. C.
1973 *Metallurgy for Engineers.* 4th ed. Arnold, London.

Rosenfeld, A.
1965 *The Inorganic Raw Materials of Antiquity.* Weidenfeld and Nicolson, London.

Rosenquist, T.
1974 *Principles of Extractive Metallurgy.* McGraw-Hill, New York.

Rostoker, W.
1986 Troubles with Cast Iron Cannon. *Archeomaterials* 1: 69–90.
1987 White Cast Iron as a Weapon and Tool Material. *Archeomaterials* 1: 145–148.

Rostoker, W. and E. R. Gebhard
1981a Iron Smelting at Isthmia. *J. Hist. Met. Soc.* 15.1: 41–44.
1981b The Reproduction of the Rooftiles for the Archaic Temple of Poseidon at Isthmia, Greece. *J. Field Arch.* 8: 211–227.

Rostoker, W., B. Bronson, and J. Dvorak
1984 The Cast Iron Bells of China. *Technology and Culture* 25.4: 750–767.
1989 Smelting to Steel by the Japanese Tatara Process. *Archeomaterials* 3: 11–25.

Rostoker, W., B. Bronson, J. Dvorak, and G. Shen
1983 Casting Farm Implements, Comparable Tools and Hardware in Ancient China. *World Arch.* 15.2: 196–210.

Rostoker, W., M. B. Notis, J. R. Dvorak, and B. Bronson
1985 Some Insights on the "Hundred Refined" Steel of Ancient China. *MASCA J.* 3.4: 99–103.

Rostoker, W. and J. R. Dvorak
1986 Additional Studies on the Smithsonian Bloom. In *Proc. 24th International Archaeometry Symp.*, edited by J. S. Olin and M. J. Blackman, pp. 279–289. Smithsonian Institution Press, Washington, D.C.

Rostoker, W. and D. Rostoker
1989 The Wedgewood Temperature Scale. *Archeomaterials* 3: 169–172.

Routledge, R.
1891 *Discoveries and Inventions of the 19th Century.* 9th ed. Geo. Routledge & Sons, Ltd., London.

Routledge, W. S. and K. Routledge
1910 *With a Prehistoric People: The Akikiyu of British East Africa.* Arnold, London.

Rubtsov, N. N.
1975 [1962] *History of Foundry Practice in USSR.* Indian National Scientific Documentation Centre, New Delhi.

Samuels, L. E.
1980a *Optical Microscopy of Carbon Steels.* Am. Soc. Met., Metals Park, Ohio.
1980b The Metallography of a Wrought Iron Anchor from the Bark Endeavor. *Metallography* 13: 357–368.

Sambhasiva Iyer, V. S.
1901 Reports on Prospecting Work in Parts of Chitaldroog and Tumkur Districts during the Field Seasons of the Years 1896–1899. *Records, Mysore Geol. Department* 2: 102–116.
1902 Notes on Economic Mineral Products of the Chitaldroog and Shimoga Districts. *Records, Mysore Geol. Department* 3: 240–260.

Sanders, C. A. and D. C. Gould
1976 *History Cast in Metal.* Cast Metals Institute-American Foundrymen's Society Joint Publication. Des Plaines, Illinois.

Sanyal, H.
1968 The Indigenous Iron Industry of Birbhum. *Indian Econ. and Social Hist. Rev.* 5.1: 101–108.

Sasaki, M.
1982 A Seven-Branched Sword Made by the Hundred Refined Method. *Tetsu-To-Hagane* (Journal Iron and Steel Institute Japan) 68.1: 178–184. [in Japanese]

Sassoon, H.
1964 Iron Smelting in the Hill Village of Sukur, North-Eastern Nigeria. *Man* 64: 174–178.

Sayre, E. V., G. Harbottle, R. W. Stoenner, W. Washburn, J. S. Olin, and W. Fitzhugh
1982 Dating an Iron Bloom Associated with the Voyages of Sir Martin Frobisher. *Nuclear Chemical Dating Techniques.* American Chemical Society Symposium Series 176 (22): 441–451.

Schallenberg, R. H. and D. A. Ault
1977 Raw Materials Supply and Technological Change in the American Charcoal Industry. *Technology and Culture* 18: 436–466.

Schenck, H.
1983 Analysis of Cast Iron from Catoctin, Maryland. *MASCA J.* 2.1: 107–111.

Schmidt, P. and D. H. Avery
1978 Complex Iron Smelting and Prehistoric Culture in Tanzania. *Science* 201: 1085–1089.

Schubert, H. R.
1942 The First Cast-Iron Cannon Made in England. *J. Iron and Steel Inst.* 146: 131–140.
1951–1953 Early Refining of Pig Iron in England. *Trans. Newcomen Soc.* 28: 59–73.
1957 *History of the British Iron and Steel Industry from c. 450 BC to AD 1775.* Routledge & Kegan Paul, London.

Schurmann, E.
1958 Die Reduktion des Eisens im Rennfeuer. *Stahl und Eisen* 78.19: 1297–1308.

von Schwartz, C.
1899 Correspondence on Syed Ali Bilgrami's Iron Industry in Hyderbad. *J. Iron and Steel Inst.* 56.2: 89–99.
1901 Über die Eisen- und Stahlindustrie Ostindiens. *Stahl und Eisen* 21: 209–211, 277–283, 337–341, 391–399.

Scott, J. G. and J. P. Hardiman
1900 *Gazetteer of Upper Burma and the Shan States,* pt. 1, vol.2. Superintendent of Government Printing, Rangoon.

Scott, H.
1971 Letters to Sir Joseph Banks, 1790–1801. In *Indian Sci. and Technology in the Eighteenth Century.* Impex, Delhi.

Scrivenor, H.
1854 *History of the Iron Trade from the Earliest Times to the Present Period.* 2nd ed. Longman, Brown, Green, and Longman, London.

Sherby, O. D. and J. Wadsworth
1983–84 Damascus Steels—Myths, Magic and Metallurgy. *Stanford Engineer,* Fall–Winter.
1985 Damascus Steels. *Sci. Am.* 252 (Feb): 112–120.

Shockley, W. H.
1904 Notes on the Coal- and Iron-Fields of Southeastern China. *Trans. Am. Inst. Mining Met. & Eng.* 34: 841–862.

Sisco, A. G. and C. S. Smith, trans.
1956 *Réaumur's Memoirs on Steel and Iron.*. University of Chicago Press, Chicago.

Smith, C. S.
1957 Decorative Etching and the Science of Metals. *Endeavour* 16.64: 199–208.
1960 *A History of Metallography: The Development of Ideas on the Structure of Metals Before 1890.* University of Chicago Press, Chicago.
1968 *The Science of Steel, 1532–1786.* MIT Press, Cambridge, Massachusetts.

Smith, C. S. and R. J. Forbes
1957 Metallurgy and Assaying. In *A History of Technology,* vol. 3, edited by C. Singer, E. J. Holmyard, A. R. Hall, and T. J. Williams, pp. 27–71. Clarendon Press, Oxford.

Smith, C. S. and M. T. Gnudi, trans.
1942 *The Pirotechnia of Vannocio Biringuccio.* MIT Press, Cambridge, Massachusetts.

Smith, R. H., R. Maddin, J. D. Muhly, and T. Stech
1984 Bronze Age Steel from Pella, Jordan. *Current Anthropology* 25.2: 234–236.

Snodgrass, A. H.
1980 Iron and Early Metallurgy in the Mediterranean. In Wertime and Muhly 1980: 335–374.

Solyom, G. and B. Solyom
1978 *The World of the Javanese Keris.* East-West Center, Honolulu.

de Sousberghe, L.
1955 Forgerons et fondeurs de fer chez les Ba-Pende et leurs voisins. *Zaire* 9.1: 25–31.

Spencer, J. R.
1963 Filarete's Description of a Fifteenth Century Italian Iron Smelter. *Technology and Culture* 4: 201–206.

Staff, Henan Provincial Museum
1978 *Han Period Stack Mold Casting of Cast Iron: The Excavation and Study of the Wenxian Cast Iron Mold Kiln.* New China Publishing House, Beijing. [in Chinese]

Stead, J. E.
1918 Iron, Carbon and Phosphorus. *J. Iron and Steel Inst.* 97: 389–415.

Stech Wheeler, T. and R. Maddin
1976 The Techniques of the Early Thai Metalsmith. *Expedition* 18.4: 38–47.

Stodart, J.
1818 A Brief Account of Wootz or Indian Steel. *The Asiatic J.* 5: 570–571.

Stodart, J. and M. Faraday
1820 Experiments on the Alloys of Steel, Made with a View to Its Improvement. *Quarterly J. of Sciences, Literature and the Arts.* 9: 319–330.

Stone, G. C.
1934 *A Glossary of the Construction, Decoration and Use of Arms and Armor.* Southworth, New York.

Straker, E.
1931 *Wealden Iron.* G. Bell, London.

von Stromer, W.
1977 Innovation and Growth in the Late Middle Ages: The Invention of the Wire Mill as a Stimulator. *Technische Geschichte,* VDI-Verlag Gmb H., Dusseldorf: 89–120.

Stuhlmann, F.
1910 *Handwerk und Industrie in Ostafrika.* L. Friederichsen, Hamburg.

Sturdy, J.
1693 On the Iron-Works in Lancashire. *Phil. Trans. Royal Soc.* 17: 695–699.

Sun, E.-Z. and S.-C. Sun, Trans.
1966 *T'ien-Kung K'ai-Wu by Sung Ying'Hsing, Chinese Technology in the 17th Century.* Pennsylvania State University Press, University Park, Pennsylvania.

Swank, J. M.
1892 *History of the Manufacture of Iron in All Ages.* American Iron and Steel Association, Philadelphia.

Swarup, D. and R. C. Misra
1944 Indigenous Smelting of Iron in Mirzapur District, U.P. *Quarterly J., Geol. Mining and Met. Soc. India* 16.4: 127–131.

Takahashi, I.
1986 Enterprising *Tatara* Ironmaking of the Pre-Modern Period in Japan. *Bull. of the Metals Museum*, Special issue, vol. 11: 76–80. Japan Institute of Metals, Sendai.

Tanimura, H.
1980 Development of the Japanese Sword. *J. Metals* Feb: 63–73.

Taussig, F. W.
1930 [1892] *The Tariff History of the United States*. 8th ed. G.P. Putnam, New York, New York.

Tavernier, J. P.
1688 *Collection of Travels Through Turkey, into Persia and the East-Indies*. London.

Tegengren, F. R.
1923–24 The Iron Ores and Iron Industry of China, pt. 2. *Memoirs of the Geological Survey of China*, Series A, No. 2. Peking.

Temin, P.
1964 *Iron and Steel in Nineteenth-Century America, An Economic Inquiry*. MIT Press, Cambridge, Massachusetts.

Terekhova, N. N.
1974 Technology of Cast Iron Production in Ancient Mongolia. *Sovetskaia Arkheologiia* 1974.1: 69–78. [in Russian]

Tessmann, G.
1913 *Die Pangwe, Völkerkundliche Monographie eines West-Afrikanischen Negerstämmes*. E. Wasmuth, Berlin.

Tholander, E.
1971 Evidence of the Use of Carburized Steel and Quench Hardening in Late Bronze Age Cyprus. *Opuscula Atheniensia* 10: 15–22.
1975 Comments on Medieval Osmond Iron. *J. Hist. Met. Soc.* 9.2: 68–70.

Thurnwald, R.
1932 *Economics in Primitive Communities*. Oxford University Press, London.

Thwing, C. B.
1920 Application of Pyrometry to the Ceramic Industries. *Pyrometry—Am. Inst. Mining Met. and Eng. Symp*: 516–521.

Todd, J. A. and J. A. Charles
1978 Ethiopian Bloomery Iron and the Significance of Inclusion Analysis in Iron Studies. *J. Hist. Met. Soc.* 12.2: 63–87.

Trendall, A. F. and R. C. Morris
1983 *Iron Formation: Facts and Problems*. Elsevier, Amsterdam.

Trinder, B. S.
1973 *The Industrial Revolution in Shropshire*. Phillimore, London.
1982 *The Making of the Industrial Landscape*. Dent and Sons, London.

Tsuna, M. and C. S. Smith
1983 *Kodo Zurako-Smelting of Copper*. Burndy Library, Norwalk, Connecticut.

Türkdoğan, E. T.
1978 Blast Furnace Reactions. *Met. Trans.* 9B: 163–179.

Turner, T.
1893 The Production of Wrought Iron in Small Blast Furnaces of India. *J. Iron and Steel Inst.* 44: 162–172 with disc., 173–181.

Tylecote, R. F.
1962 *Metallurgy in Archaeology: A Prehistory of Metallurgy in the British Isles*. Edward Arnold, London.
1965 Iron Smelting in Pre-Industrial Communities. *J. Iron and Steel Inst.* 203 (April): 340–348.
1968 *Solid Phase Welding of Metals*. St Martin's Press, London.
1969 Recent Work on Early Ironmaking Sites in the Stamford Area. *Bull. Hist. Met. Group* 4.1: 24–27.
1975 The Origin of Iron Smelting in Africa. *West African J. Arch.* 5: 1–9.
1976 *A History of Metallurgy*. The Metals Society, London.
1977 The Downpatrick Bloom. *J. Hist. Met. Soc.* 11.2: 83.
1980 Furnaces, Crucibles and Slags. In Wertime and Muhly 1980: 183–228.
1981 The Medieval Smith and his Methods. In *Mediaeval Industry*, edited by D. W. Crossley, pp. 42–50. Council for British Archaeology Research Report No. 40, Oxford.
1986 *The Prehistory of Metallurgy in the British Isles*. Institute of Metals, London.
n.d. Oxidation Enrichment Bands in Wrought Iron. Paper presented at the Etruscan Conference, Piombino. ca. 1980.

Tylecote, R. F., J. M. Austin, and A. E. Wraith
1971 The Mechanism of the Bloomery Process in Shaft Furnaces. *J. Iron and Steel Inst.* 209: 342–363.

Tylecote, R. F. and E. Owles
1960 A Second Century Smelting Site at Ashwicken. *Norfolk Arch.* 32.2: 142–162.

Tylecote, R. F. and R. Thomsen
1973 The Segregation and Surface Enrichment of As and P in Early Iron Artifacts. *Archaeometry* 15.2: 193–198.

Untracht, O.
1975 *Metal Techniques for Craftsmen*. Doubleday, Garden City, New York.

Ure, A.
1856 *A Dictionary of Arts, Manufactures and Mines*, vol. 1: 1061–1114. D. Appleton, New York.

Varoufakis, G.
1982 The Origin of the Mycenaean and Geometric Iron on the Greek Mainland and in the Aegean Islands. In *Early Metallurgy in Cyprus, 4000–500 B.C.*, edited by J. D. Muhly, R. Maddin, and V. Karageorghis, pp. 315–324. Department of Antiquities, Republic of Cyprus, Larnaca.

Vetters, H.
1966 Ferrum Noricum. *Anzeiger der Österreichischen Akademie der Wissenschaften—Philosophisch Historische Klasse* 103.25: 167–185, and plates.

Voysey, H. W.
1832 Description of the Native Manufacture of Steel in India. *J. Asiatic Soc. Bengal* 1: 245–247.

Wagner, D. B.
1984 Some Traditional Chinese Iron Production Techniques Practised in the 20th Century. *J. Hist. Met. Soc.* 18.2: 95–104.

Wagner, D. B.
1985 *Dabieshan, Traditional Chinese Iron-Production Techniques Practised in Southern Henan in the Twentieth Century*. Scandinavian Institute of Asian Studies, Monograph Ser. 52. Copenhagen.

Waldbaum, J.
1980 The First Archaeological Appearance of Iron and the Transition to the Iron Age. In Wertime and Muhly 1980: 69–98.

Walzel, R.
1951 The 2000-Year Tradition of the Austrian Iron and Steel Industry. *J. Iron and Steel Inst.* 168: 364–373.

Wang, C. C.
1923–24 Supplementary Notes on the Iron Ores of Shansi. *Mem. Geol. Surv. China*, Ser. A, No. 2: 435–443.

Wang, C. T. and Y. T. Wang
1930 A Study of General and Economic Geology along the Chengtai (Shansi) Railway. *Geol. Bull.* 15: 53–118.

Wang, Z.
1982 *Han Civilization* (trans. by K.C. Chang). Yale University Press, New Haven, Connecticut.

Ward, H. D.
1972 "Best Yorkshire" from West Yorkshire. *J. Iron and Steel Inst.* 210: 396–405.

Watson, B.
1969 *Records of the Historian: Chapters from the Shi Chi of Ssu-Ma Ch'ien*. Columbia University Press, New York.

Watson, E. R.
1910 *Iron and Steel Work in the Province of Bengal*. Constituent Monograph Series, Calcutta.

Webster, G.
1955 A Note on the Use of Coal in Roman Britain. *J. Soc. of Antiquaries* 35: 199–216.

Wedgwood, J.
1782 An Attempt to Make a Thermometer for Measuring the Higher Degrees of Heat . . . *Philosophical Transactions of the Royal Society* 72, Pt. 2: 305–326.

Wembah-Raschid, J. A. R.
1973 *Ironworking in Ufipa*. National Museum of Tanzania, Dar-es-Salaam.

Wertime, T. A.
1962 *The Coming of the Age of Steel*. University of Chicago Press, Chicago.

Wertime, T. A. and J. D. Muhly, eds.
1980 *The Coming of the Age of Iron*. Yale University Press, New Haven, Connecticut.

Wheeler, T. S., J. D. Muhly, and R. Maddin
1976 A Steel Chisel of the Fourth Century BC from Al Mina in Syria. *Levant* 8: 107–111.

Whitaker, P. and T. H. Williams
1969 An Examination of a North Indian Sword (Firangi) in the Collections of the Pitt Rivers Museum, Oxford. *Bull. Hist. Met. Group* 3.2: 39–45.

White, J.
1982 *Discovery of a Lost Bronze Age, Ban Chiang*. University Museum, Philadelphia.

White, J. R.
1980 Historic Blast Furnace Slags: Archeological and Metallurgical Analysis. *J. Hist. Met. Soc.* 14.2: 55–64.
1981 New Light on Early American Ironmaking; The Eaton Furnace. *J. Hist. Met. Soc.* 15.2: 88–93.

Wilkinson, H.
1837 On the Cause of the External Pattern or Watering of the Damascus Sword Blade. *J. Royal Asiatic Soc.* 4: 187–193.

Willen, D., W. Soedel, and V. Foley
1976 The Story of Weyland the Smith. *J. Hist. Met. Soc.* 10.2: 84–86.

Williams, A. R.
1977 Roman Arms and Armour: A Technical Note. *J. Archaeol. Sci.* 4: 258–264.

Williams, A. R. and K. R. Maxwell-Hyslop
1976 Ancient Steel from Egypt. *J. Arch. Sci.* 3: 283–305.

Wingrove, J.
1970 Identification of Iron Oxides. *J. Iron and Steel Inst.* 208: 258–264.

Wood, C.
1893 Discussion of Turner's Paper on the Production of Iron in Small Furnaces in India. *J. Iron and Steel Inst.* 44.2: 177–180.

Wood, N.
1978 *Oriental Glazes, Their Chemistry, Origins and Re-creation*. Pitman/Watson Guptill, London.

Wright, J. H.
1972 Metallurgical Examination of a Roman Iron Beam from Catterick Bridge, Yorkshire. *Bull. Hist. Met. Group* 6.1: 24–27.

Yang, K.
1982 *Development of Ironworking in Ancient China*. Peoples Press, Shanghai. [in Chinese]

Yule, H.
1842 Notes on the Iron of the Kasia Hills, for the Museum of Economic Geology. *J. Asiatic Soc. Bengal* 11: 853–857.

Yule, H. and A. C. Burnell
1903 *Hobson-Jobson*. John Murray, London.

Zacharias, S. and H. G. Bachmann
1983 Iron Smelting in Western Africa: Ivory Coast. *J. Hist. Met. Soc.* 17.1: 1–4.

Zhengzhou, Henan Provincial Museum
1978 Excavation of the Iron Smelting Site at Ancient Yingchen, Chengchow. *Wen Wu* 1978.2: 23–43. [in Chinese]

Zschokke, B.
1924 Du Damassée et des lames de Damas. *Rev. Métallurgie* 21: 658–669.

# Index

M



N



O



P



Q



R



S



V



W



Z

# ROCK QUALITY, SEISMIC VELOCITY, ATTENUATION AND ANISOTROPY

BALKEMA – Proceedings and Monographs in Engineering, Water and Earth Sciences

# ROCK QUALITY, SEISMIC VELOCITY, ATTENUATION AND ANISOTROPY

NICK BARTON

Taylor & Francis
Taylor & Francis Group
LONDON / LEIDEN / NEW YORK / PHILADELPHIA / SINGAPORE

*Taylor & Francis is an imprint of the Taylor & Francis Group, an informa business*



Typeset by Charon Tec Ltd (A Macmillan Company), Chennai, India
Printed and bound in Great Britain by CPI Bath Press (CPI-Group), Bath



Published by: Taylor & Francis/Balkema
P.O. Box 447, 2300 AK Leiden, The Netherlands
e-mail: Pub.NL@tandf.co.uk
www.balkema.nl, www.taylorandfrancis.co.uk, www.crcpress.com

*Library of Congress Cataloguing-in-Publication Data*

Barton, Nick, 1944–
Rock quality, seismic velocity, attenuation, and anisotropy/Nick Barton.
p. cm.
ISBN 0-415-39441-4 (hardcover: alk. paper)
1. Soil-structure interaction. 2. Earthquake engineering. I. Title.
TA711.5.B37 2006
624.1′51—dc22

2006005909

ISBN10: 0-415-39441-4 (Hbk)
ISBN13: 978-0-415-39441-3 (Hbk)

# Table of contents

# Preface

This book traces an accelerating path through an important part of the earth sciences, describing seismic behaviour and rock mechanics interpretation at many scales, to illuminate what lies beneath the earth's immediate surface. Although geophysics, and the rock mechanics and engineering geology of discontinuous media share the same medium, they have had a mostly separate development – with little cross-referencing in the multitude of journals. Regrettably, we seldom see geophysics colleagues at our rock conferences. This book attempts to bridge this void in strategic locations.

Seismic velocity, seismic quality (the inverse of attenuation), and anisotropy are some of the very basics of geophysics, and they depend absolutely on the rock and fluid properties, the rock mass structures, the jointing, the fracturing, the microcracks and the other pore space. These are some of the fundamentals of earth science. All contribute to the resultant dynamic stiffnesses, and to the fluid pressure micro-flow reactions, whether at dam foundation depths, tunnel depths, reservoir-well depths, or earthquake depths. All components of the anisotropic, dynamic, stiffness-velocity-permeability half-space, respond together in a logical pattern. Attempting to understand this pattern is a major objective of this book.

The assumed 'shared earth' response is revealing itself with increasing speed. Despite the very small strains and displacements involved in seismic wave loading there are inevitable, encouraging parallels, to the rock mechanics of larger strains and displacements. This makes seismic response more understandable and more logical for a wider group of professionals, with contributing areas of expertise.

In synthetic modelling in geophysics, there is now much interest in the rock joint or rock fracture compliances that may hold part of the secret of fractured reservoir description. These same properties, when inverted, are used over much larger displacements, in rock stability and deformation modelling. Remarkably, the dynamic compliance and static stiffness of fractures and joints have mostly had a compartmentalized development in the different disciplines. A dynamic, micro-strain-based *normal compliance* of $10^{-13}\,\mathrm{mPa}^{-1}$ derived from shear-wave anisotropy measurement in the sub-surface, is of recognisable magnitude when inverted, to compare with the pseudo-static 'macro-strain' *joint normal stiffness* (i.e. 10,000 MPa/mm or 10 MPa/micron) obtained from incremental loading tests on similar rock joints at similar high stress levels.

The level of rock stress, the joint wall roughness, and the joint wall compressive strength, which are also important components of aperture and permeability, provide estimates of these physical properties, not just the diagonal members of a stiffness matrix. Here we have a classic reason for a *disconnect* between part of the earth sciences, which can be bridged with advantage.

Attenuation and rock quality, another area of disconnect, can *also* be linked, but not quite so simply as taking the inverse of attenuation and calling it seismic quality. The universally used seismic quality Q of geophysics, that we will often call $Q_{seis}$, shows some qualitative and quantitative connections to *rock mass quality*, also called Q, and widely used in rock engineering since the 1970's. The *rock mass quality* (Q), which we will often call $Q_{rock}$, is composed of several attenuation-causing parameters, that are directly equivalent to block size, inter-block friction and a rough measure of effective stress and permeability.

There are clear, broad links between $Q_{rock}$ and $Q_{seis}$, due to the discovery of a mutual connection to the empirically derived and stress-dependent deformation modulus of rock masses. This connection is despite the fact that only micro-strains, micro-displacements, and micro-flows (squirt) occur with the passage of dynamic waves. Rock *mass* behaviour is non-linear and scale-dependent. Load-deformation curves have different gradients at different stress levels. Dynamic waves seem to sense this non-linearity, and they apparently sense *some* of the scale effect too.

This book is dedicated to making some of these cross-discipline empirical connections, in a simple non-mathematical way, so that the people who see a lot of rock in their daily endeavours (geologists, engineering geologists, rock mechanics and rock engineers), and those who see, and interpret, and model complex seismic results, from earthquakes, from fractured petroleum reservoirs, and from laboratory rock physics reservoir simulations, can more easily communicate in the common anisotropic stiffness-velocity-permeability half-space that is earth science.

Communication in words and diagrams, rather than through complex formulae and matrices. At least half of the people working in the earth sciences are not as good at mathematics as the other half may have assumed.

## Acknowledgements

First and foremost this book is an acknowledgement to the many thousands of earth scientists working with geophysical interpretation of the near-surface, the sub-ocean, and the seismic shallow crust. Their dedication and interesting publications have made this book a possibility. This volume is a well-illustrated documentation of just some of their excellent work. The journey through their contributions has been one of increasing excitement.

Efforts have been made to reproduce the physical essence of reviewed work with suitable choice of author's figures. Ricardo and Marcelo Abrahão have excelled in the expert redrawing of such figures, and are sincerely thanked for their painstaking work. The writer's summaries of key aspects of reviewed work are interspersed with personal and rock mechanics based interpretations with which authors need not be in full agreement.

Material contributions, in the form of inaccessible articles, figures and data, and some valuable discussions and improved insight, have kindly been provided by Dr. Enru Liu, Dr. Eda Quadros, Dr. Baotang Shen, Dr. Axel Makurat, Prof. Stavros Bandis, Dr. Karstein Monsen, Prof. Michael King, Dr. Stuart Crampin, Dr. Heloise Lynn, Harald Westerdahl, Dr. Sonja Maultzsch, Dr. Paul Chapman, Dr. Rudi Lubbe, Dr. Tor Arne Johansen, Dr. Barry New, Dr. Saul Denekamp and Dr. Tore Lasse By, who enthusiastically introduced the writer to cross-hole seismic tomography in 1986.

Part I of this book was mostly completed while the writer was Visiting Professor in the University of São Paulo Polytechnic (USP). The writer's kind neighbour in the Mining Department, Prof. Lineu Ayres da Silva, was indirectly responsible for the five years extension involved in starting and completing Part II of this book. A recently purchased volume by Kearey and Vine, 1996 lay open on his desk. A plate tectonics section of a plunging sub-ducting crust with labels 'low Q', 'high Q' caught the writer's rock-engineering attention. What did this 'Q' mean? Some of the complex answers, and a simple one showing promise, will be found in Part II.

My final acknowledgements are firstly to Pat Coughlin, who has ensured a smooth-running and expert manuscript production over a long period of endeavour. This started with the deciphering of handwriting and ended with countless explanations of Microsoft's hidden logic. The enthusiastic team at Taylor & Francis, Germaine Seijger and Lukas Goosen and the Charon Tec team have produced a work to be proud of. The reader can be the judge of this. Finally my thanks and apologies to a tolerant and loving wife Eda, who also ensured some key insights into rock-fluid interactions.

## Permissions to Reproduce Figures

The nature of this book, specifically a wide-reaching literature review, involving some 830 references from some forty different journals and publishing houses, has made obtaining permissions to reproduce figures a daunting and sometimes impossible task regarding author-permissions, due to the several hundreds of first authors, and thousands of multiple authorships. There are instances where we have been unable to trace or contact the copyright holder. If notified, the publisher will be pleased to rectify any errors or omissions at the earliest opportunity. Many key authors are retired, regrettably some have died, including Bengt Sjögren, who's published work from 1979, 1984 and 2000 was an important source for key figures in several chapters of Part I. The most prominent authors have kindly given permission for multiple reproduction of figures from my limited selection from their important contributions. All publishers as listed below, have kindly given their permission for multiple reproduction of the numerous figures reproduced in this reference volume. Their joint permissions, and those of contacted authors, *and* the contribution of all authors that could not be contacted for whatever reason, are gratefully acknowledged. Their excellent work, reproduced in this book, is a sincere acknowledgement of their contributions to geophysics.

*Acoustical Society of America (ASA)*: Journal of the Acoustical Society of America: Figure 13.42

*American Association of Petroleum Geologists (AAPG)*: Figure 15.36

*American Geophysical Union (AGU):* Journal of Geophysical Research: Figures 2.8, 3.1, 5.33, 5.34, 5.35, 10.14, 10.21, 10.25, 10.27, 10.28, 10.33, 10.37, 10.38, 10.41, 10.43, 10.44, 10.47, 10.48, 10.52, 10.53, 10.55, 10.58–10.60, 11.1, 11.6, 11.7, 11.8, 11.9ab, 11.10–11.21, 11.24–11.30, 11.31a, 11.32, 11.33, 11.35, 11.36, 11.38, 11.40–11.42, 11.48, 11.49, 11.52, 11.54–11.64, 11.66–11.71, 12.11, 12.22, 12.23, 13.2, 13.5a, 13.25, 13.29, 13.32, 13.33, 13.46, 14.16, 14.25, 14.26, 15.8, 15.11, 15.14, 15.18, 15.63. Figure Part II; Tables: 10.5, 10.6, 11.2, 15.2, 15.3, 16.5, 16.6

*American Institute of Mining, Metallurgical and Petroleum Engineers (AJME)*: 16.42, 16.68

*American Institute of Physics (AIP)*: Figure 10.21

*American Physical Society (APS)*: Physical Review E: Figure 10.64

*American Society of Civil Engineering (ASCE)*: Journal of Geotechnical Engineering: Figure 2.15

*American Society of Mechanical Engineers (ASME)*: Transactions of the American Society of Mechanical Engineering: 12.6; Journal of Applied Mechanics: 2.9

*Blackwell Publishing:* Geophysical Prospecting: Figures 1.3, 1.5, 1.7, 1.8, 1.10, 1.11, 3.9, 4.3, 5.2–5.4, 5.10, 5.11, 6.11, 6.17, 8.12, 9.2, 10.65, 10.67, 13.24, 13.25, 13.36–13.41, 13.44, 13.48, 13.61, 14.15, 15.5, 15.6, 15.22, 15.28, 15.39, 15.40–15.42, 15.47, 15.48, 15.51–15.53, 15.55, 16.20–16.22; Geophysical Journal International (Geophys. J. Int.): 10.22–10.24, 15.1a, 15.3, 15.4, 10.67; Other sources: Figures Part II, 11.1, 11.2, 11.18; Table 11.1

*Cambridge University Press*: Figures 11.3, 13.1, 13.2, 13.5 and 14.4

*Centek Publishers,* Luleå: Figure 16.13

*Comité Francais de Géologie de l'Ingénieur et de l'Environnement (CFGI)*: Paris: Figures 5.6, 5.7, 8.5; Tables 8.1, 8.2

*Coyne et Bellier*: Figures 7.7, 6.19, 6.21

*Elsevier:* International Journal of Rock Mechanics and Mining Sciences and Geomechanics Abstracts: Figures 2.1, 3.2, 3.8, 4.7ab, 4.13, 4.14, 4.17, 4.20, 5.29, 5.30ab, 6.9, 6.20, 7.18, 7.20, 7.25, 7.26, 7.31, 7.32, 8.2–8.4, 9.6, 15.9, 13.53–13.55, 13.58, 15.17, 16.2, 16.4, 16.6, 16.7, 16.9, 16.12, 16.16, 16.17, 16.44, 16.46, 16.69, 16.73, 16.74; Table 4.1; Engineering Geology: Figures 5.17, 5.19, 15.26, 14.39; Journal of Applied Geophysics: 14.15, 15.5a, 15.56, 15.57; Table 16.8; Tectonophysics: Figures 11.31b, 11.53, 16.64, 16.65, 16.76. Other sources: Figures 1.1, 1.6, 2.18, 4.12, 4.21, 5.13, 10.57, 11.5, 11.34, 15.23; Tables 2.2, 5.2, 11.1

*European Association of Geoscientists and Engineers (AEGE)*: First Break: Figure 15.31; Other sources: Figures 10.2, 10.3, 10.10; 10.20, 10.21, 10.31, 10.36, 13.24, 14.37, 14.38, 15.27, 15.31, 15.37, 15.38, 15.43, 15.45, 15.46, 15.54; Table 13.2

Geophysical Research Letters: Figures: 4.9, 10.52, 11.51, 11.56, 15.44, 16.19; Other sources: Figures 9.7, 12.7, 12.8, 11.39, 11.46, 11.54, 13.11, Table 11.3

*Geological Society of America (GSA)*: Geology: Figures 3.13, 10.6, 16.11, 16.56, 16.63; Figure 1.4

*Geological Society*: The Quarterly Journal of Engineering Geology: Figures 3.7, 3.10, 5.15, 5.16; Other sources: 2.12, 11.47, 13.56, 13.57, 15.16, 16.23

*Imperial College,* London: Figure 16.6

*Imprime Adosa,* Madrid: Figure 3.3, 5.1, 5.8, 5.9, 8.16

*Institut du Bâtiment et des Travaux Publics; Annales d'ITBTP*: Figure 6.20

*Institut Français du Pétrole (JFP)*: Oil & Gas Science and Technology: Figures 3.5, 14.32, 14.33

*International Association for Engineering Geology and the Environment (IAEG)*: Figures 3.3, 5.1, 5.6–5.9, 8.5, 8.16; Tables 8.1, 8.2, 16.7

*International Commission on Large Dams (ICOLD),* Paris: Various sources: Figures 3.6, 6.7, 7.4, 9.1, Table 6.3.

*International Society for Rock Mechanics (ISRM)*: ISRM News Journal: Figures 7.2, 7.3, 6.18, 8.21–8.23, Table 12.1

*Ishikawa Soil Incorporated Association*: Figure 2.13

*Japan Tunnelling Association (JTA);* Tunnels and Underground: Figure 8.6

*John Wiley & Sons*: Figures 7.30, 6.23

*Kansas Society of Petroleum Engineering*: Figures 13.2, 13.5ab

*Laboratório Nacional de Engenharia Civil (LNEC),* Lissabon: Figures 2.2, 6.1, 6.15, 6.22, 6.23

*Nagra;* Nagra Bulletin: Figure 9.1

*National Academy Press*, Washington: Figures 6.2, 6.3, 6.8, 7.1, 7.8, 9.11; Table 6.2

*Norwegian Petroleum Society (NFP)*: Figures 14.29, 14.30, 15.36

*Office of Nuclear Waste Isolation (ONWI)*, Columbus: 16.10, 16.14, 16.15, 16.29–16.33, 16.46, 16.67

*Österreichischen Gesellschaft für Geomechanik (ÖGG)*, Felsbau: Figure 6.4

*Oyo Corporation*: Figure 2.12

*Royal Astronomical Society (RAS)*: Quarterly Journal of the Royal Astronomical Society: Figures 11.5, 11.37, 11.50

*Schlumberger*: Oilfield Reviews: Figures 12.24–12.26, 14.15, 15.1b, 15.19, 15.36, 15.1b; Other sources: 4.10

*Seismological Society of America (SSA)*: Bulletin of the Seismological Society of America: Figures: 10.39, 10.40, 10.46, 10.52, 10.61, Table 10.7

*SGE Editoriali,* Padova: Figure 2.13

*SKB,* Stockholm: Figure 7.23

*Society for Mining, Metallurgy and Exploration (SME)*: Various sources: Figures 2.4–2.7, 5.1a, 6.12, 7.12–7.15, 7.28, 15.7, 15.25, 16.27, 16.75

*Society of Exploration Geophysicists (SEG)*: Geophysics: Figures 2.11, 2.19, 2.21, 3.11, 3.16, 4.15, 7.27, 10.1, 10.4–10.11, 10.13, 10.15ab, 10.16–10.19, 10.29, 10.30, 10.34, 10.35, 10.52, 10.64, 10.66, 10.68–10.72, 11.22, 11.23ab, 11.43, 11.48, 12.27, 13.3, 13.4, 13.6–13.8, 13.11–13.13, 13.17–13.23, 13.26–13.31, 13.34, 13.35, 13.42–13.45, 13.50–13.52, 14.1–14.3, 14.6–14.15, 14.18–14.24, 14.28, 14.31, 15.11, 15.29, 15.30, 15.60, 16.64. Tables: 10.10, 14.1–14.3, 15.1; The Leading Edge: Figures: 12.1a–d, 12.2a–d, 13.10, 13.14a–b, 13.15, 13.16, 14.33–14.36, 14.38, 15.15, 15.24, 15.35, 15.60; Canadian Journal Exploration Geophysics: Figures 10.63, 15.12–15.14, 15.32; Other sources: Figures 12.3, 12.30, 14.1, 15.5c, 15.10, 15.29, 15.44, 15.54, 15.65abc, 15.66, 16.6

*Society of Petroleum Engineers (SPE)*: SPE Journal: Figures 13.2, 13.5ab, 14.32, 14.33; Other sources: Figures 12.12, 12.13, 12.29

*Southern Africa Institute of Mining and Metallurgy (SIAMM)*: Figure 15.46

*Springer Science and Business Media*: Rock Mechanics: Figures 2.10, 16.10, 16.26, 16.41, 16.54ab, 16.57ab, 16.58; Pure and Applied Geophysics – Pageophysik: 7.22ab, 10.12, 10.49ab, 10.50ab, 10.51, 10.52, 10.54; Other Sources: Figures 10.42, 13.1, 16.60; Table 3.1

*Stanford Rock Physics & Borehole Geophysics (SRB)*: Figures 13.2, 13.5ab

*Swedish National Science Council*: Figure 1.45

*Tapir Academic Press,* Trondheim: Figure 15.25

*Thomas Telford:* Geotechnique: Figures 12.5, 12.9, 12.10, 15.2, 16.2, 16.8, 16.53, 16.75
Other sources: Figure 4.4, Tables 1.2, 1.3

*University of California Berkeley*: Figure 16.46

*Wilmington*: Tunnel & Tunnelling International: Figure 9.13

**PhD Theses:**
S. Bandis, 1980, University of Leeds (Fig. 16.3?, 16.16, 16.18, 16.40, 16.47, 16.52, 16.66, Tbl. 16.2, 16.3);
T. Cadoret, 1993, University of Paris (Fig. 13.2, 13.5ab, 13.20);
D. Han, 1986, Stanford University (Fig. 13.2, 13.4, 13.5e);
K. Iwai, 1976, University of California Berkeley (Fig. 16.46);
D.L. Jizba, 1991, Stanford University (Fig. 13.5d);
Y.-Q. Liu, 2003, University of Edinburgh (Fig. 14.15b);
R. Lubbe, 2005, Oxford University (Fig. 16.20, 16.23);
N. Lucet, 1989, University of Paris (Fig. 13.2, 13.5ab);
E. Quadros, 1982 (Msc), University of São Pualo (Fig. 16.6);
A. Shakeel, 1995, Imperial College, Univ. London (Fig. 13.58);
J.C. Sharp, 1970, University of London (Fig. 16.6);
C. Slater, 1997, University of Edinburgh (Fig. 15.20, 15.34, 15.35);
S.R. Tod, 2002, University of Cambridge (Fig. 15.44);
J. Yan, 2003, University of Edinburgh (Fig. 13.14);
J. Yuan, 2001, University of Edinburgh (Fig. 14.27).

# Introduction

## The multi-disciplinary scope of seismic and rock quality

Seismic, sonic and ultrasonic measurements are utilised by a large number of geo-science, geo-engineering and geo-resource disciplines. Their use is so widespread in the earth-sciences, that it should be of no surprise to us that such techniques are also used to register such diverse subjects as osteoporosis in cows, and the control of 'crispiness in breaded fried chicken nuggets'. The latter was a thesis in Biological Systems Engineering.

Since rock engineers tackle different problems from petroleum engineers and geophysicists, who in turn tackle different problems from tectonophysicists, there has been an understandable yet regrettable compartmentalisation between the disciplines. Both practitioners and researchers in each of these major fields, generally go to different conferences and read and publish in different journals, as there are 'too many' choices of each in each discipline, even in each speciality where we earn our living. The luxury of cross-discipline interaction, occasionally experienced with great interest and resulting stimulation, is usually defeated by time, cost and also in part, by technical-language barriers, and even mathematics.

An interesting example of partial 'compartmentalization' is *stiffness* and *compliance*. Each have followed almost separate development since the late 1960s in rock mechanics, and since the early 1980s in geophysics. Each are essential to each subject; for numerical modelling of stability and deformability in rock engineering; for improved interpretation of attenuation, anisotropy and shear wave splitting in the geophysics of fractured petroleum reservoirs. Yet the dynamically measured, micro-deformation fracture compliances in geophysics (in the normal and shear directions), are numerically close to the inverse of incrementally-loaded joint stiffnesses in rock mechanics, at least when rock quality is high.

The frequently illustrated material in this book has been assembled as a result of an interest in a variety of civil, mining, petroleum, geophysics and earth-science fields. The common denominator has been rock mass and rock joint behaviour as presumably impacting the seismic interpretation. An interesting and very large selection of seismic velocity and seismic quality related data, from practitioners working in widely varied disciplines, has been assembled. Much has obviously been left out or not yet seen. Much is still under development.

The chapters of Part I are mostly civil engineering related with strong links to the interpretation of rock conditions at both laboratory and field-scale, with their impact on engineering of tunnels and dams and planned nuclear waste repositories. The chapters of Part II go deeper both figuratively and literally, and consider much larger scale uses of seismic attributes, from hydrocarbon reservoirs and the use of multiple dynamic energy sources, to the interpretation of mid-ocean spreading-ridges, to crustal conditions interpreted from natural earthquake hypocentres.

The phenomen of seismic anisotropy, known already in the nineteenth century to give lower stiffness perpendicular to layering than parallel, is now in widespread use for investigating fractured rock at depth. Features of the rock mass, though of sub-seismic-wave size, can be detected at many kilometers depth, due to shear wave splitting, giving polarization parallel and perpendicular to dominant jointing. Different time delays for the fast and slow shear wave components vary with fracture properties and with frequency, giving frequency-dependent anisotropy.

Efforts have been made to seek out and to reproduce in brief, with helpful figures, the seismic measurements and interpretations which have a clear or potential *rock quality* content, at whatever scale. Clearly the term 'rock quality' conceals various techniques and scales of measurement, and varied interests in 'rock quality' per se. A rock mass with *high velocity* and *high rock quality* (i.e., exhibiting low attenuation) would make life less profitable for machine bored tunnellers due to slow progress and frequent cutter-changes. Aggregate producers would need more drilling and explosives per ton, and would seek other quarries. The very existence of hydrocarbon reservoirs and their productivity would be severely prejudiced if either 'rock quality' or 'seismic quality' was too high. Others would welcome good 'rock quality' characteristics, for example producers of dimension stone and clients expecting cheap drill-and-blasted tunnels requiring little rock support.

## Revealing hidden rock conditions

At the beginning of most rock engineering projects we are operating 'blind', and any help to 'see' what may lie below our dam foundation, or ahead of our tunnel, saves schedules, budgets and sometimes lives as well. The beauty of seismic, sonic or ultra-sonic investigations is that they can be applied over a virtually unlimited range of scales, to 'see' micro-cracks closing under stress in the laboratory, or to 'see' fluctuations in effective stress across a regional fault caused by changes of reservoir level, and to monitor the effects of water-flooding in a fractured petroleum reservoir.

Already in 1917, Fessenden had proposed (and patented) the use of a cross-hole seismic technique to locate ore bodies. The scale of investigation can be increased by orders of magnitude to 'earth-scale', when illuminating the seismic structure of the earth's crust, and further again to depths of 5000 km or more, to the solid iron core of the earth, as a result of global-station analyses following large earthquakes.

Sjøgren, 1984, gave the civil engineering (near-surface) profession a particularly useful guide in the use of shallow seismic refraction techniques for those involved in shallow sub-surface projects. The fundamental principle is that seismic waves propagate with significantly different velocities in different near-surface geotechnical and geological strata, due to the seismic visibility of weathered, low-stressed materials in general. This also means that the velocities tend to increase rapidly with depth, which must not be misinterpreted as meaning better quality per se. Intermediate high-speed layers, or hidden low velocity layers obviously disturb this simplified picture, and velocity anomalies and incorrect depth interpretations result unless separate analysis i.e., downhole vertical seismic profiling (VSP), or coring is performed.

Fundamental difficulties in the context of rock engineering (and in all other disciplines too) are that the means of access, superficial or along boreholes, are often limited by the geometry of the problem, by the (urban or sub-sea) location, and by the cost. The freedom to choose optimal experimental layouts is therefore limited. As pointed out by Cosma, 1995, this may cause blind zones, even in the immediate vicinity of the observation points.

In the case of soil or weathered rock horizons, seismic velocity interpretation readily distinguishes the water table from a lithological boundary by inspection of the shear or transverse wave velocity ($V_s$). If this remains constant across the region of changing water content, while $V_p$ changes, a groundwater surface is indicated, since the shear waves do not respond to changing water content due to the lack of shear stiffness. If $V_s$ also changes, a geotechnical or geological layer will have been crossed. Typical ranges of $V_p$ for a variety of near-surface sediments and rocks are reviewed in Chapter 1.

One of the historic and important applications of refraction seismic in *civil engineering*, has been at dam sites, which were investigated in great numbers, especially in the 1950s, 60s and 70s. Rock quality, permeability, and deformation modulus were of fundamental importance. Associated hydropower tunnels such as headrace and tailrace tunnels have been the subject of countless thousands of kilometres of seismic refraction spreads, not to mention all the power house foundations and high pressure penstock locations.

The seismic spreads at the ground surface should if possible be set out in optimal directions to investigate suspected sub-surface anomalies. Since the ray paths are essentially following sub-horizontal paths, steeply dipping or vertical features such as faults or deeply weathered zones can be readily located and given a characteristic seismic signature. Localised P-wave velocities of 2 or 3 or 5 km/s have distinct engineering implications for near-surface tunnelling or foundation stripping. Their interpretation in relation to rock type (uniaxial strength and porosity) and in relation to the depth of measurement, or to stress level and stress-induced anisotropy, will be reviewed in detail in this book, with the help of a quantitative *rock mass quality* description.

## Some basic principles of P, S and Q

The P-wave is a longitudinal wave, in which the direction of particle motion coincides with the wave propagation. It is often termed the first arrival or compressional wave. By contrast, the lower velocity transverse S-wave has particle motion in the plane perpendicular to the direction of wave propagation. An S-wave is of two possible basic types: the SH-wave in which particle motion is parallel to a boundary, usually the ground surface, and the SV-wave which has particle motion perpendicular to both the wave propagation direction, and to the particle motion of the SH-wave.

When passing through anisotropically fractured petroleum reservoirs, a shear wave will likely split into fast (qS1) and slower (qS2) polarized components, giving clues about the fracturing character and perhaps the principal stress direction. The latter coupling may be more complex than convention suggests however, due to adverse stress-closure-permeability behaviour in reservoir rocks, unless they are strong enough to tolerate tens of megapascals of effective normal stress across their 'open' joints or fractures. Slight shearing and dilation may actually be needed on conjugate joint or fracture sets, to explain permeability and production from fractures in weaker reservoir rocks, and to explain the 'surprising' maintenance of permeability deep into the crust.

There is a 'problem' of frequency dependence for all the component velocities of P- and S-waves, but in fact in the problem lies the more accurate interpretation. There are exciting current developments in these dispersive, frequency-dependent interpretations of velocities and attenuation, and in their relation to anisotropy, where rock mechanics knowledge of 'joint stiffnesses', or their dynamic micro-strain-based near-inverses: the geophysicist's 'fracture compliances', are proving extremely important supplements to the earlier focus on the elliptic aspect ratios of micro-cracks, and the larger-scale – and smaller magnitudes – of the aspect ratios of almost closed fractures.

## Q and Q

Seismologists have had a long tradition of utilising a quality factor Q-*seismic* (with numerous sub-sets such as the basic Qp, Qs, and Qc , the latter from the coda or tail-end of a dynamic wave sequence). Q-*seismic* was popularized by a famous Knopoff, 1964 paper with the briefest possible title: 'Q'. We will see the possibility of a Q-*seismic* relation with another quality descriptor called the 'Q-*value*', from rock engineering, not directly, but via a mutual apparent relation to the stress-dependent pseudo-static deformation modulus: surprisingly not to the dynamic modulus, at least not in the top kilometre or so.

Q-*seismic* is a dimensionless factor whose inverse ($Q^{-1}_{seis}$) indicates, if simply stated, the percentage loss of energy of a single wave length due to various (and sometimes disputed) mechanisms of attenuation in the rock mass at many possible scales. Reduction in wave amplitude is the most obvious effect. The attenuation is caused by scattering from geo-structures of different scales, and by absorption in intrinsic micro-mechanisms like normal and shear micro-displacements across microcracks and joints, therefore involving friction to some degree, and relative micro-movement of fluids between the pore-space, the micro-cracks and the jointing or fracturing.

As a result of the passage of the very slightly deforming seismic waves there will be a lot of references to 'squirt flow' losses in Part II of this book, in connection with anisotropic attenuation, which is one of several properties of the fluid conducting structures of fractured or naturally jointed hydrocarbon reservoirs.

In parallel but previously almost unrelated endeavours, a prominent engineering geologist (Deere, 1964) developed a simple empirical rock quality factor RQD, related with the degree of jointing or fracturing in drill-core. In the 1970s, with no knowledge of $Q_{seis}$, the rock quality Q-value was developed, which includes RQD as one of the six parameters. The rock engineering *rock quality* Q-value describes the degree of jointing (as relative block size) and important 'internal' joint properties like roughness and clay-filling (giving the inter-block friction coefficient). It also incorporates estimates of the permeability and the stress-to-strength ratio.

Frequent use will be made of the Barton et al., 1974 and Barton 2002 rock quality Q-value and $Q_c$-value in various parts of this book. It provides a simple link to seismic velocity, and it probably has the potential for explaining some attenuation mechanisms as well. The rock quality Q-value has a six orders of magnitude scale of quality (from 0.001 to 1000), and it predicts a two to three orders of magnitude range of deformation modulus. Completely unjointed, massive rock masses, with $Q \approx 1000$, will clearly show almost no attenuation. At many kilometres depth, $Q_{seis}$ values are of similar magnitude. Completely decomposed, near-surface, faulted rock with $Q \approx 0.01$–$0.001$ will obviously give complete attenuation (i.e. effectively lower than the theoretically lowest possible $Q_{seis}$ and highest possible $Q^{-1}_{seis}$ – each probably beyond measurement limits).

It is expected that graphs of *Q (seismic quality factor*) versus *Q (rock quality factor)* in rock masses (as opposed to lab-samples), can show strong correlations in the future, when geophysics data is reported in parallel with rock quality data. Each of the 'Q-factors' will be described in greater detail later in this book. We will also see the 'problem' of frequency-dependence, and the 'problem' of anisotropy, but both these *problem areas* are obviously concealing the potential for improved interpretation of the structures beneath the earth's surface, both shallow and deep.

## Limitations of refraction seismic bring tomographic solutions

Refraction seismic methods have been used for at least fifty years, but have some fundamental limitations that include masking of lower velocity layers under higher velocity materials, such as basalts above petroleum reservoirs, and difficulties with multiple velocity layers in close proximity. Some unexpectedly costly tunnelling has resulted from mistakes in interpretation, due to such features. However, as with most limitations, there are various solutions, and geophysicists have been extremely creative, and also willing to modify and apply techniques from other well-funded fields like medicine.

While P-wave and S-wave measurement between two points can be expressed as average wave velocities (or give a rather unhelpful 'average' picture of a patients brain), there is the possibility of using more comprehensive multiple source *and* receiver positions in separate multiple-boreholes, thereby giving positional (2D or 3D) tomographic imaging. A tumour in an unfortunate patient, and real-time scanning of brain-wave activity, as illuminated in medicine, have their engineering-scale equivalents. A fault zone delaying a tunnel, and four-dimensional fluid-migration-imaging in a producing reservoir would be approximate, large-scale geophysics equivalents. The most basic imaging analogy has been practiced for many years by geophysicists, who use earthquake sources and global monitoring stations to deduce the structure of the whole earth. So perhaps *geo*-physicists actually helped to inspire medical imaging of the human body?

In intermediate-scale, near-surface civil engineering, the strategic positioning of pairs of boreholes across complex zones or faults can be used for optimal characterization of these features, if they appear to be a threat to progress of a tunnel, or to dam foundation integrity. In special cases cross-hole tomography measurements may lead to the avoidance of collapse, as more reliable decisions can be taken concerning the need for strengthening by pre-grouting, or the need for special pre-installed ground support, or perhaps even ground-freezing. Tunnels with inadequate overburden or severe water leakage potential such as inundation by rivers or lakes, or local inflows that would allow pore pressure draw-down compaction in soft clays beneath important buildings, can also benefit greatly from seismic-based decisions for special treatment of the ground.

*Part I* which occupies the first third of this book, will be found to contain mostly civil-engineering and tunnel engineering treatments of the velocity-quality links that are helpful when interpreting near-surface conditions.The complementary laboratory testing that has often accompanied geophysics investigations of the near-surface, will also have emphasis on lower stress. Because of this, the effect of weathering and alteration and excavation on seismic attributes, will each be emphasised. Despite the obvious challenges of seismic interpretation in fractured and faulted petroleum reservoirs at many kilometers depth, or of mid-ocean ridge investigations beneath three kilometers of ocean, many geophysicists insist that obtaining high resolution images from ground level to just 50 m depth, is still one of the major challenges of modern geophysics. This happens to be the layer of the subsurface closest to most of our civil engineering endeavours, from tunnels, to dams, to the foundations for high buildings.

*Part II* of this book tackles greater depths, greater scales, and more subtle geophysical detail, as benefits this rapidly developing field. Geophysics has been in 'rapidly developing' phases many times in the past. The latest phase is due to many parallel developments, not least an acceptance of the benefits of three-dimensional surveys, of monitoring reservoir changes over time (4D), each requiring the ever-developing power of modern computers for the complex processing of huge amounts of digital data. Investment in geophysics is growing further, due to the inestimable advantages of improved information. The continued search for reliable earthquake precursors, and the pressures to find more hydrocarbons in more heterogeneous reservoirs, and improve the recovery from those already being depleted, are each driving the developments in this remarkable field. In the future, more geophysical investments may also be used to aid in the search for potable water, which already far exceeds the price of gasoline in many locations.

# Nomenclature

| | |
|---|---|
| $\beta$ | angle subtended between a discontinuity and the major principal stress $s_1$ |
| $\gamma$ | rock mass density (t/m$^3$) |
| $\gamma$ | Thomsen S-wave anisotropy parameter (horizontally polarized waves) |
| $\varepsilon$ (also e) | crack density |
| $\varepsilon$ | Thomsen P-wave anisotropy parameter |
| $\zeta$ | Thomsen S-wave anisotropy parameter (vertically polarized waves) |
| $\lambda m^{-1}$ | frequency of joints (or fractures) per meter (also F m$^{-1}$) |
| $\mu$ | shear modulus |
| $\sigma_c$ | uniaxial compression strength (MPa) |
| $\sigma_{h\,min}$ | minimum horizontal component of stress |
| $\sigma_{H\,max}$ | maximum horizontal component of stress |
| $\sigma_r$ | radial stress around an excavation in rock |
| $\sigma_v$ | vertical component of principal stress |
| $\sigma_1\ \sigma_2\ \sigma_3$ | principal stresses |
| $\sigma_\theta$ | tangential stress around a (circular) opening |
| $\sigma_{\theta\,max}$ | maximum tangential stress |
| $\sigma_{\theta\,min}$ | minimum tangential stress |
| $\tau$ | shear stress (in a direct shear test) |
| $\phi$ | friction angle of joint, fracture, filled discontinuity, fault (geomechanics) |
| $\phi$ | fractional porosity (rock physics) |
| $\phi_b$ | basic friction angle, flat unweathered surfaces, low stress |
| $\phi_c$ | critical state line defining $s_1 = 3s_3$ |
| $\phi_{peak}$ | peak friction angle of a joint, fracture |
| $\phi_r$ | residual friction angle of a joint, fracture, fault |
| $\Delta$ | change in value (e.g. $\Delta$e, $\Delta$E applying to changes in joint or fracture apertures) |
| $\Delta_v$ | vertical component of deformation |
| $\Delta_h$ | horizontal component of deformation |
| $\psi$ | axial modulus |
| ANDRA | Agence Nationale pour la gestion des Déchets Radioactifs |
| AR | advance rate (TBM, actual weekly, monthly rate) |
| AVO | amplitude variation with offset |
| AVOA | amplitude variation with offset and azimuth |
| $B_n$ | dynamic normal compliance (of single joint or fracture) |
| $B_t$ | dynamic shear compliance (of single joint or fracture) |
| BB | Barton-Bandis constitutive model for rock joints, used with UDEC as UDEC-BB |
| BEM | boundary element method of numerical modelling |
| BGS | British Geological Survey |
| BHA | bottom hole assembly |
| BHC | borehole compensated sonic logging tool |
| BHTV | borehole televiwer |
| BISQ | Biot and squirt flow model |
| BP | British Petroleum |
| BWIP | Basalt Waste Isolation Project, Hanford, Washington, USA |

| | |
|---|---|
| c | cohesion of intact rock, joint, fracture, or rock mass |
| CBTF | Conoco Borehole Test Facility |
| CC | cohesive component of rock mass (from Q-value) |
| CDR | compensated dual resistivity log |
| CSFT | coupled stress/shear flow test/temperature, for HM, HTM testing of joints |
| CSM | Colorado School of Mines, Idaho Springs experimental mine facility |
| md, $\mu$d | Darcy-based unit of permeability (md, $\mu$d for milli-, micro-darcies) |
| DEM | distinct element modelling |
| D-H-M | dynamic-hydraulic-mechanical coupling |
| $\Delta e$ | change of hydraulic aperture (joint, fracture: interpret from flow test) |
| $\Delta E$ | change of (mean) physical aperture (joint, fracture: interpret from deformation) |
| $E_{dyn}$ | dynamic axial or Young's modulus from $V_P$ and $V_S$ measurement |
| $E_e$ | modulus of elasticity (pseudo-static *unloading* stiffness: plate load test) |
| $E_{F\ dyn}$ | (as $E_{dyn}$ but field-scale, based on seismic measurements, shortened to $E_F$) |
| $F_{L\ dyn}$ | (as $E_{dyn}$, lab-scale, based on ultrasonic measurements, shortened to $E_L$) |
| $E_{mass}$ | pseudo-static modulus of deformation (also D, $E_d$ and M) from *loading* stiffness of rock mass |
| e | hydraulic aperture of a joint or fracture ($k_{intrinsic}$ laminar flow, defined as $e^2/12$) |
| e (also $\varepsilon$) | crack density |
| E | mean physical aperture of joint or fracture (empirical JRC-estimated, or BB-model) |
| EDA | extensive dilatancy anisotropy |
| EDZ | excavation disturbed/damaged zone ( typically around tunnels) |
| *Mini-EDZ* | *'alteration zone'* typically around boreholes or wells |
| EOR | enhanced oil recovery |
| F $m^{-1}$ | frequency of fractures (or joints) per meter |
| FEM | finite element method of numerical modelling |
| FC | frictional component of rock mass (from Q-value) |
| FLAC | two-dimensional continuum code for modelling small or large deformations in rock or soil |
| $FLAC^{3D}$ | 3D continuum code for modelling small or large deformations in rock or soil |
| FM, FMS | formation micro-scanner |
| FRACOD | fracture mechanics boundary element code for modelling fracturing process in rock |
| FZI | flow zone indicator |
| GRM | generalized reciprocal method |
| HDR | hot dry rock |
| HPHT | high pressure high temperature (well) |
| HRSN | high resolution seismic network, Parkfield, California |
| HSP | horizontal (in-tunnel) seismic reflection profiling |
| HTI | as TIH, transversely isotropic, horizontal axis of symmetry |
| HTM | hydro-thermal-mechanical (coupling) (also MHT) |
| i | with + or − implies dilation or contraction when loaded in shear |
| $I_{50}$ | point load index for 50 mm size samples |
| IPT | Institute of Technological Research (São Paulo) |
| ISONIC | sonic while drilling tool |
| ISRM | International Society of Rock Mechanics |
| $J_a$ | rating for joint alteration, discontinuity filling in Q-calculation |
| JCS | joint wall compression strength (MPa) |
| $J_n$ | rating for number of joint sets in Q-calculation |
| $J_r$ | rating for joint surface roughness in Q-calculation |
| JRC | joint roughness coefficient (dimensionless: range 0 to 20) |
| $J_v$ | volumetric joint count (sum of frequencies for different sets) |
| $J_w$ | rating for water softening, inflow and pressure effects in Q-calculation |

| | |
|---|---|
| K,k | permeability (intrinsic: units of length$^2$, engineering: units of m/s) |
| K | bulk modulus (also $K_{bulk}$) |
| $K_{int}$ | intermediate principal permeability |
| $K_{max}$ | maximum principal permeability |
| $K_{min}$ | minimum principal permeability |
| $K_n$ | normal stiffness (of joint or fracture: strongly non-linear, sample dependent) |
| $K_{n\ dyn}$ | dynamic normal stiffness (of joint or fracture) |
| $K_o$ | ratio of rock stresses $s_{h\ min}/s_v$ |
| $K_S$ | shear stiffness (of joint or fracture: non-linear, sample dependent, scale dependent) |
| $K_{s\ dyn}$ | dynamic shear stiffness (of joint or fracture) |
| L | Lugeon unit of water injection (l/min/m of borehole/1MPa excess pressure $\approx 10^{-7}$ m/s) |
| *Lg* | coda waves, tail of seismogram |
| LOFS | life of field seismic |
| LSS | long-spaced sonic tool |
| LWD | logging while drilling |
| M | deformation modulus (pseudo-static *loading* stiffness: plate load test. Also $E_{mass}$, D) |
| $M_{1,2}$ | dynamic elastic moduli at frequencies $f_1$ and $f_2$ |
| MAR | mid-Atlantic ridge |
| MHF | massive hydraulic fracturing |
| MIT | Massachusets Institute of Technology |
| MPBX | multiple position borehole extensometer |
| MWD | measurement while drilling |
| n | effective stress coefficient (Biot) |
| n% | porosity of matrix |
| NAFZ | North Anatolian Fault Zone, Turkey |
| ND | natural directivity |
| NGI | Norwegian Geotechnical Institute, Oslo, Norway |
| NMO | normal moveout |
| NPF | Norsk Petroleumsforening (Norwegian Petroleum Society) |
| OC | over-closure of joints, mechanical or thermal |
| O/R | open/rock-to-rock sections of shearing joint, opposite rotation |
| OBC | ocean bottom cable |
| OBS | ocean bottom seismometers |
| P | volumetric stress |
| $P_g$ | direct (P-) wave (crustal scale studies) |
| Pn | refracted (P-) wave (crustal scale studies) |
| Pr | support pressure, radial capacity of support in a tunnel |
| PR | penetration rate (TBM, uninterrupted boring) |
| Q | rock mass quality rating ('Q-value' range $10^{-3}$ to $10^3$, dimensionless) |
| $Q_{rock}$ | rock mass quality rating, distinguish from $Q_{seis}$, seismic quality, inverse of attenuation |
| $Q_c$ | seismic quality of coda wave |
| $Q_E$ | seismic quality in extensional resonance mode |
| $Q_e$ | seismic quality component (Young's mode of elastic excitation) |
| $Q_k$ | seismic quality component (bulk mode of elastic excitation) |
| $Q_o$ | seismic quality, *Lg coda at 1 Hz* |
| $Q_P$ | seismic quality of P-wave (through given medium) |
| $Q_s$ | seismic quality component (shear mode of elastic excitation) |
| $Q_S$ | seismic quality of S-wave (through given medium) |
| $Q_c$ | rock mass quality rating (Q or $Q_{rock}$ normalized by $\sigma_c/100$) |
| $Q_o$ | Q (or $Q_c$ or $Q_{rock}$) calculated with $RQD_o$, oriented in the loading or measurement direction |

| | |
|---|---|
| $Q_{seis,}$ | seismic quality factor ('Q'), inverse of attenuation, also for $Q_P$ or $Q_S$, or the coda wave $Q_c$ |
| $Q_{tbm}$ | rock-machine quality factor for TBM tunnel boring machines based partly on Q-value |
| QVO | Q(seismic) versus offset |
| r,R | Schmidt hammer rebound % on wet joint surfaces, dry intact samples, respectively |
| REV | representative elemental volume |
| RMR | rock mass rating developed by Bieniawski |
| RQD | rock quality designation developed by Deere (modified core recovery %) |
| $RQD_o$ | RQD oriented in the loading or measurement direction |
| RQI | reservoir quality index |
| SAFZ | San Andreas Fault Zone |
| SCV | Site Characterization and Validation, SKB project in Stripa mine, Sweden |
| S(fr) | steel fibre reinforced sprayed concrete |
| S/C ratio | ratio of subsidence to compaction magnitudes, above and within reservoirs |
| SKB | Swedish Nuclear Fuel Co. (Stockholm) |
| SRF | rating for faulting, strength/stress ratios, squeezing, swelling: 6th parameter in Q-value |
| TBM | tunnel boring machine |
| 3DEC | three-dimensional distinct element code for modelling jointed rock masses |
| TIH | transversely isotropic, horizontal axis of symmetry (also HTI) |
| TIV | transversely isotropic, vertical axis of symmetry |
| TSP | (in-tunnel) seismic reflection profiling |
| TSX | tunnel sealing experiment |
| UCS | uniaxial compressive strength of rock cylinder |
| UDEC | universal distinct element code, for modelling jointed, fractured rock in 2D |
| (3DEC) | three-dimensional distinct element code, for modelling jointed, fractured rock in 3D) |
| URL | Underground Research Laboratory, Manitoba, Canada |
| $V_P$ | P-wave seismic velocity (km/s) |
| $V_S$ | S-wave seismic velocity (km/s) |
| VSP | vertical seismic profiling |
| WAP | wide aperture profile |
| WIPP | Waste Isolation Pilot Plant, New Mexico |
| w.r.t. | with respect to (index only) |
| ZEDEX | Zone of Excavation Disturbance EXperiment, SKB project, Äspö, Sweden |
| $Z_N$ | dynamic normal compliance (of joints or fractures) ($= 1/K_{n\ dyn}$) |
| $Z_T$ | dynamic shear compliance (of joints or fractures) ($= 1/K_{s\ dyn}$) |

## Cross-discipline differences and connections

- effective stress = total stress minus pore pressure in *geomechanics*
- differential stress = shear stress caused by $\sigma_1 - \sigma_3$ application in *geomechanics*
- differential pressure = confining pressure minus pore pressure in *rock physics*
- compliance = $(\text{dynamic stiffness})^{-1}$, compliance $\leqslant (\text{pseudo-static stiffness})^{-1}$
- $Q_{seis} = 1/\text{attenuation}$, $Q_{rock} < Q_{seis}$, but $Q_{rock}$ provides estimate of $E_{mass} \sim Q_{seis}$

# Part I

# 1 Shallow seismic refraction, some basic theory, and the importance of rock type

*'Nature has left us an incomplete and often well-concealed record of her activities, and no 'as constructed' drawings!' (Stapledon and Rissler, 1983)*

*'Tenders for the Tay pipeline crossing did not allow time for boreholes to locate bedrock. Seismic refraction took one day to confirm that the trench would not encounter rock. The pipeline was laid in sediments.' (Gardener, 1992)*

*'The time may come when the various relations between geophysical parameters and rock properties can be usefully combined into a single classification system.' (Darracott and Orr, 1976)*

## 1.1 The challenge of the near-surface in civil engineering

Refraction seismics is by far the oldest method used in exploration seismology, with its origin traced to R. Mallet from 1848. Shallow refraction seismic measurements using first arrival, compressional P-wave velocities close to the surface often give a remarkable picture of near surface conditions due to some fortuitous interactions of physical phenomena. Firstly, weathering and the usual lack of significant stress near the surface has allowed joint systems, shear zones and faults to be exaggerated in both their extent and severity. Secondly, stress levels are low enough to allow joints and discontinuities to be seismically visible due to their measurable apertures.

So-called acoustic closure occurs at greater depths than those usually penetrated by conventional hammer seismic, unless rock strengths are rather low (e.g., New and West, 1980; Hudson et al., 1980). (At this juncture, we need to differentiate between two 'J.A. Hudson' authors, one in geophysics, the other in rock engineering, and both very prominent in their chosen fields. We will occasionally refer to 'rock' Hudson in Part I, and later in Part II to 'seismic' Hudson).

Since micro-fractures and rock joints are sensitive to stress levels, the more closed state of the discontinuities that are perpendicular to the major stress, and the more open state of those that are parallel will give the rock mass anisotropic stiffness. Consequently the rock mass will frequently display anisotropic seismic velocities. By implication, hydraulic conductivities and deformation moduli that show anisotropic distributions will be, at least in part, detectable by seismic measurements. Anisotropy will also be caused by layered inter-beds, foliation and schistocity, and of course by a dominant joint set. Simple examples of (azimuthal) anisotropy, applicable in civil engineering, will be given in Chapter 3, while larger-scale examples of anisotropy detection will be described in much greater detail, and from various fields of the earthsciences, in Chapters 13, 14 and 15 in Part II.

Despite the obvious challenges of seismic interpretation in fractured and faulted petroleum reservoirs at many kilometers depth, or of mid-ocean ridge investigations beneath three kilometers of ocean, many geophysicists insist that obtaining high resolution images from ground level to just 50 m depth, is still one of the major challenges of modern geophysics. This happens to be the layer of the subsurface closest to most of our civil engineering endeavours, from tunnels, to dams, to the foundations for high buildings.

Undoubtedly, the '0 to 50 m' challenge is mainly due to the extreme variability of the near-surface, resulting from the contrasting geological materials and weathering grades that are often present. There is also a velocity gradient that is extreme compared to anything found at greater crustal depths, where consolidation effects smooth out some of the differences. The first 5 m of unconsolidated dry beach sand may see velocity increase from 150 m/s to 300 m/s, (Bachrach et al., 2000), giving a gradient of 30 s-1, which may be an order of magnitude higher than the gradient over the next 50 to 100 m, where weathered and jointed rock may typically be found.

There are an infinite number of challenges in the near-surface. Some of the worse may be karst phenomena in limestones, or the 'inverse' problems of *core-stone* anomalies in the case of sparsely jointed but deeply weathered granites and gneisses. These features have caused tunnelling surprises in numerous countries, with nearly as numerous arbitrations as a result. Although completely weathered Grade V is an expected feature beneath the Grade VI soil in tropical terrains, Grade V saprolite sometimes confusingly swaps places with the usually deeper, and almost unjointed Grade I or II. (Saprolite is a weak, water sensitive, weathered in-place, sometimes beautifully structured and coloured relic of the rock).

If this *reversal* of weathering grades appears in a tunnel arch *beneath* massive, high velocity core-stones, or if there is a generally very undulating rock surface, with frequent tunnel penetrations into weathered materials, there can be major delays. A tunnel collapse is difficult to avoid when water is present, unless preparations have been made, as a result of the more frequent exploratory drilling demanded when seismic anomalies such as these are suspected.

Pre-injection ahead of the tunnel face, and heavier tunnel support, would be the very basic requirements in a drill-and-blasted tunnel. (This is one of the purposes of the 'Q-system' of rock mass characterization and tunnel support selection). In the case of a TBM (tunnel boring machine) excavation, a change to a *closed mode* in the case of a hybrid machine with earth-pressure-balance (EPB) would be needed, especially if the weathered depressions in the bedrock contained water, as is usually the case.

Best advice of all, as *a direct result of a seismic refraction survey*, would be to drive a deeper tunnel from the start. It is easy to imagine subway station construction under such heterogeneous conditions. It could be extremely time-consuming, and even dangerous. The cost of deeper access to the stations, via longer escalators, would be a small price to pay for much reduced tunnelling and station costs.

Sjøgren, 2000 suggested the following list of essential information expected from near-surface seismic surveys, performed for civil engineering geotechnical investigations:

- The velocities of the overburden layers, including the upper, less consolidated rock layers.
- The thickness of the various overburden velocity layers, and the total depth to the main refractor.
- A detailed determination of the velocity distribution in the main refractor.
- An estimate of the uncertainty of the velocity and depth determinations.
- An analysis of the (velocity-) depth structure.
- An assessment of velocities in vertical and lateral directions in relation to the geology.
- Seismic results in relation to results from other investigations, if available.
- Conclusions and recommendations resulting from the investigation that are of importance to the project.

Although reflection methods have eventually dominated the field of exploration seismics due to the various needs involved with deeper exploration, there is 'universal' use of shallow refraction seismic in sub-surface investigations for civil engineering projects around the world, due to its apparent simplicity and low cost. Furthermore, refraction seismics can be used to remove (from the more deeply focussed reflection data), the 'adverse' effect of the first meters or tens of meters of the heterogeneous weathered layer, where differences in the original rock quality may cause tens of meters of sub-surface 'topography' in the case of on-land exploration.

## 1.2 Some basic aspects concerning elastic body waves

It is usually assumed that the strains associated with the passage of a seismic wave are of minute, sub-micron magnitude, and except in the neighbourhood of the source, the strains are generally assumed to be elastic. Based on this assumption, the velocities of propagation of seismic waves are determined by the appropriate elastic moduli and densities of the materials passed through. The general form of the classic equations linking these three quantities is $V = (E/\gamma)^{1/2}$. Compressional bodywaves (primary or P-waves) propagate by alternating compression and dilation (Figure 1.1 a) in the direction of the waves.

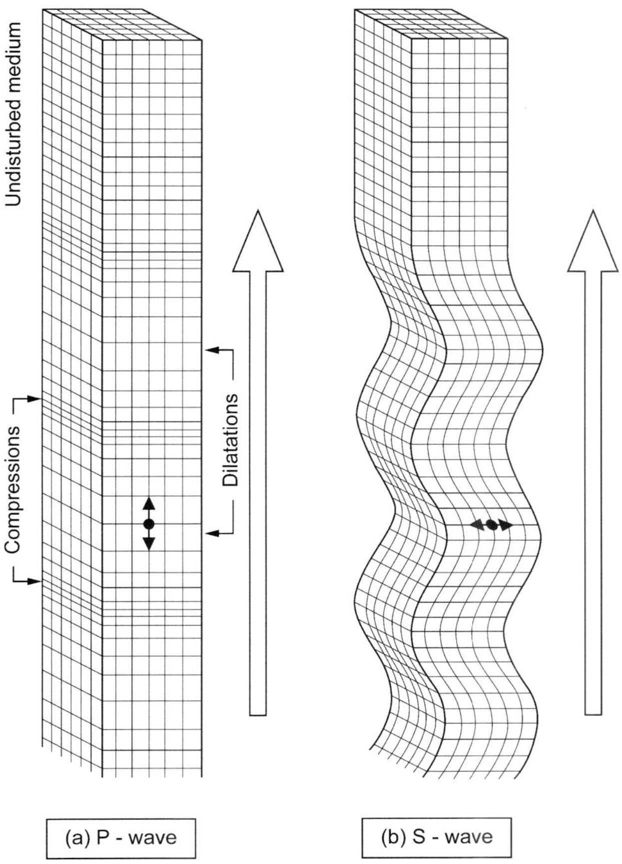


*Figure 1.1* Elastic deformations and particle motions associated with the propagation of body waves: a) P-wave, b) S-wave. Based on Bott, 1982.

The oscillating uniaxial strain involved in the case of a confined body, means that the axial modulus ($\psi$) controls the velocity of propagation, thus:

$$V_p = \left(\frac{\psi}{\gamma}\right)^{1/2} \tag{1.1}$$

Shear bodywave waves, termed secondary, transverse or S-waves propagate by a sinusoidal pure shear strain (Figure 1.1 b) in a direction perpendicular to the direction of the waves. The shear modulus ($\mu$), which is given by the ratio of shear stress ($\tau$) divided by the shear strain ($\tan\theta$), will therefore control the (lower) velocity of propagation, thus:

$$V_s = \left(\frac{\mu}{\gamma}\right)^{1/2} \tag{1.2}$$

The third important elastic modulus influencing the conversion between dynamic properties is the bulk modulus (K), defined as the ratio of the volumetric stress (P) and the volumetric strain ($\Delta v/v$). Since the three moduli are linked by the equation $\psi = K + 4/3\mu$, it follows that $V_p$ can also be expressed as:

$$V_p = \left[\frac{K + 4/3\mu}{\gamma}\right]^{1/2} \tag{1.3}$$

This equation therefore demonstrates the fundamentally faster nature of $V_p$ in relation to $V_s$. The ratio of these two dynamic properties are also linked by the dynamic Poisson's ratio for the material, as will be shown in the next section, which contains some standard equations.

### 1.2.1 Some sources of reduced elastic moduli

In the case of micro-cracked, fractured, or jointed rock masses, there is a correspondingly reduced set of moduli in relation to the undisputed elastic nature of the intact matrix, because of micro (and probably elastic) displacements in normal and/or shear directions across and/or along the micro-cracks, fractures or joints. These represent an important part of the source of attenuation of the seismic waves in the dry state, due both to various scales of wave scattering and due to the intrinsic micro-deformations. Added losses are incurred if these micro- or macro-discontinuities are partly saturated, since there is communication with the pores and eventual pore fluid, and minute flows may be initiated to equilibrate pressures. These micro-imbalances will only be equilibrated when the frequency is sufficiently low.

The above mechanisms mean that dynamic properties, such as the velocities, Poisson's ratio and attenuation tend in practice to be dispersive, or frequency dependent. They are also of course *rock quality* and *environment-dependent*, in the broadest possible meanings of these words. As rock quality declines, or the surface is approached, there develops a serious discrepancy between the dynamic or elastic properties of the intact matrix and the dynamic properties of the (partly discontinuous) medium. The ratio between the dynamic properties of the (partly discontinuous) medium and the *static* deformation properties, such as the (rock mechanics) deformation moduli and joint stiffnesses (the inverse of compliances), may rise into double figures in this complex region, where velocity-depth gradients are often extreme.

At depth, under high confinement, and if rock quality is high, it is assumed that there will be only small discrepancies between the dynamic properties of the matrix and the dynamic properties of the rock mass. The 'static' crack and joint stiffnesses, being so high, will be close to the dynamic (inverted) crack and joint *compliances*.

There is controversy however, about the ratio of the dynamic normal and shear compliances, and the (inversed) ratio of the 'static' normal and shear stiffnesses. There is even controversy over whether friction is a valid attenuation mechanism, at the level of these micro-displacements.

In the rock mechanics of 'static', 'macro-deformations', we are familiar with a significant mismatch between the high normal stiffness, and the much lower (and scale-dependent) shear stiffness. Concerning the ratio of dynamic compliances, geophysicists seem not to be so sure, a dilemma that also probably affects whether *friction* is, after all, to be a valid attenuation mechanism, as assumed in much of the geophysics literature, virtually up to the present day.

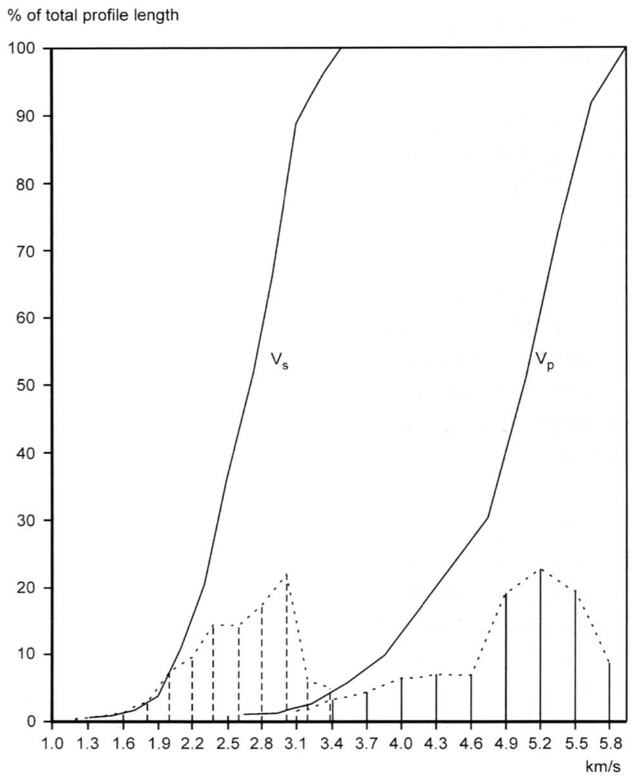


*Figure 1.2* Mean $V_p$ and $V_s$ statistics from 4 km of seismic profiles in metamorphic and igneous rocks. Sjøgren, 1984.

## 1.3 Relationships between $V_p$ and $V_s$ and their meaning in field work

The advantages of using both P-wave and S-wave data to interpret seismic results in hard rock was strongly emphasised by Sjøgren, 1984. This has been reinforced by the successively easier acquisition of multi-component, multi-channel data, and rapidly developing PC analysis capabilities. In addition to many other sets of data, some of which will be referred to later, Sjøgren, 1984 presented average $V_p/V_s$ ratios from 93 rock sections from 5 different sites in igneous and metamorphic rocks. These are reproduced in Figure 1.2.

P-wave velocities ranged from 3.3 to 5.7 km/s, and S-wave velocities from 1.6 to 3.4 km/s. On average, $V_p/V_s$ ratios were 1.89 in the rock mass with lower velocities (heavier jointing) and 1.80 in the rock mass with higher velocities (sparser jointing).

These two ratios of $V_p/V_s$ imply *rock mass quality Q-values* of roughly 1 to 10, and 10 to 100 respectively, according to the following *near-surface, hard rock* $V_p$–Q relationship (Barton, 1991).

$$V_p \approx 3.5 + \log_{10} Q \tag{1.4}$$

The rock mass quality Q-value, mentioned earlier in the introduction, is composed of parameter pairs (RQD/$J_n$, $J_r/J_a$ and $J_w$/SRF – see Appendix A for descriptions and ratings). These effectively describe *relative block size, the inter-block friction coefficient*, and *an active stress*. This rock quality term will be utilised in various places in this book, not least as a possible interpretation of *seismic quality Q* (the inverse of attenuation). (Improvements to equation 1.4 will be developed in Chapter 5, to allow for its application to weaker and more porous rock types, and to adjust it for depth or stress effects).

A rock quality Q-value of <1 tends to be heavily jointed, containing some clay, while values >1, are tending towards better quality, with wider spacing of joints, less joint sets, and no clay. (Q may reach values of about 1000 to 2000 in the case of massive, unjointed rock masses, confined at depths of say 500 m or more).

The ratio $V_p/V_s$ depends on dynamic Poisson's ratio ($\nu$) according to the following:

$$\frac{V_p}{V_s} = \sqrt{\frac{2 - 2\nu}{1 - 2\nu}} \tag{1.5}$$

From equation 1.5 one can derive the value of dynamic Poisson's ratio as follows:

$$\nu = \frac{(V_p/V_s)^2 - 2}{2(V_p/V_s)^2 - 1} \qquad (1.6)$$

The ratio $V_p/V_s$ is about $\sqrt{3}$ for hard (zero-porosity) rocks, for which $\nu \approx 0.25$. However, in the case of unconsolidated sediments, the ratio $V_p/V_s$ can even reach values of 20 to 40 for near surface material, for which $\nu$ is commonly greater than 0.45. Later in this chapter, high values of (dynamic) Poisson's ratio for a near-surface fault zone will also be seen, for similar reasons to the above.

A rock quality interpretation, linking these dynamic parameters, can be added here, by taking Sjøgren's (1984) P- and S-wave results from 4.1 km of seismic profiles for hard but sometimes weathered metamorphic and igneous rocks (Figure 1.2). The mean value of $V_p/V_s = 1.89$ in the more heavily jointed rocks (perhaps a rock quality $Q \approx 1$–$10$), and the mean value of $V_p/V_s = 1.80$ in sparsely jointed rocks (perhaps a rock quality $Q \approx 10$–$100$), can be used to calculate dynamic Poisson's ratios of 0.30 and 0.28 respectively.

As lower rock quality Q-values are approached in shear zones and faulted zones (e.g. rock quality $Q \approx 0.1$), the ratio $V_p/V_s$ increases to about 2.0, corresponding to a calculated value of dynamic Poisson's ratio of about 0.33. The corresponding Q-value (from equation 1.4, using minimum $V_p$ data from Figure 1.2) is indeed about 0.1. Extremely low Q-values, for example $Q = 0.01$–$0.001$ (when $V_p \approx 1.5$–$2.5$ km/s) will be needed before dynamic Poisson's ratio values become as large as 0.45 (as indicated for near-surface shear zones, in a later section of this chapter).

Further basic equations linking $V_p$, $V_s$, dynamic Poisson's ratio ($\nu$), density ($\rho$) and dynamic Young's modulus $E_{dyn.}$ are as follows (Darracott and Orr, 1976):

$$V_p = \left( \frac{E_{dyn.}(1 - \nu)}{\rho(1 + \nu)(1 - 2\nu)} \right)^{1/2} \qquad (1.7)$$

$$V_s = \left( \frac{E_{dyn.}}{\rho \times 2(1 + \nu)} \right)^{1/2} \qquad (1.8)$$

$$E_{dyn.} = V_s^2 \times \rho \times \frac{3(V_p/V_s)^2 - 4}{(V_p/V_s)^2 - 1} \qquad (1.9)$$

In the case of massive rocks of low porosity, the *static* and *dynamic* values of the elastic constants (e.g., the elastic moduli $E_{stat.}$ and $E_{dyn.}$) are quite close, while for heavily fractured and clay bearing zones, large differences between $E_{stat}$ and $E_{dyn}$ are seen (e.g. Cosma, 1995). Rock mass quality descriptions such as Q or RMR or RQD, which are described in more detail later, correlate better with static moduli than with dynamic moduli. Numerous relations between these moduli will be given in Chapter 6.

## 1.4 Some advantages of shear waves

In addressing the challenge of resolving the 0–50 m resolution problem, Dasios et al., 1999 reported multi-component investigations at four shallow sites (thick clays, clay/sand sequences over chalk, mudstone overlying granodiorite bedrock, and landfill), using a combination of both compressional and shear wave seismic.

The authors of course admit that there is a higher level of effort required to conduct multi-component seismic, requiring a three-component source configuration, and three-component geophones, but otherwise conventional multi-channel seismic recording systems, and PC-based processing software. Obviously the surveys are more difficult and more time consuming than compression-wave refraction or reflection, but the level of geophysical information is that much more useful.

They varied the acquisition geometry to optimize results. They found that under all the conditions, shear-waves penetrated with less attenuation than compressional waves, also being unaffected by water saturation. Shear-wave reflections from shallow interfaces were in some cases less affected by noise compared with the equivalent compressional-wave reflections.

They offered the following simple explanation of why shear-waves offer better vertical resolution than compressional-waves, particularly in shallow, unconsolidated sediments. The dominant reason is that the shear-wave velocities in such cases, are only a fraction (sometimes less than one fifth) of the compressional-wave velocities. This results in very small wavelengths, despite the fact that the dominant frequency of shear wave data is generally lower than is the case for compressional wave data. In order to obtain the same level of resolution with P-waves, energy of very high dominant frequency has to be generated, and this is correspondingly more attenuated in the low seismic Q sub-surface.

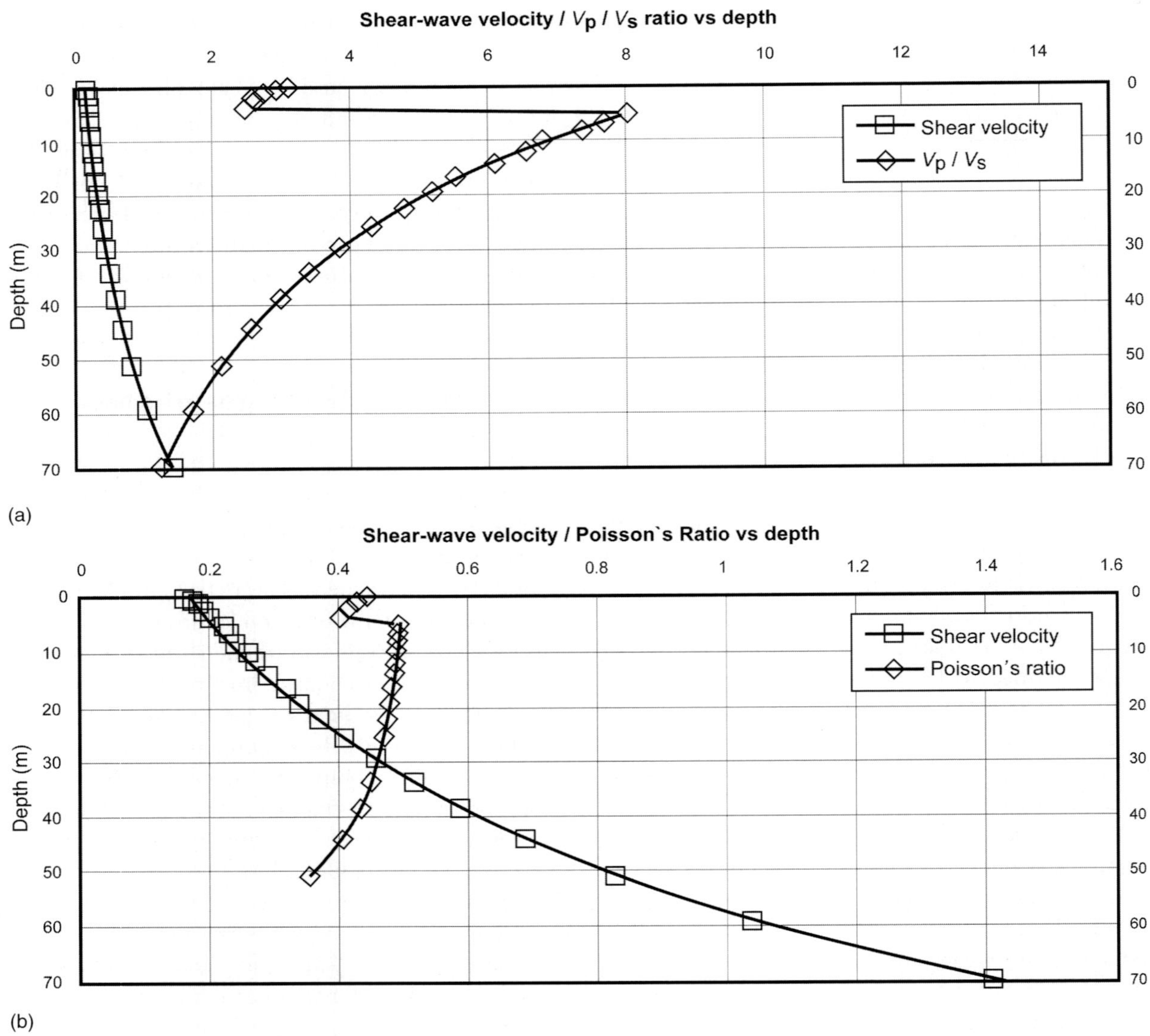


*Figure 1.3* a) Shear-wave velocities (km/s) and $V_p/V_s$ ratios versus depth. b) Shear-wave velocities (km/s) and dynamic Poisson's ratio versus depth for a clay-over-mudstone-above-basement sequence, with an interpreted water table at 4.7 m depth. Results of multi-component seismic at one of four shallow sites described by Dasios et al., 1999.

Since shear-waves are not attenuated at the water table, and are little affected by changes in fluid saturation, they can more easily detect lithological changes with correspondingly less ambiguous velocity contrasts. The authors found that under conditions of full water saturation, P-wave velocity contrasts between lithologies were small, whereas the shear-wave velocities reflected the true lithological changes.

In this brief summary, the results from a site they investigated in Crewekerne, Dorset will be reproduced. Their results give a good illustration of the subtle interaction of $V_p$, $V_s$ and the dynamic Poisson's ratio, in the presence of a water table, and also show the effect of increasing depth in a uniform sedimentary rock. At their site, thin clays overlayed mudstone, with a basement of granodiorite at 70 m depth.

Figure 1.3 shows multi-component plots of shear-wave velocity versus the ratio of $V_p/V_s$, and of shear-wave velocity versus (dynamic) Poisson's ratio, each as a function of depth to the basement rock at 70 m depth.

Analysis of the P-wave first arrivals gave velocities of 496 m/s and 1,766 m/s for the unsaturated and saturated layers, and indicated a depth of 4.7 m for the top of the saturated zone. The dynamic Poisson's ratio showed a small decrease close to the surface as a result of the consolidation, then a sharp increase as a result of the water saturation, followed by a steady decrease with depth within the uniform water-saturated mudstones.

As expected from the theoretical calculation of the dynamic Poisson's ratio, there is a certain accentuation of the above trends for the case of the ratio $V_p/V_s$, except that this ratio reduces faster at shallow depth (rather than when deeper), due to the strong $V_p$ gradient. The authors point out that the P-wave velocity was not available at the greatest depths, due to attenuation, so the data was extrapolated to 70 m. There were indications of shear-wave anisotropy in the uppermost meters of clay, but whether due to desication fractures or some form of layering is not certain.

Although outside the usual range of $V_p/V_s$ (about 1.4 to 2.0) for water saturated rocks, it is of interest to see details of the development of $V_p/V_s$ ratios in unconsolidated (e.g., subsea) sediments, by noting the progression from soft soils through compacted soils, to rocks. Hamilton, 1979 gave comprehensive $V_p$-depth, $V_s$-depth and $V_p/V_s$-depth data for silty clays, turbidites and mudstones to 1 km depth. $V_p$ values increased slowly from about 1.5 to 2.3 km/s as porosity reduced with increased depth, while $V_s$ increased rapidly from only .05 km/s close to the surface, to 0.15 km/s at depth. $V_p/V_s$ *ratios* therefore reduced very rapidly from double figures down to about 2.5 at 1000 metres depth. The dynamic values of Poisson's ratio decreased, as a result, from about 0.49 near the surface to 0.41 at 1000 m depth.

In connection with these high values of Poisson's ratio for *sediments*, it is significant to note the relatively high values of (dynamic) Poisson's ratio that tend to be recorded in shear zones and fault zones at much shallower rock engineering projects. Gardener, 1992, calculated the values of Poisson's ratio from $V_p$ and $V_s$ measurements at the Transfynydd power station in Wales, estimating 0.45 for the shear zones, where the $V_p$ velocity range was 1.6–2.7 km/s.

The higher values of Poisson's ratio for shear zones have pseudo-static parallels to the special feature of heavily jointed rock masses, which can show 'expansion ratios' or pseudo-Poisson's ratios far in excess of 0.5, and even in excess of 1.0 as (shear) failure is approached (Barton and Bandis, 1982). Elastic continuum theory is of course 'violated' by the shear displacements tending to occur on the failing joint surfaces.

## 1.5 Basic estimation of rock-type and rock mass condition, from shallow seismic P-wave velocity

When first investigating the bedrock for suitability for near-surface tunnelling or other relatively shallow construction in rock, the preliminary use of shallow refraction seismic is very typical, where surface access (including noise) do not present major problems. As a minimum, the information gives a $V_p$ – depth profile of inestimable value for further planning of the sub-surface investigation, in particular the optimal siting of boreholes for future core-logging and permeability testing.

Figure 1.4 reproduces four examples of shallow refraction results from Sjøgren 1984, demonstrating the helpful information about the location, width and depth of zones of lower velocity. Later in this chapter, and in subsequent chapters, we will be seeing the many ways of interpreting such velocities in terms of rock quality and degree of fracturing, each tempered by the effect of rock type, density, porosity, depth (or stress level), and of course the possible anisotropy (or directional dependence) of the result in relation to an anisotropic jointing frequency, and horizontal stress anisotropy.

The seismic refraction survey provides numerous depth to bedrock and quality of bedrock assessments at a small fraction of the cost and time needed for drilling. Depths are given at the impact points (hammer or shallow explosive source) and at the detector points (geophones or 3D seismometers), so a close spacing of detectors gives the equivalent of a large number of soundings or borings.

Sjøgren, 1984, gives the example of 5 m detector and 25 m source separations for a 10 m deep bedrock investigation. A 100 m profile gives the equivalent of 250 m of soundings, and a complete distribution of relative quality beneath the profile. With the 10 m source and 50 m detector separations needed for a deeper survey to 50 m depth, the equivalent of 650 m of soundings per 100 m profile is given.

The knowledge and experience of the geophysical team is essential in setting out optimal profiles in relation to the geology and structural geology, in particular in relation to anisotropic, layered media, and in relation to fault and shear zones. 'Correct' interpretation of the calculated information cannot be divorced from the geology, since a given velocity ($V_p$ or $V_s$ or dynamic Poisson's ratio) is not unique to any one material but part of a scale or gradation in the specific geological profile at the site, and reflects

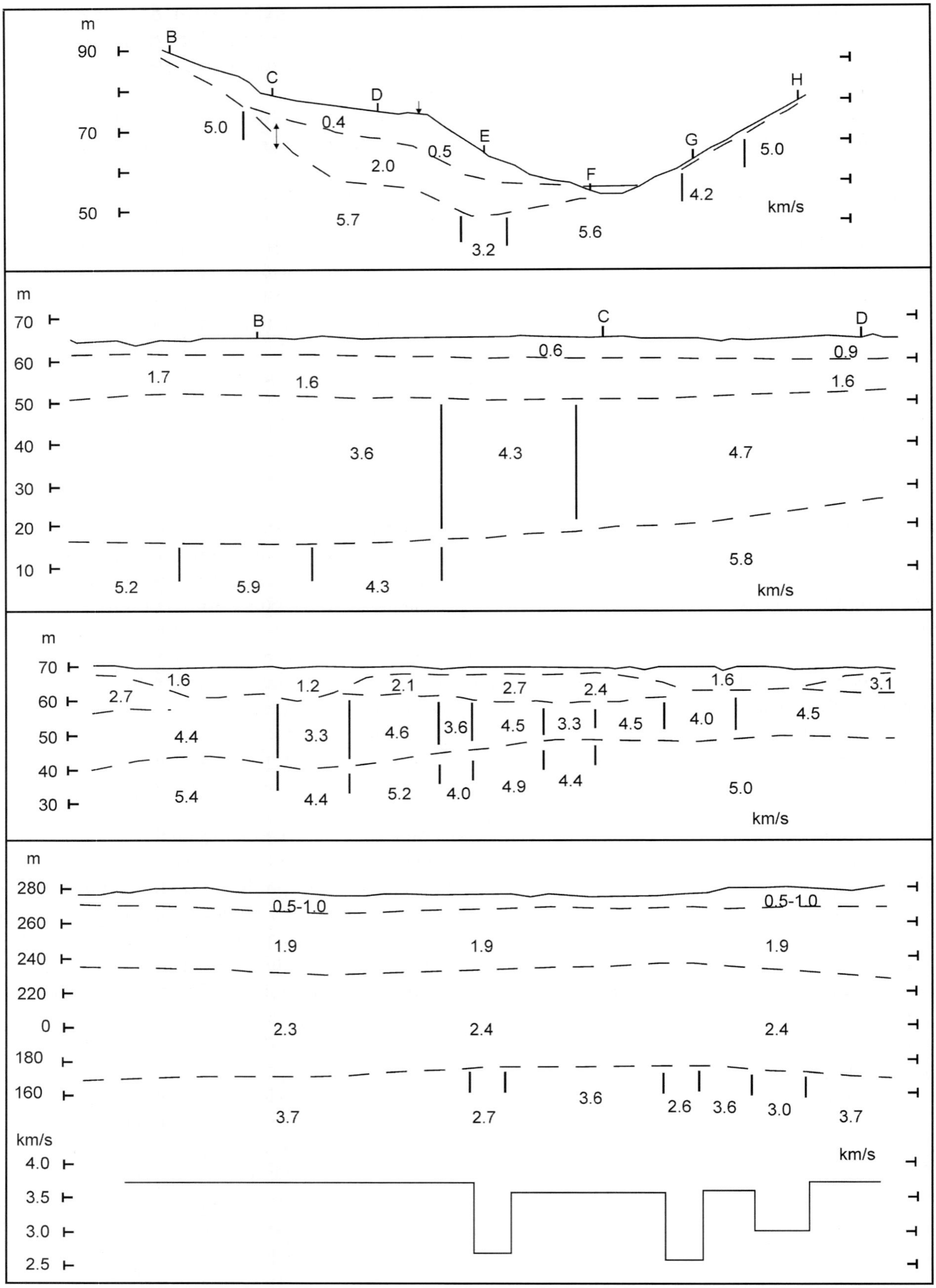


*Figure 1.4* Seismic refraction results illustrating the wealth of potential information obtained concerning near surface conditions. Sjøgren, 1984.

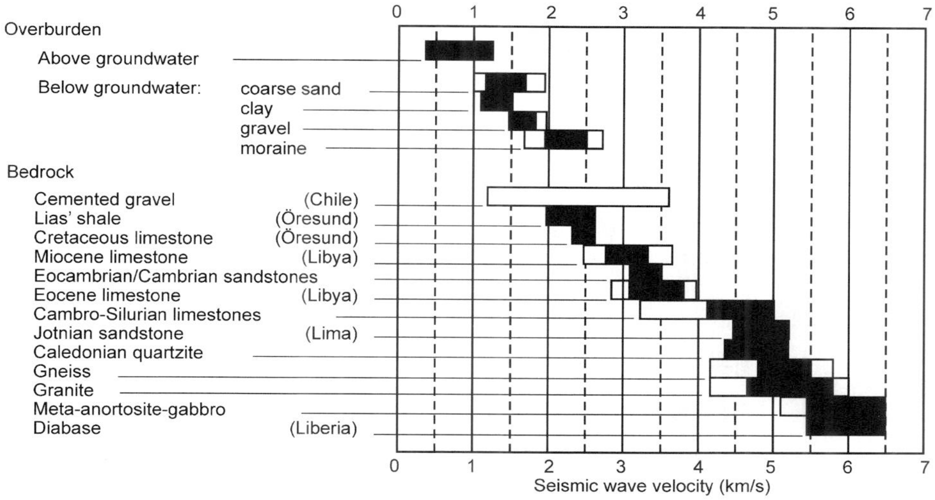


*Figure 1.5* Typical ranges of $V_p$ for sediments and for little weathered, moderately fractured rocks. Sjøgren, 1984.

various 'environmental' factors acting on each rock domain, as will be demonstrated in subsequent chapters.

The later geological and rock quality interpretation of core recovered from boreholes drilled close to the seismic profiles is the domain of engineering geologists, who besides identifying rock type, will perform careful logging of RQD, joint or fracture spacing, joint roughness and discontinuity mineral filling identification (or testing). The performance of rock quality characterization of drillcore is also standard practice for civil engineering and many mining projects, using the Q-value (Barton et al., 1974, Barton, 2002) and RMR (Bieniawski, 1989) as a minimum. Although these two methods have similarities, and common goals, there are differences, and care is needed in converting Q to RMR and visa versa, e.g. Barton, 1995.

Typical ranges of velocities for relatively competent (little weathered moderately fractured) rocks are given in Figure 1.5. Much lower velocities, covering most of the lower diagonal space between 1 km/s and 6 km/s are seen with extremes of weathering, jointing and fault related fracturing. The following is an example of the effects of weathering for just one rock type, from Sjøgren, 1984:

Table 1.1 Typical range of $V_p$ for gneiss (Sjøgren, 1984).

| | |
|---|---|
| 500 m/s | Soil (above water table) |
| 1700 m/s | Highly weathered biotite gneiss |
| 2800 m/s | Weathered biotite gneiss |
| 3500 m/s | Jointed biotite/granitic gneiss |
| 4900–5400 m/s | Sound biotite gneiss |

A similar range of values from the SSDS Project granites in Hong Kong (Gardener, 1992) gives a useful qualitative impression of variations caused by weathering and jointing in the same rock type.

Table 1.2 Typical range of Vp (km/s) for granite (Gardener, 1992).

| | |
|---|---|
| Decomposed granite (soil) | 1.6–1.8 |
| Fracture zones | 2.8–3.5 |
| Jointed granite | 3.5–4.5 |
| Intact granite | 4.5–6.5 |

At the hazardous second Severn Estuary crossing between England and Wales, tidal currents are so strong that 85% of the crossing had continuous rock outcrops between low and high tide. Sonar buoys and bottom drag cable gave the following relatively tight ranges of velocities for five rock types that were confirmed with boreholes, enabling the rocks to be identified across the site.

Table 1.3 P-wave velocities at the Second Severn Crossing (Gardener, 1992).

| Rock description | Average velocity (km/s) | Velocity range (km/s) |
|---|---|---|
| Triassic mudstone | 2.1 | 1.7–2.3 |
| Triassic siltstone | 2.4 | 2.2–2.6 |
| Triassic sandstone | 2.6 | 1.8–3.1 |
| Carboniferous siltstone and sandstone | 3.0 | 2.5–4.4 |
| Carboniferous sandstone | 4.0 | 3.0–4.4 |

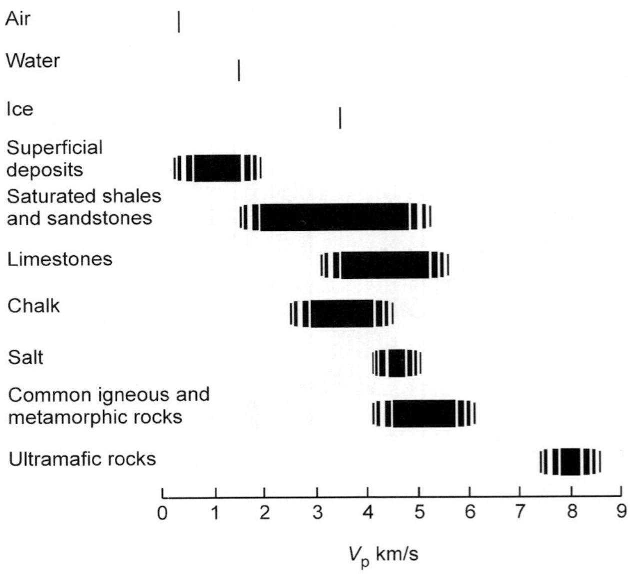


*Figure 1.6* Typical ranges of $V_p$ for common rock types. Griffiths and King, 1987.

Table 1.4 A list of typical $V_p$ and $V_s$ values from Press, 1966.

| Material | $V_p$ (m/s) | $V_s$ (m/s) |
|---|---|---|
| Alluvium | 500–2100 | |
| Clay | 1100–2500 | |
| Sand | 200–2000 | |
| Glacial Till | 400–1700 | |
| Sandstone | 1400–4500 | |
| Shale, Slate | 2300–4700 | |
| Limestone | | |
| Soft | 1700–4200 | |
| Hard | 2800–6400 | |
| Crystalline | 5700–6400 | |
| Dolomite | 3500–6900 | |
| Granite, Granodiorite | 4600–6000 | 2800–3200 |
| Diabase | 5800–6000 | |
| Gabbro | 6400–6700 | 3400–3600 |
| Basalt | 5400–6400 | 2700–3200 |
| Schist | 4200–4900 | 2500–3200 |
| Gneiss | 3500–7500 | 3300–3700 |
| Water | 1450 | |
| Air | 335 | |

Table 1.5 P-wave velocities for different horizons in a river bank terrace (Sen and Bandyoadhyay, 1990).

| | |
|---|---|
| 0.7–0.8 km/s | Clayey and silty soil with pebbles |
| 1.7–1.8 km/s | Pebbles/cobbles in silty sand matrix |
| 2.1–2.2 km/s | Cobbles/boulders in silty sand matrix |

Table 1.6 P-wave velocities for phyllites at a Himalayan dam site (Dhawan et al., 1983).

| | |
|---|---|
| Overburden/weathered phyllites | 925–1200 |
| Unweathered phyllites | 2520–4500 |

Griffiths and King, 1987, also give typical $V_p$ ranges for common rock types. These are reproduced in Figure 1.6, as a source of cross-referencing. Fractured, faulted and heavily jointed zones extend the six major ranges for these rocks far to the left on occasion. Note the extremely high velocities of the dense, ultramafic rocks, which lie outside the common range of 1 to 6 km/s.

A comprehensive set of *in situ* seismic $V_p$ values, and some $V_s$ values for common rock types, is also shown in Table 1.4. The data are given by Press, 1966.

The wide ranges of velocity for sandstone, shale, limestone and dolomite are mainly due to the wide ranges of porosity (and density) for these materials. The surprisingly high range for gneiss is due to the wide range of mineralogical composition (and density) for this rock.

The marked variation of velocities that are measured in superficial deposits (0.5 to 2.0 km/s in Figure 1.5) are partly caused by location either above or below the water table, as shown by Sjøgren's 1984 data set. The list given in Table 1.4 shows 0.2 to 2.0 km/s just for the case of sand, mostly for this reason.

The following is perhaps a good example of the influence of particle size in river born sediments.

The last line of Table 1.5 (for cobbles and boulders) differs significantly from the range 1.3 to 1.9 km/s for 'river boulders' given by Dhawan et al., 1983, presumably due to differences in porosities. In the latter case, the 'silty sand matrix' is presumably absent. These last authors also give data for phyllites, which do not appear on the foregoing figures or tables of $V_p$ data.

## 1.6 Some preliminary conversions from velocity to rock quality

Due to the seismic 'visibility' of jointing in the upper 25 to 30 metres or so, Sjøgren et al., 1979 and Sjøgren, 1984 and others, have been able to record significant correlations between $V_p$, RQD and joint frequency. These authors compared the results from a total of 113 km of P-wave profiles from fifteen sites, with the results of 2.85 km of core-logging from seventy four drill holes at eight of the hard rock sites. (They also had 5 km of S-wave surveys at five of the sites).

The range of rocks occurring at the measured locations, mostly in Norway, were: amphibolite, granite, gneiss, meta-anorthosite, pegmatite, porphyry, quartzite and mylonite. The authors were careful to emphasise that the correlations they derived between P-wave velocity and jointing descriptions such as mean RQD and mean frequency F($m^{-1}$) were relevant only to unweathered igneous and metamorphic rocks, and generally for the upper 20 to 30 metres.

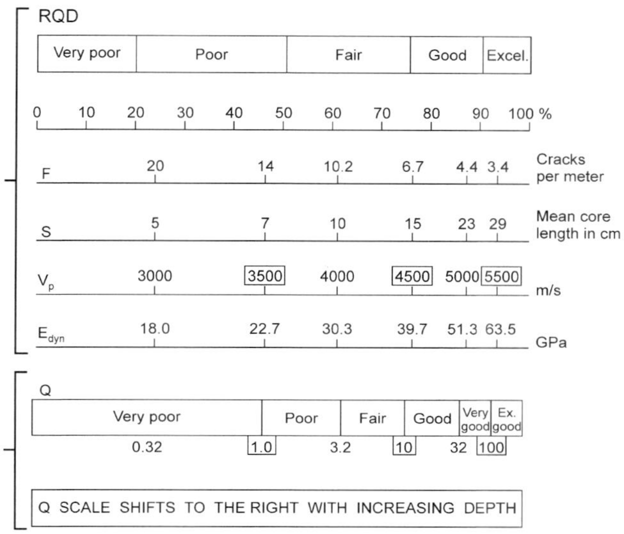


*Figure 1.7* Mean values of physical and dynamic properties for hard, unweathered igneous and metamorphic rocks, based on shallow refraction seismics. Sjøgren et al., 1979. Q-Scale added by Barton, 1995.

Figure 1.7 reproduces Sjøgren et al., 1979 data in the readily absorbed format used by the authors. Mean values for all the parameters apply. The $E_{dyn}$ modulus is the same as the symbol (ψ) used to denote the *uniaxial* or axial modulus in equation 1.1.

Since their measurements were shallow, the effect of stress-induced joint closure was minimised. They also effectively removed other sets of variables by generally recording correlations for hard and almost unweathered igneous and metamorphic rocks. The usual variables of depth, porosity, uniaxial compressive strength and density were therefore largely removed.

A hard rock, near surface correlation of $V_p$ and the rock quality Q-value was derived by Barton, 1995, on the basis of trial and error fitting to cases known to the writer and also Q-logged. (These will be described in Chapter 5). In Figure 1.8, the important effects of porosity, uniaxial strength and depth are ignored, as for the Sjøgren et al., 1979 data. Note that the RQD and F/m mean data have been smoothly extrapolated beyond both ends of the Sjøgren et al., 'data-base' represented by Figure 1.7.

## 1.7 Some limitations of the refraction seismic velocity interpretations

The seismic refraction method has some important limitations. One is that the horizontal or sub-horizontal ray

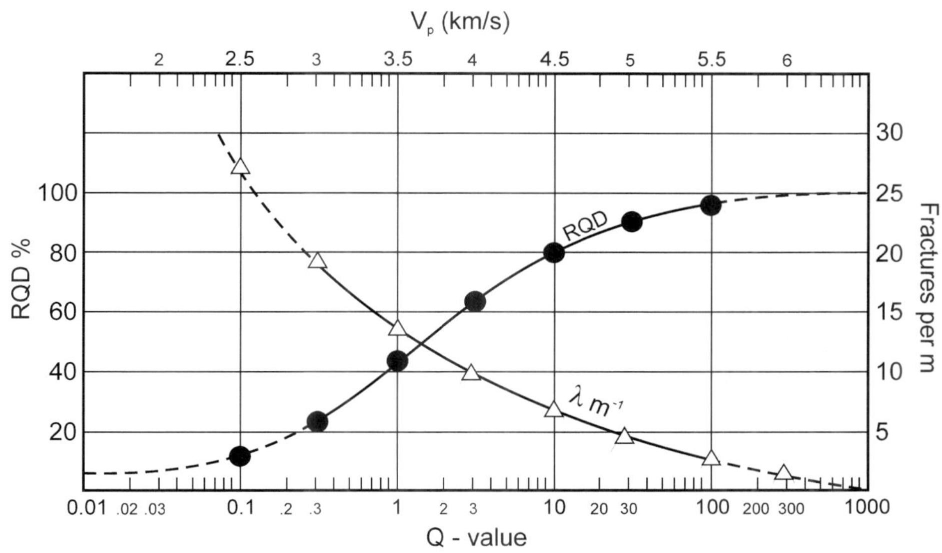


*Figure 1.8* RQD and $Fm^{-1}$ trends from Sjøgren et al., 1979. Q-scale, (as also in Figure 1.7) added by Barton, 1995. These results and approximate correlations to seismic-frequency $V_p$ are relevant for hard, low porosity, unweathered, near-surface rock masses.

paths record only the upper part of each seismic layer. A thin high-speed layer can mask underlying material, while a low-velocity intermediate layer will not be recognised for similar reasons. Depth calculations to underlying refractors will be erroneous. The hidden low velocity zones can of course be detected by up-hole shooting from a borehole to the seismic spread (i.e. VSP), or by inspection and index testing of core, if available.

A useful review of refraction seismic methods and commonly used methods of interpretation is given in Whiteley, 1990. Citing Sjøgren, 1984 that low velocity zones are consistently interpreted as being shallower than the borehole confirmation, Whiteley, 1990 went on to compare three interpretation methods of the simple yet frequently occurring situation shown in Figure 1.9. The three solutions were obtained by independent geophysics practitioners. It is clear that control drilling of identified features must be made before making important decisions such as minimum depth of cover over a sub-sea tunnel.

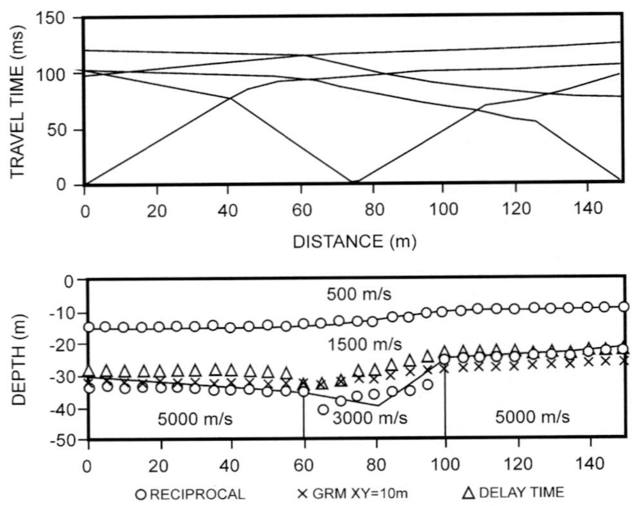


*Figure 1.9* Three independently derived depth estimates for a low velocity zone detected by refraction seismics. Whiteley, 1990.

Stapledon and Rissler, 1983 emphasised the following potential shortcomings of the seismic method, as related to near-surface investigations for civil engineering projects.

1. Minor geological defects such as weathered seams or minor faults may govern the engineering behaviour of a site, especially if their orientation is unfavourable. The seismic method 'generally is unable to detect such defects'. Is this pessimism justified or is it entirely correct?
2. P-waves are first to arrive at the geophones 'and must take the shortest path through the best rock'. Will this mean that they do not represent the average local quality but the best local quality?

Both shallow and deep refraction seismic are subject to velocity anomalies, causing time-distortion and therefore depth anomalies, as we have seen in the above shallow example. Although the following case of Armstrong et al., 2001 is strictly speaking a deeper reservoir case, it provides such a nice illustration of the hazards of depth interpretation that it will be referred to in this first chapter.

As the authors point out, the overburden above a hydrocarbon prospect is often more or less horizontal or perhaps with gently dipping sedimentary layers. It is generally paid little attention, in relation to the focussed investigation of the reservoir target at depth. Clearly, the overburden at typical shallow refraction sites for civil engineering projects may be far different from these apparently uniform conditions. Yet even the 'simple' site may contain velocity anomalies, which reduce the image quality. The time horizons suffer 'push-down' beneath slow-velocity anomalies, and 'pull-up' beneath fast velocity anomalies.

The geological 'reality' shown on the left in Figure 1.10 contrasts nicely with the equivalent post-stack seismic time section, shown on the right. There is significant distortion of layer 'horizons', and perhaps surprisingly persistent follow-through to depth. Armstrong et al., 2001 describe a method for compensating for these discrete overburden velocity anomalies. (Their warning about possible miss-calculation of reserves, or miss-positioning of infill wells, or by-passing of incremental reserves in relation to petroleum engineering, can clearly apply in other contexts to shallow civil engineering projects, where for example there are buried sediment-filled channels that have been known to plague certain tunnel projects, due to high water storage, and 'constant' inflow pressure).

Also concerning the possibility of depth estimation error, it was emphasised by Bradford and Sawyer, 2002, in the context of shallow seismic *reflection* measurements, that larger depth and layer thickness estimates would occur when using conventional velocity analysis, if in the presence of the extreme velocity gradient close to the dry-saturated transition, (usually fairly close to the surface in temperate climates).

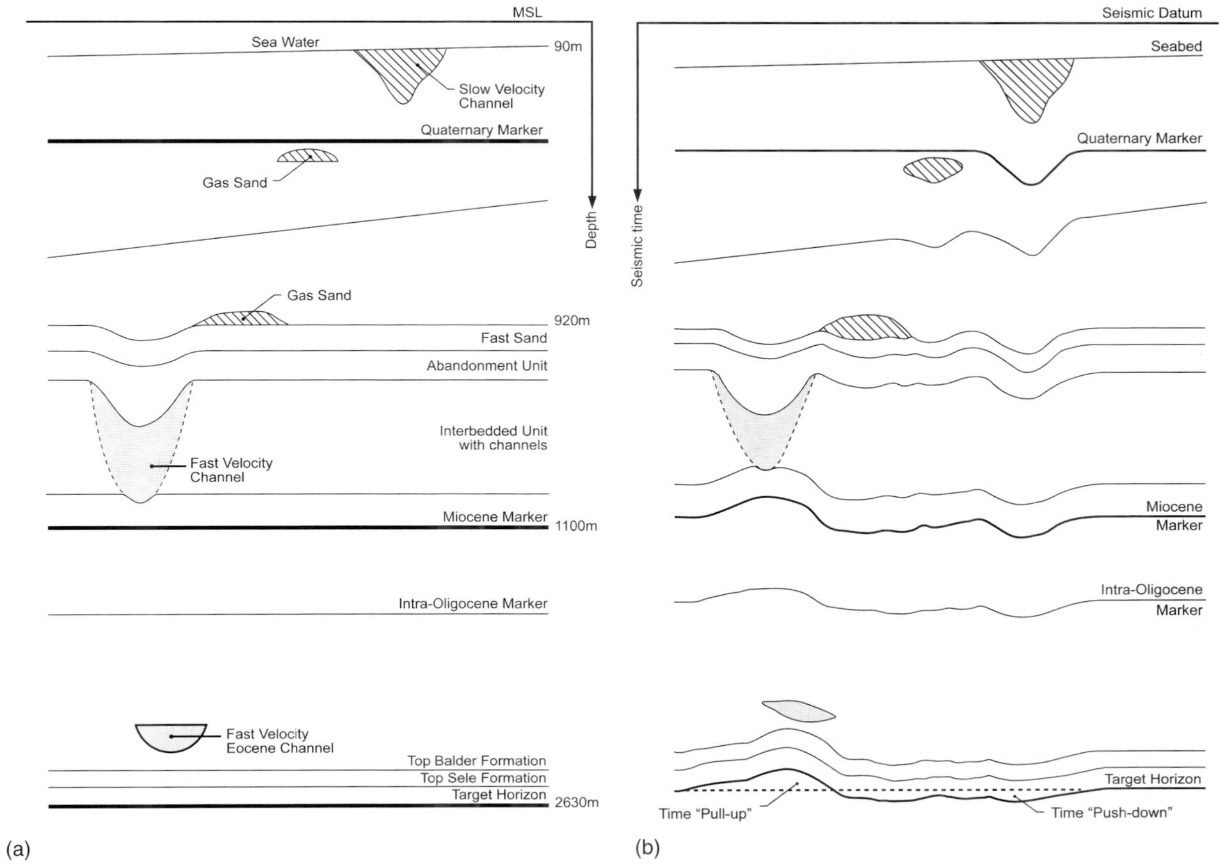


*Figure 1.10* Schematic cross-section through an overburden containing velocity anomalies of different geological age. The idealized vertical section is shown on the left, and the contrasting, time-horizon distortions, post stacking, are shown on the right. Armstrong, et al., 2001.

The authors indicated (as also seen earlier), that the P-wave velocity could increase by a factor of four or more at this transition, in the case of unconsolidated sediments changing from dry, through partial saturation, to full saturation – which may occur just above the water table, due to the action of capillary forces. A large velocity gradient (e.g. 400 m/s to 1600 m/s from 1 to 10 m depth) apparently violates many of the assumptions made in conventional reflection data processing schemes.

In a recent paper, Sjøgren, 2000, evaluated several standard methods for interpreting travel time curves. He utilised the ABC method (originating from 1931), the ABEM correction method (early 1950s, detailed in Sjøgren, 1984), the mean-minus-T method (mid-fifties, also adopted in the ABEM method), and the Hales method (1958), in order to critically evaluate a more recent (1980) generalized reciprocal method (GRM) of Palmer, in particular Palmer, 1991.

Sjøgren, 2000 expressed concern about the usefulness of the GRM method for near-surface geotechnical investigations, where details of the various overburden layers *are* required since they may have important consequences for the subsequent geotechnical design.

A relatively more recent technique for modelling of travel times and travel time inversion in refraction seismics is the so-called Eikonal solver. In principle, this involves the calculation of travel times on a regular velocity grid. Early versions, originating from the late 1980s were restricted to a plane topography for the recording surface.

Lecomte et al., 2000, describe a first order Eikonal solver that can incorporate the exact topography of the surface terrain, and any arbitrary lateral variation of velocity. There is no restriction on the velocity contrast. In effect, the model is built up layer by layer, with the refractor imaging, and the velocity mapping being performed for each identified refractor at a time, as seen

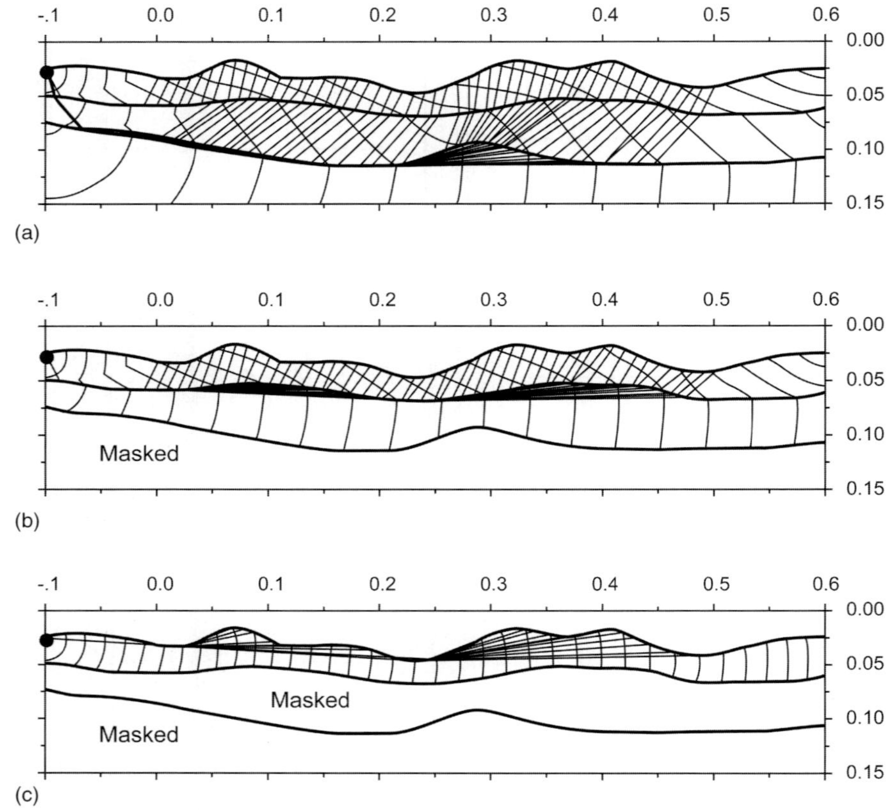


*Figure 1.11* An arbitrarily chosen model for demonstrating a new method of refraction seismic inversion. Lecomte et al., 2000 used this synthetic model for demonstrating some basic elements of the Eikonal solver, which calculates the travel time of the fastest wave at any point of a regular velocity grid, using the head waves generated in refraction seismics. The three stages of modelling shown here are: a) Wave-fronts and raypaths at receivers along the surface, when considering the *whole model* (but minus the acoustic wave velocity in the air, which was omitted for clarity. b) Wave-fronts and raypaths at receivers along the surface, when *masking the deepest layer*. c) Wave-fronts and raypaths at receivers along the surface, when *masking the two deepest layers*.

schematically in one of their illustrative figures, reproduced here in Figure 1.11.

## 1.8 Assumed limitations may hide the strengths of the method

In later chapters the numerous factors influencing seismic velocities such as joint frequency, porosity, rock (and joint wall) strength, density, depth, stress, stress anisotropy, degree of saturation, type of saturating fluid, will each be reviewed. This will be done in order to emphasise that the seismic method has numerous complications, but also inestimable advantages, and that some of the assumed shortcomings can be due to misinterpretations (often an over-simplification, or perhaps even a pre-conceived opinion about 'seismic' limitations).

Consider for example the jointed chalk marl at the Chinnor Tunnel in the UK, where jointing in this weak material were seismically closing at about 15 metres depth, to give a stable 1.6 km/s field velocity, despite changes in the degree of jointing ('rock' Hudson et al., 1980). This can be contrasted with the jointed gneiss at the Gjøvik cavern in Norway, which gave a continuous rise in velocity from *3.5 to 5.5 km/s in the first 50 metres depth* due to increased stress, yet had almost unchanged rock quality. The joint frequency, RQD and rock quality Q-values did not show an improvement with depth (Barton et al., 1994). Both the above observations could be interpreted as 'limitations' of the method. In fact they are demonstrating specific and quite logical physical laws of behaviour.

The latter is an example of the need to interpret seismic velocity with knowledge of depth and/or stress level, since a rock quality Q-value increase from perhaps 1 to 100 might otherwise be assumed in these first 50 metres, based on a 'hard rock', 'stress-less' interpretation (i.e. equation 1.4, and Figure 1.8).

The interpretation cannot be divorced from consideration of what is actually occurring in a rock mass as depth increases, i.e. joint normal stiffness increases, joint aperture reduces, joint frequency reduces (usually), and the clay content in the joints reduces (usually). The list can be lengthened considerably by adding that the stress increases, the deformation modulus increases, the permeability reduces, the pore pressure increases. Since there may be changes of lithology with depth increase, the provision 'usually' should really be added to *each* of the above.

When considering the possibility of relating seismic velocity to a rock quality descriptor such as the rock quality Q-value, another quite basic problem must also be considered. A velocity of 2.5 km/s for massive chalk marl of high porosity, as in the better parts of the UK-France Channel Tunnel, will have entirely different engineering consequences to that of a regional fault-zone of the same 2.5 km/s velocity, crossing a Japanese high-speed rail tunnel excavation, and delaying progress by months. In this best quality chalk marl, near-world record speeds of (TBM) tunnel boring were achieved.

The natural velocity of the unjointed rock under in situ conditions (Sjøgren et al., 1979), and the contrast seen in low velocity zones is the main index of difficulty, since *an order of magnitude reduction* in the rock quality Q-value will generally accompany each 1.0 km/s reduction in seismic velocity, according to the simplified model shown in Figure 1.8.

Low velocity and potentially high permeability zones will be the natural focus of attention in most sub-surface civil engineering projects. However, in a TBM (machine-bored) tunnelling project, there will be serious delays if there is too much high velocity rock, due to the slow progress made in hard, sparsely jointed rock. This would give a 'reversed' $V_p$–rock quality indication, due specifically to poor borability and the need for frequent cutter changes.

A Q-value based '$Q_{tbm}$' rock-machine quality factor has been developed for this specific problem (Barton, 2000), which also allows for the fact that more jointing is good for progress – up to some limit, when other problems may arise. (More tunnel support is needed, there could be gripper-setting problems, the cutter head could even be jammed).

In the case of drill-and-blasted tunnels, the drillability and blastability components of the cycle time for one round of advance would also be demonstrating a 'reversed' $V_p$-quality behaviour. However, the more dominant effect of a reduced need for time consuming rock support would be consistent with the normal 'high velocity = high quality' concept. Since it is more logical that $V_p$ increases should accompany rock quality or Q-value increases, such cases of 'reversed' quality must be treated as separate, method-specific cases.

## 1.9 Seismic quality Q and apparent similarities to Q-rock

A fundamental feature of the propagation of stress waves in all materials is the absorption of energy and the resulting change in shape of the transient waves. Non-linear friction has traditionally been assumed to be one of the dominant attenuation mechanisms (Kjartansson, 1979), but as will be discussed in Part II, there are a large number of mechanisms that can explain the different degrees of attenuation in different rock masses, including scattering due to structure, and *squirt flow*.

Geophysicists commonly characterise seismic attenuation by the seismic quality factor Q, which of necessity has been termed $Q_{seis}$ throughout this book, to distinguish it from the rock mass quality Q. Intuitively $Q_{seis}$ is related to, but no relation of, the rock mass quality Q of Barton et al., 1974, which is a 'quality number' also widely used in civil and mining engineering in the last several decades.

The seismic quality $Q_{seis}$ is often defined as the maximum energy ($E_{max}$) stored during a cycle, divided by the energy lost ($\Delta E$) during a cycle:

$$Q_{seis} = \frac{2\pi}{\Delta E} \times E_{max} \tag{1.10}$$

For dry rocks, $Q_{seis}$ has been claimed to be independent of frequency over a reasonable frequency range (McKenzie et al., 1982). However, as will be seen in Part II, $Q_{seis}$ is a remarkably sensitive indicator of anisotropy, and *is* frequency dependent in the case of saturated or partly saturated conditions. When attenuation ($Q_{seis}^{-1}$) is high, $Q_{seis}$ obviously has a low value. Chapter 10 in Part II, will address $Q_{seis}$ in detail, and also draw some tentative parallels between 'Q and Q', via a common link to deformation properties.

Intuitively, the rock mass quality Q (of Barton et al., 1974) which has a high value in high-modulus, high-velocity rock masses (i.e., Q = 100, $V_p$ = 5.5 km/s,

$E_{mass} \approx 45$ GPa) would seem likely to have high energy storage and low energy loss under such conditions, and therefore qualify for a high value of $Q_{seis}$ and a low value of attenuation ($Q_{seis}^{-1}$). Conversely, in a rock mass characterized by many joint sets with clay coatings and fillings on many joints (i.e., $Q = 0.1$, $V_p = 2.5$ km/s, $E_{mass} = 5$ GPa), both low energy storage and high energy loss per cycle would be expected. ($Q_{seis}$ is low and attenuation $Q_{seis}^{-1}$ is correspondingly high.)

Further discussion of $Q_{seis}$ is given in Chapter 4, based on recent seismic attenuation tomography (see *alternative tomograms*). However, a much fuller treatment of $Q_{seis}$ is given of both lab-scale and *in situ* scale attenuation phenomena, in Part II, chapter 10, and in relevant parts of Chapters 13 and 15.

# 2 Environmental effects on velocity

In this chapter, the effects of near-surface weathering on the seismic velocity $V_p$ will be reviewed. This automatically introduces the separate, but closely integrated effects of density, porosity, uniaxial compressive strength, and the depth and degree of saturation. Depth and stress effects will only be superficially reviewed here; that discussion belongs in later chapters of Part I dealing with anisotropy and rock-burst or stress-slabbing in deep excavations, and is of course a fundamental aspect of all the deep or high pressure seismic results reviewed in Part II. The review will be *loosely organised into sub-sections,* on density, porosity, uniaxial strength and water content, but *overlap* will inevitably occur within each sub-section. Weathering and depth effects are inseparable from the general presentation of reviewed data. The key result of the inter-relationships in this near-surface environment, is a velocity-depth gradient even in one rock type, that can easily climb into two and three figures, for example 2 km/s increase in the space of 20 m. The dual effects of rock mass strength and quality improvement, and of vertical *and horizontal* stress increase, are usually responsible.

## 2.1 Density and $V_P$

The strong influence of density on P-wave velocity, and the stabilisation of density below the weathered zone are nicely demonstrated in Figure 2.1 (Ikeda, 1993). The marked fluctuations in velocity at depth were interpreted by the author as due to high *shear* stresses, which were interpreted from hydraulic fracturing conducted in the same holes. (Presumably the elastic isotropic estimate of $\sigma_{H\ max}$, based on the '3P-Q' model was several times the magnitude of the measured $\sigma_{H\ min}$). Note the typical rapid increase in velocities in the 25 to 75 m depth zone, which partly mirror density increases and are partly related to joint closure and less frequent jointing.

In the case of a range of rock types including marl and peridotite, Kujundzíc and Grujíc, 1966 found a linear relation $V_p = 4.75\gamma - 7.3$ (r = 0.88) for $V_p$ in km/s, and density $\gamma$ in gm/cm$^3$. Seismic velocities ranged from 2.3 to 6.5 km/s, and densities from 2.1 to 3.0 gm/cm$^3$.

Early Bulgarian experiences with seismic registration of weathering effects are provided by Iliev, 1966. Fresh and weathered monzonite were shown to have the following ranges of properties, and linear relationships between $V_p$–n% and $V_p$–$\gamma$ (gm/cm$^3$). (See Figure 2.2)

| | $V_p$ km/s | E (GPa) | $\gamma$ (gm/cm$^3$) | n% |
|---|---|---|---|---|
| Fresh monzonite | 5.0 | 50 | 2.61 | 2 |
| Weathered monzonite | 1.4 | 6 | 2.34 | 12 |

The linear $V_p$–n% and $V_p$–$\gamma$ relationships conceal a non-linear $V_p$–uniaxial compressive strength trend. When a reduction in $V_p$ due to $\sigma_c$ and due to n% are assumed, the strength of the porosity relationship as above, needs modification. Iliev, 1966, noted that a weathering coefficient could usefully be defined as the ratio $(V_{po}-V_{pw})/V_{po}$ where sub-scripts (o) and (w) signify fresh and weathered. The coefficient approaches values of 0 and 1 at opposite ends of the weathering scale.

Many of the long span bridges in Japan have been constructed in soft rock such as Tertiary mudstones and sandstones, or weathered Tertiary granites. The long span bridges of the Honshu - Shikoku Bridge system described by Yamamoto et al., 1995, and Ishikawa et al., 1995, had foundation sizes in the 50 to 100 m range but nevertheless had contact pressures as high as 1 to 2 MPa.

For this reason, Japanese authorities devised comprehensive routines for geological and geotechnical investigations. Seismic methods, in situ deformability, strength and classification schemes were used extensively, especially when trying to extrapolate the results of in situ testing at more convenient onshore sites, to the actual undersea locations of the pier foundations.

The data given in Figure 2.3 shows in situ seismic-velocity-based rock classes, porosities, densities and degree of saturation, each intimately linked. This remarkable, and useful diagram covers all the sub-titles of this chapter, but has been included at the beginning, together with density, to indicate the *integrated* nature of these measures of rock mass quality (or lack of quality). Other useful data sets include $V_p$ versus deformation modulus, seismic rock class versus shear strength, and deformation modulus versus RQD and uniaxial strength. Most of the data was obtained from measurements in medium to lower quality, weathered granites.

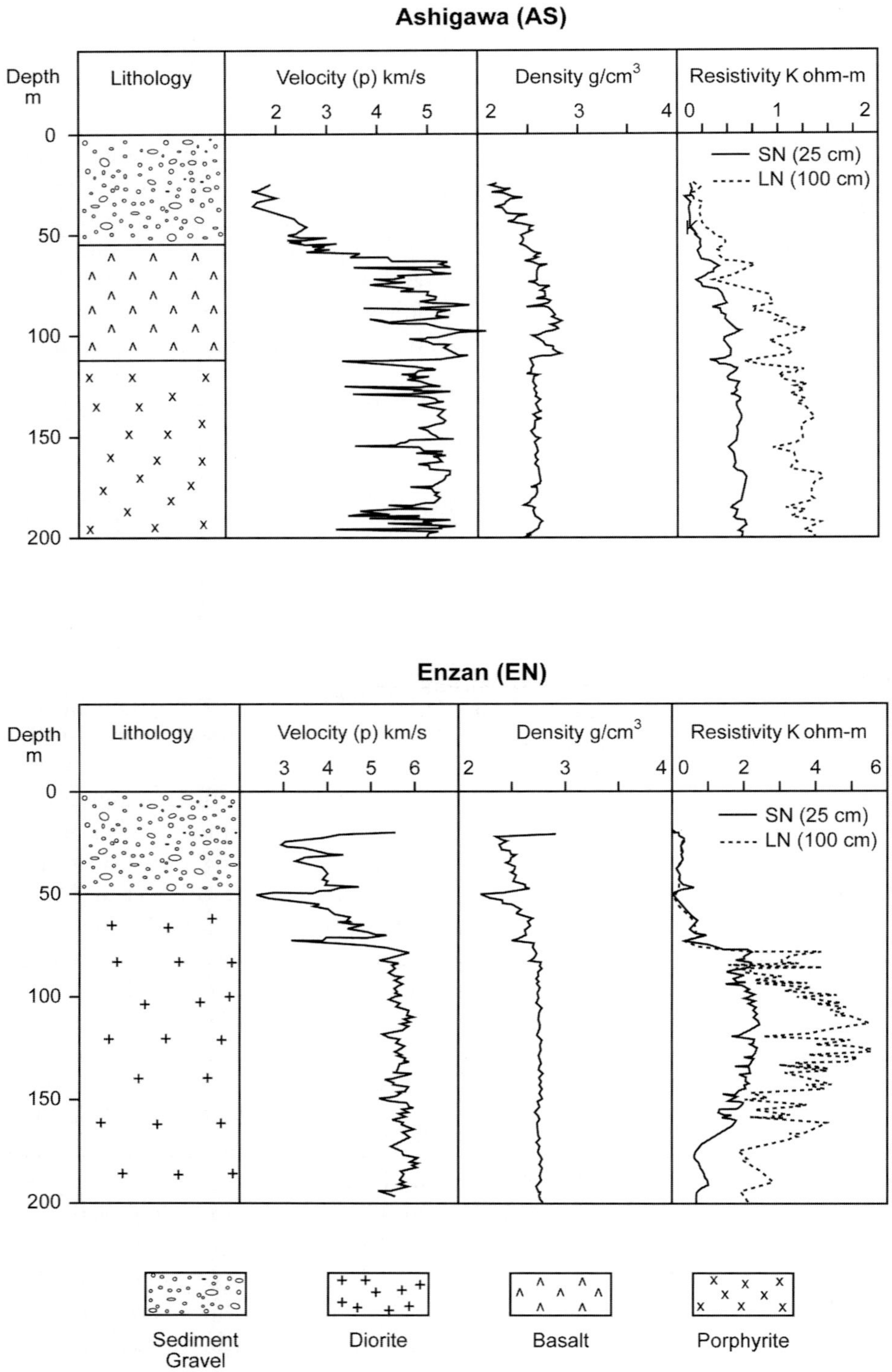


*Figure 2.1* Influences of weathering, depth of measurement and density on Vp and resistivity. Ikeda, 1993.

The authors used an extended version of the Tanaka and Japan Highways classification (which involved the six classes A, B, $C_H$, $C_M$, $C_L$ and D where subscripts mean high, medium and low), and included three classes at the lowest end of the scale ($D_H$, $D_M$, $D_L$) for rock masses with velocities in the range 1.5 to 2.7 km/s. Table 2.1 shows the scheme adopted, which cross-correlates with deformation modulus, density, porosity and resistivity.

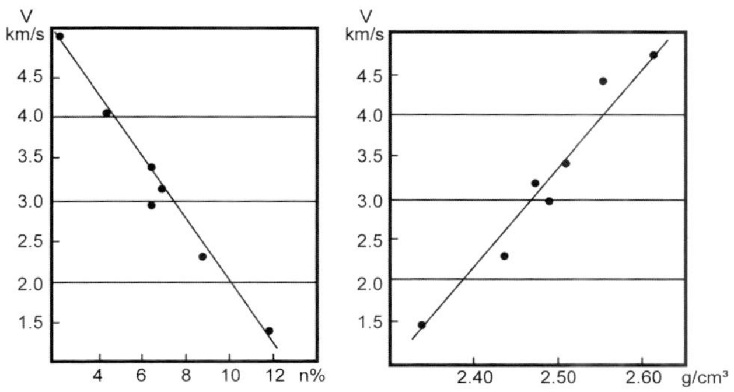


*Figure 2.2* Effects of weathering on $V_p$ of monzonite are seen in linear $V_p$-n% and $V_p$-γ relationships. Iliev, 1966.

A large collection of *laboratory* $V_p$–γ ($gm/cm^3$), and $V_p$–n% data is given by Kelsall et al., 1986, for the case of basalts from California and dolerites from S.W. England. Data that fall *outside* the general trend for the intact rock are ascribed to *fissured* and persistently microcracked rock, shown by the black data points in Figures 2.4 and 2.5.

The lower seismic velocity of the fissured samples is accentuated by the air-dried state of these samples. When plotted on a log-linear scale, the uniaxial strength is seen to broadly correlate with the air-dry $V_p$ value. The data set given in Figure 2.6 goes to

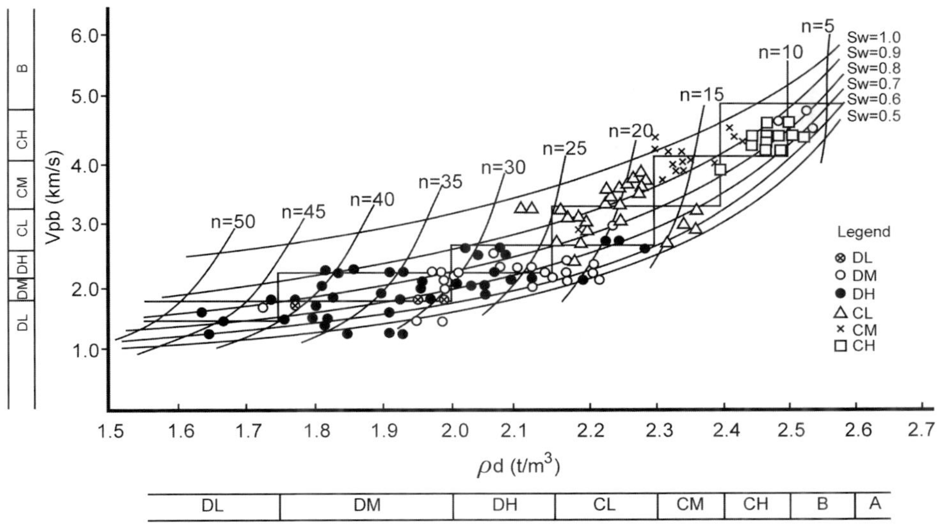


*Figure 2.3* Inter-relationships between rock class, Vp, porosity, density and degree of saturation at the Honshu - Shikoku Bridge project in Japan. Ishikawa et al., 1995.

Table 2.1 Extended (low velocity) Japanese classification scheme used at Honshu-Shikoku Bridges, showing cross-correlation of parameters. Note extreme range of densities due to weathering. Yamamoto et al., 1995.

| Rock class | $V_{pr}$ (km/s) | Rt (Ω • m) | $E_{sb}$ (GPa) | $\rho_c$ ($10^3 kg/m^3$) | $\rho_R$ ($10^3 kg/m^3$) | $n_c$ (%) | $n_R$ (%) |
|---|---|---|---|---|---|---|---|
| $D_L$ | 1.5–1.8 | 1–4 | 0.05–0.3 | 1.7–2.1 | 1.55–1.75 | 33–58 | 50–64 |
| $D_M$ | 1.8–2.2 | 4–7 | 0.3–0.8 | 2.1–2.3 | 1.75–2.0 | 19–33 | 37–50 |
| $D_H$ | 2.2–2.7 | 7–12 | 0.8–1.5 | 2.3–2.5 | 2.0–2.15 | 11–19 | 27–37 |
| $C_L$ | 2.7–3.3 | 12–20 | 1.5–3.0 | 2.5–2.55 | 2.15–2.3 | 6–11 | 20–27 |
| $C_M$ | 3.3–4.0 | 20–50 | 3.0–6.0 | 2.55–2.6 | 2.3–2.4 | 3.5–6 | 15–20 |
| $C_H$ | 4.0–4.8 | 50–120 | 6.0–12.0 | 2.6–2.65 | 2.4–2.5 | 2.0–3.5 | 11–15 |

Note: $V_{pr}$: P-wave velocity of rock mass; Rt: resistivity of rock mass; $E_{sb}$: deformation modulus from pressure meter; $\rho_c$: density of core; $\rho_R$: density of rock mass; $n_c$: porosity of core; $n_R$: porosity of rock mass.

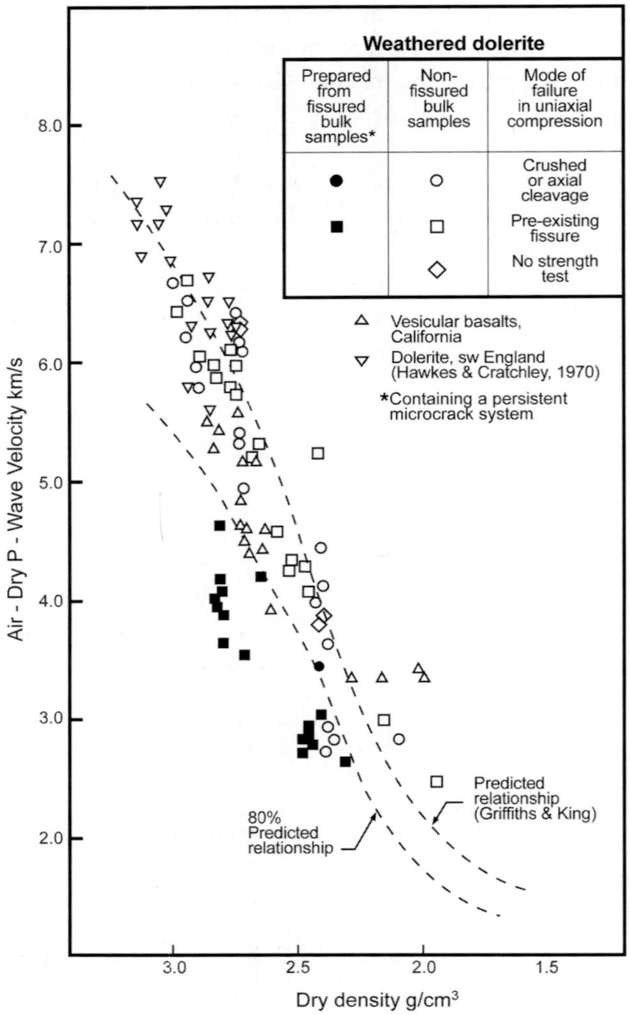


*Figure 2.4* Effects of dry density on $V_p$ for air-dry samples of dolerites. Note increased density of fissured samples, presumably indicating a subtle change in composition. Kelsall et al., 1986.

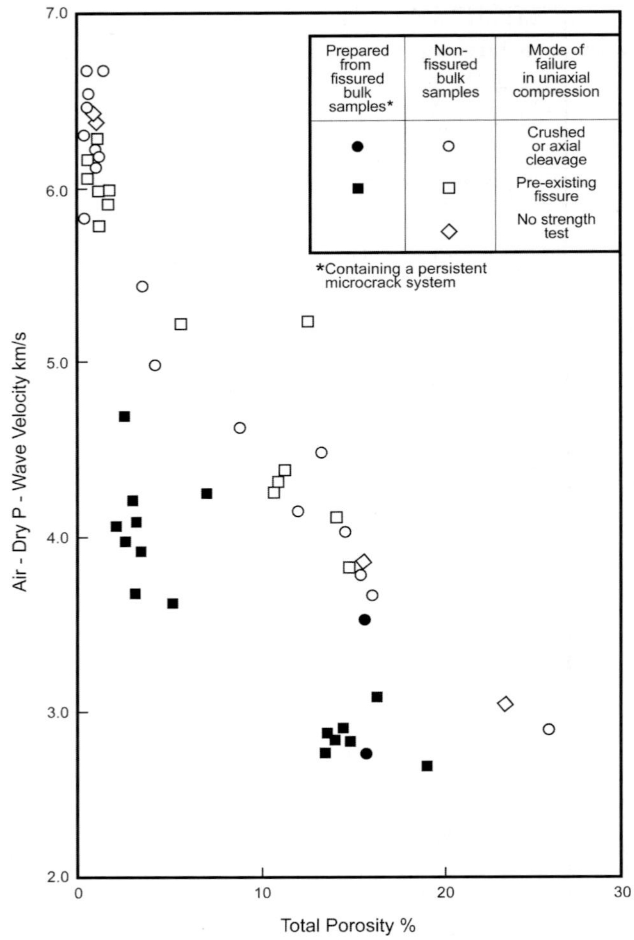


*Figure 2.5* Effects of porosity variations on $V_p$ for air-dry samples of dolerites. Note increased density of the fissured samples, presumably indicating a change in composition. Kelsall et al., 1986.

unusually high levels of strength (500 MPa) and velocity (7.5 km/s); the latter a direct function of the *high density* (2.9–3.1 gm/cm$^3$) of the dolerites. Note the ratio of $V_p$ (air-dry) to $V_p$ (saturated) given in Figure 2.7. The samples with the pre-existing fissures show greatest contrast in these velocities, due to the positive effect of wave transmission through water filled fissures (or joints).

Before leaving this section on (mostly) $V_p$ and density trends, caused by weathering, it is instructive to also look at extreme $V_p$ values due to exceptionally high densities, both from natural causes and from the influence of high stresses (1000 MPa or 10 kbars). The velocity of a variety of high density rocks such as dunite and serpentinite are shown in Figure 2.8. For densities in the range of 2.5 to 4.5 gm/cc, velocities ranged from 6 to more than 9 km/s at these extremely high pressures (Birch, 1961).

Velocity (and of course density), have been used with success for identifying minerals from host rocks. Salisbury et al., 2000, used seismic imaging of known ore bodies in central and eastern Canada, together with high pressure laboratory tests, using what they termed a 'crack closure pressure' of 200 MPa confining pressure. They drew various envelopes in velocity-density space, to distinguish commonly occurring sulphide ores, and

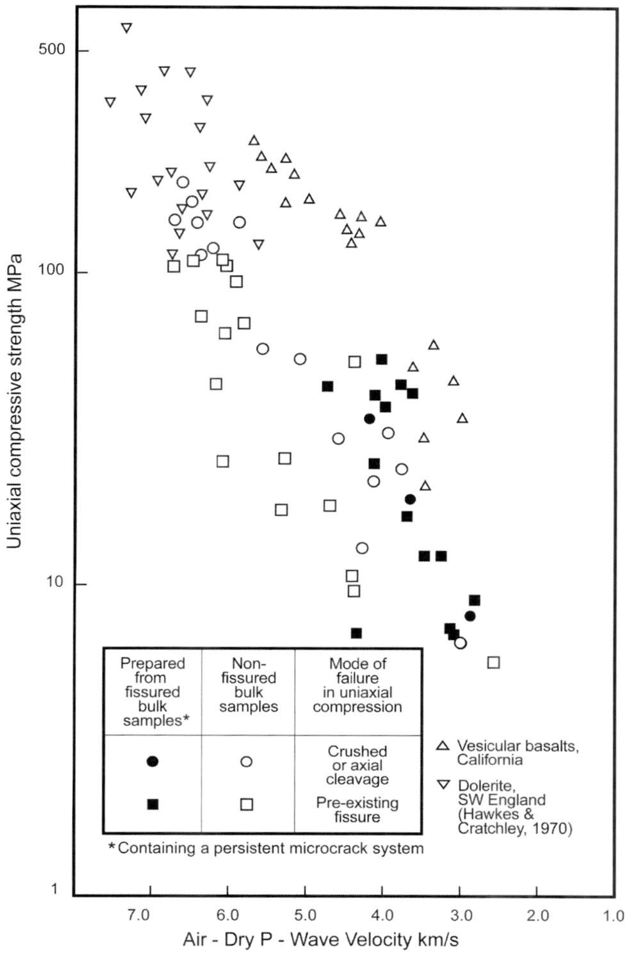


*Figure 2.6* Vp–$\sigma_c$ relation for high strength and weathered rocks. Kelsall et al., 1986.

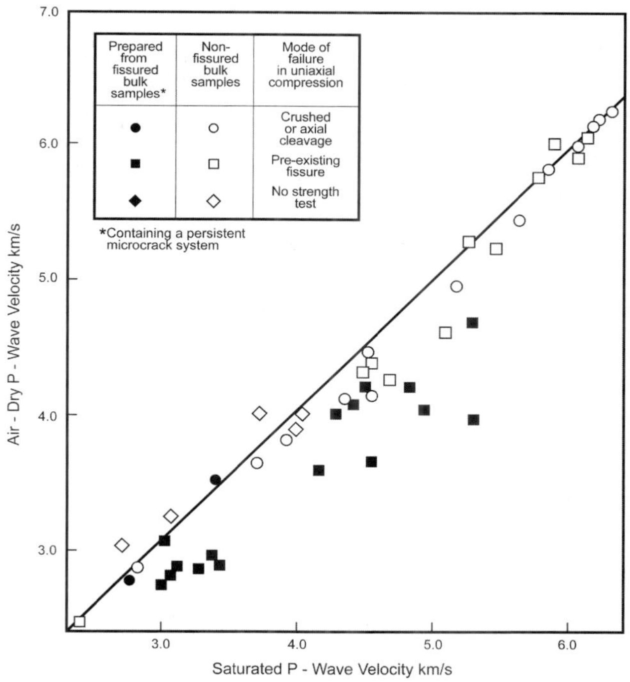


*Figure 2.7* Air-dry and saturated $V_p$ values for intact and fissured samples of dolerite. Kelsall et al., 1986.

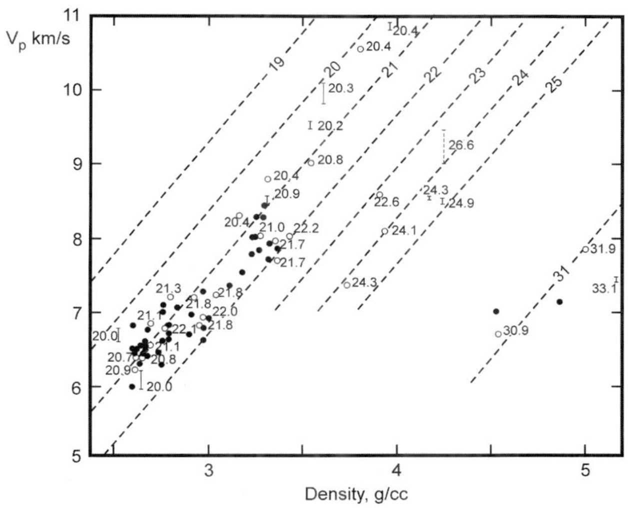


*Figure 2.8* Extreme Vp-density data for crustal rocks at 1000 MPa confinement. Birch, 1961. (Numbers: next to open circles = mean atomic weights; on diagonal lines = constant mean atomic weights (approx.)).

typical silicate host rocks. We may select some contrasting combinations:

| | | |
|---|---|---|
| Pyrite: | $V_p = 8.0$ km/s | density $= 5.0$ gm/cm$^3$ (the extreme member) |
| Pyrrhotite: | $V_p = 4.7$ km/s | density $= 4.6$ gm/cm$^3$ |
| Chalcopyrite: | $V_p = 5.5$ km/s | density $= 4.1$ gm/cm$^3$ |
| Serpentinite: | $V_p = 5$ to 7 km/s | density $= 2.4$ to 2.9 gm/cm$^3$ (host rock) |

At the shallowest depths of the earth's crust, namely the soil cover, specific depth-density-$V_p$ relationships are also evident. Brandt, 1955, developed a theory for the influence of pressure and porosity (and saturation) on the seismic velocity in porous granular media. His elasticity-based, Hertz contact theory predicted that $V_p$ should be proportional to the *1/6 power* of the effective stress. He then compared (in Figure 2.9) this $V_p$-depth gradient with test data for soil, clay and gravel measured by Nasu, 1940. The slopes of the test data plotted

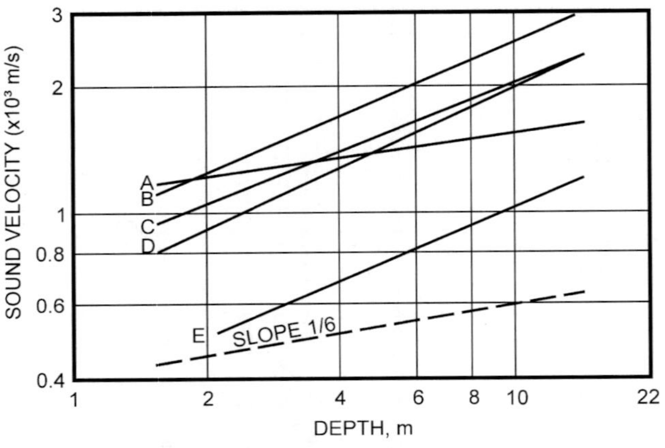


*Figure 2.9* Effect of depth of burial on $V_P$ values of soils in relation to theory. The gradient 1/6 is drawn from the origin, signifying only the $V_P$-depth relation, not the velocity magnitude, which can vary widely depending on the composition of the soil. Brandt, 1955.

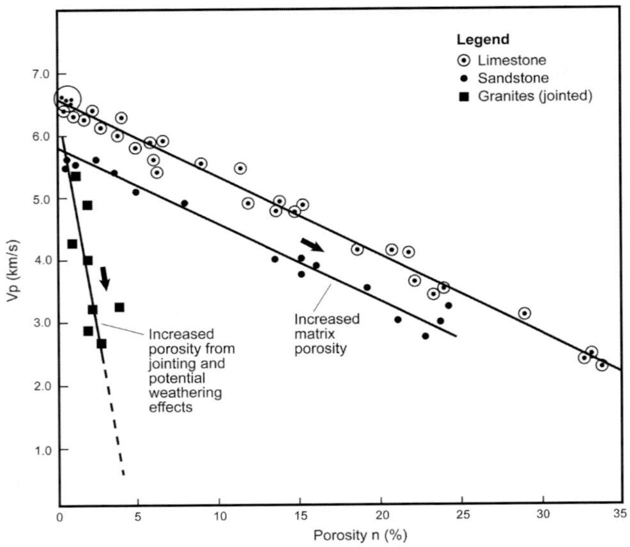


*Figure 2.10* $V_P$ – porosity data for limestones, sandstones and (jointed, weathered) granites. Fourmaintraux, 1975.

on log ($V_P$)–log (depth) scales, ranged from 1/2 to 1/7, bracketing his theoretical gradient prediction of 1/6.

In practice this data and the accompanying theory can help to explain the virtual seismic 'disappearance' of heavily stressed, faulted gouge at great depth or at large induced stress levels. Such was experienced, for example, ahead of a stuck TBM in an 800 m deep tunnel where cross-hole tomography was designed to investigate a known fault (Contract report, NGI, 1998).

## 2.2 Porosity and $V_P$

There is a wealth of data in the literature concerning the effect of the rock matrix porosity on the P-wave velocity. This is found in most abundance in the rock physics investigations to do with petroleum reservoirs, and will be reviewed in Part II, Chapter 13, and elsewhere.

In general, an approximate inverse proportionality is found between velocity and porosity, but there are many subtle variations bought about by, for example clay-content in sandstones. *High pressure* data (from Chapter 13) suggests a fairly strongly curved, concave relation for porous (30 to 80%) marine sediments, with velocities falling rapidly at first, with a plateau of about 1.5 km/s beyond extreme porosities of about 50%. Something nearly approaching linearity is seen in the case of sandstones. A simple, illustrative set of experimental data, applicable in civil engineering is that provided by Fourmaintraux, 1975, which is reproduced in Figure 2.10. Remarkable linearity is shown in these three cases.

The strong influence of the porosity of the matrix in rocks such as limestone and sandstone, and the linear nature of the $V_P$–n% inverse relationship is clearly demonstrated. In the case of the granites, where joint porosity and presumably weathering of the matrix, are the chief sources of porosity, the reduction of velocity is even more marked. The uniaxial compressive strength is also strongly related to matrix porosity in the case of porous rocks such as limestones and sandstones. It may therefore be logical to allow for the influence of both $\sigma_c$ and n% when seeking an integrated $V_P$– rock quality – deformability chart, to be developed in subsequent chapters.

Wilkens et al., 1984, found that the percentage of *clay content* in sandstones had a marked effect on the P-wave velocity of dry samples. Although this particular data set was related to petroleum reservoirs, and confining pressure was consequently very high (50 MPa), the data will be included in this chapter, as it gives a good illustration of the adverse influence of clay-content, which is a particularly relevant aspect of the near-surface, due to weathering effects on some constituent minerals.

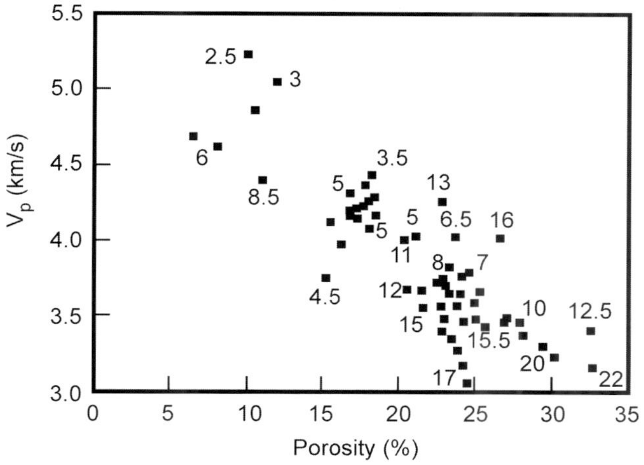


*Figure 2.11* Effect of % clay content on the variation of $V_P$ for dry sandstones, for given values of porosity. (Note: high pressure data, for illustration of relative effects. See Part II, Chapter 13 for high pressure rock physics data). Wilkens et al., 1984.

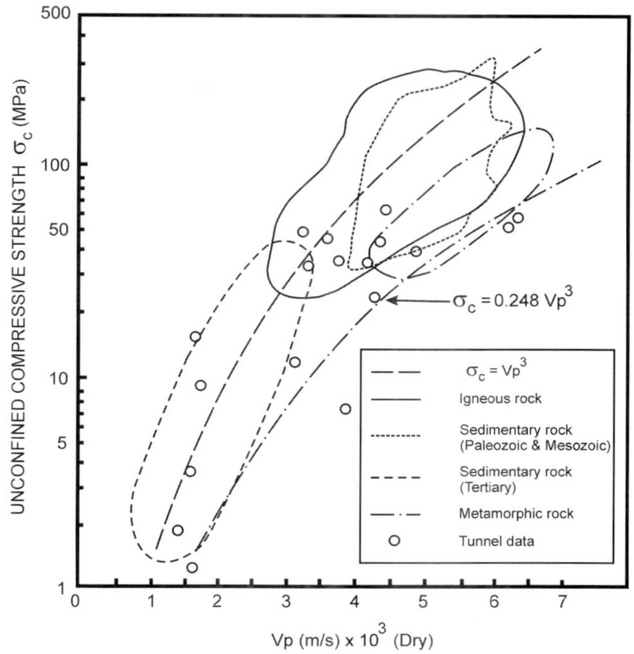


*Figure 2.12* $\sigma_c$–$V_P$ trends from Ohkubo and Teresaki, 1977, with data from a tunnel site in basalt, tuffs and agglomerates. Won and Raper, 1997.

Clearly the higher velocities given in the figure will not be so closely approached in the case of *near-surface* clay-bearing sandstones, but the relative effects of clay-content *are* illustrative.

For a given porosity, say 20%, $V_P$ was shown to range from 3.5 to 4.5 km/s due to clay content reducing from 15% to 5% (approx.). Figure 2.11 shows $V_P$–n% data for dry sandstones, with the % of clay displayed next to the data.

## 2.3 Uniaxial compressive strength and $V_P$

Classification of uniaxial strength by means of seismic velocity alone is obviously suspect since porosity, density and grain size will also be important to differing degrees. However, if envelopes are used to separate the major rock groups, then $\sigma_c$–$V_P$ relationships become somewhat clearer, as illustrated by Ohkubo and Teresaki, 1977, and Won and Raper, 1997, in Figure 2.12. The open circles are data for basalt, tuffs and agglomerates from investigations at a tunnel and highway cutting site in Australia. Note the trend lines ($\sigma_c \approx V_P^3$ and $\sigma_c \approx 0.25\,V_P^3$) whose 4:1 range is still insufficient to encompass the range of data produced in the Japanese study (Ohkubo and Teresaki, 1977 Oyo Corporation. Technical Note RP-479).

Evangelista and Pellegrino, 1990, referred to extensive Japanese data assembled by Ogawa, 1986, in also citing the potential link between porosity and uniaxial compressive strength. Figure 2.13 shows the enormous influence that porosity has on uniaxial strength, indicating a bi-linear trend in a semi-log plot. The influence of porosity on density, and the influence of uniaxial strength on stiffness, means that several inter-related physical properties play their role in increasing or decreasing seismic velocity. Microcracking, jointing, stress level and degree of saturation (including type of fluid) add to the complexity, as will be extensively demonstrated in later chapters, both in Part I and Part II.

Several hundred uniaxial compression strength tests on flysch sandstones were conducted by Pininska, 1977, in three orthogonal directions. The following general trend can be seen in their $\sigma_c$–$V_P$ plot:

| | | | | | | |
|---|---|---|---|---|---|---|
| $V_P = 2.0$ | 3.0 | 4.0 | 5.0 | 5.5 | (6.0) est. | km/s |
| $\sigma_c = 10$ | 20 | 40 | 80 | 100 | (160) est. | (MPa) |

However, the scatter of data was very large, and one could refer to ranges of the above velocities of as much as ±1.5 km/s in the *enclosed region* of the above tabulation. The doubling of strength for each 1 km/s increase

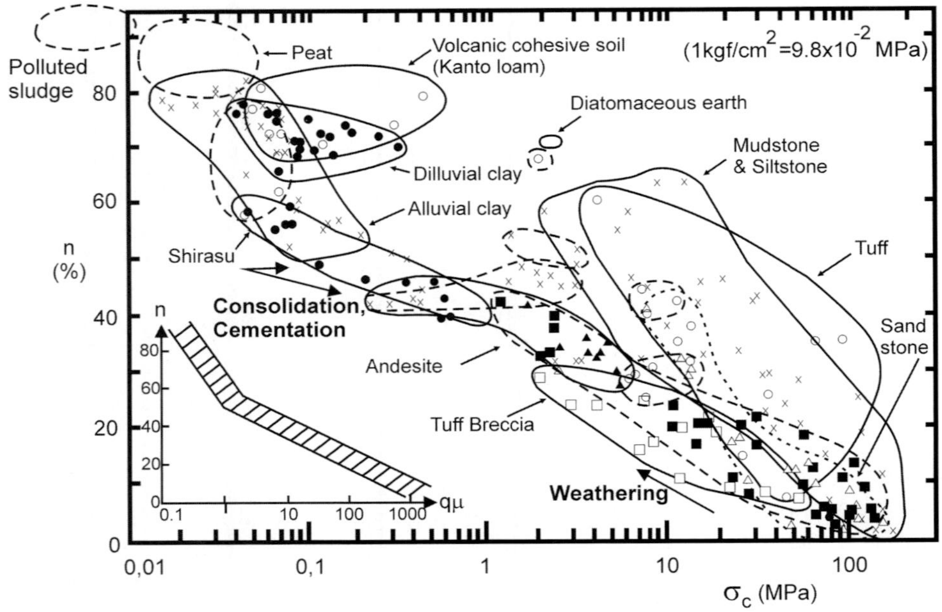


*Figure 2.13* The inter-relationship between porosity and uniaxial compressive strength. Ogawa, 1986 from Evangelista and Pellegrino, 1990.

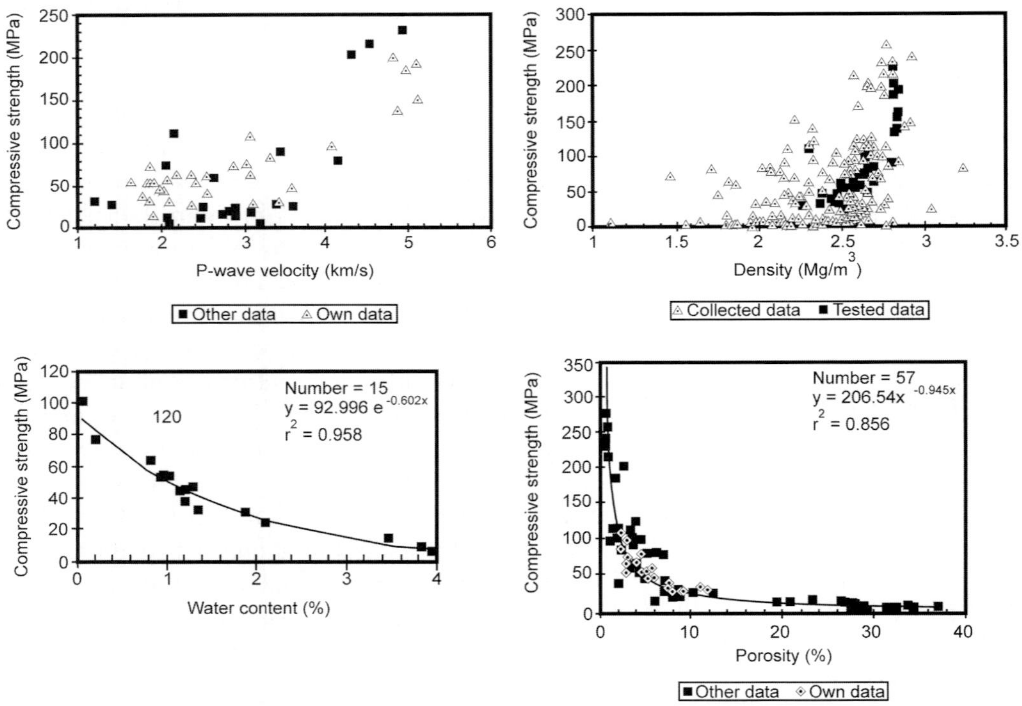


*Figure 2.14* Poor correlation of $V_p$, $\gamma$ and $\sigma_c$ is evident for shale, due in part, to the similar densities of component minerals. Lashkaripour and Passaris, 1995.

in $V_p$ is a good mean trend. We can add this to another general trend, namely that the rock mass quality Q-value increases 10-fold for each 1 km/s increase in $V_p$ for the case of hard, low porosity rocks at shallow depth.

Index tests such as the point load test and Schmidt hammer test (with density included in the interpretation) are known to correlate reasonable closely with uniaxial compressive strength. For this reason, Wei and Liu, 1990,

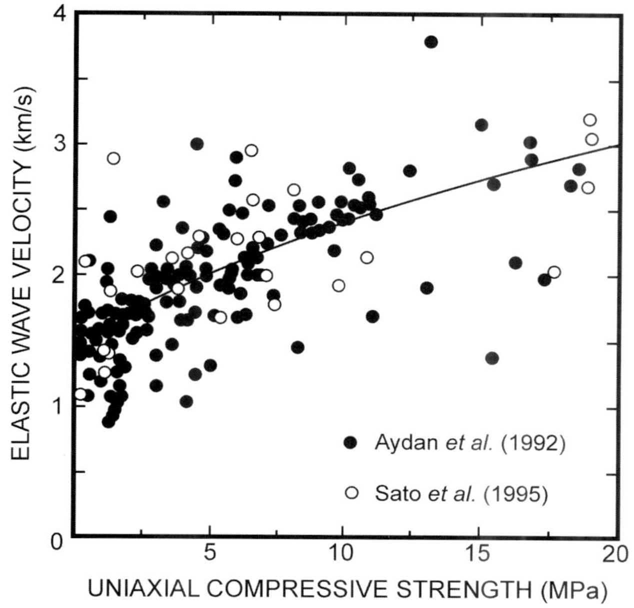


*Figure 2.15* $V_p$-$\sigma_c$ data for Tertiary mudstones and sandstones from Japan. Aydan et al., 1992 and Sato et al., 1995.

used $V_p$, Schmidt rebound value (R) and point load test ($I_{50}$) to evaluate the weathering grade of four igneous rocks. Their database was very large (1069 $I_{50}$ tests, 499 $V_p$ tests, and 1330 Schmidt hammer tests) and gave, as one might expect, strong correlation to the pre-determined weathering grades.

In the case of shale, with its mixed content of similar density minerals, i.e. quartz ($\gamma = 2.66\,gm/cm^3$), illite and montmorillonite ($\gamma = 2.61\,gm/cm^3$), the correlation of compressive strength and *density* is inevitably poorer than for most rocks (Lashkaripour and Passaris, 1995). Since seismic velocity usually correlates well with density, it is perhaps inevitable that *in the case of shales, P-wave velocity is not a sensitive indicator of compressive strength,* as shown by the large spread of data in Figure 2.14 from the same authors.

Water content on the other hand correlates extremely closely with uniaxial strength for shales, e.g., $\sigma_c \approx 90e^{(-0.5w)}$ (w = water content), based on the mean of data from two coalmines. The porosity and compressive strength were linked in a strongly non-linear manner.

At the extreme low end of the $V_p$-$\sigma_c$ spectrum ($\sigma_c < 1$ MPa to 10 MPa), the laboratory data for Tertiary mudstones and sandstones from Japan given in Figure 2.15 shows roughly an order of magnitude increase in strength (1 MPa to 10 MPa) for 1 km/s increase in velocity (i.e., 1.5 km/s to 2.5 km/s). The 'decade' rule-of-thumb, referred above, was again demonstrated for much of the data. The Japanese data shows slightly lower strengths (i.e., 5 MPa compared to 10 MPa) for a velocity of 2 km/s, though as with Pininska, 1977, the scatter of data is large when porosity variations are not shown.

## 2.4 Weathering and moisture content

Effects of weathering on the physical and seismic properties of four rock types from a dam site (quartz diorite) and from three quarries (andesite, basalt and dacite) were reported by Saito, 1981. This very comprehensive study, involving hundreds of samples with different weathering grades and porosities, gives a very useful picture of some key trends between strength, hardness, porosity, degree of water saturation and P-wave velocity. These behavioural trends are fundamental to an understanding of the in situ behaviour, where the addition of joints to the cracks and pores tested here, adds another layer of complexity.

Saito, 1981, collected numerous block samples of the different rocks and weathering grades and cast these in regular shaped concrete blocks, before coring cylindrical samples for his tests. Schmidt (N-hammer) tests were made on these larger blocks. Figure 2.16a illustrates and describes the typical weathered zones (1 to 5), and an idea of the ranges of compression strengths (dry samples) and porosities are given in Figure 2.16b and 2.16c. The very different porosities of the three volcanic rocks compared to the crystalline quartz diorite are well reflected in the clear separation of the $V_p$ values shown in Figure 2.16c.

The extended $V_p$ range of <1 km/s to almost 6 km/s was the result of the huge range of porosities (57% to 1%). When only uniaxial strength and velocity were plotted (Figure 2.16d) the fundamental differences in porosity were not seen due to the relatively high strength of the three volcanic rock types. Figure 2.16e shows how Saito's Schmidt (N) hammer rebound data related to $V_p$ in a quite linear manner, not showing the same 'plateau' effect seen with $V_p$ versus $\sigma_c$. This is an encouraging aspect of this ultra-simple test method, which was adopted for registering the compressive strength of (fresh or weathered) rock-joint walls (JCS) in the shear strength criterion of Barton and Choubey, 1977.

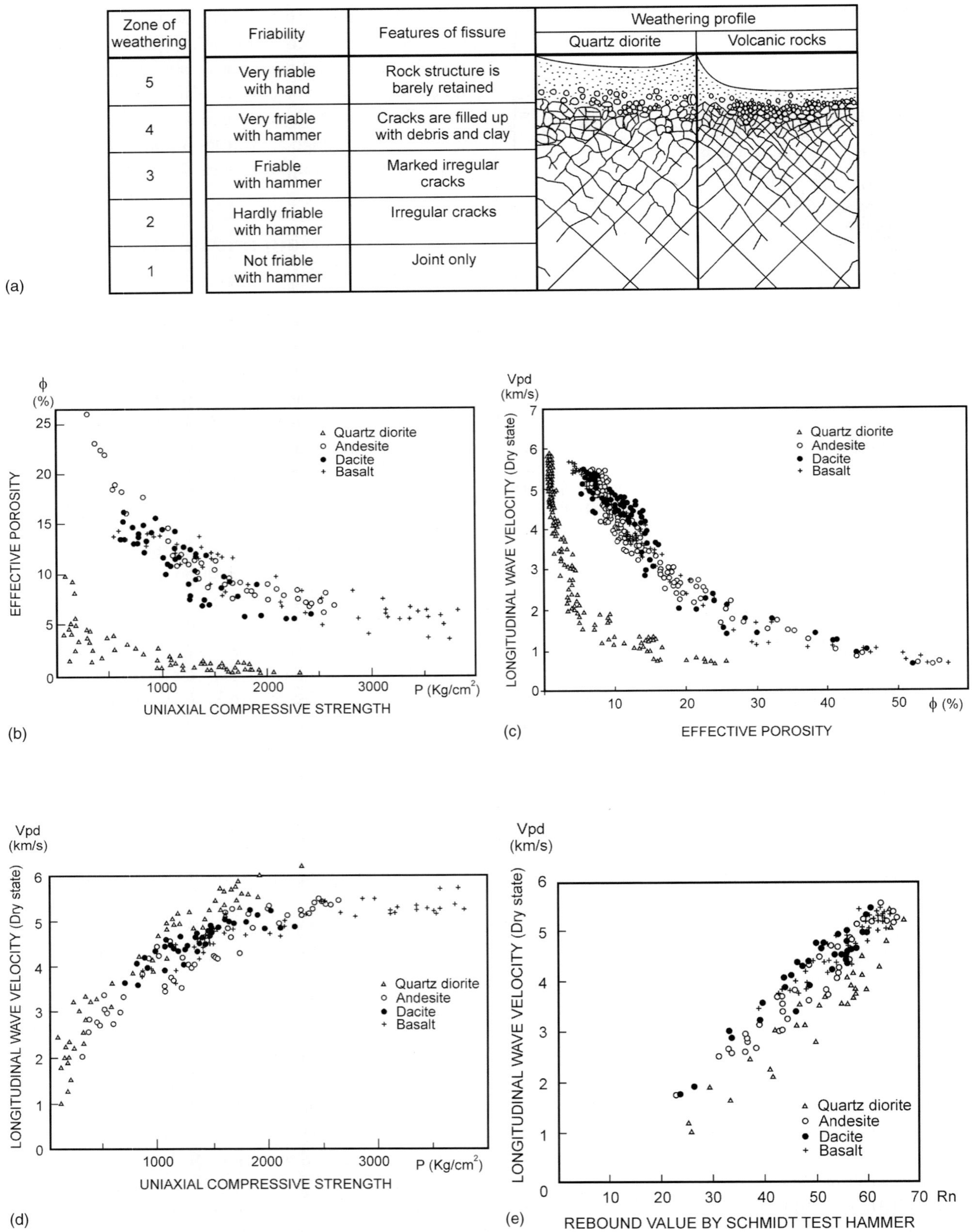


*Figure 2.16* (a to e) Effects of weathering at four sites in Japan cause huge ranges of porosity, strength and P-wave velocity. Saito, 1981.

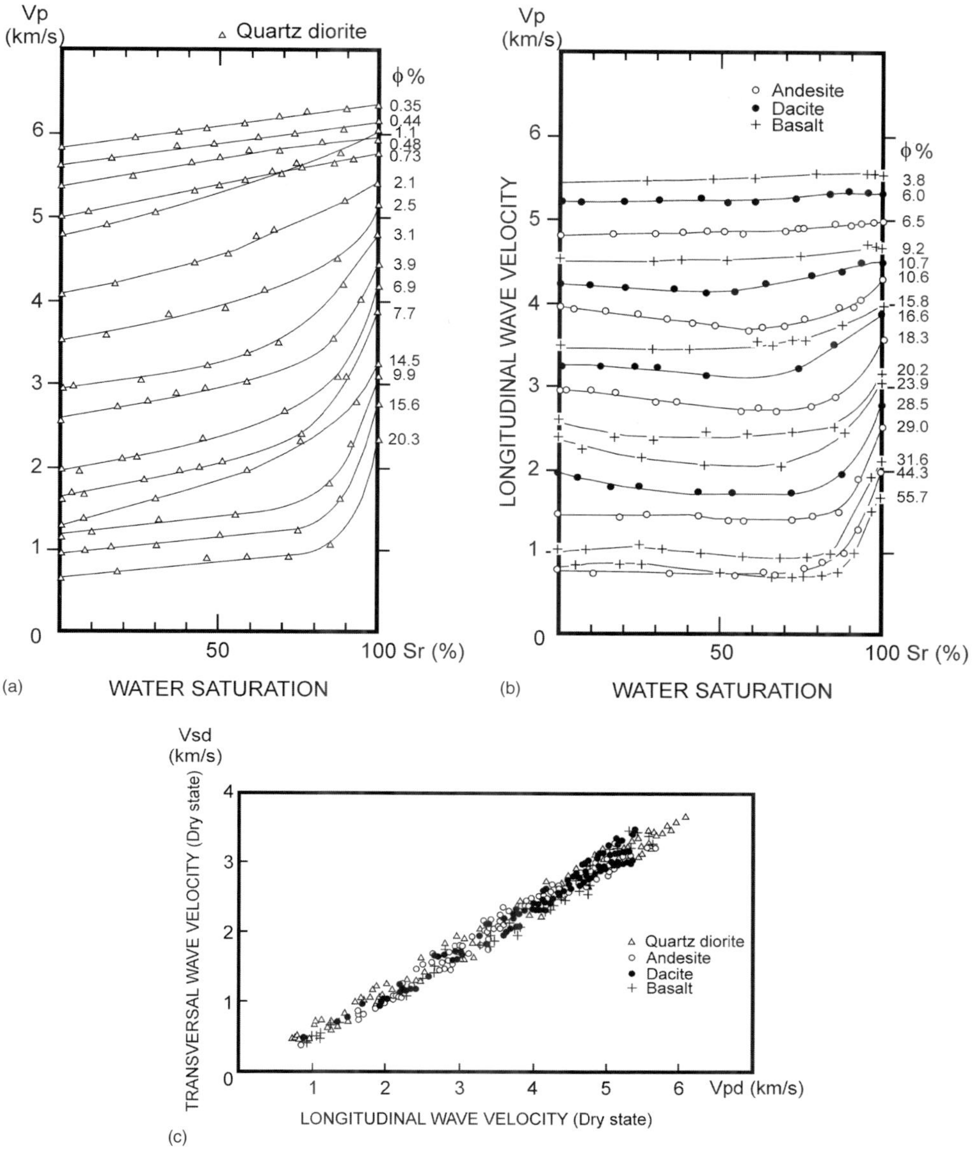


*Figure 2.17* Top: a), b) effect of water saturation on $V_p$. Bottom: c) Dry $V_s/V_p$ trend over a wide range of $V_p$. Saito, 1981.

The significant differences of behaviour caused by porosity reappear when *degree of water saturation* and its effect on $V_p$ are shown *side-by-side* in Figure 2.17a and 2.17b. The higher porosities corresponding to higher weathering grades show very strong (even 200–300%) increases in $V_p$ from initial low values, as saturation exceeds about 85%. Much less sensitivity to saturation (just a weak linear effect) was seen for the fresher, low porosity, high $V_p$ samples, where $V_p$ increased from just 5 km/s to 5.5 km/s with saturation, in the case of a low porosity sample.

The $V_p/V_s$ ratios that Saito derived from many hundreds of data are shown in Figure 2.17c. These particular data are for *dry* samples. $V_p/V_s$ ratios are seen to reduce from about 2.0 at low velocity to about 1.6 at high velocity, broadly following the trends discussed in Chapter 1.

An example of the effect of saturation is given a simple theoretical basis by Grainger et al., 1973. Their evaluation of chalk foundation qualities at the proposed site for a proton accelerator facility in Norfolk, England revealed one anomalous result, when low quality chalk (grade V), which was normally sampled above the water table ($V_p \approx 0.7$ km/s) showed a velocity of 1.95 km/s at one location. The grade V chalk was described by the authors as a structureless melange of angular fragments set in a matrix of deeply-weathered, remoulded chalk.

A *matrix version* of the time average equation of Wyllie et al., 1956, was used by the authors to explain this anomaly as follows:

$$\frac{1}{V} = \frac{\phi}{V_{fl}} + \frac{1-\phi}{V_{sd}} \tag{2.1}$$

where $V_{fl}$ = velocity in fluid,
$V_{sd}$ = velocity in solid, and
$\phi$ = ratio of the path length in the fluid to the total path length (i.e. the 'porosity')

The authors assumed $V_{sd} = 2.3$ km/s for the intact chalk fragments, and first assumed dry conditions with $V_{fl} = V_{air} = 0.33$ km/s, when substituting the measured value for grade V = 0.7 km/s in equation 2.1, to give $\phi = 0.29$. Returning to the assumed saturated conditions, using the calculated 'porosity' of 0.29 and $V_{fl} = V_{water} = 1.44$ km/s, the authors calculated a P-wave velocity for the saturated chalk of 1.97 km/s, close to that measured by the shallow refraction seismic.

An interesting variant of the above time average equation is illustrated for the case of *jointed rock*, by the manner in which joints are assumed to change the seismic velocity. McDowell, 1993, presented the classic equation of Wyllie et al., in the form:

$$\frac{L}{V_p(\text{rock mass})} = \frac{nw}{V_p(\text{joint filler})} + \frac{L - nw}{V_p(\text{rock matrix})} \tag{2.2}$$

where L is the path length
n is the number of joints
w is the average width of the joints

In practice the velocity through dry jointed rock is lower than given by the above, even when the velocity through air ($V_p = 330$ m/s) is taken into account. This is because an air-filled joint will tend to act as an acoustic barrier, except round its ends or across points of contact. The actual travel distance has increased, but (L) has not been corrected in the above equation.

In the case of water-filled ($V_p = 1.44$ km/s) or clay-filled joints, the formula is likely to be more correct, due to the improved coupling across the joint walls. The saturated condition was successfully modelled by the *matrix version* of the time average equation, in the above *structureless melange of angular fragments set in a matrix of deeply-weathered, remoulded chalk,* analysed by Grainger et al., 1973.

## 2.5 Combined effects of moisture and pressure

We have seen from Saito's 1981 data, the importance of the degree of saturation on $V_p$ in the absence of pressure (Figure 2.17a,b). In fact, the *combined* effect of degree of saturation and stress level have significant influence in rock engineering projects due to the common 'environmental changes' that are introduced when we excavate a tunnel, slope or foundation, causing changes of stress (especially unloading) and changes in the pore pressures and drainage routes in the so-called 'excavation disturbed zone' (EDZ), which will be reviewed in detail in Chapter 7.

The excavation process (blasting, boring, ripping, etc.) causes stress redistribution and the release of radial stresses. Monitoring of $V_p$ in such zones sometimes shows areas of increase, but more usually significant reductions in velocity, especially when the rock mass is significantly jointed or damaged by the high tangential stresses, and by the excavation process itself, particularly if by less careful versions of drilling-and-blasting.

It is usually assumed that the release of radial stress and the formation of new fractures by blast gasses are the two chief causes of velocity (and modulus) reductions in the EDZ. However, the results of tests on the *effect of saturation* in weathered and micro-cracked materials (Saito, 1981 and Kelsall et al., 1986) seen in Figures 2.7 and 2.17, obviously suggest a third mechanism of velocity reduction, namely *drying out.* The presence or absence of stress and the dry or saturated state create the largest 'environmental' changes to $V_p$ besides weathering state ($n\%$, $\gamma$, $\sigma_c$).

Nur and Simmons, 1969, classic experiment with repeated measurements of $V_p$ on a sample of Chelmsford granite ($n \approx 1\%$) showed a reduction of $V_p$ from 5.4 km/s (when the sample was saturated) to a value of about 3.9 km/s after four days at room temperature, while drying out. The rapid change in the first 5 hours, even though the rock is of low porosity, is seen in Figure 2.18.

Nur and Simmons, 1969, data also show very strong sensitivity to confining stress, especially in the case of dry samples, which would mean that an unloaded tunnel wall ($\sigma_r = 0$) in a dried-out rock mass would tend to show significantly lower velocities than if still saturated.

If velocity reductions appear to exceed what one would expect in relation to reasonable modulus reductions in an EDZ (from Chapter 7), then drying out seems to be a distinct possibility. The tabulations below, that also belong with high stress data from Part II, show the potential strength of such effects, in comparing dry and saturated samples. (Data extracted from Nur and Simmons tabulations, and rounded).

The very fine grain size in the Solenhofen limestone (0.01 mm) compared to the millimetre-size, or several millimetre-size grains of the other rocks, and its complete lack of crack porosity is the reason for the almost complete *lack of pressure sensitivity* for this rock. Micro-cracks are presumably the chief cause of the above sensitivities to pressure and degree of saturation in the case of the crystalline rocks (upper half of Table 2.2). A large-scale parallel would be the effect of 'environment' (stress and degree of saturation) on jointed rock which we will see in other data sets in later chapters, in particular the data connected with EDZ experiments (Chapter 7).

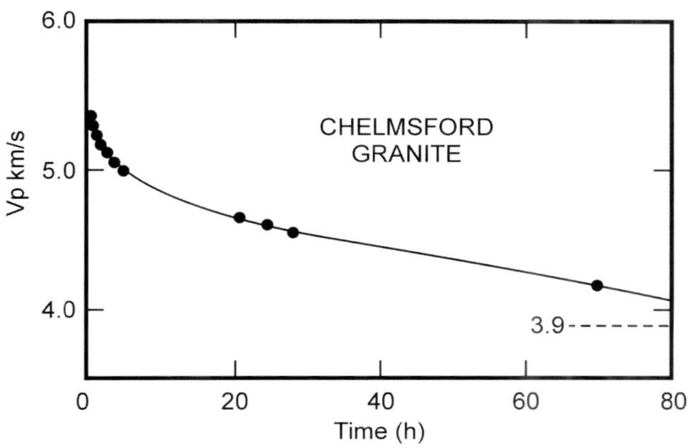


*Figure 2.18* Slow air-drying of a saturated sample of granite reduces $V_p$ by 1.5 km/s. Nur and Simmons, 1969.

To conclude this section on the *combined effects of moisture and pressure,* some data sets will be 'borrowed' from future topics in this book, namely the higher pressure world of Part II, relevant to petroleum reservoirs and earthquake related tectonophysics.

An idea of the eventual non-linear nature of $V_p$-stress data, is given in some early King, 1966, experiments with hydrostatic loading of sandstones, shown in Figure 2.19 with classic *psi* and *ft/sec* units. The water-saturated and dry states show classic 'knee' shapes, and velocities that begin to converge at high stress, due to closure of microcracks. The improved coupling with water, still gives the highest velocity in the saturated state. The maximum pressures in King's experiments were about 35 MPa.

The strong effect of extreme confining pressure, especially when these pressures go far beyond the uniaxial strength of the rocks, is typically illustrated by classic 'knee' shaped $V_p$–$\sigma_3$ curves. Figure 2.20 shows a variety of behaviours from high pressure laboratory test results on dry samples, given by Wepfer and Christensen, 1991. A compressible shale (3.0 to 5.7 km/s) and a porous sandstone (2.2 to 4.0 km/s) show strongest effects of confining pressure, while low porosity sandstone, dolomite and limestone show only 200 to 300 m/s increases. The authors refer to velocity hysteresis; the effect of pressure in closing cracks in the stress range 0–200 MPa (0–2 kb) is

Table 2.2 Confining pressure and dry/saturation effects on the $V_p$ (km/s) of some hard rocks (Nur and Simmons, 1969).

| | | | Confining Pressure (MPa) | | | | |
|---|---|---|---|---|---|---|---|
| $V_p$ km/s | Rock type | Porosity | 0 | 5 | 10 | 20 | 40 |
| dry | Casco granite | 0.7% | 3.3 | 4.2 | 5.1 | 5.7 | 6.0 |
| saturated | | | 5.3 | 5.8 | 6.0 | 6.1 | 6.3 |
| dry | Westerly granite | 0.9% | 3.8 | 4.5 | 5.0 | 5.3 | 5.6 |
| saturated | | | 5.5 | 5.6 | 5.7 | 5.8 | 5.9 |
| dry | Troy granite | 0.2% | 4.5 | 5.7 | 5.9 | 6.2 | 6.3 |
| saturated | | | 5.7 | 6.2 | 6.2 | 6.3 | 6.4 |
| dry | Webatuck dolomite | 0.7% | 5.0 | 5.9 | 6.4 | 6.7 | 6.9 |
| saturated | | | 6.4 | 6.6 | 6.7 | 6.8 | 6.9 |
| dry | Solenhofen limestone | 4.7% | 5.6 | 5.6 | 5.6 | 5.6 | 5.6 |
| saturated | | | 5.6 | 5.6 | 5.6 | 5.7 | 5.7 |
| dry | Bedford limestone | 12.3% | 2.6 | 2.8 | 3.0 | 3.4 | 3.8 |
| saturated | | | 4.5 | 4.6 | 4.7 | 4.8 | 4.8 |

not matched by equal crack opening when unloading occurs in this region, and velocity hysteresis results.

## 2.6 Combined effects of moisture and low temperature

This chapter on 'environmental' effects on $V_p$ would not be complete without reference to the influence of low *temperature and ice formation* on $V_p$. Construction in permafrost, and monitoring of the ground-freezing method, for tunnelling though unstable water bearing areas under environmentally-sensitive areas such as city streets, could each benefit from seismic monitoring to determine the progression or regression of the ice front.

Data given by Timur, 1968, show velocity increases upon freezing that vary from about 20 to 50% for many saturated porous rocks, as compared to their P-wave velocities at room temperature. In general, the largest increases are for the most porous rocks. A shale showed only 8% increase. Dry rock samples are hardly affected by cooling below 0°C.

The enormously contrasted temperature-velocity graphs for the dry and saturated states, for the first few

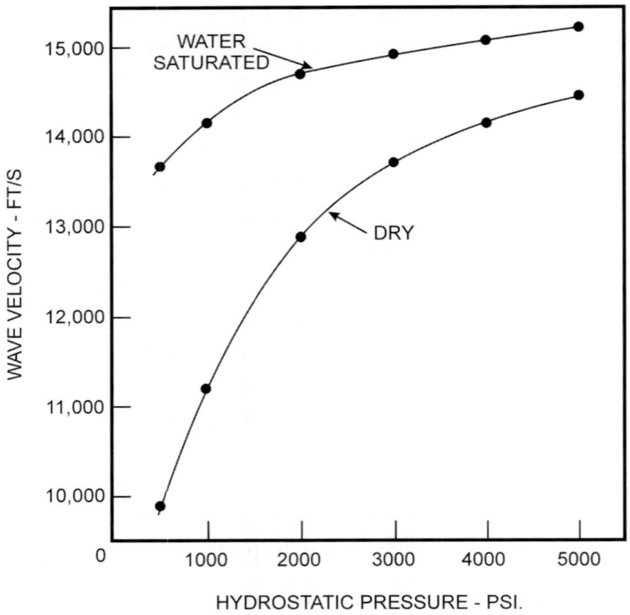


*Figure 2.19* Effect of dry and water-saturated states on $V_p$-versus-stress, for a sandstone, King, 1966. (Note: an intermediate curve for kerogen, lying just below the water-saturated curve, has been removed since not relevant to Part I).

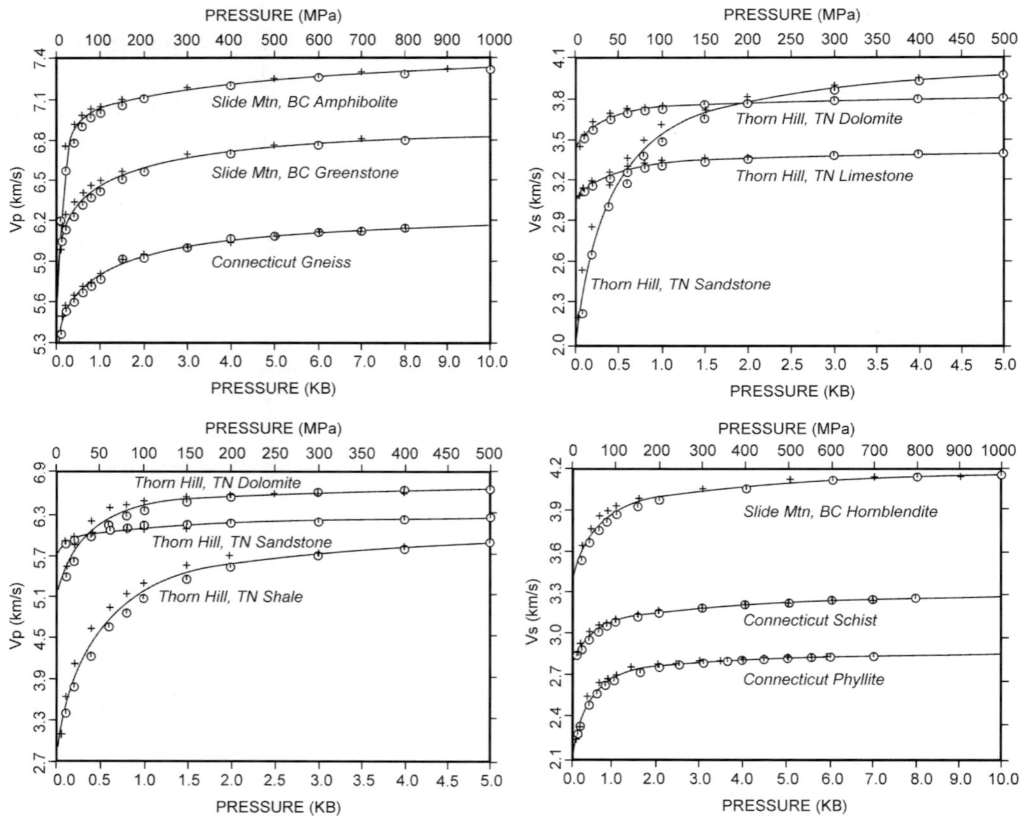


*Figure 2.20* High pressure effects on $V_p$ (500 to 1000 MPa) for a variety of rock types. Wepfer and Christensen, 1991.

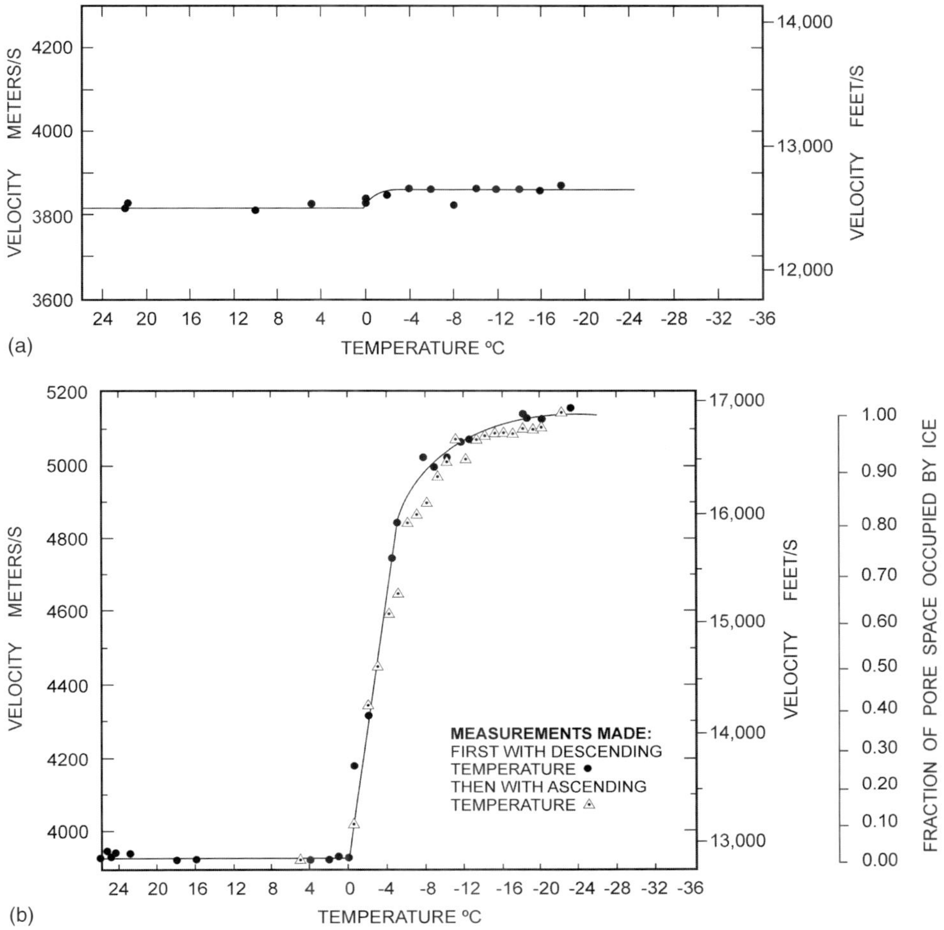


*Figure 2.21* Contrasting effects of low temperature on $V_p$ for Berea sandstone in the dry and wet state, with $\sigma_1 = 31.3$ MPa in each case. Timur, 1968.

degrees below 0°C, is nicely demonstrated in Figure 2.21. This contrast is due to the different rates of freezing in pore volumes that have different *area/volume ratios*. Surprisingly perhaps, the author explains that the smallest pores actually freeze later due to less favourable area/ volume ratios. In the 'macro-discontinuity' world of jointed rock, one would expect that the smaller, finer tips of cracks and joints would freeze first, due to the more stationary conditions, making 'ice-wedging' such an effective mechanism of weathering in mountainous terrain, and in more northerly and southerly climates.

# 3 Effects of anisotropy on $V_p$

In this chapter the 'simple' approach to anisotropy caused by micro-cracks or jointing will be taken, considering principally P-wave, azimuthal anisotropy, and anisotropy caused by stress difference. Besides micro-cracks that may be aligned due to tectonic history or due to existing or applied stress anisotropy, there will be fundamental reasons for velocity anisotropy in foliated, schistose, layered or inter-bedded rocks with unequal layer stiffness. When jointing and faulting are included, with the special effects of stress anisotropy on these larger scale features, the potential causes of velocity anisotropy will be numerous. Although velocity anisotropy complicates interpretation, at the same time it also provides important information for a rock engineering project, and of course for a fractured petroleum reservoir, if correctly interpreted. It will be seen that the classic alignment of a dominant joint set with the maximum horizontal stress direction is often a cause of a double-anisotropy effect. Both the near-surface and high stress treatment of P-wave anisotropy, as introduced in this chapter, will be supplemented in Chapter 14, by studies at considerably greater depth, principally in fractured reservoirs. In Chapter 15 the anisotropy information found in *shear waves* will finally be the focus of attention, as a lot more information is contained in waves that polarize in parallel (fast $qS_1$), and (slow $qS_2$) perpendicular directions relative to the discontinuities. These shear wave components show dispersive, frequency dependent levels of anisotropy, caused, in principle, by the dimensions, density and stiffnesses of the fracturing and jointing. There are also those who attribute the shear-wave anisotropy at depth mostly to micro-cracks.

## 3.1 An introduction to velocity anisotropy caused by micro-cracks and jointing

It is reported that Maurycy Rudzki, the first Professor of Geophysics at a university in Cracow, stated his intention to do research on the propagation of seismic waves in anisotropic media, in 1896. (Helbig and Szaraniec, 2000). It has also been recognised since early in the 1900s that compressive stress affects the elastic properties of rock, and that change of properties are related to the behaviour of the micro-cracks under load. In Chapter 2 it was seen how micro-cracked and fissured samples were particularly sensitive to the degree of saturation, since they are seismically much more visible when dry and unloaded, than when saturated and strongly loaded.

Nur and Simmons, 1969, reported important results of stress-induced anisotropy, noting that the largest velocity change took place in the direction of the applied stress. Prior to loading, isotropic velocity was usually recorded. Nur, 1971, showed how the observed velocity anisotropy caused by stress effects on crack closure could be modelled, in fair to good agreement with experimental results.

Figure 3.1 shows the relative effects of hydrostatic stress (0 to 50 MPa), and uniaxial stress (0 to 40 MPa) on the compressional wave velocity $V_p$. Here we are also looking ahead into levels of stress appropriate to Part II. Under uniaxial stress, the velocity increase parallel to the stress direction is much greater than the velocity increase perpendicular to the stress, due to preferential closure of those micro-cracks that are aligned more or less perpendicular to the applied stress. The effect may be enhancing in situ velocity anisotropy effects since $\sigma_{Hmax}$ often tends to be parallel or sub-parallel to major jointing, and velocity parallel to these joints is highest, independent of the above 'intact' rock effects. Similarly to these micro-cracks, minor joint sets in situ tend to be closed by the major principal stress, giving a further reason for stress-induced velocity anisotropy at larger scale.

Holt et al., 1997, suggested that stress dependent velocity (caused by micro-cracks) seen in cores taken from great depth may be mainly a result of coring damage caused by the release of anisotropic stresses. This stress dependent behaviour is particularly pronounced at low stresses compared to the virgin stress, such as in a triaxial test performed below the original stress state. Above the previous stress state, the sensitivity to stress change was less. Holt et al., 1996 and 1997, observed that there was little or no stress dependence when no cracks were formed in the recovery process, nor was there stress dependence when the rock was loaded (or unloaded) near the original stress state. In a limited stress regime around the original stress state, the rock behaved as a linear elastic substance.

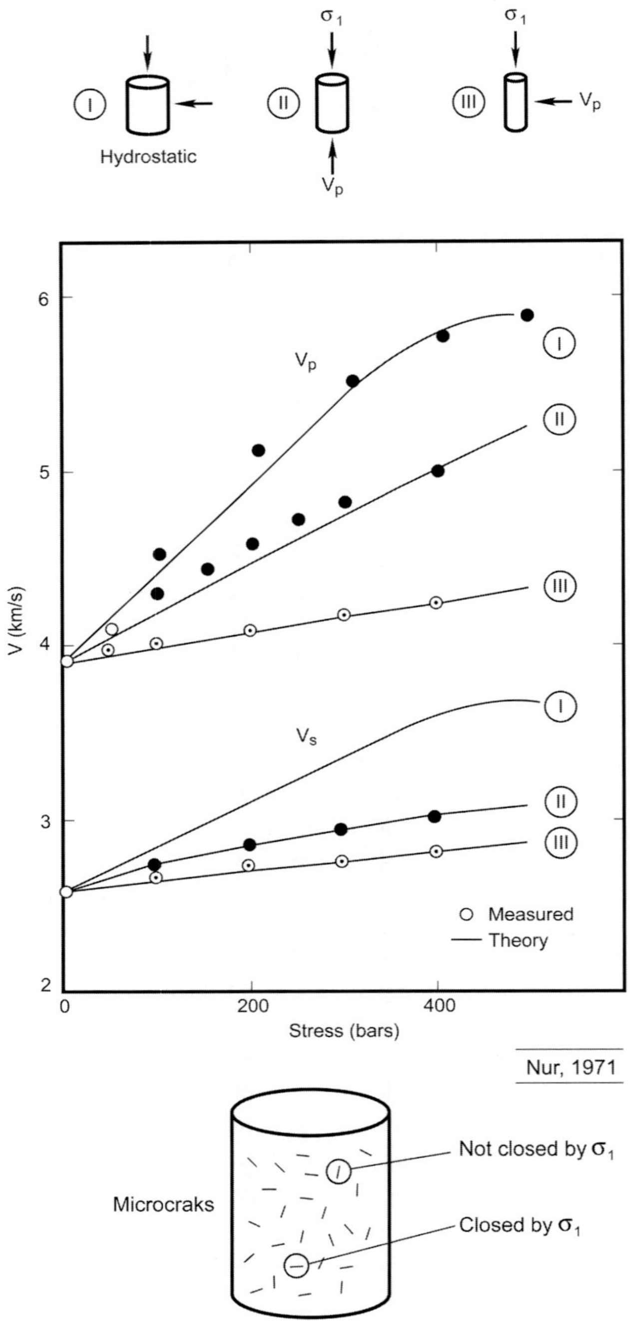


*Figure 3.1* Velocity anisotropy under stress (I hydrostatic, II measured parallel to uniaxial, III measured perpendicular to uniaxial), due to micro-crack closure perpendicular to stress. Nur, 1971.

A comprehensive in situ and laboratory study was reported by Engelder and Plumb, 1984, using shallow boreholes at in situ sites that were free of joints and above the water table. Dried core from the same holes was also utilised. Systematic measurements were made along different azimuths to check for anisotropic velocities as a result of anisotropic in situ stresses. The stress-relieved cores showed azimuth dependent velocity reductions of as much as 20% in granite and as low as 1% in limestone. In the case of the granite, the maximum anisotropy was consistent with the in situ stress orientation. Significantly, the difference between $V_p$ (in situ, stressed) and $V_p$ (core, unstressed) was usually less than 0.5 km/s when the in situ stress difference ($\sigma_1$–$\sigma_3$) was limited to about 10 MPa. These velocity differences tended to be between 0.5 and 1.5 km/s when the stress difference ($\sigma_1$–$\sigma_3$) was as large as 20 to 40 MPa in situ.

The dilation and brittle fracturing that occurs when rock is highly stressed was found by Rummel et al., 1978, to be a significant source of P-wave anisotropy. They utilised a biaxial loading arrangement with fast-reacting servo-control, to study the development of dilation adjacent to the shear failure surfaces developed in granite. They found that the P-wave velocity increased continuously in the direction of maximum compression in the pre-peak region. In the post-failure region the P-wave velocity decreased almost reversibly with reducing compression.

By comparison, the minimum and intermediate principal stress directions suffered a marked reduction of P-wave velocity (recorded as travel time increases), after fracturing was initiated. As they pointed out, enhanced permeability would be a related phenomen of such dilation: this would presumably occur mostly in the $\sigma_1$ direction. The authors mentioned the need to be aware of the possibility for increasing velocity anisotropy, when interpreting field seismic data in crustal regions where large tectonic stresses are assumed to be operating.

Failure processes in intact rock (Berea sandstone) up to and beyond the brittle ductile transition, with simultaneous monitoring of axial and lateral P-wave and S-wave signatures were reported by Scott et al., 1993. Confining pressures of 20 to 138 MPa were used. Slight P-wave anisotropy at the start of each test (due to a weak bedding fabric) were strongly enhanced at increasing axial strains, as micro-cracks tended to close perpendicular to $\sigma_1$, and open parallel to $\sigma_1$. At failure in the brittle regime, the shear fracture formation caused a small break in the P-wave signal, followed by constant $V_p$ (axial) and $V_p$ (lateral) velocities as shown in Figure 3.2a.

During ductile deformation at much higher stresses, the P-wave anisotropy continued to increase, presumably due to a more pervasive micro-crack and grain crushing development. By comparison, the shear fractures

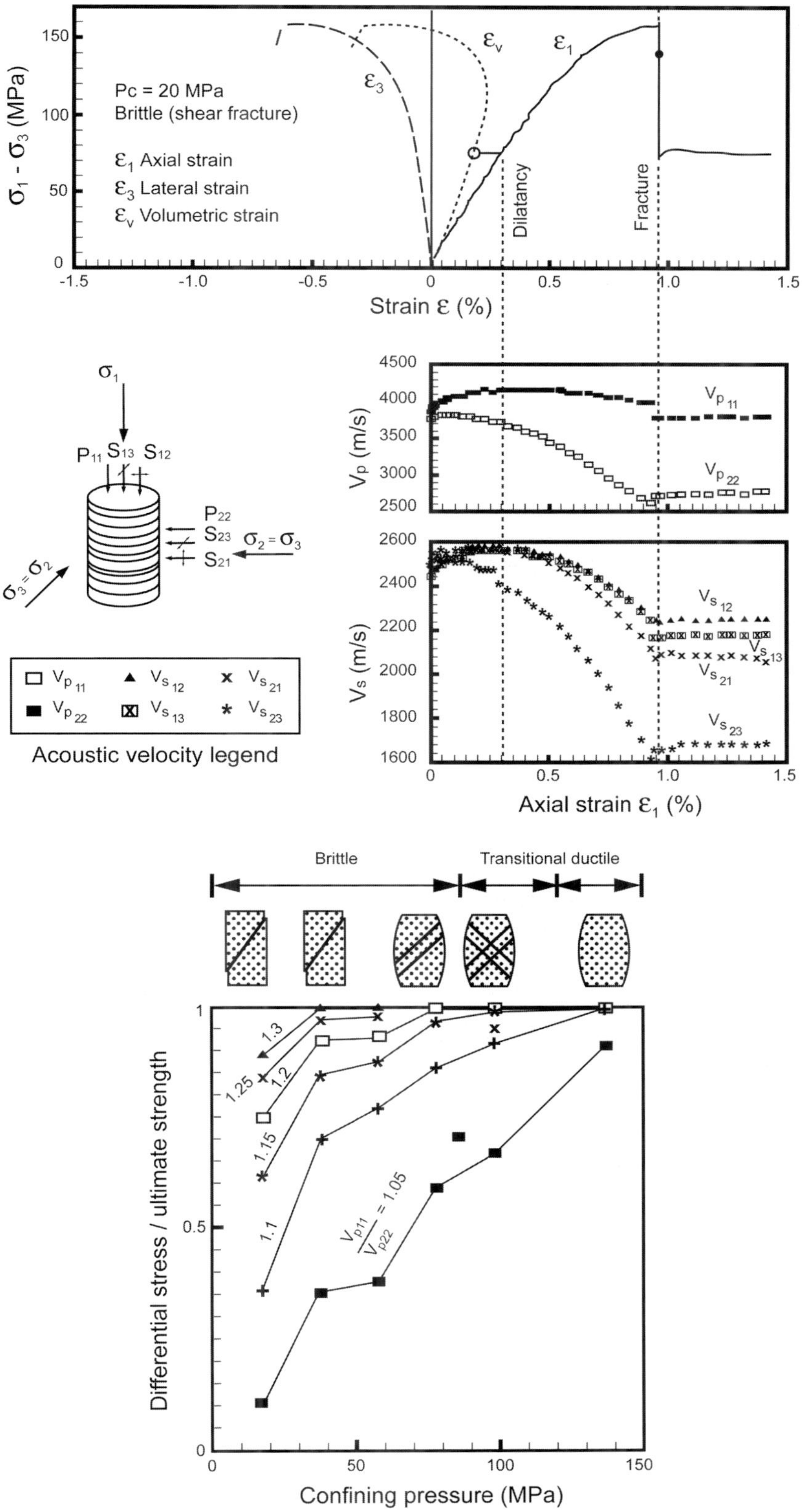


*Figure 3.2* a) P- and S-wave anisotropy as a function of confining pressure level for Berea sandstone samples. Note the effect on the velocities of the onset of dilatancy, and the fracturing event. b) As the ratio of differential stress to ultimate strength rises, the P-wave anisotropy is seen to increase, but high confinement removes this anisotropy. Scott et al., 1993.

developed at lower confining pressure were only surrounded by a limited zone of micro-cracks.

Scott et al., show an interesting plot of P-wave velocity anisotropy in relation to confining pressure and differential stress level (Figure 3.2b) that nicely demonstrates the increasing anisotropy of $V_p$ (axial) and $V_p$ (lateral) close to failure (at least 30% drop in $V_p$ lateral) and the reduction of this anisotropy at high stress levels.

## 3.2 Velocity anisotropy caused by fabric

Intact specimens of rock that exhibit strongly anisotropic or orthotropic tendencies such as slate, show significant velocity differences when measured parallel to foliation (e.g., 5.2 km/s) and perpendicular to foliation (e.g., 4.2 km/s). Duellmann and Heitfeld, 1978, show that this anisotropy varies smoothly as the angle of incidence to the foliation is varied from 0° to 90°, as shown for loading and unloading cases in Figure 3.3. The minor velocity hysteresis seen on unloading is presumably due to load-deformation hysteresis of the fabric, or of eventual micro-cracks. It is a slightly stronger effect when loading-unloading occurs perpendicular to the fabric, as one might expect.

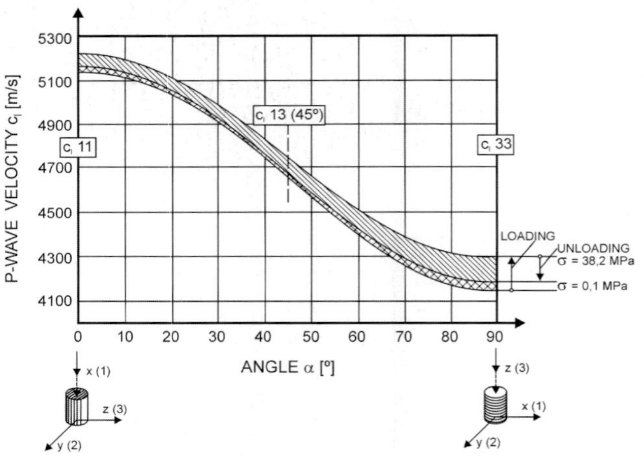


*Figure 3.3* Velocity anisotropy of intact samples of slate due to cleavage. Duellmann and Heitfeld, 1978.

A extensive collection of laboratory data that show the clear effect of the measurement direction in relation to the foliation (0°, 45° or 90°) was given by Tsidzi, 1997. He used the ultrasonic pulse transmission technique to derive $V_p$ data for intact samples of amphibolite, gneiss, hornfels, phyllite, schist, slate and quartzite (the latter only weakly or very weakly foliated). Both dry and unsaturated conditions were tested. The effects of loading were not reported. Tsidzi, 1997, suggested that 'strongly', 'moderately' and 'weakly' foliated rocks could be expected to show velocity anisotropies of 40–20%, 20–6% and 6–2% respectively. In Table 3.1, some results have been selected from the much larger set of data given by the author.

Strongly foliated gneiss from the Nagra project in Switzerland showed even stronger anisotropy, giving, in the dry state, a parallel-to-schistocity $V_p$ value of 4.4 km/s, and only 3.1 km/s perpendicular to this direction (Hesler et al., 1996). Figure 3.4 shows that both these extremes were achieved at the lowest axial stress of about 2 MPa, while the application of more than 25 MPa appeared to largely remove the velocity anisotropy; the slow perpendicular direction *converging* to the fast parallel direction above this stress level.

This convergence is in direct contrast to the micro-crack related divergent (increasing anisotropy) behaviour shown by Nur's (1971) results, in Figure 3.1. As noted in Chapter 2, the effect of saturation is to remove much of the effect of load increase on velocity, and the same appears to be the case for anisotropy caused by fabric. Figure 3.4 shows only slight velocity anisotropy in the case of saturated samples of gneiss, though the data set is limited.

An important contribution to the understanding of the *three-dimensional anisotropy* of *dense* shales was reported by Zinszner et al., 2002. They used ultrasonic techniques in the laboratory, to measure the multi-directional P-wave

Table 3.1 A selection of $V_p$ anisotropy data, showing the effect of foliation, schistocity and cleavage, and the dry or saturated state, when the velocity measurement direction is parallel, at 45°, or perpendicular to the particular planar fabric. Tsidzi, 1997.

| Rock Type | β = 0° | β = 45° | β = 90° | β = 0° | β = 45° | β = 90° |
|---|---|---|---|---|---|---|
| Condition | Dry | Dry | Dry | Saturated | Saturated | Saturated |
| Gneiss (SW) | 5102 | 4211 | 3956 | 5918 | 5237 | 5081 |
| Phyllite (F) | 6010 | 5130 | 5090 | 6050 | 5417 | 5307 |
| Schist (SW) | 6641 | 5802 | 5151 | 6706 | 5932 | 5378 |
| Slate (F) | 5913 | 5074 | 4893 | 5745 | 4722 | 4283 |

velocity across a 6×6×6 cm, 18-sided truncated cube of the Tournemire shale, whose format is illustrated in Figure 3.5a, using a slate example. Their interpretation of qP velocities is shown in the form of a Wulff's stereogram, in Figure 3.5b. As may be noted, the minimum velocity of approx. 3,200 m/s is recorded perpendicular to the bedding (Z-axis), while the maximum of approx. 4,250 m/s is parallel to the bedding (X, Y, etc.)

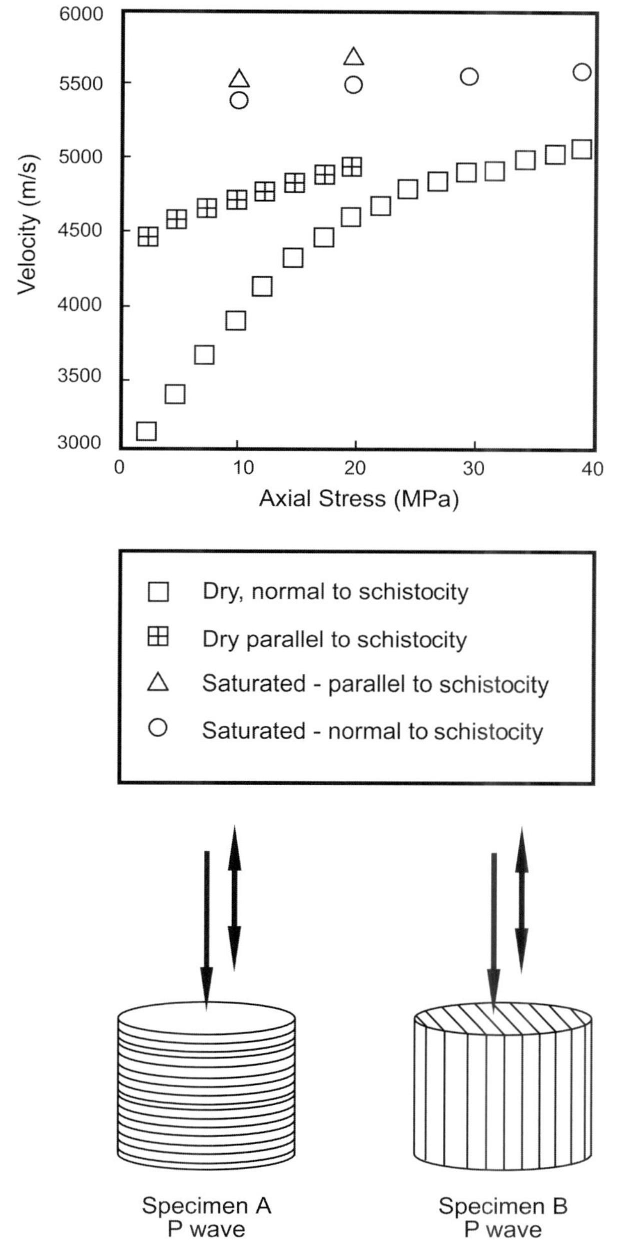


*Figure 3.4* Effect of schistocity in a strongly anisotropic gneiss, loaded to 40 MPa, parallel or perpendicular to the fabric. Hesler et al., 1996.

The authors also gave the results of velocity measurements under uniaxial stress levels from 0 to 20 MPa, which show a remarkable *lack* of stress sensitivity: the $V_p$ – stress curves giving almost horizontal straight lines between the seven different stress levels applied. However, the directional effect was marked, possibly accentuated by a *tendency* for slight shear in 'diagonal' directions of loading relative to the bedding. The lowest velocities were in the ZXX, YZY, ZXZ, YZZ and ZYZ (sub-perpendicular to bedding) directions giving velocities of only 1,700 to 1,800 m/s, while in the XXX, YYX, YYY, XXZ, YYX and YYZ (sub-parallel to bedding) directions, velocities were as high as 4,200 to 4,300 m/s.

Under the level of compression applied, and with presumed careful preservation of the samples, the ZZZ

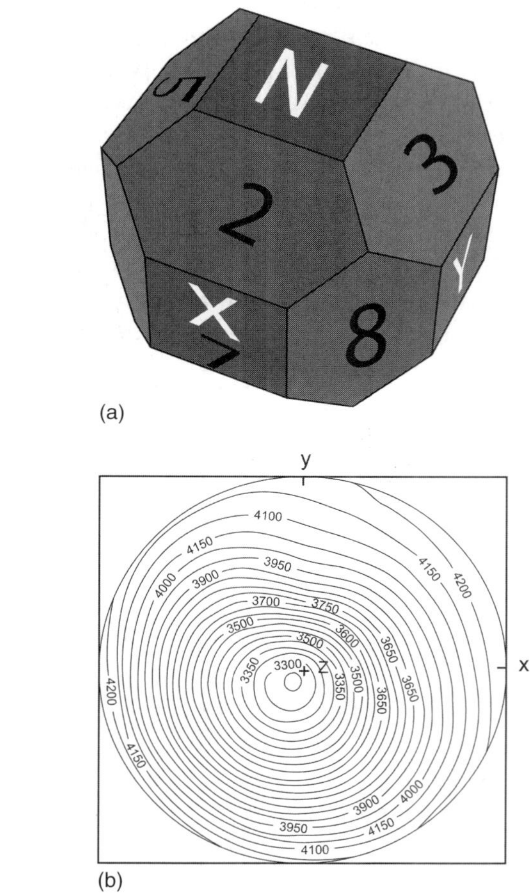


*Figure 3.5* Ultrasonic measurements on a truncated cube of dense shale recovered from the Tournemire experimental tunnel, south of Aveyron, in France. a) Truncated cube model of slate, showing axes. b) Interpolation of qP velocity measurements for a sample of the shale, from the west gallery. Zinszner et al., 2002.

direction gave intermediate velocities in this case, roughly 3,200 to 3,300 m/s. These were similar to *in situ* P-wave velocities calculated from seismic tomography, where in the vertical direction they recorded 3,125 m/s, with some reduction to 2,950 m/s in a tectonically disturbed area near a sub-vertical fault.

## 3.3 Velocity anisotropy caused by rock joints

Masuda, 1964, gave a simple but illustrative example of the effect of jointing and joint direction on the anisotropic velocity of blocks of granite at the Kurobe IV dam site in Japan. Figure 3.6 shows P-wave velocities in the dry and saturated state, for three orthogonal measurement directions. Velocity anisotropy was significant and sometimes amounted to 20% or even 25% difference in velocity. The slowest direction was of course when crossing the joints, the fastest when parallel. The loading state of the blocks was not referred to, but judging by the extreme effect of the dry or wet state, possibly the blocks were under low or zero load when these velocity measurements were made.

A massive granitic site in Cornwall, England, with one set of predominant jointing striking ESE–WNW, (note rotated axis), caused seismic velocities to be quite anisotropic, with maximum velocities of 5.5 km/s parallel with this jointing and minimum values of 5.1 km/s more or less perpendicular to the jointing (New, 1985). The velocity rosette shown in Figure 3.7 is a convenient way of representing the anisotropy, but the possible reasons for some of the other features on the rosette, for example the marked reduction between 30° and 40° (not exactly perpendicular to the 120–130° joint orientations) was not given. Perhaps the principal stress had rotated some 20° to 30° to 140–150°, giving a low velocity perpendicular to $\sigma_h$ (minimum), or shear stress effects were involved.

Noting the complexity of describing jointed rock masses and their physical anisotropy in relation to deformability and seismic velocity, Oda et al., 1986, developed a crack tensor technique which they compared with laboratory tests on artificially jointed samples, and with in situ tests on jointed granite. The artificial samples of gypsum plaster were cast with artificial, deformable cracks made of deformable greased paper. In Figure 3.8a, it will be noted that the cracks have either a random distribution or an ordered 'N-S' distribution. The squared velocity ratio $(V/V_o)^2$ which is the measured ultrasonic velocity normalised by that of the intact sample ($V_o$), showed corresponding isotropic or anisotropic distributions.

| Direction | Velocity (km/s) Dry | Wet |
|---|---|---|
| a | 2.92 | 4.47 |
| b | 3.39 | 4.90 |
| c | 3.50 | 4.90 |
| a | 2.84 | 3.66 |
| b | 3.32 | 4.37 |
| c | 3.41 | 4.53 |
| a | 2.77 | 3.94 |
| b | 3.26 | 4.82 |
| c | 3.32 | 4.92 |
| a | 3.55 | 4.50 |
| b | 3.58 | 4.63 |
| c | 3.63 | 4.64 |

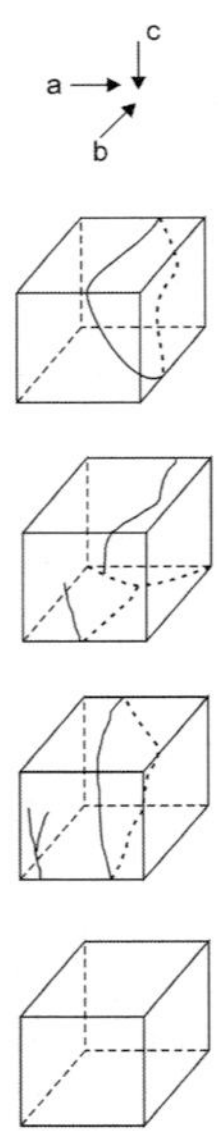


*Figure 3.6* Effects of measurement direction (and saturation) on $V_p$ values across jointed blocks of rock, at Kurobe IV dam site in Japan. Masuda, 1964.

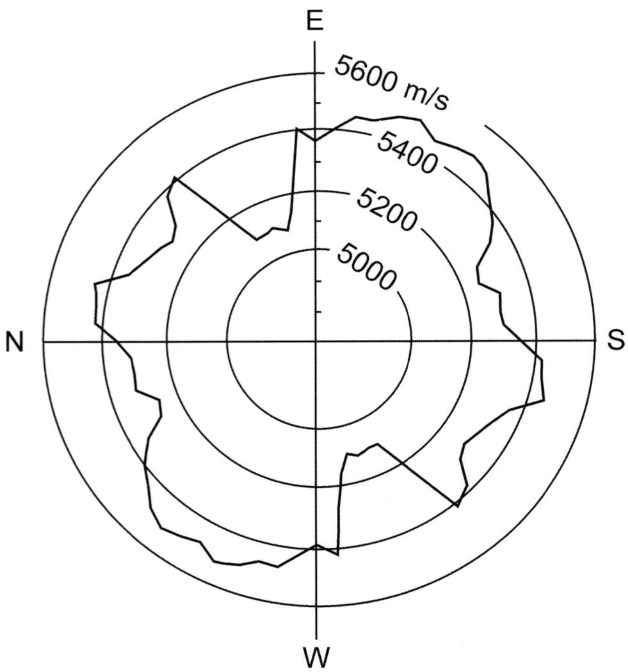


*Figure 3.7* $V_p$ anisotropy at a massive granite site in Cornwall, England. $V_{p(max)}$ was parallel to the single set of joints. New, 1985.

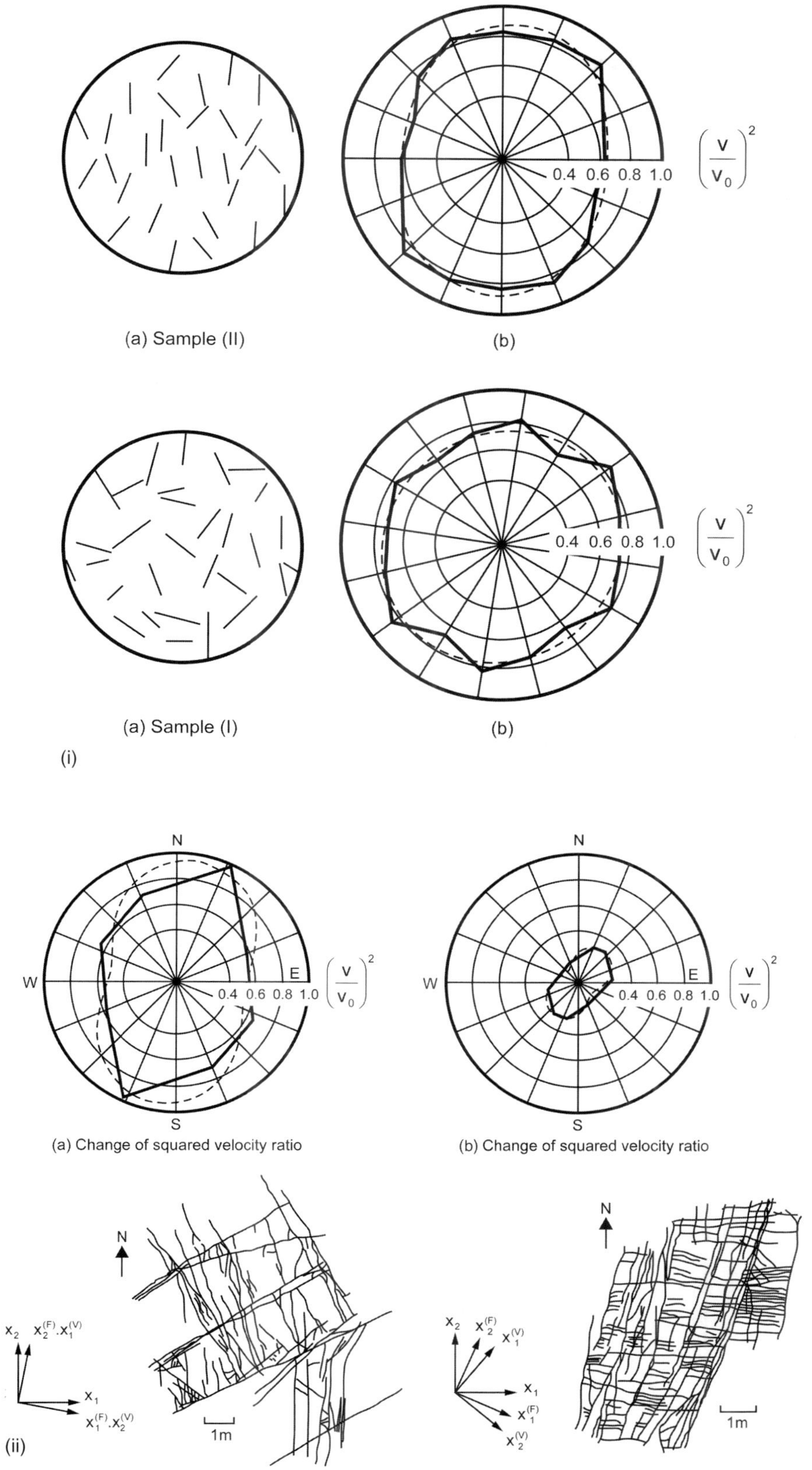


*Figure 3.8* Velocity anisotropy of gypsum samples with flaws, and of two jointed granite sites in Japan. Oda et al., 1986.

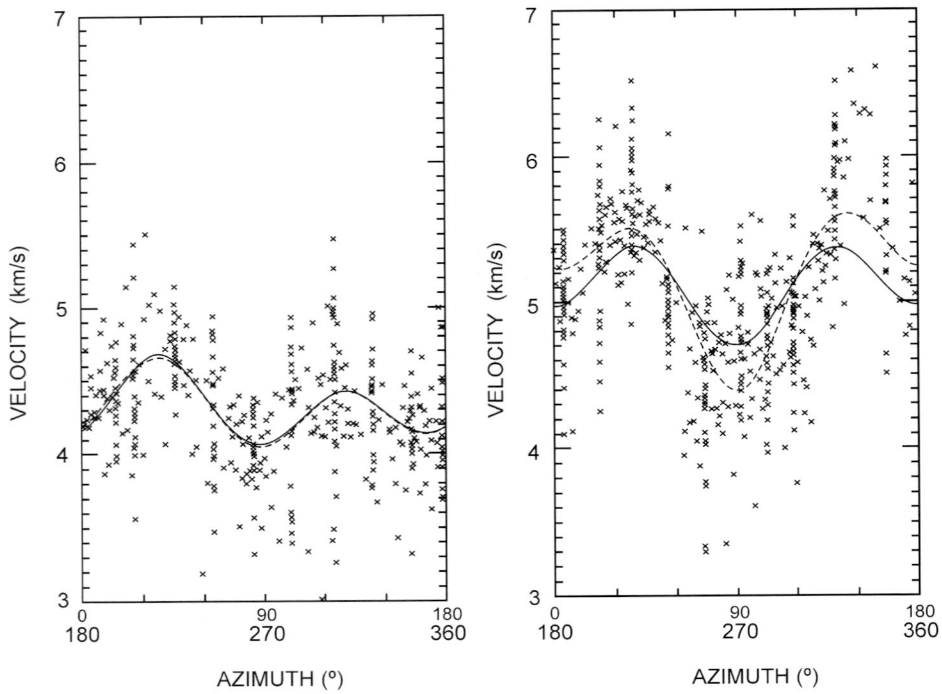


*Figure 3.9* Azimuthal $V_p$ anisotropy in jointed limestones at a 'dry' (left) and saturated site (right). Bamford and Nunn, 1979.

The graphic results of these authors' *in situ* tests are shown in Figure 3.8b. Both granite sites were anisotropic and the intensity of jointing differed, as clearly shown by the magnitude of the squared velocity ratio ($V_o$ for granite samples was 4.5 km/s). The authors' crack-tensor calculation showed remarkably good agreement with the seismic anisotropy measurements. The orientations of the velocity distributions are clearly dominated by the two 'fast' velocity directions sub-parallel to the relevant joint set directions.

As seen in Figure 3.8a, small uniformly distributed random cracks cause seismic velocities to be isotropically reduced in relation to an uncracked matrix. Attenuation is increased isotropically, and $V_p/V_s$ ratios are also changed. In contrast, most jointing shows some overall alignment, and ensures anisotropic seismic response.

An analysis of the seismic refraction tests at regularly jointed sites in limestones (Bamford and Nunn, 1979) given by Crampin et al., 1980, indicated that the velocity anisotropy (shown in Figures 3.9a and 3.9b) was also very sensitive to the degree of saturation of the joints. The two maxima (at about 40° and 120°) were clearly related to two sets of joints that intersected at about 80°.

Details of similar seismic refraction tests to those referred to in Figure 3.9 are given by Nunn et al., 1983. They used radial (20° interval) geophone spreads at a total of four sites in chalk in Lincolnshire, England. The chalk was not exposed at the seismic measurement locations, but two quarries in the area were mapped to obtain joint orientation data. Strong velocity anisotropy was measured at three of the four sites (Figure 3.10), and maxima at between 5° and 25° were found to correspond with dominant near-vertical joints which were perpendicular to the axis of a major monocline, which had a predominant direction of 15° ± 7°.

The velocity anisotropy of all four sites is compared in Figure 3.10a, and a comparison of velocity anisotropy and resistivity anisotropy for site CFR is given in Figure 3.10b. For the case of site RGQ, $V_{max.}$ and $V_{min.}$ were 2.85 and 1.75 km/s, giving a total velocity anisotropy $(V_{max.} - V_{min.})/V_{max.} = 0.38$, i.e., approximately ±20% around the mean of 2.25 km/s. Resistivity anisotropy (Figure 3.10b) was greater than seismic anisotropy for the case compared, possibly due to the strong contribution of a fluid bearing joint set.

An in-depth investigation of anisotropy caused by persistent sub-vertical jointing at a geothermal site in the USA (Beaver County, Utah) was described by Leary and Henyey, 1985. The authors analysed *in detail* why, if a significant number of vertical joints remained open (due

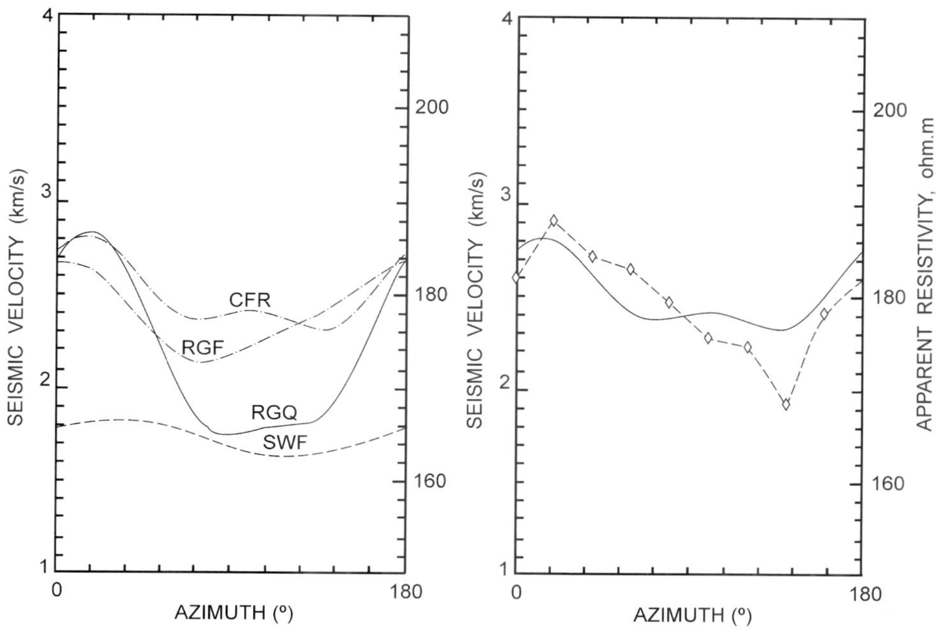


*Figure 3.10* a) Azimuthal $V_p$ anisotropy at jointed limestone sites in Lincolnshire, England. b) A comparison of $V_p$ and resistivity anisotropy at one of the four sites (see diamond symbols), is also given. Nunn et al., 1983.

to horizontal stress anisotropy), the elastic properties and hence the seismic velocities would be anisotropic.

Compressional waves travelling perpendicular to the joints would obviously be slowed more than those travelling parallel to the joints. However, the authors cautioned that minor geologic structure and mineral fabric could also influence the measured velocity anisotropy. Following earlier work by Garbin and Knopoff, Crampin, and McGonigle and Bamford, Leary and Henyey, 1985, gave simplified equations for the effect of cracks (or joints) on seismic velocity, and the effect of the dominant direction of the cracks (or joints). The following two equations are given for the dry and saturated states. Ignoring higher order terms:

*1) For dry cracks*:

$$V_P^2 = V_{P0}^2(1 - 71/21\varepsilon - 8/3\varepsilon\cos 2\theta + \varepsilon/21\cos 4\theta) \quad (3.1)$$

*2) For saturated cracks*:

$$V_P^2 = V_{P0}^2(1 - 8/21\varepsilon + 8/21\varepsilon\cos 4\theta) \quad (3.2)$$

where: $\varepsilon = Nr^3/V$ is the crack (or joint) density of N cracks of radius r in a volume V, and $\theta$ is the angle of incidence relative to the crack plane normals, where $\varepsilon \ll 1$. $V_{po}$ is the velocity without cracks.

At the geothermal site in question, the authors used 22 clusters of shots (sources) each located within 1 km of the wellhead, and used geophones downhole at depths ranging from 30 m to about 700 m. The source clusters were at about 160 m, 280 m and 370 m from the wellhead along six radial lines. The close-in shots were far enough from the wellhead that casing or tube waves did not obscure first arrivals.

The results of these tests are shown in Figure 3.11. They demonstrate both azimuthal velocity anisotropy and *velocity-depth* effects. The two sets of data shown in the figure, represent average P-wave velocity for seismic waves originating 370 m from the wellhead, which were received at *two depth ranges* in the well (0–300 m, shown as triangles and 460–520 m, shown as squares).

The 1.5 to 2.0 km/s increase in velocity is surprisingly large for an average depth increase of only about 350 m. However, Barton et al., 1994, showed a similar velocity increase even in the first 50 m at the Gjøvik cavern site in Norway, due to several MPa increase in stress in rock that had more or less unchanged frequency of jointing and RQD and rock quality Q, in these first 50 metres. In other words, the increased stress acting on the joints (3 to 5 MPa in this case),

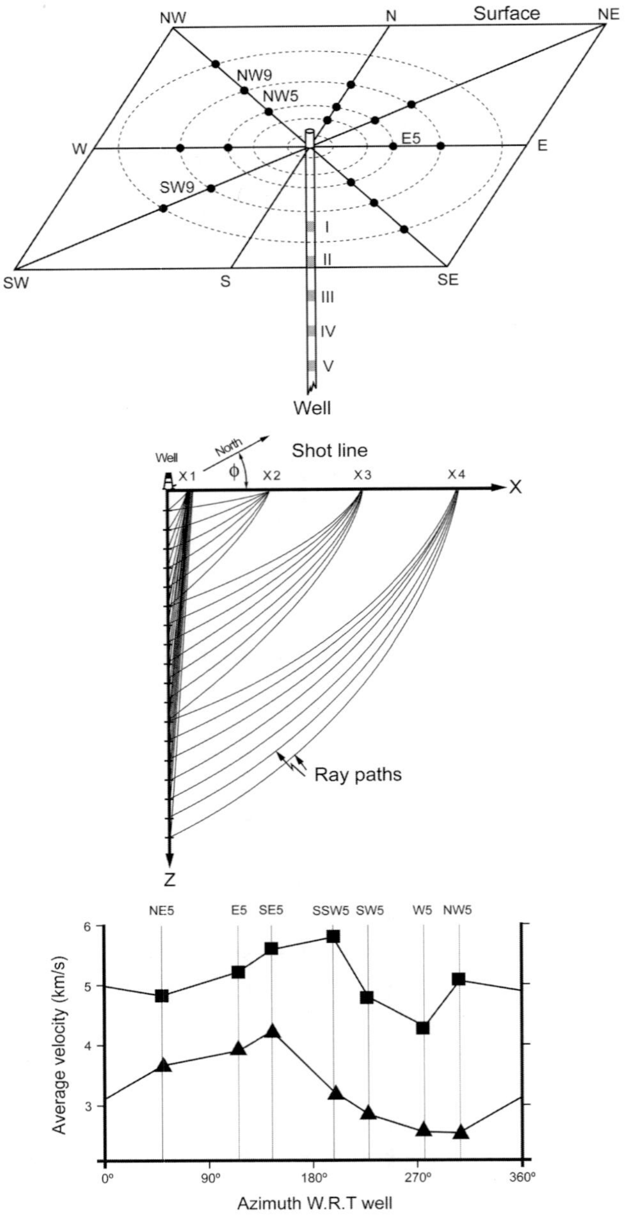


*Figure 3.11* Effects of joint set anisotropy on velocities, with depth effects superimposed, based on an 'areal well shoot' or 3-D VSP measurements at a borehole. $V_p$ anisotropy results are shown in the lower diagram, for receivers at depth ranges of 0–300 m (lower, average-velocity curves) and 460–520 m (higher, average-velocity curves). Leary and Henyey, 1985.

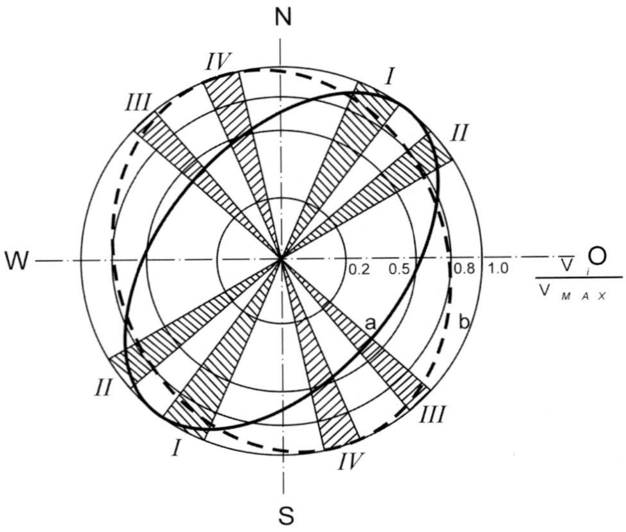


*Figure 3.12* Ultrasonic and longer wave length seismic investigation of four joint sets in dolomitic limestones, which had greater frequency (sets I and II), or lesser frequency (sets III and IV). The strengths of the velocity anisotropy of the different frequencies of jointing are distinguished in a logical manner. Lykoshin et al., 1971.

rather than reduced joint frequency, was the reason for the velocity increase of 2 km/s at a site saturated nearly to the surface. The important effect of stress on joint closure and seismic 'visibility' is not treated in the above equations (3.1 and 3.2). This topic will be treated fully in Chapter 5.

When jointing intensity is quite different between the different sets of joints, the use of different seismic wave lengths may be important in distinguishing the behaviour in different orientations. Lykoshin et al., 1971, describe the use of ultrasonic measurement with wave lengths of 0.8 to 0.1 m, and the seismic method with wave lengths of 8 to 15 m, for *distinguishing the velocity anisotropy* caused by the closely spaced joints ($S \approx 0.1$ to 0.2 m), from the velocity anisotropy caused by the much wider spaced joints.

The results of this interesting, and quite early dual-scale velocity anisotropy measurement are shown in Figure 3.12. Sets I and II correspond to the joints with the closest spacing, and Sets III and IV which had lower frequency, were separately logged by this method.

A helpful diagram of jointing, 'broken down' into its component sets, was presented by Olson and Pollard, 1989. This is shown in Figure 3.13. The figure was used by Schoenberg and Sayers, 1995 to illustrate their building of a stiffness matrix for a rock mass. This they did by summing the *compliance* tensor of an unjointed background rock and the compliance tensors for each set of parallel or aligned joints. This they inverted into the form of a *stiffness* tensor, which they suggested was more

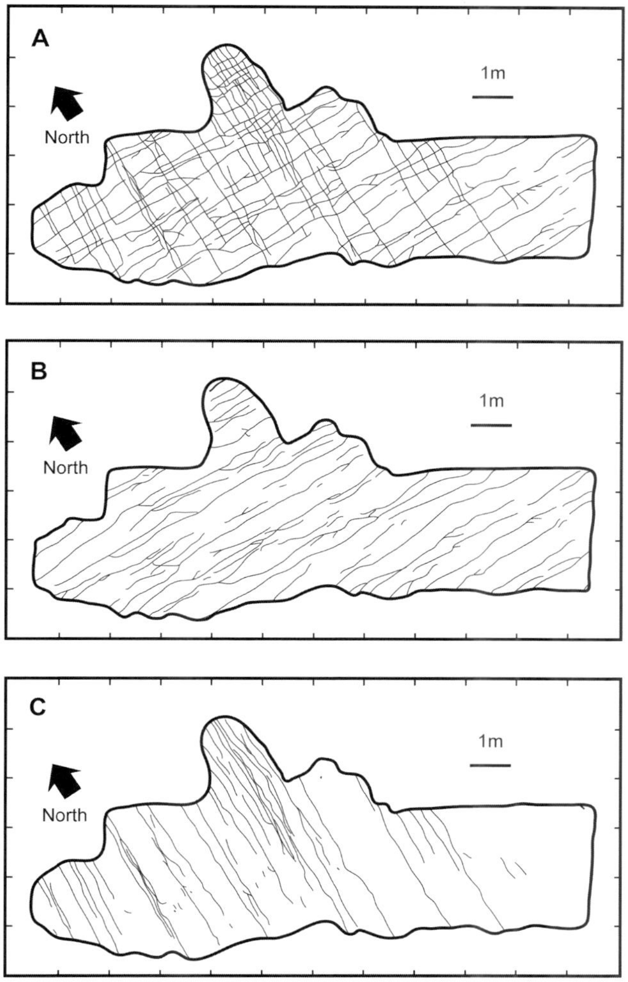


*Figure 3.13* a) Joint traces from a 1 m thick bed of limestone, 'dis-aggregated' in b) and c) into their two component sets. Olson and Pollard, 1989.

useful in the consideration of elastic wave propagation through rock masses, than the compliance, which is the inverse of stiffness (as commonly used in rock mechanics). Schoenberg and Sayers went on to apply their rock mass stiffness matrix to the interpretation of *shear-wave* anisotropy, which is strictly the topic of Chapter 15.

An aspect that will reduce the predicted anisotropy when applying the Schoenberg and Sayers stiffness matrix, is the fact that they assumed equal shear and normal compliances, based on seismic imaging of some perhaps not ideally suited, roughened 'lucite-sheet' models. (These will also be discussed in Chapter 15).

Compliance is the inverse of stiffness. In the 'macro-displacement' world of rock mechanics, there is a one to two order of magnitude difference between the normal and shear stiffnesses, i.e. Kn >> Ks, even for the case of small laboratory samples of rock joints, whether under low or high normal stress. This is because Kn and Ks are entirely different physical *deformation processes*, involving the normal deformation of a stiff compressed joint, and the less stiff shearing deformation along the same joint. The 'macro-displacement' stiffness anisotropy increases further with increased block size. (Barton and Bandis, 1982).

However, it is at present uncertain to which degree this fundamental *rock mechanics* aspect influences the *dynamic rock physics* aspect of micro-displacements that are presumably elastic in nature. There is controversy on this aspect, and even on whether friction along joints is a valid *attenuation* mechanism, as assumed for so long. We will attempt to resolve some of these questions in the more comprehensive chapters of Part II.

## 3.4 Velocity anisotropy caused by interbedding

The commonly occurring interbedding (alternation) of sedimentary strata of different stiffness, such as sandstone, shale and mudstone; layers which also have different porosity, density and uniaxial strength, causes anisotropy of all the major mechanical parameters and also affects all the components of velocity. (In relation to petroleum reservoirs, the reader is directed to Part II, Chapter 14, for a fuller discussion of this fundamental topic). In this section, some observations of the effects of inter-bedding on near-surface civil engineering projects will be presented.

Fine layering of sedimentary strata means that the dominant wavelength of a seismic pulse is long compared to the thickness of individual layers. The medium will then exhibit effective (and real) anisotropy, with a vertical symmetry axis in the case of horizontal layering. In the presence of hydrocarbons this layered medium may show substantial attenuation and velocity dispersion. The combination of effective anisotropy and attenuation means $Q_{seis}$ anisotropy and anisotropic velocities (Carcione, 2000) as we shall see later.

Oberti et al., 1979, reported a very instructive set of *in situ* near-surface measurements that involved down-hole sonic logging, cross-hole logging and comparison with deformation moduli determined at different depths below plate loading tests. The latter were performed parallel and perpendicular to the strata, and could therefore be compared with the anisotropic velocities.

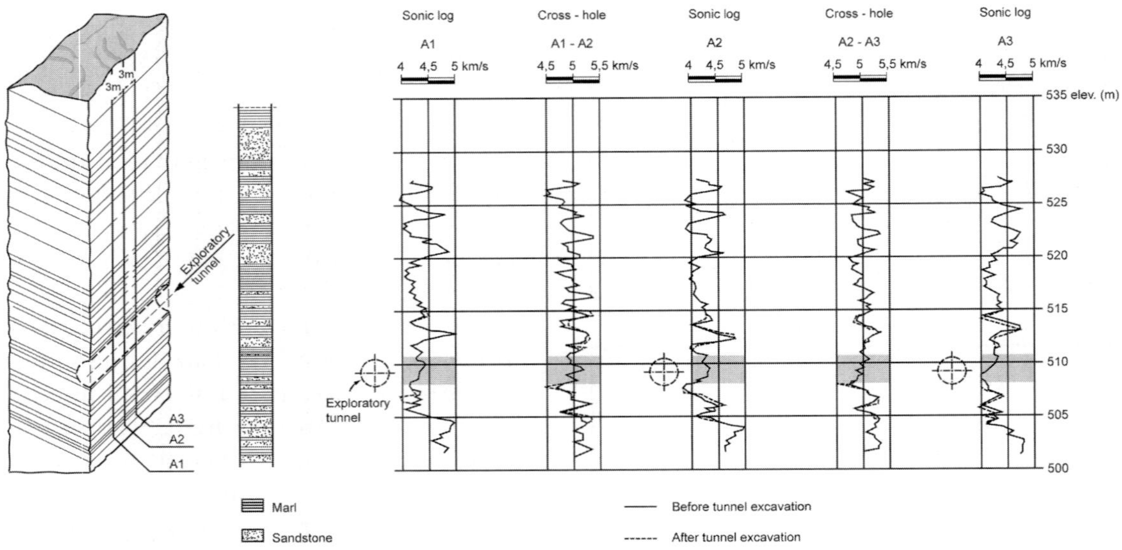


*Figure 3.14* Seismic cross-hole and downhole investigations of marl-sandstone interbedded strata, at a dam site in Italy. Oberti et al., 1979.

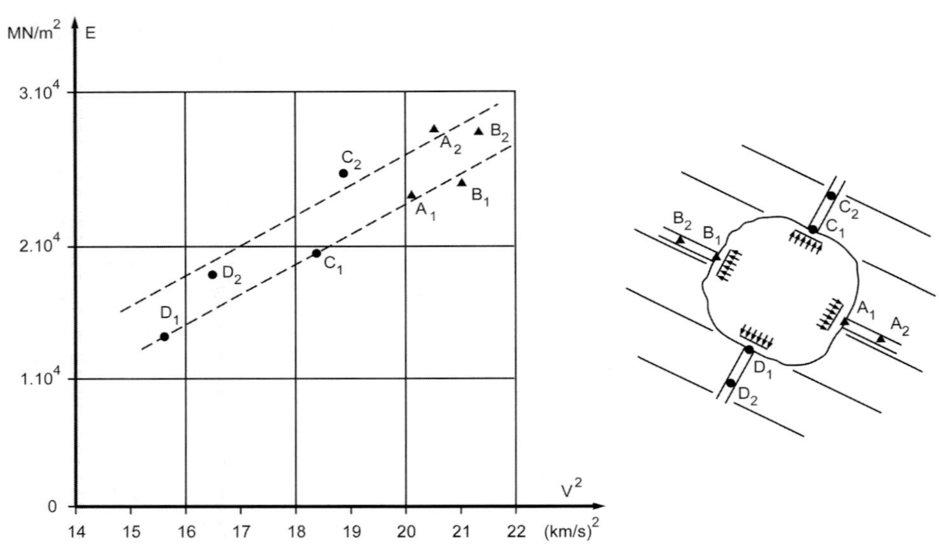


*Figure 3.15* Correlated anisotropy for the velocity and 'static' deformation moduli, as recorded in the interbedded marl-sandstone sequences shown in the previous figure. Oberti et al., 1979.

The rhythmically layered sandstone and marl, with a dip of 27°, formed the foundation for an arch-gravity dam in the Apennines in Italy.

Figure 3.14 illustrates the geological sequence and location of boreholes. The exploratory tunnel used for the plate loading tests, shown in Figure 3.15, was at 30 to 35 m depth, and ran parallel to the strike of the inter-bedded strata.

The three boreholes (A1 to A3) were parallel and spaced at 3 m centres. Sonic and cross-hole logs are shown sequentially in Figure 3.14. The mean velocity anisotropy in this orthotropic rock mass was 4.3 km/s (perpendicular to the layers) and 5.0 km/s (parallel to the layers). Differences can be noted between the higher velocities in the sandstones and the lower velocities in the marl.

Figure 3.15 shows a comparison of the sonic measurements performed in central boreholes beneath each plate loading location, where deformations were also recorded with extensometers, so that 'static' deformation moduli could be calculated at different depths. The lower static moduli and lower velocities of the disturbed near-surface

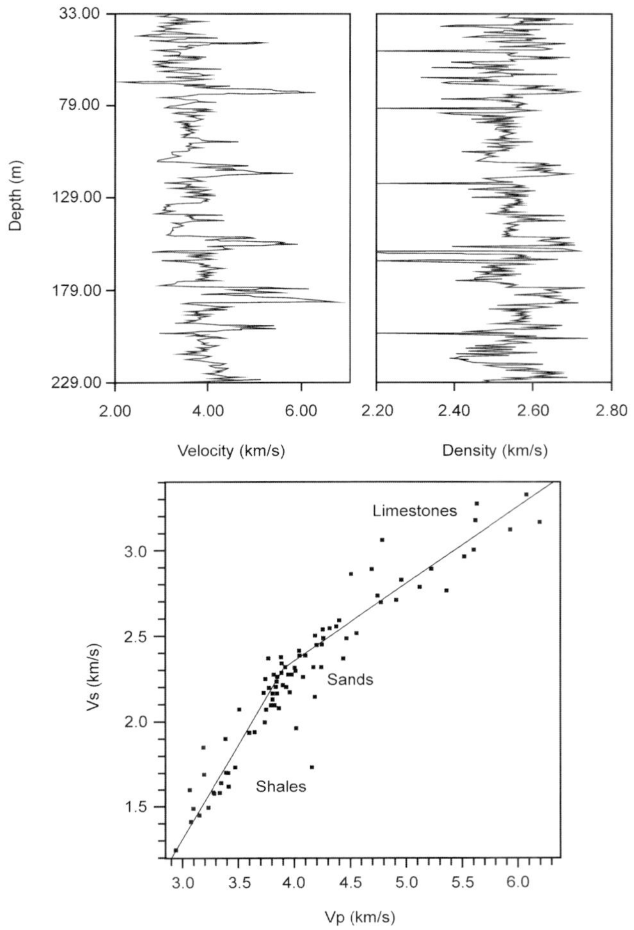


*Figure 3.16* Inter-bedded limestones, shales and sandstones in a 230 m deep well, Sams, 1995. Resolution of detail in finely interlayered sequences.

rock are evident, especially that of the marl in the invert, where moisture content perhaps had increased.

As can be seen in Chapter 5, the deformation moduli and velocities measured in these tests correlate quite closely with the '$V_p$–Q–M' model, where the rock quality Q-value, or the velocity, are both seen to correlate with the 'static' deformation modulus (M), provided that appropriate corrections are made for porosity, rock strengths and depth.

A case intermediate between the near-surface and petroleum reservoir interpretation was presented by Sams, 1995, for inter-bedded limestone, shale, sandstone sequences in a 230 m deep (and subsequently additional) research well. The combined use of a borehole compensated (BHC) sonic logging tool, a compensated formation density tool, and a Formation Micro Scanner (FM) was capable of resolving much of the detail of finely interlayered rock sequences. This research will be re-visited in several contexts in Part II, particularly regarding frequency dependent and stress dependent attenuation, described in Chapter 10.

## 3.5 Velocity anisotropy caused by faults

A final category of anisotropy that will be described in this chapter is that caused by major faulting. An instructive example is provided from Japan, where Ikeda et al., 1981, describe some of the extensive Japanese high-speed railway (Shinkansen) tunnel investigations in major fault zones. They show characteristic variations in velocity in a major 300 m wide faulted zone in the Rokkô Tunnel, with three zones of velocities as low as 2.2 km/s (Figure 3.17a, b, c).

Extensive investigation galleries enabled the authors to investigate the effect of the angle between the fault boundary and the seismic investigation line. When the angle is very acute, and if the fault is also dipping at a shallow angle ($\varphi$ in Figure 3.17b), a false high velocity ($V'$) may be registered, or it becomes impossible to register the fault.

The authors also assembled seismic data from 100 rail tunnels, with emphasis on fault zones, and heavily fractured rock. In Figure 3.17c, $V_c$ is the higher velocity of the *competent* rock surrounding the heavily fractured zone, which is given a velocity classification A to F in the table shown in this lower figure. The heavily fractured zone has the lower velocity V. They observed that the clay core of fault zones could have velocities as low as 0.8 to 2.5 km/s.

It is of interest to note that the two parallel fault zones depicted in Figure 3.17a, which have a minimum velocity of 2.2 km/s, created *many months of delay in tunnelling*, as can be judged by the profusion of investigation adits into this regional Otsuki Fault zone. In another rock type, on the other side of the globe, the same seismic velocity of about 2.0 to 2.5 km/s allowed tunnelling progress (with TBM) of up to 1500 metres per month, the reason being the high porosity of an entirely different, easily excavated, and relatively stable rock type, when not heavily jointed. The UK-France Channel Tunnel *chalk marl* had a porosity $n \approx 27\%$, and an easily cuttable strength of only 4 to 9 MPa.

The Q–$V_p$ relationship ($V_p \approx 3.5 + \log_{10} Q$) for hard, non-porous, near-surface rocks presented in Chapter 1, would *normally* predict a Q-value as low as 0.003 to 0.01 (exceptionally poor) for such low velocities: as if the low

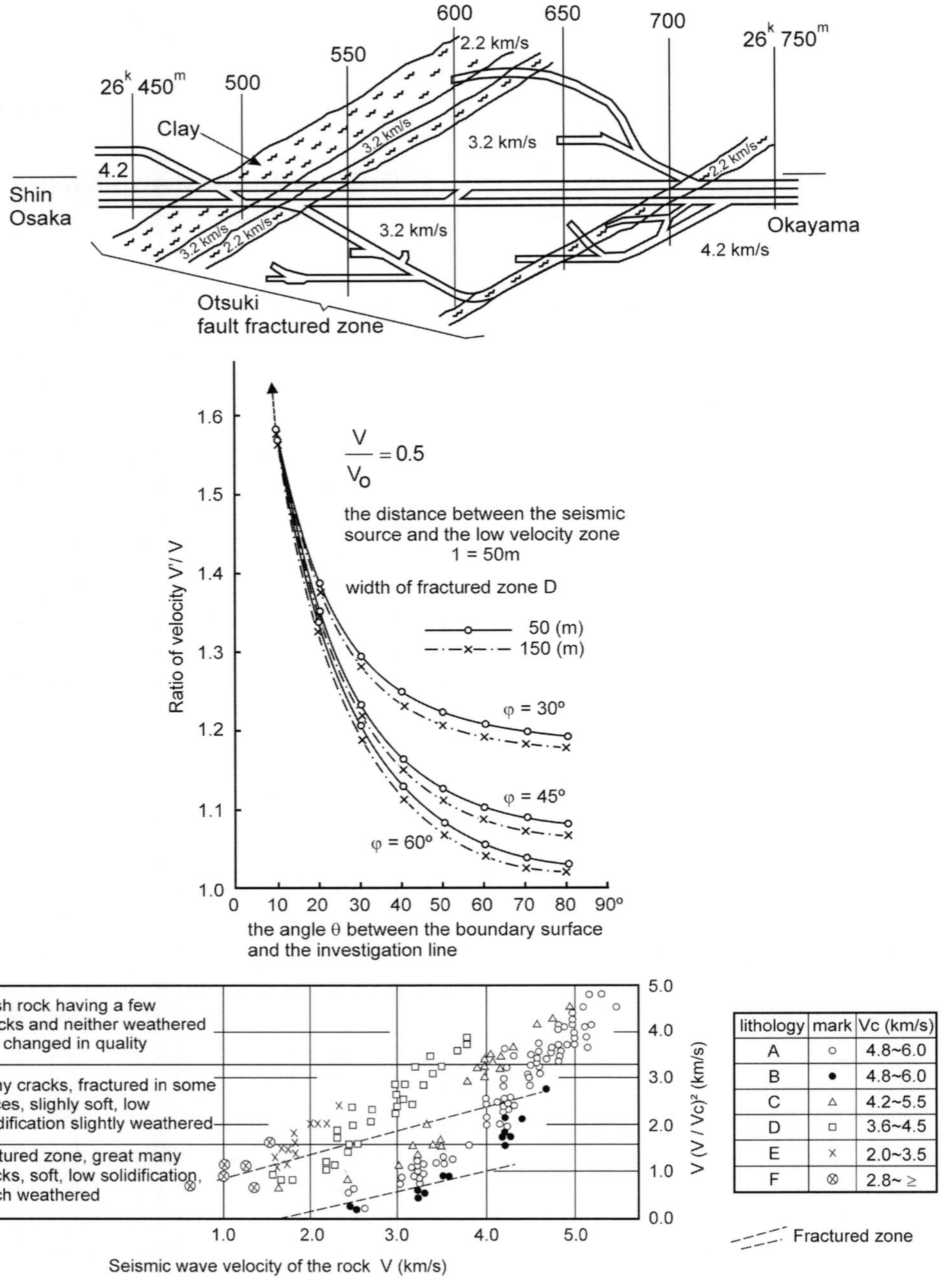


*Figure 3.17* Seismic investigations of fault zones at Japanese rail tunnels. Ikeda et al., 1981. a) Plan view of fault zone crossing the tunnels. b) False high velocity (V′) caused by too acute angle of the seismic investigation line relative to the fault zone. Width of fractured zone has small effect. c) Integrated results from 100 rail tunnels giving a velocity ratio expression for the low fault zone velocity (V) in relation to the surrounding competent rock ($V_c$).

velocity represented a fault zone. But in the chalk marl in question, Q-values were in the range 10 to 20 where these record tunnelling speeds were achieved. The missing corrections for a low uniaxial strength (i.e. $\sigma_c \approx 5$ MPa) and a high porosity (i.e. n ≈ 27%) will be described in Chapter 5. They are essential for integrating rock quality Q and $V_p$ – in softer rocks.

# 4 Cross-hole velocity and cross-hole velocity tomography

Cross-hole and between-gallery seismic work has been performed for many years at major dam projects, particularly at the sites of arch dams, where the deformation moduli of the rock foundation and valley walls are of most concern. Unfortunately, the large number of dams constructed from the 1960s to the early 1980s did not have the benefit of tomographic imaging, in which not just the average velocity between source and receiver, but also the approximate location and velocity could be displayed, following inversion of the multi-source-multi-receiver-position data. The use of pairs of boreholes (or multiple holes), for direct access to a 'hidden plane' (or planes), on which representative velocities and their location could be calculated has many advantages for assessing the severity of fault zones, the need for treatment of the ground, or even in some cases the avoidance of bad ground. In this chapter, a wide variety of cross-hole seismic tomographic imaging of the sub-surface will be reviewed, from tunnels and caverns, to mining pillars, blasting-effect analysis, excavation disturbed zone mapping, and analysis of grouting efficiency.

## 4.1 Cross-hole seismic for extrapolation of properties

In the initial stages of site description for a civil engineering project, geological mapping of major structures may be followed by imaging of these features, using large scale reflection techniques. As emphasised by Cosma et al., 2001, subsequent access in a very limited number of holes will normally suggest VSP measurements, with sources at the ground surface. When the construction phase is begun, access via a larger number of boreholes, even those drilled from shafts or tunnel walls will allow a combination of smaller scale VSP, and direct cross-hole tomography, giving *velocity and location*. Later in this chapter we will see some exceptional applications of 'close-in' seismic tomography.

A classic example of cross-hole and between-gallery seismic is that shown in Figure 4.1 from the 220 m high Mratinje Dam in Yugoslavia, as reported by Kujundzíc, 1979. This figure shows the separate sites of the deformation tests for determining E (the dynamic elastic modulus), D (the deformation modulus) and $V_p$ (the local value of $V_p$ at the deformability test site).

Some of the cross-hole, between gallery and borehole-to-gallery velocity measurements are shown in more detail in Figure 4.2 (from Ivanovíc et al., 1970), where the 'fans' of velocity can be readily observed. By relating the velocity at each test site to the moduli, the larger scale cross-hole results could be used to extrapolate the expensive and time-consuming tests to other parts of the foundation. In Chapter 6 we will see some of the inter-relationships that have been developed between $V_p$, $E_{intact}$ and D, for comparison with $E_{dynamic}$ which can be derived from $V_p$, $V_s$ and density, as indicated in Chapter 1.

Possible pitfalls when performing *cross-hole seismic* measurements in low velocity layered media such as clays, which presumably will also affect *cross-hole seismic tomography* in similar media are illustrated in Figure 4.3. McCann et al., 1975, indicate that there is an apparent decrease in the velocity of the high velocity layers with increasing separation of the boreholes. First arrivals at the common depth of 7.4 m showed velocities of 2.18, 1.97, 1.83, 1.81 and 1.78 km/s with borehole separations increasing successively from 2.9 m to 15.1 m. The high frequency direct first arrival received at small borehole separations was replaced by a long dispersed waveform at the largest separations. Attenuation of the higher frequency, higher velocity part of the wave at increasing distance was apparently occurring. The authors used a high voltage 'sparker' source in their measurements. The strongly attenuating properties of the clay were presumably the cause of this result.

Various seismic wave characterisation methods were compared at a rock anchor foundation site by Ebisu et al., 1992. Figure 4.4 shows P-wave data interpreted from seismic refraction, downhole logging, cross-hole and surface exploration. The discrepancies between the systems should serve as a warning that many factors need to be

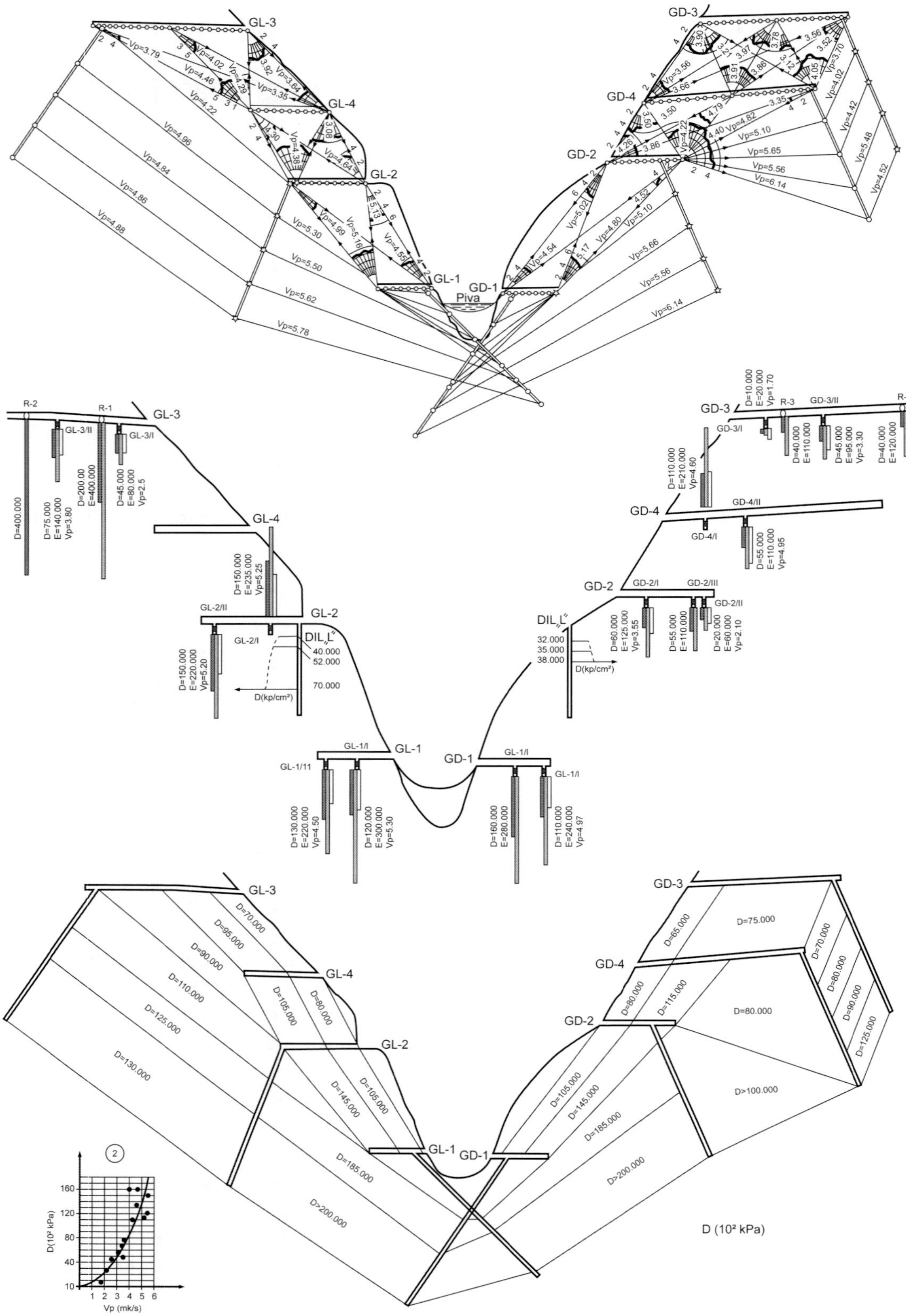


*Figure 4.1* Cross-hole and gallery-to-gallery seismic tests at the Mratinje Dam in Yugoslavia, for extrapolating deformability tests. Kujundzíc, 1979.

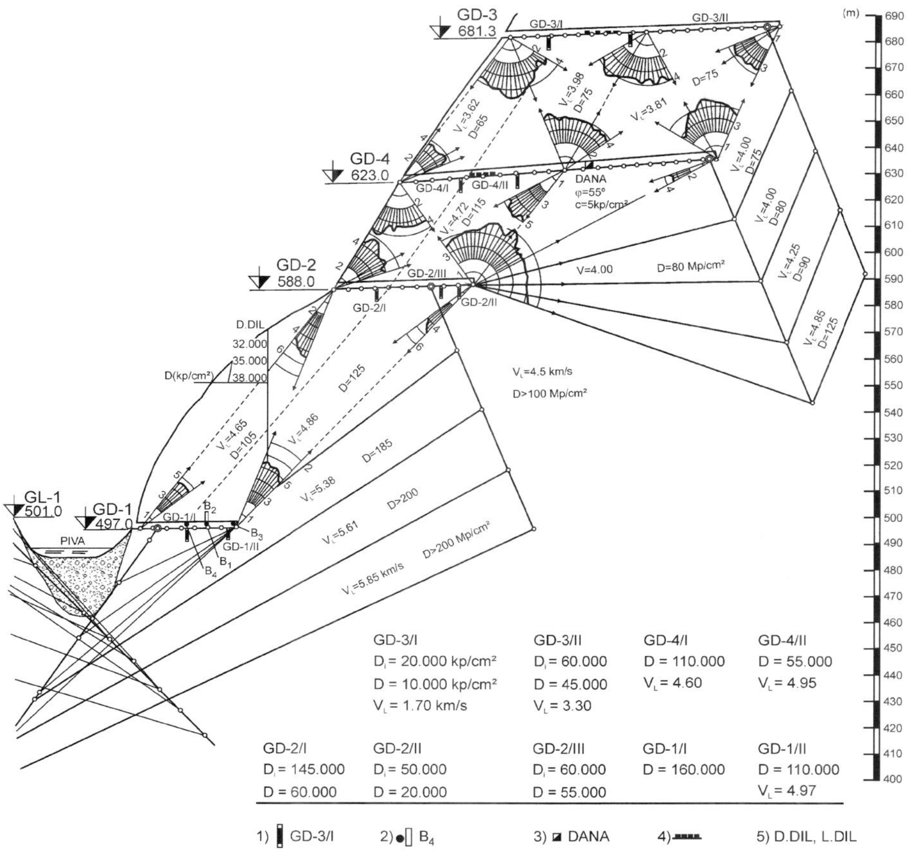


*Figure 4.2* Classic example of the use of gallery and cross-hole seismics for extrapolating quality and deformation modulus values at the Mratinje Dam, Yugoslavia. Ivanovíc et al., 1970.

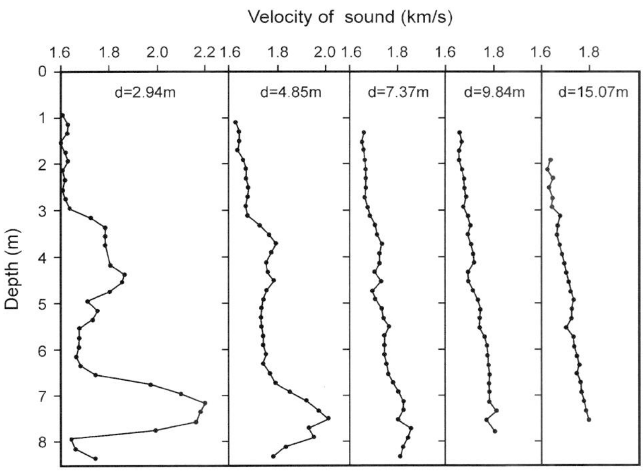


*Figure 4.3* Dispersion effects in layered Oxford clay give apparent reduced $V_p$ for stiff layer with increased cross-hole separation. McCann et al., 1975.

evaluated when interpreting the results of a suite of tests. Ebisu et al. prefer the use of the surface wave velocity $V_r$ to correlate with modulus of deformation. They show on a log-log plot of modulus versus $V_r$, that a consistent trend is established. A modulus of 0.3 GPa corresponds to $V_r \approx 0.5$ km/s, 1 GPa corresponds to $V_r \approx 0.8$ km/s.

Usually, when comparing cross-hole and downhole velocity measurements, the downhole sonic probe is considered to give a small-scale, and usually higher velocity than the averaged cross-hole result. However, the small-scale excavation damage zone (EDZ) that may also accompany *a borehole in incompetent rock*, may presumably be the reason for sometimes measuring a lower velocity at the small scale. In Figure 3.14 in the chapter on anisotropy, the sonic log of Oberti et al., 1979, generally showed about 0.5 km/s lower velocity than the cross-hole result, where the hole spacing was 3 metres. The sonic log also showed greater sensitivity to the layering (marls and sandstone) in this case.

The large reduction of the velocity in the 'within-the-borehole' measurement shown in Figure 4.4 (Ebisu et al., 1992) was not discussed by the authors, but is perhaps an expression of damage caused by the drilling/flushing process in these near-surface weathered materials at the rock anchor foundation in Japan.

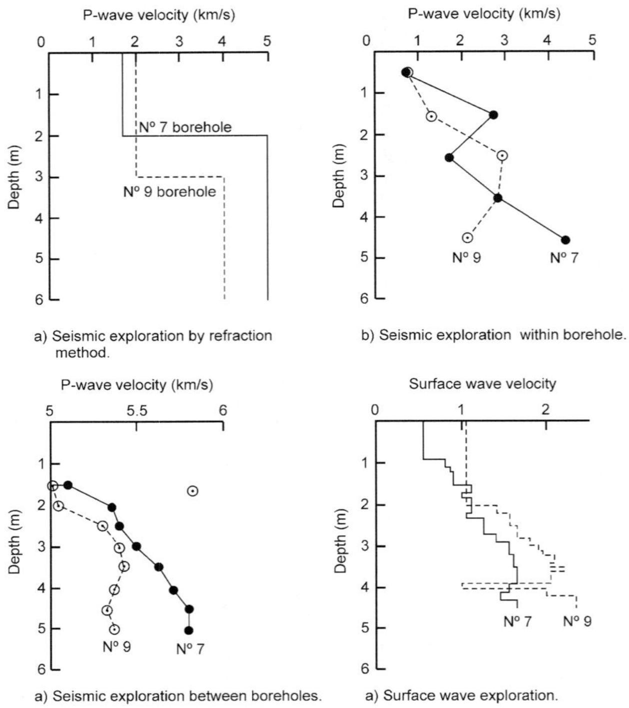


*Figure 4.4* Contrasting P-wave velocities at a rock anchor foundation, using four methods of measurement. Ebisu et al., 1992.

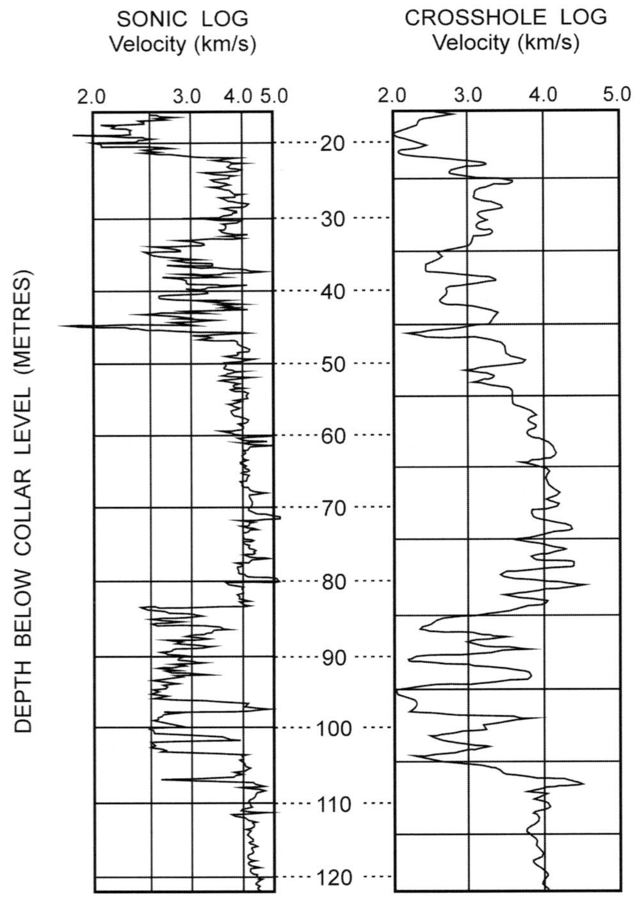


*Figure 4.5* Comparison of a sonic log and a cross-hole (mean velocity) log. Whiteley, 1990.

A more 'normal' comparison between a downhole sonic log and a cross-hole log is that shown in Figure 4.5 from Whiteley, 1990. Hole spacing was 40 m. The latter shows a smoothed, average behaviour. While general trends are seen to be remarkably similar, details between the two logs clearly differ due to the change of scale and location.

## 4.2 Cross-hole seismic tomography in tunnelling

The system of seismic data analysis used in tomographic studies was probably adapted from the medical profession, although the use of 'superficial' seismic sources (earthquakes) for inversed imaging of the internal structure of the earth seems to be a possible source of inspiration. The efficient data handling and graphic presentation techniques represented by *the tomographic method*, were rapidly adopted in rock engineering projects since roughly the mid-1980s, and to an increasing degree in petroleum engineering.

The simple principle of the method is that a string of receivers (hydrophones or 3D accelerometers) suspended (or pushed into) a borehole at (e.g. 2.5 or 5 m) regular spacing, are used to receive the seismic signals from a source of dynamic energy in an adjacent hole. The artificial source can be a 1 gm detonator cap, a downhole hammer, a sparker or a piezoelectric high frequency source, which is moved successively down the sender hole. The same spacing of sources and receivers (e.g. 2.5 or 5 m), is normally used. The inversion of the travel times of the multiply crossing ray paths, into velocities, or into other seismic attributes, is organised in principle into a regular grid in which average solutions for the local velocity (or other seismic attribute), are produced.

Tomographic plots of velocity, amplitude and velocity difference are commonly employed. Most frequently, the method is used from single pairs or multiple pairs of boreholes drilled from the surface or from the face of a tunnel, to image a pending (or already intersected) fault zone. Increasingly in recent years the method is also being used in mining to delineate highly stressed and burst prone areas, which seem to be most closely associated with steep *gradients* of velocity, where high shear stresses may be present. The seismic tomography method can be used remotely and safely in hostile environments, to image highly stressed regions of a mine or overstressed rock around a deep tunnel. (The interesting use of 'passive' sources such as acoustic emission (AE) will be illustrated briefly in Chapter 7, where *average* velocities can be calculated.)

Figure 4.6 show some potential layouts for the borehole arrays. A moving source, for example mining equipment, can also be used to obtain a tomographic image, if a suitable array of receivers is in place, and if measurements are repeated at regular intervals over a suitable length of time. Westman et al., 1996, utilised a long wall shearer in an Appalachian coal mine in the USA, and sampled this source at ½ to 1 minute intervals during mining shifts, while the shearer was moving. Their receivers were geophones fixed to rock bolts in the mine entry roofs, close to the mining face. They produced attenuation tomograms that changed with time as mining advanced in response to high stress anomalies, stress release phenomena, changed degrees of jointing and stress induced fracturing.

The assumption is often made that P-waves have travelled directly from source to receiver, and a straight line tomography program is used. Curved ray path tomography is preferred to allow for velocity anisotropy and for refraction (McDowell, 1993).

By, 1987, described a comprehensive layout of vertical boreholes for cross-hole seismic tomography, which was performed in Oslo for a difficult, faulted section of the twin tube, 13 m span Fjellinjen road tunnels (Figure 4.6a). Some 20 m of soft clay underlying downtown Oslo had to be protected from groundwater pressure drawdown. At one location, the rock cover over the arch consisted of only 3 to 5 m of crushed alum shale (damaging to concrete), beneath 20 to 30 m of soft clay, in a major regional fault zone. Selection of freezing for one of the tunnel tubes was made on the basis of the seismic results, which were based on cross-hole measurements from five boreholes of 60 m depth and a total of eight cross-hole sections.

In contrast to this layout of vertical holes, Hope et al., 1996, working in chalk, used single holes drilled radially into the wall of a pilot tunnel, and the upper and lower walls of a tunnel, to give two triangular shaped spreads (see Figure 4.6b). They obtained a distribution of velocities ranging from 1.8 to 2.5 km/s between the crown positions (2 m intervals) and the borehole, and 1.9 to 2.3 km/s between the invert positions (2 m intervals) and the borehole. Lower velocity zones were consistent with additional jointing associated with a listric fault cutting through the chalk. This was verified after benching down and widening of the pilot tunnel had been performed, to create a cylindrical oil storage cavern. The initial refraction seismic survey from the pilot tunnel had indicated a range of velocities of approximately 2.3 to 2.6 km/s, representing generally uniform conditions. Laboratory samples of the 10 MPa chalk had indicated a mean P-wave velocity of 2.4 km/s at natural water content (13 to 14%), and 2.5 km/s when fully saturated.

An example of the potential benefits obtained from cross-hole seismic tomography at a near-surface cavern site is shown in Figure 4.7a and b. The measurements were performed for the Gjøvik Olympic cavern site investigation in Norway in 1990. The position of the planned, 62 m span, 140,000 $m^3$ cavern was adjusted in order to penetrate as little as possible of the lower velocity, near-surface zone ($V_p \approx 4.1$ to 4.3 km/s).

This was proved in later cavern logging to have rock quality Q-values as low as 2 to 5 at the shallow end of the cavern. This quality results from a low to moderate RQD (frequent smaller pieces of core <10 cm long), up to four joint sets ($J_n = 12$–15), and with some alteration of the joint walls ($J_a = 2$). Positive aspects were considerable joint roughness ($J_r = 2$ to 3), and surprisingly high horizontal stresses. (See Appendix A for Q-parameter ratings.)

These moderate velocities fit the hard rock, near-surface relation $V_p \approx 3.5 + \log_{10} Q$ presented in Chapter 1 quite closely, for the relevant shallow conditions (approx. 25 m depth). At the other, deeper end of the cavern, Q-values also fell to 2 or 3. Significantly, this rock quality Q was *lower than the* $V_p$ *values would have indicated*, with this shallow seismic relationship. The fundamental need for depth or stress adjustments in a Q–$V_p$–M (static modulus of deformation) relationship, are discussed in Chapter 5.

The details of NGI's cross-hole tomography, analysed in more detail in Chapter 5, indicate a continuous rise in velocity down the 60 m deep boreholes (approximately from 3.5 km/s to 5.5 km/s), despite more or less constant joint frequency, RQD and rock quality Q-values down the lengths of the recovered rock cores.

This is a good example of stress effects on in situ $V_p$ values, since hydraulic fracturing stress measurements had shown $\sigma_{h\ min}$ (and the elastic theory estimate of $\sigma_{H\ max}$), to be about 3 and 5 MPa respectively, at cavern depth, i.e. equivalent to depths of ≈100–200 metres, if vertical stress alone had been responsible for the rise in $V_p$.

Shifting to another category of seismic tomography applications for tunnelling, it is interesting to note that deviated boreholes are quite frequently used in combination with sea-bottom hydrophones to obtain

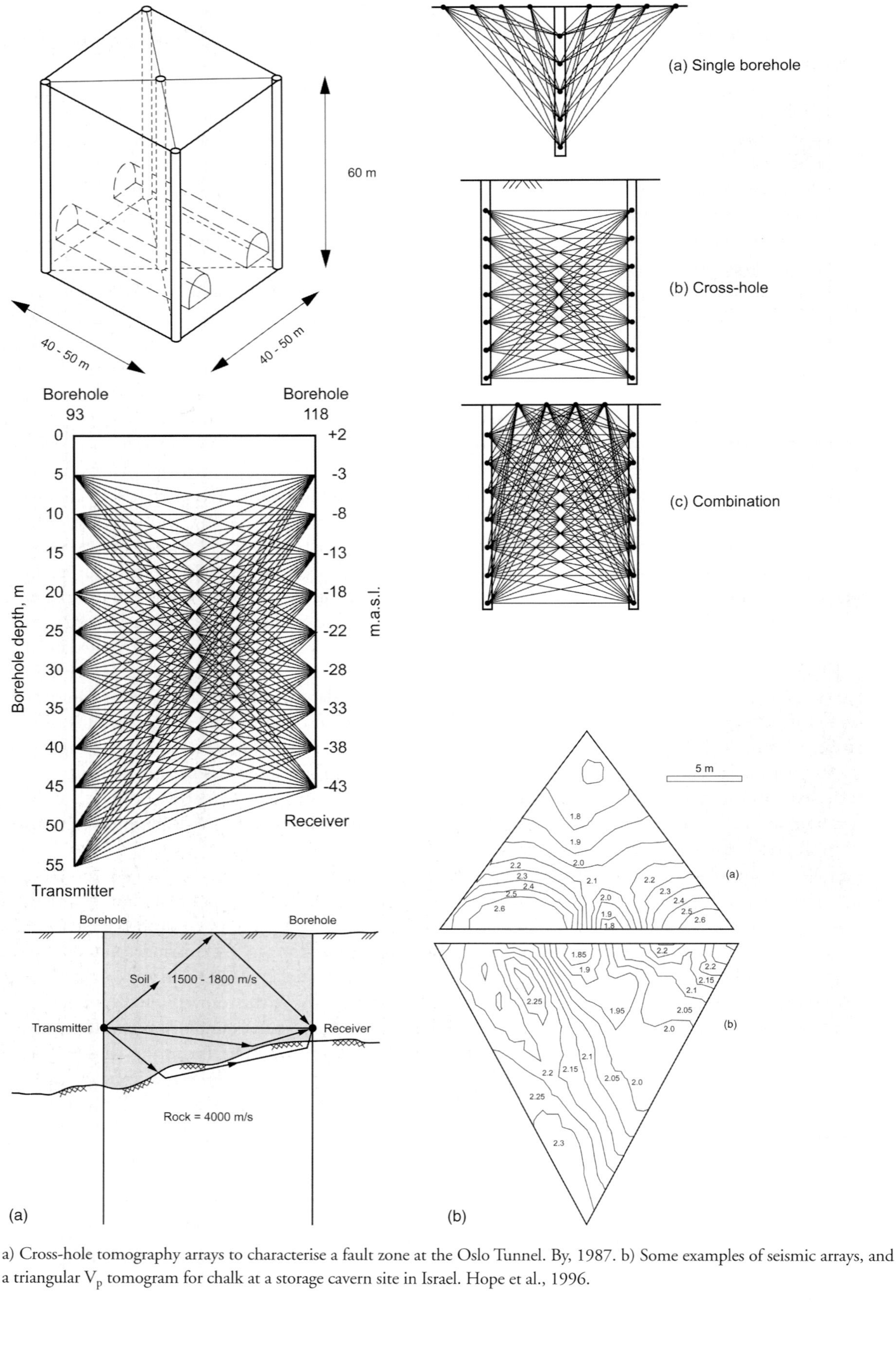


*Figure 4.6* a) Cross-hole tomography arrays to characterise a fault zone at the Oslo Tunnel. By, 1987. b) Some examples of seismic arrays, and a triangular $V_p$ tomogram for chalk at a storage cavern site in Israel. Hope et al., 1996.

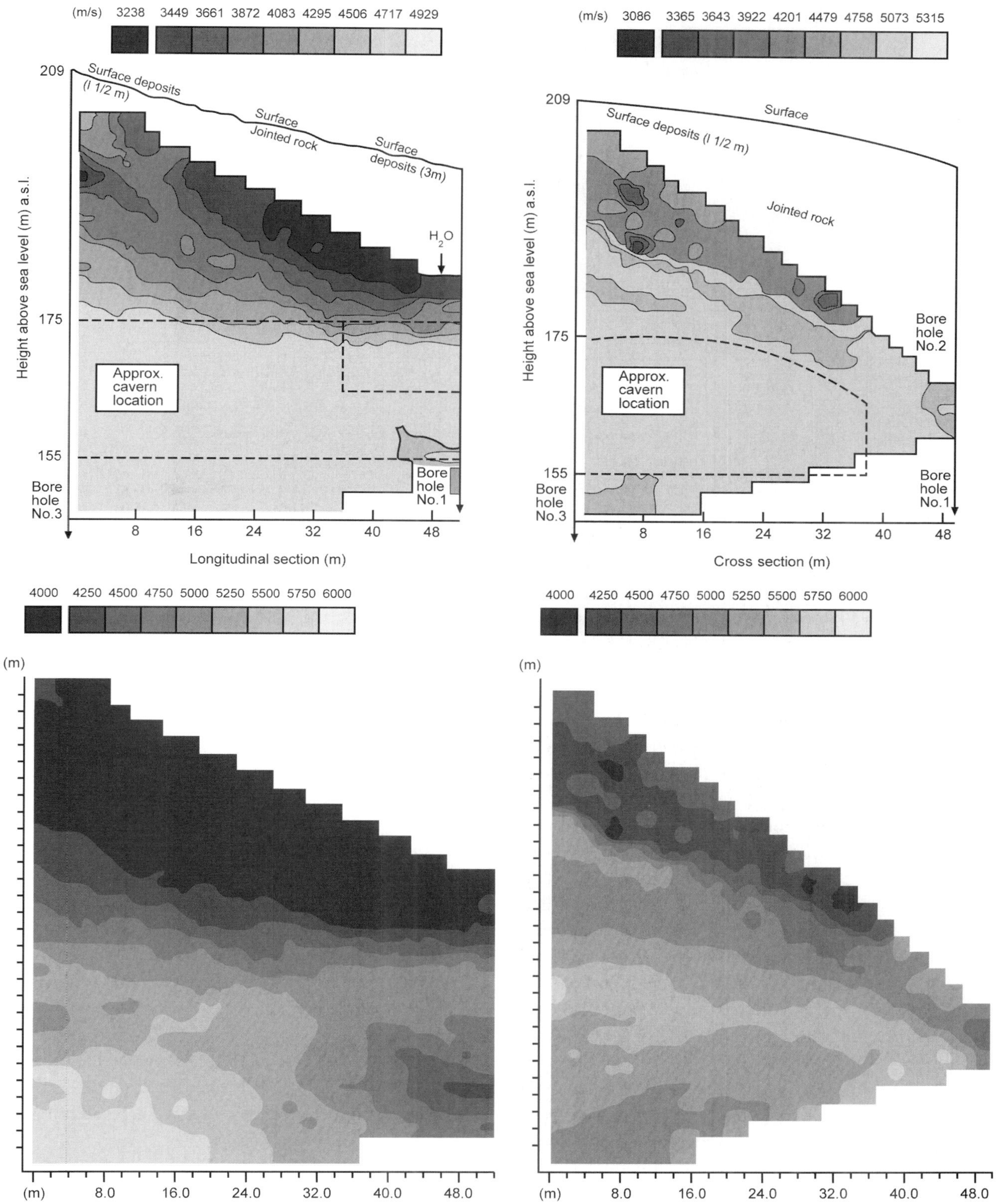


*Figure 4.7* Cross-hole velocity tomography performed by NGI for the Gjøvik cavern site investigation, Norway. Note the use of different velocity scales, to improve the velocity information a) above the cavern, b) at cavern depth. Barton et al., 1994.

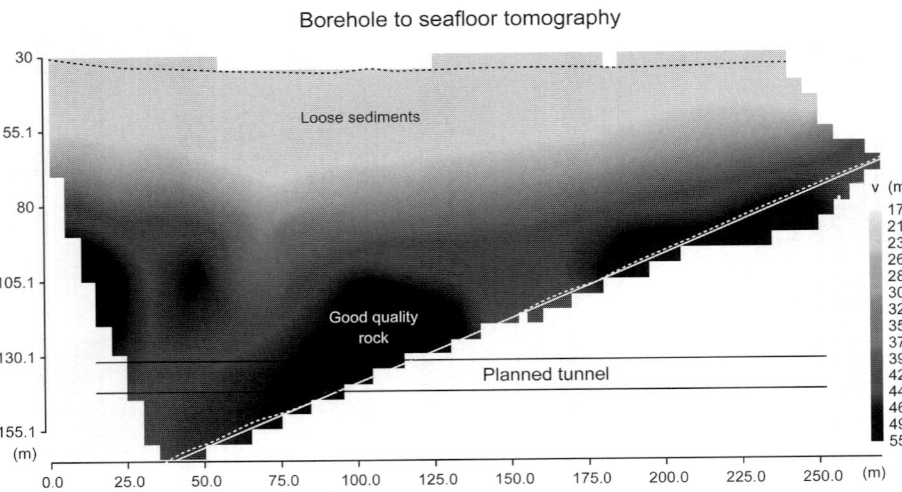


*Figure 4.8* Sub-fjord borehole to seabed seismic tomography, using differently inclined and deviated 250 to 300 m holes drilled from the land and from an intervening island, and seabed geophones. (Westerdahl and Cosma, priv. comm.1998.)

information about major faults known to be sub-parallel to these deep fjord depressions. There are numerous sub-fjord and sub-sea tunnel sites in Norway, that have utilised seismic tomography for the near-land part of their routes, with sub-sea refraction measurements for the less accessible kilometres of these tunnels.

A typical case is shown in Figure 4.8, for planning the 1997–1999 tunnelling under the Oslo Fjord, near Drøbak in southern Norway (pers. comm. Westerdahl, NGI and Cosma, Vibrometric, 1998). In this particular case the 50 to 70 m thick, fjord-bottom sediments, caused some reduction in resolution. The fault zones were correctly predicted and later encountered in the tunnel, but some details of their structure could not be detected as well as expected.

Because of uncertainties about the likely quality of difficult pre-grouting into a boulder and clay filled depression in the bedrock, against a back-pressure from more than 100 m of seawater, a deeper by-pass tunnel was excavated to maintain schedule. Penetration of the major zone was prepared with more extensive (but partly unsuccessful) pre-grouting, followed by a time-consuming freezing, when the full scope of the situation was understood. The quality of pre-grouting (or that of the freezing process) could probably have been detected by means of seismic tomography monitoring.

Excellent examples of the use of cross-hole seismic tomography, ('geotomography') in *mountain* tunnelling, are given by Chang and Lee, 2001, who refer to several tunnels in South Korea, surveyed by these and other methods. The authors point out that severe topographic changes and gradational weathering in mountainous terrain, make the use of conventional travel-time refraction seismic hard to use, as long geophone arrays may receive shortest path direct waves earlier than the refracted head waves. There is also gradational, *progressive* weathering, rather than distinct layering, with less clear development of head waves.

Consequently, the authors report widespread use of tomographic inversion techniques for their tunnels through steep terrain, and use not only of hole-to-hole, but also hole-to-surface and surface-to-surface configurations of sources and receiver strings. The authors also show the parallel use of downhole logging, with $V_p$ and $V_s$ based interpretation of the three elastic moduli and velocity-depth gradients. The additional use of rock quality RMR and Q-value core logging, and the *extrapolation* and *intrapolation* of properties afforded by the geophysics, provides a quite reliable basis for tunnel support strategies, with key attention to fault zones and portal areas.

Other near-surface uses of cross-hole seismic tomography that can be mentioned in this section are of course the possibilities of using geotomography at dam sites. It is easy to imagine the benefits of *correctly located* low and high velocity zones in a dam foundation such as that illustrated in Figures 4.1 and 4.2, where at that time, (in the 1960s and 1970s), only average velocities between holes could be determined to extrapolate deformation moduli across the foundations.

At dam sites that are located in limestones and marbles, solution cavities can prove extremely difficult to find and treat with conventional drilling and injection. Deep, sediment-filled scour-holes in dam site canyons, extending foundation depths many tens of metres could also be mapped more successfully with cross-hole seismic tomography.

By et al., 1988, described the use of the technique at a dam site in northern Norway. Small, concentrated low velocity zones of about 3 km/s were observed at numerous locations in the marble-mica schist dam foundations, using cross-hole measurements between seven inclined boreholes. The information formed a fence-like picture across the foundation, enabling decisions to be made about modified grout-curtain designs.

Cross-hole seismic tomography from boreholes drilled from the surface can also be performed at greater depth, in order to extrapolate core-logging data to (presently) inaccessible locations, as a means of optimising layouts for mines or nuclear waste repositories, or research-related underground laboratories.

An early example of cross-hole seismic tomography at the Underground Research Laboratory (URL) site in Manitoba, Canada is given by Wong et al., 1983. They first conducted a form of 'cross-hole VSP' by fixing a transmitter in one hole at 100 m depth, while the receiver was moved by 2.25 m intervals in a vertical hole 16 m away. In this case both P and S arrivals were inverted, giving average values of $V_p$ and $V_s$ of 5.5 and 3.1 km/s, and a deduced dynamic Young's modulus of 65 GPa and a dynamic Poisson's ratio of 0.245.

Due to the relatively coarse resolution achieved when they subsequently conducted cross-hole seismic tomography with a 175 m borehole spacing, they felt the need to plot the so-called residual velocity, to accentuate information levels at the location of one of the now quite well known, and much researched fracture zones. This intersected borehole URL-6 at 275 m, and was proved to be the same zone at 300 m depth in borehole M2a, due to the consistently low velocity between these two locations. The calculation of *residual velocity* shown in Figure 4.9 was designed to remove background velocity and velocity-depth trends. The authors used piezoelectric transducers as high frequency sources, in order to improve resolution.

The authors expressed the residual velocity as:

$$V_p = V'_p - 5.5 - 0.002\,(z - 100) \qquad (4.1)$$

with depth expressed in metres (for $z \geqslant 100$ m). They were thus reducing the measured velocity by 0.2 km/s at 200 m depth, by 0.4 km/s at 300 m depth, making an assumed velocity – depth gradient of $2\,s^{-1}$. A lot of discussion concerning such gradients will be found in Chapter 11, relevant to continental velocity-depth structures, and those derived from mid-ocean, spreading ridge investigations of new basalt crust.

At the UK Nirex Ltd Sellafield site in NW England, where it had been proposed to locate a low and intermediate level nuclear waste repository following many

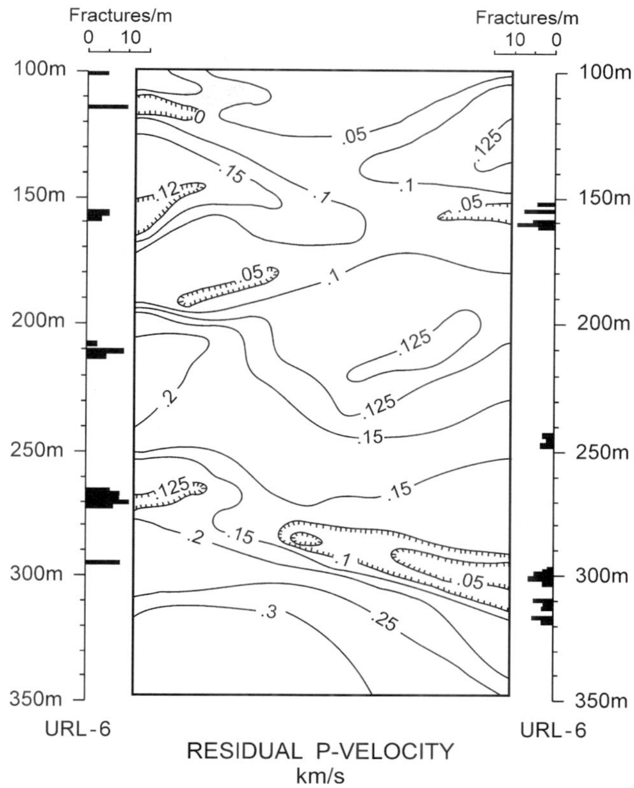


*Figure 4.9* Residual $V_p$ tomogram between two boreholes at the Underground Research Laboratory (URL), Canada. These were spaced 175 m apart. The calculation of *residual velocity* was designed to remove background velocity (5.5 km/s) and a measured velocity-depth gradient of $2\,s^{-1}$ seen from equation 4.1. Wong et al., 1983.

years of investigations of the site, cross-hole seismic tomography was performed between some of the 1000 to 1200 metre deep boreholes. The layout of the holes and the results of two of the tomograms are shown in Figure 4.10.

Stress levels at 1200 metres depth were as high as 40–50 MPa in the direction BH2 to BH5, and about 20 to 30 MPa in the direction BH2 to BH4. Because of the fault structures sub-parallel to $\sigma_{H\,(max)}$ seen crossing the BH2 to BH4 tomogram, and in view of the lower horizontal stress acting in this direction, one would expect lower average velocities in this tomogram than between BH2 and BH5. This proves not to be the case, and one must therefore question whether the between-borehole distance of 600 m is giving similar attenuation problems as shown in Figure 4.3 earlier in this chapter.

The other possible conclusion to draw from the velocity tomograms is that perhaps a deep zone with different joint orientation caused the lower velocity region at 800 to 1100 metres depth. Whatever the explanation, the

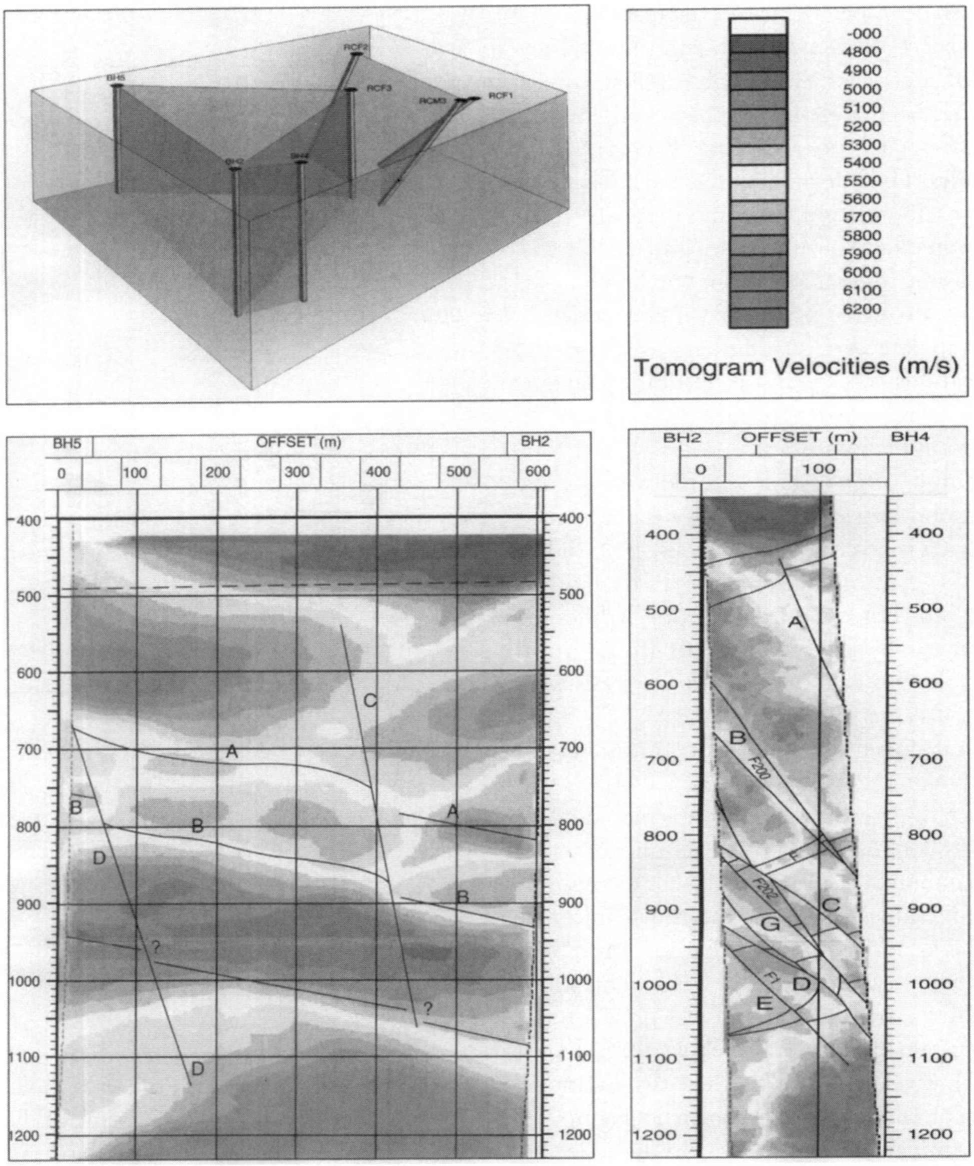


*Figure 4.10* Deep (1000–1200 m) cross-hole tomography at the UK Nirex Ltd Sellafield site. (Schlumberger GeoQuest, Nirex Report S/94/007, by kind permission.) (See Color Plate 1)

frequently occurring velocities of about 5.2 to 6.2 km/s relate quite poorly with the mean weighted Q-value of about 3 (range 1 to 10) which were logged by the NGI team of engineering geologists. (Barton et al., 1992a).

This discrepancy is most likely due to stress effects on $V_p$ in the jointed (ignimbrite and welded tuff) rock mass (Barton, 1995). This will be discussed further in the next chapter, where effects of depth and stress in jointed media are reviewed in more detail. Correlations are finally developed between depth and velocity for a given rock quality Q, also incorporating uniaxial compressive strength and matrix porosity.

## 4.3 Cross-hole tomography in mining

Phenomenological results of stress change causing velocity change will be presented in this section, prior to the in depth review of stress effects on velocities in jointed media to be given in Chapter 5. Although the cases reviewed are from mining, it may be useful to start with an interesting high pressure tomography experiment from the laboratory, described by Scott et al., 1994.

The effects of their high pressure loading of an intact cylinder of Berea sandstone with a steel indentor, was monitored by 20 acoustic sensors arranged in a ring

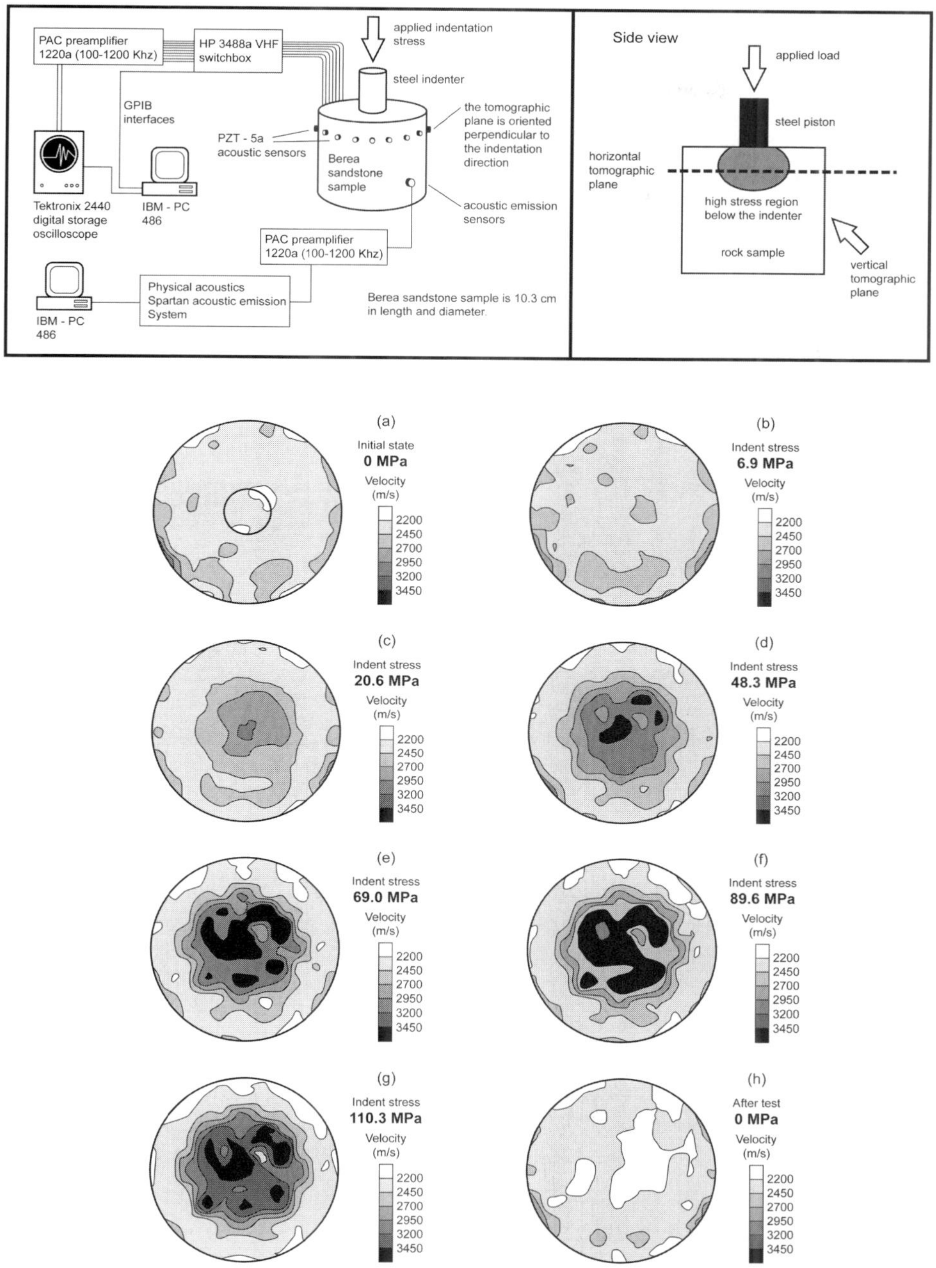


*Figure 4.11* Ultrasonic tomography to monitor the loading on an indentor on Berea sandstone. a) Experimental set-up, showing acoustic sensors, load application, and data acquisition. b) Cross-section, showing tomographic plane. c) Acoustic tomograms for different indentor stress levels a to h. Scott et al., 1994.

around the sample. The sample had a porosity of 18%. One hundred and thirty ray-paths were analysed to calculate the velocity in 97 individual elements. The experimental set-up is shown in Figure 4.11a,b and the tomographic images for eight load increments (including final unloading) are shown in Figure 4.11c.

It was found that the mean velocity of 2.3 km/s for the sample increased to 3.0 km/s a short distance under the indentor at 20.6 MPa applied stress, and finally to 3.55 km/s at 110 MPa applied stress. The rate of velocity increase declined at higher stresses, presumably due to the already reduced pore space. However, the sample appeared to have remained nearly in the elastic state, and the velocity after final unloading was very similar to that before loading, except for some increase in the area showing the lowest velocity. Slight damage was assumed.

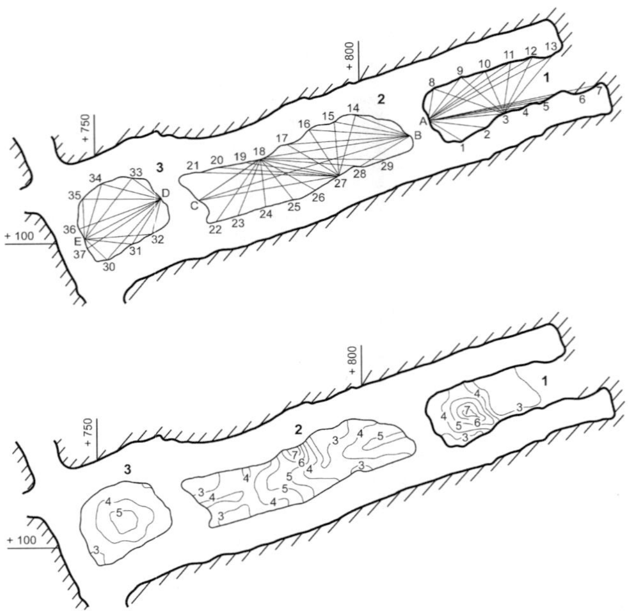


*Figure 4.12* Cross-pillar seismic tomography showing ray paths and $V_p$ values (km/s) across pillars in the Masua Mine in Italy. The host rock was dolomitic limestone, and the orebody was mineralized limestone. Barla, 1993.

A second experiment involving a vertical plane of measurement, revealed velocity increases as before, but the levels achieved differed from those in the horizontal plane (they were lower) indicating differential stress induced anisotropy.

Following this laboratory demonstration of pressure effects on $V_p$ in intact (but porous) rock, we can examine some cases from mining where monitoring of stress changes was carried out at much larger scale.

Barla, 1993, describes the use of seismic tomography across three pillars in the ore-body of the Masua Mine in Italy. While there was a general tendency of high velocity (up to 7 km/s) in central parts of the pillars, and lower velocities (3–4 km/s) on the outsides of the pillars, there was however some variation, and in one perhaps highly mineralised zone, the velocities were highest at one edge of a pillar, as shown in Figure 4.12.

Friedel et al., 1995, 1996a and 1996b, used seismic velocity tomography both in a coal mine (Foidel Creek, Colorado), and in the deep Homestake gold mine, in South Dakota, USA for monitoring of apparent stress changes and stress gradients as a result of mining. At the coal mine reported by Friedel et al., 1996a, they monitored *velocity* and *velocity changes* in two yield pillars alongside active longwall panels.

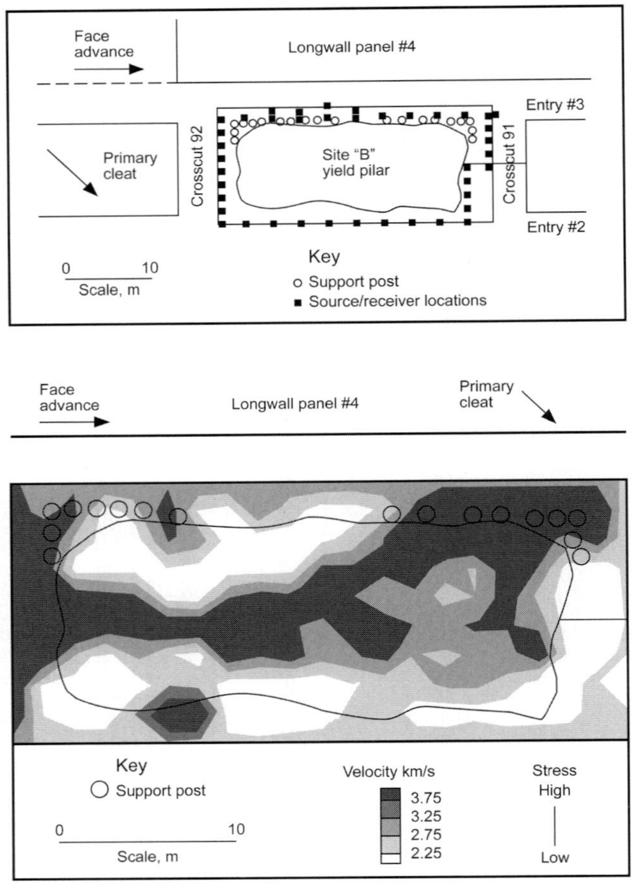


*Figure 4.13* Cross-pillar seismic tomography across a coal pillar, showing the relation between $V_p$ and the perceived stress level in 'yield pillar A'. Friedel et al., 1996a.

Figures 4.13 and 4.14 show the tomographic test set-up in each case, and below this the velocities (<2.25 to >3.75 km/s) and velocity changes (> −1.5 to >2.5 km/s) as a result of the adjacent longwall panel advance. The one day of advance (some 8 metres) caused reductions in velocity, presumably just as required for a yield pillar function. The local reductions in velocity probably reflect the adverse effect of an increase in the vertical pillar stress ('$\sigma_1$'), which would cause loosening and reduced velocity in horizontal directions (as monitored), where the rock was not well confined.

Gas and coal outbursts in Polish mines in the Lower Silesian coal basin, and the difficulty or impossibility of obtaining test samples due to the fineness of the discontinuities, led Poldolski et al., 1990, to use time-lapsed tomographic imaging to monitor velocities and related areas of high stress. The authors describe a 70 ton roof fall (and 2600 $m^3$ ejection of methane) and how

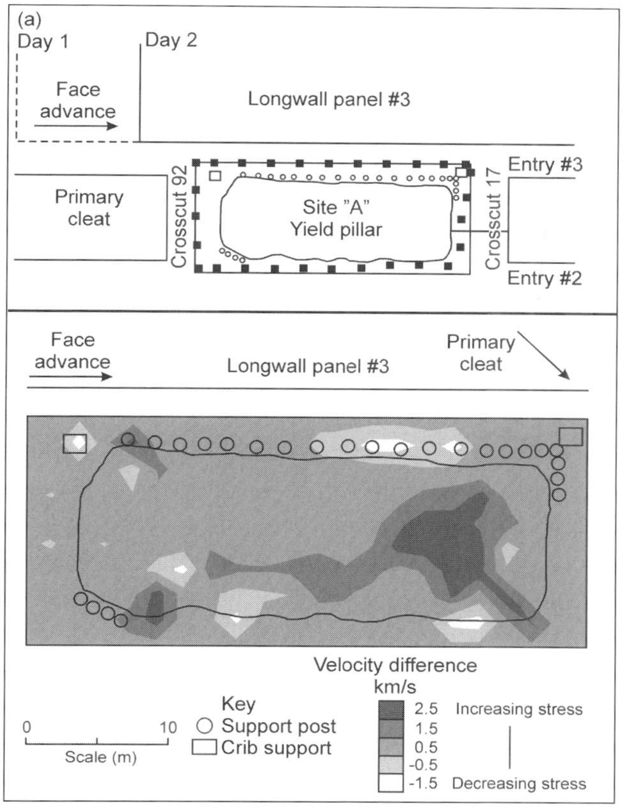


*Figure 4.14* $V_p$ as a monitor of increasing stress in 'yield pillar B', adjacent to a longwall-mining advance in coal. Friedel et al., 1996a.

high seismic velocities in the same area correlated with increased volumes of coal ejection and degassing from blast holes drilled for shooting the longwall face. The link between high stress and high velocity – prior to failure, is clear.

Friedel et al., 1995, reported monitoring between two levels of the deep Homestake gold mine in the USA. Their results indicated a sensible correspondence between low velocity zones and back-filled areas, ore chutes, and so on. High velocity gradients were interpreted as locations of potential rock burst. We shall see more examples of the effects of high stress on velocities, when reviewing the work that has been done in excavation disturbed zones (e.g. Cosma et al., 2001) in Chapter 7, and also see the possibilities of using acoustic emission (AE) as a remote method of monitoring high stress gradient problem areas.

Cross-hole seismic tomography has also been in use to delineate the detailed structure of orebodies, beyond what can be achieved by intermittent core drilling. A good example was described by Wong, 2000, from the McConnell orebody, near Sudbury, Ontario. This is a steeply dipping sulphide, crossed by a number of deviated boreholes, as illustrated in Figure 4.15a. The author described the use of a non-destructive piezoceramic vibrator source which was successively lowered down each of the inclined water-filled holes, with a hydrophone string of detectors in the nearest neighbouring hole. Figures 4.15b and 4.15c show a schematic of the equipment, and 1/5 th of the ray-paths between two of the adjacent holes. The (approximately reproduced) tomogram shown in Figure 4.15d indicated a clearly delineated orebody velocity of about 4.0 to 4.5 km/s, compared to the 5.9 to 6.5 km/s of the host rock.

The so-called pixel dimensions for the tomographic imaging and interpretation were only 1.5 m × 2.5 m. Dominant frequencies were in the 3–4 kHz range. As a first approximation, a straight ray-path assumption was made to speed the interpretation. As the author pointed out, actual raypaths were likely to curve due to refraction in a non-uniform geologic medium, and could be degraded by false features or artefacts. Checking of the tomogram structure, using independent means, including the recovered core, was therefore advised.

## 4.4 Using tomography to monitor blasting effects

Several investigators have used seismic velocity tomography to follow the effects of loosening and void formation caused by blasting. Cumerlato et al., 1988, performed seismic tomographic analysis of pre-blast and post-blast quarrying effects in dolomite, in a lime quarry in the USA, using a modified refraction seismic technique. Figure 4.16 shows pre-blast and post-blast velocity distributions, and clear advantages of a modified blast hole loading factor for controlling fracturing. High velocity zones ($V_p$ = 4.5–6.0 km/s) were reduced to low velocity (0 to 3.5 km/s) when blasting performance was unfavourable, due to all the crushing and void formation.

Maxwell and Young, 1993, used a velocity difference image technique for analysing the effect of an explosive detonation in a borehole in granite. The experimental set-up is shown in Figure 4.17a and b. The velocity difference images, examples of which are shown in Figure 4.17c and d, are computed from before-and-after-blasting time-delays, along common ray paths. The authors observed extension of the lower velocity zone away from the blast hole, sub-parallel to the trace of assumed

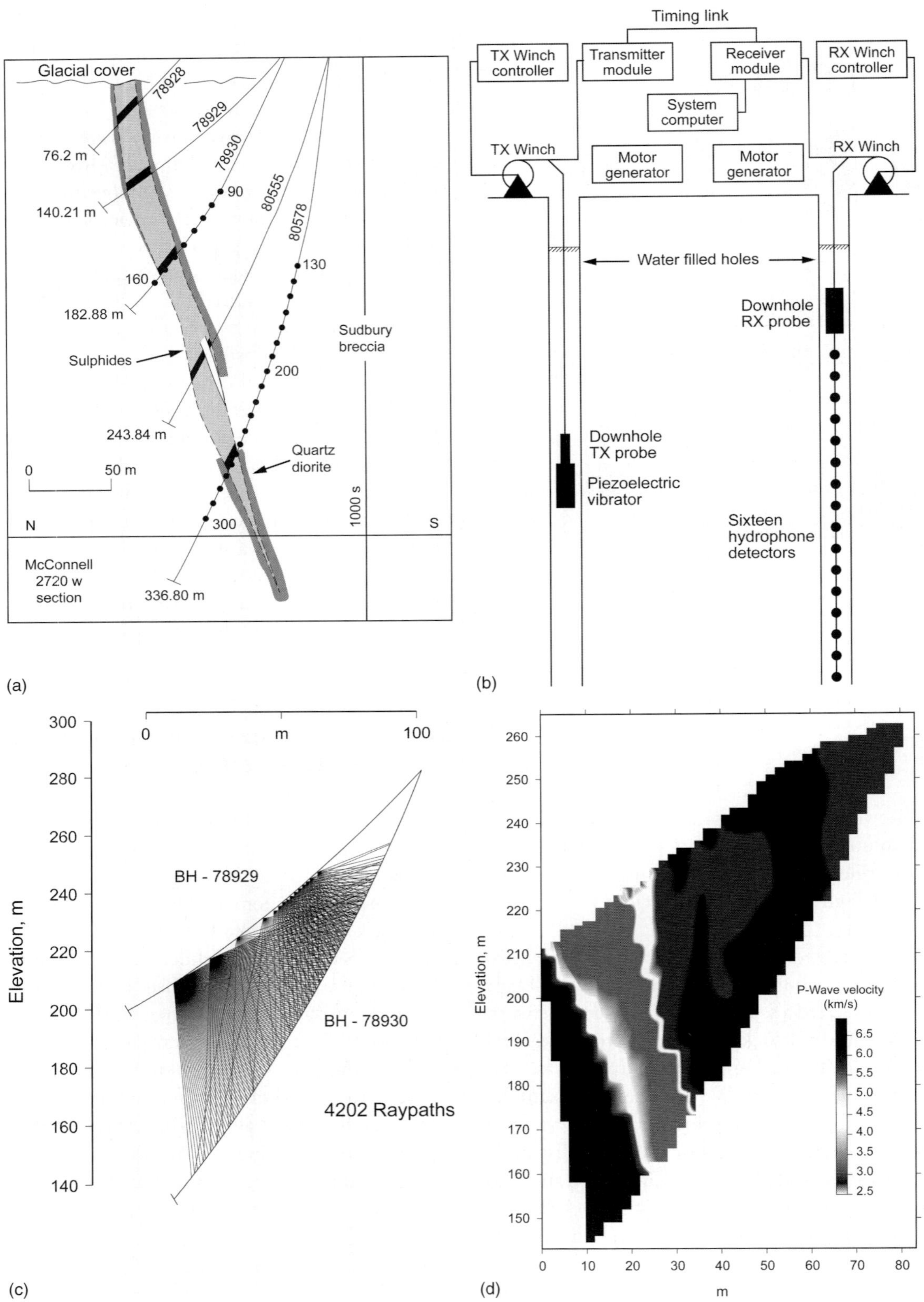


*Figure 4.15* a) Sulphide orebody delineated by boreholes. b) Schematic of equipment and acquisition geometry (shown vertical). c) One-fifth of the total 4,200 raypaths for one pair of holes. d) Approximate velocity tomogram, showing the lower velocity of the sulphide orebody ($V_p$ = 4.0–4.5 km/s), compared to the host rock ($V_p$ = 5.9–6.5 km/s). Wong, 2000.

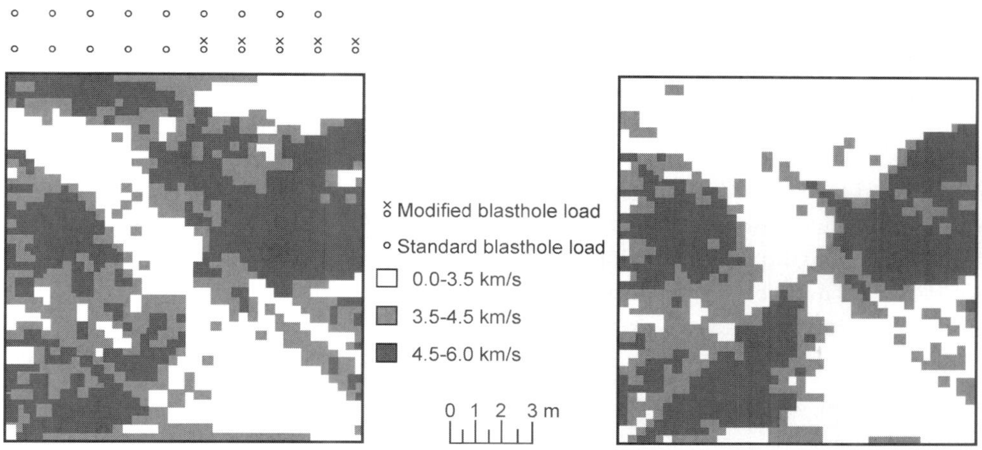


*Figure 4.16* Pre-blast (left) and post blast velocity tomograms. Cumerlato et al., 1988.

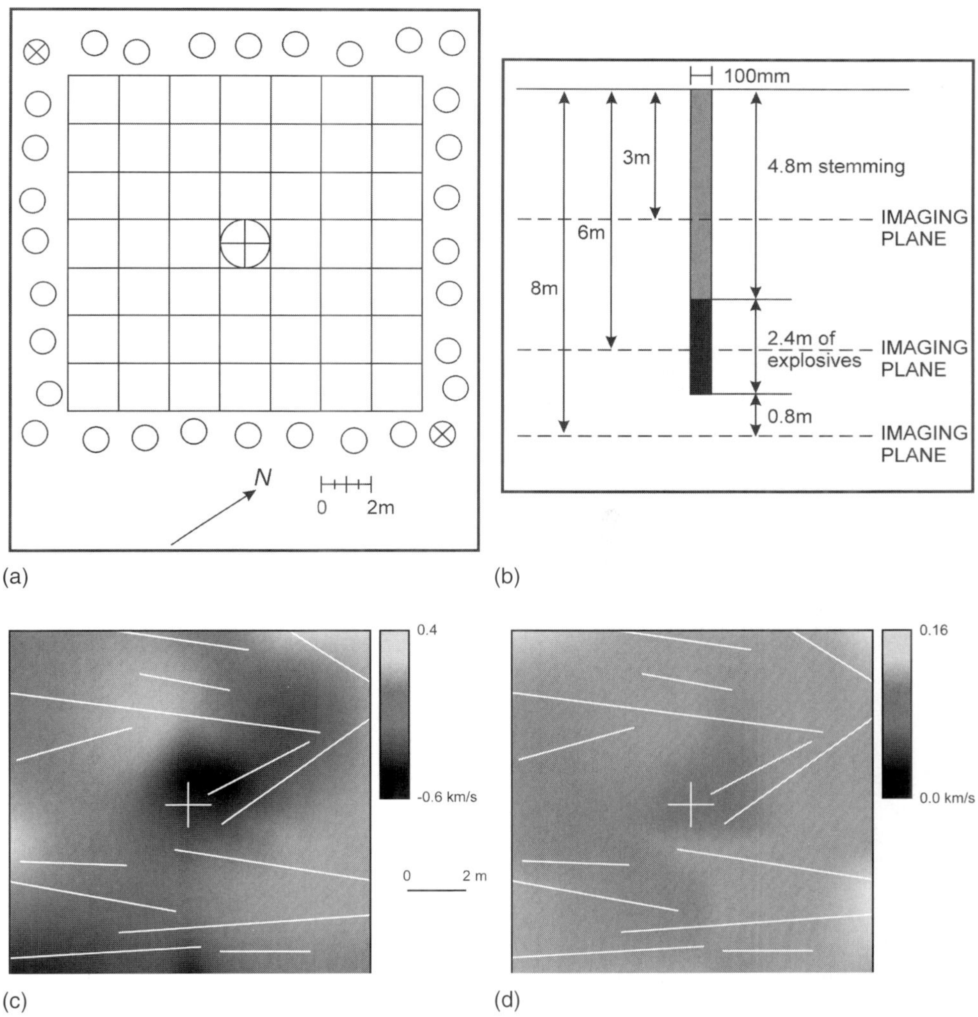


*Figure 4.17* a,b) Cross-hole tomography set-up, for monitoring blasting effects in a borehole in granite. c,d) Velocity difference tomograms showing reduced velocity caused by blasting. Error tomogram on right. Maxwell and Young, 1993.

joint planes. Secondary changes may have been associated with changes in the water table.

Seismic tomography for controlling blast fragmentation results for mine areas where in situ leaching was planned, were described by Thill et al., 1992. The results shown in Figures 4.18 and 4.19 were obtained from cross-hole measurements by the US Bureau of Mines at the experimental Edgar Mine in Idaho Springs, Colorado.

They found good correlation between pre-blast and post-blast velocities that corresponded to areas where blast induced fracturing had occurred. These low velocity zones (for example Figure 4.19) showed increases in velocity when fluid was injected. These were also the areas where lost circulation occurred when drilling was performed. The joint aperture changes and new fractures presumably created poor acoustic coupling so were readily detected as velocity reductions, later to be partly recovered when there was resaturation in the area.

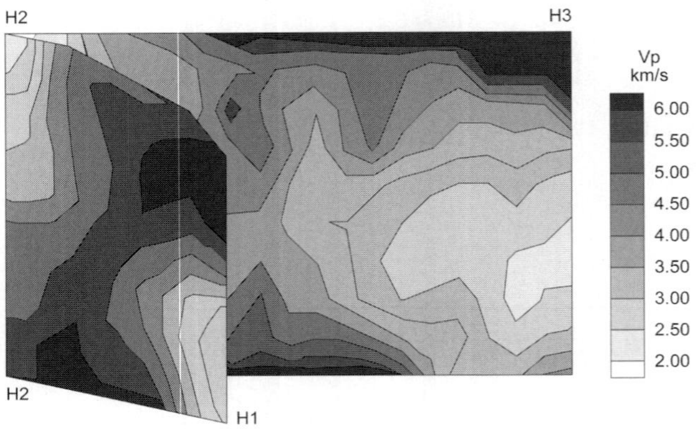


*Figure 4.18* Pre-blast 'fence' tomogram at the USBM Edgar Mine in Colorado, USA. Thill et al., 1992.

## 4.5 Alternative tomograms

Thill et al., 1992, show a method they developed of constraining the seismic parameters (e.g., velocity) at a common borehole axis, when two non-coplanar tomograms were to be joined in a 'fence' type presentation, as illustrated in Figure 4.18. One result that is not immediately obvious is the $V_p/V_s$ ratio tomogram shown in Figure 4.19. Why the post-blast high velocity (undamaged) areas should have the highest $V_p/V_s$ ratios (i.e. 1.88 or higher) while fractured zones with low velocities have the lowest ratios, must presumably be due to the changed saturation, since if each area were saturated the opposite result would be expected, as we saw in Chapter 1.

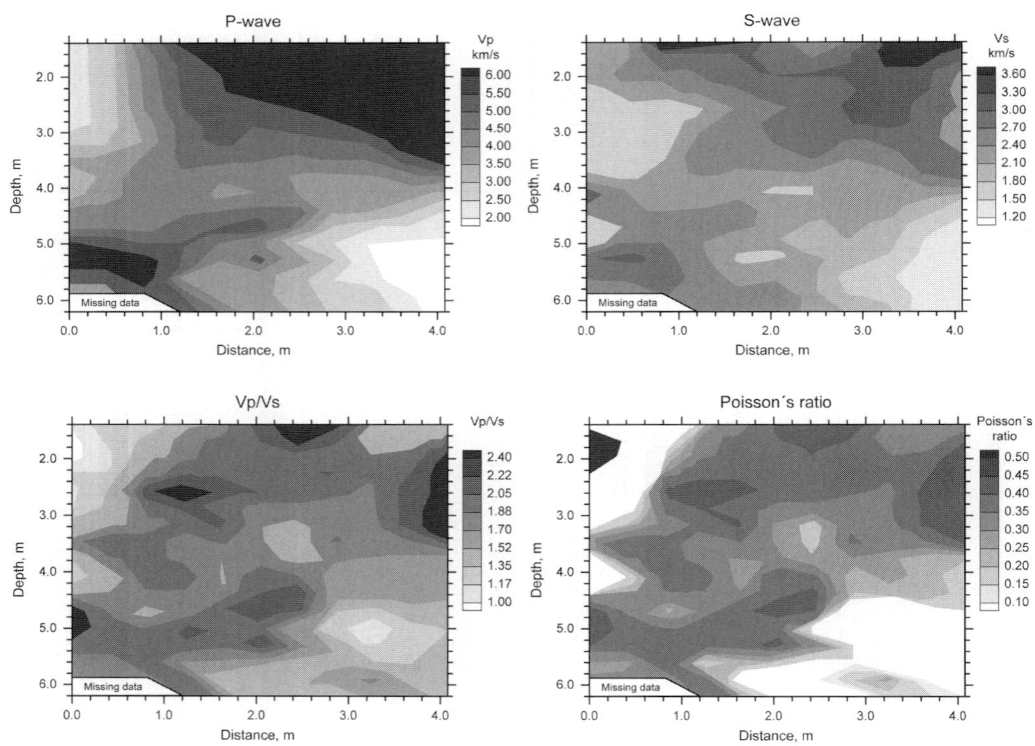


*Figure 4.19* Post-blast tomograms at the stope leaching site, Edgar Mine, Colorado, USA. Thill et al., 1992.

Another factor also seen in Chapter 1, is the basic, and theoretically determined correspondence between high $V_p/V_s$ ratios and high dynamic Poisson's ratios. The general form of both these tomograms is seen from Figure 4.19 to be similar, following the theoretical basis given in Chapter 1.

In a nice example of the capabilities of alternative tomographic descriptions of a site, using different seismic wave form analyses, Watanabe and Sassa, 1996, give three tomographic plots of the same experimental mining site (the Kamioka Mine in Japan). The geological setting and source and receiver locations are shown in Figure 4.20a and b. The P-wave velocity tomogram shown in Figure 4.20c indicates high velocities, even in locations where 'fractures', 'basic dike' and 'fault' are shown, suggesting high stresses and reduced sensitivity to jointing and faulting. The authors therefore utilised amplitude attenuation tomography (Figure 4.20d) and pulse broadening tomography (Figure 4.20e), which correlated better with the geologic structures.

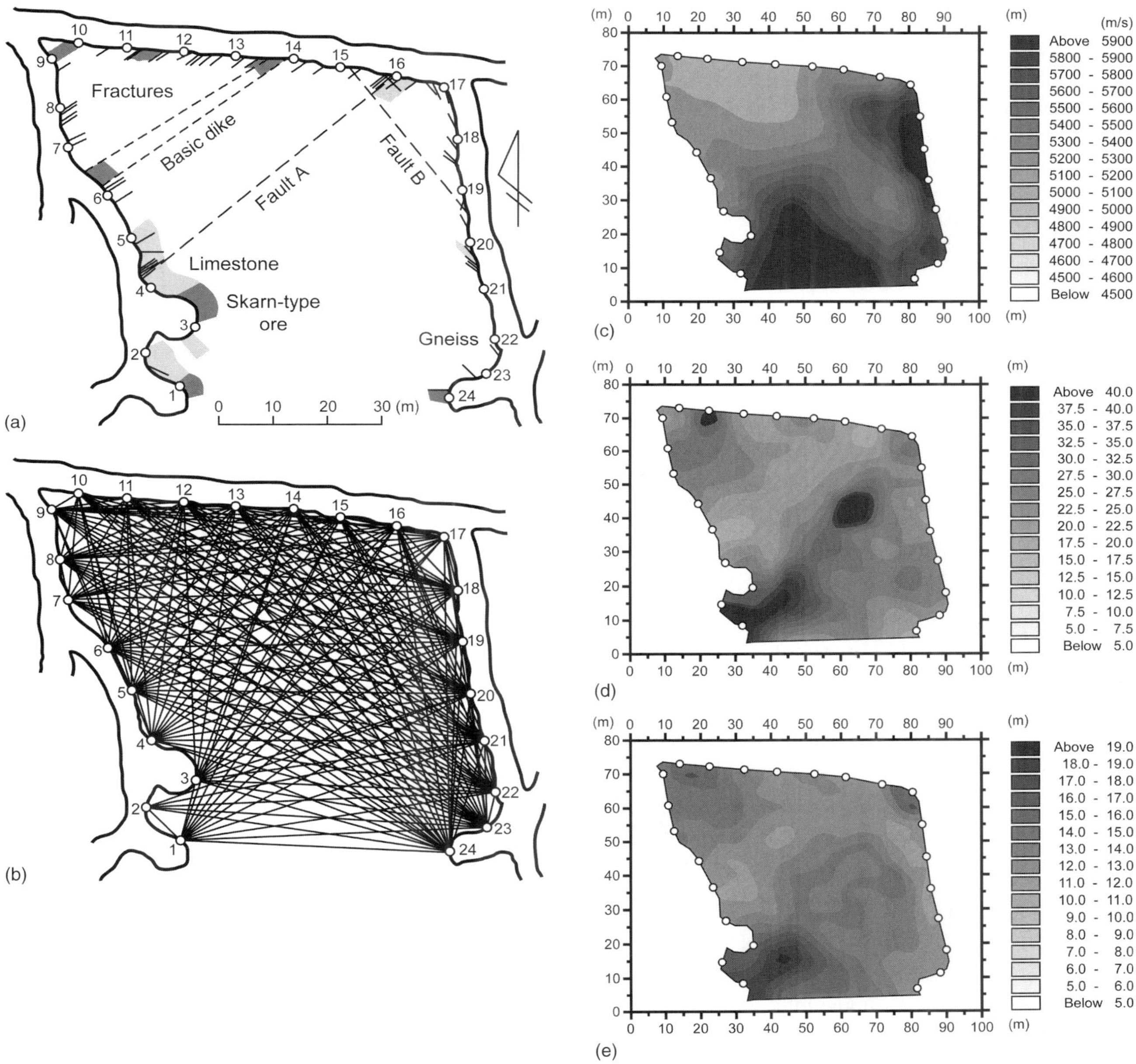


*Figure 4.20* a) to e). Three tomograms comparing P-wave velocity, amplitude attenuation and pulse broadening methods of analysis at the Kamioka Mine in Japan. Watanabe and Sassa, 1996.

Amplitude attenuation was calculated for the first arrival P-wave, and was based on the fact that amplitude decreases by geometrical spreading, and is a function of the distance between the source and receiver. The attenuation coefficient ($\alpha$) is given by:

$$\alpha = \frac{\pi f}{QV} \tag{4.2}$$

where (Q) is the seismic quality factor, (f) is the frequency and (V) is the velocity.

Watanabe and Sassa, 1996, suggested that the seismic Q-value was an inherent parameter of the medium that was *independent of frequency* in the seismic wave frequency range. At the same mine site they listed the following $V_p$ and Q-seismic values for cores.

Table 4.1 Seismic velocity and Q-seismic of rock cores measured in the laboratory (Watanabe and Sassa, 1996).

| | Velocity (m/s) | Q-seismic |
|---|---|---|
| Gneiss | 5700 | 79 |
| Limestone | 5470 | 22 |
| Skarn | 4900 | 28 |
| Basic dyke | 5170 | 36 |

The gneiss showed the highest $V_p$ (5.7 km/s) and the highest Q-seismic value (79). In situ, the gneiss between points 5 and 20 (Figure 4.20b) had an average velocity of 5.53 km/s and a Q-seismic value of 16.

The amplitude attenuation tomography shown in Figure 4.20d gives the Q-seismic values. A dark colour corresponds to high Q-seismic values of low attenuation. Light colours correspond to the low Q-seismic values associated with the fault. Soft limestone and areas oozing water reportedly also showed lower Q-seismic values (i.e., about 10 or less). There seems therefore to be more evidence here of a fairly close implicit relationship between the Q-value (the rock mass quality rating of Barton et al., 1974) and the Q-seismic value. This will be explored in greater detail in Part II, Chapter 10.

The final tomographic plot shown in Figure 4.20e is called pulse-broadening tomography. The dark colour denotes a large broadening factor, or low attenuation. The pulse broadening technique is based on the fact that the wavelength lengthens and the frequency reduces as a seismic wave travels in the rock mass. The rise time or pulse width of the first arrival P-wave is used.

## 4.6 Cross-hole or cross-well reflection measurement and time-lapse tomography

Although strictly outside the scope of Part I of this book, which deals mostly with civil and mining engineering topics, an exception will be made here, concerning a description of the obvious benefits of using cross-well seismology in the petroleum industry. Paulsson et al., 1993, recommended not only cross-hole tomography but also well-to-well reflection measurements, as shown diagrammatically in Figure 4.21, to obtain a better understanding of the (increasing-as-time-goes-by ?) heterogeneity of reservoirs.

They demonstrated how repeated (time-lapsed) surveys could be used to follow the progress of enhanced oil recovery (EOR) programmes, such as steam injection, and also pointed out the advantages of the downhole location of both source and receivers, since the attenuating weathered (or soft-sediment) layer is no longer limiting the high frequencies that can be recorded.

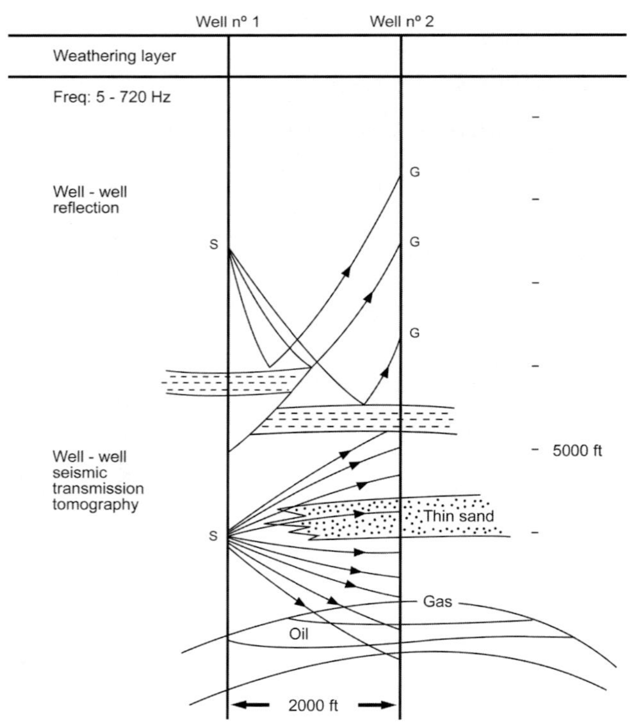


*Figure 4.21* Cross-well transmission and reflection tomography for petroleum reservoir definition. Paulsson et al., 1993.

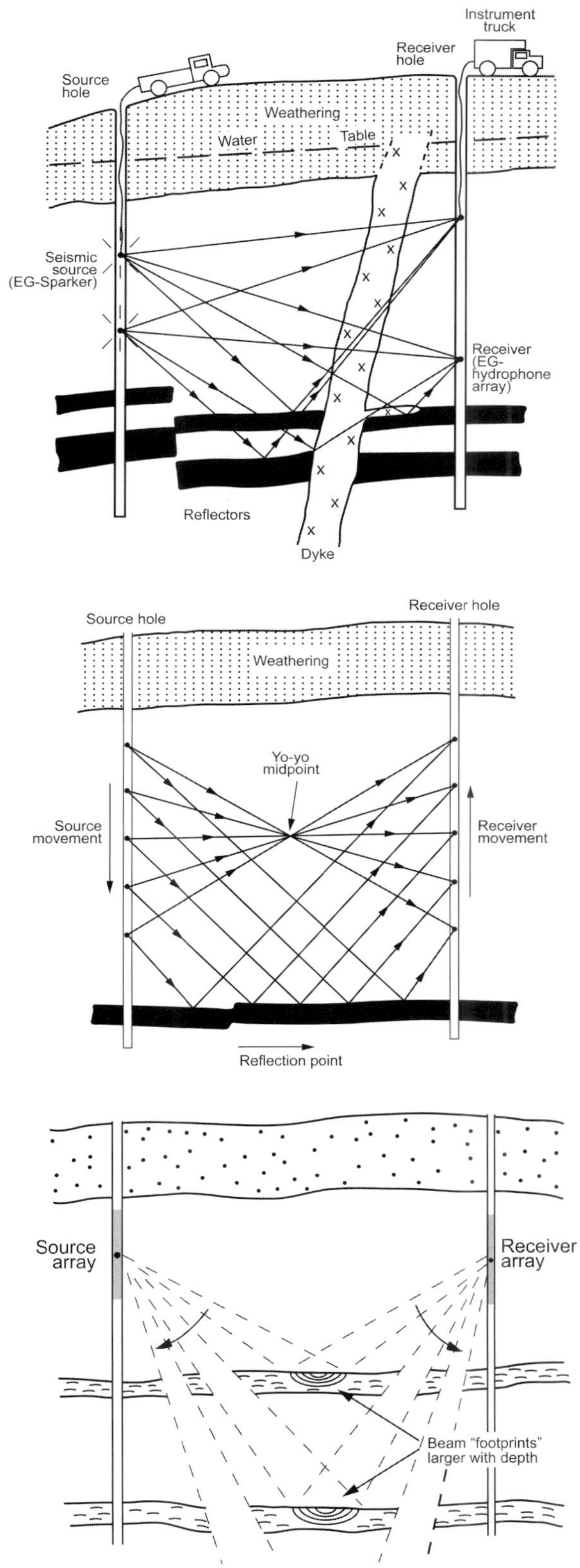


*Figure 4.22* High resolution, cross-hole reflection imaging technique, showing the yo-yo and beam steering methods, for delineating shallow coal seams in Australia. Whiteley, 1990.

Their results showed the strong correlation between oil saturation and velocity (e.g., 1.5 km/s at 20% oil saturation to 2.7 km/s at 60% oil saturation). They also noted the high velocity zone that developed when injecting cold water, due to resaturation, a result that could presumably locally reverse in a jointed reservoir, if effective stress reductions and joint aperture increases exceeded the otherwise positive effect of resaturation on $V_P$.

In a keynote lecture at the 6th IAEG Congress in Amsterdam, Whiteley, 1990, gave particular emphasis to the high resolution, cross-hole reflection imaging technique. The three diagrams given in Figure 4.22a, b, and c illustrate the basic field set-up and two of the methods (yo-yo and beam steering) for imaging targets from multiple positions.

In this particular application from Australia, interest was focussed on the location of shallow coal seams and of unfavourable structural features that would affect mining operations. A modified marine sparker source was lowered in one hole, and an array of closely spaced marine hydrophones were located in an adjacent hole, which could be up to 150 metres away. Borehole depths were up to 300 metres.

# 5 Relationships between rock quality, depth and seismic velocity

Efforts to relate rock quality and seismic velocity have been made at intervals, during the development and integration of rock engineering and engineering geology. Included in this review will be RQD, joint or fracture frequency ($Fm^{-1}$), and the Q-value (the 'static' rock mass quality rating). Their various relationships to P-wave velocities obtained from shallow refraction seismic, and also from down-hole sonic logging, will be explored. The correlations obtained have had emphasis on hard rocks, with or without weathering, without the complication of matrix porosity variations, or large ranges of strength and density. These preliminary empirical correlations between RQD and velocity ratio, and between $Fm^{-1}$ or the Q-value and $V_p$, must necessarily include the effect of depth or stress level on $V_p$, for them to be more widely applied. On the basis of numerous reviews of deeper seismic measurements, a method is developed in this chapter, that includes matrix porosity and rock strength besides all the rock mass attributes of jointing, faulting, weathering and clay. To these are added the all important influence of depth or stress level, causing gradual or rapid closure of many or all of the joint sets.

## 5.1 Some preliminary relationships between RQD, F, and $V_p$

Relationships between $V_p$ (lab, therefore intact) and $V_p$ (field, therefore jointed) have been suggested as a seismic measure of degree of jointing for many decades. Deere et al., 1967, found that the ratio $V_{field}/V_{lab.}$ when squared, was numerically very close to the value of RQD (expressed as a ratio rather than a percentage), at least for near-surface measurements. (RQD is defined as the % of core that has core sticks >10 cm long, for selected structural domains, or for specific lengths of core). The following simple table shows the central trend of this relationship, which however shows considerable scatter. (It should be noted that seismic refraction velocities in the field are being compared with the higher frequency, typically ultrasonic measurements of the laboratory.)

$$RQD\% \approx 100\left(\frac{V_{field}}{V_{lab}}\right)^2 \qquad (5.1)$$

Table 5.1 Relationship between rock quality, RQD and velocity index, Deere et al., 1967. ($V_F$ = field value of $V_p$, $V_L$ = laboratory value of $V_p$).

| Quality description | RQD (%) | Velocity index $(V_F/V_L)^2$ |
|---|---|---|
| Very poor | Less than 25 | 0–0.25 |
| Poor | 25–50 | 0.25–0.5 |
| Fair | 50–75 | 0.5–0.75 |
| Good | 75–90 | 0.75–0.9 |
| Excellent | Over 90 | Over 0.9 |

Other authors, reviewed by McDowell, 1993, have suggested the following evaluation of rock quality, as expressed by RQD.

Table 5.2 Seismic evaluation of Rock Mass Quality (see McDowell, 1993). The ratios are field-seismic/lab-ultrasonic.

| Quality description | RQD (%) | Joint frequency ($m^{-1}$) | $\frac{V_F}{V_L}$ | $\left(\frac{V_F}{V_L}\right)^2$ |
|---|---|---|---|---|
| Very poor | 0–25 | >18 | 0–0.4 | 0–0.2 |
| Poor | 25–50 | 15–18 | 0.4–0.6 | 0.2–0.4 |
| Fair | 50–75 | 8–5 | 0.6–0.8 | 0.4–0.6 |
| Good | 75–90 | 5–1 | 0.8–0.9 | 0.6–0.8 |
| Excellent | 90–100 | 1 | 0.9–1.0 | 0.8–1.0 |

The above sets of relationships are only approximate, as too few factors that obviously affect $V_p$ values for the rock mass are actually 'captured' in the RQD value alone. RQD on its own is an insufficient descriptor of the rock mass quality. However, as a single parameter it is very effective in heavily jointed rock masses, where it is particularly sensitive to the state of weathering or alteration, since the frequent occurrence of clay-fillings in the accentuated near-surface jointing, gives appropriately low values of RQD, for example 10 to 20%. (Note that 'incompetent' rock pieces that can be broken by hand are excluded, i.e. local RQD = 0%, following Deere's recommendations.)

When joints are tightly closed by effects of depth or horizontal stress, $V_F$ will more closely approach the value of $V_L$ without RQD being affected. Therefore whether RQD, or the Bieniawski, 1989 RMR, or the Barton et al., 1974 Q-value are used to correlate with seismic velocities, a depth or stress correction is required for use below the superficial penetration of surface refraction measurements, in other words when depths exceed some few tens of metres. In fact a depth correction may be needed at shallower depths, but with RQD, RMR and Q typically increasing rapidly in the first tens of meters, a reliable depth correction is problematic, since three variables are changing at once (quality, depth, velocity).

Turk and Dearman, 1986, proposed a *seismic fissurization* index K that was based on the difference between P-wave velocity of a dry, intact sample loaded to $^1/_2\sigma_c$ (half the uniaxial strength), and the velocity of the dry rock mass.

$$K = \frac{V_{\sigma/2} - V_{mass,dry}}{V_{mass,dry}} \qquad (5.2)$$

They showed that K was sensitive to increased porosity caused by weathering (e.g., for fresh or weathered andesites: n = 1.9%, K = 0.21, and for n = 9.1%, K = 0.68 respectively). When they analysed data from King et al., 1978, concerning $V_p$ and joint frequency measurements from above the water table in andesite, K was shown to vary from about 0.1 to 0.3 with increased joint frequency, while the in situ velocity (for unsaturated conditions) varied from about 6 km/s to 5 km/s. This data and King et al., 1978, source data are shown in Figure 5.1a and b.

Karmis et al., 1984, also investigated the effect of fractures (saw-cut or tensile) on the seismic velocity. When $F(m^{-1})$ was plotted against the velocity ratio ($V/V_0$), a linear relationship was given. In approximate terms, the following was found:

| $Fm^{-1}$ | $V/V_0\%$ |
|---|---|
| 18 | 50 |
| 9 | 75 |
| 4.5 | 90 |

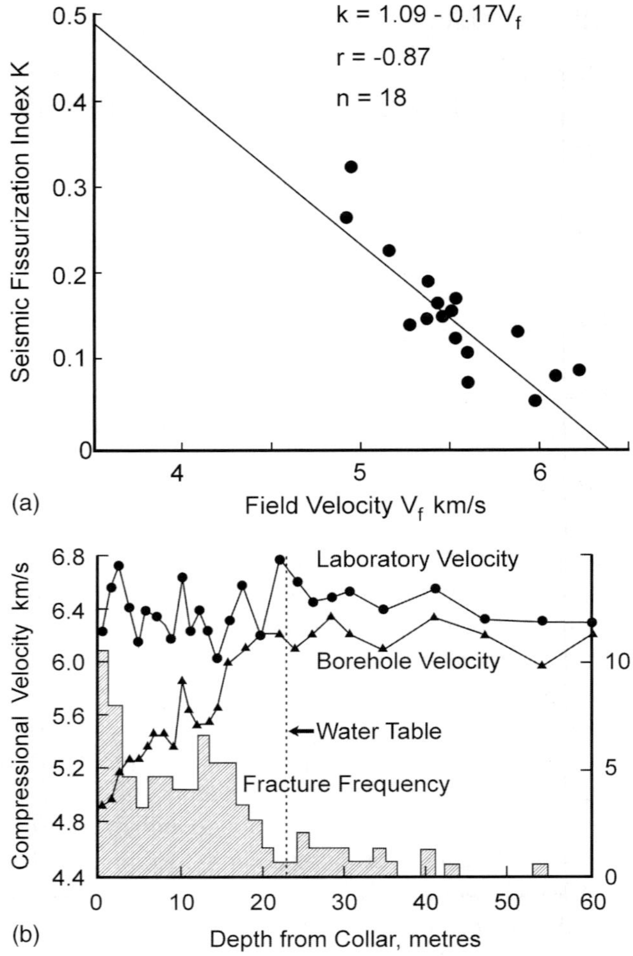


*Figure 5.1* Fissuration index K in relation to in situ velocity for dry rock masses. Turk and Dearman, 1986, with andesite data from above the water table, from King et al., 1978.

One of the most thorough analyses of seismic refraction measurements in mostly hard, jointed rock environments was that given by Sjøgren et al., 1979 and Sjøgren, 1984. The authors' experience from some 113 km of P-wave surveys (15 sites) and 5 km of S-wave surveys (5 sites) were compared with the results of 2.85 km of core from 74 drill holes at 8 of the hard rock sites. The range of rocks occurring at the measured locations, mostly in Norway, were: amphibolite, granite, gneiss, meta-anorthosite, pegmatite, porphyry, quartzite and mylonite.

The authors were careful to emphasise that the correlations they derived between P-wave velocity and jointing descriptions such as mean RQD and mean frequency $F(m^{-1})$ were relevant only to *unweathered igneous and metamorphic* rocks, and generally for the upper 20 to

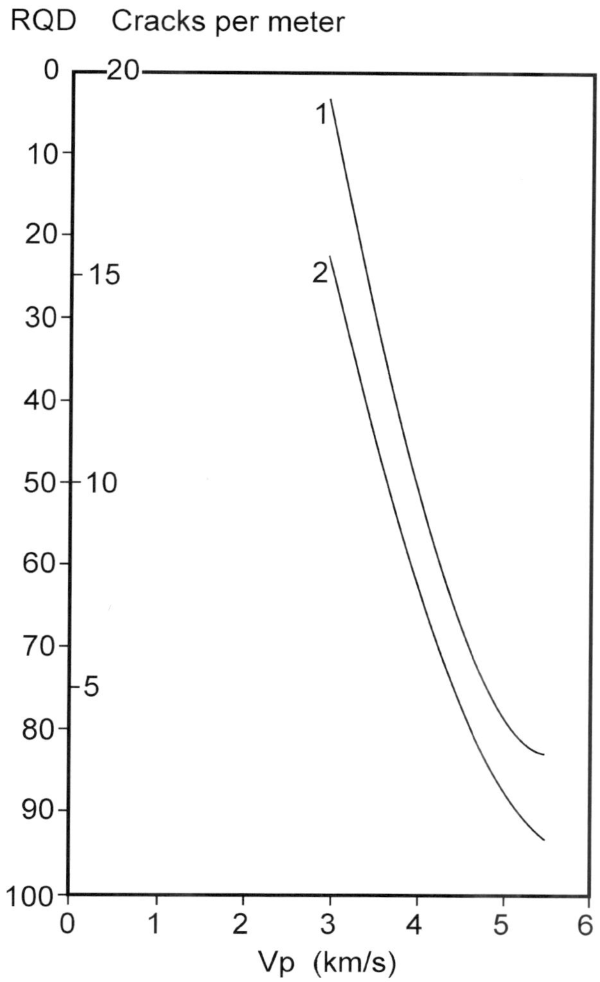


*Figure 5.2* Curve 1 = joints per metre, curve 2 = RQD, as a function of $V_p$, from shallow refraction seismic in hard, unweathered rocks, mostly from Norway. Sjøgren et al., 1979.

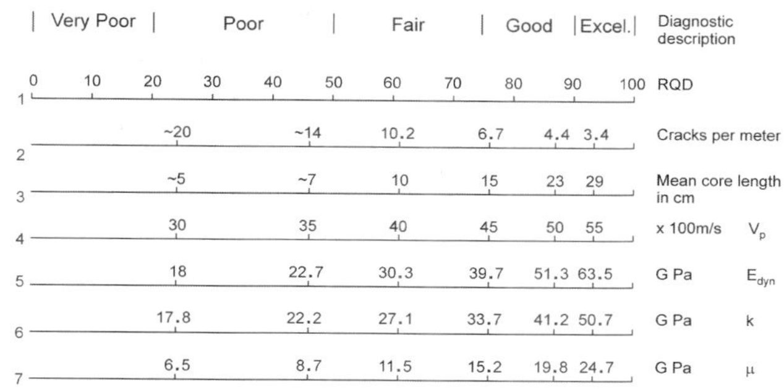


*Figure 5.3* Mean values of physical and dynamic properties for hard, unweathered igneous and metamorphic rocks, based on shallow refraction seismics. Sjøgren et al., 1979.

30 metres. They also emphasised the differences caused by weathering.

Figure 5.2 shows mean numbers of cracks (joints) per metre ($F^{-m}$) for a given velocity in more or less unweathered Scandinavian rocks. The general trend for mean RQD values versus velocity are also given for this 'unweathered' data set. Deviation from these average curves were reported to be about 1 crack/m at the higher velocities, and 1.5–2.0 cracks/m for the lower velocities. Corresponding dispersions of RQD values were 2–3% and 5–6% respectively.

The 74 drill holes were as close as possible to, or on the seismic lines, and directed to be as perpendicular as possible to the tectonic structure or foliation. $Fm^{-1}$ values are therefore close to the maximum and RQD values close to the minimum, when considering a line sample through the rock mass.

The general joint frequency and RQD trends for these unweathered jointed rocks, including resulting dynamic moduli are summarised in Figure 5.3. (In this figure k = bulk modulus, and $\mu$ = shear modulus. This data is reproduced at larger scale in a subsequent comparison with Q-values.)

Sjøgren et al., 1979, discussed various factors that could alter the proposed mean joint frequency and mean RQD versus $V_p$ trends shown in Figures 5.2 and 5.3. They pointed out that the 'natural' velocity of the unjointed (or most massive) rock from site to site could vary due to rock type, mineralogy, etc. (One could also add to this list the inter-related technical terms: porosity, density and uni-axial compressive strength.) Besides these fundamental causes for variation, the effects of weathering and depth of measurement were obviously of particular influence.

For this reason the authors addressed most of their attention and derived most of their data from the depth zone of up to 20 or 30 metres. When they conducted subsequent tunnel measurements, they found that a 30 to 50 m depth resulted in a general increase in velocity of about 5 to 15%, greatest for the lower velocity. When $V_p$ was lower than 3 km/s they had observed 'considerably greater' increases with depth, and also a common *reduction of the widths* of the low velocity zones with increased depth (40 to 60% was quoted).

In a later publication, Sjøgren, 1984, gave his earlier example of hard rock correlations between mean RQD, mean joint frequency per metre ($F^{-m}$) and mean P-wave velocity (shown in Figure 5.2) with an additional curve 3 related to the mean trend of RQD in Permian and Triassic sandstones (Figure 5.4). Obviously these sets of measured data cannot all fit the simple relation of Deere et al., 1967, that $(V_F/V_L)^2 \approx RQD/100$.

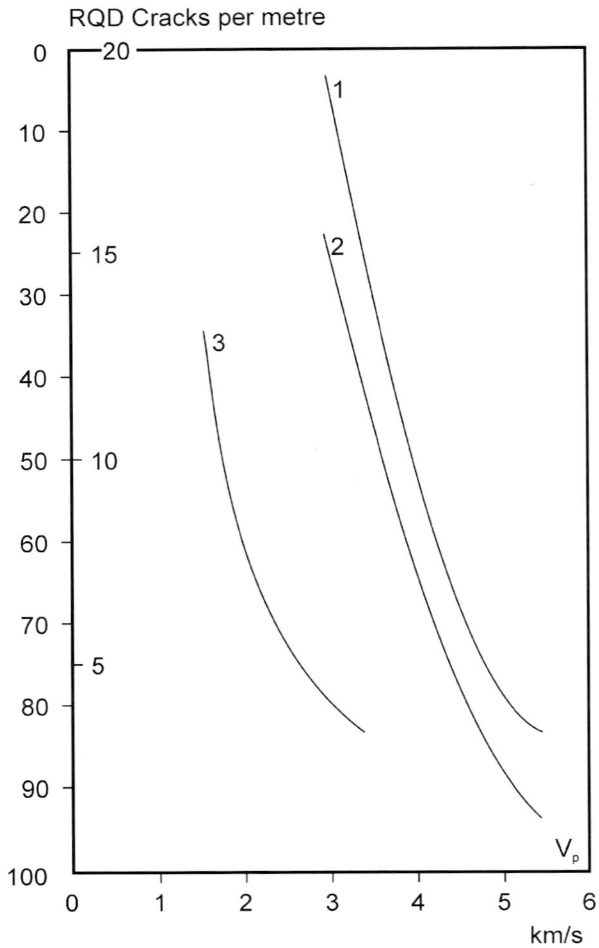


*Figure 5.4* Mean RQD and $Fm^{-1}$ as a function of $V_p$ for the previous hard rock sites (curve 1 = $Fm^{-1}$, curve 2 = RQD), and for Permian and Triassic sandstones. Sjøgren, 1984.

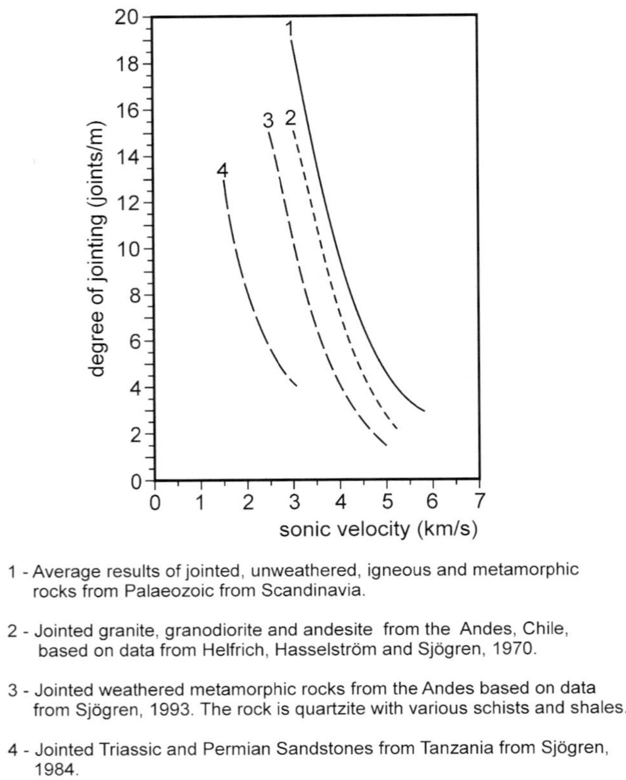


*Figure 5.5* Data from Sjøgren and co-workers for $Fm^{-1}$ versus $V_p$ for various rock conditions, with increased weathering effect and/or reduced strength, from right (#1) to left (#4). Palmström, 1996.

The large reductions in velocity (1.5–2 km/s) are clearly a function of increased porosity (and density) and uniaxial strength (or deformation modulus E). When there is a tendency for weathering, or for matrix porosities higher than normal for hard rocks, then the *same joint frequency will be recorded at lower seismic velocities*. The four curves from Sjøgren and co-workers, shown in Figure 5.5 represent at the one extreme (curve No. 1) the same data as given in Figure 5.2 for hard, unweathered rocks at shallow depth. The degree of weathering increases, rock strength eventually reduces, and the matrix porosity increases while progressing from curve No. 1 to curve No. 4 in Figure 5.5.

The data were assembled from Sjøgren and co-workers, and are derived from measurements in Scandinavia, in the Andes and in Tanzania. In relation to empirical correlation possibilities, corrections for weathering, porosity and rock strength (or density) are probably needed to explain the range of data.

Extensive fracture frequency and velocity data were presented by Niini and Manunen, 1970. The data in Figure 5.6 were derived from 55 vertical or steeply inclined holes drilled into the upper 15 metres of bedrock, along 100 km of seismic traces made for the 120 km long Helsinki water supply tunnel.

The possible complication of increased stress, from tectonic causes or from 15 to 30 m of additional soil cover, has apparently meant that high fracture frequencies were recorded even when velocities were as high as 4.5 km/s. Possibly all fractures in the core, and the natural joints, were each included in 'fractures/m', since these numbers for $Fm^{-1}$ are significantly larger than the Sjøgren data sets. Fracture frequencies were unusually high (F = 26/m for 0–5 m rock depth, F = 21/m for 5–10 m rock depth) and seismic velocities were strongly affected as a result.

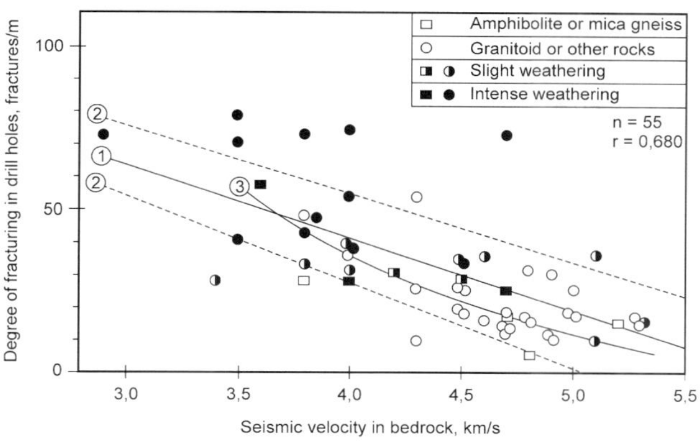


*Figure 5.6* Joint frequency and velocity trends from the Helsinki water tunnel. Niini and Manunen, 1970. There appears the possibility that both joints and artificial fractures were counted, in view of the extremely high $Fm^{-1}$ magnitudes, for a given velocity.

The authors gave the following ranges of results for *intact rock samples* obtained from 31 drillholes in the granites and from 2 drillholes in the mica gneisses.

| | $V_p$ (field) km/s | $V_p$ (lab) km/s | $E_{dyn}$ (lab) GPa |
|---|---|---|---|
| Granites | 3.4–5.3 | 5.8–5.5 | 61–82 |
| Mica gneisses | 3.8–5.2 | 5.5–5.8 | 85–96 |

An unusual set of data for *fractured zone widths* was also described by Niini and Manunen, 1970. Zone widths were shown in relation to velocities measured within the low velocity zones and also outside the zones. Figure 5.7 shows that fracture zone widths were larger, as one would expect, if the velocity outside the zone was also low. Narrowest zones tended to have lowest internal velocities, and highest external velocities.

Developments in logging joint and fracture frequency effects using a downhole acoustic tool that could log in water-filled or dry boreholes were described by King et al., 1978. Figure 5.8a and b show good examples of the relationship between the larger scale downhole velocities and the laboratory velocities from a mining location in andesites and pegmatites. Laboratory conditions of humidity and stress (7 MPa) were matched to the mining stope conditions as far as possible. The data shown in Figure 5.8 can also be taken as a useful example of excavation disturbed zone (EDZ) and drainage effects, although since joint frequency also increases towards the mine opening, the combined effects of $Fm^{-1}$ and % saturation cannot be separated.

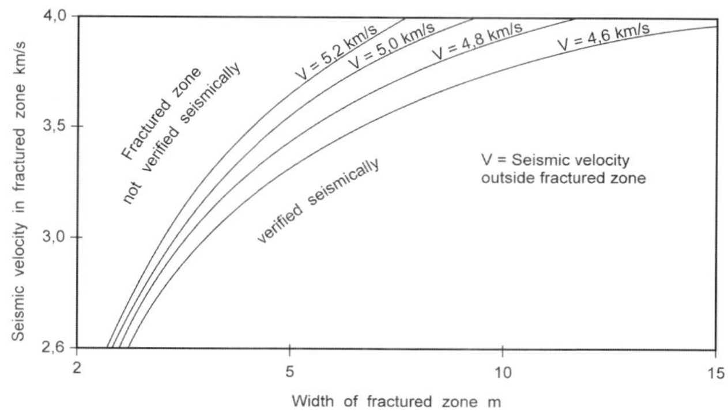


*Figure 5.7* The separation of velocities *within fractured zones*, and outside the fractured zones, from studies for the 100 km long Helsinki water tunnel. Niini and Manunen, 1970.

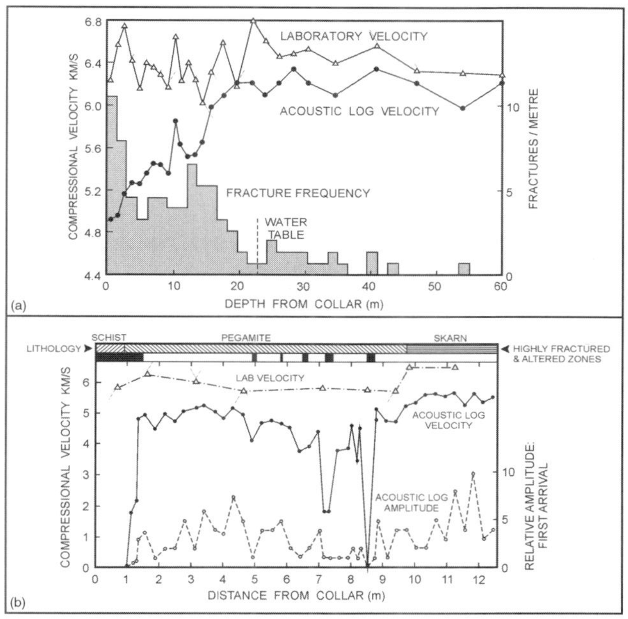


*Figure 5.8* Examples of joint frequency effects on downhole acoustic log $V_p$ values, with comparison to laboratory velocities under the same stress levels as *in situ*. King et al., 1978.

The highly fractured and altered zones were, of course, strongly correlated to velocity and amplitude reductions. The same authors also made some useful assessments of the effect of logged fracture (or joint) frequency ($Fm^{-1}$) on velocities in cored 60 m deep boreholes (Figure 5.8), and assembled other near-surface and underground data to investigate correlation of the squared velocity ratio $(V_F/V_L)^2$ with joint frequency ($Fm^{-1}$). Care was taken to discount the drilling induced fractures. The authors gave the following correlation (see Figure 5.9).

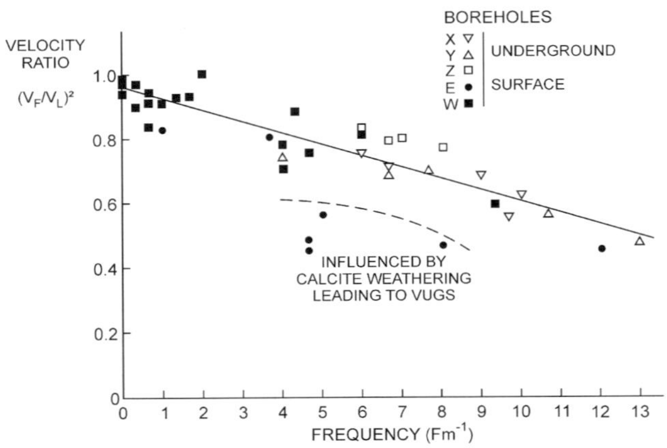


*Figure 5.9* A collection of near-surface and underground data, showing velocity ratio squared $(V_F/V_L)^2$ as a function of fracture frequency. King et al., 1978.

$$(V_F/V_L)^2 = 0.96 - 0.036F \quad (5.3)$$

(Correlation coefficient = 0.93)

## 5.2 Relationship between rock quality Q and $V_p$ for hard jointed, near-surface rock masses

Due to the seismic 'visibility' of jointing in the upper 25 to 30 metres, Sjøgren et al., 1979, and Sjøgren, 1984, were able, as we have seen, to record significant correlations between $V_p$, RQD and joint frequency. Since their measurements were shallow, the effect of stress-induced joint closure was minimised. They also effectively removed other sets of variables by generally recording correlations for hard and almost unweathered igneous and metamorphic rocks. The variables of depth, porosity, uniaxial compressive strength and density were therefore largely removed. This has some advantages for what follows.

A hard rock, near surface correlation of $V_p$ and rock quality Q-value can also be derived on the same basis as above, if effects of porosity, uniaxial strength and depth are first ignored, as for the Sjøgren et al., 1979, data shown in Figures 5.2, 5.3 and 5.4.

On the basis of the Norwegian Geotechnical Institute's cross-hole seismic tomography measurements at the Gjøvik (62 m span) cavern site in Norway, shown in Chapter 4, and based on Q-logging of the same boreholes, a preliminary model for a $V_p$-Q-value relationship was developed by Barton, 1991. This was subsequently confirmed by analysis of several other seismic, cross-hole and Q-logs including a useful set from the Xiaolangdi multi-purpose hydroproject in China, where plate loading tests also provided 'static' deformation modulus measurements that showed broad correlation with the Q-values independently logged by the writer.

Various hard rocks such as granites, gneisses, volcanic ignimbrite and competent sandstones were also tested in projects in Norway, England and Hong Kong, where first-hand information on Q-logging had been obtained.

The proposed relationship, which was briefly introduced in Chapter 1, was as follows:

$$V_p \approx 3.5 + \log_{10} Q \quad (5.4)$$

This empirical relation is plotted in two different ways together with Sjøgren et al., 1979, trends for RQD and $Fm^{-1}$ (derived from Figure 5.2). Figures 5.10a and b show the result. When the rock quality Q-value is 1.0, midway between the extremes of 0.001 and 1000, $V_p = 3.5$ km/s. The velocity changes by roughly 1 km/s (upwards or downwards) for each ten-fold change in rock quality Q-value.

This model has now been tested on sites in several countries where rock quality Q-logging of core has been performed. The fit to measured data is quite good, provided that depths are shallow (i.e. down to 25 m, near the usual limit of shallow refraction seismic surveys performed where depth of weathering is relatively limited). A further necessary condition is that the rocks are non-porous and reasonably hard (i.e. typically with uniaxial strengths of 100 MPa or more). This model for hard rocks, and a modified one for soft porous rocks to be developed later, can be used for initial interpretation of seismic data.

The table of data given in Figure 5.11, from Sjøgren et al., 1979, can also be expanded to include the Q-value scale, as shown at the bottom of this figure. As also noted in the figure, depth or stress effects, discussed in detail later, will mean that the Q-scale must be shifted more and more to the right in relation to $V_p$, as depth increases. The same of course will actually apply to the RQD and $Fm^{-1}$ scales. For the above reasons, the suggested correlations must be strictly applied to near-surface seismic data (shallow refraction seismic or shallow boreholes) in hard, non-porous, largely unweathered rocks, but of course can apply to heavily fractured and sheared zones, (i.e., faults) containing clay.

It is interesting to note the perceived inter-relationships between eight methods of rock mass classification using the Chinese descriptive classes 'soft rock', 'hardpan',

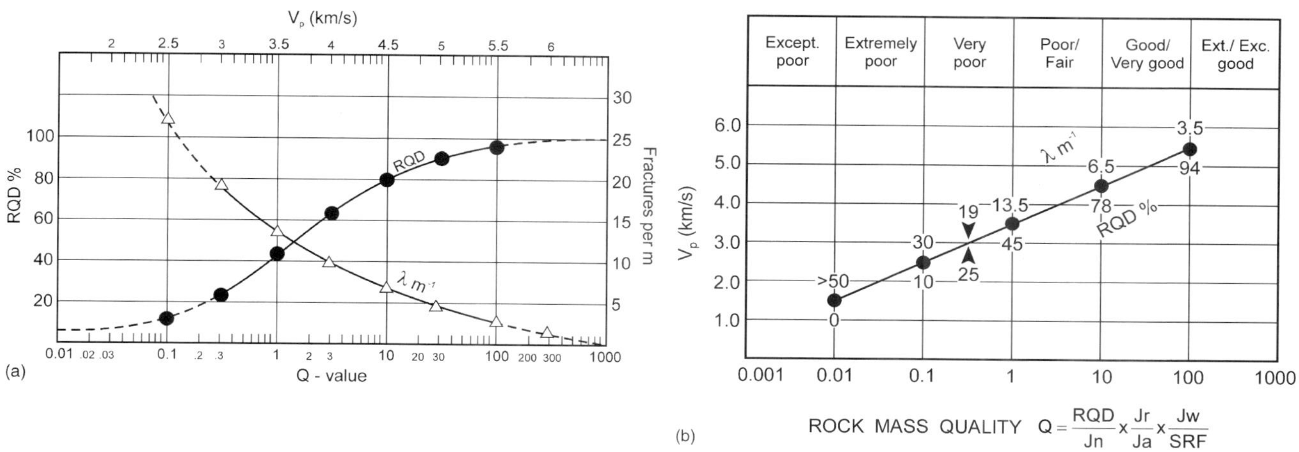


*Figure 5.10* a) Mean RQD and $Fm^{-1}$ trends for hard, near-surface, low porosity rock masses, from Figure 5.2 Sjøgren et al., 1979, with an estimated extrapolation of the 'extremes' (dashed-lines), by the writer. The lower rock quality Q-scale, added by the writer, is also only relevant for hard, unweathered, low porosity, near-surface rock masses. b) Note rearrangement of Q-$V_p$ relationship, with appended RQD and $Fm^{-1}$ data from Sjøgren et al., 1979, after Barton, 1995.

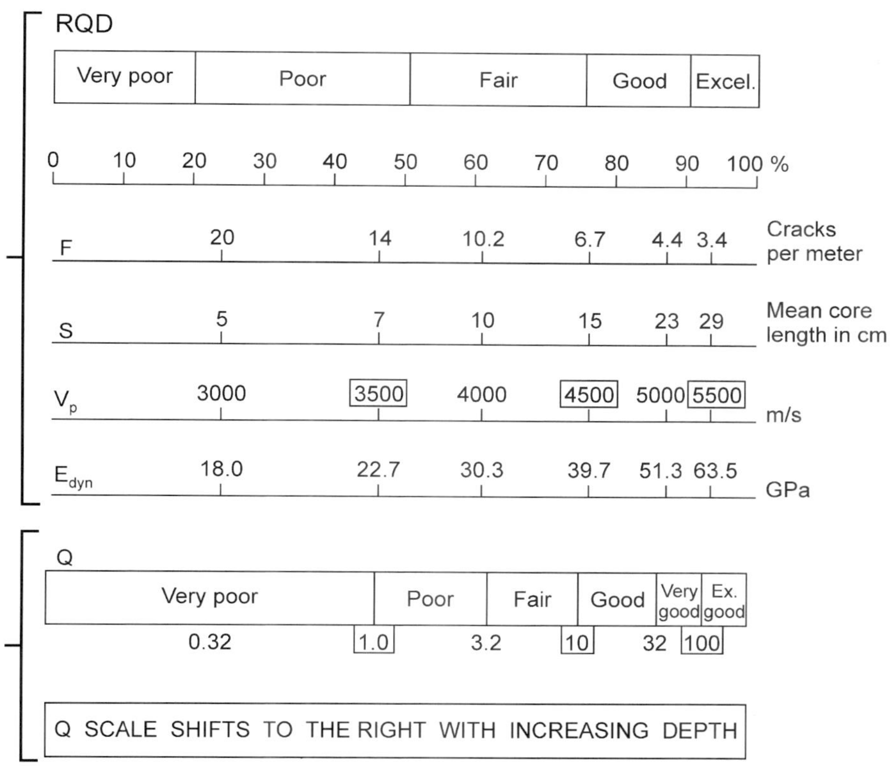


*Figure 5.11* Mean RQD, $Fm^{-1}$, $V_p$ and $E_{dyn}$ data for hard, near-surface, low porosity rock masses, from Sjøgren et al., 1979. The rock quality Q-value scale was added by the writer. Note the need for a shift of the Q-scale to the right, with increasing depth.

'normal soil' etc. that was given by Chen, 1982, using the seismic velocity ranges as a reference. Approximate similarity to the empirical model ($V_p \approx 3.5 + \log Q$) that was developed ten years later is indicated in Figure 5.12.

Comparison of rock quality Q-values logged in boreholes (or mapped at the surface) and seismic velocity measurements are not yet very common in the rock mechanics literature, though data is available at numerous

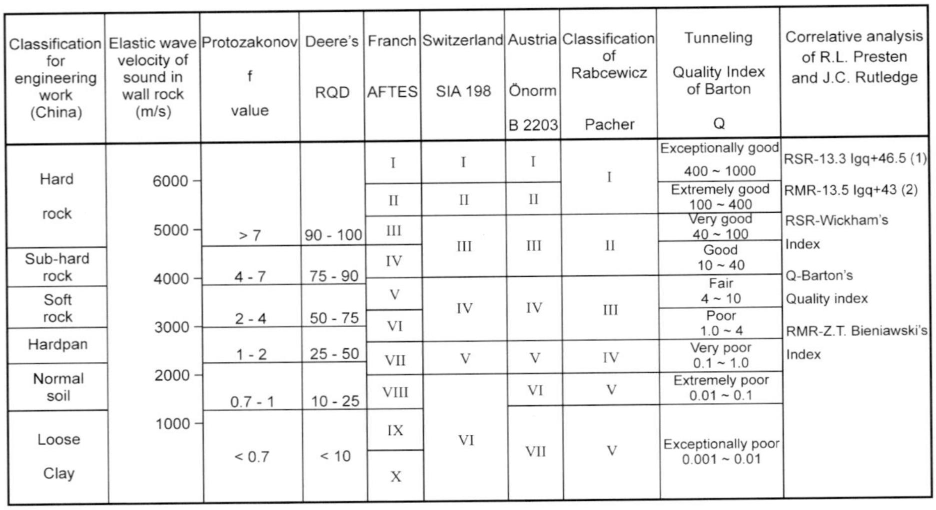


*Figure 5.12* Inter-relationships between various rock mass classification schemes, $V_p$, RQD, and the rock quality Q-value. Chen, 1982.

sites. Chan, 1993, describes the engineering geological investigations performed for the purpose of optimising the location and orientation of a 24 metre span cavern to be used for refuse transfer, at Mount Davis in Hong Kong.

General mapping of tuffaceous outcrops using the rock quality Q-system (70% $Q > 10$, 25% $Q = 10$ to 2, 5% $Q = 2$ to 0.1) was supplemented by inclined drill holes and surface refraction seismic measurements. Figure 5.13 compares the results of core-logging (Q range $\leqslant 1$ to $\geqslant 100$) and an adjacent seismic profile, showing near-surface velocities of 4.5, 4.7 and 5.9 km/s at about 25 m depth.

Velocities of only 1.1 and 3.2 km/s were recorded in a weakness zone. Based strictly on the simple Q–$V_p$ relationship for low porosity, hard rocks at shallow depth ($V_p \approx 3.5 + \log_{10}Q$), the following wide range of Q-values can be anticipated from the above velocities:

$V_p = 5.9, Q \approx 250$ *very good quality*
$V_p = 4.7, Q \approx 16$ *good quality*
$V_p = 4.5, Q \approx 10$ *good quality*
$V_p = 3.2, Q \approx 0.5$ *very poor quality*
$V_p = 1.1, Q \approx 0.004$ *exceptionally poor quality*

These appear to be reasonable values for massive rock, good quality rock, and fault zones respectively. However, a basic problem with such comparisons is that the shallow refraction seismic is not penetrating to the same depths as the core-logging. The broad, low velocity

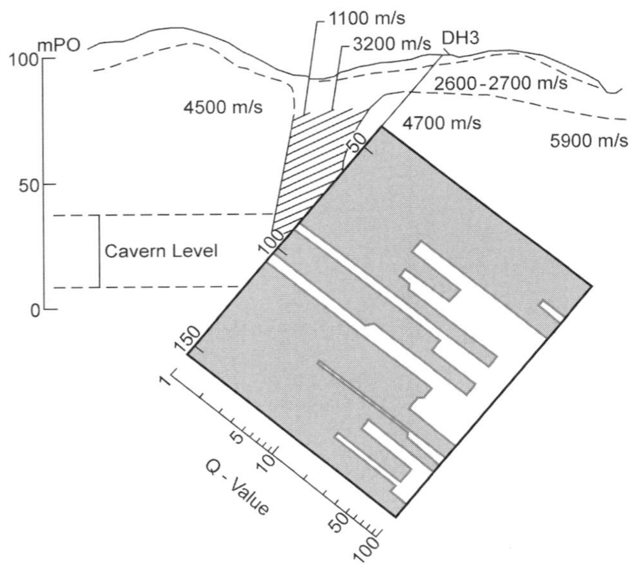


*Figure 5.13* Comparison of refraction seismic velocities and core-logged rock quality Q-values at a cavern site in Hong Kong tuffs. Chan, 1993.

zone shown in Figure 5.13 is clearly quite complex at cavern level (50 to 70 m depth) since it has rock quality Q-values of less than 1 (two locations), and Q-values of about 10, 20 and 70 in other parts of the zone. The shallow seismic picks up $V_p = 1.1$ ($Q \approx 0.004$?) and $V_p = 3.2$ ($Q \approx 0.5$?) but not the higher quality slabs (?) of rock apparently existing at greater depth within the same fault zone.

## 5.3 Effects of depth or stress on acoustic joint closure, velocities and amplitudes

Bertracchi et al., 1966, gave some early Italian experiences of cross-hole and downhole sonic logging, and noted a consistent tendency for increased velocity with depth (usually 30 to 60 m/s increase per metre) in the depth range 5 to 25 m. However, related core logging results were not given. This increase in velocity gives an extreme gradient of 30 to 60 $s^{-1}$, which is about an order of magnitude greater than in the subsequent few hundred meters.

Depth effects on seismic velocity were also reported by Cecil, 1971, from a survey of Swedish tunnels. Velocities at 50 to 60 m depth in high quality rock were up to 17% higher, while for low quality, heavily jointed rock they were up to 38% higher. By comparison Sjøgren et al., 1979, reported 5 to 15% increase at 30 to 50 m depth compared to that at the surface. In both the above cases the comparative rock qualities at the different depths are a factor of uncertainty.

There are also cases in the literature where depth effects are, apparently, absent. Bertacchi and Sampaola, 1970, conducted repeated measurements of seismic velocity at four Italian dam sites, using a combination of downhole sonic logging and cross-hole measurements. The deepest measurements (to 100 m depth) were conducted at the 160 m high Alpe Gera dam, founded in serpentine with marked foliation. Measurements were repeated over a four-year period and showed invariance with time, and independence of reservoir storage level. In view of the dependence of $V_p$ on effective stress, it is tempting to assume that the measurements were concentrated downstream of a successful grout curtain and effective drainage. Entirely different experiences are seen at other dam sites, as we shall see in a later chapter.

An illustrative set of depth-related data, which nevertheless is inconclusive, is that given by Mouraz Miranda and Mello Mendes, 1987, in Figure 5.14. The objective was to explain the reducing rates of penetration and increased specific energy used when diamond core drilling and downhole hammer drilling in a 22 metre deep profile of weathered granites. Since all the indices of quality used were increasing with depth (i.e. hardness, RQD, density), it is inevitable that a very large increase in $V_p$ (1.0 to 4.5 km/s) should have been registered.

Hudson et al., 1980 ('rock' Hudson), demonstrated some fundamental effects concerning the influence of *rock strength*, *joint spacing* and depth (or *stress*) on seismic velocity in a weak porous rock. This was due to some fortuitous circumstances at a test tunnel in chalk. Their studies were performed in an experimental machine-bored tunnel in the Lower Chalk at Chinnor in Oxfordshire, England, and also in the laboratory. All the studies were performed in chalk from above the water table, but with a natural water content of about 17 to 20% and a density of 2.2 $gm/cm^3$. Firstly, Figure 5.15a shows the influence of *intact strength* on $V_p$ values, using penetration tests at ch. 79 m in the tunnel. A similarly strong effect of joint or discontinuity spacing on $V_p$, from the same location in the tunnel is shown in Figure 5.15b.

The most interesting result was the effect of stress level (or tunnel depth) on seismic velocity. Velocity increased from typical values of 1.1–1.3 km/s in the first 30 metres of tunnel, up to 1.5–1.6 km/s between 40 and 60 m inside the tunnel, where overburden had increased to some 15–20 m. This increase in $V_p$ might appear to be as expected, but it actually occurred despite an *increase* in the frequency of joints and discontinuities in the chalk, as one progressed further into the tunnel (Figure 5.16). Often, $V_p$-depth data can be ambiguous because velocity increases occur at depth, due also to less frequent jointing. Here the two effects were, by unusually good fortune, separated.

New and West, 1980 working on the same problems, also performed loading tests on artificially fractured or sawn interfaces for various rocks, and found that for several different surface roughnesses in the case of the chalk from the Chinnor Tunnel, a stress of about 0.4 MPa was required for 'acoustic closure' (Figure 5.17). Significantly, this stress also corresponded to the overburden stress where *in situ* $V_p$ values had shown a certain flattening out. The maximum *in situ* $V_p$ values of about 1.6 km/s (influenced by a fracture frequency as high as ten per meter), may be compared with $V_p$ values for intact blocks of 1.95 km/s (range 1.6 to 2.2 km/s).

Before leaving the Chinnor Tunnel, it may be of interest that Hudson, Jones and New, 1980, also mentioned very low P-wave velocities (0.6–1.0 km/s) for badly fractured areas of the chalk, and quoted permeability values of $10^{-6}$ to $10^{-4}$ m/s, or approximately 10–1000 Lugeons. According to a rule-of-thumb ($L \approx 1/Q_c$ developed in Chapter 9), *strength-normalized* rock quality $Q_c$ -values might then be expected to range from about 0.1 to 0.001, in the absence of complications caused by clay sealing.

This range of qualities (where $Q_c = Q \times \sigma_c/100$) is broadly what might be expected from rock quality Q-logging in badly fractured areas of this weak rock, since if one assumes values of uniaxial compressive

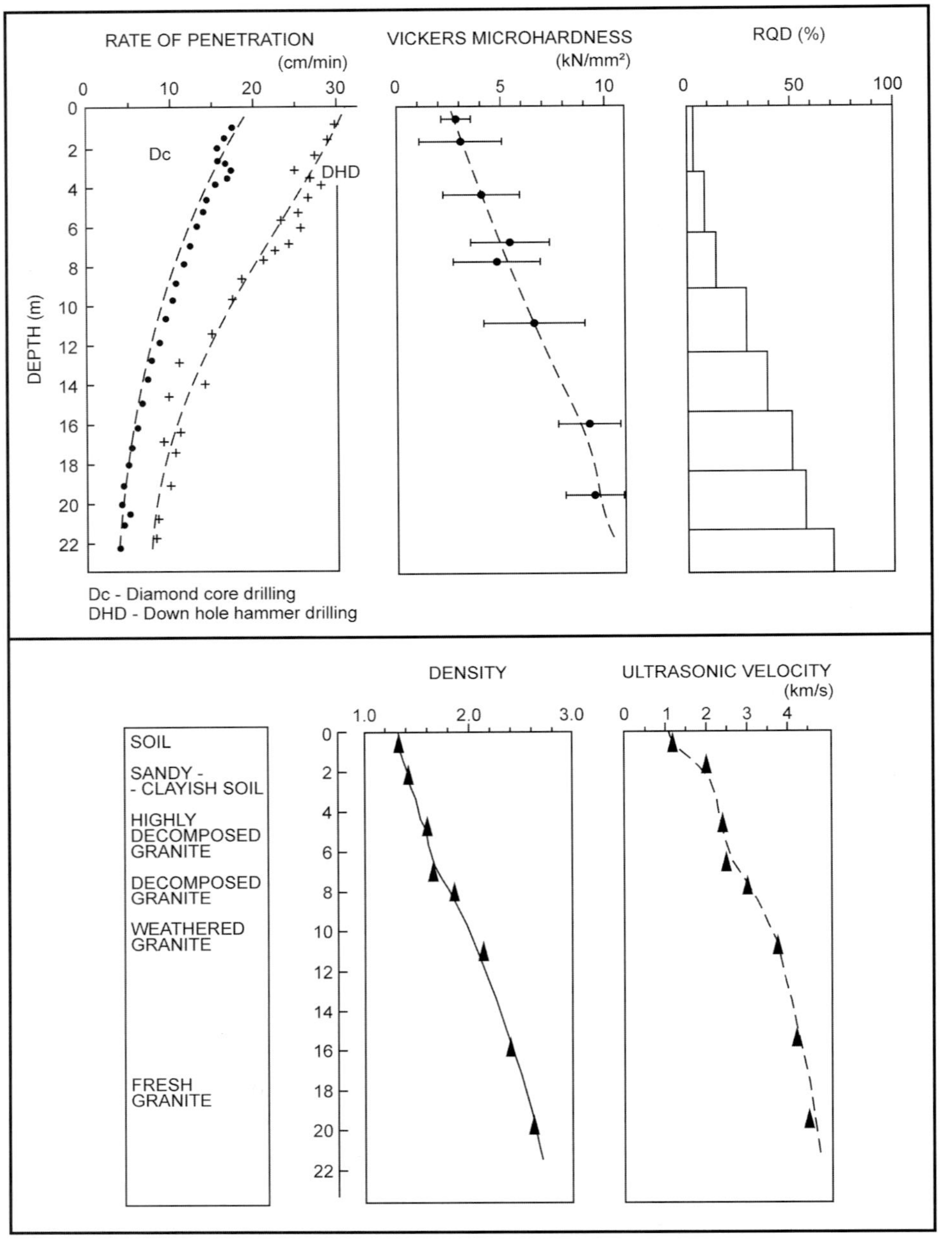


*Figure 5.14* Simultaneous increases in hardness, RQD, and density give a cumulative effect on $V_p$ (+450%) and drilling rate (−75%). Mouraz Miranda and Mello Mendes, 1987.

strength ($\sigma_c$) for the chalk from about 10 MPa down to 1 MPa (when weathered), the implied rock quality Q-values would be respectively 1 and 0.1, closely resembling 'well-jointed', and 'heavily jointed and weathered' rock, respectively.

New and West, 1980, also discussed the question of stress level (or tunnel depth) on joint closure. They questioned the applicability of the Deere et al., 1967 relation (i.e., $(V_p \text{ field}/V_p \text{ lab})^2 \approx \text{RQD}/100$: see beginning of this chapter), since they argued that stress could 'acoustically close' joints, and joint frequency as expressed by RQD would then prove to have little effect on the velocity. Their experiments on artificial flat surfaces in contact, to simulate smooth joints in various rocks, show the approach of 'acoustic closure' at about 2 to 4 (?) MPa normal stress in Figure 5.17. However, these artificial surfaces were probably not responding quite in the usual non-linear manner, in terms of stress-closure, such as

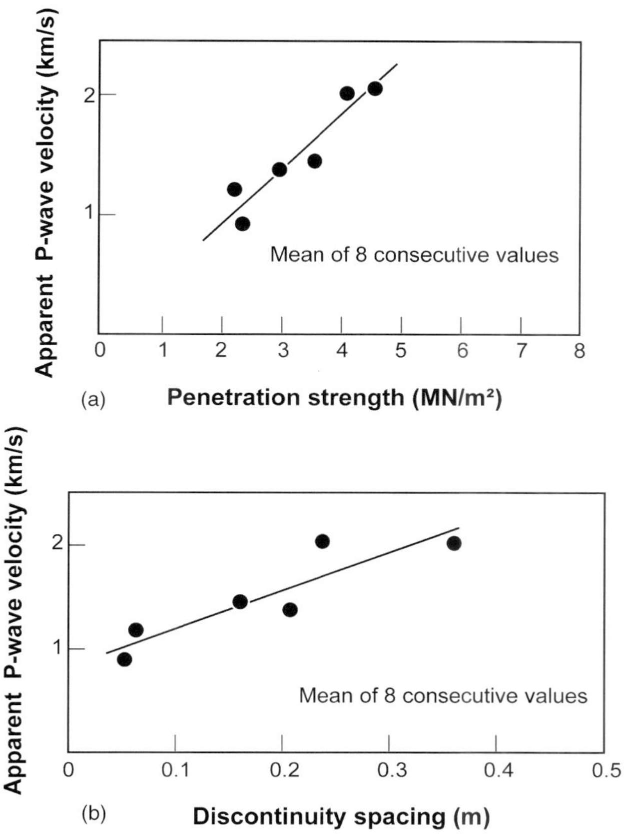


*Figure 5.15* Penetration strength and discontinuity spacing show a strong influence on the P-wave velocity for partly saturated, porous chalk from the Chinnor Tunnel. Hudson, Jones and New, 1980.

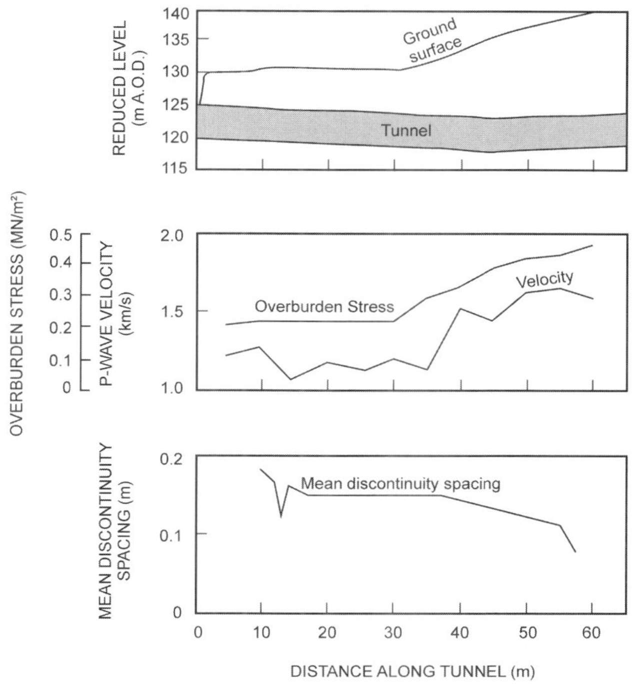


*Figure 5.16* Seismic measurements at the Chinnor Tunnel in England. Increase of overburden stress caused $V_p$ to increase (as expected), but this was despite an unusual *increase* in joint frequency with greater depth. 'Absolute proof' of a stress-related increase in $V_p$ is evident from this case, since the matrix did not change. Hudson, Jones and New, 1980.

reported by Bandis et al., 1983, and they only represented the behaviour of 'joints' that were perpendicular to the principal stress.

In Figure 5.18, New and West, 1980, show the insensitivity of $V_p$ to joint frequency changes in a sandstone, as measured along the Kielder aqueduct tunnel, in northern England. However the principal joints were reportedly vertical only (i.e. one set that had responded to the effects of 100 m overburden and previous geological history, and were perhaps tight and closed).

In a mudstone section of the same tunnel, calcite filled joints and some heavily jointed and faulted zones showed generally strong effects on velocity, as seen in Figure 5.19. In general, a rock mass with several joint sets is likely to show stress sensitivity to greater depth or stress level than the authors are implying, and if the rock is very weak and porous, volume changes will occur at greater stress giving velocity increases. In a rock like chalk marl, joints may close easily with moderate stress, as shown by the Chinnor experience, but large stress increases will inevitably 'reactivate' sensitivity to stress as compaction occurs.

'Acoustic closure' studies by Westerman et al., 1982, were based on artificially sawn, ground and then acid-etched surfaces of a corallian limestone. Perhaps because of the partial 'non-mating' of the etched depressions in each surface they found that normal stress levels as high as 20 MPa were required to reach the seismic velocity (5.6 km/s) of the intact rock. The parabolic-type increase in velocity from 3.9 km/s was quite smooth, and corresponded to a reducing attenuation, the latter stabilising beyond normal stresses of about 10 MPa. The samples were dry as far as can be understood from the results obtained.

The ratio of stress to strength, for example the ratio JCS/$\sigma_n$ used in the shear strength criterion of Barton and Choubey, 1977 may be useful for explaining so-called 'acoustic closure', since it is not only the stress level, or depth, but also the rock joint stiffness or strength that determines the contact area needed for the less attenuating

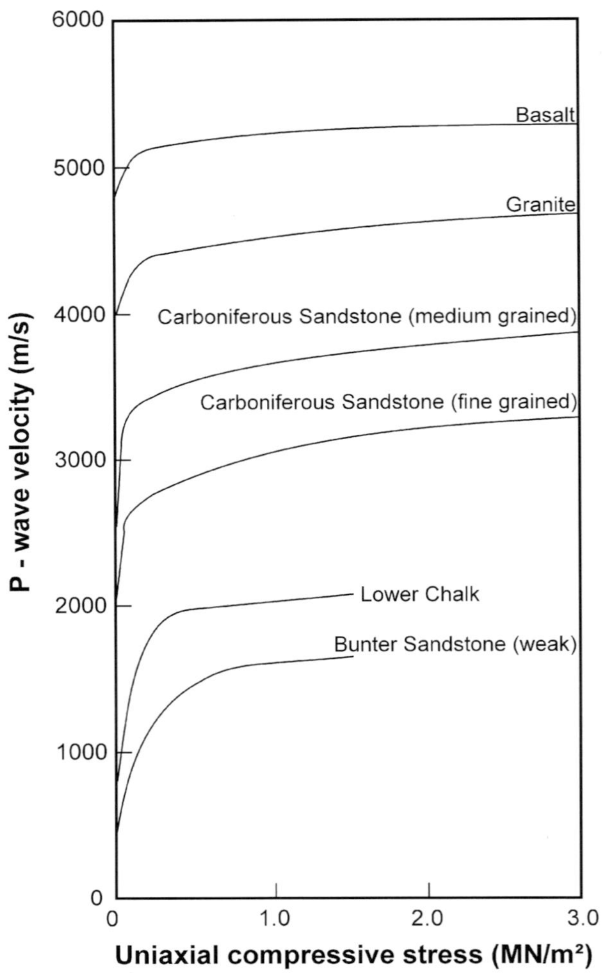


*Figure 5.17* $V_p$ increases for flat, dry artificial rock surfaces under normal stress. New and West, 1980.

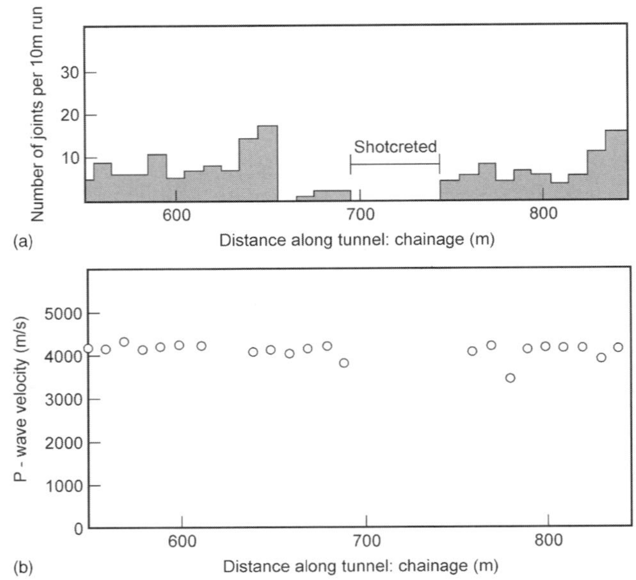


*Figure 5.18* Insensitivity of $V_p$ to vertical joint frequency in sandstone at the Kielder Tunnel, with 100 m overburden. New and West, 1980.

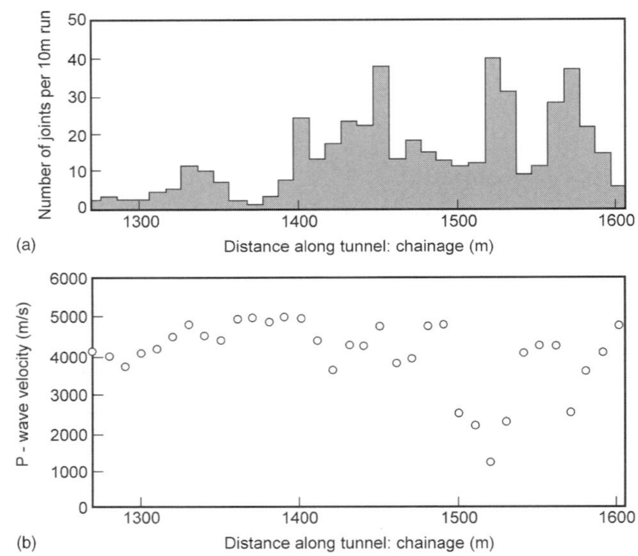


*Figure 5.19* Sensitivity of $V_p$ to frequencies of calcite-filled joints in mudstone section of Kielder Tunnel. New and West, 1980.

seismic transmission across joints. The ratio of JCS (joint wall compression strength and $\sigma_n$ (effective normal stress) is closely related – or perhaps nearly identical – to the ratio of $A_1$ (assumed contact area)/$A_0$ (actual contact area). The ratio $A_0/A_1$ is very small for the case of hard unweathered rock joints near the surface, which continue to show velocity increase for many hundreds of metres, while $A_0/A_1$ is much larger for the case of a soft porous rock mass like chalk marl, which can show 'acoustic closure' at stresses as low as 0.4 MPa according to Hudson, Jones and New, 1980.

Tanimoto and Ikeda, 1983, found that $V_p$ was approximately proportional to the normal stress applied to simulated joints over the range 3 to 20 MPa, but that $V_p$ dropped rapidly below stress levels of 3 MPa. A cut-off aperture of about 40 μm separated their experimental results, with apparently no influence of $Fm^{-1}$ on $V_p$ below this aperture. The above effects and new experimental data are the main reason why Tanimoto and Kishida, 1994, and others are advocating the use of compression wave amplitude as a supplement to velocity data, for better sensitivity to jointing, particularly for the higher stress levels than those usually employed in shallow refraction seismic.

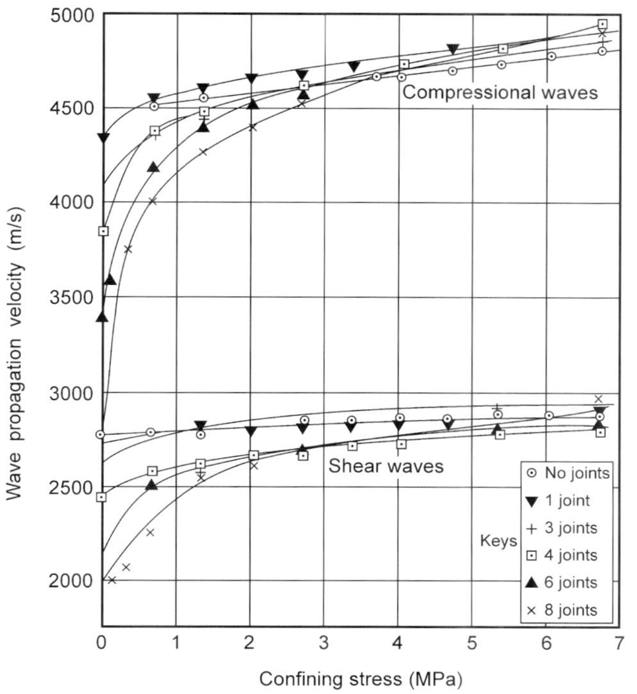


*Figure 5.20* Tests on artificially fractured columns of marble, showing $V_p$ and $V_s$ as a function of normal stress, in the dry state. Stacey, 1977.

Laboratory tests using artificially jointed columns of marble that could be axially loaded and submerged in water were described by Stacey, 1977. These very interesting and instructive tests showed that the commonly used P-wave velocity was not the most sensitive parameter as regards rock quality. 'Rock quality' in this set of experiments was simulated by having 0, 1, 2 or up to 8 tension fractures distributed along the 0.46 m length of marble under dry or wet conditions, or with clay in the fractures. The basic P-wave and S-wave velocity responses to different levels of normal loading (0 to 7 MPa) in the dry state are shown in Figure 5.20.

The number of fractures corresponds to F values ($m^{-1}$) of about 2, 6, 9, 13 and 17 per metre. The major increases in *P-wave velocity* (e.g., 2.8 to 4.4 km/s) occur in the first 2 MPa of normal loading, thereafter less rapid increases are seen, but the rise is consistent and nearly constant up to the maximum applied stress of 7 MPa. The number of fractures had much less influence on velocities at stresses above 2 to 3 MPa, in the dry state. Under zero stress levels the number of fractures had the greatest influence. When, by contrast, the tension fractures were wet and clay-coated, their number reportedly had 'negligible' effect *on P-wave velocity*.

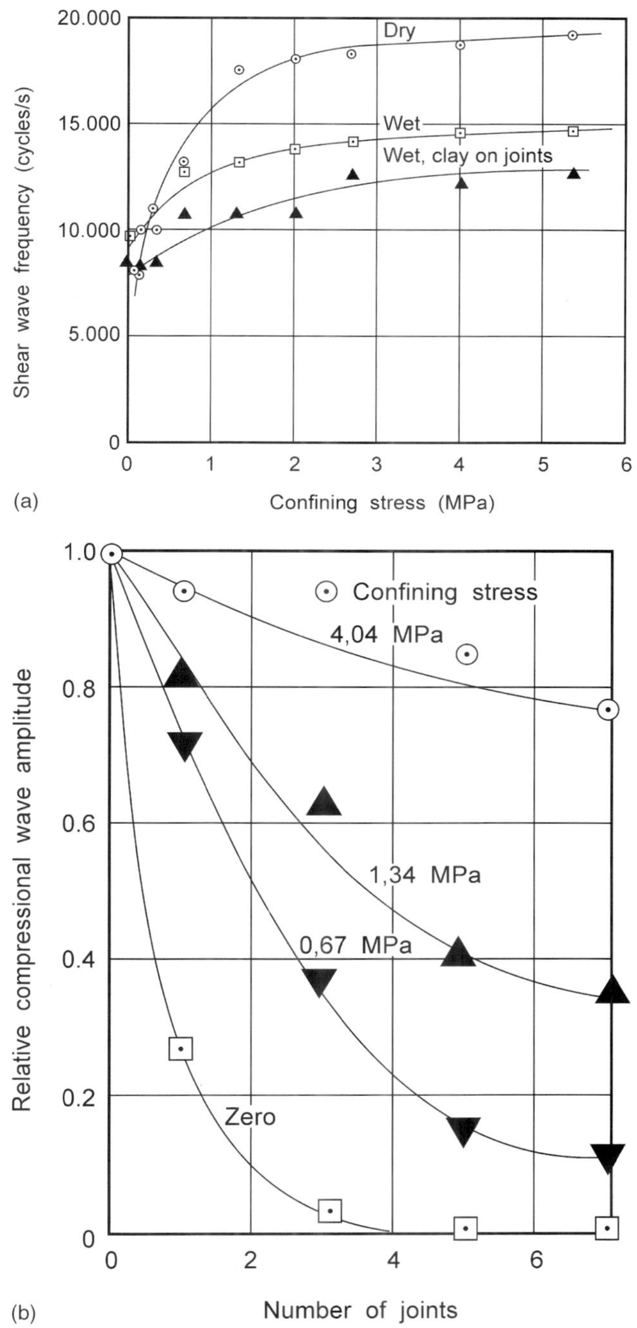


*Figure 5.21* Tests on artificially fractured columns of marble. a) S-wave frequency. b) relative P-wave amplitude. Stacey, 1977.

Stacey, 1977, went on to demonstrate that *shear wave frequency* was also very sensitive to the degree of joint closure caused by stress. Frequency increases from about 8 kHz to about 19 kHz were indicated for the case of well compressed, dry tension fractures, with less sensitivity in the case of wet conditions, or with wet clay fillings. These results are shown in Figure 5.21a.

(There appears to be a possible link here to shear wave splitting, polarization and anisotropy interpretation, for the case of fractured reservoirs with gas or liquid saturation – see Chapters 13 and 15.)

Furthermore, the *compressional wave amplitude* showed great sensitivity to both stress level and the frequency of the tension fractures, as demonstrated in Figure 5.21b. Compressional wave amplitude is of course an indicator of the level of *attenuation*, which was shown to be maximum in the case of *unloaded, multiple-fracture* models, and minimum for the case of *well confined, single fractures*. Attenuation is treated in detail in Part II, Chapter 10.

A related though more comprehensive set of experiments was reported by Tanimoto and Ikeda, 1983, using numerous 50 mm diameter cylinders of various rocks (rhyolite, sandstone, granite, granodiorite and tuff). The rock cylinders had five different lengths, making it possible to represent a 'line sample' through and imagined rock mass, with 0, 10, 20, 30 or 40 joints per metre. Filter paper, dry or saturated, was used to produce the desired 'joint apertures' and moisture contents. Figure 5.22 shows a schematic of the various 'line samples' created by cylinders of different length. In the case of the natural joint sample, both normal and shear load was applied.

In Figure 5.23, the authors showed that the ratio $k = V_p$ (jointed)/$V_p$ (intact) was hardly influenced by the frequency or number of the types of 'joints'. However, they qualified this by indicating that it was their apertures of *less than 0.04 mm* (or 40 μm) that caused the frequency of fractures to have little influence on the *P-wave* velocity. In the context of numerous seismic surveys underground that they had included in their evaluations, they suggested that apertures were considerably wider than 40 μm, therefore allowing fracture or joint frequency ($Fm^{-1}$) to influence the velocity ratio $V_J/V_i$ as follows:

$$V_{jointed} = V_{intact}\left(\frac{5}{F+4}\right)^{1/2} \tag{5.5}$$

No reduction in P-wave velocity is predicted with 1 joint per meter, but with 5, 10 or 20 per meter, the ratio k is predicted to show successive reductions to 0.75, 0.6 and 0.5 (to the nearest decimal places). So a hard crystalline rock with $V_{intact}$ of 5.5 km/s, might show reductions to 4,100, 3,300 and 2,500 km/s respectively. This appears inherently representative of experiences in fractured zones, at least in the near-surface.

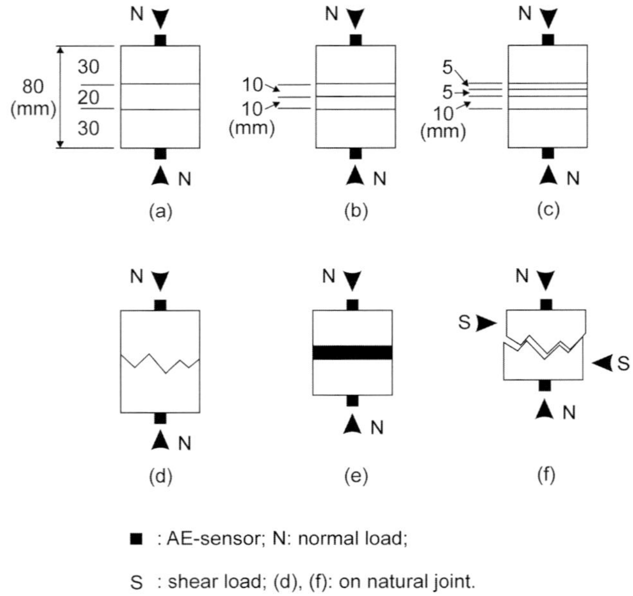


*Figure 5.22* Schematic of the multi-cylinder 'line samples', individual 'filled-joints' and a natural joint, that were studied by Tanimoto and Ikeda, 1983.

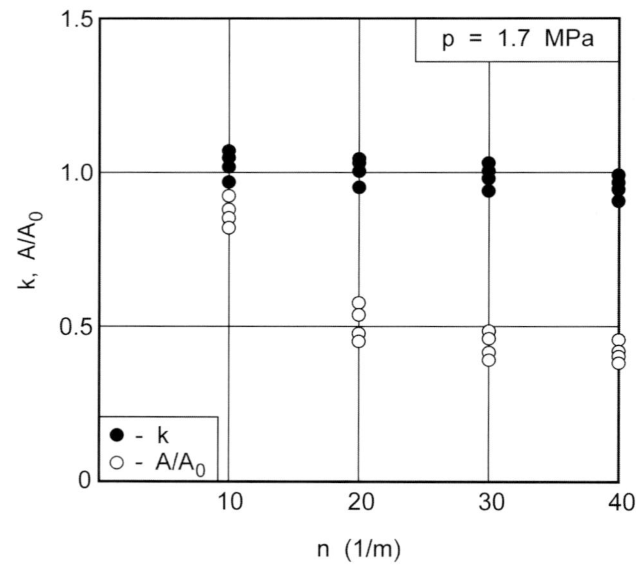


*Figure 5.23* Joint frequency (n), velocity ratio (k), and amplitude ratio $A/A_0$, indicating little influence of 'joint' frequency (or $Fm^{-1}$) on the P-wave velocities, but a strong influence on the amplitude ratio. Tanimoto and Ikeda, 1983.

One may perhaps conclude that heavily jointed rock masses have a degree of acoustic coupling that is not as complete as in the (aperture <40 μm) laboratory experiments, due to variously oriented joints and lower near-surface stresses, various rock-to-rock contact areas, and various ranges of weathering, mineral coatings and fillings. The contact area ratio $A_0/A_1$ referred to earlier in this chapter will obviously change completely in the presence of clay filling under high compressive stresses.

As in the case of Stacey, 1977, Tanimoto and Ikeda, 1983 found that *compressive wave amplitude* gave a very sensitive measurement of the number of 'joints'. The amplitude ratio $A/A_0$ ($A_0$ for intact rock) is shown in Figure 5.23 (open circles), and shows excellent sensitivity to number of 'joints'. These authors also found that *apertures less than 0.01 mm (10 μm) did not have influence on the wave propagation*, even when the normal stress was as low as 1–2 MPa. Physical apertures (E) of this size (in contrast to hydraulic apertures (e) which are smaller due to roughness effects), are probably rare in the upper 20–30 metres of rock masses where refraction seismic is carried out, so this result is probably consistent with experience in the field.

(One may speculate whether such a finding could have application in the interpretation of so-called 'open' fractures in petroleum reservoirs: would these need to be of greater aperture than this order-of-magnitude, before they could cause shear wave splitting?)

A combination of permeability testing and index testing of the relevant joints (i.e. roughness JRC, and wall strength JCS) using the methods described by Barton et al., 1985, for converting between hydraulic and physical joint apertures would indicate the rough order of magnitude of the 'necessary' hydraulic apertures to satisfy this possible 'rule-of-thumb' that E must be $\geqslant 10\ \mu m$, for having influence on wave propagation.

A look ahead to Chapters 15 and 16 where joint properties are discussed in detail, would suggest that hydraulic apertures of about 10, 1.8 and 0.3 μm might be operating with $E = 10\ \mu m$, if *joint roughnesses* were respectively 2.5 (quite smooth and nearly planar), 5 (near-planar but some small undulations) and 10 (non-planar with marked inclined asperities). These latter would hardly be considered as 'open' joints, and in a reservoir situation are perhaps (as suggested by the Tanimoto and Ikeda, 1983 results) not capable of influencing seismic waves, nor, by implication, shear wave splitting.

Tanimoto and Ikeda, 1983, also investigated the effect of larger apertures (or thicknesses of discontinuity fillings) using more sheets of filter paper to give a range of 1 to 3.4 mm thickness. There was strong sensitivity of $V_p$ to stress level (0.3 to 3 MPa) and to the total cumulative joint aperture and moisture condition.

Fratta and Santamarina, 2002, also used columns of blocks under stress to demonstrate velocity-stress sensitivity. They studied the effects of varying thicknesses of kaolinite gouge materials, finding that *shear* wave velocity gave a very sensitive indication of the effect of gouge thickness at even lower stress levels, equivalent to about 1 to 10 m, typical of the weathered zone. For example at 0.25 MPa normal stress, equivalent to about 10 m depth, the conditions a) no gouge, b) 0.5–1 mm of gouge, c) 2.0 mm of gouge, d) 2.5 mm of gouge, gave S-wave velocities of about 850 m/s, 750 m/s, 600 m/s and 450 m/s respectively. The strongly non-linear (convex) $V_s$ – normal stress curves, showed velocities ranging, respectively, as low as 450 to 250 m/s at an equivalent depth of only 1 meter.

In hard rocks, such as the numerous cases reported by Sjøgren et al., 1979 and Sjøgren, 1984, there is a significant in situ correlation between $V_p$ and joint frequency (and RQD), due presumably to the fact that this 'cut-off aperture' (whatever it may be in different rocks) has not been reached at the moderate (20–30 metres) penetration of shallow refraction seismic surveys. Tanimoto and Ikeda, 1983, found that $V_p$ was proportional to $\sigma_n$ in the stress range 3–20 MPa, but dropped sharply for $\sigma_n < 3$ MPa. By chance, or similar physics, in the field case records used to design the empirical $V_p$–Q–M chart to be shown later (Barton, 1995), $V_p$ and depth are also found to be linearly related from about 200 m to 1000 m depth (5 to 25 MPa), with $V_p$ falling rapidly for depths in the range 25 m to 100 m, i.e., for $\sigma_n$ or $\sigma_v$ (or $\sigma_h$) $<2.5$ MPa.

### 5.3.1 Compression wave amplitude sensitivities to jointing

One of the most thorough studies of the effect of joint parameters on seismic signatures was reported in a subsequent study by Tanimoto and Kishida, 1994 and Kishida, 1999, which was built on these earlier investigations of Tanimoto and Ikeda, 1983. The advantages of compressive wave amplitude compared to $V_p$ for sensing joint frequencies at the higher stress levels was emphasised again, and convincingly demonstrated experimentally. The authors also compared (conventional) seismic velocity tomography (SVT) with seismic amplitude tomography (SAT). The latter reportedly corresponded more closely to jointing observed with a borehole scanner, as compared to the more frequently used (SVT). The studies were made in boreholes at a dam site, using an exploratory adit for further confirmation.

The extensive laboratory studies reported by Tanimoto and Kishida, 1994, were based on cylindrical rock specimens, with a total of 86 cylinders representing sandstone, rhyolite, shale, granite, granite porphyry and slate. In the case of 'joint samples', natural joints were moulded and replicated with hard mortar or plaster, to reproduce the same roughness as in nature. Since the authors were very concerned about the correct measurement of roughness they utilised a non-contact laser scanner micrometer mounted on a 3D, movable stage roughness profiler.

They show in Figure 5.24 a relationship between the joint roughness coefficient JRC (from Barton and Choubey, 1977) and '$M_s$'. The latter is derived from a power spectrum versus frequency relation analysed by the so-called 'maximum entropy method', which the authors selected in preference to the 'fast Fourier transform' method conventionally used. The inter-relationship between '$M_s$' and JRC shown in Figure 5.24 is needed to interpret the effect of joint roughness variations used in their seismic velocity and seismic amplitude ($A/A_0$ results shown collectively in Figure 5.25 and 5.26. Each horizontal pair of figures compares the sensitivity of $V_p$ and amplitude ratio $A/A_0$ to the following:

1. 'joint' frequency (25, 37.5 or 50 $m^{-1}$) and normal stress level (0 to 3 MPa)
2. 'joint' aperture (0, 1.05 or 2.10 mm) and normal stress level (0 to 3 MPa)
3. joint roughness (see $M_s$–JRC relation) and normal stress level (0 to 3 MPa) using joint replicas
4. shear displacement of joint replicas (0 to 7 mm) for three normal stresses (0.2 to 0.6 MPa)
5. dilation (0 to 1 mm) during joint shear at three normal stresses (0.2 to 0.6 MPa) for JRC $\approx$ 13
6. aperture change during normal loading of joint (JRC $\approx$ 8 to $\approx$ 13).

These two figures demonstrate why *shallow* seismic refraction measurements which operate at low stress levels *are successful* in distinguishing joint frequency and aperture (achieved here with moist filter paper), but why *amplitude measurements* appear to be *much more desirable* than velocity measurements at higher stress levels, if joint frequencies and character are still to be distinguished. This emphasises the value of attenuation measurement, or of seismic Q. It reflects richly of the mechanisms involved with seismic detection of joints.

Better 'joint' inter-locking from increased stress, or an absence of dilation and shear, gave increased amplitude as expected, and the opposite occurred (loss of energy), in the case of shearing or joint dilation 'post peak'.

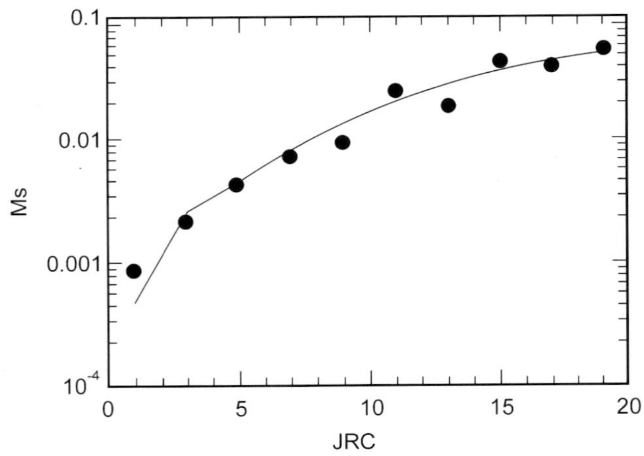


*Figure 5.24* Correlation of the parameter $M_s$ used by the authors, with JRC for the natural joints. Tanimoto and Kishida, 1994.

These comprehensive results leave one in no doubt about the potential advantages of amplitude measurement compared to velocity measurement. Even such details as joint roughness, which obviously also relates to acoustic coupling and contact area, demonstrates that amplitude measurement has clear sensitivity to the mechanics involved in the joint plane. The ratio $A/A_0$ tends to get smaller (increased attenuation) as roughness increases, which fits with the picture of joint closure difficulties when joints are rough. (See extensive treatment of rock joint behaviour in Chapter 16).

The above authors would probably be the first to agree that further studies with real joints would be an advantage, and of course that further field studies using velocity and amplitude tomography, together with comprehensive joint surveys, would be necessary for refining the interpretation of these useful techniques. Since SAT and SVT methods reportedly show good correspondence when filling materials and wider apertures are found (i.e., near surface), an understanding of the inter-relationships can perhaps best be obtained in this low stress, weathered, or partly weathered zone, since wide ranges of both $V_p$ and ($A/A_0$) are seen, and the rock mass quality Q-value also varies strongly in this region.

Some useful indicators about rock joint closure mechanics can also be obtained from a study reported by Nihei and Cook, 1992. They utilised a combination of acoustic emission (AE) monitoring and P- and S-wave monitoring of artificial tension fractures in sandstone.

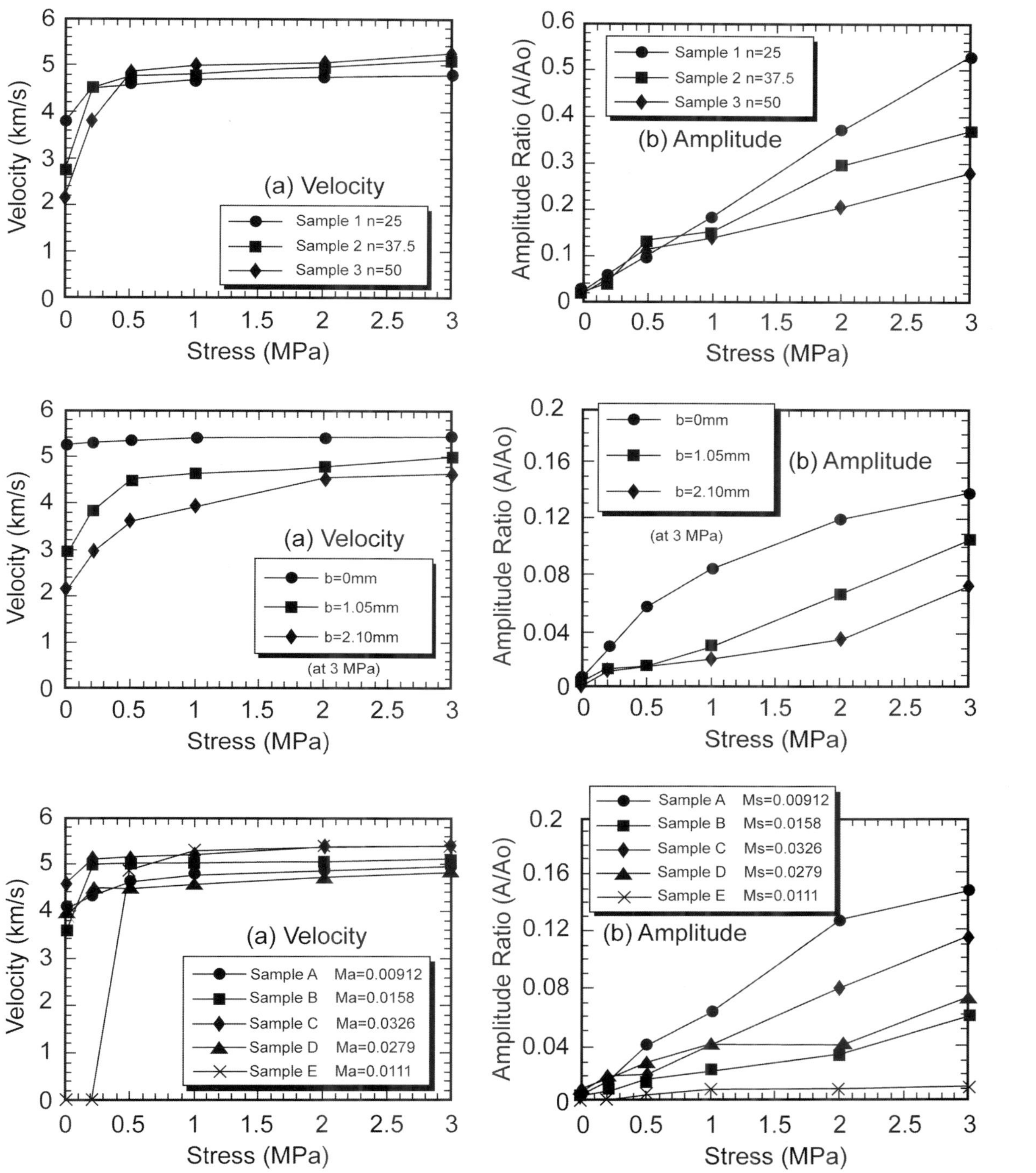


*Figure 5.25* Parallel comparison (see a and b pairs) of $V_p$-monitoring, and amplitude ratio monitoring, of *loading effects* (0 to 3 MPa) with: *top*) varied 'joint' frequency, *centre*) varied 'joint' filling thickness, *bottom*) varied joint roughness for the natural joints. Tanimoto and Kishida, 1994.

These rough fractures were loaded to 24 MPa in *mated* and *unmated* conditions. As a result they produced widely different levels of acoustic emission (Figure 5.27), especially on the first cycle of loading.

The velocity-stress behaviour shown in Figure 5.28 indicated a consistent pattern of behaviour, with increased contact area at higher stress causing marked increases in both the P-wave and S-wave velocities. The

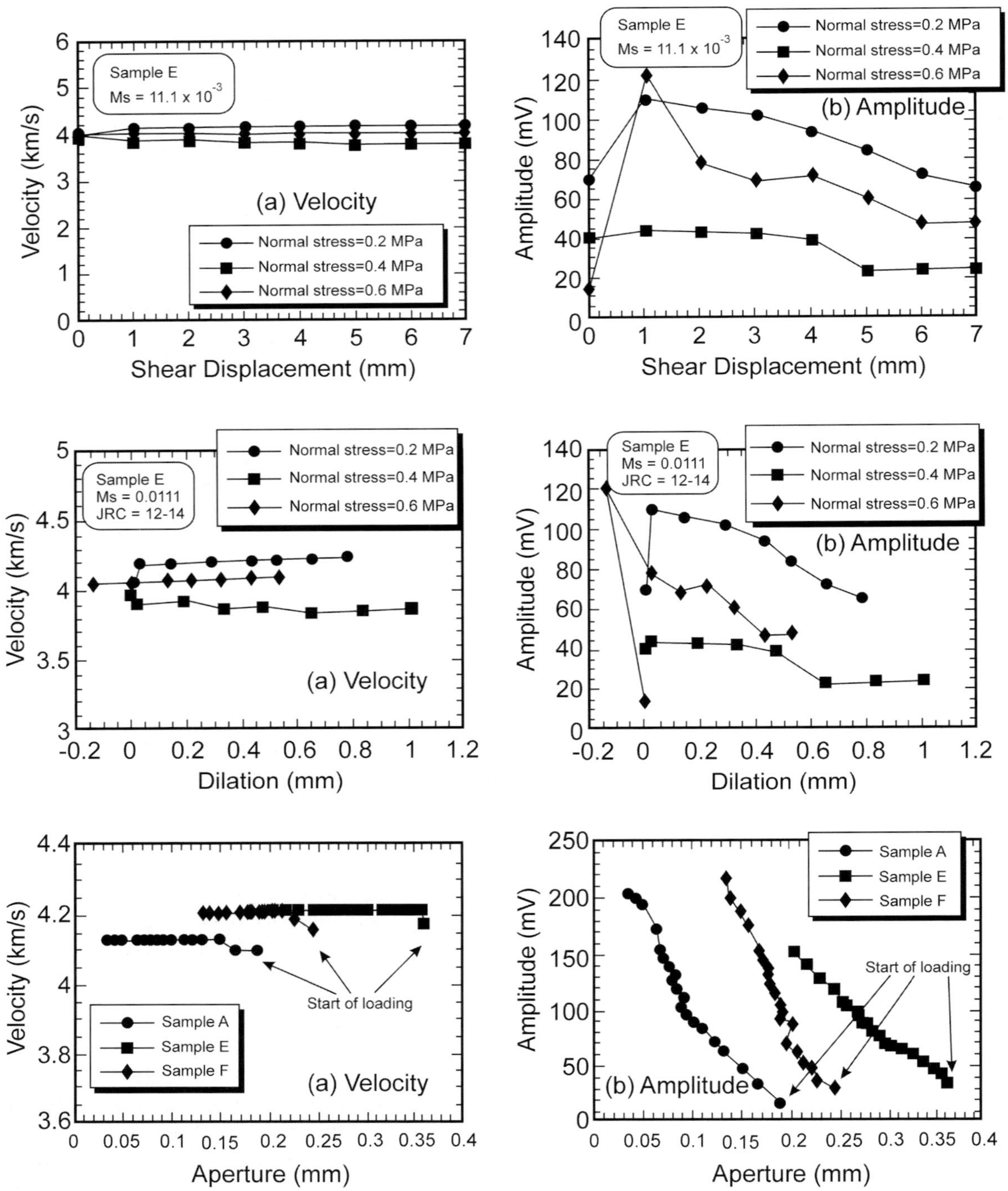


*Figure 5.26* Parallel comparison (see a and b pairs) of $V_p$-monitoring, and amplitude ratio monitoring of a) *shear displacement* along a natural joint, b) *dilation* during shear, c) *aperture* changes with three joints of different roughness. Tanimoto and Kishida, 1994.

mated fracture more closely approached the intact sample in terms of velocity, showing rapid increases in $V_p$ from about 2.8 km/s (at 2.5 MPa) to 3.4 km/s (at 10 MPa), and thereafter a slower increase of velocity. The unmated fracture showed nearly parallel behaviour, but at velocities of about 0.2 km/s slower, both for P-waves and S-waves.

It is apparent that rough-walled rock joints with their typically high JRC values (often 15 to 20) are more difficult to close acoustically, so rough joints in

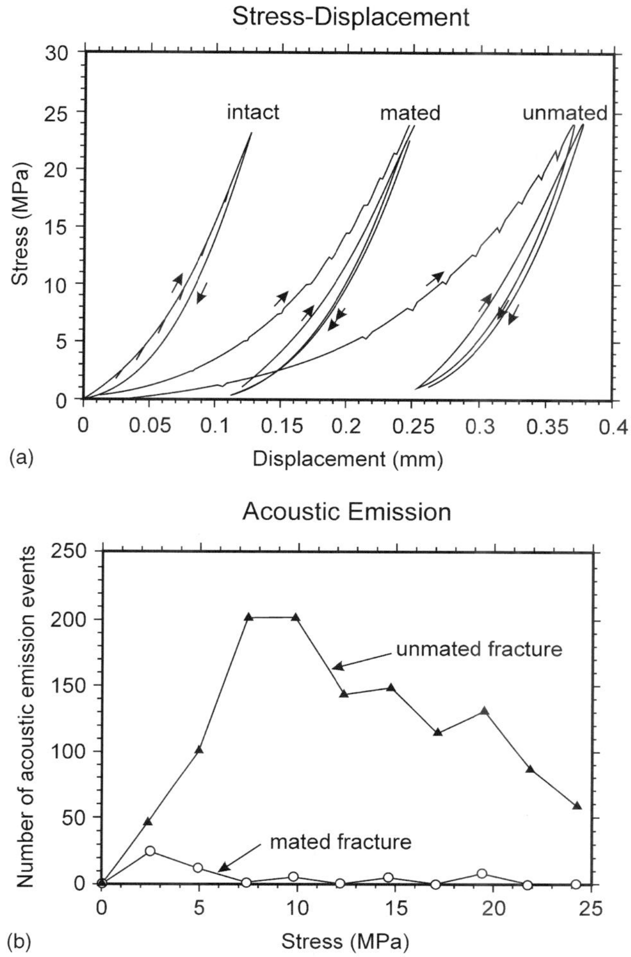


*Figure 5.27* a) Stress-deformation cycles comparing intact, mated, and unmated tension fractures. b) The number of acoustic emission events associated with loading the mated and unmated fractures to 24 MPa. Nihei and Cook, 1992.

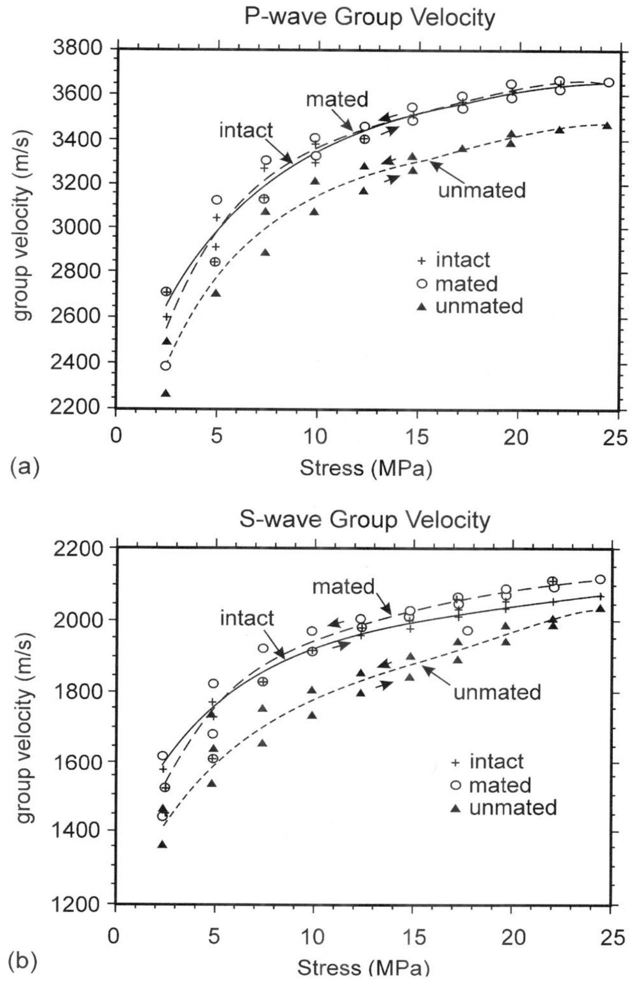


*Figure 5.28* A comparison of P-wave and S-wave velocities when loading the intact, mated and unmated samples from about 2.5 to 25 MPa. Strong stress sensitivity is shown, due particularly to the dry state. Nihei and Cook, 1992.

hard rocks could be seismically 'visible' to significant depths (e.g., 10 MPa or more, corresponding to several hundreds of metres overburden). There are however other factors involved, such as mechanical *over-closure, and thermal over-closure*, (Barton, 2004), which would give tighter apertures than 'expected' from the present depth of burial or exposure. (See Chapter 16).

The above closure aspects would contrast with the evidence from some of the experimental work on flat surfaces in weak rock reviewed earlier, for example that of New and West, 1980, which would suggest much lower levels of stress sensitivity in the case of smoother rock joints, except when closer to the surface.

In massive granite at the Underground Research Laboratory (URL) in Manitoba, Canada, micro-seismic EDZ and stress-induced failure sensing reported by Talebi and Young, 1992, showed P-wave velocities ranging from 5.6 to 5.9 km/s, and S-wave velocities ranging from 3.3 to 3.4 km/s for the depth range 310 to 400 metres (approximately) down the 4.6 m diameter shaft. (The ratio of $V_{p(mean)}/V_{s(mean)}$ was exactly 1.70 in this massive granite.)

Velocities increased by about 0.1 km/s for every 30 m increase in depth, (gradient = $3.3\,s^{-1}$), based on the 1 m deep measurements using numerous shallow boreholes drilled into the walls of the shaft. If stress-induced fracturing had been involved, a linkage between increased depth and increased velocity would be less certain, as dilation and reduced velocity might equally well occur, with such shallow holes. (Core recovered from different depths at URL showed reduced compressive strength and modulus with increased sampling depth, due to increases in stress-induced microcracking.)

### 5.3.2 Stress and velocity coupling at the Gjøvik Cavern site

The foregoing review of depth and stress effects on P and S-wave velocities measured across loaded joint assemblies, or at tunnel sites in natural settings, can be concluded by a brief review of the stress effects on velocities that were documented at the Gjøvik Olympic cavern site in Norway. Figure 5.29 is a reminder of some of the cross-hole seismic tomography discussed in Chapter 4.

Figure 5.30a shows the stress measurement results at the cavern site, based on hydraulic fracturing and hydraulic jacking of existing joints, performed by Tunbridge, of NGI. Both $\sigma_{h\ min}$ and the estimated $\sigma_{H\ max}$ show rather high values, many times greater than the calculated vertical stress gradient. The jointed gneiss at the site had steeply dipping conjugate jointing with up to four joint sets, but fortunately these were joints with significant roughness and undulation.

Based on the evidence shown in the velocity tomograms (Figure 5.29 and other tomograms from Chapter 4), a maximum velocity rise of some 2 km/s (3.5 to 5.5 km/s) occurred in the depth range 10 to 60 metres, over which range, horizontal stresses may have increased by 2 MPa to 4 MPa, depending on direction relative to $\sigma_{H\ max}$ and $\sigma_{h\ min}$ (Figure 5.30a).

The above rise of velocity occurred in a rock mass with a typical Q-value of about 8 to 10, with a joint frequency that ranges from about 3 to 10 per metre, and an RQD that ranges from about 60 to 100%. There is no evidence from the four core logs of consistently increasing rock quality with depth, so the velocity increase of nearly 2 km/s occurred mostly as a presumed result of the increase in horizontal stress. The gneissic rock mass at the site was generally saturated to within a few metres of the surface, and Lugeon-type permeability tests in the four investigation boreholes indicated values in the range 1 to 0.01 Lugeons, mostly 0.1. An inversion of the median Lugeon value is close to the mean Q-value, the significance of which will be clarified in Chapter 9.

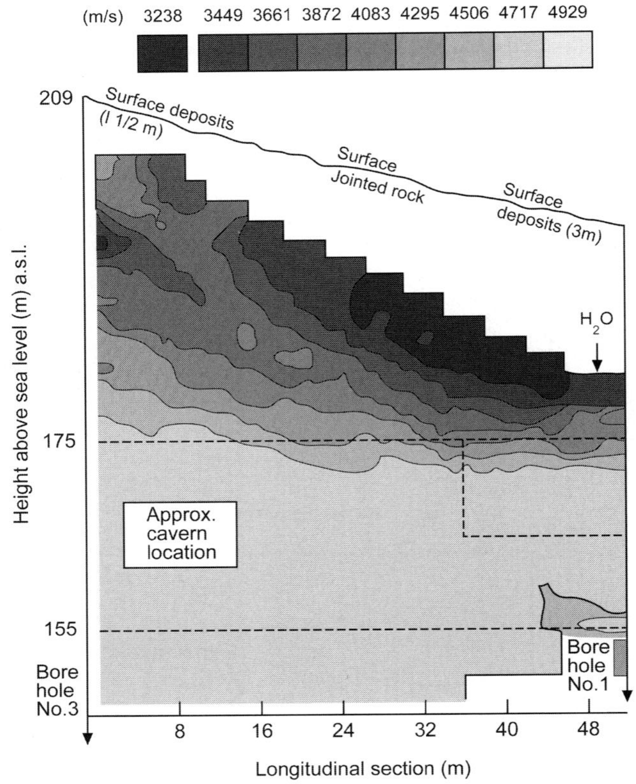


*Figure 5.29* Cross-hole seismic tomography at the Gjøvik cavern site, showing strong increases in velocity with depth, actually caused most by high horizontal stress levels (3 to 6 MPa) in the upper 50 meters. There were velocity-depth gradients as high as $80\,s^{-1}$ in the upper 20 m, and as high as $40\,s^{-1}$ over the first 50 m. Remarkably, the rock quality parameters: RQD, $Fm^{-1}$ and Q, did not improve beyond about 5 m depth. Barton et al., 1994.

## 5.4 Observations of effective stress effects on velocities

Dam construction represents a significant local source of rock mass loading, in which the total vertical stress initially is increased without necessarily changing the pore pressure, since the reservoir takes time to fill. When grouting of the foundations is very thorough, this assumption is of course suspect and a more complicated picture may arise. When reservoir impounding begins there are likely to be significant reductions in effective stress near the dam, and fluctuations of the latter, usually on a seasonal basis.

Savich et al., 1983 noted the above effects when the 270 m high Inguri arch dam was being constructed. A

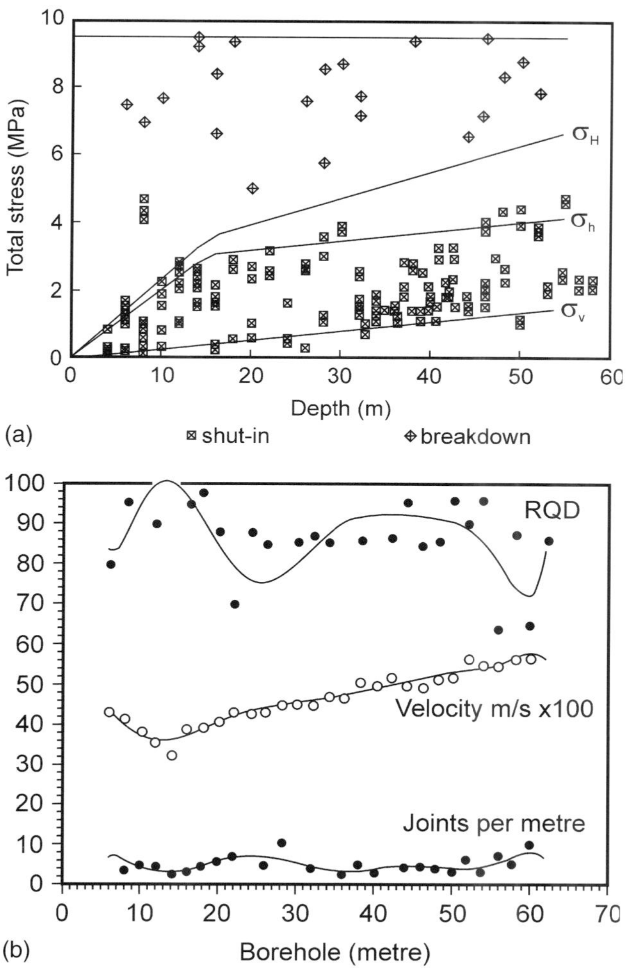


*Figure 5.30* a) Hydraulic fracturing based stress estimation at Gjøvik. b) Velocity depth trend next to one of the boreholes where seismic tomography was performed. RQD and $Fm^{-1}$ (and also Q-value) did not show improvement with depth. Kristiansen, 1991 pers. comm., Barton et al., 1994.

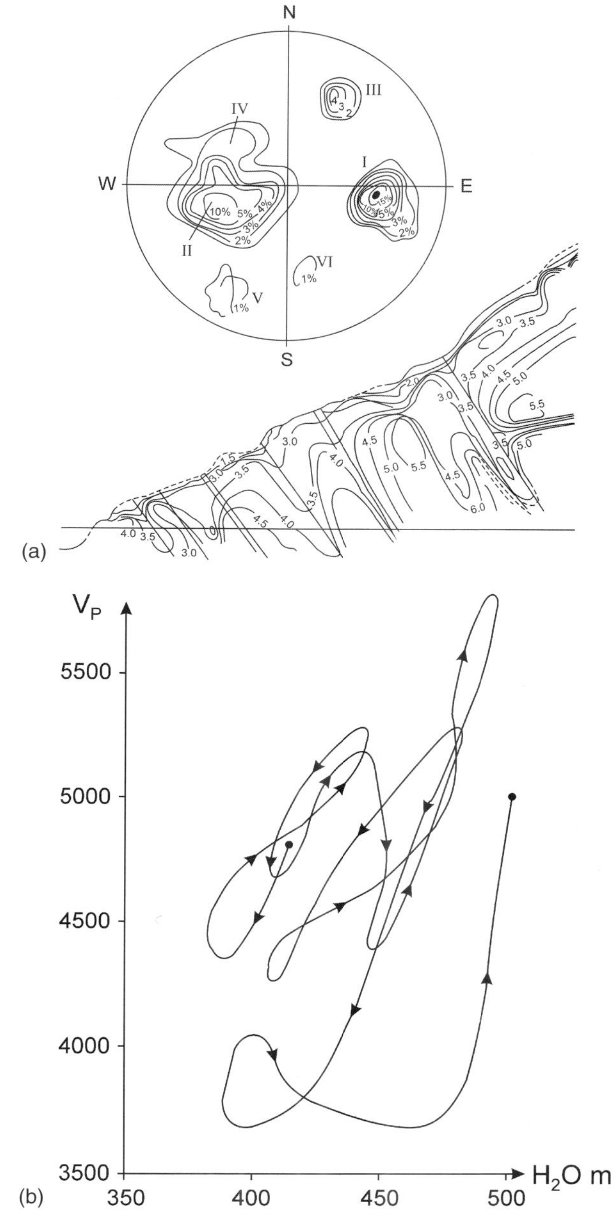


*Figure 5.31* $V_p$ distribution in Inguri dam abutment, and records of $V_p$ changes with construction, reservoir filling and seasonal fluctuations. Savich et al., 1987.

sharp change in deformation was noted when impounding of water began. Significantly, $V_p$ values reduced ($V_s$ values also reduced), and there was a reduction in resistivity, each implying reduced effective stresses due to the rise of joint water pressures in spite of the grouting and drainage measures. The most intensive reductions in $V_p$ occurred during periods of water level rise, with weaker reductions, when the water level was constant. Interestingly, the 'conventional' effective stress type linkage with $V_p$ was modified, and radically changed in subsequent long term monitoring.

Savitch et al., 1987 gave updated and longer term syntheses of the Inguri dam response, following another four years of performance monitoring, since their last referred article (Savitch et al., 1983). Figure 5.31a shows the deeply penetrating lower velocity contours in the left abutment and a joint pole concentration plot. Figure 5.31b shows the complex coupling of $V_p$ and reservoir level response, with predominant gradients showing $V_p$ *increasing* with reservoir level (H).

A nine-year record of $V_p$, percentage change of $V_p$, dam load and reservoir level fluctuation shown in Figure 5.32b confirms the general, but slightly delayed increase in $V_p$ with reservoir level, following the initial reduction in $V_p$ which appeared to be effective stress related. A more detailed look at $V_p$–H response, with smoothed-out behavioural trends (Figure 5.32b) shows a small delay.

Savitch et al., 1987 interpret the near-synchronous behaviour (after first reservoir filling) as closure of joints or cracks due to the reservoir load on the dam. However, at higher pressures perhaps a widening of large cracks occurs. Reservoir draw-down rate appears to be important since if the draw-down rate exceeds the 'permeability velocity', a sharp decrease in $V_p$ occurs, corresponding to the classic effective stress model.

Before leaving this chapter on effective stress-depth effects, we will review an interesting set of cases presented by Moos and Zoback, 1983. Seismic velocity measurements in four wells to roughly 1 km depth, can serve as a bridge to the much deeper seismic velocity-depth profiles that are given in Chapter 11.

Four wells, varying in depth from 0.6 to 1.2 km were surveyed with borehole televiewer, and sonic logged, which involved measuring vertical travel times over a 1.2 to 1.5 m interval. One of the wells (the Crystallaire well, termed XTLR, was drilled 4 km from the San Andreas fault, in crystalline rocks, in the Mojave Desert area. This well was the subject of three separate seismic investigations using sonic logging (from below the water table), vertical seismic profiling (VSP) using an air gun source, and nearby seismic refraction.

The three different frequencies were respectively 2 MHz, 75 Hz and 10 Hz, and these meant three fundamentally different wave lengths in relation to the fractures. Wave lengths were respectively 20 to 60 cm (i.e. similar to the fracture spacing), 60 m (using a 30 m geophone interval) and several hundreds of metres. In addition, laboratory samples were taken at intervals down the core, and were subjected to appropriate confining pressures, based on an assumed effective stress gradient of 16.7 MPa/km, designed to correspond to the *effective* overburden stress. These four sets of data are compared in Figure 5.33. The majority of the fractures (or joint sets) were steeply dipping.

Overall there was remarkable agreement between the three in situ methods, but the ultrasonic measurements on the intact cores gave significantly higher velocities; about 2 km/s higher at shallow depth and about 1 km/s higher at intermediate depths. There was a relatively high degree of fracturing in this well, as shown in Figure 5.34, which also shows the *detailed* sonic log, and its 'divergency' from the lab (intact) data, as the surface is approached.

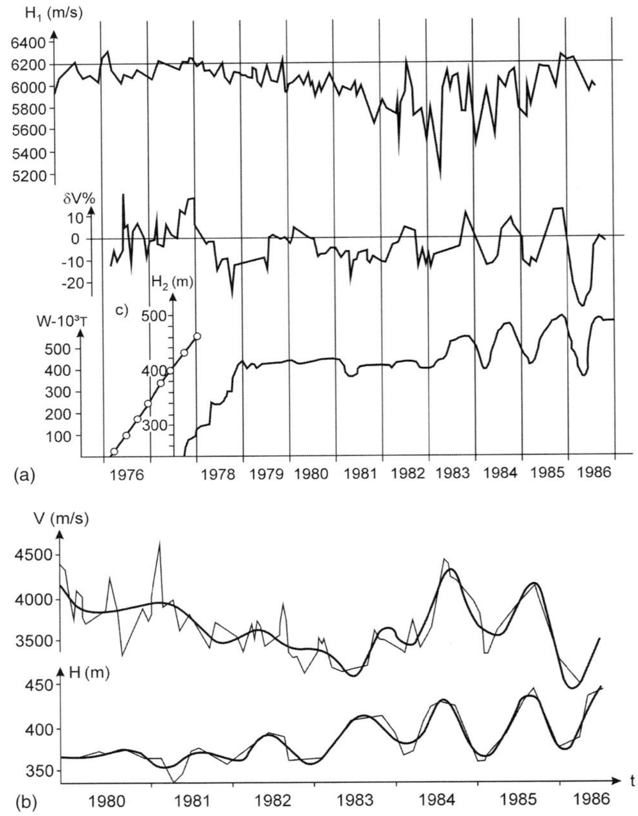


*Figure 5.32* Ten years of recording P-wave velocity and reservoir fluctuations at Inguri dam. *Top*) Ultrasonic, borehole P-wave velocity fluctuations, 'relative velocities', and reservoir levels, including filling. *Bottom*) Seismic P-wave velocity fluctuations show slight 'inertial delay' relative to the seasonal reservoir fluctuations. Savich et al., 1987.

Another well reported by Moos and Zoback, 1983, showed reduced depth-dependent velocities in relation to the above, due to the somewhat lower fracture frequency. This was the MONT-1well, from the Monticello reservoir in South Carolina, USA, drilled through granodiorites. Joint or fracture frequencies and P-wave velocities are shown together with $V_s$ and $V_p/V_s$ ratios in Figure 5.35. The solid circles representing the ultrasonic tests on the laboratory samples were in this case very close to the in situ sonic-log data for this sparsely fractured well.

These two contrasting wells showed increases of average $V_p$ from about 3.5 to 5.2 km/s (XTLR, $Fm^{-1} = 1$ to 4) and 5.4 to 6.1 km/s (MONT-1, $Fm^{-1} = 1$ to 2), over the respective 100 to 850 m and 50 to 1100 m depth ranges. Naturally, almost every fracture zone (or joint swarm) was a zone of low velocity.

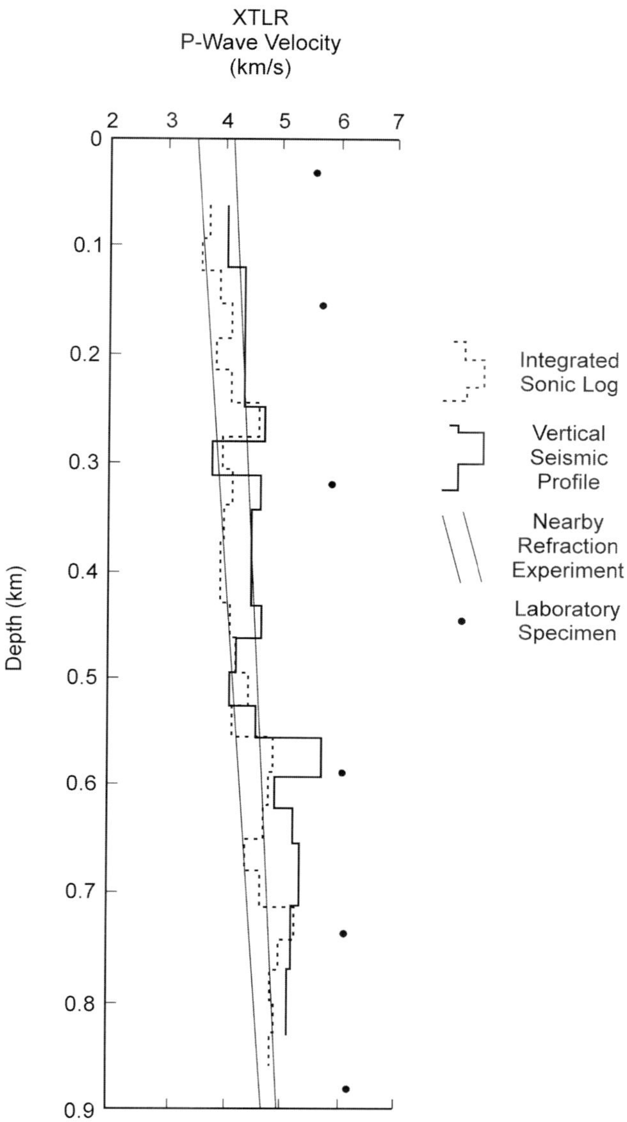


*Figure 5.33* Comparison of velocity-depth trends obtained from three different frequencies of field measurement (smoothed sonic log, VSP, refraction seismic) with ultrasonic tests on intact laboratory samples, tested at appropriately increasing effective confining stress levels. XTLR well in crystalline rocks, 4 km from San Andreas fault zone. Moos and Zoback, 1983.

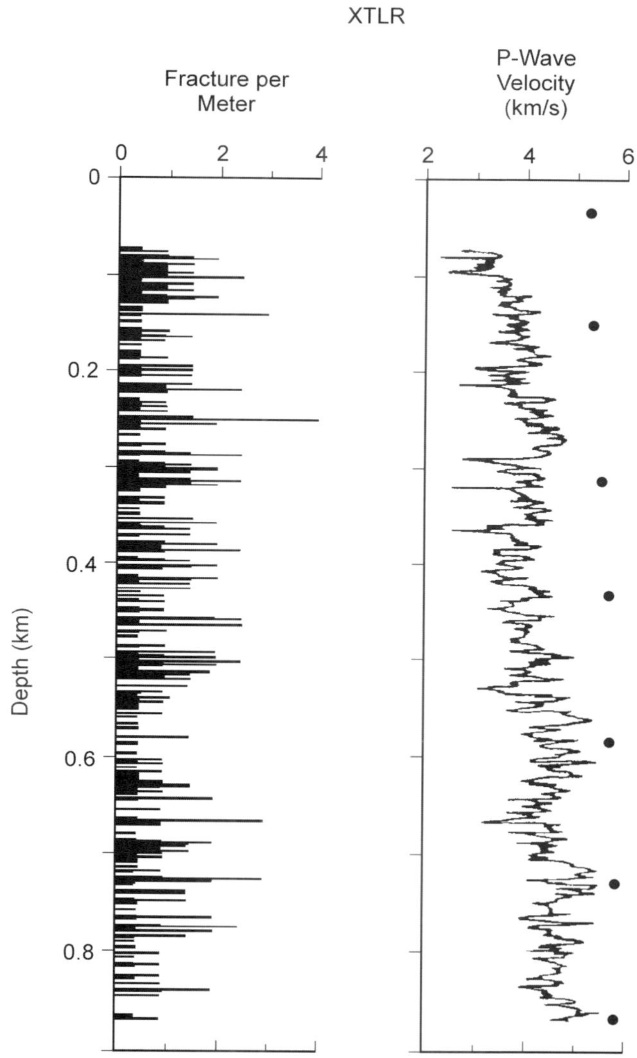


*Figure 5.34* Details of joint frequencies in the XTLR well, showing generally small variations with depth except in upper and lower 50 m, together with the generally increasing trend of the sonic log velocities with depth. Moos and Zoback, 1983.

Besides the description of joint frequency, there was relatively little information in Moos and Zoback's interesting article from which to judge the number of joint sets or the joint character. However, an attempt can be made to match these quite different velocity-depth gradients with a deconvoluted version of Figure 5.36. This is shown in Figure 5.37, and will be explained shortly.

The average velocity-depth gradients shown in Figures 5.33 (XTLR well) and in Figure 5.35 (MONT-1 well) are as follows:

$$\text{XTLR} \quad \left(\frac{V_p}{\text{depth}}\right)_{\text{mean}} = \frac{5.2 - 3.5}{0.75} \approx 2.3\ \text{s}^{-1}$$

$$\text{MONT} - 1 \quad \left(\frac{V_p}{\text{depth}}\right)_{\text{mean}} = \frac{6.1 - 5.4}{1.05} \approx 0.7\ \text{s}^{-1}$$

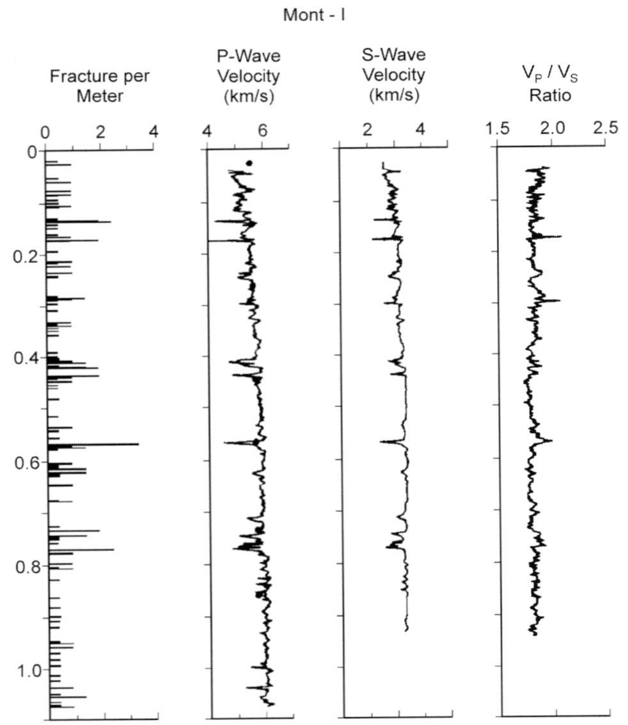


*Figure 5.35* Velocity-depth trends were less marked in the MONT-1 well, in less jointed granodiorites, where laboratory sample velocities were also close to the field result. Moos and Zoback, 1983.

Two other wells reported by Moos and Zoback, 1983, showed little, or in one case reversed (negative $V_P$/depth gradient) due to the increasing frequency of fractures from about 350 to 600 m in the well. Velocities in the top 200 m averaged about 5.5 km/s. When little $V_P$-depth sensitivity is shown, it suggests few fractures. In fact the range of average $V_P$ for a well named MONT-2 was as high as 5.9 to 6.1 km/s despite medium fracture density. This suggests (in Q-value terms) that there might have been only one set of joints, giving a high Q-value, and further that the joints and rock had characteristics allowing for stiff, possibly smooth-walled apertures.

## 5.5 Integration of velocity, rock mass quality, porosity, stress, strength, deformability

It has been shown in previous sections how the P-wave velocity is sensitive to each of the factors listed in the above sub-title. To this we must also add moisture content (for the matrix) and ground water level (for the rock mass). The assumption will be made in the following development that seismic velocity measurements will most frequently be made in saturated rock masses. The

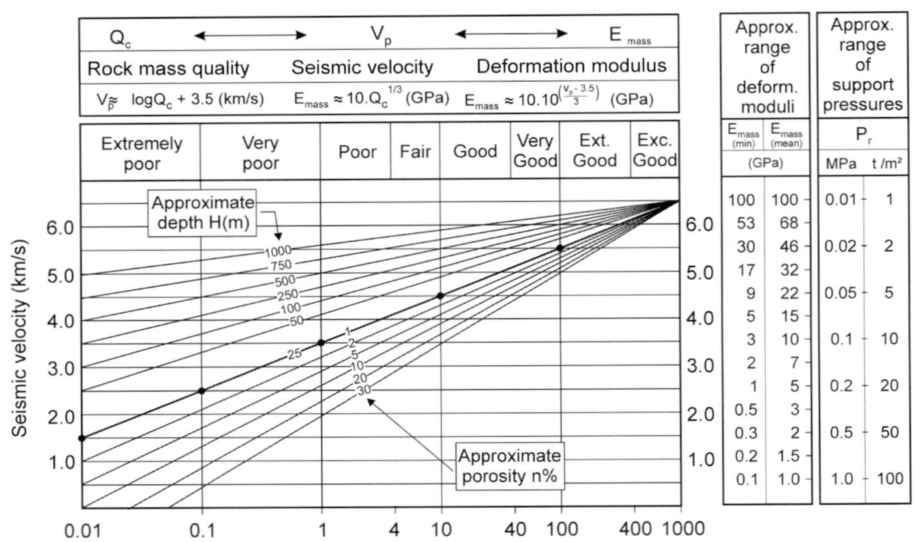


$$Q_c = \left[\frac{RQD}{J_n} \times \frac{J_r}{J_a} \times \frac{J_w}{SRF}\right] \frac{\sigma_c}{100}$$

*Figure 5.36* Integration of rock quality Q-$V_p$-$E_{mass}$ in a model that incorporates depth, porosity and rock strength adjustments. Note that $E_{mass}$ (or M) represents the *static* modulus of deformation, from plate loading tests and from back-analysis of measured deformations. $V_p$ is the seismic velocity measured from refraction seismic, and from cross-hole seismic tomography, in the case of greater depths. Barton, 1995.

correlations developed will be based on this assumption, and systematic errors will of course arise if drainage causes drying out of the matrix and/or joint water.

Since velocity–rock quality correlation is a complex task, no hesitation must be made in adding some degrees of sophistication to the simple model $V_p = 3.5 + \log_{10}Q$ proposed earlier (Figure 5.10). Avoidance of mathematics suggests the use of a graphic method for converting the formulation for hard, unweathered, low-porosity, near-surface rock masses (i.e. typical Sjøgren et al., 1979, data) to conditions towards the other end of the seismic and rock quality scale, e.g., low strength, weathered, high porosity, highly stressed (or unstressed) rock masses.

The development shown in Figure 5.36 which was introduced by Barton 1995 and 1996a, has opposing corrections for porosity and depth (i.e. stress) since these cause opposing influences on velocity. In addition, an adjustment for uniaxial compression strengths different from a typical hard rock 100 MPa (or more) is made by the following simple normalisation of the rock quality Q-value:

$$Q_c = Q \times \frac{\sigma_c}{100} \quad (\sigma_c \text{ expressed in MPa}) \tag{5.6}$$

This correction is necessary because the *rock quality* Q-value was originally developed in 1974, for correlation with tunnel and cavern rock reinforcement and support needs (i.e. rock bolts and sprayed concrete, respectively). The rock quality Q-value only uses the ratio strength/stress ($\sigma_c/\sigma_1$ in the SRF factor – see Appendix A), when major principal stress levels (and their concentration as maximum tangential stress = $3\sigma_1 - \sigma_3$) are causing stress-related fracturing.

It is probable that in a tunnel excavation disturbed zone (EDZ), the potentially large values of SRF (that reduce the Q-value directly) can also be used in principal to predict the measured reductions in velocity and deformation modulus that are typically recorded in the walls of deep shafts and tunnels (e.g. Barton and Bakhtar, 1983 who back-calculated moduli of <3.5 GPa in the outer 3 m, or one radius, of a 1,600 m deep shaft in steeply bedded, highly stressed, jointed quartzites, compared to >65 GPa at a depth of two shaft diameters, using multiple position borehole extensometers, or MPBX).

In highly stressed cases, $\sigma_c$ will tend to be *high* for hard massive rocks subject to dynamic and sometimes explosive stress-slabbing, and $\sigma_c$ will be *low* for soft rocks that are subject to a slower, 'plastic-deformation' type of squeezing. In equation 5.6, $\sigma_c/100$ corrects the

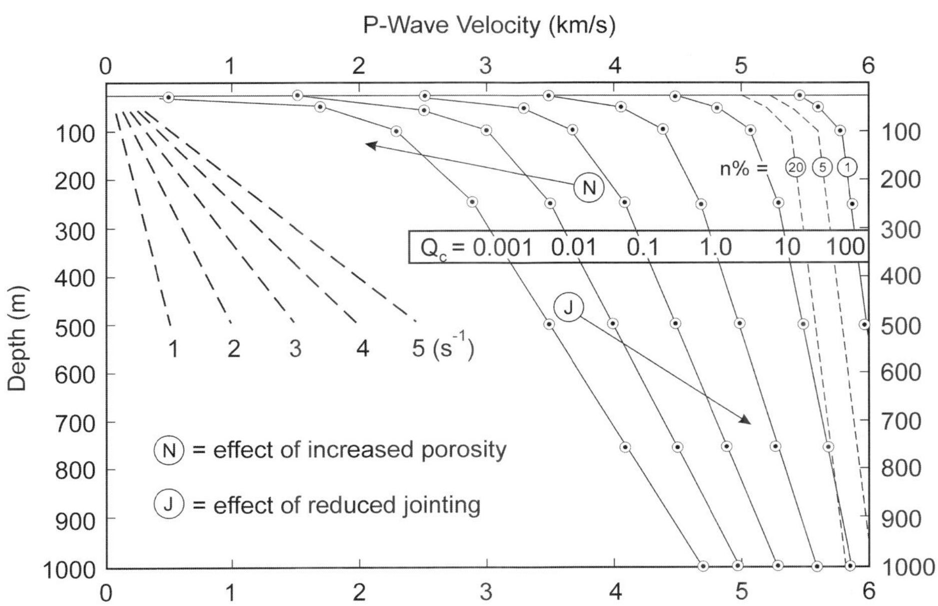


*Figure 5.37* Conversion of the Barton, 1995 rock quality $Q_c$-$V_p$ model of Fig. 5.36, into a more familiar velocity-depth format. Note that there will be a tendency for 'curve-jumping' (i.e. 'Q-jumping'), as a near-surface rock quality improves at greater depth. This will be due to the reduced effects of weathering, and due to a tendency for reduced jointing frequency. Note the contrasting directions of the 'N' and 'J' arrows shown in the figure, together with the $s^{-1}$ (km/s/km or m/s/m) gradients.

Q-value to an approximately suitable value of $Q_c$ to correlate more closely with the rock mass static deformation modulus (M, or $E_{mass}$) and with the seismic velocity $V_p$, particularly for the case of softer rocks.

The 'opposed corrections' given by the two sets of diagonal lines in Figure 5.36 are designed to do the following:

1. A strongly non-linear initial correction for depth gives greater sensitivity to 'acoustic joint closure' for weaker and lower quality rock masses.
2. A weakly non-linear porosity correction also gives larger changes of velocity for the weakest rocks.

The chart, which should be considered as an approximate engineering guideline, was developed with the hard rock ($V_p = 3.5 + \log_{10}Q$) relationship (Figure 5.10) as a 'core' (see diagonally-aligned black discs in Figure 5.36). Development for soft rocks occurred by a process of trial-and-error fitting of Q, $V_p$, $\sigma_c$, n and depth data from known sites in jointed chalks, (Saul Denekamp, personal communication), jointed chalk marl from the UK end of the Channel Tunnel, sandstones, mudstones, shales, welded tuff and ignimbrite. Data from Israel, England, Japan and China were included. Depths ranged from about 25 meters to more than 1000 metres in the case of tuff/ignimbrite from UK Nirex's Sellafield site, where cross-hole tomography and laboratory tests could be compared with NGI's detailed rock quality Q-logging (and joint index testing) of 9 km of drill core (see Chapter 4).

The first two empirical relationships listed in the top of Figure 5.36 were derived from extensive field test data for hard rocks (Barton, 1996). Testing with soft rocks has shown that the modified $Q_c$ term gives satisfactory fit, which is improved when the porosity and depth corrections are also made. Thus we have the following approximations for general use. Note that both $V_p$ and deformation modulus (M or $E_{mass}$) are predicted to increase with depth (see detailed discussion in Barton, 2002a).

$$V_p \approx \log_{10} Q_c + 3.5 \text{ (km/s)} \tag{5.7}$$

$$M(\text{mean}) \approx 10Q_c^{1/3} \text{ (GPa)} \tag{5.8}$$

Although there is little data for deformation modulus measurement at 'undisturbed' depths of hundreds of metres, it will be noted that the predicted static moduli (Figure 5.36, right-hand column) become closer to seismically derived dynamic E moduli (from $V_p$, $V_s$ and density, see Chapter 1). If truly undisturbed static modulus testing could be achieved, the normal discrepancy between static deformation modulus and dynamic E modulus might be lessened, despite the fundamentally different levels of strain involved in each case.

An illustration of application of the Q–$V_p$–M method to very soft rocks can be given here, based on Q-logging of tunnels in chalk marl (Terlingham Tunnel, Beaumont Tunnel, and UK sector Channel Tunnels, and selected marine drill core: PB1 to PB8). Details of the rock quality Q-logging are given by Barton and Warren, 1996c.

A weighted mean value of Q = 8 was obtained from the so-called 'precedent study' of nearby tunnels in chalk marl, and from marine drillcore. This Q-value was found to compare closely with the overall mean Q-value recorded in the contractor-owner (Trans Manche Link – EuroTunnel) TBM face logs from the 20–24 km subsea chainage near the English south coast, where an overall mean value of Q = 9 was obtained from the three machine bored tunnels. In this zone, significant tunnelling problems were caused by the (salt) water leakage and overbreak, hindering PC-element ring-building, and causing problems with electrical equipment on the TBM.

The mean $\sigma_c$ value for the chalk marl was 6 MPa. Thus from equation 5.6 we have:

$$\bar{Q} = 8,\ \bar{\sigma}_c = 6 \text{ MPa},\ \bar{Q}_c = 8 \times \frac{6}{100} = 0.48$$

This $Q_c$ value intersects the central diagonal line (equation 5.7) in Figure 5.36 at $V_p$ = 3.2 km/s. Correction for average porosity (n = 27.7%) results in a reduction of 1.6 km/s giving 3.2–1.6 = 1.6 km/s. Tunnel depths of, for example 40 m, bring this value up to about 2.0 or 2.1 km/s. Offshore geophysics carried out during several campaigns indicated P-wave velocities generally in the range 2.0 to 2.6 km/s for the UK chalk marl. A Q-value of about 20 is needed to explain the upper velocity of 2.6 km/s using the above method. This is in line with the otherwise generally good rock mass Q-values, registered outside the 4 km zone with much overbreak.

The predicted moduli of M ≈ 0.5 GPa (minimum) and 3 GPa (mean) compare with (disturbed) but unjointed laboratory scale moduli of 0.64 GPa (mean) and a range of 0.15 to 4.2 GPa. Deformation measurements in the tunnels were interpreted as indicating an *in situ* modulus of about 1 to 2.5 GPa for a range of rock qualities, quite consistent with the above.

The correction $Q_c = Q \times \sigma_c/100$ can also be applied in the direction of very hard rock, to adjust the value of $Q_c$ above that of the logged Q-value. As we shall see in Chapter 7 on EDZ phenomena, this gives a useful hard rock correlation between measured velocities and observed Q-values from greater depths. We will examine predicted and measured velocities and moduli from greater depths in Chapter 7.

The sloping lines for depth adjustments that are shown in Figure 5.36 can be converted to velocity-depth gradients, with a format more familiar to geophysicists, as shown in Figure 5.37. Here one may note the marked linearity at depths beyond the first 250 m, and the marked non-linearity in the upper 100 m in particular.

With awareness of the different velocity scales, we can compare the velocity-depth trends of the XTLR well near the San Andreas fault (Figure 5.34) with the overall gradients of a rock mass quality of $Q_c = 0.1$ in Figure 5.37. This *suggests* a well tectonised rock mass with several joint sets, and possibly with clay coatings and smooth or slickensided joints. Unfortunately, the detailed condition of core seems seldom to be described in published articles, so whether this 'picture' is realistic is uncertain.

At the extreme end of the rock mass quality spectrum, one may refer to the seismic studies of a several million cubic meter post-glacial landslide, at Koefels in Austria, which covers a valley over an area of some 10 km$^2$. Brueckl and Parotidis, 2001, found that the landslide mass could be approximated by a $V_p$-depth model which we can simplify to $V_p \approx 400D^{0.32}$ m/s. So at 50 m, 100 m, 200 m and 300 m (the average depth), the velocities would, in round figures be approximately as low as 1,400, 1,750, 2,300 and 2,500 m/s, compared to a bedrock of 5.2 km/s.

As a first approximation one might view the $Q_c$ -$V_p$ – depth model of Figure 5.37 and consider a faulted-crushed-rock Q-value of 0.01, which would reduce to a $Q_c$ – estimate of 0.001 if weakened rock of say, 10 MPa uniaxial strength was involved. The relevant $V_p$ – depth curve gives a certain geometric similarity to the above velocity increases down to 300 m depth, but needs a minor 'parallel shift' to somewhat lower values.

Such could be 'achieved' by a) a minor downwards adjustment of velocity, due to a porosity correction ($-V_p$) for slightly altered rock, or b) by a larger downwards adjustment due to the larger mass porosity of the failed materials, e.g. 20%, but then necessarily from a higher $Q_c$ value as a starting point. Detective work, preferably aided by local knowledge of the site, may result in improved insight, where a hint concerning the deformation modulus of the slide masses (about 5 GPa) might also need to be used.

# 6 Deformation moduli and seismic velocities

Although it was not the original intention in this book to explore in detail the 'static' *deformability* of rock masses, a superficial treatment is necessary, in view of the common use of seismic measurements to extrapolate near-surface deformation or elastic modulus measurements. We saw in the last chapter how the loading of individual joints, or of multiple jointed columns of rock in the laboratory created changes in the P-wave and S-wave velocities and amplitudes. Detailed monitoring of in situ loading tests using sonic logging in boreholes beneath plate loading tests, such as shown in Chapter 3 (Figure 3.15), also shows correlation with the moduli obtained. However, due to the inevitable damage in the unloaded zone around the sites of the tests, the lowest moduli (and lowest velocities) are usually registered closest to the loading plates, flatjacks or pressure chambers, while higher velocities and moduli are registered at greater depth. This is probably due to the more uniform stresses *outside* the excavation disturbed zone, or EDZ, where shear stresses are less due to the radial stress ($\sigma_r$) being higher. There is increasing discrepancy between dynamic and 'static' deformation moduli in the near-surface, low quality, weathered zone, that may unfortunately need to be used as a dam or large building foundation.

In this chapter there will be detailed treatment and examples of the multitude of 'moduli' that have proliferated with the combined use of loading (and unloading) 'static' moduli, and the attempts to extrapolate or compare these with the 'dynamic' moduli, both from laboratory samples and *in situ* measurement.

## 6.1 Correlating $V_p$ with the 'static' moduli from deformation tests

The method that is perhaps most common for relating seismic velocity and in situ modulus measurements, is what we may call the *seismic characterisation method*. The rock mass at, or beneath, the site of a deformability test is logged by downhole, cross-hole or shallow refraction seismic. On the basis of the local $V_p$ value the measured moduli are extrapolated (or interpolated) to other parts of the foundation using larger scale cross-hole, between-gallery, or VSP (hole to surface or tunnel perimeter) velocity measurements. Classic examples of this were shown in Figure 4.1 and 4.2.

A useful basic comparison between the 'static' modulus of deformation $E_d$ (or M in Figure 5.36) and the so-called modulus of elasticity ($E_e$) is shown in Figure 6.1. The modulus of elasticity ($E_e$) is traditionally obtained from the gradient of the *unloading* curves, which are often similar, and supposedly have elastic character due to the frequent closed state of the stress-deformation loops. These results were given in the 1st ISRM Congress by Kujundzíc and Grujíc, 1966. Three of their figures are reproduced in Figure 6.1.

The total deformation measured at the highest load level, after several loading cycles, is the usual basis for the calculation of $E_d$ (Figure 6.1a). Thus defined, $E_d$ may change with stress level, while $E_e$ is considered a constant. Partly for this reason, the inequality $E_e > E_d$ can be quite large, as it depends also on the stress level, which in turn is usually based on the size and type of dam (or other structure) to be located on the particular rock foundation.

The basic inequality of $E_d$ and $E_e$ is clearly shown in Figure 6.1b and c. Kujundzíc and Grujíc, 1966, assembled most of their data from tests on limestone foundations at Yugoslavian dam sites carried out in the 1950s and 1960s. As can be seen from the data, the largest inequalities (ratios of 1.5 to 2.5) are seen at the lowest moduli values (<5 GPa). The values of $E_e$ and $E_d$ followed approximately the following trend:

$$E_e \approx \frac{15.50\ E_d}{6.45 + \sqrt{E_d}} \tag{6.1}$$

where $E_e$ and $E_d$ are expressed in GPa. Let us look at two examples to illustrate the trends of $E_e > E_d$, with differences reducing when higher quality rock masses are involved. Assuming $E_d$ of 5 and 50 GPa, we see from equation 6.1 that the elastic moduli are predicted to be 8.9 and 57.3 GPa.

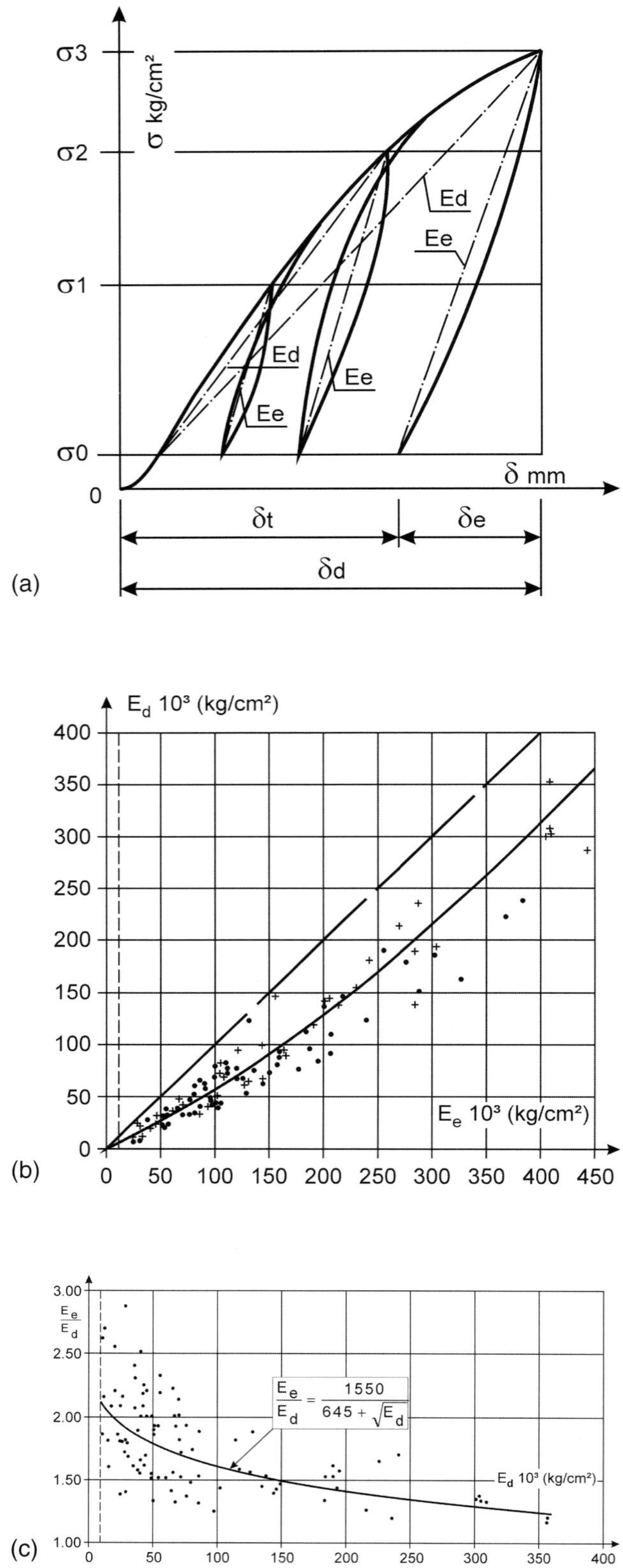


*Figure 6.1* Definition of $E_d$ and $E_e$, and examples of their differences, from limestone dam foundations in Yugoslavia. Kujundzíc and Grujíc, 1966.

Note that these larger differences with $E_e > E_d$ at lower qualities, are largely caused by an increasing hysteresis in the load-unload behaviour of the individual rock joints, as rock mass quality deteriorates, typically due to weathering and unloading effects near the surface. In Chapter 16 we will also see the different shapes of these rock mass loading curves caused by predominance of joint closure (concave type) over shear (convex type). A combination of both components may give a semblance of 'linearity'. (Barton, 1986).

(The above inequality $E_e > E_d$ is in the 'static' loading, ultra-low-frequency sphere of geophysics. This behaviour is *not* directly related to another key inequality in geophysics, namely that the (static) joint *stiffnesses* become less than the inverse of the (dynamic) joint *compliances*, as rock mass quality deteriorates. These aspects are discussed in Chapters 15 and 16 in relation to shear-wave splitting).

Grujíc, 1974, also reported results of $V_p$ measurements in the galleries at a Yugoslavian dam site, as shown in Figure 6.2. The ability to use $V_p$ measurements to extrapolate flatjack or plate loading test results is obviously an important part of the economy and thoroughness of major foundation studies. In most cases, working hypotheses are established that relate deformation moduli with the seismic velocities. In the case of the Mratinje dam on limestone foundations, Grujíc found the following relationship:

$$E_d = 9.1 + 4.8\,(V_p - 3.6)\ \text{GPa} \tag{6.2}$$

With $V_p = 4.5$ km/s, (an implied nominal 25 m depth rock quality $Q_c = 10$), this equation suggests a static deformation modulus $E_d$ of 13.4 GPa. This is midway between the *minimum* and *mean* M-values shown in Figure 5.36. The tendency for 'disturbed zone', near-valley-side reductions in $V_p$ and presumed reductions in $E_d$ (or M) can be seen in Figure 6.2.

The basis for the above correlation between $V_p$ and `$E_d$, were the so-called *polar* measurements of $V_p$ surrounding the flatjack (and therefore loaded) locations. These are illustrated in plan and in vertical section, in Figure 6.3. The upper line in Figure 6.3c shows $E_e$ (=E), and the lower line shows $E_d$ (=D), and their correlations with velocity. These obviously relate to velocities of the *flatjack-loaded* rock mass, giving a range of $V_p$, together with rock quality variations, from 1.5 to 5.5 km/s. The rocks involved were variously jointed and weathered Triassic limestones.

A very thorough review of inter-relationships between seismic velocity measurements and E-moduli and

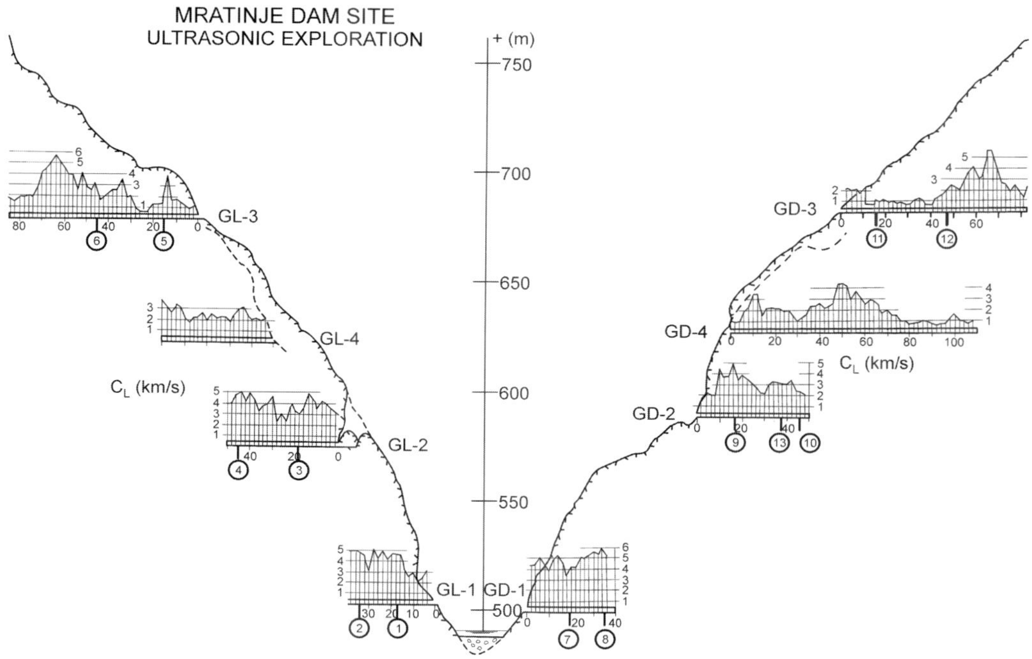


*Figure 6.2* Example of $V_P$ measurements in exploration galleries at a dam site in Yugoslavia. These are used to economically extrapolate deformation modulus test results to other parts of the dam foundation. Grujíc, 1974.

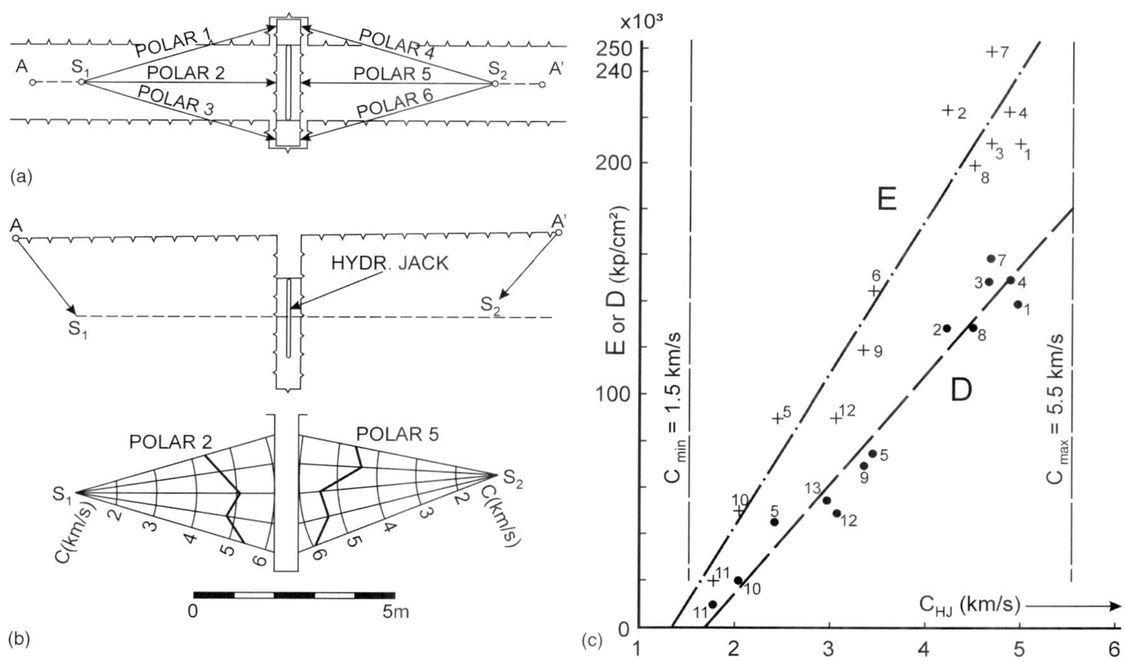


*Figure 6.3* Flatjack deformation tests for determining the moduli D(or $E_d$), and E (or $E_e$), and correlation with velocity in the Triassic limestones. Grujíc, 1974.

D-moduli obtained from plate loading tests in Italy was given by Ribacchi, 1988. These are shown in Figure 6.4. The particular emphasis of this review article was rock mass deformability, and complexities caused by anisotropy, which the author nevertheless concluded, was greatest with samples that had been recovered for laboratory tests. The results of in situ investigations at eleven sites were given. These involved several varieties

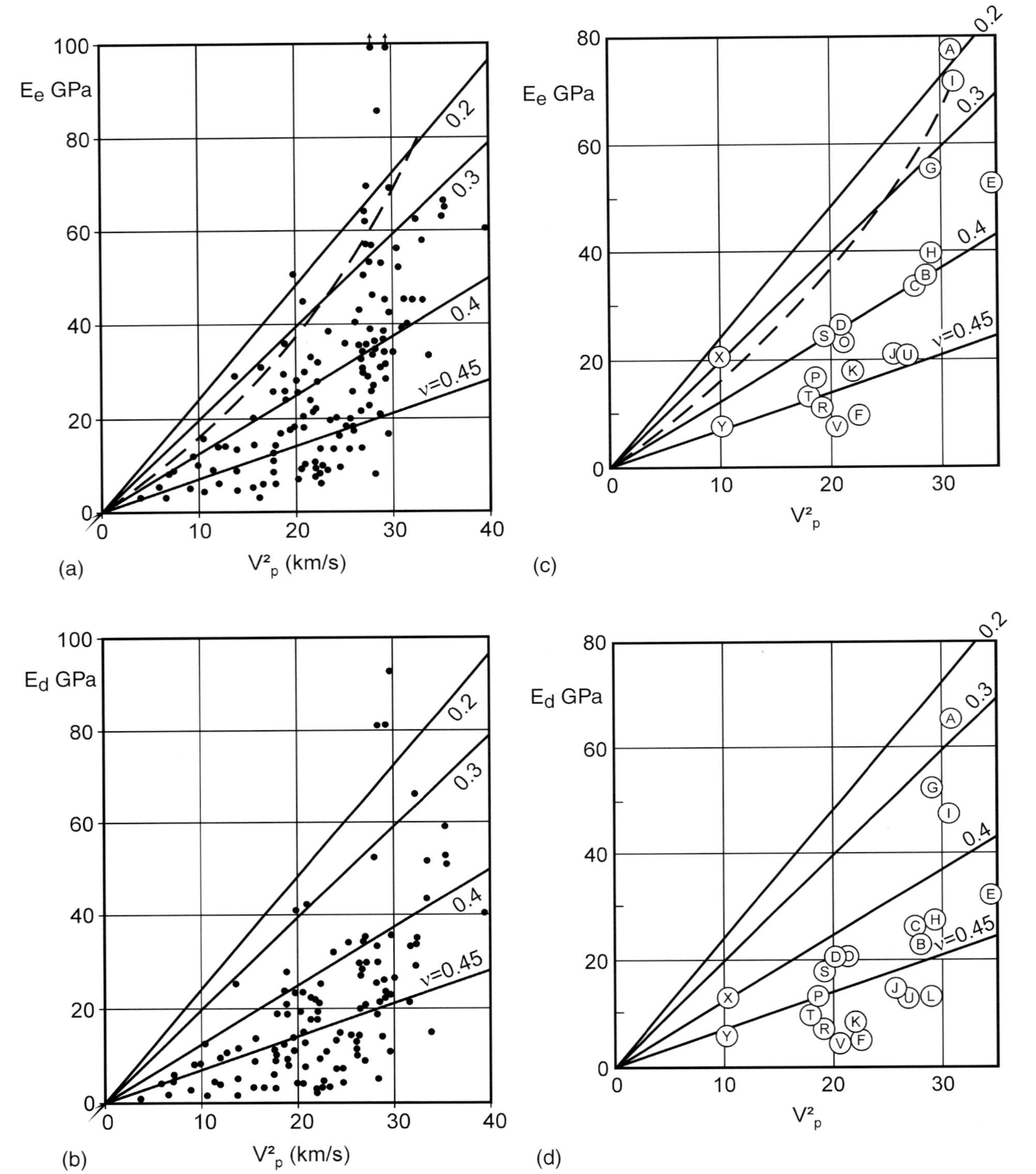


*Figure 6.4* a,b) Results of in situ deformation tests at eleven sites, showing $E_e$ and $E_d$ correlated with $V_p^2$. c,d) Mean data from each site shows reduced scatter. Note dynamic Poisson's ratio gradients Ribacchi, 1988.

of limestones, schists, gneisses, granodiorites, mylonites, dolomite, a high porosity (n = 30%) calcarenite, and inter-bedded sandstones and mudstones. Both the plate loading tests and the seismic velocity measurements were conducted perpendicular and parallel to layering, bedding, or schistocity, as appropriate.

The author cited examples of sites that showed good correlation between modulus and typical indices of jointing (RQD, and joint frequency $Fm^{-1}$). The scatter of more than 100 data points for modulus E or D versus $V_p^2$ in Figure 6.4 is quite large for the whole sample, but reportedly showed good correlation for limestones and dolomites. Figures 6.4a and b show all data points for $E_e$ and $E_d$, while somewhat reduced scatter for the mean data at each site is shown in Figures 6.4c and d. Ribacchi's data from a limestone and dolomite site are shown in Figures 6.5a and b. Here, modulus is plotted on a log scale, and velocity is not squared as above, thereby showing much narrower scatter. Trend lines for these and other limestones (Yugoslavian) are shown in Figure 6.6.

The deformation moduli M(mean) and M(min.) tabulated in the inset to Figure 5.36 (in Chapter 5) are also encompassed by the above trend lines. The log-linear equation given in Figure 5.36 simplifies to:

$$M = 10^{(V_p - 0.5)/3} \text{ (GPa)} \tag{6.3}$$

and serves as a mean value for the in situ E-moduli ($E_e$ shown in Figure 6.5a), while it is closer to an upper-bound for the in situ deformation moduli ($E_d$ shown in Figure 6.5b). The M(min.) values given in Figure 5.36 independently encompass the minimum $E_d$ trend lines given in Ribbacchi's Figure 6.6.

Kikuchi et al., 1982, also presented a very comprehensive set of modulus and velocity data, this time from investigations at Japanese dam sites. They established a rock grading system based on a combination of seismic refraction data, deformation moduli from plate load tests, Mohr Coulomb parameters from in situ shear tests, and rock rebound hammer (modified to 30 mm impact diameter for soft rocks). Their data extended from weathered or soft rocks with uniaxial strengths as low as 2 MPa up to extremely hard rocks with uniaxial strengths as high as 300 MPa. The *in situ* P-wave velocities ranged from 0.4 up to 5 km/s.

Figure 6.7a and b show examples of their $V_p$–rock grade, and $V_p$-D (modulus of deformation) correlations, while Figure 6.7c gives the well-correlated log E–$V_p$ relationship. The elastic (E) modulus and the deformation (D) modulus were defined as tangent and secant gradients at the maximum loading of 7 MPa. Their best-fit trend lines for E and D compare quite closely with envelopes 2 or 3 and 1, respectively in Figure 6.6 (from Ribacchi, 1988).

Broad correlations between seismic P-wave velocities and in situ loading test results at dam sites are also demonstrated by Navalón et al., 1987. Investigations at

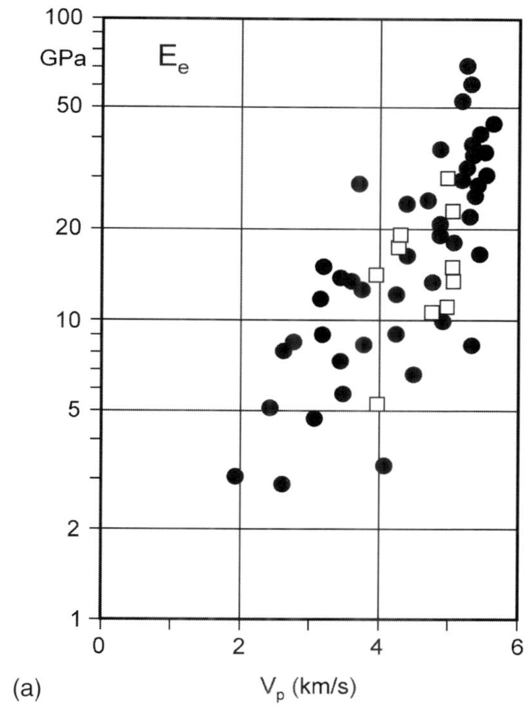


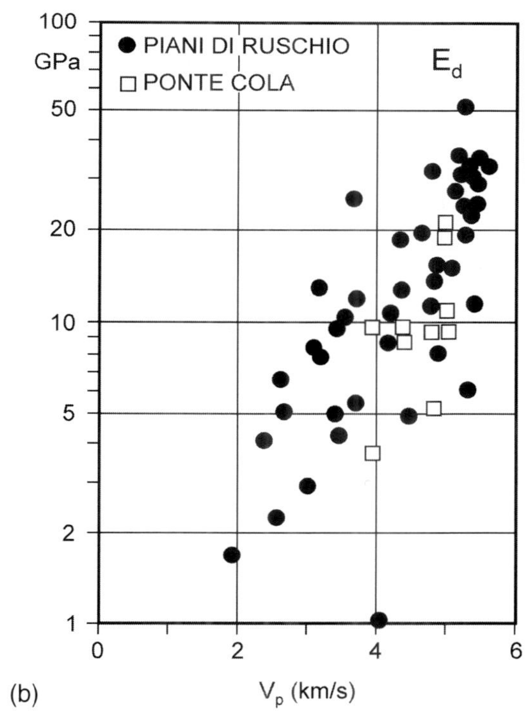


*Figure 6.5* $V_p$ versus logarithm of deformation moduli $E_e$ and $E_d$, for two Italian dam sites in limestone and dolomite. Ribacchi, 1988.

two similar sites consisting of crystalline limestones and marly limestones with innumerable clayey interlayers of millimetre thickness produced the results shown in Table 6.1 (M = deformation modulus).

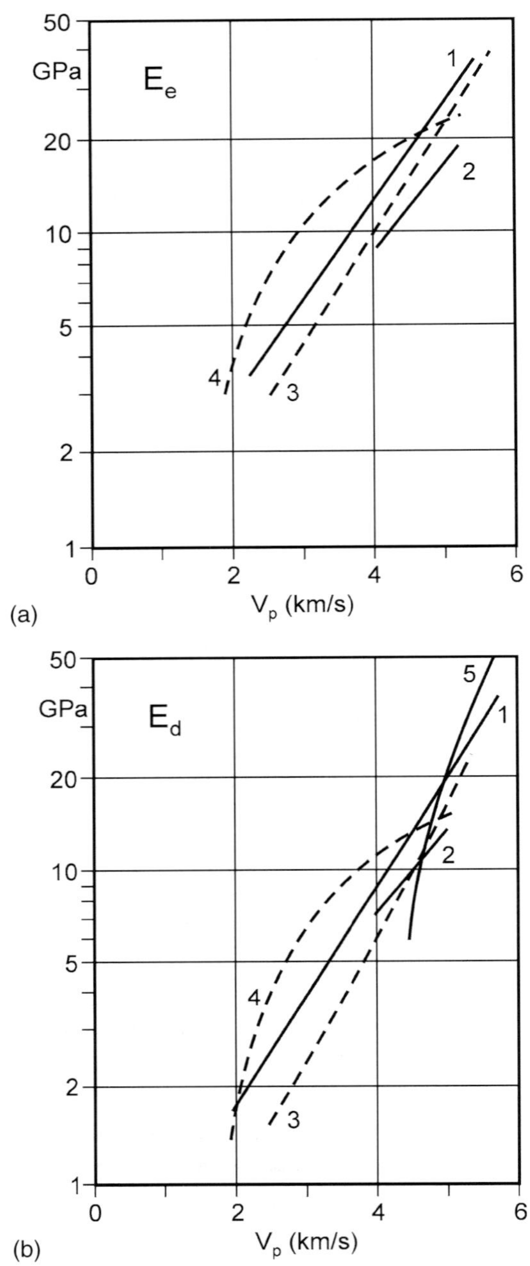


*Figure 6.6* $V_p$ versus logarithm of deformation moduli $E_e$ and $E_d$, from Italian, Yugoslavian and American *in situ* test data. See Ribacchi, 1988 for references.

These three depth zones were defined at each of the two dam sites, and were relevant to foundation testing in the left and right abutments and in the riverbed at each site. Similar geology was involved in each case.

The above low velocity data broadly fits the simple empirical equation developed in Chapter 5 (Figure 5.36):

$$\text{D or M} \approx 10Q^{1/3} \text{ GPa} \tag{6.4}$$

when $V_p$ values are converted to Q-values using the 'hard rock' relation $V_p \approx 3.5 + \log_{10}Q$. However, the 'deep zone' data with $V_p \approx 3.5$ km/s have surprisingly high moduli (approx. 14 to 20 GPa) which are higher than that predictable from equation 6.4. For some reason the moduli but not the velocities appear to be increased by the 'deep zone' boundary condition (i.e. by higher stress).

The process of rock mass loading and even failure under various sizes of plate loading (0.28 to 0.45 m in diameter) was monitored by Savitch et al., 1974, using simultaneous $V_p$ measurement in the first one metre beneath the plates. The testing was performed at the Inguri arch dam in Russia, which is founded on medium and intensely jointed limestones. The range of deformation moduli was 1 to 15.7 GPa.

Increments of loading were applied, to final levels of about 10, 20, 30 or 50 MPa (depending on rock quality) which caused characteristic rises and sharp falls of $V_p$-values (Figure 6.8a, b, c) for various rock qualities, which were sampled by arranging the loading tests at different depths below the surface. The $V_p$ monitoring was performed cross-hole, to 1 m depth below the plates, in boreholes that were located close to the edges of the plates, across a diameter.

Figure 6.8 shows small arrows on each set of curves. Those arrows pointing downwards signify 'failure' as evaluated by the seismic results, while the arrows below the curves signify failure as assessed by the accelerating

Table 6.1 Deformation moduli and velocities from numerous plate load tests. Navalón et al., 1987.

| A) | $V_p$ (km/s) | M (GPa) | B) | $V_p$ (km/s) | M (GPa) | C) | $V_p$ (km/s) | M (GPa) |
|---|---|---|---|---|---|---|---|---|
| Decompression zones | 0.7–2.0 | 0.4–3.9 | Intermediate zones | 1.8–2.2 | 3.4–4.9 | Deep zones | 3.3–3.6 | 17.6–19.7 |
| | 0.9–2.3 | 0.7–5.9 | | 2.6–2.7 | 7.4–8.3 | | 3.5 | 14.2–16.2 |
| | 1.0–2.4 | 0.9–6.4 | | 2.6 | 7.3 | | 3.5 | 16.2 |
| | 1.5–2.5 | 2.0–6.4 | | 2.6 | 7.4 | | | |

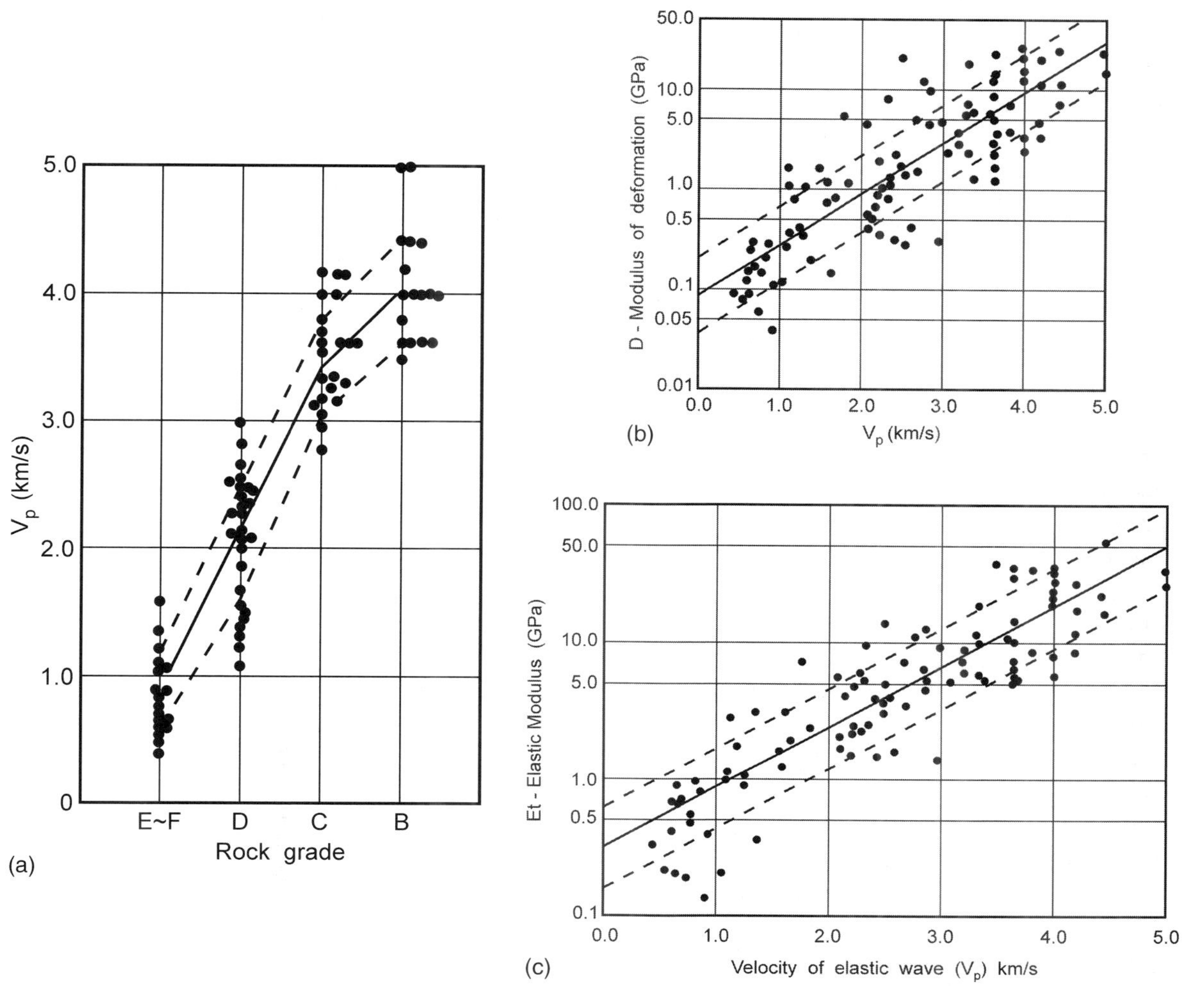


*Figure 6.7* Correlations of refraction seismic velocities with weathering grade, deformation modulus and elastic modulus from Japanese dam sites. Kikuchi et al., 1982.

deformation. Consistent relationships between deformation modulus, velocity and the ultimate strengths of the rock masses are shown by the following tabulation:

Table 6.2 Plate load tests taken to ultimate failure. Savitch et al., 1974.

| Test No. | D (GPa) | $V_p$ (km/s) | $\sigma_c$ (mass) MPa |
|---|---|---|---|
| 3 | 0.9 | 1.3–2.2 | 2–3 |
| 6 | 1.0 | 2.2–2.4 | 3–5 |
| 4 | 2.8 | 3.6–3.9 | 19–24 |
| 1 | 3.4 | 3.6–3.8 | 22–23 |
| 7 | 15.4 | 3.9–4.2 | 42–78 |

Coal, which in some ways resembles jointed rock at reduced scale, also shows the expected increases in velocity with stress increase (uniaxial or triaxial), but Shea and Hanson, 1988, identified two other phases of behaviour as well which resemble those we have just seen in Savitch et al., 1974 plate load tests.

Phase I represented the rapid rise in P-wave velocities due to 'closing of layer cavities' at quite low levels of loading (MPa assumed). Attenuation also decreased in this phase as shown in Figure 6.9a Phase III marked the decline of these two trends, while Phase II represented the increase in attenuation that probably signified the creation of micro-cracks. There was hardly any increase in P-wave velocity in this phase, although the S-wave velocity continued to increase.

Figure 6.9b shows that triaxial conditions (1.7 MPa confinement) caused consistently reducing attenuation in Phase I and II, then a sudden increase in attenuation and reduction in velocity as failure approached in

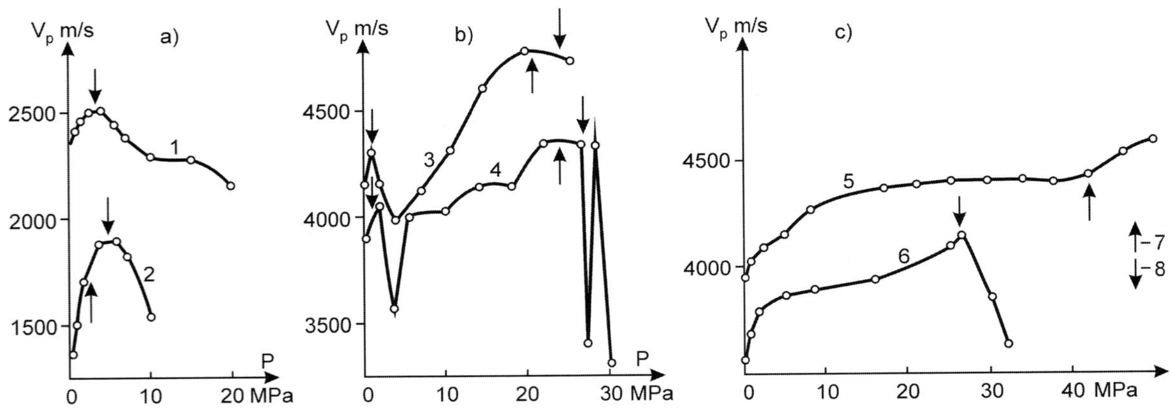


*Figure 6.8* Effect of stress on $V_p$ values beneath plate load tests at the Inguri arch dam in Russia. Savich et al., 1974.

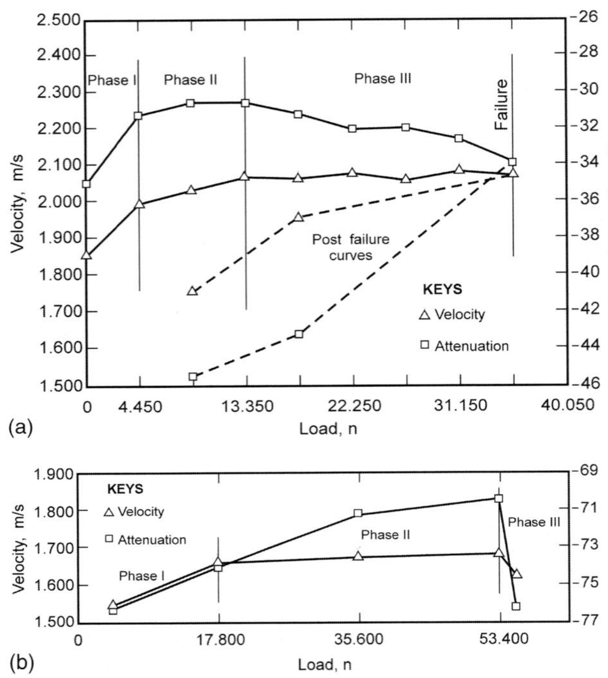


*Figure 6.9* a) Uniaxial, and b) triaxial test of coal, showing velocity and attenuation changes caused by cleat and microcrack behaviour. Shea and Hansen, 1988.

Phase III. Broadly similar (but inverted) behaviour of the P-waves was also seen.

## 6.2 Dynamic moduli and their relationship to static moduli

In this section, which principally concerns dynamic moduli and dynamic Poisson's ratio, the use of $V_P$ and $V_s$ measurements to derive the four standard dynamic elastic properties of rock masses will be discussed. The validity or otherwise of these dynamic, small-strain, elastic moduli to rock engineering design will also be addressed. The four standard equations, two of which were introduced earlier in Chapter 1 are reproduced here together, for ready reference (ISRM, 1998).

Dynamic Young's modulus $E_{dyn}$:

$$E_{dyn} = \rho V_s^2 \frac{3(V_P / V_s)^2 - 4}{(V_P / V_s)^2 - 1} \tag{6.5}$$

$$= \rho V_P^2 \frac{(1 + \nu_d)(1 - 2\nu_d)}{1 - \nu_d} \tag{6.6}$$

Dynamic Poisson's ratio $\nu_d$:

$$\nu_d = \frac{1}{2}\frac{(V_P / V_s)^2 - 2}{(V_P / V_s)^2 - 1} \tag{6.7}$$

Dynamic shear modulus $\mu$:
(also with symbol G)

$$\mu = \rho V_s^2 = \frac{E_{dyn}}{2 + 2\nu_d} \tag{6.8}$$

Dynamic bulk modulus $K_{bulk}$:

$$K_{bulk} = \rho(V_P^2 - 4/3\ V_s^2) \tag{6.9}$$

$$= \frac{E_{dyn}}{3 - 6\nu_d} \tag{6.10}$$

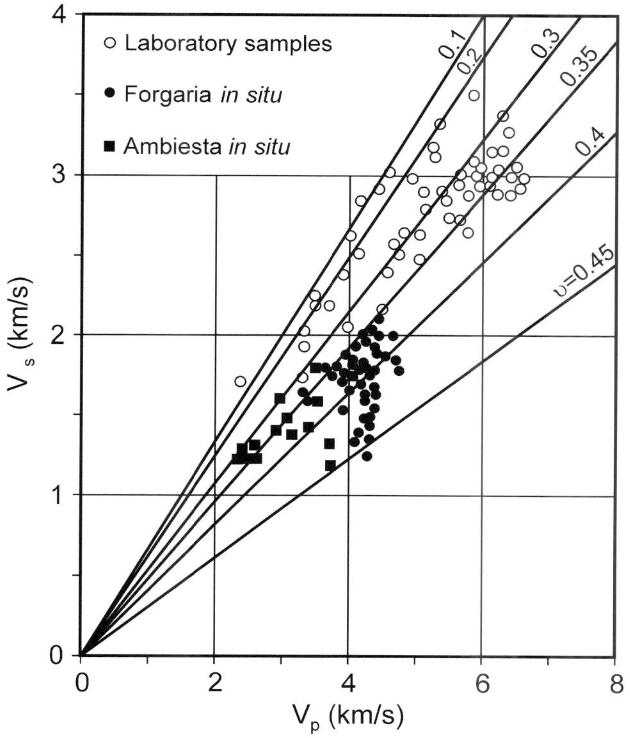


*Figure 6.10* Contrasting laboratory and field data for the $V_p$ and $V_s$ values of limestones, with calculated dynamic Poisson's ratios. Note the over-riding tendency of higher dynamic Poisson's ratio in the case of the lower velocity *in situ* data. Ribacchi, 1988.

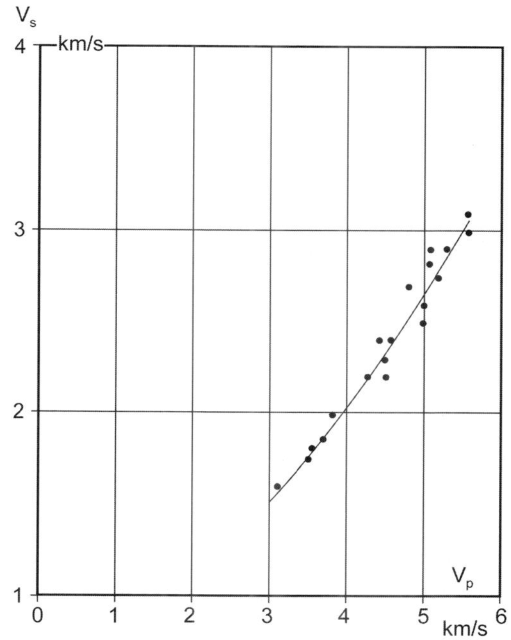


*Figure 6.11* Ratio of $V_p$ and $V_s$ at a hard igneous rock site in Norway. $V_p/V_s \approx 1.8–1.9$. The full range of dynamic Poisson's ratios was from 0.15 to 0.39, with higher values when velocity was lower. Sjøgren et al., 1979.

It will be noted that if the dynamic Poisson's ratio is estimated (rather than derived from $V_p$ and $V_s$), the three dynamic moduli can theoretically be estimated from $V_p$ measurement alone. As pointed out later, this can cause significant inaccuracies, and such values given in the literature should be treated with caution.

The manner in which $V_s$ and $V_p$ values are distributed in relation to the general quality of the rock mass was illustrated in Chapter 1 (Figure 1.7 from Sjøgren, 1984). Further sets of data, both from laboratory samples and from comparable field data (Ribacchi, 1988) are illustrated in Figure 6.10.

Ribacchi's data shows a particularly clear demarcation between the lower dynamic Poisson's ratio in the case of the higher velocity laboratory data, and the opposite trend for the lower velocity field data. High dynamic Poisson's ratio are a clear sign of the influence of jointing. In shear zones and faulted rock, high values of $\nu_d$ are common.

During 'static' flatjack biaxial loading tests of rock masses, 'static Poisson's ratios' (or *lateral expansion coefficients*) in excess of 0.5 may even be measured. In special cases, values in excess of 1.0 have been registered. This occurs as shear failure is approached, in the case of biaxially loaded model rock masses, having two conjugate fracture sets that are under significant levels of shear stress. (Barton and Hansteen, 1979, Barton, 1993a, and Barton, 2004b).

Measurements of $V_p$ and $V_s$ at an unweathered site in Norway are shown in Figure 6.11, and indicate ratios of $V_p/V_s$ of about 1.8 to 1.9. The corresponding dynamic Poisson's ratios were found to lie in a narrow range, more than 80% of the values were between 0.26 and 0.32. The mean value of 0.28 in fact lay close to the maximum RQD and minimum joint frequency trend for this particular site, and corresponded to a ratio $V_p/V_s$ between 1.8 and 1.85. In the rock masses investigated, the authors found that the full range of dynamic Poisson's ratios was from 0.15 to 0.39.

Deere et al., 1967, addressed the important differences between field measurements of $E_{F\ dyn}$ and laboratory measurements of $E_{L\ dyn}$ of the intact rock, by utilising their observation that $(V_{Field}/V_{Lab})^2$ resembled RQD/100. (This was referred to in Chapter 5, Table 5.1.) This was based on the Onodera, 1963, suggestion of using the field/lab velocity ratio $V_F/V_L$ as a measure of rock quality.

The resulting method for estimating $E_{L\ dyn}/E_{F\ dyn}$ is shown in Figure 6.12. As can be noted from the spread of

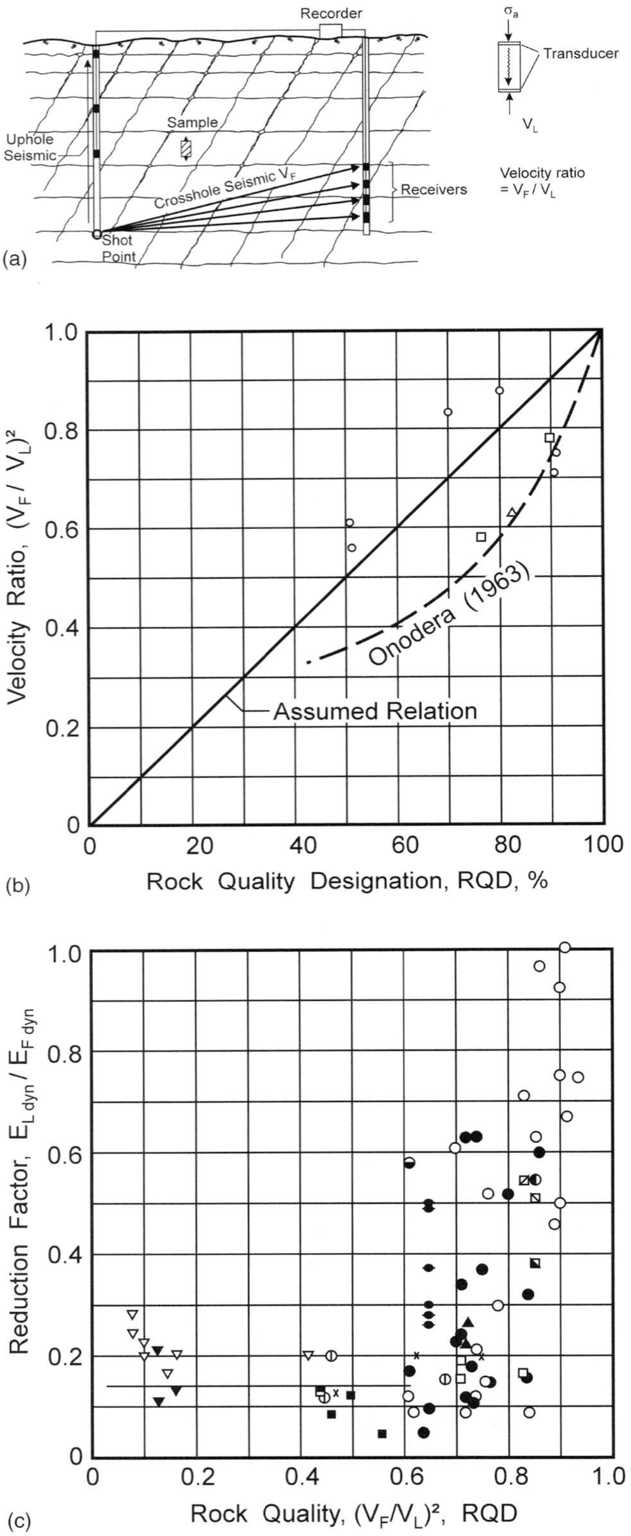


*Figure 6.12* Utilising the velocity ratio (squared) or RQD%/100, to estimate the ratio of the dynamic modulus for intact samples ($E_{L\,dyn}$), compared to the dynamic field modulus ($E_{F\,dyn}$). Deere et al., 1967.

data, the inverted knee-shaped trend does *not* give a good method for estimating the lowest values of the ratio, as RQD and $(V_F/V_L)^2$ vary too much in the region $E_{L\,dyn}/E_{F\,dyn} \approx 0.1$–$0.3$. In higher quality rock masses however, a more consistent trend is observed, some of the data from the USA even indicating $E_{F\,dyn} \approx E_{L\,dyn}$, for values of $(V_F/V_L)^2$ as high as 0.9, or RQD = 90% or more.

When commenting on the differences between static and dynamic moduli, Wang and Nur, 1992, make the usual observation concerning the different strain amplitudes involved (perhaps $10^{-3}$ and $10^{-6}$ respectively) but also point out that the different moduli are thereby directly caused by the presence of pores, cracks (and joints). These are deformed in static tests, but hardly deformed by dynamic waves, but when stresses are very high and pores, cracks and joints are (almost) closed, the static and dynamic moduli are likely to be very close.

The possibility of estimating $E_{dyn}$ values from $V_p$ measurement alone, by just estimating the value of the dynamic Poisson's ratio ($\nu_d$) instead of measuring both $V_p$ and $V_s$ was referred to earlier. This would involve using equation 6.6 instead of the correct method utilising equation 6.5. In warning against this short-cut, Stacey, 1977, assembled what he called 'reliable' $E_{dyn}$ values from the literature. These values are shown in Figure 6.13a. (For the numbered references see Stacey, 1977).

Stacey also assembled a large number of $E_e$-$E_{dyn}$ data (Figure 6.13b) and $E_d$-$E_{dyn}$ data (Figure 6.13c). These two figures show the apparently irrelevant nature of $E_{dyn}$ in comparison to the standard rock engineering methods of testing deformation modulus for design purposes. The doubt nevertheless remains that most of our large scale methods are testing an excavation disturbed zone, or EDZ, rather than the undisturbed state, with its higher and maybe more isotropic in situ stresses.

There is a contrary factor that most rock masses observed or tested, are actually on a major unloading curve due to erosion or due to rock excavation, and may have correspondingly higher joint stiffnesses, than if loaded up without this prior unloading.

The mismatch of the static and dynamic moduli in jointed rock masses, except where rock qualities are very high and strains are very small, is probably a universal rule, unless extremely small strains are actually involved in the 'static' loading.

In the last two decades, it has apparently been recognised in soil engineering that strain levels associated with normal foundation designs are rather small, for example 0.01 to 0.1%, and therefore stiffnesses may be successfully described by the correlations obtained from in situ

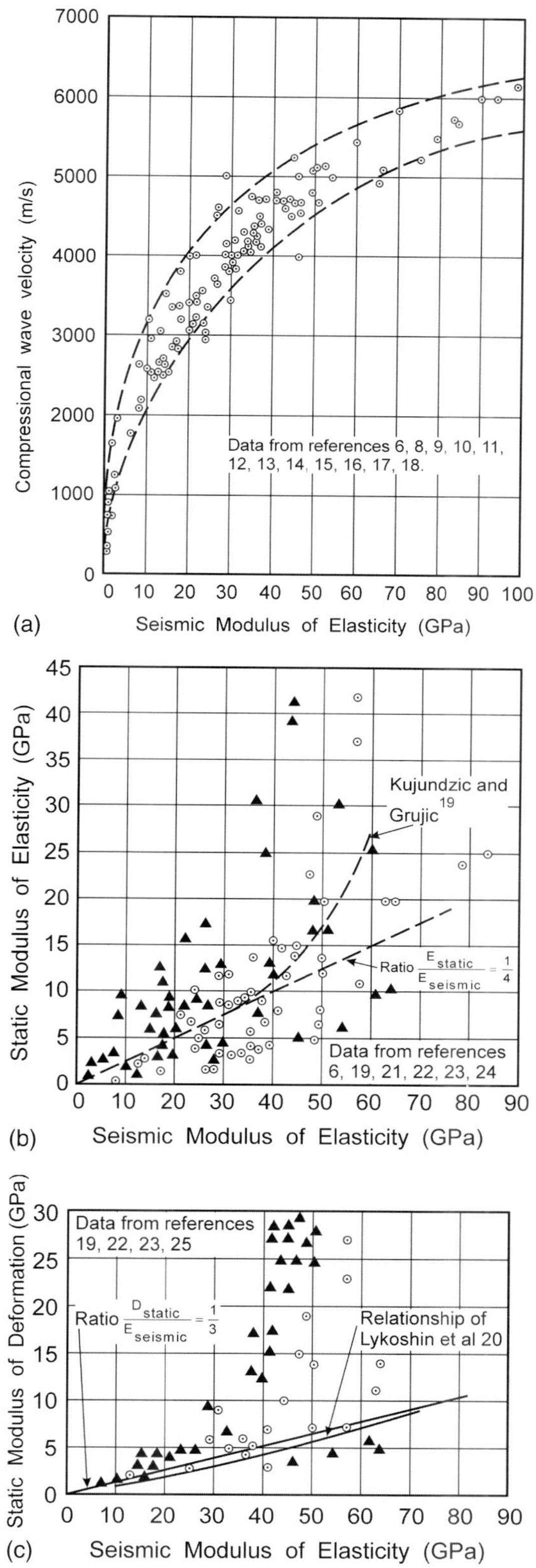


*Figure 6.13* An extensive collection of 'reliable' $E_{dyn.}$ data, and of the ratios $E_e/E_{dyn.}$ and $E_d/E_{dyn.}$. Stacey, 1977. (See individual references in article).

seismic measurements (Matthews et al., 1997). Such measurements also have the great advantage of registering the stiffness of the ground at in situ stress levels and in the undisturbed condition.

A corollary to the above is the predicted high deformation moduli for jointed rocks at depth, shown in Figure 6.14 (Barton, 1995). A pre-condition here is that the rock mass is undisturbed and strains are small, corresponding more closely with the higher seismic velocities seen at greater depths in the same rock masses. When a tunnel or test adit is constructed at considerable depth, the EDZ effect will alter the above conditions in a complex way, to a degree that depends on rock quality and the care with which the adit-excavation has been performed.

One must expect a certain seismic velocity gradient, a deformation modulus gradient, a deformation gradient and even a permeability gradient and pore pressure gradient, and finally a possible gradient of saturation.

The natural complexity of a site may also tend to increase the range of moduli and moduli ratios in relation to those measured in one particular lithology. The Latiyan Dam site in Iran was founded on granites, pegmatites, migmatites and gneiss, and weathered layers of each of these. Lane, 1964, compares four of the basic moduli commonly obtained at dam sites:

- laboratory $E_{L\ dyn}$ (from laboratory $V_p$, $V_s$ and Poisson's ratio)
- field $E_{F\ dyn}$ (from field $V_p$, $V_s$ and Poison's ratio)
- field deformation modulus (D, $E_d$ or M) (from plate or flatjack *loading*)
- field elastic modulus ($E_e$) (from plate or flatjack *unloading*)

and also gives ratios of each, as obtained in three exploratory tunnels in the dam foundations. The surprisingly large ratios of these moduli, for rather poor rock mass qualities (Q mostly < 0.1?) are shown in Table 6.3.

Link, 1964, also gave a wide-reaching comparison of dynamic and static moduli from projects (usually dams) in many countries. He found that the ratio of $E_{F\ dyn}/D$ from field tests (category 6 in Table 6.3), ranged from about 1 to 16, with most values of the ratio lying in the range 3 to 7. A large number of the static loading tests were pressure chamber tests from Central European dam sites, and the author pointed out that the seismic measurements (from $V_p$, $V_s$ and dynamic Poisson's ratio) were also of quite large scale. As we see from the above table, ratios of $E_{F\ dyn}/D$ can be even larger than the above when the rock quality is poor, due to the basic inequality

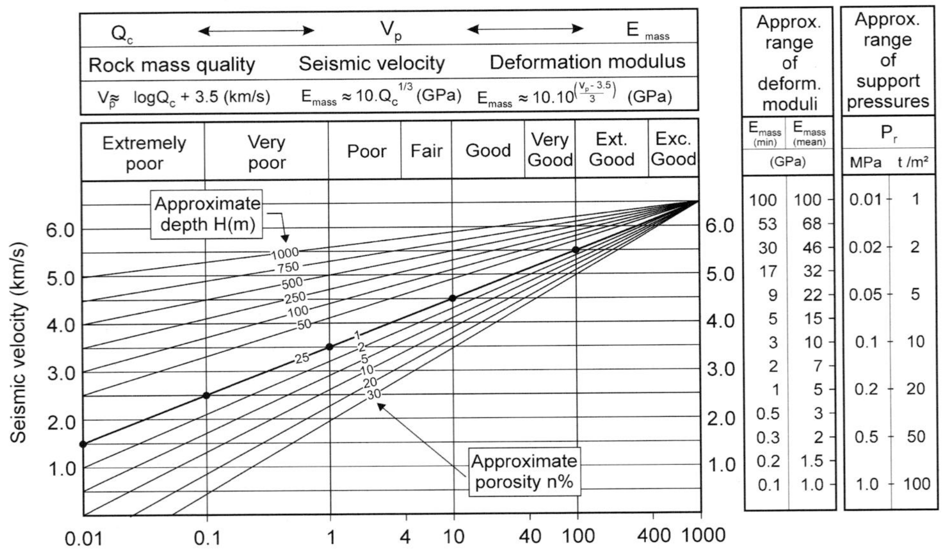


$$Q_c = \left[\frac{RQD}{J_n} \times \frac{J_r}{J_a} \times \frac{J_w}{SRF}\right] \frac{\sigma_c}{100}$$

*Figure 6.14* High values of *static* deformation modulus $E_{mass}$ (or M) for the rock mass (also referred to as D, or $E_d$ by some authors), are predicted where stresses and rock quality are high. Here $E_{mass}$ approaches $E_{F\ dyn}$. Barton, 1995.

Table 6.3 Ranges of moduli and ratios of moduli at three complex sites (after Lane, 1964).

| Modulus | | Tunnel 1 | Tunnel 3 | Tunnel 4 |
|---|---|---|---|---|
| 1 Laboratory Dynamic ($E_{L\ dyn}$) | (GPa) | 35.8 | 45.4 | 35.8 |
| 2 Field seismic ($E_{F\ dyn}$) | (GPa) | 17.7 | 25.4 | 16.4 |
| 3 Average modulus of deformation field loading (D or $E_d$ or M, depending on author) | (GPa) | 1.8 | 1.3 | 1.3 |
| 4 Average modulus of elasticity field unloading ($E_e$) | (GPa) | 4.8 | 4.6 | 3.8 |
| 5 Ratio laboratory dynamic to field seismic ($E_{L\ dyn}/E_{F\ dyn}$) | | 2.02 | 1.78 | 2.18 |
| 6 Ratio field seismic to modulus of deformation ($E_{F\ dyn}/D$) | | 10 | 20 | 12 |
| 7 Ratio field seismic to unloading modulus of elasticity ($E_{F\ dyn}/E_e$) | | 3.7 | 5.6 | 4.4 |
| 8 Ratio laboratory dynamic to modulus of deformation ($E_{L\ dyn}/D$) | | 20 | 36 | 27 |
| 9 Ratio laboratory dynamic to jacking modulus of elasticity ($E_{L\ dyn}/E_e$) | | 7.4 | 9.9 | 11.5 |

of the two *static* unloading/loading moduli: $E_e/E_d$ (or $E_e/D$), shown, for example by Kujundzíc and Grujíc, 1966.

Link, 1964, made special reference to an extremely high value of $E_{F\ dyn}$ measured (interpreted) under the lower slopes of the Vajon limestones. The value of 140 GPa was considered the result of high overburden, and/or residual stresses.

Graphic presentation of the inequalities between $E_{F\ dyn}$ and $E_e$ were given by Kujundzíc and Grujíc, 1966. Figure 6.15a and b show the significant inequality of $E_{dyn}$ and $E_e$ for the case of the limestones tested in their Yugoslavian dam site tests. A general trend was noted as follows:

$$E_e \approx \frac{E_{dyn}}{(5.3 - 0.05E_{dyn})} \qquad (6.11)$$

where $E_e$ and $E_{dyn}$ are expressed in GPa.

The lower the dynamic modulus the larger the ratio of the dynamic/static moduli. When the inequality of $E_e$ and $E_d$ is also considered, the very large ratios of $E_{dyn}/E_d$ of 10 to 20 given by Lane, 1964, and Link, 1964, are more readily understood.

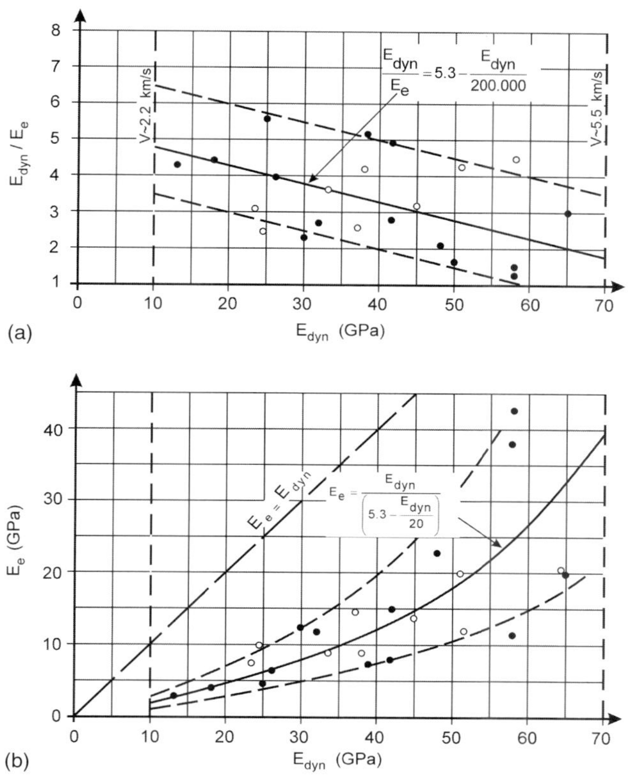


*Figure 6.15* Inequality between $E_e$ (static unloading) and $E_{dyn.}$ (or $E_{F\ dyn}$), for limestones tested at Yugoslavian dam sites. Kujundzíc and Grujíc, 1966.

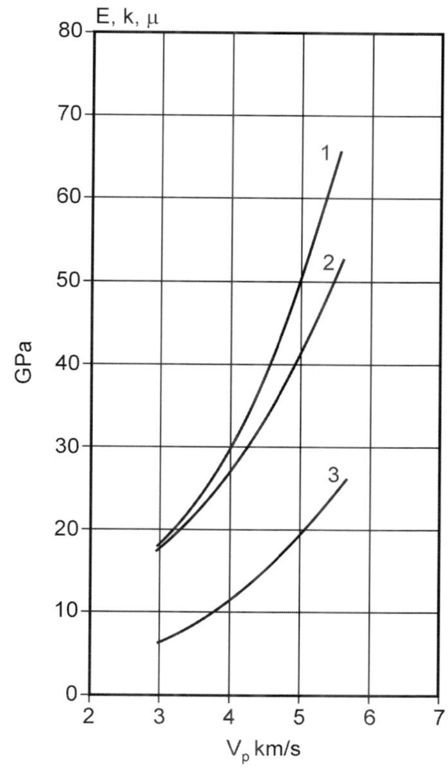


*Figure 6.16* Average curves for the field-scale dynamic elastic (E: curve 1), bulk (k: curve 2), and shear (μ: curve 3) moduli, in relation to refraction seismic $V_p$, based on 80 examples from igneous and metamorphic areas. Sjøgren, 1984.

## 6.3 Some examples of the three dynamic moduli

Sjøgren et al., 1979, gave average curves for the three dynamic moduli (Young's, bulk and shear), reproduced in Figure 6.16. These were based on 80 examples from three igneous and metamorphic rock areas (1 = dynamic E modulus, 2 = bulk modulus K, and 3 = shear modulus G). Each are shown in relation to measured values of $V_p$ from refraction seismic, which ranged from 3 to 5.5 km/s.

Deviations from the curves occur with changes in the dynamic Poisson's ratio ($\nu_d$). This is because of the fundamental linkage between $V_p$ and $V_s$:

$$\frac{V_p}{V_s} = \left(\frac{1 - \nu_d}{1/2 - \nu_d}\right)^{1/2} \qquad (6.12)$$

and because of the elastic equations (6.5 to 6.10) linking $V_p$ and $V_s$ with the dynamic elastic ($E_{dyn}$), bulk ($k_{\ bulk}$) and shear (μ) moduli. Concerning $E_{dyn}$ and μ, high Poisson's ratios give values below the curves in Figure 6.16 and low Poisson's ratios give values above the curves.

The authors found that the dispersion of the dynamic Poisson's ratio values always increased at lower velocities (e.g., dispersion ≈ 0.02 at 5.5 km/s, and ≈ 0.065 at 3.5 km/s). In Figure 6.17, their calculated values of dynamic Poisson's ratios (from equation 6.7) are shown integrated with $V_p$ magnitudes, with calculation of equivalent values of $E_{dyn}$ and $k_{\ bulk}$. The reduced moduli where even higher values of dynamic Poisson's ratio are calculated, can be readily imagined.

In this connection it is of interest to refer to the results of $V_p$ and $V_s$ logging of shallow boreholes. Results given by Chang and Lee, 2001 from a Korean tunnelling project, show the elastic 'constants' for a 30 m deep borehole. (Figure 6.18a). The $E_{dyn}$ moduli of 0.07 and 0.23 GPa are less than the bulk moduli (k) of 0.15 and 0.38 GPa in the residual soils in the top 10 m of the hole, where the dynamic Poisson's ratios are as high as 0.43 and 0.40 respectively, and $V_p$ only 0.5 km/s.

It is not until the soft rock is reached at 24 m depth, that $E_{dyn}$ of 5.92 GPa becomes greater than k of 5.20 GPa, with $V_p$ increased to 1.85 km/s, and dynamic Poisson's

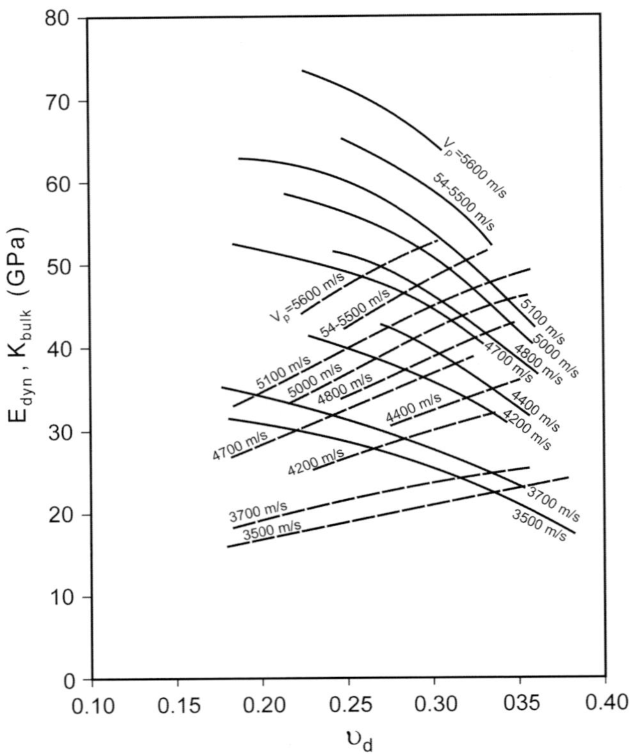


*Figure 6.17* Dynamic moduli $E_{dyn}$ and $k_{bulk}$, as a function of the calculated dynamic Poisson's ratio, for different $V_p$ values. The solid curves refer to the appropriate magnitudes of $E_{dyn}$, and the dashed-curves refer to the $k_{bulk}$ values. Note the reducing values of $E_{dyn}$ at higher values of the dynamic Poisson's ratios. Sjøgren et al., 1979.

ratio reaching a 'rock-like' 0.31. With the harder rock at 30 m depth, the respective values have changed to 24.07 GPa, 18.43 GPa, 3.45 km/s and 0.28.

Figure 6.18b shows a similar set of results for a deeper, mountain tunnel borehole, where $V_p$ varied from 5.2 km/s beyond 30 m depth, to 5.5 km/s beyond 150 m depth. Here $E_{dyn}$ increased from 50 GPa to beyond 60 GPa, with corresponding increases in rock density, and reductions in Poisson's ratio.

## 6.4 Use of shear wave amplitude, frequency and petite-sismique

In Chapter 5, we saw how S-wave *frequency* was used by Stacey, 1977, to monitor the closure-under-stress of dry, wet and clay-coated joints. P-wave *amplitude* was also found by Stacey 1977, Tanimoto and Ikeda, 1983, and Tanimoto and Kishida, 1994, to be particularly sensitive to joint closure and joint shearing mechanisms.

Here we will make a brief, premature digression to large-scale, high stress earthquake environments, to illustrate one of the many uses of shear wave monitoring. Chen et al., 1993, showed in research related to earthquake source mechanisms, that stable sliding and 'stick-slip' could be differentiated by interpretation of continuous shear-wave monitoring. *Continuous sliding* between rock interfaces under pressure reportedly caused a *continuous decrease in shear wave amplitude*, while stick-slip behaviour was detectable by a rapid drop in amplitude prior to or during 'slip', and by an increase in amplitude during each 'stick' period.

The authors suggested this method as an alternative to the classic but possibly unreliable observation that the ratio of $V_p/V_s$ showed a large decrease in earthquake source regions prior to earthquakes (Agarwal et al., 1973). (Anomalies in the amplitude ratio $A_s/A_p$ have also been reported before earthquakes.) A well known but tragic fact is that Chinese seismologists succeeding in predicting the 1975 Liaoning earthquake due to the decreased $V_p/V_s$ ratio, but failed one year later to predict the 1976 Tangsham earthquake which reportedly killed almost one million people.

In rock engineering projects, particularly arch dam sites investigated by French consultants Coyne et Bellier, direct use has been made of the correlation between the 'frequency of the second arrival' (the S-wave frequency) and the in situ static modulus of deformation (variously described as $E_d$ D, $E_{mass}$ and M in the foregoing text and referenced figures).

Schneider, 1967, and Londe, 1979, were instrumental in developing and publicising the 'petite-sismique' method, which was subsequently given some impetus by Bieniawski, 1978, with the additional test data and seismic data from some South African projects. At this time a linear relation between in situ static modulus (D or M) and S-wave frequency (f) was proposed (D = 0.054f-9.2), as shown both in Figures 6.19 and 6.20.

Subsequently the method, termed SCARABEE (Systeme Complet d'Analyse des Roches d'Appui de Barrages et d'Excavations) was standardised and expanded by Coyne et Bellier and Geodia, 1995, in France. Use was made of a normalised value of the Schmidt rebound (÷40), a normalised value of the dynamic modulus (÷44 GPa) and of the S-wave frequency (÷750 Hz), with a standard gain (×45 dB), to give a more comprehensive picture of rock mass conditions. The correlation between D and f is now considered to be of the non-linear form: ($D \approx 0.17 f^{1.7}$), as shown in Figure 6.21, from Coyne et Bellier, 1998 and Carrère, 1999.

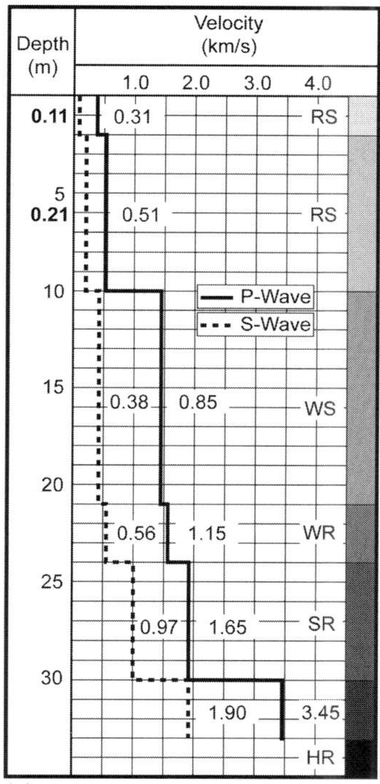


| $\gamma$ | $\nu$ | G | E | K | $\mu$ | E | K |
|---|---|---|---|---|---|---|---|
| gm/cm³ | Poission | kgf/cm² | kgf/cm² | kgf/cm² | GPa | GPa | GPa |
| 1.90 | 0.43 | 230 | 660 | 1520 | .02 | .07 | .15 |
| 1.90 | 0.40 | 840 | 2340 | 3830 | .08 | .23 | .38 |
| 2.00 | 0.38 | 2890 | 7940 | 10800 | .29 | .79 | 1.08 |
| 2.20 | 0.35 | 6900 | 18550 | 19900 | .69 | 1.86 | 1.99 |
| 2.40 | 0.31 | 22580 | 59180 | 52030 | 2.26 | 5.92 | 5.20 |
| 2.60 | 0.28 | 93860 | 240720 | 184320 | 9.34 | 24.07 | 18.43 |
| | | | | | | | |

(a)

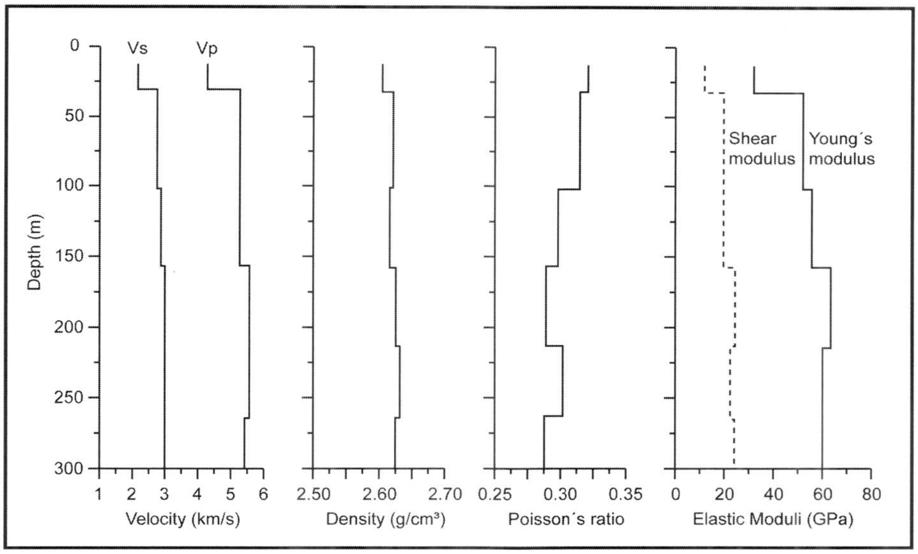


(b)

*Figure 6.18* The dynamic elastic moduli compared from down-hole $V_p$, $V_s$ and density logging, a) for a shallow 30 m deep hole, b) for a 300 m deep mountain tunnel borehole. Chang and Lee, 2001. Note GPa, and symbol conversion, G to $\mu$.

## 6.5 Correlation of deformation moduli with RMR and Q

The extensive *in situ* deformation modulus data presented by Ribacchi, 1988, which was reproduced earlier in this chapter, represents a useful source of field data with which to test how well the standard RMR-deformation modulus and Q-deformation modulus relationships relate to large scale data. The mean deformation modulus ($E_d$) data presented in Figure 6.4 for twenty different sites (as $E_d$ versus $V_p^2$) is presented *again* in Figure 6.22, this time with modulus in relation to RMR, shown as the lower scale.

The source of the solid curve of $E_d$-versus-RMR given in Figure 6.22 is shown in Figure 6.23, where both the

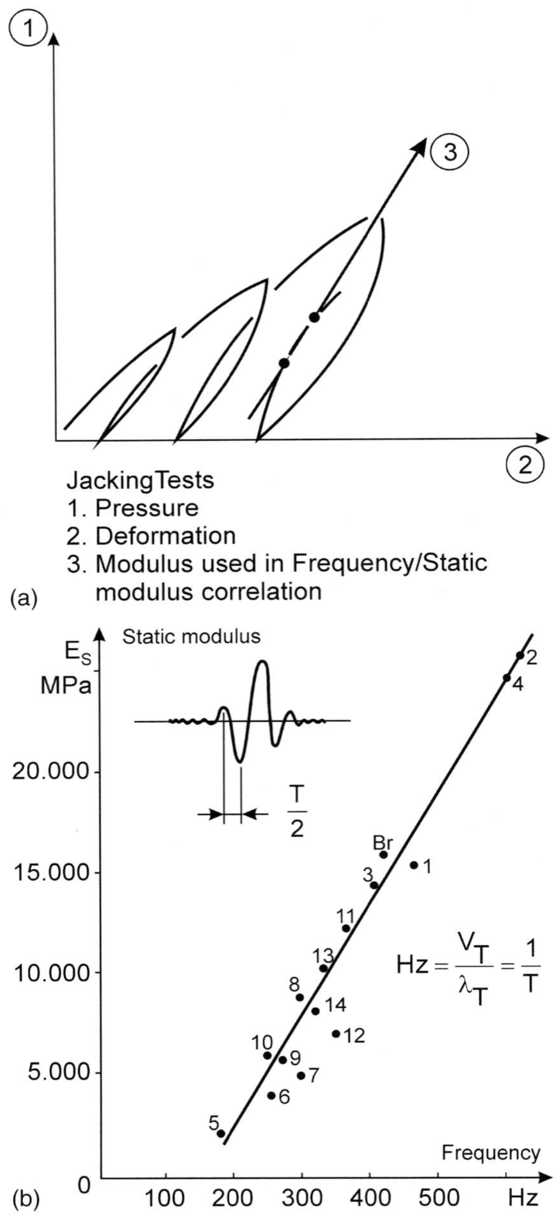


Figure 6.19 The originally assumed, linear correlation between S-wave frequency and tangent modulus of deformation, in the so-called 'Petite sismique' method of modulus estimation. Londe, 1979.

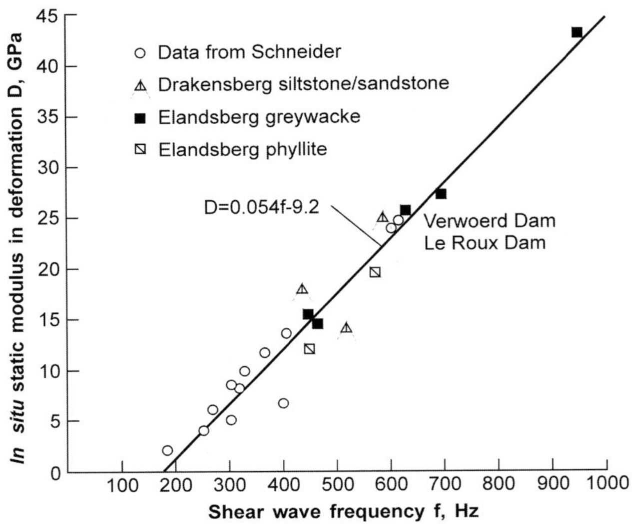


Figure 6.20 'Petite sismique' method of modulus estimation, with additional, linearly distributed data from South African hydropower projects, from Bieniawski, 1978. Earlier data from Schneider, 1967 is also shown.

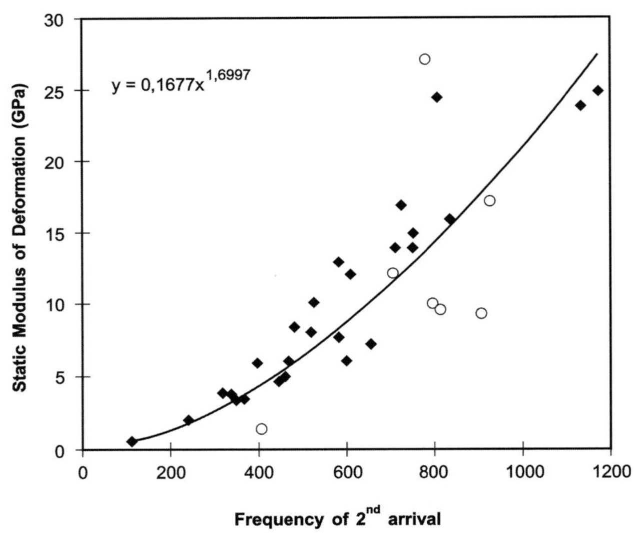


Figure 6.21 Updated, non-linear relationship between the static modulus of deformation and the shear wave frequency, from Coyne et Bellier, 1998, Carrère, 1999.

(Seraphim and Pereira, 1983) equation and the Q-system equation for 'static' deformation modulus prediction, are each shown.

It can be seen from Figure 6.23, that the Seraphim and Pereira, 1983, relation is as follows:

$$E_d = M = 10^{((RMR-10)/40)} \tag{6.13}$$

The Q-system based equation for deformation modulus prediction gives a similar trend:

$$E_d = M = 10Q_c^{1/3} \tag{6.14}$$

where $Q_c = Q \times \sigma_c/100$, so $Q_c \approx Q$ in a typical hard rock situation. This equation is the source of the

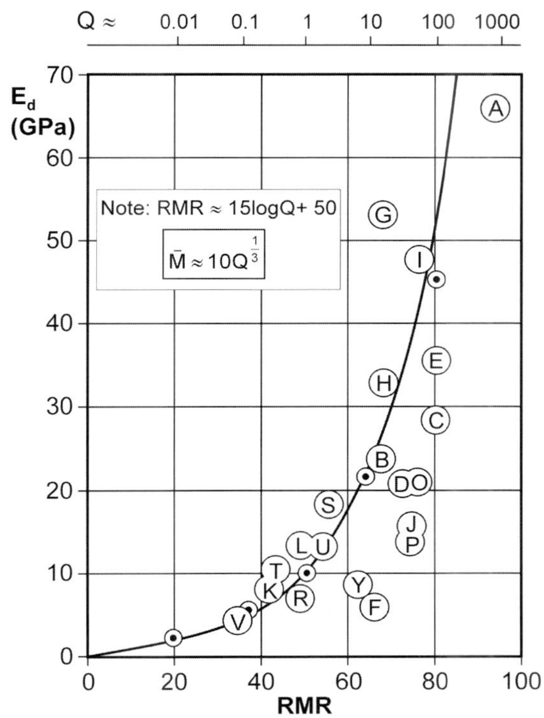


*Figure 6.22* RMR – $E_d$ fit to Ribacchi, 1978 data using the empirical Seraphim and Pereira, 1983 equation. The Barton, 1995 RMR to Q conversion: (RMR ≈ 15 log Q + 50) has been added to give the uppermost Q-scale, together with 'circled-dots' representing the $M = 10Q_c^{1/3}$(GPa) relation.

'circled-black-dots' in Figure 6.22, which lie on, or just below, the 'RMR-curve'.

The solid and dashed lines shown in Figure 6.23 are also seen to almost coalesce for Q-values less than 1 and RMR less than 50. If we further assume that Q and RMR can be related by the following simple relation (Barton, 1995):

$$\text{RMR} \approx 50 + 15 \log_{10} \text{Q} \tag{6.15}$$

then we can supply a Q-value scale to the top of Figure 6.22, where in approximate terms:

| RMR | = | 5 | 20 | 35 | 50 | 65 | 80 | 95 |
|---|---|---|---|---|---|---|---|---|
| Q | = | 0.001 | 0.01 | 0.1 | 1.0 | 10 | 100 | 1000 |

For lower strength rocks, the estimate of static deformation modulus is suitably lower using the Q-based equation 6.14 compared to the RMR-based equation 6.13, and actually more in line with the extremely low moduli that are usually measured, in the case of soft porous rocks. From earlier in this chapter one may note from Figure 6.7b (Kikuchi et al., 1982, Japanese tests) that one third of the deformation moduli lie in the range 0.05 to 1.0 GPa, with correspondingly low P-wave velocities (mostly 0.5 to 2.5 km/s).

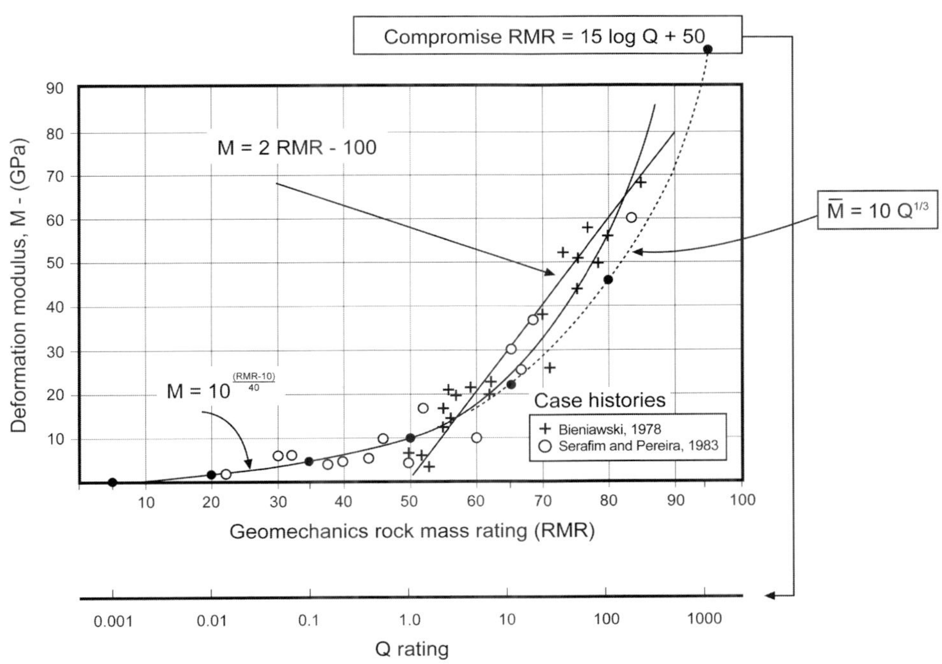


*Figure 6.23* A comparison of the old and the new RMR – deformation modulus relations (Seraphim and Pereira, 1983, and Bieniawski, 1989), and the more recent formulation involving Q or $Q_c$. Barton, 1995.

An example application of equation 6.14 may be useful here. Let us suppose that the conventionally logged Q-value (for tunnel support selection) in a weak sandstone of $\sigma_c \approx 2$ MPa is as low as 0.01 due to heavy jointing and low frictional strength. When we correct Q = 0.01 to $Q_c$, a value of $Q_c = Q \times \sigma_c/100$, or $Q_c = 0.0002$ is obtained. Substitution in equation 6.14 gives a predicted $E_d$ or M value of 0.6 GPa. The RMR equation 6.13 gives a predicted $E_d$ or M value of 1.8 GPa, when Q = 0.01 or RMR = 20. Such an estimate may be too high for soft rocks. As may be noted from Figure 5.36, in Chapter 5, an additional correction for porosity is also available in the 'Q-diagram', to improve fit to Tertiary type, weak, porous sedimentary rocks.

In the case of an exceptionally poor quality faulted zone or extremely weathered zone with $Q \approx 0.001$, with crushed rock pieces having a strength of only 1 MPa (almost a soil) we would need to use a $Q_c$-value of 0.00001 in equation 6.14, which would correspond to an exceptionally low $E_d$ or M value of only 0.2 GPa. This is close to the 'lowest possible' $M_{(min.)}$ value tabulated in Figure 6.14, and corresponds apparently (and perhaps realistically) to a P-wave velocity as low as 0.5 km/s.

In Figure 6.24, showing Ribacchi, 1988 data from Figures 6.5b and 6.6b, the following relation between $\overline{M}$ and $V_p$ is plotted. (This relation was given in Figure 6.14, see inset at top):

$$\overline{M} \approx 10.10^{((V_p - 3.5)/3)} = 10^{((V_p - 0.5)/3)} \qquad (6.16)$$

The objective was to again test the relation against the extensive data and data trends for limestones and dolomites given by Ribacchi, 1988. In this figure, the predicted $M_{(mean)}$ values given by this equation are seen to represent upper-bound envelopes in relation to the $E_d$ data given in the figure. The $M_{(min.)}$ tabulation given in Figure 6.14 appears to represent a lower-bound envelope in the case of the five trend lines given in Figure 6.22.

It is important to try to understand why such high velocities as, for example, 4.0 to 4.5 km/s can be associated with $E_d$ values as low as 5 to 10 GPa as can be seen in Figure 6.24b. Although the assumed porosities of the limestones and dolomites might affect $E_d$ values more than $V_p$ values, there is much evidence, for example from Figure 2.10 (Chapter 2, Fourmaintraux, 1975) that this cannot be so.

Possibly the most likely explanation, which has been raised earlier in this chapter, is that the large scale deformation modulus test as practised at dam sites (plate load,

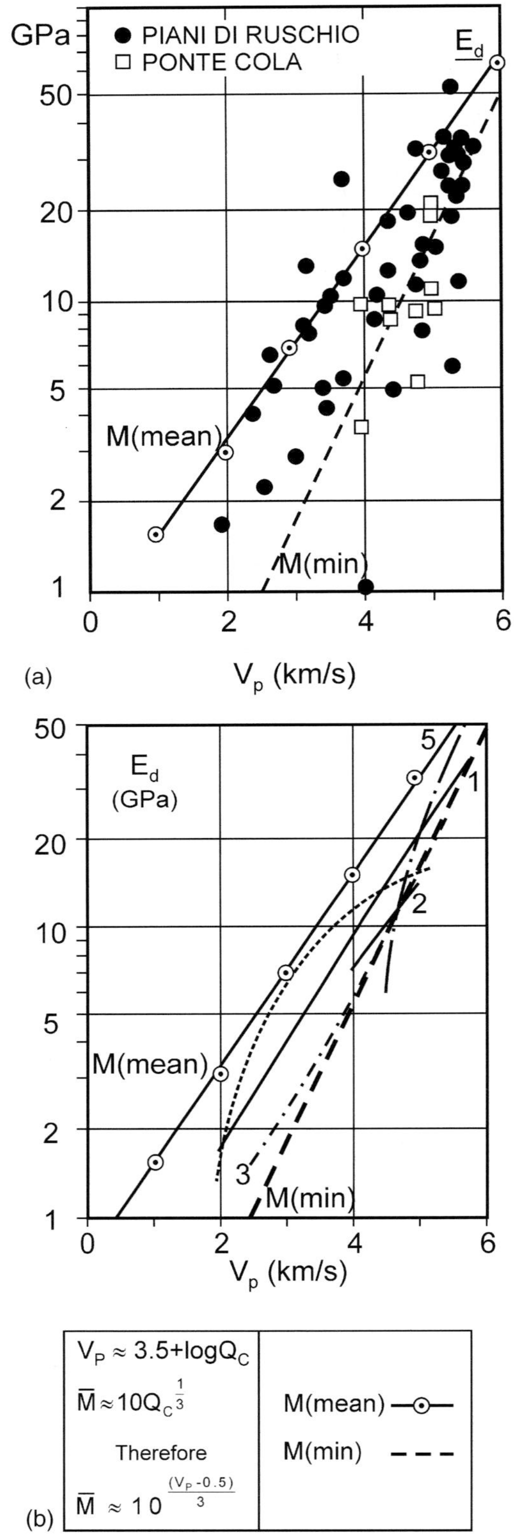


*Figure 6.24* Plate load data for limestones and dolomites (see Figs. 6.5b and 6.6b) from Ribacchi, 1988, and a comparison of the M-$Q_c$ model. ($M_{mean}$ is the solid line: left, $M_{min.}$ is the dashed line: right).

flatjack or occasionally pressure chamber) is nearly always testing an excavation disturbed zone (EDZ), in a loading direction parallel to the radial stress ($\sigma_r$), while the velocity measurement may be averaging velocities over a larger volume, and may tend to be recording velocities parallel to the $\sigma_\theta$ (tangential stress) direction, which is a much higher, maximum local principal stress, compared to the minimum radial stress. The latter obviously approaches zero, due to the effect of excavating a test adit.

The stress anisotropy ($\sigma_\theta >> \sigma_r$) around an excavation, which can explain higher values of $V_p$ parallel to $\sigma_\theta$ than parallel to $\sigma_r$, could therefore explain the higher-than-expected $V_p$ values, and the higher-velocity-than-expected $E_d$–$V_p$ trend lines seen in Figure 6.24b.

In this connection, it is significant that the empirical equations that were introduced in Chapter 5:

$$V_p \approx 3.5 + \log_{10} Q_c \text{ and } \overline{M} \approx 10Q_c^{1/3}$$

were developed independently (at different times) from different data sets. The first equation was developed from core logging and refraction seismic, and from deeper cross-hole tomography data. The second equation was developed from large scale modulus of deformation data where either the Q-value was known from independent logging or from approximate conversion from logged RMR values, using equation 6.15. The implied connection between M and $V_p$ (elimination of $Q_c$ between the above equations) to give equation 6.16, is a 'pure' link, which ignores the potential complication that anisotropic stresses ($\sigma_\theta > \sigma_r$) may cause anisotropic velocities in the EDZ test zones.

# 7 Excavation disturbed zones and their seismic properties

The existing surface of exposed rock that may be found where soil cover is absent can be considered as nature's disturbed zone. Steep mountainsides in rocky terrain, steep gorges or valley sides where dams may be founded or glaciated terrain are typical examples. As we have seen earlier the special coincidence of low stress, weathering effects and possibly more frequent jointing (with joint apertures above the limits of 'acoustic closure') make such zones extremely 'visible' in seismic refraction or sonic logging of shallow boreholes. Velocities may be up to several km/s lower in the near-surface zone than at greater depth, in some cases even when joint frequencies remain unchanged. In an attempt to get away from the weathered zone, investigators of dam sites and deep foundations may often construct adits or shafts for conducting deformability and other geomechanical tests. Unfortunately, however much care is taken, even to the extent of non-blasting methods, a disturbed zone results. The removal of stressed rock and its usual replacement by air at atmospheric pressure (a convenient 'definition' of tunnelling), results in a radial stress ($\sigma_r$) that approaches zero at the excavation walls. The tangential stress ($\sigma_\theta$) may assume many values (including negative) depending upon the existing stress anisotropy, joint orientation, rock strength and the disturbance caused by the excavation method.

## 7.1 Some effects of the free-surface on velocities and attenuation

The problem with the EDZ is that many of the geomechanical properties we are most interested in investigating at large scale are themselves affected by the process of obtaining access. Only a borehole-size intrusion may be considered nearly free of damage, but it will cause local disturbance to the stresses, that may influence permeability measurements and may also influence preliminary load cycles from borehole dilatometers. The 'skin-effect' around an excavation adit is far from 'skin-deep', and may extend many metres or to one or several radii from the excavation wall, especially in the case of softer rock that is highly stressed.

We must assume both a damaged zone and a disturbed zone that together have been given the nomenclature EDZ (excavation damage and disturbed stress zone). $ED_1D_2Z$ would be more specific, as with e.g. line drilling of an experimental tunnel, as at URL in Canada, we have only $ED_2Z$.

An unusual and instructive geophysical monitoring of the Dneiper ship lock excavation, which reached a depth of more than 20 metres, is shown in Figure 7.1. The effect of loosening caused by blasting, stress relief (and presumably inadequate slope reinforcement) is shown very

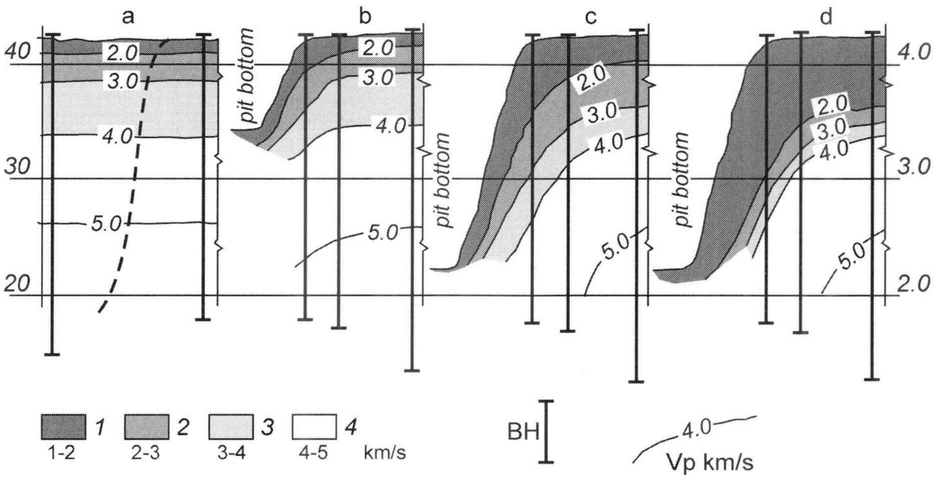


*Figure 7.1* Free-surface effect, and slope excavation (and degradation) effects on P-wave velocities. There is a 1 year delay between measurements c) and d). Savich et al., 1983.

clearly. There is a one-year delay between diagrams (c) and (d). Savich et al., 1983, refer to a 200–300% reduction in velocity, a 75 to 85% reduction in deformation modulus, and a 1 to 20 times increase in 'joint voids'.

This can be considered an excavation disturbed zone (EDZ) study related to slopes, in which velocity changes were related both to loosening and presumably to water drawdown. The same basic mechanisms may be at work around tunnels, in which EDZ effects are important for may reasons, including increased support needs, reduced deformation modulus, and frequently increased leakage (or permeability) in the case of larger or less stable tunnels. These effects may be remedied by high pressure pre-grouting with micro or ultra-fine cements. This will be discussed later. In the case of very stable rock, and small tunnels, inflow may reduce in relation to expectation. (Olsson, 1992).

Price et al., 1970, reported the results of seismic refraction measurements at three sites in the UK where bolting, anchoring and grouting of rock slopes was being designed. Their measurements at Edinburgh Castle in *complexly* jointed basalt foundations, suggested a 16% increase in $V_p$ as a result of bolting and anchoring (6 to 12 m deep) followed by grouting.

At Cheddar Gorge, which cuts through massive, but bedded limestones, an outer layer of about 5 m thickness showed a velocity of 2.9 km/s, while deeper into the walls of the gorge, the velocity was 5.5 km/s.

The vertical, pseudo columnar jointing in dolerite at Stirling Castle showed ratios of $V_{field}/V_{lab}$ as low as 0.1 due to stress relieved joints and weathering near the surface. Slope 'EDZ' can clearly exceed the EDZ associated with tunnelling, due to the added influence of weathering. The authors also described early trials with a downhole seismic probe for easier access to inaccessible locations behind steep slopes. Cross-hole measurements clearly indicated the presence of stress-relieved joints behind the slope faces.

The excavation efficiency of large dragline excavators in open cast coal mines as a function of efficient blasting practice was investigated by Young et al., 1985, using what they called 'seismic spectroscopy' to quantify the degree of brokenness of the rock before and after blasting. They utilised seismic attenuation and anisotropy diagrams to evaluate the efficiency of the blasting process. Certain frequencies were found to be attenuated due to the effect of fracture size and porosity (and void space). The principle of the method is shown in Figure 7.2.

The attenuation spectrum of the broken rock mass differed fundamentally from that of the unbroken, more homogeneous rock. As shown in this most interesting Figure 7.3, attenuation was larger and more *irregular* for

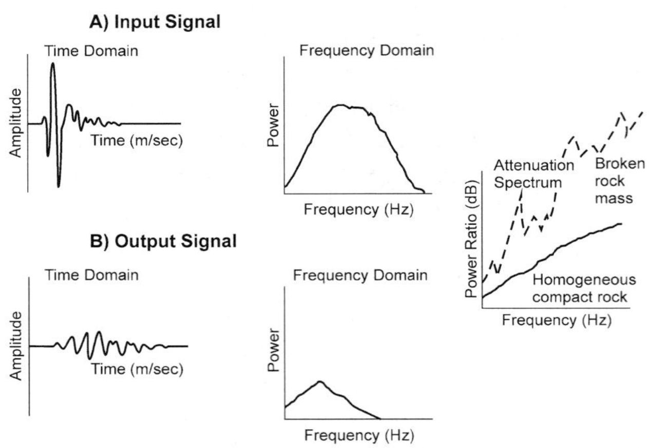


*Figure 7.2* Seismic attenuation compared in broken rock, and in compact rock masses. Young et al., 1985.

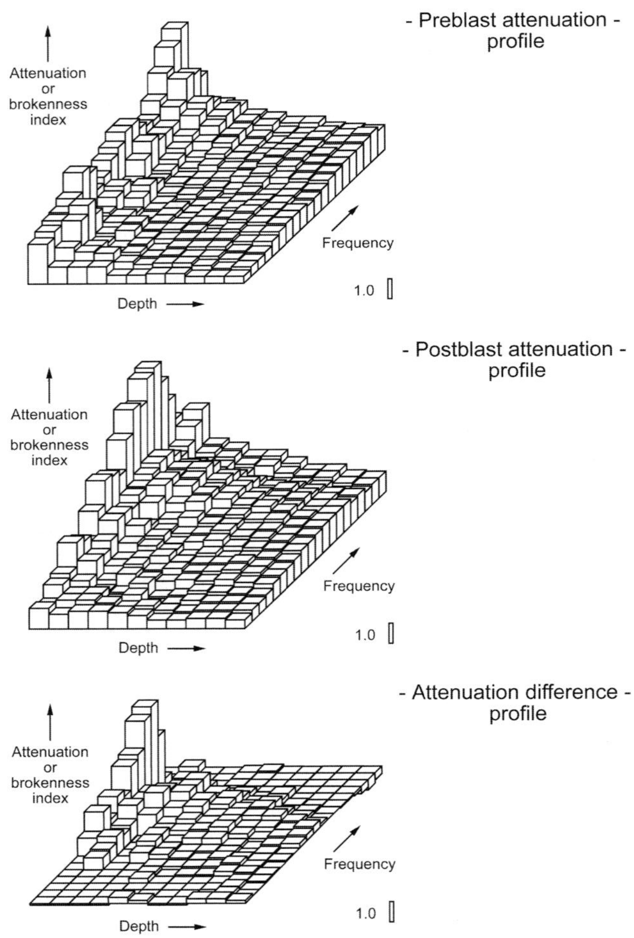


*Figure 7.3* Effect of blasting and depth below surface on the seismic attenuation. Attenuation is greatest near-surface, and at intermediate frequencies. Young et al., 1985.

shallow depths (where fractures open more easily) and more *regular* and limited at greater depths, where confinement limits fracture or joint opening. Associated compressional wave velocities typically changed from 1.1 and 1.8 km/s (at the surface/at greater depth) prior to blasting, down to 0.6 and 1.1 km/s respectively, after blasting.

These excavation disturbance effects were accompanied by rotations of both attenuation anisotropy axes and velocity anisotropy axes, presumably as a result both of disturbance to pre-existing joint patterns and blast-induced fracturing in the proximity of a high wall.

## 7.2 EDZ phenomena around tunnels based on seismic monitoring

In the years before the 1980s, reports of EDZ investigations in the rock mechanics literature were usually in connection with hydropower projects. An impressively early *model* for subsequent investigations was provided by Hasselstrøm et al., 1964, at a dam site in Sweden, who compared cross-hole and downhole sonic logging results in an investigation gallery, as shown in Figure 7.4.

Velocities were seen to fall from about 5.5 to 3.5 km/s in the outer 1 metre of their 1.5 × 2 m gallery. The

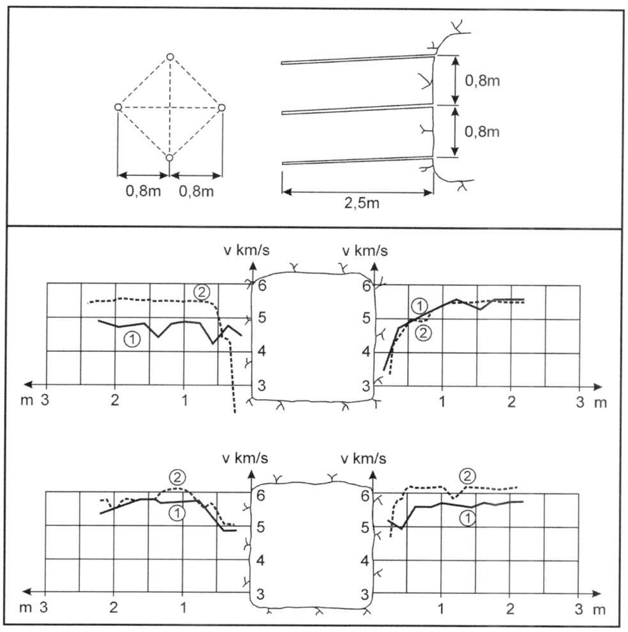


*Figure 7.4* An early EDZ measurement at a Swedish dam site gallery that compares cross-hole (1) and sonic log (2) results. Hasselstrøm et al., 1964.

authors cited the same reasons for the velocity reduction that we hear in more recent times at nuclear waste investigation sites such as Hanford (King et al., 1984), URL (Maxwell and Young, 1996), Äspö (Emsley et al., 1996) and Stripa (Olsson, 1992). Fracture formation, joint disturbance, stress redistribution and possible desiccation of the existing joint system were all listed by Hasselstrøm et al., in 1964, and are equally relevant (and complicated) today.

A classic EDZ investigation in relation to pressure tunnel design was reported by Kujundzíc et al., 1970. They performed a trial chamber test for investigating post-stressing effects on the concrete liner of their 5 m diameter, circular tunnel. In the course of this study, they utilised numerous grouting boreholes (32 in all) for conducting cross-hole seismic along the tunnel axis at eight different radial positions. Their results are shown in Figure 7.5.

They visualised the existence of three zones around the tunnel: 1) the loosened zone (with lowest velocities); 2) the stress bearing ring (with highest tangential stresses and velocities); 3) the uninfluenced zone (with declining velocities and background stresses). Their mean results ($V_p$ = 3.5 km/s at the tunnel wall, $V_p$ = 5.5 km/s at 1 m radius, and $V_p \approx$ 4.5 km/s in the undisturbed zone) shown in the centre of Figure 7.5 can be interpreted by means of the $V_p$-stress effect model as discussed in Chapter 5 (Figure 5.36).

Significant sophistication was added to the analysis of disturbed zone phenomena by Russian engineers, who analysed a variety of effects, including anisotropic velocities using ultrasonic and seismic methods. They emphasised the need to consider the use of different wave lengths (see Chapter 3, Lykoshin et al., 1971, Figure 3.12).

In relation to the EDZ logging of a shaft in diabase, these same authors used a time-average equation to estimate the joint void ratio (e) as a function of depth, as shown in Figure 7.6:

$$e = \frac{V_a(V_r - V_m)}{V_m(V_r - V_a)} \tag{7.1}$$

where $V_a$ was given the value 345 m/s assuming air-filled joint voids, $V_r$ is the intact rock velocity and $V_m$ the mean velocity at the depth of measurement. For example, we can substitute the P-wave velocities 0.345, 5.0 and 3.5 km/s respectively to obtain an estimated

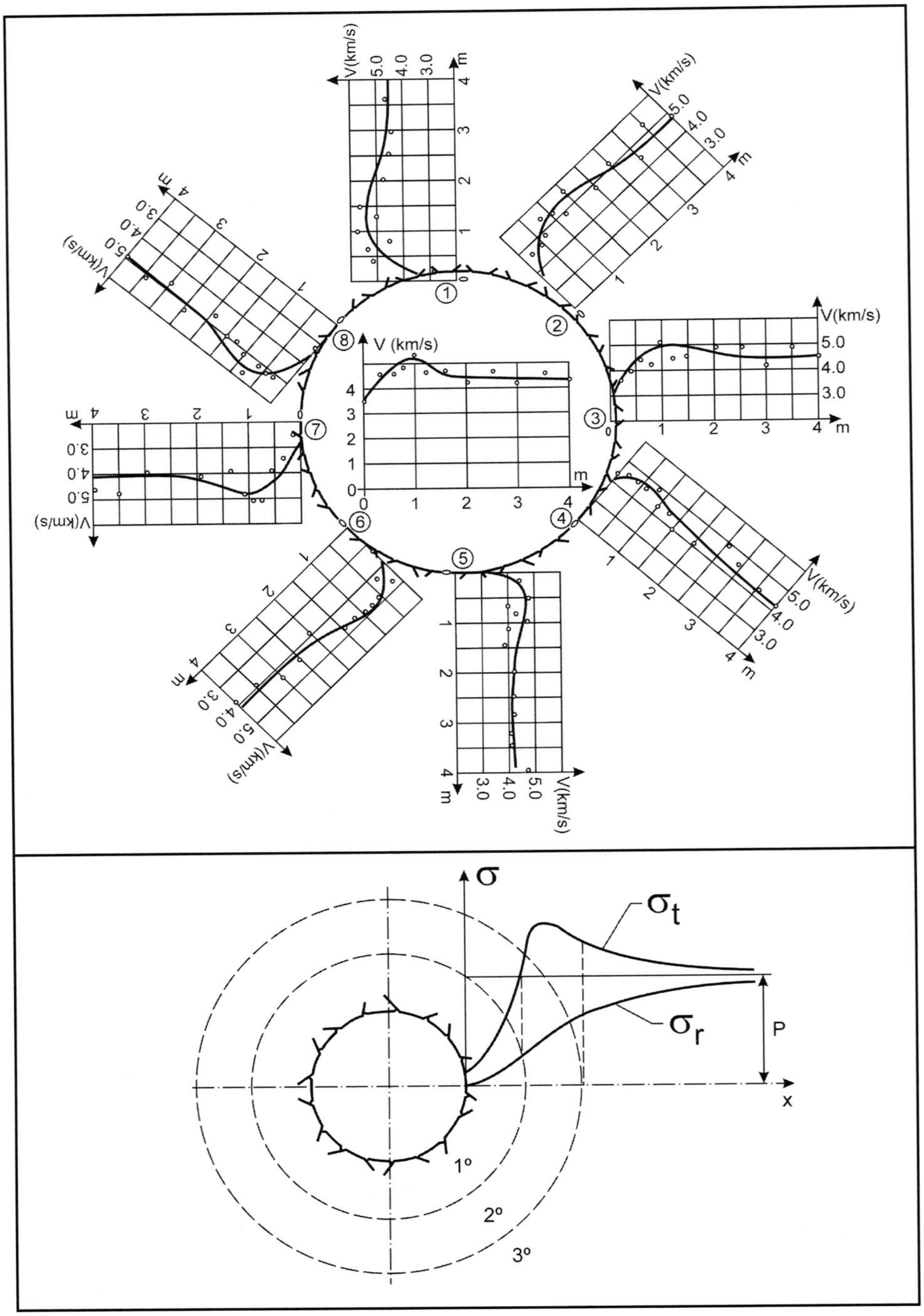


*Figure 7.5* Pressure tunnel investigations of seismic velocity as a function of radius, and therefore as a function of tangential stress, using cross-hole measurements. Kujundzíc et al., 1970.

joint void ratio of 0.03 (or 3%) at an imagined shallow depth into the wall of an excavation.

At greater depth when $V_m$ has increased, for example, to 4.5 km/s, the calculated joint void ratio will have reduced to 0.008 (or 0.8%). If we suppose that greater depth into the excavation wall (or the results of high stress) cause $V_m$ eventually to equal $V_r$, then clearly the void space will theoretically vanish. Lykoshin et al., 1971, recommended the use of ultrasonic logging with the minimum transmitter-receiver distance (e.g., 0.1 m) to obtain the relevant velocity distributions in such studies.

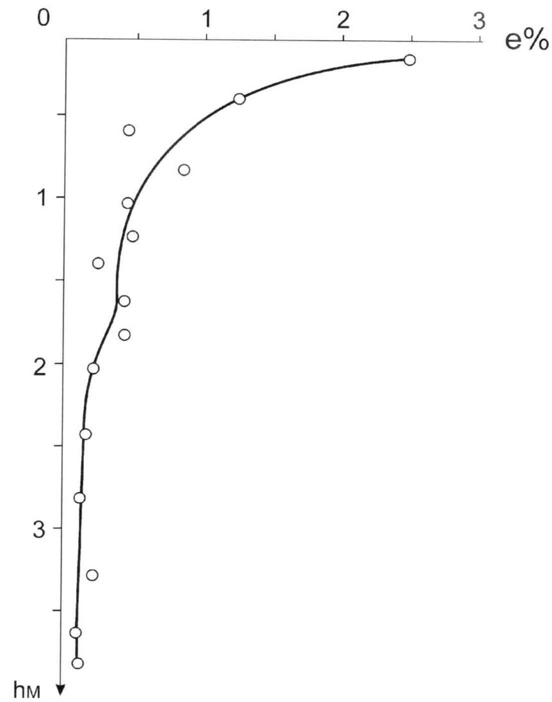


*Figure 7.6* Calculations of joint void ratio (%) in the EDZ surrounding a shaft in diabase, using a time average equation. Lykoshin et al., 1971.

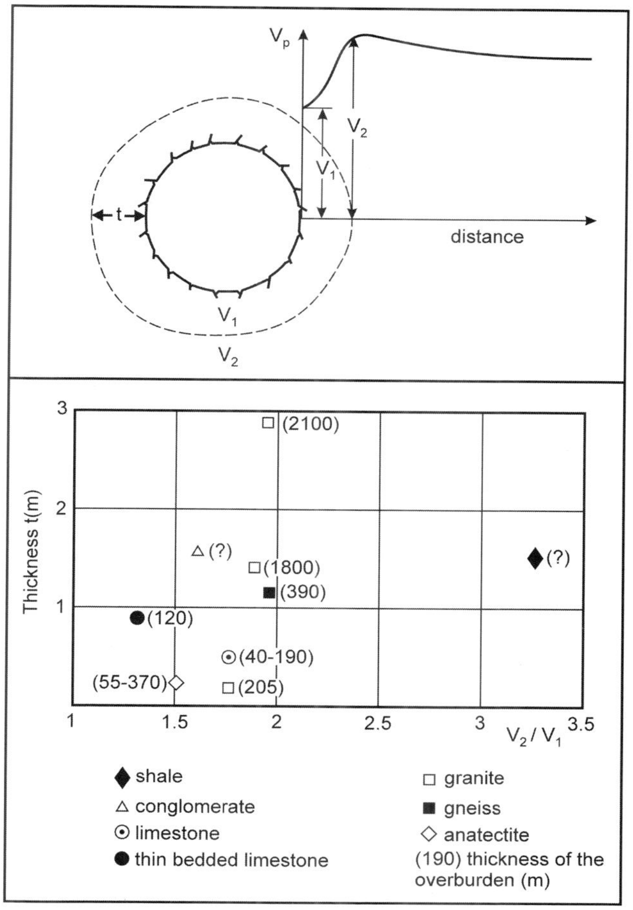


*Figure 7.7* Tunnel EDZ measurements assembled by Capozza, 1977 for a variety of rock types. Tunnel depths (m) are shown in parentheses.

The simple time-average equation given above is a particularly useful theoretical means of investigating the void space created in an EDZ, since we can also investigate the theoretical effects of drying out, by substituting the $V_a$ value of 0.345 km/s (as in equation 7.1) in place of a $V_w$ value for water of 1.44 km/s. However, the resultant values of $V_r$ for the intact rock and $V_m$ for the jointed rock will also tend to change due to drying, most rapidly in the case of $V_m$ (because of the influence of $V_a$) and less rapidly in the case of $V_r$.

The example of reducing $V_p$ with time due to gradual drying of an intact sample of granite (Nur and Simmons, 1969, Figure 2.18, Chapter 2) could be referenced at this juncture. Lykoshin et al., 1971, also gave a more complex expression for the velocity of eventual joint filling materials, based on the P-wave velocities $V_a$ and $V_w$, and on three other component velocities of the rock and rock mass.

Capozza, 1977, reviewed the results of a large number of EDZ-style investigations of tunnels. These were made using both seismic and cross-hole seismic techniques. Figure 7.7 shows the ratio of velocities $V_2/V_1$ in the first few metres depth in numerous tunnel walls. Values in parentheses show the depth of the tunnel in metres. It is typical for $V_p$ to be reduced by 25 to 50% close to the tunnel walls. The 3 m thick 'skin' observed in the tunnel with 2100 m of overburden is presumably a result of stress slabbing in the granite.

Bernabini and Borelli, 1974, described a variety of early seismic tests performed at hydroelectric projects in Italy in the 1950s, 1960s and early 1970s. In hydropower tunnels in gneiss and granite, they measured EDZ effects using seismic refraction techniques along four lines in each tunnel (two in the arch, and two in the lower walls). There were marked reductions of velocity (at least 50 to 60%) in the measurements made just 100 metres from the tunnel entrance (for example 4.8 to 2.2 km/s) as can be seen in Figure 7.8. Further from the tunnel entrance (300 m and 500 m) the higher quality and higher velocity rock showed less marked reductions in velocity due to excavation disturbance.

Dhawan et al., 1983, described seismic refraction measurements at a dam site in the Himalayas and attempts to correlate results with 1100 metres of core logging and in situ testing in the same quartzitic and slaty phyllites. The in situ testing was performed in boreholes and in an investigation drift in the dam foundations, shown in Figure 7.9.

The values of deformation modulus, ranging from about 0.5 to 3 GPa from plate jacking and flat jack tests, seemed to correlate poorly with the wider range of seismic velocities (1.5 to 4.0 km/s), possibly due to the discrepancy of the plate jacking loading direction (radial) in relation to the perpendicular-to-tangential-stress direction followed by most of the seismic refraction ray-paths. The flat-jack loading could possibly have been performed in the location of minimum tangential stress, if horizontal slots, rather than vertical slots were cut in the wall, and if horizontal along-the-valley stresses were maximum principal stresses in the location of the tests.

The disturbed zone surrounding the drift showed a thick low velocity zone (1.4 km/s) of about 3 m thickness in the first 10 metres of the drift. This exceeded the diameter of the drift (2 metres). Further inside the drift, in less weathered material, the higher velocity material ($V_p$ = 2.85 km/s) came closer to the walls of the drift, as shown in Figure 7.9.

In the same year, Bonapace, 1983 described rock mechanics testing for the design of an extremely high head (1260 m) inclined pressure shaft. Deformation modulus testing at different scales culminated in a radial

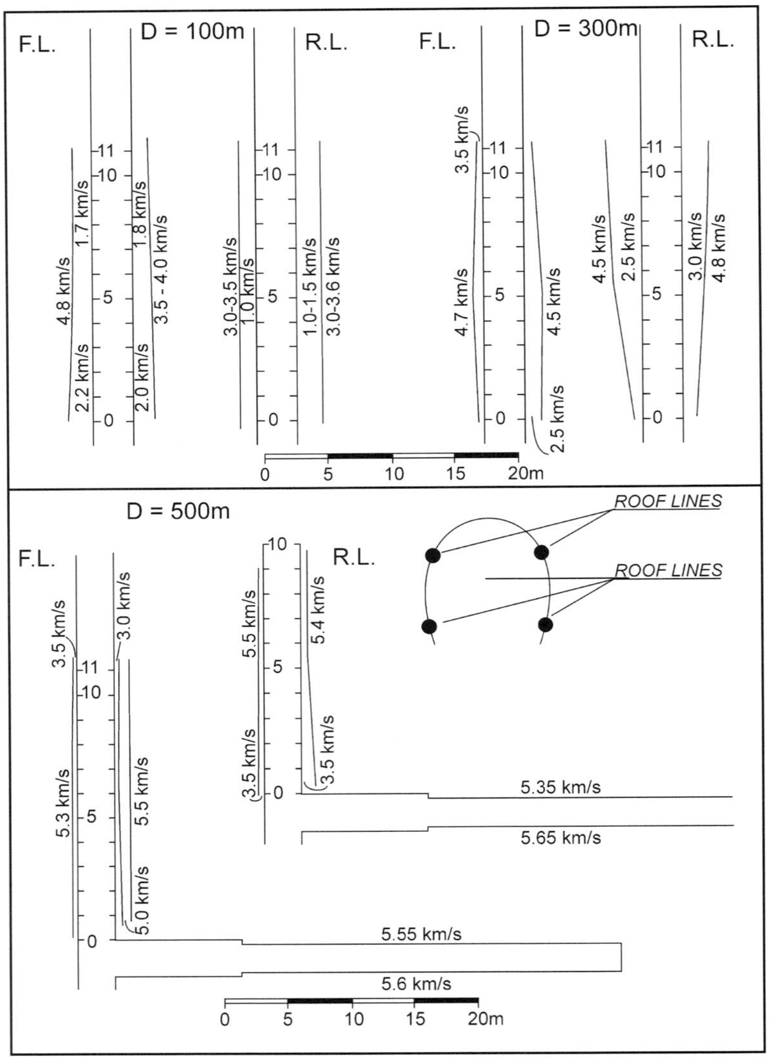


*Figure 7.8* Seismic refraction profiles along tunnels in gneiss and granite give a crude indication of EDZ. Berabini and Borelli, 1974.

jacking test chamber (diameter = 3 metres) which gave the lowest modulus of 11.5 GPa. In order to extrapolate this result along the shaft alignment, 4 m deep boreholes were drilled at 10 m intervals, in order to measure seismic velocities and correlate these with the measurements at the radial jacking location.

These EDZ-style measurements in the 4.8 m diameter shaft wall, shown in Figure 7.10, gave velocities as low as 1 to 1.5 km/s in the outer 1.0 metre of the drill-and-blasted shaft, but velocities no lower than 2 km/s in the outer 0.5 metre of the TBM excavated section. In general, velocities were 1 km/s higher in the case of the TBM excavation, over the depth range of 1 to 4 metres from the shaft wall.

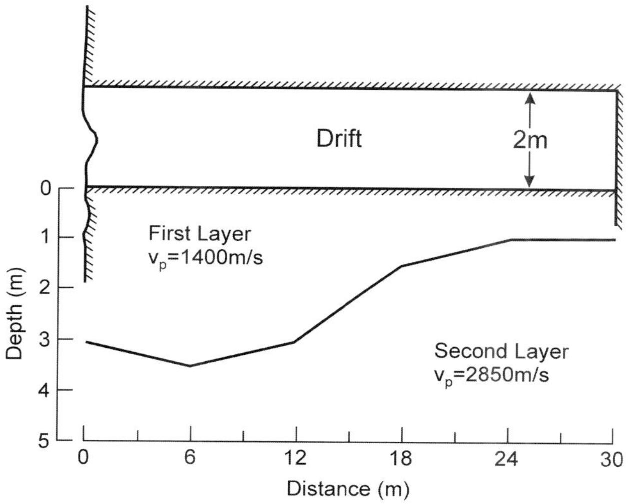


*Figure 7.9* EDZ effect accentuated by the initial weathered zone at an investigation adit at a dam site in the Himalayas. Dhawan et al., 1983.

Downhole logging by Stead et al., 1990, in a potash mine in Saskatoon, Canada also showed distinct EDZ effects as shown in Figure 7.11a and b. Holes in the roof showed velocity anomalies connected with the opening of clay seams (or the presence of clay seams), while anomalies and reductions in velocity in the walls of mine entries were reportedly associated with stress-induced fracturing of the potash. Freshly mined openings showed only slight reduction in velocity in the future EDZ zone, as fracturing took time to develop. The P-wave velocity clearly appeared to be more sensitive to these anomalies than the S-wave velocity, judging by the two figures.

Holcomb, 1988, found that mining of excavations in the bedded salt of the WIPP site in New Mexico had greatest effect on the *attenuation* of compressional waves. The EDZ could be detected to a depth of some 3 metres (about 1/2 D). The mining induced radial stress relief and increased tangential stress would tend to cause under-saturation due to dilation of the salt. The reduction in *compression wave amplitude*, which increased with time after excavation, was a stronger indicator of EDZ than the minor reduction in P-wave velocity (<0.5 km/s).

In softer rocks such as chalk the EDZ effect can be even more marked in terms of percentage change in velocity. At a storage cavern site in Eocene chalk with one predominant set of vertical joints of about 1 m spacing and a second, less well developed set, McDowell et al., 1992, showed mean background velocities of 2.34 km/s reducing to 1.47 to 1.56 km/s in the outer 1 to 2 metres of the wall of the pilot tunnel.

Borehole investigations at the same site had shown significant increases with depth; $V_P$ was 1.25 km/s from

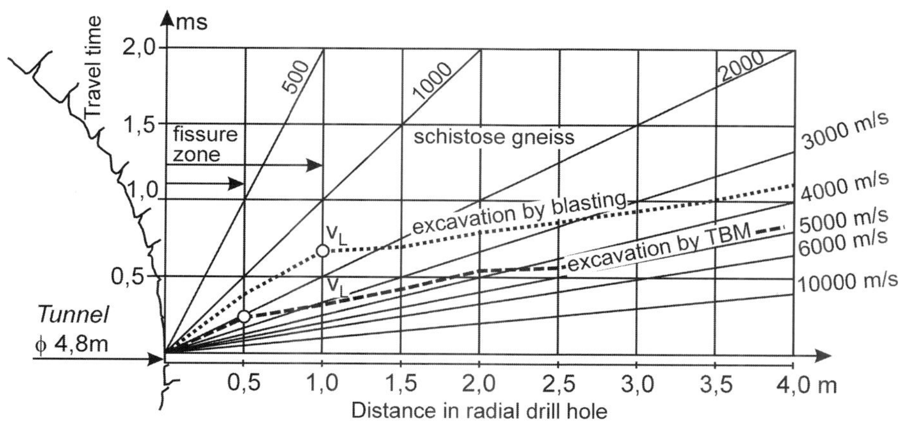


*Figure 7.10* Contrasting EDZ effects on $V_P$ from drill-and-blast and TBM excavated shaft. Note the travel time-distance-velocity plotting format. Bonapace, 1983.

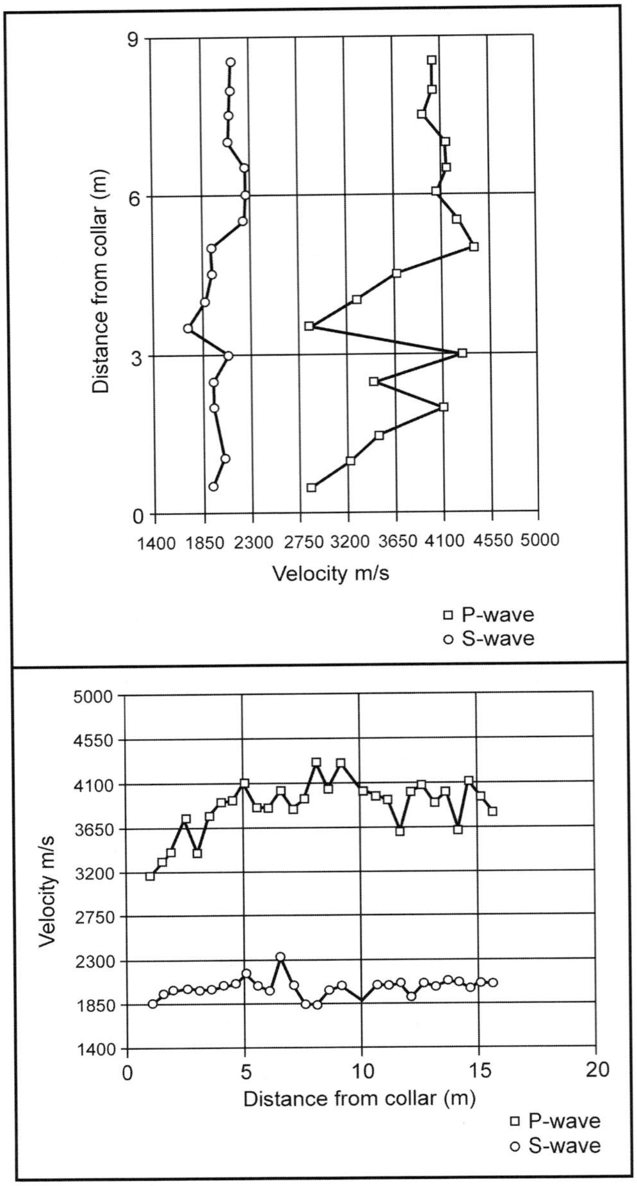


*Figure 7.11* EDZ effect and clay seam velocity anomalies in a potash mine roof. $V_P$ is seen to be more sensitive to the EDZ than $V_S$. Stead et al., 1990.

0–30 m, about 2.2 km/s at 70 m depth, and 2.6 km/s at 90 m depth.

(This closely follows the $Q_c = 0.1$ velocity–depth curve in Figure 5.37, with a 10% porosity correction, but requires an upward correction to an appropriate rock quality Q-value: e.g. to $Q = 1$ if $\sigma_c$ was 10 MPa.)

A corresponding increase in S-wave velocities was not recorded in the chalk; values were mostly 1.4 km/s, with small variations ($\pm$0.3 km/s) related to lithology and RQD variations. This suggests a stress-related increase in $V_P$, rather than systematically reduced jointing. In this relatively low velocity chalk, the authors experienced the seismic restriction caused by shotcreting of a fault zone. This stiffer, high strength layer caused a thin, high velocity surface layer to be registered.

## 7.3 EDZ investigations in selected nuclear waste isolation studies

Starting in the late 1970s, but increasingly from the 1980s and since that time, there have been a growing number of nuclear waste disposal studies at potential or generic sites, and at underground laboratory facilities. Some of the facilities have become 'household' names in rock mechanics circles, starting with Stripa (in Sweden), Grimsel (in Switzerland), URL – underground research laboratory (in Canada), BWIP – basalt waste isolation plant (in the USA), Äspö (in Sweden) and several other prominent facilities. In this review of selected EDZ data, significant results from four of these facilities will be described here. In Part II, Chapter 16, further results from additional sites in the United States will be described, due to their seismically-relevant *jointed-rock-mechanics* aspects.

### 7.3.1 BWIP – EDZ studies

King et al., 1984, and King et al., 1986, reported on a series of cross-hole seismic measurements performed in a flow-entablature, columnar-jointed basaltic rock mass. The columns were regular but sinuous, 0.15 to 0.36 m in thickness, dipping 70 to 90°, with frequent low angle, discontinuous cross-jointing. The measurements were made between four horizontal boreholes drilled 12 metres into the wall of a drill-and-blasted underground opening, at 46 m depth, and theoretically above the water table. The objective was to investigate the effect of blast damage and stress redistribution, i.e., two of the assumed chief components of the EDZ or excavation damage and disturbed zone.

A vertical separation of boreholes 1 and 2 of 3 m, and a horizontal separation of boreholes 3 and 4, also of 3 m (in the form of a cross) allowed both the vertical, horizontal and diagonal paths to be investigated, thereby crossing different joint sets (predominantly sub-vertical columnar) at different angles. Figure 7.12a and b show the test set-up.

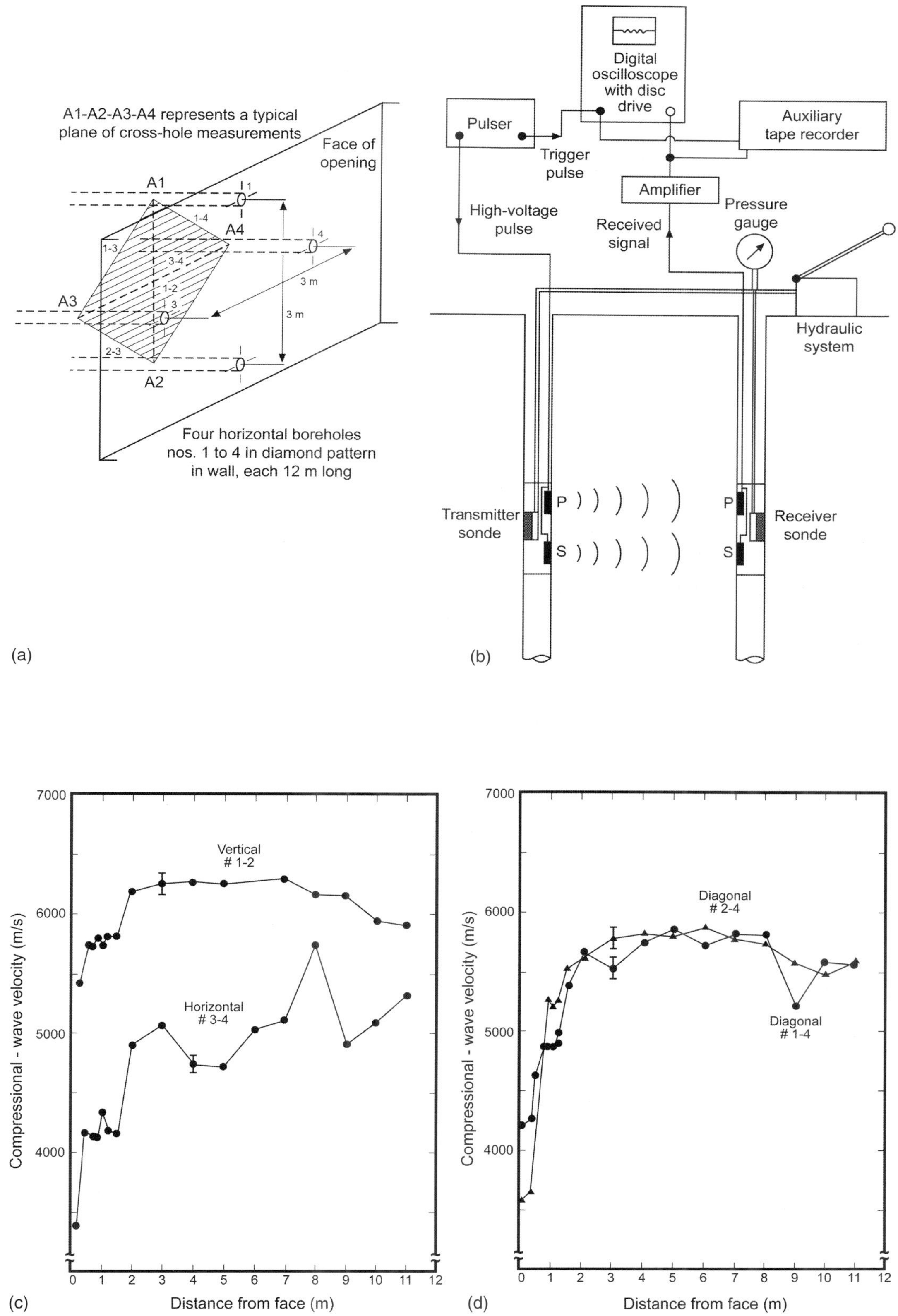


*Figure 7.12* Experiment for measuring EDZ effects in a tunnel wall (or 'face'), at the Basalt Waste Isolation Project (BWIP). a) Test set-up, b) test principle, c) and d) four of the selected ray paths and the effect of the EDZ on the P-wave velocities. King et al., 1984.

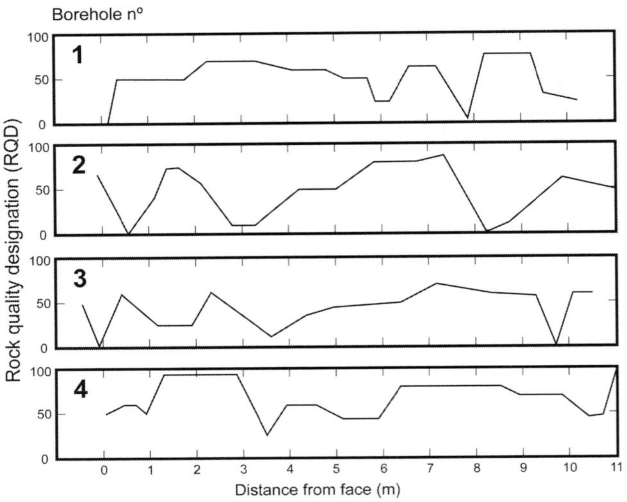


*Figure 7.13* RQD variation in the boreholes in basalt showed no marked EDZ effect. King et al., 1984.

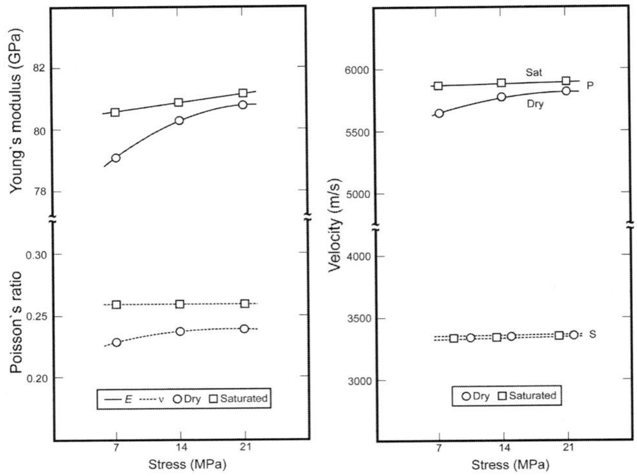


*Figure 7.14* Effects of stress on laboratory samples of the basalt, in the dry and saturated states. King et al., 1984.

The large contrasts in $V_p$ values for the vertical path (#1 to #2) and for the horizontal path (#3 to #4) close to the opening (1.5 to 2.0 km/s difference) are indicated in Figure 7.13c. A significant reduction in velocity of 1 to 2 km/s, most in the diagonal directions shown in Figure 7.13d, is seen in the outer 2 to 3 metres. There is also some indication of a tangential stress concentration effect; the background (far-field) velocity of about 5.4 to 5.8 km/s appears to be elevated by about 0.5 m/s from about 4 to 8 m depth in the wall.

Signs of loosening effects of the sub-vertical joints are evident in the horizontal velocity reductions. The additional effects of potential stress anisotropy and joint anisotropy are clearly illustrated in King et al., 1984, 1986, results. The authors registered no consistent trend in RQD values with depth. Figure 7.13 does not suggest that increased jointing caused the marked velocity reduction.

The velocity reductions are a product of blast-damage, stress relief (and redistribution) and possible reduction in moisture content. The authors noted water flow from some of the horizontal holes during the tests and originally assumed more or less saturated conditions. The horizontal and (to a lesser extent) the diagonal seismic measurement paths crossed the more open columnar joints, and these features clearly opened most as a result of excavation, giving the strongest reductions in velocity (55% to 65%).

King et al., 1986, also investigated the effects of stress (7 to 21 MPa) and saturation on $V_p$, on $V_s$, on the dynamic modulus, and also on the Poisson's ratio of ten intact specimens. Comparison of Figure 7.12 and Figure 7.14 indicate a higher velocity in the vertical in situ planes than in the intact laboratory samples (6.3 km/s compared to 6.0 km/s). In the dried-out state, the parameters showing most response to stress level were $V_p$ and the dynamic Young's modulus.

The authors also analysed in some detail the seismic quality factor which also has the symbol Q. This seismic quality (which we have termed $Q_{seis}$ to avoid confusion) is *inversely* proportional to the attenuation coefficient, for a given frequency. Only in the vertical direction was it significantly increased *in situ*, perhaps in response to the tangential stress increase. King et al., 1986 laboratory test data for the same rock also showed the positive effect of stress on $Q_{seis}$, with the S-waves showing the strongest reduction in attenuation and the highest increase in $Q_{seis}$, as clearly shown in Figure 7.15.

In a subsequent paper, Zimmerman and King, 1985, used the known effects of saturation on dynamic moduli. Theoretically, shear modulus is unaffected, while bulk modulus increases linearly to equal the uncracked rock when s = 100%. They suggested that the degree of saturation and the joint or crack density were each contributing to the measured reductions in cross-hole velocities $V_p$ and $V_s$, as shown in Figure 7.16a and b.

These results are presented in order to emphasise the possibility of drying out of some of the joints, despite water flow from some of the holes. The theoretical analysis of crack density does not appear to be supported by the RQD measurements in general (Figure 7.13) but is perhaps an expression of joint void ratio changes (equation 7.1) with the joints closest to the tunnel wall showing the largest voids and therefore suggesting an apparent (but false) increase in joint density.

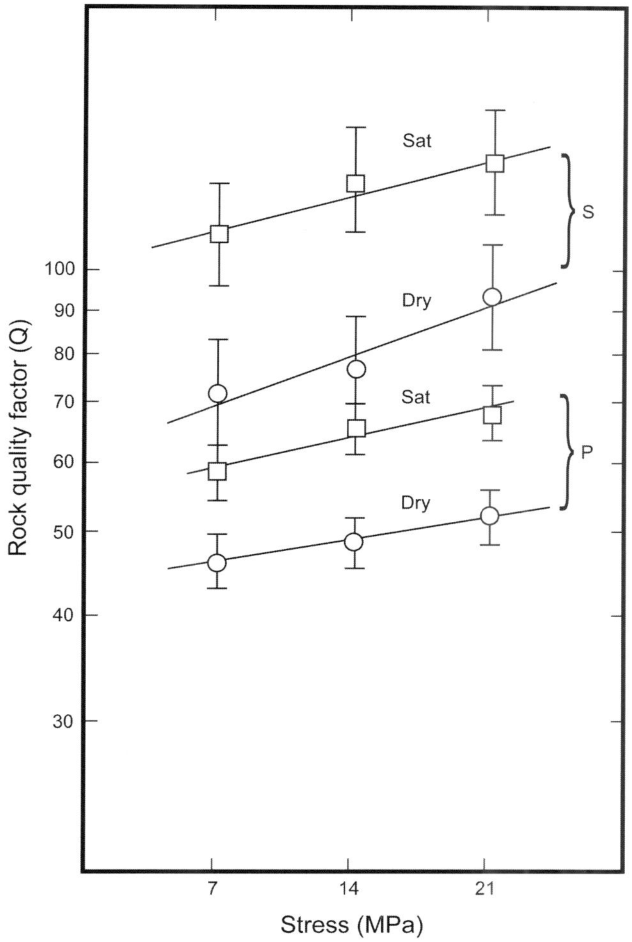


*Figure 7.15* The effect of stress on the seismic quality Q, the inverse of attenuation, for the basalt samples. Note that the authors' use of the term 'rock quality factor' is left unchanged in this reproduction of their drawing. (As will be discussed in detail in Part II, Chapter 10, there are indeed certain numerical similarities between *rock quality* Q and *seismic quality* Q). King et al., 1984.

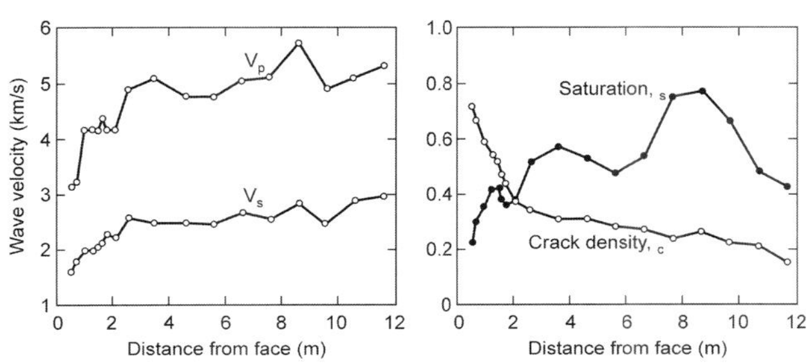


*Figure 7.16* $V_p$ and $V_s$ as a function of depth in the tunnel wall (or 'face'), saturation level and crack density variations in the EDZ in BWIP basalt. Zimmermann and King, 1985.

### 7.3.2 URL – EDZ studies

The underground research laboratory (URL) in Manitoba, Canada was the subject of numerous and very informative rock mechanics and geophysical studies. Novel instrumentation was used for monitoring responses to test tunnel excavation through the limited areas of natural fracturing. However, the dominance of massive, unjointed, highly stressed granite resulted in particular focus on stress-related EDZ, with down-hole sonic logging and acoustic emission monitoring, together with parallel laboratory tests and numerical modelling studies. Several phases of investigation of EDZ effects, and several scales of investigation were accomplished in this mostly massive, highly stressed granite. (Martin et al., 1995).

We can start at the smallest scale, by looking at the effect of the high in situ stresses, and mode of excavation, on the state of micro-cracking in cored samples. The French group, ANDRA (Homand-Etienne and Sebaibi, 1996) selected core samples from radial boreholes drilled in the walls of a drift where excavation was either by *normal blasting* or by *smooth blasting*. Micro-cracking, partly induced by the excavation and partly pre-existing, indicated $V_p$ reductions of about 1 to 1.5 km/s in the outer 0.8 m of the normally blasted tunnel, and reductions of about 0.5 km/s in the outer 0.5 m of the smooth blasted excavation, as shown in Figure 7.17. In each case the core samples recovered from various depths into the tunnel wall had been machined into cubic specimens prior to the velocity measurements.

The effects of highly anisotropic, sub-horizontal stresses at the URL were studied in a unique test tunnel ($ED_2Z$ only), at the 420 m level. (Figure 7.18). Excavation was by line drilling and reaming, followed by mechanical breakout to avoid blast damage. Principal stresses of approximately 60, 45 and 11 to 15 MPa, caused classic 'break-out' resembling that in a borehole.

The originally intended circular, 3.5 m diameter tunnel, was excavated parallel to $\sigma_2$ in order to maximise the potential for stress-induced fracturing. The isotropic-elastic theoretical tangential stresses of 165 MPa ($3\sigma_1 - \sigma_3$) in the 'rotated' 11o'clock roof and 5o'clock floor, and −15 MPa ($3\sigma_3 - \sigma_1$) in the side walls, caused prominent V-shaped notches of rock failure. Associated micro-cracking and stress changes were imaged tomographically (Maxwell and Young, 1996) as will be shown shortly.

The walls of boreholes drilled in this carefully excavated $ED_2Z$, provided important information about stress-related disturbance. This was measured directly

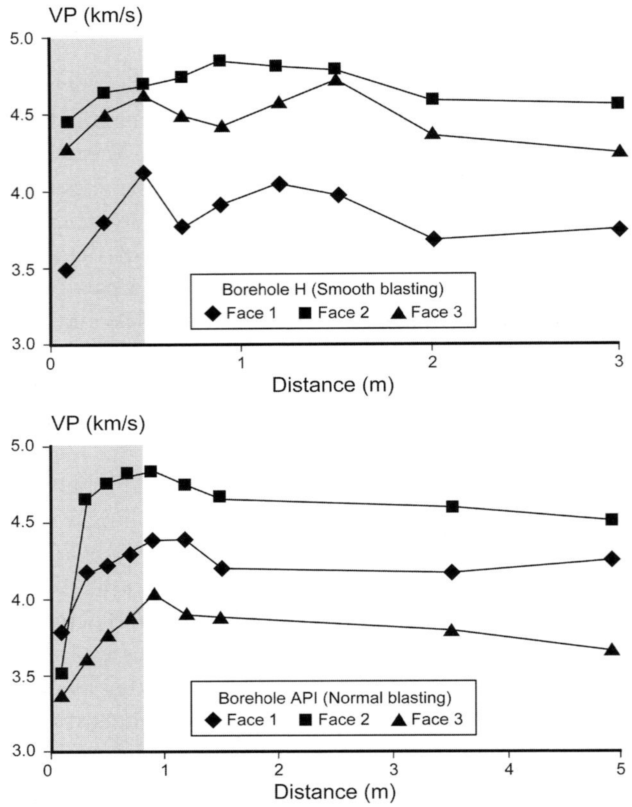


*Figure 7.17* Core samples recovered from 3 m long radial holes at a drift in the highly stressed URL, where both *smooth-blasting,* and *normal blasting* were used, show evidence of micro-cracking having originated in the differently configured EDZ. Homand-Étienne and Sebaibi, 1966.

in 1 m deep boreholes using a micro-velocity probe with 10 cm separation of the transducers. Maxwell and Young, 1996, installed their probe in holes drilled both into the *tensile region* in the test tunnel wall, and into the *compressive region*. Since their measurements of velocity were in a radial direction, micro-cracking was registered in the 'fast' direction and in the 'slow' direction, the latter perpendicular to the micro-cracking.

Figure 7.19a shows how both the P-wave and S-wave velocities reduce by some 10% in the outer 1 m of the tunnel wall in the tensile region of tangential stress. The unstressed core from the same hole showed additional damage and lower velocities due to the loss of radial stress (Figure 7.19b).

Carlson and Young, 1993, and Maxwell and Young, 1996, also recorded acoustic emission (AE) locations during careful (mine-by) extension of the 420 m deep test tunnel. Calculated compressional wave velocities showed quite strong anisotropy in this massive granite. This was obviously caused by the principal stress anisotropy (60, 45 and 11 to 15 MPa). The lower hemisphere $V_p$ stereonet shown in Figure 7.20 gives a clear indication of the link between stress anisotropy (caused by the very different maximum and minimum tangential stresses), and P-wave velocity. The authors' velocity data showed a clear EDZ effect in the first 1 metre depth of tunnel wall with velocities reducing from about 5.8 km/s to 5.3 km/s in the case of the vertical direction. The acoustic emission results confirmed that rock failure was initiating just inside the tunnel wall, orthogonal to the $\sigma_1$ direction. The authors

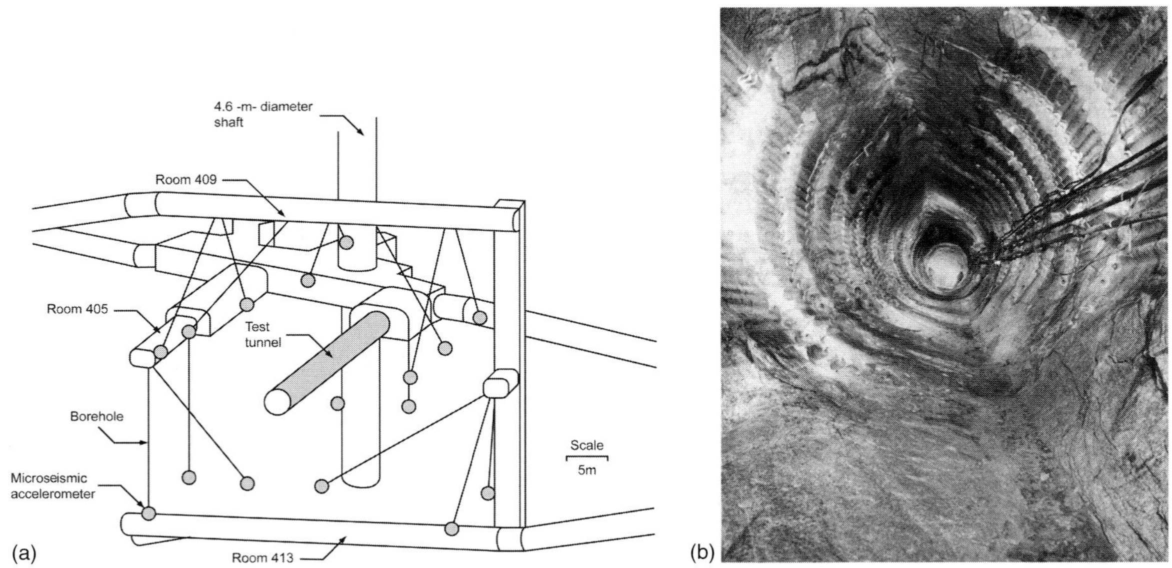


*Figure 7.18* A unique line-drilled and hand-mined test tunnel at URL, showed classic break-out related to the sub-horizontal $\sigma_1$ orientation. Maxwell and Young, 1996. Note location of AE sensors.

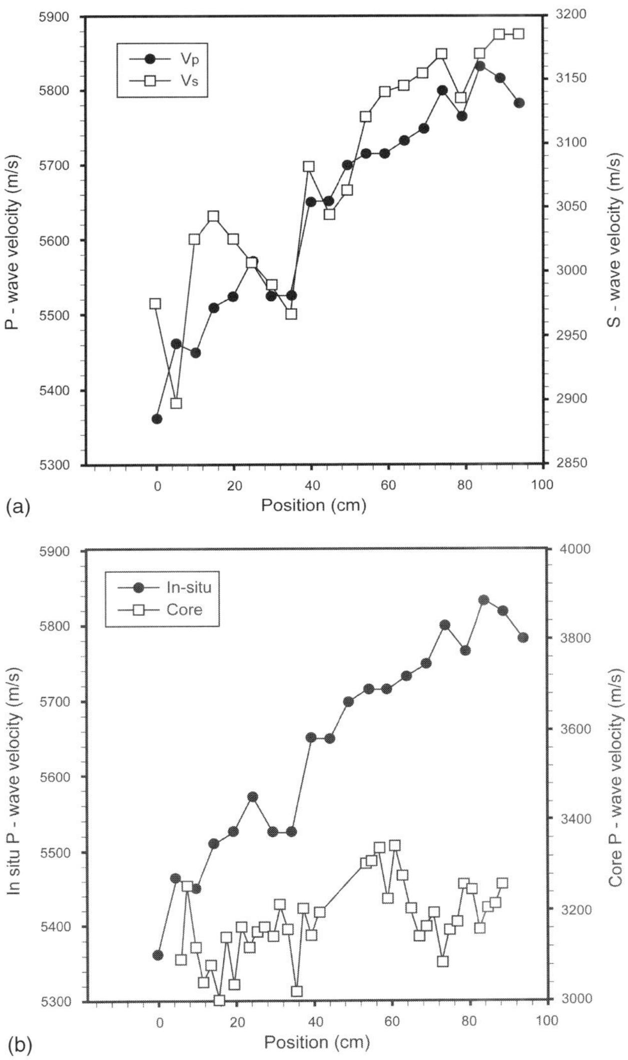


*Figure 7.19* EDZ effect registered by downhole sonic probe at URL test tunnel. a) $V_p$ and $V_s$ *in situ* down a 1 m deep horizontal hole b) P-wave velocity *in situ* compared with that of unstressed, but micro-cracked core from the same hole. Maxwell and Young, 1996.

considered that the $V_p$ anisotropy (10 to 14%) was due to open-crack porosity in addition to the micro-crack fabric.

The location of the microseismic sensors in relation to the drilled-and-hand-mined test tunnel can be seen in Figure 7.18. The AE events shown in Figure 7.21 are seen to cluster both where tangential stresses were highest and where seismic velocity ($V_p$) gradients were steepest. Relatively decreased velocities were seen in the two regions that were under tensile tangential stress.

It is of particular interest to note the 'broadness' of the high velocity regions, which presumably reflect an increase in deformation modulus due to the particular alignment of the maximum tangential stresses. The relevant isotropic elastic stress distribution calculated with a boundary element program is also shown in Figure 7.21.

In subsequent work at the URL for an experimental tunnel sealing experiment, Young and Collins, 1999 were able to demonstrate AE-interpreted reductions in average P-wave *and* S-wave velocities in the highly stressed zones caused by post-excavation of larger diameter 'dog-collars' (or sealing-bulkheads) for forming separate concrete and bentonite seals. The particular tunnel; Room 425, was elliptical in shape to reduce the previously

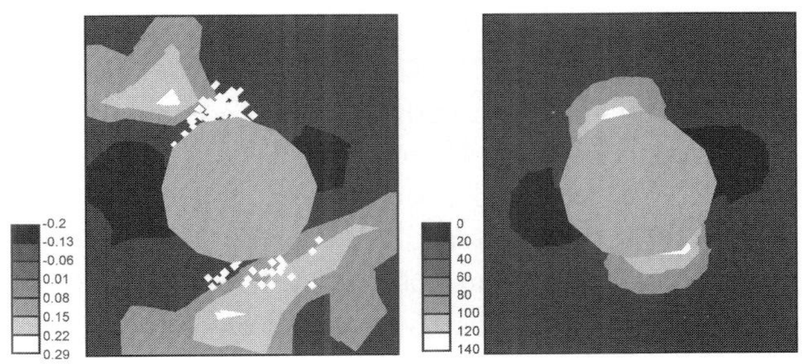

*Figure 7.21* AE events, and the interpreted regions of increased average velocity, are reasonably consistent with the elastic continuum model of the stress distribution. Maxwell and Young, 1996.

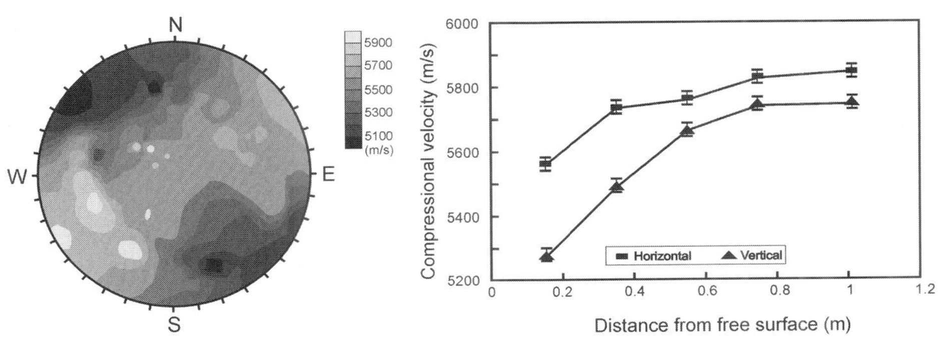


*Figure 7.20* Principal stress-orientated velocity anisotropy (lower hemisphere plot) and EDZ effects on $V_p$ at the same URL test tunnel, Canada. Carlson and Young, 1993.

referred maximum tangential stresses to a tolerable 105 MPa.

The microseismic instrumentation available to monitor this tunnel sealing experiment (TSX) was extensive, consisting of 16 grouted-in-borehole triaxial accelerometers, operating in the 0.1–10 kHz band, and providing 3D coverage of an outer 100 m × 100 m × 50 m volume surrounding the tunnel. Two higher resolution AE arrays for recording in a 10 m × 10 m × 10 m volume around the collars, consisted of 24 ultrasonic transducers operating in the 50–250 kHz band. Each of these arrays had 16 receivers and 8 transmitters.

Young and Collins, 2001 described the way their AE-based monitoring of temporal changes in $V_p$ and $V_s$ was used to estimate the theoretical change in crack density (c) and saturation (s) along any particular ray path. They also refer to Maxwell et al., 1998 who used these techniques to monitor (c) and (s) along radial boreholes at the line-drilled mine-by test tunnel shown in Figure 7.18.

The theoretical assumption was that the saturation (s) did not affect the dynamic shear modulus ($\mu$), but that the dynamic bulk modulus (k) increased linearly with (s), even equalling that of uncracked rock when $s = 1$.

Concerning the crack density (c), defined as the number of cracks per unit volume multiplied by the cube of the mean crack radius, they assumed that the dynamic Young's modulus $E_{dyn}$ and the dynamic Poisson's ratio in the *damaged* material, when normalized by the *undamaged* material, decreased *exponentially* with (c). Young and Collins, 2001 showed a small reduction in (s) from 0.485 to 0.455, and a small increase in crack density (c) from 0.198 to 0.206, during a 16 months monitoring period, for a ray-path from the tunnel perimeter to 3.5 m into the wall of the tunnel.

The development of a rock fracture adjacent to the heat-producing curing of the concrete part of the bulkhead, and its effect on AE monitoring of velocities, is also described by Young and Collins, 2001 and Young and Baker, 2001. This fracture caused a temporary loss of AE signal, both for average P-wave velocity and P-wave amplitude monitoring. There was subsequent recovery of both signals caused by remedial grouting. The approximate 90% drop in P-wave amplitude caused by the fracture was about 30% recovered by the effect of grouting.

A steady, slow rise in $V_p$ (40 to 50 m/s increase was noted during one of the monitoring periods), was associated with the slow hardening of the concrete. All ray-paths through the concrete bulkhead, when averaged, suggested 0.18 m/s and 0.13 m/s *per day* increases in $V_p$ and $V_s$. When constant density was assumed, these increases suggested a theoretical 3.5% *per year* increase in $E_{dyn}$ for the concrete.

The focus on microseismic and acoustic emission monitoring when excavating variously shaped tunnels in these highly stressed, massive granites has naturally aroused curiosity about both the exact location of failure around an opening (how close to the wall is initiation?), and curiosity concerning the type of fracturing responsible for the microseismic events.

Cai et al., 1998, together with mining-rock-mechanics colleagues Kaiser, and Martin, who had most responsibility for URL rock mechanics experiments, collectively address the dominance of tensile fracturing in brittle rocks, and the apparently unrealistic source sizes provided by shear-failure based models.

Contrary to failure around tunnels or boreholes in weaker, less brittle materials, which seems to be dominated by log-spiral-type single-set or conjugate-set shear failure (e.g. Addis et al., 1991, Barton, 2004), the rock mass fracture process near underground openings in hard brittle rocks appears to be dominated by extension cracking, as extensively reviewed by Cai et al., 1998. The authors therefore argue that the focal mechanism will differ from that of natural earthquakes, where slip on pre-existing faults may dominate.

Out of the 3,500 events recorded when excavating the line-drilled test tunnel depicted in Figure 7.18, some 800 events were located within the volume of a subsequently excavated 3.5 m thick slice, that was line-drilled from the floor of the tunnel. These events mostly clustered in the region, shown in Figure 7.22a, where 2D deviatoric stress (actually a shear stress) was larger than the crack initiation threshold, which the authors suggested was about 70 MPa, or $(0.4 \pm 0.1)\sigma_c$. This was also where the 'notch' failures occurred.

When the authors estimated the crack sizes using traditional shear models, they appear to have obtained unrealistically large crack radii. They argue that the ratio of the S-wave energy to the P-wave energy is an important indicator of the type of focal mechanism. The S-wave energy usually dominates, as the energy radiated in P-waves is only a small fraction of that of S-waves. It appears from their review that when $E_s/E_P < 10$, the cracking process involves a dominant tensile failure component, whereas if $E_s/E_P > 20$ to 30 the cracking process is dominated by shear failure. Reportedly, many mining-induced seismic events of large moment magnitude also have high $E_s/E_P$ ratios, and can be analysed realistically by shear models.

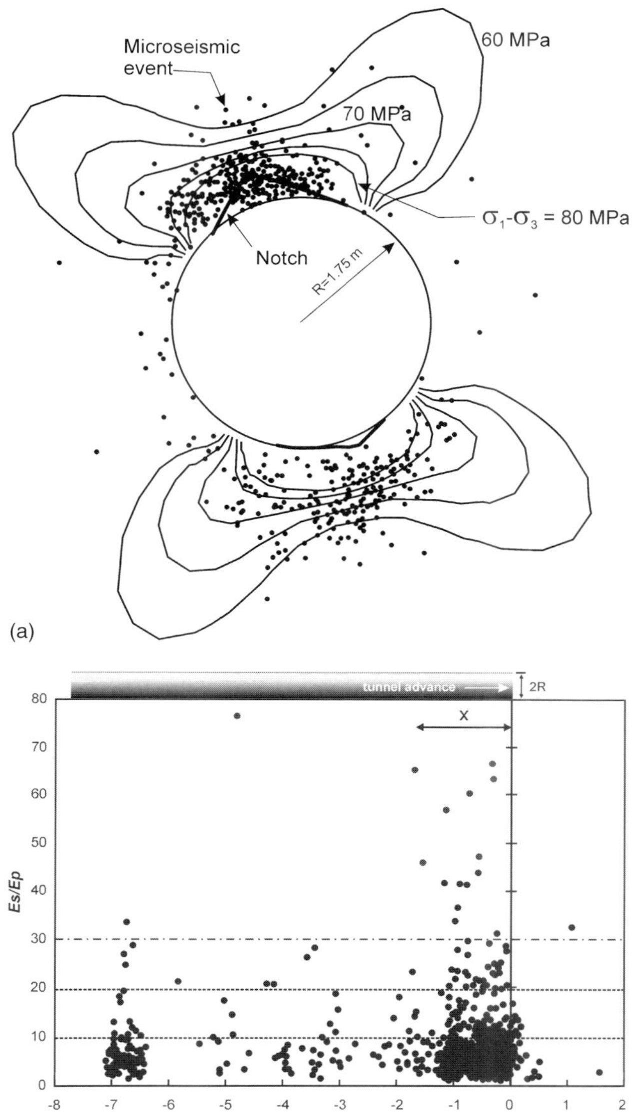


*Figure 7.22* a) Some 800 microseismic events recorded in the 3.5 m thick slice that was line-drilled in the floor of the URL mine-by tunnel. The notch formation is also shown, together with a 2D calculation of the deviatoric ($\sigma_1$–$\sigma_3$) stress contours from an isotropic elastic model. b) Microseismic event location relative to the advancing face of the test tunnel, and the relevant ratio of S-wave and P-wave energies ($E_s/E_P$). The 78% of ratios <10 suggested dominance of tensile cracking sources. Cai et al., 1998.

Their URL data for the 800 or so events clustered within the line-drilled 3.5 m thick slice, showed that $E_s/E_P$ ratios were most frequently between about 6 and 12, and 78% of events had ratios <10. Figure 7.22b shows the recorded ratios of $E_s/E_P$ and their location in relation to the advancing face of this carefully excavated test tunnel. In the region where X/2R < 2, i.e. within two diameters of the advancing face, where the 3D stress distribution is changing to a 2D distribution, there are greatest numbers of likely tensile-dominated events, and a small number with such high S-wave/P-wave energy ratios ($E_s/E_P$), that shearing events may be suspected. Martin et al., 1997 reported that the 'slabbing' associated with the notch formation (Figure 7.22a), started at 0.14 to 0.28 × 1 diameter from the face (X/2R = −0.14 to −0.28).

Earlier during the URL work, Maxwell and Young, 1996, had reported an interesting case of 'passive-source' (i.e. AE) tomography from South Africa. *Concurrence of AE events with high stress* conditions ahead of the mining face in the South African Blyvoor gold mine were again associated with high velocities (for example 5.8–5.9 km/s) in the P-wave tomogram. Therefore using passive source (AE) tomography, the velocity image could potentially be used to map problem areas. The majority of small magnitude rock bursts in the mine were located in regions of high *velocity gradient*, between a low-velocity failed zone and a high-velocity, highly-stressed zone. Logic would perhaps indicate that this was a region of high shear stresses.

### 7.3.3 Äspö – EDZ studies

The Äspö hard rock laboratory (HRL) is the second major location in Sweden for investigating nuclear waste disposal problems such as the excavation disturbance zone. (The first was at the disused Stripa mine, where many international teams cooperated in SKB's facility.) At the 420 m level in Äspö HRL, extensive seismic and radar investigations were performed, principally in order to compare the depth of excavation damage zones in immediately adjacent drill-and-blast and TBM sections of tunnel. The site of the ZEDEX (zone of excavation disturbance experiment) is illustrated schematically in Figure 7.23.

Some of the results of the radial cross-hole tomography performed by Cosma and Enescu, 1996a,b, are shown in Figure 7.23. The borehole radar and seismic cross-hole tomography produced comparable locations for some major joints, which correlated with core logging in the relatively good quality granite and diorite (rock mass quality Q-value mostly 10 to 40, weighted mean = 23).

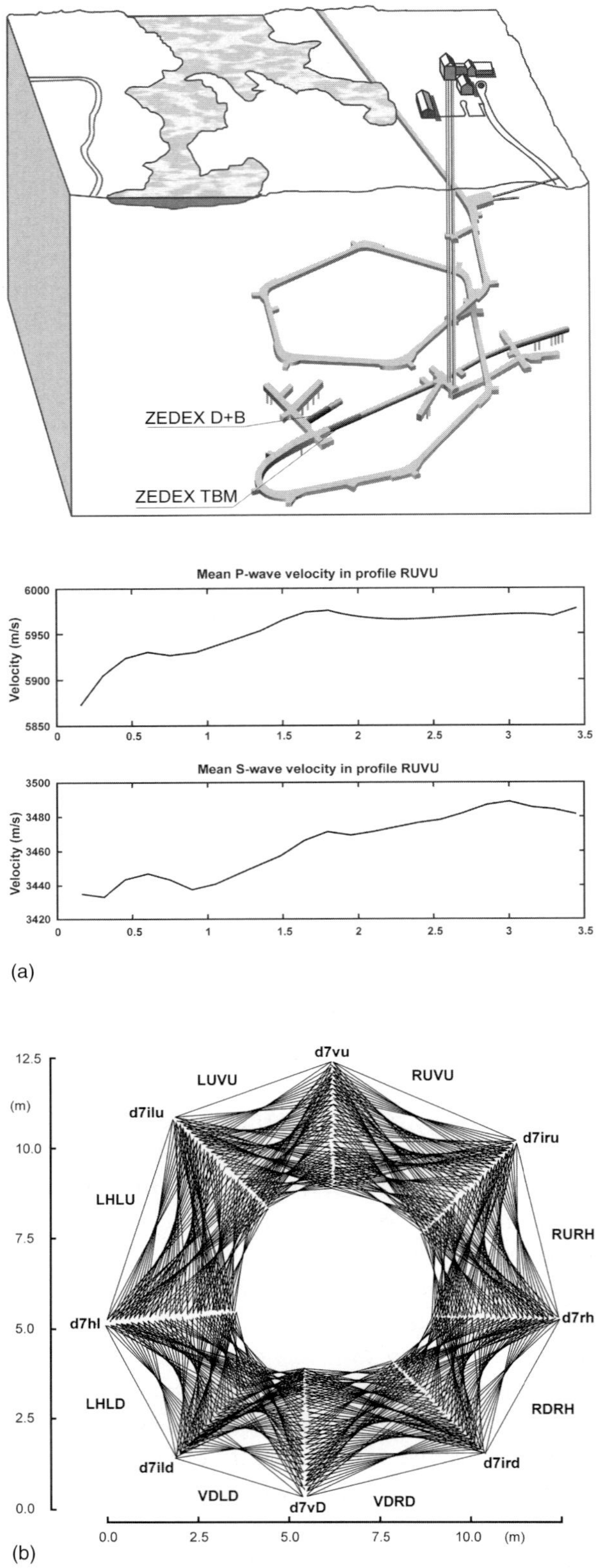


*Figure 7.23* The Äspö ZEDEX site for TBM and D + B evaluation of EDZ effects, with cross-hole seismic tomography results. Cosma and Enescu, 1996a,b.

There was a very mild EDZ effect on $V_p$ and $V_s$ due to the high stresses and the partly discontinuous jointing. P-wave velocities measured across horizontal planes were 6.06 ± 0.1 km/s for both the TBM and drill-and-blast tunnels. In the vertical planes, differences in the two cases were observed, 5.96 ± 0.15 km/s around the TBM drift and 6.26 ± 0.1 km/s around the drill-and-blast drift. Prior to excavation, the three-dimensional principal P-wave velocities were 6.06 (vertical), 5.99 (horizontal, NW) and 5.90 (horizontal, NE) km/s respectively.

Principal stresses at the ZEDEX site were approximately 32, 17 and 10 MPa. Emsley et al., 1996, showed that $V_P$ (fast) was parallel to $\sigma_3$ (vertical), while $V_P$ (slow) was perpendicular to $\sigma_1$ due to anisotropy caused by jointing. In this case $V_P$ (intermediate) was parallel to $\sigma_1$. In fact, both jointing and the stress-anisotropy caused the moderate anisotropy in seismic velocities at Äspö.

In the ZEDEX damage zone studies at Äspö, Bauer et al., 1996, also recorded the changes of velocity in 3 m long radial boreholes drilled in the wall and floor of the TBM and drill-and-blasted drifts, using sonic logging. Reductions in $V_P$ from the background value of 5.9 to 6.0 km/s were recorded in the first 0.25 m into the TBM tunnel walls, and up to 1 metre into the walls of the drill-and-blasted tunnels.

It is of interest to note the 'flat' response seen in Figure 7.24, where tangential stress changes around the tunnel are not 'seen' by the seismic velocity, due presumably to the good quality of the rock (Q-value ≈ 22–24) and the existing high stresses, which were high enough to have acoustically closed the joints, but not high enough to cause excavation induced micro-cracking, as shown earlier at the URL (Homand-Etienne and Sebaibi, 1996).

The rock quality $Q_c$-$V_P$-M inter-relationships reproduced in Figure 7.25 can be utilised to demonstrate some concurrence with the above data. Some 800 metres of drill core logged by NGI at the ZEDEX site showed a weighted mean rock quality Q-value of 24 ('good' quality). If we utilise the $\sigma_c/100$ normalisation of Q (Chapter 5) using $\sigma_c \approx 200$ MPa as measured, $Q_c$ equals approximately 50. $V_P$ is predicted to range from about 5.8 to 6.2 km/s for an equivalent depth range (relevant to the principal stress range) of about 400 to 1200 metres, rather close to the measured values reported above.

The intact rock laboratory E-modulus was 69 GPa, and a simple UDEC model showed concurrence with the small measured deformations of 1 to 2 mm when a

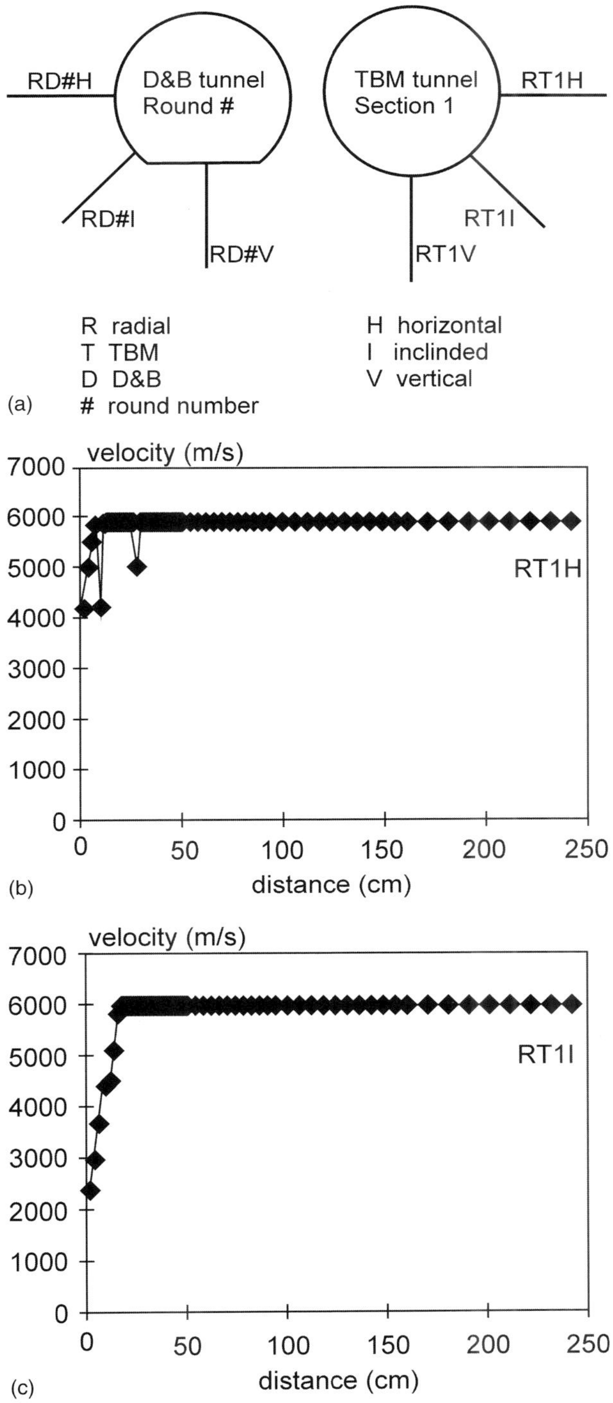


*Figure 7.24* EDZ measurements of $V_p$ in 3 m long radial holes in the TBM section of the Äspö ZEDEX project. Bauer et al., 1996.

deformation modulus of 60 GPa was used. The equation relating mean deformation modulus M with $V_p$ shown in Figure 7.25 (see inset), suggests deformation moduli ranging from 58 to 79 GPa, for the predicted velocity range of 5.8 to 6.2 km/s. Calculated dynamic moduli around the drill-and-blast drift ranged from 76 to 79 GPa. Presumably the situation $M \approx E_{dyn}$ has some implied relation to 'acoustically closed' jointing.

### 7.3.4 Stripa – effects of heating in the EDZ of a rock mass

The EDZ studies performed in relation to nuclear waste isolation problems focuses much attention on the disturbed properties in this zone. We have seen many examples of increased disturbance from the frequently measured reductions in seismic velocity and deformation modulus, and these reductions, taken in conjunction with increased joint void ratios, generally lead us to expect enhanced permeability in the EDZ. This does not always occur however.

In the Stripa SCV (Site Characterisation and Validation) experiment (Olsson et al., 1993), inflow to the test tunnel was less than almost all the hydraulics modellers had predicted, based on dedicated borehole permeability testing, using holes drilled along the future periphery of the tunnel.

Several potential reasons for this discrepancy were described by the researchers involved, including dissolved air coming out of solution at the reduced pressures, blast gas invasion of the joints, and Poisson expansion in the third (axial) dimension causing increased axial (i.e. normal) stress on dominant joints crossing the tunnel. Rock mechanics modelling with two- and three-dimensional distinct element codes (UDEC-BB and 3DEC) had shown insufficient shear (mostly <1 mm) for dilation-enhanced permeability changes, due to the small size (2 × 3 m) of the test tunnel, which had quite discontinuous jointing. This property was modelled with 'numerically glued' joint ends. (Barton et al., 1992b).

The planned use of the geosphere as a potential disposal volume for nuclear waste has meant that the heating (and cooling) effect in the rock exposed in the floor or walls of test tunnels has been the subject of much research. Large diameter disposal boreholes for high level waste canisters will also be in the EDZ of the excavations, and will create their own smaller EDZ around the large boreholes. What effects can we expect on local rock stresses, on seismic velocities, and on permeability, due to the production of considerable heat over long time-spans in the early disposal period, followed by the cooling period?

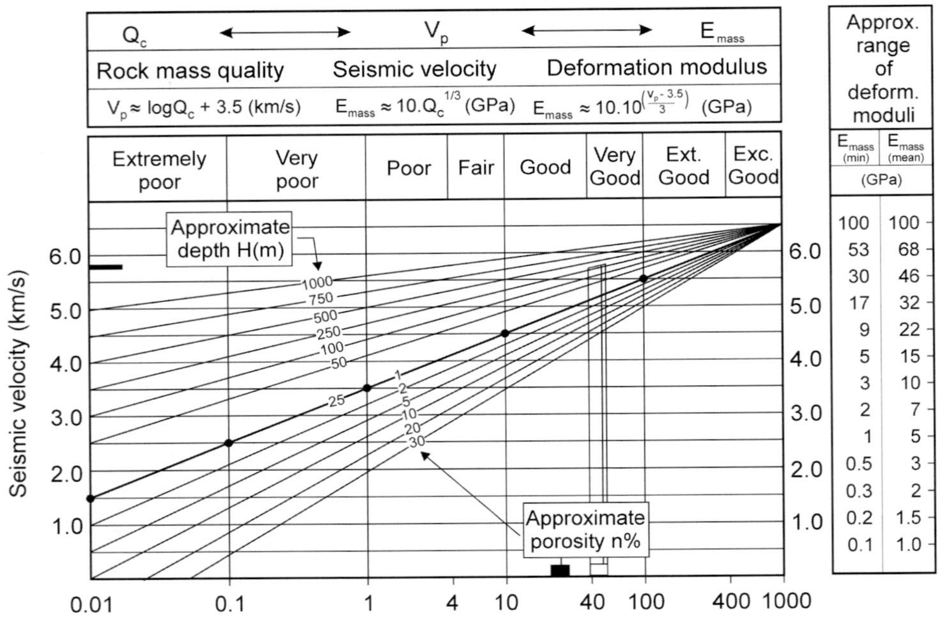


$$Q_c = \left[\frac{RQD}{J_n} \times \frac{J_r}{J_a} \times \frac{J_w}{SRF}\right] \frac{\sigma_c}{100}$$

*Figure 7.25* Rock quality Q-value observations from 800 m of core logging, corrected to $Q_c$ by the high value of $\sigma_c \approx 200$ MPa, gives a realistic prediction of $V_p$ and deformation modulus at 450 m depth, when compared with ZEDEX data.

A drift in the Stripa mine in Sweden, used for borehole heater tests, showed some interesting effects of thermally induced strains. These caused increased seismic velocities between drained monitoring holes in the jointed quartz monzonite. A schematic of the experimental set-up shown in Figure 7.26a, indicates the relative locations of the ultrasonic monitoring holes, which were drilled to 10 metres depth, twice as deep as the heater borehole.

Paulsson and King, 1980, showed the increases of seismic velocity for cross-hole measurements (M8–M6), as a function of time after heater turn-on and turn-off. Pre- and post-heater velocities were generally similar, as shown in Figure 7.26b, and a velocity anomaly at about 3 m depth was 'smoothed-out' by the heating but 'returned' when the rock was cooled.

Across the mid-plane of the heater, at 4.2 m depth, the four sets of cross-hole velocities (Figure 7.26c) showed more or less parallel behaviour, except for an extra strong reduction in velocity for M7–M9 after heater turn-off. The authors give interesting similar-trending curves for monitored displacement and stress change about 1 m from the heater (Figure 7.26d). They do not have a confirmed explanation for the anomalous net reduction in velocity at the end of the test (M7–M9). However, it might be related to the improved closure of joints at elevated temperature, even under constant normal stress, as measured by Hardin et al., 1981, and discussed in more detail by Barton et al., 1985.

Upon cooling, the less rough, interlocked joints may 'spring-open' more than their closed neighbours, to avoid tensile stress development. This could then cause a marked reduction in seismic velocity if the open joint or joints, happened to cross the path of the seismic array. Related local increases in joint conductivity, and reduced shear strength, of any 'open' and more planar joints, could be an added uncertainty in nuclear waste related disposal scenarios, as emphasised by Barton and Makurat, 2006.

Further details of this Stripa heater experiment (Paulsson and King, 1980) were subsequently reported in a comprehensive analysis by Paulsson et al., 1985. The full duration of the test was 750 days with 398 days of heating. The long period of cooling generally returned seismic velocities to values *lower* than before the heating, suggesting permanent changes (such as local excessive joint opening as hypothesised above). A significant quantity of water expelled during the heating signified a

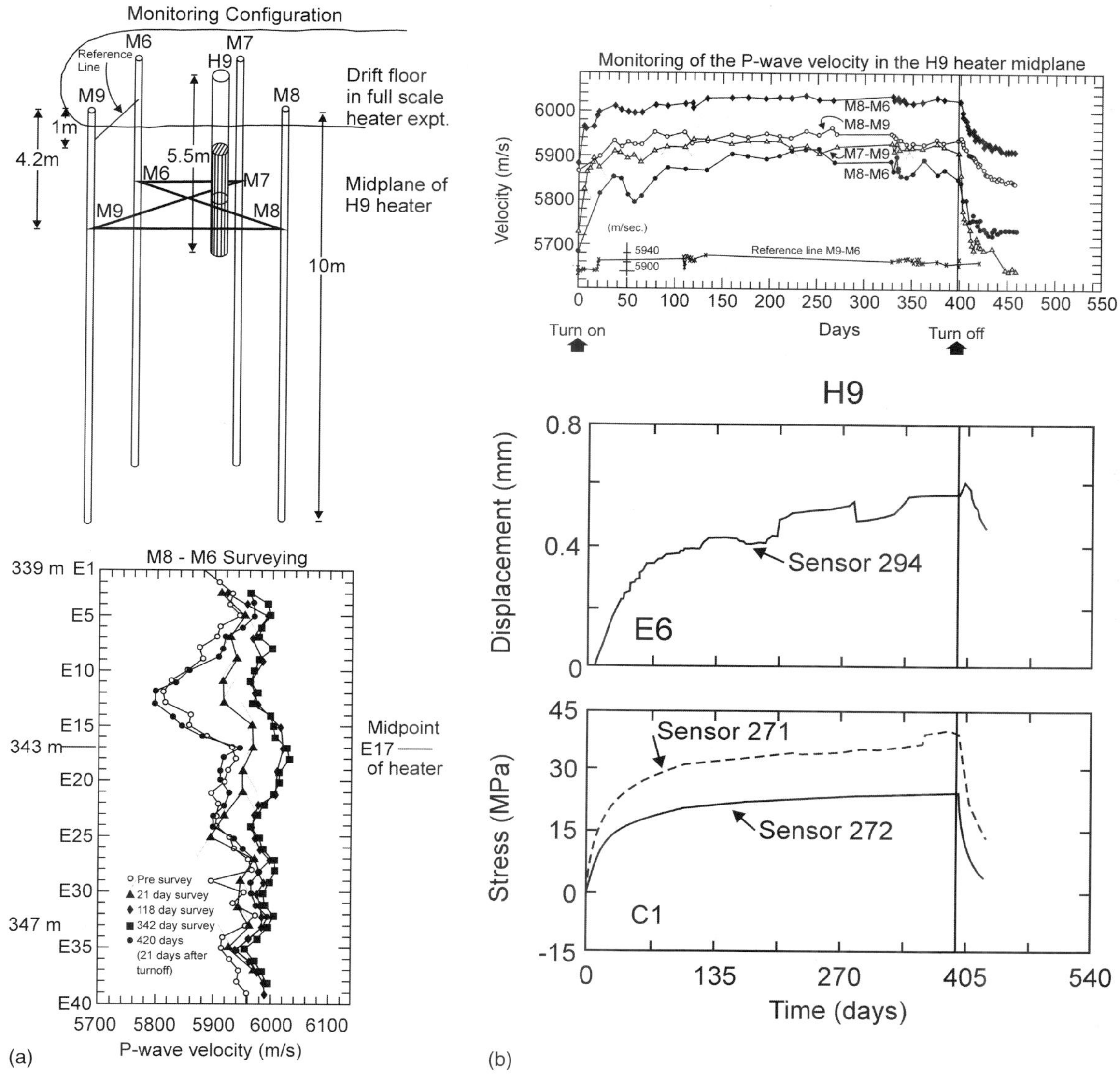


*Figure 7.26* Heater experiments at Stripa, with velocity changes matched by stress and displacement records. Paulsson and King, 1980.

general closing of the joints. Temperatures were over 100°C in a small region around the heater and water was expelled also from distant boreholes where perhaps the low initial permeability was less reduced.

The initial increase in velocity with temperature was linear and varied from 2 to 4 m/s/°C. The average joint frequency in the test area, analysed from 224 m of core, was 8.3/m, and an elastic continuum analysis conducted prior to the test had indicated larger stresses and local displacements than were actually measured, presumably due to the compliance of all these joints. This effect was also presumably experienced in a 'heated-mine-by' experiment in the Climax Mine, in the USA.

The full record of P-wave and S-wave velocities over the 750 days duration of the test is shown in Figure 7.27a and b. The largest velocity changes caused by the heating, amounting to 0.2–0.3 km/s were interpreted as occurring in the direction of the minimum horizontal stress, which is logical since the calculated thermal stress was as much as 55 MPa in, presumably, the direction of maximum horizontal stress.

The effect of the heating in an in situ experiment such as that described by Paulsson and King, 1980, and Paulsson et al., 1985, is to change both the stresses in the rock and the degree of saturation, particularly close to the source of heat. In an effort to understand and

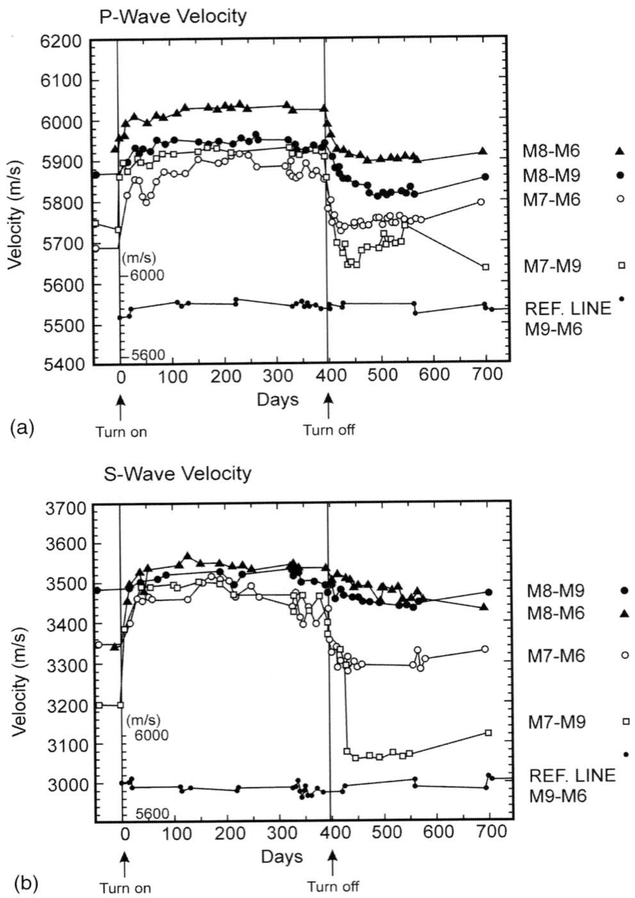


*Figure 7.27* Complete P- and S-wave velocity record for one of the Stripa heater tests. Maximum $V_P$ increase was parallel to minimum stress. Paulsson et al., 1985.

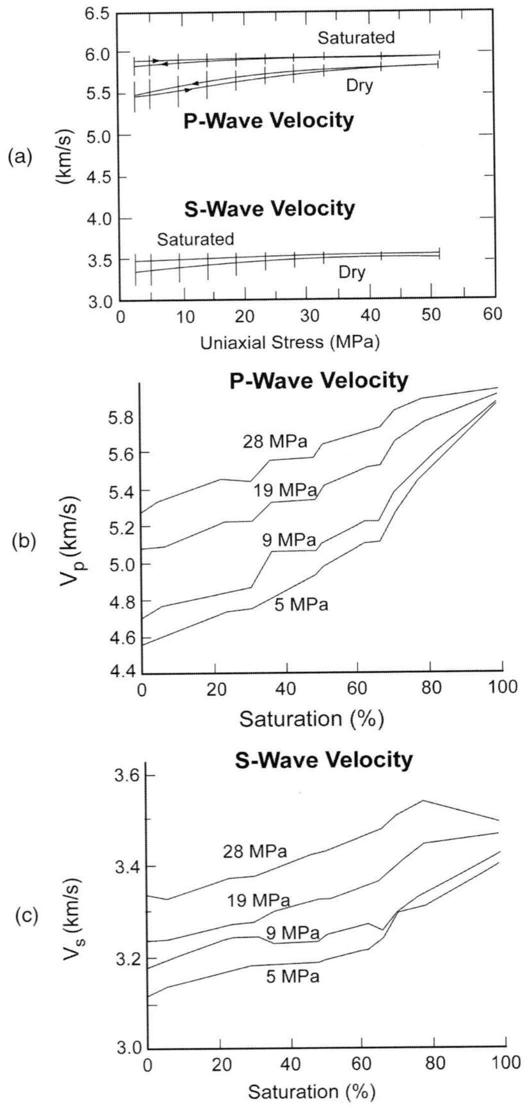


*Figure 7.28* Uniaxial tests on intact but micro-cracked Stripa granite show the important influence of stress and degree of saturation on $V_P$, and to a lesser extent, $V_S$. Paulsson, 1984.

isolate some of these effects, Paulsson, 1984, conducted uniaxial tests (0–52 MPa) on granitic (quartz monzonite) samples from the same Stripa test drift. The samples were intact but because of micro-cracks showed strong effects of saturation levels on P-wave velocities.

Figure 7.28a shows the effects of axial stress on axially measured P and S-wave velocities in the dry and saturated states. The standard deviation of results from eleven specimens is shown by the length of the crosses.

A clearer indication of the important effect of degree of saturation on $V_P$ at different stress levels is given in Figure 7.28b. Lesser effects are seen on $V_s$ in Figure 7.28c. This particular sample showed stronger effects than the general trend given in Figure 7.28a, and may presumably have been due to increased porosity from low aspect-ratio micro-cracks, since the initial velocity when dry was only 4.6 km/s (at 5 MPa), and rose to more than 5.8 km/s with full saturation.

## 7.4 Acoustic detection of stress effects around boreholes

Plona et al., 1997, utilised the effects of compressive stress concentrations around boreholes to investigate if principal stress orientations could be determined by acoustic measurements from within boreholes, at stress levels lower than needed for break-out. They referred to the triaxial tests on sandstone reported by Sammonds et al., 1989, reproduced in Figure 7.29, to emphasise the possibility of differentiating the high tangential stress from the low tangential stress region around a

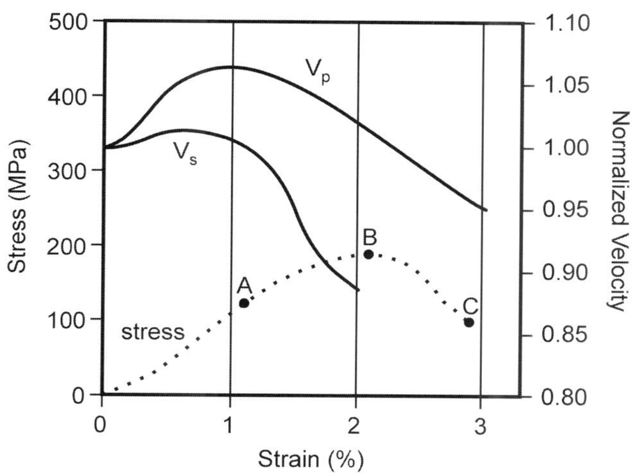


*Figure 7.29* High pressure triaxial tests on sandstones showing the strong coupling of stress with velocity changes. Sammonds et al., 1989.

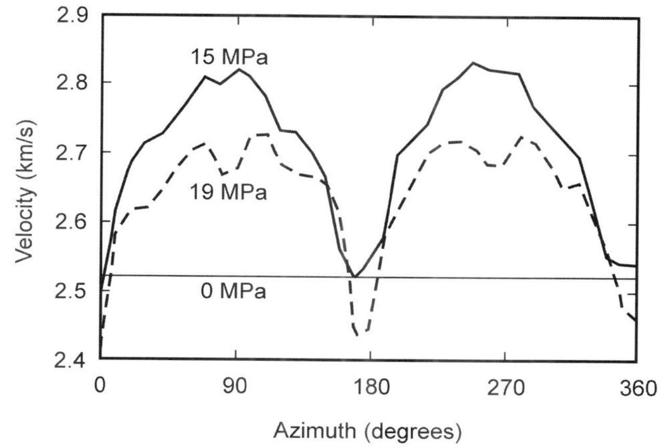


*Figure 7.31* High frequency, axial P-wave monitoring around a 10 cm borehole in a uniaxially loaded cube of sandstone. Plona et al., 1997.

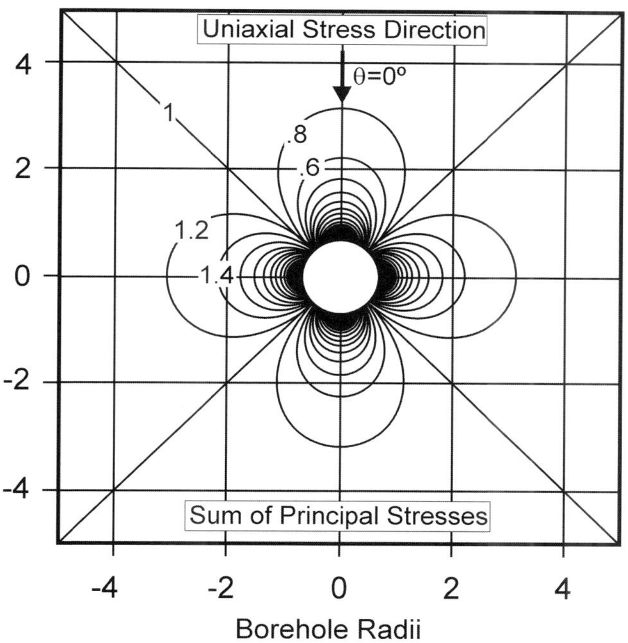


*Figure 7.30* Theoretical elastic stress distribution around a uniaxially loaded borehole. Jaeger and Cook, 1977.

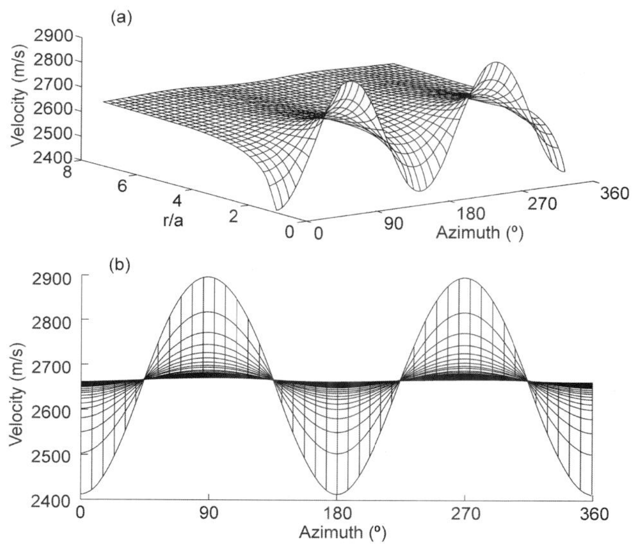


*Figure 7.32* Acousto-elastic calculation of $V_p$ anisotropy around a uniaxially loaded borehole in sandstone ($\sigma_H$ = 10 MPa) Plona et al., 1997.

borehole. An elastic model of the latter is shown in Figure 7.30 for the case of uniaxial loading (Jaeger and Cook, 1977).

Plona et al., 1997, used a 50 cm cube of sandstone with a central 10 cm diameter borehole loaded uniaxially to 21 MPa, to investigate the potential of axial acoustic refraction monitoring at numerous azimuth locations around the borehole. Their principal results are shown in Figure 7.31.

Break-out reportedly started at the 15 MPa stress level where velocity maxima were registered across a diameter. The decline of velocity seen at 19 MPa was due to mechanical damage in the same diametrically opposite $\sigma_{\theta\ max}$ locations. In Figure 7.32, the authors show the results of an acousto-elastic model for a borehole in sandstone, loaded with a boundary stress of 10 MPa. The general similarity of model and experiment is striking.

One may wonder whether these effects are taken into account in the general interpretation of sonic logging down boreholes, since several hundred m/s variations in velocity are seen in a 'simple' unjointed sandstone, with admittedly, an extreme induced tangential stress anisotropy of 30 MPa for $\sigma_{\theta\ max}$, and (-) 10 MPa for $\sigma_{\theta\ min}$ (based on the isotropic elastic tangental stress 'rule-of-thumb' 3A-B, 3B-A). Significant azimuthal velocity anisotropy effects would seem to be possible in wells where there was significant horizontal stress anisotropy, plus structure in the form of vertical or sub-vertical jointing that would be variously opened or closed by the respective effects of $\sigma_{\theta\ min}$ and $\sigma_{\theta\ max}$. Is it possible that the sonic logging 'misses' such effects of velocity anisotropy around the horizontal plane, due to principally vertically and steeply inclined ray paths?

# 8 Seismic measurements for tunnelling

Tunnel face seismic tomography using a pair of boreholes that was performed by NGI at the Oslo Fjellinjen Tunnel, in the late 1980's, was used to help the contractor plan for ground freezing in one tube, while careful multiple drift excavation was performed in the other tube. The 2 × 13 m span motorway tunnel beneath the Oslo downtown harbour district, passed through a wide regional fault zone with 2 to 4 m of crushed alum shale over the arch, underlying 20–30 m of soft, sensitive marine clays. The advantages of the ahead-of-the-event data far outweighed the relatively short stoppages and modest cost involved.

There are now several companies around the world who are offering the use of reflection techniques for mapping marked reflectors out to many hundreds of meters ahead of an advancing tunnel face. Here, both sources and receivers are placed within the tunnel. By the nature of the reflection result, it is difficult to determine if the quality will be better or worse. Others have proposed, and demonstrated, the advantages of combining this with in-tunnel seismic refraction, with an in-tunnel source and both internal and external (mountain side) receivers. One can then obtain reliable velocity and rock quality predictions ahead of the face, both in front of, and behind the reflectors, which will thereby correlate better with possible pending tunnelling difficulties.

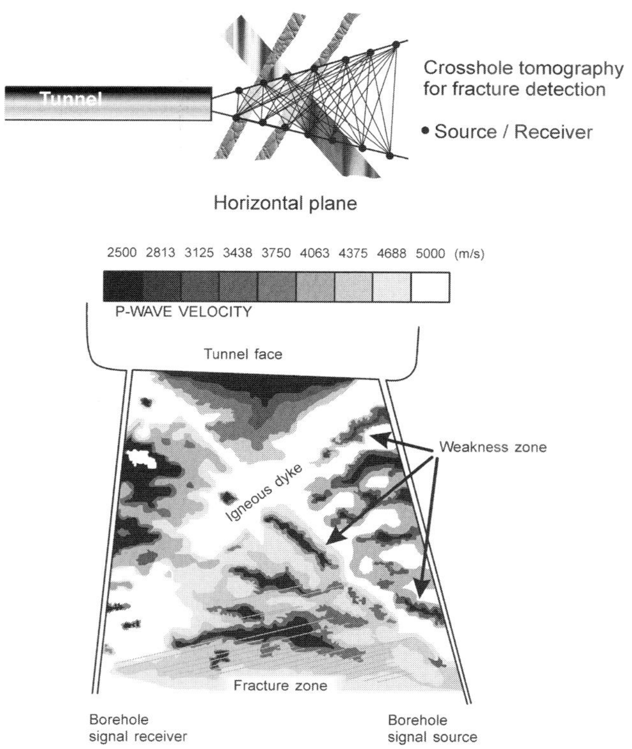


*Figure 8.1* Cross-hole seismic tomography performed by NGI at the face of the Oslo Tunnel in 1987, when approaching a major fault zone. Nord et al., 1991.

## 8.1 Examples of seismic applications in tunnels

Concerning the obvious need to have information 'ahead-of-the-event' in a tunnel, Nord et al., 1991, commented that the present 'lack of information has only been accepted due to the high cost of obtaining it'. The authors went on to analyse the duration and frequency (in metres) of down-time at some hard rock projects. They concluded that probe drilling to 50 metres, and 1/2 to 1 hour for the seismic probing measurements would likely be sufficient. It does not take long to percussion-drill to 50 m with modern hydraulic jumbos, when 1 to 4 m per minute rates of penetration are achieved.

We will now go backwards in time to an impressive early example of the use of geophysical surveys in tunnels, as given by Scott et al., 1968, for the Straight Creek pilot bore of 4.0 m diameter, driven under the continental divide in Colorado, USA in 1963 and 1964. (Figure 8.2). Both seismic and resistivity measurements were made at regular and irregular intervals along the bore, in order to sample each class of rock. Five rock classes were defined, based principally on joint spacing (<3 cm to >0.9 m), mineral alteration (%) and presence of fault gouge, foliation or schistocity.

The rock types themselves (granite, diorite, gneiss, migmatite and schist) did not appear to determine rock class. The seismic velocity of the deep layer beneath the

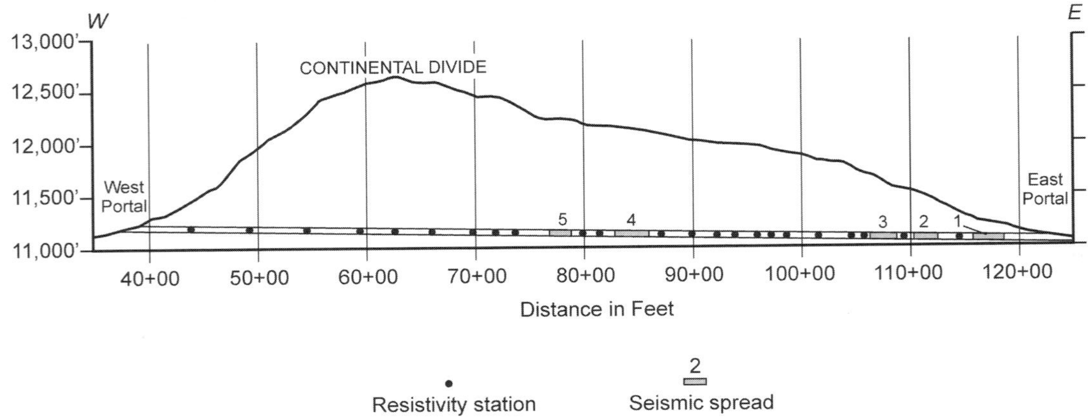


*Figure 8.2* Cross-section of Straight Creek pilot bore showing geophysical test locations. Scott et al., 1968.

excavation-disturbed low-velocity layer (caused by stress redistribution and gravity induced loosening in the arch) was correlated to a number of key construction parameters by the authors, and good correlation was indicated in many cases. The following *deep layer velocities* were measured at the five seismic spreads shown on Figure 8.2: #1 5.2 km/s, #2 5.1 km/s, #3 4.8–6.1 km/s, #4 4.2 km/s and #5 6.0 km/s. Later, the much lower *shallow layer velocities* will be introduced into the discussion.

Figure 8.3a and b compare the thickness of the low-velocity layer and the so-called 'tension arch', defined by the authors as the depth where no further measured dilation of the rock in the arch could be detected. The range of seismic velocities shown (14000 ft/s ≈ 4.3 km/s, 20,000 ft/s = 6.1 km/s) correspond to the range of rock classes 5 to 1, and obviously reflect the confinement effect from the mostly 200–500 m of overburden.

The worst quality (class 5) corresponded to seismic spread #4 in Figure 8.2 ($V_p = 4.2$ km/s), which had a local overburden of about 300 metres, and obviously was strongly affected by the depth or stress level. Our earlier hard rock relation $V_p \approx 3.5 + \log_{10} Q$ with depth correction, (Figure 5.36), suggests a possible Qc – value of about 0.1 assuming low porosity, hard rock is involved. At the surface a $V_p$ value of 2.5 km/s is implied for this rock mass quality i.e. severely faulted ground which appears to correspond to the authors' description of class 5 rock. (A disturbed zone of at least one diameter was indicated for such rock.)

Scott et al., 1968, showed good correlations between $V_p$ and construction data such as steel set spacing (Figure 8.3b), and rate of construction (Figure 8.3d). They also showed a significant correlation of tunnel support type with *electrical resistivity* (Figure 8.4a) which is itself partially correlated with the deep layer velocity (Figure 8.4b).

Already at this early date (the mid 1960s), the authors envisaged a time when geophysical probing ahead of tunnels would have enough correlations to conditions in tunnels of different size, and 'in all environments', that a full diameter bore could be driven directly, without the need for time-consuming pilot bores.

It is of interest to speculate that the great difficulties encountered when driving the full-scale, twin bore Straight Creek tunnel were caused by, among other factors, an unexpectedly large scale effect caused by the 4 m to 12 m tunnel size difference, and perhaps insufficient appreciation of the effect of stress on the seismic velocities at that time. There is also the possibility of adverse interaction between the twin tubes, with 'plastic zone' overlap, a problem of relevance when assessing risk in twin-bore TBM tunnelling, where conditions are very unfavourable for any type of tunnelling. (Barton, 2004c).

The attempted use of steel sets in fault zones at Straight Creek, probably allowed too much (scale dependent) loosening of the ground, with the low velocity layers even more affected than in the pilot bore of 4 m span. The seismic velocities of the *shallow layers* at the five seismic spreads listed above (#1 to #5, Figure 8.2) were respectively: 3.0 km/s, 2.3–2.7 km/s, 2.3–3.1 km/s, 1.3–1.6 km/s (worst case, class 5) and 2.3 km/s.

Perhaps more attention should have been paid to these extremely low EDZ velocities (and to the thicknesses of such zones in the worst rock) which caused almost insurmountable problems in the main bore, which took several year to complete. The Q-system support pressure database (Barton et al., 1974) includes Straight Creek main bore as almost the highest recorded tunnel support

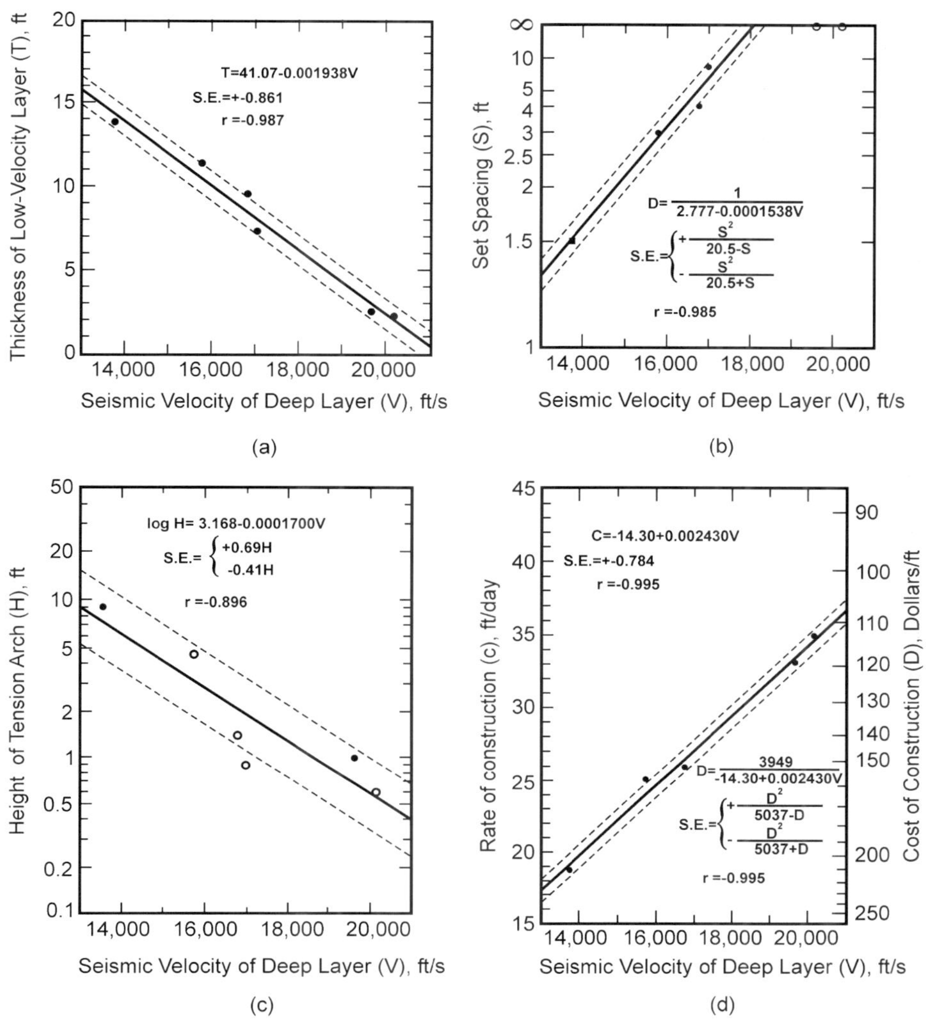


*Figure 8.3* Disturbed zone thickness and some support and construction rate details based on $V_p$ measurement at the Straight Creek pilot bore. Scott et al., 1968.

capacity of at least 300 tnf/m$^2$, and an estimated rock mass 'quality' Q-value as low as 0.001. Multiple perimeter drifts filled with concrete of some 2 to 3 m thickness were needed in some zones.

Ikeda, 1970, assembled a comprehensive set of technical data from some 70 rail tunnels in Japan. He first classified *rock types* into classes A to F, as reproduced in Table 8.1. Examples of Class A rocks were metamorphic rocks such as gneiss, quartzite, etc., while examples of Class E were Pleistocene rocks such as mudstone and volcanic ejecta.

He then listed typical ranges of seismic velocities ($V_p$) under *rock conditions* (Classes 1 to 7), shown in Table 8.2 for each of the previous rock type classes A to F. The two examples under *Class 3 rock condition* would be A/3 = 4.0 to 4.6 km/s, E/3 = 2.6 km/s. Based on the classification of rock conditions (Classes 1 to 6), he gave tunnel support loads in the range 0.01 to 0.3 MPa (1 to 30 tons/m$^2$ range for Classes 1 to 6), and spacing of the steel arch support in the range 1.5 to 0.75 m, and concrete thickness in the range 0.3 to 0.9 m, for 30 m$^2$ and 60 m$^2$ tunnel sections from his 70 case records. These data are reproduced in Figure 8.5 a, b, c and d. This is a valuable set of early case records and their technical description, using seismic velocities.

A somewhat finer division of rock types than the original Japanese Railways classification of Ikeda, 1970 has been used in more recent years by the Japanese Highway Authority. This is reproduced in Figure 8.6 (from Barton and Itoh, 1995), showing the addition of

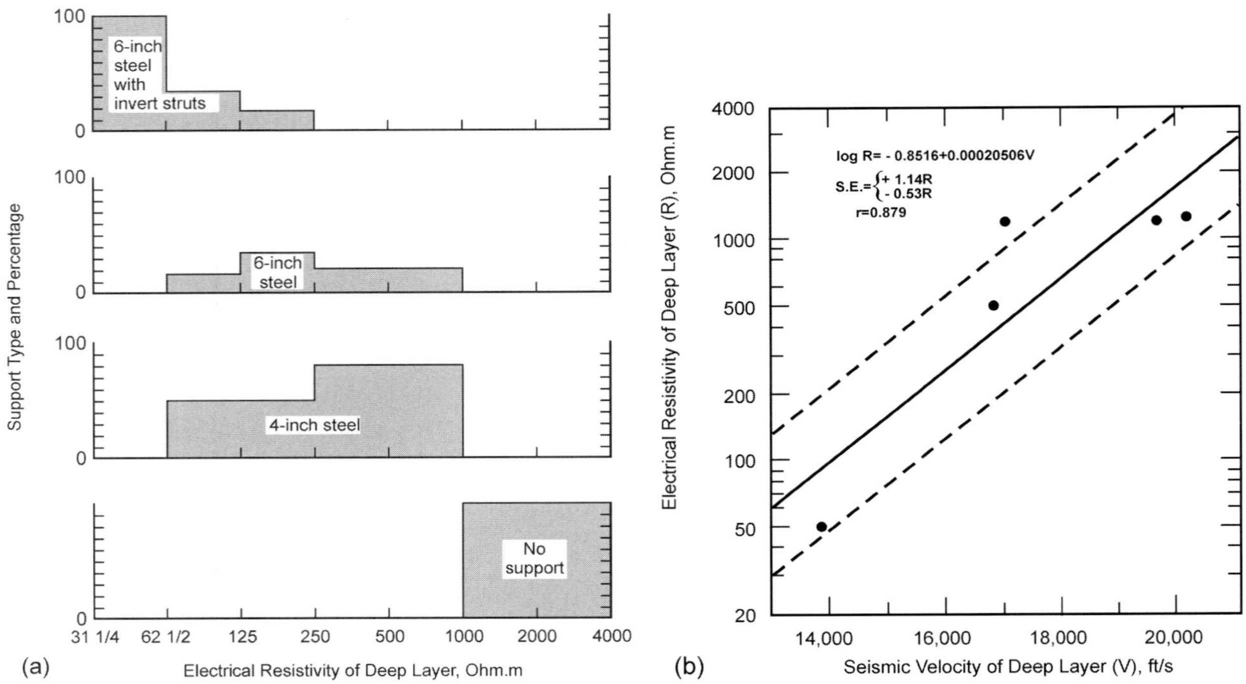


*Figure 8.4* Some cross-correlation between support type, resistivity and undisturbed (deep layer) velocity. Straight Creek pilot bore. Scott et al., 1968.

Table 8.1 Rock *type* classes (A to F) of Ikeda, 1970 based on lithology and geology.

| Rock qualities | | Names or rocks |
|---|---|---|
| A | a) Paleozoic rocks Mesozoic rocks | clayslate, sandstone, greywacke, conglomerate, chert, limestone, 'schalestein', etc. |
| | b) Plutonic rocks | granite, granodiorite, diorite, gabbro, peridotite, etc. |
| | c) Hypabyssal rocks | granite porphyry, quartz porphyry, porphyrite, diabase, etc. |
| | d) Volcanic rocks (some part) | rhyolite and andesite of Mesozoic era, basalt, etc. |
| | e) Metamorphic rocks | gneiss, hornfels, schist, phyllite, quartzite, etc. |
| B | a) Metamorphic rocks having conspicuous schistocity | |
| | b) Paleozoic and Mesozoic rocks having fine bedding planes | |
| C | a) Mesozoic rocks (some part) | shale, clayslate, sandstone, tuff breccia, 'schalestein', etc. |
| | b) Palaeogene rocks (some part) | silicified shale and sandstone, tuff breccia, welded tuff, etc. |
| | c) Volcanic rocks (greater part) | rhyolite, dacite, andesite, basalt, dolerite, etc. |
| D | a) Tertiary rocks | mudstone, shale, sandstone, conglomerate, tuff, tuff breccia, welded tuff, agglomerate, etc. |
| E | a) Pleistocene rocks Neogene rocks | mudstone, siltstone, sandstone, sand and gravel rock, tuff, terrace, talus, fan, volcanic ejecta, agglomerates, etc. |
| F | a) Alluvium rocks Diluvium rocks | clay, silt, sand, sand and gravel, loam, volcanic ejecta, fan, talus, terrace, etc. |

Table 8.2 Rock condition classification using seismic P-wave velocities based on 70 Japanese rail tunnel case records analysed by Ikeda, 1970. (Note writer's interpretation of some of author's words under 'Remarks' and 'Notes').

| Classification of rock conditions | Rock qualities A | B | C | D | E | F | |
|---|---|---|---|---|---|---|---|
| 1 | 5.0 | | 4.8 | 4.2 | | | |
| 2 | 5.0–4.4 | | 4.8–4.2 | 4.2–3.6 | | | |
| 3 | 4.6–4.0 | 4.8–4.2 | 4.4–3.8 | 3.8–3.2 | 2.6 | | good |
| 4 | 4.2–3.6 | 4.4–3.8 | 4.0–3.4 | 3.4–2.8 | 2.6–2.0 | | medium |
| 5 | 3.8–3.2 | 4.0–3.4 | 3.6–3.0 | 3.0–2.4 | 2.2–1.6 | 1.8–1.2 | |
| 6 | 3.4 | 3.6 | 3.2 | 2.6 | 1.8 | 1.4–0.8 | |
| 7 | | | | | 1.4 | 1.0 | bad |

Remarks:

1) Always go to a poorer class when ground water will inflow into tunnel.
2) Rocks with expansive nature have no relation to this classification.
3) Raise 1 or 2 classes when the Poisson's ratio of the weathered rock is better than 0.3.

Notes:

1) The numbers 1–7 are the rock condition ratings.
2) The numerals show the velocities of elastic wave in the rock (km/s).
3) See Table 8.1 for the rock qualities A through F.

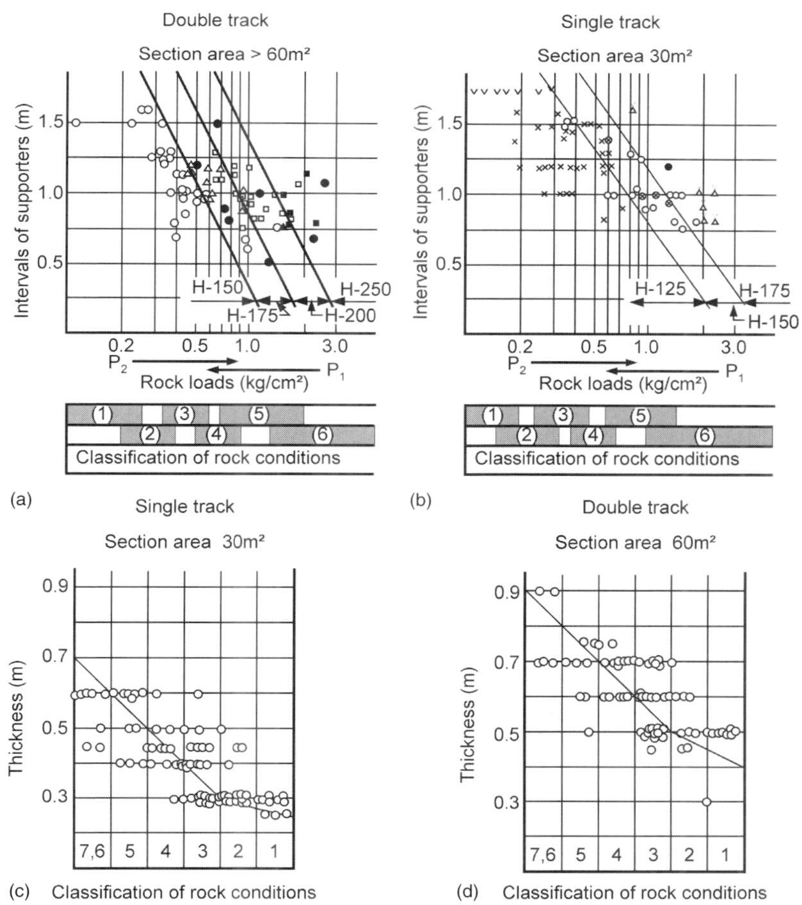


*Figure 8.5* Relationships between support intervals (steel sets), concrete lining thickness and support pressure, each as a function of rock condition classes 1 to 7, which were defined by $V_p$ ranges. Ikeda, 1970.

a preliminary rock quality Q-scale at the base of the figure. A comparison of the velocity ranges shown in Figure 8.6, with Table 8.2 shows, of course, broad similarity in view of the common, relatively young geology, that has to be tunnelled by the two state authorities.

Note that the appended Q-scale is based on the simple shallow, hard rock equation ($V_p \approx 3.5 + \log Q$), with no immediate attempt to correct for uniaxial strength through $Q_c$, nor to adjust for significant matrix porosities. The 'fit' of $V_p$ with Q will therefore be increasingly in error in the case of the lower velocity, weaker, and more porous rocks, which require individual interpretation of these (usually) inter-related factors, which collectively have a strong $V_p$-reducing effect.

This means that a low velocity Tertiary sandstone with a shallow-seismic refraction velocity of 2.5 km/s, may have an initially implied $Q_c$ value of only 0.1, while the implied $Q_c$ value is 1, if $n \approx 15\%$. (Figure 5.36). A significantly higher apparent Q-value (due to $\sigma_c << 100$ MPa) is implied, concerning the usual factors describing the structure of the rock mass, like RQD, $J_n$ (number of joint sets), $J_r$ and $J_a$. However if the rock is incompetently weak, RQD is zero (according to the definition of Deere et al., 1967), but a minimum of 10 in the Q-calculation to avoid an impossible Q = 0. Furthermore, and most importantly, there may be an unfavourably large ratio of stress-to-strength in the context of tunnelling, giving the need for an elevated SRF (see Appendix A).

The differences seen above are due to a fundamental difference between *characterization* (with no excavation involved), and a *classification*, which has become known in tunnelling, as a scheme for selecting appropriate rock support and reinforcement, via RMR (Bieniawski, 1989), or Q.

An account of pre-investigations and experiences while driving Norway's first sub-sea road tunnel to the west-coast island of Vardø, was given by Palmström, 1982. Seismic refraction profiles totalling almost 13 km covered a 500 m wide zone across the 1.5 km wide sound and gave depth to bedrock with an accuracy of ±0.5 m in relation to 36 soundings in the sea bed.

The presumed and actual weakness zones encountered during driving of the 9.4 m span tunnel are shown in Figure 8.7. The rock cover (mean 40 to 50 m) and shallow sea depth (20–40 m) are also shown. Up to 4 joint sets were logged in the quartzitic sandstones, siltstones and shales, which had frequent clay coatings on the generally steeply dipping joints, giving poor stability, especially in the low velocity (e.g. 3.2–3.3 km/s) crushed zones shown

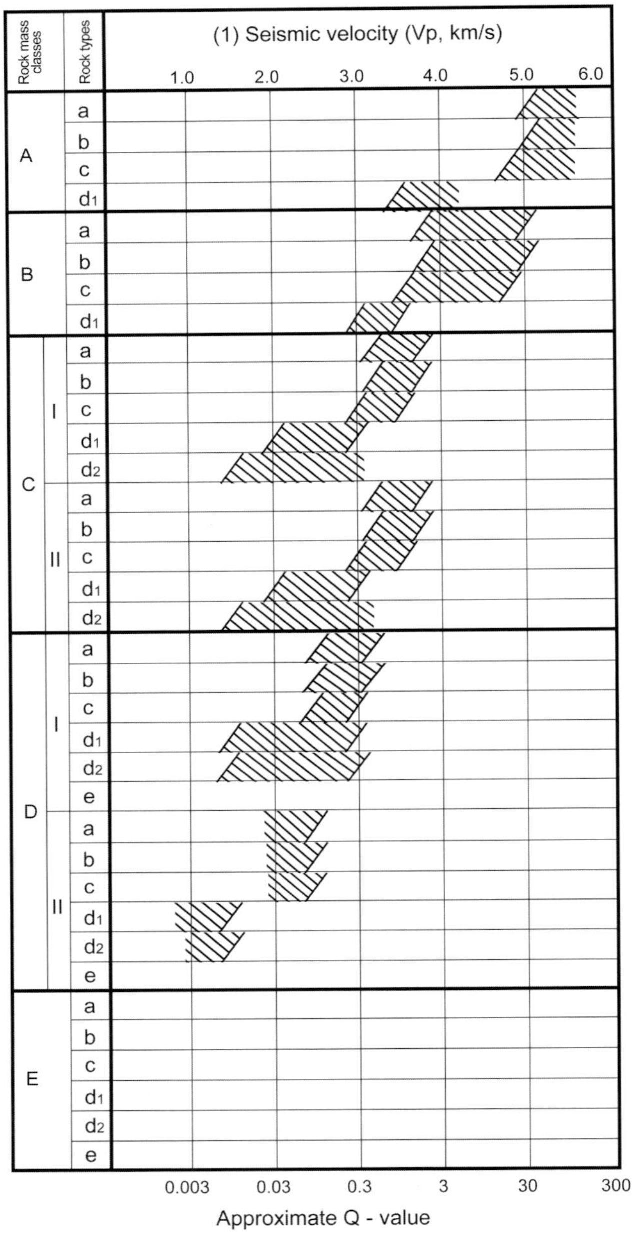


*Figure 8.6* The Japanese Highway rock-mass-class and rock-type tabulation, together with the anticipated seismic P-wave velocities from shallow refraction seismic. Approximate rock mass quality Q-scale from Barton and Itoh, 1995.

in Figure 8.7. The overall distribution of $V_p$ is also shown in Figure 8.8, for a rock mass with a mean logged Q-value, outside the weakness zones, of the following:

$$Q = \frac{90}{12} \times \frac{1}{4} \times \frac{1}{1} = 1.9$$

Palmström, 1982, recorded a 560 m length of concrete lining (350 m placed at the face), 2500 m$^3$ of shotcrete

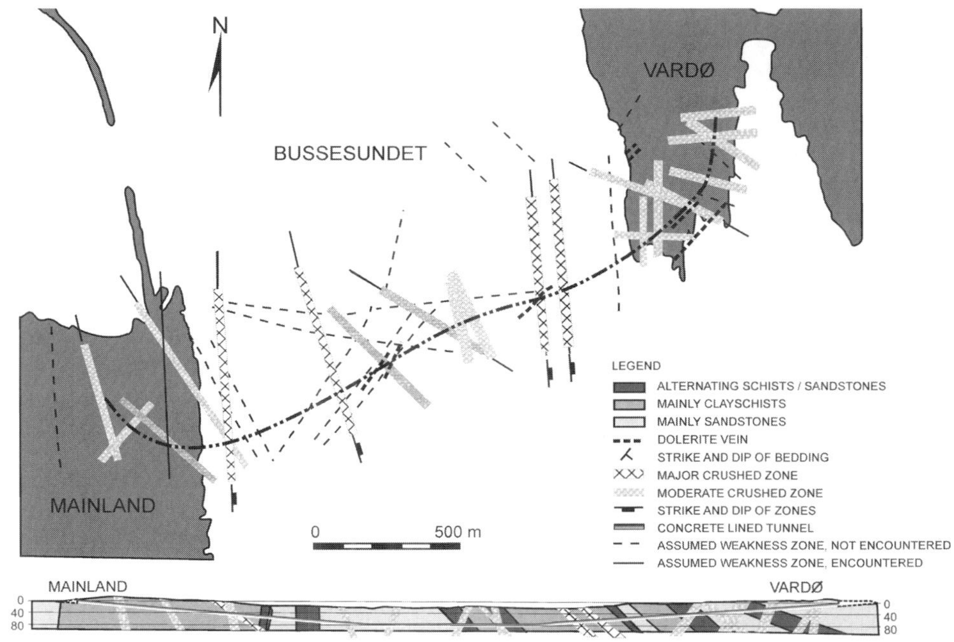


*Figure 8.7* Norway's first sub-sea tunnel to Vardø, showing assumed and encountered weakness zones. Palmström, 1982.

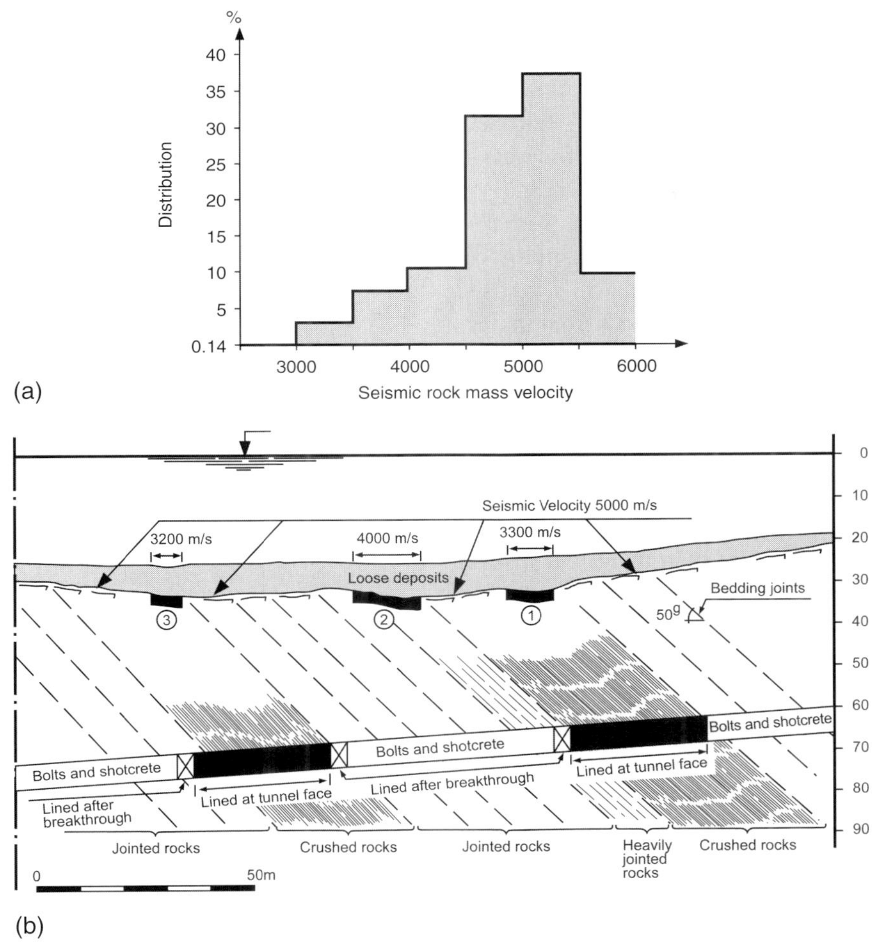


*Figure 8.8* P-wave velocity statistics for Vardø sub-sea tunnel. Lower velocity zones and corresponding support methods. Palmström, 1982.

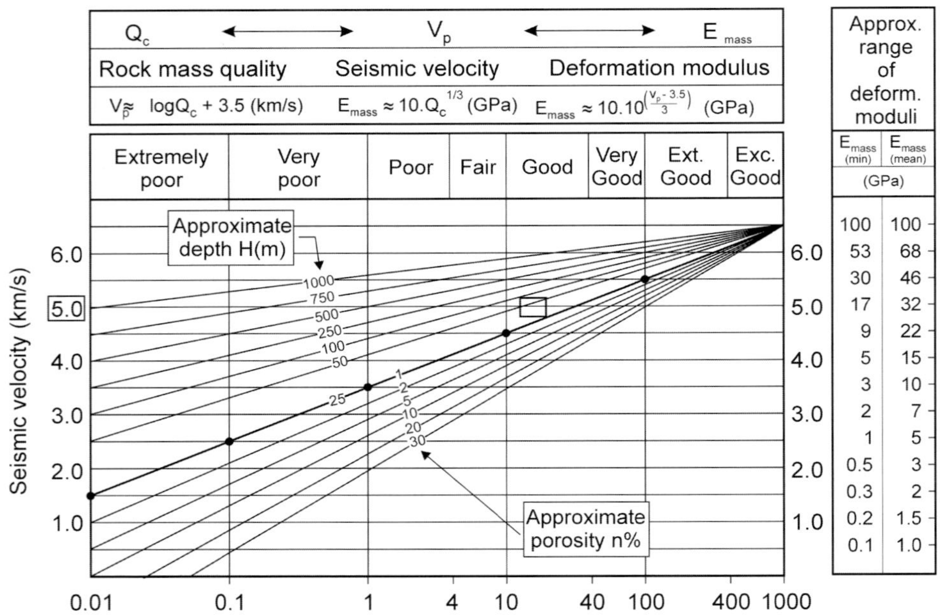


$$Q_c = \left[\frac{RQD}{J_n} \times \frac{J_r}{J_a} \times \frac{J_w}{SRF}\right]\frac{\sigma_c}{100}$$

*Figure 8.9* $V_p$-$Q_c$-M correlations with depth and porosity correction. Barton, 1995.

and 168,000 rock bolts giving an average tunnelling progress of 17 metres/week and a cost of USD 8000 per metre. The weighted mean value of $V_p$ outside the weakness zones was close to 5.0 km/s which, according to Figure 8.9 (Barton, 1995) implies a Q-value, perhaps equal to 15 if we assume 50 m of cover and a nominal 1% porosity for these generally hard rocks.

This estimate of rock mass quality from $V_p$ compares poorly to the quoted value of about 2, which might of course be in error. Another reason for this discrepancy might be stress-related, or an effect of the hard beds of quartzitic sandstones, *masking* the lower velocity, weaker beds of siltstones and shales.

The low-velocity zones ($V_p$ = 3.2–3.3 km/s) also created greater tunnelling difficulties than the values would suggest, implying an artificially elevated velocity in relation to rock quality. This will be discussed further.

There is obviously a broad correlation between tunnel support costs (as a percentage of total costs) and *seismic velocity*, as shown for example by Nilsen, 1998, for the case of half a dozen sub-sea tunnels (Figure 8.10). Support costs may rise from 50% to at least 75% of total costs, when the P-wave velocity reduces from 5.5 to 4.5 km/s. As pointed out by Nilsen, the most important factor is the quality of pre-investigation and follow-up, and an understanding of the need for good investigations by the owner.

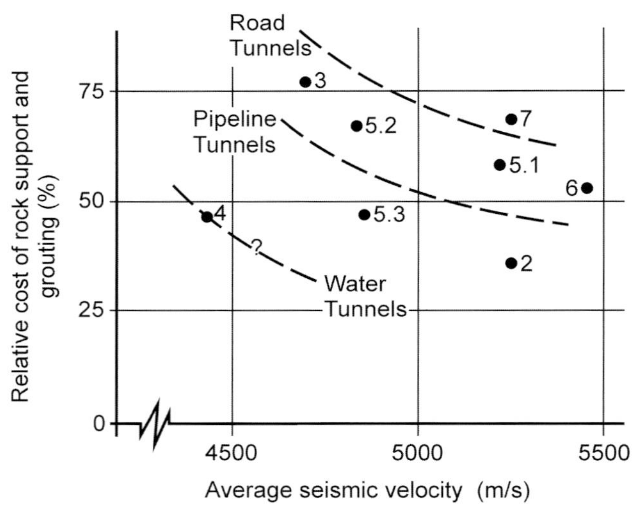


*Figure 8.10* Support costs (as % of total cost) versus $V_p$, for some road, pipeline and water tunnels. Nilsen, 1998.

Errors in interpretation of seismic measurements, for example due to horizontally interlayered stiff (meta-sandstones) and softer shales, have on occasion caused unwelcome surprises, i.e., with 'false' velocities, apparently as high as 4.5 to 5.0 km/s, nevertheless requiring immediate cast concrete lining up to the tunnel face. In one such case, the depth effect upon $V_p$ shown in Figure 8.9, may have been responsible for some of the 'false' velocity increase in

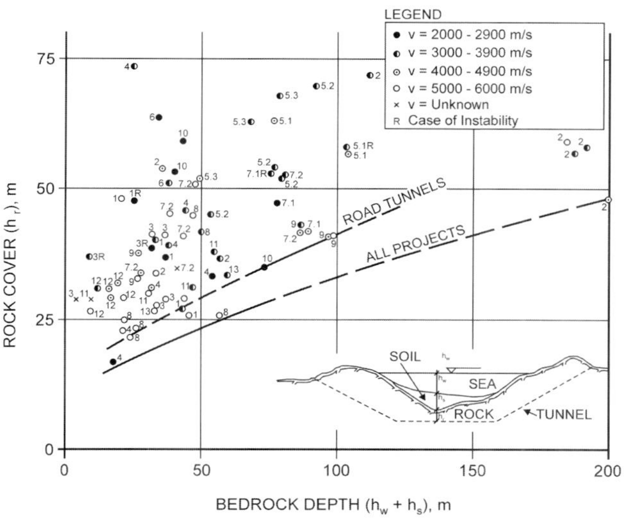


*Figure 8.11* Rock cover in relation to seismic velocities at sub-sea tunnel sites. Nilsen, 1998.

a generally extremely dry sub-sea tunnel (Barton and Monsen, NGI contract report concerning North Cape Tunnel, 1997, for Norwegian Road Authority).

The logged rock mass quality Q-values of only 0.01 and the need for heavy support right up to the face were 'inconsistent' with the general range of $V_p \approx$ 4.5–5.0 km/s. In this case the rock load and the sea depth may each have contributed, due to the undrained situation.

Nilsen, 1998, also gave rock cover versus bedrock depth (water + sediment) data for numerous sub-sea tunnels together with associated P-wave velocities. As expected, the highest velocities were usually associated with the lowest rock covers, while the zones with lowest velocities (as low as 2 km/s) tended to have higher designed rock cover, as shown in Figure 8.11.

Sjøgren et al., 1979, performed shallow refraction seismic investigations at sites where subsequent subsurface excavations were planned or actually carried out. They were therefore able to show several cases where seismic velocity showed a broad correlation with tunnel support measures such as shotcrete and concrete lining.

Although excavation span is listed in Figure 8.12, the authors did not give depth, for the caverns or tunnels listed. Significant depth is probably the reason why velocities between 3.8 and 4.7 km/s (which might correspond to hard rock Q-values of about 2 to 15 at 25 m depth) apparently were associated with such major stability problems at Rendalen (cases 5 to 8), where different sections of the headrace tunnel required from

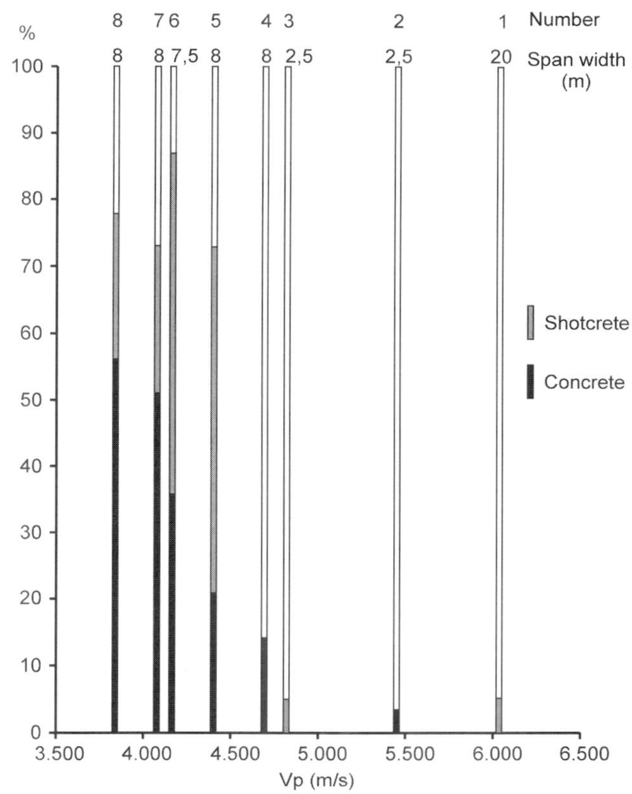


*Figure 8.12* Mean P-wave velocities at some tunnel and cavern sites in Norway, with shotcrete and concrete lining frequencies. Sjøgren et al., 1979.

15 to 55% concrete lining and from 20 to 50% shotcrete lining (with rock bolts).

The *mean* velocities given in Figure 8.12 are of course a bit misleading, as it is the lower values in any given distribution of velocities that require the rock support. For example the few shotcrete lined sections of the Mongstad oil storage caverns (case 1, Figure 8.12) would certainly have had a lower, local velocity than the mean value of 6.0 km/s for these massive, foliated meta-anorthosites.

As the authors point out using an illustrative statistic from the Vardeåsen site in Norway, the high velocities (4.8 to 6.2 km/s) completely dominate the usual range of velocities from these relatively unweathered Scandinavian sites, and it is the much smaller number of tectonic zones (shear zones, faults), dykes and joint swarms with velocities from about 2.5 to 3.5 km/s that cause the construction problems, especially when high inflows of water occur.

If we utilise the *shallow-depth, hard rock, low porosity* Q-Vp conversion given below:

$$Q \approx 10^{(V_p - 3.5)} \tag{8.1}$$

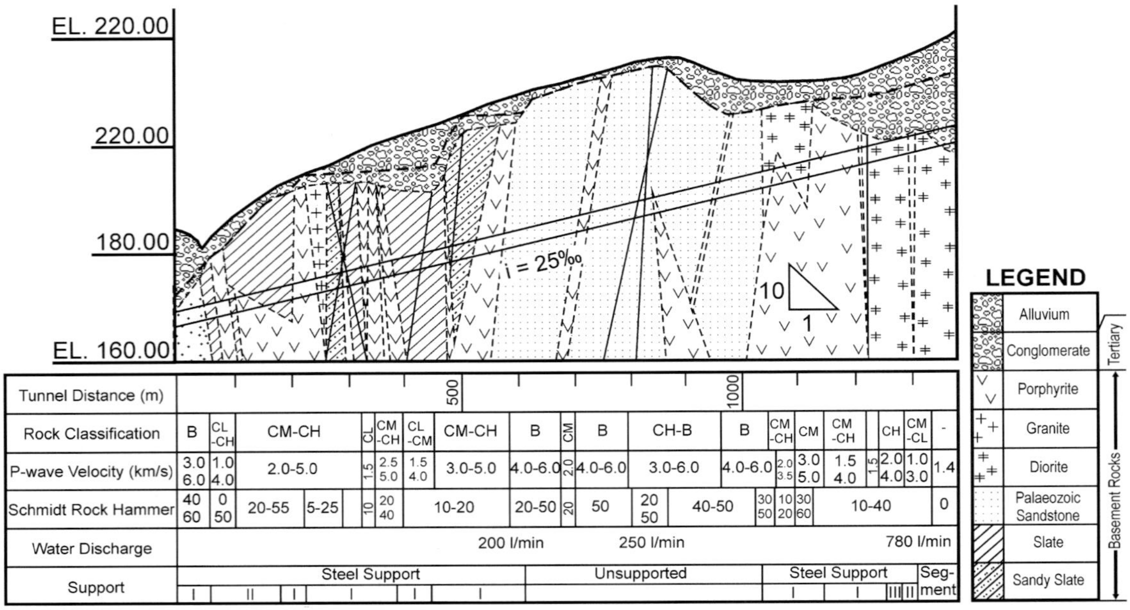


*Figure 8.13* Geological profile and rock classification details for Tunnel A. Mitani et al., 1987.

the above velocity ranges for good rock and fault or fracture zones can be approximately converted as follows:

1) $V_p = 4.8$ to $6.2\,km/s$ ($Q \approx 20–500$)
2) $V_p = 2.5$ to $3.5\,km/s$ ($Q \approx 0.1–1$)

As can be seen from Figure 8.9, a velocity as high as 6.2 km/s may for reasons of depth of measurement (or unusually high rock strength) only imply a Q-value of say 25, so conversions using this equation must always be related to the more comprehensive Q–$V_p$–depth–porosity model reproduced in Figure 8.9.

At the Veas sewage treatment plant near Oslo, Norway, where seismic surveys were performed by Geoteam (Sjøgren, 1984), and logging of excavations in the nodular limestone were performed by NGI colleagues, the reported correlations between $V_p$ and rock quality Q were as follows:

| $V_p$ (km/s) | Rock quality Q |
|---|---|
| 3.0 | 1 to 5 (with support) |
| 4.4 | 20 (no support needed) |

For hard rock at shallow depth, with negligible porosity (central trend, Figure 8.9), the above velocities would have suggested somewhat lower Q-values than the above (0.3 and 8 respectively). However, the nodular limestone, consisting of inter-bedded and well cemented shale and limestone layers, has some porosity (5 to 10% could be estimated) and its uniaxial compressive strength ($\sigma_c$) is less than the nominal 100 MPa for hard rocks. We can therefore use the porosity (n%) and $Q_c = Q \times \sigma_c/100$ corrections in the $Q_c$–$V_p$ chart, which then help to explain the somewhat higher Q-values.

Cecil, 1971, warned of another source for the possible lack of correlation between seismic velocity and rock quality, when considering the presence of thin clay coatings in otherwise widely spaced jointing. The latter cause potential stability problems but may hardly change RQD or $V_p$ values. On the other hand, the $J_r/J_a$ terms in the Q-system may capture the correct stability problems by reducing the Q-value, but Q will then not correlate so well with the velocities. These potential pitfalls in the general, near-surface correlation for hard rocks (equation 8.1) need to be carefully evaluated from case to case.

## 8.2 Examples of the use of seismic data in TBM excavations

A good example of the application of seismic velocity measurement to interpret TBM penetration rates is given by Mitani et al., 1987. They investigated the rate of advance of two small diameter TBM tunnels (∅ = 2.6 m) by measuring $V_p$ and Schmidt hammer rebound values for the wide variety of rocks encountered. Tunnel A (shown in Figure 8.13) was mostly driven in sandstones, slates, porphyry and weathered conglomerates, with generally strongly developed jointing and poor stability.

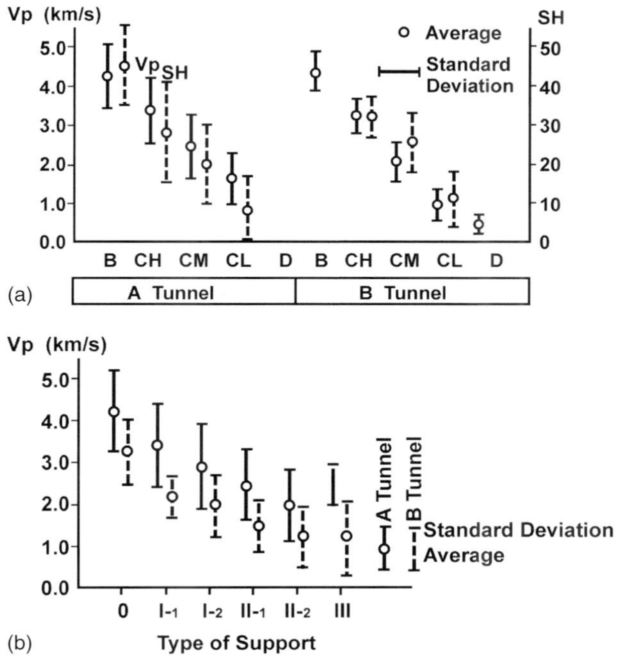


*Figure 8.14* a) $V_p$ (solid bars), and Schmidt hammer rebound %, in two Japanese TBM tunnels of 2.6 m diameter, b) correlation of $V_p$ to steel rib support spacing. Mitani et al., 1987.

Tunnel B was driven in more homogeneous but variously weathered granites.

Figure 8.14a shows $V_p$ in relation to Schmidt hammer rebound %, while Figure 8.14b shows the degree of rock support in relation to the $V_p$ values. (Rock support consisted of unsupported sections, or various grades of steel rib, steel channel or I-beam, with successively reducing spacing, i.e., 1.5, 1.2, 1.0, 0.8 m as $V_p$ reduced. Penetration rates, ranging from 0.6 to 5 metres per hour are shown correlated with seismic velocities in Figure 8.15.

Utilisation of geomechanics and seismic testing to correlate with TBM penetration rates (PR = net-instantaneous, and AR = effective advance rate), was described by Sampaola et al., 1978. Figure 8.16 shows a quite sensitive correlation between progress in m/hr and rock mass class A, B or C (which represented statistically homogeneous zones). The tunnel was only 6.4 m in diameter, and was bored in granites of variable quality caused by alteration and variations in jointing frequency. The depth range for the tunnel was not given by the authors. The TBM appears to have been a little under-powered in relation to the strongest, least jointed rock mass class.

The set of data given by Sampaolo et al., 1978, can be reproduced approximately as shown in Table 8.3.

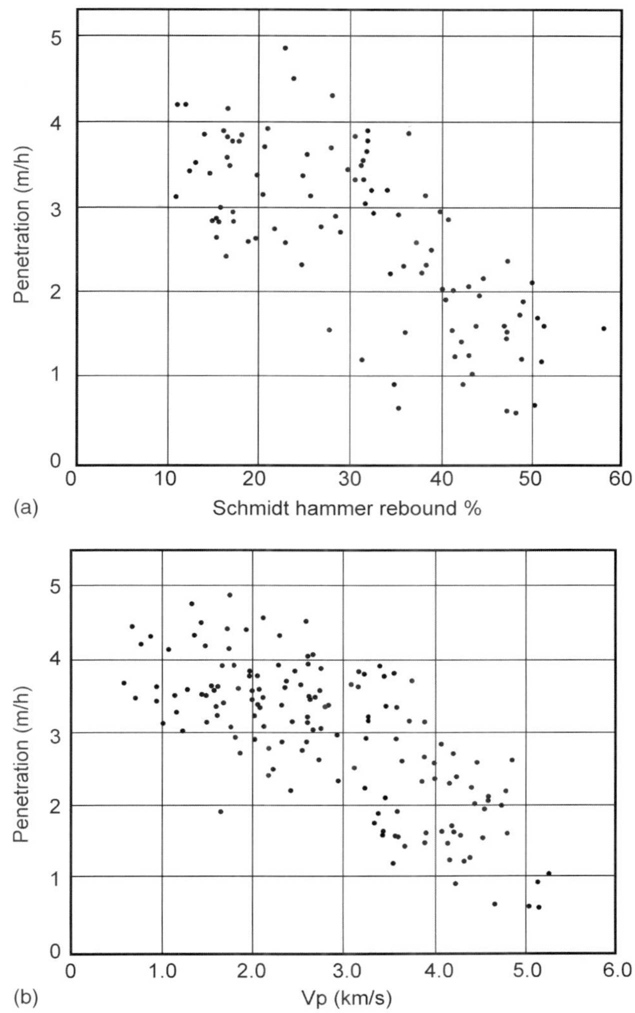


*Figure 8.15* Net penetration rates as a function of Schmidt hammer rebound and seismic velocity for 2.6 m diameter TBM tunnels in Japan. Mitani et al., 1987.

As can be noted by the above, the range of assumed rock qualities (approx. Q ≈ 0.1 to 15) do not penalise advance rates by any time-consuming support needs, so effective rates of advance for this hydropower tunnel were inversely 'correlated' to two of the geomechanics measures given above, and were therefore effectively inversely correlated with Q-values. This is consistent with the $Q_{TBM}$ model of Barton, 2000.

## 8.3 Implications of inverse correlation between TBM advance rate and $V_p$

In view of the intended aim of correlating, where possible, the seismic velocity and the rock quality, it is appropriate to consider in more detail the *inverse correlations* seen in the two previous case records. We have seen from Figure 8.15 the strong correlation between support needs and seismic velocities, which follow the *normal trend* of increased support with lower $V_p$-values, and with lower assumed Q-values.

However, it will be noted that net penetration rates (PR), correlate *inversely* with $V_p$ values, in other words, the increased degrees of jointing and reduced strength (also seen in the Schmidt hammer results) help to increase the penetration rate. The same trend is seen in Table 8.3 for a larger TBM (∅ = 6.4 m) boring in granites.

The documented trends of degree of jointing and rock strength on drilling or boring rate seen in the above examples are summarised in diagrammatic form in Figure 8.17. The same type of inverse correlation with $V_p$ values can be envisaged by converting the rock mass quality Q-class to a $V_p$-class in the upper portion of the figure. Faster boring will correlate with lower $V_p$ values up until some limit, as suggested by the two descending portions of the drilling rate trends.

Until the above support/stability limit is reached, the net penetration rates (PR) seen in the comprehensive data of Mitani et al., 1987, may be considered to have the approximate upper and lower bound values given in Table 8.4, in relation to $V_p$ and assumed, shallow depth (nominal 25 m), hard rock Q-values.

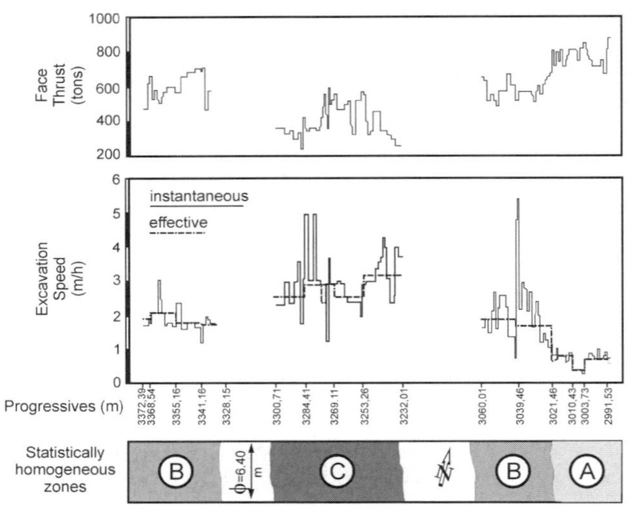


*Figure 8.16* Correlations between rock mass class A, B, C and excavation speed for a TBM driven hydropower tunnel. Respective classes had $V_p > 5$ km/s, 3.5–4.5 km/s, <3.5 km/s, and $\sigma_c$ >150 MPa, 50 MPa and <8 MPa. Sampaolo et al., 1978.

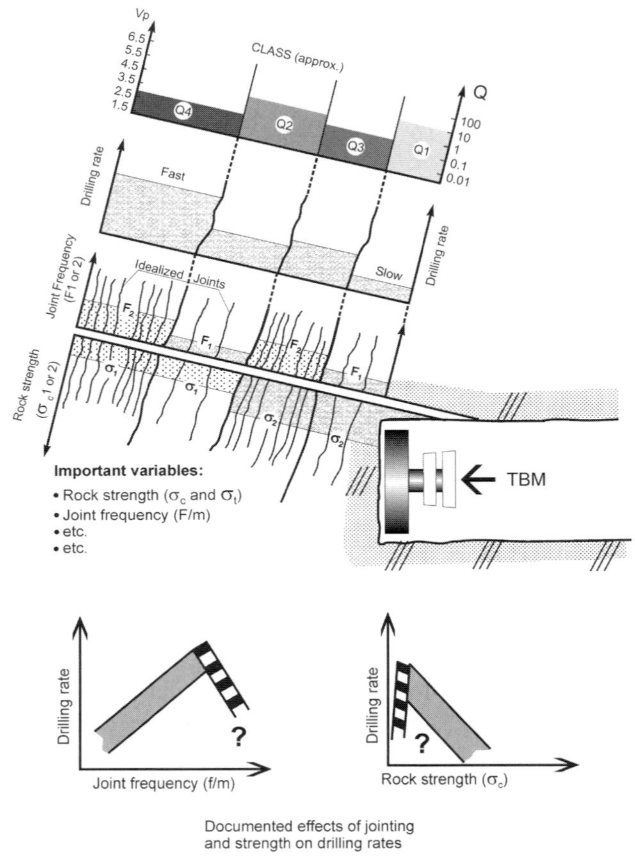


*Figure 8.17* Conceptual inverse correlation of boring or drilling rate with Q-value or $V_p$-value. Modified from Barton, 1996b.

Table 8.3 Correlations between advance rates and seismic velocities (Sampaola et al., 1978), with last column added by writer, using central trend for 100 m depth, or 50 m depth, from Figure 8.9.

| Zone | Alteration | $V_p$ (km/s) | $\sigma_c$ (MPa) | Effective advance rate (m/hr) (AR) | Net rates m/hr (PR) | F $m^{-1}$ | $Q_{est.}$ (Barton, 1995) |
|---|---|---|---|---|---|---|---|
| A. Sound granite | little or no alteration | ⩾5.0 | ⩾150 | 0.4–0.6 | 1.0 | <2 | ⩾8 or ⩾15 |
| B. Jointed granite | medium degree of alteration | 3.5–4.5 | 50 | 1.8–2.0 | 2.5–5 | ⩽2, ⩽5 | 0.07–2, 0.2–4 |
| C. 'Cataclastic rock' | high or very high alteration | ⩽3.5 | ⩽8 | 2.5–3.0 | 4–5 | >5 | ⩽0.07, ⩽0.2 |

Table 8.4 Upper and lower-bound PR–$V_p$–Q trends from Mitani et al., 1987 with additional interpretation.

| | Net penetration rate (PR) | | | | | | | |
|---|---|---|---|---|---|---|---|---|
| | Upper-bound | | | | Lower-bound | | | |
| PR (m/hr) | 1.0 | 2.0 | 3.0 | 4.0 | 1.0 | 2.0 | 3.0 | 5.0 |
| $V_p$ (km/s) | 6.0 | 5.0 | 4.0 | 3.0 | 4.5 | 3.0 | 2.0 | 1.5 |
| Q | 300 | 30 | 3 | 0.3 | 10 | 0.3 | 0.03 | 0.01 |

## 8.4 Use of probe drilling and seismic or sonic logging ahead of TBM tunnels

The last decade of developments with double-shield TBM's that use the PC-element liner for thrust when resetting the grippers, have made it possible to tunnel through a range of geological conditions, generally with less delays. However, the complexity of the machines has drawbacks in commissioning and learning-curve time, and a simpler design could in principle be used, if prior information of ground conditions were available through probe drilling and geophysical measurements. The concrete element liner, while convenient from many points of view, may be an expensive solution if most of the rock is actually of very good quality, requiring only light support.

Nord et al., 1991, indicated that TBM advance rate could be optimised in mixed ground conditions, by always selecting the right tunnelling mode in response to advance information. The other needs for advance information relate of course to support needs and to the possibilities of water inrush or caving ahead of the face. There are numerous TBM around the world that are stopped for long periods (some few, even permanently), due to inadequate information about pending adverse rock conditions. (Barton, 2000)

In Figure 8.18, the concept of advance seismic velocity information for subsequent rock quality class and tunnel support class estimation is presented. Ideally this should be made a routine operation, made by fast percussion drilling, such that support components such as steel arches, rock bolts or shotcrete can be immediately available, and applied with appropriate timing , at and behind the tunnel face.

A more complete concept is illustrated in Figure 8.19. Here, displacement monitoring is also performed in an effort to roughly confirm the $V_p$–Q correlation. Since special depth or anisotropic stress effects make the Q-$V_p$

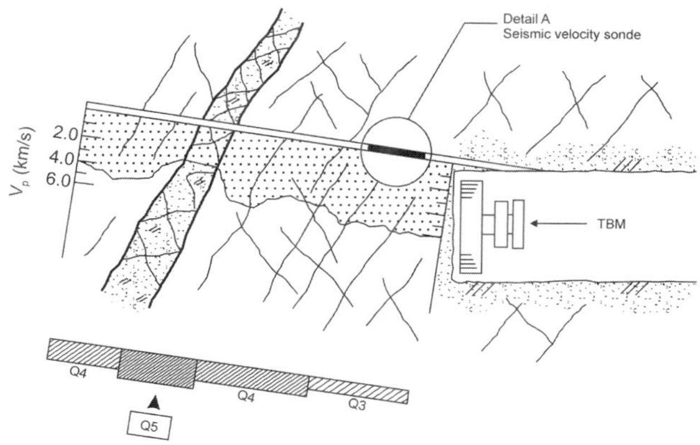


*Figure 8.18* Conceptual drawing of sonic probe and conversion of data to rock mass Q-values and rock support classes. Barton, 1996b.

linkage complex in many cases, the approximate expectation that final displacement (Δ, millimetres) is approximately given by:

$$\Delta \approx \frac{\text{SPAN(m)}}{\text{Q}} \tag{8.2}$$

may be an invaluable correction to the $V_p$-Q correlation, which may have a 'stress problem', concealing the actual poor quality. (See Barton, 2000, for detailed correlations between Δ, tunnel dimension, Q-value and other stress-related factors).

Unfortunately, the fact that the rock mass Q-value logged in a TBM tunnel may be higher than that logged in a drill-and-blasted tunnel will affect the measured deformation as described above. (Q will appear to be higher and Δ will be smaller). Likewise, if refraction seismic measurements were performed along the wall of the TBM tunnel, the values of $V_p$ obtained would also tend to be higher than in the equivalent drill-and-blasted tunnel for at least two reasons:

1) Reduced level and depth of damage in wall of TBM tunnel
2) Higher tangential stresses closer to the tunnel wall.

Such aspects will influence other details of the behaviour of the rock mass, due to coupled behaviour. For example there will be a tendency for lower permeability and less drainage around the TBM tunnel, which, for reasons of more complete saturation might also increase the seismic velocity. However, a seismic velocity probe ahead of the tunnel will *not see the difference* between the TBM

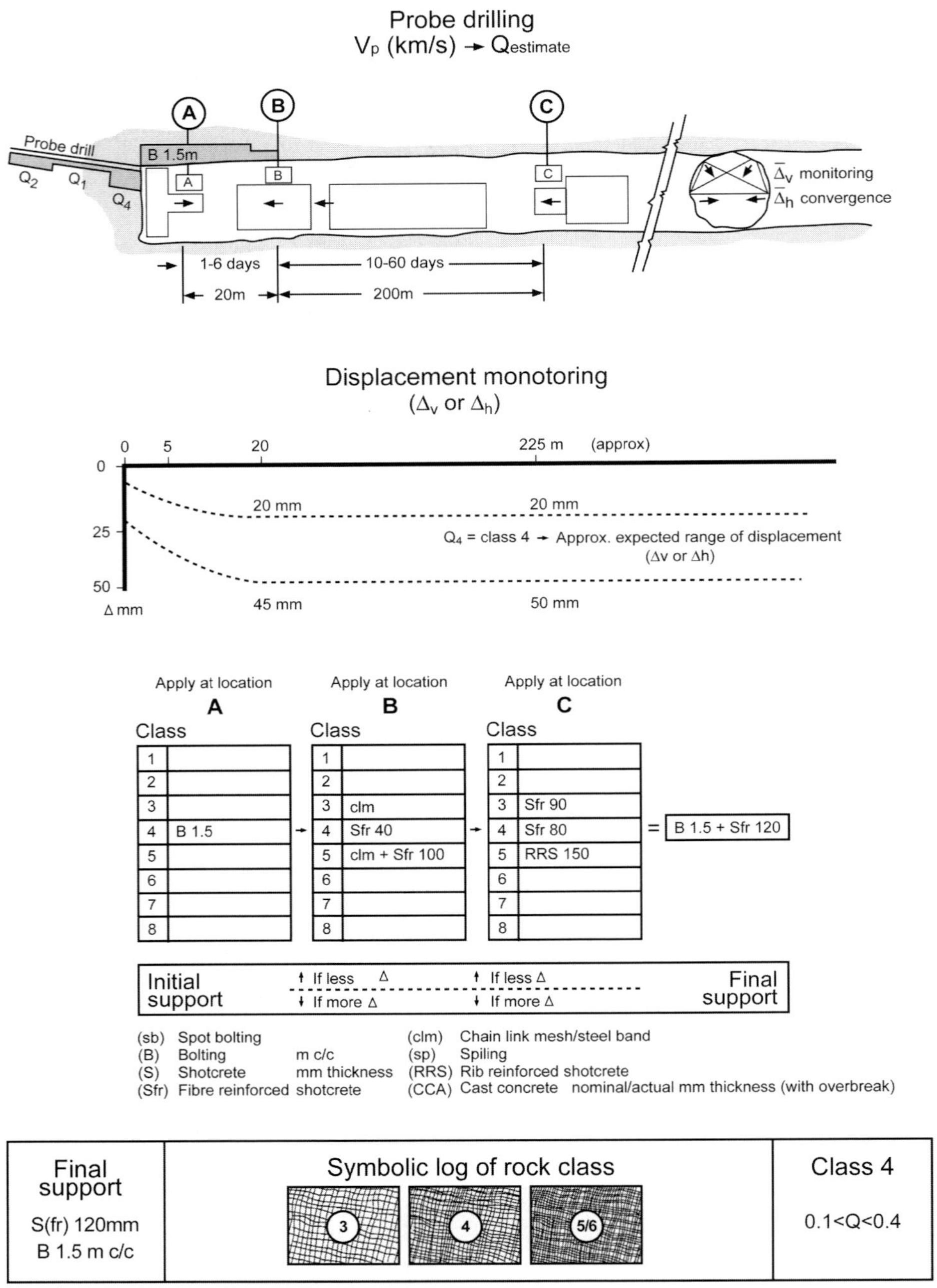


*Figure 8.19* Conceptual use of sonic probe for rock mass Q-value estimation, with displacement monitoring for confirmation of support needs. Barton, 1996b.

tunnel and the drill-and-blast tunnel, unless the measurement was made too close to the face of the tunnel.

## 8.5 In-tunnel seismic measurements for looking ahead of the face

The Japanese, with an extremely active tunnelling industry, in combination with far from ideal geological conditions, appear to have taken a lead in the use of seismic for 'probing ahead' of their tunnels, in particular TBM tunnels, where the consequences of delays are more critical, due to both the investment level and the normal expectation of fast tunnelling. Following on from Mitani et al., 1987 analyses of the relation between TBM progress and seismic velocity, reviewed earlier, we will briefly refer to some

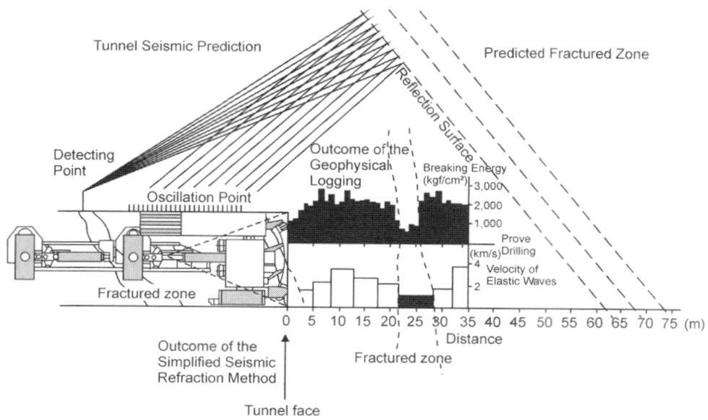


*Figure 8.20* Seismic reflection for identifying a change of conditions, combined with the results of sonic logging in a probe hole. From Kajima Corp. Nishioka and Aoki, 1998.

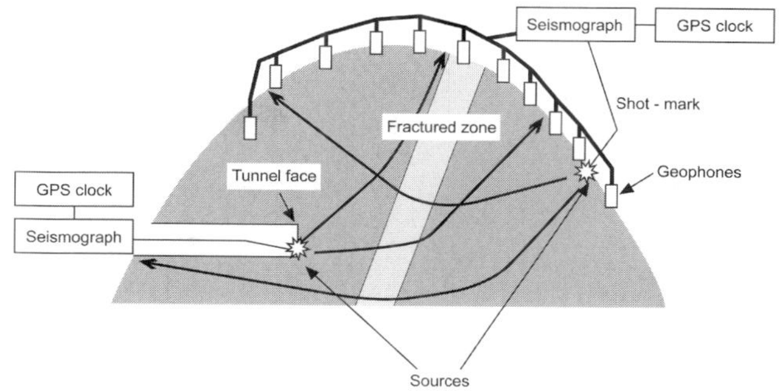


*Figure 8.21* Seismic refraction principle, for both in-tunnel and surface sources, with both in-tunnel and surface receivers. Hayashi and Saito, 2001.

more recent advances concerning seismic 'probing ahead'.

The advantage of being fore-warned of changed conditions by means of in-tunnel reflection methods is nicely illustrated in Figure 8.20. Also shown is the use of a probe hole, and a sonic logging of this hole, with a low velocity fractured zone correlating with an interval of reduced 'breaking energy', possibly actually referring to compressive strength, in view of the units used. Clearly, the reflector further ahead, presents the possibility of either reduced or increased rock mass quality.

A radically reduced $V_p$ and Q-value and a decision for pre-injection, pre-reinforcement and perhaps overboring may save weeks or months of delay and cost only one or two days in 'lost' production. Such an investment appears worthwhile.

Hayashi and Saito, 2001 described an interesting approach to seismic surveys for tunnelling that is a logical extension of conventional high-resolution surface refraction seismic, namely the use of sources and receivers also at, and close behind, the tunnel face. The concept is shown schematically in Figure 8.21. GPS clocks are needed to synchronise the sources within the tunnel and the receivers at the surface.

The authors pointed out that the already developed reflection method of HSP or TSP (horizontal or tunnel seismic profiling, e.g. Sattel et al., 1992), with both source and receiver in the tunnel, locates seismic reflectors ahead of the tunnel face, as desired, but that the reflector distributions are not related to rock quality directly. It is also difficult to determine if the rock quality will get better or worse at a given reflector, and they suggest that there may be inaccuracies of location in view of the unknown actual velocity field.

The authors suggested that this situation could be rectified by using the in-tunnel refraction estimates of velocity distributions ahead of the face, so that the accuracy of reflector positions could be increased. With the necessary velocity distribution ahead of the face, the rock mass could be characterized both up to and beyond the better-located reflectors.

The authors justified their method by demonstrating the steadily improving match to a hypothetical mountain velocity model, as numbers of in-tunnel sources were increased. The model, and two stages of improvement, are reproduced in Figure 8.22.Theoretical travel times were calculated by ray-tracing, and were considered like observed data.

The authors applied their proposed method to a tunnel under construction, in Mesozoic slates, sandstones and chert, with an overburden varying from 100 to 300 m. Figure 8.23 shows the detailed surface refraction seismic model of the mountain terrain, which was produced before tunnel construction. A general velocity along the tunnel route of about 4.0 to 4.2 km/c was indicated at this stage. The subsequently installed in-tunnel sources and receiver are also indicated, together with the string of surface receivers down the 'opposite' mountain face.

An in-tunnel reflector method that was being used in this tunnel, had *imaged* a clear reflector some distance ahead of the face when the face was at 439 m. The in-tunnel source refraction method subsequently utilised with only one in-tunnel source, predicted a sharply declining velocity ahead of the face. In fact a face collapse occurred at chainage 544 m, 105 m ahead of the measurement location, and a 300 m wide zone of weak rock, with velocity as low as 3.7–3.8 km/s was indicated by the second method.

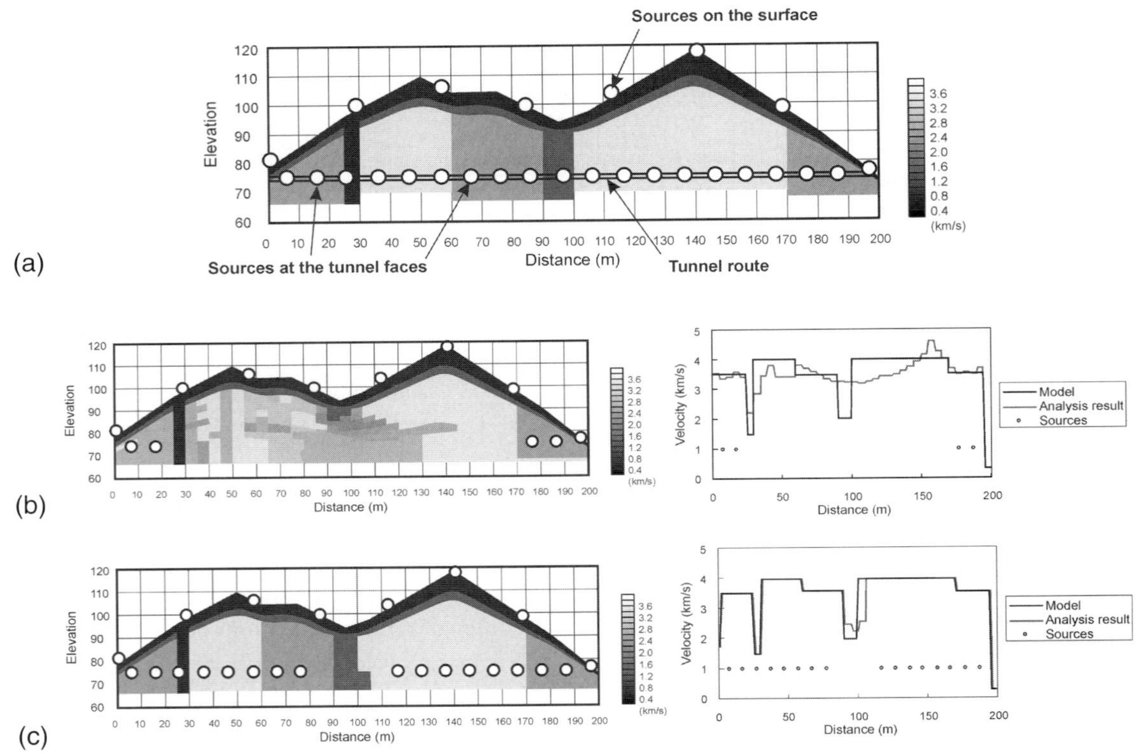


*Figure 8.22* a) A velocity model with ideal number of in-tunnel sources and limited surface sources. b) Two sources within the tunnel portals, and limited surface sources, improves the reconstructed velocity model, with improved definition and location of the low velocity zones. c) A series of mostly systematic tunnel sources, but with a central gap, gives greatly improved match to the 'actual' case. Hayashi and Saito, 2001.

## 8.6 The possible consequences of insufficient seismic investigation due to depth limitations

In the final section of this chapter a TBM case record will be briefly referred, in which the depth of tunnel (mostly 700 to 900 m), and mountain-side screes and loose deposits, had hindered the correct interpretation of the actually very adverse structural conditions. The case also illustrates the problem with stress effects, or virtual 'seismic closure', on this occasion due to compaction and near-invisibility of actually very troublesome faults.

Figure 8.24 indicates the initial geometric difficulties of this valley-parallel, TBM-driven headrace tunnel, for the Pont Ventoux hydroelectric project in north-west Italy. This drawing of a valley-parallel fault swarm was developed several years after the original investigation, and shows the limited surface seismic that was attempted at the time, together with some insufficiently deep boreholes.

The case was summarized in Barton, 2004c as follows: 'Unpredicted fault swarm parallel to valley-side, together with very high (and fault-eroding) water pressures, at depths of 700–900 m. TBM tunnel (diameter 5 m) eventually ran sub-parallel to individual faults, causing delays of at least half a year for each 1 m wide fault (AR ≈ 0.005 m/hr). TBM finally abandoned; new contractor for D + B from other end of tunnel.'

The tunnel was plagued by these sub-parallel valley-side faults for at least 2 years, with up to 6 months fault-related delays on several occasions, until TBM tunnelling was abandoned. An attempt to detect the continued presence of a particular fault (see sketches from daily geological log in Figure 8.25) using *seismic tomography* between two divergent pilot boreholes, proved to be unsuccessful, due to the presumed confining and densifying effect of the high stresses (from 800 m of overburden) on the fault-zone materials.

In Figure 8.26, the rock quality $Q_c$-based velocity-depth model is shown again, this time with some appended comments concerning the possibly elevated velocities of highly confined fault zones. Such zones, despite their $Q_c$ values as low as 0.01 or even 0.001, can nevertheless exhibit a stiffness and compactness at

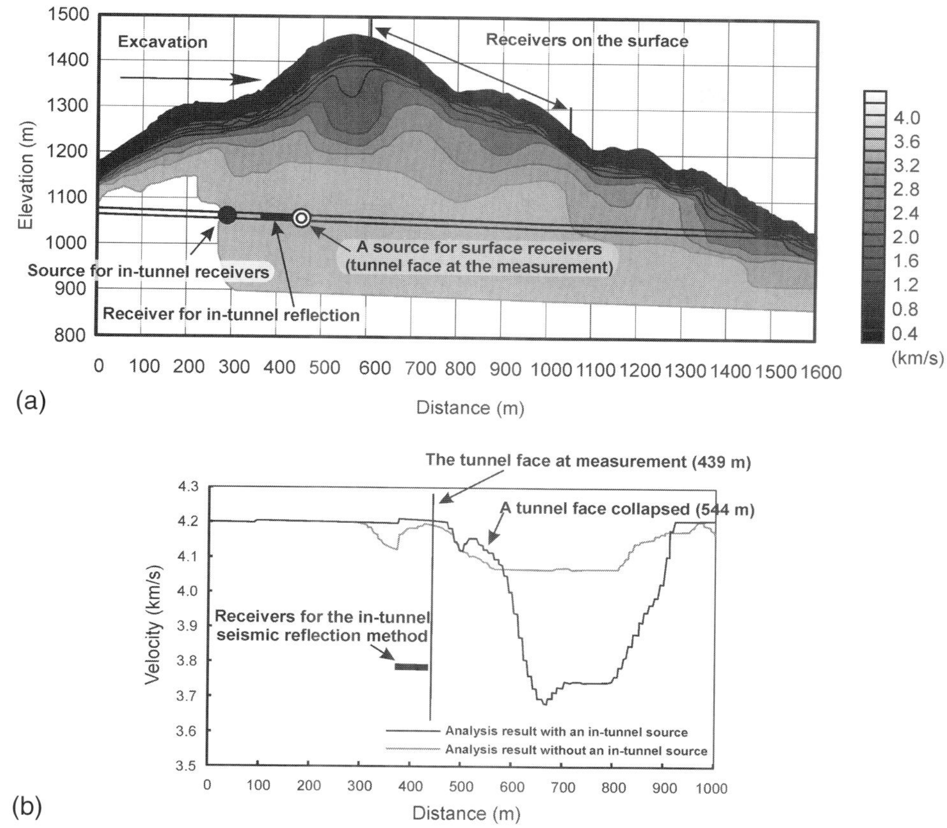


*Figure 8.23* a) Pre-construction surface seismic result, showing the in-tunnel sources for in-tunnel and surface receivers. b) An in-tunnel reflector method had indicated a reflector ahead of the face and a small reduction in velocity was assumed. The in-tunnel source refraction method subsequently predicted a sharply declining velocity ahead of the face, and a face collapse occurred about 105 m ahead of this location. Hayashi and Saito, 2001.

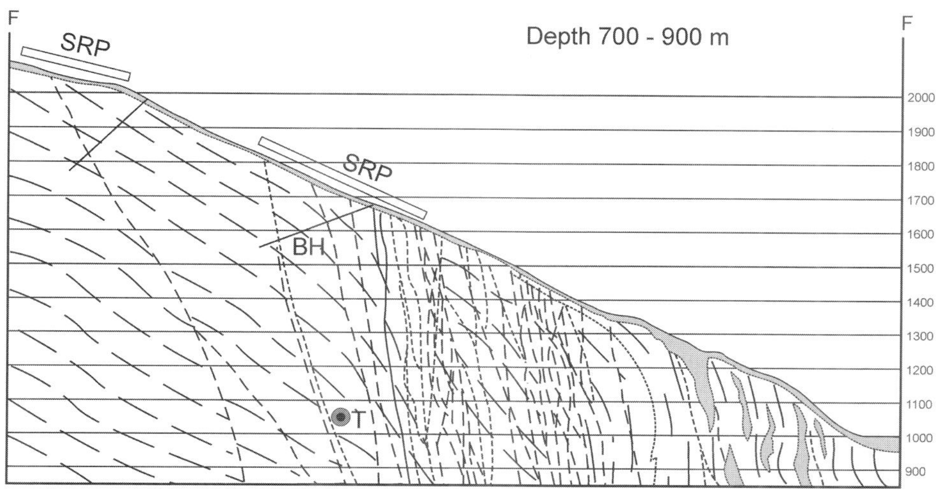


*Figure 8.24* Original seismic refraction profiles and inadequate borehole depths, are compared with the geologist's later re-assessment of the actual valley-parallel fault swarm, that had a dramatic effect on the fate of the TBM, and the final decision 'to drill-and-blast' from the other end of the tunnel. Pont Ventoux, Italy.

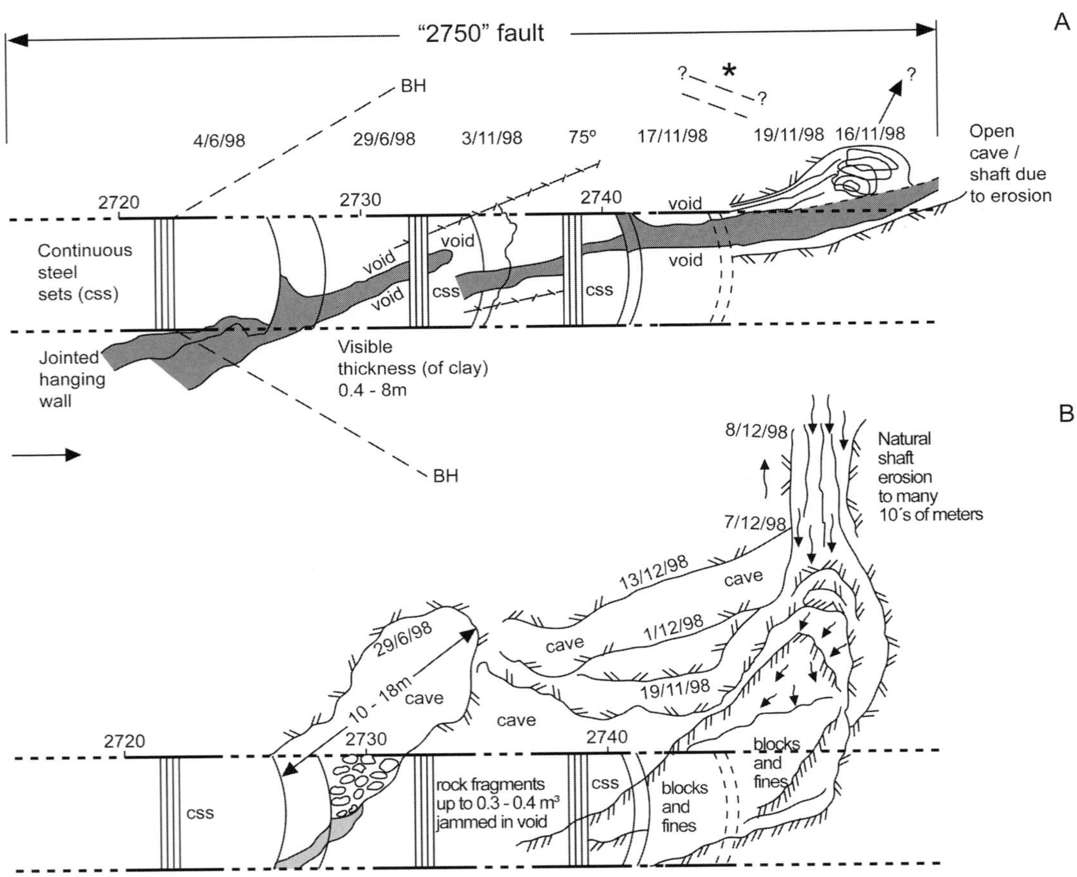


*Figure 8.25* Plan and elevation views of the '2750' fault at Pont Ventoux, based on a super-position of the site geologists sketches of the developing situation. This was a case of an unexpected combination of high stress, high permeability, and high fault frequency, and the eroding power of high pressure water. Barton, 2004c. Attempts to 'probe ahead' using seismic tomography between two diverging boreholes, proved not to be as successful as expected, due to the relative 'invisibility' of the assumed extension of the fault.

depth, that makes them nearly 'invisible'. But in fact there is hope that they will still show a recognisable contrast to the even higher velocities of surrounding rock, as suggested in the labels in this figure.

There is an interesting support for the above logic from the first case record referred to in this chapter, namely the continental divide Straight Creek pilot tunnel in Colorado, described by Scott et al., 1964. It may be recalled that the in-tunnel seismic refraction had been differentiated into 'deep layer' and 'shallow layer' velocities.

The following *deep layer velocities* were measured at the five seismic spreads shown on Figure 8.2: #1 5.2 km/s, #2 5.1 km/s, #3 4.8–6.1 km/s, #4 4.2 km/s and #5 6.0 km/s. The seismic velocities of the *shallow layers* (i.e. the loosened, near-surface-of-the-tunnel layers) at the five seismic spreads listed above, were respectively: 3.0 km/s, 2.3–2.7 km/s, 2.3–3.1 km/s, 1.3–1.6 km/s (worst case, class 5) and 2.3 km/s.

The worst quality (class 5) corresponded to seismic spread #4 in Figure 8.2 (deep layer $V_P = 4.2$ km/s), which had a local overburden of about 300 metres, and obviously was strongly affected by the depth or stress level. At the surface a $V_P$ value of 2.5 km/s is implied for this rock mass quality i.e. severely faulted ground which appears to correspond to the authors' description of class 5 rock. (A disturbed zone of at least one diameter was indicated for such rock.).

So in fact one may conclude that the deep layer velocities bore no resemblance to the eventual *major tunnelling difficulties* experienced when excavating the 12 m span twin tunnels. The deep layer velocities were either undisturbed (but highly stressed), or were perhaps subject to additional tangential stress compaction effects.

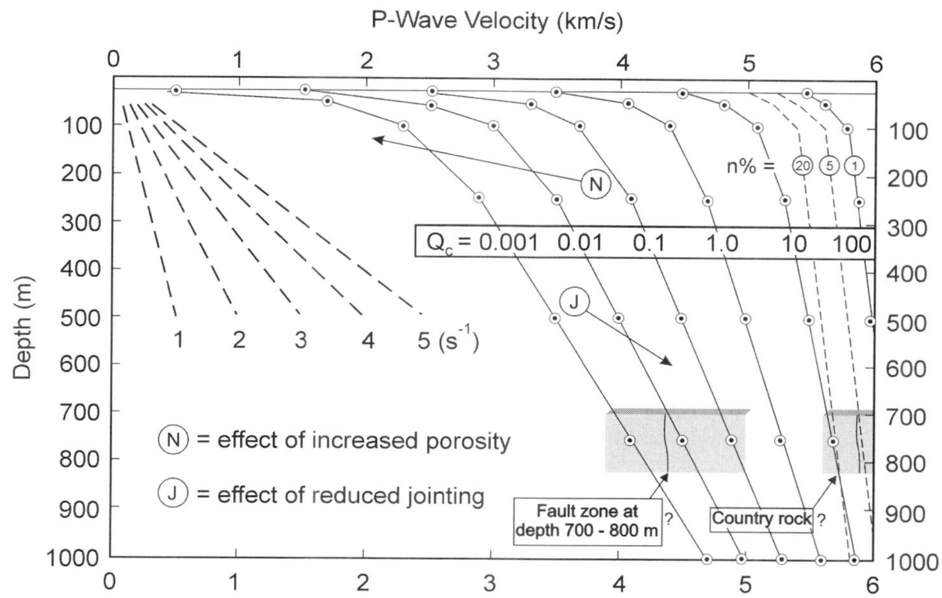


*Figure 8.26* Rock mass quality $Q_c - V_P$ – depth model, showing the potentially elevated P-wave velocities of nevertheless seriously-delaying fault zones, if encountered at great depth, as at the Pont Ventoux head-race tunnel. Contrast to the even more elevated velocity of the surrounding 'country rock' can nevertheless be expected. Barton, 2004c.

The more relevant, extremely low EDZ velocities from the pilot bore investigations, truly representing the poorer rock classes, were actually what caused the almost insurmountable problems in the main bore, which took several year to complete. Multiple perimeter drifts filled with concrete, making some 2 to 3 m effective wall thickness were needed in some zones.

One may also note from the Hayashi and Saito, 2001 case record, reproduced in Figure 8.23, that the face collapse at 544 m chainage, actually occurred in a $V_P \approx 4.1$ km/s rock mass. This is 'illogical', without the depth or stress effect that masks, in velocity terms, the true low quality. The 300 m overburden at this collapse location, would from Figure 8.26, suggest a near-surface $V_P$ of about 2.5 km/s – i.e. relevant to a serious fault zone, or extremely poor rock.

Finally, one may note the adverse effects of low Q-values on TBM progress, shown in Figure 8.27, specifically because of fault-zones. Velocity measurements at depth may *not* suggest such low values of Q. The TBM may nevertheless be delayed.

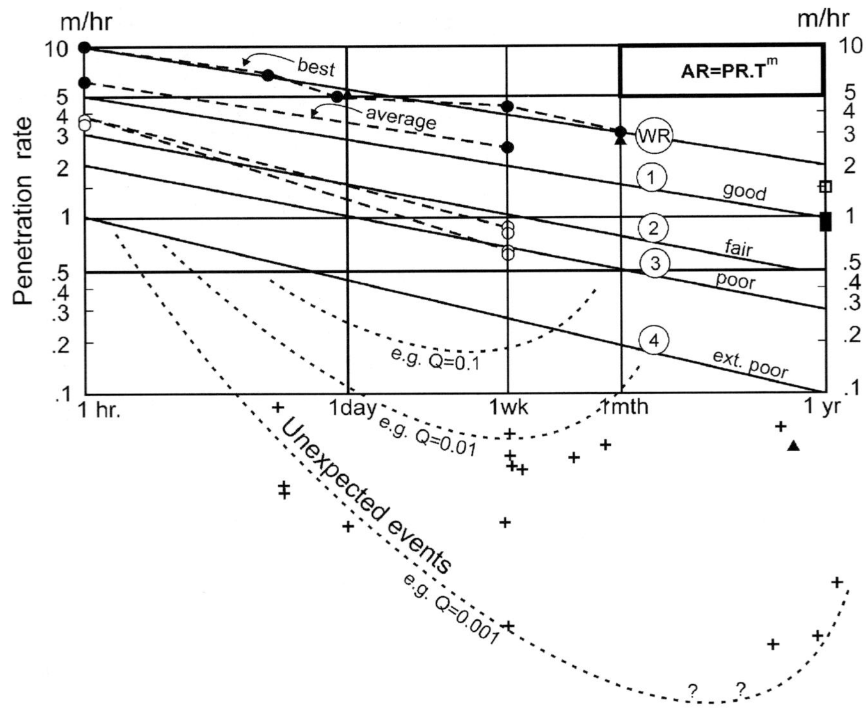


*Figure 8.27* The typical performance trends derived from analysis of 140 TBM tunnels, with 'unexpected events' strongly tied to low rock mass quality Q-values. Barton, 2000. It is probable that extremely low *actual* Q-values might show a deceivingly 'high' range of P-wave velocities, in the case of imaging ahead of deep tunnels.

# 9 Relationships between $V_p$, Lugeon value, permeability and grouting in jointed rock

Since matrix porosity, and in particular joint porosity, each affect the permeability of a rock mass it is perhaps logical that the seismic P-wave velocity should show some degrees of correlation with the permeability. When the joint void space is artificially increased by a particular type of 'permeability' testing (i.e., high pressure Lugeon injection tests), stronger correlation with velocity can be expected, due to the lowering of effective stress. This may apply most strongly in the case of rock masses of poor quality that are easily deformed around the injection boreholes, during the Lugeon tests, but the possible presence of clay is a complicating factor. The unit Lugeon is defined as the number of litres per minute flowing from each metre of a double-packered section of borehole, under an excess injection pressure, above the groundwater pressure, of 1 MPa.

Since a rock mass generally contains joints and microcracks, which are both a source of water and compliant, the application or existence of an anisotropic (effective) stress distribution may preferentially have closed those oriented at an obtuse angle to the major stress, while keeping those at acute angles or sub-parallel to the major stress, 'open'. There are then grounds for expecting both anisotropic permeability tensors, and a corresponding anisotropic velocity, with both maxima tending to be parallel or sub-parallel to the major stress. Since sub-vertical jointing may dominate in the same way that horizontal stress anisotropy may dominate, the anisotropy will tend to be related to azimuth. However, there is a potential source of error here. The permeability test holes must *intersect* the 'open' structure to register their higher permeability. The test holes needs to be drilled in the 'slow' direction, parallel to the minimum stress. The lower Q-value given by crossing all the 'open' joints, should correspond to the higher permeability.

## 9.1 Correlation between $V_p$ and Lugeon value

In Chapter 2, the strong effect of saturation on P-wave velocity was convincingly demonstrated by the extensive data of Saito, 1981 (Figure 2.17) and by the use of the time average equation for dry and saturated chalk (Grainger et al., 1973). The approximate 'porosity' which appears in the time average equation may contain air or water, and this porosity obviously affects the overall velocity, i.e., whether $V_{air} = 0.33$ km/s or $V_{water} = 1.44$ km/s is involved. So we have a theoretical starting point for a saturated velocity. The key question is whether this helps in predicting possible permeabilities. Does the saturated velocity give any indication of actual flow resistance?

Extensive sets of in situ measurements of rock foundation moduli, permeability and seismic velocity were assembled by the Comité National Français, 1964, from numerous dam site investigations. For the special case of two sites in jointed granite (from France's Massif Central), a strong correlation was evident between $V_p$ and the Lugeon test results.

Figure 9.1 shows an approximately linear distribution of data on a semi-log plot of $V_p$ versus the Lugeon value.

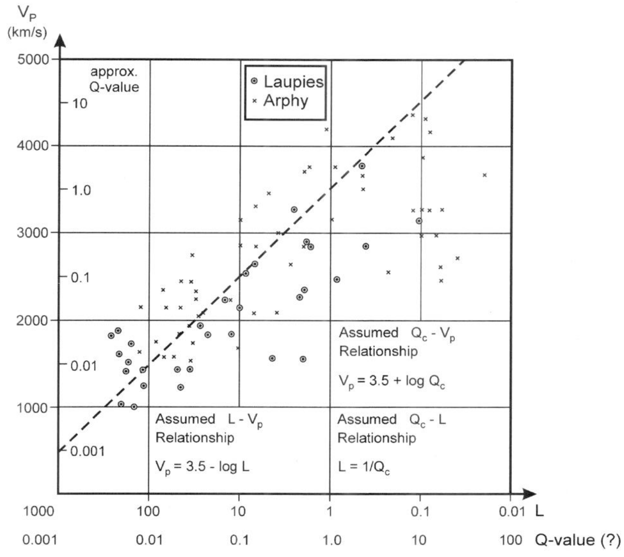


*Figure 9.1* Evidence for a correlation between $V_p$ and Lugeon value at two granitic dam foundations (Comité National Français, 1964) (Q-value scales have been added by the writer.)

If we make the assumption that shallow refraction seismic, or relatively shallow cross-hole measurements of velocity were used, then we can tentatively investigate the relation $V_p \approx 3.5 + \log_{10} Q_c$ (the diagonal line in Figure 9.1) as a means of relating Q-value and Lugeon value.

In very approximate terms we can see from the data that not only proportionality, (i.e., $L \propto 1/Q$) but equality $L \approx 1/Q$ is evident in the approximate range 100 to 1.0 Lugeon. The scatter of velocities and Q-values is seen to be about one order of magnitude, in other words neither velocity measurement, not Q-logging must be substituted for the testing. However, $L \approx 1/Q$ (Lugeon) might be utilised in extrapolation exercises, or to identify non-conforming behaviour.

We may therefore tentatively write:

$$L(\text{Lugeons}) \approx \frac{1}{Q} \tag{9.1}$$

as a useful approximate for fitting to data in some rock masses, and for explaining deviation (i.e. channelling, in other cases). There is an upper-bound $V_p$-L trend in the French data that exactly parallels the $Q \approx 1/L$ envelope; as can be clearly seen in Figure 9.1.

Two campaigns of core drilling at a shallow tunnel site in Wales; first with vertical holes, then with 45° inclined holes to intersect more of the steep structure, gave mean Lugeon results of 12 and 28. If we assume $Q \approx 1/L$, then Q values of 0.08 and 0.036 are derived. Completely independent Q-logging of the relevant boreholes (8 from phase 1 vertical holes, and 13 from phase 2 inclined holes) by the writer, gave weighted mean Q-values from many hundreds of observations of the six Q-parameters that were as follows:

Q (BH 1 to 8) = 0.11 ('higher' Q, lower permeability: L = 12)

Q (BH 13 to 21) = 0.08 ('lower' Q, higher permeability: L = 28)

The tunnel itself showed an overall weighted mean Q-value of 0.05, i.e. it was rather unstable rock. Downhole $V_p$ logging in BH 1 to 8 gave a mean $V_p = 2.6$ km/s for the same depth range that was core logged. This converts to a predicted Q-value of 0.12, almost the same as logged. By using the following 'hard rock' method of conversion, based on $V_p \approx 3.5 + \log_{10} Q$ and $L \approx 1/Q$, and eliminating Q we obtain:

$$L \approx 10^{(3.5-V_p)} \tag{9.2}$$

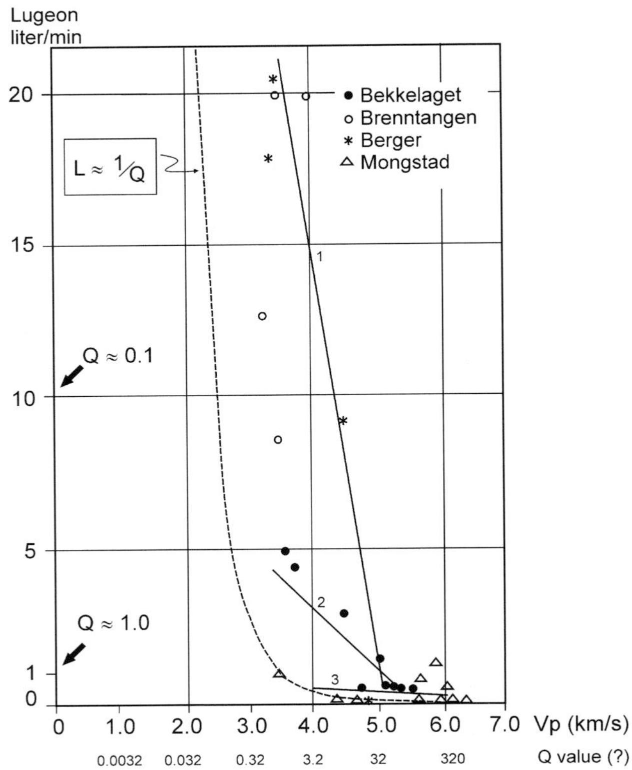


*Figure 9.2* Correlation between $V_p$ and Lugeon values at four hard rock sites in Norway. After Sjøgren et al., 1979. (Q-values scale added by the writer. Stippled curve given by equation 9.1).

which suggests a Lugeon value of 8 when $V_p$ is 2.6 km/s. The measured value in the relevant holes was 12.

The above logging data from clay-bearing metasediments shows remarkable similarities to the $L \approx 1/Q$ model, and also shows the potential anisotropy of the Q-value due to different joint sampling frequency with hole orientation. Lower Q-values, higher Lugeon values, and lower seismic velocities will tend to be measured when perpendicular to major structure. The opposite occurs when paralleling major structure. Of course there can be exceptions to this basic concept caused primarily by an eventual rotated stress-anisotropy, that no longer matches the joint patterns: a less likely scenario.

Sjøgren et al., 1979, gave correlations between Lugeon tests and seismic velocities for several locations from four of their investigated hard rock sites in Norway. A total of 29 data points are given in Figure 9.2. They defined 1 Lugeon in the usual way, and mentioned the constant pressure of 1 MPa. It is not known if this standard excess pressure was reduced closer to the surface, but if not, this could be the reason for some of the

Table 9.1 Approximate correlations between measured transmissivities and seismic parameters based on measurements in five boreholes in marl. Albert, 2000.

| Transmissivity | $[m^2/s]$ | $10^{-14}$ | $10^{-12}$ | $10^{-10}$ | $10^{-8}$ | $10^{-6}$ | $10^{-4}$ | $10^{-2}$ |
|---|---|---|---|---|---|---|---|---|
| $V_p$ | [m/s] | 5500 | 5117 | 4733 | 4350 | 3967 | 3583 | 3200 |
| $V_s$ | [m/s] | 3000 | 2700 | 2400 | 2100 | 1800 | 1500 | 1200 |
| $V_p/V_s$ | | 1.72 | 1.84 | 1.96 | 2.09 | 2.21 | 2.33 | 2.45 |
| Dynamic shear modulus | [GPa] | 24 | 20 | 16 | 12 | 7 | 3 | 0.1 |
| Dynamic E-modulus | [GPa] | 60 | 50 | 40 | 30 | 20 | 10 | 0 |
| Poisson's ratio | | 0.28 | 0.30 | 0.32 | 0.35 | 0.37 | 0.39 | 0.41 |

unexpectedly high values (e.g., 10 to 20 Lugeons) seen in rock with velocities from 3.5 to 4.5 km/s. However, in general their results showed the expected lower Lugeon results at higher velocities, as in the French data.

The rule of thumb $1/Q \approx$ number of Lugeons was tested against the Sjøgren et al., 1979, data using the assumed, near-surface, hard rock relationship $V_p \approx 3.5 + \log_{10} Q$ (Barton, 1995). It appears from Figure 9.2 that the '1/Q' curve is a suitable lower bound to some of the velocity-Lugeon data. Accuracy could potentially be improved if Lugeon values were correlated with depth zones, and if these depth zones had been given, a more correct $V_p$–Q relationship could have been selected from Figure 8.9 (see Chapter 8).

Curiously, some of the points plotted by Sjøgren et al., 1979, exactly fit the dotted line ($Q \approx 1/L$) relationship, as also experienced in Figure 9.1, possibly indicating conforming or non-conforming data, as the case may be.

In poorer quality rock masses, a Lugeon test is a form of rock mass deformability test, with unusually sensitive ('aperture cubed') registration of joint deformation, due to the tendency, under laminar flow conditions, for flow rates to be proportional to $e^3$, where e is the hydraulic aperture. However, in very stiff rock masses this may not apply, especially if channel flow dominates due to outwash channels in filled joints, or if joint jacking (slight opening) occurs due to low $K_o$ ($=\sigma_v/\sigma_h$) values, or if jacking occurs due to over-dimensioned water pressures close to the surface.

From extensive work in marl formations in Switzerland, Albert, 2000, indicated quite strong relations between selected seismic parameters and transmissivity measurements in five deep boreholes at Wellenberg, a potential nuclear waste repository site. Good correlations with transmissivity were obtained with $V_s$, dynamic shear ($\mu$) and E-moduli, $V_p$, $V_p/V_s$ and dynamic Poisson's ratio.

Table 9.1 shows the author's approximate correlations between the seismic parameters (using linear scales) with the logarithmic transmissivity scale. (The transmissivity is the product of the permeability and the thickness of the measured aquifer or 'aquiclude'.) Most of the measurements reported by Albert, 2000, were in the transmissivity range $10^{-5}$ to $10^{-12}\,m^2/s$. A 1 m long test section assumption would convert these to the more familiar engineer's m/s units.The borehole depths ranged from about 400 m to 1800 m, and included faulted and brecciated rock.

One must assume that the 'hydraulic tests and fluid logging' were of the non-deforming type, unlike the civil engineering Lugeon testing (often for evaluating groutablity), discussed elsewhere in this chapter. Compaction effects on this relatively weak rock at borehole depths up to 1800 m presumably have affected permeabilities more than the seismic velocities. The smaller steps in velocity at the lower transmissivities resemble the effect of *depth* in the $Q_c$-M-$V_p$ engineering model (Figure 8.9). The successive reductions in dynamic E-modulus with increased transmissivities (and $Q_c$-values?) have a certain similarity to this rock engineering model, bearing in mind the $E_{dyn} > M$ inequality.

Examples of correlations between seismic refraction surveys and drilling and tunnelling results are given by Sjøgren, 1984. This example is given in this chapter due to permeability links. Figure 9.3 shows successive stages of an investigation, and confirmation during construction, for a water supply tunnel beneath the Skien river in Norway. Four seismic refraction profiles are shown in the top figure. Three low velocity zones were indicated beneath the river, the largest of which ($V_p = 2.5$ km/s) was proved by an inclined borehole to be a partly consolidated breccia and loose alum shale (core loss averaged 75% in this zone).

The Lugeon value in this zone was 14, which might correspond to a $Q_c$ value of about 1/14 (=0.07). This is close to the value of Q that could be predicted from Figure 8.9, using a nominal porosity for the zone of 5%, and the 50 m depth shown in Figure 9.3c. At this

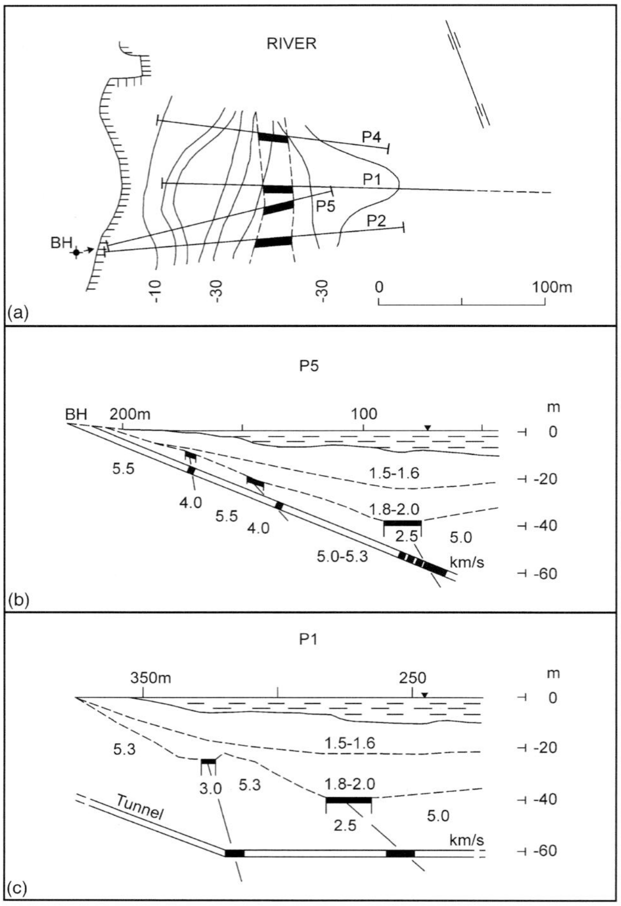


*Figure 9.3* Correlations between refraction seismic velocities and borehole and tunnelling experiences through the same zones. Sjøgren, 1984.

depth (and with n = 5%) Q = 0.07 corresponds to $V_p$ = 2.5 km/s, as measured by chance or good physics. The tunnel was driven through the same zone, in the direction of profile 1 (Figure 9.3c). Grouting was necessary for the $V_p$ = 3.0 km/s zone (perhaps Q ≈ 0.3 and L ≈ 3 from equations 9.1 and 9.2). Probe drilling and heavy reinforcement was used through the 12 m wide fractured alum shale zone, which had $V_p$ = 2.5 km/s (and L = 14, and Q ≈ 0.07?).

## 9.2 Rock mass deformability and the $V_p$-L-Q correlation

Dam sites throughout the world are investigated by means of borehole water injection tests, typically using double packers, and injection-pressures related to depth below the surface, but usually limited to about 0.25 or 0.5 or 1.0 km/cm$^2$ per metre depth. (The choice is related to national or regional practice.) The number of Lugeons is expressed by the well-known relation L = litres/min/m/1 MPa excess pressure. Most of the flow losses (and joint deformation) occur close to the borehole in such a test, which differs greatly from the careful, low pressure pumping (or extraction) tests favoured in permeability testing (Quadros, 1995).

By good fortune or correct physics, the modulus of deformation (M) that was shown in Figure 8.9 is proportional to $Q^{1/3}$ or to $Q_c^{1/3}$ in the case of rocks weaker or stronger than our nominal $\sigma_c$ = 100 MPa. Similarly, it is well known that flow rate is more or less proportional to $e^3$ in jointed rock masses (where e = equivalent hydraulic aperture of the joints, and intrinsic permeability can be expressed as $e^2/12$). The smaller value of (e) approaches the physical aperture (E) when e ≈ 1.0 mm, and this inequality (E/e ⩾ 1) is related to joint roughness JRC (Barton et al., 1985, Barton and Quadros, 1997).

Around the injection borehole we may assume that the natural joint apertures are deformed significantly, especially when maximum injection pressures of 0.025 up to 0.1 MPa per metre depth are used. The latter European injection pressure limit at dam sites is about two times the assumed vertical total stress. When $k_o$ (= $\sigma_h/\sigma_v$) is <1.0 and causes lower minimum stress than these figures, some slight hydraulic jacking of some of the joints is an obvious consequence in the initial radii around the boreholes.

The following basic assumptions will be made concerning this all-important joint deformation region around the injection holes:

1. The Lugeon value (L) which is recorded as volumetric flow rate (litres/min) will tend to be proportional to the cube of the new apertures that have been created, i.e., $\alpha(\Sigma\ E^3)$. There is some evidence from grouting results (over-coring or excavation) that the most permeable and well-connected joints open most at the expense of others in the same set. The resulting Lugeon value will often be dominated by the $E_{max}$ value and we can ***roughly*** approximate here that $L \propto E^3_{max}$, since the smallest micron-size apertures will contribute only minutely.
2. The locally gapped joint will have an aperture $E_{max}$ that is approximately inversely proportional to deformation modulus M.
3. The calculation of a 'double' Boussinesque elastic foundation calculation for the radially distributed deformation of *each side of the joint*, with realistic input for dimensions, supports this.

Therefore we have the following possible inter-relationships between maximum apertures, Lugeon values, deformation moduli and Q-values, which in turn are linked to seismic velocities:

$$L \propto E_{max}^{3}$$
$$E_{max} \propto M^{-1}$$
$$M \propto Q_c^{1/3}$$
$$\text{Therefore } L \propto Q_c^{-1}$$

(Note '$\propto$' implies 'approximately proportional to' in the above proportionalities).

These simple proportionalities therefore suggest that the number of Lugeons may indeed be proportional to $1/Q_c$, unless other mechanisms than local joint deformation are responsible for the flows, for example outwashed chlorite fillings, severely canalised flow due to basalt flow-top weathering, uncontrolled hydraulic joint jacking, and so on. The data that follow the lower-bound trend $L \approx 1/Q$ in Figures 9.1 and 9.2 are therefore consistent with this theoretical model, and explain why $V_p$ and L (Lugeon) can show a degree of correlation.

In Figure 9.4, the above inter-relationships have been expressed in the form of a nomogram, using the basic structure derived in Chapter 5 (Figure 5.36). Five illustrative 'type curves' have also been added, to show what might be typical $V_p$-Q-M-L data for fault zones, weak porous rock, hard jointed rock and hard massive rock. For simplicity (and continuity of the curves), it has been assumed that porosity only develops in the shallow, near-surface weathered zone in each case ($H < 25$ m).

In the case of a *hard porous* rock, there will be a velocity correction caused by porosity at all depths, and this will mean that the central curve shown in Figure 9.5 will give the correct velocity, roughly midway between the 'hard rock' reference curve (top) and the porosity correction curve (bottom).

The steepening gradient of the type curve is in this case affected by the assumed Q-value increase at greater depth. If, for some reason, this does not occur, then the 'type curve' could be a straight vertical line, as for the 'hard massive' rock shown in Figure 9.4. Any porosity correction would merely reduce the 'height' (i.e., reduce the maximum velocity) of such a line. The $Q_c$ correction factor is the same as that developed in Chapter 5 (Figure 5.36). The nomogram can be illustrated by the following 'coupled' example.

At the Chinnor Tunnel in chalk marl, Hudson et al., 1980, ('seismic' Hudson) referred to very low velocities (0.6 to 1.0 km/s) for badly fractured/jointed areas of the chalk marl and quoted permeabilities from Lugeon type tests of $10^{-4}$ to $10^{-6}$ m/s in these areas. If we assume for simplicity that 1 Lugeon $\approx 10^{-7}$ m/s then the very high Lugeon values obtained of 1000 to 10 imply $Q_c$ values of 0.001 to 0.1 according to equation 9.1 (where Q has been replaced by $Q_c$).

These low $Q_c$ values can be converted to 'tunnel support' Q values of 0.02 to 2 if we assume a mean $\sigma_c$ value of 5 MPa for the chalk marl. This range is in line with expectations for the heavily jointed rock mass at Chinnor. (The term 'tunnel support Q-value' is used to remind of the original development of the Q-system for selecting tunnel support: Barton et al., 1974.)

Although one should in general resist the temptation to convert 1 Lugeon to $10^{-7}$ m/s (approx.) as if 'rock mass permeability', because deformability of the medium is very likely in the case of Lugeon testing in weaker rocks, it is nevertheless of interest to note that the 1000 to 0.001 Lugeon scale shown in Figures 9.4 and 9.5, would convert to $10^{-4}$ m/s to $10^{-10}$ m/s. This resembles the wide range of permeability often encountered where thousands of well tests are assembled in one plot.

However, channelled flows in weathered basalt flow tops may exceed $10^{-2}$ m/s, and some massive igneous rock may have permeabilities as low as $10^{-12}$ m/s, due to lack of joint connectivity and lack of micro-cracks. (The latter may appear only after sampling from strongly anisotropic virgin stress states, as discussed in Chapter 3).

A further example of Q-$V_p$-L correlation can be developed from the columnar basalt foundations of the Segunda Angostura dam site in Argentina. Classification of the site together with preliminary testing were reported by Di Salvo, 1982.

mean RMR = 63
mean Q = 8.5

(These are close to the Barton, 1995 suggested inter-relationship $RMR \approx 15 \log_{10} Q + 50$)

$V_p$ (downhole) = 4.5 km/s below 'decompressed zone'
$V_p$ (downhole) = 2.0 km/s in the 'decompressed zone'

The higher velocity suggests $Q_c \approx 10$, based on the relation $V_p \approx 3.5 + \log_{10} Q$ for hard rock. A uniaxial strength for the basalt of e.g. 125 MPa, would mean $Q \approx 8$.

L (Lugeon) in decompressed zone = 16, suggesting $Q_c \approx 0.06$ based on $L \approx 1/Q_c$.

A $Q_c$-value of 0.06 suggests a $V_p$ value of 2.3 km/s, i.e., very close to the measured velocity.

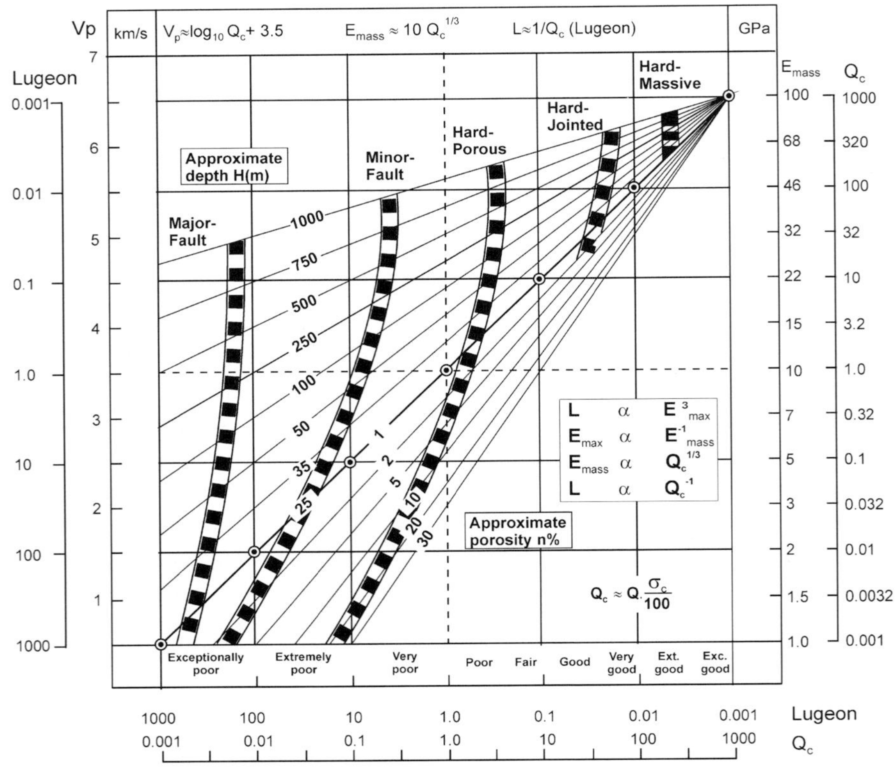


*Figure 9.4* Potential inter-relationships between $V_p$, Q, M and L, with corrections for depth, porosity and compression strength. Barton, 1999.

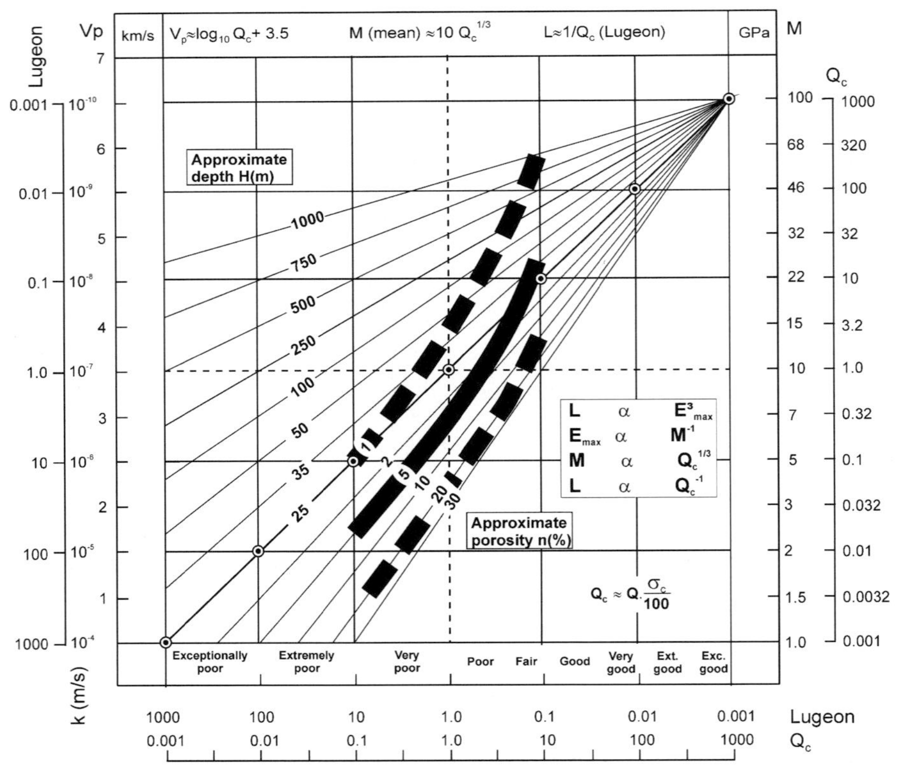


*Figure 9.5* Hard porous rock of 10–20% porosity. Example type curve for estimating $V_p$-Q-M-L data.

## 9.3 Velocity and permeability measurements at in situ block tests

During the 1970s and 1980s, a series of large scale, in situ block tests were performed by Pratt and co-workers in the USA, in order to evaluate test methods and instrumentation suitable for nuclear waste disposal projects which were being planned at that time. The block tests were designed to give large-scale properties (1 to 3 metres scale) under controlled loading conditions (using flatjacks), and some were at elevated temperatures (using borehole heaters).

The effects of stress application on velocity and permeability in jointed granite were first investigated on a large scale by Pratt et al., 1977, who used a flat-jack loaded block measuring 3 × 3 × 3 m, which contained three sub-parallel, vertical joints. The rock was an anisotropic, but quite massive granite, and the site was in Wyoming, USA. The authors investigated velocity changes as a function of applied stress (0 to 9 MPa) applied either parallel or perpendicular to the jointing (so-called E-W or N-S velocities, respectively).

Results for different measurement lengths, including 0.15 m long laboratory samples, are shown in Figure 9.6. The lab samples, which may have experienced microcracking on release by drilling, show the strongest $V_p$ response to stress increase. Although the 3 m cubed block was released on all four vertical sides, the 'contact' with 'virgin' rock stresses along its intact base may presumably be the reason for less response of $V_p$ to stress along these in situ measurement lengths of 1.0 and 2.85 metres. The block also remained nearly saturated, compared to the lab sample '0.15 m, D' (D = dry, S = saturated).

More details of the in situ response of rock mass velocity to increasing stress are given in Figure 9.6c. Increased wetting of the surface of the block caused the small (0.1 km/s) increases in velocity seen between the pairs of curves 4 and 9, 1 and 8 respectively. The authors finally presented a composite plot (fully coupled behaviour) of joint displacement (mm), flow rate along joint $J_1$ ($cm^3/s$) and velocity, each as a function of stress. Figure 9.6d shows that increased closure of the joint after about 2 to 3 MPa normal loading, caused a plateau on the permeability-stress curve, and a sharp reduction in the sensitivity of $V_p$ to further stress increase, especially beyond 5 MPa loading. This is consistent with stress-$V_p$ data reviewed in Chapter 5, and broadly in line with the non-linear effect of depth on velocity shown in Figures 5.36, 9.4 and 9.5.

Another in situ block test, this time in jointed sandstone in Colorado, USA was reported by Swolfs et al., 1981. The block was 2 $m^3$ in volume and contained a near-vertical joint. The joint was calcite filled, and appeared to be about 1.5 mm wide at the surface. The P-wave velocity of the surrounding jointed rock of about 1.5 km/s appeared to be independent of joint frequency and orientation. This is surprising in view of the presumably drained state of the test site (Figure 9.7a). However, 'moist' laboratory samples had about the same value of $V_p$.

In situ stresses of about 1 MPa were relieved by line drilling of three sides of the block. The long side of 2.3 metres and 1.2 m depth was parallel to the joint. This resulted in $V_p$ and $V_s$ changing from 1.5 and 0.8 km/s to 0.9 and 0.5 km/s respectively. Calculated values of $E_{dynamic}$ thereby changed from 3.3 to 1.2 GPa, assuming a rock density of 1.97 $gm/cm^3$, because the sandstone has a high porosity of 25%. The uniaxial strength was about 11 MPa, and static Young's modulus was 2.3 GPa, based on laboratory samples. The block was loaded uniaxially (normal to the joint) and biaxially, using multiple flatjacks in each of the three slots.

The effect on P-wave and S-wave velocities is shown in Figures 9.7c and d. Pre-excavation velocities (shaded lines) were reached at about 1 MPa. This is exactly the stress acting when undisturbed velocities were measured. An anomalous increase in joint deformation was also recorded above this same stress level of about 1 MPa.

The authors also applied shear stresses to the joint by activating the flatjacks at the end of the block, while holding a constant normal stress across the joint (0.7 or 1.4 MPa). Since the block was attached at its base, joint shearing was limited (even at the top surface of the block) to about 0.7 mm, which represents pre-peak strength. Dilation was negligible (10 μm), and is perhaps the reason why $V_p$ and $V_s$ slightly increased during application of shear stress to 3.0 MPa, probably mostly in response to the simultaneous application of normal stress of 0.7 or 1.4 MPa (Figure 9.7a). If significant dilation had occurred during increased shearing, a reduced velocity would presumably have resulted. The small velocity response to moderate stress change seems to be a feature of relatively unjointed, porous rock.

The authors also performed a permeability test using injection in a central hole that intersected the joint. They calculated a permeability of $3.7 \times 10^{-7}$ m/s.

There are several interesting coincidental values of the reported tests that we can compare with the $Q_c$-$V_p$-M-L model (Figure 9.4). If we follow the ambient

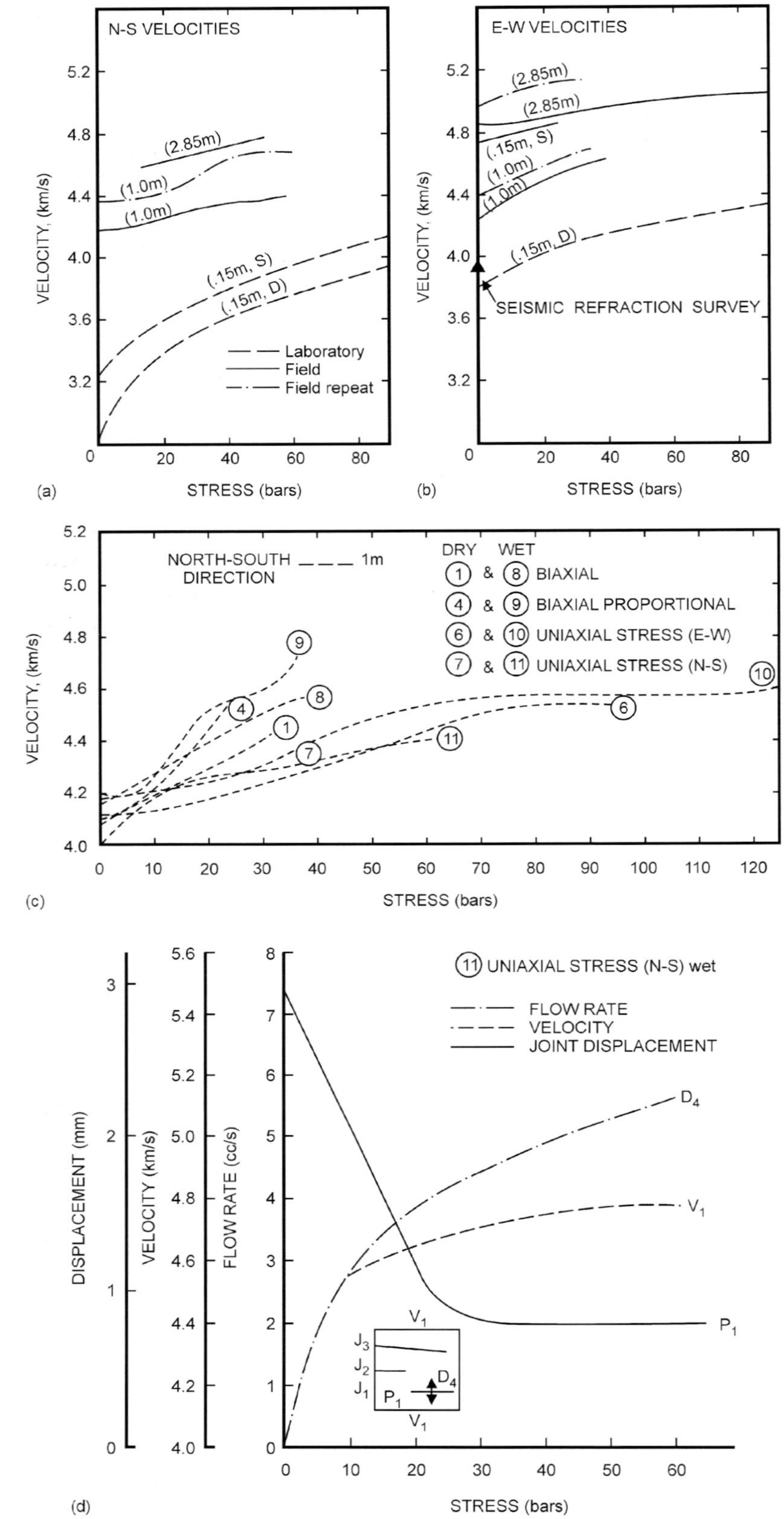


*Figure 9.6* $V_p$ changes caused by loading a 3 × 3 × 3 m block of granite containing vertical joints, and laboratory tests of the same rock. a,b) Velocity-stress behaviour for three types of loading conditions, and for three measurement sizes. c) Nomogram linking effects of uniaxial joint closure stresses with joint J1 deformation $D_4$, velocity across jointed block, and flow rate along part of joint J1. Pratt et al., 1977. Note tendency for acoustic closure beyond 5 MPa.

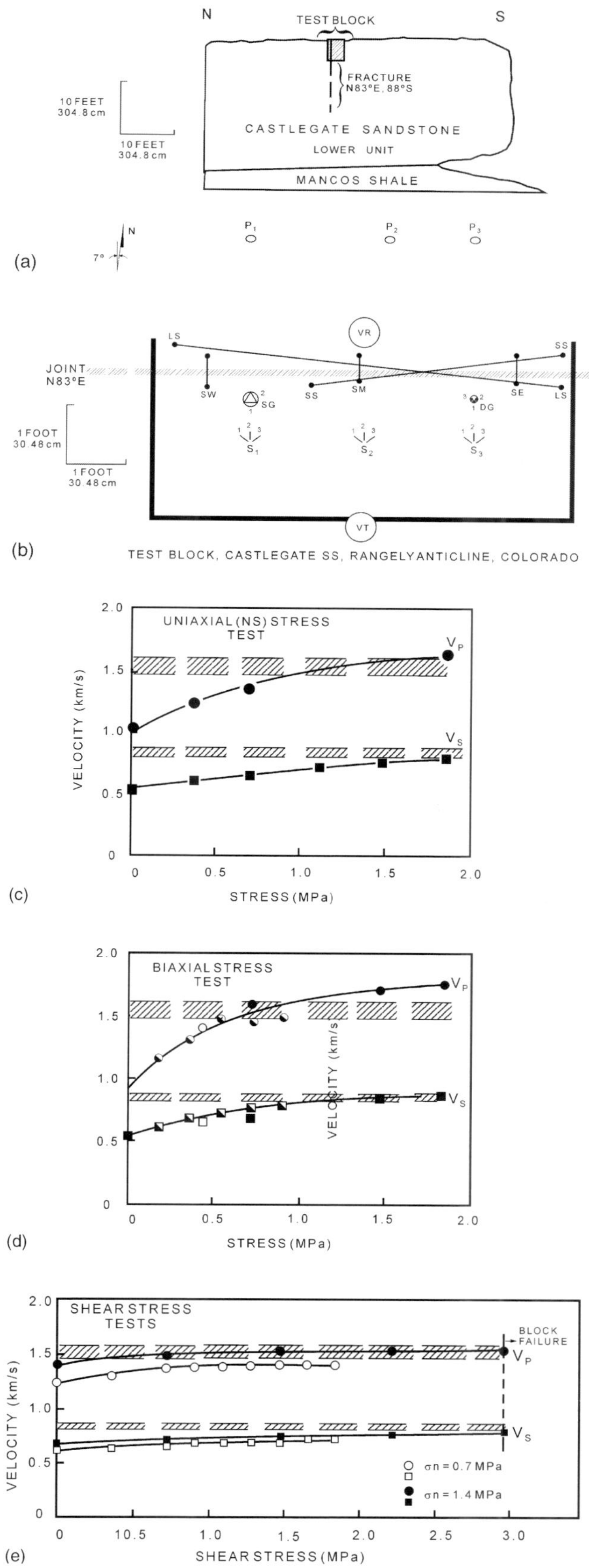


*P-wave velocity* of 1.5 km/s at the ambient stress of about 1 MPa (equivalent to about 20–25 metres of overburden) in the lower left-hand corner of the Q-$V_p$-M-L chart, we find a $Q_c$ value of about *0.4* at 25% porosity. Independent of this, the $1/Q_c$ model for *Lugeon estimation* suggests a back-calculated $Q_c$-value of *0.27*. This is very close to the velocity-based estimate. The low uniaxial strength of 11 MPa means that the Q-value can be estimated as about (0.4 or 0.27) 100/11 = 3.6 or 2.5. These are close on a six-order of magnitude Q-scale.

The estimation of deformation modulus (M) can be based directly on $V_p$ according to Figure 9.4. Thus we see that 2 GPa is estimated, which is close to the laboratory value of 2.3 GPa, and to the $E_{dynamic}$ estimates of 1.2 GPa (unloaded) and 3.3 GPa (undisturbed, loaded to approximately 1 MPa). In this case this deformation modulus estimate is based on $V_p$ (Figure 9.4, right-hand column of M values derived from):

$$M \approx 10^{(V_p - 0.5)/3} \tag{9.3}$$

and this gives a more accurate estimate of 2.1 GPa when $V_p = 1.5$ km/s. The relevant modulus value is also obtained using the direct equation between M and Q:

$$M \approx 10\, Q_c^{1/3} \tag{9.4}$$

which again gives an estimated 2.1 GPa, when using $Q' = 0.01$. We refer to Q′ as 'Q-prime' since it has not been corrected for porosity. The real $Q_c$ value needs the porosity correction, and final correction for the ratio $\sigma_c/100$, to reach the assumed rock mass quality Q, which we estimated from both velocity measurement and independent Lugeon testing as ranging from about 2.7 to 3.6.

Further checks on rock mass quality can be made the direct way by using the authors' descriptions of the jointing; three sets, spaced at 0.6, 0.9 and 0.3 metres, with the most prominent set filled with about 3 mm of calcite. Via the volumetric joint count of Palmström, 1983, we can calculate $J_v = 6.1$, and RQD ≈ 95%.

*Figure 9.7* a,b) Loaded block test in (drained) unit of *in situ* sandstone containing a vertical joint, loaded on three sides by flat-jacks. c,d) $V_p$ – and $V_s$ – stress trends for uniaxial and biaxial loading, compared with pre-slot velocities – shaded. e) Effect of joint shearing on $V_p$ at two different normal stress levels. Swolfs et al., 1981.

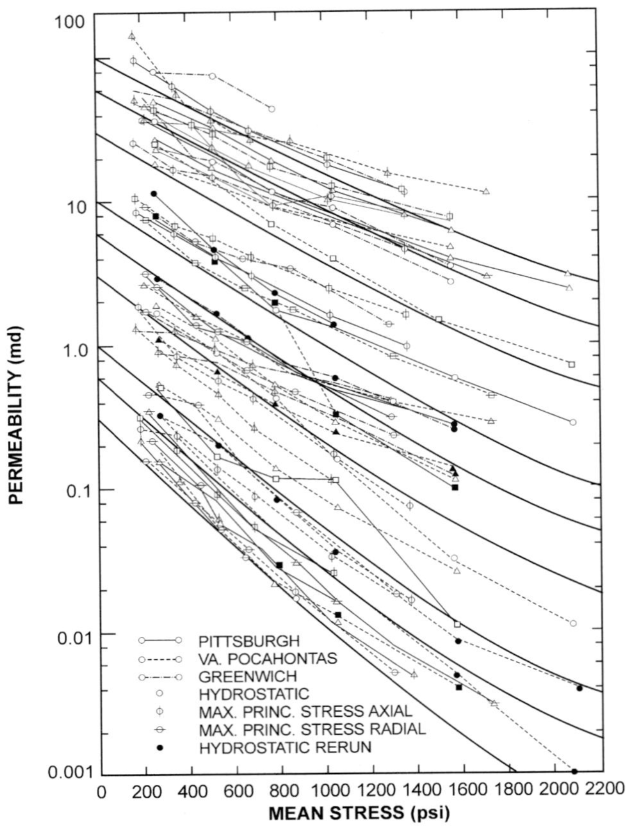


*Figure 9.8* Permeability – stress coupling for three bituminous coals, due to the detailed cleating or jointing: an extreme analogue for jointed rock masses. Somerton et al., 1975.

The independently estimated Q-value is therefore approximately as follows:

$$Q = \frac{RQD}{J_n} \times \frac{J_r}{J_a} \times \frac{J_w}{SRF}$$

$$= \frac{95-100}{9} \times \frac{1-1.5}{6} \times \frac{1}{1} = 1.7 - 2.8$$

All of the above estimates are very close, considering the logarithmic (six orders of magnitude) Q-value rock quality scale. We have thus demonstrated that Q, $V_p$, M and L are inter-related, and that we may be able to include the Lugeon value in this inter-relation, if care is taken to eliminate irrelevant non-deforming, channel flow cases. The implication is that depth or stress level, also an axis in Figure 9.4, also plays an important role in these mutual inter-relationships.

Using an analogue material for heavily jointed rock, namely coal, one can also see how there is great potential sensitivity between velocity, stress level and permeability,

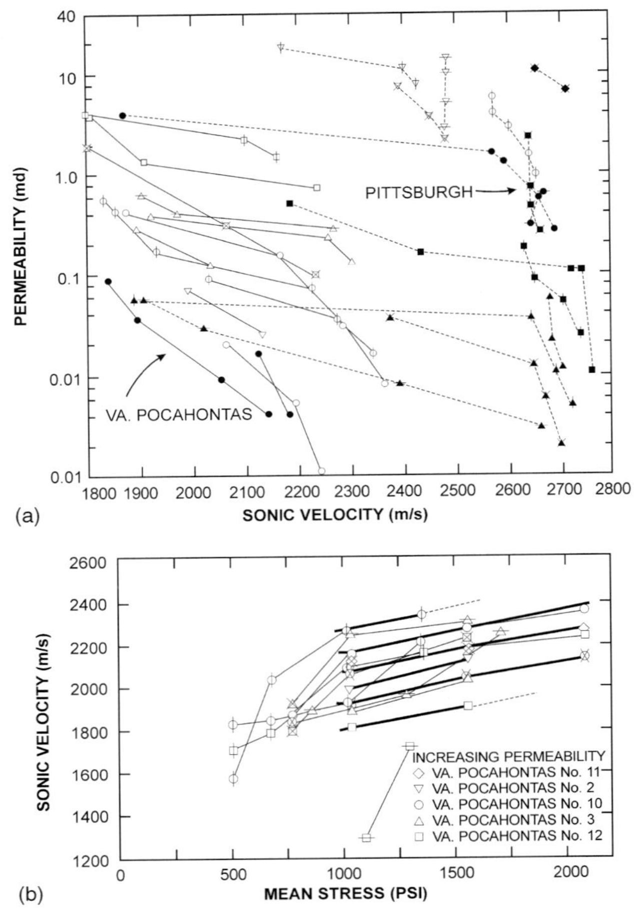


*Figure 9.9* a) Permeability-$V_p$ coupling for two of the bituminous, cleated coals. b) Velocity-mean stress coupling for one of the cleated coals. Somerton et al., 1975.

which will also be present in jointed rock masses at large scale, when *in situ* effective stress states are altered by large scale pumping or injection experiments.

Three bituminous coals having large differences in hardness and degree of jointing (cleats, etc.) showed almost equally great sensitivity to applied stress level, despite their five order of magnitude range of permeabilities (0.1 to 100 millidarcys). Somerton et al., 1975, applied mean stresses over the range 1 to 14 MPa and noted between two and three orders of magnitude reduction in permeability (Figure 9.8).

Simultaneous monitoring of ultrasonic velocity showed increases of velocity of about 0.3 to 0.6 km/s (from 1.8 km/s when stress-free) for each order of magnitude reduction in permeability. This is shown in Figure 9.9a together with the $V_p$-stress behaviour of one of the coals in Figure 9.9b. Both these figures indicate greatest changes in $V_p$ and permeability at the lowest stress levels (and lowest velocities), just as found in rock masses, due to improved acoustic coupling across joints.

## 9.4 Detection of permeable zones using other geophysical methods

Since the mid-eighties, researchers working at nuclear waste related rock laboratories such as Stripa, Äspö, and Grimsel have utilised both seismic and radar tomography to characterize major fault zones. Their studies have generally helped to explain why these relatively small volumes of fractured (or heavily jointed) rock are responsible for such large percentages of the total flow of water. At the Grimsel site, Martel and Peterson, 1991, found that seismic velocity tomography delineated major geologic structures better than radar attenuation or radar slowness tomography. However, they point out that anomalies on tomograms can reflect a wide range of features (rock types, alteration, changed porosity) besides different degrees of jointing or fracturing. Knowledge of geological background data is therefore helpful in making better interpretations.

Injection of brine for tracing flow paths has been successful in many projects. Martel and Peterson, 1991, found that radar attenuation difference tomograms were more reliable in locating brine than slowness tomograms, at the Grimsel US/BK site. Radar and seismic signals are sensitive to different physical parameters (mechanical stiffness and electro-magnetic wave conductivity, respectively). For this reason, the respective tomograms highlight different features of the rock mass. Radar may delineate permeable zones (porosity caused by pore space or by joint apertures) in slightly different locations to the low seismic velocity zones associated with clay filled discontinuities. The one will usually lie parallel to the other, since higher permeability may be associated with the heavily jointed zones that are often found adjacent to faults. This was a phenomenon that at first made geophysics teams question each other's coordinates at the Stripa SCV (Site Characterisation and Validation) site (Olsson., 1992).

Green et al., 1989, described the use of vertical seismic profiles and cross-hole seismic surveys at the Camborne School of mines 2 km deep hot dry rock geothermal project. The objective was to show that the micro-seismicity generated during stimulation experiments corresponded to regions of enhanced permeability. The majority of micro-seismic events were interpreted as minute (10 to 50 microns) shear displacements on a set of near-vertical joints that were not aligned to $\sigma_{H(max)}$. The joint roughness was assumed to create some increase in permeability despite the assumed small shear displacements.

During a stimulation experiment in which the reservoir was kept 'inflated' by a well-head pressure of 6 MPa and a flow rate of 9 litres/s, the velocity showed small reductions in the depth zone between 2100 to 2500 metres. The seismic data suggested that the permanently stimulated cracks and joints were dilating as the pore pressure increased, even though the pore pressure was only 20% of that required for jacking (30 MPa). A certain degree of joint aperture increase can be expected from the elevated pore pressure, whether or not significant shearing was occurring.

Aoki et al., 1991, describe the use of cross-hole seismic measurements to compare with (and verify) the directional distributions of hydraulic diffusivity. In the case of tests in a heavily jointed rhyolite, the lower velocity zones (3–4 km/s) between two of the boreholes corresponded quite closely to the location of highly permeable zones between these boreholes at 6 and 13 metres depth, as seen in tomographic plots of cross-hole test data.

It is well known that low resistivity measurements correlate with zones of increased water content and frequently with higher permeability. At a site in South Korea, where the writer logged a series of boreholes in weathered granites, the opportunity arose to compare these independently derived Q-parameter statistics with resistivity tomograms that were given to the writer after his draft report was delivered.

It was found that sections of the boreholes with increased joint frequency (low RQD, high $J_n$) did not always correlate with low resistivity and vice versa, as was reasonably to have been expected. The parameters that *did* show a consistent correlation with low resistivity were the low values of $J_w$ (estimated, for example, from iron staining or apparent aperture) and the high values of $J_a$ (for example from sand or silt fillings *and* due to clay fillings).

The latter gives low resistivity due to the ionic effects of the clay, since water content (and permeability) are clearly lower in such discontinuities than in those that are sand or silt filled. There is therefore in fact a potential source of error in judging the meaning of low resistivity zones.

This end of the rock mass quality spectrum is also unfortunately the region where the usual link of low $V_p$, low rock mass quality Q-value and high permeability also may break down, due to the 'adverse' effect of clay on permeability. We will see in Chapter 16 that a *partial solution* has been found for this clay-based phenomenon, by rearrangement of the function of two Q-parameters, namely the reversal of $J_r$ with $J_a$ in a simple term called '$Q_{H_2O}$'. This depth-dependent model appears to provide a realistic, preliminary estimate of permeability to many kilometres depth.

## 9.5 Monitoring the effects of grouting with seismic velocity

A simple illustration of the benefits of seismic velocity monitoring at grout injection jobs was given by By, 1988, using cross-hole (average velocity) measurements at a dam abutment in Norway. Blast damage from reflected waves caused by a nearby quarry for rockfill was suspected to have caused shearing and dilation along adversely dipping foliation planes, giving $V_p$ measurements as low as 0.5 km/s in the drained, 5 to 10 metres depth zone.

Following extensive surface injection (Figure 9.10) $V_p$ locally increased to between 2 and 6 km/s. However, the depth interval of 7 to 15 metres did not show acceptable velocities (only 2 to 3 km/s) and this was confirmed by additional water leakage tests. A strong depth-velocity effect was observed at the site (velocities rising from 2 to 5 km/s from about 10 to 25 m depth) presumably related both to *apparent* rock quality improvement at depth, and to a post-stressing effect from the increasingly confined grout at greater depth.

Rodrigues et al., 1983, also refer to the correlation between seismic velocity and reduced permeability achieved by grouting at the Cabril dam site founded on granite, in Portugal. The dam had been grouted 30 years previously and had already quite a high P-wave velocity (range 4.2 to 5.5 km/s), partly as a result of this earlier foundation treatment. The new round of grouting increased $V_p$ by 2 to 20% and reduced the permeability, as registered by Lugeon testing, by anything from 40 to 100%. A larger grout take (where there was presumably a larger rock mass 'porosity'), also corresponded to the locations where the largest increases in $V_p$ were registered, following the grouting.

It was noted by Grujíc, 1974, at the 220 m high Mratinje dam (shown in Figures 4.1 and 4.2) that effective consolidation grouting could be performed when $V_p$ was in the range 2.5 to 3.5 km/s (i.e., approximately $Q \approx 0.1$ to 1.0 or 10 to 1.0 Lugeon according to the relation $L \approx 1/Q$ Lugeon. Velocities above 4.0 km/s ($Q \approx 3$? or $K \approx 0.3$ Lugeon?) could not be improved upon by the grouting. Such results emphasise the reasons for combined use of high injection pressures and micro or ultrafine cements, if e.g. 0.3 Lugeons (or rather 4.0 km/s), should be improved upon. Barton, 2004a.

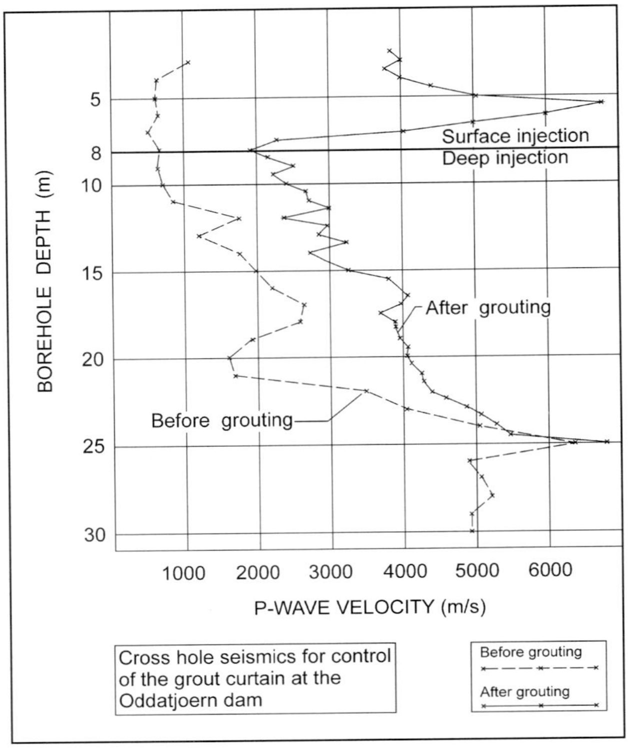


*Figure 9.10* Superficial and deeper-layer grouting at a dam site abutment, where nearby (and too close) quarry blasting had caused suspected shearing along the dipping foliation planes, resulting in a (drained) P-wave velocity of only 0.5 km/s. Note the dramatic improvements in the (assumed) foundation properties as a result of grouting. By, 1988.

At the 270 m high Inguri arch dam in Georgia, Savitch et al., 1983, used the seismic velocity criteria shown in Figure 9.11 for judging the success of grouting. One can first interpret that very high pressures must have been used here, since it is implied that velocities as high as 4.5 km/s could be improved by grouting.

However, the depth effect on $V_p$ (e.g., Figure 9.4) is probably playing a role here. A $V_p$ value of 4.5 km/s implies $Q_c = 10$ in near-surface, hard un-weathered rocks. However at the 270 m high dam, deep injection grouting and deep $V_p$ monitoring (say at 100 to 200 m depth) might have caused a depth (or stress) related enhancement that was equivalent to a much lower rock

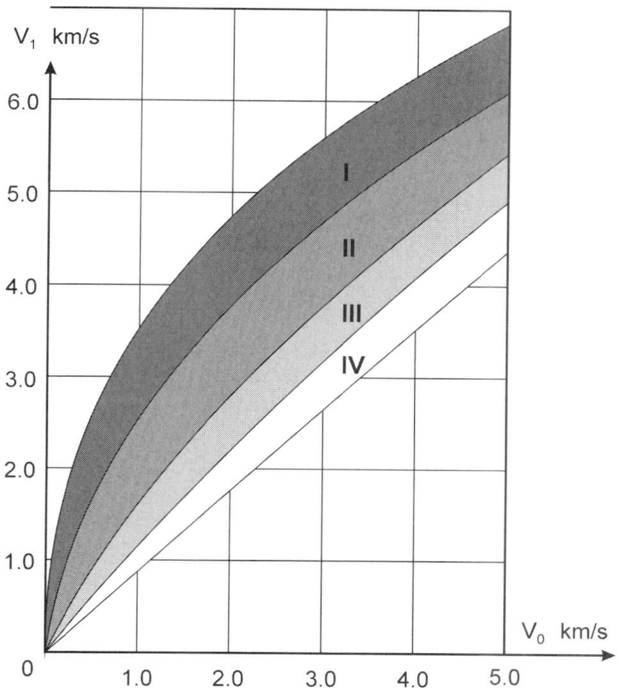


*Figure 9.11* Grouting efficiency (I = excellent, II = good, III = Satisfactory, IV = unsatisfactory) based on velocity monitoring at the Inguri arch dam. Savich et al., 1983.

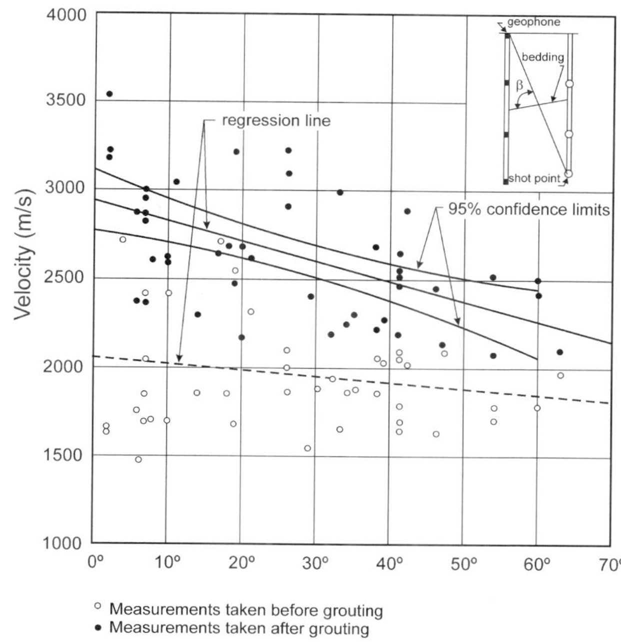


*Figure 9.12* Before, and after grouting $V_p$ measurements in sandstones and marls, showing increased velocity, and increased anisotropy. Capozza, 1977.

quality of $Q_c = 1$, or even less, which was likely to be injectable. Savich et al., 1983, results are therefore readily understandable when the $V_p$-Q-depth effect is taken into account.

Grouting at the Zavoj hydro electric project in Yugoslavia was monitored by cross-hole velocity measurements and by cross-hole seismic tomography performed between two galleries. Slimak et al., 1991, showed three-dimensional contour plots of velocity distributions before and after grouting, and a difference tomogram showing the net gain in velocity, as a result of the grouting. The authors emphasised the efficiency of such measurements in checking the effectiveness of a large-scale injection programme.

Unfortunately, the authors did not give the results of Lugeon testing before and after the grouting. The velocity increase of only 0.25–0.75 km/s and the relatively high velocity (mostly 4.0–5.5 km/s) before grouting suggest only a moderate result. One may speculate that these velocities are also affected by stress level, since 'near-surface' Q-values of 3 to 100 derived from the above velocities (if measured in the upper 25 metres) would imply low Lugeon values (0.33 to 0.01 Lugeon) if one accepts $L \approx 1/Q$ as a useful lower bound estimate, as discussed earlier. These are hardly values that would justify grouting. Presumably some of the velocity increase caused by grouting is due to the increased stress, and some due to reduced volume of joint apertures and better seismic coupling. Velocity monitoring alone may therefore not guarantee a good (i.e., low leakage) grouting result.

Bernabini and Borelli, 1974, describe a variety of early seismic tests performed at hydro electric projects in Italy in the 1950s, 60s and early 70s. In 'stratified rocks' they show about 35% increase in seismic velocity caused by successful grouting with cement. However, they observed that the scatter of data did not change; the stratified rock maintained its anisotropic character, just the mean velocity was increased (1.96 to 2.65 km/s).

The before-and-after grouting measurements of $V_p$ referred to by Bernabini and Borelli, 1974, are reproduced in more detail by Capozza, 1977. The inter-bedded sandstones and marls experienced an average velocity increase of 0.7 km/s (2.0 to 2.7 km/s) as a result of the grouting. Since the cross-hole measurements performed before and after grouting gave a range of ray-path angles (β) in relation to the gently dipping bedding, it was possible to show the influence of angle β° on the results.

Figure 9.12 from Capozza, 1977, shows not only the higher velocity after grouting but also the increased anisotropy, which was closer to that of the unweathered formation at depth. Presumably, the weathered and jointed sandstone layers between the marl were more easily injected, giving this increased anisotropy.

Wenhua, 1991, described the use of seismic velocity measurements to monitor the effects of grouting on the deformation modulus of jointed and faulted power-house foundations, at the 1750 MW Gezhouba hydro-electric project on the Yangtze River in China. The fracture zone of concern originally had a modulus of deformation as low as 0.1 GPa and a permeability greater than 10 Lugeons, and was affected by small, karstic voids and cracks. Velocity measurements *in the faulted and permeable zone*, showed values of 2.5–3.2 km/s after grouting, compared to 1.5–2.0 km/s before the treatment. The average values for the whole foundation were 3.18 km/s before grouting and 4.74 km/s after grouting which imply an effective Q-value increase from (very approximately) 0.5 to 17 or a Lugeon value reduction from perhaps 2 to 0.06 (using equations 9.1 and 9.2). This implied low Lugeon value is so low that a stress related effect on the 4.74 km/s velocity achieved after grouting is suspected.

As suggested earlier, the stress related effect may be a combination of depth (greater than the reference 25 m, Figure 9.4) and post-stressing by high pressure grouting, which could give a 'locked-in' stress after curing at least in vertical or sub-vertical planes. In horizontal planes any potential 'locked-in' stress caused by local 'lenses' of grouting, would need to be over a limited volume, if arching were to give a local stress greater than $\sigma_v$. 'Artificially' high $V_p$ values could then be registered, which might exaggerate the true effect of the grouting, which is primarily designed to reduce permeability, but has several other positive effects (i.e., increased modulus, shear strength, etc.).

## 9.6 Interpreting grouting effects in relation to improved rock mass Q-parameters

Since we have indicated a general potential relationship between Lugeon value, Q-value, and measured velocity, and have also obviously noted an apparent change in Lugeon value and velocity as a result of grouting, it is of interest to investigate the potential physical effects of grouting on Q-parameters. The tunnelling situation can be used in this exercise, as pre-injection is a commonly needed measure to reduce problems ahead of a tunnel face, either in permeable, leaking rock masses, or beneath environmentally sensitive areas, where groundwater draw-down cannot be tolerated.

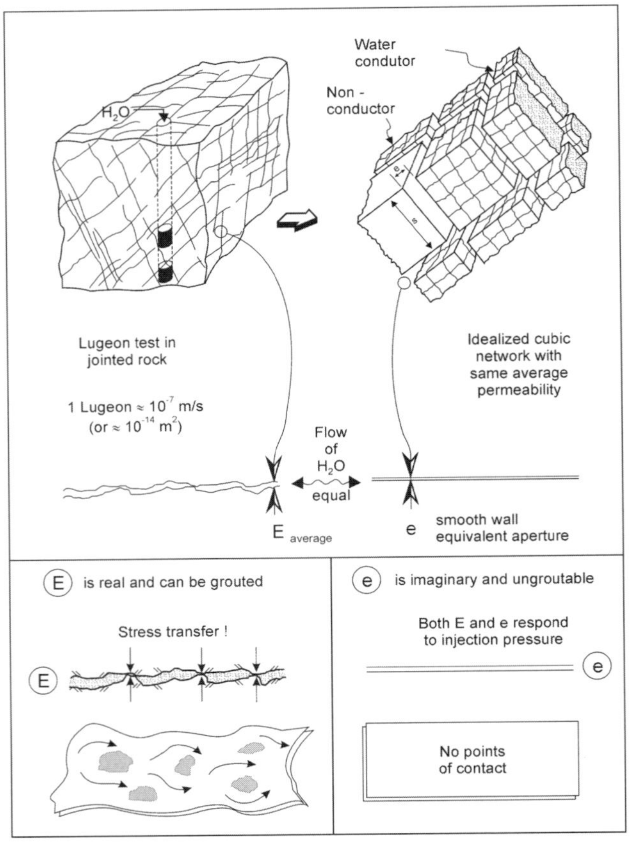


*Figure 9.13* Top left: a depiction of a Lugeon or water injection test in a rock mass with three joint sets, and the Snow, 1968 idealized cubic network, consisting of a Poisson-distributed, and limited number, of conducting 'smooth parallel plates' with equal permeability. The lower diagrams emphasise the joint-roughness-related, inequality of the physical joint apertures (E), and the theoretical hydraulic apertures (e). Barton, 2004a.

Most tunnel engineers experience that correctly carried out pre-grouting reduces leakage, and that it apparently increases deformation modulus and probably shear strength, since tunnels that are pre-injected show each of these implied characteristics, meaning improved stability, less deformation, and lessened support needs. The same is probably true in dam foundations, minus the support needs.

A helpful, if very idealized figure, concerning the 'available' joint porosity for potential grout penetration, is given in Figure 9.13. The top right-hand diagram is based on Snow, 1968 with the addition of non-conducting joints between Snow's idealized cubic

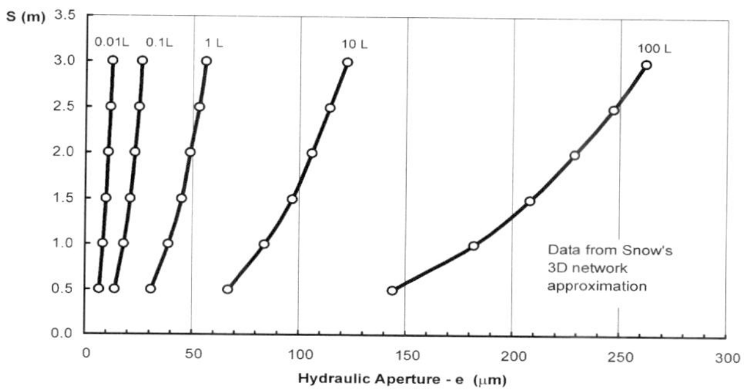


| 0.01 Lugeon | | 0.1 Lugeon | | 1 Lugeon | | 10 Lugeon | | 100 Lugeon | |
|---|---|---|---|---|---|---|---|---|---|
| S (m) | e (μm) | S (m) | e (μm) | S (m) | e (μm) | S (m) | e (μm) | S (m) | e (μm) |
| 0.5 | 6.7 | 0.5 | 14 | 0.5 | 31 | 0.5 | 67 | 0.5 | 144 |
| 1.0 | 8.4 | 1.0 | 18 | 1.0 | 39 | 1.0 | 84 | 1.0 | 182 |
| 1.5 | 9.7 | 1.5 | 21 | 1.5 | 45 | 1.5 | 97 | 1.5 | 208 |
| 2.0 | 10.6 | 2.0 | 23 | 2.0 | 49 | 2.0 | 106 | 2.0 | 229 |
| 2.5 | 11.5 | 2.5 | 25 | 2.5 | 53 | 2.5 | 114 | 2.5 | 247 |
| 3.0 | 12.2 | 3.0 | 26 | 3.0 | 56 | 3.0 | 122 | 3.0 | 262 |

*Figure 9.14* The evaluation of equation 9.8 in graphic and tabulated format, for typical Lugeon values between 0.01 and 100, equivalent to assumed isotropic rock mass permeabilities of approx. $10^{-9}$ to $10^{-5}$ m/s, and average spacings for the water-conducting joints of 0.5 m to 3.0 m.

network. The physical (aperture E), compared to the theoretical (apertures e), available for grouting are depicted in the lower diagrams.

A further visualization of the size of the theoretical hydraulic apertures available for grouting, if the rock mass had three equal joint sets, is given in Figure 9.14, which was derived in Barton, 2004a, from equations in American units from Snow, 1968. Based also on the hydraulic theory of Louis 1967:

1. Permeability of one smooth parallel plate:

$$k = \frac{e^2}{12} \tag{9.5}$$

2. Permeability of *1 set* of parallel plates:

$$K_1 = \frac{e^2}{12} \times \frac{e}{S} \tag{9.6}$$

3. Permeability of '*the conducting rock mass*' (3 sets):

$$K_{mass} \approx \frac{2e^3}{12S} \tag{9.7}$$

Since 1 Lugeon $\approx 10^{-7}$ m/s, and $10^{-7}$ m/s $\approx 10^{-14}$ m$^2$, a *laminar flow* 3D interpretation of Lugeon tests can be expressed as follows:

$$e \approx (6LS \times 10^{-8})^{1/3} \tag{9.8}$$

where (e) and (S) in millimeters, L is the average Lugeon value, and each apply to the local *domain, rock type, or borehole depth.*

Although average physical apertures (E) are significantly larger than (e), they are hardly of different orders of magnitude. It is therefore clear that there will be difficulties of grouting a $V_p = 4.5$ km/s rock mass (measured at nominal 25 m, shallow refraction seismic depth), if we assume the approximate validity of equation 9.2: $L \approx 10^{(3.5-V_p)}$, suggesting a Lugeon value of only about 0.1 in this case.

It is simple to understand from grouting case records that the higher the Lugeon value, or *the lower the velocity before grouting*, the better the potential improvement (e.g. Figure 9.11). A useful demonstration of this is the practical case of trial grouting of a dam abutment in Figure 9.15, from Quadros and Correa Filho, 1995. Three boreholes permeability-tested before grouting, were re-drilled in roughly the same location following the grouting, so that the before-and-after permeabilities could be compared. The lower-left diagram, treating just the individual borehole results, suggested that the following before-and-after results could be expected. (Only industrial cement was used in these tests).

| Before | After |
|---|---|
| $k = 10^{-3}$ m/s | $k = 10^{-7}$–$10^{-8}$ m/s |
| $k = 10^{-5}$ m/s | $k = 10^{-6}$–$10^{-7}$ m/s |

In the tunnel situation, the need for reduced tunnel support following pre-grouting, can be documented, if prognoses of required support using an 'ungrouted Q' are accepted as realistic. This claim has been supported by recent rail tunnels for the Norwegian Jernbaneverket in the Oslo area. Tunnels were driven under built-up areas founded on clays, using over-lapping pre-injection 'umbrellas' established every 3 to 4 rounds, by performing a regular, high pressure (5 to 10 MPa), single-stage, 24 hours-duration pre-grouting routine, over many kilometres if tunnel. (Moen, 2004).

Since tunnel deformation is closely linked to SPAN/Q (Barton et al., 1994, Barton, 2002) and support needs are linked directly to Q, the inescapable conclusion (which would also be arrived at by velocity monitoring

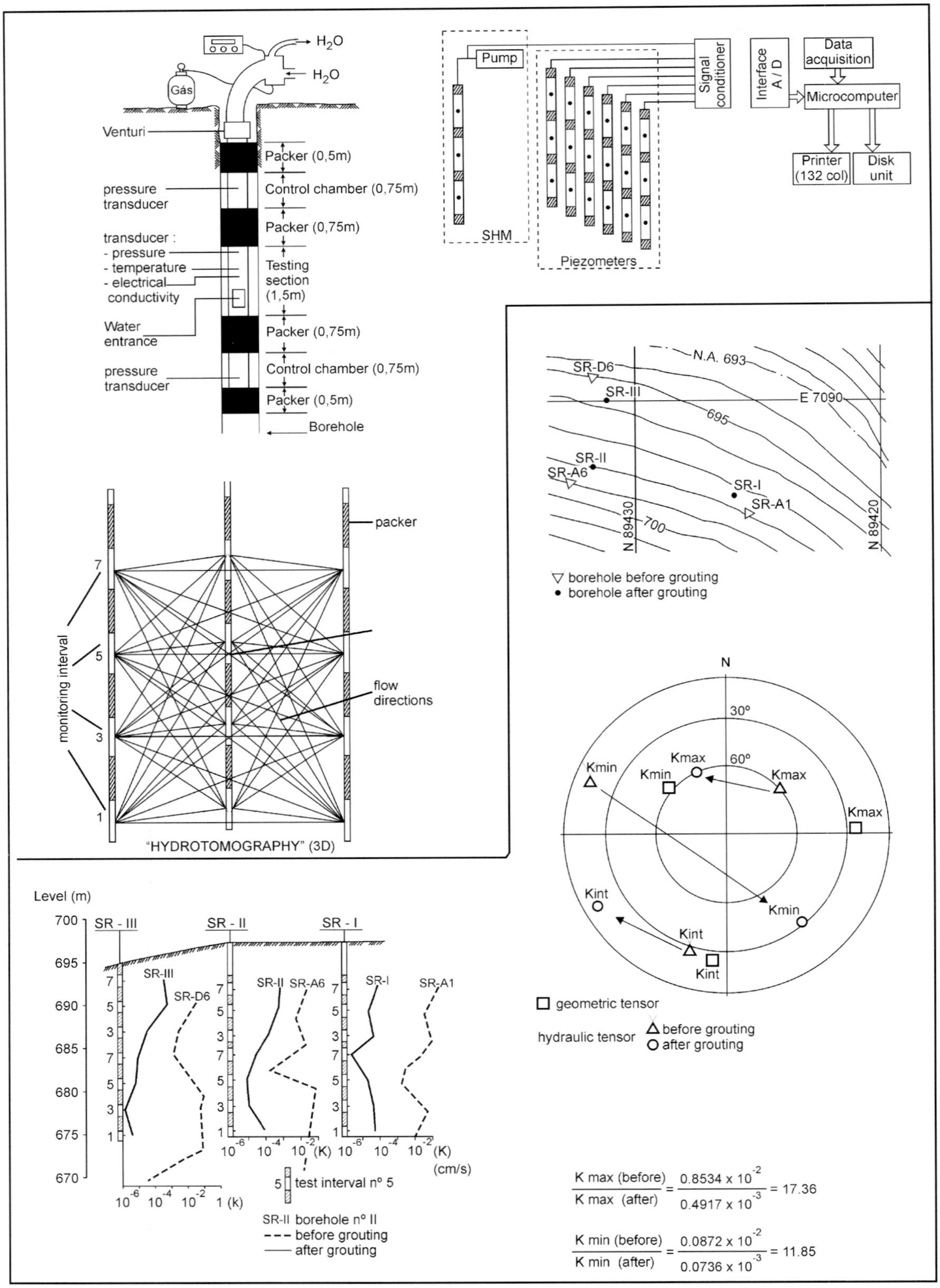


*Figure 9.15* Permeability testing at a dam abutment in Brazil, using before-and-after testing of the effect of grouting, and both single-hole and 3D measurements. Note the rotation of permeability tensors, and their reduced magnitude, suggesting progressive sealing of joint sets. This suggests a possible scenario for individual Q-parameter improvements. Quadros and Correa Filho, 1995.

and deformability testing) is that the effective Q-value has itself been increased by the pre-injection.

The Q-value (Barton et al., 1974, Barton and Grimstad, 1994) is determined from the modified core recovery RQD (i.e., counting competent pieces of core $\geqslant$ 100 mm in length as recovery). In addition to RQD, Q is calculated from the number of joints sets ($J_n$), the roughness ($J_r$) and the degree of alteration ($J_a$) of the least favourable set, and from the water inflow ($J_w$) and stress/strength condition (SRF). From Figure 9.4, a velocity increase of 1 km/s from say 3.5 to 4.5 km/s at a dam site, or in a wet, jointed zone ahead of a large tunnel, will *imply* that the Q-value has increased from 1 to 10 as a result of grouting.

Following equation 9.1, a drop in Lugeon value from 1.0 to 0.1 is also implied, and using equation 9.3: $M \approx 10Q_c^{1/3}$, the modulus of deformation may be predicted to have increased from 10 GPa to at least 20 GPa. Are these changes possible to explain via changes in the six component Q-parameters? The answer is definitely yes, but the exact answer will always be unknown.

We can speculate that the following orders of magnitude of *effective* rock mass quality improvement (each very modest), *may* occur in practice during grouting ahead of a tunnel or at a dam foundation:

1. RQD of say 30%, increases to say 60%, due to grouting of the most prominent set of joints that were most permeable. This occurs largely as a result of hydraulic joint jacking.
2. $J_n$ of say 15 (four sets) is effectively reduced to 9 (three sets) for the same potential reasons as above. (This is a *very* conservative argument).
3. $J_r$ of 1.5 (rough, planar) changes to 2 (another set) or to 4 (discontinuous), also for the same reasons.
4. $J_a$ of 2 (weathered) changes to 1 (another set) or to +0.75 (cemented), also for the same reasons.
5. $J_w$ of 0.5 (high pressure inflow) changes to 0.66 (small inflow) due to preferential sealing of the most permeable set. (This is also conservative).
6. SRF of 1 (unchanged). (In the case of a minor fault, even SRF might change).

We therefore have the following potential 'before' and 'after' scenarios:
From:

$$Q = \frac{RQD}{J_n} \times \frac{J_r}{J_a} \times \frac{J_w}{SRF} \tag{9.9}$$

$$Q_1 = \frac{30}{15} \times \frac{1.5}{2} \times \frac{0.5}{1} = 0.8 \tag{9.10}$$

$$Q_2 = \frac{60}{9} \times \frac{2-4}{1-0.75} \times \frac{0.66}{1} \approx 9 - 23 \tag{9.11}$$

The effective Q-value has increased in terms of round figures, by a factor of 10 to 30, which is broadly consistent with the increased $V_p$ and M values, and with the reduced Lugeon value and rock support needs. It should be noted in particular, that when the seldom reported or measured 3D permeability is analysed in before-and-after-grouting scenarios, a rotation of the permeability tensors (and reduction of their magnitude) is seen (Figure 9.15, from Quadros and Correa Filho, 1995). This is the tentative justification for suggesting, as above, that the least favourable joints – and those causing the lower before-grouting *velocities* – are those that are (*first*) sealed by the grout.

In this particular example we can estimate the following 'hard rock, shallow near-surface' results for before and after grouting, based on $V_p \approx 3.5 + \log_{10} Q$, $L \approx 1/Q$, $M \approx 10Q^{1/3}$:

Table 9.2 Potential effects of grouting according to empirical predictions.

| | Before grouting | After grouting |
|---|---|---|
| Q | 0.8 | $9 \rightarrow 23$ |
| $V_p$ | 3.4 (km/s) | $4.5 \rightarrow 4.9$ (km/s) |
| L | 1.3 (Lugeon) | $0.1 \rightarrow 0.04$ (Lugeon) |
| M | 9.3 (GPa) | $21 \rightarrow 28$ (GPa) |

As with some of the cases reviewed earlier, this apparently good grouting result would need to be attributed to hydraulic joint jacking and perhaps to the use of micro-cements. In relation to the Inguri arch dam (Figure 9.11), Savich et al., 1983, would allocate the result ($V_p = 3.4 \rightarrow 4.5 \rightarrow 4.9$ km/s) to class II (good grouting result).

The interaction of *rock mechanics*, *rock hydraulics* and *rock dynamics* through application of seismic monitoring and rock quality description has many applications for rock engineers. The 'core' interactions ($V_p$, Q, L and M) illustrated above and in Figure 9.4 can also be expressed in alternative ways. By 'extracting' the uniaxial strength ($\sigma_c$) of a rock from $Q_c(= Q \times \sigma_c/100)$ we

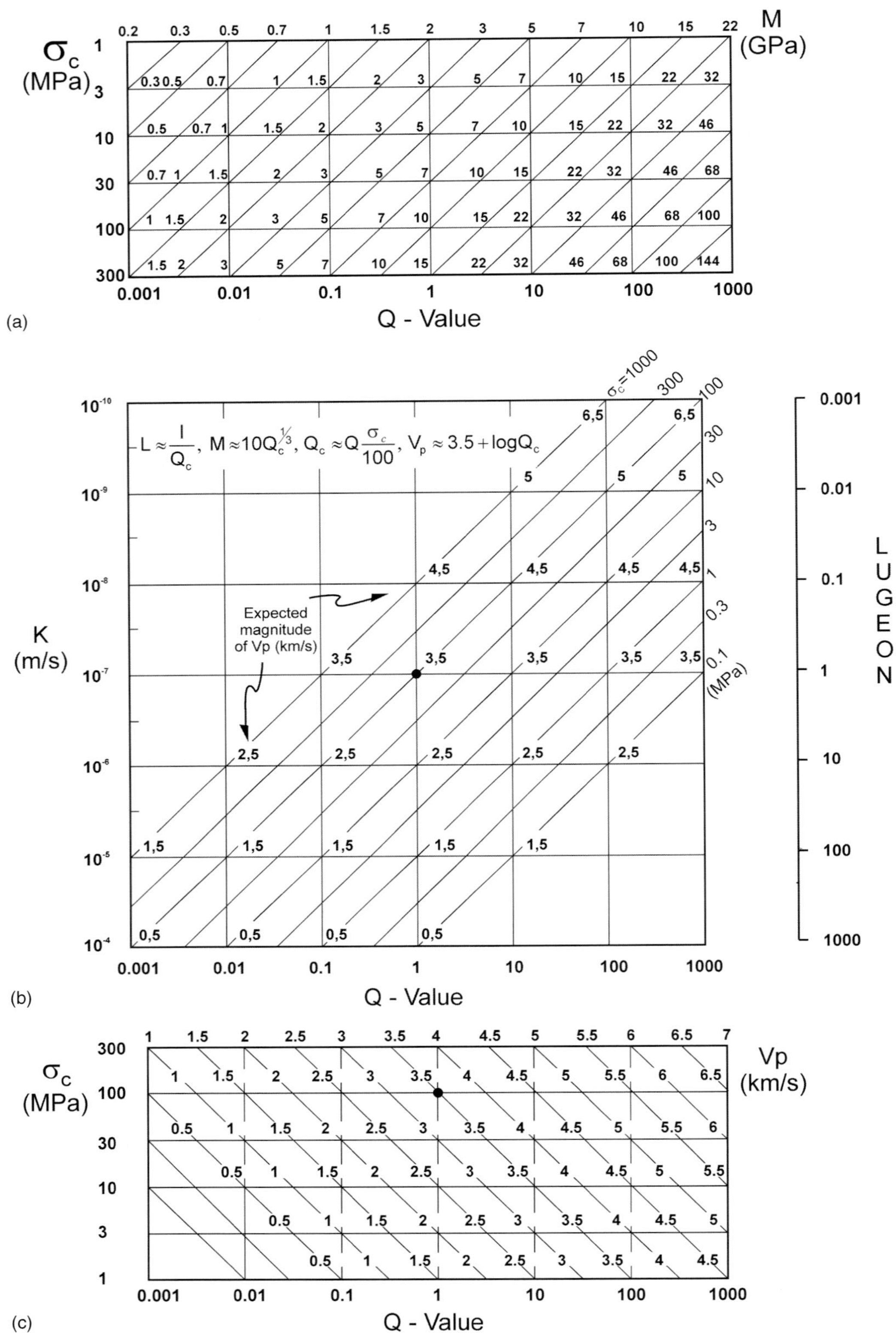


*Figure 9.16* Alternative nomograms for estimating typical interactions between $\sigma_c$, Q, M, $V_p$ and L. Note that porosity and depth (or stress) effects have been ignored for simplicity of presentation. (All predictions for nominal n = 1% porosity, and typical refraction seismic depth ≈ 25 m.)

can make $\sigma_c$ one of the principal variables as shown in the nomograms for M, L and $V_p$ shown in Figure 9.16.

The numbers distributed within Figure 9.16a are estimates of the static deformation modulus (M), while the numbers distributed within Figure 9.16b and c, are the estimates for P-wave velocity. The units of M are GPa and $V_p$ are km/s respectively. The common x-axis in each diagram is the Q-value of the rock mass, i.e., the rock mass quality and not the seismic quality, although as discussed in more detail in the next chapter, these numbers, each having the same Q symbol, are inevitably related quite closely.

# II Introduction to Part II

The subduction zone diagram showing 'extremely low Q', 'low Q' and 'high Q' reproduced on the back cover of this book, the 'familiar' Vp-depth trends of mid-ocean ridge seismic investigations, and the continent-wide seismic velocities also showing 'familiar' increase with depth were each strong reasons for delving deeper into the subject of 'seismic velocity and rock quality', which was the original title planned for this book. Possible parallels with engineering scale phenomena were evident, in something resembling a 'fractal' earth.

Of course the subduction zone 'Q' values proved to be attenuation related. However the possibility that a commonly used rock mass quality parameter, also called 'Q' (range 0.001 to 1000) and seismic quality *Q* (range 1 to 5,000), where attenuation = $Q^{-1}$, could in some way be related, proved to be one of the incentives for deeper research into seismic phenomena, and are the reasons for developing Part II material.

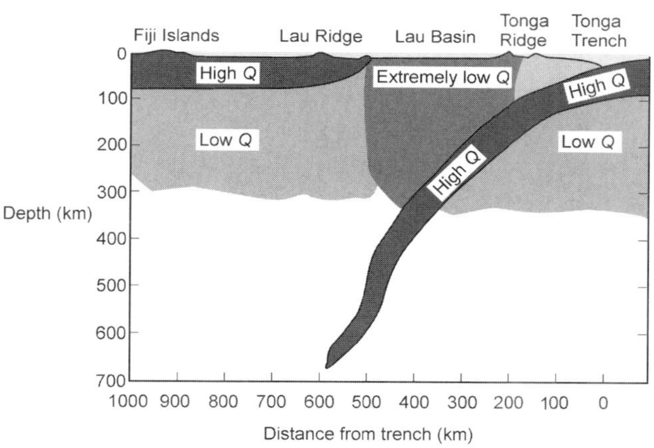


*Fig. PART II* Schematic section, after Barazangi and Isacks 1971 and Kearey and Vine 1996, of the Tonga arc, with inferred seismic Q variations. A possible relationship between seismic Q and rock mass quality Q is one of the first objectives of Part II.

Part II contains a wide sampling of interesting large scale continental and sub-ocean seismic behaviour, also high pressure laboratory rock physics tests designed to improve understanding of both crustal and reservoir variation-with-depth phenomena. A broad sampling of *in situ* reservoir related topics is given in later chapters, such as borehole stability and their seismic effects, and fractured reservoir investigations, involving P-wave anisotropy, S-wave splitting and polarization, and poro-elastic modelling of the dispersive and anisotropic nature of fractured reservoir simulations. An attempt has been made to bridge between engineering geological and geophysical scales of depth and time, in this continued investigation of 'seismic velocity and rock quality'.

The last chapter addresses geomechanics understanding of joint and fracture behaviour, in particular permeability-stress performance, with a view to suggesting alternative interpretations of aligned fracture orientations that actually involves multiple sets. Deep well behaviour in which impermeable and permeable fracture sets are separated by the determination of either dominance of normal stress or dominance of shear stress, represents a more correct understanding for maintenance of permeability in the face of high effective reservoir stresses. The 'parallel to $\sigma_{H\ max}$' assumption for aligned single sets of conducting fractures from shear-wave anisotropy may be an over-simplification, and is often in conflict with geomechanics test data and theory.

# 10 Seismic quality Q and attenuation at many scales

In this chapter the term *'Q'* used in the title in the classic paper of Knopoff, 1964 will be distinguished from the engineering rock mass quality Q-value of Barton et al., 1974, by reference to the seismic quality as 'seismic Q', $Q_{seis}$, or $Q_p$ or $Q_s$ if the compressional wave or shear wave components have been distinguished. In fact, as we shall see, there are obvious connections between $Q_{seis}$ and the rock quality Q-value; a heavily jointed clay-bearing rock mass with low Q-value (probably less than 0.1) will inevitably cause great attenuation and have a correspondingly low $Q_{seis}$ (perhaps less than 5), while an almost unjointed massive rock mass with very high Q-value (e.g. 100–500) will inevitably cause little attenuation and have correspondingly high $Q_{seis}$, depending on whether shallow or at great depth.

Knopoff, 1964, introduced his review of seismic Q (or $Q_{seis}$) by stating 'Were it not for the intrinsic attenuation of sound in the earth's interior, the energy of earthquakes of the past would still reverberate through the interior of the earth today. The chaos resulting from this awesome prospect is a speculation which lies outside the scope of this paper.' We can conclude that $Q_{seis}$ and any of the physical reasons for $Q_{seis}$ that are captured in the Q-value rating (Appendix A) are fundamental to our well-being, even though low values of both may cause problems when tunnelling or when preparing a large dam foundation.

## 10.1 Some basic aspects concerning attenuation and $Q_{seismic}$

Using the definition of $Q_{seis}$ given by Knopoff, 1964, as a starting point, we may refer to the familiar electrical circuit theory for energy loss:

$$\frac{2\pi}{Q_{seis}} = \frac{\Delta E}{E} \qquad (10.1)$$

In this definition, $\Delta E$ is the amount of energy dissipated per cycle of a harmonic excitation in a certain volume, and E is the peak elastic energy in the system in the same volume. Common sense would suggest that $Q_{seis}$ can never be less than $2\pi$; however values below this magnitude are quite frequently recorded near the surface, including negative values which presumably may reflect interpretation difficulties of some sort.

At the time of Knopoff's review it was customary to assume that $Q_{seis}$ was substantially independent of frequency. His assumptions of 'a homogeneous sample' and 'at low frequencies' are clearly important in view of what is now understood about potential dissipation mechanisms in microcracked rock samples or in rock masses with sets of bedding planes and/or joints. Laboratory experiments on many homogeneous solids had shown that up to moderately high frequencies, the dimensionless quantity $Q_{seis}$ was virtually independent of frequency. This preliminary conclusion indicated that the mechanism by which energy was removed from elastic waves in solids was not the same as the mechanism for attenuation in liquids, where attenuation is frequency dependent.

Some typical values of $Q_{seis}$ for longitudinal excitation of various solids, selected from Knopoff, 1964, are reproduced below.

In this very selective list, the attempt is made to link $Q_{seis}$ to the relative stiffnesses of these materials. In reality the satisfactory-looking 'order' seen here is more scattered. One may comment already that the sandstone,

Table 10.1 Some examples of $Q_{seis}$ for longitudinal or bending excitation of various solids, selected from Knopoff, 1964, sorted by magnitude.

| Material | $Q_{seis}$ |
|---|---|
| Steel | 5000 |
| Copper | 2140 |
| Silica | 1250 |
| Glass | 490 |
| Diorite | 125 |
| Limestone | 110 |
| Lead | 36 |
| Sandstone | 21 |
| Shale | 10 |
| Celluloid | 7 |

since presumably not jointed, was likely to have been very weak and porous, in view of later values for sandstones that we will review.

Interestingly, and as a less serious aside, the 2003 Paramont film 'The Core', about an improbable voyage towards the centre of the earth to 'fix' an electromagnetic hazard, contains an opening sequence where the soon-to-be-seconded professor (Dr. Joshua Keys, played by Aaron Eckhart) has written the following Q-quality factors on the blackboard of the University of Chicago lecture room:

Shale 20 to 70
Limestone 45 to 90
Granite 40 to 230
Also: $Q \rightarrow \infty$: no attenuation

The writer was informed about this sequence by a lawyer who wondered if it was 'my' Q. These $Q_{seis}$ magnitudes are of course entirely feasible extensions of the above list of $Q_{seis}$ magnitudes selected from Knopoff. As we will see, the numbers for any rock will change with degree of microcracking, with pressure, with the dry or saturated state, with weathering, and with the degree of jointing and faulting when at larger scale. Due to each of the above, the frequency band of the dynamic loading will also affect the result, due to factors to be explored in this chapter.

It is now known that higher frequencies are attenuated at a higher rate than lower frequencies. Thus in a constant seismic Q (or rock quality Q) region of the rock mass, the amplitude of high frequency waves will decrease faster than that of the low frequency waves. Although most of the early evidence suggested that the seismic quality factor Q was frequency-*independent* over a wide range of frequencies (e.g. $10^{-2}$ to $10^{7}$ Hz for the case of shales, MacDonal et al., 1958), this is now generally rejected, as a result of laboratory tests conducted under different frequencies, under smaller more realistic strain levels and over wider ranges of confining pressures.

More recent data from in situ well tests conducted over wide ranges of frequency show the fundamental frequency dependence of seismic Q, due to the range of scales of the various attenuation mechanisms. Some of these newer sets of data will be reviewed in this chapter.

Although Knopoff, 1964 assumed that the attenuation of elastic waves in dry (intact) rock was independent of frequency, it is a different matter when microcracks and joints and water saturation (or partial saturation) are added. Attenuation increases and becomes frequency dependent, due to the inertial forces of the fluid in the microcracks and joints, and due to scattering.

Walsh, 1966, proposed a frictional-dissipation-at-crack-surfaces model to explain the simpler attenuation in dry rock. There were parallels with his observation of hysteresis when loading and unloading rock in uniaxial compression.

Concerning attenuation in intact rock, Walsh envisaged the following. Among the large number of cracks of all orientations and lengths, some are open and some are closed at any given pressure. As a compressional wave traverses the rock, (micro-scale) sliding on one crack-face past the other will occur on cracks which are barely closed and which have favourable orientation with respect to the wave propagation. This crack-face motion is opposed by friction, and some of the elastic energy of the wave is dissipated.

As the wave traverses the material, the normal stress between the crack-faces increases, and thus the frictional shear stress also increases. As the wave passes, the direction of the frictional shear stress is reversed, and again work must be done against friction as the crack returns to its equilibrium position. Clearly, micro-scale deformations are implied here.

Numerous mechanisms have been proposed to explain attenuation of seismic waves in rock and in rock masses. Johnston et al., 1979 listed the following in their landmark paper:

- Matrix anelasticity
- Frictional dissipation due to relative motions at grain boundaries and across crack surfaces (cf. Walsh, 1966)
- Fluid flow causing relaxation due to shear motions at pore-fluid boundaries
- Relative motion of the matrix frame with respect to the fluid inclusions in the case of fully saturated rock (cf. Biot, 1956a)
- Squirt phenomena (cf. Mavko and Nur, 1975 and O'Connel and Budianski, 1977)
- Gas pockets squeezing when only partial saturation
- Geometrical effects due to small pores, larger irregularities, thin beds (this category obviously extends to major discontinuities, faults, rock boundaries, dykes etc.)

Attempts to illustrate some of the smaller scale mechanisms of intrinsic attenuation are reproduced from Johnston et al., 1979, in Figure 10.1. The fluid flow attenuation mechanisms really fall into two

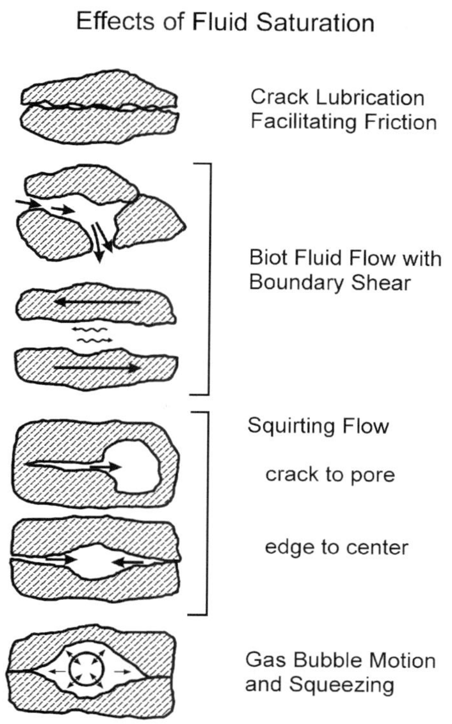


*Figure 10.1* Schematic illustrations of several of the proposed intrinsic attenuation mechanisms. Johnston et al., 1979.

frequency-dependent categories: the inertial resistance which is important at ultrasonic frequencies, and squirt flow which is more prominent at lower frequencies.

It appears that friction across thin cracks and grain boundaries may be the dominant attenuation mechanism (at small scale), if strain levels are sufficient. Increasing pressure decreases the aperture and effective number of cracks, and thereby reduces attenuation. Water wetting and saturation reduces the friction coefficient, thereby increasing attenuation. We will see much more detail of these aspects in the next section, and a detailed treatment of the effect of strain levels and pressure on attenuation mechanisms.

### 10.1.1 A preliminary discussion of the importance of strain levels

A useful summary of key concepts concerning elastic (contra plastic) strain as a function of frequency, and relaxation mechanisms as a function of frequency, was given by Batzle et al., 2005. Two diagrams presented in their EAGE Madrid poster are reproduced here, in

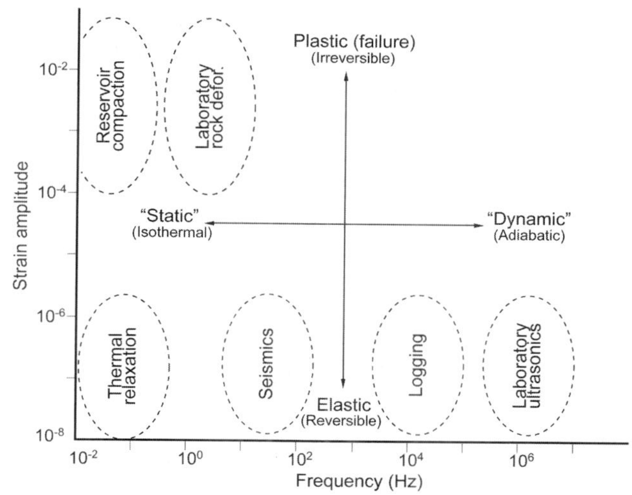


*Figure 10.2* A generalized chart concerning strain magnitudes and frequencies, for various deformation processes. After Batzle et al., 2005.

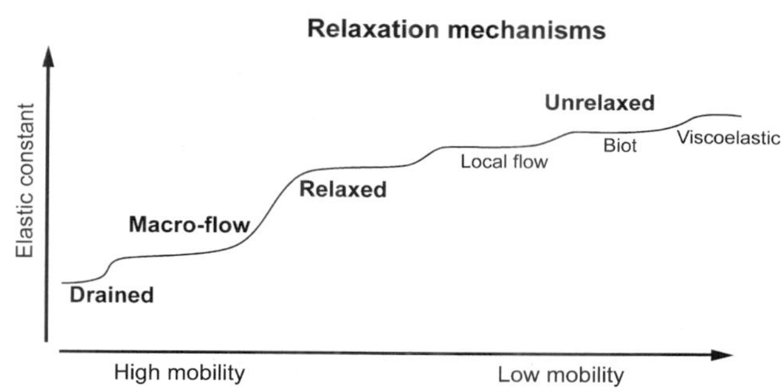


*Figure 10.3* Conceptual diagram of elastic constants for different relaxation mechanisms, with a frequency scale, and an indication of whether the fluid involved is of low or high mobility. After Batzle et al., 2005.

Figures 10.2 and 10.3. These supplemented their extended abstract. They give a useful perspective on the orders of magnitudes involved in these fundamental earth-science topics. The authors emphasised that moduli or velocities measured in one amplitude or frequency domain were usually not valid in other domains, since different deformation mechanisms would likely be operating.

Concerning the 'elastic ↔ plastic' 'static ↔ dynamic' *cross-plot* in Figure 10.2, one may put forward a 'jointed-rock-mechanics' viewpoint that in the presence of the usual heterogeneities of jointed, stressed rock *masses*, there will be a tendency for the four ellipses below the horizontal 'static-dynamic' axis to stretch their long axes upwards, into *larger strain territory*. The reason for this opinion is that dynamic *joint compliances* derived from the registration of seismic anisotropy, reviewed in later chapters, have inverted magnitudes that are, perhaps surprisingly, partly within experimental ranges of pseudo-static *joint stiffness* data, from rock mechanics 'macro-deformation' testing.

It is difficult to believe, on this basis, that joint micro-displacements involved in developing the characteristic deformation-load units of compliance ($m.Pa^{-1}$) from *in situ* seismic inversion, could be as small as 'sub-atomic', as a prominent physicist has suggested. Such an opinion probably stems from consideration of the effect of *microstrain* on *microcracks* (in intact samples), giving 'too small' influence to mobilize conventional concepts of friction, as we shall see shortly.

In the *in situ* reality, a rock mass consists of joints and discontinuities with both length dimensions, and spacings, many, many orders of magnitude larger than the rock physicists microcracked 'intact' samples. The nano-strains to sub-microstrains presumably experienced in a rock mass during the passage of seismic waves (depending on distance from source and its magnitude) may then, through discrete micro-displacement in the rock mass, experience attenuation due to the larger scale features as well.

Possibly the response of the joint to the dynamic pulse, is to initiate response from the *current* operating normal and shear stress-deformation gradients. Dynamic micro-excursions above and below these gradients would then occur, with an increased (or nearly equal) slope presumably depending on the quality of the joint walls and surrounding rock.

A useful geophysics concept, in this context, is that 'rock quality is defined as whether $E_{dynamic}$ is *more than or equal* to $E_{static}$'. Clearly, as shown in Chapter 6, when rock is hard and joints are fresh, the *inequality* of these two moduli is small – in which case the inverse of 'dynamic' joint or fracture compliance is also likely to be similar to the 'static' joint or fracture stiffness. When on the other hand rock (and joint) quality is poor, which in the rock mechanics world would be when rock quality $Q_c$ was low ( and joint wall compression strength JCS also low, perhaps even with clay-smear-or-filling), there will then be a big inequality, with $E_{dynamic} >$ or $>> E_{static}$, and a presumably corresponding inequality of 1/dynamic compliance > or >> 'static' stiffness, or as we shall see in Chapter 16, $1/Z_N \geq K_n$, (and $1/Z_T >$ or $>> K_s$).

In both the above scenarios: high or low rock quality, and the existing stress-deformation gradients, be they steep or shallow, will likely determine the 'static' *starting* point for the similarly steep, or much increased gradients of 1/Z. It is surely logical to assume that the four ellipses depicted in Figure 10.2 stretch more into higher strain territory, the lower the rock quality, with the likelihood of some slight, irreversible deformation in the case of seismic, low frequency motion, especially near the rock surface, where stresses and deformation resistance are low.

According to the dispersion relation of Kjartansson, 1977, cited by Mavko and Nur, 1979, the attenuation $Q^{-1}$, or inverse seismic quality Q can actually be used to explain the difference between the static and dynamic moduli of intact rock (See Chapter 6 for general results for rock *masses*). It is well known that the dynamic modulus can be at least double that of the static modulus even in intact rock, if flat pores or microcracks are present. Part of the difference in moduli may be due to fluid stiffening, in addition to the above dynamic-compliance/static-stiffness differences required when going up to *in situ* scale.

According to Kjartansson, 1977, the ratio of moduli (M) at different frequencies (f) can be expressed as:

$$\frac{M_1}{M_2} \approx \left(\frac{f_1}{f_2}\right)^{2/\pi Q} \tag{10.2}$$

It is assumed here that $Q_{seis}$ remains roughly constant over the frequency band of interest. Taking a 'static' value of $f_2 = 0.01$ Hz and a dynamic $f_1 = 10^5$ Hz and $Q_{seis} = 1000$, 100, 50 and 20, we find predicted ratios of $M_1/M_2$ of 1.01, 1.1, 1.2 and 1.7 respectively. Obviously, the more flaws (pores, microcracks) that are present in a rock sample, the lower will be the seismic quality Q, and the higher the predicted ratio of $M_1/M_2$ (the dynamic/static ratio of moduli).

The above is consistent with the idea of a broadly related $Q_{seis}$ and rock quality Q, since dynamic moduli diverge more from the static moduli, as rock (mass) quality reduces. The above 'intact rock' difference is accentuated when larger scale is considered, since the rock joints will usually have lower values of the 'static' normal and shear stiffnesses, than the inverse of the dynamic compliances of the same joints.

### 10.1.2 A preliminary look at the attenuating effect of cracks of larger scale

A useful insight into the effect of changes in crack porosity (and number of cracks) on the seismic quality factor $Q_{seis}$, was given by Remy et al., 1994. We will utilise this in this introductory section, before reviewing intact laboratory data concerning seismic Q. The authors' laboratory investigations involved *sixteen* freeze-thaw cycles (20°C to −20°C) over a period of *sixteen* days, in order to simulate part of the first appearance of weathering effects. The rock investigated was a thin-bedded (<1 cm), Jurassic limestone from Lorraine in France. Cylinders (5 cm diameter, 10 cm length) and cubes (5 cm sides) were used, having a bulk density of 2.1 gm/cm$^3$ and porosity of 22%. The bedding planes were perpendicular to the axes of the cylinders, and parallel to the top surface of the cubes. The repeated cycles of freezing (5 hrs), frozen (6 hrs), thawing (5 hrs), thawed (8 hrs) and corresponding changes of P-wave velocity are shown in Figure 10.4.

P-wave velocities were higher when frozen (e.g. 4.7 km/s) than when thawed (e.g. 3.4 km/s) due to the higher wave velocity in ice (3.8 km/s). Maximum velocities were reached at the end of the freezing. As shown in Figure 10.5a and b, $V_P$ (frozen state) fell with each cycle, while $V_P$ (thawed) fell most rapidly on the first two cycles. It is important to note that the creation of new cracks caused under-saturation of the initially water-saturated samples, which were jacketed, and immersed in a solution of methanol.

The two marked drops in $Q_{seis}$ values signify cracking episodes, the second of which was perpendicular to the bedding planes (during the 8th cycle). Physical evidence for the cracking was seen from hydrostatic loading tests on the cubic samples, where definition of the total volumetric crack porosity (the sum of the components of each axis) was recorded. This parameter increased successively, and uniformly, during sixteen cycles of freezing and thawing, and clearly intimately affected the reduction in $Q_{seis}$.

The reduced values of $V_P$ and $Q_{seis}$ with successive accumulation of crack-related damage have direct parallels in rock mass quality changes (i.e. reduced rock mass Q-value due to the fact that RQD reduces, $J_n$ may increase, $J_w$ reduces and, subsequently $J_a$ increases as a result of weathering. See Appendix A for descriptions of the Q-parameters of rock quality (Barton et al., 1974).

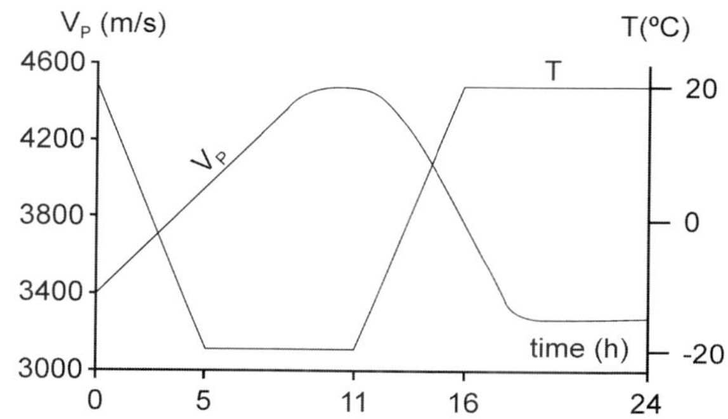


*Figure 10.4* A unit freeze-thaw-time cycle of 24 hours applied to thin-bedded limestones, and its basic effect on $V_P$. Remy et al., 1994.

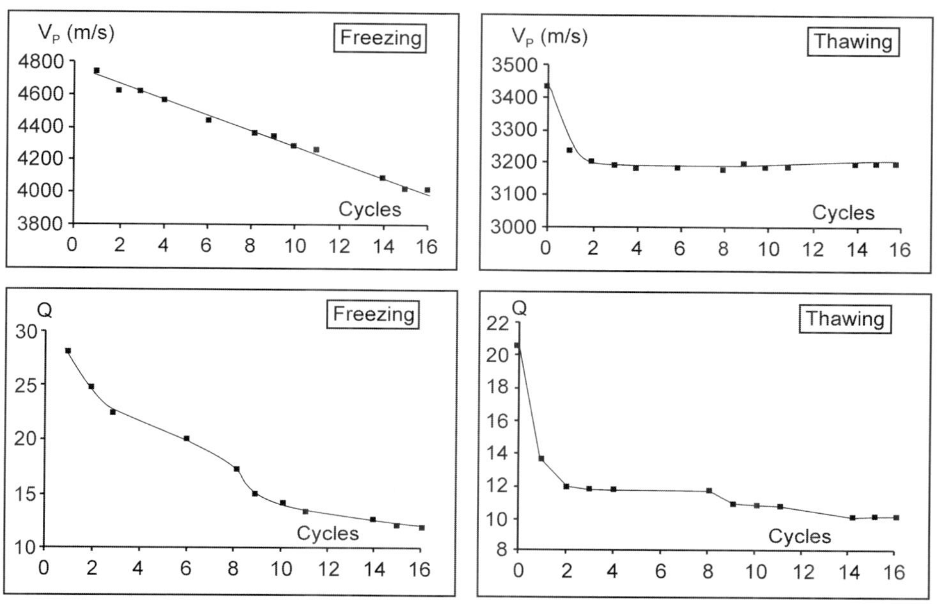


*Figure 10.5* a) Velocity $V_P$ versus number of freezing and thawing cycles. b) Seismic Q versus number of freezing and thawing cycles. (Remy et al., 1994).

In the literature there are numerous references to the relatively low values of $Q_p$ (and $Q_s$), in near-surface jointed or altered rock (e.g. 50 or less) and the higher values for intact samples of rock (e.g. 10 to 250) and the higher still values for deep igneous and metamorphic basement rocks (e.g. 100 to several 1000s), sills (e.g. 2000) and salt (e.g. 1000).

## 10.2 Attenuation and seismic Q from laboratory measurement

A compilation of $Q_{seis}$ values for specific groups of rocks, without distinction between different frequencies or degrees of saturation shows, inevitably, a wide scatter as shown in Figure 10.6, from Bradley and Fort 1966. This shows only porosity as the plotted variable, and consequently a range of $Q_{seis}$ from less than 10 (for porous sandstones) to nearly 900 (for low porosity igneous and metamorphic rocks). There is a general trend of $Q_{seis}$ inversely proportional to porosity, but this is compromised by too many hidden mechanisms of attenuation.

Early investigations of the effect of the degree of water saturation in reducing $Q_{seis}$ for porous rocks, and its strong pressure sensitivity as the rock reverts from dry to different degrees of saturation, are shown in Figure 10.7, from Gardner et al., 1964 data, reproduced by Johnston et al., 1979. As we shall see later, when considering both the P-wave and S-wave related attenuations, the ratio of $Q_s/Q_p$ proves to be an even better indicator of the degree of saturation.

In each case, pressure, with its microcrack-closing ability, causes a rise in velocity and a reduction in attenuation. $Q_{seis}$ therefore rises. According to the model of attenuation developed by Johnston et al., 1979, the relative contributions of friction and fluid flow on the overall attenuation are as shown in Figure 10.8.

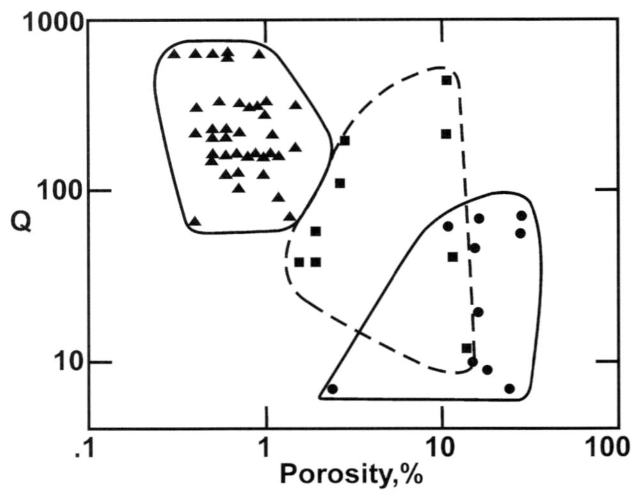


*Figure 10.6* $Q_{seis}$ as a function of porosity for igneous and metamorphic rocks (triangles), limestones (squares), and sandstones (circles). A wide range of frequencies and degrees of saturation contribute to the scatter of data. From Bradley and Fort 1966, reproduced by Johnston et al., 1979.

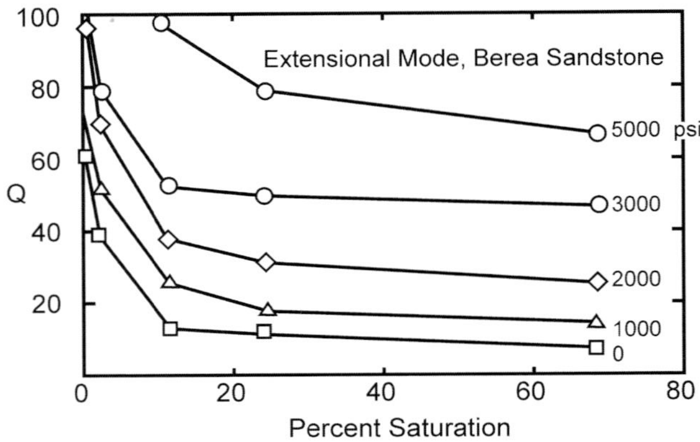


*Figure 10.7* Saturation and pressure dependence of $Q_{seis}$. From Gardner et al., 1964 data.

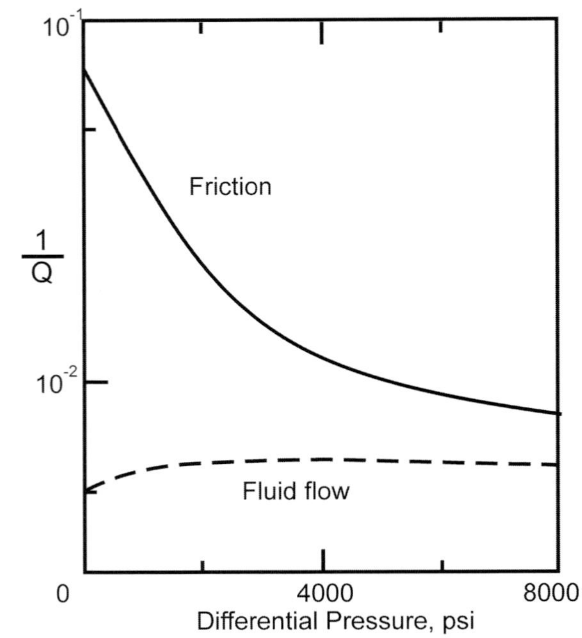


*Figure 10.8* Relative contribution of friction-based and fluid flow-based attenuation for a brine saturated Berea sandstone, according to the model of Johnston et al., 1979.

As pointed out by Johnston et al., 1979, since the porosity and permeability (of these intact specimens) is relatively unchanged by the range of pressures applied, there is limited effect on the fluid-flow contribution to attenuation. Such would presumably not be the case if a jointed specimen or a jointed rock mass was involved, where pressure sensitivity of the permeability and secondary porosity would be marked, and non-linear, thereby giving a strong rise in $Q_{seis}$ with the reduced attenuation and velocity increase.

### 10.2.1 A more detailed discussion of friction as an attenuation mechanism

According to the models of Johnston et al., 1979, the relative effects of frequency and pressure can be combined to elevate the total $Q_{seis}$ (specifically $Q_p$) of the Berea sandstone. At low pressures, the friction mechanism dominates and is almost independent of frequency. With increasing pressure and low frequencies $Q_p$ climbs beyond 100, but as frequency increases there is a reduction of $Q_p$ due to the contribution of squirt flow and so-called shear relaxation. Eventually, at very high frequencies, $Q_p$ declines sharply again due to scattering. This general scheme of predicted behaviour is illustrated in Figure 10.9.

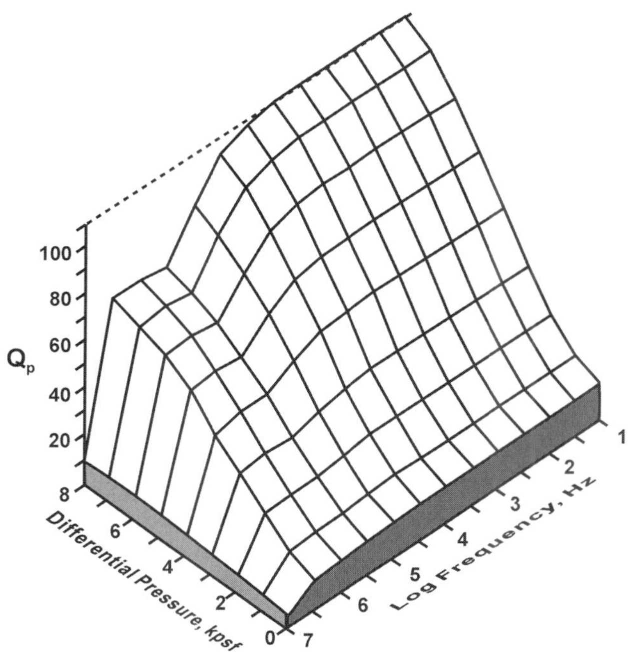


*Figure 10.9* Total $Q_p$ predicted for brine-saturated Berea sandstone, from Johnston et al., 1979.

The question of whether friction is a viable source of seismic attenuation; along microcracks, across crack-tips, (and also along joints and filled discontinuities, and within the multiple surfaces of faults), will now be addressed again, with the benefit of more understanding of the effects of strain levels, provided by Winkler and Nur, 1982. With its title: 'Seismic attenuation: effects of pore fluids and frictional sliding', one would certainly expect that *both* mechanisms were still to be emphasised as potential sources of attenuation.

In their conclusions the authors however, state the following: 'Since the conditions required for sliding friction to be observed (large strains and small confining pressures) generally do not apply to seismic wave propagation in the earth, we conclude that simple frictional sliding is not a significant attenuation mechanism in situ.'

Their conclusion was drawn, at least partly, on the basis of extensional resonance tests, conducted on long, thin (intact) bars of homogeneous rock, such as sandstone, which were contained inside a long pressure vessel, and made to oscillate with an electro-magnet, while supported rigidly at their mid-point. Figure 10.10 shows the results of resonance decay measurements, giving both

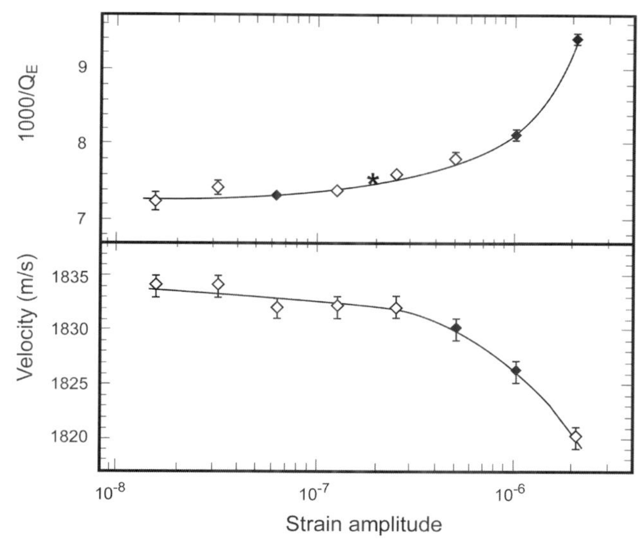


*Figure 10.10* Variation of attenuation ($1000/Q_E$) and velocity with strain amplitude, based on extensional resonance decay measurements on long (intact) bars of sandstone, suspended at their mid-point in a pressure vessel, and excited by an electro-magnet at one end, with a phonograph pick-up at the other end. (Note $Q^{-1}$ sensitivity of 19%, and velocity sensitivity of only 0.7% to the 2-order of magnitude strain amplitude variation). Winkler and Nur, 1982.

Table 10.2 Effect of strain amplitudes on extensional seismic quality, showing negligible effect on velocity. Selected data from Winkler and Nur, 1982.

| Material | Strain amplitude | $Q_E$ | Velocity |
|---|---|---|---|
| Sierra white granite | $1.44 \times 10^{-6}$ | 185 | 3,629 m/s |
| | $4.15 \times 10^{-8}$ | 204 | 3,637 m/s |
| Berea sandstone | $2.10 \times 10^{-6}$ | 103 | 1,937 m/s |
| | $2.30 \times 10^{-8}$ | 140 | 1,955 m/s |
| Lucite | $1.43 \times 10^{-6}$ | 23.4 | 2,108 m/s |
| | $3.04 \times 10^{-8}$ | 23.2 | 2,108 m/s |

velocity and $Q^{-1}$ as a function of strain amplitude. The authors used frequencies from 500 to 9000 Hz, and studied the effects of confining pressure, degree of saturation, strain amplitude, and frequency.

The changes in attenuation and velocity they observed with increasing strain amplitude (Figure 10.10) *were* interpreted as evidence of frictional sliding at grain contacts. But since this amplitude dependence supposedly disappeared at strains and confining pressures that they considered were typical of seismic wave propagation in the earth, they consequently inferred that frictional sliding was not a significant source of seismic attenuation in situ.

They referred to other problems with the frictional attenuation mechanism. Savage, 1969 had pointed out that for typical strain amplitudes of seismic waves, and for reasonable microcrack dimensions, the computed slip across crack faces would be less than the inter-atomic spacing. They assumed, probably correctly, that this extremely small interaction would not be described by conventional models of macroscopic friction. In addition, they referred to the widely held assumption that frictional attenuation caused nonlinear wave propagation, which had not apparently been observed at the low strain amplitudes typical of seismic waves.

Winkler and Nur provided a useful summary of the strain amplitude dependence of extensional attenuation $Q_E$ for intact samples of several rock types, and some man-made materials. (The Massilon sandstone result is shown in Figure 10.10). Only the materials (i.e. rocks) that contained potential (micro) sliding surfaces, indicated strain amplitude dependence, and the author's tests showed that (intact) rock samples almost lost this dependence with moderate increases of pressure, as shown in Figure 10.11.

Noting the effect of *confining pressure* on extending the strain amplitude 'limit' for $Q_E$ sensitivity, to about

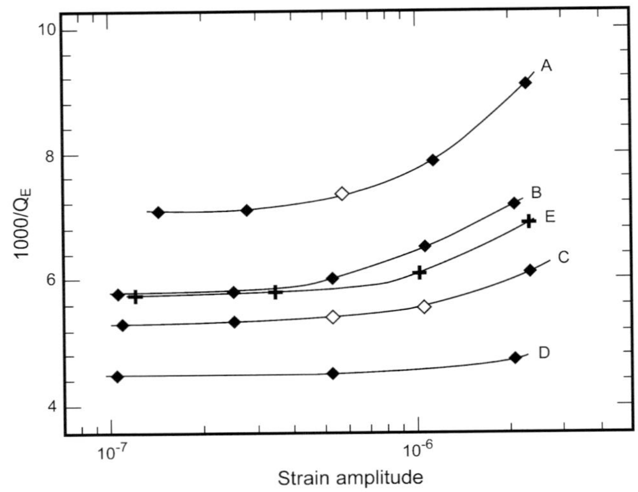


*Figure 10.11* Effect of confining pressure in reducing the strain amplitude-dependence of extensional attenuation $Q_E$, for dry Berea sandstone. Curve A = 1 MPa, B = 2 MPa, C = 3 MPa, D = 5 MPa, E = 5 − 3 = 2 MPa (helium pore pressure of 3 MPa). Note therefore the closeness of curves B and E. Winkler and Nur, 1982.

$10^{-6}$, as shown when comparing Figures 10.10 and 10.11, the authors suggested that it was significant that this strain level was in the same range as that needed for cusped stress-strain loops to become elliptical (Brennan and Stacey, 1977). They then posed the question: why is a strain of $10^{-6}$ or larger needed to cause frictional attenuation? (*At least for the case of intact rock specimens, excited in the extensional mode?*).

They explored the answer to this question by suggesting that displacements across crack surfaces should *at least* be comparable to *inter-atomic* spacings of about $10^{-10}$ m. They then equated a shear strain ($\varepsilon$) to a maximum displacement ($d$) across a crack of length ($L$), suggesting $d \approx \varepsilon.L$. With ($d$) $\approx 10^{-10}$ m, ($\varepsilon$) $\approx 10^{-6}$, a crack length ($L$) of $10^{-4}$, or 0.1 mm is implied. The authors considered this to be a realistic upper-bound for microcrack sizes in rock, so concluded that at strains below $10^{-6}$, sliding displacements would generally be too small for *friction* to describe the (sub-micron) interaction.

The obvious corollary to this is to pose the question: what about all the larger cracks, i.e. intra-bedding joints, tectonic joint sets, major clay-filled discontinuities, and multiple internal interfaces in fault zones, all of which have large, or extremely large ($L$). With potential length dimensions of 0.1 m 1.0 m and 10 m for the 'smallest' three of the above five categories of *discontinuity*, and assumed spacings of the same order of magnitude (for convenience of estimation), an unchanged

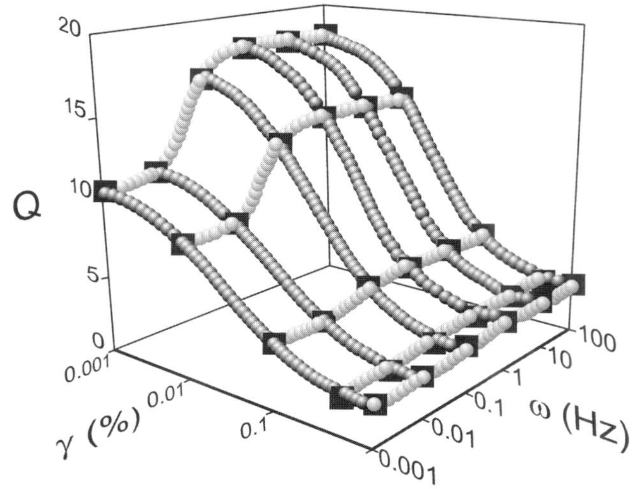


*Figure 10.12* Seismic Q as a function of angular frequency and shear strain level, measured on cylinders of near-surface clay, in a resonant column apparatus. Marmureanu et al., 2000.

continuum-based shear strain of $10^{-6}$ generated close to a given seismic source, might well imply maximum (close to the source) *displacement discontinuity events* of the order of 0.1, 1.0 and 10 μm for these three joint/discontinuity types, if the continuum strain was converted to intermittent discontinuous shearing events with the same frequency as their length scale.

Can such events be the source of dynamic joint *compliances* in geophysics, that have recognisable (nearly same order) magnitudes and units, as the MPa/mm pseudo-static stiffnesses of rock joints that are familiar to rock mechanics engineers? Shear strains decaying to one or two orders of magnitude *less* than $10^{-6}$, further from seismic sources, are surely still capable of providing displacement discontinuities of sufficient magnitude for frictional attenuation to be a valid mechanism in rock *masses*, as opposed to intact bars of homogeneous rock.

While on the subject of the importance of strain level and frequency on $Q_{seis}$, it is of interest to look at soils, nicely illustrated by the results of Marmureanu et al., 2000, using resonant column equipment. They tested cylindrical samples from surface soil layers, applying torsional and longitudinal vibrations, in studies connected with seismic risk mitigation. Figure 10.12 shows seismic Q as a function of shear strain level (γ%) and frequency, almost showing independence from frequency over a typical engineering seismology range of interest, i.e. about 5 to 100 Hz. The angular and shear strain dependence of soil, giving non-linear behaviour, was emphasised in their focus on earthquake hazard estimation.

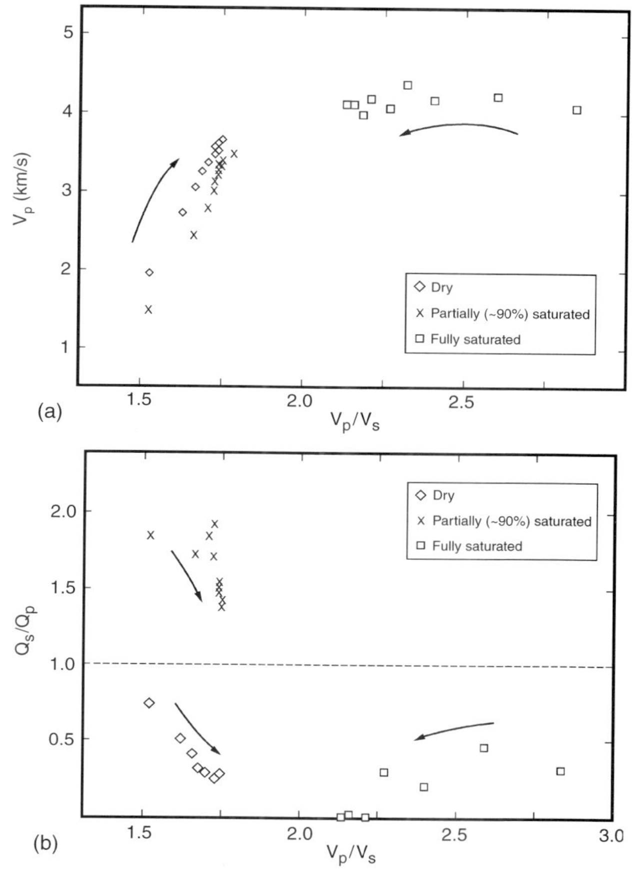


*Figure 10.13* Cross-plots of $V_p$ versus $V_p/V_s$, and $Q_s/Q_p$ versus $V_p/V_s$, showing the distinctive effects of the dry, partly saturated, or fully saturated states, when using these parameter ratios. Winkler and Nur, 1982.

### 10.2.2 Effects of partial saturation on seismic Q

Figures 10.13 a and b, show a useful summary of some of Winkler and Nur, 1982 work on the effects of the dry, partly saturated, or fully saturated state on the P-wave velocity and its variation with $V_p/V_s$. A 'companion' set of data for the moisture-detecting ratio $Q_s/Q_p$ versus $V_p/V_s$ is also shown.

The S-wave attenuation increases with saturation ($Q_s$ reduces), thus making the ratio $Q_s/Q_p$ a particularly sensitive indicator of the degree of saturation, since P-wave attenuation, though increasing with initial saturation levels, eventually reduces to less than the S-wave attenuation: thus the ratio $Q_s/Q_p$ reduces to low levels, since $Q_p$ has increased. The separation of data into 'environmental compartments' is very interesting, and also useful for in situ interpretation.

As we have seen above, seismic attenuation in partly saturated rock with pore space and microcracks provides interesting insights into the frequency dependence of seismic Q and $Q^{-1}$ for this small-scale element of rock mass. The early work of Knopoff, 1964, showed that attenuation in dry rock was independent of frequency even over a wide range, because of the assumed velocity-independence of sliding against friction across crack faces. When a sample is fully saturated, attenuation becomes frequency-dependent because of energy losses from viscous dissipation, which depends on shearing velocity.

As discussed by Walsh, 1995, when attenuation is plotted against saturation for the same rock, there are peaks of attenuation at low saturation (e.g. 1%) and at high saturation (e.g. above 60 to 90%), depending on whether loading is in pure compression or in shear, where the peak attenuation comes only at greater saturation. It is believed that the attenuation peak at very low saturations is the result of viscous losses in fluid trapped in microcracks, which are filled first due to stronger capillary forces where apertures are very small.

Over a wide range of saturation from a few percent to some 50%, no change of attenuation occurs, but as continuous saturated regions arise, fluid pressures rise in response to the pore volume reduction caused by the harmonic compressive wave, and energy loss becomes frequency dependent due to viscous dissipation. However, at very high frequencies, no fluid transfer occurs and the saturated region responds elastically. At very low frequencies, flow occurs, but if viscous stresses are low, dissipation of energy may be negligible.

The bell-shaped frequency dependence of seismic $Q^{-1}$ occurs at the intermediate frequencies, when viscous dissipation is not negligible. This is illustrated in Figure 10.14. The narrow peaks of attenuation in both shear and hydrostatic compression were suggested by Walsh, 1995, to be the result of the 'squirt' phenomenon (Mavko and Nur, 1975).

At low frequency, pore fluids influence the attenuation due to their lack of rigidity, compressibility and density, while at higher frequencies, attenuation occurs due to viscous and inertial forces. Nur, 1973, interpreted temporal velocity anomalies as evidence of dilatant strain and varying pore water saturation in the crust prior to certain earthquakes. Mavko and Nur, 1979, showed that even a small amount of water can dramatically enhance the attenuation, when very flat pores (or joints) are present. This is because high pressure gradients cause (micro) flow at the contact between wet and dry pore space.

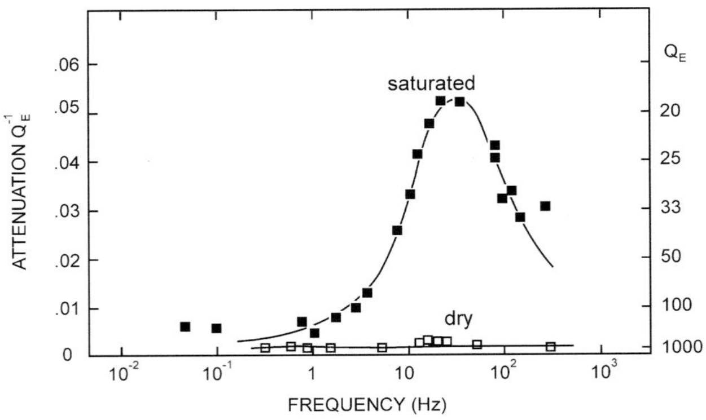


*Figure 10.14* Attenuation as a function of frequency in dry rock (open squares) and rock fully saturated with water. The bell-shaped curve is characteristic of viscous damping. After Paffenholz and Burkhardt, 1989; and Walsh, 1995.

P and S waves passing through a medium exert oscillatory stresses which can be resolved into normal and shear components in the plane of each pore space. Attenuation can be demonstrated both for the normal component and for the shear component. The ratio $Q_p/Q_s$ is $>1$ for dry rocks and is $<1$ when almost fully saturated, as we have seen earlier. It appears from the model of Mavko and Nur, 1979, that the state of saturation of the flat cracks (or rock joints) rather than the overall saturation of the rock is the most important factor for the attenuation.

## 10.3 Effect of confining pressure on seismic Q

Resonant bar techniques for the sonic frequency range, and pulse transmission techniques for the ultrasonic frequency range were used by Lucet and Zinszner, 1992, to demonstrate that not only frequency range, but also *confinement* can affect the seismic quality $Q_{seis}$. Their 3 to 7 kHz and 500 kHz testing with some 30 rocks that included limestones and sandstones, included confinement to 45 MPa and water saturation. Pore pressure was fixed at 1 atmosphere. Care was taken to select core from adjacent samples in the same homogeneous quarried block, so that the effect of different frequencies could be truly compared.

Figure 10.15a shows a set of results for a sandstone, in which sonic and ultrasonic attenuation as a function of increasing confining pressure are (in this case) similar. The vertical scale of $1000 \times Q^{-1}$ shows that $Q_p$ (or $Q_E$), increased from about 6 to nearly 100 as a result of confinement. This can be seen by 'inserting' seismic Q

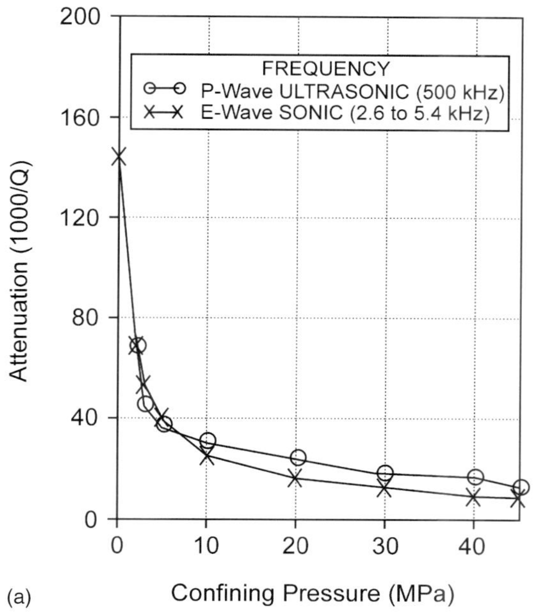


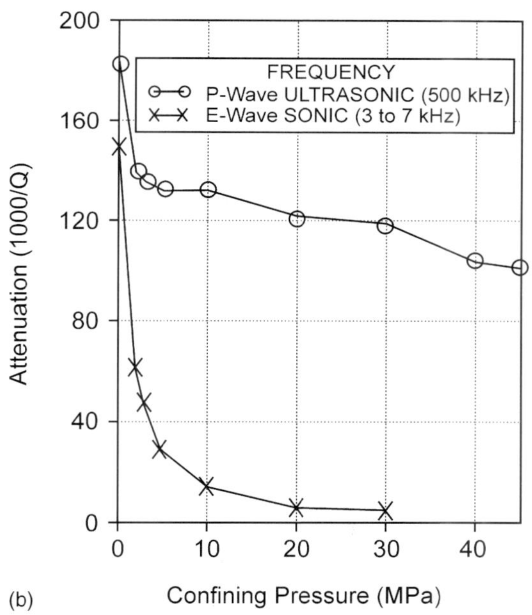


*Figure 10.15* Sonic (resonant bar extensional mode), and ultrasonic measurements, a) on a saturated sandstone, and b) on a saturated crinoidal limestone, as a function of confinement. Lucet and Zinszner, 1992.

values down the right-hand axes at convenient arithmetic intervals, giving $Q_{seis}$ values of 5, 10, 20, 50 and 100. To one with a rock mechanics background, a resemblance to E-modulus increases with confinement is seen in both sets of sonic data, with units of GPa. More of this will be seen later.

In the case of a crinoidal limestone shown in Figure 10.15 b, there is clear separation of the attenuation according to frequency. The ultrasonic $Q_p^{-1}$ (attenuation) is significantly higher, or $Q_p$ numerically much smaller (6 to 10) than for the sonic tests, where $Q_E$ ranges from 7 to 100 or more, as confinement is increased. The authors interpreted these differences as being due to scattering of waves due to 'density' heterogeneities in the case of the limestone. Another limestone which was fine-grained showed less dramatic separation of behaviour as a result of frequency differences, and almost negligible effect of confining pressure. Seismic Q values were in this case a more or less constant 50 (ultrasonic) and a more or less constant 100 (sonic), over the full confining pressure range.

In an important series of tests on two sandstones, Prasad and Manghnani, 1997, investigated not only the effects of effective stress change, but also pore pressure changes on the P-wave velocity and attenuation $Q_p^{-1}$ Their experimental set-up, which is simply and clearly illustrated, has been reproduced in Figure 10.16. This figure defines Pc and Pp, and the difference Pd = Pc - Pp is found in subsequent figures showing their results.

The two sandstones investigated, Berea and Michigan, had bulk densities of 2.28 and 2.36 gm/cm$^3$, and corresponding porosities of 21.2% and 16.9%, respectively, causing the higher velocities in the Michigan sandstone. The Berea sandstone had visible bedding planes and weakly cemented angular grains with microcracks. We can therefore select this sandstone for reproducing some of the author's important results.

These results, and equivalent ones for the rounded-grained and less porous Michigan sandstone, enabled the authors to differentiate the pore pressure dependence of the two sandstones. Referring to the classic effective stress equation:

$$P_e = P_c - nP_p \tag{10.3}$$

where n is the effective stress coefficient (Biot, 1962, Todd and Simmons, 1972), the authors found that both the Berea sandstone and the Michigan sandstone had values of n that reduced from about 0.78 and 0.62 respectively, when the confining pressure was high. These results applied to experiences in interpreting $V_p$. In the case of $Q_p$, equivalent results were 1.10 and 0.86, reducing to 0.81 and 0.71, respectively. In other words, Vp and $Q_p$ measured at elevated pore pressures and elevated confining pressures are governed by effective stress coefficients significantly less than the classic n = 1 obtained for more permeable media. The authors Prasad and Manghnani cited differences in the type of

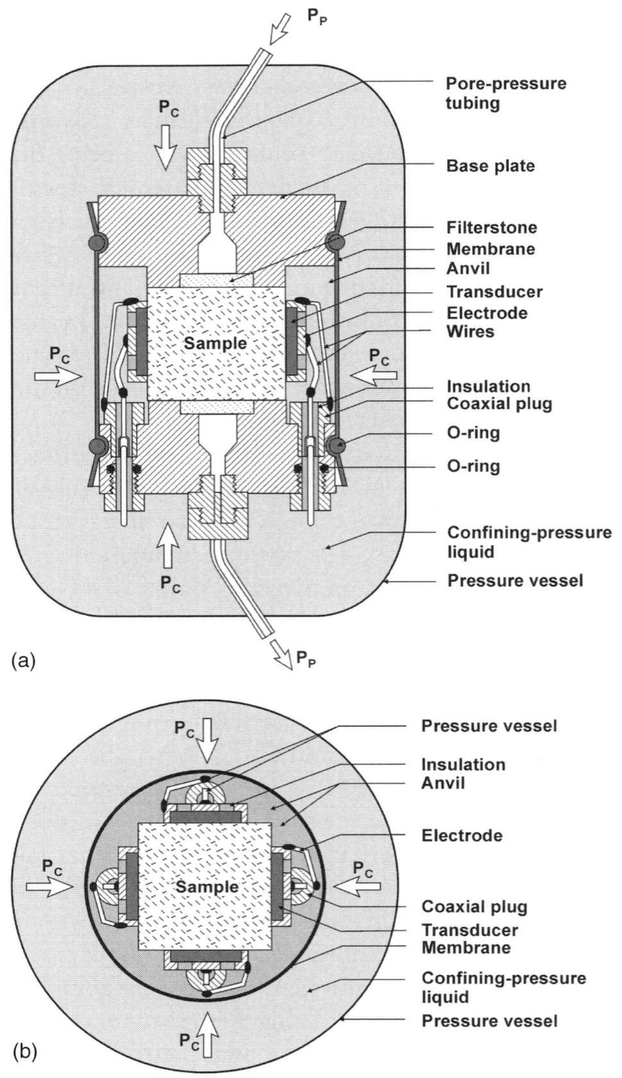


*Figure 10.16* Schematic diagram of the ultrasonic pulse transmission experiments of Prasad and Manghnani, 1997.

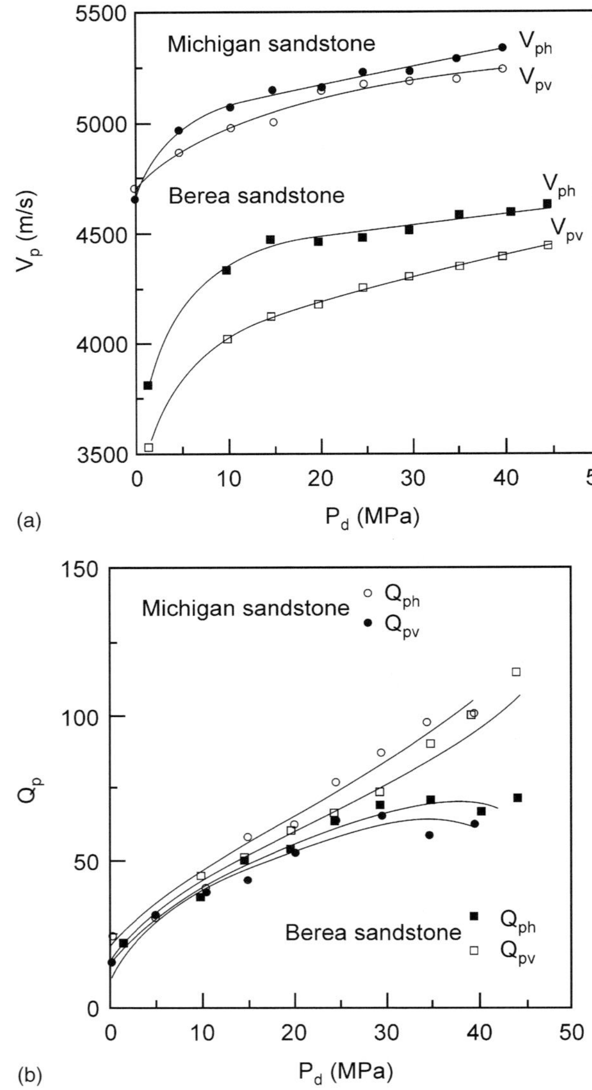


*Figure 10.17* Vp and $Q_p$ as a function of effective confining pressure for two sandstones, in this case at a pore pressure of 1 atmosphere. Prasad and Manghnani, 1997.

contact areas between the grains in the two sandstones as the reason for the differences in the pore pressure dependencies of the two sandstones.

Before leaving the *above* results of confining pressure on $Q_p$ (there will be more data in some other chapters), it may be of interest to mention a finding, now a conviction, first noted when writing the chapter dedicated to *rock physics* results (Chapter 13). It was finally recognised that the variation of $Q_p$ with confining pressure resembled the well known rock mechanics effect of triaxial confinement on the E-modulus of rock samples. When the latter is expressed in GPa, quite remarkable likeness to seismic $Q_p$ was noticed. This can be seen in broad terms in Figures 10.15, 10.17, 10.18 and 10.19. The likeness of $E_{mass}$ in GPa and *in situ* $Q_p$ has continued to be seen in field data reviewed.

Almost all rock mechanics modulus data, from laboratory testing representing near-surface to kilometre depths, and from in situ testing at dam sites and deep tunnel deformation back-analysis, show moduli within the extreme range of 1 to 150 GPa, most commonly 5 to 75 GPa. In exceptionally weathered, weak, or clay-bearing conditions, moduli can reduce to 0.1 GPa, where 'total' attenuation in less than a wave length no doubt occurs.

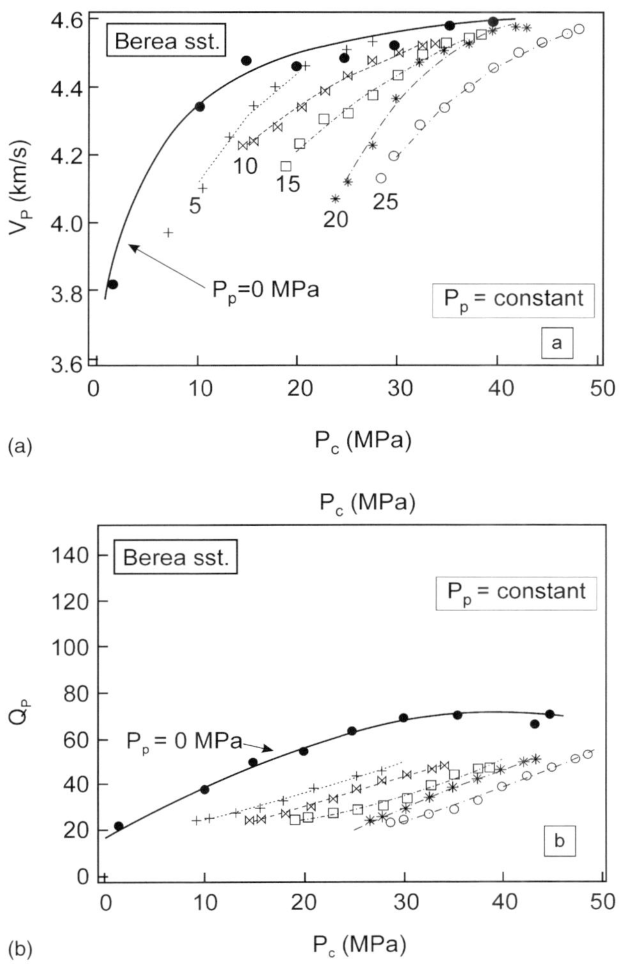


*Figure 10.18* Changes in Vp and Qp in Berea sandstone: in both cases with pore pressure $P_P$ constant. Prasad and Manghnani, 1997.

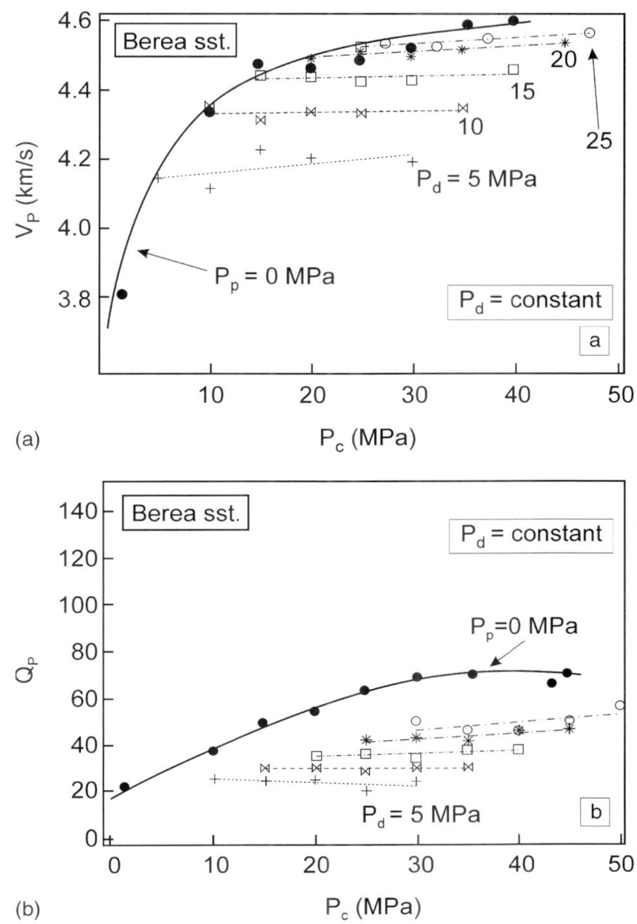


*Figure 10.19* Changes in $V_p$ and $Q_p$ in Berea sandstone: in both cases with differential (or effective) pressure $P_d$ constant. Prasad and Manghnani, 1997.

### 10.3.1 The four components of elastic attenuation

Before looking at the (non-linear) effects on attenuation of samples loaded towards fracturing, at the end of this section on laboratory tests with confining pressure, it is appropriate to refer to Batzle et al., 2005 laboratory testing of intact, porous samples in a so-called *forced-deformation apparatus.* These authors' tests neatly demonstrated the relative magnitudes of the elastic attenuation components.

According to Nur and Winkler, 1979, the different modes of elastic attenuation ($1/Q_k$ = bulk, $1/Q_p$ = compressional, $1/Q_e$ = Young's and $1/Q_s$ = shear) are related to each other through inequalities.

1. $1/Q_s < 1/Q_e < 1/Q_p < 1/Q_k$ (for low $V_p/V_s$ with partial saturation)
2. $1/Q_s = 1/Q_e = 1/Q_p = 1/Q_k$
3. $1/Q_s > 1/Q_e > 1/Q_p > 1/Q_k$ (for high $V_p/V_s$ with full saturation)

The Batzle et al., 2005 *forced-deformation apparatus* was capable of applying frequencies from 0.3 Hz to 2,000 Hz, with strain amplitudes below $10^{-7}$. Note that the latter is *very* low. Micro-valves were used to control fluid movement into or out of the samples, in response to the dynamic loading. Both brine-saturated and partly brine-saturated states were investigated.

Batzle et al., 2005 found that opening or closing their sample boundaries to fluid, using special micro-valves, caused a significant change in the velocity and dispersion values, when at full saturation. Two sets of their experimental results are reproduced in Figure 10.20. The authors noted with the open boundary, that low frequencies caused the rock-fluid conjunct to behave as if partially saturated. The fluid movement

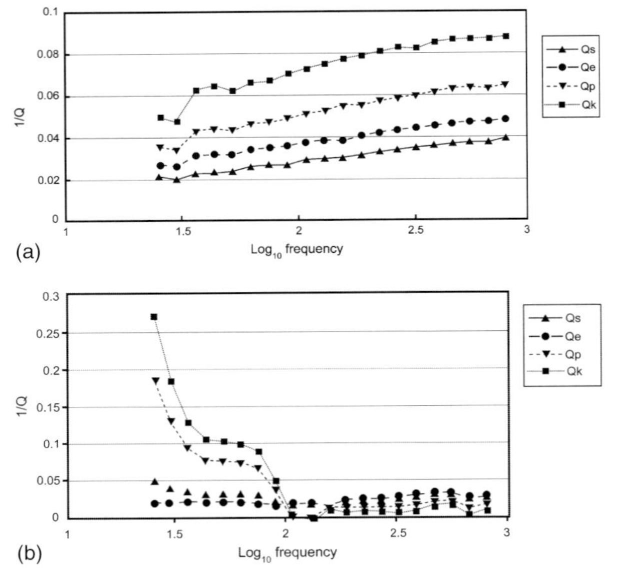


*Figure 10.20* The relation between the four different modes of elastic attenuation with a) partial brine saturation, where attenuation due to the bulk modulus dominates, and b) 100% brine saturation and an open boundary, where at low frequencies $1/Q_K$ and $1/Q_p$ dominate, as fluid can flow in and out of the samples. Batzle et al., 2005.

across the boundary was absent at high frequencies, due to the lack of time to reach (pressure) equilibrium. Beyond 100 Hz, saturated samples (with open boundaries) showed a low cumulative elastic attenuation.

Batzle et al., 2005 cited the fundamental coupling of attenuation, velocity and frequency, from the illustrative Cole and Cole, 1941 developments in dielectrics, which were applied to attenuation measurements by Spencer, 1981. The authors added the effect of fluid mobility and partial saturation, and indicated the typical measurement window, in Figure 10.21

The authors also addressed the more complex question of attenuation and frequency dependence, or dispersion, in samples of shale, finding that dispersion had strong directional dependence. Because of the low permeability and inhibited fluid motion, *yet observed dispersive results* because of strong attenuation, they suggested that interactions among clay particles and between the clays and bound water may be responsible.

They also showed that viscous fluids like heavy oil had their own internal viscous losses, which could contribute to overall rock attenuations. The conclusion from their studies of the multiple components of (elastic) attenuation was that attenuation-related attributes extracted from seismic data have to take such controlling parameters into account.

Problematic here is that the fracturing or joint set developments in a rock mass will often be concentrated in the higher modulus layers, whose internal attenuation

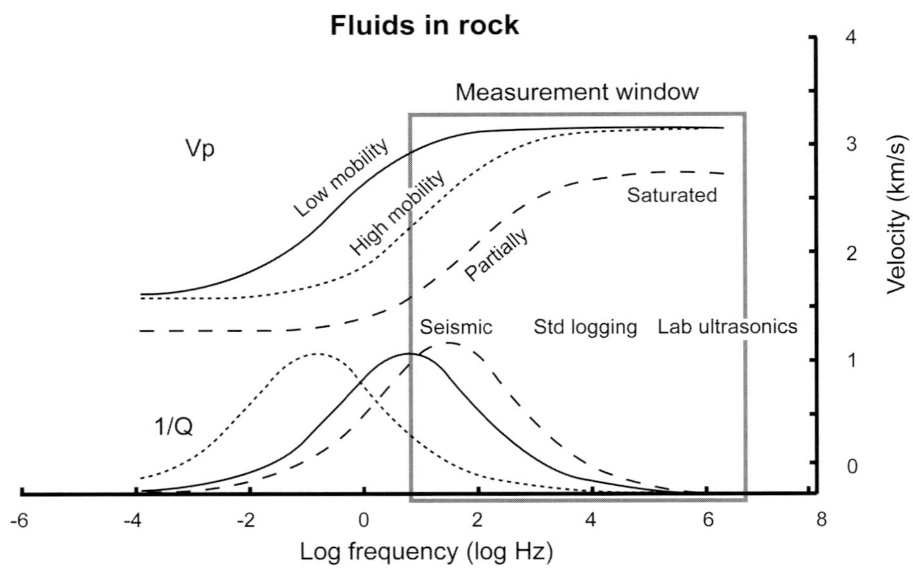


*Figure 10.21* The Cole-Cole relation coupling velocity, attenuation and frequency from the field of dielectrics. Cole and Cole, 1941. This was applied to attenuation in rock by Spencer, 1981. Batzle et al., 2005 also indicated the approximate dispersive effect of low or high fluid mobility, and of partial saturation.

components have thereby changed, or are different, from surrounding rock. Fluids of different viscosity in the differently fractured layers will add to the challenge of inverting data.

### 10.3.2 Effect on $Q_P$ and $Q_S$ of loading rock samples towards failure

An important contribution was made to our understanding of the influence of fracturing on attenuation, with simultaneous velocity effects, by Wulff et al., 1999. The authors made a careful study of the seismic effects of microfracturing during constant, low strain rate uniaxial compression testing, up to and beyond the point of microfracturing. They tested tuffaceous sandstone and granite samples, both related with Hot Dry Rock projects in Japan.

As the authors pointed out, attenuation was not *directly* related to the strength and elastic moduli, but to mechanisms such as fluid flow, friction and scattering due to microcrack and crack density effects. Testing only dry specimens, they concentrated on interpreting the relative roles of scattering and friction.

They reviewed several studies of attenuation in dry rock, (slate, sandstone, gabbro), that indicated good agreement of observations with the intrinsic attenuation mechanism of *frictional sliding*, developed by Walsh, 1966, and good agreement with the semi-empirical pressure-dependent theory of Johnston et al., 1979, that is also based on Walsh, 1966. They also cited studies of *scattering* attenuation in micro-fractured marble where the *scattering* attenuation theories of Hudson, 1981 and 1990 (the first-order scattering model), did not predict sufficient attenuation in relation to test results. Wulff et al., 1999 found that crack density *squared* was needed in the Hudson, 1981 model, rather than a linear relation to crack density, to explain their own results.

The authors' tests on blocks of tuffaceous sandstone and granite, measuring 100 × 100 × 250 mm, were conducted under 'room-dry' conditions, following four weeks of drying at room temperature. New cracks were assumed to be dry 'or at least not to have absorbed enough water molecules to permit fluid flow during the time of the experiment'.

Several of their very interesting results are reproduced here. Figure 10.22 shows the separate effects of axial strain (with associated development of microcracking) on P- and S-wave velocities, and on $Q_p$ and $Q_s$, for two samples of tuffaceous sandstone (t2-l, and t1-f). The wave propagation was *perpendicular* to the

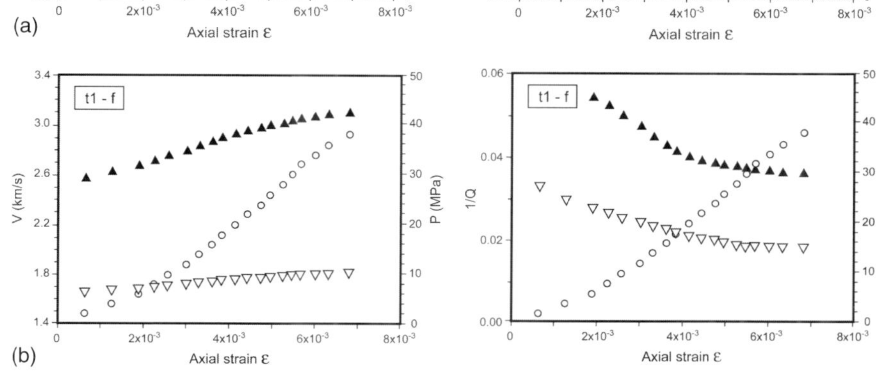


*Figure 10.22* Effects of uniaxial stress-strain: symbol (o), and the associated microcracking, on $V_p$, $V_s$, $Q_p$ and $Q_s$ for two tuffaceous *sandstone* samples (100 × 100 × 250 mm, n = 18.6%, $\gamma$ = 2.05 gm/cm$^3$) from Japan. Note sample failures at 25 and 39 MPa. a) The upper pair of results (sample t2-l), have wave propagation (400 kHz) perpendicular to the loading direction, and therefore perpendicular to dominant microcracking. b) The lower pair (sample t1-f), have wave propagation (also 400 kHz), parallel to the loading direction, causing increased velocity and less attenuation with increased load. (solid symbols: $V_p$ and $Q_p$). Wulff et al., 1999.

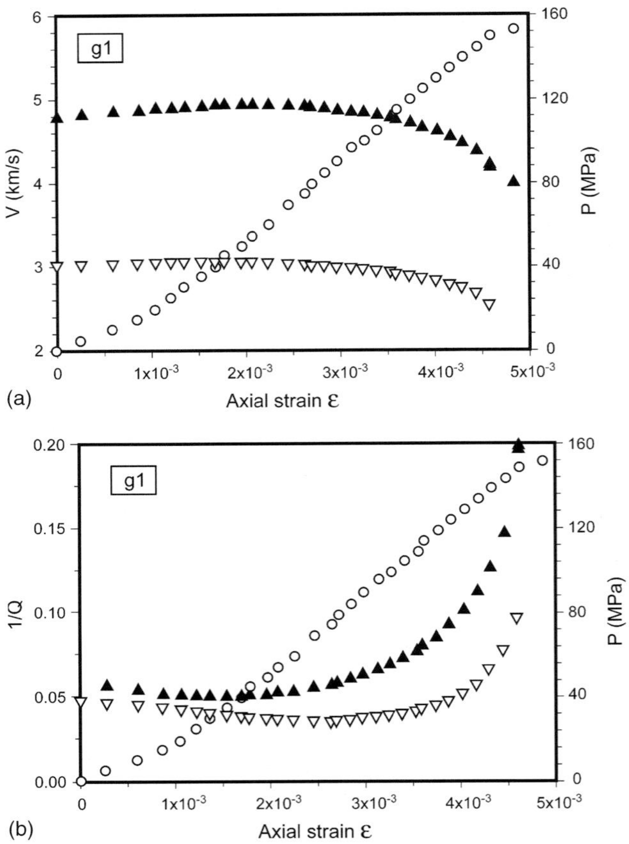


*Figure 10.23* Effects of uniaxial stress-strain: symbol (o), and the associated microcracking, on $V_p$, $V_s$, $Q_p$ and $Q_s$ for a *granite* sample (100 × 100 × 250 mm, e.g., average grain size 1.3 mm, n = 1.4%, $\gamma = 2.62\ gm/cm^3$) from Japan. Note failure at about 150 MPa. Wave propagation (400 kHz), is perpendicular to the loading direction, and therefore perpendicular to dominant microcracking, causing the reduced velocities and increased attenuation (reaching a minimum $Q_p = 5$). (solid symbols: $V_p$ and $Q_p$). Wulff et al., 1999.

loading direction in Figure 10.22a, and *parallel* to the loading direction in Figure 10.22b. In each case the axial load – axial strain curves are given by (o) symbols. Equivalent results for a granite specimen, with wave propagation perpendicular to the loading direction, are shown in Figure 10.23.

The increase of seismic $Q_p$ from about 18 to 30 (and $Q_s$ from about 30 to 50) in the case of the measurement *parallel* to the loading direction, are both potentially recognisable as 'deformation modulus' results, if the latter were expressed in GPa. On the other hand, the measurements of seismic Q made *perpendicular* to the loading direction, showing both $Q_p$ and $Q_s$ reducing from about 18 to 7 or 10 were actually registering a 'Poisson expansion' effect. In the case of the tests on granite shown in Figure 10.23, seismic Q was also measured *perpendicular* to loading, so registered similar reductions to $Q_p$ and $Q_s$, following a slight increase in $Q_s$ during the first half of the loading.

The authors investigated theoretical crack densities (crack number density × crack radius cubed), based on the theories of Hudson, 1981 and 1990. They interpreted a non-linear increase in attenuation with crack density as being due to pressure-increased crack sizes in addition to crack density increase. The frequency dependence of the P-waves, proportional to approx. $f^{2 \text{ or } 3}$, suggested attenuation by scattering and possibly by friction. They therefore investigated the (Rayleigh) scattering attenuation predicted by the Hudson 1981 model, and found that only when using the largest plausible crack dimensions could they explain the total attenuation, if scattering alone was responsible.

The attenuation mechanism caused by frictional sliding along the tapered tips of microcracks, according to Mavko, 1979, was also investigated, giving a good fit to the total attenuation of two of the four samples, based on their assumptions. The mechanism is independent of frequency, which the authors found consistent with the fact that the measured total attenuation was less frequency dependent than expected if scattering was the only mechanism. Figure 10.24a compares the scattering attenuation calculated with the Hudson, 1981, model, based on maximum plausible crack sizes, with measured data, and Figure 10.24b shows the calculated attenuation due to crack-tip friction, following Mavko, 1979. The authors concluded that attenuation in the rocks investigated was probably by a combination of frictional attenuation and scattering.

A laboratory study involving 'flaws', this time naturally existing, concerning the cavities in carbonate rock, such as vugs or karsts, was described by Hackert and Parra, 2003. These cavities cause scattering attenuation like the fracturing seen above, but quantification is difficult due to the unknown scale and structure of the cavities. The authors described the use of X-ray computerized tomography scans to obtain the exact vug structure of two cores. They then used 3D finite-difference modelling to determine the P-wave scattering attenuation at ultrasonic frequencies.

$Q_{seis}$ in the saturated states were as low as 8 and 15 near the source frequency of 250 kHz. The two cores had respective total porosities of 32.1% and 16.6%, with CT-computed vuggy porosities of 13.4% and 4.5%. The respective dry-state P-wave velocities were 3.97 and

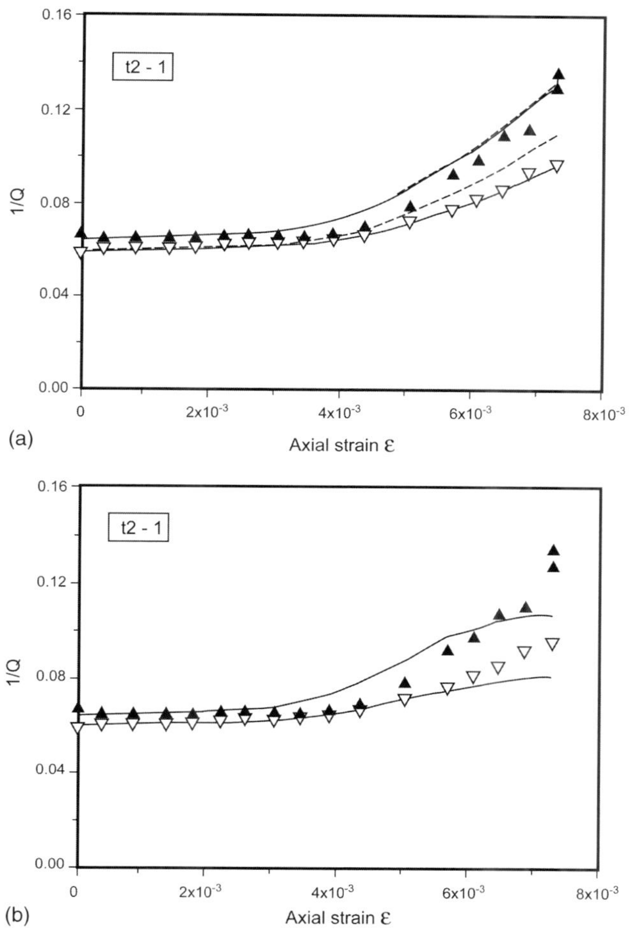


*Figure 10.24* a) Calculated scattering attenuation for *sandstone* sample t2-l, using the Hudson, 1981 method, assuming crack sizes of 600 μm. Solid lines represent the model with randomly oriented cracks, and dashed lines represent cracks oriented in the loading direction. b) Calculated attenuation due to friction, using the Mavko, 1979 model, based on estimated crack densities for randomly oriented cracks. Wulff et al., 1999.

4.25 km/s, and dry densities 1.85 and 2.20 gm/cm$^3$. The authors observed that if the *vugs* had been *karsts* 1000 times larger (about 5 m), then the attenuation would have been seen at seismic frequencies in the range 100 to 500 Hz.

## 10.4 The effects of single rock joints on seismic Q

We will end this section on laboratory tests concerning seismic Q, with an appropriate *transitional stage*, namely the seismic Q behaviour of laboratory samples that are

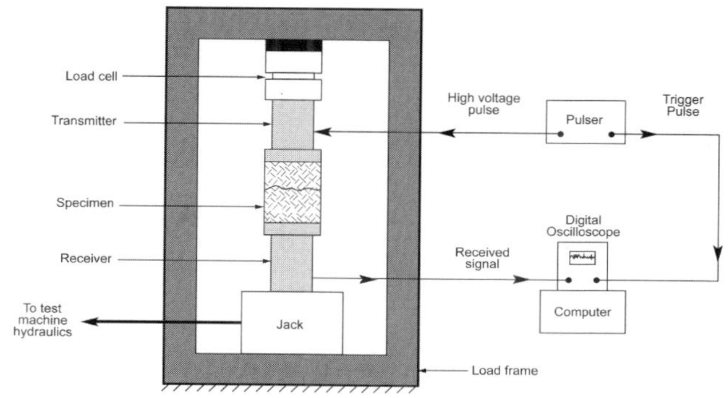


*Figure 10.25* Basic test set-up, for conducting normal loading and dynamic testing of joint samples, with add-on facilities for hydraulic testing and contact area estimation using Woods Metal. Pyrak-Nolte et al., 1990.

'divided' by single natural joints or fractures. Landmark work was done in this area by Laura Pyrak-Nolte and colleagues Neville Cooke and Larry Myer, with important links to the hydraulics of joints or fractures via Paul Witherspoon. This pioneering research, originating from the University of Berkeley and from Lawrence Berkeley Laboratory, followed on from the rock mechanics developments of Goodman twenty years previously, and the theoretical geophysics of Schoenberg, and represents one of the few and important links between rock mechanics, hydraulics and geophysics.

Most of Pyrak-Nolte's and colleagues' better known work was focussed on the behaviour of just three samples of joints in quartz monzonite from Stripa Mine 'granite' in Sweden. Even the much described sample numbers E30, E32 and E35 are sometimes referred to by geophysicists. These robust samples were subject to numerous tests, on numerous occasions, and have given the profession important insight into 'fully-coupled' earth science behaviour. We will review different aspects of this work in this *and* later chapters.

An understanding of the basic principles for their tests is given in Figure 10.25. Besides the dynamic testing under normal load, as indicated, there was the possibility to measure permeability by linear (sector-to-sector) flow across the circular joint specimens, which had a diameter of 52 mm. There was also a facility to inject non-wetting molten Woods Metal into heated joint samples, which upon cooling, gave a measure of the area of the joint available for flow, at the given normal stress level. Joint roughness, such as JRC was not described, but a test result was referred to by Pyrak-Nolte et al., 1987a, where negligible effect of temperature (95°C) on aperture was indicated. (This differs from some other experiences of temperature effects on joint apertures where there is appreciable roughness, e.g. Barton et al., 1985, Barton, 1999, Barton and Makurat, 2006).

The *hydraulic* apertures of the three joints E 32, E 30 and E 35 can be interpreted as reducing from approximately 3, 5 and 26 μm to approximately <1, <1 and 6 μm, as the joints were closed by measured amounts of approximately 4, 9 and 22 μm (Pyrak-Nolte et al., 1987a). Thus $\Delta E/\Delta e$ was in the low range of 1.1–2.25, suggesting quite planar joints, in relation to the Barton et al., 1985 model for the JRC-controlled measured inequality $E \geqslant e$, or $\Delta E \geqslant \Delta e$ usually seen. A preliminary estimate suggests that JRC may have been in the range of only 2 (two cases?) to 4, which would readily explain the relatively high stiffness and small closures under stress, exhibited especially by two of the three Pyrak-Nolte et al., joint samples (E30 and E32).

Figure 10.26a shows joint (or fracture) deformation versus normal stress for one of a series of load-unload cycles. The equivalent 'specific stiffness' (the inverse of tangent slopes) for these three load-deformation events, is shown in Figure 10.26b. These roughly 1,000 to 30,000 MPa/mm normal stiffnesses are of the same order of magnitude as the results for fresher, i.e. stiffer joint samples in various hard rock types, tested by Bandis, 1980. (See Bandis et al., 1983 and joint stiffness data reproduced in Chapter 16).

Pyrak-Nolte et al., 1990 performed dynamic tests both across the three joint samples and across *intact* samples taken from adjacent core. (Stripa 'granite'/quartz monzonite, $\gamma = 2.65\,\text{gm/cm}^3$, Young's modulus $E = 60\,\text{GPa}$). The three pairs of samples had equal length and diameter (77 × 52 mm).

Figure 10.27 shows three sets of ultrasonic ('0' to 1.5 MHz) P-wave amplitude spectra, conducted in the dry state in this case, for the three pairs of 'companion' samples (jointed, and adjacent intact). The comparison of 'intact' and jointed response gives a very instructive image of the effect of the more open and deformable E 35 joint (top), on P-wave transmission, showing selective filtering of highest frequency. (Note the lower normal stiffness of E 35 in Figure 10.26b, compared to the very stiff and presumably well interlocked E 32 sample). Sample E 30, with intermediate stiffness, gives intermediate response.

This fairly long, but necessary introduction to these important tests, brings us to the subject of seismic Q. Pyrak-Nolte et al., 1990 calculated seismic Q by taking the ratio of the spectral amplitudes of the intact and jointed 'companion' samples, in comparison to the spectral amplitudes obtained from a non-attenuating cylinder of aluminium of identical dimensions, and loaded in an identical manner to the same loads.

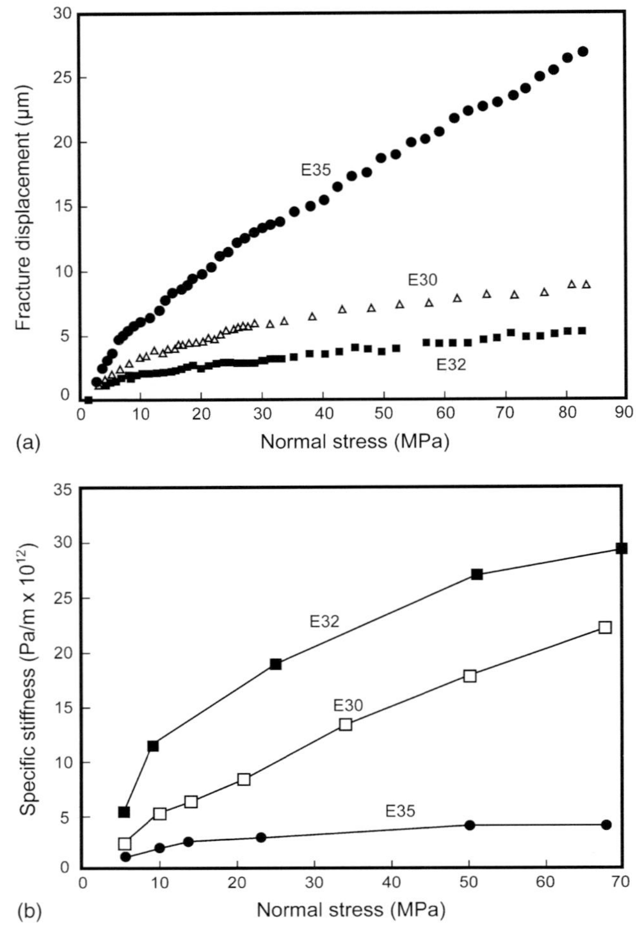


*Figure 10.26* a) One set of load-deformation results for the three quartz monzonite (Stripa granite) joint samples. b) Specific stiffness (inverse tangent slopes) as a function of normal stress, for the three joint samples tested dry. Pyrak-Nolte et al., 1987a.

The authors, following Johnston et al., 1979, compared the spectral amplitudes of the different samples, starting with the dispersive wave equation:

$$A = A_0\, e^{\varpi x / 2Qc} \qquad (10.4)$$

where
$\omega$ = frequency
$X$ = travel path length
$Q$ = inverse of attenuation
$c$ = phase velocity of wave
$A_0$ = amplitude at $x = 0$

The ratios of the spectral amplitudes of the seismic pulses transmitted through the companion rock samples

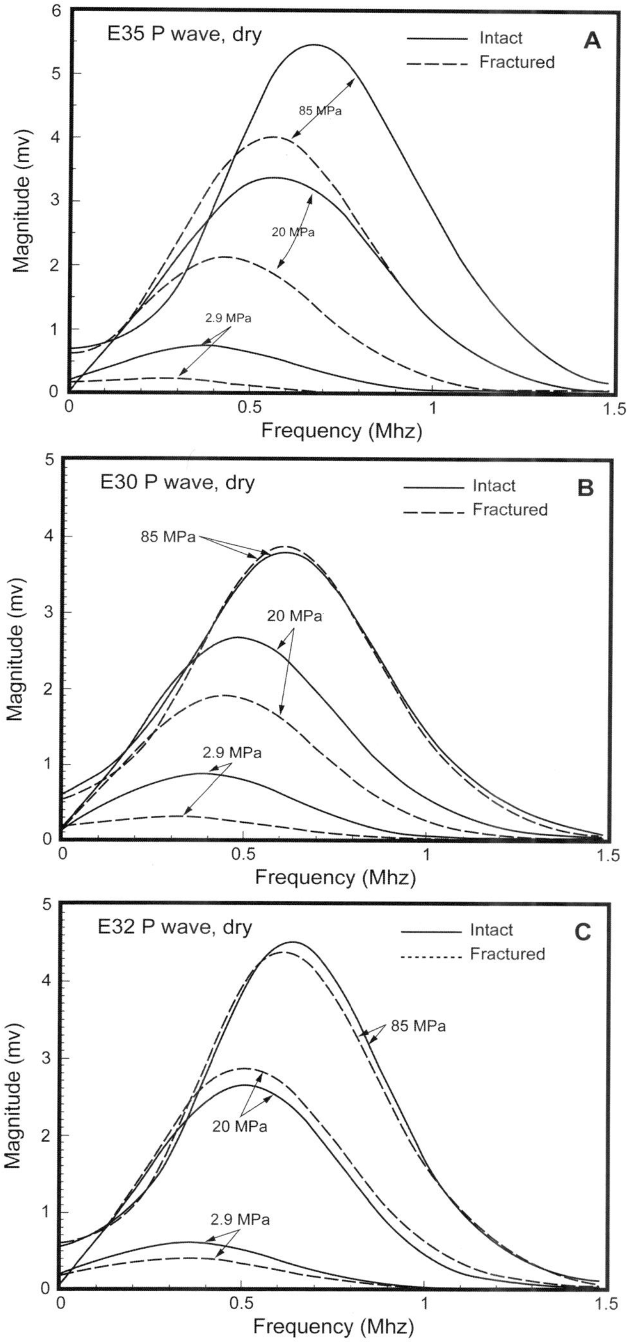


*Figure 10.27* Comparison of intact and jointed sample response to ultrasonic P-waves up to 1.5 MHz frequency. The magnitude spectra show the positive, magnitude-increasing, effect of higher normal stress. Sample E 35 has least normal stiffness, due to its 25 μm of closure under normal stress to 85 MPa. It demonstrates the maximum filtering of higher frequencies, compared to the high stiffness sample E 32, which only closed some 5 μm under 85 MPa stress. Pyrak-Nolte et al., 1990.

Table 10.3 Seismic $Q_p$ and $Q_s$ calculated by Pyrak-Nolte et al., 1990, using equation 10.4, for both jointed and intact, and dry and saturated states, at two levels of normal (axial) stress. The results for the most deformable E 35 joint sample are selected, as the 25 μm joint closure with 85 MPa stress increase is considered realistic for non-planar joints.

| Specimen type and test condition | $Q_p$ 2.9 MPa | $Q_p$ 20 MPa | $Q_s$ 2.9 MPa | $Q_s$ 20 MPa |
|---|---|---|---|---|
| Sample No. | E 35 | E 35 | E 35 | E35 |
| Jointed-dry | 7 | 14 | 12 | 23 |
| Intact-dry | 12 | 39 | 32 | 71 |
| Jointed-wet | 9 | 30 | 28 | 39 |
| Intact-wet | 15 | 51 | 41 | 56 |

compared to the aluminium ($A/A_1$), were then calculated (by Johnston et al., 1979) as:

$$\ln\frac{A}{A_1} = \frac{\pi f x}{Qc} + \ln\frac{A_0}{A_{01}} \tag{10.5}$$

where $f = \omega/2\pi$

The authors gave a comparison of seismic Q calculated from this equation, for both the dry and saturated states. A selection of their interesting results is reproduced in Table 10.3.

The marked reduction in attenuation when loading the jointed specimen at 20 MPa instead of 2.9 MPa is typical of *in situ* response. $Q_p$ increases by a factor of 2 when dry, and by a factor of 3 when wet. It is interesting to note however, that the *less attenuating* intact specimen shows $Q_p$ increasing by a factor of *at least* 3, both when dry and when wet. The seismic waves are of course transmitted perpendicular to the microcracks most likely to be closed by the axial stress.

Pyrak-Nolte et al., 1990 made an alternative seismic Q calculation, because of the non-linearity of the spectral ratio data. By assuming that $A_0/A_{01} = 1$ in equation 10.5, they were able to re-arrange the equation and express seismic Q as a function of frequency. Figure 10.28a shows the result of applying equation 10.6 to the data from the dynamic tests on jointed sample E 30. This joint had an intermediate level of normal stiffness in relation to E 35, and to the least deformable, stiffest joint E 32 (Figure 10.26).

$$Q = -\frac{\pi f x}{c \ln A/A_1} \tag{10.6}$$

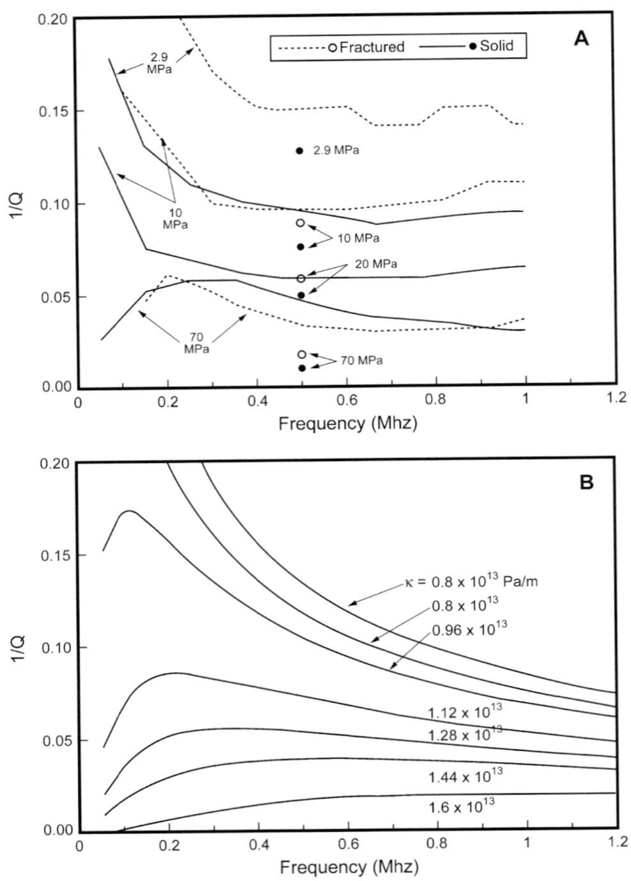


*Figure 10.28* a) The frequency-dependent attenuation calculated for the medium-stiff sample E 30 (dry), with comparison to the lesser attenuation of the intact (dry) companion sample. Normal stress levels were 2.9, 10 and 70 MPa. Small circles were the result of calculation using equation 10.5, at a specific frequency of 0.5 MHz. b) Analytical solutions assuming increasing values of dynamic normal stiffness, as described in the text. Pyrak-Nolte et al., 1990.

Figure 10.28a shows the dispersive results of the seismic Q calculation, using this equation for the tests conducted on E 30, when in the dry state. The small circles in the figure were calculated using equation 10.5, at a specific frequency of 0.5 MHz. The similar shape of the curves for the jointed and intact samples was interpreted by the authors as evidence for similar (closure-under-stress) behaviour of both the joint and the microcracks most likely to be (partly) closed by the axial stress.

The theoretical curves shown in Figure 10.28b were developed by the authors, by assuming a joint or fracture density of 1 per 77 mm, as tested, in an otherwise non-attenuating medium (clearly a simplification). The single fracture was represented by a displacement discontinuity (based on the theory of Schoenberg, 1980), with dynamic normal stiffness varying from $6.4 \times 10^{12}$ to $1.6 \times 10^{13}$ Pa/m. In rock mechanics units, this is more easily understood as 6,400–16,000 MPa/mm, in fact typical for the *pseudo-static* normal stiffness of fresh rock joints at high stress levels. (See Chapter 16.) The theoretical curves show good correspondence to the interpreted data using the spectral amplitude ratio method described above.

Before leaving these interesting studies for the time being, the earlier referred suspicion that deformation modulus (in GPa) is similar to seismic Q, will be addressed again. If we 'insert' a virtual $Q_{seis}$ scale down the right-hand axes in Figure 10.28, we obtain 'simple' magnitudes for seismic Q of 5, 10, 20, and 100 for $1/Q_{seis}$ values of 0.2, 0.1, 0.05 and 0.01. The jointed sample shows the lowest $Q_{seis}$ values when measured at the 'lower' frequencies of 0.2 MHz, and values varying from 5 to about 20 as normal stress increases from 2.9 to 70 MPa while the solid control sample, at 0.5 MHz shows $Q_{seis}$ of 8, 15, 20 and 90 as stress was raised from 2.9 to 10, 20 and 70 MPa. These results remain in the typical range of moduli (when expressed as GPa), but the low stress value of $Q_{seis}$ seems to be lower than expected.

In-seam seismic measurements in coal have been used for a number of years to indicate the state of stress and fluid drainage in this fine-structured, deformable, low velocity material, which in some ways resembles a miniature (*and property-scaled*) roughly cubically-jointed rock mass. A set of laboratory test data, including effects of the dry to fully saturated state, with confining pressures *from near-surface*, up to mine-relevant levels (2 to 40 MPa) was given by Yu et al., 1993, using a transversely isotropic Permian coal from Tower Colliery, Wollongong, Australia.

Some of the key results of these comprehensive studies are reproduced in Figures 10.29 and 10.30. The strong effect of water saturation, which tends to fill the flat (low aspect ratio) cracks and cleats in the coal, is evident in all the data. This miniature 'rock mass' also displayed the classic anisotropic effects of lower $V_p$ perpendicular to bedding, and higher $V_p$ parallel to bedding, with 45° wave transmission giving intermediate values. S-waves were little affected by saturation, which is also a traditional result, when S-wave splitting is not involved. In Chapter 15 we shall see that polarized split shear-waves are affected by degree of saturation and even fluid type, due to changed joint or fracture compliances.

At higher confining pressures, the water content (one of the contributing causes of attenuations) was reduced

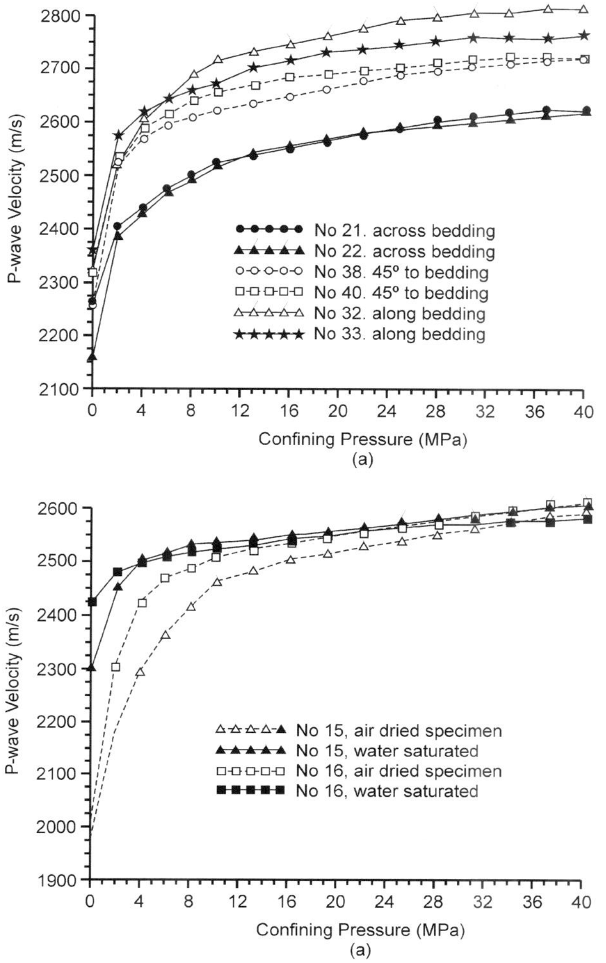


*Figure 10.29* $V_p$ as a function of confining pressure (pore pressure = atmospheric), for dry and saturated macro-bedded coal specimens. The specimens were loaded and dynamically tested in three orientations relative to bedding. Yu et al., 1993.

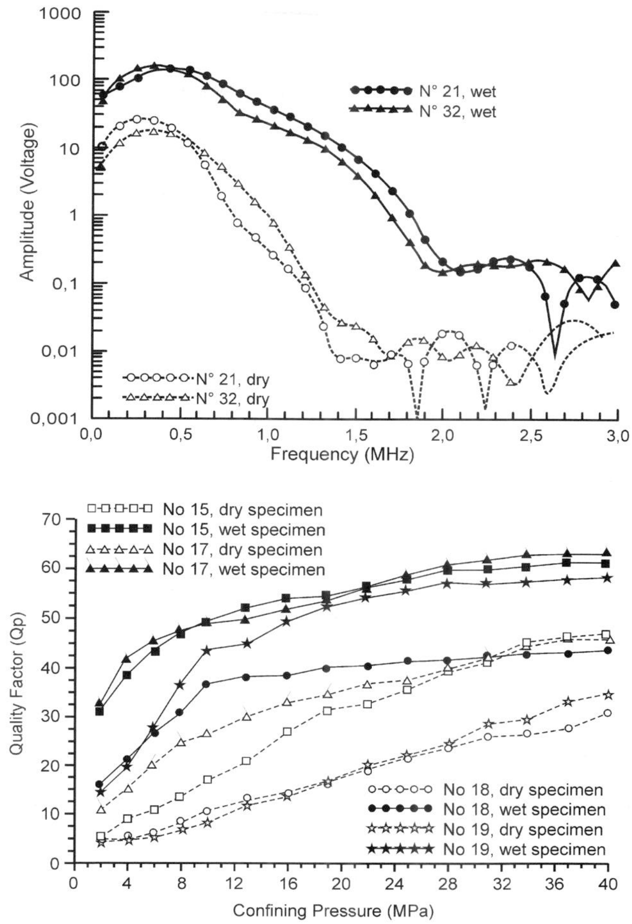


*Figure 10.30* a) The spectral amplitude behaviour across and parallel to the coal's bedding, showing the strong effect of saturation on wave transmission. b) $Q_p$ for dry and saturated coal specimens, as a function of confining pressure (pore pressure = atmospheric). Yu et al., 1993.

by partial closure of the fine cubic structure of cracks and cleats. $V_p$ (dry) and $Q_p$ (dry) approached the values of $V_p$ (saturated) and $Q_p$ (saturated) as confining pressure was increased. The seismic Q values were lowest for the four dry specimens, while there was much less attenuation for the four saturated samples. The authors emphasised the fact that the *decrease* in attenuation for fully-saturated specimens of coal, with its low aspect ratio cracks, differed diametrically from the usual result for sandstones. This could be questioned based on earlier results, but the point is made that cracks close easier than equant pore space.

Yu et al., 1993, also emphasised the relatively low values of *dynamic* E-moduli for the coal (e.g. about 8 GPa: at low stress?) and the relatively high values of the dynamic Poisson's ratio (e.g. about 0.4) in relation to typical intact rocks. The effect of higher stress in the coal reducing the difference in seismic Q between dry and saturated conditions was assumed to be because of both increased frictional resistance along the cracks, and due to the reduced water content caused by the closing cracks. Again we see the general trend of seismic Q increasing with stress in a similar manner to deformation modulus.

Thanks to these excellent laboratory Q-studies, the scene is now set for going into the field, to see fractures and rock joints (not forgetting the 'ever-present' microcracks), in their *in situ* seismic Q environment. First we will look at some near-surface seismic Q, including some quite shallow studies in reservoir-type sediments.

This will be followed by seismic Q at great depth connected with earthquakes and continental-scale studies. Finally we will return to 'medium' petroleum reservoir depths at the end of this chapter on seismic Q, where the potential economic rewards of understanding $Q_{seis}$ are pressing further developments. Understanding petroleum reservoir behaviour, first from more rock physics (laboratory) contributions, then from *in situ* anisotropy effects, (i.e. shear wave splitting), forms much of the material in Chapters 13 to 15.

## 10.5 Attenuation and seismic Q from near-surface measurements

Seismic *reflection* and refraction techniques used to illuminate major features beneath the earth's surface obviously depend on the existence of seismic wavelets. There may not always exist sufficient seismic impedance contrasts between rock boundaries to cause reflection or refraction. But *absorption* is continuous, and significant information can accumulate on the progressing wavelets. Ecevitoglu and Bingol, 1999 pointed out that the absorption information may be crucial as far as the rock's consolidation, porosity, fractures, and fluid contents are concerned. Near surface measurements of seismic Q are however complicated by the presence of near-surface weathering layers, and of course by faulting. The seismic energy will be strongly attenuated, and waveforms may also be distorted.

### 10.5.1 Potential links to rock mass quality parameters in jointed rock

With our progression from laboratory to field scale, *implicit* links between $Q_{seis}$ and rock mass quality Q (Barton et al., 1974, see Appendix A), can apparently be seen, due to the logical results of near-surface, seismic Q that rapidly increase with depth. This also mirrors the way that rock mass deformation modulus, calculated from the rock mass Q-value, also increases with depth, to match *in situ* measurements of this fundamental rock mass parameter. This seems to be because the features of a rock mass that are described by a rock mass characterization method, such as rock mass quality Q, contains elements of the in situ medium deemed potentially important for both the intrinsic and scattering attenuation mechanisms.

When moving to *in situ* scale, joint spacing (captured in RQD) and the number of joint sets ($J_n$) are clearly going to have an influence on scattering losses, just as microcracks and induced rock sample cracking and individual joints are seen to influence scattering and intrinsic losses in laboratory samples, as shown by Wulff et al., 1999, and Pyrak-Nolte et al., 1990.

It is also intuitively reasonable to suspect that the condition of joints – their degree of interlock as determined both by stress level and by roughness ($J_r$), and the presence or absence of mineral coatings or clay fillings ($J_a$) – will have potential influence on (micro) permeability and therefore on potential squirt losses, when there is a variable degree of saturation, as is frequently the case in the near-surface.

Joint characteristics are also expected to influence eventual frictional losses, if magnitudes of *continuum-based shear strain* (actually discrete micro-displacement discontinuities), are of sufficient magnitude. The fact that *seismically determined in situ* dynamic rock joint *compliances*, the inverse of stiffnesses, have immediately recognisable (i.e. expected) magnitudes, and are expressed in the same (but inverted) units as in the rock mechanics of discontinua (e.g. Bandis et al., 1981, 1983, Pyrak Nolte et al., 1990), is further justification for looking also *beyond* microcrack-scale, for the contributions of the jointing of the *rock mass*, to both scattering and intrinsic losses.

In the last section on laboratory testing of the dynamic response of joints under load, the results of Pyrak-Nolte et al., 1987a and 1990, showed *'static'* normal stiffnesses for the most deformable joint (E35, Figure 10.26b) that varied from 5,000 MPa/mm at 5 MPa normal stress, to about 30,000 MPa/mm at 70 MPa normal stress, each in the dry state. These authors' ultrasonic joint measurements, showed for the same sample E35, a *dynamic* normal stiffness varying from 'only' 4,000 to 7,600 MPa/mm (dry), and from 9,500 to 15,000 MPa/mm (saturated), at comparable low normal stresses of 2.9 and 6.0 MPa respectively. At normal stress levels of 70 MPa, the authors showed dynamic normal stiffnesses of 32,000 MPa/mm when dry, and 59,000 MPa/mm when saturated.

Only the latter is higher than the *'static'* stiffness, showing similar relative increases as the $E_{dynamic}$ to $E_{static}$ inequality that reduces strongly with high quality rock joints and rock masses.

Of importance for *in situ* uses of seismic Q, the authors Pyrak-Nolte et al., 1987b noted that *velocities* soon reached the level of the intact rock, when using high frequencies. They showed that this 'seismic closure' could occur at lower stress levels when the joints were less stiff (as for their sample E 35, Figure 10.26b), and at higher stress when the joints were stiffer. Most importantly, even when the effect of a joint on *velocity* was almost 'erased' by stress and high frequencies, the change of amplitude of the transmitted wave, i.e. the spectral amplitude basis for seismic Q-estimation, remained very strong.

Kang and McMechan, 1994 showed *near surface* data from N. Texas where the smallest $Q_p$ value of 36, and the smallest $Q_s$ of 23, were relevant to the highly variable surface weathered zone. According to these authors, very few in situ measurements of scattering in the upper few metres to tens of metres were available at that time.

'The near surface velocity/density structure may be more variable than previously thought: some of this variability may have been hidden in previous measurement that did not explicitly separate intrinsic and scattering effects.' This of course is supported, implicitly, by engineering experiences: see for example Chapter 1 and the rapidly changing (laterally and with shallow depth) refraction seismic velocities of Sjøgren et al., 1979. Such would also imply rapidly changing rock mass qualities, deformation moduli, and by implication, seismic Q, since in the *near-surface*, $V_p \approx 3.5 + \log Q$ km/s, where Q is in this case the rock mass quality of Barton et al. 1974. (See Chapter 5 for rock quality and velocity variation at shallow depth).

A hydraulically conductive, gently dipping fracture zone at SKB's study site at Finnsjön, north of Stockholm, was imaged using the seismic reflection method. Amplitude decay curves as a function of distance, given by Juhlin, 1995, showed that a seismic quality $Q_{seis}$ of 10 fitted the data, assuming an average frequency content of 150 Hz and a P-wave velocity of 5.5 km/s. The value of $Q_{seis} = 10$ was assumed to be relevant to the upper 100 metres of this granodioritic rock. Juhlin, 1995, considered the result to be consistent with higher $Q_{seis}$ values of 30 and 50 at depths of between 200–1100 m in crystalline rocks of comparable character. Again note the similarity to rock mass deformation modulus expectations, when the latter is expressed in GPa.

In relation to the Q-value of *rock quality*, a P-wave velocity of 5.5 km/s at 100 m depth in a crystalline, hard, low porosity rock (Figure 5.36, Part I) suggests a $Q_c$ value of about 40, and when compression strength of say 200 MPa are allowed for, the rock mass quality Q-value would be about 20. This is a very typical rock quality value for a good quality but jointed crystalline bedrock with two to three joint sets. It might have Q-parameters as follows (see Appendix A).

$$Q \approx \frac{100}{6} \times \frac{2}{1} \times \frac{0.66}{1} \approx 20$$

Greater frequency of jointing in a fracture zone would probably reduce this value to 1 or less. This is consistent with independent Q-logging results at SKB's Swedish nuclear waste investigation sites, performed on 4,000 m of core by the writer in 2003.

Shaw et al., 2004, reported near-zero offset VSP investigation of $Q_{seis}$ in a 50 to 600 m deep section of a well through a Faroe Islands Upper and Middle basalt series, typical of other North Atlantic basalt formations. The source used was a 150 cubic inch air gun fired in a pit under 2 m of water. The receiver was a clamped, three-component geophone, with spatial intervals of 10 m. The authors were able to assess the errors in the $Q_{seis}$ assessment, by testing with slightly different receiver separations of 280, 290, 300, 310 and 320 m. Their results, expressed as $Q_{seis}$ versus mid-point depths from 200 to 450 m, showed $Q_{seis}$ increasing rapidly from about 10 to 50 in the upper third of the well, and levelling off at about 60 at greater depths. The effect on seismic Q of lower stress in the upper levels of the basalts is implied in these and other studies. The results are also typical of rock mass deformation modulus variation with depth, where rock quality Q might typically vary from about 2 to 20, based on Q-logging of numerous basalts.

Payne et al., 2005 described a seismic (sparker P-wave) experiment at a shallow borehole test site in N.E. England in variously jointed Cretaceous chalk. Cross-well seismic was performed between three wells, at a frequency band width of 500 to 3,000 Hz. Spectral modelling was performed to provide $Q_{seis}$ estimates for a shallow 30–36 m deep highly jointed zone, with permeability of about 1 darcy ($\approx 10^{-5}$m/s), and for a deeper (36–50 m), less jointed interval, which had an implied permeability close to that of the matrix of about 1 millidarcy, or $10^{-8}$m/s. The respective $Q_p$ values were 20 and 60.

To help assess whether the higher attenuation in the highly jointed zone was mostly caused by scattering rather than by intrinsic mechanisms, the authors used a discrete particle numerical model, as described by Toomey and Bean, 2000. (Although several numerical

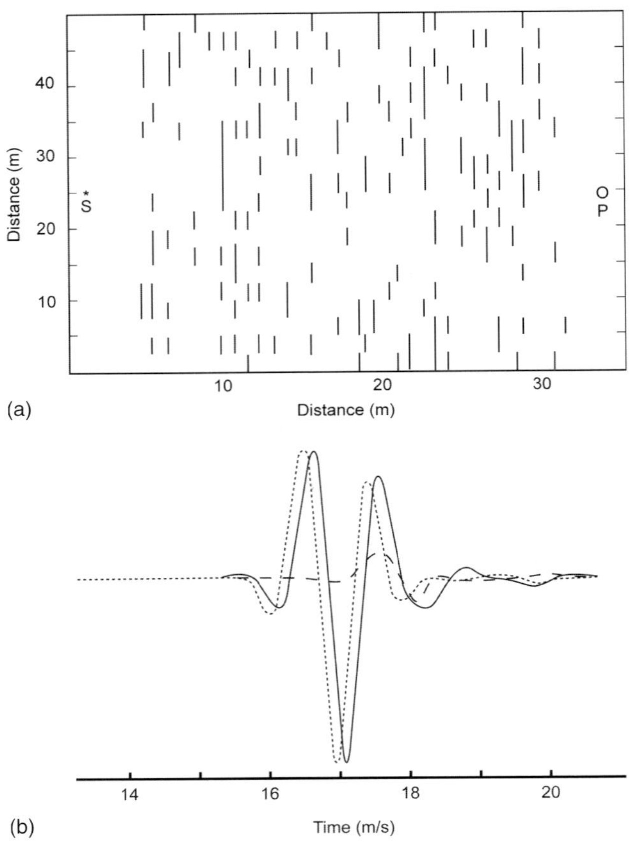


*Figure 10.31* a) Numerical simulation, based on Toomey and Bean, 2000 model, of a random, vertically fractured section, with source and receiver at opposite sides, representing part of the cross-hole experiment in jointed chalk. b) Amplitude of received waves, as affected by fractures with two different compliance assumptions. Reference case without fractures is the solid line. Payne et al., 2005.

models will be described in Chapter 15, we may examine this model here, in the context of seismic Q).

A horizontal compressional wave was applied from one side of a random generation of vertical fractures having compliant bonds, with the receiver at the other side. The compressional wave amplitude at the receiver is shown in Figure 10.31, and indicates three amplitude-time curves.

When the (dynamic) compliance $Z_N$ was as large as $4 \times 10^{-12}$ m/Pa (or dynamic $K_n$ as small as 250 MPa/mm), there was merely a small time shift of the signal in relation to the homogeneous unfractured case (see dotted, parallel curve). The authors indicated that this implied that the apparent seismic Q of this numerical assembly of particles-containing fractures was then 'close to infinite'.

(This was perhaps surprising, as dynamic stiffnesses as low as 250 MPa/mm do not imply either high stress or non-attenuating conditions. Perhaps the discontinuous nature of the modelled 'fracturing' had something to do with this 'close to infinity' seismic Q).

When on the other hand, the fracture compliance was increased by an order of magnitude to $4 \times 10^{-11}$ m/Pa (or when the dynamic $K_n$ was as small as 25 MPa/mm), a marked reduction in wave amplitude was registered (see third, attenuated curve in Figure 10.31b). An apparent seismic Q of 15 was calculated for this attenuating case. Scattering within the highly jointed zone was therefore considered a plausible mechanism, presumably because a dynamic $K_n$ as low as 25 MPa/mm (and its inverted compliance value) were thought to be reasonably representative of the shallow joints or fractures in the chalk.

As will be shown in Chapter 16, such low values of normal stiffness indeed imply very low normal stress, quite consistent with the 'macro-deformation' and comparatively 'static' normal closure testing results for natural, fresh or partly weathered joints reported by Bandis, 1980 and Bandis et al., 1983. 'Static' (extreme low frequency) testing, perhaps at a rate of only $10^{-3}$ Hz, displays strong non-linearity in relation to normal stress level, when loading is from zero to about 60% of the joint wall strength JCS – usually represented by many tens of MPa of normal stress. The so-called 'initial normal stiffness' described by Bandis, has values of the same low magnitude as assumed above for the dynamic modelling.

An interesting example of low, near-surface seismic *Q* structure, with comparison to refraction seismic structure, was given by Ecevitoglu and Bingol, 1999. They introduced a new methodology to rapidly compute and graphically map seismic *Q*, arguing that absorption measurements are tedious, subject to noise, and not common in everyday geophysics in the near-surface.

It appears from a necessarily brief (extended abstract) description that the direct wave and all the refracted waves were each considered as the first breaks.

Figure 10.32 shows an example of their tomography-like seismic Q distribution from the near-surface to 50 m depth. The high Q anomaly of 20, at 45 m depth, was found independently in a conventional seismic refraction interpretation with an upper layer of $V_p$ of 1.97 km/s, and a second layer of 4.26 km/s, resulting in a 45 m depth for the refractor. 'The exact location we have found independently from seismic Q imaging'.

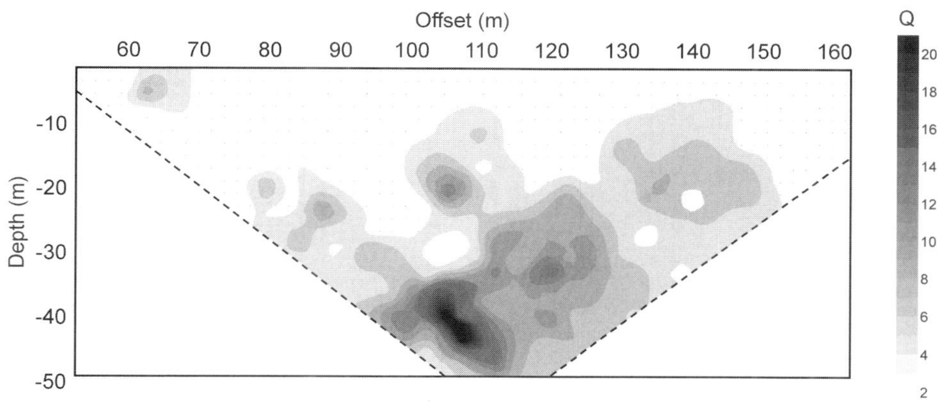


*Figure 10.32* Seismic Q imaging of an anomaly at 45 m depth with correspondence to a $V_p$ of 4.26 km/s from independent refraction seismic imaging of this second layer of higher velocity. Ecevitoglu and Bingol, 1999.

### 10.5.2 Effects of unconsolidated sediments on seismic Q

Extremely low values of $Q_p$ in unconsolidated sediments such as a value of 4 between 60 and 100 m depth in sands and gravels (Gibbs and Roth, 1989), and values between 2 and 6 for the case of artificial, glycerol-saturated, random packs of glass beads and coarse sands (Molyneux and Schmitt, 2000), emphasise the character of these unconsolidated and unlithified low $Q_{seis}$ media.

If energy dissipation is small, the seismic quality $Q_{seis}$ (also called the internal friction or dissipation factor) was previously defined as:

$$\frac{2\pi}{Q_{seis}} = \frac{\Delta E}{E} \tag{10.7}$$

where E is the elastic energy stored at maximum stress and strain and $\Delta E$ is the energy loss per harmonic excitation cycle. $Q_{seis}$ can however apparently be smaller than $2\pi$ (i.e. $\Delta E > E$), but alternative definitions of $Q_{seis}$ seem to be needed if larger dissipation (i.e. exceptionally low $Q_{seis}$) is measured or assumed.

In the case of shallow seismic investigations in sediments, it is likely that $Q_{seis}$ has a frequency-dependent component because near-surface layers of sediment tend to be unconsolidated and may contain fluid. This was verified by Jeng et al., 1999, who measured $Q_s$ values as low as 2 to 5 using different sources of energy, and found these lowest values corresponded to the lowest frequencies used of about 50 Hz. A roughly 5 times higher frequency (250 Hz) resulted in about 6 to 8 times higher $Q_s$-values (16 to 30 approx.). Less attenuation (higher $Q_{seis}$) is observed at higher frequencies because it is less easy to accelerate the pore fluid along the pores (due to inertial forces) than to compress the pore fluid, as in the flat-pores model of Mavko and Nur, 1979.

Jeng et al., 1999 used artificial source and receiver pairs, and a frequency-dependent Q estimation, in contrast to the conventional spectral ratios with constant Q assumption. When examining the triaxial-geophone data with varied (2 m interval) offsets, the frequency-dependent and frequency-independent assumptions reportedly gave 'dramatic variation' of Q.

The authors carried out experiments at three different sites in Taiwan, but concentrated their attention at the Yuan-Lin site in the foothills of central Taiwan, where two different sources were available. The surface of the site had a 2 to 3 m thick layer of alluvium and unconsolidated sediments, overlying a 200 m thick gravel formation. Their data showed $Q_p$ values linearly increasing from between 1 and 3, to between 10 and 16, as frequency was increased from 50 to 300 Hz. There was marked instability, and therefore lack of linearity, at frequencies beyond 300 Hz. The frequency components for the power law $Q = kf^n$ were 1.11 and 0.93 for the P and S waves, respectively.

Their investigation using the conventional frequency-independent assumption for Q, and geophone intervals of 5, 7.5, 10, 12.5, 15 and 17.5 m gave average Q tending to increase from about 10 to 13 over this range of geophone intervals. The modified frequency-dependent approach at the same location, gave Q values varying approximately linearly from about 2 to about 18, as frequency was increased from 57.5 to 575 Hz. The authors conclude that for weathered loose layers $Q_{seis}$ smaller than $2\pi$ is obviously possible, despite the classic formulation of energy loss per harmonic cycle,

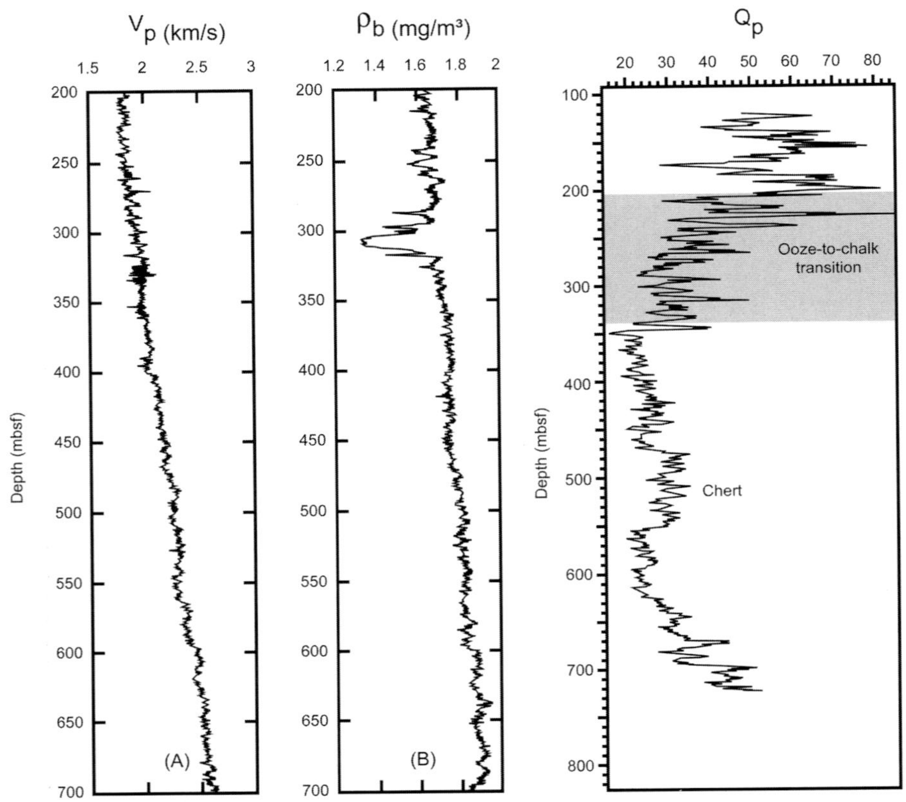


*Figure 10.33* Downhole logs of $V_p$, density and $Q_p$ for ooze and transition to increasing layers of chalk, in sub-ocean studies made during the Ocean Drilling Program, at the Ontong Java Platform, in the western Pacific. Frazer et al., 1997.

compared to the elastic energy stored at maximum stress and strain.

The authors Frazer et al., 1997, working in the Ocean Drilling Program at the sub-ocean Ontong-Java Platform carbonate sections in the western Pacific, emphasised that attenuation measurements made at ultrasonic frequencies in the laboratory, often on disturbed sediment samples, or estimated from seismic experiments over long wave lengths, may reveal little about the geologic/depth evolution of sediment attenuation. Compaction of loose grains, through diagenesis, to sedimentary rocks at depth, can be a fragile environment to sample, especially when the shear-wave velocity is lower than the borehole fluid velocity.

The authors used the Schlumberger long-spaced sonic (LSS) tool which has two sources and two receivers in a special arrangement. Every 6 inches (0.15 m), four microseismograms were recorded, with three source-receiver spacings. The frequencies involved were about 5 to 25 kHz, lying between laboratory and seismic measurements.

The actual $V_p$-density log for one of the holes investigated is shown in Figure 10.33. The velocity increased almost linearly from 1.8 km/s at 200 m below sea level, to 2.6 km/s at 700 mbsl. The reduced density appeared to be at the base of the ooze-to-chalk transition, where more chalk was present. In the transition zone the chalk first appeared in distinct, several centimetre thick layers, separated by ooze layers. The thickness and proportion of the chalk increased towards the base, where there was more calcium carbonate cement.

The seismic quality $Q_p$ for the same hole is also shown in Figure 10.33. In the loose, high-porosity sediments the attenuation was assumed to be mostly due to fluid motion relative to the framework of loosely packed grains. With greater depth of burial, the number of points of contacts and their load increases, and friction was assumed to become a more important mechanism of attenuation. The authors showed porosities as high as 60 to 70% in the ooze, from 60 to 75% in the transition, and thereafter reducing in the chalk from 60 to 45%. Presumably the effects of layering/bedding and perhaps jointing through the thin chalk beds (?), and related fluid flows, contributed to the higher attenuation at the base of the transition, where $Q_p$ was as low as 20. An increase in $Q_p$ was seen where harder chert occurred at greater depth.

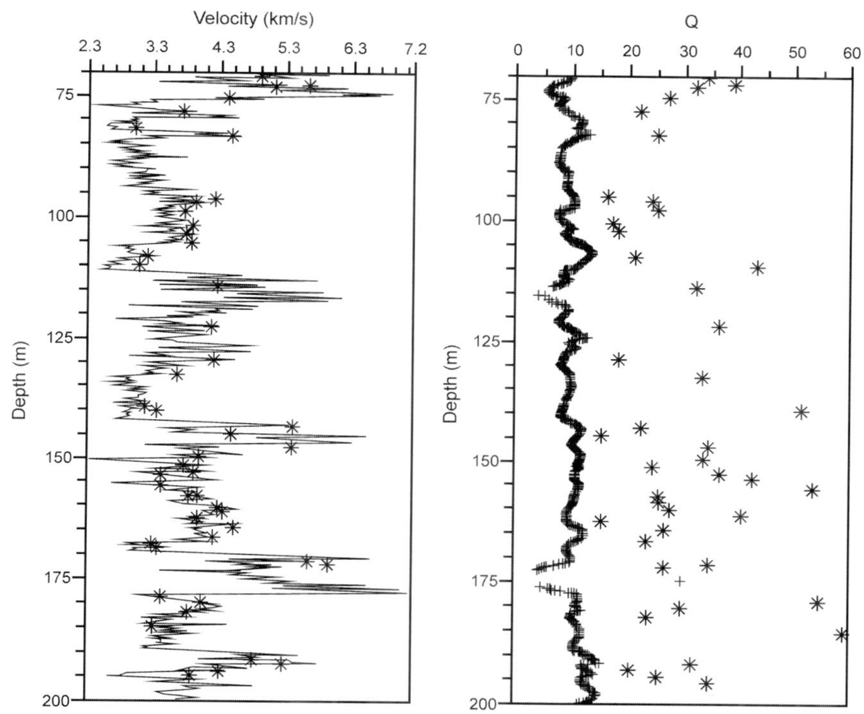


*Figure 10.34* Laboratory tests, shown by stars, and sonic borehole logs of $V_p$ and $Q_p$ showing the marked effect of frequency on $Q_p$. Sams et al., 1997. The sonic logging gave the lowest estimates of seismic Q of all the methods investigated.

### 10.5.3 Influence of frequency variations on attenuation in jointed and bedded rock

When a dynamic load is applied to a rock at low frequencies, the fluid in compliant large aspect ratio cracks will tend to be squeezed into the pores and cracks that are less compliant. As we have seen, the geophysics profession has termed this mechanism 'squirt flow' (e.g. Mavko and Nur, 1975; Palmer and Traviolia, 1980; Jones, 1986; Dvorkin et al., 1995). At higher frequencies inertial effects cause the fluid in the compliant cracks to be less mobile, and there is lower attenuation, making seismic Q higher.

Laboratory tests which offer the flexibility of using different saturating fluids having different viscosities, in fact show that there is a peak attenuation when the product of frequency and viscosity is between 1 and 10 (units of Hz.Pa.s) (e.g. O'Connel and Budiansky, 1977). It appears that cracks or joints with aspect ratios of about $10^{-3}$ to $10^{-4}$ cause most of the attenuation (Jones, 1986).

The assumption of frequency-independent *intrinsic* Q and frequency-dependent *scattering* Q implies that when the total Q for S-waves ($Q_s$) is smaller than for P-waves ($Q_p$), the *intrinsic* Q is dominant; when it is larger ($Q_s > Q_p$), *scattering* Q is dominant over intrinsic Q (Kang and McMechan, 1994). In fact, scattering and intrinsic Q contributions are separable by assuming that they have different frequency dependencies, as we shall see in particular from Chapter 13.

A local (micro) fluid flow mechanism is found to be the only mechanism that can account for widely observed variations of compressional and shear wave attenuation with frequency, both in partially saturated and fully saturated rocks. However, evidence for frequency-dependent attenuation from field experiments is apparently less conclusive: a few cases were reviewed by Sams et al., 1997.

Sams et al., 1997, made a very important contribution in this area of frequency effects, by investigating a sequence of saturated sedimentary rocks (a finely layered sequence of limestones, sandstones, siltstones and mudstones) using four boreholes drilled to 250–280 m depth at the Imperial College test site in NE England. They acquired many data sets at widely different frequencies:

- VSP experiments (30–280 Hz)
- Cross-hole experiments (200–2300 Hz)
- Sonic logging (8–24 kHz)
- Laboratory measurements (0.3–0.9 MHz)

P-wave velocities for core samples at equivalent depths, and *ultrasonic* $Q_p$ estimates measured on core samples (each shown by stars) are compared with the sonic log results in Figure 10.34. The good correspondence in the

Table 10.4 Frequency dependence of $V_p$ and $Q_p$ (Sams et al., 1997).

| Type | Freq. range | Median Vp | 1000/Qp | i.e. Qp |
|---|---|---|---|---|
| Core | 500–900 kHz | 3.95 | 37.0 | 27.0 |
| Sonic | 8–24 kHz | 3.48 | 96.5 | 10.4 |
| VSP | 30–280 Hz | 3.20 | 32.0 | 31.3 |
| Cross-hole | 200–2300 Hz | – | 63.6 | 15.7 |

case of Vp, and the poor correspondence in the case of $Q_p$ at these two different frequencies, is readily seen.

The dependence of median $V_p$ and median $Q_p$ measurements on frequency is shown in Table 10.4.

These contrasting results for the frequency dependence of $V_p$ and attenuation (expressed as 1000/Q), are plotted in Figure 10.35 with curve fitting based on the authors' modelling, assuming squirt flow, following Jones, 1986. This modelling suggests that different sizes of 'soft pore space', i.e. bedding and intra-bedding joints are sensed by the different frequencies, as a single aspect ratio fits lower frequency VSP and cross-hole data, but not necessarily the sonic data.

Sams et al., 1997 estimates of the amount of stiff porosity (i.e. conventional matrix pores) apparently confirmed the results of Mavko and Jizba, 1991, and showed that the soft porosity responsible for squirt-flow attenuation occupied only a fraction of the total pore space.

Sams et al., 1997, found that the dominant aspect ratios causing most attenuation were 8.1 to $8.8 \times 10^{-4}$, even though a much wider distribution of crack (and joint) geometries is obviously present in this and other rock masses. We can estimate as an example that if the mean intra-bed joint apertures were 0.05 mm, the implied lengths of these would be only about 6 cm. Perhaps bed-limited jointing, specifically that under least effective normal stress, could be responsible for most of the attenuation, in view of these relatively 'large' aspect ratios. Sams et al., concluded that the marked frequency dependence demonstrated by their measurements *'points to the amount of information about the rocks that we should be able to obtain from broad-band seismology once we have fully understood the processes that are operating'.*

It is unfortunate that readily obtainable rock quality descriptions such as RQD and the Q-value are not presented together with these important geophysics results. If the physical components of these rock masses were logged in the conventional manner of engineering geologists, possibly it would be easier to understand the *potential* roles of scattering and intrinsic attenuation in these results.

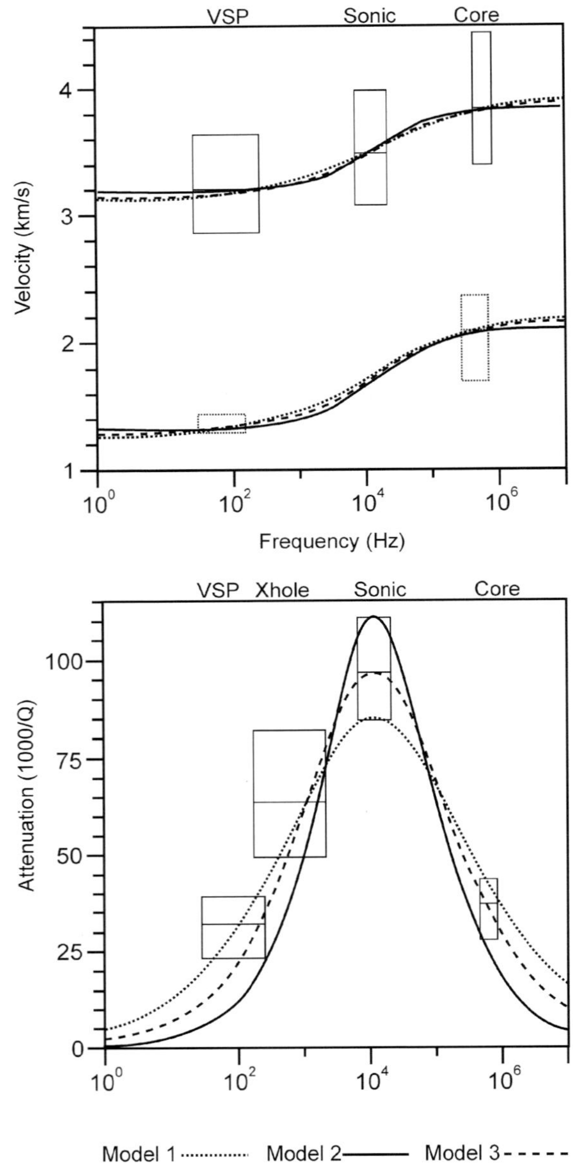


*Figure 10.35* The dependence of P-wave attenuation, $V_p$ and $V_s$ on the frequency of measurement in a finely layered sequence of limestones, sandstones, siltstones and mudstones. Sams et al., 1997. The curves relate to the author's modelling of squirt flow losses, using the model of Jones, 1986. A range of aspect ratios was used to represent the 'soft pores', or assumed cracks, intra-bed jointing, and bedding planes.

Concerning the four sets of Sams et al., 1997, field data, presenting 1/Q versus frequency for VSP, cross-well, sonic and ultrasonic (shown in Figure 10.35), Vogelaar and Smeulders, 2005 recently showed that the levels of attenuation measured in these field experiments in the relatively shallow experimental borehole site, exceeded by far, the theoretical prediction of Biot, 1956a,b, comparing just the viscosity-based damping of the Biot theory. This comparison is shown in Figure 10.36.

In efforts to improve the fit of theoretical approaches to the four sets of attenuation-frequency data presented by Sams et al., 1997, Vogelaar and Smeulders, 2005 modelled a periodically layered porous medium, where the repeating layers 1 and 2 had pore fluids with different properties. They applied Biot's poroelasticity equations, *together* with elements of White et al., 1975, who separated strain due to the fast compressional wave and strain due to the slow compressional wave, the former obeying the wave equation, and the latter the diffusion equation.

They showed that the resulting numerical model solution based on White's local flow model, demonstrated an attenuation rising to 0.05 (Q = 20), more than an order of magnitude higher than the Biot theory, at a frequency of between 20 and 100 Hz. They stated that an extension of the White model to higher frequency, made it capable of predicting the levels of attenuation seen in the field data. The three factors they investigated showed firstly a maximum attenuation at a specific frequency, secondly the maximum attenuation occurred at some specific percentage of gas, and thirdly that increasing the gas fraction caused the attenuation peak to shift towards higher frequencies.

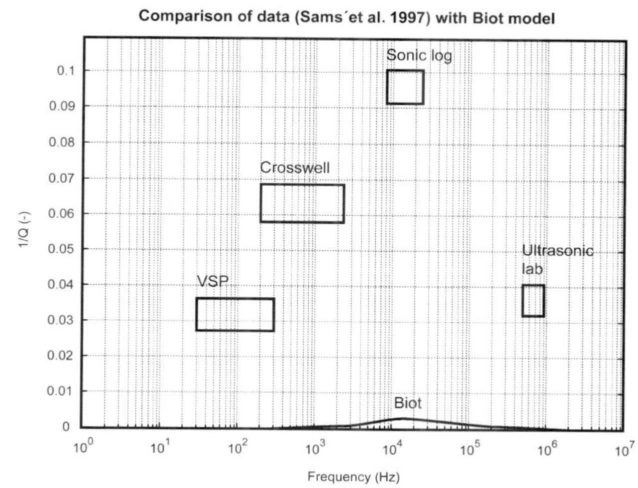


*Figure 10.36* Comparison of Sams et al., 1997 field data of attenuation versus frequency, with modelled data from the viscosity-based damping of the Biot theory. Vogelaar and Smeulders, 2005.

## 10.6 Attenuation in the crust as interpreted from earthquake coda

Since the 1960s, the seismic quality of the crust as interpreted from attenuation of earthquake waves has been the focus of much attention. This research was guided by attempts to find reliable ways of interpreting the precursors of earthquakes. The source, in place of surface-explosives or borehole piezoelectric devices, was the earthquakes themselves, and their after-shocks. We will trace some of the earlier measurements, and progress from surface recordings to some of today's down-hole recording of earthquake sources.

### *10.6.1 Coda $Q_C$ from earthquake sources and its relation to rock quality $Q_C$*

Coda waves are the tail of a seismogram (after the arrival of major wave types such as P, S and surface waves) recorded at a certain distance from an earthquake epicentre. (Aki and Chouet, 1975). Seismic coda waves of local earthquakes appear to be produced by backscattering of waves from numerous randomly distributed heterogeneities. The longer the waves travel, the greater the variety of heterogeneities they encounter. The later portions of a seismogram may therefore be the result of some kind of averaging of many samples of the heterogeneities of the intervening crust (Aki, 1969).

The spectral contents of the early part of a local earthquake seismogram depend strongly on the travel distance and on the nature of the wave path to the recording station. The coda excitation also depends on the local geology of the station site, and can be 5 to 8 times larger on sediment than on granite (Aki, 1969). Most coda measurements are made in the 20 to 200 seconds time window. As we shall see later, there are obvious advantages of in-borehole seismometer location, at kilometre depths, to help minimize 'site effects'.

The attenuation related to the rate of decay of the coda is termed $Q_c^{-1}$ in the geophysics literature, so seismic $Q_c$ has by chance, an identical symbol to rock quality $Q_c$ (Barton, 1995), that is used to describe rock mass quality Q (from Barton et al., 1974). This original, widely used term for rock mass quality, was at this time also normalized by uniaxial compressive strengths greater or lesser than 100 MPa, to the form $Q_c = Q \times \sigma_c/100$. This was done to improve fit to velocity and modulus of deformation data. As we have noticed, there appear to be some numeric similarities between seismic $Q_p$ at shallow depth (i.e. <1 km), and the estimated deformation modulus ($E_{mass}$, or symbol M), when expressed in GPa.

It may be of interest to observe that the *inverse* of rock mass quality ($Q_c^{-1}$) is roughly proportional to the rock mass permeability or Lugeon value for central ranges of rock quality, when without the complication of clay-sealing of joints that is common near the surface. (The Lugeon value L ≈ $1/Q_c$, where 1 Lugeon ≈ $10^{-7}$ m/s, Barton, 1999, 2002). By implication, less attenuation (high coda $Q_c$) would correspond not only to high rock qualities $Q_c$, but also to lower permeability, and higher deformation modulus.

### 10.6.2 Frequency dependence of coda $Q_C$ due to depth effects

The coda $Q_c$ was found to increase with frequency, though according to Aki and Chouet, 1975, this did not necessarily mean that the coda $Q_c$ of crustal material was frequency dependent. The above frequency effect was thought to be due to the dependence of coda $Q_c$ on depth, since waves were scattered from different parts of the earth's crust. As the primary waves from an earthquake spread out, they leave behind a pool of scattered energy which quickly becomes homogeneous inside the pool because of high diffusivity.

Since a large volume surrounding the earthquake source is 'sampled', the seismic coda $Q_c$ has been considered as a potential measure of the assumed changing rock properties, due to accumulation of stress and strain in the hypocentral zone (e.g. Chen and Long, 2000).

Temporal variations of coda $Q_c$ before or following earthquakes have been reported in some cases, while unfortunately in other cases, no changes have been noted. Several of these case records will be reviewed in this section of Chapter 10. The apparent frequency dependence of coda $Q_c$ waves can be explained if the coda waves at 1 Hz are primarily composed of surface waves scattered from shallow heterogeneities, while coda waves at 20 Hz are primarily back scattering body waves from deeper heterogeneities in the high $Q_{seis}$ lithosphere. At the two sites investigated by Aki and Chouet (western California and Japan), the coda $Q_c$ ranged from 50 to 200 at 1 Hz in the shallowest crust (resembling possible deformation moduli in GPa), to about 1000 to 2000 at 20 Hz in the deeper crust.

The latter, by chance, resembles the rock mass $Q_c$ value itself, assuming that jointing is effectively closed or almost absent. On the other hand there is no possibility of ever acquiring a reliable measure of deformation modulus at extreme depth, without compressing ultra-small, unjointed laboratory samples, as done by tectonophysicists in the past. Perhaps, unknown to the writer, the E moduli at 20 km depth or 300 (to 500) MPa effective (to total) confining stress, could reach much higher values than the most typical 50 to 75 GPa seen at an order of magnitude smaller depths, for the case of hard crystalline rocks. As we shall see in Chapter 16, the differential stress ($\sigma_1 - \sigma_3$) tolerated by small rock samples can be increased by a factor of 5 to 10, by similar magnitudes of confining stress to the above. (See Figures 16.57a and b).

Aki's and Chouet's observations were interpreted as showing the *combined effect* of variation of coda $Q_c$ with depth and the frequency-dependent composition of coda waves. The average coda $Q_c$ over the depth range 0 to 12 km was 300 in the case of the Stone Canyon site, some 15 km to the north of the San Andreas Fault, where average magnitude 1 earthquakes were analysed by these authors.

Figure 10.37 shows the regional variations of the coda attenuation $Q_c^{-1}$ in the range of frequencies from 1 to 20 Hz derived from the California Stone Canyon earthquake events, and from the Japanese Tsukuba Oishiyama earthquake events. Attenuation $Q_c^{-1}$ reduces and coda Qc increases with increasing frequency. At Tsukuba, where earthquakes are deeper, attenuation $Q_c^{-1}$ was lower and coda $Q_c$ therefore higher, especially at the highest frequency of 24 Hz. ($Q_c^{-1}$ ≈ 0.001, $Q_c$ ≈ 1000).

Carpenter and Sanford, 1985, used spectra from 130 digitally recorded micro-earthquakes (M = −0.9 to 0.3) to compute the apparent seismic Q for upper crustal rocks near Socorro, in the Rio Grande Rift, New Mexico. Most of the seismic wave attenuation due to intrinsic absorption and scattering was computed over the frequency range of 3 to 30 Hz. Their apparent seismic Q values were found to increase with event distance, for the eight recording stations used in the study. This increase was modelled with a varying thickness, low seismic Q, low-velocity layer, lying above a relatively high seismic Q, high-velocity half space.

As illustrated in Figure 10.38, the waves from the more distant earthquakes would have a greater fraction of their total ray path in the deeper rocks, therefore indicating that seismic Q was greater at depth. The seismic $Q_p$ and $Q_s$ values were found to be less than 50 directly beneath the sites at 0.3 to 2 km. This again

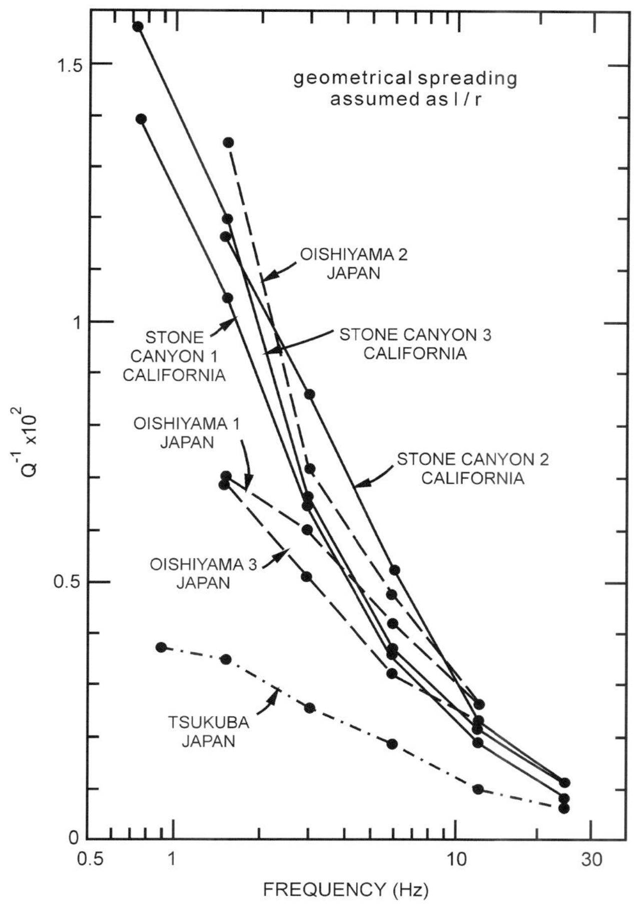


*Figure 10.37* Regional variations of $Q_c^{-1}$ for the frequency range 1 to 20 Hz, derived from analysis of earthquake coda waves. Aki and Chouet, 1975. (A potential seismic $Q_c$ scale on the right side of the figure would range from a minimum of 67 at about 1 Hz, through 100, 200 and 1000 at 25 Hz).

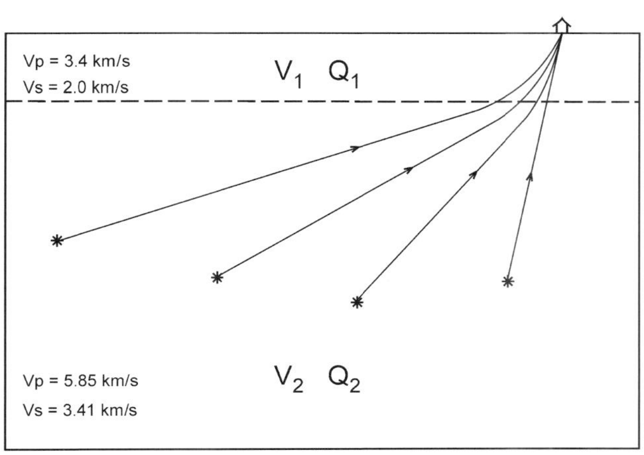


*Figure 10.38* The two-layer model used to interpret event-distance effects on seismic Q. Carpenter and Sandford, 1985.

resembles rock mass deformation moduli magnitudes, expressed in GPa.

A stronger seismic Q gradient near the surface was accounted for by lumping all of the low seismic Q material into this one layer. Significantly, the near-surface S-wave $Q_s$ values for stations resting on apparently competent Pre-Cambrian and Paleozoic rock were quite low (<25) and generally less than for stations resting on Tertiary tuffs. The authors interpreted this as being due to a greater incidence of open and water-filled fractures in the otherwise more competent rock. Such would also be consistent with lower rock quality Q-values, due in particular to higher, near-surface values of SRF, causing lower Q-values and lower rock mass deformation moduli. (See Appendix A for Q-parameter ratings for describing different rock mass conditions).

$Q_p/Q_s$ ratios ranged from 0.34 to 1.39, and a decrease of this ratio was generally measured with increasing distance. Carpenter and Sanford took this to imply varying degrees of saturation in the upper crustal rocks. Near the surface, fully saturated rocks have $Q_p > Q_s$, while at depth, partially saturated or dry rocks may have $Q_s > Q_p$. (Winkler and Nur, 1982).

In a gas-producing region of Uzbekistan in 1.4 km thick Tertiary sediments, Clouser and Langston, 1991, determined values of $Q_p$ (10 to 70) and $Q_s$ (10 to 25) based on a spectral ratio method of analysing aftershocks from the 1984 Gazil earthquake. There was conjecture that these 10 to 20 km deep thrust-faulting events could have been induced by *gas extraction.*

Clouser and Langston, 1991, investigated various $Q_p$-$Q_s$ relations, comparing some theoretical straight line relations with some in situ measurements. Figure 10.39 shows the stratigraphic section through this 1.4 km of sediments, and increasing P and S wave velocities down to the basement. The $Q_p$ and $Q_s$ relations are shown in Figure 10.40, together with black dots representing other authors' data for comparable sedimentary rock and rock sequences. The various initials against the $Q_p$-$Q_s$ curves are from their six monitoring stations SP-1, GSN, OFT, GAZ, TSV and K31. Intuition would now suggest that these lower $Q_p$ and $Q_s$ magnitudes, varying as they do from about 5 to 70, may represent the lower frequency, nearer-surface sampling of the jointed crust.

Clouser and Langston, 1991, evaluated the following two equations:

$$Q_s = \frac{4}{3}\left[\frac{V_s}{V_p}\right]^2 Q_p \qquad (10.8)$$

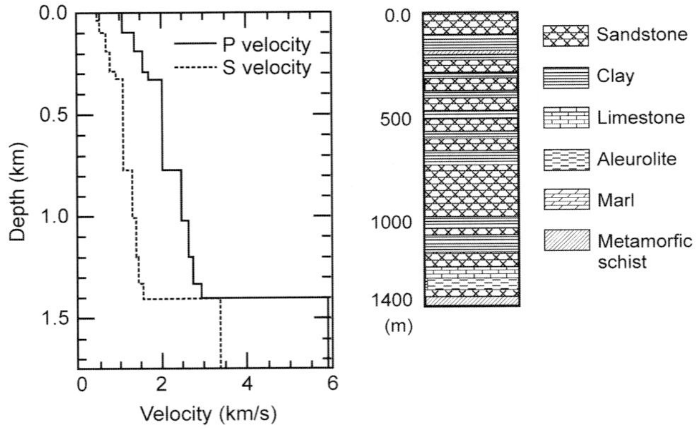


*Figure 10.39* Stratigraphic section through the 1.4 km Tertiary basin investigated by Clouser and Langston, 1991.

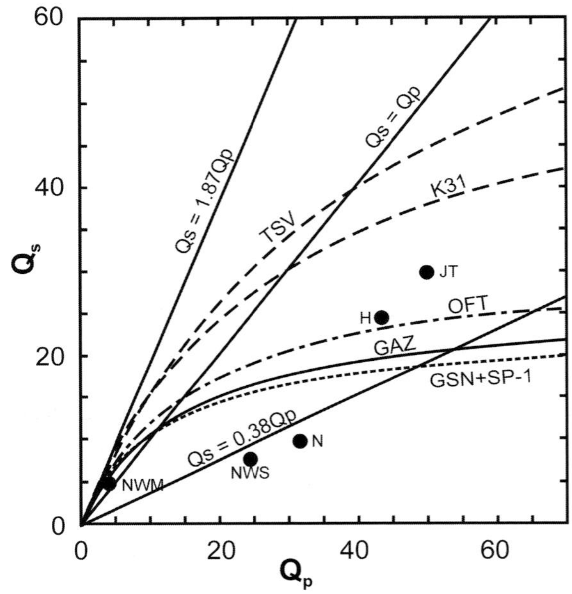


*Figure 10.40* $Q_p$-$Q_s$ relations determined for each station, from spectral ratio slopes and travel times. Black circles are laboratory and in situ data from other authors. Clouser and Langston, 1991.

$$Q_s = \frac{V_p}{V_s} \times Q_p \tag{10.9}$$

Using their average $V_s/V_p$ and $V_p/V_s$ ratios of 0.534 and 1.87 respectively, the above equations gave the lower and upper limit curves shown in Figure 10.40, namely $Q_s = 0.38\ Q_p$ and $Q_s = 1.87\ Q_p$. Their measurements using spectral ratios, and accounting for differences in basin thickness between each station, showed $Q_s$ varying from 10 to 70.

### 10.6.3 Temporal changes of coda $Q_c$ prior to earthquakes

Initially encouraging indications of a temporal change in coda $Q_c$ before two large Chinese earthquakes were reported by Jin and Aki, 1986. This was based on measurements of coda decay rate for small local earthquakes in the general area that is 'preparing' for a major earthquake. The authors referred to the significant difference of duration of coda $Q_c$ from local earthquakes, which may last several minutes in a stable area like Norway, but which die out quickly in seismically active places like California. (Presumably, significant differences in attenuation, seismic Q and the rock quality Q-value would also be registered in such contrasting regions).

During the three-year period preceding the Tangsham earthquake in China (M = 7.8, 1976), the coda $Q_c$ was about 3 times *lower* than it was before or after this period, i.e. attenuation was greater in the three years before the event. Apparently a comparable change occurred at the time of the Haicheng earthquake, also in China. The authors emphasised that *first appearances* suggested a change of focal depth. Tsujiura, 1978, reportedly found that coda $Q_c$ for Japanese earthquakes were higher when focal depths were at 100–160 km, than when at 40–80 km. Roecker et al., 1982, found coda $Q_c$ to be four times higher when sampling 400 km depth than when sampling less than 100 km depth.

The authors concluded that a coda $Q_c$ change by a factor of 2 to 3 would imply a change in focal depth of about 100 km. They also discussed the possibility of changes in predominant frequency affecting the coda $Q_c$:

$$Q_c = Q_o f^m \tag{10.10}$$

where $0 < m < 1$. For the coda Qc to change by a factor of 2 to 3 the frequency must change significantly also. However, their analyses showed mean variations of frequency of only 20%. They therefore concluded that the observed coda Qc change was due to a 'change in the property of the earth medium', namely the opening of cracks or dilation. For the case of the referenced Tangsham earthquake, the actual low coda Qc value was 71 (in the period 1973 to 1976) and 200 (in the period 1969 to 1972). The P-wave velocity was reportedly also anomalously low in the period 1973–1975 just preceding the earthquake. However, Jin and Aki, 1986, preferred using coda Qc, since it covered an entire ellipsoidal region determined by the lapse time, whereas monitoring of the Vp/Vs anomaly requires that locations of source and receiver make the wave path go directly through the anomalous source region.

Peng et al., 1987, used several thousand seismograms of small earthquakes in the Mammouth Lakes area in the USA to measure values of Qc from the decay of the earthquake coda. They added a certain degree of sophistication to the interpretation, by observing two opposite trends at different distances from the source:

a) in the region near the main epicentre, measurements of coda attenuation ($Q_c^{-1}$) were higher for earthquakes that occurred *after* the main shock, than coda attenuation $Q_c^{-1}$ of those that occurred prior to the main shock (i.e. $Q_c$ was lower after the main shock, due to *greater* attenuation).
b) further away from the main epicentre, measurements of coda attenuation $Q_c^{-1}$ were lower after the main shock than before it (i.e. $Q_c$ was higher after the main shock, due to *reduced* attenuation.

Could this perhaps be due to the reduced shear stress at distance, and more 'damage' closer to the source? Such would also give lower rock qualities: quite logically closer to the source following the main shock.

Peng et al., 1987, reviewed numerous cases of temporal changes of coda attenuation ($Q_c^{-1}$) before and after large earthquakes. For brevity we will make a list without individual references, and refer the reader to the above authors. To aid interpretation with respect to coda quality $Q_c$ as opposed to attenuation $Q_c^{-1}$, we will list changes to coda $Q_c$ in the list that follows. The changes are therefore inverted compared to the Peng et al., 1987 list:

1. 30% reduction in coda $Q_c$ *before* Hawaii earthquake, 1975 (M = 7.2)
2. 20% reduction in coda $Q_c$ *before* 3 large Kuril-Kamchatka earthquakes, (M = 8.0)
3. 30% reduction in coda $Q_c$ (at 6 Hz) *before* Petatlan earthquake (M = 7.6)
4. anomalous, low coda $Q_c$ *before* E. Vamanashi earthquake, 1983 (M = 6.0)
5. 300% reduction in coda $Q_c$ for 3 years *before* Tangsham earthquake, 1976 (M = 7.8)
6. 200% reduction in coda $Q_c$ for period *before* Haicheng earthquake, 1975 (M = 7.3)

Following this list an *increase* in coda $Q_c$ is mentioned for a 2 to 3 year period before the Misasa earthquake in Japan, 1983 (M = 6.2), and the authors also record the possibility that coda $Q_c$ changes *without an obvious relation* to a major earthquake.

### 10.6.4 Possible separation of attenuation into scattering and intrinsic mechanisms

Based on their own Mammouth Lakes data and on other case records, Peng et al., 1987 proposed that coda $Q_c$ reduced within the aftershock zone but increased outside this zone, when the main shock occurred. The observed temporal changes to coda $Q_c$ suggested that *heterogeneities* were responsible for the *scattering* component.

Presumably such features as lithological boundaries, clay-filled discontinuities and branch faults, could cause scattering while the intrinsic component of attenuation might be microcrack and joint related (i.e. thin cracks that were bearing high fluid pressures, that were therefore sensitive to small stress changes. Peng et al., 1987 emphasised that separation of the intrinsic and scattering components of coda $Q_c$ introduced a severe non-uniqueness in determining these parameters. However, they suggest that probably a small number of strong scatterers dominated the coda in a seismically active zone.

Concerns about the relative magnitudes of intrinsic attenuation ($Q^{-1}$ intrinsic) and scattering ($Q^{-1}$ scattering) has led to methods for separating these effects by integrating the S-wave energy for successive time windows, as a function of earthquake hypocentral distance. The method is described by Hoshiba et al., 1991. Figure 10.23 shows how it was applied to Central California, Long Valley and Hawaiian earthquakes, giving a separation of $Q_s^{-1}$ scattering and $Q_s^{-1}$ intrinsic, whose sum is equal to $Q_s^{-1}$ total. Interestingly, when adding two larger $Q_s$ intrinsic and $Q_s$ scattering components, the sum $Q_s$ total is *less* than either of the above, since attenuation is being accumulated, not $Q_{seis}$.

In general, for frequencies $\leqslant$6.0 Hz, scattering $Q_s^{-1}$ was greater than intrinsic $Q_s^{-1}$, whereas above 6.0 Hz the opposite applied. In all three regions, scattering $Q_s^{-1}$ was strongly frequency dependent, decreasing proportionally to frequency increase, or even faster. Intrinsic $Q_s^{-1}$ was considerably less frequency dependent. A concurrent suggestion of a depth-dependent intrinsic $Q_s^{-1}$, which increases with depth, apparently can reduce the discrepancy between theoretical predictions and observations (Zeng et al., 1991).

Sato and Fehler, 1998, (Seismic wave propagation and scattering in the Heterogeneous Earth, Springer Verlag & AIP Press), who collected regional seismic data from around the world, also showed the separation of the scattering and intrinsic attenuation. Clearly, and in almost all cases, the lower frequencies give the greatest attenuation

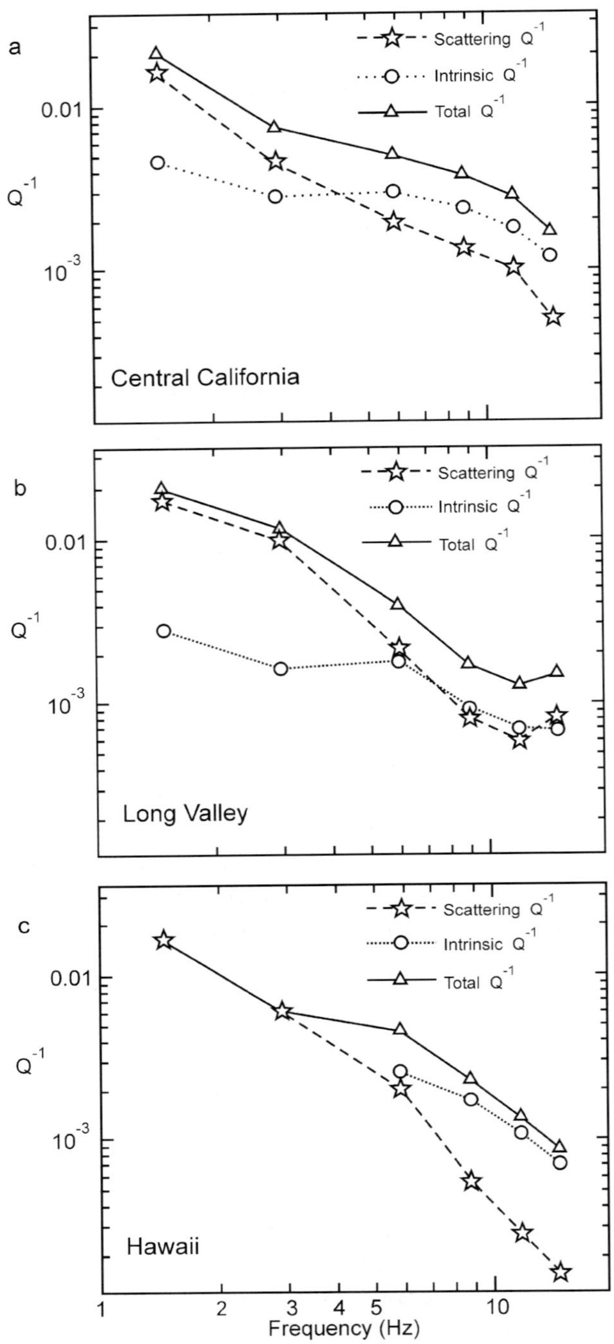


*Figure 10.41* Plots showing total $Q_s^{-1}$, together with scattering $Q_s^{-1}$ and intrinsic $Q_s^{-1}$, for earthquakes that occurred between 1987 and 1990. Focal depths are between 5 and 20 km and magnitudes are between 1.5 and 3.6. Mayeda et al., 1992.

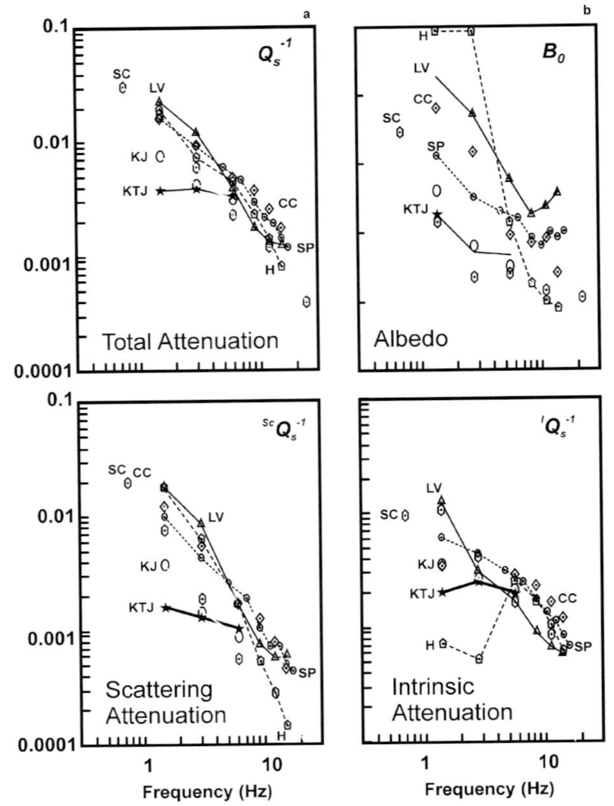


*Figure 10.42* Regional seismic data assembled by Sato and Fehler, 1998, showing the relative magnitudes of the intrinsic and scattering attenuation and its frequency dependence, even between 1 and 10 Hz. KTJ = Kanto-Tokai, Japan; KJ = Kanto, Japan; LV = Long Valley, California; H = Hawaii; SC = Southern California; SP = Southern Spain; and CC = Central California.

both from scattering and from intrinsic mechanisms. This seems to suggest the extreme importance of the structural geology (joint sets, faulting), and of the conducting properties of these larger scale, nearer the surface features. Figure 10.42 shows the relative magnitudes of these total, scattering and intrinsic attenuations. (The so-called 'Albedo' ($B_o$) is defined as the ratio $Q_{sc}^{-1}/Q_{total}^{-1}$).

Mayeda et al., 1991, had suggested that the more complex regions of Hawaii and Long Valley required models which incorporated heterogeneously distributed, non-isotropic scatterers in a layered medium with depth-dependent *intrinsic* $Q_s^{-1}$. They suggested that their results for scattering $Q_s^{-1}$ indicated a length scale of heterogeneity at least comparable to the wavelength for the lowest frequencies studied, of the order of a few kilometres, presumably implying fault-size features, or perhaps lithological contacts.

### 10.6.5 Changed coda Q during seismic events

As we have seen, and perhaps confusingly for earthquake precursor analysis, some investigators have recorded decreases in the seismic coda Qc value *prior* to major earthquakes, while others have recorded decreases *following* the occurrence of a major earthquake.

Beroza et al., 1995, using nearly identical (doublet) earthquakes in a pre-seismic, co-seismic and post-seismic search for temporal changes in coda $Q_c$, found a *nearly stable attenuation throughout* the Loma Prieta, California sequence. The main shock in October 1989 ($M = 6.9$) nevertheless reportedly resulted in an almost total stress drop, and effects on coda Q had clearly been expected. In earlier studies, increases in coda $Q_c$ by as much as 50% had been cited as precursors to large earthquakes. Unfortunately, changes in coda $Q_c$ by as much as 50% had also been reported even in the absence of a large seismic event.

To address possible effects of source size, geometric spreading and earthquake mechanisms and location, most researchers estimate coda $Q_c$ from *late coda*, the part of the seismic wave arriving after twice the S-wave arrival time. The relatively long paths taken by waves in the late coda are assumed to sample a large volume of the crust with a variety of takeoff angles from the source.

Using earthquake doublets, Beroza et al., 1995, employed a technique for measuring coda $Q_c$ that was assumed to be insensitive to geometric spreading, location and source mechanism, because these factors are common to the two (doublet) events. Furthermore, both the early and late coda can be used; but the former is usually sensitive to the focal mechanism.

The stability of coda $Q_c$ throughout the Loma Prieta sequence was in sharp contrast to other studies (e.g. Peng et al., 1987), who reported larger pre-cursory changes in coda $Q_c$ for many (but not all) of the earthquakes that were analysed, as we saw earlier.

Previously it had been assumed that the coda samples a large volume. However, Chen and Long (2000) showed that temporal variations in the seismic coda $Q_c$ could in fact be explained by hypocentre migration over small distances compared to the radius assumed to be sampled by the coda. These authors used data from an earthquake swarm in Georgia, USA (Norris Lake Community).

They first noted the apparent temporal variation in coda $Q_c$ during a five month period. Figure 10.43 shows that reductions occurred in the three orthogonal directions (vertical, NS, EW) *during the period of maximum seismic activity.* Over 4000 shallow (<1.2 km) earthquakes were recorded altogether. However for this analysis, 108 earthquakes of similar size were chosen that were uniformly distributed during the period of the swarm activity. About 2 weeks before the peak of seismic activity, the coda $Q_c$ began to reduce.

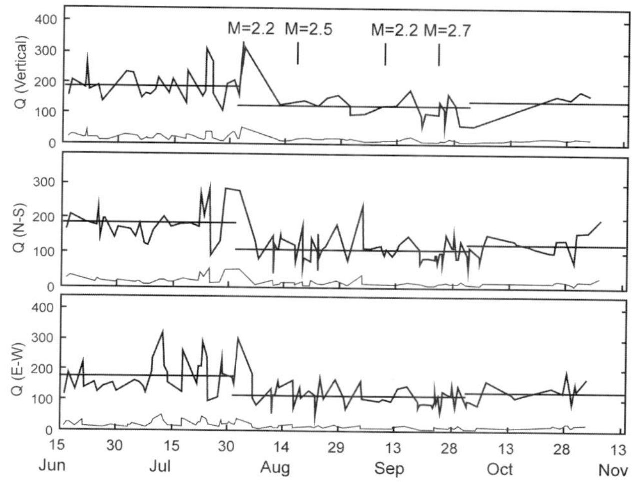


*Figure 10.43* Temporal variation of coda $Q_c$ for a swarm of earthquakes with peak activity in August and September. Chen and Long, 2000.

Chen and Long, 2000, found that their coda $Q_c$ values were related to the locations of the earthquakes. Early events were mainly located in high coda $Q_c$ areas, while later earthquakes occurred preferentially in low coda $Q_c$ regions, in other words there was a shifting of hypocentres. (Rock quality or attenuation levels could presumably change also in a given location, due to fracturing events, but hypocenter migration was considered the main effect). The coda attenuation $Q_c^{-1}$ increased from about 0.005 ($Q_c = 200$) to 0.010 ($Q_c = 100$) from the early period to after the peak activity, due to this assumed shift of hypocentres.

A short distance (<1 km) migration of hypocentres generated this halving of magnitude of $Q_c$, possibly related to migration from 'geologically uniform' gneiss to an area with 'mafic dikes and more complicated topographic relief'. These dikes were interpreted as strong, and inhomogeneously distributed scatterers. The authors concluded that the normal assumption that coda scatterers (i.e. large discontinuities) are uniformly distributed may be one reason for earlier misinterpretations of temporal variations in coda $Q_c$. An assumed homogeneity of velocity structure would be another.

The strong spatial variation of coda $Q_c$ (and by implication of the rock quality $Q_c$-value also) does not eliminate the possibility that intrinsic attenuation and scattering may also be affected by more subtle changes in rock mass properties and the effects of stress change. However, in rock engineering, we would certainly expect that spatial variations of rock mass quality $Q_c$ would tend to be greater than those caused by stress change, especially in view of the moderate stress drops that occur as a result of earthquakes (i.e. often just fractions of, or a few MPa).

Following *rock quality* $Q_c$ values a little further, one can see from Figure 5.36 (Part I, Chapter 5), that higher rock quality $Q_c$ values imply higher P-wave velocities and higher deformation moduli. Such areas would likely be under highest stress, and have least permeability, so a spatial migration of hypocentres to lower seismic *coda* $Q_c$ areas, with lower *rock mass quality* $Q_c$ and lower moduli and higher permeabilities, is entirely logical.

Perhaps the 'complicating' factor of frequency dependence of seismic Q reviewed earlier, is another way of recognising a scale effect. Certainly the evaluation of rock mass quality $Q_c$ could also be considered scale-dependent, since the inclusion of larger volumes of the rock mass (including faults) will inevitably adversely affect all the six Q parameters, resulting in lower overall rock mass quality and the strong likelihood of greater attenuation.

Hellweg et al., 1995, also used the Parkfield Dense Seismograph Array (of the US Geological Survey) to estimate coda $Q_c$ from up to 42 recordings for each earthquake that occurred. Coda $Q_c$ was determined in two frequency bands (4 to 8 Hz, and 8 to 16 Hz), from a tight cluster of 26 seismic events between December 1989 and January 1994. Despite magnitude (M 4.7 and M 4.6) events in 1992 and 1993, they found that $Q_c$ had not systematically changed. Figure 10.44 however, shows various interesting trends of the data, including a certain anisotropy regarding azimuth (graph c), and a clear distance dependence (graph d) which may be related, as observed earlier, with deeper sampling of the earth's crust as distance increases, and therefore higher coda $Q_c$ values.

Concerning the coda $Q_c$ magnitude and time period, Hellweg et al., 1995, suggest that there had been no systematic change in the coda $Q_c$. One may however observe that if allowed to plot a least-squares (or other) best fit to the time/date data in graph (e), a certain reduction of $Q_c$ with time would be observed. However this would apparently be invalid as the events cannot actually be compared with each other directly, as they are determined from different length windows regarding each seismogram. The authors proposed that $Q_c$ should always be measured from the same length window starting at the same lapse time regardless of the source location.

As a contrast to the coda $Q_c$ obtained from seismically active areas, Kvamme and Havskov, 1989, determined the coda $Q_c$ in Southern Norway, finding values at 10 Hz frequency to vary from 780 to 1530, for source-to-station distances varying from 15 to 300 km.

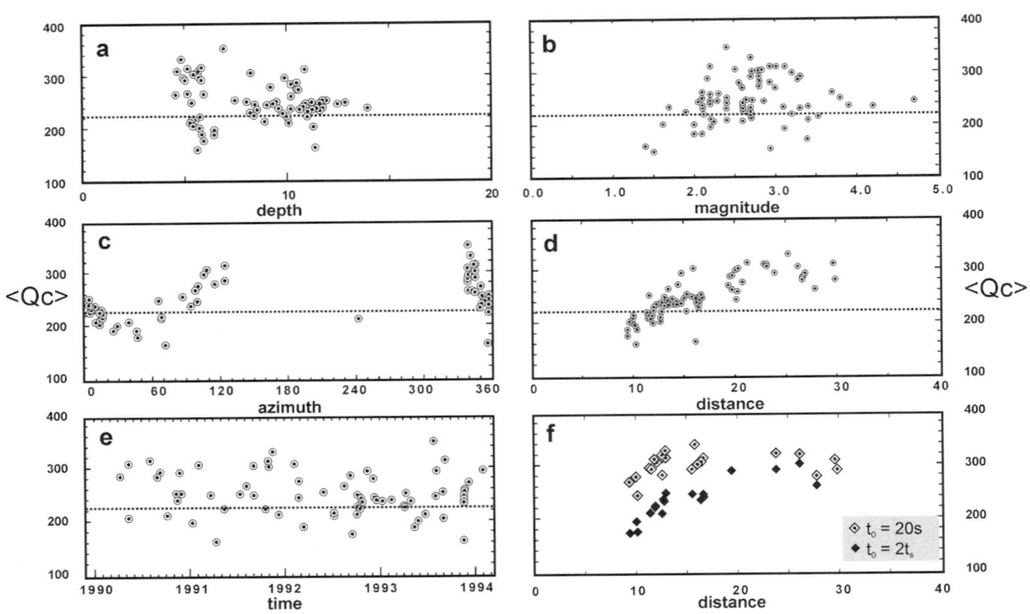


*Figure 10.44* Coda Qc from the 4–8 Hz band: a) coda $Q_c$ dependence on depth, b) coda $Q_c$ dependence of earthquake magnitude, c) coda $Q_c$ dependence on azimuth, d) coda $Q_c$ dependence on epicentral distance (km assumed), and e) coda $Q_c$ as a function of time. Dotted lines show the average of all coda $Q_c$ values, f) coda $Q_c$ calculated for selected events with epicentres less than 30 km. Filled diamonds are measured from a 30 s window which starts at $2t_s$. Hollow diamonds are measured from a 30 s window which starts at a lapse time of 20 s. Hellweg et al., 1995.

A certain increase in coda $Q_c$ with window length was interpreted as increased $Q_c$ with depth, as in other studies reviewed here. They considered the Norwegian measurements of coda $Q_c$ to be similar to values found in another shield area (Canada), but observed stronger frequency dependence as possible evidence of stronger scatterers in Southern Norway. Some of the paths shown in Figure 10.45, certainly cross some major regional zones of weakness (i.e. the Oslo fjord), and regional faulting.

They used window lengths of 5, 20, 30 and 40 s and observed variations of coda $Q_c$ with frequency:

$$Q_c = Q_o f^m \tag{10.11}$$

and found ($m$) to be 1.15 for most of the Norwegian data. Comparison of their own and other frequency dependent coda $Q_c$ are reproduced in Figure 10.46.

Hiramatsu et al., 2000, reported temporal changes in coda $Q_c^{-1}$ in the Tamba region of Japan, to the northeast of the main rupture zone of the 1995 M 7.2 Hyogoken Nanbu earthquake. This region has the densest distribution of Quarternary active faults in Japan, with very high seismic activity for several decades.

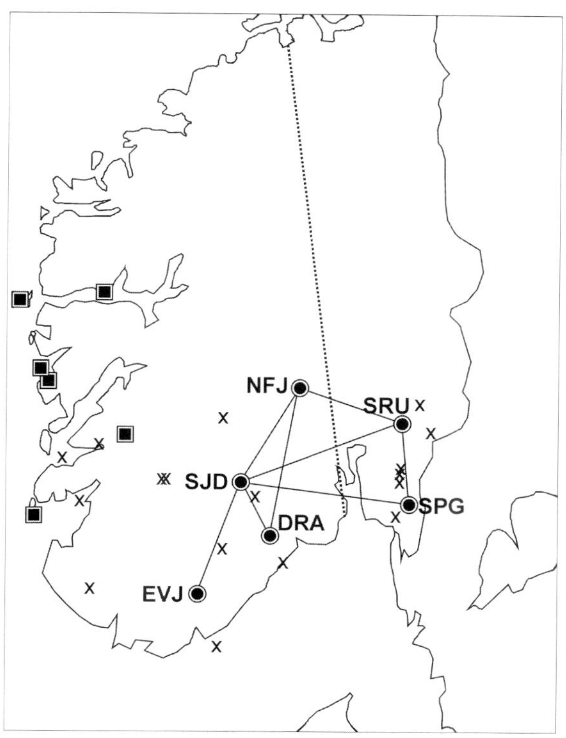


*Figure 10.45* Location of Southern Norway profiles for Kvamme and Havskov, 1989, spectral ratio analyses. The dotted line direction gave seismic $Q_p = 575$ from measurements by Kanestrøm and Haugland, 1971.

The authors used local earthquakes recorded from 1987 through 1996, and concentrated their analyses on high-quality data from the depth range 45 to 15 km. Values of coda $Q_c^{-1}$ were averaged over three recording stations for each earthquake. They divided the data into two periods: 8 years before and 2 years after the major (M = 7.2) 1995 earthquake.

The average value of coda $Q_c^{-1}$ increased after the major earthquake, especially for the lower frequency bands between 1.5 and 4 Hz, as illustrated in Figure 10.47, and in Table 10.5. The authors emphasised that no change in focal mechanism was reported, citing the fact that changes in epicentres, focal depths, or focal mechanism can cause false temporal changes in coda $Q_c^{-1}$. Even in a small 1 × 1 × 1 km volume, the value of $Q_c^{-1}$ increased after the main shock at frequencies below 5 Hz, suggesting that changed epicentres were not the cause of the increased attenuation (and reduced seismic quality). The average depth remained in the 9 to 10 km range.

The authors considered that a numerically estimated change of shear stress at 10 km depth of only 0.02 MPa due to the Hyogo-ken Nanbu earthquake, was the cause of the increased attenuation. The sensitivity of $Q_c^{-1}$ to shear stress change was estimated to be ≈10 $(MPa)^{-1}$ at around 3 and 4 Hz frequency, which the

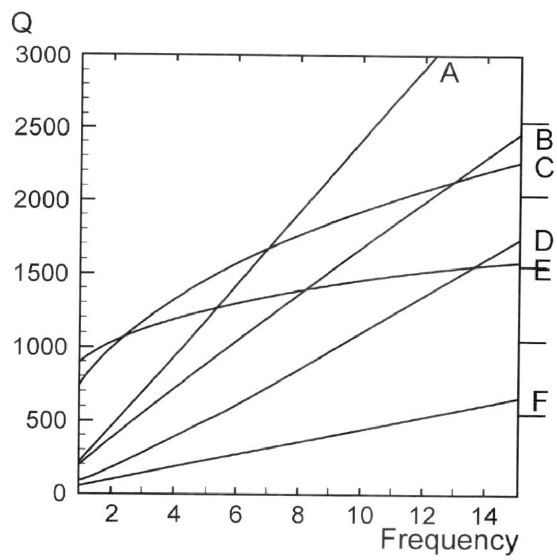


*Figure 10.46* Coda $Q_c$ as a function of frequency for different regions. A–Aleutian Islands, m = 1.05, B–Carolina, USA, m = 0.94, C–New England, m = 0.40, D–Southern Norway, m = 1.15, E–Canadian shield, m = 0.20, F–former Montenegro region, Yugoslavia, m = 1.00. (See Kvamme and Havskov, 1989 for references).

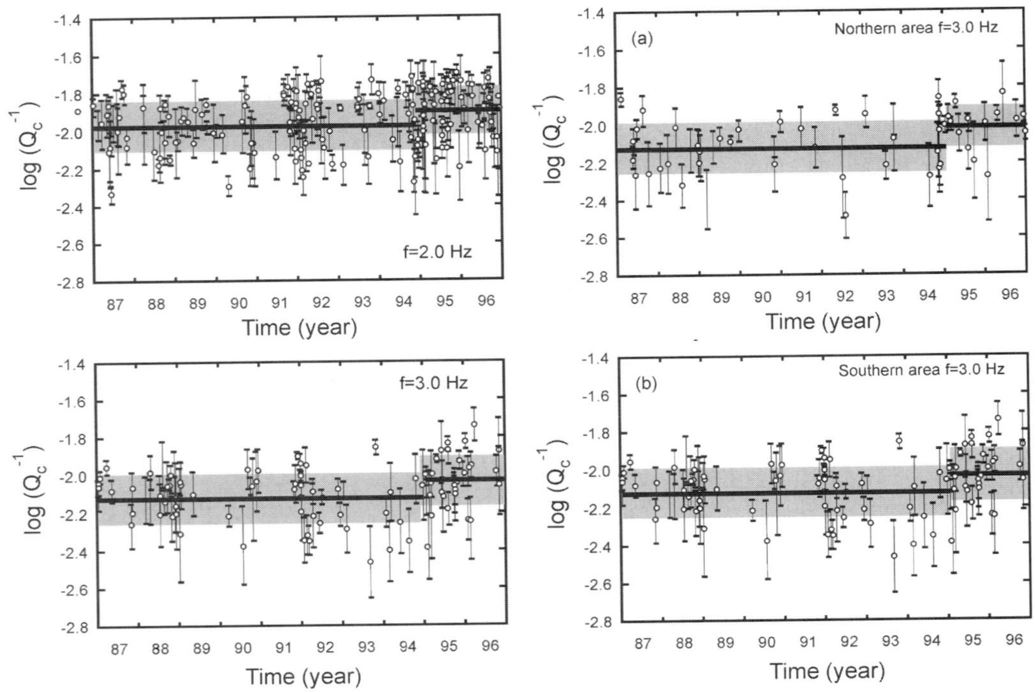


*Figure 10.47* Distinct increases in attenuation (coda $Q_c^{-1}$) following the January 1995 Hyogo-ken Nanbu (M = 7.2) earthquake. Hiramatsu et al., 2000.

Table 10.5 Increase of average $Q_c^{-1}$ and decrease of average $Q_c$ following the 1995 Hyogo-ken Nanbu earthquake. (Hiramatsu et al., 2000).

| Frequency (Hz) | $Q_c^{-1}$ before | $Q_c^{-1}$ after | $Q_c$ before | $Q_c$ after |
|---|---|---|---|---|
| 1.5 | 0.012 | 0.015 | 81.3 | 67.6 |
| 2.0 | 0.011 | 0.013 | 91.2 | 77.6 |
| 3.0 | 0.0076 | 0.0093 | 131.8 | 107.2 |
| 4.0 | 0.0054 | 0.0062 | 186.2 | 162.2 |

authors suggested was much larger than the stress sensitivity of seismic velocity. However they also referred to the fault or fracture dimensions of micro-earthquakes (M = 3) as about 400 m, consistent with the characteristic length of scatterers of 300–600 m estimated from the wavelengths of 3 to 4 Hz which had the greatest influence on $Q_c^{-1}$ increases.

They found in addition, that the frequency dependence of coda $Q^{-1}$ varied with time. Using the proportional to power of frequency expression:

$$Q_c^{-1} = Q_o^{-1} f^{-n} \tag{10.12}$$

where $Q_o^{-1}$ is the $Q_c^{-1}$ value at 1 Hz and n is a constant, they used the least squares method to determine values of ($n$) for the activity periods I to V (shown in Figure 10.48). All ten frequency bands (from 1.5 to 24 Hz) were analysed. As clearly shown in this figure, there was a clear reduction in the n-value in the years leading up to the event, followed by a marked increase in the n-value, especially after the major event. As suggested by Kvamme and Havskov, 1989, (and others), a greater density of scatterers could be the logical cause of this greater sensitivity to frequency, following the major earthquake.

### 10.6.6 Attenuation of damage due to acceleration

Mandal et al., 2001 interpreted 110 aftershocks following the $M_w$ 6.4 Chamoli earthquake in the Garhwal Himalaya as propagating up-dip along a thrust plane from 20 to 2 km depth, the main shock having occurred at 15 km depth. The region had five earthquakes exceeding magnitude 6, and twelve exceeding magnitude 5 in just the 20th century, which was presumed to have caused a high level of shallow crustal heterogeneity. They analysed 48 of the local earthquakes with magnitudes varying from 2.5 to 4.8, with recordings at nine digital stations, with three-component seismometers, covering an area with a 150 km radius. Sub-surface recording is not mentioned.

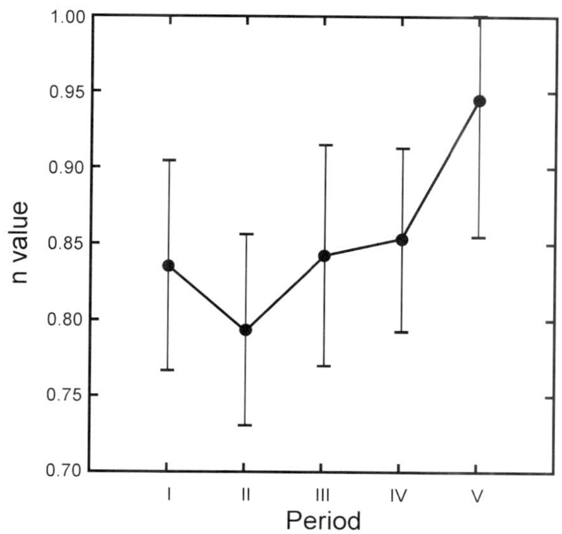


*Figure 10.48* Frequency dependence of $Q_c^{-1}$ increases over the same 10-year period. Hiramatsu et al., 2000. The sharp rise after the event in period IV was referred to in their figure.

The estimated Qo value ($Q_c$ at a frequency of 1 Hz) at the various stations was on average only 30 at 1 to 3 Hz, indicating a strongly attenuating crust. Higher $Q_c$ at higher frequencies were interpreted as propagation of back-scattered body waves through deeper parts of the lithosphere where less heterogeneities were expected. South of the main central thrust belt is a less rigid, slightly metamorphosed sedimentary wedge, while north of the thrust are more rigid highly metamorphosed crystalline rocks. The frequency dependence of the Garhwal Himalaya coda $Q_c$ also gave a quite 'steep' curve in relation to Figure 10.46 data from other seismically active regions. $Q_c = 30 f^{1.2}$ was typical.

The author compared the Garhwal Himalaya $Q_o$ of 30, and its frequency component (n) of 1.21, with the Indian Peninsula $Q_o$ of 550 and n = 0.84. From the above high attenuation, the authors, in 2001, infer that the acceleration (from a hypothesised 7.4 magnitude earthquake at Chamoli, with 1 g at the source), would decay to 50% at 20 km distance, and to 7% at 100 km, suggesting less significant damage beyond 100 km if such should occur.

In a later paper, Mandal et al., 2004 report on the low coda $Q_c$ interpreted from the devastating, intraplate $M_w$ 7.7 earthquake that occurred in Gujarat in 2001, named the Ghuj earthquake. On this occasion the distribution of aftershocks defined a 45° dipping zone of some 40 × 60 km in area. The presence of high seismicity, in a zone V (potential M8 earthquake zone), despite the intraplate location of this Kutch area, is evidenced by the two 7.7 magnitude quakes of the last 200 years. The authors suggested, as at Garhwal, that these were the reasons for a high level of shallow crustal heterogeneity.

The coda $Q_c$ interpreted from 200 local earthquakes of magnitudes varying from 3.0 to 4.6, in an area of 140 km radius, suggested $Q_o = 102\ f^{0.98}$. On this occasion, based on the attenuation curve estimated for $Q_o = 102$, they estimated a ground acceleration at 240 km distance (from an assumed 1 g source) of 0.13 g, almost agreeing with a 0.11 g accelerograph recording at similar distance. To be read in conjunction with this locally attenuating, seismically active crust, is the very variable velocity-depth model for the uppermost 6 km of the region given by Mandal et al.:

0.9 km 2.2 km/s
0.2 km 4.8 km/s (basalt)
2.6 km 3.5 km/s (Jurassic sediments)
0.1 km 4.8 km/s
2.2 km 5.9 km/s

### 10.6.7 Do microcracks or tectonic structure cause attenuation

When considering the mechanisms of attenuation of the crust from a great earthquake, the decay of wave amplitude with distance can clearly take many forms: conversion of elastic energy into heat, scattering caused by heterogeneities such as changes of rock types, changes of velocities, sub-surface topography, and cracks and faults. The authors Mandal et al., 2004 also cite 'sliding along grain boundaries', but in the opinion of the writer this seems insignificant in relation to the, admittedly surface-magnified, ground deformation, with widespread uplift, deep cracks, liquefaction, ejection of water and sand, even at 275 km distance, and collapse of 50 high-rise buildings in a city (Ahmedabad) 240 km distant from the epicentre. There was damage even at 1,600 km distance on the SE coast of India – probably due to the high seismic Q of the intervening crustal plate.

It seems intuitively *impossible* that micro-friction along microcracks, joints and faults, as an attenuation mechanism does not exist in the crust. An earlier opinion from the 1980s, based mostly on the basis of high frequency, sub-micro-strain oscillation tests on intact rods of sandstone, and continuum concepts of strain, deduced that friction was not a source of attenuation in the earth. Some geophysicists hold this viewpoint even today.

Due to the higher stiffness of grain boundaries and microcracks – they have 'high' aspect ratios – it would also seem that the ubiquitous rock joint, and its probable several sets, could be a likely candidate for microfriction attenuation combined with squirt flow losses. Attenuation caused by friction is certainly easy to envisage in the case of an earthquake induced, 8.5 m rupture at 23 km depth, reducing to about 1 m near the surface, where there was uplift of the southern side of a major fault. Neighbouring rock masses resisting such motion are clearly absorbing frictional energy, suffering microdisplacement discontinuities in the process.

If joint and fault compliances that fit attenuation data somehow 'acquire' realistic magnitudes and units of $Pa^{-1}$ m (or the more familiar MPa/mm of stiffness) during even minor man-made seismic or sonic exploration, how can friction-caused attenuation, of however small magnitude, actually be avoided in the case of natural seismic events? Such 'displacement-discontinuity' events in a jointed rock mass are surely a part of the intrinsic attenuation. That scattering occurs from these features as well, seems hardly justification for excluding compliance effects from intrinsic mechanisms, as implied in some interesting analyses of fault-related attenuation.

From another seismically active region, the authors Akinci et al., 2004, present highly attenuated data from the North Anatolian Fault Zone in Turkey, and compare this with a sediment influenced region of Southern Germany. The NAFZ is an intra-continental transform fault boundary between the Eurasian Plate in the north, and the Anatolian block in the south. This major strike-slip fault extends through many segments, for about 1400 km, from a triple plate junction in eastern Turkey to the Aegean in the west.

Figure 10.49 and 10.50 show the results of the authors P-wave $Q_p$ analyses for the western and eastern portions of the NAFZ, with $Q_p$ versus distance data, and $Q_p$ versus seismic event magnitude. Since there is little sediment beneath the stations in either region, the low $Q_p$ results possibly apply to crystalline basement, a lot of it of Triassic age, although 'site effects' from a more attenuating near-surface rock mass beneath the recording stations, should perhaps not be neglected here. Most of the seismic Q results lie between about 4 and 40, suggesting, on the basis of an intuitive $Q_{seis} - Q_{rock\ quality}$ relation, that fracturing/jointing and faulting must be extensive in this tectonically disturbed region.

The range of data for $Q_p$ shown in Figure 10.49, fortuitously or for scientific reasons, *exactly* matches the range of expected rock mass deformation moduli for the following conditions:

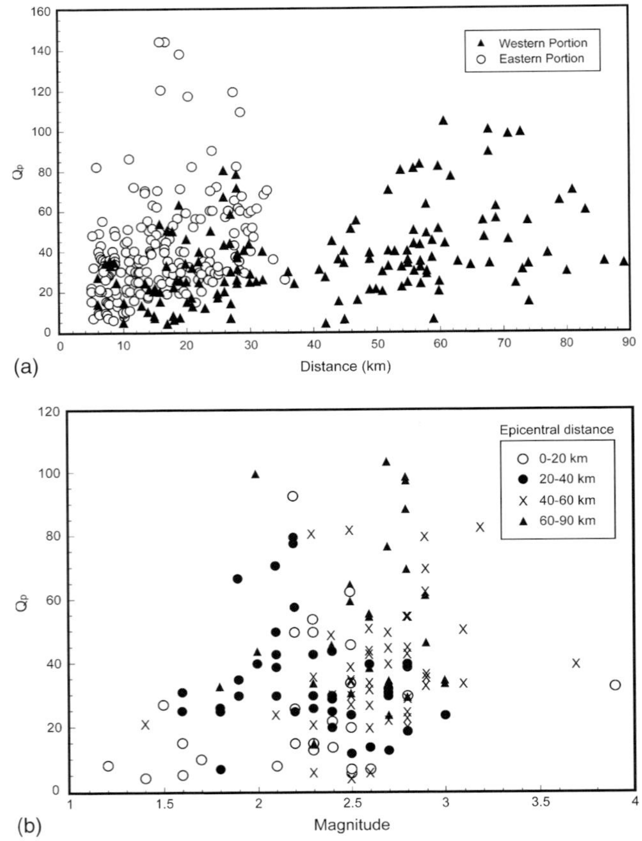


*Figure 10.49* $Q_p$ interpreted from seismic events, in terms of a) distance and b) magnitude, in the western and eastern portions of the North Anatolian Fault Zone. The NAFZ stretches 1400 km from eastern Turkey to the Aegean. Lack of sediments at all the receiver stations suggested to the authors that the low $Q_p$ applied to crystalline basement rocks of Triassic age. Akinci et al., 2004. Sub-site attenuation also seems a possible source of the lowest values. Significant levels of jointing/fracturing/faulting must be assumed in view of the low values of $Q_p$.

1. Near-surface (e.g. immediately below station): rock with a compression strength more than 10 MPa and a rock mass Q-value in excess of 1 (heavily jointed), giving rock mass $Q_c$ = 0.1. Figure 10.54, and 13.60 in Chapter 13 show $E_{mass}$ = 5 GPa (as per minimum seismic $Q_p$ in Figure 10.49a).
2. At 1 km depth, rock with a compressive strength of 100 to 300 MPa, and a rock mass Q-value of 100 (jointed but massive) to 1000 (almost without joints – or completely closed joints) giving Qc from

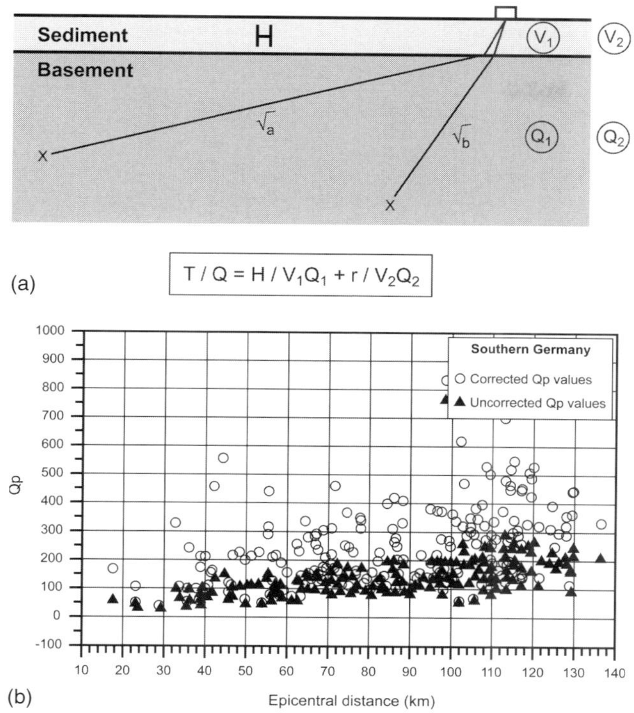


*Figure 10.50* a) Separation of sediment and basement $Q_p$ values by proportioning of the respective velocities and distances, following Hough et al., 1988. b) Effect of correction on the corrected $Q_p$ values for the Southern German seismic data. Akinci et al., 2004.

100 to 3000. Figure 10.54, and 13.60 in Chapter 13 show $E_{mass}$ from approx. 50 to 140 GPa (according to empirical equations: specifically 46 to 144 GPa).

3. Note the great body of seismic $Q_p$ data between 10 and 40, suggesting domination of near-surface, sub-station, heavily jointed and sometimes faulted rock mass. Refer to 'minor fault' curve in Figure 10.54, at depth from 50 to 1000 m: Qp data is matched by moduli of 10 to 40 GPa.

By way of comparison of a seismically active region with one that was less so, the authors also presented their $Q_p$ results for a region of southern Germany. As indicated in Figure 10.50, they made a correction for the low Q sedimentary cover of some 2 km thickness, which had a low range of $Q_p$ from 6 to 10, thereby revealing the corrected $Q_p$ for the 8 to 10 km of basement of between 100 and 500. The fact that the uncorrected 2 km deep data match the above, while the 8 to 10 km of basement exceeds the above $Q_p$ – $E_{mass}$ (potential) link, can perhaps be ascribed to the need for an extreme extrapolation of deformation moduli to an order of magnitude greater depth than ever directly measured. Linked $Q_p$-$E_{mass}$ values in a range of 100–400 (also GPa) would appear reasonable, but this is pure speculation of course.

### 10.6.8 Down-the-well seismometers to minimise site effects

Knowledge of attenuation magnitudes in the upper few kilometres of the earth's crust is clearly an essential ingredient in modern seismic hazard analysis. Earlier studies using only surface seismometers were limited by relatively high noise levels and by the strong attenuation at shallow depth, preventing high frequency signals from being recorded.

The installation of borehole seismometers in more recent years has greatly improved knowledge of the near-surface attenuation of some individual sites. Abercrombie, 1998 pointed out that seismograms recorded at the earth's surface are contaminated by both seismic and man-made noise. The frequency range of observed signals tends to be limited to below a few tens of Hertz at most surface sites. This makes it difficult to link observations of attenuation in the real earth with the much smaller scale and much higher frequency laboratory studies. The ideal situation for recovering uncontaminated measurements is installation of wide-bandwidth seismograms down deep boreholes, where there is low background noise. The problem of strong attenuation in the near-surface is thereby avoided.

As part of the Californian earthquake prediction programme, a region of the San Andreas fault near Parkfield, California was equipped with an array of seismic instrumentation in boreholes averaging about 250 m depth, with a more extensive downhole array both at 1.5 km and almost 3 km in the Cajon Pass deep well, some 1.5 km NE of the fault. Deep instruments reportedly failed soon after installation in the 5150 ft. deep Varian well (Malin et al., 1987). In this area, quite different basement rocks are juxtaposed by the fault, with higher velocities to the SW than to the NE.

Deployment of closely spaced seismometers has reportedly shown extensive variation in recorded amplitude, frequency content, and coda duration over short distances, meaning that the 'site effect' (the near-surface, near-receiver rock), can have a strong effect on earthquake recordings. Amplification in the low velocity, low density near surface, scattering, and resonance within shallow layers, plus attenuation of high frequency energy, also play a role in earthquake damage.

Seismic Q (i.e. $Q_p$ unless $Q_s$ is specified) is now known to be as low as 10 in the upper 100 m of many sites, in varying rock types. Attenuation at shallow depths in fact appears to exhibit little dependence on rock type. This, to a degree, is also basically true for the rock mass quality Q. According to reviews by Abercrombie, 1998, and Abercrombie (2000), the instrumented deeper boreholes show seismic Q increasing with depth to about 100 between 1 and 2 km, and reaching about 1000 at greater depths. Adams and Abercrombie, 1998, found seismic Q ≈ 1000 below 2.5 km depth at Cajon Pass. Studies using both direct and coda waves down-hole-recorded at over 2 km depth, have shown seismic Q to be high (≈1000) at seismogenic depths in California. Hough et al., 1998, found seismic Q ≥ 1000 at 5 km depth at Anza, in California.

A useful comparison of the *relative magnitudes* of attenuation in the Varion and Cajon Pass wells given by Abercrombie, 1998, is reproduced in Figure 10.51. Despite the difference in rock types, the seismic Q profiles are essentially similar. However at greater depths, the thicker and lower velocity Franciscan rocks to the NE of the San Andreas fault at Parkfield, appear to give lower $Q_p$ and $Q_s$ than the higher velocity Mesozoic crystalline basement rocks at Cajon Pass. Abercrombie's analyses of the spectral ratios at different recording depths, having linear slopes on a log-linear plot, suggested that the *near-surface* seismic Q was almost *independent* of frequency, in the frequency range of about 2 to 100 Hz.

A velocity model crossing the fault at right angles is shown in Table 10.6, based on Abercrombie, 2000. $Q_p$ and $Q_s$ estimates down the Varion well on the NE side of the fault are given in Table 10.7.

Earthquake sources reportedly lie within the steep velocity gradient separating the higher velocity SW side of the fault from the lower velocity NE side. This is consistent with micro-seismic and rock burst observations reported in mining and deep tunnelling projects referred to in Chapter 7. Greatest AE activity occurs where velocity gradients are highest, as in such parts of deep gold mines, according to interpretation of seismic AE arrays.

Abercrombie, 1997 calculated that an earthquake recorded at the Cajon Pass wellhead, with a hypocentral distance of 15 km, would suffer 90% attenuation in the upper 3 km, 80% in the upper 1.5 km, and 50% in the upper 300 m, at frequencies of a few Hertz and above. Equivalent calculations using the data in Figure 10.51, for the same earthquake hypocentral distance at the Varian well, suggested that less than 50% of the attenuation would occur in the upper 1 km: the relatively lower and deeper seismic Q magnitudes at this well constituting a form of 'site effect', with most attenuation occurring at greater depth than 1 km.

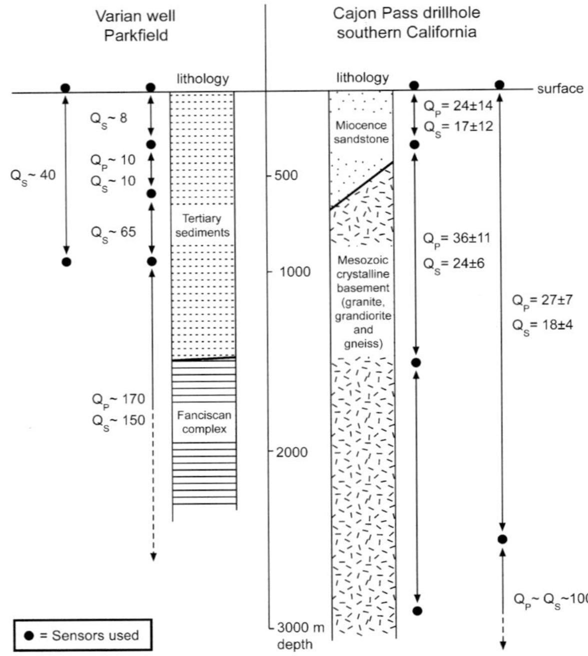


*Figure 10.51* A comparison of attenuation-depth profiles for $Q_p$ and $Q_s$ in the Varian well, Parkfield, and in the Cajon Pass well, S. California. Both sites show very low $Q_{seis}$ near the surface, increasing with depth. As observed elsewhere in this chapter, similarity to depth-modulus trends for jointed rock masses is striking. Abercrombie, 1998.

Table 10.6 A velocity and attenuation model for either side of the San Andreas Fault (SAF), with average seismic Q inversion results for the upper 5 km, assuming a $Q_p$ value of 1000 below 5 km (Ambercrombie, 2000).

| Depth | $V_p$ (SW) | SAF | $V_p$ (NE) |
|---|---|---|---|
| 0–2 km | 4.7 | SAF | 4.0 |
| 2–5 km | 5.9 | SAF | 5.15 |
| 5–10 km | 6.1 | SAF | 5.35 |
| >10 km | 6.5 | SAF | 6.0 |
| | $Q_p$ (SW) | | $Q_p$ (NE) |
| 0–5 km | 199 | 49 | 99 |
| 5–12 km | 1000 | – | 1000 |
| | $Q_s$ (SW) | | $Q_s$ (NE) |
| –5 km | 236 | 78 | 84 |
| 5–12 km | 1000 | – | 1000 |

[SAF = San Andreas Fault: $Q_p \approx 50$, $Q_s \approx 80$ (top 5 km)].

Table 10.7 $Q_p$ and $Q_s$ estimates at various depth intervals in the Varian (VAR) well from Ambercrombie (2000) and Jongmans and Malin, 1995.

| Depth interval (m) | $Q_s$ | $Q_p$ |
|---|---|---|
| 0–298 | 7–10 | 20 |
| 298–938 | 31 | 30 |
| 572–938 | 53–94 | 55 |
| 0–938 | 18, 33–45 | 33 |

Analysis of seven earthquakes recorded at 2.5 km depth in the Cajon Pass well, reported by Learey and Abercrombie, 1994, found a weak increase in $Q_s$ with frequency (≈500 at 10 Hz, increasing to ≈1200 at 100 Hz). They found that *intrinsic attenuation* was the dominant mechanism, because of the ratio $Q_{scatter} > 10\ Q_{intrinsic}$, at all observed frequencies.

(Relevant here, perhaps, is a contribution to the question of frequency-dependence by Van Der Baan, 2002. He used wave localization theory to show that 'constant' Q might be due to apparent attenuation due to scattering losses, if the earth displayed fractal characteristics over a certain range of scales, thereby creating an absorption band. The author drew attention to the usual constant Q at low frequencies, and some form of positive power law dependence at higher frequencies. Constant Q for example between periods of 1 hour and 10 seconds ($0.3 \times 10^{-4}$ to 0.1 Hz), is followed by an increase for f > 1 Hz. The author suggested that it may be very difficult to distinguish between intrinsic and scattering attenuation using only frequency-dependent Q measurements. He also claimed that, in most cases, the common assumption that the scattering and intrinsic effects could be separated, by allowing only for frequency-dependent *scattering*, could be invalid. His numerical modelling indicated that in the case of non-perfect fractals, a constant Q only occurred within a given frequency band, and that Q became proportional to $f^{\alpha}$ for higher frequencies, as in fact observed in most cases.)

Adams and Abercrombie, 1998, analysed more than 100 earthquakes recorded *at a range of depths* in the Cajon Pass borehole. They employed a multiple time lapse method to determine the relative contributions of intrinsic and scattering attenuation concerning $Q_s$ and its frequency dependence. They confirmed only a weak frequency dependence ($Q_s \approx 800$ at 10 Hz, increasing to ≈1500 at 100 Hz, similar to the above). These authors also found intrinsic attenuation to be the dominant mechanism, compared with scattering, at seismogenic depths.

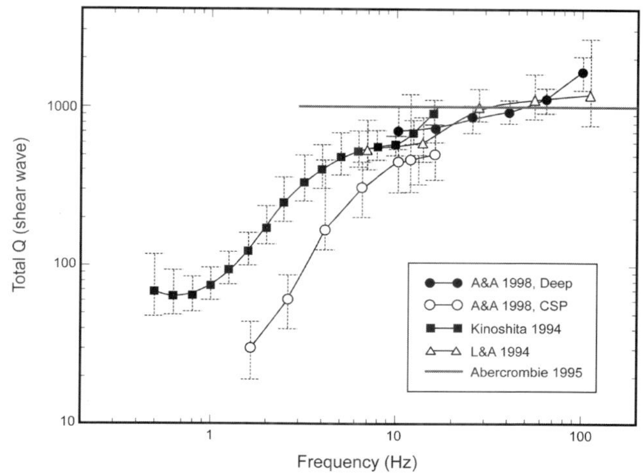


*Figure 10.52* A compilation of $Q_s$ estimates and their frequency dependence between 1 and 10 Hz. Abercrombie, 1998, from Adams and Abercrombie, 1998, Leary and Abercrombie, 1994, and Kinoshita, 1994.

Abercrombie, 1998 cited several studies from other tectonic areas in the Western USA and from Japan, which showed frequency dependence of $Q_s$ between frequencies of 1 to 10 Hz, with a levelling off of frequency dependence above 5 or 10 Hz. The lesser frequency dependence above 5 to 10 Hz (see compilation in Figure 10.52), is suggested as a possible reflection of changes in the nature of the crust at scale lengths of a few hundred metres (an REV effect?), or that it could be due to an artefact of the models, which at the outset, assume body or surface waves that are isotropically attenuated (no lateral or depth variation) – reasonable with at-depth recordings, but not with many surface recordings.

The level of current tectonic activity, a thinner warmer crust, the presence of large crustal faults characterized by low velocity zones, were each referred to, as possible reasons for the marked frequency-dependence between 1 and 10 Hz. There has been much made of the strong and continued frequency dependence indicated by earlier recordings of $Q_p$ and $Q_s$ from seismic coda, but most likely such records were not *recorded* at various depths *below* the surface, and therefore had numerous potential 'site effects'.

Abercrombie, 2000, compared recordings of nine local earthquakes at seven different depths down the Varian A1 well (0, 24, 298, 572, 877, 907 and 938 m) and at the other borehole network in the Parkfield area (the high resolution seismic network – HRSN). The layout of these measurement locations in relation to the various branches of the San Andreas fault is shown in Figure 10.53. In the

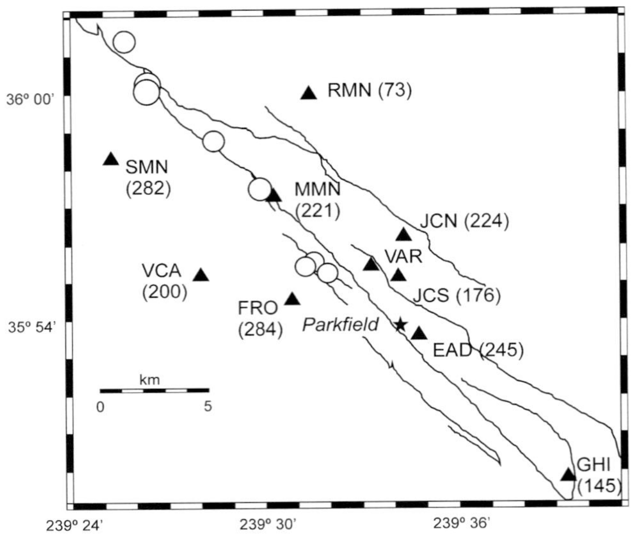


*Figure 10.53* Parkfield, California high resolution seismic network (triangles) and Varian array (VAR). Earthquake epicentres used in the Abercrombie, 2000 analysis of depth dependent and spatially variable attenuation, across the San Andreas fault are shown as circles. The station depths follow the station initials.

analyses performed, Abercrombie, 2000, assumed that attenuation was exponential and frequency-independent

The San Andreas fault zone was confirmed as being a strongly attenuating zone with $Q_p$ averaging 50 over the depth range 0 to 5 km. Seismic Q appears to increase most rapidly at shallow depths, as is also the case for $V_p$, for the rock quality Q or $Q_c$, and therefore for the rock mass deformation modulus. The attenuation to the deepest instrument (0.9 km) on the lower velocity NE side of the fault was comparable to the attenuation to the instrument at only 200 m depth on the higher velocity SW side of the fault. Abercrombie emphasised the value of multiple-depth monitoring and lateral arrays, in improving the level of characterization.

If $Q_p$ or $Q_s$ were linked to rock quality terms, such as the rock mass deformation modulus, even better understanding might perhaps be achieved of the physical nature of the rock masses close to and within the fault. The range of $Q_p$ from the top 300 m to 2 km depth, seen in Figure 10.51 (10 to 180) is a near match to entirely feasible deformation moduli (expressed in GPa), as we saw for the Turkish data.

### 10.6.9 Rock mass quality parallels

The potential reasons behind the strong attenuation in near-surface rock masses are numerous, and can be listed from Abercrombie, 1998 discussion of this topic, as follows:

1. Fracture densities at outcrops and in cores suggest perhaps an order of magnitude decrease at 500 m or more depth (this of course is variable).
2. The presence of fractures (or joints) in the upper kilometres, and the moderate pressures, suggest that friction may be a dominant mechanism of intrinsic attenuation.
3. Analysis of seven earthquakes recorded at 2.5 km depth in the Cajon Pass well, had reported $Q_{scatter} > 10\ Q_{intrinsic}$, at all observed frequencies.
4. Joints and fractures are also major scatterers of seismic energy, and reduction of scattering losses with depth would be expected due to their reduced frequency and greater closure with higher stress.
5. Mining induced seismic events (Spottiswood, 1993) at 2 to 3 km in South Africa show $Q_{seis}$ of about 1000 through 'solid rock' (i.e. probably more massive quartzites), while close to stopes through highly fractured ground (see the velocity EDZ of Chapter 7) could be as low as 20.

The above mechanisms of attenuation, and the depth-dependent seismic Q data, continue to support the idea that seismic Q gives a strong reflection of *rock mass* characteristics, with low values of seismic Q corresponding to the poorer, more jointed, more open structure that is typical of shallower rock, and perhaps present beneath recording stations that are supposedly without 'site effects', i.e. those founded *at the surface,* on crystalline basement rocks.

It seems increasingly reasonable to assume that the typical features of rock masses that make them variable media for engineering construction (foundations for large buildings and dams, tunnels and rock caverns), make them also 'variable media' and variably attenuating, in terms of seismic Q. Rock mass quality is important, perhaps at all depths where jointing is present or slightly open, because of implicit deformation moduli links to $Q_p$ variation with depth in the upper 2–3 km. Up to this 'mining depth', empirical rock mass data is still acquirable, or already existing.

One may speculate about what the rock mass Q-value distribution would be at the various depths in the Varian well, if Q-logging had been performed on recovered core, as done regularly in (usually shallower) nuclear waste related projects, such as the 9 km of Q-histogram core logging, sometimes to 1.6 km depth, performed at UK Nirex's Sellafield site (Barton et al.,

Table 10.8 Three progressively worsening rock mass qualities and their predicted (*near-surface, hypothetical recording station*) properties. Consult tables in Appendix A for explanation of the selected ratings.

| RQD | $J_n$ | $J_r$ | $J_a$ | $J_w$ | SRF | Q | $\sigma_c$ (MPa) | $Q_c$ | $V_p$ (km/s) | M GPa | K (m/s) |
|---|---|---|---|---|---|---|---|---|---|---|---|
| 30 | 15 | 1 | 4 | 1 | 5 | 0.1 | 50 | 0.05 | 2.2 | 3.7 | $2 \times 10^{-6}$ |
| 60 | 12 | 1.5 | 2 | 0.66 | 2.5 | 1 | 100 | 1 | 3.5 | 10 | $10^{-7}$ |
| 90 | 6 | 2 | 1 | 0.66 | 1 | 20 | 150 | 30 | 5.0 | 31 | $3.3 \times 10^{-9}$ |

Note: Near-surface: $V_p \approx 3.5 + \log Q$ km/s, $M(=E_{mass}) \approx 10\ Q^{1/3}$ GPa, $K \approx 1/Q_c \times 10^{-7}$ km/s (Barton, 2002). A significant degree of anisotropy can be provided if desirable or relevant, by using oriented $RQD_o$ and values of $J_r$ and $J_a$ perpendicular to the loading or (dynamic) testing direction. The effects of anisotropic stresses or the effects of increased depth, and the effects of matrix porosity on $V_p$ and M can be handled using the equivalent depth and porosity corrections in Figure 10.54.

1992), and the Q-histogram logging performed at four 1000 m deep SKB wells in Sweden at Forsmark and Simpevarp in 2003.

One cannot help but wonder whether the low seismic Q, also recorded elsewhere at crystalline sites, such as the hundreds of $Q_p$ values lying between 4 and 40, recorded at the NAFZ (Figure 10.49a) from 5 to 30 km distant events, are also a typical reflection of a strongly jointed, tectonically disturbed rock mass, also present beneath the recording sites, and sampled by the lower frequency wave lengths.

It is very easy to imagine a potential range of rock quality Q, composed of the following typical Q-parameter ratings, in such site locations (see Appendix A for description of the ratings used to quantify the six parameters):

$$Q = \frac{RQD}{J_n} \times \frac{J_r}{J_a} \times \frac{J_w}{SRF} = \frac{45-90}{9-15} \times \frac{0.5-2}{2-4} \times \frac{0.66-1}{1-2.5} \qquad (10.13)$$

(i.e. three or four sets of joints, spacing typically 15–50 cm, one or more sets possibly slickensided or smooth-undulating, with weathered or clay-smeared joint walls, dry or partly saturated with water, with typical shallow (<25 m, or <100 m) near-surface, low-stress characteristics.)

From the above example, we obtain $Q = 0.1 - 10$ (quality described as 'very poor' – 'poor' – 'fair'). This calculated range shows, obviously, that $Q_{\text{rock quality}} < Q_{\text{seismic}}$, but when $Q_{\text{rock quality}}$ is used to estimate the rock mass deformation modulus $E_{mass}$, values of from about 2 to 30 GPa would be obtained with an appropriate 10 to 100 MPa range of UCS, and the specified 25 to 100 m depth.

Further insight into the meaning of these Q-parameters and their link to engineering parameters may be given by recording the fact that the first two terms $RQD/J_n$, describing joint structure (the *potential scattering* component), has a maximum range from 100/0.5 = 200 (massive rock without joints), through 45/9 = 5 (closely spaced joints in three sets), to 10/20 = 0.5 ('earth-like', crushed rock). This pair of parameters alone, has by chance, a certain 'familiarity' in relation to relevant seismic Q magnitudes reviewed earlier.

Sophistication, related to shear strength, is added by the next two Q-value terms $J_r/J_a$, describing, with some level of accuracy, the friction coefficient (a potential *intrinsic attenuation* component). (see Appendix A for rating descriptions, and for a graphic presentation of $J_r/J_a \approx \mu$).

Table 10.8 shows some hypothetical constructions of typical near-surface Q-parameter ratings for potential *recording-site* qualities. Such Q-parameter ratings have been applied on thousands of engineering sites world-wide, and the rock engineering profession is as familiar with '$Q = 1$' as the geophysics profession is familiar with '$Q_p = 10$'. Interestingly, and frustratingly, both numbers can remain constant when actually composed of different contributions from the various components.

In rock engineering it is therefore good practice to quote all six assumed parameter values. In geophysics it would be excellent if at least the near-surface scattering (caused by $RQD/J_n$) and intrinsic components (caused by $J_r/J_a$ and $J_w$) could be separated, as increasingly seen with more deeply-acquired data. The hidden ubiquitous microcracks have a 'multiple role' in increasing *both* the attenuation components to varying degrees, but this is possibly masked by significant *near-surface* jointing, since of potentially stronger effect. Microcracks also have a role in increasing the effect of weathering, thereby reducing RQD and $J_w$, and increasing $J_a$, and possibly SRF – due to loosening, all of which , it is suggested, cause a reduction in *both* of these Q values ($Q_{seis}$ and $Q_{rock}$).

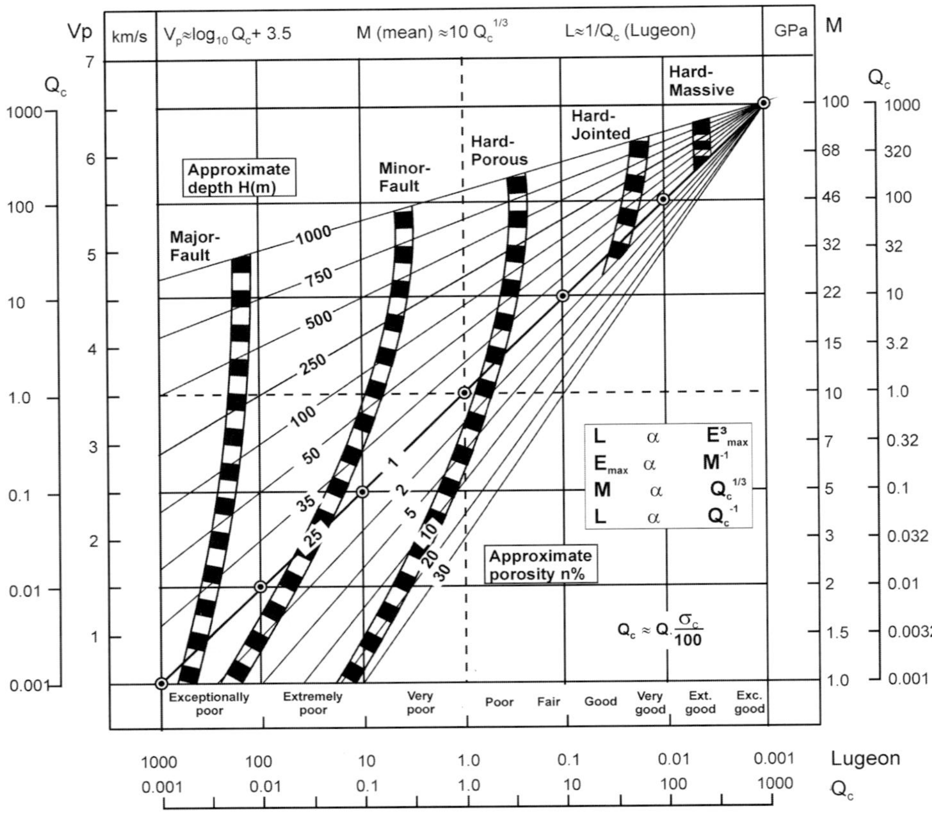


*Figure 10.54* Rock engineering parallels to seismic Q. (See Chapters 5 and 9). The rock mass quality Q, with normalization to $Q_c$ to account for weak or strong rock, appears to follow the trends of seismic Q values (with rock quality $Q_c < Q_{seis}$), Deformation modulus M ($=E_{mass}$) expressed in GPa appears to match seismic $Q_p$ quite closely. Strongly attenuating fault zones, almost 'invisible' to seismic velocity, are seen to 'maintain' an apparently attenuating level of rock quality $Q_c$ at 1 km depth. It has been observed that shallow Californian earthquakes are never found in regions with $V_p > 6.3$ km/s. With appropriate stress correction, this implies some degree of jointing in the neighbourhood of causative faults.

While microcracks are vitally important for attenuation in laboratory samples, they should be less so in the near-surface rock mass, where jointing and weathering-induced porosity may readily dominate attenuation. Some microcracks have also become macro-cracks in this zone. Joint sets in petroleum reservoirs at 2 to 5 kilometres depth remain a major source of attenuation (also polarizing shear waves), but microcracks presumably still contribute to the local attenuation caused by the fluids. Any remaining pressure-resistant pore space and higher aspect-ratio microcracks at 5 to 10 km depth, may contribute to the strongly declined attenuation, but there will still be scattering from major faults and eventual rock boundaries. Jointing, usually assumed to have ceased to exist at great depth, may in fact be present in the form of minor faulting, thereby explaining the maintained (but low) permeability. See the last sections of Chapter 16 for discussion of these aspects.

## 10.7 Attenuation across continents

As introduction to this section, concerning attenuation across continents, it is appropriate to reproduce a well-known diagram of plate tectonics. That reproduced in Figure 10.55 is from Isacks et al., 1968, as presented in Kearey and Vine, 1996. It shows the classic subduction of the lithosphere into the asthenosphere at opposite edges of plates.

### 10.7.1 Plate tectonics, sub-duction zones and seismic Q

The two-dimensional cross-section of the main features of an oceanic subduction zone, shown in Figure 10.56, is reproduced from Kearey and Vine, 1996. The Benioff zone shown in this figure is the source of deep

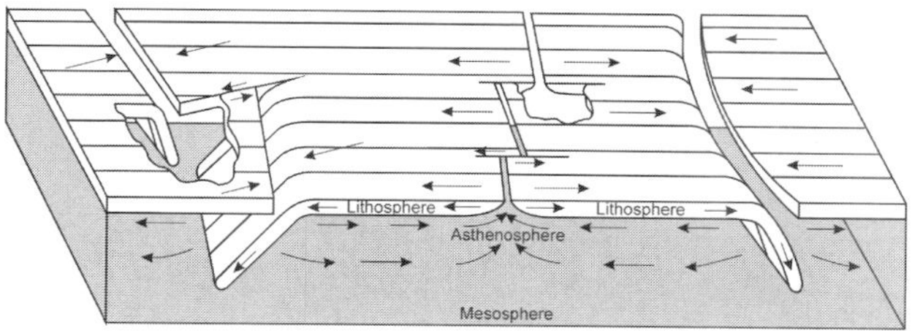


*Figure 10.55* Block diagram summarising the main features of plate tectonics motions. The arrows shown in the asthenosphere represent possible complimentary flow in the mantle, in relation to the lithosphere. From Isacks et al., 1968, reproduced in Kearey and Vine, 1996.

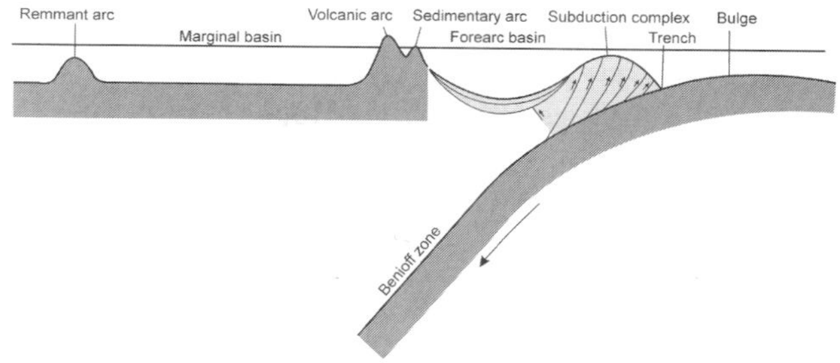


*Figure 10.56* Cross-section showing main features of an oceanic subduction zone. From Kearey and Vine, 1996.

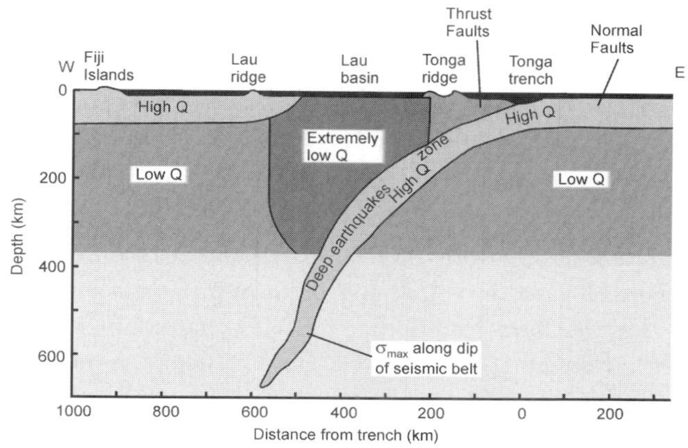


*Figure 10.57* Schematic section crossing the Tonga trench showing the high and low seismic attenuation regions, and the high seismic Q tongue of sinking (and thrusting) lithosphere, which is the source of deep earthquakes. Bott, 1982. The continued application of seismic Q, to hundreds of kilometres depth, is evident, and there is clearly structural/tectonic logic in the location of many of the adjectives: extremely low, high etc.

earthquakes, with the maximum stress direction following the dip of this zone, as indicated in more detail in Figure 10.57.

The 'extremely low Q' region shown in the figure ($Q_P$ of 50 or less) lying in the mantle above the deep earthquake zone appears to be a common feature above such seismic activity. From a rock mechanics point of view, it is reasonable to suppose that this is a zone of high shear and tensile stress, with tectonic disturbance of the strata as a result.

Within the original theory of plate tectonics, plates were considered to be internally rigid and to act as extremely efficient stress guides. Supposedly a stress applied to one margin of a plate was transmitted to its opposite margin with no deformation of the plate's interior (Kearey and Vine, 1996). The plates may be 80 to 150 km thick and thousands of kilometres in width.

In fact there are obviously some locations where intraplate deformation does occur, such as the thrust faulting in intraplate mountain belts. In view of the friction to be overcome, from an engineering viewpoint it would seem clear that the thrust at one side of the plate must be more than at the opposite margin, and the variable modulus of deformation, albeit high, would then seem to have relevance in the overall behaviour.

It has long been suspected, e.g. Molnar and Oliver, 1969, that high values of $Q_{seis}$ correspond to regions of high strength and high velocity, while low values correspond to low strength and low velocity, possibly associated with high temperature and adjacent volcanic activity. A long time ago, Daly, 1940, had correctly defined crustal strength as 'enduring resistance to shear stress with a limiting value'.

The Benioff zone, which is the source of numerous earthquakes even down to a depth of as much as 700 km, often dips at about 45°. The structure of such zones is known from analysis of the different seismic arrivals. High seismic Q travel paths suffer little attenuation and represent stronger (and stiffer) rock. In the case of the Tonga trench which has ocean depths of some 8 km, seismic waves passing up the length of the Benioff zone appear to pass through a region of high seismic Q (about 1000) while those travelling to lateral recorders pass through a more normal region of low seismic Q (about 150).

The zone of very high attenuation (or extremely low seismic Q of about 50) identified in the uppermost mantle above the down-going slab, is a region about 300 km wide (Barazangi and Isacks, 1971). It is implied that this is much weaker (and less stiff) than elsewhere. In the Basin and Range province of the Western USA where there is also anomalously low $Q_{seis}$ in the upper mantle, extensive normal faulting has occurred in a tensional stress regime.

In the foregoing summary of estimates of $Q_{seis}$ in the region of subduction zones, we have seen values of $Q_{seis}$ ranging from high values in deep Benioff zones ($\approx$1000), low value in more normal (shallower) regions ($\approx$150), and extremely low values in the uppermost mantle above the down-going slab ($\approx$50); presumably in the region of the pull-apart-basin.

### 10.7.2 Young and old oceanic lithosphere

Noting that the coda $Q_c$ decay rate was supposedly independent of the source-receiver path, yet reduced by an order of magnitude from stable continental upper crust to active tectonic areas, Jin et al., 1985, investigated coda $Q_c$ from 22 local earthquakes recorded on islands located in young and old *oceanic lithosphere*.

The authors found that coda $Q_c$ values increased with frequency proportional to $f^n$, where n ranged from 0.46 to 0.61, the higher values being for younger oceanic lithosphere. This was similar to continental regions where the active, tectonic, western side of the USA also show higher n values. Crustal coda $Q_c$ values at a frequency of 1 Hz reportedly increase from about 140–200 in the western US to around 1300 in the stable central US, as we shall review shortly.

Figure 10.58 shows a large collection of data for coda $Q_c^{-1}$ (attenuation) versus frequency of measurement, which includes the island data of Jin et al., 1985 (termed 'this study') and numerous other authors' data assembled by R.S. Wu). See Jin et al., 1985 for a complete listing of the sources of data. Most of the geographic areas are marked on the figure. The AKU and GUA data are from Akureyri (Iceland) and the Mariana Islands of Guam.

The coda $Q_c$ data shown in Figure 10.58 demonstrate largest differences at low frequencies, while all the trends converge at higher frequencies. Coda $Q_c$ tended to increase beyond 1000 when frequency was greater than about 20 Hz. Jin et al., 1985 proposed the equation:

$$Q = Q_o \left( \frac{f}{f_o} \right)^n \qquad (10.14)$$

where $f_o$ is 1 Hz, and $Q_o$ is coda $Q_c$ at 1 Hz. They found a good linear relation between log Q (i.e. coda $Q_c$) and log f. Gradient n was 0.346 for the older GUA (Guam)

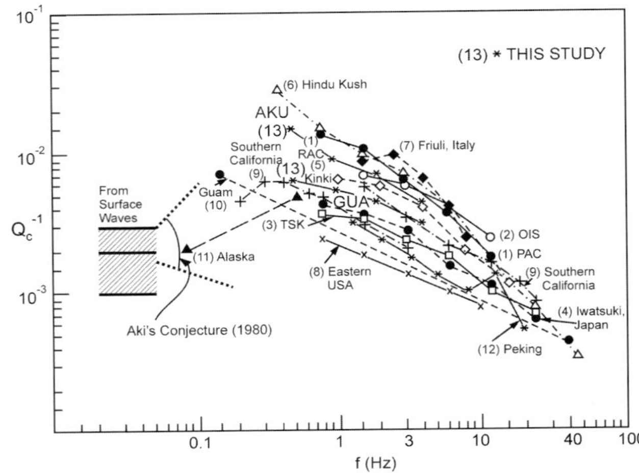


*Figure 10.58* Coda $Q_c^{-1}$ from coda analyses cited by Jin et al., 1985, spanning several sites in the USA, Alaska, Japan, Guam, China, Afghanistan, Iceland and Italy. See Jin et al., 1985 for individual references. Note low-frequency, low seismic $Q_c$ trend, from suspected shallow crust sampling.

data and n was 0.61 for the younger AKU Iceland data, as indicated graphically in Figure 10.58

Their conclusions from a careful analysis of the data trends were that:

1. At frequencies from about 0.5–1 Hz, the average coda $Q_c$ values increased from the youngest oceanic lithosphere to the oldest by a factor or 2.
2. The regional change on the continents follows the same trend, but the difference could be as much as a factor of 10.
3. The coda $Q_c$ values from the old oceanic lithosphere were therefore low in comparison to those obtained from stable parts of the continents.

### 10.7.3 Lateral and depth variation of seismic Q and seismic velocity

Solomon and Toksöz, 1970, were among the early researchers who noted the significant lateral variations in attenuation. They referred to Oliver and Isacks, 1967 experiences of $Q_{seis}$ changing by an order of magnitude over lateral distances of a few 10s of kilometres. Demonstrable regional differences in crust and upper mantle $Q_{seis}$ are also found in the continental USA, for example 200 in the Western seismic belt and 1000 in the east-central US, as will be reviewed shortly.

Variation of $V_p$ and $V_s$ with depth and *lateral location* in the crust, and corresponding $Q_p$ and $Q_s$ increases

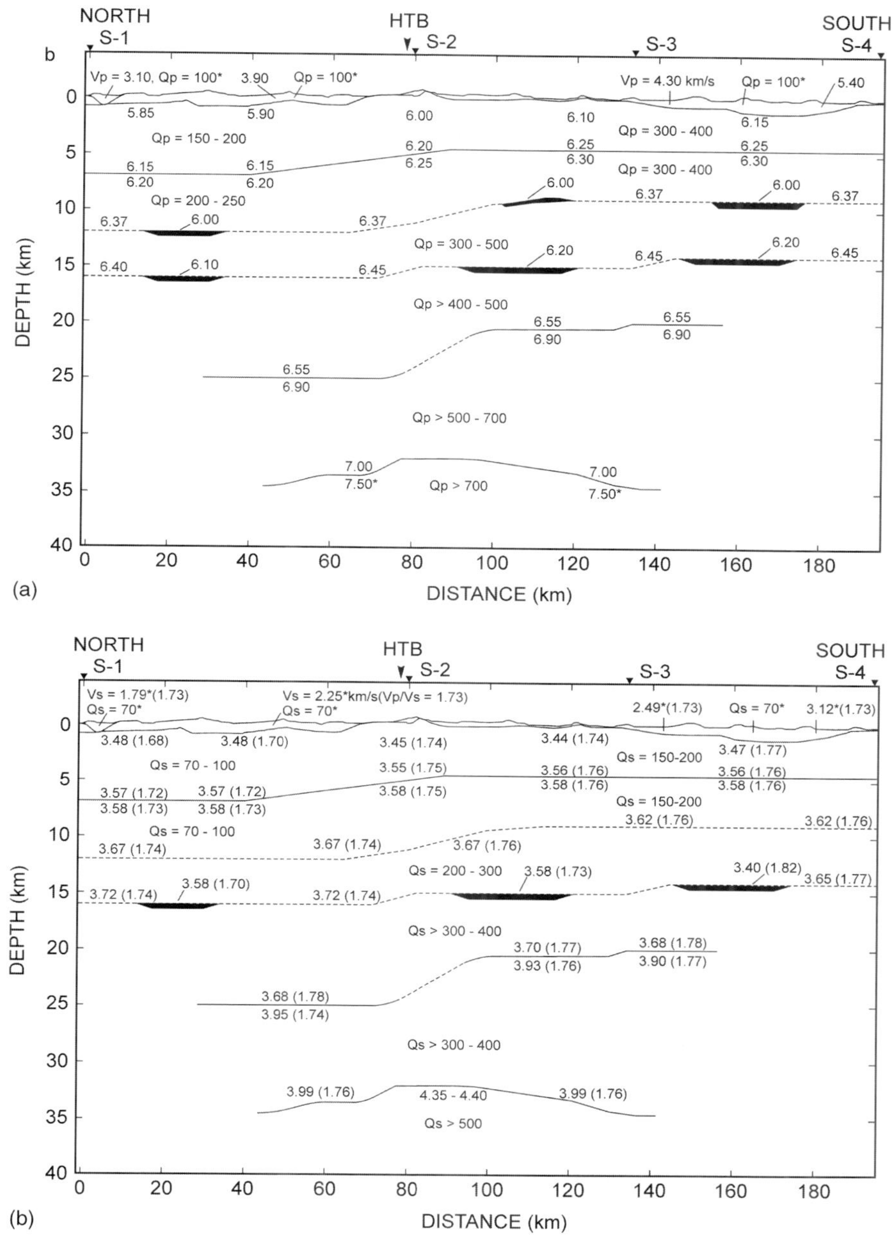


*Figure 10.59* Deep seismic refraction results showing depth variation of a) $V_p$ and seismic $Q_p$, b) $V_s$ (and $V_p/V_s$) and seismic $Q_s$. Iwasaki et al., 1994.

with depth, are shown in Figure 10.59a and b, from Iwasaki et al., 1994, from an extensive seismic refraction experiment conducted on a 194 km N-S line across the Kitakami massif of E. Northern Honshu in Japan. A Jurassic accretionary complex lies to the north, and pre-Silurian and Silurian-lower Cretaceous marine sediments lie to the south.

In these studies, the uppermost crust was covered with a thin (0.5 to 1.0 km) 'surface layer' with $V_p$ ranging from 3.1 to 5.4 km/s. The velocity structure below this showed *lateral variation* (as must surely be expected over a 190 km profile), with successive increases in $V_p$ and seismic $Q_p$ as depth increased to 35 km. The Moho occurs at about 32 to 34 km depth at the base of the selected profile.

Inspection of the variation of $V_p$ and seismic $Q_p$ as depth increases was typically as in the simplified table shown on the next page.

Note the possibility of tentative extrapolation to 2 (and 5) km depth at the top of Figure 10.54. The implication of the data from Iwasaki et al., 1994, is of a significantly jointed upper 1 km, with low rock quality Q-values. Beyond 2 km depth we move outside the area of empirically derivable deformation moduli, so the seismic $Q_p$-$E_{mass}$ (GPa) similarity breaks down because of lack of empirical data for $E_{mass}$.

$Q_p$ beneath two of the world's four major continental rifts (the Rio Grande, SW USA, and the East African rift) were estimated from 1000 km and 600 km linear arrays during the 1980s. With $Q_p$ for the crust taken as 480 to 650, $Q_p$ values for the Rio Grande rift were 95 and 100 respectively. The value of $Q_p$ for the more strongly upwarped asthenosphere beneath the East African rift was only 27 by comparison. In general terms the asthenosphere is believed to have lower velocity, higher attenuation and the possible presence of partial melt, which significantly reduces the shear modulus. This reportedly causes a larger perturbation in P-wave velocity than density, which in turn signifies partial melt (Halderman and Davis, 1991).

Table 10.9 Typical $V_p$-$Q_p$ inter-relation, with depth as the important variable. (Derived from Iwasaki et al., 1994 data, reproduced in Figure 10.59)

| $V_p$ (km/s) | 3 to 4 | 5.9 | 6.2 | 6.3 | 6.4 | 6.5 |
|---|---|---|---|---|---|---|
| $Q_p$ | 100 | 150 | 200 | 300 | 400 | 500 |
| Depth (km) | 1 | 2 | 5 | 8 | 12 | 15–20 |

### 10.7.4 Cross-continent Lg coda Q variations and their explanation

We will conclude this section on seismic quality and attenuation, with continental broad-scale coda $Q_c$ variation first across Eurasia, then South America and the United States. Mitchell et al., 1997 showed that at a frequency of 1 Hz, the coda Q (from the so-called Lg coda) varied between 200 and about 1000, with lowest values in the orogenic belt formed by the collision of African/Arabian and Indian plates with the Eurasian plate (Figure 10.60). Low values are also found in the Arabian Peninsula ($Q_c$ = 350–500), which is a region of recent uplift, extension and volcanism.

High coda $Q_c$ (>800) were found, as expected from earlier results, beneath three shield areas (East European, Siberian and Indian) and beneath the oldest portion of the Altaid belt. Mitchell et al., 1997, suggested that the different coda $Q_c$ magnitudes across Eurasia were proportional to the length of time elapsed since the most recent episode of large scale *compressional* tectonic activity.

They interpreted low coda $Q_c$ as resulting largely from hydrothermal fluids generated by tectonic activity or heating, residing in permeable portions of the Eurasian crust. Crustal $Q_c$ appears to increase with time, as fluids are lost to the surface or absorbed by metamorphism.

Companion papers that appeared in the same number of Pure and Applied Geophysics, described recently interpreted continental structures of $Q_o$ (the *Lg* coda

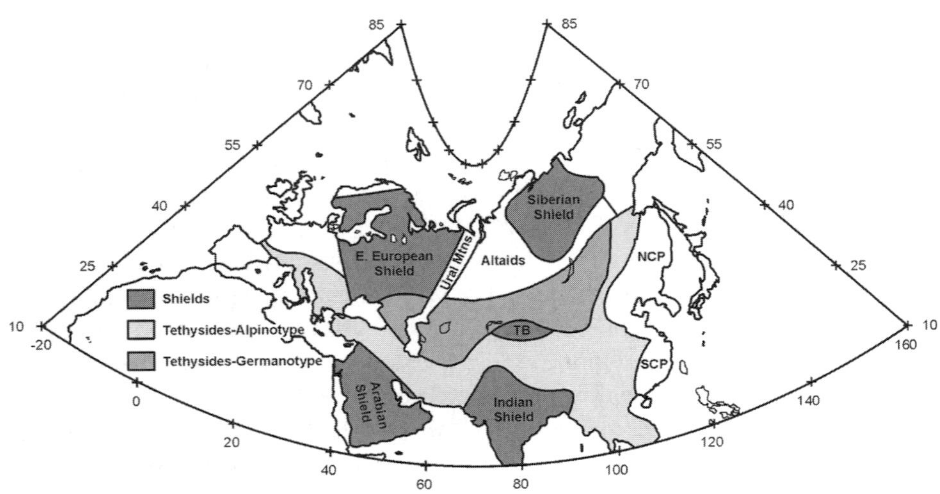


*Figure 10.60* Simplified tectonic map of Eurasia. A tomographic plot of $Q_c$ at 1 Hz was given for each 3° by 3° cell. Mitchell et al., 1997.

at 1 Hz) for South America (DeSouza and Mitchell, 1998), and North America (Baquer and Mitchell, 1998). These authors used 389 seismic recordings in the case of S. America, and 218 in the case of N. America, to produce back-projection tomography, by inversion, giving regionalized maps of $Q_o$ and of its frequency dependence ($\eta$) at 1 Hz.

The *Lg* phase was explained by Mitchell and Hwang, 1987, as being prominent on regional short-period seismograms, where in stable (high Q) continental regions, it can be observed to distances as great as 4000 km, and forms the basis of magnitude scales for small earthquakes, as recorded over regional (i.e. large) distances. The main *Lg* phase is followed by a coda, the main duration of which can also be used to determine the magnitude of regional events. However, the later part of the coda may reportedly not be coherent across arrays of seismograph stations, indicating that part of the coda is due to scattering.

According to these most recent 1998 studies, the seismically active South American Andean Belt was typified by low $Q_o$ (250–450), in a similar manner to the low $Q_o$ (250–300) region west of the Rocky Mountains (the Basin and Range province and active Californian coastal regions). In South America there were broad regions of very high $Q_o$ (700–1100) spanning the central Brazilian shield, and the Amazonian and Paraná Basins, whereas in North America the highest $Q_o$ region was the Northern Appalachians and some of the central lowlands (650–750). The Gulf Coastal Plain and the southern portion of the Atlantic Coastal Plain had intermediate values (400–500), while the Atlantic Shield in South America also had intermediate values (450–700), these last possibly related to the tectonic and igneous activity that occurred during the break-up of Gondwanaland.

The authors of these continental studies suggested that the low $Q_o$ in the Andes, particularly in two belts across the southern and northern Andes, was probably related with higher upper mantle temperatures, or that there were more, deep hydrothermal fluids in these belts. Fluids in the upper crust, and the energy loss they represent, were also cited as the likely reason for low $Q_o$ in the region west of the Rocky Mountains, with variations in that region caused by variable amounts of fluids in faults, joints and rocks of variable permeability. They cited recent studies that showed that a *shear velocity* transition from high to low velocity, lay further west at 'intermediate' depths between 25 and 100 km, than at greater depths than this.

Several of the above (1998) $Q_o$ ranges for regions of the United States showed some differences to the earlier study of Singh and Herrmann, 1983, where a broad belt in the eastern region below the Great Lakes, had values varying from 1000 to a maximum of 1300. The lowest belt of $L_g$ coda $Q$ along the western coast (Oregon, Washington, California) was given as 200 in these earlier studies.

### 10.7.5 Effect of thick sediments on continental Lg coda

Baquer and Mitchell, 1998 emphasised the role of thick deposits of Mesozoic and younger e.g. Cretaceous sediments, typically sandstones and shales, in significantly reducing $Q_o$ in various regions of North America, while *older* sedimentary rocks did not. They also cited the 'positive' effect of dolomites and limestones in maintaining high $Q_o$, and of fluids that had been lost with time. Earlier work by Mitchell, 1995, had suggested that seismic Q was influenced in a 'positive' (less attenuating) direction by the time elapsed since the most recent major episode of tectonic activity. This seemed to be supported by these most recent measurements, and by the Eurasian studies.

In earlier studies, Mitchell and Hwang, 1987 had investigated in some detail, whether the lateral variations in $L_g$ attenuation across the United States, could be explained by known variations in the thicknesses of shallow sedimentary layers. They stated that many of the features of the coda at frequencies near 1 Hz could be approximately duplicated in synthetic seismograms produced by *plane-layer models*, which included layers of low-velocity surface sediments.

As they pointed out, soft and unconsolidated sediments could be characterized by very low velocity and low $Q$ values. However, deep sedimentary basins bounded by sharp discontinuities could influence *Lg* by scattering (or even wave blockage, as described by Baumgardt, 1985), and thick sediments of low $Q$ could cause rapid attenuation due to intrinsic absorption. Their assumption, based on earlier studies, was that *Lg Q* and coda $Q$ were approximately equal, and could show *regional variations* of greater than 1:6 across the United States.

Two aspects to be investigated were how far regional variations could be explained by sediment of different age (as we have seen in the latest studies), and why there could be low and laterally varying values of Q in the

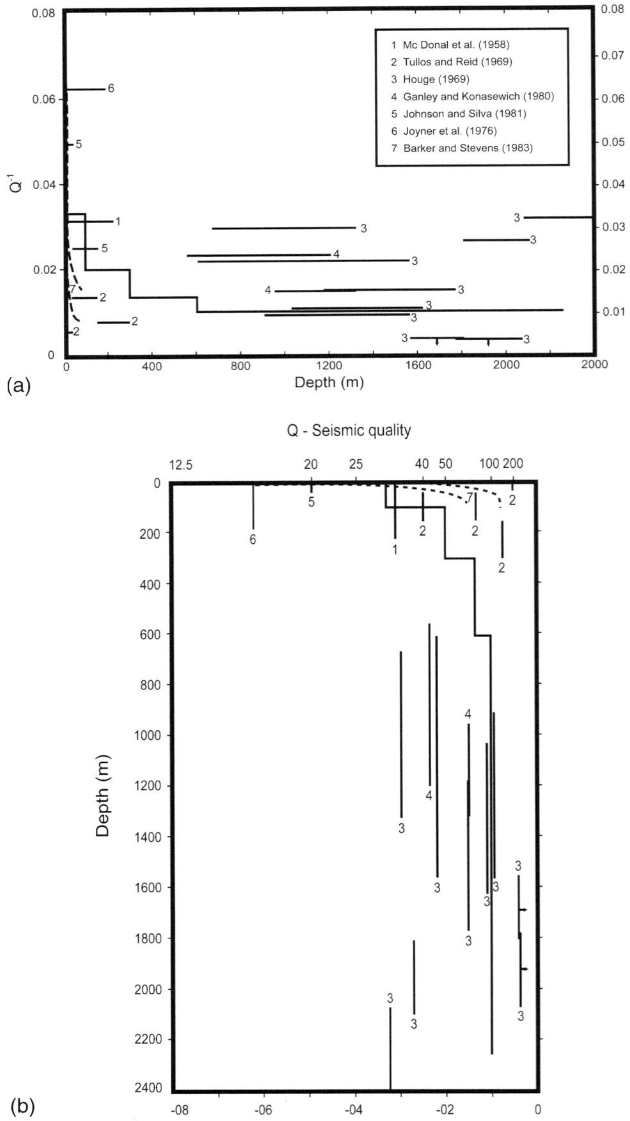


*Figure 10.61* a) An assembly of $Q^{-1}$ data for sandstone and shale sediments as a function of depth. (See Mitchell and Huang, 1983 for complete references). 1: Pierre shale, 2: Gulf Coast sediments, 3: various VSP data sets, 4: unknown, 5: unconsolidated sediments, 6: San Francisco Bay sediments, 7: El Centro area sediments. b) Reversal and rotation of data by the writer to match Figure 10.62 $V_p$ – $Q_c$ format. (See Chapter 5 for derivation).

crystalline crust, where deep accumulations of sediment were absent, as in the western USA. Mitchell and Hwang assembled $Q_{seis}$ values for sandstones and shales, excluding limestones due to their high $Q_{seis}$ values.

Figure 10.61a shows this sedimentary $Q^{-1}$ data, with reversal and rotation by the writer (Figure 10.61b) to match the plotting format of rock mass quality $Q_c$ versus depth, reproduced in Figure 10.62. The implicit 'geometric' similarity to the velocity – depth structure of the *jointed rock mass* Q-$V_p$-M model (Barton, 1995), is striking. Possibly the rock mass quality $Q_c$ value is about 1/10 of the coda $Q$, which would then give an equivalent velocity scale along the top of Figure 10.61b, stretching from about $< 3.5$ km/s (due to porosity effects) to about 5.8 km/s (porosity 'compensation' by 2 km depth). A rock quality $Q_c = 1$ suggests $V_p = 3.5$ km/s (less, with significant porosity), while a rock quality $Q_c = 200$ suggests $V_p = 5.8$ km/s (less, with significant porosity, but partly compensated by the depth effects of $>$2 km of sediment). Conversion of rock quality $Q_c$ to modulus of deformation again indicates the potential match of seismic $Q_p$ and $E_{mass}$, when given in units of GPa.

As Mitchell and Huang noted, there was clear evidence for lower seismic $Q$ at depths of less than 400 m. For the modelling of sediment-layer effects, the authors used the stepped trend in Figure 10.61a, where the following was suggested:

- 0–100 m $Q \approx 30$
- 100–300 m $Q \approx 50$
- 300–600 m $Q \approx 75$
- $>$600 m $Q \approx 100$

In their synthetic modelling the authors assumed frequency *independence* of seismic $Q$ below 1 Hz. The authors found that their low seismic $Q$ thick-sediment models, while applicable and explanatory of *Lg* coda $Q$ in those regions with seismic $Q$ ranging from 400 to 1300, were not applicable to the lower $Q$ of the western United States. Surprisingly, the authors, at that time, did not mention jointing and faulting in (the seismically more active) western United States, as a likely cause for lateral and depth variation of the generally low $Q$ in this region of mostly crystalline basement.

## 10.8 Some recent attenuation measurements in petroleum reservoir environments

As will also be observed in many of the $V_p$, $V_s$, $Q_p$, and $Q_s$ sets of data to be presented in Chapter 13, the attenuation is a relatively more sensitive indicator of the degree of saturation than velocity. In particular, the $Q_s/Q_p$ ratio in sedimentary rocks is much more sensitive to the degree of saturation than the $V_p/V_s$ velocity ratio. Seismic Q has therefore become increasingly important in hydrocarbon exploration. For similar reasons, Winkler and Nur, 1979, suggested using these relatively

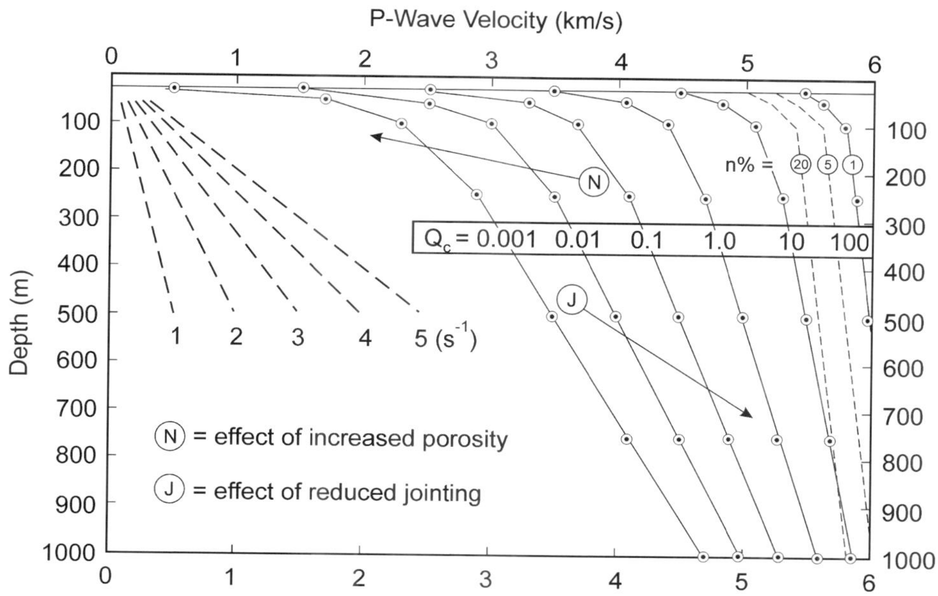


*Figure 10.62* Rock quality $Q_c$-$V_P$ depth model for comparison to previous data set. Making a gross approximation of rock quality $Q_c \approx 1/10$th of *Q*, an approximate velocity scale of <3.5 km/s to <5.8 km/s (less, due to sediment porosities) is suggested in Figure 10.61b.

large changes in seismic Q with saturation as an earthquake predictor, since pre-earthquake dilatancy could affect the degree of saturation in fault zone rocks. These saturation/partial saturation effects were reviewed in section 10.2, but derived only from intact bars of dynamically excited sandstone,

Anomalously low seismic Q values at depth, adjacent to a 5 km deep well where VSP was performed, were reported by Keehn and Kanasewich, 1987. The spectral ratio method was used to obtain values of seismic Q of 10 in a Lower Triassic sandstone, and a second low seismic Q of 12 at about 4,000 m depth. (Figure 10.63). The authors considered that they were observing the effects of scattering caused by intrabed multiples, together with intrinsic attenuation associated with the sandstone lithology.

The strongly attenuating zone between 1,930 and 2,320 m depth was associated with an almost uninterrupted Lower Triassic sandstone layer. The authors referred to studies showing that increased content of sand, rather than shale, were responsible for high attenuation. For some reason, the possibility of a fractured reservoir in this location was not directly referred to, possibly due to confidentiality.

The likelihood that oil bearing rock in a jointed reservoir was being described, is however evident from their conclusions. 'The existence of the two low-Q zones may be attributed to one or more of the following factors which contribute to high attenuation: the presence of fluids of higher viscosity than those in the

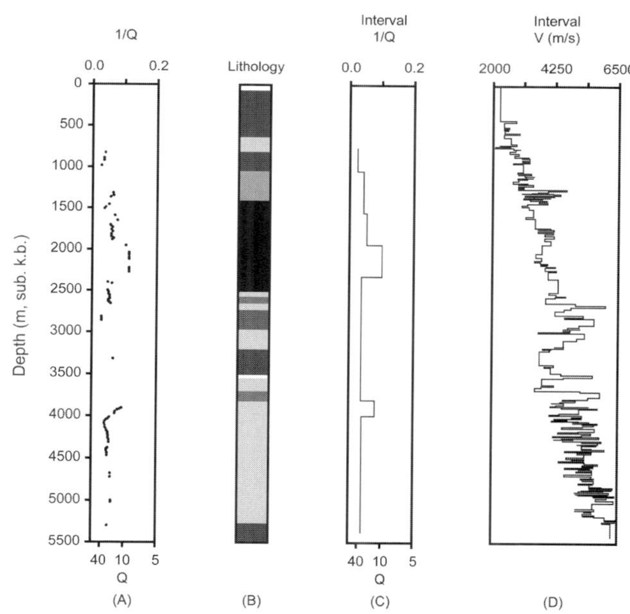


*Figure 10.63* Depth plot of 1/Qseis and Q, smoothed with a 200 m running average, together with the interval P-wave velocities. The well is on Melville Island, in the Canadian Arctic. Keehn and Kanasewich, 1987.

remainder of the hole, pores with smaller aspect ratios, more pore space, and a higher degree of saturation'.

Cross-well tomography data of Quan and Harris, 1997 shown in Figure 10.64, emphasises the generally high level of attenuation in transversely isotropic ('layer cake') sedimentary series. Seismic Q values were between about 20 and 90. The attenuation coefficient (α) in the central log, is given by the relation $Q_{seis} = \pi f/\alpha \times V$, where V is the wave speed, and f is the frequency. It is apparent from the figure that there were certain similarities, in this case, between the cross-well velocity structure and the seismic quality Q. The 8 to 17 K ft/sec velocity scale converts to 2410 m/s – 5120 m/s. Each of these values ($V_p$ and $Q_{seis}$) showed a clear differentiation between shale, limestone and clay, and there was a certain indication of a less jointed (or less porous) area in the overlying chalk.

The cross-well Q seismic data shown in Figure 10.64 are unfortunately specific only to the frequency range

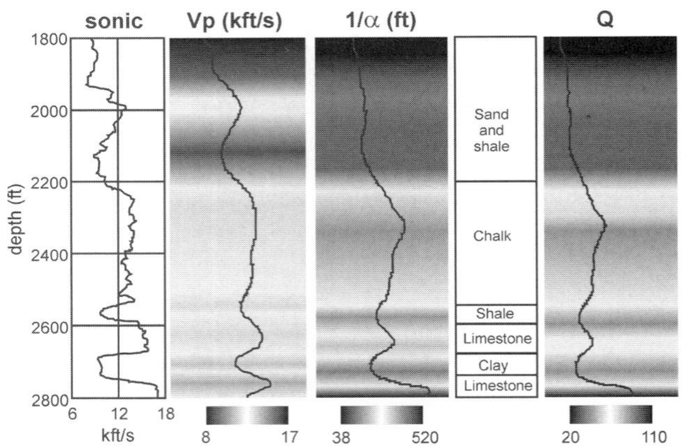


*Figure 10.64* Sonic log, cross-well $V_p$, and cross-well attenuation and seismic Q from roughly 600–900 m depth, in the BP Devine Test Site (Quan and Harris, 1997), reviewed by Pride et al., 2003. Centre frequency was 1750 Hz.

employed, as emphasised by the important set of attenuation data from the Imperial College test site in NE England, reproduced earlier, in Figures 10.34 and 10.35, from Sams et al., 1997. The differences in attenuation between sonic, cross-well, VSP and ultrasonic measurements in the same formation were significant

The four boreholes utilised by Sams et al., 1997, were drilled to about half the depth of the above, to just a few hundred metres depth in a layered sequence of limestones, sandstones, siltstones and mudstones. The variability of 1/Q for each type of seismic survey was due both to rock heterogeneity (i.e. implicit rock quality Q-value variations) and frequency variations.

The authors Hustedt and Clark, 1999, drew attention to the fact that the seismic attenuation factor Q is an important parameter in the processing and interpretation of seismic data, both because of the *detrimental* effect it has on the data, and because it can itself be an indicator of rock properties.

Hustedt and Clark referred to the QVO (Q versus offset) technique that had recently been introduced by Dasgupta and Clark, 1998. This could be used for extracting Q from routine marine surface seismic reflection data. As they explained, in exploration analyst jargon: 'The QVO method applies the well-known spectral-ratio method to a true-relative-spectrum-processed, NMO-corrected, CMP gather.'

They compiled QVO-derived seismic Q-values in relation to interval velocities, from a variety of hydrocarbon exploration settings, as shown in Figure 10.65. Q-values ranged from 50 to 700–800, suggesting that some of

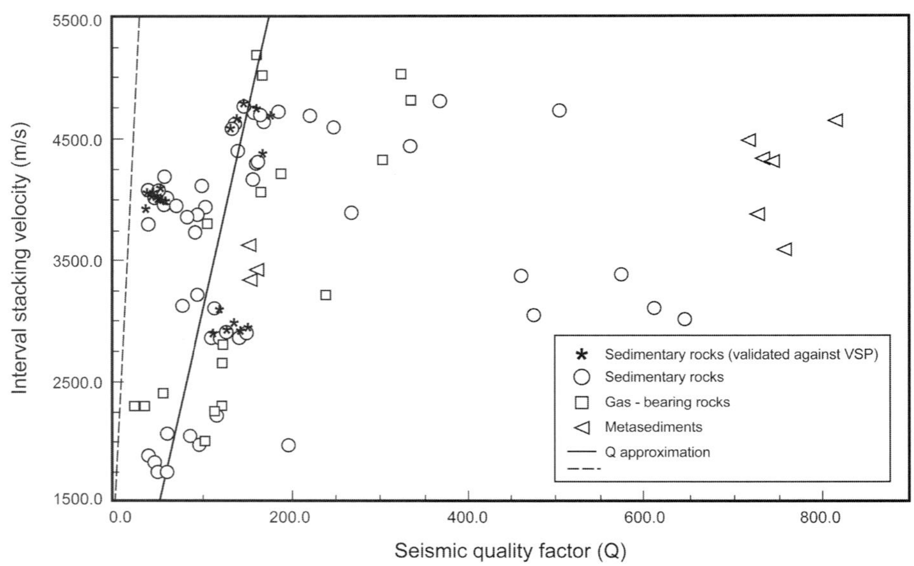


*Figure 10.65* A compilation of QVO (Q versus offset) data for seismic quality Q versus interval velocity. Hustedt and Clark, 1999.

the 'rules-of-thumb' attempting to relate an approximate Q to the interval velocity, may be inappropriate. The potential 'central trend' of the data – a steepening curve – was unfortunately defeated by a dense 'clump' of low seismic Q (40 to 50), yet displaying medium high velocity, of about 3.9 to 4.1 km/s.

In the context of jointed rock (possibly inapplicable at the undefined, but presumed several kilometres depths), a certain $V_p - Q$ coupling, as in Figure 10.64, would be understandable. In relation to the 'curving trend' of much of the data in Figure 10.65, a slightly better fit than the straight lines could be $Q_p \approx (V_p\text{-}500)^{0.6}$, implying the 'disappearance' of a measurable $Q_p$ at $V_p$ as low as 0.5 km/s.

### 10.8.1 Anomalous values of seismic Q in reservoirs due to major structures

Dasgupta and Clark, 1998, reported seismic Q values of 46 and 130 for top chalk and base chalk, from some North Sea data. In Figure 10.66 $Q_p$ is shown falling from 100 to a minimum value of about 50 in the *anticlinal crest*. The implied reservoir values (13 to 33) caused by *gas effects* agreed with trends shown in laboratory studies (See Chapter 13).

A low value of seismic Q for a fault zone (or possibly several faults) encountered in a well in the North Sea ($Q_{seis}$ averaging 45), was described by Worthington and Hudson, 2000. Figure 10.67 shows the roughly 1000 m to 2000 m depth trace of the fault in the seismic migrated time section, reproduced from Harris et al., 1997, and its approximate intersection with the well. The

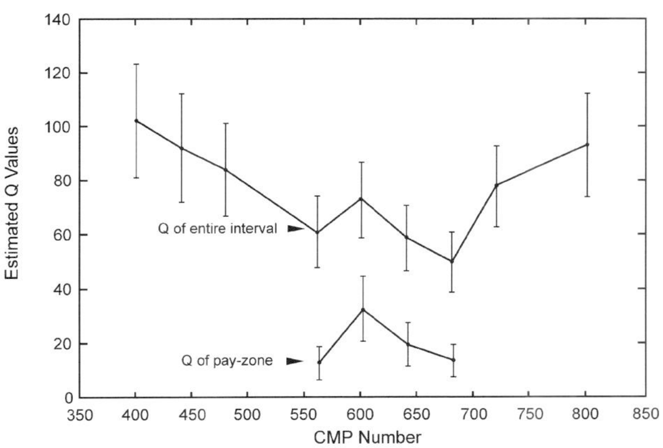


*Figure 10.66* $Q_p$ values interpreted in a North Sea UK sector anticlinal crest, with further reduced values in a gas bearing pay zone. (Dasgupta and Clark, 1998).

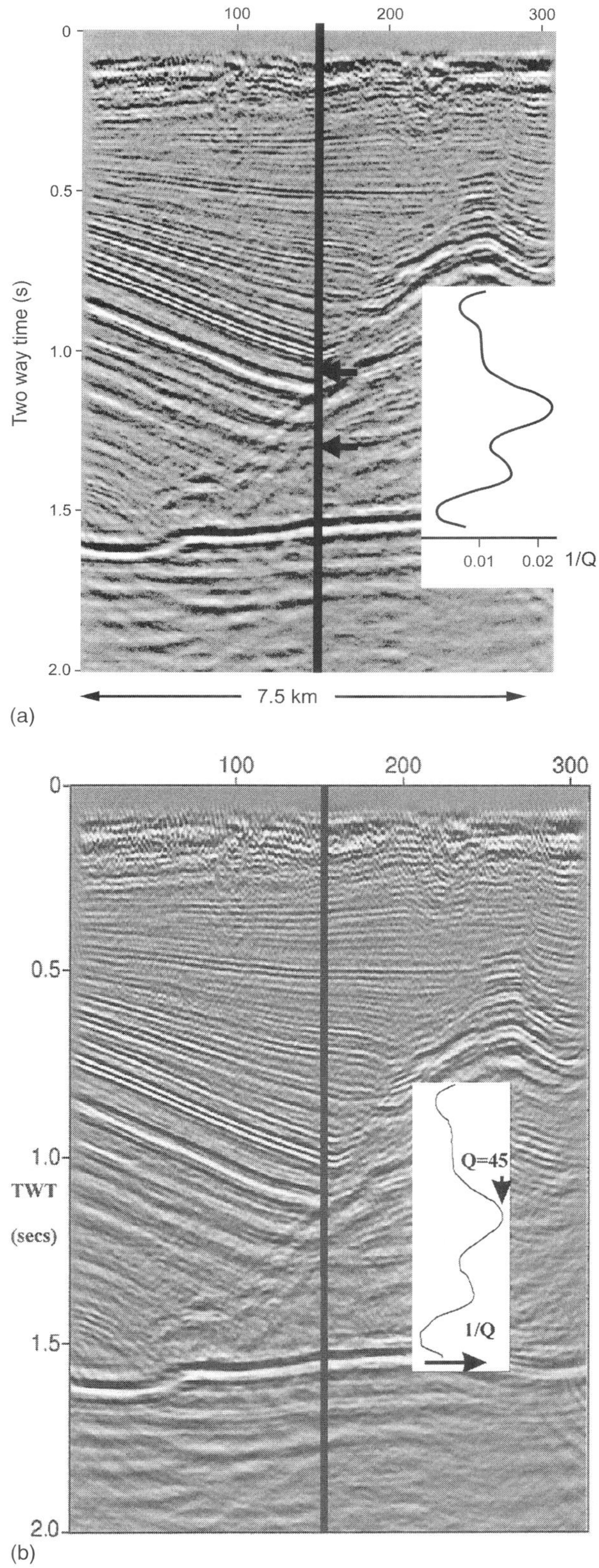


*Figure 10.67* Anomalously low seismic Q related with fault zones in a North Sea reservoir. Worthington and Hudson, 2000, from Harris et al., 1997 data.

fault zone caused an abrupt increase in attenuation, relative to the Triassic and Lower Jurassic age sandstones, siltstones and claystones that were predominantly encountered in the well.

Worthington and Hudson described their modelling of the effects of a down-going P-wave between 1000 and 2000 m depth, by assuming that a fault or several faults, intersected the transmission path. By using a compliance model of a major discontinuity with not completely conforming opposite faces, they showed the need for a remarkable, but actually very realistic, inequality of the normal and shear *compliances*.

We will examine these important parameters (whose inverse is dynamic stiffness), in detail in Chapters 15 and 16, seeing the similarity of their inverted magnitudes, to the normal stiffnesses of joints, clay-filled discontinuities, or faults that are more familiar in the macro-displacement world of rock mechanics. This subject was also addressed earlier in this chapter, concerning the important work of Pyrak-Nolte et al., 1990 related to dynamic and static loading tests on joints.

### 10.8.2 Evidence for fracturing effects in reservoirs on seismic Q

Evidence for the subtle effects on seismic Q of fracturing in petroleum reservoirs, was given by Parra et al., 2002, who described field characterization at the Buena Vista Hills reservoir, in California. They described the use of seismic Q derived from high-resolution cross-well seismic data, to detect vertical, joint-like tectonic fracturing dominating in the Antelope Shale and en echelon, sigmoidal, vein fracturing that was restricted to the Brown Shale, where the joint-like fractures also occurred, but with less frequency. The Brown Shale displayed both low $Q_{seis}$, and low P-wave velocity.

The sand-shale sequences were too finely layered to be detected by sonic logging, with layer thicknesses and bed thicknesses ranging from fractions of centimetres to tens of centimetres to meters. The Antelope Shale formation, containing the highest densities of jointing, consisted of thin, siliceous, clay-free shale beds, with intercalated thin laminae of clayey sand and carbonates. A 290 m core interval contained nearly 750 sand laminæ. The cores averaged 28% porosity, but had only 70 *micro*-darcies permeability due to the dominance of siliceous shale. Since only 5% of the rock consisted of the sands, hydrocarbon production was assumed to be due to the fracturing, which indicated permeabilities of 2.5 to 5 *milli*-darcies.

The en echelon vertical fractures were very short 0.4 to 8 cm, frequently occurring features, perpendicular to the bedding in the Brown Shale. The joint-like features that were more dominating in the Antelope Shale, were nearly vertical, and also had modest heights averaging only 13 cm, due to the bedding thickness limits. Joint densities were from 0 to 2 per meter. There were also less frequent larger fractures and micro-faults, also perpendicular to the bedding.

The cross-well tomography, with sources and receivers at 1.5 m (5 feet) intervals over 457 m (1500 feet) of the reservoir formation, showed expected, distinctly layered velocity trends in the range of about 2.4 to 3.4 km/s over a selected depth range of 3,900 ft (1,190 m) to 4,600 ft (1,400 m).

Figure 10.68 indicates the P-wave velocity and computed $Q_{seis}$ for an interval in the Antelope Shale. Also shown is the core-plug permeability. It is particularly interesting to note the good 'geometric correlation' of $V_p$ and $Q_{seis}$ in the fractured part of the formation with lower core-plug permeability. This was the Brown Shale, which contained both styles of fracturing. In the lower parts of the formation with more sand, there was a marked increase in core-plug permeability, but $Q_{seis}$ remained low, even below 20, probably due to intrinsic squirt-flow attenuation in the sand and carbonate beds, where the P-wave velocity was markedly higher. The Q seismic data is reproduced at more exaggerated scale in Figure 10.69. This shows the strong influence of the two styles of jointing that were described.

The authors also conducted poro-elastic numerical modelling, based on a Biot squirt flow attenuation mechanism. They demonstrate in Figure 10.70 the effect of frequency and azimuth angles on computed attenuations, referring to the actual vertical fracture azimuths of 0° to 30°.

It is interesting to note from their modelling of the Brown Shale, that there was little attenuation of seismic waves propagating from sources *at the surface*, where $Q_{seis}$ was a surprising 1000. In contrast, they found that the en echelon and joint-like features in this shale were strongly attenuating to seismic waves propagating *parallel to stratification*, and perpendicular to the fractures ($Q_{seis}$ of 20), in the frequency range of sonic and cross-well seismic.

The lower Antelope Shale with its frequent sand beds, indicated a higher attenuation of seismic waves propagating from the surface, with $Q_{seis}$ typically 100, while waves propagating *parallel to stratification* and

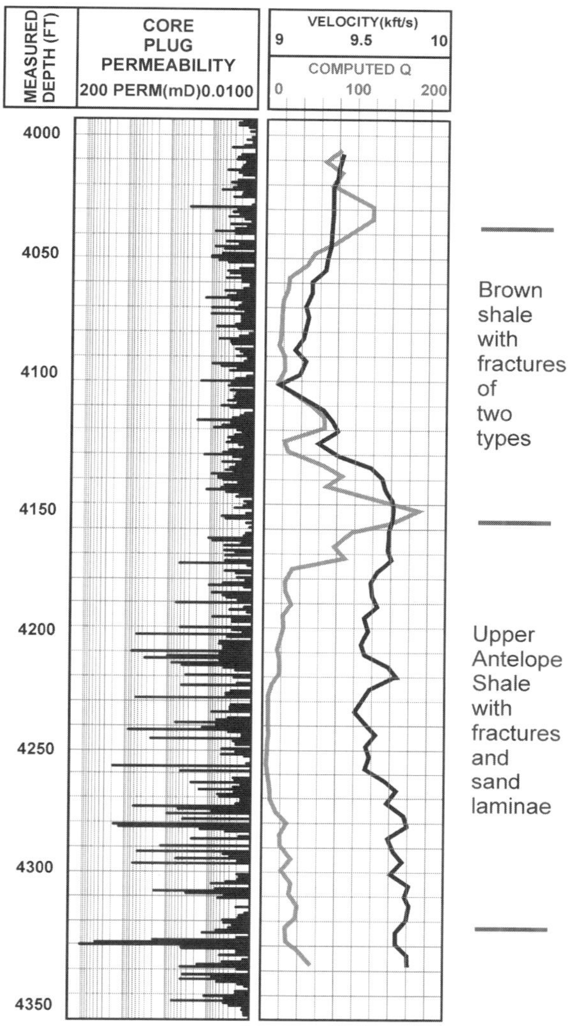


*Figure 10.68* $V_p$ and $Q_{seis}$ results computed from velocity dispersion data (1 to 10 kHz), for a selected 1,220 to 1,330 m deep section of the en echelon fractured and infrequently jointed Brown Shale, and the lower, well jointed, sand-laminae bearing Antelope Shale, Buena Vista Hills field, California. Also shown is the core-plug permeability, which is highest where sand is more frequent in the Antelope Shale. The uppermost low $Q_{seis}$ zone corresponds to the Brown Shale. Parra et al., 2002.

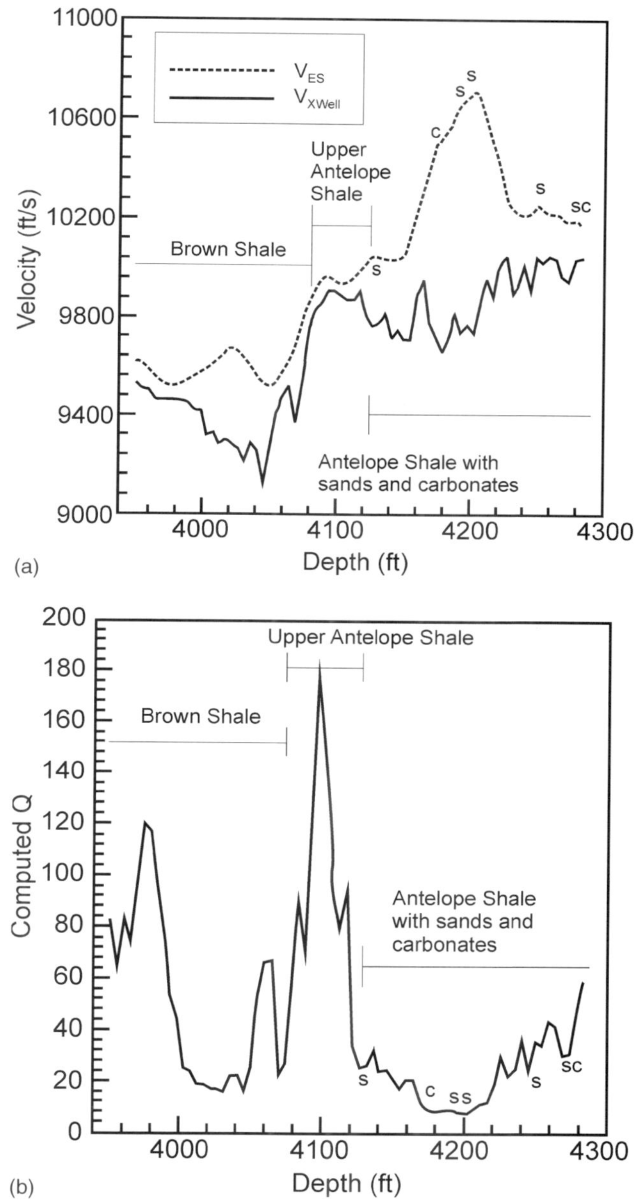


*Figure 10.69* Low seismic Q in jointed Antelope shale, and in *en echelon* fractured Brown shale, plotted at exaggerated scale. Buena Vista Hills studies described by Parra et al., 2002.

perpendicular to 'open' fractures, indicated seismic Q values of 18 to 20. For their modelling, they had selected permeabilities perpendicular to the fracture systems of 50 μd, similar to the matrix, while parallel to the fractures they selected 5 md. (Note: 1 darcy = $10^{-8}$ cm$^2$, or $10^{-12}$ m$^2$, which is roughly equivalent to an engineering unit of $10^{-5}$ m/s for water).

The uses of seismic Q appear to be expanding rapidly, as time goes by, due to its greater sensitivity to some physical properties than seismic velocity. Rossi et al., 2005 described the use of attenuation and velocity tomography, using an array of ocean bottom seismometers (OBS), crossed by a dense pattern of shot lines, on the western continental margin of Svalbard, on the lower part of the continental slope, close to an active mid-ocean ridge.

Data was acquired within the EU Hydratech Project. An important boundary condition was a reflector, marking the boundary between gas-hydrate and free-gas

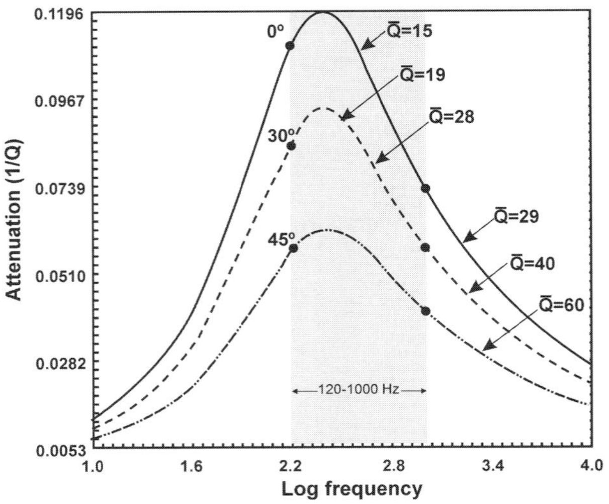


*Figure 10.70* They computed $Q_{seis}$ values of 15, 19 and 28 in the frequency range of 120 to 1000 Hz using an assumed squirt flow length of 3 cm for the modelled fractures. (Parra et al., 2002).

bearing sediment zones below the reflector. $Q_p$ showed a strong decrease from around 200 to values declining below 50 and even below 25, below this reflector, corresponding to the sediments containing free-gas. The high values of $Q_p$ corresponded to the expected gas-hydrate zone, and its probable 'strengthening of the solid frame' of the rock. The authors *generally* found good spatial agreement between the $Q_p$ and $V_p$ variations, both vertically and laterally. The corresponding $V_p$ values were about 1.75 to 1.8 km/s above the reflector in the gas-hydrate zone, declining rapidly to 1.5–1.6 km/s in the free-gas zones below.

The authors also noted that the small graben-causing faults in the area, correlated with the $Q_p$ and $V_p$ reductions. Interestingly, the $Q_p$ values began to decline just above the reflecting boundary, where $V_p$ was still high. The authors considered that this might have been due to interference in the frequency-shift calculations, caused by the liquid to gas phase change. One other anomalous result when comparing the two parameters was the almost full recovery of $V_p$ at greater depth, while $Q_p$ increased only to about 75.

At greatest depth there was a further unexplained $Q_p$ of about 20, while the velocity attained its highest value of 1.9 km/s. The consistent rise in velocity in the deeper levels resembled typical velocity-depth (stress-closing-structure) trends seen in particular in Chapter 11. Possibly a strongly developed structure, under significant effective stress, would be capable of showing an 'increased $V_p$, reduced $Q_p$' reaction.

### 10.8.3 Different methods of analysis give different seismic Q

On several occasions, different methods of estimating seismic Q have been compared by the same authors. We will briefly review two such cases, in order to emphasise both the difficulties that can sometimes arise, and the potential errors involved. As an observer of geophysical results, rather than a practitioner, it is not possible to judge whether the levels of error reported here, are a serious threat to the use that the geophysics community is currently making of seismic Q, in all its various forms.

Badri and Mooney, 1987 used several processing methods, in both time and frequency domains, to compute the seismic quality factor Q for water-saturated, unconsolidated sediments. The methods used included measurements of the spectral amplitude ratio, peak-to-peak and first-peak amplitude ratio, rise time, pulse broadening, and the Futterman causal attenuation operator for attenuating signals.

The authors used compressional seismic waves generated from explosive sources ranging in size from 1 to 64 mg of silver azide, at a depth of 7.6 m below the 70% saturated, silty sandy clays, near Wendover, Utah. The hydrophone receivers were spirally distributed at distances ranging from 25 to 200 m from the source. The computed seismic Q values showed a remarkable variation, as indicated below. The authors suggested that the spectral amplitude ratio method was probably the most reliable, as it is applicable independent of the source.

1. Spectral amplitude ratio method, with five explosive source sizes, over frequencies of 450–725 Hz. Average Q = 23.
2. Peak amplitude ratio method. Average Q = 123.
3. Rise-time method. Range of Q = 50–207.
4. Pulse broadening technique. a) Quarter-cycle measurement: range of Q = 25–158. b) Half-cycle measurement: range of Q = 26–114.
5. Futterman causal attenuation operator. Range of Q = 200–300.

Intuitively, in view of the unconsolidated material and shallow depth involved, one could be permitted to assume that the three lowest seismic Q values listed above (23, 25 and 26), were likely to be the most realistic, and the author's preference for the spectral amplitude ratio method, seen in much of the literature, seems likely to be the most reliable.

Toverud and Ursin, 2005 went further than the above, and compared eight methods of determining seismic Q, using (almost) zero-offset VSP for three separate zones, obtained from a well off the Norwegian coast. The source was deployed at 4 m depth at 40 m of horizontal offset from the well. Depths analysed ranged from 2907–3907 m, using 10 m intervals. A minimum frequency close to zero up to 90 Hz was indicated.

The authors evaluated eight different attenuation models, using a least squares model-fitting approach. They used the geometric ray approximation approach of Ursin and Arntsen, 1985, for point source, vertical wave propagation in a 1D viscoelastic medium, with plane-wave reflection coefficients. A formula for the complex velocity was assumed, with inversion of the attenuation parameters at three different depth intervals, to obtain the parameters in three homogeneous layers: (2907–3335 m was 427 m of mainly shales, with some limestone and marls; 3335–3650 m was 315 m of marl/limestone, with some shale; 3650–3907 m was 257 m of sandstone, siltstone and shale).

Table 10.10 Comparison of eight methods for estimating seismic Q. Toverud and Ursin, 2005. *Minimum normalized misfits are shown.* (For brevity, results for two of the three layers are selected here. Velocities have been rounded to the nearest 10 m/s. Variations were less than 5 m/s).

| Model | Layer | $Q_p$ | $V_p$ m/s |
|---|---|---|---|
| Kolsky-Futterman | 1 | 32 | 3.14 |
| | 3 | 36 | 2.98 |
| Power law | 1 | 35 | 3.14 |
| | 3 | 39 | 2.98 |
| Kjartansson | 1 | 31 | 3.14 |
| | 3 | 37 | 2.98 |
| Müller | 1 | 34 | 3.15 |
| | 3 | 40 | 2.98 |
| Azimi's second law | 1 | 36 | 3.15 |
| | 3 | 44 | 2.98 |
| Azimi's third law | 1 | 34 | 3.14 |
| | 3 | 40 | 2.98 |
| Cole-Cole | 1 | 22 | 3.14 |
| | 3 | 25.5 | 2.99 |
| Standard linear solid (SLS) | 1 | 27.5 | 3.14 |
| | 3 | 32 | 2.99 |

The results for the second layer (marl/limestone with some shale) gave considerably higher Q and more variation. This zone had the highest content of limestone and marl, which perhaps explains both the higher values (Q = 90–160), and the large range. The apparent good correspondence of the models for layers 1 and 3 was a function of the use of the *minimum* (normalized) *misfit* in Table 10.10. An example of their normalized misfit analysis, for the Kolsky-Futterman model, is shown in Figure 10.71. They considered that this performed slightly better than the other models, except in the middle layer.

In reality, and in Figure 10.72, the authors show that there is actually a lot of difference between the models concerning their frequency-dependence. As may be noted, the Kolsky-Futterman and Kjartansson models assume almost negligible frequency dependence.

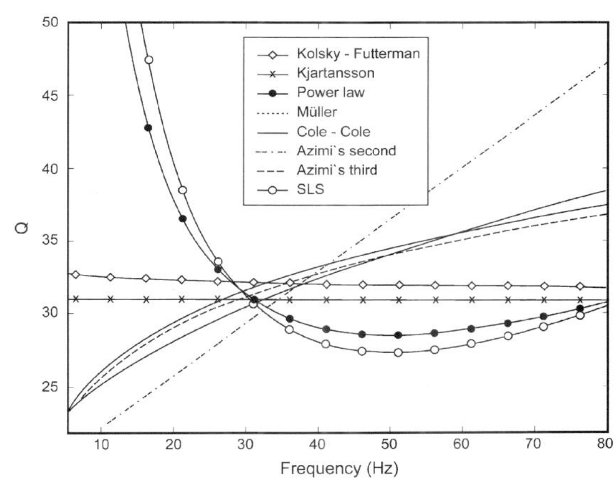


*Figure 10.72* A comparison of the eight models for interpreting seismic Q, for the case of layer 1. Large differences between models are indicated concerning frequency dependence. Toverud and Ursin, 2005.

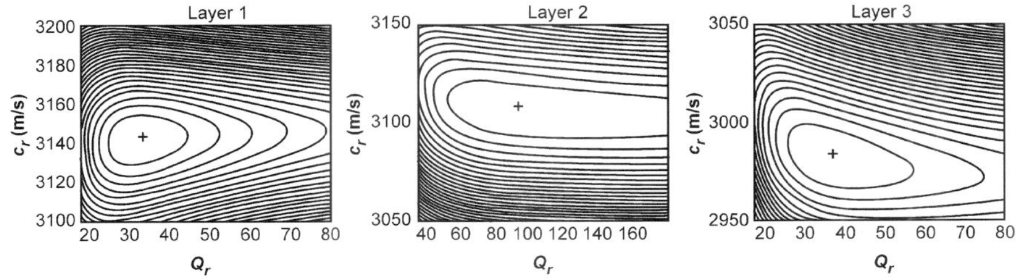


*Figure 10.71* An example of Toverud and Ursin, 2005, normalized misfit analysis, using the Kolsky-Futterman model, for the three layers of VSP data analysed for seismic Q.

# 11 Velocity structure of the earth's crust

This chapter summarises the velocity structure of the continental crust, the continental margins, and the sub-ocean spreading ridges, where zero-age crust is forming. The velocity-depth models of large scale (100 m to 1 km to 50 km) naturally represent an extension of near-surface experiences from Part I. However there are some important parallels and points of basic similarity, especially beneath 3 km of ocean. Although mainly concerned with velocity-depth trends and their reasons, there are also some parallel seismic Q results, where separation into Chapter 10 would have been undesirable.

## 11.1 An introduction to crustal velocity structures

The text of Part I of this book was dominated by civil engineering scale velocity-depth and rock quality trends. In Part II, Chapter 11 we will now concentrate on the velocity-depth trends of the continental and oceanic crusts. However, when contemplating assembling a review of the velocity structure of the crust, a complete section to the inner core of the earth, as summarised in Figure 11.1, is clearly an important starting point for delineating the outer boundaries of behaviour.

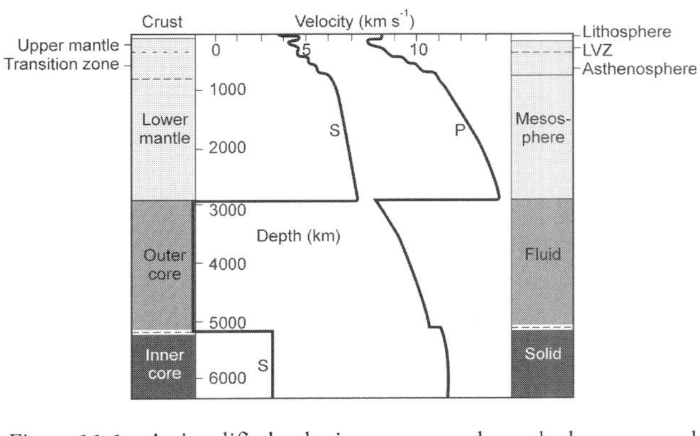


*Figure 11.1* A simplified velocity structure through the crust and mantle down to the inner core. Kearey and Vine, 1996, from Hart et al., 1977.

Figure 11.1, reproduced from Kearey and Vine, 1996, shows a 'familiar' increase in $V_p$ and $V_s$ through the crust and upper and lower mantle. However, due to the great pressure and much higher densities of the rocks involved, the magnitude scale *for* $V_s$ (4 to 7 km/s) now looks more 'familiar' than the exceptional 8 to 13 km/s range of the P-wave velocities. The depths of nearly 3000 km of course exceed by up to five orders of magnitude, the near-surface phenomena reviewed in Part I.

When reaching the outer core and assumed fluid conditions beyond 3000 km, the S-wave is shown as falling to zero due to the loss of shear strength, and the P-wave also shows a dramatic fall of some 5 km/s to 'only' 8 km/s, increasing thereafter to about 10 km/s, before reaching the inner core with its increased density, where the S-wave makes a return, albeit to less than 4 km/s.

In 1909, Mohorovičič interpreted a first arrival P-wave of magnitude 5.7 km/s within 200 km of an earthquake epicentre, and another first arrival P-wave of 7.9 km/s at greater distance from the epicentre, as evidence for a velocity discontinuity. This is now termed the Moho. Figures 11.2a, b show travel time-distance gradients consistent with the Moho velocity discontinuity both for the case of the thicker sub-continental crust (shown here as 54 km) and for the case of the much thinner sub-ocean crust (shown here as about 12 km). The direct wave ($P_g$) and the refracted wave ($P_n$) show different gradients.

At first, there was suspicion of another velocity discontinuity at intermediate (sub-ocean) depth, based on Conrad, 1925. A velocity increase from 5.6 to 6.3 km/s is shown in Figure 11.2b. It is now known that the Conrad discontinuity is not always present and a gradational increase with depth is generally seen. We shall see much evidence of these gradational increases in velocity in this chapter.

Information about the uppermost parts of the earth's crust is now available from direct sampling in ultra-deep boreholes, and indirectly from experimental data on velocities measured over ranges of temperature and pressure consistent with crustal conditions. Pressure (total stress) increases at a rate of about 30 MPa per

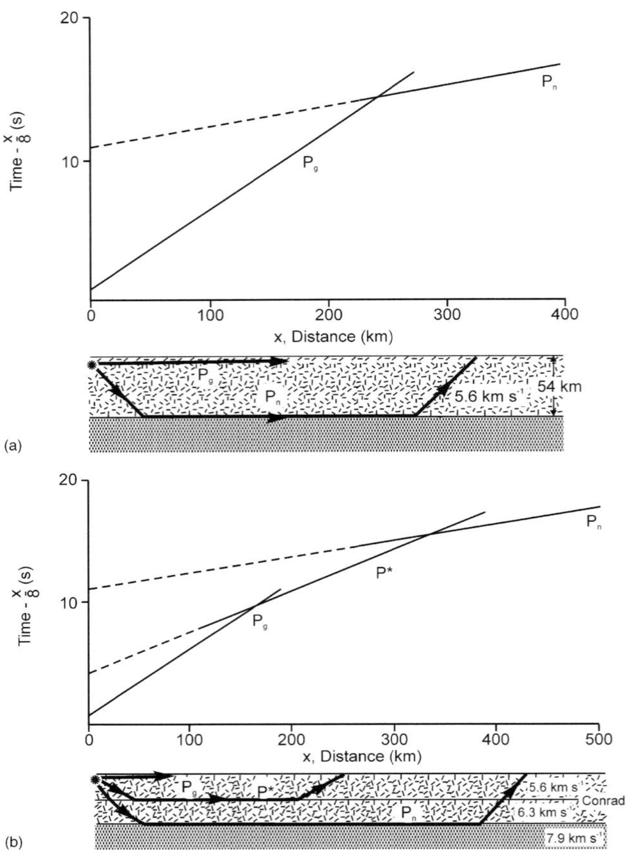


*Figure 11.2* a, b Diagrams of time-distance gradients that demonstrate the sub-continental and sub-ocean evidence for the Moho. From Kearey and Vine, 1996.

kilometre due to the high average density, but there is an initial pore pressure increase to perhaps 20 km or more. At shallower depth, it is common to assume about 16 or 17 MPa/km increase of the *effective* stress (i.e. 26 minus 10 = 16 MPa/km) in crystalline rock, and closer to 10 or 12 MPa/km in less dense hydrocarbon reservoir sediments, neglecting over-pressured zones. Temperature increases at a rate of about 25°C per kilometre up to the Moho, which usually varies from 20 to 80 km depth beneath continents.

The *uppermost* 5 km of the crust shows a rapid increase in deformation modulus and density, as pore space and joints are closed. However, the thermal expansion partly balances the increase in seismic velocity, and P-wave velocities above about 6.5 km/s do not appear to be common. The P-wave velocity range in the *lower* crust, from about 6.5 to 7.6 km/s is explained by chemical transformation to more dense phases, e.g. basalt to garnet granulite to ecologite, or by the presence of higher density

Table 11.1 A simplified, classic model for the seismic structure of oceanic crust (from Bott, 1982, reproduced by Kearey and Vine, 1996).

| | P-wave velocity (km/s) | Average thickness (km) |
|---|---|---|
| Water | 1.5 | 4.5 |
| Layer 1 (sediment) | 1.6–2.5 | 0.4 |
| Layer 2 | 3.4–6.2 | 1.4 |
| Layer 3 | 6.4–7.0 | 5.0 |
| | *Moho discontinuity* | |
| Upper mantle | 7.4–8.6 | |

gabbroic anorthositic rocks which therefore give a higher range of velocity (Kearey and Vine, 1996).

The oceanic crust of the earth is much thinner than the continental crust, and is usually about 6 to 7 km thick, beneath an average water depth of 4.5 km. Table 11.1 reproduced from Kearey and Vine, 1996, and based on Bott, 1982, gives velocities for a simplified layered model, while Figure 11.3 shows more detail and a more gradual velocity increase (dash line) based on inversion techniques.

The type of measurements traditionally required to obtain such information, so-called reversed deep-sea refraction, are illustrated in Figure 11.4. Here we can see the typical layer 1, 2, 3 and Moho separations of velocity, that were state-of the-art prior to more extensive investigations (and investments), from the end of the 1970s.

Measurements of the type illustrated were extended from continental shelf to deep sea, giving as in Figure 11.5, a good illustration of the relative thickness of continental and ocean crust, which are due to great differences in age, as we shall see later in this chapter.

Of course there are also anomalies to complicate the simple picture of increasing velocity (and seismic Q) with depth, as implied so far. A seismic low-velocity zone at depth in the crust is widely accepted as evidence of a region of partially molten rock. It can explain the occurrence of low seismic Q and large negative gradients of both velocity and seismic Q with depth. It was also considered by Mavko and Nur, 1975 as a likely zone of relaxation that could be responsible for transient deformation following large earthquakes on plate boundaries. Their 'melt squirt' mechanism – possibly involving flow of molten rock between cracks of different orientation to the changed stress field, apparently gives a relaxation time of the right magnitude (a few years) to explain transient deformations that may follow the 'elastic rebound' phase of deformation, following

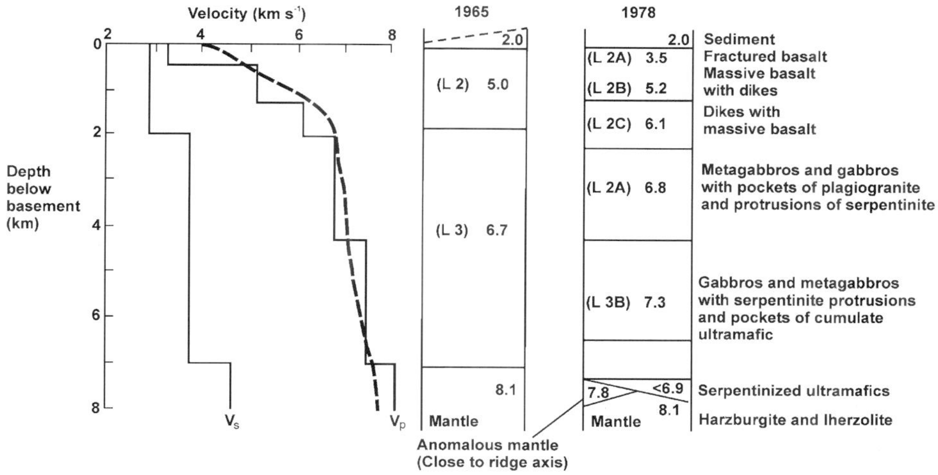


*Figure 11.3* Simplified models, dating from 1965 and 1978, of supposed layering in the oceanic crust. On the left, with the benefit of improved inversion techniques, is a Spudich and Orcutt, 1980, and Harrison and Bonatti, 1981, interpretation of a more gradational increase in velocity with depth.

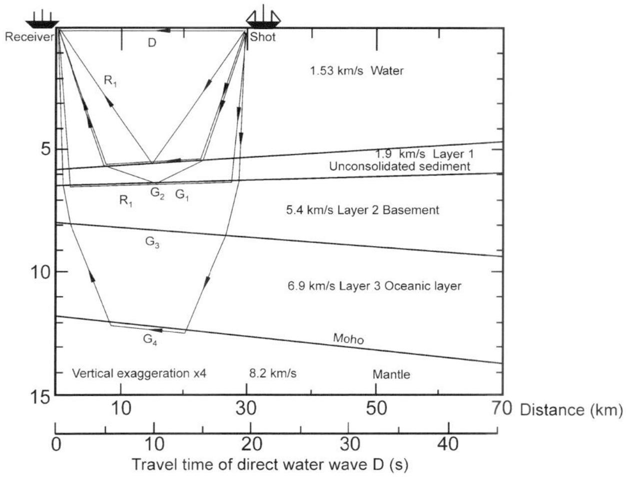


*Figure 11.4* Reversed deep-sea refraction, using two ships and explosive charges. From Bott, 1982, based on Talwani, 1964.

the accumulation of surface deformations that may be approximately 'cancelled' during a large earthquake.

A multidisciplinary investigation of the tectonostratigraphic terrain that compose the Alaskan lithosphere by Beaudoin et al., 1992a, revealed low-velocity (<6.4 km/s) rocks extending to a depth of approximately 27 km. In this case, little complexity was suggested, with seismic layering typically as shown in Figure 11.6

The bedrock composition was metasediments, metagranitic rocks and granitic plutons. Principal mineralogical compositions were quartz, plagioclase and mica, which reportedly have similar average compressional wave velocities. Therefore although the geology was complex, the seismic structure was simple. The average velocity-depth gradients for the investigated terrain were as shown in Figure 11.7. Here the field data is compared with relevant laboratory data.

To conclude this introduction to crustal velocities we will return to greater depth, by first considering the velocity and seismic Q structure within and above descending crustal material, followed by a glimse of the deeper velocity and seismic Q trends. One of the most typical subduction zones in the world is the north-eastern Japan arc. The oceanic Pacific plate subducts downwards into the mantle at a convergence rate of about 10 cm/yr and at an angle of 30° and steeper at greater depth. Many shallow earthquakes occur beneath the Pacific ocean along the upper boundary of the Pacific plate. Intermediate-depth and deep earthquakes are generated within the subducted Pacific plate. Beneath Japan, shallow earthquakes also occur in the upper crust of the continental plate. Active volcanoes are distributed on the land area, parallel to the trench axis.

A modern interpretation of this north-eastern Japan convergent margin is shown in Figure 11.8 from Hasagawa et al., 1994. In the base of the mantle wedge, low-velocity, low seismic Q zones are distributed in parallel to the dip of the high seismic velocity, high seismic Q subducting plate. Decompression melting within the ascending flow of hot mantle material from depth produces low seismic velocities and high seismic attenuations.

The lower portion of the crust and mantle wedge are governed by creep or flow, and are weak and incapable of supporting high stress. According to the review by

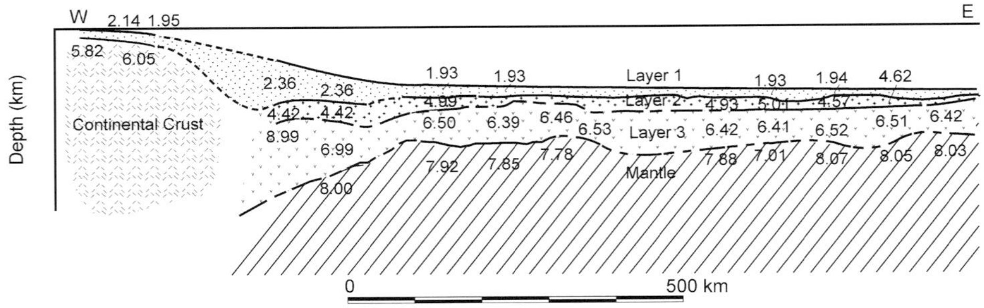


*Figure 11.5* Refraction lines from Argentina and 1000 km eastwards into the Atlantic Ocean. From Ewing, 1965, reproduced in Bott, 1982.

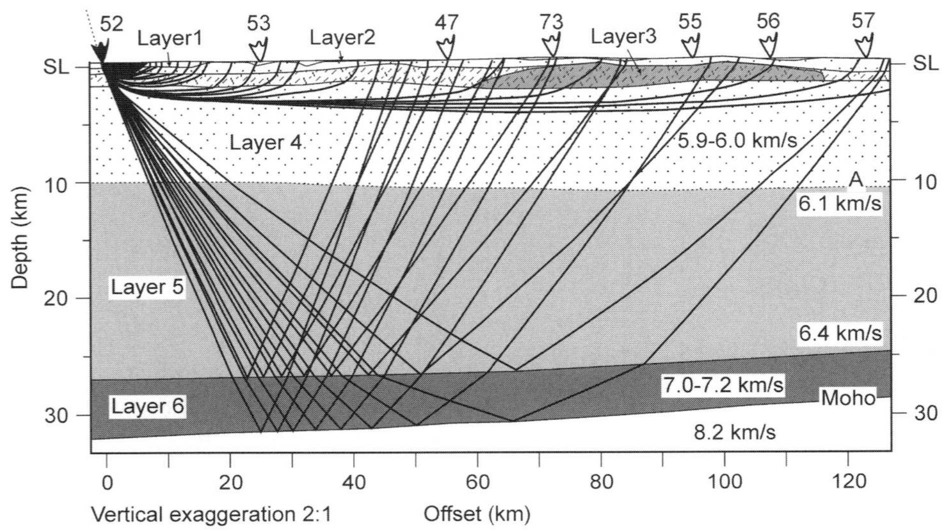


*Figure 11.6* Typical ray diagram (for shotpoint 52) for the East Central Alaska crust (Beaudoin et al., 1992a).

Hasagawa et al., 1994, horizontal compressive stress caused by the convergence between the subduction plate and the overlying continental plate is supported mostly by the upper 15 km of the crust. This is a strong seismic zone, resulting in shallow, thrust-fault earthquakes. Stress concentration will also arise beneath the volcanic regions where the seismogenic zone is locally thin.

P-wave velocities plotted on a depth scale of 0 to 1200 km for the western USA, determined from spectral amplitudes of seismic body waves, given by Archambeau et al., 1969, appear to 'start' at about 6 km/s rapidly reaching about 8 km/s through the crust with intermittent increases to almost 12 km/s at 1200 km depth. Their studies also suggested a frequency-dependent value of seismic Q, with magnitude increasing with frequency.

Seismic Q values inferred from the 'anelastic dissipation' of compressional body waves and surface waves are shown to increase from about 150 to 8000 between the depths of 100 km and 1200 km. These extreme depth trends for $V_p$ and $Q_p$ are shown in Figure 11.9. Strong variations in the upper 20 km of solid crust are suggested in this large-scale data, within the $Q_p$ range of about 130 to 1000. There is an inferred fall of $Q_p$ from about 1000 to 150 through the Moho, between depths of about 40 and 80 km, followed by a rapid increase of $Q_p$ to 200 km depth, and a slower increase of $Q_p$ values to about 8000 at 1200 km depth.

## 11.2 The continental velocity structures

A definitive, updated summary of the seismic velocity structure and composition of the earth's continental crust has been provided by Christensen and Mooney, 1995, who gave a global review based on 560 determinations by

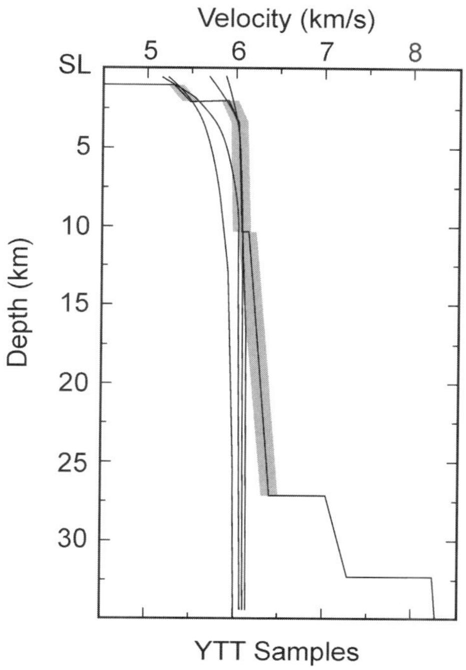


*Figure 11.7* Average $V_p$-depth data (stepped line) compared with temperature corrected laboratory data (curved lines). Beaudoin et al., 1992a.

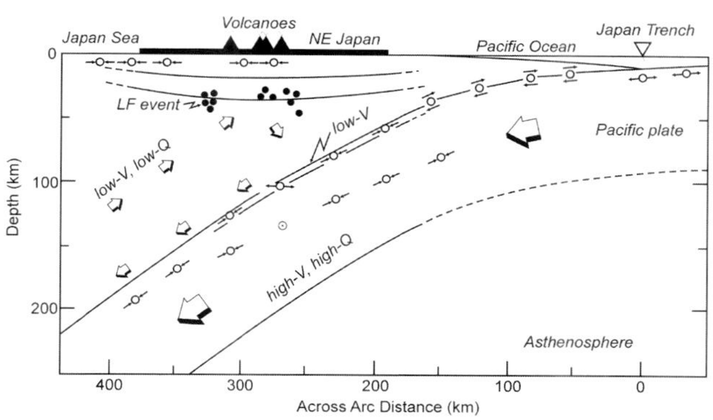


*Figure 11.8* Schematic cross-section of crust and upper mantle in the NE Japan convergent margin. Numerous open circles show focal mechanisms. Solid circles show low frequency micro-earthquakes in low velocity (and low seismic Q) zones beneath volcanoes. Hasagawa et al., 1994.

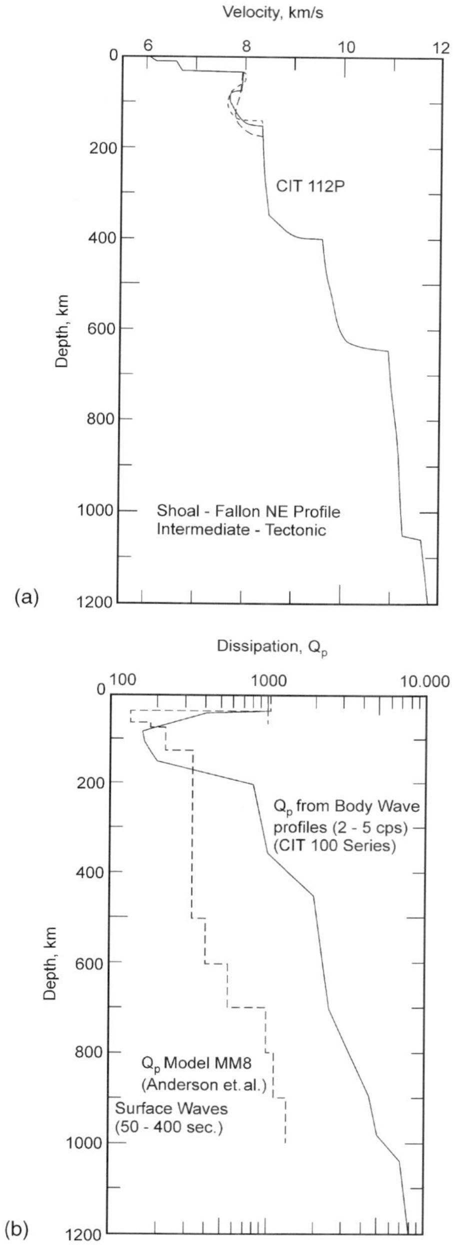


*Figure 11.9* a, b Ultra-deep $V_p$ and $Q_p$ structure interpreted for western USA, by Archambeau et al., 1969.

more than 100 investigators. The geographic locations are illustrated in Figure 11.10. The data reviewed and selected by these authors covers the years 1950 to 1993. In the refraction seismic methods applied, the apparent seismic velocities are directly measured, while the depths of the refracting horizons are calculated from the uppermost layer down to the deepest layer. The depth determinations generally have larger percent errors than the velocities.

The average crustal thickness, weighted to correctly represent the total global areas of each major crustal type is 41.1 km, while the thinnest is in Ethiopia (Afar Triangle: 16 km), and the thickest is in China (Tibetan Plateau: 72 km). The average compressional wave velocity is 6.45 km/s. By chance, this is close to the 'focal point' in Figure 5.36, satisfying the intact, massive rock quality Q-value = 1000 'limit' of 6.5 km/s, for an undefined, average mineralogy.

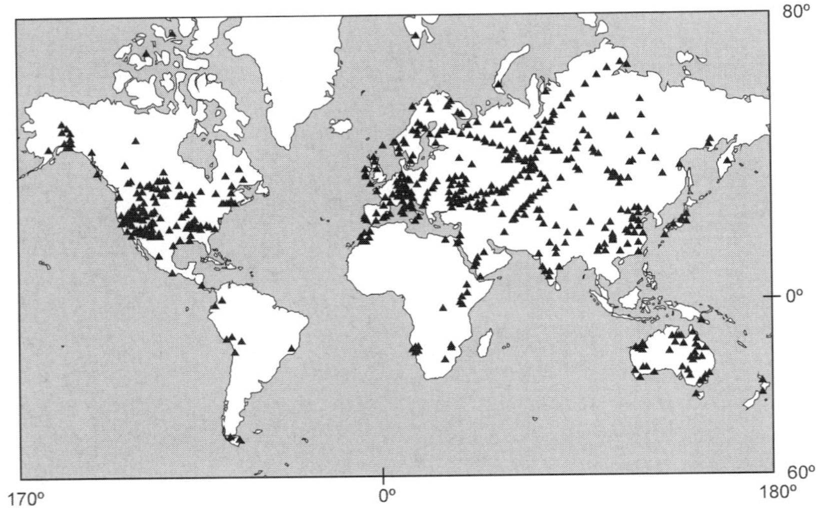


*Figure 11.10* Locations of 560 seismic velocity-depth measurements. Christensen and Mooney, 1995.

Average compressional wave velocities of common crustal rocks show excellent correlations with density. Based on tests of 3000 cores of igneous and metamorphic rocks, taken to 1 GPa confining pressure (similar to 35 km depth), Christensen and Mooney estimated a mean 2830 kg/m$^3$ density for the continental crust. The seismic velocity data was considered to be accurate to 3%, or about ±0.2 km/s, while the depths were considered to be accurate to 10%.

Figures 11.11a, b and c show the very clear trends of crustal depth, average crustal velocity and upper mantle velocity (the normal P-wave termed $P_n$ by tectonophysicists).

These three worldwide compilations suggest a 10 to 20% thicker continental crust than previous estimates (due to under-representation of shields, platforms and orogens while the average velocities lie within previous Eurasian, North American and global estimates of 6.30 to 6.55 km/s (for $V_p$, continental crust) and 7.7 to 8.6 km/s (for $P_n$ of uppermost mantle).

Histograms of average velocities for 5, 10, 15, 20 and 25 km depths are reproduced from Christensen and Mooney, 1995, in Figure 11.12.

Shallow crustal velocities of less than 5.0 km/s, corresponding to sedimentary rocks have not been included. (This applies to the upper 10 km.) The histograms at 5 and 10 km are sharply peaked at 6.0 to 6.2 km/s, typical of crystalline upper crust. Possibly a minimum of 'effective' jointing is found at these high pressures of 140 to 280 MPa. In other words in engineering terminology, we would be talking of a very massive tight structure, probably with Q-values of rock mass quality (Barton et al., 1974; Barton, 2002) of 250–500, combined with the effect of exceptionally high confining stress.

If sedimentary rocks were included, we would likely be operating with a rock mass quality $Q_c$ value ($Q_c = Q \times \sigma_c/100$ plus porosity adjustment) less than the above, depending on the effects of $\sigma_c < 100$ MPa and porosity > 5% on the seismic velocity (see Figures 5.36 and 5.37 in Chapter 5).

In Table 11.2, mean velocities for five principal tectonic provinces as a function of depths of 5, 10, 15, 20 and 25 km are given. The uppermost (lowest $V_p$ values) reflect a great range of lithologies, and presumably some residual (i.e. tightly closed) jointing.

The velocity-depth gradients for these five tectonic provinces, and for an average continental crustal model, are compared in Figures 11.13a and b. The almost linear gradient between 5 and 25 km for the average crust displays a gradient of about $0.6/20 = 0.03\,s^{-1}$, while the gradient between 5 and 10 km for the five tectonic regions is approximately $0.5/5 = 0.1\,s^{-1}$. The reduced gradient at greater depth is due to the expansion effect caused by increased temperature.

Based on a very extensive (3000 cores) laboratory study, stretching over some ten years, Christensen and Mooney, 1995, were able to distinguish anisotropic (mineral/fabric orientation related) velocities for a wide range of crustal rock types. The results for a confining pressure of 1 GPa (≈35 km depth) are reproduced in Figure 11.14. However the authors pointed out that it

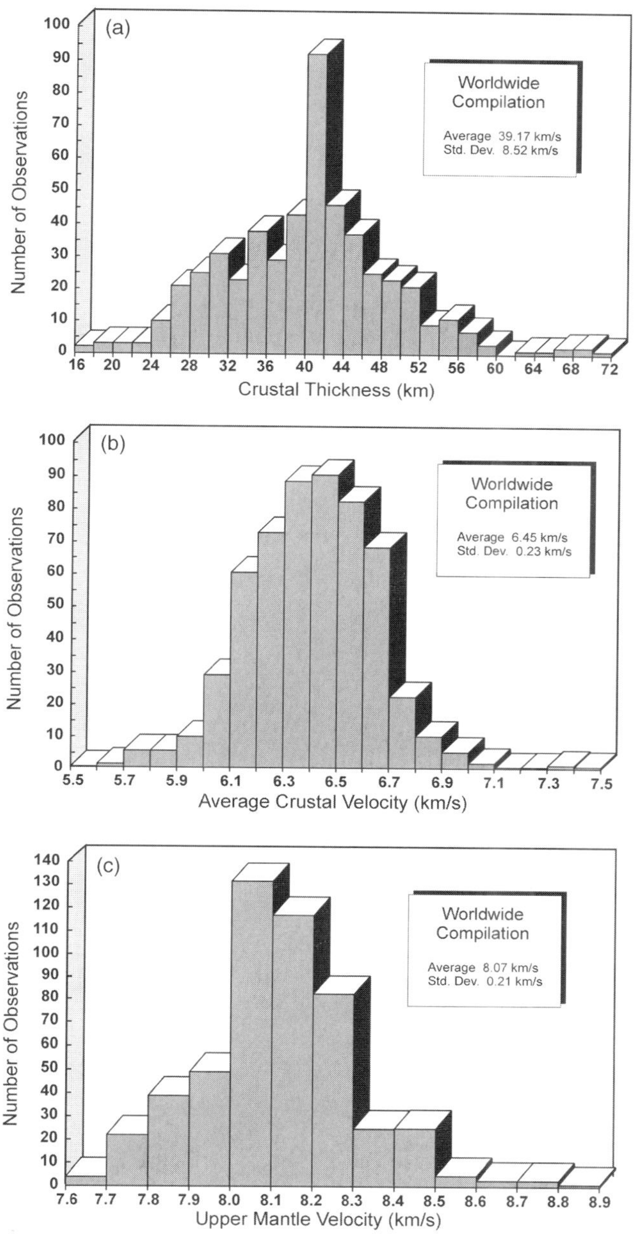


*Figure 11.11* a) Histogram of crustal thickness, from 560 measurements. b) Histogram of average continental crust velocity ($V_p$). c) Histogram of uppermost mantle velocity beneath the continental crust ($P_n$). Christensen and Mooney, 1995.

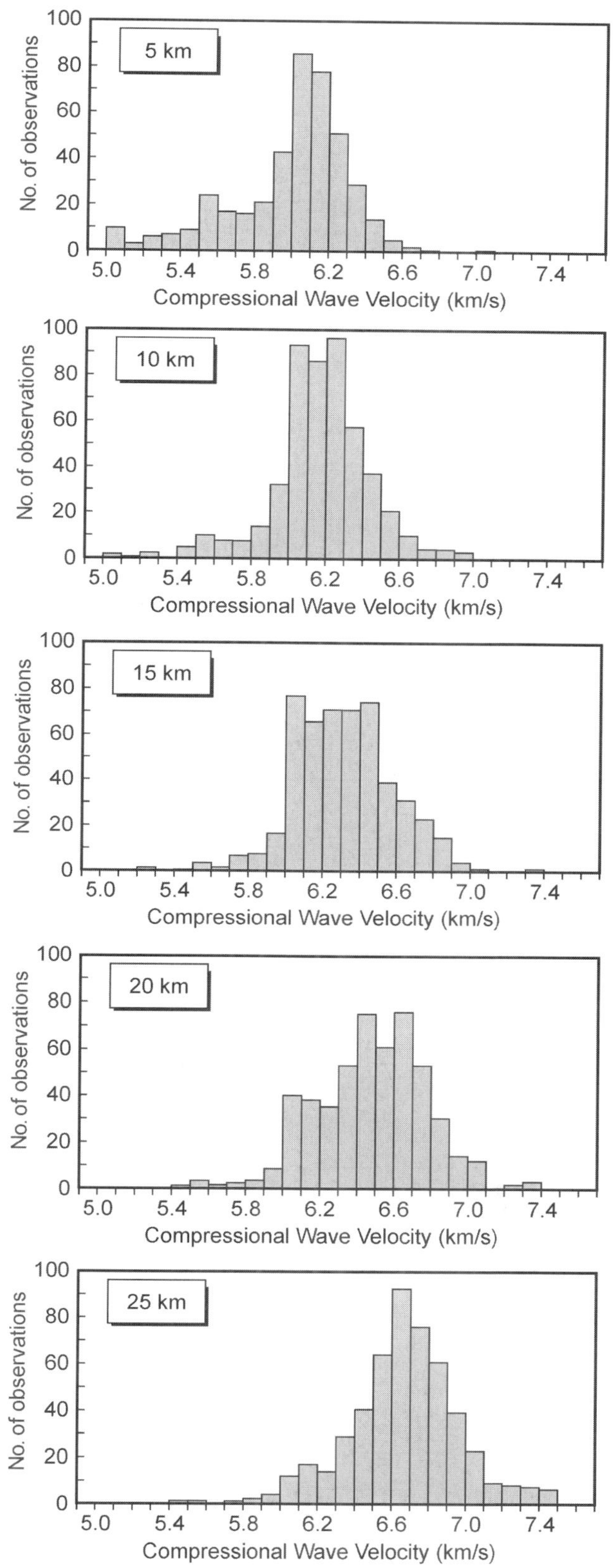


*Figure 11.12* Histograms of crustal velocity at 5 km depth intervals. Christensen and Mooney, 1995.

was not possible to take into account the effects of larger scale anisotropy in their crustal averages, since the presence of *crustal anisotropy* had only recently been documented.

They expected maximum anisotropy in upper crustal metamorphic rock with abundant phyllite and higher

Table 11.2 Velocities for five principal tectonic provinces. Christensen and Mooney, 1995.

| Crustal property | Orogens | Shields and platforms | Continental Arcs | Rifts | Extended crust | Average crust |
|---|---|---|---|---|---|---|
| $V_p$ at 5 km | 5.69 ± 0.67 | 5.68 ± 0.81 | 5.80 ± 0.34 | 5.64 ± 0.64 | 5.59 ± 0.88 | 5.95 ± 0.73 |
| $V_p$ at 10 km | 6.06 ± 0.39 | 6.10 ± 0.40 | 6.17 ± 0.34 | 6.05 ± 0.18 | 6.02 ± 0.45 | 6.21 ± 0.27 |
| $V_p$ at 15 km | 6.22 ± 0.32 | 6.32 ± 0.26 | 6.38 ± 0.33 | 6.29 ± 0.19 | 6.31 ± 0.32 | 6.31 ± 0.27 |
| $V_p$ at 20 km | 6.38 ± 0.34 | 6.38 ± 0.26 | 6.55 ± 0.28 | 6.51 ± 0.23 | 6.53 ± 0.34 | 6.47 ± 0.28 |
| $V_p$ at 25 km | 5.53 ± 0.39 | 6.53 ± 0.27 | 6.65 ± 0.28 | 6.72 ± 0.35 | 6.69 ± 0.30 | 6.64 ± 0.29 |

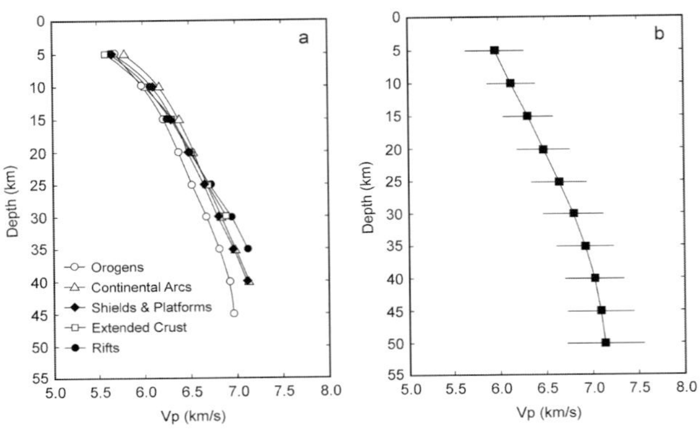


*Figure 11.13* a) Average velocity depth trends for five tectonic provinces, compared to b) the average crust. Christensen and Mooney, 1995.

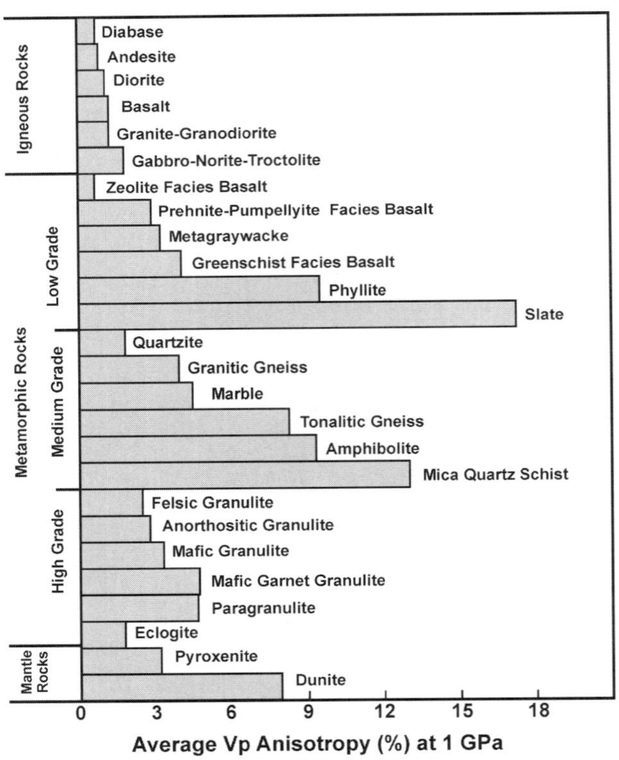


*Figure 11.14* Average anisotropies for laboratory samples at 1 GPa confinement. Christensen and Mooney, 1995. Phyllite, slate and schist dominate as expected, even at these high pressures.

grade slate, and in deeper crustal sections of amphibolite and mica schist. To this one could perhaps add the possibility of azimuthal velocity anisotropy, even at depth, resulting from regions of strong horizontal stress anisotropy, in e.g. thrust belts.

In general, these results are based on rock cores taken in three mutually perpendicular directions. The authors' reported that the change in anisotropy with depth was minimal for most of the rock types. Anisotropies reach 9.5%, 13.0% and 17.2% on average for the mica quartz schist, phyllite and slate respectively. Even at upper mantle depths, azimuthal-dependent Pn velocities are consistently shown, particularly along (and across) the axes of continental rift structures. When one considers the added effect of near-surface (upper 5 km) jointing that may be parallel or sub-parallel to dominant fabric, anisotropy will presumably tend to increase on average as the surface is approached.

Average velocities at 20 km depth equivalent and at 309°C (using average heat flow assumptions) for each rock type are shown in Figure 11.15. The majority of rocks lie between velocities of 6.0 and 7.0 km/s.

As has been noted, the effect on velocity of increased heating in a given laboratory test sample tends to counteract the effect of increased pressure due to thermal expansion effects. However a single rock type 'taken from 5 km to 50 km' does not show the average continental crust velocity-depth gradient, which is brought about by a combination of lithological changes, mineralogical changes and temperature-induced expansion.

Within the mid-crustal depths of 10 to 25 km, where amphibolite facies rocks are likely to comprise the bulk of the crust, there is a gradual change in composition from granitic gneiss and tonalitic gneiss to mafic mineral assemblages rich in amphibolite. At greater depth, the garnet content increases. These gradual changes give the crust its composite average velocity-depth gradient, in relation to the single rock type, laboratory trends shown in Figure 11.16.

The contrasting trends shown in Figure 11.16 are again from Christensen and Mooney, 1995, who must

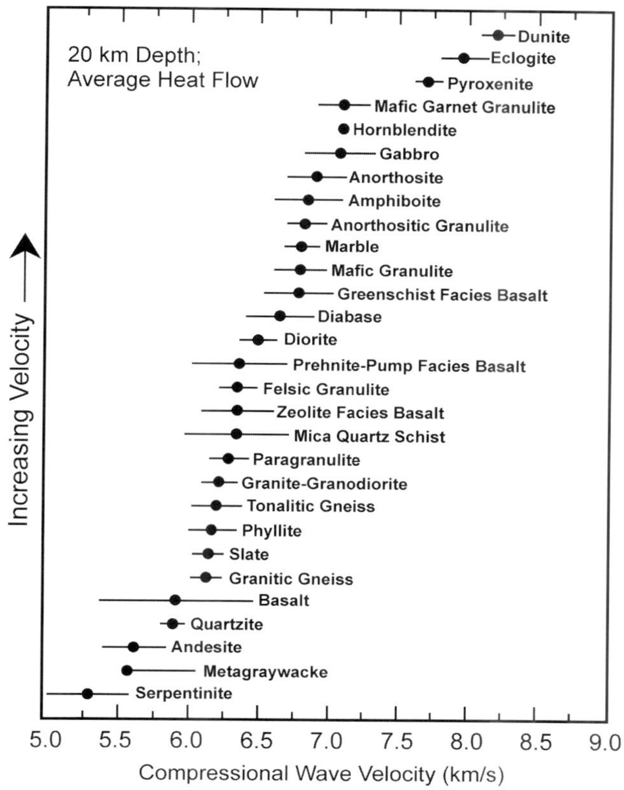


*Figure 11.15* Average laboratory velocities for each rock type, at 20 km equivalent depth, and at a temperature of 309°C. Christensen and Mooney, 1995.

be commended for their extraordinarily far-reaching review, from which we have borrowed many figures in this chapter on crustal velocities.

A new global model for the earth's crust, based on seismic refraction data published in the period 1948–1995 was provided by Mooney et al., 1998. The model was based on 5° × 5° tiles (that measure 550 × 550 km at the equator). In each tile, crustal properties were described by seven layers: 1) ice, 2) crater, 3) soft sediments, 4) hard sediments, 5) crystalline upper crust, 6) middle crust, 7) lower crust.

The source location (mid-profile) of the numerous seismic refraction profiles for this monumental (2000 cases) study by Mooney et al., 1998 and others, are given in the world map in Figure 11.17. The triangles refer to the locations within continents, and on margins, where a velocity-depth function could be extracted from a published interpretation. The mid-point of a major profile corresponds to the triangle location. In about 10% of the cases the shear velocity-depth profile was also reported.

There were 2592 tiles of 5° × 5°, and more than 2000 available sets of field measurements of oceanic and continental crust. Primary continental and oceanic crustal types and mean $V_p$ as a function of depth are shown in Figures 11.18a and b. Note the predominance of average velocities 6.0, 6.1 and 6.2 km/s for the upper crust (range 5.7 to 6.3 km/s) and mostly 6.6 km/s for middle crust (range 6.4 to 6.7 km/s).

On continents, the P-wave velocity averages 2.0 to 3.0 km/s in unconsolidated soft sediments, and 4.0–5.3 km/s in the consolidated (hard) layer. A comparison with a 'site-specific' (continental USA) vertical-section, from Kearey and Vine, 1996 is also shown in Figure 11.18c, for comparison with the continental 'crustal types' model in Figure 11.18a. One-dimensional crustal models of $V_p$, $V_s$ and density to 40 km depth for the whole globe, continental crust and shelf, and the oceanic crust are given in Figure 11.19.

Interesting insights into the local nature of crustal reflections, and of course excellent velocity-depth data from sonic logging and VSP have been obtained from the KTB deep drilling project in Germany, where results of 9.1 km of borehole logging and core analysis were available in Harjes et al., 1997. The thirteen authors of this paper related some interesting experiences about the nature of the strongest reflectors, which tended not to be lithological boundaries alone, but fluid-filled fractures and cataclastic fracture zones, sometimes associated with such boundaries.

The most pronounced and discrete reflections were reportedly found in the compositionally homogeneous amphibolite unit, and originated from hydraulic fracture zones at 4.0 and 4.8 km. Other reflections correlated with fluid-filled fracture zones in gneiss-amphibolite contrasts, so uniqueness could not be determined. One may speculate that the hydraulic fracture zones had become propped in some way, perhaps due to sheared-dilated sections (non-parallel to maximum stress), as discussed by Barton, 1986, and extensively reviewed in Chapter 16.

The predominantly gneiss and amphibolite sequences shown in Figure 11.20 showed lower $V_p/V_s$ ratios in the gneiss than in the amphibolite (due mainly to quartz content differences). In general, decreases of the $V_p/V_s$ ratios were caused by decrease of $V_p$ rather than by an increase of $V_s$, which the authors liken to typical behaviour in fractured, or jointed, and porous rocks.

The mean trends and individual results of sonic and VSP measurements down this unusually deep borehole, are shown to at least 8.5 km depth in Figure 11.20. The strong $V_p$ (and $V_s$) velocity-depth gradient shows the

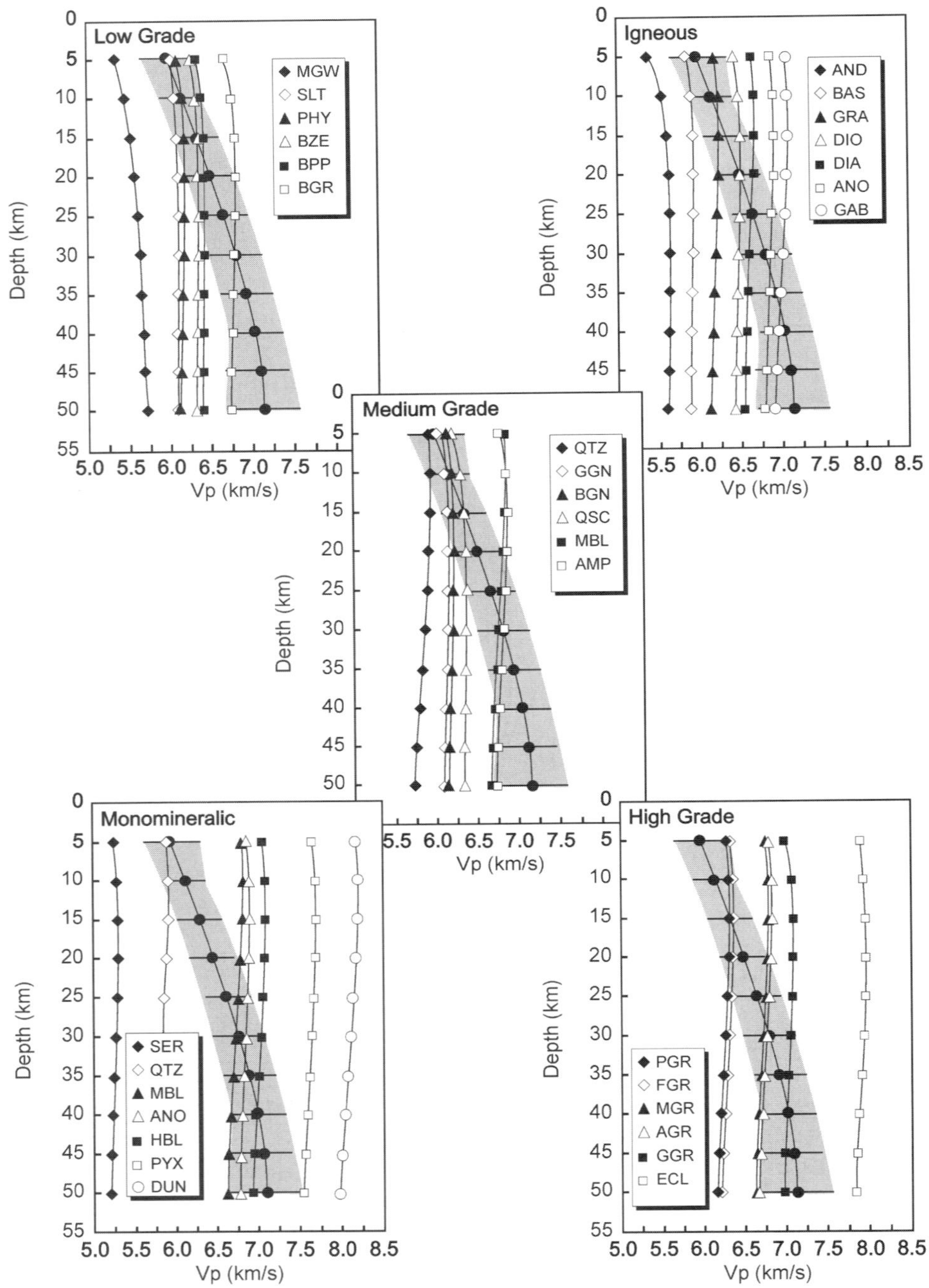


*Figure 11.16* Average continental crustal velocities (shaded) compared to average laboratory velocities at simulated depths and temperatures. Christensen and Mooney, 1995.

classic knee-shape, with average $V_p$ increasing from about 3 km/s close to the surface to about 5.3 km/s at 500 m depth and to about 5.7 km/s at 1000 m (based on mean VSP data). Comparison with the velocity-depth-Q (rock mass quality Q or $Q_c$) curves shown in Figures 5.36 and 5.37, suggests that this upper 1 km of paragneisses (with some amphibolite) may have Q-values (or $Q_c$ values) as low as 1, which signifies a good deal of jointing. The calliper log measurements also shown in Figure 11.20a indeed support the idea of borehole walls with a good deal of joint-related 'overbreak', which normally occurs when there are 2 or more joint sets (i.e. $J_n$ probably in excess of 4 or 6 in the Q-system of rock mass quality description, combined with unfavourable

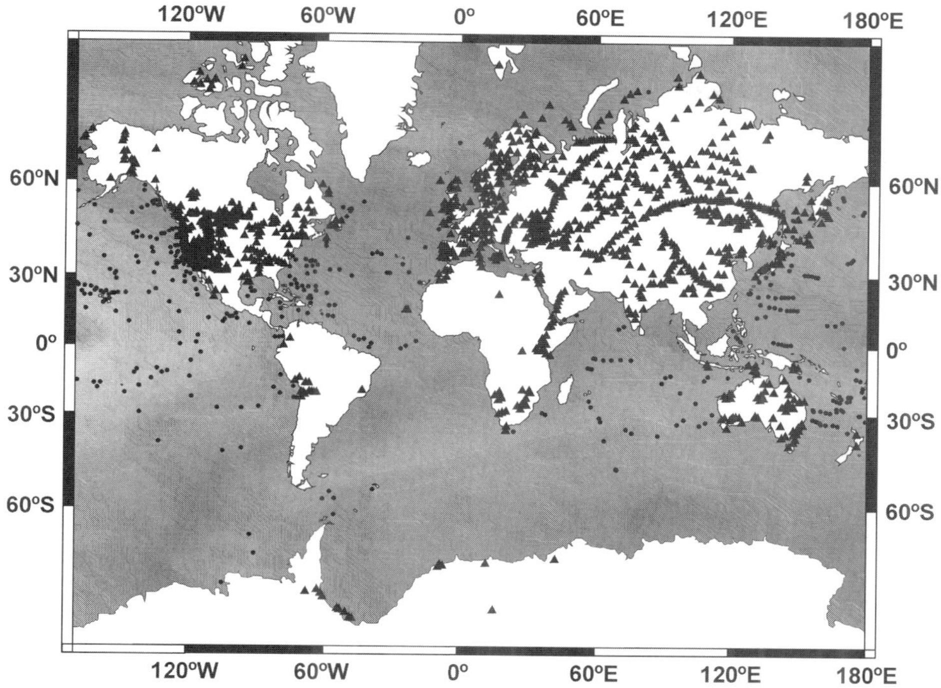


*Figure 11.17* Location map for the Mooney et al., 1998 global crustal model seismic refraction profiles. The triangles refer to the locations within continents, and on margins, where a velocity-depth function was extracted from a published interpretation. In about 10% of the cases the shear velocity-depth profile was also reported.

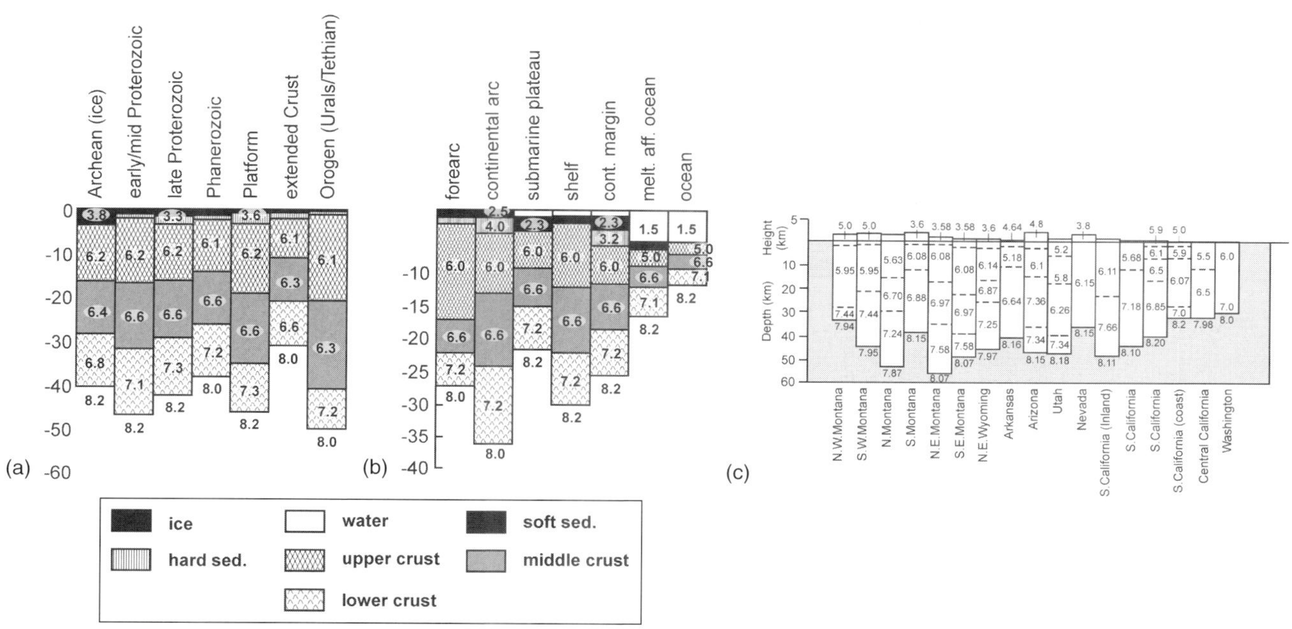


*Figure 11.18* $V_P$-depth structures for primary crustal types. a) Continental. b) Oceanic. Mooney et al., 1998. c) Continental USA velocity-depth section. Kearey and Vine, 1996.

anisotropic stress and presumably water pressure see: Appendix A for Q-parameter ratings).

Moving to an entirely different geology and location, and into an artificial 'seismic' environment, namely the Nevada Test Site in the western USA, one may note that nuclear 'events' of 155 to 1300 kilotons equivalent yield were used for forward modelling of surface velocity data that was recorded within 15 km of the underground nuclear test explosions. Barker et al., 1991, used a plane-layered structural model of the porous, low-density volcanic sequences beneath the Pahute Mesa to derive the velocity-depth structures shown in Figure 11.21. These velocity structure models were needed to determine the effective source functions of the underground explosions.

The upper part of these velocity-depth trends show broad similarity to the $V_p$-depth-$Q_c$ (normalised rock mass quality) trends shown in Figure 5.37. Clearly the porous, jointed volcanics are at the lower end of the $Q_c$ range, or alternatively represent a 'porous version' of the trends shown by $Q_c = 1$ to 10 in Figure 5.37.

Before leaving continental velocity structure, we may look at two near-surface extremes, namely *sea ice*, or

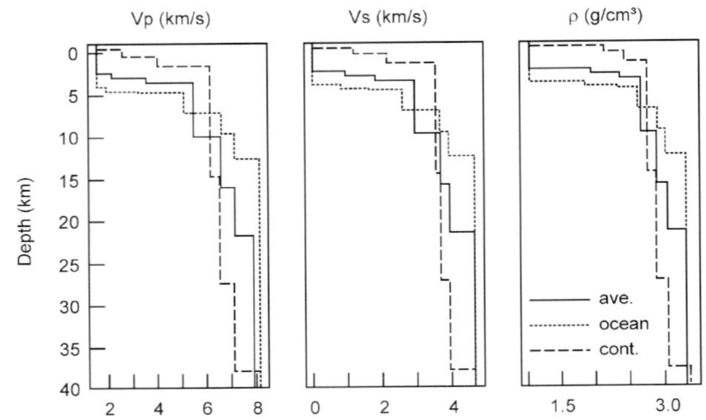


*Figure 11.19* Depth-velocity-density profiles from a crustal model CRUST 5.1. Mooney et al., 1998. The predominance of oceans causes the average and oceanic crust velocities to be low in the upper 3 km.

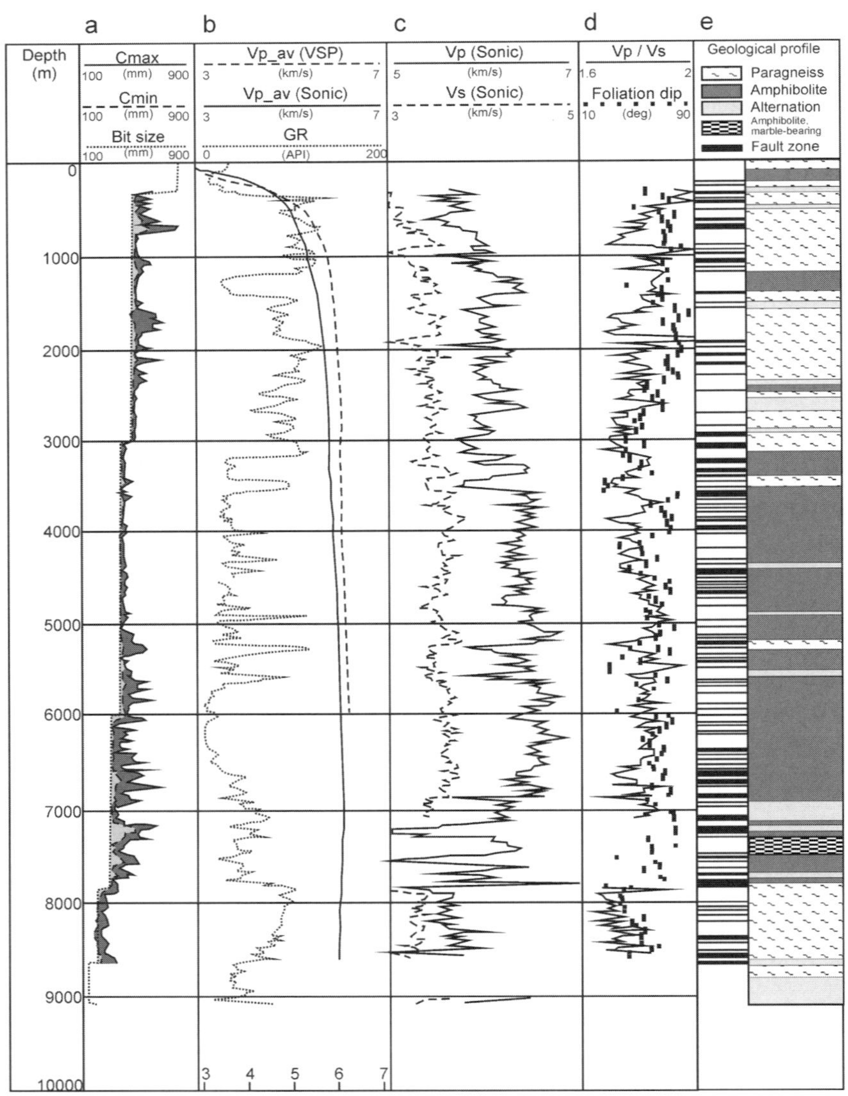


*Figure 11.20* Borehole measurements and geological profile of the KTB super-deep well. Note (a) shows calliper log measurements and hole diameter. Note the ubiquitous nature of faulting at all depths. Harjes et al., 1997.

*glacial ice* and *beach sand*. These occur just above sea level and they have two aspects in common. They each display high gradients of velocity, but from different starting points.

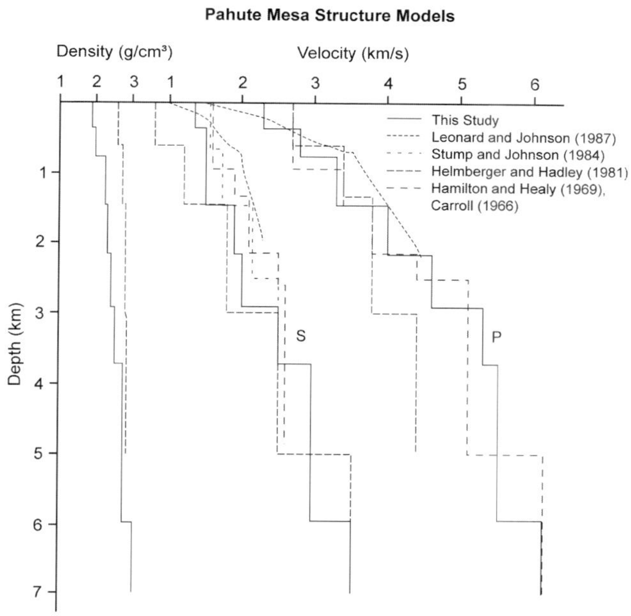


*Figure 11.21* Velocity-density-depth trends for the Nevada Test Site Pahute Mesa. Barker et al., 1991.

While thin, floating sea ice, typically $\leq 10$ m thick constitutes an approximately constant velocity layer, an ice accumulation such as the Ross Ice Shelf, Antarctica of many hundreds of metres thickness, displays a sharply declining velocity gradient with depth.

A multi-layered upper 50 to 100 m called *firn* is responsible for the steep velocity gradient. This includes snow cover which becomes *firn* after one melt season, and eventually becomes glacial ice when permeability to liquid water drops to zero with subsequent burial.

Investigations using seismic reflection and seismic refraction profiles, reported by Beaudoin et al., 1992, were located on the 200 to 850 m thick Ross Ice Shelf as shown in Figure 11.22.

The principal results of $V_p$ versus depth are reproduced in Figure 11.23. Compressional wave velocities in the near-surface ranged from 500 m/s at the surface to 2000 m/s at 10 m depth, a gradient of $150\,s^{-1}$. From 10 m to 70 m depth, the velocity increased from approximately 2.0 to 3.8 km/s, which represents a gradient of $30\,s^{-1}$.

In this region, metamorphism of the firn is governed by recrystallisation. Below about 70 m, any further compaction of the ice is by deformation of existing air pockets, with little effect on velocity (though possibly giving an orthotropic distribution). Of the four compression wave

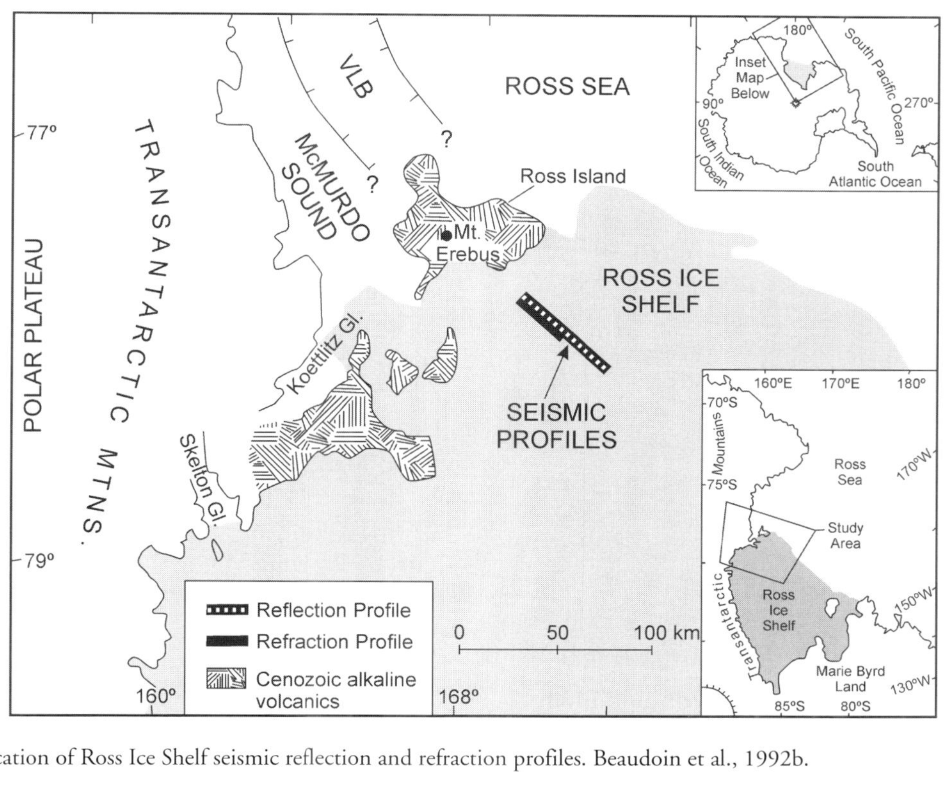


*Figure 11.22* Location of Ross Ice Shelf seismic reflection and refraction profiles. Beaudoin et al., 1992b.

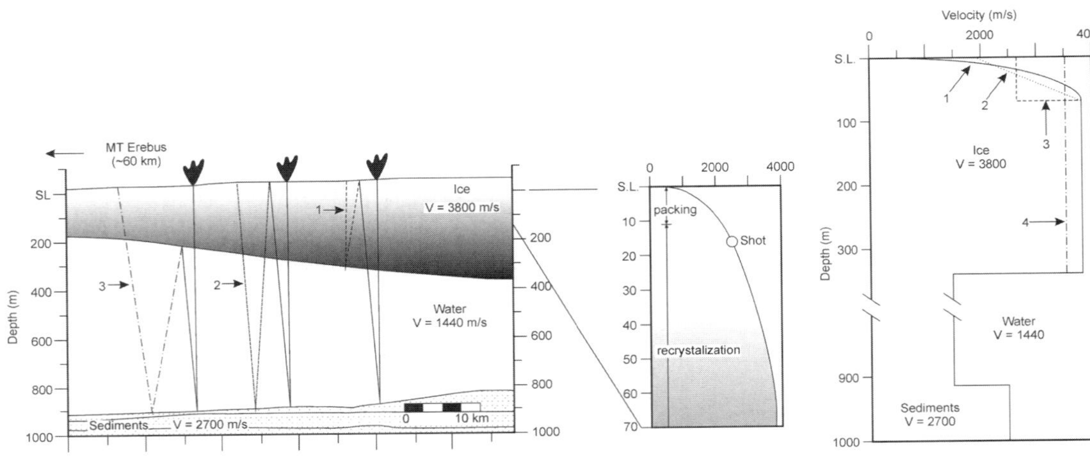


*Figure 11.23* a) Cross-section of ice and water. b) Near-surface $V_p$-depth gradient caused by firn c) Overall $V_p$-depth profile, and chosen models 1 to 4. Beaudoin et al., 1992b.

velocity models shown in Figure 11.23c, No. 1 was consistent with the observed data. Below the ice, 570 m of water with a velocity of 1.44 km/s reached down to subsea sediments with a velocity of 2.7 km/s.

Conducting the shallowest possible high-resolution seismic reflection and refraction experiment in the upper 2 m of a sea-*beach sand,* Bachrach and Nur, 1998, measured a minimum P-wave velocity of 0.04 km/s. They used only a 0.1 m distance between the shot and receiver. They calculated a theoretical minimum possible value of 0.013 km/s, considering the top few centimetres of dry sand as a suspension of sand in air.

The effective elastic modulus ($M_{eff}$) and the velocity of the air-quartz mixture were calculated from the following equations:

$$\frac{1}{M_{eff}} = \frac{\phi}{M_{air}} + \frac{1-\phi}{M_{quartz}} \quad (11.1)$$

$$M = K + \left(\frac{4}{3}\right)G \quad (11.2)$$

$$\rho_{eff} = (1-\phi)\rho_q + \phi \times \rho_{air} \quad (11.3)$$

$$V_p = \sqrt{\frac{M_{eff}}{\rho_{eff}}} \quad (11.4)$$

The following parameters were used: bulk modulus $K_{quartz} = 36.6$ GPa, shear modulus $G_{quartz} = 45$ GPa, $\rho_q$ = quartz density, $\rho_{air}$ = air density, $\phi$ is the porosity of the mixture, assumed at the critical value of 40%.

Their estimates of theoretical velocity, *and* their low measured values give velocities that are actually much lower than the velocity of sound in air.

## 11.3 The continental margin velocity structures

Velocities at continental margins, such as that obtained in the Atlantic margin seismic experiment described by Holbrook et al., 1994, naturally show some of the highest lateral variations of velocity, plus the familiar $V_p < 6.5$ km/s to 10 or even 20 km depth beneath the continental material. Figures 11.24 a and b show velocities and geological interpretation side by side, for a 240 km section off the East coast of the US. Short black and white lines are reflectors.

The multi-channel data was acquired using a 177 litre airgun array and 6 km long streamer, and coincident wide-angle data from ten ocean bottom seismic instruments. These seismic results along the US East Coast continental margin show the presence of a huge, high velocity (7.2–7.3 km/s) igneous body of as much as $2.7 \times 10^6 km^3$ in volume. This East Coast Margin Igneous Province (ECMIP) probably extends seaward, making it one of the worlds really large igneous provinces. The high velocity (in relation to thickness) is nicely demonstrated in the two further versions of the velocity-depth-distance sections, from Kelemen and Holbrook, 1995, shown in Figure 11.25.

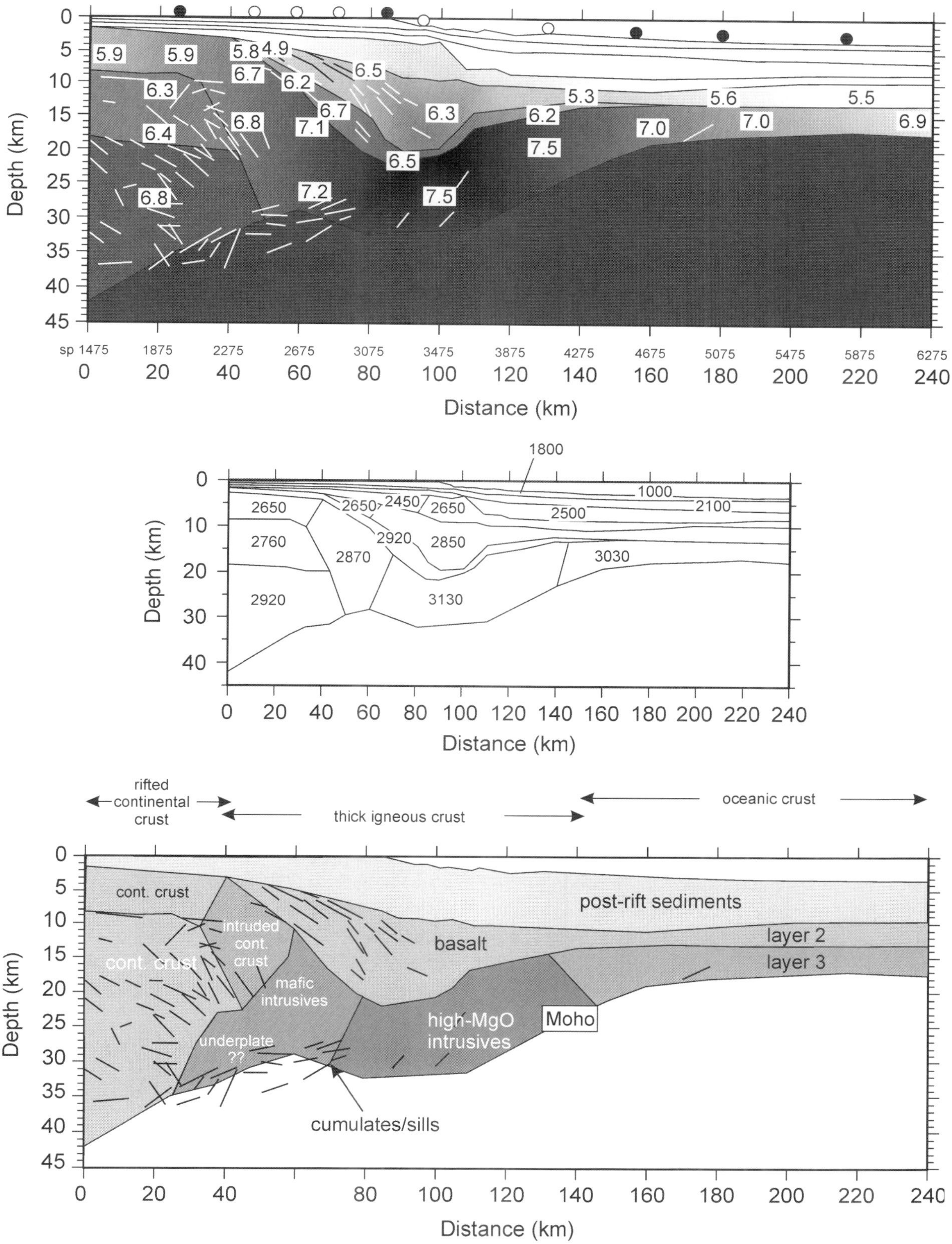


*Figure 11.24* US East Coast continental margin velocities (a), densities in kg/m$^3$ (b), and geology (c). Holbrook et al., 1994. Note that black and white lines are reflectors.

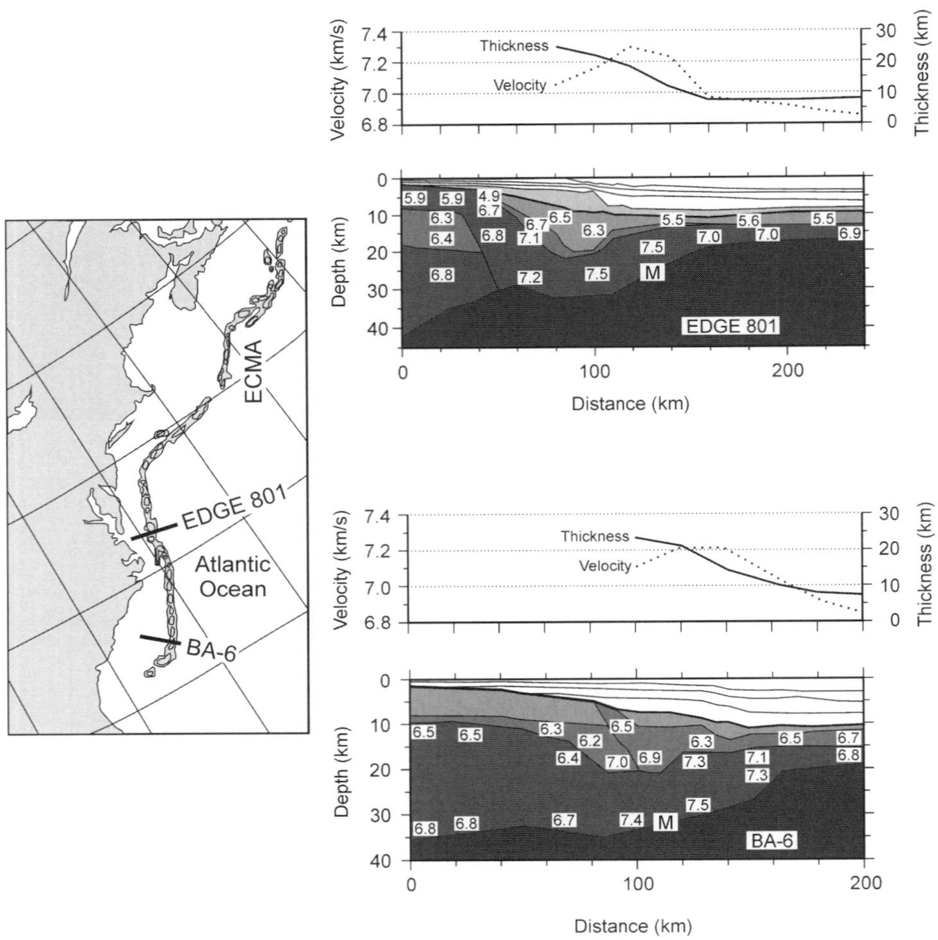


*Figure 11.25* Location and velocity-depth trends of two sections (BA-6, and EDGE-801) through the US East Coast margin, showing the velocity-thickness anomaly. Kelemen and Holbrook, 1995.

### 11.3.1 Explaining a velocity anomaly

In an effort to understand the likely composition of the rock in this huge magnetic and seismic velocity anomaly, Kelemen and Holbrook, 1995, assembled numerous high pressure laboratory $V_p$ data to try to differentiate the 25 km thick high velocity crust from the general 8 km thick ($V_p = 6.9$ km/s) sub-ocean crust.

Figure 11.26a shows a multiple linear regression fit to 188 garnet-free, igneous and metamorphic rocks. Measured $V_p$ at 25°C and confining pressures from 0.6 to 1.0 *GPa*, are compared with the bulk composition by weight of $SiO_2$ or MgO in the samples. The empirical relation obtained was:

$$V_p = 8.054 - 0.024\,(\%SiO_2) + 0.029\,(\%MgO) \tag{11.5}$$

(where $V_p$ is km/s).

This equation was subsequently corrected to lower-crustal temperatures (≈400°C) using an assumed $dV_p/dT$ gradient of (−) 0.0005 km/s/°C, by subtracting 0.2 km/s.

Figure 11.26b shows calculated $V_p$ for rocks crystallised from mantle melts as a function of the pressure of partial melting in the mantle. This was estimated by Kelemen and Holbrook, 1995, using their relation:

$$V_p = 6.712 + 0.16\,P_{melting}(GPa) + 0.661\,F_{melting} \tag{11.6}$$

where $V_p$ = km/s and $F_{melting}$ is the melt fraction of the parental melt, using reported $SiO_2$ and MgO contents, and the temperature corrected (−0.2 km/s) version of equation (11.5).

According to Kelemen and Holbrook, 1995, the goodness of fit of equation 11.6 did not substantially improve when other oxides like FeO, CaO, $Al_2O_3$, $Na_2O$ etc. were entered, because these compositional

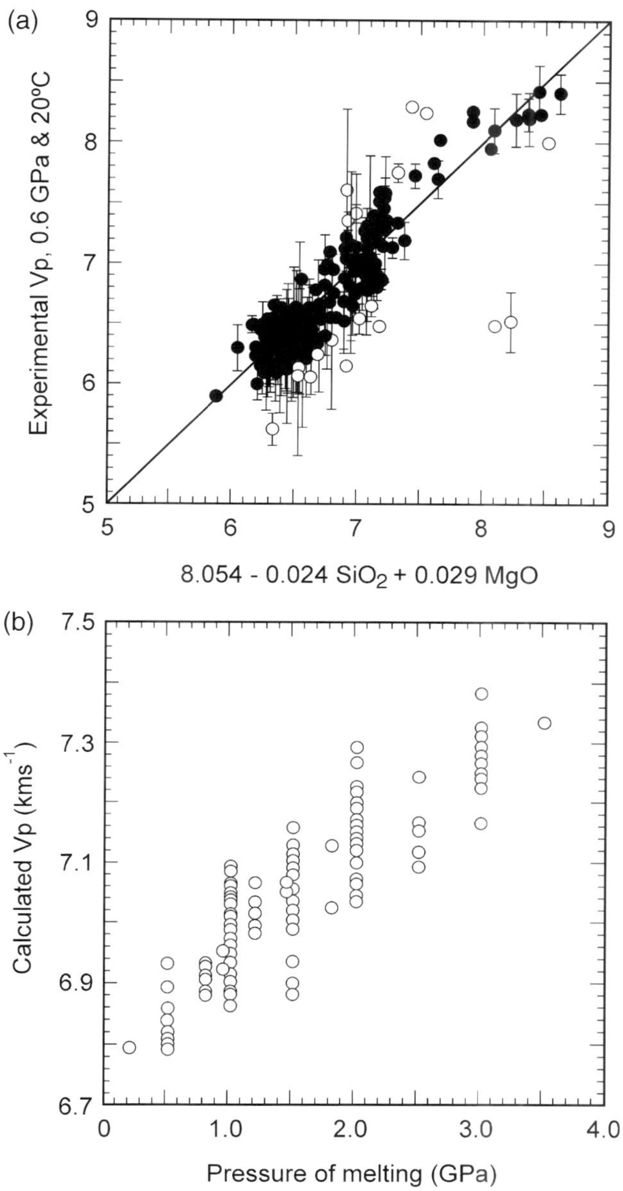


*Figure 11.26* a) $V_p$ and mineral composition for 188 igneous and metamorphic rocks at confining pressures of 0.6 to 1.0 GPa. b) Empirical calculation of $V_p$ versus pressure of partial melting. Kelemen and Holbrook, 1995.

variables were closely correlated with $SiO_2$ and MgO in the experimental set of data shown in Figure 11.26a.

According to their analyses, the high velocity body could have been produced from partial melting of mantle peridotite, using lower estimates of melt fractions (<10%) but higher average pressures (⩾2.0 GPa) than that producing normal mid-ocean ridge basalt. They surmised active upwelling of the asthenosphere at faster rates than lithospheric spreading rates, to produce the necessary high pressure conditions.

Gravity anomalies at the surface may be an expression of non-hydrostatic stresses at depth, implying that significant deviatoric stresses may exist. In the case of the Hawaiian Islands, gravity anomalies associated with flexure of the crust on either side of the Hawaiian ridge are associated with average velocity reductions in Layer 2 of some 0.8 to 0.9 km/s within the flexural arch, some 155 km from the ridge (Brocher and ten Brink, 1987).

Elastic and elasto-plastic flexural models for the region give predicted stress drops of 80 MPa in the upper lithosphere. These authors compare this with a similar confining pressure drop necessary to reduce velocities in porous basalts by 0.5 km/s in the laboratory.

The lateral velocity variations to the north and south of the Hawaiian ridge, produced partly as a result of this flexure, are shown in Figure 11.27. The models go from ocean floor to the bottom of Layer 2. We shall see many more models of oceanic velocities in the next section of this chapter.

Tomography was used by Hole et al., 2000, to invert earthquake and air gun travel time data in the San Francisco Bay area, to obtain 3D seismic velocity and earthquake hypocentres. Most hypocentres were relocated up to 2 km from their catalogue locations, and the 3D approach was also important for mapping lateral velocity contrasts (subvertically through most of the crust) where major strike-slip faults were present. These lateral velocity variations correlated well with known surface geology differences.

Strong velocity contrasts of 0.3 to 0.6 km/s were observed in the *middle crust* when crossing the San Andreas fault. Weaker contrast (0.1 to 0.3 km/s) existed at other depths, and across two other faults. The relocated seismicity hypocentres on the active strike-slip faults defined steeply dipping planes beneath the surface expression of each fault.

Figure 11.28 compares a Hole et al., 2000, 3D based velocity tomogram with a 2D refraction model of Holbrook et al., 1996.

Throughout the Hole et al., 2000, San Francisco Bay study area, *no earthquake was found to occur in regions with* $V_p > 6.3$ *km/s*, and usually the various inversion tests produced a maximum velocity of 6.2 km/s. They surmised that the base of seismicity may be thermally controlled by a deeper brittle-ductile transition in the relevant Franciscan rocks. A simpler, shallow depth Q-M-$V_p$ model argument (Barton, 1995, 2002) might be that fracturing, that has to be present, has kept $V_p$

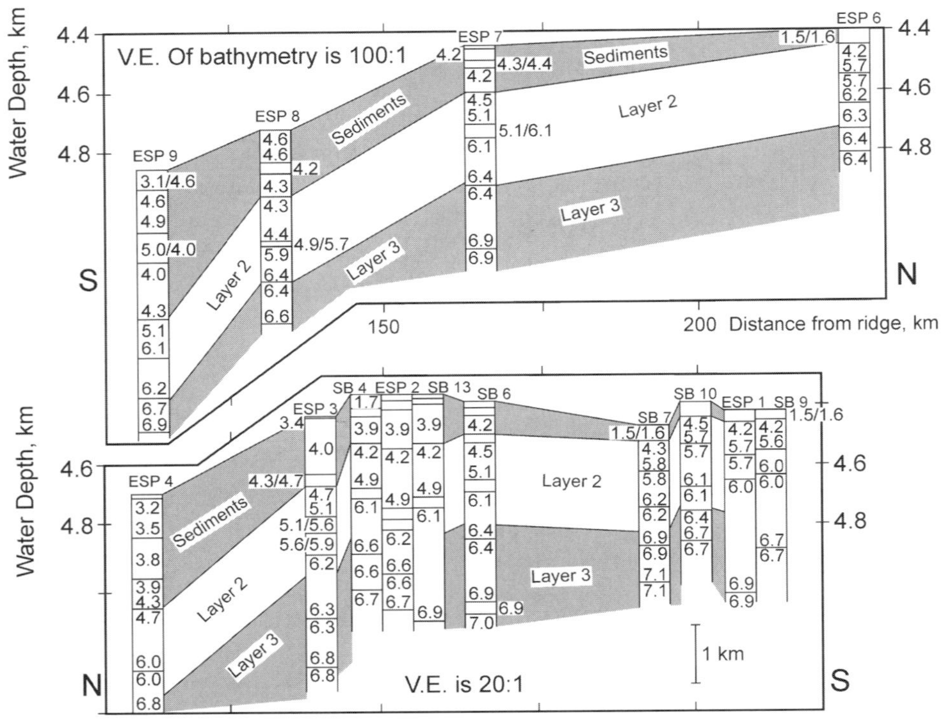


*Figure 11.27* Velocity-depth solutions from N and S of the Hawaiian ridge. Brocher and ten Brink, 1987.

below the 6.5 km/s 'limit' for intact, strong, highly stressed rock masses. However there is of course a need to extrapolate the Q-M-$V_p$ model to greater depths to be related even to shallow earthquakes.

The amount of detail in depth-velocity structures for onshore and offshore southern California (adjacent to the Pacific and North America plate boundary) was recently increased with new 3D $V_p$ and $V_p/V_s$ models using P and S-P travel times from local earthquakes and from controlled sources (Hauksson, 2000). A 15 km horizontal grid-spacing, and an average vertical grid spacing of 4 km, down to 22 km gave new insight into the heterogeneity of crustal structure in this earthquake-prone region. The near-surface increase in P-wave velocity, from the surface to 8 km depth was found to be rapid and had a logarithmic shape for stable blocks, but was slower and had a more linear slope for sedimentary basins (Figures 11.29a, b, c).

Ratios of $V_p/V_s$ varied widely in the upper 5 km and often fell outside the typical ratio of 1.7 to 1.8 generally seen at lesser depths. Values as high as 1.9 to 2.0 were seen in sedimentary basins and in locations below an offshore channel (Santa Barbara). High $V_p/V_s$ ratios may be related to the high fluid content of near-surface cracks and joints, in other words, the joint sets that still remain partly open.

The author was able to recalculate the hypocentres of recent earthquakes using the improved three-dimensional velocity models, which clearly differentiated the sedimentary basins from the nearby mountains. Hauksson also compared his 3D $V_p$-depth profiles (using double standard deviations) with laboratory $V_p$ measurements of triaxially loaded samples from McCaffree et al., 1998.

It is of interest to note the 'reluctance' of the *in situ* data in Figure 11.31 to exceed $V_p = 6.5$ km/s – which in the $V_p$-Q-M model of Barton, 1995, is the supposed limit for completely unjointed rock masses, or rock masses with neither primary or secondary porosity and 'normal' composition (i.e. granites, gneisses etc.). Presumably the stress levels at 5 or more kilometres depth are sufficient to completely close (in a joint-normal direction) the apertures of any joints, since stresses of the order of 130 MPa and more are close to the expected JCS values of joint walls in the schists, intrusives and gneisses. (JCS = joint wall compressive strength, Barton and Choubey, 1977.) On the other hand with shearing deformation along the joints, apertures and permeability

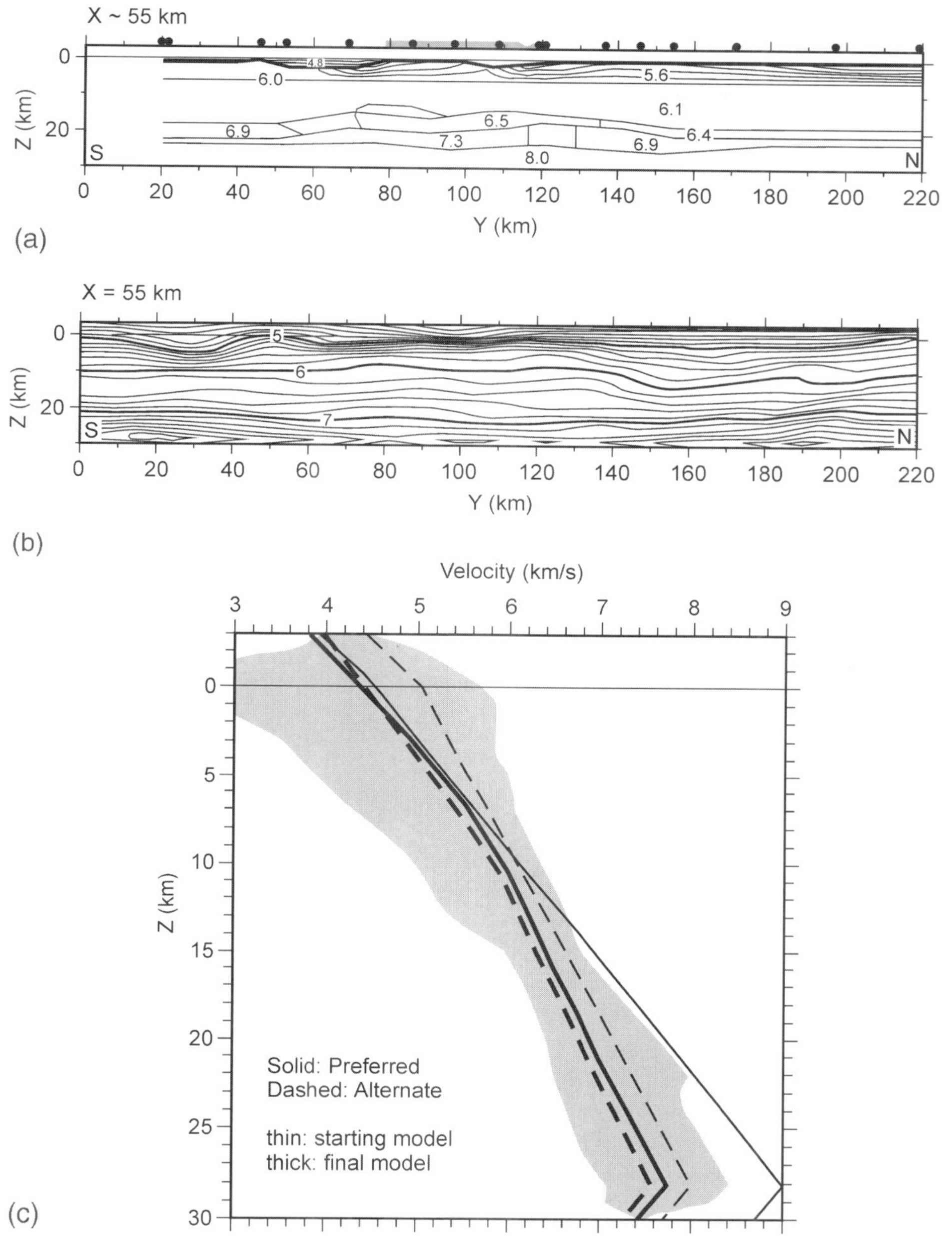


*Figure 11.28* Comparison of a) 2D and b) 3D velocity tomograms for San Francisco Bay area crust. c) Range of velocity-depth models providing solutions to the inverted travel times. Hole et al., 2000.

could remain of finite magnitude at these (and greater) depths, as suggested by the work of Zoback and colleagues, reviewed in Chapter 16.

The 3D velocity data shows mean values of 5.4 and 5.7 km/s at one kilometre depth. In the Q-M-$V_p$ model, (see Figures 5.36 and 5.37), a significant amount of jointing and/or alteration along the joints would be suggested, with a $Q_c$ (rock quality) value on either side of 1, if we make no allowance for porosity $n > 1\%$ or $\sigma_c < 100$ MPa, or $> 100$ MPa. The relatively low velocities of some of the laboratory samples of schist and intrusives (as low as 4.5 to 5.0 km/s at 1 km depth equivalent) does suggest that n% and $\sigma_c$ adjustments would be needed to fine-tune $Q_c$ rock quality estimates.

There are other important details regarding the velocity model for the San Gabriel ranges discussed above

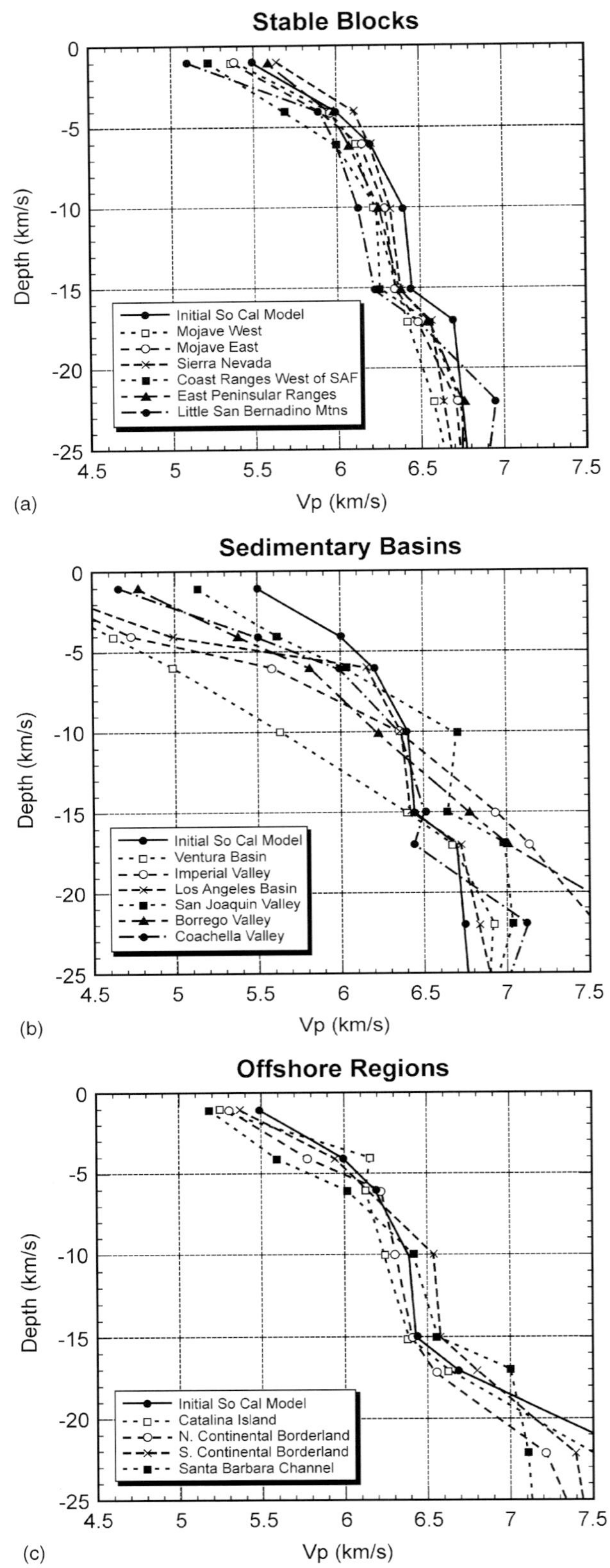


*Figure 11.29* Velocity-depth data for a) stable blocks, b) sedimentary basins and c) offshore regions off southern California. Hauksson, 2000.

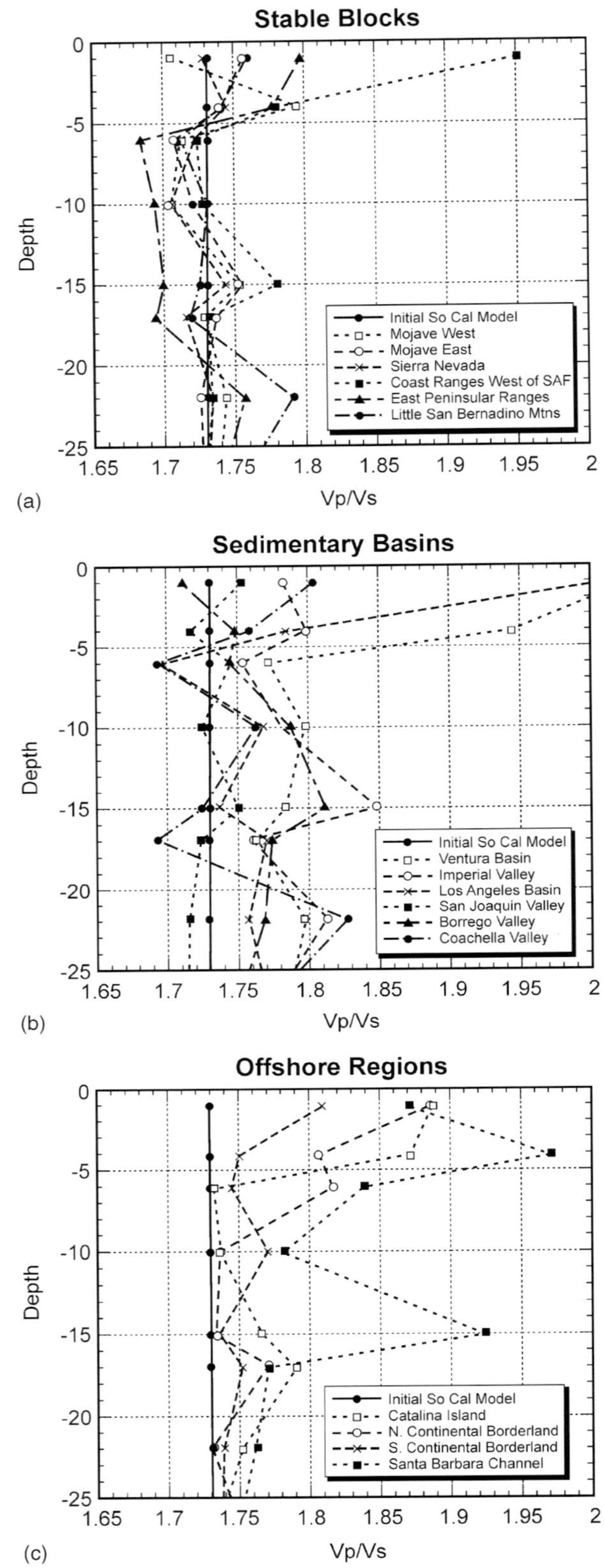


*Figure 11.30* $V_p/V_s$ ratio trends for a) stable blocks, b) sedimentary basins, and c) offshore regions of southern California. Hauksson, 2000.

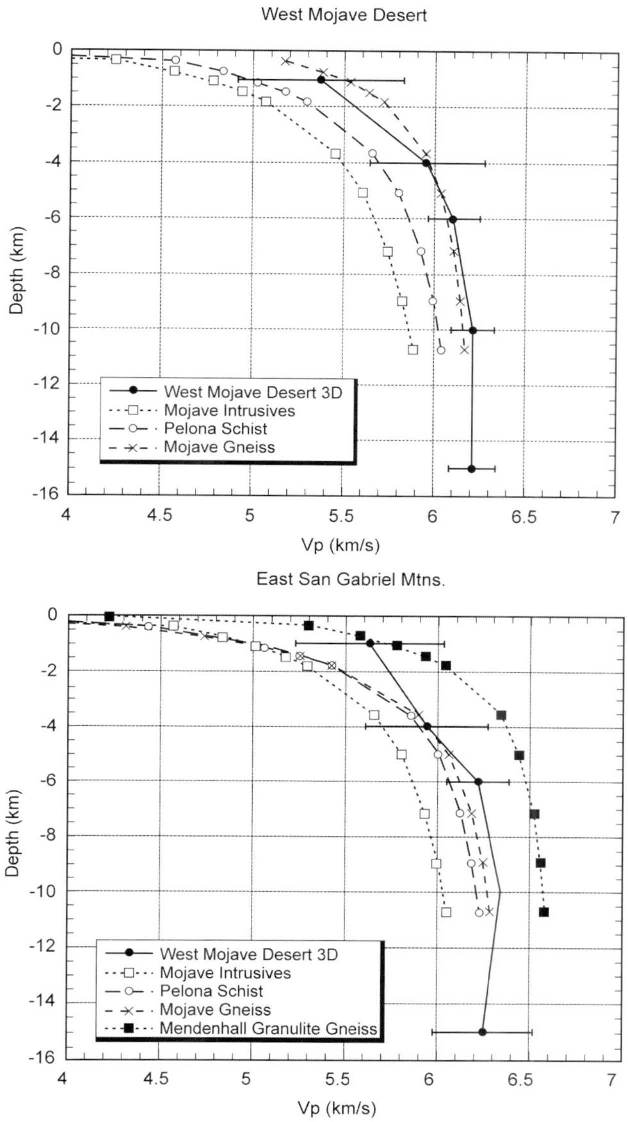


*Figure 11.31* Comparison of velocity-depth trends for two onshore regions of southern California (solid lines: '3D') with laboratory tests (all dotted lines). Hauksson, 2000 and McCaffree et al., 1998.

(Figure 11.31b). Hauksson, 2000, warns that for *shallow earthquakes*, the seismic waves from the hypocentres to the recording stations are travelling through the schist with subhorizontal ray paths, while rays from *deep earthquakes* may have steep angles of incidence. Thus in both the Mojave (Figure 11.31a) and San Gabriel (Figure 11.31b) terrains, the rays from the hypocentres will sample the average velocity of the schist, because, in relation to the foliation dip angles (45° to 90° for the Pelona schist) the rays have all possible azimuths and a large range of take-off angles.

## 11.4 The mid-Atlantic ridge velocity structures

The object of systematic geophysical inversion techniques is to derive structures which fit a given set of observations. For many years, sub-oceanic marine seismic refraction profiles were interpreted as a small number of layers separated by planar interfaces, with a constant velocity assumption for each layer. A major advantage of layer solutions was that they could be developed using a desk calculator.

As equipment (e.g. sonobuoys and repetitive sources such as airguns) and computing capacity improved, homogeneous layering assumptions from the 1960s, e.g. Ewing, 1963, were first replaced by much finer layering and then in the mid-1970s by continuous gradients in velocity (e.g. Kennett and Orcutt, 1976). These authors showed that the seismic data do not require uniform layering as a solution, but they did not exclude the possibility of homogeneous layering. A typical set of their solutions, with bounds compared with the layered solution is shown in Figures 11.32a, b.

The first geophysical *downhole logging* data for oceanic crustal material is reportedly that of Kilpatrick, 1979. He found that the predictions of low velocities from refraction seismic were borne out by downhole sonic logging. In situ sonic velocities were typically from 1.5 to 4.8 km/s in the upper 200 m of oceanic Layer 2A. Calculated porosities of 13 to 41% were unexpectedly high. Formation damage away from the drilled holes was considered to be minor, as electrical resistivity away from the hole showed a lack of radial variation. The measurements were made in hole 396B (leg 46) of the Deep Sea Drilling Project, near the mid-Atlantic ridge.

The reasons for the high porosities were interpreted as being due to a combination of sediments, rubble, and solid basalt in contrast to the compact nature of basalt samples used in laboratory tests, which often has $V_p$ between 5.5 and 6 km/s and porosities from only about 2 to 8%. Open fractures and voids were assumed to exist on a scale larger than the laboratory samples, giving high permeability throughout the drilled section.

A decade of Lamont-Doherty Geological Observatory sonobuoy data led Houtz and Ewing, 1976, to conclude that the P-wave velocity of the sub-ocean crust at and near ridge crests actually exhibited an increase in velocity with age. Numerous results from the Atlantic and Pacific shown in Figure 11.33 showed an obvious link between $V_p$ and age, up to some 40 million years. Deeper and older layers did not show systematic increase in velocity.

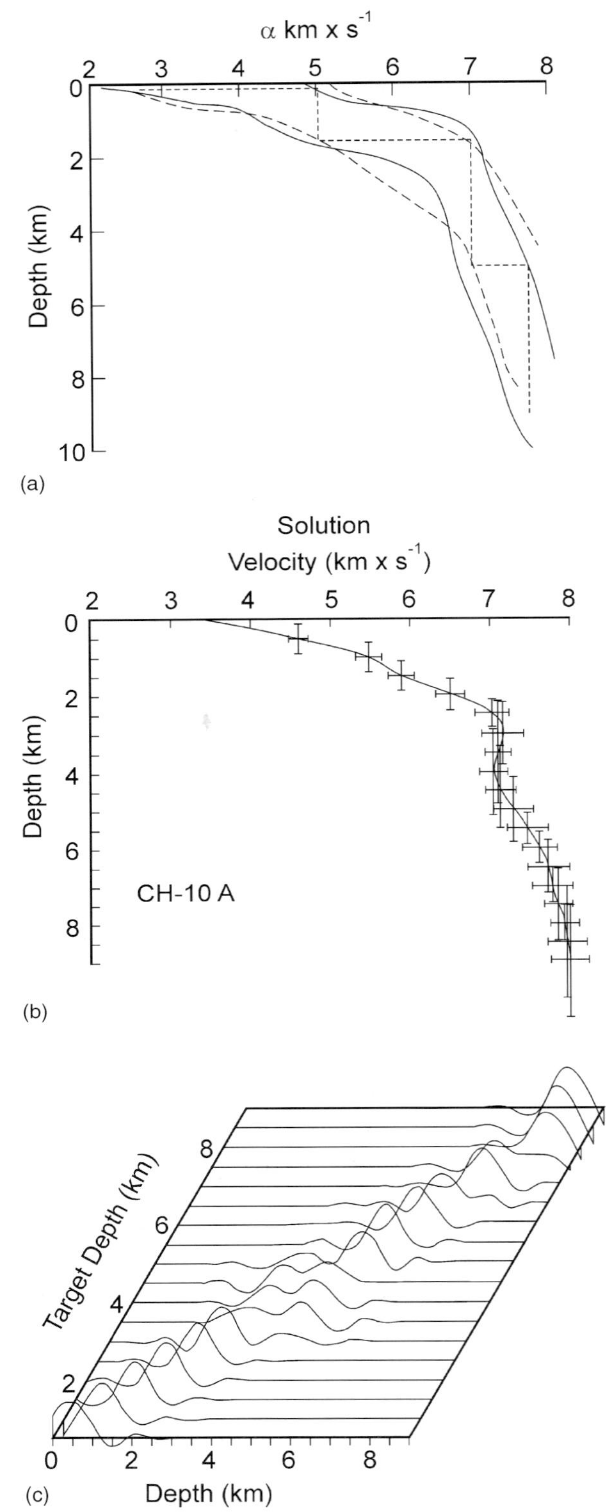


*Figure 11.32* a) Velocity-depth bounds from inverted marine refraction profiles, compared to Layer 2 and 3 constant $V_p$ models. b) Mean value and error bars for same profile CH-10A. c) Resolving kernels for CH-10A. Kennett and Orcutt, 1976.

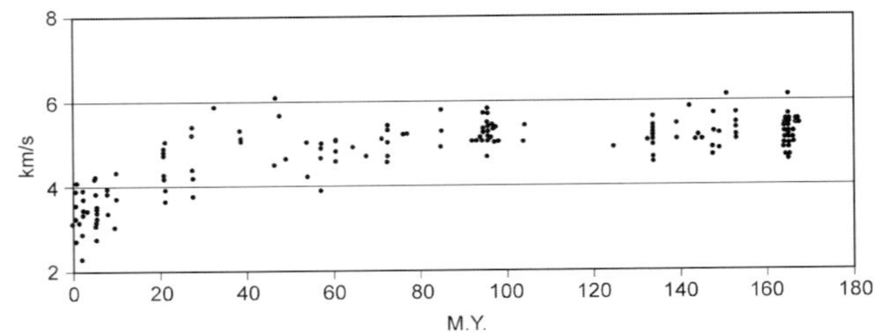


*Figure 11.33* Measured velocities as a function of age for Atlantic and Pacific sites. Houtz and Ewing, 1976.

The rocks concerned were generally pillow basalts or vesicular, and could be weathered or massive and fresh. The authors argued that it was difficult to envisage any diagenetic change with the relatively low (effective) overburden pressures, so filling of voids and cracks (and presumably joint sets) with products of hydrothermal mineralization must presumably be one of the mechanisms involved in the increasing velocity with age.

Different calculated porosities for 4 m.y. and 62 m.y. crust as a function of depth, given by Whitmarsh, 1978, and reproduced in Figure 11.34, suggested zero porosity beneath 1.5 km of Layer 2A oceanic crust. The two rectangles, the asterisk and the three dots were from limited borehole data available at that time. Filling of cracks by hydrothermal minerals with increasing age was cited as the likely mechanism.

The early (and continued) difficulty of obtaining samples of oceanic crust to several kilometres depth, led Salisbury and Christensen, 1978, to 'reconstruct' the intact rock $V_p$, $V_s$ and dynamic Poisson's ratio structure, along a traverse through an on-land (Bay of Islands, Newfoundland) ophiolite complex. The $V_p$ and $V_s$ data shown in Figures 11.35 and 11.36 were derived from hydrostatically confined and water saturated intact samples. They suggested that the velocity structure should be indistinguishable from normal oceanic crust, but with the notable difference that the structure (voids, joints, fractures etc.) were not of course sampled. The velocities, especially in the upper 1 km, therefore represented maxima.

The authors showed from earlier studies (Christensen and Salisbury, 1972), the strong link between velocity and density for oceanic crust basalts. At high porosity, with a density of only 2.5 g/cm$^3$, $V_p$ tended to be only about 4.5 km/s, and from then on showed a linear increase to about 6.5 km/s by the time the density had reached 3.0 gm/cm$^3$.

Clearly, seismic velocity and density are mutually dependent properties of a rock, and each are strongly

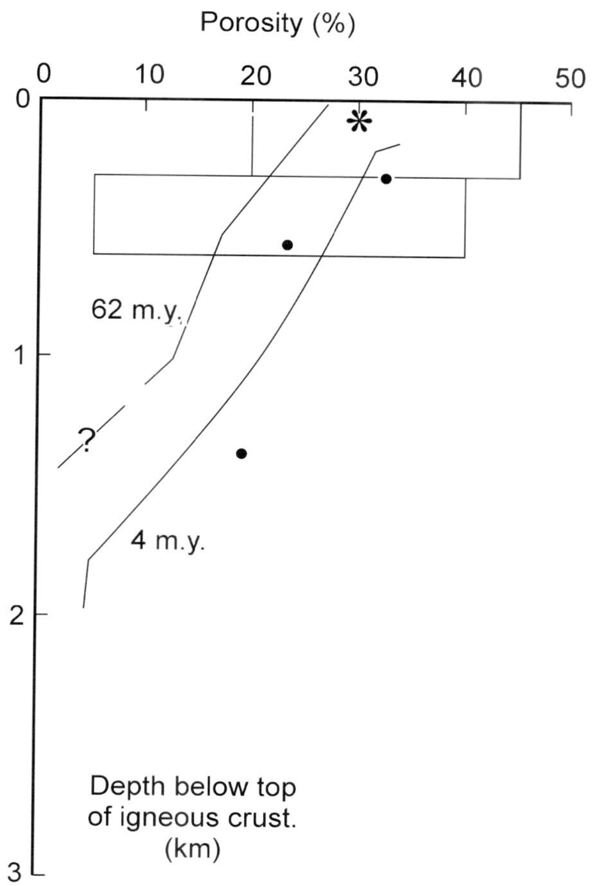


*Figure 11.34* Estimates of porosity-depth relations for different ages of oceanic crust (basalt) given by Whitmarsh, 1978.

linked to porosity, uniaxial compressive strength and Young's modulus. A major collection of density-$V_p$ measurements for a wide range of sedimentary, metamorphic and igneous rocks is shown in Figure 11.37, from Ludwig et al., 1970, and P.J. Barton, 1986. The studies with oceanic crust basalts cited above ($V_p$ = 4.5 km/s with $\gamma$ = 2.5 gm/cm$^3$, $V_p$ = 6.5 km/s with $\gamma$ = 3.0 gm/cm$^3$) clearly fit this huge data set.

The thin line in the centre of the scatter in Figure 11.37 is the mean velocity-density relationship, while the heavy boundaries contain the great majority of data. At any given value, a density variation of about ±0.2 to 0.3 gm/cm$^3$ and a velocity variation of ±0.5 to 1.0 km/s are seen. The thick vertical bar, corresponding to a density of 2.8 gm/cm$^3$ is typical for crystalline continental crust, with $V_p \approx$ 5.7 to 7.0 km/s. Closed circles represent sedimentary rocks and open circles represent metamorphic and igneous rocks, relevant in these oceanic crust studies.

While density is a bulk property independent of direction, seismic velocities can be anisotropic due to the effects of microcrack alignment (e.g. in recovered, stress relieved samples) or due to fabric anisotropy. We may consider that density is related to the 'hard porosity' of the rock, while 'soft porosity' in the form of jointing (i.e. the rock quality Q-value) gives higher or lower velocities depending on great depth or shallow depth respectively.

### 11.4.1 A possible effective stress discrepancy in early testing

Extensive laboratory testing of oceanic basement rocks from deep drilling in the mid-Atlantic ridge by Hyndman and Drury, 1976, highlighted the discrepancy between laboratory seismic properties and in situ, bulk velocities obtained from seismic refraction.

One of the problems was obviously sampling bias, and the other may have been the early tendency to test at much too high effective stresses. The 'intact' basalt cores contain only disconnected vesicules and were earlier assumed to be under a high effective stress due to both sea load and crustal load. The actual effective stress acting on deep-ocean shallow crust, is most likely to be relatively low, due to a highly permeable assemblage of basalt, sediment layers and joints/fractures and larger voids.

The typical samples for laboratory tests were minicores of 2.5 cm diameter drilled transversely to the recovered pieces of 6 cm diameter vertical core. The drillers operating onboard the Glomar Challenger drilling ship generally recovered only 20% of the sections drilled in mid-ocean upper-crustal rocks, so there was a further source of sampling bias.

Due to the controversial opinion of appropriate stress levels for testing laboratory samples, discussed above, confining pressures of 17 MPa and even up to 207 MPa were used by Hyndman and Drury, 1976, for on-board velocity measurements. A thin metal sheet jacketed the sample so the confining pressure was 'close to the external pressure, and the internal pore pressure was small.' Results of these high pressure velocity-density studies are reproduced in Figure 11.38.

Hyndman and Drury, 1976, found that the vesicular porosity of their relatively fresh, young basalts, caused Poisson's ratio to *increase* with increased velocity (i.e. from 0.28 at $V_p$ = 5.8 km/s, to about 0.3 at $V_p$ = 6.4 km/s, due to the vesicules affecting $V_p$ more than

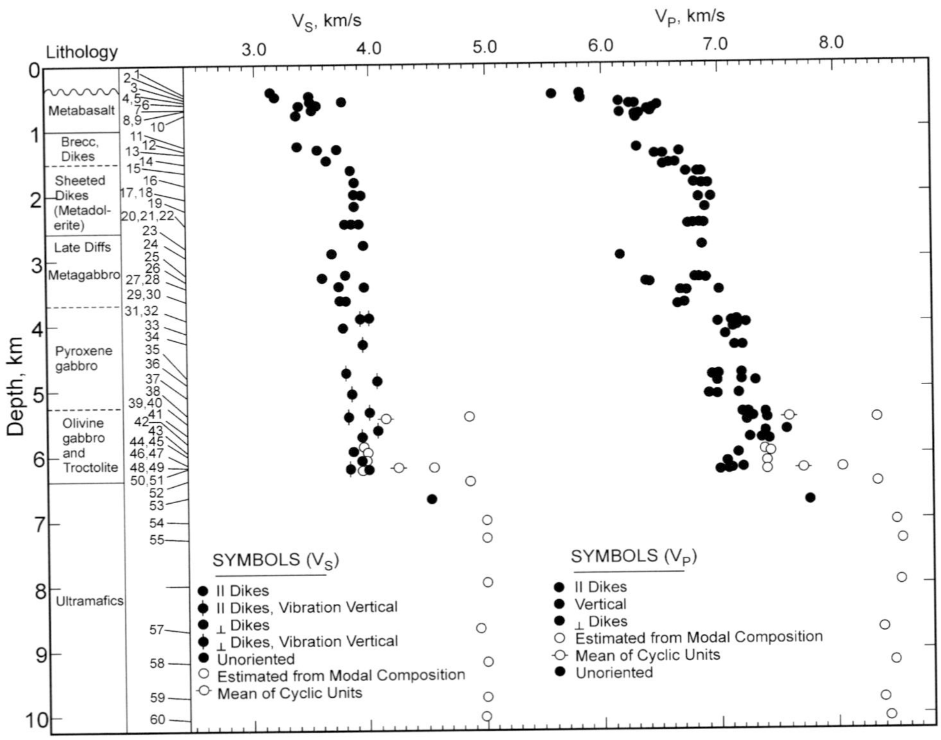


*Figure 11.35* $V_p$ and $V_s$ data from hydrostatically confined laboratory samples, plotted as a function of depth in the Blow-Me-Down massif of the Bay of Islands ophiolite complex Newfoundland. Salisbury and Christensen, 1978.

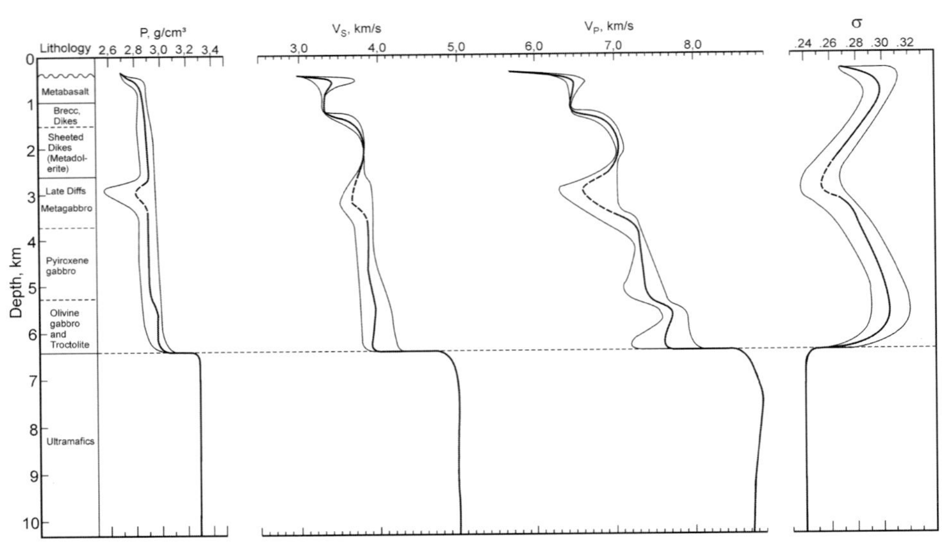


*Figure 11.36* Smoothed envelopes of the same $V_p$ and $V_s$ test data, plus density and dynamic Poisson's ratio, for the ophiolite Blow-Me-Down massif samples. Salisbury and Christensen, 1978.

$V_s$. In contrast, Christensen and Salisbury, 1972, testing older and shallower depth basalts found Poison's ratio *reducing* with increased velocity, due to the greater effect of grain boundary weathering on $V_s$ than $V_p$. The large scale joints, fractures and voids in situ are also likely to cause increased Poisson's ratio with reducing velocities, in contrast to the decrease caused by equi-dimensional (vesicular) pore spaces.

Even in 100-m.y.-old sea floor the reduction in velocity caused by weathering appeared to only extend to

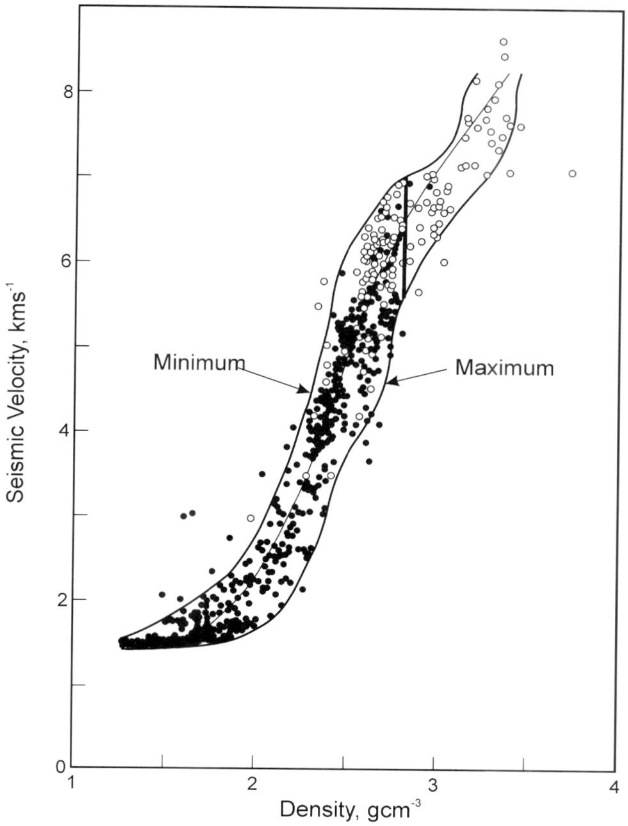


*Figure 11.37* An extensive set of laboratory P-wave and density data for sedimentary (closed circles), metamorphic and igneous rocks (open circles). Ludwig et al., 1970, and P.J. Barton, 1986.

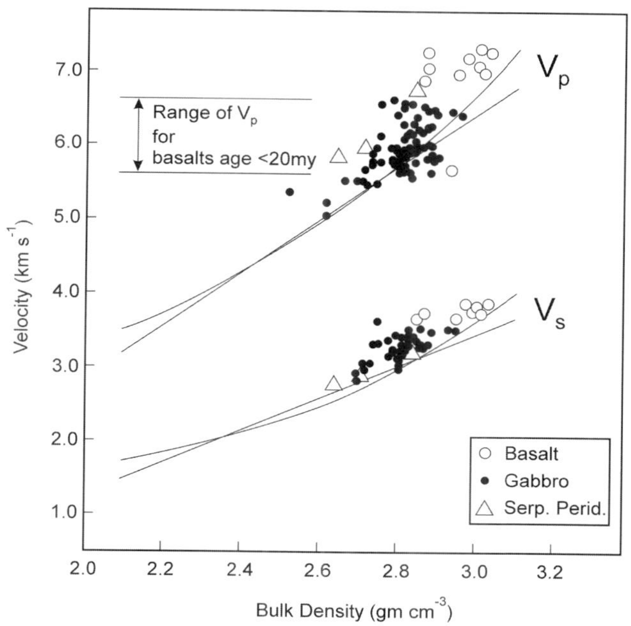


*Figure 11.38* $V_P$-$V_S$-density data from high pressure triaxial tests of three rock types. Hyndman and Drury, 1976.

about 50 m depth. The mean laboratory velocity for basalts younger than 20 m.y. and generally from deeper than 50 m, appeared to be frequently in the range 5.9 to 6.0 km/s. Hyndman and Drury, 1976 showed laboratory velocities (of about 5.5 to 6.5 km/s) obtained with 50 MPa confining pressure, next to the refraction-seismic inferred velocity depth profiles. However, laboratory data for samples recovered from about 3 m to 60 m depth, given by Hyndman, 1979, gave velocities from as low as 4 km/s to 5.2 km/s just below the ocean floor, up to 5.4–6.5 km/s at 60 m depth.

These were also presumably tested at excessively high confining stress levels. The 2.5 to 3.5 km/s *in situ* velocities were showing an *apparent* discrepancy of about 3 km/s relative to the intact rock, but in fact some of this difference was presumably due to the excessively high confining pressures applied to the intact samples.

It is clear that most of the velocity discrepancy was due to the dominance of larger voids and fractures, and the effect these had on the effective stresses. In a bulk sense, the effective stresses are much lower than those assumed or actually acting on the intact, but vesicular basalt.

Presumably the vesicules may originally be gas-filled at a pressure at least as high as the surrounding water pressure into which they were injected. Whether the matrix porosity is too low to allow either inward saturation with water (while under the ocean) or escape of some of the excess gas pressure (when brought to atmospheric pressure) is perhaps still a point of controversy.

### 11.4.2 Smoother depth velocity models

A new picture of the seismic structure of the oceanic crust began to emerge at the beginning of the 1980s, with the work of Spudich and Orcutt, 1980a. It was found that velocity models in which velocity varied smoothly with depth generally explained wave amplitude variations better than the earlier 'thick, homogeneously layered' models. Some indications of this were apparent some 10 years earlier, but in this compilation, Spudich and Orcutt, 1980a included many sites to confirm the new trends.

The three homogeneous layers 2A, 2B and 2C suggested by Houtz and Ewing, 1976, were now considered too simplified, as finer structure, with significant lateral variations, showed a mix of velocity gradients, but generally within the range 1 to 2 $s^{-1}$.

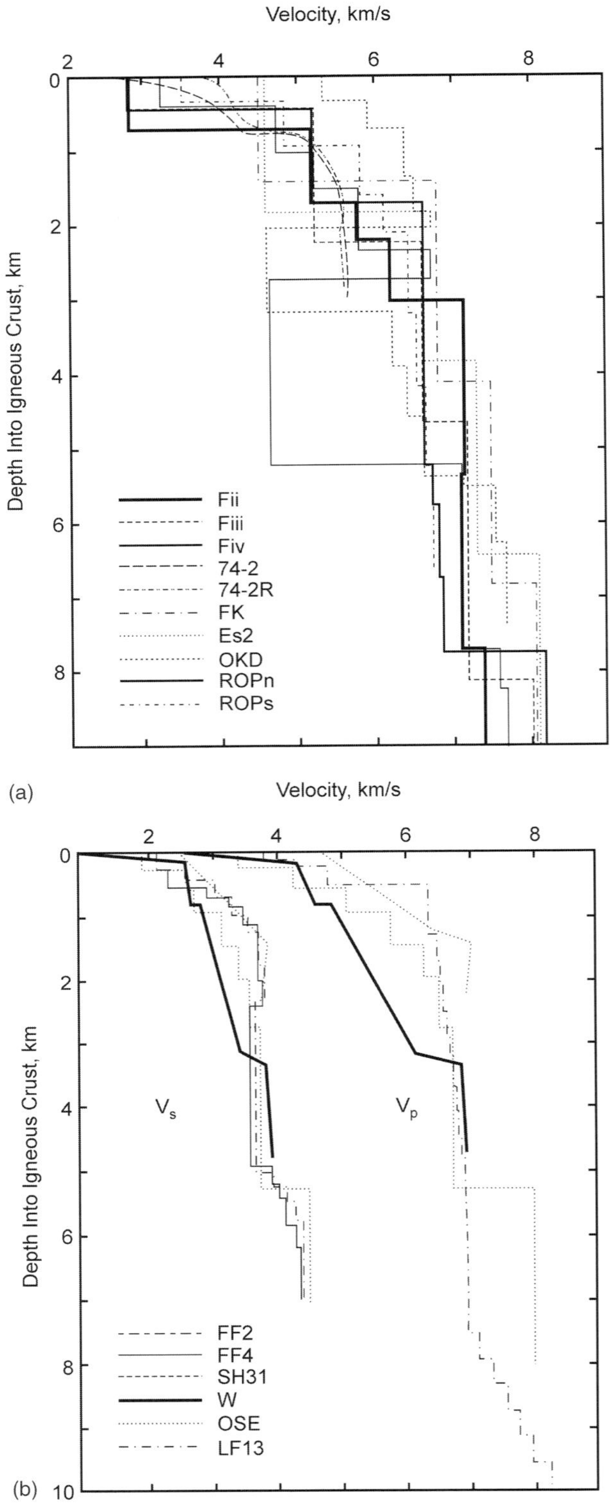


*Figure 11.39* a) Mid-ocean ridge and near-ridge (i.e. younger) $V_p$-depth profiles derived by synthetic seismogram modelling. b) $V_p$-depth profiles derived from sites older than 20 m.y. Spudich and Orcutt, 1980a. For references to each profile, see their paper.

Figure 11.39a shows selections of the velocity depth models assembled by Spudich and Orcutt, 1980a. The upper diagram is for ten ridge or near-ridge sites, which contrast significantly (at shallow depth) with the higher velocities of the six older than 20 m.y. sites shown in Figure 11.39b.

Spudich and Orcutt discussed possible reasons for the quite steep velocity gradients (approximately 1 to 2 $s^{-1}$) of oceanic 'layer 2' as being caused by finer structure, which was highly variable laterally (i.e. it varied with increased or decreased age). However the finer structure was difficult to resolve with the currently existing resolution of explosion seismology.

Drilling of the shallow crust from submersibles, that had begun in the 1970s, had shown that the shallow crust was permeated with numerous sediment and/or water filled fissures, i.e. voids larger than laboratory sampling scale. Spudich and Orcutt, 1980a argued that if 'layer 2' were, *on average*, composed of the Mid-Atlantic Ridge basalt identified by Hyndman and Drury, 1976:

$$V_p = 5.94\,\text{km/s}$$

$$V_s = 3.26\,\text{km/s}$$

$$\gamma = 2{,}80\,\text{Mg/m}^3$$

$$n = 7.8\%$$

then the addition of another 10% of porosity in the form of larger water-filled cracks or fissures, could give a $V_p$ range from 5.5 to 2.6 km/s using current crack and spherical pore models. This range could encompass nearly the entire range of layer 2A and 2B velocities observed by Houtz and Ewing, 1976.

The aspect ratio of cracks and fissures, whether they were water filled or sediment filled, and whether they could close in response to increased effective stress was, naturally, the subject of much discussion. It was also assumed by now that alteration of the older basalt could have resulted in progressive infilling and cementation of the cracks, thus explaining the increased velocities with age.

### 11.4.3 Recognition of lower effective stress levels beneath the oceans

On the subject of effective stresses in the uppermost permeable sub-ocean crust, Todd and Simmons, 1972,

Table 11.3 Interpretation of upper ocean crust P-wave velocities east of Guadalupe Island. Profile FFZ of Spudich and Orcutt, 1980b. See Figure 11.40.

| Layer | Depth of crust (km) | $V_p$ km/s | $\gamma$ (Mg/m$^3$) |
|---|---|---|---|
| 3 | 0.40 | 4.6 | 1.99 |
| 4 | 0.43 | 5.0 | 2.15 |
| 5 | 0.60 | 5.0 | 2.15 |
| 6 | 0.63 | 5.75 | 2.43 |
| 7 | 0.79 | 5.75 | 2.43 |
| 8 | 0.85 | 6.20 | 2.60 |
| 9 | 1.14 | 6.20 | 2.60 |
| 10 | 1.22 | 6.20 | 2.60 |
| 11 | 1.40 | 6.30 | 2.64 |
| 12 | 1.47 | 6.30 | 2.64 |
| 13 | 1.72 | 6.42 | 2.68 |
| 14 | 1.90 | 6.90 | 2.86 |

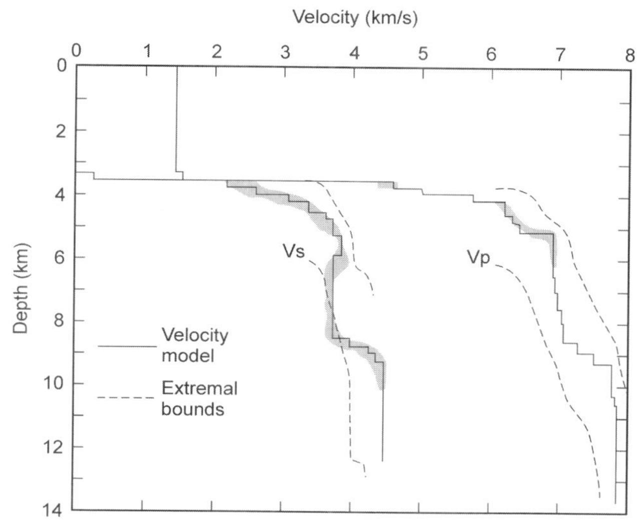


*Figure 11.40* Best fitting $V_p$ and $V_s$ for profile FF2, described by Spudich and Orcutt, 1980b.

and Spudich and Orcutt, 1980b, seem to have been among the first to argue that it must have been the rapidly increasing effective stresses that were acting on the shallow sub-ocean crust that was causing the velocity increase. In other words, with $P_o$ as the water pressure at the ocean floor (often 3 or 4 km depth of water), the pore pressure $P_p$ acting at depth z into the ocean crust will be:

$$P_p = P_o + \gamma_w g z \tag{11.7}$$

while $P_e$ the external stress at the same depth in the crust will be:

$$P_e = P_o + \gamma_r \times g z \tag{11.8}$$

Therefore the effective stress $P_p - P_e$ will be given by:

$$P_e - P_p = (\gamma_r - \gamma_w) g z \tag{11.9}$$

This obvious cause of an effective stress gradient in the crust was cited by these authors as a reason for measuring velocities in laboratory tests down to zero effective stress, rather than the practice (at that time) of measuring velocity at elevated triaxial stress states with zero pore pressure.

Based on an interpretation of the sharp velocity gradients shown in Table 11.3, (where a water depth of 3.4 km has been subtracted) and based on Figure 11.40, Spudich and Orcutt, 1980b, argued that a vesicular ('spherical') porosity of 18% would match velocities in the top 200 m of the basalt crust (i.e. 4.6 km/s) and that reductions of porosity to 2% could explain the increased velocity down to 0.6 km/depth (i.e. 5.75 km/s). While discussing the likelihood of fractures and voids as well as matrix porosity, they focused more on matrix-type porosity changes than on effective stress-induced closure of joints or fractures.

### 11.4.4 Direct observation of sub-ocean floor velocities

Three new experiments carried out on the Mid-Atlantic Ridge (MAR), near latitude 23°N, were described by Purdy, 1987. The uppermost few hundred metres of the oceanic crust were tested using a fixed ocean floor hydrophone receiver, and a controllable explosive source that was towed within a few tens of metres of the rugged bottom topography.

These 1 to 2 km refraction lines produced direct observation of the $V_p$ structure of the upper 200 to 300 m of the young igneous crust. One of the experiments was carried out over the site of Hole 648B of the Ocean Drilling Programme on a small volcano within the median valley of the MAR. This was close to a 'zero age' location, the two others were 14 km apart, above 7 m.y. old crust. The latter gave higher layer 2A velocities than the 'zero age' location.

The sea floor velocity at 'zero age' was observed to be 2.1 km/s, overlying an initial 4 s$^{-1}$, roughly 200 m deep, linear velocity-depth gradient. The crust at this location consisted of fresh basalt lavas, with laboratory measured velocities in excess of 5.8 km/s (However these tests, carried out as early as 1978, were conducted at a confining pressure as high as 50 MPa. Later tests in 1984 were also at high 40 MPa confining pressures, suggesting about 6.0 km/s for the intact basalt, if 'artificially' confined at too high pressure).

On the basis of the above in situ/laboratory comparison, Purdy, 1987, suggested that the 3.7 to 3.9 km/s difference in velocities must be due to the presence of large-scale porosity, and various models were discussed. It was inferred that from 30 to 50% porosity might be needed using conventional arguments about pore shape of that period. However, Purdy also referred to the Todd and Simmons, 1972, and Spudich and Orcutt, 1980b emphasis of the actual importance of effective stress (subtraction of pore pressure from total stress), which had been used for a long time in soil mechanics, following Terzaghi's theory of effective stress developed earlier in the 20th century. Naturally, with exceptional water pressures of around 30 to 40 MPa, it is understandable that the theory of effective stress was apparently late in being adopted in this hostile sub-ocean environment.

Purdy considered the possibility ('an elegant solution') that increasing differential [sic] (effective) stress could be responsible for the $4\,s^{-1}$ velocity-depth gradient, and presumed a 4 to 5% per 100 m porosity reduction, as seen in the first 200 to 300 m of sub-sea layer 2A. In fact experience from engineering (e.g. tunnelling) projects close to the surface, does suggest very high velocity gradients when rock quality Q and $Q_c$ values (Barton, 1995; 2002) are a) low and b) are rapidly increasing with depth, such as Q = 0.1 followed by Q = 1 and very quickly Q = 10 etc. i.e. nominal near-surface 'jumps' in $V_p$ from 2.5 to 3.5 to 4.5 km/s with a superimposed stress increase effect on $V_p$ as well. (See later comments on the question of gradients of velocity, and 'curve jumping', i.e. increases of rock quality Q-values with depth.)

The reality below the near-surface zone is that both effective stress increases and clay gets compacted, and there is less clay as depth increases. Therefore one progresses rapidly from low rock quality Q or $Q_c$ to higher Q or $Q_c$ values quite quickly, with obvious consequences for increased $V_p$.

The differences between in situ velocity measurement in the shallow oceanic crust and the higher matrix velocities measured at suitable (low) effective stress levels, is obviously caused not only by moderate changes to the matrix porosity but by stress-sensitive (low aspect ratio) jointing and fracturing. White and Whitmarsh, 1984, found that sub-vertical, water-saturated conjugate joint sets at 30° to 60° to the ridge axis of mid-Atlantic Ridge oceanic crust could explain their observed horizontal (azimuthal) velocity anisotropy of up to 0.4 km/s. There was apparently a negligible seismic influence of pervasive ridge-parallel fractures on this anisotropy, which the authors explained by their infilling by sediment, or hydrothermal precipitation, or by their closure under the high ridge-normal principal horizontal stresses.

In the case of this mid-Atlantic crust of 1.1 to 3.4 m.y., White and Whitmarsh reported that the top of the basement had a velocity of approximately 3.7 km/s which increased on average at 1.0 to $1.2\,s^{-1}$ in the uppermost 2.5 km, giving 6.0 km/s at about 2 km depth. The uppermost 200–300 m showed higher gradients than this.

In relation to the $V_p$-Q-porosity-depth model (Figure 5.36 in Part I), such velocities would suggest Q-values of about 4 to 6 if the matrix porosity was about 5 to 10%. If we assume a mean 'uniaxial' compressive strength $\sigma_c$ of about 200 MPa, then a Q-value of about 2 to 3 is suggested, i.e. significantly jointed, perhaps with the following general character (see Appendix A):

$$Q \approx \frac{50}{12} \times \frac{2}{4} \times \frac{0.66}{0.5} \approx 3 \tag{11.10}$$

At 1000 m depth a velocity of the order of 5.5 to 6.0 km/s would be predicted, if the rock at this depth had unchanged character.

Christensen, 1984, investigated pore pressure effects on basalts and dolerites and verified the strong effect of pore pressure variation on the velocities and on dynamic Poisson's ratios. The latter increased significantly as a result of increased pore pressure. He discussed the possibilities of over-pressured zones due to seals caused by rapid accumulation of low permeability clays and shales, and also theorised that release of water accompanying low grade metamorphic reactions in basalts could result in excess pore pressure and resulting changes (reductions) in seismic velocities, and increases in Poisson's ratios. The authors noted that the *pore pressure coefficient* was less than 1, and was not a constant for a given sample but depended on confining pressure and on pore pressure.

### 11.4.5 Sub-ocean floor attenuation measurements

Reportedly the first *direct* measurements of Upper Oceanic Crust compressional wave *attenuation* were

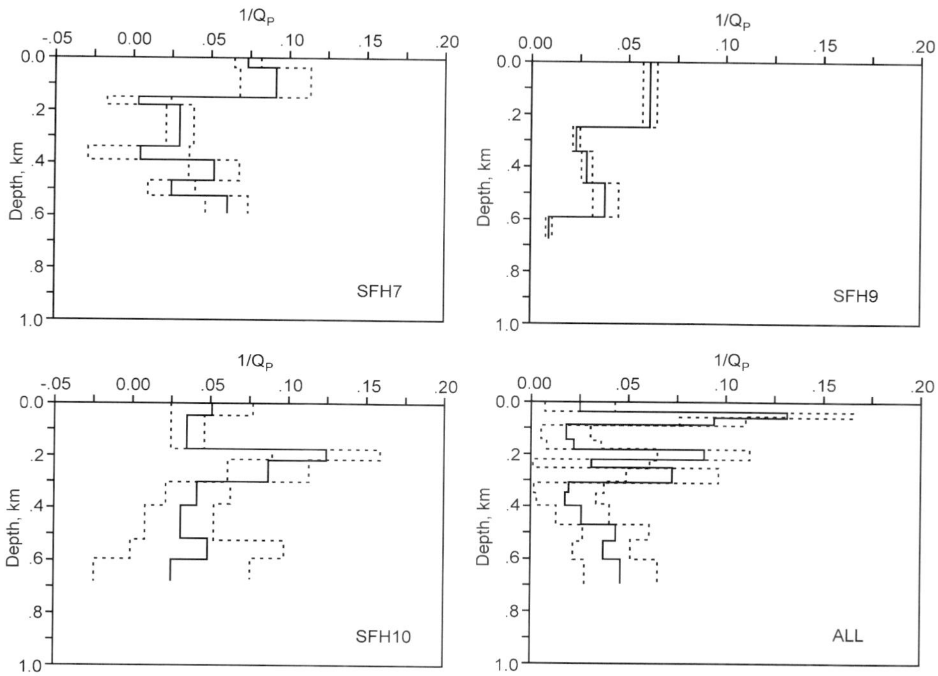


*Figure 11.41* Selected sub-ocean $Q_p$ profiles given by Jacobsen and Lewis, 1990. Variable attenuation, shows only partial consistency of increasing $Q_p$ with depth.

described by Jacobsen and Lewis, 1990, using seafloor hydrophones and large (56 to 116 kg) explosive sources. The site was on 0.4 m.y. old crust, 13 km SE of the Juan de Fuca Ridge. At the same site a seafloor velocity of 2.7 km/s increased uniformly to 5.6 km/s at 679 m depth, with gradients as high as $4.6\,s^{-1}$ at the surface and $4.1\,s^{-1}$ at depth. Values of seismic $Q_p$ obtained by Jacobsen and Lewis, 1990, varied from 4 to 275, but mostly clustered between about 10 and 20 in the upper 100 m, which was significantly lower than earlier estimates based on synthetic seismograms.

They did not find a consistent increase of $Q_p$ with depth, but several sets of data for $1/Q_p$ did show such a trend of $1/Q_p$ reducing with depth. (Figure 11.41) The variations presumably might be connected with variable degrees of fracturing or cooling joints, and partial closure with effective stress increases. Their results showed that $Q_p$ was linearly related to frequency between 15 and 140 Hz, but frequency-independent components of attenuation were also evident.

Pujol and Smithson, 1991, who analysed seismic wave attenuation from VSP measurements in the Columbia Plateau basalts, found values of $Q_p$ of about 50 (with $V_p = 5.0$ km/s) that were close to the value of $Q_p$ of 40 found in Eastern North Sea basalt by Rutledge and Winkler, 1987. As has been argued in Chapter 10, there is an implicit relation between seismic Q and the rock mass quality Q-based deformation modulus $E_{mass}$, or M, when this is expressed in GPa. The above $V_p$ of 5 km/s suggests a near-surface $Q_{rock} = 32$. Shallow, sub-ocean seismic Q of 10 to 20 might imply a significant degree of 'structure', if equivalent $Q_{rock}$ values were, say less than 5.

Elsewhere, *shallow ocean crust basalts* have shown $Q_p$ values of between 20 and 50. Dry samples of oceanic basalts from layer 2, tested at (artificially elevated) confining pressures of between 40 and 100 MPa have given $Q_p$ in the range 5 to 85. Differences are attributed to crack content, degree of alteration and matrix porosity.

These values are lower than the $Q_p$ values normally obtained for sound basalts, where values of between 100 and 600 can often be obtained (Wepfer and Christensen, 1990). Wepfer and Christensen, 1987, reporting the first laboratory measurements of $Q_p$ for dry and water-saturated *oceanic basalts* under appropriate pressure and temperature conditions, showed $Q_p$ varying from 8 to 100 at ultrasonic frequencies. The range was dependent on the state of alteration and porosity.

The sudden steps up, and down from, very high *in situ* $Q_p$ values like 200–300, even negative $1/Q_p$ steps, leads one to question whether the early ship-board triaxial test routines had an element of (local) correctness, meaning that some volumes of intact basalt can perhaps be subject to high 30 MPa plus-rock-depth confinement loads, interspersed by a majority of permeable and low effective-stress-loaded permeable blocks.

### 11.4.6 A question of porosities, aspect ratios and sealing

Shaw, 1994, using Kuster and Toksöz, 1974 theory, postulated that thin cracks preferentially close at shallow depth while lower aspect-ratio cracks do not. However, all crack populations were assumed to decrease with depth. It was pointed out that even 0.1% porosity in the form of thin ($\alpha = 0.001$) cracks significantly affect seismic velocities as these close, but this hardly affects overall porosities if thicker cracks and voids remain open.

Near the surface (depth A in Figure 11.42), where both crack populations were assumed to be open, Shaw, 1994, estimated a *Poisson's ratio* of 0.28. At intermediate depth (B), he postulated that only the thicker ($\alpha = 0.1$) cracks and voids were open, resulting in an anomalously (and as observed) low Poisson's ratio of 0.24. At greater depth (C), all cracks were assumed to be sealed, returning the velocities to that of the host rock, and Poisson's ratio was again about 0.28. In older crust, hydrothermal deposition caused thin cracks to seal first. Thicker cracks

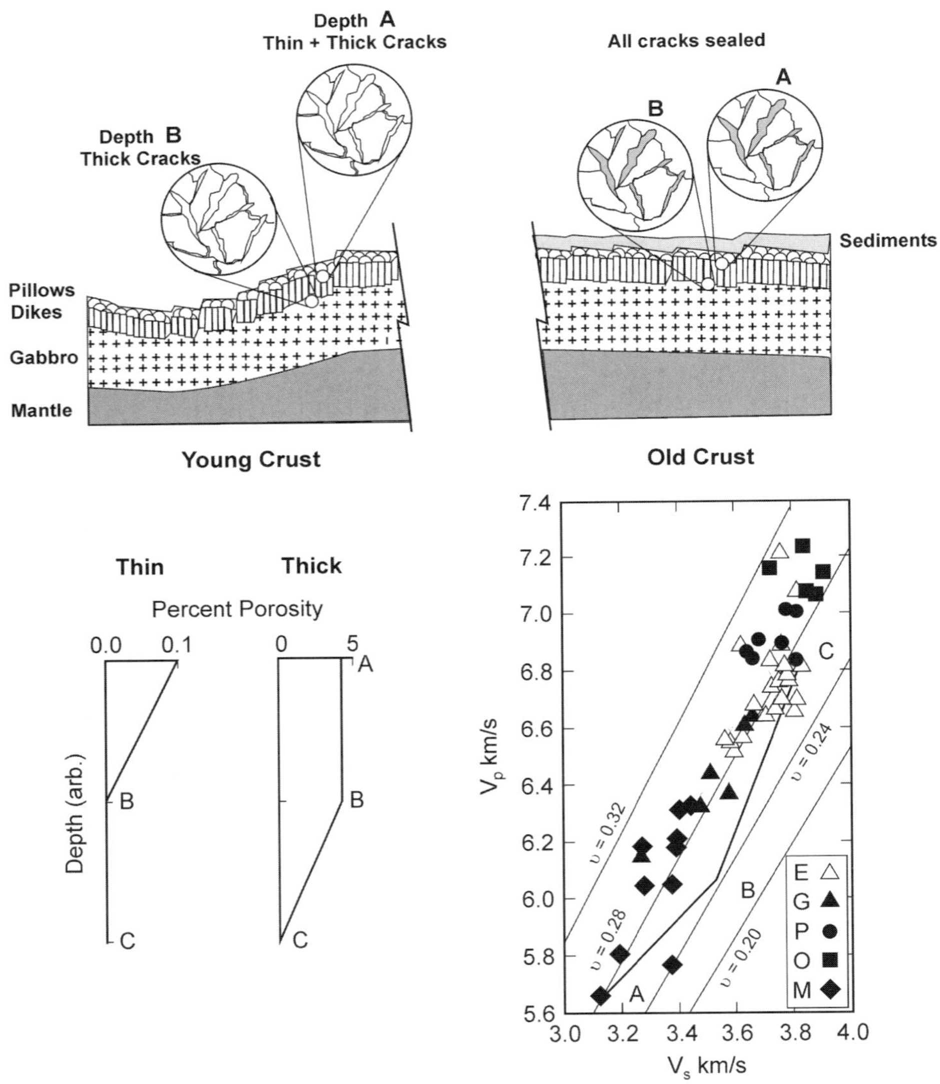


*Figure 11.42* Top: a) *For young crust*: thin and thick cracks at depth A; thin cracks are sealed at depth B, leaving only the thick cracks. b) *For old crust*: all cracks are sealed. The above causes a Poisson's ratio anomaly at depths of about 0.8 to 1.5 km, as shown in diagram d) in relation to $V_p$ and $V_s$ data. Shaw, 1994.

could remain unclosed and unsealed until the crust was older, which then restored Poisson's ratio to laboratory values.

### 11.4.7 A velocity-depth discussion

The strong focus on velocity-depth data in these investigations of the oceanic crust, in particular the supposedly 'anomalous' velocities and gradients discovered in the mid-ocean fracture zones, should lead us to consider the fundamental non-uniqueness of velocity-depth relations, as emphasised in a thought provoking article by Al-Chalabi, 1977. Figure 11.43a shows a smooth velocity-depth function such as:

$$V_z = V_o + k.z \tag{11.11}$$

with actual small-scale fluctuations commonly seen in a sonic downhole log. In this simple equation which is attributed to Slotnick, 1936. $V_o$ is the (P-wave) velocity at the surface and $V_z$ is the velocity at vertical depth z.

As pointed out by Al-Chalabi, 1977, the fluctuations shown in Figure 11.43a, which represent actual variability (and borehole effects) may not be seen in seismic work, when the seismic wavelength is greater than the scale of the fluctuations.

The actual variations of sonic velocity with depth can be described by an extremely wide range of 'parameter value combinations', and these may well be parameters of convenience, as Al-Chalabi pointed out. In the rock quality Q-system the 'convenient' parameters are clearly those considered in the formulation of Q, rather than additional parameters not thought of.

The non-uniqueness of the parameters in velocity-depth functions, and the lack of physical significance of any specific value of a given parameter had been overlooked up to then, according to Al-Chalabi, 1997.

An investigation of the velocity-depth gradients that are synthesised in the $V_p$-Q-value-porosity-depth model of Barton, 1995 follows from Figure 5.36 (Part I), using a plotting format that can readily be compared with the oceanic crust fracture zone data of Layer 2A and 2B.

Figure 11.44 shows the results which were extracted directly from Figure 5.36 for the case of six specific Q or $Q_c$-values ranging from 0.001 (intensely fractured, thick clay-bearing discontinuities) up to 100 (quite massive, unweathered competent rock mass with few widely spaced joints, principally one set only). See Appendix A for relevant – but non-unique – parameter ratings.

The very steep $V_p$-depth gradients typically seen close to the ocean floor, in the first few hundreds of meters of the new crust, could also be analysed with this near-surface based empirical method, developed mostly from civil engineering and engineering geological projects. Note

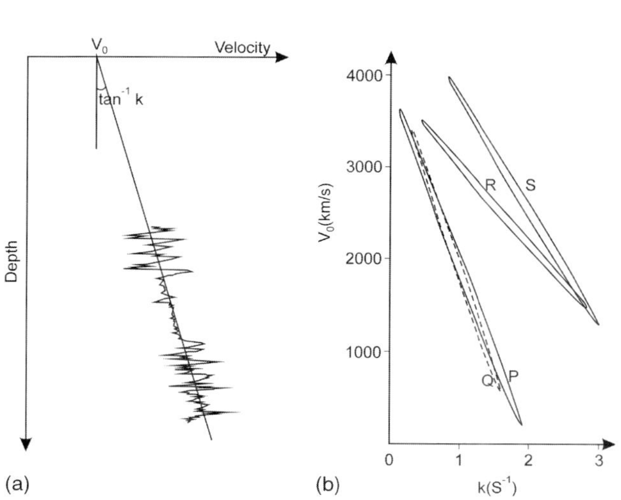


*Figure 11.43* a) A linear velocity-depth fit to a unit showing fluctuations at sonic log scale. b) Surface velocity-gradient trends for a unit logged at different depths in four wells. Al-Chalabi, 1997.

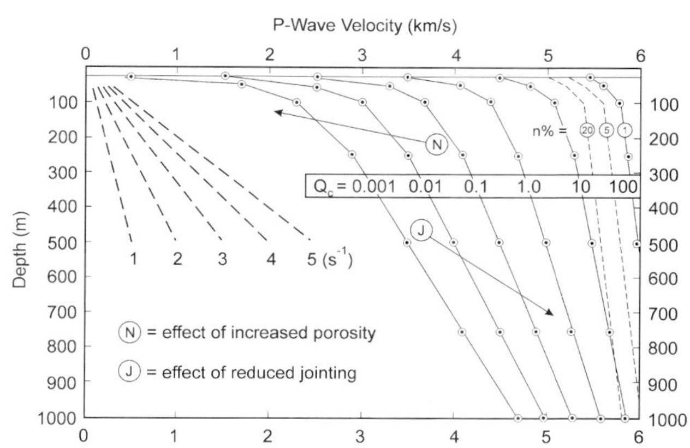


*Figure. 11.44* $V_p$-depth trends for six specific rock quality $Q_c$ values, showing the assumed minor effect of porosity (n) when $Q_c$ is high. Influence of n% (the 'hard' porosity) increases with lower $Q_c$-values. Gradient $k = km/s/km = s^{-1}$ is shown on the left-hand side. Trends of $V_p$-depth-$Q_c$ were derived from Figure 5.36 (Barton, 1995). Note similarity with oceanic crust fracture zone data for Layers 2A and 2B. The $Q_c$-value represents mainly the 'soft porosity', i.e. jointing.

the effect of porosities ⩾ 1%, which increase strongly in influence as one moves from $Q_c = 100$ to the lower rock qualities (and larger 'soft porosities') towards the left-hand-side of Figure 11.44.

In practice there will be a tendency for increased porosities close to the more weathered surface (arrow *N* trend), while with *increasing* depth, trend *N* will be reversed and trend *J* may dominate (i.e. reduced joint frequency, mineral healing, increased Q-value and $Q_c$ value, meaning that 'curve-jumping-to-the-right' will be necessary). This empirically-based, near-surface method could perhaps help to explain 'anomalously high' gradients through Layers 2A and 2B. Both trend *N* and trend *J* stimulate such an effect.

Although 'parameter value combinations' in the rock quality Q-system (Barton et al., 1974, Barton, 2002) are definitely 'non-unique' (as per Al-Chalabi, 1997), a physically plausible situation is described by this empirically-based choice of increasing 'hard porosity' close to the surface, and reducing 'soft porosity' at depth (higher $Q_c$-values).

In a similar manner to the above, we can extract $V_p$-depth *gradient* (k) data from Figure 11.44, and express it in the simple form given by equation 11.11. The results are shown in Figure 11.45.

Four sets of data are shown in Figure 11.45:

A = mean gradient, 25 to 100 m (extreme)
B = mean gradient, 100 to 500 m
C = mean gradient, 500 to 1000 m
D = mean gradient, 25 to 1000 m

In the case of D, giving the overall gradient from $V_0$ to $V_{1000}$, the separate effect of increased porosity is shown, which moves curves successively to the right. In each of these four cases, a uniaxial compressive strength of 100 MPa (nominal) has been assumed (giving $Q_c = Q$ in Figure 5.36). Higher values of $\sigma_c$ than 100 MPa, due to lack of weathering and low porosity would obviously give higher surface velocities and a lower gradient k ($s^{-1}$), thereby plotting to the left of these four sets of '100 MPa hard rock' lines. The opposite would be the case with $\sigma_c < 100$ MPa (i.e. with younger rocks).

### 11.4.8 Fracture zones

The low velocity mid-Atlantic fracture zone studies reviewed in this section, show gradients of 3.0 to 3.5 $s^{-1}$ for the upper 0.5 to 0.8 km, and seabed velocities as low as 1.9 to 2.7 km/s. Reference to Figure 11.45 suggests that curves B and D with suitably increased porosities (>nominal 1%) would fit such data very well.

Data from mid-Atlantic Ridge fracture zone anomalies, discovered during the 1970s and 1980s, were assembled by Detrick et al., 1993. (e.g. Figure 11.46) These emphasise the extreme heterogeneity of their thickness and internal structure. In general, they consist

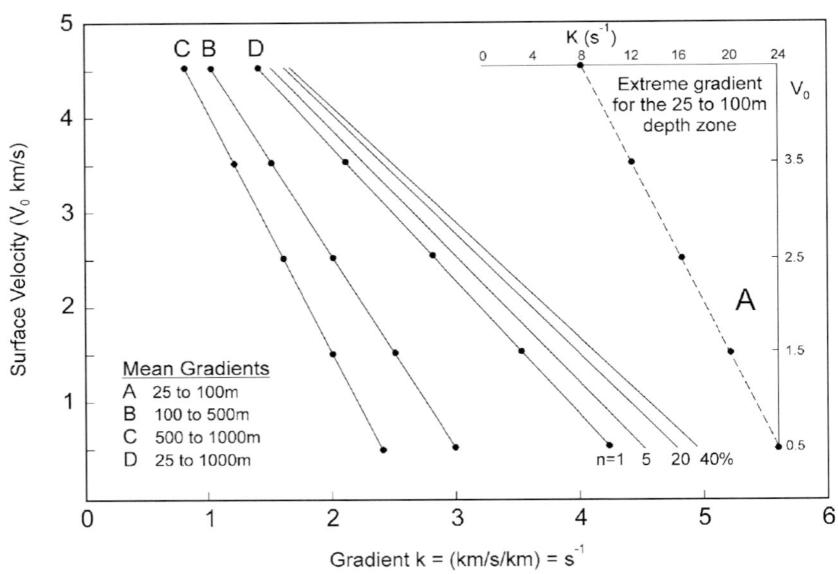


*Figure 11.45* Analysis of $V_p$-depth gradients (k, $s^{-1}$) as a function of surface velocity $V_o$ magnitudes, from the $V_p$-Q-value-depth-porosity model, Figure 5.36, from Barton, 1995). Based on the Al-Chalabi, 1997 plotting format, shown in Figure 11.43b.

of thin intensely fractured and hydrothermally-altered basaltic sections, overlying a rather shallow Moho. The sites of some of these investigations are shown in Figure 11.46a, and a typical structural cross-section is shown in Figure 11.46b.

Velocity-depth trends for four of the large Atlantic fracture zones are shown in Figure 11.47. Initial velocity-depth gradients to 2 km depth appear to vary from about 2 to 3 $s^{-1}$, though even steeper gradients are seen in the uppermost 100 to 200 metres.

## 11.5 The East Pacific Rise velocity structures

Following the forgoing summary of advances in understanding of mid-ocean ridges and fractured zones for the case of the mid-Atlantic ridge, we will now retrace some of the steps made in studies of the East Pacific rise. Many of the names of researchers will be familiar after reading or perusing the last section.

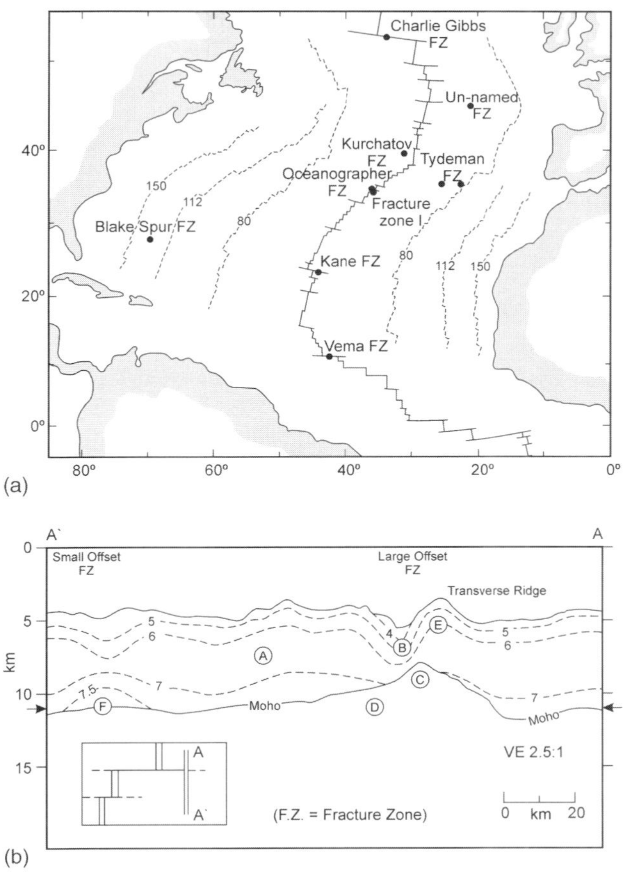


*Figure 11.46* a) Simplified tectonic map of North Atlantic fracture zones. b) Generalised velocity-depth structural cross-section of a large Atlantic fracture zone. Detrick et al., 1993.

Ewing and Purdy, 1982, both well known for their mid-Atlantic ridge studies, assumed a linear velocity-depth gradient in the upper 500 to 800 m of young (0 to 4 m.y.) oceanic crust on the flanks of the East Pacific Rise. The data shown in Figure 11.48 indicates an average gradient of between 3.0 and 3.5 $s^{-1}$ for the upper 0.5 to 0.8 km of oceanic crust, with seabed velocities ranging from as little as 1.9 to 2.7 km/s. Ewing and Purdy suggested that an even higher gradient might exist in the upper 400 to 500 metres.

The evidence of very low velocities in the upper-most oceanic crust was reportedly consistent with visual/submersible and photographic evidence of *pervasive fracturing* in mid-ocean ridge crustal regions, where the basalt layer was exposed, and it was consistent with drilling and logging results that showed high porosity (Hyndman and Drury, 1976; Kirkpatrick, 1979).

Spudich and Orcutt, 1980a, had reasoned that a 10% porosity in the form of large fissures, added to a measured matrix porosity of about 8%, could readily produce a P-wave velocity of 2.6 km/s. For the case of rubble zones, Hyndman and Drury, 1976, had estimated a porosity of about 20%.

The $V_p$-depth data interpretation shown in Figure 11.48 indicated to Ewing and Purdy, 1982 that 'the percentage of cracks and voids' diminished rapidly with depth, giving a $V_p$ of about 5.2 km/s at 800 m depth. If we enter the $V_p$-Q-porosity-depth diagram shown in Figure 11.44, at a velocity of 2.5 km/s, and at a *matrix* porosity of 8%, we see a rock quality $Q_c$ value of about 1 (typical of weathered, heavily jointed rock).

At 800 m depth, with assumed unchanged rock mass quality (but with higher effective stress), the P-wave velocity is predicted to be 5.3 km/s, almost the same as above, but without the linear-trend assumption. In other words, the effect of increased depth may have largely removed the porosity component created by the tectonic and thermal fracturing and jointing, but need not have removed (and indeed could not have removed) the matrix porosity of a competent volcanic rock which already had intruded into a pore pressure regime as high as 30 MPa, resulting from a 3000 m ocean depth.

Ewing and Purdy, 1982 considered that their observed data showed a significantly lower gradient of about 1 $s^{-1}$ below 800 m, which would give a velocity of 6 km/s, appropriate for the 'solid unweathered basalt', at about 1.5 km depth. They reckoned that this might be a reasonable maximum depth of significant fracturing

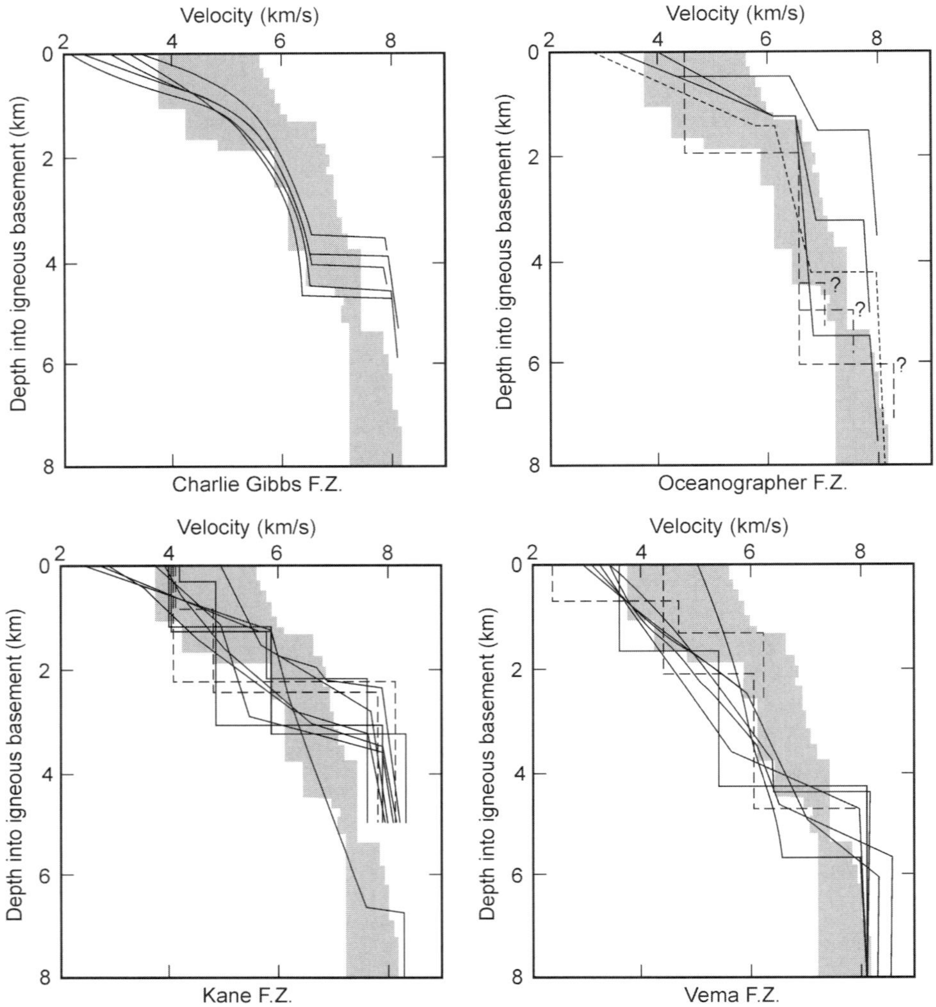


*Figure 11.47* Velocity-depth data from four of the large Atlantic fracture zones. Shaded areas are for 'normal' oceanic crust from White, 1984.

although it could be argued that initial fracturing extended to greater depth, and some healing has already occurred there by compaction and/or cementation'.

Purdy, 1982, concurrently reported laterally homogeneous velocity-depth behaviour for two areas separated by 110 km on the flanks of the East Pacific Rise. The non-linear $V_p$-depth curves shown in Figure 11.49, which are averages for two areas of 20 to 40 km lateral extent, show, in this case, an *inverse relation* between age and velocity, since the youngest crust has highest velocity. Furthermore there was 100 m of sediment overlying the older of the two sets of crustal data, which would tend to add to the recorded velocity, yet it remains lower than the youngest $V_p$-depth curve.

As a point of curiosity, the initial parts of the curves to the 'knees' at 400 and 600 m have gradients of about 4.2 and 4.4 $s^{-1}$. With reference to the Q-$V_p$-depth-porosity-strength chart, in Figure 5.36, taking the nominal 25 m line as 'surface', the measured surface velocities of 3.05 and 2.4 km/s shown in Figure 11.49 suggest rock quality $Q_c$ values of about 0.5 and 0.08 (i.e. 'very poor' and 'extremely poor' engineering tunnelling qualities), following which at 400 and 500 m depths, Figure 5.36 predicts velocities of about 4.7 and 4.6 km/s, close to those measured. (Note that the empirical $V_p$-Q-depth-porosity-strength relationship was determined by trial and error, from land-based refraction seismic in jointed and faulted rocks, and from both shallow and very deep cross-hole tomography measurements, each with Q-logging of relevant core).

The less steep gradients of about 2 $s^{-1}$ over the next 500 m depth to 1 km were the result of measured velocities of about 5.8 and 5.4 km/s. These compare to predicted velocities of about 5.5 and 5.3 km/s, from

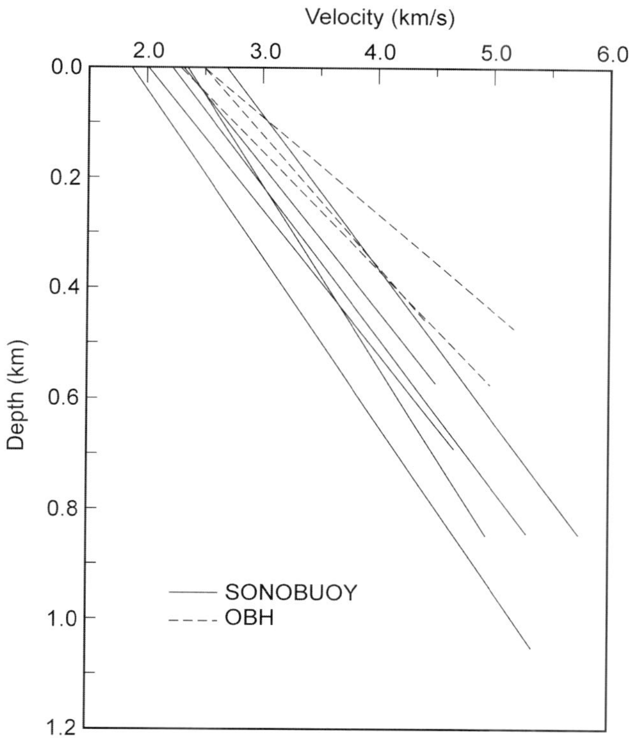


*Figure 11.48* Linear $V_p$ – depth assumptions for the shallowest fi km of East Pacific Rize oceanic crust (ROSE area), from Ewing and Purdy, 1982. They utilised a time-distance inversion method suggested by Dorman and Jacobsen, 1981, which required linear $V_p$ – depth gradients in each layer. (OBH: ocean bottom hydrophone).

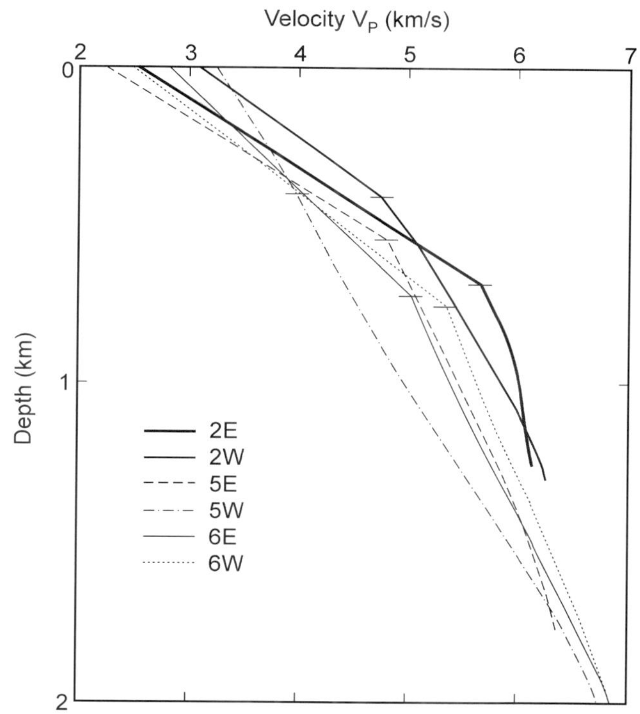


*Figure 11.49* Velocity-depth behaviour for 0.5 m.y.: crust (2E 2W), and 4 m.y.: crust (5E 6E 6W), from seismic refraction data on the flanks of the East Pacific Rise. Purdy, 1982.

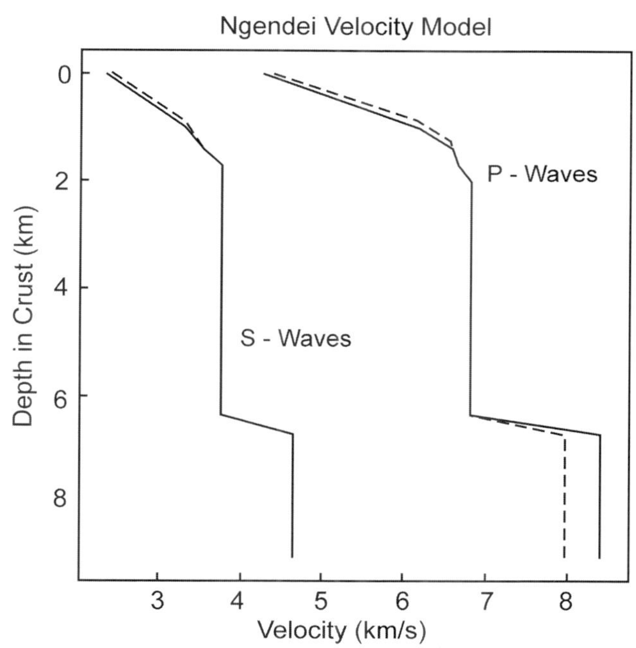


*Figure 11.50* Anisotropic velocity versus depth model which satisfies the Ngendei, South Pacific data. Solid line: NNE velocity, dashed line: ESE velocity. Shearer and Orcutt, 1986.

Figure 5.36. Differences in porosity and uniaxial strength between the youngest 0.5 m.y. and older 4 m.y. crust could be used to further distinguish and interpret the relative degrees of jointing, causing 'curve jumping' with increased depth, due to the likelihood of changed rock quality $Q_c$ values with age.

Due to assumed aligned cracks, Shearer and Orcutt, 1986 found that travel times were affected by azimuth, in measurements performed during the Ngendei expedition to the South Pacific. They estimated 0.2 km/s difference in P-wave velocities and 0.1 km/s difference in S-wave velocities in the upper 1.5 km, caused by azimuth. In the upper mantle, from about 7 km depth below the sea bed (Figure 11.50), the difference in P-wave velocity was about 0.45 km/s, but a nearly isotropic S-wave was indicated.

They interpreted the crustal anisotropy by a model involving aligned cracks parallel to the original spreading ridge, resulting in a fast direction *perpendicular* to the fossil spreading direction. The upper mantle anisotropy was consistent with there being aligned olivine crystals, in which the fast direction was *parallel* to the fossil spreading direction.

The problems posed by zero-age oceanic crust with $V_p = 2$ km/s, compared to about 6 km/s for intact basalt continued to provide challenges for theoreticians and practitioners working on the origin, formation and structure of mid-oceanic crust. Studies resembling Mid-Atlantic Ridge theories about hard and soft porosity (low aspect ratios) and preferential mineral sealing, were also performed with East Pacific Rise data.

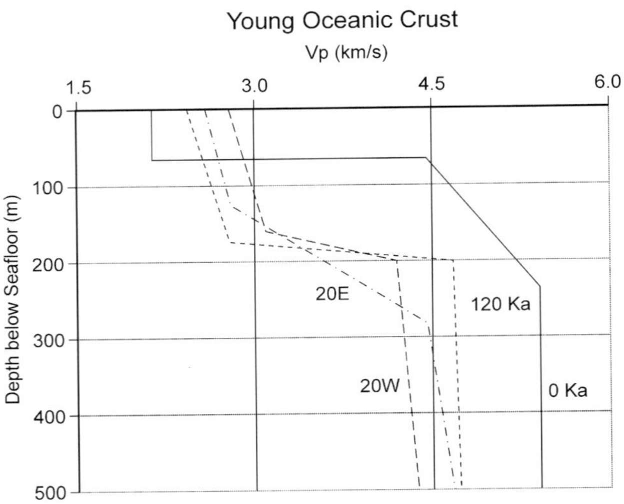


*Figure 11.51* $V_p$-depth structure for zero-age crust, 20 ka crust (east and west) and 120 ka crust near the East Pacific Rise axial summit graben. Christeson et al., 1991 and Purdy et al., 1991.

## 11.5.1 More porosity and fracture aspect ratio theories

Low aspect ratio cracks, and their reduced frequency of occurrence *and* reduction in aperture with depth, and probable sealing with hydrothermal minerals in the case of older oceanic crust, were some of the variables confronting those researching the variable structure of mid-ocean crusts.

Using theories termed extended-Walsh and extended-Kuster-Toksöz, Berge et al., 1992, utilised a range of crack aspect ratios ranging from extremes of 0.5 to 0.001, for depth zones ranging from 0 to 500 m below the sea floor, and succeeded in matching the $V_p$-depth trends for 0, 20 and 120 ka (1 ka = 1000 years) oceanic crust from Christeson et al., 1991, and Purdy et al., 1991. These are shown in Figure 11.51. The method of Berge et al., 1992, was one of data fitting, not forward prediction.

Berge et al., 1992 theorised that for 120 ka material with $V_p = 2.5$ km/s, porosity should lie between 24 and 34%. Slower ($V_p = 2.2$ km/s) zero-age crust was less well-bounded; a porosity of between 26 and 43% was predicted. The extended Walsh model used by Berge et al., 1992, required porosity-depth distributions for the various crustal ages, as shown in Figure 11.52.

Wilkens et al. 1992 managed to match Ocean Drilling Program/Deep Sea Drilling Project (ODP/DSDP9 Hole 504B and Hole 418A $V_p$ – depth data, to 500 m depth, by modelling cracks of small aspect ratio that, in a 'fast' model became sealed if of sufficiently small aspect ratio, and in a 'slow' model did not seal. Deeper in the profile they 'closed an increasing volume of lower aspect ratio pores'.

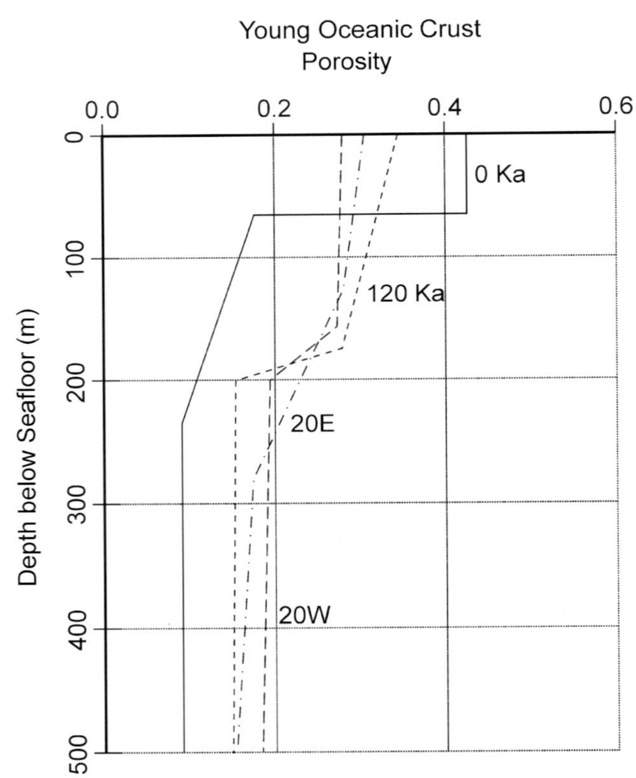


*Figure 11.52* Theoretical variation of porosity for matching $V_p$-depth data, with extended Walsh crack aspect ratio fit. Berge et al., 1992.

### 11.5.2 First sub-Pacific ocean core with sonic logs and permeability tests

At the beginning of the 1980s, in a sub-ocean Deep Sea Drilling Project borehole in the eastern equatorial Pacific ocean, in the Costa Rica Ridge area, it was possible for the first time to correlate core (but usually of low % recovery) with downhole sonic logs, borehole televiewer logs, and permeability test results. This was first performed to a depth of 1 km, through layers 2A, 2B and 2C. A schematic section and downhole logging results from Newmark et al., 1985, is shown in Figure 11.53.

Based on the vertical borehole logging (i.e. biased *against* vertical structure) the upper 50 metres contained numerous horizontal to sub-horizontal fractures, thick basalt flow units, and thin interbeds of pillow structures. In general the uppermost 100 m was an aquifer of rubbly pillow basalts, breccias and a few massive flows, and greatest variability and largest velocity gradients occurred here. The next 0.5 km was composed of pillow basalts, flows and breccias with an abundance of minerals and alteration products. Basalt dikes were typical in the lower 350 m. Velocities, porosities and permeabilities varied approximately as shown in Table 11.4.

The fact that average recovery of core was only 20% suggests many vertical and sub-vertical discontinuities were not sampled. Several of the well logs suggested the presence of zones of intense fracturing and open porosity, but the reducing permeabilities with depth clearly supported the general observation of increased mineral sealing with depth, and presumably increasing effective stress effects as well. (Of course the second leads to the first, if finest fractures are preferentially sealed).

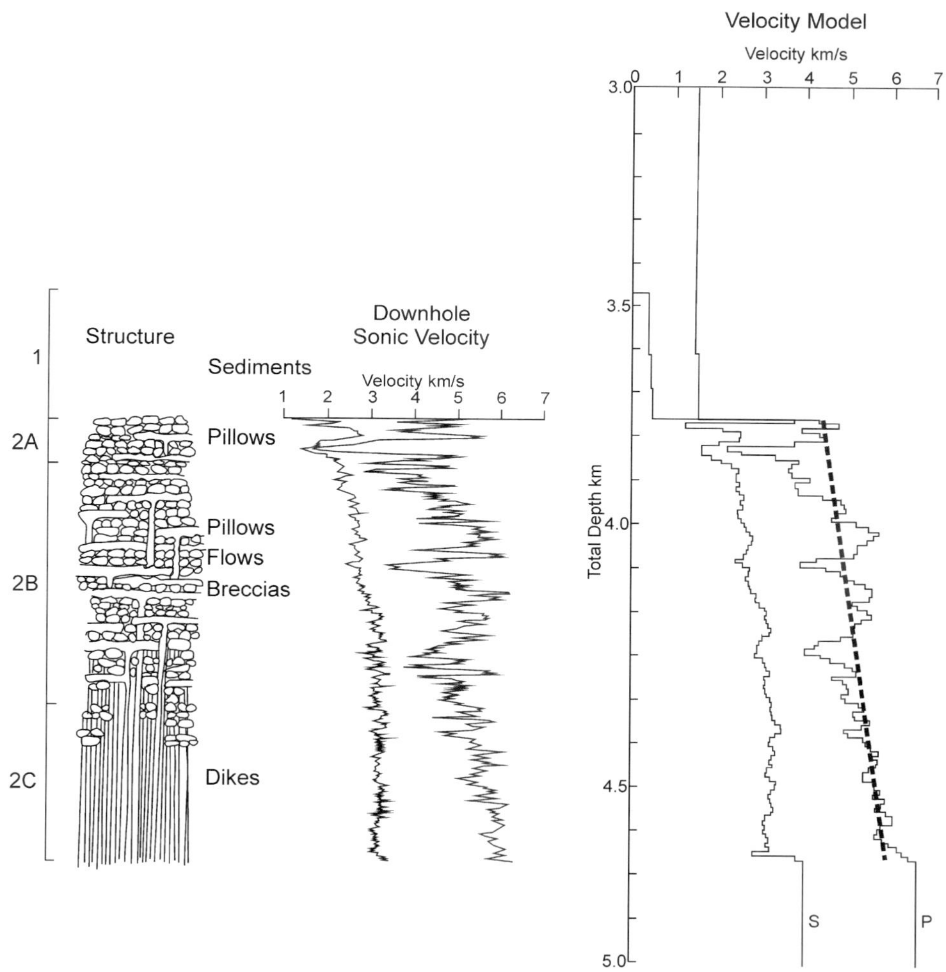


*Figure 11.53* Downhole (504B) sonic velocities and schematic structure of 1 km of oceanic 2A, 2B and 2C crust, from the equatorial eastern Pacific (Costa Rica Ridge area). Newmark et al., 1985. Note sediment and rock velocity contrast.

Table 11.4 Approximate velocities, porosities and permeabilities from downhole measurements in the top 1 km of Hole 504B, equatorial Pacific ridge. After Newmark et al., 1985.

| Zone | $V_p$ | n% | k ($cm^2$) | ≈K (m/s) | $V_p$-depth gradients |
|---|---|---|---|---|---|
| Upper 100 m | 3.7 | 4–10 | $10^{-9}$–$10^{-10}$ | $10^{-6}$–$10^{-7}$ | Steepest |
| Middle 550 m | 4.8 | – | $10^{-11}$–$10^{-12}$ | $10^{-8}$–$10^{-9}$ | Moderate |
| Lower 350 m | 5.6 | 1.5–4 | $10^{-12}$–$10^{-13}$ | $10^{-9}$–$10^{-10}$ | Moderate |

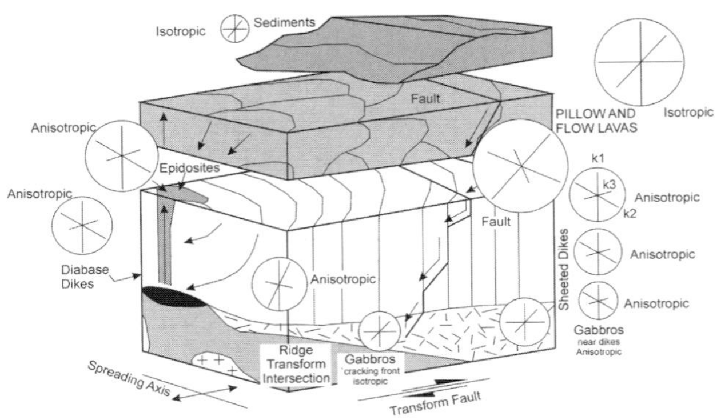


*Figure 11.54* Principal permeability directions in layer 2B oceanic crust, showing anisotropic principal permeability magnitudes $K_1$, $K_2$ and $K_3$. Interpretation based on observations of fracturing and mineralization of an (on land) ophiolite in Cyprus. van Everdingen, 1995.

The in situ *bulk* permeability of extrusive volcanic rocks measured in the Deep Sea Drilling Project (DSDP) drillhole 504B (Costa Rica Rift area), was subsequently quoted as $4 \times 10^{-14} m^2$ (or about $4 \times 10^{-7}$ m/s) at 20°C, which corresponds to that of a well-jointed rock mass with a rock quality Q-value ≈ 0.25, based on a conversion of the above permeabilities to a Lugeon value of about 4 (see Chapter 9). In contrast, permeabilities of unfractured basalt, diabase and gabbro may lie in the range $10^{-16}$ to $10^{-23} m^2$ (or about $10^{-9}$ to $10^{-16}$ m/s).

Systematic investigations of jointing characteristics in ophiolitic (on land) remnants of oceanic lithosphere were used by van Everdingen, 1995, to infer the possible joint structure effects on permeability in layer 2B of the oceanic crust. A compilation of permeabilities measured or inferred for sub-sediment pillow lavas (from about 250 to 900 metres beneath the sea floor) and for the underlying sheeted dike complex (from 900 to 1600 m) showed a range from $10^{-10}$ to $10^{-18} m^2$ (or $10^{-3}$ to $10^{-11}$ m/s). This happens to be comparable to the usual maximum range of measured land-based permeability-depth measurements, e.g. to 1 km depth in igneous and metamorphic rocks. (Barton, 2002).

The Troodos ophiolite estimates of van Everdingen, 1995, suggested a decrease in joint aperture and trace length with depth, and very marked anisotropy. The interpreted principal permeability directions in layer 2B oceanic crust given by van Everdingen, 1995, are illustrated in Figure 11.54. These land-based measurements appeared to have been at least partly based on the apertures indicated by epidote, quartz and later calcite fillings.

Based on the usual inequality of hydraulic aperture (e) and the (rough-walled) average physical aperture (E) (i.e. E ≥ e, Barton et al., 1985), the above method of estimating apertures could explain the higher estimates of permeability (e.g. $10^{-3}$ m/s), which would also be reflecting the negative effective stress episodes that would necessarily occur during hydrothermal penetration of fluids. Such was probably not typical of the effective stress conditions in operation when the above DSDP permeability measurements were made, since the permeabilities were of only moderate magnitude.

(If effective stresses were locally negative during the DSDP permeability testing – a hazardous boundary condition for drilling of wells – then the resulting larger apertures, e.g. 1 mm or more, actually giving much higher permeabilities, would then have satisfied the assumed 'mineral-filled opiolitic boundary condition' of e = E.)

A combination of deep sea crustal permeability measurements and interpretations of mineral-filled 'frozen' apertures from the Troodos ophiolite are shown in Figure 11.55.

Similar, ophiolite observations of fossil flow porosities and permeabilities, based on cubic law calculations using mineral filling thicknesses (e.g. Norton and Knapp, 1977), were reported by Nehlig and Juteau, 1988. With apertures of 1, 2 and even 5 mm (e.g. of epidote), it is clear that artificially high estimates of 'permeability' are made, such as many estimates of $10^{-10} m^2$, or about $10^{-3}$ m/s. These are exceptionally high. If many such

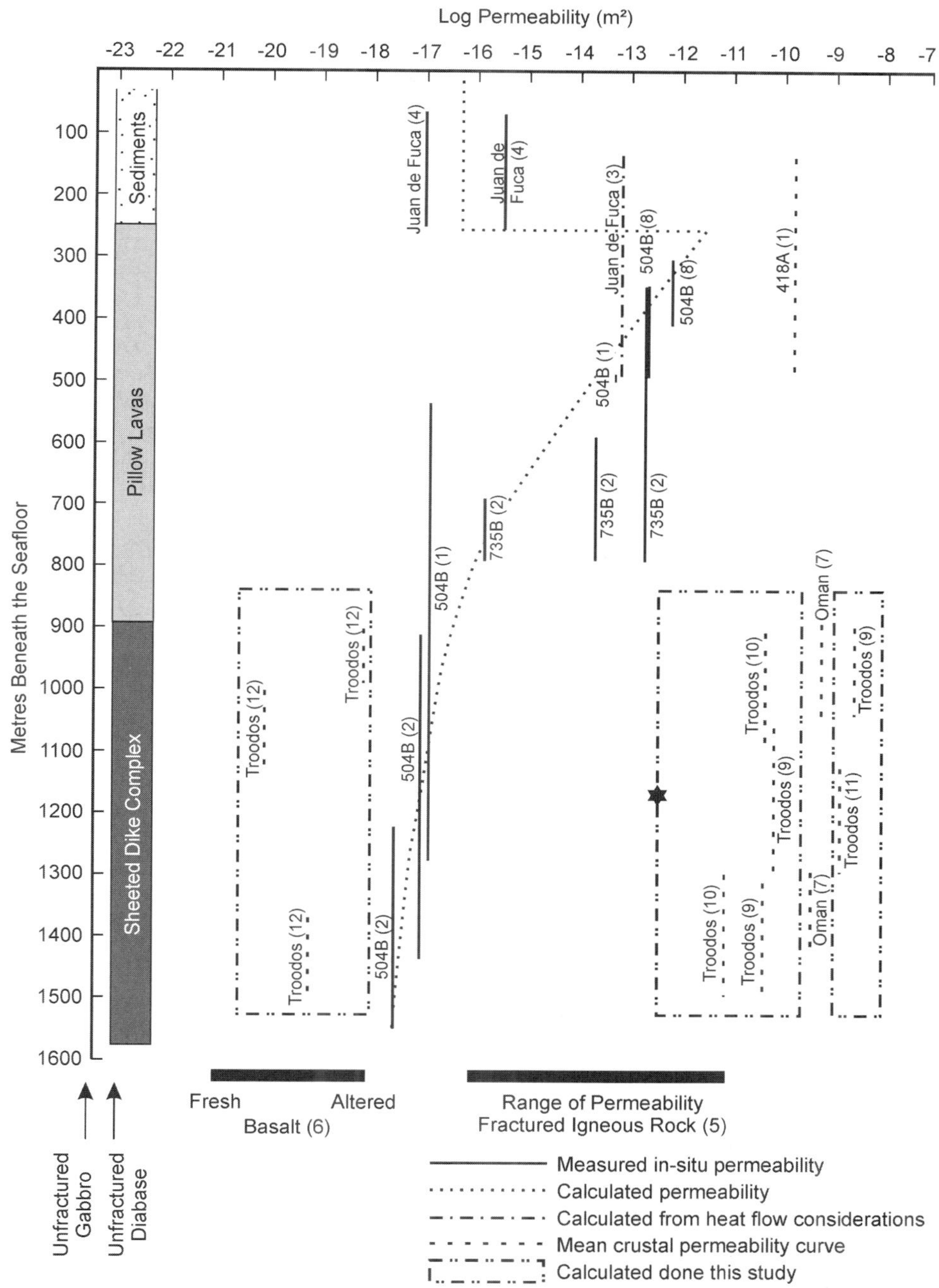


*Figure 11.55* An interpretation of possible permeability trends in the first 1600 m of ocean crust, based on parallel-plate modelling, with matrix addition, plus measured permeabilities from various sources. See van Everdingen, 1995, for references.

apertures were caused during negative effective normal stress episodes, as seems likely in sill and dike intrusions, they would not reflect 'virgin' permeabilities, as existing prior to the hydrothermal fluid injections. On the basis of this reasoning, the 'fossil' apertures observed in recovered core may not accurately reflect the porosity available at the time of hydrothermal fluid injection.

### 11.5.3 Attenuation and seismic Q due to fracturing and alteration

Swift et al., 1998a described the *seismic attenuation,* for the upper *1.8 km* of Hole 504B (Costa Rica Ridge area: see upper 1 km of permeability data in Table 11.4 from Newmark et al., 1985). About 60% of the total observed

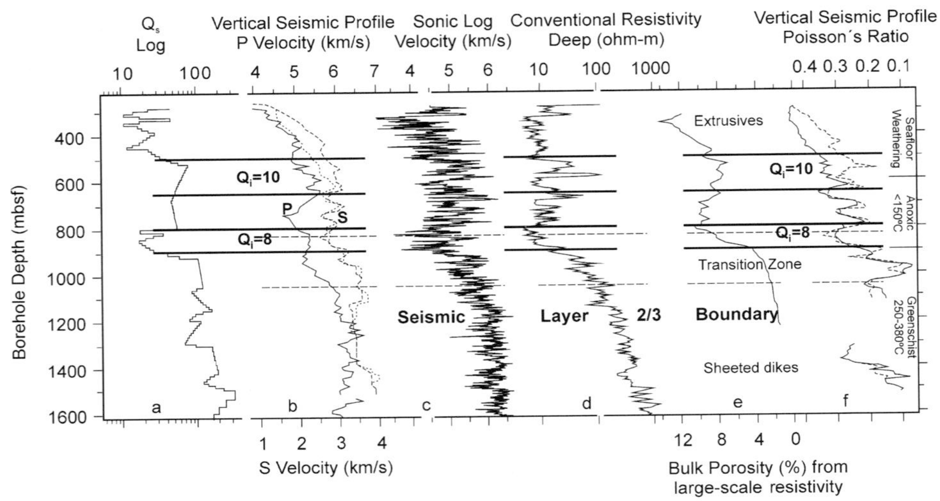


*Figure 11.56* VSP and sonic log measurements in upper 1.6 km of Hole 504B. VSP Poisson's ratio log also shown. $Q_i$ (intrinsic) values are from Swift et al., 1998a with $Q_s$ (scattering) values at 10 m intervals from Goldberg and Yin, 1994. Note the very high dynamic Poisson's ratios at shallow depth, derived from the VSP $V_p$ and $V_s$ values, due to thin flows, pillow lava, and breccia.

amplitude decay apparently occurred in the pillow basalt, due to geometrical spreading and impedance contrasts, and much of the remaining amplitude decrease was concentrated in two layers, about 500–650 m and 800–900 m below the ocean floor, as shown in Figure 11.56.

Attenuation in these layers was described by low seismic Q of 10 and 8 respectively, due to *intrinsic* attenuation mechanisms. In the case of the *upper* zone the authors believed that *alteration-mineralogy* may have been responsible for the attenuation, as there was no crack-related reason for the high attenuation. In the past this may have been a zone of high porosity prior to pore-space filling by zeolites.

In the case of the *lower* zone with seismic Q of only 8, the authors mention that this 800–900 mbsf interval coincided with features in core descriptions and logs that suggested the presence of an intracrustal deformation zone, or a sub-horizontal fault. Tectonically disturbed rock, a local minimum in resistivity, a maximum in inferred porosity, a decrease in Young's modulus, increased fracturing in the wall of the borehole, low aspect ratio cracks containing fluid, are among the various descriptions of this low seismic Q zone. It could have been helpful to also know the rock mass quality Q, or at least RQD, among all these mostly *qualitative* descriptions.

Attenuation studies in upper ocean crust cited by the authors, indicated a certain age relationship, with attenuation greatest and seismic Q lowest in the youngest rock, as reviewed by Swift et al., 1998a.

- $<1$ Ma $Q_{seis} \approx 20$ to 60
- Near ridge axis $Q_{seis} \approx 10$ to 20 (near sea floor)
- Within first 0.4 Ma $Q_{seis} \approx 50$ to 60 (near sea floor)
- $>$few Ma $Q_{seis} > 300$ (1 to 2 km depth)
- Lab samples of basalt at 50 MPa confinement: $Q_{seis} \approx 55$ to 120
- Lab samples of diabase dikes at 100 MPa confinement: $Q_{seis} \approx 70$ to 370.

Other attenuation data from the upper crust in these mid-ocean ridge structures, show seismic $Q_p$ estimates varying from 35 to 80 in general (Harding et al., 1989; Vera et al., 1990; Wilcock et al., 1992) and dropping as low as 11 to 20 in the uppermost crust (Christeson et al., 1994). Kappus et al., 1995 found that their reflectivity synthesis computed with seismic $Q_p$ varying from 50 to 100, fitted original data very well.

The high attenuation, low seismic Q zone described by Swift et al., 1998a, had sonic log velocities down to 1 km/s *lower* in this fractured interval. The $Q_i$ *intrinsic* attenuation and $Q_s$ *scattering* attenuation interpretation, the corresponding sonic and resistivity lows and the dynamic Poisson's ratio determined from VSP, are each reproduced in Figure 11.56.

Swift et al., 1998b also referred to the large-scale resistivity measurements at Hole 504B, giving parallel estimates of bulk porosity reduction with depth.

The increase in Poisson's ratio is caused by disproportional reduction in S-wave velocity compared to P-wave velocity, which theoretical studies by Shearer, 1988, have shown should occur with relatively thin cracks or joints, having aspect ratios less than 0,005 in an otherwise isotropic solid.

Swift et al., 1998b gave an interesting comparison of laboratory data (open circles in Figure 11.57) obtained apparently from 100 MPa confinement, with $V_p$, $V_s$ and $\nu_{dyn}$ trends obtained from VSP. The 100 MPa confinement applied in earlier laboratory studies was an incorrect simulation of an actual much lower *effective* stress gradient, as discussed earlier in this chapter. Possibly some effect of a correct effective stress gradient might have been observed on the intact samples, if this had been applied. We can see that, in relation to this presumably unrealistic 100 MPa confinement, there is up to 2 km/s deviation (reduction) of in situ P-wave velocities due to structure, at the top of the hole. The difference is more than 3 km/s in the case of S-waves, presumably due to the water-filled structure close to the ocean floor.

Swift et al., 1998b suggested that large-scale, well-oriented vertical fractures (i.e. a joint set) formed tectonically, did not have a detectable effect on P-wave velocities. Presumably this is an expression of the effect of tight closure by stress, which has also been observed in the case of an obvious fault ahead of a (stuck) TBM tunnelling machine. (Seismic velocity tomography performed from diverging holes ahead of the particular tunnel face at about 800m depth, failed to detect the low velocity that would 'normally', (nearer the surface), be an obvious feature of such a fault: see Figures 8.25 and 8.26 in Part I.

### 11.5.4 Seismic attenuation tomography across the East Pacific Rise

Some results of the first three-dimensional tomographic study of crustal seismic attenuation across the East Pacific Rise near 9°30′N, dating from 1988, was reported by Wilcock et al., 1995. The rather unique layout of 480 explosive charges distributed over an ocean-bottom 16 × 16 km grid is reproduced in Figure 11.58. Solid symbols are ocean bottom receivers, which included analogue and digital hydrophones and seismometers. Bathymetric contours (m) are also shown. The East Pacific Rise near 9°30′N is a fast spreading ridge, characterized by a sharp upper-crustal to mid-crustal velocity inversion some 1.5 to 2 km below the seafloor, presumed to be the roof of an axial magma lense. Small mid-crustal (i.e. 3 to 5 km deep) magma chambers appear to be a common feature of these fast-spreading ridges. Since a narrow lense of partly melted rock would solidify rapidly in a cooling environment dominated by hydrothermal

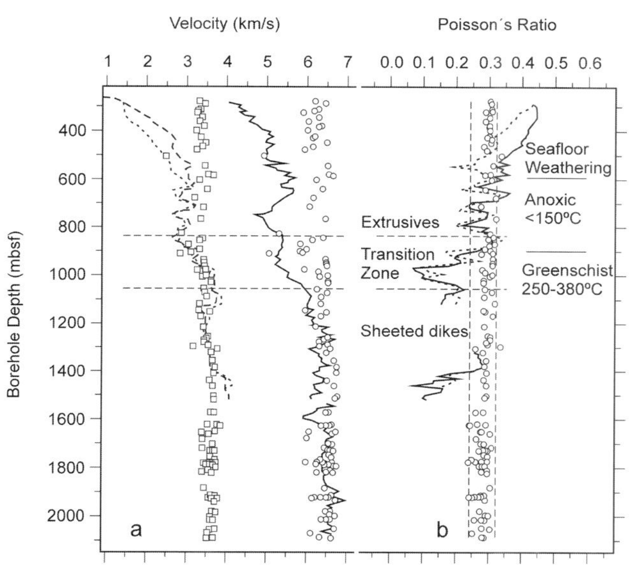


*Figure 11.57* Comparison of VSP (in situ) velocity structures in Hole 504B on the Costa Rica Rift, with numerous researcher's laboratory data (open circles), which show little effect of depth with the (artificial) applied effective stress of 100 MPa. Swift et al., 1998a.

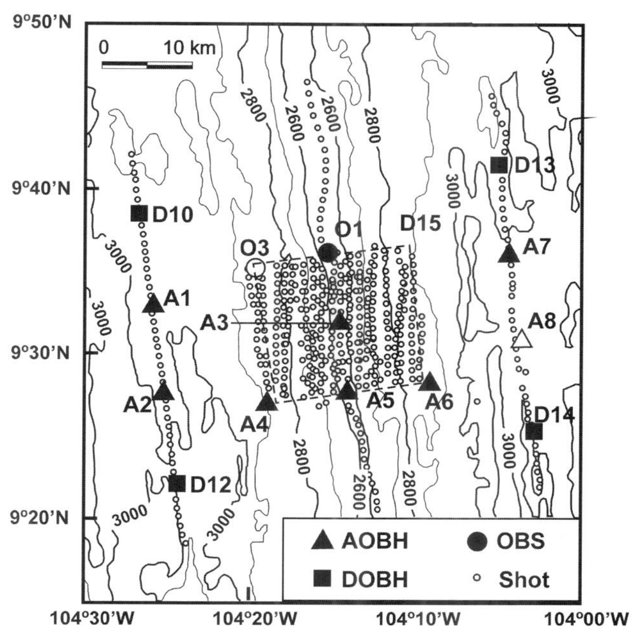


*Figure 11.58* A 16 × 16 km grid with 480 explosive charges (open circles) and the ocean bottom receiver array (sold symbols). Bathymetric contours (m) are also shown. Wilcock et al., 1995.

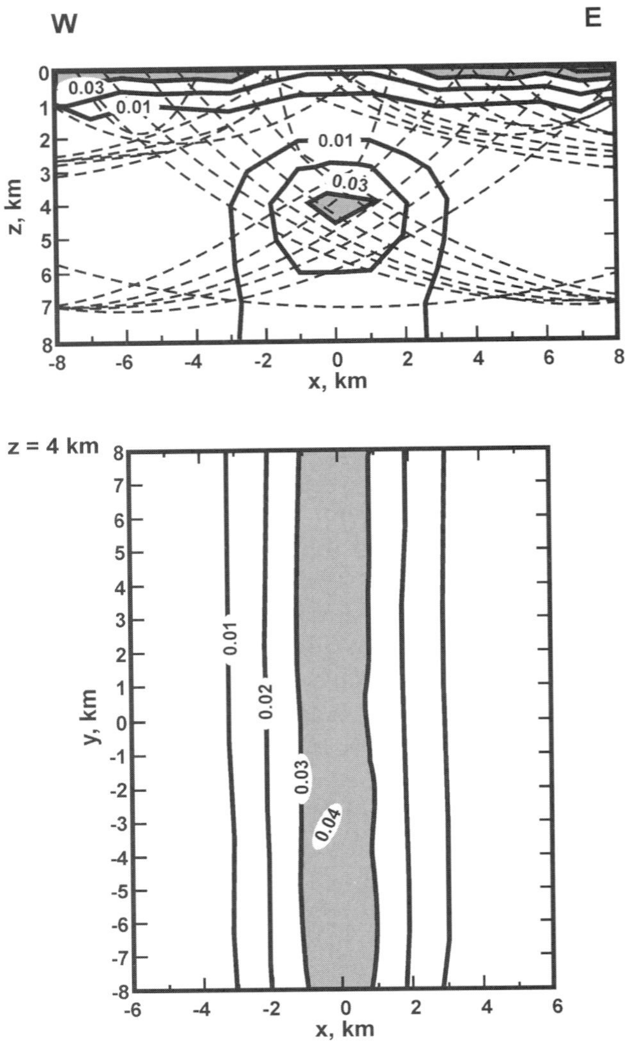


*Figure 11.59* Results of the inversion for a vertical cross-section and an along-axis slice, showing the lower crustal, sub-ocean ridge, $Q_p^{-1}$ attenuation structure. The along-axis result is at 4 km depth. The $Q_p$ values range from 25 to 100. Wilcock et al., 1995.

circulation, it is generally considered that the supply of magma from below must be relatively steady-state and uniform, deep beneath the axis of the ridge.

The attenuation expressed as $Q^{-1}$, (or $Q_p^{-1}$) for a central vertical section, and for a horizontal cross-section at 4 km depth is shown in Figure 11.59. The four attenuation ($Q^{-1}$) values give estimates of seismic Q = 25, 33, 50 and 100 from the central lense (with an assumed few percent of melt) to 2 and 3 km off-ridge distances.

In the *upper 1 km* by contrast, seismic Q averages about 35 off-axis, increasing to 65 near the axis rise. One may speculate that this might be due to a horizontal stress enhancement above the lense, because 20 km off axis, seismic Q remains at only 45, despite greater age at increasing distance from the ridge.

Inversions for individual receivers showed that seismic Q increased from average values of 40–50 in the upper 1 km, to at least 500–1000 at depth greater than 2 to 3 km. Results appear to be in agreement with other studies of attenuation in young oceanic crust. In an on-bottom refraction study near the centre of this tomography experiment, Christeson et al., 1994, measured seismic Q in the upper layer 2A of only 11 to 22.

Wilcock et al., 1995, cited three dominant attenuation mechanisms in stable continental crust and marine sediments as: *intrinsic attenuation* as a result of Coulomb friction along cracks (Walsh, 1966), *the flow of pore water* (Biot, 1962) and *scattering*. They suggest that these mechanisms, together with scattering from the rough seafloor, may also be important in young, igneous oceanic crust. Laboratory measurements on oceanic basalts (Wepfer and Christensen, 1990) suggested that attenuation would also increase with porosity, degree of alteration and water saturation.

One may append here the more specific and presumed 'micro-shearing 'and micro-flow' terms to qualify the above 'friction' and 'pore water flow', since there are physicists who visualise only sub-atomic magnitudes of deformation across microcracks with the passage of seismic waves, and others who even deny that friction can be involved in attenuation.

The subject of attenuation is indeed controversial, but there are clear indications, presented in Chapter 10, and further discussed in Chapters 13, 15 and 16, that micro-shear and micro-normal deformations along/across attenuating cracks and joints (i.e. displacement discontinuities), as interpreted from seismic anisotropy field measurements, are closely following the stiffness (or compliance) magnitudes seen in the 'static' macro-deformation (stress-closure and shear-displacement-dilation) testing of joints that is more common in rock mechanics. Dynamic compliances are often smaller than the inverse of 'static' stiffnesses, but only marginally so in the case of *normal* stiffness in rock masses of good quality. They seem to be of the same order of magnitude, or even closer. Dynamic shear stiffnesses *may be* up to two orders of magnitude stiffer than static shear stiffnesses. This seems hardly enough difference to prevent friction from being involved in attenuation.

Since fractures are preferentially oriented parallel to mid-ocean rise axes, it has been postulated that attenuation should be higher for waves propagating perpendicular to the rise (Macdonald, 1982). Unfortunately there were insufficient axis-parallel paths in the tomographic investigation of Wilcock et al., 1995, to distinguish the two orientations.

### 11.5.5 Continuous sub-ocean floor seismic profiles

As time went by, investigations of the mid-oceanic ridge areas become even more extensive with the added possibility to compare new results with ever more numerous earlier studies. An integrating report of this nature by Kappus et al., 1995, also described a high-resolution seismic velocity profile of the uppermost 500 m of East Pacific Rise crust at 13°N, along a 52 km segment of the ridge crest. The continuous profile, synthesised from 70 individual 1-D models spaced at 750 m, showed remarkable lack of variation. The 53 km segment was however more than 100 km from the nearest first order transform fault and over 10 km from the nearest second order discontinuity.

The main features revealed were a thin low velocity layer 2A consisting of about 80 m of (nearly) constant velocity rock (2.45 ± 3% km/s at the sea floor) followed by a steep gradient through 150 m of rock to the base of layer 2A. The thickness of the 4 km/s iso-velocity contour was mostly 130 ± 20 m, increasing to 180 m towards the north. This implied a maximum $V_p$-depth gradient of about $(4.0 - 2.45)/0.05 \approx 31$ through a 50 m section compares with the also very steep gradients at shallow depth in jointed (sub-continental) rock masses shown in Figure 11.44, as derived from the Q-$V_p$-depth-porosity-strength model of Barton, 1995, 2002).

Figure 11.60 shows the velocity of the top layer 2A (mean 2.45 km/s) and the thickness of the 4 km/s *iso-velocity* contour (mostly 130 ± 20 m). Velocities at the top of layer 2A and at the top of layer 2B are shown in Figure 11.61. A reflection deeper in the crust (triangles) at a velocity of 5.5 to 6.1 km/s is also shown.

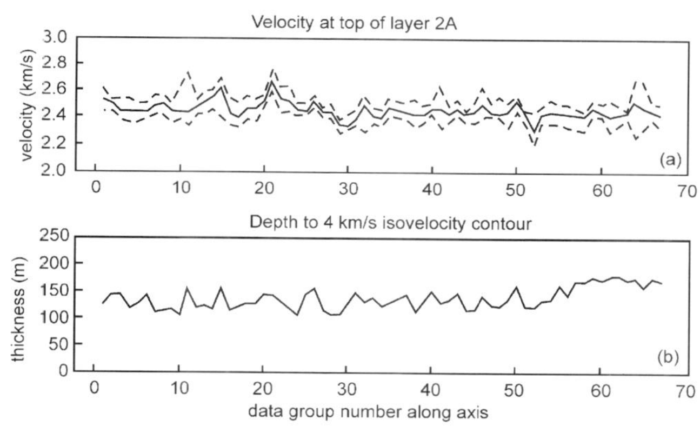


*Figure 11.60* a) Velocity at top of layer 2A (mean 2.45 km/s) and b) thickness of layer 2A where the 4 km/s *iso-velocity* contour is found (130 to 180 m). Kappus et al., 1995.

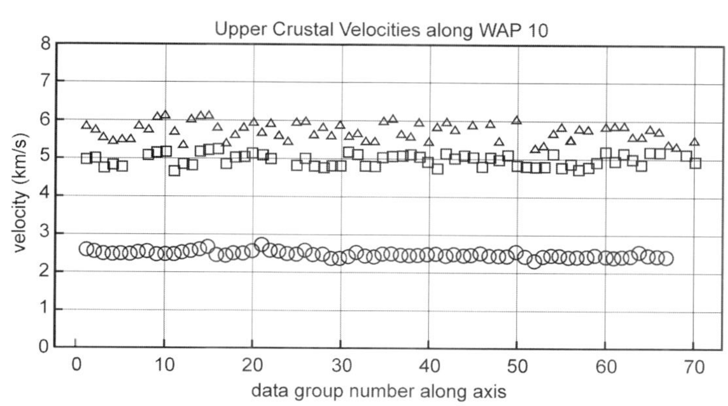


*Figure 11.61* Velocities at top of layer 2A (circles) and at top of layer 2B (squares). A reflection deeper in the crust (triangles) was interpreted as the lid of a magma chamber. Kappus et al., 1995.

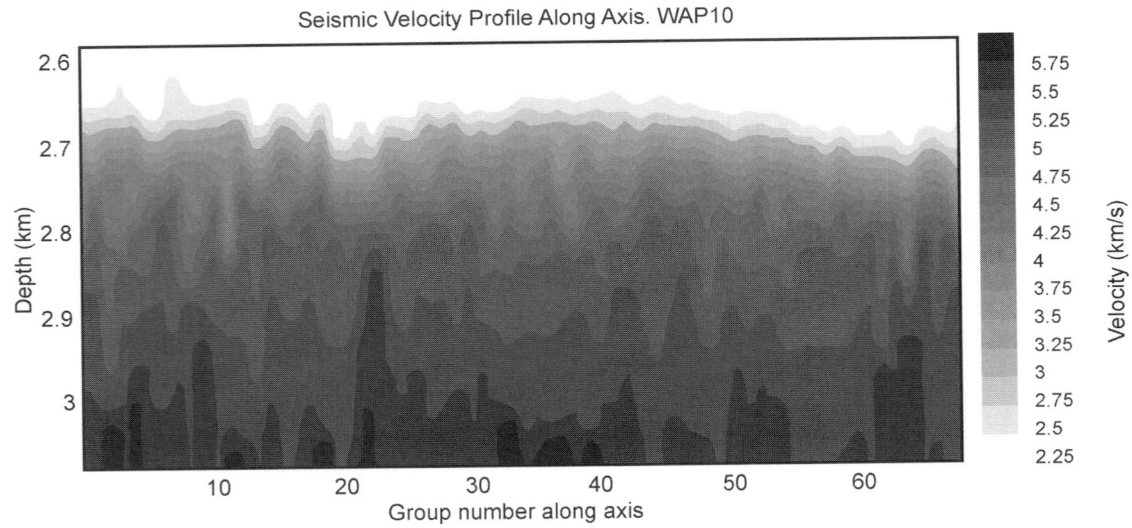


*Figure 11.62* Seismic velocity structure of the upper 500 m of a 52 km long segment along the mid-oceanic ridge crest of the East Pacific Rise. Kappus et al., 1995.

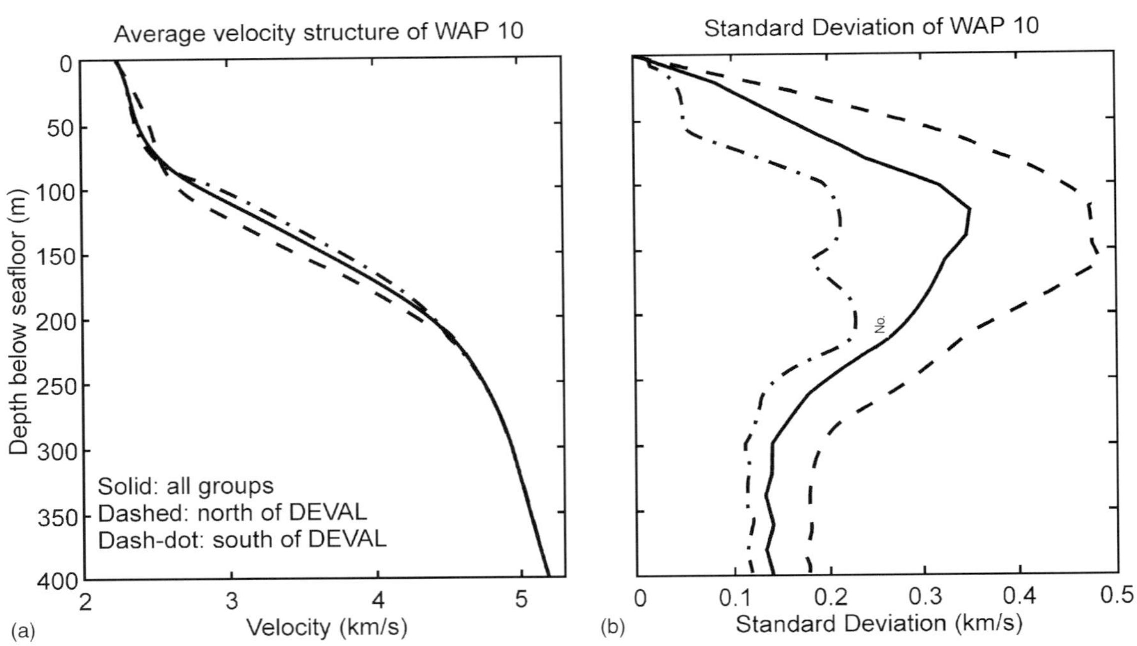


*Figure. 11.63* Average P-wave velocity-depth curve and standard deviation for 70 profiles covering a 52 km segment of the ridge crest, to a depth of 400 m. Kappus et al., 1995.

The relative uniformity of much of the 52 km long segment (measured at 70 locations) is further emphasised by the strikingly beautiful contoured velocities shown in Figure 11.62, which show rapid increases in velocity at first followed by slower increases due to longer depth intervals.

The base of layer 2A was assumed to be the lower part of the steep velocity gradient at about 230 m below the seafloor. This transition to an entirely different (2B) gradient is seen more easily in Figure 11.63, which shows the average velocity depth behaviour (solid lines) and the mean of N and S parts of the segment (dashed lines). A starting model of velocity versus depth, and various iterations is shown in Figure 11.64.

For the purpose of estimating gradients, the average velocity-depth data is reproduced in Table 11.5. This data

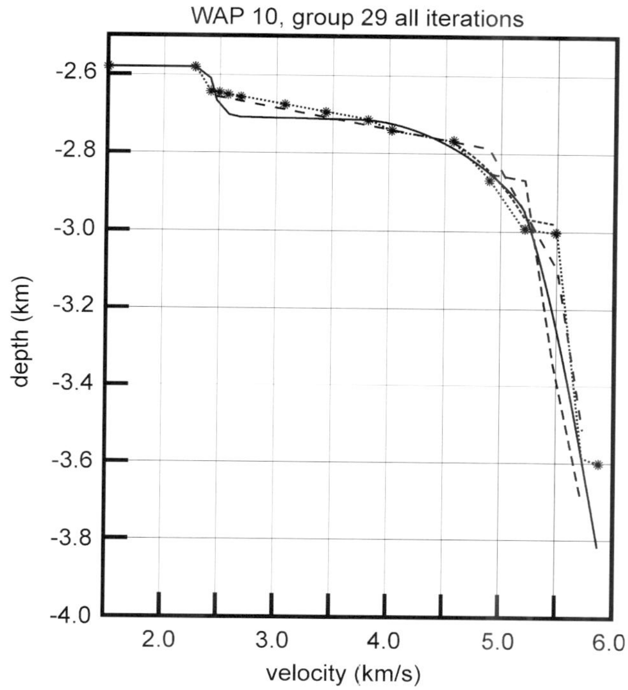


*Figure. 11.64* One of the starting models for $V_p$-depth inversion, to 1 km depth below the ocean floor. Successive iterations are shown, the fourth with asterisks. Kappus et al., 1995.

has been plotted among the 'soft porosity' (joint-related) curves of the Qc-$V_p$-depth-porosity-strength model, reproduced for easier comparison in Figure 11.65.

The Table 11.5 data plotted in Figure 11.65, shows strong evidence of structural (and matrix porosity) effects, and much rock quality improvement with depth increase (i.e. curve-jumping), in the upper 250 m. Interestingly the $V_p$ data below this depth suggests typical jointed rock Q-values (Barton et al., 1974; Barton 2002) in the range 2 to 8 down to 1000 m depth, with data paralleling the trends of $Q = 1$ and $Q = 10$ suggesting effective stress-joint-closure effects. A uniaxial strength of 100 MPa has been assumed in this example. If the basalt was closer to 200 MPa strength, a uniform shift to half as high Q-value (i.e. from about 5 to 2.5) would be involved. A plausible Q-parameter construction to explain such values of rock quality beyond 250 m depth could be the following:

$$\frac{60}{12} \times \frac{3}{4} \times \frac{0.5}{0.5} = 3.8$$

*Attenuation* in the upper crust of these mid-ocean ridge structures is clearly quite strong, with seismic $Q_p$ estimates varying from 35 to 80 in general (Harding et al., 1989; Vera et al., 1990; Wilcock et al., 1992). Kappus et al., 1995 found that their reflectivity synthesis computed with seismic $Q_p$ varying from 50 to 100, fit their original data very well.

Table 11.5 Average velocity-depth data from 70 profiles along a 52 km segment of the East Pacific Rise ridge crest. Kappus et al., 1995.

| Depth below seafloor, m | P-wave velocity km/s | Standard deviation km/s |
|---|---|---|
| 0 | 2.25 | 0.000 |
| 40 | 2.38 | 0.135 |
| 80 | 2.57 | 0.240 |
| 120 | 3.17 | 0.350 |
| 160 | 3.82 | 0.323 |
| 200 | 4.36 | 0.289 |
| 240 | 4.71 | 0.217 |
| 280 | 4.88 | 0.158 |
| 320 | 5.00 | 0.140 |
| 360 | 5.10 | 0.140 |
| 400 | 5.19 | 0.138 |
| 440 | 5.26 | 0.141 |
| 480 | 5.36 | 0.172 |

The low velocities of seafloor, age zero, mid-oceanic ridge crest materials from numerous studies in the period 1976 to 1994, reviewed by Kappus et al., 1995, had the following values in km/s: 2.5, 3.1, 2.1, 3.5, 2.35, 2.2, 2.0, 2.45, 2.4 and 2.7 (read from the zero-age end of eventual ranges of velocities). The mean value of 2.5 km/s implies a near-sea floor rock quality Q-value of only 0.1, as also roughly indicated in Figure 11.65. Velocity-depth gradients for layer 2A as a whole appear to have ranged from 3.5 to 5.5 $s^{-1}$, though they do not appear to have been quoted in many of the papers referenced by Kappus et al., 1995.

These low velocity, zero-age, crustal values have to be contrasted to laboratory velocities for young basalts of at least $5.6 \pm 0.4$ km/s (e.g. Hyndman, 1976). Information from drill holes (Alt et al., 1986; Nehlig and Juteau, 1988) reinforce the idea that the low velocities are strongly linked to structure (i.e. discontinuities, joints, fractures) and to matrix porosity, since there is evidence of strong circulation of hydrothermal fluids, mixing of cold and hot fluids and alteration, which could be intense at some levels.

Collier and Singh, 1998, utilised wide-aperture seismic reflection data with much improved vertical resolution (shots and receivers placed every 100 m), and

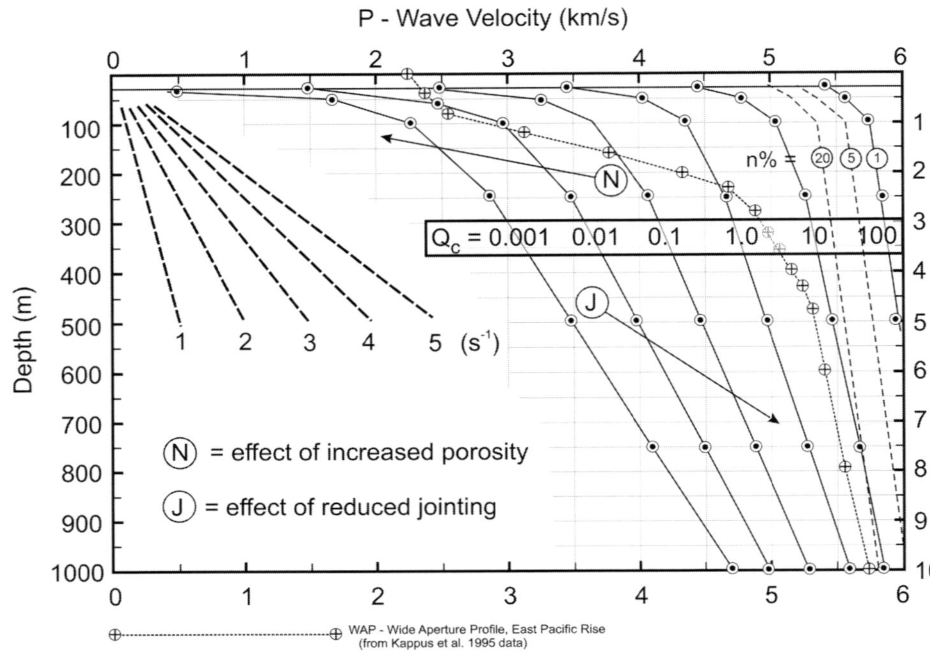


*Figure 11.65* Comparison of Kappus et al., 1995, mean Wide Aperture Profile data with the rock engineering $Q_c$-$V_P$-depth-porosity-strength model derived from Barton, 1995, 2002. A UCS value of 100 MPa gives Q = $Q_c$.

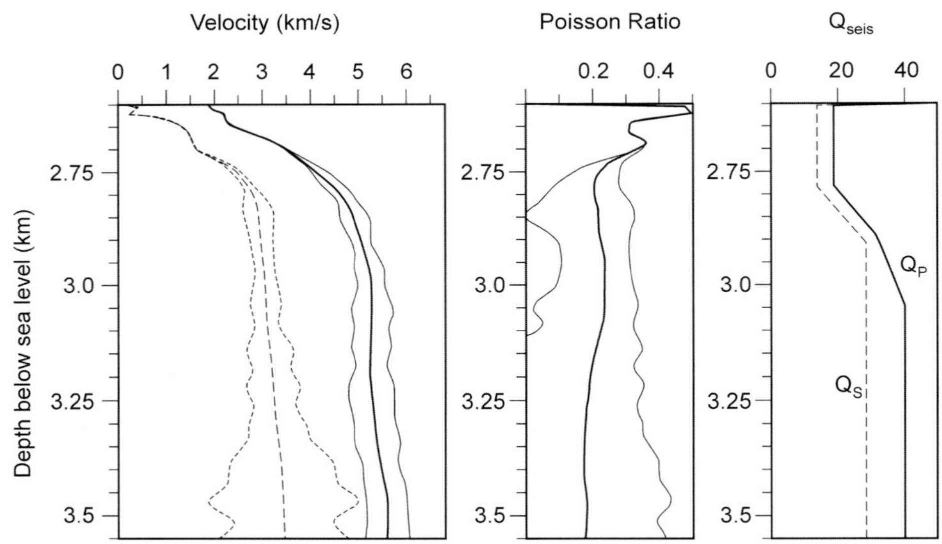


*Figure 11.66* Waveform inversion results from measurements on the East Pacific Rise near 14°S. Thinner lines show the one standard deviation error band. Collier and Singh, 1998.

applied full wave form inversion to interpret sub-ocean crustal structure beneath the East Pacific Rise near 14° S. They provided evidence of extremely high dynamic Poisson's ratios as high as 0.48, with a sharp drop to 0.25 within 200 m of the ocean floor, across the 2A/2B transition. A very low $V_s$ in the upper 50 to 100 m thick layer-2A ($V_P$ = 1.9 km/s, $V_s$ = 0.4 km/s) was responsible for the initial high value of 0.48. Their results included an estimate of seismic $Q_P$ in the low range of 18–30 across layer 2A, with increases at greater depth, as shown in Figure 11.66.

Their calculations suggested a porosity in excess of 30% in layer 2A, which reduced to 6–7% at the top of the 2A/2B transition, and further reduced to

Table 11.6 $V_P$-depth data as a basis for rock quality $Q_c$ estimation. From Figure 11.66 data.

| Depth below seafloor | $V_P$ (approx.) | $Q_c$ (approx.) | Gradient ($s^{-1}$) |
|---|---|---|---|
| 10 m (nominal) | 2.0 | 0.04 | – |
| 75 | 3.0 | 0.04 | 15 |
| 150 | 4.0 | 0.2 | 13 |
| 250 | 5.0 | 3 | 10 |
| 800 | 5.5 | 4 | 2 |

about 5% at a depth of 600 m below the seafloor, within layer 2B.

Hydrothermal alteration seems to be mostly responsible for the reducing bulk porosity and for preferentially sealing low aspect-ratio cracks. They also cited the higher porosity of extrusives, i.e. pillow lavas, compared to the deeper intrusives, which consist of dikes and sills. The increasing lithostatic (effective) pressure also preferentially closes the cracks or joints with lowest aspect-ratios, which became predominant across the 2A/2B transition. As they and others have emphasised, this can have a dramatic initial effect on $V_P$ gradients, until 'seismic closure' is achieved at sufficiently high effective stress. Table 11.6 gives the approximate indications of $V_P$-depth gradients.

Referring to Figures 11.44 and 11.65, we see that, in rock engineering terms we need to 'curve-jump' from initially very low rock quality $Q_c$-values, consistent with extremely fractured and altered conditions near the surface, to a typical poor quality jointed medium ($Q_c$ = 3 to 4) at greater depth in layer 2B.

## 11.6 Age effects summary for Atlantic Ridge and Pacific Rise

Finally, this chapter will conclude with a broad review of age effects for both mid-Atlantic Ridge and Pacific Rise data. A very wide ranging assembly of seismic velocity data for uppermost oceanic crust (layer 2A), by Carlson, 1998, (with one and a half pages of referred authors) suggested that most of the age-dependent increase in seismic velocities occurred 'rapidly' with velocities nearly doubling in <10 million years.

The apparently rather heterogeneous sets of data (Figure 11.67) when synthesised by Carlson, 1998, using mean and 9-point interval median values, showed much clearer trends of accelerated velocity increase with early age, and clear distinction from typical layer 2B velocities

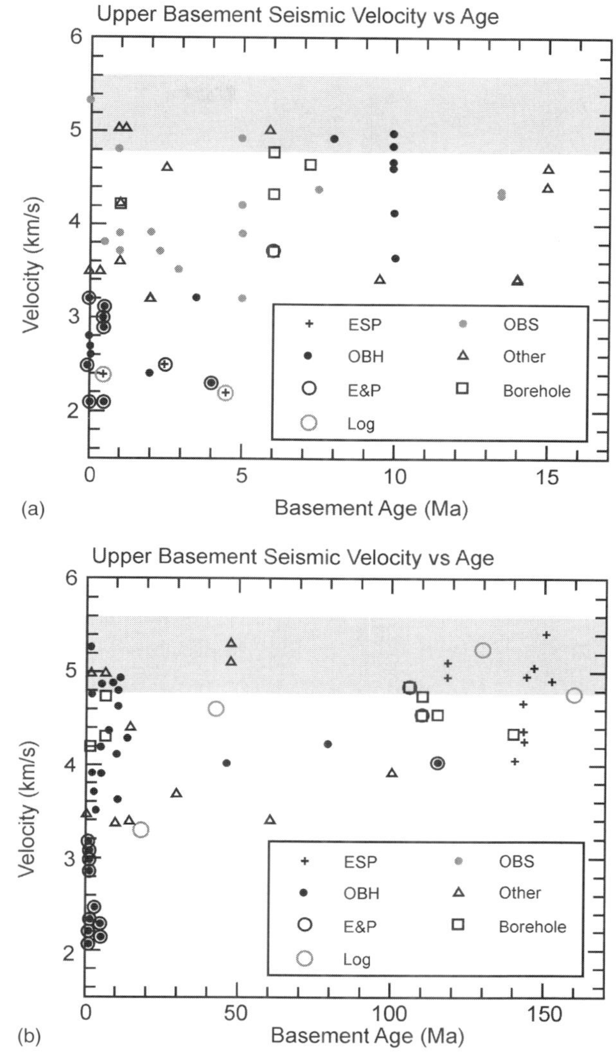


*Figure 11.67* Upper crustal velocities in the Pacific and Atlantic oceans for a) 0–150 m.y. b) detail of first 15 m.y. Typical layer 2B velocities (5.2 +/− 0.4 km/s) shown by dark band, from Houtz and Ewing, 1976 and Houtz, 1976. (OBS = ocean bottom seismometer, OBH = ocean bottom hydrophone, log = downhole, borehole = borehole seismic surveys). Carlson, 1998 and his numerous cited authors.

(shown by dark band). Layer 2A appears to persist as a low velocity capping of the ocean crust, even when >15 m.y. old, as shown in Figure 11.68.

The trend for increased velocities as age increases is also shown clearly by the statistics for <1 m.y., 1–5 m.y. and ⩾5 m.y. in Figure 11.69, also from Carlson, 1998. The link between hydrothermal alteration and seismic velocity increase, due to deposition of minerals first in

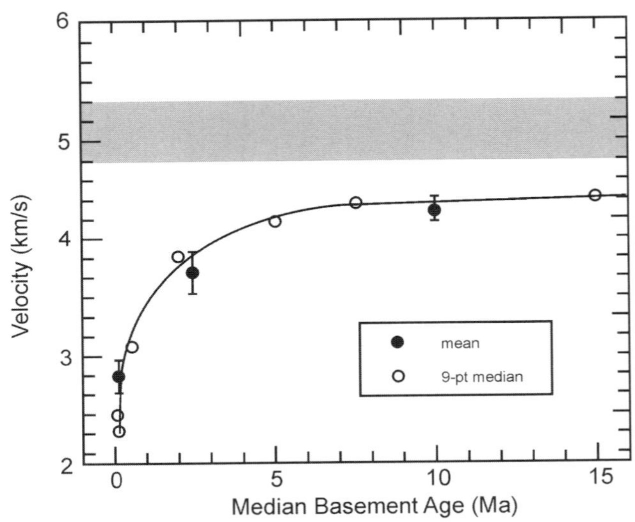


*Figure 11.68* Mean velocities (black circles), and 9-point median velocities (open circles), as a function of median basement age. Mean velocities are from age intervals 0–1, 1–5 and 5–20 Ma. The data demonstrates that layer 2A persists as a low velocity (and low rock quality) layer, capping the crust. Hatched area is typical layer 2B. Carlson, 1998 and his numerous cited authors.

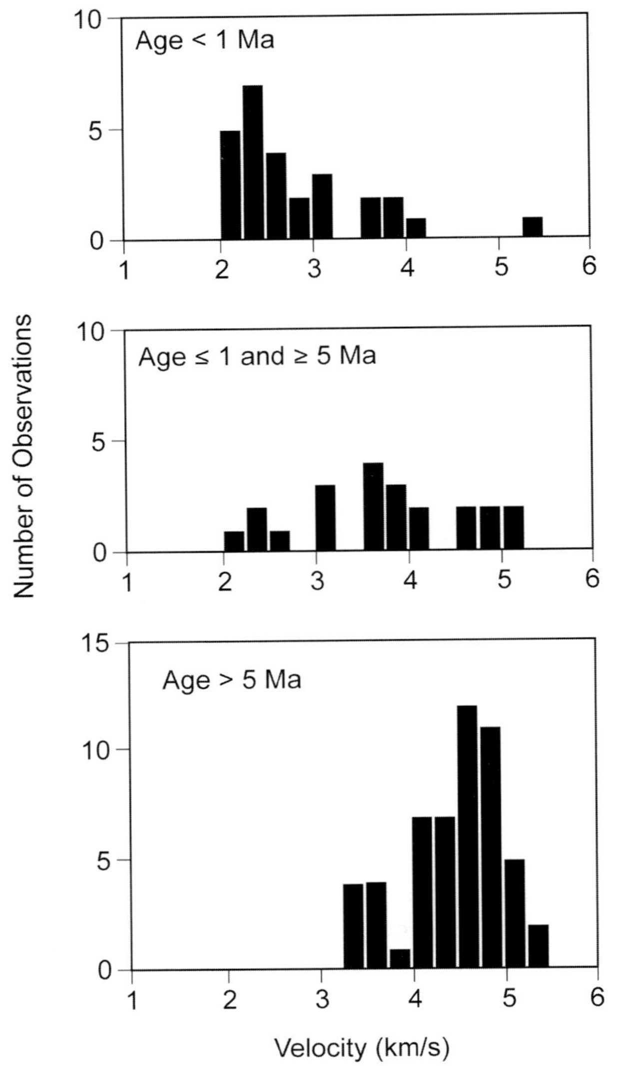


*Figure 11.69* Distribution of seismic velocity in the upper crust (layer 2A) from age intervals 0–1, 1–5, >5 Ma. By comparison, layer 2B velocities generally range from 4.5 to 5.6 km/s. Carlson, 1998 and his numerous cited authors.

the thinnest cracks and joints, gives logical support to the notion of most rapid alteration when the hydrothermal activity has been most vigorous.

However, a potential problem of interpretation exists as pointed out by Carlson, 1998, because hydrothermal void filling causes a simultaneous increase in velocity and reduction in hydraulic conductivity, therefore supposedly reducing heat flow to the ocean flow. Yet the heat flow has continued for far longer than the early period of velocity increase. When one considers the cubic flow law, suggesting that mineral sealing of small apertures will have less effect on hydraulic conductivity than sealing of large apertures, it seems reasonable to envisage that the sealing of smallest apertures first will not compromise the increase in seismic velocity and the maintenance of a reasonably high heat flow.

Further interesting details of hydrothermal circulation and mineralization, as an explanation for seismic velocity increases with age, were given by Grevmayer et al., 1999. Their work was concentrated in Layer 2A, on the eastern flank of the East Pacific Rise at 14°S, along a 720 km by 25–40 km wide corridor, with only thinly sedimented seafloor of up to 8.5 Ma age. For 0.5 Ma and 8 Ma crust, they derived P-wave velocities of 2.9 and 4.3 km/s, respectively. Over the last 8.5 m.y. there has been local ridge spreading at an average annual rate of 85 mm/year in this area.

Their forward modelling of 17 (split) profiles on 0.5 to 8.3 m.y.-old crust for three age ranges is shown in Figure 11.70. When velocity is plotted versus age as in Figure 11.71, a rapid then gradually slowing rise in velocity is seen, which is similar of course, to the wider-reaching review of Carlson, 1998, shown in summary in Figure 11.68. Grevemeyer et al., data indicate a continuous decrease of the velocity gradient in layer

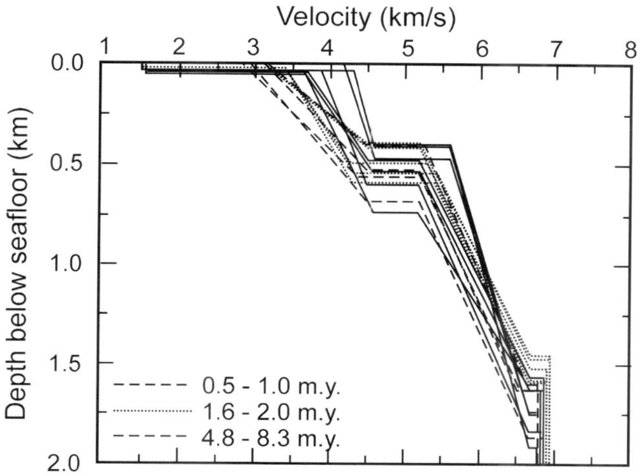


*Figure 11.70* $V_p$-depth models of upper oceanic crust from inversion of 17 OBH split profiles, on 0.5 to 8.3 m.y. crust, from the East Pacific Rise at 14°S. Note the three age groupings. Grevemeyer et al., 1999.

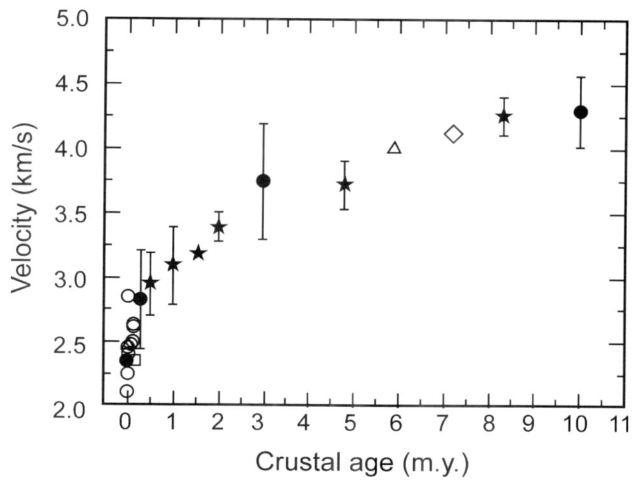


*Figure 11.71* Layer 2A velocity as a function of age (at top of layer). New results (star-symbols) and various results from referred studies, including Carlson, 1998 (solid circles). Grevemeyer et al., 1999.

2A, with age up to 10 m.y. Their data only just reaches the plateau seen in Figure 11.68, beyond about 8 m.y.

Grevemeyer et al., 1999, argued convincingly that hydrothermal mineral filling of open void spaces was the reason for age dependent velocities, and that velocities in layer 2A remained constant in crust older than ≈10 m.y. An 8 m.y. crust at the top of layer 2A, showed 4.3 km/s which is high for the Pacific. Even 110 m.y. crust and older, usually showed between 4.0 and 4.3 km/s.

Interesting parallels to the above hydrothermal mineral filling of fractures, can be gleaned from civil engineering, where the sealing of jointed rock by pre-grouting with fine-grained micro-cements ahead of tunnels, or the use of industrial (coarser grain size) cements in dam foundations, are common ways of both sealing and improving properties. Quadros and Correa Filho, 1995, measured rotations (and magnitude reductions) of the three *permeability tensors*, when conducting multiple-borehole 3D hydrotomography before and after grouting. (Holes were redrilled after grouting, for the second round of permeability testing).

These tensor or principal value rotations are interpreted as due to sealing of the most permeable and least stressed joint sets. This process can also be interpreted by small changes in five or six of the Q-value parameters (ratings in Appendix A), which has been shown by Barton, 2002, to cause some dramatic potential improvements in the rock mass properties such as modulus of deformation, seismic velocity, and frictional and cohesive strength. These joint-grouting property-improvement aspects will be described at the end of this chapter.

### 11.6.1 Decline of hydrothermal circulation with age and sediment cover

It is well established that hydrothermal circulation naturally declines with the age of the crust. Anderson et al., 1985, and Evans, 1994, deduced on-axis (zero age) upper crustal permeability of about $6 \times 10^{-12}\,m^2$ ($\approx 6 \times 10^{-5}$ m/s), this decreasing to about $7 \times 10^{-14}\,m^2$ ($\approx 7 \times 10^{-7}$ m/s) within 6 m.y. Seismic velocities for crust of the same age were 2.2 km/s and 4.0 km/s according to the Grevemeyer et al., 1999, data. The same authors surmised that permeability may reduce to about $10^{-14}\,m^2$ ($\approx 10^{-7}$ m/s) or less, by the time the crust is old enough to have reached a 4.3 km/s 'plateau'.

In thinly sedimented areas, sealing/plugging of crustal pore spaces appears to extend for 7 m.y., perhaps up to 15 m.y. However, in regions with significant sediment cover, the previously open seawater convection cooling system is hindered, and temperatures rise, thereby accelerating the formation of secondary minerals and porosity sealing.

Rohr, 1994, used these arguments to explain 4.3 km/s velocities in only 1.5 to 2.0 m.y. crust at the Jan de Fuca Ridge. So Grevemeyer et al., 1999 concluded that basement temperature, which is a function

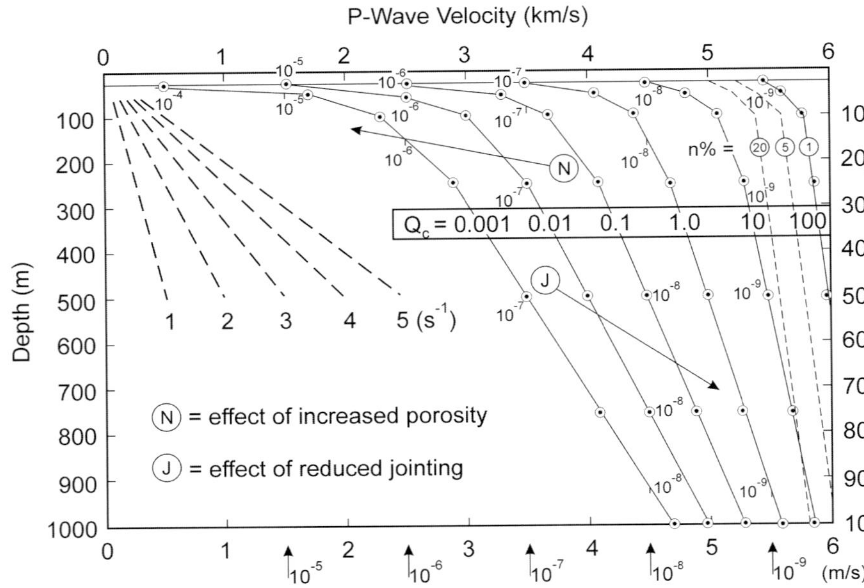


*Figure 11.72* The rock quality $V_P$-depth-$Q_c$ curves with rule-of-thumb estimates for the approximate permeability (m/s) caused by the 'soft' porosity (i.e. the jointing). As in the case of pre-grouting with micro-cements ahead of tunnels, one can argue for improvement of various Q-parameter ratings through sealing-mineralization, provided that clay and layer-lattice minerals are not involved. Harder minerals will have cumulative positive effects on Q-values, and therefore on permeability and other linked rock engineering parameters. 'Curve-jumping' or 'Q-jumping' can therefore occur. Based on Barton, 2002.

of heat flow, sediment thickness and sediment permeability, governed the evolution of the seismic properties of upper ocean crust.

Since massive, non-vesicular basalt has a P-wave velocity of about 7 km/s (Wilkens et al., 1988), then mature sealed ocean crust, with velocities nearer 4 km/s, must have a residual porosity consisting of vesicules, fractures, breccia and inter-pillow voids. Most importantly, it is actually under a low effective stress, just as at the earth's surface, despite (and because of) 30 to 40 MPa hydrostatic pressure, and because of the shallow sediment and rock cover.

In the rock quality world of Q and $Q_c$ ($Q_c = Q \times \sigma_c/100$) there is a rough rule-of-thumb (Barton, 2002) that due to different degrees of 'soft' porosity caused by jointing, the Lugeon value ($1L \approx 10^{-7}$ m/s or $10^{-14}$ m/s) is approximately inversely proportional to $Q_c$. This applies in a central range of rock quality, uncomplicated by clay sealing of joints, which reduces both Q and permeability together, thereby defeating the inverse proportionality.

In Figure 11.72, shown earlier in connection with velocity-depth gradients, the above rule-of-thumb permeabilities have been marked at appropriate depths along the Q or $Q_c$ 'isoquality' curves. It is relatively easy to invoke this 'jointed rock' model as an analogue to explain a surface or near-surface permeability, as referred above, of about $10^{-5}$ m/s ($10^{-12}$ m$^2$), with increases to $10^{-6}$ and $10^{-7}$ m/s ($10^{-13}$ and $10^{-14}$ m$^2$), as age allows the effective rock mass quality to rise, due to mineralization sealing in some of the 'soft' porosity. At the nominal 25 m deep 'surface' drawn in Figure 11.72, $Q_c$ changing from 0.01 through 0.1 to 1 (i.e. 'curve jumping' due to improving quality with age) implies permeability reducing from $10^{-5}$ through $10^{-6}$ to $10^{-7}$ m/s.

We can express these analogues for rock mass quality improvement with age, due to sealing-mineralization (as opposed to clay-lubrication) in another way in the $V_P$-$Q_c$-M-L interaction nomogram in Figure 11.73.

Here we have marked the above analogue for permeability reduction with age (roughly $10^{-5}$ to $10^{-6}$ to $10^{-7}$ m/s) at the nominal near-surface depth of 25 m (central diagonal), and at 50 m. Also shown is the implied improvement of modulus of (static) deformation for the rock mass: 2 to 5 to 10 GPa, based on the simple empirical relation (units: GPa):

$$M(\text{or } E_{mass}) \approx 10Q_c^{1/3} \tag{11.12}$$

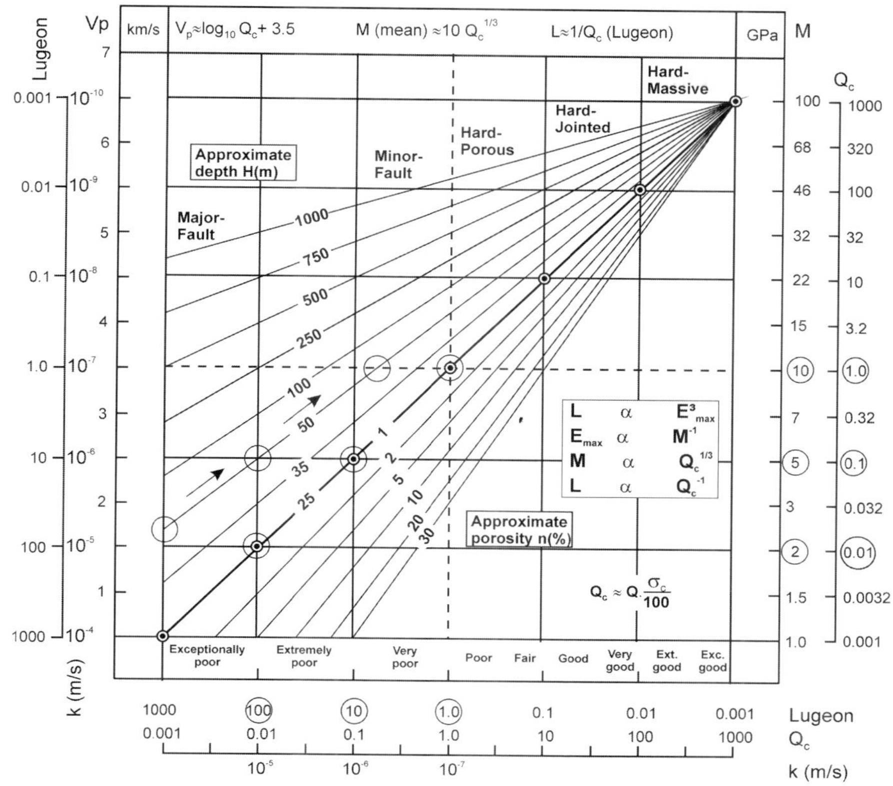


*Figure 11.73* The rock quality $Q_c$-$V_p$-M-L interaction nomogram, with appended 'circles' to mark the engineering consequences of 'curve-jumping' on permeability, modulus and velocity. After Barton, 2002.

### 11.6.2 The analogy of pre-grouting as a form of mineralization

In this final section of Chapter 11, a possible rock engineering analogy to the hydrothermal mineral sealing of ocean-floor basalts will be demonstrated, using the analogy of high pressure (5 to 10 MPa) pre-injection pressures used in an 'umbrella' of numerous (20 to 40) boreholes, which are commonly drilled and injected ahead of leaking tunnels, or where there is environmental sensitivity at the surface and significant inflows to the tunnel cannot be allowed.

The argument for improved rock mass quality due to the sealing (by micro-cement and micro-silica) of successive joint sets, is based on the evaluation of individual Q-parameter descriptions and ratings. Each small improvement, like the reduction of the number of effective joint sets due to sealing (i.e. $J_n$ reduces, see Appendix A), may have cumulative effects on rock qualities Q and $Q_c$. There exists in situ proof of improved qualities due to observations of improved stability, less deformation, less need for rock support (bolting and shotcreting quantities are reduced), and of course reduced or negligible inflows.

Permeability K (m/s) may reduce from $10^{-5}$ to $10^{-8}$, $10^{-6}$ to $10^{-8}$ or $10^{-7}$ to $10^{-8}$ m/s, the relative degree of improvement being related to the 'severity' of pre-treatment conditions. Higher pressure injection, from 5 to 10 MPa excess pressures (above joint water pressure), can cause permeabilities to reduce to between $10^{-8}$ and $10^{-9}$ m/s. There is an interpreted 1 to 5 litres of grout per 1 $m^3$ of (locally-injected) rock mass in successful, high-pressure grouting (Barton, 2004a). In other words there is the effect of some joint deformation close to the injection holes, made permanent by the subsequent hardening.

Such would also be expected during hydrothermal episodes of injection, the subsequent 'fossil' mineral fillings therefore somewhat exaggerating the *pre-injection* apertures and permeabilities, despite some subsequent pressure adjustment prior to crystallization. Sealing of major channels is 'a problem' in both scenarios, because any continued flow will tend to hinder crystallization/hardening in the two processes.

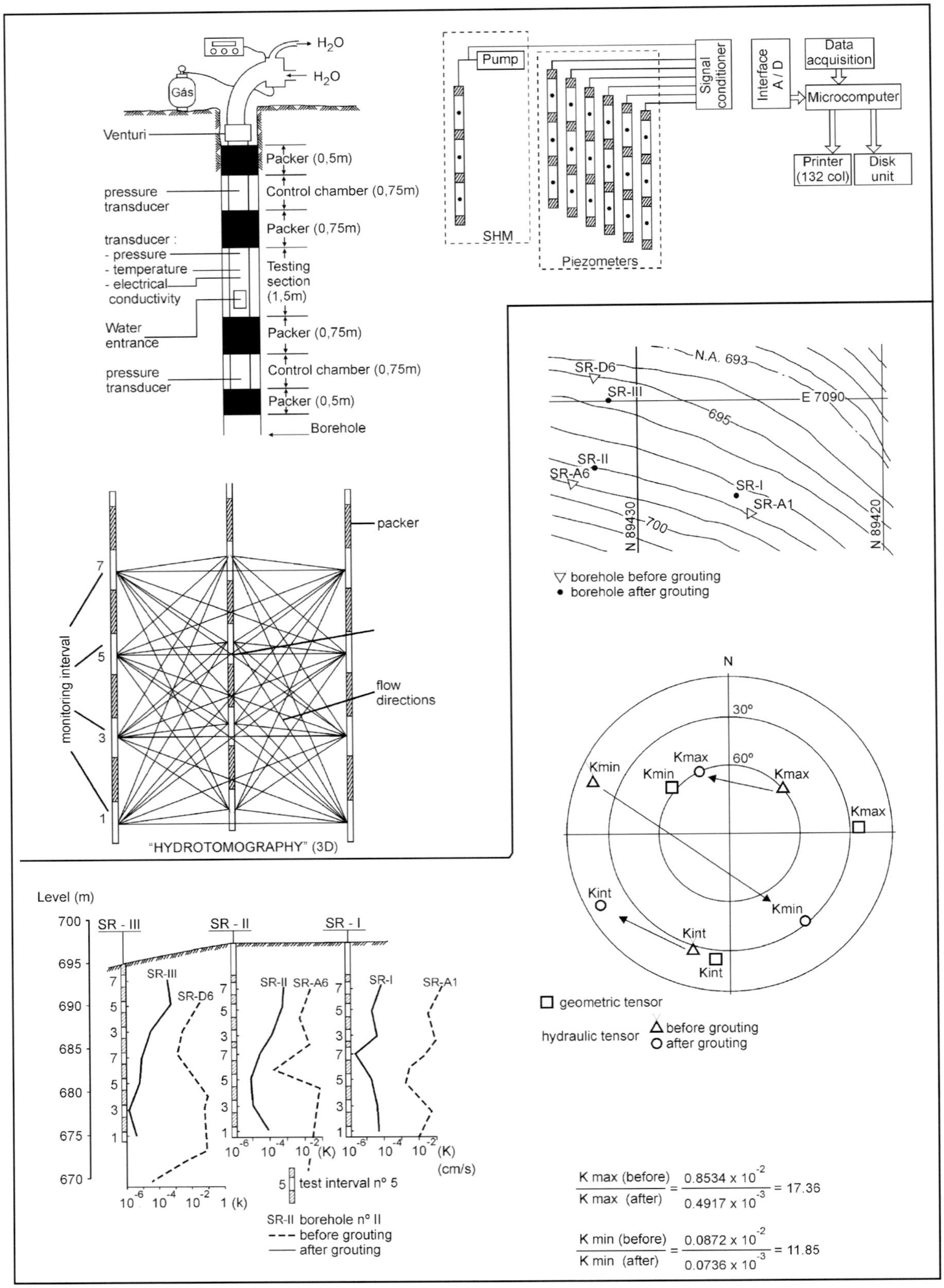


*Figure 11.74* Three-dimensional permeability monitoring (hydro-tomography) performed before and after grouting, showing rotation of permeability tensors, and reduction in principal magnitudes. Quadros and Correa Filho, 1995; Barton and Quadros, 2003.

Table 11.7 A hypothetical model of potential Q-parameter improvements as a result of both pre-injection with micro-cement (in tunnelling) and through (hard, resistant) mineral sealing in sub-ocean crust.

| Improvement of rock mass properties with pre-grouting, as analogy to mineralization | | | |
|---|---|---|---|
| Effective RQD | Increases | e.g. 30 to 50% | |
| Effective $J_n$ | Reduces | e.g. 9 to 6 | |
| $J_r$ | Increases | e.g. 1 to 2 | (changed set)* |
| $J_a$ | Reduces | e.g. 2 to 1 | (changed set)* |
| $J_w$ | Increases | e.g. 0.5 to 0.66 | (perhaps $J_w = 1$ is achieved) |

SRF would reduce only near surface, e.g. 2.5 to 1
(*it may be appropriate to qualify with the word 'perhaps' in these cases)

Before pre-grouting $Q \approx \frac{30}{9} \times \frac{1}{2} \times \frac{0.5}{2.5} \approx 0.3$ (i.e. prior to mineralization)

After pre-grouting $Q \approx \frac{50}{6} \times \frac{2}{1} \times \frac{0.66}{1} \approx 11$ (i.e. after mineralization)

| Before pre-grouting | After pre-grouting | Empirical equation |
|---|---|---|
| $Q \approx 0.3$ | $Q \approx 11$ | $Q \approx \frac{RQD}{J_n} \times \frac{J_r}{J_a} \times \frac{J_w}{SRF}$ <br> $Q_c = Q \times \frac{\sigma_c}{100}$ |
| $V_p \approx 3.0$ km/s | $V_p \approx 4.5$ km/s | $V_p = 3.5 + \log Q_c$ km/s |
| $L \approx 3$ ($3 \times 10^{-7}$ m/s) | $L \approx 0.1$ ($10^{-8}$ m/s) | $L \approx 1/Q_c$ |
| $M \approx 7$ GPa | $M \approx 22$ GPa | $M \approx 10Q_C^{1/3}$ GPa |

As referred to in Chapter 9, it has also been shown by Quadros and Correa Filho, 1995 and Barton and Quadros, 2003 that permeability tensors can rotate and reduce in magnitude as a result of grouting. This is shown in Figure 11.74. It appears to be evidence of successive sealing of the joint sets, starting presumably with the set that is under least normal stress (probably nearly parallel to $\sigma_{max}$), or with the most permeable set (or sets) perhaps caused by (conjugate) shearing and dilation. This presumably could also govern the chronological order of deposition in hydrothermally opened fractures and major inter-connected pore space in the sub-ocean basalts.

The mechanism by which rock mass quality and rock mass velocity increases, as a result of both successful pre-grouting and 'successful' mineralization of sub-ocean crust (as it gets hotter due to sediment sealing), could be as in Table 11.7 (see Appendix A for descriptions and ratings of the six parameters).

The hypothesised improvements in Q-parameters are very conservative particularly concerning the number of joint sets that may be sealed. More likely, the three sets represented by $J_n = 9$, could be reduced to two sets ($J_n = 4$), or even to one set ($J_n = 2$). On the other hand, the grout is an 'inferior' fill in relation to hard rock, and may not then provide the expected increases in velocity and modulus, shown by the empirical equations in Table 11.7. So a greatly increased *apparent* Q-value, would not then give realistic velocity improvements.

The exception would be the case of injection into weak rocks with comparable compressive strengths and densities (roughly 20–30 MPa and 1.5 gm/cm$^3$). A beautiful example of an igneous intrusion into what was, or has since become a weaker rock is shown in Figure 11.75. This could act as a reminder that the injected hydrothermal fluids, and of course magma, if subsequently very stiff compared to the surrounding rock, will tend to be jointed due both to cooling and deformation, thereby maintaining a level of permeability at depth.

*Figure. 11.75* An igneous intrusion (dike) tends to have elevated permeability due to the number of joint sets (typically four: $J_n = 15$). High-modulus mineralized veins may be fractured by subsequent tectonic deformation, helping to maintain some permeability despite the 'sealing' process. (see Plate 2)

# 12 Rock stress, pore pressure, borehole stability and sonic logging

By their very nature, hydrocarbon-bearing rocks rely on pore-space and permeability for the possibility of having recoverable reserves that can be produced at a well. The necessary entrapment beneath a shale-sealed anticline or by the juxtaposed impermeable layer caused by fortuitous faulting, are two basic scenarios for the present-day existence of the reserve. The necessary migration of the hydrocarbons from source rocks into potential entrapment structures, without escape to the atmosphere, adds to the adverse statistics of hydrocarbon exploration. Too close to the surface the sealing properties of shale, salt or clay-smear in faults, may have been compromised by lack of plasticity and too high permeability. Too deep, the pore space and permeability of the reservoir may be compromised, giving a reduced reserve and the need for permeability enhancement and gradient enhancement, or a decision for non-development.

Following seismic interpretation of potential hydrocarbon-bearing structures, the need for expensive exploratory drilling and well-testing follows. Besides reservoir access for production testing, the hole is used for sonic logging and selected side-wall and regular core recovery, to better define the properties of the different lithologies, seals and reservoir rocks. Fortunately, or unfortunately according to viewpoint and tools available, rock reacts to the drilling of boreholes with a complex coupling of rock stress and strength magnitudes (plus the anisotropies of each), not forgetting the necessary subtraction of pore pressure for conversion of the three principal stresses to *effective stresses*. The appropriate selection of wellbore 'temporary support' in the form of mud pressure, using variable mud weight, determines the state of the borehole wall in the different lithologies, prior to setting and cementing the casing.

Due to various opinions about an 'alteration zone' around the wells, there is now widespread acceptance of the need for logging while drilling (LWD) with monopole and dipole tools, to obtain 'early' velocity responses, which may differ significantly from subsequent wireline logging. The differences are probably due to stress-fracturing, increased permeability, and accelerated mud-filtrate invasion. The frequent development of a near-wellbore, tangentially-distributed *discontinuum*, in cases of insufficiently mud-supported rocks, may perhaps have been overlooked, when modelling infiltration.

## 12.1 Pore pressure, over-pressure, and minimum stress

Before addressing the details of seismic signatures of reservoir rocks, caused by numerous environmental effects such as pressure, temperature and fluid type in Chapter 13 (at laboratory *rock physics* scale) and in Chapters 14 and 15 (at reservoir-scale), it is appropriate to consider the components and modifiers of the most fundamental of reservoir parameters, namely the *effective stress magnitude*. The rock stress and its variations with direction, depth and location, and the pore pressure and sometimes over-pressure which are influenced by compaction and also by fluid type, are the major boundary conditions. Their relative magnitudes affect both the laboratory test simulations, the drilling programme, the production planning, and the reservoir production and depletion, possibly for 50 years or more in a large reservoir.

### *12.1.1 Pore pressure and over-pressure and cross-discipline terms*

Without attempting too much detail, since 'pore pressure analyst' has become a speciality career choice for numerous petroleum engineers, it is worthwhile following the helpful philosophy of Bruce and Bowers, 2002, and mentioning various cross-discipline differences for describing the effects of pore pressure and over-pressure. Reproducing selected diagrams from their practical article, we see

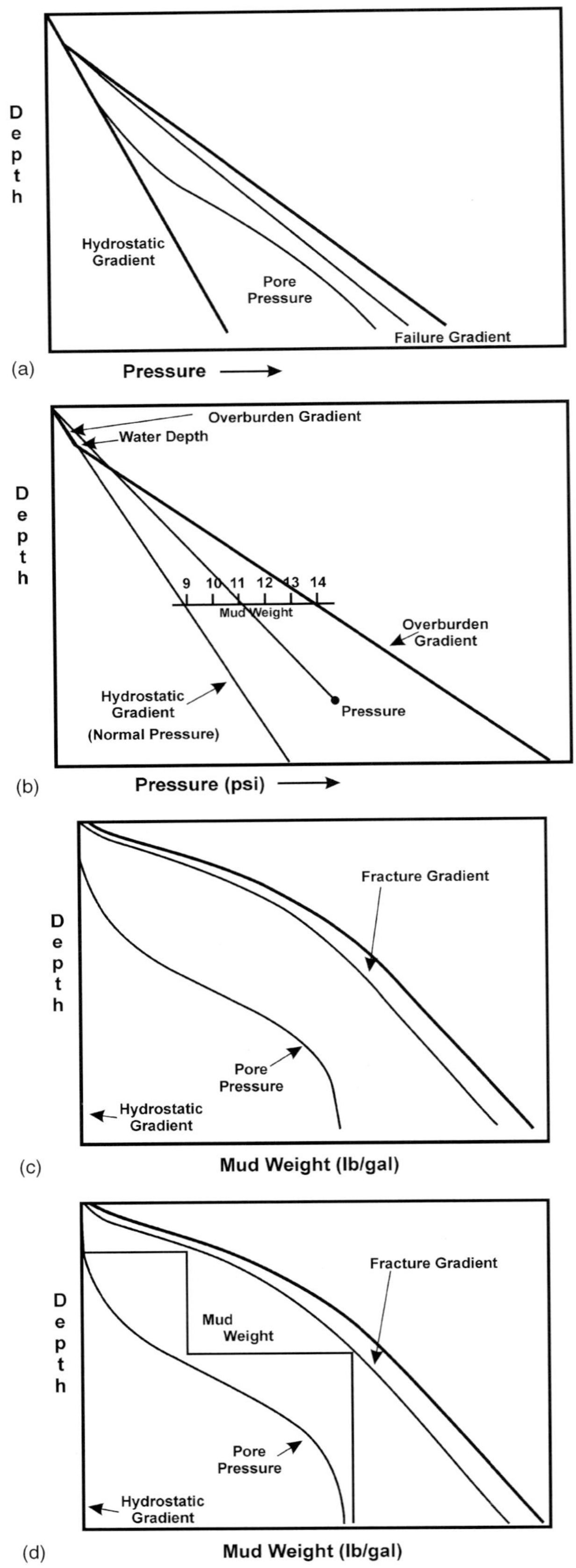


the following details in Figure 12.1a to d, and Figure 12.2 a to d.

Most fundamentally, over-pressure and the *presence* of petroleum products, both increase the pore pressure. So the effective stress is reduced, which will have the effect of causing a reduction in P-wave velocity, as illustrated in Figure 12.3 from Bowers, 2002.

Over-pressure commonly occurs where low permeability layers (as in shale) prevent fluid from escaping as rapidly as pore space compacts. Excess pressure in relation to hydrostatic then builds up as newly deposited sediments cause squeezing of the trapped pore fluids, which could be water, oil or gas or even two or three of these close together.

To conclude this brief section on over-pressure effects, we may refer to more sophisticated considerations that over-lap with environmental effects discussed in the next chapter. Carcione and Gangi, 2000, reported results from their modelling of seismic attributes of gas generation and over-pressure. Their model for basin-evolution showed that pore pressure effects were seismically visible when the effective pressure was less than about 15 MPa, with oil-to-gas conversion of only 2.5%. Here they differentiated live oil containing gas, from the dead oil-free gas which may become seismically visible.

They found that a small conversion of oil to gas was sufficient to make the pore pressure equal to the confining pressure. The large changes of predicted velocity, as shown in Figure 12.4, were the result of the fact that the dry rock moduli were strongly affected by low effective pressures.

### 12.1.2 Minimum stress and mud-weight

A complication for drillers when drilling through inter-bedded lithologies, or where hole stability is marginal due to limited rock strength or local 'structure', is their

*Figure 12.1* a) The basic gradients when there is over-pressure, considering $\sigma_{min}$ and the possibility of failure (hydraulic fracturing) by too high mud pressure. b) Depth-pressure gradients, showing equivalent mud-weights (in American units of lbs/gallon), where sea water = 8.54 lbs/gallon. [1 lb/gal. = 0.0519 psi/ft] c) Change from pressure-depth axis, to mud-weight-depth format (as preferred by drillers for obvious reasons). d) To prevent hydraulic fracturing by high mud-weights, as needed where there is overpressure, casing will be set to protect the overlying units from fracturing. Bruce and Bowers, 2002.

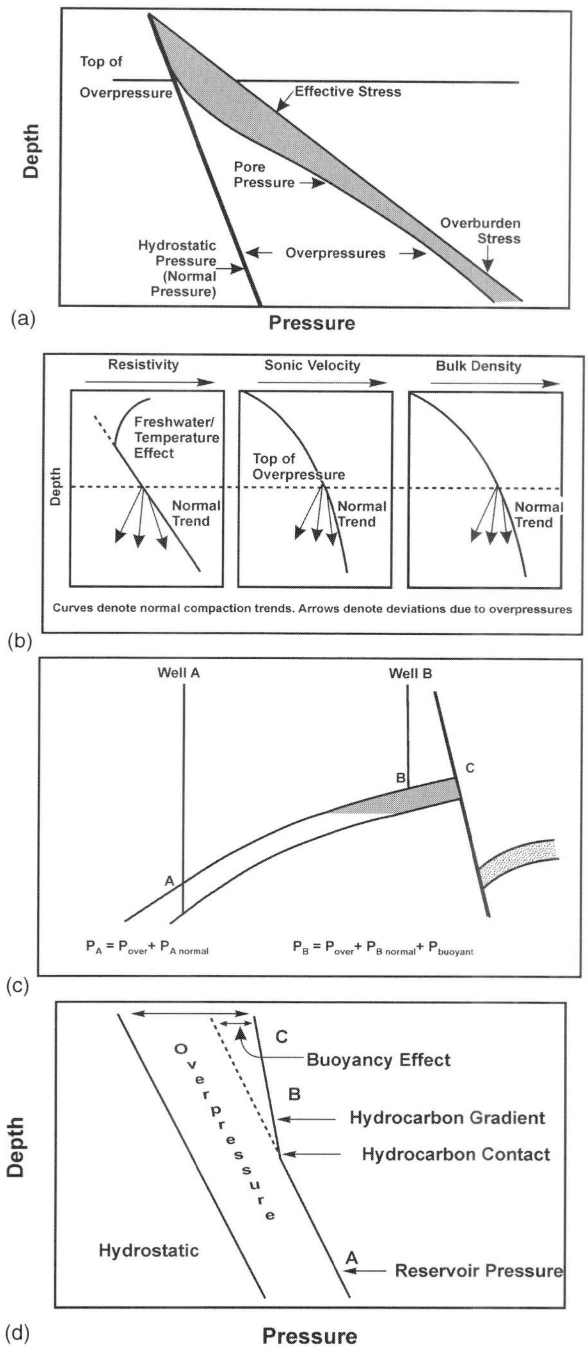


*Figure 12.2* a) Overpressure in relation to normal hydrostatic pressure and its potential effect on effective stress-depth trends (in relation to $\sigma_{vertical}$). b) Resistivity, velocity, density, depth trends with various degrees of deviation possible where over-pressure changes the effective stress. c) Pressure at well B is the sum of normal (hydrostatic) pressure, plus over-pressure, plus a buoyancy effect caused by the reduced density of the petroleum. d) The pressure components given in terms of depth-pressure gradients. Bruce and Bowers, 2002.

choice of mud weight. This heavy fluid acts as a temporary support of the walls of the well during drilling, until replaced by steel casing that is grouted into intimate, impermeable contact to the rock. Well testing, possible minimum rock stress estimation by mini-hydrofracing and subsequent production occurs through shaped-charge perforations of the casing in the reservoir intervals, and also in the cap-rock, if minimum stress is to be measured, to determine the minimum stress difference.

Inter-bedded 'brittle' layers (like sandstone as opposed to 'sand'), and 'plastic' layers (like shale or salt) will likely exhibit fluctuating minimum principal stress. Since the shale and salt-rocks may have insufficient shear strength to tolerate a significant stress difference, the *minimum rock stress* (the minimum of three principal rock stress directions) in the shale (or salt) will often exceed the minimum stress in the reservoir sandstone by up to several MPa. An example from measurements described by Barton, 1986, is shown in Figure 12.5.

The shale (and the salt) may, if encountered at sufficient depth during drilling, require the support of an active mud-weight to prevent creep or squeezing, as graphically illustrated by Bradley, 1978, for the case of deviated wells, in Figure 12.6. So the drillers choice of mud-weight, or the setting of casing, becomes critical where support of the well is needed adjacent to a reservoir rock like sandstone or fractured limestone or chalk, which would tend to have a *minimum* rock stress *less than* that of these weaker, sealing 'plastic' layers. The reservoir horizons could potentially fracture, or have a permeable joint under lower normal stress than the mud-weight needed to keep the plastic materials from squeezing and jamming the drill-string. Invasion of mud (lost circulation), into any reservoir horizon is obviously very undesirable.

## 12.2 Stress anisotropy and its intolerance by weak rock

The local measured variations of minimum principal stress shown in Figure 12.5 are due to intolerance of stress difference in the weaker shale, as compared to the sandstone. As discussed earlier, such a mechanism also occurs in the case of salt. Swolfs, 1977 provided a graphic comparison of tolerance of stress difference ($\sigma_v - \sigma_{h\ min}$) for different rock types. Figure 12.7 shows data for $\sigma_v$ (overburden, calculated) minus $\sigma_{h\ min}$ (minimum horizontal principal stress) versus vertical stress. Rocks such as granite, stronger sandstones and

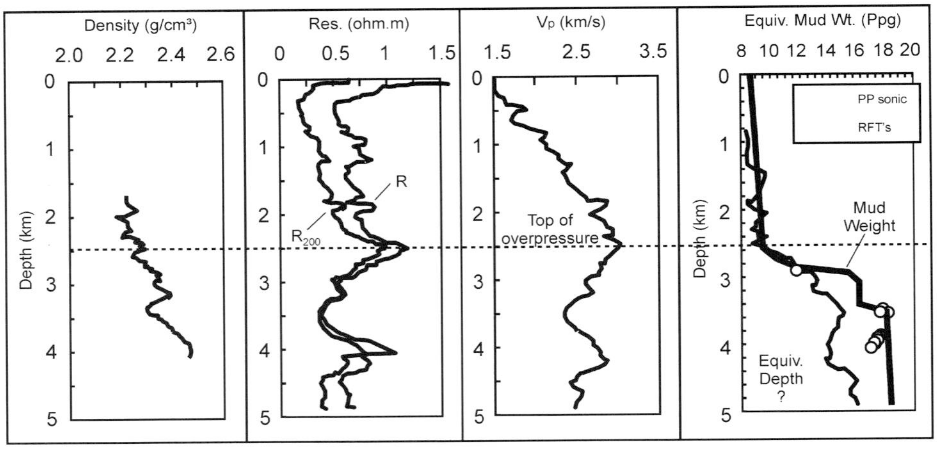


*Figure 12.3* Example of high-pressure well from Bowers, 2002, where velocity (and resistivity) undergo reversals, due to 'under compaction'.

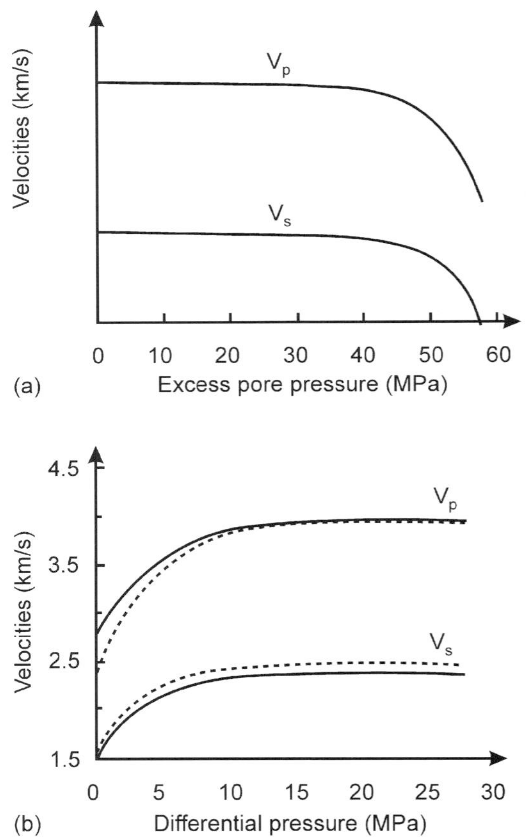


*Figure 12.4* A model of low frequency (25 Hz) P- and S-wave velocities versus excess pore pressure in the case of deep gas resources. The seismic visibility begins due to the sensitivity of $V_p$ and $V_s$ to reduced effective stress, especially in the dry state (dotted lines). From Carcione and Gangi, 2000.

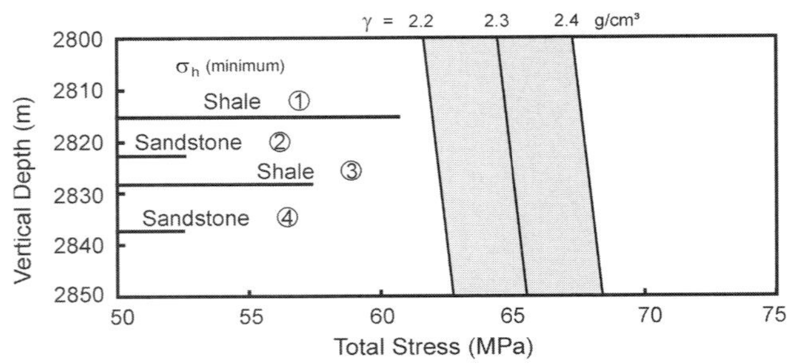


*Figure 12.5* Minimum principal (rock) stress in inter-bedded reservoir shales and sandstones, as recorded by mini-hydrofracing tests, to determine optimal stimulation by MHF (massive hydraulic fracturing), Barton, 1986.

stronger shales tolerate differential stress much better than weaker shales and salt rocks.

Swolfs, 1977 compilation of vertical (over-burden) and minimum stress data for these typical North American reservoir rocks (and of some harder rocks like granite and gneiss) to depths of 5 km, shows greatest stress anisotropy near the surface, and in the harder rock types. Figure 12.8 shows a range of $K_o$ from 0.35 to 2.9. Bedded salt in particular, plus weak rocks like Tertiary oil shale and Tertiary tuff showed $\sigma_{h\ min}/\sigma_v$ close to 1.0.

From the point of view of stress-azimuth dependent *anisotropic velocity*, the added influence of stress-oriented jointing and the character of these joints will likely be more important than stress anisotropy effects on the matrix alone, due to the nature of 'soft' porosity (low aspect ratio) jointing. When shear wave splitting

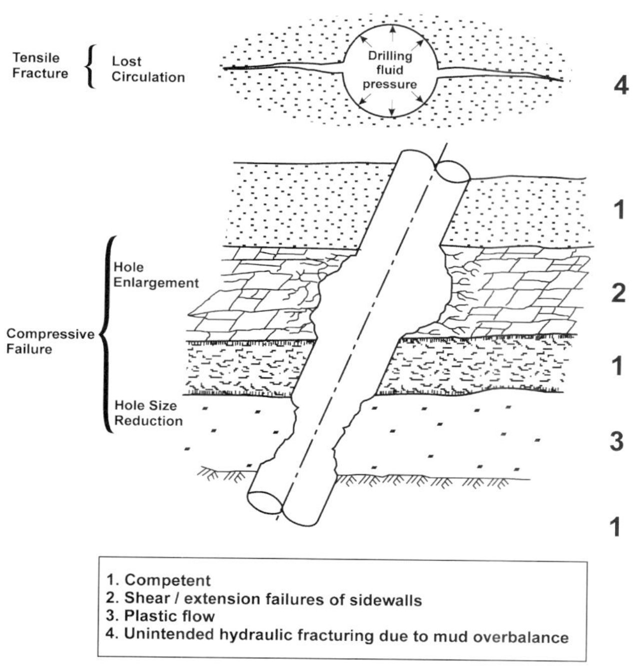


*Figure 12.6* The driller's dilemma: avoidance of lost circulation (hydraulic fracturing or jacking), with the simultaneous need to support the rock walls of the well before casing is in place. Bradley, 1978.

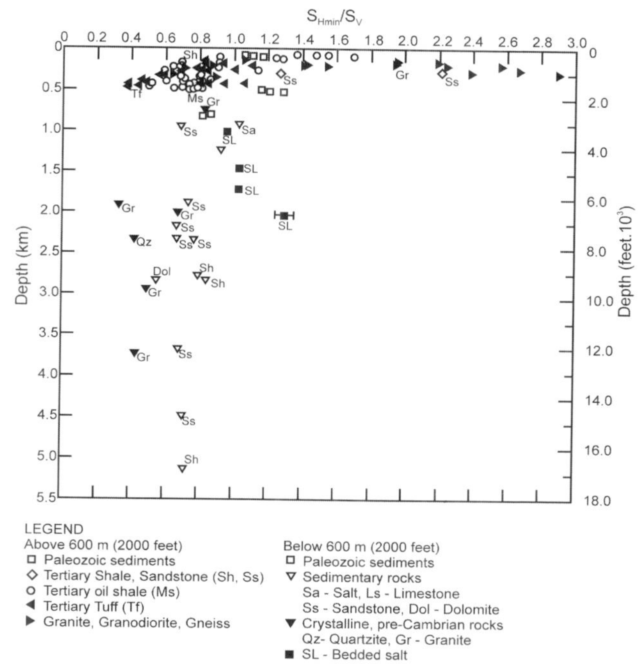


*Figure 12.8* Compilation of North American reservoir (and harder rock) and $\sigma_{h\ min}/\sigma_v(=K_o)$ ratios as a function of depth. Swolfs, 1977.

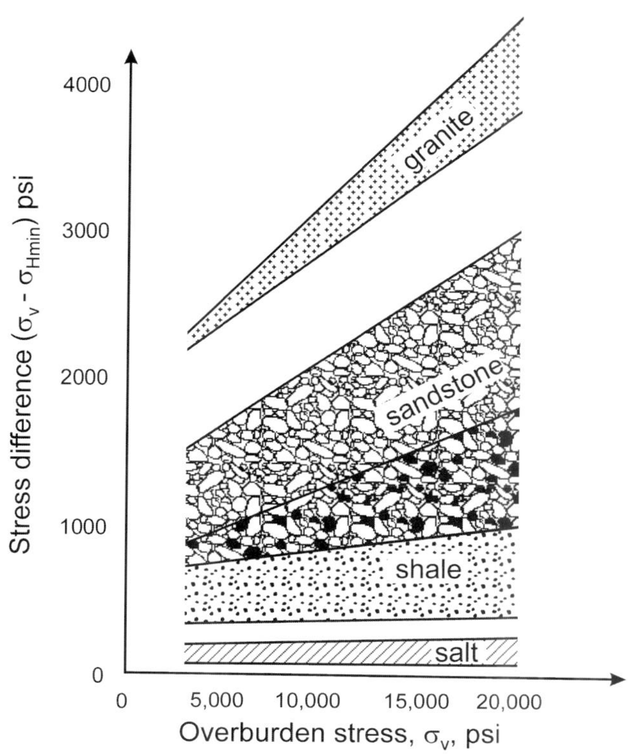


*Figure 12.7* Different degrees of stress difference (or shear stress) tolerance of reservoir rocks, compared to granite. Swolfs, 1977.

and joint shearing mechanisms are treated in Chapters 15 and 16, it will be found that the influences of anisotropic stress on the jointing will be of over-riding importance, since jointing may not always be parallel to the major stress direction, as traditionally expected.

### 12.2.1 Reversal of $K_o$ trends nearer the surface

As one approaches the surface, inter-bedded rock types resembling reservoir sequences, as illustrated in Figure 12.9, show the reverse of the previously discussed shear stress intolerance. Hydraulic fracturing tests reported by Barton, 1981 and Barton, 1986 generally showed low $K_o$ ratios ($\sigma_{h\ min}/\sigma_v$) in the weaker materials like shale and siltstone, and maxima in the sandstones.

This reversal of $K_o$ trends (at a certain, unknown depth) if a general effect, may have certain ramifications when comparing stress-induced velocity anisotropy (and velocity 'oscillation') near-surface and at greater depth, for example when evaluating the applicability of 'research-borehole' seismic testing. There could be possible consequences for the relative magnitudes of attenuation, as low $K_o$ (in addition to lower stress levels near

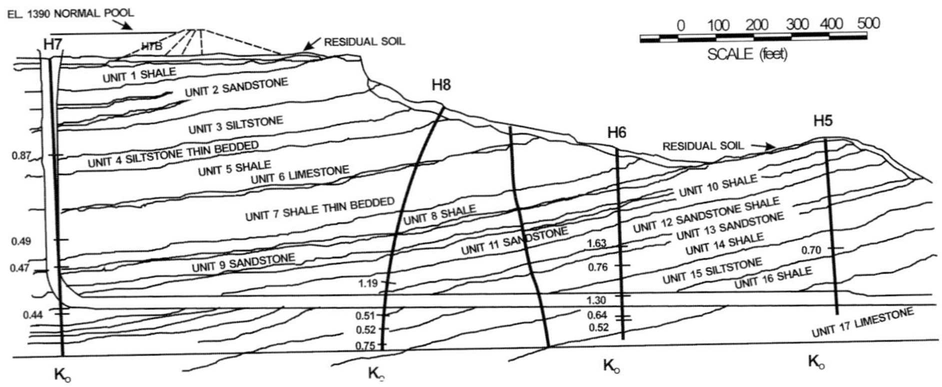


*Figure 12.9* A general tendency for lower $K_o$ ratios ($=\sigma_{h\ min}/\sigma_v$) in weaker rocks like shale and siltstone, according to hydraulic stress measurements at 100 to 300 m depth. Barton, 1981; Barton, 1986.

the surface) will tend to enhance attenuation, as seen in Chapter 10, as compared to the reduction of attenuation at depth due to higher stress and $K_o$ values closer to 1.0 i.e. less anisotropy.

Thus $Q_p$ and $Q_s$ values could be expected to be markedly lower and exhibit more anisotropy near-surface than at depth, as compared to rock sequences that did not show this $K_o$ reversal. Whether such a trend has been recognised in seismic signatures, for these specific reasons, is uncertain.

A possible modulus-related explanation for this $K_o$ reversal was suggested by Barton, 1986. Figure 12.10 shows imaginary stress-strain curves for loading-unloading (i.e. sedimentation-erosion) curves for a stiff pseudo-elastic sandstone and for a non-elastic, hysteretic shale. Imaginary deep-burial (but-on-the-unloading-curve) moduli of $E_2$ (shale) $>$ $E_1$ (sandstone) and shallow-burial (but-on-the-unloading-curve) moduli of $E_3$ (sandstone) $>$ $E_4$ (shale) would give an explanation for the above depth-related $K_o$ reversal. Whether the potentially different (unloading) moduli at these different depths would cause a part of an inhomogeneous velocity and attenuation structure, in addition to that caused by stress anisotropy and joint structure azimuth variations, remains an open question.

Another open question, is perhaps whether the 'oscillating' minimum effective stress ($\sigma_{min}$ – pore pressure) in such alternating reservoir rocks is a significant reason for variable attenuation. In theory, it should cause 'oscillations' of $V_p$ and $V_s$ over-and-above those caused by the alternating media, with their different intrinsic properties like density, porosity, modulus and degree of fracturing or jointing. Possibly all the above cyclic properties of $\sigma_c$, n% and $K_o$, and the different compliances of the bedding planes and joints, could each play their role in the velocity-depth and attenuation (intrinsic and scattering) behaviour, and as we shall see in Chapter 15, influence shear wave splitting and anisotropy.

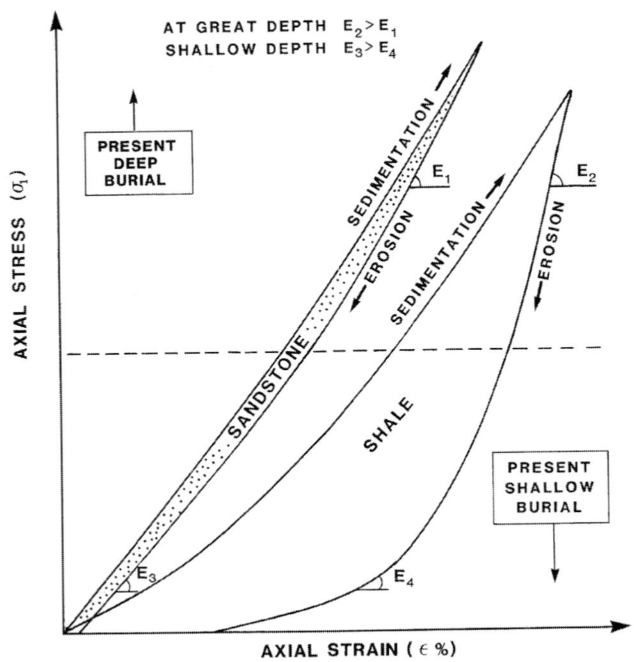


*Figure 12.10* Imaginary stress-strain curves for loading-unloading (i.e. sedimentation-erosion) curves for a stiff pseudo-elastic sandstone and for a non-elastic, hysteretic shale. Imaginary deep-burial (but-on-the-unloading-curve) moduli of $E_2$ (shale) $>$ $E_1$ (sandstone) and shallow-burial (but-on-unloading-curve) moduli of $E_3$ (sandstone) $>$ $E_4$ (shale) would give an explanation for depth-related $K_o$ reversal. Barton, 1986.

## 12.3 Relevance to logging of borehole disturbed zone

The hydrocarbon reservoir exploration and production industry has found that the subject of borehole failure modes is an important 'complication' concerning the interpretation of sonic-logging of wells. There are now acoustic dipole and monopole logging devices that can be used in a *logging while drilling* LWD mode, that acquire responses from more than one hundred wave forms, in order to delineate formation fracturing response, and virgin conditions further from the walls of the wells, before additional 'alteration' has occurred from stress and/or mud-filtrate invasion, as often seen in subsequent wireline logging.

There are possibilities for local velocity (and seismic Q) *enhancement* due to tangential stress increase in the case of competent rock like limestones, or low porosity sandstones. In the case of over-stressed, fracturing ('dog-earing') sections of rock, and especially in the case of incompetent rocks like shales, *reduction* of velocity (and seismic Q) will occur, due to the mini-EDZ that form as a result of drilling. In the radial direction, stress reduction, which may be azimuthally varying, will tend to locally reduce the velocity, and thereby also the seismic Q. It is impossible to support each lithology with the ideal mud-weight, so some suffer the consequences, just as occurs in a tunnel where the contractor might be trying to reduce costs by under-supporting.

In Part 1, Chapter 7, a range of effects that the EDZ (excavation disturbed zone) could have on seismic velocity inhomogeneity around *tunnels* was reviewed. Due to the smaller size of wells and the use of mud for hole support, the recognition of the behavioural data that can be extracted from anisotropic stress effects on small-scale EDZ round wells, seems to have come much later to the petroleum industry, than in tunnel engineering. This comment does not of course apply to borehole ellipticity, a much-used historical indicator of the minimum horizontal stress axis.

The comment does apply to what lies *behind* the ellipticity. In tunnels it is all too easy to see, and occasionally even be killed by, the effects of structure-induced wedge release, or stress-fractured 'lenses' of rock. We are also able to install multiple-position borehole extensometers (MPBX), both in tunnels and in shafts, to measure the anisotropic radial-distribution of deformation, thereby giving deformation moduli as a function of direction and radial depth (e.g. 2.5 to 70 GPa variation of deformation modulus with radial depth, at

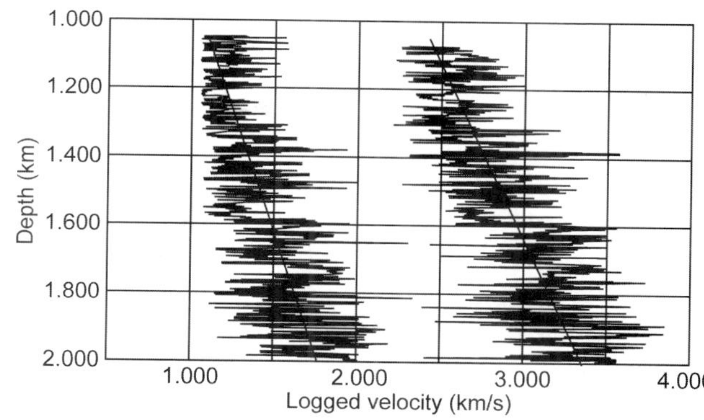


*Figure 12.11* Shear and compressional sonic logs in 'compacting shale' between 1000 m and 2000 m depth in the South China Sea. Miller et al., 1994. 'Substantial variation' about the assumed (central) trends is taken by the writer as a possible indication of over-stress in numerous layers that, by the nature of the variation, were presumably insufficiently mud-supported.

1.6 km depth in steeply bedded quartzites: Barton and Bakhtar, 1983). Velocity variations and permeability variations as a function of position and radial depth around the tunnel or shaft, can also be determined, thereby relating these parameters to eventual stress anisotropy. (See Chapters 7 and 8)

Plona et al., 1997, reported high frequency P-wave monitoring around a uniaxially loaded block of sandstone containing a borehole, and Figures 7.31 and 7.32 (and Section 7.4) gave indications of the significant velocity anisotropy (10 to 15%) due to unequal tangential stress concentration. They also suggested that the in situ detection of these azimuth-dependent velocity differences could be used to predict principal stress orientations, at stress levels below those needed for break-out. This was possibly signalling the early progress of important industry developments in dipole, logging-while-drilling, which will be illustrated later.

Suspicion that wellbore effects must be important for logging results can be high-lighted by reference to Figure 12.11, taken from the sonic logging of a 'compacting shale sequence' in the South China Sea, reported by Miller et al., 1994. Should we really believe that the 'substantial variation' around the trends is a true reflection of different 'virgin' lithological properties? Are fluctuations of ±500 to 1000 m/s about the mean 'real', or could they be due to *accentuation* of velocity changes, due to positive (+m/s) or negative (−m/s) mini-EDZ effects, as discussed, and illustrated above? Are the logging tools registering formation data, or wellbore effects?

There is a strong likelihood that mini-EDZ in the weaker, less well mud-supported zones, have reduced, radial-dependent velocity, due to failure and deformation in the over-stressed zones. Stronger inter-beds could show an opposite trend due to tangential stress enhancement of the velocities. Mini-EDZ penetrating several diameters can in fact be *detected, and circumvented* by deeper sensing, shear-wave based, dipole logging tools. However the probable *discontinuum* reality (caused by log-spiral shearing, to be illustrated shortly), is only referred to as e.g. 'shale alteration'. There is also the question of fabric and jointing and bedding plane effects, not least on the progress rates for mud-filtrate invasion. This seems not to be addressed, as yet.

Fjær and Holt, 1999 addressed the possible effects of the theoretical, isotropic, *elastic* stress distribution around a wellbore on sonic logging results. They suggested that with conventional long-spaced sonic tools, the zone of changed velocity due to the reduction in radial stress in relation to back-ground stress, would be mostly within a ½ well radius, and would not be detected, unless specially designed tools were used. They thought that the larger (and possibly also smaller) P-wave in the tangential directions would not be measurable by a conventional sonic logging tool.

However, the authors referred to Plona et al., 1998 who had suggested use of a multi-pole logging tool capable of distinguishing between velocities in different azimuthal directions, close to the well. Plona and colleagues work, already referred to in Part I, is a central rock mechanics-based part of the newer well-logging developments that will be described shortly.

Concerning processes important for *well logging interpretation,* we must consider modes of well behaviour, going strictly beyond elastic isotropic stress-distribution concepts. As we shall see shortly, the evidence to be presented for deeper penetration of mini-EDZ, presumably related with the 'shale alteration' referred to by Brie et al., 1998, is much in line with current industry interest. There turns out to be a serious potential contrast in logging results, when comparing the later (e.g. 1–2 weeks later) result of wireline logging, with the few hours delay represented by LWD, or logging while drilling.

The more recent *shear-wave anisotropy* based logging, images a volume of up to several borehole diameters away from the wall, also *beyond* the stress-related fracturing and mud-filtrate invasion or 'shale alteration', thereby giving presumed 'virgin' formation attributes as well.

## 12.4 Borehole in continuum becomes borehole in local *discontinuum*

In mining and rock mechanics, there are well known, distinctive differences between stress-induced failure of hard dilatant brittle rocks like quartzites and hard sandstones (extensional splitting and crushing), and intermediate strength and less dilatant rocks, in which the traditional 'dog-earing' takes on a different shape due to log-spiral shear failure. These two basic modes were contrasted by a person with a broad experience of civil and petroleum related fields; Maury, 1987 (ISRM commission on failure modes). The two basic modes were linked to possible dilatant/non-dilatant rock failure behaviours by Barton, 1987.

In Figure 12.12, various modes of *continuum failure* (of the previously intact matrix) have been assembled for comparison. They include a recently reported 'tabular slot' type of breakout, probably related with *compaction band* development during laboratory-scale drilling in porous blocks of sandstone.

The more plastic model material, driven to the extreme of 'well closure' conditions, showed undrained strength mobilization and residual strength development reaching to many diameters. There was also a compacted zone. One may speculate that subsequent recovery (re-drilling) of such a zone would still leave permanent effects in relation to velocity reduction and seismic Q reduction. Local, dramatic changes of seismic properties should not deceive those interpreting well logging.

The variable azimuth drilling described by Addis et al., 1990 was performed with the important three-dimensional 'detail' of drilling while under 3D stress states, as happens in reality. The same result is not obtained when loading a block with a *pre*-drilled hole. The deep log-spiral development in the weak (0.5 MPa) cemented-sand tested by Addis et al., 1990, typically occurred when the major principal stress was about *ten-times* higher than the uniaxial strength, with the minor and intermediate principal stresses of 60% or 80% of the maximum.

The borehole failure at the bottom of Figure 12.12 is probably a typical mode for not just non-dilatant, but actually *contracting-with-shear* clay-rich materials. Encountered in tunnelling situations, such fault-related, clay-rich materials may flow to form a weakly-inclined 'delta', flowing many tens of meters from the previous tunnel face (e.g. a Nathpa Jakri HEP headrace tunnel access drift, NE India). A case of 7,000 $m^3$ of fault-zone

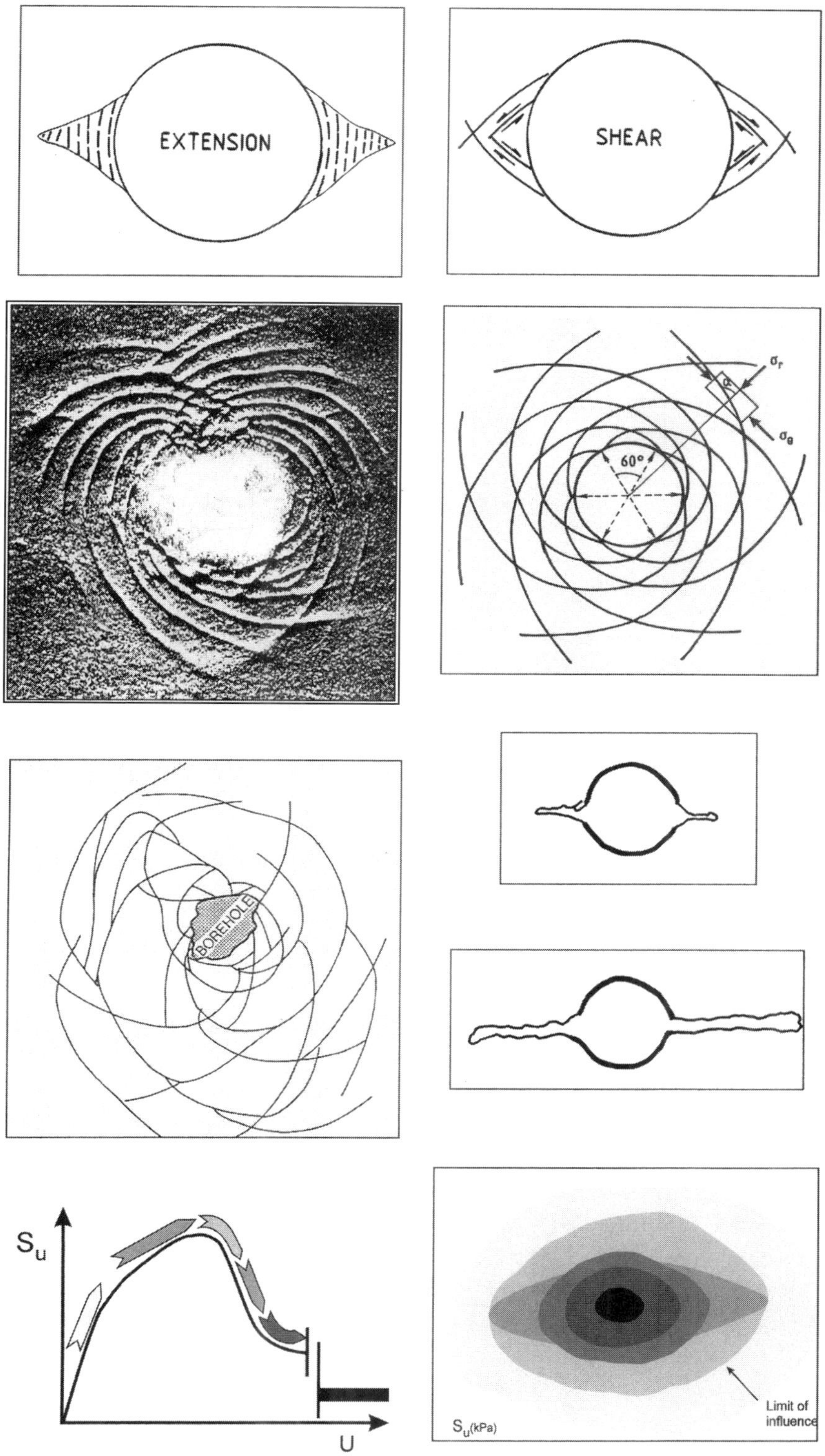


*Figure 12.12* A wide collection of failure modes for circular openings. *Upper pair:* from Maury, 1987 (ISRM commission on failure modes). *Second pair:* from NGI joint industry project and Addis et al., 1990. *Third pair:* left - deep log-spirals from NGI study by Rawlings, 1991, right – slotted breakouts sketched from Haimson, 2003. See Lee and Haimson, 2006 for recent results. Bottom pair: from NGI joint industry project: Bandis, 1988. Undrained shear strengths (in kPa) from centre: 0–5 (f), 5–10 (p), 10–30 (p), 70–75 (c), 40–50 (s), and 50–60 (v). (f: flow, p: plastic, c: compacted, s: softening, v: virgin). See Zoback et al. 2003 for detailed *theoretical analysis* of break-out phenomena.

Table 12.1 Failure and deformation modes typical in tunnels (after Barton, 2004b) may also apply at greatly reduced scale to the mini-EDZ that are likely to occur locally, where insufficient mud-weight has been applied.

| Description | Mode of behaviour |
|---|---|
| 1. Hard, massive, brittle rocks that dilate during failure even when stresses are high. Stress-induced failure may be delayed as 'strength corrosion' occurs. | Extension failure, thin-walled stress-slabbing, dynamic ejection, bursting. The symmetric 'dog-eared' fall-out due to the anisotropic stresses may have a 'nose'. |
| 2. Hard or medium hard, bedded and jointed rock that can shear and dilate along structural planes, while under moderate to high stress levels. | Anisotropic response. Shear stress dissipates by slight shear on bedding planes and joints. Deformations are moderate. Block falls can occur. |
| 3. Soft, massive, non-brittle rocks that may, or may not dilate during shear failure. Typical for young e.g. Tertiary rocks such as the mudstones and siltstones in Japan. | Failure may occur by log-spiral shear development and *tangential strain*. Radial deformations are large, and pressure on support is high. Multi-diameter influence.* |
| 4. Very soft, plastic rocks (and clays) that contract when sheared under significant stress levels. | Post peak strength loss reaches an extreme of virtual 'flow', with in-rush of failed material. |

(*Twin tunnels need pillars 4 to 5 times their span c.f. Japan, Taiwan, in order to minimise excavation interaction)

debris, filling a 70 m$^2$ tunnel for 100 m has also been seen by the writer, in the Pinglin tunnels in Taiwan. Water pressures may of course be fundamental catalysts for such dramatic and sometimes tragic events.

The 'slotted' failures described by Haimson, 2003, were obtained in polyaxially-loaded blocks of Berea sandstone, with 23 mm diameter hole-drilling while under stress, parallel to the intermediate vertical stress. Haimson's samples with 17% porosity developed 'conventional' dog-earing, while the weaker 22% and 25% porosity samples developed increasing lengths of symmetric 'slotting' (see simple holes-with-slots traced from the experimental results, reproduced in Figure 12.12). The uniaxial strengths of the three sandstones were 53 MPa, 42 MPa and 22 MPa respectively. The depicted 'slotting' occurred during drilling, with principal applied boundary stresses of 50, 60 and 90 MPa, i.e. up to *four-times* the uniaxial strength of the weakest sandstone.

Table 12.1, from Barton, 2004b, summarizes a quite complete range of failure and deformation behaviours for tunnels, with potential relevance to mini-EDZ development around boreholes, and possible consequences for sonic logging, and for bed-property enhancement or degradation, as discussed above.

As emphasised by Barton, 2004b, the actual modes of physical behaviour experienced by boreholes and tunnels, are unlikely to be predictable when modelling with conventional Mohr-Coulomb type (c + tan $\varphi$) shear strength criteria, because rock tends to fail first by loss of continuity at *small strain*, caused by loss of local tensile or cohesive strength, followed by frictional mobilization at *larger strain*. In modified Mohr-Coulomb terms it is a case of 'c *then* tan $\varphi$', not 'c + tan $\varphi$'. Numerical models that are programmed, or manually-steered, to dissipate cohesion while mobilizing friction are capable of matching observed behaviour. The BEM (Boundary Element) fracture mechanics code FRACOD (Shen et al., 2002) seems to model log-spiral fracture development, and over-stress dissipation, in a particularly realistic way.

If stress anisotropy (and stress magnitudes in relation to rock strengths) are sufficient to cause break-out, then clearly the principal horizontal stress direction is easy to predict from calliper logs, it being at right-angles to the largest well diameter in the case of a vertical well. This 'simple' situation may be disturbed by break-out caused by structure, i.e. steeply dipping jointing causing wedge-shaped fall-outs assisted by local bedding.

Such mechanisms are a sign of a mini-EDZ that is in progress around the well, and one that could be used in dedicated, azimuthal, short-base, well-logging. The influence of stress fracturing on velocity anisotropy, including but beyond the pre-failure states investigated by Plona et al., 1997, means that there will be anisotropically distributed fracturing around eventual over-stressed part of a well, that had not been adequately supported by mud pressure.

To emphasise (and probably exaggerate) the likelihood of possible serious differences in EDZ-potential down a well in alternating hard/soft rocks, one may refer again to some of the borehole stability studies reported by Addis et al., 1990. These were performed in the late 1980s at the Norwegian Geotechnical Institute, in a joint oil-industry study. Figure 12.13 shows one of the

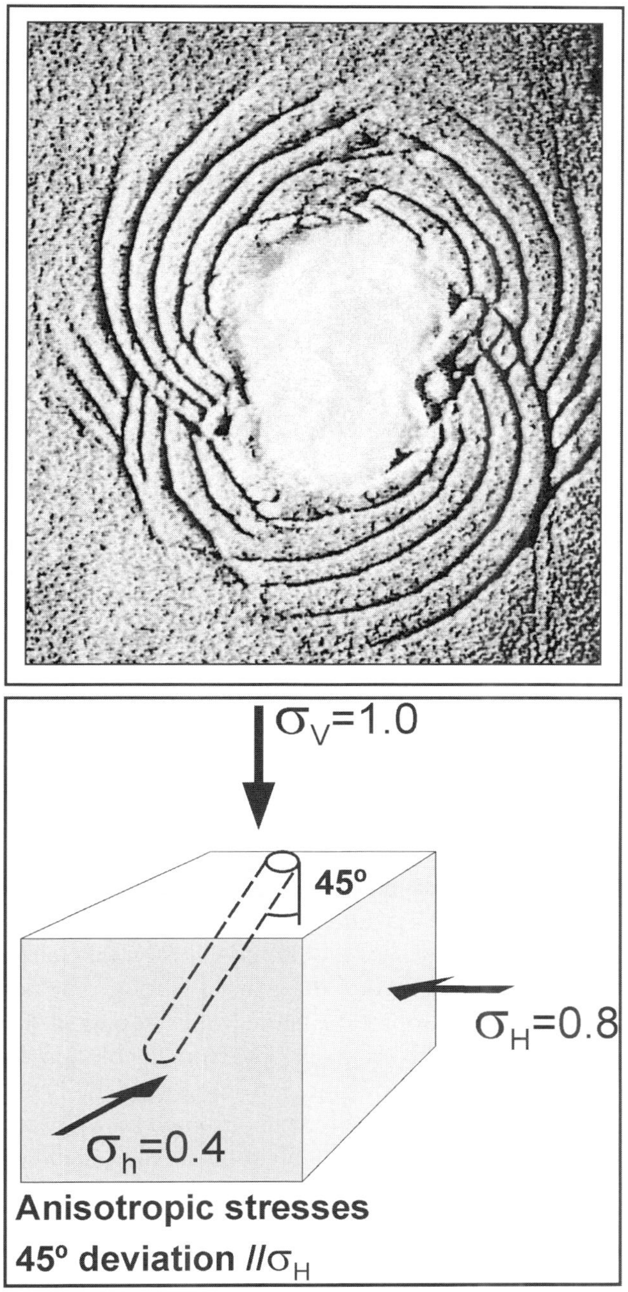


*Figure 12.13* One of a series of log-spiral-type failures around heavily stressed model boreholes that were drilled at different deviation angles in relation to a moderate 3D stress anisotropy. Addis et al., 1990, and NGI contract report, 1990.

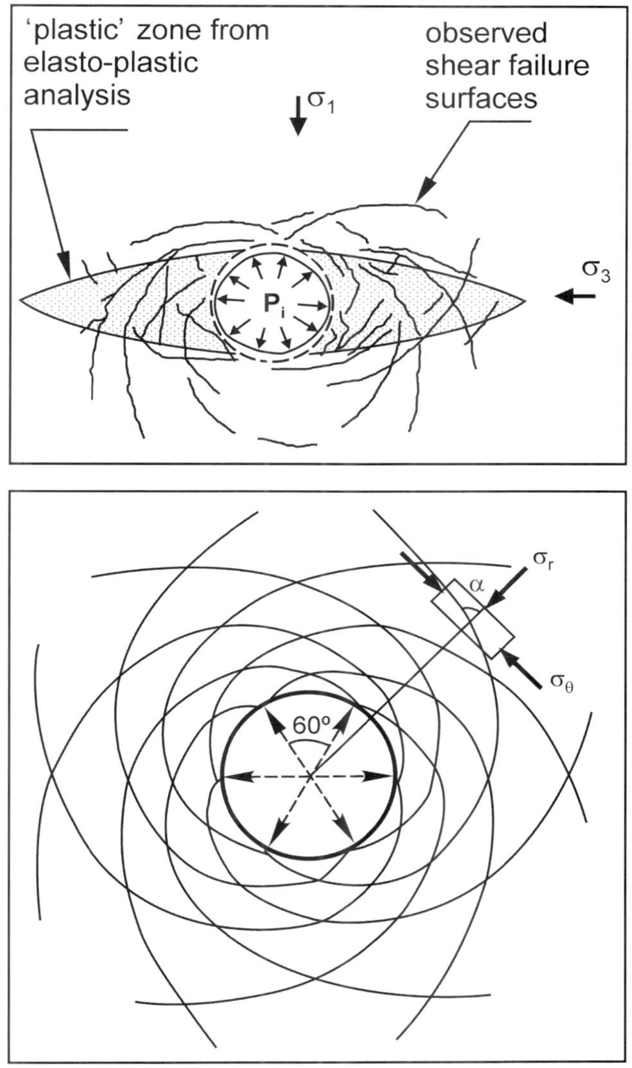


*Figure 12.14* Closed-form elasto-plastic analysis of an internally-supported well, compared to the physical-model reality of log-spiral shear-failure surfaces, that help to dissipate the highest stresses further into the rock mass. NGI contract report, 1990.

'symmetric', double-cusp, intersecting log-spiral modes of shear failure that was exposed around the borehole, following polyurethane-foam hole-stabilization, while still under stress. Cubic 50 × 50 × 50 cm blocks of model sandstone ($\sigma_c$ = 0.5 MPa) were used for 3D application (via flatjacks) of three equal or unequal principal stresses, with drilling performed into the modelled block of rock while under stress, either as a 'vertical' well, or at different deviation angles with respect to any of the principal stresses, as shown in this case.

Application of an analytical closed-form elasto-plastic isotropic continuum model to a heavily stressed, drilling-mud-supported well, shown in Figure 12.14 suggested that a narrow, deeply-penetrating, elliptical 'plastic' zone would develop, where the Mohr-Coulomb shear strength criterion was exceeded. The reality – as far as a physical model represents reality – was for one of earth science's favourite continuum theories to be ignored, and instead we witnessed the development (in every case), of a 1 to 2 diameter wide zone of log-spiral failure surfaces, making in effect a *discontinuum*, which actually allowed the wall-rock to dissipate the highest shear stresses further into the surrounding rock, just as occurs in tunnelling through over-stressed 'squeezing' rock. A 'radial' strain is observed in the opening – there is *measured radial closure* – but the reality is *tangential strain* due to the log-spiral shearing. This mechanism has been well documented by Aydan et al., 1992.

These three-dimensional failure surfaces 'followed the borehole', and also curved in front of the face of the hole, whenever the holes were drilled into the highly stressed blocks of rock simulant. This was seen following sectioning. Such an EDZ, developed in an insufficiently mud-supported weakness zone, or in a bed of shale adjacent to a reservoir sandstone (with insufficient $\sigma_{hmin}$ to tolerate higher mud weight), would tend to locally accentuate both the expected low seismic velocity, and an expected low seismic Q value, that could be obtained from seismic logging.

Before leaving rock failure mechanisms in the *intact matrix* around boreholes, a hybrid (intact plus jointed) result will be shown from modelling with the FRACOD fracture mechanics BEM code of Shen, described in Shen et al. 2002. The two models shown in Figure 12.15, performed by Shen for the case of a deep excavation under high stress, had deliberately sparse jointing. Different levels of 'log-spiral-type failure are demonstrated, depending on the 'disturbance' to stress distributions, caused by jointing, which dissipates some of the highest, near-wall tangential stresses.

The resultant 'dispersed' fracturing (as opposed to jointing) would presumably be a strong source of scattering attenuation – and velocity reduction. Note that the principal applied stress levels in the case of this strong rock were only about 35–38% of the uniaxial strengths. A brittle sandstone of 50 MPa UCS would be acted on by an equivalent ratio of effective stress beyond about 1200–1300 m depth, with standard density and pore pressure assumptions, considering a $\sigma_{H\ max}$ value no larger than the vertical effective stress.

Of particular interest to the frequent *velocity discrepancy* between LWD and later wireline logging of the same formation (see later examples), is the possible development of fracturing over time, and the result this has on the 'factor of safety' against further shear failure. Figure 12.16 shows two episodes of fracture development over time from the same FRACOD model, and demonstrates a distinct rotation of the dark grey-scale, from a 'NNE-SSW' to 'E-W' azimuth. The grey-scale represents F. of S. $< 1.0$.

## 12.5 The EDZ caused by joints, fractures and bedding-planes

Due to the influence of deformation of 'soft' as opposed to 'hard' porosity, a borehole for hydrocarbon exploration that penetrates variably jointed and faulted ground, will actually experience variable small amounts of *deformation*, due to different degrees of joint closure, joint opening, and joint shearing. There will also be the pseudo-elastic response, due to both *loading* (at the diametrically-opposite $\sigma_{\theta\ max}$ locations) and potential *unloading* (at the diametrically-opposite $\sigma_{\theta\ min}$ locations) of the matrix as well as the joints, the latter usually dissipating some of the theoretical (isotropic, elastic) peaks of maximum and minimum tangential stress.

This process will occur even with a constant mud-weight, since the mud – unlike rock bolts in a tunnel – cannot prevent joint movements of unequal magnitude at different points around the opening, although the mud may help to make them very small.

Figure 12.17 shows four idealized, two-dimensional models of circular hole excavations in anisotropically loaded 'blocky-rock', with four successively reduced block sizes. There are 250, 1000, 4000 and 10,000 blocks in the respective models. Two of the coarsest block geometries are shown in the top of the figure, while all four of the joint-shearing fields are given. The reducing block sizes can be glimpsed within the two lower models. These models represent 'crushed zones' at borehole scale, i.e. gross approximations to fault-zone-crossing stability problems.

Figure 12.18 demonstrates analytical Mohr-Coulomb formulations used by Shen to produce the general result for zones of shearing on differently inclined, conjugate or perpendicular joint sets. The four-sector EDZ, at roughly 45°, 135°, 225° and 315° is bisected by the two applied boundary principal stresses, e.g. a 'N-S' $\sigma_H$ and an 'E-W' $\sigma_h$. Shear displacements occur on the joints in these four sectors, even when stresses are *isotropic*, as shown by UDEC-MC or UDEC-BB modelling of tunnels through horizontally and vertically jointed rock.

A more realistic model of 'fractured rock' (such as might be found in the neighbourhood of fault zones), actually representing slightly random jointing in tuff, is shown in Figure 12.19. The model represents a bored

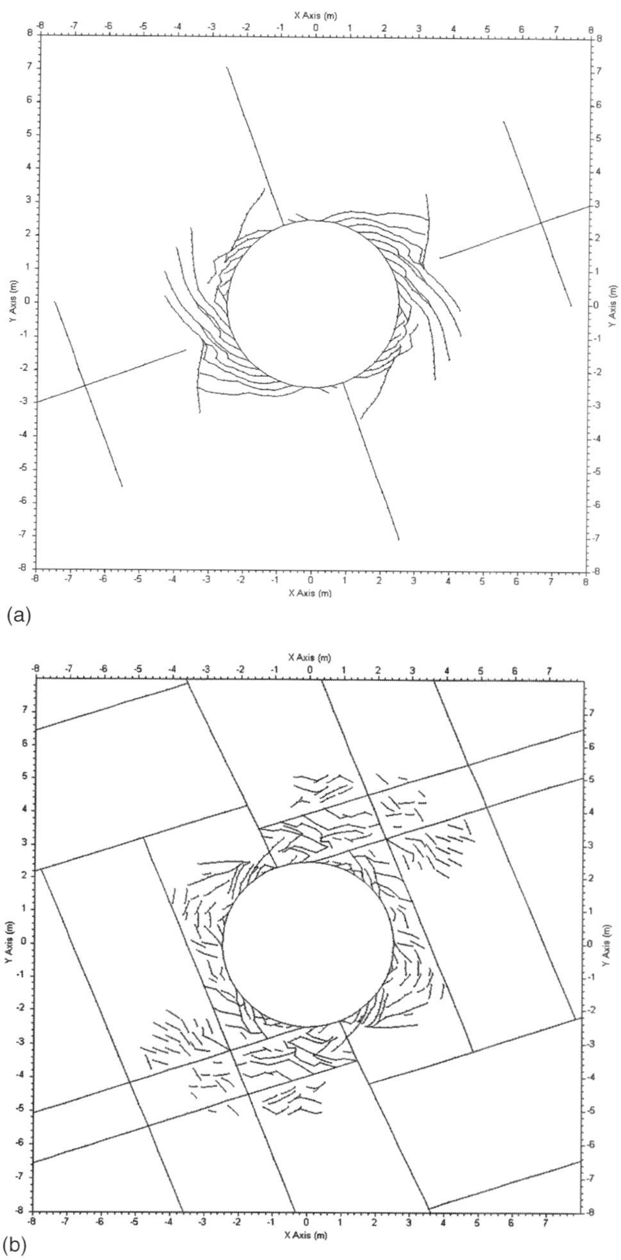


*Figure 12.15* Fracture mechanics BE modelling of circular openings in high over-burden environments, using the FRA-COD code developed and demonstrated by Shen. The modelled rock mass was hard and of high modulus (UCS = 160 MPa, M = 50 GPa), with assumed shear strength of c = 40 MPa, $\varphi = 35°$ and appropriate fracture toughnesses at the modelled effective stress levels of 30 to 60 MPa. a) EDZ with sparse jointing and $\sigma_1$ ('NS') of 55 MPa and with $\sigma_2$ ('EW') of 40 MPa. b) EDZ with significant jointing and $\sigma_1$ ('EW') of 60 MPa and with $\sigma_2$ ('NS') of 30 MPa. Note: deformations do not track subsequent fall-out/loosening. Nick Barton & Associates contract report, 2005.

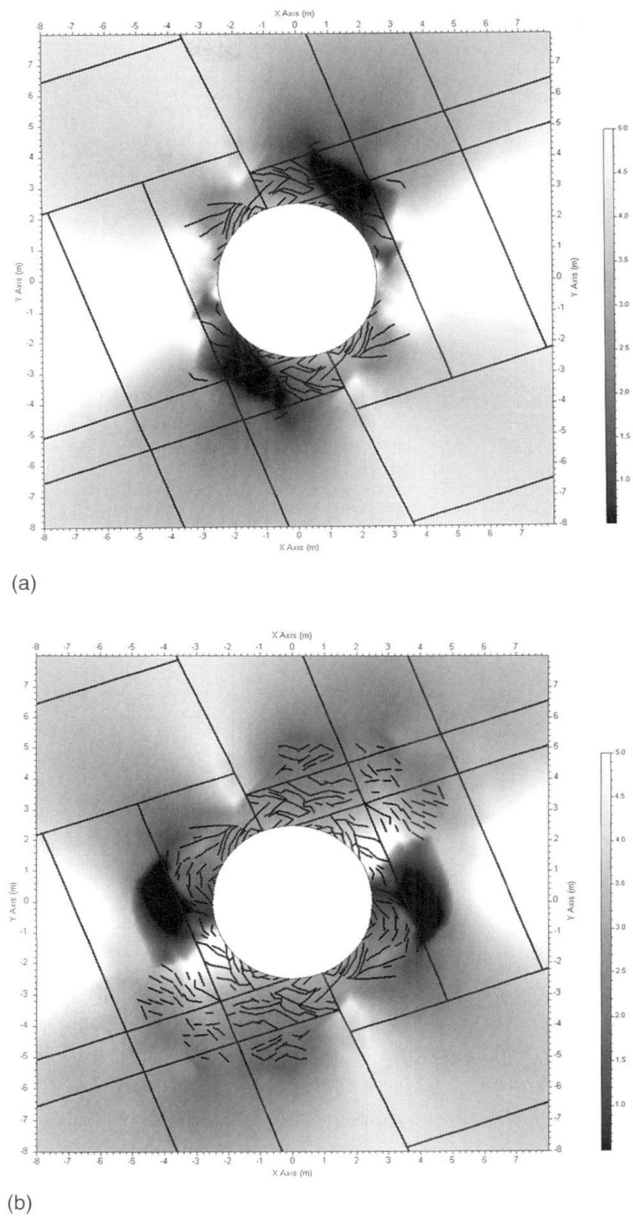


*Figure 12.16* Fracture development over time in a FRACOD model of a circular opening in a jointed zone. Note the 'rotation' of the diametral-pair of red regions, which represent low factors of safety against shear failure. Further fracturing dissipates and displaces the low F. of S. zones, suggesting that more fracturing could occur across the 'E-W' diameter. Changed seismic response over time is easy to imagine, also a mud-filtrate invasion speed that could be highly non-uniform, due to developing permeability in the partly connected *discontinuum.* Nick Barton & Associates 2005 contract report. See Figure 12.15 for input data. (See plate 3).

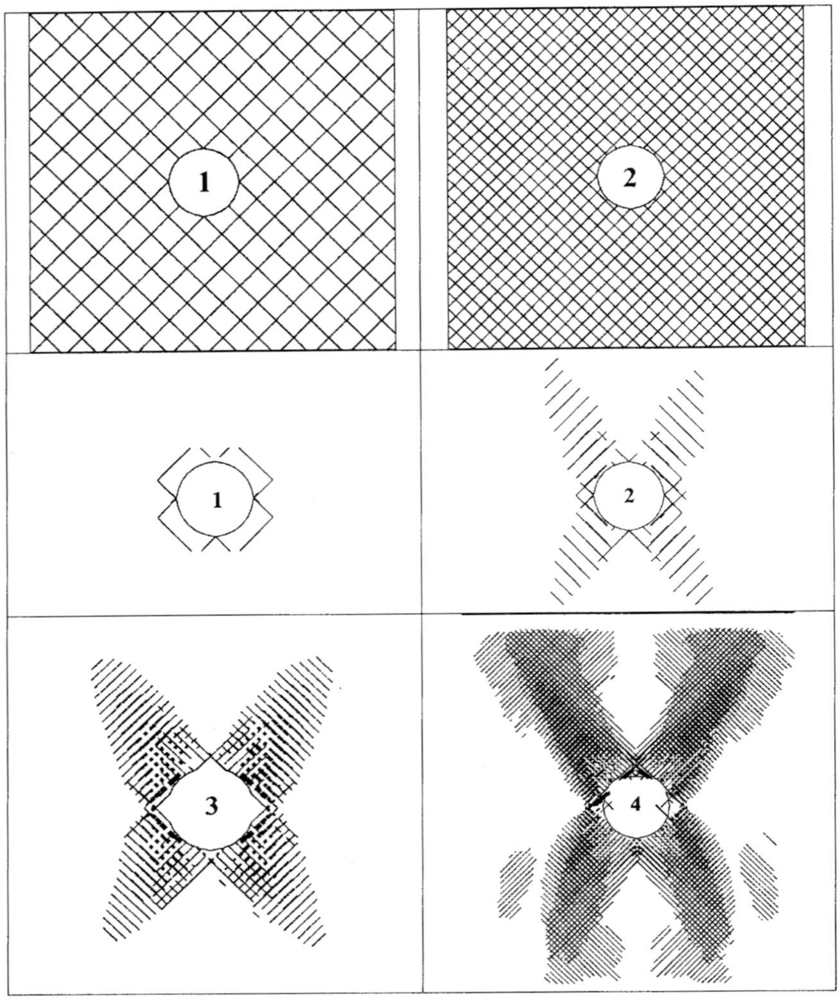


*Figure 12.17* Shear deformation zones developed around circular openings, as a result of slip along both joint sets, when excavated under anisotropic stress, and when with limited joint strength, e.g. due to clay-smear or general smoothness. Cundall's distinct-element UDEC-MC code (with Mohr-Coulomb sub-routine), was used for the analyses, with four successively reduced block sizes. Note the increasing multi-diameter EDZ, as block size and therefore rock quality reduces. Line thickness depicts joint shear magnitude. Shen and Barton, 1997.

*Figure 12.18* Theoretical, Mohr-Coulomb based solutions for shear displacement zones involving slip on conjugate joint sets. Note that the largest joint-shearing EDZ is for the 60° case, in which shear stress is maximized relative to the joint directions. The smallest joint-shearing EDZ is for the 90° case, in which the shear stress is minimized relative to the joint directions. Shen and Barton, 1997.

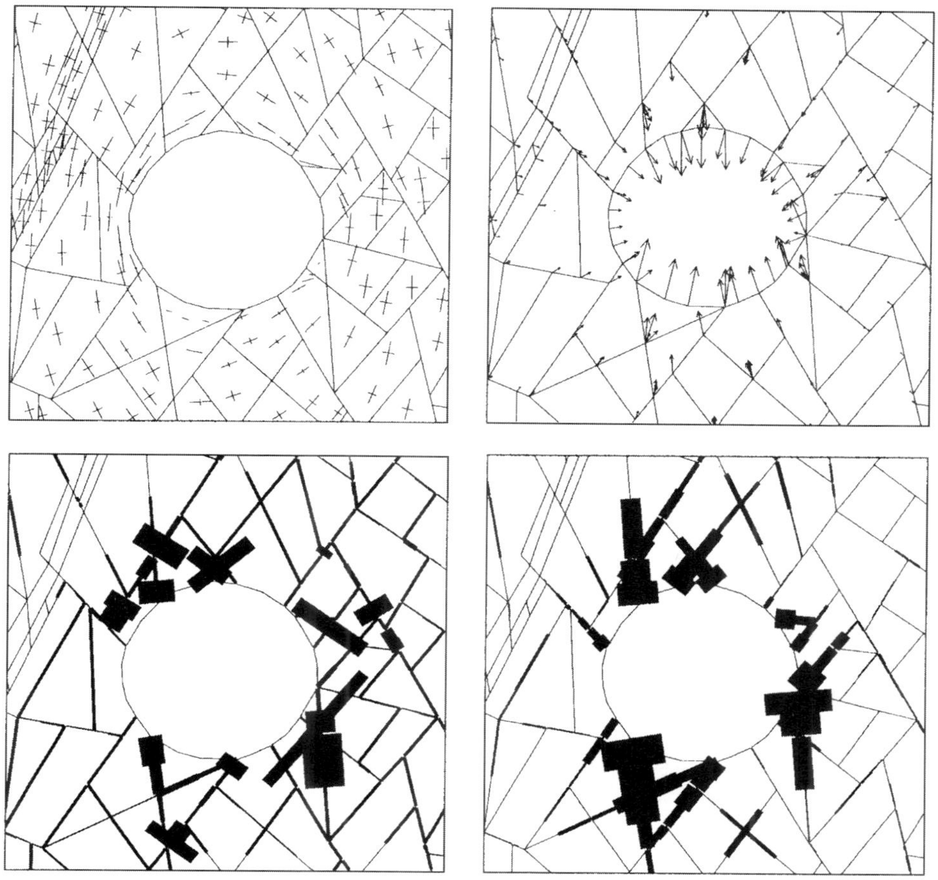

*Figure 12.19* UDEC-BB modelling of a TBM tunnel EDZ in jointed tuff, at a simulated 650 m depth. Note far-field $\sigma_h = \frac{1}{2}\sigma_v$. This model is assumed to represent a generic case of heavily fractured (faulted) and sheared rock close to a vertical borehole. Clockwise (with 8 m full-scale excavation diameter): principal stresses (max. 30 MPa), deformations (max. 11 mm), joint shearing (max. 8 mm), and joint (and block-corner channel) conducting apertures (max. 2 mm). Hansteen, 1991 (NGI contract report), and Barton, 2000.

excavation at 650 m depth, and had the predicted, close-to-the hole EDZ behaviour. (Note that the model's loaded boundaries extended some 10-diameters from the hole, and the hole was also close to a simulated clay-bearing fault.) Four basic EDZ components are shown:

- Principal stresses ('N-S'), and redistributed tangential stress
- Displacement vectors: maximum parallel to 'N-S' stress maximum
- Joint shearing (clockwise or anticlockwise) and proportional to line thickness
- Joint (and block corner channel) conducting apertures

The modelled rock was actually jointed tuff, with input data from index testing of numerous joints from each set, as recovered from deep boreholes at the UK Nirex Sellafield's site in NW England. A second model, this time representing a horizontal bored tunnel at 400 m depth through inter-bedded sandstone and shale, is shown in Figure 12.20.

If we scale the roughly 5–15 mm range of deformations from 8 m tunnel size in hard rock, to a nominal, and convenient, 16 cm well size, with correspondingly reduced rock strengths, the magnitudes become 0.1 to 0.3 mm. The deformations, although small (i.e. sub-millimetre size when the hole is stable), are probably up to a few orders of magnitude larger than the *displacement discontinuities* that are sufficient to generate joint compliance changes (normal and shear), that are of recognisable (inverted) magnitude, in relation to the joint *stiffnesses* familiar in rock mechanics. The lower

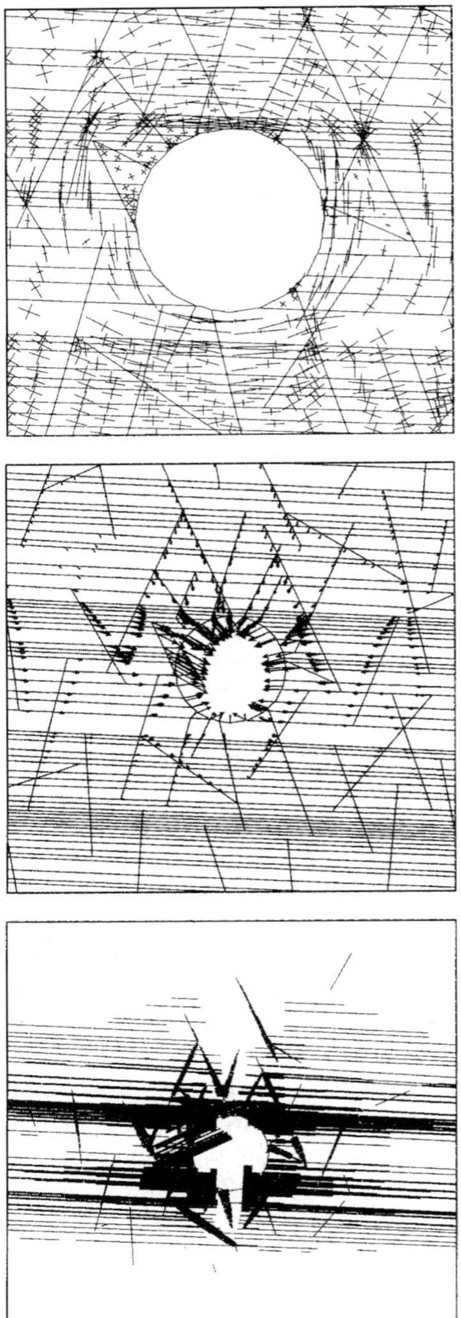

*Figure 12.20* UDEC-BB modelling of a TBM tunnel EDZ at 300 m simulated depth, in inter-bedded sandstone and shale. Note far-field $\sigma_h = \sigma_v$. This model is assumed to represent a generic, shallow horizontal well. From top: principal stresses (max. 50 MPa), deformations (max. 15 mm), joint shearing (max. 8 mm). Note that 'buckling', or cracking of thin beds, does not occur in UDEC, unless the necessary failure surfaces are discretized before-hand. Chryssanthakis, 1991 (NGI contract report), and Barton, 2000.

dynamic compliances will signal the reduced velocity and the reduced seismic Q, via the reduced dynamic (and static?) moduli.

Since both isotropic and anisotropic stresses, but especially the latter, can (indeed must) cause a *mini-disturbed zone* around a well, it is likely that this could influence dipole sonic logging down the well. In the case of the sedimentary rock model, with shale beds, at a moderate 300 m depth, the joint shearing and deformation EDZ are each of several hole-diameters in extent.

Is it not possible that the *mini-EDZ* illustrated here, can be the root cause of much of the intense *fluctuation* seen in the sonic logging results presented in Figure 12.11? Low $V_p$ and low seismic Q zones could be depressed further by an accentuated EDZ. Conversely, high $V_p$ and high seismic Q could be further elevated by the effect of high tangential stresses close to the borehole wall, where there was no failure or joint deformation (or joint presence) to speak of.

Consider the following cross-discipline *parallels*. When assessing the rock quality of the walls and arches of tunnels, the *observed rock*, which is the visible part of the tunnel-scale EDZ, is *classified* (using the Q-system: see Appendix A), in order to select appropriate rock reinforcement (grouted rock bolts) and tunnel support (sprayed, steel-fibre-reinforced concrete). The latter is the equivalent of the borehole mud pressure, and is badly needed in a complete load-bearing ring, in the rapidly deteriorating and deforming clay-bearing zones, in order to control deformation and prevent local tunnel collapse.

Outside the tunnel EDZ, the rock mass would be *characterized* as a better quality rock mass. If one performed both sonic logging and azimuthal dipole sonic logging – a development described by Brie et al., 1998 (see later), the borehole *mini-EDZ* can be *classified* by the one tool, while the hydrocarbon-bearing or reservoir sealing formation away from the immediate influence of the hole, can be *characterized* with higher velocity by the other tool. The same arguments as above apply to seismic Q. As we shall see, at least part of this rock mechanics logic is being applied in modern well logging, with multi-wave-form acquisition.

A layered model, consisting of a 'thinly-bedded' model sandstone and micaceous inter-layers that was expertly fabricated by Bandis while working at NGI, is shown in Figure 12.16, to emphasise again that if one starts with a discontinuum, the possibility of deformation is increased. In contrast to the jointed UDEC models shown earlier, this steeply inclined bedding allowed a buckling mode of deformation to develop, causing an elliptical yielding zone, with a long-axis of several diameters. It is easier to imagine this case when rotated through 45°. A mud-pressure capable of preventing the buckling action can readily be imagined causing unwanted hydraulic fracturing, as depicted in Figure 12.6. Such buckling mechanisms are common in thinly-bedded coal-measure rocks, and presumably can compromise the integrity of horizontal wells, if drilled where beds are very thin.

In the foregoing, the possible effects of the *weakness* of formations, in the face of drilling with insufficient mud-pressure, have mainly been addressed. Accompanying potential shear-failure, there is a tendency for dilation, unless stress levels are high enough to suppress this expansion. The local and bulk dilation is of course part of the 'alteration' zone phenomenon, and is a basic reason for an initial velocity reduction (e.g. under LWD), followed intuitively by increasing permeability, therefore accelerated mud-filtration weakening, with a possible spiral into lower velocity territory, by the time of wireline logging. The permeability of a damaged borehole wall is unlikely to remain constant in the interval between LWD and wireline logging.

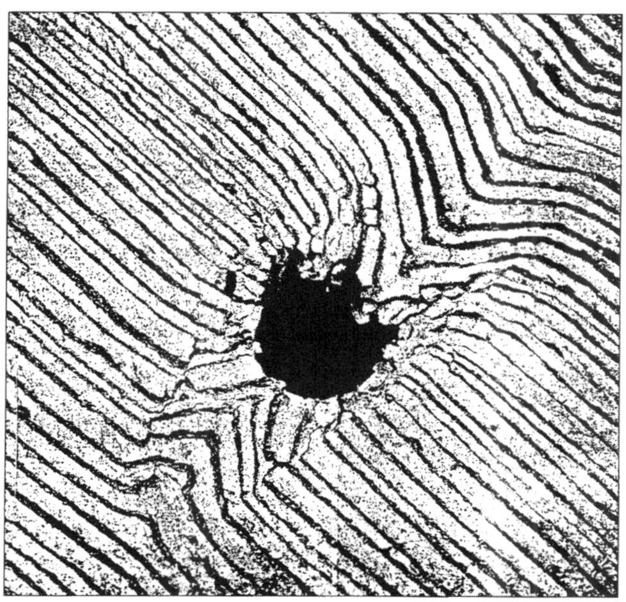

*Figure 12.21* A buckling mode of deformation when penetrating horizontal or steeply dipping thinly-cycled beds, would also tend to compromise the 'correct' value of logging-based $V_p$ and seismic Q, in the case of dipole azimuthal-logging. Model fabricated by Bandis, 1987 (pers. com.).

## 12.6 Loss of porosity due to extreme depth

Loss of porosity with pressure or depth is well documented in sandstones recovered from sedimentary basins such as the North Sea and the US Gulf Coast. These basins typically have porosity versus depth curves with mean porosity of about 23 to 27% at 2 km depth, decreasing to 3 to 7% at 7 km depth. This change in porosity signals changes (increases) in mechanical strengths and of course increases in seismic velocities, and, all other things being equal, would suggest reduced attenuation or higher values of seismic Q. In a downhole or reservoir environment, inhomogeneity caused by structure, for example more intense fracturing or jointing in the resulting stiffer sandstone might have a reversed effect on the above assumption of reduced attenuation with depth.

An assessment of the onset of the brittle ductile transition of sandstones reported by Scott and Nielsen, 1991, has relevance here in the case of increased depths of burial, and possible components of behaviour around deep wells, in view of the approximate doubling of the field effective stresses in the tangential stress arch around an initially non-yielding borehole wall, i.e. when still responding 'elastically' with horizontal stresses of similar magnitude, giving $\sigma_{\theta\ (\text{max and min})} \approx 3\sigma_1 - \sigma_3$.

The authors Scott and Nielsen, subjected sandstones with a wide range of porosities (6.0%, 13.9%, 19.9%, 20.9%, 22.0% and 27.6%) to triaxial confinement (0.1, 5, 10, 30, 50, 70, 90, 110 and 130 MPa), and also reviewed tests at much higher confining stresses. All their tested sandstones had similar grain size, shape, orientation and composition, and all were cored from fresh unweathered blocks, without strong bedding plane anisotropy.

Figure 12.22 shows how the higher porosities were readily driven to the brittle-ductile boundary; solid circles representing ductile (non-dilatant) cataclastic flow, while open circles represented brittle, dilatant, shear fracture.

On the shear strength development at reservoir stress levels are shown in Figure 12.23 subject of extreme burial effects, Chuhan and Bjørlykke, 2002 contrasted the normal mechanical compaction of sandstones down to 2–2.5 km depth, with grain crushing phenomena at greater depths. The 'limits' of normal mechanical compaction appears to correspond to the 'knee' seen in velocity-effective stress tests in the laboratory up to about 25 MPa. They mentioned the precipitation of quartz cement at greater depth, where temperature was higher,

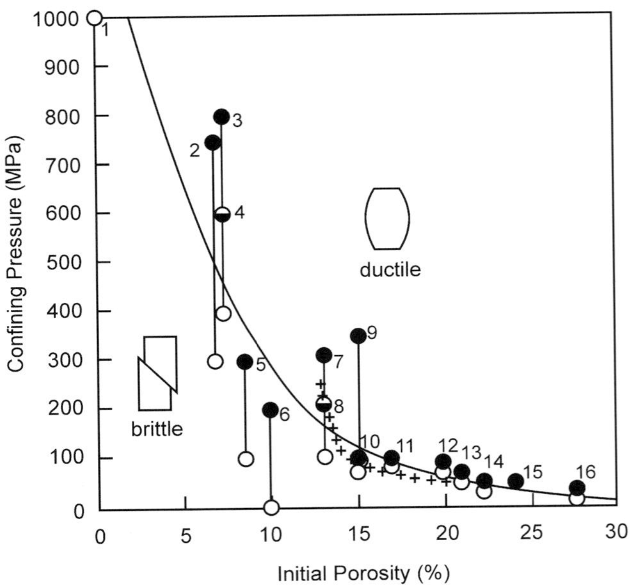


*Figure 12.22* High pressure triaxial tests on sandstones of various porosities, showing the approximate limit of brittle behaviour. Scott and Nielsen, 1991 and cited authors. Note the extreme confining pressures. Solid circles represent cataclastic flow, open circles represent brittle shear fracture.

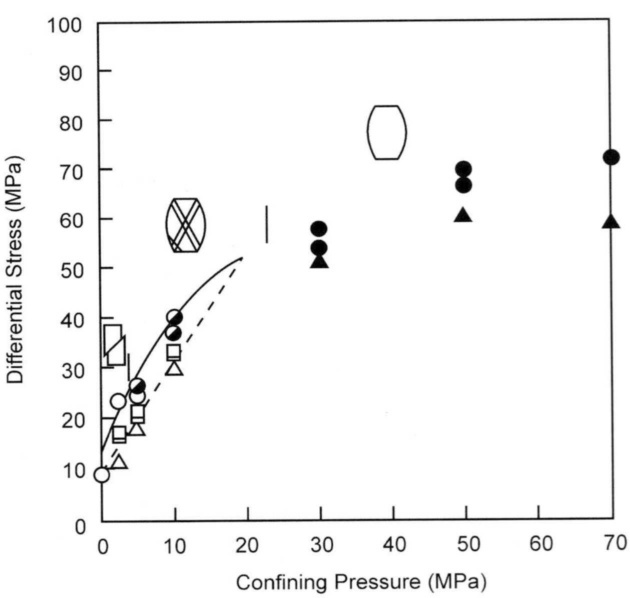


*Figure 12.23* Shear strength envelope separating brittle and ductile behaviour for sandstones of various porosities. Scott and Nielsen, 1991 and cited authors. Solid symbols represent cataclastic flow, open symbols represent brittle shear fracture, and half-black represents transitional behaviour.

causing enough strengthening of the grain framework to prevent further compaction.

According to Chuhan and Bjørlykke, grain *fracturing* is commonly seen in North Sea reservoirs, in coarse-grained sandstones, but rarely in fine-grained sandstones. Some deeply buried sandstones (>5 km) at Haltenbanken showed evidence of more extensive *grain crushing*, because grain coatings had apparently delayed quartz cementation. Deeply buried reservoir sandstones (>6 km) from the Azerbaijan area, that were subject to rapid subsidence and a low geothermal gradient had very little quartz cement and showed evidence of intense grain fracturing. This was assumed to have occurred when the *effective* stress was about 40–50 MPa, at 4–5 km depth.

## 12.7 Dipole shear-wave logging of boreholes

Brie et al., 1998, in a multi-authored, major contribution to 'new directions in sonic logging', provided a fascinating glimpse of the newer techniques of well logging, and of the challenging conditions that were met in different types of formations. Their article confirms many of the foregoing suspicions that what we have termed *mini*-EDZ, are indeed a source of concern in certain formations, and more importantly, that these deeper 'alteration zones' can be detected and seismically characterized, *and avoided,* with the help of novel dipole transmitter tools, that generate *flexural waves.* Flexural waves are shear waves that are polarized into *fast* and *slow* directions, and penetrate several hole-diameters into the formation, thereby revealing potential stress-induced 'alteration', and/or drilling mud-induced alteration. As we shall see in much greater detail in Chapter 15, shear wave measurements seem to be the most useful mode for fractured reservoir interpretation at many scales.

The amount of sophisticated equipment that well drillers may now connected to 'follow' the drill-bit, using a down-hole mud motor in the *bottom-hole assembly* (or BHA), is impressive for its sophistication and necessarily compact nature. It resembles in some ways the several hundreds of meters of generally much less sophisticated *back-up* that are set to be pulled (on rails), behind a large tunnel boring machine. Consider this paragraph from Brie et al., 1998, concerning an interval of drilling down an Angolan (West African) offshore exploratory well:

*'A long single-bit run was conducted over a 7-day period that covered a depth from 1000 to 8000 ft. The wellbore*

*was deviated 20° in this interval. The BHA consisted of a PowerPak mud motor, a CDR Compensated Dual Resistivity tool, a PowerPulse MWD (measurement while drilling) telemetry system – for real-time transmission – and an ISONIC sonic-while-drilling tool. The ISONIC tool, placed above the PowerPulse system, was approximately 104 ft (32 m) away from the bit. At the average rate of drilling, the LWD (logging while drilling) measurements were made fewer than four hours after the formation was first cut. Wireline sonic logging was run after the 7-day drilling run was completed, and then only after circulating the well for several hours.'*

Brie et al., 1998 emphasised that LWD *logging while drilling* sonic logs, acquired in freshly drilled rock, show 'remarkable differences', compared with wireline measurements that are usually taken many days, or a week or two, after the drilling has exposed the formation. They pointed out that both well surveys bring important but different information about wellbore rock properties. The cause of these fundamental differences is summarized in a helpful diagram from their publication, shown in Figure 12.24.

Since the 'altered zone' around the borehole may continue to develop during the week or so that may separate the two types of logging, the later wireline log may be influenced by a reduced modulus in an annular zone around the borehole, particularly in soft formations. Water uptake in this zone (from mud-filtrate), as well as the initial over-stress, will also reduce the modulus in shales and shaly sands, according to Brie and co-authors, lending support to our initial question concerning the real meaning of the South China Sea log shown earlier in Figure 12.11. A so-called 'bi-compressional arrival' may be registered – a phantom arrival too fast to be a shear wave – actually caused by trapping of the wave-front by the low-modulus damage zone.

Figure 12.25 illustrates the novel dipole transmitter principle. During logging, flexural waves are generated by sequential firing in two perpendicular directions (see bottom of tool), first along the tool x-axis, then along the tool y-axis. These induced shear waves are split and polarized into fast ($qS_1$) and slow ($qS_2$) directions, respectively parallel and perpendicular to dominant fracturing or to the principal (horizontal) stress direction (if the tool is in a vertical hole, and more distant response is analysed). The shear waves are registered by the dipole receiver pairs shown at the top of the tool.

The important property of shear-wave splitting is illustrated in Figure 12.26. In the example shown, the shear wave splitting is caused by the y-axis aligned fracturing, with fastest velocity (as also with P-waves), in the direction parallel to structure, possibly also aligning with the major principal stress direction. (In Chapter 15, numerous examples of this exceptionally fortuitous property of S-waves will be given, from fractured reservoir analysis). The slow direction (as also with P-waves) is perpendicular to the fracturing – which could be microcracks, cracks, joints or faults – according to the scale of example considered. Shear-wave splitting, and polarization is one of the most valuable of all seismic anisotropy properties for fracture and fracture-fluid investigations, perhaps matched by the anisotropic and dispersive attenuation of P- and S-waves (see Chapters 14 and 15).

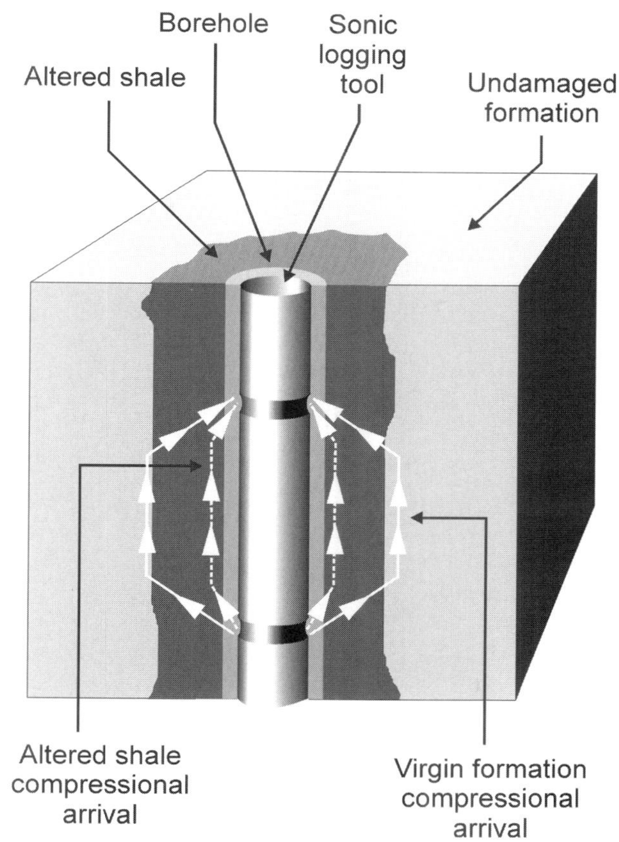


*Figure 12.24* Stress-related damage, possibly compounded by subsequent water or mud-filtrate weakening, have the potential effect of reducing the modulus in the several days or one week delay between LWD (logging while drilling a few hours behind the drill-bit), and the subsequent wireline sonic logging. Entrapment of wave-fronts in the lower modulus damage zones results in 'bicompressional arrivals': i.e. second arrival compressional waves. Brie et al., 1998.

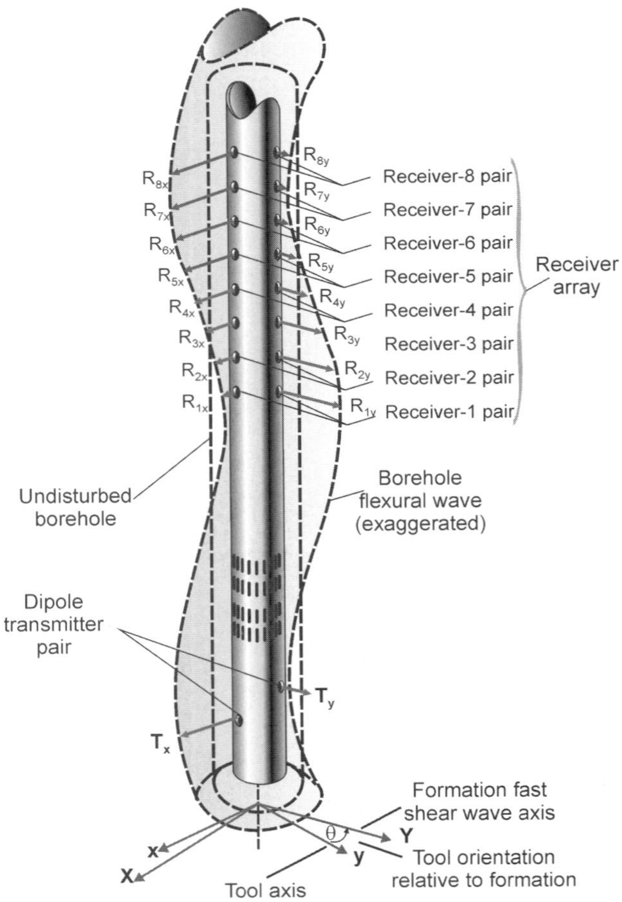


*Figure 12.25* Dipole transmitter tool for sequential firing in x- and y-directions, to generate flexural (shear) waves that become split and polarised due to dominant fracturing or stress, thereby becoming fast (with longer wavelength) parallel to structure, and slow (with shorter wave length) perpendicular to structure. Brie et al., 1998.

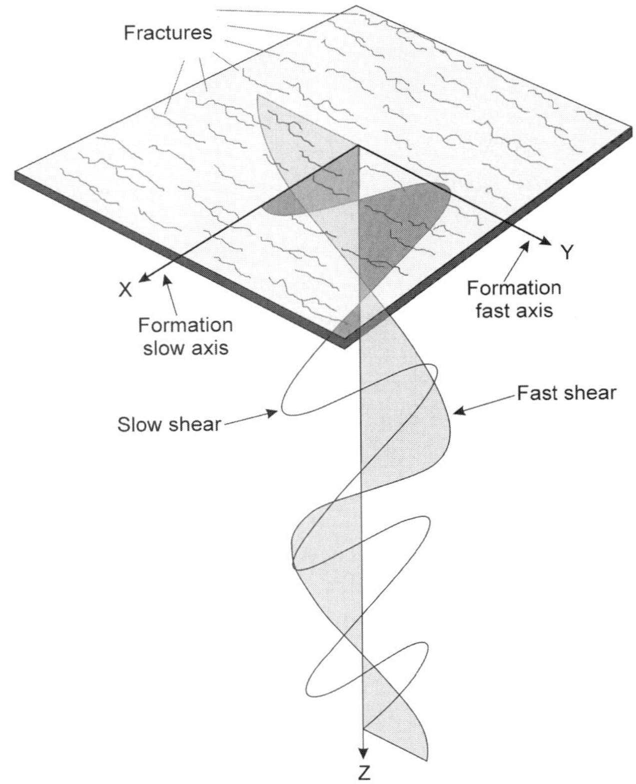


*Figure 12.26* An illustration of the shear-wave splitting and polarization phenomenon, from Brie et al., 1998. Note the longer wave length of the fast shear-wave.

The two orthogonal dipole transmitters, and the multiple receiver pairs, which are aligned in orthogonal directions, measure the components of *slowness* in any direction within planes *perpendicular* to the borehole. In fact four *sets* of waveforms are recorded. By minimizing (with *Alford* rotation), the cross-receiver amplitude-based energies, the rotated direction of the fastest shear waves becomes the *fast-shear tool azimuth.* A magnetometer relates this direction to true north.

The logging tool results are converted to graphic read-out of the two contrasting velocities, together with the energy anisotropy, the slowness anisotropy, and the time anisotropy. Large energy differences, with the minimum staying low, signal strong anisotropy. Both the acoustic time anisotropy and the slowness anisotropy are sensitive to properties deeper within the formation than the surface effects of drilling.

A significant finding regarding the distinction between stress-related anisotropy around boreholes, and fracture related anisotropy, is shown in Figure 12.27 from Winkler et al., 1998 (his co-authors Sinha and Plona were also contributors to Brie et al., 1998). Borehole stress concentrations (in competent rock like the tested Berea sandstone) caused *radial stress gradients* that were different in the two principal directions, thereby causing frequency dependent (dispersive) effects such that the shear-wave that is polarized parallel to the major stress, is fastest *only at low frequency* ($<25$ Hz).

The fact that shear wave anisotropy allows the investigation of a volume of the formation up to several diameters from the borehole axis, means that it can sense jointing, and stress-induced fracturing, that are missed by conventional logging tools. This means that it is particularly useful for registering the additional jointing and fracturing that tends to be present on either side of a fault. (Endo et al., 1997)

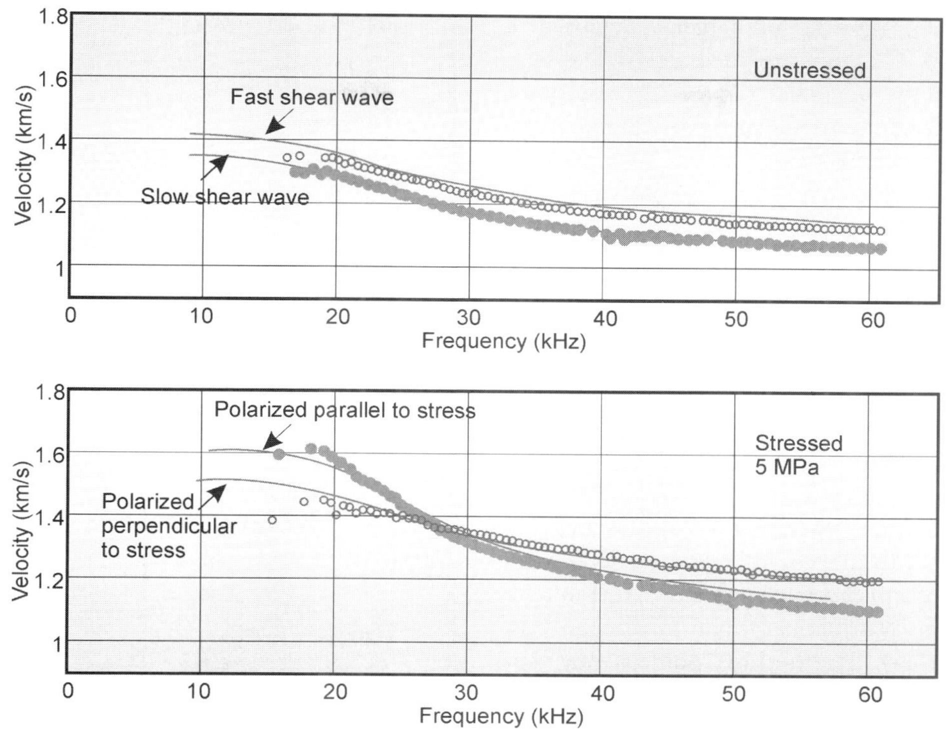


*Figure 12.27* Dipole *dispersion cross-over* of fast and slow shear waves, demonstrated by modelling, and by testing unloaded and loaded blocks of Berea sandstone containing a borehole. Caused by unequal radial stress gradients in the two principal stress directions, when *uniaxially* loading one of the blocks of sandstone containing an instrumented borehole. Winkler et al., 1998. This study should ideally be performed on a borehole drilled while the block is under stress, e.g. following Addis et al., 1990, Haimson, 2003.

A case was described from an Egyptian oil-producing well drilled in basement granites. With the shear-wave anisotropy logging, the fault signature was clear, with the fast *azimuth* starting to change slowly when about 20 m from the fault, with a rapid change of nearly 65° across the fault, followed by a return to an intermediate *azimuth* some 30 m beneath the fault, and a final return to background azimuth at greater distance. The particular fault had high permeability.

A final mention will be made of two other very useful applications of LWD with dipole shear-wave anisotropy analysis, which is available almost in real-time (some hours behind the drill-bit). The first is the application to drilling of *horizontal* well sections, designed to intersect a maximum amount of structure, and simultaneously avoid the less favourable parallel to $\sigma_{H\ max}$ hole direction. The second is early warning while drilling in formations with rapidly changing pore pressure. LWD is then an important aid in choosing appropriate mud pressures. When porosities are no longer maintaining a normal trend of increased compaction with depth, the slowness data will tend to diverge from the expected trend (Brie et al., 1998).

By the beginning of the 21st century, it has become more common to use dipole sonic logs to help interpret AVO or amplitude variation with offset anomalies, and to help tie in ocean bottom seismic surveys (OBS) with four-component (4C) acquisition. These techniques will be briefly described with case examples, in Chapter 14. The use of both wireline dipole logging in vertical holes, and pipe-conveyed dipole tools for deviated and horizontal wells has now given reservoir geophysicists improved means of *calibrating* the responses of their rock physics based reservoir models, against small-scale measurements.

### 12.7.1 Some further development of logging tools

More recent descriptions of mechanical damage and anisotropy evaluation using a new modular dipole sonic logging tool developed by Schlumberger, were given by Pistre et al., 2002. This tool provided a complete seismic characterization in radial, axial and azimuthal directions. Improved monopole and cross-dipole technology, together with 13 axial levels, each with 8 azimuthal sensors, with each of the 104 receivers individually digitised, resulted in 104 waveforms per transmitter firing. Five transmitters are used. As the authors emphasised, radial rock property variations arise because of non-uniform stress distributions and mechanical and chemical near-wellbore 'alteration' due to the drilling process. The development of a *discontinuum* close to the wellbore is not mentioned, but perhaps implied in the word 'alteration'.

The authors emphasised that radial gradients of acoustic slowness arose due to wave speeds being a function of stress. (To this could be added the velocity reducing effects of a *discontinuous* medium, due to rock failure and various possible deformation mechanisms, as illustrated earlier). The authors pointed out that anisotropy could also be caused by intrinsic shale anisotropy, possibly combined with external differential stresses. Inversion for these anisotropic, radial and azimuthal rock properties could now be achieved from the various acoustic behaviours, and from use of broadband dispersion curves.

A particular feature of the new tool was that profiling of radial variations in *compressional slowness* could be achieved through monopole acquisition with a wide range of transmitter-receiver spacings, from very short to very long. Radial inversion of *shear slowness* was quantified through inversion of the broadband dispersions of the dipole *flexural* (Figure 12.25) and Stonely wave modes. When radial gradients were detected, in e.g. sandstones, there could be an increased risk of sanding, while lack of radial property gradients suggested mechanically intact rock, far from failure. The question of tangential-stress *enhanced* velocities and moduli, capable in principle of adding to the velocity *oscillation* seen in Figure 12.11, was not mentioned. (As shown in Chapter 5, Figure 5.36, the modulus that is influenced by cracking or jointing, shows strong stress-sensitivity).

An example of part of a log display from an exploration/development well, is reproduced (in grey-scale) in Figure 12.28, from part of a very comprehensive, colourful log display given by Pistre et al., 2002. This shows the result of the dipole *flexural* wave split into fast (left) and slow (right) shear directions. The six-tone curves represent *slowness* (expressed as μs/ft) at different radii from the wall of the borehole. The upper half of the profile shows a large radial gradient, from high *slowness* near the wellbore, to a lower *slowness* at a radius of 24 in, or 63 cm. At the base of the profile the individual curves tend to overlay, indicating a sounder rock with negligible radial gradient of shear velocity.

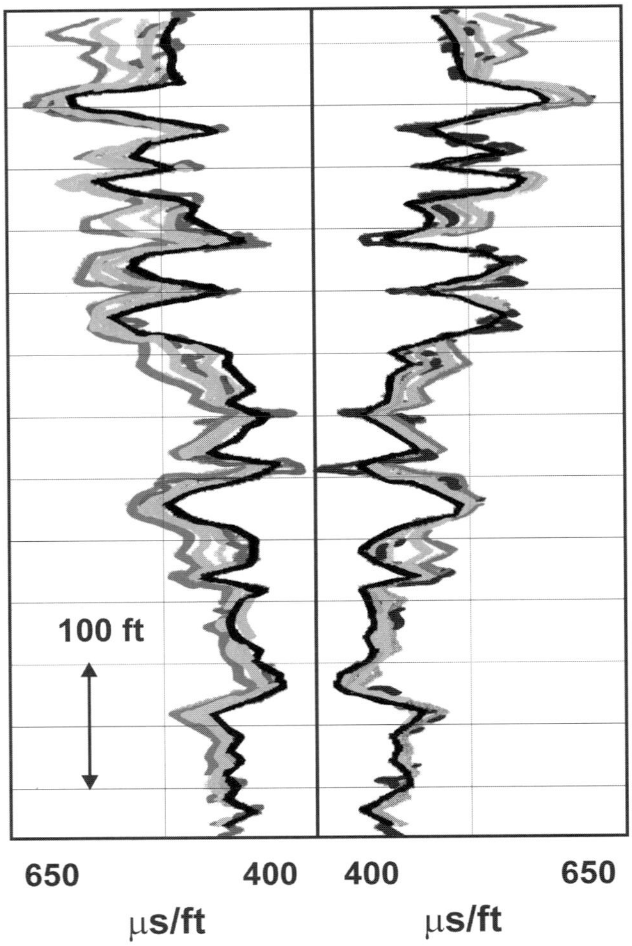


*Figure 12.28* Part of a dipole-generated *flexural* wave monitoring log, here split into fast (left) and slow (right) shear directions by stress and/or fracturing. Differences are expressed here as *slowness*, for different radial depths behind the wellbore wall. Pistre et al., 2002.

## 12.8 Mud filtrate invasion

The related themes discussed and illustrated in the forgoing pages:

- intensely fluctuating sonic log velocities in interbedded (sandy?) shales

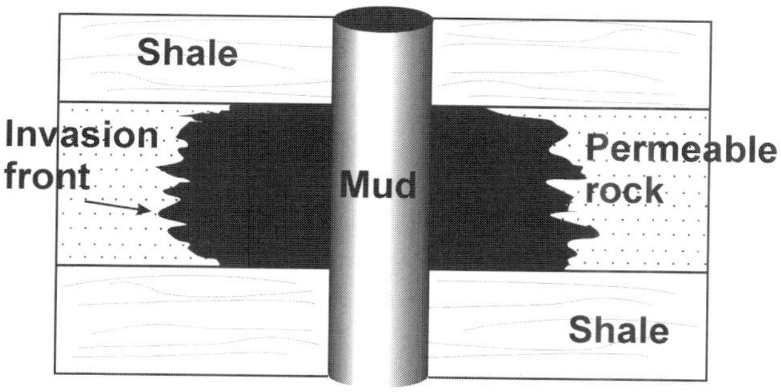


*Figure 12.29* Conceptual mud-filtrate invasion in a permeable rock formation. After Chi et al., 2004.

- mini-EDZ around experimental boreholes
- log-spiral shear surfaces observed in physical and numerical models
- deformation (shearing, opening) of joints close to numerically modelled wells
- damage zones confirmed by bicompressional arrivals in real wells

suggest that mud-filtrate invasion modelling may perhaps need to also consider flow through local *discontinua*, such as interconnected log-spiral shear surfaces in softer rocks, or less well- connected tensile and shear fractures in harder rocks, and flow along bedding planes and joints in general cases without over-stress.

Chi et al., 2004 mention the fairly common finding, based on LWD and subsequent wireline logging differences, that near-wellbore formations are often altered by stress, stress release and an assumed mud-filtrate invasion, as envisaged, schematically in Figure 12.29. However, a relatively uniformly-paced invasion, based on porosity-permeability conversion may perhaps be compromised by an actual *permeability enhancement,* based on the central thesis of this book that rock quality, i.e. degree of jointing, stress-induced fracturing, deformation moduli, velocity and permeability, are quantitatively linked.

According to the authors, the 'alterations' cause the physical properties in the near-wellbore region to be different from those of the uninvaded rock formation. In addition, stress concentration may cause formation anisotropy, and an azimuthally varying radial variation of velocities. As they point out, in well-consolidated hard rock formations, mechanical damage is less pronounced than in soft formations, so mud-filtrate invasion would then be more localized.

Formation properties inferred from wireline logging measurements may not reflect the true properties, so their opinion was that a realistic description of the invaded zone would be important for processing and interpretation of logs. The problem is caused by the invaded (or 'altered') zone being deeper than that illuminated by the logging tool, meaning that the velocities will not reflect those of the formation, but of the damaged zone, therefore requiring corrections.

They used multilayer velocity models to interpret well measurements, for example the following layer depths and velocities for the case of a *slow* and *fast* formation:

| Layer | Radius | $V_p$ slow m/s | $V_p$ fast m/s |
|---|---|---|---|
| 1 | 0 | 1500 | 1500 |
| 2 | 0.10 | 2300 | 4390 |
| 3 | 0.18 | 2350 | 4512 |
| 4 | 0.26 | 2400 | 4634 |
| 5 | 0.34 | 2450 | 4756 |
| 6 | 0.42 | 2500 | 4878 |

These differed in a realistic manner from a more commonly assumed sharp-interface model. In some ways this small-scale gradational model mirrors the early controversy concerning stepped or gradational sub-ocean spreading-ridge velocity modelling, also applying to an unchanged rock type, in that case basalt, and applying to kilometre-scale depths. (Chapter 11) Both phenomena are reflections of the changing degree-of-fracturing, and of its interaction with the local effective stress.

Chi et al., 2002, emphasised that synthetic seismograms often did not correlate with measured seismograms, when correlating seismic data with acoustic logs. It appeared that the standard approach here was to 'correct' the acoustic logs via a Biot-Gassmann *fluid substitution,* to 'free sonic logs from mud-filtrate invasion effects'. Doing this, it was assumed that the measured velocities were those of the invaded zone, saturated with mud filtrate. By 'displacing the saturation fluid' theoretically, new velocities were obtained, and taken as the *virgin formation velocities.* However, there is a potential problem here, if the near-wellbore is stress-fractured, despite 'removal' of the saturating fluid.

In the example they described, surprisingly, only of the order of 2 to 3% velocity changes resulted from the Biot-Gassman fluid substitution, showing that a reduction of $V_s$ and an increase for $V_p$ had occurred in relation to true formation properties. This magnitude of change seems immaterial in relation to the total effect of potential wellbore damage, clearly mostly caused by 'rock mechanics' effects (i.e. stress induced fracturing of some form).

In addressing the needs of a multi-physics approach to the complex question of formation evaluation, based on time effects registered by well logging (LWD followed by wireline), Torres-Verdin et al., 2003 utilised a variety of modelling approaches. In Figure 12.30 they show the result of a four-days mud-infiltrate invasion model, using constant-permeability-with-radius assumptions, but with a 1:10 ratio of $K_{vertical}$ and $K_{horizontal}$ in the formation of interest. The figure shows the calculated formation *resistivity* distribution, due to the unequal rates of invasion in the central formation.

This anisotropy modelling gives an improved vision of possible realities, but due to wellbore damage, modellers and analysts should perhaps also consider a *radius-dependent* distribution of permeability, related with the radius-dependent velocity caused by the (also radius-dependent) degree of stress-induced fracturing and stress redistribution. Such will only be necessary of course, when the formation characteristics, in relation to the stress levels and mud-pressure applied, cause unwanted 'alteration' of properties, both immediately, and exacerbated by time effects. Infiltration rates seem likely to be accelerated at smaller radius, due to the possible presence of a miniature, failure-related *discontinuum*. (Revealing velocity-versus-radius measurements in tunnels were reviewed in Chapter 7, see for instance the circular tunnel cross-hole seismic results in Figure 7.5).

Briggs et al., 2004 in research at MIT, compared LWD and wireline data that was collected over the same interval in the same well. Both monopole and dipole logs were measured by the wireline tool, which was run some 10 days following completion of the drilling. The LWD tool provided dipole measurements.

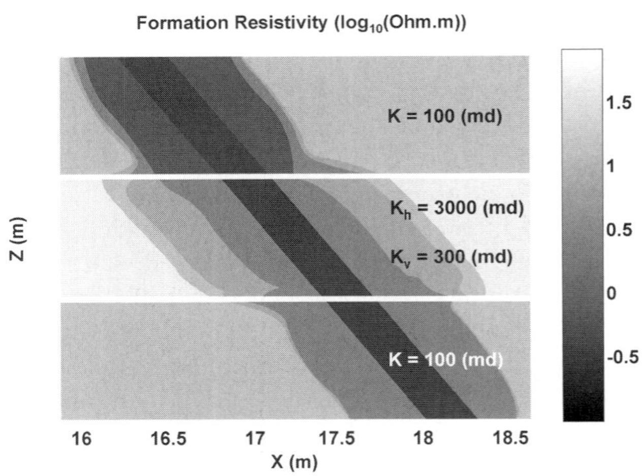


*Figure 12.30* Modelling of four-days mud-infiltrate invasion, using constant-permeability-with-radius assumptions, but with a 1:10 ratio of $K_{vertical}$ and $K_{horizontal}$ in the formation of interest. Torres-Verdin et al., 2003.

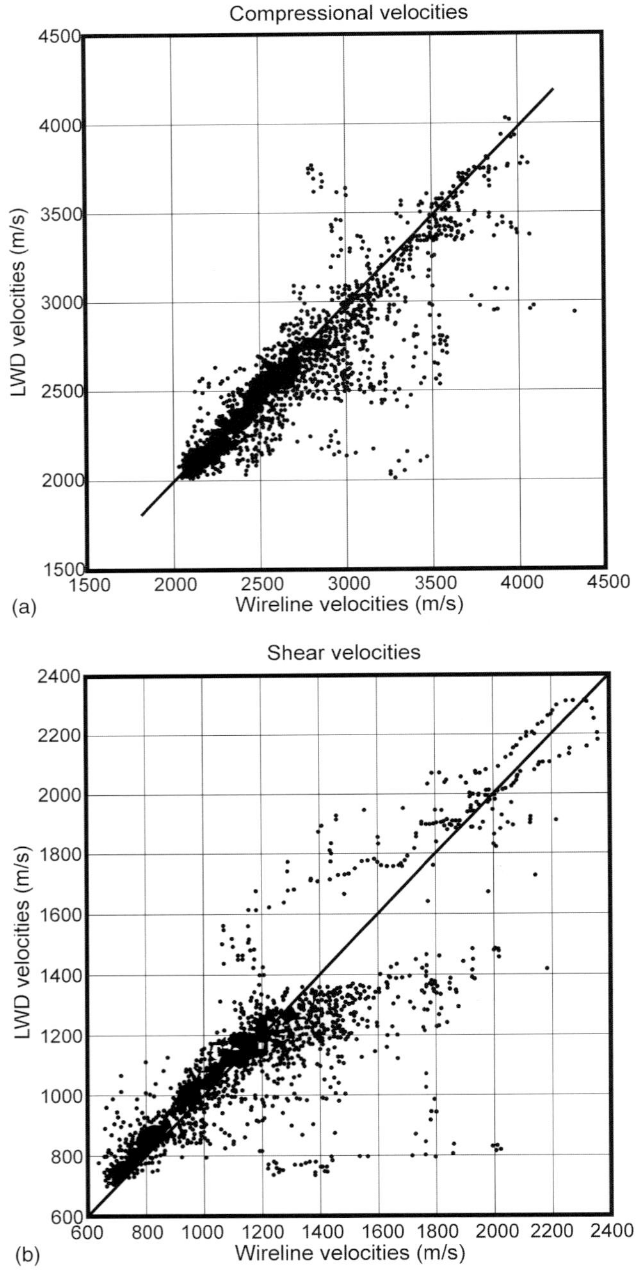


*Figure 12.31* Scatter plots comparing higher frequency (shallower-viewing, more disturbed) LWD tool velocities, with the lower frequency (deeper-viewing, less disturbed) wireline tool velocities. Briggs et al., 2004. In the case of the higher velocities, the earlier LWD measurement possibly views a less cracked and/or less invaded borehole wall, both tools registering relatively high velocities.

The authors found that for this well, there was on average about 5% discrepancy in shear wave data, and about 3% in compressional wave data. These trends suggested higher average velocities from the subsequent, deeper-viewing wireline logs. A short section of the formation showed velocity differences of just over 10%, with an opposite trend. Significantly, both tools measure in different frequency ranges, and had different *offsets* between source and first receiver.

The authors pointed out that as a general rule-of-thumb, a tool sees 1 inch into the formation for every foot separating the source and first receiver. Furthermore, low frequencies (1–3 kHz) see 2 to 3 borehole diameters, while the higher frequencies see less than one borehole diameter. Consequently, the higher frequency LWD tool was assumed to see the formation nearer to the borehole wall than the lower frequency wireline tool. This would mean that the damage zone would tend to be seen by the LWD tool, while the virgin formation would tend to be seen by the lower frequency wireline tool. However, the deeper penetration of split shear waves nevertheless detects fracturing and anisotropy in the formation, outside the damage zone, as we saw earlier.

The overall scatter plots of LWD versus wireline P- and S-wave velocities shown in Figure 12.31 support this *radius-bias*, with the wireline velocities sometimes 1 km/s faster in $V_P$, and 0.5 to 1.0 km/s faster in $V_S$. The authors emphasised that in poorly consolidated zones that would be susceptible to damage by drilling, the slower velocities should be accounted for, either by making a sufficient correction, or by using lower frequencies and/or larger offset, in order to see deep enough into the formation to register undisturbed velocities.

Material reviewed in this chapter has demonstrated that geomechanics/rock mechanics wellbore stability studies have an important place in improved understanding of well-logging anomalies (e.g. Fjaer et al. 1992). However, the existence of a miniature, but potentially up to several diameter EDZ *discontinuum*, which cannot be modelled in conventional Mohr-Coulomb based continuum modelling (Barton, 2004b), due to incorrect *addition* of the cohesive and frictional components, is perhaps the root cause of the phrase 'alteration zone' being used to describe the complex, time-dependent interactions occurring in the over-stressed, near-wellbore zone. The use of logging tools that illuminate to greater depth, due to low frequency, but that can also give information about the discontinuous zone, are clearly of importance for improved understanding of this *cross-discipline* region that surrounds wellbores.

The mini-EDZ theme will be terminated with two photographs from a large 'borehole', namely one of the world's first TBM tunnels, excavated by a 7 ft (2.1 m) diameter steam-driven machine credited to Beaumont,

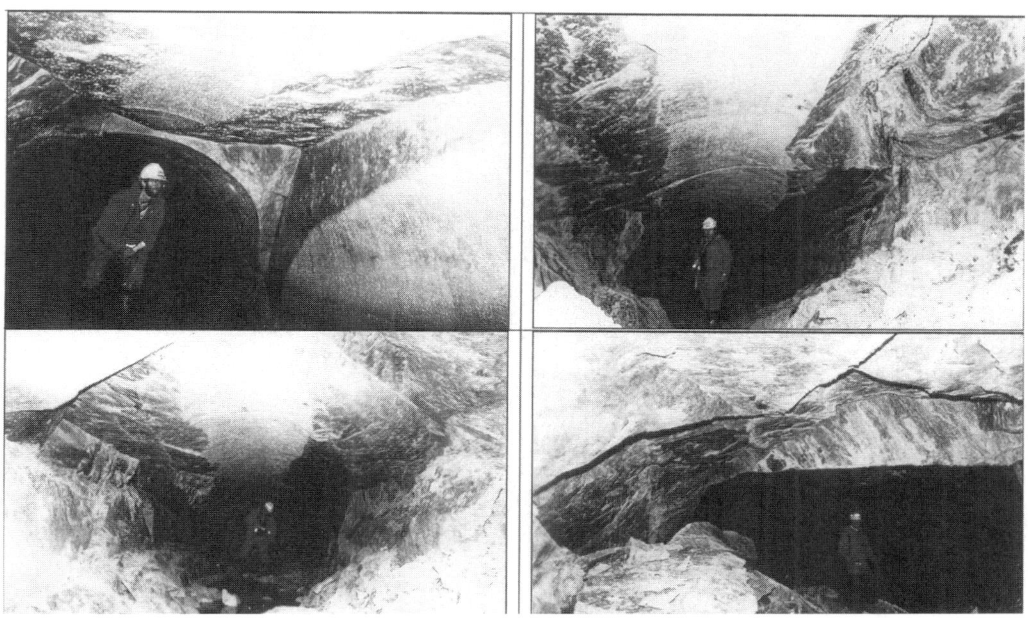

*Figure 12.32* One of the world's first TBM tunnels, from 1880, credited to Beaumont. Excavation in chalk marl of UCS = 4 to 9 MPa, close to the Channel Tunnel between England and France. Note structurally controlled and (vertical) stress-controlled breakout to at least one-radius on diametrically opposite sides of the tunnel. Barton and Warren, 1996.

for 'pilot-drilling' towards France, in an early (1880) effort to examine the feasibility of a sub-sea link between England and France.

The four pictures shown in Figure 12.32 are very informative for this chapter. They show: a) structurally controlled, three joint set fall-out, which would be a source of calliper-log 'noise' in other (vertical well) circumstances, disturbing the assumption of hole-ellipticity parallel to $\sigma_h$ min. Photos b) and c) show progressive stress-controlled 'break-out' at least doubling the effective diameter, caused in this case by tunnel loading under a 50–60 m chalk cliff along the S. England Folkestone Warren coast line. The weak (UCS = 4 to 9 MPa) chalk marl at tunnel level, has failed in combined shear and compression, due to a vertical stress of only about 25% of UCS, possibly with some bedding plane influence in the 'break-out'.

The *discontinuum* lies on the tunnel floor and behind what we see at the tunnel wall. In a 'mud-filtrate invasion' scenario, the fluid would not penetrate in an isotropic or linear manner, but faster into the haunches, or 'E-W'. In a further possible illustration of unwanted hole collapse, this time due to 'over-pressure', photograph d) shows failure due to successive bedding plane opening under the sea section of this tunnel. The original circular 2.1 m diameter hole has 'migrated' with time, into the collapsing, bedded rock above the crown. Could this be the occasional fate above and below horizontal sections of wells waiting for casing installation, in case of thinly bedded strata with over-pressure?

## 12.9 Challenges from ultra HPHT

The drive to discover more petroleum, which is especially relevant at the time of publication (2006) with oil prices above 70 US $ per barrel, is stimulating the exploration of deeper petroleum reserves, with all the associated difficulties of high temperatures, high fluid pressures, and high rock stresses. Some of this is so-called 'infrastructure-led exploration', searching deeper or laterally *and* deeper from developed fields with their existing production facilities.

Since a 1990 accident report (the Piper Alpha platform disaster in the UK sector of the North Sea), there have existed formal definitions of *high pressure high temperature* (HPHT) wells. With time this has been extended to the following three-tier system for classifying these extreme conditions.

Table 12.2 Definitions of maximum HPTP pressure and temperature tiers (after Willson, 2006).

| Tier | Reservoir pressure | Reservoir temperature |
|---|---|---|
| Tier I | 15,000 psi (103 MPa) | 350°F (177°C) |
| Tier II | 20,000 psi (138 MPa) | 400°F (204°C) |
| Tier III | 30,000 psi (207 MPa) | 500°F (260°C) |

The Tier II and Tier III categories are termed *ultra*-HPHT and *extreme* HPHT for obvious reasons. The current record for the offshore environment is reportedly Mobile Bay, off the coast of Alabama, at 138 MPa and 215°C. There are now several North American deep gas reservoirs both onshore and offshore that would classify as Tier III. As might be expected, there are operating temperature limits for 'conventional' LWD logging while drilling components, and for steerable downhole drilling motors. Even Tier I temperature limits may be reached at well depths in excess of 6 or 7 km, meaning that real-time data may no longer be available while drilling. However, subsequent wireline logging can presently be performed at temperatures up to the Tier III 'limit' of 260°C.

As we shall see in Chapter 13 on rock physics phenomena at extreme levels of confining stress, the high state of compaction at great depth means that the porosity and seismic velocity of typical reservoir rocks may exhibit little sensitivity to changing depth. The prediction of fluid pressures and fracture gradients (e.g. Figure 12.1) is then more difficult.

A related problem of extreme rock pressure is that the Terzaghi, 1943 theory of effective stress:

$$\sigma' = \sigma - p \tag{12.1}$$

may no longer give a correct description of the magnitude of the three effective principal rock stress components that would normally define the likely stability or instability of a deep well. At great depth, the rock *skeleton* may bear a greater proportion of the total rock stresses due to an *effective stress parameter* $\alpha$ that is less than 1.0, following the Biot, 1956b generalized theory of poroelasticity:

$$\sigma' = \sigma - \alpha p \tag{12.2}$$

A 'moderate' example from Hettema and de Pater, 1998 is a clay-rich sandstone with an *unstressed* porosity of 20%, which demonstrated an $\alpha$-value of 0.9 under zero differential pressure, reducing to 0.6 at 36 MPa differential pressure. A more extreme example is a limestone tested by Laurent et al., 1993, with α-values reducing from 0.8 to 0.2 as porosity decreased from 23% to 4.5%. Such effects may increase the effective stress components by tens of MPa, which leads, as also in TBM tunnelling (Barton, 2000) to reduced drilling penetration rates as the rock is stronger. However it also leads to the opposite effect of a potentially increased likelihood of stress-induced fracturing or (log-spiral) shearing, depending on rock type. (In the world of TBM tunnelling there are extremes of 0.1 m/hr and 10 m/hr penetration rates, the former due to extremely hard rock at great depth in an Idaho mine, with a TBM giving insufficient thrust.

Practical measures for reducing the double impact of both high effective stresses, and the additional thermoelastic effect of high drilling mud *temperatures* were proposed by Maury and Guenot, 1995, by introducing mud cooling systems. When the circulating mud is cooler than the formation, thermoelastic contraction means lower tangential stresses, with the dual effect of reducing the likelihood of *compressive* stress-induced fracturing, but an increased likelihood of mud pressure induced tensile cracking, due to the reduced minimum tangential stresses on opposite sides of the well, and rotated 90° from the potential *compressive* fracturing locations.

The tensile cracking allows mud loss, but some is returned when the temperature subsequently rises. The apparent wellbore 'ballooning' represents the mud loss, prior to the subsequent gain of fluid. The subject of mud temperature management is of great current interest for ultra deep wells, for extending fracture gradients, as discussed by Pepin et al., 2004 and others, in recent literature.

As pointed out by Willson, 2006 in a helpful technical review of the subject ('Feeling the heat, can't stand the pressure?'), manipulating the mud temperature during the drilling process, before formation temperature is re-established, can also be used 'in the opposite direction' to positively influence the hydraulic fracture (mud-loss) gradient if there is no risk of formation well-bore *compressive* failure. This would be done by *increasing* the mud temperature. However, in the case of shale, an initially undrained condition when exposed by drilling means that the greater thermal expansion coefficient of the contained water will have greater influence on the stability than the rock's response to increased or lowered mud temperatures (Li et al., 1998).

There are reportedly particular problems when penetrating salt rocks at high temperature and pressures (Willson and Fredrich, 2005), due to the particular sensitivity of the creep rates of salt to high temperature. There is a so-called 'undefined mechanism' of creep at lower temperatures, and a 'dislocation climb mechanism' at high temperatures, which can result in orders of magnitude increased rates of 'creep'. Already between 60°C and 230°C there is a reported 200× increase. Clearly this is of particular concern to petroleum companies who are developing reserves beneath thick bedded salt formations or next to salt dome structures. A review of drilling problems, and an emphasis of the need for high pressure and high temperature creep tests for salt is given by Maia et al., 2005.

# 13 Rock physics at laboratory scale

King, 2005 recently summed up the major exploration-related goals of rock physics research. They are 'to understand how lithology, porosity, confining stress and pore pressure, pore fluid type and saturation, anisotropy and degree of fracturing, temperature, and frequency influence the velocities and attenuation of compressional P- and S-waves in sedimentary rocks'. At the end of this list the author added 'and vice versa', presumably emphasising the interactive and complex nature of the reality. Isolation of just two or three of these variables for intense, high quality investigation, clearly gives just a small increment to the overall understanding.

His review article ended by emphasising the remaining challenges: relationships between attenuation, anisotropy, fractures and fluid flow – and determining these relationships across the frequency spectrum of core, log and seismic measurements. In this chapter we will review some of the important *increments* in this understanding, mostly by first treating velocity-related experiments, then attenuation-related experiments. Description of various matrix behaviours will be followed by description of laboratory tests that include jointing or induced fracturing.

There was early recognition (e.g. Birch and Bancroft, 1938), that seismic velocities of rocks were strongly influenced by microcracks, and that seismic attributes representative of the intrinsic mineralogy and porosity, could only be obtained by applying pressure to the rocks. Much of the rock physics understanding of reservoir rock behaviour (of both matrix and joints), has therefore to be achieved at *elevated pressure*, and the importance of *temperature* is also well recognised, but less frequently an experimental variable.

Mavko, 2002 pointed out that because of the growing complexity of recently discovered oil fields, a major shift had taken place in the use of seismic methods in the last decade of the 20th century. Interpretation of the increased *spatial variability* due to heterogeneous distributions of porosity, clay content, fracture density and permeability, was now more strongly based on rock physics understanding. Gradually, more order had been discovered in relations that once seemed scattered, such as velocity-versus-porosity, porosity-versus-permeability, $V_p$/Vs-versus-saturation and lithology. The author discussed the seismic signatures of cementation, sorting, shaliness, lithology, fluid content, and compaction.

Both the well-established porous matrix behaviours, and several of the newer areas of knowledge such as joint and fracture behaviour, will be reviewed in this chapter, before moving to reservoir-scale in Chapters 14 and 15. Of necessity the selection of material is subjective, but designed to be informative, and broad-based, within the confines of one chapter. A lot of supplementary data will be found concerning seismic Q, to add to the introductory material of Chapter 10. In this chapter it is presented together with the velocity data.

A 'shallow perspective' of this subject, certainly in relation to depth of observation, will be found in Chapter 2, based mostly on civil engineering related investigations. A certain relation to many of the high-pressure velocity-*parameter* trends presented in the present chapter will be noted. The reader is referred to some particularly interesting assemblies of low-stress data. The particular effects of variable weathering and fissuring common to the near-surface is documented, which is absent in reservoir-related studies, with the possible exception of related effects from *alteration* close to fault zones, that may exhibit hydrothermal alteration of their various mineral assemblages.

## 13.1 Compressional velocity and porosity

In the petroleum industry there is a history of at least half a century for inferring porosity from well logs, and for indicating pore fluid type. Among the oldest, and most popular expressions was that put forward by Wyllie et al., 1956, based on time-averaging in the solid and fluid phases. Using $T_p$ to represent the total travel time of a P-wave, and $T_s$ and $T_f$ to represent the travel time components expected in the *solid* and *fluid* phases respectively, it is assumed that:

$$T_p = T_s + T_f \tag{13.1}$$

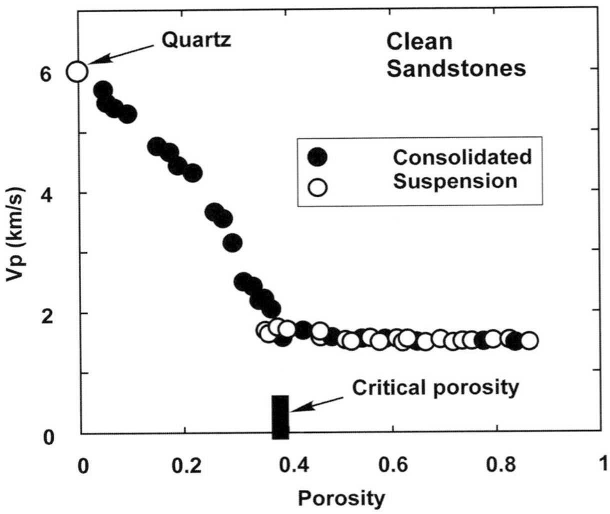


*Figure 13.1* P-wave velocity versus porosity for compact clean sandstones, and for a suspension of component grains, behaviour that is separated by the critical porosity $\varphi_c$ concept. Nur et al., 1991, and Mavko et al., 1998.

Inserting the fractional porosity ($\phi$) of the rock, and $V_p$ for measured velocity, the component velocities $V_{ps}$ and $V_{pf}$ contribute as follows by substitution in equation 13.1:

$$\frac{1}{V_p} = \frac{1-\phi}{V_{ps}} + \frac{\phi}{V_{pf}} \tag{13.2}$$

A close variant of this Wyllie et al., 1956 time-average prediction, is the popular velocity-to-porosity equation of Raymer et al., 1980, for $\phi < 0.37$:

$$V_p = (1-\phi)^2 V_{ps} + \phi\, V_{pf} \tag{13.3}$$

However, as pointed out recently by Dvorkin and Nur, 1998, in an appropriately titled note: 'Time-average equation revisited', there is actually only a limited theoretical scope for such time-average equations, requiring arrangement of the two components in layers normal to the direction of propagation, with a wave length small compared to the thickness of an individual layer.

Nur et al., 1991 championed the concept of a *critical porosity* that separates both the mechanical and acoustic state of a rock and its component grains. At lower porosities than $\phi_c$ the mineral grains are load-bearing, giving correspondingly higher velocities, while at porosities greater than $\phi_c$ a fluid phase will be load-bearing, the particles being in suspension, with the obvious consequences of a very 'flat' velocity-porosity response. An example of this concept, for the case of clean sandstones, is given in Figure 13.1.

A 'reservoir rock' selection of P-wave velocity versus porosity data for saturated samples, is reproduced in Figure 13.2 for ready reference. This limited $V_p$-$\phi$ selection, from Mavko et al., 1998 appendices, is also credited to original authors in the figure caption. Chalk, two sets of limestone data of widely different porosity ranges, dolomite, and two sets of sandstone data are reproduced. As we shall see shortly, explaining some of the wide scatter in the data requires specification, as a minimum, of effective stress level, degree of cementation, and clay-content. Examples of such discrimination are shown in Figure 13.3, from Dvorkin and Nur, 1998.

## 13.2 Density, $V_s$ and $V_p$

The age-depth relationships derived for numerous wells in sandstone-shale units by Faust, 1951 (see beginning of Chapter 14), had a certain grouping of velocities with age, due to variations of porosity and the resulting densities. Hard porosity in the form of pores tended to decrease with age and depth, while (soft) porosity in the form of joints tended to increase with age due to tectonic influences, but reduced strongly with depth. Only the hard porosity will have a *significant* effect on density.

Gardner et al., 1974, showed simple velocity-density trends for reservoir rocks, which on a log-log plot were almost linearly distributed. Figure 13.4 shows the trends for sandstone/shale/limestone/dolomite, where the central relationship (dashed line), as in equation 13.4 is:

$$\gamma = 0.23V^{1/4} \tag{13.4}$$

where $\gamma$ = bulk density in gm/cm$^3$, and velocity $V_p$ is ft/s. (Note: density × velocity = impedence = gm/cc × m/s)

We can visualise more familiar (km/s) velocities and densities by estimating $\gamma$ when $V_p$ is 3, 4 or 5 km/s (9,840, 13,120 or 16,400 ft/s). The three results for $\gamma$ according to this equation are approximately 2.29, 2.46 and 2.60 gm/cm$^3$ respectively. When $V_p$ = 6 km/s a density of 2.72 gm/cm$^3$ is predicted. Each of the above is a realistic reflection of both mineralogy and porosity differences, as density and velocity increase. Several sets of seismic data that included the fundamental effect of density were presented in Chapter 2, but these were not specifically related to reservoir rock, and were usually at 'near-surface' stress levels, related to civil engineering projects, and often had the density- and velocity-reducing influences of weathering.

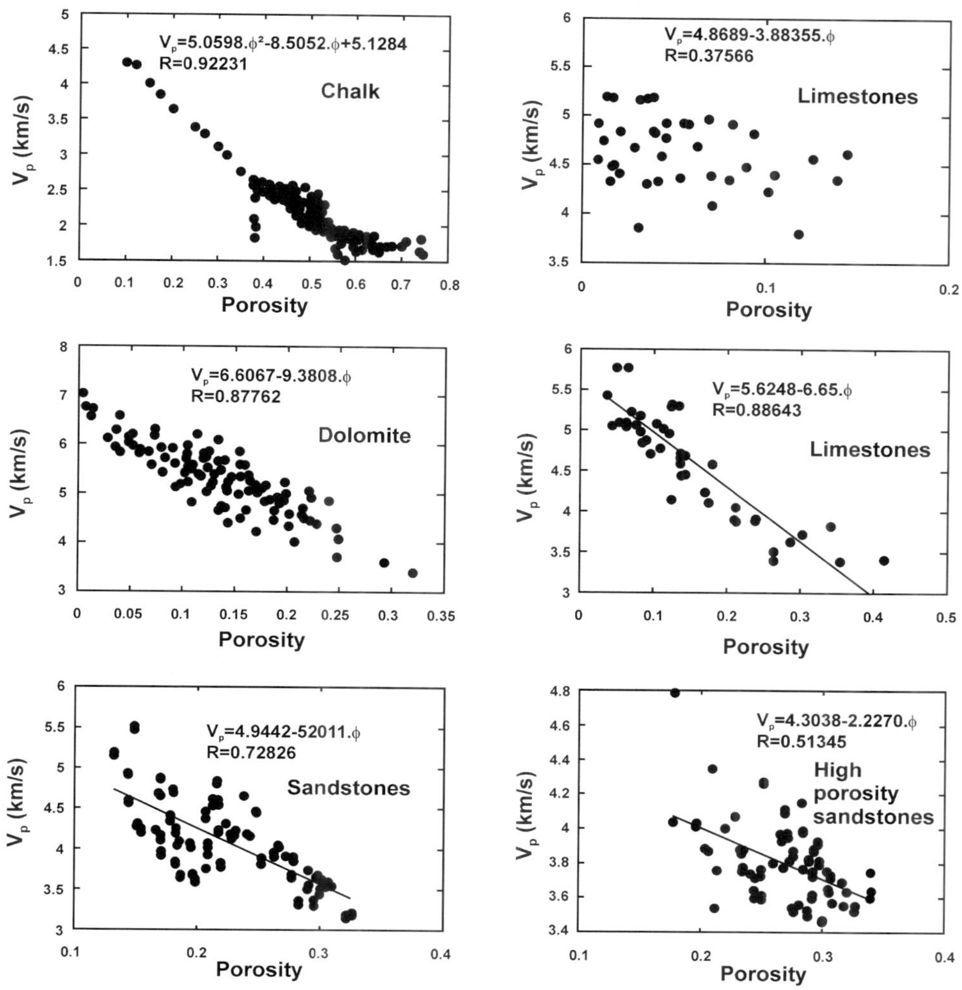


*Figure 13.2* An assembly of $V_P$-porosity data for saturated chalk, two limestones of widely different porosity, dolomite, and two sandstones of medium and high porosity. a) Chalk: from sonic log and porosity log, Urmos and Williams, 1993. b), c) Limestones: ultrasonic, 10 to 50 MPa effective pressures, Cadoret, 1993, Lucet, 1989, Yale and Jamieson, 1994. d) High porosity sandstone: ultrasonic, 35–40 MPa effective pressures, Strandenes, 1991. e) Sandstones (with yet-to-be-defined clay-content): ultrasonic, 30–40 MPa effective pressures, Han, 1986. f) Dolomites: ultrasonic, 10–35 MPa effective pressures, Geertsma, 1961, Yale and Jamieson, 1994. After Mavko et al., 1998.

A 'reservoir rock' selection of P-wave velocity versus density data, is reproduced in Figure 13.5a for ready reference. This limited $V_P$-$\gamma$ selection, from Mavko et al., 1998 appendices, is also credited to original authors in the figure caption. Chalk, limestone, dolomite, and three sets of sandstone data are reproduced. Again for reference purposes, both the shear-wave and compression-wave velocities, for the same six groups of rocks, are shown in Figure 13.5b.

The relatively 'ordered' density-$V_P$ trends for the chalk, limestone and dolomite reflect the simpler mineralogy. The contrast to the widely scattered density-$V_P$ data for the three groups of sandstones is evidence of the variable mineralogy of 'sandstones', with 10–15% variation in density possible for the same velocity, particularly in the case of the tight gas sandstones, which gives a correlation coefficient of only 0.39. In contrast to these variations, the $V_s - V_P$ trends are consistently uniform, as befits characterization by seismic waves.

In Chapter 14, an early introduction to the age and depth effects on velocity is reproduced from Faust 1951, who analysed well survey results from some 500 petroleum wells in the USA and Canada. Faust used data from about 300 kilometres of well sections. The great majority of data was for mixed *shale/sandstone* sections.

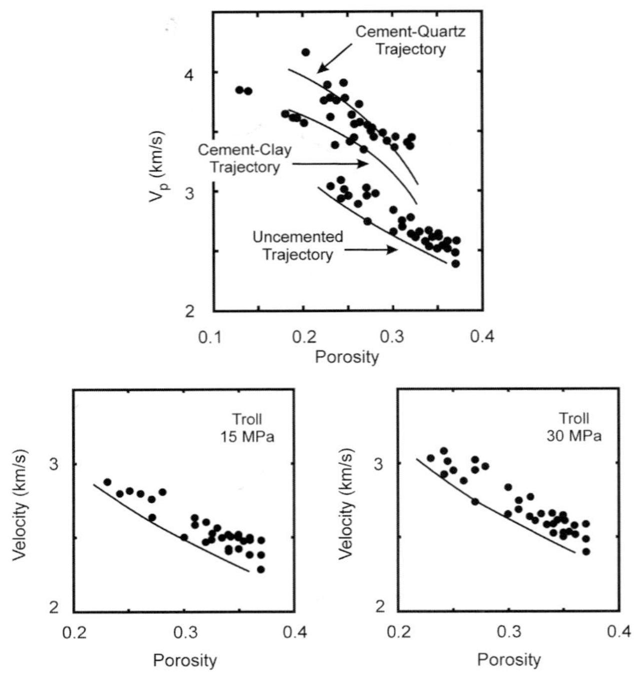


*Figure 13.3* Specification of sandstone condition: quartz-cemented, clay-cemented, uncemented, and specification of effective stress level (applied on uncemented Troll sand), helps to sort P-wave velocity data that displays 'unexplained' scatter. Dvorkin and Nur, 1998. The sub-envelopes beneath the Troll sand data are from Dvorkin and Nur, 1996.

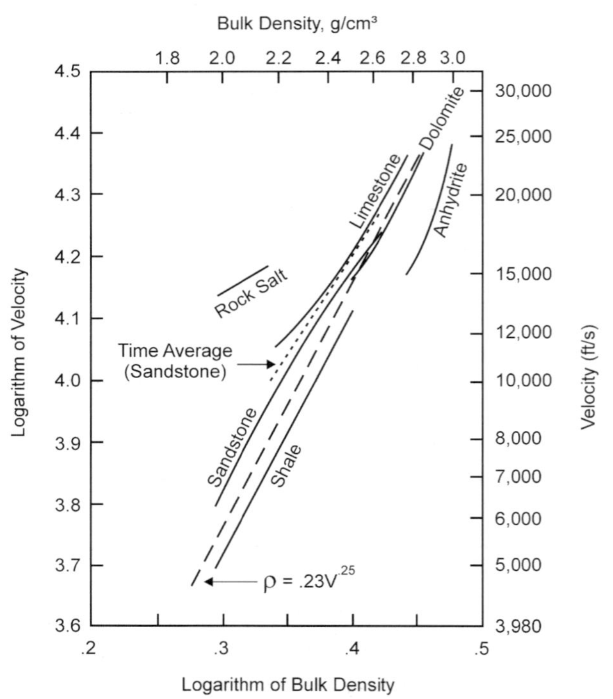


*Figure 13.4* Log-log trends for $V_p$ (ft/s), and γ. Gardner et al., 1974.

A non-systematic comparison of shale and sand (sandstone) velocities revealed an average discrepancy of only 350 ft/sec, or 106.7 m/s in velocity between these two, frequently inter-bedded units, the sandstone having the highest velocity by this small average margin. We will present the *also* remarkably close *$V_p$ versus $V_s$ trends* for water-saturated sandstones and shales, from Castagna et al., 1993, as reproduced in Mavko et al., 1998, at the appropriate location in Chapter 14 (Figure 14.4), to emphasise the remarkably similar $V_p$ and $V_s$ signatures of these two 'dissimilar' lithologies, when in a compacted state. The necessity of using impedence (gm/cc × m/s), attenuation, and anisotropy, for *seismically* distinguishing these two most essential reservoir 'partners' is clear.

## 13.3 Velocity, aspect ratio, pressure, brine and gas

An important early paper in the area of theoretical modelling of porous rock behaviour was presented by Toksöz et al., 1976, who examined the numerous factors affecting seismic velocities of intact samples of porous rocks with emphasis on sandstones. They developed theoretical formulations to represent the solid matrix, and assumed *spherical* to *oblate spheroidal* pores, of widely varying aspect ratios, to match numerous laboratory data. As one would expect, they found that small aspect ratios (flatter voids) caused greatest reductions to elastic moduli and velocities.

They also predicted and confirmed that the properties of the saturating fluid (gas, oil or water) produced greater effects on the compressional velocities than on the shear velocities. The P-wave velocities were predicted, correctly, to be higher when the rock was saturated with water, than when dry or gas-saturated.

When *fitting* their theoretical model to P- and S-wave velocities that were *measured* at different pressures, they required pore shape spectra ranging from spheres to very fine cracks (aspect ratios from 1 to $10^{-5}$) for sandstones, limestones and granites, both under dry and saturated states. As igneous rocks have low porosities, the pore shape has great influence on the elastic and seismic properties, and dry and water-saturated behaviours are markedly different, as was also seen in Chapter 2.

Compressional velocities were highest with brine saturation and lowest with gas saturation, but the difference declined with increasing pressure. Poisson's ratios for gas saturated rocks were lower than for those with

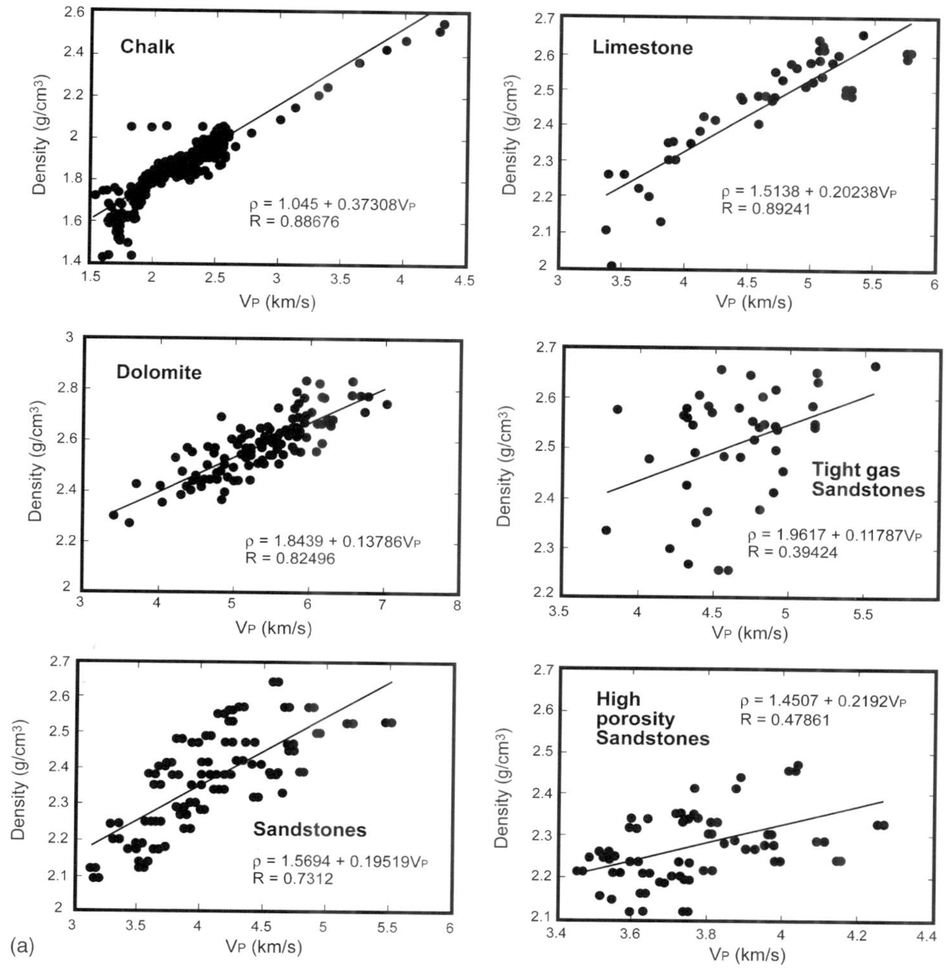


*Figure 13.5a* An assembly of *$V_p$-density data* for saturated chalk, limestone, dolomite, and three sandstones of low, medium and high porosity. a) Chalk: from sonic log and porosity log, Urmos and Williams, 1993. b) Limestone: ultrasonic, 10 to 50 MPa effective pressures, Cadoret, 1993, Lucet, 1989, Yale and Jamieson, 1994. c) Dolomites: ultrasonic, 10–35 MPa effective pressures, Geertsma, 1961, Yale and Jamieson, 1994. d) Tight gas sandstones: ultrasonic, effective pressures 40 MPa, Jizba, 1991. e) Sandstones: ultrasonic, 30–40 MPa effective pressures, Han, 1986. f) After Mavko et al., 1998. f) High porosity sandstone: ultrasonic, 35–40 MPa effective pressures, Strandenes, 1991. After Mavko et al., 1998.

brine saturation, and this difference persisted to great depths according to their model.

Figure 13.6, from Toksöz et al., 1976, is a good example of their modelling predictions, showing the relative predicted effects on $V_p$ of brine-filled and gas-filled cracks in a 16% porosity sandstone model. The fluid occupying the smaller aspect ratio cracks has more influence on velocities at low pressures, due to the greater pressure sensitivity of the fine cracks.

Figure 13.7 shows the same authors' predicted Poisson's ratio variations, with varying degrees of brine or gas saturation, and also as a function of differential pressure to 50 MPa. Greatest sensitivity, as one would expect, was shown when pressure or depth was smaller.

From summaries of the numerous experimental and theoretical trends given by Toksöz et al., 1976, Table 3.1 was developed.

The effects of the dry or brine-saturated states, and the influence of effective stresses, as predicted theoretically by Toksöz et al., 1976 so long ago, were nicely illustrated by more recent testing by King and Marsden, 2002, who tested numerous sandstones both dry and brine saturated. Ultrasonic P- and S-wave measurements were made on ten sandstones with porosities less

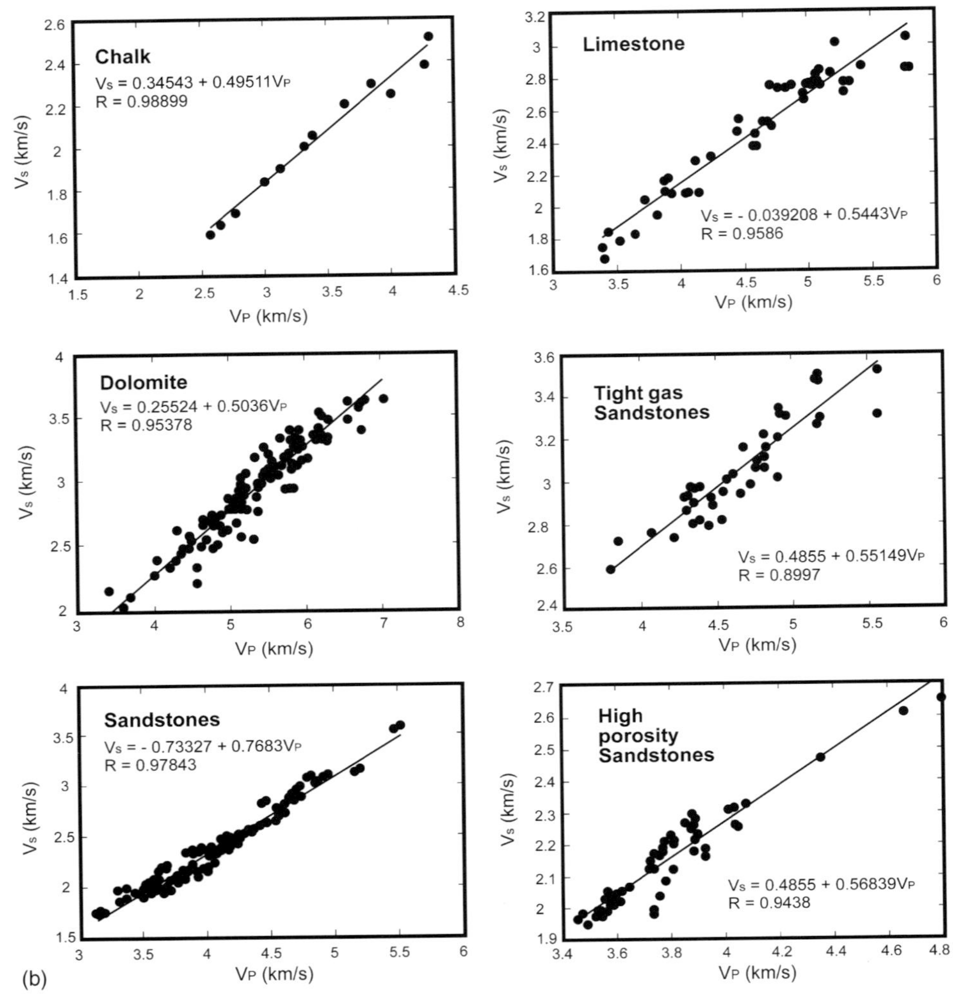


*Figure 13.5b* An assembly of $V_s$ - $V_p$ *data* for the same saturated chalk, limestone, dolomite, and three sandstones of low, medium and high porosity, that are presented in Figure 13.5a. a) Chalk: from sonic log and porosity log, Urmos and Williams, 1993. b) Limestone: ultrasonic, 10 to 50 MPa effective pressures, Cadoret, 1993, Lucet, 1989, Yale and Jamieson, 1994. c) Dolomites: ultrasonic, 10–35 MPa effective pressures, Geertsma, 1961, Yale and Jamieson, 1994. d) Tight gas sandstones: ultrasonic, effective pressures 40 MPa, Jizba, 1991. e) Sandstones: ultrasonic, 30–40 MPa effective pressures, Han, 1986. f) After Mavko et al., 1998. f) High porosity sandstone: ultrasonic, 35–40 MPa effective pressures, Strandenes, 1991. After Mavko et al., 1998.

than 10%, and thirty-four specimens with porosities in the range 20 to 30%, under hydrostatic effective stresses up to 60 MPa (in the case of the stronger, lower porosity samples), and up to 40 MPa in the case of the higher porosity set.

Figure 13.8 shows the $V_p$ and $V_s$ results for the dry and saturated specimens to a common hydrostatic stress of 40 MPa. Equations relating $V_p$ and $V_s$ are shown in the figure.

King and Marsden, 2002, also presented their $V_p$ and $V_s$ results as a function of the effective stress (10, 20, 40 or 60 MPa). Figure 13.9 shows both the $V_p$-$V_s$-effective stress trends, and the dry-saturated trends. The addition of a 'Gassman predicted' result for the saturated state, given by the authors, has been omitted from these figures for the sake of clarity.

## 13.4 Velocity, temperature and influence of fluid

Although far from exhaustive, a limited set of data for temperature effects on hydrocarbon-saturated samples will now be presented to illustrate some of the geophysical changes that can be used to monitor producing, stimulated reservoirs. Nur, 1989, referred to the new 'four dimensional seismology', in other words the ability to monitor in three dimensions the effect of time during

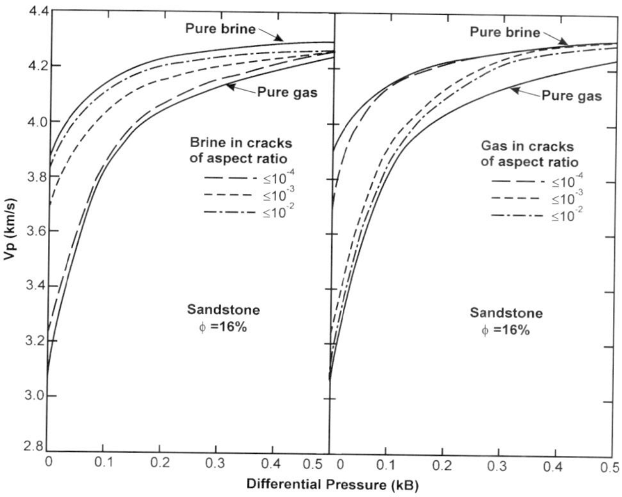


*Figure 13.6* Theoretical variations of $V_p$ with differential pressure (0 to 50 MPa) and different aspect ratio cracks, for various brine or gas saturation levels. Toksöz et al., 1976.

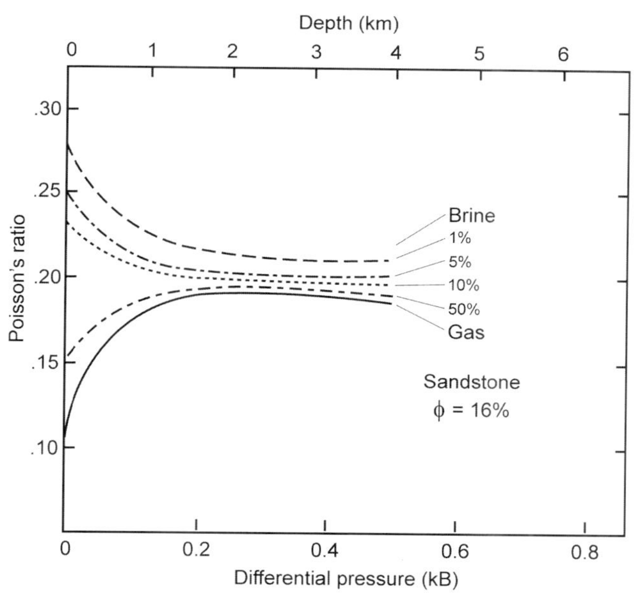


*Figure 13.7* Toksöz et al., 1976 model predictions for Poisson's ratio as a function of pressure or depth, and as a function of the degree of saturation with brine, gas or a mixture of the two.

various flooding methods. The basis for such an ability would be the strong dependence of velocity on temperature, (water-flooding causing local cooling, steam flooding the opposite), plus the significant influence of the relative hydrocarbon and brine saturations.

The six sets of results shown in Figure 13.10 show an easily detectable effect of temperature, with greatest effect when 100% oil saturation, and least effect (almost zero effect) when 100% gas saturated or 100% brine

Table 13.1 Typical relative effects of environmental conditions for porous reservoir rocks (derived from Toksöz et al., 1976).

| Lower velocity ($V_p$) if | Higher velocity ($V_p$) if |
|---|---|
| Low water saturation | High water saturation |
| Dry or gas saturated *(if flatter pores)* | Dry or gas saturated *(if rounder pores)* |
| Some immiscible gas (in brine) | Saturated with brine |
| Higher porosity | No immiscible gas |
| Over-pressured | Lower porosity |
| Shallow depth | Under-pressured |
| Thin pores | Greater depth |
| After several cycles of freezing | Rounded pores |
| Room temperature | Frozen |
| Extremely high temperature | Low or moderate temperature |
| **Lower velocity ($V_s$)** | **Higher velocity ($V_s$)** |
| If water saturated | If dry or gas saturated (and highly porous) |

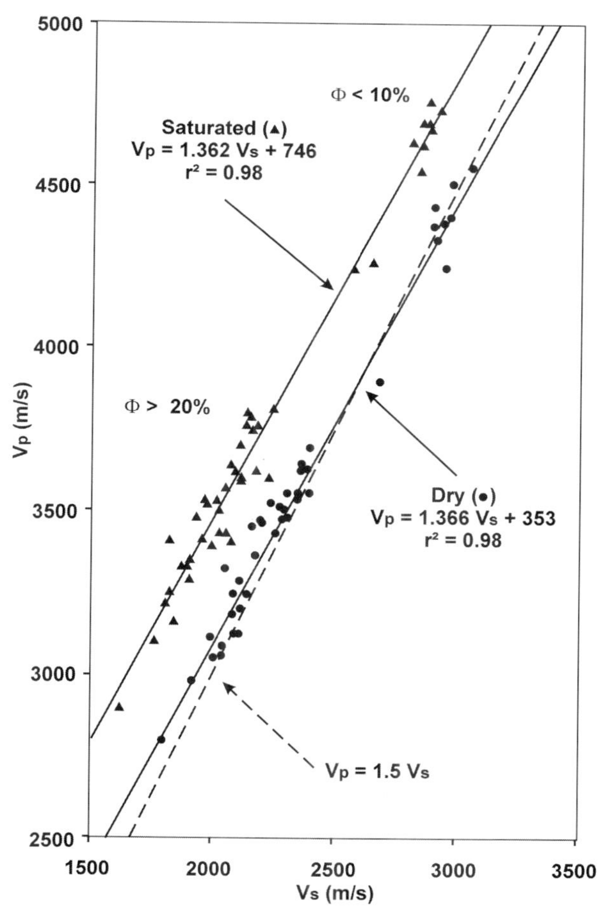


*Figure 13.8* $V_p$ and $V_s$ as a function of condition (dry or brine-saturated), at a common hydrostatic effective stress of 40 MPa. King and Marsden, 2002.

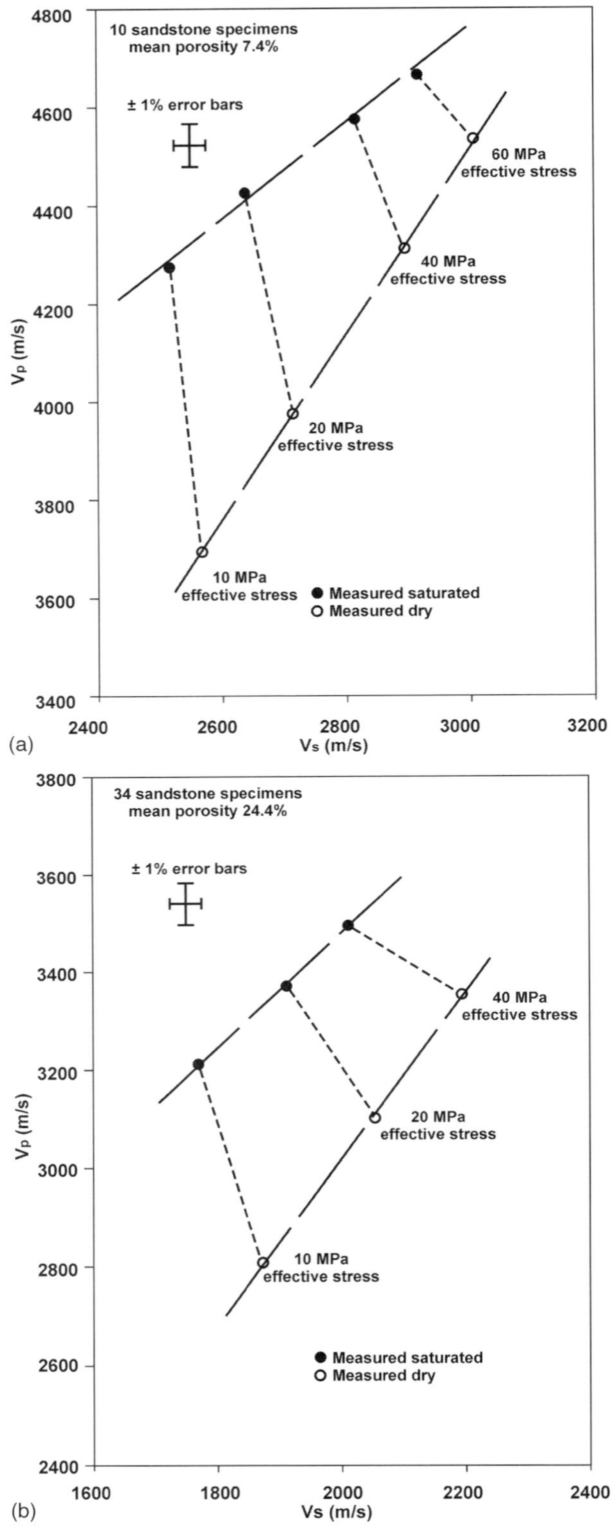


*Figure 13.9* $V_p$ and $V_s$ as a function of effective hydrostatic stress, for a) *ten* lower porosity sandstones, b) *thirty four* high porosity sandstones. Dry and brine-saturated results are shown. Redrawn from King and Marsden, 2002.

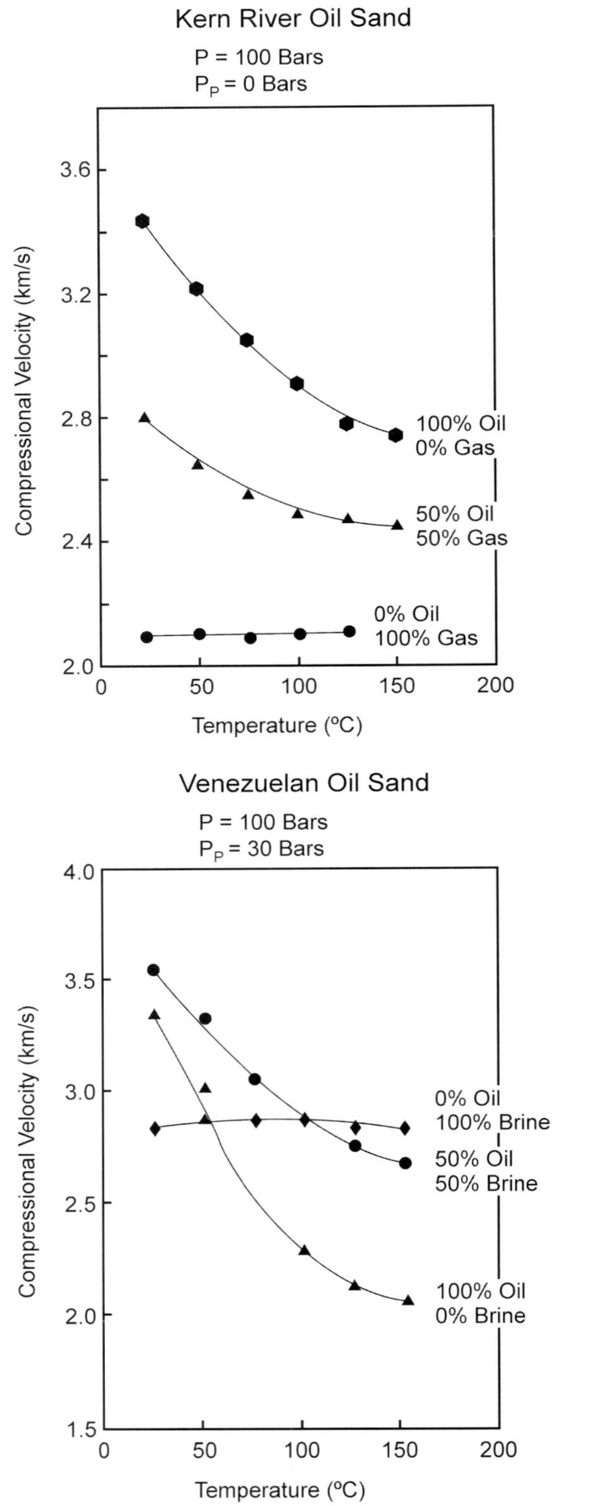


*Figure 13.10* $V_p$ behaviour as a function of relative hydrocarbon saturation and temperatures up to 150°C. Top: Kern River oil sand: P = 100 bars, $P_P$ = 0 bars, Bottom: Venezuelan oil sand: P = 100 bars, $P_P$ = 30 bars. Nur, 1989.

saturated. This temperature effect on velocity reduces to about half strength when 50% oil/50% gas or 50% oil/50% brine are present. In other words, when the oil is removed from these sands, the velocities successively become independent of temperature, with roughly half the effect when 50% oil remains.

Nur, 1989, also referred to a pilot steam flood experiment in viscous tar sands in Canada, conducted by Amoco. Changes in travel-time caused by heating effects reducing $V_p$ were readily detected. A differential travel-time plot revealed the areas closest to the wells (and partly between the wells) where velocity had been reduced by the change in viscosity. Laboratory tests of the tar sand showed an S-shaped reduction in velocity from $V_p = 2.4$ km/s at 25°C, to a final plateau of about 1.1 km/s at 150°C. (Den Boer and Matthews, 1988).

In their review of acoustic velocity and attenuation in porous rocks, Winkler and Nur, 1995 refer to the work of Wang and Nur, 1990 who measured the temperature dependence of compressional and shear wave velocities in sandstones that were either dry, saturated with water, or saturated with crude oil. Figure 13.11 shows the quite widely differentiated results, due again to the temperature dependence of the viscosity of the oil. The phenomenon was described as follows by Winkler and Nur. At low temperature the higher viscosity means that the oil cannot flow easily, so the dynamic measurement is on the high-frequency, high velocity, unrelaxed side of the local-flow mechanism. As temperature increases viscosity reduces, so fluid flows more easily, and velocity therefore decreases since measurement is on the relaxed side of the absorption/dispersion mechanism.

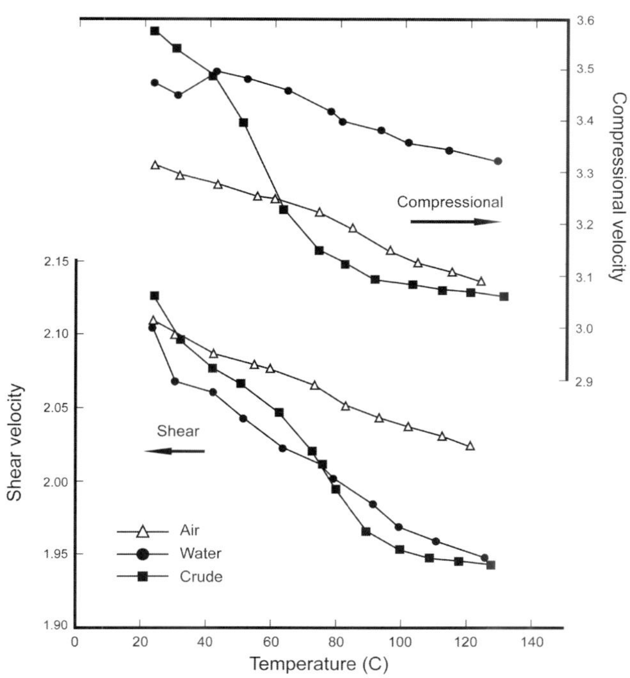


*Figure 13.11* Compressional and shear wave velocities in Boise sandstone, as a function of temperature and saturating fluid. The samples were dry, or saturated with water or saturated with crude oil. A common effective stress of 15 MPa was applied. Wang and Nur, 1990 in Winkler and Murphy, 1995.

## 13.5 Velocity, clay content and permeability

A comprehensive series of laboratory tests reported by Klimentos, 1991, were designed to investigate the influence of clay content on the P-wave velocities of saturated sandstones under varying confining and pore fluid pressures up to 40 MPa. Forty-two samples of sandstones were investigated, having the following range of characteristics, in order to see their combined and individual effects on velocity:

- Porosity: 2 to 36%
- Permeability: 0.001 to 306 mD
- Clay content: negligible to 30%

The principal results are shown in Figure 13.12, with sorting according to clay-content, shown in Figure 13.13.

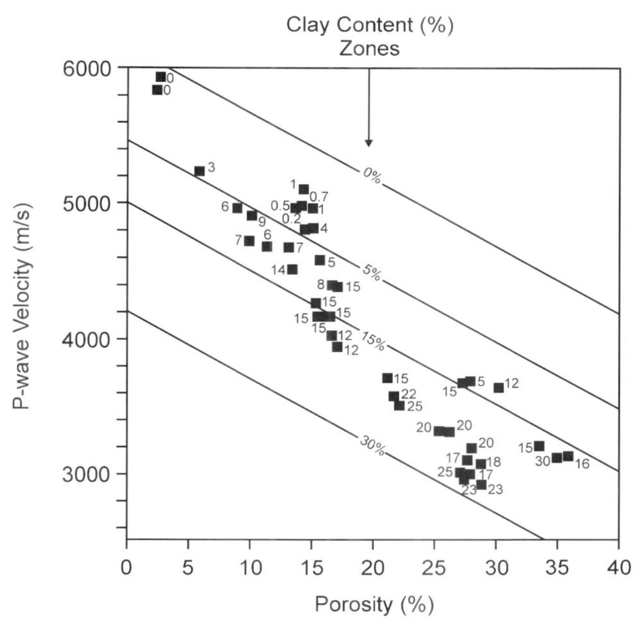


*Figure 13.12* P-wave velocity at 1 MHz frequency and 40 MPa confining pressure, showing clay contents and porosities. Klimentos, 1991.

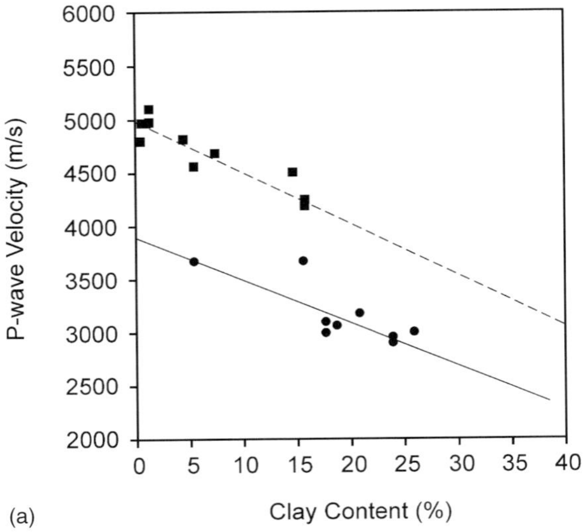


(a)

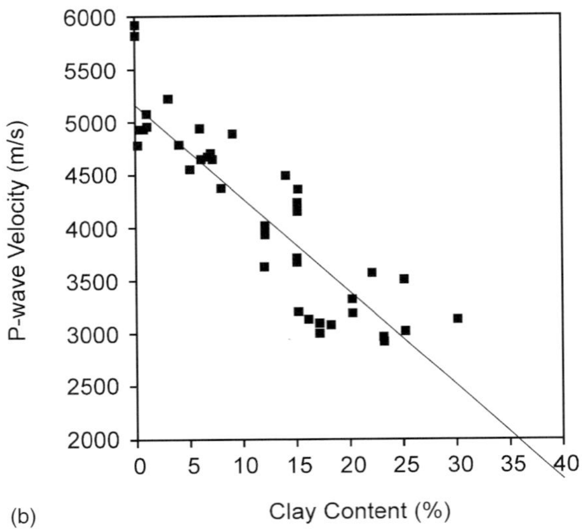


(b)

*Figure 13.13* a) P-wave velocity as a function of clay content for porosities of 6 to 36%. b) P-wave velocity as a function of clay content with average 15% porosity (squares) and 28% porosity (circles), each at 1 MHz and 40 MPa confining pressure. Klimentos, 1991.

From Figure 13.13, where porosity differences were used to distinguish the effect of clay content. Klimentos, 1991, gave the following multivariable linear regression equations for estimating $V_p$:

1. at ultrasonic frequencies (1 MHz):

$$(r = 0.96)\ V_p = 5.66 - 6.11\phi - 3.53c + 0.0007K \qquad (13.5)$$

2. at seismic frequencies:

$$(r = 0.93)\ V_p = 5.27 - 5.40\phi - 2.54c + 0.001K \qquad (13.6)$$

where $\phi$ = porosity (fraction), c = clay content (fraction) and K is permeability in millidarcies.

The effect of permeability alone on the P-wave velocity is seen to be negligible. In fact a misleading, weak increase in $V_p$ with increasing permeability arises due to the velocity-clay content and clay content-permeability interrelations. Clearly these equations can be reformulated to give estimates of (matrix) permeability.

Klimentos and McCann, 1990, also drew attention to the complex nature of permeability – depending as it did on porosity, pore size distribution, inter-connectedness of the pores, and tortuosity – the latter two presumably being especially compromised by clay content. They also posed as an open question, what the relative effects would be of frequencies of 10 Hz to 1 kHz (as used in seismic exploration) or frequencies of 10 to 20 kHz (as used in borehole logging), when clays were present in the *in situ* sandstones.

## 13.6 Stratigraphy based velocity to permeability estimation

Gutierrez et al., 2002, also addressed the question of clay-content, referring to the initially undifferentiated data set of Han, 1986 for sandstones. Figure 13.14 shows the undifferentiated $V_p$ versus porosity data of Han, 1986: *top-left*, and the *stratigraphy-guided*, *clay*-content differentiation: *top-right*, that makes the nearly 2 km/s variation in $V_p$ at one porosity, understandable.

Jan, 2003 presented a finer clay-content break-down of Han's data, including some additional data, which is included in Figure 13.14c and d. Both the P-wave and S-wave velocities were measured at 40 MPa confining pressure and 1 MPa pore pressure.

In order to emphasise the potential of some *stratigraphy-guided* lab-to-field velocity-porosity-permeability correlations, Gutierrez et al., 2002 presented a well log for La Cira-Infantas Oil Field (LCI). This is reproduced in Figure 13.15. The left-hand velocity log applied to the highly variable (also laterally) clean-to-shaly, loosely consolidated, Tertiary sandstones, lying in a highly faulted, asymmetrical anticline. Well-logged $V_p$ ranged mostly from 2.5 to 4 km/s, while core-based porosity was mostly between 10 and 30%, and there was a four order-of-magnitude range of core-based permeability. The core-based results are plotted in Figure 13.15b and c.

The authors plotted a series of velocity-porosity diagrams, starting with an 850 m interval with a few thousand data points for the undifferentiated deposition

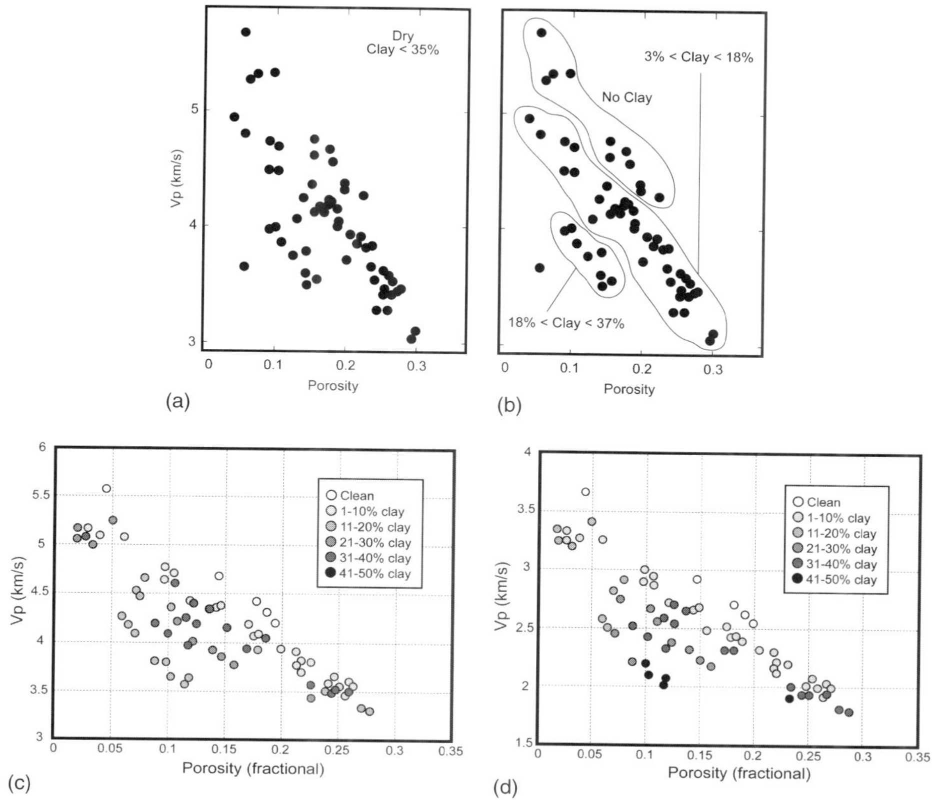


*Figure 13.14* a) Unsorted $V_p - n\%$ data for sands with <35% clay. b) Logical differentiation of $V_p - n\%$ trends when grouped by clay-content. Gutierrez et al., 2002. c) and d) A more detailed presentation of Han, 1986, data given by Yan, 2003.

cycles of the Tertiary basin. Successively smaller deposition cycles were then considered, first down to a specific operational zone, then so-called 'fining-up', giving successively higher $V_p$-n% correlation coefficients, reflecting the more uniform sedimentary environment and diagenetic nature of the smaller cycles. An essentially linear plot was shown for one of the fining-up cycles: in a 5 m section $V_p$ was 3.5 km/s at n = 10%, and 2.5 km/s at n = 30%, with correlation coefficient r = 0.973.

The authors found that due to *a large fraction of silt* (whose mineralogy is close to that of clean sand), the clay-content in 100% shale intervals could be as low as 20–30%. Poor sorting in the shale caused a reduction in total porosity, which caused an increase in the velocity in relation to existing models, and in relation to the data sets in Figure 13.14, which show lower velocities with these amounts of clay.

The concept of sorting data into common categories, using *stratigraphy* and other *matching techniques*, was also the theme in Prasad, 2003, who showed that by grouping and sorting rocks into hydraulic units, it was easier to establish relationships between *velocity* and *permeability*, through a more relevant match of porosity and permeability with common sediment compaction and cementation history.

This work followed the permeability-porosity matching of Amaefule et al., 1993, who showed the importance of separation into hydraulic units. Possibly we can draw a parallel here to the common separation of rock mass qualities into classes (Q = 1 − 4: poor, Q = 4 − 10: fair, etc), for different structural domains, which is the basis for prediction of similar behaviour, such as particular reinforcement needs in a tunnel, a particular range of deformation moduli or P-wave velocities.

Amaefule et al., 1993 and Prasad, 2003 used the following simple 'classification' relations:

$$RQI = 0.0314\ (k/\varphi)^{1/2} \qquad (13.7)$$

$$FZI = 0.0314/\varepsilon\ (k/\varphi)^{1/2} \qquad (13.8)$$

RQI is known as the *reservoir quality index*, with permeability (k) in units of millidarcies, and (φ) is the

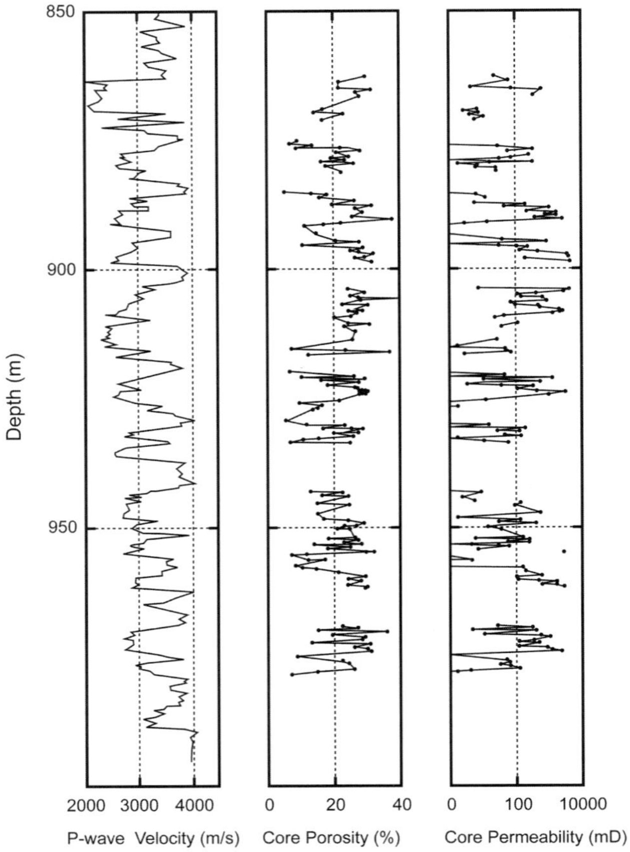


*Figure 13.15* a) A *well log* from the heterogeneous Tertiary sands of the LCI field, showing $V_p$, with stratigraphy-guided *core data* for b) porosity and c) permeability. Gutierrez et al., 2002.

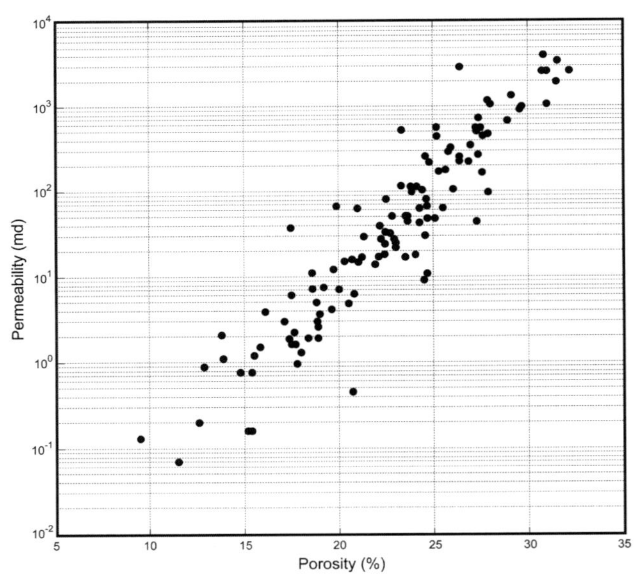


*Figure 13.16* Porosity-permeability correlation from core of a second well in the LCI Tertiary sands, which allowed sonic-logging data to be used for permeability prediction. Gutierrez et al., 2002.

fractional porosity. The term (ε) is the *void ratio*, given by φ/(1-φ). It thus links FZI the *flow zone indicator* with RQI, using the ratio of pore volume to solid volume. Rocks with FZI values within a narrow range belonged to one hydraulic unit: they had similar flow properties. A semi-log plot of porosity versus permeability showed similar FZI values plotting together.

Prasad first tested this older method, showing unsorted data (left), then data sorted by FZI (right) in Figure 13.17. This demonstrated that the FZI concept could be extended to seismic parameters, giving a strong correlation between *velocity* and *permeability*, when using the appropriate FZI grouping. She used a laboratory-test data base, which included porosity, permeability, velocity and attenuation data from tests at similar confining pressures. Figure 13.18 shows a much larger set of data with $V_p$ – k correlation through FZI.

### 13.6.1 Correlation to field processes

Prasad, 2003 also provided a practical illustration of the effect of depth-of-sediment on porosity, velocity and permeability development to 500 m below sea floor, using marine logging results from 'Site 977, ODP Leg 161 (Shipboard Scientific Party, 1996). In the parallel diagrams reproduced in Figure 13.19, the scattered down-hole log data is shown beneath the smoothed trend lines A-B in each case. Curves D-D' represent the hypothetical trend if porosity is 'frozen' by a cementation episode from point D. Note the FZI-predicted maintenance of high permeability, and the 'non-correlating' increase in velocity (but one that can be explained by FZI).

Curves C-C' represent the hypothetical effects of an 'opposite' trend – an influx of pore-filling materials from point C. There is a rapid reduction in porosity. This can be used to give an estimated reduction in velocity followed by an FZI-predicted reduction in permeability. Prasad, 2003 *calculated* the permeability scatter-curves directly from assumed FZI values of 0.1 (left curve), and 0.25 (right curve), using the log data for porosity and velocity. Note the similar scatter-shape of porosity data, and FZI-calculated permeability, due presumably to the use of $(\text{porosity})^{-0.5}$ and $(\text{void ratio})^{1.0}$ in the FZI estimate.

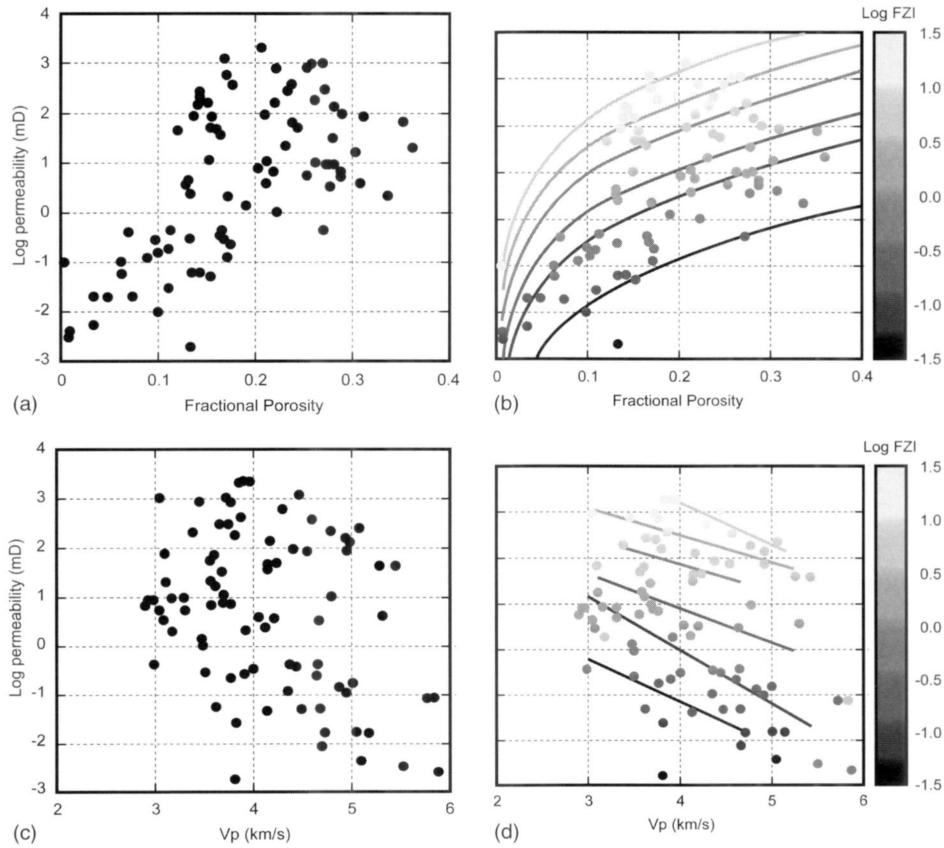


*Figure 13.17* a) Unsorted *log k* versus $\varphi$ data on the left, and b) sorted data using the expression for FZI with measured matrix parameters on the right c) Unsorted *log k* versus $V_p$ data on the left, and d) sorted data using FZI on the right. Prasad, 2003.

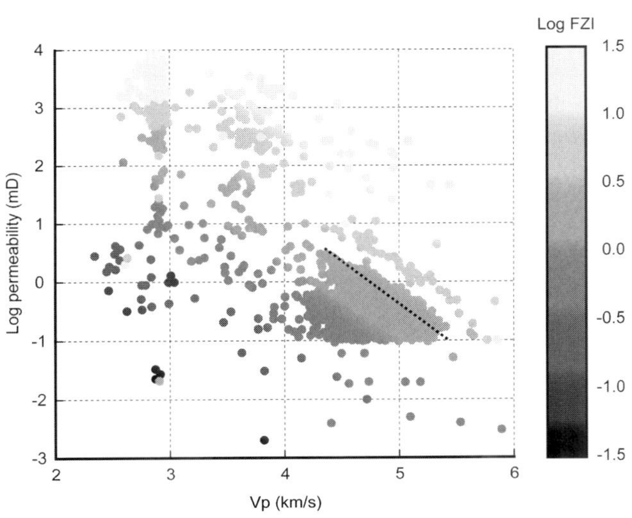


*Figure 13.18* Permeability estimated from velocity data, using FZI values from a larger data base of reservoir sandstones, marine reservoir sand and a tight sandstone. Prasad, 2003.

## 13.7 Velocity with patchy saturation effects in mixed units

Knight et al., 1998, showed that uniform or smooth variations of velocity with degree of saturation were strictly a function of an *assumed or actual homogeneous* distribution of saturation due to lithological uniformity. They investigated the more complex (and common) effects of having *mixed* lithological units, which tended to create a heterogeneous or *patchy*, saturation distribution, with different signatures during imbibition and drainage. With a more complex distribution of saturations due to lithology differences, it was only when close to saturation of 100%, that there was a consistent steep rise in velocity. (From another field, and for harder rock types, see Saito, 1981, in Figure 2.17a, b in Chapter 2 of this book).

The authors found that pore-scale and sample-scale fluid distribution effects, and of course capillary effects, caused different $V_p$ response (in degree but not general

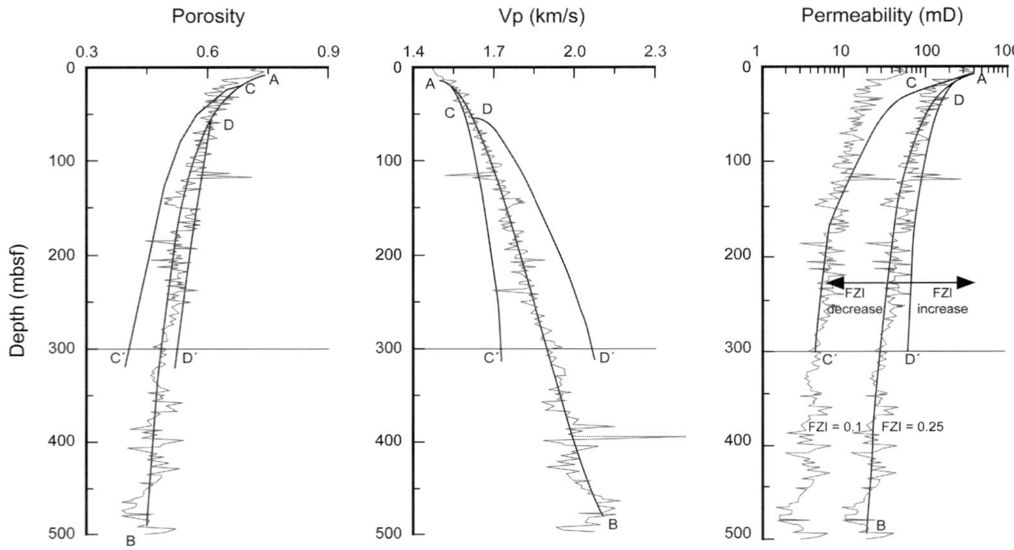


*Figure 13.19* An illustration by Prasad, 2003 of the possibility of predicting permeability from log data of porosity and permeability, using a relevant, logging-based FZI-value. Marine logging results from 'Site 977, ODP Leg 161 (Shipboard Scientific Party, 1996). Hypothetical mineralization from D (curve D-D') , and hypothetical pore-filling from C (curve C-C').

style) when draining as compared to imbibing. The drainage process creates a more heterogeneous distribution of saturation.

Local full saturation of the crack-like regions of the pore space tend to stiffen these regions in relation to high frequency, but at low frequency these 'patches' can drain to the less saturated pore space. The phenomenon appears to be shifted in frequency at 'macroscopic-patch' scale. Such results clearly impact poroelastic modelling with different frequencies, which is discussed in Chapter 15.

Knight et al., 1998, found that the pore fluid relaxation time increased as the size of the volume occupied by the fluid increased. They argued that the size of a patch (or partly drained rock joint?) may be orders of magnitude larger than a compliant pore. So if pore fluid is arranged in patches, the apparent ($V_p$) stiffening of partially saturated rock in response to a dynamic wave may occur even at low frequencies.

Examples of frequency-dependent differences in $V_p$-saturation response, for the case of carbonate samples, are shown at lower frequency (1 kHz) in the first example, in Figure 13.20a. Presumably the lower frequency allows the wave-induced pressure changes in the pore fluid to dissipate, so that the pore fluid pressure is very close to that of the high-compressibility gas in the dry pore space, as described by Knight et al., 1998.

As a result, the pore fluid lying in thin, compliant pores can flow freely into the dry pore space, in a squirt-flow type of attenuation response. It does not therefore allow reinforcement of the compliant part of the rock, so velocities are low.

At high frequencies, (Figure 13.20b), this pressure equilibrium cannot occur because the pore fluid relaxation time is greater than the seismic wave period. Pore fluid in the thin compliant pores is then effectively 'trapped', and it therefore reinforces the otherwise compliant pore spaces, resulting in higher apparent modulus and velocity.

Knight et al., 1998 described a comprehensive investigation with a controlled distribution of (10) lithologic units, an assumed state of capillary equilibrium, and calculated the saturation level in each unit from corresponding capillary pressure curves. Their contrasting calculations for shaley and clean sand, shown in Figure 13.21, demonstrate the effect of saturation heterogeneity.

The effect of 'patchiness' (of saturation), compared to homogeneous conditions is nicely demonstrated in Figure 13.22. The two parallel lines defining 'patchy' $V_p$-saturation response, are two different theoretical solutions given by Knight et al., 1998, in which just *two different lithologies* have been combined.

In Figure 13.23, the 'extremes' created by mixing the ten lithological units are shown. (Properties were given in Figure 13.21a). The smooth, conventional result (Figure 13.23a) was obtained by a pore-scale mixture of the ten sand-to-shaley-sand units, while the 'multi-stepped' response shown in Figure 13.23b was obtained using a patchy mixture of the ten lithological units.

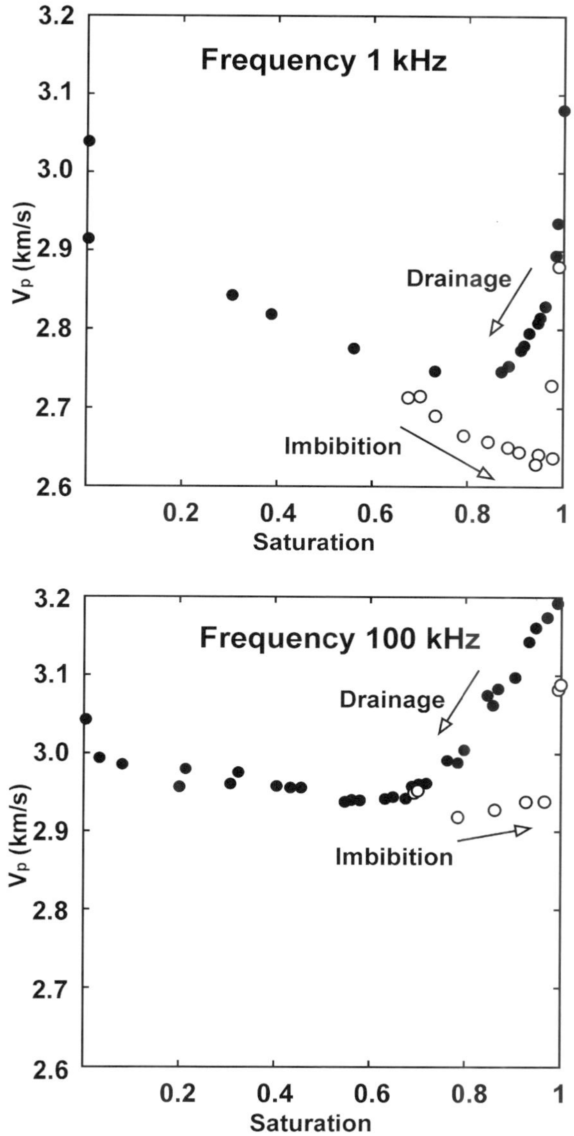


*Figure 13.20* P-wave velocity responses to a) lower and b) higher frequency in the case of a carbonate sample of 30% porosity studied by Cadoret, 1993. Note 'hardening' at high frequency, and the different imbibition and drainage responses with changing levels of water/air saturation, as described by Knight et al., 1998.

## 13.8 Dynamic Poisson's ratio, effective stress and pore fluid

Carcione and Cavallini, 2002, described modelling in relation to Poisson's ratio (the dynamic value), as a function of 'differential pressure' and pore fluid type. (The 'differential pressure' is the hydrostatic confinement minus pore pressure, and is referred to as effective stress in soil and rock mechanics.)

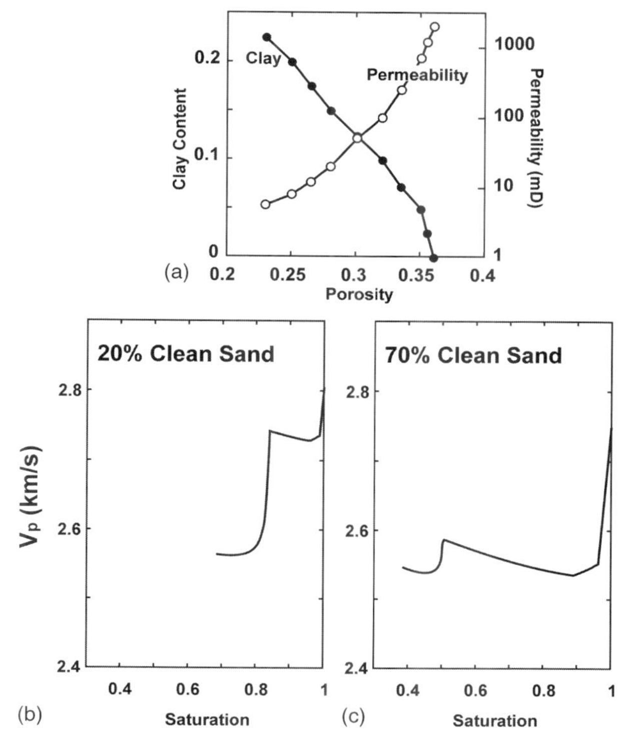


*Figure 13.21* a) Properties of 10 lithological units, showing clay content and permeability versus porosity. b) and c) P-wave velocity (calculated) versus saturation for mixtures of shaley and clean sand. Knight et al., 1998.

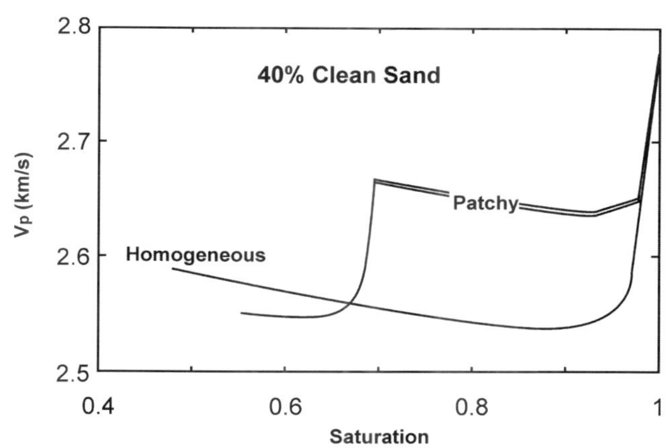


*Figure 13.22* P-wave velocity versus saturation in a 40% mixture of clean sand and shaley sand. The two lithologies were combined in a patch arrangement and as a homogeneous mixture. Knight et al., 1998.

The authors highlighted the fact that Poisson's ratios are anomalously high for cases of over-pressure, where *effective stress* can approach the fracturing (negative) side of the usual lithostatic and pore pressure gradients (as approached in Figure 12.1 of chapter 12).

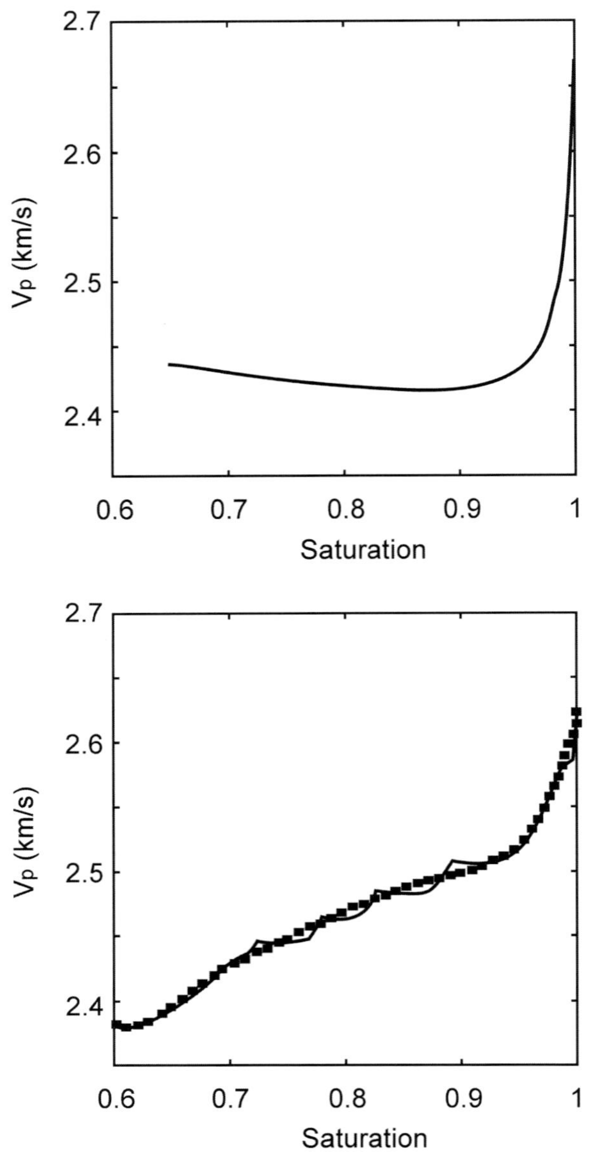


*Figure 13.23* P-wave velocity versus saturation for a) a pore-scale 'homogeneous' mixture of ten units, b) a patchy mixture of ten units, from sand to shaley-sand. Knight et al., 1998.

Dynamic Poisson's ratio, as we have seen in Part I, can be calculated from:

$$\nu = \frac{1}{2}\left(1 - \frac{1}{a-1}\right), \quad \text{where } a = \left(\frac{V_p}{V_s}\right)^2 \tag{13.9}$$

Carcione and Cavallini emphasised that it was the *aspect ratio* of the cracks and pores and the nature of the saturating fluid that determined $\nu$. Rock containing mainly stiff, equi-dimensional pores does not show major variations of $\nu$ with effective stress. However, closure of low aspect ratio (compliant) microcracks and pores will tend to increase the bulk modulus K more than the shear modulus $\mu$, having the following effect on ($\nu$) :

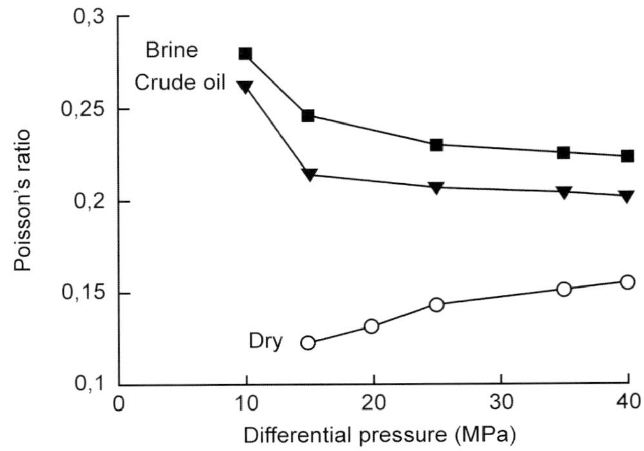


*Figure 13.24* Some experimental results for dynamic Poisson's ratio, as a function of effective stress (hydrostatic confinement minus pore pressure) and fluid type. Khazanehdari et al., 1998, as reproduced by Carcione and Cavallini, 2002.

$$\nu = \frac{1}{2}\left(1 - \frac{1}{1/3 + K/\mu}\right) \tag{13.10}$$

Therefore, in dry rocks, dynamic Poisson's ratio increases with differential or effective stress. However, in saturated rocks the compliant pores have become stiffened in relation to high frequency waves, so ($\nu$) changes less as effective stress increases. However, at low effective stress, or when pore pressures are very high, the effective stress sensitivity is marked, and ($\nu$) increases.

Some experimental results that the authors quoted, from tests by Khazanehdari et al., 1998, are shown in Figure 12.24, and indicated the increasing sensitivity at low effective stresses. Carcione and Cavallini were particularly interested in the responses of ($\nu$) close to the hydraulic fracturing limit when sealed over-pressured beds were under-compacted, and where there could be excess pressure due to oil-to-gas conversion, as investigated by Carcione and Gangi, 2000. As they pointed out, at zero effective stress, $V_s$ is (locally) zero as the rock mass is hydraulically fractured and load is born by the fluid. However $V_p$ is not zero, therefore the ratio (a) in equation 13.9 tends to infinity and $\nu \rightarrow 0.5$.

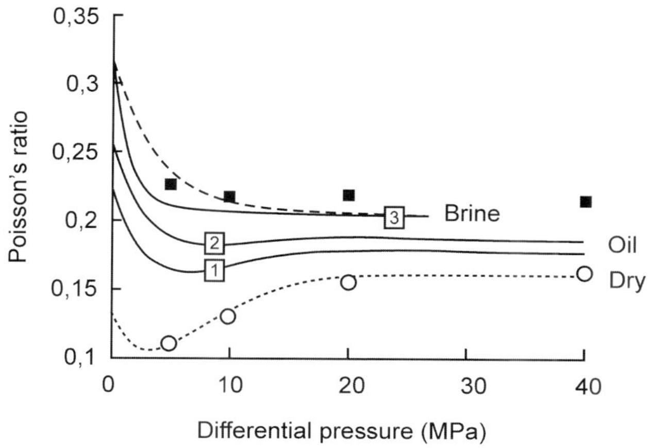


*Figure 13.25* Dynamic Poisson's ratio $\nu$ versus differential stress (hydrostatic confinement minus pore pressure). The squares and open circles are from Winkler, 1985. Curves marked 1, 2 and 3 were calculated results for full oil saturation, 50% oil saturation and full brine saturation, respectively. Carcione and Cavallini, 2002.

Carcione and Cavallini used the theory of Carcione and Gangi, 2000, to develop theoretical trends for Poison's ratio increase when approaching zero effective stress, showing in Figure 13.25, similar trends to the available experimental results.

## 13.9 Dynamic moduli for estimating static deformation moduli

In engineering fields that involve design for building foundations, dams or tunnels, or indeed wellbore stability for hydrocarbon exploration and production, it is well documented that static moduli of deformation can be significantly lower than the dynamic moduli predicted from P- and S-wave velocities. (See Chapter 6, for extensive rock engineering based comparisons of these quantities.) Such differences will also tend to be greater in jointed *in situ* rock, as opposed to microcracked lab samples.

Such differences in moduli are of obvious interest to the petroleum industry. Tutuncu et al., 1998, expressed the opinion that knowledge of non-linear elastic properties (that are largely responsible for the differences between dynamic and static moduli) is essential for optimal drilling, effective well completions and efficient reservoir management. For example, when applying distinct element modelling to the Ekofisk reservoir subsidence in the mid 1980s, (Barton et al., 1986), it was appreciated that there were obvious uncertainties about the choice of moduli of deformation for the various layers of

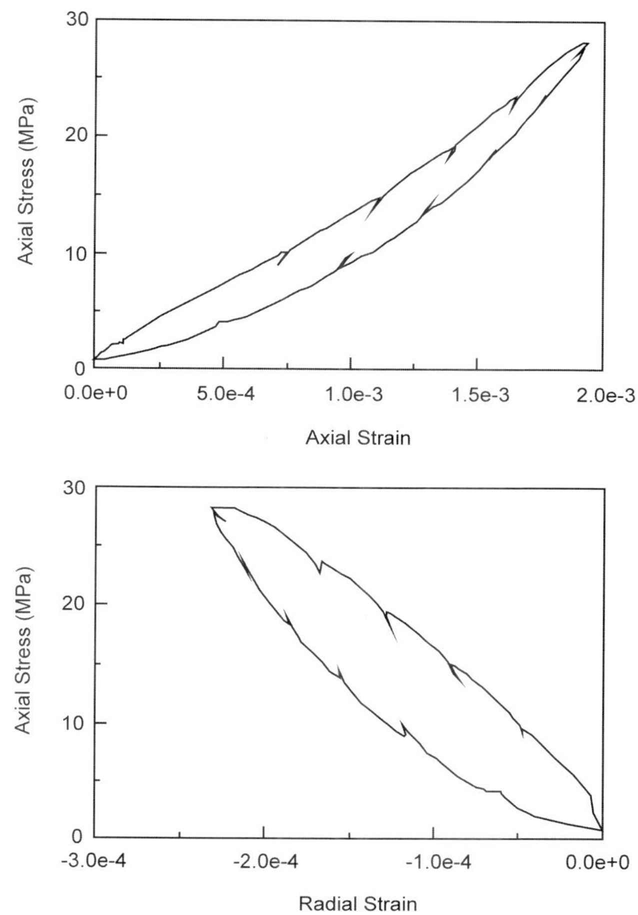


*Figure 13.26* Non-linearity and hysteresis observed in uniaxial cycling (with constant differential stress). The rock was a tight gas sandstone, tested dry. Tutuncu et al., 1998.

the 3 km of overburden, which were inevitably velocity-based at that time. Only the reservoir itself (chalk) was core-sampled and laboratory tested.

Based on laboratory observations of the elastic *non-linear* behaviour for sandstones as illustrated in Figure 13.26 a and b, Tutuncu et al., 1998, showed that the frequency of measurement was all important for the geophysical estimate obtained, since $E_{ultrasonic} > E_{log} > E_{low\ freq.} > E_{static}$. Their ultrasonic laboratory measurements were conducted at 1 MHz, 180 kHz, 100 kHz and 50 kHz, and their low frequency measurements at 2 kHz to 1 Hz, and their static measurements at 0.05 Hz to 0.001 Hz.

To understand the frequency dependence of the various 'dynamic' moduli, it was necessary to see the effect of frequency on the velocities of the P- and S-waves. Firstly, for the case of dry porous sedimentary rocks, it was generally concluded by the authors that dynamic

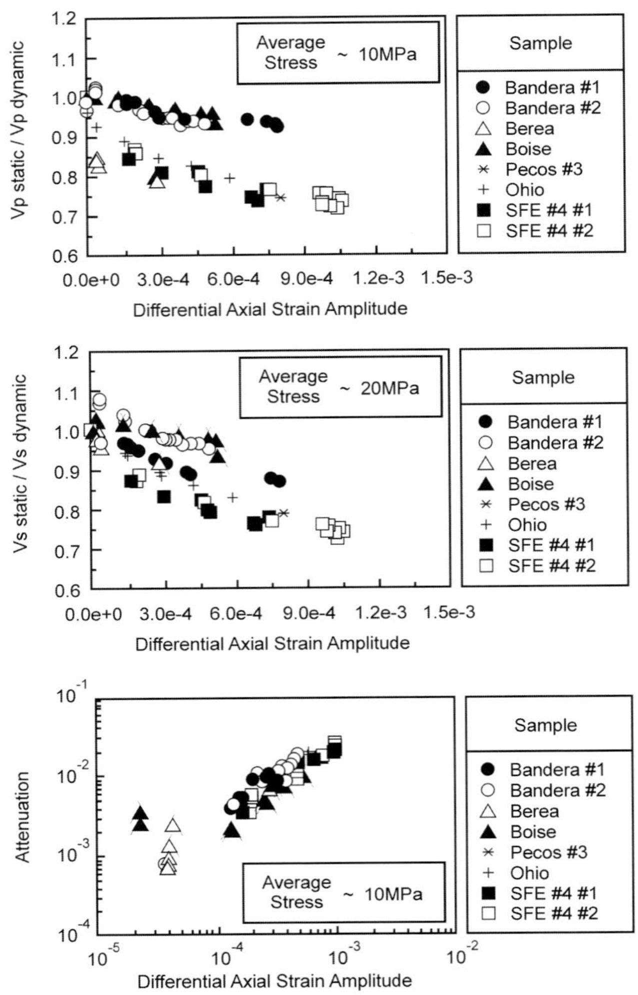


*Figure 13.27* Ratios of $V_p$ *static/*$V_p$ *dynamic*, and $V_s$ *static/*$V_s$ dynamic, and attenuation, each as a function of strain magnitude. Tutuncu et al., 1998.

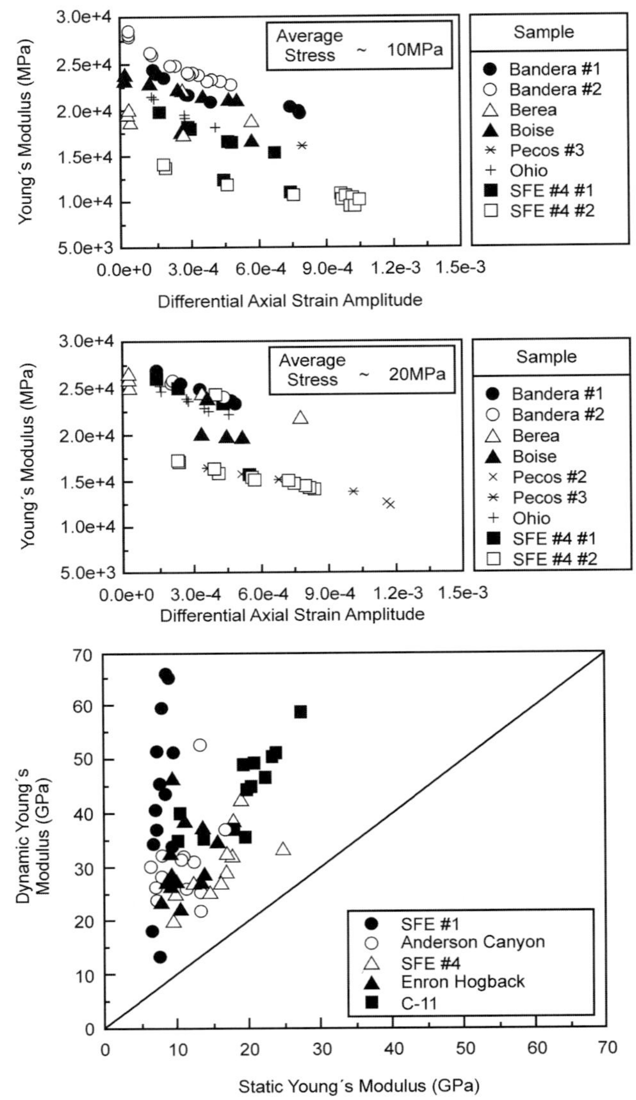


*Figure 13.28* Young's modulus as a function of strain amplitude at two different confining stress levels, and a comparison of E *dynamic*/E *static*. Tutuncu et al., 1998.

tests gave results that were independent of frequency below ultrasonic frequencies. However, when ultrasonic frequencies were approached (>0.1 MHz) the wave length could become comparable to the grain size and scattering became an important attenuation mechanism, which increased with frequency and when heterogeneities were present.

In fluid saturated porous rocks, $V_p$, $V_s$ and attenuation depend on frequency even when scattering is negligible. Since $V_p$ increases with frequency more than $V_s$, the resulting $V_p/V_s$ increase may give a pronounced increase in the dynamic modulus estimate. Under the more rapidly oscillating loads, the fluid in the pores and grain boundary cracks are not allowed sufficient time to (micro-) flow or 'squirt', and the rock acts as if unrelaxed so the properties measured will be undrained and stiffer. By contrast, low-frequency measurements give sufficient time for fluid transfer and squirt to occur from microcracks, so that a relaxed, drained or less stiff behaviour is registered. Squirt flow appeared to be the dominant mechanism for attenuation and velocity dispersion at frequencies from 100 Hz to 10 kHz.

When strain amplitude was increased (as in static measurements), the good agreement between wave propagation models and experimental data broke down, and large discrepancies were experienced between measured and predicted velocities and attenuations.

One of the main objectives of the Tutuncu et al., 1998, study was to develop a methodology to predict hard-to-measure static (low frequency) moduli and attenuation from the relatively easy to measure dynamic moduli. They therefore conducted stress cycling measurements at various stresses and cycling frequencies. The fact that $V_{ultrasonic} > V_{log} > V_{static}$ is emphasised by their plots of $V_p$ 'static'/$V_p$ dynamic versus differential strain amplitude. A typical set of $V_p$, $V_s$, Young's modulus and attenuation data plotted in this format is shown in Figure 13.27 a, b and c.

For the above general reasons, when comparing dynamic Young's modulus to 'static' Young's modulus, Tutuncu et al., 1998, were able to show large discrepancies, since the 'static' measurements (0.05 to 0.001 Hz) were at so much higher strains than low strain dynamic measurements. Results are shown in Figure 13.28, which resemble rock engineering data shown in Chapter 6.

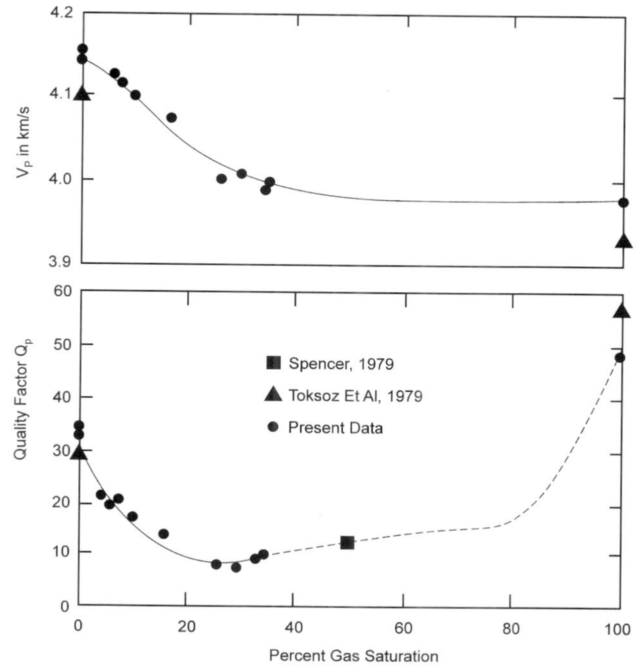


*Figure 13.29* Compressional wave velocity and seismic quality factor $Q_p$ as a function of gas/brine saturation for sandstone at 10.3 MPa effective stress. Frisillo and Stewart, 1980.

## 13.10 Attenuation due to fluid type, frequency, clay, over-pressure, compliant minerals, dual porosity

In this section we will trace parts of the development of attenuation as a means of improved characterization of reservoir rocks. The dispersive, frequency-dependent nature of seismic Q, and the greater sensitivity of the ratio of $Q_s/Q_p$ to fluid and partial saturation than $V_p/V_s$, as already reviewed briefly in Chapter 10, will be illustrated by interesting cases reported in both past and recent literature.

### 13.10.1 Comparison of velocity and attenuation in the presence of gas or brine

Some of the subtle differences between velocity and attenuation (in the form of seismic Q) were shown by Frisillo and Stewart, 1980 tests with *variable* gas/brine saturations. The tests on Berea sandstone ($n = 19.7\%$, $k = 376$ mdarcy) shown in Figure 13.29 give $V_p$ and $Q_p$ on the vertical axes, and percent gas saturation on the horizontal axes. Zero percent gas saturation corresponds to 100% brine saturation.

Frisillo and Stewart's data represented by black circles, is shown supplemented by some reasonably consistent data (also for Berea sandstone) reported by Spencer, 1979 (squares), and by Toksöz et al., 1971 data for dry or full saturation (triangles). The expected reduction in $V_p$ by reduced brine saturation (and increased gas saturation) was matched *initially*, by greater attenuation (a $Q_p$ of 33, reducing to approximately 9). At the far end of the saturation scale, when the samples become 'room dry' and reached 100% saturation with nitrogen, the attenuation reduced sharply, and $Q_p$ reached a value of about 50. Clearly this is related to the eventual absence of squirt flow with increased dryness.

### 13.10.2 Attenuation when dry or gas or brine saturated

In 1979, Toksöz et al. presented ultrasonic laboratory data on dry and water- or brine-saturated rocks, investigating how the attenuation varied with 'differential' pressure. Pore fluid pressure ($P_f$) and confining pressure ($P_c$) on their specimens was controlled independently, $P_c - P_f$ giving their quoted 'differential' pressures. In a companion paper, reviewed in Chapter 10, Johnston et al., 1979, presented the assumed mechanisms of attenuation, and formulated theoretical models that fitted this laboratory data. The present data was obtained at ultrasonic frequencies (0.1 to 1.0 MHz), using a Berea sandstone with 16% porosity.

Toksöz et al., 1979 used a frozen rock (limestone), showing very small attenuation and very high $Q_{seis}$, as a

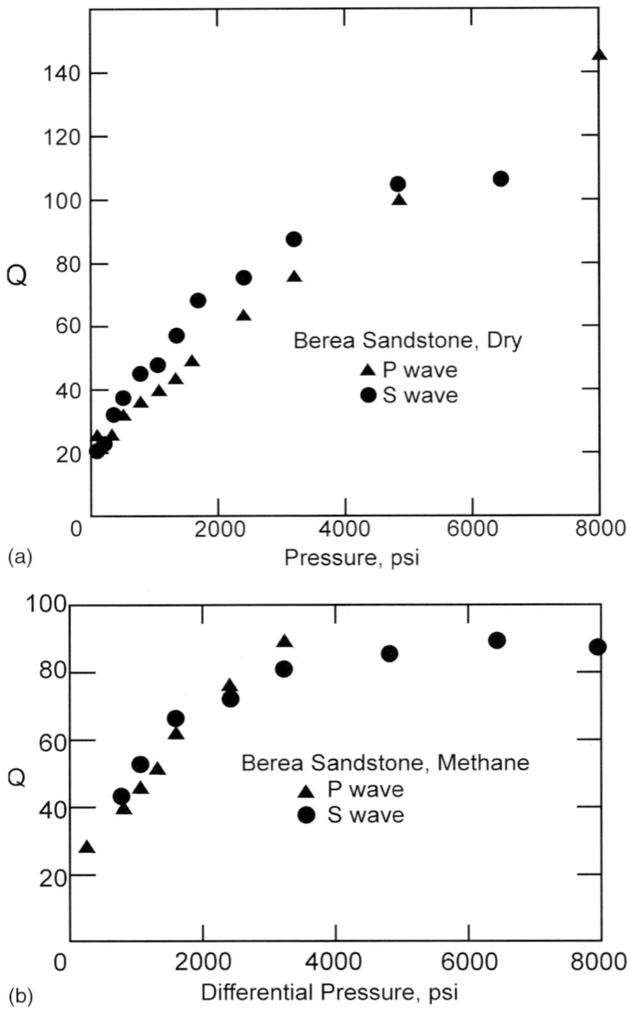


*Figure 13.30* $Q_p$ and $Q_s$ values as a function of confining (or differential) pressure, in a) dry ($P_f = 0$) and b) methane saturated ($P_f = 0.465\ P_c$) Berea sandstone. Toksöz et al., 1979.

reference standard. Their subsequent detailed studies of environmental effects such as dry, methane-saturated, or brine-saturated, were conducted on Berea sandstone.

At zero pressure, the P-wave velocities of their dry and brine-saturated Berea sandstone were about 3.3 and 3.8 km/s respectively, rising rapidly to about 4.1 and 4.2 km/s at 3000 psi (about 21 MPa). At the highest differential pressure used there was little difference between the dry and brine-saturated $V_p$ in relation to the differentiation of the dry or brine-saturated condition. These and similar velocity trends for methane and for other brine concentrations were used in their calculation of seismic Q, which are shown in Figure 13.30 and 13.31, for the case of dry, methane-saturated, and two NaCl concentrations.

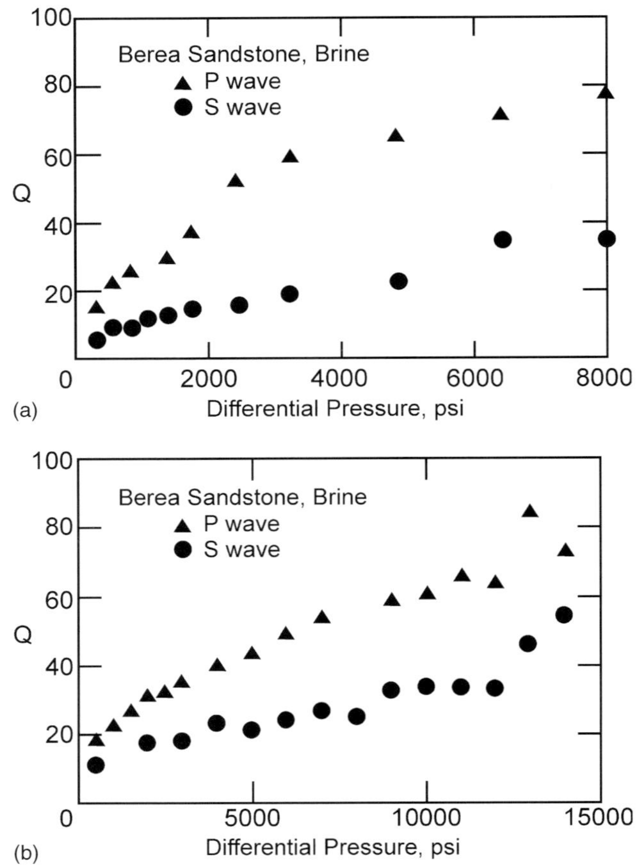


*Figure 13.31* $Q_p$ and $Q_s$ values as a function of differential pressure ($P_f = 0.465\ P_c$) for brine-saturated Berea sandstone, with NaCl concentrations of a) 67.2 ppm and b) 161.3 ppm respectively. Toksöz et al., 1979.

The anomalous high value of $Q_p$ at highest differential pressure in Figure 13.30a was assumed to be due to some pore collapse and locking of grain boundaries. Several important trends can be seen in the Q data. Firstly, $Q_p$ and $Q_s$ were both higher (less attenuation) in the dry or methane saturated states, than in the case of brine saturation. Furthermore, $Q_s$ was more often larger than $Q_p$ in the case of the dry and methane-saturated sandstone. In contrast, there was a consistently wide separation of $Q_p$ and $Q_s$ of some 100 to 150% ($Q_p > Q_s$) in the case of the brine saturation.

### 13.10.3 Effect of frequency on velocity and attenuation, dry or with brine

Winkler, 1983a, provided a remarkably detailed collection of velocity and attenuation data for three sandstones when tested in triaxial compression, over a range of frequencies (400 to 2000 kHz) in a dry or brine-saturated state. Effective stresses were increased in the sequence 2.5, 5, 10, 20 and 40 MPa in the case of the dry samples.

The three sandstones had the following basic physical properties:

| | Massilon sst. | Berea sst. | Boise sst. |
|---|---|---|---|
| Porosity % | 24,6 | 20.3 | 24.9 |
| Permeability (mD) | 1425 | 107 | 286 |
| Grain size (μm) | ≈200 | 150–200 | 150–200 |
| Quartz % | 94 | 80 | 46 |

(Note: 1 Darcy = $10^{-12}$ m$^2$ or $10^{-5}$ m/s, approximately)

The three 'sandstones' were of Mississippian age (first two) and Pliocene (last), with geological descriptions quartose sandstone, greywacke and arkose, respectively. Grain densities varied only from 2.68, 2.65 to 2.63 gm/cm$^3$, respectively.

The frequency-dependent, saturation-dependent and effective stress-dependent variations of Vp and attenuation (1/Q) measured by Winkler in this study, are reproduced in Figure 13.32 (a to f ).

The dry samples all showed *negative velocity dispersion*, meaning velocity decreasing with increasing frequency, while the attenuation increased as the third to *fourth* power of frequency. This was taken as evidence of scattering within the pore spaces between the grains. (Inter-grain scattering in aluminium reportedly also results in $f^4$ attenuation.)

The brine-saturated 'sandstones' mostly showed slight, positive, velocity dispersion (at least at the lower confining pressures) while attenuation varied with only the first or second power of frequency. This change in attenuation-frequency dependence was taken as evidence of local fluid-flow loss mechanisms. The saturated rocks always showed much stronger attenuation (lower $Q_p$ ) than the dry samples.

All the data sets showed the strong influence of effective stress, which had greatest influence on attenuation when the samples were brine-saturated and at the lowest levels of effective stress, as we have seen earlier in this chapter.

Parallels to *compliant* joints affecting rock mass behaviour (Chapters 15 and 16), were the observations by Winkler, 1983b, that the nature of the grain contacts was all important. Compression and dilation of relatively compliant contacts (and strong pressure dependence) in the case of the three sandstones could readily induce fluid flow into and out of the contact regions.

Interestingly, the reported behaviour of fused glass-bead samples was entirely different, as the sintered contacts between the glass beads were very stiff and lacked pressure dependence, and would not therefore generate local (micro) fluid flows (squirt) which would have increased the attenuation.

Figure 13.33 shows this contrasting $V_p$ and attenuation behaviour for the fused glass-bead samples, which had spherical 177–210 μm beads (initially) that were fused by heating to give a porosity of 26.6% and a permeability of 'several' Darcies (1 Darcy = $10^{-12}$ m$^2$), i.e. it was significantly more permeable than the three 'sandstones', yet could not generate squirt-related attenuation.

### 13.10.4 Attenuation for distinguishing gas condensate from oil and water

On the basis of the differentiation of $Q_p$ values, listed below, Klimentos, 1995, suggested a way of distinguishing *gas and condensate* from *oil and water* in sandstone reservoirs, but at the same time questioned whether the method could also be used in carbonate reservoirs.

1. in perfectly dry rocks, $Q_p$ is very high
2. in fully liquid saturated rocks $Q_p$ is at an intermediate level
3. in partially saturated rocks $Q_p$ is low

Ranges of $Q_p$ and $Q_s$ for sandstone reservoirs were reportedly as follows based on well log, i.e. sonic frequencies:

- Gas and gas-condensate bearing sandstones $5 < Q_p < 30$ $15 < Q_s < 30$
- Oil bearing sandstones $8 < Q_p < 100$ $15 < Q_s < 50$
- Water bearing sandstones $9 < Q_p < 100$ $15 < Q_s < 50$

Cross-over of P-wave and S-wave attenuation (or cross-over of $Q_p$ and $Q_s$, due to the increase in attenuation of P-waves by gas (lower $Q_p$) and the absence of effects on the S-wave attenuation (as above), was the basis of the

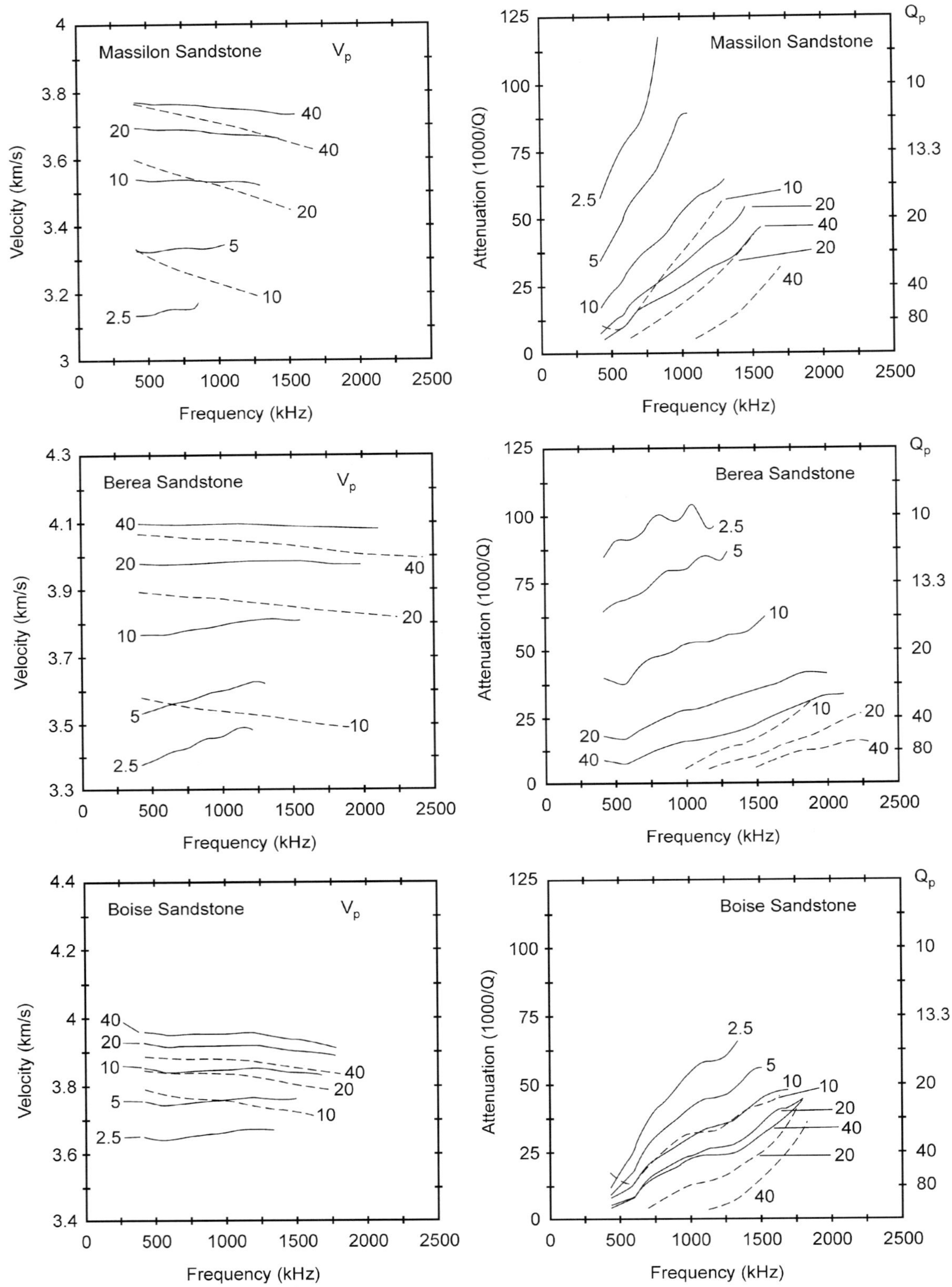


*Figure 13.32* Compression wave velocities and attenuations ($1000/Q_p$) as a function of frequency, and whether dry (dashed line) or brine saturated (solid line), for three 'sandstones'. Numbers on the curves represent the applied effective pressures (MPa), Winkler, 1983a. Note interpreted $Q_p$ scale added on right-hand axis for convenience.

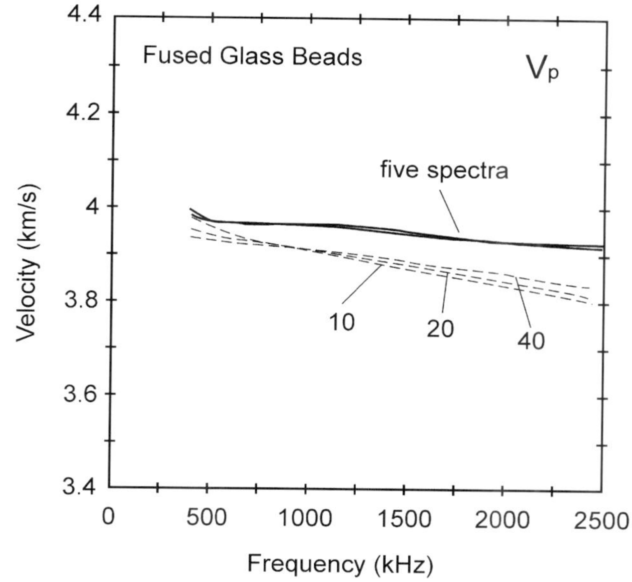


*Figure 13.33* $V_p$ and attenuation as a function of frequency, and effective stress, for dry (dashed line), or brine-saturated (solid line) fused glass beads. Winkler, 1983b.

method suggested. Typical sets of well data such as the following were cited:

| | | | $V_p$ km/s | $Q_p$ | n% |
|---|---|---|---|---|---|
| 1. | Gas-bearing sandstone | (2.5 km depth) | 4.0 | 5 | 15 |
| 2. | Water-bearing sandstone | (2.5 km depth) | 4.2 | 40 | 15 |

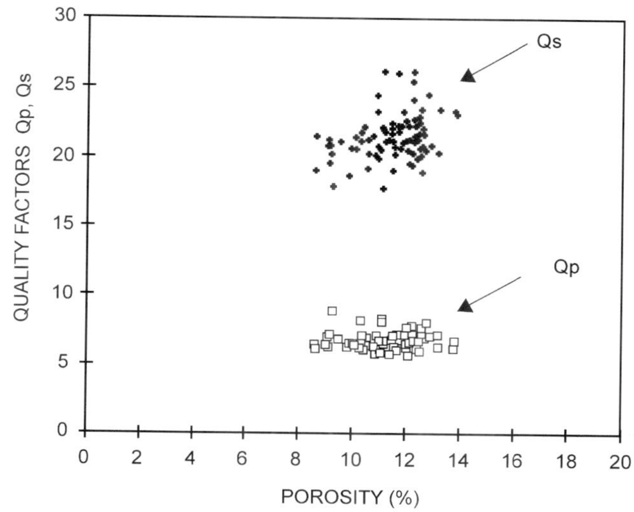


*Figure 13.34* Indication of a gas zone in a well, due to $Q_s > Q_p$. Klimentos, 1995.

A specific set of data for $Q_s$ and $Q_p$ as a function of porosity in a *gas zone* in one of the wells analysed is shown in Figure 13.34. The low $Q_p$ value for the gas-bearing sandstone suggests the presence of partial saturation with a liquid phase, and possibly higher pore pressure as well.

### 13.10.5 Attenuation in the presence of clay content

It has previously been established that compressional wave velocities are inversely proportional to the clay content of sandstones (see previous section), with softening of the sandstone matrix and reduced dynamic deformation modulus as a result. (Han et al., 1986). In the study of Klimentos and McCann, 1990, the first systematic study of intrapore clays on compressional wave *attenuation* seems to have been made. These authors noted that there was a general lack of correlation between porosity and permeability for clay-bearing sandstones, but in general low permeability was associated with high clay content, and high permeability with low clay content.

Although the Biot, 1956a and Biot, 1956b theory accounted well for attenuation in clay-free sandstones, it apparently failed by an order of magnitude to account for the attenuation effect of the clay. Klimentos and McCann suggested, as others since then, that this strong clay-related attenuation was due to viscous interaction between the clay particles and the pore fluid. Since the permeabilities were strongly dependent on the

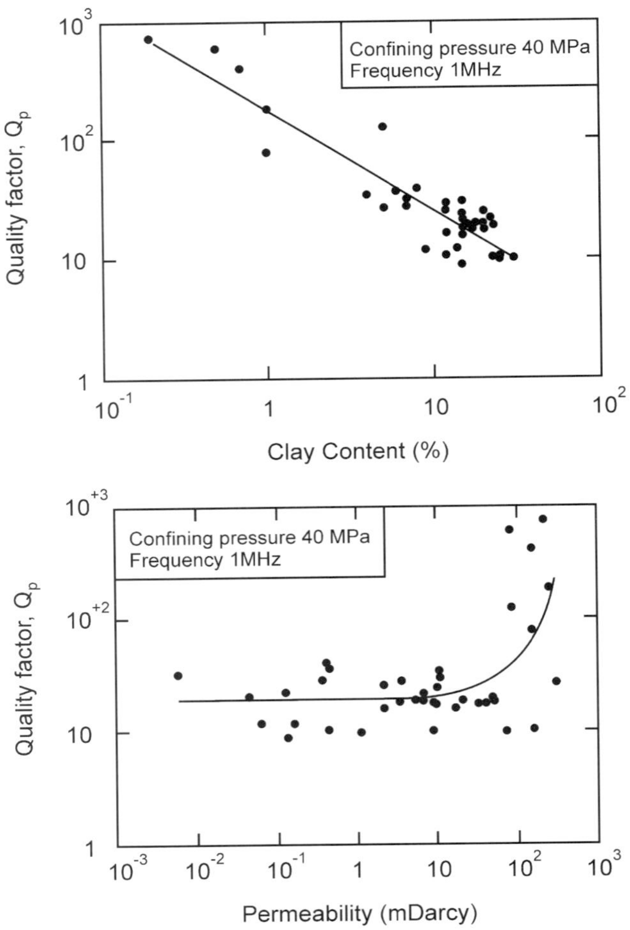


*Figure 13.35* Top: Seismic quality $Q_p$ versus volumetric clay content for 39 sandstones with porosities from 6 to 36%. Bottom: Permeability (which is dominated by clay content) versus $Q_p$. Confining pressure was 40 MPa in each case. Klimentos and McCann, 1990.

clay-content, the dependence of the attenuation on the permeability arose from the over-riding influence of clay-content. One may also add here that any possible thin layering of clay-rich horizons could, in an *in situ* environment, cause an increase in the shear compliance. Clay-bearing discontinuities tend to have low ('static') shear stiffness and high (dynamic) shear compliance.

Klimentos and McCann, 1990, measured the attenuation of compressional waves in 42 sandstones under a confining pressure of 40 MPa. The frequency range was limited to 0.5–1.5 MHz, using a pulse-echo method. The objective was to investigate the role of porosity, clay content and permeability on the attenuation. Intrapore clays were naturally found to be important in causing attenuation and in modifying the permeability.

Figure 13.35 shows the strong influence of clay content (% by volume) on $Q_p$, and a less clear relation of $Q_p$ with permeability. The data set excluded the few samples with zero clay content, as they could not be plotted on the logarithmic scale.

$Q_p$ and clay content showed excellent correlation. The best fit statistical relationship was:

$$Q_p = 179C^{-0.843} \tag{13.11}$$

where C was the percent clay content by volume. The correlation coefficient was 0.91.

### 13.10.6 Attenuation due to compliant minerals and microcracks

A number of important trends of behaviour regarding potential mechanisms of attenuation behaviour in the presence of compliant minerals and joints under the effect of confinement, have been revealed by researchers at Imperial College, using ≈260 m deep research boreholes, located in Northern England. The near surface sedimentary series of rocks (sandstones, siltstones and limestones)have been extensively investigated, both in laboratory, and with multi-frequency field surveys.

The water-saturated microcracked clean sandstones, and their in situ counterparts (bedded/jointed sandstones) exhibited strong pressure dependence/depth dependence respectively, with strong decreases in attenuation, and increases in $Q_p$ at higher pressures. (Best and Sams, 1997; Best, 1997).

This effective stress dependence was attributed to reduced squirt flow as pressure rose, due to partial closure of joints and bedding planes. Other sandstones and siltstones at the site had varying proportions of clay and kerogenic organic matter, which seemed to reduce sensitivity to pressure; $Q_p$ was quite low even at high pressure, which was attributed to 'clay squirt flow', Best 1997. Data from Best et al., 1994 and Best, 1997, combined in Figures 13.36 a and b, shows systematic reduction of $Q_p$ (from 80 plus to about 20) as the percentage of compliant minerals in sandstones and siltstones increased from a few percent to nearly 80%.

Best and Sams, 1997 speculated that clay squirt flow would be an important mechanism at both seismic and sonic frequencies, if larger scale geologic features were involved, such as inter-bedded permeable and impermeable layers.

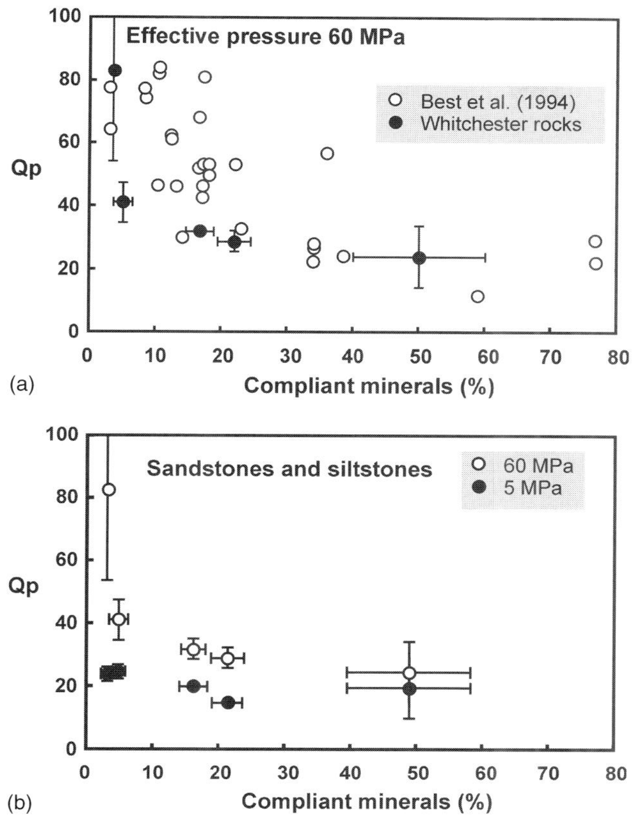


*Figure 13.36* Ultrasonic data for siltstones and sandstones; $Q_p$ versus compliant mineral content, showing the effect of confinement. Best, 1997 and Best et al., 1994.

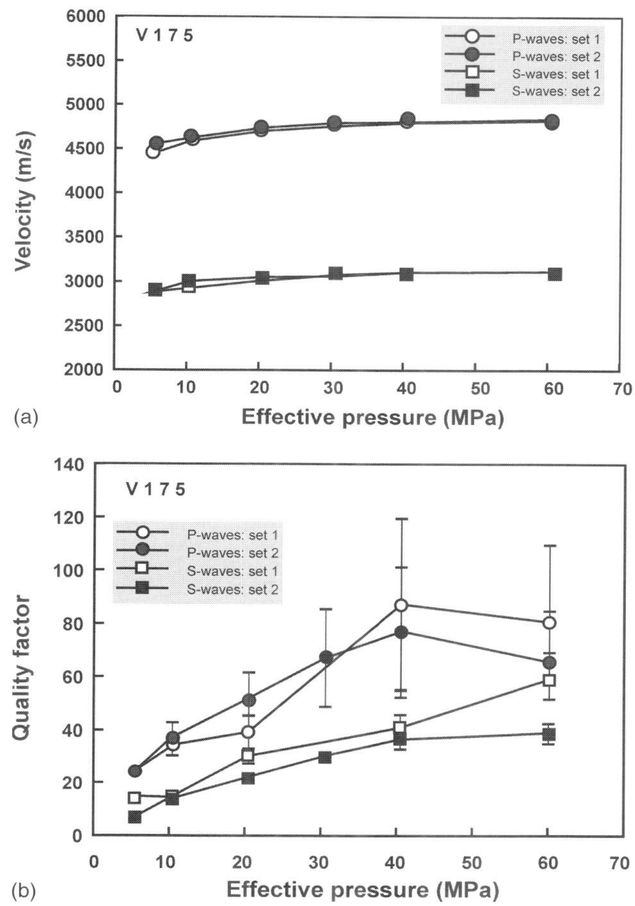


*Figure 13.37* Comparative $V_p$, $V_s$ and $Q_p$, $Q_s$ responses of a saturated, microcracked sample of sandstone to confining pressures varying from 5 to 60 MPa. Best, 1997.

From a rock mechanics/rock engineering point of view it would seem important to measure the deformation properties imparted by these increased contents of compliant minerals. In the Q-value world of rock mass engineering quality, increased compliant mineral content would mean reduced uniaxial strength $\sigma_c$ and therefore reduced $Q_c$ ($=Q \times \sigma_c/100$) and modulus.

There were clearly significant differences in the elastic deformation properties of the siltstones, sandstones and limestones as intact materials. Best, 1997, mentioned a maximum range of instantaneous sample shortenings of 0.74% (siltstone), 0.18% (sandstone) and 0.05% (limestone) as a result of applying the 60 MPa confining pressure. Moduli of 8.1, 33.3, and 120 GPa are implied. P-wave velocities for the laboratory samples were about 3.4 to 3.7, 3.6 to 4.6 and 6.0 to 6.3 km/s respectively.

In the case of a clean, saturated sandstone pervaded by microcracks, Best and Sams, 1997 and Best, 1997, showed that $Q_p$ was a more sensitive indicator of the effect of confining pressure than $V_p$. The comparative ultrasonic responses of velocity and attenuation to confining pressures of up to 60 MPa are compared in Figure 13.37.

The sensitivity shown alerts one to the possible complication of sampling-induced microcracking (i.e. related to the initial stress release when coring), which is ever-present in laboratory samples acquired from anisotropic-stress environments, if tested at artificially low confining pressures.

At 5 MPa confining pressure (close to the estimated in situ confining pressure at the borehole site), $Q_p$ was $24 \pm 2$, increasing to $83 \pm 29$ at 60 MPa. The grain contact microcracks were apparently closing beyond about 40 MPa, with a consequent reduction in squirt flow related attenuation.

The most interesting set of data from the point of view of rock mass quality (as opposed to rock matrix quality), and in relation to attenuation behaviour is shown in Figure 13.38. This strictly field data (belonging to

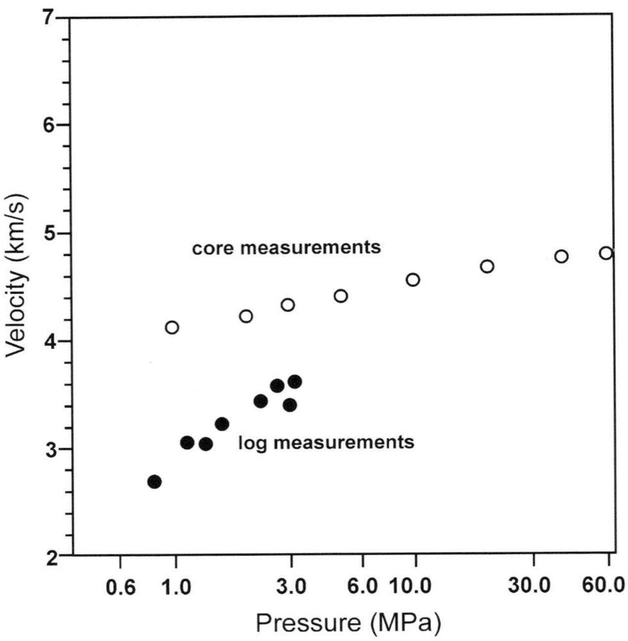


*Figure 13.38* Comparison of core (Whitchester sandstones) and downhole sonic log measurements, at equivalent pressures and depths. Best and Sams, 1997.

Chapter 14) is presented here for the sake of continuity. Presumably an 'altered zone' (stress-fracturing and mud-filtrate invasion: Chapter 12) that tends to affect sonic logging data at reservoir depths, would be only a very minor factor at this shallow 260 m deep well.

The figure shows the stress dependence of the micro-cracked laboratory specimen (mild) and the stronger response to pressure of the bedded/jointed in situ structures. Best and Sams, 1997 did not comment on the relative bedding/jointing frequency at the different depths in the borehole. However, the pressure/depth-velocity gradient shown in Figure 13.38 can be estimated to be approximately:

$$\frac{3.6 - 2.7}{3.0 - 0.8}\frac{(\text{km/s})}{(\text{MPa})} \approx \frac{0.9}{0.09} \approx 10\ \text{s}^{-1}$$

In the above, the pressure increment 2.2 MPa has been converted to an approximate 90 m depth (=0.09 km) to give the conventional $V_p$-depth gradient in $s^{-1}$ units. This steep gradient suggests quite a low rock quality Q-value, representing the soft porosity in this bedded sedimentary strata, which responds strongly to pressure increase at shallow depth. In Chapter 11, Figure 11.72, an empirical $V_p$-$Q_c$-depth trend was shown, that suggests, for velocity-depth gradients to 250 m that are in excess of $5\ s^{-1}$, that $Q_c \leqslant 0.1$.

A typical finely-layered sequence of limestones, sandstones, siltstones and mudstones, with bedding and perhaps a set of bed-limited joints, could be expected to have a 'poor-quality' rock quality Q-value, lowered by bedding planes (counted as one set, as is customary, and possible cross-bedding, giving a moderate RQD. Softer inter-beds would effectively increase $J_a$ (See Appendix A for a description of these terms.)

We may estimate $Q \approx 20/4 \times 1.5/4 \times 0.66/1$, or about 1, certainly to the nearest order of magnitude. Together with a rough estimate of uniaxial strength of closer to 10 to 40 MPa than either 1 or 100 MPa, the resulting $Q_c$ value would be very roughly 0.1–0.4, making velocity of about 3.5 km/s at about 250 m depth easily understood, if the matrix porosity was about 10% (refer to Figure 11.73).

A medium low rock quality Q-value, typical for sedimentary inter-bedded strata, and a correspondingly lower rock quality $Q_c$ value, would give significantly reduced rock mass static modulus of deformation $E_{mass}$ (or M), Barton, 1995, from the relation:

$$M = 10 \times Q_c^{1/3} \tag{13.12}$$

Values of M would be in the range 4 to 7 GPa, for the case of $Q_c = 0.1$–$0.4$. This 'static' modulus actually has the appearance of correlating, in broad terms, with the lowest values of seismic $Q_p$ obtained at lowest frequency. In Chapter 10 we saw frequency-dependent seismic Q values as low as 9 to 12 for this in situ site (Figure 10.35), from sonic log based (8 to 24 kHz) measurements.

### 13.10.7 Attenuation with dual porosity samples of limestones

Assefa et al., 1999, conducted ultrasonic (0.7 to 0.85 MHz) compressional and shear wave attenuation measurements on forty, 5 centimetre-sized samples of *water- or oil-saturated* oolitic limestones, at 50 MPa effective hydrostatic confining pressures ($\sigma' = 55 - 5$ MPa pore pressure = 50 MPa). They found that attenuation reached a maximum value in the samples which had the most fully developed 'dual-porosity'. This dual porosity consisted of inter-particle macro-pores (dimensions up to 0.3 mm) and micro-pores (with dimensions of 5 to 10 microns). They interpreted this increased attenuation (low $Q_{seis}$) as evidence of a squirt-flow mechanism, as

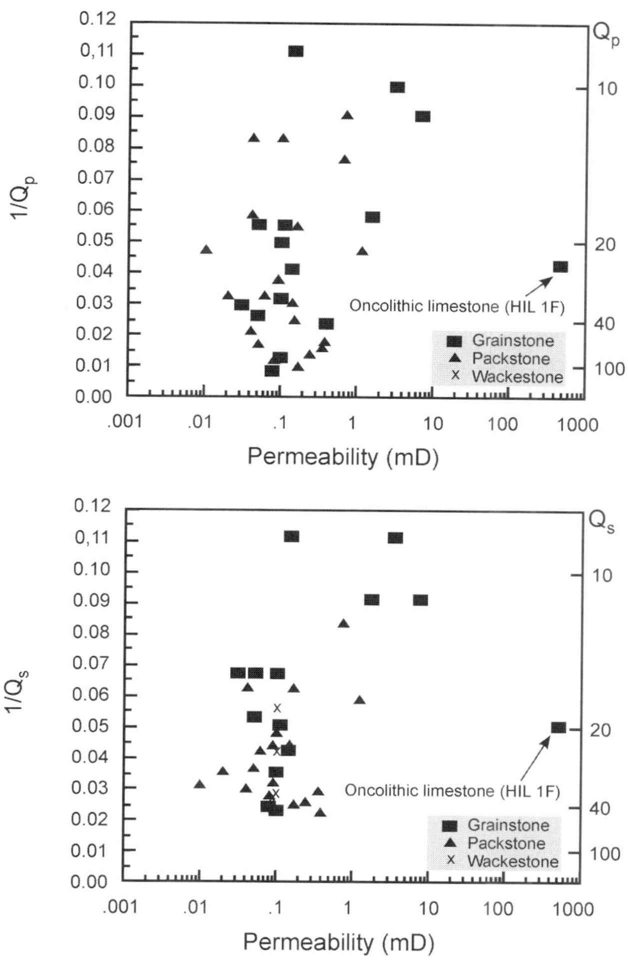


*Figure 13.39* Attenuations ($1/Q_p$ and $1/Q_s$) for water saturated limestones, as a function of (helium) porosity (range 3 to 17 %) and a roughly three-orders-of-magnitude permeability scale. Assefa et al., 1999.

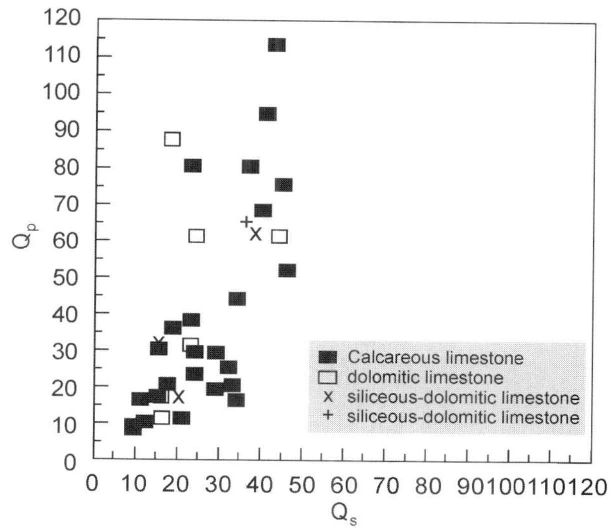


*Figure 13.40* Compressional-wave seismic quality factor ($Q_p$) versus shear-wave quality factor ($Q_s$), showing a dominance of $Q_p$ over $Q_s$ when there was *less* attenuation. Assefa et al., 1999.

found in shaley sandstones. They also suggested that conventional dual porosity (i.e. joints and pores) present in the case of, for instance, in situ limestones, could similarly cause seismic attenuation due to squirt flow. Modelling these dual porosity aspects will be addressed in Chapter 15.

Klimentos and McCann, 1990, and others had previously shown how attenuation in sandstones was dependent on the pore-filling minerals, particularly the clay content. They had shown how the attenuation was significantly higher in clay-rich sandstones, than in clean clay-free sandstones of identical porosity. Klimentos, 1995, had later measured compressional- and shear-wave attenuation from sonic wave forms in three gas and oil reservoir wells, and was able to show that $Q_p/Q_s$ was about 1/3 in gas bearing sandstone, while $Q_p/Q_s$ was about 5 in water/oil saturated sandstone. The combination of $Q_p/Q_s$ and $V_p/V_s$ with other well logs enabled differentiation of gas-bearing from oil-bearing reservoirs, as we have seen in other research.

Assefa et al., 1999, found that their 'dual porosity' (or bimodal porosity) limestone specimens showed higher attenuation (lower seismic Q) when permeability and (total) porosity were also larger. Their results are shown in Figure 13.39. Both distributions of pore size were important, and the attenuation was shown to be the sum of Biot-type fluid flow and squirt flow to/from the larger, moderately interconnected inter-granular pores, which sometimes contributed about 90% of the total porosity.

Assefa et al., 1999 also presented results for the ratio of $Q_p$ and $Q_s$, sorted by mineralogical differences in their samples. Figure 13.40 shows the generally larger magnitude of $Q_p$ consistent with the general effect of saturation, as referred to above. They also compared $Q_{seis}$ values ($Q_p$ and $Q_s$) for oil-saturated and water-saturated, showing in Figure 13.41 how the water saturated specimens generally showed less attenuation (higher $Q_p$ in particular).

The authors posed the question of whether the ultrasonic data (0.7–0.85 MHz) for these small 'intact' dual-porosity limestone samples, were of any value to geophysicists trying to interpret propagation through dual-porosity (porous and jointed) limestones in the field, at frequencies in the 50 Hz to 30 kHz ranges, as used in seismic and sonic log surveys.

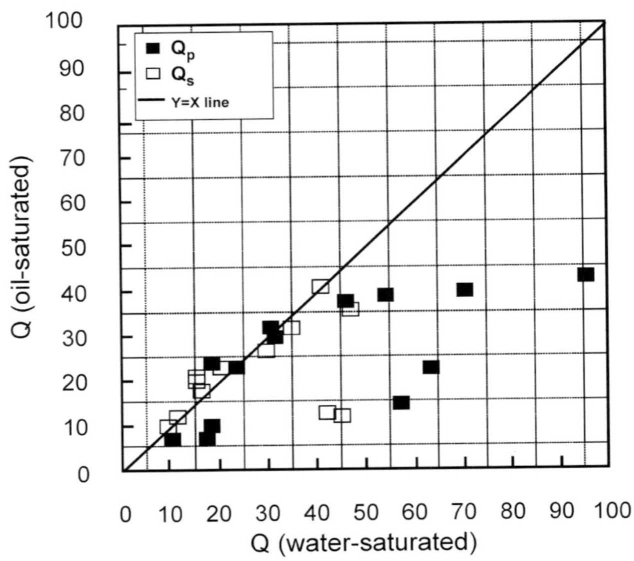


*Figure 13.41* Comparison of oil-saturated and water-saturated seismic quality factors, for bi-modal porosity limestones, at 50 MPa effective hydrostatic stress. Assefa et al., 1999.

Concerning *in situ* jointed, dual-porosity chalk (with higher porosity than the limestones), Newman and Worthington, 1982, measured $Q_p$ and $Q_s$ values as low as 4.0 and 5.2, and 3.5 and 5.9, in two near-surface fissured (jointed) chalks, using seismic frequencies. Assefa et al., suggest that when interpreting the propagation of low frequency seismic waves, the potential for squirt flow attenuation in a large scale dual porosity system such as jointed limestones will clearly be present.

Possibly this is why a certain correlation is being noticed between *seismic Q* and the 'static' modulus of deformation, expressed in GPa and readily estimated from *rock quality Q* (equation 13.12). The components of the rock quality Q-value reflect many potential attenuation-causing factors, e.g. RQD/$J_n$ for scattering due to relative block size, $J_r/J_a$ concerning frictional and conductive properties of the joints expected to be relevant for squirt flow (also including mechanisms in clay), $J_w$ as a direct link to permeability, and SRF related to increased attenuation where stress is low, and reduced attenuation where stress is high.

### 13.10.8 Attenuation in the presence of over-pressure

Establishing wells in deep sea environments, where there may be over-pressured zones due to rapid sedimentation of alternating sands and shaly sediments, presents a potential hazard due to the risk of so-called *shallow water flows* (SWF). Sands in the Gulf of Mexico can present problems for these reasons, at water depths of between 400 and 2100 m, and depths below the mudline of as much as 1200 m. As the name implies, SWF can also be a hazard in shallow water drilling, where effective stresses and compaction of sediments can be minimal, and progressive instability during drilling at a new well can potentially engulf neighbouring wells.

Prasad, 2002, used rock-physics principles involving *velocities and attenuation, to* study this problem. Older data on sands tested at very low effective stresses (in Prasad, 1988), were added to by new data, by performing 1 MHz pulse generator testing of jacketed, lightly confined, clean beach sands of grain size 250–550 μm. Due to very low values of shear wave velocity at low effective stresses, for example 400 m/s at 1 MPa, while equivalent compression wave velocities were closer to 1,800–1,900 m/s, there was an exponential increase in the ratio of $V_p/V_s$ to values beyond 5 and 10, and even beyond 100 at negligible effective stress. This trend, which became very noticeable below 2 MPa, is shown in Figure 13.42a, with a log scale for effective stress shown in Figure 13.42b.

Prasad showed that there was a dramatic change in the S-wave signals when testing at extremely low effective stress levels, with high *attenuation* of the shear waves at the lowest pressures, indicating the sand was close to a state of suspension, with low shear strength. The amplitude of the S-waves decreased dramatically below 1 MPa. In contrast, P-wave attenuation reduced marginally at very low effective stress. Due to the decreasing S-wave velocity, Poisson's ratios increased rapidly to just below 0.5 at negligible effective stress levels.

The diagnostic use of the velocities and seismic Q values for registering over-pressure, and the presence of gas can conveniently be 'tabulated' as follows, based on the author's summary text:

*Detecting over-pressure*

Reduction in $V_p$ may be ambiguous, as both over-pressure and gas reduce $V_p$

$V_p/V_s$ increases with over-pressure

$V_p/V_s$ and Poisson's ratio both increase exponentially, when sediment approaches a state of suspension

*Detecting gas*

Reduction in $V_p$ may be ambiguous, but $V_s$ will be unaffected by presence of gas

$V_p/V_s$ decreases with gas saturation

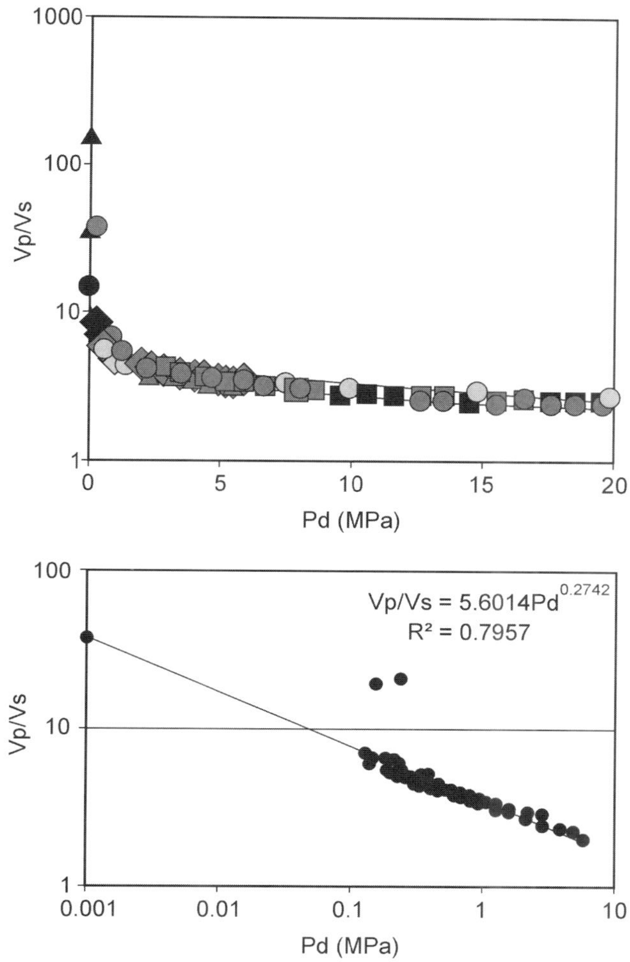


*Figure 13.42* Ratio of $V_p/V_s$ showing exponential increase at low effective stresses (Note: *differential* pressure Pd is *effective* stress, defined as $P_{overburden} - P_{pore}$). Prasad, 2002. An empirical prediction from work by Hamilton, 1971 is glimpsed beneath the recent data in the upper diagram.

Carcione and Gangi, 2000, added another important aspect to the understanding of over-pressure effects on seismic attenuation, by considering and modelling, the relative effects of pore-space compressibility, and the compressibility *and thermal expansion coefficient* for the fluid mixture filling the pore space. Their model consisted of reservoir sand that was buried at a constant sedimentation rate, under a geothermal gradient which was constant both in time and depth. Their model (Figure 13.44) showed realistic reductions of velocity and quality factor with reducing differential (or effective) pressure, especially when less than 15 to 20 MPa, as commonly observed in numerous studies reviewed in this book.

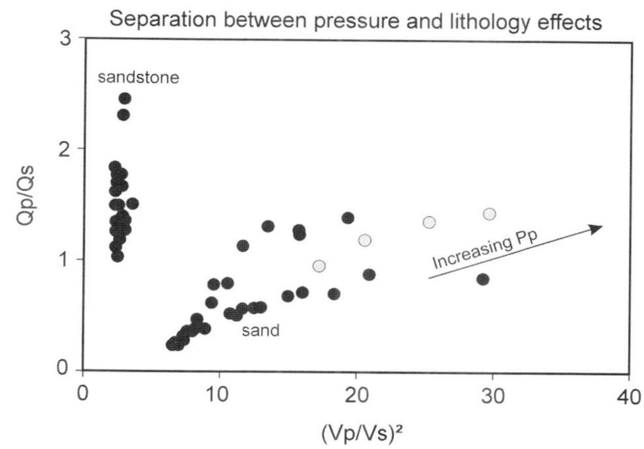


*Figure 13.43* Distinguishing between sand and sandstone is very clear using the ratio of $Q_p/Q_s$ in combination with $(V_p/V_s)^2$. The effect of increasing pore pressure, and reducing effective stress is suitably accentuated. Prasad, 2002.

They pointed out that rocks saturated with fluids of high compressibility and low thermal expansion coefficient were generally *under-pressured*, while rocks saturated with fluids of low compressibility and high thermal expansion coefficients were generally *over-pressured*. The latter could therefore be seismically 'visible'. Of course at high differential pressures, the velocities (and quality factors) became almost constant. The authors' model was able to predict pore pressure from seismic properties, if reliable wave velocities and quality factors could be obtained.

## 13.11 Attenuation in the presence of anisotropy

A further combination of interesting rock physics test data and sophisticated modelling abilities were described by Carcione, 2000, concerning petroleum source rock containing kerogen and different amounts of water. The objectives of the study were to obtain a model for source rocks that would be capable of relating seismic anisotropy (of velocity and attenuation) to kerogen content, pore pressure and water saturation. The author succeeded in demonstrating that *anisotropic velocities and attenuation* could be used as strong indicators of kerogen content and maturation, which depends on pressure change.

Some test data, from the North Sea Kimmeridge Shale, is shown in Figures 13.45a and b. The author's model, shown as solid curves, was based on a viscoelastic transversely isotropic medium composed of illite and smectite

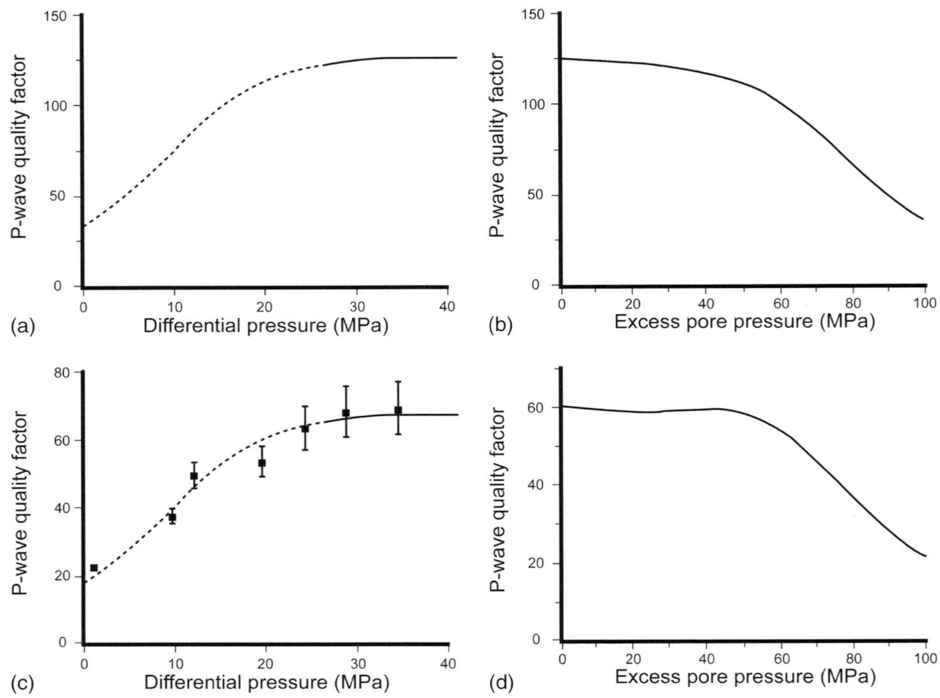


*Figure 13.44* Modelling of low frequency (25 Hz): a) and b), and ultrasonic (1 MHz): c) and d) P-wave seismic Q *versus* differential pressure, or *versus* excess pressure, for a water-saturated model sandstone reservoir. Experimental squares for bedding-parallel Q of Berea sandstone, from Prasad and Manghnani, 1997. (Dotted lines correspond to 0 to 2 km, where the rock is normally pressured, and the continuous lines to the range 2 to 8 km where the rock is over-pressured.) Carcione and Gangi, 2000.

and organic matter. The data for $V_p$ and $V_s$ versus kerogen content, was presented from measurements in bedding normal (0°) and bedding-parallel (90°) directions. The marine Kimmeridge Shale was from the Draupne Formation, located at between 3480 and 3580 m depth, in the central Viking Graben of the North Sea.

The modelled seismic $Q_p$ and $Q_s$ trends, shown in diagrams d) and e), demonstrated an attenuation anisotropy that was higher than the velocity-based stiffness anisotropy. The largest anisotropies were at 18% kerogen content for the case of attenuation, and 30% kerogen content for the case of stiffness or velocity. In Figure 13.45, diagrams e) and f), Carcione demonstrated the modelling of $V_p$ and $V_s$ for the case of a fixed kerogen content of 35%, and several water saturation levels ($S_m$ = 0 to 0.7), again for the case of bedding normal (0°) and bedding parallel (90°) directions. A key variable in these plots was the strong effect (in the case of the bedding-normal (0°) direction of measurement), of an excess pore pressure of up to 50 MPa, giving as expected, successively lower velocities, with greatest reductions for the highest saturation levels.

The different maturation stages of this source rock were modelled by evaluating the kerogen to oil conversion and the excess pore pressure, with fracturing estimated if a change of pore pressure of as much as 48 MPa occurred, compared to the calculated lithostatic minus hydrostatic pressures of 82 MPa and 34 MPa. A kerogen content of 35% was assumed here. The sonic log result for $V_p$ showed a typical reduction from about 4 km/s to a fairly constant value of only 2.6 km/s for the 100 m thick section of this valuable over-pressured source rock.

### 13.11.1 Attenuation for fluid front monitoring

In 4-D, time-lapse seismic monitoring of reservoir processes, several 3-D seismic surveys of the same reservoir locations made at different times are compared. Differences in reflection amplitude or impedance indicate changes in the reservoir. There is an increasing move to have permanent down-hole sources and receivers for obtaining greater detail of the movement of fluids during production and injection (Ziolkowski, 1999), and for enabling the periodic performance of high frequency cross-hole imaging (e.g. see Chapter 14).

Wulff and Mjaaland, 2002 studied time-dependent fluid-front 4-D seismic effects, using a large scale *laboratory* test, in which a block of lower Triassic sandstone of 17% porosity, was successively submerged in a

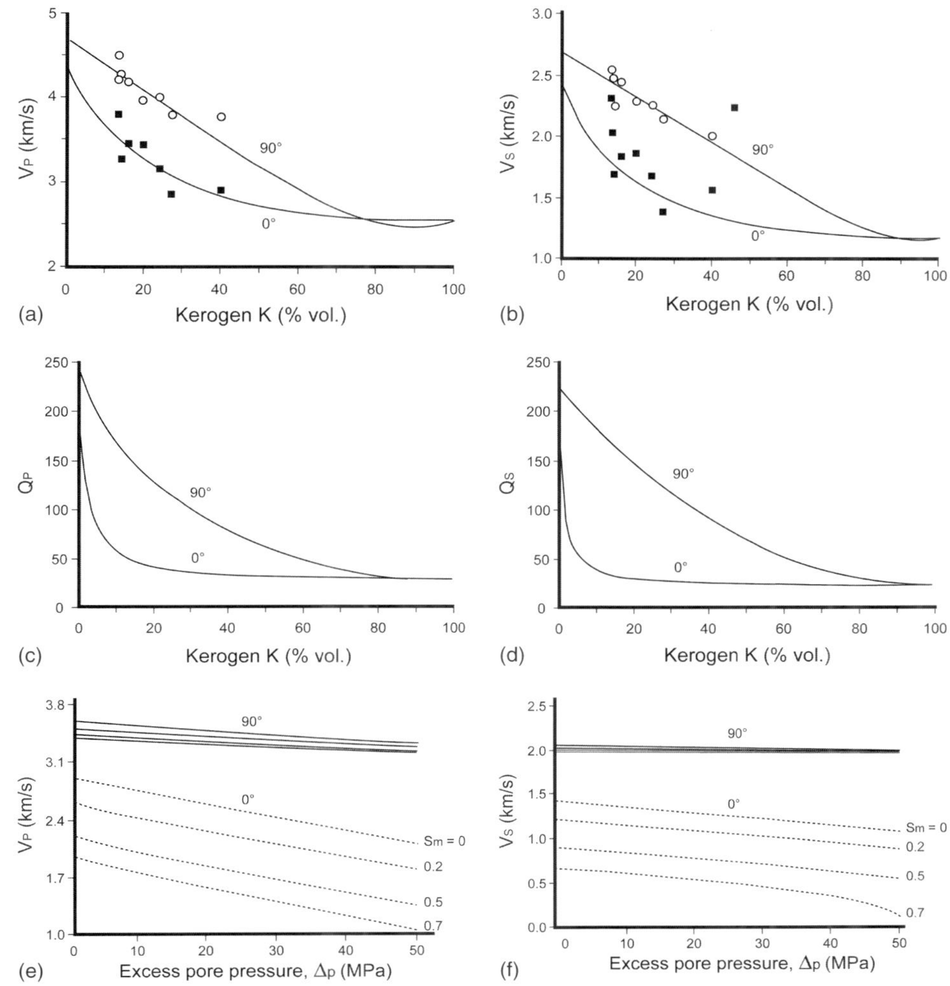


*Figure 13.45* Experimental velocity data for Kimmeridge Shale source-rock samples from three and a half kilometres beneath the North Sea, with comparison to viscoelastic, transversely isotropic model results. Anisotropic velocity, attenuation, and velocity dependence on excess pore pressure was demonstrated, when measured or modelled in bedding-normal (0°) and bedding parallel (90°) directions. Carcione, 2000.

water tank in four stages, with immediate and long term (300 days) seismic monitoring of the effect of the intermittent 'water-flood' and capillary effects. They used P- and S-wave transducers, with centre frequency of 500 kHz. Transmitted and reflected waves were used for the monitoring, with six transducers glued to the top and (submerged) bottom of the block.

The water-flooding caused the velocity, amplitude and frequency of the transmitted waves to diminish significantly, with reversal upon final drying of the block. The authors evaluated both the attenuation coefficient ($\alpha$), where $1/Q = \alpha V/\pi f$, and the attenuation 1/Q, relative to the signals obtained for the dry rock. Seismic Q was also estimated by the spectral ratio method. At maximum saturation of the block, the spectral ratio method indicated a very high maximum $1/Q_p$ of 0.5, and $1/Q_s$ of 0.3. The authors therefore preferred to use the attenuation calculated from ($\alpha$). This gave $1/Q_p$ of 0.065 and $1/Q_s$ of 0.07. Attenuation had increased very rapidly, such that maximum attenuation was reached prior to full saturation, followed by a quite fast decline in the three following months. This was assumed to be due to the decline in local fluid flow (squirt) effects which were dominant prior to full saturation, prior to full flooding.

The authors assumed that the velocity reductions were due to water adsorption effects causing a reduced modulus (strictly a laboratory, as opposed to an *in situ* reservoir effect, unless gas replaced by water-flooding could be considered?). They concluded that improved interpretation of reservoir processes required not only P- and S-wave measurement, but also amplitude and attenuation measurement.

## 13.12 Anisotropic velocity and attenuation in shales

Anisotropy in shale has not been frequently studied due to difficulties with sample disturbance when handling fissile materials. The set of data from Johnston and Christensen, 1995, (Figure 13.46), shows the effect of $V_p$ measurement direction, and the effect of confining pressure, and is a useful summary of the effects of preferred clay mineral orientation.

Data sets reviewed by Johnston and Christensen, 1995 showed $V_p$ anisotropy of 20 to 30%, and $V_s$ anisotropy of 32 to 35%. The maximum velocity was always parallel to bedding and the minimum perpendicular to bedding, just as it is for the case of rock joints, as reviewed in Chapter 3. Figure 13.47 shows the complete distribution of velocities, with the vertical/horizontal axes showing respectively the perpendicular to bedding and parallel to bedding magnitudes (units km/s).

Domnesteanu et al., 2002, measured the anisotropic velocity and attenuation of fully saturated shales under *overpressured* conditions, Figure 13.48, apparently for

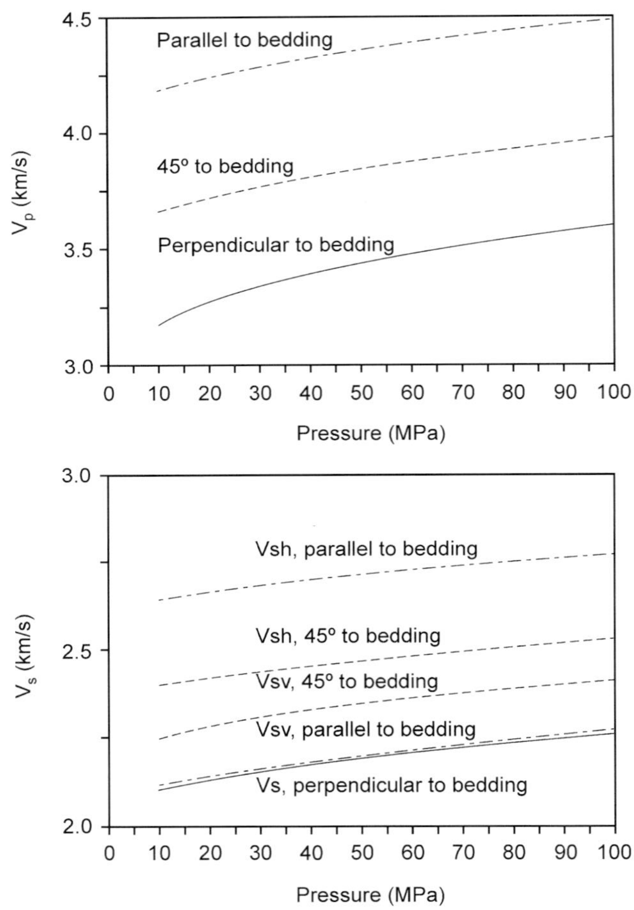


*Figure 13.46* Effects of confining pressure and direction of velocity measurement for shale. Anisotropy is caused by the preferred orientation of clay particles in the fabric of the shale. ($V_{sh}$ is the velocity of the shear wave vibrating parallel to bedding, and $V_{sv}$ is the velocity of the shear wave vibrating perpendicular to bedding). Johnston and Christensen, 1995.

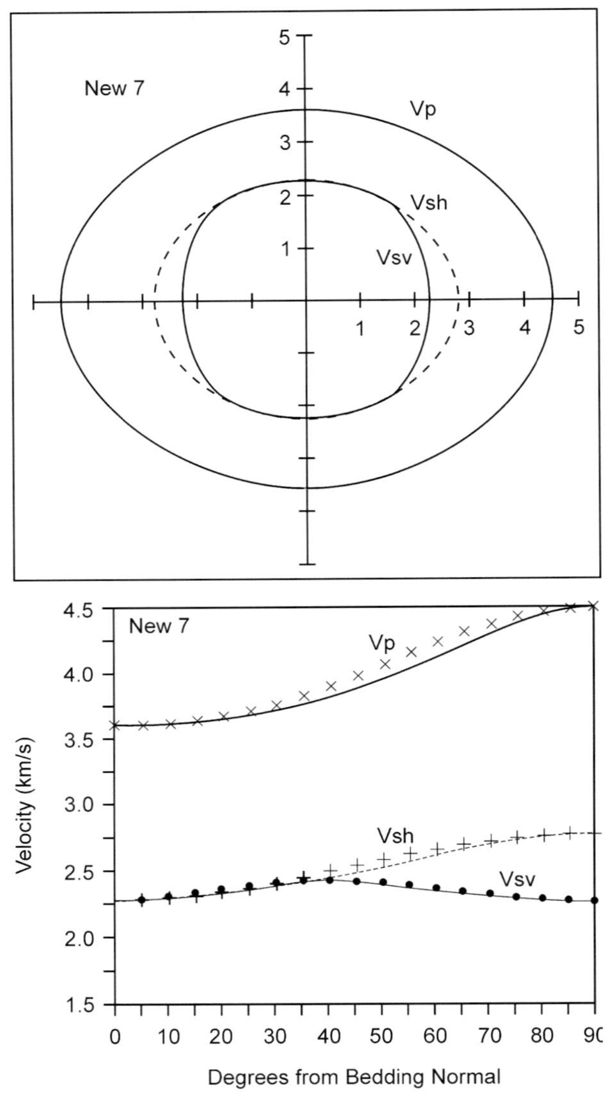


*Figure 13.47* Velocity anisotropy caused by the preferred orientation of clay particles in the fabric of the shale: sample New 7. Note *shear-wave splitting* when making an increasingly acute incident angle to the direction of bedding. ($V_{sh}$ is the velocity of the shear wave vibrating parallel to bedding, and $V_{sv}$ is the velocity of the shear wave vibrating perpendicular to bedding). Johnston and Christensen, 1995.

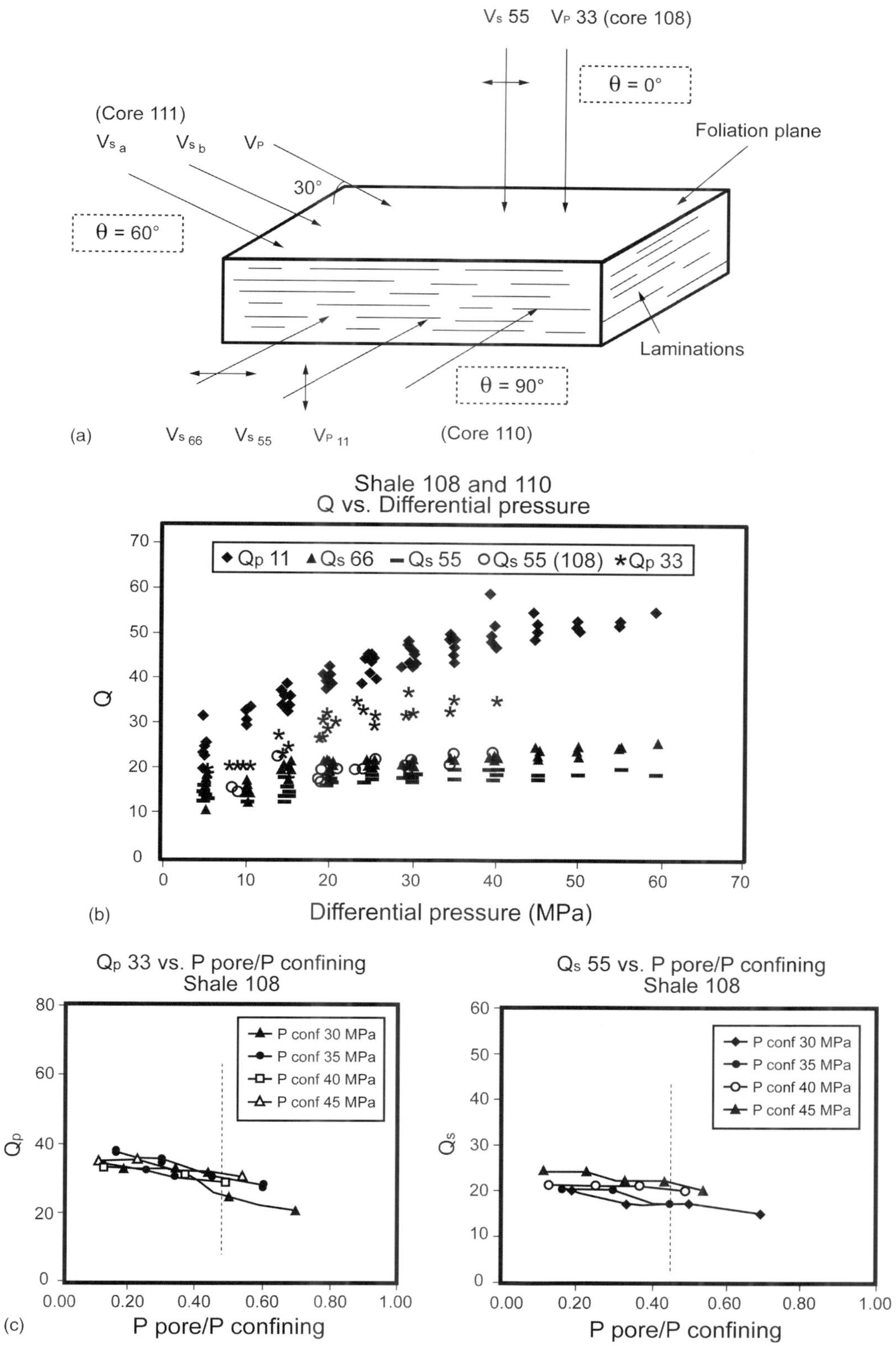


*Figure 13.48* a) Propagation and vibration directions with respect to foliation, and relevant to shale sample number. b) Seismic qualities for P- and S-waves through over-pressured shales, as a function of differential pressure, and as a function of propagation direction relative to the foliation. c) One set of examples of P-wave and S-wave attenuation as a function of over-pressure, where $P_P/P_c > 0.46$ corresponds to over-pressure. Domnesteanu et al., 2002.

the first time in the laboratory. They used an ultrasonic reflection technique. The shale cores were from the North Sea. Confining and pore pressures were applied that were relevant to in situ conditions, giving *differential* pressures from 5 to 60 MPa. (In view of the low permeability of the shale, the use of the rock mechanics adjective for $\sigma_1 - \sigma_3 =$ *differential* stress, will not be questioned, as the actual *effective stress is* uncertain, due to test rate dependency).

The seismic signature of the shale was explored by taking the samples through cycles of over-pressured and normally pressured states, whilst increasing the overall confining pressure. The authors found that each incremental increase in pore pressure caused the shale to expand slightly, counteracting the opposite effect of increasing the confining pressure. It was found that the shales behaved elastically at confining pressures higher than 35 MPa. As expected, the compressional and shear wave velocities and seismic quality factors increased with increasing *differential pressure* (i.e. with reduced over-pressure.

The plane of circular symmetry was parallel to the foliation (see Figure 13.48a). The degree of sample anisotropy was found to be related to the depth of origin of the shale. Both $V_s$ *and* $V_p$ decreased with increasing over-pressure. Nevertheless, the *ratio* $V_p/V_s$ decreased with increasing *differential* pressure. As expected there was a general increase in seismic $Q_p$ and $Q_s$ with differential pressure, but with lowest values showing least increase with stress. Under a fixed *differential* pressure, an *increasing* pore pressure was found to reduce $Q_p$ by about 16%, or up to 8 units. An anisotropic pattern of wave attenuation is evident from Figure 13.48b.

The authors found that the relative proportions of Biot 'fluid-past-frame' attenuation, and local squirt flow attenuation were different in the plane parallel to foliation, and in the plane perpendicular to the layering. Squirt flow, localized between compliant and non-compliant pores, was considered a predominant loss in the bulk modulus, and a small loss in the shear modulus. It was naturally considered to be more active in the plane of the foliation, than perpendicular to it. Biot flow was considered a predominant loss in the shear modulus, and a very small loss in the bulk modulus. Since arising from fluid movement in the open pores, it was considered to be related to the macro-permeability.

The authors found that at differential pressures <20 MPa, compressional waves propagating perpendicular to the layering ($V_{p33}$) were attenuated by squirt flow, and hence more attenuated than the compressional waves propagating parallel to the layering ($V_{p11}$). $Q_p$ reduced with overpressure, regardless of propagation direction, while $Q_s$ was highly dependent on propagation direction, and depended less on pore pressure than $Q_p$. Perhaps predictably, the results suggested a strong link between the rock framework, the pore geometry and connectivity, and therefore the response of pore fluid to the propagation of seismic waves, in specific directions.

### 13.12.1 Attenuation anisotropy expressions $\varepsilon$, $\gamma$ and $\zeta$

At a seismic Q workshop in Madrid, Best et al., 2005, described an ultrasonic pulse-echo technique for investigating both velocity and seismic Q anisotropy of P- and S-waves in finely inter-bedded reservoir-type rocks. They had modified the earlier version of the equipment, so as to be able to study shear wave anisotropy, using a 360° rotating S-wave transducer, for observing the shift of arrival time. This will be described here because of obvious relevance to attenuation in finely bedded reservoir rocks, which will be partly addressed in the next section.

Their studies were performed at effective stress levels of 5 and 40 or 50 MPa, on vertically or horizontally aligned samples of Carboniferous sandstone, siltstone or limestone obtained from the Imperial College experimental borehole site in northern England.

Following Thomsen (1986) *velocity* anisotropy expressions for $\varepsilon$, $\gamma$ and $\zeta$ in weakly transversely isotropic media, and because of the excellent stability of their pulse-echo data, they gave equivalent expressions for attenuation ($Q^{-1}$), (in place of $V_p$, $V_{sh}$ and $V_{sv}$) and derived relevant results for these new parameters:

$$\varepsilon(Q^{-1}) = \frac{Q_p^{-1}(H) - Q_p^{-1}(V)}{Q_p^{-1}(V)} \tag{13.13}$$

$$\gamma(Q^{-1}) = \frac{Q_{sh}^{-1}(H) - Q_{sh}^{-1}(V)}{Q_{sh}^{-1}(V)} \tag{13.14}$$

$$\zeta(Q^{-1}) = \frac{Q_{sh}^{-1}(H) - Q_{sv}^{-1}(H)}{Q_{sv}^{-1}(H)} \tag{13.15}$$

where V denoted vertical, H horizontal, and the sub-scripts p, sh and sv denoted P-waves, S-waves with horizontal polarization, and S-waves with vertical

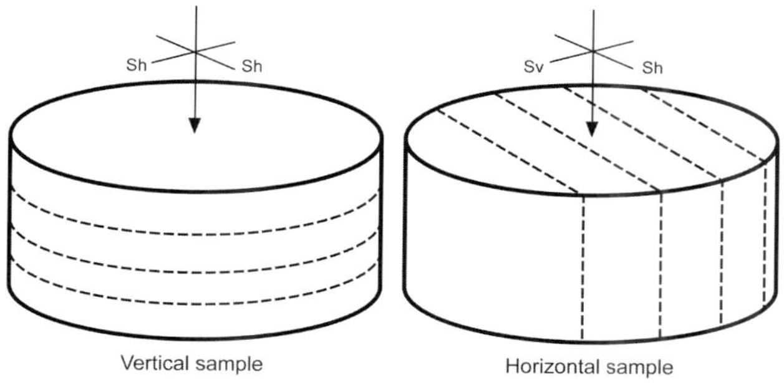


*Figure 13.49* Orientation of laboratory (sub-core) samples, with bedding features (dashed lines) either perpendicular or parallel to the shear wave transmission. Shear wave polarization is shown by the Sv and Sh components. Best et al., 2005.

Table 13.2 $Q^{-1}$ anisotropy parameters for thinly bedded sandstone samples. Best et al., 2005.

| Lithology | Pressure (MPa) | $\varepsilon(Q^{-1})$ % | $\gamma(Q^{-1})$ % | $\zeta(Q^{-1})$ % |
|---|---|---|---|---|
| Sandstone | 5 | −11.8 (+/−3.3) | 0.0 (+/−1.2) | −27.3 (+/−1.0) |
| Sandstone | 40 | −42.5 (+/−5.1) | 75.0 (+/−4.6) | −7.7 (+/−1.9) |

polarization (see Figure 13.49). The authors introduced the parameter $\zeta$, to describe the anisotropy between the fast and slow S-wave polarizations in the H sample. (The subjects of *shear wave splitting* and *polarization* are treated in detail in Chapter 15).

A brief sampling of Best et al., 2005 results for sandstone, using these new expressions, is given in Table 13.2. The samples exhibited visible sub-millimetre, fine horizontal layering of clay/organic matter. At an equivalent 500 m depth (for the case of low density, porous rock), the effective stress of 5 MPa indicates that there is significant $Q_p^{-1}$ anisotropy ($\varepsilon$), none for $Q_{sh}^{-1}$ ($\gamma$), and strong anisotropy for $Q_{sv}^{-1}$ ($\zeta$), i.e. consistent with transverse anisotropy.

At 40 MPa, roughly equivalent to effective stresses at 4,000 m depth in porous sedimentary rock masses, there was a dramatic increase in anisotropy for $Q_p^{-1}$ and $Q_{sh}^{-1}$ (the latter perhaps surprising), while the $Q_{sv}^{-1}$ anisotropy reduced, as would be expected at higher stress. A second sample showed significantly different results, except in the case of $Q_{sv}^{-1}$ behaviour (reduction with increased stress).

The anisotropy of seismic Q will be addressed in much more detail in Chapter 15, due to the importance of shear wave splitting for revealing attenuating and fluid-conducting joint structure, in hydrocarbon reservoirs.

## 13.13 Permeability and velocity anisotropy due to fabric, joints and fractures

A very interesting laboratory study of velocity and permeability anisotropy, for the case of tight gas sandstones containing sub-vertical, conjugate-type jointing, was reported by Dürrast et al., 2002. The individual joints were either 'open', open-and-mineralized, or mineralized. The sandstones were of very low matrix permeability (<30 μd), and had a sedimentary layering consisting of fine clay layers. Production from a reservoir in such tight sandstones naturally depends on the jointing (i.e. the natural fractures).

The authors were able to study the three-dimensional P-wave velocity of *spherical* samples machined from the core, in at least 100 directions, to obtain $V_p$ symmetry without prior assumptions.

Some selected results of the P-wave anisotropy measurements, plotted stereographically on lower-hemisphere Schmidt net projections (perpendicular to the core axis), are shown in Figure 13.50a. Their results for *six* spherical samples, under the first condition of *zero* confinement, showed a range of maximum $V_p$ (saturated) of 4.6–5.2 km/s and minimum $V_p$ (dry) of 3.7–4.0 km/s, with the difference plots (saturated minus dry) showing a maximum range of 0.5−1.1 km/s for the six samples, under zero confinement. The sedimentary layering and cross-bedding tended to dominate the $V_p$ distribution, with jointing (open or mineralized) having less effect on $V_p$, but tending to change the symmetry of the $V_p$ distribution, giving a more monoclinic symmetry.

Permeability *and* P-wave velocity measurements were performed on *spherical* samples of the same sandstones, some containing both the sedimentary fabric *and* the steeply dipping jointing. $V_p$ under four levels of confinement, up to 100 MPa are shown in Figure 13.50b.

These tests demonstrated several important trends of behaviour. Naturally, in the absence of jointing, the higher permeability and velocity values were recorded in the plane of the nearly horizontal layering, perpendicular to the core axis. Figure 13.51a shows the permeability in three orthogonal directions, as a function of confining pressure, with results dominated by the successively reducing porosities of the five selected samples. Note the low values of the vertical (Z-axis)

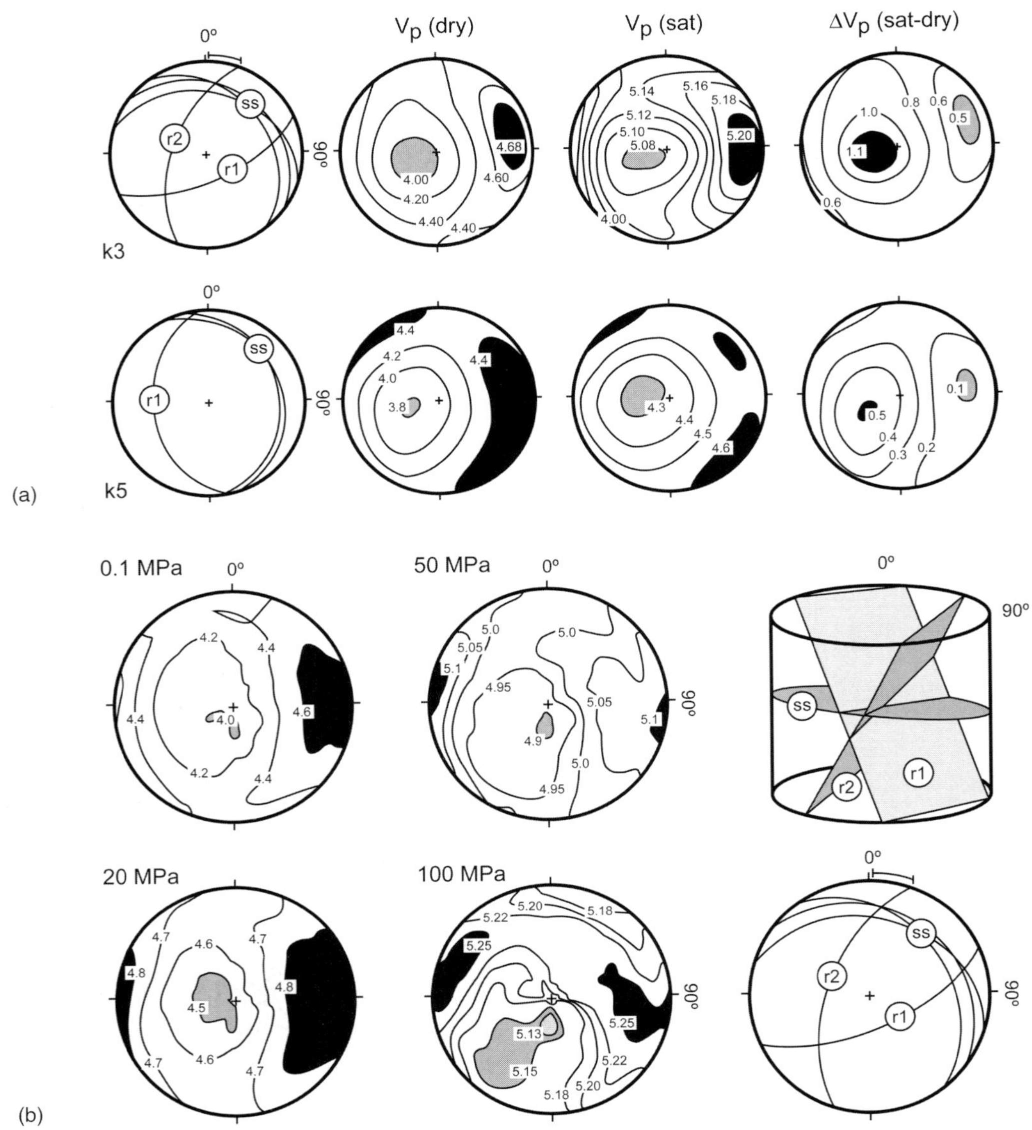


*Figure 13.50* a) Three-dimensional P-wave velocity measurements on *spherical* samples in more than 100 directions, in the dry and saturated states. Two of the six results (k3 and k5) for the tight gas sandstones are shown. These illustrated cases were under zero confinement. b) Confined tests with sedimentary layering (ss) and sometimes jointing ($r_1$ and $r_2$), provided an anisotropic, confining pressure-dependent mix of effects on velocities. (Note: sample porosities are listed in Figure 13.51). Dürrast et al., 2002.

permeabilities in each case, due to the thin dark-clay layers obstructing vertical flow.

However where discrete jointing occurred, with sub-vertical orientation, as in the case of sample k5, the *highest* permeabilities were recorded parallel to the core axis, i.e. *vertically.* This clear trend is shown in Figure 13.51b. With confining pressure below 50 MPa, there was a *mono-clinic* symmetry of the P-wave velocity distribution caused by the combination of sedimentary layering, jointing, and cross-bedding, while at higher confinement, the sedimentary layering was virtually the only remaining feature, resulting in a more *transversely isotropic* symmetry.

The authors commented on the significance of the surface roughness of the joints providing a significant fluid path, even at higher confining pressures. From 50 to 80 MPa confining pressure, corresponding to depths of several kilometres, the permeabilities parallel to the fractures were up to *nine* times higher than those parallel to the sedimentary layering. In view of the possibility of conjugate shearing of such joint sets, one may speculate that in a reservoir environment, with anisotropic stress, there would be a possibility of relative *maintenance* of joint permeability despite high effective stresses. These possibilities, and the influence of joint-surface related

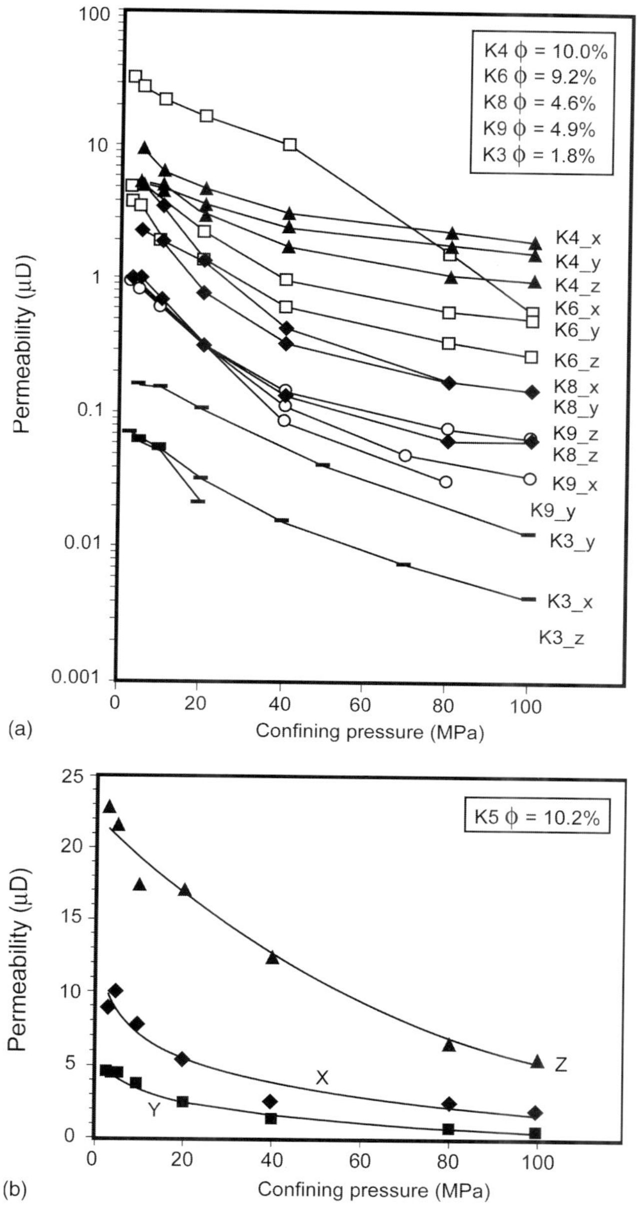


*Figure 13.51* a) Orthogonal permeabilities of five spherical samples, having successively reducing porosities. Vertical permeability (Z) is least due to dominance of sub-horizontal dark-clay layers. b) Orthogonal permeabilities of jointed sample k5, which had a porosity of 10.2%. Note effect of sub-vertical jointing on the higher vertical (Z) permeability in this case. Dürrast et al., 2002.

properties, like roughness (JRC), and wall-strength (JCS), in assisting permeability maintenance (unless each are too low), are mentioned in Chapter 14 and addressed in detail in Chapter 16.

The large, spherical-sample test equipment used in the above studies, was also described in a companion paper, by one of the above co-authors, from the Institut Français du Pétrole, see Rasolofosaon and Zinszner, 2002. The new equipment, based on that developed by Arts et al., 1996, now allows independent application of pore pressure and confining pressure, while measuring P-wave velocities in multiple directions as we have seen. These two authors described interesting comparisons of the anisotropic 3D permeability *and* elasticity tensors of various reservoir rocks. The permeability tensors were obtained by a tracer injection and X-ray technique.

They found that in some cases, the elastic property anisotropy, and the hydraulic anisotropy were closely related in terms of symmetry directions – this occurred when the two mechanisms shared the same cause, such as layering or jointing. Good agreement was seen between the two types of anisotropy for a North Sea sandstone of 16% porosity (Figure 13.52), while in the case of a dolomite of 23% porosity, there were marked differences due to the influence of small-scale disconnected fissures. Cases where there was no correlation, highlighted the challenges faced in estimating permeability and monitoring fluid flow from seismic measurements in the field. In an earlier study with the same equipment, Rasolofosaon et al., 2000 had shown a comparison of crack and fabric analysis with inversion of the multi-directional ultrasonic data.

### 13.13.1 Seismic monitoring of fracture development and permeability

An important experimental testing facility at Imperial College, described by King et al., 1995, Shakeel, 1995, and King, 2002, allows the application of extremely high (hundreds of MPa) *polyaxial* stress states to small (40 to 50 mm) cubical specimens of rock. Figure 13.53 illustrates the loading frame and principal loading and velocity measurement directions, with definition of the nine components of P- and S-wave velocity, and the vibration directions of the S-waves. Pietzo-electric transducers were used to produce and detect the pulses of compressional waves (450–800 kHz), and either of two shear waves (350–750 kHz), polarized at right-angles, and propagating in one of the principal stress directions.

By holding the minimum principal stress very low (e.g. 2 or 3 MPa), and increasing $\sigma_1$ and $\sigma_2$ in unison, to high levels, it was possible to create a set of closely spaced extension/tension fractures perpendicular to the minimum stress direction. Shakeel and King, 1998 and

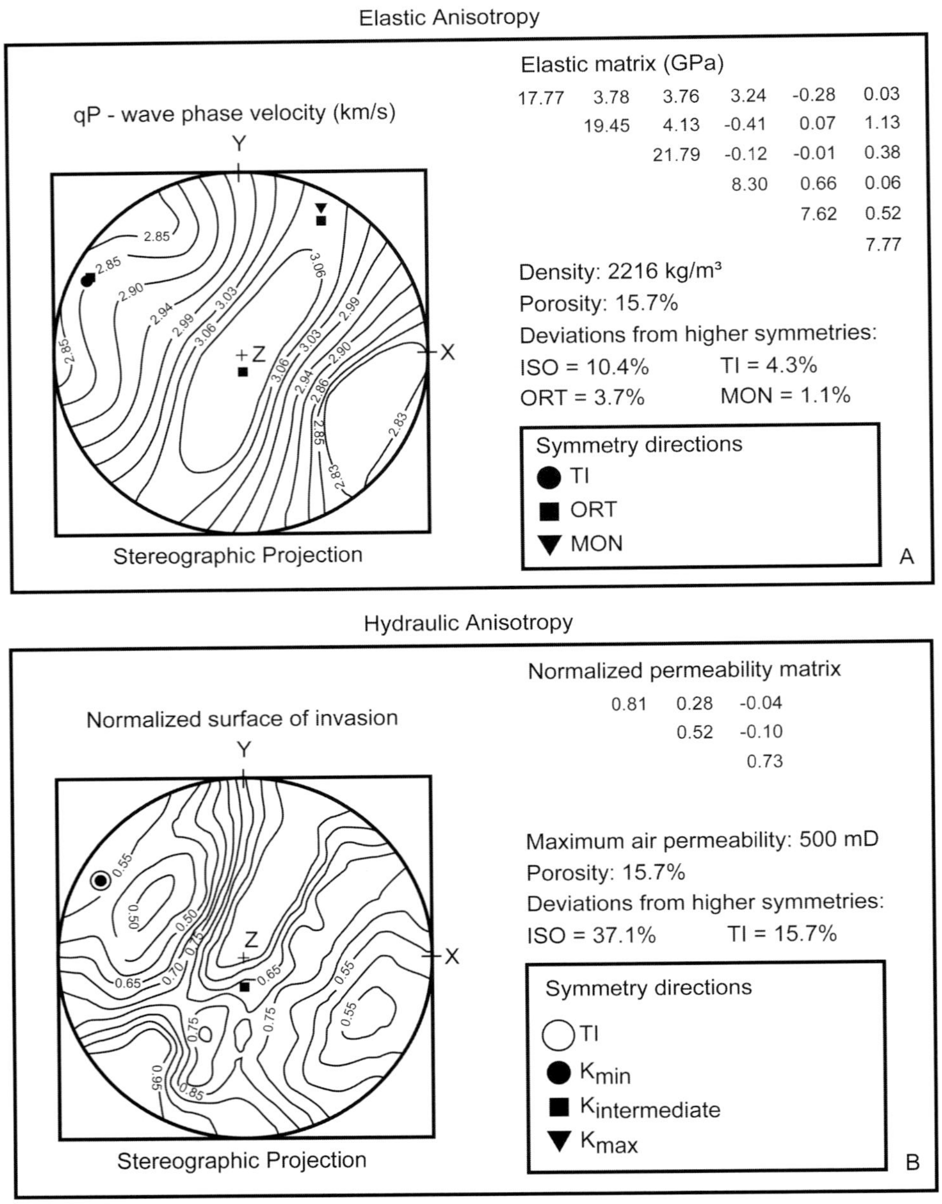


*Figure 13.52* Comparison of elastic anisotropy, measured on a spherical North Sea sandstone sample, and hydraulic anisotropy, measured by a tracer-injection X-ray technique in the same rock sample. ISO: isotropic, TI: transversely isotropic, MON: monoclinic symmetry, ORT: orthorhombic symmetry. Rasolofosaon and Zinszner, 2002.

King, 2002 described dynamic hydro-mechanical (ultrasonic and flow) measurements both before fracturing, during fracturing, and in load-unload cycles after fracturing.

Figure 13.54 reproduces some of the very interesting experimental results of principal P-wave and S-wave velocities that were recorded during the initially almost biaxial loading of a 6% porosity, and almost isotropic sandstone up to fracturing. Diagrams a) and b) show $V_{p1}$ and $V_{s1}$ increasing steadily parallel with the high 'biaxial loading. At approximately 100 MPa, the set of 'biaxial' extension fractures started to develop (point F), and the perpendicular velocities $V_{p2}$ and $V_{s2}$ indicated rapidly declining velocities due to the presence of the new fractures.

Figure 13.55 indicates what happened when the newly fractured cube was reloaded with equal *hydrostatic* stress. There was a more rapid increase in $V_{p2}$ and $V_{s2}$ as the fracture set was (nearly) closed, while parallel with the fractures, the velocities $V_{p1}$ and $V_{s1}$ behaved almost as

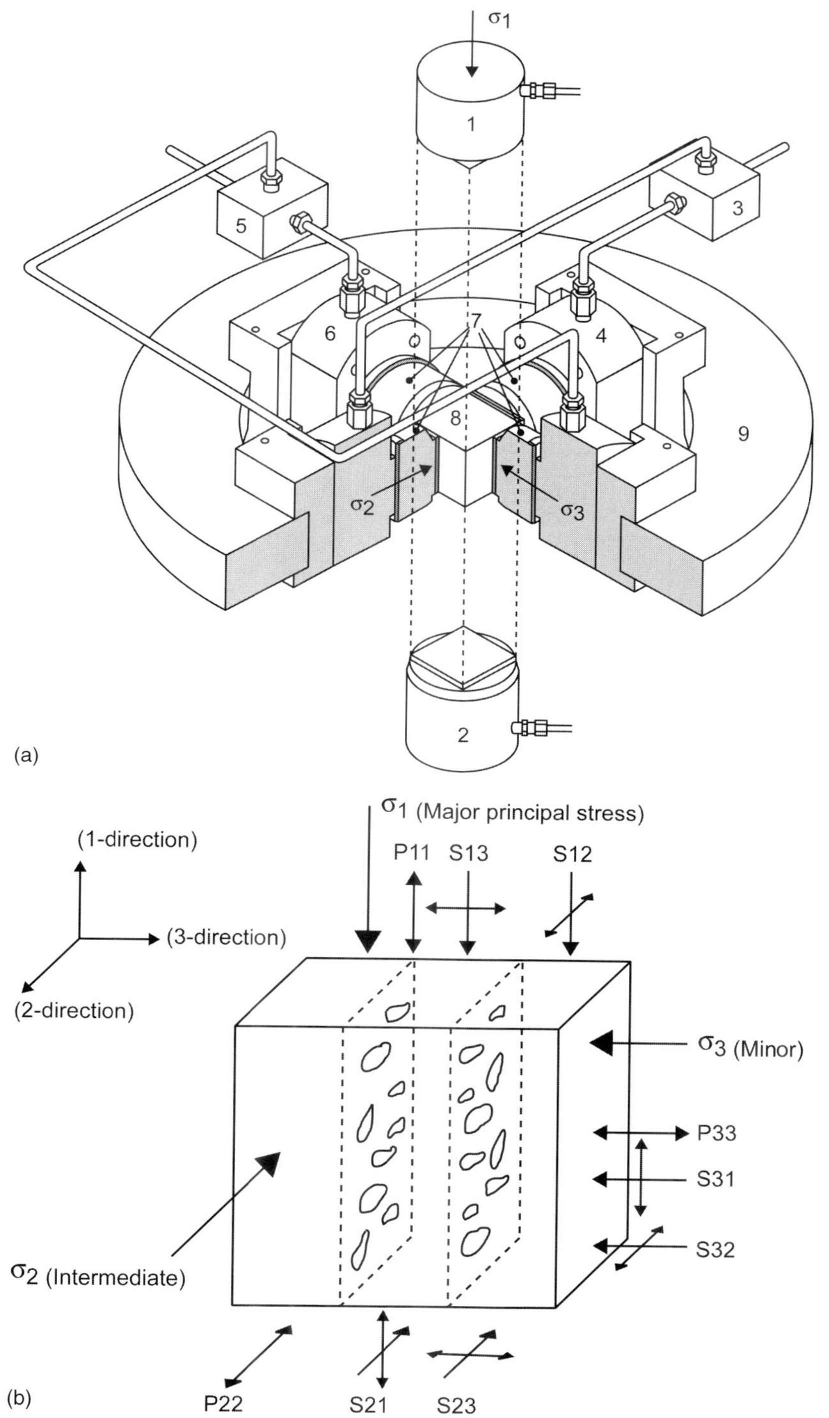


*Figure 13.53* a) Section through the 3D static and dynamic loading facility at Imperial College. Permeability can also be measured in the presence of a high pore pressure. b) Principal loading and velocity measurement directions, with definition of the nine components of P- and S-wave velocity, and the vibration directions of the S-waves. King et al., 1995, King 2002.

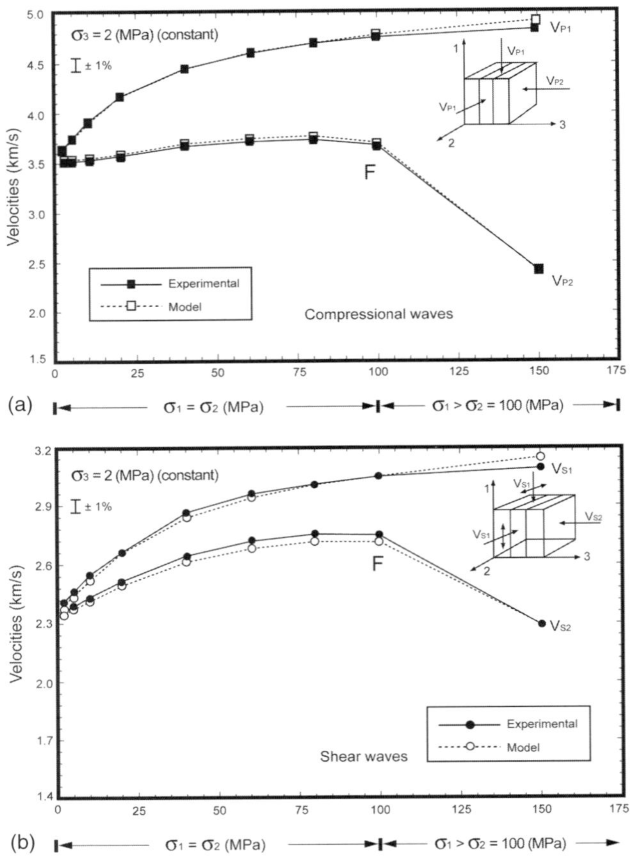


*Figure 13.54* Ultrasonic P- and S-wave components from tests performed across a 6% porosity, isotropic cube of Crossland Hill sandstone, when loaded in a polyaxial facility. Pre- and post-fracturing results are shown, and compared with excellent modelling results, from Shakeel, 1995. Measurable permeability (>1 md) parallel to the pending fractures, did *not develop* until velocities reduced at fracturing point F. King, 2002.

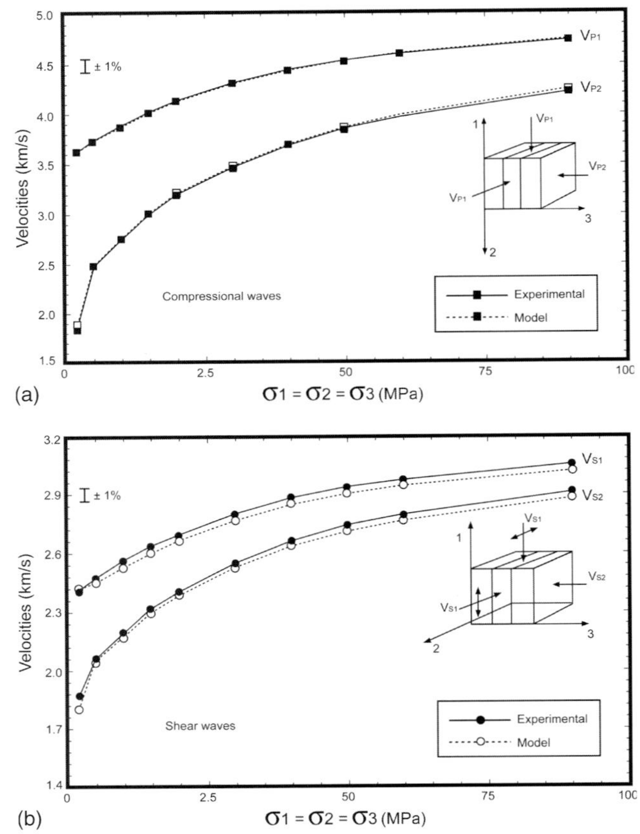


*Figure 13.55* Ultrasonic P- and S-wave components from tests performed across a 6% porosity, isotropic cube of Crossland Hill sandstone, when re-loaded in a polyaxial facility in parallel and perpendicular-to-fracturing directions, upon subsequent hydrostatic loading to 90 MPa. Note the almost unchanged velocities measured parallel to the fracture set that was developed. King, 2002.

before. The *very* closely matching analytical modelling results by Shakeel, 1995, using a method developed for a transversely isotropic cracked medium by Nishizawa, 1982 are virtually identical with the experimental result, in terms of velocities.

King et al., 1997 reported several interesting results concerning the *attenuation* of various of the nine velocity components (calculated by the spectral ratio technique), in an equivalent series of three test cycles on a cube of Penrith sandstone of 13% porosity. Figure 13.56a shows selected results for the '3' direction, perpendicular to the pending fracture set, which presumably started to initiate at point F, when the major stress was increased beyond 100 MPa to initiate fracturing. Note the reduction of seismic Q from about 20 before fracturing, to less than 5, somewhat resembling a reduction in deformation modulus, when expressed in units of GPa.

Figure 13.56b shows the hydrostatic loading test phase for the same set of fractures, with a low stress seismic Q increasing from 10 to about 29 with hydrostatic (therefore normal) loading to 70 MPa. The 'before-fracturing' result is also shown for comparison, indicating seismic Q increasing from 20 to 40, as a result of loading to 70 MPa. These again resemble increase in moduli, expressed in units of GPa. In Figure 13.57, the attenuation for P- and S-waves in the same perpendicular '3' direction, are plotted against permeability measured *parallel* to the fracture.

King, 2002 described the less demanding development of a set of parallel (cleavage) fractures in slate, simply by axially loading a cylindrical sample in a conventional *triaxial* cell. These easily formed, smooth, planar cleavage

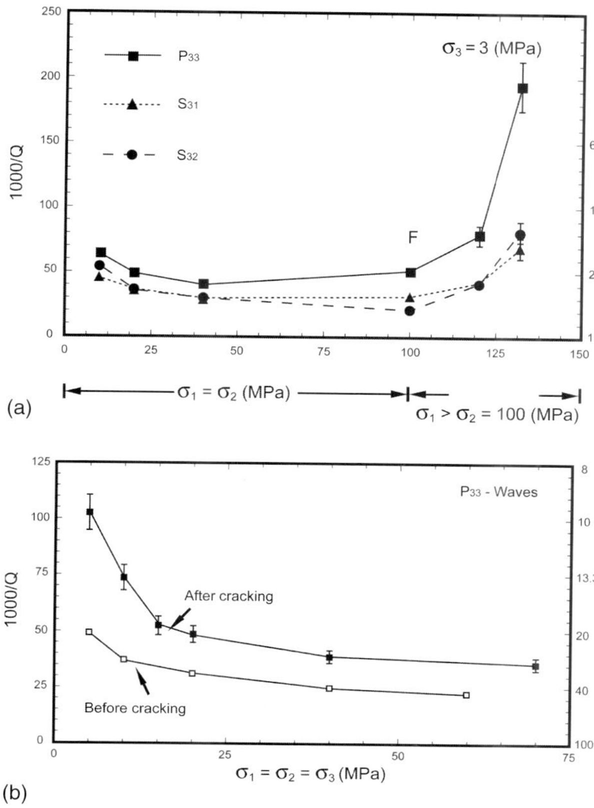


*Figure 13.56* a) Fracturing test cycle on a cube of Penrith sandstone of 13% porosity, with *attenuation* results for the '3' direction, perpendicular to the pending fracture set, which presumably started to initiate at point F. b) Hydrostatic loading test phase for the same set of fractures, with comparison to 'before-fracturing' phase. King et al., 1997. *Note: seismic Q values have been added on the right-hand axes, as the geophysicists tradition for expressing attenuation as '1000/Q' is perhaps hiding a physically viable mechanism related to attenuation, namely a certain modulus increase due to normal loading, of unknown but similar magnitude to seismic Q, when the former is expressed as GPa.*

fractures had a spacing of about 1 mm, while in the sandstone, fractured in the polyaxial cell, the *rougher* fractures were spaced many mm to 1–2 cm apart, based on a photograph of a fractured cube, provided in King, 2002.

A particularly interesting, and possibly challenging result was obtained from the dynamically monitored permeability measurements. King and his colleagues had found that the low permeability (<1 md) sandstone, started to develop measurable permeability once the velocities $V_{p2}$ and $V_{s2}$ started to reduce (at point F in Figure 13.54), indicating crack development.

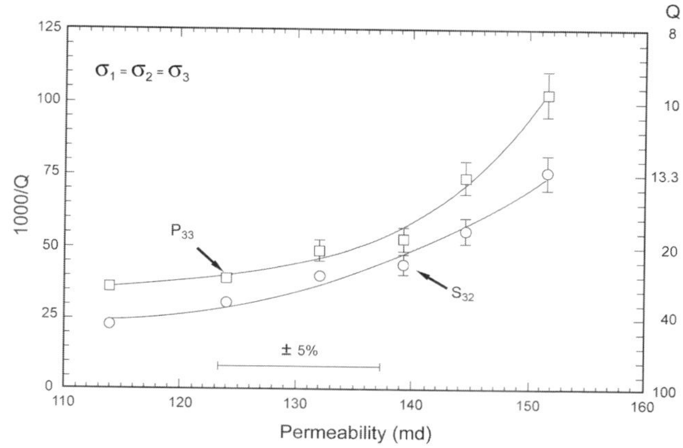


*Figure 13.57* Attenuation for P- and S-waves in the '3' perpendicular to fracture set direction for the same cube of Penrith sandstone of 13% porosity, plotted against the permeability to flow *parallel* to the fracture set. Results apply to the hydrostatic loading case, as in Figure 13.56b. King et al., 1997.

In the subsequent hydrostatic loading, the velocities tended to converge indicating near-closure, but the permeabilities of three similarly fractured sandstones, shown in Figure 13.58 reduced much more slowly with stress increase than the smooth cleavage fractures in the slate. The shear wave velocity $V_{s2}$ showed a continuous rise.

King, 2002 reported 'considerable hysteresis' in $V_{s2}$ – permeability behaviour, with subsequent crack-closing cycles. An example is shown in Figure 13.58b. King was of the opinion that the reason for the significant differences in *velocity – permeability* behaviour between the smooth and the rough fractures was 'unclear', but of course cited the difference in roughness. For some reason, rough fractures closing due to stress increase contributed to increased velocity, but suffered a less-than-expected reduction in permeability.

The reason for the different behaviour of the rough fractures compared to the smooth, may be that E (physical aperture) > e (hydraulic aperture), for the case of rough fractures (or joints), and $E \approx e$ for smooth fractures (or joints). (Barton et al., 1985). This would mean faster physical closure than hydraulic closure for rough joints.

The greater inequality with increased roughness JRC, of the average physical aperture (E), compared to the theoretical, smooth-wall hydraulic aperture (e), described more fully in Chapter 16, means that a fracture or joint with an assumed (low) aspect ratio, say $10^{-4}$, may have a permeability equivalent to an even smaller aspect ratio

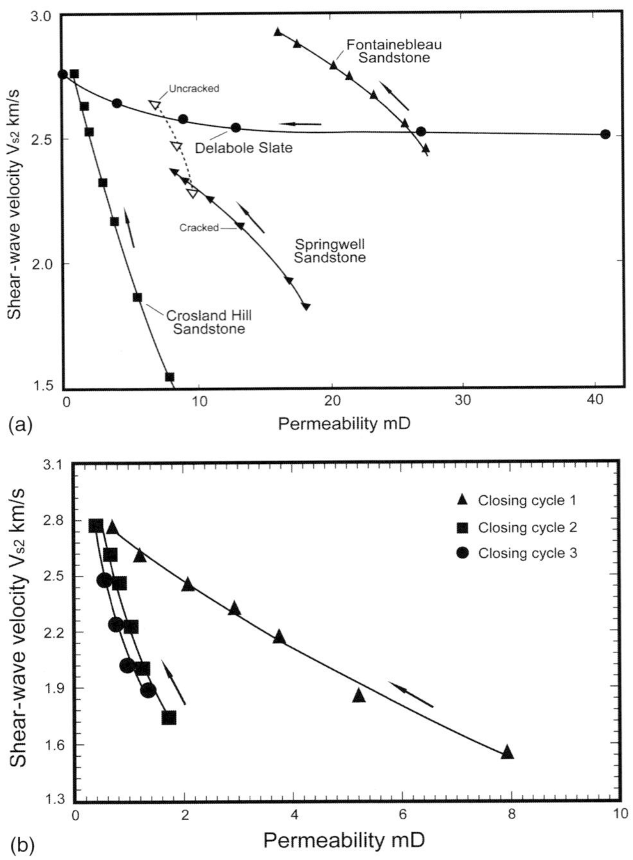


*Figure 13.58* Shear-wave monitoring of fracture closure and permeability reduction, due to increased normal loading in the Imperial College *D-H-M polyaxial* cell. a) Three cubic samples of different sandstones, each with a set of parallel, rough extension fractures, display quite different behaviour to smooth (cleavage-parallel) extension fractures in slate. b) Stress-closure *cycles* for the fractures in Crossland Hill sandstone, show strong hysteresis on the first cycle. King, 2002 and Shakeel, 1995.

e.g. $10^{-5}$. This means that a 100 μm physical aperture may have a 10 μm hydraulic aperture.

So the physical aperture (E) of a *rough joint* or *rough fracture*, with a typical JRC = 10–15 for a well controlled extension fracture, closes much faster than the hydraulic aperture under stress, perhaps helping to explain the strong velocity response and the weaker permeability response. When on the other hand the fracture has the lowest JRC of 1, typical for cleavage joints in slate, there will be no inequality between E and e. JRC = 1 is the practical limit to the following relation, implying 'table-top' planarity and smoothness. The empirical equation of Barton et al., 1985 was as follows:

$$e = \frac{E^2}{JRC^{2.5}} \quad (13.16)$$

where JRC, the joint roughness coefficient of Barton and Choubey, 1977, is fully explained in Chapter 16. With E = 100 μm, and JRC = 15, e = 11.5 μm. With JRC = 1, E = e = e.g. 10 μm. Such a fracture offers no resistance to closure: it closes to a significant degree under even a very low normal stress.

Possibly the above result is one of the reasons that in a jointed or fractured reservoir, the performance of 4D seismic can be quite successful if effective stresses are not too high, as the velocity (and particularly the attenuation) are relatively sensitive indicators of small permeability changes. By the nature of jointed reservoirs, there are unlikely to be commercially viable hydrocarbon-bearing fractures or joints with very low JRC values, as joint closure under stress would preclude both permeability and 'storage', if such was needed due to low porosity matrix.

Hydraulic fractures, or widely opening joints, due to the cooling and pressure-drive effect of water-flooding, could be expected to show a 'first-closing' cycle showing large hysteresis, when heat flow returns (see Bandis et al., 1983 ambient tests).

Velocity anisotropy due to a set of fractures, parallel or perpendicular to principal loading axes, as beautifully illustrated in the above polyaxial experiments is, inevitably, only a part of the information needed when trying to interpret field data from a jointed or fractured reservoir. Gibson and Toksöz, 1990 addressed the important question of crack orientations ranging from 0° to 90° in relation to the stress axis. Cracks of small aspect ratio (roughly $10^{-4}$) were modelled, applying the theory of Walsh, 1965 concerning crack closure under stress applied obliquely to the crack normals. The variation of the *quasi* compressional velocity with orientation, and the important link to permeability variation with crack orientation and stress level were each addressed.

Some difficulty in closely matching the sparse experimental results was indicated, including matching the more rapid closure at low stress. The reality of fracture sets having different properties affecting the stress-closure behaviour, and the possible influence of intersecting or interconnecting fractures also altering the stress-closure behaviour were discussed, but these difficulties could not of course be modelled analytically.

In this complex rock mechanics area, there were by that time promising advances in distinct element modelling

Table 13.3 'Rule-of-thumb' for order of magnitude estimates of seismic $Q_p$, based on the empirical scheme linking the *jointed rock mass quality* Q, $V_p$, $E_{mass}$ – and *in very approximate terms* – the rock mass permeability. (Barton, 1999). Water at 20°C assumed in mD to m/s conversion

| | | | | | | | | | | | |
|---|---|---|---|---|---|---|---|---|---|---|---|
| $V_p$ km/s | 1.5 | 2.0 | 2.5 | 3.0 | 3.5 | 4.0 | 4.5 | 5.0 | 5.5 | 6.0 | 6.5 |
| $Q_{seis}$ (est.) | 2 | 3 | 5 | 7 | 10 | 15 | 22 | 32 | 46 | 68 | 100 |
| K mD | 1000 | | 100 | | 10 | | 1 | | 0.1 | | 0.01 |
| K m/s | $10^{-5}$ | | $10^{-6}$ | | $10^{-7}$ | | $10^{-8}$ | | $10^{-9}$ | | $10^{-10}$ |

with the two-dimensional UDEC and three-dimensional 3DEC codes of Cundall, allowing dynamic wave transmission and attenuation through intersecting joint sets of any desired orientation or block size. Various numerical modelling methods are discussed in Chapter 15.

## 13.14 Rock mass quality, attenuation and modulus

This chapter will be concluded by bringing an empirical rock mechanics scheme into consideration, as justification for the tentative, repeated conclusion, first launched in Chapter 10, that seismic Q very much resembles the static modulus of deformation expressed in GPa, *strictly for the case of jointed or fractured rock*, for which a good data base exists. The special fracturing tests of King and his colleagues at Imperial College, have given repeated support for this simple hypothesis. Deformation moduli are almost always in the range of 5 to 100 GPa, and seismic Q seems also to be most frequently in this range in the case of near-surface rock masses (to 1 kilometre depth?), as frequently suggested by attenuation data presented in Chapters 10 and 13.

The empirical expression for static deformation modulus: $E_{mass} = 10\ Q_c^{1/3}$, and the expression linking velocity and rock quality: $V_p = 3.5 + \log Q_c$ can be merged, by elimination of $Q_c$ into the form:

$$E_{mass} = 10^{(V_p - 0.5)/3}\ (\text{GPa})\ (\approx Q_{seis}?) \qquad (13.17)$$

The $V_p$ scale on the *left-hand side* of Figure 13.59 can be followed across the diagram ('ignoring' the rock mass quality $Q_c$), all the way to the *right-hand side* deformation modulus estimates, giving (in Table 13.3) proposed $V_p$ – $E_{mass}$ (GPa) *or seismic Q* 'first-pass' estimation of (better than) order of magnitude values of seismic Q, specifically where jointed or fractured rock masses are involved. For good measure, a very rough estimate of permeability is also given, based on the Lugeon value ($1\ L \approx 10^{-7}$ m/s, and with 1 Darcy $= 10^{-12} m^2 \approx 10^{-5}$ m/s, for *water* at 20° C, we have 1 Lugeon ≈ 10 millidarcies).

There is something quite familiar about these estimates of *seismic Q* in relation to $V_p$, based on the large number of cases in fractured rock, reviewed in Chapter 10, and in relevant parts of Chapter 13. Note that $V_{p2}$ in Figure 13.54a is 2.5 km/s at the end of the cracking phase, just as the reported seismic Q of a second sandstone was 5 at the same stage of loading, following the same method of fracture development.

Beyond the 'jointed rock' modulus limit of roughly 150 GPa, the question of seismic Q value estimation is problematic, but remarkably it starts to resemble the rock mass Q and $Q_c$ value, i.e. 500, 1000, 2000 – the latter representing *completely intact hard rock* at moderate, deep and kilometre-depths, or relevant to hard, very hard and extremely hard unjointed rock at moderate depths of say ½ to 1 km.

In case of knowledge of the approximate uniaxial compressive strength ($\sigma_c$) of the rock where the P-wave velocity measurement is made, the improved linkages between ($\sigma_c$) and deformation modulus (and therefore an *approximate seismic Q*) can be applied, as shown in Figure 13.60. Example: $V_p$ measured at 2 km depth in fractured sandstones = 5.0 km/s. Uniaxial strength of sandstones ≈ 50 MPa. Estimate of 'static' deformation modulus is 25 GPa. A first estimate of seismic Q, specifically $Q_p$ is therefore 25.

The possible relative differences in *frequency* effects on $V_p$ and seismic Q is of course a source of additional error in this simple method. Porosity effects not captured by direct effects on ($\sigma_c$) may be subtracted from the estimates of modulus, using the graphical adjustments suggested in Figure 13.59. These are empirical by nature and have an insignificant data-base at high rock quality Q-values, since high n% seldom accompanies high Q-values.

In Chapter 7 concerning excavation disturbed zones, we reviewed the very thorough seismic studies of the Basalt Waste Isolation Project (BWIP) conducted by King et al., 1984 and 1986, and several other researchers. The review included EDZ effects on P- and S-wave velocities in the columnar basalt, using cross-hole seismic measurement at different depths into the tunnel wall in horizontal, diagonal (inclined), and vertical directions.

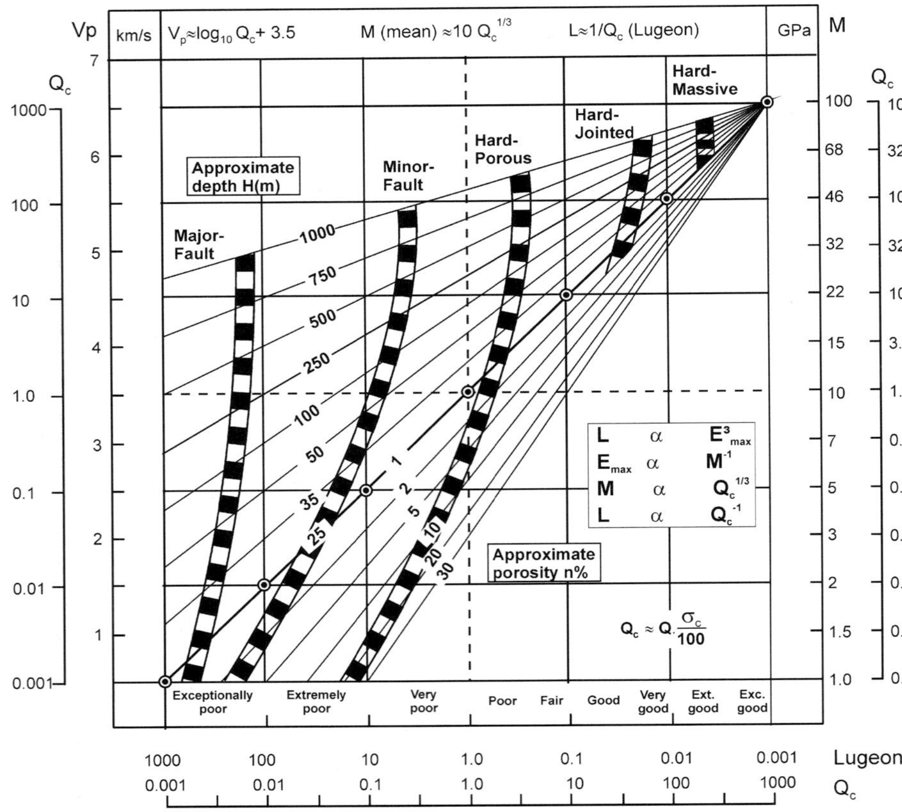


*Figure 13.59* For the case of *jointed or fractured rock*, there is a strong resemblance of the magnitude of seismic Q, to the 'static' deformation modulus expressed as GPa. The figure shows the inter-relationships developed from an empirical expression linking the rock mass quality Q-value with the 'static' deformation modulus $E_{mass}$ (or M) and velocity $V_p$. Permeability, in Lugeons ($1\,L \approx 10^{-7}$ m/s ≈ 10 millidarcies for water at 20°C) is also roughly linked, in the absence of clay-sealing of the joints. In this chart a nominal uniaxial compressive strength of 'hard' rock, namely 100 MPa forms the basis of the graphics, but using the $Q_c = Q \times \sigma_c/100$ relation, other compressive strengths can be applied, as shown in Figure 13.60. Barton, 1999a.

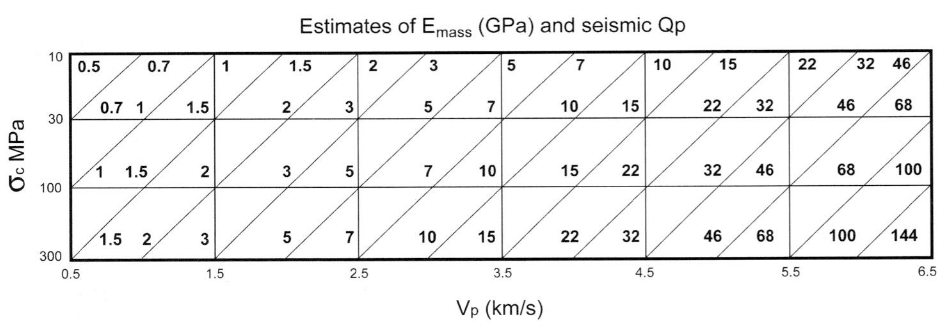


*Figure 13.60* A rough method for estimating *seismic Q* from P-wave velocity and uniaxial compressive strength linkages to 'static' deformation modulus. By implication, and also in practice, the trend when moving *downwards-and-to-the-right* is for increased depth and reduced porosity, thereby reaching values of $E_{mass}$ in the higher range of 50 to 100 or more typical for crystalline rocks or hard limestones or well-cemented sandstones.

A very interesting result of relevance to the possible link between seismic $Q_p$ and the static deformation modulus determined from $Q_{rock}$ is shown in Figure 13.61. This is not laboratory data for $Q_p$, but it is normalized by laboratory data, on the assumption that the *relative* $Q_p$ value *in situ* cannot be more than that of an *intact sample* of the same rock at the same stress level. King et al., 1986 normalized their *in situ* estimates of $Q_p$, using an assigned maximum value of 50, where the *highest* value of $Q_{seis}$ was recorded, 9 m from the tunnel face, in

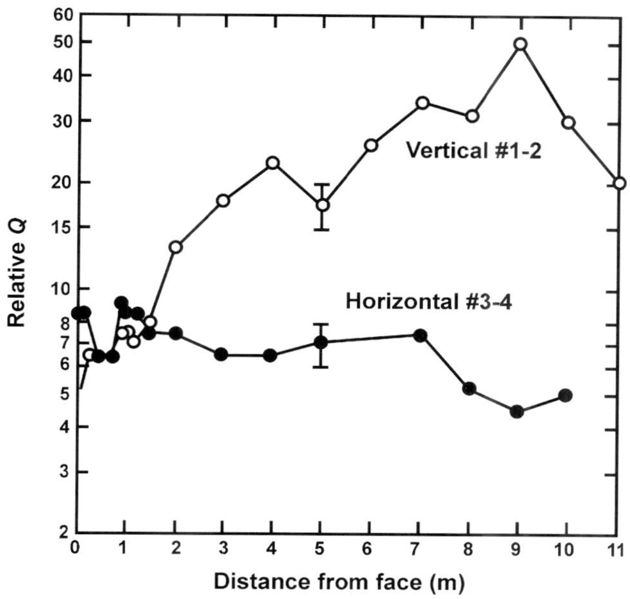


*Figure 13.61* *Relative* $Q_p$ values interpreted from seismic cross-hole measurements in the face of a shallow (46 m) tunnel. Measurements in horizontal (strongly jointed) and vertical (sparsely jointed) directions, in thin-column basalts of 0.15 to 0.36 m thickness at BWIP. King et al., 1986.

a vertical direction (not crossing any columnar joints, and probably few if any horizontal cross-column joints). The value of 50 corresponded to the laboratory test value at appropriate low stress levels.

Note the horizontal ray-path *relative Q* of about 5 to 8, measured in a direction crossing the maximum number of joints (sinuous six-sided basalt columns of 0.15 to 0.36 m thickness), and possibly under the influence of a low horizontal stress, typical of many near-surface columnar basalts. The depth of the tunnel was only 46 m. In the vertical direction, with theoretical vertical tangential stress effects perhaps more clearly imprinted, due to the relative lack of *horizontal* joints, there is an unmistakable similarity of the increase in *relative Q* to the deformation modulus increase (and magnitude, when expressed as GPa) that one often interprets from tunnel and shaft extensometer arrays.

For example Barton and Bakhtar, 1983, back calculated static deformation moduli from extensometer measurements in a deep (1.6 km) shaft in steeply bedded and jointed quartzites, obtaining moduli from 2.5 to 60 GPa from near the shaft wall to one diameter depth. There is also an unmistakable similarity of these numbers to expected $Q_p$ ranges, as also in Figure 13.61.

# 14 P-waves for characterising fractured reservoirs

The ability to detect the presence of viable hydrocarbon-bearing structures, with sufficient porosity, and with tolerable matrix and mass permeability, are among the challenges of the petroleum geologist, petrophysicist, and geophysicist, whose joint role may last far into the reservoir engineers' production phase. In this chapter we will give examples of some of the basic ways of 'seismically illuminating' reservoirs at larger scale, sometimes extending over 100's of $km^2$ to depths of 5 km. (Small-scale dipole and monopole sonic logging was described in Chapter 12, because of its intimate connection to rock stress, borehole stability and mud pressure). The larger scale techniques illustrated in this chapter range from cross-well tomography, VSP 2D and 3D multi-azimuth walk-away surveys, 4C multi-component surveys, AVO and AVOA for detecting fracture orientation, and 4D repeated surveys for tracking reservoir changes over time. One of the causes of such changes is the use of water flooding, which causes various coupled mechanisms, besides an advancing oil/water contact. 4D can also be used for monitoring the effects of compaction, and subsidence. In general, apart from a brief treatment of C-waves or converted P-S waves, we will leave a detailed description of the use of shear wave splitting and polarization for Chapter 15. This remarkable method for characterization of fractured reservoirs is a suitable finale, and a good introduction to the need for more geomechanics (in Chapter 16), for improved understanding of fractured reservoir behaviour.

## 14.1 Some classic relationships between age, depth and velocity

The most fundamental and earliest means of interpreting possible reservoir conditions at kilometre depths was the use of seismic P-wave velocity well surveys, using a simple VSP concept. There was early recognition of velocity increase with depth, with the following smooth velocity-depth function attributed to Slotnick, 1936:

$$v_z = v_o + Kz \tag{14.1}$$

where $v_o$ is the (P-wave) velocity at the surface and $v_z$ is the velocity at vertical depth z.

There was also early recognition of a quite systematic trend linking velocity to the *geological age*, in combination with the present depth of occurrence. The interesting empirical method of Faust, 1952 will be summarized as a 'geological' and 'stress-effect' introduction to hydrocarbon reservoir investigation.

Faust discovered that the greatest rate of velocity increase occurred at *shallow depth in the oldest units*, which is fundamental early proof of the importance of dual porosity. The likelihood of more joints in the stiffer, older units make these units more sensitive to stress change. However, with only $V_p$ as a dynamic indicator of conditions, acoustic closure represented a limit to the sensitivity, especially for the weaker, younger reservoir rocks.

Faust, 1951, used data from almost 1 million feet (or about 300 kilometres) of well sections, in 500 petroleum well surveys, mostly from the USA and some from Canada. The great majority of data was for mixed shale/sandstone sections. A non-systematic comparison of shale and sand (sandstone) velocities had revealed an average discrepancy of only 350 ft/sec, or 106.7 m/s in velocity between these two, frequently inter-bedded units, the sandstone having the highest velocity by this small average margin.

The ironic similarity of velocities for these two basic dissimilar as 'chalk-and-cheese' units, is a reminder of the potential 'non-uniqueness' of P-wave velocity, and the recognised need for alternative interpretation methods, such as impedance and attenuation, to distinguish the different lithologies and their fluid-bearing signatures.

The potential closeness of velocities for shale and sandstone is actually surprising, in view of the greater tolerance of the stronger sandstone to stress anisotropy, often resulting in several MPa greater minimum stress

Table 14.1 Mean depth-velocity data from Tertiary and Pennsylvanian shale-sandstone units, selected from Faust, 1951, from wells in the USA only.

| Mean depth Z (ft) | Mean velocity of Tertiary (ft/s) | No. of wells of Tertiary | Mean velocity of Pennsylvanian (ft/s) | No. of wells of Pennsylvanian |
|---|---|---|---|---|
| 1 025 | 6 800 | 71 | 9 420 | 18 |
| 2 500 | 7 660 | 63 | 11 110 | 14 |
| 3 500 | 8 160 | 63 | 11 720 | 18 |
| 4 500 | 8 670 | 64 | 12 230 | 18 |
| 5 500 | 9 220 | 54 | 12 650 | 14 |
| 6 500 | 9 520 | 53 | 12 710 | 7 |
| 7 500 | 9 860 | 48 | 13 320 | 5 |
| 8 500 | 10 220 | 31 | 13 390 | 4 |
| 9 500 | 10 670 | 23 | 13 020 | 4 |
| 10 500 | 11 090 | 13 | 14 030 | 3 |
| 11 500 | 11 300 | 2 | 14 500 | 2 |

in the shale, which is frequently a fluid barrier for the hydrocarbon-bearing sandstone.

Faust minimised lithological variations by averaging all measurements at the same depth and age. At the outset he assumed that velocity (V) could be expressed as a function of depth (Z), age (T) and lithological variables (L). Since L was considered too problematic (i.e. how to compare the separate units limestone, shale and sandstone) shale and sand were accepted as representing equivalent sections, as they alternated too frequently in relation to the usual interval of down-hole velocity measurement, which then was as much as 500 feet (or 152 m). Table 14.1 gives an extract of the data for the two most frequently represented geologic ages, namely the Tertiary and the Pennsylvanian, which follows the Permian. The oldest units were Devonian (15 wells) and Ordovician (3 wells).

Although reproducing very 'smoothed' data due to the averaging process, Faust, 1951, was careful to point out that the maximum deviations from the velocity averages could be great. Nevertheless both minimum and maximum values also demonstrated increased velocity with depth. Comparison of same-age same-depth data from different regions of the USA, reportedly showed little systematic deviation of the mean from one area to another. However the Devonian of the Appalachian Basin and the Eocene and Cretaceous of SW Texas had velocities more typical of data 3000 ft (915 m) deeper. This was thought to be due to unusually high degrees of cementation.

Figure 14.1 is derived from the extensive data that was partly sampled in Table 14.1. It shows the quite systematic average trends of velocity, age and depth. The greatest rate of velocity increase occurs at 'shallow' depth in the oldest units, which is presumably a function of more joints that therefore make the older units more sensitive to stress change. The steepest velocity-depth gradients occur, of course, in the 300 m to 1 km depth range, 1000 ft being the approximate limiting depth of measurement.

The trend of the data is given by Faust, 1951, as:

$$V = \alpha_m Z^{1/n} \tag{14.2}$$

which was reportedly similar to a current weathering correction method known as the 'Blondeau weathering'. Z is depth in feet, $\alpha_m$ is a constant with units of velocity (that proves to be age-dependent) and n is a constant, independent of age.

Faust plotted the same data in log-log format, as shown in Figure 14.2. The velocity at each depth could then be represented as:

$$\log V = \log \alpha_m + \frac{1}{n} \log Z \tag{14.3}$$

To the extent that equation 14.2 is correct, $\alpha_m$ represents the velocity for a particular geological age at 1 foot depth (0.3 m). The relevant geological ages and the 'zero depth' velocity constants ($\alpha_m$) are given in Table 14.2. The constant n = 6, applies for all the curves in Figure 14.2.

By plotting $\alpha_m$ against age, shown in Figure 14.3, a linear log-log plot was obtained of the form:

$$\alpha_m = \alpha T^{1/6} \tag{14.4}$$

where T is geologic age in years and α is a constant. Equations 14.2 and 14.4 could therefore be combined as:

$$V = \alpha (TZ)^{1/6} \tag{14.5}$$

In Faust, 1951, α is given as 125.3, and is numerically equal to velocity in feet per second when TZ = 1. From the literature of the time, values of α of 121.5 and 127.2 were also quoted by Faust. A mean value of 124.7 from these three results suggests that α = 125 would be an acceptable constant.

An example will now be given to illustrate this simple, early technique.

| | |
|---|---|
| Assume | Pennsylvanian = 220 $\times 10^6$ years (Table 14.2)<br>$\alpha_M$ = 3,047 ft/s = 929 m/s (Table 14.2) |
| Evaluating equation 14.5 at | Z = 1,000; 1,500; 2,000; 3,000 ft, we obtain:<br>V = 9,712; 10,391; 10,900;<br>11,663 ft/s = 2.96; 3.17; 3.32; 3.56 km/s |

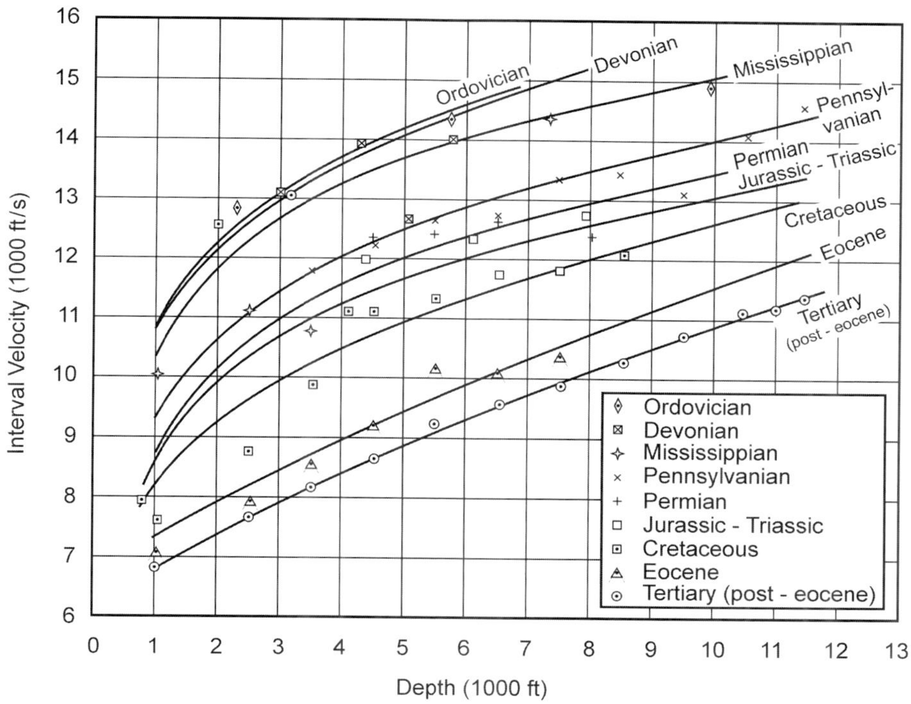


*Figure 14.1* Velocity-depth trends for in situ reservoir rocks of various geological ages (linear scales). Faust, 1951.

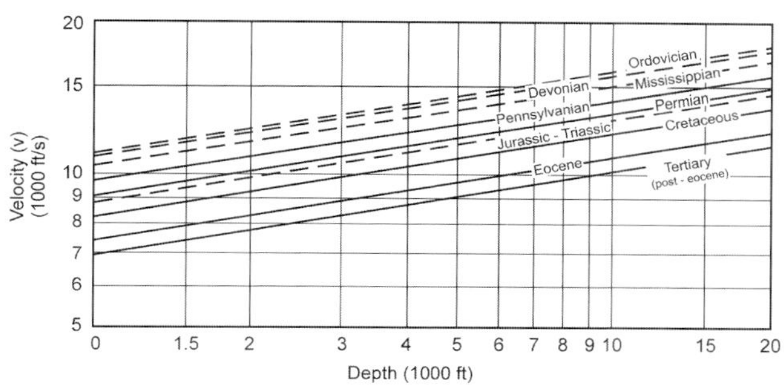


*Figure 14.2* Velocity-depth trends for in situ reservoir rocks of various geological ages (log scales). Faust, 1951.

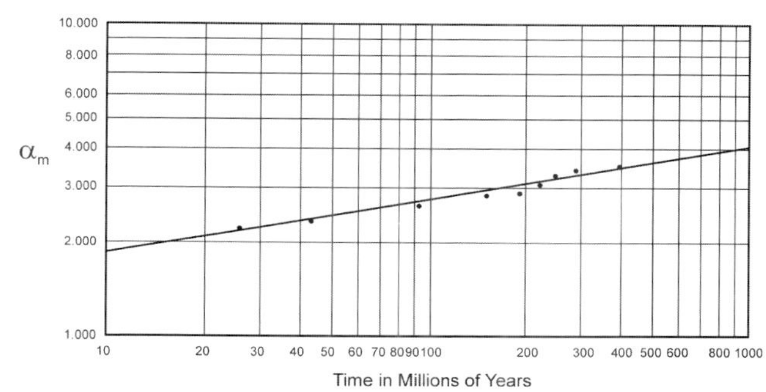


*Figure 14.3* Mean values of $\alpha_m$ (Velocity in ft/s at nominal 1 ft depth) for different ages. Faust, 1951.

Table 14.2 Geologic age and 'zero depth' velocity constants. Faust, 1951.

| Geologic age | Geologic time | Velocity constant $\alpha_m$ ft/s | Velocity constant $\alpha_m$ m/s |
|---|---|---|---|
| Tertiary | $26 \times 10^6$ yrs | 2 190 | 668 |
| Eocene | $43 \times 10^6$ yrs | 2 332 | 711 |
| Cretaceous | $93 \times 10^6$ yrs | 2 607 | 795 |
| Jurassic-Triassic | $152 \times 10^6$ yrs | 2 823 | 861 |
| Permian | $192 \times 10^6$ yrs | 2 866 | 874 |
| Pennsylvanian | $220 \times 10^6$ yrs | 3 047 | 929 |
| Mississippian | $245 \times 10^6$ yrs | 3 235 | 986 |
| Devonian | $284 \times 10^6$ yrs | 3 380 | 1 030 |
| Ordovician | $390 \times 10^6$ yrs | 3 439 | 1 048 |

These predicted velocities at 305 m, 457 m, 610 m and 915 m depth, are apparently much less sensitive to stress increase than jointed rock would be to sub-horizontal wave paths, from refraction seismic or cross-hole tomography. Perhaps this is because we are looking at an inter-bedded sandstone-shale unit of Pennsylvanian age, with the predominant measurement direction in a sub-vertical direction, with shot points only some 1,000 ft (305 m) from the well, and with downhole receivers to record the approximately 0.09 to 0.08 s average arrival time, limited to each 1,000 ft depth interval (in the case of the Pennsylvanian age sediments).

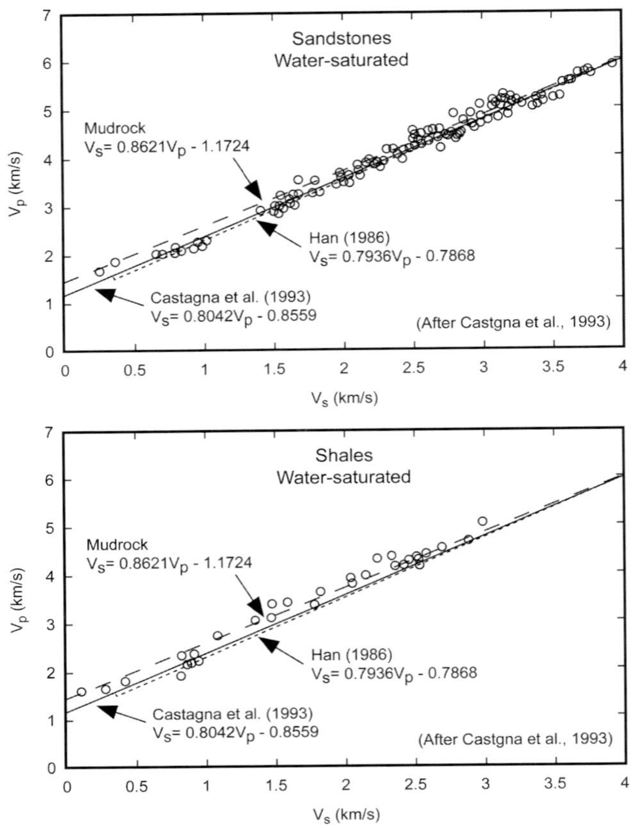


*Figure 14.4* Remarkable similarities of $V_s$ versus $V_P$ trends for water saturated sandstones and shales, according to ultrasonic testing in the laboratory at effective pressures of up to 40 MPa. Castagna et al., 1993, as reproduced in Mavko et al., 1998.

The natural suspicion that the 'coarse' VSP-style well logging available at this time may have been the cause of the poor differentiation of the shale and sandstone in terms of their relative P-wave velocities, as referred to at the beginning of this review of Faust, 1951, is definitively *not supported* by confined ultrasonic laboratory data for shales and sandstones. Figure 14.4, from work by Castagna et al., 1993, and reproduced by Mavko et al., 1998, shows that the three given empirical relationships for $V_P$ and $V_s$ seem to fit *both* data sets very closely. Han's easy to remember 1986 relationship:

$$V_s = 0.79V_p - 0.79 \quad (\text{km/s}) \tag{14.6}$$

was shown by Mavko et al., 1998 to give a very good fit to a wide variety of (water saturated) shaly sands, from 4% to 39% porosity, with clay volume fractions spanning 0 to 55%, and over a confining pressure range of 0 to 40 MPa, equivalent to depths up to roughly 3 km.

## 14.2 Anisotropy and heterogeneity caused by inter-bedded strata and jointing

Fine layering of alternating porous and impermeable strata is obviously one of the basic sedimentary systems that contribute to the existence of potential reservoir rocks in sedimentary basins. Sandstones with their usually elevated porosities, may be the recipients of the hydrocarbons, and the variously aged shale facies may contribute both as the deeper and or laterally occurring source rocks, and as the elevated seals or cap-rocks. Some form of deformation of the recipient strata, through folding or faulting, is usually necessary to ensure that a trap is formed. Petroleum may therefore be found within anticlinal structures containing favourable jointing, if capped by impermeable shales or salt rock. Frequently it is the larger scale faulting that is responsible for the porous reservoir rocks and sealing layers to be juxtaposed, or sealed by shale-smear if sufficiently plastic. The final requirement is a favourably tilted dip direction.

Fine layering of sedimentary strata means that the dominant wavelength of a seismic or sonic pulse is long compared to the thickness of individual layers. The medium will nevertheless exhibit effective (and real) anisotropy, with a vertical symmetry axis in the case of horizontal layering. In the presence of hydrocarbons this layered medium may show substantial attenuation and velocity dispersion, which will be compounded with the additional (or separate) presence of jointing or fracturing.

The authors Helbig and Thomsen, 2005 emphasised that anisotropy, and the associated new techniques: primarily shear wave splitting and polarization, bought with them a new exploration concept: *neither* exploring for the presence of reservoir rock, *nor* for the presence of hydrocarbons, but for the presence of crack or fracture *permeability*. In 4D repeated surveys, *changes* of fracture permeability due to flooding and production has also become a modern goal in the use of seismic and attenuation anisotropy.

Following Lynn, 2004, a 'thin bed' is $3/8$ of a wave length, the limit for a discrete reflection both from the top and bottom of the bed. Wave scattering, attenuation and dispersion occur when the ordered heterogeneities have scale lengths of about 0.3–0.01 of the wavelength, while the smallest scale of ordered heterogeneity, less than 0.01 of the wavelengths, *may be the cause of most of the azimuthal and offset dependent velocity*. Conventional seismic wavelengths are much longer than the scale lengths of either of the features that govern dual-porosity flow in a reservoir.

As pointed out by Williams and Jenner, 2002, the earth does not care which tools or frequencies we use; it still knows it is anisotropic. Strong P-wave velocity anisotropy is being observed in every geologic environment, with the possible exception of basins under primary deposition and burial. P-wave azimuthal anisotropy was previously ignored, and left to the research and technology group, but is now known to be one of the most significant properties of the acquired seismic data. Unfortunately, in the marine environment, fully populated offsets in each azimuth class are less common than on land, but Williams and Jenner emphasised that even narrow azimuth data gave an opportunity to see the effects of azimuthal anisotropy.

Hand-in-hand with the basic anisotropy caused by sedimentary layering, and deformation processes, is marked *heterogeneity*, occurring at many scales and for many reasons. As Nur, 1989 pointed out, reservoirs are much more heterogeneous than anybody likes to believe, and as time goes by more and more reservoirs are re-classified as severely heterogeneous, due to a multitude of dynamic flow-related cyclical events during their formation, and due to fracturing and faulting in subsequent geologic eras, each of which become better understood as time, and seismic developments, advance. An appropriate quotation from Lynn, 2004, also has relevance here. 'Fractures are like cockroaches. There is no such thing as one cockroach. If you see one, a whole family of all scale lengths is hiding nearby.'

In a survey of one hundred fractured (i.e. jointed, not MHF) oil reservoirs from around the world, it was found convenient to divide the reservoirs into four groups (Allan, 2002):

1. Type I: little matrix porosity and permeability – fractures provided both storage capacity and fluid-flow pathways.
2. Type II: low matrix porosity and permeability – matrix provides some storage capacity – fractures provide fluid-flow pathways.
3. Type III: *micro-porous* reservoirs with high matrix porosity but low matrix permeability – matrix provides the storage capacity – fractures provide the fluid-flow pathways.
4. Type IV: *macro-porous* reservoirs have high matrix porosity and permeability – matrix provides both storage capacity and fluid-flow pathways – fractures merely enhance permeability.

The author warned that Type I and Type II reservoirs could be easily damaged by excessive production rates (due presumably to the rather strong sensitivity of permeability to unnecessarily high effective stress levels), but many performed well under unassisted primary recovery when managed properly. In Type III reservoirs the recovery factor was dependent on lithology, wetability, and fracture intensity. The choice of appropriate EOR (enhanced oil recovery) was essential for optimal exploitation. In Type IV reservoirs, the recovery factor was most sensitive to the selected drive mechanism.

The sophistication of investigation methods, using multi-component and multiple-frequency methods, gives the capability of revealing heterogeneity and fracturing at many scales, even if indirectly, thanks to some remarkable and fortuitous dynamic wave properties.

A good analogy to the developing heterogeneity of a better understood reservoir, is all the adverse *faulting* gradually revealed by successive drilling at a potential nuclear waste disposal site, perhaps causing its eventual rejection, after many years of costly investigations. With huge quantities of petroleum in place at a potential or existing reservoir, rejection is seldom an option, and better understanding through improved seismic and enhanced production techniques, are the obvious ways forward.

### 14.2.1 Some basic anisotropy theory

For reference purposes we need to summarize some basic elements of isotropic and anisotropic behaviour, since various categories of anisotropic behaviour will now be treated in somewhat more detail than in earlier chapters.

A linear elastic isotropic medium requires only two constants to specify the stress-strain behaviour, either Young's modulus (E) and Poisson's ratio ($\upsilon$), or alternatively Lamé's constants ($\lambda$) and ($\mu$), where ($\mu$) is the shear modulus. These pairs of parameters can be derived from each other using standard equations of elasticity (see Chapter 1, and refer also to the Rock Physics Handbook by Mavko et al., 1998, and Birch, 1961).

Matrix format (6 × 6) is used to represent the three-dimensional elastic tensors, which in the isotropic case, requires only two parameters, as follows:

$$\begin{pmatrix} \lambda + 2\mu & \lambda & \lambda & 0 & 0 & 0 \\ & \lambda + 2\mu & \lambda & 0 & 0 & 0 \\ & & \lambda + 2\mu & 0 & 0 & 0 \\ & & & \mu & 0 & 0 \\ & & & & \mu & \end{pmatrix} \quad (14.7)$$

where the term $\lambda + 2\mu$ can be expressed as $c_{11}$, $\lambda$ as $c_{12}$, and $\mu$ as $c_{44}$.

Figure 14.5 shows the contrasting elastic tensors for three classes of *anisotropy*:

- Transversely isotropic with vertical axis of symmetry (TIV) typical of fine layering in shales. (Five independent constants, since $c_{66} = \frac{1}{2}(c_{11} - c_{12})$. Tensor elements 1,3 and 2,3 are the same term $c_{13}$)

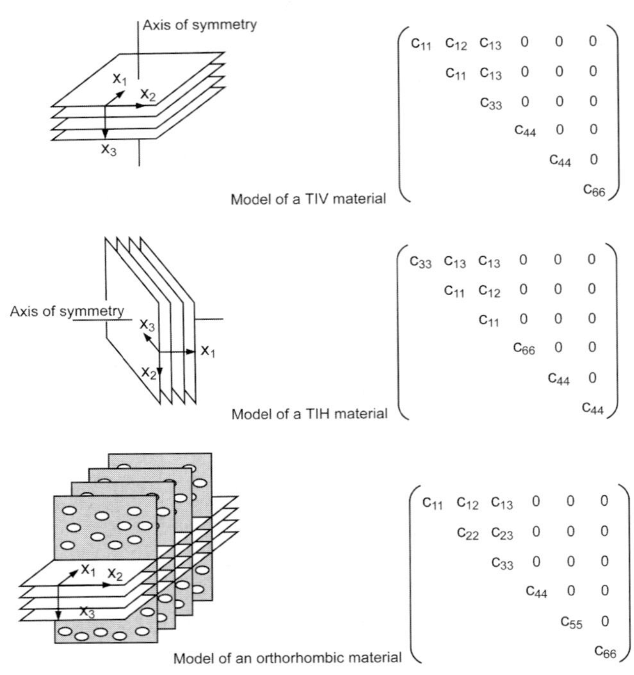


*Figure 14.5* Elastic tensors in 6 × 6 matrix format, for two *single* 'geological' classes of symmetry (TIV and TIH), combined as more realistic orthorhombic material in the third diagram, and requiring nine instead of five independent constants. Typical of bedded or layered rock with anisotropic horizontal-stress related, aligned vertical jointing, fracturing and/or microcracking.

- Transversely isotropic with horizontal axis of symmetry (TIH) perpendicular to the vertical layers, but typically caused by stress-related aligned, vertical jointing, fracturing, and/or microcracks. (Also five independent constants: tensor elements 1,2 and 1,3 are the same term $c_{13}$)
- Orthorhombic symmetry typical of horizontally bedded rock containing a set of vertically aligned joints, fractures and/or microcracks. (Nine independent constants).

Symmetry, as above, is an important property, if found, as the wave fronts will be similarly symmetric. When processing orthorhombic data (which can have a second perpendicular fracture set and still be termed orthorhombic), *if fast and slow directions have been identified*, then azimuth sectoring can be applied in these directions, and according to Lynn, 2004, most 'isotropic' processing codes can function fairly well on these azimuth-sectored data.

## 14.3 Shallow cross-well seismic tomography

For consistency of presentation we may start this brief review of reservoir characterization, by referring to an interesting *shallow* 3D tomographic imaging of both seismic velocity and attenuation, given by Brzostowski and McMechan, 1992. This was based on a 1989 3D survey shot by Peko Oil USA, and by Texaco in SE Oklahoma. Figure 14.6 (a to d) shows the surface geology and $V_p$ tomograms for the first, second and third constant-thickness slices centred at depths of 32, 96 and 106 m. The corresponding seismic $Q_p$ tomograms are shown in Figure 14.7.

As is readily noted from this interesting set of data, the alluvium (e.g. unit $Q_{al}$) had both low velocity (≤2.25 km/s) and low $Q_p$ (≤6) at 32 m average depth, while the lithified sandstones had velocities between 2.5 and 3.5 km/s at 32 m average depth and $Q_p$ values of about 7 to 10. Clearly the $V_p$ values for all units and the corresponding $Q_p$ values increase with depth. However, more lateral variation is shown in the $Q_p$ tomogram. As we have seen both in Chapter 1 and in Chapter 11, this lateral variability is also common at much larger scales.

The suggested link between $V_p$ and $Q_p$ developed at the end of Chapter 13 (Figure 13.60) suggests $Q_p$ values lower than 2 to 3, in the case of weak materials in the presence of velocities lower than 2.25 km/s. Where velocities climb to 3.5 km/s in much of layer 3 (centred

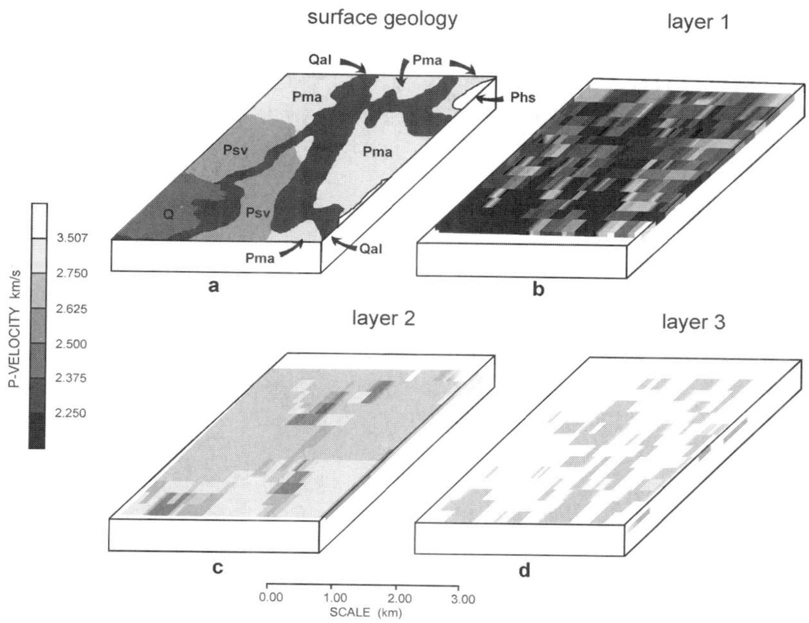


*Figure 14.6* 3D P-wave velocity tomograms for mid-depths of 32, 96 and 106 m depth in quaternary sediments, alluvial stream beds and sandstones. Brzostowski and McMechan, 1992.

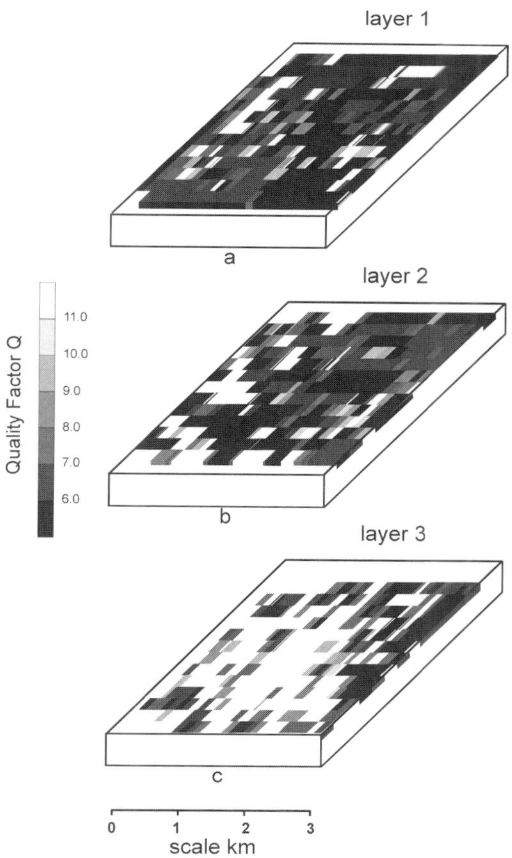


*Figure 14.7* 3D $Q_p$ tomograms corresponding to Fig. 14.6 slices. Brzostowski and McMechan, 1992.

at 106 m depth), the measured $Q_p$ was predominantly from 5 to >10, and Figure 13.60 suggests $Q_p$ of at least 5 to 7, in the case of moderately weak materials, such as 'shallow' sandstones.

Since both $V_p$ and seismic Q contain a certain level of information about the pore space and degree of jointing or fracturing, and the saturating fluid, the trial combination of these parameters in the form of tomographic plots of $Q_{seis}/V_p$, possibly offers the possibility of a closer estimate of transport properties.

In an interesting study using tomographic imaging between a total of six wells with known lithology, but unknown transport properties, Liao and McMechan, 1997 showed what might be delineated by the combination of tomographic images of $V_p$, seismic Q, and their ratio $Q_{seis}/V_p$. Figure 14.8 shows the 210–400 m deep shale/sandstone sequence. The Gypsy sands were high porosity (mean 20%), high permeability (mean 560 mD), water-saturated, clean *sand channels* lying within a silty flood-plain.

High-resolution cylindrical piezoelectric *bender* transducers were used as both sources and receivers, over a band width of 300–1300 Hz. From the diamond-pattern of six wells, the data between wells 1, 5 and 7 were selected for $Q_{seis}$ and $V_p$ tomographic analysis. Overall full-depth, three-hole, results are shown in Figure 14.9, while a detailed 90 m section between wells 7 and

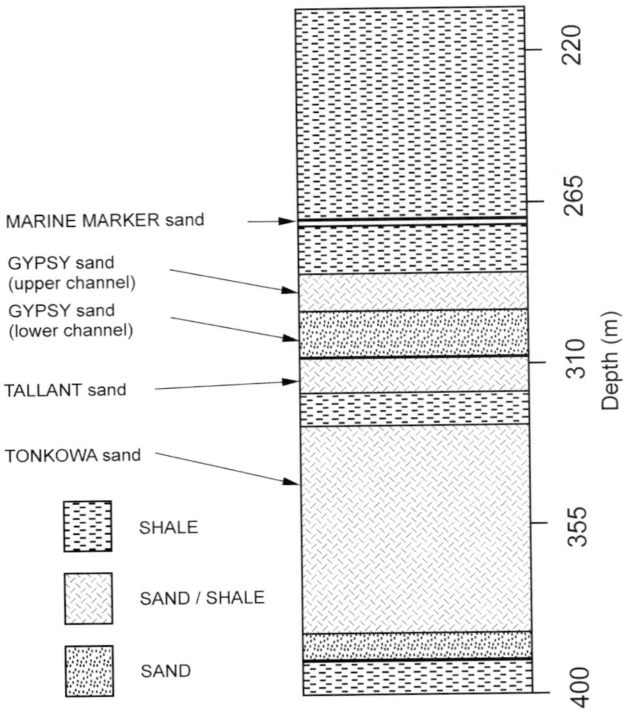


*Figure 14.8* Sand, shale and sand-shale lithology at the three-well tomographic site described by Liao and McMechan, 1997.

5 is shown in Figure 14.10. This has a marked shale-rich upper 40%, and a sand-rich middle-to-lower section.

The typical in situ values of the parameters were as follows:

| | $V_P$ | $Q_{seis}$ | $Q_{seis}/V_P$ |
|---|---|---|---|
| sandstone | 3.0 km/s | 45 | 15 |
| shale | 3.6 km/s | 30 | 8.3 |

The low velocity was a consequence of the high porosity (mean 20%) of the channel sand. The relatively low attenuation ($Q_{seis}$ as high as 45) was reportedly due to its full water saturation, and high porosity. The authors considered it likely that high values of the ratio $Q_{seis}/V_P$ would correlate with the most porous and most permeable zones, and low values with reduced flow such as the shale-rich layers. They considered that a thin and a thick shale barrier did show appropriately low values of the ratio (i.e. light colours in Figure 14.10, panel c).

The authors cited Castagne et al., 1993, who had suggested that sandstones often have higher velocity and show less attenuation than shales in the same environment. In these channel sands the low velocity (3 km/s) is clearly a function of lack of cementation and their high porosity, while the full-saturation makes

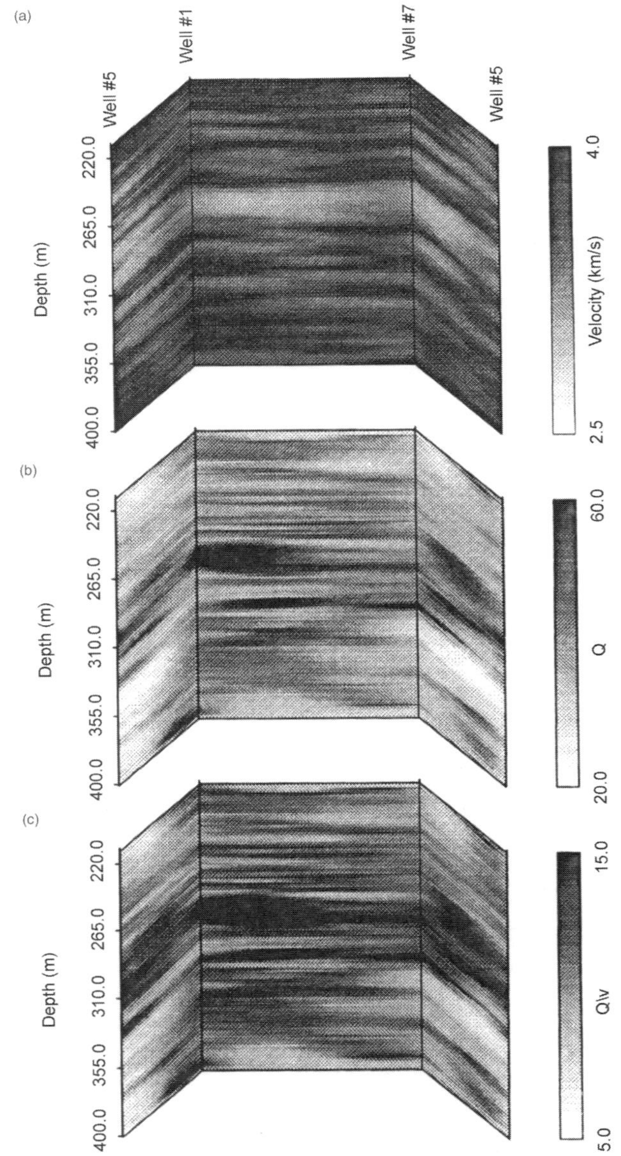


*Figure 14.9* *Fence* diagrams of seismic tomography analysis between wells 1, 5 and 7 to full depth. The wells were 99 m, 141 m and 105 m apart. Liao and McMechan, 1997. Note $V_P$, $Q_{seis}$ and $Q_{seis}/V_P$ tomograms. Liao and McMechan, 1997. Reproduced by kind permission.

them less attenuating (no squirt flow losses), giving a relatively high $Q_{seis}$ of 45. This combination (ratio = 15) is many times higher than the modest ratios of $Q_P/V_P$ suggested in the *jointed rock* model suggested in Figure 13.60, at the end of Chapter 13.

Perhaps, on reflection, this contrast in $Q_{seis}/V_P$ ratios, being so marked, could delineate the difference between unconsolidated sands and jointed sandstones, and between weak plastic shales, and the less desirable fissured/jointed, or indurated variety.

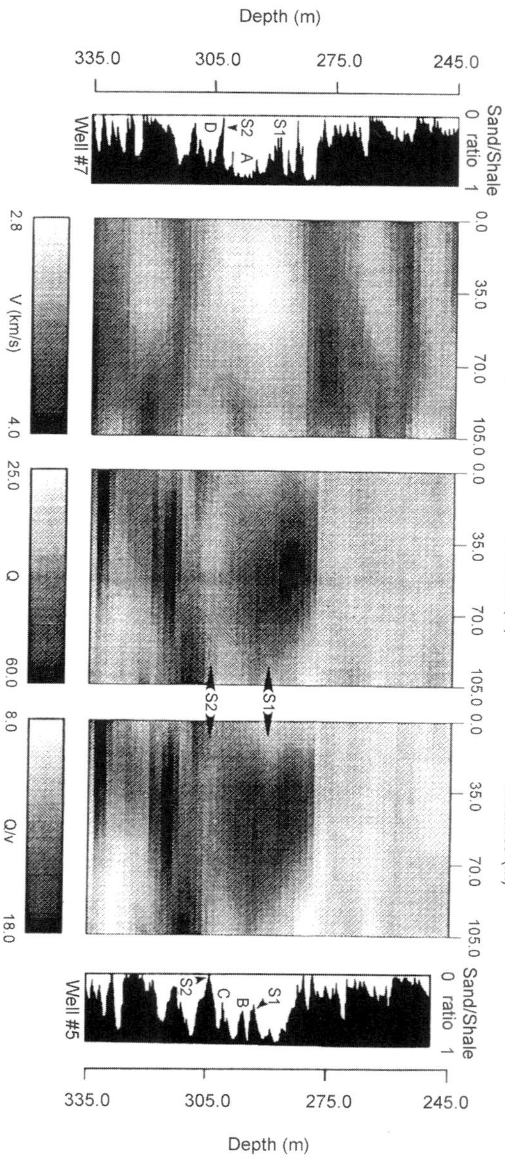


*Figure 14.10* Selected tomographic images of $V_p$, $Q_{seis}$, and their ratio $Q_{seis}/V_p$, between well 7 (left) and well 5 (right), for contrasting shale-rich (top), and sand-rich (lower-middle) zones within a 90 m section between the wells. When dominated by shale, ratios of $Q_{seis}/V_p$ are low, and when dominated by sand they are high. Sand/shale proportions for each borehole are shown. Liao and McMechan, 1997. Reproduced by kind permission.

### 14.3.1 Shallow cross-well seismic in fractured rock

The question of resolution levels necessary for fracture detection was illustrated by Majer et al., 1997. Working at the Conoco Borehole Test Facility, their goal in shallow (15 to 35 m) water saturated limestones, was to image gas filled fractures in wells with 50 to 100 m separation. The authors had started out using standard low resolution VSP and surface reflection methods, which yielded little information on the individual fractures, but did indicate P-wave anisotropy due to a N70°E trending set of joints.

Cross-well, high frequency (1 to 10 kHz) surveys using piezoelectric sources, were later used with success, to image an individual fracture zone in the limestone, before, during, and after air injection. The air followed the hydrologically-predicted pathway, and helped to seismically 'illuminate' the zone. Subsequent slant-hole coring verified the presence of a conducting, vertical fracture $<1$ m from the seismically located position. The wave-lengths used were of the order of fi to 1 m, and the fracture aperture reportedly as large as 1 mm.

The P-wave velocity of the limestones determined from earlier near-offset VSP varied from about 3.8 to 4.2 km/s between the shallow depths of 16 and 26 m, which suggests a rock quality Q-value of 2 to 5 from the empirical relation $V_p \approx 3.5 + \log Q$ derived in Part I, relevant to nominal 25 m depth, '100 MPa, low-porosity' rock. Q = 2–5 is typical for rock masses with three sets of joints, moderate block size, and with possible weathering of the joint walls: i.e. from Appendix A: $Q \approx 90/9 \times 2/2 \times 0.66/2.5 \approx 2$ to 3.

A *seismic* Q of the order of 5–20 times higher than $Q_{rock}$, as suggested from numerous cases reviewed in Chapter 10, suggested an 'order of magnitude' estimate of $Q_{seis} \geqslant 10$–25, at this shallow, jointed site.

The method suggested in Figure 13.60, and Table 13.3, linking $Q_{seis}$ to the 'static' deformation modulus estimate, suggests a $Q_{seis}$ value of about 13–14 from $Q_{rock} = 2$ to 3, if the uniaxial compressive strength of the limestone was around 100 MPa. Alternatively, utilising Figure 13.60, for evaluating the possible effect of lower uniaxial strengths, we see that $Q_{seis}$ for this jointed (i.e. fractured) limestone might be about 12 to 16 with the given $V_p$ range of 3.8–4.2 km/s.

### 14.3.2 Cross-well seismic tomography with permeability measurement

A new poroelastic analytical model called '*super – k*' was described by Yamamoto, 2003, who was working not in petroleum provinces, but on a limestone *aquifer*. Pointing out that extraction of permeability from velocity and

Table 14.3 Seismically derived, frequency-dependent cross-well data for $V_p$ and Q, with the seismically derived (*super-k*) prediction of permeability and porosity. Limestone section at 462 m depth. Yamamoto, 2003.

| Frequency (kHz) | $V_p$ (m/s) | Q | k (d) | φ |
|---|---|---|---|---|
| 2 | 3545 | 14.0 | 33.7 | 0.34 |
| 4 | 3580 | 18.1 | 33.7 | 0.35 |
| 8 | 3616 | 31.0 | 35.0 | 0.36 |
| 12 | 3652 | 49.0 | 35.7 | 0.34 |

Note: k from pumping test = 33.2 d
k from packer test = 36.3 d
porosity from neutron log: φ = 39%

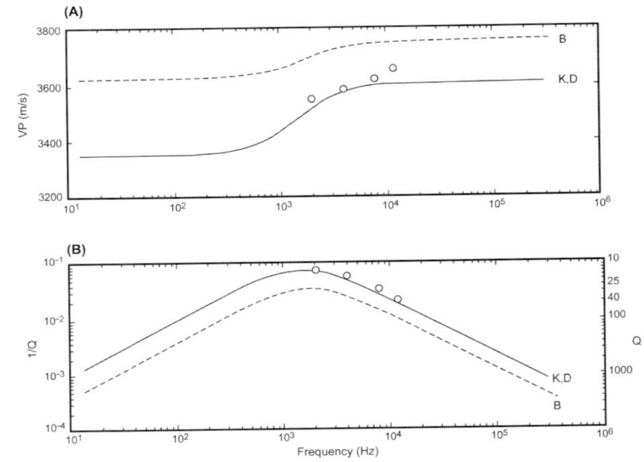


*Figure 14.11* Comparison of models (K = *super – k,* B = Biot, and D = BISQ) with cross-well velocity and attenuation data from 300–480 m depth at a permeable aquifer site. Data from Yamamoto and Kuru, 1997. Yamamoto, 2003.

attenuation data had been one of the important unsolved problems in seismic exploration, the author was able to demonstrate good correspondence between acoustically-imaged permeability and hydraulic pumping (extraction) and packer (injection) tests. An unusual close, 11 m horizontally-spaced cross-well set-up was used for deriving 4 kHz velocity and attenuation tomograms. Data was also acquired at five other frequencies (1, 2, 3, 5, and 0.250 kHz). Permeability was measured in four packered intervals down the same 300–480 m deep test section.

His model reportedly coincided numerically with the combined Biot and squirt flow mechanism (the BISQ model of Dvorkin et al., 1994), when permeability was >100 md, and when the frequency was <100 kHz (the so-called *super-k* regime, where the pore fluid is always relaxed). The limestone where the author performed these studies had extremely high permeability (10–250 d). He described a large wave attenuation and velocity dispersion in this highly permeable limestone. This in situ data is reproduced in Figure 14.11.

The author divided the 180 m × 11 m imaged cross-section into seven, mostly 25 m long sub-sections, for the purpose of velocity and attenuation inversion. His analytical inversion model for permeability involved a quadratic equation for a non-dimensional permeability-frequency parameter, with attenuation as a polynomial constant.

It may be noted that although the porosity and permeability were of exceptional magnitude, the seismic velocities of 3.5 and 3.6 km/s obtained at the two lowest frequencies of 2 and 4 kHz, were matched with seismic Q values of 14 and 18, which are quite close the *jointed rock* model potentially linking $V_p$ and $Q_p$, if the assumption is made that the uniaxial strength of the limestone is high, and that the high porosity and permeability is due to solution channels. (If porosity was distributed it would presumably resemble chalk more than limestone). Significantly, the *mismatch* with the $Q_{seis}$ *modulus-model* (Figure 13.60) increases strongly at higher frequencies.

### 14.3.3 Cross-well seismic in deeper reservoir characterization

A special issue of *Geophysics* was devoted to cross-well methods in 1995, due to the large number of unsolicited papers on the subject, following strong oil industry interest in use of cross-hole tomographic methods since the late 80's, for imaging below the resolution of surface seismic, for such purposes as steam injection front imaging. An important discovery from the early period of cross-well research, was that very high frequency could be propagated over distance of several hundreds of metres, when both source and receiver were down-hole, in deep boreholes.

The additional discovery of very high bandwidth with downhole receivers and sources apparently confirmed the phenomenon that seismologists also have to live with, if without downhole instruments (Chapter 10), that most of the surface-based frequency loss is due to attenuation in the near-surface low-Q zone. Rector, 1995, mentioned an order of magnitude improvement in subsurface resolution, with many investigators now using secondary arrivals, such as reflections, to obtain very high resolution images. As at the surface, where refraction imaging was gradually replaced by reflection imaging, the same occurred with VSP, and now was occurring when receiver *and* source were downhole.

Mathisen et al., 1995 gave an example of the use of *time-lapse* cross-well data in heavy oil sands, using a cemented multiple-receiver cable in one hole. Twenty seven cross-well surveys were acquired between two wells, during a 3½ month period, before, during, and after a 34 days steam injection cycle. $V_p$ and Poisson's ratio tomograms were used to track the changes caused by both the temperature and viscosity reduction. Stratification with a dip of 20° controlled the flow of steam, which caused progressive reductions in P-wave velocity of up to 90 to 270 m/s adjacent to and above the injectors. Poisson's ratio tomograms showed a corresponding decrease (0.1), in the same areas. Time-lapse S-wave tomograms demonstrated great stability with time, showing only very slight change (i.e. shear waves not registering fluid viscosity changes), and therefore demonstrating that the reservoir 'rock' was stable despite the fluid changes.

## 14.4 Detecting finely inter-layered sequences

Remaining at shallow depth for the moment, namely a 250 m deep research well through finely inter-layered limestone, shale, and sandstone sequences in the Imperial College Borehole Test Site, one may note the important work by Sams, 1995, who showed how the combined use of a borehole compensated (BHC) sonic logging tool, a compensated formation density tool, and a Formation Micro Scanner (FM) could resolve much of the detail of finely interlayered rock sequences.

Figure 14.12 shows the success of multiple borehole logging tools in resolving some details of finely inter-layered rocks.

Sams, 1995, pointed out that the minimum detectable layer thickness using standard sonic tools may be no less than 15 cm. In the limit when the layers become thin with respect to the wavelength, the particular layer will be transparent. Standard logs will tend to under-estimate the effects of layering on measurements of attenuation and anisotropy. Consequently, intrinsic attenuation and dispersion may be incorrectly assessed.

Any distorted frequency dependence of the intrinsic attenuation would subsequently adversely affect estimates of the physical rock properties and of the contained fluids. This is because the intrinsic attenuation is

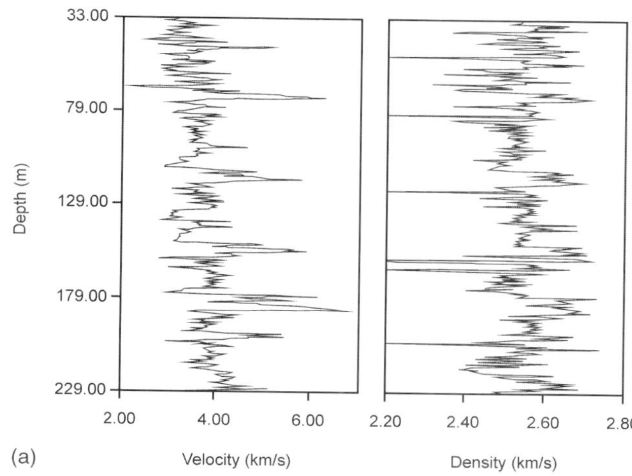

(a)

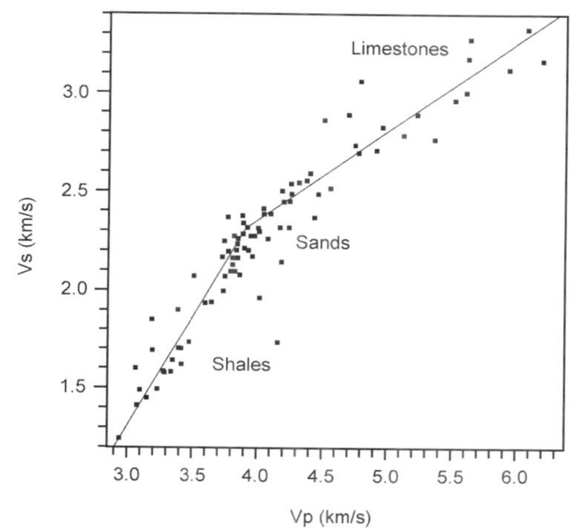

(b)

*Figure 14.12* a) The results of multiple borehole logging tools in resolving the velocities and densities of finely inter-layered rocks. b) The relative velocities of the shales, sandstones and limestones from *laboratory tests* of $V_p$, and an assumed $V_s$ correlation. Sams, 1995.

important for remote detection of fluids, such as kerogen rich shales and their state of maturation.

As discussed at length in Chapter 12, the probability of a variable, mini-EDZ in the walls of a well that penetrates inter-bedded strata of different stiffnesses, may mean accentuated fluctuation of velocities when sonic logging, *if the well is of sufficient depth.* Both $V_p$ and an associated $Q_{seis}$, are intuitively likely to move in the direction of lower values *if* the mini-EDZ takes the form of shear fracturing and additional joint and fissure displacements and loosening, as it apparently can in shale, in the so-called 'alteration' zone.

Both parameters may move in the opposite direction of higher values, when the tangential stress concentration (e.g. $3\sigma_H - \sigma_h$) makes a positive contribution to the 'tightness' of both a stiff matrix (like limestone) and any local joint structure in these stiffer facies.

The velocity-facies definition given in Figure 14.12, with implicit and definitive separation of shales, sandstones and limestones, leads one to suggest the potential for *facies* or *strength* based estimates of $Q_{seis}$, using such differentiated velocities. With appropriate ranges of uniaxial strengths and velocities for the three rock types (shale, sandstone, limestone), one might arrive at 'representative' $Q_{seis}$ values of 6–7, 10–15, and 20–40 for the less frequent limestone layers, using the *modulus-model* of Figure 13.60.

The key question is then whether the inter-bedded *mix of facies* will create a 'weighted average' response, concerning the high frequency, best definition of layering when well logging. In Chapter 10, where $Q_{seis}$ was treated in detail, one may note from Figure 10.34, from the multi-frequency investigations of Sams et al., 1997 in these same experimental boreholes, that the *8–24 kHz sonic logging* gave the lowest $Q_P$ range of about 6–14, with a mean $Q_P$ of 10.4 and a mean $V_P$ of 3.48 km/s. (See Table 10.4). These fit the *modulus-model* predictions.

Figure 14.13, a virtual copy of Figure 10.35, is repeated here with the important $Q_P$ scale firmly in place, to supplement the somewhat misleading '1000/Q' typically found in numerous publications. If $Q_{seis}$ is accepted as potentially being 'a rock mass property' (i.e. pores and joints and saturating fluid included), it apparently helps to give a very believable range of $Q_P$ of about 5 to 12, with sonic velocities ranging from 3 to 4 km/s as shown in the figure.

The lower frequencies of *cross-hole of 0.2–2.3 kHz*, gave a mean $Q_P$ of 15.7. Interestingly, the lowest frequency 30–280 Hz VSP, giving poorest definition of the fine inter-layering, gave a mean $Q_P$ of 31.3. At first sight it would appear possible that this is because of the dominant effect of higher velocity layers (limestone) and the relative invisibility of the low velocity layers to longer wave-lengths, thereby 'biasing' the weighted-average picture that applies most fully at higher frequency when 'all' layers are illuminated. However, in relation to velocity, this argument does not hold, as Figure 14.13 shows VSP velocities to be somewhat lower (range of approx. 2.8 to 3.6 km/s). The higher core velocities, ranging from 3.4 to 4.4 km/s, are for matrix without bedding or cross-bedding joints, naturally giving higher $Q_P$, they do not fit the *modulus-model* well since they are not representative of the jointed, bedded rock mass.

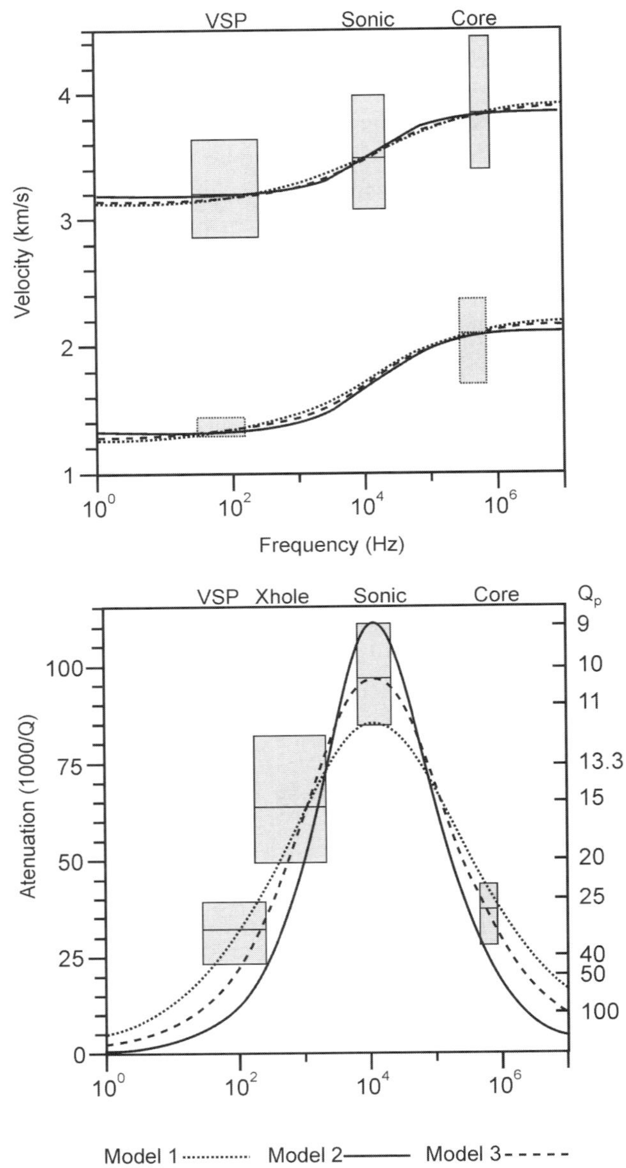


*Figure 14.13* Sams et al., 1997 multi-frequency results for four scales of seismic illumination. The highest frequency sonic logging, giving a form of weighted-average of the numerous rock layers present, seems to be quite well matched by the *modulus-model* of Figure 13.60.

### 14.4.1 Larger scale differentiation of facies

Larger scale differentiation of facies described by Kabir and Verschuur, 2000, is shown in Figure 14.14. In this synthetic Picrocol data set, which nevertheless was based on real data shot over a salt dome, the model assumed constant velocity layers, which was a simplification.

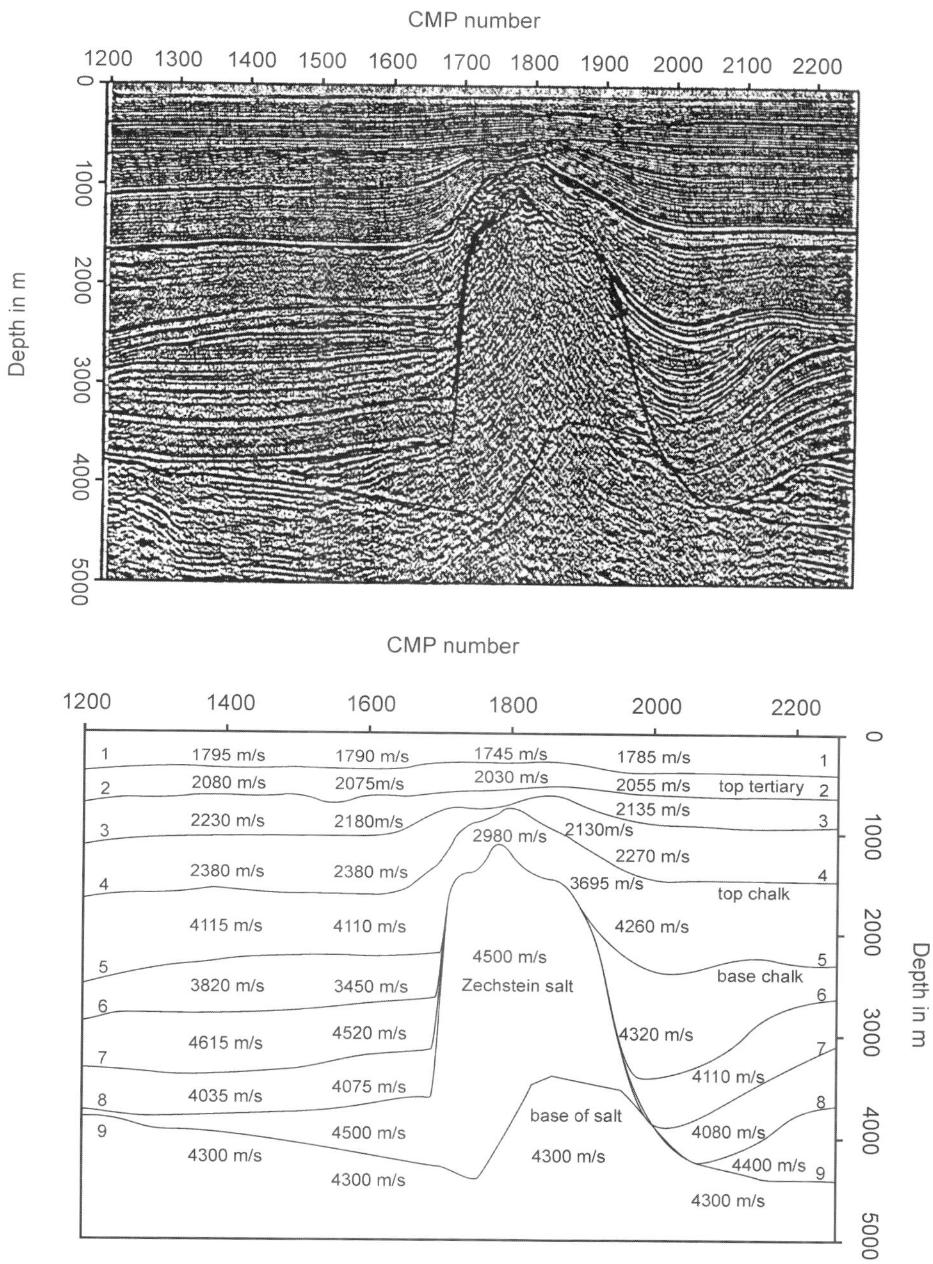


*Figure 14.14* Estimated velocity models based on an IFP Picrocol data set from a southern North Sea salt dome and surrounds. Note lateral velocity variations in individual layers. Kabir and Verschuur, 2000. Reproduced by kind permission.

Nevertheless each layer was described with an average velocity and a vertical velocity gradient. Overburden stress and fluid pressure were assumed to increase linearly with depth. The authors discussed the sometime need for lateral velocity gradient in addition to vertical velocity gradient, in interpreting variations. They suggested the use of well velocities to improve depth conversion and geological interpretations.

Fracturing or jointing caused by anticlinal structures can be a key to effective drainage of oil and gas from a porous but low permeability matrix such as tight sandstone or chalk. Depending on the efficiency of the cap-rock, often shales, there may also be over-pressured shale layers in the reservoir. Fracturing or jointing causing reduced velocities may be seen as a seismic time-sag. This can be accentuated by the presence of gas rather than oil, and can also be accentuated by the effect of over-pressure.

## 14.5 Detecting anisotropy caused by fractures with multi-azimuth VSP

In the same way that we had to reject earlier assumptions about elastic isotropic rock mechanics behaviour, the sedimentary rocks containing hydrocarbons have also proved to be neither isotropic nor homogeneous, but heterogeneous and anisotropic. Besides detecting azimuthal velocity anisotropy, signalling aligned fracturing and/or anisotropic rock stress, there are now needs for spatial resolution of variable structure, and also for resolution of temporal changes. A more recent challenge is the resolution of azimuthal variation in attenuation as an indicator of horizontal permeability anisotropy.

All seismic data are now known to vary with the offset (angle in relation to the well), and with the azimuth. Since all seismic data also varies with frequency, there is an increasing recognition that *3D multi-component, multi-mode and multi-azimuth* acquisition may be required and economically justified. The seismic wavelength at which the measurement is made, determines what seismic attributes can be measured, and whether the rock looks homogeneous and isotropic or heterogeneous and anisotropic.

According to Lynn, 2004 when the cause of the ordered anisotropy replicates itself across different scale lengths, or exhibits fractal tendencies, the anisotropy measured in modern dipole well logs at kilohertz frequencies, may match the anisotropy trends that are detected at another scale, with 10–100 kH reflection seismic. A given set of vertically aligned fractures will therefore cause anisotropy with *low frequency* measurement, signal distortion with *mid-frequency* measurement, and lead to reflections by the *highest frequency* waves.

The practice of VSP (vertical seismic profiling), walkaway VSP, multi-azimuth VSP, and numerous other techniques related to VSP are so fundamental in exploration that a diagram for illustrating the technique is probably superfluous. Nevertheless, since a figure is worth a thousand words: Figure 14.15 shows the principle of e.g. walkaway VSP, with surface shot points shown along a single azimuth, and downhole receiver positions. There is also a technique for performing VSP for horizontal well sections, using geophones clamped to the formation, and hydrophones suspended in the borehole fluid. Subtraction of seismograms for the down-going and the up-going pulses leaves an enhanced reflection. (Christie et al., 1995).

There is a growing trend to instrument some petroleum wells on a 'permanent' basis, so that 4D seismic can be used 'more easily', to monitor changes bought about by different water-flood and production practices, using *rock physics* principles to assist in the interpretation. As we saw in Chapter 10, the *in-well* installation practice apparently started in the late 80's and early 90's in the case of permanent installations in deep holes adjacent to the San Andreas fault in California.

### 14.5.1 Fracture azimuth and stress azimuth from P-wave surveys

The original goal of a seismic study reported by Lynn et al., 1999a, for the US Department of Energy, was to evaluate and map fracture azimuth and relative fracture density throughout a naturally-fractured gas reservoir interval. P-waves travelling in the plane parallel to aligned vertical fractures (thus giving transverse isotropy with a horizontal axis – known as HTI media), have higher velocity than the P-waves travelling perpendicular to the fractures. The latter direction corresponds to the high attenuation direction, due to the lower frequency of the P-waves, and therefore a lower seismic Q. The advantages of using of both travel time and amplitude, or travel time and frequency, for detecting permeability was emphasised.

As the authors pointed out, the detection of relative crack density and crack orientations was not the final goal, rather it was the *detection of azimuthal anisotropy that was attributable to structure that controlled the fluid-flow properties at reservoir scale.* With sufficient fracture density, connectivity and permeability anisotropy were likely. A threshold value of anisotropy would probably be involved. The detection of *permeability* anisotropy could be considered as one step beyond the detection of vertically aligned fractures, and/or the detection of unequal horizontal stresses.

The authors emphasised the importance of knowledge of the orientation of maximum horizontal stress, due to the *commonly assumed* strong correlation between directionality of reservoir flow and the local, present-day orientation of the maximum horizontal stress. Typically, oriented four-arm calliper logs show a long axis that is oriented parallel to the minimum horizontal stress

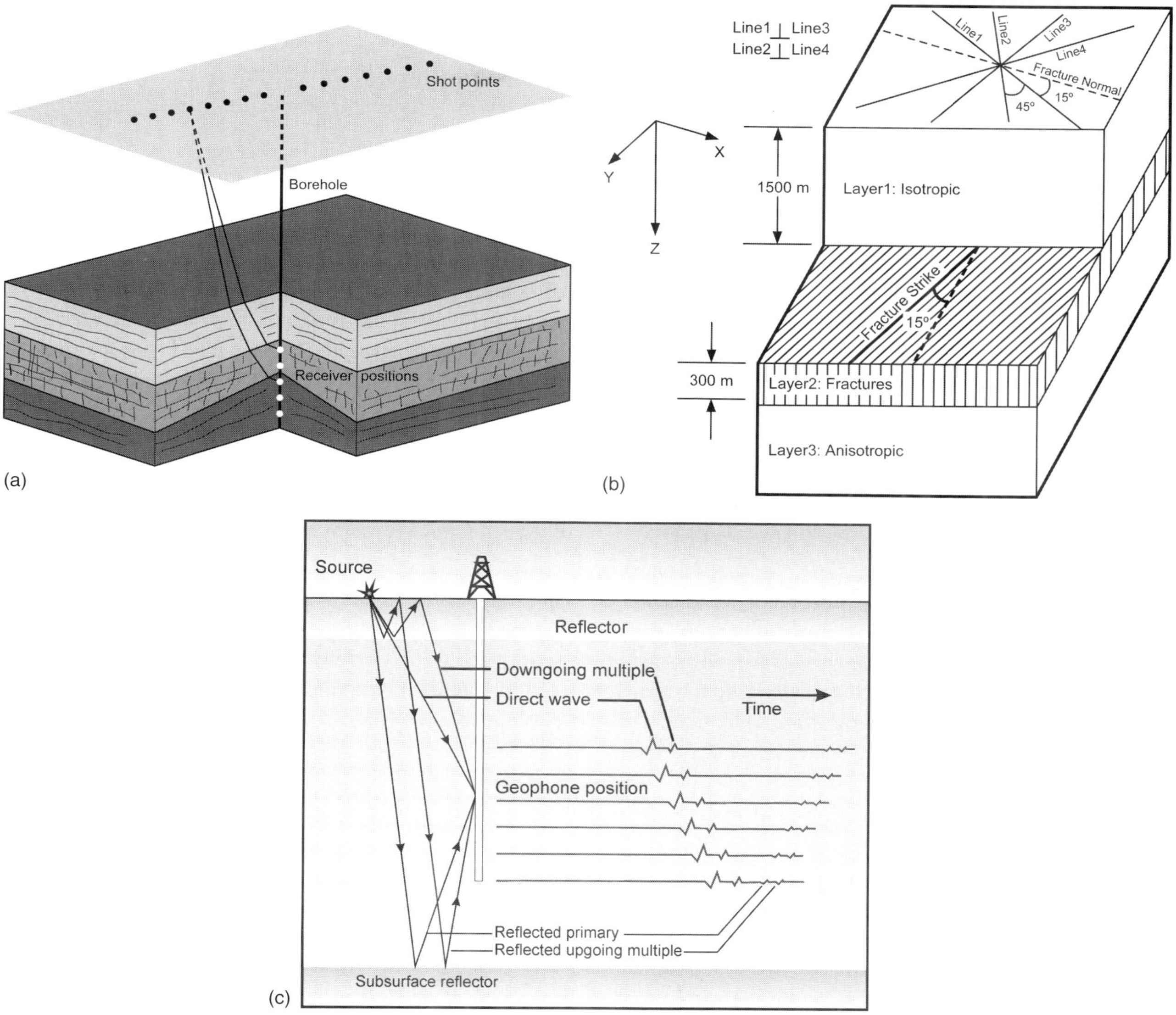


*Figure 14.15* a) Simple illustration of the principles of *walk-away* vertical seismic profiling (VSP). The usual practice will be the use of shot points along multi-azimuth lines, preferably symmetrically distributed, such as in two perpendicular arrays related to the dominant fracturing direction (e.g. Grimm et al., 1999), or as illustrated by b) with a more complete double-orthogonal coverage for the case of the hypothetical fractured reservoir (Liu, 2003). A *walk-around* lay-out at constant offset or radius, for improved characterization of fractured reservoirs has also been used (Horne, 2003). Use of wide aperture layouts, extending on the surface to at least the target depth is recommended, e.g. Lynn et al., 1999. c) *Zero*, or limited *offset* VSP with the direct and reflected wave-paths explained (Christie et al., 1995).

direction, if there is stress-induced break-out. (However, as mentioned earlier in this chapter, and discussed in more detail in Chapters 15 and 16, there may be joint- or fracture-related reasons for a careful evaluation of this commonly-held viewpoint.)

Interestingly for later discussion of this topic, the authors mention the 'problem' of a NW-oriented production trend at a neighbouring well, yet with a WNW-oriented fracture strike (based on oriented core and impression-packer results). The WNW fracture direction also agreed with the maximum stress indicated by break-out analysis from calliper logs. A plausible explanation of *enhanced dilation* of the WNW fracture set, due to structural flexure of the anticlinal structure, mentioned improved *connectivity* as a possible explanation of the (roughly 20°?) rotated NW production trend.

Analyses at the Rulison Field by Lynn et al., 1999b, using just *two azimuthal bins* of data, also showed a 'discrepancy' of orientations: a similar present day stress as above (N 70–80° W), but principle P-wave velocity directions of N30W and N60E. The authors continued to assume that the direction of $\sigma_{H\,max}$ would be the direction of open cracks or fractures, but did not discuss the possibility of the interaction of e.g. two sets of joints or fractures, or the possibility of conjugate shearing as an explanation of 'open' fractures.

This topic is treated in detail in Chapter 16, following analysis of the shear strength of rock joints and fractures, and analysis of what has been learned by Zoback and co-workers, about water conducting joint (or minor fault) directions relative to $\sigma_{H\,max}$ directions. Their findings should have special relevance in petroleum reservoirs where rock strengths are limited, yet fractures are assumed to be 'open', despite $\sigma_{h\,min}$ magnitudes of tens of MPa.

In Chapter 16, (and Barton, 2005), a further reason is discussed for such minor angular discrepancies, namely a dilation-related contrary-rotation of fluid lenses contra rock-to-rock contacting asperities, when *non-planar joints* (or fractures), are actually under significant shear stress, and therefore significantly 'open'. This geometric effect could possibly rotate shear wave splitting mechanisms.

Interesting observations on the subject of 'open' fracture orientations were made by Laubach et al., 2002. They cited comparisons of *measured* stress directions and orientations of open, flow-controlling fractures that showed that open fractures in the sub-surface were not necessarily parallel to maximum compressive stress ($\sigma_{H\,max}$). Fractures perpendicular to this direction could also be 'open' if partially filled with mineral cements, and for this reason, sealed fractures parallel to $\sigma_{H\,max}$ were numerous.

They pointed out that a determining factor for fluid flow was the degree of mineral cements deposited within the fractures, either 'at the time of fracturing' (*synkinematic*), or as *post-kinematic* cements precipitated after fractures ceased opening. They suggested, from experiences in both compressional and extensional provinces, with production data from 2,400 to 6,400 m depth, that the divergence between $\sigma_{H\,max}$ and 'open' fractures demonstrably contributing to flow was 'from a few degrees to 90 degrees'.

Grimm et al., 1999 also chose a relatively simple (two-azimuth bin) method of estimating the spatial distribution of gas-producing natural fractures in a tight gas reservoir. This relatively large P-wave survey was divided into two volumes, with ray-paths parallel and perpendicular (+/−45°) to the assumed dominant fracture direction (based on structural geology). This method was used in order to construct azimuthally differentiated, and azimuthally limited seismic attributes, that would highlight the formation's fracture and matrix response, respectively.

A rough measure of fracture anisotropy was given by the resulting P-wave interval velocities, measured in two such azimuthal *bins*. (Two-azimuthal binning is however blind to anisotropy near 45° to the principal chosen directions). The authors found that when the anisotropy exceeded 5 to 10% using this simpler approach, there was good agreement with the principle directions obtained by more sophisticated methods. There was correspondingly robust correlation and identification of zones of high fracture density and permeability.

The authors reported that *reflectivity* and *frequency* were also anisotropic, and actually had *better correlations with gas productivity than P-wave velocity anisotropy*. Ray paths travelling parallel to fractures giving increase in frequency may have been due to squirt flow mechanisms parallel to the fractures. The authors used the neural network technique (with 85% success), to infer commercial *prospectivity* over undrilled areas, using the available geological, geophysical and engineering attributes.

The authors advised full-azimuth 3-D P-wave surveys for such analyses, with maximum offsets equal to, or greater than target depth, using azimuthally isotropic source and receiver arrays. Naturally they also recommended processing in as many azimuths as allowed by cost, with independent velocity analyses in each azimuth.

An admirably detailed integration of *structural orientation* data with near-offset VSP at the Conoco Borehole Test Facility, was given by Queen and Rizer, 1990, based on the surface outcrop joint orientations, rock fabric orientation (from point-loading), and from the results of joint orientation using a borehole televiewer and oriented core. Figure 14.16 shows the principal orientation data at these different scales. The two joint sets seen at the surface proved not to be as consistent as expected with the three sets of jointing in the sub-surface. The quite consistently oriented point-load fracturing traces could not apparently be related to any observable microcrack directions or to fabric, and would then seem likely to have some relation to *residual stresses* in the samples.

The seismic anisotropy measurements consisted of near-offset VSP in twenty three 30 m intervals from 183 to 853 m depth, and azimuthal VSP uniformly spaced at 15° intervals around a 120° quadrant, with nine sources on a 290 m radius centred on the Conoco 33–1 well.

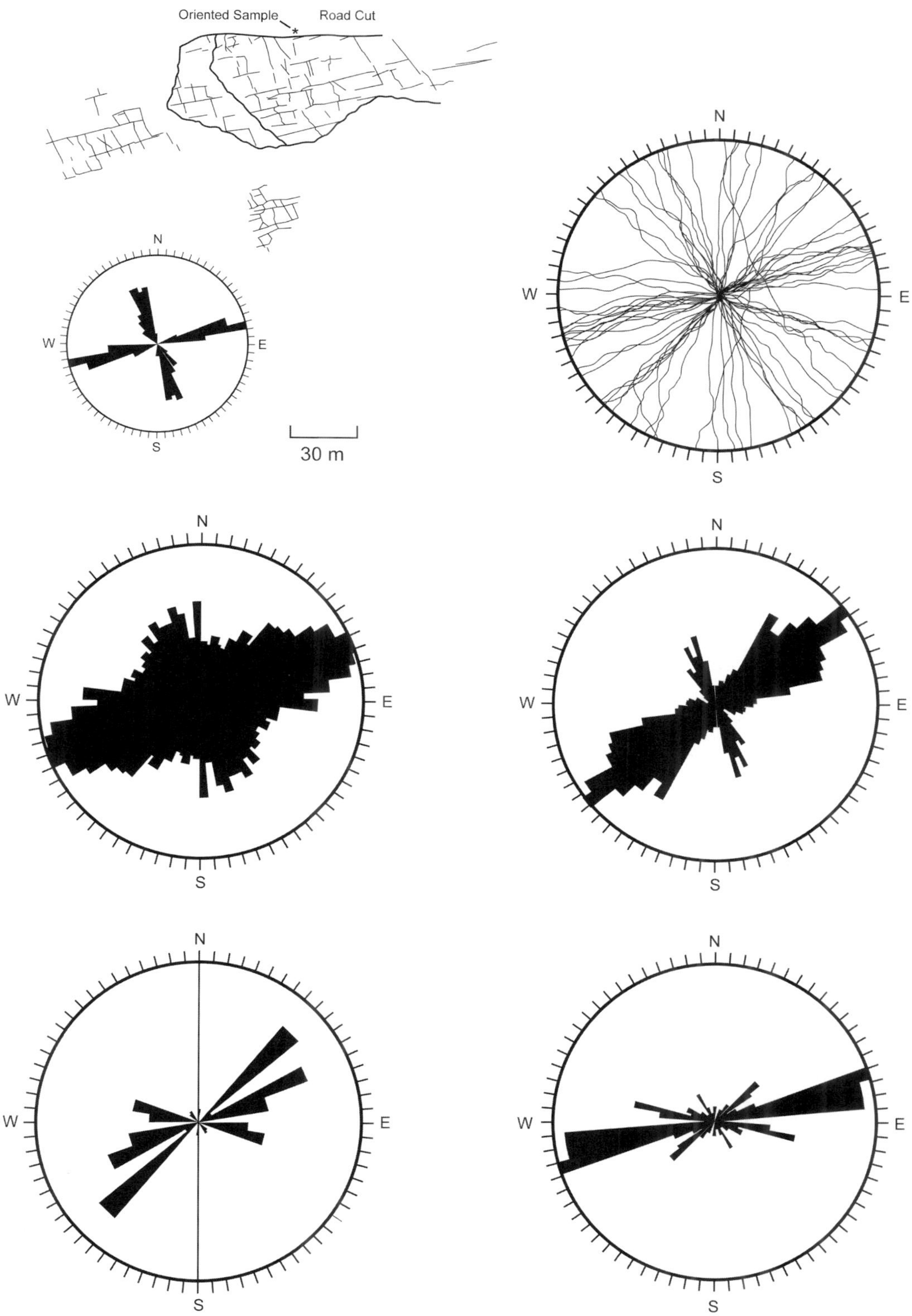


*Figure 14.16* *Top left*: surface outcrop and joint rosette. *Top right*: superimposed tensile-fracture traces from point-load (steel ball) indentation testing of the limestone. *Middle: left and right*: cumulative induced-fracture lengths and orientations from 35 tests on oriented surface samples of the limestone, and from 9 tests on oriented indurated shale core from 734 m depth down the Conoco 33–4 well. *Bottom left:* Joint rosette concerning accumulative length of sub-surface jointing of given strike, interpreted from BHTV and oriented core (with bias against non-vertical jointing). *Bottom right:* Polar histogram for all levels and azimuths of all the nine-component VSP data, with lengths weighted by travel time. Queen and Rizer, 1990.

The results of shear wave splitting and polarization, from this multi-component survey (strictly also belonging in Chapter 15), suggested closest correspondence with the 'middle' set of (ENE) fractures, which was close to the BHTV-interpreted break-out analysis indication of the (perpendicular) major horizontal stress.

Alternatively, the ENE direction could be the theoretical and therefore practical *resultant* of shear wave splitting influences from the other sets, which strike to each side (NE and ESE). The point load fracturing, perhaps responding to a residual stress (since no microscopically-visible oriented microcracks, nor aligned fabric were seen), was also in the conforming ENE direction. The authors were of the opinion that their data, although consistent, did not have enough resolution for detailed engineering analysis of fluid flow.

### 14.5.2 Sonic log and VSP dispersion effects and erratic seismic Q

The authors De et al., 1994, from Chevron, compared P- and S-wave velocities and seismic quality factors ($Q_p$ and $Q_s$) using vertical seismic profiling (VSP) and sonic log measurements in five wells, which were situated in California, Texas and Alberta. The expected bias (VSP transit times were greater than sonic log times) were attributed to normal *velocity dispersion*, due to which higher frequency (sonic) waves travel at higher velocities than lower frequency seismic waves. Differences in average P-wave travel times ranged from 2.5% to 7% in the different wells, giving velocity differences between the two methods that were consistently in the direction predicted by dispersion.

The authors discussed additional potential causes of the systematic velocity differences, citing local *stress concentrations* around the boreholes, altered zones and velocity anisotropy or lateral inhomogeneity. In fact stress concentrations around the wells, will cause a magnification of (tangential) stress, in the same direction as major principal stress, and diminution of (tangential) stress in the perpendicular direction. If these effects are strong enough in relation to rock strength, shear surfaces may develop, first giving break-out, subsequently a *log-spiral-fractured* discontinuum. Various forms of borehole wall disturbance are possible, as emphasised in Chapter 12.

De et al., 1994, calculated seismic quality factors ($Q_p$ and $Q_s$) both from a velocity *dispersion formula* and from *spectral ratios*. When the two $Q_{seis}$ values agreed, they concluded that velocity dispersion resulted solely from absorption. They showed logs of $Q_p$ down a 1000 m deep borehole that showed an average difference of $Q_p$ (dispersion formula) minus $Q_p$ (spectral ratio method) of 13. Typical ranges of $Q_p$ were 10 to 50 (again reminding one of the expected magnitude of deformation moduli, as expressed in GPa).

They found that VSP (i.e. *seismic*) $Q_p$ were systematically smaller than those from *sonic* $Q_p$, therefore suggesting a bias between VSP and sonic log $V_p$ that could not be explained by intrinsic attenuation alone under a constant $Q_{seis}$ assumption. Besides these observations, the authors observed that individual values of $Q_p$ or $Q_s$ changed 'erratically with depth' unless depth averaging was used. The authors suggested that the validity of seismic Q 'does not always correlate with the data quality or with the rocks themselves'.

Since there is a potential relation between rock quality $Q_c$ and seismic $Q_p$ via the empirically derived 'static' modulus of deformation (Figures 13.59, 13.60), it must be emphasised that the rock mass quality Q-value also usually varies rapidly down any given drill-core, unless rock mass conditions are unusually uniform. So on that basis it would not then be surprising to see 'erratically' varying $Q_p$ or $Q_s$ with depth.

An example of rock mass (core-logged) Q-value variations and an alternative rock mass rating RMR from Bieniawski, 1989, down a recovered core is shown in Figure 14.17. It is unfortunate that such logs of rock quality are never (?) a part of the geophysicists' reporting of seismic attributes. There is after all, a strong empirical relationship between $Q_{rock}$ and $V_p$, and there is an implicit similarity in the variable values of $Q_{rock}$ with both rock quality (obviously), and also with depth, as we saw in Chapter 10.

## 14.6 Dispersion as an alternative method of characterization

Seismic *attenuation* has come to be recognised as potentially very sensitive to reservoir properties. This is because of its sensitivity to fractures, joints or bedding planes, and in turn, due to sensitivity to changes of effective stress. Attenuation levels are also sensitive to the saturating fluid and petro-physical properties. However, as emphasised by Hackert et al., 2001, field measurements of in situ attenuation are complicated due to reflections, geometrical spreading losses and *varying formation stiffness*.

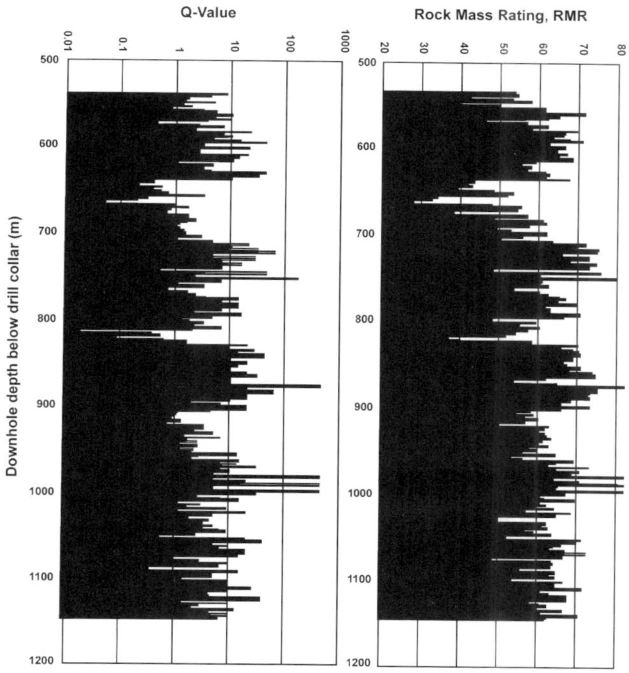


*Figure 14.17* Examples of *rock mass quality variability* in terms of Q-values and RMR, down a 600 m section of the UK Nirex Ltd borehole RCF1, at Sellafield. This was the planned site for a 500–700 m deep shaft, to develop a *Rock Characterisation Facility*, for nuclear waste disposal feasibility studies. NGI contract report, 1994. The rock was tuff and ignimbrite, of (obviously) varying quality due to jointing and faulting. These Q-values, modified to $Q_c$ using UCS, correlate with $V_p$, $E_{mass}$ and depth.

It has recently been discovered that *dispersion,* or the well known dependence of seismic velocity on frequency, can be used for reservoir characterization, as dispersion is mathematically related to seismic attenuation. Sams et al., 1997 (reviewed earlier in this chapter, and also in Chapter 10), compared velocities and attenuation from four frequency regimes: vertical seismic profiling (VSP), cross-well measurements, sonic logging and core measurements. They established dispersion and attenuation profiles as a function of these different frequency measurements. A distinct peak in attenuation versus frequency was observed at about 20 kHz, related to a crack relaxation mechanism (Aki and Richards, 1980) with realistic crack aspect ratios according to the crack model of Jones, 1986.

Hackert et al., 2001, used a similar multiple-frequency (*cross-well*: 1 kHz and *sonic log:* 10 kHz) investigation to create logs of intrinsic dispersion and attenuation for a shale formation in the Buena Vista Hills reservoir in California. They found that high dispersion (and low $Q_{seis}$) values correlated with thin sand and carbonate beds within the shale. These beds were at least ten times as permeable as the host shale formation. A velocity and density log for the part of the formation (3950 to 4300 ft depth) is shown in Figure 14.18.

The main Antelope Shale (lower half of log) contained more than 100 thin sand and carbonate layers: the most visible ones marked with (S) and (C). These contrasted with the uppermost Brown shale, and Upper Antelope shale that contained no sand or carbonate layers.

High frequency measurements differ from low frequency measurement due to *elastic scattering* and *intrinsic attenuation*. Based on work by Marion et al., 1994, and Brown and Seifert, 1997, the authors of this reservoir study gave the following equation for rationalising their two-frequency approach:

$$\underset{(\approx 10\text{kHz})}{V_{\text{sonic}}} = \underset{(\approx 1\text{kHz})}{V_{\times\text{well}}} + \Delta V_{sc} + \Delta V_i \tag{14.8}$$

where $\Delta V_{sc}$ = change of velocity due to *elastic scattering,* and $\Delta V_i$ = change of velocity due to *intrinsic attenuation.* The former is related to geometric effects including the natural anisotropy (layering) of the medium, while the latter is caused by inelasticity of the rock matrix and/or viscous losses of saturating fluids in compliant pores. They also defined a term ($V_{ES}$) from the above terms:

$$V_{ES} = V_{\text{sonic}} - \Delta V_{sc} \tag{14.9}$$

As can be noted in Figure 14.19, $V_{ES}$ is significantly higher than $V_{\times\text{well}}$, particularly in the anisotropic sand and carbonate section.

Hackert et al., 2001 conducted forward modelling of their cross-well set-up in this layered medium, *first* assuming $Q_{seis}$ values roughly proportional to the *dynamic elastic moduli,* as follows:

Shale $Q_{seis} = 34$
Sandstone $Q_{seis} = 66$
Dolomite $Q_{seis} = 88$

They used the difference between the $V_{ES}$ and $V_{\times\text{well}}$ curves to derive the predicted $\Delta V_i$, which yields the intrinsic $Q_{seis}$, using a method described by Aki and Richards, 1980. The results of this analysis indicated $Q_{seis}$ values in the thin sand and fractured carbonate beds of between 10 and 50 (appearing to potentially match our *modulus-model* logic), while the bulk shale had $Q_{seis} > 50$ in general, which appears not to be well matched by

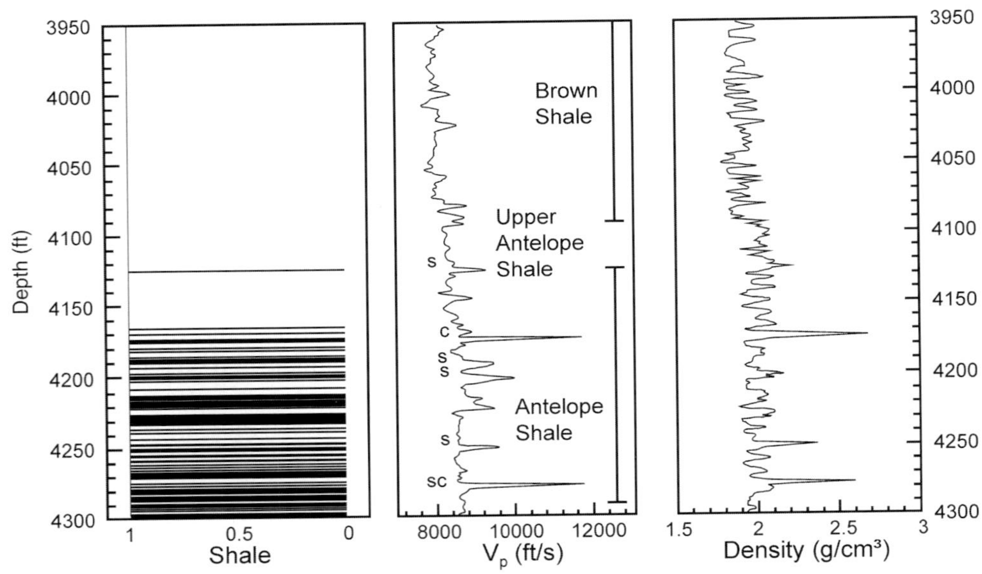


*Figure 14.18* P-wave velocity and density logs in relation to the location of non-shale layers (sand and carbonate) in the Antelope shale. Hackert et al., 2001.

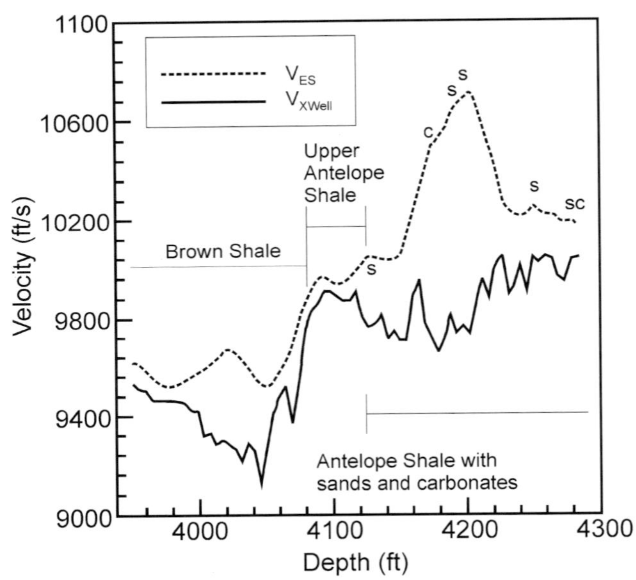


*Figure 14.19* Cross-well velocity data compared to computed $V_{ES}$. Hackert et al., 2001 computed $Q_{seis}$, finding lower $Q_{seis}$ with multiple sand and fractured carbonate layers as shown earlier in Figure 10.69b.

*modulus-model* logic, suggesting other attenuation mechanisms than fracturing and contained fluid effects.

## 14.7 AVO and AVOA using P-waves for fracture detection

AVO (amplitude variation with offset) and AVOA (amplitude variation with offset and azimuth) will be briefly described in this section. Azimuthally dependent variation of P-wave amplitude can be related, both theoretically, and clearly in practice, to the presence of fracturing, and gives reasonable estimates of the orientation of the fracturing, particularly if only one set is involved, or if one set is dominant. Two studies related by Pérez et al., 1999a and 1999b will be briefly summarized, as they illustrate the method, and they also compare results with shear-wave related studies, using C-wave (P to S converted waves), and the shear-wave splitting mechanism.

Although the use of shear waves are theoretically favoured for fracture set detection, using the shear wave splitting and polarization mechanism to be described in Chapter 15, there has apparently been a certain reluctance to use shear waves, due to more expensive acquisition and more expensive processing routines. For these reasons, the use of P-waves for fracture set detection and estimation of orientation, has attracted a lot of interest, even though P-wave travel times need to be detected in many directions to obtain the necessary information.

AVO analysis using P-waves, is based on some principles that can best be illustrated by the practical example, given by the authors. If seismic data acquisition is conducted parallel to the fracture orientation, the fractures will have minimal influence on the reflection properties, regardless of the angle of incidence, or offset. This is because the P-wave particle motion is parallel to the fractures. If the line is instead oriented more perpendicular to the fractures, at larger angles of incidence than zero, the reflection coefficients will be affected strongly. In fact at the largest angles of incidence, especially when

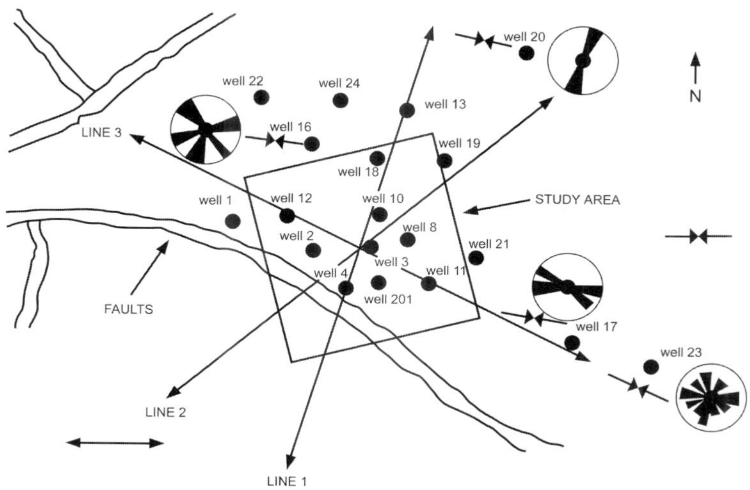


*Figure 14.20* Survey geometry for Pérez et al., 1999a and 1999b analyses of 3D AVO and converted (C) wave P-S shear wave splitting analyses. Small inward-facing arrows indicate $\sigma_{H\ max}$ direction interpreted from break-out orientation logs, from wells 16, 17, 20 and 23. The (unconventional) rose diagrams, which have been blackened for clarity, indicate the various fracture orientations, and their density, based on FMS logs at these boreholes.

perpendicular, the P-wave velocity is expected to also be affected by the acoustic properties of the fluids filling the fractures. So in the presence of anisotropy, the reflection amplitude will vary with offset, due to changed angle of incidence, and also will change with azimuth (AVOA).

The authors used data acquired from three, intersecting 10 km, three-component seismic lines, with three different azimuths. The three lines intersected at one of the wells, where there were results of FMS (formation micro-scanner) and caliper logs for estimating both the dominant fracture strike and the direction of $\sigma_{H\ max}$. The latter was oriented NW-SE, and the authors assumed that the fracture set in approximately this direction (one of several sets), would be the one most likely to be 'open', and therefore most detectable.

In this study the far offset extended to 3600 m, a bit more than the depth of the target zone: a 35 m thick fractured limestone at approximately 3000 m depth. This had a P-wave velocity of only 3513 m/s, and an S-wave velocity of 1890 m/s, with a density of 2.5 gm/cm$^3$. In other words it would seem to have been extremely well fractured or over-pressured.

A geophysical term NMO (normal moveout) is relevant here, as the so-called NMO *stretch* decreased frequency at far offsets, which therefore affected the amplitudes. This distortion was most significant at shallow depth and large offset. During processing, the authors observed a higher frequency (and velocity), along the offset line that had an azimuth parallel to the assumed dominant ('open') fracture set, which agrees with other studies.

The authors opinion at this time was that there were few studies that related AVO attributes to fracture or crack parameters. However they were convinced that P-wave *AVO gradients* were affected by fracture-induced azimuthal anisotropy. Analysis involved calculating reflection coefficient curves in two azimuths, one parallel, the other perpendicular to fracture orientation (implying the need to know the expected result). They found that the *AVO gradient* was larger for the lines perpendicular to fractures than for the line parallel to fractures: this direction showed a somewhat higher reflection coefficient.

In a companion paper, Pérez et al., 1999b compared their local 2D AVO study of the fracture effects at the intersection of their three azimuthal lines (with tie-in to well data at this point), with more comprehensive 3D azimuthal AVO analysis, and with analysis using converted (C) P-S waves, using shear wave splitting and polarization (see Chapter 15).

The converted P-S waves have the advantage that they can be generated by compressional ( i.e. explosive) sources, yet are expected to contain the same information as pure S (or SS) waves. These comprehensive studies were finally used to assist reservoir engineers in exploiting the fractured limestone layer at 3 km depth, using horizontal wells oriented perpendicular to the densest (or the assumed more 'open'?) fracture set.

Figure 14.20 shows the layout of the three survey lines relative to the wells, with details of some fracture-strike orientations from FMS logs. Figure 14.21 shows the results of two of the three fracture direction analyses,

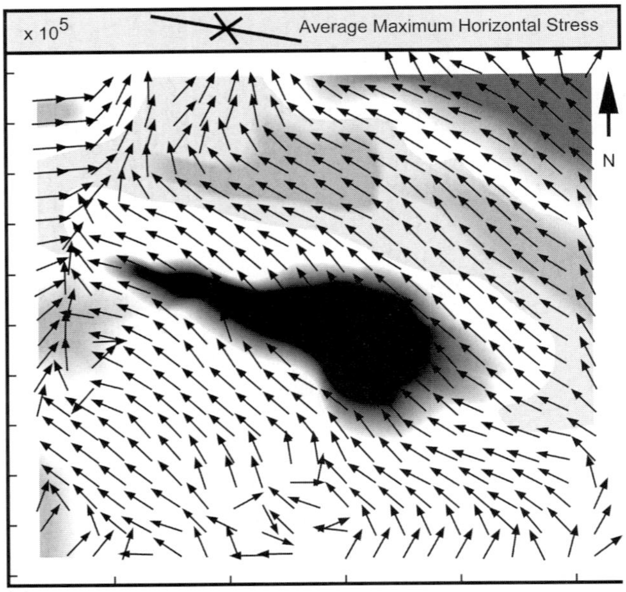


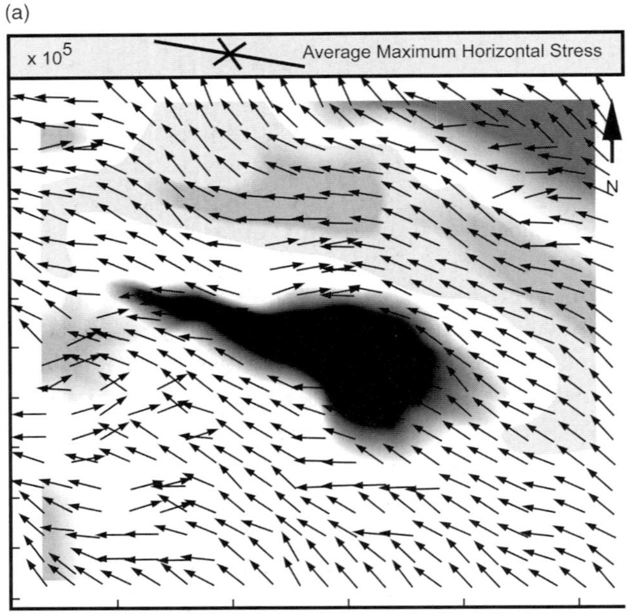


*Figure 14.21* Fracture orientations within the investigation area, projected to the 35 m thick fractured limestone at 3 km depth. Interpretation from a) Converted P-S based shear wave analysis, b) 3D azimuthal AVO analysis. Note that the result using converted P-S waves was relatively closer to $\sigma_{H\ max}$. Pérez et al., 1999b.

which were described by Pérez et al., 1999b. The upper one is from the converted (C) wave P-S analysis, using the shear wave splitting and polarization method ($qS_1$ parallel, and $qS_2$ perpendicular to jointing – see Chapter 15).

The second fracture orientation plot was derived from the 3D azimuthal AVO analysis. The 'independence' of the plots was guaranteed by the sequential timing of the surveys. The authors estimated that the azimuth of the *maximum AVO gradient* was 56°, based on a Rüger, 1996 formula for calculating the reflection coefficients in TIH (transversely isotropic media with a horizontal symmetry axis due to a vertical fracture set). Consequently, since this is assumed to be perpendicular to fracture orientation, the 3D AVO based fracture azimuth estimate was 146°, which was shown to be as much as *36°* from the interpreted $\sigma_{H\ max}$ direction. When based on the converted P-S shear wave splitting, the deviation was less.

This brings us to a topic introduced later, when reviewing water-flood analyses summarized by Heffer, 2002. Heffer's results from water flood case records, may suggest to those with rock mechanics background, the possibility of conjugate-shearing of joint sets that are *intersected* by the $\sigma_{H\,max}$ direction. This topic is treated in more detail in both Chapters 15 and 16, as it both contradicts the 'standard industry' assumption of 'open fractures parallel to $\sigma_{H\ max}$', at the same time helping to explain real angular discrepancies between dominant 'open' fracture azimuths, and the perpendicular-to-break-out based $\sigma_{H\ max}$ direction. (Refer to Figure 16.71 for visualization of this shear mechanism, and the potentially important *fluid lense rotation* phenomenon).

The authors Pérez et al., 1999a and 1999b felt that their analyses of fracture orientation trends from 3D AVO and from converted P-S (and also from 3D NMO ellipticity: not shown here due to much 'poorer' stress-direction fit), had 'captured the regional orientation of maximum horizontal stress in the field'. They also concluded, significantly, that even though the reservoir had several fracture systems, most of the applied methods picked 'the one that is closer to the maximum horizontal stress'. See also Tingay et al., 2005 (Fig. 8c)

As we shall see in Chapter 16, this may not be a necessary condition for maximum permeability direction – most likely this will be found sub-parallel to *the dominant set of a conjugate pair* that has suffered more shear and dilation than its neighbouring set. At shallow depth however, the standard 'industry assumption' (also in civil engineering) is more correctly focussed on maximum permeability (and $V_p$) being parallel to the $\sigma_{H\ max}$ direction. This is because this joint direction has verifiable apertures and permeability *near the surface.*

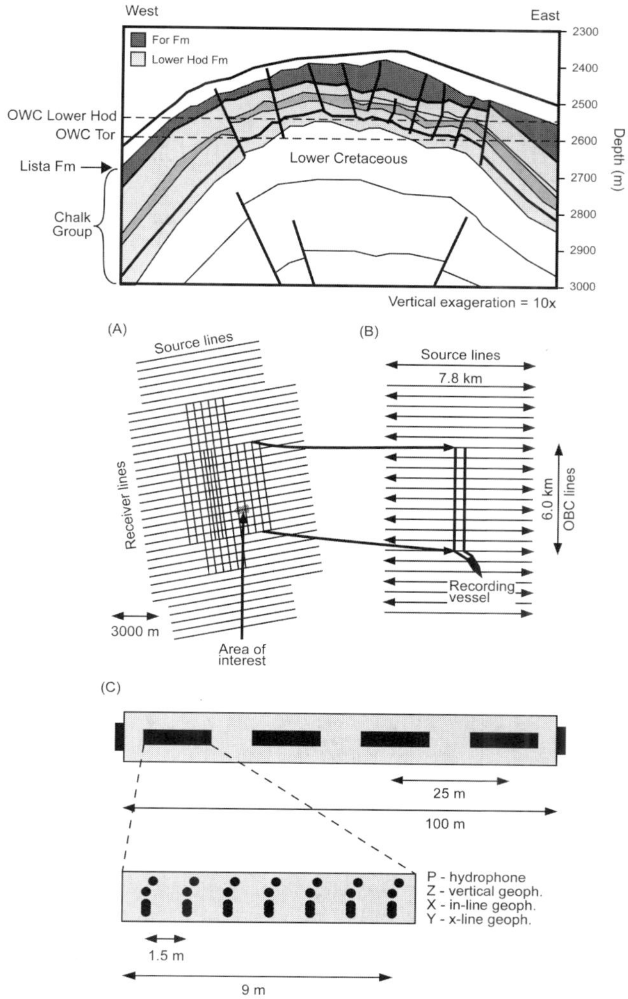


*Figure 14.22* a) Exaggerated vertical cross-section of faulted anticline defining the Valhall reservoir. The small-scale fracture sets are thought to be related to the domal faulting, which has principal directions of NW-SE and NE-SW. b) Schematic of orthogonal 3D-4C ocean bottom cable (OBC) data acquisition over Valhall. Source lines were 7.8 km long at 600 m spacing, with shots every 25 m over pairs of receiver lines. Each receiver cable of 6 km length, also at 600 m spacing, had receiver arrays at 25 m spacing, each with seven four-component sensors at 1.5 m spacing in each group. Note: patch shooting, over pairs of receiver lines, using air guns at 7.5 m depth. c) Details of a single, multi-component receiver unit, which are uniformly spaced along the cables, Hall and Kendall, 2003.

### 14.7.1 Model dependence of AVOA fracture orientation

As described later in this chapter, a 3D-4C (three-dimensional, four-component) ocean bottom seismic (OBS) survey was performed at the Valhall field in 1998. (See details in Figure 14.22, from Hall and Kendal, 2003). This, and previous 2D-4C studies had shown the ability of shear waves, derived from converted P-S waves, to penetrate through a gas cloud and illuminate the target fractured zones, otherwise invisible to P-waves in these locations.

As part of the seismic studies using the new 3D OBS data at Valhall, Hall and Kendall, 2000 and subsequently Hall and Kendall, 2003, addressed the problem of *model-dependence in AVOA analyses* of the multi-azimuth OBS-acquired, vertical P-wave part of the data (i.e. component Z in Figure 14.22c).

The authors suggested that the fracture strike direction could be *ambiguous*, since the azimuthal variation in the *near-offset* AVO gradient, could be positive or negative, relative to the fracture direction. They found, unfortunately, that the direction of the most positive AVO gradient could correspond to either the fracture-normal, or the fracture strike direction, *depending on the character of the fracturing*, i.e. depending on whether brine-filled or gas-filled – which were the 'end-member' models.

They therefore recommended forward modelling in order to constrain the interpretation of AVOA. The objective was to predict the near-offset AVO gradient anisotropy for equivalent medium models, based on the different elastic properties of the different fracture models, and compare this with observed AVOA in the data. The effective medium models were based on the *additional fracture compliance* terms ($Z_N$ and $Z_T$) allowing poroelastic modelling of the (micro) hydraulic connectivity between fractures and pore space (as discussed in more detail in Chapters 10 and 15).

Log-based data for the fractured chalk were as follows: $V_p = 3134$ m/s, $V_s = 1534$ m/s, density 2300 kg/m$^3$. Permeabilities related to production were an order of magnitude higher than core data, due to the significant amount of fracturing. Hall and Kendall, 2003 mentioned that *azimuthal anisotropy* was also observed from the *dipole sonic logging* in the producing horizon, using the flexural shear-wave splitting principle, as described in Chapter 12.

Their comparison of model and field data suggested that a large crack density (about 0.1) and thin fractures, were needed to match the high anisotropy, and this matching also suggested liquid-filled rather than gas-filled fractures, unless aspect ratios were $<0.00025$. Low matrix permeability was also suggested. They considered a most important finding was that the high

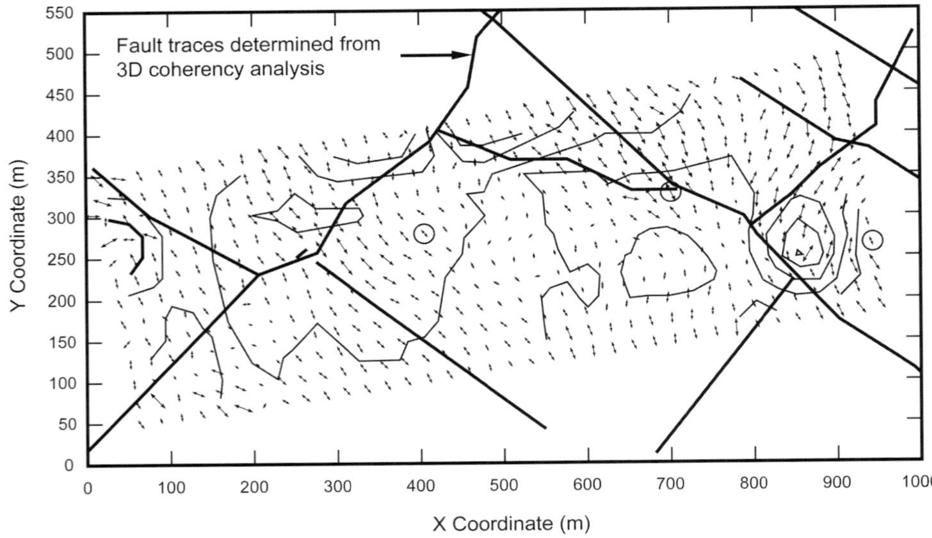


*Figure 14.23* Fracture pattern map, assuming that the high levels of observed anisotropy could only be modelled if the *most positive* near-offset AVO gradient was in the direction *perpendicular* to dominant fracturing. Hall and Kendal, 2003.

levels of observed anisotropy could only be modelled if the *most positive* near-offset AVO gradient was in the direction *perpendicular* to dominant fracturing. Hall and Kendal 2003 emphasised that in other situations, the reverse could be true, i.e. the direction of the *most negative* AVO gradient could be in the direction *perpendicular* to the dominant fracturing.

The principal fracturing directions interpreted from the AVOA studies described, were NNW-SSE to N-S, the former nearly resembling the 'compartmental-like' NW-SE component of faulting, as shown in Figure 14.23.

### 14.7.2 Conjugate joint or fracture sets also cause anisotropy

From experiences of domal-structure jointing at Ekofisk, one may pose a tentative question: could the above images of dominant, fracturing directions actually be images of the similar strike of conjugate, steeply dipping sets? If so, then an important mechanism of compaction and production-maintenance could also be at work, namely down-dip shearing, despite the one-dimensional strain boundary condition. (Barton et al., 1988).

The modified apertures caused by slight shear-induced dilation, could perhaps have influence on the above aspect ratio assumptions, where the inequality of physical aperture (E) and conducting aperture (e) (Barton et al., 1985), actually results in *two possible aspect ratios*, whose individual relevance may depend on the mechanism being modelled, i.e. squirt flow attenuation would actually be governed by a different aspect ratio to that determining the fracture compliances – if these mechanisms were being modelled in poroelastic models that required *aspect ratio* as input.

The standard assumption that maximum permeability in a rock mass tends to be parallel to a dominant set of vertical joints or fractures, which themselves usually trend parallel to the maximum horizontal stress, is a simply understood concept for which there is also theoretical support (e.g. Sayers, 1990). The arguments clearly extend also to the P-wave and S-wave velocities.

The concept and the theory are defensible, when the rock is of sufficient strength in relation to the effective normal stress, to provide (partly) 'open' joint apertures in the major stress-parallel direction. This generally applies at least to near-surface rocks, and it may apply at reservoir depths in the case of harder rock types.

However, if stress-closure modelling that is based on empirical rock mechanics data, can demonstrate virtual closure of the set under discussion, at effective normal stress levels of several tens of MPa, then other mechanisms are likely to be operating, in order to explain a viable production rate, assuming that matrix permeability is *insufficient* to explain production, as for instance at both the Valhall and Ekofisk jointed chalk reservoirs.

The conjugate shear mechanism, and an important *detail* of this mechanism for the common case of non-planar joint or fracture surfaces, is introduced in Chapter 15, and further quantified in Chapter 16, with

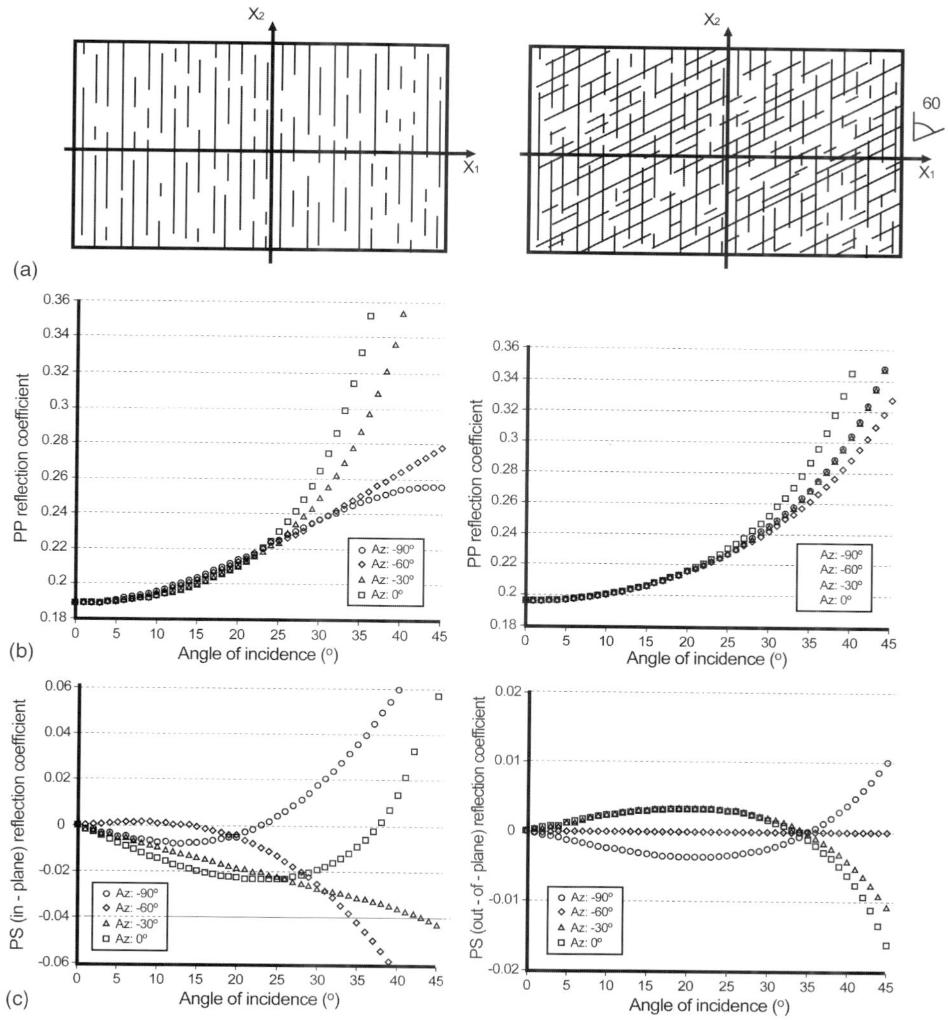


*Figure 14.24* a) Two synthetic models of jointed or fractured reservoirs, with one set, or two intersecting sets of joints. The modelling of overlying shale ($V_p$ = 3700 m/s) and reservoir *matrix* ($V_p$ = 5877 m/s and density 2.4 gm/cm$^3$), was supplemented by compliance additions to account for the joints, but with neglect of complex joint-set interaction (as modelled in UDEC). b) Azimuthal P-wave reflection coefficients as a function of angle of wave incidence at four selected azimuths for models 1 and 2. c) Equivalent coefficients for split shear waves: P-SV for model 1 and P-SH for model 2. Note greater azimuthal sensitivity of a single set of joints, and greater azimuthal separation of the reflection coefficients for the case of polarized shear waves. Sensitivity of P-waves to azimuth is limited, below angles of incidence of 25°. Chen et al., 2005.

particular reference to the 'shearing critical crust' concept of Zoback and co-workers. So far little recognition of the importance of this mechanism is evident in reservoir geophysics literature.

Sayers and Dean, 2001, and also Chen et al., 2005, addressed the question of the effect of additional fracture sets, in particular *non-orthogonal* vertical sets, on the AVO response. The first authors showed sinusoidal-type P-wave reflection coefficient trends, when plotted versus azimuth (0 to 360°), with variations depending on the assumptions for the ratios of fracture normal and shear compliance, and whether these ratios were equal or unequal for the different sets.

Some of the theoretical assumptions concerning the ratio of compliances being 1.0 for the case of 'open, gas-filled' fractures seem not to match engineering concepts, due to the entirely different mechanisms involved in closure and shear. However a lower ratio of $Z_N/Z_T$ (or Bn/Bt) for the case of clay filled fractures seems reasonable.

Chen et al., 2005 presented reflection coefficients as a function of both angle of incidence and azimuth for two very 'tangible' images of single-joint set and two-joint set models, as illustrated in Figure 14.24. The authors set up synthetic jointed reservoir models with the model 1 and model 2 jointing, intending to have a fracture density of 0.1 in each case.

### 14.7.3 $V_P$ anisotropy caused by faulting

Concerning anisotropy in the neighbourhood of *faults*, Williams and Jenner, 2002, warned of the possibility for multiple fracture directions, with the fast velocity no longer equal to the matrix or bulk rock velocity. The normal elliptical $V_{fast}$ and $V_{slow}$ distribution is then replaced by superimposed multiple ellipses, which have the effect of reducing the observed velocity, and the previous directionality with a single set of joints or fractures will be lost. Due also to rapid changes in fracture frequency, rapid changes in velocity are also seen – in fact responding to the rapid changes in rock mass quality Q close to, and across faults, as frequently observed in tunnelling, and when logging fault-zone core for rock quality purposes.

The authors showed a velocity difference tomogram of $V_{fast} - V_{slow}$, with velocity differences of up to 150 m/s in the neighbourhood of known faulting, due not to bulk velocity changes, but due to changes in fracturing frequency and directions. Since even 2-D offset lines cross the earth's azimuthal velocity field, a case was made for dense sampling near fault and fracture zones. Mostly, azimuthally varying P-wave velocities are attributed to fractures, but the authors suggested that stress-aligned microcracks, as always suggested by Crampin, may also be responsible, and that the fractures did not need to be 'open'.

### 14.7.4 Poisson's ratio anisotropy caused by fracturing

A useful contribution to the question of seismic anisotropy caused by fracturing is found in a description of geothermal development in the Olkaria field of the Kenya rift. This incipient caldera structure has had a number of injection episodes. It was monitored with a seismic network comprising 18 stations. Simiyu, 2000 gave $V_p$, $V_p/V_s$, and $\nu$ data from 2613 well-located micro-earthquakes, 45 quarry blasts and 25 calibration shots. Velocity-depth data, extending beyond 4 km depth, for six geothermal fields in the area (identified by initials) are reproduced in Figure 14.25. The well distributed stations made possible the presentation of less frequently seen, *azimuthally dependent* dynamic Poisson's ratio data, as shown in Figure 14.26.

For the case of geothermal reservoirs, $V_p/V_s$ and $\nu$ are more controlled by $V_p$ than $V_s$ due to the various fluid-phases in the reservoir. $V_p/V_s$ ratios increase when going from a vapour-saturated (*low pore pressure*) condition to a liquid-saturated (*high pore pressure*) condition. Water-dominated geothermal fields tend to have higher values of $\nu$ (i.e. 0.2–0.3), while steam-dominated fields have lower values of $\nu$ (i.e. 0.15–0.2). NW-SE and NE-SW trending structures associated with reservoir fluid channelling were suggested by the anisotropic distribution of $\nu$, according to Simiyu, 2000.

## 14.8 4C four-component acquisition of seismic including C-waves

There is a multitude of technical jargon in the geophysical industry, not least because of the advanced developments, the large number of practitioners, and the complex processes of seismic data gathering, and methods of interpretation. One of the simpler sources of confusion is double use of C – which singly used refers to *converted* waves (i.e. P-wave converting to S-wave mode, at the sea floor), and the 4C term meaning four-component seismic recordings, typically using ocean-bottom-cable (OBC).

A helpful diagram, illustrating each term, and also presenting an important 'new' seismic acquisition method, was found in Yuan, 2001, who analysed four-component seafloor data to determine the possible presence of vertical (TIV) or horizontal (TIH) transverse isotropy, in the presence of mode-converted shear waves.

The illustrated acquisition technique consists in principle of implanting four-component (4C) sensors into the seabed. These consist of one hydrophone, one vertical geophone, one in-line horizontal geophone, and one cross-line horizontal geophone. (See also Figure 14.22c). Conventional air gun arrays (P-wave generating sources) are used, towed by a shooting vessel, while a recording vessel stays above the receiver array. The receiver (OBC) array must of course be relocated, to give multiple-azimuth data if not already installed in multiple azimuth directions. There is a potential tendency for 3D OBC to give 'patchy coverage' in offset and azimuth (e.g. Hall and Kendall, 2003), though not in the case of the extensive survey illustrated in Figure 14.22.

An interesting proposal by Thomsen, 2002, based on imminent plans at BP, was to 'fuse' 4D (repeated surveys for reservoir management), with 4C. Although 4C technology had proved an economic success, e.g. for imaging beneath gas clouds, and helping to delineate anisotropic structure, it had not according to Thomsen caused a *revolution*. He also considered that 4D, though very useful

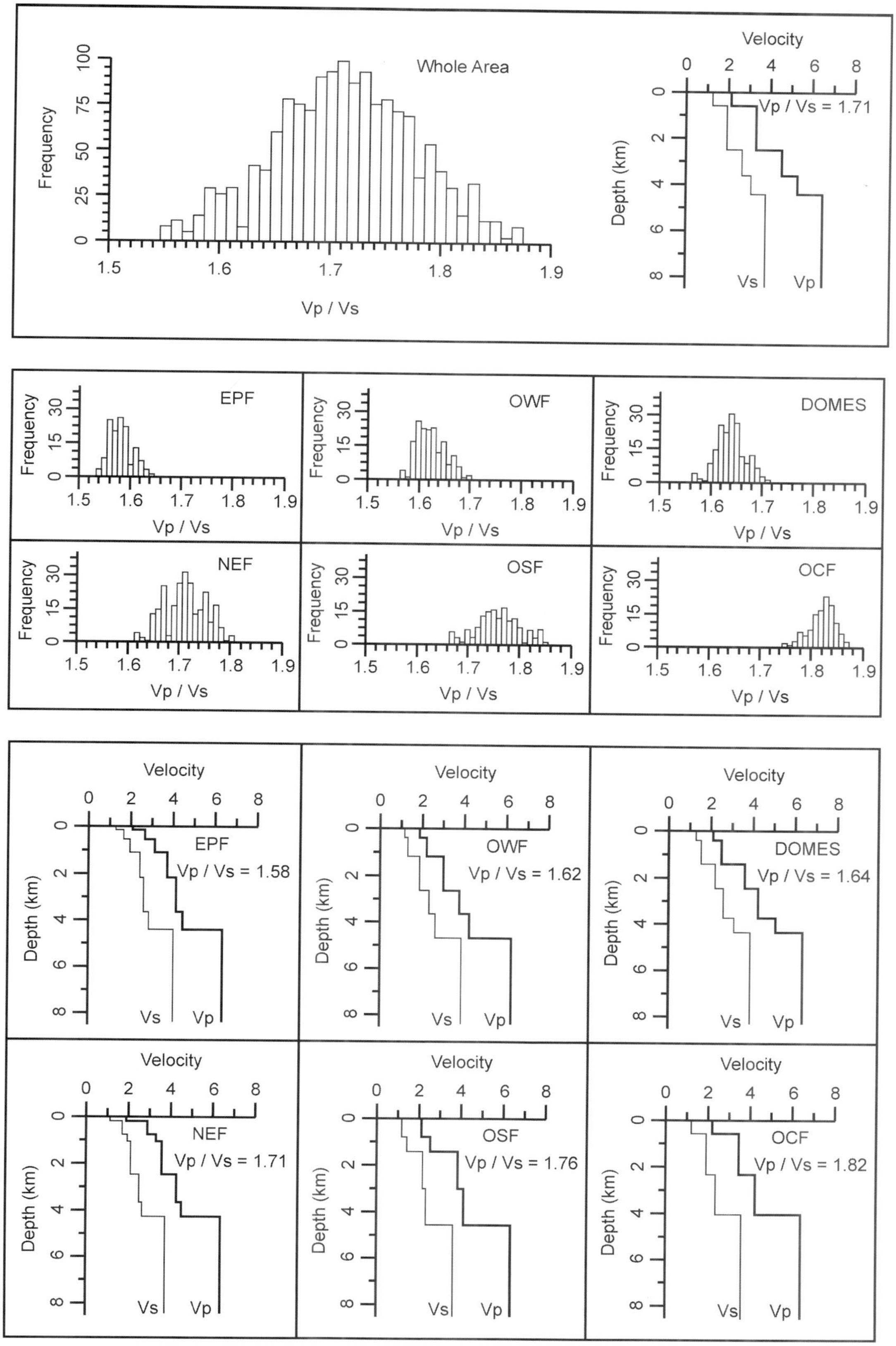


*Figure 14.25* $V_p$ and $V_s$-depth trends and $V_p/V_s$ histograms for six geothermal fields in the Kenya Rift. Simiyu, 2000.

for enhancing reservoir management, was expensive as each new survey cost as much as the previous one.

Thomsen's vision was to make subsequent 4D surveys much cheaper than the first, so that re-shoots could be done frequently. The key was to accept the investment of permanently installed receivers in the seafloor, so that the marginal cost increase for making 4D receivers also 4C, would be born by the economic gains made from 'dozens' of 4D re-shoots. Management needed to become confident in the gains to be made from frequent 4D4C surveys.

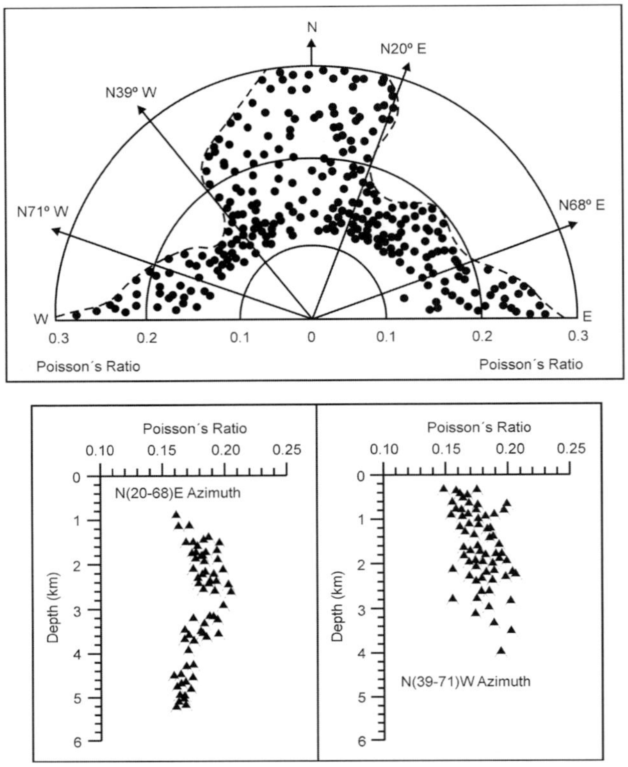


*Figure 14.26* Dynamic Poisson's ratio as a function of azimuth and depth. Simiyu, 2000.

Such a philosophy, based on huge investment until prices were driven down by marketplace economics, could revolutionize the practice of reservoir management according to Thomsen, with 'history matching' replaced by 'parameter estimation' and *prediction* of performance. The consequence of frequent full-field 4D4C re-shoots, providing full-field estimates of all required reservoir parameters, would result in much more efficient exploitation of reserves, a production increment coming sooner, and also possibly larger, than discovery and exploitation of new fields.

Regarding penetration of gas clouds, referred to by Thomsen, Granger et al., 2000 referred to the fact that even *2D*-4C surveys at the Valhall reservoir in the North Sea, had been able to penetrate through a gas cloud using the converted or C-waves, since the resultant shear-wave component could illuminate a target otherwise invisible to P-waves. In 1998, a *3D*-4C seismic survey was performed at Valhall with the intention of improving the structural imaging of the crest of the field, and for establishing the potential for jointly using P-wave and C-wave data for reservoir characterization, lithology prediction, and for stress and fracture orientation.

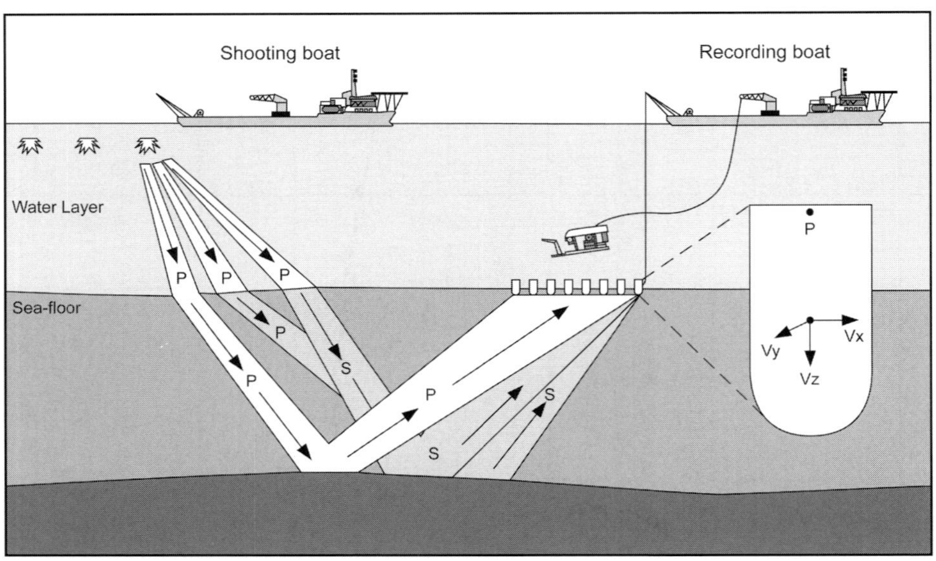


*Figure 14.27* Configuration of four-component (4C) seafloor seismic, for P-wave and mode-converted S-wave acquisition, using ocean-bottom-cables (OBC). Companie Général de Géophysique (CGG) diagram, from Yuan, 2001.

The P-wave part of this work was briefly reviewed earlier, see Figures 14.22 and 14.23.

## 14.9 4D seismic monitoring of reservoirs

In recent years, many more oil companies have been utilising Ocean Bottom Cable (OBC) acquisition methods to make repeated three-dimensional measurements over time, so-called four-dimensional or 4D seismic. (e.g. Bull-Gjertsen, 1998). This has made it possible to track reservoir depletion phenomena, such as changing pore pressure, and especially to track water-injection fronts in water flood treatments. Experiences from several North Sea oil and gas reservoirs have shown that the higher velocity of reservoir zones saturated with water allows operators to register where the oil-water contact is moving, and where oil may be by-passed by the water flooding.

In this section we will see a brief selection of uses for 4D seismic, including the mapping of oil saturation changes over time, the monitoring of water-flooding, and the detection and quantification of reservoir compaction, and even near-surface, sub-sea subsidence.

Reservoir monitoring with 4D seismic in its most basic form is the inversion of seismic data to obtain (dynamic) reservoir properties, which can subsequently be used to predict pore pressure change at a distance from the wells due to the assumed, laboratory-sample-based effective stress sensitivity of the reservoir rocks. While there is high sensitivity to effective stress in shallow reservoirs, the typical 'plateau' that may be reached at high effective stresses (i.e. beyond roughly 25 to 50 MPa, depending on rock type), suggests that the fluid compressibility effects may become more important at greater depth.

### 14.9.1 Possible limitations of some rock physics data

There is unfortunately a basic complication that the more compressible grain-boundary cracks with their low aspect ratios may be partly the result of stress unloading when drilled and bought to the surface (Holt et al, 1996, 1997, Nes et al., 2000), and it is such samples that are the basis for much of the collective assumption of a given stress sensitivity.

There is in addition the possibility that joint sets that are under lower levels of effective stress may still be contributing to some of the (anisotropic) sensitivity to effective stress change, and they certainly contribute, through finite compliances, to attenuation, as already seen in Chapter 10. This emphasises the need for both velocity and attenuation monitoring in 4D seismic.

In addressing stress sensitivity questions related to 4D seismic, MacBeth, 2004, developed semi-empirical pressure-sensitive bulk-moduli and shear-moduli formulations using the concept of *excess compliance,* which was designed to capture all categories of weakness in the rock, specifically in numerous types of sandstones. He found that the bulk and shear modulus pressure-sensitivities (i.e. sensitivities to effective stress change), lay between 1 and 10% per MPa.

This meant that a 5 MPa pore pressure decline in a reservoir could produce from 5 to 50% increase in bulk modulus, depending on sand/sandstone type. Rocks of higher porosity tended to show higher pressure sensitivities, and the reverse behaviour was also shown, as intuitively expected. MacBeth pointed out several potential problem areas related with using ultrasonic core measurements for interpreting changes of reservoir seismic velocities in 4D seismic monitoring. Some of these are summarized below:

1. In saturated rock, higher velocities are generally measured at higher frequencies.
2. Saturated samples containing microcracks would project a lower stress sensitivity with laboratory ultrasonics, than with seismic waves.
3. Cores loaded back to their original in situ stress state do not recover their original velocities, and due also to the possible microcracking caused by sampling, may show increased sensitivity to effective stress relative to in situ.
4. On the other hand, jointing is not a part of the laboratory sample response.
5. Cores may be selected from the most productive and competent part of the reservoir.
6. Subsequent core-to-log correlations, and interpretation using long wave-length seismic averages may cause underestimation of pressure sensitivity.

Most of the rocks studied by MacBeth were predicted to have only 1% per MPa sensitivity to stress change (at 24 MPa effective stress). However, elevated temperature testing may be needed; see Chapter 16.

### 14.9.2 Oil saturation mapping with 4D seismic

An impressive and very clear indication of the utility of 4D time-lapse seismic analysis of producing reservoirs was provided by Eide et al., 2002. The authors described quantitative mapping of changes in oil saturation over a 30 km$^2$ area in the northern part of the Gullfaks field in the North Sea. A base survey dating from 1985 before the start of production, was supplemented by *monitoring surveys* in 1996 and 1999, for assisting well planning, and these reportedly had a large economic impact on production.

The oil saturation mapping project utilised a rock physics model to compute changing density from P-wave and S-wave velocities at the three survey times, with porosity, percentage of clay, saturation and pressure effects extracted from a flow simulator. Synthetic seismic amplitude and amplitude difference volumes were generated by forward modelling, based on time-dependent elastic parameters.

The change in reflection strength at the top of the reservoir horizon, and the change in oil saturation in the upper cells of the reservoir simulation model, using segments with high correlation of seismic changes and saturation changes, were used to generate maps of saturation change. Statistical analysis indicated that the relationship between the *change in reflection strength* and the *oil saturation change*, depended on the height of the original oil column.

Coloured maps of saturation change from 1985 to 1996, were contrasted with those from 1996 to 1999, showing as expected, significantly larger areas and stronger changes (i.e. $\Delta S_o$ approximately as high as 0.35 to 0.5 in 4 to 5 regions) in the first 11-year period, compared to the less well distributed, small areas and small changes in the second, 4-year period. Probability maps of large, medium and low saturation changes based on a stochastic simulation were also developed.

## 14.10 4D monitoring of compaction and porosity at Ekofisk

Landrø and Stammeijer, 2002, described two seismic methods for monitoring compacting reservoirs, based on seismic data acquired in 4D surveys at Ekofisk. One was based on *prestack* travel time changes, the other on *post-stack* travel time *and* amplitude changes. Velocity-porosity relationships were not required. The authors emphasised that 4D time-shifts in a compacting reservoir were a combined result of increased velocity due to compaction, and reduced layer thickness, i.e. reinforcing effects. They referred to measurements of time-shifts of as much as 12–16 ms, between 1989 and 1999, related to an

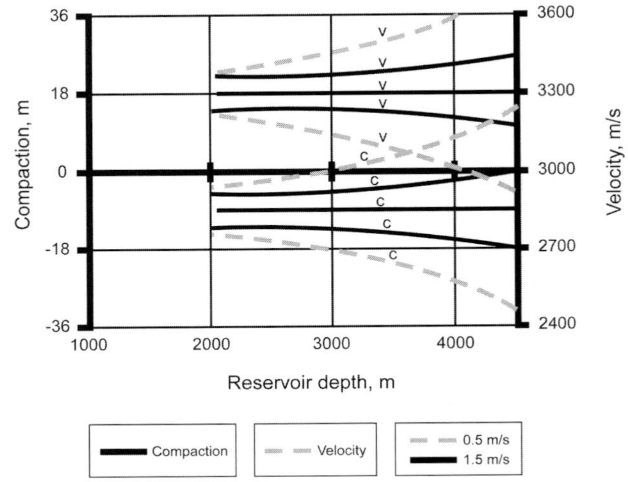


*Figure 14.28* Uncertainty in estimated velocity and compaction magnitudes, as a function of reservoir depths, with assumed excellent 4D data yielding a time-shift error estimate of only 0.5 ms (solid-lines). With poorer data, and 1.5 ms error (dotted-lines). Maximum offset assumed limited to 3000 m. Compaction: *black*, Velocity: *grey*. Landrø and Stammeijer, 2002.

estimated 6 m of additional compaction at 3 km depth. The authors related time shift, thickness change, and velocity change with the following normalized relation:

$$\frac{\Delta t}{t} \approx \frac{\Delta z}{z} - \frac{\Delta v}{v} \qquad (14.10)$$

The authors assumed a 4D time-shift error of 0.5 ms at near-offset, and 2 ms at far-offset, and a reservoir at 3000 m depth, with 9 m of compaction. A major challenge here, was to discriminate between compaction and velocity changes, both in the 300 m thick reservoir, and in the 3 km of overburden. Figure 14.28 shows a set of their compaction and velocity-change estimates for a more general case. As reservoir depth increased, the magnitude of uncertainty increased for both components.

Smith et al., 2002 described some of the detailed operating problems caused by the Ekofisk subsidence, including the difficulty of extrapolating seismic depth conversion from known horizon-depths at existing vertical wells, to predict depth structure *away* from the wells. Two hundred wells in the main field area, giving excellent depth control, contrasted with limited down-flank control, above this large domal structure. The authors found that simply scaling seismic interval velocities to well interval velocities, suggested larger structural variation than expected. Extrapolated depths

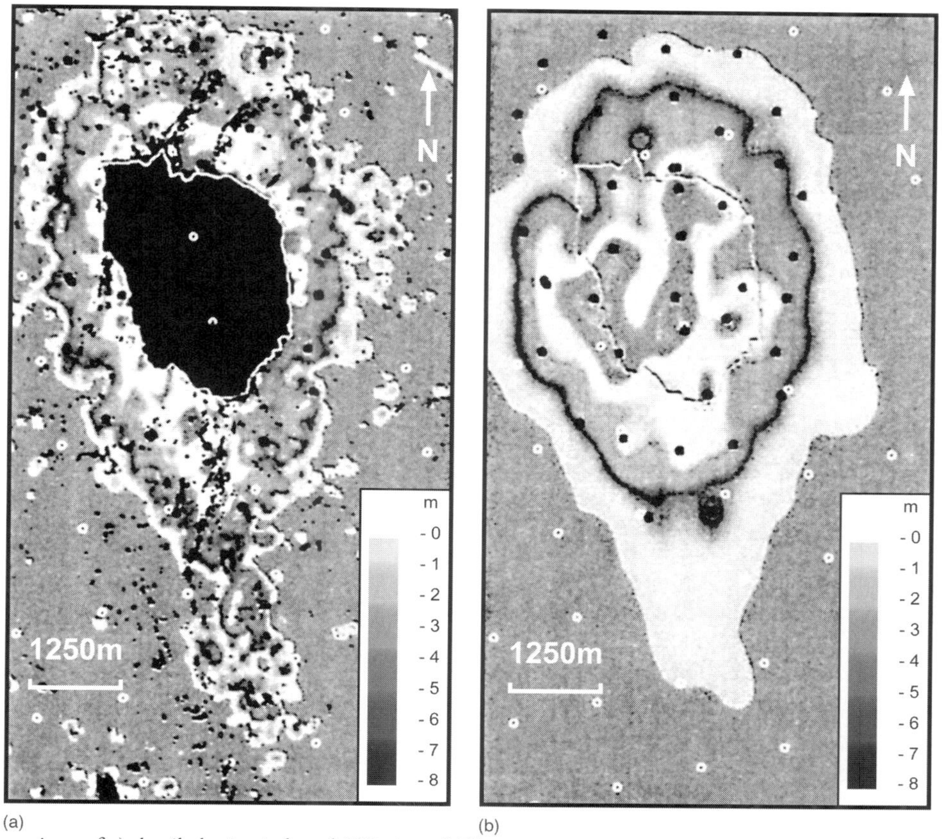


*Figure 14.29* Comparison of a) detailed seismic-based (4D time-shift) compaction interpretation (with adjustment for the velocity reduction caused by a subsiding overburden), with b) geomechanics-based one-dimensional strain compaction model, that included porosity reduction due to weakening effect of water saturation, seen in the next figure. Smith et al., 2002. Note gas cloud effect in centre of seismic model. (See Plate 4). Reproduced by permission from NPF.

down-flank were *deeper* than nearby exploration wells. The authors therefore developed a method for improving the seismic depth structure match to the horizontal sections of wells that penetrated far down-flank.

The authors mentioned that a straightforward time-lapse comparison between the 1989 3D seismic survey, with the newer 3D seismic survey in 1999, had given unrealistically large values for compaction and subsidence. Subsequently a geomechanics-based model was developed, based on vertical uniaxial strain behaviour for (intact) chalk, that related porosity changes to layer thickness changes. The authors mentioned use of a linear porosity-velocity relationship for the chalk, to compute depth conversion from velocity change. An example of the good general fit between the seismic time-shift compaction model, and the geomechanics-based compaction model, is reproduced in Figure 14.29. (see also Plate 4).

The water-weakening porosity-reduction model used at Ekofisk is shown in Figure 14.30, from a companion paper by Guilbot et al., 2002. The water injection programme, designed to compensate for the big pressure draw-down following about 20 years of production, had side effects, well known from early rock mechanics testing, of reducing the strength of the chalk.

Figure 14.30 shows how this water-weakening was tied to assumed porosity reduction, due to accelerated pore-collapse, based on tests of 'water-flooded' triaxially confined chalk. In fact the water weakening effect, although not positive in terms of arresting settlement of the sea-bed, has given a strong boost to the compaction drive mechanism, resulting in exceptional recovery of the reserves.

The extensive casing damage to numerous wells at Ekofisk is one set of evidence of discontinuous behaviour, due to stretching of the overburden and differential bedding plane slip. A new source of evidence for discontinuous behaviour during the compaction at Ekofisk can be seen in the results of the 4D seismic. A 1989 to 1999 'time lapse' comparison (Geo, 2001),

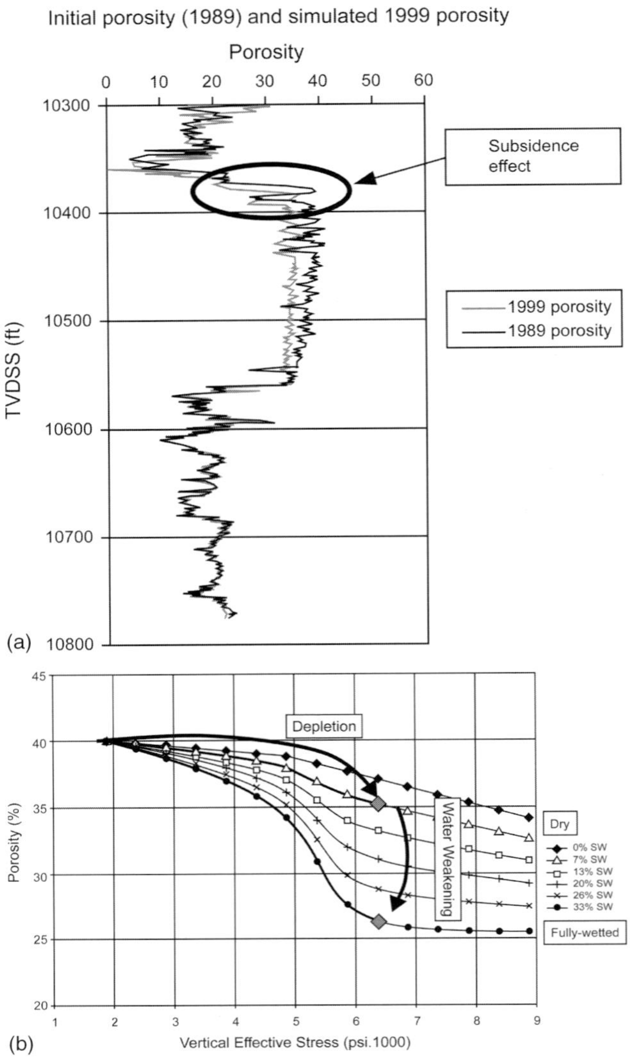


*Figure 14.30* The water-weakening model for estimating porosity reduction as a function both of vertical effective stress, and degree of saturation. Guilbot et al., 2002.

appears to show fault related discontinuities in the estimated 'tomogram' of compaction magnitudes, and also a somewhat larger compaction (>8 m in the central 1.5 × 2.0 km), as compared to the 'smoother' continuum calculations of the geomechanical model reported by Smith et al., 2002.

### 14.10.1 Seismic detection of subsidence in the overburden

It is interesting to note from Smith et al., 2002 that subtle overburden velocity changes had occurred due to the 'stretching' of the overburden in response to the 10 years of incremental compaction. Apparently the above-mentioned, initial over-estimates of compaction given by the initial 4D comparison of 3D-1989 and 3D-1999 data, did not account for this reduction in velocity.

Large-scale, 2D axi-symmetric modelling with 3 × 10 km 'layered-jointed-and-faulted' models with UDEC-MC (Barton et al., 1986, 1988), add support to possible reasons for a velocity reduction mechanism, as they showed a lot of locations in the over-burden with horizontal shearing, due to velocity-based moduli-contrasts between different layers. These events could presumably affect velocities determined from VSP, at most offsets. The combination of bedding plane shear and the stretching caused by the overburden possibly *subsiding nearly as much as the compaction,* were presumably also the inherent source of numerous casing collapses mentioned by Ekofisk authors in the past. At the Wilmington field, where hundreds of casings collapsed, seismic-magnitude events were recorded due to bed slip, one of which measured 24 cm.

Stretching causing slight opening of numerous sub-vertical, bedding limited joints, and a general horizontal stress reduction on sub-vertical faults, would presumably only affect velocities determined from wide-offset (large aperture) VSP.

At Ekofisk there was inevitably a mismatch of *3D isotropic continuum* based subsidence modelling (mostly giving S/C ratios of about 0.6, and *2D discontinuum modelling*, which suggested S/C ratios as high as 0.75, increasing to beyond 0.85 as compaction progressed. (Barton et al., 1988). The 'simpler' 2D discontinuum modelling matched the steep 1985 subsidence bowl better than the 3D continuum models, which is understandable considering the *150* km$^3$ of (obviously discontinuous) rock involved.

This total volume of deforming rock, has required billions of dollars of extra investment, due to the late 1980's need to jack all platforms and risers 6 m, protect the central storage tank with a 100 m diameter concrete wall, and finally establish new platform facilities away from the central subsidence during the 1990's. However, the compaction is fortuitously giving much more in return, with greatly increased recovery.

The huge volume of rock that is deforming, happens to be some $10^{15}$ times larger than the core-plugs used for uniaxial strain testing. Possibly the high stress levels are responsible for a *relative lack of scale effect*, when these small tests are used for compaction modelling. Alternatively, the effects of joint deformation in the compaction, are in some way 'cancelling' a possible scale effect.

### 14.10.2 The periodically neglected joint behaviour at Ekofisk

Ekofisk is a classic jointed (or fractured) reservoir, having too low matrix permeability for production without the jointing. Yet jointing tends to be 'ignored' when modelling, due to the large scales involved (possibly as much as 15 km$^3$ of chalk, with 300 m thickness). A more complex aspect that is inevitably missing from large-scale geomechanical models, which due to size constraints, are usually *isotropic elasto-plastic continuum* in nature, is the actual jointed nature and possible coupled behaviour of the Ekofisk joints, and their special behaviour during compaction, especially when stimulated by HTM effects during water-flooding (i.e. cooling, reduced shear strength, greater permeability).

The joints are typically in the form of conjugate sets of steeply dipping joints, making small block sizes of e.g. 15 to 40 cm, whose roughly common strike reportedly 'rotates' around the pear-shaped flanks of the anticlinal structure. The rotation is associated with the rotating directions of maximum horizontal stress, also shown by Teufel and Farrall, 1992, which is reproduced in Figure 14.31.

Two-dimensional distinct element (jointed) modelling with UDEC-BB performed for the Norwegian Oil Directory, soon after the seabed subsidence was discovered (e.g. Barton et al., 1986, 1988), had shown that *down-dip shearing* could occur with various joint geometries and with the higher porosities. The one-dimensional compaction modelling was performed using the effect of a 20 MPa reduction in pore pressure experienced prior to water flooding, causing a 20 MPa increase in effective stress. In subsequent more comprehensive UDEC-BB modelling at NGI, performed by Gutierrez in the 1990's, 24 MPa was used, confirming the joint-shearing trends of the earlier modelling. The horizontal stress acting in all these models was, in effect, the rotating major horizontal stress shown in Figure 14.31.

Input data for the 1985 models were obtained from tests on jointed core recovered from Ekofisk, using joint index testing (e.g. joint wall roughness and tilt tests), direct shear tests, and coupled-shear-flow-temperature biaxial tests (CSFT, Makurat et al., 1990), using hot Ekofisk oil or carbonate-equilibrated sea water. The

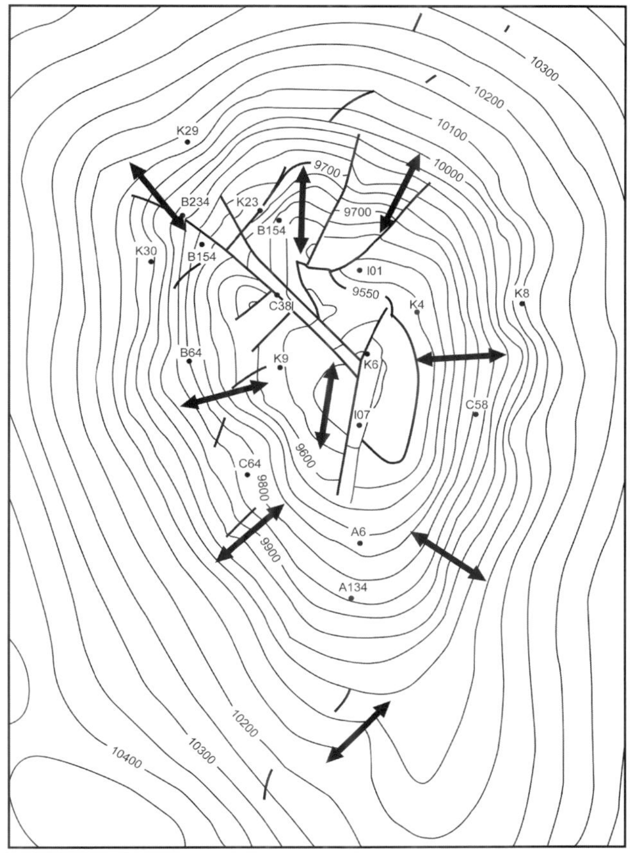


*Figure 14.31* The anticline-related rotation of the maximum horizontal stress directions at reservoir levels at Ekofisk. Teufel and Farrell, 1992 and Bruno and Winterstein, 1994. These stress rotations relate also to the rotating strike of the steeply-dipping conjugate joint directions, which give a rotating trend for principal permeability directions. A rotating trend for P-wave anisotropy, and for polarized shear waves would also be expected. (See Chapter 15 for rotation effects measured in the overburden).

Barton-Bandis joint strength and stiffness parameters derived from JRC, JCS and $\varphi_r$, were scaled from assumed in situ block sizes, for each set of conjugate joints.

The less than obvious *discontinuum* component of the compaction mechanism, was due to matrix contraction providing 'space' for the down-dip shear, despite the one-dimensional ('roller-boundaries') constraint. Significantly, the down-dip shear mechanism caused the development of a higher $k_0$ (ratio of horizontal to vertical stress), than in the case for '1-D' compaction of unjointed chalk, thereby helping to stabilize the mechanism. (Some of this modelling is shown in the joint behaviour related geomechanics material, treated in Chapter 16).

It seems highly likely that the multitude of (*inevitably*) shearing joints in this pear-shaped 9 × 14 km and 300 m thick reservoir, may be an important reason for the rapid weakening by water during water flooding, due to the huge increases in surface area exposed to water, caused by the multiple joint surfaces, with strength loss and increased shear, following mechanisms interpreted at larger scale by Heffer 2002, reviewed shortly.

The 'several-micron' sized apertures modelled in the 1980's, were apparently almost maintained by shearing due to joint-surface non-planarity, even without subsequent 1990's and present cooling effects from water-flooding. Sufficiently permeable jointing (and fractured rock near faults), will also be potential causes of faster water break-through to producer wells. Faster break-through of the water front close to faults was referred to by Guilbot et al., 2002.

These authors reported that about 9 ms of time-shift had occurred at the top of the reservoir in the 10-year period between the two 3D surveys, but mentioned a range of 0–20 ms in the water-flood areas. Strongest compaction (i.e. largest time-shift) was usually interpreted where reservoir thickness and porosity was highest. As a possible corollary to this, the above joint-shearing mechanism was also found to occur with most strength where modelled porosity was high (e.g. 40%) rather than low (e.g. 25%).

There are undoubtedly multiple reasons for the higher compactions where porosity and thickness was greatest, and also multiple reasons for the continued reservoir compaction, despite (or because of?) extensive water-flooding from 1990. Barton, 2002b suggested the following:

- accelerated pore collapse
- joint shearing compaction
- joint permeability maintenance (due to dilation counteracting effective normal stress increases)
- water-weakening (accelerated through a vast area of joint planes)

Maury et al., 1996 have pointed out the inevitability of compaction increase when water-flooding water-sensitive reservoir rocks. Strength reduction actually applies to the majority of rocks, according to the experimental data reviewed by Barton, 1973, but perhaps is more extreme for a very porous weak material like chalk.

From a change-of-seismic-velocity point of view, the effect of each of the above mechanisms must be extremely complicated, possibly only pore collapse causing potentially increased velocity, assuming break-down of the rock frame is not itself a source of velocity reduction: probably it is initially.

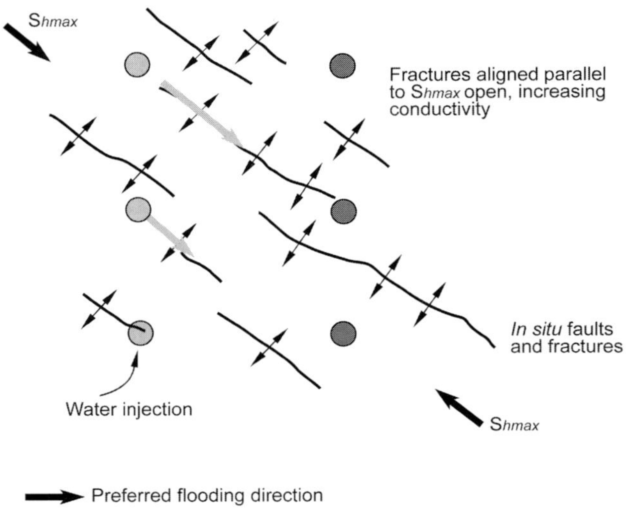


*Figure 14.32* The conventional and frequently occurring fracture and fault opening parallel to the $S_{h\ min}$ direction with water flooding. Heffer et al., 1997, Heffer, 2002.

## 14.11 Water flood causes joint opening and potential shearing

In water-flooding, for *stimulating* petroleum production, indeed for *driving* production, there is both a local increase in pore pressure at the injector wells, and a reduction in temperature, causing some contraction of the matrix, both of which help to dilate and indeed create, fractures. Gutierrez and Makurat, 1997, have demonstrated these effects in fully-coupled MHT modelling.

Heffer, 2002, referred to earlier studies of water-flood effects in eighty reservoirs, both fractured and unfractured (Heffer et al., 1993, 1995), and discussed the important question of *flood directionality* in relation to the major horizontal stress level. The conventional and expected mechanism of fracture or fault opening *exactly* in the direction of $S_{h\ min}$, as shown in Figure 14.32, was not in fact so common as one might expect.

Figure 14.33 shows the important statistical spread of observations, from 33 cases of reservoirs known to be fractured, and from 47 cases thought to be unfractured, at least *prior* to the water-flooding.

There are in fact subtle and important indications in these two 'joint rosettes' (in this case *fault or fracture*

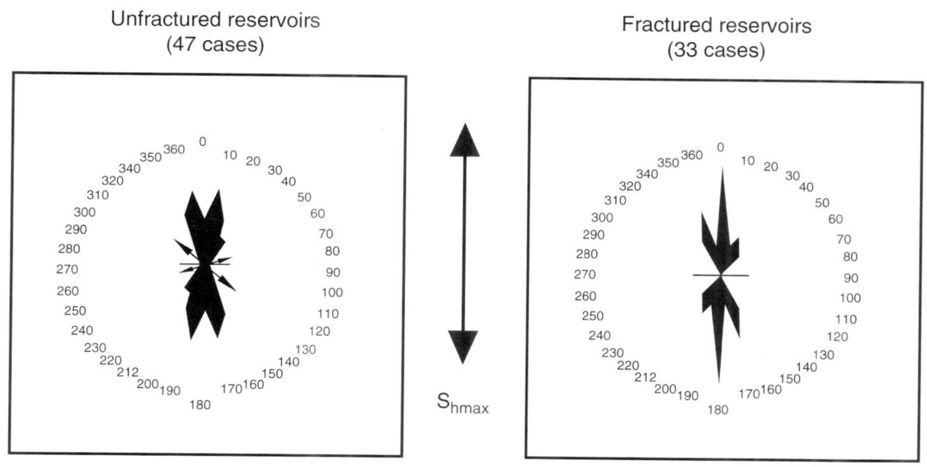


*Figure 14.33* Statistical data for water flood directionality effects in relation to $S_{h\ max}$ directions in 47 unfractured reservoirs, and in 33 fractured reservoirs. Heffer et al., 1997, Heffer, 2002. See also extensive data sets in Tingay et al. 2005.

rosettes), that the unloading caused by the two water-flood effects mentioned above, are in many cases stimulating shear-failure (of 'intact' rock), or shear displacement, and therefore further dilation, of existing joint or fracture sets, or faults.

The direction of $\sigma_{H\ max}$ shows a frequent tendency to have *bisected* the geologic features that are the basis for the rosettes shown in Figure 14.33. The mechanism may (often?) be one of conjugate shear, in which case showing strong parallels to the findings of Townend and Zoback, 2000 concerning the frequency of water conducting features being under shear stress. These topics are addressed in detail in Chapter 16.

## 14.12 Low frequencies for sub-basalt imaging

Many prospective ocean margins are covered by large areas of basalts. These tend to be extremely heterogeneous, and scatter and attenuate the seismic energy of conventional seismic reflection surveys, making it difficult to obtain seismic images from deeper reflectors. Ziolkowski et al., 2003, argued that since high frequencies were scattered more than low frequencies, it would be logical to emphasise low frequencies, by using much larger air guns, towing the source and receivers at greater depths than usual, i.e. 15–20 m. This was done in 2001, over an area of the NE Atlantic margin that holds promise of very large accumulations of hydrocarbons in the Mesozoic and Palaeozoic sediments, which are covered by higher velocity Cenozoic flood basalts.

The new sub-30 Hz data was shown to suffer significantly less attenuation due to scattering, and revealed deep reflections not previously seen in conventional surveys. One of the problems identified in numerous previous attempts had been the scattering and attenuation from the rough basalt surface, and absorption due to faults and joints, plus interference from inter-bedded units of e.g. claystone and siltstone.

Strong sea-surface reflections had also been a problem. Basalts of this region are very heterogeneous at scale lengths of tens of metres, approximately an order of magnitude less than the seismic wavelength when of $<10$ Hz. So use of the long wave-lengths could avoid the problems of thin inter-layering and lateral inhomogeneity.

The authors generated synthetic seismograms, using an available sonic well log in basalt, and assumed a single deep reflector. They used the following model:

1. 0–400 m $V_p = 1500$ m/s
2. 400–1000 m $V_p = 2500$ m/s
3. 1000–1800 m $V_p = 3500$–$5500$ m/s (strong oscillation in basalt flows, greater extremes)
4. 1800–5000 m $V_p = 3000$ m/s (assumed sedimentary basin)
5. 5000–6000 m $V_p = 4000$ m/s

The modelled basalt was 800 m thick. Their subsequent field survey, using large air guns towed at 15 m depth, were able to image deep reflectors in a much clearer manner than achieved in a conventional survey.

We may conclude this section with an exotic use of reflection seismic imaging, described by Dypvik et al.,

1996 and Tsikalas et al., 1998, namely the imaging, in the Barents Sea, of the 40 km diameter Mjølner impact structure, originating some 145 million years ago, which involved an 850–1400 km$^3$ disturbed volume of rock, possibly caused by 'only' a 0.9 to 3 km diameter meteor. A transient crater of 16 km diameter and 4.5 km depth, and ejecta of some 175 km$^3$, were interpreted from the seismic images. Large amplitude tsunami waves, and energy release of some $10^{20}$ or $10^{21}$ Joules, equivalent to an 8.3 magnitude earthquake were estimated by the authors.

## 14.13 Recent reservoir anisotropy investigations involving P-waves and attenuation

Mapping the azimuthal velocity anisotropy of P-waves, to reduce risk of drilling low-productivity wells in relatively unfractured parts of the UK offshore Claire Field, west of the Shetland Islands, was described by Smith and McGarrity, 2001. Their company, BP, had performed multi-azimuth walk-away ('float-away') VSP in an offshore well to attempt a calibration with oriented core data and with FMI logs. Three 4 km walk-away lines were performed, using a downhole triaxial accelerometer sensor array, both in the overburden (1486–1535 m) and in the fractured reservoir (1850–1906 m). They described the obvious importance of 'open conductive fractures' with producing-well intersection of these features as the goal.

The authors cited azimuthal variation in the *shear modulus* of the fractured rocks as the reason for the P-wave velocity anisotropy. 'P-waves travelling obliquely to the fractures can be influenced by the shear properties of the rock'. This is perhaps an alternative way of acknowledging the theoretical contribution of the shear stiffness (and normal stiffness) of the fractures to the anisotropy, which also causes shear wave splitting, as we shall see in Chapter 15.

The authors also mentioned the more conventional dominance of one fracture set orientation, with an orthogonal subset, and mentioned variation of *fracture density* in this unequal two-set system, as a reason for variations in the degree of seismic anisotropy. We will investigate *fracture density* in detail in the next chapter, mostly in connection with the interpretation of shear-wave anisotropy.

A schematic classification of fracture types in the Claire reservoir, shown in Figure 14.34, suggesting considerable complexity, reduced nevertheless to the predominance of

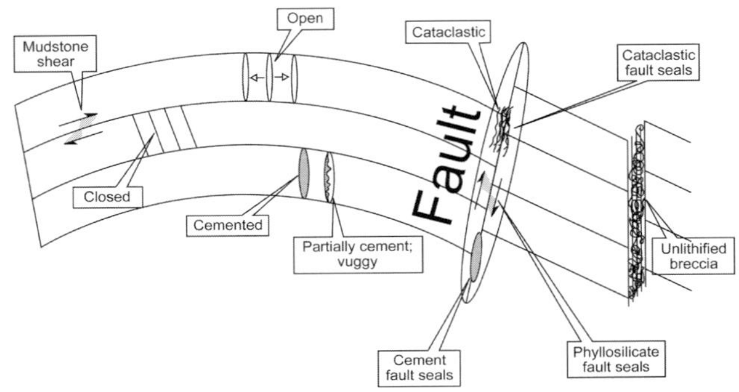


*Figure 14.34* Fracture-type schematic for the Clare field reservoir, given by Smith and McGarrity, 2001.

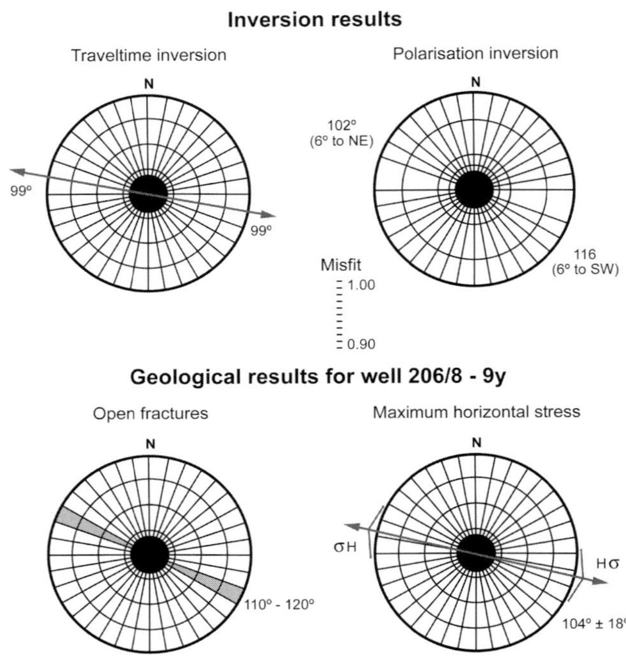


*Figure 14.35* Polar plots showing orientation from core, FMI fracture orientation, and maximum stress orientation (90° to calliper-log elongation?), compared with travel time inversion ($V_p$ anisotropy) and polarization inversion, using postulated velocity models to fit the direction of wave front motion at VSP downhole receivers. Smith and McGarrity, 2001

(almost) $\sigma_H$-parallel 'open' fractures as the clear source of preferred orientation for both the travel time and the polarization inversion, as shown in Figure 14.35. The magnitude of the (P-wave) velocity anisotropy, typically 6%, actually ranged from 0 to 15%, and the authors were able to demonstrate reasonable correlation of high-flow wells with higher (P-wave) velocity anisotropy and low-flow wells with lower anisotropy.

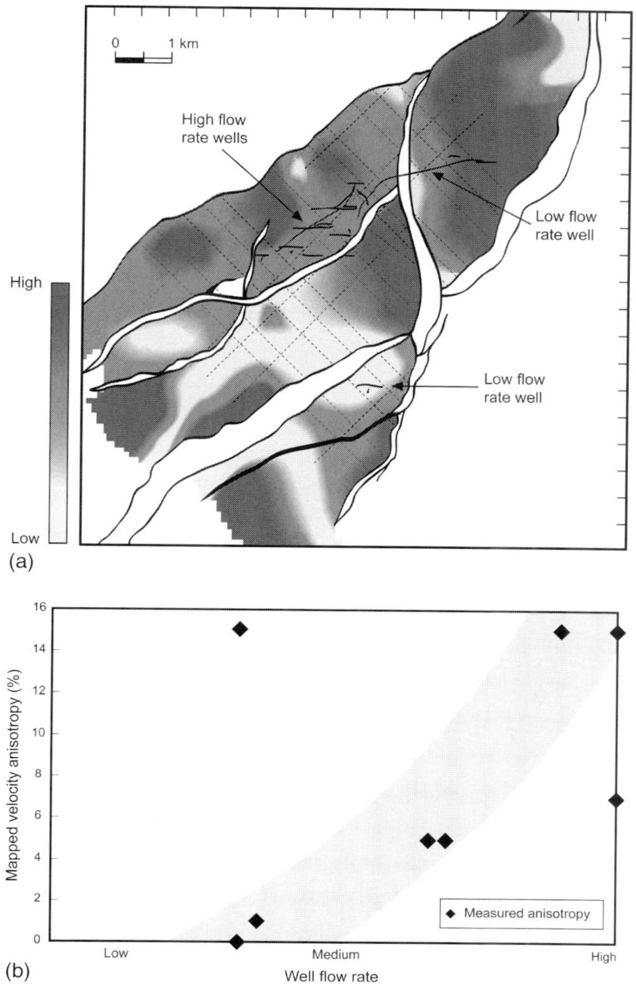


*Figure 14.36* a) Density of shading represents degree of P-wave velocity anisotropy at reservoir (unit V) level. b) Well flow rates versus estimated P-wave velocity anisotropy. Smith and McGarrity, 2001.

The very rough correlation between velocity anisotropy and well flow rate, are shown in Figure 14.36b. It was therefore suggested that conventional P-wave seismic acquired by towed arrays in marine environments may help characterize fractured reservoirs, thereby maybe avoiding the need for expensive multi-component shear-wave acquisition.

Following on from the above *azimuthal anisotropy of P-waves* at the Claire Field, as discussed by Smith and McGarrity, 2001, a later study by Maultzsch et al., 2005, with BGS colleagues from Edinburgh, addressed the question of *attenuation anisotropy* at this field. They also used the multi-azimuth walk-away VSP, and demonstrated that the fractured, oil-saturated reservoir also showed a consistent azimuthal variation in attenuation.

These authors first demonstrated the $1/Q_{seis}$ related modelling abilities of the Chapman dynamic poroelastic matrix-and-aligned-fractures model, which is reviewed in Chapter 15. A simulation of P-wave attenuation in the case of a modelled rock mass with vertical fracturing is shown in the polar diagram reproduced in Figure 14.37, showing in b), a maximum attenuation $1/Q = 0.15$ for the vertical direction of fractures.

First they measured a zero-offset $Q_{seis}$, using data from all source offsets and azimuths. They obtained $Q_{seis} = 18$ for the reservoir which was fractured (again a very plausible match to deformation modulus in GPa), and $Q_{seis} = 35$ to 40 for the overburden, which they assumed was relatively unfractured. Secondly they analysed attenuation as a function of azimuth for each offset, both in the reservoir and in the overburden, utilising two tool settings. In the reservoir, minimum attenuation lay between N70°E and N100°E. But at the shallower receiver setting in the overburden the

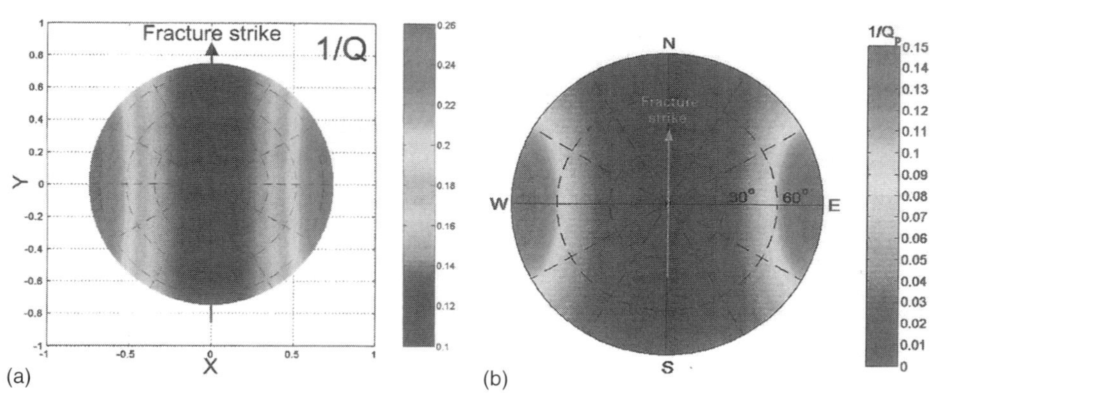


*Figure 14.37* Two examples of attenuation modelling with a set of vertical aligned fractures, using Chapman's dynamic poroelastic matrix-and-fracture-set model. See Chapter 15 for a description of this model. (Reproduced by kind permission, Maultzsch pers. comm. 2005). (see Plate 5).

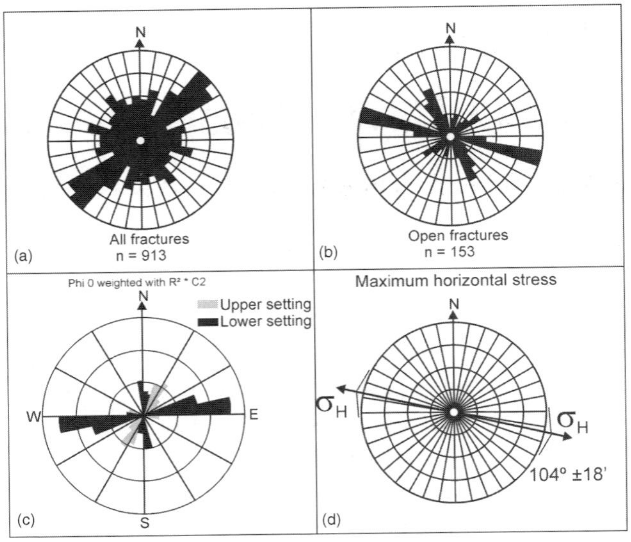


*Figure 14.38* a) Fracture rose diagrams for all fractures, and b) open fractures at the Claire Field. c) measured minimum attenuation direction from multi-azimuth walk-away VSP. d) major horizontal stress estimate. (Maultzsch et al., 2005, and Smith and McGarrity, 2001).

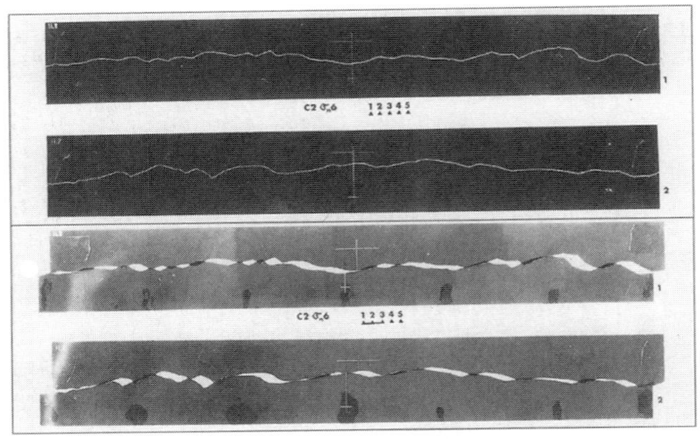

*Figure 14.39* A sheared, rough fracture analogue for 'open' fractures, that shows the dilated parts (white) as having a different average orientation compared to the black, contacting (partly crushed, load-bearing) parts of the asperities. Barton, 1973.

minimum attenuation was scattered between wider azimuths, as shown in Figure 14.38a.

Rose diagrams of all the fractures and of the open conducting fractures, from others' logging of cores and borehole images are reproduced in Figure 14.38a and b. Maultzsch et al., 2005, also provide, in Figure 14.38c, their measured azimuth of minimum attenuation, which although close to a match with the open fractures, is *actually some 20° oblique to these.* Curious to see where the major horizontal stress is oriented, we can add this as in Figure 14.38d, from Smith and McGarrity, 2001, just reviewed. It is close to parallel to the open fractures, yet the minimum attenuation was 20° different.

A possible explanation is that the 'open fractures' are open due to limited (but sufficient) shear dilation episodes in the past, when $\sigma_{H\,max}$ was perhaps not parallel to these joints. From a rock mechanics point of view, and armed with several experiences of measuring a considerable $\sigma_{h\,min}$ magnitude at reservoir depths using hydraulic fracturing, one must question whether there can be 'open' fractures in a petroleum reservoir at right angles to $\sigma_{h\,min}$, unless:

a) the rock is unusually strong and that joints are rough,
b) there exists a close-to-fracturing pore pressure,
c) that *shearing has occurred,*
d) there is a suitable quantity of hard mineralization to 'bridge' and maintain an earlier porosity.

Fractures can presumably be conceived as 'open' if they have good connectivity and reasonable apertures (for example 10 to 100 μm conducting apertures?) with respect to the gradients that are operating.

If limited shearing and therefore dilation has occurred in the past, and if there are limited influences from a second set of fractures, then Figure 14.39 can be a possible explanation for the apparent 20° rotation in the above case. Note the possible influence of a rotation of the fluid-bearing parts of the fractures (black), in relation to the contacting parts taking the load. This model will be developed further in the last two chapters of this book.

# 15 Shear wave splitting in fractured reservoirs and resulting from earthquakes

In this chapter the effect of structural anisotropy on *shear wave splitting and polarization phenomena* will be treated in some detail, due to its extreme importance in helping to characterize jointed or fractured reservoirs, and due to the improved insight it is giving into earthquake phenomena. The structural anisotropy may be stress aligned, and there are then logical ties to the principal permeability or drainage directions. Dominant jointing and natural fractures are of increasing interest to petroleum companies, both for production and for aiding stimulation, where matrix permeability is low but hydrocarbon storage high. This vitally important structural feature is notoriously poorly sampled by vertical core and well-bore scanning, since itself often sub-vertical or vertical. The 'miracle' of shear-wave splitting (with assistance from azimuthal AVO P-wave surveys) has provided the means of detecting the presence of these compliant, fluid-bearing fractures. Contrary to the classic wisdom of porous media fluid substitution theory, *in situ* fractures also seem capable of signalling to the *shear-waves*, whether they contain gas or brine, through subtle velocity reduction of the *slow* S-wave, due to the fluid-compressibility-altered fracture compliance magnitudes. Geophysicists utilise an unfortunately ambiguous way of describing *fracture density*: as number per unit volume times radius cubed. This ambiguity, meaning that millions of microcracks or a hand-full of fractures can give the same magnitude, nevertheless seems to have a remarkable proportionality to shear-wave anisotropy, but clearly this can be altered by changed compliances. The need to define a specific volume of fractures for reservoir understanding is urgent. Consequently, recent numerical dual-porosity poro-elastic modelling developments have become increasingly important for exploring the frequency-dependent velocity and attenuation resulting from the various potential scales of anisotropy. Case records both from seismology and petroleum engineering will be used to show recent trends in analysis of seismic survey data. Relevant rock mechanics experience with coupled stress-deformation-flow testing of rock joints will be referred to in several contexts, where deemed appropriate. There will also be a strong focus on the possible links between the joint or fracture *shear and normal compliance* used by geophysicists, and the macro-deformation, and inverted *stiffnesses* used for many years in rock mechanics models of jointed media. The need for compatible measures of *volume-defined* fracture densities and *in situ* values of compliance (as opposed to those obtained from hand-sized joint samples or roughened plates of Lucite) is necessary for further development.

## 15.1 Introduction

In view of the very widely accepted knowledge that vertical and sub-vertical jointing is extremely common in most rock masses (due to such diverse effects as cooling, bed-flexure and tectonic influences of horizontal stress), it is *unfortunate* to say the least, that vertical boreholes are usually the first, and seemingly also the second choice, for sampling and gaining access to the sub-surface. As pointed out by numerous authors, and also quantified by R. Terzaghi, 1965, the *sampling bias* caused by the mismatch of borehole diameter and horizontal spacing of vertical structure, *and* the vertical borehole itself, is extreme.

Our vertical boreholes provide such a poor sample of the *jointed* sub-surface, that P-wave azimuthal anisotropy on the one hand, and shear-wave splitting caused by the presence of vertical or aligned structure on the other hand, are truly god-given means for rectifying our poor sampling strategies. If the 'economy' of a vertical well as opposed to a steeply inclined one, and the subsequent 'cost' of an extensive seismic survey were combined, there would perhaps be more reason for rapidly deviating our boreholes at least 10° or 15°, in order to sample the increasingly understood relevance of vertical and sub-vertical structure on hydrocarbon production. As emphasised by Laubach, 2003, a central challenge of sub-surface fracture characterization is obtaining data on essential fracture attributes where direct observation is (remarkably) unlikely.

Recently, five boreholes inclined just 30° from the *horizontal*, with a combined length of 1.4 km were used to investigate the steeply folded geology and structure along the route of a rail tunnel in Norway. The value and relevance of the core logging and permeability measurements was thereby increased 10- or 100-fold in relation to the equivalent length of vertical holes. Of course this is not a recipe for deep oil-field wells, but a fairly early deviation of 10° or more would greatly improve understanding of both the overburden jointing and the reservoir jointing. The potential anisotropy of the overburden cannot be ignored in seismic inversion.

As Barkved et al., 2004 pointed out in their review of multicomponent data, countless reservoirs have been discovered, characterized, and monitored by P-waves. However, P-waves cannot solve every seismic imaging or reservoir description problem. With the addition of S-waves, (usually in the form of converted PS-waves, see Chapter 14), oil and gas companies have found an enormous quantity of new reserves that could not have been found with P-waves alone. The new reserves have been more effectively exploited by better identification of fracturing, and therefore better placement and deviation of the wells. Not infrequently, shear-wave technology provides information where shallow gas has obscured P-wave imaging over central parts of a field. A particularly good example is the Ekofisk reservoir.

Shear waves bring additional knowledge to a seismic study due to the different rock mass properties that are sampled. The traditional view is that shear-wave velocity remains unchanged whether a formation contains gas, oil or water. However, as we shall see, because of the effects of fluid compressibility on the normal stiffness of fractures (or its inverse: compliance), the shear waves passing through a fractured or jointed medium, *will* give the ability to distinguish between oil (low compressibility) and gas. (Van der Kolk et al., 2001).

There is possibly a small point of controversy here, concerning whether the fractures and shear-waves can *both* be vertical, when theoretically only shear compliance is sensed. It is uncertain if shear compliance will be sufficiently affected by the fluid compressibility. A more certain effect of fluid compressibility – i.e. the ability to distinguish between oil and gas is when normal compliance is involved in the case of sub-vertical fractures (or sub-vertical shear-waves) giving a finite incidence angle (Sayers, 2002b).

Shear waves were once considered just noise and had to be filtered out. Now, with proper multicomponent recording, S-waves can deliver important information both concerning rock and fracture properties, and concerning fluid type in the case of shear-wave splitting.

## 15.2 Shear wave splitting and its many implications

A landmark paper on the relatively new technique of shear wave splitting, summarising the first ten years of developments, was given by Crampin and Lovell, 1991. Following theoretical developments of Keith and Crampin, 1977, and the suggestions of Crampin, 1978, shear wave splitting was positively identified above small earthquakes by Crampin et al., 1980, and many times subsequently. Shear wave splitting had already been noted in a number of sedimentary hydrocarbon basins, with Lynn and Thomsen, 1986, and Willis et al., 1986, reporting at the 56th SEG (Society of Exploration Geologists) meeting in Houston.

A simple schematic of the shear wave splitting principles, given by Crampin and Lovell, 1991, is reproduced in Figure 15.1a. When a steeply inclined shear wave meets sub-vertical, aligned discontinuities, the shear wave splits into two components, which have different arrival times and different, usually orthogonal polarization. Both these aspects are shown in Figure 15.1.

A more complete version of this simple diagram, from Barkved et al., 2004, is also reproduced in Figure 15.1. This shows, depending upon ones preference for microcracks or joints, the simplest source of polarized shear waves: a set of vertical joints or fractures. These, like microcracks, cause the transmitted shear-wave to split into a fast and slow component, registered as time delay, due to the attenuating effect of fracture compliance on the S-wave component that has particle motion *perpendicular* to the fracture strike. The difference in travel-time between the fast and slow waves (termed $qS_1$ and $qS_2$ elsewhere in this book), is strongly related to fracture density, and also to fracture compliance, since both have an attenuating effect.

An often referred aspect of shear wave splitting and polarization is that the faster of the two polarized components is parallel or sub-parallel to the direction of maximum horizontal stress and/or to the preferentially-oriented fluid-filled microcracks, cracks or sets of sub-vertical joints or fractures. The shear-wave velocity anisotropy is often in the range 0.5 to 5%, sometimes

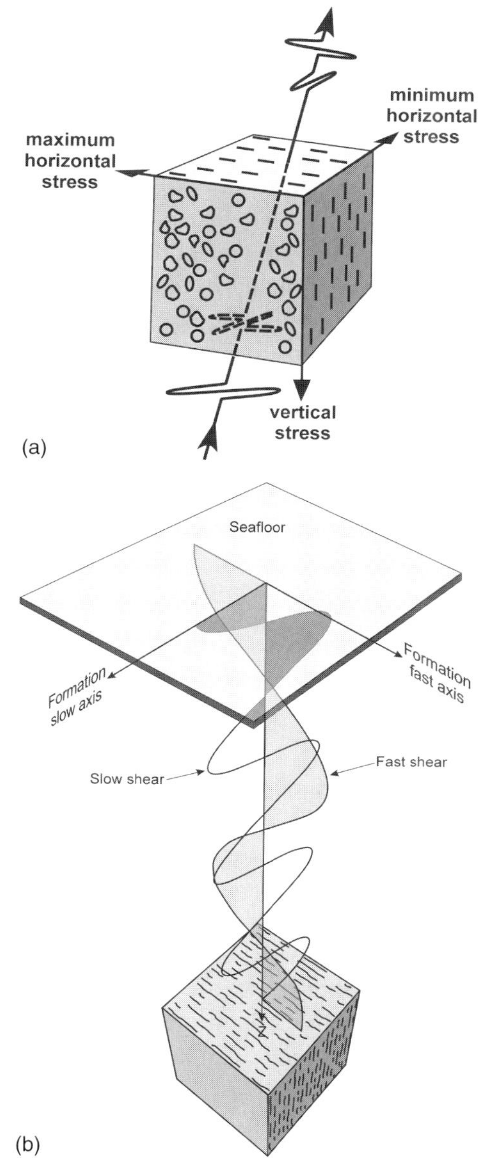


*Figure 15.1* a) Simple schematics of shear-wave splitting principles, as a direct or indirect result of the principal stress and the associated stress-aligned, fluid-filled cracks. Crampin and Lovell, 1991. An updated version of this classic diagram, from the BGS Anisotropy Project in Edinburgh, shows larger 'cracks', signifying not just aligned microcracks, but perhaps an aligned joint set, also assumed to be roughly parallel to maximum stress. b) A more comprehensive diagram of the principles of shear-wave splitting, from Barkved et al., 2004. The fast $qS_1$ particle motion is polarized in the average direction of fracture strike, while the slow $qS_2$ particle motion is polarized perpendicular to the average fracture strike. Note the 'formation fast axis' substitution for major horizontal stress. They may of course be synonymous.

much more. The distribution of stress-aligned cracks or inclusions is referred to by many authors as 'extensive-dilatancy anisotropy' (or EDA), following Crampin's publications.

Fracture (or joint) characterization using seismic methods is considerably more complicated than for the isotropic case, due to the existence of the three distinct body waves, which propagate with different velocities and polarizations. Following general convention, these waves are referred to as the qP (quasi-compressional), $qS_1$ and $qS_2$ (fast and slow quasi-shear) modes, or simply $S_1$ and $S_2$.

At the time of the Crampin and Lovell, 1991 review of the first ten years of recordings (which accelerated rapidly with the development of digital, three-component geophones in the late 1980s) it was still not known exactly what caused the shear-wave splitting, because aligned pore space, layering and consistently aligned micro-cracks and joint (or fracture) sets could each cause the phenomenon.

Shear-wave splitting had been identified by this time, in a wide range of rock types. It was considered remarkable that, with all the different scales and characters of aligned fluid-filled cracks, inclusions or fractures in sedimentary, metamorphic and igneous rocks (with their wide differences in strength and modulus), the differential shear-wave anisotropy varied only within narrow limits (0.5 to 5%). Since that time, with increasing application at well-fractured reservoirs, this range has been greatly exceeded, as we shall see later.

The magnitude of shear-wave anisotropy appeared already to correlate well, with the amount of hydrocarbon production where there was larger-scale jointing or fracturing. There was also evidence, e.g. from Lewis, 1989, that the delay between the split shear waves appeared to decrease with increasing depth, yet an accumulative delay with increasing depth had been expected.

Crampin and Lovell identified several difficulties with shear wave splitting interpretation in the case of *earthquakes*. Because of the relative steepness needed for the incident wave to make an acute angle to (typical) sub-vertical structure, there was a need for the recording site to be within a so-called 'shear wave window'. The epicentral distance from the recording sites needed therefore to be considerably less than the focal depth of the earthquake.

Another limitation with interpretation of shear wave splitting from earthquakes was the combination of the large scale and the generally complex geology and tectonic structure between the source and the recordings.

The shear waves would likely pass through a range of rock types with different ages. Velocity, fracturing style and individual fracture-set properties would therefore likely vary, and each split shear wave could therefore split again, giving multiple splitting, with the influence of the joint structures near the recording site as one of the prominent results. Multiple splitting obviously makes time delay estimation more difficult.

Crampin and Lovell emphasised that the fluid filled features (cracks, inclusions, fractures) being the most compliant (i.e. with least *stiffness*, as referred in rock mechanics), would potentially be most sensitive to pore pressure changes or deformation, and thereby modify the way that the shear waves pass through the rock mass.

### 15.2.1 Some sources of shear-wave splitting

Crampin and Lovell, 1991, listed five possible scales of azimuthal anisotropy that could cause shear wave splitting:

1) aligned crystals
2) direct stress induced anisotropy
3) lithological anisotropy (e.g. aligned grains)
4) structural anisotropy (fine layering)
5) stress-aligned crack-induced anisotropy

Subsequently, much evidence for larger-scale fracture- or joint-set alignment effects on shear wave splitting have been obtained, which is a very important 'addition' for shear wave splitting, in view of the importance of these structural alignments for hydrocarbon production from fractured (naturally jointed) reservoirs. The larger scale features dominate drainage potential from the matrix to the joints, and thence to the wells, while the smaller scale EDA(extensive dilatancy anisotropy), popular with Crampin in many subsequent papers, would logically dominate drainage from the pores to the microcracks.

However, Crampin and Lovell emphasised that it was the stress-aligned vertical fluid-filled cracks that were most likely to cause more uniform splitting. That the two shear wave components will be oriented (polarized) parallel (or sub-parallel), and perpendicular (or sub-perpendicular) to the present most permeable fracture set directions, that *may* themselves be parallel (or sub-parallel) to the present major horizontal stress is the logical extension from crack-scale to joint or fracture scale. But as we shall see in Chapter 16, there are other joint or fracture alignments (caused by shearing), that are of major importance too.

Temporal changes of shear wave splitting were referred to already in Crampin and Lovell's '10 year review' paper. These had occurred before and after earthquakes, and after hydraulic pumping in granite. Changes in the aperture of the fluid-filled, aligned features (cracks, joints or fractures) would presumably have changed the aspect ratio of these features, and changes in $K_n/K_s$ ratios would be a macro-deformation rock mechanics consequence of such changes.

In the case of earthquake source investigations, it is unfortunate that shear-wave polarizations tend to be determined by the anisotropic joint or fracture structure close to the recorder, which of geometric necessity is likely to be much nearer the surface than the source. However, in the petroleum industry with its more limited depths, down-hole recording as in VSP, can focus on the structural domain of interest. For this reason, many of the important developments up to the time of Crampin and Lovell's 1991 '10-year' review, had been made with VSP configurations.

Two further important practical details concerning the potential uses of shear-wave splitting technology were emphasised by Crampin and Lovell, 1991. Firstly that the shear-wave-train likely contained many times the information carried by the P-wave-train, and that this was in the wave forms themselves rather than just in arrival times. Secondly that the multiplicity of source-to-geophone ray paths required to analyse P-wave arrival times, was not needed to obtain information from *shear waves*.

The problem of relative scale was also emphasised by Crampin and Lovell, 1991. The likely dimensions of the different phenomena (from extension-dilation cracks perhaps of microns to fractions of millimetre size, to fractured reservoir features of many metres size) were each much smaller than the wave-lengths of most shear waves (i.e. tens of metres in reflection experiments to several kilometres in teleseismic shear waves). It was emphasised that multi-offset, multi-azimuth, three-component VSP was likely to be the best way to attempt to analyse the geometrical aspects. Some of these investigations will be reviewed later.

In Chapter 10 concerning the phenomenon of attenuation, the inverse trends of velocity and attenuation magnitudes were seen. High velocities only occur where attenuation is low or $Q_{seis}$ is high. Links to rock mass qualities Q and $Q_c$ through deformation moduli were also cited, i.e. high velocities only occur where rock qualities Q and $Q_c$ are high. $Q_{seis}$ was likened, numerically, to the rock engineering static deformation modulus in the first kilometre (when expressed in GPa), and to the $Q_c$ value itself at extreme depth, since both seem to depend on the degree of jointing, clay-content, and rock hardness, despite the micro-displacements involved in seismic loading.

Crampin and Lovell made the interesting observation that the reciprocal relationship between velocity and attenuation (e.g. Crampin, 1981) was one of the reasons why the leading split shear wave was such a stable phenomenon; because it was travelling in the fast direction and was less attenuated than the slower split shear wave. The oriented rock mass quality $Q_o$ described by Barton, 2002, also has its maximum value in this same direction, i.e. sub-parallel to dominant structure, due mainly to higher RQD in this direction. (The use of oriented RQD, termed $RQD_o$, gives Q the directional dependency $Q_o$, together with the changed $J_r/J_a$ ratio representing the frictional strength of the joints across the sampling direction. See Appendix A for description of these terms).

Crampin and Lovell, 1991, concluded their '10-year review' by claiming that 'progress in understanding shear-wave propagation is the most fundamental advance in seismology for some decades'. They also speculated on the future uses of shear-wave splitting, including monitoring of hydrocarbon production and monitoring the stability of major civil engineering works in rock. The former is now much used.

## 15.3 Crack density and EDA

The geophysicist's crack density is defined as:

$$e = \frac{Na^3}{V} \tag{15.1}$$

where a is the crack radius, *cubed* due to the argument that this relates to the energy of elastic deformation associated with the crack. (O'Connell and Budansky, 1974). N is the number of such cracks in volume V. According to the authors Leary et al., 1990, crack density is often in the range of $e \approx 0.01$ to 0.05 in widely different geological and tectonic regions. This opinion seems to stem from the articles of Crampin, several of which will be reviewed here.

Let us examine some orders of magnitudes to get a 'feel' for this actually rather ambiguous geophysics parameter:

1) *Microcracks*: Assume $10^6$ microcracks of radius 100 microns in a $10 \times 10 \times 10$ cm rock sample. This gives $e = 10^6 \times (10^{-4})^3/(10^{-1})^3 = 0.001$. If the 'microcracks' were of 1 mm size, the crack density would be 1.0. The 'crustal range' is contained within these 'extremes'. (The microcracks would need to be 330 to 465 μm to lie in the mid-range of $e = 0.01$ to 0.05) with the above frequency.
2) *Fractures (or joints)*: Assume 10 joints of 1 m radius in a 10 m³ block of rock *in situ*. We have: $e = 10 \times 1^3/10 = 1.0$. (This represents a well-connected rock mass if for instance there were on average 3.3 joints in each principal direction of the 2.15 m 'cube'. If there was only one aligned set, the spacing would be close to 20 cm, if two conjugate sets, the spacing of each would approximate 40 cm. This fracture density is significantly *less* than much of the jointed chalk at Ekofisk, as we shall see later. If there was only *one* such joint per *1000* m³ (10 m on a side), the 'crack density would be the same as our more extreme microcrack example with $e = 0.001$. (We would need 10 to 50 joints of 1 m radius per 10 m cube to lie in the mid-range of $e = 0.01$ to 0.05).
3) *Fault swarm*: Assume 10 medium-sized faults of 464.5 m radius in 1 km³. In this case $e = 10 \times 464.5^3/10^9 = 1.0$. If we veer to the other extreme, we would require only one minor fault of 100 m radius in 1 km³ to give $e = 0.001$. (We would need 10 to 50 minor 100 m faults per 1 km³ to lie in the mid-range of $e = 0.01$ to 0.05)

To a non-geophysicist, it is difficult at first viewing, to see why so much seismic interpretation is related to this parameter when it is so ambiguous. To check again:

1) *microcracks.* $e = 10^7 \times (10^{-4})^3/(10^{-1})^3 = 0.01$ *(ten million @ 100 μm/10 cm cube)*
2) *fracture* $e = 10 \times 1^3/10^3 = 0.01$ *(ten @ 1 m/10 m cube)*
3) *minor fault* $e = 10 \times 100^3/1000^3 = 0.01$ *(ten @ 100 m/1 km³)*

These three very unequal scenarios with their equal 'crack density' would inevitably have totally different mechanical and fluid-conducting properties. Yet surprisingly, they are supposed to generate equal shear wave anisotropy, as we shall see.

The Leary et al., 1990 review of the 1988 Chapman conference (and 3rd International Workshop on Seismic Anisotropy) papers on 'Seismic Anisotropy of the Earth's Crust', gave a useful broad-brush reflection of some of the earlier opinions on fracture and crack induced anisotropy and its measurement, prior to the great advances in data acquisition and computer processing, that occurred mostly during the 1990s, that has given 'anisotropy' a central role in earthquake interpretation and fractured reservoir exploration and subsequent production monitoring.

Aligned fracturing may be detected and monitored over a huge range of length scales, using polarized shear waves. Dimensions may range from crustal dimensions of 10–100 km, through 1–1000 m reservoir scale fractures and faulting, to millimetre and micron-sized microcracks. The potential of fluid-filled microcracks to react to crustal stress and strain led Crampin et al., 1984 to propose the 'extensive-dilatancy-anisotropy' (EDA) concept. The EDA hypothesis is that crustal fluids prop open a population of high-compliance voids or inclusions that are nevertheless capable of remaining open against the least principal stress. The implication of 3D principal stress anisotropy at depth is that EDA cracks will tend to be aligned in a vertical plane, striking parallel to the major horizontal stress. With this configuration, Crampin 1978 had reasoned that a microcracked but otherwise isotropic crust was transversely isotropic, with a horizontal symmetry axis.

Shear waves polarized parallel to the microcracks, and also travelling parallel to them, hardly sense the presence of the cracks, and travel at almost the wave speed of the unfractured matrix. However, the shear waves travelling in this same parallel-to-structure direction, with polarizations *normal* to the microcracks, sense the reduced shear stiffness caused by the cracks and are slowed, in a similar manner to shear waves that might be travelling along the (perpendicular) axis of symmetry. As a result of the differential wave speeds, the shear waves travelling parallel to the aligned microfractures, with their two different polarizations, separate in time in proportion to the length of travel path and the density of the crack population. The majority of polarized shear wave observations show the fast wave polarized parallel or sub-parallel to the accepted local or regional maximum stress field.

EDA can in principle refer to a wide range of crack sizes, including fluid-conducting fractures. However, it seems that because many of the observed seismic properties can be simulated by propagation through distributions of *microcracks*, Crampin appears to have focussed on this scale of EDA for many years, while others have now demonstrated dispersive effects caused by fractures of entirely different dimensions, using double-porosity or triple-porosity poroelastic models, with pore-microcrack-fracture-fluid interaction.

The 'problem' is that the delayed, approximately orthogonal shear-wave arrivals are also consistent with a feasible density of aligned vertical microcracks. However, the relative stiffness of microcracks, having much higher aspect ratios than inter-locked fractures or joints, means that they cannot respond in the same way as fractures, to a given change in fluid pressure – at least according to classic geophysics teaching. Furthermore there is the *crack density* definition, which is a source of confusion (for non-geophysicists).

### 15.3.1 A discussion of 'criticality' due to microcracks

According to Crampin, 1993b, several oil companies were already reporting shear wave splitting in almost all their three-component reflection surveys in sedimentary basins. Furthermore, perhaps contrary to Crampin's expectations, the splitting, as he observed, was assumed to be due to large fractures within fractured reservoirs. Shear wave splitting was also visible in reflections from layers above the reservoirs, apparently suggesting to Crampin that exclusive dependence of splitting on large fractures seemed unlikely.

In fact as we will see later in this chapter, shear wave splitting is also seen to follow the saucer-shaped subsidence bowl far above a compacting North Sea jointed chalk reservoir, with correlation to the exact location of increased sea depths. Joint-stretch in the overburden seems likely to be the cause, but if *unconsolidated* sediments were the actual source of the splitting/polarization, microcracks or even macro-cracks in the sediments would need to be invoked to explain the polarization match to the subsidence bowl.

It is not quite clear why, but Crampin, 1993a was of the opinion that fluid-filled microcracks were the most compliant elements of the rock mass. If this opinion was because of the assumption that larger scale (and lower aspect-ratio) fractures and joints would be closed at depth, then indeed the usually *less compliant* microcracks could remain as perhaps the most compliant element of the rock mass. But here we run into difficulties concerning the interpretation of 4D repeated surveys over producing reservoirs.

Rock physics provides evidence of effective stress sensitivity of e.g. S-waves, P- and S-wave attenuation, and of their anisotropy and dispersive nature. (Chapter 13). When changes of oil saturation in producing parts of a reservoir are monitored, or water/oil front progression is monitored (see examples in Chapter 14), it is surely the fractures that are the primary targets for the diagnostic seismic waves. Do the pore-to-microcrack responses to production-caused pore pressure, occur fast enough to be 'in-step' with the responses in the *probably* more compliant fractures and joints? After all, the fractures have far higher permeability in general, than the matrix. As we have seen in the earlier critique of 'crack density', this parameter does not seem to help in answering this question, because of its basic ambiguity.

In his paper 'Arguments for EDA', Crampin, 1993a also addressed the meaning of the typical 1% to 5% differential shear-wave anisotropy, reportedly measured in a wide range of rock types, but as we shall see, there are many exceptions in fractured areas. Crampin considered that the 1% to 5% was also equivalent to the generally limited range of effective crack densities, which he assumed were usually $0.01 \leqslant e \leqslant 0.05$. Crampin noted that the percentage of differential shear wave anisotropy was usually about $e \times 100$, for a $V_p/V_s$ ratio of about 1.7 (1.732 was quoted).

Crampin, 1993a avoided mentioning specific rock-volume dimensions regarding *crack density* ($e$) in the following paragraph, which will be quoted in full, as it reveals an unexpected way of viewing rock mass stability and the fluid-bearing nature of rock masses, which is difficult to match with rock mechanics and hydraulics experiences.

*'A crack density of $e = 0.05$ (shear-wave anisotropy about 5%) is equivalent to a crack of diameter 0.7 in each unit cube. A crack density of $e = 0.1$ has a crack diameter of 0.93 in each unit cube (check: $e = 1 \times 0.465^3/1 = 0.100$), and this is clearly near the critical crack density at which an intact rock fragments, as very close cracks begin to coalesce to form through-going fractures. Thus the upper limit of crack densities ($e = 0.05$, with occasional excursions to 0.1) is probably due to the limit of the number of fractures for an intact rock mass to remain intact. If this limit is exceeded the rock mass fragments and the pore fluid would disperse and, once dispersed, the cracks would tend to close and crack-healing occur which would lead to a lower crack density of open cracks. The only other occasion when substantially larger crack densities (up to $e = 0.4$) have been claimed for field observations are reported by Crampin et al., (1980a) for observations on the surface of limestone pavements, where the pavements are characterized by massive jointing and the rock cannot be considered as intact.'*

In a later article, Crampin, 2000 relates $e = 0.055$ to 'fracture criticality', and states that 'almost all rocks are (then) marginally close to the 'critical percolation threshold in a stressed fluid-saturated solid, when shear strength is lost and fracturing occurs'. It has to be emphasised that these are foreign concepts to rock engineers: extensively jointed rock masses with far higher *crack densities* have perfectly adequate strength to tolerate e.g. tunnelling.

Rock stress and joint roughness helps retain both strength and permeability. If there are two or three joint sets contributing to '$e = 0.055$', there is actually little cause for concern about 'loss of shear strength', or 'dispersion of fluid'. Three joints of 52 cm *diameter*, intersecting (or avoiding each other) in a 1 $m^3$ block of rock also gives $e = 0.055$. This would be a rather stable rock mass compared to near-surface experiences. When blasting one would see most of the remaining 'half-pipes' of the blast holes, but with occasional 'block-corner' overbreak beyond the half-pipes.

If the same 1 $m^3$ block was (almost) divided into *eight* component smaller cubes by three typical, near-surface, mutually perpendicular joints, each of 1 m diameter, the *crack density* would have increased to: $e = 3 \times 0.5^3/1 = 0.375$. We regularly construct tunnels in such rock, and if stress and joint roughness are adequate, and there is no clay, there may be no need for immediate rock support of the exposed perimeter of such a tunnel. That the tunnel may *leak* is a different problem. It does not signal 'criticality', rather 'normality'. In a reservoir, 'leakage' is desirable when in the right direction.

After all, where will the fluid go, and where will the rock blocks go when surrounded by neighbouring fluid and neighbouring rock blocks at 3 km depth? They will continue to contribute their coupled roles in supporting the weight of the over-lying 3 km of rock, and they will continue to provide (limited) void-space for oil storage, and allow percolation towards producing wells. There is surely nothing *critical* about this, just *normality*.

### 15.3.2 Temporal changes in polarization in Cornwall HDR

An interesting phenomenon was observed by shear-wave splitting when monitoring the Cornwall hot-dry-rock HDR project. Crampin and Booth, 1989 reportedly found that there were 7° to 10° rotations in the polarization directions as a result of deep-well injection of (cold)

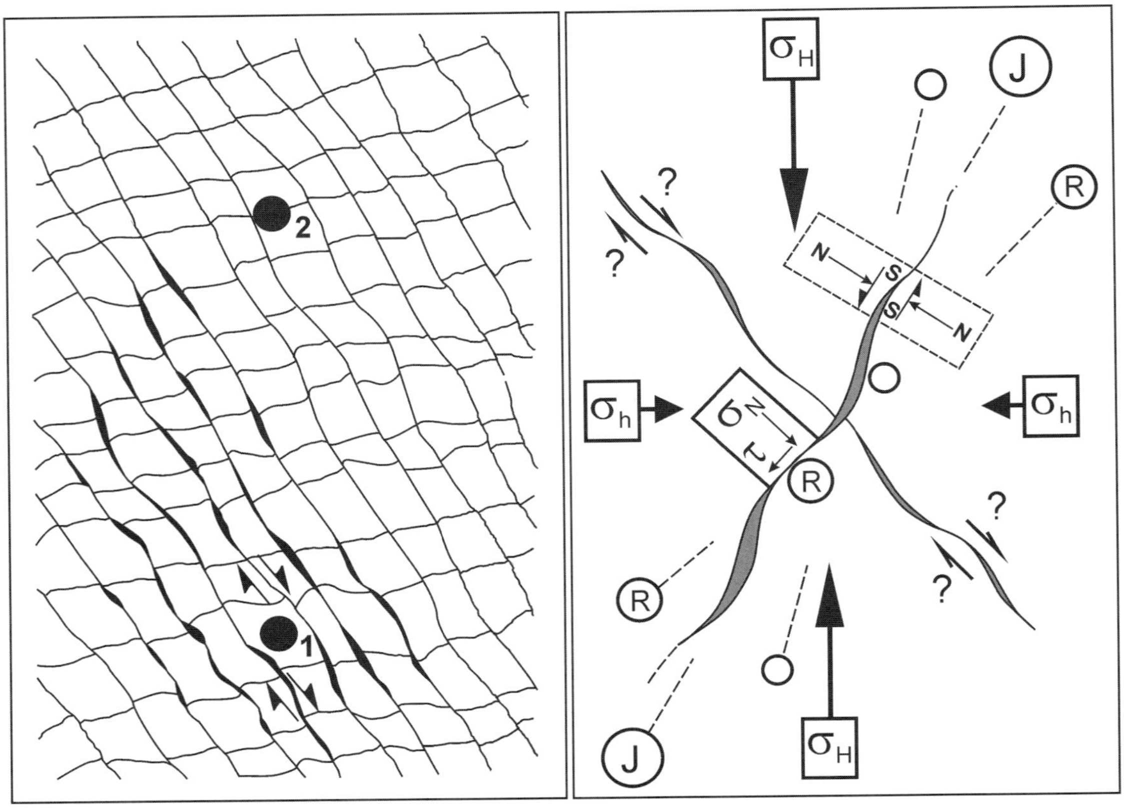


*Figure 15.2* (a) The Cornwall joint-shearing mechanism visualized by Barton, 1986 as a result of cold water injection and effective stress reduction in a joint set *not* perfectly aligned parallel to the major stress, and under a differential (shear) stress. (b) An exaggerated diagram of (conjugate, or single-set) joint shearing, showing the possibility of *opposite rotation* of the rock-to-rock contacts (R), and open (O) fluid-filled sections during shearing of *the usually non-planar* joints needed to give joint permeability in reservoirs. This shearing-mechanism could possibly be the source of polarization rotation, both here and elsewhere, where 'deviation' with respect to $\sigma_{H\,max}$ has been noted.

water into the jointed and (possibly) microcracked granite, which was also AE monitored to several kilometres depth, during the pumping. Batchelor and Pine, 1986.

They considered that this consistent change in polarization was a stable, reliable and significant result. Although the title of the Crampin and Booth paper suggests an 'explanation' that could have been consistent with the writer's joint-shear-dilation hypothesis (Barton, 1986, see below), Crampin 1993a actually suggests the following mechanisms: *'The results suggest that before the dilation of the incipient joints by pumping, the fluid-filled EDA cracks are aligned parallel to the in-situ stress field. After the joints have been dilated by pumping and the stress field modified, the EDA cracks close to the joints are realigned parallel to the joints.'*

There follows a reference to opening and closing of microcracks and sub-critical crack growth, that seem not to relate directly to any well explained *joint-dilation* mechanism. Was the dilation considered to be a *normal* expansion of the apertures, or was there sufficient, inclined, differential stress, for *shearing* to occur as a result of the pumping? The answer would have a direct effect on how the stress field was modified.

Figure 15.2*a*, illustrates an alternative interpretation of the Cornwall polarization rotation of 7° to 10°, referred to above. The jointing at depth was under some shear stress, as it was *not* aligned parallel to the major horizontal stress. With cold water injection there is both contraction and effective normal stress reduction, allowing for some slight shearing that is likely to be part of the source of microseismic 'clouds' that propagated to greater depths down the presumably dilating, slightly shearing jointing. (Batchelor and Pine, 1984 and Barton, 1986). The hypothesis that slight joint shearing can lead to polarization rotation is illustrated in Figure 15.2*b*. The details of this mechanism are described in Chapter 16.

Crampin, 1993a emphasised many times the (reportedly) remarkable 'uniformity' of the mostly 1% to 4% (but occasionally 10%, and even 30%) shear-wave anisotropy, reportedly from a multitude of rocks and from a wide range of scales of investigation, i.e. near-surface: 10's of meters, sedimentary basins: 1–2 km, mixed geology above earthquakes: 5–15 km, and whole crust: 30 km. Focus on the 1% to 4% range, if generally justified, is implying limited EDA dimensions of a few microns to a few meters.

*Crack density* does not need to be 0.015–0.045 to 'explain' a frequent shear-wave anisotropy of 1.5% to 4.5%. Almost 'all rock' is unlikely to be pervaded by tens of millions of microcracks per hand-specimen – if it was there would certainly be a 'critical crust'. Rather rock masses are often pervaded by one two or three joint sets – and it is these nearer-the-surface, very common structural-geological features, that *may* be imprinting their shear-wave splitting result on the recorded waves.

### 15.3.3 A critique of Crampin's microcrack model

Because *many* of the observed seismic properties can be simulated by propagation through distributions of *microcracks*, Crampin appears to have focussed on this scale of EDA in the last decade. Others have now demonstrated dispersive effects caused by fractures of many different dimensions, using double-porosity or triple-porosity poroelastic models, with pore-microcrack-fracture-fluid interaction (see later in this chapter).

The 'problem' is that the delayed, approximately orthogonal shear-wave arrivals are also consistent with an (occasionally) feasible, high density of aligned vertical microcracks. However, the relative stiffness of microcracks, having *much higher aspect ratios* than inter-locked fractures or joints, means that they cannot respond in the same way as fractures, to a given change in fluid pressure, despite Crampin's apparent claims to the contrary.

This classic aspect ratio stiffness argument, originating at least from Walsh, 1966 seems to be discounted or ignored by Crampin. He appears to overlook the ambiguity of *crack density*, and trust in the assumed attributes of the physically somewhat unlikely APE model reproduced in Figure 15.3, when making statements such as: 'fluid-saturated microcracks are highly compliant'... ...'large cracks would be stiff and much less compliant' ... ... 'there is very little evidence that the splitting is caused by larger cracks or fractures' ... ... 'the detailed geometry of fluid-saturated microcracks in almost all rocks and reservoirs' ... ... 'the underlying reason for the calculability (of APE) and predictability (1.5% to 4.5%) is that the rock mass is so heavily (micro) cracked that it can be considered as a critical system'.

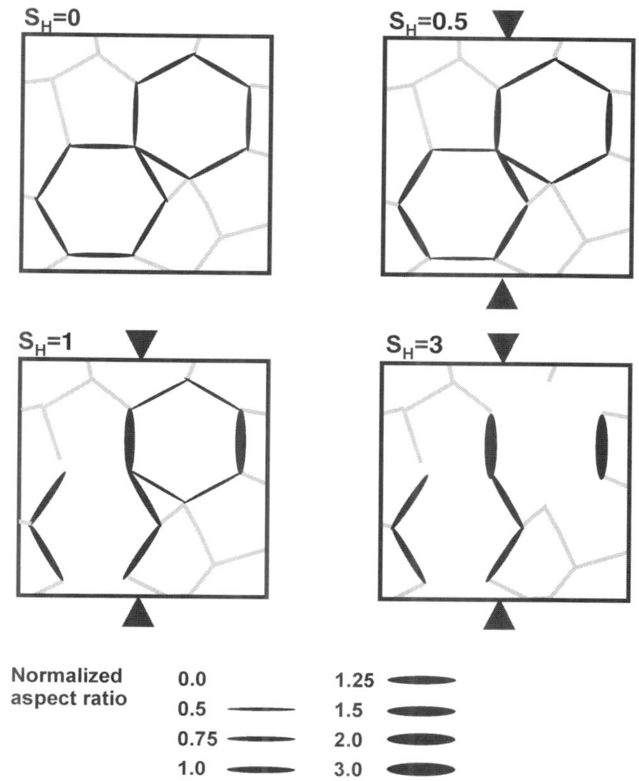


*Figure 15.3* Schematic illustration of the Zatsepin and Crampin, 1997 APE model. The actually 3D, assumed 'hexagonal boundary cracks' are vertical. Note the four values of increasing maximum horizontal differential stress normalized to the critical value at which cracks first begin to close. Note the remarkable, assumed increases in normalized crack aspect ratios, with increased '$S_H$' stress. Aspect ratios were apparently chosen to give a porosity of 5%.

### 15.3.4 90°-flips in polarization

Figure 15.3 illustrates the anisotropic, poro-elastic or APE model of Zatsepin and Crampin, 1997, that appears to have featured more and more in the interpretations and conclusions found in Crampin's numerous recent articles concerning analysis of shear-wave splitting in both earthquake and reservoir scenarios. The dramatic increases in aspect ratio, the second stiffest component of rock porosity (after equant porosity, and before joint or fracture porosity) is supposedly able to develop in any rock, presumably because of the over-looked resistance of the intervening crystals/grains of perhaps 50 to 150 GPa Young's modulus, that obviously will resist such assumed expansion.

How could such a lateral volume increase, and therefore stress increase, from countless trillions of expanding, aligned microcracks be absorbed in a rock mass without a general reversal of the $\sigma_{H\ max}$ and $\sigma_{h\ min}$ directions? Surely an *occasional* mechanism such as this is the *actual cause* of '90°-flips' in the polarization, as occasionally observed/suspected in earthquake source zones. (Crampin et al., 2002, Crampin and Peacock, 2005).

The hypothesised extreme fluid pressures that supposedly can exceed $\sigma_{h\ min}$, and even approach $\sigma_{H\ max}$ in magnitude (Figure 15.4, after Crampin et al., 2002), are surely geotechnical impossibilities? If the conceptual, hexagonal, vertical microcracks of the APE model (Figure 15.3) were miraculously present in e.g. a soft mudstone, then, and only then, could such an aspect-ratio expansion be envisaged, with or without the help of high pore pressure. They would be likely to cause a '90°-flip in polarization', due to the resistance of the surrounding material.

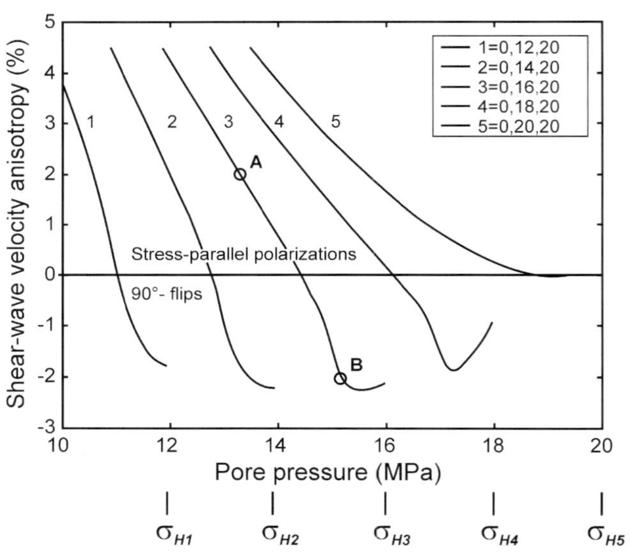


*Figure 15.4* Crampin et al., 2002 explanations for '90°-flips' in polarization, as apparently observed above small earthquakes in Iceland. Their model supposes high fluid-pressures within 1 or 2 MPa of a 'critical stress'. This figure shows modelled variations of shear-wave anisotropy with increasing fluid pressure, for five different sets of principal axes of stress. They presume that anisotropy becomes negative for pore-fluid pressures close to $\sigma_{H\ max.}$ (Inset shows five assumptions for '$s_h$, $s_H$, and $s_v$').

The '90°-flips' explanation shown in Figure 15.4 supposes that the assumed high (actually extreme, and surely impossible?) pore pressure causes the faster split shear-waves that were previously parallel to $\sigma_{H\ max}$ to do a 90°-flip and become the slower wave parallel to $\sigma_{h\ min}$. It is not explained, nor is it clear in any way, how a pore pressure can build up to beyond $\sigma_{h\ min}$ without dissipation due to hydraulic fracturing, nor how the pore pressure could thereafter even approach $\sigma_{H\ max}$.

A possible 90°-flip *would* appear to be possible on occasion with extreme $\sigma_{H\ max}$ loading, causing the usually unlikely and untenable lateral expansion of aligned microcracks, as depicted in Figure 15.3. What seems to be a geotechnically very unlikely process (the EPA model) perhaps has merit in just such earthquake-related cases of 'axial-overload'.

## 15.4 Theory relating joint compliances with shear wave splitting

The geophysicists' progression from seismic propagation in isotropic media, to anisotropic layered media to transversely isotropic layered media containing one set of vertical fractures (a joint set), later increased to two sets of perpendicular fractures, and subsequently to non-orthogonal vertical sets, and later still to non-vertical sets, has resulted in a progression of theoretical papers in the geophysics literature, containing, inevitably, an increasing content and complexity of 6 × 6 compliance and stiffness matrices.

Here we will summarize an intermediate stage in this progression, from Schoenberg and Sayers, 1995, by way of introducing, as simply as possible, some necessary theoretical aspects. These authors addressed the problem of how the presence of a single joint or fracture set would affect the elastic moduli of the fractured rock, since it is the elastic moduli and the density that determine the behaviour of seismic waves, assuming a linear, loss-free, elastic behaviour.

In simplest possible terms, appropriate to the non-mathematical treatment in this book, the effective elastic compliance tensor $S_{ijkl}$ of a rock containing fractures, is designed to relate the average *strain* $\varepsilon_{ij}$ over a representative volume V, to the average *stress* components $\sigma_{ij}$. One therefore can write:

$$\varepsilon_{ij} = S_{ijkl}\,\sigma_{ij} \tag{15.2}$$

The compliance matrix that includes the effect of a set of vertical fractures, has to include the *excess compliances* caused by the additional presence of the fractures. The authors assumed that this could be simply expressed as the sum ($S$) of the compliance of the isotropic back-ground rock ($S_b$) and the *excess compliance* matrix associated with the fractures ($S_f$). The latter is composed of the effects of a fracture-normal compliance $Z_N$, and of a fracture-shear compliance $Z_T$ related with linear slip deformation. Based on the assumptions made, the single set of vertical fractures produces an unusually simple compliance matrix:

$$S(f) = \begin{bmatrix} Z_N & 0 & 0 & 0 & 0 & 0 \\ 0 & 0 & 0 & 0 & 0 & 0 \\ 0 & 0 & 0 & 0 & 0 & 0 \\ 0 & 0 & 0 & 0 & 0 & 0 \\ 1 & 0 & 0 & 0 & Z_T & 0 \\ 0 & 0 & 0 & 0 & 0 & Z_T \end{bmatrix} \tag{15.3}$$

From this matrix, Schoenberg and Sayers, 1995 developed the $6 \times 6$ compliance matrix ($S$) for the combined transversely isotropic fractured medium, involving the various combinations of Lamé constants ($\lambda$ and $\mu$) shown in Chapters 1 and 14. The simple addition of the three fracture compliance terms ($Z_N$, $Z_T$ and $Z_T$) was made in the same (1,1 5,5 and 6,6) locations in the combined compliance matrix. They then inverted this to form the elastic *stiffness* matrix to represent the *combined* isotropic back-ground and vertically fractured medium. They noted that this stiffness matrix was of exactly the same form as that developed previously by Crampin, 1984 for the case of an isotropic medium containing a vertical array of 'penny-shaped' fractures, although Crampin also had second-order terms for fracture density.

Of importance to the *shear-wave splitting context* of this chapter, are the stiffness matrix terms relating to the *fast shear wave* propagating parallel to the fractures, given as the $C_{44}$ term, and the *slow shear wave* propagating perpendicular to the fractures, given by the $C_{55}$ term. For the case of the vertically propagating waves through the vertical fractures, there is no fracture compliance term in $C_{44}$, only the Lamé constant $\mu$ or $\mu_b$ for the background rock. (This was inverted in the case of $S_{44}$ in the compliance matrix, as $1/\mu$ or $1/\mu_b$).

For the case of the slow shear wave, the $C_{55}$ term was given as $\mu_b(1\text{-}\delta_T)$, with $\delta_T$ (limits: $0 \leq \delta_T < 1$) expressed as $(Z_T\,\mu_b)/(1 + Z_T\,\mu_b)$. In other words there is no $Z_N$ term, which requires dipping fractures, or non-vertical wave propagation for $Z_N$ to be mobilized. (In the compliance matrix the $S_{55}$ term was simply $1/\mu_b + Z_T$).

### 15.4.1 An unrealistic rock simulant suggests equality between $Z_N$ and $Z_T$

Schoenberg and Sayers, 1995 followed their vertical fracture set compliance matrix treatment, by citing the results of shear compliance ($Z_T$) and normal compliance ($Z_N$) interpretation from loaded, roughened Lucite ('Plexiglas') plates used 'to simulate a fractured medium', reported by Hsu and Schoenberg, 1993. Schoenberg and Sayers apparently derive their assumption that $Z_N \approx Z_T$ for the case of dry (gas saturated) cracks from this laboratory work. (Note equivalent use of symbols $B_n$ and $B_t$).

(It is argued in Chapter 16 that interlocked rock joints usually have quite different magnitudes of normal and shear *stiffness*, with the expectation that *at least some of these differences* remain when considering micro-displacement and the (presumed) elastic response of joints to dynamic wave loading. However, this is at present a subject under debate and therefore controversial).

Hsu and Schoenberg, 1993 interpreted the normal and shear compliances from the results of static and dynamic loading tests on numerous (60 or 200), roughened Lucite ('Plexiglas') plates, loaded normally. Their extremely close spacing (0.7 mm) and their continuity means that *fracture densities* were actually in the 'impossible' range of 7.5 to 25, up to three orders of magnitude too high to represent fractured (or microcracked) rock.

Table 15.1 indicates that these artificial-material tests were taken to quite high normal stress levels in relation to the assumed strength of Lucite. One can imagine that under the highest loads, something approaching a

Table 15.1 Shear and normal compliance differences obtained from normally-loaded Lucite plates with rough surfaces. Schoenberg and Sayers, 1995. Derived from Hsu and Schoenberg, 1993.

| Stress (MPa) | $(Z_T - Z_N)/(Z_T + Z_N)$ |
|---|---|
| 6 | 0.1736 |
| 12 | 0.1177 |
| 18 | 0.0332 |
| 24 | (−) 0.0035 |

'welded' state could develop for the contacting micro-asperities, since the 'roughening' would reduce the initial contact area. If a 'locally-welded' state developed, this would perhaps help to explain the similar (actually almost identical) values of $Z_T$ and $Z_N$ that they obtained.

It is recommended that the phenomena observed in highly stressed roughened plates of Lucite, with their extreme fracture densities, should *not* be used to simulate rock joint response to dynamic or static loading, because the suggested 'equality' of $Z_N$ and $Z_T$ has permeated some of the geophysics literature, and may be far from realistic for all but the smallest laboratory specimens.

Schoenberg and Sayers, 1995 used a 'difference/sum' $(Z_T - Z_N)/(Z_T + Z_N)$ format to compare the magnitudes of $Z_T$ and $Z_N$. They used the experimental components $c_{11}/\rho$, $c_{33}/\rho$, $c_{44}/\rho$ and $c_{66}/\rho$ from Hsu and Schoenberg, 1993, for calculating the results given in Table 15.1. One may 'solve' for specific examples of $1/K_n$ and $1/K_s$, thus finding the ratios of difference/sum.

Let us first assume realistic small-scale pseudo-static values of $K_n = 1000$ MPa/mm, and $K_s = 10$ MPa/mm, and argue (*for the sake of preliminary assessment),* that $K_n \approx 1/Z_N$, and $K_s \approx 1/Z_T$. Thus $(Z_T - Z_N)/(Z_T + Z_N) = (0.1 - 0.001)/(0.1 + 0.001) = 0.98$. Clearly the low numbers for the ratio difference/sum, given in the table, imply much closer values of the two compliances. Suppose $Z_T = 0.1$ mm/MPa as before, while $Z_N = 0.05$ mm/MPa. This would suggest a difference/sum ratio 0.05/0.15, still not close enough to the Hsu and Schoenberg 1993 highly-stressed Lucite data. Significantly, with the increased pressure, $Z_T$ and $Z_N$ were approaching equality.

With assumed much closer values of $Z_T = 0.1$ mm/MPa, and $Z_N = 0.099$ mm/MPa, one obtains a ratio of $0.001/0.199 = 0.0050$. The 'welded' asperity argument for the highest stress levels would not on this basis seem unreasonable. If approximate equality to the inverse of stiffness was still tested, the relevant values of $K_n$ and $K_s$ would be 10.1010 MPa/mm and 10.0 MPa/mm, clearly unrealistically close in rock mechanics terms, in view of the entirely different displacement mechanism involved: the former involving 'closure' (in a stiffening direction), the latter involving 'slip' (not necessarily stiffening, possibly softening, in the absence of roughness).

Viewing the obviously different mechanisms involved in mobilizing the 'spring stiffnesses' in the normal and shear directions in the various materials depicted in Figure 15.5, it is hard for an experimentalist to understand how equality could be expected between S and n, or $Z_N$ and $Z_T$, since in pseudo-static testing and its

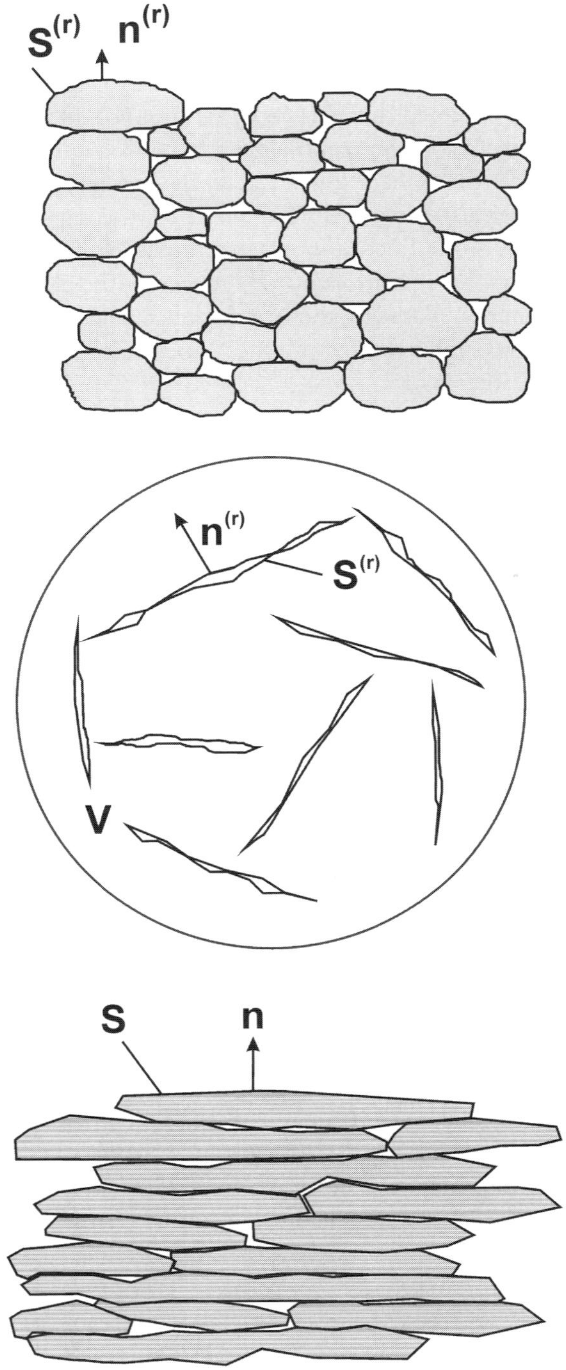


*Figure 15.5* Discontinuities of different scale, between sand grains, in a microfractured or jointed rock mass, and between clay particles in shale, as depicted by geophysicists concerned with the inequality (or unlikely equality) of the shear and normal compliances. S and n, or $S^{(r)}$ and $n^{(r)}$, are depicting the shear and normal compliances (of the $r^{th}$ discontinuity) in the three or four different scales of media. (Sayers, 2002a, 2002b, Liu et al., 2000)

empirical modelling, the mechanisms involved in closing and shearing are entirely different, and are modelled in entirely different ways as a result. Some of this difference would seem likely to remain at the micro-scale of dynamic wave loading.

The question is whether heavily loaded, roughened Lucite plates can be accepted as realistic models for a set of stressed rock fractures. Clearly they should not be accepted, or accepted only with great caution, in view of the non-brittle 'plastic' behaviour of Lucite. It is therefore urged that one should re-evaluate the relevance of the '$Z_T/Z_N \approx 1.0$' theory, and adopt, *in the first instance*, data from tests on actual rock joints, for example those described by Pyrak-Nolte et al., 1990 (see later), or those derived (implicitly) from Chaudry, 1995 and King et al., 1995 polyaxial tests. The latter studied principal stress-developed rock fractures in highly stressed cubes of isotropic sandstone (see Chapter 13).

Even though both these data sets are likely of too small scale, with respective dimensions of 52 mm and 41 mm, they are extremely likely to be closer to relevant reservoir jointing than artificial Lucite surfaces.

### 15.4.2 Subsequent inequality of $Z_N$ and $Z_T$

Later work by Sayers, 2002a, was based on elastic anisotropy expressions involving both second-rank and fourth rank tensors for describing the seismic effect of microcrack and grain boundary contact *normal and shear compliances* in sandstones. By matching his theoretical predictions to polyaxial loading on blocks of sandstones, it was found that even at pre-fracturing microcrack scale, the shear compliance was deduced to be greater than the normal compliance, based on mismatch of theory with measurement.

We may draw some parallels between the two scales of discontinuous phenomena addressed by Sayers 2002a and 2002b, each caused by present or past stress anisotropy. As Sayers, 2002a and other authors have pointed out, grain boundaries and microcracks may cause anisotropic behaviour in the presence of *stress anisotropy*, in otherwise 'isotropic' rock materials. The sandstone depicted in Figure 15.5a, will exhibit stress-induced P-wave anisotropy and S-wave anisotropy due both to preferential stress-alignment of the microcracks, and due to the likely difference in the magnitudes of the normal ($Z_N$) and shear ($Z_T$) compliances between the microcrack faces and between grain contacts.

The same will apply to the fractured or jointed medium of much larger scale, where horizontal stress anisotropy will often have caused the alignment of the natural joints or fractures. (We have termed the natural features: rock *joints* in most of this book, to distinguish them from *artificially induced fractures*, as developed around boreholes or tunnels or mine openings, or fracturing induced by MHF stimulation of tight reservoirs. This preferred terminology is following the recommendation of ISRM, 1978 and seems logical in view of the number of artificial (man-made) fractures that have to be described in these shared, *rock engineering* disciplines.

At the larger scale, Sayers, 2002b, deduced that $Z_T > Z_N$, when analysing the results of the Chaudry, 1995 polyaxial tests on 41 × 41 × 41 mm cubes of sandstone, which had a fracture-development cycle. (See descriptions of these interesting tests in Chapter 13). This compliance inequality is of course consistent with the experience of $K_s < K_n$, concerning the pseudo-static shear and normal *stiffnesses* of joints and fractures, where stiffness is the rough inverse of compliance. The magnitude of $K_n$ proves to be less than, but quite close to $1/Z_N$ in good quality (unweathered) hard rock, while in the shearing direction, $K_s < 1/Z_T$, sometimes $\ll 1/Z_T$ (see Chapters 13 and 16).

### 15.4.3 Off-vertical fracture dip or incidence angle, and normal compliance

Sayers, 2002a pointed out that off-axis velocity measurements would be needed to actually quantify the mismatch $Z_T > Z_N$, so he was actually not able to determine this explicitly with the principal axes polyaxial testing of Chaudry, 1995 and King et al., 1995. Sayers, 2002b therefore went on to address the problem of vertical shear wave propagation in jointed media with *off-vertical dips*, using the example of two conjugate sets with oppositely oriented dip angles, as depicted in Figure 15.6. Unlike the vertical wave propagation through the vertical joints, with dependence of shear-wave splitting on only the shear compliances, as discussed earlier, the shear wave components $qS_1$ and $qS_2$ depend here on both the shear *and* normal compliances, since the incident angles are no longer parallel to the joint planes.

In such cases where both shear *and* normal compliances are involved, normal compliance is reportedly *reduced* (i.e. stiffened) by fluids of non-zero bulk modulus. Gas and

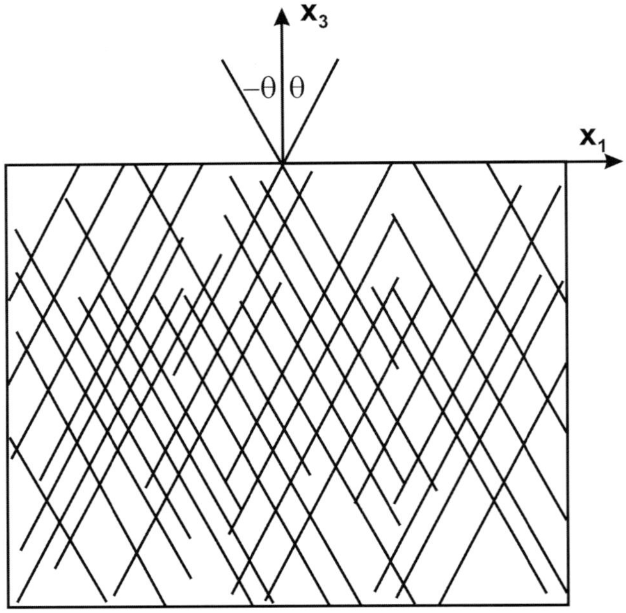


*Figure 15.6* Two conjugate sets of fractures with dips of θ and −θ. This model was used by Sayers, 2002b to investigate shear wave splitting phenomena when incident shear waves were no longer parallel to vertical fracturing.

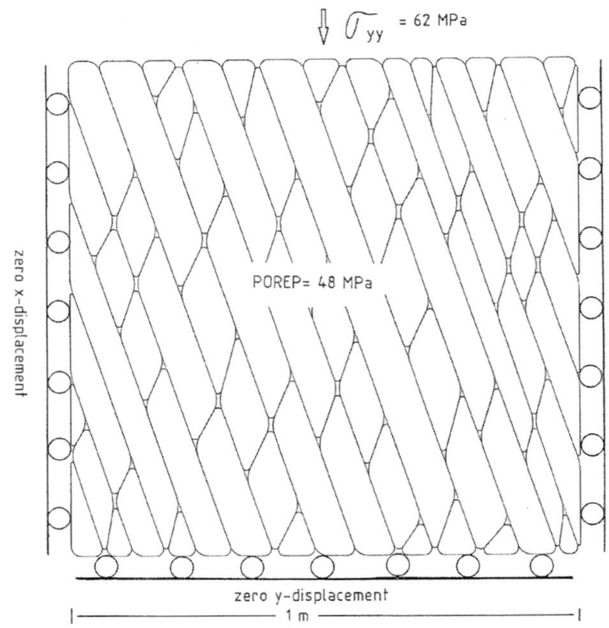


*Figure 15.7* A cross-discipline example of unequal conjugate jointing used to represent a high-porosity section of the Ekofisk reservoir, modelled by the distinct element code UDEC-BB, with non-linear, effective stress and size-dependent shear and normal stiffnesses. Barton et al., 1986.

oil should be distinguishable by respectively greater and less shear wave anisotropy, as the stiffening effect of the oil makes the fracture normal stiffness less contrasted to the back-ground medium. For dipping joints or fractures, there proves to be a significant *decrease* in shear wave anisotropy if the fluid has a higher bulk modulus, making the normal *stiffness* of the fractures greater. The average of the two shear wave velocities is therefore also increased.

Bakulin et al., 2001, also pointed out that both geophysical and geological data acquired over naturally jointed reservoirs often revealed the presence of *multiple* sub-vertical fracture sets, which made the effective medium *monoclinic.* They developed a model for handling two sets of *unequal,* and non-orthogonal vertical joint sets, which yielded the azimuths and compliances of both sets of joints, as well as the P- and S-wave velocities of the assumed isotropic background medium. From a second model, consisting of a single set of *micro-corrugated* joints, they stated that monoclinic symmetry stemmed this time from coupling between the normal and tangential 'slip'. This coupling caused shear wave splitting dependence on the fluid content of the fractures.

Experience from local core logging and photography of jointed core from the Ekofisk reservoir by the writer in 1985, and inspection of jointing in the Lagerdorf chalk quarry in German, also supports the notion of a possible dominant set when conjugate jointing is present. Figure 15.7 illustrates a specific realization of jointing that was simulated when modelling the local-scale (1 × 1 m, 2D) compaction mechanism principles, in this massive 30 km$^3$ North Sea chalk reservoir. The behaviour of the joints was based on measured values of wall strength (JCS) and roughness (JRC), which give pseudo-static values of both normal and shear stiffness (the latter may be significantly lower than the dynamic values). In rock mechanics modelling the non-linearity of these parameters is also an issue, with extreme stiffening of $K_n$ at high stress, and less stiffening of $K_s$ with stress, which in pseudo-static loading is also scale (block-size) dependent (Barton and Bandis, 1982).

Sayers 2002b, again claimed that the ratio of $Z_N/Z_T$ (or $B_N/B_T$ in his terminology) was approximately equal to 1.0 for the case of 'gas-filled open fractures'. This was possibly based on Hsu and Schoenberg, 1993 results for loading tests on 'roughened plates of Lucite', to quite high normal stress levels in relation to the strength of the Lucite. If so, the result should be regarded with great caution, as highly-stressed, roughened Lucite plates, with three-orders of magnitude too high *crack densities*, surely bear little similarity to normally-stressed, interlocked joints in brittle rock, as discussed earlier.

The presence of a joint-filling fluid with non-zero bulk modulus in place of gas would, as suggested by Sayers, 2002b cause a lower value of $Z_N/Z_T$ (or $B_N/B_T$), but not it is suggested, just lower in relation to the above artificially high ratio of 1.0, but rather in relation to a ratio of $Z_N/Z_T$ that was already significantly lower than 1.0, as seems intuitively likely due to the different mechanisms involved (namely 'closure' in the case of $Z_N$, and 'slip' in the case of $Z_T$. (See Figure 15.5).

Interesting supportive field data was cited by Sayers, 2002b, concerning the differences in observed velocities and shear wave splitting intensity, when in the presence of gas as opposed to oil. (This will be reviewed later in this chapter). We may recall that with *non-vertical* fractures and with vertically propagating waves, both the shear *and* normal compliances are assumed to be involved in the shear wave splitting process. The field data from Oman (Guest et al., 1998, van der Kolk et al., 2001) for a fractured (supposedly nearly vertically fractured) carbonate reservoir showed higher shear wave anisotropy (due to a lower qS2) over the gas cap volume, than where there was oil. As pointed out by van der Kolk et al., 2001, this *in situ fractured media* result is in disagreement with the prediction of Gassmann's theory for porous media, in which the shear modulus should be independent of the fluid. The shear modulus of the matrix governs the fast shear wave, with no $Z_N$ or $Z_T$ influence.

If the *in situ* fractures were actually not perfectly vertical, as seems to be implied by Sayers, 2002b, then the normal compliance would also be involved in the intensity of the shear wave splitting. Because the normal compliance is reduced in the presence of a fluid with non-zero bulk modulus (i.e. oil), the ratio of $Z_N/Z_T$ (or $B_N/B_T$) would reduce. According to Sayers, this should cause a significant decrease in shear wave splitting and an increase in shear wave velocity with the increased fluid bulk modulus, and therefore stiffened fractures.

Without attempting to describe complex details, it may be pointed out from recent work by Gurevich, 2003, that the compliance matrix for a fluid-saturated, *porous-and-fractured* medium is not equivalent to the compliance matrix of any *solid* medium with a single set of parallel fractures. This is due to the wave-induced (micro) flow of fluids between pores and fractures. Such attenuating (micro) flow is now being modelled with double-porosity poro-elastic models, which are reviewed later in this chapter.

### 15.4.4 Discussion of scale effects and stiffness

In Chapter 16 we will see the inequality of the *pseudo-static* shear and normal stiffness ($K_s < K_n$) of virtually all joints (and filled discontinuities), and the likelihood of changing ratios of $K_n/K_s$ as pore pressure changes affect the effective normal stress. These joint *stiffnesses* have been used in rock mechanics finite element modelling since the late 1960s, (Goodman and Duncan, 1968), and finite difference distinct element (DEM) modelling since the 1970s (Cundall, 1971). The important question is whether geophysicists will be able to apply the accumulated knowledge from these macro-displacement, *inversed* equivalents of the geophysicists' micro-strain *compliances*.

*If* the stiffnesses and compliances of these two disciplines can be related, despite the probable different orders of magnitude of dynamic and 'static' deformations involved, then the more researched parameters $K_n$ and $K_s$ could perhaps give shear wave splitting 'even more information' than presently assumed, since one often sees *equal* normal and shear compliances assumed in papers describing shear wave splitting analyses, implying that information may be limited at present, at the less accessible dynamic micro-scale of displacements.

Major stress-aligned, fluid-filled or sediment-filled inclusions, (and cracks, joints or fractures) are the diverse sources of *mechanical anisotropy* that can be presumed to be common to all rocks. They may be the guaranteed sources of pseudo-static normal to shear stiffness ratios considerably in excess of 1, most likely from about 5 to 50 as shown later in this chapter, and reinforced in Chapter 16. They alone actually guarantee anisotropic behaviour in the 'static' (macro-deformation or macro-strain) regime.

The fact that the same (but inverted) units for compliance and stiffness are used by the different ('static' and dynamic) professions, is some guarantee that deformation, like 'squirt', is taking place, however small this may be.

The question that remains to be answered is whether, for instance, the ratio of these compliances or inverted dynamic stiffnesses for *application in situ*, can be in any way based on *dynamic* tests on three joints of about 50 mm diameter, as tested in important work at Berkeley (Pyrak-Nolte et al., 1990), to be reviewed next, or whether there could be some hidden relation to the much more common and easier *pseudo-static* tests on e.g. at least 200 samples of 100 mm length (as tested by

Barton and Choubey, 1977 and Bandis, 1980), or whether in fact *in situ block sizes* of decimetres or meters in size, should be the basis for *in situ* dynamic estimation – by back-analysis.

Strictly for the case of pseudo-static loading, the *in situ block sizes* were suggested by Barton and Bandis, 1982, for up-scaling (to lower *in situ* values), of the JRC (roughness) and JCS (wall-strength) components of strength and stiffness. The natural block size was suggested as the limit for reducing these components, based on *pseudo-static* biaxial loading tests on 400, 1000 and 4000 interlocked, fractured blocks, generated in brittle rock-like model materials mostly consisting of dense, weakly-cemented fine sand, as described by Barton and Hansteen, 1979.

There seems no physical reason why a 50 mm diameter core containing a joint (from a disused mine in Sweden), should have a strong link to *in situ* reservoir scale, just because this may be a convenient sampling size also for reservoir joints. That would be fortuitous indeed. There is clearly still less intuitive reason to use tests on roughened plates of 'Plexiglas' in this extrapolation. However, as there is so little data in this area, we must utilise what is available, and acknowledge the following very useful experimental contribution from Berkley in the late eighties.

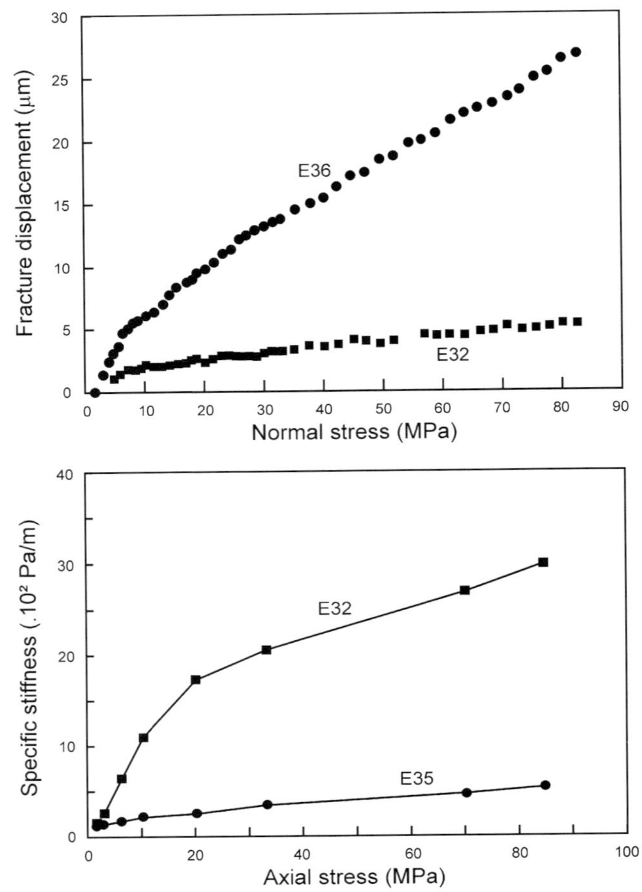


*Figure 15.8* Normally loaded stress-deformation, and specific stiffness trends for two of the three jointed samples tested by Pyrak-Nolte, as reported in Pyrak-Nolte et al., 1990. The static and dynamic stiffnesses for samples E 32 and E 35 are contrasted in Table 15.2.

## 15.5 Dynamic and static stiffness tests on joints by Pyrak-Nolte

Several aspects of the important work of Pyrak-Nolte et al., 1990, and of other published work with contemporaries at Berkeley, were reviewed in the laboratory-scale rock physics discussions in Chapter 13. The static and dynamic normal stiffness of three natural joints, their permeability, and the effect of normal stress on velocity, attenuation and seismic Q were each addressed. In this chapter on shear wave splitting, we will consider particularly the data the authors provided on the *dynamic normal and shear stiffnesses*. This can be inverted to compare with *joint compliance* data given in this chapter, and also in Chapter 16, where more rock mechanics pseudo-static data is introduced.

The load-deformation data given by Pyrak-Nolte, 1995 reproduced in Figure 15.8, shows the 52 mm diameter jointed quartz monzonite sample E 35 (from the Stripa mine in Sweden), closing by more than 25 μm, compared to only about 5 μm closure for sample E 32 when loaded to 80 MPa. The latter is remarkably stiff behaviour. The contrasting behaviour *possibly* indicates respectively 'rough' and more 'planar' joint surfaces, (or differences in the degree of inter-lock), causing the widely different specific stiffness trends for these two samples. (Joint roughness quantifications, and interlock descriptions seem not to have been reported).

The diameter of the jointed samples E 32 and E 35 was only 52 mm, which may have influenced the relatively high values of $K_{s(dyn)}$, and therefore also the relatively low values of $K_{n(dyn)}/K_{s(dyn)}$ seen in Table 15.2. This is commonly the case in rock mechanics pseudo-static testing of rock joints too: small samples have higher shear stiffness, normal stiffness may be little influenced (if correctly sampled *and* interlocked). Therefore the ratio of normal/shear

Table 15.2 Estimates of *dynamic* normal and *dynamic* shear stiffnesses derived by fitting to recorded response spectra, selected from Pyrak-Nolte et al., 1990 data sets. Note the stiffer *water-saturated* behaviour. The seismic quality $Q_p$ values for these two jointed samples are also appended. (The unit MPa/μm is suggested to aid interpretation. Note that the stiffnesses are high to very high, with 1 MPa/μm = 1000 MPa/mm).

| STRESS $\sigma_n$(MPa) | Sample E 32 $K_{n(dynamic)}$-*dry* | ($\times 10^{12}$ Pa/m) or MPa/μm $K_{n(dynamic)}$-*saturated* | Sample E 35 $K_{n(dynamic)}$-*dry* | ($\times 10^{12}$ Pa/m) or MPa/μm $K_{n(dynamic)}$-*saturated* |
|---|---|---|---|---|
| 2.9 | 15.0 ($Q_p$ = 9) | 35.0 ($Q_p$ = 17) | 4.0 ($Q_p$ = 7) | 9.5 ($Q_p$ = 9) |
| 10 | – | 80.0 | 11.5 | 20.0 |
| 20 | – ($Q_p$ = 24) | 100.0 ($Q_p$ = 36) | 20.0 ($Q_p$ = 14) | 25.0 ($Q_p$ = 30) |
| 70 | 120.0 | – | 32.0 | 59.0 |
| **STRESS $\sigma_n$(MPa)** | **Sample E 32 $K_{s(dynamic)}$-*dry*** | **($\times 10^{12}$ Pa/m) or MPa/μm $K_{s(dynamic)}$-*saturated*** | **Sample E 35 $K_{s(dynamic)}$-*dry*** | **($\times 10^{12}$ Pa/m) or MPa/μm $K_{s(dynamic)}$-*saturated*** |
| 2.9 | 3.5 | – | 1.9 | – |
| 10 | 9.5 | – | 4.8 | – |
| 20 | 17.0 | – | 6.2 | – |
| 70 | 55.0 | – | 7.4 | – |
| **STRESS $\sigma_n$(MPa)** | **Sample E 32 $K_{n(dyn)}/K_{s(dyn)}$ (dry)** | **See *high* stiffness behaviour in Fig. 15.8b** | **Sample E 35 $K_{n(dyn)}/K_{s(dyn)}$ (dry)** | **See *low* stiffness behaviour in Fig. 15.8b** |
| 2.9 | 4.3 | | 2.1 | |
| 10 | – | | 2.4 | |
| 20 | – | | 3.2 | |
| 70 | 2.2 | | 4.3 | |

Note: $Q_s$ values were generally larger than the above $Q_p$ values: see Table 10.3, Chapter 10.

Table 15.3 $K_{n(dyn)}/K_{s(dyn)}$ *ratios* for sample E 30, which was of intermediate stiffness compared to E 32 and E 35. Sample diameter was also 52 mm. Derived from Pyrak-Nolte et al., 1990 data sets.

| Normal stress (MPa) | Ratio of $K_{n(dyn)}/K_{s(dyn)}$ |
|---|---|
| 1.4 | 2.2 |
| 2.9 | 1.6 |
| 6.0 | 1.6 |
| 10.0 | 1.3 |
| 20.0 | 2.0 |
| 33.0 | 2.7 |

stiffness tends to be low when very small sample sizes are used. There is a lot of data from *near-surface* rock engineering projects where sample size is 600–1000 mm, and laboratory tests tend also to be in a higher 100–300 mm range compared to the high stress tests on (inevitably) smaller samples. Some of this data is reviewed in Chapter 16. Note the lack of any tendency for $K_{n(dyn)} = K_{s(dyn)}$, or its inverse $Z_N = Z_T$, as deduced by Schoenberg and Sayers, 1995, from dynamic tests on dry, roughened 'Plexiglas' plates.

Dynamic stiffness ratios for a joint of intermediate stiffness (E30) were also reported by Pyrak-Nolte et al., 1990. This is given separately in Table 15.3.

It may be noted from the above tables that the range of $K_{n(dyn)}/K_{s(dyn)}$ ratios for these three small jointed samples was 1.3 to 4.3 with a mean of 2.5. This means that

Table 15.4 Comparison of static and dynamic normal stiffness data for the Pyrak-Nolte et al., 1987a jointed sample E 35, which was the most deformable sample (possibly due to greater roughness). These dynamic stiffnesses were determined by curve matching.

| Normal Stress (MPa) | Static stiffness $K_{n(static)}$ ($\times 10^{12}$ Pa/m) | Static stiffness $K_{n(static)}$ MPa/mm | Dynamic stiffness $K_{n(dyn)}$ ($\times 10^{12}$ Pa/m) | Dynamic stiffness $K_{n(dyn)}$ MPa/mm | Ratio $K_{n(dyn)}/K_{s(dyn)}$ |
|---|---|---|---|---|---|
| 2.9 | 1.0 | 1000 | 4.5 | 4500 | 4.5 |
| 10.0 | 2.2 | 2200 | 8.0 | 8000 | 3.6 |
| 33.0 | 3.3 | 3300 | 25.0 | 25000 | 7.6 |

the ratio of the inversed *compliances* $Z_N/Z_T$ for these 52 mm diameter samples ranges from 0.2 to 0.8, with a mean of 0.4. The remaining question is what relevance this convenient 'core-sized' data has to *in situ* reservoirs in general, with a typical spread of jointed block sizes from perhaps 200 mm to 5,000 mm, and mostly sedimentary rock as opposed to Stripa 'granite', or hard quartz monzonite, which had $E_{static} = 60$ GPa, and $V_p = 4.9$–6.1 km/s over a stress range of 84 MPa, (Pyrak-Nolte et al., 1987b).

Pyrak-Nolte et al., 1987a reported an earlier set of (dry) dynamic/static normal stiffness measurements, performed on one of the same jointed samples (E 35). In view of the relative scarcity of such dynamic/static data for rock joints in the literature, this is reproduced in Table 15.4.

Units of stiffness more familiar in rock mechanics (MPa/mm) are also given in this table, as this unit is easier to visualize in practical terms (e.g. 100 MPa for 0.1 mm closure) than that used in geophysics, where the inverse *compliance* is also typically reported in 'complex' units of $10^{-12}$ m.Pa$^{-1}$ (with 'm' sometimes omitted altogether). It is even more 'practical' to consider the unit MPa/μm for the first static stiffness listed in this table. One thereby comes much closer to the elastic, *presumably* sub-micron, dynamic wave pulse effects on micro-closure, and on micro-fluid movement during squirt-flow attenuation.

In connection with the above ratios of $K_{n(dyn)}/K_{s(dyn)}$, it may be of interest to refer to the equivalent data for *artificial surfaces* in *aluminium*, reported by Pyrak-Nolte et al., 1992, based on interface wave experiments. An aluminium cylinder of 293 mm diameter was sawn in two, planed smooth, and then sandblasted with 300 μm grit. This artificial, planar-but-roughened interface, which can perhaps be likened to the roughened 'Plexiglas' surfaces studied by Hsu and Schoenberg, 1993, showed predicted (curve-fitting-based) ratios of $K_{n(dyn)}/K_{s(dyn)}$ ranging from 0.90 at 13.2 MPa, to 0.47 at 32.9 MPa normal stress.

The aluminium results are therefore quite different from dynamic rock joint behaviour. The actual magnitudes of dynamic normal and shear stiffness were somewhat higher than for the (obviously rougher) rock joints tested by Pyrak-Nolte, for instance 22.5 and 25.0 $\times 10^{12}$ Pa/m, in the case of the normal stress of 13.2 MPa. (In alternative units these translate to 22500 and 25000 MPa/mm, or 22.5 and 25.0 MPa/μm, i.e. very stiff behaviour, perhaps due to 'asperity-weld'.

### 15.5.1 Discussion of stiffness data gaps and discipline bridging needs

In comparison with the above *static* and *dynamics* data for one joint sample in igneous rock, we will see in Chapter 16 some rock mechanics data from Bandis, 1980, with numerous *static* normal stiffness data for rock joints in limestone, sandstone and siltstone (and other non-sedimentary rocks). We will see $K_n$ (static) values varying from 250 MPa/mm at 10 MPa normal stress, to 31500 MPa/mm at 40 MPa normal stress. (Priv. comm. Bandis, 2005). The lowest value was for a rough, *weathered* joint (JRC = 15, JCS = 44 MPa) in limestone, the latter a smoother joint (JRC = 7.6, JCS = 160 MPa) in *almost unweathered* limestone. *Unstressed* apertures, prior to standard stress-cycling to achieve an approximation to *in situ* (consolidated) conditions, were about 0.5 and 0.2 mm respectively.

Concerning shear stiffness, which is of relevance when discussing the possible magnitude of shear *compliance* in shear wave splitting, we will see in Chapter 16 a whole range of possible static *shear stiffnesses* that are seen to be *inversely* related with the sample size. The extent to which dynamic *shear compliances* are related (much more weakly, but perhaps *directly*) to sample size, seems to be one of the remaining unsolved areas in this important area of seismic detection of anisotropy, and the subsequent goal of *interpretation of permeability*.

Some pieces of the 'dynamic-permeability' jigsaw are 'complete', but there are missing bridges between dynamic and static testing, between high stress and low stress testing, and across the related small sample to large sample void. The 'completed' parts of the jigsaw are the abilities to predict both (pseudo-static) stiffnesses $K_n$ and $K_s$ and the less tangible physical (E) and hydraulic (e) apertures, for any size of jointed rock blocks. This ability is based on index tests or estimates of the rock strength (specifically the joint wall compressive strength JCS), and the roughness (specifically the joint roughness coefficient JRC). These parts of the 'jigsaw' will be illustrated in Chapter 16, in case of future use when the three missing 'bridges' identified above, are robust.

### 15.5.2 Fracture stiffness and permeability

Concerning one of the three 'bridges' discussed above, the three joints in Stripa 'granite' (quartz monzonite) that were so extensively tested and modelled by Pyrak-Nolte et al., 1990, and Pyrak-Nolte, 1996, (see also Chapter 13 review), gave interesting evidence of a strong connection between the normal stiffness (termed 'fracture specific stiffness' by the author), and the permeability, as recorded by flow rate per unit head, during 'sector-to-sector' flow across the otherwise sealed joint perimeters. Figure 15.9 shows the strong relationship between flow rate per unit head (proportional to permeability), and the joint normal stiffness, which changes due to the effect of normal stress, as illustrated in Figure 15.8.

The data shows that joints that support less flow obviously have higher stiffness, due also to the effect of normal stress. The author suggested that the joint stiffness was inversely related to the cube root of the flow, and that fluid flow and the seismic response of a joint could therefore be inter-related through the normal stiffness of the joint in question.

Pyrak-Nolte also found that the joints that *attenuated* seismic waves most at a given stress level, also supported more flow, which is logical. This implies an implicit link between the permeability and the rock quality Q-value and more specifically between the permeability and the modulus of deformation, when *not* reduced by clay, since $Q_p$ and the *pseudo-static* modulus of deformation M (or $E_{mass}$), expressed in GPa, were seen to display some remarkable, empirical similarities in Chapters 10 and 13.

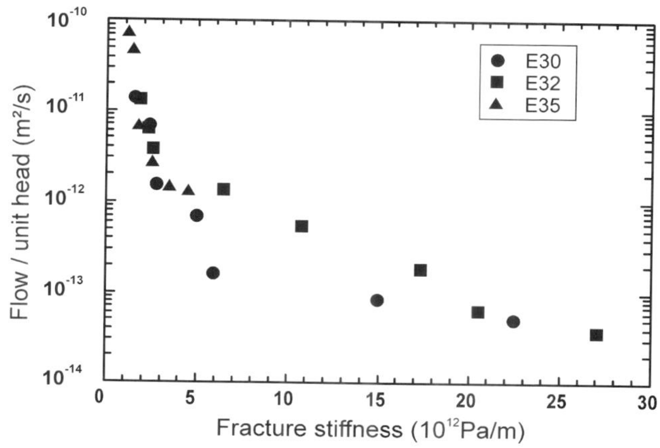


*Figure 15.9* Effect of joint normal stiffness on the flow rate per unit head (which is proportional to permeability. Data for three joints in quartz monzonite, with linear (not radial), sector-to-sector flow, crossing core of 52 mm diameter. Pyrak-Nolte et al., 1987a, Pyrak-Nolte, 1996.

## 15.6 Normal and shear compliance theories for resolving fluid type

Interest in fracture-induced seismic anisotropy has rapidly evolved from the earlier estimation of 'only' fracture orientation, (with an assumed indication of major horizontal stress), to fracture intensity, and the attempted prediction of fluid saturation and permeability anisotropy. To make this advance, the sensitivity of fracture compliance to fluids has to be understood. There are several models addressing this important question.

The authors Liu et al., 2000, investigated various possible representations of fractures, for use in dynamic modelling of fractured reservoirs. These consisted of an *in-plane* distribution of small cracks, an *in-plane* distribution of contacts, and a thin layer with material infill. They were able to derive specific expressions for the fracture compliances $Z_N$ and $Z_T$. Their different models indicated strong sensitivity of the ratio $Z_N/Z_T$ to the bulk modulus of the fracture infill material, with the most rapid change in the *compliance ratio* (and values closer to 1.0) occurring when the infill bulk modulus approached zero, such as for gas-filled fractures. However with realistically small fracture aspect ratios (i.e. $< 0.001$), much lower values than 1 were suggested, as shown in Figure 15.10.

Liu et al., 2000, and Liu and Li, 2001 presented equations for the compliance ratio ($Z_N/Z_T$), which incorporated the aspect ratio of cracks, and the ratio of bulk moduli for fluid and matrix. One of their equations, based on earlier work by Schoenberg and Hudson,

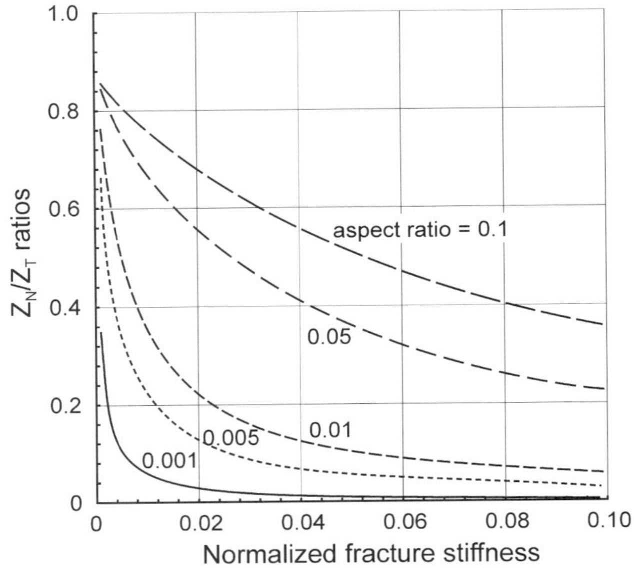


*Figure 15.10* Liu et al., 2000 model of *in-plane* small cracks, for estimation of the ratio $Z_N/Z_T$ versus the normalized bulk modulus ($\lambda_f$) of the fracture infill, via components of Hudson's models. The matrix was a so-called Poisson's solid with equal Lamé constants (bulk and shear moduli $\lambda = \mu$).

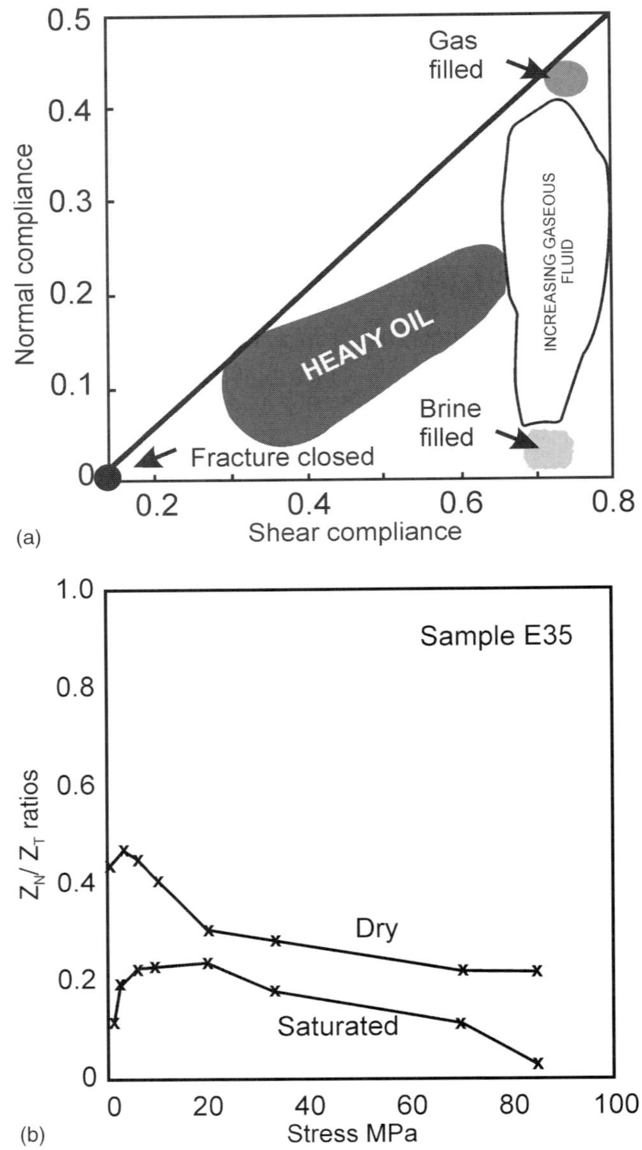


*Figure 15.11* (a) A fracture compliance based prediction of fluid type (brine, heavy oil or gas) based on fracture aspect ratios and on the ratio of fluid and matrix bulk moduli. Liu and Li, 2001. (b) Laboratory evidence for the importance of fluid (dry or saturated) on the ratio $Z_N/Z_T$ as a function of stress, from Pyrak-Nolte et al., 1990.

suggested that $Z_N/Z_T \approx 1$ if the fractures were dry, and $Z_N/Z_T \approx 0$ if the fractures were filled with liquid. Their interesting theory, which however is not directly supported by the dynamic rock mechanics experience with rock joints reviewed in the previous section, is also illustrated in Figure 15.11a. Their equation, on which the figure is based, is reproduced here for reference:

$$\frac{Z_N}{Z_T} = \frac{7}{8}\left[1 + \frac{5}{2\pi}\left(\frac{a_c}{c_c}\right)\left(\frac{k_f}{k}\right)\right]^{-1} \qquad (15.4)$$

where $a_c$ and $c_c$ are the 'long and short axes of the elemental cracks in the fracture planes', which they also refer to as the aspect ratios, and $k_f$ and k are the bulk moduli of the fluid and matrix.

Figure 15.11a suggests, as does the above equation, that $Z_N$ and $Z_T$ are zero (i.e. stiffnesses $K_n$ and $K_s$ are infinite, or very high) when a fracture is 'closed'. Of course if the fracture is no longer present this is logical, but not if stresses are merely high and contact areas are high, but significantly less than 100%. Such might be typical for reservoir fractures in weaker materials. Perhaps for the highest-porosity chalk, showing general compaction due to pore collapse, fractures would truly be of infinite stiffness, since effectively absorbed in the matrix collapse. Based on the different bulk moduli of gas, brine and heavy oil, Liu and Li, 2001 allocated different $Z_N$ versus $Z_T$ 'regions' in Figure 15.11a.

As partial support for the predicted rends, Liu and Li, 2001, cite the dynamic stiffness data from Pyrak-Nolte et al., 1990, and invert the ratios (see Tables 15.2 and 15.3) to give the $Z_N/Z_T$ versus normal stress trends reproduced in Figure 15.11b. They note the increased ratio of $Z_N/Z_T$ from the saturated to the dry state.

Converting this figure to rock mechanics terminology, the dynamic stiffness ratios $K_n/K_s$ are predicted to vary from about 4 to 45 with increasing stress (when saturated), and from about 2 to 4 with increasing stress (when dry).

The authors also referred to the Lucite-plate samples tested by Hsu and Schoenberg, in which honey-saturated (sic) tests showed even greater sensitivity to the dry and saturated states, and surprisingly close magnitudes of $Z_N$ and $Z_T$ ($Z_N/Z_T$ close to 1.0) when their samples were dry, as reviewed earlier. As we have seen, the 'intense fracturing' (more like thin inter-bed studies) created by the Hsu and Schoenberg, 1993 tests on either 60 or 200 (!) 0.7 mm thick, roughened (10 μm) Lucite plates, created an extreme *crack density* (almost without its equal) of 7.5 to 25, up to *three* orders of magnitude greater than *in situ* microcracking or fracturing. Possibly this factor is partly to blame for the unrealistic $Z_N \approx Z_T$ result, that has permeated into several geophysics publications.

### 15.6.1 In situ compliances in a fault zone inferred from seismic Q

The low value of seismic Q for a fault zone (or possibly several faults) encountered in a well in the North Sea ($Q_{seis}$ averaging 45), described by Worthington and Hudson, 2000, represented an abrupt increase in attenuation, relative to the Triassic and Lower Jurassic age sandstones, siltstones and claystones that were predominantly encountered in the well.

Worthington and Hudson modelled the effects of a down-going P-wave between 1000 and 2000 m depth, by assuming that one or up to several faults, intersected the transmission path. They used a theoretical compliance model developed by Hudson et al., 1997, to demonstrate the need for a very large but (to rock mechanics thinking), realistic inequality of the shear and normal *compliances*, suggesting the need for $Z_N$ (or Bn) $= 4.4 \times 10^{-14}$ m.Pa$^{-1}$, and $Z_T$ (or *B*t) $= 1.1 \times 10^{-9}$ m.Pa$^{-1}$. The equations used to estimate these compliances are quite complicated, and are reproduced here to emphasise the geophysical 'way of thinking' regarding micro-deformation compliances. (According to Worthington and Lubbe, 2004, the following two equations each had a small error in the first *and* second terms in brackets. In place of the 2000 versions with $(4/\pi) \times (\alpha/\omega a)$ and $(8/\pi) \times (\alpha/\omega a)$, the equations are as follows:

$$B_n = r^w \left(\frac{4\mu}{\pi a}\right)\left(1 - \frac{\beta^2}{\alpha^2}\right)\left\{1 + \frac{2\left(r^w\right)^{1/2}}{\sqrt{\pi}}\right\} + \frac{\kappa' + (4/3)\mu'}{\Delta} \tag{15.5}$$

$$B_t = r^w \left(\frac{8\mu}{\pi a}\right)\frac{\left(1 - \dfrac{\beta^2}{\alpha^2}\right)}{\left(3 - \dfrac{2\beta^2}{\alpha^2}\right)}\left\{1 + \frac{2\left(r^w\right)^{1/2}}{\sqrt{\pi}}\right\} + \frac{(\mu')}{\Delta} \tag{15.6}$$

Here, α, β are the P-wave, S-wave velocities, μ is the rigidity of the rock, $r^w$ is the proportion of the fault surface area that is estimated to consist of welded contact (assumed <0.2), *a* is the mean radius of the contact areas, μ′and κ′are the rigidity and bulk modulus of the fault fill, and Δ is the mean aperture of the fault.

For purposes of comparison, one may first *invert* these presumed micro-deformation magnitudes of geophysics, to the much more easily understood 'physical' macro-deformation units of rock mechanics, namely shear and normal stiffnesses of $K_n \approx 20{,}000$ MPa/mm (and 20 MPa/μm), and $K_s \approx 1$ MPa/mm. These by chance, or due to *more inter-related* micro-and-macro physical processes at larger scale and at lower frequency, are similar to the values one may need to use on different occasions in rock mechanics pseudo-static modelling.

Heavily stressed rock joints, perhaps equivalent to the 'welded' portions of the modelled fault or faults, have normal stiffnesses, as we shall see in Chapter 16, that are of this high magnitude, while clay-filled discontinuities will tend to have much lower values for obvious reasons of low frictional strength. The low values of shear stiffness that are implied by the authors' modelling, are in fact very reasonable values for blocks within faults of large dimension, that are assumed to dominate in large scale deformation processes, such as overburden subsidence over compacting reservoirs.

For example, Barton et al., 1986 and Barton, 2002b, describe ('static') $K_s$ as low as $10^{-2}$ MPa/mm, needed to realistically model large scale discontinua in the Ekofisk reservoir subsidence, thereby obtaining a better match to reality than continuum models. Such low values were selected from a Barton, 1982 assembly of large-scale shear stiffness data (see Chapter 16). The extremely compliant values of pseudo-static shear stiffness were due to the assumed large effective 'block sizes' of the major deforming elements of a conservatively estimated 150 $km^3$ deforming volume of overburden. (The 300 m thick pear-shape reservoir in jointed chalk, in area roughly 9 × 14 km, has a depth of roughly 3 km).

The approx. 1.0 MPa/mm, for the dynamic shear stiffness derived by Worthington and Hudson for the sub-North Sea fault, is equivalent to a static shear stiffness under 20 − 25 MPa effective stress (1.5 km gives about 37 − 15 = 22 MPa effective stress) of roughly 10 m effective block sizes. (See Figure 16.14 in Chapter 16).

An empirical expression for the macro-deformation, pseudo-static shear stiffness, used in rock mechanics, is given in Chapter 16, and is evaluated with various input data assumptions, including the effect of saturated or dry conditions. In principle, the simple empirical equation for shear stiffness, from Barton and Choubey, 1977, is based on the peak shear strength equation divided by the displacement to peak. In other words a constant gradient of load versus shear deformation is assumed. In the case of normal stiffness, the empirical hyperbolic equation of Bandis, 1980, also given in Bandis et al., 1983, gives the basis for normal stiffness estimation.

Both these stiffness equations are components of the Barton-Bandis constitutive model, and utilise the relevant joint wall roughness coefficient JRC, and the joint wall strength JCS, as measured or estimated input data. In the case of shear stiffness, the residual friction angle $\phi_r$ is also needed. In the case of faults, we may consider JRC ≈ 0 (zero effective roughness), and $\phi_r$ may be as low as 10° to 20°, depending on the mineralogy of the clay or possible 'shale-smear'. (For a comprehensive review of the sometimes very low *in situ* pseudo-static shear strengths of clay-filled discontinuities, often tested at large scale at dam sites, but at low to moderate stress levels, see Barton, 1973b).

A return to the important themes of *dynamic compliance* and pseudo-static stiffness will be found in Chapter 16, to better explore the similarities and dissimilarities, as the case may be. At this stage of comparison one can assume that the pseudo-static values of stiffness are lower (in the normal direction), and much lower (in the shear direction) than the equivalent dynamic values of stiffness. In view of the extreme lack of present data for the dynamic values, and the wealth of data for the pseudo-static values, it may well prove useful to 'bridge-the-gap' between the disciplines and 'quantify' (<) and (<<). This will be attempted.

## 15.7 Shear wave splitting from earthquakes

Monitoring shear waves caused by the effects of earthquakes requires a certain consideration of the geometry of the situation. There is a so-called *shear wave window*, which is the area on the earth's surface above the earthquake, where ray-paths subtend angles of incidence less than about 35°. This derives from the requirement of angles of incidence less than $\theta = \sin^{-1}(V_p/V_s)$. (Nuttli and Whitmore, 1962.) For a Poisson's ratio of 0.25, θ is about 35°. Outside this window the shear-wave waveforms are severely distorted.

Crampin and Peacock, 2005 mention the added distortion if recordings are made on irregular topography. They also mention the 'helpful' surface effect that a low velocity layer may effectively widen the shear-wave window by refracting the upcoming shear-waves towards vertical incidence. The effective window may then be as much as (2 x) 45° to 50°.

The polarizations of the leading, faster, split shear-waves within the shear-wave window above many recorded earthquakes are typically observed to be scattered by ±10° to ±20° about a direction parallel to the direction of maximum horizontal compression. (Crampin and Booth, 1985). There can be many reasons for this scatter besides general heterogeneity with a 'mixed' geology along the path length. Furthermore, and unfortunately, the location of the main source of shear wave splitting is not known, which is why in-hole instrumentation at up to several kilometres depth is so attractive (see Chapter 10), to remove the attenuating near-surface layers, which perhaps could also be the source of somewhat differently oriented structure and/or major principal stress.

### 15.7.1 Shear-wave splitting in the New Madrid seismic zone

Rowlands et al., 1993, using a network of more than thirty three-component digitally recording seismometers in the New Madrid seismic zone in the central USA, recorded shear-wave splitting, which they attributed to EDA cracks. A compressed summary of the seismic zone, an example of shear-wave splitting, and of the split shear-wave polarizations, together with the velocity structure through the local crust is given in Figure 15.12.

The majority of the seismic events were between 3 and 15 km depth, mostly in the Precambrian basement. The shear waves propagating upwards encountered three high-to-low velocity interfaces and one low-to-high interface. The aligned polarizations lay in the ENE-WSW direction (average N62°E), reportedly coincident with the direction of the regional maximum horizontal stress documented by Zoback and his colleagues.

This good correlation suggested to the authors that the anisotropy causing the shear-wave splitting was likely to be caused by the presence of fluid-filled, stress-aligned vertical *extensive dilation anisotropy* (EDA) cracks. The time delays (120 to 180 ms) and associated shear wave travel times (3.8 to 4.6 s) suggested a shear wave anisotropy over the whole path of 3 to 4%, consistent with a Crampin, 1993a, review of these ubiquitous phenomena.

Rowlands et al., 1993, found evidence that the anisotropy causing the polarization alignments may have been present only in the top 5 km, rather than evenly distributed. They also reported that the basal Palaeozoic deposits ($V_p = 4.83$ km/s, see table in Figure 15.12) were considered to be fractured, while the thick shale formation might have introduced a thin-layer anisotropy.

### 15.7.2 Shear-wave splitting at Parkfield seismic monitoring array

Liu et al., 1993 also confirmed shear-wave splitting at a majority of the recording stations at the Parkfield borehole seismic network along the San Andreas fault zone, obtained during 18 months of recordings from previous campaigns by co-authors Evans, Booth and Crampin, reported in 1984 and 1985, but recorded earlier than this. At three of the stations, distant 1 to 5 km from the fault, the polarizations were consistently normal or subnormal to the fault strike and *parallel* to the direction of the regional maximum horizontal stress.

At station MM, immediately adjacent or within the fault zone (see Figure 15.13, reproduced from Liu et al., 1993), first motion ($qS_1$) was polarized parallel to the fault strike, i.e. in this case *perpendicular* to the regional maximum principal stress of about N30°E.

There was some evidence of *temporal variation* of the shear wave time delays in connection with a magnitude $M_L = 4$ San Andreas fault earthquake; time delays at Station MM (7 ms/km) were roughly twice those at station VC (4 ms/km). The authors concluded that the relatively greater shear wave splitting observed at station MM suggested that the fluid-filled fractures within the fault zone were more extensive than in the surrounding crust, which is logical.

The fault parallel polarization of the leading split shear wave at station MM was taken by the authors to indicate that either the stress was highly irregular in the immediate vicinity of the fault, or that the fault-related fractures were aligned by fault shearing rather than by the regional principal stress.

The Liu et al., 1993, analyses were performed with co-authors' data, also recorded between January 1989 and July 1990. All the earthquakes during which shear waves were clearly visible above the P-wave coda, showed evidence of shear-wave splitting. In a few cases, only one of the anisotropic shear-wave polarizations was recorded, due to attenuation of the slower component. The dominant frequency was 20 Hz, yielding elliptical motion. Various different polarization alignments of some stations (Figure 15.13b) were thought to be possible due to local topography, or due to changes in angle of incidence due to refraction across hard rock-to-sediment interfaces.

A marginal *decrease in the time delay* following the $M_l = 4$ earthquake of May 1989, followed by an irregular increase was noted from stations MM and VC. Similar *decreases in time delay* at the time of other larger earthquakes ($M = 3.5$, $M = 6.0$) were cited. The average time delay at MM on the fault zone was about twice as large as those at station VC 5 km away, on the southwest block of the SAF. A greater density of microcracks and fractures in the fault zone was cited, and change of stress affecting the geometry of these fluid-filled features was assumed as a possible cause for the temporal changes.

A migration of focal depth for about 100 days after the $M_l = 4$ event was first cited as a possible cause of the decrease in time delay. However, for earthquakes with focal depths above 7 km depth, there was a pronounced increase in time delays. So the conclusion was drawn that the *anisotropy of the fault zone* was concentrated above 7 km.

Since the polarizations at VC, VR and ED (Figure 15.13b) were approximately perpendicular to the SAF, and parallel to the regional principal stress direction (N30°E), they were considered to be consistent with

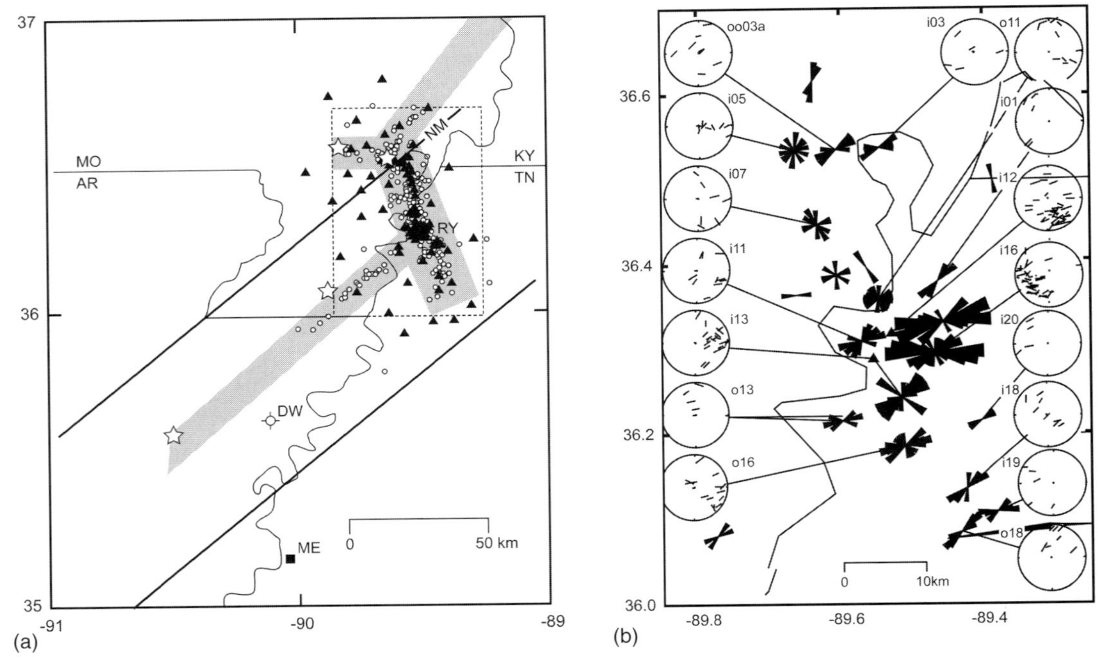


| Layer | Thickness (km) | $V_p$ (km/s) | $V_s$ (km/s) | Density (g/cm$^3$) | Geology |
|---|---|---|---|---|---|
| 1 | 0.65 | 1.8 | 0.6 | 2.2 | Unconsolidated marine clastic sediments |
| 2 | 1.85 | 6.02 | 3.56 | 2.65 | Palaeozoic shales and carbonates |
| 3 | 2.50 | 4.83 | 3.20 | 2.60 | Palaeozoic clastics and carbonates |
| 4 | 12.0 | 6.17 | 3.57 | 2.75 | Precambriam basement |
| 5 | 10.0 | 6.60 | 3.82 | 2.90 | Upper crust |

(c)

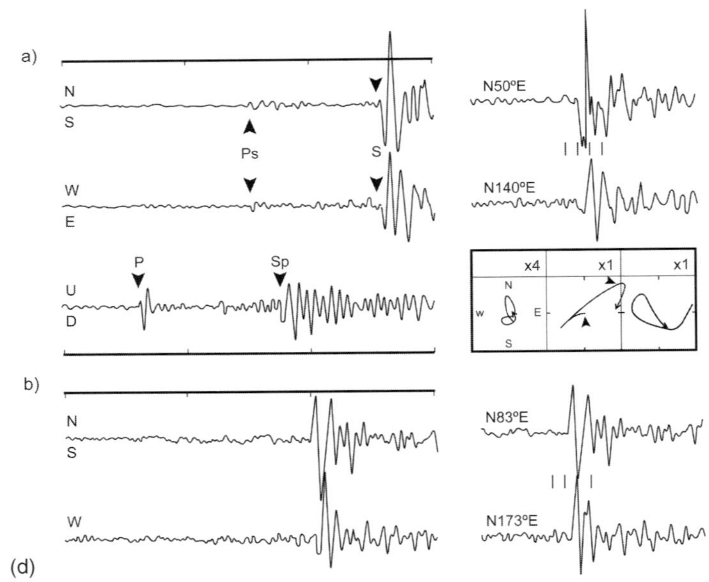


*Figure 15.12* A compressed summary of Rowlands et al., 1993, measurements and analyses of shear-wave splitting at the New Madrid seismic zone. (a) seismic map, (b) shear-wave polarization examples, (c) table showing velocity structure and geology, (d) a shear wave splitting example (focal depth 9.1 km, epicentral distance 5.0 km. Traces on left are the original, 3 seconds duration. Traces on right are the horizontal components rotated into the fast and slow split shear-wave polarizations.

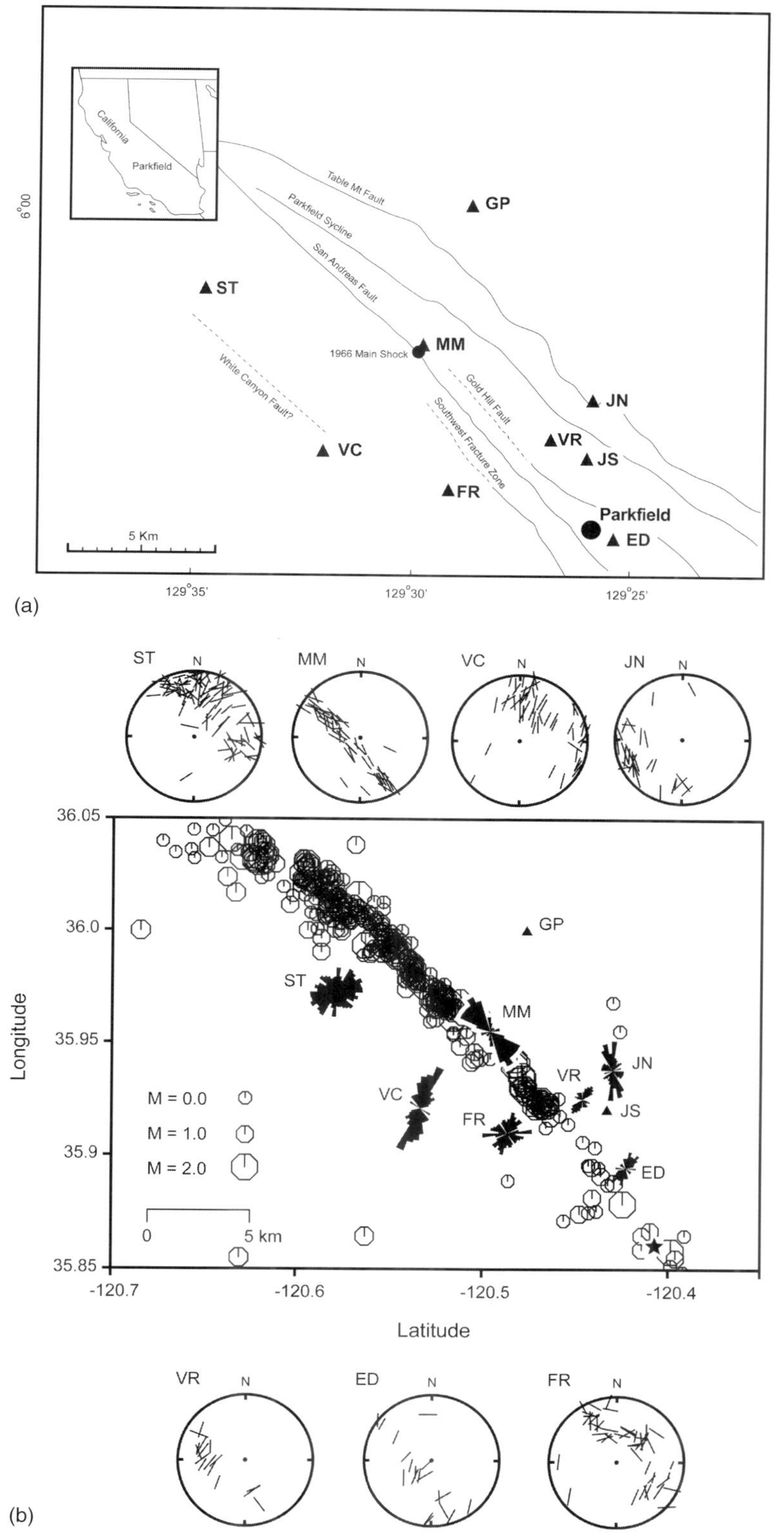


*Figure 15.13* (a) San Andreas fault and associated shallow thrust faults, with Parkfield high resolution seismic network (HRSN). (b) Epicentres of local small events within the shear-wave window. (Star in SE corner was an $M_l = 4$ event at 8.3 km depth.) Lower hemispheres show polarizations of leading split shear-waves, beneath 7 of 9 stations. Liu et al., 1993.

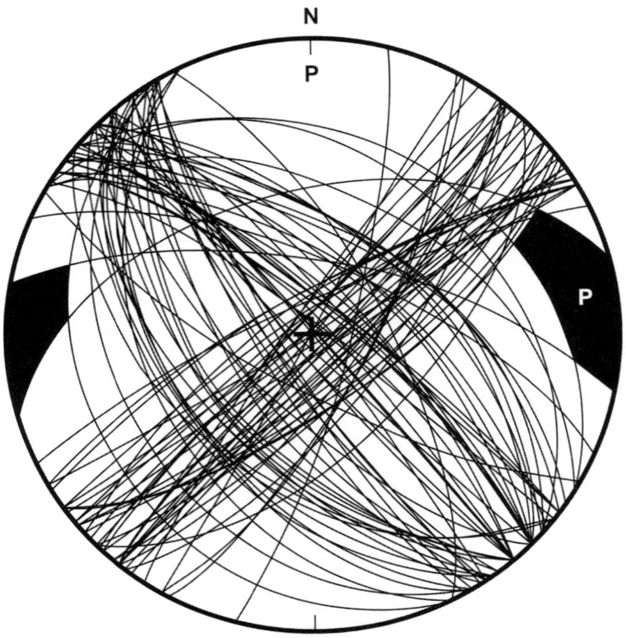


*Figure 15.14* Superimposed fault plane solutions for earthquakes close to station MM. 95% of the shear waves propagated within 15° of vertical. Nishioka and Michael, 1990, cited by Liu et al., 1993.

the extensive-dilation anisotropy (EDA) concept of stress-aligned fluid-filled cracks.

Nearer the fault, where the leading shear wave is polarized parallel to the fault, the indication was of *fault parallel cracks* instead of the adjacent *regional stress parallelism* of the other stations. The internal structure of the fault gouge and transition zone was assumed to be the reason for this rotation. In the immediate location of the fault zone, fault plane solutions for 68 events close to station MM (Figure 15.14) indeed show a different, and quite narrowly defined maximum stress axis between N15°W and N10°E, i.e. rotated in relation to the regional stress along N30°E.

### 15.7.3 Shear-wave splitting recorded at depth in Cajon Pass borehole

Liu et al., 1997 also analysed 51 local earthquakes recorded at 2.5 km depth in the Cajon Pass scientific borehole, to assess shear-wave splitting. Time delays between the split shear-waves of up to 44 *ms* per km were identified for 32 of the events. Their analyses suggested that the anisotropic structure near Cajon Pass had *orthorhombic* symmetry. 'Stress-aligned' fluid-filled microcracks and pores were assumed, i.e. the standard Crampin EDA assumption, but in fact the polarization direction of N 13° W was reportedly *not* consistent with the numerous stress measurements and borehole break-out analyses of the surroundings, and of the uppermost 3 km of the borehole. These had suggested N 57° E ± 19°. (The strike of the San Andreas Fault, 4 km distant, is N 60° W.) In the upper 1000–1820 m at Cajon Pass the polarizations of split shear waves had earlier indicated a nearly consistent N 70° E ± 10°. The horizontal stress from inferred *focal mechanisms* around Cajon Pass was reported as N 17° W, which was much closer to the polarization direction N 13° W of the authors, using 'below 2.5 km depth' seismic data.

The authors commented that the behaviour of shear-waves in the vicinity of fault zones is complicated, with the leading shear-wave polarizations often exhibiting *fault-parallel* alignments near the fault, but alignments with the regional stress field away from the immediate fault zone. (This would suggest respectively fault-aligned and stress-aligned joint sets, as the possible source of shear-wave splitting.) The fact that stress directions inferred from shear-wave polarizations at 2.5 km depth were different from those inferred nearer the surface did not suggest to the authors that *joint orientations could be different*, but that microcrack-controlling stress directions were different. The authors concluded from their study that the San Andreas Fault was probably driven by deep, regional tectonic stresses.

### 15.7.4 Stress-monitoring site (SMS) anomalies from Iceland

From another seismically active region, Crampin, 2003 described the preliminary establishment of a stress-monitoring site in Iceland, using state-of-the-art borehole instrumentation to monitor shear-wave splitting between a controlled-source well and two receiver wells. Such sites were designed to identify the nearly negligible changes of stress, which in appropriate circumstances could monitor the build-up of stress (or crustal adjustments) before earthquakes and volcanic eruptions.

Figure 15.15 reproduces an interesting series of observations from the SMS at Húsavik, Iceland, from 8th to 24th August, 2001, also reported in Crampin, 2003. The 12-hours (per day) histograms of seismicity

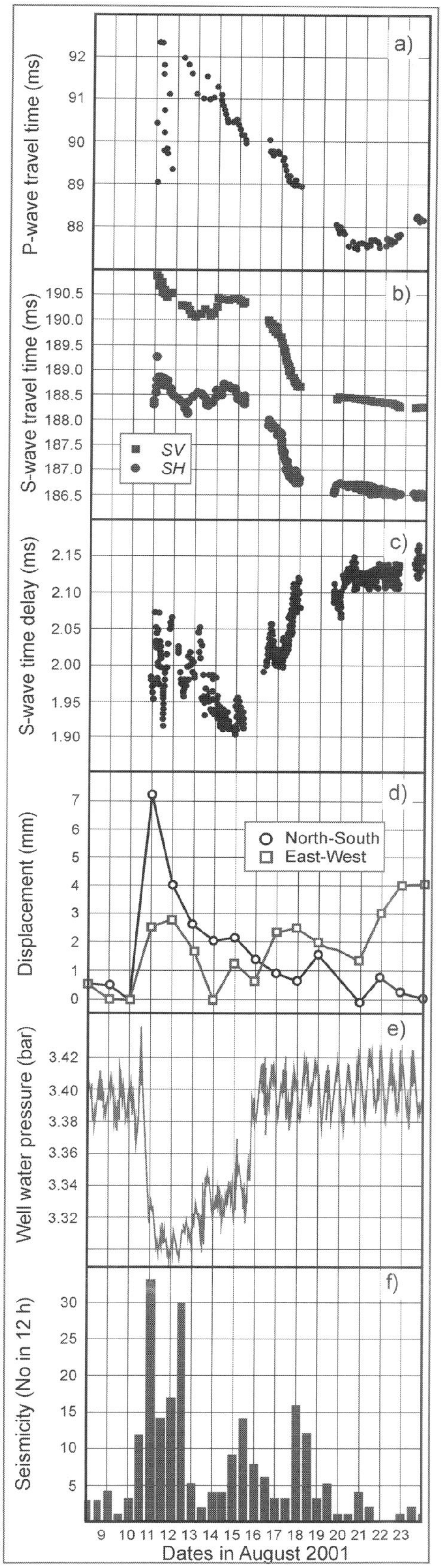


recorded within 100 km of Húsavik show peak activity coinciding with an anomalous 1 m water level drop at 33 m depth in a well on nearby Flatey Island, immediately above the Húsavik-Flatey Fault. There was also correspondence with GPS displacement anomalies of many millimetres magnitude around Húsavik, and with various changes in prior and *subsequent* P-wave and S-wave travel times and delays.

The shear-wave splitting was the most sensitive parameter besides ground deformation, showing a 10% variation in time delays. Significantly, Crampin, 2003 also mentioned that with the controlled source and cross-borehole seismic represented by SMS borehole equipment, one was 'free of the 90°-flips' that may occur in source zones, (due to extreme pore pressures according to Crampin, or due to extreme loading in the $\sigma_{H\ max}$ direction, causing an extreme 'Poisson expansion' due to Crampin APE-crack expansion, according to the writer: see earlier discussion).

### 15.7.5 SW-Iceland, station BJA shear wave anomalies

Figure 15.16 reproduces the key results of a four-year study in Iceland, reported by Volti and Crampin, 2003 for seismic recording Station BJA in south-west Iceland, during the period 1st January 1996 to 31st December, 1999. The middle and top diagrams show the variation of *time-delays with time*, for raypaths in Band-1 and Band-2, which separate observations making solid angles of ±15°−45°, and ±15° to the average 'crack' plane. The time-delays in *ms* per km are normalized to a 1 km path length.

The vertical lines through the time-delay points are error-bars, based on errors in hypocentral distance. The irregular curves are nine-point moving averages. The inclined lines in Band-1 (middle diagram) are least-squares estimates starting just before the minima of the

*Figure 15.15* Some coordinated observations of changes brought about by presumed, seismically induced crustal adjustments, recorded at the Stress-Monitoring Site (SMS) at Húsavik from 8th to 24th August, 2001. The 12-hours (per day) histograms of seismicity recorded within 100 km of Húsavik (diagram f), show peak activity coinciding with a variety of P- and S-wave changes (diagrams a to c), to ground movements (d), and to water-well level changes (e). Crampin, 2003.

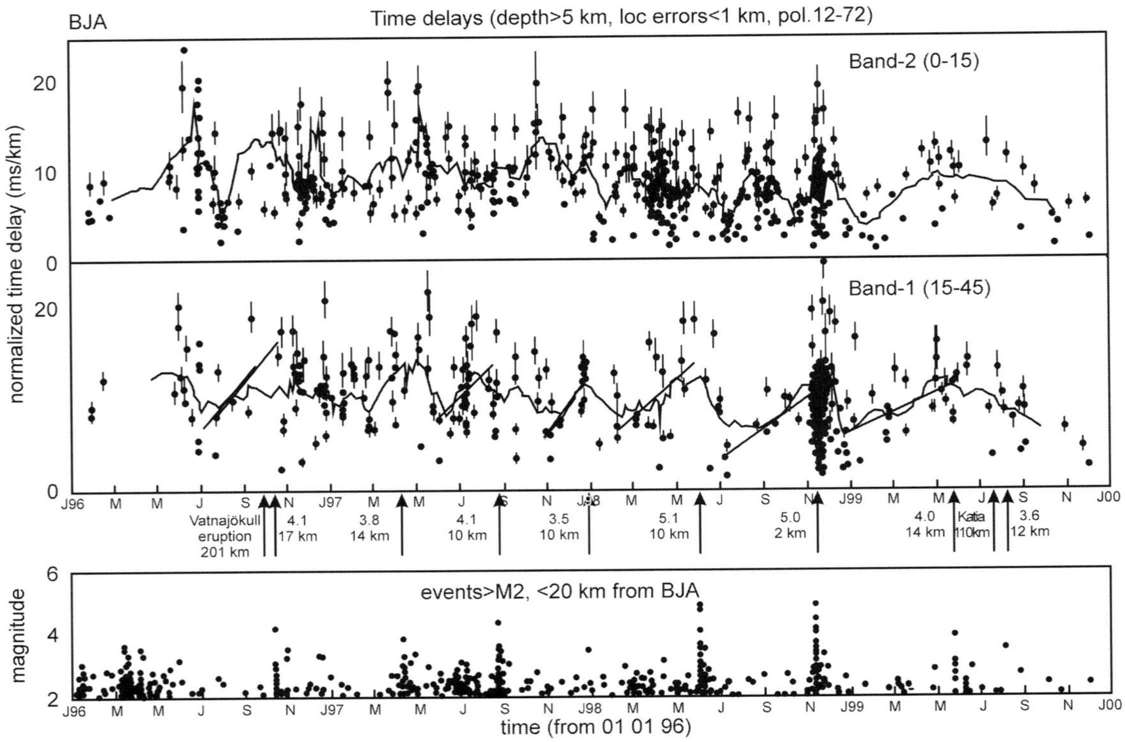


*Figure 15.16* Correlation of time delay changes with timing of seismic events in Iceland. The inclined lines are least-squares estimates of time-delay increases, which end with each of the larger events. Volti and Crampin, 2003.

nine-point moving average, and ending *when an earthquake or large eruption* takes place.

The arrows indicate the time of these events, with date and epicentral distance. The bottom diagram shows the magnitudes of earthquakes greater than M = 2, within 20 km of this recording station.

It was Crampin's opinion that the changes observed in shear-wave splitting, both before the earthquakes and volcanic eruptions, were due to changes in the rock mass, rather than associated with changes in the immediate source zones. Crampin claimed that the shear-wave splitting showed '*that almost all in-situ crack distributions verge on fracture-criticality*'. In the opinion of the writer, the Iceland data is of great interest because of the extreme sensitivity of quite different scales of 'crack', namely joints and natural fractures, to minute changes of effective stress and/or fluid pressure.

As argued previously, this claim *that almost all in-situ crack distributions verge on fracture-criticality* does not appear valid, if one assesses that the greatly expanding microcracks in the APE model (Figure 15.3) is an improbable model for rock masses. The *crack density* definition is ambiguous, and as applied by Crampin in numerous publications, implies that even hand-specimens 'of almost all rocks', would be pervaded by tens to hundreds of millions of microcracks, since the $(radius)^3$ term with suitably small microcracks (e.g. 10 or 100 μm) guarantees completely unrealistic numbers of microcracks to generate, for example, $e = 0.045$ (according to Crampin a *typical* maximum *crack density*), as needed for generating the *typical* maximum 4.5% shear-wave anisotropy.

In fact, as we shall see, there are reported crack densities and shear wave anisotropies far higher than suggested by Crampin, when measuring shear wave polarization through jointed or fractured reservoirs. When the *crack density* calculation is applied to the often observed scale of several meters (in wells and deep tunnels), quite reasonable numbers of larger fractures or joints are predicted, and these features with their extremely low aspect ratios are, according to fundamental principles of geophysics and rock mechanics, surely *more* compliant than the microcracks apparently favoured by Crampin.

At great depth, as stresses are more isotropic and very high, there may be no microcracks or open fractures to speak of, as they may be completely closed. It is then

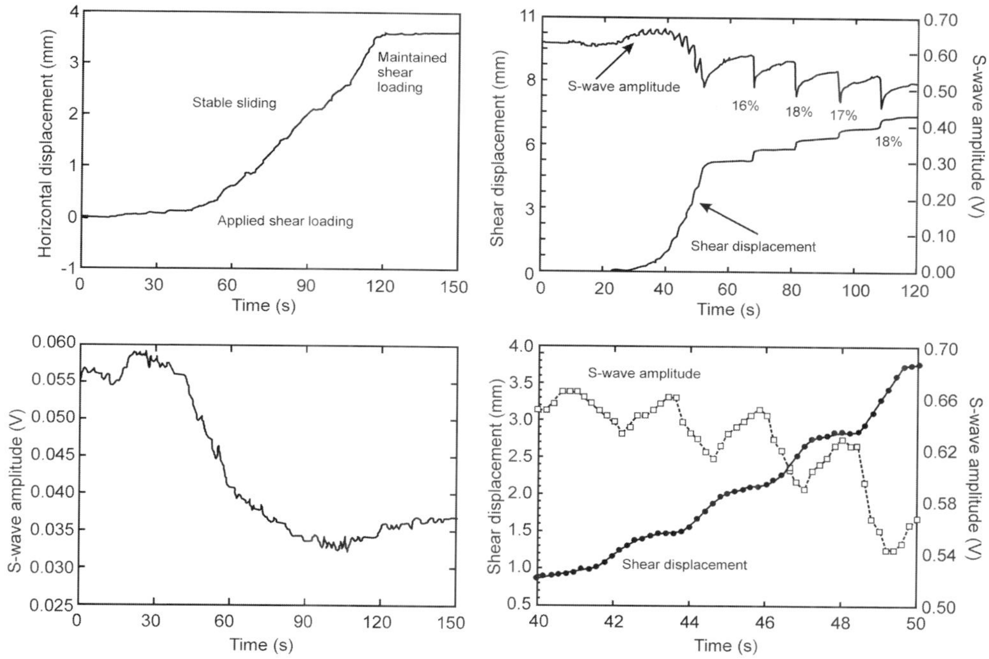


*Figure 15.17* *Left-hand diagrams:* Stable sliding of several millimetres on normally loaded (6 MPa) planar-but-roughened surfaces in Stripa 'granite' (quartz monzonite), caused a marked reduction (up to 50%) in the shear-wave amplitude. *Right-hand diagrams:* Stable sliding under a normal stress of about 10 MPa, followed by stick-slip effects, showed cyclic reduction of the shear-wave amplitude in abrupt steps for each slip event. The lower diagram shows an expanded view of the 'continuous' sliding, showing actual (but inaudible to AE) faster-slip episodes, which also caused quite rapid reductions of shear wave amplitude. Chen et al., 1993.

that the ambiguous, undefined location of the source depth of the shear wave splitting comes into play, and one can assume with some confidence that the source of the splitting will tend to be shallower than perhaps desired, where conditions of *anisotropy* are more favourable for shear-wave splitting.

### 15.7.6 Effects of shearing on stiffness and shear wave amplitude

In relation to the possible use of shear waves in detecting earthquake precursors, it is interesting to note Chen et al., 1993 investigations of the *amplitude of shear waves* under the influence of stable sliding and stick-slip. There were strong indications of reduced shear wave amplitude, both under stable sliding, and during stick-slip events, with build-up of amplitude when 'stuck' and rapid reductions in amplitude just before and during slip. Figure 15.17 shows some key results. There was reportedly no change in velocity during the stick-slip process, which could help to explain why $V_p/V_s$ reductions prior to some earthquakes are not necessarily reliable precursors for all cases.

The authors noted that the reduced shear-wave amplitude was probably associated with reductions in (specific) shear stiffness of the joint or joints undergoing shear. As shown by Boitnott et al., 1992, and as also modelled with the 'JRC-mobilized' concept of the Barton-Bandis joint constitutive model (Barton, 1982: see Chapter 16), there is likely to be an 'invisible' shearing of micron-size, prior to *measurable* sliding events, which could be the reason for the precursory shear-wave amplitude reduction noted by the authors. This reduction is registered before sliding is detected, and will logically occur before any dilation is registered, with the eventual mobilization of roughness in the 'sliding-up' phase, when peak strength is approached.

### 15.7.7 Shear-wave splitting at a geothermal field

Shear-wave splitting at one of the largest geothermal sites in the world, the Cerro Prieto Geothermal Field (CPGF), was described by González and Munguía, 2003. Data from both weak and strong-motion earthquakes was used, with seismic recording at about a dozen stations, in and surrounding the production zone of the CPGF. Eight of their seismic stations showed that the faster shear waves were polarized in a range of directions from N 14°W to N 17° E. At the four remaining stations the polarization trends were between N 25° E and N 67° E. For the entire area N-S was the best average.

Surprisingly, in view of the need for *joint connectivity* at a geothermal site, the authors followed the EDA convention, and assumed that it was *microcracking* that caused most of the shear-wave splitting. So when there was variation, they assumed that *variation of stress direction* was responsible, rather than for instance variation of sheared conjugate joint set directions. One of the areas of deviation corresponded to the epicentre of a swarm of seismic activity, which would also have suggested shearing of larger structural features.

In the production zone of the field, the widest spread of orientation (up to 90° variation according to their equal area rose diagrams), and trending to the west, was actually registered, suggesting to the writer something that might resemble conjugate joint shearing, as illustrated in Figure 15.2. Good connectivity in a production zone, satisfying a 'constant' N-S average maximum stress direction, can readily be provided by conjugate jointing, preferably with some non-planarity if high permeability is to be generated due to slight dilation. See the extensive discussion of this topic in Chapter 16.

### 15.7.8 Shear wave splitting during after-shocks of the Chi-Chi earthquake in Taiwan

Before concentrating on reservoir measurements, it is useful to record the following earthquake after-shock monitoring result from Taiwan. Seismic velocity and $Q_{seis}$ anisotropy were recorded in the case of near-surface monitoring of after-shocks, following the destructive Chi-Chi earthquake in Taiwan, in 1999. Seismograms recorded at a 200 m deep station gave clear indication of upgoing split shear-waves with fast and slow components, indicating 8% velocity anisotropy below the top 200 m.

The authors Liu et al., 2005, estimated $Q_{seis}$ values in the 2–15 Hz frequency band, and found values of 61 to 68 in the fast direction, and 43–52 in the slow direction. As commented upon many times, such values closely resemble possible deformation moduli (M, expressed in GPa), which can be readily estimated by the joint-property-based rock mass quality Q. ($M \approx 10\ Q_c^{1/3}$, see Figure 15.33.) If the rock quality immediately below the 200 m recording depth (that was assumed to have caused the shear wave splitting), was actually poor, resemblance to dynamic moduli of deformation (expressed in GPa) would be a more appropriate suggested approximation.

The authors cited microcrack alignment and their response to in situ stress as the reason for the anisotropy. However, based on the scale-dependence of dispersion seen in recent dynamic poroelastic matrix-with-fractures modelling, reviewed later in this chapter, it would seem that the anisotropy is more likely to be caused by preferential sub-vertical joint set anisotropy, which gives anisotropy at the low seismic frequencies, as opposed to microcrack anisotropy that is *dominant* from >1 kHz. At higher frequencies, there is insufficient time in each cycle of wave motion to allow significant movement of fluid, giving a relatively unrelaxed state with less attenuation, as we shall see in the model of Chapman, 2003.

### 15.7.9 Shear-wave splitting under the Mid-Atlantic Ridge

An interesting combination of shear-wave splitting analysis from local earthquakes, and P-wave anisotropy measurement with controlled sources, was used by Barclay and Toomey, 2003, to interpret the anisotropy at a 35° N sector of the Mid-Atlantic Ridge, under an ocean depth mostly in excess of 2 km. The anisotropy was attributed to a shallow distribution of vertical, fluid-filled cracks, aligned parallel to the trend of the axial valley. (Figure 15.18a). The vertical cracks give rise to so-called hexagonal anisotropy, with a horizontal symmetry axis normal to the crack planes.

Most of the shear wave delay was attributed to the shallowest 500 m (seismic layer 2A: see Chapter 11). Here the shear-wave anisotropy was from 8 to 30% in this highly fissured layer. The authors considered that isolated fluid-filled cracks at depths from 500 m to 3 km were too tight to be detected by the P-wave portion of their survey, but may have contributed to the shear-wave delays. The authors' analyses were restricted to shear-waves arriving within the shear-wave window (0° to $\sin^{-1} V_p/V_s$). The average value of $V_p/V_s$ was given as 2.9 for seismic layer 2A (the shallowest 400 m).

The authors reported that the time delays ranged from 35 to 180 *ms*, with an average delay of 90 *ms*. This was reportedly similar to other studies with micro-earthquakes (e.g. 100–300 *ms* in Iceland, 100–230 *ms* in Hawaii, 10–125 *ms* at the San Andreas Fault). The axial valley floor was reportedly heavily fractured and

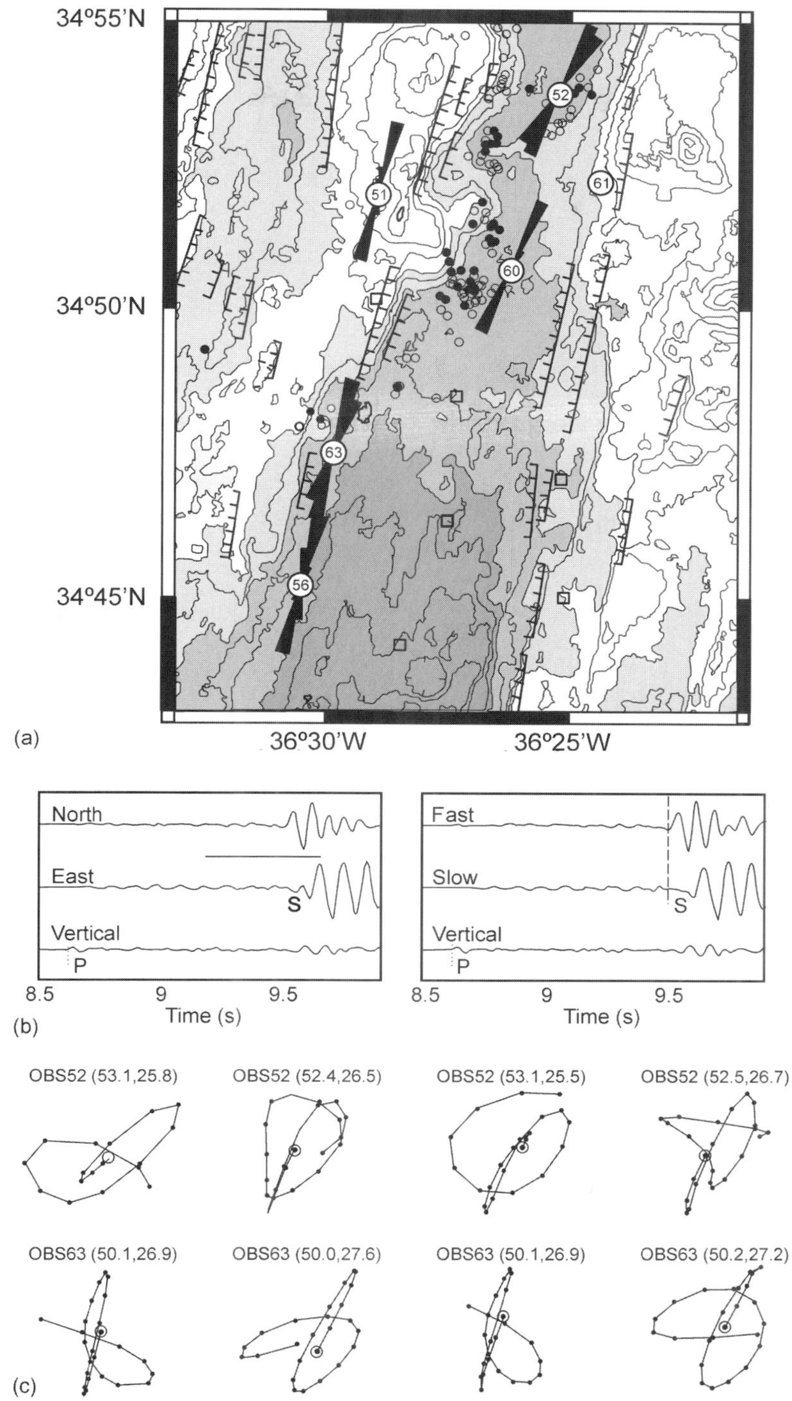


*Figure 15.18* (a) Bathymetric contours along a Mid-Atlantic Ridge axial valley at 35°N, showing rose diagrams of the fast polarizations. Each rose histogram has been normalized to the size of its most populated (azimuthal) bin. Solid circles denote the 65 earthquakes. Squares denote the six OBS instruments, which were spaced 4 km apart, covering an area of approximately 18 × 10 km. (1830 air gun shots were recorded during the first day, followed by micro-earthquake recording for 43 days.) Fault scarp (lineations) are also shown. (b) Examples of the three-component seismograms before and after rotation to the fast and slow directions. (c) Selected examples of shear-wave splitting in horizontal particle motions, with sampling points every 7.81 *ms*. The open circles are the origins. Each trace is 203 *ms* long. Selected from Barclay and Toomey, 2003.

fissured, and also had faults with strike in the assumed maximum stress direction.

The focal depths of the micro-earthquakes were limited to 3–4 km, and the authors emphasised that resolution of depth-dependent structure was not possible, but they suspected shallow structure, based on evidence from the shallow concentration of P-wave anisotropy, which reportedly decreased from >4% at 500 m depth to zero below 1.5 km depth. (P-wave anisotropy was defined as $100 \times (V_{p\ max} - V_{p\ min})/V_{p\ average}$, whereas, following Crampin, 1989, S-wave anisotropy was defined as $100 \times (V_{s\ max} - V_{s\ min})/V_{s\ max}$).

## 15.8 Recent cases of shear wave splitting in petroleum reservoirs

Detection of fracturing by long wave-length seismic waves, where vertical wells have given a limited sampling of the possible vertical or sub-vertical structure, is seemingly a constantly developing field. The economic pressure for discovery of new fields, perhaps more heterogeneous than their predecessors, and with less favourable matrix permeabilities, is focussing increased attention on fracture definition and understanding. It is also assuming growing importance as so many fields are in a mature phase and need production assistance, with fracturing a key to maximizing the final recovery. Optimal early production strategies are also pressing for seismic monitoring assistance.

In contrast to the above seismic studies that mostly relied on earthquakes as the natural source for their inversion to anisotropic properties, the hydrocarbon-bearing regions are today, by their very nature, stable areas of commercial fluid accumulation. It may be possible to use 'natural' sources such as AE only later in the life of a compacting reservoir, such as Ekofisk or Valhall, where for instance, shearing has been detected.

The type of dynamic source will depend on whether the reservoir is on land or off-shore. On land it may range from explosives, to arrays of heavy trucks with vibrators, which can provide repeated P-wave or S-wave sources. Offshore, sub-sea-surface air-guns or explosives provide the P-wave source, with conversion to PS-waves at the sea-floor interface, and subsequent processing as S-waves. A recent application of the suction-anchor principle for sea-floor generation of S-waves (SS for emphasis compared to converted PS) is currently being tested in Norway (Westerdahl, 2005 priv. comm.).

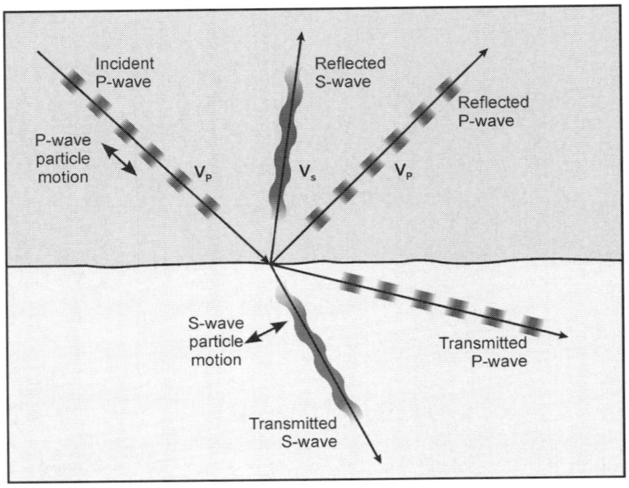


*Figure 15.19* Conversion of a P-wave source (from a sub-surface air-gun in water), to PS (converted) S-waves, showing the different reflected and transmitted angles. The converted waves reflect at sub-surface interfaces according to Snell's Law. An S-wave always reflects more vertically than a P-wave because of its lower propagation velocity. This asymmetry complicates the acquisition and processing of converted-wave surveys. Barkved et al., 2004.

### 15.8.1 Some examples of S-wave and PS-wave acquisition methods

Barkved et al., 2004 discussed the significantly different demands of recording and processing S-waves (which give a fully three-dimensional wave field), and the simpler recording, and processing of P-waves. The 'graphics' of conversion of P-waves generated by the typical sub-surface, or sub-sea explosive or air-gun source, to S-waves, was previously shown in Chapter 14 (Figure 14.27). In Figure 15.19, the different angles subtended at an interface by the converted, *reflected* S-wave, and by the converted, *transmitted* S-wave are emphasised. The particle motion perpendicular to the wave propagation direction for the case of the converted (PS) S-waves, means that they are constrained to the plane of reflection, and in the figure, would have particle motion in the plane of the page.

Three-component sensors (one vertical, two horizontal) are required to record the fully three-dimensional S-wave field, with one horizontal component aligned in the direction of wave propagation. When combined with a hydrophone that is sensitive to fluid pressure changes, the (marine) acquisition system is referred to as *four-component* (4C) technology.

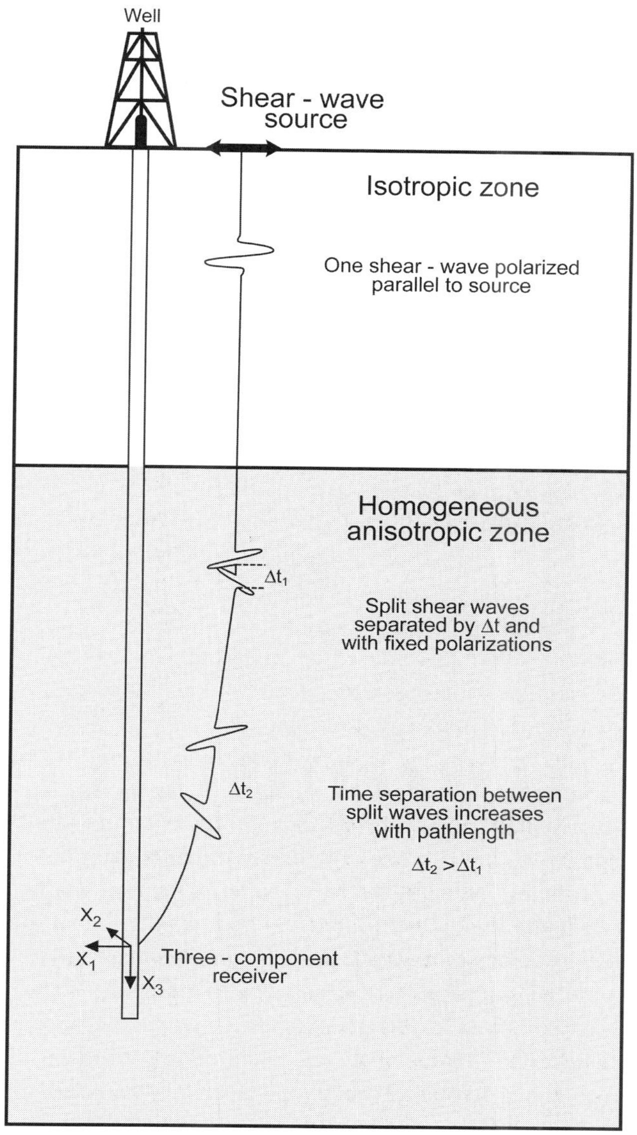


*Figure 15.20* Schematic of walk-away VSP, showing the shear wave splitting phenomenon as polarization occurs in an anisotropic zone, caused by sub-vertical fractures. It also indicates the increased time delay between the fast and slow $qS_1$ and $qS_2$ split shear waves, prior to their acquisition by three-dimensional or four-component sensors in the well. Slater, 1997.

On land the shear-wave source can be a direct S-wave (termed SS-wave, for distinction from converted PS). A useful representation of the subsequent shear-wave splitting principle, when for instance performing walk-away VSP with a shear-wave source, is illustrated in Figure 15.20, from Slater, 1997, whose Caucasan reservoir analyses will be reviewed later.

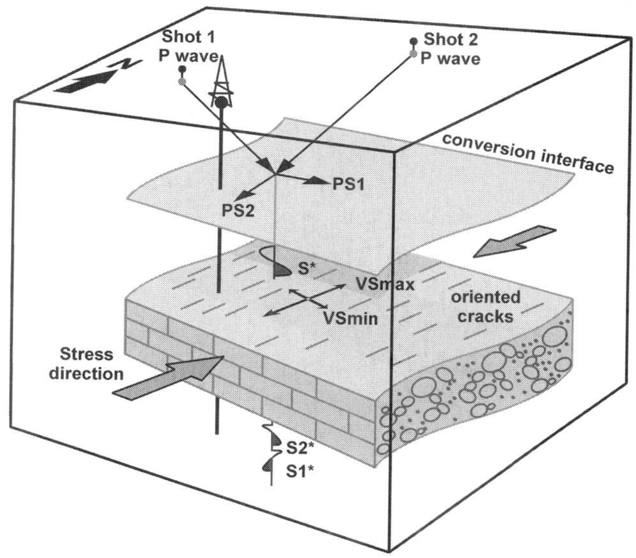


*Figure 15.21* Split shear waves S1 (or $qS_1$) and S2 (or $qS_2$ ) from polarization, due to stress aligned fracturing. The three-component wave fields were generated by far-offset (60° or 120°) P-waves. Stenin et al., 2002.

Variable azimuth VSP studies were described by Stenin et al., 2002, who analysed the converted PS-wave three-component wavefields, in order to detect and characterize the fractured intervals in the Archangelsk region of Western Russia. They used far-offset shot points, for generating P-waves, subsequently converted to PS waves as illustrated in Figure 15.21, as a supplement to conventional VSP with variable azimuth sources. The two P-wave sources with 120° or 60° separation are shown in the figure.

Converted components PS1 and PS2, and $qS_1$ and $qS_2$ polarized waves, caused by the assumed stress aligned fracturing are also shown. The velocity anisotropy of $V_{s\,max}$ and $V_{s\,min}$ caused by the oriented fracturing, and by higher compression in the interpreted stress-parallel direction, is also indicated. They reportedly confirmed their seismic anisotropy analyses with core data from offset wells that could sample the fracturing.

Horne, 2003, emphasised the additional advantages of *walk-around* VSP, (a circular path of multiple sources at fixed offset) shown in Figure 15.22a, for improved fracture definition. In this connection he pointed out that an incident P-wave, when converted to a PS wave that passes through a plane of (horizontal) symmetry, such as that caused by a vertical set of fractures, will

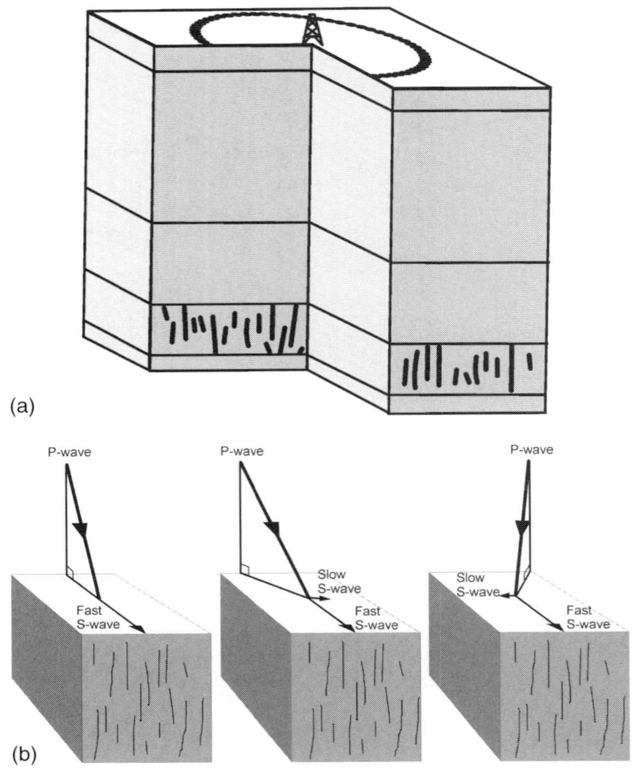


*Figure 15.22* (a) Walkaround VSP with (b) polarity reversal for converted PS waves as they pass at different angles in relation to strike, through a plane of symmetry. Horne, 2003.

show polarity reversal when incident on 'either side' of the fracture strike. This is illustrated in Figure 15.22b.

In all the acquisition systems, the overburden is obviously involved in wave transmission, but has tended in the past to be 'ignored', in the sense of assuming a vertical symmetry axis, with no aligned fracturing. Tjaaland et al., 2001 emphasised that the overburden may also display anisotropy in the form of azimuthal-dependent velocity. In traditional seismic processing the overburden was assumed to be isotropic and elastic.

Due to the actual anisotropy of the overburden in relation to the assumptions, errors will be introduced when the seismic data are inverted to obtain reservoir parameters. They conducted synthetic modelling with realistic overburden anisotropy to compare results with the isotropic overburden assumption, and showed that travel times, and P- and S-wave velocities were each affected. If attenuation in the overburden was ignored in the analyses, the S-wave velocity could show errors of 20% or more.

### 15.8.2 Classification of fractured reservoirs

3D seismic surveys, using compressional waves to generate shear-wave reflections (converted PS-waves), as above, have found a very important role in the identification and characterization of fractured reservoirs. With shear-wave splitting and polarization due to the presence of the (relatively) compliant fracture properties, the detection of basic structural information such as fracture density, strike and (due to symmetries) dip, can be estimated, and as we shall see, some indications of both fluid-type and permeability may also be obtained.

It is naturally believed that stress variations and orientations can also be determined, though here we have seen some variations and therefore suggested possible geomechanics reasons (shearing) for occasional diversion from this assumption. These will be detailed more in Chapter 16. Such deviations from the major stress direction would be caused by prior deformation along the more conducting, larger-scale features, or by today's (or yesterday's) water-flood treatments.

The shear-wave splitting *component* from two sets of fractures that lie on either side of the major stress direction is clearly another reason for *apparent* deviation from $\sigma_{H\ max}$, if (or because) one of the sets dominates. Presumably we should also be open to the possibility that if the fracture or joint set *stiffnesses* differ, then the effect of a basic *crack density* difference between the sets could be compromised, i.e. altered to a degree.

Nelson, 1985, 2001 suggested a fractured reservoir classification that considered the dual contribution of both the matrix and the fractures on the *porosity* and the *permeability*. The total porosity was composed of the relative amounts of matrix and fracture porosity, and the relative levels of permeability were caused by the matrix and the fracturing combined. This simple scheme is shown in Figure 15.23.

Type I reservoirs are heterogeneous and anisotropic, where fractures dominate in terms of both porosity and permeability. In Types II and III more reserves are stored in the matrix, but fractures *control* in Type II, and *assist* in Type III. In Type IV reservoirs, fractures still cause anisotropy and can even create barriers, and they are assumed to provide no additional porosity or permeability. Nelson 2001 suggested that unproductive wells are often a result of not recognising the significance of fractures, or joints, early in the development of a field. When wells were drilled based on the assumption of evenly distributed, matrix-controlled production (a

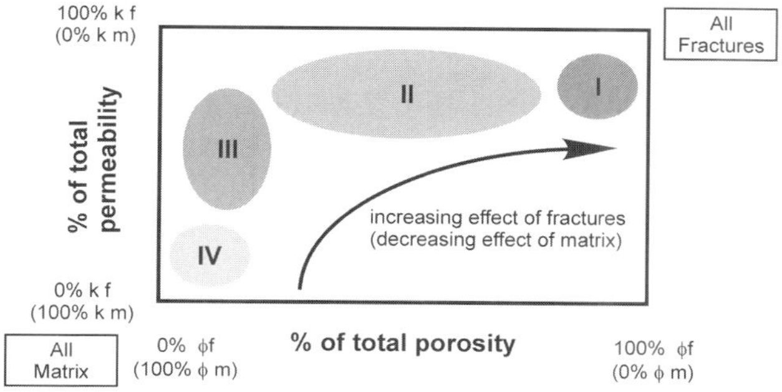


*Figure 15.23* Fractured reservoir classification of Nelson, 2001, based on the relative contributions of matrix and fracture porosity and permeability, to the overall permeability and productivity of the reservoir.

form of 'fracture denial'), it was presumably inevitable that the term 'bright-spot' would be the fore-runner to full-blow fracture appreciation.

Laubach et al., 2000 proposed a fractal-based solution to the remarkable fact, that despite the growing interest in fractured reservoirs, seemingly worldwide, the wells are always vertical and the target structures probably vertical too (see also Ch. 14). As a consequence, cores and well-logs often give little or no useful information about the fractures. (A fortunate exception is the conjugate fracturing found in anticlinal structures, where samples of both the oppositely-dipping joint sets can hardly be avoided.) The authors referred to the use of *sidewall cores* to help pin-point zones having high fracture intensity; not by reaching out to the poorly sampled fractures, but by using *microfracture* and *diagenesis* data to infer the presence of the macro-fractures.

Figure 15.24 shows a comparison of the very consistent microfracture strikes (nearly EW), compared with the broader but consistent orientations of *macrofractures* both from the same well and from the same formation combined as one data set. One may speculate both about the probable orientation of the maximum horizontal stress, and about the possibility that the 'symmetric' range of strikes for the macrofractures means they are conjugate joints and possibly sheared features, perhaps with extra good conductivity as a result (see Chapter 16).

Also shown in Figure 15.24 is the fractal (up-scaling similarity) trend of spacing and aperture. According to the authors: 'Microfracture proxies for large fractures are the surrogates that can provide complete, reliable, bed-by-bed evidence of fracture attributes.' The authors found that in many rocks there was a diagenesis event contemporaneous with the fracture development. The large aperture fractures in their experience, had mineral bridges and

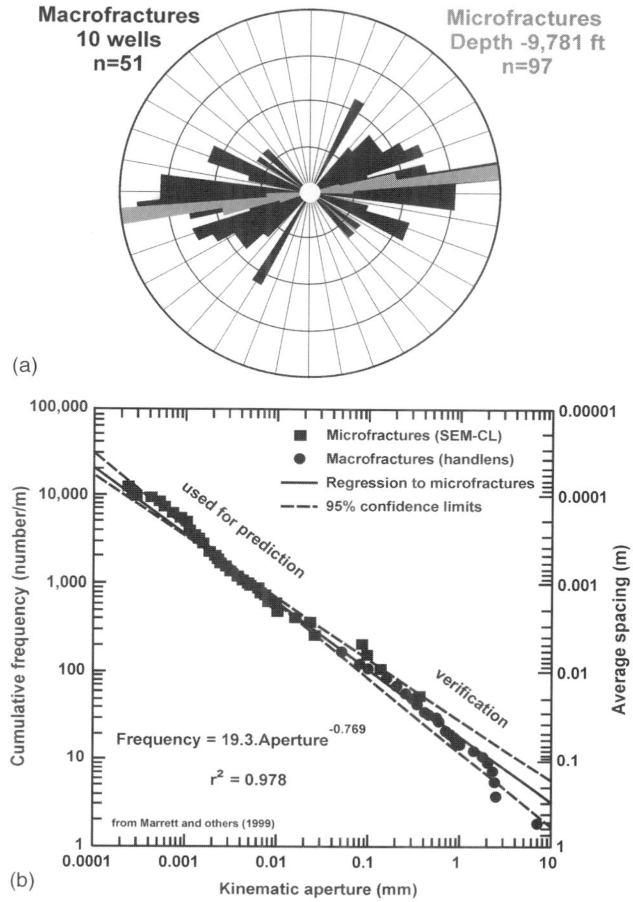


*Figure 15.24* (a) East Texas well data for microfracture orientations, with comparison to the (symmetric) spread of macrofractures from the same well and local formation. (b) Scaling patterns for aperture and spacing of microfractures through to macrofractures. Laubach et al., 2000.

abundant preserved porosity. This meant that the cements did not fill the large fractures, but those below a certain characteristic size were completely filled. These were the target for side-wall cores due to their abundance. Intermediate size fractures (the emergent threshold) were those filled-to-partly open, and by the nature of Figure 15.24b were less frequent. They termed the original apertures kinematic apertures. Cases subsequently sealed were suffering from clogging by post-kinematic cement.

Further observations on the subject of 'open' fracture *orientations* were also made by Laubach et al., 2002. They cited comparisons of *measured* stress directions and orientations of open, flow-controlling fractures. These showed that open fractures in the sub-surface were not necessarily parallel to maximum compressive stress

($\sigma_{H\ max}$). Fractures perpendicular to this direction could also be 'open' if partially filled with mineral *synkinematic*, or *post-kinematic* cements. They emphasised that *sealed* fractures parallel to $\sigma_{H\ max}$ were numerous. It was suggested, from experiences in both compressional and extensional provinces, with production data ranging from 2,400 to 6,400 m depth, that the divergence between $\sigma_{H\ max}$ and 'open' fractures demonstrably contributing to flow, was 'from a few degrees to 90 degrees'.

### 15.8.3 Crack density and shearing of conjugate sets at Ekofisk might enhance splitting

In earlier parts of this chapter concerning shear-wave splitting, the approximate rule-of-thumb relation between *crack density* and shear wave anisotropy was quoted from the paper 'Arguments for EDA', by Crampin, 1993a. Crampin addressed the meaning of the typical 1% to 5% differential shear-wave anisotropy, reportedly measured in a wide range of rock types. He considered that the 1% to 5% was also equivalent to the generally limited range of effective crack densities, which he assumed were usually in the range $0.01 \leqslant e \leqslant 0.05$. Crampin noted that the percentage of differential shear wave anisotropy was usually about $e \times 100$, for a $V_p/V_s$ ratio of about 1.7 (1.732 was quoted).

One may speculate whether these apparently lower *crack densities* implied by many earthquake studies, are a direct reflection of the sampling of 'average rock' (mostly imprinted on the shear-waves closer to the surface), and logically derived from low seismic Q and seismicity-prone provinces (see Chapter 10). In contrast, crack densities interpreted in fractured reservoirs from split shear-wave data, may represent a 'biased sample', i.e. may be caused by a rock mass that is more jointed or fractured than the norm at this depth.

In connection with *near-surface crack density* influences, it is of interest to note the common experience referred to by Crampin, 1993a for relatively large time delays between split shear-waves to be set up in the top tens to hundreds of meters. The variations that can occur near the surface have been termed *natural directivity* or ND.

To an engineering geologist, this phenomenon would probably imply increased occurrence of jointing, and to a rock mechanic with some geophysics interpretation added, the larger time delays would also be expected to be related to the reducing normal and shear stiffnesses of the near-surface jointing. Neither profession, it is suggested, would be thinking of changes in the frequency of microcracks in the top tens to hundreds of meters. A common experience is also that the near-surface joint orientations *do reflect* the major horizontal stress direction, and consistency with permeability tensor principal directions may also be expected, as demonstrated by 3D hydrotomography by Quadros et al., 1999.

A convenient example of much higher *crack densities* ($e = N.a^3/V$) than apparently detected in the earthquake studies of Crampin and others will be given here, to introduce the alternative viewpoint that *crack density* may be more appropriately applied to describe the *frequency of jointing or fractures* that typically compose the fluid-bearing, tangible, visible structure making up the fractured reservoir, as opposed to a focus on microcracks. Following Leary et al., 1990, that a given number of crack populations $N_i$ of radii $a_i$, will give a total crack density that is the sum of the densities $e_i$, we can consider the *crack density* for producing parts of the Ekofisk reservoir in the North Sea.

Phillips Petroleum geologist's core logging (H. Farrell, pers. comm. 1985), of *conjugate* steeply-dipping jointing in the porous, highly productive sections of the reservoir, indicated about 10 to 12 dominant (>1 m long) set no. 1 joints crossing a '1 m window', with oppositely dipping set no. 2 joints showing about 4 to 6 shorter joints (30–50 cm) in this same volume. In this very real case, with obviously the desirable fluid flowing towards the producing wells, we can estimate a much larger crack density than apparently suggested by earthquake studies (reportedly 0.015–0.045, Crampin, 1993a).

A *mid-range* (2D) representation of the above joint description could be estimated roughly as follows:

$$\sum e = e_1 + e_2 \approx \frac{11 \times 0.5^3 + 5 \times 0.2^3}{1.0} \approx 1.4$$

The rock mass depicted in 2D in both Figures 15.7 and 15.25, is far from 'intact', but it tolerated an original effective vertical stress of the order of $\sigma_v' \approx 62 - 48 = 14$ MPa, with a lower effective horizontal stress. During the first 20 years of production this effective vertical stress had built up to about 38 MPa, due to the 24 MPa pore pressure reduction prior to large-scale waterflooding (which was followed by 6 m jack-up of all platforms, and final relocation of central platforms). In a sense the 'fragmentation' worries of Crampin at high crack densities was seemingly being demonstrated, but in fact it was the matrix pore-space that was collapsing in the highest porosity chalks, supplemented by a shearing mechanism.

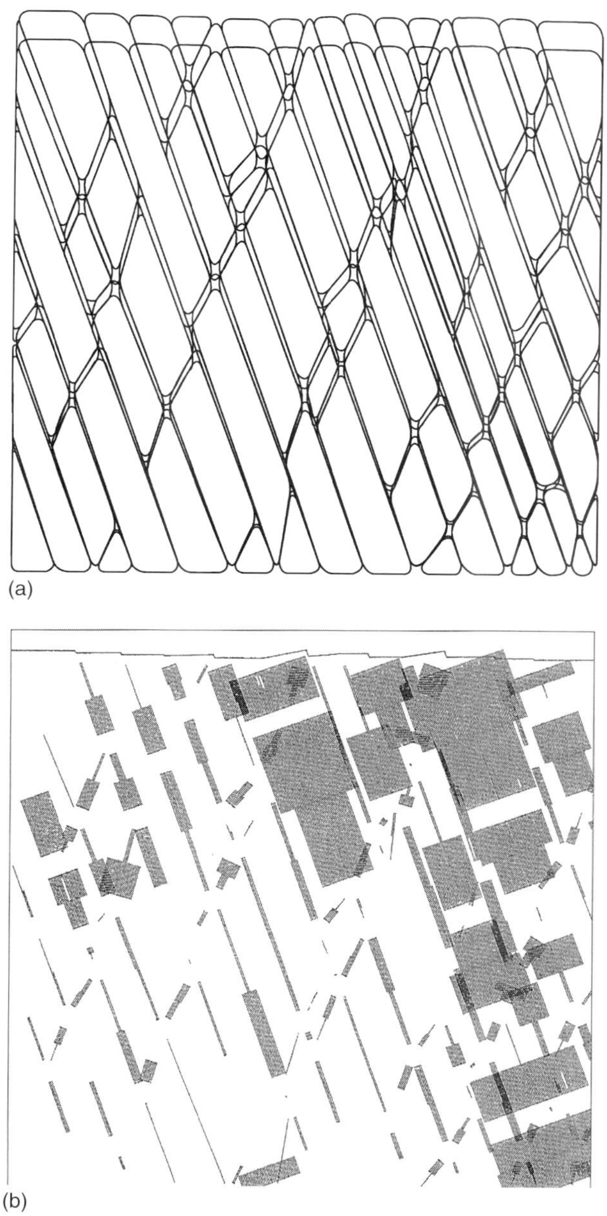



*Figure 15.25* (a) A UDEC-BB model of a compacting, conjugately-jointed 1 × 1 m 'element' of the Ekofisk chalk, which has a depicted *crack density* of the order of 1.4. The modelled compaction of this idealized high-porosity 'element' of the reservoir was 4.8%. (b) Joint shearing (shown proportional to line thickness) was modifying the compaction in relation to that of an unjointed porous matrix. M. Christianson UDEC-BB modelling: Barton et al., 1986, 1988.

When discretely modelled (as in Figure 15.25), the pore pressure reduction (in 1985) was limited to 20 MPa. A corresponding increase of the minimum horizontal effective stress, plus deformation effects, means that the highly jointed rock mass was well confined. It would be correctly described as *interlocked* rather than 'fragmented', and it is clearly an exceptionally good reservoir rock mass regarding fluid-flow, with planned production possibly to 2050, an 80-year reward to Phillips Petroleum, for not abandoning the North Sea, due to the promising result of their final drillhole, when exploratory drilling in the late 1960s.

Domal carbonate or chalk reservoirs of high porosity, with steeply dipping, as opposed to flat-lying jointing, can apparently be successful producers because of a remarkable joint shearing mechanism, despite the one-dimensional (vertical) strain constraint. Matrix shrinkage under the large increase of effective stress, 'makes space' for down-dip joint shearing. The latter helps to maintain joint aperture due to shear-induced dilation, and apparently may even provide a pseudo-confinement effect (increased $k_0$: ratio of horizontal to vertical stress), which would make the jointed reservoir somewhat stiffer (in a vertical direction), than the unjointed rock (Barton et al., 1985, 1986). Significantly, in view of the fact that our shearing theories were not at first believed, Albright et al., 1994 mention Ekofisk exhibiting *'shear fracture microseismicity, possibly indicating that subsidence is caused by a combination of pore collapse and shear sliding'*. They state that subsidence surpassed early model estimates based on pore collapse, indicating other mechanisms at work. By implication shearing was also occurring at fault scale. (See further discussion of this shearing mechanism at joint, or fracture scale, in Chapter 16.)

In this connection one should refer to possible doubt concerning the use of laboratory 1D-compaction data for modelling pore-collapse versus porosity. One should recall that the modelling of compaction applies to some *30 km³* of reservoir chalk, about *twelve orders of magnitude* larger than the laboratory samples. Some level of scale effect may be present in the matrix, as certainly found for the case of joint strength, though with strongly reduced scale effect at high effective stress levels. The *relative contributions* of the matrix and the joints (and faults) to the over-all compaction is therefore inevitably uncertain where such large volumes are involved. The operator's core-logging geologists reportedly detected slickensides on conjugate joint sets, when drilling new holes during the 1980s for pressure maintenance, using equilibrated sea-water injection. Slickensides had reportedly not been detected in earlier characterization of the Ekofisk field in the late 1960s, where production started in 1971.

The most porous chalk (n = 40%) that was first modelled by the NGI team in 1985, with a modulus of only

0.33 GPa, showed maximum shearing of 3.9 mm, with an average of approximately 0.39 mm for all the joints. Later modelling by Gutierrez, with a higher pore pressure reduction of 24 MPa, showed up to 10 mm maximum joint shear. Presumably such deforming features would be seismically visible in practice (i.e. a strong source of shear wave splitting), just as the reproduced direct shear tests on rough fractures (Barton, 1973) shown in Figure 15.26 are 'visible', due to the dilation that is caused.

Water flooding may have stimulated this shearing mechanism due to intuitive, preferential weakening at highly stressed joint-wall contacts, *prior* to water-softening of the less permeable matrix. (Barton, 2002b). However, the Ekofisk rock mass does not 'fragment' or suffer 'dispersion' of pore fluid because of high *crack density*, although the weakening effect of the water *is* tending to cause 'rubble-isation' of the matrix-plus-joints, but this is outside the scenario envisaged by Crampin, 1993a. The oil undoubtedly flows more easily towards the producing wells in response to the compaction drive and due to the originally well-developed *crack-density-determined* connectivity and resulting permeability. Loss of matrix (and joint) strength due to water-saturation, gives an additional production effect.

The authors Saenger and Shapiro, 2002 used an explicit finite difference scheme to model elastic waves in their variously 'cracked' models. Interestingly, considering the above independently estimated *crack density* $e = 1.4$ for the Ekofisk conjugate jointing, derived by the writer from a geologist's description of the jointing, is the fact that Saenger and Shapiro find a value of $e = 1.43$ to represent the *critical crack density.* They liken this level of crack density to a medium with '*only finite sized pieces of solid, and there is no continuum through which an elastic wave can propagate*'. To one from a different background (rock mechanics), this would almost seem like an unintended definition of the typical near-surface jointed rock masses through which we are frequently performing seismic refraction, core drilling, and driving tunnels, with frequent need for rock support.

From another chalk reservoir, in Texas, relative levels of shear-wave anisotropy detected by VSP above the Austin Chalk reportedly correlated with production (Crampin, 1993a), as one would certainly expect at Ekofisk. A tendency for near-surface layers to reflect this same anisotropy was also noted for the Austin Chalk. Dimming of the amplitudes of the slow shear-wave, (due to greater attenuation i.e. lower $Q_{seis}$), correlated with fractured zones, which were verified by horizontal drilling. The detection of the fractured zones was vital for good production, due to the low matrix permeability of the respective chalks, despite their high porosity.

Leary et al., 1990 referred to exceptional 2.44 km/s and 1.83 km/s fast and slow shear-wave velocities interpreted parallel and perpendicular to the inferred alignment of fractures, in an oil-bearing stratum in the same Austin chalk. The shear-wave velocity anisotropy reported by

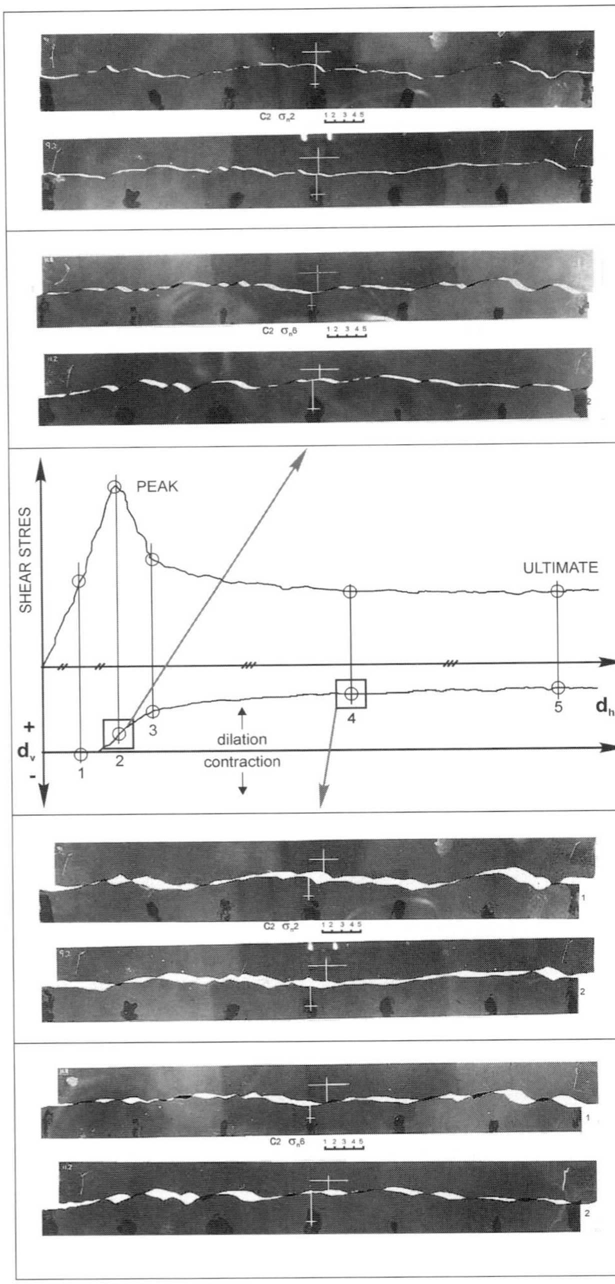


*Figure 15.26* Reconstruction of the measured shear-dilation path from direct shear tests of rough tension fractures. $\sigma_n$ 2 and $\sigma_n$ 6 depict stress level. Barton, 1973a.

Johnston, 1986 of about 30%, would have implied a *crack density* of the order of 0.3, based on the Crampin '100*e*' relation, which presumably requires modification if there are more than one joint or fracture set contributing to the *crack density*. The ratio $V_p/V_s$, and the relative levels of compliance of the contributing joints (not directly mentioned by Crampin due to microcrack focus), will also alter this '100x' relation.

If the crack density is contributed to by two sets of oppositely dipping conjugate joints, then following Sayers, 2002b, we can expect both shear *and* normal compliance contributions (from both sets), to the slowness of the slow shear-wave, and presumably therefore a lesser need for an extreme (interpreted) *crack density* to 'explain' a high value of shear-wave anisotropy.

### 15.8.4 Links between shear wave anisotropy and permeability

Horne et al., 1997, presented what they term 'a global optimization' to address the problem of ray-tracing, to invert the observations of shear-wave splitting from two 'near-offset' VSP data sets, in which two sources were employed along diametrically opposite azimuths, about the wellhead.

Their shear-wave splitting analyses were based on shallow depth VSP and logging data for the Conoco Borehole Test Facility (CBTF), where an 18-layers interpretation of velocities, densities and thicknesses was in use by researchers. This layered model was split into five anisotropic zones that corresponded to the observed discontinuities in the shear-wave splitting estimates, as obtained from the diametrically opposed VSP surveys. Figure 15.27 shows the basic elements of the velocity density model, and the five selected zones.

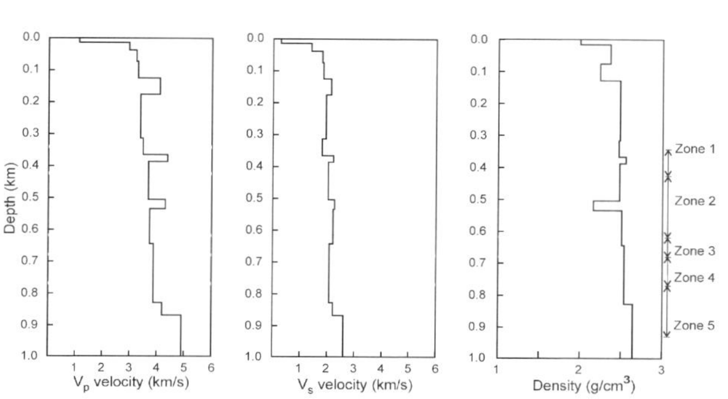


*Figure 15.27* The velocity and density model for the CBTF well. Five anisotropic zones were identified from the shear-wave splitting analysis. Horne et al., 1997.

Horne et al., 1997 described each anisotropic zone using three parameters: the crack density, the aspect ratio and the crack content, and a variety of advanced operations which we need not be concerned with in this simplified review. The authors' results supported earlier studies that also concluded that the observed anisotropy at the CBTF well was likely to be due to a sub-vertical fracture set dipping approximately 18° (to the SE). They showed that the fracture dip could be obtained from the shear-wave anisotropy when using opposite-azimuth VSP. The authors' representation of the velocity anisotropy structures of qP, $qS_1$ and $qS_2$ wave forms for the case of vertical cracks (or joints) and 20° dipping cracks (or joints) is shown in Figure 15.28.

In this last case reviewed we saw how the use of opposed-azimuth VSP could be used to interpret the *dip*

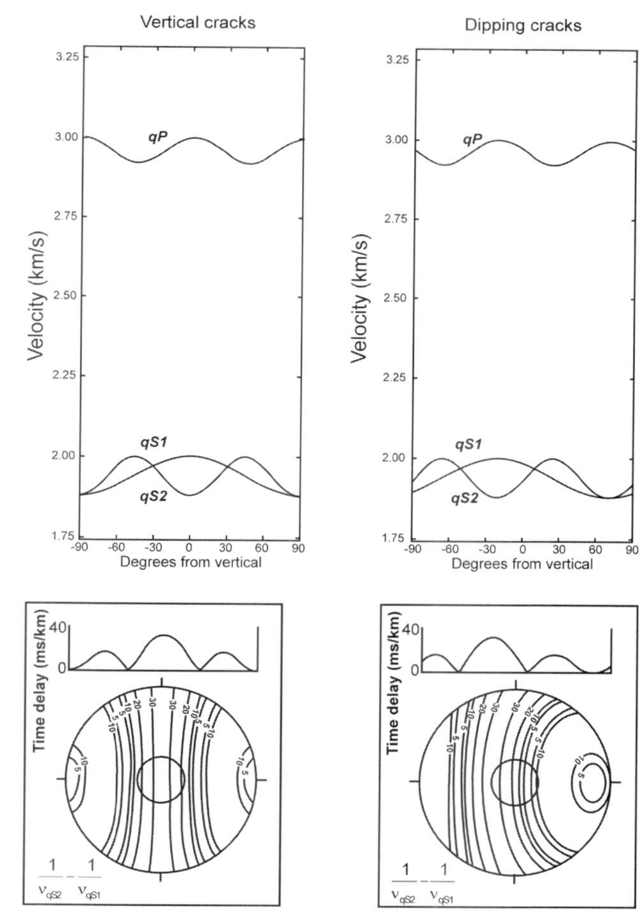


*Figure 15.28* A comparison of anisotropic velocity components, if vertical (left) or 20° dipping cracks (right). Lower plots show time delay variations over a hemisphere of propagation directions. Horne et al., 1997.

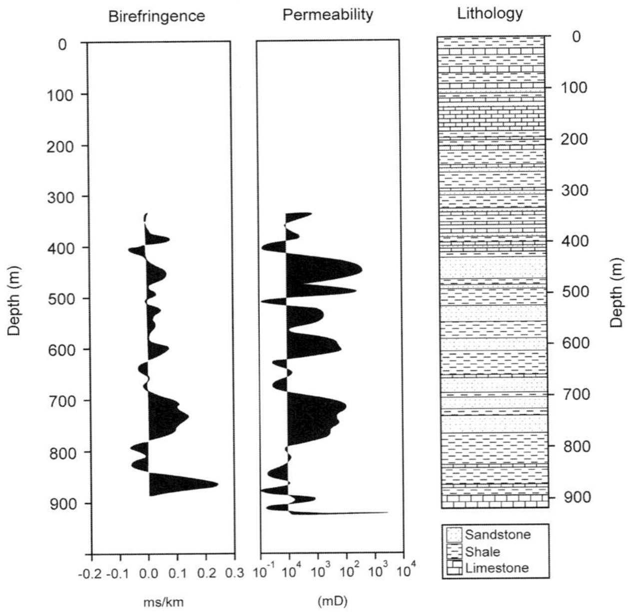


*Figure 15.29* Azimuthal connection is indicated between the degree of seismic anisotropy ('birefringence') and the permeability. Several anisotropic domains are indicated. (Horne and MacBeth, 1996, as reproduced in MacBeth and Li, 1999.)

of fracture sets or joint sets. The more usual application of shear wave splitting is to determine the azimuth.

Figure 15.29, again from the Conoco test site, shows the Horne and MacBeth, 1996, linkage between seismic anisotropy (quite often termed birefringence) and the recorded permeability. We will recall that there were five domains (or zones) of anisotropic wave-splitting at this site. MacBeth and Li, 1999 also emphasised that the relationship between the stress field and a fracture model may require investigation from case to case. It appears that the hydraulically connected fractures may be those most readily detected, which implies that the link between permeability and anisotropy is being detected also.

However, the generally assumed influence of a nearly parallel major horizontal stress, and therefore minimum normal stress must be assessed carefully, as we shall see in Chapter 16. It may follow present convention in shallow wells, while a (possibly conjugate) shearing mechanism may be needed at depth, unless rock strength, or partial mineralization 'bridging' are sufficient to keep fractures 'open', despite a significant effective normal stress.

For vertical joints or fractures, the most pronounced anisotropy is registered when the source-receiver propagation paths are parallel and perpendicular to the strike of the fractures. The response will be a complicated function of aspect ratio (which includes a normal stress component), fracture fill and/or porosity. A minimum of three source-receiver lines are capable of detecting fracture strike, if the particular set of fractures gives directional velocity variations in relation to the strength of other sets. The response is strongest when fractures are gas filled, as we shall also see later from a case record from Oman. In addition, a lower normal effective stress perpendicular to the major set of fractures will give a stronger velocity contrast. The gas to oil ratio in the reservoir is therefore an important component of the interpretation.

The authors pointed out that the new generation of vertical cables, seabed seismic sensors and walk-away (and/or 3D) vertical seismic profiling was leading to high-resolution anisotropy estimation, specifically in the offshore environment. The common goal was to detect this commercially important azimuthal anisotropy. If the alignment of fracture sets could be deduced, this would assist in the optimal lay-out of producing wells, and perhaps subsequent injection wells for water-drive.

Azimuthal anisotropy detection techniques can help resolve both the large fault-scale structures, and the expected inter-block scale fracture (or joint) set structures. The advantage of this ability to interpret structure more fully, at several scales, can be envisaged in the complex stratigraphy shown in Figure 15.30.

MacBeth and Li expressed the opinion that azimuthal anisotropy determination at fracture set scale, could fill the gaps between the fracture characterization from core, from borehole logs, from outcrop analogues, and that inferred from 3D seismic above complexly faulted and fractured reservoirs. The conceptual scale-dependency of fault throw magnitudes, after Yielding et al., 1992 and MacBeth and Li, 1999, is sketched in Figure 15.31.

Interesting from a rock mechanics viewpoint was the suggestion by MacBeth and Li that 'sub-vertical wave propagation through vertical fractures could provide direct knowledge of the conditions influencing tangential movement across the fracture faces'. They suggested that this could lead to discrimination between the solid or liquid content of the fractures, evaluation of porosity, and perhaps geometric aspects of the surface topography. One may speculate if the numerically modelled joint shearing shown earlier for an element of the Ekofisk jointing (Figure 15.25) could be seismically detected, likewise the rock-to-rock (R) and fluid-bearing open (O) parts of shearing joints depicted in Figure 15.2b.

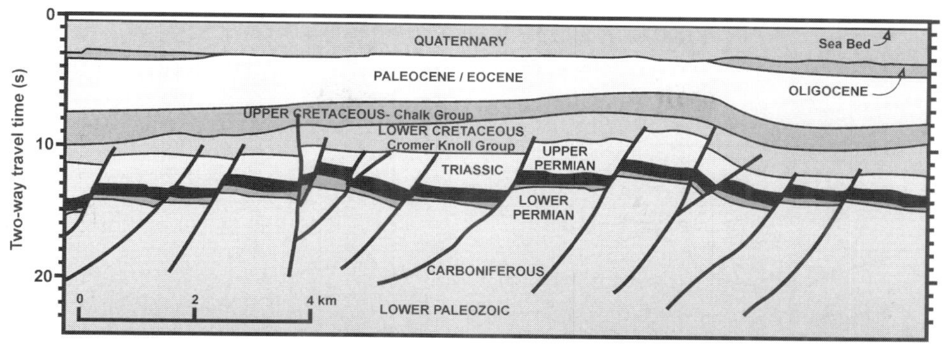


*Figure 15.30* A seismic interpretation that shows complex stratigraphic and structural relationships (London-Brabant Massif, North Sea). Seismic azimuthal anisotropy analysis offers potential for resolving major fracture set azimuths within the faulted blocks. MacBeth and Li, 1999.

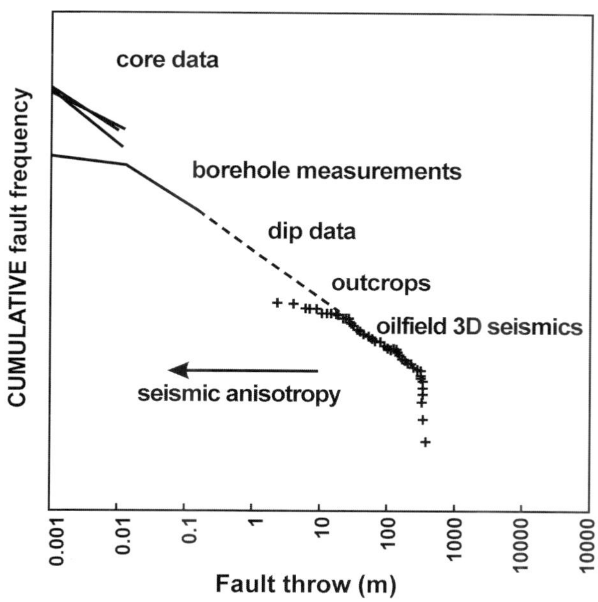


*Figure 15.31* MacBeth and Li, 1999 suggested that seismic anisotropy analysis could fill a gap between the fracture characterization from core and that from borehole logs and outcrop analogues, and that inferred from 3D seismic. From Yielding et al., 1992.

### 15.8.5 Polarization-stress alignment from shallow shear-wave splitting

Liu et al., 1993 had previously addressed shear-wave sources from the *shallow* (50 m deep) multi-azimuth, reversed VSP at the Conoco Borehole Test Facility, where earlier analyses had demonstrated that fracture parameters could be estimated from cross-hole, reversed VSP, and borehole data, as we saw in Chapter 14.

The authors found that the shear wave data could be interpreted as showing both velocity anisotropy (azimuthal variation in travel times), and attenuation anisotropy (azimuthal variation of amplitudes). The common central azimuth N70°E, which was also the strike of the dominant, steeply dipping macro- and micro-fractures, happened also to be the direction of maximum horizontal stress at shallow depth in the area. (Figure 15.32a, b and c). At this shallow near-surface location there was clearly excellent stress-fracture alignment, as expected.

The nearby surface exposures suggested two approximately perpendicular fracture sets at the CBTF. Liu et al., 1993 managed to infer the multiple fracture sets using shear-wave polarization, with sufficient azimuthal coverage, in this case 160° of data. Their equal area polar projections show two maxima in approximately orthogonal directions, which almost exactly parallel the strikes of the two fracture sets. Figure 15.32a shows the outcrop jointing and Figure 15.32b the velocity structure. The target for these analyses was a fractured limestone from 16 to 29 m depth, sealed above and below by impermeable shales. Its velocity of about 4 km/s does indeed suggest well fractured rock (but a small velocity-depth gradient across this layer is perhaps likely).

Reference to Figure 15.33, which reproduces key rock mass quality $Q_c$-$V_p$-depth charts, suggest a rock quality in the range $2 < Q_c < 4$, when a porosity adjustment of presumably some few per cent is made. This typical jointed rock quality, could be 'generated' by a logical combination of e.g. RQD $\approx$ 45–90, $J_n \approx 9$ (3 sets of joints; two sub-vertical, plus bedding), $J_r \approx 2$ (smooth but undulating joints), $J_a \approx 2$ (slight weathering), $J_w \approx 0.66$ (wet, some water flow), SRF $\approx 2.5$ (near

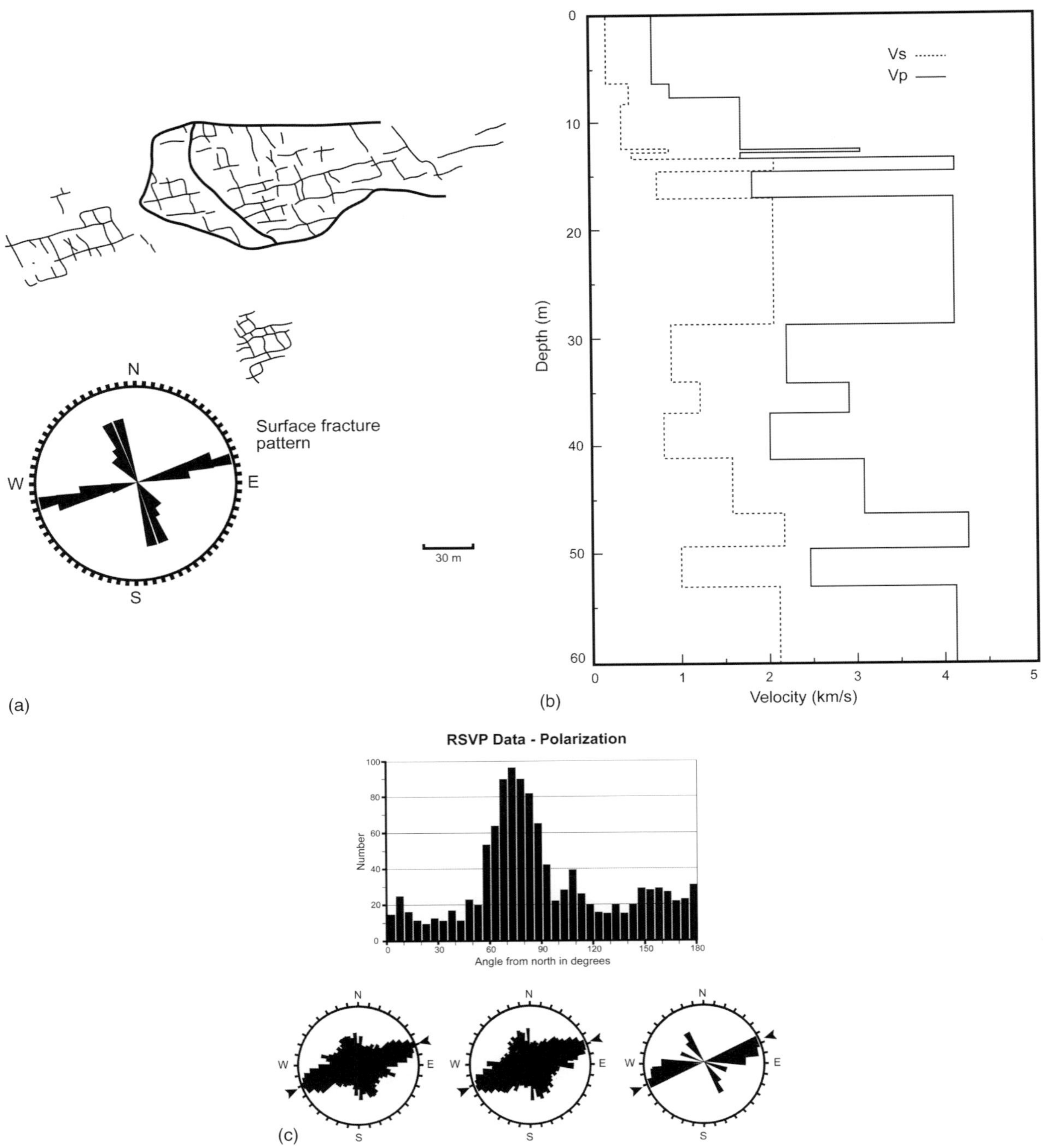


*Figure 15.32* (a) Outcrop jointing, (b) assumed velocity structure, (c) polarization of shear-wave arrivals. Arrowheads show actual dominant fracture strike. Liu et al., 1993.

surface, some looseness in relation to greater depth). (See Appendix A for full Q-system rating tables.)

$$Q \approx \frac{45-90}{9} \times \frac{2}{2} \times \frac{0.66}{2.5} \approx 1.3 - 2.6$$

If the uniaxial strength of the limestone was e.g. 100 MPa, then $Q_c$ ($=Q \times \sigma_c/100$) would be $\approx 1$ to 3, which is mutually consistent with the measured velocity of about 4 km/s.

A simple approach is to check along the '4 km/s diagonal' in the simplified near-surface (nominal 25 m

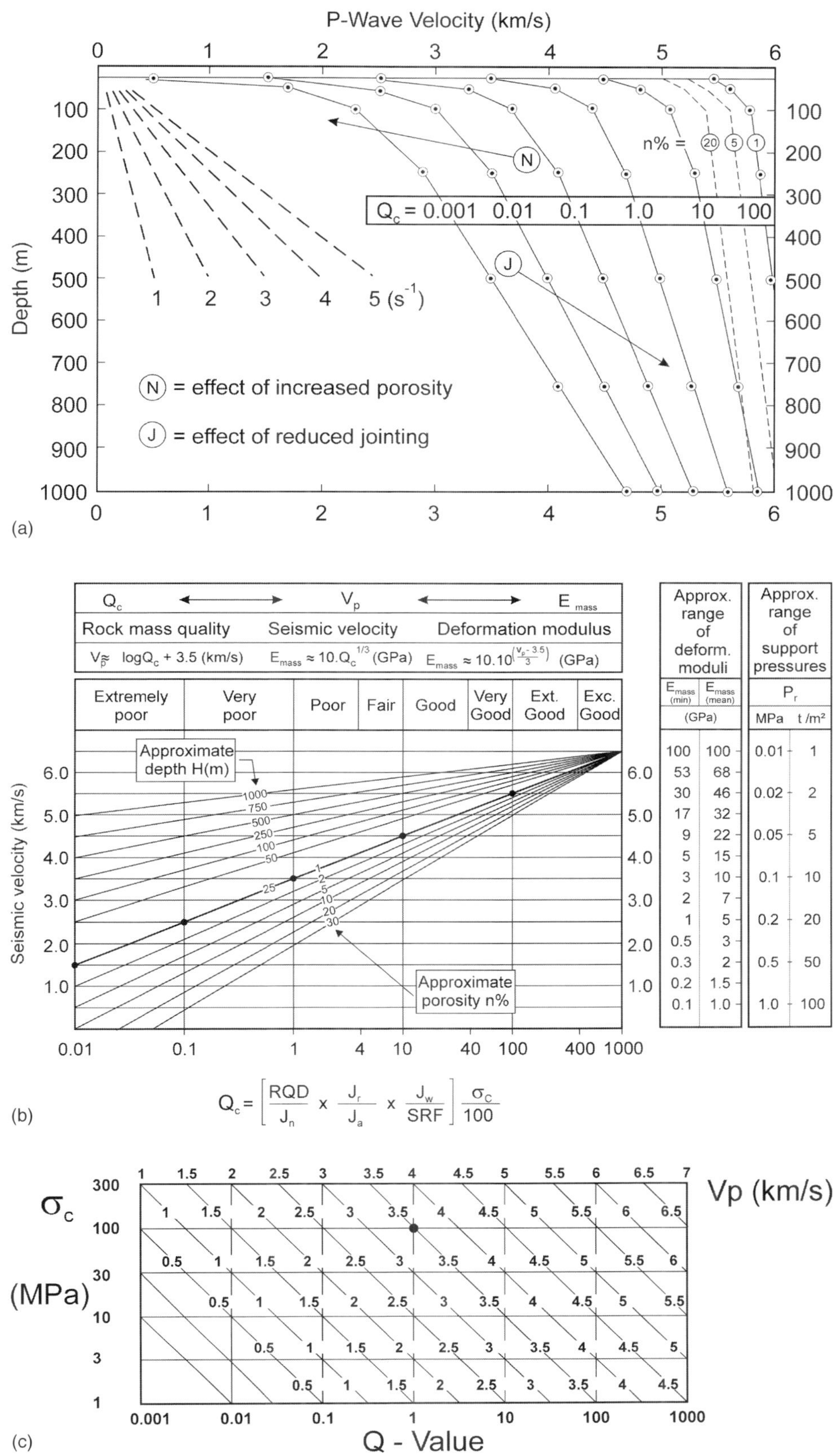


*Figure 15.33* (a) Rock mass quality $Q_c$-$V_p$-depth gradients, (b) $Q_c$-$V_p$-depth-porosity-modulus, (c) simplified Q-$V_p$-$\sigma_c$ chart (nominal 25 m depth). Based on Barton, 1995 and Barton, 2002a. Note: support pressure in b) refers to tunnels.

depth) model shown in Figure 15.33c. Note that in the area of $Q \approx 2$ to 3, the implied uniaxial compressive strength is around 100 MPa, which appears realistic for a limestone. If in reality the rock was closer to say chalk, then higher matrix porosity, lower uniaxial strength, and slightly less jointing than assumed above (giving a higher rock quality Q-value, could still be invoked to derive a suitable P-wave velocity of about 4 km/s.

Great azimuthal consistency is shown by this Liu et al., 1993 data set, due perhaps to the shallow depth, which allows the 'open' joints or fractures to indeed follow the *conventional* concept with likely parallelism with the $\sigma_{H\ max}$ direction. As pointed out earlier, and analysed in more detail in Chapter 16, such parallelism may be compromised at reservoir depths in less competent rock than limestone, due to effective normal stress induced joint 'closure'. A (conjugate) shearing mechanism might then be required for sufficient 'openness' of the conducting joints. Alternatively, suitable amounts of mineral filling for 'bridging' might be needed, neither too much (= sealing), nor too little (= stress-closure).

A shear wave polarization *component* (from e.g. two similar conjugate joint sets) that was still roughly consistent with the $\sigma_{h\ max}$ direction, might be assumed from *much deeper* sets of perpendicular break-out analysis, from caliper logs. But here there would be the possibility of a lot of 'noise' in the data, possibly from joint-induced break-out *at other angles*, more consistent with the possible conjugate, partly sheared (minor-faulted) jointing. In addition to this possible 'anomaly' in relation to conventional thinking, there is also the possible 'rotation effect' caused by the joint shearing per se, giving different orientation for the fluid-bearing (O) and stress-bearing (R) parts of a non-planar, shearing joint (Figure 15.2b).

The above aspects will be addressed in more detail in Chapter 16, where important supporting case records will be found for the 'shearing anomaly'. It is possible that parallelism of one *conducting* set of joints to $\sigma_{h\ max}$ is *in fact the anomaly*, when no longer close to the surface, in view of the stress-induced joint closure phenomenon, unless the reservoir rock is rather strong, like limestone, or a hard sandstone but not chalk, or poorly cemented sands, which would have sufficient matrix flow to make shear wave splitting a less relevant mechanism.

### 15.8.6 Shear-wave splitting in argillaceous rocks

In a doctorate study at the University of Edinburgh, Slater, 1997, also investigated if there was a relationship between seismic anisotropy and productivity. He used VSP walk-away data that was shot at two North Caucasus oil fields. Due to the usual low permeability of clay rocks, hydrocarbon reservoirs in argillaceous rocks are not commercially viable unless they are strongly fractured. Prior to the commercial use of shear-wave splitting, primarily since the early 1990s, argillaceous reservoirs were reportedly found almost accidentally, while drilling towards prospects in other rock units.

Slater, 1997 investigated the azimuthal anisotropy to check if the strong *azimuthal variation in productivity* was caused by variations in fracture intensity. In one of the reservoirs, walk-away VSP at two of the wells indicated velocities increasing as the direction of propagation moved away from the vertical direction. The reservoirs were in layered clays.

In Figure 15.20, introducing the topic of shear wave splitting, a schematic of the walk-away VSP was shown, from Slater 1997. Shear wave splitting and polarization was occurring in the near-surface anisotropic zone, caused by sub-vertical fractures. Also indicated is the increased time delay between the fast and slow $qS_1$ and $qS_2$ split shear waves. Later in this chapter, we will see many more examples of these phenomena, and also see models capable of simulating the squirt-flow causes of *dispersion* or frequency dependence at different fracture-size scales.

Slater, 1997, examined the typical transverse isotropy of the lower strata in these clay reservoirs, which display isotropy about a vertical axis of symmetry as shown in Figure 15.34. The VSP analyses demonstrated strong increases in velocity when the direction of propagation moved away from the vertical direction.

The very low velocities calculated from VSP at wells 85 and 87 at one of the Caucasan oil fields are shown in Figure 15.35. The layered sandstones, limestones and clays, and the Maicop Clay, *do not follow* the rock quality $Q_c$-depth-$V_p$ trends shown elsewhere in this book, except in the top few hundred metres where azimuthal anisotropy indicates jointing that would tend to show closure with depth as in the $Q_c$-$V_p$-depth model of Figure 15.33. Perhaps below this anisotropic zone, the change to clay, and the reported high pore pressures cause under-compaction, despite depth increase. A lack of fit with the writer's Q-$V_p$-depth-soft-porosity model would then be logical, unless a correction for over-pressure was made, and an 'apparent depth' estimated.

An important finding from careful analysis of much earlier multi-offset VSP in the Paris sedimentary basin, indicated non-parallel shear wave polarizations at different azimuths and offsets. Polarization by sub-parallel cracks or fractures was not sufficient to explain these

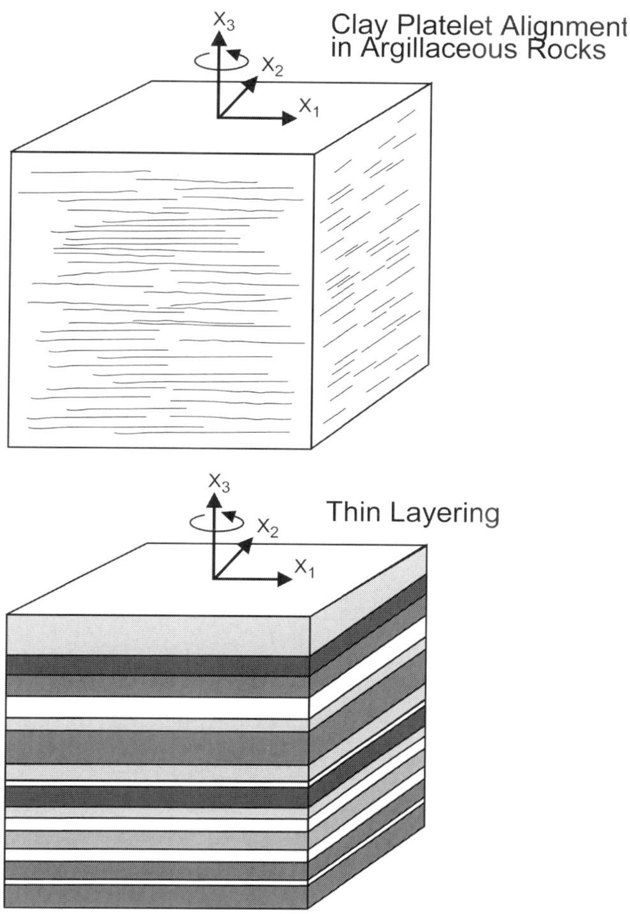


*Figure 15.34* Examples of transversely isotropic layering in argillaceous/clay reservoirs, where VSP analyses demonstrated strong increases in velocity when the direction of propagation moved *away from the vertical* direction. See Slater et al., 1993 and Slater, 1997, for details of the 'anisotropic cuspidal phases' discovered in these studies.

data, and Bush, 1990, demonstrated that the anomalous behaviour could be the result of a combination of matrix anisotropy due to layering (termed 'azimuthal isotropy', with vertical axis of symmetry, as in Figure 15.34), and crack anisotropy. A paper by Bush and Crampin, 1991, showed the consistency of this combination of mechanisms at five of the offsets.

### 15.8.7 Time-lapse application of shear-wave splitting over reservoirs

Reportedly the first time-lapse survey designed to measure possible *changes in shear-wave splitting* above a *petroleum reservoir*, was reported by Roche et al., 1997. The Vacuum field is a fractured dolomite reservoir in New Mexico, and the survey was a repeated 3C (converted wave) survey, i.e. 3C × 3C, or 9C survey. The argument was made that because the dolomite was quite hard it should be a poor candidate for conventional time-lapse studies, (i.e. its joints probably had high JCS or joint wall compression strengths, and perhaps high joint roughness JRC: see Chapter 16 for the influence of JCS and JRC on normal stiffnesses, the approximate inverse of dynamic compliance). The joint parameters used for *pseudo-static* estimation of both normal and shear stiffness (i.e. JRC and JCS), are relevant here due to the pseudo-static nature of the planned $CO_2$ flooding, and its effect on the increased pore pressure and therefore reduced effective stresses.

Before the $CO_2$ injection, the distribution of the shear-wave splitting *anisotropy parameter* ($\gamma$) varied between +/−8%, but averaged about 4%. Note: ($\gamma$) as given by Thomsen, 1986 is defined from the elastic stiffness matrix as $(c_{44} - c_{55})/2c_{55}$, where $qS_1 = \sqrt{(c_{44}/\rho)}$, and $qS_2 = \sqrt{(c_{55}/\rho)}$. After the injection of $CO_2$, the pore pressure at the injector well increased from 10.6 to 17.0 MPa, giving a maximum $\Delta P_p$ of 6.4 MPa. This was sufficient to give a P-wave velocity *decrease* of only 4%. The shear-wave anisotropy parameter ($\gamma$) on the other hand changed by up to 14% (from +4% to −10%). These changes were successfully modelled by Angerer et al., 2001. The polarized fast and slow shear-wave velocities did not rotate during these time-lapse changes, they apparently exchanged places.

According to a review of Winterstein et al., 2001, including 23 wells in six locations in NW USA, and not specifically related to fractured reservoirs, *shear-wave splitting anisotropy* as defined above, was commonly in the range $0 \geqslant \gamma \geqslant 21\%$. The general trend of the anisotropy data suggested a certain consistency in one layer or domain, with an abrupt transition to different values. One could argue that such could be caused by sedimentary beds of different geological ages with differently oriented joint sets, and also by local stress rotation near fault zones.

Some time-lapse surveys using P-waves, with its relatively more simply processed information, were described in Chapter 14. At the Ekofisk reservoir, the gas-cloud was seen in Figure 14.29 to obscure P-waves from a large portion of the central, and most porous part of the reservoir. In the context of the newer S-wave technology, Barkved et al., 2004 referred to *the world's first time-lapse, marine, multicomponent (3D/4C) survey*, as that performed in September 2002 at the Ekofisk jointed-chalk reservoir

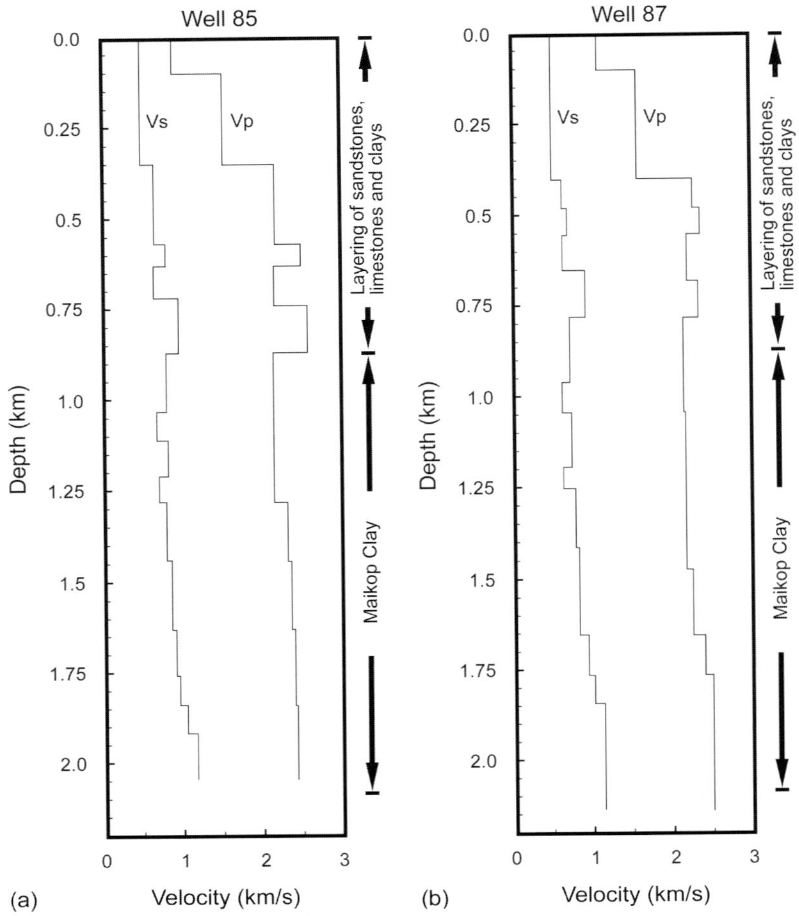


*Figure 15.35* Very low shear wave and P-wave velocities at two wells, as interpreted from VSP. The geological description of the 0.75 km of overburden is also seen. There appears to be over-pressure in the Maikop Clays. Slater, 1997. Alternatively, the reduced sensitivity to effective stress may be due to low pore compressibility for the clays. Holt et al. 2005.

in the North Sea. This baseline was subsequently compared with a monitoring survey acquired in December 2003. In each case, seabed cables were used to acquire data with a wide range of azimuths. (Van Dok et al., 2004).

With 'only' about $3 \times 10^8\,m^3$ of oil out of a total of about $1.1 \times 10^9\,m^3$ produced by 2003, and production expected to 2050, it is clear that the 'belated' application of shear-wave technology still has an important role to play. From each survey, converted PS-waves were analysed to determine the principal directions of fast and slow shear-waves. At every sensor location, recorded traces were binned (collected) into 10° azimuth sectors, and then stacked, giving 36 traces of each component at every receiver location. Even this limited time-lapse of 15 months indicated some small *changes in the direction* of the fast shear wave, and in the difference between fast and slow shear velocities. The differences were not consistent across the field.

According to Barkved et al., 2004 (with 'Ekofisk-author' Van Dok as one of the co-authors), the reasons for the small changes detected by the S-waves 'had yet to be understood'. The small-scale joint-shearing mechanism identified in distinct element (UDEC-BB) studies for the Norwegian Petroleum Directorate (Barton et al., 1985, 1986), that was discussed earlier, was later 'confirmed' by slickensided conjugate joint faces, in core recovered from subsequent wells (post 1985) for water-flooding and production.

According to Phillips Petroleum Co. geologists this slickensiding had not previously been noted, and nor was it noted in the older jointed cores made available to NGI for laboratory direct shear and coupled shear-flow-temperature (CSFT) tests. The shear mechanism may seem surprising in view of the 1D-strain ('roller-boundaries') boundary condition, since a 9 × 14 km reservoir of 300 m thickness can hardly expand laterally during compaction: this occurs more in the stretching, subsiding, overburden as 'seen' by shallow shear-wave splitting at Valhall, to be reviewed shortly.

A *possible* explanation for the small changes of polarization direction and of shear-wave anisotropy at Ekofisk, can perhaps be found in the conjugate (or single) shear mechanism that was illustrated in Figure 15.2b. This mechanism was also invoked earlier in this chapter, as a possible explanation of a larger polarization rotation at the Cornwall hot dry rock geothermal project. With potential 'opposite-rotation' of fluid lenses and rock-to-rock contact areas (Barton, 2005), there could be subtle domination of effects from the primary relative to the secondary conjugate joint set, and if the 'O-R' mechanism can be detected by shear waves, *and if the strike of the two conjugate sets is not equally oriented*, then a small rotation could be explained.

Variation about the field, with the 'radially' trending jointing and rotating principal stress (Figure 14.31) would easily explain variation of such trends. Others might quote EDA-(micro)-cracks and stress rotation as the possible cause. More subtle mechanisms may be at work, and additional complications in the neighbourhood of fault zones are almost inevitable.

At the Valhall Field, quite close to Ekofisk, BP installed a *permanent* seabed cable array, covering 45 $km^2$ area, to monitor changes using regularly repeated 3D multicomponent seismic surveys, to help determine the best reservoir drainage strategy. The plans for this installation were alluded to in Chapter 14. As is well known, and as will also be indirectly demonstrated in Chapter 16 using Barton-Bandis joint closure-permeability modelling, it is all too easy to produce too fast thereby prejudicing the permeability of the rock joints and fractures, close to producing wells, where pore pressure reduction may cause 'too high' effective normal stress on the producing fractures. A slower production helps to maintain the vital permeable routes through the reservoir, especially where matrix porosity is superior to its permeability.

In such modelling, the joint-roughness-dependent conversion between mechanical aperture (following pore pressure-induced effective stress increase) and the conducting aperture, needs to be differentiated. Fortunately, the loss of mechanical aperture (E) occurs more rapidly than the loss of the smaller hydraulic aperture (e), as shown by Barton et al., 1985, Barton and Quadros, 1997. (See Chapter 16 for review of $E \geqslant e$ data sets).

A very interesting application of shear-wave splitting and polarization was described by Olofsson and Kommedal, 2002. They referred first to the significant

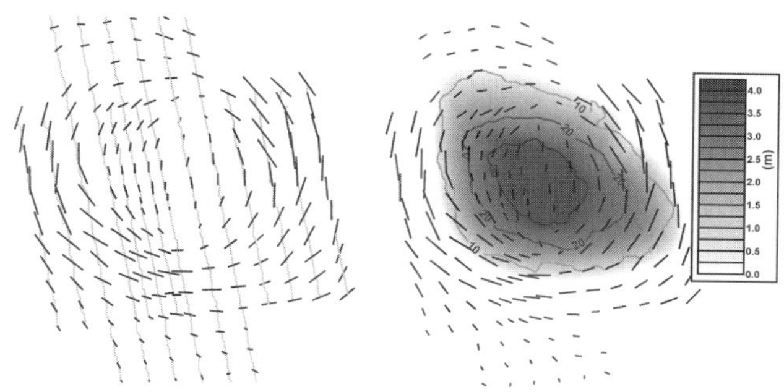


*Figure 15.36* Shear-wave splitting and polarization results for the shallow overburden above the compacting Valhall reservoir. Lines show the $qS_1$ direction, with their *length* corresponding to the $qS_1 - qS2$ time delay or 'lag'. The 'rotation' presumably corresponds to the relative 'visibility' of sub-vertical (bedding-limited?) joints caused by 'stretch' in all directions. Olofsson and Kommedal, 2002, also Gaiser and Van Dok, 2003 and Barkved et al., 2004.

time delay of the slow shear wave in relation to the fast wave in the reservoir, where both the polarization directions, and the time delay correlated with the geological model of fracturing. In this particular paper they presented the first results of shear-wave splitting in the *shallow* overburden, indicating a remarkable, and very convincing match to the assumed 'stretch' of sub-vertical jointing caused by subsidence. (Mention of these mechanisms was made in Chapter 14, for the case of Ekofisk overburden velocity reduction). Figure 15.36 shows the result of their shallow overburden shear-wave polarization, with lines showing the $qS_1$ direction, with their *length* corresponding to the $qS_2$ time delay or 'lag'.

Barkved et al., 2004 also commented on the above near-surface Valhall result, and stated the following: 'The actual mechanism causing the shallow shear-wave splitting is not known. Azimuthal anisotropy is usually associated with fracturing, stress or lithology. In this case the amount of anisotropy is small at the centre of the field, where the subsidence is largest, but the anisotropy is large on the flanks and small again farther from the centre. This strongly points to shear-wave splitting being sensitive to changes in stress or strain.' By modelling changes of *triaxial* stress in (continuum) layers above similar compacting reservoirs, Herwanger and Horne, 2005 produced similar, but quite circular models of shear-wave polarization.

The referred authors have not apparently focussed on *intra-bed* jointing as the likely source of the partial 'squareness' of some of the strongest anisotropy (i.e. the

'NNE-trending' and longest lines). The depth giving this possible dominant 'imprint' to the polarization and velocity anisotropy is of course not known. Large-scale (axisymmetric, 10 km radius) distinct element modelling of the *Ekofisk* overburden response to modelled compaction, using numerous coarsely 'bedded-jointed-and-faulted (2D) UDEC models (Barton et al., 1985, 1986), showed distinct 'joint' opening and some 'bedding' shear in the overburden, due to the stretch caused by the subsidence. These effects worsened with increased compaction profiles, the subsidence/compaction S/C ratio exceeding 0.85 as compaction approached 10 m.

A 3D version of such discontinuum modelling (with the 3DEC code, also developed by Cundall), would obviously have shown similar reactions from other perhaps perpendicular 'joint' sets. This 'joint' opening occurred most strongly where bending was strongest, and *least* both centrally and further out beyond the flanks. It is therefore suspected that the 'stress or strain' referred to by Barkved et al., 2004 could rather be termed intra-bed *joint-opening effects,* since the strength and location of these phenomena are likely to match the subsidence-bowl shape at Valhall. Where the polarization is 'diagonal', (i.e. 'NE-SW' or 'NW-SE' relative to the 'N-S' page), presumably the components to polarization (i.e. joint compliance) from both sets could be operating. (See also polarization at 300 m depth, in Barkved and Kristiansen, 2005.)

### 15.8.8 Temporal shear-wave splitting using AE from the Valhall cap-rock

An unusual petroleum reservoir case record concerning temporal variation of attenuation, was described by Carter and Kendall, 2005. This concerned the utilisation of micro-seismic events generated in the siltstone (above-shale?) caprock, above Norway's Valhall jointed chalk reservoir, in rock at about 2 km depth. Subsidence, observable at the sea-floor (see previous review), caused this AE activity, which was recorded 300 to 500 m away, by a vertical string of six 3-component geophones, placed in an abandoned well near the crest of the anticlinal structure.

Over a period of 56 days, continuous recording gave 572 events, 324 of which were located reliably (Dyer et al., 1999). Carter and Kendall performed shear-wave splitting analysis, comparing relative frequency content of the fast ($qS_1$) and slow ($qS_2$) shear waves. Their

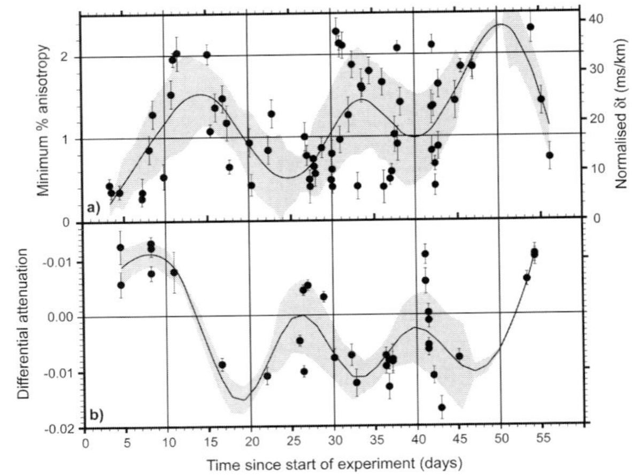


*Figure 15.37* (a) Temporal variation in *minimum* % anisotropy and b) temporal variation in differential attenuation, from shear wave splitting analysis using AE recorded in siltstone caprock, above the compacting Valhall jointed chalk reservoir. Carter and Kendall, 2005.

results were surprising in two ways: 1) that anisotropy appeared to be temporal, 2) that sometimes $qS_2$ was richer in higher frequencies than $qS_1$.

These two unusual results are illustrated in Figures 15.37 and 15.38. The authors defined a *differential attenuation,* as the difference in energy loss per cycle experienced by $qS_1$ and $qS_2$. This approximates the $1/Q_{seis}$ that a homogeneous constant $Q_{seis}$ material would require, in order to produce the observed difference in frequency content between $qS_1$ and $qS_2$.

One may speculate that opening of sub-vertical, bed-limited jointing, with changing joint-wall contact character, could be responsible for such variation over time. According to Barkved and Kristiansen, 2005, sea floor subsidence at Valhall exceeds 5.4 m, approximating 0.25 m/year. The larger neighbouring Ekofisk field has suffered significantly larger subsidence, at an early rate of about 0.45 m per year when detected in the mid 1980s, and about 10 m of total compaction with large numbers of sometimes repeated casing damage, by the turn of the century. Since the shear-wave phenomena may be relying on vertical jointing, the question would be whether bed-slip could be affecting the successive opening and closing of bed-limited jointing.

In the case of the Wilmington field, under Long Beach, California, a significant seismic event occurred, with a 20 cm presumed bed-slip, at one stage during more than 10 m of subsidence. Dussault, 2001 described the sand/shale interfaces above the Ekofisk reservoir as those most prone to causing casing damage, with episodic

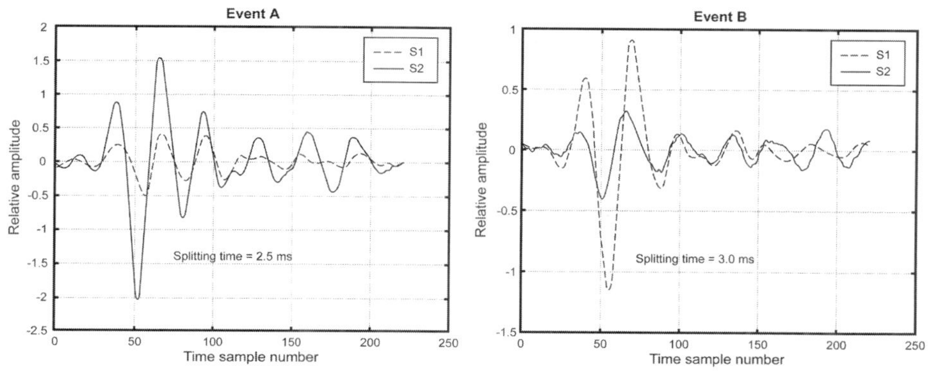


*Figure 15.38* Shear wave splitting analysed from AE events from siltstones in the Valhall field caprock. *Event A* exhibits the expected maximum attenuation of the *slow* wave S2. *Event B* exhibits maximum attenuation of the *fast* wave S1. Carter and Kendall, 2005.

microseismic stick-slip events close to the top of the reservoir. Perhaps these ongoing bed-slip events were influencing the steep or vertical bed-limited jointing, giving stimulus to $qS_1$ to $qS_2$ reversals. The phenomenon might possibly be related to the '90°-flips' discussed earlier in this chapter, which were suspected by the writer as due to principal effective stress 'flips'.

### 15.8.9 Shear-wave splitting and fluid identification at the Natih field

A series of interesting papers were published concerning a large scale 3D shear-wave experiment performed over the 1 km deep fractured carbonate (chalky limestones) Natih field in Oman. The structure of the field was described as 'not very complex' by Potters et al.,1999. The matrix permeability was a very low 1 to 30 mD. Hard shales overlie the 300 m of carbonates, and these Fiqa shales exhibited some interesting anomalies as we shall see. The fractures in the carbonates were *nearly* vertical (a detail that seems to be important if both shear *and* normal compliances are to be involved, following Sayers 2002b, reviewed earlier). However the authors of the two reports reviewed did not emphasise this aspect.

Outcrop mapping included fractures too large to be generally detected by core or FMS analysis, meaning that vertical wells were the usual poor samplers of typical sub-vertical structure. The outcrop, of necessity 50 km distant, had joint character as shown in Figure 15.39a (scale not given by Van der Kolk et al., 2001), but possibly the same as the '1 m' scale given for the reservoir in Figure 15.39b. The two fracture or joint sets had either NE-SW strike (the dominant direction), or NW-SE strike. Note the consistent dip signs on both drawings, despite the 'vertical or sub-vertical' assumption. These dips, even if minimal, seem to be important for using shear waves to distinguish between gas and oil in the fractures, following Sayers, 2002b.

Extensional and shear fractures were noted in each of the principal strike directions. The dominant NE-SW extensional set were continuous over hundreds of meters, and were responsible for the dominant permeability direction, as established by tracer tests. Along the crest of the shell-shaped 6 × 10 km *antiform*, which terminates at a major fault zone, the bed-curvature being increased, there was evidence of the NW-SE striking extensional set also participating in the fluid-flow network.

The scope of the nine-component three-dimensional (9C3D) survey/experiment was impressive, with 10,800 3C geophones, up to 1000 vibrator positions per day, and 22 million traces recorded on 2,000 tapes during the 32 days, and 28 km$^2$ of field work. The survey basic grid size was only 25 × 25 m. Shear-wave anisotropy exceeding 15% was registered over about half of the field. The *average* well flow rate in areas of large time delay was higher than that in areas with low S-wave anisotropy, but significantly the local fracture swarms giving individual wells high productivity were too small to be detected seismically. Potters et al., 1999 suggested that the *absence* of strong seismic anisotropy did not however preclude the presence of fractured zones.

Recalling the earlier critique of fracture density in this chapter, it is encouraging to note the authors' reference to the '*well-known ambiguity that a given amount of shear anisotropy can be caused by an infinite number of combinations of fracture densities and sizes. Since different combinations have different rheological (and flow) properties, this*

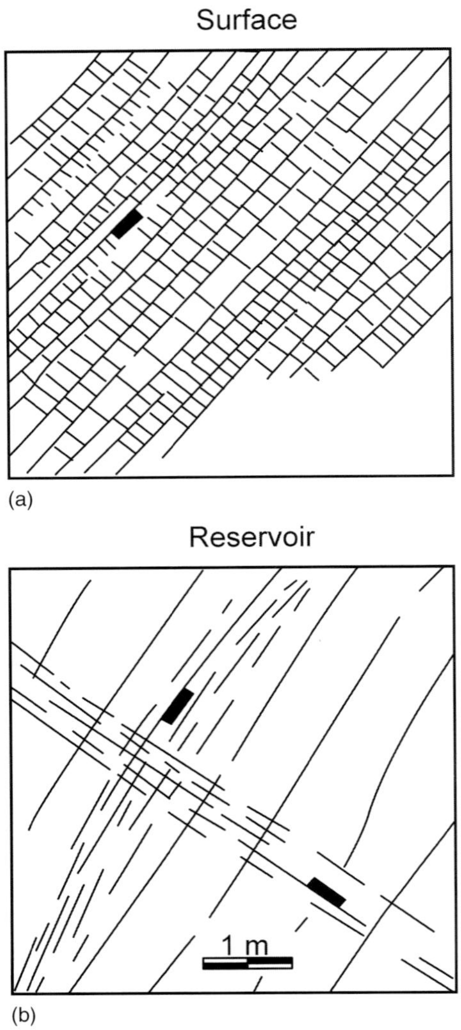


*Figure 15.39* A partly contrasting, and partly consistent information from the surface exposure jointing (50 km distant), and from the 1 km deep reservoir jointing at the Natih chalky-limestone reservoir in Oman. This was the site of a large shear-wave experiment in 1991. Van der Kolk et al., 2001.

*problem needs to be solved possibly by incorporating a theoretical or empirical fracture size distribution*'. The latter sounds like a good solution, since the ubiquitous microcracks seemingly favoured by Crampin, do not seem to have utility in the face of the need (and existence) of the permeable fracture networks of interest especially to oil companies.

The authors also addressed another uncertainty. '*Flow is strongly influenced by fracture aperture, a property which is, at best, only weakly expressed in the reservoir's mechanical properties*'. Here we may interject a possible future geomechanics-based solution, since the 'mechanical properties' referred to are in the case of shear-wave anisotropy, a function of the dynamic shear stiffness (or compliance) for the case of vertical incident waves, and a function of both the shear and normal dynamic stiffnesses for the case of non-vertical waves (or non-vertical fractures), following e.g. Sayers, 2002b.

As we shall indicate in Chapter 16, there are direct links between the rock mechanics of pseudo-static rock joint shear and normal stiffness behaviour, and their stress-aperture behaviour, both hydraulic and mechanical, with $e \leqslant E$. This is due to their common prediction by the Barton-Bandis constitutive laws, using joint characterization parameters JRC and JCS for wall roughness and wall strength respectively (Barton and Choubey, 1977, Bandis et al., 1981, 1983).

In Chapter 16 we will indicate the common mismatch between the dynamic and static normal stiffnesses (roughly reflecting the differences between dynamic and static moduli), which are therefore roughly predictable. The dynamic shear stiffness (or compliance) sensed by the slow shear-wave in the case of *vertical* incidence and *vertical* fractures obviously also carries information of relevance to the stress-closure-aperture-permeability behaviour of the dominant joints. By estimating the roughness parameter JRC from core or well-bore images, the (effective, confined) joint wall strength JCS could be estimated, which then allows estimation of the mechanical aperture at the given effective stress levels. Conversion to an estimate of hydraulic aperture is the final stage.

Van der Kolk et al., 2001, concluded the series of articles by Shell and their collaborators in Oman concerning the Natih 3D shear-wave experiment, by presenting, reportedly for the first time, evidence that *shear waves were sensitive to fluid type* (gas or oil), in fractured or jointed media. Regions of gas were characterized by *slow* shear waves, and this had particular consequences for two phenomena described by the authors.

Firstly, the shear-wave splitting map of the Natih reservoir exhibited much larger splitting values (i.e. anisotropy) over the gas cap on the reservoir. This increase in anisotropy was due to the *decrease in the slow shear-wave*, which senses both the fractures and the fracture filling fluid. A second phenomenon was the shear-wave data from directly above the reservoir. The thick Fiqa shale also exhibited a low shear-wave velocity anomaly, but associated with a gas chimney.

Van der Kolk et al., 2001 used a semi-dynamic *effective medium* model, to help to explain some of the

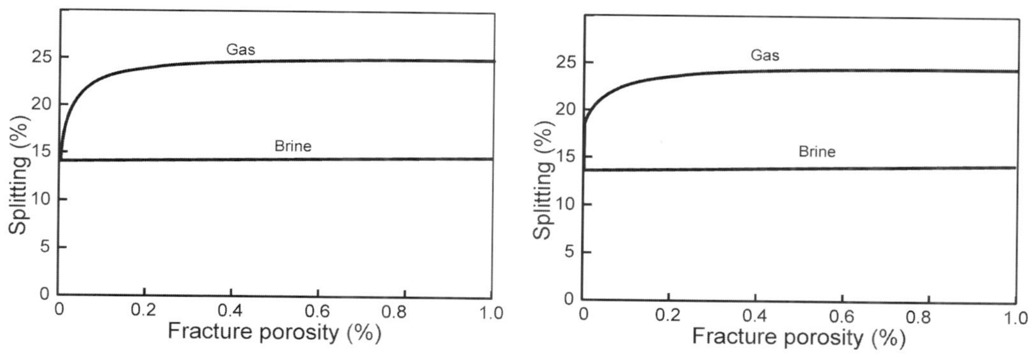


*Figure 15.40* An effective medium *BOSK* model prediction of the shear-wave splitting magnitude for a set of 1 m long and 10 m long fractures, as a function of fracture porosity, i.e. directly related to aperture. Van der Kolk et al., 2001. Note splitting % = $(V_{s1} - V_{s2})$ $V_{s1} \times 100$.

observed phenomena at the Natih field. This model was based on a combination of the work of Budianski and O'Connell from the 1970s concerning the elastic moduli and visco-elastic properties of cracked fluid-saturated solids, together with extensions for arbitrary crack orientation statistics from Sayers and Kachanov in the 1990s. They termed the combined model the *BOSK theory.* With this model, 'squirt' losses were also accounted for between the pores and fractures. This model correctly predicted that the slow shear-waves, polarized perpendicular to the fractures, were more influenced by the fluid type occupying the fractures, than were the compressional-waves. (Classic Gassmann, 1951 theory anticipates a sharp *reduction* in the P-wave velocity in the presence of gas in porous media, and supposes that the S-wave velocity is relatively unaffected by the type of fluid).

Figure 15.40 shows a *BOSK* model prediction of the degree of shear-wave splitting as a function of *fracture porosity*, for a fixed *fracture density* of 0.2 (where $e = N a^3/V$). To avoid the ambiguity of *e*, discussed at length earlier in this chapter in connection with EDA, the authors specified reservoir-like fracture lengths of a) 1 m and b) 10 m in these two *BOSK* realizations. Note that there are subtle differences only where *fracture porosity* is extremely small. The fracture *porosity* at Natih was about 0.1%. It is noted that the *BOSK* model, as for many other effective medium models, (to be briefly reviewed soon), apparently makes no distinction between the mechanical (E = actual) assumed crack aperture and the hydraulic aperture (e). In jointed rock (E) actually controls stiffness and deformation moduli. In rock mechanics, we do not often use the crack aspect ratio. On the other hand, (e) controls the intrinsic permeability, given as $(e)^2/12$, and both apertures (e and E) probably influence attenuation, since they effectively define two different aspect ratios, therefore influencing the assumed squirt-flow losses and the assumed stiffness.

The e $\leqslant$ E inequality* is therefore a potential source of error for the case of effective medium modelling of rough-walled, tightly compressed (i.e. deeply buried) cracks, joints or fractures, when using only *one aspect ratio.*

The crest of the Natih reservoir structure, with its higher (50% or more) shear-wave splitting had higher local fracture densities than the flanks, but this was not a sufficient reason for this higher value. Using the *BOSK* theory, van der Kolk et al., 2001 were able to investigate the effect of fluid viscosity (causing dispersive, frequency-dependent behaviour). Figure 15.41a shows the modelled effect of *gas replacing brine* on the *vertical* P- and S-wave velocities, as a function of fracture density.

The modelling results indicated that with a fracture porosity of 0.025%, the shear wave splitting was about 50% higher in gas-filled, compared to brine-filled fractures. With lower fracture densities than 0.15, the 'classic

* The two joint apertures will frequently differ by a factor of about 2 to 5, most for higher roughness JRC, and highest normal stress level. This inequality was demonstrated at the Technical University of Trondheim. Heimli, 1972, used a pre-instrumented *intact core*, subsequently split axially, so that different experimentally set values of E = 0.05, 0.1, 0.2 mm could be known with certainty. These were subsequently compared to the smaller hydraulic apertures (e) back-calculated from flow tests. The e $\leqslant$ E inequality has since been confirmed many times, and is explained by wall roughness effects, eventually quantified by JRC. See Figures 16.6 and 16.7 in the next chapter. Data sets for (e) and (E) and an empirical JRC-based model are shown. (Barton, 1972, Barton et al., 1985, Barton and Quadros, 1997).

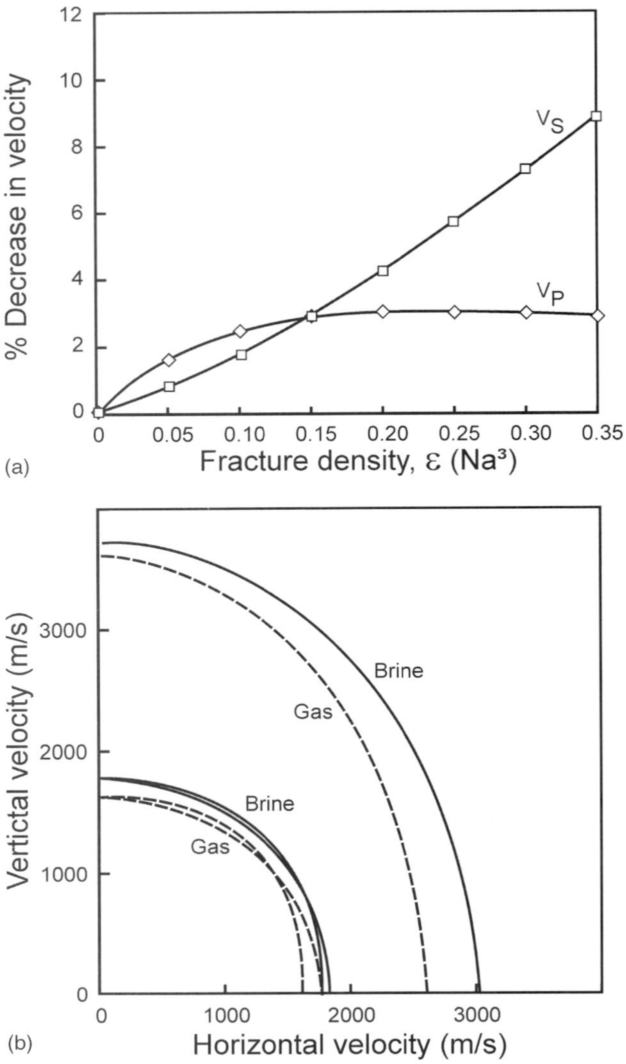


Figure 15.41 (a) A *BOSK* effective medium prediction of the effect on *vertical* P- and S-wave velocities, of gas replacing brine. (Note Vs refers here to $V_{s2}$, the slow shear wave polarized perpendicular to the fractures.) Two sets of vertical fractures, in an isotropic, low porosity matrix were modelled, with a wide range of fracture densities. (b) P-wave and polarized S-wave surfaces for gas and brine filled fractures, showing the modelled influence of wave direction relative to two sets of vertical fractures. The fractures are oriented 45° to either side of these figure axes. Van der Kolk et al., 2001.

behaviour' of P-waves more affected by fluid type than S-waves was seen. When P-waves *crossed* the fractures, rather than paralleling them, the predicted sensitivity to fluid type was found to be stronger in P-waves than S-waves, agreeing with the traditional behaviour.

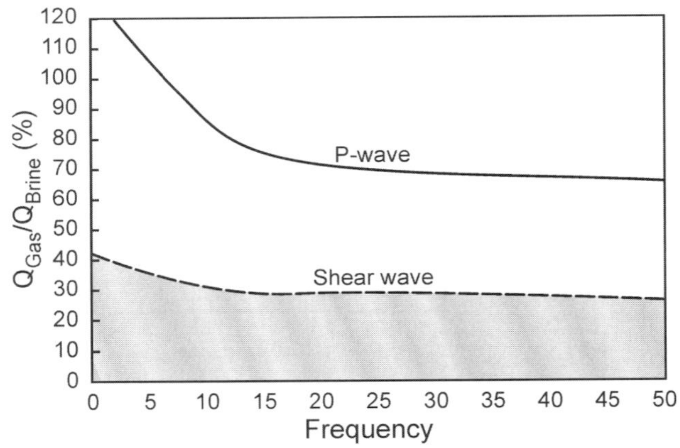


Figure 15.42 The ratio of seismic $Q_{gas}/Q_{brine}$ in modelled fractured media, using the *BOSK* model. Shear waves were much more attenuated than compressional waves by the introduction of gas. Van der Kolk et al., 2001.

The authors concluded that *vertically* moving P-waves, and the fast shear wave, were hardly affected by whether it was gas or brine in the fractures. It was the slow S-wave that was polarized perpendicular to the fractures that registered the fluid type, i.e. whether of low or high compressibility.

The *BOSK* model was also used by van der Kolk et al., 2001 for studying the effect of frequency, and the relative effects on attenuation of gas or brine. The ratio of seismic Q for gas compared to the seismic Q for brine ($Q_{gas}/Q_{brine}$) indicated greater attenuation of the S-waves with the introduction of brine. This result is shown in Figure 15.42.

The authors made some important and undoubtedly correct conclusions, which are nevertheless contrary to standard exploration practices, and will therefore be quoted in full: '*Both the observations and the* (BOSK) *modelling suggest that the S2 shear-wave propagation depends on the fluid type in the fractures. For propagation parallel to and polarization perpendicular to the plane of the fractures, the observed effect is exactly opposite to what is predicted by Gassmann fluid substitution in a porous matrix. We can conclude therefore that Gassmann theory is not sufficient to model fluid replacement in (heavily) fractured media. (Matrix porosity effects must still be included in the calculations, of course). Many standard exploration practices are based on Gassmann substitution, e.g. direct hydrocarbon indicators (DHIs) such as bright spots, flat spots and sometimes AVO effects. The results obtained above suggest that these techniques may be invalid in fractured media and new extensions should be explored*'.

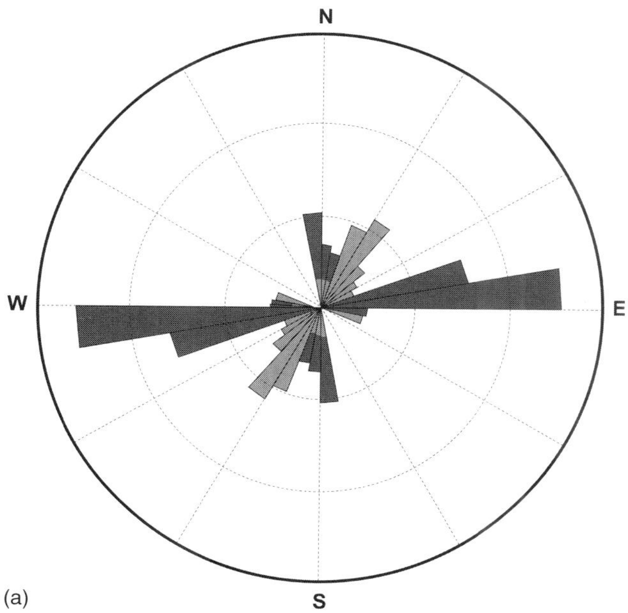

(a)

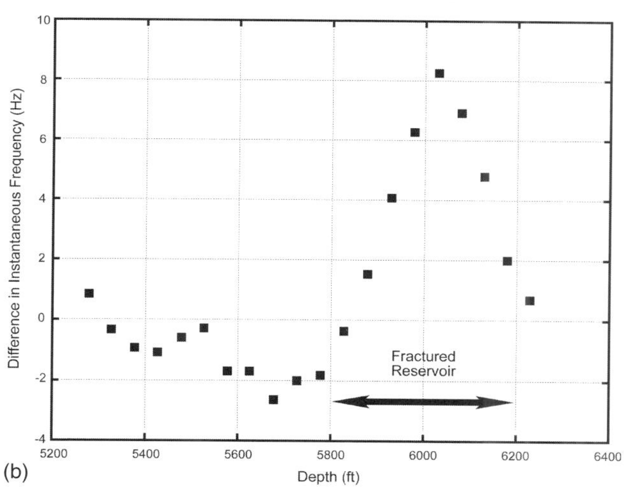

(b)

*Figure 15.43* (a) Rose diagram showing the consistent P-wave anisotropy directions in the reservoir (black) and the overburden (grey). (b) Differences in frequency between the fast and slow shear-waves, indicating a consistently lower frequency for the slow shear waves in the fractured reservoir from 5,800 to 6,200 feet. Maultzsch et al., 2005.

At the end of Chapter 14, an analysis of the Claire Field by Maultzsch et al., 2005 concerned P-wave attenuation anisotropy derived from analysis of field VSP data. In this case, with no gas cloud, the *fractured reservoir* displayed much stronger anisotropy than the *overburden* (Figure 15.43a), the authors drew attention to the advantages of using *relative* attenuation, rather than *absolute* attenuation. Their concept of relative attenuation or $1/Q_{seis}$ was derived from comparison of spectra recorded for different azimuths, but with identical source-receiver offsets. Absolute attenuation is typically recorded by comparison of spectra recorded along an individual ray path. The authors utilised the field measurements of travel time anisotropy, and were able to invert for fracture strike, fracture intensity (usually ambiguous), *and* scale length (presumably making the intensity formulation non-ambiguous).

When attempting to measure the absolute values of attenuation with the traditional spectral ratio and the instant frequency method of Dasios et al., 2001, they found that the instantaneous frequencies of the events fluctuated strongly: the heterogeneous rock mass giving zones of apparent negative $Q_{seis.}$ The authors argued that since the fast and slow *shear-waves* had similar wave lengths, they therefore probably sampled the same heterogeneities. They therefore analysed the differences in the *instantaneous frequencies* of the two waves. Despite the scatter, the differences in the two sets of frequencies were consistent with their location in either the overburden or in the fractured reservoir.

Figure 15.43b, showing these differences in instantaneous frequency, indicates increased attenuation of the slow shear wave, which had lower frequency in the reservoir layer (5,800 to 6,200 ft), but essentially the same frequencies in the overburden. One could perhaps speculate on another less dominant fracture direction in the overburden, in view of the mostly not quite zero differences in instantaneous frequencies.

## 15.9 Dual-porosity poro-elastic modelling of dispersion and fracture size effects

The ability to model various aspects of jointed rock behaviour has existed for many years, and is a complex and constantly expanding field. No attempt can therefore be made to give an exhaustive treatment in a single section of a single chapter. Since this section *will* address dual porosity poro-elastic modelling, with several examples, we will first summarize the different modelling needs and capabilities developed in *rock mechanics*, which is an increasingly close neighbour of geophysics, particularly in recent years when micro-deformation depending on joint stiffnesses, and micro-flow simulation depending on apertures, has become important for interpreting seismic response.

It is probably fair to say that rock mechanics modelling advanced very far in the last several decades of the 20th century, for the dual reasons of rock engineering and civil engineering needs, and due to the relative accessibility of input data, specifically from drill core, shallow exploration adits, and extensive use of shallow refraction seismic, plus sampling from within tunnels during construction. This has been a 'constant' for many decades, although description (and use) of the *joint properties* has improved dramatically, with perhaps less isotropic continuum modelling than previously.

In contrast, geophysics data acquisition abilities (and quantities of data) have exploded in just the last decade or two, with digital recording of increasingly complex 3D 4C receiver arrays, both on-land and especially off-shore. This development continues unabated. There is now an enormously increased need for realistic numerical models for geophysical interpretation, particularly in the area of structural anisotropy, fractures sizes and properties, and the frequency dependence of their seismic responses.

### 15.9.1 A brief survey of rock mechanics pseudo-static models of jointed rock

By way of an extremely brief history, one can mention the finite element modelling with joint elements that was developed at the end of the 1960s and promoted for rock engineering use by Goodman et al., 1968 and others. These authors, and Goodman, 1970 were perhaps the first to define the (pseudo-static) joint normal and shear stiffnesses needed for input to their 1D joint-slip elements for *discontinuous* finite element modelling. The deformation and stability of slopes, dam foundations and tunnels were the primary focus in rock engineering, and fluid flow and effective stress analyses were of course required too. These early 2D FEM models had deformable joint elements requiring *linear* estimates of normal and shear stiffness and of course frictional and eventual cohesive strength. Three dimensional FEM modelling, for example for dam-and-foundation interaction studies, with more limited numbers of (major) joint and fault planes was also performed during the 1970s and subsequently, but was obviously very time-consuming.

In the early 1970s Cundall, 1971 developed a 2D finite difference distinct element model for randomly or regularly jointed rock masses, making it easier to model large assemblages of blocks, with the ability, if required, to follow large deformations by tracking edge and corner contacts. Initial rigid block calculations in μDEC were subsequently replaced by simply deformable, then fully deformable blocks in UDEC, with the ability to model dynamic loading, micro- or macro-deformations, and fluid flow within the joints, but not in the matrix, where only pore pressures were modelled. Here we see one of the limitations, which of course is no longer there when only modelling flow in (un-jointed) porous media.

UDEC first had Mohr Coulomb linear joint strengths and linear stiffnesses, then a Cundall continuously yielding law, followed by non-linear shear strength and stress dependent shear and normal stiffnesses, following the block-size-sensitive Barton-Bandis constitutive model, in which most of the input data can be derived from simple index tests performed on drill-core, giving joint parameters JRC, JCS, and $\phi_r$. Block-size determined the scaling of these two joint roughness and compressive strength parameters, and the core-logged Q-value gave a stress-dependent deformation modulus. As we have seen in Part I, (and in Figure 15.33) the P-wave velocity from shallow refraction seismic could be used to estimate the *static* deformation modulus, and to extrapolate or interpolate such data from borehole to borehole.

This non-linear 2D model was termed UDEC-BB. Flow and fluid pressures were modelled by converting the physical joint apertures (E) developed at any time during the modelled joint deformation, into the (usually) smaller hydraulic apertures (e), using a linear-laminar flow assumption with joint permeability given by $k = e^2/12$. The conversion between (E) and (e) was found to depend on roughness JRC (Figure 16.7).

Subsequently 3DEC was developed by Cundall, together with colleagues at Itasca, giving the ability to model three-dimensional assemblages of jointed, deformable blocks, with any desired moduli and joint or fault properties. Linear strength and stiffness laws were followed for the joint sets and faults, to reduce calculation time. More recently, the pore space defined by the complex, three-dimensionally deforming, intersecting joint sets was fully defined, allowing flow modelling and dynamic effective stress modelling to be performed, also in three dimensions. Of course there are more dedicated 3D fracture flow models like FRACMAN (Dershowitz/Golders) and NAPSAC (AEA Harwell), but these are lacking comprehensive joint deformation modelling.

There are now several numerical models in use in rock mechanics for also modelling rock failure or cracking of the matrix blocks, caused by over-stress. These can accommodate a more limited number of pre-existing joints. Prominent among these are the particle flow codes of Cundall ($PFC^{2D}$ and $PFC^{3D}$), and the non-linear fracture mechanics code of Shen (FRACOD). The latter in particular, seems capable of realistic modelling of log-spiral type break-out around tunnels or boreholes, without recourse to 'manual' *degradation* of cohesion, and *mobilization* of friction, as needed when using an inappropriate 'c + tan ϕ' formulation. This linear Mohr Coulomb law is acceptable, or at least much used, for particulate clays and sand modelling in soil mechanics, and is reasonable for pre-existing planar joints, though probably best for faults. It is incorrect for previously intact hard rock, where 'c *then* tan ϕ' is a more correct formulation than 'c + tan ϕ', due to the widely different strains involved in cohesive failure of a brittle material and subsequent frictional sliding along the failure surfaces. (See discussion by Barton, 2004b).

### 15.9.2 A very brief review of slip-interface, fracture network and poro-elastic crack models

Concerning the dynamic modelling of cracked rock in geophysics, we can quote Tod, 2002 who is a prominent new contributor to this field: '*There are many theories available in the literature, resulting from a range of theoretical backgrounds that provide a description of an effective medium appropriate to describing the properties of a matrix material permeated with cracks on a length scale far less than the wavelength of seismic waves. While many of these theories agree qualitatively and are capable of describing a number of observed features, each has its particular shortcomings.*'

For the modelling of cracks in geophysics, we will not go further back in history than to mention Schoenberg 1980, and Hudson, 1980. These classic developments assumed, for greater simplicity (there was enough mathematics without fluids), that there was no exchange of fluid, either between the fractures or cracks themselves, or between these and the rock matrix. The importance of fluid had of course been known for a long time, and was modelled within the pore space by Gassman 1951 and within the cracks by O'Connel and Budiansky, 1977, and by many others since then. Thomsen, 1995, showed how seismic anisotropy was enhanced by transfer of fluid between fractures and equant porosity, with perfect pressure equalization at very low frequency, and reduced equalization at higher frequencies, giving the 'unrelaxed' behaviour modes.

#### 15.9.2.1 Schoenberg slip interface concept

Schoenberg 1980 modelled elastic wave behaviour using linear slip interfaces. These allow reflection, transmission, conversion, and delay to take place at the modelled interface, with the magnitudes depending on the specific stiffness, the frequency content, and the angle of incidence. The assumption is that when an elastic wave propagates across a fracture, there is a displacement discontinuity that is linearly related to the normal or shear force generated. The seismic particle displacement is *discontinuous*, while the seismic stresses are assumed to be *continuous*. In the linear slip model, the displacement discontinuity vector Δu was assumed to be linearly related to the traction $t$ as follows:

$$\Delta u = \begin{bmatrix} Z_T & 0 & 0 \\ 0 & Z_N & 0 \\ 0 & 0 & Z_T \end{bmatrix} t \qquad (15.7)$$

In geophysics it became customary to talk of fracture *compliances*, with inverted nomenclature and units (e.g. MPa/mm for stiffnesses, and $m.Pa^{-1}$ for compliances). Naturally, the compliances or stiffnesses used in geophysics refer to the dynamic properties of the joints or fractures, which generally have somewhat greater stiffness (or lower compliance) than the pseudo-static values commonly used in rock mechanics modelling. These differences have been mentioned many times, and will be quantified further in Chapter 16.

#### 15.9.2.2 Hudson effective medium concept

Hudson, 1980, in contrast to Schoenberg, utilised a 'method of smoothing' for the effect of modelled cracks, which was capable of representing the elastic parameters of a 'cracked' material, in the form of an *effective medium*. This allowed calculation of the effect of incident dynamic waves of long wavelength. Subsequently, Hudson et al., 1996 extended the model to allow for the cracks to be connected via the porosity of a rock matrix. In this extended case, cracks could be deformed by an incident wave in a manner that depended on their aspect ratio and on their orientation with respect to the incident wave. Clearly the modelling of intrinsic attenuation mechanisms such as squirt flow, and its frequency dependence, was transformed by this extended capability.

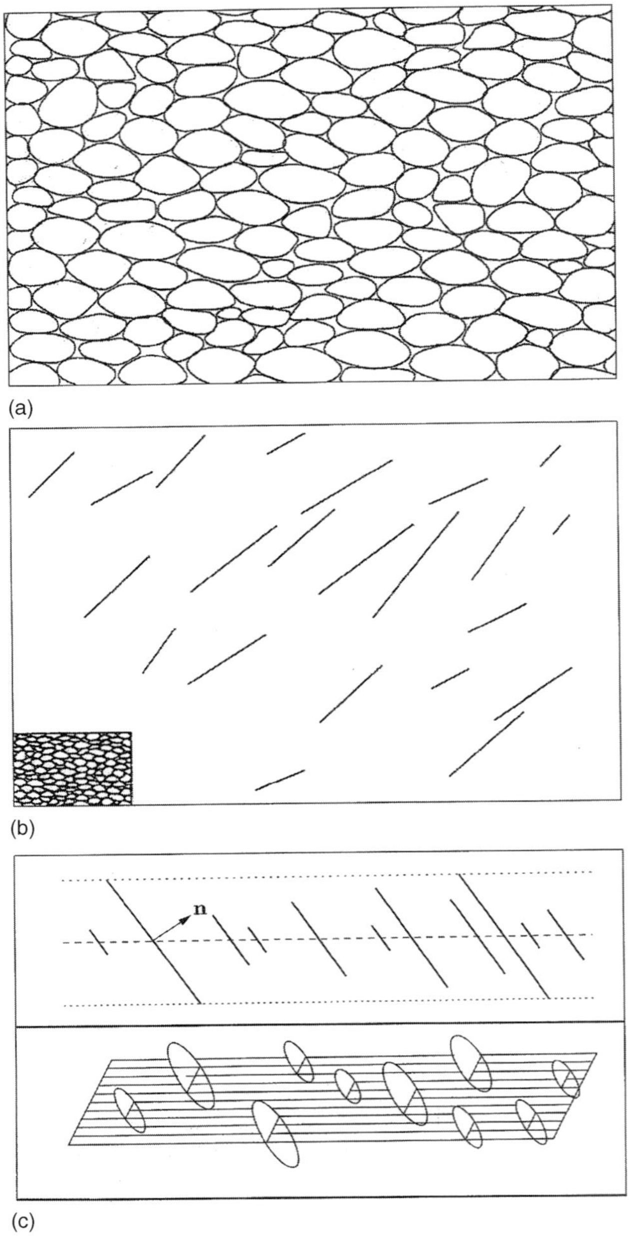


*Figure 15.44* (a) Schematic of the pore space created by the misfit of (sand) particles, (b) a possible distribution of cracks that are subject to a sub-parallel stress, c) oblique cracks in a fault zone applied in a new model. Tod, 2002, and Tod et al., 2002.

### 15.9.2.3 Tod crack density decay model

Tod et al., 2002, extended this model further, by allowing for a continuous but independent distribution of both crack orientation and aspect ratios, and by allowing each to depend on the applied stress and on the fluid pressure. At high frequencies, the cracks behaved, as expected, as isolated (stiff) features, while at low frequencies they behaved as 'poorly-drained', since squirt losses at lower frequencies cause a 'soft' behaviour because of some (micro) drainage.

The model developed by Tod effectively allows the crack density to decay with an increase in applied stress, from an initial value representing the unstressed state. However, the model is purely elastic, and relaxes to its original state during unloading. Crack density was designed to decrease with increasing compressive stress, from e.g. 0.1 at zero effective stress to e.g. 0.05 at 200 MPa. Only cracks with normals lying in the $\sigma_h$ minimum direction were assumed to remain open.

The model is capable of capturing the changed anisotropy caused by fluid pressures and applied stresses, which impact the aspect ratios of the cracks. However the pore space, illustrated in Figure 15.44a is non-compliant. Wave speeds approach that of the matrix at high pressures. Shear waves proved to be more sensitive to pressure change than the compressional waves, as known from other studies.

The non-compliant pores of the Tod model, transfer fluid to the physically unconnected cracks, therefore giving dispersive, or frequency-dependent velocities and attenuation. The theory predicts that there is a pressure at which anisotropy reaches a maximum value, before the conducting properties are reduced, by increasing stress, eventually to that of the uncracked matrix. Tod also extended his modelling to the case of faults with parallel cracks aligned at an angle to the main fault plane (Figure 15.44c).

Since these authors, and also Chapman, 2002 and 2003 (see later), utilise aspect ratios that can in practice be reduced to very low values by high stress, the inequality of physical (E) and hydraulic (e) apertures will affect interpretation, since fluid volumes in actual cracks or joints are greater than hydraulic capacity, since permeability $k \approx e^2/12$. In dynamic modelling seen so far, this distinction does not appear to have been treated.

However, the improved model described by Tod et al., 2002, (together with Cambridge Professor 'seismic' Hudson: there is a prominent Professor J.A. Hudson in rock mechanics too), provides a new level of sophistication, since crack aspect ratio, crack density, crack orientation and responses to applied stress and fluid pressures, are each incorporated into the permeable, but non-compliant pore space between the matrix 'grains'. The new formulations and the interdependencies of crack orientation, aspect ratio and applied (effective) stress gives an elastic loading and unloading behaviour, but

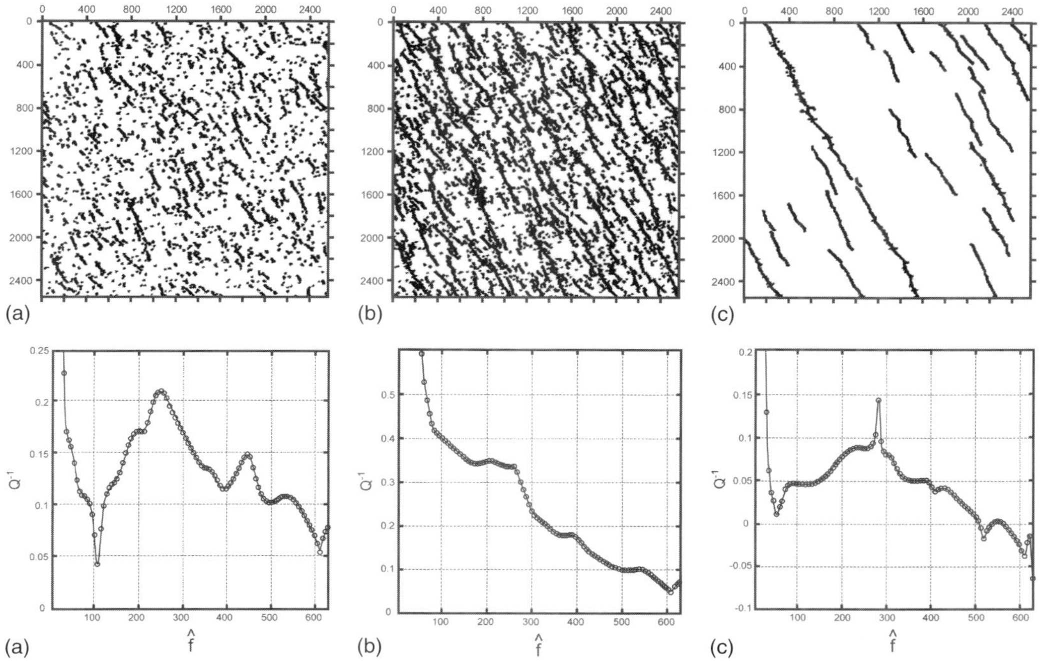

*Figure 15.45* Three examples of the nucleating branching fracture networks, showing their changing attenuation and frequency dependence. Source wavelets of 30 Hz applied at centre of each model. Vlastos et al., 2003a. Reproduced by kind permission.

with no hysteresis, and no stress history, in obvious contrast to the larger-deformation modelling of the rock mechanics UDEC-BB code.

### 15.9.2.4 Vlastos-Narteau automaton model

Interesting new developments in modelling capabilities in geophysics were also demonstrated by Vlastos et al., 2002 and Vlastos et al., 2003a. The first authors Vlastos and Narteau, utilised the 2D finite difference method termed the multi-scale 'cellular automaton model' developed by Narteau, 2001 to study progressive changes of attenuation in nucleating, growing, branching and coalescing arrays of fractures. They studied the effects of scattering attenuation in what evolves, through shear fracturing, into an *anisotropically* fractured network, with fractures of increasing length and reduced frequency as shear localization develops.

Examples of the successive stages of fracturing that can be generated are shown in Figure 15.45. These are successive 'snapshots' generated by a particular realization of this dynamic network model.

In the seismic attenuation modelling described by the authors, only the scattering attenuation component is modelled. This is related to the structural heterogeneities (i.e., the growth and coalescence of fractures), but does not include squirt or intrinsic attenuation losses. Nevertheless, scattering attenuation is also found to be frequency dependent, showing $Q_{seismic}$ values as low as $1/0.6 = 1.7$ for case (b) in Figure 15.45.

In the numerical fracture models illustrated in Figure 15.45, background values of $V_p$ and $V_s$ were 3.3 km/s and 2.0 km/s respectively. Density was 2.2 kg/m$^3$. Vlastos et al., 2002 and Vlastos et al., 2003a used what are considered by the writer to be unrealistically *equal* normal and shear compliances ($Z_N = Z_T = 5.6 \times 10^{-10}$ GPa$^{-1}$: presumably GPa$^{-1}$.m?), which (may) correspond to rock mechanics normal and shear stiffnesses ($K_n$ and $K_s$) of about 1.8 MPa/mm. In general terms, due to the higher dynamic modulus of rock masses, it is likely that the micro-deformation, dynamic compliances should be much lower (i.e. much stiffer in rock mechanics terminology) than the macro-deformation values typically measured in 'static' loading tests on joints, as frequently performed in rock mechanics, and the shear compliance should perhaps be higher (i.e. the shear stiffness lower).

Presumably with rock mechanics experiences of e.g. $K_s/K_n \approx 1/10$ or an approximated $Z_T/Z_N \approx 10$, even greater attenuation, and greater anisotropy would have been indicated. Clearly there is promise for future links between fracture frequency, fracture character and seismic attenuation measurements.

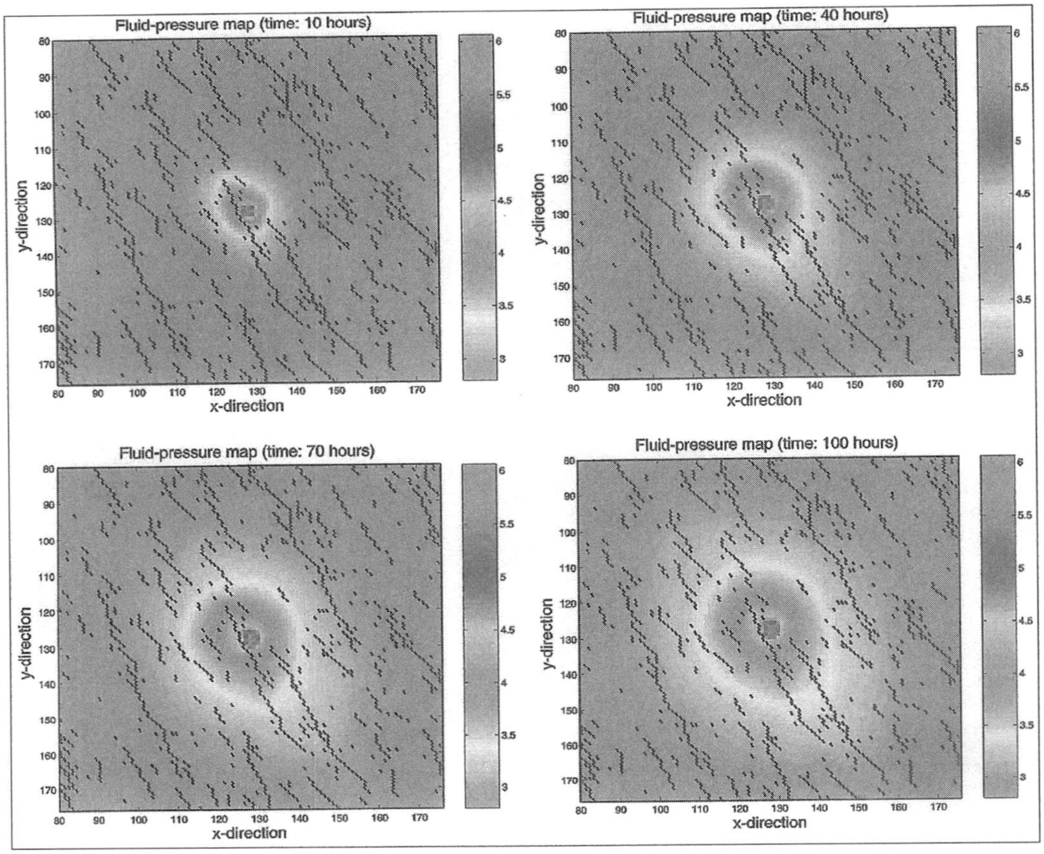


*Figure 15.46* 'Snapshots' of the anisotropic, elliptical, pore pressure propagation with time, in the Vlastos et al., 2003b, fractured network model. Reproduced by kind permission.

Vlastos et al., 2005 presented a more comprehensive version of the Narteau 'stochastic-deterministic' fracture network modelling, with a 12-stage evolution of fracturing. Application of the spectral ratio method to the scattering attenuation, resulted in increasing then declining attenuation, giving seismic Q values reducing from 100 to 7 to 2.5 (with the maximum density of fracturing), and increasing again through 6, 10 and 100 as the fractures were fewer and longer, as typified by stage c) of Figure 15.45.

It is clear that the wide variety of fracture-scale lengths illustrated in the most well-connected networks were within the so-called 'percolation threshold', and were *largely* responsible for the low $Q_{seis}$ values in these 'central' models, typified by case b) in Figure 15.45.

Vlastos et al., 2003b, with co-authors from BGS and the universities of Berkeley, Edinburgh and Cergy-Pontoise in France, and Lin et al., 2004, used a 2D fluid flow model and a pre-existing, anisotropically fractured numerical network, similar to the above, to study the seismic signatures of a central (borehole) fluid injection into a simulated 2560 × 2560 m model. They show interesting 'snapshots' of the anisotropic, elliptical pore pressure progression that develops with time.

Figure 15.46 reproduced in grey-scale, shows the pore pressure development 10, 40, 70, and 100 hours after the start of injection. They explained that the ellipticity of the pore pressure field, besides being due to the presence of fracturing, was due to assuming a diffusivity along the fractures, many orders of magnitude greater than the background.

As the authors pointed out, injection decreased the effective stress, thereby increasing the compliance (or reducing the stiffness) of the modelled fractures. On this occasion a Schoenberg, 2002, non-linear compliance formulation was used. An increased fracture opening and/or pore throat size, could have effectively decreased the stiffness of the rock and rock mass in terms of squirt flow, thereby reducing velocities and increasing intrinsic attenuation.

The authors came to the surprising conclusion that the P-waves were not sensitive to pore pressure changes, as opposed to the S-waves and coda waves that showed high sensitivity. Could the reasons for this be found in their assumptions for compliances, in relation to rock mechanics experiences of the approximate inverse: the stiffnesses? The authors reportedly set compliances as follows (but presumably the unit should be $Pa^{-1}$.m, as otherwise the stiffnesses would be many orders of magnitude too high):

At zero stress: $$Z(_{To}) = 5.681 \times 10^{-9} GPa^{-1}$$
$$Z(_{No}) = 2.8409 \times 10^{-10} GPa^{-1}$$

If we assume a typing error, and intended units of $Pa^{-1}$.m, then rock mechanics stiffnesses, *if* they had been equal to the inverse, would have been:

$$K_s \approx 0.176\,MPa/mm$$
$$K_n \approx 3.52\,MPa/mm$$

These give a recognizable ratio $K_n/K_s$ of 20. However, the equivalent $K_n$ value, as will be seen Chapter 16, is exceptionally low, and the shear stiffness also low. Both are representative of an effective normal stress of perhaps less than 0.1 MPa. This could be the explanation of the lack of P-wave sensitivity to pore pressure change, since the modelled rock mass was, perhaps inadvertently, given the equivalent of a high confining stress. At 'infinite' stress, the authors used $Z(_{To})/5$ (i.e. $K_s \times 5$) and $Z(_{No})/2$ (i.e. $K_n \times 2$), changes that would be greatly exceeded in rock mechanics terms, when the original stiffnesses were so low, due to the assumed low stress.

Real rock joints display strongly non-linear pseudo-static stiffness at lower stress levels, when subjected to mechanical loading causing macro-deformation. Furthermore, *near-surface rock masses* clearly display a strong dependence of P-wave velocity on effective stress level (and therefore pore pressure), at least in the first several hundreds of metres, and sometimes to depths of one or even two kilometres if the rock has high modulus, or if there is over-pressure. The rock quality $Q_c$ -$V_p$ - porosity – depth relations which are empirically based (e.g. Figure 15.33 and Chapter 15), confirm the relevance of low $K_n$ at low stress, with a strong stiffening at high stress. So the authors' conclusion regarding $V_p$ not sensitive to pore pressure changes, may inadvertently be a function of the 'equivalent-to-low-stress' input data, and the application of high (fracture-closing) stress levels.

#### *15.9.2.5 Chapman triple-porosity poro-elastic model*

Chapman, 2002 and 2003, developed a dual-porosity poro-elastic model, based on the following important observation that typical laboratory samples, clearly unfractured, nevertheless display dispersion, anisotropy, stress sensitivity, and dependence on fluid type and degree of saturation, as we have seen in numerous contexts in earlier chapters. Chapman's argument for developing his new model was that, when the fractures or cracks were removed from *preceding* models, a linear-elastic material remained, contrary to observation.

Chapman et al., 2001 contrasted his more recent poro-elastic double (actually triple) -porosity model with the anisotropic poro-elastic model of Zatsepin and Crampin, 1997. Although the driving process of this earlier model was the migration of fluid along inter-crack pressure gradients (and subsequent preferential crack closing), the calculation of the induced velocity changes actually relied on crack models which assumed that fluid could *not* move at the time scale of a seismic wave. As pointed out, this could be a good approximation for high frequency laboratory ultrasonic experiments, but not for low frequency e.g. 100 Hz field data. The restriction that fluid should not move also ruled out the (correct) modelling of attenuation.

Chapman had earlier presented a poroelastic theory that modelled the effects of squirt flow at the grain scale, which gave a good match to experimental data. Chapman's new theory combined two or more length scales explicitly: the grain scale and a set of fractures of any desired characteristic length (e.g. stress-aligned dominant jointing of many potential scales: 0.01, 0.1, 1, 10 metres). He termed the latter 'meso-scale' anisotropic fractures. Naturally, he also modelled the all-important interaction between this meso-scale and the grain-scale equant matrix porosity and ellipsoidal microcracks. Because there may be three or more scales involved, it has been termed a 'triple-porosity' model in this book.

An important feature introduced by the inclusion of fractures was that dispersion occurred at lower frequency ranges than those over which the micro-structure dispersion occurred. He found that the larger the size of the fractures, the lower was the relevant frequency band.

Figure 15.47 gives a clear demonstration of the ability of this model of equant porosity with elliptical microcracks (first without meso-scale fractures), to model frequency dependent velocities (dispersion) in the case of an isotropic model *without* fractures. Figure 15.47 a and b, show that the dispersion is predicted

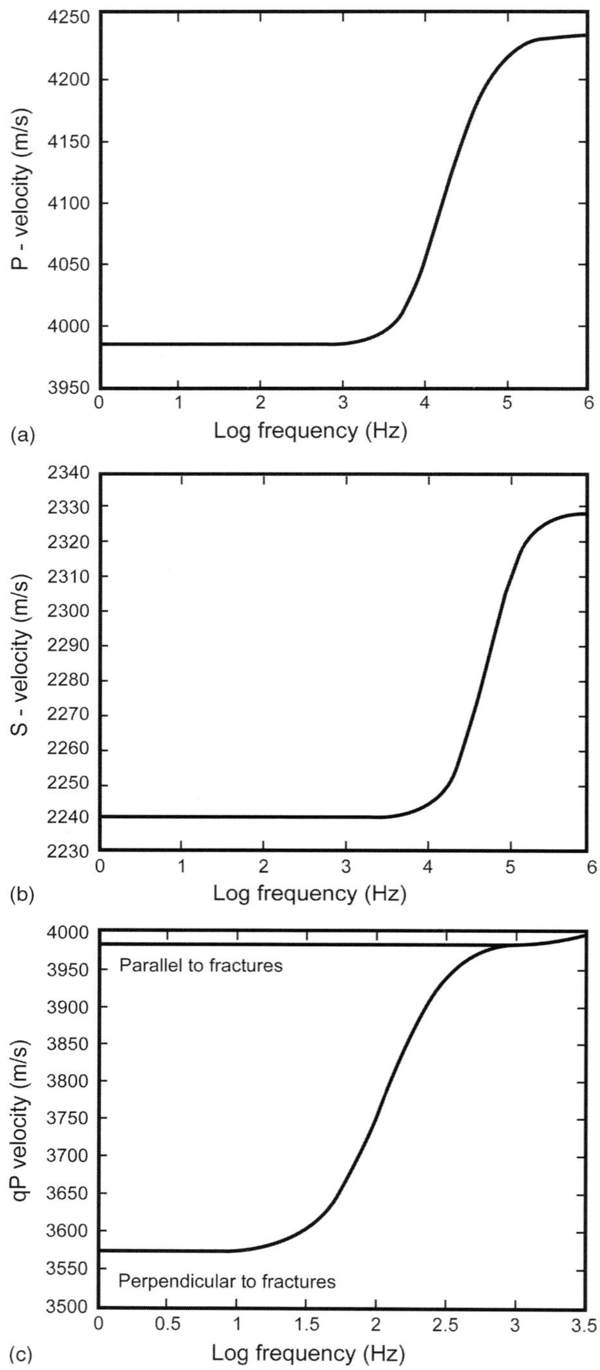


*Figure 15.47* (a) (b) Chapman model of dispersion with no meso-fractures. (c) dispersion and anisotropy when 10 cm fractures are added. Chapman, 2003.

over a limited frequency band, roughly between 1 kHz and 100 kHz. The particular model had moduli of 17.5 GPa, density 2300 kg/m$^3$, a microcrack density of 0.1 and 10% porosity.

Next, Chapman added fractures of 10 cm length, at a density of 0.05, making the model *transversely isotropic.* Figure 15.47c shows the dispersion of the qP-wave velocities travelling parallel and perpendicular to the model fractures. Clearly, the parallel waves are unaffected by the fractures, but squirt flow gives dispersion for the perpendicular waves. There is a significant reduction in velocity at low frequencies, but no reduction in velocity above about 1 kHz, due to the *apparent* stiffness of the fluid inclusions. Introduction of even small 10 cm long fractures had caused dispersion to begin at lower frequencies.

Figure 15.48 shows modelled shear wave velocity anisotropy [defined as before, as $100 \times (qS_1 - qS_2)/ qS_1$] for the case of a mix of microcracks of 200 μm size, 10% equant porosity, and meso-fractures of up to 1 m size. (See visualization of the model in Figure 15.48a).

Shear wave propagation was at 20° incidence to the fractures. Anisotropy reduced with increasing frequency. In contrast, Figure 15.48b had no equant porosity (0%) and crack density is reduced from 0.1 to 0.02. A marked reduction (and smaller variation) of shear wave anisotropy can be noted. The absence of the pores, means that fluid can no longer flow out of the microcracks with the passage of the seismic waves, so anisotropy and attenuation are reduced. Note that the shear-wave anisotropy is only weakly following the '100*e*' rule-of-thumb of Crampin, since the two examples should then show 10% and 2% respectively.

When modelling the attenuation, there was a singularity (zero attenuation) in the case of the pure shear wave, when the frequency was only 40 Hz, and the angle of incidence was 90°. This was believed to be due to the perfectly aligned fractures† this aspect being avoided in the new Tod, 2003 model, which utilises a 'nearly aligned' crack set.

When Chapman modelled with a much higher frequency (3 kHz), the qP, qS and pure shear S waves were

† On the subject of modelling P-wave anisotropy in the case of perfectly aligned fractures, Willis and Rao, 2005 pointed out that a P-wave can be cancelled out due to the time delay of the wave, when travelling between two fractures, thus creating a significant notch (or even a null) in the spectral ratios of reflected seismic traces. This happens when the P-wave length is about twice the fracture spacing. However, they also suggest that the frequency location of the notches themselves could be used to determine the fracture spacing and losses, due to scattering attenuation from the fractures.

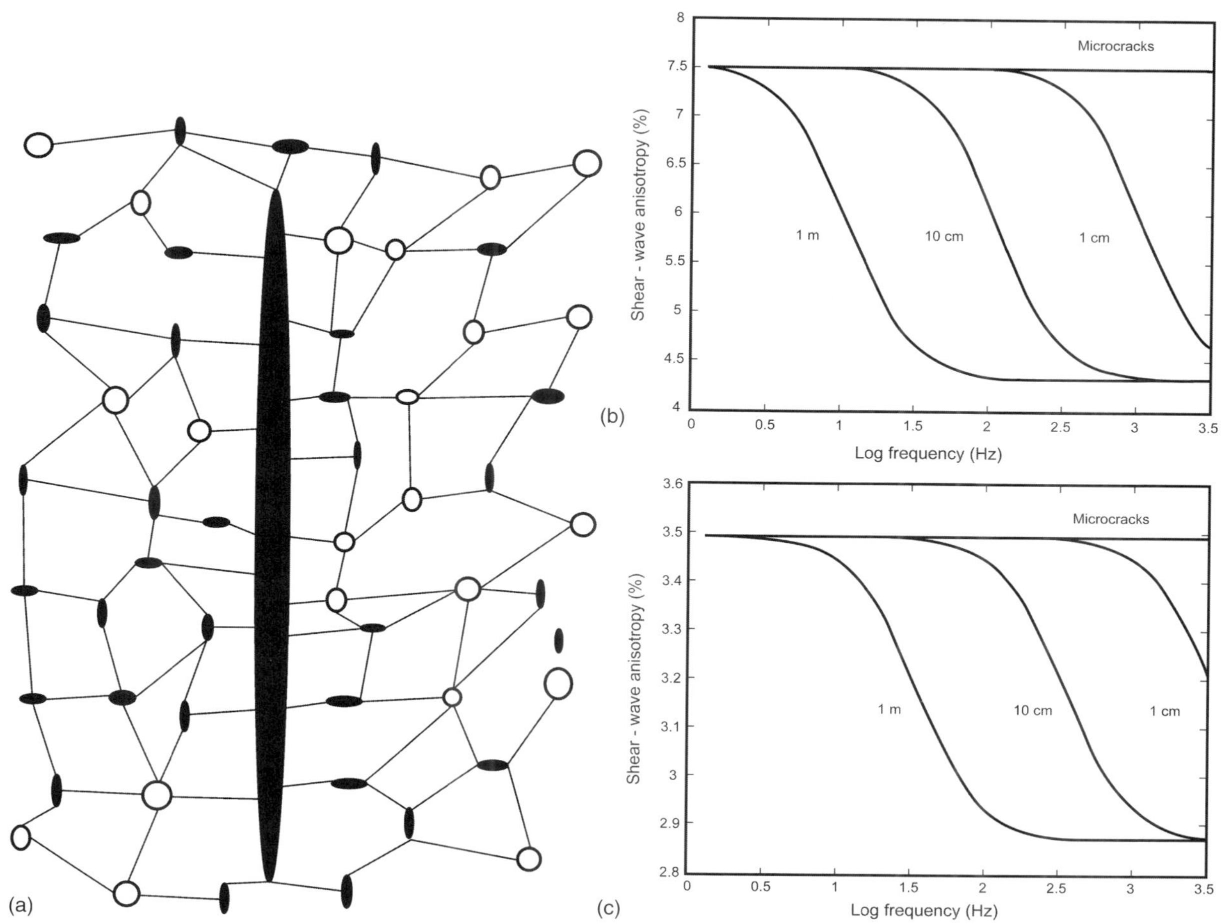


*Figure 15.48* (a) Visualization of Chapman's model of connected equant pores, microcracks and fractures with low aspect ratios, (b) S-wave anisotropy with 10% porosity, and ellipsoidal microcracks and fractures of various sizes, giving a crack density of 0.1. (c) similar to (b) but with 0% porosity, and 0.02 crack density, which caused reduced attenuation, and reduced S-wave anisotropy. Chapman, 2003.

all attenuated, with (1000/Q) in the range 18 to 24 (approx.), i.e. $Q_{qP}$ and $Q_{qS}$ values in the range 42 to 56, which is typical for $Q_p$ and $Q_s$ of moderately fractured rock closer to the surface (see Chapter 10).

This brings one to the interesting question of stress sensitivity. By setting smaller aspect ratios for the meso-fractures compared to the microcracks, one can in some way mimic the effect of higher stress, and thereby cause less attenuation (higher $Q_{seis}$), as the fluid cannot 'squirt' or flow so easily, in response to the passage of the seismic waves.

Setting a certain aspect ratio is like creating a physical aperture (E), whereas the assumed *permeability* of this crack of 'fixed' aspect ratio is actually a function of its hydraulic aperture (e). Because of joint or fracture roughness (JRC), the usual inequality: $E \geqslant e$ occurs, with $e \approx E$ only when roughness is absent, and JRC = 1. In reality, a chosen aspect ratio will have less permeability than assumed, due to joint roughness.

Because of varying degrees of roughness for most joint types, and the usual inequality of these two apertures, dispersion will likely begin at lower frequencies than might be modelled; there will be extra resistance for fluid flow (in relation to the modelled aspect ratio), as roughness increases. Yet it will contain the assumed volume of fluid. The significant inequality of E and e is demonstrated in the explanation of some rock mass 'groutability' terminology, in Figure 15.49, and by UDEC-BB modelling of tunnels, in Figure 15.50.

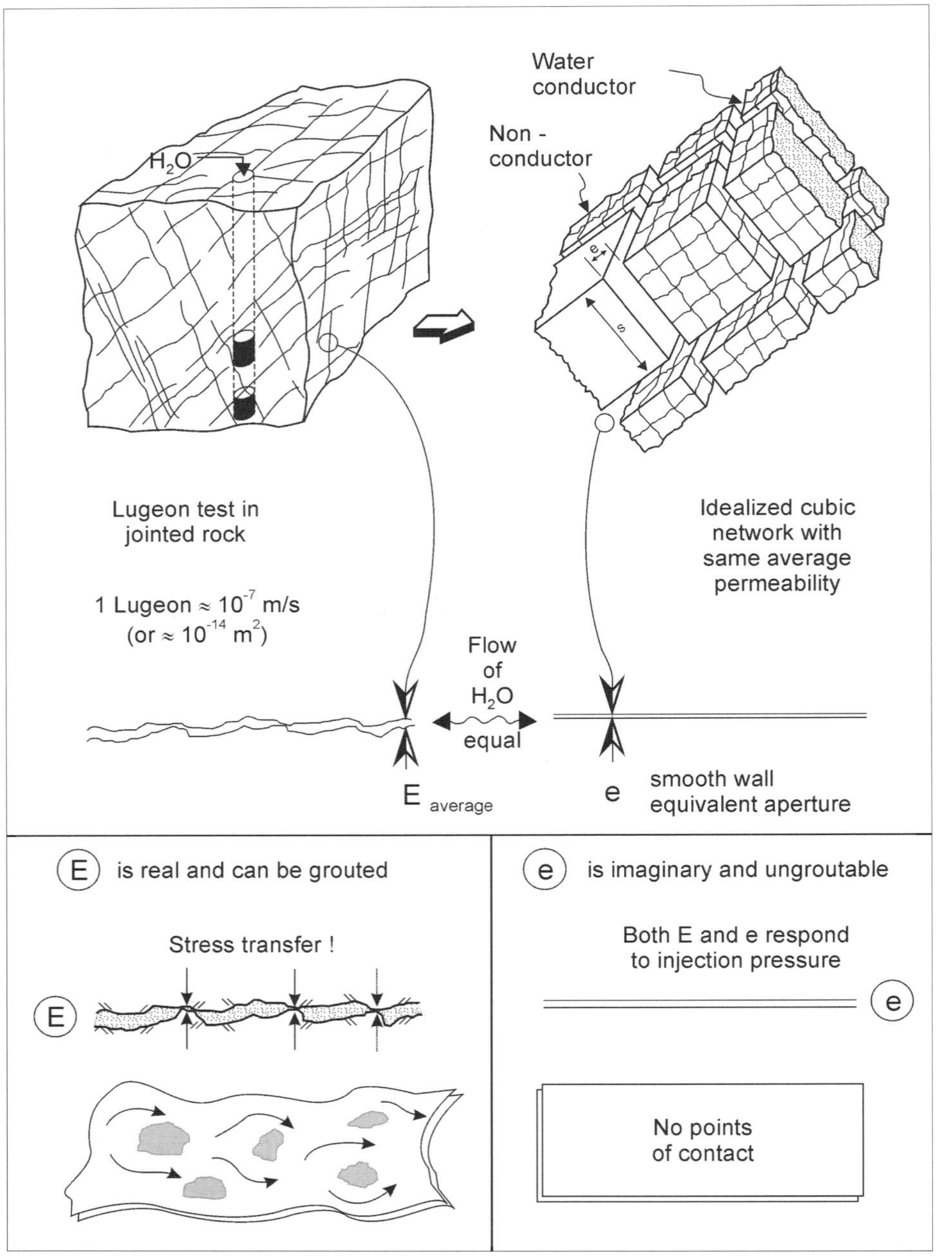


*Figure 15.49* Physical demonstration of joint aperture concepts (E) and (e) regarding groutability. Barton and Quadros, 2003.

#### *15.9.2.6 Maultzsch-Chapman fracture size estimation*

Maultzsch et al., 2003, with colleague Chapman from the Edinburgh BGS Anisotropy Project, broached the question that seems to represent an important ambiguity in many earlier models in geophysics, namely how best to interpret the commonly used 'crack density' formula:

$$e = \gamma a^3 \text{ or } e = \frac{Na^3}{V} \qquad (15.8)$$

where a = crack radius, and N/V is a number per unit volume ($\gamma$ = 'number density').

This means, as they point out in Figure 15.51a, that the same crack density can be caused by a few large fractures or many small cracks. For instance, a fracture density of 0.05 means that N/V (where $e = Na^3/V$) can be expressed in many different ways, such as:

a) 50 × 1 mm cracks/1 cm cube
b) 5 × 1 cm cracks/10 cm cube

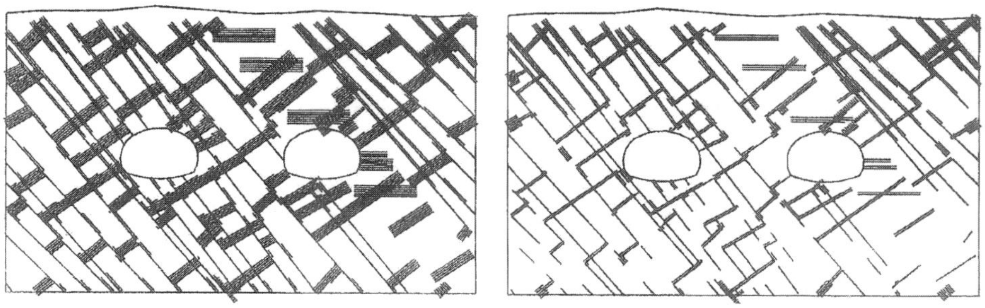


*Figure 15.50* The inequality of physical aperture (E) and hydraulic aperture (e), demonstrated in a distinct element UDEC-BB model of tunnels in Oslo. Makurat and Barton, 1988, NGI contract report. Makurat et al., 1990a.

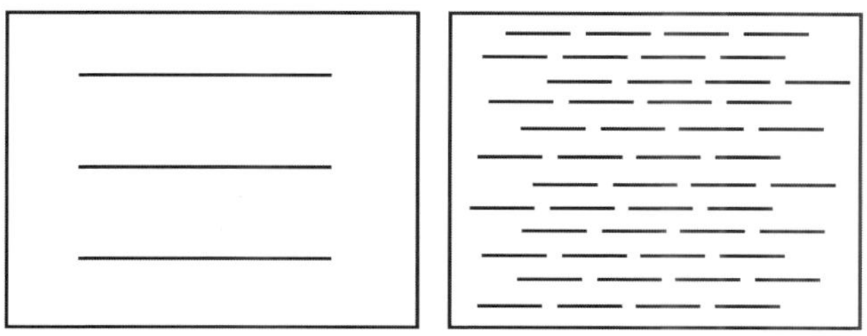

*Figure 15.51* The problem of the *crack density* definition $e = Na^3/V$ used in geophysics is that both the illustrated representations of fractures of different size can have the same numerical value of *e*. From Maultzsch et al., 2003.

c) 50 × 10 cm fractures/1 m cube
d) 1 × 1 m fracture/2.7 m cube

While the crack density can be roughly estimated from the time delay of split shear waves, since a shear wave anisotropy of approximately 100*e* is expected according to Crampin, the result is seemingly ambiguous, since fracture compliances will also play a role in the time delay and anisotropy estimate. This ambiguity is hardly satisfactory when trying to understand the structure of a fractured reservoir.

This geophysics *crack density* concept is in stark contrast to what engineering geologists are *expected* to do during field logging, namely attempt to present the spacing statistics of the different sets of joints (with a mean $\lambda\,m^{-1}$ for each set), and where possible, also present the length statistics of the joints belonging to the different sets, and whether both or only one end of the joint in question is visible in the exposure. (ISRM, 1978).

Quantification as joint spacing and length and number of sets, as used by engineering geologists, would be the theoretical ideal, but with compromise necessary due to the 'invisible' nature of the targets in geophysics. Mixed scale fracturing obviously needs to be modelled without ambiguity concerning fracture size, since it is of fundamental importance when attempting to assess fractured reservoir structure, as also argued by van der Kolk et al., 2001 in the case of the Natih field in Oman.

When an engineering geologist is logging exposed rock masses or core, he also uses one or more rock mass classification methods, such as RQD, Q, and RMR).‡ Naturally, being able to see the rock surface exposure, tunnel wall or core, is obviously of inestimable value for giving this more correct description of reality. The microcrack scale is of course excluded from general field observations.

Maultzsch et al., 2003, investigated the length-scale frequency-dependence of the Chapman model, taking account of fractures up to 10 m size in Figure 15.52.

‡ The first two terms of the rock mass quality Q-estimate are one step simpler than estimating $\lambda\,m^{-1}$ for each set, as above. For RQD, one records the percentage of core pieces longer than 10 cm, dividing this by the rating ($J_n$), for the number of joint sets (see Appendix A). These first terms in the Q-value estimate ($RQD/J_n$), give a close approximation to relative block size, and the degree of freedom for block movement. This is fundamental for rock mass stability and inter-connectivity to fluids. It is tempered by the next pair of parameters $J_r/J_a$, which closely represent the coefficient of friction, with contrasting contributions from joint roughness and clay filling, in the first instance for the least favourable joint set. The combined value $Q' = RQD/J_n \times J_r/J_a$, often termed Q-prime, clearly has strong links to seismic attenuation and anisotropy.

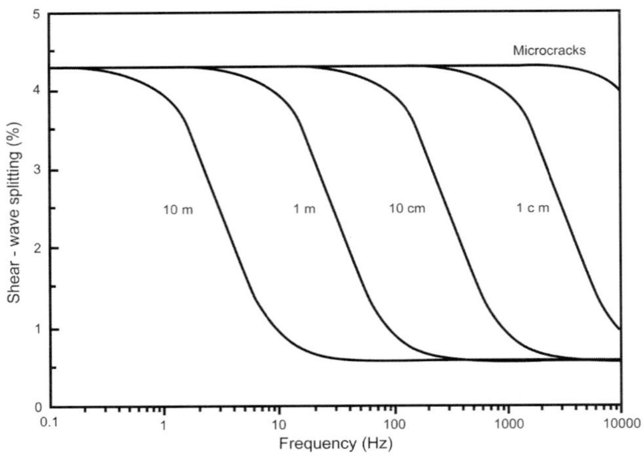


*Figure 15.52* A further example of the scale and frequency dependence of shear wave splitting anisotropy, according to the Chapman model, that connects pores, microcracks and formation-scale fractures. In this example, from Maultzsch et al., 2003, the waves are propagating at 30° to the fracture planes, and are split to give the dispersive anisotropies shown.

Here the percentage anisotropy of the polarized $qS_1$ and $qS_2$ shear wave velocities is shown. It is implied that shear wave splitting anisotropy can be registered from wave splitting at numerous scales of crack/fracture sizes. However it is understood that attenuation in a well fractured reservoir may prejudice the arrival of the slow $qS_2$ wave.

For any given fracture radius, the shear-wave anisotropy is seen to *decrease* as frequency *increases*. The larger the size of fractures, the lower the frequency range where velocity dispersion and frequency dependence of anisotropy occurs. This property was subsequently used by the authors for *inversion to fracture size.*

As also in Figure 15.48, the modelled results of shear wave splitting shown in Figure 15.52 suggest that the EDA concept with focus only on microcracks as the source of shear-wave splitting, may need to be re-evaluated. Microcracks, as modelled, appear to give a 'constant' potential source of shear-wave splitting independent of frequency. In the examples shown in the two figures, anisotropies of 7.5%, 3.5% and 4.25% are shown. The 'meso-fractures' of 1 m and 10 m (radius) apparently have an equally strong role in shear-wave anisotropy as microcracks, but only at the seismic frequencies appropriate to earthquake studies. There is therefore no good reason for only assuming that microcracks are responsible for shear-wave splitting in seismically disturbed rock masses, such as along the instrumented sections of the San Andreas Fault, as reviewed earlier in this chapter.

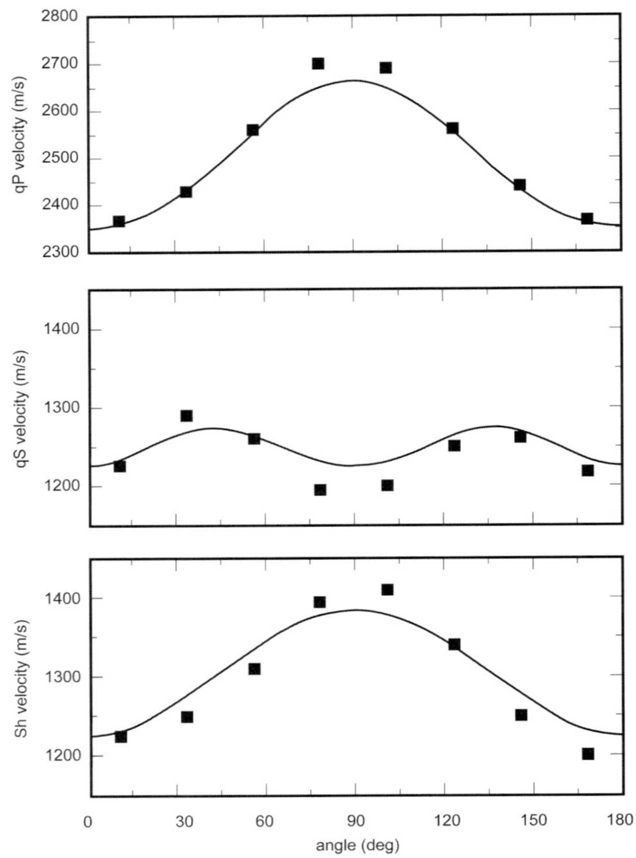


*Figure 15.53* Comparison of Chapman model to the physical saturated sandstone-and-fracture model of Rathore et al., 1995. From Maultzsch et al., 2003.

Maultzsch et al. 2003 demonstrated the application of the Chapman model with two examples. One was a novel physical model of known fractures and pore structure located in IKU, Trondheim, Norway, made by embedding thin metal discs into a sand-epoxy matrix, with subsequent acid leaching of the discs to create cracks of known geometry and orientation (Rathore et al., 1995). This physical model will be described in greater detail at the end of this chapter. It appears likely to have aperture characteristics satisfying E = e, (i.e. without contact or apparent stress transfer, and with limited, if any, roughness). The Maultzsch et al., 2003 match to these laboratory determined qP, qS and Sh velocity components with azimuth, was excellent, and is shown in Figure 15.53.

### 15.9.3 Applications of Chapman model to Bluebell Altamont fractured gas reservoir

Since the Chapman, 2002 and 2003 dynamic equivalent medium model handles squirt flow in a triple-porosity poroelastic medium with porosity, microcracks and a fracture set of any desired fracture length or density, it is logical that it becomes a useful tool for demonstrating both the frequency dependence and azimuthal variation of the attenuation ($1/Q_p$) of P-waves.

Maultzsch et al., 2002 used the Chapman model, and various fracture sizes to demonstrate azimuthal dependent $Q_{seis}$ (strictly $1/Q_p$) and also frequency dependence, each for four different fracture sizes. This synthetic model had a common fracture density of 0.05, and fracture sizes (radii) of 1 mm, 1 cm, 10 cm, and 1 m. Figure 15.54 shows the predicted results. Note the tendency for 10 cm and 1 m fractures to give highest attenuation at lower seismic frequencies.

Maultzsch et al., 2003 applied the Chapman model to nine-component VSP field data, from the Bluebell Altamont Field in the Uinta Basin, Utah in the USA (Lynn et al., 1999a). This was a tight, fractured gas reservoir in 'sandstones' (or lenticular sands encased in shales and carbonates), from which production was primarily controlled by the size, orientation and concentration of natural fractures, i.e. intersecting joint sets.

The near-offset VSP had the source located 550 feet west of the well, and the three components receivers were placed at depths from 2800 feet to 8650 feet, with 50 ft spacing. The Green River reservoir formation was located from 6687 to 8591 feet deep (approx. 2000 to 2600 m). A P-wave and two orthogonal S-wave sources were used, yielding a nine-component data set.

Shear wave splitting had long been recognised, with the fast shear waves giving a consistent (slightly rotating with depth) N46°W to N40°W (average N43°W) polarization. This direction was identified by the authors with the fracture strike, but, actually as we shall see later it is closer to $\sigma_H$ max but nevertheless caused by fractures of large scale. Significantly, the time delay between the fast and the slow shear waves showed a sharp *increase with depth* at the reservoir level, indicating the presence of the fractures. Maultzsch et al., 2003 tested the Chapman model, following the Liu et al., 2003 methodology for interpreting the frequency dependent anisotropy at this fractured reservoir.

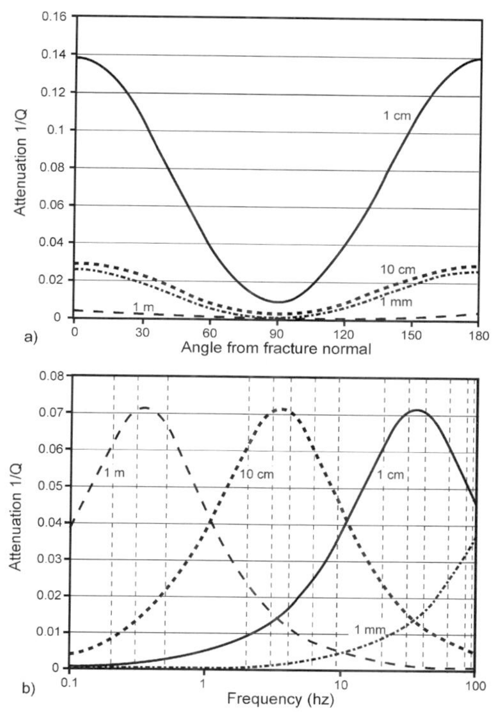


*Figure 15.54* (a) Variation of P-wave attenuation with angle from the fracture normal, for different modelled fracture radii (*frequency* = 30 *Hz* in example). (b) Variation of P-wave attenuation with frequency and fracture radius. Chapman, 2002 and 2003 model, in Maultzsch et al., 2002.

The time delays between the two split shear waves, showed a systematic variation with frequency, and of course with depth. The time delay was largest (32 to 38 ms) at lowest frequencies (5–15 Hz), and smallest (30 to 34 ms) at higher frequencies (20–40 Hz). This data was used for *inverting for the theoretical fracture density and fracture radius.*

The roughly 2000 to 2600 m deep reservoir had $V_p$ = 4877 m/s, $V_s$ = 2575 m/s, density 2.60 and porosity 9.4%. For the modelling, an aspect ratio of 0.0001 was chosen for the fractures (i.e. 100 μm per 1 metre).

Figure 15.55a shows the results of Maultzsch et al., 2003 modelling of the relative error between the measured and the model-computed time delays. This was for a range of fracture densities, fracture sizes and frequencies. The black section of the 'bend' in the figure shows minimum error, for a fracture density of 0.035 and a fracture radius of about 3 m. Reportedly, there was evidence from Lynn et al., 1995, of fracture lengths of about 2 to 3 m, based on borehole images. The

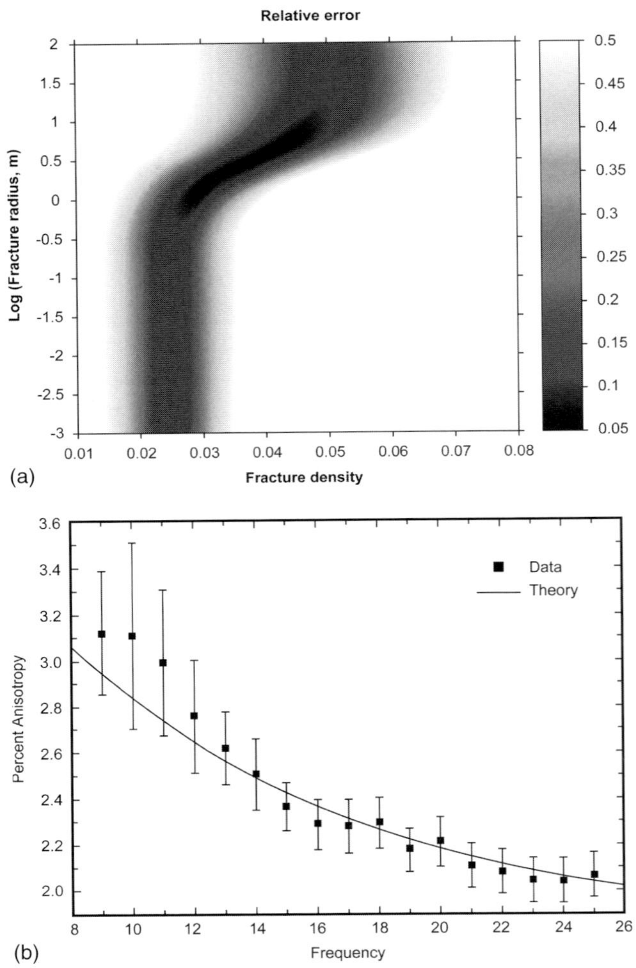


*Figure 15.55* (a) Relative error between VSP measurements and Chapman model results, shows minimum at realistic fracture radius a ≈3 m, and fracture density of 0.035 or about one fracture/10 m cube. (b) Percentage anisotropy versus frequency match is excellent. Maultzsch et al., 2003.

consistently modelled polarization angles were 43° (the same as the average N43°W polarization that was measured), and the *percentage anisotropy as a function of frequency*, shown in Figure 15.55b, was a reasonably good match; and excellent at frequencies above about 13 Hz.

Liu et al., 2003a, also described methods of analysis needed for interpreting multi-component VSP from the fractured gas Bluebell-Altamont reservoir in the Uinta Basin in Utah. This was the subject of the above Maultzsch et al., 2003 application of the Chapman, 2003 dynamic model of pores, ellipsoidal microcracks and aligned meso-fractures, which was used to interpret both fracture strike and fracture dimensions at this field.

Liu and co-authors emphasised the importance of frequency on the magnitude of the S-wave anisotropy, as demonstrated by the Chapman model. They also emphasised that in the Bluebell-Altamont Field VSP, the polarized *fast* S-waves actually showed no apparent variation with frequency (unless at very low frequency). However the time delays between the split shear waves decreased as the frequency increased, due presumably to apparent stiffening of the squirt phenomenon.

At the surface there were two joint sets, one striking N22° to 32°W, and the other N60° to 77°E, within about 5° of perpendicular, on average, as described by Lynn et al., 1999. However, the maximum horizontal stress (at reservoir level), was estimated as N40° to 45°W based on perpendicular borehole elongation using four-arm calliper logs in two adjacent boreholes. Geologically recent natural hydraulic fracturing (gilsonite dikes) were oriented at a consistent N40° to 45°W as well.

The extensive analyses reported by Lynn et al, 1999a, included shear wave splitting, and azimuthal P-wave response due to the fractures (the latter was reviewed in Chapter 14). Liu et al., 2003a concentrated on an analysis of the frequency dependence of the S-wave anisotropy, looking at data both above and within the 1980 to 2590 m deep reservoir. The techniques of rotation, band pass filtering, minimising off-diagonal energy and so forth, are beyond the confines of our simplified treatment.

Significantly, except for the very low frequency band below 10 Hz, the polarizations were generally between 40° to 45° for the three frequency band-widths between 10 and 40 Hz, over the whole depth interval 853 to 2636 m, both above and within all the reservoir, (see Figure 15.56a). This agreed well with the average N43°W of the interpreted major stress, but did *not quite agree with the dominant jointing*, as actually proposed in Liu et al., 2003a and Maultzsch et al., 2003.

Three distinct time-delay intervals were detected (see Figure 15.56b and Table 15.5). Within the three depth intervals there was a superimposed frequency dependence, as shown in Figure 15.56b. The steep time delay-frequency data for the reservoir (interval III) is reproduced at larger scale in Figure 15.57a, and the different gradients give corresponding estimates of frequency dependent shear-wave anisotropy percentages in Figure 15.57b.

According to Liu et al., 2003a, there were *at that time* very few reports of frequency–dependent anisotropy in the literature, from the exploration geophysics community. However, for earthquake data, Marson-Pidgeon

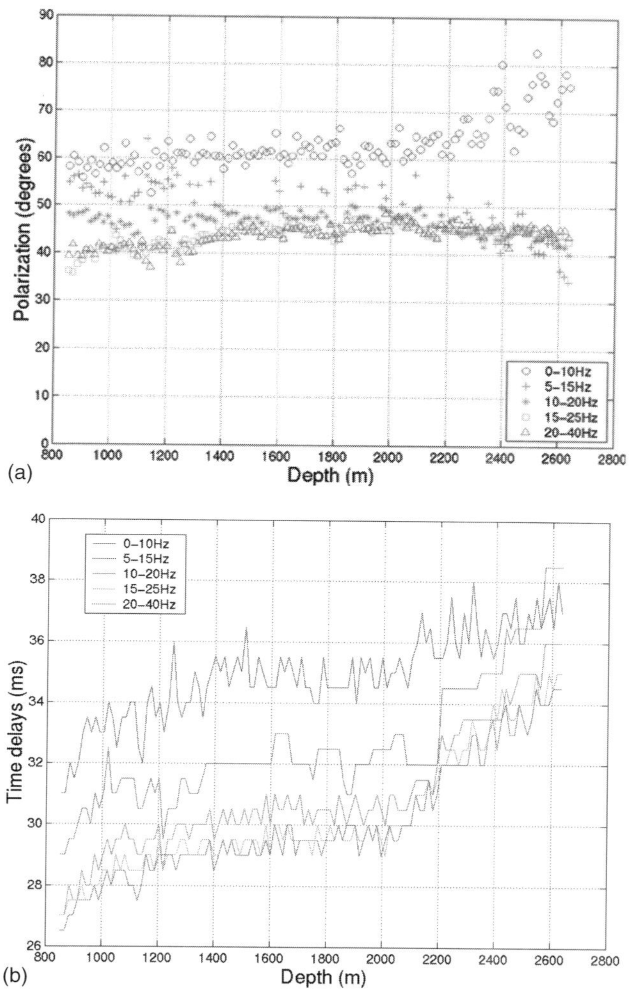


*Figure 15.56* (a) Except at low frequency (0–10 Hz), a reasonably constant polarization at 40° to 45° is shown. (b) Time delays show three intervals: *gradient*, flat, *gradient*, implying *anisotropy*, isotropy, *anisotropy*. Bluebell-Altamont Field anisotropy interpretation, from Liu et al., 2003a, with kind provision of files from Liu, pers.com. 2005. (see colour Plate 6).

Table 15.5 Time delay-depth gradients for interpreting seismically isotropic and anisotropic (fractured) depths at the Bluebell-Altamont Field. Liu et al., 2003a.

| | |
|---|---|
| Interval I, 853–1219 m | Time delays linearly increase with receiver depth increase: i.e. seismically *anisotropic*, as shear wave splitting is occurring |
| Interval II, 1219–2072 m | Time delays almost constant: i.e. seismically isotropic, as no further shear wave splitting |
| Interval III, 2072–2636 m (reservoir interval) | Time delays abruptly start increasing: i.e. strong *shear wave anisotropy* of 3 to 4%, due to splitting. Attributed to fracturing. |

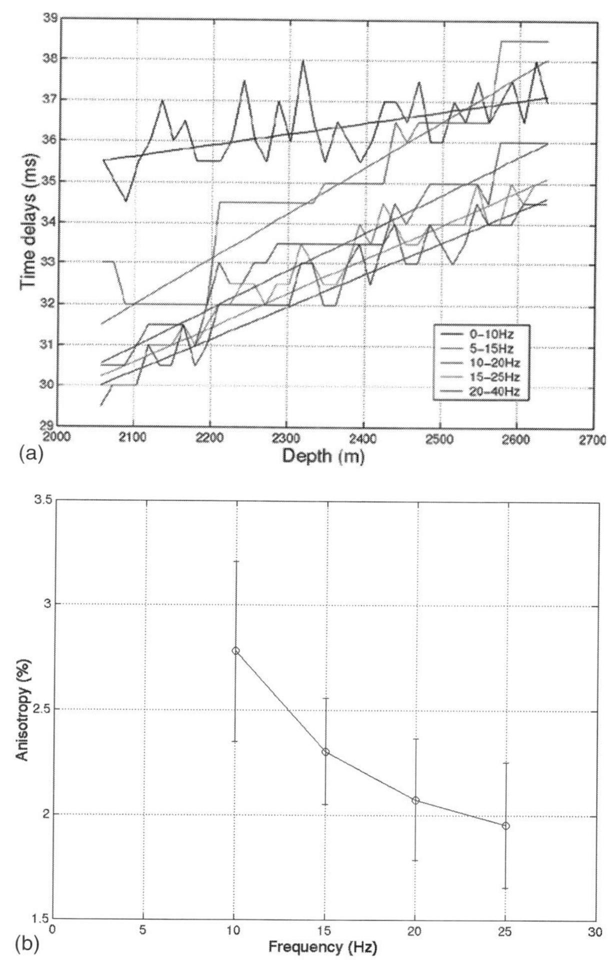


*Figure 15.57* (a) Steep time delay gradient in reservoir interval, with frequency dependence. (b) Interpretation of anisotropy percentage as function of frequency. Bluebell-Altamont Field anisotropy interpretation, from Liu et al., 2003a, with kind provision of files from Liu, pers.com. 2005. (see colour Plate 7).

and Savage, 1997, had reportedly shown a systematic decrease in time delay with increasing frequency, as also shown by the reservoir data in Figure 15.57.

Liu and co-authors suggested that if a proper mechanism (or mechanisms) for this frequency dependence could be understood, as implicitly shown in the Chapman, 2003, dynamic model of triple-scale porosity reviewed earlier, then meso-scale fractures and their fluid flow properties could be understood. Naturally, it is logical to refer to Figure 15.52 and see again from the example of Chapman triple-porosity modelling, the

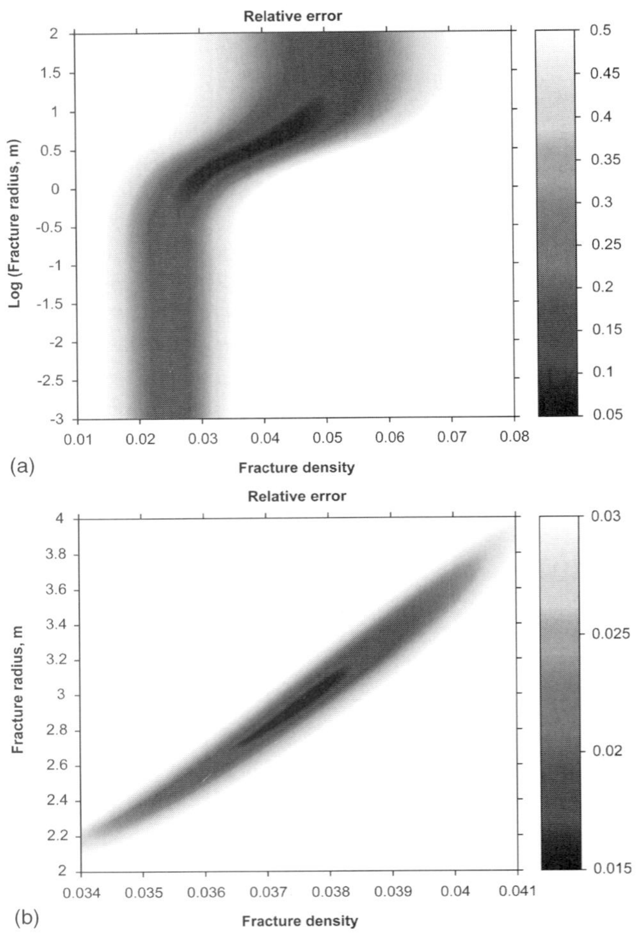


*Figure 15.58* (a) The relative error between the predicted and measured time-delay/depth, evaluated over four frequency values for a range of possible fracture densities and sizes, for comparing with multi-component shear wave VSP data acquired in the Bluebell-Altamont field in Utah. (b) The rms error zoomed around the minimum, where the error is less than 5%. (Chapman model application, by Maultzsch et al., 2003 and Liu et al., 2003b). (see colour Plate 8).

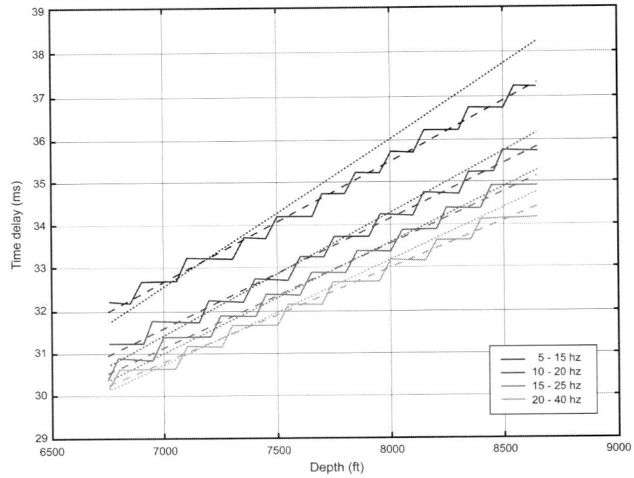


*Figure 15.59* Time-delays estimated from synthetic data for 4 frequency bands (solid lines). The real data is shown with dashed lines, showing generally excellent fit. Liu et al., 2003b.

tendency for an increased relevance of *larger-scale fractures* or joints, for explaining shear-wave anisotropy at the lower frequencies.

In a companion paper Liu et al., 2003b, also gave the Maultzsch et al., 2003 prediction of *fracture density* (appropriately re-named, from *crack density*) and mean fracture radius, obtained with the Chapman model. Figure 15.58 shows the more complete result of this analysis of the Bluebell-Altamont Field anisotropy, with interpreted fracture density of 0.04 and fracture radius of 3 m. These imply, from $e = Na^3/V$, about one of these major fractures per 9 m cube of rock.

The estimated time delays from the Chapman model (stepped) were finally compared with mean trends from the reservoir (dotted lines). (Figure 15.59). The slight discrepancy ('a matter for future investigation') could perhaps, among other more complex factors, be due to the inequality of physical aperture (E) and hydraulic (smooth parallel-plate) aperture (e), when modelling the aspect-ratio-dependent treatment of squirt flow losses and frequency dependence in the Chapman model. In other words a given aspect ratio assumption contains a fluid volume equivalent to thickness (E), but permeability in the 'real world' is governed by (e) since $K = e^2/12$ if laminar. The effect of $E/e \geqslant 1$ is caused by joint or fracture roughness (Barton et al., 1985).

If looking for more reasons for partial lack of fit between model and one may note that Olsson and Barton, 2001, showed from coupled *shear-flow* experiments by Olsson, (reviewed in Chapter 16), that the $E/e \geqslant 1$ inequality with pure normal closure is modified in a subtle way when there is joint *shearing*. There would then also be greater relative mobilization of shear stiffness, compared to normal stiffness, which is not of course modelled in the Chapman poro-elastic model, but could be the 'reality' *in situ*. There is also the possibility of the shear-related 'O/R' rotation shown in Figure 15.2b at the beginning of this chapter.

Figure 15.60, which summarises the various azimuthal relationships pertaining to the Lynn et al., 1999a Bluebell-Altamont field investigations in Utah, in fact

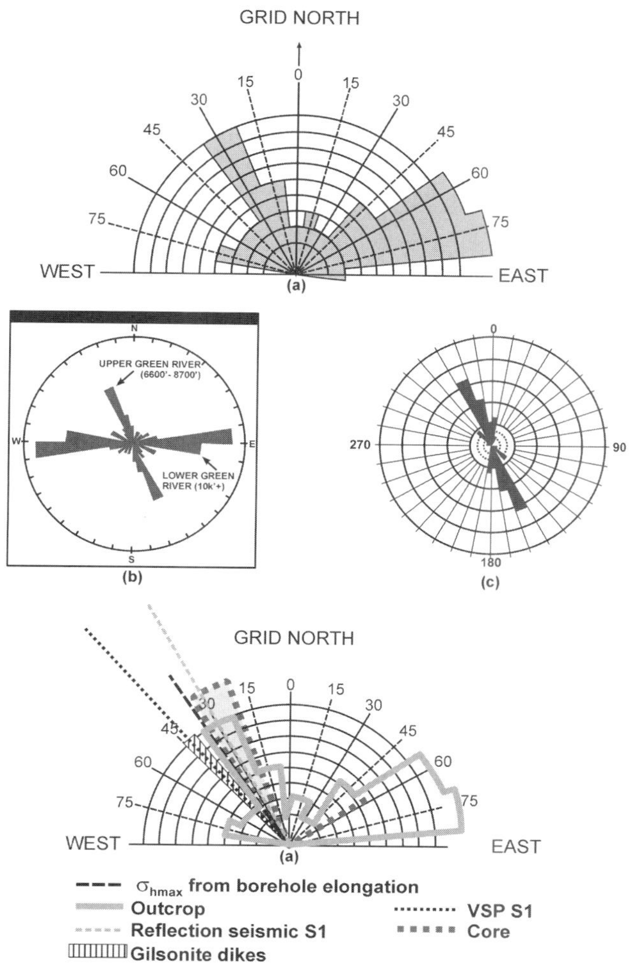


*Figure 15.60* Azimuthal fracture orientations from outcrop and core, compared to interpreted stress and interpreted seismic data, for the Bluebell-Altamont field. Based on Lynn et al., 1995 and Lynn et al., 1999a.

indicates a 15° discrepancy between *reflection* seismic interpretation of S1, and the polarized shear-wave interpretation of S1. This actual *small discrepancy* of the polarization directions for the supposed 3 m fractures, in relation to the oriented fractures in the core (a possible 15° to 25° discrepancy) could be due to experimental/acquisition errors (Lynn, 2005 pers. comm.), but might also be due to a phenomenon such as illustrated in Figure 15.2b.

Liu and co-authors emphasised that the Chapman model reverted to a grain-size squirt flow model when without fractures. But with the introduction of a fracture set, two characteristic frequencies then existed: the traditional lab-scale squirt frequency and a lower frequency that depended on fracture sizes. This meant that dispersion (frequency dependent velocity) could occur down to seismic frequencies, as we have seen in Figure 15.52.

### 15.9.4 The SeisRox model

There are an increasing number of numerical models that can represent the effect of a single set of fractures (typically vertical) on the otherwise transversely isotropic behaviour of a layered reservoir. Two contrasting reservoir idealizations were presented by Johansen et al., 2004 and are shown in Figure 15.61. A reminder of the theoretical velocity components, kindly provided by Johnsen (pers. comm., 2005) is shown in Figure 15.62.

a) an isotropic medium without fractures (Figure 15.61a)
b) an transversely isotropic material with one set of fractures giving a horizontal symmetry axis (Figure 15.61b)

Johansen et al., 2004, coupled a rock physics model with a finite difference scheme for visco-elastic seismic modelling (termed SeisRox), to represent the coupling between fluid-filled fractures and pores. They demonstrated a stiffening behaviour (increased velocities), as frequency was increased. The contrasting effects of their models of pores only (Model A) and communicating pores and fractures (Model B), are shown in Figure 15.61.

The properties assumed for their fractured reservoir layer (Table 15.6), gives insight into the level of modelling detail. Note the rather large aspect ratios for the assumed cracks, which fall short of actual fracture (or joint set) simulation in this particular case.

To emphasise the links between stiffness, velocity and attenuation, we may refer to the published article (Johansen et al., 2004) in which the vertical axes of Figures 15.61a and b were given the alternative nomenclature 'real part of stiffness' in place of velocity, and 'imaginary part of stiffness' in place of attenuation. The four sets of curves of behaviour for the six 'V' and 'Q' components were unchanged, in this alternative nomenclature, yet in place of a velocity scale of 2.0 to 4.5 km/s, they gave a stiffness scale of 5 to 50 GPa, and in place of an attenuation scale (1/Q) of 0 to 0.20 (or minimum Q = 5) they gave a stiffness scale of 0 to 6 GPa (the 'imaginary part of the stiffness').

It is of interest at this juncture, to refer to Figure 15.33b, for the purpose of cross-discipline connections, suitably tempered by the dynamic situation. This diagram

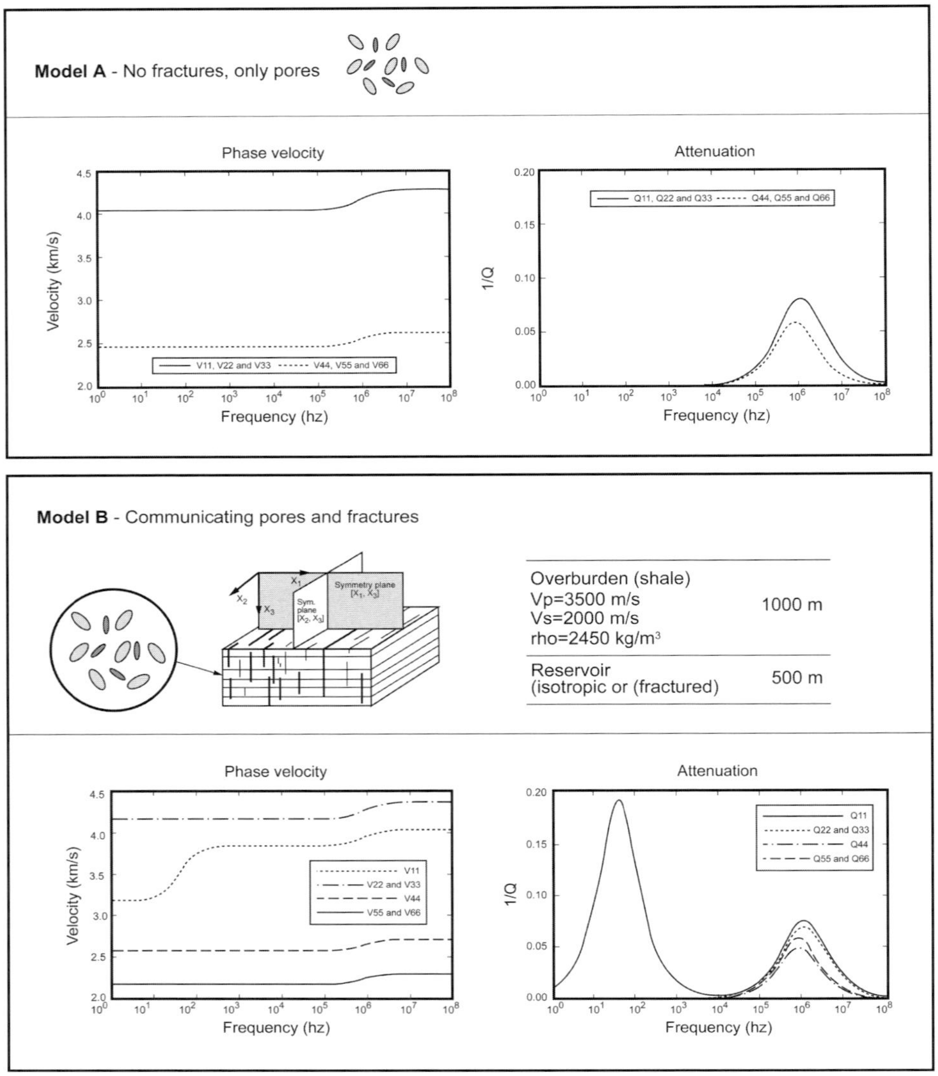


*Figure 15.61* Modelled wave speeds and attenuation as a function of frequency, using the SeisRox model, for (a) an isotropic medium with random pores, (b) communicating pores and oriented vertical cracks in a transversely isotropic medium with horizontal symmetry axis. Johansen, pers. comm., 2005.

shows the empirical link between P-wave velocity from refraction seismic, and the *static* deformation modulus M (as opposed to the *dynamic* value obviously just referred). The relationship has 'common ground', namely, the rock quality Q or $Q_c$ value, which can however be dispensed with, when the proposed direct link from seismic $V_p$ and static modulus (M) is used (Barton, 2002a).

It may be noted, for purpose of comparison, that the above 2.0 to 4.5 km/s range of velocities, with their equivalent (dynamic) 'real part of stiffness' of 5 to 50 GPa, have a predicted static moduli (M) range of 3 GPa to 22 GPa, respectively. The maximum attenuation (1/Q) scale value of 0.20, giving the particularly low $Q_{seis}$ value of 5, and an equivalent (dynamic) 'imaginary part of stiffness' of 6 GPa, following Johansen et al., 2004, would (falsely) give a low $Q_c$ estimate of about 0.2 ('very poor rock quality') if one temporarily ignored the $E_{dynamic} > M_{static}$ norm, and used the empirical link $M = 10\ Q_c^{1/3}$ shown in Figure 15.33b.

### 15.9.5 Numerical modelling of dynamic joint stiffness effects

In the foregoing double-porosity models we have witnessed the use of crack representation by aspect ratios.

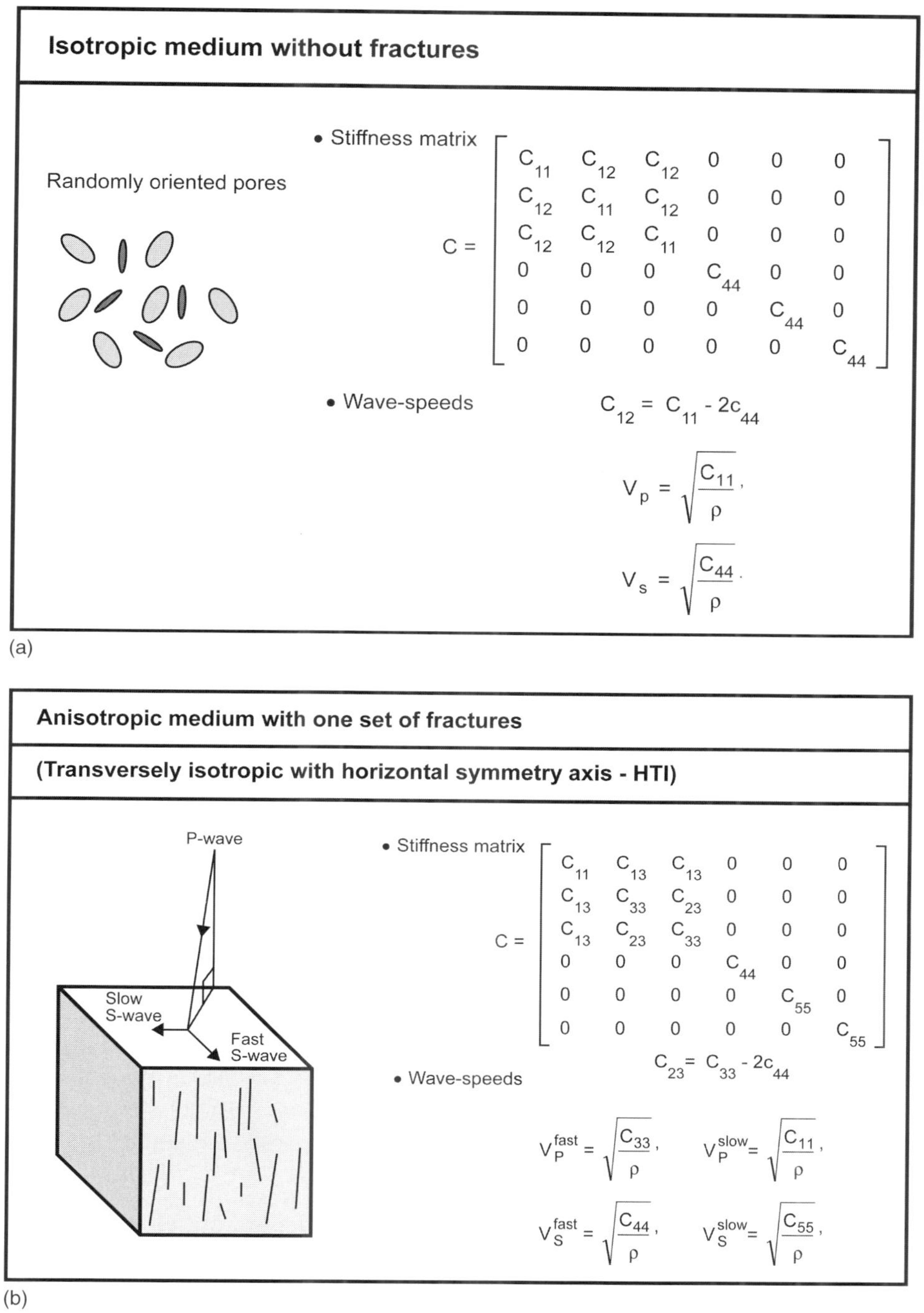


*Figure 15.62* The stiffness matrices showing the velocity components modelled in Figure 15.61. Johansen, pers. comm., 2005.

Important insight into the influence of (dynamic) joint stiffness and joint spacing effects on seismic dispersion is provided by the modelling of Monsen, 2001. The motivation for his study was the interpretation of high frequency investigations of rock masses, in which the wave length may be less than the joint spacing, for instance in cross-hole seismic at frequencies of several kHz. Monsen first explored the influence of *single fractures*, and assumed dynamic normal stiffness values (100, 1,000 and 10,000 MPa/mm) equivalent to widely different normal stress levels, from e.g. close to surface to kilometre depths, as this range indicates (see Chapter 16).

Table 15.6 Parameters used to compute effective visco-elastic properties of reservoir rock. (Johansen et al., 2004).

| | |
|---|---|
| Porosity (%) | 15 |
| Permeability (mDa) | 50 |
| Clay content (%) | 15 |
| Viscosity (cP) | 1 |
| Fluid density (kg/m$^3$) | 1000 |
| Fluid velocity (m/s) | 1500 |
| Aspect ratio quartz-related pores | 0.15 |
| Aspect ratio clay-related pores | 0.027 |
| Aspect ratio cracks | 0.01 |
| Crack density | 0.1 |
| Effective density (kg/m$^3$) | 2383 |

The soft-fracture near-surface simulation, showed the single fracture to be a good reflector, with a wide range of low to medium transmission coefficients (T) until frequencies were lower than 1 kHz, when T became unity. The stiffest, deepest simulation, showed transmission coefficients of mostly 1.0, except at frequencies higher than 100 kHz. In effect the time delay here was close to zero except at the highest frequencies. There was naturally more delay and attenuation as the dynamic fracture normal stiffness was reduced.

The effect of multiple parallel fractures on the group velocity, as compared to a 'sedimentary rock' matrix velocity of just below 2490 m/s is illustrated in Figure 15.63. The same range of dynamic normal stiffness as above was used (i.e. from near-surface to several kilometres equivalent depths), and three alternative spacings of 0.1, 1.0 and 10 m were used. The reduced velocity with small fracture spacing is of course a familiar trend in near-surface rock mass quality (Q) relations to $V_P$, with limited effect on velocity (and rock quality), when the spacing is as high as 1 or particularly 10 meters.

However, as also suggested by the results in Figure 15.63, when equivalent depth is of kilometre to several kilometres magnitude, a fracture spacing as small as 0.1 m has small, or negligible effect on the *perceived* velocities with dynamic stiffnesses of $K_n$ = 1,000 and 10,000 MPa/mm, unless at frequencies beyond 10 kHz. This is a good illustration of the benefit of high frequency cross-well measurements, in particular in view of the long transmission distances of high frequency waves at depth, when joints and fractures are stiff, as also referred to in Chapter 14.

When modelling wave propagation through jointed or fractured rock with distinct element codes like

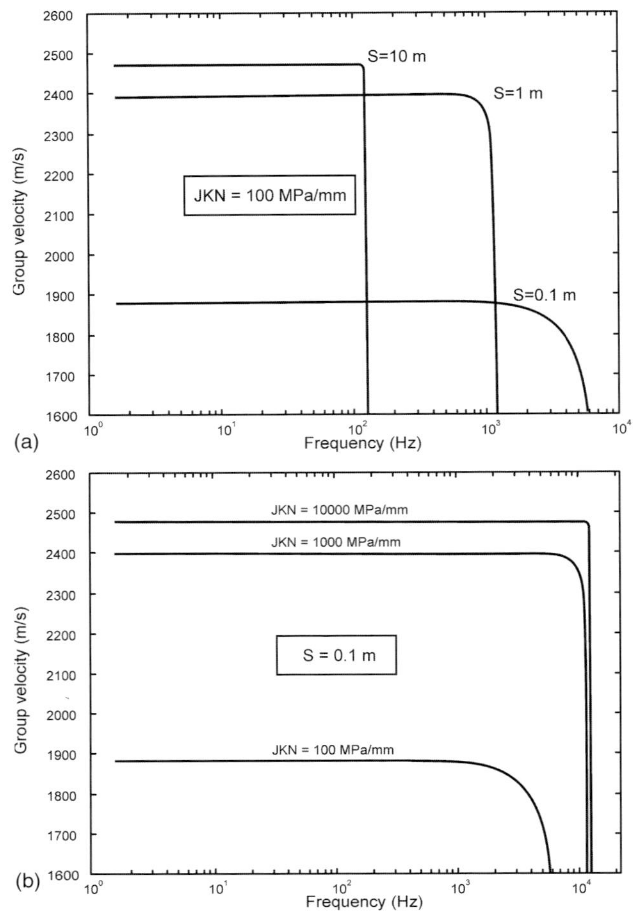


*Figure 15.63* (a) Group velocity versus frequency for three fracture spacings, at a constant (near-surface, and very low) dynamic normal fracture stiffness of 100 MPa/mm. This behaviour corresponds quite accurately with near-surface rock quality (Q) versus *seismic* velocity expectations. (b) Group velocity versus frequency with a constant fracture spacing of only 0.1 m, and the full range of dynamic normal stiffnesses of 100, 1,000 and 10,000 MPa/mm. The benefit of cross-well, high frequency surveys is demonstrated quite well by these scoping studies. After Monsen, 2001.

UDEC, the provision of realistic input data concerning joint or fracture dynamic stiffnesses is fundamental for a representative result. In Figure 15.64, we see the effect of a more realistic 'jointed rock' (in 2D), on the attenuation of wave amplitude, on velocity reduction, and on frequency change. The centre of each model (intact or jointed) is the source location, with a measurement point at the boundary in each case. (Non-reflecting/absorbing boundaries are of course used). In the cases

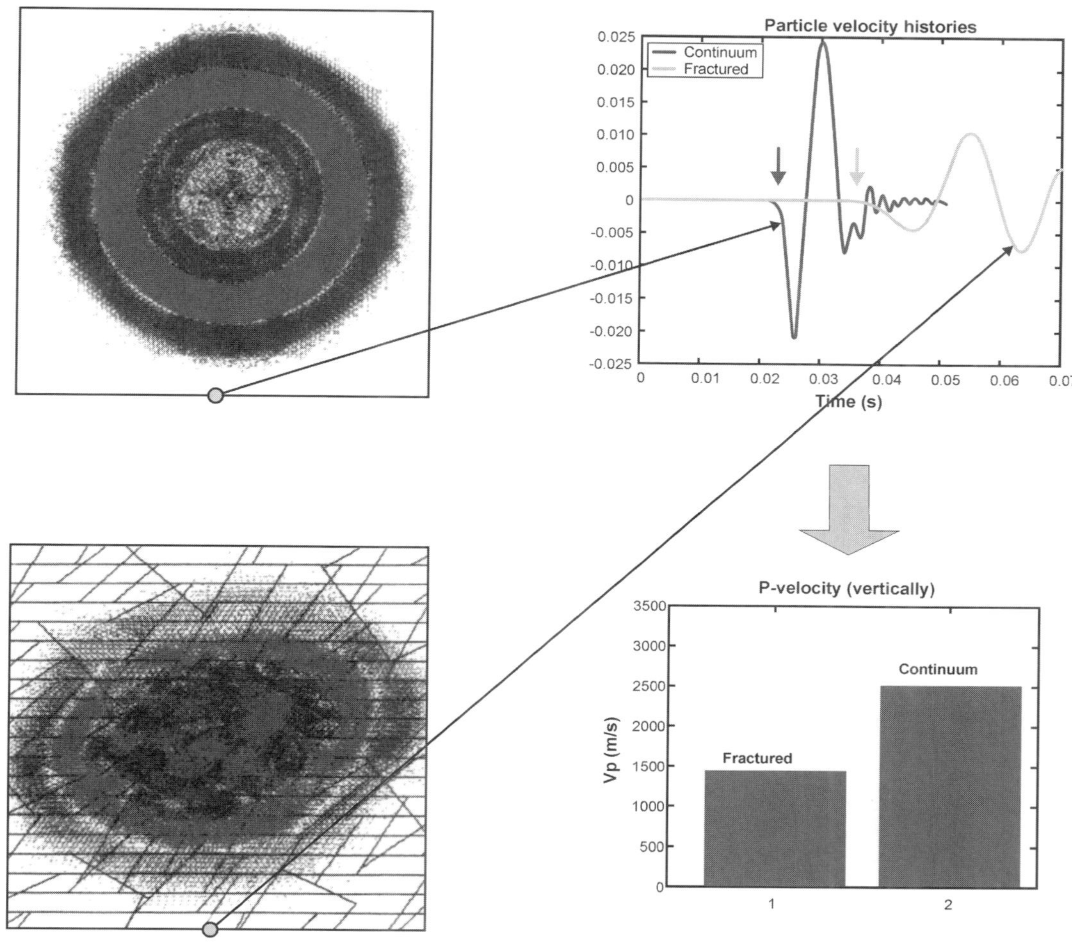


*Figure 15.64* An illustration of wave propagation differences between intact and realistically jointed (but 2D) models, with dynamic normal and shear stiffness input estimates appropriate to the dynamic wave form. Note the multiple effects of (vertical) wave transmission through the layered and jointed medium: i.e. the delay, the amplitude reduction (i.e. attenuation), the velocity reduction and the frequency reduction. (Monsen, 2005 priv. comm).

illustrated, the dynamic stiffnesses assumed were within the range studied by Monsen, 2001. There is at present no matrix flow modelling in UDEC.

### 15.9.6 A 'sugar cube' model representation

A practical 'sugar-cube' model with three orthogonal joint or fracture sets was proposed by Brown et al., 2002, as a means of integrating seismic data to production data. The authors combined the ideas of Oda, Kachanov, Schoenberg and Sayers to produce the rather obvious representation of a multiply-jointed rock mass, where three, more or less mutually perpendicular joint sets is almost a rule rather than an exception. Such a representation for the *permeability* of rock masses at dam sites was used by Snow, 1968, and is utilised for grouting predictions in rock engineering (e.g. Figure 15.49).

Brown et al., 2002 assumed that both the fracture permeability tensor and the fracture compliance tensor would 'diagonalise' in the same principal coordinate system. The model predictions for P-wave and S-wave phase velocities, contrasting the matrix velocity with the effect of either a drained or brine-filled fracture network is shown in Figure 15.65.

The authors noted that when the compliance of the fractures was large, as in the field, where fractures tend to be weaker, the effect of fluid changes on the velocity would tend to be dominated by effective stiffness increases due to e.g. brine saturation compared to drained or gas filled. The authors drew parallels to the Natih field experiences described by van der Kolk et al., 2001, which were reviewed earlier in this chapter.

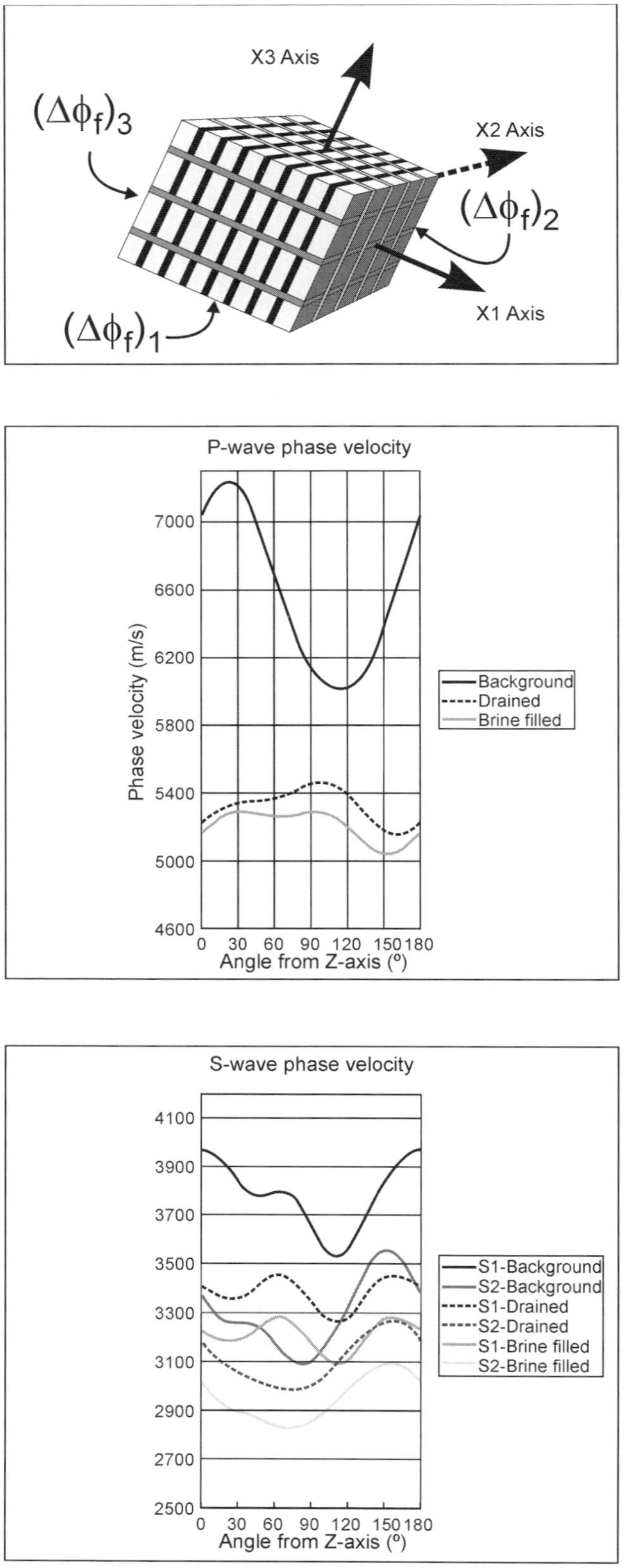


*Figure 15.65* A 'sugar-cube' model of compliance and permeability, for integrating seismic and flow measurements. Brown et al., 2002.

## 15.10 A porous and fractured physical model as a numerical model validation

The technique for constructing synthetic, fractured sandstones by Rathore et al., 1995, was referred to earlier, as a means of basic validation of numerical models. The (future) cracks are placed as metallic discs in a porous sand-epoxy mixture, and subsequently leached out, leaving empty voids. They thus have a known aspect ratio, size, position and orientation and are a close approximation to basic aspects of dual-porosity theoretical models. (Both however have the potential weakness in relation to reality, that stress and displacement is not transferred across the crack faces, so E = e, since roughness may be absent).

The crack geometry studied by Skjærstein and Fjær, 2000, consisted of 1425 circular parallel cracks per $dm^3$, with all cracks of 8.25 mm diameter and 20 $\mu$m aperture. The crack density (number of cracks per unit volume × crack radius cubed) was 0.1. A representation of the crack-modelling principles is shown in Figure 15.66a.

The authors pointed out that the (Rayleigh wave) scattering was strongly dependent on the wavelength-to-crack size ratio, and therefore on the frequency for a given crack size. Since the velocity anisotropy was not equally dependent on frequency at high ratios of wavelength-to-crack-size, the big discrepancy between $V_{min}/V_{max}$ and $A_{min}/A_{max}$ should decrease with decreasing frequency.

With the crack size studied by Skjærstein and Fjær, 2000, the P-wave velocity anisotropy was greatly reduced by saturation, while in earlier models studied by Rathmore et al., 1995, there was significantly less reduction in velocity anisotropy with saturation, due probably to the smaller size of cracks that were used, giving different conditions of pore pressure equilibration.

Skjærstein and Fjær, 2000, testing first dry then saturated cracks, found that the attenuation was strongly dependent on the incidence angle, as indicated in Figure 15.66b and in Figure 15.67. The attenuation, which was caused by scattering, distorted the wave forms and changed their frequency content. The angular dependence of attenuation was affected by the frequency (e.g. 100, 250 or 500 kHz). In the case of the 100 kHz excitation pulse applied across the 16-sided prismatic, fracture-filled sample, the wavelength-to-crack diameter ratio was in the range of 1.7 to 3.0.

Under the experimental conditions investigated by Skjærstein and Fjær, 2000, the *attenuation anisotropy* was

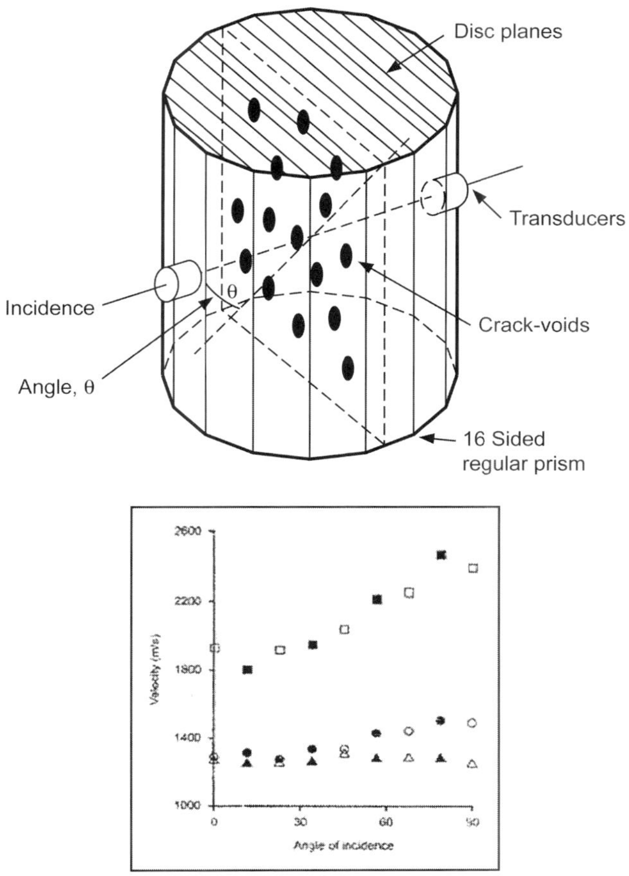


*Figure 15.66* (a) The prismatic, sixteen-sided, crack-bearing synthetic sandstone used by Skjærstein and Fjær, 2000. (b) A comparison of P-wave and polarized S-wave ($qS_1$ and $qS_2$) results using variable angle of incidence, for the two cases of i) through-going spot-welded fractures: open symbols, and ii) small parallel cracks: closed symbols. Squares are P-waves, circles/triangles are S-waves polarized parallel/quasi-normal to the cracks. Fjær, 1997.

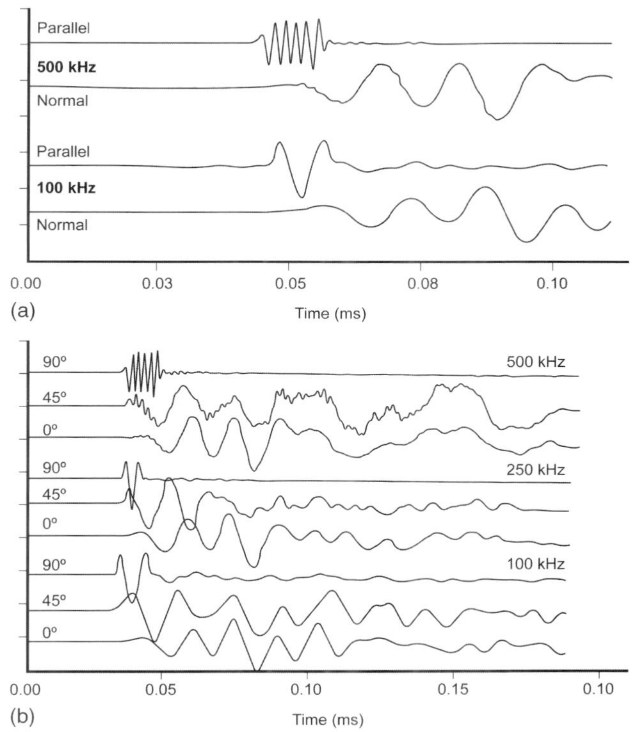


*Figure 15.67* (a) Recorded P-wave signals propagating parallel or normal to dry cracks in a synthetic, crack-bearing porous sandstone (D/t = 8.25/0.02 mm/mm) from Fjær, 1998. (b) Recorded P-wave signals when using three different excitation frequencies, in a dry crack-bearing synthetic, porous sandstone. A five-cycle pulse was used at 500 kHz. These samples had 1425 circular parallel cracks/$dm^3$ with the same D/t measurement as above. Skjærstein and Fjær, 2000

much more visible than the velocity anisotropy. Also at higher frequencies, where the P-wave velocity anisotropy vanished, the attenuation anisotropy remained high and clearly visible. Likewise, when saturated, causing a significant reduction in the P-wave velocity anisotropy, the attenuation anisotropy remained high. The authors recorded $V_{min}/V_{max} = 0.80$, and $A_{min}/A_{max} = 0.01$ for the same wave frequency.

In Figure 15.66b, we may note one of the 'complicating' factors concerning the diagnostic use of P- and S-wave polarization. Fjær, 1997, presented the P-wave and polarized S-wave results in parallel for i) through-going but spot-welded cracks, and ii) small parallel cracks. The closed symbols in this figure represent 5.5 mm diameter cracks of 20 μm aperture, while the open symbols represent the through-going but spot-welded cracks, where the contact spot diameters were 1.4 mm and the open parts the same 20 μm. Each case studied was dry. These two special (and rather theoretical) geometries were chosen to represent a Hudson, 1981, type of crack distribution, and a White, 1983, model with spot-contact points on the larger fracture planes.

When the wave-length was many times larger than the crack diameters, the cracked material effectively appeared as a homogeneous medium with reduced stiffness, while at higher frequencies, the smaller wave length-to-crack diameter ratio effectively made the medium appear more heterogeneous. Shear wave splitting with fast and slow components then showed greater contrast.

# 16 Joint stiffness and compliance and the joint shearing mechanism

This final chapter is designed to act as a cross-discipline reference point between rock mechanics and engineering geological behaviour in the 'static' world of slow-and-macro deformation processes, and the geophysicists 'dynamic' world of fast-and-micro deformation and attenuation processes. That there are important links between the two in terms of joint or fracture compliance and its inversion: stiffness, and in terms of rock quality, deformation modulus, and seismic quality, has been established in various contexts in the chapters of Part II. In particular, this last chapter attempts to extend current thinking regarding fractured reservoirs, 'open' joints, and assumed $\sigma_{H\ max}$ parallelism, to also embrace the possibility, even probability, that multiple-joint-sets and shear-dilation-conductivity coupling are needed, due to the inevitable tendency for joints in less competent rocks to close at reservoir depths, even when under the influence of 'only' $\sigma_{h\ min}$. Shear wave splitting and polarization from two sets of conjugate (or $\sigma_{H\ max}$ straddling) joints can also appear to satisfy 'open joints were parallel to $\sigma_{H\ max}$' assumptions. The important improvement would be that rock mechanics theory is not violated, when assuming that unsheared joints in weaker reservoir rocks can be conductors, when actually under high levels of effective stress. Shearing to great depth has been verified in other areas of the earth-sciences, and the need for adoption in the geophysical interpretation of petroleum reservoirs is discussed, with simple joint-model illustrations. The 'critical shearing crust' is interpreted here in terms of the non-linear Barton-Bandis shear strength and coupled behaviour constitutive model. An evaluation of the importance of joint roughness at several scales, and the need for dilation-corrected stress transformation in the earth sciences is also treated.

## 16.1 Some important non-linear joint and fracture behaviour modes

In the Chapter 15 review of some of the recent dynamic, poroelastic, multiple-porosity modelling, mostly only one set of 'perfectly' aligned, or partially aligned, sets of cracks, fractures (or rock joint analogues) was seen. The mathematical complexity of multiple sets clearly presents numerical problems, but no doubt such will be solved in due time, as it was, with simpler boundary conditions, in rock mechanics, due to the need for modelling hydraulically communicating, deforming, multiple joint sets, but without the dynamic interaction with grain-size pore space.

The deliberate addition in Figure 16.1, of a secondary 'crack set' to the Tod, 2002 conceptual model of a variably oriented single set of cracks, fractures or joints, is to emphasise the obvious: that fractured (or jointed) reservoirs often have at least two sets of joints or fractures, that can assist in the drainage of the matrix towards the wells. Primary drainage pathways are usually considered

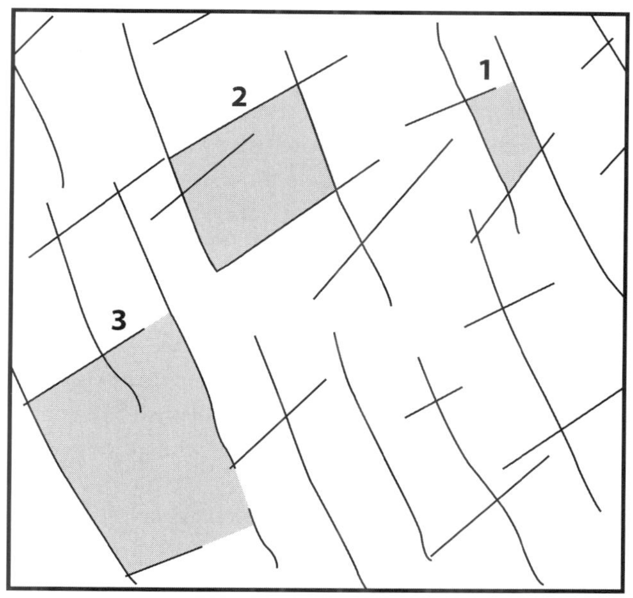


*Figure 16.1* A secondary set of joints (i.e. natural fractures) have been added to the Tod, 2002 schematic, to emphasise that in many cases there will be additional intersecting sets of joints in fractured reservoirs. Joints may also be under a state of shear stress, rather than the implied normal-stress-only.

to be along the set that approximates the $\sigma_H$ maximum direction, as we have seen, with some variations, in the shear wave splitting analyses reviewed in the last chapter. This conventional assumption may be modified however, if the rock is weak and porous, since conducting joints might be too tightly closed, or not exist, at depths of kilometres, unless there was shearing. This aspect has dominated the selection of material for this chapter.

In case of significant *non-alignment* of the present-day $\sigma_{H\ maximum}$ direction with the dominant set (or sets) of joints, then shear-dilation-conductivity coupling becomes a potential mechanism for effective reservoir drainage from matrix-to-joints-to-producing well. The range of effective block sizes illustrated in Figure 14.1 (1, 2 and 3 are examples) will then also come into consideration, since block size will determine how much shear-dilation-permeability coupling has been possible, especially with elevated pore pressure.

Scaling of the small-scale roughness $JRC_o$ to effective roughness $JRC_n$ for the appropriate block size, following the Barton-Bandis constitutive model, suggests that case 3 with the largest block size, can have presented the lowest shear stiffness (and shear strength) to the historic anisotropic stress that may have caused shearing in the past. A sheared-joint situation would presumably tend to enhance the velocity anisotropy (both for azimuthal P-waves, and for split shear-waves).

A cross-disciplinary reference to the (macro-deformation) hyperbolic normal-closure behaviour of rock joints, and to the scale effect on shear strength caused by relative block-size, are key items in this 'sheared-earth' thinking. These macro-deformation aspects of jointed rock behaviour are shown at the top of Figure 16.2, taken from Bandis et al., 1983, and 1981. This *macro-deformation* behaviour is shown here, in view of the influence it could have on the *micro-deformation compliances* of joints that were 'lying close' to the respective rock mass *stiffness* curves. Of course their will be differences between the inverted magnitudes of dynamic compliance and 'static' stiffness, but differences may be minor when rock quality is reasonably high.

The assumed stress-deformation behaviour of three categories of rock mass, as visualised by Barton, 1986, are shown below the normal-closure (N) and shear-displacement (S) components, in Figure 16.2c, and in Table 16.1. There is large scale, up to 8 $m^3$ biaxial flat-jack loading, and uniaxial plate-loading verification of these three categories of non-linear and hysteretic behaviour. (The linear load-deformation line in case B appears to be due to two partly opposed non-linear mechanisms).

Numerical discrete element modelling of these three categories of jointed structures, was performed by Chryssanthakis et al., 1991, with UDEC-BB, incorporating the Barton-Bandis non-linear and scale-dependent joint behaviour laws, to confirm this type of rock mass stress-deformation behaviour. Type A and Type C can be considered typical of many reservoir jointing scenarios (greatly simplified), with sedimentary bed-limited jointing in Type A, often with a vertical major principal stress, but with an intermediate horizontal principal stress that could cause fracture-set alignment. Type C, representing conjugate jointing, is of tectonic origin, as for instance in the anticlinal Ekofisk reservoir in jointed chalk. The deformation, joint-shear and principal-stress plots from this numerical modelling are shown in Figure 16.3.

The one-dimensional application of the Barton-Bandis normal closure modelling, shown in Figure 16.4, from Barton et al., 1985, follows the multiple load-unload consolidation cycles performed on numerous natural rock joints by Bandis, 1980. In the discrete element UDEC-BB modelling performed in rock mechanics analyses of e.g. slopes, tunnels and caverns for civil engineering use, we automatically follow the 'fourth cycle' which is assumed to approximate the 'undisturbed' in situ condition. There is naturally an exaggerated *hysteresis* (and closure) in the first cycle of a stress-closure test on a sampled (= disturbed) joint. In the context of the micro-displacements experienced by joints with the passage of seismic waves, it is clearly the consolidated state one is concerned with.

Hatchell and Bourne 2005, have recently highlighted the importance of *hysteresis* in the interpretation of time-lapse time shifts surrounding depleting, compacting reservoirs. Depletion induced time shifts have opposite sign inside and outside the reservoir. Laboratory testing of velocity-effective pressure needs to also be focussed on unloading and on strain behaviour.

Models that use linear slip theory, following earlier Schoenberg work, and therefore require fracture compliances rather than aspect ratios, are gaining ground in geophysics, as we have seen from some of the modelling reviewed in Chapter 15. Here we will consider a remark by Gurevich, 2002, who describes the theoretical construction of a poro-elastic model with aligned fractures that obey linear-slip theory, and are characterized by 'excess' normal and tangential fracture compliances, following Schoenberg and Sayers, 1995. Gurevich commented on the fact that several authors had expressed, intuitively, the likelihood of a link between normal and

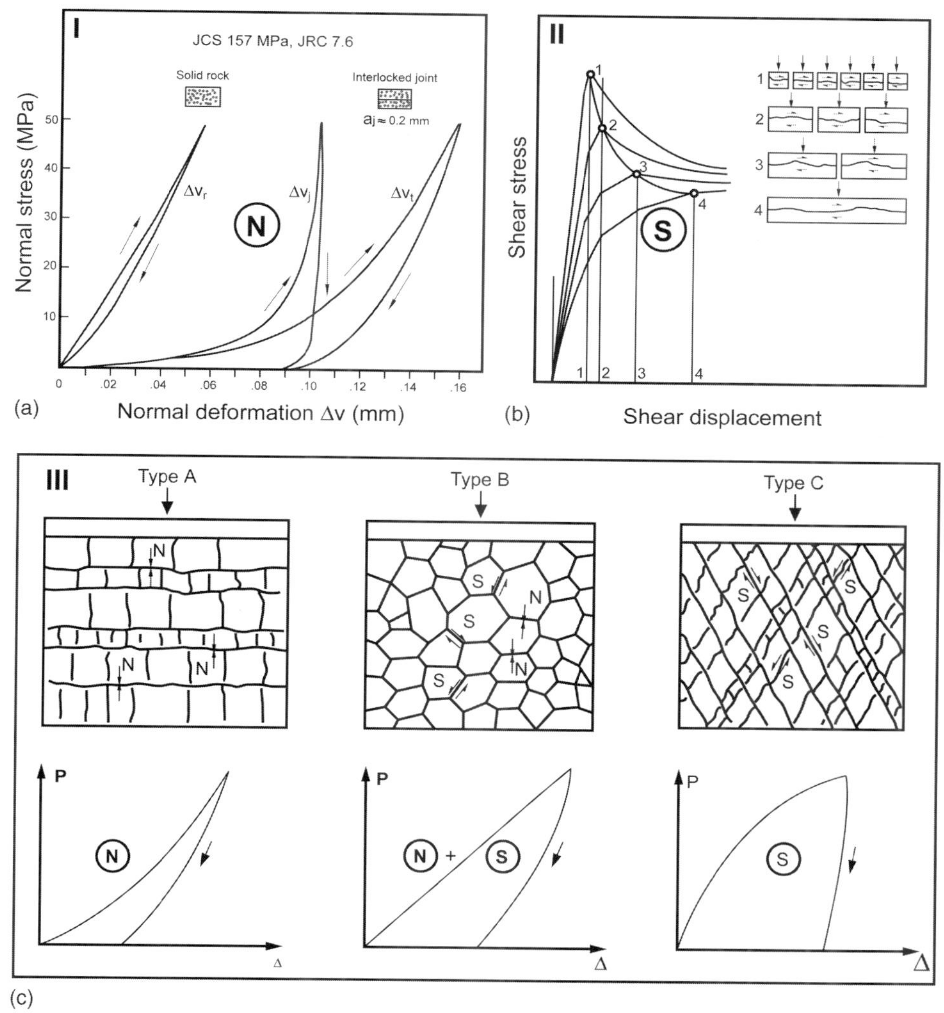


*Figure 16.2* a) Non-linear, hysteretic normal-closure behaviour of rock joints. b) Shear strength dependency on block size, also showing strong non-linearity and two reasons for strong shear stiffness scale effects. c) Conceptual pressure-deformation curves due to N and S components, or their combination. From top left: Bandis et al., 1983, Bandis et al., 1981, and Barton, 1986.

Table 16.1 Three characteristic load-deformation behaviour-modes for rock masses.

| Type | Dominant mode | Shape | Hysteresis | Lateral expansion* | Poisson $\nu$ |
|---|---|---|---|---|---|
| A | normal | concave | small | small | low |
| B | normal + shear | linear | moderate | moderate | medium |
| C | shear | convex | large | large | high |

* Under conditions of *one-dimensional strain* the relative magnitudes of these differences in behaviour will be reduced. Reduction in pore pressure, and compaction of the matrix, would allow a Type C mechanism in a compacting reservoir to develop further, as for instance in the Ekofisk reservoir.

tangential fracture compliances and the porosity of the host medium.

From the rock mechanics world of inverted compliance, i.e. stiffness (where $K_n$ and $K_s$ stand for normal and shear stiffness in units of MPa/mm), there is also an *implied* linkage to porosity, since the term JCS, the *joint wall compressive strength*, and its ratio with effective normal stress ($JCS/\sigma_n'$) is an important component of the equation for shear stiffness and peak shear strength (Barton and Choubey, 1977), and of the hyperbolic equation for normal stiffness (Bandis, 1980). The value of JCS is obtained from rock mechanics index testing, using a Schmidt (L-) impact hammer, the rebound of which (10 to 95% approx.) can be linked to uniaxial compression strength $\sigma_c$ (or UCS), also using rock density, following Miller, 1965. Clearly porosity will therefore affect this conversion of rebound to JCS (or to matrix UCS). Hence the connection to stiffness, and therefore perhaps compliance.

The Schmidt hammer is used as this is almost the only way to record the reduced strength of the few millimetres of (eventually weathered, or altered) joint walls. In other words JCS $\leq \sigma_c$, unless there has been mineralization at some time, as shown in the photographs (Figure 16.5, a and b) of a wave-cut platform in dolomite, from Kimmeridge Bay in Southern England. Note the complete cementation of all joint sets in Figure 16.5c, which probably removes the need for a roughness and wall-strength based classification method.

## 16.2 Aspects of fluid flow in deforming rock joints

Several recent workshops have focussed strongly on seismic anisotropy, suggesting an active recognition of the

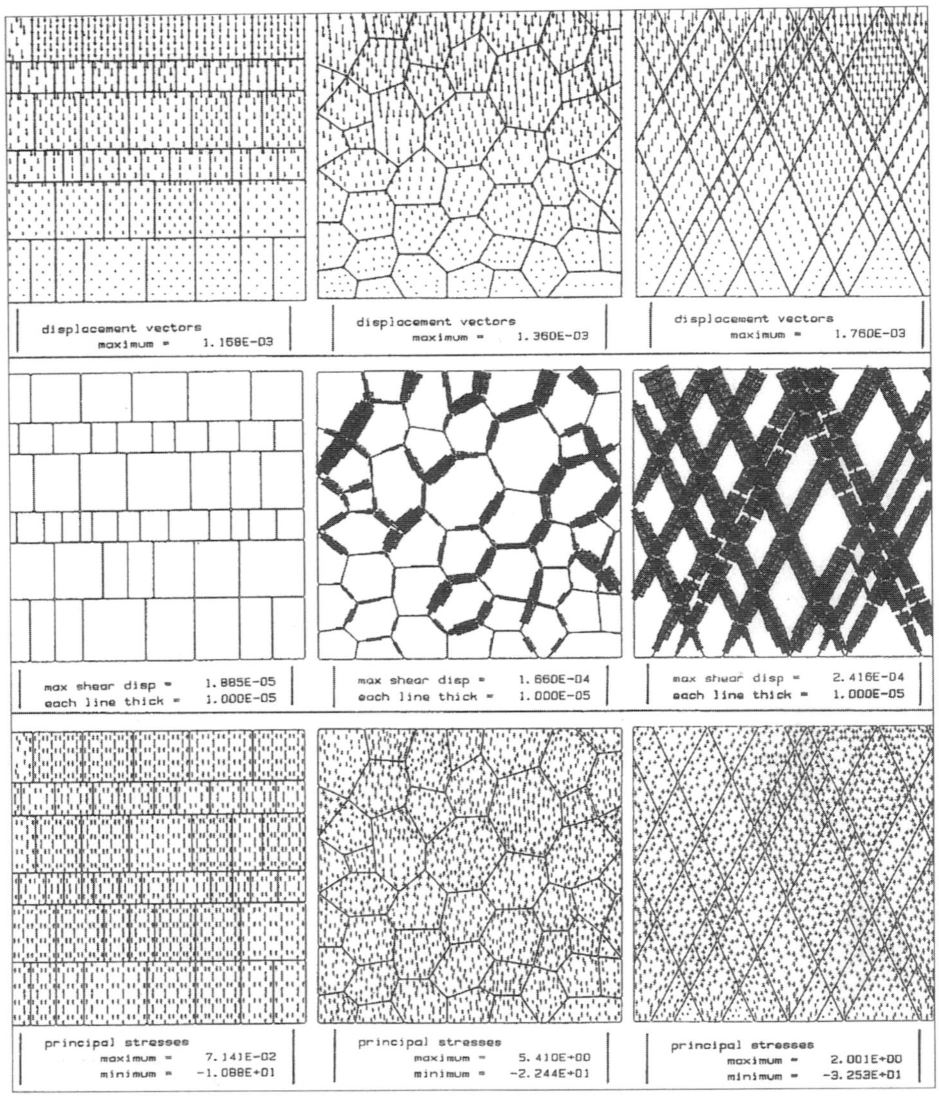


*Figure 16.3* Distinct element modelling with UDEC-BB code, with the Barton-Bandis joint behaviour sub-routine, demonstrates the relative magnitudes of the shear and normal deformation components. Top: deformation, Centre: joint shearing, Bottom: principal stresses. Chryssanthakis et al., 1991.

over-riding importance of dominant fracture sets (or joint sets in engineering geology terms) for drainage of reservoirs. This is particularly important where matrix permeabilities are low, yet storage in the form of porosity, obviously high enough for commercial development, sometimes to an exceptional degree. The Ekofisk field in the North Sea, is expected to have at least 80 years of production from this well jointed reservoir.

### 16.2.1 Coupled stress-flow behaviour under normal closure

Coupled stress-flow-deformation joint behaviour has long been of concern in rock mechanics, where there is a 'standard' test called the coupled shear flow test (CSFT), developed by Makurat at NGI, in Norway,

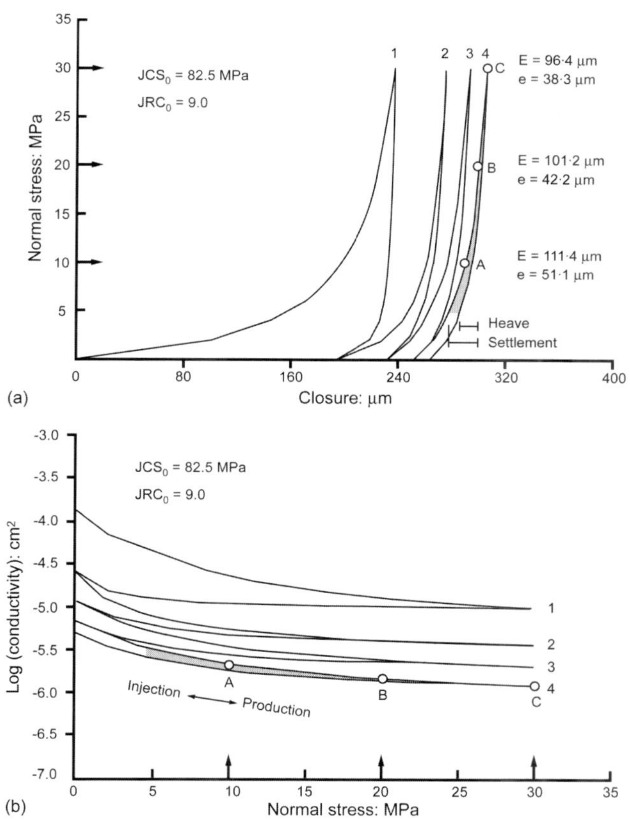


*Figure 16.4* Normal closure and conductivity cycles and *hysteresis* from the BB model. An imaginary injection-production cycle is highlighted in gray. Barton et al., 1985. (The importance of *hysteresis* in time-lapse interpretation both inside and outside a compacting reservoir was recently emphasised by Hatchell and Bourne, 2005.)

(Makurat et al., 1990), and first used for extensive studies of the conjugate joints (or fractures) in the Ekofisk chalk, to investigate oil and brine flow under normal and shear deformation, also with the effect of elevated temperature. Various aspects of this work with laboratory-scale testing of reservoir joints and reservoir-scale deformation modelling in the chalk were described by Barton et al., 1986.

Although dispersive (frequency-dependent) attenuation and velocity are dynamic concepts, involving *micro-responses* in the interior of any rock joint, it seems intuitively likely that the *macro-responses* of joints to coupled deformation and flow can represent 'end-members' for the dynamic loading and squirt attenuation concepts of geophysicists.

The normal and shear *compliances* (units of m.GPa$^{-1}$) presently utilised in geophysics, have their rock mechanics inverse values, termed normal and shear *stiffness* (units of GPa/m). Their (inverted) magnitudes are remarkably similar, especially in the case of $K_n$, suggesting some yet to be determined relationship, albeit perhaps in an 'end-member' role.

Under normal closure (i.e. the perpendicular incident wave case), the hydraulic aperture (e) of a rock joint, as measured in a (macro) flow test, is related to the larger (mean) physical aperture (E) by the roughness-related empirical equation (Barton et al., 1985):

$$e \approx \frac{E^2}{JRC^{2.5}} \qquad (16.1)$$

(where $E \geqslant e$. The joint roughness coefficient JRC, varies from 1.0 for smooth joints: giving, implicitly $e = E$, to about 20 for extremely rough tension fractures).

There seems every reason for these 'macro-roughness' and 'macro-flow' concepts to apply as boundary conditions, in the case of intrinsic attenuation, at least in broad terms. The fundamental importance of joint or fracture (macro) roughness is demonstrated by empirical data in Figure 16.6, and Figure 16.7 shows the form of this empirical relation.

An illustrative example of the physical mismatch of E and e, and the practical effect this can have on e.g. the groutability of a near-surface rock mass is illustrated in Figure 16.8. This 'cubic' model of Snow, 1968, with additions by Barton, 1986 is a convenient vehicle for planning the choice of suitable grout particle sizes ($d_{95}$), with the rule-of-thumb $E > 4\ d_{95}$. When there is *insignificant* mismatch of these quantities E and $d_{95}$ due to high stress or due to the low permeability of tight joints, then

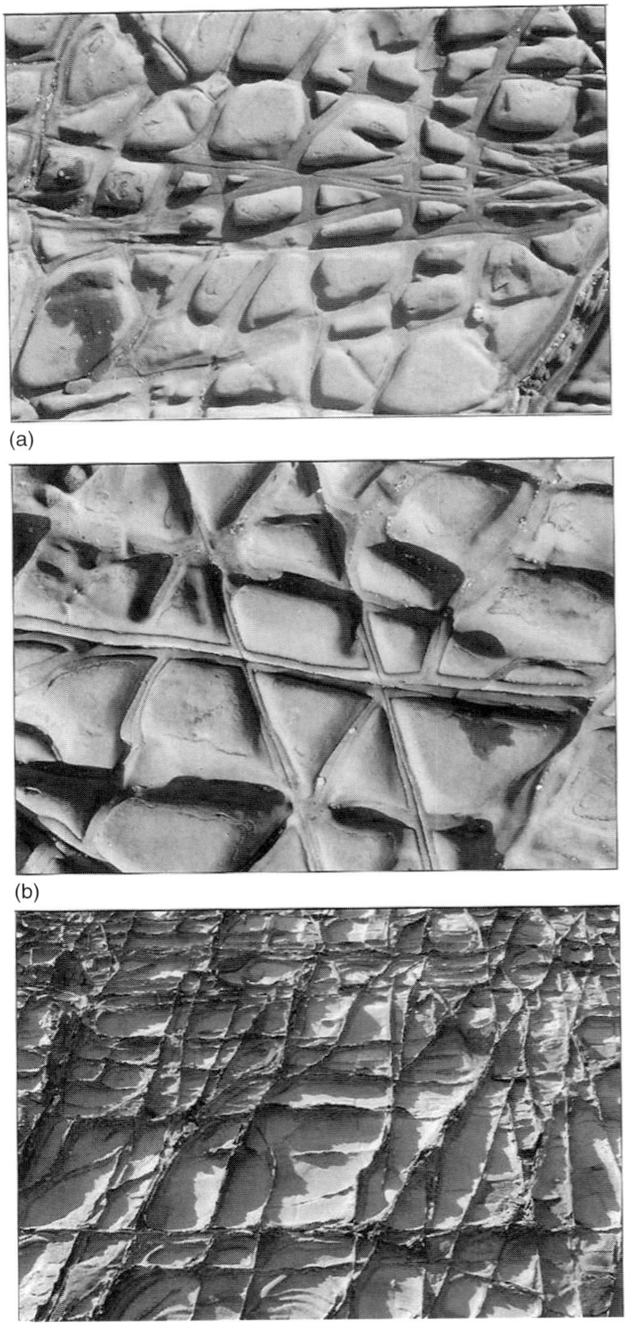
(a) (b) (c)

*Figure 16.5* A wave-cut platform in a jointed dolomite bed. These beds occur at intervals in the Kimmeridge shale, outcropping in Kimmeridge Bay, Dorset, England. The joints show a) implied JCS $< \sigma_c$ due to weathering and preferential wave erosion, and b) implied (local) JCS $> \sigma_c$, due to subsequent mineralization of dominant conducting joints. c) A fine example of joint cementation which may prevent the use of normal joint characterization techniques. Widemouth Bay, near Bude, Devon, England. (See Plate 9).

higher (locally deforming) injection pressures are used e.g. Barton, 2003. Note the interpreted increase of the spacing of *water-conducting* joints or fractures with depth, based on a statistical analysis of the Lugeon water injection test, which was depicted diagramatically in Figure 15.49, in Chapter 15.

The above very simple *single*-parameter approach to characterizing the mismatch of the mechanical or physical aperture (E) and the usually smaller hydraulic aperture (E), does not seem to be known in the geophysics literature reviewed. There is of course data from an increasing number of research projects that has supported, and continues to support, the fundamental mismatch of these quantities, with early sources of such data listed in Figure 16.6.

Key experimental research in this area includes Witherspoon et al., 1979, Barton et al., 1985, Gale, 1987, Gentier 1986, Pyrak-Nolte et al., 1990, Makurat et al., 1990, and Cook, 1992. A more recent review by Renshaw, 1995, and aperture modelling contributions from Sisavath et al., 2003 and Liu, 2005 emphasise the interest in this 'detail', which in fact is fundamental in controlling the world's supply of oil, gas and water. Liu also considered that naturally fractured formations under reservoir conditions would have very small apertures (0.1–10 $\mu$m), very rough fracture surfaces, and a large proportion of the fracture surfaces in contact with each other.

This opinion, which is shared if only normal closure (i.e. interlock) is considered, turns out to be the least productive condition, exceeded only by more planar joints in weaker rock, which would hardly conduct fluids if tightly closed. With somewhat lesser roughness and slight shear, a fundamental enhancement to these (too) limited apertures is possible, as we shall see below.

### 16.2.2 Coupled stress-flow behaviour under shear deformation

When there is shear displacement (i.e. a discontinuity in the shear stress-load behaviour), experiments show that the roughness 'iso-curves' of Figure 16.7, are cut nearly at right angles in the E/e versus e space. Some of these trends are seen in Figure 16.6, from a slightly sheared, heated block test (see 'N–S', 'E–W' etc.). Esaki et al. 1995, confirmed these 'perpendicular' trends, and Olsson and Barton, 2001, formalised the results, following further CSFT testing by Olsson (Figure 16.9) for the case of the mismatch of (e) and (E) that occurs when there is shear

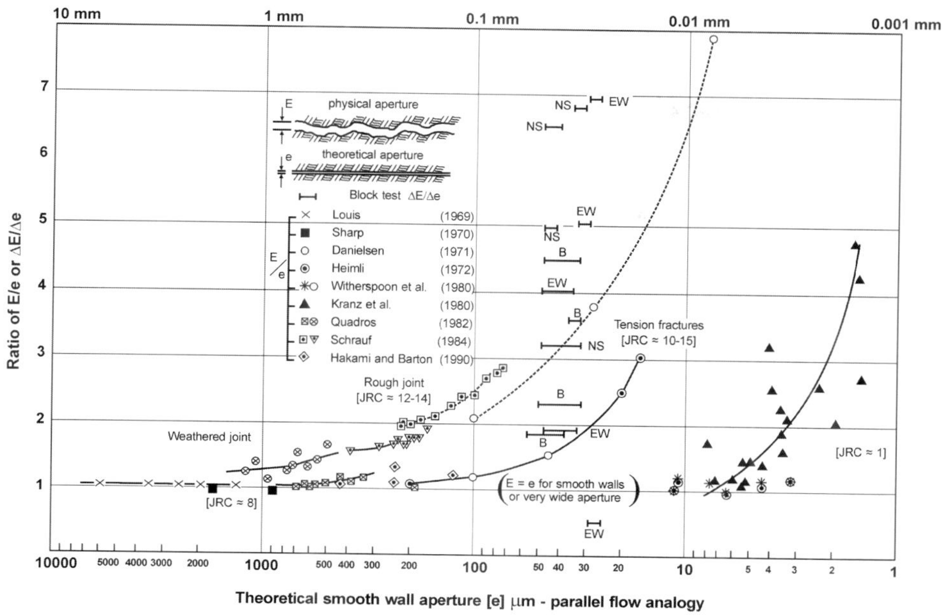


*Figure 16.6* Mismatch of (e) and (E) due to joint or fracture roughness. Barton and Quadros, 1997, (updated from Barton et al., 1985)

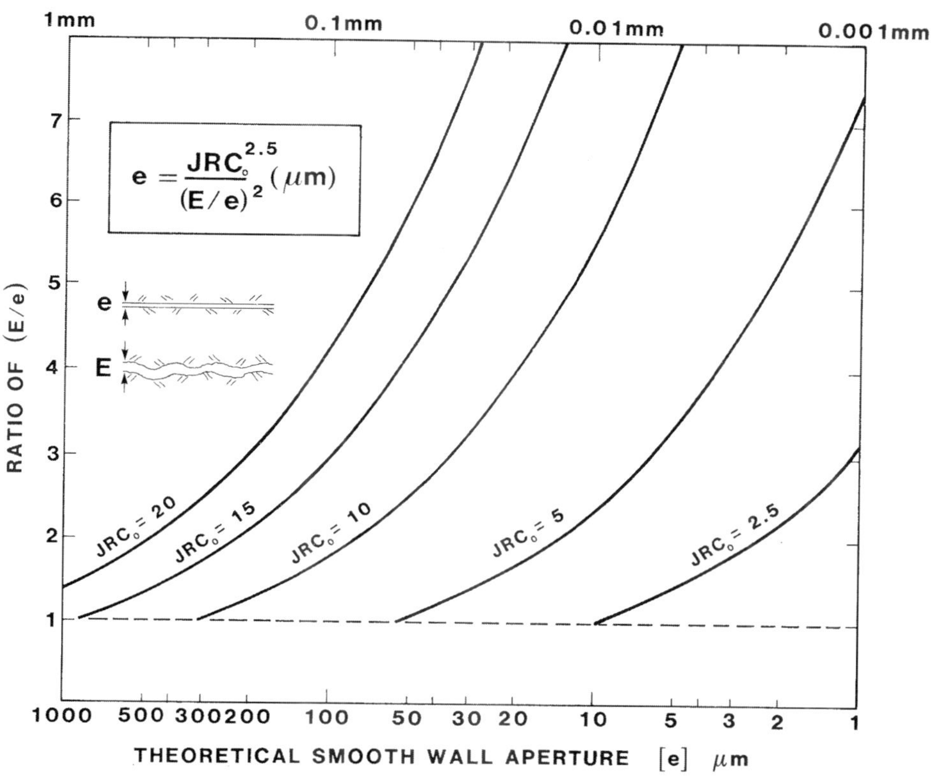


*Figure 16.7* An empirical model for the mismatch of E and e, from Barton et al., 1985.

(and some gouge production). The following empirical relation was found to apply:

$$e = E^{1/2} \times JRC_{mob} \qquad (16.2)$$

The $JRC_{mobilized}$ concept of joint or fracture shear strength mobilization is explained with a low stress example in Figure 16.10. The early mobilization of friction, followed later by dilation and roughness mobilization and

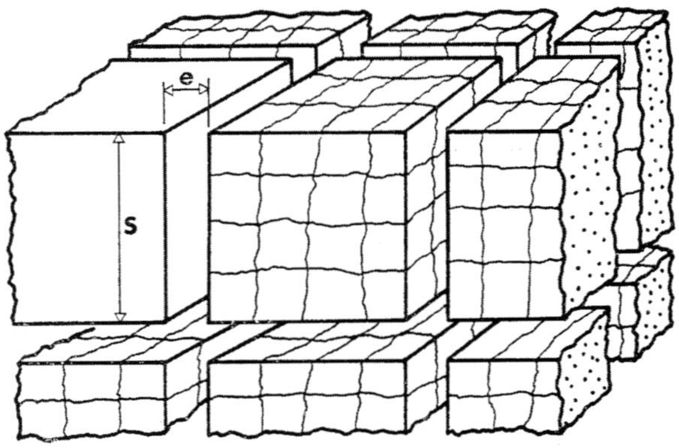


ONE SET

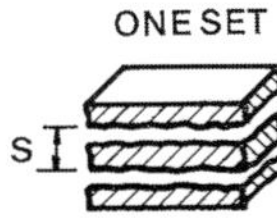


$n = \frac{E}{S}$

THREE SETS

POROSITY OF ROCK MASS $n = \frac{3E}{S}$ (E > e)

| Depth zones | S(m) | e(μm) | E(μm) |
|---|---|---|---|
| 5-15m | 0.3 | 150 | 218 |
| 15-25m | 0.4 | 110 | 186 |
| 25-45m | 0.6 | 80 | 159 |
| 45-60m | 0.7 | 60 | 138 |

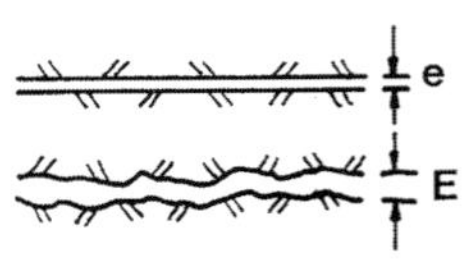


$$E = \sqrt{e.\ JRC^{2.5}}$$

Grout-Take Estimates / 1m³ rockmass

| Depth zone | 5-15m | 15-25m | 25-45m | 45-60m |
|---|---|---|---|---|
| Grout (litres) | 2.2 | 1.4 | 0.8 | 0.6 |

*Figure 16.8* A cubic-network approach to near-surface permeability at dam sites (after Snow, 1968), with the addition of aperture differentiation and roughness discrimination by Barton, 1986.

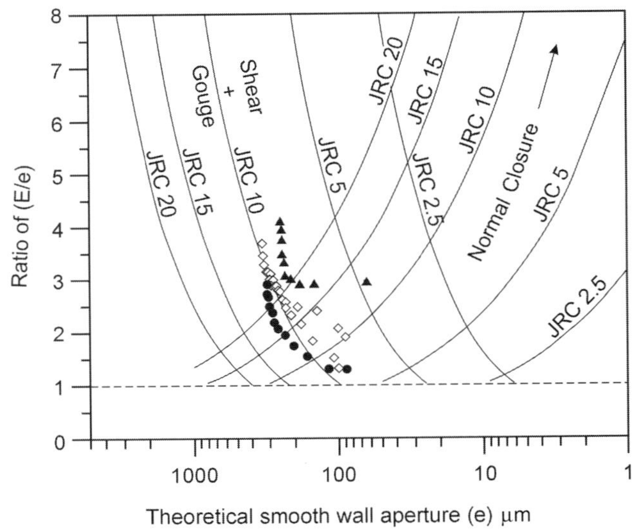


| | | |
|---|---|---|
| Normal Closure | $e = \frac{JRC_0^{2.5}}{(E/e)^2}$ | $\}\ e = E^2/JRC_0^{2.5}$ |
| Shear+ Gouge | $e = \left(\frac{E}{e}\right) JRC^2$ | $\}\ e = \sqrt{E} \cdot JRC_{mob}$ |

*Figure 16.9* Data points are from CSFT (coupled shear-flow tests) by Olsson, in Olsson and Barton, 2001, showing 'perpendicular' trends in relation to the E-e-JRC model, due to shearing and gouge formation. The 'sloping-to-the-right' curves are relevant to shear behaviour, and are similar to trends shown by Esaki et al., 1995. In other words shearing, dilation, and possible gouge formation causes data to climb the curves (for $JRC_{mob}$), shown to the left.

final roughness degradation, was quantified by Barton, 1982, and forms an important basis for the shear-part of the Barton-Bandis constitutive model.

Note the location of 'dilation begins', corresponding to the initial mobilization of roughness. In the 'roughness destroyed' post-peak area there is a physical reduction in JRC, in other words retesting would show a reduced peak strength for the joint in question. The early mobilization of friction (prior to dilation), can be envisaged as a potential contributor to the intrinsic frictional attenuation mechanism, but would presumably involve (almost) recoverable micro-slip.

The other part of intrinsic attenuation involving fluid flow: potential viscous shearing, or 'squirt', will clearly be affected by minor (or larger) amounts of (pre-) shearing. Figure 16.11 shows a set of experimental data (solid black points), from a self-weight, very low stress 'CSFT' from Maini and Hocking, 1977. The 'twin' curves in the other three diagrams were obtained using the shear, dilation and flow coupling given by the Barton-Bandis model, based on the $JRC_{mob}$ concept shown in Figure 16.10. Note the large change of measured and modelled permeability due to slight dilation, for the case of these very planar (low JRC) cleavage joints.

A more recent coupled shear-flow test performed at higher stress by Olsson, using controlled normal stiffness loading in a large direct shear box apparatus (see Olsson and Barton, 2001), is shown in Figure 16.12. The term 'krm' in the individual figures stands for applied normal stiffness, with 'krm = 0 kN/mm' representing a conventional, constant normal load, zero-stiffness test.

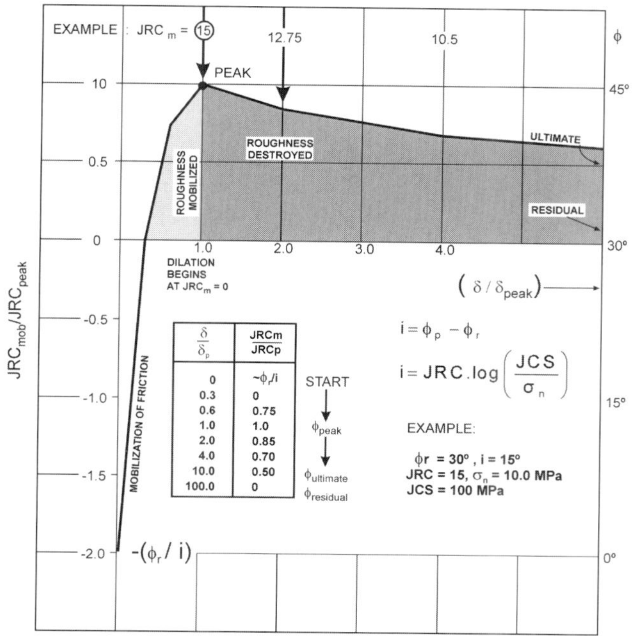


*Figure 16.10* The $JRC_{mobilized}$ concept of Barton, 1982. This shows shear strength development for rock joints expressed as a dimensionless $JRC_{mob}/JRC_{peak}$, and reduction of this ratio with dimensionless $(\delta/\delta_{peak})$ displacement. Peak shear strength is given by $\tau = \sigma_n \tan [JRC \log (JCS/\sigma_n) + \phi_r]$, from Barton and Choubey, 1977. This diagram also gives a direct method for predicting the theoretical (maximum) permeability development with displacement and dilation. For this one uses the JRC-based $E \geq e$ inequality, and the initial-aperture-enhancing dilation. (For gouge-reduced permeability estimation, one uses the $JRC_{mobilized}$ relation, as shown in Figure 16.9).

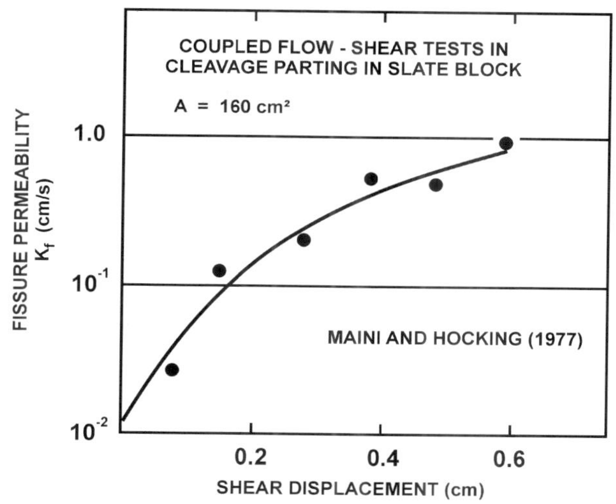


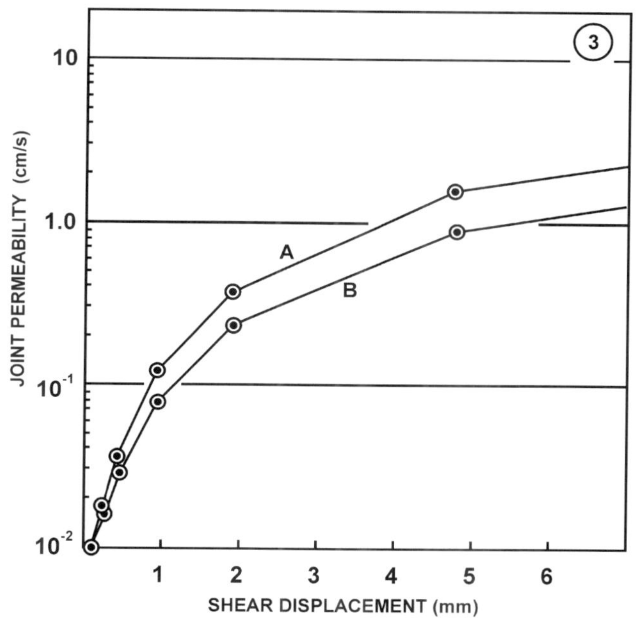


*Figure 16.11* An early coupled shear flow test by Maini (in Maini and Hocking, 1977), and an attempt to match this by estimating the shear-induced increase in joint aperture due to full application of the dilation effect, using the Barton-Bandis model. (Barton, 1982).

A CSFT test by Makurat (Makurat et al., 1990) is shown in Figure 16.13. This was selected as it shows that with gouge production caused by shearing damage, there will be a discrepancy between the dilation-based permeability modelling (assuming full application of the dilation-enhanced aperture), and the experimental result. The measured permeability increase with shearing was less than expected from the measured dilation, due to gouge production compromising the effect of the increased aperture.

In two of the above cases, the theoretical effect of the modelled dilation on permeability is shown. In the low stress case (Figure 16.11), a reasonably good match to the dramatic, two-orders-of-magnitude increase in permeability with less than 0.6 mm of shear is suggested. The tests by Olsson (Figure 16.12) show a smaller, but significant increase (e.g. one order of magnitude increase with 2 mm of shear) but with reduced rate at larger shear deformation.

From the point of view of attenuation potential in the reservoir situation, the presence of 'open' fractures caused by limited shear, may be an important additional effect for dynamically induced (micro) fluid flow, and squirt losses, during the passage of a wave front. Sheared fractures may represent some of the 'open' fractures

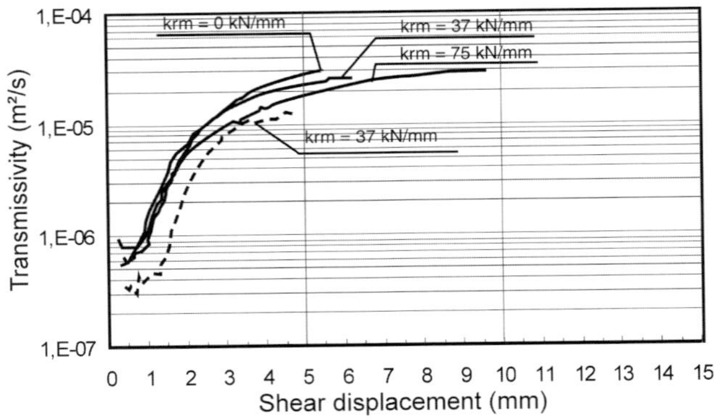


*Figure 16.12* CSFT tests by Olsson using controlled normal stiffness during the shearing. One order of magnitude increase in permeability with 2 mm of shear is indicated. (Olsson and Barton, 2001).

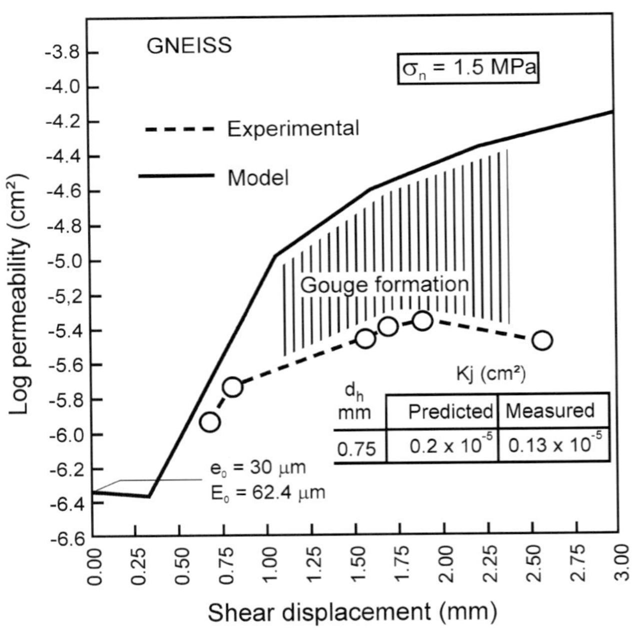


*Figure 16.13* CSFT test by Makurat, showing the experimentally measured increase in permeability during 2.5 mm of shearing on a joint in weathered gneiss, compared to the modelled, much larger permeability increase, if one assumes no blockage to flow by gouge production. Makurat et al., 1990b and Bandis et al., 1985.

often referred to, and they could be an important source of *in situ* anisotropy, as described in Chapters 14 and 15. However, with higher effective normal stresses, and with the likely presence of some gouge, making for a tortuous flow path and generally tighter apertures (i.e. smaller aspect ratios, despite shearing), flow *resistance* is likely to be significant.

## 16.3 Some important details concerning rock joint stiffnesses $K_n$ and $K_s$

Normal stiffness $K_n$ is defined as the normal stress increment required for a small closure of a (usually very tight) joint or fracture, at a given level of effective stress. It is therefore clear that it is usually of significantly larger value than the shear stiffness $K_s$ which is the stiffness in shear, usually taken as the average slope up to the shear strength-displacement peak.

Since compliance in both normal and shear directions is becoming of such importance in recent geophysics modelling, some of the parallel, but macro-displacement aspects of the inversed parameters, $K_n$ and $K_s$ will be further explored in this section.

It is believed that when rock quality is good, and joints are hard, there will be great similarity in magnitude between $K_n$ and its dynamic 'inverse' $1/Z_N$, where $Z_N$ is the dynamic normal compliance. This is probably because of the similarity, in good rock conditions, between $E_{dynamic}$ and $E_{static}$, as investigated in Part I. Between $K_s$ and its dynamic 'inverse' $1/Z_T$, where $Z_T$ is the dynamic shear compliance, there is likely to be a significant difference, due to the *lesser* likelihood that block size will be the determinant factor, as it is for the magnitude of the pseudo-static shear stiffness $K_s$.

When rock is of poor quality, and joints are altered or clay-bearing, there will be greater discrepancies between each of the above three pairs of parameters, but $K_n$ and $1/Z_N$ should remain within an order of magnitude of each other. These aspects have been discussed in slightly different context, in a number of chapters in Part II.

Barton, 1972, Barton and Hansteen, 1979, Bandis, 1980, Bandis et al., 1981 and Barton, 1982 have each emphasised the scale-dependent nature of shear stiffness $K_s$. An assembly of various sets of data is reproduced in Figure 16.14, from Barton, 1982.

By means of characterizing joint samples recovered in core, using the JRC (joint roughness coefficient) and JCS (joint wall compressive strength) parameters, it is possible with Barton-Bandis joint modelling to simply estimate the above scale-dependent shear stiffnesses. An example, using two different laboratory (core) sized samples, is shown in Figure 16.15 for an assumed *block-size* range of 100 mm to 10 m. The three diagonal pairs of curves, represent shear stiffnesses at three effective normal stress levels. Techniques for estimating the joint characteristics JRC and JCS, developed by Barton and Choubey, 1977, are illustrated in graphical format later in this chapter,

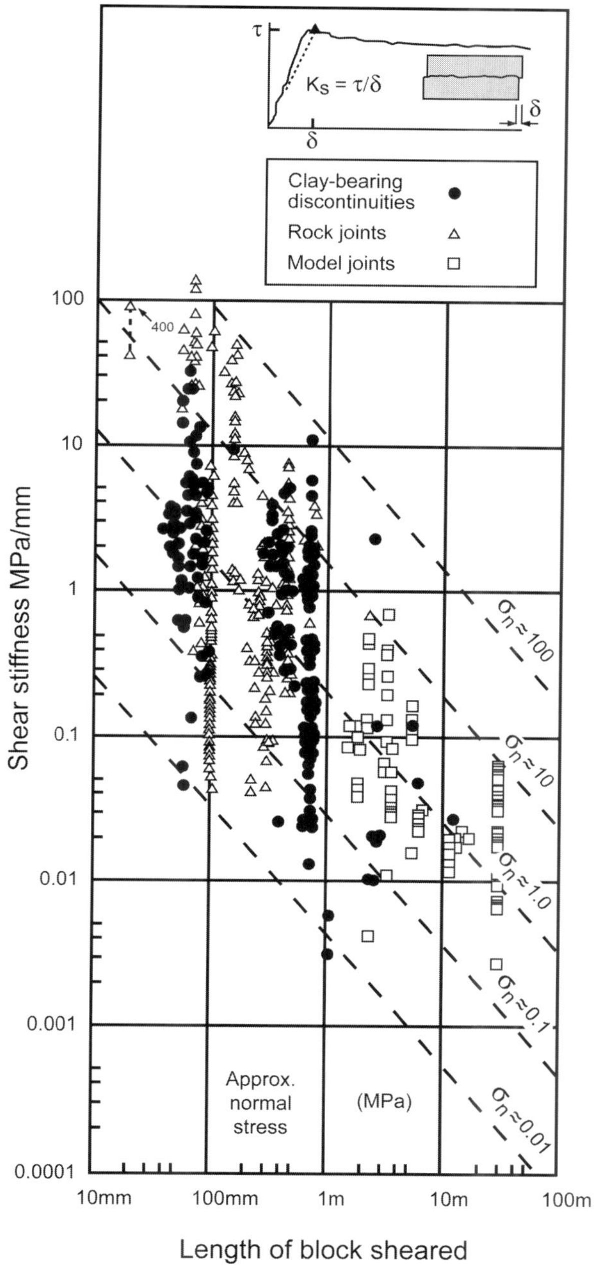


*Figure 16.14* Experimental 'large strain' data for the shear stiffness ($K_s$) of rock joints, clay-bearing discontinuities, and model tension fractures. A *very strong* scale dependence, and a *strong* (effective normal) stress dependence are each implied. These are fundamental properties of rock masses in 'macro-deformation' processes. Barton, 1982.

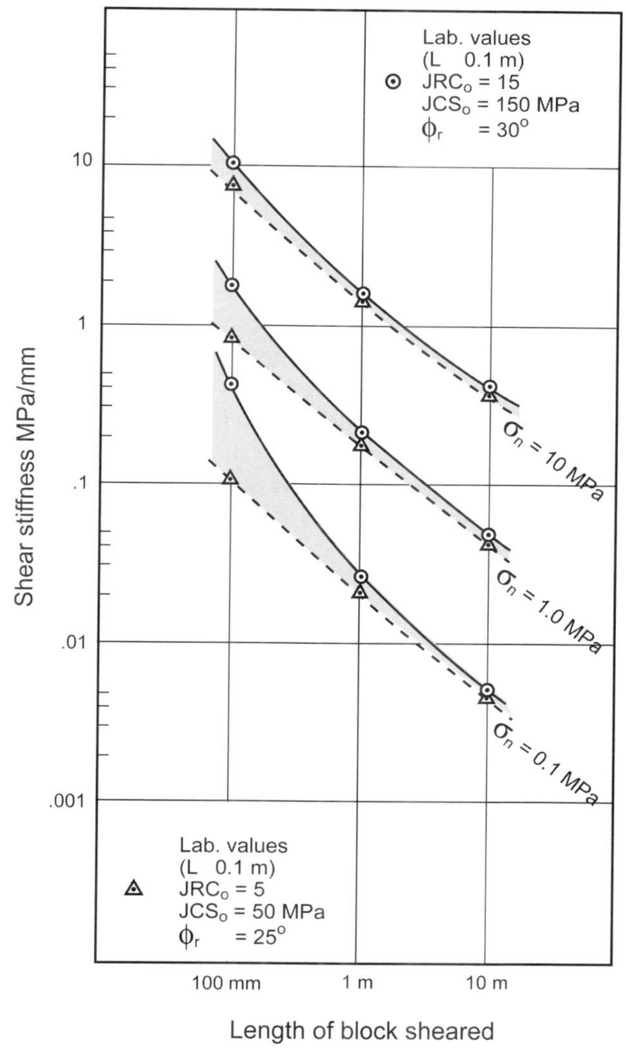


*Figure 16.15* Estimation of peak shear stiffness based on extrapolation of two different joint characters (as observed at core-scale) using Barton-Bandis scaling, from Barton, 1982. The *extent of direct relevance* of these 'macro-displacement' stiffnesses to seismic 'micro-displacements', using the stiffness-compliance inversion, with adjustment for the $E_{dynamic} \geq E_{static}$ inequality, is presently *more* uncertain, when moving in the direction of larger block sizes.

when exploring the effects of stress on modelled joint apertures.

Normal stiffness ($K_n$), by comparison to $K_s$, may not be scale-dependent, although it is undoubtedly 'sample-dependent', i.e. if a larger, more continuous, water-bearing, feature is sampled, it will likely have lower normal (and shear) stiffness than a less continuous, probably unweathered and usually rougher joint surface (Barton, 1990b).

As shown in extensive studies by Bandis, 1980, and Bandis et al., 1983, the normal stiffness of joints $K_n$ is especially sensitive to sample disturbance (i.e. inevitable unloading when recovering samples), and the first load cycle as depicted in Figure 16.16, shows much lower initial stiffness and much larger hysteresis. Lower normal stiffness is also seen following shearing, due to a certain mismatch of the previously interlocking roughness. The reduced contact area causes the reduced stiffness, with typically ½ to ⅙ reduction, depending on stress level (Bandis et al. 1983). Adverse effects on the dynamic compliances $Z_N$ and $Z_T$ are likely too.

The contrasting normal stiffnesses from first, second and third cycles of loading for fresh joints in hard rock are shown in Figure 16.17. Corresponding curves for joints in weaker rocks or weathered rocks are markedly less steep, and show much more hysteresis on the first cycle.

The relative effects of different rock types on normal stiffness including the weaker sandstones, siltstones and limestones are illustrated in Figure 16.18, from Bandis, 1980. Here we see the total (rock + joint) deformation $\Delta V_t$.

### 16.3.1 Initial normal stiffness measured at low stress

A major tabulation of *initial* normal stiffness ($K_{ni}$), normal stiffness and maximum closure ranges for first, second and third load cycles for numerous rock joint samples with widely different JRC and JCS values was assembled by Bandis, 1980. In Table 16.2, just the average values have been reproduced.

As will be noted for *moderately weathered* and *fresh* sandstones, siltstones and limestones, the *third* cycle, have $K_{ni}$ values in the respective ranges as follows:

| | | |
|---|---|---|
| Sandstone | 27 to 37 | MPa/mm |
| Siltstone | 23 to 54 | MPa/mm |
| Limestone | 99 to 118 | MPa/mm |

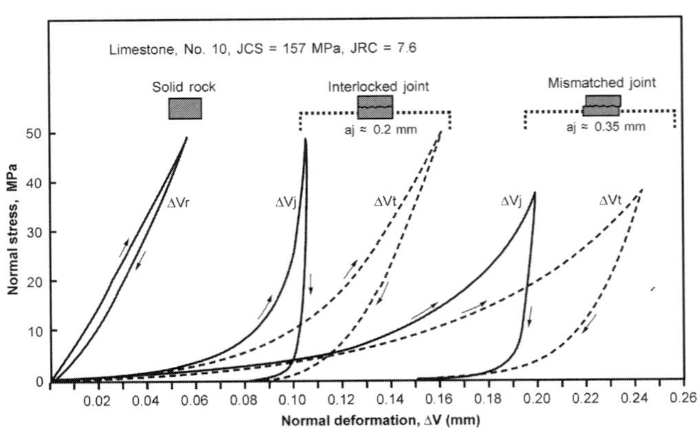


*Figure 16.16* Normal stiffness testing of rock joints $\Delta V_t$ = total deformation $\Delta V_J$ = net joint closure. First cycle of load-unload. Bandis, 1980; Bandis et al., 1983.

It is clear from inspection of data in Figure 16.17, and elsewhere, that these average third cycle, *initial* normal stiffness values of roughly 30 to 120 MPa/mm, correspond to effective normal stresses in the lowest

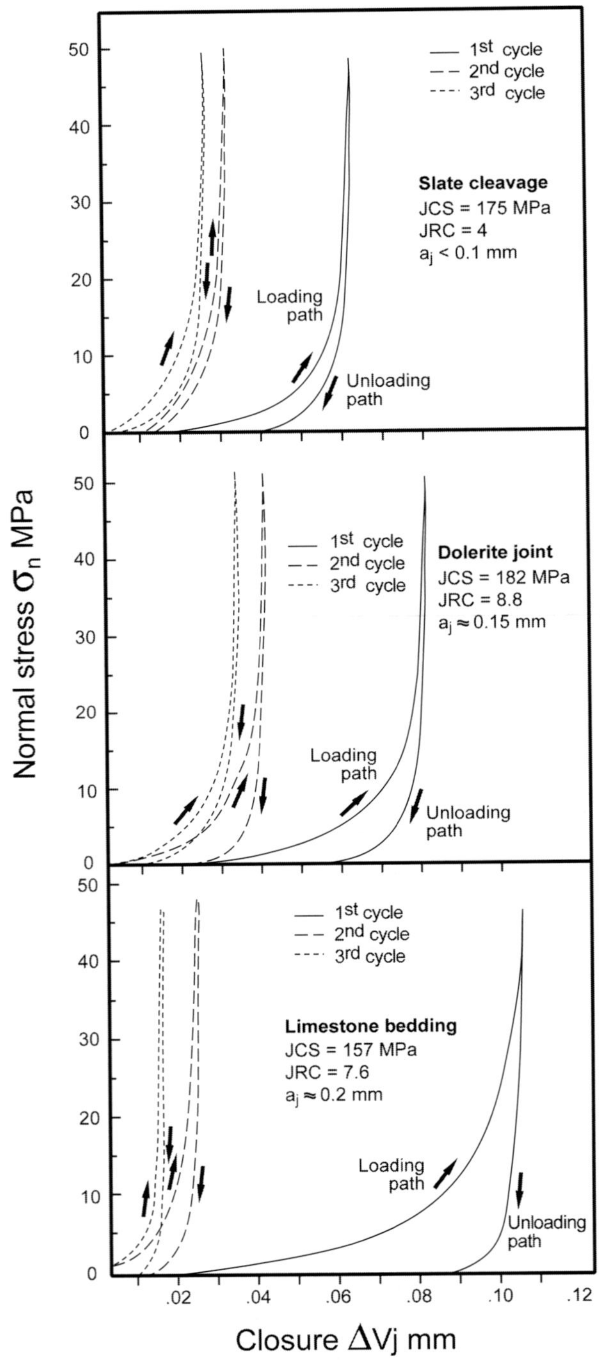


*Figure 16.17* Three load-unload ($\Delta V_J$) cycles for joints in hard unweathered rock. Bandis et al., 1983. These will be used for estimating normal stiffness dependence on normal stress, at the highest stress levels.

possible range of perhaps 0 to 1 MPa, and at the same time correspond to the *pre-consolidated condition*, since measured after three load-unload cycles.

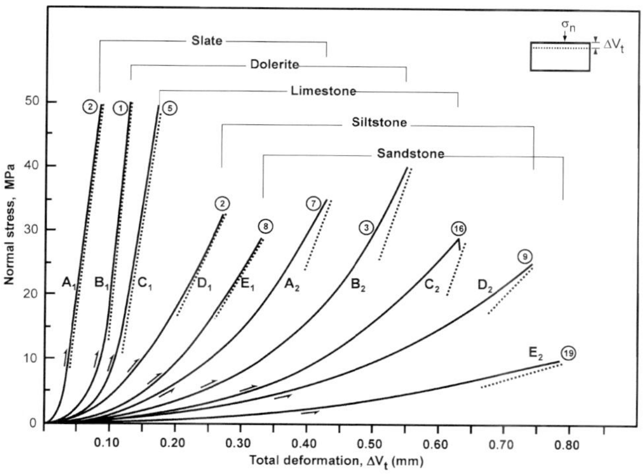


*Figure 16.18* Comparative total deformation ($\Delta V_t$ = rock + joint) for different rock types, based on the most deforming first cycle of loading. Bandis, 1980.

### 16.3.2 Normal stiffness at elevated normal stress levels

From 0 to 5 MPa, there is a rapid rise of the average (pseudo-static) normal stiffnesses to between 250 and 500 MPa/mm, and between 0 and 10 MPa, 400 to 800 MPa/mm is more common for the average normal stiffness of the fresher joints, and about 200 MPa/mm for the (JCS = 40–50 MPa) joints in weaker rock. The *average* normal stiffnesses between 0 and 50 MPa are of the order of 2,000 to 3,000 MPa/mm for the hardest joints.

However, considering local or *incremental stiffnesses*, at tens of MPa stress levels (such as 10 to 40 MPa), normal stiffnesses for these unweathered hard joints with high JCS may rise to tens of thousands, but may only rise to about 1,000–2,000 MPa/mm in the case of the weaker rocks (e.g. with JCS = 40–50 MPa).

Before proceeding to a detailed treatment of possible stiffness ratios, we will present normal stiffness data kindly prepared by Bandis from his 1980 doctoral studies in the University of Leeds. This can be compared, where

Table 16.2 Bandis, 1980 data for $K_{ni}$ (3 cycles) and maximum closure $V_m$ (3 cycles).

| Rock type and weathering state of joints | JCS (MPa) | No. of joints | $V_m$ and $K_{ni}$ data: 1st cycle $K_{ni}$ MPa/mm | 1st cycle $V_m$ mm | 2nd cycle $K_{ni}$ MPa/mm | 2nd cycle $V_m$ mm | 3rd cycle $K_{ni}$ MPa/mm | 3rd cycle $V_m$ mm |
|---|---|---|---|---|---|---|---|---|
| **SLATE** | | | | | | | | |
| Fresh | 175 | 3 | 35.0 | .039 | 181.1 | .027 | 266.2 | .027 |
| Mod. weathered | 142 | 1 | 13.1 | .106 | 69.0 | .046 | 235.5 | .039 |
| Weathered | 77 | 3 | 12.7 | .331 | 32.6 | .146 | 63.6 | .118 |
| **DOLERITE** | | | | | | | | |
| Fresh | 167–182 | 2 | 24.2 | .101 | 67.1 | .053 | 111.2 | .067 |
| Weathered | 60–76 | 3 | 10.8 | .484 | 50.0 | .150 | 80.4 | .109 |
| **LIMESTONE** | | | | | | | | |
| Fresh to slightly weathered | 152–170 | 11 | 18.8 | .091 | 85.3 | .042 | 117.9 | .030 |
| Mod. weathered | 94–120 | 5 | 30.1 | .116 | 51.7 | .060 | 98.9 | .040 |
| Weathered | 35–53 | 5 | 8.5 | .373 | 44.2 | .115 | 56.9 | .079 |
| **SILTSTONE** | | | | | | | | |
| Fresh | 105 | 5 | 18.9 | .135 | 43.9 | .072 | 53.5 | .063 |
| Mod. weathered | 67 | 2 | 10.8 | .310 | 20.8 | .205 | 22.9 | .184 |
| Weathered | 44 | 3 | 10.4 | .514 | 27.8 | .104 | 35.0 | .113 |
| **SANDSTONE** | | | | | | | | |
| Fresh to slightly weathered | 68–95 | 8 | 12.8 | .170 | 23.4 | .075 | 37.2 | .054 |
| Mod. weathered | 64–58 | 9 | 9.3 | .240 | 17.9 | .101 | 26.7 | .089 |
| Weathered | 22 | 4 | 3.1 | .469 | 11.3 | .131 | 15.6 | .080 |

possible with data in Table 16.4 from Pyrak-Nolte et al., 1990 that was reviewed in Chapter 15.

To conserve space in Table 16.3 normal stiffness values for *weathered* joints are excluded, in view of their lesser relevance for reservoir conditions. A rough (JRC = 15), *weathered* (JCS = 44 MPa) joint in limestone showed $K_n$ values of 250, 1,275 and 2,550 MPa/mm at normal stress levels of 10, 25 and 40 MPa. More planar (JRC = 6), weathered joints (JCS = 25–45 MPa) in sandstone and siltstone displayed normal stiffnesses of between 200 and 900 MPa/mm at normal stress levels of 10 and 25 MPa.

Several of the pseudo-static values of normal stiffness from the tests of Bandis, 1980 shown in Table 16.3 had higher values of pseudo-static normal stiffness than the dynamic normal stiffnesses of this more deformable joint in quartz monzonite. Another series of tests by Pyrak-Nolte and co-workers, using a stiffer (E 32) sample, and

Table 16.3 Normal stiffnesses in (MPa/mm) measured by Bandis, 1980 in *pseudo-static* loading. Aperture (a) represents the approximate unstressed aperture prior to testing. (Bandis, 2005 pers. comm.)

| | Limestone | Limestone | Sandstone | Sandstone | Siltstone |
|---|---|---|---|---|---|
| Normal | F | SW | F | MW | F |
| Stress | JRC = 10 | JRC = 7.6 | JRC = 12 | JRC = 7.5 | JRC = 9 |
| | JCS = 160 MPa | JCS = 160 MPa | JCS = 68 MPa | JCS = 44 MPa | JCS = 105 MPa |
| | a = 0.25 mm | a = 0.2 mm | a = 0.25 mm | a = 0.25 mm | a = 0.15 mm |
| (MPa) | (MPa/mm) | (MPa/mm) | (MPa/mm) | (MPa/mm) | (MPa/mm) |
| 10 | 1,200 | 850 | 470 | 350 | 350 |
| 25 | 6,300 | 10,500 | 12,750 | 12,750 | 3,100 |
| 40 | 15,750 | 31,500 | – | – | – |

Table 16.4 Comparison of static and dynamic normal stiffness data for the Pyrak-Nolte et al., 1987b jointed sample E 35, of 52 mm diameter, which was the most deformable of three joint samples, possibly due to greater roughness.

| Normal Stress | Static stiffness $K_{n(static)}$ | Dynamic stiffness $K_{n(dyn)}$ | Ratio |
|---|---|---|---|
| MPa | MPa/mm | MPa/mm | $K_{n(dyn)}/K_{n(static)}$ |
| 2.9 | 1,000 | 4,500 | 4.5 |
| 10.0 | 2,200 | 8,000 | 3.6 |
| 33.0 | 3,300 | 25,000 | 7.6 |

Table 16.5 Selected *dynamic* normal and *dynamic* shear stiffnesses from tests on two joints in quartz monzonite. Pyrak-Nolte et al., 1990.

| STRESS | Sample E 32 | (all stiffnesses in MPa/mm) | Sample E 35 | (all stiffnesses in MPa/mm) |
|---|---|---|---|---|
| $\sigma_n$ (MPa) | $K_{n(dynamic)}$-*dry* | $K_{n(dynamic)}$-*saturated* | $K_{n(dynamic)}$-*dry* | $K_{n(dynamic)}$-*saturated* |
| 2.9 | 15,000 | 35,000 | 4,000 | 9,500 |
| 10 | – | 80,000 | 11,500 | 20,000 |
| 20 | – | 100,000 | 20,000 | 25,000 |
| 70 | 120,000 | – | 32,000 | 59,000 |
| $\sigma_n$ (MPa) | $K_{s(dynamic)}$-*dry* | $K_{s(dynamic)}$-*saturated* | $K_{s(dynamic)}$-*dry* | $K_{s(dynamic)}$-*saturated* |
| 2.9 | 3,500 | – | 1,900 | – |
| 10 | 9,500 | – | 4,800 | – |
| 20 | 17,000 | – | 6,200 | – |
| 70 | 55,000 | – | 7,400 | – |

the same E 35 sample as above, provided both dry and saturated results. These are repeated in Table 16.5 in compatible (MPa/mm) units, for ready comparison.

## 16.4 Ratios of $K_n$ over $K_s$ under static and dynamic conditions

Due to equipment capacity limitations, when Bandis, 1980 tested both the normal *and* shear stiffnesses of the same joint sample, involving tests in a shear box rather than in a high-capacity loading frame, significantly lower levels of normal stress than any of the above were applied. Values of $K_s$ ranging from 1 to 7 MPa/mm were obtained, where normal stress values were in the range of approx. 1 to 5 MPa, for the 100 mm long samples.

It seems reasonably certain that $K_s$ *static* is likely to be some orders of magnitude lower than $K_s$ *dynamic* at comparable normal stress levels. Future testing to develop an improved data base is obviously required. Significantly we do not yet seem to know the effect of *in situ* block sizes on $K_s$ *dynamic*. Block size has a dramatic, and well documented reducing effect on $K_s$ *static* (e.g. Barton, 1972, Bandis et al., 1981, Barton, 1982), as also seen in Figures 16.14 and 16.15.

When the ratio of $K_n/K_s$ (both static) could be compared directly on the same 100 mm long samples, mostly at the low normal stress of 1 MPa, Bandis found that the ratio varied from extremes of 5.7 (dolerite) to 132 (slate), but with most ratios of $K_n/K_s$ in the narrow range of 11 to 15.

Table 16.6 gives the ratios of $K_{n(dyn)}/K_{s(dyn)}$ for the dry samples listed in the Table 16.5. Note that the ratios of dynamic stiffness are for 52 mm diameter specimens. Under *pseudo-static* testing, the size of sample plays a significant role in reducing the shear stiffness, thereby increasing the ratio $K_n/K_s$ to much larger values than the range given in Table 16.6. Under dynamic testing this appears not to be the case.

It may be noted from the above tables that the range of $K_{n(dyn)}/K_{s(dyn)}$ ratios for these three, 52 mm diameter jointed samples in quartz monzonite was only from 1.3 to 4.3 with a mean of 2.5. This means that the ratio of the inverted *compliances* $Z_N/Z_T$ for these small samples ranges from 0.2 to 0.8, with a mean of 0.4. The remaining question is what relevance this convenient 'core-sized' data has to *in situ* reservoirs in general, with a typical spread of jointed block sizes from perhaps 200 mm to 5,000 mm, and mostly sedimentary rock as opposed to Stripa 'granite', or hard quartz monzonite.

Table 16.6 Ratio of *dynamic* $K_n$/*dynamic* $K_s$ for the three Pyrak-Nolte et al., 1990 joint samples in quartz monzonite. The authors' data set is partially incomplete. All samples were of 52 mm diameter.

| STRESS | Sample E 30 | Sample E 32 | Sample E 35 |
|---|---|---|---|
| $\sigma_n$ (MPa) | $K_{n(dyn)}/K_{s(dyn)}$ (dry) | $K_{n(dyn)}/K_{s(dyn)}$ (dry) | $K_{n(dyn)}/K_{s(dyn)}$ (dry) |
| 2.9 | 1.6 | 4.3 | 2.1 |
| 10 | 1.3 | – | 2.4 |
| 20 | 2.0 | – | 3.2 |
| 33 | 2.7 | – | – |
| 70 | – | 2.2 | 4.3 |

### 16.4.1 Frequency dependence of fracture normal stiffness

The dynamic stiffness data for the three joints reported by Pyrak-Nolte et al., 1990, were acquired over a limited ultrasonic frequency range (mostly 0.1 to 1 MHz). Different aspects of these tests were reviewed in Chapters 10 and 15, and have been summarised in part in this chapter, for comparison with static data. An important 'detail' is the fact that the dynamic stiffness, showing roughly four to eight times greater magnitude than the static values for the 'soft' joint in Table 16.4, is actually dependent on the frequency.

Pyrak-Nolte and Nolte, 1992 addressed this important aspect, suggesting that the frequency-dependence may be a simple consequence of the fracture or joint-wall geometry. They referred to a ratio of dynamic/static normal stiffness 'of typically three', and likened this ratio to the difference of the dynamic to static moduli. As we have seen in Part I, the ratio is rock quality dependent, and the Table 16.4 result for the more deformable joint (E 35), is consistent with this picture: the 'softer' the joints, the greater will be the difference. This applies to an even greater extent if the shear stiffness of the joints is also involved in the respective static and dynamic loading directions.

Pyrak-Nolte and Nolte 1992 applied the displacement-discontinuity theory of Schoenberg, involving assumptions of discontinuous displacement, but continuous stress across the fractures. An assumption of inverse proportionality between the discontinuity displacement and the specific stiffness of the fracture is also involved. The authors assumed that the different parts of any given fracture in intimate interlock due to normal stress, will

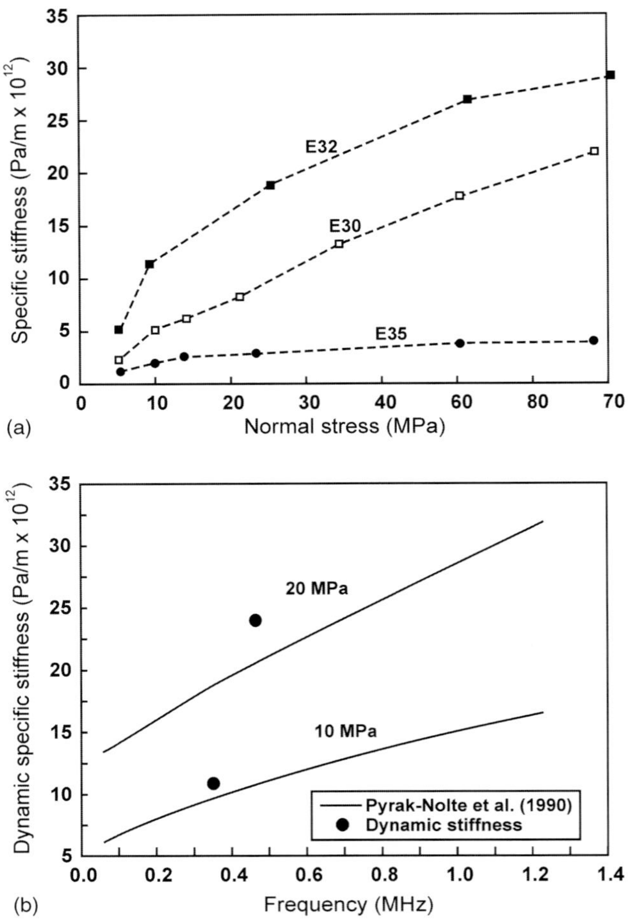


*Figure 16.19* a) Static normal stiffness trends for three joints in quartz monzonite (Pyrak-Nolte et al., 1987a). b) Dynamic stiffness as a function of frequency, for normal stress levels of 10 and 20 MPa. Pyrak-Nolte and Nolte, 1992.

have different *characteristic frequencies*, due to the distribution of voids and highly stressed rock-to-rock contacts. This will mean that different parts of the fracture will pass lower frequencies, while other parts will pass higher frequencies. They therefore assumed that the *local* transmission coefficients would depend on the *local* static stiffnesses.

The authors used a so-called *stratification percolation* model to simulate fracture aperture distributions and to match (with excellent precision) experimental amplitude-frequency curves from the tests by Pyrak-Nolte et al., 1990 referred to earlier. The result of their dynamic stiffness-frequency modelling is shown in Figure 16.19b: the two curves apply to normal stress levels of 10 MPa and 20 MPa, and were considered an improvement on the earlier frequency-*independent* values of 11,000 and 24,000 MPa/mm (shown by black dots). For the sake of completeness, Figure 16.19a shows one of the experimental data sets for the *static* normal stiffness of the three joints, which was reviewed in Chapter 10.

Pyrak-Nolte and Nolte, 1992 observed that the *dynamic* stiffness is equal to the *static* stiffness at low frequencies. This is clearly an important 'end-member' when considering the interpretation of fracture stiffnesses at much lower seismic exploration frequencies. It suggests that fracture properties, as defined in detail in Bandis 1980 static data (i.e. JCS and JRC, and resultant behavioural trends) will have *more to say* when comparing static and dynamic behaviour at seismic frequencies, than when comparing static and dynamic behaviour at laboratory ultrasonic frequencies. The *in situ* scale of fracturing and block sizes, and the use of lower frequencies, suggest that one should remain open to the possibility that static behaviour can be a useful guide to possible joint-parameter influences at seismic exploration frequencies. Note from Figure 16.19 that laboratory ultrasonic frequencies 'automatically' cause $K_n$ (dynamic) >> $K_n$ (static). At frequencies below 0.1 MHz, somewhat closer dynamic to static magnitudes are suggested.

### 16.4.2 Ratios of static $K_n$ to static $K_s$ for different block sizes

Since there are several variables involved in joint stiffness, beside the influence of wall strength JCS and roughness JRC, we will in the following simplify by assuming *either* small or moderate *in situ* block dimensions of 100 mm (i.e. bed-limited blocks) *or* 1 m (i.e. moderately jointed reservoirs). Block volumes might well rise to 10's, 100's or 1000's of $m^3$, in sparsely jointed rock masses. Since $K_n$ is not believed to be *scale-dependent* (but *sample-dependent*), while $K_s$ is clearly very scale dependent, it can be expected that *static* ratios of $K_n/K_s$ will tend to rise strongly with increasing block size. $K_s$ is very scale dependent because both *peak shear strength* and *displacement to peak are* scale dependent. It could be described as doubly-dependent on scale.

Comparison of the reviewed $K_n$ trends, with the $K_s$ *ranges* seen at (effective) normal stresses from 1 to 10 MPa in Figure 16.14 (usually about 1 to 10 MPa/mm for 100 mm samples, and 0.2 to 2 MPa/mm for 1 m block sizes), indicates that ratios of $K_n/K_s$ are most likely to be in the following *very approximate* ranges, based on data reviewed earlier:

1) 1 to 10 MPa: 100 mm size: $K_n/K_s$ = (200–800)/(10) = 20–80
2) 1 to 10 MPa: 1 m size: $K_n/K_s$ = (200–800)/(2) = 100–400
3) 10 to 40 MPa: 100 mm size: $K_n/K_s$ = (1000–*20000*)/(10–25) = 100–*800*
4) 10 to 40 MPa: 1 m size: $K_n/K_s$ = (1000–*20000*)/(2–4) = 500–*5000*

*(Italics implies uncertainty in incremental, high stress, normal stiffnesses)*

This means that the *pseudo-static* $K_n/K_s$ ratios may be in the low hundreds to low thousands at typical reservoir effective stresses, in the case of weaker or harder rocks, respectively. Here we assume that $K_s$ remains between about 2 and 20 MPa/mm, at large (1 m block-size), and small-scale (0.1 m block-size) respectively, as suggested at stress levels of about 10–20 MPa in Figure 16.14.

At the higher (effective) normal stress levels, $K_n$ is seen to rise much faster than $K_s$, so it is therefore that ratios of $K_n/K_s$ may climb to a range of about 500 to 5,000, in the context of in situ block sizes, as opposed to joints in small, highly stressed laboratory samples. It goes without saying that the reasons for the 'mismatch' of these macro-deformation stiffnesses, is because they are defining such different processes: $K_n$ approaches the stiffness (and stability) of the intact rock, while $K_s$ approaches the failure direction. (This is emphasised in Figure 15.5).

These two different stiffnesses are truly the *hard* and the *soft* stiffnesses of a rock mass, just as the matrix pore space, and the joints, are respectively the *hard* and the *soft* porosities of a rock mass. However, we have seen that *ultrasonic, laboratory-scale* measurements suggest less *softness*, since the ratios of *dynamic* $K_n/K_s$ are much lower than the static ratios just listed. The most important question that remains to be solved is whether seismic exploration frequencies, combined with much larger *in situ* fracture sizes and block sizes, will cause larger *dynamic* $K_n/K_s$ ratios than those listed in Table 16.6.

If the significantly 'anisotropic' or unequal static joint stiffnesses are in any way responsible for an (inverted) 'anisotropy' or non-equality of $Z_N/Z_T$, then they are surely a strong reason for the potential 'visibility' of water/oil/gas conducting joints to shear waves, as their potential weakness in shear and their anisotropy of stiffness ($K_n/K_s$) may be so marked compared to the more isotropic intact rock.

One could suggest, based on the above, that joints (or 'meso-scale' fractures) should be more seismically visible than stress-aligned microcracks, or the 'extensive dilatancy anisotropy' (EDA), if visibility depended on compliance contrasts, unless for some reason macro-deformation stiffness ratios ($K_n/K_s$) have *no influence* as 'starting points' for the micro-deformation dynamic compliance ratios ($Z_N/Z_T$). In good quality rock when $E_{static}$ is close to $E_{dynamic}$, there should in fact be close correspondence of $K_{n(static)}$ and $1/Z_N$ (dynamic), but apparently from what we have seen thus far, there may not be good correspondence of $K_{s(static)}$ and $1/Z_T$ (dynamic), unless block size *and* frequency have influence.

### 16.4.3 Field measurements of compliance $Z_N$

Important recent contributions to the problem of determining field-scale joint or fracture compliances were made by Lubbe and Worthington, 2005 and Lubbe, 2005. These were not reviewed in earlier chapters as they are most directly comparable with the foregoing rock mechanics work concerning *in situ* joint stiffnesses, and the scale effect known to act on the (static) shear stiffness $K_s$. These topics belong here in Chapter 16.

Lubbe and Worthington utilised three 40 m deep boreholes drilled in the floor of a Carboniferous limestone quarry near Bristol, in southwest England. Core analysis, wire-line logging, and local fracture mapping provided the important characterization data, for future comparison to other sites. Although using much lower frequency sources than in the laboratory (2000 Hz from a sparker), these cross-hole investigations were of course at significantly higher frequency than that of seismic exploration. The stiffness-frequency trends shown in Figure 16.19 should therefore be remembered.

The authors found that the rock mass had a P-wave anisotropy of 10% which they attributed to the predominantly *horizontal*, partly open fractures. P-wave velocities were often in the range of 5 to 6 km/s. An average of 5250 m/s for the fractured regions was quoted. The average density was 2,600 kg/m$^3$. For a near-surface location, this velocity suggests a rock mass quality $Q_c$ value of about 55, and a deformation modulus of about 38 GPa, assuming a UCS of a nominal 100 MPa. Lubbe et al., reported an intact seismic Q as high as 60, and an *in situ* value as low as 25 – the implied minimum. Again the similarity of $Q_{seis}$ to the potential static E-modulus or rock mass deformation modulus, expressed as GPa is striking: see Figure 13.60.

By using the upgoing or downgoing waves between borehole source and receivers, Lubbe and Worthington were able to estimate (dynamic) fracture normal compliances within the range of $2.5 \times 10^{-13}$ to $3.5 \times 10^{-12}$ m/Pa. These we can invert to dynamic normal stiffnesses in the range 4000 to 286 MPa/mm, or reversing the order and adopting a more 'friendly' unit, between 0,3 to 4 MPa/$\mu$m. Assuming an average of two continuous fractures per meter from 5 to 20 m depth, between all three boreholes, they estimated a mean fracture compliance of $1.25 \times 10^{-12}$, which converts to 800 MPa/mm in more easily understood rock engineering units. Later in the paper a 'reasonably robust average' of 1000 MPa/mm was suggested, following some numerical scoping.

As they pointed out, their compliances were an order of magnitude higher (therefore the stiffnesses lower) than obtained from laboratory experiments at higher stress and on smaller and stiffer samples. However, they referred to other field data showing low $K_{n(dynamic)}$ values of only 500 MPa/mm (dynamic compliance = $2 \times 10^{-12}$ m/Pa), for columnar joints in basalt, as interpreted by Myer et al., 1995 from Hanford Basalt cross-hole tests performed by King et al., 1986. (In Chapter 13 we also noted a very low seismic Q in the same horizontal measurement direction at this site, i.e. crossing the basalt columns, which was probably related to low levels of horizontal stress at this shallow test tunnel site: see Figure 13.61).

Examination of Figure 16.19 and Tables 16.4 and 16.5 in fact shows the *stiffer* end of the above range of *in situ* data to be the same as the 'softer' E 35 sample of Pyrak-Nolte, when tested at only 2.9 MPa normal stress, while the least stiff *in situ* result of 286 MPa/mm, is obviously closer to Bandis, 1980 *static* $K_n$ data shown in Table 16.3.

Lubbe and Worthington considered that the vertical fractures at the quarry (Figure 16.20) appeared to be stiffer than the measured bedding planes, resulting in horizontal velocities that were close to those for the unfractured rock. There appears to the writer the possibility that the quite high horizontal P-wave velocities of typically 5.3 to 6.2 km/s could be due a horizontal stress concentration beneath the floor of the quarry. If so, this would give an 'equivalent depth' correction to the empirical $Q_c - M - V_p$ relationships shown in Figure 13.59, thereby 'not requiring' so high $Q_c$ values. This would actually be more in line with the quite jointed appearance of the quarry seen in Figure 16.20, where a rock mass Q-value of about the following magnitude would normally be expected (see Appendix A for ratings):

$$Q = \frac{90}{9} \times \frac{2}{1} \times \frac{0.66}{1} = 13 \quad (E_{Mass} \approx 24\,GPa)$$

*Figure 16.20* Oxford University Earth Sciences Department website photograph of the Carboniferous limestone quarry near Bristol that was used for careful *in situ* investigations of normal compliance. (Lubbe et al., 2005). Note dipping bedding planes and sub-vertical joints, presumably seldom sampled by the vertical boreholes in the floor of the quarry. The cross-hole seismic was performed between three vertical holes spaced at 7 m. These are in the foreground. (www.earth.ox.ac.uk)

The authors also discussed shear compliances at the end of their paper, referring to Worthington and Hudson, 2000, fault studies. 'Values of shear compliance of the order of $10^{-10}$ to $10^{-9}$ m/Pa can be predicted and tentatively compared to field observations'. These can be inverted to *in situ* fault-related dynamic normal stiffnesses of only 10 and 1 MPa/mm, giving a somewhat closer match to the scale-dependent *static* shear stiffnesses plotted in Figure 16.14 and calculated in Figure 16.15. Such *in situ*, modelling-based estimates would be about one order of magnitude stiffer in relation to the static stiffness values for *in situ* block sizes of say 10 m, assuming '*in situ*' effective stress levels above and below a nominal 10 MPa. If on the other hand, *in situ* block sizes were only 0.5 m, as feasible in the neighbourhood of a fault, such 'soft' dynamic values (also obtained at lower seismic exploration frequencies), would match static values very closely.

### 16.4.4 Investigation of normal and shear compliances on artificial surfaces in limestones

Lubbe and Worthington, 2005 also referred to Lubbe, 2005 laboratory data for *artificial fractures* machined from Jurassic and Carboniferous limestones, tested dry under 5 to 60 MPa normal stress, at ultrasonic frequencies between 0.6 and 1.0 MHz. Using the displacement discontinuity theory of Schoenberg, 1980, values of $K_n$ (dynamic) ranging from 22,220 to 200,000 MPa/mm were derived, when inverted from the reported $4.5 \times 10^{-14}$ to $0.5 \times 10^{-14}$ m/Pa compliance units.

These dynamic stiffnesses are similar, to somewhat stiffer, than the Pyrak-Nolte data for *natural joints* of 52 mm size reviewed earlier. Lubbe, 2005 found that the ratio of normal to tangential compliance for the 50 mm diameter 'fractured' samples, was approximately 0.4 in the case of smoother surfaces, and increased from 0.3 to 0.8 in the case of corrugated surfaces that experienced 'asperity-crushing' beyond 30 to 40 MPa normal stress. The ratio of 0.4 for the smoother surfaces is the same as the *mean* value reported by Pyrak-Nolte et al., 1990 from their tests on 52 mm joint samples, as reviewed in Chapter 15.

Lubbe et al., 2005 reported details concerning the type of artificial fractures that were prepared for these tests. Samples were either cut with a 'fine-toothed' diamond saw and then polished, or, in the case of just the Jurassic limestone, samples were cut and grooved to produce a 'corrugated fracture interface' with specific voids and contact areas. The two 'fracture' types, which were all tested dry, are shown in Figures 16.21 and 16.22, together with the respective, and contrasting, compliance-normal stress behaviour. In these figures we have added stiffness scales on the right-hand axes, to aid comparison with the laboratory stiffness data reviewed earlier in this chapter.

The Jurassic limestone samples had porosities of approximately 13%, while the Carboniferous limestone

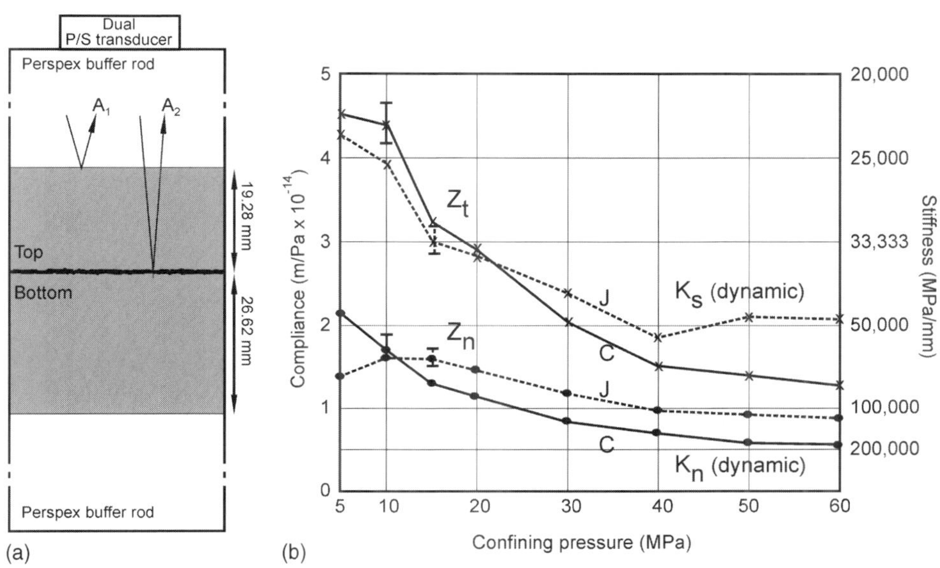


*Figure 16.21* a) Samples of Jurassic (J) and Carboniferous (C) limestone, with 'fractures' prepared with a fine-toothed diamond saw. b) Shear and normal compliance versus normal stress, with inversion to dynamic stiffnesses shown on the right-hand axis. Lubbe et al., 2006.

samples had porosities of only 2.4%. Densities were 2710 and 2660 kg/m$^3$ respectively.

The lower stiffnesses seen in the case of the corrugated fracture surface, giving artificial (initial) limits to contact areas, has some resemblance to the reduced (static) normal stiffnesses recorded by Bandis, 1980 for slightly sheared joints, shown in Figure 16.16 (see 'mismatch' result). A marked acceleration in the axial shortening of the sample was noted by the authors between stress levels of 30 and 40 MPa, which also corresponded with the *increment* of dynamic normal stiffness seen in Figure 16.22. It is also consistent that a *decrement* of dynamic shear stiffness should also occur at this same stress level, resembling a 'roughness-destroyed' portion of pseudo-static shearing simulation (the $JRC_{mobilized}$ concept), shown in Figure 16.10.

The authors discussed the marked difference of compliance behaviour exhibited by their two types of artificial fracture. The 'grooved' sample showed a steady *reduction* in both normal and shear stiffness up to 30 or 40 MPa, exactly the reverse of the stiffening behaviour of the sample that actually had a more 'natural' contact area development. They proposed, very reasonably, that a gradual weakening (of the asperities) was occurring in the grooved case. They also reported a layer of powdered rock development beyond 40 MPa. Finally they doubted that the theoretical use of the $Z_N/Z_T$ compliance ratio was justified, as an indicator of potential fluid saturation (and fluid type), as reviewed and also doubted in Chapter 15 (see Figure 15.11).

It seems, also from earlier review of the Perspex plates model of Hsu and Schoenberg, 1993, and of a cut 'fracture' in aluminium, that the less 'rock-joint-like' the sample, the more likely it is that the ratio of $Z_N/Z_T$ will approach 1.0. In the case of the above grooved limestone interfaces, the ratio rose markedly, towards 0.8, beyond a stress level of 40 MPa.

In the rock mechanics world of *pseudo-static* testing, including laboratory and field scales of blocks, the more 'rock-mass-like' the sample becomes (i.e. when moving into the *in situ* environment), the *higher* the ratio of $K_n/K_s$. If this behaviour has a certain influence, even only minor, on exploration seismic behaviour, (at orders of magnitude lower frequency than all the laboratory compliance testing reviewed in this book), then a lower ratio of $Z_N/Z_T$ would be expected, far from the neighbourhood of 1.0, as seen in many laboratory-based or modelling-based geophysics publications.

Concerning the actual effect of water saturation compared to the dry state, which we could tentatively extrapolate to the hydrocarbon equivalent of 'brine-saturated compared to gas', we will see later in this chapter that *both* the *pseudo-static* $K_n$ and $K_s$ magnitudes will reduce somewhat when water saturated. Devoid of any attenuation in the form of fluid-changed scattering or squirt losses, these reductions can simply be explained (and quantified) by reduced JCS (joint wall strength) with moisture for the case of a reduced $K_n$, and reduced JCS and $\phi_r$ (residual friction angle), for the case of a reduced $K_s$.

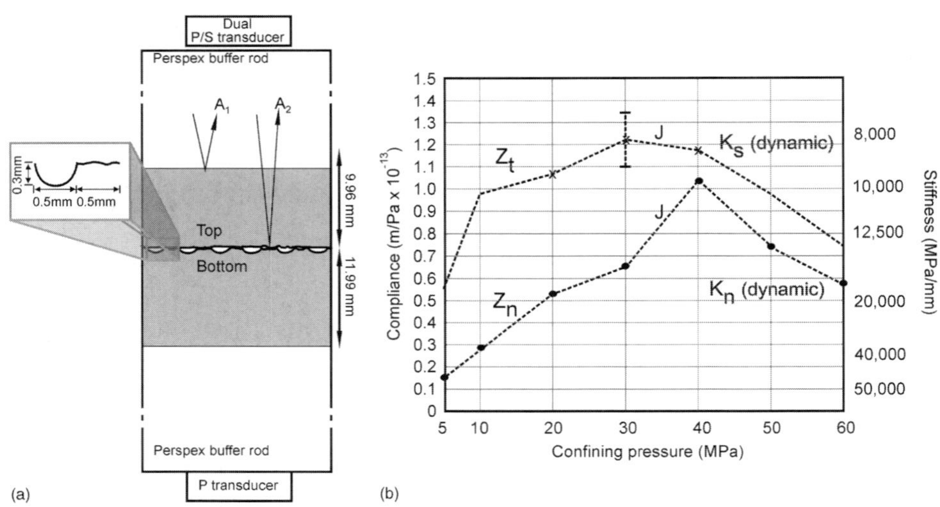


*Figure 16.22* a) Samples of Jurassic (J) limestone, with 'fractures' prepared by cutting and grooving, to produce a 'corrugated fracture interface'. b) Shear and normal compliance versus normal stress, with inversion to dynamic stiffnesses shown on the right-hand axis. Lubbe et al., 2006.

Although the micro-displacement phenomena mobilized by dynamic waves may not be *fully* affected by such pseudo-static reductions in stiffness, there is of course increasing evidence and logic, in reduced dynamic stiffnesses in the case of weakened joint or fracture surfaces. After all, such reductions will also tend to reduce $E_{static}$ (or M), *and* $E_{dynamic}$, assuming that frequencies are not too high to prevent micro-flows of fluid.

### 16.4.5 The Worthington-Lubbe-Hudson range of compliances

Lubbe, 2005 considered that the main conclusion from his research on compliance, could be summarised in Figure 16.23. This was kindly provided for reproduction in this book. The careful three-author title to this sub-section is due to the joint contribution of these researchers, and is also based on the figure's earlier appearance in Worthington and Lubbe, 2004. The data

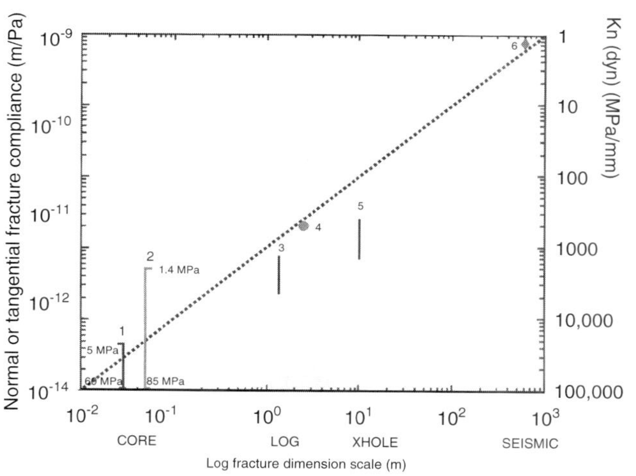


*Figure 16.23* Scaling of fracture compliance from laboratory and field data. The data numbered 1, 3 and 5 was produced by Lubbe from compliance measurements on artificial fracture surfaces in limestone, reviewed above, from the limestone quarry data reviewed earlier, and from a highly fractured shale-rich field site. The latter two included sonic logging and cross-hole interpretations, assuming a given mean bedding-plane fracture spacing. The other data are explained in the numbered text that follows. The horizontal axis of the figure was diagrammatic, since 'it is not possible to measure any parameter that accurately represents the size of the fracture'. (Worthington and Lubbe, 2004, Lubbe, 2005, Figure 7.1 and Lubbe, 2005 pers. comm.).

numbered 1, 3 and 5 extending from ultrasonic to sonic to cross-hole frequencies, represented Lubbe's D.Phil. contribution to this important compliance-*size* trend.

We will summarise the differently numbered data sets, and also convert compliance units to dynamic stiffness units, for easier reference to both pseudo-static and dynamic test data, where the 'Pa/m' format (or MPa/mm) is used, instead of the geophysicists' somewhat cumbersome 'm/Pa', with the need of multiplication by a very small number like $10^{-12}$.

1. Laboratory compliance tests, using artificial 50 mm diameter surfaces in Jurassic and Carboniferous limestone. The vertical bar bounds the 5 to 60 MPa normal stress range. The complete range of both normal and shear compliances was $0.5 \times 10^{-14}$ to $4.5 \times 10^{-14}$ m/Pa, or 200,000 to 22,220 MPa/mm (equivalent to a very stiff-sounding 200 to 22.2 MPa/$\mu$m). See Figures 16.21 for the differentiation of $Z_N$ and $Z_T$ results. (Data from the 'grooved' samples: Figure 16.22, are not incorporated in Figure 16.23).
2. The compliance interpretations of Pyrak-Nolte et al., 1990 for dynamic tests on three natural joints in quartz monzonite, of 52 mm diameter. Normal stress levels ranged from 1.4 to 85 MPa. Compliances ranged from $5.3 \times 10^{-13}$ (the lowest $Z_T$ value) to $10^{-14}$ m/Pa (the highest $Z_N$ value). In stiffness units, these correspond to 1,900 MPa/mm to 100,000 MPa/mm (or 1.9 to 100 MPa/$\mu$m).
3. This sonic-log based field data was from the Carboniferous limestone quarry at Tytherington, depicted in Figure 16.20. The approx. 23 kHz compressional wireline measurements were analysed by Lubbe, first using a $Z_N = Z_T$ assumption and the theories of Schoenberg, 1980 and Pyrak-Nolte et al., 1987b. Allowance for a 20° variation in fracture dips resulted in an estimate of $Z_N = 4.8$ $(\pm 2.6) \times 10^{-13}$ m/Pa, or 2,080 MPa/mm (range 4,405 to 1,350 MPa/mm). Lubbe also used the Schoenberg and Sayers, 1995 *excess compliance* theory, using the vertical velocity reduction accorded to fracturing ($\pm$500 m/s) to give a $2.5 \times 10^{-13} \leqslant Z_N \leqslant 5.0 \times 10^{-13}$ m/Pa estimate of normal compliance. This range converts to a $K_n$ (dynamic) range of 4,000 to 2,000 MPa/m respectively. Lubbe used a third calculation method, dispensing with the $Z_N = Z_T$ assumption, and basing his estimates on Hudson and Crampin theories to interpret the 10% P-wave anisotropy as an assumed 0.035

*fracture density* for the horizontal 'open' joints. This resulted in a similar $2.6 \times 10^{-13} \leqslant Z_N \leqslant 5.2 \times 10^{-13}$ m/Pa, or 3,850 to 1,920 MPa/mm dynamic normal stiffness. As one may note from the location of these larger scale, lower frequency data in Figure 16.23, only the low stress laboratory ultrasonic data lies in the same magnitude bracket.

4. The data point labelled '4' in Figure 16.23 was the Myer et al., 1995 normal compliance estimate for Hanford Basalt columnar jointing, derived from cross-hole measurements. The normal compliance estimate was $2 \times 10^{-12}$ m/Pa, or a dynamic normal stiffness of 500 MPa/mm. (In Worthington et al., 2001, the respective values are $5 \times 10^{-12}$ m/Pa, or 200 MPa/mm). Both are of similar magnitude to the laboratory (100 mm) scale pseudo-static $K_n$ data obtained at higher (10 MPa) stress, as measured by Bandis, 1980 (see Table 16.3).
5. The vertical bar labelled '5' in Figure 16.23 is from a highly fractured shale-rich field site called Reskajeage, where Lubbe, 2005 again used Hudson and Crampin theories to relate a 22% P-wave anisotropy to a 0.07 *fracture density*. Normal compliance estimated for single fractures were as follows: $6.6 \times 10^{-13} \leqslant Z_N \leqslant 8.3 \times 10^{-13}$ m/Pa, which converts to a dynamic normal stiffness range of 1,510 to 1,208 MPa/mm. A second method of estimation, again involving the $Z_N = Z_T$ assumption and Schoenberg and Pyrak-Nolte theories, resulted in the following range of compliances: $1.9 \times 10^{-12} \leqslant Z_N \leqslant 2.4 \times 10^{-12}$ m/Pa, meaning dynamic normal stiffnesses ranging from 525 to 410 MPa/m. Thirdly, Lubbe, 2005 followed the methodology of Pyrak-Nolte et al., 1990 and Myer et al., 1995, modelling the observed attenuation from the P-wave transmission amplitudes in a highly anisotropic 'open' fractured zone, deriving a normal compliance of $4.5 \times 10^{-13}$ m/Pa, or a dynamic normal stiffness of 2,230 MPa/mm, both reportedly consistent with cross-hole travel time data. Lubbe remarked that the Reskajeage compliances were higher (and stiffnesses therefore lower) than at the limestone quarry site, due to the greater continuity of hydraulically conducting 'open' fractures.
6. The data point labelled '6' in Figure 16.23 was from Worthington and Hudson, 2000 and related to a computed *shear* compliance estimated for a major fault zone from 1 to 2 km depth, intersecting a North Sea well at an angle of 50°. (see Figure 10.67). The shear compliance estimate was $1.1 \times 10^{-9}$ m/Pa, or a dynamic shear stiffness of 0.9 MPa/mm. The theories of Hudson and Schoenberg were used to derive a range of possible low stiffnesses, as reviewed in Chapters 10 and 15. The frequency band was 10 to 150 Hz.
7. Finally, the dotted black line in Figure 16.23 was based on the theoretical prediction by Hudson et al., 1997 for *tangential* compliance. For fault-like features it also applied to *normal* compliance if the 'fracture' was filled with weak material, assuming a 10% area in 'welded contact'. Lubbe, 2005 therefore considered that the internal consistency of this spread of data demonstrated that *compliance increased with the assumed scale of the fractures*, dynamic stiffness thereby reducing with size. The experience from *pseudo-static* testing, and also from structural-geological logic, obviously supports such a thesis, and has long been a part of *shear stiffness* prediction in relation to block size, since Barton, 1972.

Lubbe, 2005, perhaps incorrectly, felt that the very low interpreted shear compliance for the North Sea fault zone might be highlighting 'the theoretical possibility of $Z_N/Z_T$ being close to zero', since he considered that the normal compliance of a fracture at 2 km depth 'would be vanishingly small'. Clearly this is an exaggeration, as the effective stress at 2 km depth in a petroleum reservoir may be 'only' 20–25 MPa. However, the important point was emphasised that the ratio $Z_N/Z_T$ could be *very small*, thereby indirectly linking the dynamic behaviour to the pseudo-static behaviour. In subsection 16.4.2, potential pseudo-static ratios of $K_n/K_s$ for typical 1 m *in situ* block sizes were derived, ranging from 100–5000. These are large ratios, but they are finite. $Z_N/Z_T$ never vanishes, as perhaps implied in Figure 15.10, at very small aspect ratios. That $Z_N/Z_T$ becomes very small at large scale is however easy to understand. Worthington and Lubbe, 2004 presented a slightly different version of Figure 16.23, describing their analysis of a cross-hole tomography investigation in Cornwall, which was published by Herwanger et al., 2004. They estimated the compliance contribution of a dominant, dipping fractures zone, which showed velocities with a range of about 3.0 to 4.5 km/s, and an estimated *fracture density of 0.07*. Using the Schoenberg and Sayers, 1995 *excess compliance* concept, they estimated total values of $Z_N$ and $Z_T$ for the fracture zone of $6.5 \times 10^{-12}$ and $7.0 \times 10^{-12}$ m/Pa. Due to a heterogeneous distribution of fracture frequency in the zone,

Table 16.7 An assembly of (macro-deformation, pseudo-static) normal and shear stiffnesses from laboratory and in situ tests on clay-filled discontinuities, from Brazilian dam foundations. After Infanti and Kanji, 1978.

| Clay-filling Thickness (mm) | $K_s$ MPa/mm | $K_n$ MPa/mm | Ratio $K_n/K_s$ |
|---|---|---|---|
| 50–100 | 0.01–0.1 | 0.1–0.5 | 6 (5 to 10) |
| 12–20 | 0.1–0.6 | 0.5 – 2.0 | 3.6 (3.3 to 5) |
| <1 mm | >1.0 | >5.0 | 5 |

they estimated a possible range of $7.0 \times 10^{-13}$ m/Pa if 10 fractures/m, and $7.0 \times 10^{-12}$ m/Pa if only 1 fracture/m. These invert to dynamic stiffnesses of 1,430 and 143 MPa/m respectively. Their estimated range could be added as a '5b' set of data in Figure 16.23 (immediately to the right-side of line '5', thereby slightly extending the trends for lower stiffness at larger assumed size, as sampled by 'lower' frequency.

### 16.4.6 Pseudo-static stiffness data for clay filled discontinuities and major shear zones

While addressing the subject of likely large-scale compliances and dynamic stiffnesses, civil engineering contributions to the subject of pseudo-static stiffnesses of clay-filled (i.e. major) discontinuities can be cited, from Brazilian dam foundation testing. Table 16.7 shows macro-deformation normal and shear loading tests from the 1970s, reported by Infanti and Kanji, 1978. The applied effective normal stress levels were from only 0.5 to 2.5 MPa, representing 'near-surface' conditions in a geophysics perspective.

A much more comprehensive assembly of shear stiffness data from *in situ* testing at Brazilian, and other dam sites was subsequently reported by the same authors, in Infanti and Kanji, 1990. The general trends for their extensive shear stiffness data are summarised in Figure 16.24. Note the dominance of values of shear stiffness of less than 1 MPa/mm, but at stress levels that are an order of magnitude less than reservoir effective stresses. Much of the data range corresponds to the black-discs shown at 0.7 m size in Figure 16.14.

An interesting *coupled-process* investigation of large scale fracture zones at the Underground Research Laboratory (URL) in Manitoba, based on extensive borehole-based measurements, was reported by Martin et al., 1990. Hydraulic pressure changes caused by

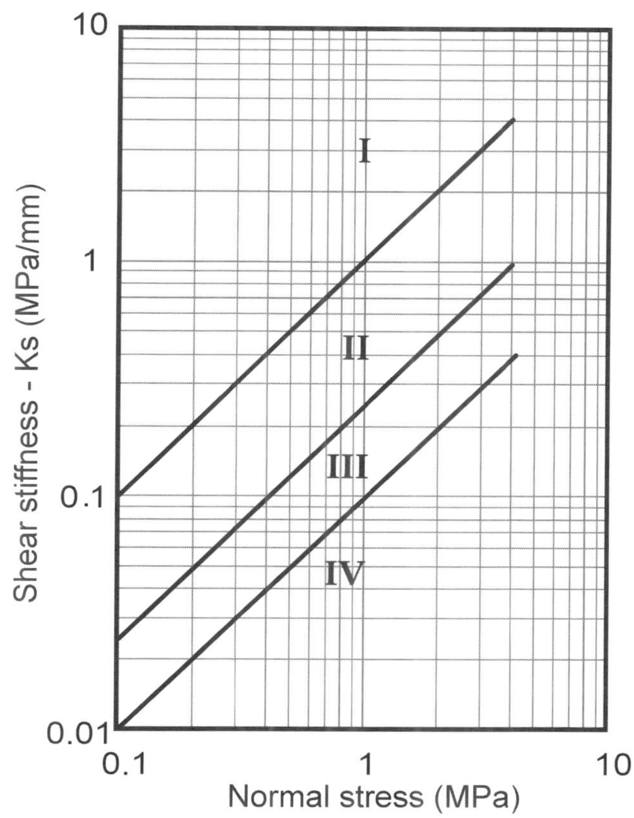


*Figure 16.24* General trends from *in situ* tests at dam sites, concerning the (pseudo-static) shear stiffness of normal-and-shear loaded rock joints and clay-filled discontinuities, as a function of effective normal stresses in the near-surface range of 0.1 to about 4 MPa. I) Rough unfilled joints, or with thin clay film, $\delta_{peak} < 1$ mm. II) Joints without filling, or with thin clay film, $\delta_{peak}$ 1.0–2.5 mm. III) Filled joints (thickness up to 12 mm) with $\delta_{peak}$ 2.5–5.0 mm. IV) Smooth joints with fillings >20 mm, and $\delta_{peak} > 20$ mm. Infanti and Kanji, 1990.

injection or pumping could be monitored by so-called 'Pac-ex' combined double-packer-extensometer units installed in boreholes intersecting the fracture zones. Interpretation of the normal-deformation effective-stress change allowed estimation of large scale pseudo-static normal stiffness from measurements in a total of ten boreholes, with numerous measurements concentrated in 'Fracture zone 2', which was a major dipping feature intersected over hundreds of meters extent at the URL.

The data sets describing the *in situ* normal stiffnesses were extensive, due to the range of normal stress levels measured in different locations adjacent to and within the fracture zone, and due to the fact that the pore pressure could be altered at will. In a so-called cataclastic and very permeable section of FZ 2, where the effective stress was determined to be as low as 0.2 to 2 MPa, the measured *normal stiffnesses* ranged from only 2 to 6 MPa/mm. Elsewhere, where effective normal stresses were as high as 20 MPa, the *normal stiffness* ranged between about 20 and 500 MPa/mm. Flow-rate interpreted *hydraulic apertures* were as large as 400 μm in the cataclastic zone, and as small as approx 20, 40 and 60 μm where normal stiffnesses and stresses were an order of magnitude, or more, higher. Reasonably good correspondence to Barton-Bandis coupled modelling was reported by the authors.

The low values of normal and shear stiffness reviewed here, represent very attenuating, heavily fractured, and sometimes clay-bearing rock masses. In terms of attenuation (1/Q) they would represent cases of complete energy loss, not per cycle, but in just part of a cycle, i.e. below the traditionally defined $Q_{seis} \geqslant 2\pi$ 'limit'.

As shown earlier in this chapter, ratios of pseudo-static $K_n/K_s$ perhaps of several hundred to several thousand, are suggested if *clay-free* fracturing or jointing is to be represented. But with a dominance of *clay-filling* due to faulting as above, ratios of $K_n/K_s$ perhaps as low as about 5, and therefore ratios of $Z_N/Z_T$ as high as about 0.2, would be suggested. (Refer also to the discussion on fault-scale stiffnesses given in Chapter 15, section 15.6.1).

Of course it is not known whether the existing 'static' stress-deformation conditions acting on the joints or fractures in question, could be considered to have significant influence on the stiffer dynamic micro-displacement 'excursions'. Such cyclic excursions presumably will tend to occur with suitably increased gradient, *above* the existing in situ stress-deformation gradient (or loading path). When this is already 'soft' due to clay-filling, higher (dynamic) compliances seem reasonable to expect. Interestingly, Lubbe, 2005 refers to Japanese work concerning faults under high stress that were assumed to potentially suffer *inelastic* behaviour in the solid wall-rock material, during transmission of elastic waves within the shallow sub-surface. 'Reverse' stiffening, as in the grooved sample of Lubbe (Figure 16.22) could then affect the compliance/stiffness response.

### 16.4.7 Shear stress application may apparently affect compliance

Nakagawa performed an interesting laboratory test which demonstrated that the application of *static shear stress* on a planar fracture actually caused normal incidence P⟶S conversion. This led Schoenberg and Nakagawa, 2002 to demonstrate that linear slip theory and off-diagonal terms in the Schoenberg fracture compliance matrix, could be used to relate the dynamic traction on the experimental fracture with the dynamic displacement discontinuity across the fracture.

They found that due to an applied shear stress, a fracture compliance matrix had a single off-diagonal term that coupled the normal component of dynamic stress with the tangential component of displacement, and vice versa. The off-diagonal component ($Z_{NV}$) appeared from their laboratory tests to be *linearly related to the shear stress component.*

In a partly related field, where joint stiffnesses are of concern for the modelling of deformation in rock masses as a result of, for example, tunnel or slope excavation, we generally assume a *linear* initial shear stress-displacement relation. If the joints are displaced in shear (as depicted in Figure 16.16 – see 'mismatch'), there will tend to be a linear relation between *normal stiffness* and normal stress, following Bandis, 1980 and Bandis et al., 1983. Perhaps from these physical results we can approximate the off-diagonal response referred to above.

An example of the Barton-Bandis shear strength-displacement-dilation modelling of joints or fractures of different sizes, is shown in Figure 16.25. One may note the modelling of a delayed dilation as block size increases. Each of the responses to shear stress are scale-defined. The likelihood of minor cross-jointing *in situ*, to thereby define blocks of a certain size, will be found to give different macro-shear stiffnesses, due to the 'double-effect' of scale on both peak shear strength and displacement to peak; the latter also tending to determine the initial slope.

The extent to which this 'macro-deformation' behaviour will 'steer' dynamic micro-deformation behaviour is of course uncertain, but the scale effect seems unlikely to be totally ignored. As a corollary, why should small laboratory samples give more 'correct' values of compliance and stiffness than in situ block sizes? Possibly the Worthington-Lubbe-Hudson data shown in Figure 16.23 is tending to resolve this dilemma.

In view of the Nakagawa laboratory observation that application of shear stress causes P⟶S conversion on a fracture, one may speculate whether the 'open' fractures that are the dream of well production, are also under sufficient shear stress through limited mis-alignment with major principal stress, that the shear wave splitting phenomenon is also affected in an equivalent manner to the above wave conversion.

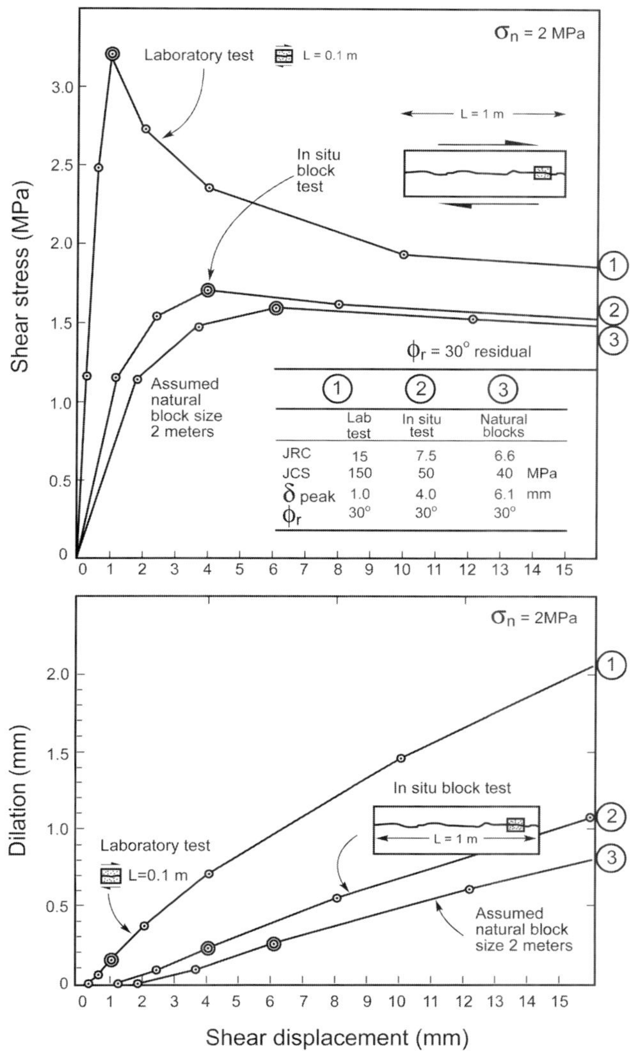


*Figure 16.25* Example of manually calculated shear stress-displacement-dilation curves for different block sizes, using the $JRC_{mobilized}$ concept of Barton, 1982. Scaling of parameters JRC and JCS shown in the table, follows developments described by Bandis et al., 1981.

Figure 16.10 showed the dimensionless JRC mobilized concept that is used to predict the various 'macro-deformation' shear strength-displacement-dilation behaviours depicted in Figure 16.25. In the initial part of the curve where 'mobilization of friction' is written (i.e. a reservoir joint under limited shear, or differential stress), one may imagine an impinging dynamic wave causing very slight, *cyclic*, *μm-magnitude* (?) shear displacements. Since such shear waves are known to 'mobilize' a shear compliance of recognisable magnitude with recognisable units ($m.Pa^{-1}$) during *dynamic excursions* in this area, it is surely justified to continue the classic assumption that 'friction' is indeed one of the sources of seismic attenuation ($Q^{-1}$) in a rock mass.

## 16.5 Effect of dry or saturated conditions on shear and normal stiffnesses

The effect of the environmental conditions (gas, brine, or oil) on the anisotropy-causing *compliances* of rock joints or fractures, at levels of effective normal stress relevant to reservoir conditions, have obviously not yet been investigated to the level required, now that *shear wave splitting* and *fracture diagnostics* are becoming more common, and more important in hydrocarbon *exploration* and *time-lapse monitoring*. There is the added complication that the application of high, reservoir-related stress levels has demanded the use of rather small (roughly 40 to 50 mm) sample sizes in the very few laboratories who have investigated (dynamic) joint compliances. We have reviewed readily available data, under various contexts, in Chapters 10, 13 and just now in this chapter.

With the exception of temperature and oil-saturation effects, there is a large body of relevant test data from *earlier* rock mechanics investigation that can be used to deduce the approximate normal and shear stiffness magnitudes for rock joints. Joints and fractures developed in different rocks obviously have different rock strength, wall-alteration and roughness (UCS, JCS and JRC) characteristics. Although not direct models for compliance, the different characteristics can perhaps be of value in guiding the extrapolation of the presently *very* small, and *very* small-scale compliance data set.

The variable shear strength and shear and normal stiffnesses of rock joints have been of concern for a long time (for almost 40 years, at publication time), due to the need for basic rock stability calculations, and for subsequent coupled flow-deformation analyses.

It is therefore logical to look at this earlier evidence for the effect of *dry* (assume equal to gas-saturated) or *water-saturated* conditions on joint or fracture properties. This line of investigation is followed here, under full recognition that the 'macro-deformation' behaviour of rock joints has to be applied with caution to the tentative extrapolation of micro-deformation dynamic/seismic wave loading effects, being mindful of the usual $E_{dynamic} \geqslant E_{static}$ inequality.

Before referring back to 'environmental' test data from the 1960s and 1970s, it is helpful to address the components used in the most comprehensive-yet-simple-to-apply joint model. The Barton-Bandis criterion is widely used and referenced outside geophysics literature, where it is unknown, and is part of a well established rock mechanics numerical modelling procedure (UDEC-BB). A useful feature of the model is the ease with which input data can be acquired from index tests on drill-core.

The shear-related components JRC, JCS and $\phi_b$ were introduced by Barton, 1973a, and were refined by the addition of $\phi_r$ and simple index test methods by Barton and Choubey, 1977. This development was based on numerous tilt-shear and direct shear tests on 130 fresh and partly weathered joint samples. The just developed JRC-JCS-$\phi_r$ joint strength parameters were subsequently utilised by Bandis, 1980 in very extensive studies of both shear and normal stiffness and scale effects on shear strength. These three strength parameters, and the scaling of the first two, have been referred to elsewhere in this chapter. Here we will provide some details of the parameters themselves.

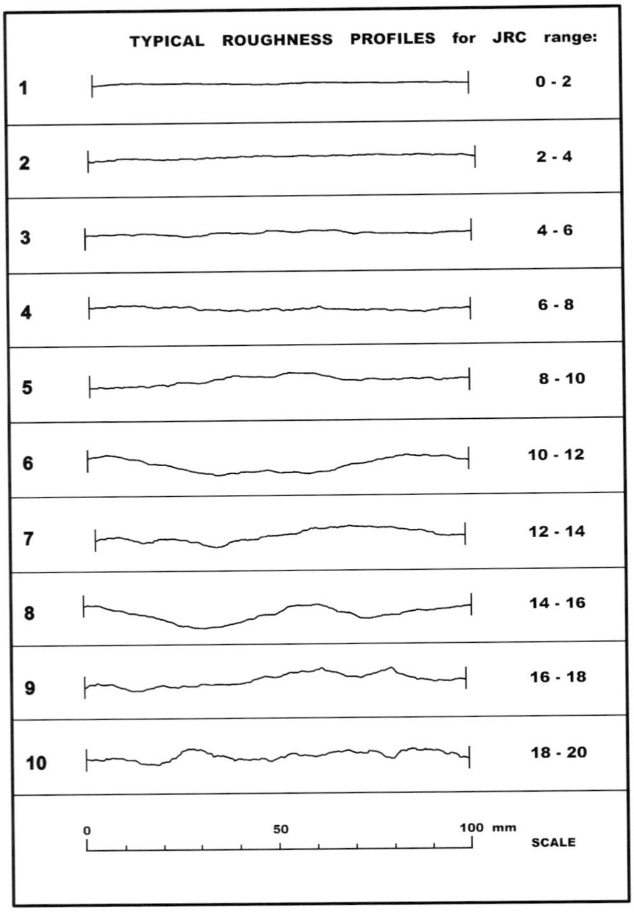


*Figure 16.26* Laboratory-scale joint roughness profiles with their measured JRC values, from Barton and Choubey, 1977. Note that the roughness is defined by *one* parameter for simplicity. The lab-scale $JRC_o$ value is reduced when larger block sizes are involved, using a $JRC_o$ to $JRC_n$ scaling equation, for block sizes of length $L_n$ in place of $L_o$. Scaling equations are given later in the text.

### 16.5.1 Joint roughness coefficient (JRC)

It is reasonable to believe that the joint roughness coefficient (JRC), which varies from 0 to about 20 (for completely plane to extremely rough joints) is *not* significantly affected by the dry or wet condition, since it is essentially 'geometry'. Examples of laboratory-scale (100 mm) profiles and their JRC values (now termed $JRC_o$) are shown in Figure 16.26. Examples of $JRC_n$ profiles of longer joints, as may be used in larger-scale shear strength estimation, are shown in Figure 16.27.

In softer, younger rocks, JRC may perhaps reduce *slightly* during the process of compaction caused by pore pressure reduction when producing from a reservoir. This effect, which can hardly be evaluated, is a function of the changing effective stress/strength ratio: specifically $\sigma_n'/JCS$.

The idea behind JRC is that it is extremely easily determined, as indicated by several index test methods, shown diagrammatically in Figure 16.28. These include self-weight tilt tests which can be performed on joints recovered from core, as for instance performed on joints from the Ekofisk chalk reservoir in the mid-eighties. Also shown is the roughness-profile gauge 'comb' method, which one enterprising manufacturer, without the writers knowledge, has called 'the Barton comb'. The amplitude (a) of asperities per measurement length (L) can also be used, with a/L scaling to $JRC_n$ (Barton and Bandis, 1990).

In practice, the index tests illustrated in Figure 16.28 are performed on several samples of each joint set that is to be modelled (e.g. in a UDEC-BB distinct element model). Makurat et al., 1990 gave examples of such applications, from one of the earlier tunnel modelling exercises with UDEC. An extract from this modelling, concerning the different apertures (e and E), their distribution with depth, and effects of tunnel excavation, was illustrated in Figure 15.50.

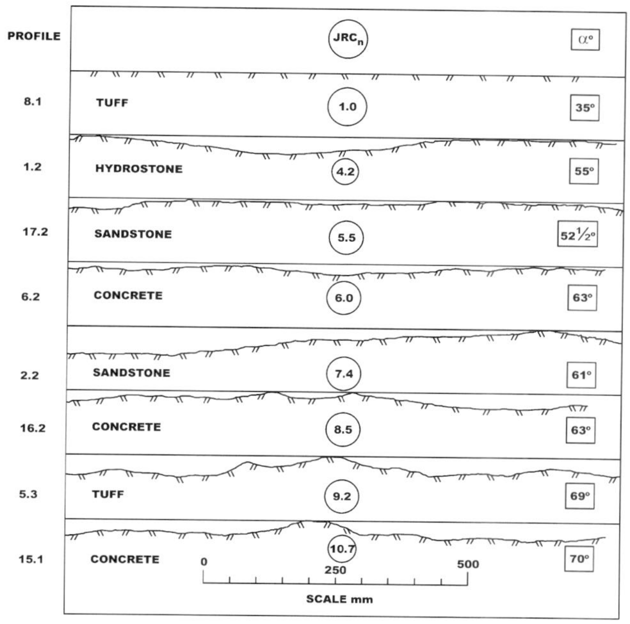


*Figure 16.27* Large-scale (1 m) laboratory-determined roughness profiles, and their $JRC_n$ values as determined from self-weight *tilt testing*, from Bakhtar and Barton, 1984.

### 16.5.2 Joint wall compression strength (JCS)

The joint wall compression strength (JCS) is known to generally reduce with water saturation compared to the dry state. This is because of the much researched effect of moisture on the uniaxial compressive strength ($\sigma_c$), and on the Point load tensile strength ($I_{50}$), which emanates from the late 1960s. When correctly performed, JCS measurement by Schmidt hammer is performed on clamped, water saturated joint samples, specifically to record eventual moisture-reduced wall strength, with JCS $\leqslant \sigma_c$ in general, as illustrated in Figure 16.5a.

It will be noted from the peak shear strength criterion of Barton-Choubey (for laboratory scale estimation), and from the Barton-Bandis (larger scale) version, to be given as equation 16.5, that when $JRC_o$ or $JRC_n$ are *zero*, due to complete planarity and smoothness (but without complications of work-hardened *polishing*), the ratio $JCS/\sigma_n'$ no longer influences the shear strength. This assumes of course that the remaining value of shear strength ($\sigma_n' \tan \phi_r$) correctly captures the essence of $\phi_r$ for the rock concerned. This is the function of the following ratio r/R.

### 16.5.3 Basic friction angle $\phi_b$ and residual friction angle $\phi_r$

The third and fourth components of the shear strength criterion are the *basic* friction angle ($\phi_b$) of unweathered artificial, planar, dry rock surfaces, and the *residual* friction angle ($\phi_r$) applying to flat, non-dilatant, saturated, well-sheared surfaces that may be weathered, i.e. $\phi_r \leqslant \phi_b$. Since it is very difficult to reach this minimum strength value in standard direct shear tests, an ultra-simple empirical equation was developed to enable *estimation* of $\phi_r$ from the easily measured $\phi_b$ (often close to 30°), using Schmidt hammer rebound (r) on the saturated joint walls to compare to the rebound (R) on the dry artificially cut rock surfaces. With r $\leqslant$ R, the residual strength is estimated as follows:

$$\phi_r = (\phi_b - 20°) + 20(r/R) \qquad (16.3)$$

The back-calculation of JRC from tilt tests or direct shear test, and the JRC-labelled profiles in Figure 16.26 are based on this method of estimating $\phi_r$. Since moisture generally reduces both JCS and $\phi_r$, there are inevitable consequences of saturation both on the shear stiffness (which is given by JRC, JCS and $\phi_r$), and on the normal stiffness (which is given by JRC and JCS).

Barton, 1973a and Barton and Choubey, 1977, gave extensive reviews of the effects of saturation on $\phi_b$, $\phi_r$, and $\sigma_c$. One of the first sources of data for these reviews was Horn and Deere, 1962, who showed that massive crystal structures such as quartz and calcite may cause the 'coefficient of friction' (flat, artificial surfaces, i.e. $\phi_b$ in the context of these tests) to *increase* with saturation, while layer-lattice structures such as mica and chlorite caused the 'coefficient of friction' ($\phi_b$) to reduce.

Barton, 1973a, tabulated numerous test results from the *dry* and *saturated* states for the three categories 1) $\phi_b$ (or $\phi_r$) 2) $\sigma_c$ (or $\sigma_t$: the tensile strength), and 3) the overall shear strength (which incorporates effects on $\phi_b$, $\phi_r$ and JCS). Rock types included sandstones, siltstones, limestones, chalk, and shale, and several igneous rocks such as granites, basalts, dolerites, porphyries, metamorphic gneisses and slates. This review showed that:

1. $\phi_b$ may *typically* range from 26° to 34°, while $\phi_r$ can be several degrees less, depending on the degree of weathering (since r < R). In general terms, moisture ('wet' compared to 'dry') caused from 0° to 4° (generally 1° to 3°) reduction in $\phi_b$ and $\phi_r$, since pure crystal structures (massive/layer-lattice)

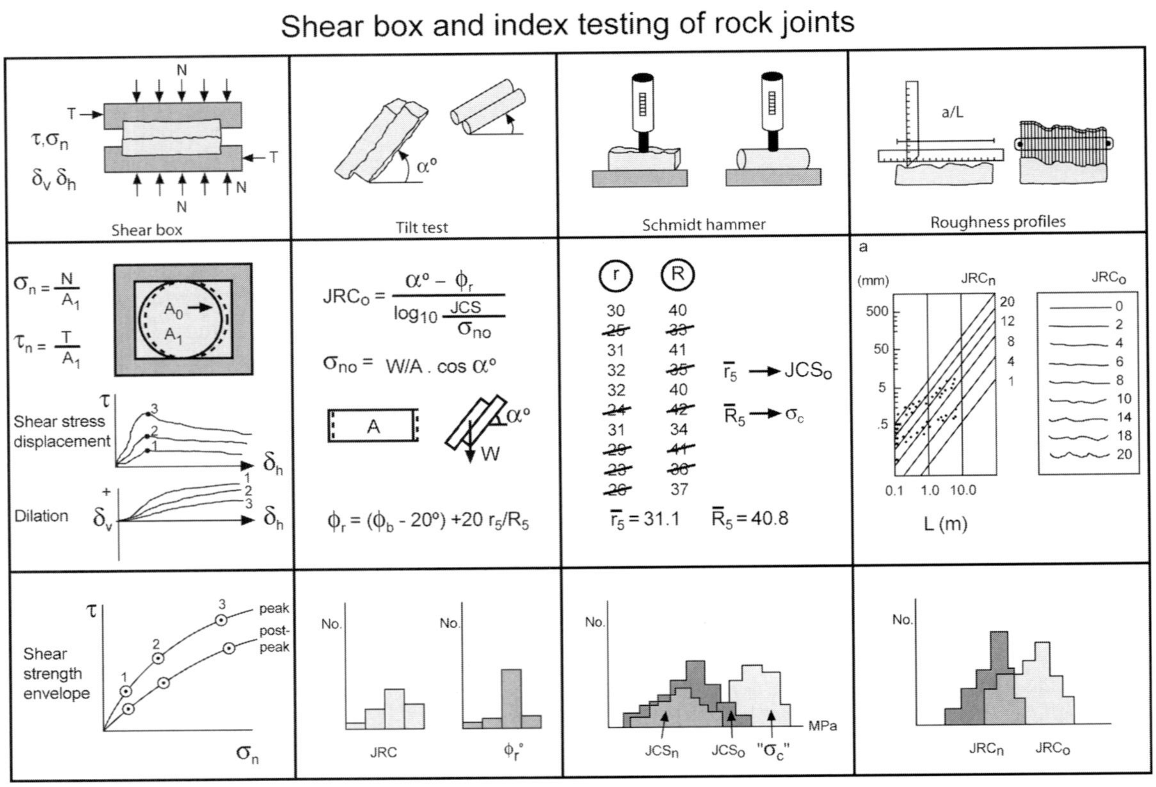

*Figure 16.28* Diagrammatic illustration of joint characterization methods, from Barton, 1999a. a) The left column illustrates direct shear box testing, where shear force T is designed to act 'in-line' in practice, to avoid over-turning moments. Note shear strength-displacement-dilation and shear strength envelope results for three hypothetical joint samples. b) The second column illustrates self-weight *tilt tests*, performed on the joints for back-analysis of JRC, and performed on (un-polished) core sticks or sand-blasted flat-sawn surfaces (both unweathered) for input of $\phi_b$ to the empirical equation for estimating $\phi_r$. c) The third column shows Schmidt hammer tests for estimating JCS and $\sigma_c$, and for input of the *degree-of-weathering* ratio $r_5/R_5$ (see worked example showing use of highest 50% in each case) for input to the $\phi_r$ estimation. d) The fourth and final column illustrates amplitude/length (a/L) measurement, and 'comb' roughness profiling, with respective application in the a/L diagram for $JRC_n$ estimation, or for $JRC_0$ profile-matching at laboratory scale, using Figure 16.26. e) At the bottom of three of the columns, the statistical distribution of each of the key joint strength-and-stiffness parameters is illustrated, together with a reminder of the likely scale-effects on JRC, JCS and the uniaxial compressive strength $\sigma_c$. The core photograph is a fine example of contrasting JRC values from an inter-locked joint and a minor fault in welded tuff, with respective $JRC_0$ values of about 15 and 1.

Table 16.8 Examples of the effect of water saturation compared to dry conditions, on the basic friction angles of flat, unweathered rock surfaces. From review by Barton, 1973a.

| Rock type (flat, smooth, non-polished surface) | $\phi_{b\ dry}$ (degrees) | $\phi_{b\ wet}$ (degrees) |
|---|---|---|
| sandstone | 26–35 | 25–33 |
| siltstone | 31–33 | 27–31 |
| limestone | 31–37 | 27–35 |

are seldom the only component of rock joint samples as opposed to mineral samples. However, an exceptional 9° reduction for slate (dominant layer-lattice) was noted due to water saturation. Three examples are given in the Table 16.8 extract from Barton, 1973a.

2. In general terms moisture ('wet' compared to 'dry') caused from 10% to 50% (generally 20 to 40%) reduction in uniaxial strength ($\sigma_c$ or UCS), and a similar reduction in point load or tensile strength ($\sigma_t$). (Water injection into a reservoir like chalk can therefore have serious consequences – which may be overridingly positive especially if planned for). Water saturation causes reduced UCS, reduced JCS, reduced $\phi_b$ and reduced $\phi_r$, therefore reduced joint normal stiffness, reduced joint shear stiffness and reduced peak and residual shear strength.

As shown in Figure 16.28, the value of JCS can be estimated from Schmidt hammer tests on saturated, fresh or weathered joint surfaces, using a density-*uniaxial compression strength* conversion, from Miller, 1965. See also ISRM, 1978. It is logical to expect that JCS will also be adversely affected by moisture, as is the case for UCS, since JCS specifically involves asperity failure/ stiffness when contributing to either the shear or normal stiffness in the Barton-Bandis models for these two components. Most likely JCS will be *more* sensitive to weakening by water than the rock matrix, if there is any trace of weathering, and because of the increased surface area and microcracking associated with jointing.

### 16.5.4 Empirical equations for the shear behaviour of rock joints

We can illustrate the above effects of the dry or saturated state on shear stiffness, by evaluating examples. (See Barton, 1982 for derivation of the following empirical equations, two of which were derived in Bandis et al., 1981 from Bandis, 1980 scale-effect experiments).

*A) Shear stiffness*

$$K_s = \frac{\tau_{peak}}{\delta_{peak}} \tag{16.4}$$

*B) Shear strength*

$$\tau_{peak} = \sigma'_n \tan\left[JRC_n \log\left(\frac{JCS_n}{\sigma'_n}\right) + \phi_r\right] \tag{16.5}$$

*C) Displacement to peak strength*

$$\delta_{peak} \approx \frac{L_n}{500}\left(\frac{JRC_n}{L_n}\right)^{1/3} \tag{16.6}$$

($\delta_{peak}$, $L_n$ in metres)

*D) Large-scale joint roughness*

$$JRC_n \approx JRC_o\left(\frac{L_n}{L_o}\right)^{-0.02\,JRC_o} \tag{16.7}$$

*E) Large-scale joint wall strength*

$$JCS_n \approx JCS_o\left(\frac{L_n}{L_o}\right)^{-0.03\,JRC_o} \tag{16.8}$$

*F) Initial unstressed aperture*

$$E_o \approx \frac{JRC_o}{5}\left(0.2\frac{\sigma_c}{JCS_o} - 0.1\right) \tag{16.9}$$

where $L_o$ = lab-scale sample length (usually 100 mm)
$L_n$ = in situ block size (generated by the intersecting joint set)
$E_o$ = millimetres

We will assume the following input data:

| | | |
|---|---|---|
| $\phi_r$ dry = 30° (e.g. gas) | | $\phi_r$ wet = 27° (e.g. brine) |
| $JCS_o$ dry = 70 MPa | | $JCS_o$ wet = 50 MPa |
| $JRC_o$ = 8 | $L_o$ = 100 mm | $L_n$ = 500 mm |
| | $\sigma'_n$ = 10 MPa, 20 MPa | |

Thus according to the expected effects of moisture on two of the three strength components, there is a

Table 16.9 Barton-Bandis modelling of the possible effects of water saturation on the shear stiffness of joints sampled at two different scales.

| A. Dry cases | B. Wet cases |
|---|---|
| 1. $JRC_n \approx 8\left(\frac{500}{100}\right)^{-0.02\times 8} = 6.2$ | 1. $JRC_n$ (assume unchanged) |
| 2. $JCS_n \approx 70\left(\frac{500}{100}\right)^{-0.03\times 8} = 47.6$ MPa | 2. $JCS_n \approx 50\left(\frac{500}{100}\right)^{-0.03\times 8} = 34.0$ MPa |
| 3. $\delta_{peak} \approx 0.001\left(\frac{6.2}{0.5}\right)^{1/3} \approx 2.3$ mm (0.0023 m) | 3. $\delta_{peak}$ (assume unchanged) |
| 4. i) $\tau_{peak} = 10\tan\left[6.2\log\left(\frac{47.6}{10}\right) + 30°\right] = 6.8$ MPa | 4. i) $\tau_{peak} = 10\tan\left(6.2\log\left(\frac{34.0}{10}\right) + 27°\right) = 5.8$ MPa |
| ii) $\tau_{peak} = 20\tan\left[6.2\log\left(\frac{47.6}{20}\right) + 30°\right] = 12.7$ MPa | ii) $\tau_{peak} = 20\tan\left(6.2\log\left(\frac{34.0}{20}\right) + 27°\right) = 10.8$ MPa |
| 5. i) $K_s \approx \frac{6.8}{2.3} \approx 3.0$ MPa/mm at $\sigma_n' = 10$ MPa | 5. i) $K_s \approx \frac{5.8}{2.3} \approx 2.5$ MPa/mm at $\sigma_n' = 10$ MPa |
| ii) $K_s \approx \frac{12.7}{2.3} \approx 5.5$ MPa/mm at $\sigma_n' = 20$ MPa | ii) $K_s \approx \frac{10.8}{2.3} \approx 4.7$ MPa/mm at $\sigma_n' = 20$ MPa |

Table 16.10 Barton-Bandis modelling of the adverse effect of water saturation on the conducting apertures of a joint at two different normal stresses.

| | $\sigma_n' = 10$ MPa | $\sigma_n' = 20$ MPa |
|---|---|---|
| 1. $JCS_o$ dry = 70 MPa | e = 5.1 μm | e = 1.6 μm |
| | E = 30.3 μm | E = 16.8 μm |
| 2. $JCS_o$ wet = 50 MPa | e = 3.9 μm | e = 1.2 μm |
| | E = 26.5 μm | E = 14.4 μm |

moderate 13 to 15% reduction in *shear stiffness* from the dry to saturated cases, in the examples in Table 16.9, in which moderate assumptions for effects of saturation have been made.

A similar exercise for normal stiffness is more easily performed using the BB numerical model. A cursory application shows that the initial normal stiffnesses (where $K_{ni}$ is the stress-closure gradient at very low stress), are in the region of 28 to 32 MPa/mm with the above input data. Normal stiffness rises by a factor of at least thirty at 20 MPa normal stress, on cycle #1.

With the foregoing assumptions for input data ($JCS_o$ dry = 70 MPa, $JCS_o$ wet = 50 MPa, it is found from BB modelling that the initial (first load cycle) normal stiffnesses are relatively little affected by the dry→wet condition. However, the final apertures, being so small, are significantly affected as a result of the assumed reduction in wall strength JCS due to moisture.

The *incremental normal stiffnesses* can easily rise by a factor of 10 to 100 when joints in weaker rock are 'imaged' at an effective normal stress of 10 or 20 MPa, as may apply in a reservoir. As we also saw earlier, from Bandis normal stiffness testing, $K_n$ can rise to $10^3$ or $10^4$ MPa/mm if effective stresses become close in magnitude to the saturation-reduced JCS values. This is logical, since the joints in question will be almost closed. Thus, fluid saturated joints in weaker rocks may have *very high* ratios of $K_n/K_s$. We will return to this theme of 'almost closed' joints in greater detail later.

Concerning the anisotropic shear wave splitting and polarization into the fast $qS_1$ and slower $qS_2$ wave speeds, the former parallel and the latter perpendicular to the dominant joint or fracture sets, one may assume that it is the relative low value of $K_S$ (or the high value of shear compliance $Z_T$), rather than the high value of $K_n$, that is most responsible for shear wave splitting. It would also appear reasonable to assume that very tight (low-permeability) joints, with high normal stiffness, will tend to give faster $qS_2$ wave speeds. Shear-wave anisotropy, and attenuation anisotropy might therefore tend to be increased. Theoretical treatment in this area of 'compliance-anisotropy' was given in Chapter 15.

## 16.6 Mechanical over-closure, thermal-closure, and joint stiffness modification

Unfortunately, when joint samples are recovered from a borehole for laboratory testing, or exposed in a tunnel or mine adit for in situ testing, the object of our investigations is 'cold', in relation to its likely formation temperature. The laboratory may also be 'cold' in relation to the sample's *recent* environment. Several pieces of experimental evidence from heated lab-tests, from heated block tests and from heated 'mine-by' tunnelling experiments, suggest that the rock joint normal stiffness, and therefore the apparent rock mass deformation modulus *changes*, when the rock is heated in the presumed direction of higher joint formation temperatures. What *appears* as a measured reduction in stiffness and modulus, is actually producing a stiffer medium. Clearly this process is also going to affect normal compliances, particularly when rock quality is high, and $E_{static}$ is closer to $E_{dynamic}$.

Both indirect circumstantial evidence, and directly measured evidence for these phenomena, were reviewed by Barton and Makurat, 2000, in an NGI contract report ('Hydro-Thermal and Mechanical Hysteresis and Over-Closure Effects in Joints and Rock Masses caused by Mechanical and Thermal Loading and Unloading'). Some published results of such tests were given by Barton et al., 1985, discussed in Barton, 2004b, and reported by Barton and Makurat, 2006.

One of the best documented cases, and also the first of very few hydro-thermo-mechanical *in situ* tests in rock mechanics, was described by Hardin et al., 1981, and further analysed by Barton, 1982. The 8 $m^3$ *in situ* test set-up, allowed biaxial (normal) loading, or uniaxial loading in 'N–S' and 'E–W' directions, i.e. shearing. Flat-jack loading on four (or two) sides of the sawn-in-place block of quartz monzonite provided the desired stress levels. There was also a line of borehole heaters to raise the temperature by some 60°C. A simplified schematic is shown in Figure 16.29.

The *mean* joint spacing in the area of the block test is indicated. The aspect of the test to be described here is the permeability test holes, drilled parallel to the long diagonal, foliation joint. Jointed core extracted from these holes is drawn in Figure 16.30, and shows the measured joint roughness traces, *and* the tilt-test failure angles, giving $JRC_n$ estimates of 8.0, 8.3 and 7.9 due to the 200 to 300 mm long samples, while the mean $JRC_o$ was 13.

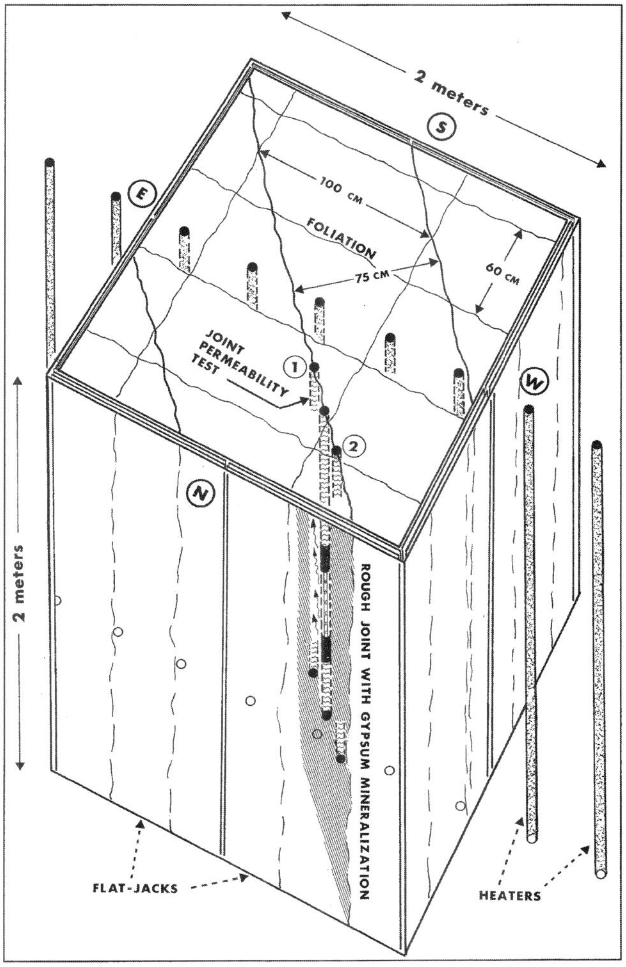


*Figure 16.29* Idealisation of average local jointing in the 2 × 2 × 2 m TerraTek/CSM heated block test in quartz monzonite. This test gave the first well instrumented *in situ* data on MHT coupling, suggesting that joint apertures closed under the effect of increased temperature, despite carefully controlled normal stresses. Note the flat-jack loading of four vertical boundaries, and the line of borehole heaters. Hardin et al., 1981.

The flat-jack loading test cycles are shown in Figure 16.31. Maximum stress was generally 1000 psi, or 6.9 MPa. The very first unload-load-unload-load cycles are shown in Figure 16.32, plotted in terms of normal stress versus hydraulic aperture (e), where permeability during low-pressure-gradient laminar flow was assumed to be given by $K = e^2/12$. The vertical axis of the figure shows the $e = 38\,\mu m$ *hydraulic aperture* before the block was line-drilled to make space for the boundary flat-jacks. This unloading caused the hydraulic aperture to increase to $e = 61\,\mu m$. Following a minor

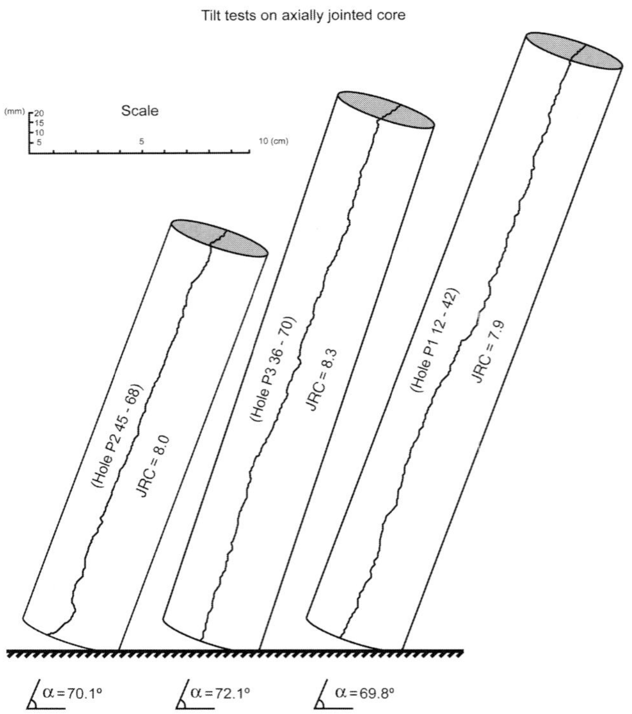


*Figure 16.30* Three of the axially jointed core from the block test permeability test holes, with as-recorded roughness profiles, correct tilt angles at shear failure (70.1°, 72.1° and 69.8°) and back-calculated $JRC_n$ roughness coefficients of 8.0, 8.3 and 7.9 at the given length scales. Barton, 1982.

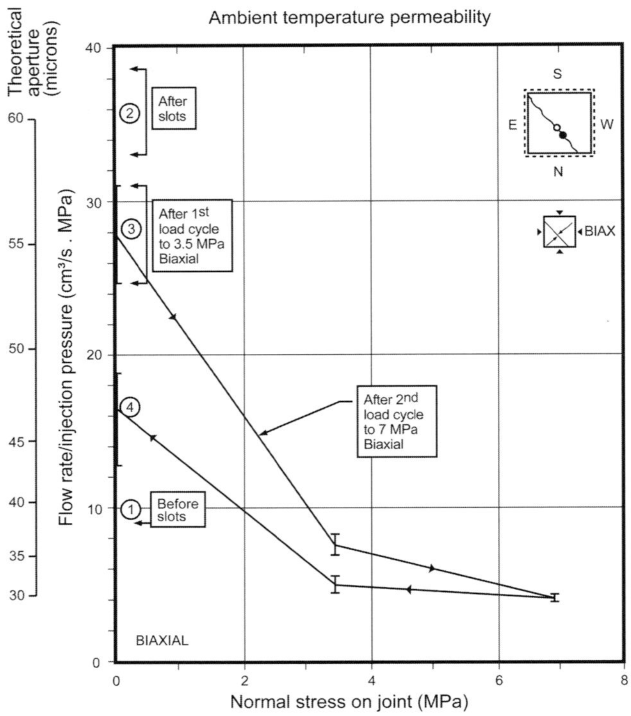


*Figure 16.32* Hydraulic aperture (e) versus normal (applied) stress for the first ambient temperature, biaxial load-unload cycle. Note apertures before and after drilling of the flat-jack slots. Hardin et al., 1981. (See *normal stiffness* calculation, and possible *normal compliance* interpretation, in text.)

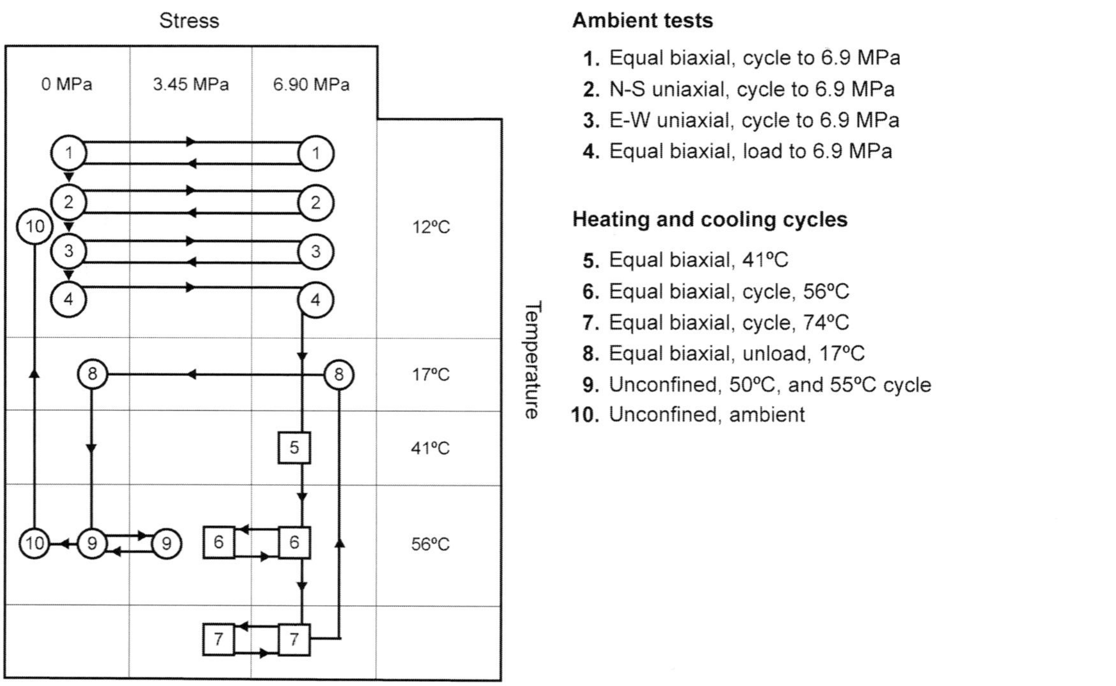


*Figure 16.31* The main boundary-stress-temperature test paths, for the HMT heated block test. Hardin et al., 1981.

0–3.5–0 MPa 'dress-rehearsal' test cycle, the first ambient 0 to 6.9 to 0 MPa *biaxial* test is shown, with $\sigma_1 = \sigma_2$, $\sigma_n = 6.9$ MPa and $\tau = 0$ MPa.

### 16.6.1 Normal stiffness estimation

The typical increasing normal stiffness seen from $\sigma_n = 3.45$ to 6.9 MPa, mirrors laboratory testing. If one had assumed e = E, then the *normal stiffness* in this increment would have been $3.45/(34.4 - 30.0) \times 10^{-3}$ MPa/mm = 784 MPa/mm. However, the physical apertures can be estimated to have reduced from 144.8 μm to 135.2 μm, making $\Delta E \approx 9.6$ μm. This estimate is based on $JRC_o = 13$ and application of the empirical model $e = E^2/JRC_o^{2.5}$, shown in Figure 16.7. The interpreted normal stiffness is then about 360 MPa/mm, which is exactly in line with $K_n$ laboratory test results reviewed earlier in this chapter.

*If we make the assumption* that the rock quality was sufficiently high for $E_{static} \approx E_{dynamic}$, this $K_n$ value of 360 MPa/mm would *imply*, by inversion, a normal compliance $Z_N$ of about 0.0028 m/GPa, or $2.8 \times 10^{-12}$ m/Pa. In fact the rock mass quality *was* very good, as the following estimation of Q-parameters shows (see Appendix A for ratings).

$$Q = \frac{100}{9} \times \frac{3}{1} \times \frac{1}{0.5} \approx 67 \quad Q_c = Q \times \frac{\sigma_c}{100}$$

$$Q_c \approx 67 \times \frac{200}{100} \approx 130 \quad \text{('very good')}$$

Chapter 13, Figure 13.59 relating $Q_c - V_p - M -$ depth – porosity, suggests a *static modulus* of deformation as high as 70 GPa, when taking into consideration the favourable mine-adit-based (and applied) stress level, equivalent to 300 m depth, and a nominal (hard rock) 1% porosity. This figure also suggests a seismic refraction $V_p$ close to 6 km/s for such good rock, located next to the cross-hatched 'hard-jointed', 'hard-massive' boundary. One could then reasonably assume that the measured *normal stiffness* $K_n \approx 360$ MPa/mm derived from the permeability test ( under a 3.45 to 6.9 MPa increment of normal stress), was almost equivalent, in this case, to a *normal compliance* $Z_N \approx 2.8 \times 10^{-12}$ m/Pa. The roughly 1 m size result would 'plot' immediately above the (Hudson) diagonal line in Figure 16.23.

The joint characterization of the permeability test hole cores, gave $JRC_o = 13$ and $JCS_o = 90$ MPa. One could estimate that these values would both be at the upper end of the roughness and strength scales for petroleum reservoir rocks, and perhaps relate best to the stronger carbonates. The uniaxial strength of the present igneous rock was however, 'too high'. The JCS $< \sigma_c$ mismatch was due to slight alteration of the rough joint walls.

### 16.6.2 Thermal over-closure of joints and some implications

When temperature was applied in the above block test, unexpected *over-closure* of the diagonal test joint was experienced. Figure 16.33 shows a three-parameter (coupled-process) plot of (e) versus ($\sigma_n$) versus temperature (°C), and the following result.

The added joint closure means that this cooled (sampled), unloaded joint is displaying an apparent reduction in normal stiffness. This will be estimated shortly. In the reality of e.g. a warm-shallow, or hot-deep petroleum reservoir, joints are likely to be better interlocked than when we unload and cool them for (initial) laboratory testing. In fact it would be better to keep them hot

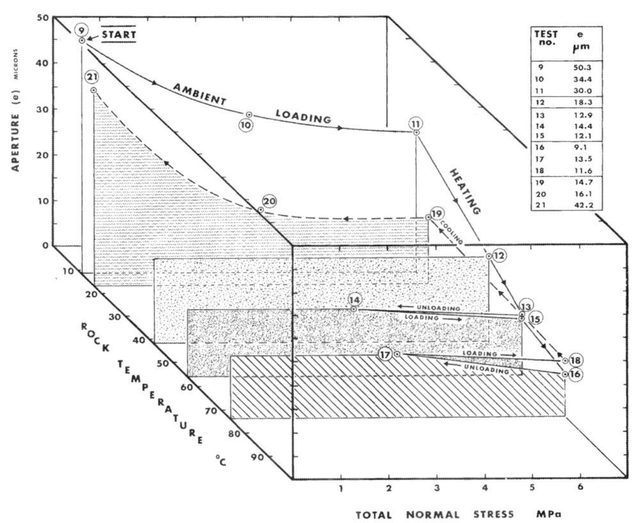


*Figure 16.33* Hydraulic aperture (e), versus normal stress ($\sigma_n$), versus average rock temperature (T°C) in the permeability test volume of the TerraTek heated block test, CSM mine, Colorado. Note aperture (e) reductions from 30.0 μm to 18.3 μm, to 12.9 μm and finally to 9.1 μm as a result of temperature rise, despite constant applied stress. This gives an *apparent* reduction in the normal stiffness in this test, but in the warmth of a deep petroleum reservoir, would have allowed joints to remain stiffer since their formation. Barton, 1982. see Plate 11.

Table 16.11 Effect of temperature on joint (hydraulic) apertures at the heated block test.

| Test No. | 11 | 12 | 13 | 16 |
|---|---|---|---|---|
| $\sigma_n$ (MPa) | 6.9 | 6.9 | 6.9 | 6.9 |
| e ($\mu$m) | 30.0 | 18.3 | 12.9 | 9.1 |
| Temp°C | 12° | 41° | 55° | 74° |

Note: water viscosities corrected for T°C, before calculating (e).

before testing, *and* test them hot too, to get the most relevant response, whether for simulating *in situ* normal stiffness, or dynamic compliance. The key parameter affected by thermal-closure is the thermal expansion coefficient. Joints must be included, and tested hot.

The explanation for this phenomenon was assumed to be quite simple (Barton, 1982). Namely that the joint in question, and perhaps the huge majority of joints developed in the crust, were formed at variously elevated temperatures. They were thereby given a primeval 'finger-print' of 3D-roughness, which was influenced by all the minerals (or grains) forming the joint walls. When cooled, various subtle changes would occur, causing reduced fit.

In the case of igneous rocks, many of the earliest jointing episodes would be at the cool side of the brittle-ductile transition. When such a joint, any joint, is encountered today, (e.g. at shallow depth in a mine, or if drilled at depth and bought to the surface causing unloading and further cooling?), the 3D roughness *finger-print*, though very recognisable, would be subtly altered in its finer details. These *details* (very important at the micron-scale) would be a combination of:

a) a slight expansion from (anisotropic?) stress relief when sampled from depth
b) a sampling damage, but neglected for sake of simplicity
c) a *general* thermal contraction effect, but this might allow the equally altered *finger-prints* from each wall to still fit perfectly
d) locally inhomogeneous contraction superimposed on c), due to non-equality of the thermal contraction coefficients of the constituent minerals and grains along the joint walls

Factors a) and d) are good reasons for less than 'perfect fit' of the primeval joint walls in a testing situation, when initially testing at *ambient temperature*. Factors a) and b) alone, without temperature considerations, are enough to explain the need for several load-unload-load cycles to move beyond the strong hysteresis 'always' experienced on the first load cycle, as emphasised by Barton, 1971, Bandis, 1980 and Bandis et al., 1983, and demonstrated in Figures 16.17.

To illustrate the order of magnitude involved, from the above coupled HMT block test, one may utilise the ambient $K_n$ value of approximately 360 MPa/mm as a starting point. This value was established under ambient conditions, with a final stress increment of 3.5 to 6.9 MPa, therefore reaching the same stress level as applied in the heated part of the test. The constant 6.9 MPa was carefully monitored during heating of the block, to avoid thermal expansion-caused increases in flat-jack pressure. The successive reductions of conducting aperture were from 30 $\mu$m to 18.3 $\mu$m to 12.9 $\mu$m to 9.1 $\mu$m, for temperature increases from 12°C to 41°C to 55°C and finally to 74°C. The process took about 1 month: in geophysics terminology about $10^{-7}$ Hz.

As shown earlier, if one had assumed e = E, then the normal stiffness in the previous ambient increment would have been equal to 3.45 MPa/(34.4 – 30.0) $\mu$m $\times 10^{-3} \approx$ 780 MPa/mm. But since $\Delta E \approx$ 14.0 $\mu$m, based on $JRC_o = 13$ and the empirical model $E^2 \approx e^2 . JRC_o^{2.5}$, the interpreted ambient normal stiffness was about 360 MPa/mm, which was in line with (ambient) $K_n$ test results reviewed earlier in this chapter. The estimations of thermally-changed (static) normal stiffnesses (and therefore *potentially* changed dynamic compliances) are set out in Table 16.12, for transparency.

The assumption is made, for purposes of calculation, that each thermally-induced reduction in aperture occurred at the end of the last load increment. The successive closures are therefore allowed to have an *accumulative* effect on normal stiffness, as if each test started at 12°C and 3.45 MPa normal stress. In fact, since a strict definition of stiffness is stress increase divided by aperture reduction, one could claim that the joint had 'zero' stiffness, since no stress increment (rather a thermal increment) was applied.

If this (almost reversible) thermal over-closure phenomenon is of general application in the case of the less planar jointing, it implies that when the joint in question is 'sampled at depth' by seismic anisotropy determination, in the familiar warmth of a petroleum reservoir, the elevated stress and temperature will likely result in lower compliance, due to its previously imprinted greater stiffness. However, the present temperature at depth is unlikely to be the same as the temperature when the joint was formed.

Table 16.12 Estimation of thermally-induced, apparent reductions of (static) normal stiffness, from the heated block test. For $K_n$ estimation assume each test starts at 12°C and 3.45 MPa, and ends at the specific elevated temperature.

| Temperature increment | Hydraulic aperture (e) μm | Δe (μm) | Physical aperture E (μm) estimated | ΔE (μm) | Kn (MPa/mm) |
|---|---|---|---|---|---|
| 12 to 12°C (ambient) | 34.4 to 30.0 | 4.4 | 144.8 to 135.2 | 9.6 | 360 |
| 12 to 41°C | 30.0 to 18.3 | 11.7 | 135.2 to 105.6 | 29.6 | 88 |
| 41 to 55°C | 18.3 to 12.9 | 5.4 | 105.6 to 88.6 | 17.0 | 61 |
| 55 to 74°C | 12.9 to 9.1 | 3.8 | 88.6 to 74.5 | 14.1 | 49 |

The joint will also be open to changes bought about by the partial cooling caused by e.g. *water flooding*. The above mechanisms suggest a 'better than expected' aperture and permeability enhancement as a result of water flooding. Efforts to reduce the natural warming of this water on the way down the injection well, could be beneficial too, but avoidance of matrix contraction and oil-bypass (unstable fingering) might be a contrary requirement.

This *hydrothermomechanical* HTM coupling 'detail' (in fact a quite important experimental detail), has not yet been absorbed into numerical modelling. It has not even been coded in the Barton-Bandis joint model, mainly because data is presently limited, as the phenomenon is little known, researched, or acknowledged in the rock mechanics (modelling) community. Barton and Makurat, 2006 have addressed the problem once more, in an attempt to bring *over-closure* (in rock mechanics) into focus, as it has been in soil mechanics for decades, but regarding a somewhat different mechanism (i.e. hard, over-consolidated clays, with their altered matrix properties due to unloading from higher historic effective stress levels.).

### 16.6.3 Mechanical over-closure

There is another unresearched aspect, which may have application to shear wave splitting phenomena (and assumed anisotropy), and that is *mechanical* over-closure. Rough joints or fractures that are historically loaded to a higher effective normal stress, and later sheared (or shear-wave loaded) under today's lower effective normal stress, (due for example to an episode of over-pressure), will have a higher shear strength, a higher shear stiffness, and a higher normal stiffness, than if conventionally tested at 1:1 $\sigma_n$ levels.

Over-consolidation ratios of 4:1 and 8:1 were shown by Barton, 1973, to cause several degrees higher frictional strength in relation to 1:1 control samples, in the case of rough tension fractures which remained 'over-closed' following removal of the pre-stress. This caused difficulties with *over-stable* modelled rock slopes, in 40,000-block tension fracture models. Barton, 2004b, discussed this phenomenon as being due to 'JRC at right angles', (i.e. an effective JRC in a perpendicular sense relative to the joint plane). This causes effective locking of asperities, if roughness is sufficient, and the stress reduction significant. When JRC is sufficiently low, such as less than 10, possibly only less than 6, the phenomenon is not expected (refer to Figure 16.26 100 mm roughness profiles in this context).

The writer has experience of a very rough tension fracture, generated to make a sample for demonstrating tilt testing, actually tolerating a tilt angle of 180°, i.e. going from horizontal (no shear, pure normal stress from the self weight of a carefully placed upper half of the block) to 90° (vertical) to 180° (upside-down). In other words even the small pre-stress from self weight loading, amounting to about 0.02 MPa, was sufficient to give an apparent tensile strength to the fracture, due to asperity inter-lock.

Barton, 1973, discussed similar experiences reported by a colleague at Imperial College, who registered an error when preparing a direct shear test. An over-closure episode prior to shear testing (M. de Freitas, pers. com. 1970) caused a jointed sample to be too strong to shear, even though the normal stress was already reduced to the correct level. The sample required mechanical wedging to open it, when extracted from the over-loaded DST apparatus.

## 16.7 Consequences of shear stress on polarization and permeability

In Chapters 14 and 15, when addressing P-wave and S-wave anisotropy in reservoirs, several case records were reviewed that showed *some angular discrepancy* between the various *indicators* of predominant joint strike (FMS, oriented core), the assumed $\sigma_{H\ max}$ direction (90° from borehole calliper-log long axis), and the principal axes of either (or both) P-wave and S-wave anisotropy. The latter is obtained from shear-wave polarization when splitting into fast and slow $qS_1$ and $qS_2$ components. Figures 14.16, 14.35 and 15.60 are three of the cases.

In this section of Chapter 16, we will explore in much greater detail, some of the reasons for believing that *shearing* (as well as mineralization 'bridging') may be a necessary mechanism for explaining hydrocarbon (especially oil) production, from highly stressed joints or fractures.

Vertical, dominant fractures perpendicular to the minimum horizontal stress direction, as often sought and sometimes 'confirmed', may be a more complex aspect of geomechanics than usually assumed. Contrary to the assumption of joints or fractures loaded by 20 to 40 MPa 'minimum' stress actually being 'open', there may be:

1. insufficient strength (specifically $JCS_{confined}$) for *conducting* joints in the assumed direction
2. there may be two sets of joints contributing their seismic 'components' on either side of the assumed $\sigma_{H\ max}$ direction
3. there may be conjugate shearing on one or more joint set and therefore permeability enhancement (or maintenance).

## 16.7.1 Stress distribution caused by shearing joints, and possible consequences for shear wave splitting

The *in situ* heated block test of 8 m$^3$ volume, described in the previous sub-section, was subsequently the object of distinct element modelling by numerical modelling colleagues at NGI in Oslo. Some of this work is reported in Chryssanthakis et al., 1991. The modelling was a form of validation of the distinct element code UDEC, with the Barton-Bandis joint model as a subroutine, termed UDEC-BB in this non-linear form, or UDEC-MC when with simpler linear Mohr Coulomb joint modelling. An example of the complex stress distribution in even just a four-block (simplified) model is shown in Figure 16.34. The flat-jacks were simulated as fluid-filled boundaries. Note the complex stress distribution, even with equal (biaxial) $\sigma_1 = \sigma_2 = 5.4$ MPa

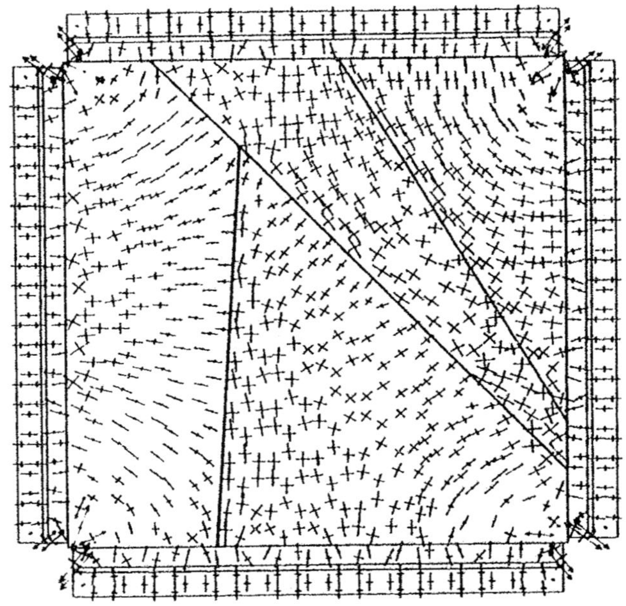

*Figure 16.34* A UDEC-BB model of the Terra Tek/CSM heated block test. Note the complex stress distribution even in an 'equal-biaxial' loading situation, with only four (assumed) interlocking blocks. Input data for the different joints were: JRC = 10 or 13, JCS = 90 or 120 MPa, $\phi_r$ = 25° or 28°. Chryssanthakis et al., 1991.

(approx.) loading. Normal stresses across the joints are clearly in equilibrium, but the stress in the perpendicular direction can be dramatically different, with at least 10 MPa difference, from one side of a joint to the other side.

One may visualise that the passage of shear waves through such a jointed (fractured reservoir) medium could perceive the joints as both displacement discontinuities and stress discontinuities, even though the classic Schoenberg 1980 assumption is for *stress continuity* across the *displacement discontinuity*.

If this stress discontinuity complication (in the joint-parallel direction) was generally in operation, there could then be four potential contributors to shear wave splitting, needing acceptance or rejection by the theoreticians who debate, conclude and publish such things:

1. the possible stress discontinuity (even when joints were under no shear stress)
2. the potential stiffness anisotropy ($K_{n\ dyn} > K_{s\ dyn}$, $Z_N \neq Z_T$)
3. the excess compliances/stiffnesses (the presumed source of ($qS_1 - qS_2/qS_1$ anisotropy) with different contributions according to incident angle

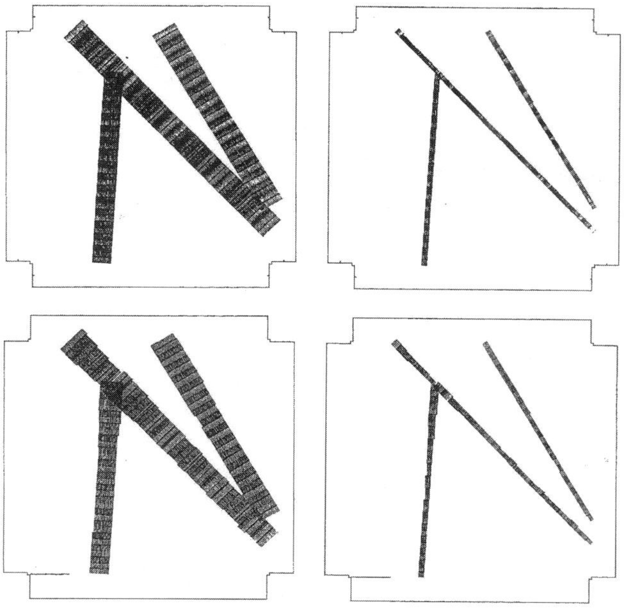

*Figure 16.35* Equal biaxial ($\sigma_1 = 5.4$ MPa, $\sigma_2 = 5.4$ MPa) loading effect on the maximum values of the physical (E) and hydraulic (e) apertures. Top: 142 μm, 35 μm, Bottom: with dilation-limited shearing, E = 169 μm, e = 49 μm. Chryssanthakis et al. 1991.

4. the presence of a variable thickness film of water between the opposed, contacting sides of non-planar slightly sheared joints.

If there has been pre-shear (in a 'critical crust' scenario), then presumably the parts of the joints that were dilated and fluid filled would have changed local stiffness in relation to the rock-to-rock stress-transferring parts, thereby contributing differently to attenuation, P-, and S-wave anisotropy.

In Figure 16.35, the modelling of physical (E) and hydraulic (e) apertures is shown, based on the assumed JRC = 10 or 13, JCS = 90 or 120 MPa, $\phi_r = 25°$ or 28° input data. The upper pair of (E) and (e) models are under equal biaxial stress, while the lower pair are under N–S ($\sigma_1$ only) loading. Slight non-uniformity of aperture is shown, though shearing was very limited, for important reasons which will be discussed below, and treated later.

The *maximum* joint apertures shown in Figure 16.35 were as follows:

$\sigma_1 = \sigma_2 = 5.4$ MPa  e = 35 μm
E = 142 μm
$\sigma_1 = 5.4$ MPa, $\sigma_2 = 0$  e = 49 μm
E = 169 μm

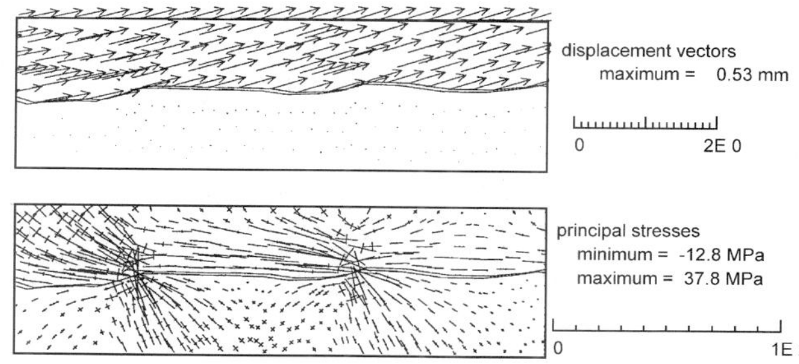


*Figure 16.36* Multi-linear modelling of joint roughness with UDEC, showing shear-displacement vectors of upper sample, and the asperity-controlled principal stresses caused by the simulated shearing. Gutierrez and Barton, 1994.

The limited shearing corresponds also to the reality of this *in situ* test. There was no more than 0.25 mm shear, despite exceeding the shear strength according to conventional (Mohr) stress transformation of stresses. This is because the conventional stress transformation equations, for converting principal stresses into shear and normal stress components, fail to take account of the contribution of dilation to what are actually *non-coaxial stresses and displacements*. The mobilized dilation angle needs to be added into the cos 2 β , sin 2 β terms of the classic transformation equations. Barton, 1986.

If we move now to a more direct simulation of joint shearing, in a simulated DST direct shear test, as opposed to a uniaxial/biaxial test, we avoid the problem of the stress transformation error, as the principal stresses are applied parallel and perpendicular to the joint plane.

Gutierrez in Gutierrez and Barton, 1994, used the discrete element code UDEC to simulate the roughness of rock joints in a direct, unusually realistic manner, by *constructing* multi-linear sectors to represent some of the measured 100 mm long roughness profiles of Barton and Choubey, 1977 (from Figure 16.26). One of the UDEC models is shown in Figure 16.36. Note the dilation-related inclination of the deformation vectors for the 'upper' sample, the bottom being fixed. The principal stresses applied were parallel and perpendicular and there was simultaneous flow of fluid simulated from left to right, along the joint void. This particular joint simulation had $JRC_o = 5$. The constant normal stress was only 1 MPa.

It is very easy to note the entirely different principal stress vectors *above and below* the shearing joint. The only locations where there are more similarities in magnitudes, but not directions, are the concentrated regions of stress transfer at the two major asperities, which resemble the more exaggerated case of a sheared undulating

*Figure 16.37* Interlocked and sheared joints in 'wavy' columnar basalt, demonstrating the role of asperities and dilation on aperture distribution. Columbia River Basalts, Washington State, USA. See Plate 12.

joint in basalt, from the Columbia River basalt sequence in the Western USA, shown in Figure 16.37.

On the basis of such rock stress related phenomena, one may then pose the following questions to geophysics theoreticians. *In the case illustrated by the shearing joint model of Gutierrez, will shear wave splitting occur more readily where the principal stresses 'plunge' into another direction (as at the stress-transferring asperities), or will shear wave splitting be able to occur across the $\sigma_1$ stress discontinuity, as generally seen in Figure 16.36, due to the general presence of a 'fluid-filled' lense between these same locations?*

A painstaking 'diagnostic' of the simulated shearing joint was performed by Gutierrez. This is shown in Figure 16.38. In the diagrams we see the following distributed phenomena, along the 100 mm long simulated rock joint:

a) individual contact lengths (mm) for 0, 1 and 2 mm shear
b) individual contact angles for 0, 1 and 2 mm shear
c) stress to strength ($\sigma_n$/JCS) ratios for 0, 1 and 2 mm shear
d) contact apertures for 0, 1 an)d 2 mm shear

There is probably a correlation of high stress to strength ratios where there are small contact lengths and high contact angles. The highest (compressive) stress concentration was 38 times the applied (average) normal stress ($\sigma_n$), and the maximum tensile stress was 13 times higher than $\sigma_n$. Apertures were mostly 'closed', or 'opened' to about 2 mm. (There appears to be a 'mm for m' misprint in diagram d, where $10^{-3}$m and $10^{-6}$m should be the two approximate extremes).

Finally, to round off this present discussion of the possible consequences of an actual local *stress-discontinuity* at joint surfaces, especially when under shear stress, one may consider the inter-bedded bituminous shale and dolomite 'Kimmeridge clay' source rocks depicted in Figure 16.39a and b. The views are taken looking upwards from the base of the 10 m high cliffs at Kimmeridge Bay in Dorset.

Although a certain consistency can be noted in the orientation of the two (actually up to three) sub-vertical joint sets in the harder dolomite beds, it is easy to imagine the relative complexity of the joint stress distributions, in view of the stress discontinuity trend discussed above. Interpretation of shear wave splitting in such a fractured reservoir environment, perhaps with non-aligned horizontal stress anisotropy, can be imagined as somewhat demanding.

### 16.7.2 The strength-deformation components of jointed rock masses

The potential for shear-wave splitting as a means of monitoring temporal changes in the geometry of the aligned,

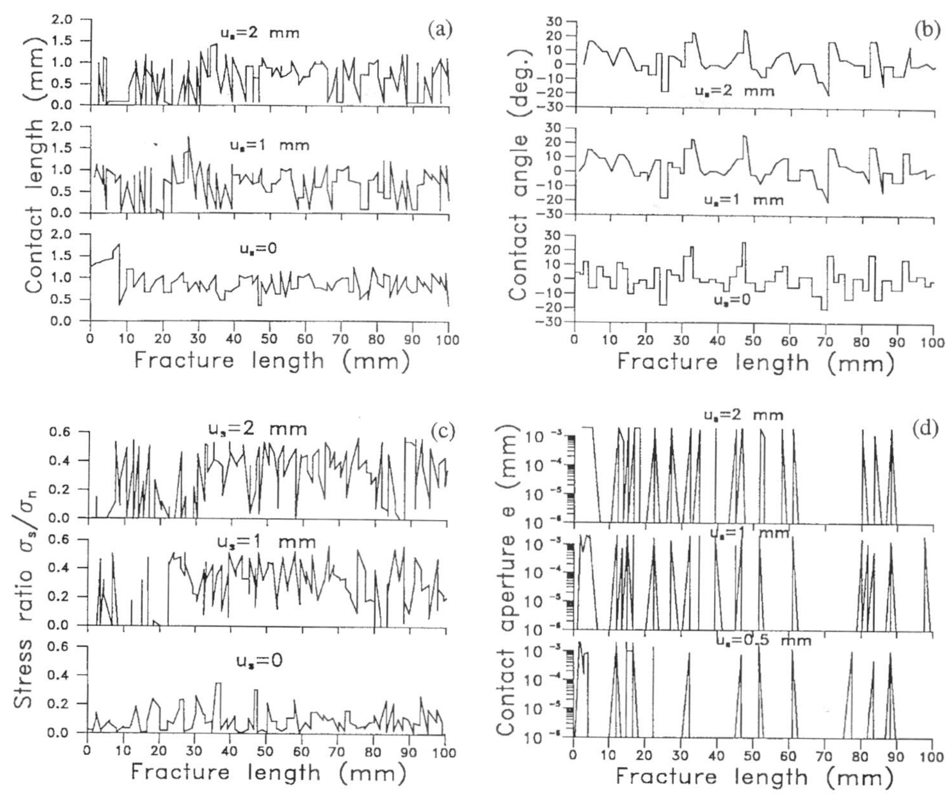


*Figure 16.38* Numerical diagnostic of the sheared UDEC joint, showing contact lengths, contact angles, contact apertures and stress ratios, with 0 or 0.5 mm, 1.0 mm and 2 mm of shearing. Gutierrez and Barton, 1994.

*Figure 16.39* Vertical view through the inter-bedded bituminous shales and dolomite bed, in Kimmeridge Bay, showing a) complex, b) ordered sub-vertical joint patterns. Dorset coast, S. England. *See Plate 13.*

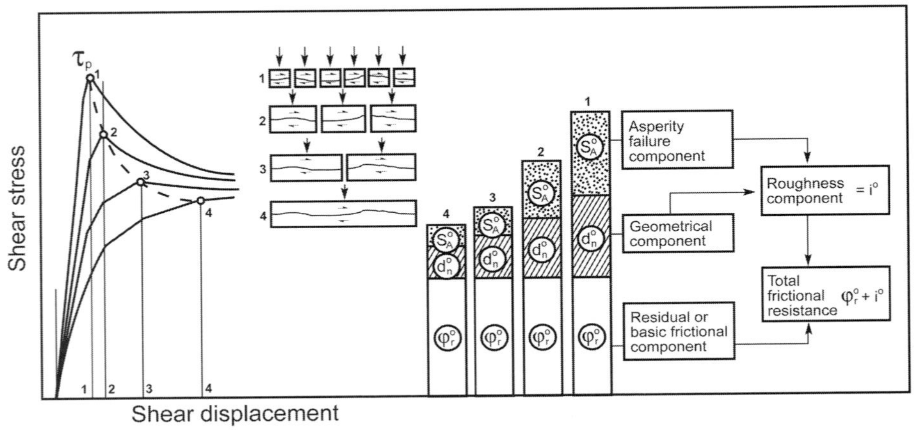


*Figure 16.40* Depiction of the effect of block size on the shear strength components of joints, from Bandis 1980, and Bandis et al., 1981.

fluid-filled features responsible for the splitting, can be further visualised by reference to cases where shear deformation occurs in the course of production from a petroleum reservoir. Due to the usual non-planarity of joints (despite their obvious depiction as straight lines or planes in the literature), there will usually be non-linear normal stress-closure-opening behaviour, and non-linear shear-dilation behaviour, Some of the details of joint behaviour need to be explored further, if stress anisotropy causes shearing during production, and if seismic monitoring is to be used for diagnosing the effects of new production measures.

There is the potential for shearing when producing from conjugately jointed reservoirs when a vertical principal stress roughly intersects the two steeply dipping sets of conjugate joints in the reservoir (cf. Ekofisk, as described in Barton et al., 1986). This case will be discussed later in this chapter.

The potential for shear also exists when water flooding into a jointed reservoir with non-aligned, anisotropic principal stress, for enhancing production. A related case is the injection of cold water into a geothermal reservoir, to establish hydraulic connection to wells on the far side of the reservoir.

In the first case a gradual but major *reduction* in pore pressure (at least 20 to 24 MPa, prior to water-flooding) had caused an equivalent effective stress increase, following some 15 years of *production.* The porous chalk matrix apparently contracted sufficiently to make space for down-dip shearing, despite one-dimensional compaction, (see Figure 15.25).

In the second case the *injection* caused an *increase* in pore pressure, and sufficient reduction in effective normal stress for the unequal principal stresses (70 and 30 MPa, Pine and Bachelor, 1984) to cause slight shearing. Evidence for this was obtained from acoustic emission monitoring, which showed migration of micro-shearing to successively greater depths, when attempting to establish hydraulic connection between wells, for extraction of thermal energy. A schematic of such a case was illustrated in Chapter 15, Figure 15.2a, and temporal effects on shear-wave anisotropy were documented.

The representative block size, given by the number of joint sets and their spacing and orientation, plus the surface character and length of the joints, combined with the effective normal stress levels acting on each set, collectively contribute to the possible response of a rock mass to shearing, when under the influence of an anisotropic, and non-aligned principal stress field. Figure 16.40, from Bandis, 1980, shows the dramatic potential effect of the *individual block size,* on shear strength, and displacement to peak, especially when joint roughness is significant.

Barton and Choubey, 1977, had anticipated the effect of block size on displacement to peak ($\delta_{peak}$), as shown conceptually in Figure 16.41. Despite identical $JRC_n$ for joints AA and BB, the altered block size was expected to reduce $\delta_{peak}$. Clearly the bulk modulus of the second case would nevertheless be lower than for the massive rock mass, as there are so many more joints involved in the deformation process.

Confusingly perhaps, it may be the 'better quality' rock masses, with wider joint spacing, and with greater joint continuity, that could be most prone to (previous) shearing deformation along the joint sets, when under the influence of a non-aligned, anisotropic, (differential) stress field, caused by some historic tectonic adjustments, of which there is no shortage.

A heavily fractured rock mass, though obviously with lower (bulk) modulus, may resist significant shearing

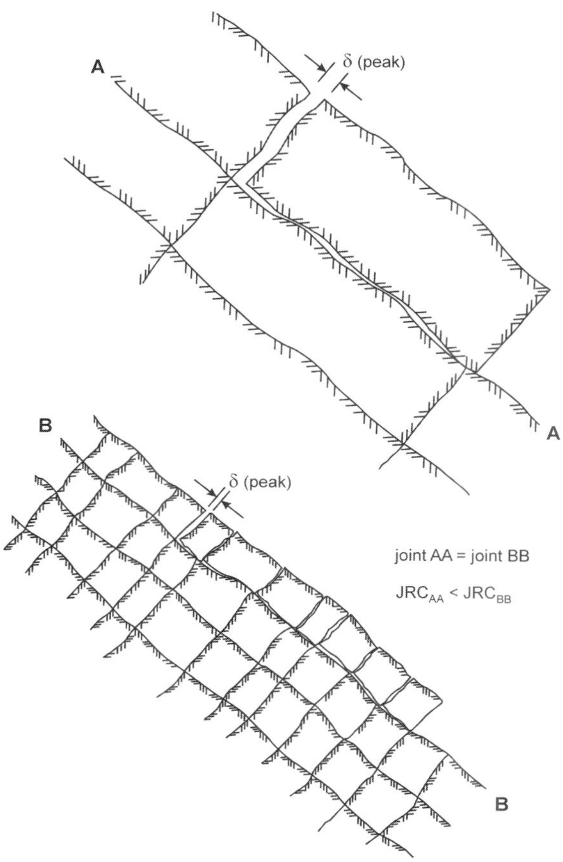


*Figure 16.41* Scale effect on δ (peak) caused by increasing block size. Barton and Choubey, 1977. The scaling of $\delta_{peak}$ is one of the components of $K_s$ that causes low stiffness as block size increases.

along individual joints, due to the higher collective shear resistance of smaller block sizes. When the block sizes are small, blocks can rotate slightly, thereby sampling the steeper joint-surface asperities that contribute to higher strength. (Barton and Bandis, 1982).

Some physical modelling evidence of the influence of block size is shown schematically in Figure 16.42a, from Barton and Hansteen, 1979, as also analysed further by Bandis et al., 1981. The (differential) stress-deformation behaviour of the three different models is shown in Figure 16.42b. Back-analysis showed that the elemental tension fractures, common to each model, whether forming 250, 1000, or 4000 interlocked blocks, had a '$JRC_n$' value averaging 20, from numerous 'large-scale' direct shear tests. The tension-fractured assemblies, made with a dynamic 'double-guillotine' device (Barton, 1972), each fabricated of identical rock-like, brittle material, had back-calculated JRC values for their first (photogrammetry-measured) shearing events of 21.6, 25.1 and 26.7 when tested in biaxial shear, i.e. shear strength rising as block size reduced.

The mass 'Poisson ratios' estimated from these 'static' biaxial shear tests are seen to rise dramatically as differential stress rises. Values exceed the continuum 'limit' of 0.5, due to the influence of micro-shearing along numerous fracture surfaces. Presumably, in the context of a sheared reservoir situation, such could be detected by the seismic anisotropy response. Shear wave anisotropy would be enhanced most by the more easily shearing (and therefore more conducting) *primary set of fractures.* (Note that these would *not* be parallel to the customary $\sigma_{H\ max}$ direction).

In these physical tension fracture models, the first parallel set of fractures developed on the double-guillotine table, were continuous, and remained continuous, following the development of the secondary, intersecting set, which due to *up-stepping* and *down-stepping* when crossing the pre-existing fractures, would impart an anisotropic strength and stiffness to the assembly. It is reasonable to believe that in the context of conjugate joint sets, this dominance of a primary set is common. It was certainly the case in the jointed chalk of the Ekofisk field, as reviewed in Chapter 15, Figure 15.7.

### 16.7.3 Permeability linked to joint shearing

In a deep petroleum reservoir, and strictly from a geomechanics viewpoint, one can hardly imagine that 'open' joints can be found at typical 2 to 4 km reservoir depths, *unless* there are one or more of the following conditions: mineral 'bridging' holding parts of fractures open, high-strength rock and rough-surfaced joints, or joint shearing (possibly where there is less roughness).

This viewpoint is held because of a probably too high effective minimum stress ($\sigma_{h\ min}$) of perhaps 20 to 40 MPa magnitude, in relation to probable similar magnitudes of $JCS_n$ in the case of the common weaker reservoir rocks. Shearing, as in weak Ekofisk chalk, and perhaps at a larger number of reservoirs than the petroleum industry suspects, is linked of course to the potential for dilation, and therefore potential permeability maintenance (in the face of the high normal stresses), and perhaps even permeability enhancement.

Example values of the three basic components of joint or fracture shear strength and stiffness (JRC = roughness, JCS = joint wall compressive strength, $\phi_r$ = residual

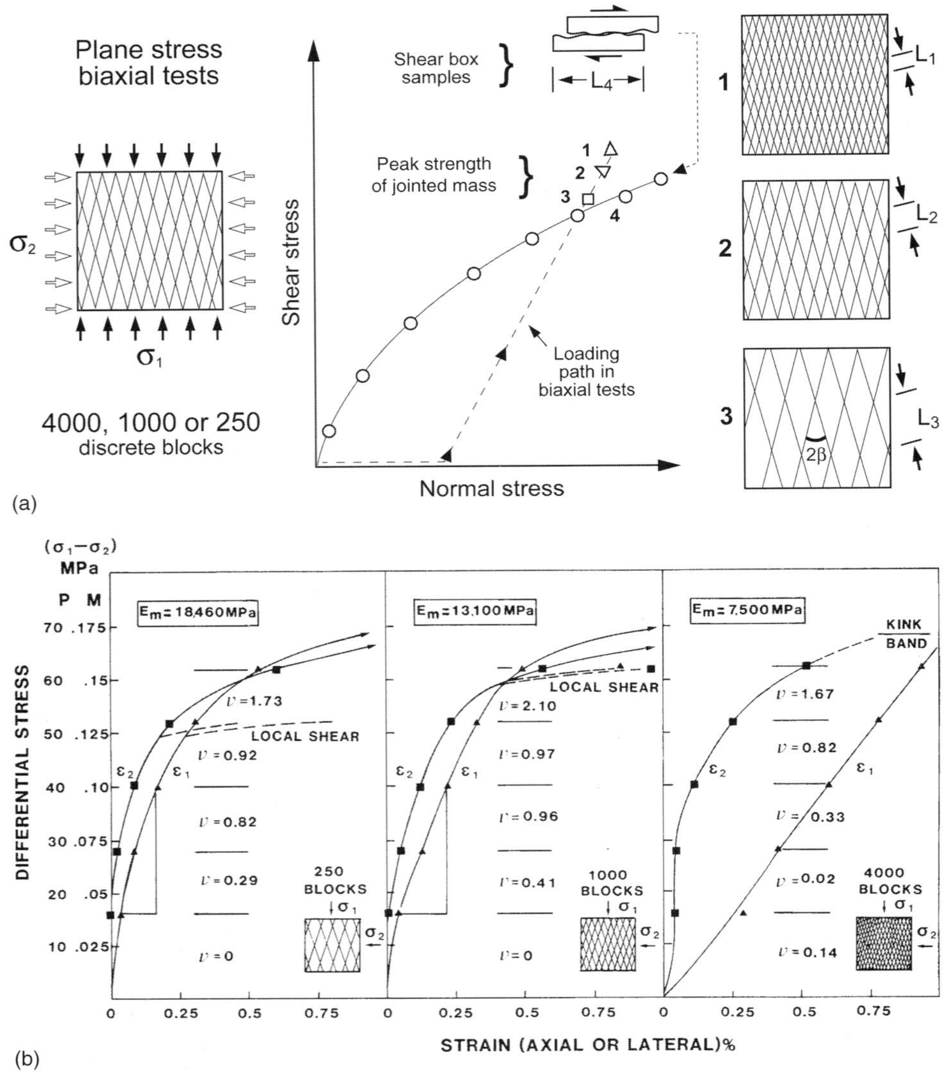


*Figure 16.42* a) A comparison of the shear strength of interlocked assemblies of tension-fractured, idealised, model rock masses, with 250, 1000 or 4000 blocks. b) The respective differential stress-versus-strain curves in the $\sigma_1$ and $\sigma_2$ directions for the same models. Note entirely different behaviour of the most heavily fractured model with 4000 blocks, due to greater freedom for (micro) block-rotation. Barton and Hansteen, 1979.

friction angle) are shown contributing to *peak shear strength* in Figure 16.43. Estimated *peak dilation angles* are also shown on each strength envelope, at appropriate normal stress levels. Clearly as shearing continues beyond peak, rates and angles of dilation will steadily reduce. However the joint (minor fault?) aperture has probably accumulated some vital void space for fluid conduction, minus areas of contact with crushed material and finer gouge causing local blockage, as we saw in Figure 16.13.

Barton-Bandis modelling of individual joint responses as either *block size* or normal stress changes are shown in Figure 16.44, from Barton et al., 1985. The relative amounts of predicted *dilation*, with the given input data assumptions, are shown in the middle diagrams. In the lowest of the three sets of predicted behaviour curves, the theoretical *maximum* change in permeability with the shear-induced dilation is shown. Here, gouge production (and partial joint blocking) is ignored. Note the assumption of an initial hydraulic aperture ($e_o$) of 25 μm in each case.

The dilation curves show how the initial physical aperture ($E_o$) changes. The sum of $E_o + \Delta E$ ($=E_1$) is converted to $e_o + \Delta e$ ($=e_1$) to estimate permeability from the assumed cubic law $K = e^2/12$.

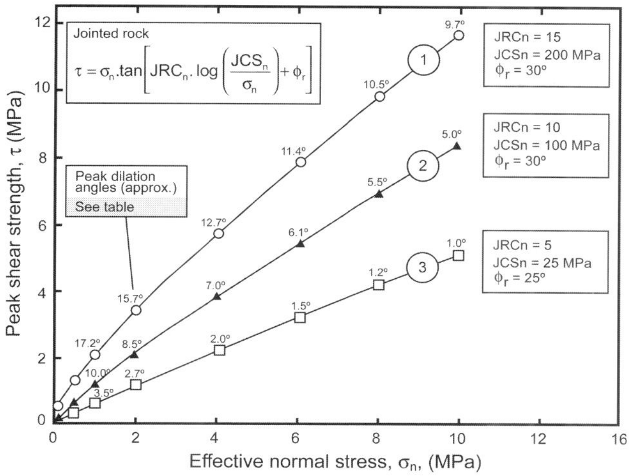


Estimates of peak dilation angles ($d_n$°)

| Curve n° | Effective normal stress (MPa) | | | | | | | |
|---|---|---|---|---|---|---|---|---|
| | 0.1 | 0.5 | 1.0 | 2.0 | 4.0 | 6.0 | 8.0 | 10.0 |
| 1 | 24.7 | 19.5 | 17.2 | 15.7 | 12.7 | 11.4 | 10.5 | 9.7 |
| 2 | 15.0 | 11.5 | 10.0 | 8.5 | 7.0 | 6.1 | 5.5 | 5.0 |
| 3 | 6.0 | 4.2 | 3.5 | 2.7 | 2.0 | 1.5 | 1.2 | 1.0 |

*Figure 16.43* The peak shear strength equation for *in situ scale* rock joints, and example strength envelopes. The subscripts on JRC: ($JRC_n$) and on JCS: ($JCS_n$) imply large scale, block-size-dependent values in general. In the two uppermost strength envelopes shown, input values are more typical of laboratory samples, where roughness tends to dominate behaviour. The lower envelope might apply to a minor fault, or to a weak, or weathered rock joint.

## 16.7.4 Reservoir seismic case records with possible shearing

Seeking more detail about the fracture directions and stress direction at the Bluebell-Altamont Field, one may note from Lynn et al., 1999, that the borehole elongation data indicated a range of possible major *stress* orientations from N20° to 50°W. Fracturing could presumably have influenced this result. Three methods of *fracture* orientation (outcrops, FMS log and core samples) indicated a spread between N20° to 35°W, with the second set less well delineated at about N60° to 80°E.

The Maultzsch et al., 2003, Liu et al., 2003a and Liu et al., 2003b analyses of shear wave anisotropy, reviewed in Chapter 15, actually suggest polarization more related to the major stress direction than the dominant fracture direction. However, the variation of anisotropy at lower frequencies suggests that their results *must* be fracture related rather than stress-aligned microcrack related (EDA).

In Figure 15.60, we could see the various sources of orientation data from Lynn et al., 1999. There was clearly great consistency in the direction of the N20° to 35°W joint set, which was presumed by Maultzsch et al., 2003, Liu et al., 2003a and Liu et al., 2003b to be the source of the $qS_1$ polarization. Yet diagram b) in Figure 15.60 showed 'VSP S1' at about N43° to 45°W, some 10° to 20° different from the fracture orientation obtained from core data, yet almost parallel to the gilsonite dykes.

One may speculate that a possible explanation for the potential discrepancy between $qS_1$ polarization and the fracture orientation could be due to non-planarity of the fractures combined with slight shearing. It could also be due to unavoidable, minor sources of error, in data acquisition, as suggested by Lynn (pers. comm. 2005).

The Chapman modelling reviewed in Chapter 15, had suggested 6 m diameter fractures at a frequency of only one per 9 to 10 m cube of rock mass, i.e. implying quite large block sizes, but this would be dependent on the spacing of the N60° to 80°E set.

With large block size and non-coaxial major principal stress one could have a situation in the reservoir (especially if fluid pressures had been high) that, *in principle*, could resemble the shear-dilation mechanism shown in Figure 16.45. These are rough tension fractures that have a prototype scale of many metres length (due to up-scaling of a weak brittle model material). The measured profiles of roughness have been sheared following the shear-dilation (X, Y coordinates) measured during direct shear tests at a) low normal stress (top: $\sigma_n$ 2) and b) at high normal stress (bottom: $\sigma_n$ 6). Two different degrees of pre-peak and post-peak shear and dilation are shown.

The black (overlapped) contacting areas that transfer shear and normal stress have, on average, a significantly different orientation to those (white) areas that are dilated, and which in a 'slightly-sheared' reservoir situation would contain most fluid. Could it be that the rotated mean direction of these 'fluid lenses', obviously distributed in three dimensions, could be more responsible for the shear wave polarization than the *average* joint plane, or would it be the rock contact areas?

These large, rougher than normal tension fractures show a 10° to 20° effective rotation of the lenticular apertures in relation to the mean fracture planes. They might represent the areas where squirt attenuation was least active, due to higher permeability.

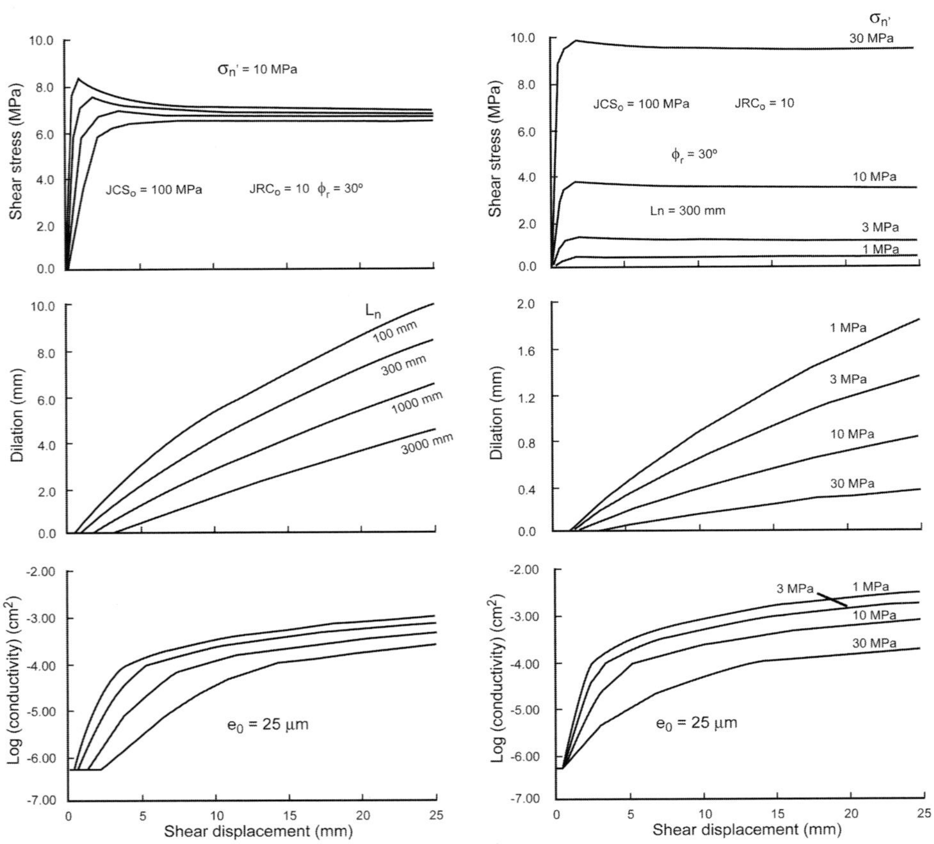


*Figure 16.44* Barton-Bandis models of shear strength-displacement-dilation-permeability coupling. Curves generated by Bakhtar with a programmable calculator. (Barton et al., 1985.) The permeability estimate is the theoretical, dilation-produced maximum, devoid of considerations of gouge production. The latter requires a modified treatment of roughness for (E) to (e) conversion, using $JRC_{mobilized}$, as shown by Olsson and Barton, 2001. (See Figure 16.9).

The oppositely rotated *rock-to-rock* contact areas with their extremely high stress, are less likely planes for reflection. So if the lenses were fluid-filled, an apparent rotation between the reflectiveness to S-wave and the squirt attenuation source areas *might* be the result, in relation to the mean plane of reference, some 10°–20° between the two.

Previous figures in this chapter showed that shear strengths, shear displacements, dilations and shear stiffnesses, were each affected adversely by increasing block size. (Barton and Bandis, 1982). Due to reductions of effective roughness JRC and wall strength JCS, and *increased* displacement to peak, each caused by larger block size, the *in situ* (pseudo-static) joint shear stiffness even at 10 to 20 MPa effective normal stress, may be close to 1 MPa/mm for the case of say 2 m *in situ block sizes*. Such a low pseudo-static stiffness is two or more orders of magnitude *softer than* the 'Hudson compliance' diagonal (Figure 16.23). Is this also significant?

### 16.7.5 The apertures expected of highly stressed 'open' joints

As introduction to the question of 'open' fractures and their orientation, we refer to another case record of reservoir seismic anisotropy investigation, this time involving a large (43 square mile) 3D full-azimuth, full-offset P-wave survey in a fractured carbonate gas field in Texas, as described by Lynn et al., 2000. The deep target formations were carbonate and chalk formations at 14,000 to 15,500 ft (about 4,280 to 4,730 m). The faster P-wave velocity direction was parallel to the strike of the local structure, and interpreted as also parallel to the local maximum horizontal stress direction. As the authors state: *'The interpreted open fractures are those approximately parallel to the local maximum horizontal stress, and are considered to be the fluid-flow pathways or permeable conduits'.*

Clearly, at four-and-a-half km depth, one is far removed from the near-surface, where $\sigma_{H\ max}$ oriented

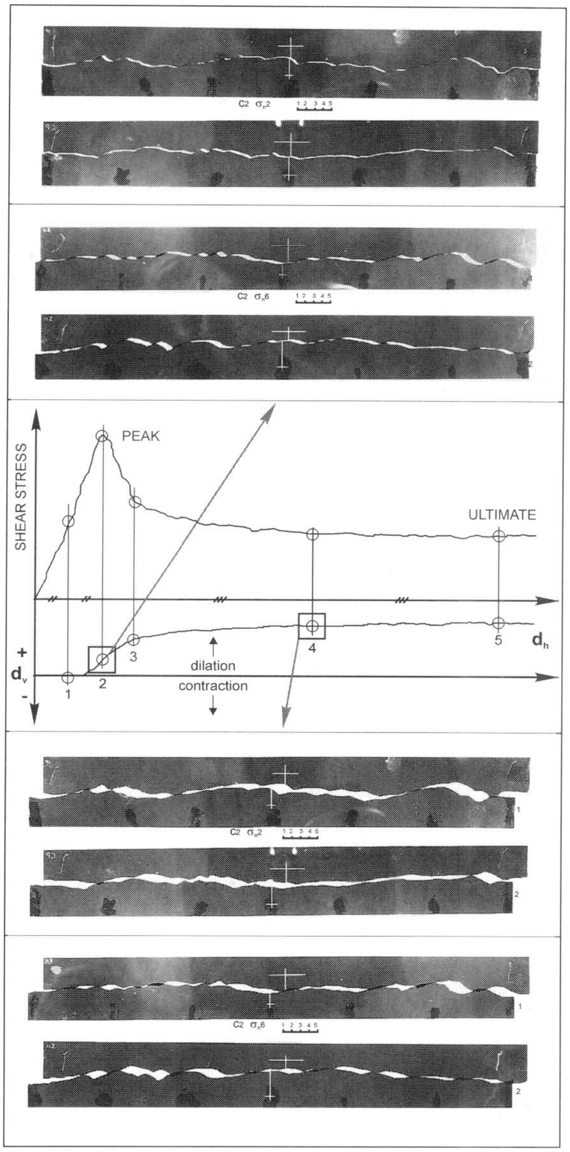


*Figure 16.45* Reconstructed shear-and-dilation events for model tension fractures with the illustrated roughness profiles, cut out as plastic 'replicas'. Note the 'opposite' rotations of the (potentially) fluid-bearing lenses, which are down-dipping to the right, and the up-slope to the right rock-to-rock contact areas which are of 'double' thickness therefore signifying the production of crushed particles. The lenses will therefore have debris/gouge at their extremities. Barton, 1973a.

joints are certainly very typical. Such cases were reviewed in Part I of this book, and were the source of marked azimuthal P-wave anisotropy. In contrast to the near-surface, the adverse effect of many tens of MPa minimum effective stress would have suggested to a rock mechanic that the more 'open' joints should be *at an angle* to the major stress. Such a scenario for the 'open' joints has been emphasised and convincingly demonstrated in deep well analysis, by Colleen Barton, Zoback and Moos, 1995, and this and later studies by Zoback and his colleagues will be reviewed later in this chapter.

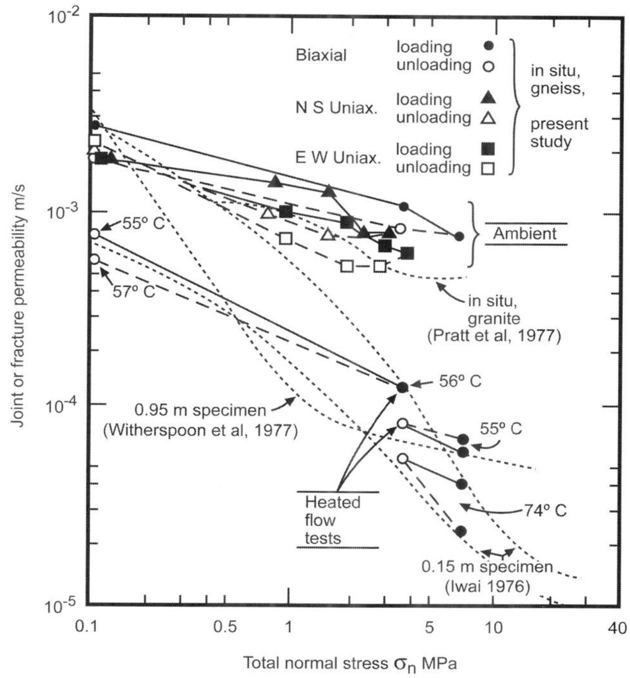


*Figure 16.46* A collection of large scale stress-permeability tests on joints in hard crystalline rocks, both at ambient and elevated temperature, showing 'linear' log-log-permeability-stress trends, with various gradients. Barton, 1982. (Pratt et al., 1977, Witherspoon et al., 1977, Iwai, 1976)

Before considering some convincing evidence for the possible/probable *non-alignment* of 'open' sub-vertical fractures with the major principal stress direction ($\sigma_{H\,max}$), from these and other authors, one must consider the case of fractures that *are* orientated in this classically assumed 'mode 1' direction. In such cases, the minimum effective stresses may tend to keep the 'open' fractures actually more closed than open, unless the rock is rather strong, and/or the fractures are 'bridged' by mineral cementation.

Even for joints in *hard* crystalline rocks, coupled (MH) stress-permeability tests of joints at mostly in situ scale (1 m diameter jointed cylinders, and 2 to 8 $m^3$ jointed in situ blocks), demonstrate a marked reduction in permeability with normal stress. Figure 16.46 from Barton, 1982, shows 'linearity' on a log-log plot of permeability versus normal stress. One earlier ambient

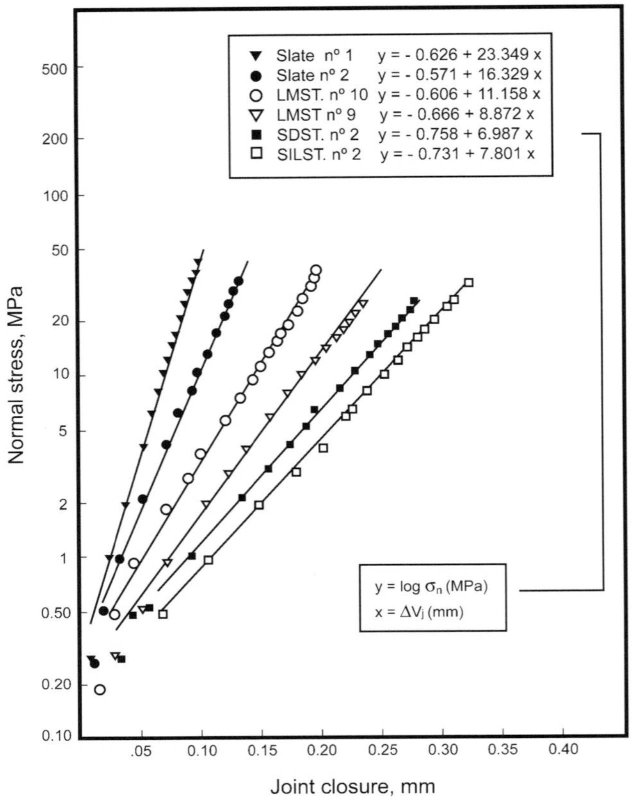


*Figure 16.47* Log-linear relation showing the effect of normal stress on the *closure* of physical aperture for various rock joints, in mostly sedimentary rocks, as tested by Bandis, 1980.

temperature and one *heated* block test (described in Figure 16.29 to 16.33), constitute the main body of data assembled in this figure. There was experimental evidence from these tests of even smaller conducting apertures when temperature was elevated. This would add to the downward trend of permeability, with increased depth. Hydraulic apertures in the lower right corner of the figure are down to a few microns in size. It seems *unlikely* that such would be considered 'open' in the context of oil producing reservoirs.

The physical *closures* of joint apertures in (mostly) sedimentary rocks, as a result of applied normal stress, were shown by Bandis, 1980 to *decline linearly* against the log of normal stress. This relevant result is shown in Figure 16.47. The large closures seen in the weaker rocks like siltstone, would be expected to nearly 'close' such joints, and with conversion from physical aperture (E) to hydraulic aperture (e), small residual permeabilities would be expected.

The laboratory normal stress-conducting aperture data shown in Figure 16.48 was assembled by Makurat (pers. comm. 2006), from numerous tests performed in the CSFT apparatus developed by Makurat at NGI (Makurat et al., 1990, Makurat, 1996, Makurat 2006). As is readily observed, there are a large number of test data showing extremely small apertures, often well below 10 μm, when even moderate normal stress levels in relation to JCS values were applied. The data for tuff and ignimbrite joints from the UK Nirex Ltd planned Sellafield Rock Characterization Facility, were derived from tests at up to 30 MPa normal stress. Although this is equivalent to effective stress levels at about 1.5 km depth in a petroleum reservoir, the ratio of applied $\sigma_n$/JCS was less than 0.15, due to the high strength of the tuff. The fact that the joints were quite planar, with $JRC_o$ mostly from 3 to 6, was an important contributor to their small conducting (and mechanical) apertures under stress.

It was noted by Makurat that tests on weak reservoir rocks like chalk and shale, often showed higher conducting apertures than expected, in relation to quite high ratios of $\sigma_n$/JCS, such as 0.7, 1.4 and even 2.0. This was interpreted as due to sampling damage, as it is easy to 'lose' material (prior to reaching the laboratory environment), due to recovery by drilling and subsequent transport. Loose grains and cracked pieces can be more easily lost when strength is in the 2 to 5 MPa range, as for several of the JCS values of joints in the chalk, sandstone and shale.

The hoped for 'open' joints in a fractured reservoir can hardly be expected to be significantly open, unless the joints or fractures are rather rough, or are 'bridged' (but not sealed) by hard minerals, thereby resisting complete closure during hydrocarbon production. Of course if the fluid is over-pressured and is very close to the minimum total rock stress, so that the effective stress ($\sigma_{h\,min} - P_f$) is small, then closure again would be limited until much later in production.

If we consider 3 km depth, and a fairly typical minimum total stress ($\sigma_{h\,min}$) magnitude of 40 to 50 MPa (compared to an assumed density-based total vertical stress ($\sigma_v$) of 60 to 70 MPa), there may be a hydrostatic-based pressure of about 30 MPa for the 3 km deep reservoir. It is then clear that the effective normal stress of 10 to 20 MPa will be holding joints parallel to $\sigma_{H\,max}$ nearly closed, since they are undisturbed, and have no way of losing material, as above.

We can deviate briefly to the micro-world of elliptical cracks and consider that the theoretical closure pressure

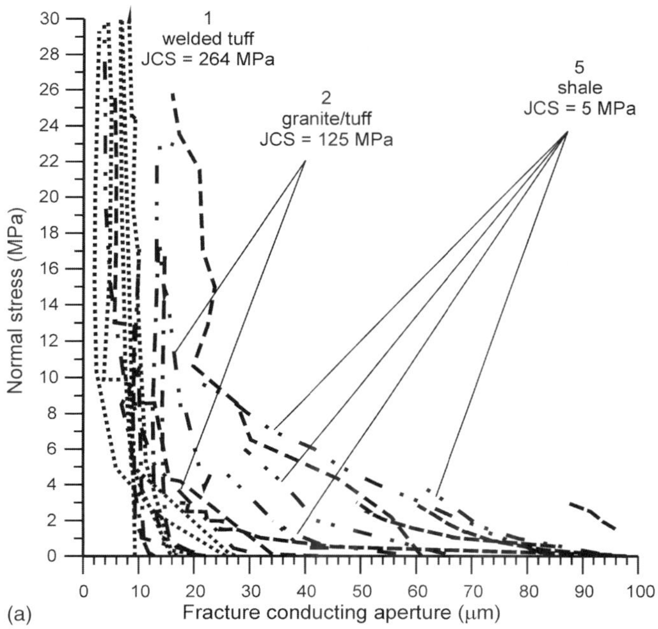


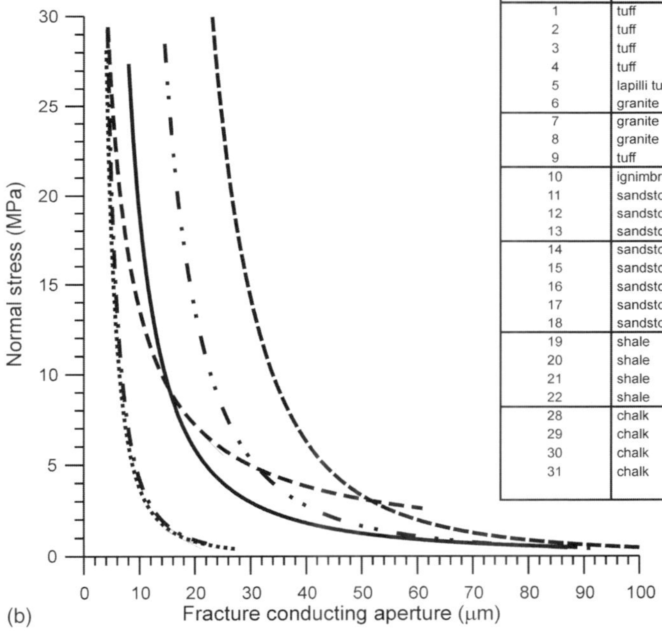


| Sample no. | rock type and fracture properties | | | line code |
|---|---|---|---|---|
| | Rock type | JCS (MPa) | JRC | |
| 1 | tuff | 264.6 | 3.0 | 1 |
| 2 | tuff | 264.6 | 3.0 | |
| 3 | tuff | 264.6 | 4.0 | |
| 4 | tuff | 264.5 | 4.2 | |
| 5 | lapilli tuff | 198.2 | 3.5 | |
| 6 | granite | 162.0 | 6.4 | |
| 7 | granite | 125.0 | 1.9 | 2 |
| 8 | granite | 125.0 | 3.8 | |
| 9 | tuff | 124.0 | 5.7 | |
| 10 | ignimbrite | 87.1 | 6.3 | 3 |
| 11 | sandstone | 86.8 | 4.0 | |
| 12 | sandstone | 67.2 | 4.9 | |
| 13 | sandstone | 60.7 | 4.7 | |
| 14 | sandstone | 35.5 | 10.0 | 4 |
| 15 | sandstone | 35.5 | 7.0 | |
| 16 | sandstone | 35.5 | 8.0 | |
| 17 | sandstone | 35.5 | 14.0 | |
| 18 | sandstone | 35.5 | 12.0 | |
| 19 | shale | 5.0 | 3.0 | 5 |
| 20 | shale | 5.0 | 3.0 | |
| 21 | shale | 5.0 | 3.0 | |
| 22 | shale | 5.0 | 3.0 | |
| 28 | chalk | 2.1 | 5.6 | 6 |
| 29 | chalk | 2.1 | 7.1 | |
| 30 | chalk | 2.1 | 9.4 | |
| 31 | chalk | 2.1 | 9.0 | |

*Figure 16.48* a) Normal stress-conducting aperture behaviour from CSFT tests in natural joints in tuff, ignimbrite, granite, sandstones, shale and chalk. b) Mean behaviour for each rock type, using power-law extrapolation. Makurat, 1996, Makurat pers. comm. 2006.

P, for the case of a stiffer elliptical crack of aspect ratio (α), under hydrostatic stress (Walsh, 1965) is:

$$P = 3K_s \times \alpha \frac{1 - 2\nu}{2\left(1 - \nu^2\right)} \tag{16.10}$$

in which $K_s$ = bulk modulus of solid material (e.g. about 40 GPa for clean sandstones), and $\nu$ = Poisson's ratio of solid material (0.17 for clean sandstones). The closure pressure for a microcrack of aspect ratio $10^{-3}$ is then approximately 40 MPa. Using this theory, King and Marsden, 2002, had assumed that microcracks with

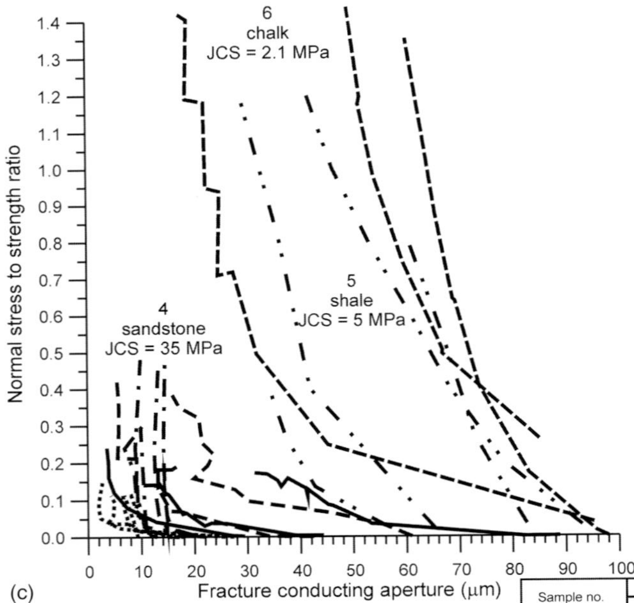


| Sample no. | rock type and fracture properties | | | line code |
|---|---|---|---|---|
| | Rock type | JCS (MPa) | JRC | |
| 1 | tuff | 264.6 | 3.0 | 1 |
| 2 | tuff | 264.6 | 3.0 | |
| 3 | tuff | 264.6 | 4.0 | |
| 4 | tuff | 264.5 | 4.2 | |
| 5 | lapilli tuff | 198.2 | 3.5 | |
| 6 | granite | 162.0 | 6.4 | |
| 7 | granite | 125.0 | 1.9 | 2 |
| 8 | granite | 125.0 | 3.8 | |
| 9 | tuff | 124.0 | 5.7 | |
| 10 | ignimbrite | 87.1 | 6.3 | 3 |
| 11 | sandstone | 86.8 | 4.0 | |
| 12 | sandstone | 67.2 | 4.9 | |
| 13 | sandstone | 60.7 | 4.7 | |
| 14 | sandstone | 35.5 | 10.0 | 4 |
| 15 | sandstone | 35.5 | 7.0 | |
| 16 | sandstone | 35.5 | 8.0 | |
| 17 | sandstone | 35.5 | 14.0 | |
| 18 | sandstone | 35.5 | 12.0 | |
| 19 | shale | 5.0 | 3.0 | 5 |
| 20 | shale | 5.0 | 3.0 | |
| 21 | shale | 5.0 | 3.0 | |
| 22 | shale | 5.0 | 3.0 | |
| 28 | chalk | 2.1 | 5.6 | 6 |
| 29 | chalk | 2.1 | 7.1 | |
| 30 | chalk | 2.1 | 9.4 | |
| 31 | chalk | 2.1 | 9.0 | |

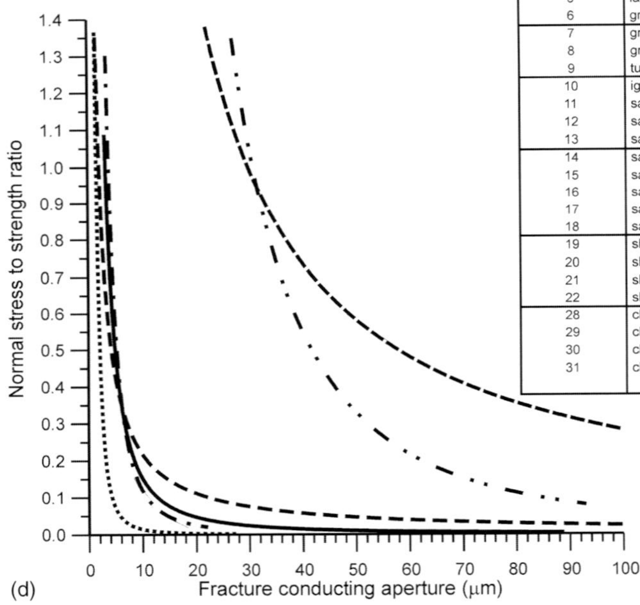


*Figure 16.48* c) Normalised stress-to-strength ratio ($\sigma_n'$/JCS) versus conducting aperture plots, which emphasises the different rock strength magnitudes. d) Power-law fit to mean data for a given rock type, with extrapolation to higher stress-to-strength levels. Makurat, 1996, Makurat pers. comm. 2006.

smaller aspect ratios would not be open at applied effective stresses greater than 40 MPa.

If up-scaling is performed to much longer cracks and discontinuous joints, with aspect ratios of e.g. $10^{-5}$ or less, the above theory for 'elliptical' cracks will presumably break down, as rock joints with some wall roughness appear not to close completely (Bandis et al., 1983). The normal stiffness merely becomes extremely high, but complete closure is virtually prevented by asperity roughness, like a joint roughness coefficient $JRC_o$ that

| Barton Bandis Joint Model | | NORMAL CLOSURE CALCULATION | | | | | |
|---|---|---|---|---|---|---|---|
| INPUT PARAMETERS | | SNORM | CYCLE 1 | CYCLE 2 | CYCLE 3 | CYCLE 4 | CYCLE 5 |
| JRC | 5 | LOAD | 20 | 20 | 20 | 10 | 0 MPa |
| JCS | 25 | UNLOAD | 0 | 0 | 0 | 0 | 0 MPa |
| SIGMAC | 30 | APERTURE | 0,140 | 0,030 | 0,020 | 0,016 | 0,015 mm |
| | | KNP | 4,60E+03 | 2,80E+04 | 4,50E+04 | 1,40E+04 | 3,00E+01 |
| CALCULATED PARAMETERS | | | | | | | |
| LOAD | | KNI | 4,76 | 16,42 | 24,13 | 29,18 | 30,42 MPa/mm |
| | | VMI | -0,140 | -0,030 | -0,020 | -0,016 | -0,015 mm |
| | | AJ | 0,210 | 0,061 | 0,041 | 0,034 | 0,033 |
| | | BJ | 1,499 | 2,030 | 2,100 | 2,125 | 2,130 |
| UNLOAD | | KNI' | 16,42 | 24,13 | 29,18 | 30,42 | 23,64 MPa/mm |

| INPUT PARAMETERS | | | FULL SCALE PARAMETERS | | |
|---|---|---|---|---|---|
| JRCo | 5,00 | | JRCn | 4,26 | |
| JCSo | 25,00 | MPa | JCSn | 19,64 | MPa |
| Lo | 0,10 | M | DPEAK | 2,03 | mm |
| Ln | 0,50 | M | ROUGH | 1,25 | |
| PHIr | 27,00 | DEG | LFACT | 0,01 | |
| SIGMA n | 10,00 | MPa | PHIr RAD | 0,47 | Rad |
| SIGMAC | 30,00 | MPa | PPEAK | 28,25 | Degrees |
| APERTURE | 0,000 | mm | KS | 2,650 | MPa/mm |

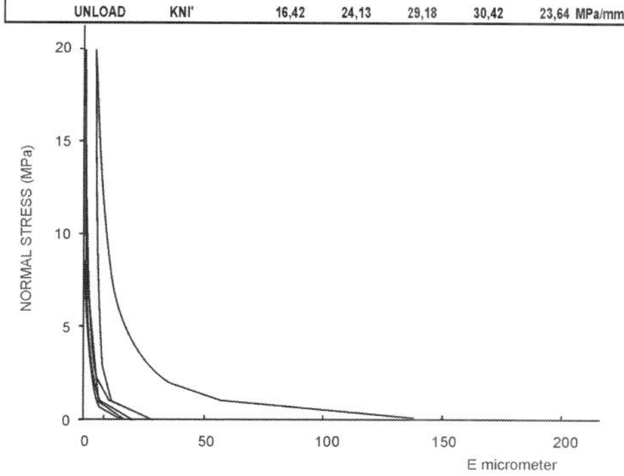


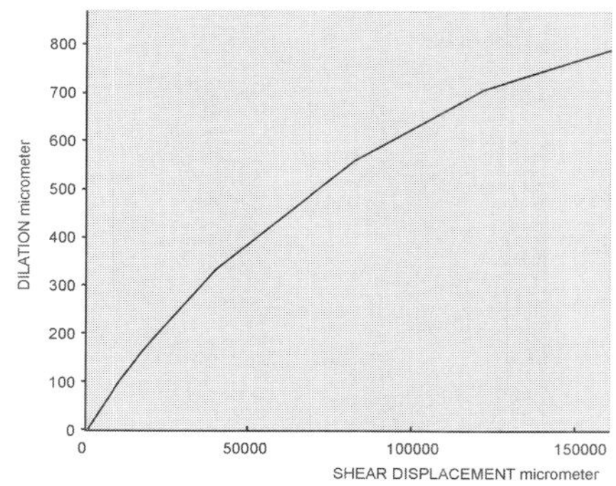


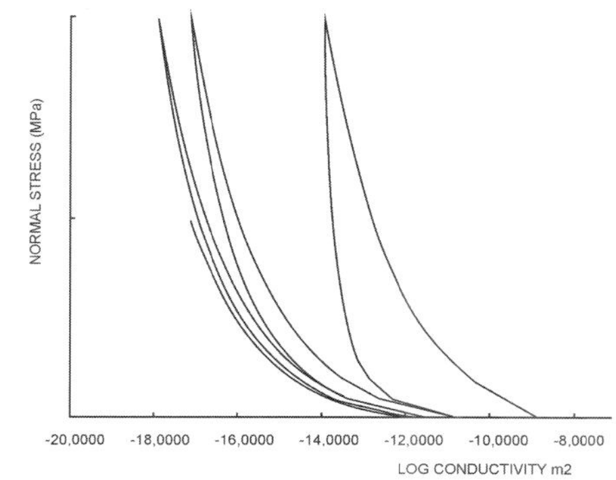


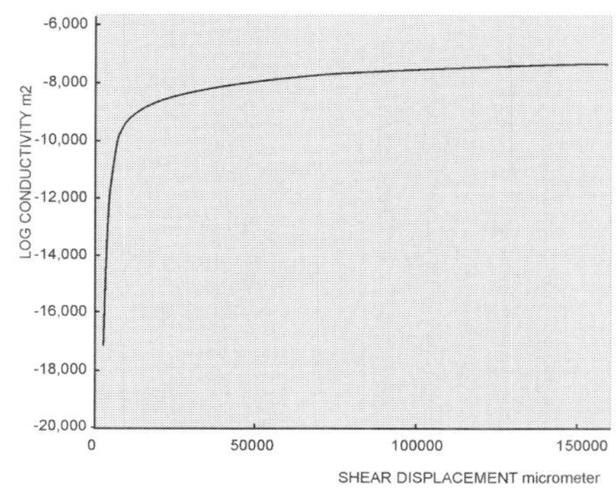


*Figure 16.49* BB normal closure-permeability and shear dilation-permeability modelling. Case A 'sandstone' with JRC = 5, JCS = 25 MPa, $\phi_r = 27°$ and $\sigma_c = 30$ MPa. Note extremely small apertures and low permeability in the left-side plots (= normal closure) and the marked dilation and rapid permeability enhancement in the right-side plots (= shear).

is acting *at right-angles*. However, we would be reaching micron size (e.g. 1 to 5 μm or less) in such cases, *unless* previous shearing had occurred.

It would appear that the 'conventional wisdom' of the most conducting joints being parallel or sub-parallel to the major stress direction, which is clearly proved in the *near-surface,* by means of three-dimensional permeability testing (e.g. Quadros et al., 1999), needs to be re-assessed in the context of petroleum reservoirs, in weaker sedimentary rocks at great depth. Will a joint set with apertures of a few microns or less, really satisfy the description 'open joints', and will they cause shear wave splitting when they have such a high area in contact as is necessary for $\sigma_h$ minimum of 10, 20 or 30 MPa to be transferred across their relatively weak joint walls? If the 'open' apertures are actually mainly 'tape-worm-like' channels, within 'closed' joint planes, would shear wave splitting and recognisable fast and slow wave speed anisotropy have been registered by the geophysicists?

### 16.7.6 Modelling apertures with the BB model

In order to investigate this further, the stress-closure ($\sigma_n - \Delta E$) and stress-permeability ($\sigma_n - \Delta K$) cycles resulting from Barton-Bandis modelling of three hypothetical, but realistic joint characteristics were investigated. The input data assumptions are listed in Table 16.13, and Figures 16.49 to 16.51 and Table 16.14 show selected results of this parameter study.

The first two of these BB modelling figures shows that the modelling of both cases originates from an initial

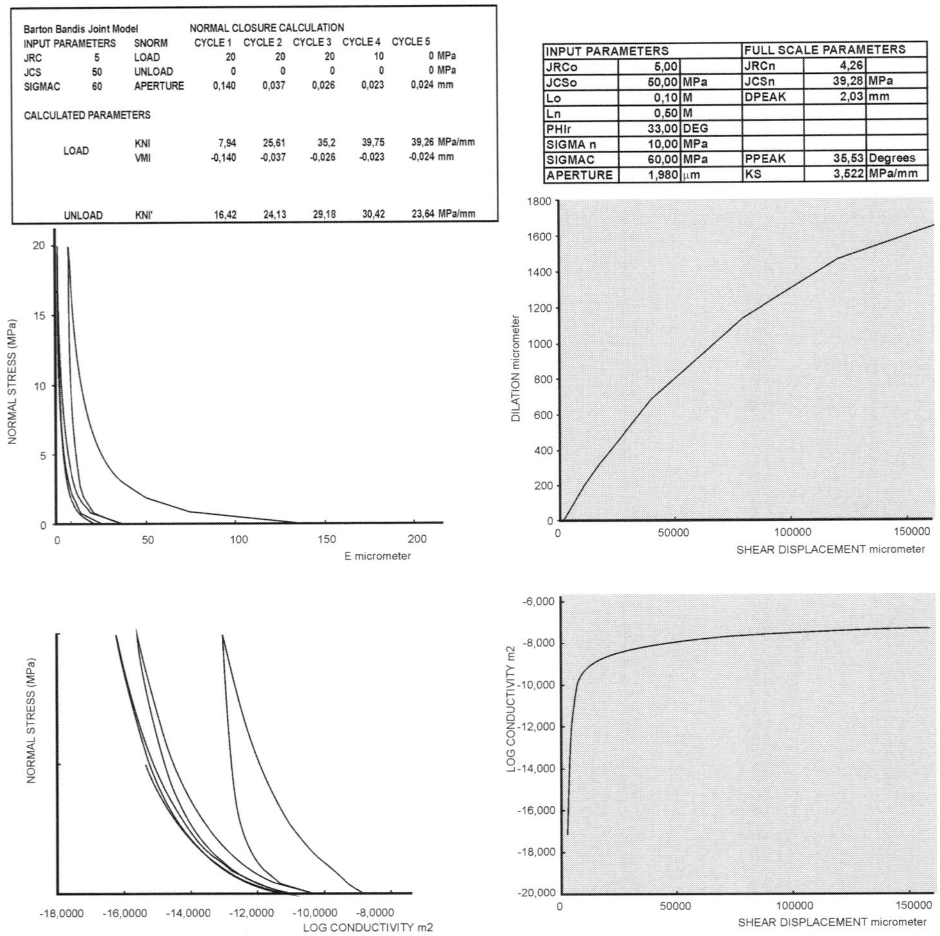


*Figure 16.50* BB normal closure-permeability and shear dilation-permeability modelling. Case B 'carbonate' (low roughness) with JRC = 5, JCS = 50 MPa, $\phi_r = 33°$ and $\sigma_c = 60$ MPa. Note extremely small apertures and low permeability in the left-side plots (= normal closure) and the marked dilation and rapid permeability enhancement in the right-side plots (= shear).

Table 16.13 Input data assumptions for investigating the physical and conducting apertures likely to be available with an 'open joints parallel to $\sigma_{H\,max}$ assumption'. Unless joints are rough, and the rock is hard, this orientation proves unlikely to give the assumed 'open' joints at reservoir effective stress levels. Note subscripts $JRC_o$ and $JCS_o$ representing lab-scale parameters, which are assumed to be most representative for *normal closure* modelling.

| | |
|---|---|
| Case A | $JRC_o = 5$ (medium rough, slightly undulating) |
| | $JCS_o = 25$ MPa (medium weak rock e.g. sandstone) |
| | $\sigma_c = 30$ MPa $\phi_r = 27°$ |
| Case B | $JRC_o = 5$ (medium rough, slightly undulating) |
| | $JCS_o = 50$ MPa (medium strength carbonate rock) |
| | $\sigma_c = 60$ MPa $\phi_r = 33°$ |
| Case C | $JRC_o = 10$ (rough, slightly undulating) |
| | $JCS_o = 50$ MPa (medium strength carbonate rock) |
| | $\sigma_c = 60$ MPa $\phi_r = 33°$ |

*unstressed* aperture ($E_o$) of 0.14 mm (or 140 μm). This was estimated from the following empirical equation (Barton, 1982) based on Bandis, 1980, tests of numerous natural joints.

$$E_o(\text{mm}) \approx \frac{JRC_o}{5}\left(0.2\frac{\sigma_c}{JCS_o} - 0.1\right) \quad (16.11)$$

This equation makes allowance for the experimental result that initial apertures are larger when the small scale joint roughness coefficient ($JRC_o$) is greater than about 5, as in Figure 16.51 which has $JRC_o = 10$. The ratio of $\sigma_c/JCS_o$ allows for the increased aperture when there is weathering when $JCS_o < \sigma_c$. Clay-fillings or clay-coatings, as tackled in the rock quality Q-system, are not considered in Barton-Bandis modelling.

| Barton Bandis Joint Model | | NORMAL CLOSURE CALCULATION | | | | | |
|---|---|---|---|---|---|---|---|
| INPUT PARAMETERS | | SNORM | CYCLE 1 | CYCLE 2 | CYCLE 3 | CYCLE 4 | CYCLE 5 |
| JRC | 10 | LOAD | 20 | 20 | 20 | 10 | 0 MPa |
| JCS | 50 | UNLOAD | 0 | 0 | 0 | 0 | 0 MPa |
| SIGMAC | 60 | APERTURE | 0,280 | 0,090 | 0,068 | 0,059 | 0,056 mm |
| CALCULATED PARAMETERS | | | | | | | |
| LOAD | | KNI | 13,5 | 20,17 | 23,4 | 25,41 | 26,33 MPa/mm |
| | | VMI | -0,277 | -0,063 | -0,039 | -0,032 | -0,029 mm |
| UNLOAD | | KNI' | 16,42 | 24,13 | 29,18 | 30,42 | 23,64 MPa/mm |

| INPUT PARAMETERS | | | FULL SCALE PARAMETERS | | |
|---|---|---|---|---|---|
| JRCo | 10,00 | | JRCn | 7,25 | |
| JCSo | 50,00 | MPa | JCSn | 30,85 | MPa |
| Lo | 0,10 | M | DPEAK | 2,42 | mm |
| Ln | 0,50 | M | | | |
| PHIr | 33,00 | DEG | | | |
| SIGMA n | 10,00 | MPa | | | |
| SIGMAC | 60,00 | MPa | PPEAK | 36,55 | Degrees |
| APERTURE | 29,500 | μm | KS | 3,067 | MPa/mm |

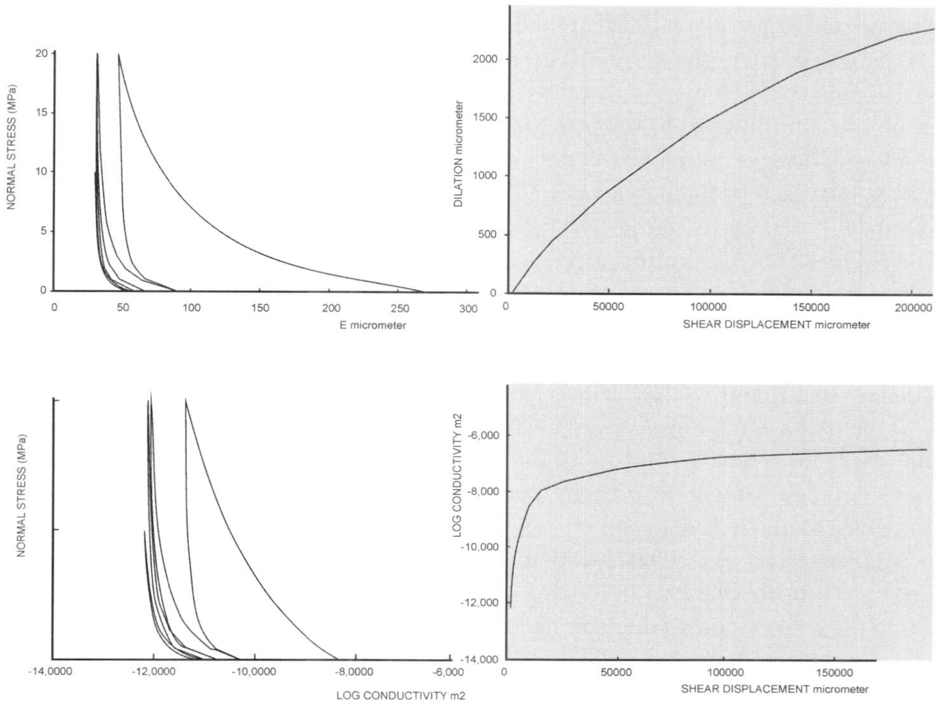


*Figure 16.51* BB normal closure-permeability and shear dilation-permeability modelling. Case C 'carbonate' (high roughness) with JRC = 10, JCS = 50 MPa, $\phi_r = 33°$ and $\sigma_c = 60$ MPa. Note improved apertures and permeability in the left-side plots (= normal closure) and the stronger dilation and rapid permeability enhancement in the right-side plots (= shear).

Table 16.14 Estimates of physical (E) and conducting (e) apertures, for two assumed reservoir rock joint scenarios, with low and moderate strength rock, and low roughness. Predicted apertures are too small to be considered 'open', with this limited roughness. 3 = 3rd cycle, 4 = 4th cycle of loading.

| | Case A | Case B |
|---|---|---|
| Input data | $JRC_o = 5$ | $JRC_o = 5$ |
| | $JCS_o = 25$ MPa | $JCS_o = 50$ MPa |
| At $\sigma'_n = 10$ MPa | $E_3 = 0.90\ \mu m$ | $E_3 = 2.26\ \mu m$ |
| $\sigma'_n = 20$ MPa | $E_4 = 0.37\ \mu m$ | $E_4 = 1.03\ \mu m$ |
| At $\sigma'_n = 10$ MPa | $e_3 = 0.01\ \mu m$ | $e_3 = 0.09\ \mu m$ |
| $\sigma'_n = 20$ MPa | $e_4 < 0.01\ \mu m$ | $e_4 = 0.02\ \mu m$ |

Both models with the least rough joints were consolidated three times to 20 MPa, then to 10 MPa on the 4th cycle, as if there was a certain overpressure. They show extremely small minimum apertures ($E_3$, $E_4$) on the third and fourth cycles of consolidation (actually 'reloading', as in the Bandis experiments, to reach an almost stable assumed 'in situ' aperture).

The effective normal stresses applied were to simulate $\sigma_{H\ min}$ with an 'open joints parallel to $\sigma_{H\ max}$ assumption', as referred to by many geophysicists with presumably the agreement of the field reservoir engineers. Hard rock and rough joints (or 'bridging' with hard minerals) seem to be needed for this orientation to satisfy the 'open' joints requirement, judging by the very small apertures predicted for these two least-rough-joints cases.

*One set* of conducting fractures will give the rock mass an (anisotropic) mass permeability equal to $K_m$ (following Louis, 1967):

$$K_m = K_r + K_j \frac{e}{L} \tag{16.12}$$

where $K_r$ = permeability of rock matrix

$K_j$ = joint permeability

e = average hydraulic aperture

L = average spacing of conducting joints

Since in the highly stressed ($\sigma_n' = 20$ or 10 MPa) examples (cases A and B), $e_3$ and $e_4$ are in the range <0.01 to 0.07 μm, it is clear that both the $K_j = e^2/12$ term, and the e/L term will give minute contributions to the bulk permeability $K_m$, and one will depend on the matrix permeability $K_r$ (and porosity), in each of these two cases. On the other hand when JRC = 10 (as in Figure 16.51, the apertures and permeabilities are considerably larger, and 'open' joints in the direction of $\sigma_H$ maximum can then be imagined.

There is however, a dilemma exposed by the $JCS_o$ assumptions in Tables 16.13 and 16.14, concerning which magnitude of the *confined joint wall strength* $JCS_o$ one should actually select for reservoir joint modelling. Clearly a positive effect of confinement on strength (represented by the expanding Mohr circle diameter $\sigma_1 - \sigma_3$), should help to increase apertures somewhat, but would likely be insufficient to give 'open' character in the case of the less rough joints in the weaker, porous reservoir rocks.

The problem for a weak reservoir rock, like porous sandstone or chalk, is that very high confinement, as when at several kilometres depth, may no longer be positive for strength development, due to the onset of plastic deformation and eventual pore-collapse trends. The Mohr strength envelopes tend to curve rapidly towards a maximum shear strength with increased confinement (i.e. have strongly non-linear internal friction angles), and therefore reach a maximum value of $\sigma_1 - \sigma_3$ at the point where the Mohr envelope becomes horizontal, usually corresponding to a 'critical state' line defined by a gradient $\sigma_1 = 3\sigma_3$ suggested by Barton, 1976. This will be shown later.

At higher confinement a 'cap' or declining trend occurs. Loss of porosity or reduction in void ratio for a rock like porous chalk, plus a reducing rate of gain in strength, occurs at successively lower effective mean stresses as the porosity rises from e.g. 20 to 30 to 40% and more. In contrast, hard rocks, and especially very hard rocks, show great benefit of confining stresses equivalent to many kilometres depth, by not crossing the 'critical state' line until hundreds of MPa confinement, or up to tens of kilometres of equivalent depth in exceptional cases. These aspects will be discussed further in the next section.

### 16.7.7 Open joints caused by anisotropic stress, low shear strength, dilation

In contrast to the normal closure modelling of apertures discussed above, the *shear-dilation modelling* that is also shown in Figures 16.49 to 16.51, indicates that significant apertures will be developed after some few mm of shearing in all three cases, even with the weaker of the 'reservoir' rocks. So repeated shearing in a 'critical crust' scenario (as interpreted by Townend and Zoback, 2000, and others who support the 'geomechanical' school of thought), could easily develop the apertures needed to make the above case A, and case B 'hydrocarbon reservoirs' productive, with truly 'open' sheared joints and less need for a permeable matrix, than seen in the normal stress-closure examples.

In the case of shearing under an arbitrary effective normal stress level, the peak shear strengths ($\tau$) can be estimated from the following equations, which were presented earlier. Block-size induced scale effects, reducing $JRC_o$ and $JCS_o$, are seen to reduce peak shear strength. They will also affect the subsequent shear resistance reductions, towards *residual* strength, the latter taking perhaps 1 meter, or more, of shear displacement.

$$\tau = \sigma_n' \, JRC_n \left( \log\left(\frac{JCS_n}{\sigma_n'}\right) + \phi_r \right)$$

$$JRC_n \approx JRC_o \left(\frac{L_n}{L_o}\right)^{-0.02\,JRC_o}$$

$$JCS_n \approx JCS_o \left(\frac{L_n}{L_o}\right)^{-0.03\,JRC_o}$$

It will be noted that in contrast to the *normal closure* modelling of the three cases A, B and C in the previous section, there is here a necessity for $\phi_r$ estimation, since shearing is to be modelled. In Table 16.13, $\phi_r = 27°$ was assumed for the case A, weak 'sandstone', and $\phi_r = 33°$ for cases B and C, representing stronger 'carbonates' with two different joint roughness magnitudes. In situ block sizes $L_n$ of 1.0 m were assumed here in each case, in order to demonstrate the orders of magnitude of possible strength reduction, due to reductions of $JRC_o$ and $JCS_o$ according to the scaling equations of Bandis et al., 1981. The experimental form of the $JRC_o$ and $JCS_o$ reduction with block size is illustrated in Figure 16.52.

Note that the dilation curves shown in Figures 16.49 to 16.51, were *non-conservatively* linked directly to

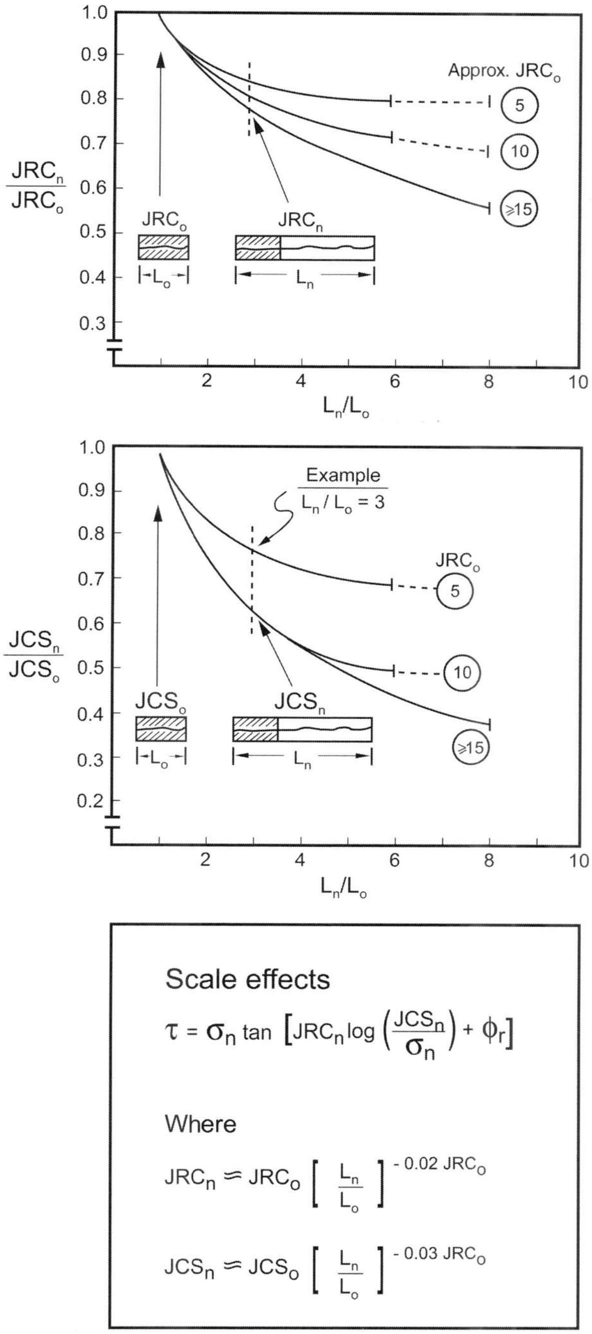


*Figure 16.52* Experimentally determined reductions in JRC and JCS, based on analysis of the extensive scale-effect investigations of Bandis 1980. Note the use, and important influence, of small-scale roughness $JRC_o$ in this scaling. Bandis et al., 1981.

aperture increase, following Barton et al., 1985. The reality from case to case is difficult to estimate, without using the Olsson and Barton, 2001, JRC $\rightarrow$ $JRC_{mob}$ modified, E to e conversion (Figure 16.9), for *roughly*

Table 16.15 Estimation of shear strength reduction due to block size, for the three reservoir rock scenarios. Assumed mean block size = 1.0 m. (Note the friction coefficient format: useful for comparison with data from the 'critical shearing crust' section that follows).

| *Case A* (assume $\sigma'_n$ = 10 MPa) | *Case B* (assume $\sigma'_n$ = 10 MPa) | *Case C* (assume $\sigma'_n$ = 10 MPa) |
|---|---|---|
| $JRC_o$ = 5 | $JRC_o$ = 5 | $JRC_o$ = 10 |
| $JRC_n$ = 4.0 | $JRC_n$ = 4.0 | $JRC_n$ = 6.3 |
| $JCS_o$ = 25 MPa | $JCS_o$ = 50 MPa | $JCS_o$ = 50 MPa |
| $JCS_n$ = 18 MPa | $JCS_n$ = 35 MPa | $JCS_n$ = 25 MPa |
| $\phi_r$ = 27° | $\phi_r$ = 33° | $\phi_r$ = 33° |
| $\tau_o$ = 5.5 MPa | $\tau_o$ = 7.4 MPa | $\tau_o$ = 8.4 MPa |
| $\mu_o$ = 0.55 | $\mu_o$ = 0.74 | $\mu_o$ = 0.84 |
| $\tau_n$ = 5.3 MPa | $\tau_n$ = 7.0 MPa | $\tau_n$ = 7.1 MPa |
| $\mu_n$ = 0.53 | $\mu_n$ = 0.70 | $\mu_n$ = 0.71 |

*estimating* gouge-production effects on *reduced* permeability enhancement with shear and dilation. This will result in less extreme increases in permeability with shear unless the rock is hard, and/or the roughness low and the rock hard enough to resist 'wear'.

There may be an approximately 50% reduced conducting aperture caused by gouge, according to Makurat's CSFT (coupled shear flow test) results. (Makurat et al., 1990). This will be depend partly on the $\sigma'_n$/JCS ratio which describes one of the important components of damage. Damage during shear will also depend on JRC, as steeper asperities tend to get more damaged, though also cause more dilation. (Barton and Choubey, 1977 developed a damage coefficient on the basis of the above). These *opposed* tendencies obviously have complex consequences on the permeability development with shear.

As discussed before, both smaller scale and larger scale joint or fracture non-planarity, will cause several possible interpretations of '*fracture orientation*', in relation to shear wave splitting and polarization. Interpreted anisotropy results that are actually non-parallel to $\sigma_{H\,max}$ but not greatly different, might sometimes mislead analysts into thinking that the productive fractured reservoirs they are interpreting, have 'open' fractures approximately 'parallel' to $\sigma_{H\,max}$, when an alternative interpretation is possibly more realistic in geomechanics terms. (Here we recall from Chapter 15 that two conjugate sets can provide components suggesting $\sigma_{H\,max}$ parallelism.).

As demonstrated in the above BB-modelling exercises, that parallelism of 'open' fractures to the $\sigma_{H\,max}$ direction is *unlikely* in weaker rock unless joints are quite rough, or production very poor, despite the fracturing (i.e. if the joints were nearly closed they would satisfy the geomechanics criterion, but not the production requirement). In stronger (e.g. carbonate) rock, with greater roughness, there may be naturally larger apertures, which would give 'open' fractures also without the benefit of historic or recent tectonic shear deformations.

## 16.8 Non-linear shear strength and the critical shearing crust

Hydraulic fracturing based stress measurements performed for a pressure tunnel design in 1980 in Georgia, USA while working in the USA, and acoustic emission events recorded by others when injecting water into the Cornwall Geothermal Project (Pine and Batchelor, 1984), were the stimulation for the sheared-fracture/sheared-joint sketches given in Figures 16.53a and 16.53b, which we will use to introduce the 'critical shearing crust' concept. These are 'forced' events, caused by pore-pressure increase. Nature appears capable of something similar without human intervention.

When there is a significant mismatch of horizontal stress magnitudes, and pre-existing joints that are not parallel to $\sigma_H$, there is a good statistical chance for joint shearing, when attempting hydraulic break-down in a minifrac stress measurement (Barton, 1981). In a leaking pressure tunnel-liner situation, the same event is possible at much larger scale.

In this section, the important findings of Colleen Barton and Zoback and co-workers will be reviewed regarding the convincing evidence for the existence of significant shear stress on joints or fractures that were *hydraulically conducting* in a deep boreholes in crystalline rock. Data from several wells are now available, thanks to the analyses of the Zoback team. By careful analysis of fracture orientation and inflow logging, they also showed that *insignificant* shear stresses were acting on joints that were *not* conducting.

In this work, they compared their stress estimates with the 'standard' $0.6 \leqslant \mu \leqslant 1.0$ joint friction model that has been popular in the USA since Byerlee, 1978. (The 'Byerlee law' suggested a generally applicable $\mu \approx 0.85$ for the coefficient of friction). Before reviewing their *flow or no-flow* data, we will look at the actual non-linearity of shear strength, in all its forms. Their data on conducting joints and fractures fits the Byerlee 'band' of shear strength, but the fit can be improved by considering non-linear friction.

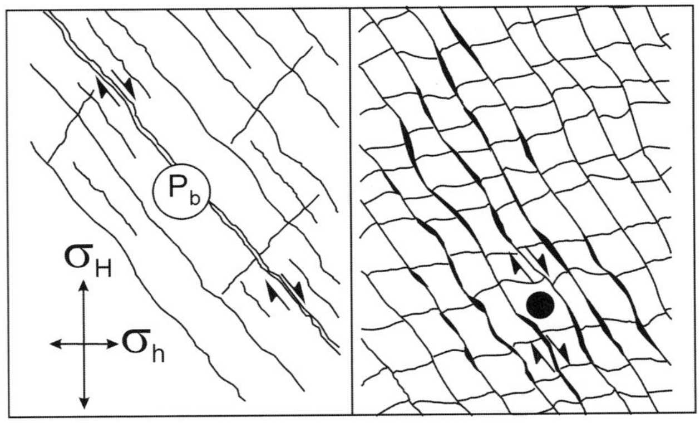


(a)

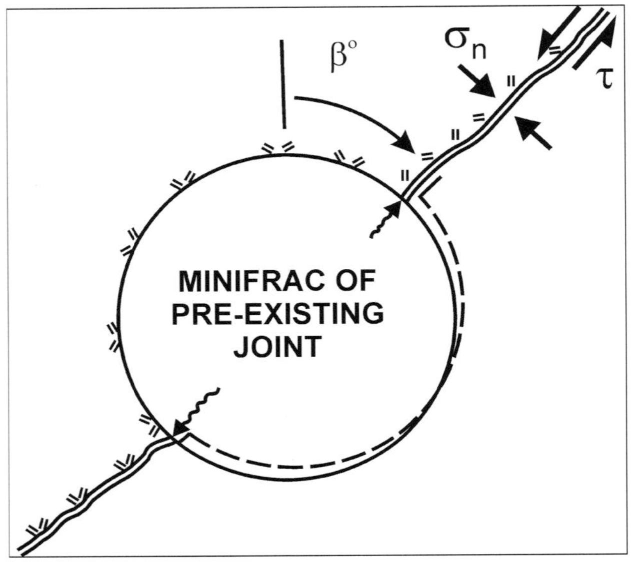


(b)

*Figure 16.53* a) The nature of conducting 'open' joints or hydraulic fractures caused by pore pressure and shear stress induced shearing. Barton, 1986. b) The suggested mechanism of hydraulic fracture 'breakdown' by shear mobilization, given by Barton, 1981, for the case of pre-existing joints where stress measurements (or leaking pressure tunnels) are involved.

### 16.8.1 Non-linear strength envelopes and scale effects

In an attempt to understand the *actual non-linearity* of shear strength envelopes, both for rock joints, for fractured triaxial samples at high stress, and for intact rock itself, Barton, 1976 assembled and analysed a large number of test data from the literature. The following contrasting sets of (non-linear) shear strength data, consist of high pressure triaxial shearing of (small)

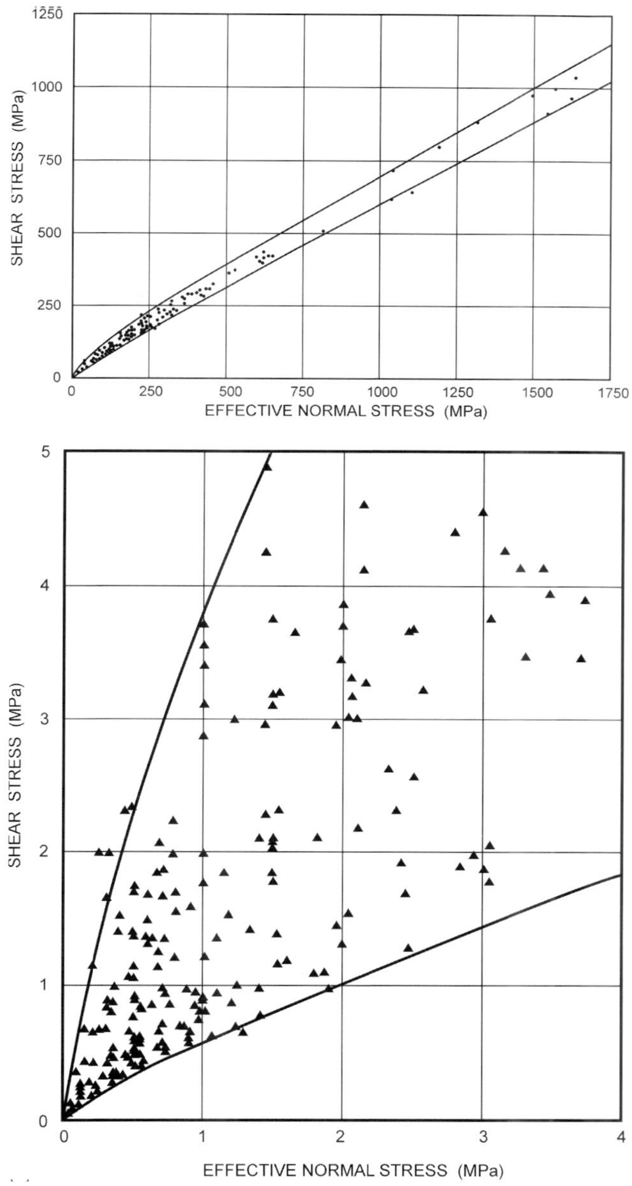


*Figure 16.54* a) Shear strength of *induced fractures* in small, extremely high pressure triaxial tests following failure of the centimetre-size intact samples. (Data from Byerlee, 1978 and others). b) Shear strength from direct shear tests on natural rock joints (one test per sample), at low rock engineering stress levels. Typical sample lengths of 100 mm to 1000 mm. Barton, 1976.

'faulted' samples, shown in Figure 16.54a, and a low pressure set of data shown in Figure 16.54b. The latter were obtained from direct shear tests on natural rock joints, and included the test results for the 130 fresh and slightly weathered joint samples tested by Barton and Choubey, 1977.

Returning to these and further data sets in 1990, at the *First International Workshop on Scale Effects in Rock Masses* in Loen, Norway, and armed with some important scale-effect research by Bandis, 1980, as summarised in Bandis et al., 1981, the writer (Barton, 1990b) sketched approximate 'scale effect' bounds on the significantly updated data sets.

Figure 16.55 shows the assumed 'shifts' in the data boundaries. It was estimated that in case a) due to the intensely high stresses to 800 MPa, there would be a very limited scale effect (assuming faults were excluded). The engineering data at stress levels mostly below 4 MPa shown in Figure 16.55b, were known to display a severe scale effect. In other words tests on 100 mm samples of a given joint were always of higher strength than 1000 mm samples, unless roughness was virtually absent.

Subsequently, the generally expected *non-linearity* of the shear strength envelopes for rocks, rock joints and crushed rock were assembled by Barton, 1999a, to compare the various empirical strength criteria that have the form:

$$\tau = \sigma_n \tan(X \log_{10} Y + Z) \tag{16.13}$$

where X represents the effect of roughness, e.g. JRC, and Y represents the strength/stress ratio, i.e. joint wall strength JCS/$\sigma_n'$ in the case of engineering stress levels, which will need the confined strength $(\sigma_1 - \sigma_3) >$ JCS, i.e. the strength/stress ratio $(\sigma_1 - \sigma_3)/\sigma_n'$, in the case of shearing at high stress levels (Barton, 1976). The schematic arrangement of the different strength envelopes is shown in Figure 16.56.

Two of the sources for the schematic envelope for intact rock in this figure, are reproduced in Figure 16.57. These show the tendency of the numerous, mostly ultra high-stress data to cross the 'critical state' line (defined by Barton, 1976), in a more-or less horizontal orientation, i.e. beyond the brittle-ductile transition, in the advancing 'plastic' state. Note that $Z = \phi_c = 26.6°$ in this model, as indicated by the expanding and then declining Mohr circles in Figure 16.58.

The general non-linearity of the Barton-Choubey (small-scale) and Barton-Bandis (general scale) constitutive equations for the shear strength of rock joints, can be reduced to the traditional, but actually less geotechnically representative Mohr Coulomb parameters c and $\phi$, by the simple equations shown in Figure 16.59. The particular case shows how to derive c and $\phi$ between arbitrary effective normal stress levels of $\sigma_{n3}$ and $\sigma_{n4}$.

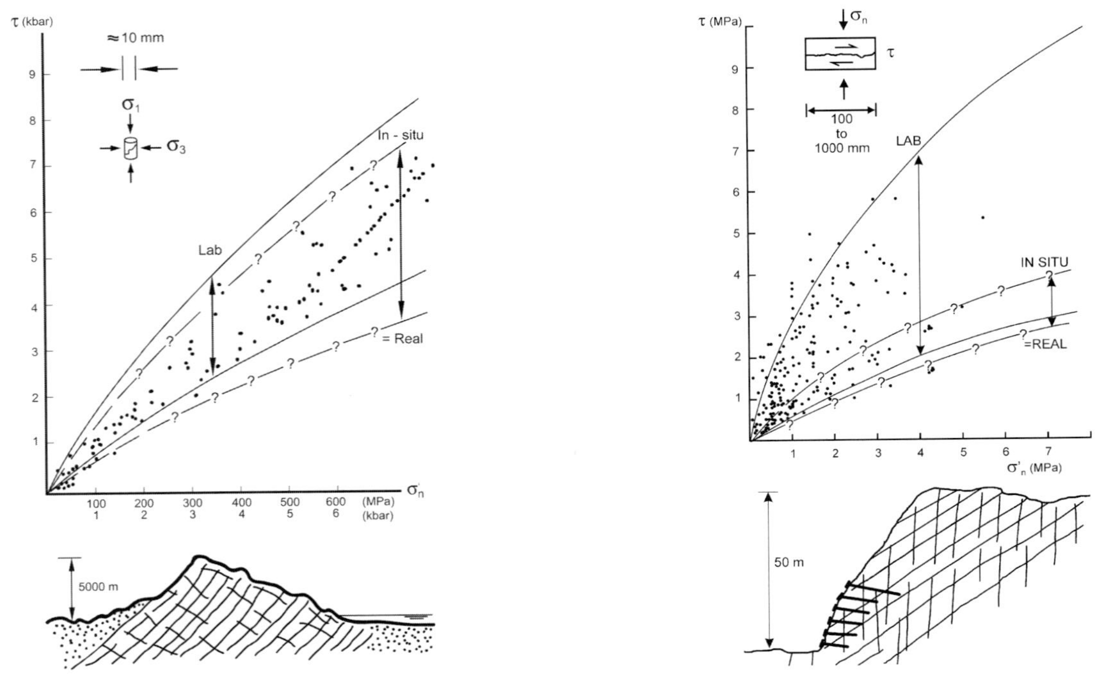


*Figure 16.55* a) Triaxial, high stress and b) direct shear (low stress) data for induced fractures and joints respectively. Scale effect curves are estimates only. Barton, 1990b. Adapted from Barton, 1976 with additional data from Byerlee, 1978.

The general strengthening-by-confinement that we have seen in the above shear strength envelopes, will obviously apply *strongly* in the case of deep wells in hard crystalline rock, both for the intact rock, and for the faulted discontinuities or sheared joints. We will address this aspect again when reviewing deep permeability data.

Strengthening-by-confinement is likely to be of only *moderate* strength in the case of hydrocarbon reservoirs in *weaker* sedimentary rock, due to their limited tolerance of very high confinement, before 'state change', such as pore collapse (in chalks), or general porosity reduction. (Refer to Figures 12.22 and 12.23 for the case of sandstones). These *medium weak* rocks would obviously not support 20 to 50 MPa effective stresses, *and mostly* maintain porosity, without some strengthening due to confinement.

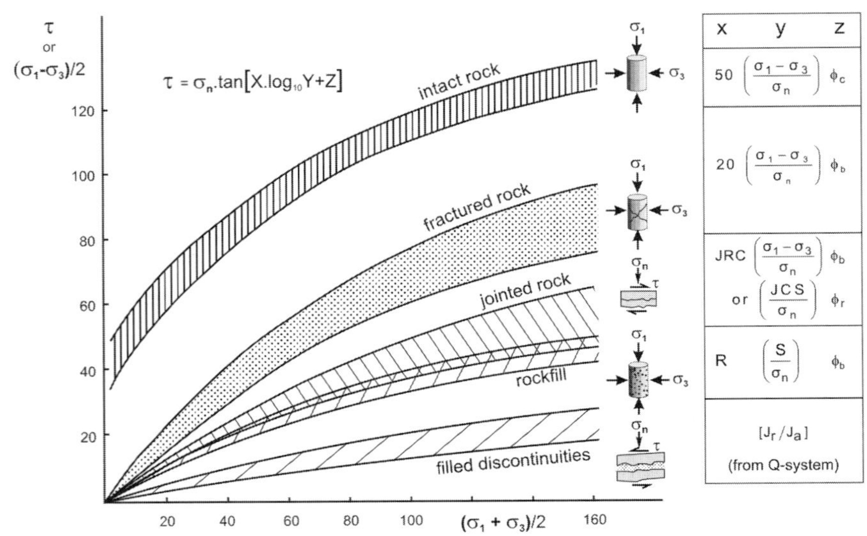


*Figure 16.56* Schematic representation of the non-linearity of strength envelopes for various 'earthscience' materials, showing a comparison of the empirically derived roughness (X) and strength/stress (Y) ratios. Barton, 1999a.

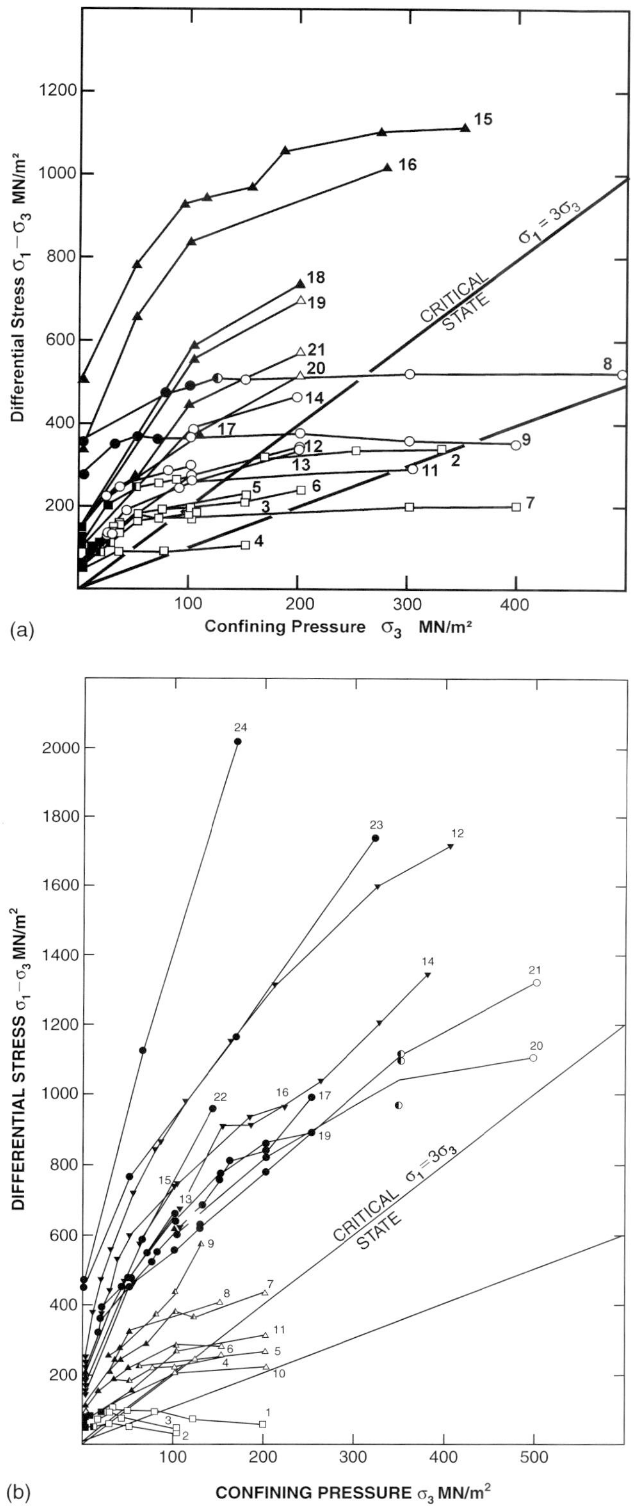


*Figure 16.57* Shear strength data for intact rock at a variety of mostly high confining pressures, from the numerous sources reviewed by Barton, 1976.

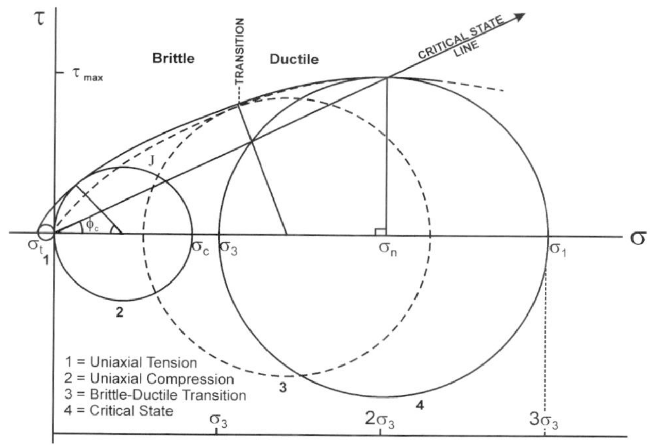


*Figure 16.58* A general model for representing the shear strength of intact rock, showing the succession of key Mohr circles that also span the brittle-ductile transition and reach the 'critical state line', defined by Barton, 1976 as $\sigma_1 \approx 3\sigma_3$. From this point, strength envelopes may have a tendency to show reduced strength with increasing confinement, in the 'cap' region.

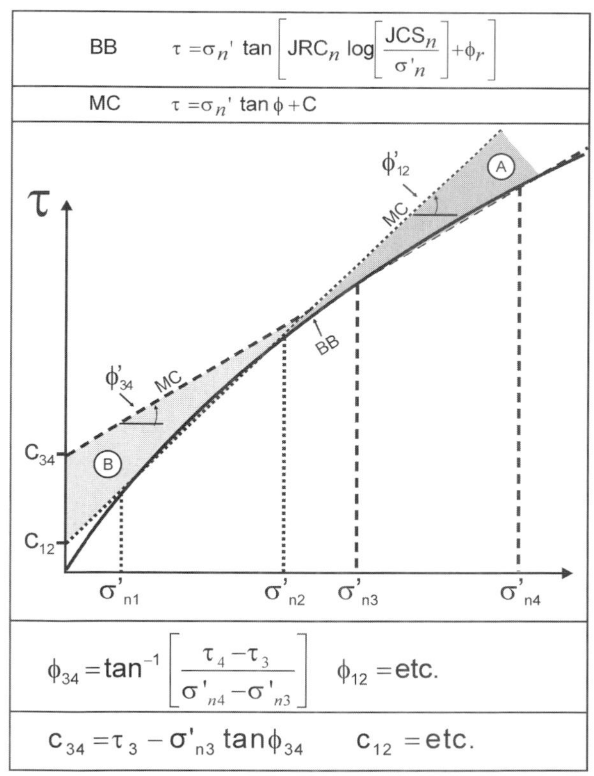


*Figure 16.59* Conversion of the non-linear Barton-Bandis shear strength equation to Mohr Coulomb parameters c and $\phi$ at a specific range of stress. Note use of large-scale rock joint characteristics, i.e. *in situ* block size $L_n$ with $JRC_n$ and $JCS_n$, compared to $L_o$ with $JRC_o$ and $JCS_o$ for lab tests.

That there are limits to the strengthening-by-confinement in the case of *weaker* rock, can be seen from the 'weakest' strength envelopes in Figure 16.57, which are seen in diagram b) to cross the 'critical state line' at comparatively low confinement. There might also be the occasional 'adverse' effect of water flooding on e.g. the Ekofisk chalk, where the confined strength was indeed exceeded in the most porous sections, due to draw-down causing pore collapse, but with excellent compaction drive production as a result of the subsequent further weakening by water.

Before leaving the failure of intact rock it is of interest to see the dilation and brittle fracturing that occurs when rock is highly stressed. This was found by Rummel et al., 1978, to be a significant source of *P-wave anisotropy.* They utilised a biaxial loading arrangement with fast-reacting servo-control, to study the development of dilation adjacent to the shear failure surfaces. They found that the P-wave velocity increased continuously in the direction of maximum compression in the pre-peak region. In the post-failure region, the P-wave velocity decreased almost reversibly with reducing compression.

By comparison, the minimum and intermediate principal stress directions suffered a marked reduction of P-wave velocity (recorded as travel time increases), after fracturing was initiated. (Enhanced permeability would be a related phenomenon of such dilation: this would presumably occur mostly in the $\sigma_1$ direction). The authors mention the need to be aware of the possibility for *increasing velocity anisotropy*, when interpreting field seismic data in crustal regions where large tectonic stresses are assumed to be operating.

Figure 16.60 shows the complete stress-strain curves for their granite specimens, which were tested in a 1.0 *GPa* pressure vessel. These triaxial tests were on cylindrical samples. The markedly non-linear shear strength envelope resulting from these highly confined samples is shown in the same figure, together with the frictional strength of the resulting fractures, which are also non-linear, just as they were in Byerlee, 1978.

The frictional strength of the resulting fractures, follows the non-linear shear strength model of Barton, 1976 closely, with an equivalent JRC of 20, and the use of confined strength ($\sigma_1 - \sigma_3$) in place of JCS, as indicated in Figure 16.56. This was also the case when matching equivalent sets of data for various rock types, given by Byerlee, 1978. The friction coefficient (approx. 0.7) for the Rummel et al., 'fractures' compares well with the 'critical shearing crust' well data of Zoback and co-workers, as we shall see shortly, although $\mu = 0.7$ is

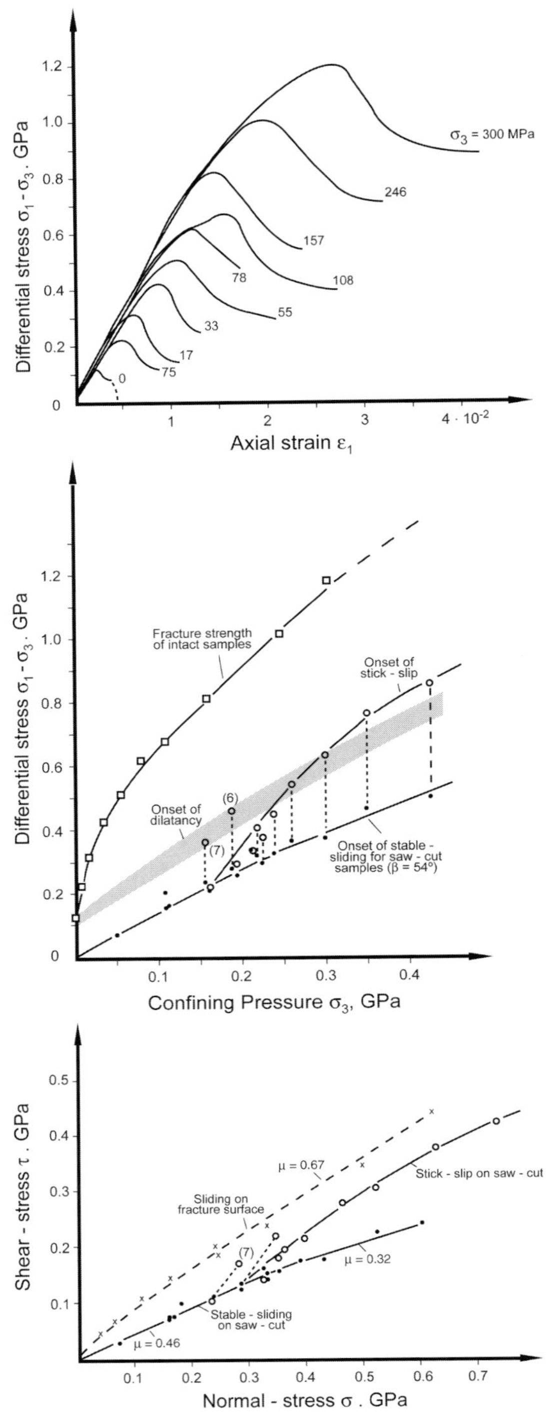


*Figure 16.60* Shear stress – axial strain data for high pressure triaxial testing of intact granite, showing strength envelopes for subsequent sliding on the 'minor faults', and stick-slip using some artificial saw cut surfaces of the granite. Note the reducing non-linearity of the strength envelopes for these three categories. Rummel et al., 1978.

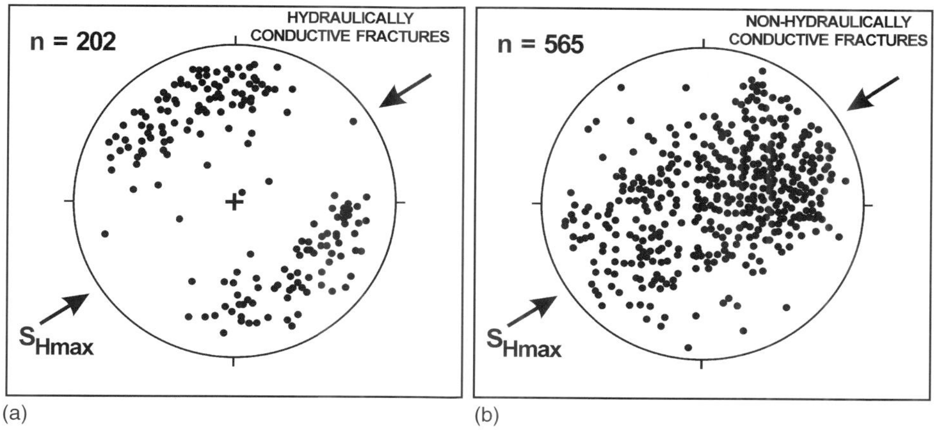


*Figure 16.61* Distinctive pole populations of joints or fractures that were conductive, and those that were not, from analysis of Cajon Pass Scientific Drillhole results. Colleen Barton et al., 1995.

lower than some of their data. Note the lower friction coefficients of the saw-cut surfaces, which resemble minor faults at residual strength.

## 16.9 Critically stressed open fractures that indicate conductivity

A significant contribution to our understanding of the geomechanics acting at *deeper levels* below the earth's surface, was made by Zoback and Colleen Barton and co-workers, regarding the delineation of fracture directions that appear to be conducting directions, and those that do not. The 'simple' yes or no concerning conducting directions is fundamental evidence, also for geophysicists and petroleum engineers.

Here we will review several cases that support the philosophy of 'open' joints *actually* often being those that are under a significant state of shear stress, rather than being parallel or sub-parallel to $\sigma_{H\ max}$ and without shear stress. Authors in geophysics and even those working in the petroleum industry, who are each dependent on 'open' joints: both for shear wave anisotropy, and for petroleum production, appear to have favoured the 'parallel to $\sigma_{H\ max}$' model.

Figure 16.61 shows the differentiation of fracture orientations a) for those that were hydraulically conductive, and b) for those that were hydraulically non-conductive. These different populations were recorded by Colleen Barton et al., 1995 in the Cajon Pass Scientific Drillhole, which was drilled to 3.5 km depth into *granites and granodiorites*, 4 km from the San Andreas Fault near San Bernadino in California. The hydraulically active fractures were located by temperature anomalies, and mapped by televiewer, to an accuracy of ±1.0 m from each temperature anomaly.

The above authors had noted break-out direction anomalies over large depth intervals, in relation to the perpendicular-to-$\sigma_{H\ max}$ direction norm. These proved to be associated with slip on joints (or minor faults). Their plots of normalised mobilization of friction: $(\tau/\sigma_v)$ versus $(\sigma_n - P_p)/\sigma_v$ shown in Figure 16.62 are a convincing demonstration of the importance of shear stress (rather than just minimum normal stress) for explaining the 'openness' of joints, and therefore their conductive capacity.

One may speculate that in such granites and granodiorites, the values of JRC and JCS that one might typically estimate, would probably be high enough even for joints that were under $\sigma_{h\ min}$ normal stress, to be marginally conductive. However, shearing-enhanced permeability would obviously greatly dominate in signal strength, as apertures would be very small in the case of the 'closed' joints.

Such a proposition can be tested. A quick 1-D Barton-Bandis modelling, starting with equation 16.11, with $JRC_n = 5$ or 10, $JCS_n = 100$ MPa, and an effective normal stress of 75 MPa suggests *conducting apertures* as small as 2.5 and 15 μm. Apertures appear to reach such 'residual' levels after a few tens of MPa, due to the combined effect (in this crystalline rock case), of high wall strength and significant joint wall roughness.

Comprehensive supplemental data to the above deep borehole interpretation was more recently reported by Townend and Zoback, 2000, and Zoback and Townend, 2001. These very extensive data sets are reproduced in

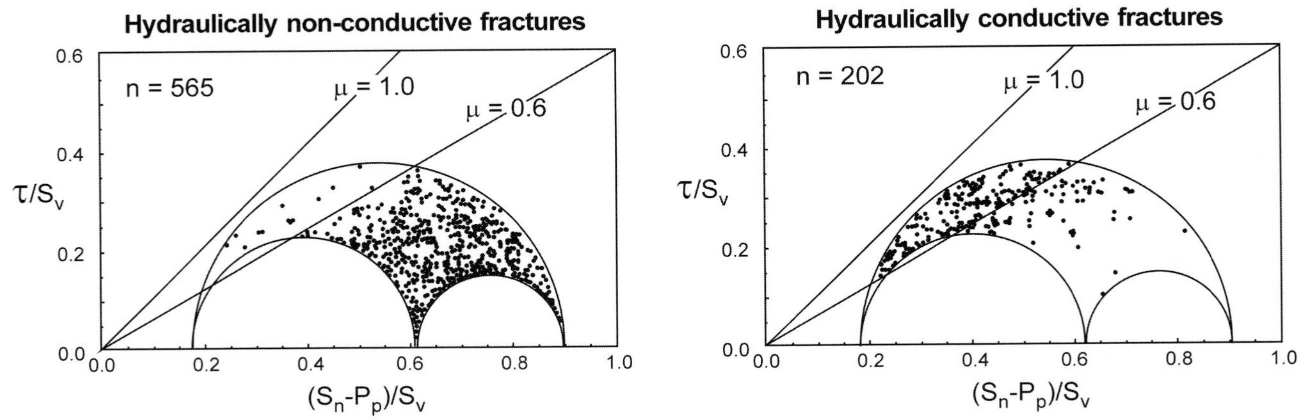


*Figure 16.62* Mohr stress representation of the shear and normal stress components, acting on the Cajon Pass fractures, (i.e. sheared joints or minor clay-free faults?), according to their conductive or non-conductive characteristics. Note authors' friction coefficient limits, as per Byerlee, 1978, ($\mu = 0.6$ to 1.0). These could apparently be extended/modified to 0.4 to 0.9 in practice. Colleen Barton et al., 1995.

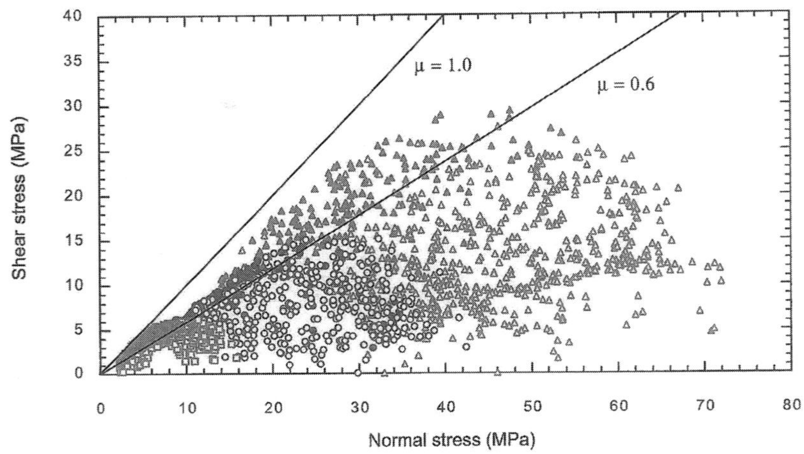


*Figure 16.63* Normal and shear stresses for fractures identified as hydraulically conducting (closed symbols) or non-conducting (open symbols). Cajon Pass (triangles), Long Valley (circles), Nevada Test Site (squares). Townend and Zoback, 2000, with data also from Colleen Barton et al., 1995. (Zoback, 2006 pers. comm., by kind permission.) See Plates 14, and 15.

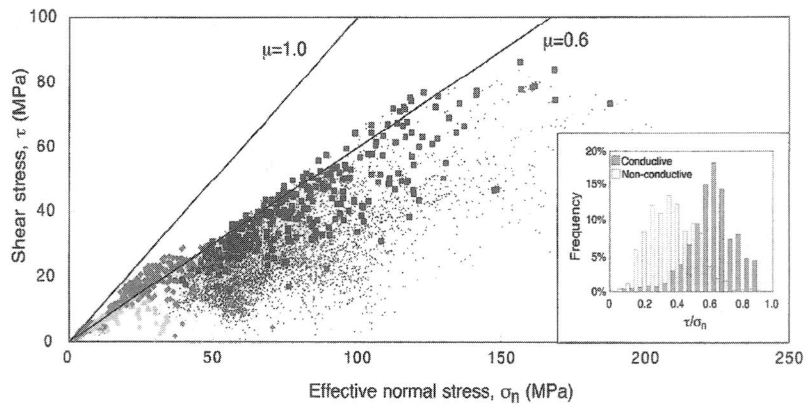


*Figure 16.64* Normal and shear stresses for fractures identified as hydraulically conducting or non-conducting, using borehole imaging. Cajon Pass (red diamonds and dots), Nevada Test Site (green circles and dots), Long Valley (yellow triangles and dots), KTB (Germany – blue squares and dots). Inset shows $\mu = \tau/\sigma_n$ for combined data set. Zoback and Townend, 2001, with data from Ito and Zoback, 2000, and from Colleen Barton et al., 1995. (Zoback, 2006 pers. comm., by kind permission.)

Figures 16.63 and 16.64 by kind permission. Note the widely different stress magnitudes in these two figures.

Figure 16.63 shows conductive (closed symbols) and non-conductive (open symbols or dots) from Cajon Pass (triangles), Long Valley (circles) and Nevada Test site (squares). It is clear from this data that either due to joint character or due to stress level, or a combination of both, the interpreted friction coefficients can range from (at least) $\mu = 0.4$–$1.0$ (Yucca Mountain: at normal stresses below 16 MPa), to $\mu = 0.5$–$0.8$ (Long Valley: 10 to 40 MPa), to $\mu = 0.5$ to $0.9$ (Cajon Pass: 14 to 70 MPa).

The quite wide range of $\mu$ for Nevada Test Site (i.e. Yucca Mountain) can be due both to the lower stress levels (i.e. shallower depth), and to the wide range of joint roughnesses. Joint roughnesses have been interpreted to be from about 1 to 15 for $JRC_o$, when assuming, for simplicity, a representative JCS of 100 MPa, and $\phi_r$ of 30°. These estimates are based on *in situ* roughness recordings in the exploratory TBM tunnels, and on a series of medium scale direct shear tests.

Values as low as $\mu = 0.4$, shown in Figure 16.63, implying a 22° angle of friction, suggest some clay-fillings. In fact even lower values are indicated in the Nevada Test Site data set, but these are non-conducting, as expected for clay-filled discontinuities.

A clearer picture of the frictional content of the stress data from these six deep boreholes, is reproduced in Figure 16.65a and 16.65b, from Townend and Zoback, 2000, and Zoback and Townend, 2001. The crosses shown in the figures are the error bars for the different sets of data. The three dotted lines representing $\mu = 1.0$, 0.6 and 0.2 were derived by these authors from the Jaeger and Cook, 1979, equation:

$$\frac{S_1 - P_f}{S_3 - P_f} = \left(\sqrt{\mu^2 + 1} + \mu\right)^2 \qquad (16.14)$$

The authors Townend and Zoback, 2000, and Zoback and Townend 2001, do not give specific explanations for the likely reasons for the above ranges of friction coefficient (notably 0.6 and 1.0). As interpreted, these are the levels of mobilized friction tolerated by the analysed joint planes, at the time of analysis. A man-made change of pore pressure would not change the differential stress, but would change the effective mean stress or the effective normal stress, thereby altering the shear resistance of the joint planes in question, and their friction coefficients, in view of non-linearity with stress.

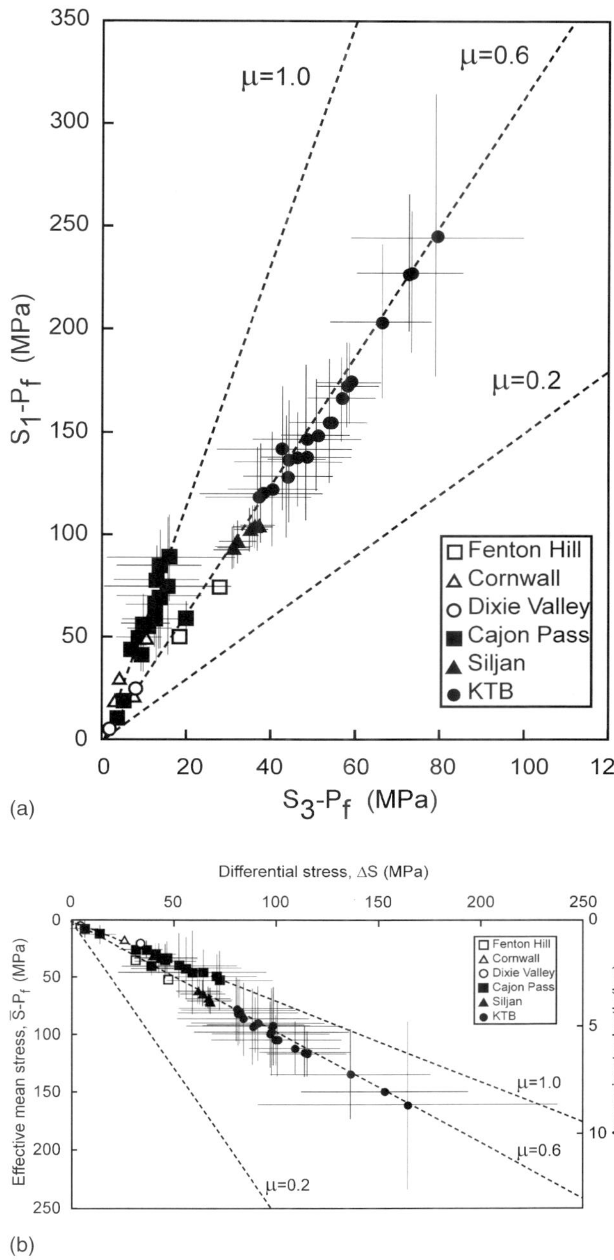


*Figure 16.65* a) Maximum effective stress versus minimum effective stress, for six sets of deep borehole data.
b) Differential stress versus effective mean stress. See Townend and Zoback, 2000, and Zoback and Townend, 2001, and other referred authors who are the sources for the various data sets.

It is notable that the Cajon Pass rocks showed consistently high values of (resistance to) $\mu \approx 1$. As shown earlier in this chapter, it is *likely*, in the case of joints and fractures, that this is due to the interaction of a moderate to high roughness $JRC_n$, medium to high $\phi_r$

(say 30° approx.), and tolerable stress/strength ratios $\sigma_n'/JCS_n$. The high strength rock will give high values of $JCS_n$, or more specifically: high confined strengths $(\sigma_1 - \sigma_3)$, if joint walls had *minimal* alteration.

We can 'assemble' estimates of input data for application in the high-stress, large block-size, version of the Barton-Bandis criterion (Barton and Bandis, 1990):

$$\tau = \sigma_n' \tan\left(JRC_n \log\left(\frac{(\sigma_1 - \sigma_3)_n}{\sigma_n'}\right) + \phi_r\right) \quad (16.15)$$

where $(\sigma_1 - \sigma_3)_n$ is an estimate of the confined strength of the rock, with assumed absence of alteration along the joint or fracture walls. If there was alteration and reduced wall strength, one would expect a less 'confined' estimate of $JCS_n$, i.e. of magnitude $< (\sigma_1 - \sigma_3)_n$ for the triaxially confined intact rock. Of course *in situ*, the rock is poly-axially confined.

The following assumptions will be made to explore if the toleration of transformed shear and normal stress in a ratio as high as 1, i.e. $\mu \approx 1$, can be 'explained' with a representative range of parameter values, as follows:

| | |
|---|---|
| $JRC_n = 8$ or 10 | $\phi_r = 29°–31°$ |
| $JCS_n = 200$ MPa or $(\sigma_1 - \sigma_3)_n = 400$ MPa | $\sigma_n' = 30–60$ MPa |

This is an effective stress range similar to the easily estimated vertical effective stress at (hydrostatically saturated) depths of *2 and 4 km*, assuming a simple average crystalline rock density of 2500 kg/m$^3$ (i.e. $2 \times 2.5 - 2 \rightarrow 30$ MPa, $4 \times 2.5 - 4 \rightarrow 60$ MPa). As will be demonstrated, unless *in situ* block sizes around the analysed wells are quite small, in fact significantly less than 0.5 m, then impossibly high, lab-scale $JRC_o$ values would be needed to generate mobilized friction as high as $\mu = 1$. One would need to greatly exceed the '$JRC_o$ limit' (about 20), illustrated in Figure 16.26.

As one may note from the estimates in Table 16.16, a range of $\mu = 0.65$ to 0.9 (approx.) can be reasonably 'explained', but specifically from *peak shear strength* estimation. If we engage in further analysis of maximum possible *confined* strengths $(\sigma_1 - \sigma_3)_n$ by looking at more of the available triaxial strength data (Figure 16.57), we see the possibility of even higher strengths than 400 MPa at appropriate depths to our example, *if the rock is very strong under triaxial* conditions. (Unfortunately for the 'parallel to $\sigma_{H\,max}$ model' weaker reservoir rocks do not have such benefits from confining stress.).

Table 16.16 Estimates of potential tolerance of mobilized frictional strength, to explore the meaning of $\mu = 1$ for e.g. Cajon Pass data from Zoback and Townend, 2001.

| Case A (minor alteration of joint walls) | Case B (hard unaltered joint walls) |
|---|---|
| JRCn = 8, | JRCn = 10, $(\sigma_1 - \sigma_3)_n$ = |
| JCSn = 200 MPa, $\phi_r = 29°$ | 400 MPa, $\phi_r = 31°$ |
| $\sigma_n' = 30$ MPa $\mu \approx 0.72$ | $\sigma_n' = 30$ MPa $\mu \approx 0.91$ |
| $\sigma_n' = 60$ MPa $\mu \approx 0.65$ | $\sigma_n' = 60$ MPa $\mu \approx 0.82$ |

Substitution of 500 MPa or 600 MPa for $(\sigma_1 - \sigma_3)_n$ in the Case B example in Table 16.16, again at effective stresses of 30 and 60 MPa respectively, would raise the Case B estimates of *peak* $\mu$ from 0.91 to 0.94 (with 600 MPa), or from 0.82 to 0.87 (with 500 MPa). Here we have ignored the inevitable small size of the samples used to generate the Figure 16.57 data (down to approx. 10 mm size cylinders), and have therefore assumed limited scale effects at these intensely high pressures.

### 16.9.1 The JRC contribution at different scales and deformations

The only other conceivable contributor to higher shear strength values might appear to be roughness $JRC_n$. However, with block sizes of 0.5, 1.0, or even 5.0 m, the empirical scaling laws would demand quite exceptional $JRC_o$ values for the small-scale roughness, and frankly, little possibility that 'the critical shearing crust' was ever mobilized. The *optimistic* value of $JRC_n = 10$, applied in Case B in Table 16.16, is already actually 'too high', since it implies $JRC_o$ values (for nominal 100 mm scale), as high as 20 or 21, even when the assumed *in situ* block size round the analysed wells is limited to 0.5 m. This is shown, by regression, from the inverse application of the previously described block-size scaling of roughness:

$$JRC_o \approx JRC_n \left(\frac{L_n}{L_o}\right)^{-0.02\,JRC_o} \quad (16.16)$$

The standard set of ten $JRC_o$ values and their roughness profiles, from Barton and Choubey, 1977 were reproduced in Figure 16.26. The ten selected samples were mostly granite, aplite and gneiss, plus basalt, hornfels, slate and calcareous shale. The roughest sample no. 10 was an artificial tension fracture in a weak soapstone.

The exceptional roughness needed to explain a tolerance of $\mu$ as high as 0.94, namely: $JRC_o = 20$ if $L_n$ is 0.5 m, therefore $JRC_n = 10$, together with the highest defensible confined strength of 600 MPa ($JCS_n = (\sigma_1 - \sigma_3)_n$), is still below the interpreted values of transformed *in situ* principal stress anisotropy, of mostly $\tau / \sigma'_n = \mu = 1$. This was for the Cajon Pass research borehole data of Zoback and Townend, who had interpreted effective mean stresses up to almost 60 MPa, at 3.5 km depth. The only possible explanation seems to be that there was a small average block size, even at several kilometres depth, in this tectonically disturbed area.

Two additional perspectives concerning roughness are provided here, in view of the important potential contribution of large-scale roughness in explaining high values of interpreted, transformed anisotropic stresses. The even larger-scale roughness profiles shown together with appropriate $JRC_n$ values, in Figure 16.27, were obtained from extensive characterization of fractures developed in sandstone, tuff, concrete and hydrostone samples, with sheared areas of about 1.0 m × 1.10 m. Characterization included roughness profiling at different scales, Schmidt hammer testing, tilt testing of the diagonally 'jointed' samples (weighing some 2.5 tons), and final biaxial shear testing in a 1 $m^3$ loading frame at TerraTek, in Salt Lake City, as reported by Bakhtar and Barton, 1984. As will be discussed at the end of this chapter, great difficulty was experienced in shearing some of the rougher samples. This difficulty may have parallels to the Zoback and Townend deep borehole interpretations.

Bandis, 1980, performed a remarkable series of scale-effect tests, using repeatedly cast replicas of natural joints, which were tested at different sampling lengths, that could then be compared with each other since the material strengths and compositions were identical. He presented a set of eleven roughness profiles, with equivalent lengths of between 5 and 15 m, together with back-calculated $JRC_n$ values. These are shown in Figure 16.66. The various profile lengths in this up-scaled format, represent typical medium to hard rock strengths.

Inspection of the major undulations on some of these long profiles, and on some of the 1 m profiles shown in Figure 16.27, suggest that *significant shearing* of joints involved in the 'critical shearing crust' scenario, would likely be limited to $JRC_n$ values somewhat below 10. Otherwise there would be a combination of too much volume increase, developing too much increase in normal stress and stiffness, thereby causing too high shear strength for the mechanism to be viable.

*Figure 16.66* Profiles and interpreted large-scale roughness coefficients ($JRC_n$), from Bandis, 1980 scale-effect investigations with model materials cast as joint replicas. Details are also given by Bandis et al., 1981.

### 16.9.2 Does pre-peak or post-peak strength resist the assumed crustal shear stress?

A further factor to be considered when evaluating and attempting to geomechanically 'explain' the interpreted $\mu$-mobilized values close to 1.0, is that it is clearly not logical that the conducting joints identified in these deep borehole studies by Zoback and his colleagues, should be exactly at *peak strength*. This point or eventual plateau on the shear strength–displacement curve is in many ways a *singularity*, involving some few millimetres of shear in the case of *in situ* block sizes, and perhaps a millimetre or less in typical lab samples. A wide-ranging review of such displacement data, from Barton, 1982 is reproduced in Figure 16.67.

The data is divided into three size categories (roughly 5 to 30 cm = lab-scale, 0.3 to 3 m = *in situ test* scale, >3 m = exotic tests and up-scaled models). Even for small samples there is a wide range of 'strains-to-peak', when expressed as a displacement-to-peak divided by sample length, and expressed as a percentage.

A very interesting point arises when considering whether the shear-stressed joints involved in the 'critical shearing crust' are at their peak, pre-peak, or post-peak shear strength. Reference to the $JRC_{mob}$ concept, which was shown in Figure 16.10, and to the shear strength-displacement-dilation worked examples shown in Figure 16.25, reveals that the 'necessary' permeability enhancement due to the beginning of dilation, actually occurs when roughness starts to be mobilized, and may already be active after a few millimetres of shearing.

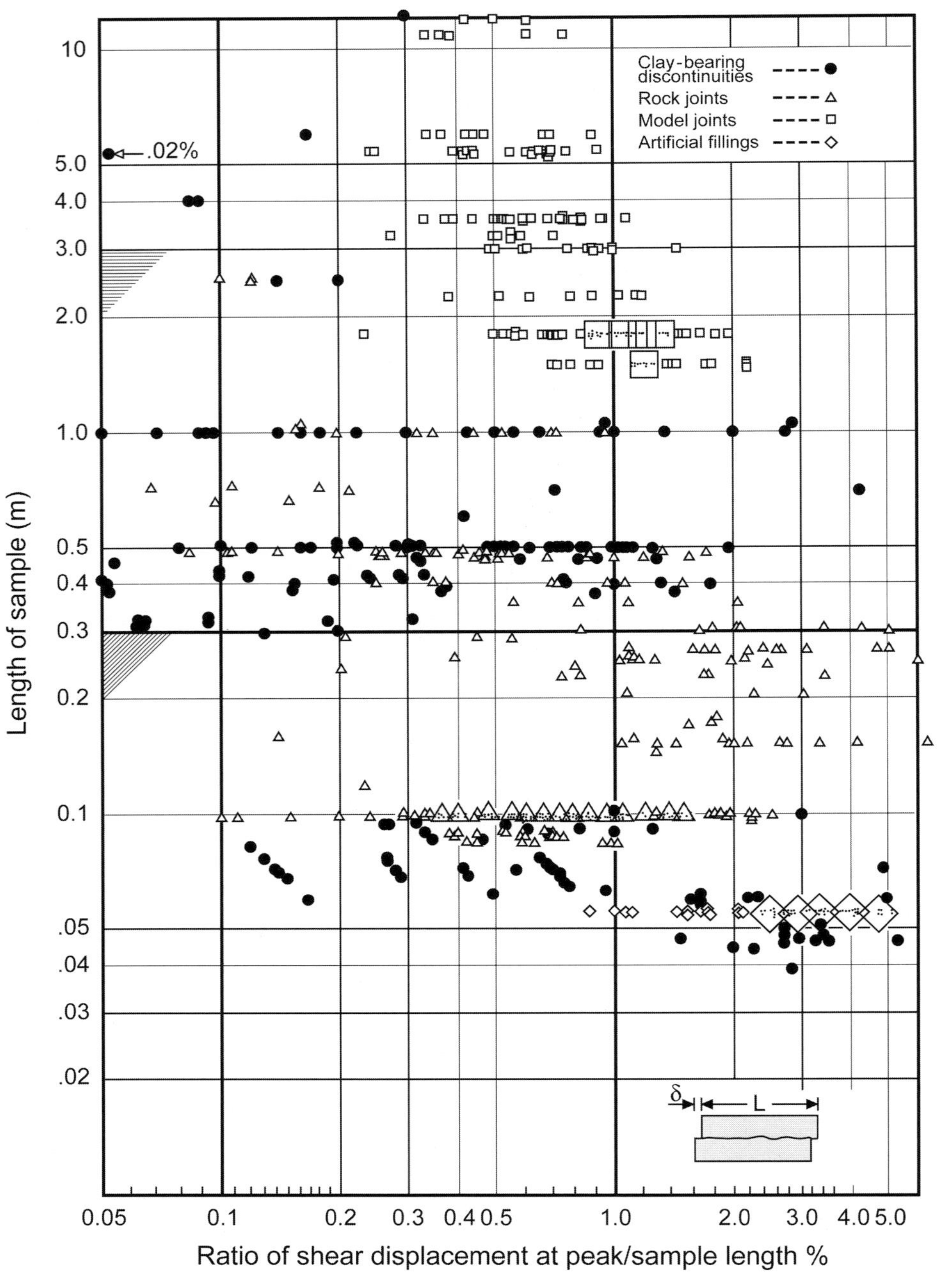


*Figure 16.67* Extensive test data review of displacement-to-peak ($\delta_{peak}$), from a wide variety of sample/block sizes, from laboratory and *in situ* direct shear tests. Barton, 1982.

As can be seen, the larger the modelled block size the 'later' this dilation and assumed permeability enhancement occurs.

In view of the need for combined shear stress and permeability, to qualify for registration in the mobilized-friction-with-flow diagrams (Figures 16.61 to 16.65), the joints need to be sheared, but they actually do not need to be sheared very far. If they are rough and undulating, too much dilation would prevent further shear, due to a massive build-up of normal stress.

When there is *faulting*, or hydrothermal alteration, with *clay-filled* or *clay-coated* discontinuities present in the immediate vicinity, the relevance of the parameters $JRC_n$ and $JCS_n$ in the Barton-Bandis joint constitutive equation appear, at least at first sight, to be limited. The reality may be no remaining roughness of note

($JRC_n \rightarrow 0$), but with the possibility of a residual *(i) value*, to account for the added shear resistance given by any remaining undulation, with effective inclination angle *(i)* in relation to the mean plane. This could be simply accounted for by applying the Patton, 1966 equation, possibly with a cohesion intercept (c) added for the case of (obviously cohesive) clay-fillings:

$$\tau = \sigma_n' \tan(\phi_r + i) + c \tag{16.17}$$

As was shown by a wide review of the shear strength of clay-filled discontinuities, the residual friction angles ($\phi_r$), and therefore overall frictional strength relevant to such features ranges, from below 10° to about 25°, depending on mineralogy and degree of over-consolidation (Barton, 1973b).

If one attempts to use the Barton-Choubey strength criterion: $\tau = \sigma_n \tan[JRC \log(JCS/\sigma_n) + \phi_r]$ to see if it fits clay-related experimental data, it will be noticed that when $\sigma_n$ (strictly $\sigma_n'$ for effective stress) exceeds JCS, (e.g. an over-consolidated fault gouge of 5 MPa strength with a local effective normal stress of 20 MPa), the log ($JCS/\sigma_n'$) term becomes negative. Contraction-with-shearing, in place of dilation is then actually predicted by the model, (as for normally consolidated clay, or heavily loaded over-consolidated clay), but not if JRC = 0. Clearly it would be relevant to also apply a suitably reduced value of $\phi_r$ relevant to clay-smeared surfaces, to obtain a meaningful fit to data. *Predictions* of fault strength with this 'rock joint and rock fracture' constitutive equation are of course of limited reliability, and there is little option besides testing or back-analysis.

In further relation to the interpreted mobilized friction coefficients of 0.6 to 1.0 shown in Figures 16.61 to 16.65, one may refer to direct shear test results on *tension fractures* developed in weak, brittle model materials, as reported by Barton and Hansteen, 1979. (These materials could be formulated to give various frictional and cohesive strengths, and consisted of fine sand, fine Ballotini glass spheres, $Pb_3O_4$ red lead, and small proportions of gypsum and water, with curing at high enough temperatures to cause disassociation of some of the water: $Ca.CO_3.2H_2O \rightarrow Ca.CO_3.\frac{1}{2}H_2O$ thereby reducing the compressive strength: Barton, 1971).

It may be noted from the normal stress magnitudes given in Figure 16.68 that there is a model/prototype (M/P) stress *and* strength scaling of 400, assuming a strong 175 MPa prototype rock. The length scale from Buckingham's $\pi$-theorem for dimensionless products,

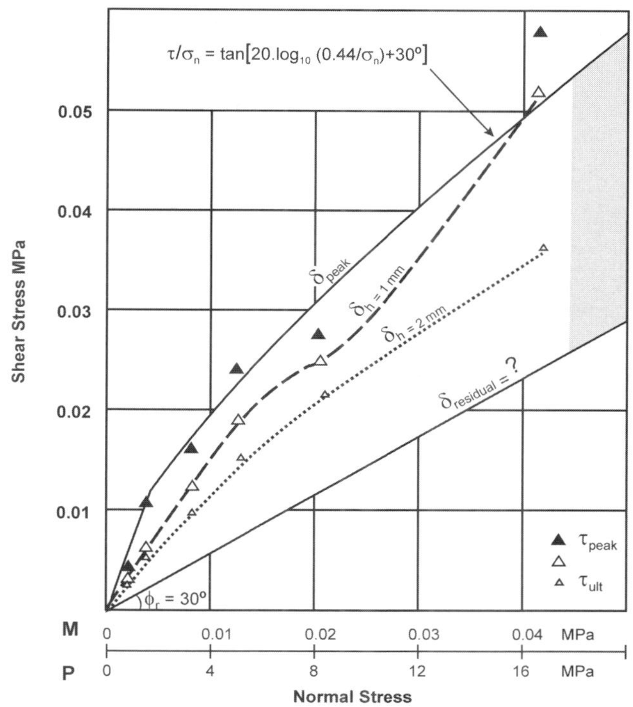


*Figure 16.68* Shear strength envelopes (peak, $\delta_h$ = 1%L, and $\delta_h$ = 2%L) and assumed residual strength for direct shear tests on model tension fractures of 100 mm length. Barton and Hansteen, 1979. The material was a brittle, very fine-grained, sandstone-like material. The stress and strength scale: $\psi$ = 400, and the length scale: $\lambda$ = 320, was due to lower model density of 2.0 gm/cm$^3$. Note that $\mu$ = 0.6 and $\mu$ = 1.0 correspond to the 'residual' strength envelope (0.58 = tan 30°) and to the mean of the $\delta_h$ = 1 mm and 2 mm strength envelopes, representing the shear strength remaining (at different stress levels), following a mean 0.45 m of shearing at full scale, for these simulated 30 m long fractures. However these samples were free to dilate, with constant normal stress applied.

is then $\lambda = \psi \times \gamma_m/\gamma_r$, where $\gamma_m$ and $\gamma_r$ are the respective densities of model material and rock. With respective values of 2.0 gm/cm$^3$ and 2.5 gm/cm$^3$ (and $\psi$ = 400), one can estimate a length scale $\lambda$ = 320. The samples therefore represent approximately 30 m long, rough undulating major fractures.

Due to this scaling, the given shear displacement magnitudes of 1 and 2 mm as recorded in the 100 mm long direct shear tests correspond to about 0.3 m and 0.6 m at full scale. The larger of these two shear displacements has reduced the shear strength of these rough, high strength simulated fractures ($JRC_o$ = 20,

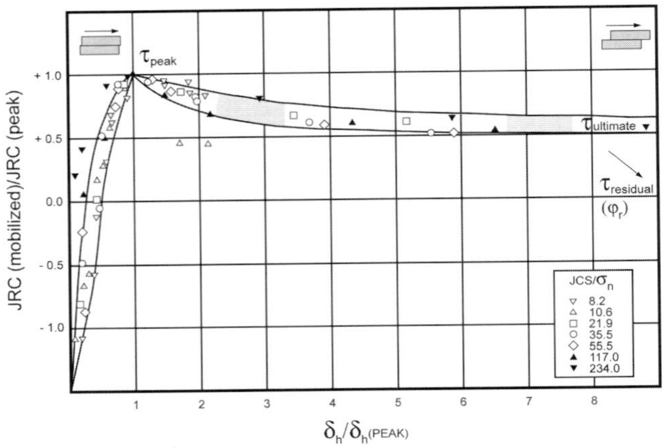


*Figure 16.69* a) A dimensionless representation of the complete shear stress-displacement behaviour of model tension fractures shown in conventional format, in Figure 16.70. The shear test data has been normalised using the $JRC_{mobilized}$ concept (Barton, 1980, 1982).

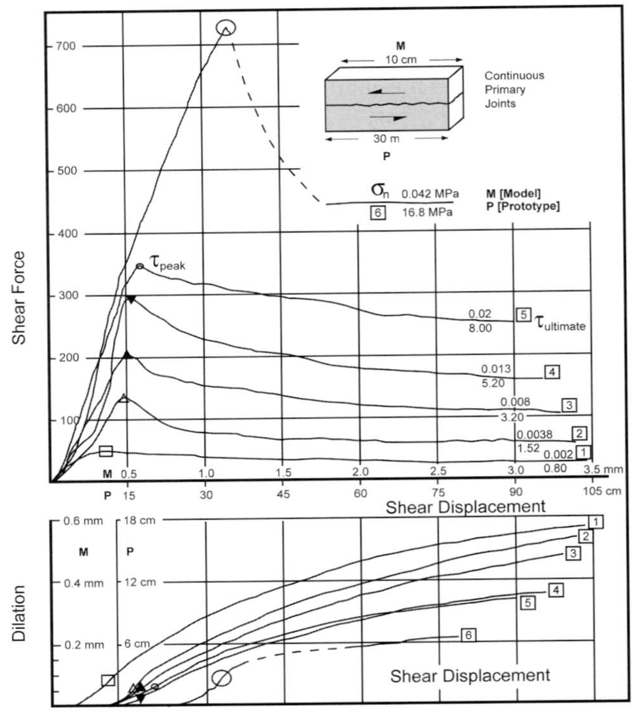


*Figure 16.70* Direct shear and dilation data for model tension fractures representing 30 m long, rough-undulating fractures in hard rock. This is the source of DST data for the dimensionless representation in Figure 16.69. Barton, 1993.

$JCS_o = 0.44 \times 400 = 176$ MPa) to a minor fault character of $\mu \approx 1.0$ at a simulated, medium normal stress of 8 MPa, and to $\mu = 0.8$ at a simulated maximum normal stress of 16 MPa, but each free to dilate.

The various data sets from Zoback and Townend, reproduced in Figures 16.64 and 16.65, are obviously composed of variable rock strengths and variable joint/fracture/minor fault roughnesses, since the different data belong, obviously, to a wide variety of rock types, joint characters, and minor fault types. These will inevitably have suffered various deformation magnitudes, controlled by a variety of non-linear shear strength envelopes, representing different degrees of pre-peak, close-to-peak and post-peak shearing, as suggested by the model tension fracture envelopes, shown in Figure 16.68.

In relation to post-peak shearing, it is helpful to invoke the $JRC_{mobilized}$ concept again. (See Figure 16.10). Figure 16.69 shows a dimensionless set of shear strength displacement data, representing the whole range of behaviour shown in conventional stress-displacement format in Figure 16.70. $JRC_{mob}$ can be estimated at any desired location along shear stress-displacement curves, by evaluating the mobilized shear strength ($\tau_{mob}$) at the point of interest. It is estimated by a simple re-arrangement of the Barton-Choubey peak shear strength equation:

$$JRC_{mob} = \frac{\tan^{-1}\left(\dfrac{\tau_{mob}}{\sigma_n}\right) - \phi_r}{\log \dfrac{JCS}{\sigma_n'}} \tag{16.18}$$

It will be noted that the 'ultimate' strength, after some 1 m of simulated shearing (or 3% of the length of the fractures) is still well above the true residual behaviour, as indicated in Figure 16.69. This is also found when testing rock joints. It is not possible to reach true residual strength in the normal confines of a (linear) direct shear device. It is likely that Zoback and co-workers' frictional interpretation of deep borehole 'critical crust' behaviour is showing $\mu = 0.4$ to 1.0 coefficients due not only to different $JRC_n/JCS_n/\phi_r$ and $\sigma_n'/JCS_n$ ratios, but also due to different degrees of shearing, both pre- and post-peak.

## 16.10 Rotation of joint attributes and unequal conjugate jointing may explain azimuthal deviation of S-wave polarization

We have previously investigated hypothetical joints in typical reservoir 'sandstone' and reservoir 'carbonate'

(JCS = 25 MPa and 50 MPa), and seen the dramatic closure of aperture that was predicted to occur if these joints were consolidated/closed by 10 or 20 MPa or higher effective normal stress, representing an assumed range of (effective) $\sigma_h$ min, acting on sub-vertical joints or fractures in the reservoir.

Although down to a certain depth, an increased $JCS_n$ could be assumed, by utilising a confined compressive strength $(\sigma_1 - \sigma_3)_n$ for the joint walls, it has been demonstrated with triaxial test data that this increase is limited in the case of weaker porous rocks. It may not be sufficient to explain 'normal-closure' apertures more than a few microns, according to modelling predictions. This is especially so if roughness is limited.

With the advent of improved digital data acquisition, and strong interest in the use of shear waves for interpreting reservoir fracturing, it has become increasingly important to correctly interpret what a certain level of anisotropy and attenuation at a certain frequency means, when analysing the fast and slow shear waves $qS_1$ and $qS_2$. How far the industry has come in accepting the need for a shearing mechanism interpretation on some occasions, or a non-aligned set of fractures, or that *two intersecting fracture* sets are actually the source of the measured components, is difficult to judge, even from most recent publications.

We should probably consider five possibilities:

- $\sigma_{H\,max}$-aligned 'open' joints due to sufficient wall strength and roughness
- $\sigma_{H\,max}$-aligned 'open' joints due to hard-mineral 'bridging-but-not-blocking'
- 'open' joints at some angular deviation from the $\sigma_{H\,max}$ direction
- apparently '$\sigma_{H\,max}$-aligned-open-joint-set' that actually is two conjugate sets
- a generally non-planar joint wall topography in all cases

The third and fourth of these cases will be addressed here. In the handful of cases reviewed in earlier chapters, the anisotropic P-waves, or the polarized, split qS-waves do indeed orient themselves in *approximately* the 'correct' (conventional) direction in relation to the calliper-logged-perpendicular-to-break-out assumption for the $\sigma_{H\,max}$ orientation. However, we noted, e.g. from Lynn et al., 1999, and Maultzsch et al., 2003 data, that the line-up of assumed fracturing and assumed major stress directions was not 'perfect'. In fact there could be 10° to 20° discrepancies, if one can rely on the data acquisition accuracy of each set of measurements. There are many possible sources of error (Lynn, 2005, pers. comm.), so this is uncertain.

The logic and data of Zoback and Colleen Barton and Townend and their colleagues regarding the probability of shearing is surely incontrovertible. Although we perceived certain difficulties with $\mu$ as high as 1.0 at several kilometres depth, the 'critical-shearing crust' phenomenon was easy to verify on a *local scale*, using for instance non-linear shear strength and Barton-Bandis coupled behaviour modelling.

Although presented by these authors in the context of mostly multiple-kilometre deep boreholes in predominantly *crystalline* rock, it is the opinion of the writer that it is the petroleum industry that will have the greatest benefit of recognising the possibility (and sometimes the necessity), of a shearing mechanism, often conjugate in form. At Ekofisk in the North Sea, a model-predicted, and subsequently core-sampled confirmation of a conjugate shearing mechanism (causing slickensided joints), was probably fundamental for the two sets of 60° dipping joints to be able to continue to drain the porous, low permeability chalk matrix. (Nowadays there is an additional component of compaction drive, due to water weakening of the matrix and joints)

There would probably have been no development at Ekofisk if there had been just one set of joints parallel to $\sigma_{H\,max}$. They would have been closed before discovery (i.e. therefore never discovered), and 'further closed' by a 20 to 24 MPa increase in effective normal stress, (prior to water flooding), if an obstinate owner had decided for development. Of course other solutions (MHF and proppants) would perhaps have been found effective.

So the details of fracturing, in terms of *number* of joint or fracture sets, may be important for production, and therefore ideally should be detectable by seismic interpretation. In addition, there is the simultaneous importance of the *non-planarity* of most joints and fractures. At one extreme, one has the cleavage joints in slates that give 'billiard-table' planarity and smoothness, and clearly have no place in explaining petroleum production. They would 'fail to produce' on two counts: their apertures would be too minute when acted on by reservoir levels of effective *normal* stress, and in a 'shearing earth' scenario, where they would have difficulty tolerating $\mu = 0.4$ when wet, they would nevertheless not dilate when sheared. So permeability would be unlikely on both counts. At the other extreme there could be marked non-planarity of the one or two key joint sets, as illustrated in the next figures.

Intermediate between these extremes of planarity or non-planarity are all the other numerous cases, which are vitally necessary for petroleum production because a *compromise between the extremes* is needed for productivity, if shearing is to be possible historically. There will be no shearing (or only pre-peak shearing), if non-planarity is too great, as suggested earlier when reviewing the two sets of $JRC_n$ profiles.

Figure 16.71 shows several features that can explain an interpreted 'open fracture set' orientation that differs somewhat from the calliper assumed direction. The undulating (non-planar) nature of conjugate joints that have become slightly sheared, or even conjugate *minor* faults, through successive shearing in response to $\sigma_H - \sigma_h$ (or $S_1 - S_3$) and perhaps fluctuating pore pressure, has been exaggerated for clarity.

As has been noted previously in both Chapters 15 and 16, from the reconstructed shearing of model fractures by Barton, 1973a, there is in Figure 16.71 an *opposite* 'rotation' of the fluid filled parts of the joints/minor faults, shown with open O-symbol, in relation to the rock-to-rock shear and normal stress transfer sections of the asperities, which lie on the opposite 'slopes', and are denoted by the R-symbol.

Open (O) and contacting (R) sectors, lie on either side of the assumed mean plane of the joint (J). Note that shearing of one of the conjugate sets has been assumed to take 'priority', thereby displacing the secondary set, which may also shear, but to a lesser degree. This is a frequent and almost inevitable consequence of 'conjugate shear'.

Considering the potential for shear wave splitting; if we can make the assumption that the average orientation of the fluid-filled sections (O) of the joints or minor faults have a *different* influence on the splitting/polarization mechanism of shear waves, than the average orientations of the rock-to-rock sections of the joints (R), then we have immediately come 10° or more, further away from the $\sigma_H$ (or $S_1$) direction. The 'contrary' fluid rotations can be visualised from the 'frozen' evidence of quartz injections into a *shear stress field*, causing actual shearing, as illustrated in Figure 16.72.

As indicated in the extensive caption to Figure. 16.71, the influence of a second set of potentially shearing fractures/minor faults may be of importance in the eventual polarization directions of the split qS-waves. If the dominant joint or fracture set physically displaces the secondary set (as noted in the UDEC-BB modelling of Ekofisk joints, Barton et al., 1986), then we have the possible influence of higher shear resistance and therefore higher shear stiffness for the secondary

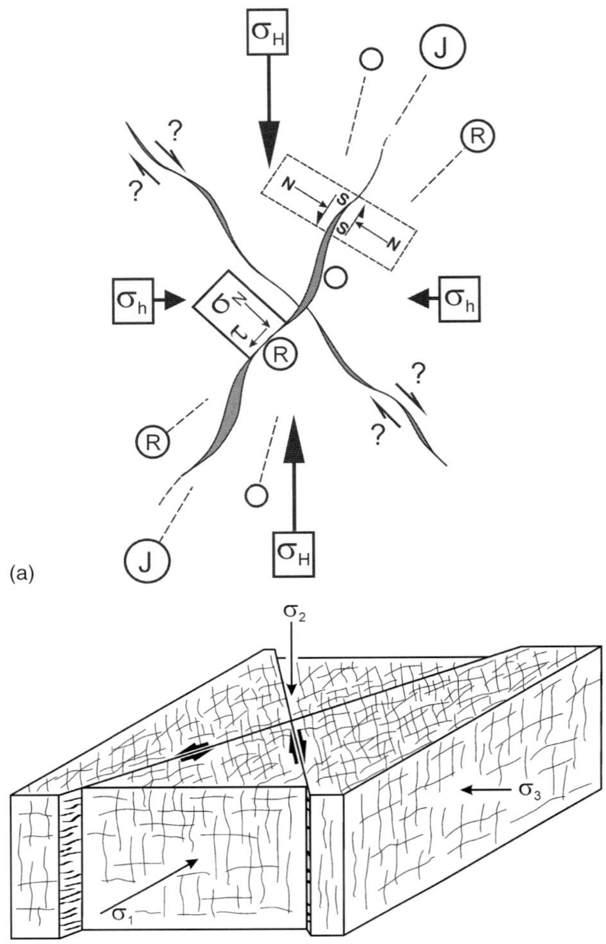


*Figure 16.71* Idealised and exaggerated joint/fracture/minor fault undulations, for a conjugate pair of 'joints', with one dominant and displacing the secondary set. Note contrary 'rotations' of 'R' and 'O' in relation to mean plane 'J'. Shear wave splitting is assumed to occur *with different strength* from the slip-prone 'R' sections, with their finite shear and normal stiffnesses (and $K_n \gg K_s$), compared to that from the more fluid-bearing 'O' direction. (Theoretical opinion is sought here!) Reflection may occur more strongly from the more fluid-bearing 'O' direction. Rotation of the polarization of the $qS_1$ and $qS_2$ waves *towards* the $\sigma_{H\,max}$ direction would require the partial help of an opposite component of polarization from the secondary set of joints, which might be less conducting and thereby have reduced lengths of fluid lenses. It may be assumed that the pseudo-static shear stiffness $K_{S2}$ of the minor, displaced set would be increased in relation to $K_{S1}$ by the 'off-set' mechanism. This results in an actual cohesive intercept in the sense of shear resistance of the secondary set, further stimulating shear in the dominant set.

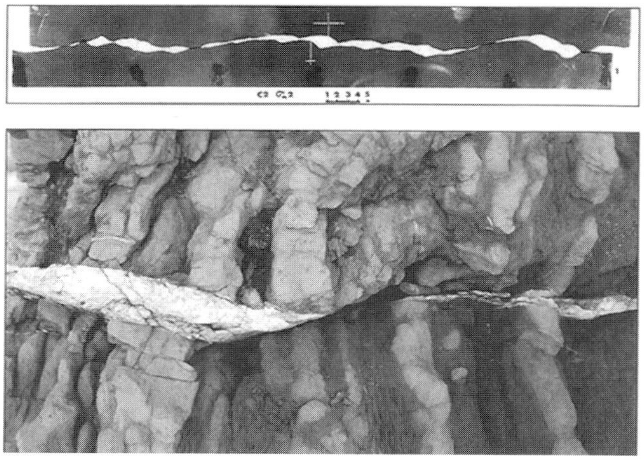

*Figure 16.72* a) Reconstructed shearing of a rough fracture at low stress, showing the effective contrary rotation of fluid-filled and rock-to-rock sections. b) A 'frozen' quartz filling of a minor fault showing the same contrary rotation of 'fluid lenses' and rock-to-rock contacts. Høvik, Norway.

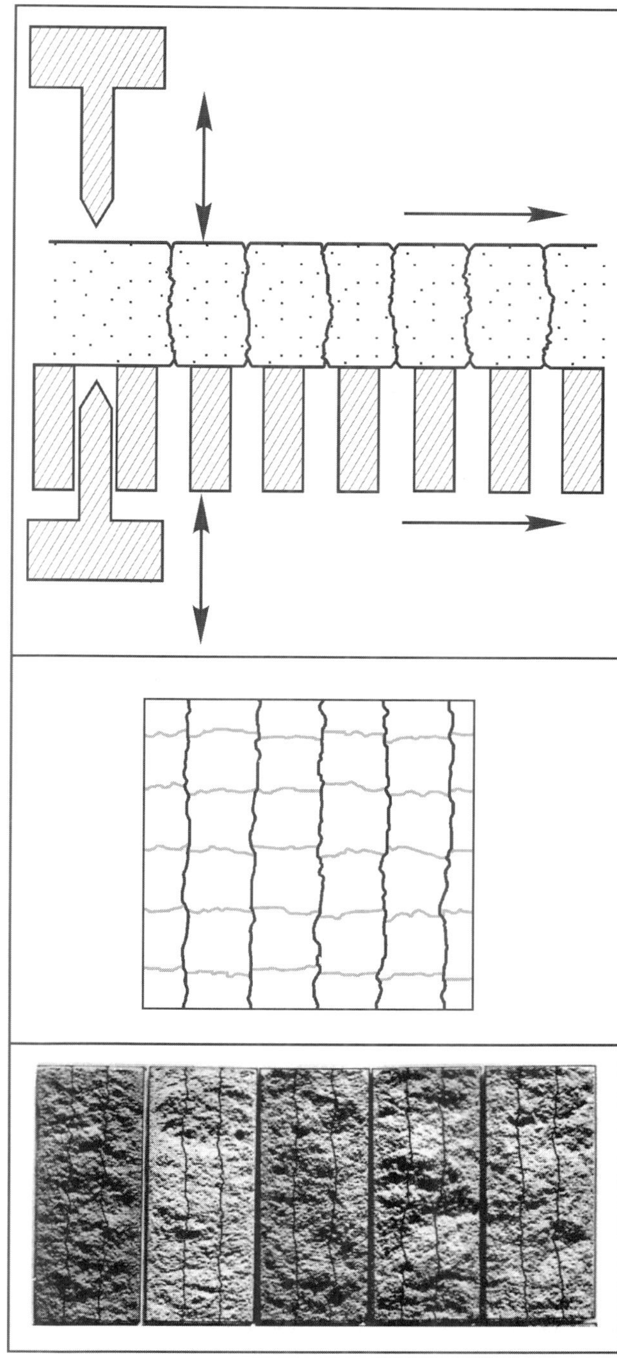

*Figure 16.73* a) Double-bladed guillotine for developing parallel sets of tension fractures, and for creating intersecting sets by azimuthal rotation of the whole sample. b) Continuous joints of set no. 1, and stepped/offset joints of set no. 2, as often observed in nature. c) Roughness profiles of the rough model tension fractures, as measured by photogrammetry. Barton, 1972a.

set. Presumably this could influence (reduce) shear compliance in the secondary direction to some degree, even though compliance is a dynamic micro-displacement parameter.

Figure 16.73 shows a double-bladed guillotine method for creating primary parallel fracture sets, and intersecting sets of secondary parallel fractures in 25 mm thick slabs of weak brittle model materials. These types of interlocking 'fractured' models were used in pseudo-static biaxial shear testing, as illustrated previously in Figure 16.42. They were also used to create 'two-dimensional' physical models of steep, 'jointed' rock slopes (Barton, 1971) with up to 40,000 blocks, and also to physically model the excavation of large span, 'near-surface' caverns, in various anisotropic stress fields, and with various intersecting 'jointing' patterns (Barton and Hansteen, 1979). In each case the progressive deformation fields were monitored by photogrammetry.

Figure 16.74 shows the results of direct shear tests of three classes of model joint that have parallels in nature. Peak and ultimate (larger deformation) strength envelopes are shown for a) primary (set no. 1), b) primary cross-cut set no. 1 (*PCJ*), and c) secondary (stepped, castellated) set No. 2.

Clearly the secondary set has a true *cohesion* intercept, and in a real situation at reservoir depth, would lie somewhere between the shear strength envelopes for the intact material and the dotted 'J' curve in the Mohr circle-based strength envelope diagram (Figure 16.58).

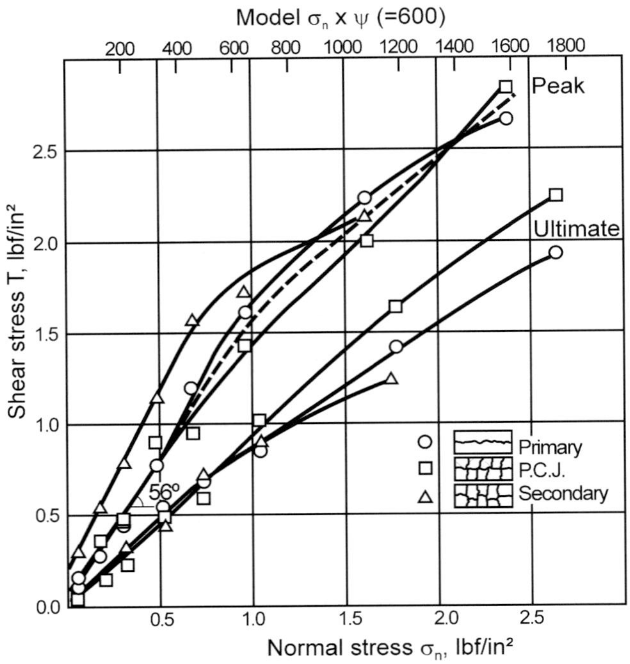


*Figure 16.74* Direct shear test strength envelopes for primary, primary but cross-jointed and secondary joints, indicate that there will be higher shear stiffnesses, and therefore potentially lower compliance for the less dominant directions. Barton, 1972a.

The offset joints have shear strengths that are 1.5 to 3 times higher than the primary joints, with greatest difference at lowest stress.

This *offset-by-shear* mechanism will obviously cause the shear stiffness of this secondary set to be higher than that of the primary set, and perhaps cause the dynamic *compliance* of the secondary set to be lower than that of the primary set. In the case of conjugate sets of vertical reservoir jointing, this could be another potential source of azimuthal deviation of the shear-wave polarization, *away* from the likely intersecting $\sigma_{H\,max}$ direction.

The modelled (30 m long prototype) *primary* joints gave measured ratios of $K_n/K_s$ at prototype, scaled-up stress levels as follows (Barton, 1972a).

| $\sigma_n$ | 3 | 7 | 11 | MPa |
|---|---|---|---|---|
| $K_n/K_s$ | 250 | 75 | 73 | |

At that time, in the late sixties, there was not the incentive to further investigate stiffnesses, apart from the interesting problem that large anisotropy (or inequality) of these two fundamentally different (pseudo-static) stiffnesses reportedly caused *numerical stability* problems in the FEM joint element modelling (Goodman et al., 1968) that preceded the modelling of jointing in UDEC and UDEC-BB, that started with the developments of Cundall, 1971.

## 16.11 Classic stress transformation equations ignore the non-coaxiality of stress and displacement

In Figure 16.71a, there is discrepancy between the applied shear stress orientation (S) parallel to the mean plane (J), compared to the subsequent direction of shear displacement when roughness is mobilized, and a *non-co-axial* dilation begins. The resolved shear and effective normal stresses ($\tau$ and $\sigma_n'$) acting on the rock-to-rock contacting parts (R) of the non-planar dominant joint, emphasise this non-coaxiality. The non-planarity of most joints, perhaps many minor faults, (and sections of major faults), actually causes non-coaxial stress-and-displacement (i.e. violation of the coaxial stress and strain principle of St. Venant). This means that the 'global' application of the Mohr Coulomb equation and following classic stress transformation equations is actually erroneous to some varying degree.

$$\tau = \frac{1}{2}(\sigma_1 - \sigma_3)\sin 2\beta \qquad (16.19)$$

$$\sigma_n = \frac{1}{2}(\sigma_1 + \sigma_3) - \frac{1}{2}(\sigma_1 - \sigma_3)\cos 2\beta \qquad (16.20)$$

($\beta$ is the acute angle between $\sigma_1$ and the plane in question. Note that equation 16.20 is sometimes quoted erroneously in literature, with a (+) sign between the two parts of this equation. A joint at 30° to the $\sigma_1$ direction *cannot* have $\sigma_n > \tau$ as implied in this particular case: clearly $\tau > \sigma_n$).

In theory the plane should be imaginary (thereby not rotating the local principal stress directions), it should also not shear, and it should certainly not dilate, causing another error.

The problem with the classic stress transformation equations was experienced first hand by Bakhtar and Barton, 1984, who were attempting to biaxially shear 1.0–1.1 m long diagonal fractures developed in 1 $m^3$ blocks in a large test frame at TerraTek, Salt Lake City, as referred to earlier in this chapter. (See large-scale roughness profiles in Figure 16.27). In fact the same problem

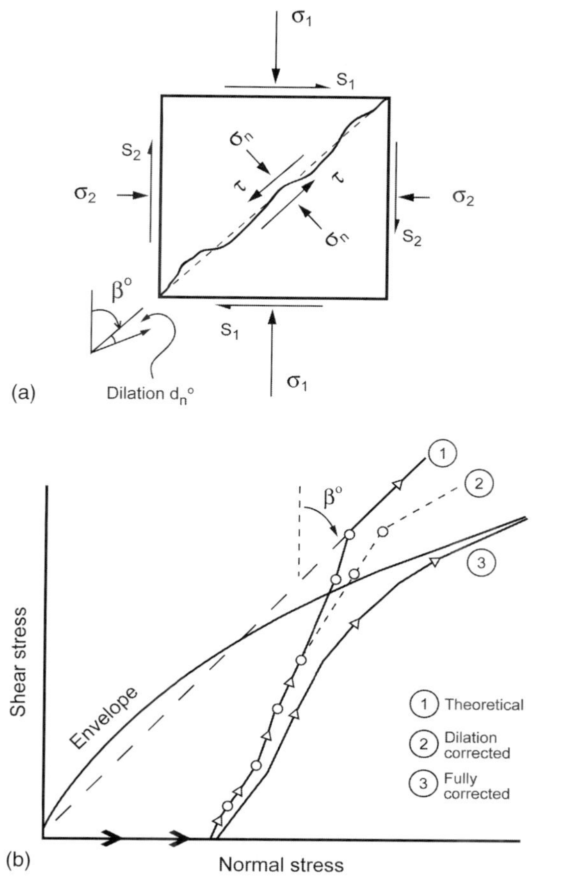


*Figure 16.75* a) Large scale biaxial shear tests of 1 $m^3$ blocks with diagonal, well characterized joints or fractures. Note non co-axial 'stress and strain' due to dilation angle $d_n$. b) Calculated, scale-adjusted shear strength envelope and 1) *theoretical,* 2) *dilation corrected* and 3) *fully corrected* loading paths that explain the difficulty of shearing due to a break-down of the classic stress transformation equations, when not modified to account for the mobilized dilation angle. Bakhtar and Barton, 1984; Barton, 1986; Barton, 1999a.

was experienced earlier, but not recognised, in the *in situ* block test depicted in Figure 16.29. The limited (<0.25 mm) shear on this occasion was naturally assumed to be due to the 'attached' base of the block.

Ten 1 $m^3$ samples with varying degrees of fracture roughness had been prepared and thoroughly characterized, including the performance of large scale tilt-testing, following Barton and Choubey, 1977. The experimental setup is schematised in Figure 16.75a. Great care was taken in minimising side friction ($S_1$ and $S_2$ symbols in the figure), using double flatjacks and a thin Teflon sandwich separated by a fluid film of molybdenum disulphide grease.

The discrepancy between the carefully estimated full scale strength envelope, and the assumed loading path (no. 1) is shown in Figure 16.75b. Correction for side-friction and use of modified stress transformation equations, finally explained the 'inexplicable' high stresses needed to achieve shearing. In fact most of the ten samples could not be significantly sheared, due to the classic neglect of the problem of dilation. The following equations were needed to explain these experimental difficulties:

$$\tau = \frac{1}{2}(\sigma_1 - \sigma_3)\sin 2(\beta + d_{n\ mob}) \qquad (16.21)$$

$$\sigma_n = \frac{1}{2}(\sigma_1 + \sigma_3) - \frac{1}{2}(\sigma_1 - \sigma_3)\cos 2(\beta + d_{n\ mob}) \qquad (16.22)$$

$$\text{where } d_{n\ mob} = \frac{1}{2}\text{JRC}_{mob}\log\left(\frac{\text{JCS}_n}{\sigma_n'}\right) \qquad (16.23)$$

(The mobilized roughness concept of Barton, 1982 was illustrated in Figure 16.10. The conventional equation for *peak* dilation angle, uses $JRC_o$ or $JRC_n$ and not $JRC_{mob}$.)

***Example:*** Close to shear failure of the diagonally fractured block (Fig. 16.75):

1) $\sigma_1$ had been *increased* to 28 MPa, $\sigma_2$ was *reduced* to 0 MPa
2) $\beta = 45°$
3) Assume $d_{n\ mobilized} = 10°$

With the 'classic' assumption of *no dilation*, equations 16.19 and 16.20 give: $\tau = \sigma_n = 14.0$ MPa. (since sin 90° = 1.0, cos 90° = 0). Shear failure of the fracture had been expected/predicted even before $\mu = 1$ was reached, as assumed here.

With modified 'dilating' equations 16.21 and 16.22, the sin 110° and cos 110° terms (=0.939, −0.342) indicate the actual difficulty in shearing, since $\tau$ = 13.16 MPa (lower than assumed), and $\sigma_n$ = 18.79 MPa (higher than assumed), dilation causing an effect similar to an increasing angle $\beta$.

The new calculation of the *actually applied* ratio of shear and normal stress (with an assumed mobilized dilation angle of 10°, was *too low.* With *applied* $\mu$ only = 13.16/18.79 = 0.70, the strength of the joint was too high for shearing more than a fraction of a millimetre, i.e. pre-peak. Both these changes ($\tau\downarrow$, $\sigma_n\uparrow$) in the 'actual' compared to the 'assumed' *boundary stresses* are also illustrated by the contrasting stress paths shown in Figure 16.75b. They help to explain the dangerous burst of one of the real-life flat-jacks at 28 MPa pressure, nearly causing injury, and displacing pictures around the laboratory walls due to the high pressure oil burst. Only two of the surfaces illustrated in Figure 16.27 could be sheared easily.

Since these experimental difficulties, further descriptions and analysis of the problem have been given by Barton, 1986, and Barton, 1999a, including convincing evidence from the analysis of a large series of *biaxial* hydraulic-jack based shear box tests, where the applied loads $S_1$ and $S_2$ had to be resolved into the assumed components $\tau$ and $\sigma_n$ acting on the 45°/45° oriented joint planes in the centre of the shear box. (This is not required in a standard shear box, with perpendicular and parallel application of forces in the principal planes).

The main author of these biaxial shear box tests was convinced that the JRC-JCS criterion of Barton and Choubey, 1977 needed to be modified. In fact the consistent 'error' in relation to a 1:1 gradient between 'predicted' and 'measured' (as compared to an almost perfect 1:1 relation in standard shear boxes), was caused by unintended neglect of a dilation adjustment in the classic transformation equations, as described above.

On the basis of the above, Barton, 1999a argued that 'classic' stress transformation in all potentially dilating geotechnical materials (e.g. rock joints, rock-fill, dense sand, over-consolidated clay), may need to be considered as violation of the St. Venant principle of co-axial stress and strain.

One may speculate that the earlier referred Cajon Pass analysis of Zoback and Townend, indicating in this case, a tolerance of $\mu = 1$, could also be subject to the possible need of a mobilized dilation correction, since this research borehole is located in the tectonically active San Andreas fault area, with the possibility of a more tectonized 'broken' rock mass with smaller block-sizes and perhaps rougher joints. As demonstrated earlier in this chapter, it was extremely difficult to justify such high tolerance of shear stress, using all known means of maximizing shear strength.

The $d_{n\ mobilized}$ equation 16.23, means that one must consider roughness (and therefore block-size), *and* the magnitude of shear displacement. This is because the magnitude of $JRC_{mobilized}$ depends numerically, upon the degree of shear displacement. So the mobilized dilation angle will depend on the present displacement, in relation to the initial, usually interlocked condition of the joint.

A long way beyond the $\delta_{peak}$ singularity (where the strength and dilation angle are maxima), there will be little need of a correction for dilation. So for mature sections of faults one can apply the standard transformation with almost no error, whereas in newer regions of faulting, a roughness/dilation estimate may well be required for *more correct* interpretation. Research in this area, starting in the laboratory, is urgently needed.

## 16.12 Estimating shallow crustal permeability from a modified rock quality Q-water

Townend and Zoback, 2000, and Zoback and Townend, 2001, argued convincingly for acceptance of a 'critical crust' concept, with *hydrostatic* rather than lithostatic-related pore pressure throughout the brittle depth of the crust. Their analyses of deep well data were reviewed earlier in connection with the importance of the mobilized ratio of shear and normal effective stress, as a strong indicator of water-conducting fractures.

From the $JRC_{mobilized}$ concept we argued that permeability enhancement could even be with *pre-peak* shear displacements of just millimetre size. However the sometimes high ratios of *resisted* $\mu$ ($= \tau/\sigma_n'$) of up to 1.0 obviously suggests a *close-to-peak* condition, since we showed that it is difficult to construct such high *in situ* full-scale shear strengths. *Post-peak* seems unlikely in the case of the *highest* values of resisted $\mu$, while the values below 0.6 to 0.7, implying mobilized friction angles less than 31°–35°, could well be representing post-peak strengths, or minor faulting. Clearly even lower values of $\mu$ of 0.4 to 0.5 (i.e. mobilized friction angles of only 22° to 27°), as seen in Figures 16.62 to 16.65, have to be minor faulting, but without the 'complication' of sealing gouge, since they were also recorded as conducting features, by the relevant authors.

Zoback and Townend also assembled the deep borehole permeability measurements that are reproduced in Figure 16.76. In the case of the KTB hole in Germany, permeability test intervals varied between a few tens of metres to 3.5 km, with bulk permeabilities between about $5 \times 10^{-15}$ to $10^{-17}$ m$^2$, or *approximately* $5 \times 10^{-8}$ to $10^{-10}$ m/s if the water was close to 20°C.

In a shallower series of tests (<1 km depth) at the Monticello reservoir, Zoback and Hickman, 1982, (with data also shown in Figure 16.76), reported permeabilities ranging from $10^{-15}$ to $10^{-16}$ m$^2$, or $10^{-8}$ to $10^{-9}$ m/s. Thermal models of borehole temperature suggested

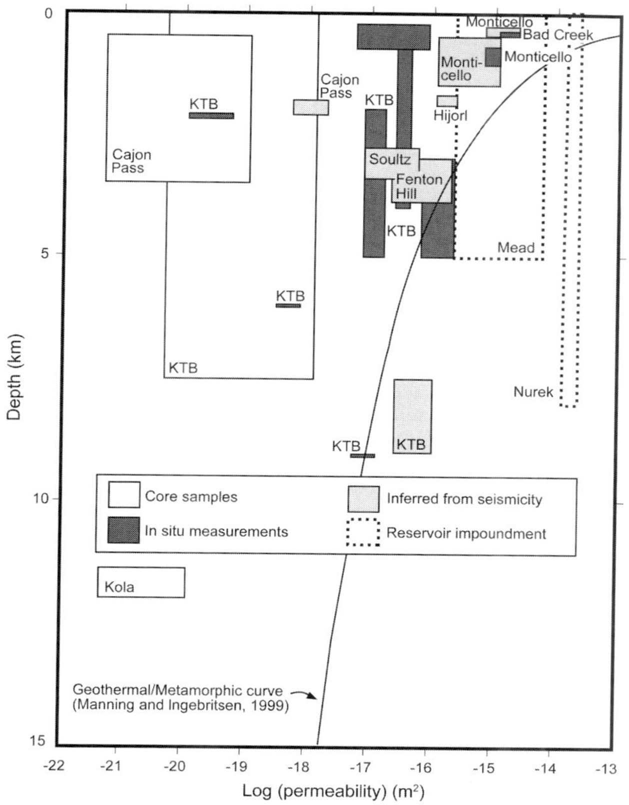


*Figure 16.76* Zoback and Townend, 2001, assembly of deep crustal permeability from well-known deep borehole projects such as Cajon Pass, KTB, Kola. Intact samples (with K< $10^{-18}m^2$) are given on the left. See original paper for references to other projects.

$10^{-14}$ ($10^{-7}$ m/s) for the 0 to 2 km depth interval at Kola, and $10^{-17}m^2$ ($10^{-10}$ m/s) for the 6 to 8 km depth interval. The above, and similar data, led Zoback and Townend, 2001, to conclude that the upper crust had permeabilities of $\approx 10^{-17}m^2$ ($10^{-10}$ m/s) to $10^{-16}m^2$ ($10^{-9}$ m/s) over 1 km and 10 km (depth) scales.

### 16.12.1 The problem of clay-sealed discontinuities

With this interesting evidence for the maintenance of some significant permeability to great depth, presumably related with even deeper operation of the 'critical shearing crust' mechanism, it was of interest to know if there was a possibility that a rock mass description scheme like the Q-system, could 'explain' such diverse ranges of permeability, and the partial maintenance to greater depth.

As we have seen earlier, nearer the surface, the degree of potential *connectivity* in a rock mass may be reflected in the *relative block-size* ratio of RQD/$J_n$, because of the great importance of the number of joint sets $J_n$. However when clay was responsible for low RQD values, this relation would likely be compromised. One began to speculate whether the second pair of shear-strength related parameters, $J_r/J_a$ could also have a role in 'explaining' relative magnitudes of permeability for individual joint sets?

We will start by describing the actual *limitation* of what was proposed in Barton, 2002a, where an extremely simple approximation between rock quality and the Lugeon value was proposed for central 'jointed rock' qualities in the integrated Q-diagram, reproduced here for easier reference, as Figure 16.77.

$$L \approx 1/Q_c \qquad (16.24)$$

where L = Lugeon (1 Lugeon $\approx 10^{-7}$ m/s $\approx 10^{-14}m^2$ at 20°C), and where $Q_c = Q \times \sigma_c/100$, where $\sigma_c$ = uniaxial compression strength of the rock.

As preliminary examples we see K $\approx 10^{-14}m^2$ ($\approx 10^{-7}$ m/s) when $Q_c = 1$ (a quite heavily jointed rock mass, e.g. 3 to 4 joints sets, near-surface), and we see K $\approx 10^{-16}m^2$ ($\approx 10^{-9}$ m/s) when $Q_c = 100$ (sparsely jointed, quite massive rock, typical at hundreds of metres depth and beyond). So far the approximation is reasonable, but as hinted earlier in this book, the relation $L \approx 1/Q_c$ inevitably breaks down when there is clay-filling along the joints or fault zones.

As may be noted from Figures 16.76 and 16.77, the above deep well data (often $\approx 10^{-9}$ to $10^{-10}$ m/s, or 0.01–0.001 L), lies in a partly 'consistent' position on the Q (or $Q_c$) – $V_p$ – L chart concerning the high quality end of the diagram, showing 'hard jointed' and 'hard massive' black-and-white curves. However, extrapolation of the diagram to much greater depth would be needed, by extending the '1000 m' depth diagonal upwards, into higher velocity and higher modulus territory.

Recent compilations of permeability data from the Äspö site in Sweden, which is a quite well-jointed site in diorite and granites, shows a range of K = $10^{-4}$ to $10^{-10}$ m/s based on analysis of numerous tunnel probe holes from 50 to 400 m depth. There was also a recorded 260:1 permeability anisotropy around the spiral access tunnel (Vidstrand, 2003). An interesting scale effects analysis by the same author showed K varying from about $10^{-5}$ m/s to $10^{-8}$ m/s at the 100 m scale, and K varying from about $10^{-4}$ m/s to $10^{-10}$ m/s at the 10 m

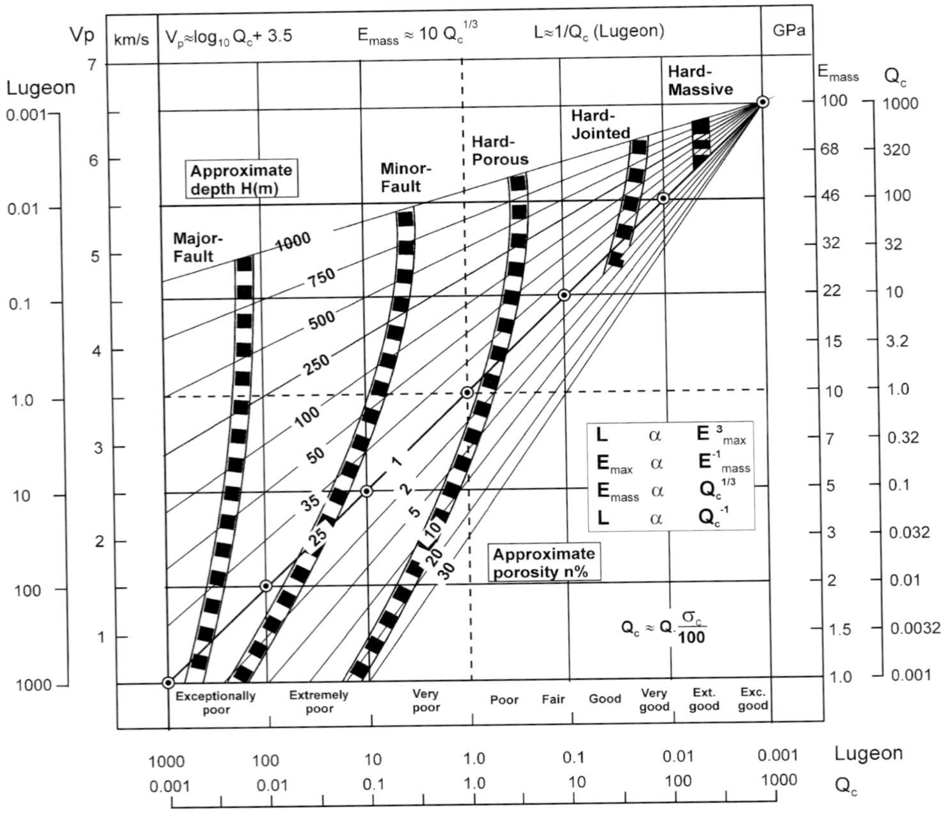


*Figure 16.77* The jointed rock mass quality Q-diagram, showing potential integration of the UCS-normalised $Q_c$ value, $V_p$, $E_{mass}$ and Lugeon value. (1 Lugeon $\approx 10^{-7}$ m/s). (See Chapter 9).

scale, Below the 1 m scale, values reduced to intact matrix values of $10^{-12}$ m/s as a minimum.

The above six orders of magnitude corresponds to the full range of permeabilities shown in Figure 16.77. The simple (too simple) correlations between rock quality, velocity and permeability, presently represented in Figure 16.77, are reproduced in Table 16.17.

In Figure 16.78 two sets of equations are shown in the table above the figure. The uppermost set of equations correspond to the $L \approx 1/Q_c$ inverse relation discussed above, and show the standard 'definition' of rock mass quality Q, here normalised by the UCS of the rock. The lower part of the table shows recent work by the writer for rationalizing the permeability – Q-value linkage, which has inconsistencies where clay-filled joints or discontinuities are concerned. One may note from the following:

$$Q_{water} = \frac{RQD}{J_n} \times \frac{J_a}{J_r} \times \frac{J_w}{SRF} \tag{16.25}$$

Table 16.17 Some suggested inter-relations between rock mass quality, P-wave seismic velocity, and permeability. Note that 'faulted rock' qualities, with $Q < 0.1$, implying clay-filled (and probably sealed) discontinuities, have been excluded.

| | Extremely/ very poor | Poor | Good | Very good | |
|---|---|---|---|---|---|
| $Q_c$ | 0.1 | 1 | 10 | 100 | |
| Lugeon | 10 | 1 | 0.1 | 0.01 | |
| K (m/s) | $10^{-6}$ | $10^{-7}$ | $10^{-8}$ | $10^{-9}$ | (approx.) |
| $V_p$ | 2.5 | 3.5 | 4.5 | 5.5 | (km/s) |

that the friction-based $J_r/J_a$ ratio has been *inverted* to $J_a/J_r$ to allow improved consistency where clay-filling is concerned. With clay-filling on dominant jointing, $Q_{water}$ now increases, and the inverse permeability reduces. A further step to enable general stress or depth dependent permeability to be modelled, is normalization by 100/JCS; i.e. when the joint wall compression strength is less than 100 MPa, there will be an increase

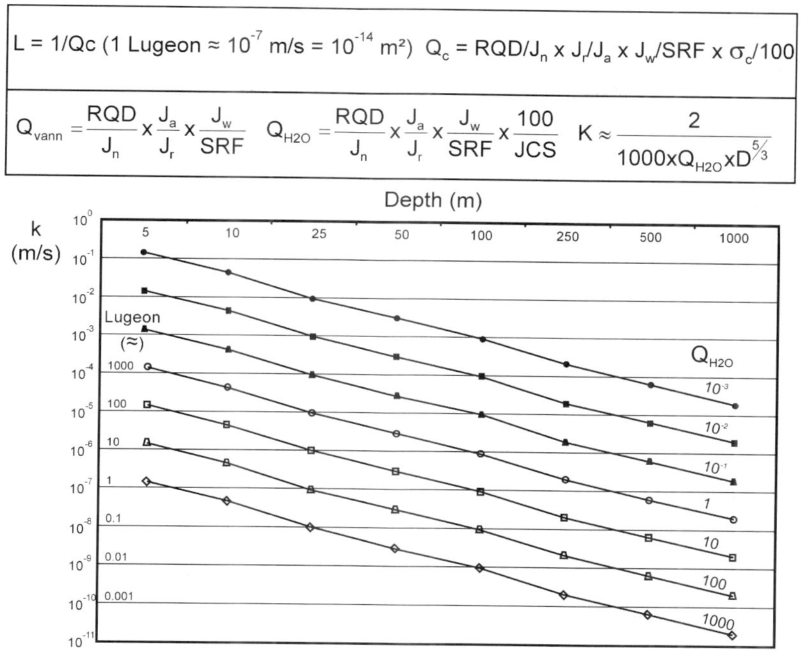


*Figure 16.78* The permeability-depth trends that are predicted by the $Q_{H_2O}$ modification of the rock quality Q formula. The trends are preliminary, but show the potential for a comprehensive rock mass description method to give a first estimate of possible permeability values.

in 'quality', and the inverse, giving permeability, will reduce further:

$$Q_{H_2O} = \frac{RQD}{J_n} \times \frac{J_a}{J_r} \times \frac{J_w}{SRF} \times \frac{100}{JCS} \qquad (16.26)$$

Finally, there is a preliminary depth-permeability-$Q_{H_2O}$ relation, based on analysis of shallow-borehole Lugeon testing at a rail tunnel project in the Oslo area, and at a metro tunnel project in São Paulo.

$$K \approx \frac{2}{1000 \times Q_{H_2O} \times D^{5/3}} \qquad (16.27)$$

The inclined lines in Figure 16.78 show this relation in graphic form, with an extended permeability range $10^{-1}$ to $10^{-11}$ m/s. It is relatively easy (perhaps too easy), with the above formulations, to 'construct' realistic permeability estimates at depth, with any desired jointed or faulted rock mass quality/character (see Appendix A for details of the Q-parameter ratings). So far, the method is appropriate to 'jointed and faulted' rock problems, and application in highly porous rock types, with significant matrix permeability *must* be avoided.

Example 1.
Clay-bearing, well-jointed rock at 100 m depth, with a low UCS of 10 MPa.

$$\left(\text{regular Q-value} = \frac{50}{9} \times \frac{1.5}{4} \times \frac{0.66}{1} = 1.4, \text{ i.e. 'poor'}\right)$$

$$Q_{H_2O} = \frac{50}{9} \times \frac{4}{1.5} \times \frac{0.66}{1} \times \frac{100}{10} = 98$$

$$K \approx \left(\frac{2}{1000 \times 98 \times 100^{5/3}}\right) = 9 \times 10^{-9} \text{ m/s}$$

(Quite low permeability due to clay coatings, and compressible joint walls).

Example 2.
Sparsely jointed, rough undulating joints, quite massive rock, with high UCS = 200 MPa, with high water pressure at 1000 m depth.

$$\left(\text{regular Q-value} = \frac{100}{3} \times \frac{3}{1} \times \frac{0.5}{0.5} = 100, \text{ i.e. 'very good'}\right)$$

$$Q_{H_2O} = \frac{100}{3} \times \frac{1}{3} \times \frac{0.5}{0.5} \times \frac{100}{200} = 5.5$$

$$K \approx \left(\frac{2}{1000 \times 5.5 \times 1000^{5/3}}\right) = 4 \times 10^{-9} \text{ m/s}$$

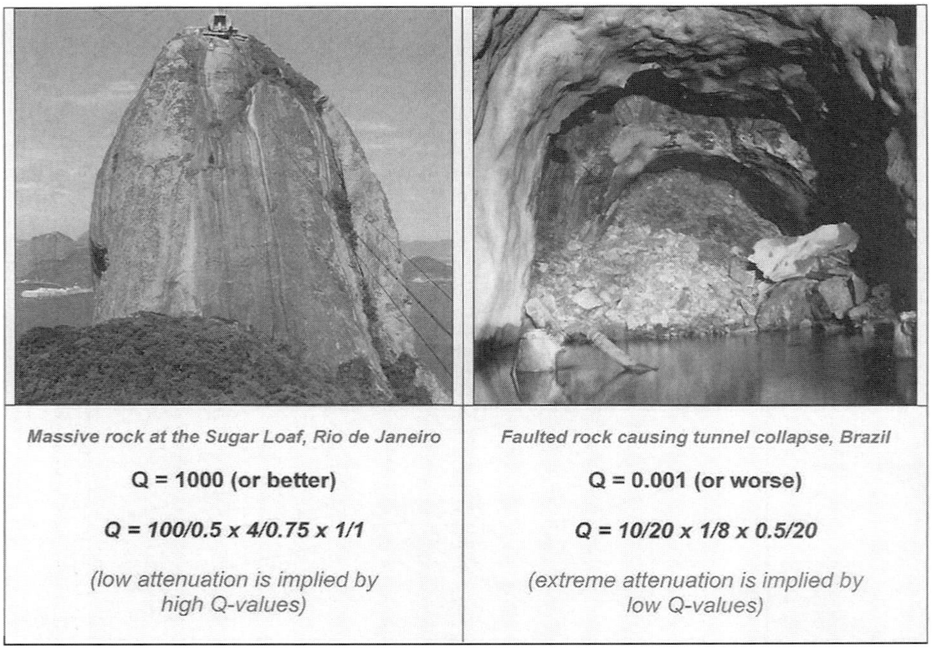


*Figure 16.79* a) An example of a massive rock mass with rock quality $Q \approx 1000$ and a deformation modulus in excess of 100 GPa. A very low attenuation is implied. b) The fault-collapse blocking the tunnel on the right would give almost the lowest rock quality $Q \approx 0.001$, and a modulus of deformation lower than 1 GPa. It is perhaps 'off-the-scale' regarding the conventional definition of $Q_{seis,}$ and would need to be under stress to allow spectral analysis of *measurable* amplitudes. Its Q-value would then be higher too. See Plate 16.

(Quite low permeability due to limited jointing and high stress level, which cancels out the conducting effect of hard, rough joint walls. Near-surface, significant permeability).

Example 3.
Negligibly jointed, low permeability, massive rock at 5 km depth (here we extrapolate beyond the curves of Figure 16.78, and base the estimate on the equations).

$$\left(\begin{array}{l}\text{regular Q-value} = \dfrac{100}{1} \times \dfrac{4}{1} \times \dfrac{1}{0.5} \\ \quad = 800, \text{ i.e 'extremely good quality'}\end{array}\right)$$

$$Q_{H_2O} = \frac{100}{1} \times \frac{1}{4} \times \frac{1}{0.5} \times \frac{100}{200} = 25$$

$$K \approx \left(\frac{2}{1000 \times 25 \times 5000^{5/3}}\right) = 5.5 \times 10^{-11} \text{ m/s}$$

(extremely low permeability due to lack of jointing and high stress level)

By way of illustrating rock mass Q-values and their extreme range, and also rounding off this wide review of 'the rock beneath our feet', a pair of contrasting rock conditions are illustrated in Figure 16.79. By chance both are from Brazil, one a famous landmark, the other a particularly difficult tunnelling project. The different qualities are typified rather well by the need for a cable car to reach the top of '$Q = 1000$' rock, while a boat was needed for hazardous tunnel inspections of several similar '$Q = 0.001$' fault zones, due to the flooded state of the tunnel between each fault blockage.

As we have seen in several chapters the seismic Q values, the inverse of attenuation, might have similar numbers, remarkably, as these pseudo-static moduli (when expressed in GPa) namely 100 and 1.

While permeability might be even lower than 0.001 Lugeon in the Sugar Loaf (i.e. $<10^{-10}$ m/s, or $<10^{-17}$ m$^2$), roughly corresponding to the Figure 16.77 prediction, the fault zone quality of 0.001 would be very unlikely to give $10^{-4}$ m/s or $10^{-11}$ m$^2$ close to the surface, as implied in Figure 16.77, because a lot of sealing clay is obviously present. If devoid of clay, Q could remain quite low, and permeability quite high.

Application of '$Q_{H_2O}$' from Figure 16.78 to this faulted case, with an assumed 0.1 MPa strength-of-clay JCS estimate near-surface, and 1 MPa at 1000 m, would give a predicted range of permeability of about $10^{-6}$ m/s at the surface – i.e. wet conditions, while at 1000 m depth, a value closer to $10^{-9}$ m/s would be suggested. These estimates appear reasonable.

# 17 Conclusions

## Introduction

In a great range of applications stretching from tectonophysicists' interest in microcracks, frequency dependent attenuation and earthquake source mechanisms, civil engineers' concerns with low modulus dam foundations, petroleum engineers' interest in shear-wave anisotropy and the permeability of fractured reservoirs, or tunnelling engineers' concerns with the approaching difficulties of a low velocity regional fault zone, the common use of seismic measurements unites many fields of earth science.

The richly illustrated material in this book has been assembled as a result of an interest in a variety of civil, mining, petroleum, geophysics and earth science fields. The common denominator has been rock mass and rock joint behaviour and their impact on the seismic interpretation of the sub-surface.

The geophysics of the sub-surface and the rock mechanics of jointed media often focus on related aspects of the same infinitely variable material. Yet different scales and different frequencies have caused these disciplines to have a mostly separate development, with limited cross-referencing in the multitude of journals. Bridging this void in some strategic locations is the major objective of this book. Two of the strongest bridges will be seismic quality and rock quality, and fracture compliance and stiffness, as applying in particular to fractured reservoirs.

*Part I* of this book is mostly focussed on civil engineering, and the links between seismic interpretation of rock conditions at laboratory and field-scale, and their impact on rock quality interpretations for tunnels, deep foundations, dams, planned nuclear waste repositories, and mines.

*Part II* of this book focuses on greater depths, greater scales, and more subtle geophysical detail, as befits this rapidly developing field. The chapters of Part II treat attenuation and anisotropy in detail, and topics range from the use of shear wave splitting to interpret anisotropic fractured hydrocarbon reservoirs, to the interpretation of mid-ocean spreading-ridges, and crustal conditions as interpreted between earthquake source zones and the near-surface.

## PART I

### Chapter 1 Shallow seismic refraction and importance of rock type

#### 1.1

Many geophysicists insist that obtaining high-resolution images from ground level to just 50 m depth is still one of the major challenges of modern geophysics. This happens to be the layer of the subsurface closest to most of our civil engineering endeavours, from tunnels, to dams, to the foundations for high buildings.

#### 1.2

Shallow refraction seismic measurements – using first arrival, compressional P-wave velocities close to the surface can give a remarkable picture of near surface conditions due to some fortuitous interactions of physical phenomena. Weathering and the usual lack of significant stress near the surface has allowed joint systems, shear zones and faults to be exaggerated in both their extent and severity. Stress levels are low enough to allow joints and discontinuities to be seismically visible due to their measurable apertures. Acoustic closure occurs at greater depths than those usually penetrated by conventional hammer seismic, unless rock strengths are rather low.

#### 1.3

Micro-fractures and rock joints are sensitive to stress levels. The more closed state of the discontinuities that are perpendicular to the major stress, and the more open state of those that are parallel will give the rock mass anisotropic stiffness. The rock mass will therefore frequently display anisotropic seismic velocities. Hydraulic conductivities and deformation moduli that show anisotropic distributions should therefore be easily detected by near-surface seismic refraction measurements.

### 1.4

The intimate interaction of the dynamic Poisson's ratio and the compressional P-waves and transverse S-waves, gives strong correlation with the physical condition of sediments, due to the propagation of P-waves in fluids and the lack of propagation of S-waves due to the absence of shear strength in fluids. Soft, saturated sediments that display Poisson's ratios of almost 0.5 have negligible S-wave velocities and large ratios of Vp/Vs as a result. This ratio eventually falls to below 2.0 as Poisson's ratio falls and lithification occurs, as seen at greater depth. Shear zones or faults display high values of Poisson's ratio and higher ratios of Vp/Vs for similar reasons, despite the relatively low value of Vp in such zones.

### 1.5

Shear-waves offer better vertical resolution than compressional-waves in shallow, unconsolidated sediments. Shear-wave velocities in such cases are only a fraction of the compressional-wave velocities. This results in very small wavelengths, despite the fact that the dominant frequency of shear wave data is generally lower than for compressional waves. To obtain the same level of resolution with P-waves, energy of very high dominant frequency has to be generated, and this is correspondingly more attenuated in the low seismic Q sub-surface.

### 1.6

Shear-waves are not attenuated at the water table, and are little affected by changes in fluid saturation. They may therefore more easily detect lithological changes with correspondingly less ambiguous velocity contrasts. Under conditions of full water saturation, P-wave velocity contrasts between similar lithologies may be small, whereas shear-wave velocities may reflect true lithological changes.

### 1.7

Rock mass quality can be described by the Q-value (here frequently termed $Q_{rock}$), which is a logarithmic scale of quality from about 0.001 (for major, clay-filled fault zones) through 1.0 (for well-jointed rock) to 1000 (for massive, unjointed bodies of rock). Static rock mass deformation moduli might vary from 0.1 GPa, through 10 GPa to 100 GPa for the same range of qualities, more exact values depending on the strength and porosity of the rock material, and on the level of stress or depth. $Q_{rock}$ also has relationship to the seismic-frequency based magnitude of Vp, with 0.5, 3.5 and 6.5 km/s corresponding, roughly, to the above rock mass qualities, when measured near-surface.

### 1.8

The Q symbol is also used by geophysicists as a measure of the attenuation or energy loss when seismic waves propagate through a medium. Low energy storage and high energy loss per cycle, signifying low $Q_{seismic}$ values and high attenuation (equal to the inverse $Q^{-1}_{seis}$), will obviously correspond to low rock quality Q-values, and vice versa, suggesting potential correlation between these two Q symbols. As will be shown in Part II, correlation may be most valid through a 'shared earth' parameter: the stress-dependent rock mass deformation modulus.

### 1.9

The seismic refraction method has some important limitations. Horizontal or sub-horizontal ray paths record only the upper part of each seismic layer. A thin high-speed layer can mask underlying material, while a low-velocity intermediate layer will not be recognised for similar reasons. Depth calculations to underlying refractors will then be erroneous. Hidden low velocity zones can be detected by up-hole shooting from a borehole to the seismic spread (i.e. reversed VSP), and of course by inspection and index testing of core, if boreholes are available.

### 1.10

The overburden at typical shallow refraction sites for civil engineering projects may not present uniform conditions. Even a simple site may contain velocity anomalies, which reduce the image quality. The time horizons suffer *push-down* beneath slow-velocity anomalies, and *pull-up* beneath fast velocity anomalies.

## Chapter 2 Environmental effects on velocity

### 2.1

Numerous factors influence seismic velocity. Joint frequency, porosity, rock (and joint wall) strength, density, depth, stress, stress anisotropy, degree of saturation, and type of saturating fluid are among the primary influences. Velocity generally increases with depth and with rock stress increase, due to increased joint normal stiffness, joint aperture reduction, joint frequency reduction, and reduction of clay content in the joints. The increased depth also causes deformation modulus to increase, and in general, the permeability reduces due to the increase in effective stress. There may also be changes of lithology with increased depth. Seismic interpretation must therefore be neither separated from the geology, nor from the structural geology.

### 2.2

Surface weathering effects alter numerous properties of both the rock material and the rock mass. Reduced density, reduced compression strength and increased porosity correlate both separately and collectively with strongly reduced *matrix* P-wave velocity. Because of the increased void space in *weathered* materials, there is a marked reduction in velocity as the degree of water saturation reduces below 100%. There is a less pronounced reduction in velocity with reduced saturation, when rocks are *unweathered*. Dry rock shows the greatest range of P-wave velocities.

### 2.3

Linear trends for $V_p$ versus the inverse of porosity (1/n%), and linear $V_p$ versus density ($\gamma$) relationships, may conceal a *non-linear* $V_p$–uniaxial compressive strength ($\sigma_c$) trend for the matrix. Weathering reduces all three parameters, and therefore also the velocity. The approximately inverse proportionality between velocity and porosity is influenced by many subtle variations bought about by such factors as clay-content in sandstones. Where joint or crack porosity increases due to weathering, and the matrix porosity increases too, the reduction of velocity may be several km/s, causing very steep velocity-depth gradients near the surface.

### 2.4

A large majority of *matrix* velocity-strength data lies between the trend lines $\sigma_c \approx V_p^3$ and $\sigma_c \approx 0.25^3_p$, with $V_p$ expressed in km/s. However, this 4:1 range is still insufficient to encompass some high porosity, low strength data, and is also insufficient to encompass some high strength igneous rock data. $\sigma_c \approx 2V_p^3$ may be needed as an upper limit. A doubling of rock strength for each 1 km/s increase in $V_p$ is also a good mean trend at intermediate velocities. The effectively concave velocity-strength trend implied by the above, is replaced by a linear $V_p$ – Schmidt (N) hammer rebound trend, with $V_p \approx 0.1 \times N$ as a first approximation.

### 2.5

At the low end of the $V_p$-$\sigma_c$ spectrum ($\sigma_c < 1$ MPa to 10 MPa), laboratory data for intact samples of Tertiary mudstones and sandstones show roughly an order of magnitude increase in strength (1 MPa to 10 MPa) for 1 km/s increase in velocity (i.e., 1.5 km/s to 2.5 km/s). Concerning the *rock mass*, another general trend is that rock mass quality Q-values increase 10-fold for each 1 km/s increase in $V_p$ for the case of hard, low porosity rocks at shallow depth.

### 2.6

In the case of shale, with its mixed content of similar density minerals (quartz $\gamma = 2.66\ gm/cm^3$, illite and montmorillonite $\gamma = 2.61\ gm/cm^3$ ), the correlation of compressive strength and density is inevitably poorer than for most rocks. P-wave velocity is therefore not a sensitive indicator of compressive strength for shales.

### 2.7

When slopes or foundations are excavated, there is usually a change of stress (mostly unloading) and a reduction in pore pressure. Improved drainage and increased permeability are also likely. Monitoring of $V_p$ in such zones usually shows significant reductions in velocity, especially when the rock mass is significantly jointed or damaged by the excavation process. Depending on the type of excavation, there may also be shear stress development due to the excavation, which may accentuate such effects. Weathering, and frost damage near the surface, may accentuate these velocity reductions in susceptible materials, if left unprotected or insufficiently supported. Velocity will then tend to decrease over time, signifying reduced resistance to shear stress.

## Chapter 3 Effects of anisotropy on $V_p$

### 3.1

Even in the unstressed state, and even if stresses are isotropic, the presence of microcracks, fabric, bedding or jointing will give anisotropic distributions of seismic velocity if these features are themselves anisotropically distributed, as is obviously the case for fabric and sedimentary structures.

### 3.2

Intact specimens of rock that exhibit strongly anisotropic or orthotropic tendencies such as slate, show significant velocity differences when measured parallel to foliation (e.g., 5 km/s) and perpendicular to foliation (e.g., 4 km/s). This anisotropy varies smoothly as the angle of incidence to the foliation is varied from 0° to 90°. Intact samples of quartzite, amphibolite, hornfels, gneiss, phyllite, schist, and slate exhibit 'weak', 'moderate' and 'strong' degrees of foliation. They can be expected to show matrix velocity anisotropies of 2–6%, 6–20% and 20–40% respectively.

### 3.3

Micro-cracks that are closed by stress give higher velocity in the direction of the applied stress. Fabric such as schistocity will also give velocities that are strongly dependent on stress levels, when the loading is normal to fabric and the rock is dry. The same reasoning applies to joint sets that are closed by the stress or depth effect. When however loading is parallel to predominant microcracks, fabric or jointing, velocities are higher and there is less sensitivity to stress level, either in the dry or saturated states.

### 3.4

It has been suggested that stress dependent velocity caused by microcracks seen in cores taken from great depth, may be mainly a result of coring damage caused by the release of anisotropic stresses. Stress dependent behaviour is particularly pronounced at low stresses compared to the higher virgin stress. Above the previous stress state, the sensitivity to stress change is less. There is little or no stress dependence when no cracks are formed in the recovery process, or when the rock is loaded or unloaded near the original stress state. In a limited stress regime around the original stress state, the rock matrix may behave as a linear elastic material.

### 3.5

Vertical or sub-vertical joint sets showing dominant strike and continuity of the primary set, cause anisotropic P-wave velocities. The degree of velocity anisotropy may range from 10% for high velocity rock masses to as much as 40% for lower velocity rock masses. The magnitude of anisotropy is accentuated if the dominant set is also parallel or sub-parallel to an anisotropic distribution of the principal horizontal stress magnitudes $\sigma_{H\,max}$ and the perpendicular $\sigma_{h\,min}$. Near-surface jointed chalk has shown velocity contrast in perpendicular directions as extreme as $V_{max.}$ of 2.85 km/s and $V_{min.}$ of 1.75 km/s, giving a total velocity anisotropy $(V_{max.}-V_{min.})/V_{max.} = 0.38$ or 38%.

### 3.6

The commonly occurring interbedding of sedimentary strata, such as sandstone, shale and mudstone, represent layers that also have different porosity, density, modulus and uniaxial strength. These contrasts of the major mechanical parameters cause variable velocity anisotropy for perpendicular and parallel wave propagation. The fine layering of sedimentary strata means that the dominant wavelength of a seismic pulse is long compared to the thickness of individual layers. The medium will exhibit a vertical symmetry axis in the case of horizontal layering. (For detailed treatment of anisotropy, see Part II).

## Chapter 4 Cross-hole velocity and cross-hole velocity tomography

### 4.1

Traditional cross-hole or between-adit velocity measurements at dam sites gives only average velocities for use in extrapolation of deformation measurements. Cross-hole seismic tomography allows both the width and location of low velocity zones to be estimated, and has revolutionized sub-surface seismic measurements for tunnels, caverns, deep foundations (and petroleum reservoir monitoring: see Part II).

### 4.2

Severe topographic changes and gradational weathering in mountainous terrain, make the use of conventional travel-time refraction seismic hard to use, as long geophone arrays may receive shortest path direct waves earlier than the refracted head waves. Gradational, *progressive* weathering, rather than distinct layering, causes less clear development of head waves. The use of tomographic inversion techniques for tunnels through steep terrain, is therefore an attractive alternative, and may include tunnel to surface imaging, where fault zones are expected.

### 4.3

There is an apparent decrease in the velocity of high velocity layers with increasing separation of boreholes. The high frequency direct first arrival received at small borehole separations may be replaced by a long dispersed wave-form at the largest separations. Attenuation of the higher frequency, higher velocity part of the wave at increasing distance may occur in strongly attenuating rock masses.

### 4.4

When comparing cross-hole and downhole velocity measurements, the downhole sonic probe is considered to give a smaller-scale, and usually higher velocity magnitudes than the averaged cross-hole result. However, the small-scale excavation damage zone (EDZ) that may accompany a borehole in incompetent rock, may be the reason for sometimes measuring lower local velocities at the small scale. Cross-hole measurement generally shows a smoothed, average behaviour. While general trends are similar, the details between the cross-hole and sonic logs differ, due to the change of scale and sampling location.

### 4.5

When seismic velocity tomography is performed at different scales around loaded rock samples, across loaded mine pillars, or through rock masses under high in situ stresses, a clear effect of rock stress on seismic velocity is demonstrated. In hard jointed rocks near the surface, the P-wave velocity may increase by 1 or even 2 km/s as a result of a stress increase of only 5 MPa, and this may occur where there is no rock mass quality improvement. In soft jointed rock, acoustic closure may mask the subsequent effects of a stress increase, unless higher stress causes compaction of the pore space. Care needs to be taken in the interpretation of higher velocities measured at depth. These may or may not be associated with improved rock quality at depth.

### 4.6

When the cross-hole tomography method is used to image highly stressed and burst prone areas in mining, steep velocity *gradients* may be found associated with such zones. Attenuation tomograms that change with time as mining advances, may be due to high stress anomalies, stress release phenomena, changes of joint aperture and stress induced fracturing. High shear stresses may be present where steep velocity gradients occur. Seismic velocity tomography can also be used to follow the effects of loosening and void formation caused by blasting.

### 4.7

P-wave *amplitude* and S-wave *frequency* measurements in the laboratory have shown superior sensitivity to stress change and to the effects of joint frequency change, as compared to P-wave *velocity*. Amplitude attenuation tomography and pulse broadening tomography may therefore correlate better with variable geology or variable structure when in situ stresses are very high, or where deep mining is involved.

## Chapter 5 Relationships between rock quality, depth and seismic velocity

### 5.1

The simplicity of first arrival P-wave velocities from shallow refraction seismic, and the easy access for core drilling have stimulated various correlations between Vp, RQD, joint frequency and the rock mass quality Q-value. For *hard, low porosity rock* a useful rule-of-thumb is that $V_p \approx 3.5 + \log Q$ km/s, where Q is the rock mass quality (range of Q = 0.001 to 1000).

### 5.2

When rock has a uniaxial strength ($\sigma_c$) significantly lower (or higher) than the nominal 100 MPa for hard rocks, the Q-value needs to be modified to $Q_c = Q \times \sigma_c/100$ in the above $V_p$–Q relationship. When rock matrix porosity (n) is significantly more than the nominal 1% for hard rocks, $V_p$ must be corrected by a porosity-related reduction in velocity. When depth is significantly greater than the nominal 25 m for shallow refraction seismic, a depth or stress related increase in velocity occurs, which can also be estimated in the case of rock masses dominated by 'soft' porosity, or jointing. A non-linear correlation of $V_p$ with *static* deformation modulus has also been developed, which can be utilised both with, or without, knowledge of the rock mass quality Q.

### 5.3

Data from sites in chalk, chalk marl, sandstones, mudstones, shales and tuff have been used to develop empirical $Q_c$ – velocity – depth corrections, which are non-linear in the case of depth or stress level, but nearly linear in the case of porosity.

### 5.4

For near-surface measurements in harder rocks, the ratio of P-wave velocities $V_{field}/V_{lab.}$, when squared, has often proved to be numerically close to the value of RQD (with RQD expressed as a ratio rather than a percentage). RQD is defined as the % of core that has core sticks >10 cm long. It applies to a given core run, or to selected structural domains, or to specific rock types, or to cycles of rock types, where there is some continuity. Since stress can 'acoustically close' joints in weaker rock, joint frequency as reflected in RQD may then prove to have little correlation with the velocity. This is where other seismic attributes than P-wave velocity become important.

### 5.5

Increasing depth will usually cause both the vertical *and* horizontal stress to increase. Besides stress increase, both RQD and $Q_{rock}$ typically increase rapidly in the first tens of meters, making a reliable depth correction problematic, since the three variables quality, depth, and velocity are often all changing at once. Sensitivity of velocity to stress is greatest in these first few MPa of stress increase. The velocity-depth relation is non-linear, and has a steep gradient. Velocity increase with depth can occur in harder rock masses, even when rock mass quality (Q-value, RQD, or joint spacing) remains constant.

### 5.6

A measured 20 metres deep profile of weathered granites, showing improvement with depth of all the indices of quality (hardness, RQD, density, etc.), demonstrated a very large increase in $V_p$ from 1.0 to 4.5 km/s. The associated velocity-depth gradient of 175 $s^{-1}$ emphasises the potential value of rock quality description. Velocity-depth gradients as high as 80 $s^{-1}$ in the upper 20 m, and as high as 40 $s^{-1}$ over the first 50 m were recorded at a cavern site in jointed gneiss with only slight weathering. Unusually, the rock quality parameters: RQD, F $m^{-1}$ and Q, did not improve at this site between 5 m and 60 m depth. A maximum velocity rise of some 2 km/s (3.5 to 5.5 km/s) occurred in the depth range 10 to 60 metres, in which horizontal stresses were interpreted from stress measurements to have increased by 2 MPa to 4 MPa, depending on direction relative to $\sigma_{h\,min}$ and $\sigma_{H\,max}$.

### 5.7

Fracture zone widths in the upper tens of meters tend to be larger if the velocity outside the fracture zone is also low. Narrowest zones tend to have lowest internal velocities, and highest external velocities. There is commonly a reduction of the widths of low velocity zones with increased depth.

### 5.8

When apertures are less than approx. 0.04 mm (or 40 μm), the frequency of such fractures appears to have little influence on the *P-wave* velocity. This is implied by experiments with 'line-samples' in the form of multiple-jointed columns of carefully machined rock cylinders. Seismic surveys underground also suggest that apertures need to be considerably wider than 40 μm, for fracture or joint frequency (F $m^{-1}$) to influence the velocity ratio $V_{jointed}/V_{intact}$.

### 5.9

No reduction in P-wave velocity is predicted with as little as 1 joint per meter. However, with 5, 10 or 20 joints per meter, a hard crystalline rock with $V_{intact}$ of 5.5 km/s, is predicted to show reductions to 4,100, 3,300 and 2,500 km/s respectively. Strong sensitivity of $V_p$ to stress level in the range 0.3 to 3 MPa is implied for the case of filled discontinuities, including sensitivity to the total cumulative joint aperture and moisture condition. $V_p$ is found to be *proportional* to $\sigma_n$ in the stress range 3–20 MPa, but drops rapidly when $\sigma_n < 3$ MPa. These experimental findings are similar to the empirical, field-data based $Q_{rock} - V_P$ – depth model.

### 5.10

*Compressive wave amplitude* gives a more sensitive measurement of the density of jointing than P-wave velocity. The amplitude ratio $A/A_0$ ($A_0$ for intact rock) shows excellent sensitivity to the density of jointing. Apertures less than 0.01 mm (10 μm) apparently do not have influence on the wave propagation, even when the normal stress is as low as 1–2 MPa. Physical apertures (E) of this size, in contrast to hydraulic apertures (e) which are smaller due to roughness effects, are probably rare in the upper 20–30 metres of rock masses where refraction seismic is carried out, so this result is probably consistent with experience in the field.

### 5.11

Hydraulic apertures of about 10, 1.8 and 0.3 μm are implied with E = 10 μm, if *joint roughnesses* are respectively 2.5 (quite smooth and nearly planar), 5 (near-planar but some small undulations) and 10 (non-planar with marked inclined asperities. The smallest of these apertures would hardly be considered as 'open' joints.

### 5.12

*Shear-wave* velocity gives a very sensitive indication of the effect of gouge thickness at low stress levels, equivalent to about 1 to 10 m, typical of the weathered zone. At 0.25 MPa normal stress, equivalent to about 10 m depth, conditions varying from no gouge to 2 mm of gouge, may give S-wave velocity reductions of as much as 50%.

### 5.13

Careful experiments have demonstrated that shallow seismic refraction measurements that operate at low stress levels are likely to be successful in distinguishing joint frequency and aperture. Amplitude measurements appear to be much more desirable than velocity measurements at *high* stress levels, if joint frequencies and character are to be distinguished. This emphasises the value of attenuation measurement, or of seismic Q. Reduced amplitude and increased attenuation occurs as joint roughness increases, which fits with the picture of joint closure difficulties when joints are rough. Shearing and dilation similarly reduces amplitude and increases attenuation.

## Chapter 6 Deformation moduli and seismic velocities

### 6.1

It is well known from dam site investigations that the dynamic modulus of elasticity ($E_{dyn}$) that is calculated from $V_p$, $V_s$ and density exceeds the pseudo-static modulus of elasticity ($E_e$) obtained from plate-load tests. The latter is itself usually several times larger than the static modulus of deformation $E_d$, i.e. $E_{dyn} > E_e > E_d$. In poor quality rock masses, ratios of $E_{dyn}/E_d$ may be as high as 10 to 20, due to the fundamentally different levels of strain involved, but in massive hard rocks under high stress the ratio will be quite close to unity. $E_d$ is variously termed $E_{static}$, D and M in the diverse literature.

### 6.2

The differences between static and dynamic moduli are attributed to the different strain amplitudes involved (perhaps $10^{-3}$ and $10^{-6}$ respectively). The different magnitudes of these moduli are also caused by the presence of pores, cracks and joints, which are measurably deformed in pseudo-static tests, but only very slightly deformed by dynamic waves. When stresses are very high and the pores, cracks and joints are almost closed, the static and dynamic moduli are likely to be of almost equal magnitude in the common direction of loading.

### 6.3

The modulus of elasticity ($E_e$) is traditionally obtained from the gradient of the *unloading* curves, which tend to have elastic character due to the frequently nearly closed state of the stress-deformation loops. When unloading from higher stress levels, as when back-analysing shaft or cavern deformation, higher values of both $E_d$ and $E_e$ are indicated. The rock mass becomes stiffer both in loading and unloading, due to the increased degree of joint closure as depth or stress increases.

### 6.4

The total deformation measured at the highest load level, after several load-unload cycles, is the usual basis for the calculation of $E_d$. The applied stress level is usually based on the size and type of dam, or other structure to be located on the particular rock foundation. Clearly, this limits the range of stress over which these parameters have typically been investigated. However, back-analysis of tunnels and caverns and deeper shaft deformations, and higher pressure testing for large bridge foundations, has extended the near-surface range of stress.

### 6.5

Static deformation moduli (also termed $E_{static}$, and M), can be estimated from the non-linear relationship with $V_p$, or from the rock quality Q-value, $\sigma_c$, n% and depth correlations developed in this book. In the undisturbed state, the consolidated state of the rock mass ensures a depth-dependent deformation modulus, due to the tighter interlock and higher stiffnesses of the rock joints. One should be aware that most rock masses observed or tested, are actually on part of a major unloading curve due to erosion, or due to the rock excavation involved in preparing for the test. In the case of rock masses with rougher, interlocked joints, lower joint stiffnesses may be registered, than if the jointed rock mass was loaded up without this prior unloading.

### 6.6

Since seismic velocity also senses the effects of higher joint stiffnesses, modulus can apparently be estimated from the empirical relation $M = 10^{(Vp-0.5)/3}$ GPa, even without reference to the rock mass quality $Q_c$ which is common to both. Where seismic 'closure' in relation to $V_p$ measurement occurs at shallower depth in weak rock, the above relation may cease to track the assumed, continued increase in moduli. One may then revert to the $Q_c$ value and apply the appropriate porosity $(-)$ and depth $(+)$ adjustments, in order to predict the continued, assumed modulus increase.

### 6.7

Deformability data have been measured at dam sites in weathered or soft rocks with uniaxial strengths ranging as low as 2 MPa, and at sites with extremely hard rocks with uniaxial strengths as high as 300 MPa. These relatively shallow civil engineering sites may have *in situ* P-wave velocities ranging from about 0.5 up to 5.5 km/s. Corresponding ranges of $E_e$ may be from about 0.5 to 80 GPa, and ranges of $E_d$ may be from about 0.1 to 50 GPa. At intermediate rock strengths and intermediate velocities of 3 to 4 km/s, $E_d$ and $E_e$ values tend to be from about 5 to 10 GPa and 10 to 20 GPa respectively. Porous rock, such as weaker limestones, results both in lower velocities and in lower moduli.

### 6.8

The inequality $E_e > E_d$ applies to the two rock engineering parameters that are actually in the 'static' loading, ultra-low-frequency sphere of geophysics. The inequality of the moduli is due to mechanical hysteresis, and is not related to another key inequality in geophysics, namely that the (static) joint *stiffnesses* become less than the inverse of the (dynamic) joint *compliances*, as rock mass quality deteriorates.

### 6.9

Plate loading tests taken to such high stress that rock mass failure occurs are unfortunately extremely rare. The measurement of P-wave velocity at such sites may allow tentative extrapolation to other sites through a common rock mass quality estimate. Such data are then the source of tentative rock mass strength ($\sigma_{c\ mass}$) estimation. The following selected sets of (smoothed) data reflect the potential linkages: $V_p = 2.3, 3.7, 4.0$ km/s, $E_d$ (or D) = 1, 3, 15 GPa, $\sigma_{c\ mass}$ (uniaxial-loading) = 4, 20, 50 MPa.

### 6.10

It has been recognised in *soil* engineering that strain levels associated with normal foundation designs are rather small, for example, 0.01 to 0.1%, and therefore stiffnesses may be successfully described by the correlations obtained from in situ seismic measurements. Such measurements have the advantage of registering the stiffness of the ground at in situ stress levels and in the undisturbed condition. When a tunnel or test adit is constructed at considerable depth in rock, the excavation disturbed zone (EDZ) effect will alter the above conditions in a complex way, to a degree that depends on rock quality and the care with which the excavation of the access adit and block test site has been performed.

### 6.11

The three dynamic moduli (Young's, shear, bulk), can theoretically be estimated from $V_p$ measurement alone, if the dynamic Poisson's ratio is estimated, rather than derived from $V_p$ and $V_s$. This however can cause significant inaccuracies, and such values given in the literature should be treated with caution. It is normal to register lower dynamic Poisson's ratios in the case of higher velocity laboratory data, and the opposite trend for the lower velocity field data. High dynamic Poisson's ratios are a sign of the influence of jointing. In shear zones and faulted rock, very high values (>0.4) of dynamic Poisson's ratio are common. Static values of 'Poisson's ratio' (mass-expansion-coefficient) >0.5 may be measured when loading jointed rock towards shear failure. At hard rock sites, typical *in situ* values of the three dynamic moduli $E_{dyn}$, k and $\mu$ might be 65, 50 and 25 GPa when the average P-wave velocity is as high as 5.5 km/s.

### 6.12

Quite high P-wave velocities, such as 4.5 to 5.5 km/s are sometimes reported together with static deformation moduli $E_d$ values as low as 5 to 10 GPa. The most likely explanation is that the large scale deformation modulus test, as practised at dam sites (plate load, flatjack or occasionally pressure chamber) is nearly always registering an excavation disturbed zone (EDZ) in a loading direction parallel to the (unloaded) radial stress ($\sigma_r$) direction. In contrast, the velocity measurement may be averaging velocities over a larger volume, and may be recording velocities parallel to the $\sigma_\theta$ (tangential stress) direction, or perhaps axially along the test adit wall. The tangential stress is a much higher, maximum local principal stress, compared to the minimum radial stress, which approaches zero, due to the effect of excavating the test adit. The axial direction would have intermediate stress levels, and presumably intermediate velocities as a result.

## Chapter 7 Excavation disturbed zones and their seismic properties

### 7.1

Excavation disturbed zones (EDZ) caused by tunnel, slope or open excavation may show up to several km/s reduction in P-wave velocity due to the combined effects of radial stress relief, blasting or excavation damage causing fracturing. Improved drainage and drying-out of the near-field rock mass will also contribute to the reduction in velocity. Time effects, reducing $V_p$ even more, may result from deformation or deterioration caused by inadequate rock mass reinforcement or surface protection.

### 7.2

Cross-hole seismic monitoring of a ship lock excavation, which reached a depth of more than 20 metres, caused a 200–300% reduction in velocity, a 75 to 85% reduction in deformation modulus, and a 1 to 20 times increase in (back-calculated) 'joint voids'. The combined effect of loosening caused by blasting, stress relief, presumed inadequate slope reinforcement, and a one-year delay between two of the monitoring stages, were responsible for the easily monitored degradation in properties.

### 7.3

A larger scale version of stress relief at a gorge, which cuts through massive, bedded limestones, had an outer layer of 5 m thickness with a velocity of 2.9 km/s, while deeper into the walls of the gorge, the velocity was 5.5 km/s. The potentially greater effect of weathering in nature may be thwarted by lack of multiple joint directions and permeability enhancement. Slope 'EDZ' may exceed the EDZ associated with tunnelling, due to the longer and more effective influence of surface weathering, including frost damage.

### 7.4

Seismic attenuation can be used to evaluate the efficiency of rock excavation by blasting, by quantifying the degree of brokenness of the rock before and after blasting. Certain frequencies may be attenuated more than others due to the effect of fracture size and fracture-induced voids. Attenuation will tend to be larger and more irregular at shallow depths, where existing joints and new fractures open more easily. Attenuation is more regular and more limited at greater depths, where confinement limits fracture and joint opening. Such EDZ effects may be accompanied by rotations of both the attenuation anisotropy axes and the velocity anisotropy axes, as a result both of disturbance to pre-existing joint patterns and dissipation of prior stress anisotropy.

### 7.5

The removal of stressed rock and its usual replacement by air at atmospheric pressure when tunnelling, results in a radial stress ($\sigma_r$) that approaches zero at the excavation walls. The tangential stress ($\sigma_\theta$) usually increases in diametrically opposite sectors, but may also incorporate diametrically opposite sectors of negative $\sigma_{\theta\,\min}$ if the far-field stress anisotropy is $\geqslant 1:3$. Joint set orientations relative to an anisotropic stress, adverse rock strength/stress ratios, and the disturbance caused by the excavation method (blasting, boring, or line-drilling), may each have strong influence on the physical EDZ, and on the resulting seismic EDZ.

### 7.6

Three velocity zones may be visualized around a tunnel: a 1 m or less loosened zone with lowest velocities, a stress bearing ring with highest tangential stresses and velocities, and an uninfluenced zone with reduced velocity where the background stresses are more isotropic. Illustrative mean results for the three zones may be $V_p$ = 3.5, 5.5 and 4.5 km/s respectively. Least effects on $V_p$ tend to be seen when tunnelling in saturated massive rock at greater (but pre stress-fracturing) depths, and greatest effects on $V_p$ are seen in drained or dry jointed rock at shallower depth. The hazardous stress fractured zones of tangentially-strained and radially-loosened or dynamically *loosening* rock, would show lowest velocities if measurements could be achieved.

### 7.7

Mining of excavations in bedded salt has shown greatest effect on the attenuation of P-waves. Mining induced radial stress relief and increased tangential stress tend to cause under-saturation due to dilation of the salt. A reduction in P-wave amplitude, which may increase with time after excavation, may give a stronger indicator of the EDZ than a minor reduction in P-wave velocity.

### 7.8

The effect of radial stress release close to a tunnel excavation in jointed rock, generally increases the joint apertures, which can actually be expressed as a void ratio. A time-average equation utilising the P-wave velocity through air or water-filled joint voids, together with the P-wave velocity of intact rock, can be utilised to interpret depth-dependent downhole or up-hole P-wave velocities. The result is explained in terms of an increased joint-aperture related void-ratio, with reduced depth of measurement from the excavation walls.

### 7.9

Cross-hole seismic measurements performed in a columnar-jointed basaltic rock mass at BWIP, Washington, USA were made between four horizontal boreholes drilled 12 metres into the wall of a drill-and-blasted underground opening at 46 m depth. The basalt columns were regular but sinuous, 0.15 to 0.36 m in thickness, dipping 70 to 90°, with frequent low angle, discontinuous cross-jointing. The horizontal and (to a lesser extent) the diagonal seismic measurement paths crossed the more open columnar joints, and these features clearly opened most as a result of excavation, giving the strongest reductions in velocity of 55% to 65%. The seismic quality factor $Q_{seis}$ increased in the vertical direction, in the same direction as the highest velocities recorded where there was least EDZ effect. $Q_{seis}$ in the horizontal direction remained low (5 to 8), in the same direction that showed the maximum EDZ-reduced velocity, with $V_P$ declining from 5.5 or 6.0 km/s to 3.5 km/s at the tunnel wall.

### 7.10

The underground research laboratory (URL) in Manitoba, Canada was the site of numerous geophysical studies. The dominance of massive, unjointed, highly stressed granite resulted in particular focus on stress-related EDZ, with down-hole sonic logging and acoustic emission monitoring, together with parallel laboratory tests and numerical modelling studies. At the smallest scale, the state of micro-cracking in selected core samples from radial boreholes drilled in the walls of a drift where excavation was either by *normal blasting* or by *smooth blasting*, indicated $V_p$ reductions of about 1 to 1.5 km/s in the outer 0.8 m of the normally blasted tunnel, and reductions of about 0.5 km/s in the outer 0.5 m of the smooth blasted excavation.

### 7.11

The effects of highly anisotropic, sub-horizontal stresses at URL were studied in a unique test tunnel excavated by line drilling and reaming, followed by mechanical break-out to avoid blast damage. Principal stresses of approximately 60, 45 and 15 MPa, caused classic 'break-out' resembling that in a borehole. The isotropic-elastic theoretical tangential stresses of 165 MPa ($3\sigma_1 - \sigma_3$) at 11 o'clock in the roof and 5 o'clock in the floor (with $3\sigma_3 - \sigma_1 = -15$ MPa in the side walls), caused prominent V-shaped notches of rock failure. The stress-related disturbance was measured directly in 1 m deep boreholes using a micro-velocity probe with 10 cm separation of the transducers.

### 7.12

Acoustic emission (AE) was recorded during careful (mine-by) extension of the 420 m deep test tunnel. The AE events clustered both where tangential stresses were highest and where seismic velocity ($V_p$) gradients were steepest. The acoustic emission results confirmed that rock failure was initiating just inside the tunnel wall, orthogonal to the $\sigma_1$ direction. Relatively decreased velocities were seen in the two regions that were under tensile tangential stress. Calculated P-wave velocities showed quite strong anisotropy in the massive granite, caused by the principal stress anisotropy given in the previous paragraph.

### 7.13

An experimental tunnel sealing experiment at the URL utilised AE for interpreting reductions in average P-wave *and* S-wave velocities, in the highly stressed zones that were caused by post-excavation of larger diameter 'dog-collars' or bulkheads for concrete and bentonite sealing. 3D coverage of an outer volume surrounding this tunnel sealing experiment, and two higher resolution AE arrays for recording in a 10 m × 10 m × 10 m volume around the collars, allowed AE-based monitoring of temporal changes in $V_p$ and $V_s$. These were used to estimate the theoretical change in crack density (c) and saturation (s) along any particular ray path.

### 7.14

At the Äspö hard rock laboratory (HRL) in Sweden, seismic tomography investigations were performed to compare the depth of excavation damage zones in immediately adjacent drill-and-blast and TBM sections of tunnel. Principal stresses at the ZEDEX (zone of excavation disturbance experiment) were approximately 32, 17 and 10 MPa. There was only a small EDZ effect on $V_p$ and $V_s$, due to the high stresses and the partly discontinuous jointing. P-wave velocities were mostly in the region of 6.0 to 6.3 km/s, with small reductions of velocity in a thin 'skin' next to the walls. The reductions of velocity were recorded in the first 0.25 m into the TBM tunnel walls, and up to 1 metre into the walls of the drill-and-blasted tunnels.

### 7.15

EDZ effects around the tunnel were only just detectable by seismic velocity, due to the good quality of the rock

(Q-value ≈ 22–24), and due to stresses that were high enough to have acoustically closed the joints, but not high enough to have caused excavation induced micro-cracking, as at the URL in Canada. The rock quality $Q_c$-$V_p$-M model predicted a P-wave velocity range from 5.8 to 6.2 km/s, using an equivalent depth range of 400 to 1200 metres, relevant to the principal stress range. The intact rock laboratory E-modulus was 69 GPa. The equation relating mean deformation modulus M with $V_p$, suggested deformation moduli ranging from 58 to 79 GPa, for the predicted velocity range of 5.8 to 6.2 km/s. Calculated dynamic moduli around the drill-and-blast drift ranged from 76 to 79 GPa. The particular $M \approx E_{dyn}$ situation at this site implies 'acoustically closed' jointing.

### 7.16

A drift in the Stripa mine in Sweden, used for borehole heater tests, showed increased seismic velocities between drained monitoring holes in the jointed quartz monzonite, during heating of the rock mass. The initial increase in velocity with temperature was linear and varied from 2 to 4 m/s/°C. This was the presumed result of thermally induced joint and micro-crack closure. Increased P-wave velocities were recorded in directions consistent with a thermally-induced stress increase. An initial stress anisotropy giving low stress in the same direction as the subsequent thermal stress increase tended to enhance the $V_p$ response. Greatest response was seen when degrees of saturation were also low.

### 7.17

A long period of cooling (350 days) generally returned seismic velocities to values *lower* than before the heating, suggesting permanent changes, probably connected with the hysteresis effect of joint closure and opening. Upon cooling, the less rough, interlocked joints may have 'sprung-open' more than their closed neighbours, to avoid tensile stress development. This would cause a reduction in seismic velocity if the open joint or joints crossed the path of the seismic array. A velocity anomaly at about 3 m depth was smoothed-out by the heating but returned when the rock was cooled. A significant quantity of water expelled during the heating signified a general closing of the joints.

### 7.18

The average joint frequency in the test area, analysed from 224 m of core, was 8.3/m. An elastic continuum analysis conducted prior to the test had suggested larger stresses and local displacements than were actually measured, presumably due to the thermal compliance of the joints, as noted in Part II concerning a fully coupled HTM heated block test. This thermal compliance effect was also experienced in a heated-mine-by experiment in the Climax Mine, in the USA, where deformation monitoring also failed to match predictions, since no thermal joint compliance was modelled.

## Chapter 8 Seismic measurements for tunnelling

### 8.1

Seismic velocity measurements performed from the surface prior to tunnelling need careful interpretation due to the common experiences of enhanced weathering, enhanced fault width and more open jointing experienced near the surface. Anisotropic stresses, and anisotropic fabric or structure or bedding, plus hidden lower velocity layers, add to the potential pitfalls when attempting to anticipate rock mass conditions at tunnel depth. The degree of saturation where measurements were made, compared to tunnel depth conditions is also important.

### 8.2

Seismic refraction profiles with appropriate azimuth, and borehole dip and dip-directions that also take due regard of structure and stress anisotropy can improve the quality of the sub-surface investigation. The use of only vertical boreholes drilled where there is predominantly vertical structure is a guaranteed way to obtain poor sub-surface information, if this is only to be based on core and borehole wall inspection. The application of geophysics helps to recover some of the lost information from representative core samples, especially if high-frequency cross-hole surveys are performed.

### 8.3

The common experience of improved rock quality and increased $V_p$ at depth may be checked at intervals by appropriate core-drilling. In the local absence of core, a sensible interpretation of the usual effect of depth on $V_p$ values can be made using $Q-V_p$ correlation, where Q is the rock mass quality. In very general terms at hard rock sites, a ten-fold improvement of Q-value, combined with a 1 km/s *additional* depth effect on $V_P$ might be expected at a 50 m deep tunnel. So a near-surface Q-value of 1.0 interpreted from shallow seismic refraction, where $V_P$ was 3.5 km/s, might see $V_P$ increased to 5.5 km/s at 50 m depth, but with the Q-value equal to 10 at tunnel depth, rather than Q = 100 as implied by a nearer-the-surface $V_P$ of 5.5 km/s.

### 8.4

There is evidence to suggest that not only rock overburden depth, but also water depths are important, in the case of undrained sub-aqueous tunnels that prove to be completely dry, due to low permeability over-lying rock masses such as phyllites. In such cases the total stress caused by rock load and water load may give a false impression of rock quality at tunnel depth, to the extent of 1 to 2 km/s.

### 8.5

Seismic velocity measurements performed while tunnelling have been shown to correlate in approximate terms with speed of tunnelling, support needs, and tunnel cost. Probe drilling and sonic logging of the holes can be utilised for ahead-of-the-event information on rock and hydrogeological conditions. If multiple probe holes are drilled, then cross-hole tomography can be performed to add to the tunnel contractor's preparedness for difficult zones. This has resulted in the choice of ground-freezing for safer penetration of dangerous and/or environmentally sensitive sub-urban fault zones.

### 8.6

Velocity measurements should be made sufficiently far ahead of the tunnel face, at least two to three diameters, such that additional stress concentration effects are avoided. Interpretation of local effective stress conditions will be needed when evaluating results of the velocity measurements, using $V_p-Q$ correlation. A 'false' rock quality, due to a higher P-wave velocity may be interpreted, if depth of measurement is not accounted for. This may mask the actual presence of a serious fault that could better be identified by shear waves or attenuation measurement.

### 8.7

An early example of the use of geophysical surveys in tunnels was the Straight Creek pilot bore of 4.0 m diameter, driven in the 1960's at 200–500 m depth through granite, diorite, gneiss, migmatite and schist, under the continental divide in Colorado, USA. *Deep layer velocities* were measured at five seismic spreads: 5.18 km/s, 5.1 km/s, 4.8–6.1 km/s, 4.2 km/s and 6.0 km/s. Much lower *shallow layer velocities*, representing loosening effects were respectively 3.0 km/s, 2.3–2.7 km/s, 2.3–3.1 km/s, 1.3–1.6 km/s (worst case, class 5) and 2.3 km/s. These extremely low EDZ velocities proved, in retrospect, to be more related to the insurmountable problems in the 12 m main bore, which took several year to complete. There was possibly insufficient appreciation of the effect of stress on the seismic velocities at that time. There is also the possibility of adverse interaction when there are twin tunnel tubes, with 'plastic zone' overlap (or log-spiral shear-zone overlap) in faulted or weak rock zones, a problem of relevance when assessing risk in too-close twin-bore TBM tunnelling, where conditions are unfavourable.

### 8.8

Comprehensive geological and velocity classification of rock conditions at numerous rail tunnels in Japan from the 1970's demonstrate the value of velocity-tunnel-support correlation. Recommendations were based on 70 case records from 30 $m^2$ and 60 $m^2$ tunnels. For rock-type and geological Classes 1 to 6, with their associated high to low P-wave velocities, tunnel support loads ranged from 1 to 30 tons/$m^2$, the spacing of the steel arch support ranged from 1.5 to 0.75 m, and the final concrete thickness ranged from 0.3 to 0.9 m. Today, we could place an equivalent velocity scale below the rock mass Q-value, which is used for recommending shotcrete thickness and bolt spacing.

### 8.9

When attempting to add a rock mass Q-value scale to such velocity-rock-support recommendations, the geological part of the classification is particularly important, in view of the porosity, strength and depth corrections to the $V_P - Q_c$ empirical model. Young, weak, porous rocks may have RQD = 0, and incur unfavourable SRF values due to adverse strength to stress ratios in the tunnelling situation. These two factors may alter the correlation of support-type-and-degree with the velocity, unless the velocity is EDZ-based, rather than that obtained from the fully confined (pre-tunnel) situation.

## 8.10

Sub-sea tunnelling experiences, with velocity-rock support correlation, have revealed several cases of medium and low-velocity zones actually creating greater tunnelling difficulties than the velocities would suggest, implying an artificially elevated velocity in relation to rock quality. The tunnelling situation changes the stress level in relation to the pre-tunnelling confining stress. Based on sub-sea tunnelling experiences, Norwegian studies have shown support costs rising from 50% to at least 75% of total costs, when the P-wave velocity reduces from 5.5 to 4.5 km/s. It is likely that these are 'depth-enhanced' velocities, since uncritically applied Q-values would be 100 (no support) and 10 (light support) respectively, which are unrealistic when referring to significant support costs.

## 8.11

High velocities of typically 5 to 6 km/s may dominate in refraction seismic studies at relatively unweathered hard rock sites. The much smaller number of tectonic zones (shear zones, faults), dykes and joint swarms with velocities from about 2.5 to 3.5 km/s are what cause the construction problems, especially when high inflows of water occur. In weaker, porous rock, velocities in the range 2.5 to 3.5 km/s may signify excellent stability. The contrast from back-ground velocities is clearly the measure of the stability problem, not the velocity per se. The hard rock velocities may signify rock quality Q-values of 30–300 (very/extremely good), and 0.1–1.0 (very poor) respectively, clearly reflecting the reduction of quality, and need for local, heavy tunnel support. The weak, porous rock velocities could signify Q-values as high as 10–100 at nominal 25 m depth, if corrections for 10% matrix porosity and UCS of 10 MPa were included in the assessment. The $V_P - Q_c$ empirical model can be applied to make these assessments, where $Q_c = Q \times \sigma_c/100$.

## 8.12

Despite the above pitfalls concerning the meaning of velocity when different porosities and rock strengths are involved, tunnel refraction seismic and tunnel boring machine (TBM) progress indicate a quite linear inverse relationship between the penetration rate (PR in m/hr, with uninterrupted boring), and P-wave velocity. This is also partly matched by an inverse PR and Schmidt-hammer rebound trend. Respective values might be as follows: $V_P$ = 4 km/s, 2 km/s: PR = 2 m/hr, 4 m/hr: Schmidt-hammer rebound = 40, 20. (Inverse correlation to possible Q-values of 30 and 0.3 are also implied, as in the $Q_{TBM}$ model). Harder, higher velocity rock signifies tougher boring (i.e. lower PR), but the weekly mean advance rates (AR) might be 1 m/hr and 0.5 m/hr respectively, due to the reduced tunnel support needs. While tunnel support quantities are roughly proportional to $\log_{10} Q$, PR may be inversely proportional to $\log_{10} Q$.

## 8.13

Sonic logging of probe holes, made by fast percussion drilling at from 2 to even 5 m per minute, can give advance information on seismic velocity if the holes are sonically logged. This can be used for subsequent rock quality class and tunnel support class estimation. The chosen excavation mode, and support components such as bolted steel arches, rock bolts and mesh, or shotcrete, can then be immediately available, and applied with appropriate timing, behind the advancing TBM tunnel face.

## 8.14

A seismic velocity probe ahead of a tunnel will not see the difference between a TBM tunnel and the drill-and-blast tunnel. However, if refraction seismic measurements are performed along the wall of a TBM tunnel, the values of $V_p$ obtained may tend to be higher than in the equivalent drill-and-blasted tunnel for at least three reasons. There is a reduced level and depth of damage in the wall of a TBM tunnel. Higher tangential stresses are acting closer to the TBM tunnel wall. There will be a tendency for lower permeability and less drainage around the TBM tunnel, which, for reasons of more complete saturation might also increase the seismic velocity. On the other hand, there may be a reduced value of effective stress as a result of the same reduction in permeability.

### 8.15

The reflection method of HSP or TSP (horizontal or tunnel seismic profiling) with both source and receiver in the tunnel, is capable of locating seismic reflectors ahead of a tunnel face. However, reflector images are not related to rock quality directly, but to implied *change* of quality. It is difficult to determine if the rock quality will get better or worse at a given reflector, and there may be inaccuracies of reflector location due to the unknown actual velocity field.

### 8.16

A logical extension of conventional high-resolution surface refraction seismic, is the application of sources and receivers at, and close behind the tunnel face, with receivers at the surface, if this is not too distant. GPS clocks are needed to synchronise the sources within the tunnel and the receivers at the surface. The use of in-tunnel refraction-based estimates of velocity distributions ahead of the face, can improve the accuracy of reflector positions. With the necessary velocity distribution ahead of the face, the rock mass quality can be *characterized* both up to and beyond the now better-located reflectors.

### 8.17

Velocities measured at depths of hundreds of meters or more, using in-tunnel refraction or cross-hole tomography, may bear little resemblance to the major tunnelling difficulties sometimes experienced when tunnelling. Face collapse in a $V_p \approx 4$ km/s rock mass is 'illogical', without allowing for the depth or stress effect that may mask, in velocity terms, the true low quality. A 300 m overburden at such a collapse location, suggests a near-surface $V_p$ of about 2.5 km/s, using the Q-$V_p$-depth model. This would be relevant to a serious fault zone, or extremely poor rock, and therefore more consistent with eventual collapse.

## Chapter 9 Relationships between Vp, Lugeon value, permeability, and grouting in jointed rock

### 9.1

Rock masses containing voids in the form of porosity, joints or damage zones will generally have enhanced permeability and reduced seismic velocity. Measurements at dam sites and at tunnels have indicated inverse correlation between P-wave velocity and permeability, in particular with the relatively high pressure Lugeon injection test, which may locally deform the void space created by joints or fractures. Because the rock mass quality or Q-value also correlates with the deformability, a very simple lower bound inverse relation between the Lugeon-value ($L \approx 10^{-7}$ m/s) and the Q-value is indicated. ($Q \approx 1/L$). This is not relevant however, when joint-sealing clay is present, causing *both* Q and permeability to reduce together, instead of inversely.

### 9.2

Lower Q-values and higher Lugeon values will tend to be measured when holes are oriented to cross major structure. The opposite occurs when paralleling major structure, and failing to cross the dominant jointing. The latter is typical for vertical wells prior to deviation through seismically identified oriented structure. Holes should be drilled in the 'slow' direction for registering highest permeability. Clay sealing of joints may compromise these simplified assumptions.

### 9.3

Since sub-vertical jointing may dominate in the same way that horizontal stress anisotropy may dominate, the permeability anisotropy will tend to be related to azimuth. At great depth, this model may be compromised by joint closure, and a shear mechanism may be needed to explain the maintenance of (anisotropic) permeability.

### 9.4

The void space in a rock mass that is responsible for greatest permeability, specifically the joints and any outwash channels caused by weathering or soluble minerals, can usually be injected with cement, micro-cements, or ultra-fine cements. Seismic measurements, principally in dam foundations, show remarkable increases in P-wave velocities as a result of successful grouting, when permeabilities are also favourably reduced. Part of the increase in velocity and resulting reduction in deformability may be due to the post-stressing effect of high pressure grouting, which also may set in a stressed state.

## 9.5

Most effective grouting can be expected in the joint set having greatest permeability and least magnitude of the effective normal stress. This set will thereby show greatest increase in seismic velocity as a result of the grouting. A rotation of the permeability tensor to a lesser magnitude can also be expected as a result of the grouting, and velocity anisotropy is likely to be reduced. Several of these aspects have been demonstrated using multiple borehole 3D permeability testing in Brazil.

## 9.6

The inter-related physical nature of rock mass quality (Q), deformation modulus (M), seismic velocity ($V_p$) and permeability (L) means that rock masses can be represented by type curves in nomograms linking these parameters. Although only approximate, such type curves serve to distinguish typical properties of massive rock, jointed rock, porous jointed rock, and fault zones. The mass properties of each will also be affected by depth and by matrix properties such as uniaxial compressive strength and porosity.

## 9.7

Extensive work in marl formations in Switzerland, indicated quite strong relations between selected seismic parameters and transmissivity measurements in five deep boreholes at Wellenberg, a potential nuclear waste repository site. Good correlations with transmissivity were obtained with $V_p$, $V_s$, $V_p/V_s$, dynamic shear ($\mu$) and E-moduli, and dynamic Poisson's ratio. The borehole depths ranged from about 400 m to 1800 m, and included faulted and brecciated rock. The measurements showed a very broad transmissivity range of $10^{-2}$ to $10^{-14}$ m$^2$/s. Corresponding P-wave velocities ranged from 3.2 to 5.5 km/s, $V_p/V_s$ from 2.45 to 1.72, dynamic E-modulus from 0 to 60 GPa, and dynamic Poisson's ratio from 0.41 to 0.28.

## 9.8

At the Chinnor Tunnel in chalk marl in southern England, very low seismic velocities in the range 0.6 to 1.0 km/s were registered for badly fractured/jointed areas of the chalk marl. Quoted permeabilities were $10^{-4}$ to $10^{-6}$ m/s in these areas. Assuming that 1 Lugeon $\approx$ $10^{-7}$ m/s then the very high Lugeon values of 1000 to 10 imply $Q_c$ values of 0.001 to 0.1. These low $Q_c$ values can be converted to 'tunnel support' Q values of 0.02 to 2 ('extremely poor' to 'fair'), by assuming a mean $\sigma_c$ value of 5 MPa for the chalk marl. This range of Q-values is in line with expectations for the heavily jointed rock mass at Chinnor.

## 9.9

It has been confirmed from comparison with data from carefully documented block tests performed in the USA, that the parameters Q, $V_p$, M and L are inter-related, and that the inclusion of the Lugeon value in this inter-relation is justified, if care is taken to eliminate irrelevant non-deforming, channel flow cases, and to eliminate cases where clay sealing of the joints is occurring. Depth or stress level, also plays an important role in these mutual inter-relationships.

## 9.10

Using an analogue material for heavily jointed rock, namely coal, one can see great sensitivity between velocity, stress level and permeability, which will also be present in jointed rock masses at large scale, when *in situ* effective stress states are altered by large scale pumping or injection experiments. Extensive data sets for numerous coal samples show the three key behaviour modes: permeability reduction with increased stress, permeability reduction with increased velocity, and velocity increase with increased stress. Greatest changes in $V_p$ and permeability occur at the lowest stress levels and at the lowest velocities, just as found in rock masses, due to the greater sensitivity to acoustic coupling across the cleats in the coal.

### 9.11

Researchers working at nuclear waste related rock laboratories such as Stripa, Äspö, and Grimsel have utilised seismic and radar tomography to characterise fractured zones and fault zones. Their studies have generally helped to explain why these relatively small volumes of fractured or heavily jointed rock are responsible for such large percentages of the total flow of water. Radar attenuation difference tomograms have been found more reliable in locating brine in tracer tests, than slowness tomograms.

### 9.12

Radar and seismic signals are sensitive to different physical parameters (electro-magnetic wave conductivity, and mechanical stiffness respectively). The respective tomograms therefore highlight different features of the rock mass. Radar may delineate permeable zones caused by pore space or by joint apertures, in slightly different locations to the low seismic velocity zones associated with the clay-filled sections. The one will usually lie parallel to the other, since higher permeability may be associated with the heavily jointed zones that are often found in the walls of faults.

### 9.13

Low resistivity generally correlates with zones of increased water content and frequently with higher permeability. At a site in South Korea, a series of boreholes in weathered granites were Q-logged and later compared with resistivity tomograms in a 'blind' test. Sections of the core with increased joint frequency (low RQD, high $J_n$) did not always correlate with low resistivity and vice versa. The two parameters that *did* show a consistent correlation with low resistivity were low values of $J_w$ estimated from iron staining or apparent aperture, and the high values of $J_a$ due to sand or silt fillings, *and* due to clay-fillings. The latter gives 'anomalously' low resistivity due to the ionic effects of the clay. Water content and permeability are clearly lower in such discontinuities than in those that are sand or silt filled. This represents a potential source of error in judging the meaning of low resistivity zones.

### 9.14

Dam foundation damage in the form of shearing on foliation planes, with resultant P-wave velocities as low as 0.5 km/s, was a starting point for systematic cement grouting, that saw $V_p$ locally increased to between 2 and 6 km/s. The remaining low velocities of only 2 to 3 km/s correlated with residual permeability, which was eliminated with further grouting, and consequently increased velocities. A strong velocity-depth effect was noted, related both to rock quality improvement at depth, and to a post-stressing effect from the increasingly confined grout at greater depth. New rounds of grouting at *old* dams has also demonstrated increased $V_p$ and reduced permeability. Largest grout take correspond to the locations where the largest increases in $V_p$ are registered, following the grouting.

### 9.15

It was noted in the 1970's, in the 220 m high Mratinje dam foundation, that effective consolidation grouting could be performed when $V_p$ was in the range 2.5 to 3.5 km/s. According to the Q-$V_p$-L correlations, this may correspond to $Q \approx 0.1$ to 1.0 and 10 to 1.0 Lugeon. Velocities above 4.0 km/s (implying $Q \approx 3$ and $K \approx$ 0.3 Lugeon) could not be improved upon by the grouting done at that time. Such results emphasise the reasons for combined use of high injection pressures of 5 to 10 MPa, and use of micro or ultrafine cements and microsilica, in today's pre-grouting ahead of tunnels. Systematic pre-injection where average Lugeon values are as low as 0.1 (implying $V_P$ of 4.5 km/s), may not be successful without resort to high pressure and/or use of the finest cement types.

### 9.16

A general, approximate relationship between Lugeon value, Q-value, and measured velocity, has been compared with case records showing reduced Lugeon values and monitored velocity increases, as a result of grouting. These changes also represent potential physical effects of grouting on individual Q-parameters. Conservative assumptions of individually improved Q-parameters, like increased effective RQD, reduced number of effective joint sets, and successful grouting of the least favourable joint set, combine to suggest significantly improved effective Q-values. Reduced tunnel deformation, reduced support needs, increased modulus of deformation, and increased seismic velocity are each implied, and seem to be supported by the experience of trouble-free tunnel advance following effective, high pressure pre-grouting.

## PART II

## Chapter 10 Seismic quality Q and attenuation at many scales

### 10.1

Attenuation can be simply defined as the loss of energy per cycle divided by the maximum energy per cycle in the same rock volume. The wave amplitude versus frequency diagram indicates that frequency determines the maximum wave amplitude. The inverse of attenuation, the seismic quality Q, shows high values when there is little attenuation, and low values when rock conditions and near-surface location causes strong attenuation. The standard definition: $\Delta E/E = 2\pi/Q$ suggests a minimum theoretical value of seismic Q of about 6, but lower values are reported, including possibly erroneous negative values. Several methods can be used to estimate seismic Q, which can have the forms $Q_p$ and $Q_s$ when derived from P-wave or S-wave spectral analysis, or $Q_c$ when derived from the tail of an earthquake seismogram, termed the coda, to sample the deep hypocentral region.

### 10.2

Knopoff's paper 'Q' from 1964 with long lists of seismic Q values for solids like steel, glass, lead and celluloid (5000, 490, 36 and 7), suggest that relative stiffness may be involved in some way. It was originally thought that seismic Q was independent of frequency. However, rock with microcracks, joints or fractures, and with different fluid saturation levels, is now known to cause frequency-dependent attenuation. Attenuation losses occur due to wave *scattering* from heterogeneities like joints, faults and rock boundaries, and due to *intrinsic* losses such as matrix anelasticity, friction at grain boundaries, across microcracks and across joints and fractures, related heat loss, fluid 'squirt' from fractures to microcracks to pores, and gas pocket squeezing when partly saturated. The viscous fluid related losses are particularly dependent on frequency due to the relaxed or unrelaxed state caused by low or high frequency.

### 10.3

The short period of each wave cycle, and the relatively low levels of dynamic stress, are generally assumed to cause sub-micron size reversible micro deformations and flows. There are those who have denied the early assumption of friction as an intrinsic attenuation mechanism in the earth, but this supposition was partly based on dynamic sub micro-strain work with intact bars of dynamically excited rock. Frictional losses *in situ*, due to the presence of joints and fractures, plus their contribution to losses by scattering, are now widely cited, and by various authors, when working outside the common intact-medium laboratory limits. Important here is the fact that when the effect of single joints or aligned fractures are inverted from seismic data, the respective contributions of the dynamic stiffnesses of the joints to the wave amplitude and to shear-wave anisotropy, are found to have recognisable magnitudes and the same units (Pa/m) as when working with pseudo-static loading tests. The latter *obviously* mobilize friction due to their contribution to frictional strength.

### 10.4

Review of a large body of laboratory data for seismic Q has indicated a close approximation to the pseudo-static Young's modulus ($E_{static}$) and it's increase with confining stress and sample stiffness. This was noted when substituting a non-linear seismic Q scale next to numerous sets of experimental data that showed only a linear 1/Q attenuation scale. Laboratory studies frequently show seismic Q increasing from about 10–20 to about 50–100, as confining pressure is increased from near zero to 10, 20, 40, or 70 MPa . Seismic Q magnitudes have great similarity to the variation of $E_{static}$ with confining pressure, if seismic Q is expressed as if it were GPa.

### 10.5

Intact rock samples loaded to failure show consistently increasing seismic Q when measured in the loading direction, as does $E_{static,}$ and the same samples show reducing values of seismic Q when measured in the perpendicular

(Poisson expansion) direction. Ultrasonic 0.1 to 1 MHz seismic Q values estimated during normal-stress closure tests on jointed samples, show $Q_p$ increasing from about 5 to 80, as normal stress is raised from about 3 to 70 MPa. At another extreme, cracking caused by repeated freeze-thaw cycles in limestone, have demonstrated $Q_{seis}$ steadily declining with increased cracking from about 20–25 down to 10–12, with corresponding reductions in P-wave velocity.

## 10.6

*In situ* seismic interpretations of seismic Q have revealed a certain tendency for higher seismic Q when the rock mass quality Q-value was also expected to be higher. These commonly used quality parameters are differentiated as respectively $Q_{seis}$ and $Q_{rock}$. The widely used $Q_{rock}$ value, varies from about 1 to 1000 for heavily jointed to massive almost joint free rock masses. It is composed of three ratios. The first ratio is RQD (% of core pieces >100 mm length) divided by the number of joint sets $J_n$. The second ratio $J_r/J_a$ describes roughness and alteration, and gives the friction coefficient, which includes the effect of clay-filling. Finally the estimated water pressure (or tunnel inflow) and the stress/strength ratio are evaluated. $Q_{rock} = RQD/J_n \times J_r/J_a \times J_w/SRF$. The maximum range of $Q_{rock}$ is from 0.001 for fault zones to 1000 for massive joint-free rock masses. $Q_{rock}$ provides quick estimates of seismic $V_p$, deformation modulus and permeability, and also indicates tunnel support needs in rock engineering projects.

## 10.7

The first pair of $Q_{rock}$ parameters $RQD/J_n$, representing 'relative block size', have a numerical range of 200 to 0.5. Even this has a certain similarity to the range of $Q_{seis}$ for respectively joint-free to heavily-jointed rock masses. More important is the likeness of the rock mass (pseudo-static) deformation modulus to $Q_{seis}$, because $E_{mass}$ is readily estimated from $Q_{rock}$. The great majority of rock masses, from the surface to 1 km depth, from heavily jointed to virtually joint-free rock, have deformation moduli in the approximate range 1 to 150 GPa, as derived from *in situ* testing and from back-analysis of deep tunnels and shafts. This range closely resembles the typical ranges of *in situ* $Q_{seis}$ seen in numerous shallow 0 to 2 km deep seismic data.

## 10.8

With clay and weak rock at one extreme, the deformation modulus range can extend down to 0.1 GPa, representing a strongly absorbing inelastic matrix with UCS < 1 MPa, *plus* the near-surface attenuating effect of clay-filled discontinuities. At the other extreme, with high matrix strength UCS > 200 MPa, a depth up to 1 km and negligible jointing, $E_{mass}$ may reach 150 GPa, representing very low attenuation.

## 10.9

In sedimentary, finely layered rock sequences, greater sophistication in broad-band recording has demonstrated a more complex $Q_{seis}$-deformability response than the above, with seismic $Q_p$ indicating that different values are obtained from ultrasonic 500–900 kHz testing of core (mean 27.0), sonic 8–24 kHz testing in boreholes (mean 10.4), cross-hole 200–2300 Hz testing (mean 15.7) to VSP 30–280 Hz testing from surface to borehole (mean 31.3). The rock mass was a finely layered saturated sequence of limestones, sandstones, siltstones and mudstones. The overall range of $Q_p$ from the different frequencies and depths was approximately 5 to 40. This remains in a potentially familiar range, if expressed in rock mechanics deformation modulus units of GPa.

## 10.10

Coda waves from the tail of seismograms, typically in the 20 to 200 seconds time window, after the arrival of P, S and surface waves, have traditionally been used to analyse the seismic Q signatures of both local and distant earthquakes. A large volume of rock surrounding the earthquake source is believed to be sampled by the coda, with the possibility to register changing rock properties or migrating source zones, due to the accumulation or dissipation of stress and resulting strains and possible fluid movements. Coda Qc is frequency dependant, with low frequency 1 Hz content assumed primarily composed of surface waves scattered from shallow heterogeneities, while coda waves at 20 Hz are believed to be from back-scattered body waves from deeper heterogeneities in the high $Q_{seis}$ lithosphere. Thus many surface recorded data sets show $Q_{seis}$ lower than 100 at frequencies <2 Hz, while at >10 Hz, $Q_{seis}$ is often as high as 1000 due to the combined effect of deeper crustal stress levels, and a presumed more massive rock mass.

### 10.11

The late coda is assumed to contain a variety of take-off angles from the source. The frequency dependence is presented in the form $Q_c = Q_o f^n$, with component n often increasing from pre- to post-event values. Component n may be just below to close to 1.0. A greater density of scatterers after a seismic event has been suggested as the reason for greater frequency dependence after a major event, which also suggests that frictional and partial saturation losses will be involved when new discontinuity surfaces are being developed, or extended.

### 10.12

Temporal variations of coda $Q_c$ have been reported in the periods before and following significant earthquakes. From 30% to 300% *reduction* of coda Q prior to six M 6.0 to 8.0 earthquakes have been reported, with an *increase* in coda Q in only one case, and the possibility of no obvious relation between coda Qc and a single major earthquake. Temporal variations have more recently been interpreted as due to migrating hypocentres, with an example Qc reduction from 200 to 100 interpreted as a migration during the early period of activity, to after the activity. In rock mechanics terms, this could approximate to changing from a higher modulus, higher stressed zone, to one with lower modulus and lower stress, after the event.

### 10.13

At the M 7.2 Hyogoken Nanbu earthquake near Kobe, seismic $Q_c$ was shown to have reduced following the earthquake. At respective frequencies of 1.5 and 2.0 Hz, $Q_c$ reduced from 81 to 68, and from 91 to 78, perhaps representing seismic sampling at increasing depth with higher frequency. At respective frequencies of 3.0 and 4.0 Hz $Q_c$ reduced from 132 to 107 and from 186 to 162. All the above resemble feasible deformation modulus trends, but obviously progressing to depths beyond existing $E_{mass}$ data.

### 10.14

The more recent possibility to record seismic events using down-hole seismometers, to depths of 100's of meters and even to 2 and 3 km, has made it possible at some localities, notably near the San Andreas Fault, to avoid contamination by seismic and man-made noise, and to avoid much of the 'site effect' and it's unwanted influence on recordings. Amplification in the low velocity near surface, with increased scattering and intrinsic losses, can have had a strong effect on all surface recordings. Attenuation at shallow depths *beneath* the surface layers, appears to be little influenced by rock type, with similar low values of $Q_p$ and $Q_s$ in the first 1 to 2 km down wells in widely different rock types like tertiary sediments and crystalline basement. Perhaps each of these dissimilar rocks are nevertheless stiff enough to have a partially common tectonic imprint of jointing and faulting from the same neighbouring major fault.

### 10.15

$Q_p$ values of 20, 30 and 55 from analysis of depth intervals 0–300 m, 300–940 m, and 570–940 m, and $Q_p$ increasing to 110–170 from 1 to 3 km depth, are also following the likely pattern of deformation moduli, when $Q_p$ values are expressed as if GPa. The actual addition of scattering and intrinsic losses are of course not distinguished in such a simple modulus model, but since the modulus estimate actually includes block size, inter-block friction and effective stress-to-strength ratio, it is in some way *sampling* important causes and components of both classes of attenuation. Reduced fracture density with depth, reduced friction losses on increasingly stressed fractures, and scattering from a reduced number of fractures, are reported reasons for the increasing $Q_{seis}$ values with depth. These are all recipies for increased deformation modulus too, and a common experience in deep, as opposed to shallow tunnels.

### 10.16

For cross-continent studies of seismic coda, the so-called *Lg* phase common on regional short-period seismograms is often used. It is followed by the main coda, which can also be used for determining the magnitude of regional events. The *Lg* coda at 1 Hz is commonly referred to as

$Q_o$, and has been used to describe deep crustal attenuation across continents and in component plates, orogenic belts and major sedimentary basins. Comparing South and North America on this scale, one sees lower $Q_o$ in the Andean Belt (250–450), and in the Basin and Range province west of the Rocky Mountains (250–300). By comparison, there are broad regions of very high $Q_o$ spanning the central Brazilian Shield and Amazonian Basins (700–1100), and in the eastern region in the USA below the Great Lakes.

### 10.17

Higher values of $Q_o$ reportedly reflect the length of time since the last major tectonic activity. Low $Q_o$ regions are typical of seismically active regions with higher upper mantle temperatures, or the presence of deep hydrothermal fluids, and variable amounts of fluids in major faults. Younger sediments cause local reductions of $Q_o$ due to their contained fluids, while older sedimentary rocks which have lost fluid, are less attenuating. A shallow sediment model for explaining shallow and lateral variation of attenuation shows $Q_{seis}$ values of 30, 50, 75 and 100 for sandstones and shales from 0–100 m, 100–300 m, 300–600 m and >600 m. Units of GPa are again suggested in broad terms.

### 10.18

Seismic Q has become increasingly important in hydrocarbon exploration due to the improved sensitivity to degree of saturation of the ratio $Q_s/Q_p$ compared to the velocity ratio $V_p/V_s$. The presence of low seismic Q in petroleum reservoirs can reflect the presence of over-pressure, of fracturing, and oil. Values of seismic Q as low as 10 to 40 at more than 4 km depth are good indications of favourable properties, likewise are pay-zones with over-pressured gas giving seismic Q as low as 10 to 40, when the remainder of the sequence shows seismic Q between about 50 and 130.

### 10.19

In reservoirs where cross-well tomography is performed, there is a possibility of improved definition of the structures, due to the high frequency and multiple ray-path coverage. Transversely isotropic 'layer-cake' sedimentary inter-bedding is found to be quite attenuating, with good differentiation between shale, limestone and clay. A major, dipping fault zone in a North Sea reservoir, between 1 km and 2 km depth, was interpreted from down-going P-waves as having a seismic $Q_p$ value as low as 45. Analytical modelling suggested a strong inequality between the high *dynamic* shear compliance, and the more conventional lower normal compliance. Low *pseudo-static* shear stiffness for fault-scale structures is also a necessary input to numerical distinct element models, where realistic subsidence modelling is to match steep subsidence bowls, in situations where continuum modelling fails to match such measurements.

### 10.20

In Chapter 13, there are numerous further sets of seismic Q data from laboratory-scale testing, in which the velocity and attenuation data are given side-by-side. Tests at reservoir confining stress levels and with numerous fluids (brine, gas, oil) will also be found. The rock physics data also extends into velocity and attenuation testing of the effects of bedding, foliation and induced fracturing, and the anisotropy caused by these features. In Chapter 15, the treatment of poro-elastic modelling also addresses the dispersive and anisotropic nature of seismic Q, as caused by the coupled dynamic behaviour of hydraulically connected equant pores, microcracks and aligned sets of fractures of various sizes

## Chapter 11 Velocity structure of the earth's crust

### 11.1

The *uppermost* 5 km of the crust shows a rapid increase in deformation modulus and density, as pore space and joints are closed. However, the thermal expansion partly balances the increase in seismic velocity, and P-wave velocities above about 6.5 km/s do not appear to be common here. The crust usually varies from 20 to 60 km in thickness beneath continents, while the oceanic crust is much thinner, and is usually about 6 to 7 km thick, beneath an average water depth of 4.5 km. The typical crustal P-wave velocity range is 6.0 to 6.8 km/s. In the upper 5 km, and excluding sedimentary rock with $V_p < 5$ km/s, the most typical range of velocity is 6.0 to 6.2 km/s.

### 11.2

An almost linear velocity-depth gradient between 5 and 25 km for the average crust shows a gradient of about $0.6/20 = 0.03\,s^{-1}$, while the gradient between 5 and 10 km is approximately $0.5/5 = 0.1\,s^{-1}$. Velocity-depth gradients are more than an order of magnitude steeper than this in the upper kilometre, and much steeper again in the upper 25–100 m. Gradients are even more pronounced in the upper 25 m of weathering-effected rock mass. The reduced gradient at great depth is due to the expansion effect caused by increased temperature.

### 11.3

Continental-scale velocity-depth profiles show strong *visual* resemblance to *near-surface* refraction seismic profiles to *50 m* depth in place of *50 km* depth. A uniform subtraction of about 2 km/s velocity, and a scale reduction of 1:1000 gives an almost indistinguishable result from the laterally-varying and depth-varying shallow refraction seismic obtained at a rock engineering site.

### 11.4

For many years, *sub-oceanic* marine seismic refraction profiles were interpreted as a small number of layers (Layer 2 to 2 km, Layer 3 to 7 km), separated by planar interfaces, with a constant velocity assumption for each layer. Homogeneous layering assumptions from the 1960s were first replaced by much finer layering (2A, 2B, 2C and 3A and 3B), and then in the mid-1970s by continuous gradients in velocity. The first geophysical *downhole* sonic *logging* data for oceanic crustal material was near a portion of the mid-Atlantic ridge in a leg of the Deep Sea Drilling Project. Velocities were typically from 1.5 to 4.8 km/s in the upper 200 m of oceanic Layer 2A. Interpreted porosities of 13 to 41% were unexpectedly high.

### 11.5

The reasons for high porosities were interpreted as being due to a combination of sediments, rubble, and solid basalt in contrast to the compact nature of basalt samples used in laboratory tests, which often showed $V_p$ between 5.5 and 6 km/s and porosities from only about 2 to 8%. Open fractures and voids were assumed to exist on a scale larger than the laboratory samples, also giving high permeability throughout the drilled section.

### 11.6

The P-wave velocity of the sub-ocean crust at and near ridge crests exhibit an increase in velocity with age. Numerous results from the Atlantic and Pacific mid-ocean ridge studies show an obvious link between $V_p$ and age, with an increase of $V_p$ of 2–3 km/s, in the first 40 million years. Deeper older layers do not show systematic increase in velocity.

### 11.7

Early laboratory testing of oceanic basement rocks from deep drilling in the mid-Atlantic ridge highlighted the discrepancy between laboratory seismic properties and in situ, bulk velocities obtained from seismic refraction. Confining pressures of 20, 50, and even 200 MPa were used in early on-board velocity measurements. Laboratory velocities of 5.5 to 6.5 km/s were typically obtained with 50 MPa confining pressure. The comparable, shallow depth refraction-seismic inferred velocity-depth profiles showed only 2.5 to 3.5 km/s in situ velocities, an *apparent* discrepancy of about 3 km/s relative to the intact rock, but in fact, some of this difference was due to the excessively high confining pressures applied to the intact samples.

### 11.8

It was subsequently realised that it must be the rapidly increasing (from zero) effective stresses, not the assumed *external* ocean-depth loads, that were acting on the shallow sub-ocean crust, that was causing the rapid increase in velocity with depth. In other words, the velocity gradients were similar to what is found at the earth's surface, with rock and (effective) fluid loads both increasing from zero. The theory of effective stress was apparently late in being adopted in this hostile sub-ocean environment.

### 11.9

Increasing effective stress from the ocean floor was responsible for about $4\,s^{-1}$ velocity-depth gradients, together with presumed 4 to 5% per 100 m porosity reduction, as seen in the first 200 to 300 m of sub-sea layer 2A. The differences between in situ velocity measurements in the shallow oceanic crust and the higher matrix velocities measured at suitable (low) effective stress levels, was not only caused by moderate changes to the matrix porosity, but also by low aspect ratio jointing and fracturing, which was more stress-sensitive.

### 11.10

Very high velocity gradients, similar to the above, are typically experienced at the earth's surface because the rock quality Q-value, as well as the effective stress, are both increasing rapidly with depth. We may have a near-surface $Q = 0.1$ followed rapidly by $Q = 1$ and then $Q = 10$, suggesting nominal 'near-surface' theoretical increases in $V_p$ from 2.5 to 3.5 to 4.5 km/s, with the *additional* effect of increasing depth and therefore increasing effective stress. This dual effect requires 'curve jumping' in the Q-$V_p$-depth relation.

### 11.11

A *direct* measurement of Upper Oceanic Crust P-wave *attenuation* was described in 1990, using seafloor hydrophones and large explosive sources. The site was on 0.4 m.y. old crust, and had a seafloor velocity of 2.7 km/s, which increased uniformly to 5.6 km/s at 680 m depth. Gradients as high as $4.6\,s^{-1}$ near the surface and $4.1\,s^{-1}$ at greater depth were estimated. Values of seismic $Q_p$ varied from 4 to 275, but mostly clustered between about 10 and 20 in the upper 100 m, similar again to expected deformation moduli in GPa.

### 11.12

A consistent increase of $Q_p$ with depth was not found, but several sets of data for $1/Q_p$ did show a 'stepped' trend of $1/Q_p$ reducing with depth in the first 500 m. Values of $Q_p$ were mostly from about 8 to 50 in this depth range. The sudden steps up, and down from, very high $Q_p$ values like 200–300, even negative $1/Q_p$ steps, leads one to question whether the early ship-board triaxial test routines had an element of (local) correctness. Some volumes of intact basalt could perhaps be subject to high 30 MPa rock-plus-water confinement loads, interspersed by a majority of permeable and therefore low effective-stress-loaded permeable blocks.

### 11.13

The $V_p$-Q-value-porosity-depth model uses a plotting format that can readily be compared with the oceanic crust fracture zone data of Layer 2A and 2B. Strong similarity with oceanic data is seen, probably because the rock quality Q-value specifically represents the 'soft porosity' or jointing, and models the effect on velocity of gradual joint closure with depth. The very steep $V_p$-depth gradients typically seen close to the ocean floor, in the first few hundreds of meters of the new crust, can be analysed with this near-surface based empirical method, which was developed from *near-surface* civil engineering projects. 'Curve-jumping' is needed to explain the supposedly 'anomalously high' gradients through Layers 2A and 2B. These can be modelled by assuming increased Q-values for deeper, older material.

### 11.14

East Pacific rise studies were made by many of the researchers known for their mid-Atlantic ridge studies. A linear velocity-depth gradient in the upper 500 to 800 m of young (0 to 4 m.y.) oceanic crust on the flanks of the East Pacific Rise was initially assumed, and an average gradient of between 3.0 and $3.5\,s^{-1}$ for the upper 0.5 to 0.8 km of oceanic crust was estimated, with seabed velocities ranging from as little as 1.9 to 2.7 km/s. The evidence of very low velocities in the upper-most oceanic crust was consistent with visual and photographic evidence from submersibles, of pervasive fracturing in mid-ocean ridge crustal regions, where the basalt layer was exposed. Low velocities were also consistent with drilling and logging results that showed high porosity.

### 11.15

The problems posed by zero-age oceanic crust with $V_p = 2$ km/s, compared to about 6 km/s for intact basalt continued to provide challenges for theoreticians and practitioners working on the origin, formation and structure of mid-oceanic crust. Low aspect ratio cracks, and their reduced frequency of occurrence *and* reduction in aperture with depth, and probable sealing with hydrothermal minerals in the case of older oceanic crust, were some of the variables confronting those researching the variable structure of mid-ocean crusts. It was theorised that young 120 ka material with $V_p = 2.5$ km/s, must have a porosity of between 24 and 34%. Slower $V_p = 2.2$ km/s zero-age crust was theorised to have a porosity of between 26 and 43%.

### 11.16

At the beginning of the 1980s, a sub-ocean Deep Sea Drilling Project borehole in the Costa Rica Ridge area, made it possible to correlate core, usually of low % recovery, with downhole sonic logs, borehole televiewer logs, and permeability test results. This was first performed to a depth of 1 km, through layers 2A, 2B and 2C. In the 0–100 m, 100–650 m and 650–1000 m depth zones, P-wave velocities were a 'familiar' 3.7, 4.8 and 5.6 km/s, and permeabilities were likewise a 'familiar' $10^{-6}$–$10^{-7}$, $10^{-8}$–$10^{-9}$ and $10^{-9}$–$10^{-10}$ m/s. Based on vertical borehole logging, which would be biased *against* vertical structure, the upper 50 metres was found to contain numerous horizontal to sub-horizontal fractures, thick basalt flow units, and thin interbeds of pillow structures.

### 11.17

Large scale three-dimensional tomography was performed on the East Pacific Rise sub-ocean crust at the end of the 80's, at the location of a fast spreading ridge. This was characterised by a sharp upper-crustal to mid-crustal velocity inversion some 1.5 to 2 km below the seafloor. This was presumed to be the roof of an axial magma lense, with an assumed few percent of melt. Contours of seismic Q at 4 km depth, showed values of $Q_{seis}$ of 25 and 33 nearest the 'magma', and values of 50 and 100 at 2 to 3 km off-ridge distances, in older crust. Values of $Q_{seis}$ as low as 8 and 10 have been measured at the Costa Rica Ridge area, caused by local faulting or similar features at ½ and 1 km depth in the much referred Hole 504B.

### 11.18

Inversions for individual receivers showed that seismic Q increased from average values of 40–50 in the upper 1 km, to at least 500–1000 at depth greater than 2 to 3 km. One may speculate that this abrupt increase could be due to an *undrained* increment of effective vertical stress of about 30 MPa representing the ocean load, since such very high seismic Q values appear *here* at 'shallow' (rock) depth compared to generally greater rock depths on-land. With a thin, warm sub-ocean crust, permeability may be compromised at relatively shallower depths than under tectonically deformed continental crust.

### 11.19

Age effect reviews for both mid-Atlantic Ridge and Pacific Rise data show that most of the age-dependent increase in seismic velocity occurs 'rapidly' with velocities nearly doubling in less than 10 million years. Layer 2A appears to persist as a low velocity capping of the ocean crust, even when more than 15 m.y. old. The trend for increased velocities as age increases is clearly shown by the statistics. A velocity plateau, averaging about 4.3 km/s is indicated, beyond about 7 m.y. There is a clear link between hydrothermal alteration and seismic velocity increase, due to deposition of minerals first in the thinnest cracks and joints. Hydrothermal void filling causes a simultaneous increase in velocity and reduction in hydraulic conductivity, therefore reducing heat flow to the ocean floor.

### 11.20

In regions with significant sediment cover, the previously open seawater convection cooling system is hindered, and temperatures rise, thereby accelerating the formation of secondary minerals and porosity sealing. The *on-axis*, zero age upper crustal permeability has been deduced to be about $6 \times 10^{-5}$ m/s, decreasing to about $7 \times 10^{-7}$ m/s within 6 m.y. Seismic velocities for crust of the same age are about 2.2 km/s and 4.0 km/s. Permeability may reduce to about $10^{-7}$ m/s or less, by the time the crust is old enough to have reached the approx. 4.3 km/s 'plateau'.

### 11.21

Parallels to the hydrothermal mineral filling of fractures, and increased velocities may be gleaned from civil engineering, where the sealing of jointed rock by pre-grouting with fine-grained micro-cements ahead of tunnels, or the use of industrial, coarser grain size cements in more permeable dam foundations, are common ways of both sealing and improving rock mass properties. Rotations and magnitude reductions of the three principal *permeability tensors*, when conducting multiple-borehole 3D hydrotomography before and after grouting have been documented.

### 11.22

These tensor or principal value rotations are interpreted as due to successive sealing of the most permeable and least stressed joint sets. This process can also be interpreted by small changes in five or six of the rock quality Q-value parameters (ratings in Appendix A), which may cause some dramatic potential improvements in the rock mass properties such as increased modulus of deformation, seismic velocity, and frictional and cohesive strength. There is a degree of correspondence to sub-ocean velocity increase, and their expected geomechanical effects.

## Chapter 12 Rock stress, pore pressure, borehole stability and sonic logging

### 12.1

Hydrocarbon-bearing rocks rely on pore-space and permeability for the possibility of having recoverable reserves that can be produced at a well. The necessary migration of the hydrocarbons from source rocks into potential entrapment structures, without escape to the atmosphere, adds to the adverse statistics of hydrocarbon discoveries. Too close to the surface the sealing properties of shale, salt or clay-smear in faults, may have been compromised by lack of plasticity and too high permeability. Too deep, the pore space and permeability of the reservoir may be compromised, giving a reduced reserve and the need for permeability enhancement and gradient enhancement, or a decision for non-development.

### 12.2

Besides reservoir access for production testing, a drillhole is used for sonic logging and selected side-wall and regular core recovery, to better define the properties of the different lithologies, seals and reservoir rocks. Rock reacts to the drilling of boreholes with a complex interaction of rock stress and strength magnitudes, plus the anisotropies of each, and is affected by the necessary subtraction of pore pressure for conversion of the three principal stresses to *effective stresses*. The appropriate selection of wellbore 'temporary support' in the form of mud pressure, using variable mud weight, determines the state of the borehole wall in the different lithologies, prior to setting and cementing the casing.

### 12.3

Due to various opinions about an *alteration zone* around deeper wells, there is now widespread acceptance of the need for logging while drilling (LWD) with monopole and dipole tools, to obtain 'early' velocity responses, which may differ significantly from subsequent wireline logging. The differences are probably due to stress-fracturing, increased permeability, and consequently accelerated mud-filtrate invasion. A near-wellbore, tangentially-distributed log-spiral-type *discontinuum*, in case of an insufficiently mud-weight supported weaker formation, may need to be considered when interpreting the two sets of seismic data.

### 12.4

It is important to consider the components and modifiers of the most fundamental of reservoir parameters, namely the *effective stress magnitude*. The rock stress and its variations with direction, depth and location, and the pore pressure and sometimes over-pressure which are influenced by compaction and also by fluid type, are the major boundary conditions. Their relative magnitudes affect both the laboratory test simulations, the drilling programme, the production planning, and the reservoir production and depletion, possibly for 80 years or more in a large reservoir.

### 12.5

To prevent hydraulic fracturing by high mud-weights, which are needed where there is overpressure, casing will be set to protect the overlying units from fracturing. A change from a pressure-depth axis, to mud-weight-depth format is preferred by drillers, who speak of mud-weight in lb/gal, and try to steer this between the pore pressure gradient and the fracture-gradient. Resistivity, velocity, and density depth-trends will each suffer various degrees of deviation from the norm, when there is over-pressure that changes the effective stress.

### 12.6

The fluid pressure at a well is the sum of normal hydrostatic pressure, plus over-pressure, plus a buoyancy effect caused by the reduced density of any petroleum that is present. Since over-pressure and the *presence* of petroleum products both increase the pore pressure, the effective stress will also be reduced, which will have the effect of causing a reduction in velocity.

### 12.7

Over-pressure commonly occurs where low permeability layers such as shale prevent fluid from escaping as rapidly as pore space compacts. Excess pressure in relation to hydrostatic then builds up as newly deposited sediments cause squeezing of the trapped pore fluids, which could be water, oil or gas or two or three of these close together. Models for basin-evolution show that pore pressure effects are seismically visible when the *effective pressure* is typically less than about 15 MPa, A small % conversion of *live oil* to gas is sufficient to make the pore pressure equal to the confining pressure. The large changes of predicted velocity are caused by the fact that the dry rock moduli are strongly affected by low effective pressures.

### 12.8

Minimum rock stress estimation by mini-hydrofracing, if not possible in an open-hole situation, is done by seating the double-packers on either side of shaped-charge perforations of the casing. This is done in the reservoir intervals, and also in the cap-rock interval, to determine the minimum stress difference. Interbedded 'brittle' layers like sandstone, and 'plastic' layers like shale or salt, will usually exhibit different minimum principal stress levels. This may be an additional reason for oscillating sonic log records in such interbedded strata.

### 12.9

Shale and salt-rocks may have insufficient shear strength to tolerate a significant principal stress difference. The minimum of the three principal rock stress magnitudes in the shale or salt will therefore often exceed the minimum stress in the reservoir sandstone by up to several MPa. This is desirable for hydrocarbon containment, and also for vertically limiting massive hydraulic fracture treatments. Rocks such as granite, limestones, stronger sandstones and stronger shales (which are thereby poor seals), tolerate differential stress much better than weaker shales and almost plastic salt rocks.

### 12.10

Shale or salt may, if encountered at sufficient depth during drilling, require the support of an active mud-weight to prevent creep or squeezing. The drillers choice of mud-weight, or the setting of protective casing, becomes critical where support of the well is needed adjacent to a reservoir rock like sandstone or fractured limestone or chalk, which would tend to have a *minimum* rock stress less than that of these weaker, sealing 'plastic' layers. The reservoir horizons could potentially fracture, or have a permeable joint under lower normal stress than the mud-weight needed to keep the plastic materials from squeezing and jamming the drill-string. Invasion of mud or lost circulation, into any reservoir horizon is obviously very undesirable.

### 12.11

As one approaches the surface, inter-bedded rock types resembling reservoir sequences, show the reverse of the previously discussed differential stress intolerance, because the weaker rocks are no longer 'over-stressed'. Furthermore, because of their lower deformation moduli, they attract lower stresses from a given horizontal stress field. Hydraulic fracturing tests therefore may give indications of low $K_o$ ratios ($\sigma_{h\ min}/\sigma_v$) in the weaker materials like shale and siltstone, and higher values in sandstones and limestones.

### 12.12

This reversal of $K_o$ trends at a certain depth (it was measured from 100–150 m depth, but might apply from 0 to 500 m), may have implications when comparing stress-induced velocity anisotropy and sonic log velocity 'oscillation' near-surface and at greater depth. This reversed behaviour also needs to be considered when evaluating the applicability of shallow borehole seismic testing to reservoir holes perhaps an order of magnitude deeper, with corresponding reversed $K_o$ behaviour. There may also be consequences for the relative magnitudes of attenuation, as both lower $K_o$ and lower stress levels near the surface, will tend to enhance attenuation. Thus $Q_p$ and $Q_s$ values must be expected to be lower, and exhibit more anisotropy near-surface than at depth.

### 12.13

The hydrocarbon reservoir exploration and production industry has long been aware that borehole deformation and failure modes are an important 'complication' concerning the interpretation of sonic-logging of wells. There are now acoustic dipole and monopole shear-wave producing logging devices that can be used in a *logging while drilling* LWD mode, that acquire responses from more than one hundred wave forms, some tens of meters behind the drill-bit, in order to delineate formation fracturing response, and virgin conditions further from the walls of the wells, before additional 'alteration' has occurred from stress and/or mud-filtrate invasion, as often seen in subsequent wireline logging, when the drill-string is removed.

### 12.14

There are possibilities for local velocity (and seismic Q) *enhancement* due to tangential (and diametrically-opposite) stress increase in the case of competent rock like limestones, or low porosity sandstones. In the case of over-stressed, fractured ('dog-eared') sections of rock, and especially in the case of incompetent rocks like shales, *reduction* of velocity (and of seismic Q) will occur locally, due to the mini-EDZ (excavation disturbed zone) that form as a result of drilling and possible over-stressing.

### 12.15

The mini-EDZ may mean the development of a log-spirally sheared *discontinuum,* based on physical and numerical modelling results, where the starting point was a continuum. Stress reduction in the radial direction, which may be azimuthally varying, will tend to locally reduce the velocity, and thereby also the seismic Q. It is impossible to support each lithology with the ideal mud-weight, so some suffer the consequences, just as occurs in weak zones in a tunnel where the contractor might be trying to reduce temporary support costs by under-supporting, using for example a constant single layer of sprayed concrete, where two layers were actually needed locally.

### 12.16

Due to the smaller size of wells and the use of mud for hole support, the recognition of the behavioural data that can be extracted from anisotropic stress effects on small-scale EDZ round wells, may possibly not be used in the petroleum industry, to the extent it can be used in tunnel engineering. Borehole ellipticity, a much-used historical indicator of the minimum horizontal stress axis, is the surface expression of effects *behind* the ellipticity. In tunnels it is easy to see the effects of structure-induced wedge release, or stress-fractured 'lenses' of rock. It is also possible to install multiple-position borehole extensometers (MPBX), in tunnels and in vertical shafts, to measure the anisotropic radial-distribution of deformation, thereby giving deformation moduli as a function of direction. Velocity variations and permeability variations as a function of position and radial depth around a tunnel or shaft, can also be determined, thereby relating these parameters to eventual stress anisotropy.

### 12.17

There is a strong likelihood that mini-EDZ in the weaker, less well mud-supported zones, have reduced, radial-dependent velocity, due to failure and deformation in the over-stressed zones. Stronger inter-beds could show an opposite trend due to tangential stress enhancement of the velocities. Mini-EDZ that might penetrate several diameters can be *detected, and circumvented* by deeper sensing, shear-wave based, dipole logging tools. The probable *discontinuum* caused by log-spiral shearing is often referred to only as 'shale alteration'. Fabric and jointing and bedding planes, may also affect the progress rates for mud-filtrate invasion. The geomechanics of borehole deformation and over-stress, and its coupled MHT effect on permeability, mud-filtration, and LWD-to-wireline logging differences, can be quantified in approximate terms.

### 12.19

Deeper penetration of mini-EDZ, representing 'shale alteration', may be the reason for a serious potential contrast in logging results, when comparing the 1–2 weeks later result of wireline logging, with the few hours delay represented by LWD, or logging while drilling. More recent *shear-wave anisotropy* based logging, is capable of imaging a volume of up to several borehole diameters away from the wall, therefore *beyond* the stress-related fracturing and mud-filtrate invasion or 'shale alteration', thereby giving presumed 'virgin' formation attributes as well.

## 12.20

Variable azimuth drilling in test blocks under 3D stress states, gives failure modes that cannot be obtained when loading a test block with a *pre*-drilled hole. Deep log-spiral shear failure surfaces have been demonstrated in weak cemented-sand blocks, when the major principal stress was about *eight to ten times* higher than the uniaxial strength, with the minor and intermediate principal stresses of 60% or 80% of the maximum. This level of over-stress is easily reached in deep wells in relation to shale and salt rocks.

## 12.21

With extreme weakness, the failure mode may be non-dilatant, and actually contracting-with-shear, or even flow, in the case of clay-rich materials. There are distinctive differences between stress-induced failure of hard dilatant brittle rocks, giving extensional splitting and subsequent crushing or comminution of the rock in the sharp ends of diametrically opposite 'V-shaped' corners. In the case of failure in intermediate strength and less dilatant rocks, the traditional 'dog-earing' takes on a different shape resembling localized log-spiral shear failure surfaces. Each of these modes can be demonstrated in physical simulations, and with suitable choice of numerical model, though not with conventional rock failure criteria.

## 12.22

The actual modes of physical behaviour experienced by boreholes and tunnels, are unlikely to be predictable when modelling with conventional Mohr-Coulomb type ($c + \tan\varphi$) shear strength criteria, because *intact* rock tends to fail first by loss of continuity at *small strain*, caused by loss of local tensile or cohesive strength, followed by frictional mobilization at *larger strain.* In modified Mohr-Coulomb terms it is a case of 'c *then* tan φ', not 'c *plus* tan φ'. Numerical models that are programmed, or manually-steered, to dissipate cohesion while mobilizing friction, are capable of matching physically observed behaviour. Non-linear fracture mechanics boundary-element based modelling seems to mirror reality extremely well, with log-spiral type fracture development that dissipates over-stress.

## 12.23

Different degrees of 'log-spiral-type' shear failure are demonstrated, depending on the 'disturbance' to the isotropic stress distributions, caused by different amounts of jointing or fissuring close to the hole. These geological features dissipate some of the highest, near-wall tangential stresses seen in elastic isotropic analyses. When modelling medium strong brittle rock, principal stresses of only about 35–40% of the uniaxial strengths are needed to start fracturing in the form of initial 'dog-earing'. A brittle sandstone of 50 MPa UCS would be acted on by an equivalent ratio of strength to stress beyond about 1200–1300 m depth, with standard density and pore pressure assumptions, considering a $\sigma_{H\,max}$ value no larger than the vertical effective stress. The existence of dog-earing may provide more information about formation properties than the simple registration of stress-induced ellipticity in a four-arm calliper log.

## 12.24

Due to the influence of deformation of 'soft' as opposed to 'hard' porosity, a borehole for hydrocarbon exploration that penetrates variably jointed and faulted ground, will actually experience variable small amounts of *deformation*, due to different degrees of joint closure, joint opening, and joint shearing. There will also be the pseudo-elastic response, due to both *loading* (at the diametrically-opposite $\sigma_{\theta\,max}$ locations) and potential *unloading* (at the diametrically-opposite $\sigma_{\theta\,min}$ locations) of the matrix as well as the joints, the latter usually dissipating some of the theoretical (isotropic, elastic) peaks of maximum and minimum tangential stress. This process will occur even with a constant mud-weight, since the mud – unlike rock bolts in a tunnel – cannot prevent joint movements of unequal magnitude at different points around the opening, although the mud may help to make them very small.

### 12.25

When assessing the rock quality of the walls and arches of tunnels, the *observed rock*, which is the visible part of the tunnel-scale EDZ, is *classified* (using the Q-system: see Appendix A), in order to select appropriate rock reinforcement (grouted rock bolts) and tunnel support (sprayed, steel-fibre-reinforced concrete). The latter is the equivalent of the borehole mud pressure, and is badly needed in a complete load-bearing ring, in the rapidly deteriorating and deforming clay-bearing zones, in order to control deformation and prevent local tunnel collapse. Outside the tunnel EDZ, the rock mass would be *characterized* as a better quality rock mass.

### 12.26

If one performs both sonic logging and azimuthal dipole sonic logging in a well, the borehole *mini-EDZ* can be *classified* by the one tool, while the hydrocarbon-bearing or reservoir sealing formation away from the immediate influence of the hole, can be *characterized* by the other tool, and show higher velocity and higher seismic Q as well. It appears that this familiar rock mechanics EDZ logic is effectively being applied in modern well logging, with its multi-wave-form acquisition.

### 12.27

Dipole transmitter tools are designed to generate *flexural waves*. Flexural waves are shear waves that are polarized into *fast* and *slow* directions, and penetrate several hole-diameters into the formation, thereby revealing potential stress-induced 'alteration', and/or drilling mud-induced alteration. The need for these tools confirms many of the foregoing suspicions that what we have termed *mini*-EDZ, are indeed a source of concern in certain formations, and more importantly, that these 'alteration zones' can be detected and seismically *classified*. The shear-wave analysis can also be used when *characterizing* the formation beyond the damage zone.

### 12.28

The 'altered zone' around the borehole may continue to develop during the week or so that may separate the two types of logging. The later wireline log may be influenced by a reduced elastic modulus in an annular zone around the borehole, particularly in soft formations such as shales and shaly sands. Entrapment of wave-fronts in the lower modulus damage zones results in 'bi-compressional arrivals', or second arrival compressional waves. The bi-compressional arrival is a phantom arrival too fast to be a shear wave, and actually caused by trapping of the wave-front by the low-modulus damage zone.

### 12.29

Orthogonal dipole transmitters, and the multiple receiver pairs, which are aligned in orthogonal directions, measure the components of *slowness* in any direction within planes *perpendicular* to the borehole. They use the principal of shear-wave splitting, and polarization, one of the most valuable of all seismic anisotropy properties for fracture and fracture-fluid investigations. The slow direction (as also with P-waves) is perpendicular to the fracturing – which could be microcracks, cracks, joints or faults – according to the scale of example considered. The rotated direction of the fastest shear waves becomes the *fast-shear tool azimuth*. Both the acoustic time anisotropy and the slowness anisotropy are sensitive to properties deeper within the formation than the superficial effects caused by drilling.

### 12.30

The fact that shear wave anisotropy allows the investigation of a volume of the formation up to several diameters from the borehole axis, means that it can sense jointing, and stress-induced fracturing, that are missed by conventional logging tools. This means that it is particularly useful for registering the additional jointing and fracturing that tends to be present on either side of a fault.

### 12.31

LWD with dipole shear-wave anisotropy analysis is available almost in real-time, some hours behind the drill-bit. It has proved very useful when applied to drilling of *horizontal* well sections, designed to intersect a maximum amount of structure. Simultaneously one can avoid the less favourable parallel to $\sigma_{H\ max}$ hole direction. Early warning is also given while drilling in formations with rapidly changing pore pressure. LWD is then an important aid in choosing appropriate mud pressures. The use of wireline dipole logging in vertical holes, and pipe-conveyed dipole tools for deviated and horizontal wells, has given reservoir geophysicists improved means of *calibrating* the responses of their rock physics based reservoir models, against these small-scale, but *in situ* measurements.

### 12.32

Synthetic seismograms often do not correlate with measured seismograms, when correlating seismic data with acoustic logs. Formation properties inferred from wireline logging measurements may not reflect the true properties, so a realistic description of the mud-filtrate invaded damage zone is important for processing and interpretation of the logs. A problem is caused by the invaded or 'altered' zone being deeper than that illuminated by the logging tool, meaning that the velocities will not reflect those of the formation, but of the damaged zone, therefore requiring corrections.

### 12.33

A *standard* approach is to correct the acoustic logs via a Biot-Gassmann *fluid substitution*, to free sonic logs from mud-filtrate invasion effects. It is assumed that the measured velocities are those of the invaded zone, saturated with mud filtrate. By displacing the saturation fluid *theoretically*, new velocities are obtained, and taken as the *virgin formation* velocities. Despite *theoretical* removal of the saturating fluid, one possibly should note from Chapter 15 the failure of the porous-medium based Biot-Gassmann method in the case of split shear-wave data in *fractured* as opposed to unfractured porous formations.

### 12.34

Ultra high pressure and high temperature (HPHT) wells that are increasingly relevant with high petroleum prices have several unfavourable effects on well stability. One is that the rock skeleton may bear a proportionally greater load, as the effective stress parameter ($\alpha$) becomes less than 1.0 (as per Terzaghi), when stress levels have caused reduced porosity and permeability. This higher effective stress causes slower drilling rates.

### 12.35

Mud cooling systems can be applied, which help to reduce the likelihood of compressive stress-induced failure, as matrix contraction reduces the maximum tangential stress concentration. On the adverse side, there is an increasing possibility of tensile failure and mud loss. So-called wellbore 'ballooning' represents such mud loss, but some of the mud is returned by subsequent heating. If there is no risk of compressive stress failure, then high mud temperatures can be used to control the hydraulic fracturing gradient. Shale and salt rocks pose particular challenges in HPHT environments.

## Chapter 13 Rock physics at laboratory scale

### 13.1

The major exploration-related goals of rock physics research have recently been summed up by King as: 'to understand how lithology, porosity, confining stress and pore pressure, pore fluid type and saturation, anisotropy and degree of fracturing, temperature, and frequency influence the velocities and attenuation of compressional P- and S-waves in sedimentary rocks'. At the end of this list 'and vice versa' was added, emphasising the interactive and complex nature of the reality. The remaining challenges were summed up as 'relationships between attenuation, anisotropy, fractures and fluid flow, and determining these relationships across the frequency spectrum of core, log and seismic measurements'.

### 13.2

It was recognized long ago that seismic velocities of rocks were strongly influenced by micro-cracks, and that seismic attributes representative of the intrinsic mineralogy and porosity, could only be obtained by applying pressure to the rocks. Much of the rock physics understanding of reservoir rock behaviour, of both matrix and joints or fractures, has therefore to be achieved at *elevated pressure*. The importance of *elevated temperature* is also well recognised, but is less frequently an experimental variable than it actually should be, especially if jointing is also sampled, as it should be in the case of fractured reservoirs.

### 13.3

Age-depth relationships derived from well analysis in sandstone-shale units have a certain grouping of velocities with age, due to variations of porosity and the resulting densities. Hard porosity in the form of pores tends to decrease with age and depth, while soft porosity in the form of joints tends to increase with age due to tectonic influences, but reduces strongly with depth. Only the hard porosity has a *significant* effect on density. A relatively ordered density-$V_p$ trend for chalk, limestone and dolomite is often seen, and reflects the simple mineralogy. Frequently widely scattered density-$V_p$ data for sandstones is evidence of the variable mineralogy of 'sandstones', with 10–15% variation in density possible for the same velocity, particularly in the case of tight gas sandstones. In contrast to these variations, $V_p$-$V_s$ trends are consistently uniform, as befits characterisation by seismic waves.

### 13.4

An early compilation by Faust of well survey results from some 500 petroleum wells in the USA and Canada, included data from about 300 kilometres of well sections. The great majority of data was for mixed *shale/sandstone/shale* sections. A non-systematic comparison of shale and sand (sandstone) velocities revealed an average discrepancy of only 350 ft/sec, or 106.7 m/s in velocity between these two, frequently inter-bedded units, the sandstone having the highest velocity by this small average margin. Remarkably close $V_p$ versus $V_s$ trends for water-saturated sandstones and shales, emphasise the remarkably similar seismic velocity signatures of these two 'dissimilar' lithologies, when in a compacted state. The necessity of using attenuation, and other techniques, for *seismically* distinguishing these two most essential reservoir 'partners' is clear.

### 13.5

A useful means of separating such formations is to plot impedance, or the product of velocity and density, versus $V_p/V_s$. Variation of porosity within such a diagram is a further means of distinguishing different formations, once they have been identified.

### 13.6

Theoretical modelling of porous rock behaviour has been used on many occasions for examining the numerous factors affecting seismic velocities. Theoretical formulations by Toksöz were used to represent the solid matrix, and the assumed spherical to oblate pores, using widely varying aspect ratios to match numerous laboratory data. Small aspect ratios, or flatter voids, caused the greatest reductions to elastic moduli and velocities. The properties of the saturating fluid (gas, oil or water) were found to produce greater effects on the compressional velocities than on the shear velocities. The P-wave velocities were predicted to be higher, and of course were measured as higher, when the rock was saturated with water, than when dry or gas-saturated.

### 13.7

When such theoretical models are fitted to P- and S-wave velocities that have been *measured* at different pressures, they are found to require pore shapes ranging from spheres to very fine cracks (aspect ratios from 1 to $10^{-5}$) for sandstones, limestones and granites, both under dry and saturated states. As igneous rocks have low porosities, the pore shape has great influence on the elastic and seismic properties, and dry and water-saturated behaviours are often very different as a result. Compressional velocities are predicted and measured, as highest with brine saturation, and lowest with gas saturation. These differences decline with increasing effective pressure. Poisson's ratios for gas saturated rocks are predicted to be lower than for those with brine saturation, and this difference persists to high pressures.

### 13.8

From extensive data sets and modelling we can summarize from Toksöz that $V_p$ is likely to be lower: if low water saturation, if dry or gas saturated (*in the case of flatter pores*), if in the presence of some immiscible gas (in brine), if higher porosity, if over-pressured, if at shallow depth, if containing thin pores, following several cycles of freezing, if at room temperature, if at extremely high temperature. $V_p$ is likely to be higher: with water saturation, when dry or gas saturated (*in the case of rounder pores*), when saturated with brine, when there is no immiscible gas, when of lower porosity, when underpressured, when at greater depth, when consisting of rounded pores, when frozen, when at low or moderately high temperature.

## 13.9

Water-flooding, and four dimensional seismic monitoring of its effects, relies on the strong dependence of velocity on temperature, and on the significant influence of the relative hydrocarbon and brine saturations. Both the matrix, but especially the joints, will also sense the reduced effective stress, caused by the injection pressure and by the matrix shrinkage caused by cooling.

## 13.10

Concerning the matrix behaviour in the case of oil sands, the greatest effect of temperature is seen when there is 100% oil saturation, with P-wave velocity reducing by as much as 40% (e.g. 3.4 to 2.0 km/s) from 20° to 150° C. Least and almost zero effect is seen when there is either 100% gas saturation or 100% brine saturation. As oil is removed from these sands, the velocities successively become independent of temperature, with roughly half the effect when 50% oil remains. In viscous tar sands velocity may reduce by more than 50% (e.g. 2.4 to 1.1 km/s) when steam flooding reduces viscosity due to a similar temperature rise.

## 13.11

At low temperature the higher viscosity means that the oil cannot flow easily, so the dynamic measurement is on the high-frequency, high velocity, unrelaxed side of the local-flow mechanism. As temperature increases, viscosity reduces, so fluid flows more easily, and velocity therefore decreases since measurement is on the relaxed side of the absorption/dispersion mechanism.

## 13.12

Ultrasonic measurement of the effect of clay content on the P-wave velocity of sandstones may show a 40% or greater drop in velocity (e.g. 5 km/s to 3 km/s) as clay content rises to 30% or more in a simultaneously increasingly porous sandstone. The relative effects of clay content at frequencies of 10 Hz to 1 kHz (as used in seismic exploration) or frequencies of 10 to 20 kHz (as used in borehole logging), when imaging *in situ* sandstones, is less clear, due to potential effects of anisotropy. The complex nature of permeability, which depends on porosity, pore size distribution, inter-connectedness of the pores, and tortuosity, means that permeability may be severely compromised by increasing clay content. Variable clay content also typically occurs in bedding-parallel layers, making for strongly anisotropic permeabilities.

## 13.13

When relating rock physics laboratory data to the in situ reality of frequently inhomogeneous deposition cycles, the technique of 'fining-up', to benefit from the more uniform sedimentary environment and diagenetic nature of smaller deposition cycles, gives successively higher $V_P$-n% correlation coefficients. Sorting data into common sediment-compaction and cementation-history categories, using stratigraphy and other matching techniques, also has advantages when establishing relevant relationships between permeability and porosity, and between velocity and permeability. The log of permeability may be linearly related to porosity (e.g. 30%: 1 darcy, 20%: 10 md, 10%: 0.1 md).

## 13.14

Relationships between *velocity* and *permeability*, can be developed more easily by making a more relevant match of *porosity* and permeability with common sediment compaction and cementation history. Without doing this, velocity and permeability may show poor correlation. Simple 'classification' terms can be used. RQI is known as the *reservoir quality index*, and is composed of permeability (k) in units of millidarcies, and ($\varphi$) the fractional porosity. The *void ratio* ($\varepsilon$), the ratio of pore volume to solid volume, given by $\varphi/(1-\varphi)$, links RQI and FZI, which is known as the *flow zone indicator*. Rocks with FZI values within a narrow range are found to belong to one hydraulic unit and have similar flow properties. A semi-log plot of porosity versus log permeability tends to show similar FZI values plotting together with distinct log K-$V_P$ trends. The FZI classification can be extended to seismic parameters, to obtain a stronger correlation between *velocity* and *permeability*.

### 13.15

A uniform or smooth variation of velocity with degree of saturation is strictly a function of an assumed or actual homogeneous distribution of saturation, which is possible with lithological uniformity. The more complex and common effects of mixed lithological units, may create a heterogeneous or *patchy* saturation distribution. This gives different signatures during imbibition and drainage. The drainage process creates a more heterogeneous distribution of saturation. Local full saturation of the crack-like regions of the pore space tend to stiffen these regions in relation to high frequency, but at low frequency these 'patches' can drain to the less saturated pore space. The pore fluid lying in thin, compliant pores can flow freely into the dry pore space, in a squirt-flow type of attenuation response. It does not therefore allow reinforcement of the compliant part of the rock at low frequencies, so velocities are low.

### 13.16

At high frequencies, pressure equilibrium cannot occur because the pore fluid relaxation time is greater than the seismic wave period. Pore fluid in thin compliant pores is then effectively trapped, and it therefore reinforces the otherwise compliant pore spaces, resulting in higher apparent modulus and velocity.

### 13.17

Poisson's ratios are anomalously high in cases of over-pressure, where *effective stress* can approach the fracturing (negative) side of the usual lithostatic and pore pressure gradients. The aspect ratio of the cracks and pores and the nature of the saturating fluid determine the magnitude of ($\nu$). Rocks containing mainly stiff, equi-dimensional pores do not show major variations of ($\nu$) with effective stress. In saturated rocks the compliant pores become stiffened in relation to high frequency waves, so ($\nu$) changes less as effective stress increases. However, at low effective stress, when pore pressures are very high, the effective stress sensitivity is marked, and ($\nu$) increases.

### 13.18

The sensitive response of ($\nu$) close to the hydraulic fracturing limit may be the result of sealed, over-pressured and under-compacted beds. There may also be additional over-pressure due to increased oil-to-gas conversion at zero effective stress. $V_s$ is (locally) zero as the rock mass is hydraulically fractured and load is born by the fluid. However $V_p$ is not zero, therefore $\nu$ approaches 0.5.

### 13.19

It is well documented that pseudo-static moduli of deformation can be significantly lower than the dynamic moduli predicted from P- and S-wave velocities. Knowledge of the non-linear elastic properties that are largely responsible for the differences between dynamic and static moduli is essential for optimal drilling, effective well completions and efficient reservoir management. The frequency of measurement is all important for the geophysical estimate obtained, since $E_{ultrasonic} > E_{log} > E_{low\ freq.} > E_{static}$.

### 13.20

The development of attenuation as a means of improved characterization of reservoir rocks is due to the dispersive, frequency-dependent nature of seismic Q, and the greater sensitivity of the ratio of $Q_s/Q_p$ to fluid and partial saturation than $V_p/V_s$. The expected reduction in $V_p$ by reduced brine saturation and increased gas saturation in sandstones is matched *initially*, by greater attenuation with $Q_p$ reducing from e.g. 30 to 10. At the far end of the saturation scale, when samples become 'room dry' or reach 100% saturation with nitrogen, the attenuation reduces sharply, and $Q_p$ may reach a value of 50. This is related to the eventual absence of squirt flow with increased dryness.

### 13.21

Ultrasonic (0.1 to 1.0 MHz) laboratory tests on dry and water- or brine- saturated Berea sandstone of 16% porosity, have been used on numerous occasions for investigating how velocity and attenuation vary with 'differential' pressure (confining minus pore pressure). The P-wave velocities of this sandstone, when dry *and* when brine-saturated, usually rise rapidly over the first 10 to 20 MPa, eventually reaching a plateau with little stress sensitivity. At high differential pressure such as 50 MPa, equivalent to reservoir conditions, there may be little difference between the dry and brine-saturated values of $V_p$. By comparison $Q_p$ and $Q_s$ rise more consistently, even up to high stress levels. They show less attenuation in the dry or methane saturated states, than in the case of brine saturation, where squirt losses can occur.

### 13.22

$Q_s$ is often larger than $Q_p$ in the case of the dry and methane-saturated sandstone. In the case of brine saturation, there is a consistently wide separation of $Q_p$ and $Q_s$ ($Q_p > Q_s$). This represents an important means of tracking water-gas fronts during production. The increase of $Q_p$ from e.g. 10 to 20 at low effective stress, to values as high as 60 to 80 to 100 at effective pressures of 50 MPa bears a strong resemblance to deformation modulus increases. A lower range of increase seen with brine saturation, might be due to the weakening effect of brine.

### 13.23

Tests with higher porosity sandstones (range 20–25%) tested in dry or brine-saturated states, in triaxial compression, over an effective stress range from 2.5 and 40 MPa, using a range of frequencies (400 to 2000 kHz), show important additional trends when the effect of frequency is shown together with the effect of stress level. The greater sensitivity of $Q_p$ than $V_p$ as effective stress levels rise is seen as before. However the effect of increasing frequency shows *negative* velocity dispersion for the *dry* samples, meaning velocity decreasing with increasing frequency, while in marked contrast, the attenuation increases ($Q_p$ reduces), as the third to *fourth* power of frequency. This is assumed to be evidence of *scattering* within the pore spaces between the grains.

### 13.24

*Brine-saturated* sandstones show slight, positive velocity dispersion at the lower confining pressures, while attenuation increases ($Q_p$ reduces), with only the first or second power of frequency. This change in attenuation-frequency dependence is taken as evidence of local fluid-flow loss mechanisms. The saturated rocks always show much stronger attenuation (lower $Q_p$ ) than the dry samples. The effective stress level has greatest influence on attenuation when samples are brine-saturated and at the lowest levels of effective stress. The behavioural contrast of $Q_p$ compared to $V_p$, with the former's greater sensitivity to effective stress level, and to frequency, confirms the importance of attenuation as an excellent diagnostic at reservoir stress levels.

### 13.25

Differentiation of $Q_p$ values, has also been suggested as a way of distinguishing *gas and condensate* from *oil and water* in sandstone reservoirs. In perfectly dry rocks, $Q_p$ is very high. In fully liquid saturated rocks $Q_p$ is at an intermediate level. In partially saturated rocks $Q_p$ is low. Magnitudes of $Q_p$ for sandstone reservoirs, based on well log (i.e. sonic frequencies), are usually in the following ranges: gas and gas-condensate bearing sandstone $5 < Q_p < 30$, oil bearing sandstones $8 < Q_p < 100$, water bearing sandstones $9 < Q_p < 100$. The low $Q_p$ values may be caused by low effective stress in the case of over-pressure (or by fracturing). These ranges are remarkably similar to laboratory ultrasonic data, when a range of effective stresses are applied.

### 13.26

Classic Biot theory that accounts well for attenuation in clay-free sandstones, fails by an order of magnitude to account for the attenuation effect of clay content. Strong clay-related attenuation is assumed to be due to viscous interaction between the clay particles and the pore fluid. Permeabilities are also strongly dependent on clay-content. The measurement of attenuation of compressional waves in sandstones under confining pressures of 40 MPa at ultrasonic frequencies (0.5–1.5 MHz, shows that intra-pore clay content is important in causing attenuation, and in modifying the permeability. $Q_p$ may be as low as 10 with clay contents >10%, rising up to several hundred when clay content is <1%.

### 13.27

Systematic reduction of $Q_p$ from about 80–100 to about 10–20 is seen as the percentage of compliant minerals in sandstones and siltstones increases from a few percent to nearly 80%. This is attributed to 'clay squirt flow'. This mechanism may also be important at both seismic and sonic frequencies, in the case of larger scale geologic features such as inter-bedded permeable and impermeable layers.

### 13.28

Different rock types such as siltstones, sandstones and limestones may also show a significant range of instantaneous sample deformation as a result of applying high confining pressures. Their deformation moduli are clearly different. Differences between rock types will be accentuated when bedding and jointing is also present, causing increased attenuation, and greater sensitivity to effective stress. This sensitivity may even apply to velocities, which tend to show much less sensitivity to confining pressure than seismic Q when samples are without jointing.

### 13.29

Dual porosity limestone specimens, with micro-pores and inter-particle macro-pores, show a weak trend for higher attenuation and lower seismic Q when permeability and total porosity are also larger. For example a $Q_p$ of 10 roughly correlates with 1–10 mD, while a $Q_p$ of 100 roughly correlates with 0.01–0.1 mD. Both distributions of 'pore' size are important. The attenuation can be shown to be the sum of Biot-type fluid flow and squirt flow to/from the larger, moderately interconnected inter-granular pores, which may contribute as much as 90% of the total porosity. This permeability-$Q_p$ trend is expected to strengthen when the small-scale 'dual porosity' also has the contribution of *in situ* jointing or fracturing. A requisite number of joint sets for connectivity, and well-intersection for verification, are necessary boundary conditions.

### 13.30

Ultrasonic data (0.7–0.85 MHz) for small 'intact' dual-porosity limestone samples has uncertain relevance for geophysicists interpreting propagation through dual-porosity porous and jointed limestones in the field, at the lower frequencies used in seismic and sonic log surveys (50 Hz to 30 kHz range). Dual-porosity chalk, with higher porosity than limestone, has indicated $Q_p$ and $Q_s$ values as low as 4 to 5, and 3 to 6, at *near-surface* sites, using seismic frequencies. There is potential for strong squirt flow attenuation with passage of low frequency seismic waves, in dual porosity systems such as jointed limestones. Smaller magnitudes of $Q_{seis}$ may be expected, due to the contrasting moduli at the different scales, with reduced differences at higher stress levels.

### 13.31

A degree of correlation is noticed between *seismic Q* and the 'static' modulus of deformation, expressed in GPa and readily estimated from *rock quality Q*. This modulus is stress- or depth-dependent, and may range from about 1 to 150 GPa in the upper 1 kilometre, but most frequently from 5 to 100 GPa. The components of the rock quality Q-value reflect many potential attenuation-causing factors, e.g. $RQD/J_n$ for scattering due to relative block size, $J_r/J_a$ concerning the frictional and conductive properties of the joints that are expected to be relevant for squirt flow, including loss mechanisms in clay, $J_w$ as a direct link to permeability, and SRF related to increased attenuation where stress is low, and reduced attenuation where stress is high.

### 13.32

Over-pressured zones due to rapid sedimentation of alternating sands and shaly sediments present a potential hazard when drilling, due to the risk of *shallow water flows* (SWF). Effective stresses and compaction of sediments can be minimal, and progressive instability during drilling at a new well can potentially engulf neighbouring wells, also at depths up to a kilometre. Due to very low values of shear wave velocity at low effective stresses in sands, there is an exponential increase in the ratio of $V_p/V_s$ to values beyond 5 and 10, and even beyond 100. Poisson's ratios increase rapidly to just below 0.5. There is high *attenuation* of the shear waves at the lowest pressures, as the sand is close to a state of suspension. Distinguishing between unstable sand and sandstone is very clear using the ratio of $Q_p/Q_s$ plotted versus $(V_p/V_s)^2$.

### 13.33

Saturated shales tested under *over-pressured* conditions, show anisotropic velocity and attenuation in ultrasonic laboratory testing. Velocities can be 30–40% higher parallel to bedding, where attenuation is also least. There is a general increase in seismic $Q_p$ and $Q_s$ with increasing differential pressure. Relative proportions of Biot 'fluid-past-frame' attenuation, and local squirt flow attenuation are different in the plane parallel to the layering, and in the plane perpendicular to the layering. There is a strong link between the rock framework, the pore geometry and connectivity, and therefore of the response of pore fluid to the propagation of seismic waves in specific directions.

### 13.34

The surface roughness of joints or fractures may help to maintain some permeability, even at higher confining pressures, corresponding to depths of several kilometres. Permeability parallel to jointing or fracturing may then be much higher than that parallel to eventual sedimentary layering. The additional possibility of pre-peak-strength conjugate shearing of such joint sets, due to anisotropic stress, would allow relative maintenance of joint permeability despite high effective stresses. Elastic property anisotropy, and hydraulic anisotropy may be closely related in terms of symmetry directions, when the two mechanisms share the same cause, such as layering or jointing.

### 13.35

High pressure polyaxial loading frames have been used to study the seismic signature differences between unfractured matrix and fractured matrix. Parallel fracturing has been developed in the same specimen by holding the minimum principal stress very low and increasing $\sigma_1$ and $\sigma_2$ in unison, to high levels. Velocity increasing steadily parallel to the high biaxial loading direction, contrasts with the fall in velocity in the perpendicular direction, when fracturing initiates. Subsequent reloading of the fractured sample demonstrates stronger stress-velocity dependence perpendicular to the fractures than parallel. Perpendicular to the fracturing direction, seismic Q values change from 40 to 30 as a result of fracturing, and reduce to 10 with unloading, actually resembling potential deformation modulus behaviour, when expressed as GPa. Low permeability sandstones start to develop measurable permeability when velocities start to reduce in the direction perpendicular to the stress-induced fracturing.

### 13.36

The coupled stress-permeability-velocity behaviour of smooth-planar and rough-undulating fractures is different. Rough fractures closing due to stress increase contribute to increased velocity, but suffer less than expected reduction in permeability. The reason for the different behaviour of the rough fractures compared to the smooth, may be that E (physical aperture) $>$ e (hydraulic aperture), for the case of rough, high JRC fractures (or joints), while $E \approx e$ for smooth fractures (or joints). This would mean faster physical closure than hydraulic closure for rough joints, thereby potentially explaining the stronger velocity response and the weaker permeability response to stress increase.

### 13.37

Reservoir-scale 4D seismic monitoring in fractured reservoirs is most sensitive to production-induced changes at lower effective stress levels. The velocity and particularly the attenuation, are relatively sensitive indicators of small permeability changes. By the nature of jointed reservoirs, there are unlikely to be commercially viable hydrocarbon-bearing fractures or joints with very low surface roughness JRC values, as joint closure under stress would preclude both permeability and 'storage', if such was needed due to low porosity matrix. Minerally 'frozen' stylolites in limestone and chalk are a special case, being both exceptionally 'rough', and insensitive to stress change, in comparison to rough, interlocking joints which show greatest stress sensitivity, and greatest apertures (E and e).

### 13.38

A simple empirical method is suggested for linking seismic Q values, specifically $Q_p$, with hydraulic and rock engineering properties. This is based on the fact that $Q_{seis}$ invariably resembles the *static* E-modulus in the case of intact samples, and the *static* deformation modulus in the case of jointed or fractured rock and rock masses. when $Q_{seis}$ magnitudes are expressed in GPa. Despite the dynamic micro-strain basis for spectral analysis estimation of $Q_{seis}$, the magnitude of $Q_{seis}$ and its increase with effective stress, does not so closely resemble the *dynamic* micro-strain based deformation properties such as $E_{dyn}$ as one might expect. It has been found that $Q_{seis}$ cubed and inverted gives first-order estimates of intrinsic permeability, and velocity and compression strength combined (both reflecting stiffness), give independent first-order estimates of $Q_{seis}$.

## Chapter 14 P-waves for characterizing fractured reservoirs

### 14.1

There was very early recognition in petroleum exploration, of velocity increase with depth, and early recognition of a quite systematic trend linking velocity to the *geological age*, in combination with the present depth of occurrence. The greatest rate of velocity increase was found to occur at shallow depth in the oldest units, which is fundamental early proof of the importance of dual porosity. The likelihood of more joints in the stiffer, older units means that these units are more sensitive to stress change. However, with only $V_p$ as a dynamic indicator of conditions, acoustic closure represents a limit to the stress-sensitivity of velocity, especially for the case of weaker, younger reservoir rocks.

### 14.2

An early analysis of almost 300 kilometres of well sections, in 500 petroleum well surveys, mostly from the USA, and mostly for mixed shale and sandstone sections, indicated an average P-wave velocity discrepancy of only about 110 m/s in velocity between these two, the sandstone having the highest velocity by this small average margin. The similarity of velocities for these mechanically and hydraulically dissimilar units, is a reminder of the potential 'non-uniqueness' of P-wave velocity, and the need for alternative interpretation methods, such as attenuation, and impedence, to distinguish the different lithologies and their fluid-bearing signatures. The closeness of the *in situ* velocities for shale and sandstone, also seen in rock physics experiments on the *matrix* of both rocks, is surprising, in view of the greater tolerance of the stronger sandstone to stress anisotropy, often resulting in several MPa greater minimum stress in the shale, which is frequently a fluid barrier for the hydrocarbon-bearing sandstone.

### 14.3

Fine layering of alternating porous and impermeable strata is obviously one of the basic sedimentary systems that contribute to the existence of potential reservoir rocks in sedimentary basins. Fine layering of sedimentary strata means that the dominant wavelength of a seismic or sonic pulse is long compared to the thickness of individual layers. The medium will nevertheless exhibit effective (and real) anisotropy, with a vertical symmetry axis in the case of horizontal layering. In the presence of hydrocarbons this layered medium may show substantial attenuation and velocity dispersion, which will be compounded with the additional presence of jointing or fracturing. With modern seismic techniques a new exploration concept has gradually developed, exploring not just for the presence of reservoir rock containing hydrocarbons, but exploring for the presence of permeable joint-sets and their principal direction.

### 14.4

A 'thin bed' is considered to be 3/8 of a wave length, the limit for a discrete reflection both from the top and bottom of the bed. Wave scattering, attenuation and dispersion occur when the ordered heterogeneities have scale lengths of about 0.3–0.01 of the wavelength, while the smallest scale of ordered heterogeneity, less than 0.01 of the wavelengths, may be the cause of most of the azimuthal and offset dependent velocity. Conventional seismic wavelengths are much longer than the scale lengths of either of the features that govern dual-porosity flow in a reservoir.

### 14.5

Strong P-wave velocity anisotropy is observed in every geologic environment, with the possible exception of basins under primary deposition and burial. P-wave azimuthal anisotropy, previously ignored and left to the research and technology specialists, is now known to be one of the most significant properties of the acquired seismic data. In the marine environment, fully populated offsets in each azimuth bin are less common than on land, but even narrow azimuth data gives an opportunity to see the effects of azimuthal anisotropy. As time goes by more and more reservoirs are being re-classified as severely heterogeneous, as well as fractured and anisotropic.

### 14.6

Definitions of three common classes of *anisotropy* are as follows. Transversely isotropic media have a vertical axis of symmetry, and are referred to as TIV: with horizontal axis of symmetry as TIH. The former is typified by fine layering in shales. Transversely isotropic media with a horizontal axis of symmetry, known as HIV, typically have stress-aligned, vertical jointing or fracturing, and/or microcracks. When fast and slow directions have been identified, azimuth sectoring can be applied in these directions. Even isotropic processing codes can function with such azimuth-sectored data.

### 14.7

Shallow 3D-survey based imaging of seismic velocity and attenuation, shot over several km$^2$, can be used to derive tomographic images of velocity and attenuation at specific depths in shallow reservoirs, using an iterative reconstruction algorithm. The tomograms derived for constant-thickness slices centred at increasing depths show successive increases in both P-wave velocity and $Q_p$ with increasing depth (e.g. $Q_{seis} < 5$ when $V_p < 2$ km/s, and $Q_{seis} > 10$ when $V_p > 3.5$ km/s).

### 14.8

Smaller scale, shallow cross-well tomography, has been used in reservoir sands (channel-sands) to correlate high values of the ratio $Q_{seis}/V_p$ with the most porous and most permeable zones, and low values with flow barriers, such as shale-rich layers (e.g. sands $Q_p/V_p = 45/3 = 15$, plastic shale $Q_p/V_p = 30/3.7 = 8$). Absence of jointing and high porosity and permeability apparently causes higher values of $Q_p$ (i.e. less attenuation without squirt losses), and quite low $V_p$. This is in contrast to the seismic attributes of jointed sandstones or jointed rock in general. The proposed $V_p$-UCS-deformation-modulus 'GPa-model' for $Q_p$ fits the latter and not the former. A contrast in $Q_{seis}/V_p$ ratios could therefore be used to delineate the difference between unconsolidated sands and jointed sandstones, and between weak plastic shales, and the less desirable fissured/jointed, or indurated variety.

### 14.9

At a site in the USA, the P-wave velocity of near-surface jointed limestones determined from near-offset VSP varied from about 3.8 to 4.2 km/s between the shallow depths of 16 and 26 m. This suggests a rock quality Q-value of 2 to 5 from the empirical relation $V_p \approx 3.5 + \log Q$ relevant to nominal 25 m depth, 100 MPa, low-porosity rock. $Q_{rock} = 2$ to 5 is typical for rock masses with three sets of joints, moderate block size, and with possible weathering of the joint walls: i.e. $Q \approx 90/9 \times 2/2 \times 0.66/2.5 \approx 2$ to 3. A seismic $Q_p$ value of about 13–14 can be estimated via the deformation modulus method, when $Q_{rock} = 2$ to 3, and assuming the uniaxial compressive strength of the limestone is around 100 MPa. Alternatively, using just velocity, $Q_{seis}$ can be estimated as 12–16 for this jointed limestone, based on the $V_p$ range of 3.8–4.2 km/s.

### 14.10

In this 'circular-logic' jointed rock prediction, the ratio $Q_p/V_p$, if roughly correct, would be a much lower 3 to 4, compared to the much higher ratios seen in porous, unconsolidated sediments. A 460 m deep cross-well measurement in a very permeable limestone aquifer showed $V_p$ and $Q_p$ of 3.5 km/s and 14 at 2 kHz, but at higher frequencies (12 kHz), $Q_p$ rose by 350%, while $V_p$ rose by only 3%.

### 14.11

There have been many years of oil-industry interest in cross-hole tomographic methods for imaging below the resolution of surface seismic. High frequency waves can be propagated over distances of many hundreds of metres, with minimum loss of energy, when both source and receiver are down-hole, in deep boreholes. The avoidance of near-surface (low $Q_{seis}$) losses, means that broad band-widths can be used. Very high resolution images are obtained by using second arrivals and reflection imaging.

### 14.12

At a 260 m deep research well through finely inter-layered limestone, shale, and sandstone sequences, the combined use of a borehole compensated sonic logging tool, a compensated formation density tool, and a formation micro-scanner enabled resolution of much of the detail of finely interlayered rock sequences, where the normally detectable layer thickness using standard sonic tools may be no less than 15 cm. The ability to separate shales, sandstones and limestones, based on down-hole facies recognition and velocity differences, can also be used in a tentative separation of $Q_p$ according to facies. With appropriate ranges of increasing uniaxial strengths and velocities for the three rock types shale, sandstone, limestone, one arrives at potentially representative $Q_p$ values of 6–7, 10–15, and 20–40, using the empirical $V_p$-UCS-modulus model.

### 14.13

A finely inter-bedded mix of facies as above, may tend to create a 'weighted' response in standard logging. Sonic logging (8–24 kHz) gave the lowest $Q_p$ with a mean $Q_p$ of 10 and a range of about 6–14. A $Q_p$ range estimated from the measured velocity range of 3–4 km/s, using the $V_p$-UCS-modulus method, would be about 5 to 12. The somewhat lower frequencies of cross-hole logging (0.2–2.3 kHz), gave a mean $Q_p$ of 15.7 and a range of 12 to 20. The lowest frequency VSP (30–280 Hz), giving presumably the poorest definition of the fine inter-layering, gave a mean $Q_p$ of 31.3 and a range of 25 to 45.

### 14.14

Since sedimentary rocks containing hydrocarbons have proved to be neither isotropic nor homogeneous, but heterogeneous and anisotropic, the seismic wavelength at which the measurement is made, determines what seismic attributes can be measured, and whether the rock looks homogeneous and isotropic or heterogeneous and anisotropic. All seismic data are now known to vary with offset from the well (in VSP) and with azimuth. Besides detecting azimuthal velocity anisotropy due to aligned fracturing or stress, one can now acquire spatial resolution of variable structure, azimuthal resolution of attenuation, and resolution of temporal changes, which may include azimuthal variation of attenuation in 4D monitoring, possibly due to conjugate joint shearing.

### 14.15

Since all seismic data also varies with frequency, there is increasing acceptance that 3D multi-component, multi-mode and multi-azimuth acquisition may be required and may also be economically justified. A given set of vertically aligned fractures may cause anisotropy with low frequency measurement, signal distortion with mid-frequency measurement, and lead to reflections by the highest frequency waves. The detection of the azimuthal anisotropy attributable to structure, that specifically controls the fluid-flow properties at reservoir scale, is now an important focus of attention. The detection of permeability anisotropy can be considered as one step beyond the detection of vertically aligned fractures, and/or the detection of unequal horizontal stresses.

### 14.16

There is also a growing trend to instrument selected petroleum wells on a permanent basis, especially offshore, so that 4D seismic can be used relatively more easily, to monitor changes bought about by different water-flood and production practices. Rock physics principles are used to assist in the interpretation of measured changes in velocity, amplitude and attenuation. In-well 3D accelerometer installations were applied in the late 1980's for permanent installations in deep holes adjacent to the San Andreas fault in California, where the benefit of avoiding near-surface attenuation were recognised.

### 14.17

It is commonly assumed that there is strong correlation between directionality of reservoir flow and the local, presentday orientation of the maximum horizontal stress. Oriented four-arm calliper logs typically show a long axis that is oriented parallel to the minimum horizontal stress direction, if there is stress-induced break-out. However there may be geomechanics-based reasons for carefully evaluating this commonly held viewpoint from case to case. Rock strength, joint or fracture roughness, joint closure under stress, and possible shear-displacement modes need also to be considered. Fractures perpendicular to the $\sigma_{H\,max}$ direction can also be 'open' if partially filled with mineral cements, and for this same reason, sealed fractures parallel to $\sigma_{H\ max}$ are also numerous.

### 14.18

The common assumption that the direction of $\sigma_{H\ max}$ is the direction of 'open' cracks or fractures also overlooks the possibility that two sets of joints or fractures can be involved in limited conjugate shear-displacement. With one set dominant, the orientation 'discrepancy' often reported, regarding the direction of 'open' fractures in relation to $\sigma_{H\ max}$, can be better explained. Dominant directions of fracturing can also be images of the resulting dominant strike of conjugate, steeply dipping sets, in the case of domal structures.

### 14.19

A further potential source for minor angular discrepancies, is the dilation-related contrary-rotation of fluid lenses contra rock-to-rock contacting asperities, when *non-planar* joints or fractures are under significant shear stress, and therefore significantly 'open'. This geometric effect could also potentially cause a minor rotation of shear wave splitting polarization, as argued in Chapter 15.

### 14.20

Measurements in deeper wells have indicated that seismic $Q_p$ based on seismic frequency VSP, may be systematically smaller than $Q_p$ based on higher frequency sonic logging. This is the opposite of what has been measured in shallower wells, where attenuation was least for VSP and most for sonic logging, giving lower $Q_{seis}$. The expected dispersive bias of higher frequency (sonic) waves travelling at higher velocities than lower frequency seismic (VSP) waves, remains consistent when shallow or deep. Individual values of $Q_p$ or $Q_s$ may change 'erratically' with depth unless depth averaging is used. However, rock quality Q-values down recovered core also tend to fluctuate quite strongly, and since linked to deformation modulus, $Q_p$ and $Q_s$ must also be expected to fluctuate.

### 14.21

Rock quality differences, and therefore differences in stiffness and susceptibility to failure may play a role in such fluctuations. In addition, minimum rock stress will tend to be residing in the weaker beds (i.e. shale) at shallow depth, while residing in the stiffer beds (i.e. sandstone) at depths where shale is more plastic, and therefore has a higher $\sigma_{h\ min}$. Stress concentrations around wells will cause a magnification of tangential stress, in the same direction as major principal stress, and diminution of tangential stress in the perpendicular direction. If these stress concentration effects are strong enough in relation to rock strength, shear failure surfaces may develop, first giving break-out, subsequently a possible *log-spiral-sheared* discontinuum close to the well.

### 14.22

Seismic *attenuation* has come to be recognised as potentially very sensitive to reservoir properties. This is because of its sensitivity to fractures, joints or bedding planes, and in turn, due to *their* sensitivity to changes of effective stress and to frequency. Attenuation levels are also sensitive to the saturating fluid and petro-physical properties. High dispersion (and low $Q_{seis}$) values may correlate with permeable sand and carbonate beds within shale. Such beds can be at least ten times as permeable as the host shale formation. The dependence of seismic velocity on frequency can be used for reservoir characterisation, since the dispersion is mathematically related to seismic attenuation. High frequency measurements differ from low frequency measurement due to both elastic scattering and intrinsic attenuation.

### 14.23

AVO (amplitude variation with offset) and AVOA (amplitude variation with offset and azimuth) indicate that variation of P-wave amplitude can be related both theoretically and in practice, to the presence of fracturing. Appropriate analysis of AVOA gives reasonable estimates of the orientation of fracturing, particularly if only one set is involved, or if one set is dominant. Fracture orientations can be compared to results obtained when using C-waves (P to S converted waves), and the shear-wave splitting and polarization mechanism. The converted P to S waves have the advantage that they can be generated by compressional (i.e. explosive) sources, yet are expected to contain the same information as pure S (or SS) waves, as discussed in Chapter 15.

### 14.24

Although the use of shear waves are theoretically favoured for fracture set detection, there has been some reluctance to use shear waves, due to more expensive acquisition and more expensive processing routines. For these reasons, the use of P-waves for fracture set detection and estimation of orientation, has attracted a lot of interest, even though P-wave travel times need to be detected in many directions to obtain the necessary information.

### 14.25

If seismic data acquisition is conducted parallel to the (geologically suspected) fracture orientation, the fractures will have minimal influence on the reflection properties, regardless of the angle of incidence, or offset. The P-wave particle motion is then parallel to the fractures. If the seismic line is instead oriented more perpendicular to the fractures, at larger angles of incidence than zero, the reflection coefficients will be affected strongly. At the largest angles of incidence, especially perpendicular, the P-wave velocity is also expected to be affected by the acoustic properties of the fluids filling the fractures. Thus in the presence of anisotropy, the reflection amplitude will vary with offset, due to changed angle of incidence, and will also change with azimuth (AVOA).

### 14.26

When deviation of 20° to even 40° is observed between AVO-determined dominant fracture orientation and the perpendicular-to-break-out based $\sigma_{H\,max}$ direction, and when nearly as large deviation is also obtained between shear-wave polarization and the $\sigma_{H\,max}$ direction, the possibility of conjugate-shearing of joint sets that are *intersected* by the $\sigma_{H\,max}$ direction should be considered. Although this contradicts the standard industry assumption of 'open fractures parallel to $\sigma_{H\,max}$', it helps to explain frequent angular discrepancies between dominant 'open' fracture azimuths, and the perpendicular-to-break-out based $\sigma_{H\,max}$ direction, At shallow depth, the standard industry assumption, also in civil engineering, is more correctly focussed on maximum permeability, and $V_p$, being parallel to the $\sigma_{H\,max}$ direction.

### 14.27

There is a reported problem of *model-dependence* in AVAO analyses. The dominant fracture strike direction can be ambiguous, since the azimuthal variation in the *near-offset* AVO gradient, can be positive or negative, relative to the fracture direction. The direction of the most positive AVO gradient can correspond to either the fracture-normal, or the fracture strike direction, depending on the character of the fracturing, and depending on whether brine-filled or gas-filled. Forward modelling is therefore needed in order to constrain the interpretation of AVOA. Forward modelling using fracture density, fracture aspect ratio, and fracture (*additional*) *compliance* concepts, may actually require knowledge of two different fracture apertures: the hydraulic aperture (e) that would govern squirt losses, and the physical aperture (E) that would govern compliance or stiffness and stored volume of fluid, where $E \geqslant e$. This inequality in the case of favourably rough-walled joints or fractures seems so far to have been ignored.

### 14.28

The probability for multiple joint or fracture directions in the neighbourhood of *faults*, means that the fast velocity is no longer equal to the matrix or bulk rock velocity. The normal elliptical $V_{fast}$ and $V_{slow}$ distribution is then replaced by superimposed multiple ellipses, which have the effect of reducing the observed velocity. The previous directionality with a single set of joints or fractures will be lost. Due to rapid changes in fracture frequency, rapid changes in velocity are also seen. Such is actually a response to the rapid changes in rock mass quality Q close to, and across faults, as frequently mapped in tunnelling, and when logging fault-zone core, in each case for rock quality determination.

### 14.29

There is a multitude of technical jargon in the geophysical industry. Some is exceedingly simple. Converted C waves means explosive or air gun generated P-waves converted to S-waves at an interface or at the sea floor. (Pure S-waves may be referred to as SS). The term 4C means four-component seismic recordings. These consist of one hydrophone, one vertical geophone, one in-line horizontal geophone, and one cross-line horizontal geophone. The term 4D means 3D seismic repeated at intervals for monitoring changes caused by production. Repeated 4D surveys can be made cheaper by modifying 4D receivers to also be 4C. There are now some expensive 4D4C installations that make frequent reservoir monitoring much cheaper. The consequence of frequent full-field 4D4C reshoots, as at Valhall, in the North Sea, providing full-field estimates of all required reservoir parameters, is more efficient exploitation of reserves, and a production increment obviously coming sooner than discovery and exploitation of new fields.

## 14.30

Reservoir monitoring with 4D seismic in its most basic form is the repeated inversion of changing seismic data, to obtain dynamic reservoir properties, which can subsequently be used to predict pore pressure change at a distance from the wells based on the effective stress and fluid sensitivity of laboratory samples of the reservoir rocks. History matching can be used to up-scale the rock physics matrix data, and to calibrate forward modelling of anisotropy and fracture effects. There is high sensitivity to effective stress in shallow reservoirs, but a stress-velocity plateau may be reached at high effective stresses (i.e. beyond roughly 25 to 50 MPa, depending on rock type and on fracture-surface roughness.) A velocity plateau indicates the need for up-scaling using attenuation.

## 14.31

Compressible grain-boundary cracks with their low aspect ratios may be partly the result of stress unloading when matrix samples are drilled and bought to the surface. The *in situ* velocity may not be recovered upon reloading, due to hysteresis caused also by temperature change. Saturated samples containing micro-cracks will tend to project a lower stress sensitivity with laboratory ultrasonics, than with seismic waves, due to greater relaxation, as opposed to stiffening with the ultrasonics.

## 14.32

When compacting reservoirs are 4D monitored by repeated seismic surveys, time-shifts are registered, which are a combined result of increased velocity due to compaction, and reduced layer thickness, i.e. reinforcing effects. Time-shifts of as much as 12–16 ms, between 1989 and 1999, recorded at the Ekofisk field, were related to an estimated 6m of additional compaction at 3 km depth. An uncritical time-lapse comparison between surveys, may give unrealistically large values for compaction and subsidence.

## 14.33

Extensive casing damage to numerous wells at Ekofisk is evidence of discontinuous behaviour, due to stretching of the overburden and differential bedding plane slip. This was also seen in early discontinuum models. Subtle changes are now known to occur to the overburden velocity, due to the 'stretching' of the overburden in response to the incremental compaction between surveys. There are about 150 km$^3$ of obviously discontinuous rock involved in the compaction and subsidence. Further evidence for discontinuous behaviour caused by compaction at Ekofisk can be seen in the results of the 4D seismic. Fault related discontinuities are seen in 'time lapse' tomograms of compaction magnitudes. Forward modelling of compaction details, performed in the 1980s, indicated *small-scale* down-dip shearing of conjugate jointing in the chalk, before evidence of slickensiding had been seen in newly drilled core.

## 14.34

In water-flooding, for *stimulating* and driving petroleum production, there is both a local increase in pore pressure at the injector wells, and a reduction in temperature, causing some contraction of the matrix, both of which help to dilate, and possibly shear existing joints, and perhaps create new fractures as well. The conventional and expected mechanism of fracture or fault opening *exactly* in the direction of $S_{h\ min}$ is not as common as expected.

## 14.35

There are strong indications from surveys of numerous fractured reservoirs that have been water-flooded, that the reduction of effective stress caused by the water-flood pressure, and the related contraction-cooling effects, could be stimulating shear-displacements on existing joint or fracture sets, or faults. The direction of $\sigma_{H\,max}$ shows a frequent tendency to have *bisected* the geologic features that are the basis for the joint rosettes. An implied mechanism of conjugate shear, would show strong parallels to the findings of Zoback and co-workers, concerning the frequency of water conducting discontinuities in deep wells needing to be oriented so that they are under shear stress.

### 14.36

Mapping of the azimuthal velocity anisotropy of P-waves using a downhole triaxial accelerometer sensor array, and multi-azimuth walk-away (or 'float-away') VSP, is a means of reducing the risk of drilling low-productivity wells in unfractured parts of reservoirs. Calibration with oriented core data and with FMS logs improves the likelihood that later producing-wells will intersect 'open' conductive fracture sets, where P-wave anisotropy is highest. Azimuthal variation in the *shear modulus* of the fractured rocks is cited as the reason for the P-wave velocity anisotropy. Dominance of one fracture set orientation, with a near-orthogonal subset, and variation of *fracture density* in the unequal two-set system, would be reasons for variations in the degree of seismic anisotropy, and demonstrate the benefit of 3D mapping.

### 14.37

Multi-azimuth walk-away VSP can also be used to map the *attenuation anisotropy* of a reservoir. A fractured, oil-saturated reservoir is likely to show azimuthal variation in attenuation, in a similar manner to P-wave anisotropy. As examples, we may quote $Q_p = 18$ in the fractured part of a reservoir, and $Q_p = 35$ to 40 in the overburden, which was assumed to be relatively unfractured, with minimum attenuation correspondingly scattered between wider azimuths. In the particular reservoir, minimum attenuation was some 20° to 30° oblique to both the open conducting fractures and to the $\sigma_{h\,max}$ direction, based on oriented cores and borehole images. Conjugate jointing, perhaps also pre-peak shearing, is again suggested.

### 14.38

'Open' fractures in a petroleum reservoir would seem to require that the rock is unusually strong and that joints are rough, or that there exists a close-to-fracturing pore pressure, *or that* there is a suitable quantity of hard mineralization to 'bridge' and maintain an earlier porosity. Alternatively, if some pre-peak-strength shear displacement of non-planar joints or fractures has occurred, there would be the contribution of dilation to 'openness', and the additional influence of a 10° to 20° rotation of the fluid-bearing parts of the fractures in relation to the contacting parts taking the load. This might rotate both the sources of attenuation and the sources of shear-wave polarization.

## Chapter 15 Shear wave splitting in fractured reservoirs and resulting from earthquakes

### 15.1

Vertical and sub-vertical jointing is extremely common in most rock masses. Yet vertical boreholes are usually the first, and seemingly also the second choice, for sampling and gaining access to the sub-surface. The *sampling bias* caused by the mismatch of borehole diameter and horizontal spacing of vertical structure, *and* the vertical borehole itself, is extreme and well known. If the economic savings of a vertical well, and the subsequent cost of an extensive seismic survey and its inversion were combined, there would perhaps be reason for rapidly deviating exploration boreholes at least 10° or 15°, in order to sample the increasingly understood relevance of vertical and sub-vertical structure on hydrocarbon production. On the other hand drilling and hole stability problems *might* be increased by the more frequent joint intersections.

### 15.2

Shear-wave anisotropy due to splitting and polarization caused by the presence of vertical or aligned structure, (and P-wave azimuthal anisotropy), are miraculous means, in view of the long seismic wave lengths, for rectifying these poor joint or fracture sampling strategies. The central challenge of sub-surface fracture characterization is to obtain data on essential fracture attributes where direct observation has been prejudiced by vertical wells. An early deviation of 10° or more would greatly improve understanding of both the overburden jointing and the reservoir jointing. The potential anisotropy of the overburden cannot be ignored in seismic inversion, as indicated in an increasing number of cases, especially where compaction and therefore subsidence are occurring.

### 15.3

Countless hydrocarbon reservoirs have been discovered, characterised, and monitored by P-waves. However, P-waves cannot solve every seismic imaging or reservoir description problem. The addition of S-waves, usually in the form of converted PS-waves, has given oil and gas companies an enormous quantity of new reserves that could not have been found with P-waves alone. The new reserves have been more effectively exploited by better identification of fracturing, and therefore better placement and deviation of production and water-flood wells. Multi-component recording of shear-wave attributes also provides information where shallow gas has obscured P-wave imaging over central parts of a field, such as at Ekofisk and Vallhall.

### 15.4

The basic geological source of polarized shear waves can be sets of vertical joints or fractures, or stress-aligned microcracks. These aligned features cause the vertically transmitted shear-waves to split into fast and slow components, registered as time delay, due to the attenuating effect of fracture shear compliance, on the S-wave component that has particle motion *perpendicular* to the fracture strike. Shear-waves travelling in the *parallel* direction hardly sense the presence of the cracks, and travel at almost the wave speed of the unfractured matrix. The difference in travel-time between the fast $qS_1$ and slow $qS_2$ components is strongly related to the length of travel path and to the density of the crack population. It is also related to fracture compliance. Numerous polarized shear wave observations show the fast wave polarized parallel (*or sub-parallel*), to the accepted local or regional maximum stress field. There can be several reasons for this, and also several reasons for deviation from this direction in other cases.

### 15.5

The potential of fluid-filled microcracks to react to crustal stress and strain led to the early proposal for *extensive dilatancy anisotropy* (EDA). The hypothesis was that crustal fluids prop open a population of compliant voids or inclusions that are nevertheless capable of remaining open against the least principal stress. The implication of 3D principal stress anisotropy at depth is that EDA cracks will tend to be aligned in a vertical plane, striking parallel to the major horizontal stress. With this configuration, a microcracked but otherwise isotropic crust would be transversely isotropic, with a horizontal symmetry axis.

### 15.6

The traditional view was that there were several possible *small scales* of azimuthal anisotropy that could cause shear wave splitting, such as aligned crystals, lithological anisotropy due to aligned grains, stress-aligned microcracks, and fine layering. Much evidence for the influence of larger-scale joint-set alignment effects on shear wave splitting has subsequently been obtained. These larger scale features obviously dominate drainage potential from the matrix to the joints, and thence to the wells in hydrocarbon production. The micro-scale *extensive dilatancy anisotropy* (EDA) championed by Crampin and co-workers, would logically dominate drainage from the pores to the microcracks. Relatively large time delays between split shear-waves may also be set up in the top tens to hundreds of meters of rock. These near-surface effects have been termed *natural directivity*. Principal stress aligned microcracks, or principal stress aligned intra-bed jointing, or aligned jointing from historic tectonic effects including doming and anticlines, each have shear-wave splitting potential.

### 15.7

It was earlier considered remarkable that, with all the different scales and characters of aligned fluid-filled cracks, inclusions or fractures in sedimentary, metamorphic and igneous rocks, that the differential shear-wave anisotropy varied only within narrow limits (0.5 to 5%). With increasing application at fractured reservoirs, this range, and the earlier assumed limited ranges of fracture density, have each been extended, sometimes by significant margins.

### 15.8

The traditional porous medium experience is that shear-wave velocity remains unchanged whether a formation contains gas, oil or water. However, because of the effects of fluid compressibility on the dynamic *normal* stiffness of fractures, or its inverse compliance, the polarized shear waves passing through a fractured or jointed medium, may actually have the unexpected ability to distinguish between oil and gas, specifically when incident waves and jointing are non-parallel. Regions of gas are characterized by lower magnitude *slow* shear waves, and regions of oil by higher magnitude slow shear waves. In comparison, classic Gassmann porous medium theory anticipates an S-wave velocity relatively unaffected by the type of fluid.

## 15.9

A controversial point is whether the fractures and shear-waves can *both* be vertical, where theoretically only shear compliance would be sensed, because it is uncertain if shear compliance will be sufficiently affected by these contrasting fluid compressibility effects. A more certain effect of fluid compressibility is when normal compliance is involved in the case of sub-vertical fractures or sub-vertical shear-waves, giving a finite angle of incidence. There is evidence that the delay between the split shear waves may decrease with increasing depth, yet an accumulative delay with increasing depth would naturally be expected. This perhaps suggests that stress-sensitive compliances are involved, which would match rock mechanics experience with the non-linear pseudo-static stiffness of joints or fractures.

## 15.10

In the case of *earthquake* recordings, the relative steepness required for the incident wave to make an acute angle to typical sub-vertical structure, means that there is a need for the recording site to be within the *shear wave window*. This derives from the requirement of angles of incidence less than $\theta = \sin^{-1}(V_p/V_s)$. For a Poisson's ratio of 0.25, $\theta$ is about 35°. Outside this window the shear-wave waveforms are severely distorted. The epicentral distance from the recording sites must therefore be significantly less than the focal depth of the earthquake.

## 15.11

The large depth of most earthquake sources means that shear waves will pass through a range of rock types with different ages. Velocity and individual joint-set properties are sure to vary, and each split shear wave may therefore split again, giving multiple splitting. The influence of the joint structures near the recording site gives one of the most prominent results. Down-hole instrumentation is needed if possible, in order also to minimise such site effects and the higher frequency filtering due to attenuation.

## 15.12

High velocities in rock tend to occur where attenuation is low or $Q_{seis}$ is high. This reciprocal relationship between velocity and attenuation is one of the reasons why the leading split shear wave is a very stable phenomenon, because it is travelling in the fast direction and is less attenuated than the slower split shear wave. Sometimes the slow wave is too attenuated to allow calculation of the shear-wave anisotropy. The location along the ray path where the shear-wave splitting is imprinted most strongly is not known *a priori*, and near-surface effects where jointing is stronger may be a disturbing feature. Well fractured reservoirs may over-print such effects. Lower *crack densities* implied by many earthquake studies may be a reflection of the sampling of 'average rock'. In contrast, crack densities interpreted from fractured reservoirs may represent a 'biased sample', caused by a rock mass that is more jointed or fractured than the norm, and therefore also a target for exploration and subsequent exploitation.

## 15.13

The geophysicist's crack density ($e$) is defined as number (N) of cracks per volume (V) times the crack radius (a) *cubed*. Crack density was often quoted in the range of $e \approx 0.01$ to 0.05 in reportedly widely different geological and tectonic regions. This commonly used parameter is unfortunately remarkably ambiguous. Ten million micro-cracks @ 100 μm/10 cm cube, give $e = 0.01$, while ten fractures @ 1 m/10 m cube also give $e = 0.01$, and even ten minor faults @ 100 m/1 km$^3$ give $e = 0.01$. These three scenarios, with their theoretically equal 'crack density', have very different mechanical and fluid-conducting properties. Nevertheless they *theoretically* would previously have suggested equal shear wave anisotropy. Alternative methods of forward modelling show this theory to be in error, when extremes of fracture size and compliance are involved, and due to frequency or dispersive effects.

### 15.14

The classically assumed limited range of 1% to 5% shear-wave anisotropy, was linked to a similarly limited range of assumed crack densities $0.01 \leq e \leq 0.05$. The percentage of differential shear wave anisotropy was reportedly about $e \times 100$, for a $V_p/V_s$ ratio of about 1.7 or $\sqrt{3}$. There are now known to be many exceptions to these limited levels of shear-wave anisotropy, and there are order of magnitude larger values of fracture density interpreted in fractured reservoirs. So-called fracture *criticality*, associated with higher densities, is clearly positive for good reservoir production. The rock mass does not fragment, and the shear strength is not lost, and nor does the pore fluid disperse at higher crack densities, as feared by a prominent author. This is because it is confined by a 3D stress field, and by less permeable boundaries. How-ever rock mass deformability and susceptibility to compaction may obviously increase when looking beyond matrix compaction mechanisms, as at Ekofisk.

### 15.15

Aligned fracturing may be detected and monitored over a huge range of length scales, using polarized shear waves. Dimensions may range from crustal dimensions of 10–100 km, through 1–1000 m reservoir scale fractures and faulting, to millimetre and micron-sized microcracks. The relative stiffness of microcracks, having much higher aspect ratios than inter-locked fractures or joints, means that they cannot respond in the same way as fractures, to a given change in fluid pressure, according to classic geophysics teaching. Recent poro-elastic fluid interaction modelling of the dispersive effects caused by fractures of widely different dimensions, using double-porosity or triple-porosity models, show that different fracture dimensions can be inverted from given levels of shear-wave anisotropy, based on their different response to changing frequency.

### 15.16

By the mid-eighties, some oil companies were apparently reporting shear wave splitting in almost all their three-component reflection surveys in sedimentary basins. The phenomenon was assumed to be due to fracture or joint sets within the fractured reservoirs. Shear wave splitting was also visible in reflections from layers above the reservoirs. More recently, shear wave splitting has been observed to mimic saucer-shaped and oval-shaped sub sea-bed subsidence bowls kilometres above compacting jointed reservoirs. Strength-scaled polarization patterns show excellent correlation to locations where sea depths are changing most rapidly. The amount of detected anisotropy is small above the centre of the compacting field, where the subsidence is largest, and large on the flanks, becoming small again further from the centre. Oriented joint-stretch in the overburden seems likely to be the cause, as a partially *cubic* polarization pattern resembling intra-bed jointing can be noted. If unconsolidated un-jointed sediments were the actual source of this nearer-the-surface splitting and polarization, then micro-cracks or even macro-cracks in the sediments would need to be invoked to explain the polarization match to the subsidence bowls.

### 15.17

On occasion, '90°-flips' in polarization directions are observed from presumed earthquake source zones. It has been postulated that this may be due to extreme build-up of pore pressure, causing the faster split shear-waves that were previously parallel to $\sigma_{H\ max}$ to do a 90°-flip and become the slower wave parallel to $\sigma_{h\ min}$. From a geotechnical viewpoint a 90°-flip in polarization would appear more likely with extreme $\sigma_{H\ max}$ loading, causing lateral expansion of aligned microcracks in a limited volume of rock. The volume affected by expanding, aligned microcracks may need to be limited, since the rock mass could not absorb this volume increase without a general reversal of the $\sigma_{H\ max}$ and $\sigma_{h\ min}$ directions.

### 15.18

The progression from seismic propagation in isotropic media, to anisotropic layered media, to transversely isotropic layered media containing one set of vertical fractures, later increased to two sets of perpendicular fractures, then non-orthogonal vertical sets, and finally to non-vertical sets, has resulted in a progression of theoretical papers in the geophysics literature, containing an increasing content and complexity of $6 \times 6$ compliance and stiffness matrices. Schoenberg and Sayers are prominent authors.

### 15.19

The elastic moduli and the density determine the behaviour of seismic waves, assuming a linear, loss-free, elastic behaviour. The presence of fracture sets affects the elastic moduli of the fractured rock, due to the addition of their dynamic compliance. The additional presence of fractures can be expressed as the sum of the compliance of the isotropic back-ground rock and the *excess compliance* matrix associated with the fractures. The latter is composed of the effects of a fracture-normal compliance $Z_N$, and of a fracture-shear compliance $Z_T$.

### 15.20

The simple addition of the three fracture compliance terms ($Z_N$, $Z_T$ and $Z_T$) is made in the same diagonal-term (1,1 5,5 and 6,6) locations in the combined compliance matrix. In the context of *shear-wave splitting*, the stiffness matrix term relating to the *fast shear wave* propagating parallel to the fractures, is the $C_{44}$ term, and the *slow shear wave* propagating perpendicular to the fractures, is given by the $C_{55}$ term. The Thomsen shear-wave splitting *anisotropy parameter* ($\gamma$) is defined from the elastic stiffness matrix as $(C_{44} - C_{55})/2C_{55}$ (often expressed as a percentage), where $qS_1 = \sqrt{(C_{44}/\rho)}$, and $qS_2 = \sqrt{(C_{55}/\rho)}$. Shear-wave splitting anisotropy as defined by Thomsen, is commonly in the range $0 \geqslant \gamma \geqslant 20\%$. The Crampin definition of the S-wave velocity anisotropy is a contrasting $100 \times (V_{s\,max} - V_{s\,min})/V_{s\,max}$ or $100 \times (qS_1 - qS_2)/qS_1$. For the case of the vertically propagating waves through the vertical fractures, there is no fracture compliance term in $C_{44}$, only the Lamé constant $\mu$ for the back-ground rock. This causes the fast shear-wave component to be parallel to the fractures. For the case of the *slow shear wave*, the more complex $C_{55}$ term involves $Z_T$, and not $Z_N$. The simpler *compliance* matrix $S_{55}$ term is simply $1/\mu + Z_T$.

### 15.21

Involvement of $Z_N$ in the slow shear wave velocity theoretically requires dipping fractures, or non-vertical wave propagation. When polarized shear-waves sense the different viscosity of oil or gas in the fractures, $Z_N$ is more likely to be involved than $Z_T$, which would have a less obvious dependence on fluid viscosity differences. This sensitivity is despite Gassmann's theory for porous media, in which the shear modulus should be independent of the fluid. In the case of fractured media, oil and gas are distinguishable by the reduced and increased shear wave anisotropy respectively, as the stiffening effect of the oil makes the fracture normal stiffness less contrasted to the back-ground medium. So for dipping joints or fractures, there proves to be a significant *decrease* in shear wave anisotropy if the fluid has a higher bulk modulus, making the normal *stiffness* of the fractures greater. The average of the two shear wave velocities is therefore also increased.

### 15.22

The shear compliance $Z_T$ and normal compliance $Z_N$ interpreted from loaded, roughened Lucite (Plexiglas) plates, which were used to simulate 'a fractured medium', was apparently responsible for some authors to assume that $Z_N \approx Z_T$ for the case of dry, gas saturated cracks. Seismic phenomena observed in highly stressed, finely layered (t = 0.7 mm), roughened plates of Lucite, with their extreme 'crack densities' and artificial 'fracture' surfaces, should however not be used to predict rock joint response to dynamic or static loading. The suggested 'equality' of $Z_N$ and $Z_T$ was propagated in some of the geophysics literature, but may be far from realistic for all but the smallest laboratory specimens. $Z_N$ involves 'micro-closure' in a stiffening direction, while $Z_T$ involves 'micro-slip' in a direction that may not involve stiffening.

### 15.23

An inequality of the joint or fracture compliances would be more consistent with the experience of $K_s < K_n$, concerning the *pseudo-static* shear and normal *stiffnesses* of joints and fractures, where stiffness is the rough inverse of compliance. The magnitude of $K_n$ proves to be a bit less than, but quite close to $1/Z_N$ in good quality unweathered hard rock, while in the shearing direction, $K_s < 1/Z_T$, sometimes $\ll 1/Z_T$. However $Z_T$ data from geophysics investigations is extremely limited compared to the large body of $K_s$ (pseudo-static) data that has been in use in discontinuum rock mechanics modelling since the late 1960's, first in jointed FEM studies by Goodman. If pseudo-static stiffnesses and dynamic compliances could be related, despite the different orders of magnitude of dynamic and 'static' deformations, then the more researched, stress-and-scale-dependent parameter $K_s$ and the stress-dependent $K_n$ might provide shear wave splitting analysts more information than at present. A handful of 40 to 50 mm size laboratory samples of natural joints and stress-induced fractures, and tests on Perspex plates, represent a very uncertain basis for interpreting the equality or inequality $Z_N \leq Z_T$.

## 15.24

The measured range of $K_{n(dyn)}$ / $K_{s(dyn)}$ ratios for approximately 50 mm diameter jointed samples in hard crystalline rock has been shown from limited testing to be about 1 to 4. This means that the ratio of the inverted *compliances* $Z_N/Z_T$ for such samples may range from 1 to 0.25. Convenient 'core-sized' 50 mm data from state-of-the art testing of joints in hard crystalline rock seem unlikely to have 1:1 relevance to *in situ* reservoirs in much weaker sedimentary rock, where jointed block sizes may range from extremes of perhaps 100 mm to 10,000 mm.

## 15.25

Pyrak-Nolte demonstrated at laboratory scale, that joints that support less flow tend to have higher normal stiffness, and that in principal, joint normal stiffness may be inversely related to the cube root of the permeability. The permeability and seismic response of a joint can therefore be inter-related through the normal stiffness of the joints in question. Joints that *attenuate* seismic waves most at a given stress level, due to lower normal stiffness, also supported more flow. This implies a second implicit link between the permeability, and *both* the joint index parameters JRC and JCS, since these can be used to calculate both the normal and shear stiffness, and they independently give estimates of average hydraulic and physical apertures.

## 15.26

The permeability of the rock *mass* may also be related to the rock mass modulus of deformation, when permeability and modulus are *not* reduced by clay. In simplest possible terms permeability is inversely related to $Q_{rock}$. Seismic quality $Q_{seis}$ and the *pseudo-static* modulus of deformation $E_{mass}$ expressed in GPa, can each be estimated from $Q_{rock}$, for the case of jointed rock masses.

## 15.27

The typical $10^{12}$ Pa/m unit for stiffness favoured by geophysicists, converts to 1000 MPa/mm, which is more familiar in rock mechanics. An even better experimental feel for stiffness is given by 1 MPa/μm, each of these being equivalent. In geophysics, the inverse of stiffness, termed *compliance* is also typically reported in 'complex' units such as $10^{-12}$ m.Pa$^{-1}$, instead of equivalent units of 1 μm/MPa, which is much more tangible for any rock mechanics experimentalist.

## 15.28

Depending on stress levels, rock type, and joint roughness, $K_{n(dyn)}$ may range from $10^{12}$ to $10^{14}$ Pa/m, or 1,000 to 100,000 MPa/mm or 1 to 100 MPa/μm, suggesting a mostly very small increment of dynamic displacement. $K_{n(static)}$ values from a wide range of weaker rock and joint types, may vary from as low as 100 MPa/mm to almost 50,000 MPa/mm. There is therefore a large degree of overlap between the static and dynamic stiffnesses in this *stiffest* perpendicular loading direction.

## 15.29

*Shear* compliance, as opposed to *normal* compliance is of most relevance in the case of vertical shear-waves and vertical structure, and the resulting degree of shear-wave anisotropy. Of the two *dynamic* compliances, this is the least understood component. By comparison, it is the pseudo-static shear *stiffness* that is most researched in rock mechanics. In pseudo-static testing, a whole range of possible static *shear stiffnesses* are found, that seem generally to be *inversely* related with the sample size, if there is measurable joint roughness, (and therefore possible permeability even at depth). The degree to which dynamic *shear compliances* might be related, more weakly but nevertheless directly to sample size, is one of the remaining unsolved areas in this important area of seismic detection of anisotropy, and the subsequent goal of interpreting fractured reservoir permeability.

## 15.30

Some pieces of the dynamic-permeability jigsaw are complete, but there are missing links between dynamic and static testing, and between small sample testing and the large sample reality, which causes experimental stress-magnitude problems. The completed parts of the jigsaw are the abilities to estimate both of the pseudo-static stiffnesses $K_n$ and $K_s$ and the less tangible physical (E) and hydraulic (e) apertures, for different sizes of jointed rock block, based on simple index testing. This involves estimation of joint roughness and wall strength, using respectively the un-scaled or scaled JRC and JCS components of the Barton-Bandis joint constitutive model.

## 15.31

Fracture-induced seismic anisotropy has rapidly evolved from the earlier estimation of fracture orientation, with an assumed indication of major horizontal stress, to fracture intensity, and the attempted prediction of fluid type, fluid saturation, and permeability anisotropy. To make this advance, the sensitivity of the fracture compliances to fluids has to be understood. Theoretical expressions for the fracture compliances $Z_N$ and $Z_T$ have been developed by Liu and Hudson and co-authors, which indicate strong sensitivity of the ratio $Z_N/Z_T$ to the bulk modulus of the fracture infill material, with the most rapid change in the *compliance ratio*, and values closer to 1.0 occurring when the infill bulk modulus approaches zero, such as for gas-filled fractures. One of the geophysics equations suggests that $Z_N/Z_T \approx 1$ if fractures are dry, and $Z_N/Z_T \approx 0$ if fractures are filled with liquid. With realistically small fracture aspect ratios (i.e. $<0.0001$), much lower ratios of $Z_N/Z_T$ than 1.0 are predicted, which would be more in line with rock mechanics pseudo-static experience. Zero is inadmissible for the pseudo-static ratio $K_s/K_n$, but certainly $K_n \gg K_s$.

## 15.32

Although outside this chapter's focus on shear-waves, Worthington and Hudson modelled the effects on down-going P-waves, of one or more faults intersecting the transmission path, between 1000 and 2000 m depth in a North Sea reservoir overburden. They used a theoretical compliance model to demonstrate the need for a very large inequality of the shear and normal *compliances*, suggesting the need for $Z_N = 4.4 \times 10^{-14}$ m.Pa$^{-1}$, and $Z_T = 1.1 \times 10^{-9}$ m. Pa$^{-1}$. These convert to shear and normal *stiffnesses* of $K_n \approx 20{,}000$ MPa/mm, and $K_s \approx 1$ MPa/mm. This low, back-calculated *'in situ'* shear stiffness is similar to the values used in large-scale pseudo-static modelling of compaction/subsidence in rock mechanics. In general one may assume that the *pseudo-static* values of stiffness are lower in the normal direction, and much lower in the shear direction, than the equivalent dynamic values of stiffness.

## 15.33

Stronger shear wave splitting interpreted from large time delay in locations in the neighbourhood of fault zones, such as the San Andreas fault at Parkfield, suggests that fluid-filled fractures within the fault zone may be more extensive than in the surrounding crust. Fault parallel polarization of the leading split shear wave may indicate that fault-related fractures are aligned by fault shearing rather than by the differently aligned regional principal stress that can be verified further from the fault plane. The internal structure of the fault gouge and transition zone is assumed to be the reason for this rotation. In some cases, only one of the anisotropic shear-wave polarizations is recorded, due to attenuation of the slower component. A change of stress affecting the geometry and fluid in neighbouring discontinuities following larger earthquakes, may be the reason for temporal decreases in time delay.

## 15.34

Earthquakes with shallower focal depths may show pronounced increase in time delays, suggesting that a stronger anisotropy associated with the fault zone may also be concentrated at shallower depth. Leading shear-wave polarizations exhibiting *fault-parallel* alignments near the fault, but alignments with the regional stress field away from the immediate fault zone were also verified at the Cajon Pass site. In general terms the source of shear-wave splitting may be shallower than desired, where conditions of *anisotropy* are more favourable for this mechanism.

## 15.35

Stress-monitoring sites in Iceland, using state-of-the-art borehole instrumentation to monitor shear-wave splitting between controlled-source wells and receiver wells are designed to identify the effects of nearly negligible changes of stress, which might be capable of monitoring the build-up of stress and other crustal adjustments before earthquakes and volcanic eruptions. Anomalous water level changes and GPS-measured displacement anomalies have been related to various changes in prior and *subsequent* P-wave and S-wave travel times and delays.

## 15.36

The '90°-flips' that sometimes occur in source zones, may be due to extreme pore pressure build-up according to Crampin. An alternative explanation is that they could be due to extreme loading in the $\sigma_{H\,max}$ direction, causing an extreme 'Poisson effect' due to crack expansion, thereby locally reversing the $\sigma_{H\,max}$ and $\sigma_{H\,min}$ directions. Except in an eventual clay core, a pore pressure *reduction* appears more likely than the assumed increase of pore pressure, since as shear failure of either intact or jointed rock is approached, dilation is the most likely phenomenon, unless beyond the brittle-ductile transition.

## 15.37

Major earthquake (Chi-Chi) after-shock monitoring in Taiwan indicated 8% shear-wave velocity anisotropy at a 200 m deep station due to up-going split shear-waves with fast and slow components. $Q_{seis}$ values in the 2–15 Hz frequency band were 61 to 68 in the fast direction, and 43–52 in the slow direction. Such values may closely resemble possible deformation moduli M, expressed in GPa. The anisotropy may have been caused by sub-vertical joint set anisotropy, which causes anisotropy at low seismic frequencies, as opposed to the assumed micro-crack anisotropy that is dominant from >1 kHz.

## 15.38

Shear-wave splitting analysis utilizing local 4 to 5 km deep micro-earthquakes, and P-wave anisotropy measurement with controlled sources, were used in Mid-Atlantic Ridge studies, where anisotropy was attributed to a shallow distribution of vertical, fluid-filled cracks, aligned parallel to the trend of the axial valley. Time delays ranged from 35 to 180 *ms*, with an average delay of 90 *ms*. The shear-wave anisotropy was from 8 to 30% in the highly fissured layer. Most of the shear wave delay was attributed to the shallowest 500 m in seismic layer 2A, where the average value of $V_p/V_s$ was 2.9. There was evidence for only a shallow concentration of P-wave anisotropy, which decreased from >4% at 500 m depth, to zero below 1.5 km depth. Isolated fluid-filled cracks at depths from 500 m to 3 km were too tight to be detected by the P-wave survey, but could contribute to the shear-wave delays. P-wave anisotropy was defined as $100 \times (V_{p\,max} - V_{p\,min})/V_{p\,average}$, while S-wave anisotropy was defined as $100 \times (V_{s\,max} - V_{s\,min})/V_{s\,max})$ following convention.

## 15.39

Offshore 3D seismic surveys, using compressional P-wave sources converted to PS at the sea-floor, thereafter analysed as S-waves, or use of direct S-waves (SS) generated on land, in walk-away, multi-azimuth VSP, are basic geometries for shear-wave polarization and anisotropy investigations above fractured reservoirs. Split shear waves S1 (or $qS_1$) and S2 (or $qS_2$ ) may occur due to major principal stress aligned dominant fracturing: the conventional interpretation, or perhaps due to stress bisected sets of unequal conjugate fractures, or due to fracture sets that have suffered some permeability-enhancing pre-peak shear displacement, which may also cause a deviation from $\sigma_{H\,max}$. In walk-around VSP, a circular path of multiple sources at fixed offset, an incident P-wave, when converted to a PS wave that passes through the plane of horizontal symmetry caused by aligned vertical fractures, will show polarity reversal when incident on either side of the fracture strike.

## 15.40

The overburden may also display anisotropy in the form of azimuthal-dependent velocity. Errors will be introduced when the seismic data are inverted to obtain fractured reservoir parameters, if overburden anisotropy is ignored. Shear-wave splitting and polarization occurs in both cases due to the presence of the relatively compliant fracture properties. For the reservoir, basic structural information such as fracture density, strike and dip (due to symmetries), and some indications of fluid-type (gas or oil) and permeability may also be obtained, due to the relative sensitivities of the fracture normal and shear compliances. Deviation of fracturing from $\sigma_{H\,max}$, because one of the sets dominates, or because fracture set stiffnesses differ, may compromise relative crack density interpretation.

### 15.41

Conjugate fracturing with non-vertical dips, as typically found in anticlinal or domal structures, means that samples of both the oppositely-dipping joint sets can be directly sampled, and tested, using vertical exploration wells. Nevertheless, it is a remarkable fact that exploration wells are nearly always vertical despite the target structures often being vertical too. Sidewall cores are therefore used to help locate zones having high fracture intensity; using microfracture and diagenesis data to infer the presence of the macro-fractures.

### 15.42

Large aperture fractures may require mineral bridges to have preserved conducting porosity. Fractures below a certain characteristic size may be completely filled. 'Open' fractures in the sub-surface are not necessarily parallel to maximum compressive stress $\sigma_{H\ max}$. Fractures perpendicular to this direction can also be 'open' if partially filled with *synkinematic* or *post-kinematic* mineral cements. *Sealed* fractures parallel to $\sigma_{H\ max}$ may also be numerous. Effective normal stress induced joint 'closure' caused by sufficiently high $\sigma_{h\ min}$ may occur at reservoir depths in less competent rock. There is however an important prepeak shear-displacement mechanism for maintenance of 'open' fractures that are non-parallel to $\sigma_{H\ max}$, in which conjugate sets may be *bisected* by $\sigma_{H\ max}$. This interpretation has numerous merits for explaining permeability at depth.

### 15.43

When fracture densities are as high as 1 to 2, as in well-jointed, domal chalk reservoirs, dimming of the amplitudes of the slow shear-wave, due to greater attenuation caused by lower $Q_{seis}$ tend to correlate with the most productive parts of the reservoir, as also experienced where the measurable shear-wave anisotropy is greatest. A down-dip shear mechanism may help to maintain apertures despite high effective stresses in the presence of this weaker, high porosity rock. If the high crack density is contributed to by two sets of oppositely dipping conjugate joints, one can expect shear *and* normal compliance contributions from both sets to the slowness of the slow shear-wave, which might be $<2$ km/s when porosity is high. The response is strongest when fractures are gas filled, as gas does not stiffen the normal compliance. Attenuation anisotropy should be enhanced in the case of oil-filled fractures, due to the greater contrast of their normal and shear compliances.

### 15.44

Reportedly the world's first time-lapse, marine, multi-component (3D/4C) survey, was performed in 2002 at the Ekofisk jointed-chalk reservoir in the North Sea. This baseline was subsequently compared with a monitoring 4D/4C survey acquired in 2003. Small changes of polarization direction and of shear-wave anisotropy were detected, even after this short 15 months monitoring of more than 30 years production. Azimuthal rotation of the anisotropic attenuation has recently also been detected at neighbouring Valhall, where life time seismic monitoring is established. An unequal conjugate shear mechanism could be a possible explanation for these temporal rotations, especially if the strike of the two conjugate sets was not equally oriented. Potential 'opposite-rotation' of fluid lenses and rock-to-rock contact areas due to pre-peak shearing of non-planar joints might also be detected by the shear waves. A much larger rotation of polarization azimuth was detected at a monitored HDR in Cornwall, where joint shearing was also assumed, due to plunging AE activity.

### 15.45

An important development in the dynamic modelling of the effects if fractures is that of Schoenberg, who modelled elastic wave behaviour using linear slip interfaces. These allow reflection, transmission, conversion, and delay to take place at the modelled interface, with the magnitudes depending on the specific stiffness, the frequency content, and the angle of incidence. The assumption is that when an elastic wave propagates across a fracture, there is a displacement discontinuity that is linearly related to the normal or shear force generated. The seismic particle displacement is *discontinuous*, while the seismic stresses are assumed to be *continuous*. This shares several of the interface-stiffness concepts that are basic to discontinuum modelling in rock mechanics, in the Cundall 2D and 3D codes UDEC and 3DEC.

### 15.46

Hudson utilised a 'method of smoothing' for the effect of modelled cracks, which was capable of representing the elastic parameters of a 'cracked' material, in the form of an *effective medium*. This allowed calculation of the effect of incident dynamic waves of long wavelength. Subsequently, Hudson extended the model to allow for the cracks to be connected via the porosity of a rock matrix. In this extended case, cracks could be deformed by an incident wave in a manner that depended on their aspect ratio and on their orientation with respect to the incident wave. Clearly the modelling of intrinsic attenuation mechanisms such as squirt flow, and its frequency dependence, was transformed by this extended capability.

### 15.47

Tod extended this model further, by allowing for a continuous distribution of crack orientation and aspect ratios, and by allowing each to depend on the applied stress and on the fluid pressure. At high frequencies, the cracks behave as isolated stiff features, while squirt losses at lower frequencies cause a soft behaviour. Crack density is allowed to decay with an increase in applied stress, from an initial value representing the unstressed state. The model is purely elastic, and relaxes to its original state upon unloading. Crack density is designed to decrease with increasing compressive stress, but cracks with normals lying in the $\sigma_h$ minimum direction are assumed to remain open. The model is capable of capturing the changed anisotropy caused by fluid pressures and applied stresses, which impact the aspect ratios of the cracks. The non-compliant pores of the Tod model, transfer fluid to the physically unconnected cracks, therefore giving dispersive, or frequency-dependent velocities and attenuation.

### 15.48

Chapman developed a triple-porosity poro-elastic model, based on the observation that typical laboratory samples, clearly unfractured, nevertheless display dispersion, anisotropy, stress sensitivity, and dependence on fluid type and degree of saturation. When the fractures or cracks were removed from *preceding* models, a linear-elastic material remained, in contrary to observation. Chapman combined 'meso-scale' anisotropic fractures with equant matrix porosity *and* ellipsoidal microcracks. Due to the fluid transfer between the different scales of discontinuity, dispersion occurs at lower frequency ranges than those over which the micro-structure dispersion occurs. Introduction of even small fractures causes dispersion to begin at lower frequencies, and the effect increases with increasing fracture sizes. The magnitude of shear wave anisotropy therefore shows fracture size, frequency and azimuth dependence. Attenuation is altered by setting smaller aspect ratios for meso-fractures compared to microcracks, as a result of higher stress, as the fluid cannot 'squirt' or flow so easily, in response to the passage of seismic waves.

### 15.49

Effective medium models appear at present to make no distinction between the assumed mean crack aperture (E), and the theoretical hydraulic aperture (e). In jointed rock (E) controls stiffness and deformation moduli, and the crack aspect ratio should have a similar role. On the other hand, (e) controls the intrinsic permeability, given as $e^2/12$. Both apertures probably influence attenuation, since they effectively define two different aspect ratios, therefore influencing the assumed squirt-flow losses and the assumed stiffness. Dispersion will likely begin at higher frequencies as roughness increases. Incorporating the inequality $e \leqslant E$ would be a source of improvement for effective medium modelling of *rough-walled*, tightly compressed cracks, joints or fractures, for which use of only one aspect ratio is least realistic.

### 15.50

The traditional EDA focus on microcracks as the source of shear-wave splitting, may need to be re-evaluated, in the light of these new poro-elastic double (or triple) porosity models. Microcracks appear to give a 'constant' potential source of shear-wave splitting independent of frequency. Meso-fractures, for instance of 1 m and 10 m radius, apparently have an equally strong role in shear-wave anisotropy as microcracks, but specifically at the seismic frequencies appropriate to earthquake studies.

### 15.51

Shear wave splitting through fractured reservoirs may show a sharp increase of time delay with depth at the reservoir level, indicating the presence of fractures. The time delays tend to be largest at lowest frequencies (e.g. 5–15 Hz), and smallest at higher frequencies (e.g. 20–40 Hz). However, polarization of the *fast* S-waves may show no apparent variation with frequency, unless at very low frequency. The time delays between the split shear waves may be decreasing as frequency increases, due to stiffening of the squirt phenomenon in the case of the *slow* waves. Such data have been used for inverting for the theoretical *fracture* density (an appropriate re-naming of *crack* density), and for fracture radius, using nine-component VSP data sets, with one P-wave and two orthogonal S-wave sources. Frequency-dependent anisotropy has also been demonstrated.

## Chapter 16 Joint stiffness and compliance and the joint shearing mechanism

### 16.1

Rock mechanics developments in distinct element rock mass deformation modelling, including flow within the deforming joints, can be used to illustrate an extension of the traditional geophysics concept of one set of stress-parallel 'open' joints in a reservoir. This is important because multiple sets of joints are more usual in rock masses, probably including even deep reservoirs. Multiple joint or fracture sets, such as bedding and two vertical sets also open the possibility of polarization orientations that are not parallel to the $\sigma_{H\,max}$ direction. Two conjugate vertical joint sets intersected by the $\sigma_{H\,max}$ direction can also cause shear wave splitting with polarization that *nearly* corresponds to this principal stress direction.

### 16.2

When reservoir joints or fractures are not parallel to the $\sigma_{H\,max}$ direction, they may be acted on by shear or differential stress, and this implies shear deformation, which may be needed to supplement the often too small conducting apertures which may result when in the traditional parallel to $\sigma_{H\,max}$ direction. This philosophy is based on the fact that unless joints or fractures are mineral 'bridged-but-not-blocked', or are very rough and in hard rock, both testing and modelling indicates too small apertures to be considered 'open', since conducting apertures may be $<5\,\mu m$, and often $<1\,\mu m$. Support for this geomechanics concept is provided by deep-well data showing that joints under differential (shear) stress are the conductors, with other directions apparently non-conducting.

### 16.3

The interpretation of shear wave splitting with possible multiple 'vertical' joint or fracture orientations emphasises the need for knowledge of both shear and normal compliances, if poro-elastic modelling is to extend to stress and displacement sensitive aspect ratios for the modelled joints or fractures. Presently, limited sets of laboratory data, and even more limited sonic log and cross-hole data, and a tentative extrapolation to large scale seismic interpretation of fault compliances, strongly suggest a scale effect. Increased compliance in both normal and shear directions is implied, as scale is increased or frequency reduced.

### 16.4

A comparison has been made between the presently known normal $Z_N$ and shear $Z_T$ compliance laboratory and field data, which is a dynamic and micro-deformation response, and the much larger body of laboratory and field data for the pseudo-static response of normal $K_n$ and shear $K_s$ stiffness. Data for joints, tension fractures and clay-filled discontinuities are readily available due to long term use in rock mechanics modelling of jointed rock masses. These pseudo-static measures of stiffness are given in typical units of MPa/mm, and when inverted show resemblance but generally larger magnitude, compared to the geophysicists compliance, given in typical m/Pa units.

### 16.5

When rock mass and rock joint quality is high, the dynamic modulus $E_{dyn}$ is not of much greater magnitude than the static modulus $E_{mass}$. The normal compliance $Z_N$ may then be within 1 to 1/10 of the magnitude of the inverted static normal stiffness $K_n$. A typical *laboratory-scale* $Z_N$ range of $10^{-13}$ to $10^{-14}$ m/Pa, or $K_n$ (dyn) of 10,000 to 100,000 MPa/mm is therefore found to be 1 to 10 times stiffer than typical $K_n$ (static) data of *typically* 1,000 to 10,000 MPa/mm (range maybe 100 to 50,000 MPa/mm). In the inverted worlds of geophysics and rock mechanics, $10^{-12}$ m/Pa is the same as 1,000 MPa/mm, with the identical 1 MPa/μm giving a more practical feeling for this level of micro dynamic stiffness.

### 16.6

Although the large-scale, seismically derived magnitude of shear compliance $Z_T$ from reservoirs is presently very uncertain, preliminary indications are that it may be about two orders of magnitude stiffer than large-scale pseudo-static shear stiffness $K_s$. The lower magnitude of $K_s$ is due to strongly reducing magnitudes as block size increases. A block size increase has a double effect due to reduced peak shear strength and larger displacement to reach peak strength. It is not known whether this fundamental scale effect could also influence the presumed stiffer dynamic shear stiffness, but a weaker dependence than in rock mechanics *is* suspected.

### 16.7

A scale effect on $Z_N$ or $K_n$ (dyn), as occurs with $K_s$, is not expected when merely sampling a larger portion of the same feature, However, when comparing $Z_N$ or $K_n$ (dyn) on small fractures, larger joints, or faults, a *sampling* effect must obviously be expected, since the inequality $E_{dyn} \geqslant E_{mass}$ is potentially increasing when sampling increasingly large features.

### 16.8

There is present uncertainty in the magnitudes of $Z_N$ and $Z_T$ because of limited test data at different scales. The uncertainty in the magnitudes of the pseudo-static $K_n$ and $K_s$ parameters used in rock mechanics is much less, and values can be readily predicted using a 1D model for coupled M–H joint behaviour. However, there is an inevitable lack of test data with respect to high pressure *and* large size, also in the case of pseudo-static testing.

### 16.9

A further uncertainty with all four joint parameters is that rough joints or fractures tend to be more tightly closed at high temperature, and are also more tightly closed following several load cycles if they have been sampled and unloaded, and cooled. Joints can appear 'over-closed' by stress reduction from a higher stress level, if sufficiently rough and in sufficiently strong rock. Cooling can have a similar effect. Both ambient and thermal over-closure cause higher shear strength to be registered, due to changes in the thermal expansion coefficient, when joints are included. This causes changes in aperture and permeability, and may therefore complicate the matching of reality and present theory. *During* heating, stiffness and modulus appear to be less than expected, due to the additional closure. However, following the heating, this thermal over-closure results in stiffer subsequent behaviour.

### 16.10

Despite insufficient test data from most areas, the utilization of index properties for the joints for guiding the estimation of $K_n$ and $K_s$, which is well established in rock mechanics, could provide order of magnitude estimates of the dynamic properties, or estimates of the index properties and therefore permeability, by inversion. This would presently be based on the assumed one to two order of magnitude stiffer dynamic behaviour. Index properties JCS and JRC concerning wall strength and roughness, allow estimation of the mechanical and hydraulic apertures under closure and shearing, and also estimation of the pseudo-static normal and shear stiffnesses each, thereby potentially linking dynamic data and *in situ* permeability.

### 16.11

Index data for typical reservoir rocks and joints, with variation of both strength JCS and roughness JRC, and consideration of confinement effects on strength, give prediction of very small conducting apertures of micron or less magnitude when modelling *closure* by an assumed $\sigma_{h\,min}$ of 10 to 40 MPa, unless roughness and strength are both significant. Available coupled M–H test data from the laboratory CSFT apparatus, and from *in situ* block tests, each including heating, also confirm the extremely small apertures of interlocked joints, unless rough and of high strength, or with aperture preserved by mineral bridging.

### 16.12

When modelling minor amounts of shear, excellent permeability is predicted in most cases, due to the positive influence of even pre-peak-strength shear displacements, showing the vitally important beginnings of dilation. One to two orders of magnitude increased permeability is seen in CSFT coupled *shear* testing, following only 1 or 2 mm of pre-peak or close-to-peak shear displacement. Shearing of non-planar surfaces causes the fluid-bearing lenses on 'down-slopes' to have a different average orientation than the 'up-slope' rock-to-rock contact areas. This rotation may influence the polarization direction of split shear waves, if they are more sensitive to one or the other of these average directions. Combined with two conjugate but usually unequally developed joint or fracture sets, the resultant polarization directions of the split shear waves $qS_1$ and $qS_2$, can have at least two reasons for not lining up exactly with the classically 'preferred' $\sigma_{H\ max}$ and $\sigma_{h\ min}$ directions, and for showing 4D rotations of azimuth.

### 16.13

Significant contributions from deep-well monitoring and analyses by Colleen Barton, Zoback and Townend and colleagues, has demonstrated the hydraulic flow distinction of joints or fractures that are under differential shear stress, and those that are principally under larger normal stress and insignificant shear stress, due to their mutually different orientations. In general, but quite clearly, the former are found to be water conducting, and the latter are assumed not to be water conducting, based on temperature logging. The wells are of kilometre to several kilometre depths, drilled in mostly crystalline rocks, with several wells connected with San Andreas Fault investigations. This hydraulically differentiated behaviour is despite rock strengths significantly higher than typical reservoir rocks.

### 16.14

Differential stress magnitudes based on rock stress measurements, indicate that the presently resolved ratios of shear and effective normal stress expressed as mobilized frictional strength $\mu$, are generally in the range of 0.4 to 1.0 for the case of the above water conducting fractures. On this basis fractured reservoir jointing that was parallel to $\sigma_{H\ max}$ and without mineral 'bridging' or sufficient JCS and JRC might well be very impermeable when normally pressured, but perhaps conducting when over-pressured.

### 16.15

Using non-linear Barton-Bandis shear strength, dilation and aperture interpretation, in place of the linear Byerlee $\mu = 0.6$ to 1.0 assumption for the shear strength of 'faulting', one finds that the conducting 'fractures' with $\mu$ in the range 0.6 to 1.0, require typical to high ranges of both $JRC_n$, and confined strength $JCS_n$. This means a full-scale roughness as high as 10, and full scale confined rock strengths $(\sigma_1 - \sigma_3)$ as high as 200 to 600MPa or more, to explain this higher range of *resisted* $\tau/\sigma_n'$ loading. The subscripts (n) on the index parameters signify *in situ* strength, and block sizes $L_n$ of perhaps 0.25 to 2.5m.

### 16.16

It is very difficult to model $\mu$ as high as 1.0 *in situ*, in jointed or fractured rock at kilometre depths, but easy when near-surface. The joints or fractures must be pre-peak or close to peak strength to have sufficient strength to develop implied mobilized friction angles as high as 45°. At the same time they must be sufficiently dilated to be conducting. 'Fractures' that are under the lower $\mu$ range of 0.4 to 0.6 are more likely to have the character of minor faults or have larger block sizes or some clay smear.

### 16.17

The assumption of co-axial stress and displacement in the classic stress-transformation equations seems to be in error if the joints or fractures or new fault surfaces are non-planar. Since shear displacement is implied when specifying shear strength, or a resisted value of $\tau/\sigma n' = \mu$, then dilation has also inevitably occurred at the highest range of $\mu$ from about 0.7 to 1.0, for which significant roughness or non-planarity is also implied.

### 16.18

The mobilized dilation angle $d_{n\ mob}$ estimated from the $JRC_{mobilized}$ concept, needs to be added to the angle $\beta$ used to define the joint or fracture orientation in relation to the major principal stress direction $\sigma_1$. The addition of the dilation as $\sin 2(\beta + d_{n\ mob})$ and $\cos 2(\beta + d_{n\ mob})$ in the shear and normal stress transformation equations explains the extra difficulty of shearing when dilation occurs, and might explain the high end of the interpreted values of *in situ* $\mu$.

### 16.19

An approximate, order of magnitude prediction of permeability in rock masses, can be made using the rock mass Q-value, normalized by the uniaxial strength to the form $Q_c$. This can be equated to the inverse of the Lugeon value, where 1 L is $10^{-7}$ m/s. The resulting $Q_c = 10$, $K \approx 10^{-8}$ m/s type of estimate appears to be a useful first order estimate, when $Q_c$ values range from 0.1 to 1000, implying little complication of clay smear. The range of K is then predicted to be about $10^{-6}$ to $10^{-10}$ m/s, or roughly $10^{-13}$ to $10^{-17}$ m$^2$ if water viscosity at 20°C is invoked, for simplicity.

### 16.20

Due to the problem of clay-sealing of discontinuities, and due also to the general effect of reduced permeability at significantly increased effective normal stress, a new term called $Q_{H20}$ has been developed, involving an inverted $J_r/J_a$ term, a normalized JCS, together with a simple depth-permeability equation for the 'soft porosity' represented by jointed rock. This shows promising fit to shallow civil engineering Lugeon testing, and potentially also to deep-well data, and demonstrates suitable adjustment to lower permeabilities caused by clay-filled discontinuities and increased depths.

# Appendix A – The $Q_{rock}$ parameter ratings

$$Q = \frac{RQD}{J_n} \times \frac{J_r}{J_a} \times \frac{J_w}{SRF}$$

## The six parameters defined

RQD is the % of competent drill-core sticks >100 mm in length in a selected domain (Deere et al., 1967)

$\mathbf{J_n}$ = the rating for the number of joint sets (9 for 3 sets, 4 for 2 sets etc.) in the same domain

$\mathbf{J_r}$ = the rating for the roughness of the least favourable of these joint sets or filled discontinuities

$\mathbf{J_a}$ = the rating for the degree of alteration or clay filling of the least favourable joint set or filled discontinuity

$\mathbf{J_w}$ = the rating for the water inflow and pressure effects, which may cause outwash of discontinuity infillings

**SRF** = the rating for faulting, for strength/stress ratios in hard massive rocks, for squeezing or for swelling, as appropriate

## Combination in pairs

RQD / $J_n$ = relative block size (useful for distinguishing massive, rock-burst-prone rock)

$J_r$ / $J_a$ = relative frictional strength (of the least favourable joint set or filled discontinuity)

$J_w$ / SRF = relative effects of water, faulting, strength/stress ratio, squeezing or swelling ('active stress')

*An alternative combination* of these three quotients in two groups only, has been found to give fundamental properties for describing the shear strength of rock masses – something close to ***the product of 'c' and 'tan*** $\varphi$'. By implication Q (and in particular $Q_c$) have units resembling ***MPa***. (Barton, 2002a)

## Definitions of characterization and classification as used in rock engineering

CHARACTERIZATION description of a *virgin site* (*pre-tunnelling* attributes and properties)

CLASSIFICATION description of a *non-virgin site* (*post-excavation* attributes and properties)

For example, in the excavation disturbed zone of a tunnel or steep rock slope, there will be changes in *stress, permeability, deformation modulus* and *seismic velocity*. In the case of a tunnel, it may not only be the redistributed *stresses* that have radial and tangential components. With rock joints present, four sectors of shear stress and joint displacement (roughly at 45° intervals) and the pairs of diametrically opposite maximum and minimum tangential stress, may give a complex perturbation of properties in the EDZ.

The tables appearing in the following figures are the ratings for the six '$Q_{rock}$' parameters. These have been printed as figures, in order to keep the compact style suitable for reproducing together with field logging sheets. The recommended way of recording the Q-parameter ratings is explained in the following notes, based on Barton, 2002a. Footnotes below the tables that follow, also give advice for *site characterization* ratings for the case of $J_w$ and SRF, which must not be set to 1.0 and 1.0, as some authors have suggested. This destroys the intended multi-purposes of the Q-system, which has a quite different structure compared to RMR. (Bieniawski, 1989)

## Notes on Q-method of rock mass classification

1. These tables contain all the ratings necessary for classifying the Q-value of a rock mass. The ratings form the basis for the Q, Qc and Qo estimates of rock mass quality (Qc needing only multiplication

of Q by $\sigma_c/100$, and Qo the use of a specifically oriented RQD, termed $RQD_o$ relevant to a loading or measurement direction). All the classification ratings needed for tunnel and cavern design are given in the six tables, where Q only would usually apply.

2. For correlation to engineering parameters as described in this paper, use Qc (multiplication of Q by $\sigma_c$/100). For specific loading or measurement directions in anisotropically jointed rock masses use $RQD_o$ in place of RQD in the Q estimate. This means that an oriented Qc value should contain a correctly oriented $RQD_o$ for better correlation to oriented engineering parameters.
3. Q-parameters are most conveniently collected using *histogram logging*. Besides space for recording the usual variability of parameters, for structural domain 1, domain 2 etc., it contains reminders of the tabulated ratings at the base of each histogram. Space for presentation of results for selected (or all) domains at the top of the diagram, includes *typical range, weighted mean and most frequent* (Q-parameters, and Q-values).
4. During field logging, allocate running numbers to the structural domains, or core boxes, or tunnel sections, e.g. 1 = D1, 2 = D2 etc. and write the same numbers in the allotted histogram columns, using a regular spacing for each observation such as 11, 113, 2245, 6689 etc. In this way the histograms will give roughly the correct visual frequency of all the assembled observations, in each histogram column. Besides this, it will be easy to find the relevant Q-parameters for a particular domain, core box or section of tunnel, for separate analysis and reporting. Overall frequencies of observations of each rating (or selected sets of data) can be given as numbers on separate logging sheets. Large data sets can be plotted in e.g. EXCEL when returning from the field.
5. It is convenient and correct to record rock mass variability. Therefore allow as many as five observations of each parameter, for instance in a 10 m length of tunnel or 5 m length of core. If all observations are the same, great uniformity of character is implied, if variable – this is important information. At 'the end of the day' the histograms will give a correct record of variability, or otherwise.
6. Remember that logged RQD of $<10$, including 0, are set to a nominal 10 when calculating Q, to avoid calculating $Q = 0$. In view of the log scale of Q, the histograms of RQD in the logging sheet will be sufficiently accurate if given mean values, from left to right, of 10, 15, 25, 35...85, 95, 100. The log scale of Q also suggests that decimal places should be used sparingly. The following is considered realistic 0.004, 0.07, 0.3, 6.7, 27, 240. Never report that $Q = 6.73$ or similar, since a false sense of accuracy will be given.
7. Footnotes below the tables that follow, also give advice for site *characterization* ratings for the case of $J_w$ and SRF, which must not be set to 1.0 and 1.0, as some authors have suggested. This destroys the intended multi-purposes of the Q-system, which has an entirely different structure compared to RMR (Leave blank if permeability and stress data is awaited, otherwise *estimate* Jw and SRF.)

| **1. Rock Quality Designation** | | **RQD (%)** |
|---|---|---|
| A | Very poor | 0-25 |
| B | Poor | 25-50 |
| C | Fair | 50-75 |
| D | Good | 75-90 |
| E | Excellent | 90-100 |

*Notes: i) Where RQD is reported or measured as $\leq 10$ (including 0), a nominal value of 10 is used to evaluate Q.*

*ii) RQD intervals of 5, i.e., 100, 95, 90, etc., are sufficiently accurate.*

| **2. Joint set number** | | **$J_n$** |
|---|---|---|
| A | Massive, no or few joints | 0.5-1 |
| B | One joint set | 2 |
| C | One joint set plus random joints | 3 |
| D | Two joint sets | 4 |
| E | Two joint sets plus random joints | 6 |
| F | Three joint sets | 9 |
| G | Three joint sets plus random joints | 12 |
| H | Four or more joint sets, random, heavily jointed, 'sugar-cube', etc. | 15 |
| J | Crushed rock, earthlike | 20 |

*Notes: i) For tunnel intersections, use $(3.0 \times J_n)$.*

*ii) For portals use $(2.0 \times J_n)$.*

| **3. Joint roughness number** | | **$J_r$** |
|---|---|---|
| ***a) Rock-wall contact, and b) Rock-wall contact before 10 cm shear*** | | |
| A | Discontinuous joints | 4 |
| B | Rough or irregular, undulating | 3 |
| C | Smooth, undulating | 2 |
| D | Slickensided, undulating | 1.5 |
| E | Rough or irregular, planar | 1.5 |
| F | Smooth, planar | 1.0 |
| G | Slickensided, planar | 0.5 |

*Notes: i) Descriptions refer to small-scale features and intermediate scale features, in that order.*

| ***b) No rock-wall contact when sheared*** | | |
|---|---|---|
| H | Zone containing clay minerals thick enough to prevent rock-wall contact. | 1.0 |
| J | Sandy, gravely or crushed zone thick enough to prevent rock-wall contact | 1.0 |

*Notes: ii) Add 1.0 if the mean spacing of the relevant joint set is greater than 3 m.*

*iii) $J_r = 0.5$ can be used for planar, slickensided joints having lineations, provided the lineations are oriented for minimum strength.*

*iv) $J_r$ **and** $J_a$ classification is applied to the joint set or discontinuity that is least favourable for stability both from the point of view of orientation and shear resistance, $\tau$ (where $\tau \approx \sigma_n \tan^{-1} (J_r/J_a)$.*

*Figure A1* Characterization/classification ratings for RQD, Jn and Jr.

| | 4. Joint alteration number | $\phi_r$ approx. | $J_a$ |
|---|---|---|---|
| ***a) Rock-wall contact (no mineral fillings, only coatings)*** | | | |
| A | Tightly healed, hard, non-softening, impermeable filling, i.e., quartz or epidote. | -- | 0.75 |
| B | Unaltered joint walls, surface staining only. | 25-35° | 1.0 |
| C | Slightly altered joint walls. Non-softening mineral coatings, sandy particles, clay-free disintegrated rock, etc. | 25-30° | 2.0 |
| D | Silty- or sandy-clay coatings, small clay fraction (non-softening). | 20-25° | 3.0 |
| E | Softening or low friction clay mineral coatings, i.e., kaolinite or mica. Also chlorite, talc, gypsum, graphite, etc., and small quantities of swelling clays. | 8-16° | 4.0 |
| ***b) Rock-wall contact before 10 cm shear (thin mineral fillings).*** | | | |
| F | Sandy particles, clay-free disintegrated rock, etc. | 25-30° | 4.0 |
| G | Strongly over-consolidated non-softening clay mineral fillings (continuous, but < 5 mm thickness). | 16-24° | 6.0 |
| H | Medium or low over-consolidation, softening, clay mineral fillings (continuous, but < 5 mm thickness). | 12-16° | 8.0 |
| J | Swelling-clay fillings, i.e., montmorillonite (continuous, but < 5 mm thickness). Value of $J_a$ depends on per cent of swelling clay-size particles, and access to water, etc. | 6-12° | 8-12 |
| ***c) No rock-wall contact when sheared (thick mineral fillings)*** | | | |
| KL M | Zones or bands of disintegrated or crushed rock and clay (see G, H, J for description of clay condition). | 6-24° | 6, 8, or 8-12 |
| N | Zones or bands of silty- or sandy-clay, small clay fraction (non-softening). | -- | 5.0 |
| OP R | Thick, continuous zones or bands of clay (see G, H, J for description of clay condition). | 6-24° | 10, 13, or 13-20 |

| | 5. Joint water reduction factor | Water $P_w$ ≈ (kg/cm²) | $J_w$ |
|---|---|---|---|
| A | Dry excavations or minor inflow, i.e., < 5 l/min locally. | < 1 | 1.0 |
| B | Medium inflow or pressure, occasional outwash of joint fillings. | 1-2.5 | 0.66 |
| C | Large inflow or high pressure in competent rock with unfilled joints. | 2.5-10 | 0.5 |
| D | Large inflow or high pressure, considerable outwash of joint fillings. | 2.5-10 | 0.33 |
| E | Exceptionally high inflow or water pressure at blasting, decaying with time. | > 10 | 0.2-0.1 |
| F | Exceptionally high inflow or water pressure continuing without noticeable decay. | > 10 | 0.1-0.05 |

Notes: *i) Factors C to F are crude estimates. Increase $J_w$ if drainage measures are installed.*

*ii) Special problems caused by ice formation are not considered.*

*iii) For general* ***characterization*** *of rock masses distant from excavation influences, the use of Jw = 1.0, 0.66, 0.5, 0.33 etc. as depth increases from say 0-5m, 5-25m, 25-250m to >250m is recommended, assuming that RQD /Jn is low enough (e.g. 0.5-25) for good hydraulic connectivity. This will help to adjust Q for some of the effective stress and water softening effects, in combination with appropriate* ***characterization*** *values of SRF. Correlations with depth- dependent static deformation modulus and seismic velocity will then follow the practice used when these were developed.*

*Figure A2* Characterization/classification ratings for Ja and $J_w$.

| 6. Stress Reduction Factor | | SRF |
|---|---|---|
| ***a) Weakness zones intersecting excavation, which may cause loosening of rock mass when tunnel is excavated*** | | |
| A | Multiple occurrences of weakness zones containing **clay** or chemically disintegrated rock, very loose surrounding rock (any depth). | 10 |
| B | Single weakness zones containing **clay** or chemically disintegrated rock (depth of excavation ≤ 50 m). | 5 |
| C | Single weakness zones containing **clay** or chemically disintegrated rock (depth of excavation > 50 m). | 2.5 |
| D | Multiple shear zones in competent rock (**clay-free**), loose surrounding rock (any depth). | 7.5 |
| E | Single shear zones in competent rock (**clay-free**), (depth of excavation ≤ 50 m). | 5.0 |
| F | Single shear zones in competent rock (**clay-free**), (depth of excavation > 50 m). | 2.5 |
| G | Loose, open joints, heavily jointed or 'sugar cube', etc. (any depth) | 5.0 |

Notes: *i) Reduce these values of SRF by 25-50% if the relevant shear zones only influence but do not intersect the excavation. This will also be relevant for* ***characterization.***

| ***b) Competent rock, rock stress problems*** | | $\sigma_c/\sigma_1$ | $\sigma_\theta/\sigma_c$ | SRF |
|---|---|---|---|---|
| H | Low stress, near surface, open joints. | > 200 | < 0.01 | 2.5 |
| J | Medium stress, favourable stress condition. | 200-10 | 0.01-0.3 | 1 |
| K | High stress, very tight structure. Usually favourable to stability, may be unfavourable for wall stability. | 10-5 | 0.3-0.4 | 0.5-2 |
| L | Moderate slabbing after > 1 hour in **massive** rock. | 5-3 | 0.5-0.65 | 5-50 |
| M | Slabbing and rock burst after a few minutes in **massive** rock. | 3-2 | 0.65-1 | 50-200 |
| N | Heavy rock burst (strain-burst) and immediate dynamic deformations in **massive** rock. | < 2 | > 1 | 200-400 |

Notes: *ii) For strongly anisotropic virgin stress field (if measured): When $5 \le \sigma_1/\sigma_3 \le 10$, reduce $\sigma_c$ to 0.75 $\sigma_c$. When $\sigma_1/\sigma_3 > 10$, reduce $\sigma_c$ to 0.5 $\sigma_c$, where $\sigma_c$ = unconfined compression strength, $\sigma_1$ and $\sigma_3$ are the major and minor principal stresses, and $\sigma_\theta$ = maximum tangential stress (estimated from elastic theory).*

*iii) Few case records available where depth of crown below surface is less than span width. Suggest an SRF increase from 2.5 to 5 for such cases (see H).*

*iv) Cases L, M, and N are usually most relevant for support design of deep tunnel excavations in hard massive rock masses, with RQD/Jn ratios from about 50 to 200.*

*v) For general* ***characterization*** *of rock masses distant from excavation influences, the use of SRF = 5, 2.5, 1.0, and 0.5 is recommended as depth increases from say 0-5m, 5-25m, 25-250m to >250m. This will help to adjust Q for some of the effective stress effects, in combination with appropriate* ***characterization*** *values of Jw. Correlations with depth - dependent static deformation modulus and seismic velocity will then follow the practice used when these were developed.*

| ***c) Squeezing rock: plastic flow of incompetent rock under the influence of high rock pressure*** | | $\sigma_\theta/\sigma_c$ | SRF |
|---|---|---|---|
| O | Mild squeezing rock pressure | 1-5 | 5-10 |
| P | Heavy squeezing rock pressure | > 5 | 10-20 |

Notes: *vi) Cases of squeezing rock may occur for depth $H > 350\ Q^{1/3}$ according to Singh 1993 [34]. Rock mass compression strength can be estimated from $SIGMA_{cm} \approx 5\ \gamma Q_c^{1/3}$ (MPa) where $\gamma$ = rock density in t/m$^3$, and $Q_c = Q x \sigma_c/100$, Barton, 2000 [29].*

| ***d) Swelling rock: chemical swelling activity depending on presence of water*** | | SRF |
|---|---|---|
| R | Mild swelling rock pressure | 5-10 |
| S | Heavy swelling rock pressure | 10-15 |

*Figure A3* Characterization/classification ratings for SRF.

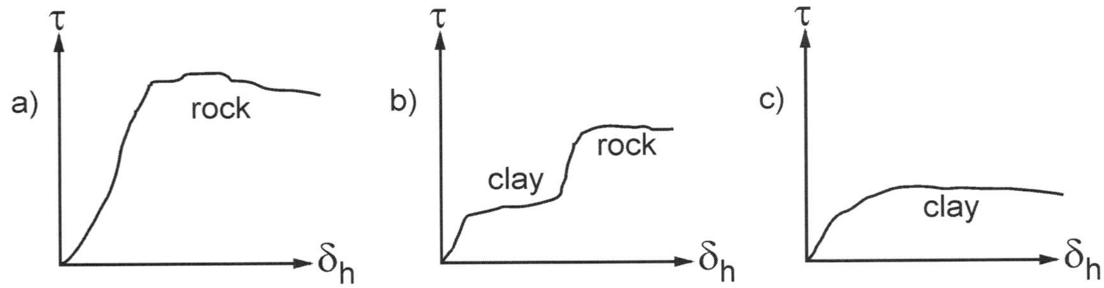


(a) Rock wall contact

| | Jr | (thin coatings) Ja = 0.75 | 1.0 | 2 | 3 | 4 |
|---|---|---|---|---|---|---|
| | | $\tan^{-1}(Jr/Ja)^o$ | | | | |
| A. Discontinuous joints | 4 | 79° | 76° | 63° | 53° | 45° |
| B. Rough, undulating | 3 | 76° | 72° | 56° | 45° | 37° |
| C. Smooth, undulating | 2 | 69° | 63° | 45° | 34° | 27° |
| D. Slickensided, undulating | 1.5 | 63° | 56° | 37° | 27° | 21° |
| E. Rough, planar | 1.5 | 63° | 56° | 37° | 27° | 21° |
| F. Smooth, planar | 1.0 | 53° | 45° | 27° | 18° | 14° |
| G. Slickensided, planar | 0.5 | 34° | 27° | 14° | 9.5° | 7.1° |

(b) Rock wall contact when sheared

| | Jr | (thin fillings) Ja = 4 | 6 | 8 | 12 |
|---|---|---|---|---|---|
| | | $\tan^{-1}(Jr/Ja)^o$ | | | |
| A. Discontinuous joints | 4 | 45° | 34° | 27° | 18° |
| B. Rough, undulating | 3 | 37° | 27° | 21° | 14° |
| C. Smooth, undulating | 2 | 27° | 18° | 14° | 9.5° |
| D. Slickensided, undulating | 1.5 | 21° | 14° | 11° | 7.1° |
| E. Rough, planar | 1.5 | 21° | 14° | 11° | 7.1° |
| F. Smooth, planar | 1.0 | 14° | 9.5° | 7.1° | 4.7° |
| G. Slickensided, planar | 0.5 | 7° | 4.7° | 3.6° | 2.4° |

(c) No rock wall contact when sheared

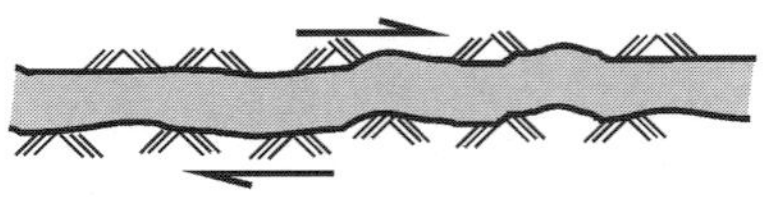

| | Jr | (thick fillings) Ja = 5 | 6 | 8 | 12 | 13 | 16 | 20 |
|---|---|---|---|---|---|---|---|---|
| | | $\tan^{-1}(Jr/Ja)^o$ | | | | | | |
| Nominal roughness of discontinuity rock walls | 1.0 | 11.3° | 9.5° | 7.1° | 4.8° | 4.4° | 3.6° | 2.9° |

*Figure A4* Graphic presentation of the meaning of Jr/Ja, representing the frictional strength of joints and clay-filled discontinuities. Note tendency for friction angle development like $\varphi + i$, $\varphi$, and $\varphi - i$, according to whether dilatant or 'normal', or contractant joint or discontinuity resistance to shearing.

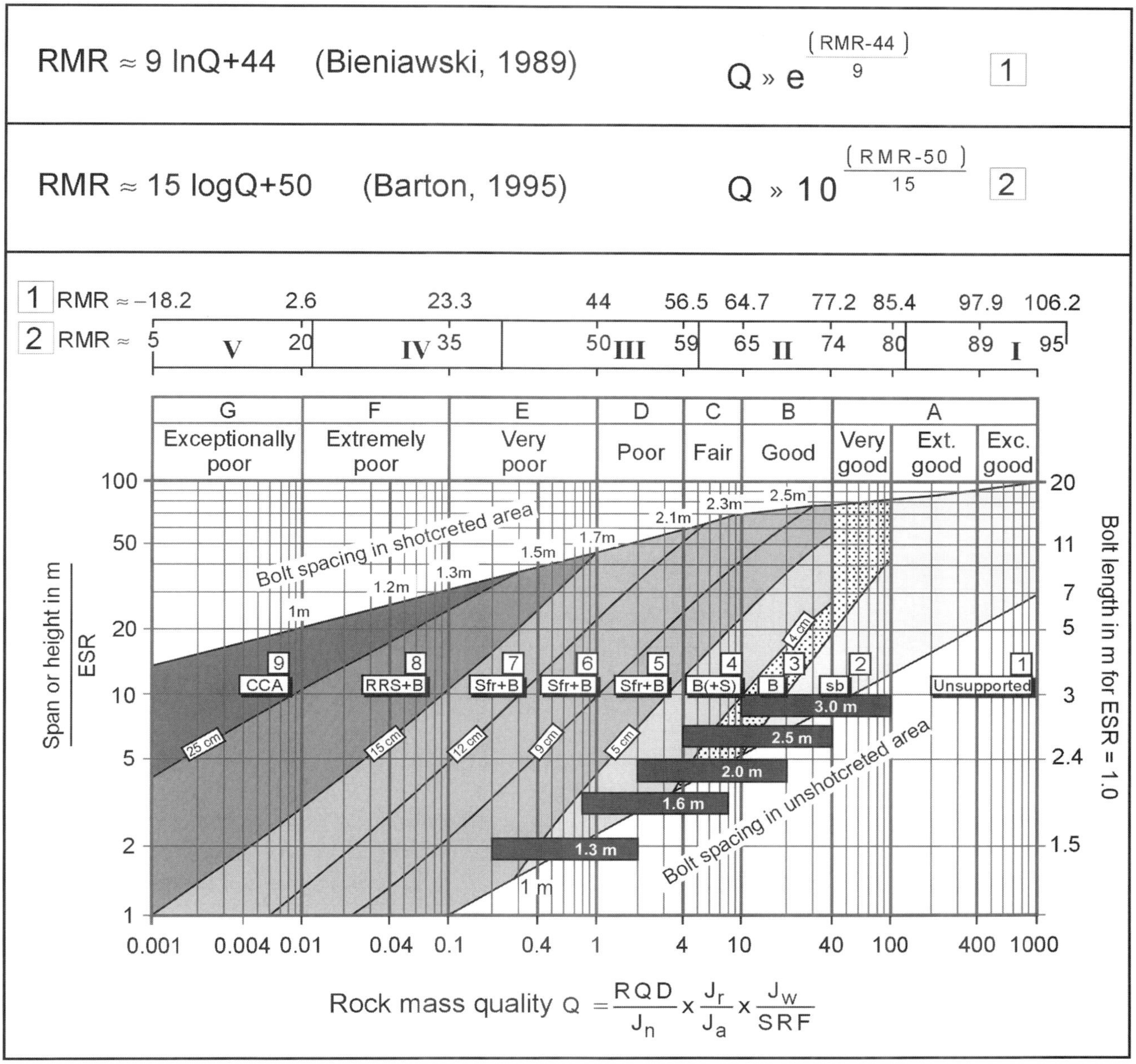


*Figure A5* The Q-system of tunnel (and rock cavern) permanent support estimation, based on Grimstad and Barton, 1993, and Barton, 2000. The other widely used 'rock mass rating' (RMR) used in engineering geology has approximate correlation to $Q_{rock}$ as shown in the equations. The version 'RMR = 15 log Q + 50' is preferred. Barton, 1995. (Note: B = systematic bolting i.e. grouted steel rebar. Sfr = fibre reinforced sprayed concrete, often called shotcrete).

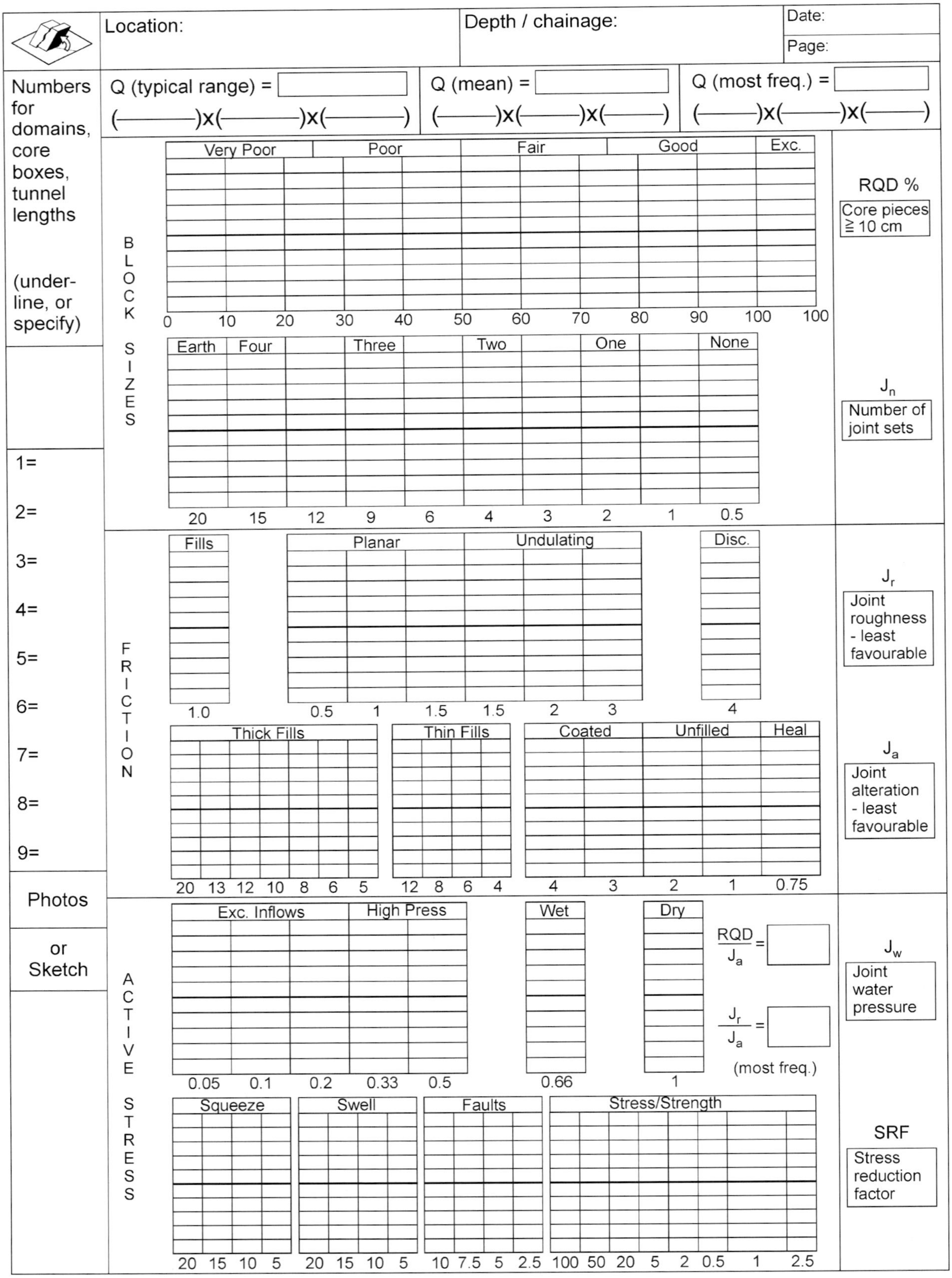


*Figure A6* The Q-parameter histogram logging sheet, for recording the number of observations of each parameter.

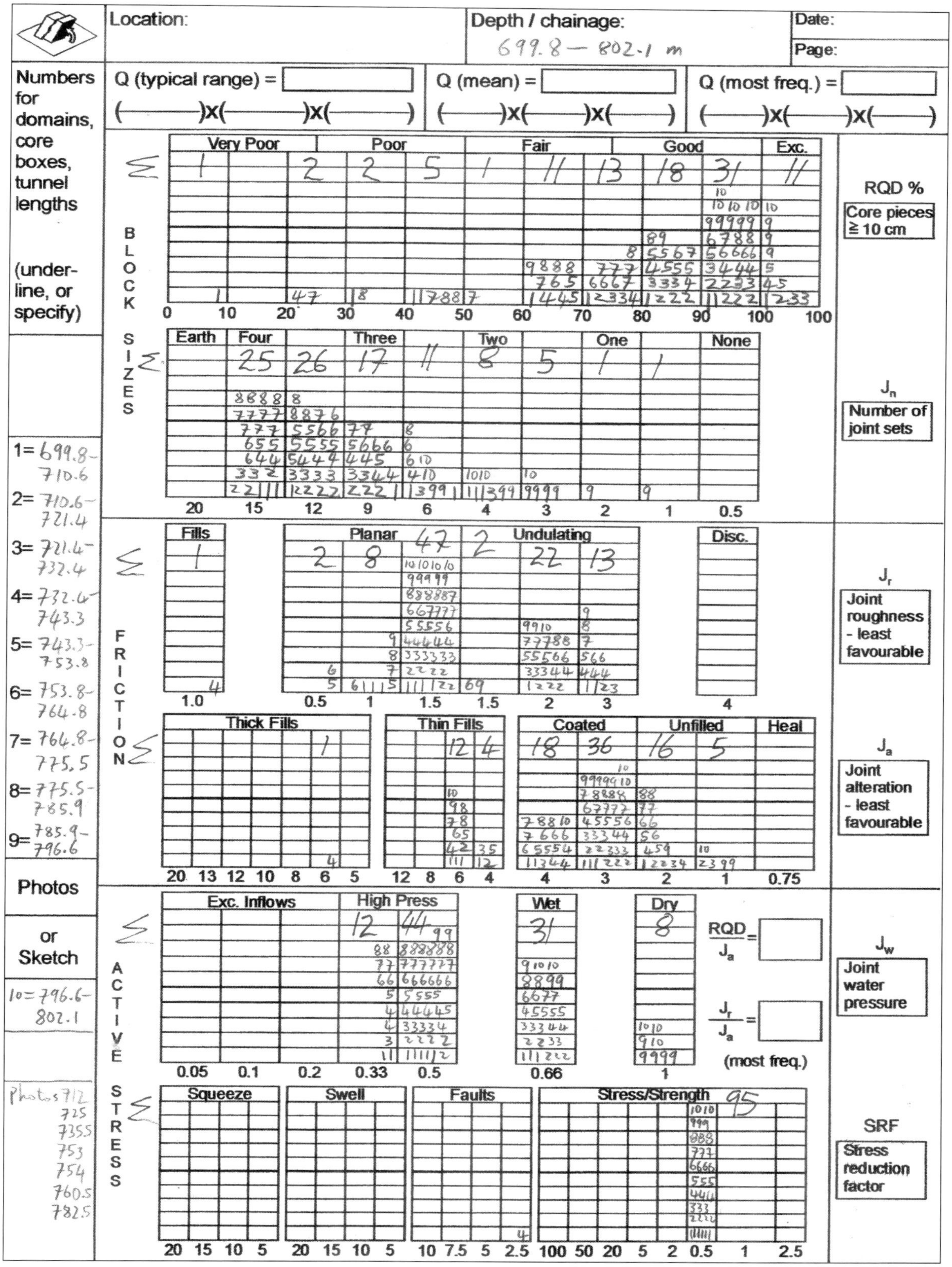


*Figure A7* Example of hand-filled Q-parameter ratings from core-logging of part of a deep borehole.

| Q - VALUES: | (RQD | / | Jn) | * | (Jr | / | Ja) | * | (Jw | / | SRF) | = | Q |
|---|---|---|---|---|---|---|---|---|---|---|---|---|---|
| Q (typical min)= | 15 | / | 15,0 | * | 1,0 | / | 6,0 | * | 0,33 | / | 1,0 | = | **0,055** |
| Q (typical max)= | 100 | / | 2,0 | * | 3,0 | / | 1,0 | * | 1,00 | / | 0,5 | = | **300,0** |
| Q (mean value)= | 83 | / | 8,6 | * | 1,8 | / | 2,9 | * | 0,63 | / | 0,6 | = | **5,91** |
| Q (most frequent)= | 95 | / | 9,0 | * | 1,5 | / | 3,0 | * | 0,66 | / | 0,5 | = | **6,97** |

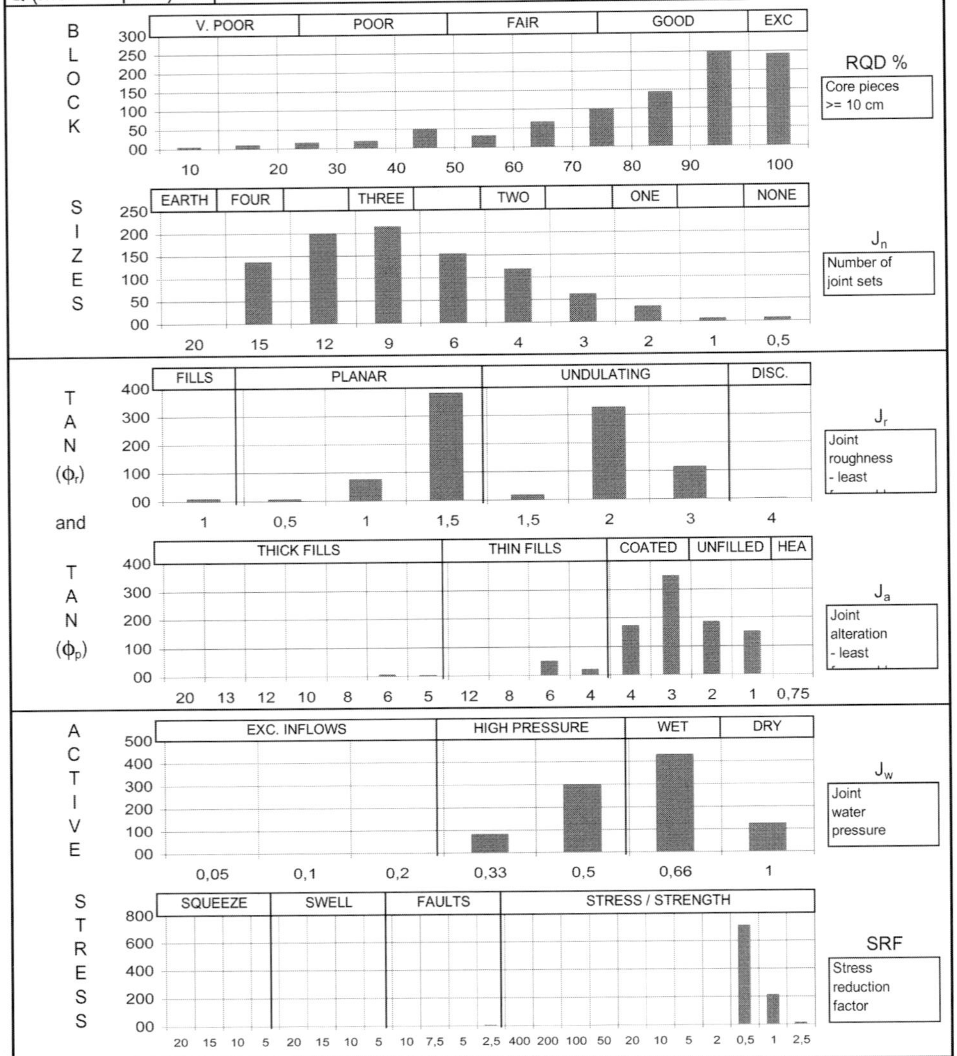


| AN EXAMPLE OF Q-PARAMETER STATISTICS | Rev. | Report No. | Figure No. A8 |
|---|---|---|---|
| Histogram-logging using the six Q-parameters, for rapid assembly of Q-statistics, following core-logging of a deep borehole. | Borehole No. : | Drawn by | Date |
| | Depth zone (m) | Checked | |
| | | Approved | |

*Figure A8* Example of EXCEL calculation of rock mass quality statistics, for the deep borehole, part of which was logged in Figure A7.

# Appendix B – A worked example

## Q-value and tunnel or cavern support

Note the overbreak caused by three joint sets in Figure B1. The joint planes (beneath the sprayed concrete) are planar and weathered or clay-coated, making for poor stability until also reinforced with rock bolts. $Q \approx 90/9 \times 1/4 \times 0.66/2.5 \approx 0.7$ ('very poor'). Estimate of permanent support requirements from Figure A5: 20 m span requires B (bolting) 1.6 m c/c (spacing) + 13 cm S(fr) (steel-fibre reinforced 'shotcrete' or sprayed concrete).

## Q-value used for geophysical estimates

1. With a simple $\sigma_c$ estimate for the gneiss of 150 MPa, $Q_{rock} = 0.66 \times 150/100 = 1.0$. Therefore at this shallow ($\leq$25 m deep) cavern site the following 'geophysical' estimates can be made, from Figure 5.36, 5.37, 13.60 and 15.33.
2. $E_{mass}$ (or M) $\approx$ 10 GPa, $V_p$ (seismic) $\approx$ 3.5 km/s, $Q_{seis} \approx 12$. (A rock mass porosity equal to a nominal 1% has been assumed).
3. From Figures 5.36 and 5.37, the empirically-derived effect of increased depth can be traced; i.e. Vp $\approx$ 5 km/s, and $E_{mass} \approx$ 32 GPa at 500 m depth.
4. By implication, with a (continued, pessimistic) assumption of unchanged rock mass quality with increased depth, the magnitude of $Q_{seis}$ (specifically $Q_p$ at seismic frequency) would be expected to be about 30.
5. The reality of improved Q-value at depth (e.g. Jr/Ja $\approx$ 2/1, and SRF = 0.5 (high stress) would mean $Q \approx 40$. With less well-developed joint sets, the rock mass quality Q-value could easily be 100, at 500 m depth.
6. With the 40–100 estimate of $Q_{rock}$, a more likely range of $V_p$ is about 5.6–6.0 km/s, with $Q_{seis}$ increased to about 60–68 at 500 m depth.

*Figure B1* The reality of near-surface construction of tunnels and caverns in rock. Note the three joint sets causing deep over-break. See Plate 17.

7. The approximate $Q_{seis}$ estimates of about *10* at 25 m depth, and about *60–70* at 500 m depth are mirrored numerous times in this book.

$Q_{rock}$ and $Q_{seis}$ appear to be approximately linked via the pseudo-static deformation modulus, which can be estimated from $Q_{rock}$, as we have seen. The possible 'explanation' for this simple (probably too simple) link is that all three parameters ($Q_{rock}$ and $Q_{seis}$ and $E_{mass}$) are reflecting the 'soft-porosity' effects of both structure and joint stiffness, with potentially several joint sets involved, not purely normal loading across one set. For some reason, rock mass dynamic stiffness as reflected in $E_{dyn}$, is too high for a good correlation.

# References

## A


Abercrombie, R.E. 1997. Near-surface attenuation and site effects from comparison of surface and deep borehole recordings. *Bull. Seism. Soc. Am.*, *87*: 731–744.

Abercrombie, R.E. 1998. A summary of attenuation measurements from borehole recordings of earthquakes: the 10 Hz transition problem. *Pure Appl. Geophys.*, *153*: 475–487.

Abercrombie, R.E. 2000. Crustal attenuation and site effects at Parkfield, California. *J. Geophys. Res.*, *105*: 6277–6286.

Adams, D.A. & Abercrombie, R.E. 1998. Seismic attenuation at high frequencies in Southern California from coda waves recorded at a range of depths. *J. Geophys. Res.*, *103*: 24,257–24,270.

Adams, L.H. & Williamson, E.D. 1923. The compressibility of minerals and rocks at high pressures. *J. Franklin Inst.*, *195*: 475–529.

Addis, M.A., Barton, N., Bandis, S.C. & Henry, J.P. 1990. Laboratory studies on the stability of vertical and deviated boreholes. *65th Annual Technical Conference and Exhibition of the Society of Petroleum Engineers*, New Orleans.

Agarwal, Y.P., Sykes, L.R., Armbruster, J. & Sbar, M.L. 1973. Premonitory changes in seismic velocities and prediction of earthquakes. *Nature*, *241*: 101–104.

Akbar, N., Dvorkin, J. & Nur, A. 1993. Relating P-wave attenuation to permeability. *Geophysics*, *58*: 20–29.

Akbar, N., Mavko, G., Nur, A. & Dvorkin, J. 1994. Seismic signatures reservoir transport properties and pore fluid distribution. *Geophysics*, *59*: 1222–1236.

Aki, K. 1969. Analysis of the seismic coda of local earthquakes as scattered waves. *J. Geophys. Res.*, *74*: 615–631.

Aki, K. & Chouet, B. 1975. Origin of coda waves: source, attenuation, and scattering effects. *J. Geophys. Res.*, *80*(*23*): 3322–3342.

Aki, K. & Richards, P.G. 1980. *Quantitative Seismology: Theory and Methods*. 932 pp. San Francisco: W.H. Freeman and Co.

Akinci, A., Mejia, J. & Jemberie, A.L. 2004. Attenuative dispersion of P-waves and crustal Q in Turkey and Germany. *Pure Appl. Geophys.*, *161*: 73–91.

Albert, W. 2000. Petrophysical parameters – an indication of rock permeability? *Nagra Bulletin*, *26*: 26–37.

Albright, J., Dangerfield, J., Johnstad, S., Cassell, B., Deflandre, J.-P. & Withers, R. 1994. Seismic surveillance for monitoring reservoir changes. *Oilfield Rev.*, January 1994, Schlumberger.

Al-Chalabi, M. 1997. Parameter nonuniqueness in velocity versus depth functions. *Geophysics*, *62*(*3*): 970–979.

Allan, J.R. 2002. Controls on reservoir performance during primary and enhanced recovery: lessons learned from 100 fractured giant fields. *American Association of Petroleum Geologists (AAPG Reservoir Geology Committee), Annual Meeting*, Houston: TX.

Alt, J.C., Honnerez, J., Laverne, C. & Emmermann, R. 1986. Hydrothermal aleration of a 1-km section through the upper oceanic crust. Deep sea drilling project hole 504B. *J. Geophys. Res.*, *91*: 10,309–10,335.

Amadei, B. & Stephansson, O. 1997. *Rock Stress and its Measurement*. 490 pp. London, New York: Chapman & Hall.

Amaefule, J.O., Altunbay, M., Tiab, D., Kersey, D.G. & Keelan, D.K. 1993. Enhanced reservoir description: using core and log data to identify hydraulic (flow) units and predict permeability in uncored intervals/wells. SPE Paper 26436, 1–16. Tulsa, OK: Society of Petroleum Engineers.

Anderson, R.N., Zoback, M.D., Hickmann, S.H. & Newmark, R.L. 1985. Permeability versus depth in the upper oceanic crust: in situ measurements in DSDP hole 504B, Eastern Equatorial Pacific. *J. Geophys. Res.*, *90*: 3659–3669.

Angerer, E., Crampin, S., Li, X.-Y. & Davies, T.L. 2002. Processing, modelling and predicting time-lapse effects of overpressured fluid-injection in a fractured reservoir. *Geophys. J. Int.*, *149*: 267–280.

Aoki, K., Shiogama, Y., Tezuka, Y., Kobuchi, T. & Masumoto, K. 1991. Evaluation of hydraulic conductivities of jointed rock mass by cross-hole permeability test. *Int. Congress on Rock Mechanics ISRM*, *7*. Aachen 1991. *1*: 423–426. Rotterdam: Balkema.

Archambeau, C.B., Flinn, E.A. & Lambert, D.G. 1969. Fine structure of the upper mantle. *J. Geophys. Res.*, *74*: 5825–5865.

Armstrong, T., McAteer, J. & Connolly, P. 2001. Removal of overburden velocity anomaly effects for depth conversion. *Geophys. Prospect.*, *49*: 79–99.

Arts, R.J., Rasolofosaon, P.N.J. & Zinszner, B.E. 1996. Experimental and theoretical tools for characterizing anisotropy due to mechanical defects in rocks under varying pore and confining pressures. Fjær, Holt and Rathore (Eds.). *Seismic Anisotropy*: 384–432. *Society of Exploration Geophysicists*.

Assefa, S., McCann, C. & Sothcott, J. 1999. Attenuation of P- and S-waves in limestones. *Geophys. Prospect.*, *47*: 359–392.

Aydan, Ö., Akagi, T., Ito, T. & Kawamoto, T. 1992. Prediction of behaviour of tunnels in squeezing ground. *J. Geotech. Eng.*, *448*(*III-19*): 73–82. JSCE.

## B


Bachrack, R. & Nur, A. 1998. High-resolution shallow-seismic experiments in sand, Part I: Water table, fluid flow, and saturation. *Geophysics, 63(4)*: 1225–1233.

Bachrach, R., Dvorkin, J. & Nur, A. 2000. Seismic velocities and Poisson's ratio of shallow unconsolidated sands. *Geophysics, 65(2)*: 559–564.

Badri, M. & Mooney, H.M. 1987. Q measurements from compressional seismic waves in unconsolidated sediments. *Geophysics, 52(6)*: 772–784.

Bakhtar, K. & Barton, N. 1984. Large scale static and dynamic friction experiments. Rock Mechanics in Productivity and Protection, *Proc. 25th Symp. Rock Mech., Evanston, IL,* Dowding, C.H. (Ed.). The American Institute of Mining, Metallurgical and Petroleum Engineers (AIMM).

Bakulin, A., Grechka, V. & Tsvankin, I. 2001. Estimation of fracture parameters of monoclinic media from reflection seismic data. *9th IWSA (Int. Workshop on Seismic Anisotropy), Abstracts, Geophysics, 66(4)*: 1294–1312.

Bakulin, A., Grechka, V. & Tsvankin, I. 2002. Seismic inversion for the parameters of two orthogonal fracture sets in a VTI background medium. *Geophysics, 67*: 292–299.

Bamford, D. & Nunn, K.R. 1979. In-situ seismic measurements of crack anisotropy in the carboniferous limestone of Northwest England. *Geophys. Prospect., 27*: 322–338.

Bandis, S. 1980. Experimental studies of scale effects on shear strength, and deformation of rock joints. *PhD Thesis,* University of Leeds, England. 385 pp., and appendices.

Bandis, S., Lumsden, A. & Barton, N. 1981. Experimental studies of scale effects on the shear behaviour of rock joints. *Int. J. Rock Mech. Min. Sci. & Geomech. Abstr., 18*: 1–21.

Bandis, S., Lumsden, A.C. & Barton, N. 1983. Fundamentals of rock joint deformation. *Int. J. Rock Mech. Min. Sci. & Geomech. Abstr., 20(6)*: 249–268.

Bandis, S.C., Makurat, A. & Vik, G. 1985. Predicted and measured hydraulic conductivity of rock joints. *Proc. of Int. Symp. on Fundamentals of Rock Joints,* Björkliden, Sweden, 269–280. Luleå: CENTEK publishers.

Baquer, S. & Mitchell, B.J. 1998. Regional variation of *Lg* coda *Q* in the Continental United States and its relation to crustal structure and evolution. *Pure Appl. Geophys., 153*: 613–638.

Barazangi, M. & Isacks, B. 1971. Lateral variations of seismic wave attenuation in the upper mantle above the inclined earthquake zone of the Tonga Island arc: deep anomaly in the upper mantle. *J. Geophys. Res., 76*: 8493–8516.

Barclay, A.H. & Toomey, D.R. 2003. Shear wave splitting and crustal anisotropy at the Mid-Atlantic ridge, 35° N. *J. Geophys. Res., 108(B8)*: 2378, EM2-1–EM2-10.

Barker, J.S., Hartzell, S.H., Burdick, L.J. & Helmburger, D.V. 1991. *J. Geophys. Res., 96(B6)*: 10,129–10,143.

Barkved, O., Bartman, B., Gaiser, J., Van Dok, R., Johns, T., Kristiansen, P., Probert, T. & Thompson, M. 2004. The many facets of multicomponent seismic data. *Oilfield Rev. Summer, 2004*, 42–56, Schlumberger.

Barkved, O.I. & Kristiansen, T. 2005. Seismic time lapse effects and stress changes: examples from a compacting reservoir, *The Leading Edge*, December 2005, 1244–1248.

Barla, G. 1993. Case study of rock mechanics in the Masua Mine, Italy. *Comprehensive Rock Engineering,* Vol. 5. Surface and Underground Case Histories. 291–334. Oxford: Pergamon.

Barton, C.A., Zoback, M.D. & Moos, D. 1995. In situ stress and permeability in fractured and faulted crystalline rock. *Mechanics of Jointed and Faulted Rock*, Rossmanith (Ed.). 381–396. Rotterdam: Balkema.

Barton, N. 1971. A model study of the behaviour of steep excavated rock slopes. *PhD Thesis*, University of London.

Barton, N. 1972a. A model study of rock-joint deformation. *Int. J. Rock Mech. Min. Sci. & Geomech. Abstr., 9*(5): 579–602.

Barton, N. 1972b. A model study of air transport from underground openings situated below groundwater level. *Proc. of ISRM Symp. on Percolation through Fissured Rock*, Stuttgart, Theme 3A, 20 pp.

Barton, N. 1973a. Review of a new shear strength criterion for rock joints, *Eng. Geol., 7*: 287–332.

Barton, N. 1973b. A review of the shear strength of filled discontinuities in rock. *Fjellspregningsteknikk, Bergmekanikk, Oslo.* 19.1–19.38 Trondheim: Tapir Press. Also NGI Publ. 105, 1974.

Barton, N., Lien, R. & Lunde, J. 1974. Engineering classification of rock masses for the design of tunnel support. *Rock Mechanics, 6(4)*: 189–236, Springer Verlag, Vienna.

Barton, N. 1976. The shear strength of rock and rock joints. *Int. J. Rock Mech. Min. Sci. & Geomech. Abstr., 13(9)*: 255–279. Also NGI Publ. 119, 1978.

Barton, N. & Choubey, V. 1977. The shear strength of rock joints in theory and practice. *Rock Mechanics, 10(1/2)*: 1–54. Vienna: Springer Verlag, Also NGI Publ. 119, 1978.

Barton, N. & Hansteen, H. 1979. Very large span openings at shallow depth: deformation magnitudes from jointed models and F.E. analysis. *Proc. of 4th Rapid Excavation and Tunnelling Conf.*, Atlanta Georgia. Maevis & Hustrulid (Eds.). *2*: 1131–1353. New York: American Institute of Mining, Metallurgical, and Petroleum Engineers, Inc.

Barton, N. 1980. Discussion of paper by Krahn, J. & Morgenstern, N.R. The ultimate frictional resistance of rock discontinuities. *Int. J. Rock Mech. Min. Sci. & Geomech. Abstr., 17*: 65–78.

Barton, N. 1981. Hydraulic fracturing to estimate minimum stress and rockmass stability at a pumped hydro project. *Proc. of Workshop on Hydraulic Fracturing Stress Measurements,* Monterey: CA.

Barton, N. 1982. Modelling rock joint behaviour from in situ block tests: implications for nuclear waste repository design. *Office of Nuclear Waste Isolation*, Columbus: OH, 96 pp. ONWI-308, September 1982.

Barton, N. & Bandis, S. 1982. Effects of block size on the shear behaviour of jointed rock. *23rd US Symp. on Rock Mechanics*, Berkeley: CA. Goodman & Heuzé (Eds.). Keynote lecture. New York: Society of Mining Engineers.

Barton, N. & Bakhtar, K. 1983. Instrumentation and analysis of a deep shaft in quartzite. *Proc. of 24th US Symp. on Rock Mechanics*, Texas A&M University.

Barton, N., Bandis, S. & Bakhtar, K. 1985. Strength, deformation and conductivity coupling of rock joints. *Int. J. Rock Mech. Min. Sci. & Geomech. Abstr.*, *22*(*3*): 121–140.

Barton, N. 1986. Deformation phenomena in jointed rock. 8th Laurits Bjerrum Memorial Lecture, Oslo. *Geotechnique*, *36*(*2*): 147–167.

Barton, N., Hårvik, L., Christiansson, M., Bandis, S., Makurat, A., Chryssanthakis, P. & Vik, G. 1986. Rock mechanics modelling of the Ekofisk reservoir subsidence. Rock Mechanics: Key to Energy Production, *Proc. of 27th U.S. Symp. Rock Mech., Tuscaloosa, AL*, Hartman, H.L. (Ed.). Society of Mining Engineers.

Barton, N. 1987. Predicting the behaviour of underground openings in rock. Manuel Rocha Memorial Lecture, Lisbon, NGI Publication 172, 1988. Also *Geotecnia 53*, July 1988 (in Portugese).

Barton, N., Makurat, A., Christianson, M. & Bandis, S. 1987. Modelling rock mass conductivity changes in disturbed zones. Rock Mechanics. *Proc. of 28th US Symp. Rock Mech., Tucson, AZ*, Farmer, I.W., Daemen, J.J.K., Desai, C.S., Glass, C.E. & Neuman, S.P. (Eds.): 563–574. Rotterdam: Balkema.

Barton, N., Makurat, A., Hårvik, L., Vik, G., Bandis, S., Christianson, M. & Addis, A. 1988. The discontinuum approach to compaction and subsidence modelling as applied to Ekofisk. *BOSS '88. Proc. of Int. Conf. on Behaviour of Offshore Structures*, Trondheim. *1*: 129–141.

Barton, N. & Bandis, S.C. 1990. Review of predictive capabilities of JRC-JCS model in engineering practice. *Rock Joints*, Barton, N. & Stephansson, O. (eds.). *Int. Conf.*, Loen, Norway. 603–610. Rotterdam: Balkema.

Barton, N. 1990a. Chapter 6, panel of experts for Xiaolangdi Project. Report No. 1 submitted to Yellow River Water & Hydroelectric Power Development Corporation, YRCC, MWR. 26–42.

Barton, N. 1990b. Scale effects or sampling bias? Closing lecture. *Proc. of First Int. Workshop on Scale Effects in Rock Masses*, Loen, Norway. Da Cunah (Ed.). LNEC.

Barton, N. 1991. Geotechnical Design. *World Tunnelling*, November 1991. 410–416.

Barton, N., Løset, F., Smallwood, A., Vik, G., Rawlings, C., Chryssanthakis, P., Hansteen, H. & Ireland, T. 1992a. Geotechnical core characterization for the UK radioactive waste repository design. *1992 Proc. of ISRM Symp. on EUROCK*, Chester, UK.

Barton, N., Makurat, A., Monsen, K., Vik, G. & Tunbridge, L. 1992b. Geotechnical predictions of the excavation disturbed zone at Stripa. *Proc. of Fourth Int. Symp. on OECD/NEA Stripa Project*, 14–16 October 1992, Stockholm, Sweden. 77–96.

Barton, N. 1993. Physical and discrete element models of excavation and failure in jointed rock. Keynote lecture presented at *ISRM Int. Symp. on Assessment and Prevention of Failure Phenomena in Rock Engineering*, 5–7 April, Istanbul, Turkey.

Barton, N. & Grimstad, E. 1994. The Q-system following twenty years of application in NMT support selection. *43rd Geomechanic Colloquy*, Salzburg. *Felsbau*, *6/94*: 428–436.

Barton, N., By, T.L., Chryssanthakis, P., Tunbridge, L., Kristiansen, J., Løset, F., Bhasin, R.K., Westerdahl, H. & Vik, G. 1994. Predicted and measured performance of the 62 m span Norwegian Olympic ice hockey cavern at Gjøvik. *Int. J. Rock Mech. Min. Sci. & Geomech. Abstr.*, *31*(*6*): 617–641.

Barton, N. 1995. The influence of joint properties in modelling jointed rock masses. Keynote lecture. *Proc. of 8th ISRM Congress*, Tokyo. T. Fuji (Ed.). *3*: 1023–1032. Rotterdam: Balkema.

Barton, N. & Itoh, J. 1995. The Q-system and NMT support techniques. (In Japanese) *Tunnels and Underground*, Japan Tunnelling Association Publication, November 95.

Barton, N. 1996a. Estimating rock mass deformation modulus for Excavation Disturbed Zone studies. *Int. Conf. on Deep Geological Disposal of Radioactive Waste*, Winnepeg, 1996, *EDZ Workshop*, 133–144. Canadian Nuclear Society.

Barton, N. 1996b. Rock mass characterization and seismic measurements to assist in the design and execution of TBM projects. *Proc. of 1996 Taiwan Rock Engineering Symp.*, Keynote Lecture, 1–16.

Barton, N. & Warren, C. 1996. Rock mass classification of chalk marl in the UK Channel Tunnels. *Channel Tunnel Engineering Geology Symp.*, Brighton, September 1995.

Barton, N. & de Quados, E.F. 1997. Joint aperture and roughness in the prediction of flow and groutability of rock masses. *Proc. of NY Rocks '97.* Linking Science to Rock Engineering. Kim, K. (Ed.). 907–916. *Int. J. Rock Mech. Min. Sci. & Geomech. Abstr.*, *34*: 3–4.

Barton, N. 1998. Quantitative description of rock masses for the design of NMT reinforcement. Keynote Lecture, *Int. Conf. on Hydro Power Development in Himalayas*, Shimla, India. Rotterdam: Balkema.

Barton, N. 1999a. General report concerning some 20th Century lessons and 21st Century challenges in applied rock mechanics, safety and control of the environment. *Proc. of 9th ISRM Congress*, Paris, *3*: 1659–1679. Rotterdam: Balkema.

Barton, N. 1999b. Rock mass characterization from seismic measurements. *SAROCKS '98, 2nd Brazilian Symp. on 5th South American Rock Mechanics Congress*, Santos, Brazil. Keynote Lecture.

Barton, N. 2000. *TBM Tunnelling in Jointed and Faulted Rock*, 173 pp. Rotterdam: Balkema.

Barton, N. 2002a. Some new Q-value correlations to assist in site characterization and tunnel design. *Int. J. Rock Mech. Min. Sci. & Geomech. Abstr.*, *39/2*: 185–216.

Barton, N. 2002b. Different scales of discontinuous behaviour in petroleum engineering. Keynote lecture, *Proc. of 3rd Brazilian Rock Mechanics Symp.*, Sao Paulo, ABMS.

Barton, N. & de Quadros, E. 2003. Improved understanding of high pressure pre-grouting effects for tunnels in jointed rock. *Proc. of 10th ISRM Congress*, South Africa, *1*: 85–91, Johannesburg: SAIMM.

Barton, N. 2004a. The theory behind high pressure grouting. *Tunnels and Tunnelling International*, 28–30 September, October. 33–35.

Barton, N. 2004b. Failure around tunnels and boreholes and other problems in rock mechanics. Letter to *ISRM News J.*, *8(2)*: May 2004, 12–18.

Barton, N. 2004c. Risk and risk reduction in TBM rock tunnelling. Keynote lecture, ARMS 2004: *3rd Asian Rock Mechanics Symp.*, Kyoto, Japan.

Barton, N. 2005. Seismic Q, rock quality, rock joint compliance and anisotropy. *Seismic Q–observations, mechanisms and interpretation. Workshop 2, 67th EAGE Conf. on European Association of Geoscientists and Engineers*, Madrid. Extended Abstract: *Q-16*, Convenors: Hargreaves. Best and Lancaster.

Barton, N. & Makurat, A. 2006. Hydro-thermo-mechanical over-closure of joints and rock masses and potential effects on the long term performance of nuclear waste repositories. *Proc. of Eurock 06*, Liege, Belgium. London: Taylor and Francis.

Barton, P.J. 1986. The relationship between seismic velocity and density in the continental crust – a useful constraint? *Geophys. J. Roy. Astr. Soc.*, *878*: 195–208.

Batchelor, A.S. & Pine, R.J. 1986. The results of in situ stress determinations by seven methods to depths of 2500 m in the Carnmenellis granite. *Proc. of Int. Symp. on Rock Stress and Rock Stress Measurement*, Stockholm, 467–478.

Batzle, M., Hoffmann, R., Kumar, G., Duranti, L., Han, D.-H. 2005. Seismic frequency loss mechanisms-direct observations. *Q–observation, mechanisms, and interpretation. Workshop at 67th EAGE Conf. on European Association of Geoscientists and Engineers*, Madrid. Extended Abstract: *Q-12*, Convenors: Hargreaves, Best and Lancaster.

Bauer, C., Homand-Etienne, F., Ben Slimane, K., Hinzen, K.G. & Reamer, S.K. 1996. Damage zone characterization in the near field in the Swedish ZEDEX tunnel using in situ and laboratory measurements. *Eurock '96*. Barla (Ed.). *2*: 1345–1352. Rotterdam: Balkema.

Baumgardt, D.R. 1985. Attenuation, blockage and scattering of teleseismic $L_g$ from underground nuclear explosions in Eurasia. Scientific Report No. 1 to AFGL, December 18, ENSCO, Inc. 90 pp.

Beaudoin, B.C., Fuis, G.S., Mooney, W.D. & Nokleberg, W.J. 1992a. Thin, low-velocity crust beneath the Southern Yukon-Tanana Terrane, East Central Alaska: results from Trans-Alaska crustal transect refraction/wide-angle reflection data. *J. Geophys. Res.*, *97(B2)*: 1921–1942.

Beaudoin, B.C., ten Brink, U.S. & Stern, T.A. 1992b. Characteristics and processing of seismic data collected on thick floating ice: results from the Ross Ice Shelf, Antarctica. *Geophysics*, *57(10)*: 1359–1372.

Berge, P.A., Fryer, G.J. & Wilkens, R.H. 1992. Velocity-porosity relationships in the upper oceanic crust: theoretical considerations. *J. Geophys. Res.*, *97(B11)*: 15,239–15,254.

Bernabini, B. & Borelli, G.B. 1974. Methods for determining the average dynamic elastic properties of a fractured rock mass and the variations of these properties near excavations. *Proc. of 3rd ISRM Congress*, Denver. *IIA*: 393–397. Washington, DC, National Academy of Sciences.

Beroza, G.C., Cole, A.T. & Ellsworth, W.L. 1995. Stability of coda wave attenuation during the Loma Prieta, California, earthquake sequence. *J. Geophys. Res.*, *100*: 3977–3987.

Bertacchi, P., Carabelli, E. & Sampaolo, A. 1966. A contribution to the use of geophysical methods for investigation of rock masses. *Proc. of 1st ISRM Congress*, Lisbon. *I*: 45–53. LNEC.

Bertacchi, P. & Sampaolo, A. 1970. Investigations on the characteristics of rock masses by geophysical methods. *Proc. of 2nd ISRM Congress*, Belgrade. *III*: Theme 8–13. Belgrade: Privredni Pregled.

Best, A., Sothcott, J. & McCann, C. 2005. Laboratory ultrasonic Q anisotropy measurements on reservoir rocks. *Seismic Q–observation, mechanisms, and interpretation. Workshop at 67th EAGE Conf.*, Madrid. Extended Abstract: *Q-18*, Convenors: Hargreaves, Best and Lancaster.

Best, A.I., McCann, C. & Sothcott, J. 1994. The relationships between the velocities, attenuations and petrophysical properties of reservoir sedimentary rocks. *Geophys. Prospect.*, *42*: 151–178.

Best, A.I. 1997. The effect of pressure on ultrasonic velocity and attenuation in near-surface sedimentary rocks. *Geophys. Prospect.*, *45*: 345–364.

Best, A.I. & Sams, M.S. 1997. Compressional wave velocity and attenuation at ultrasonic and sonic frequencies in near-surface sedimentary rocks. *Geophys. Prospect.*, 45: 327–344.

Bieniawski, Z.T. 1978. Determining rock mass deformability: experience from case histories. *Int. J. Rock Mech. Min. Sci. & Geomech. Abstr.*, *15*: 237–247.

Bieniawski, Z.T. 1989. Engineering rock mass classifications: A complete manual for engineers and geologists in mining, civil and petroleum engineering, 251pp. J. Wiley.

Biot, M.A. 1956a. Theory of propagation of elastic waves in a fluid saturated solid, 1 lower frequency range. *J. Acoust. Soc. Am.*, *28*: 168–178.

Biot, M.A. 1956b. Theory of propagation of elastic waves in a fluid saturated solid, 2 higher frequency range. *J. Acoust. Soc. Am.*, *28*: 179–191.

Biot, M.A. 1962. Generalized theory of acoustic propagation in porous dissipative media. *J. Acoust. Soc. Am.*, *34*: 1254–1264.

Birch, F. 1961. The velocity of compressional waves in rocks to 10 kilobars. Part 2. *J. Geoph. Res.*, *66(7)*: 2199– 2224.

Boitnott, G.N., Biegel, R.L., Scholtz, C.H., Yoshioka, N. & Wang, W. 1992. Micromechanics of rock friction 2: quantitative modelling of initial friction with contact theory. *J. Geophys. Res.*, *97(B6)*, 8965–8978.

Bonapace, B. 1983. Tests and measurements for the pressure tunnels and shafts of the Sellrain-Silz hydroelectric power scheme with extremely high head. *Proc. of 5th Int. Congr. Rock Mech. (ISRM), Melbourne*: *2*: D287–D292. Rotterdam: Balkema.

Bott, M.H.P. 1982. *The Interior of the Earth, Its Structure Constitution and Evolution*, 2nd Edn. London: Edward Arnold.

Bowers, G.L. 2002. Detecting high overpressure. *The Leading Edge*, February 2002, *Soc. Explor. Geol.*, 174–177.

Bradford, J.H. & Sawyer, D.S. 2002. Depth characterization of shallow aquifers with seismic reflection, Part II – Prestack depth migration and field examples. *Geophysics*, *67(1)*: 98–109.

Bradley, J.J. & Fort, Jr., A.N. 1966. Internal friction in rocks. In: Handbook of physical constants: Clark, Jr., S.P. (Ed.). GSA Pub. 175–193.

Bradley, W.B. 1978. Failure of inclined boreholes. *Trans. Am. Soc. Mech. Eng.*, *101*: 232–239.

Brandt, H. 1955. A study of the speed of sound in porous granular media. *J. Appl. Mech. ASME*, December 479–486.

Brennan, B.J. & Stacey, F.D. 1977. Frequency dependence of elasticity of rock – Test of seismic velocity dispersion. *Nature*, *268*: 220–222.

Brie, A., Endo, T., Hoyle, D., Codazzi, D., Esmersoy, C., Hsu, K., Denoo, S., Mueller, M.C., Plona, T., Shenoy, R. & Sinha, B. 1998. New Directions in Sonic Logging. *Oilfield Rev.*, 40–55. Spring 1998, Schlumberger.

Briggs, V., Rama Rao, V.N., Grandi, S., Burns, D. & Chi, S. 2004. *Research Report, MIT Earth Resources Lab.*, 12 pp.

Brocher, T.M. & ten Brink, U.S. 1987. Variations in oceanic layer 2 elastic velocities near Hawaii and their correlation to lithospheric flexure. *J. Geophys. Res.*, *92(B3)*: 2647–2661.

Brown, A. 2001. Understanding seismic attributes. *Geophysics*, *66*: 47–48.

Brown, R.L. & Siefert, D. 1997. Velocity dispersion: a tool for characterizing reservoir rocks. *Geophysics*, *62*: 447–486.

Brown, R.L., Penuela, G., Hughes, R.G., Wiggens, M.L., Civan, F. & Martinez Torres, L. 2002. Relating saturation and permeability to elastic properties of fractured rocks. *Society of Exploration Geophysicists, Expanded Abstracts*: 113–116.

Bruce, B. & Bowers, G. 2002. Pore pressure terminology. *The Leading Edge*. February 2002: 170–173.

Brueckl, E. & Parotidis, M. 2001. Estimation of large-scale mechanical properties of a large landslide on the basis of seismic results. *Int. J. Rock Mech. Min. Sci. & Geomech. Abstr.*, *38*: 877–883.

Bruno, M.S. & Winterstein, D.F. 1994. Some influences of stratigraphy and structure on reservoir stress orientation. *Geophysics*, *59(6)*: 954–962.

Brzostowski, M.A. & McMechan, G.A. 1992. 3-D Tomographic imaging of near-surface seismic velocity and attenuation. *Geophysics*, *57*: 396–403.

Budianski, B. & O'Connell, R.J. 1976. Elastic moduli of a cracked solid. *Int. J. Solids Struct.*, *12*: 81–97, Elsevier.

Bull-Gjertsen, S.Y. 1998. New knowledge about Ekofisk – four dimensional seismic gives improved reservoir control [in Norwegian]. *Teknisk Ukeblad*, *145(11)*: 34–35.

By, T.L. 1987. Geotomography for rock mass characterization and prediction of seepage problems for two main road tunnels under the city of Oslo. *Proc. of 6th Int. Congr. Rock Mech. (ISRM), Montreal*, Herget, G. & Vongpaisal (Eds.): *X(1)*: 37–40. Rotterdam: Balkema.

By, T.L. 1988. Crosshole seismics including geotomography for investigation of foundations. *Int. Symp. on Field Measurements in Geomechanics, 2.* Kobe 1987. *1*: 335–346. Rotterdam: Balkema.

By, T.L., Lund, C. & Korhonen, R. 1988. Karst cavity detection by means of seismic tomography to help assess possible dam site in limestone terrain. *Rock Mechanics and Power Plants*, Romana (Ed.). *1*: 25–31. Rotterdam: Balkema.

Byerlee, J.D. 1978. Friction of rocks. *Pure Appl. Geophy.*, *116*: 615–629.

## C

Cadoret, T. 1993. Effect de la saturation eau/gaz sur les proprieties acoustics des roches. *PhD thesis*, University of Paris.

Cai, M., Kaiser, P.K. & Martin, C.D. 1998. A tensile model for the interpretation of microseismic events near underground openings. *Pure Appl. Geophys.*, *153*: 67–92.

Capozza, F. 1979. Panel discussion: dynamic testing. *Int. Symp. on The Geotechnics of Structurally Complex Formations.* Capri 1977. *II*: 275–279. Associazione Geotecnica Italiana.

Carcione, J.M. 1992. Anisotropic Q and velocity dispersion of finely layered media. *Geophys. Prospect.*, *40*: 761–783.

Carcione, J.M. 2000. A model for seismic velocity and attenuation in petroleum source rocks. *Geophysics*, *65(4)*: 1080–1092.

Carcione, J.M. & Gangi, A.F. 2000a. Gas generation and overpressure: effects on seismic attributes. *Geophysics*, *65*: 1769–1779.

Carcione, J.M. & Gangi, A.F. 2000b. Non-equilibrium compaction and abnormal pore-fluid pressures: effects on rock properties. *Geophys. Prospect.*, *48*: 521–537.

Carcione, J.M. & Cavallini, F. 2002. Poisson's ratio at high pore pressure. *Geophys. Prospect.*, *50*: 97–106.

Carlson, R.L. 1998. Seismic velocities in the uppermost oceanic crust: age dependence and the fate of layer 2A. *J. Geophys. Res.*, *103*: 7069–7077.

Carlson, S.R. & Young, R.P. 1993. Acoustic emission and ultrasonic velocity study of excavation-induced microcrack damage at the Underground Research Laboratory. *Int. J. Rock Mech. Min. Sci. & Geomech. Abstr.*, *30(7)*: 901–907. (*34th US Symp. on Rock Mechanics*, Madison, WI.)

Carpenter, P.J. & Sanford, A.R. 1985. Apparent Q for upper crustal rocks of the Central Rio Grande Rift. *J. Geophys. Res.*, *90*: 8661–8674.

Carrère, A. 1999. Application of the SCARABEE method to re-evaluation of concrete dams safety. Geodia report. France: Coyne et Bellier.

Carter, A.J. & Kendall, J.-M. 2005. Differential attenuation of 'split' shear waves – a future tool for reservoir characterization? *Seismic Q–observation, mechanisms, and interpretation. Workshop at 67th EAGE Conf.*, Madrid. Extended Abstract: *Q-11*, Convenors: Hargreaves, Best and Lancaster.

Castagna, J.P., Batzle, M.L. & Kan, T.K. 1993. Rock physics – the link between rock properties and AVO response. In: *Offset-Dependent Reflectivity – Theory and Practice of AVO Analysis*, Castagna, J.P. & Bakus, M. (Eds.). *Investigations in Geophysics*, No. 8: 135–171. Tulsa, OK: Society of Exploration Geophysicists.

Castagna, J.P. 2001. Recent advances in seismic lithologic analysis. *Geophysics*, *66*: 42–46.

Cecil, O.S. 1971. Correlation of seismic refraction velocities and rock support requirements in Swedish tunnels. Statens Geotekniska Institut. Publ. 40. 58 pp.

Chan, T.C.F. 1993. Study of a proposed underground refuse transfer station in Hong Kong. *34th US Symp. on Rock Mechanics,* Madison: WI. Haimson (Ed.). *II*: 773–776. University of Wisconsin.

Chang, H. & Lee, T.S. 2001. Current use of geophysical site characterization techniques for civil engineering in Korea. *Proc. of 4th Int. Workshop on Application of Geophysics to Rock Engineering*, Beijing, China. *Int. Soc. Rock Mech. News J.*, *7(1)*: 45–52.

Chapman, M., Zatsepin, S.V. & Crampin, S. 2001. On frequency scaling in anisotropic poro-elasticity. *9th IWSA (Int. Workshop on Seismic Anisotropy), Abstracts, Geophysics*, *66(4)*: 1294–1312.

Chapman, M. 2002. Frequency dependent anisotropy due to meso-scale fractures with equant porosity. *EAGE 64th Conf. and Exhibition*, Florence, Italy.

Chapman, M. 2003. Frequency-dependent anisotropy due to meso-scale fractures in the presence of equant porosity. *Geophys. Prospect.*, *51*: 369–379.

Chapman, M. & Liu, E. 2004. Theories for fluid substitution in fractured carbonate reservoirs. *EAGE 66th Conf. on Exhibition*, Paris.

Chen, H., Brown, R.L. & Castagna, J.P. 2005. AVO for one- and two-fracture set models. *Geophysics*, *70(2)*: C1–C5.

Chen W.-H. 1982. Classification of soft wall rock of the Xia-Keng tunnel on the Anhui-Jiangxi railway. *Proc. of 4th Int. Congr. Int. Ass. Eng. Geol. (IAEG), New Delhi*: *V*: V.241–V.246. Rotterdam: Balkema.

Chen, W.-Y., Lovell, C.W., Haley, G.M. & Pyrak-Nolte, L.J. 1993. Variation of shear-wave amplitude during frictional sliding. *Int. J. Rock Mech. Min. Sci. & Geomech. Abstr.*, *30(7)*: 779–784. (*34th US Symp. on Rock Mechanics*, Madison: WI.)

Chen, X. & Long, L.T. 2000. Hypocenter migration as an explanation for temporal changes in coda Q. *J. Geophys. Res.*, *105*: 16,151–16,160.

Chi, S., Wu, J. & Torres-Verdin, C. 2002. Invasion correction of Acoustic Logs in a Gas Reservoir. *43rd Annual Symp. on SPWLA*, Osio, Japan. Extended Abstracts.

Chi, S., Torres-Verdin, C., Wu, J. & Alpak, O.F. 2004. Assessment of mud-filtrate invasion effects on borehole acoustic logs and radial profiling of formation elastic parameters. SPE, Paper 90159, 13 pp.

Christensen, N.I. & Salisbury, M.H. 1972. Sea floor spreading, progressive alteration of layer 2 basalts, and associated changes in seismic velocities. *Earth Planet, Sci. Lett.*, *15*: 367–375.

Christensen, N.I. 1984. Pore pressure and oceanic crustal seismic structure. *Geophys. J. Roy. Astr. Soc.*, *79*: 411–423.

Christensen, N.I. & Mooney, W.D. 1995. Seismic velocity structure and composition of the continental crust: a global view. *J. Geophys. Res.*, *100(B7)*: 9761–9788.

Christeson, G.L., Purdy, G.M. & Fryer, G.J. 1991. Structure of young upper crust at the East Pacific Rise near 9°30′ N. *Geophys. Res. Lett.*, *19*: 1045–1048.

Christeson, G.L., Wilcock, W.S.D. & Purdy, G.M. 1994. The shallow attenuation structure of the fast-spreading East Pacific Rise near 9°30′N. *Geophys Res. Lett.*, *21*: 321–324.

Christie, P., Ireson, D., Rutherford, J., Smith, N., Dodds, K., Johnston, L. & Schaffner, J. 1995. Borehole seismic data sharpen the reservoir image. *Oilfield Rev.*, Schlumberger.

Chryssanthakis, P., Monsen, K. & Barton, N. 1991. Validation of UDEC-BB against the CSM block test and large scale application to glacier loading of jointed rock masses. *Proc. of 7th ISRM Cong.*, Aachen. 693–698. Rotterdam: Balkema.

Chuhan, F.A. & Bjørlykke, K. 2002. Evidence of grain crushing during mechanical compaction of reservoir sandstones – A comparison with experimental results. *American Association of Petroleum Geologists. (SEPM: Effects of Deep Burial on Reservoir and Seal Quality), Annual Meeting*, Houston, TX.

Clouser, R.H. & Langston, C.A. 1991. $Q_p$ $Q_s$ relations in a sedimentary basin using converted phases. *Bull. Seism. Soc. Am.*, *81*: 733–750.

Cole, K.S. & Cole, R.H. 1941. Dispersion and absorption in dielectrics. *J. Chem. Phys.*, *9*: 341–351.

Collier, J.S. & Singh, S.C. 1998. Poisson's ratio structure of young oceanic crust. *J. Geophys. Res.*, *103(B9)*: 20,981–20,996.

Comité National Français 1964. La deformabilite des massifs rocheux. Analyse et comparaison des resultats. *Int. Congress on Large Dams, 8.* Edinburgh 1964. 1 (*Q28*): 287–299. Paris: ICOLD.

Cooke, N.G.W. 1992. Natural joints in rock: mechanical, hydraulic and seismic behaviour and properties under normal stress. *Int. J. Rock Mech. Min. Sci. & Geomech. Abstr.*, *29*: 198–223.

Cosma, C. 1983. Determination of rock mass quality by the crosshole seismic method. *Bull. Int. Ass. Eng. Geol.*, *26–27*: 219–225.

Cosma, C. 1995. Characterization of subsurface structures by remote sensing. Keynote lecture. *Proc. of 8th ISRM Congress,* Tokyo. Fuji (Ed.). *3*: 1013–1021. Rotterdam: Balkema.

Cosma, C. & Enescu, N. 1996a. Detailed seismic tomography in radial boreholes – D&B. Technical Note. Stockholm: SKB.

Cosma, C. & Enescu, N. 1996b. Detailed seismic tomography in radial boreholes – TBM. Technical Note. Stockholm: SKB.

Cosma, C., Olsson, O., Keskinen, J. & Heikkinen, P. 2001. Seismic characterization of fracturing at the Äspö Hard Rock Laboratory, from the kilometre scale to the meter scale. *Int. J. Rock Mech. Min. Sci. & Geomech. Abstr.*, *38*: 859–863.

Coyne et Bellier & Geodia, 1995. SCARABEE – Investigation method for the rapid assessment of rock mass quality. April 1995. France: Coyne et Bellier.

Coyne et Bellier, 1998. Application de la methode SCARABEE. Exploitation de la base èxpérimentale. France: Coyne et Bellier.

Crampin, S. 1978. Seismic waves propagating through a cracked solid: polarization as a possible dilatancy diagnostic. *Geophys. J. Roy. Astr. Soc.*, *53*: 467–496.

Crampin, S., McGonigle, R. & Bamford, D. 1980. Estimating crack parameters from observations of *P*-wave velocity anisotropy. *Geophysics*, *45(3)*: 345–360.

Crampin, S. 1981. A review of wave motion in an isotropic and cracked elastic-medium. *Wave Motion*, *3*: 343–391.

Crampin, S. 1984. Effective anisotropic elastic constants for wave propagation through cracked solids. *Geophys. J. Roy. Astr. Soc.*, *76*: 135–145.

Crampin, S., Evans, R. & Atkinson, B.K. 1984. Earthquake prediction: a new physical basis. *Geophys. J. Roy. Astr. Soc.*, *76*: 147–156.

Crampin, S. & Booth, D.C. 1985. Shear-wave polarizations near the North Anatolian fault – II interpretation in terms of crack-induced anisotropy. *Geophys. J. Roy. Astr. Soc.*, *83*: 75–92.

Crampin, S. 1989. Suggestions for a consistent terminology for seismic anisotropy. *Geophys. Prospect.*, *37*: 753–770.

Crampin, S. & Booth, D.C. 1989. Shear-wave splitting showing hydraulic dilation of pre-existing joints in granite. *Sci. Drill.*, *1*: 21–26.

Crampin, S. & Lovell, J.H. 1991. A decade of shear-wave splitting in the Earth's crust: what does it mean? What use can we make of it? And what should we do next? *Geophys. J. Int.*, *107*: 387–407.

Crampin, S. 1993a. Arguments for EDA. *Can. J. Explor. Geophys.*, *29(1)*: 18–30.

Crampin, S. 1993b. A review of the effects of crack geometry on wave propagation through aligned cracks. *Can. J. Explor. Geophys.*, *29(1)*: 3–17.

Crampin, S. 2000. Shear-wave splitting in a critical self-organised crust: the new geophysics. *70th. Annual Int. Meeting Soc. Explor. Geophys.*, Calgary.

Crampin, S., Volti, T., Chastin, S., Gudmundsson, A. & Stefánsson, R. 2002. Indication of high pore-fluid pressures in a seismically-active fault zone. *Geophys. J. Int.*, *151*: F1–F5.

Crampin, S. 2003. The new geophysics: shear-wave splitting provides a window into the crack-critical rock mass. *The Leading Edge*, June 2003.

Crampin, S. & Peacock, S. 2005. A review of shear-wave splitting in the compliant crack-critical anisotropic earth. *Wave Motion*, *41*: 59–77.

Cumerlato, C.L., Stachura, V.J. & Tweeton, D.R. 1988. Application of refraction tomography to map the extent of blast-induced fracturing. *29th US Symp. on Rock Mechanics*, Minneapolis: MN. Cundall et al. (Eds.). 691–698. Rotterdam: Balkema.

Cundall, P.A. 1971. The measurement and analysis of accelerations in rock slopes. *PhD Thesis*, University of London, Imperial College, 182 pp.


## D


Daly, R.A. 1940. *Strength and Structure of the Earth*, 434 pp., Englewood Cliffs, NJ: Prentice-Hall.

Dangerfield, J.A. & Brown, D.A. 1985. *The Ekofisk Field: North Sea Oil and Gas Reservoirs*, Kleepe, J. et al. (Eds.): 3–22, London: Graham & Trotman.

Darracott, B.W. & Orr, C.M. 1976. Geophysics and rock engineering. *Symp. on Exploration for Rock Engineering.* Johannesburg 1976. *1*: 159–164. Cape Town/Rotterdam: Balkema.

Dasgupta, R. & Clarke, R.A. 1998. Estimation of Q from surface seismic reflection data. *Geophysics*, *63(6)*: 2120–2128.

Dasios, A., McCann, C., Austin, T.R., McCann, D.M. & Fenning, P. 1999. Seismic imaging of the shallow subsurface: shear wave case histories. *Geophys. Prospect.*, *47*: 565–591.

Dasios, A., Astin, T. & McCann, C. 2001. Compressional-wave Q estimation from full waveform sonic data. *Geophys. Prospect.*, *49*: 353–373.

Davies, T.L. 2001. Multicomponent seismiology – the next wave. *Geophysics*, *66*: 49.

De Souza, J.L. & Mitchell, B.J. 1998. *Lg* Coda *Q* variations across South America and their relation to crustal evolution. *Pure Appl. Geophys.*, *153*: 587–612.

De, G.S., Winterstein, D.F. & Meadows, M.A. 1994. Comparison of P- and S-wave velocities and Q's from VSP and sonic log data. *Geophysics*, *59*: 1512–1529.

Deere, D.U., Hendron, A.J., Patton, F.D. & Cording, E.J. 1967. Design of surface and near-surface construction. In: *Rock. Failure and Breakage of Rock*, Fairhurst, C. (Ed.): 237–302. New York: Society of Mining Engineers of AIME.

Den Boer, L. & Matthews, L. 1988. Seismic characterization of thermal flood behaviour. *Unitar/UNDP Fourth Int. Conf. on Heavy Crude and Tar Sands.*

Denekamp, Saul. Personal communication.

Detrick, R.S., White, R.S. & Purdy, G.M. 1993. Crustal structure of North Atlantic fracture zones. *Rev. Geophys.*, *31(4)*: 439–458.

Dhawan, A.K., Kate, J.M. & Joshi, A.B. 1983. Geophysical exploration for a project in the Himalayas. *Proc. of 5th ISRM Congress*, Melbourne. *1*: A9–A13. Rotterdam: Balkema.

Di Salvo, C.A. 1982. Geomechanical classification of rock mass at Segunda Angostura dam, Limay river, Argentine Republic. *Int. Congress on Large Dams, 14*. Rio de Janeiro. *2(Q53)*: 519–530. Paris: ICOLD.

Domenico, S.N. 1984. Rock lithology and porosity determination from shear and compressional wave velocity. *Geophysics*, *49*: 1188–1195.

Domnesteanu, P., McCann, C. & Sothcott, J. 2002. Velocity anisotropy and attenuation of shale in under- and over-pressured conditions. *Geophys. Prospect.*, *50*: 487–503.

Dorman, L.M. & Jacobson, R.S. 1981. Linear inversion of body wave data, Part 1: Velocity structure from travel times and ranges. *Geophysics*, *46(2)*: 138–151.

Duellmann, H. & Heitfeld, K.H. 1978. Influence of grain fabric anisotropy on the elastic properties of rocks. *Proc. of 3rd IAEG Congress*, Madrid. *II(1)*: 150–162. Madrid: Imprime ADOSA.

Dürrast, H., Rasolofosaon, P.N.J. & Siegesmund, S. 2002. *P*-wave velocity and permeability distribution of sandstones from a fractured tight gas reservoir. *Geophysics*, *67(1)*: 241–253.

Dusseault, M.B., Bruno, M.S. & Barrera, J. 2001. Casing shear: causes, cases, cures. SPE Paper 72060, June 2001, *SPE Drilling and Completion*, 98–107. Tulsa, OK: Society of Petroleum Engineers.

Dvorkin, J. & Nur, A. 1993. Dynamic poroelasticity: a unified model with the squirt and the biot mechanism. *Geophysics*, *58*: 524–533.

Dvorkin, J., Nolen-Hoeksema, R. & Nur, A. 1994. The squirt flow mechanism: macroscopic description. *Geophysics*, *59*: 428–438.

Dvorkin, J., Mavko, G. & Nur, A. 1995. Squirt flow in fully saturated rocks. *Geophysics*, *60*: 97–107.

Dvorkin, J. & Nur, A. 1996. Elasticity of high porosity sandstones: theory for two North Sea data sets. *Geophysics*, *61*: 1363–1370.

Dvorkin, J. & Nur, A. 1998. Short Note: time-average equation revisited. *Geophysics*, *63*: 460–464.

Dyer, B.C., Jones, R.H., Cowles, J.F. & Barkved, O. 1999. Microseismic monitoring of a North Sea reservoir. *World Oil*, *220(3)*: 73–78.

Dypvik, H., Gudlaugsson, S.T., Tsikalas, F., Attrep, Jr., M., Ferrell, Jr., R.E., Krinsley, D.H., Mørk, A., Faleide, J.I. & Nagy, J. 1996. Mjølnir structure: an impact crater in the Barents Sea. *Geology*, *24(9)*: 779–782.


## E


Ebisu, S., Aydan, Ö., Komura, S. & Kawamoto, T. 1992. Comparative study on various rock mass characterization methods for surface structures. *ISRM Symp. on Eurock '92.* Chester, UK: 203–208. London: Thomas Telford.

Ecevitoglu, B. & Bingol, H. 1999. Q imaging from first breaks in shallow seismics. *2nd Balkan Geophysics Cong. on Exhib.*, Extended Abstract: O12-3, 110–111.

Eide, A.L., Alsos, T., Hegstad, B.K., Najjar, N.F., Astratti, D., Psaila, D. & Doyen, P. 2002. Quantitative time-lapse seismic analysis of the Gullfaks field. Geophysics in a shared earth model. *Symp.*, Norwegian Petroleum Society NPF, Kristiansand, Extended Abstract: 61–64.

Emsley, S.J., Olsson, O., Stanfors, R., Stenberg, L., Cosma, C. & Tunbridge, L. 1996. Integrated characterization of a rock volume at the Äspö HRL utilised for an EDZ experiment. *Eurock '96.* Barla (Ed.). *2*: 1329–1336. Rotterdam: Balkema.

Endo, T., Ito, H., Badri, M. & El Sheikh, M. 1997. Fracture and permeability evaluation in a fault zone from sonic waveform data. *Trans. SPWLA 38th Ann. Logging Symp.*, Houston, Paper R.

Engelder, T. & Plumb, R. 1984. Changes in *in situ* ultrasonic properties of rock on strain relaxation. *Int. J. Rock Mech. Min. Sci. & Geomech. Abstr.*, *21(2)*: 75–82.

Esaki, T., Ikusada, K., Aikawa, A. & Kimura, T. 1995. Surface roughness and hydraulic properties of sheared rock. *Int. Conf. on Fractured and Jointed Rock Masses*. Myer, Tsang, Cook & Goodman (Eds.): 93–98. Lake Tahoe, CA.

Evangelista, A. & Pellegrino, A. 1990. Caratteristiche geotechiche di alcune rocce tenere Italiane. *Conferenze di*

*Meccanica e Ingegneria delle Rocce*. Politecnico de Torino 1990. Terzo Ciclo. 2-1-2-32. Padova: SGE Editoriali.

Evans, R.L. 1994. Constraints on the large-scale porosity and permeability structure of young oceanic crust from velocity and resistivity data. *Geophys. J. Int.*, *119*: 869– 879.

Ewing, J. 1963. Elementary theory of seismic refraction and reflection measurements. In: *The Sea*, Vol. 3, Hill, M.N. (Ed.): 3–19. New York: Interscience.

Ewing, M. 1965. The sediments of the Argentine basin. *Q. J. Roy. Astr. Soc.*, *6*: 10–27.

Ewing, J.I. & Purdey, G.M. 1982. Upper crustal velocity structure in the ROSE area of the East Pacific Rise. *J. Geophys. Res.*, *87*(*B10*): 8397–8402.

## F

Faust, L.Y. 1951. Seismic velocity as a function of depth and geologic time, *Geophysics 16*(*2*): 192–206.

Fessenden, R.A. 1917. Method and apparatus for locating ore bodies. US patent 1.240,328.

Feustel, A.J., Trifu, C.-Z. & Urbancic, T.i. 1996. Rock mass characterization using intrinsic and scattering attenuation estimates at frequencies from 400 to 1600 Hz. *Pure Appl. Geophys.*, *147*(2): 289–304.

Fjær, E., Holt, R.M., Horsrad, P., Raaen, A.M. & Risnes, R. 1992. *Petroleum Related Rock Mechanics*, Elsevier.

Fjær, E. & Holt, R.M. 1994. Rock acoustics and rock mechanics: their link in petroleum engineering. *The Leading Edge*, April 1994, 255–258.

Fjær, E. 1997. Laboratory tests on artificial rocks with controlled crack parameters. *Int. J. Rock Mech. Min. Sci. & Geomech. Abstr.*, *34*: 3–4.

Fjær, E. & Suárez-Rivera, R. 1998. Fracture size determined from amplitude data. *EAGE 60th Conf. on Technical Exhibition*, Leipzig, Germany.

Fjær, E. & Holt, R.M. 1999. Stress and stress release effects on acoustic velocities from cores, logs and seismics. *SPWLA 40th Ann. Logging Symp.*, Paper WW.

Fourmaintraux, D. 1995. Quantification des discontinuités des roches et des massifs rocheux. *Rock Mech.*, *7*: 83–100.

Fratta, D. & Santamarina, J.C. 2002. Shear wave propagation in jointed rock: state of stress. *Géotechnique*, *52*(*7*): 495–505.

Frazer, L.N., Sun, X. & Wilkens, R.H. 1997. Changes in attenuation with depth in an ocean carbonate section: Ocean Drilling Program sites 806 and 807, Ontong Java Plateau. *J. Geophys. Res.*, *102*(*B2*): 2983–2997.

Friedel, M.J., Scott, D.F., Jackson, M.J., Williams, T.J. & Killen, S.M. 1995. 3D tomographic imaging of mechanical conditions in a deep US gold mine. Mechanics of jointed and faulted rock. *Int. Conf. on Mechanics of Jointed and Faulted Rock, 2.*, Vienna: 689–695. Rotterdam: Balkema.

Friedel, M.J., Jackson, M.J., Williams, E.M., Olson, M.S. & Westman, E. 1996a. Tomographic imaging of coal pillar conditions: observations and implications. *Int. J. Rock Mech. Min. Sci. & Geomech. Abstr.*, *33*(*3*): 279–290.

Friedel, M.J., Scott, D.F., Jackson, M.J., Williams, T.J. & Killen, S.M. 1996b. 3-D tomographic imaging of anomalous stress conditions in a deep US gold mine. *J. Appl. Geophys.*, *36*: 1–17.

Frisillo, A.L. & Stewart, J. 1980. Effect of partial gas/brine saturation on ultrasonic absorption in sandstone. *J. Geophys. Res.*, *85*: 5209–5211.

## G

Gaiser, J. & van Dok, R. 2003. Converted shear-wave anisotropy: new technology for fractured-reservoir management. *AAPG Int. Conf.*, Barcelona, Spain.

Gale, J.E. 1987. Comparison of coupled fracture deformation and fluid flow models with direct measurements of fracture pore structure and stress-flow properties. Rock Mechanics, *Proc. of 28th US Symp. Rock Mech., Tucson, AZ*, Farmer, I.W., Daemen, J.J.K., Desai, C.S., Glass, C.E. & Neuman, S.P. (Eds.): 1213–1222. Rotterdam: Balkema.

Gangi, A.F. 1991. The effect of pore fluids and pressures on the seismic velocities in cracked and/or porous rocks. *SEG Research Workshop on Lithology Tech, Abstracts*, 35–38.

Gardner, G.H.F., Gardner, L.W. & Gregory, A.R. 1974. Formation velocity and density – the diagnostic basics for stratigraphic traps. *Geophysics*, *39*(*6*): 770–780.

Gardener, R. 1992. Seismic refraction as a tool in the evaluation of rock quality for dredging and engineering purposes: case studies. *ISRM Symp. on Eurock '92*. Chester, UK: 153–158. London: Thomas Telford.

Geertsma, J. 1961. Velocity-log interpretation: the effect of rock bulk compressibility. *Soc. Petrol. Eng. J.*, *I*: 235–248.

Gentier, S. 1986. Morphologie et comportement hydromechanique d'une fracture naturelle dans un granite sous contrainte normale. *PhD Thesis*, Universite d'Orleans, France.

Geo, 2001. Ekofisk – old field towards new heights. (in Norwegian) *Geo- Bergverksnytt,* February 2001, 12–16. Trondheim.

Gibbs, J.F. & Roth, E.F. 1989. Seismic velocities and attenuation from borehole measurements near the Parkfield prediction zone. Central California. *Earthq. Spectr.*, *5*: 513–537.

Gibson, R.L. & Toksöz, M.N. 1990. Permeability estimation from velocity anisotropy in fractured rock. *J. Geophys. Res.*, *95*(*B10*): 15,643–15,655.

Goldberg, D. & Yin, C.-S. 1994. Attenuation of P waves in oceanic crust: multiple scattering from observed heterogeneities. *Geophys. Res. Lett.*, *21*: 2311–2314.

González, M. & Munguía, L. 2003. Seismic anisotropy observations in the Mericali Valley, Baja California, México. *Pure Appl. Geophys.*, *160*: 2257–2278.

Goodman, R.E., Taylor, R.L. & Brekke, T.L. 1968. A model for the mechanics of jointed rock. *J. Soil Mech. Found. Div., ASCE, 94*: 637–659.

Goodman, R.E. 1970. The deformability of joints. *Am. Soc. Test Materials*, Special Test Publication. *477*: 174–196.

Goodman, R.E. 1976. *Methods of Geological Engineering in Discontinuous Rocks.* 172 pp. New York: West Publishing.

Gordon, R.B. & Davis, L.A. 1968. Velocity and attenuation of seismic waves in imperfectly elastic rock. *J. Geophys. Res., 73*: 3917–3935.

Grainger, P., McCann, D.M. & Gallois, R.W. 1973. The application of the seismic refraction technique to the study of fracturing of the Middle Chalk at Mundford, Norfolk. *Geotechnique, 23*: 219–232.

Granger, P.-Y., Rollet, A. & Bonnot, J.-M. 2000. Preliminary evaluation of azimuthal anisotropy over the Valhall field using C-wave data. Anisotropy 2000: fractures, converted waves, and case studies. *Society of Exploration Geophysicists*, 772–775.

Green, A.S.P., Baria, R. & Jones, R. 1989. VSP and cross-hole seismic surveys used to determine reservoir characteristics of a hot dry rock geothermal system. *Int. J. Rock Mech. Min. Sci. & Geomech. Abstr., 26(3/4)*: 271–280.

Gregory, A.R. 1976. Fluid saturation effects on dynamic elastic properties of sedimentary rocks. *Geophysics, 41*: 895–923.

Grevemeyer, I., Kaul, N. & Villinger, H. 1999. Hydrothermal activity and the evolution of the seismic properties of upper oceanic crust. *J. Geophys Res., 104(B3)*: 5069–5079.

Griffiths, D.H. & King, R.F. 1987. Geophysical exploration. *Ground Engineers Reference Book*. Bell, F.G. (Ed.). London: Butterworths, 27/1–27/21.

Grimm, R.E., Lynn, H.B., Bates, C.R., Phillips, D.R., Simon, K.M. & Beckham, W.E. 1999. Detection and analysis of naturally fractured gas reservoirs: multiazimuth seismic surveys in the Wind River basin, Wyoming. *Geophysics, 64(4)*: 1277–1292.

Grujic, N. 1974. Ultrasonic testing of foundation rock. *Proc. of 3rd ISRM Congress*, Denver. *IIA*: 404–409. Washington, DC: National Academy of Sciences.

Guilbot, J., Smith, B. & Pirera, F. 2002. Using 4D seismic to understand the complex relationship between water injection and reservoir compaction at ekofisk field. *Geophysics in a Shared Earth Model. Symp.*, Norwegian Petroleum Society NPF, Kristiansand, Extended Abstract: 45–52.

Gurevich, B. 2002. Effect of the pore fluid on the elastic properties of a porous rock with parallel fractures. *10th Int. Workshop on Seismic Anisotropy*. Evangelische Akademie, Tutzing, Germany, Prgm. and Abstracts, 72 pp.

Gurevich, B. 2003. Elastic properties of saturated porous rocks with aligned fractures. *J. Appl. Geophys., 54*: 203–218.

Gutierrez, M. & Barton, N. 1994. Numerical Modelling of the Hydro-Mechanical Behavior of Single Fractures. Rock Mechanics Models and Measurements: Challenges from Industry, *Proc. of 1st North Am. Rock Mech. Symp., Austin, TX*, Nelson, P. & Laubach, S.E. (Eds.): 165–172. Rotterdam: Balkema.

Gutierrez, M. & Makurat, A. 1997. Coupled HTM modelling of cold water injection in fractured hydrocarbon reservoirs. *Int. J. Rock Mech. Min. Sci. & Geomech. Abstr., 34*: 3–4, Paper 113.

Gutierrez, M.A., Dvorkin, J. & Nur, A. 2002. Stratigraphy-guided rock physics. *The Leading Edge*, January 2002, 98–103.

## H

Hackert, C.L., Parra, H.O., Brown, R.L. & Collier, H.A. 2001. Characterization of dispersion, attenuation, and anisotropy at the Buena Vista Hills Field. *Geophysics, 66*: 90–96.

Hackert, C.L. & Parra, J.O. 2003. Estimating scattering from vugs or karsts. *Geophysics, 68(4)*: 1182–1188.

Haimson, B.C. 2006. Borehole breakouts in Berea Sandstone reveal a new fracture mechanism. *Pure Appl. Geophys., 160*: 813–831.

Halderman, T.P. & Davis, P.M. 1991. $Q_p$ beneath the Rio Grande and East African rift zones. *J. Geophys. Res., 96*: 10,113–10,128.

Hall, S.A. & Kendall, J.-M. 2000. Constraints on the interpretation of P-wave AVAO for fracture characterization. *Proc. of 9th Int. Workshop on Seismic Anisotropy (IWSA), Houston, TX*: Extended Abstract. Society of Exploration Geophysicists.

Hall, S.A. & Kendall, J.-M. 2003. Fracture characterization at Valhall: application of P-wave amplitude variation with offset and azimuth (AVOA) analysis to a 3D ocean-bottom data set. *Geophysics, 68(4)*: 1150–1160.

Hamilton, E.L. 1971. Elastic properties of marine sediments. *J. Acoust. Soc. Am., 28*: 266–284.

Hamilton, E.L. 1972. Compressional-wave attenuation in marine sediments. *Geophysics, 37*: 620–646.

Hamilton, E.L. 1979. $V_p/V_s$ and Poissons' ratios in marine sediments and rocks. *J. Acoust. Soc. Am., 66(4)*: 1093– 1101.

Han, D. 1986. Effects of porosity and clay content on acoustic properties of sandstones and unconsolidated sediments. *PhD Thesis*, Stanford University, USA.

Han, D., Nur, A. & Morgan, D. 1986. Effects of porosity and clay content on wave velocities in sandstones. *Geophysics, 51*: 2093–2107.

Hardin, E.L., Barton, N., Lingle, R., Board, M.P. & Voegele, M.D. 1981. A heated flatjack test series to measure the thermomechanical and transport properties of in situ rock masses. *Office of Nuclear Waste Isolation*, Columbus: OH. ONWI-260, 193 pp.

Harding, A.J., Orcutt, J.A., Kappus, M.E., Vera, E.E., Mutter, J.C., Buhl, P., Detrick, R.S. & Brocher, T.M.

1989. Structure of young oceanic crust at 13°N on the East Pacific Rise from expanding spread profiles. *J. Geophys. Res.*, *94*: 12,163–12,196

Harjes, H.-P., Brann, K., Dürbaum, H.-J., Gebrande, H., Hirschmann, G., Janik, M., Klöckner, M., Lüschen, E., Rabbel, W., Simon, M., Thomas, R., Tormann, J. & Wenzel. 1997. Origin and nature of crustal reflections: results from integrated seismic measurements at the KTB superdeep drilling site. *J. Geophys. Res.*, *102(B8)*: 18,267–18,288.

Harris, P.E., Kerner, C. & White, R.E. 1997. Multichannel estimation of frequency-dependent Q from VSP data. *Geophys. Prospect.*, *45*: 87–109.

Harrison, C.G.A. & Bonatti, E. 1981. The oceanic lithosphere. In: *The Oceanic Lithosphere. The Sea*, Emiliani, C. (Ed.). Vol. 7, pp. 21–48. New York: Wiley

Hasagawa, A., Horiuchi, S. & Umino, N. 1994. Seismic structure of the northeastern Japan convergent margin: a synthesis. *J. Geophys. Res.*, *99(B11)*: 22,295–22,311.

Hasselström, B., Rahm, L. & Scherman, K.A. 1964. Methods for the determination of the physical and mechanical properties of rock. *Int. Congress on Large Dams*, 8. Edinburgh. *1*(Q 28): 611–625. Paris: ICOLD.

Hart, R.S., Anderson, D.L. & Kanamori, M. 1977. The effect of attenuation on gross earth models. *J. Geophys. Res.*, *82*: 1647–1654.

Hatchell, P. & Bourne, S. 2005. Rocks under strain: strain-induced time-lapse time shifts are observed for depleting reservoirs. *The Leading Edge*, December 2005, 1222–1225.

Hattema, M.H.H. & de Pater, C.J. 1998. The poro-mechanical behaviour of Felser sandstone: stress- and temperature-dependent. Paper SPE/ISRM 47270, *Eurock '98 Conf.*, Trondheim, Norway.

Hauge, P.S. 1981. Measurements of attenuation from vertical seismic profiles. *Geophysics*, *37*: 1548–1558.

Hauksson, E. 2000. Crustal structure and seismicity distribution adjacent to the Pacific and North America plate boundary in southern California. *J. Geophys. Res.*, *105*: 13,875–13,903.

Hayashi, K. and Saito, H. 2001. Prediction ahead of the tunnel face by a high-resolution seismic refraction method with sources placed in the tunnel. *Proc. of 4th Int. Workshop on Application of Geophysics to Rock Engineering*, Beijing, China. *Int. Soc. Rock Mech. News J.*, *7(1)*: 28–33.

Heffer, K.J., Fox, R.J., McGill, C.A. & Koutsabeloulis, N.C. 1995. Novel techniques show links between reservoir flow directionality, earth stress, fault structure and geo-mechanical changes in mature waterfloods. *SPE 30711, SPE Ann. Tech. Conf. Exhib.*, Dallas: TX. *Also SPE J.*, *2(2)*: 91–98, June 1997.

Heffer, K. 2002. Geomechanical influences in water injection projects: an overview. *Oil Gas Sci. Technol. – Rev. IFP*, *57(5)*: 415–422.

Heimli, P. 1972. Water and air leakage through joints in rock specimens. (in Norwegian). *Fjellsprengningsteknikk, Bergmekanikk. Annual Rock Mechanics Meeting*, Oslo, Norway, 137–142, Ingeniørforlaget, Oslo.

Helbig, K. & Szaraniec, E. 2000. The historical roots of anisotropy. *Proc. of 9th Int. Workshop on Seismic Anisotropy (IWSA), Houston, TX*: Extended Abstract. Society of Exploration Geophysicists.

Helbig, K. & Thomsen, L. 2005. 75-plus years of anisotropy in exploration and reservoir seismics: a historical review of concepts and methods. *Geophysics*, *70(6)*: 15.

Hellweg, M., Spudich, P., Fletcher, J.B. & Baker, L.M. 1995. Stability of coda Q in the region of Parkfield, California: View from the U.S. Geological Survey Parkfield Dense Seismograph Array. *J. Geophys. Res.*, *100*: 2089–2102.

Herwanger, J.V., Worthington, M.H., Lubbe, R., Binley, A. & Khazanehdari, J. 2004. A comparison of cross-hole electrical and seismic data in fractured rock. *Geophys. Prospect.*, *52*: 109–121.

Herwanger, J. & Horne, S. 2005. Predicting time-lapse stress effects in seismic data. *The Leading Edge*, December 2005, 1234–1242.

Hesler, G.J., Cook, N.G.W. & Myer, L. 1996. Estimation of intrinsic and effective elastic properties of cracked media from seismic testing. *2nd NARMS '96*. Montréal, Québec. Aubertin, Hassani & Mitri (Eds.). *1*: 467–473. Rotterdam: Balkema.

Hiramatsu, Y., Hayashi, N. & Furumoto, M. 2000. Temporal changes in coda $Q^{-1}$ and b value due to the static stress change associated with the 1995 Hyogo-ken Nanbu earthquake. *J. Geophy. Res.*, *105*: 6141–6151.

Holbrook, W.S., Purdy, G.M., Sheridan, R.E., Glover III, L., Talwani, M., Ewing, J. & Hutchinson, D. 1994. Seismic structure of the U.S. Mid-Atlantic continental margin. *J. Geophys. Res.*, *99(B9)*: 17,871–17,891.

Holbrook, W.S., Brocher, T.M., ten Brink, U.S. & Hole, J.A. 1996. Crustal structure of a transform plate boundary: San Francisco Bay and the central California continental margin. *J. Geophys. Res.*, *101*: 22,311– 22,334.

Holcomb, D.J. 1988. Cross-hole measurements of velocity and attenuation to detect a disturbed zone in salt at the Waste Isolation Pilot Plant. *29th US Symp. on Rock Mechanics*, Minneapolis: MN. Cundall et al. (Eds.). A.A. Balkema, 633–640.

Hole, J.A., Brocher, T.M., Klemperer, L., Parsons, T., Benz, H.M. & Furlong, K.P. 2000. Three-dimensional seismic velocity structure of the San Francisco Bay area. *J. Geophys. Res.*, *105(B6)*: 13,859–13,874.

Holt, R.M., Fjær, E., Furre, A.-K. 1996. Laboratory simulation of the influence of earth stress change on wave velocities. In: *Seismic Anisotropy*, Fjær et al. (Eds.): 180–202. Society of Exploration Geophysicists.

Holt, R.M., Furre, A.-K. & Horsrud, P. 1997. Stress dependent wave velocities in sedimentary rock cores: why and why not. *Int. J. Rock Mech. Min. Sci. & Geomech. Abstr.*, *34*: 3–4. Paper 128.

Holt, R.M., Nes, O.-M. & Fjær, E. 2005. In-situ stress dependence of wave velocities in reservoir and overburden rocks. *The Leading Edge*, December 2005, 1268–1274.

Homand-Etienne, F. & Sebaibi, A. 1996. Study of microcracking of Lac du Bonnet granite. *Eurock '96*. Barla (Ed.). *2*: 1353–1360. Rotterdam: Balkema.

Hope, V.S., Clayton, C.R.I. & Barla, G. 1996. Class 'A' predictions of the locations of major rock discontinuities at a storage cavern site, using seismic tomography. *Eurock '96*. Barla (Ed.). *2*: 925–932. Rotterdam: Balkema.

Horn, H.M. & Deere, D.U. 1962. Frictional characteristics of minerals. *Geotechnique*, *12*: 319–335.

Horn, S. & MacBeth, C. 1996. The new generation-azimuthal variation of amplitude. Extended Abstracts, *Society of Exploration Geophysicists Workshop*, Big Sky, Montana.

Horne, S. & MacBeth, C. 1997. AVA observations in azimuthal VSPs. *59th Annual Meeting European Association Exploration Geophysicists*, Expanded Abstracts: P051.

Horne, S., MacBeth, C., Queen, J., Rizer, W. & Cox, V. 1997. Fracture characterization from near-set VSP inversion. *Geophys. Prospect.*, *45*: 141–164.

Horne, S. 2003. Fracture characterization from walkaround VSPs. *Geophys. Prospect.*, *51*: 493–499.

Hoshiba, M., Sato, H. & Fehler, M. 1991. Numerical basis of the separation of scattering and intrinsic absorption from full seismogram envelope: a Monte-Carlo simulation of multiple isotropic scattering. *Meteorol. Geophys.*, *42*: 65–91.

Hough, S.E., Anderson, J.G., Brune, J., Vernon III, B.F., Berger, J., Fletcher, J., Haar, L., Hanks, T. & Baker, L. 1988. Attenuation near Anza, California. *Bull. Seism. Soc. Am.*, *78*: 672–691.

Houtz, R. 1976. Seismic properties of layer 2A in the Pacific. *J. Geophys.Res.*, *81*: 6321–6331.

Houtz, R. & Ewing, J. 1976. Upper crustal structure as a function of plate age. *J. Geophys. Res.*, *81(14)*: 2490–2498.

Hsu, C.-J. & Schoenberg, M. 1993. Elastic waves through a simulated fractured medium. *Geophysics*, *58*: 964–977.

Hudson, J.A., Jones, E.J.W. & New, B.M. 1980. P-wave velocity measurements in a machine-bored, chalk tunnel. *Q. J. Engng Geol.*, *13*: 33–43. Northern Ireland: The Geological Society.

Hudson, J.A. 1981. Wave speed and attenuation of elastic waves in material containing cracks. *Geophys. J. Roy. Astr. Soc.*, *64*: 133–150.

Hudson, J.A. 1990. Overall elastic properties of isotropic materials with arbitrary distribution of circular cracks. *Geophys. J. Int.*, *102*: 465–469.

Hudson, J.A., Liu, E. & Crampin, S. 1996. The mechanical properties of materials with interconnected cracks and pores. *Geophys. J. Int.*, *124*: 105–112.

Hudson, J.A., Liu, E. & Crampin, S. 1997. The mean transmission properties of a fault with imperfect facial contact. *Geophysical Journal International* 129, 720–726.

Hustedt, B. & Clark, R.A. 1999. Source/receiver array directivity effects on marine seismic attenuation measurements. *Geophys. Prospect.*, *47*: 1105–1119.

Hyndman, R.D. 1976. Seismic velocity measurements of basement rocks from DSDP leg 37. *Initial Rep. Deep Sea Drill. Proj.*, *37*: 373–387.

Hyndman, R.D. & Drury, M.J. 1976. The physical properties of oceanic basement rocks from deep drilling on the Mid-Atlantic ridge. *J. Geophys. Res.*, *81(23)*: 4042–4052.

Hyndman, R.D. 1979. Poisson's ratio in the oceanic crust – a review. *Tectonophysics*, *59*: 321–333.

## I

Ikeda, K. 1970. A classification of rock conditions for tunneling. *Proc. of 1st IAEG Congress*, Paris. *II*: 1258–1265. Le Comité Français de Géologie de l'Ingénieur.

Ikeda, K., Ohshima, H. & Sakurai, T. 1981. The property and the seismic wave velocity of fractured zone. *Proc. of Int. Symp. on Weak Rock*, Tokyo. 527–532.

Ikeda, R. 1993. In situ stress heterogeneity and crack density distribution. *Int. J. Rock Mech. Min. Sci. & Geomech. Abstr.*, *30(7)*: 1013–1018. (*34th US Symp. on Rock Mechanics*, Madison, WI.)

Iliev, I.G. 1966. An attempt to estimate the degree of weathering of intrusive rocks from their physico-mechanical properties. *Proc. of 1st ISRM Congress*, Lisbon. *I*: 109–114. LNEC.

Infanti, N. & Kanji, M.A. 1978. In situ shear strength, normal and shear stiffness determinations at Agua Vermelha Project. *Proc. of 3rd Int. Cong. of the IAEG*, Madrid. Session 2. *2*: 172–183.

Infanti, N. & Kanji, M.A. 1990. *Estimating the Shear Stiffness of rock joints. Rock Joints*, Barton, N. & Stephansson, O. (Eds). *Int. Conf.*, Loen, Norway: 799–804. Rotterdam: Balkema.

Isacks, B., Oliver, J. & Sykes, L.R. 1968. Seismology and the new global tectonics. *J. Geophys. Res.*, *73*: 5855–5899.

Ishikawa, K., Ochi, H. & Tadaka, S. 1995. Rock mass classification of weathered granite and evaluation of geomechanical parameters for bridge foundation of large scale structure in the Honshu-Shikoku Bridge Authority. Rock Foundation. *Int. Workshop on Rock Foundation.* Tokyo 1995. 169–176. Rotterdam: Balkema.

ISRM. 1998. Suggested methods for seismic testing within and between boreholes. *Int. J. Rock Mech. Min. Sci. & Geomech. Abstr.*, *25*: 447–472.

Ito, T. & Zoback, M.D. 2000. Fracture permeability and in situ stress to 7 km depth in the KTB scientific drillhole. *Geophys. Res. Let.*, *27*: 1045–1048.

Ivanovíc, K., Jovanovíc, L., Markovíc, O., Kujundzíc, B. & Radosavljevíc, Z. 1970. Komplexe Untersuchungen mechanischer Charakteristiken der Felsmasse am Profil der

Bogensperre Mratinje. *Proc. of 2nd ISRM Congress*, Belgrade. *III*: Theme 8–24. Belgrade: Privredni Pregled.

Iwai, K. 1976. Fundamental studies of fluid flow through a single fracture. *PhD Thesis*, University of California, Berkeley: CA. 208 pp.

Iwasaki, T., Yoshii, T., Moriya, T., Kobnayashi, A., Nishiwaki, M., Tsutsui, T., Iidaka, T., Ikami, A. & Masuda, T. 1994. Precise P and S wave structures in the Kitakami massif, Northern Honshu, Japan, from a seismic refraction experiment. *J. Geophys. Res.*, *99*: 22,187– 22,204.

## J

Jacobsen, R.S. & Lewis, B.T.R. 1990. The first direct measurements of upper oceanic crustal compressional wave attenuation. *J. Geophys. Res.*, *95(11)*: 17,417–17,429.

Jaeger, J.C. & Cook, N.G.W. 1977. *Fundamentals of Rock Mechanics*, Halsted Press.

Jeng, Y., Tsai, J.-Y. & Chen, S.-H. 1999. An improved method of determining near-surface *Q*. *Geophysics*, *64(5)*: 1608–1617.

Jin, A., Cao, T. & Aki, K. 1985. Regional change of coda Q in the oceanic lithosphere. *J. Geophys. Res.*, *90*: 8651–8659.

Jin, A. & Aki, J. 1986. Temporal change in coda Q before the Tangsham earthquake of 1976 and the Haicheng earthquake of 1975. *J. Geophys. Res.*, *91*: 665–673.

Jizba, D.L. 1991. Mechanical and acoustical properties of sandstones and shales. *PhD Dissertation*, Stanford University.

Johansen, T.A., Jakobsen, M., Agersborg, R. & Ruud, B.O. 2004. The P-P and P-S response of a fractured reservoir. *SEG Int. Exposition and 74th Annual Meeting*, Denver, Co.

Johnston, D.H., Toksöz, M.N. & Timur, A. 1979. Attenuation of seismic waves in dry and saturated rocks: II. Mechanisms. *Geophysics*, *44(4)*: 691–711.

Johnston, D.H. 1986. VSP detection of fracture-induced velocity anisotropy. *56th Annual Int. Meeting*, Society of Exploration Geophysicists, Houston: TX.

Johnston, J.E. & Christensen, N.I. 1995. Seismic anisotropy of shales. *J. Geophys. Res.*, *100*: 5991–6003.

Jones, T. & Nur, A. 1983. Velocity and attenuation in sandstones at elevated temperatures and pressures. *Geophys. Res. Lett.*, *10*: 140–143.

Jones, T.D. 1986. Pore fluids and frequency-dependent wave propagation in a vertical transversely isotropic medium. Field experiment and model study. *Geophysics*, *51*: 1939–1953.

Jongmans, D. & Malin, P.E. 1995. Vertical profiling of microearthquake S waves in the Varian well at Parkfield, California. *Bull. Seism. Soc. Am.*, *85*: 1805–1820.

Juhlin, C. 1995. Imaging of fracture zones in the Finnsjönn area, central Sweden, using the seismic reflection method. *Geophysics*, *60(1)*: 66–75.

## K

Kabir, M.M.N. & Verschuur, D.J. 2000. A constrained parametric inversion for velocity analysis based on CFP technology. *Geophysics*, *65*: 1210–1222.

Kanestrøm, R. & Haugland, K. 1971. The Trans-Scandinavian seismic profile of 1969: profile section 3–4. In: *Deep Seismic Sounding in Northern Europe*. Vogel, A. (Ed.). Stockholm: Swedish National Science Council, 76–91.

Kang, I.B. & McMechan, G.A. 1994. Separation of intrinsic and scattering Q based on frequency-dependent amplitude ratios of transmitted waves. *J. Geophys. Res.*, *99(B12)*: 23,875–23,885.

Kappus, M.E., Harding, A.J. & Orcutt, J.A. 1995. A baseline for upper crustal velocity variations along the East Pacific Rise at 13°N. *J. Geophys. Res.*, *100(B4)*: 6143–6161.

Karmis, M., Schilizzi, P. & Agioutantis, Z. 1984. The potential and application of sonic wave methods in engineering rock characterization. Rock Mechanics in Productivity and Protection. *25th US Symp. on Rock Mechanics*, Evanston, IL. Dowding & Singh (Eds.). New York: Society of Mining Engineers, 1083–1091.

Kearey, P. & Brooks, M. 1994. *An Introduction to Geophysical Exploration*. *Geoscience Texts, Vol. 4*, Oxford: Blackwell Scientific Publications.

Kearey, P. & Vine, F.J. 1996. *Global Tectonics*, 2nd edn. 333 pp. Oxford: Blackwell Science.

Keehn, N.A. & Kanasewich, E.R. 1987. Attenuation from VSP data collected on Melville Island. *Can. J. Soc. Explor. Geophys.*, *23(1)*: 73–84.

Kelemen, P.B. & Holbrook, W.S. 1995. Origin of thick, high-velocity igneous crust along the U.S. East Coast Margin. *J. Geophys. Res.*, *100(B7)*: 10,077–10,094.

Kelsall, P.C., Watters, R.J. & Franzone, J.G. 1986. Engineering characterization of fissured, weathered dolerite and vesicular basalt. Rock Mechanics: Key to Energy Production. *27th US Symp. on Rock Mechanics*, Tuscaloosa: AL. Hartman (Ed.): 77–84. New York: Society of Mining Engineers.

Kennett, B.L.N. & Orcutt, J.A. 1976. A comparison of travel time inversions for marine refraction profiles. *J. Geophys. Res.*, *81(23)*: 4061–4070.

Khazanehdari, J., McCann, C., Sothcott, J. & Astin, T.R. 1998. The inter-relationships between velocity and attenuation from VSP survey. *Technical Proc. of 60th annual meeting of the European Association of Geoscientists and Engineers*, Leipzig. Paper No. 10–28.

Kikuchi, K., Saito, K. & Kusunoki, K.-I. 1982. Geotechnically integrated evaluation on the stability of dam foundation rocks. *Int. Congress on Large Dams, 14.* Rio de Janeiro 1982. *2(Q53)*: 49–73. Paris: ICOLD.

King, M.S. 1966. Wave velocities in rocks as a function of changes in overburden pressure and pore fluid saturants. *Geophysics*, *31*: 50–73.

King, M.S., Pandit, B.I. & Stauffer, M.R. 1978. Quality of rock masses by acoustic borehole logging. *Proc. of 3rd IAEG Congress*, Madrid. *IV(1)*: 156–164. Madrid: Imprime ADOSA.

King, M.S., Myer, L.R. & Rezowalli, J.J. 1984. Cross-hole acoustic measurements in basalt. Rock Mechanics in Productivity and Protection. *25th US Symp. on Rock Mechanics*, Evanston, IL. Dowding & Singh (Eds.). New York: Society of Mining Engineers, 1053–1061.

King, M.S., Myer, L.R. & Rezowalli, J.J. 1986. Experimental studies of elastic-wave propagation in a columnar-jointed rock mass. *Geophys. Prospect.*, *34*: 1185–1199.

King, M.S., Chaudhry, N.A. & Shakeel, A. 1995. Experimental ultrasonic velocities and permeability for sandstones with aligned cracks. *Int. J. Rock Mech. Min. Sci. & Geomech. Abstr.*, *32*: 155–163.

King, M.S., Shakeel, A. & Chaudhry, N.A., 1997. Acoustic wave propagation and permeability in sandstones with systems of aligned cracks. In: *Developments in Petrophysics*, Lovell, M.A. & Harvey, P.K. (Eds.). Geological Society of London, Special Publication No. 122, 69–85.

King, M.S. 2002. Elastic wave propagation in and permeability for rocks with multiple parallel fractures. Technical Note, *Int. J. Rock Mech. Min. Sci. & Geomech. Abstr.*, *39*: 1033–1043.

King, M.S. & Marsden, J.R. 2002. Velocity dispersion between ultrasonic and seismic frequencies in brine-saturated reservoir sandstones. *Geophysics*, *67*: 254–258.

King, M.S. 2005. Rock-physics developments in seismic exploration: a personal 50-year perspective. *Geophysics*, *70(6)*: 3–8.

Kinoshita, S. 1994. Frequency-dependent attenuation of shear waves in the crust of the Southern Kanton Area, Japan. *Bull. Seism. Soc Am.*, *84*: 1387–1396.

Kirkpatrick, R.J. 1979. The physical state of the oceanic crust: results of downhole geophysical logging in the Mid-Atlantic ridge at 23°N. *J. Geophys. Res.*, *84(B1)*: 178–188.

Kishida, K. 1999. Experimental study on the estimation of discontinuity characteristics and mechanical properties of jointed rock masses. *PhD Thesis*, Civil Engineering Department, University of Kyoto, Japan, 203 pp.

Kjartansson, E. 1977. Constant Q: wave propagation and attenuation. *EOS, Trans. AGU.*, *58*: 1183.

Kjartansson, E. 1979. Constant *Q*-wave propagation and attenuation. *J. Geophys. Res.*, *84(B9)*: 4737–4748.

Klimentos, T. & McCann, C. 1990. Relationships among compressional wave attenuation, porosity, clay content and permeability in sandstones. *Geophysics*, *55*: 998–1014.

Klimentos, T. 1991. The effects of porosity-permeability-clay content on the velocity of compressional waves. *Geophysics*, *56*: 1930–1939.

Klimentos, T. 1995. Attenuation of P- and S-waves as a method of distinguishing gas and condensate from oil and water. *Geophysics*, *60*: 447–458.

Knight, R., Dvorkin, J. & Nur, A. 1998. Acoustic signatures of partial saturation. *Geophysics*, *60*: 132–138.

Knipe, R.J., Jones, G. & Fisher, Q.J. (1998). Faulting, fault sealing and fluid flow in hydrocarbon reservoirs: an introduction. In: *Faulting, Fault Sealing and Fluid Flow in Hydrocarbon Reservoirs*, Jones, G., Fisher, Q.J. & Knipe, Q.J. (Eds.). Geological Society London Special Publication 147: *i–xxi*.

Knopoff, L. 1964. 'Q'. *Rev. Geophys.*, *2(4)*: 645–654. Publication 391, Institute of Geophysics and Planetary Physics, UCLA.

Kranz, R.L., Frankel, A.D., Engelder, T. & Scholtz, C.H. 1980. The permeability of whole and jointed Barre granite. *Int. J. Rock Mech. Min. Sci. & Geomech. Abstr.*, *16*: 225–234: Erratum, *17*: 237–238.

Kujundzíc, B. & Grujíc, N. 1966. Correlation between static and dynamic investigations of rock mass "in situ". *Proc. of 1st ISRM Congress*, Lisbon. *1*: 565–570. LNEC.

Kujundzíc, B., Jovanovic, L. & Radosavljevic, Z. 1970. Solution du revêtement d'une galerie en charge par application d'injections à haute pression. *Proc. of 2nd ISRM Congress*, Belgrade. *II*: Theme 4–66. Belgrade: Privredni Pregled.

Kujundzíc, B. 1979. Use of tests and monitoring in the design and construction of rock structures. *Proc. of 4th Int. Congr. Rock Mech. (ISRM), Montreux*: *3*: 181–186. Rotterdam: Balkema.

Kuster, G.T. & Toksöz, M.N. 1974. Velocity and attenuation of seismic waves in two-phase media: Part II, Experimental results. *Geophysics*, *39*: 587–606.

Kvamme, L.B. & Havskov, J. 1989. Q in Southern Norway. *Bull. Seism. Soc. Am.*, *79*: 1575–1588.

## L

Laine, R.G.T. 1964. Rock foundations. Diagnosis of mechanical properties and treatment. *Proc. of 8th Int. Congress on Large Dams*, Edinburgh, 1964. *1*: 141–165. Paris: ICOLD.

Landrø, M. & Stammeijer, J. 2004. Quantitative estimation of compaction and velocity changes using 4D impedance and traveltime changes. *Geophysics*, *69(4)*: 949–957.

Lashkaripour, G.R. & Passaris, E.K.S. 1995. Correlations between index parameters and mechanical properties of shales. *Proc. of 8th ISRM Congress*, Tokyo. Fuji (Ed.). *1*: 257–261. Rotterdam: Balkema.

Laubach, S., Marrett, R. & Olson, J. 2000. New directions in fracture characterization. *The Leading Edge*, July 2000, 704–711.

Laubach, S.E., Gale, J.F.W. & Olson, J.E. 2002. Are open fractures necessarily aligned with maximum horizontal stress? *American Association of Petroleum Geologists. (AAPG: Fractured Reservoirs and Multi-Phase Fluid Flow), Annual Meeting*, Houston, TX.

Laubach, S.E. 2003. Practical approaches to identifying sealed and open fractures. *AAPG (American Association of Petroleum Geologists). Bull.*, *87(4)*: 561–579.

Laurent, J., Boutéca, M.J., Sarda, J.-P. & Bary, D. 1995. Pore pressure influence in the poroelastic behaviour of rocks: experimental studies and results. *SPE Formation Evaluation*, June 1993, 117–122.

Leary, P.C. & Henyey, T.L. 1985. Anisotropy and fracture zones about a geothermal well from *P*-wave velocity profiles. *Geophysics, 50(1)*: 25–36.

Leary, P.C., Crampin, S. & McEvilly, T.V. 1990. Seismic fracture anisotropy in the earth's crust: an overview. *J. of Geophys. Res., 95(B7)*: 11,105–11,114.

Leary, P.C. & Abercrombie, R.E. 1994. Frequency-dependent crustal scattering and absorption at 5–160 Hz from Coda decay observed at 2.5 km depth. *Geophys. Res. Lett., 21*: 971–974.

Lecomte, I., Gjøystdal, H., Dahle, A. & Pedersen, O.C. 2000. Improving modelling and inversion in refraction seismics with a first-order Eikonal solver. *Geophys. Prospect., 48*: 437–454.

Lee, H. & Haimson, B.C. 2006. Borehole breakouts and in situ stress in sandstone. In situ Rock Stress. *Proc. of Int. Conf.*, Lu, Li, Kjorholt & Dahle (Eds.): 201–207. London: Taylor & Francis.

Li, X., Cui, L. & Rogiers, J.-C. 1998. Thermo-poroelastic analyses of inclined boreholes. Paper SPE/ISRM 47296, *Eurock 98 Conf.*, Trondheim, Norway.

Liao, Q. & McMechan, G.A. 1997. Tomographic imaging of velocity and Q, with application to crosswell seismic data from the Gypsy Pilot Site, Oklahoma. *Geophysics, 62(6)*: 1804–1811.

Link, H. 1964. Evaluation of elasticity moduli of dam foundation rock determined seismically in comparison to those arrived at statically. *Int. Congress on Large Dams, 8*. Edinburgh 1964. *1(Q28)*: 833–858. Paris: ICOLD.

Liu, E., Crampin, S., Queen, J.H. & Rizer, W.D. 1993. Velocity and attenuation anisotropy caused by microcracks and macrofractures in a multiazimuth reverse VSP. *Can. J. Explor. Geophys., 29*: 177–188.

Liu, E. & Li, X.-Y. 1998. Estimation of pore or fracture fluids from seismic measurements. *68th Annual Int. SEG Meeting*, New Orleans, 1624–1627.

Liu, E., Hudson, J.A. & Pointer, T. 2000. Equivalent medium representation of fractured rock. *J. Geophys. Res., 105(B2)*: 2981–3000.

Liu, E., Li, X.Y. & Queen, J.H. 2000. Discrimination of pore fluids from P and converted shear-wave AVO analysis. *Anisotropy 2000: Fractures, Converted Waves and Case Studies.* Tulsa, OK: Society of Exploraion Geophysicists.

Liu, E., and Li, X.-Y. 2001. Seismic detection of fluid saturation in aligned fractures. 9th Int. Workshop on Seismic Anisotropy. *Geophysics, 66(4)*: Abstract.

Liu, E., Queen, J.H., Li, X.Y., Chapman, M., Maultzsch, S., Lynn, H.B. & Chesnokov, E.M. 2003a. Observation and analysis of frequency-dependent anisotropy from a multicomponent VSP at Bluebell-Altamont Field, Utah. *J. Appl. Geophys., 54*: 319–333.

Liu, E., Maultzsch, S., Chapman, M. & Li, X.Y. 2003b. Frequency-dependent seismic anisotropy and its implication for estimating fracture size in low porosity reservoirs. *The Leading Edge*, July 2003: 662–665.

Liu, E., Vlastos, S. & Li, X.Y. 2004. Modeling seismic wave propagation during fluid injection in a fractured network: effects of pore fluid pressure on time-lapse seismic signatures. *The Leading Edge*, August 2004: 778–783.

Liu, E. 2005. Effects of fracture aperture and roughness on hydraulic and mechanical properties of rocks: implication of seismic characterization of fractured reservoirs. *J. Geophys. Eng., 2*: 38–47, Nanking Inst. of *Geophys. Prospect.*

Liu, Y., Booth, D.C., Crampin, S., Evans, R. & Leary, P. 1993. Shear-wave polarizations and possible temporal variations in shear-wave splitting at Parkfield. *Can. J. Explor. Geophys., 29*: 380–390.

Liu, Y., Crampin, S. & Abercrombie, R.E. 1997. Shear-wave anisotropy and the stress field from borehole recordings at 2.5 km depth at Cajon Pass. *Geophys. J. Int., 129*: 439–449.

Liu, Y., Teng, T.-L., & Ben-Zion, Y. 2005. Near-surface seismic anisotropy, attenuation and dispersion in the aftershock region of the 1999 Chi-Chi earthquake. *Geophys. J. Int., 160*: 695–706.

Liu, Y.-Q. 2003. Analysis of P-wave seismic reflection data for azimuthal anisotropy. *PhD Thesis*, University of Edinburgh.

Londe, P. 1979. Panel discussion: dynamic testing. *Int. Symp. on The Geotechnics of Structurally Complex Formations*. Capri 1977. *II*: 279–281. Associazione Geotecnica Italiana.

Louis, C. 1969. A study of groundwater flow in jointed rock and its influence on the stability of rock masses. *Rock Mechanics Res. Report*, 10, Imperial College. (Translation of PhD Thesis).

Lubbe, R. 2005. A field and laboratory investigation of the compliance of fractured rock. *PhD Thesis*, Oxford University.

Lubbe, R. & Worthington, M. 2006. A field investigation of fracture compliance. *Geophys. Prospect., 54(3)*: 319–331.

Lubbe, R., Sothcott, J., Worthington, M.H., McCann, C. & Astin, T. 2006. Laboratory estimates of fracture compliance. *Int. J. Rock Mech. Min. Sci. & Geomech. Abstr.* (In press).

Lucet, N. 1989. Vitesse et Attenuation des ondes élastiques soniques et ultrasoniques dans les roches sous pression de confinement. *PhD Dissertation*, University of Paris.

Lucet, N. & Zinszner, B. 1992. Effects of heterogeneities and anisotropy on sonic and ultrasonic attenuation in rocks. *Geophysics, 57(8)*: 1018–1026.

Ludwig, J.W., Nafe, J.E. & Drake, C.L. 1970. Seismic refraction. In: *The Sea.* Maxwell, A.E. (Ed.). *4*: 53–84. New York: Wiley.

Lumley, D.E. 2001. Time-lapse seismic reservoir monitoring. *Geophysics, 66*: 50–53.

Lykoshin, A.G., Yaschenko, S.G., Mikhailov, A.D., Savitch, A.I. & Koptev, V.J. 1971. Investigation of rock jointing by seismo-acoustic methods. *Int. Symp. on Fissuration des Roches*, Nancy, France, I–19.

Lynn, H.B. & Thomson, L.A. 1986. Reflection shear wave data along the principle axes of azimuthal anisotropy. *56th Annual Int. Meeting SEG.* Tulsa, OK: Society of Exploration Geophysicists.

Lynn, H.B., Simon, K.M., Bates, C.R., Layman, M., Schneider, R. & Jones, M. 1995. Use of anisotropy in P-wave and S-wave data for fracture characterization in a naturally fractured gas reservoir. *The Leading Edge*, August 1995: 887–893.

Lynn, H.B., Beckham, W.E., Simon, K.M., Bates, C.R., Layman, M. & Jones, M. 1999a. P-wave and S-wave azimuthal anisotropy at a naturally fractured gas reservoir, Bluebell-Altamont Field, Utah. *Geophysics, 64*: 1312–1328.

Lynn, H.B., Campagna, D., Simon, K.M. & Beckham, W.E. 1999b. Relationship of *P*-wave seismic attributes, azimuthal anisotropy, and commercial gas pay in 3-D *P*-wave multiazimuth data, Rulison Field, Piceance Basin, Colorado. *Geophysics, 64(4)*: 1293–1311.

Lynn, H.B., Weathers, L. & Beckham, W. 2000. The Mobil Onshore Texas 3D full-azimuth full-offset P-wave survey. *9th Int. Workshop on Seismic Anisotropy.* Extended Abstract.

Lynn, H.B. 2004. The winds of change: anisotropic rocks – their preferred direction of fluid flow and their associated seismic signatures. Part 1: *The Leading Edge*, November 2004, *23(11)*: 1156–1162, Part 2: *The Leading Edge*, December 2004, *23(12)*: 1258–1268.


## M


MacBeth, C. 1999. Azimuthal variation in P-wave signatures due to fluid flow. *Geophysics, 64*: 1181–1192.

MacBeth, C. & Li, X.Y. 1999. AVD – An emerging new marine technology for reservoir characterization: acquisition and application. *Geophysics, 64*: 1153–1159.

MacBeth, C. 2004. A classification for the pressure-sensitivity properties of a sandstone rock frame. *Geophysics, 69(2)*: 497–510.

MacDonal, F.J., Angona, F.A., Mills, R.L., Sengbush, R.L., Van Nostrand, R.G. & White, J.E. 1958. Attenuation of shear and compressional waves in Pierre shale. *Geophysics, 23*: 421–439.

Macdonald, K.C. 1982. Mid-Ocean ridges: fine scale tectonic, volcanic and hydrothermal processes within the plate boundary zone. *Annu. Rev. Earth Planet. Sci., 10*: 155–190.

Maia, A.C., Poiate, E.J., Falčao, J.L. & Coelho, L.F.M. 2005. Triaxial creep tests in salt applied in drilling through thick salt layers in Campos Basin, Brazil. *ISRM News J., 9(1)*: 14–24.

Maini, T. & Hocking, G. 1977. An examination of the feasibility of the hydrologic isolation of a high level waste repository in crystalline rock. Geologic disposal of high-level radioactive waste session, *Annual Meeting of the Geol. Soc. of America*, Seattle, Washington.

Majer, E.L., Peterson, J.E., Daley, T., Kaelin, B., Myer, L., Queen, J., D'Onfro, P. & Rizer, W. 1997. Fracture detection using crosswell and single well surveys. *Geophysics, 62(2)*: 495–504.

Makurat, A., Barton, N., Chryssanthakis, P. & Monsen, K. 1990. Jointed rock mass modelling. *Rock Joints*, Barton, N. & Stephansson, O. (Eds.). *Int. Conf.*, Loen, Norway. 647–656. Rotterdam: Balkema.

Makurat, A., Barton, N., Rad, N.S. & Bandis, S. 1990. Joint conductivity variation due to normal and shear deformation. *Rock Joints*, Barton, N. & Stephansson, O. (Eds.). *Int. Conf.*, Loen, Norway. 535–540. Rotterdam: Balkema.

Makurat, A., Barton, N., Tunbridge L. & Vik, G. 1990. The measurement of the mechanical and hydraulic properties of rock joints at different scales in the Stripa Project. *Rock Joints*, Barton, N. & Stephansson, O. (Eds.). *Int. Conf.*, Loen, Norway. 541–548. Rotterdam: Balkema.

Makurat, A. 1996. Fracture flow and fracture cross flow as a function of normal stress and shear displacement, Extended Abstract: *10th Kongsberg Seminar on Fractures, Fluid Flow and Transport in Fractures*, Kongsberg, Norway.

Malin, P.E., Walker, J.A., Borcherdt, R.D., Cranswick, E., Jensen, E.G. & Van Schaak, J. 1988. Vertical seismic profiling of Oroville microearthquakes: velocity spectra and particle motion as a function of depth. *Bull. Seism. Soc. Am., 78*: 401–420.

Mandal, P., Padhy, S., Rastogi, B.K., Satyanarayana, H.V.S., Kousalya, M., Vijayraghavan, R. & Srinivasan, A. 2001. Aftershock activity and frequency-dependent low coda $Q_c$ in the epicentral region of the 1999 Chamoli Earthquake of $M_w$ 6.4. *Pure Appl. Geophys., 158*: 1719–1735.

Mandal, P., Jainendra, T.S., Joshi, S., Kumar, S., Bhunia, R., Rastogi, B.K. 2004. Low coda $Q_c$ in the epicentral region of the 2001 Bhuj Earthquake of $M_w$ 7.7. *Pure Appl. Geophys., 161*: 1635–1654.

Marmureanu, G., Bratosin, D. & Cioflan, C.O. 2000. The dependence of $Q$ with seismic-induced strains and frequencies for surface layers from resonant columns. *Pure Appl. Geophys., 157*: 269–279.

Marson-Pidgeon, K. & Savage, M.K. 1997. Frequency-dependent anisotropy in Wellington, New Zealand. *Geophys. Res. Lett., 24(24)*: 3297–3300.

Martel, S.J. & Peterson, Jr., J.E. 1991. Interdisciplinary characterization of fracture systems at the US/BK site, Grimsel Laboratory, Switzerland. *Int. J. Rock Mech. Min. Sci. & Geomech. Abstr., 28(4)*: 295–323.

Martin, C.D., Davison, C.C. & Kozak, E.T. 1990. Characterizing normal stiffness and hydraulic conductivity of a major shear zone in granite. *Rock Joints*, Barton, N. & Stephansson, O. (Eds.). *Int. Conf.*, Loen, Norway. 549–556. Rotterdam: Balkema.

Martin, C.D., Young, R.P. & Collins, D.S. 1995. Monitoring progressive failure around a tunnel in massive granite. *Proc. of*

*8th Int. Conf. of Int. Soc. Rock Mech.*, Tokyo, Japan. Fuji, T. (Ed.). 627–634. Rotterdam: Balkema.

Martin, C.D., Read, R.S. & Martino, J.B. 1997. Observation of brittle failure around a circular test tunnel. *Int. J. Rock Mech. Min. Sci. & Geomech. Abstr.*, *34*: 1065–1073, Elsevier.

Masuda, H. 1964. Utilisation of elastic longitudinal wave velocity for determining the elastic property of dam foundation rocks. *Int. Congress on Large Dams*, *8*. Edinburgh 1964. *1(Q28)*: 253–269. Paris: ICOLD.

Mathisen, M.E., Vasilou, A.A., Cunningham, P., Shaw, J., Justice, J.H. & Guinzy, N.J. 1995. *Geophysics*, *60(3)*: 631–650.

Matthews, M.C., Hope, V.S. & Clayton, C.R.I. 1997. The geotechnical value of ground stiffness determined using seismic methods. *Modern Geophysics in Engineering Geology*, McCann, Eddleston, Fenning & Reeves (Eds.). 113–123. Geological Society Engineering Geology Special Publication.

Maultzsch, S., Liu, E. & Chapman, M. 2002. Modelling attenuation anisotropy and its interpretation in terms of fracture properties. *Abstract 73rd Annual Int. SEG Meeting*, Dallas: TX.

Maultzsch, S., Chapman, M., Liu, E. & Li, X.Y. 2003. Modelling frequency-dependent seismic anisotropy in fluid–saturated rock with aligned fractures: implication of fracture size estimaton from anisotropic measurements. *Geophys. Prospect.*, *51*: 381–392.

Maultzsch, S., Chapman, M., Liu, E. & Li, X.Y. 2005. Observation and modelling of anisotropic attenuation in VSP data. *Seismic Q–observation, mechanisms, and interpretation. Workshop at 67th EAGE Conference*, Madrid. Extended Abstract: Q-08, Convenors: Hargreaves, Best and Lancaster.

Maury, V. 1987. Observations, recherches et résultants récents sur la mécanismes de ruptures autour de galleries isolées. *Proc. of 6th Int. Congr. Rock Mech. (ISRM), Montreal*, Herget, G. & Vongpaisal (Eds.): *2*: 1119–1128. Rotterdam: Balkema.

Maury, V. & Guenot, A. 1995. Practical advantages of mud cooling systems for drilling. *SPE Drillings & Completions*, *10(1)*, 42–48.

Maury, V., Piau, J.-M. & Hallé, G. 1996. Subsidence induced by water injection in water sensitive reservoir rocks: the example of Ekofisk. *Paper SPE 36890, SPE European Petroleum Conf.*, Milan, Italy.

Mavko, G. & Nur, A. 1975. Melt squirt in the asthenosphere. *J. Geophys. Res.*, *80*: 1444–1448.

Mavko, G. 1979. Frictional attenuation: an inherent amplitude dependence. *J. Geophys. Res.*, *84*: 4769–4776.

Mavko, G.M. & Nur, A. 1979. Wave attenuation in partially saturated rocks. *Geophysics*, *44*: 161–178.

Mavko, G.M., Kjartansson, E. & Winkler, K. 1979. Seismic wave attenuation in rocks. *Rev. of Geophys. Space Phys.*, *17*: 1155–1164.

Mavko, G.M. & Jizba, D. 1991. Estimating grain-scale fluid effects on velocity dispersion in rocks. *Geophysics*, *56*: 1940–1949.

Mavko, G., Mukerji, T. & Dvorkin, J. 1998. *The Rock Physics Handbook: Tools for Seismic Analysis in Porous Media*, Cambridge University Press.

Mavko, G. 2002. Rock physics signatures of reservoir heterogeneities. American Association of Petroleum Geologists. (AAPG Rock Physics: *The Missing Link between Geology, Geophysics and Production*), *Annual Meeting*, Houston: TX.

Maxwell, S.C. & Young, R.P. 1993. Seismic imaging of blast damage. *Int. J. Rock Mech. Min. Sci. & Geomech. Abstr.*, *30(7)*: 1435–1440.

Maxwell, S.C. & Young, R.P. 1996. Seismic imaging of rock mass responses to excavation. *Int. J. Rock Mech. Min. Sci. & Geomech. Abstr.*, *33(7)*: 713–724.

Mayeda, K., Koyanagi, S., Hoshiba, M., Aki, K. & Zeng, Y. 1992. A comparative study of scattering, intrinsic and coda $Q^{-1}$ for Hawaii, Long Valley, and Central California between 1.5 and 15.0 Hz. *J. Geophys. Res.*, *97*: 6643–6659.

McCaffree Pellerin, C.L. & Christensen, N.I. 1998. Interpretation of crustal seismic velocities in the San Gabrial-Mojave region, Southern California. *Tectonophysics*, *286*: 253–271.

McCann, D.M., Grainger, P. & McCann, C. 1975. Interborehole acoustic measurements and their use in engineering geology. *Geophys. Prospect.*, *23*: 50–69.

McDowell, P.W., Barla, G. & Sharp, J.C. 1992. Investigation of rock mass characteristics by geophysical methods for an underground storage project site in weak jointed chalk. *ISRM Symp. on Eurock'92.* Chester, UK. 169–173. London: Thomas Telford.

McDowell, P.W. 1993. Seismic investigation for rock engineering. *Comprehensive Rock Engineering*, Hudson, J.A. (Editor in Chief). Vol. 3. *Rock Testing and Site Characterization.* 619–634. Oxford: Pergamon Press.

McKenzie, C.K., Stacey, G.P. & Gladwin, M.T. 1982. Ultrasonic characteristics of a rock mass. *Int. J. Rock Mech. Min. Sci. & Geomech. Abstr.*, *19*: 25–30.

Miller, D.E., Leaney, S. & Borland, W.H. 1994. An in situ estimation of anisotropic elastic moduli for a submarine shale. *J. Geophys. Res.*, *99(B11)*: 21,659–21,665.

Miller, R.P. 1965. Engineering classification and index properties for intact rock. *PhD Thesis*, University of Illinois, 332 pp.

Mitani, S., Iwai, T. & Isahai, H. 1987. Relations between conditions of rock mass and TBM's feasibility. *Proc. of 6th Int. Congr. Rock Mech. (ISRM), Montreal*, Herget, G. & Vongpaisal (Eds.): *1*: 701–704. Rotterdam: Balkema.

Mitchell, B.J. & Hwang, H.J. 1987. Effect of low $Q$ sediments and crustal $Q$ on $L_g$ attenuation in the United States. *Bull. Seism. Soc. Am.*, *77(4)*: 1197–1210.

Mitchell, B.J. 1995. Anelastic structure and evolution of the continental crust and upper mantle from seismic surface wave attenuation. *Rev. Geophys.*, *33*: 441–462.

Mitchell, B.J., Pan, Y., Xie, J. & Cong, L. 1997. Lg coda Q variation across Eurasia and its relation to crustal evolution. *J. Geophys. Res.*, *102*: 22,767–22,779.

Moen, P.A. 2004. The greener side of grouting in Norway, *Tunnels and Tunnelling International*, June 2004: 14–16.

Mohorovičič, A. 1909. Das bebenvom 8, X, Jahrb. *Meterol. Obs.* Zagreb, *9*: 1–63.

Molnar, P. & Oliver, J. 1969. Lateral variations of attenuation in the upper mantle and discontinuities in the lithosphere. *J. Geophys. Res.*, *74*: 2648–2682.

Molyneux, J.B. & Schmitt, D.R. 2000. Compressional-wave velocities in attenuating media: a laboratory physical model study. *Geophysics*, *65(4)*: 1162–1167.

Monsen, K. 2001. Acoustic velocity in fractured rock. *J. Geophys. Res.*, *106(B7)*: 13,261–13,267.

Mooney, W.D., Laske, G. & Masters, T.G. 1998. CRUST 5.1: a global crustal model at 5° × 5°. *J. Geophys Res.*, *103(B1)*: 727–747.

Moos, D. & Zoback, M.D. 1983. In-situ studies of velocity in fractured crystalline rocks. *J. Geophys. Res.*, *88(B3)*: 2345–2358.

Mouraz Miranda, A. & Mello Mendes, F. 1987. Rock weathering as a drillability parameter, *Proc. of 6th Int. Congr. Rock. Mech. (ISRM), Montreal*, Herget, G. & Vongpaisal (Eds.): *1*: 699–700. Rotterdam: Balkema.

Myer, L.R., Hopkins, D., Peterson, J.E. & Cook, N.G.W. 1995. Seismic wave propagation across multiple fractures. *Int. Conf. on Fractured and Jointed Rock Masses*, Lake Tahoe: CA. Myer, Tsang, Cook & Goodman (Eds.). 105–109. Rotterdam: Balkema.

## N

Nasu, N. 1940. Studies on the propagation of an artificial earthquake wave through superficial soil or sand layers, and the elasticity of soil and sand. [in Japanese]. *Bull. Earthq. Res. Inst.*, *18*: 289–304. Japan: Tokyo Imperial University.

Navalón, N., Gaztanaga, J.M. & Lopez Marinas, J.M. 1987. Cortes-La Muela hydroelectric project. The foundations of Cortes and El Naranjero dams. *Proc. of 6th Int. Congr. Rock Mech. (ISRM), Montreal*, Herget, G. & Vongpaisal (Eds.): *1*: 449–455. Rotterdam: Balkema.

Nehlig, P. & Juteau, T. 1988. Flow porosities, permeabilities and preliminary data on fluid inclusions and fossil thermal gradients in the crustal sequence of the Sumail ophiolite (Oman). *Tectonophysics*, *151*: 199–221.

Nelson, R.A. 2001. *Geologic Analysis of Naturally Fractured Reservoirs.* Boston: Gulf Professional Publishing.

Nes, O.M., Holt, R.M. & Fjær, E. 2000. The reliability of core data as input to seismic reservoir monitoring studies. *Proc. of SPE European Petroleum Conf.*, Paris, SPE Paper No. 65180. Tulsa, OK: Society of Petroleum Engineers.

New, B.M. & West, G. 1980. The transmission of compressional waves in jointed rock. *Eng. Geol.*, *15*: 151–161.

New, B.M. 1985. An example of tomographic and Fourier microcomputer processing of seismic records. *Q. J. Eng. Geol.*, *18*: 335–344. Northern Ireland: The Geological Society.

Newmark, R.L., Anderson, R.N., Moos, D. & Zoback, M.D. 1985. Structure, porosity and stress regime of the upper oceanic crust: sonic and ultrasonic logging of DSDP Hole 504B. *Tectonophysics*, *118*: 1–42.

Nihei, K.T. & Cook, G.W. 1992. Seismic behavior of fractures under normal compression. *33rd US Symp. on Rock Mechanics: Rock mechanics.* Tillerson & Wawersik (Eds.). 989–997. Rotterdam: Balkema.

Niini, H. & Manunen, T. 1970. Seismic sounding as indicator of engineering-geologic properties of bedrock in Finland. *Proc. of 1st IAEG Congress*, Paris, *II*: 753–761. Le Comité Français de Géologie de l'Ingénieur.

Nilsen, B. 1997. Hvorfor stemmer ikke terrenget med kartet – er ikke de ingeniørgeologiske forundersøkelsene gode nok? Cases of unexpected ground conditions – how much are they related to insufficient engineering geological investigations? *Fjellsprengningsteknikk – bergmekanikk – geoteknikk,* 1997. 5.1–5.10. Oslo: Tapir.

Nilsen, B. 1998. Subsea rock tunnels – some Norwegian experiences in geo-investigation and construction. *Design and Construction in Mining Petroleum and Civil Engineering*, Ayres da Silva, Quadros, Gonçalves, (Eds.). Brazil: São Paulo.

Nishioka, G.K. & Michael, A.J. 1990. A detailed study of the seismicity of the Middle Mountain zone at Parkfield, California: temporal variations as possible precursors. *J. Geophys. Res.*, *193*: 577–588.

Nishioka, K. & Aoki, K. 1998. Rapid tunnel excavation by hard rock TBMs in urban areas. In: *Tunnels and Metropolises*, Negro Jr. & Ferreira, (Eds.). 655–661. Rotterdam: Balkema.

Nolet, G. (Ed.). 1987. *Seismic Tomography with Applications in Global Seismology and Exploration Geophysics.*, D. Reidel Publishing Co. (Springer).

Nord, G., Olsson, P. & By, T.L. 1991. Geophysical ground probing in TBM Tunnelling. *Proc. of 7th ISRM Congress*, Aachen. Wittke (Ed.). *1*: 581–586. Rotterdam: Balkema.

Norton, D. & Knapp, R. 1977. Transport phenomena in hydrothermal systems: the nature of porosity. *Am. J. Sci.*, *277*: 913–936.

Nunn, K.R., Barker, R.D. & Bamford, D. 1983. In situ seismic and electrical measurements of fracture anisotropy in the Lincolnshire Chalk. *Q. J. Eng. Geol.*, *16*: 187–195. Northern Ireland: The Geological Society.

Nur, A. & Simmons, G. 1969. The effect of saturation on velocity in low porosity rocks. *Earth Planet. Sci. Lett.*, *7*: 183–193.

Nur, A. 1971. Effects of stress on velocity anisotropy in rocks with cracks. *J. Geoph. Res.*, *76(8)*: 2022–2034.

Nur, A. 1974. Tectonophysics: the study of relations between deformation and forces in the earth. *Proc. of 4th Int. Cong. on Rock Mechanics*, Denver, CO, *1A*: 243–317.

Nur, A. 1987. Seismic rock properties for reservoir descriptions and monitoring (Chapter 9). In: *Seismic Tomography*, Nolet, G. (Ed.). 203–237. D. Reidel Publishing Co.

Nur, A. 1989. Four-dimensional seismology and (true) direct detection of hydrocarbons: the petrophysical basis. *The Leading Edge*, August 1989, *8*: 30–36.

Nur, A., Marion, D. & Yin, H. 1991. Wave velocities in sediments. In: *shear Waves in Marine Sediments*, Hovem, J.M. Richardson, M.D. & Stoll, R.D. (Eds.). 131–140. Dordrecht: Kluwer Academic Publishers.

Nur, A.M. & Wang, Z. (Eds.). 1992. *Seismic and Acoustic Velocities in Reservoir Rocks.* Vol. 1–2. Geophysical reprint series No. 10. Tulsa, OK: Society of Exploration Geophysicists.

Nuttli, O. & Whitmore, J.D. 1962. On the determination of the polarization angle of the S wave. *Bull. Seism. Soc. Am.*, *52*: 95–107.


## O


O'Brien, P.N.S. & Lucas, A.L. 1971. Velocity dispersion of seismic waves. *Geophys. Prospect.*, *19*: 1–26.

O'Connel, R.J. & Budiansky, B. 1977. Viscoelastic properties of fluid-saturated cracked solids. *J. Geophys. Res.*, *82*: 5719–5736.

Oberti, G., Carabelli, E., Goffi, L. & Rossi, P.P. 1979. Study of an orthotropic rock mass: experimental techniques, comparative analysis of results. *Proc. of 4th Int. Congr. Rock Mech. (ISRM), Montreux*: *2*: 485–491. Rotterdam: Balkema.

Oda, M., Yamabe, T. & Kamemura, K. 1986. A crack tensor and its relation to wave velocity anisotropy in jointed rock masses. *Int. J. Rock Mech. Min. Sci. & Geomech. Abstr.*, *23(6)*: 387–397.

Ogawa, Y. 1986. *Geology*. Ishikawa, No. 40, Ishikawa Soil Incorporated Association.

Ohkubo & Teresaki, 1971. Oyo Corporation. *Technical Note* RP-479.

Oliver, J. & Isacks, B. 1967. Deep earthquake zones, anomalous structures in the upper mantle, and the lithosphere. *J. Geophys. Res.*, *72*: 4259–4275.

Olofsson, B. & Kommedal, J. 2002. Facing the processing challenges of the Valhall 1998 3D 4C OBS data. Extended Abstract: Geophysics in a shared earth model. *Symp.*, Norwegian Petroleum Society NPF, Kristiansand, 103–106.

Olson, J. & Pollard, D.D. 1989. Inferring paleostresses from natural fracture patterns: a new method. *Geology*, *17*: 345–348.

Olsson, O. (Ed.). 1992. Site characterization and validation – final report. *Stripa Project 92–22*. 364 pp. Stockholm: SKB.

Olsson, R. & Barton, N. 2001. An improved model for hydromechanical coupling during shearing of rock joints. *Int. J. Rock Mech. Min. Sci. & Geomech. Abstr.*, *38*: 317–329. Pergamon.

Onedera, T.F. 1963. Dynamic investigation of foundation rocks in situ. *5th US Symp. on Rock Mechanics*, Minnesota. 517–533. New York: Pergamon.


## P


Paffenholz, J. & Burkhardt, H. 1989. Absorption and modulus measurements in the seismic frequency and strain range on partially saturated sedimentary rocks. *J. Geophys. Res.*, *94*: 9493–9507.

Palmer, D. 1991. The resolution of narrow low-velocity zones with the generalized reciprocal method. *Geophys. Prospect.*, *39*: 1031–1060.

Palmer, I.D. & Taviolia, J.L. 1980. Attenuation by squirt flow in undersaturated gas sands. *Geophysics*, *45*: 1780–1792.

Palmström, A. 1982a. Problems during construction of the Vardø tunnel – a 2.6 km long submarine road tunnel. *Proc. of 4th IAEG Congress*, New Delhi, *4*: IV.231–IV.244.

Palmström, A. 1982b. The volumetric joint count – a useful and simple measure of the degree of rock mass jointing. *Proc. of 4th IAEG Congress*, New Delhi, *5*: V221–V228.

Palmström, A. 1996. Application of seismic refraction survey in assessment of jointing. *Conf. on Recent Advances in Tunnelling Technology*, New Delhi, 15–22.

Parra, J.O., Hackert, C.L. & Xu, P.-C. 2002, Characterization of fractured low Q zones at the Buena Vista Hills reservoir, California. *Geophysics*, *67*: 1061–1070.

Patton, D.F. 1966. Multiple modes of shear failure in rock and related materials. *PhD Thesis*, University of Illinois, 282 p.

Paul, R.J. 1993. Seismic detection of overpressuring and fracturing: an example from the Qaidam Basin, People's Republic of China. *Geophysics*, *58*: 1532–1543.

Paulsson, B.N.P. & King, M.S. 1980. Between-hole acoustic surveying and monitoring of a granitic rock mass. Technical note. *Int. J. Rock Mech. Min. Sci. & Geomech. Abstr.*, *17*: 371–376.

Paulsson, B.N.P. 1984. A laboratory study of seismic velocities and attentuation of host rocks for a nuclear waste repository. *Rock Mechanics in Productivity and Protection. Symp. on Rock Mechanics*, *25*. Evanston, Ill. 1063–1073. New York: Society of Mining Engineers of AIME.

Paulsson, B.N.P., Cook, N.G.W. & McEvilly, T.V. 1985. Elastic-wave velocities and attenuation in an underground granitic repository for nuclear waste. *Geophysics*, *50(4)*: 551–570.

Paulsson, B.N.P., Smith, M.E., Tucker, K.E. & Fairborn, J.W. 1993. The use of cross-well seismology to characterize and monitor a steamed oil reservoir. *Comprehensive Rock*

*Engineering*. Hudson, J.A. (ed.). Vol. 3. *Rock Testing and Site Characterization*, 651–669. Oxford: Pergamon Press.

Payne, S.S., Worthington, M.H., Odling N.E. & West, L.J. 2005. A field investigation of the permeability dependence of seismic amplitudes. *Seismic Q–observation, mechanisms, and interpretation. Workshop at 67th EAGE Conf.*, Madrid. Extended Abstract: Q-01. Convenors: Hargreaves, Best, Lancaster.

Peng, J.Y., Aki, K., Chouet, B., Johnson, P., Lee, W.H.K., Marks, S., Newberry, J.T., Ryall, A.S., Stewart, S.W. & Tottingham, D.M. 1987. Temporal change in coda Q associated with the Round Valley, California earthquake of November 23, 1984. *J. Geophys, Res.*, *92(B5)*: 3507–3526.

Pepin, G.P., Gonzalez, M.E., Bloys, J.B., Loften, J.E., Schmidt, J.H., Naquin, C.J. & Ellis, S.T. 2004. Effect of drilling fluid temperature of fracture gradient: field measurements and model predictions. Paper 04-527, *Gulf Rocks '04, 6th North American Rock Mechanics Symp. on NARMS*, Houston: TX.

Pérez M.A., Gibson, R.L. & Toksöz, M.N. 1999a. Detection of fracture orientation using azimuthal variation of P-wave AVO response. *Geophysics*, *64(4)*: 1253–1265.

Pérez M.A., Grechka, V. & Michelena, R.J. 1999b. Fracture detection in a carbonate reservoir using a variety of seismic methods. *Geophysics*, *64(4)*: 1266–1276.

Pine, R.J. & Batchelor, A.S. 1984. Downward migration of shearing in jointed rock during hydraulic injections. *Int. J. Rock Mech. Min. Sci. & Geomech. Abstr.*, *21(5)*: 249–263.

Pininska, J. 1977. Correlation between mechanical and acoustic properties of flysch sandstones. *Int. Symp. on Geotechnics of Structurally Complex Formations*, Capri 1977. *I*: 387–394. Associazione Geotecnica Italiana.

Pistre, V., Plona, T., Sinha, B. & seventeen Schlumberger associates 2002. A New Modular Sonic Tool Provides Complete Acoustic Formation Characterization. *43rd Annual Symp. on SPWLA*, Osio, Japan. Extended Abstract.

Plona, T.J., Winkler, K., Sinha, B. & D'Angelo, R. 1998. Measurement of stress direction and mechanical damage around stressed boreholes using dipole and microsonic techniques. SPE/ISRM 47234 in *Proc. of EUROCK'98*, *1*: 123–129.

Plona, T.J., Sinha, B., Kane, M.R., Shenoy, R., Bose, S., Walsh, J., Endo, T., Ikegami, T. & Skelton, O. 2002. Mechanical-damage detection and anisotropy evaluation using dipole sonic dispersion analysis. *43rd Annual Symp. on SPWLA*, Osio, Japan. Extended Abstracts.

Plona, T.J., Winkler, K., D'Angelo, R., Sinha, B., Papanastasiou, P. & Cook, J.M. 1997. Acoustic detection of stress-induced effects around a borehole. *Int. J. Rock Mech. Min. Sci. & Geomech. Abstr.*, Paper No. 290, 539–546. *Proc. of NYROCKS '97*.

Podolski, R., Kosior, A. & Piotrowski, P.C. 1990. The results of applied seismic tomography to localization of the anomalies of velocity fields in coal seams. *Int. Conf. on the Mechanics of Jointed and Faulted Rock*, Vienna, Rossmanith, H.P. (ed.): 749–755. Rotterdam: Balkema.

Potters, J.H.H.M., Groenendaal, H.J.J., Oates, S.J., Hake, J.H. & Kalden, A.B. 1999. The 3D shear experiment over the Natih Field in Oman. Reservoir geology, data acquisition and anisotropy analysis. *Geophys. Prospect.*, *47*: 637–662.

Prasad, M. 1988. Experimental and theoretical considerations of velocity and attenuation interactions with physical parameters in sands. *PhD Thesis*, Kiel University.

Prasad, M. & Manghnani, M.H. 1997. Effects of pore and differential pressure on compressional wave velocity and quality factor in Berea and Michigan sandstones. *Geophysics*, *62(4)*: 1163–1176.

Prasad, M. 2002. Acoustic measurements in unconsolidated sands at low effective pressure and overpressure detection. *Geophysics*, *67(2)*: 405–412.

Prasad, M. 2003. Velocity-permeability relations within hydraulic units. *Geophysics*, *68(1)*: 108–117.

Pratt, H.R., Swolfs, H.S., Brace, W.F., Black, A.D. & Handin, J.W. 1977. Elastic and transport properties of in situ jointed granite. *Int. J. Rock Mech. Min. Sci. & Geomech. Abstr.*, *14*: 35–45.

Press, F. 1966. Seismic velocities. *Handbook of Physical Constants*. Clark, S.G. (Ed.). 195–218 Geological Society of America.

Price, D.G., Malone, A.W. & Knill, J.L. 1970. The application of seismic methods in the design of rock bolt systems. *Proc. of 1st IAEG Congress*, Paris. *II*: 740–751. Le Comité Français de Géologie de l'Ingénieur.

Pride, S.R. & Berryman, J.G. 2003. Linear dynamics of double-porosity dual-permeability materials. II. Fluid transport equation. *Phys. Rev. E 68*, 036604, 10 pp.

Pride, S.R., Harris, J.M., Johnson, D.L., Mateeva, A., Nihel, K.T., Nowack, R.L., Rector, J.W., Spetzler, J.W., Wu, R., Yamomoto, T., Berryman, J.G. & Fehler, M. 2003. Permeability dependence of seismic amplitudes. *The Leading Edge*, June 2006, *22(6)*: 518–525. Society of Exploration Geophysicists.

Pride, S.R., Berryman, J.G. & Harris, J.M. 2004. Seismic attenuation due to wave induced flow. *J. Geophys. Res.*, *109*: 1–19. (for Section 13.3.2, end).

Pujol, J. & Smithson, S. 1991. Seismic wave attenuation in volcanic rocks from VSP experiments. *Geophysics*, *56(9)*: 1441–1455.

Purdy, G.M. 1982. The variability in seismic structure of layer 2 near the east pacific rise at 12°N. *J. Geophys. Res.*, *87(B10)*: 8403–8416.

Purdy, G.M. & Detrick, R.S. 1986. Crustal structure of the Mid-Atlantic ridge at 23°N from seismic refraction studies. *J. Geophys. Res.*, *91*: 3739–3762.

Purdy, G.M. 1987. New observations of the shallow seismic structure of young oceanic crust. *J. Geophys. Res.*, *92(B9)*: 9351–9362.

Purdy, G.M., Christeson, G.L., Fryer, G.J. & Berge, P.A. 1991. Upper crustal structure of the East-Pacific Rise from ocean bottom refraction experiments: interpretation and placement of a deep drill hole. *Eos Trans. AGU 72*: 262. (Abstract).

Pyrak-Nolte, L.J., Myer, L.R., Cook, N.G.W. & Witherspoon, P.A. 1987a. Hydraulic and mechanical properties of natural fractures in low permeability rock. *Proc. of 6th Int. Cong. Rock Mech.*, Herget, G. & Vongpaisal, S. (Eds.). V.1, 225–231, Rotterdam: A.A. Balkema.

Pyrak-Nolte, L.J., Myer, L.R., Cook, N.G.W. 1987b. Seismic visibility of fractures. *Proc. of 28th US Symp. on Rock Mechanics, Farmer, Daeman, Desai, Glass and Neumann.* (Eds.): 47–56, Rotterdam: Balkema.

Pyrak-Nolte, L.J., Myer, L.R. & Cook, N.G.W. 1990. Transmission of seismic waves across single natural fractures. *J. Geophys. Res.*, *95*: 8617–8638.

Pyrak-Nolte, L.J. & Nolte, D.D. 1992. Frequency dependence of fracture stiffness. *Geophys. Res. Lett.*, *19(3)*: 325–328.

Pyrak-Nolte, L.J., Xu, J. & Haley, G.M. 1992. Elastic interface waves along a fracture: theory and experiment. *Proc. of 33rd US Symp. on Rock Mechanics*, Tillerson & Wawersik (Eds.): 999–1007. Rotterdam: Balkema.

Pyrak-Nolte, L.J. 1996. The seismic response of fractures and interrelations among fracture properties. *1995 Schlumberger Lecture Award Paper, Int. J. Rock Mech. Min. Sci. & Geomech. Abstr.*, *33(8)*: 787–802.

## Q

Quadros, E. 1982. Determinação das caracteristicas do fluxo de água em fraturas de rochas. Dissert. De Mestrado, Department of Civil Engineering, Polytech. School, University of São Paulo, Brazil.

Quadros, E. 1995. Water percolation in rock formations in Brazil. Special lecture. *8th ISRM Congress,* Tokyo. Workshop on Rock Foundation.

Quadros, E. & Correa Filho, D.C. 1995. Grouting efficiency using directional (3-D) hydraulic tests in Pirapora Dam, Brazil. *Proc. of 8th ISRM Congress*, Tokyo. Fuji, T. (ed.). *2*: 823–826. Rotterdam: Balkema.

Quan, Y. & Harris, J. 1997. Seismic attenuation and tomography using the frequency shift method, *Geophysics*, *62(3)*: 895–905.

Queen, J.H. & Rizer, W.D. 1990. An integrated study of seismic anisotropy and the natural fracture system at the Conoco borehole test facility, Kay County, Oklahoma. *J. Geophys. Res.*, *95(B7)*: 11,255–11,273.

## R

Raestad, N. 2002. Barents Sea – geological possibilities, political complications. (in Norwegian) *Geo-Bergverksnytt*, April 2002: 12–16. Trondheim.

Rafavich, F., Kendall, C.H.St.C. & Todd, T.P. 1984. The relationship between acoustic properties and the petrographic character of carbonate rocks. *Geophysics*, *49*: 1622–1636.

Rasolofosaon, P.N.J., Rabbel, W., Siegesmund, S. & Vollbrecht, A. 2000. Characterization of crack distribution: fabric analysis versus ultrasonic inversion. *Geophys. J. Int.*, *141*: 425–453.

Rasolofosaon, P.N.J. & Zinszner, B.E. 2002. Comparison between permeability anisotropy and elasticity anisotropy of reservoir rocks. *Geophysics*, *67(1)*: 230–240.

Rathore, J.S., Fjaer, E., Holt, R.M. & Renlie, L. 1995. P- and S-wave anisotropy of a synthetic sandstone with controlled crack geometry. *Geophys. Prospect.*, *43*: 711–728.

Raymer, L.L., Hunt, E.R. & Gardner, J.S. 1980. An improved sonic transit time-to-porosity transform. *SPWLA 21 Ann. Logging Symp.*, 1–12.

Rector III, J.W. 1995. Special Issue: crosswell methods: where are we, where are we going? *Geophysics*, *60(6)*: 629–630.

Remy, J.-M., Bellanger, M. & Homand-Etienne, F. 1994. Laboratory velocities and attenuation of P-waves in limestones during freeze-thaw cycles. *Geophysics*, *59(2)*: 245–251.

Renshaw, C.E. 1995. On the relationship between mechanical and hydraulic apertures in rough-walled fractures. *J. Geophys. Res.*, *100(B12)*: 24,629–24,639.

Ribacchi, R. 1988. Rock mass deformability: in situ tests, their interpretation and typical results in Italy. *Int. Symp. on Field Measurements in Geomechanics, 2. Kobe 1987. 1*: 171–192. Rotterdam: Balkema.

Roche, E. 2001. Seismic data acquisition – The new millennium. *Geophysics*, *66*: 54.

Roche, S.L., Davis, T.L. & Benson, R.D. 1997. 3C seismic study at Vacuum Field, New Mexico. *Annual Int. Meeting Society of Exploration Geophysicists*, Expanded Abstracts: *67*: 886–889.

Rodrigues, L.F., Oliveira, R. & Correia de Sousa, A. 1983. Cabril dam – control of the grouting effectiveness by geophysical seismic tests. *Proc. of 5th Int. Congr. Rock Mech. (ISRM), Melbourne*: *1*: A1–A4. Rotterdam: Balkema.

Roecker, S.W., Tucker, B., King, J. & Hatzfeld, D. 1982. Estimates of Q in central Asia as a function of frequency and depth using the coda of locally recorded earthquakes. *Bull. Seism. Soc. Am.*, *72*: 129–149.

Rohr, K.M.M. 1994. Increase of seismic velocities in upper oceanic crust and hydrothermal circulation in the Juan de Fuca plate. *Geophys. Res. Lett.*, *21*: 2163–2166.

Rossi, G., Böhm, G., Gei, D. & Madrussani, G. 2005. Attenuation tomography: an application to gas-hydrate and free-gas detection. *Seismic Q–observation, mechanisms, and interpretation. Workshop at 67th EAGE Conference*, Madrid. Extended Abstract: Q-03, Convenors: Hargreaves, Best and Lancaster.

Rowlands, H.J., Booth, D.C. & Chiu, J.-M. 1993. Shear-wave splitting from microearthquakes in the New Madrid seismic zone. *Can. J. Explor. Geophys.*, *29*: 352–362.

Rüger, A. 1996. Reflection coefficients and azimuthal AVO analysis in anisotropic media. *PhD Thesis*, Colorado School of Mines.

Rummel, F., Alheid, H.J. & Frohn, C. 1978. Dilatancy and fracture induced velocity changes in rock and their relation to frictional sliding. *Pageophysik, 116*: 743–764, Basel: Birkhäuser Verlag.

Rutledge, J.T. & Winkler, H. 1987. Attenuation measurements in basalt using vertical seismic profile data from the eastern Norwegian Sea. *Society of Exploration Geophysicsts*, Expanded Abstracts: 711–713.

## S

Saenger, E.H. & Shapiro, S.A. 2002. Effective velocities in fractured media: a numerical study using the rotated staggered finite-difference grid. *Geophys. Prospect., 50*: 183–194.

Saito, T. 1981. Variation of physical properties of igneous rock in weathering. *Proc. of Int. symp. on Weak Rock, Tokyo, 1*, 191–196. Rotterdam: Balkema.

Salisbury, M.H. & Christensen, N.I. 1978. The seismic velocity of a traverse through the Bay of Islands ophiolite complex, Newfoundland, an exposure of oceanic crust and upper mantle. *J. Geophys. Res., 83(B2)*: 805–817.

Salisbury, M.H., Milkereit, B., Ascough, G., Adair, R., Matthews, L., Schmitt, D.R., Mwenifumbo, J., Eaton, D.W. & Wu, J. 2000. Physical properties and seismic imaging of massive sulphides. *Geophysics, 65(6)*: 1882–1889.

Sammonds, P.R., Ayling, M.R., Meredith, P.G., Murrell, S.A.F. & Jones, C. 1989. A laboratory investigation of acoustic emission and elastic wave velocity changes during rock failure under triaxial stresses. Rock at Great Depth, *Proc. of Eurock*, Maury, V. & Fourmaintraux, D. (Eds.): 233–240. Rotterdam: Balkema.

Sampaolo, A., Cello, G. & Moro, T. 1978. Preliminary results of a study relative to TBM-rock interaction. *Proc. of 3rd IAEG Congress*, Madrid. *10*: 236–237. Madrid: Imprime ADOSA.

Sams, M.S. 1995. Attenuation and anisotropy: the effect of extra fine layering. *Geophysics, 60*: 1646–1655.

Sams, M.S., Neep, J.P., Worthington, M.H. & King, M.S. 1997. The measurement of velocity dispersion and frequency-dependent intrinsic attenuation in sedimentary rocks. *Geophysics, 62(5)*: 1456–1464.

Sato, J., Itoh, J., Aydan, Ö. & Akagi, T. 1995. Prediction of time-dependent behaviour of a tunnel in squeezing rocks. *FMGM '95, 4th Int. Symp.*, Bergamo, Italy. 47–54.

Sato, H. & Fehler, M.C. 1998. *Seismic Wave Propagation and Scattering in the Heterogeneous Earth*, 308 pp. New York: Springer Verlag & AIP Press.

Sattel, G., Frey, P. & Amberg, R. 1992. Prediction ahead of the tunnel face by seismic methods. Pilot project in Centovalli Tunnel, Locarno, Switzerland. *First Break, 10*: 19–25.

Savage, J.C. 1969. Comments on velocity and attenuation of seismic waves in imperfectly elastic rock. (Gordon and Davis, 1968) *J. Geophys. Res., 74*: 726–728.

Savich, A.I., Koptev, V.I. & Zamakhaiev, A.M. 1974. In situ ultrasonic investigation of failure of limestone. *Proc. of 3rd ISRM Congress*, Denver: Co *IIA:* 418–423. Washington: DC, National Academy of Sciences.

Savich, A.I., Mikhailov, A.D., Koptev, V.I. & Iljin, M.M. 1983. Geophysical studies of rock masses. *Proc. of 5th Int. Congr. Rock Mech. (ISRM), Melbourne*: *1*: A19–A30. Rotterdam: Balkema.

Savich, A.I., Ilyin, M.M. & Yakoubov, V.A. 1987. Regularities of alteration of elastic waves velocities in footing massif of Inguri arch dam during its operation. *Proc. of 6th Int. Congr. Rock Mech. (ISRM), Montreal*, Herget, G. & Vongpaisal (Eds.): *1*: 515–519. Rotterdam: Balkema.

Sayers, C.M. 1990. Stress-induced fluid flow anisotropy in fractured rocks. *Transport Media, 5*: 287–297.

Sayers, C.M. 1998. Misalignment of the orientation of fractures and the principal axes for P and S waves in rocks containing multiple non-orthogonal fracture sets. (Abstract) *Geophys. J. Int., 133*: 459–466.

Sayers, C.M. & Dean, S. 2001. Azimuth-dependent AVO in reservoirs containing non-orthogonal fracture sets. *Geophys. Prospect., 49*: 100–106.

Sayers, C.M. 2002a. Stress-dependent elastic anisotropy of sandstones. *Geophys. Prospect., 50*: 85–95.

Sayers, C.M. 2002b. Fluid-dependent shear-wave splitting in fractured media. *Geophys. Prospect., 50*: 393–401.

Schlumberger GeoQuest, UK Nirex Ltd. Report S/94/007, by kind permission.

Schneider, B. 1967. New methods for the exploration of rock masses. (in French) *Annales ITBTP*, July/August, 235–236, Paris, France.

Schoenberg, M. 1980. Elastic wave behaviour across linear slip interfaces. *J. Acoust. Soc. Am., 68(5)*: 1516–1521.

Schoenberg, M. & Sayers, C.M. 1995. Seismic anisotropy of fractured rock. *Geophysics, 60*: 204–211.

Schoenberg, M. 1998. Acoustic characterization of underground fractures: *68th Annual Int. SEG Meeting*, New Orleans, Expanded Abstract: 1624–1627.

Schrauf, T.W. 1984. Relationship between gas conductivity and geometry of a natural fracture. *MSc Thesis*, Department of Hydrology and Water Resources, University of Arizona, 156 pp.

Scott, J.H., Lee, F.T., Carroll, R.D. & Robinson, C.S. 1968. The relationship of geophysical measurements to engineering and constructional parameters in the Straight Creek Tunnel pilot bore, Colorado. *Int. J. Rock Mech. Min. Sci. & Geomech. Abstr., 5*: 1–30.

Scott, T.E. & Nielsen, K.C. 1991. The effects of porosity on the brittle-ductile transition in sandstones. *J. Geophys. R., 96*: 405–414.

Scott, Jr., T.E., Ma, Q. & Roegiers, J.-C. 1993. Acoustic velocity changes during shear enhanced compaction of sandstone. *Int. J. Rock Mech. Min. Sci. & Geomech. Abstr., 30(7)*: 763–769. (*34th US Symp. on Rock Mechanics*, Madison WI.)

Scott, Jr., T.E., Ma, Q., Roegiers, C. & Reches, Z. 1994. Dynamic stress mapping utilising ultrasonic tomography. *1st NARMS '94.* Austin TX. Rock Mechanics. Nelson & Laubach (Eds.). 427–434. Rotterdam: Balkema.

Sen, A. & Bandyopadhyay, M.R. 1990. Subsurface exploration by seismic refraction for a pumped storage scheme in Northern India. *Proc. of 6th IAEG Congress*, Amsterdam. Price (ed.). *2*: 1025–1031. Rotterdam: Balkema.

Seraphim, J.L. & Pereira, J.P. 1983. Considerations of the geomechanics classification of Bieniawski. *Proc. of Int. Symp. on Eng. Geology and Underground Construction*, LNEC, Lisbon *1*: II33–II42.

Shakeel, A. 1995. The effect of oriented fractures on elastic wave velocities, attenuation and fluid permeability of sandstones. *PhD Thesis*, Imperial College, University of London.

Shakeel, A. & King, M.S. 1998. Acoustic wave anisotropy in sandstones with systems of aligned cracks. In *Core-Log Integration,* Harvey, P.K. & Lovell, M.A. (Eds.). 173–183. Geological Society of London, Special Publication No. 136.

Sharp, J.C. 1970. Fluid flow through fissured media. *PhD Thesis,* 181 pp. University of London.

Shaw, F., Worthington, M.H., Andersen, M.S. & Petersen, U.K. 2004. A study of seismic attenuation in basalt using VSP data from a Faroe Islands borehole. P 015, Extended Abstract: *EAGE 66th Conf. and Exhib*, Paris.

Shaw, P. 1994. Age variations of oceanic crust Poisson's ratio: inversion and a porosity evolution model. *J. Geophys. Res.*, *99(B2)*: 3057–3066.

Shea, V.R. & Hanson, D.R. 1988. Elastic wave velocity and attenuation as used to define phases of loading and failure in coal. *Int. J. Rock Mech. Min. Sci. & Geomech. Abstr.*, *25(6)*: 431–437.

Shearer, P.M. & Orcutt, J.A. 1986. Compressional and shear wave anisotropy in the oceanic lithosphere – the Ngendei seismic refraction experiment. *Geophys. J. Roy. Astr. Soc.*, *87*: 967–1003.

Shearer, P.M. 1988. Cracked media, Poisson's ratio and the structure of the Upper Oceanic Crust. *Geophys. J. Int.*, *92*: 357–362.

Shen, B. & Stephansson, O. 1993. Numerical analysis of Mode I and Mode II propagation of rock fractures. *Int. J. Rock Mech. Min. Sci. & Geomech. Abst.*, *30(7)*: 861–867.

Shen, B. & Barton, N. 1997. The disturbed zone around tunnels in jointed rock masses. *Technical Note, Int. J. Rock Mech. Min. Sci. & Geomech. Abstr.*, *34(1)*: 117–125.

Shen, B. Stephansson, O. & Rinne, M. 2002. Simulation of borehole breakouts using FRACOD$^{2D}$, In *Oil and Gas Science and Technology – Revue de l'IFP*, special issue for Int. Workshop of Geomechanics in Reservoir Simulation *57(5)*: 579–590. IFP. Rueil-Malmaison, France.

Simiyu, S.M. 2000. Geothermal reservoir characterization: application of microseismicity and seismic wave properties at Olkaria, Kenya rift. *J. Geophys. Res.*, *105*: 13,779–13,795.

Singh, S. & Herrmann, R.B. 1983. Regionalization of crustal coda *Q* in the continental United States. *J. Geophys. Res.*, *88*: 527–538.

Sisavath, S., Al-Yaarubi, A., Pain, C.C. & Zimmerman, R.W. 2003. A simple model for deviations from the cubic law for a fracture undergoing dilation or closure. *Pure Appl. Geophys.*, *160*: 1009–1022.

Sjøgren, B., Øfsthus, A. & Sandberg, J. 1979. Seismic classification of rock mass qualities. *Geophys. Prospect.*, *27*: 409–442.

Sjøgren, B. 1984. *Shallow Refraction Seismics*, London, Chapman & Hall. 268 p.

Sjøgren, B. 2000. A brief study of the generalized reciprocal method and of some limitations of the method. *Geophys. Prospect.*, *48*: 815–834.

Skjærstein, A. & Fjær, E. 2000. Angular dependent attenuation in a sandstone with parallel cracks. In: *Advances in Anisotropy – Selected Theory, Modelling and Case Studies.* Hood, J.A. (Ed.). 205–215. Society of Exploration Geophysicists.

Slater, C., Crampin, S., Brodov, L.Y. & Kuznetsov, V.M. 1993. Observations of anisotropic cusps in transversely isotropic clay. *Can. J. Expl. Geophys.*, *29*: 216–226.

Slater, C. 1997. Estimation and modelling of anisotropy in vertical and walkaway seismic profiles at two North Caucasus oil fields. *PhD Thesis*, Department of Geology and Geophysics, University of Edinburgh.

Slimak, Š.A., Ljumovíc, G.R., Lokin, P.M. & Mašala, S.R. 1991. Determination of discontinuous rock mass properties by seismo-acoustic methods. *Proc. of 7th ISRM Congress,* Aachen. Wittke (Ed.). *1*: 609–614. Rotterdam: Balkema.

Slotnick, M.M. 1936. On seismic computation with applications, I: *Geophysics 1*: 9–22.

Smith, B., Guilbot, J. & Pirera, F. 2002. Time-lapse consistent depth conversion using filtered stacking velocities and horizontal wells at Ekofisk. Extended Abstract: geophysics in a shared earth model. *Symp.*, Norwegian Petroleum Society NPF, Kristiansand, 31–34.

Smith, R.L. & McGarrity, J.P. 2001. Cracking the fractures – seismic anisotropy in an offshore reservoir. *The Leading Edge*, January 2001, 18–26.

Snow, D.T. 1968. Rock fracture spacings, openings, and porosities. *J. of Soil Mech. and Found. Div. Proc. of ASCE, SMI*. 73–91.

Solomon, S.C. & Toksöz, M.N. 1970. Lateral variation of attenuation of P and S waves beneath the United States. *Bull. Seism. Soc. Am.*, *60*: 819–838.

Somerton, W.H., Söylemezoglu, I.M. & Dudley, R.C. 1975. Effect of stress on permeability of coal. *Int. J. Rock Mech. Min. Sci. & Geomech. Abstr.*, *12*: 129–145.

Spencer, J.W. 1979. Bulk and shear attenuation in Berea sandstone: the effects of pore fluids. *J. Geophy. Res.*, *84*: 7521–7523.

Spencer, J.W. 1981. Stress relaxation at low frequencies in fluid saturated rocks: attenuation and modulus dispersion. *J. Geophys. Res.*, *86*: 1803–1812.

Spottiswoode, S. M. (1993), Seismic attenuation in deep-level mines, in *Rockbursts and Seismicity in Mines*, edited by P. R. Young, pp. 409–414, Balkema, A. A., Brookfield, Vt.

Spudich, P. & Orcutt, J. 1980a. A new look at the seismic velocity structure of the oceanic crust. *Rev. Geophys. Space Phys.*, *18*: 627–645.

Spudich, P. & Orcutt, J. 1980b. Petrology and porosity of an oceanic crustal site: results from wave form modelling of seismic refraction data. *J. Geophys. Res.*, *85(B3)*: 1409–1433.

Stacey, T.R. 1977. Seismic assessment of rock masses. *Symp. on Exploration for Rock Engineering.* Johannesburg 1976, *2*: 113–117. Rotterdam: Balkema.

Stainsby, S.D. & Worthington, M.H. 1985. Q estimation from vertical seismic profile data and anomalous variations in the central North Sea. *Geophysics*, *50*: 615–626.

Stapledon, D.H. & Rissler, P. 1983. Site exploration and evaluation. General report. *Proc. of 5th Int. Congr. Rock Mech. (ISRM), Melbourne*: *3*: G5–G25. Rotterdam: Balkema.

Stead, D., Szczepanik, Z. & Mackintosh, A.D. 1990. Borehole acoustic logging around underground potash mine openings. *Proc. of 6th IAEG Congress*, Amsterdam. Price (Ed.). *2*: 1037–1043. Rotterdam: Balkema.

Stenin, V.P., Kasimov, A.N. & Tikhonov, A.A. 2002. Cracked layers characterization using far offset VSP. *Proc. 10th Int. Workshop on Seismic Anisotropy (IWSA), Tutzing.*

Strandenes, S. 1991. Rock physics analysis of the Brent Group reservoir in the Oseberg Field. *Stanford Rockphysics & Borehole Geophysics Project*, Special Volume, 25 pp.

Swift, S.A., Kent, G.M., Detrick, R.S., Collins, J.A. & Stephen, R.A. 1998a. Oceanic basement structure sediment thickness, and heat flow near Hole 504B. *J. Geophys. Res.*, *103(B7)*: 15,377–15,392.

Swift, S.A., Lizarralde, D., Stephen, R.A. & Hoskins, H. 1998b. Velocity structure in upper ocean crust at Hole 504B from vertical seismic profiles. *J. Geophys. Res.*, *103(B7)*: 15,361–15,376.

Swolfs, 1977. *Contract Report*. TerraTek International, Salt Lake City.

Swolfs, H.S., Brechtel, C.E., Brace, W.F. & Pratt, H.R. 1981. Field mechanical properties of a jointed sandstone. Mechanical behaviour of crustal rocks, *Geophysical Monographs, American Geophysical Union*, *24*: 161–172. AGU

## T

Talebi, S. & Young, R.P. 1992. Microseismic monitoring in highly stressed granite: relation between shaft-wall cracking and in situ stress. *Int. J. Rock Mech. Min. Sci. & Geomech. Abstr.*, *29(1)*: 25–34.

Talwani, M. 1964. A review of marine geophysics. *Marine Geol.*, *2*: 29–80.

Tanimoto, C. & Ikeda, K. 1983. Acoustic and mechanical properties of jointed rock. *Proc. of 5th Int. Congr. Rock Mech. (ISRM)*, Melbourne: *1*: A15–A18. Rotterdam: Balkema.

Tanimoto, C. & Kishida, K. 1994. Seismic geotomography: amplitude versus velocity in consideration of joint aperture and spacing. *1st NARMS 94.* Austin, TX. Nelson & Laubach (Eds.). 147–155. Rotterdam: Balkema.

Terzaghi, K. 1943. *Theoretical Soil Mechanics*, New York: John Wiley & Sons.

Teufel, L.W. & Farrell, H.E. 1992. Interrelationship between in-situ stress, natural fractures and reservoir permeability anisotropy – A case study of the Ekofisk Field, North Sea. *Proc. of 1st Int. Conf. on Fractured and Jointed Rock Masses (ISRM Symp.)*, Vienna, Rossmanith, H.-P. (Ed.).

Thill, R.E., Tweeton, D.R., Jessop, J.A. & Roessler, K.S. 1992. Developments in seismic tomography for minesite investigation. Rock Mechanics, *Proc. of 33rd U.S. Symp. Rock Mech., Santa Fe, NM*, Tillerson, J.R. & Wawersik, W.R. (Eds.): 1093–1102. Rotterdam: Balkema.

Thomsen, L. 1986. Weak elastic anisotropy. *Geophysics*, *51*: 1954–1966.

Thomsen, L. 2002a. The future of 4C and 4D seismics: challenges and needs. Extended Abstract: geophysics in a shared earth model. *Symp.*, Norwegian Petroleum Society NPF, Kristiansand, 35–36.

Thomsen, L. 2002b. Understanding seismic anisotropy in exploration and exploitation. 2002 *Distinguished Instructor Short Course*, Series No.5, Society of Exploration Geophysicists and European Association of Geoscientists and Engineers, (SEG/EAGE).

Timur, A. 1968. Velocity of compressional waves in porous media at permafrost temperatures. *Geophysics*, *33*: 584–595.

Tingay, M., Müller, B., Reinecker, J., Heidbach, O., Wenzel, F. & Fleckenstein, P. 2005. Understanding tectonic stress in the oil patch: The World Stress Map Project. *The Leading Edge*, December 2005, 1276–1282.

Tjaaland, E., Kanestrøm, R., Folstad, P.G. & E. Fjær, 2001. Effects of overburden anisotropy and attenuation on seismic inversion results for reservoir parameters. *9th IWSA (Int. Workshop on Seismic Anisotropy)*, Abstracts, *Geophysics*, *66(4)*: 1294–1312.

Tod, S.R. 2002. The effects of cracks and porous flow on seismic waves in crustal rock. *PhD Thesis*, Clare College, University of Cambridge.

Tod, S.R., Hudson, J.A. & Liu, E. 2002a. Modelling frequency–dependent anisotropy due to fluid flow in bed limited cracks. *72nd Annual Int. SEG Meeting*, Salt Lake City.

Tod, S.R., Hudson, J.A. & Liu, E. 2002b. Pressure induced anisotropy of interconnected cracks. *Proc. of 10th Int. Workshop on Seismic Anisotropy (IWSA)*, Tutzing.

Tod, S.R. & Liu, E. 2002. Frequency-dependent anisotropy due to fluid flow in bed-limited cracks. *Geophys. Res. Lett.*, *29*: 15 (Paper No. 10.1029/2002GL015369).

Todd, T. & Simmons, G. 1972. Effect of pore pressure on the velocity of compressional waves in low porosity rocks. *J. Geophys. Res.*, *77*: 3731–3743.

Toksöz, M.N., Cheng, C.H. & Timur, A. 1976. Velocities of seismic waves in porous rocks. *Geophysics*, *41*: 621–645.

Toksöz, M.N., Johnston, D.H. & Timur, A. 1979. Attenuation of seismic waves in dry and saturated rocks: laboratory measurements. *Geophysics*, *44*: 681–690.

Toomey, A. & Bean, C.J. 2000. Numerical simulation of seismic waves using a discrete particle scheme. *Geophys. J. Int.*, *141*: 595–604.

Torres-Verdin, C., Alpak, O.F., Wu, J., Gao, G.-Z., Hou, J., Varela, O.J., Gambús-Ordaz, Chi, S. & Youmelin, E. 2003. A multi-physics, integrated approach to formation evaluation using borehole geophysical measurements and 3D seismic data. Society Exploration Geophysicists, *73rd Annual Int. Meeting*, Dallas, TX. Extended Abstracts.

Toverud, T. & Ursin, B. 2005. Comparison of seismic attenuation models using zero-offset vertical seismic profiling (VSP) data. *Geophysics*, *70(2)*: F17–F25.

Townend, J. & Zoback, M.D. 2000. How faulting keeps the crust strong. *Geology*, *28*: 399–402.

Tsidzi, K.E.N. 1997. Propagation characteristics of ultrasonic waves in foliated rocks. *IAEG Bulletin*, *56*: 103–113.

Tsikalas, F., Gudlaugsson, S.T. & Faleide, J.I. 1998. The anatomy of a buried complex impact structure: the Mjølner structure, Barents Sea. *J. Geophys. Res.*, *103(B12)*: 30,469–30,483.

Tsujiura, M. 1978. Spectral analysis of the coda waves from local earthquakes. *Bull. Earthquake Res. Inst.*, *53*: 1–48. Univ. Tokyo.

Tsvankin, I. & Lynn, H.B. 1999. Special section on azimuthal dependence of P–wave seismic signatures – introduction. *Geophysics*, *64*: 1139–1142.

Turk, N. & Dearman, W.R. 1986. A suggested approach to rock characterization in terms of seismic velocities. *27th US Symposium on Rock Mechanics*, Tuscaloosa, AL. Hartman (Ed.): 168–175. Littleton, CO: Society for Mining, Metallurgy and Exploration (SME).

Tutuncu, A.N. 1998a. Nonlinear viscoelastic behaviour of sedimentary rocks, Part I: effect of frequency and strain amplitude. *Geophysics*, *63*: 184–194.

Tutuncu, A.N. 1998b. Nonlinear viscoelastic behaviour of sedimentary rocks, Part II: hysteresis effects and influence of type of fluid on elastic moduli. *Geophysics*, *63*: 195–203.

Tutuncu, A.N., Podio, A.L., Gregory, A.R. & Sharma, M.M. 1998. Nonlinear viscoelastic behaviour of sedimentary rocks, Part I: effect of frequency and strain amplitude. *Geophysics*, *63(1)*: 184–194.


## U


Urmos, J. & Williams, R.H. 1993. In situ velocities in pelagic carbonates: new insights from ocean drilling program leg 130, Ontong Java. *J. Geophys. Res.*, *98(B5)*: 7903–7920.

Ursin, B. & Toverud, T. 1985. Computation of zero-offset seismic profiles including geometrical spreading and absorption. *Geophys. Prospect.*, *33*: 72–96.


## V


Van Der Baan, M. 2002. Constant *Q* and a fractal, stratified earth. *Pure Appl. Geophys.*, *159*: 1707–1718.

Van der Kolk, C.M., Guset, W.S. & Potters, J.H.H.M. 2001. The 3D shear experiment over the Naith field in Oman: the effect of fracture-filling fluids on shear propagation. *Geophys. Prospect.*, *49*: 179–197.

Van Dok, R., Gaiser, J. & Probert, T. 2004. Time-lapse shear-wave splitting analysis at Ekofisk Field. Paper G046, *66th EAGE Ann. Conf. and Exhib.*, Paris, France.

Van Everdingen, D.A. 1995. Fracture characteristics of the sheeted dike complex, Troodos ophiolite, Cyprus: Implications for permeability of oceanic crust. *J. Geophys. Res.*, *100(B10)*: 19,957–19,972.

Vera, E.E., Mutter, J.C., Buhl, P., Orcutt, J.A., Harding, A.J., Kappus, M.E., Detrick, R.S. & Brocher, T.M. 1990. The structure of 0- to 0.2-m.y.-old oceanic crust at 9°N on the East Pacific Rise from expanded spread profiles. *J. Geophys. Res.*, *95*: 15,529–15,556.

Vidstrand, P. 2003. Äspö Hard Rock Laboratory. Update of the hydrogeological model 2002. *Int. Progress Report* IPR-03-35.

Vlastos, S., Liu, E., Main, I.G. & Narteau, C. 2005. Seismic wave propagation in fractured media: scattering attenuation as an indicator of fracture network evolution. *Seismic Q–observation, mechanisms, and interpretation. Workshop at 67th EAGE Conf.*, Madrid. Extended Abstract: Q-07, Convenors: Hargreaves, Best and Lancaster.

Vlastos, S., Narteau, C., Liu, E. & Main, I. 2002. Numerical study of scattering attenuation in fractured media: the effects of scale length on multiple scattering attenuation. *Abstract 73rd Annual Int. SEG Meeting*, Dallas, TX.

Vlastos, S., Narteau, C., Liu, E. & Main, I. 2003a. Numerical study of scattering attenuation in fractured media: frequency dependence and effects of characteristic length scales. *Abstract EAGE 65th Conf. and Exhibition*, Stavanger.

Vlastos, S., Schoenberg, M., Maillot, B., Main, I., Liu, E. & Li, X.-Y. 2003b. Fluid flow simulation in a fractured medium: effects of pore pressure changes on seismic waves. *Abstract EAGE 65th Conf. and Exhibition*, Stavanger.

Vogelaar, B.B.S.A. & Smeulders, D.M.J. 2005. Low frequency attenuation in a saturated rock. *Seismic Q–observation, mechanisms, and interpretation. Workshop at 67th EAGE Conference*, Madrid. Extended Abstract: Q-05, Convenors: Hargreaves, Best and Lancaster.

Volti, T. & Crampin, S. 2003. A four-year study of shear-wave splitting in Iceland: 2. Temporal changes before earthquakes and volcanic eruptions. In: *New Insights into Structural Interpretation and Modelling*. Nieuwland, D.A. (Ed.). Geological Society of London Special Publication, 212. 135–139.

## W

Walsh, J.B. 1966. Seismic wave attenuation in rock due to friction. *J. Geophys. Res.*, *71*: 2591–2599.

Walsh, J.B. 1995. Seismic attenuation in partially saturated rock. *J. Geophys. Res.*, *100(B6)*: 15,407–15,424.

Wang, Z. & Nur, A. 1990. Wave velocities in Hydrocarbon-saturated rocks: experimental results. *Geophysics*, *55*: 723–733.

Wang, Z. & Nur, A. 1992. Seismic and acoustic velocities in reservoir rocks. Geophysics reprint series. No. 10. Wang, & Nur (Eds.). Society of Exploration Geophysicists.

Watanabe, T. & Sassa, K. 1996. Seismic attenuation tomography and its application to rock mass evaluation. *Int. J. Rock Mech. Min. Sci. & Geomech. Abstr.*, *33(5)*: 467–477.

Wei, K. & Liu, D. 1990. The zoning system for assessment of weathering states og granites. *Proc. of 6th IAEG Congress*, Amsterdam. Price (Ed.). *1*: 657–663. Rotterdam: Balkema.

Weiss, R.F., Lonsdal, P., Lupton, J.E., Banbridge, A.E. & Craig, H. 1997. Hydrothermal plumes on the Galapagos rift. *Nature, 267*: 600–603.

Wenhua, Z. 1991. Treatment of fracture zone in Gezhouba project. *Int. Congress on Large Dams, 17. Vienna 1991. 3 (Q66)*: 1667–1679. Paris: ICOLD.

Wepfer, W.A. & Christensen, N.I. 1987. Characterizing microcracks via a velocity-pressure relation. *Eos, Trans. AGU*, 68, 1503.

Wepfer, W.W. & Christensen, N.I. 1990. Compressional wave attenuation in oceanic basalts. *J. Geophys. Res.*, *95*: 17,431–17,439.

Wepfer, W.W. & Christensen, N.I. 1991. A seismic velocity-confining pressure relation, with applications. Technical note. *Int. J. Rock Mech. Min. Sci. & Geomech. Abstr.*, *28(5)*: 451–456.

Westerdahl, H. 1998. Private communication.

Westerdahl, H. & Cosma, C. 1998. Private communication.

Westerman, A.R., Reeves, M.J. & Attwell, P.B. 1982. Rock surface roughness and discontinuity closure at depth. *Developments in Geotechnical Engineering*, *32*: 63–67, Elsevier.

Westman, E.C., Haramy, K.Y. & Rock, A.D. 1996. Seismic tomography for longwall stress analysis. *2nd NARMS '96*. Montréal, Québec. Aubertin, Hassani & Mitri (Eds.). *1*: 397–403. Rotterdam: Balkema.

White, J.E. 1983. *Underground Sound*, New York: Elsevier.

White, R.S. 1984. Atlantic oceanic crust: seismic structure of a slow spreading ridge. In: *Ophiolites and Oceanic Lithosphere*. Glass, Lippard & Shelton (Eds.). 34–44. Geological Society of London.

White, R.S. & Whitmarsh, R.B. 1984. An investigation of seismic anisotropy due to cracks in the upper oceanic crust at 45°N, Mid-Atlantic ridge. *Geophys. J. Roy. Astr. Soc.*, *79*: 439–467.

Whiteley, R.J. 1990. Keynote lecture: advances in engineering seismics. *Proc. of 6th IAEG Congress*, Amsterdam. Price (Ed.). *2*: 813–825. Rotterdam: Balkema.

Whitmarsh, R.B. 1978. Seismic refraction studies of the upper igneous crust in the North Atlantic and porosity estimates for Layer 2. *Earth Plan. Sci. Lett.*, *37*: 451–464.

Wilcock, W.S.D., Solomon, S.C., Purdy, G.M. & Toomey, D.R. 1992. The seismic attenuation structure of a fast-spreading Mid-Ocean ridge. *Science*, *258*: 1470–1474.

Wilcock, W.S.D., Solomon, S., Purdy, G.M. & Toomey, D.R. 1995. Seismic attenuation structure of the East Pacific Rise near 9°30′N. *J. Geophys. Res.*, *100(B12)*: 24,147–24,165.

Wilkens, R., Simmons, G. & Caruso, L. 1984. The ratio $V_p/V_s$ as a discriminant of composition for siliceous limestones. *Geophysics*, *49*: 1850–1860.

Wilkens, R., Schultz, D. & Carlson, R. 1988. Relationship of resistivity, velocity and porosity for basalts from downhole well-logging measurements in hole 418A. *Proc. of Ocean Drilling Project, Scientific Results*, Salisbury, M.H., Scott, J.H., et al. (Eds.). *102*: 69–75.

Wilkens, R.H., Fryer, G.J. & Karsten, J. 1991. Evolution of porosity and seismic structure of Upper Oceanic Crust: Importance of aspect ratios. *J. Geophys. Res.*, *96(B11)*: 17,981–17,995.

Williams, M. & Jenner, E. 2002. Interpreting seismic data in the presence of azimuthal anisotropy; or azimuthal anisotropy in the presence of the seismic interpretation. *The Leading Edge*, August 2002, also CSEG Recorder, June 2003, 35–39.

Willis, M.E., Rethford, G.L. and Bielanski, E. 1986. Azimuthal anisotropy: occurrence and effect on shear-wave data quality. *Presented at the 56th Annual Int. Meeting*. Society of Exploration Geophysicists.

Willis, M.E. & Rao, R. 2005. Seismic attenuation losses induced by discrete vertically aligned fractures. Seismic *Q–Observation, Mechanisms, and Interpretation. Workshop at 67th EAGE Conference*, Madrid. Extended Abstract: Q-13, Convenors: Hargreaves, Best and Lancaster.

Willson, S.M. & Fredrich, J.T. 2005. Geomechanics considerations for through- and near-salt well design. Paper SPE 95621, *SPE Annual Tech. Conf. and Exhibition*, Dallas: TX.

Willson, S.M. 2006. Feeling the heat, can't stand the pressure? Risks to borehole integrity when drilling in ultra-HPHT environments. *First Break*, 24 May 2006, 37–41, EAGE.

Winkler, K. & Nur, A. 1979. Pore fluids and seismic attenuation in rocks. *Geophys. Res. Lett.*, *6*: 1–4.

Winkler, K.A., Nur, A. & Gladwin, M. 1979. Friction and seismic attenuation in rocks. *Nature*, *277*: 528–531.

Winkler, K.W. & Nur, A. 1982. Seismic attenuation: effects of pore fluids and frictional sliding. *Geophysics*, *47*: 1–15.

Winkler, K.W. 1983a. Contact stiffness in granular porous materials. Comparison between theory and experiments. *J. Geophys. Res.*, *10*: 1073–1076.

Winkler, K.W. 1983b. Frequency dependent ultrasonic properties of high-porosity sandstones. *J. Geophys. Res.*, *88*: 9493–9499.

Winkler, K.W. 1985. Dispersion analysis of velocity and attenuation in Berea sandstone. *J. Geophys. Res.*, *90*: 6793–6800.

Winkler, K.W. & Murphy III, W.F. 1995. Acoustic velocity and attenuation in porous rocks. In: *Rock Physics and Phase Relations: A Handbook of Physical Constants*, American Geophysical Union, 1995.

Winkler, K.W., Sinha, B.K. & Plona, T.J. 1998. Effects of borehole stress concentrations on dipole anisotropy measurements. *Geophysics*, *63(1)*: 11–17.

Winterstein, D.F., De, G.S. & Meadows, M.A. 2001. Twelve years of vertical birefringence in nine-component VSP data. *Geophysics*, *66*: 582–597.

Witherspoon, P.A., Amick, C.H. & Gale, J.E. 1977. Stress-flow behaviour of a fault zone with fluid injection and withdrawl. *Mineral Engineering Report. No. 77-1*, University of California, Berkeley.

Witherspoon, P.A., Amick, C.H. & Gale, J.E. 1979. Stress-flow behaviour of a fault zone with fluid injection and withdrawl. *Mineral Engineering Report No. 77-1,* University of California, Berkeley.

Witherspoon, P.A., Wang, J.S.Y., Iwai, K. & Gale, J.E. 1979. Validity of cubic law for fluid flow in a deformable fracture. *Lawrence Berkeley Lab, LBL-9557, SAC-23*, 28 pp. 1980 *Water Resour. Res.*, *16*: 1016–1024.

Wolfe, C.J., Purdy, G.M., Toomey, D.R. & Solomon, S.C. 1995. Microearthquake characteristics and crustal velocity structure at 29°N on the Mid-Atlantic ridge: the architecture of a slow spreading segment. *J. Geophys. Res.*, *100(B12)*: 24,449–24,472.

Won, G.W. & Raper, R.W.J. 1997. Downhole geophysical investigations for a proposed deep highway cutting adjacent to a rail tunnel at Murrurundi, NSW, Australia. Modern Geophysics in Engineering Geology. McCann, D.M. et al. (Eds.). 283–291. Geological Society Engineering Geology. Special Publication 12.

Wong, J., Hurley, P. & West, G.-F. 1983. Crosshole seismology and seismic imaging in crystalline rocks. *Geophys. Res. Lett.*, *10(8)*: 686–689.

Wong, J. 2000. Crosshole seismic imaging for sulphide orebody delineation near Sudbury, Ontario, Canada. *Geophysics*, *65(6)*: 1900–1907.

Worthington, M.H. & Hudson, J.A. 2000. Fault properties from seismic, *Q. Geophys. J. Int.*, *143*: 937–944.

Worthington, M.H., King, M.S. & Marsden, J.R. 2001. Determining the damping factor of sedimentary rocks required for seismically designed structures. *Int. J. Rock Mech. Min. Sci. & Geomech. Abstr.*, *39*: 801–806.

Worthington, M.H. & Lubbe, R. 2007. The scaling of fracture compliance. *Geological Society*, Special Publication, 270, 73–82.

Wulff, A.-M., Hashida, T., Watanabe, K. & Takahashi, H. 1999. Attenuating behaviour of tuffaceous sandstone and granite during microfracturing. *Geophys. J. Int.*, *139*: 395–409.

Wulff, A.-M. & Mjaaland, S. 2002. Seismic monitoring of fluid fronts: an experimental study. *Geophysics*, *67(1)*: 221–229.

Wyllie, M.R.J., Gregory, A.R. & Gardner, L.W. 1956. Elastic wave velocities in heterogeneous and porous media. *Geophysics*, *21*: 41–70.

## X

Xia, G. 2006. Fracture detection from seismic P-wave azimuthal AVO analysis: application to Valhall LOFS data. In situ rock stress. *Proc. of Int. Conf.* Trondheim, Lu, Li, Kjorholt & Dahle (Eds.). London: Taylor & Francis Group.

## Y

Yale, D.P. & Jamieson, Jr., W.H. 1994. Static and dynamic rock mechanical properties in the Hugoton and Panoma fields. *Kansas Soc. Petr. Engr.*, Paper 27939.

Yamamoto, S., Kimura, Y., Takahashi, K. & Miyajima, K. 1995. Geological investigation and evaluation of foundation bedrock for long-span bridges in Japan. Rock Foundations. *Proc. of Int. Workshop on Rock Foundations, Tokyo*, Yoshinaka, R. & Kikuchi, K. (Eds.): 177–184. Rotterdam: Balkema.

Yamamoto, T. & Kuru, M. 1997. Imaging the hydrogeological structure of the limestone aquifers of South Florida using the super cross-well tomography: a pilot experiment using wells BF-1 and BF-2: University of Miami, *Geo-Acoustics Laboratory Tech. Report No. 1038*.

Yamamoto, T. 2003. Imaging permeability structure within the highly permeable carbonate earth: inverse theory and experiment. *Geophysics*, *68(4)*: 1189–1201.

Yan, J. 2003. Improved rock physical models for integration of core, log and seismic data. *PhD Thesis*, University of Edinburgh.

Yielding, G., Walsh, J. & Watterson, J. 1992. The prediction of small-scale faulting in reservoirs. *First Break*, *10*: 449–460.

Young, R.P., Hill, J.J., Bryan, I.R. & Middleton, R. 1985. Seismic spectroscopy in fracture characterization. *Q. J. Eng Geol.*, *18*: 459–479. Northern Ireland: The Geological Society.

Young, R.P. & Collins, D.S. 1999. Monitoring an experimental tunnel seal in granite using acoustic emission and ultrasonic velocity. Rock Mechanics for Industry, *Proc. of 37th U.S. Symp. Rock Mech., Vail, CO*, Amadei, B., Kranz, R.L., Scott, G.A. & Smeallie, P.H. (Eds.): 869–876. Rotterdam: Balkema.

Young, R.P. & Baker, C. 2001. Microseismic investigation of rock fracture and its application in rock and petroleum engineering. *Proc. of 4th Int. Workshop on Application of Geophysics to Rock Engineering,* Beijing, China. *Int. Soc. Rock Mech. News J.*, *7*(*1*): 19–27.

Young, R.P. & Collins, D.S. 2001. Seismic studies of rock fracture at the underground research laboratory, Canada. *Int. J. Rock Mech. Min. Sci. & Geomech. Abstr.*, *38*: 787–799.

Yu, G., Vozoff, K. & Durney, F.W. 1993. The influence of confining pressure and water saturation on dynamic elastic properties of some Permian coals. *Geophysics*, *58*(*1*): 30–38.

Yuan, J. 2001. Analysis of four-component seafloor seismic data for seismic anisotropy. *PhD Thesis*, University of Edinburgh.

## Z

Zatsepin, S.V. & Crampin, S. 1997. Modelling the compliance of crustal rock. I Response of shear-wave splitting to differential stress. *Geophys. J. Int.*, *129*: 477–494.

Zeng, Y., Su, F. & Aki, K. 1991. Scattering wave energy propagation in a medium with randomly distributed isotropic scatterers, 1. Theory. *J. Geophys. Res.*, *96*: 607–619.

Zimmerman, R.W. & King, M.S. 1985. Propagation of acoustic waves through cracked rock. *26th US Symp. on Rock Mechanics, Rapid City* SD. Ashworth (Ed.). 739–745. Rotterdam: Balkema.

Zinszner, B., Meynier, P., Cabrera, J. & Volant, P. 2002. Vitesse des ondes ultrasonores, soniques et sismiques dans les argilites du tunnel de Tournemire. Effet de l'anisotropie et de la fracturation naturelle. *Oil and Gas Science and Technology – Rev. IFP*, *57*(*4*): 341–353.

Ziolkowski, A. 1999. Chasing 1-meter scale with multi-well, multi-level, multi-component receivers. *Offshore*, *59*(*10*): 74–76.

Ziolkowski, A., Hanssen, P., Gatliff, R., Jakubowicz, H., Dobson, A., Hampson, G., Li, X.-Y. & Liu, E. 2003. Use of low frequencies for sub-basalt imaging. *Geophys. Prospect.*, *51: 169–182.*

Zoback, M.D. & Hickman, S.H. 1982. In situ study of the physical mechanisms controlling induced seismicity at Monticello Reservoir, South Carolina. *J. Geophys. Res.*, *87*: 6959–6974.

Zoback, M.D. & Townend, J. 2001. Implications of hydrostatic pore pressures and high crustal strength for the deformation of intraplate lithosphere. *Tectonophysics*, *336*: 19–30.

Zoback, M.D., Barton, C.A., Brudy, M., Castillo, D.A., Finkleiner, T., Grollimund, B.R., Moos, D.B., Pesk, P., Ward, C.D. & Wiprut, D.J. 2003. Determination of stress orientation and magnitude in deep wells. *Int. J. Rock Mech. Min. Sci. & Geomech. Abstr.*, *40*: 1049–1076.

# Index

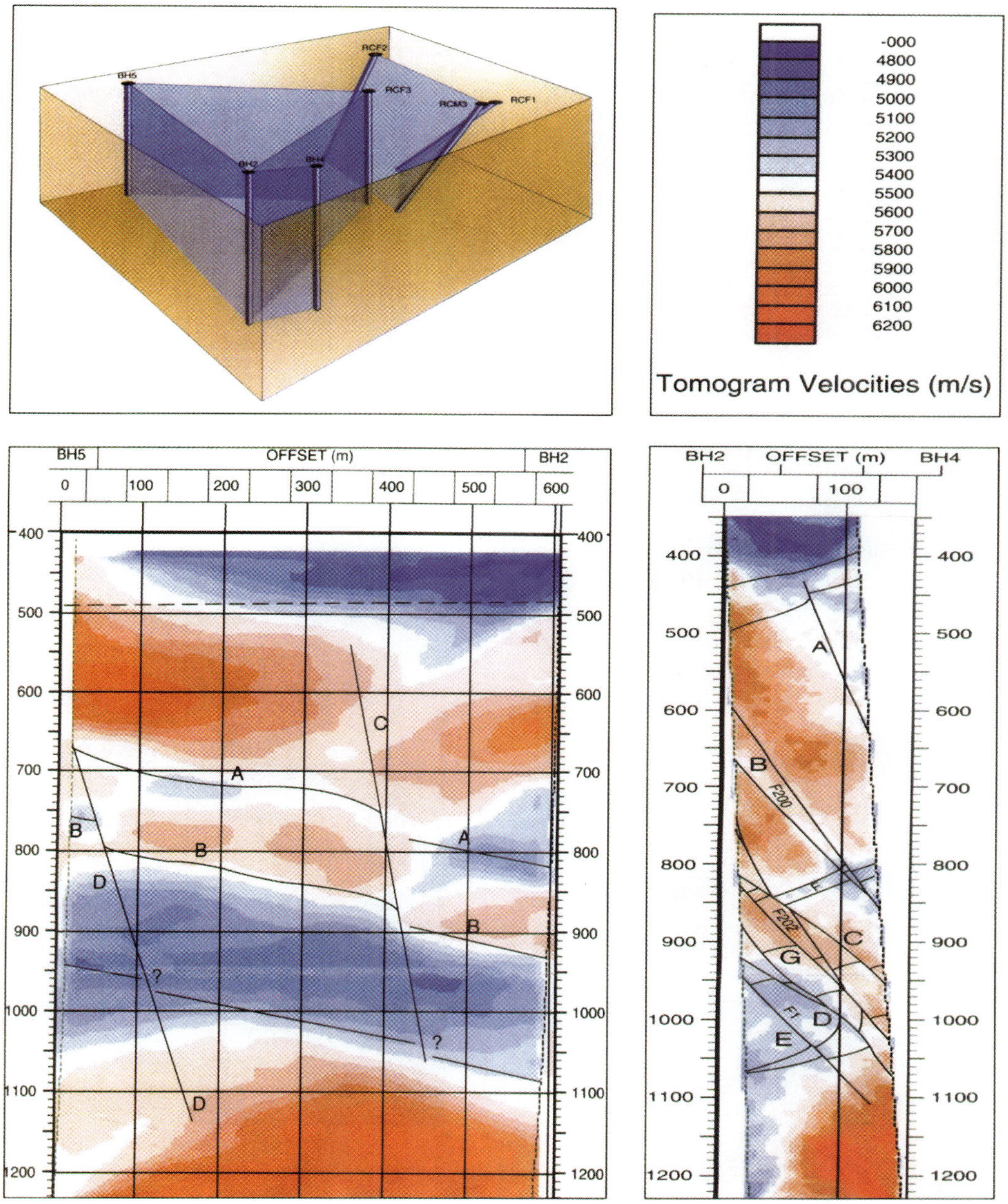


*Plate 1* Deep (1000–1200 m) cross-hole tomography at the UK Nirex Ltd Sellafield site. (Schlumberger GeoQuest, Nirex Report S/94/007, by kind permission). (Figure 4.10).

*Plate 2* An igneous intrusion (dike) tends to have elevated permeability due to the number of joint sets (typically four: $J_n = 15$). High-modulus mineralized veins may be fractured by subsequent tectonic deformation, helping to maintain some permeability despite the 'sealing' process. (Figure 11.75).

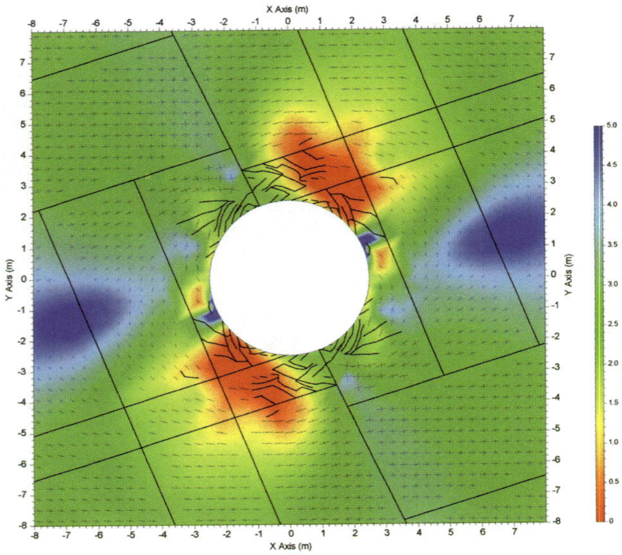


(a)

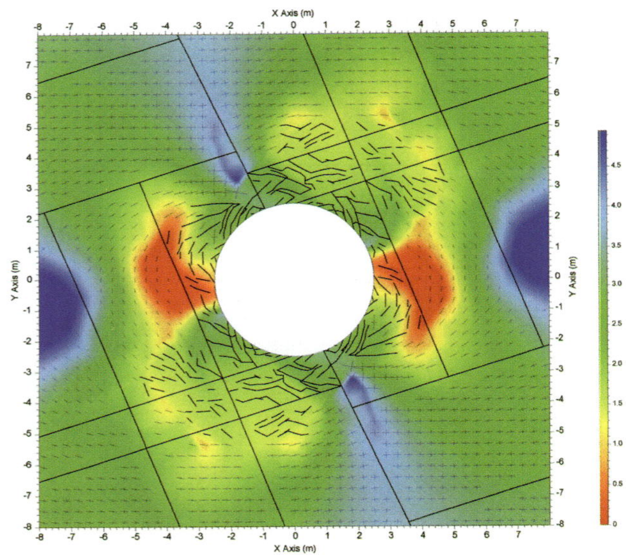


(b)

*Plate 3* Fracture development over time in a FRACOD model of a circular opening in a jointed zone. Note the 'rotation' of the diametral-pair of red regions, which represent low factors of safety against shear failure. Further fracturing dissipates and displaces the low F. of S. zones, suggesting that more fracturing could occur across the 'E-W' diameter. Changed seismic response over time is easy to imagine, also a mud-filtrate invasion speed that could be highly non-uniform, due to developing permeability in the partly connected *discontinuum*. Nick Barton & Associates 2005 contract report. See Figure 12.15 for input data. (Figure 12.16).

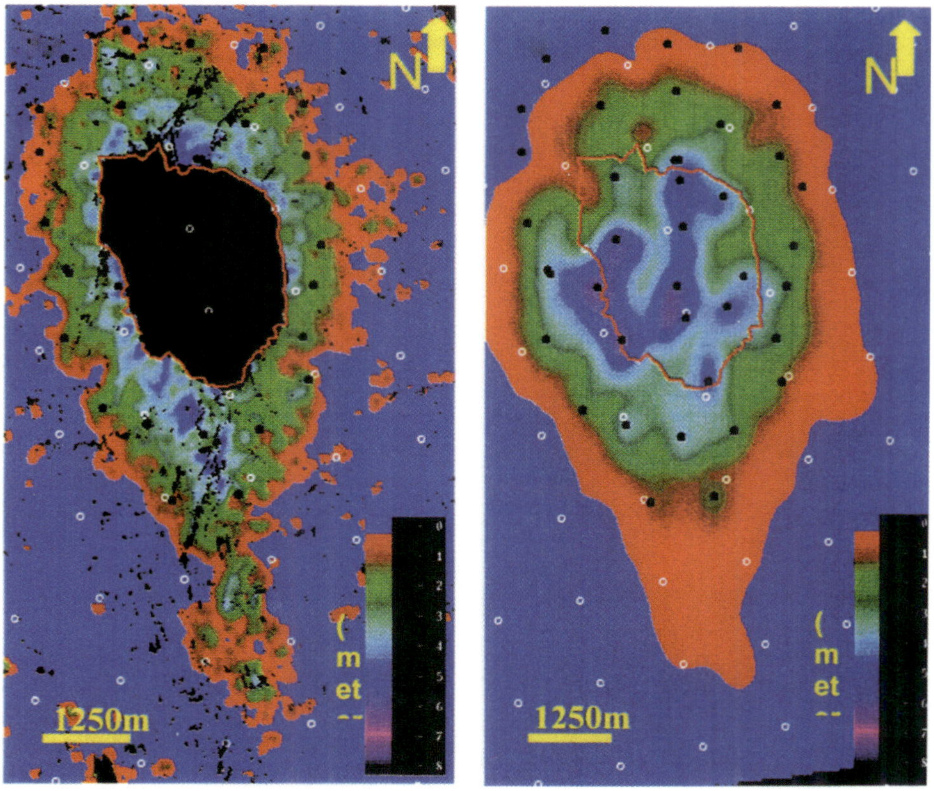


(a) (b)

*Plate 4* Comparison of a) detailed seismic-based (4D time-shift) compaction interpretation (with adjustment for the velocity reduction caused by a subsiding overburden), with b) geomechanics-based one-dimensional strain compaction model, that included porosity reduction due to weakening effect of water saturation. Smith et al., 2002. Note gas cloud effect in centre of seismic model. (Reproduced by kind permission of Norwegian Petroleum Society, NPF). (Figure 14.29).

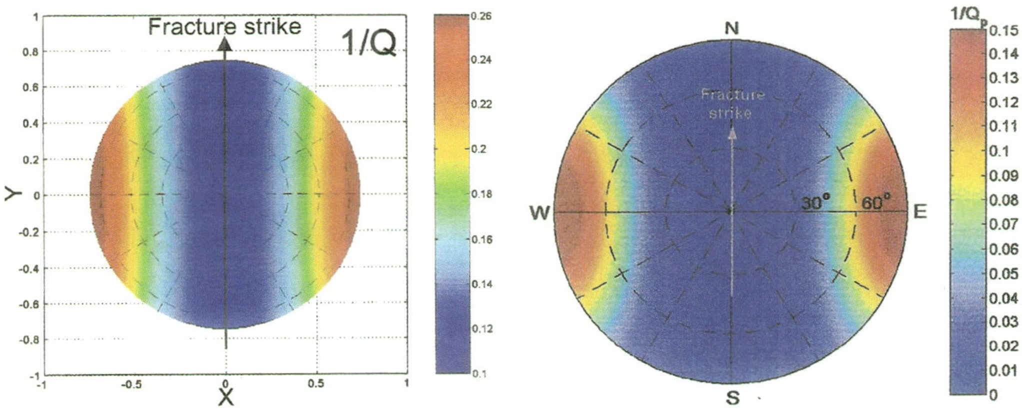


(a) (b)

*Plate 5* Two examples of attenuation modelling with a set of vertical aligned fractures, using Chapman's dynamic poroelastic matrix-and-fracture-set model. See Chapter 15 for a description of this model. Maultzsch et al., 2005. (Reproduced by kind permission, Maultzsch pers. comm. 2005). (Figure 14.37).

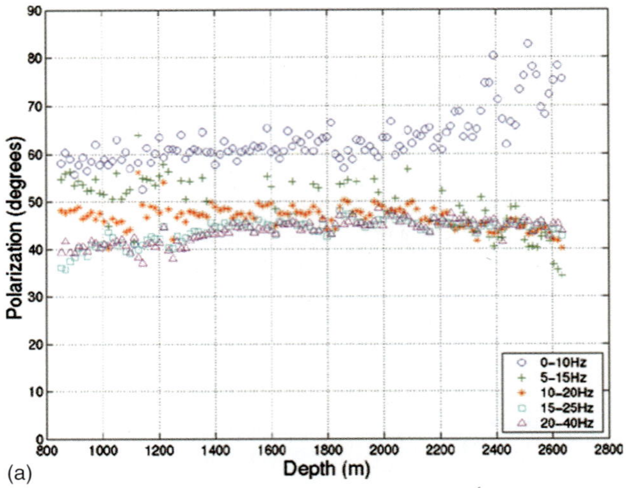


(a)

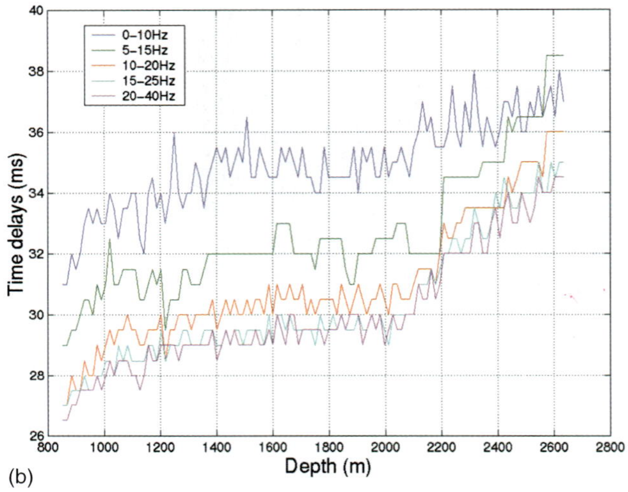


(b)

*Plate 6* Bluebell-Altamont Field anisotropy interpretation, from Liu et al., 2003a, with kind provision of files from Liu, pers.com. 2005. a) Except at low frequency (0–10 Hz), a reasonably constant polarization at 40° to 45° is shown. b) Time delays show three intervals: *gradient*, flat, *gradient*, implying *anisotropy*, isotropy, *anisotropy*. (Figure 15.56).

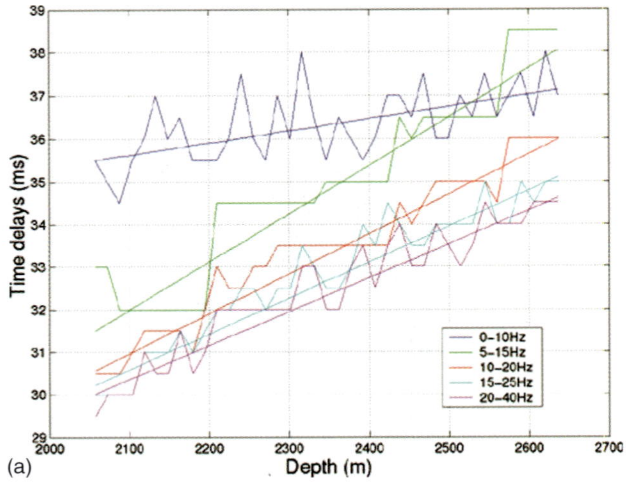


(a)

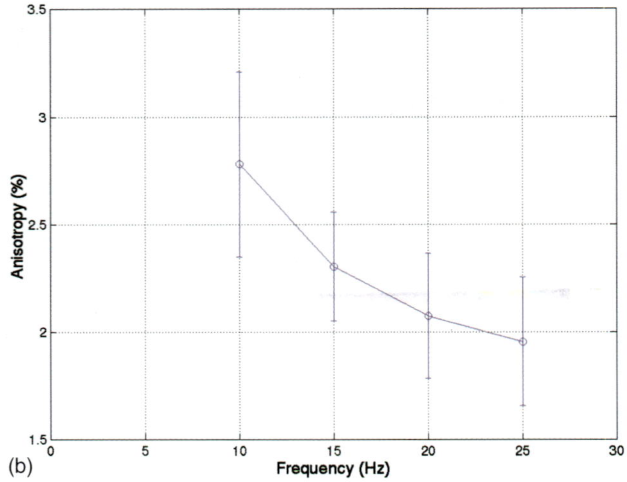


(b)

*Plate 7* Bluebell-Altamont Field anisotropy interpretation, from Liu et al., 2003a, with kind provision of files from Liu, pers.com. 2005. a) Steep time delay gradient in reservoir interval, with frequency dependence. b) Interpretation of anisotropy percentage as function of frequency. (Figure 15.57).

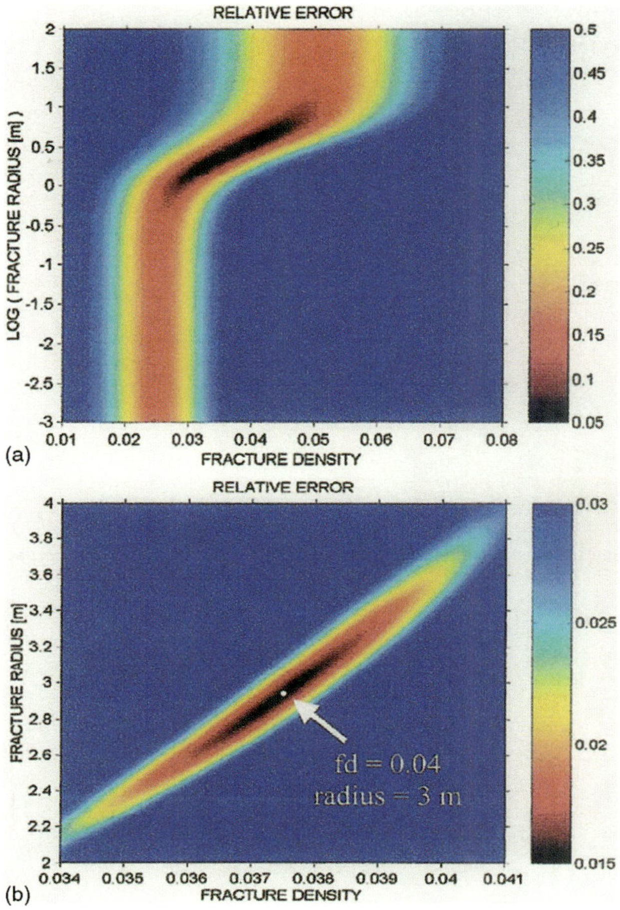


*Plate 8* a) The relative error between the predicted and measured time-delay/depth, evaluated over four frequency values for a range of possible fracture densities and sizes, for comparing with multi-component shear wave VSP data acquired in the Bluebell-Altamont field in Utah. b) The rms error zoomed around the minimum, where the error is less than 5%. (Chapman model application, by Maultzsch et al., 2003 and Liu et al., 2003b). (Figure 15.58).

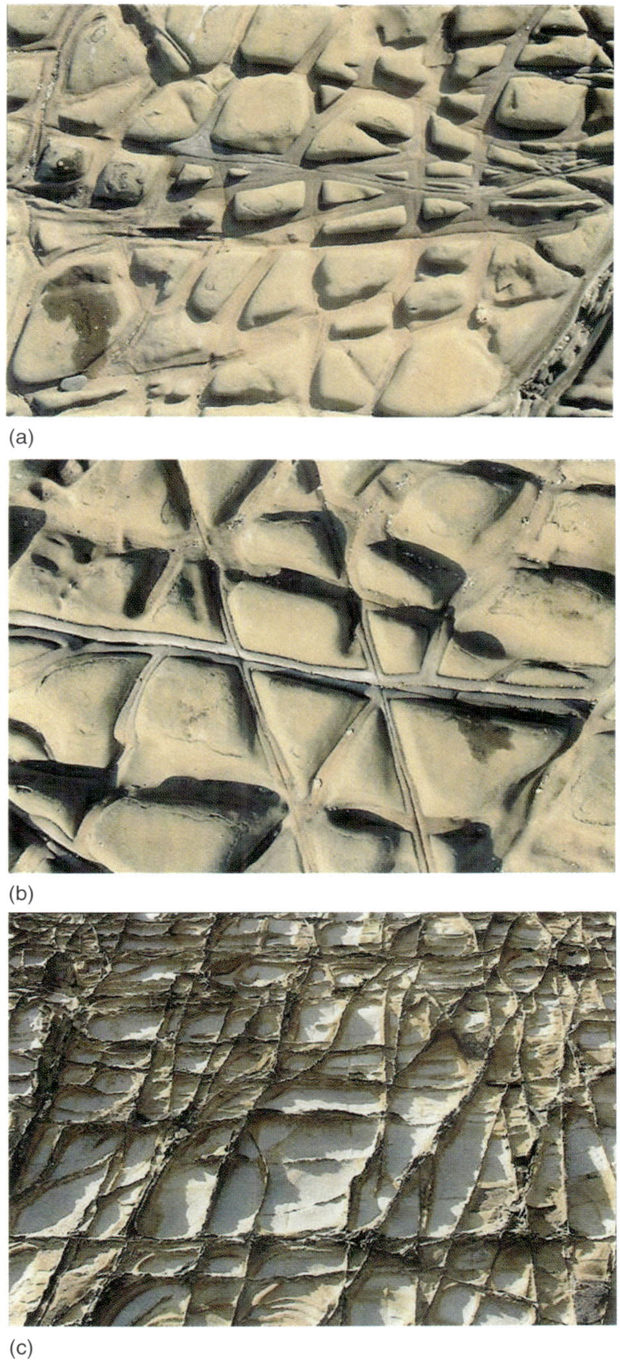


*Plate 9* A wave-cut platform in a jointed dolomite bed. These beds occur at intervals in the Kimmeridge shale, outcropping in Kimmeridge Bay, Dorset, England. The joints show a) implied JCS $< \sigma_c$ due to weathering and preferential wave erosion, and b) implied (local) JCS $> \sigma_c$, due to subsequent mineralization of dominant conducting joints. c) A fine example of joint cementation which may prevent the use of normal joint characterization techniques. Widemouth Bay, near Bude, Devon, England. (Figure 16.5).

*Plate 10* The photograph is a fine example of contrasting JRC values from an interlocked joint and a minor fault in welded tuff, with respective $JRC_0$ values of about 15 and 1. (Figure 16.28).

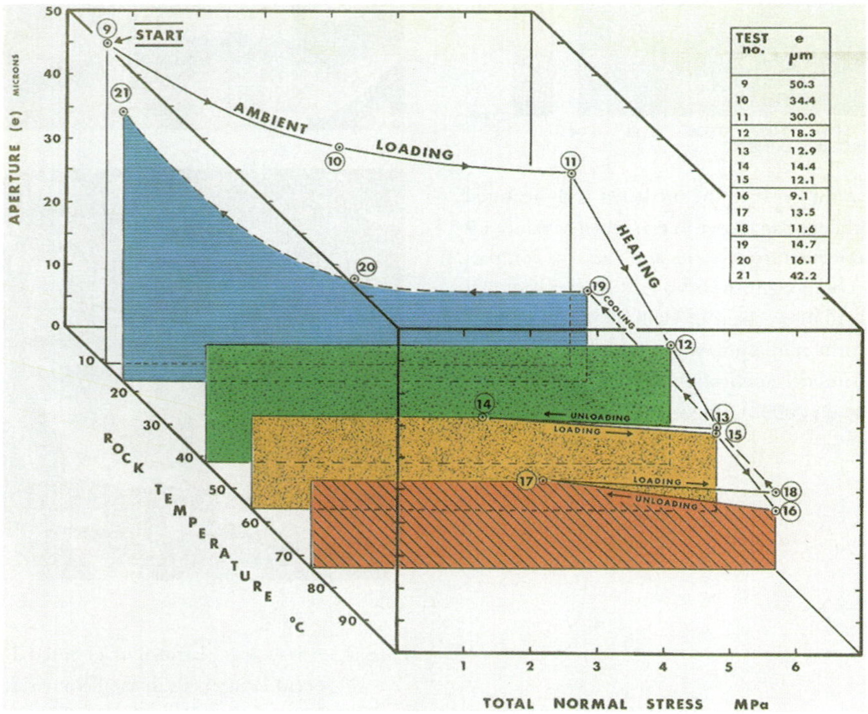


*Plate 11* Hydraulic aperture (e), versus normal stress ($\sigma_n$), versus average rock temperature (T° C) in the permeability test volume of the TerraTek heated block test, CSM mine, Colorado. Note aperture (e) reductions from 30.0 µm to 18.3 µm, to 12.9 µm and finally to 9.1 µm as a result of temperature rise, despite constant applied stress. This gives an *apparent* reduction in the normal stiffness in this test, but in the warmth of a deep petroleum reservoir, would have allowed joints to remain stiffer since their formation. Barton, 1982. (Figure 16.33).

*Plate 12* Interlocked and sheared joints in 'wavy' columnar basalt, demonstrating the role of asperities and dilation on aperture distribution. Columbia River Basalts, Washington State, USA. (Figure 16.37).

*Plate 13* Vertical view through the inter-bedded bituminous shales and dolomite bed, in Kimmeridge Bay, showing a) complex, b) ordered sub-vertical joint patterns. Dorset coast, S. England. (Figure 16.39).

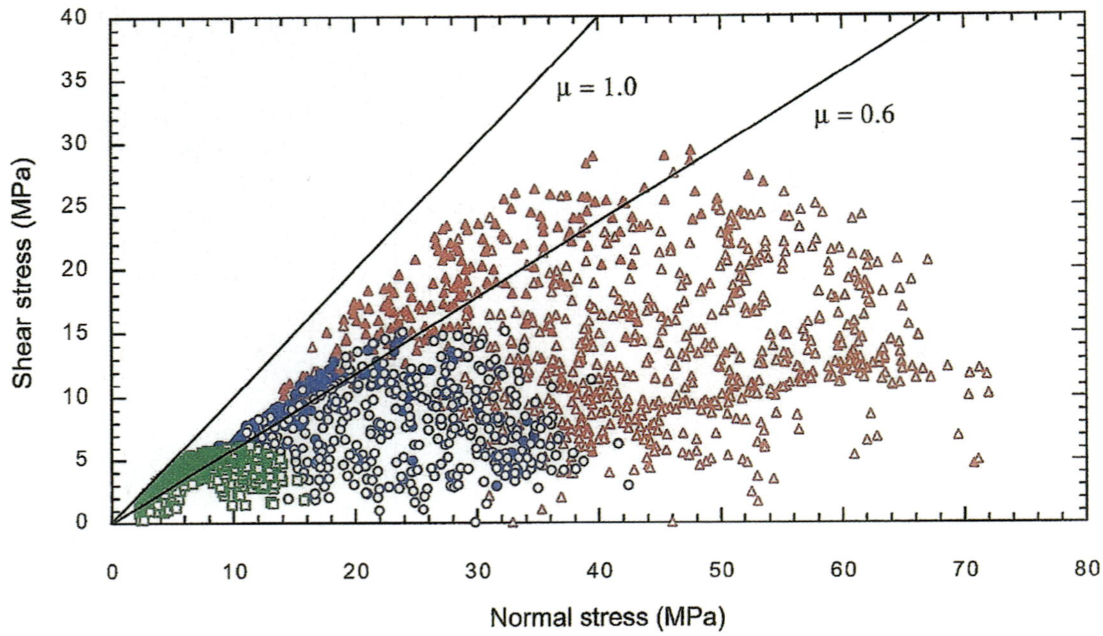


*Plate 14* Normal and shear stresses for fractures identified as hydraulically conducting (closed symbols) or non-conducting (open symbols). Cajon Pass (triangles), Long Valley (circles), Nevada Test Site (squares). Townend and Zoback, 2000, with data also from Colleen Barton et al., 1995. (Zoback, 2006 pers. comm., by kind permission). (Figure 16.63).

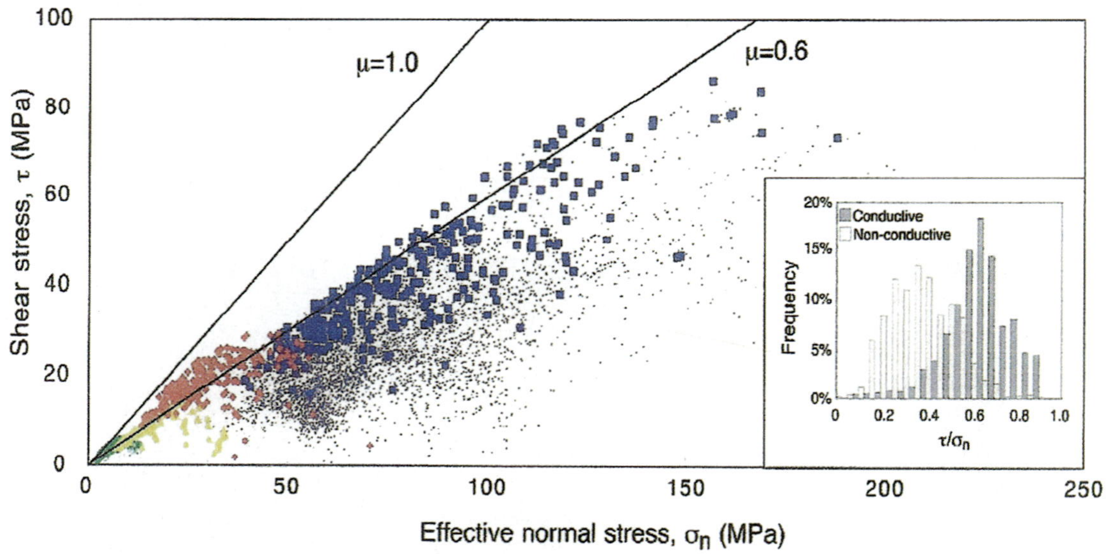


*Plate 15* Normal and shear stresses for fractures identified as hydraulically conducting or non-conducting, using borehole imaging. Cajon Pass (red diamonds and dots), Nevada Test Site (green circles and dots), Long Valley (yellow triangles and dots), KTB (Germany – blue squares and dots). Inset shows $\mu = \tau / \sigma_n$ for combined data set. Zoback and Townend, 2001, with data from Ito and Zoback, 2000, and from Colleen Barton et al., 1995. (Zoback, 2006 pers. comm., by kind permission). (Figure 16.64).

*Plate 16* a) An example of a massive rock mass with rock quality Q $\approx$ 1000 and a deformation modulus in excess of 100 GPa. A very low attenuation is implied. b) The fault-collapse blocking the tunnel on the right would give almost the lowest rock quality Q $\approx$ 0.001, and a modulus of deformation lower than 1 GPa. It is perhaps 'off-the-scale' regarding the conventional definition of $Q_{seis,}$ and would need to be under stress to allow spectral analysis of *measurable* amplitudes. Its Q-value would then be higher too. (Figure 16.79).

*Plate 17* The reality of near-surface construction of tunnels and caverns in rock. Note the three joint sets causing deep over-break. (Figure B1).